For Reference

Not to be taken from this room

Handbook of
FOOD ADDITIVES

Handbook of
FOOD ADDITIVES

An International Guide to More Than 7,000 Products
by Trade Name, Chemical, Function, and Manufacturer

Compiled by

Michael and Irene Ash

Gower

Published by
Gower Publishing Limited
Gower House
Croft Road
Aldershot
Hampshire GU11 3HR
England

Gower
Old Post Road
Brookfield
Vermont 05036
U.S.A.

Michael and Irene Ash have asserted their right under the Copyright, Designs and Patents Act 1988 to be identified as authors of this work

British Library Of Cataloguing in Publication Data
Handbook of Food Additives: International Guide to More Than 7000 Products by Trade Name, Chemical, Function, and Manufacturer
   I. Ash, Michael   II. Ash, Irene
   664.06

ISBN 0-566-07592-x

Library of Congress Cataloging-in-Publication Data
Ash, Michael
      Handbook of food additives : an international guide to more than 7,000 products by trade name, chemical, function, and manufacturer / compiled by Michael and Irene Ash.
      1040 p. 23.4 cm
      Includes bibliographical references (2 p.).
      ISBN 0-566-07592-x
      1. Food additives. 2. Food additives—Dictionaries.   I. Ash, Irene. II. Title
   TX553.A3A84 1995   94-31149
   664´.06—dc20   CIP

Typeset in Arial Narrow by Synapse Information Resources, Inc.
Printed and Bound in Great Britain by
Hartnolls Limited, Bodmin, Cornwall.

Contents

Preface

The **Handbook of Food Additives** describes approximately 5,000 trade name products and 2500 chemicals that function as food additives. This reference includes direct food additives which are defined here as substances which are: (1) intentionally added to a food to affect its overall quality or (2) expected or reasonably expected to become a part of a food as a result of any aspect of production, processing, storage, or packaging. Entries for both trade names and chemicals contain extensive information gathered from world-wide manufacturers, distributors, trade journals, government documents, and other references.

The food additive industry is growing at a rapid pace. Consumers throughout the world have created an increased demand for processed foods requiring more additive ingredients and less caloric foods requiring substitutes for fats and sugars. This has triggered a need for the development of food chemical additives that can produce more specialized effects. For example, as fat is reduced in food products, more flavors, emulsifiers, and texturizing agents are needed to mimic its properties; flavor and flavor enhancers are used to compensate for reduced sugar and sodium. The flavor industry alone totaled $5 billion in revenues for 1993.

This reference functions as a single source for information on both the trade name products and the chemicals that are used as food additives throughout the world. It includes summaries of regulatory information for the United States, Europe, and Japan presented in a tabular format. The products described in this Handbook are cross referenced in multiple ways: chemical composition, function, CAS number, EINECS number, FEMA number, and E number.

The book is divided into four sections:

Part I—*Trade Name Reference* contains almost 5,000 alphabetical entries of trade name food additive products. Each entry provides information on its manufacturer, chemical composition, CAS and EINECS identifying numbers, general properties, applications and functions, toxicology, compliance, and regulatory information as provided by the manufacturer and other sources.

Part II—*Chemical Dictionary/Cross-Reference* contains an alphabetical listing of food chemicals. Each food chemical entry includes, wherever possible, its synonyms, CAS number, EINECS number, FEMA number, and E number, formulas, chemical properties, function and application, toxicology, precautions, usage level, and regulatory information, as well as some of the manufacturers of the chemical. The trade name products from Part I that are equivalent to the chemical or contain that chemical compound as the trade name product's major chemical constituent are cross-referenced. Synonyms for these chemical entries are comprehensively cross-referenced back to the main entry.

Part III— *Functional Cross-Reference* contains an alphabetical listing of major food additive functional categories. Over 30 categories are included, e.g., acidulants, anticaking agents, colorants, emulsifiers, enzymes, flavors and flavor enhancers, nutrients and dietary supplements, surface-finishing agents, sweeteners, texturizers, thickeners, etc. Each functional category entry is followed by an alphabetical listing of the trade name products and food chemicals that have that functional attribute.

Part IV— *Manufacturers Directory* contains detailed contact information for the manufacturers of the trade name products and food chemicals referenced in this handbook. Wherever possible telephone, telefax, and telex numbers, toll-free 800 numbers, and complete mailing addresses are included for each manufacturer.

The **Appendices** contain the following cross-references:

CAS Number-to-Trade Name Cross-Reference orders many trade names found in Part I by identifying CAS numbers; it should be noted that trade names may contain more than one chemical component and the associated CAS numbers in this section refer to each trade name product's primary chemical component.

CAS Number-to-Chemical Cross-Reference orders chemical compounds found in Part II by CAS numbers.

EINECS Number-to-Trade Name Cross-Reference orders many trade names found in Part I by identifying EINECS numbers that refer to each trade name product's primary chemical component.

EINECS Number-to-Chemical Cross-Reference orders chemicals found in Part II by EINECS numbers.

FEMA Number-to-Chemical Cross-Reference orders flavor and extract chemicals that are found in Part II by Flavor and Extract Manufacturers Association (FEMA) numbers.

The Tables of Food Regulations are summaries of food additive regulatory information for the United States, Europe, and Japan and include chemical names, functions, usage levels, and limitations. This section is meant to be used as a general guide. Local food legislation should always be consulted regarding all food additive products as they vary from country to country.

This book is the culmination of many months of research, investigation of product sources, and sorting through a variety of technical data sheets and brochures acquired through personal contacts and correspondences with major chemical manufacturers world-wide as well as trade journals and reference books. We are especially grateful to Roberta Dakan for her skills in chemical information database management. Her tireless efforts have been instrumental in the production of this reference.

M. & I. Ash

NOTE:

The information contained in this reference is accurate to the best of our knowledge; however, no liability will be assumed by the publisher or the authors for the correctness or comprehensiveness of such information. The determination of the suitability of these products for prospective use is the responsibility of the user. It is herewith recommended that those who plan to use any of the products referenced seek the manufacturers instructions for the handling of that chemical.

Abbreviations

abs.	absolute
absorp.	absorption
ACGIH	American Conference of Governmental Industrial Hygienists
act.	active
ADI	acceptable daily intake (FAO/WHO)
agric.	agricultural
agrichem.	agrichemical(s)
alc.	alcohol
anhyd.	anhydrous
APHA	American Public Health Association
applic(s).	application(s)
aq.	aqueous
ASBC	Am. Society of Brewing Chemists
atm	atmosphere
at.wt.	atomic weight
aux.	auxiliary
avail.	available
avg.	average
a.w.	atomic weight
BATF	U.S. Bureau of Alcohol, Tobacco, and Firearms
BGA	Federal Republic of Germany Health Dept. certification
BHA	butylated hydroxyanisole
BHT	butylated hydroxytoluene
biodeg.	biodegradable
blk.	black
b.p.	boiling point
BP	British Pharmacopeia
br., brn.	brown
brnsh.	brownish
BVO	brominated vegetable oil
C	degrees Centigrade
CAP	Color Additive Petition (U.S.)
CAS	Chemical Abstracts Service
cc	cubic centimeter(s)
CC	closed cup
CCl_4	carbon tetrachloride
CE	Council of Europe
CFN	Council on Food and Nutrition (American Medical Association)
CFR	Code of Federal Regulations (U.S.)
char.	characteristic
chel.	chelation
chem.	chemical
CI	Color Index
CIR	Cosmetic Ingredient Review
CL	ceiling concentration
cm	centimeter(s)
cm^3	cubic centimeter(s)
CMC	carboxymethylcellulose
c.m.p.	capillary melting point
CNS	central nervous system
COC	Cleveland Open Cup
compat.	compatible
compd(s).	compound(s)
conc(s).	concentrated, concentration

consistg.	consisting
contg.	containing
cosolv.	cosolvent
cp	centipoise(s)
CP	Canadian Pharmacopeia
cps	centipoise(s)
cryst.	crystalline, crystallization
cs or cSt	centistoke(s)
CTFA	Cosmetic, Toiletry and Fragrance Association
cwt	hundred weight
DAB	Deutsches Arzneibuch
DE	dextrose equivalent
dec.	decomposes
decomp.	decomposition
deliq.	deliquescent
dens.	density
deriv.	derivative(s)
DI	deionized
diam.	diameter
dielec.	dielectric
dil.	dilute
disp.	dispersible, dispersion
dist.	distilled
dk.	dark
DMF	dimethyl formamide
DMSO	dimethyl sulfoxide
DOT	U.S. Department of Transportation
DSB	dry solids basis
EC	European Community
EDTA	ethylenediamine tetraacetic acid
EEC	European Economic Community
EINECS	European Inventory of Existing Commercial Chemical Substances
elec.	electrical
elong.	elongation
EO	ethylene oxide
EP	European Pharmacopoeia
EPA	U.S. Environmental Protection Agency
EPS	expandable polystyrene
equip.	equipment
esp.	especially
Eur.Ph.	European Pharmacopeia
exc.	excellent
F	degrees Fahrenheit
FAC	Food Advisory Committee
FAO	Food and Agriculture Organization (United Nations)
FAP	Food Additive Petition (U.S.)
FCC	Food Chemicals Codex
FDA	Food and Drug Administration (U.S.)
FD&C	Foods, Drugs, and Cosmetics
FDD	Food and Drug Directorate
FEMA	Flavor and Extract Manufacturers' Association (U.S.)
FFDCA	Federal Food, Drug, and Cosmetic Act
FG	food grade
fl	fluid
flamm.	flammable, flammability
FNB	Food and Nutrition Board

f.p.	freezing point
FPC	fish-protein concentrate
FR	Federal Register
ft	foot, feet
F-T	Fischer-Tropsch
f.w.	formula weight
g	gram(s)
G	giga
gal	gallon(s)
G-H	Gardner-Holdt
GI	gastrointestinal
glac.	glacial
gr.	gravity
gran.	granules, granular
GRAS	generally regarded as safe
grn.	green
GMP	good manufacturing practice, guanosine monophosphate
HC	hydrocarbon
HCl	hydrochloride, hydrochloric acid
HFCS	high fructose corn syrup
Hg	mercury
HLB	hydrophilic lipophilic balance
HPLC	high performance liquid chromatography
HTST	high temperature short-time pasteurization
HVP	hydrolyzed vegetable protein
hyd.	hydroxyl
hydrog.	hydrogenated
IARC	International Agency for Research on Cancer (United Nations)
i.b.p.	initial boiling point
IFRA	International Fragrance Association
IMP	inosine monophosphate
in.	Inch(es)
INCI	International Nomenclature Cosmetic Ingredient
incl.	including
incompat.	incompatible
ing.	ingestion
ingred(s).	ingredient(s)
inh.	inhalation
inorg.	inorganic
insol.	insoluble
Int'l.	International
IOFI	International Organization of the Flavor Industry
IP	intraperitoneal
IPA	isopropyl alcohol
IU	International Unit
i.v.	iodine value
IV	intravenous
JCID	Japanese Cosmetic Ingredients Dictionary
JECFA	Joint Expert Committee on Food Additives
JP	Japanese Pharmacopoeia
JSCI	Japanese Standard of Cosmetic Ingredients
JSFA	Japanese Standard of Food Additives
k	kilo
kg	kilogram(s)
l	liter(s)
lb	pound(s)

LD50	lethal dose 50%
lel	lower explosive level
liq.	liquid
lt.	light
Ltd.	Limited
m	milli or meter(s)
m-	meta
M	mega
max.	maximum
MCT	medium chain triglycerides
med.	medium
MEK	methyl ethyl ketone
mfg.	manufacture
mg	milligram(s)
MIBK	methyl isobutyl ketone
microcryst.	microcrystalline
MID	Meat Inspection Division (USDA)
min.	minute(s), mineral, minimum
misc.	miscible, miscellaneous
mixt.	mixture(s)
ml	milliliter(s)
MLD	mild irritation effects
mm	millimeter(s)
mod.	moderately
m.p.	melting point
mPa·s	millipascal-seconds
MRL	maximum residual limits
MSG	monosodium glutamate
mus	mouse
m.w.	molecular weight
N	normal
nat.	natural
need.	needles
NF	National Formulary
no.	number
nonalc.	nonalcoholic
nonflamm.	nonflammable
NTP	National Toxicology Program
NV	nonvolatiles
o-	ortho
OC	open cup
OMS	odorless mineral spirits
org.	organic
OTC	over-the-counter
o/w	oil-in-water
oz	ounce
p-	para
Pa	Pascal
PE	polyethylene
PEG	polyethylene glycol
PEL	permissible exposure level
petrol.	petroleum
pH	hydrogen-ion concentration
phr	parts per hundred of rubber or resin
pkg.	packaging
P-M	Pensky-Martens

PMCC	Pensky-Martens closed cup
POE	polyoxyethylene, polyoxyethylated
POP	polyoxypropylene, polyoxypropylated
powd.	powder
PP	polypropylene
ppb	parts per billion
ppm	parts per million
pract.	practically
prep.	preparation(s)
prod.	product(s), production
props.	properties
pt.	point
PVA	polyvinyl alcohol
PVC	polyvinyl chloride
PVP	polyvinylpyrrolidone
R&B	Ring & Ball
rbt	rabbit
RDA	recommended daily allowances
rdsh.	reddish
ref.	refractive
resist.	resistance, resistant, resistivity
resp.	respectively
r.h.	relative humidity
rhomb.	rhombic
R.T.	room temperature
s	second(s)
sapon.	saponification
sat.	saturated
SE	self-emulsifying
sec	secondary
sl.	slightly
sm.	small
SMG	succinylated monoglycerides
soften.	softening
sol.	soluble, solubility
solid.	solidification
sol'n.	solution
solv(s).	solvent(s)
sp.	specific
spec.	specification, specialty
spp.	non-specified species
std.	standard
STEL	short term exposure limit
Stod.	Stoddard solvent
str.	strength
subcut.	subcutaneous
subl.	sublimes
surf.	surface
SUS	Saybolt Universal Seconds
susp.	suspension
syn.	synthetic
t	tertiary
TBHQ	tert-butyl hydroquinone
TCC	Tag closed cup
tech.	technical
temp.	temperature

tens.	tensile or tension
tert	tertiary
thru	through
TLV	Threshold Limit Value
TOC	Tag open cup
TSCA	Toxic Substances Control Act
tsp.	teaspoon
TWA	time weighted average
typ.	typical
uel	upper explosive limits
UHT	ultra high temperature
unsat.	unsaturated
USDA	U.S. Department of Agriculture
USP	Unites States Pharmacopeia
uv	ultraviolet
veg.	vegetable
visc.	viscous, viscosity
vol.	volume
v/v	volume by volume
WFC	World Food Council
wh.	white
WHO	World Health Organization (United Nations)
w/o	water-in-oil
wt.	weight
w/w	weight by weight
yel.	yellow
ylsh.	yellowish
yr	year
#	number
%	percent
<	less than
>	greater than
@	at
≈	approximately
α	alpha
β	beta
δ, Δ	delta
ε	epsilon
γ	gamma
ι	iota
λ	lambda
ω	omega
ς	sigma
μ	micron, micrometer
μg	microgram

Part I
Trade Name Reference

A

AA USP. [CasChem] Castor oil; CAS 8001-79-4; EINECS 232-293-8; lubricant for food processing (release aid, protective coatings for vitamins, tableting); *Regulatory:* FDA approval; *Properties:* sol. in alcohols, esters, ethers, ketone, and aromatic solvs.

A.B.C. #7. [Brolite Prods.] Buttercream stabilizer that also increases yields while eliminating grit and shortening taste.

ABC-Trieb®. [BASF AG] Ammonium bicarbonate; CAS 1066-33-7; EINECS 213-911-5; baking raising agent.

Ablunol S-20. [Taiwan Surf.] Sorbitan laurate; CAS 1338-39-2; nonionic; emulsifier, emulsion stabilizer, thickener; *Properties:* oily liq.; oil-sol.; water-disp.; HLB 8.6; 100% act.

Ablunol S-40. [Taiwan Surf.] Sorbitan palmitate; CAS 26266-57-9; EINECS 247-568-8; nonionic; emulsifier, emulsion stabilizer, thickener; *Properties:* waxy solid; oil-sol.; HLB 6.7.

Ablunol S-60. [Taiwan Surf.] Sorbitan stearate; CAS 1338-41-6; EINECS 215-664-9; nonionic; emulsifier, emulsion stabilizer, thickener; *Properties:* waxy flake; HLB 4.7; 100% act.

Ablunol S-80. [Taiwan Surf.] Sorbitan oleate; CAS 1338-43-8; EINECS 215-665-4; nonionic; emulsifier, emulsion stabilizer, thickener; *Properties:* oily liq.; HLB 4.3; 100% act.

Ablunol S-85. [Taiwan Surf.] Sorbitan trioleate; CAS 26266-58-0; EINECS 247-569-3; nonionic; emulsifier, emulsion stabilizer, thickener; *Properties:* oily liq.; HLB 1.8; 100% act.

Ablunol T-20. [Taiwan Surf.] POE sorbitan laurate; nonionic; o/w emulsifier for food applics.; *Properties:* oily liq.; water-sol.; HLB 16.7; 100% solids.

Ablunol T-40. [Taiwan Surf.] POE sorbitan laurate; nonionic; o/w emulsifier for food applics.; *Properties:* oily liq.; water-sol.; HLB 15.6; 100% conc.

Ablunol T-60. [Taiwan Surf.] POE sorbitan stearate; nonionic; o/w emulsifier for food applics.; *Properties:* oily liq.; water-sol.; HLB 14.9; 100% conc.

Ablunol T-80. [Taiwan Surf.] POE sorbitan oleate; nonionic; o/w emulsifier for food applics.; *Properties:* oily liq.; water-sol.; HLB 15.0; 100% conc.

Accel®. [Quest Int'l.] Fermented specialty; flavor enhancer for a buttery background in dairy prods., prepared foods, salad dressings, bakery prods., sauces, and confections.

Accofloc® A100 PWG. [Mitsui-Cyanamid] High m.w. polyacrylamide; CAS 9003-05-8; weakly anionic; flocculant for sugar prod. (juice clarification, remelt refinery liquor clarification, scum desweetening, waste treatment); *Usage level:* 0.2-2 ppm (flocculant), 0.05-1 ppm (coagulant aid), 0.01-0.1 ppm (filtration aid); *Regulatory:* UK approved; FDA 21CFR §173.315, 173.5; *Properties:* wh. gran.; bulk dens. 5.5-6.5 g/ml; visc. 220 cps (0.1%); pH 6.5-7.5 (0.1%).

Accofloc® A110 PWG. [Mitsui-Cyanamid] High m.w. polyacrylamide; CAS 9003-05-8; anionic; potable water grade flocculant for sugar prod. (juice clarification, remelt refinery liquor clarification, scum desweetening, waste treatment); *Usage level:* 0.2-2 ppm (flocculant), 0.05-1 ppm (coagulant aid), 0.01-0.1 ppm (filtration aid); *Regulatory:* UK approved; FDA 21CFR §173.315, 173.5; *Properties:* wh. gran.; bulk dens. 0.55-0.65 g/ml; visc. 260 cps (0.1%); pH 6.5-7.5 (0.1%).

Accofloc® A120 PWG. [Mitsui-Cyanamid] High m.w. polyacrylamide; CAS 9003-05-8; anionic; potable water grade flocculant for sugar prod. (juice clarification, remelt refinery liquor clarification, scum desweetening, waste treatment); *Usage level:* 0.2-2 ppm (flocculant), 0.05-1 ppm (coagulant aid), 0.01-0.1 ppm (filtration aid); *Regulatory:* UK approved; FDA 21CFR §173.315, 173.5; *Properties:* wh. gran.; bulk dens. 5.5-6.5 g/ml; visc. 280 cps (0.1%); pH 6.5-7.5 (0.1%).

Accofloc® A130 PWG. [Mitsui-Cyanamid] High m.w. polyacrylamide; CAS 9003-05-8; highly anionic; potable water grade flocculant for sugar prod. (juice clarification, remelt refinery liquor clarification, scum desweetening, waste treatment); *Usage level:* 0.2-2 ppm (flocculant), 0.05-1 ppm (coagulant aid), 0.01-0.1 ppm (filtration aid); *Regulatory:* UK approved; FDA 21CFR §173.315, 173.5; *Properties:* wh. gran.; bulk dens. 5.5-6.5 g/ml; visc. 300 cps (0.1%); pH 7.5-8.5 (0.1%).

Accofloc® N100 PWG. [Mitsui-Cyanamid] High m.w. polyacrylamide; CAS 9003-05-8; nonionic; potable water grade flocculant for sugar prod. (juice clarification, remelt refinery liquor clarification, scum desweetening, waste treatment); *Usage level:* 0.2-2 ppm (flocculant), 0.05-1 ppm (coagulant aid), 0.01-0.1 ppm (filtration aid); *Regulatory:* UK approved; FDA 21CFR §173.315, 173.5; *Properties:*

wh. gran.; bulk dens. 0.6-0.7 g/ml; visc. 70 cps (0.1%); pH 5.8-6.5 (0.1%).

Accoline. [Vaessen-Schoemaker] Mixt. of food grade phosphates with various degrees of polymerization; additive for prep. of all comminuted meat prods. subject to heat treatment, e.g., cooked sausages; avail. in visc., nonvisc., and instant sol. types; improves binding, color, texture; *Usage level:* 3-5 g/kg final prod.; *Regulatory:* FAO/WHO, EEC compliance.

Accoline-Mix. [Vaessen-Schoemaker] Mixt. of food grade phosphates with proteins, carbohydrates, antioxidants, spices, flavor enhancers, etc.; tailor-made additive for comminuted meat prods., e.g., cooked sausages; *Usage level:* 10-30 g/kg final prod.

Acconon 300-MO. [Karlshamns] PEG 300 oleate; CAS 9004-96-0; nonionic; emulsifier, lubricant, chemical intermediate; *Properties:* Gardner 6 liq.; sol. in org. solv.; sp.gr. 0.99; dens. 8.3 lb/gal; m.p. < -5 C; 99% act.; Discontinued

Acidan. [Grindsted Prods. Denmark] Monoglyceride citric acid ester; CAS 97593-31-2; anionic; emulsifier, surfactant; *Regulatory:* EEC compliance; *Properties:* powd., flakes; HLB 11.0; 100% act.

Acid Proof Caramel Powd. [MLG Enterprises Ltd.] Double strength acid proof-type caramel; CAS 8028-89-5; EINECS 232-435-9; high intensity natural colorant; *Properties:* free-flowing powd.; 100% min. thru 40 mesh, 90% min. thru 100 mesh; pH 3.0-4.0 (1%); 4% max. moisture; *Storage:* unlimited shelf life under normal storage conditions.

Aciplex®. [Asahi Chem. Industry] Ion exchange membrane; for food industry.

actif•8®. [Rhone-Poulenc Food Ingreds.] Sodium aluminum phosphate acidic with monocalcium phosphate anhyd.; leavening agent for baking, cereals, esp. for self-rising flours; provides good baking reaction, fine texture, exceptional tenderness, good color and flavor, neutral pH, and complete freedom from aftertaste; *Properties:* wh. free-flowing powd.; 1% max. on 60 mesh, 15% max. on 140 mesh; bulk dens. 50 lb/ft³; neutralizing value ≈ 95; 6.2% Ca content; *Storage:* 12 mo min. shelf life.

actif•8® Hi-Calcium. [Rhone-Poulenc Food Ingreds.] Sodium aluminum phosphate acidic with monocalcium phosphate anhyd.; leavening agent for mfg. of self-rising flour/corn meal mixes, flour, prepared mixes, other baking prods.; *Properties:* wh. free-flowing powd.; 1% max. on 60 mesh, 15% max. on 140 mesh; bulk dens. 50 lb/ft³; neutralizing value 90 min.; 11.5% Ca content.

Actiflo® 68 SB. [Central Soya] Soy lecithin; CAS 8002-43-5; EINECS 232-307-2; amphoteric; food/standard grade rich in phospholipids; w/o emulsifier esp. suited for margarines and chocolate confections; mixing/blending aid, visc. modifier; *Regulatory:* FDA §184.1400; kosher; *Properties:* lt. amber fluid; oil-sol.; visc. 15,000 cP max; HLB ≈ 2; acid no. 25 max; 0.8% max. moisture; *Storage:* store in closed container @ 16-32 C; 1 yr shelf life.

Actiflo® 68 UB. [Central Soya] Soy lecithin; CAS 8002-43-5; EINECS 232-307-2; amphoteric; food/standard grade rich in phospholipids; w/o emulsifier for margarines and chocolate confections; mixing/blending aid, visc. modifier; *Regulatory:* FDA §184.1400; kosher; *Properties:* amber liq.; oil-sol.; visc. 15,000 cP max.; HLB ≈ 2; acid no. 25 max.; 0.8% max. moisture; *Storage:* store in closed container @ 16-32 C; 1 yr shelf life.

Actiflo® 70 SB. [Central Soya] Soy lecithin; CAS 8002-43-5; EINECS 232-307-2; amphoteric; food/standard grade rich in phospholipids; w/o emulsifier esp. suited for margarines and chocolate confections; mixing/blending aid, visc. modifier; *Regulatory:* FDA §184.1400; kosher; *Properties:* lt. amber thick fluid; oil-sol.; visc. 30,000 cP max.; HLB ≈ 2; acid no. 26 max.; 0.8% max. moisture; *Storage:* store in closed container @ 16-32 C; 1 yr shelf life.

Actiflo® 70 UB. [Central Soya] Soy lecithin; CAS 8002-43-5; EINECS 232-307-2; amphoteric; food/standard grade; w/o emulsifier esp. suited for margarines and chocolate confections; mixing/blending aid, visc. modifier; *Properties:* amber fluid; visc. 18,000 cP; acid no. 23; 0.4% moisture.

Activera™. [Active Organics] Aloe vera gel; full range of prods. for topical and ingestible use; *Properties:* avail. in liq., powd., liq. concs., and oil forms, and in drink forms.

Activera™ 1-1FA (Filtered). [Active Organics] Aloe vera gel; as a vegetable drink and juice enhancer; *Properties:* colorless cloudy liq.

Activera™ 1-1 UA (Unfiltered). [Active Organics] Aloe vera gel; used primarily as a vegetable drink for the health food industry; mixable with vegetable and fruit juices; enhances flavor of juices; *Properties:* colorless cloudy liq., almost neutral taste.

Activera™ 1-200 A. [Active Organics] Aloe vera gel; used in veg. drinks; reconstitute by adding 199 parts of deionized water to 1 part powd. by wt; *Properties:* eggshell-wh. free-flowing powd.

Adapt. [MLG Enterprises Ltd.] Blend of potassium bromate, azodicarbonamide, and other edible excipients in tablet form; dough conditioner; oxidizes flour proteins from mixer to proof and baking stages, providing greater dough tolerance; improves quality of yeast-raised baked goods; *Usage level:* 1-2 tablets; *Properties:* tablet; *Storage:* store up to 6 mos under cool, dry conditions.

ADA Tablets. [ADM Arkady] Each tablet provides 20 ppm azodicarbonamide; CAS 123-77-3; EINECS 204-650-8; provides oxidation designed to increase loaf volume and improve crumb structure; *Properties:* tablets.

Adeka Menjust. [Asahi Denka Kogyo] Alcohol and propylene glycol; preserver/quality improver for preserving fresh noodles; provides better taste and flavor.

Adeka Propylene Glycol (P). [Asahi Denka Kogyo] Propylene glycol; CAS 57-55-6; EINECS 200-338-0; pharmaceutical grade for food additives; *Properties:* APHA > 10 color; sp.gr. 1.037-1.039.

ADM Baking Powd. [ADM Arkady] Double-acting baking powd. for use in cakes and other chemically-leavened systems; combination of fast and slower acting leavening components; provides uniform and sustained gas prod.

ADM Cream Acid Salt. [ADM Arkady] A blend of leavening acids that the bakers can combine with baking soda to produce a complete baking powd.

Admul 1405. [Quest Int'l.] Sat. polyglycerol ester; nonionic; emulsifier for margarines and shortenings; *Properties:* powd.; 100% conc.; Unverified

Admul CSL 2007, CSL 2008. [Quest Int'l.] Calcium stearoyl lactylate; CAS 5793-94-2; EINECS 227-335-7; anionic; bread improver, antistaling agent; *Properties:* flake and powd. resp.; 100% conc.

Admul Datem. [Quest Int'l.] Monoglyceride diacetylated tartaric acid esters; anionic; food emulsifier used as dough conditioners in baked prods.; *Properties:* paste, powd., microbead; HLB 8.0; 100% conc.

Admul Emulsponge. [Quest Int'l.] Monoglycerides blend; nonionic; food emulsifier, aerating agent for sponge cakes, ice cream mixes; *Properties:* powd.; HLB 4.0; 100% conc.

Admul GLP. [Quest Int'l.] Lactic acid ester of monoglycerides; nonionic; food emulsifier for shortenings, toppings, and desserts; *Properties:* flake. powd.; 100% conc.

Admul GLS. [Quest Int'l.] Monoglyceride lactic acid ester; nonionic; food emulsifier for shortenings, toppings, and desserts; *Properties:* flake; HLB 3.5; 100% conc.; Unverified

Admul MG 4103. [Quest Int'l.] Monodiglycerides; nonionic; food emulsifier used in bakery prods., margarines, shortenings, creams, desserts, ice cream; *Properties:* microbead; HLB 4.0; 100% conc.

Admul MG 4123. [Quest Int'l.] Monodiglycerides; nonionic; food emulsifier used in bakery prods., margarines, shortenings, creams, desserts, ice cream; *Properties:* microbead; HLB 4.0; 100% conc.

Admul MG 4143. [Quest Int'l.] Monodiglycerides; nonionic; food emulsifier used in bakery prods., margarines, shortenings, creams, desserts, ice cream; *Properties:* microbead; HLB 4.0; 100% conc.

Admul MG 4163. [Quest Int'l.] Monoglycerides; nonionic; food emulsifier used in bakery prods., margarines, shortenings, creams, desserts, ice cream; *Properties:* microbead; HLB 4.0; 100% conc.

Admul MG 4203. [Quest Int'l.] Monoglycerides; nonionic; food emulsifier used in bakery prods., margarines, shortenings, creams, desserts, ice cream; *Properties:* microbead; HLB 4.0; 100% conc.

Admul MG 4223. [Quest Int'l.] Monoglycerides; nonionic; food emulsifier used in bakery prods., margarines, shortenings, creams, desserts, ice cream; *Properties:* microbead; HLB 4.0; 100% conc.

Admul MG 4304. [Quest Int'l.] Monoglycerides; nonionic; food emulsifier used in bakery prods., margarines, shortenings, creams, desserts, ice cream; *Properties:* paste; HLB 4.0; 100% conc.

Admul MG 4404. [Quest Int'l.] Monoglycerides; non-

ionic; food emulsifier used in bakery prods., margarines, shortenings, creams, desserts, ice cream; *Properties:* microbead; HLB 4.0; 100% conc.

Admul MG 4904. [Quest Int'l.] Monoglycerides; nonionic; food emulsifier used in bakery prods., margarines, shortenings, creams, desserts, ice cream; *Properties:* paste; HLB 4.0; 100% conc.

Admul MG 6103. [Quest Int'l.] Monodiglycerides; nonionic; food emulsifier used in bakery prods., margarines, shortenings, creams, desserts, ice cream; *Properties:* microbead; HLB 4.0; 100% conc.

Admul MG 6404. [Quest Int'l.] Monoglycerides; nonionic; food emulsifier used in bakery prods., margarines, shortenings, creams, desserts, ice cream; *Properties:* paste; HLB 4.0; 100% conc.

Admul MG 6504. [Quest Int'l.] Monoglycerides; nonionic; food emulsifier used in bakery prods., margarines, shortenings, creams, desserts, ice cream; *Properties:* paste; HLB 4.0; 100% conc.

Admul PGE 1405. [Quest Int'l.] Polyglyceryl ester; nonionic; food emulsifier with strong w/o props. and good aerating props.; *Properties:* microbead; HLB 5.0-6.0; 100% conc.

Admul PGE 1411. [Quest Int'l.] Polyglyceryl ester; nonionic; food emulsifier with strong w/o props. and good aerating props.; used in syn. creams; *Properties:* paste; HLB 5.0-6.0; 100% conc.

Admul PGMP. [Quest Int'l.] Propylene glycol palmitate; nonionic; food emulsifier for aeration of high fat content toppings, desserts, cake improvers; *Properties:* solid; HLB 2.0; 100% conc.

Admul PGMS. [Quest Int'l.] Propylene glycol stearate; CAS 1323-39-3, EINECS 215-354-3; nonionic; food emulsifier for aeration of high fat content toppings, desserts, cake improvers; *Properties:* solid; HLB 2.0; 100% conc.

Admul SSL 2003. [Quest Int'l.] Sodium stearoyl lactylate; CAS 25383-99-7; EINECS 246-929-7; anionic; o/w emulsifier; bread improver, antistaling agent; *Properties:* flake; 100% conc.

Admul SSL 2004. [Quest Int'l.] Sodium stearoyl lactylate; CAS 25383-99-7; EINECS 246-929-7; anionic; food emulsifier for starch complexing agent and dough conditioning in bread prods.; antistaling agent; *Properties:* powd.; 100% conc.

Admul WOL 1403. [Quest Int'l.] Polyglycerol polyricinoleate; nonionic; visc. modifier in chocolate; pan release agent; *Properties:* liq.; 100% conc.

Admul WOL 1405. [Quest Int'l.] Sat. polyglycerol ester; nonionic; food emulsifier for margarines and shortenings; *Properties:* microbead; HLB 5-6; 100% conc.

ADP. [Calgon Carbon] Activated carbon; CAS 64365-11-3; used for food decolorization, pharmaceutical purification; *Properties:* pulverized, 80 x 325 sieve size.

Advitagel. [Quest Int'l.] Dist. monoglyceride/emulsifier blend; nonionic; aerating agent for sponge cakes; *Properties:* paste; HLB 4.0.

Aerate Cake Emulsifier. [ADM Arkady] Hydrated blend of three highly effective emulsifiers; hydrated emulsifier system providing uniform aeration and low sp. gr. to cake batter.

AeroLite LP. [Albright & Wilson Australia] Sodium aluminum phosphate, acidic; CAS 7785-88-8; slow reaction leavening agent, primarily for baking stage; suitable for cakes, pancakes, flour premixes, frozen or refrigerated doughs; *Regulatory:* FCC, Australian compliance; *Properties:* wh. free-flowing powd.; m.w. 949.88; *Toxicology:* avoid inhaling dust, prolonged skin contact; *Storage:* store in cool, dry place to avoid caking.

Aeromix Baking Powd. [Albright & Wilson Australia] Specially formulated baking powd. based on phosphate aerators; fully-formulated baking powd. with slow dough rate of reaction for use by commercial pastry chefs; *Regulatory:* FCC, Australian compliance; *Properties:* wh. free-flowing powd.; *Toxicology:* avoid inhaling dust, prolonged skin contact; *Storage:* store in cool, dry place; close containers securely after use to prevent moisture contact.

Aerophos G. [Albright & Wilson Australia] Blend of food grade phosphate aerators based on sodium acid pyrophosphate; aerator esp. for doughnuts where med. dough rate of reaction is required; *Regulatory:* FCC, Australian compliance; *Properties:* wh. free-flowing powd.; 0.2% +250 μm particle size, 1% +150 μm; 94% acidity, 35% orthophosphate; *Toxicology:* avoid inhaling dust, prolonged skin contact; *Storage:* store in cool, dry place.

Aerophos M. [Albright & Wilson Australia] Blend of food grade phosphate aerators based on sodium acid pyrophosphate; double-acting formulated chemical aerator for self-raising flour and prepared mixes; *Regulatory:* FCC, Australian compliance; *Properties:* wh. free-flowing powd.; 0.2% +250 μm particle size, 1% +150 μm; 97% acidity, 20% orthophosphate; *Toxicology:* avoid inhaling dust, prolonged skin contact; *Storage:* store in cool, dry place.

Aerophos P. [Albright & Wilson Australia] Sodium acid pyrophosphate; CAS 7758-16-9; EINECS 231-835-0; chemical aerator used when a slow reaction is required, e.g., for baking powd., self-raising flour, and prepared mixes; *Properties:* wh. free-flowing powd.; 0.2% +250 μm particle size, 1% +150 μm; m.w. 221.94; pH 4.1 (1%); 98% acidity, 4% orthophosphate; *Toxicology:* avoid inhaling dust, prolonged skin contact; *Storage:* store in cool, dry place.

Aerophos X. [Albright & Wilson Australia] Calcium tetrahydrogen diorthophosphate monohydrate (4 parts), calcium tetrahydrogen diorthophosphate anhyd. (3 parts), calcium hydrogen orthophosphate anhyd. (1 part); chemical aerator, esp. for crumpets and biscuits, where very fast dough rate of reaction is required; nutrient; *Regulatory:* FCC, Australian compliance; *Properties:* wh. free-flowing powd.; 0.2% +250 μm particle size, 1% +150 μm; 82% $Ca(H_2PO_4)_2$; *Toxicology:* avoid inhaling dust, prolonged skin contact; *Storage:* store in cool, dry place.

Aerosil® 200. [Degussa; Degussa AG] Fumed silica; CAS 7631-86-9; anticaking and free-flow agent with high absorp. capacity; *Regulatory:* FDA 21CFR §133.146(b), 160.105(a)(d), 172.230(a), 172.480, 173.340(a), 175, 176, 177, 573.940; ≤3% for cosmetics, internal pharmaceuticals; *Properties:* fluffy wh. powd.; 12 nm avg. particle size; sp.gr. 2.2; dens. ≈ 120 g/l (densed); surf. area 200 ± 25 m^2/g; pH 3.6-4.3 (4% aq. susp.); > 99.8% assay; *Toxicology:* TLV 10 mg/m^3 total dust; LD50 > 20,000 mg/kg; may cause eye, skin, or respiratory tract irritation on overexposure; *Precaution:* incompat. with strong bases and hydrofluoric acid.

Aerosil® 380. [Degussa; Degussa AG] Fumed silica; CAS 7631-86-9; *Regulatory:* FDA 21CFR §133.146(b), 160.105(a)(d), 172.230(a), 172.480, 173.340(a), 175, 176, 177, 573.940; *Properties:* fluffy wh. powd.; 7 nm avg. particle size; sp.gr. 2.2; dens. ≈ 120 g/l (densed); surf. area 380 ± 30 m^2/g; pH 3.6-4.3 (4% aq. susp.); > 99.8% assay; *Toxicology:* TLV 10 mg/m^3 total dust; LD50 > 20,000 mg/kg; may cause eye, skin, or respiratory tract irritation on overexposure; *Precaution:* incompat. with strong bases and hydrofluoric acid.

AF 10 FG. [Harcros] Polydimethylsiloxane; nonionic; antifoam agent used for general food, poultry, and meat processing applics.; *Regulatory:* FDA §173.340, 180.1001c,d; kosher; *Properties:* wh. liq.; disp. in water; sp.gr. 1.00; dens. 8.3 lb/gal; flash pt. > 212 F (PMCC); pH 4-5 (1% aq.); 10% act.

AF 30 FG. [Harcros] Polydimethylsiloxane; nonionic; antifoam for general food, meat, and poultry processing; *Regulatory:* FDA §173.340, 180.1001c,d; kosher; *Properties:* wh. liq.; water-disp.; sp.gr. 1.01; dens. 8.4 lb/gal; flash pt. (PMCC) > 212 F; pH 4-5 (1% aq.); 30% act.

AF 70. [GE Silicones] Silicone compd.; defoamer in fermentation, corn oil mfg., deep-fat frying, esterification of veg. oil, yeast processing; direct and indirect food additive; *Regulatory:* kosher; *Properties:* liq.; sol. in aliphatic, aromatic, and chlorinated hydrocarbons; sp.gr. 1.01; dens. 8.4 lb/gal; visc. 1500 cps; flash pt. (OC) 315 C; 100% act.

AF 72. [GE Silicones] PEG-40 stearate, sorbitan stearate, and silica; nonionic; food-grade antifoam agent, surfactant for fermentation, brine systems, chewing gum base, fruit processing, instant coffee, cheese whey, jelly, pickle, potato, rice, sauce, soft drink, veg., yeast, and wine processing, sugar refining, syrups; *Regulatory:* kosher; *Properties:* wh. fluid; sl. sorbic acid odor; sol. in water; sp.gr. 1.01; dens. 8.4 lb/gal; visc. 1000 cps; 30% silicone; 44.2% solids.

AF 75. [GE Silicones] PEG-40 stearate, sorbitan stearate, and silica; nonionic; food-grade antifoam emulsion, surfactant for fermentation, brine systems, chewing gum base, fruit processing, instant coffee, cheese whey, jelly, pickle, potato, rice, sauce, soft drink, veg., yeast, and wine processing, sugar refining, syrups; *Regulatory:* kosher; *Properties:* wh. emulsion, sl. sorbic acid odor; water-sol.; sp.gr. 1.02; dens. 8.4 lb/gal; visc. 3000 cps; 10% silicone; 13.75% solids.

AF 100 FG. [Harcros] Silicone antifoam; antifoam for food, edible oils, meat and poultry processing; *Regulatory:* FDA §173.340, 180.1001c,d; kosher; *Properties:* translucent liq.; sol. in aliphatic, aromatic, and chlorinated solvs.; insol. in water; sp.gr. 1.01; dens. 8.4 lb/gal; pour pt. < 0 F; flash pt. (PMCC) > 600 F; 100% conc.

AF 8805 FG. [Harcros] Silicone antifoam; nonionic; antifoam for general food, poultry, and meat processing, agric.; *Regulatory:* FDA §173.340, 180.1001c,d; kosher; *Properties:* wh. liq.; water-disp.; sp.gr. 1.00; dens. 8.3 lb/gal; flash pt. (PMCC) > 212 F; pH 4-5 (1% aq.); 5% act.

AF 8810 FG. [Harcros] Silicone antifoam; nonionic; antifoam for general food, poultry and meat processing; *Regulatory:* FDA §173.340, 180.1001c,d; kosher; *Properties:* wh. liq.; sp.gr. 1.00; dens. 8.3 lb/gal; flash pt. (PMCC) > 212 F; pH 4-5 (1% aq.); 10% act.

AF 8820 FG. [Harcros] Silicone antifoam; nonionic; antifoam for general food, meat, and poultry processing; *Regulatory:* FDA §173.340, 180.1001c,d; kosher; *Properties:* wh. liq.; water-disp.; sp.gr. 1.00; dens. 8.3 lb/gal; flash pt. (PMCC) > 212 F; pH 4-5 (1% aq.); 20% act.

AF 8830 FG. [Harcros] Silicone antifoam; nonionic; antifoam for general food, meat and poultry processing; *Regulatory:* FDA §173.340, 180.1001c,d; kosher; *Properties:* wh. liq.; water-disp.; sp.gr. 1.00; dens. 8.3 lb/gal; flash pt. (PMCC) > 212 F; pH 4-5 (1% aq.); 30% act.

AF 9000. [GE Silicones] Silicone compd.; defoamer in many nonaq. direct and indirect food additives (corn oil mfg., deep-fat frying, esterification of veg. oils); *Regulatory:* kosher; *Properties:* sol. in aliphatic, aromatic, and chlorinated solvs.; sp.gr. 1.01; dens. 8.4 lb/gal; visc. 2500 cps; flash pt. (OC) 315 C; 100% silicone content.

AF 9020. [GE Silicones] Dimethicone aq. emulsion; nonionic; defoamer for food-processing systems (fermentation, brine systems, wine, yeast, etc.); *Regulatory:* kosher; *Properties:* wh. emulsion; disp. in warm or cold water with mild agitation; sp.gr. 1.01; dens. 8.4 lb/gal; visc. 3500 cps; fermentation, brine systems, chewing gum base, fruit processing, instant coffee, cheese whey, jelly, pickle, potato, rice, sauce, soft drink, veg., yeast, and wine processing, sugar refining, syrups; 20% silicone, 28.75% solids.

AF HL-36. [Harcros] Nonsilicone antifoam; for fermentation, processing beet sugar and yeast, distillation; *Regulatory:* FDA §173.340(a)(3); kosher; *Properties:* clear liq.; water-disp.; sp.gr. 1.00; dens. 8.3 lb/gal; flash pt. (PMCC) > 300 F; 100% act.

AFP 2000. [Solvay Enzymes] Acid fungal protease obtained by controlled fermentation of *Aspergillus niger* var.; CAS 9014-01-1; EINECS 232-752-2; enzyme for hydrolysis of proteins incl. casein, hemoglobin, gelatin, soya, fish, and other plant and animal proteins under acid conditions; prevents haze in fruit juice; used for fermentation, fish and soya processing, and for protein modification; *Usage level:* 0.01-0.1% on wt. of substrate; *Properties:* lt. tan to wh. powd., free of offensive odor; readily water-sol.; *Storage:* store in sealed containers under cool, dry conditions; store at 5 C for extended storage life.

Agar Agar NF Flake #1. [Meer] Agar agar; CAS 9002-18-0; EINECS 232-658-1; emulsifier, stabilizer, gellant for foods, chiffon pies, meringues, pie fillings, icings, toppings, cookies, cream cheeses, yogurt, chocolate milk drinks, sherbets, jelly candies, sweets, poultry/meat canning, jams, pet foods; antistaling for breads; *Properties:* water-sol.

Ajax®. [Rhone-Poulenc Food Ingreds.] Monocalcium phosphate monohydrate FCC; CAS 10031-30-8; EINECS 231-837-1; leavening agent for baking powd., self-rising flour, pancake flour, prepared mixes; ingred. in bread improvers, mfg. of cookies, crackers; acidulant in bread; acidulant and buffer in dry beverage mixes; *Properties:* brilliant wh. free-flowing cryst.; 93% thru 100 mesh, 12% thru 200 mesh; m.w. 252.2; bulk dens. 71 lb/ft³; neutralizing value 80; pH 3.7 (1%); 15.9-17% assay (Ca).

Akocote 102. [Karlshamns Food Ingreds.] Partially hydrog. soybean and cottonseed oils; standard nonlauric hard butter for use in confectionery and bakery coatings, drops, and candy centers; *Properties:* solid; m.p. 97-101 F; 0.05% moisture.

Akocote 106. [Karlshamns Food Ingreds.] Partially hydrog. soybean and cottonseed oils with mono- and diglycerides; standard nonlauric hard butter for use in confectionery and bakery coatings, drops, and candy centers; *Properties:* solid; m.p. 104-107 F; formerly Capkote 3-6.

Akocote 109. [Karlshamns Food Ingreds.] Partially hydrog. soybean and cottonseed oils; nonlauric fat for confectioner's and bakery coatings, drops, candy centers; *Properties:* liq., flake; drop pt. 109 F.

Akocote 112. [Karlshamns] Partially hydrog. soybean and cottonseed oils; standard nonlauric hard butter for food applics.; *Properties:* m.p. 110-114 F.

Akodel 95. [Karlshamns Food Ingreds.] Partially hydrog. vegetable oil (palm kernel, coconut); lauric fat for confectioner's coatings, drops, and centers; *Properties:* liq.; drop pt. 95 F.

Akodel 102. [Karlshamns Food Ingreds.] Partially hydrog. vegetable oil (palm kernel, coconut, palm); lauric fat for confectioner's coatings, drops, and centers, vegetable dairy, icings; *Properties:* liq., solid, flake; drop pt. 102 F.

Akodel 108. [Karlshamns Food Ingreds.] Partially hydrog. vegetable oil (palm kernel, coconut, palm); lauric fat for confectioner's coatings, drops, and centers; *Properties:* liq., flake; drop pt. 108 F.

Akodel 112. [Karlshamns Food Ingreds.] Partially hydrog. vegetable oil (palm kernel, coconut, palm); lauric fat for confectioner's coatings, centers, icing stabilizer, vegetable dairy prods.; *Properties:* liq., flake; drop pt. 112 F.

Akodel 118. [Karlshamns Food Ingreds.] Partially hydrog. vegetable oil (palm kernel, coconut, palm); lauric fat for confectioner's coatings, centers, icing stabilizer, vegetable dairy prods.; *Properties:* liq., flake; drop pt. 118 F.

Akofame. [Karlshamns Food Ingreds.] Partially hydrog. soybean oil; CAS 8016-70-4; EINECS 232-410-2; high stability oil used for frying, color/flavor carrier, antidusting applics.; for export trade; *Properties:* liq.

Akofil A. [Karlshamns Food Ingreds.] Partially hydrog. soybean oil; EINECS 232-410-2; specialty shortening for bakery filling, sandwich cookie fillings; *Properties:* liq., plastic; m.p. 96 F.

Akofil N. [Karlshamns Food Ingreds.] Partially hydrog. soybean oil; EINECS 232-410-2; specialty shortening for bakery filling, sandwich cookie fillings; *Properties:* liq., plastic; m.p. 99 F.

Akoleno D. [Karlshamns Food Ingreds.] Partially hydrog. cottonseed oil; CAS 68334-00-9; EINECS 269-804-9; nonlauric fat for candy centers and caramel; *Properties:* solid; m.p. 101-105 F; Discontinued

Akoleno S. [Karlshamns Food Ingreds.] Partially hydrog. soybean oil; EINECS 232-410-2; standard nonlauric filling fat for candy centers and caramel; *Properties:* solid; m.p. 97-101 F; iodine no. 80 max.; formerly Caplite S.

Akoleno SC. [Karlshamns Food Ingreds.] Partially hydrog. soybean and cottonseed oils; standard nonlauric filling fat for candy centers and caramel; *Properties:* solid; m.p. 94-97 F; formerly Caplite 3-4.

Akolizer C. [Karlshamns Food Ingreds.] Hydrog. cottonseed oil; CAS 68334-00-9; EINECS 269-804-9; crystallization enhancer, m.p. modifier; *Properties:* liq., flake, bead; drop pt. 142 F; iodine no. 5.

Akolizer P. [Karlshamns Food Ingreds.] Hydrog. palm oil; crystallization enhancer, m.p. modifier; *Properties:* liq., flake; drop pt. 140 F; iodine no. 2.5.

Akolizer PKC. [Karlshamns Food Ingreds.] Partially hydrog. vegetable oil (palm kernel, cottonseed) with lecithin; crystallization enhancer in confectioner's coatings, candy centers; *Properties:* liq., flake; drop pt. 115 F; iodine no. 5.

Akolizer RSC. [Karlshamns Food Ingreds.] Partially hydrog. vegetable oil (rapeseed, cottonseed, and soybean); crystallization promoter, m.p. modifier; *Properties:* liq.; drop pt. 145 F; iodine no. 5.

Akolizer S. [Karlshamns Food Ingreds.] Hydrog. soybean oil; EINECS 232-410-2; crystallization promoter, m.p. modifier; *Properties:* liq., flake, bead; drop pt. 155 F; iodine no. 5.

Akolizer SC. [Karlshamns Food Ingreds.] Partially hydrog. vegetable oil (soybean and cottonseed); artificial color and flavor variety avail.; m.p. modifier; *Properties:* liq., flake, bead; drop pt. 125 F; iodine no. 50.

Akomax E. [Karlshamns Food Ingreds.] Shea and palm oils; nonlauric specialty fat as tempering fat, for use in coatings, drops, and centers; compat. with cocoa butter; *Properties:* liq., solid; m.p. 36 C; iodine no. 33-35.

Akomax R. [Karlshamns Food Ingreds.] Shea, palm, and illipe oils; nonlauric specialty fat as tempering fat, for use in coatings, drops, and centers; compat. with cocoa butter; *Properties:* liq., solid; m.p. 34 C; iodine no. 35-36.5.

Akopol E-1. [Karlshamns Food Ingreds.] Partially hydrog. soybean oil with sorbitan tristearate; fractionated nonlauric hard butter for food applics., nontempering confectioner's coatings, vegetable dairy prods.; *Properties:* liq., solid; drop pt. 95-99 F.

Akopol R. [Karlshamns Food Ingreds.] Partially hydrog. soybean and cottonseed oils; fractionated nonlauric hard butter for use in nontempering confectioner's coatings, vegetable dairy; *Properties:* Lovibond 3.0 R max. liq./solid/flake; m.p. 97-101 F.

Akopuff. [Karlshamns Food Ingreds.] Partially hydrog. vegetable oil (soybean, cottonseed), water, salt, citric acid, lecithin, artificial flavor; margarine/puff paste for layer doughs (i.e., French and puff pastry); *Properties:* plastic; m.p. 124 F.

Akorex. [Karlshamns Food Ingreds.] Partially hydrog. soybean oil; EINECS 232-410-2; for applics. requiring high stability oil; used as candy centers, color/flavor carriers, in frying, lubricants, spray coatings, vegetable, dairy, and antidusting applics.; *Properties:* liq.; drop pt. 20-23 C; iodine no. 84-87.

Akorex B. [Karlshamns Food Ingreds.] Partially hydrog. soybean and cottonseed oils; high stability oil for antidusting, candy centers, color/flavor carrier, frying, lubricant, spray coating, vegetable, and dairy applics.; *Properties:* liq.; drop pt. 19-22 C.

Akorex C. [Karlshamns] Partially hydrog. canola oils; used as spray oil, in bakery prods., nutritional snacks; *Properties:* iodine no. 84.

Akorine 2A. [Karlshamns] Partially hydrog. palm kernel and cottonseed oils; fractionated lauric hard butter for confectionery and pastel coatings and centers; *Properties:* m.p. 92-95 F; iodine no. 5-8; formerly SP-2A.

Akorine 3. [Karlshamns] Partially hydrog. palm kernel and soybean oils; fractionated lauric hard butter for confectionery and pastel coatings and centers; *Properties:* m.p. 96-99 F; iodine no. 6 max.; formerly SP-3; Discontinued

Akorine 4. [Karlshamns] Partially hydrog. palm kernel oil; CAS 68990-82-9; fractionated lauric hard butter for confectionery and pastel coatings and centers; *Properties:* m.p. 94-97 F; iodine no. 3 max.; Discontinued

Akorine 9F. [Karlshamns] Partially hydrog. palm kernel and cottonseed oils with sorbitan tristearate; fractionated lauric hard butter for use in confectionery and pastel coatings, drops, and centers; *Properties:* drop pt. 97 F; iodine no. 8 max.; Discontinued

Akowesco 1. [Karlshamns Food Ingreds.] Palm kernel oil; CAS 8023-79-8; EINECS 232-425-4; fractionated lauric hard butter for confectionery and pastel coatings and centers; *Properties:* solid; m.p. 92-96 F; iodine no. 6-8; formerly SP-1 Plus.

Akowesco 2. [Karlshamns Food Ingreds.] Partially hydrog. palm kernel oil; CAS 68990-82-9; fractionated lauric hard butter for confectionery and pastel coatings and centers; *Properties:* solid; m.p. 92-95

F; iodine no. 4 max.; formerly SP-2.

Akowesco 5. [Karlshamns Food Ingreds.] Partially hydrog. palm kernel oil; CAS 68990-82-9; fractionated lauric hard butter for confectionery and pastel coatings and centers; *Properties:* solid; m.p. 94-97 F; iodine no. 3 max.; formerly SP-5.

Akowesco 7. [Karlshamns Food Ingreds.] Partially hydrog. palm kernel oil; CAS 68990-82-9; fractionated lauric fat for peanut-flavored drops, coatings, and centers; *Properties:* solid; drop pt. 98 F.

Akowesco 45 AC. [Karlshamns Food Ingreds.] Partially hydrog. palm kernel oil; CAS 68990-82-9; fractionated lauric fat for confectioner's and pastel coatings, drops, and centers; *Properties:* solid; drop pt. 96 F.

Akowesco 90. [Karlshamns Food Ingreds.] Partially hydrog. palm kernel oil with lecithin; fractionated lauric fat for confectionery centers; *Properties:* liq., solid; drop pt. 88 F.

Akowesco Plus Series. [Karlshamns] Palm stearine added at custom levels.

Akwilox 133. [Am. Chem. Services] Brominated soybean oil; CAS 68952-98-7; food additive in soft drinks for visc. adjustment; *Properties:* lt. amber liq.; bland odor, taste; sp.gr. 1.33; 100% act.

Albriphos™ Blend 75-25. [Albright & Wilson Am.] Blend of food grade sodium tripolyphosphate and sodium hexametaphosphate; used in curing pickles to stabilize the loss of natural fluids in cured meat prods.; *Properties:* gran.; 2% max. on 20 mesh, 60% min. on 100 mesh; pH 8.7-9.2 (1%); 59-61% P_2O_5; *Toxicology:* nonhazardous, nontoxic.

Albriphos™ Blend 90-10. [Albright & Wilson Am.] Blend of food grade sodium tripolyphosphate and sodium hexametaphosphate; used in curing pickles to stabilize the loss of natural fluids in cured meat prods. and during cooking, thawing, and reheating of cooked poultry prods.; *Properties:* gran.; 2% max. on 20 mesh, 65% min. on 100 mesh; pH 9.0-9.8 (1%); 57-59% P_2O_5; *Toxicology:* nonhazardous, nontoxic.

Albriphos™ Blend 928. [Albright & Wilson Am.] Blend of food grade sodium tripolyphosphate, sodium hexametaphosphate, and sodium acid pyrophosphate; sequestrant for calcium, magnesium, and iron in food prods.; used in curing pickles to stabilize the loss of natural fluids in cured meat; *Properties:* gran.; 2% max. on 14 mesh, 55% min. on 100 mesh; pH 8.2-8.6 (1%); 58% min. P_2O_5; *Toxicology:* nonhazardous, nontoxic; *Precaution:* do not handle curing pickles contg. Blend 928 in iron or galvanized equip. or piping.

Albrite Diammonium Phosphate Food Grade. [Albright & Wilson UK] Diammonium phosphate; CAS 7783-28-0; EINECS 231-987-8; buffering agent in food; yeast nutrient; *Regulatory:* FCC, UK compliance; *Properties:* wh. crystals; sol. 69 g/100 g water @ 20 C; m.w. 132.06; bulk dens. 1.0 g/ml (loose); pH 8.0 (10 g/l sol'n.); 98% min. assay.

Albrite Dicalcium Phosphate Anhyd. [Albright & Wilson UK] Dicalcium phosphate food grade; CAS 7757-93-9; EINECS 231-826-1; source of calcium

and phosphorus in foods for infants, invalids, and geriatric patients and in animal feeds; *Regulatory:* EEC, UK compliance; *Properties:* wh. impalpable powd., odorless, tasteless; m.w. 136.06; 39-42% calcium as CaO; *Storage:* protect packages from water and contamination.

Albrite Monoammonium Phosphate Food Grade. [Albright & Wilson UK] Ammonium phosphate; CAS 7722-76-1; EINECS 231-764-5; buffer, dough conditioner, and yeast nutrient in food preps.; yeast nutrient in wine-making; *Regulatory:* FCC, UK compliance; *Properties:* wh. crystals; sol. 37 g/100 g water @ 20 C; m.w. 115.03; bulk dens. 1.05 g/ml (loose); pH 4.6 (10 g/l sol'n.); 98% min. assay.

Albrite MSP Food Grade. [Albright & Wilson Australia] Monosodium phosphate food grade; CAS 7758-80-7; EINECS 231-449-2; acidic buffer for foodstuffs; processed cheese emulsifier; mfg. of starch phosphates; *Regulatory:* FCC, Australian compliance; *Properties:* wh. free-flowing powd.; 0.2% +250 μm particle size, 1% +150 μm; m.w. 119.98; pH 4.4 (1%); 99% act.; *Toxicology:* avoid inhaling dust, prolonged contact with skin; *Storage:* store in cool, dry place to avoid caking.

Albrite Phosphoric Acid 85% Food Grade. [Albright & Wilson Australia] Phosphoric acid; CAS 7664-38-2; EINECS 231-633-2; acidulant for foods and beverages; *Regulatory:* FCC, Australian compliance; *Properties:* clear sl. visc. liq.; m.w. 98.00; dens. 1.69 g/ml; f.p. 23 C; 85.2% act.; *Toxicology:* corrosive; avoid contact with skin, eyes, clothing; *Storage:* store in warm area; avoid freezing.

Albrite SAPP Food Grade. [Albright & Wilson Australia] Sodium acid pyrophosphate; CAS 7758-16-9; EINECS 231-835-0; for nonleavening applics., e.g., potato processing, general purpose buffering and acidifying agent, meat processing, cheese emulsifying salt; *Properties:* wh. free-flowing powd.; 0.2% +250 μm particle size, 1% +150 μm; m.w. 221.94; pH 4.1 (1%); 98% acidity, 5% orthophosphate; *Toxicology:* avoid inhaling dust, prolonged skin contact; *Storage:* store in cool, dry place.

Albrite STPP-F. [Albright & Wilson Australia] Sodium tripolyphosphate, food grade; CAS 7758-29-4; EINECS 231-694-5; buffer, texturizer for processed meat, fish, and cheese; dispersant for fats; color retention aid; *Regulatory:* FCC, Australian compliance; *Properties:* wh. free-flowing powd.; 3% +210 μm particle size, 39% +75 μm; m.w. 367.86; dens. 1.0 g/ml; pH 9.7 (1%); 94% act.; *Toxicology:* avoid inhaling dust, prolonged contact with skin; *Storage:* store in cool, dry place to avoid caking.

Albrite STPP-FC. [Albright & Wilson Australia] Sodium tripolyphosphate, food grade; CAS 7758-29-4; EINECS 231-694-5; buffer, texturizer for processed meat, fish, and cheese; dispersant for fats; color retention aid; *Regulatory:* FCC, Australian compliance; *Properties:* wh. coarse free-flowing powd.; 1% +500 μm particle size, 65% +75 μm; m.w. 367.86; dens. 0.55 g/ml; pH 9.7 (1%); 92%

act.; *Toxicology:* avoid inhaling dust, prolonged contact with skin; *Storage:* store in cool, dry place to avoid caking.

Albrite TSPP Food Grade. [Albright & Wilson Australia] Tetrasodium pyrophosphate food grade; CAS 7722-88-5; EINECS 231-767-1; buffer, texturizer for processed meat, fish, and cheese; sequestrant and stabilizer in dairy foods; *Regulatory:* FCC, Australian compliance; *Properties:* wh. free-flowing powd.; 1% +210 μm particle size, 39% +75 μm; m.w. 265.95; dens. 0.7 g/ml; pH 10.1 (1%); 97% act.; *Toxicology:* avoid inhaling dust, prolonged contact with skin; *Storage:* store in cool, dry place to avoid caking.

Alcolec® 30-A. [Am. Lecithin] Lecithin with corn flour carrier; emulsifier and conditioner for baked goods, yeast buns, pie crusts, cakes, cookies; facilitates dough handling, contributes to longer shelf life, improves blending and mixing, reduces sticking; *Usage level:* 1-2%; *Regulatory:* kosher approved.

Alcolec® 140. [Am. Lecithin] Refined soybean lecithin containing high level of phosphatidylcholine; CAS 8002-43-5; EINECS 232-307-2; o/w emulsifier; produces liposomes as fat extender/substitute or for encapsulation of ingreds. such as yeast; for cakes, bread, frozen dough, cake mixes, instant milk, cocoa powd., soup and sauce mixes, egg replacer, chocolate, emulsions; *Regulatory:* kosher approved; *Storage:* store in cool, dry place.

Alcolec® 495. [Am. Lecithin] Lecithin; CAS 8002-43-5; EINECS 232-307-2; nonionic; o/w emulsifier, wetting agent for aq. and oil-base systems; approved for food use; enhanced water dispersibility; *Regulatory:* FDA 21CFR §184.1400, GRAS; kosher approved; *Properties:* amber med. visc. fluid; visc. < 10,000 cP; HLB 5-6; acid no. < 25; 100% conc.; *Storage:* 1 yr shelf life in original, unopened containers.

Alcolec® BS. [Am. Lecithin] Single bleached lecithin FCC; CAS 8002-43-5; EINECS 232-307-2; nonionic; w/o emulsifier, wetting and dispersing agent, stabilizer, release and lubricating agent, foam suppressant, solubilizer, and emollient; choline source; *Regulatory:* FDA 21CFR §184.1400, GRAS; kosher approved; *Properties:* Gardner 14 max. liq.; acid no. 32 max.; 1% max. moisture; *Storage:* 1 yr shelf life in original unopened container.

Alcolec® C-150. [Am. Lecithin] Lecithin and cocoa butter (40%); controls visc. and yield value for chocolate mfg.; reduces quantity of cocoa butter required; *Regulatory:* FDA compliance; *Properties:* plastic; 50% phosphatidylcholine, 0.4% moisture; *Storage:* store in cool, dry place below 25 C.

Alcolec® Extra A. [Am. Lecithin] Lecithin; CAS 8002-43-5; EINECS 232-307-2; nonionic; emulsifier for aq. and oil-base systems, food industry; higher phosphatide content; *Properties:* liq.; 100% conc.

Alcolec® EYR SM. [Am. Lecithin] Maltodextrin (70%), soybean lecithin (30%) high in phosphatidylcholine (20%); functional replacement for egg yolk in baking applics (cookies, muffins, nonflowable batters);

contains no cholesterol; *Regulatory:* kosher approved; *Properties:* wh. to pale yel. free-flowing powd., neutral odor, bland taste; disp. in water; 3% moisture; *Storage:* store under cool, dry conditions.

Alcolec® F-100. [Am. Lecithin] De-oiled lecithin FCC; CAS 8002-43-5; EINECS 232-307-2; food emulsifier; instantizing agent for milk powd., cocoa beverage powds., dessert powds., protein powds., gravy mixes, cake mixes, etc.; choline source; *Usage level:* 0.2-0.9% (as instantizer); *Regulatory:* FDA 21CFR §184.1400, GRAS; kosher approved; *Properties:* lt. tan/yel. powd., bland odor and taste; acid no. 36 max.; 1% max. moisture; *Storage:* store below 25 C; 1 yr shelf life in original, unopened container.

Alcolec® FF-100. [Am. Lecithin] De-oiled lecithin FCC; CAS 8002-43-5; EINECS 232-307-2; food emulsifier; instantizing agent for milk powd., cocoa beverage powds., dessert powds., protein powds., gravy mixes, cake mixes, etc.; choline source; *Usage level:* 0.2-0.9% as instantizer; *Regulatory:* FDA 21CFR §184.1400, GRAS; kosher approved; *Properties:* lt. tan/yel. fine powd., bland odor and taste; acid no. 36 max.; 1% max. moisture; *Storage:* store below 25 C; 1 yr shelf life in original, unopened container.

Alcolec® Granules. [Am. Lecithin] Lecithin FCC; CAS 8002-43-5; EINECS 232-307-2; nonionic; emulsifier, stabilizer, diet supplement; instantizing agent for milk powd., cocoa beverage powds., protein powds., gravy mixes, cake mixes, etc.; choline source; *Regulatory:* FDA 21CFR § 184.1400, GRAS; kosher approved; *Properties:* lt. tan/yel. gran., bland odor and taste; sp.gr. 0.5; acid no. 36 max.; 97% act., 1% max. moisture; *Storage:* store below 25 C; 1 yr shelf life in original, unopened container.

Alcolec® HWS. [Am. Lecithin] Hydroxylated lecithin in a wheat starch carrier; improves pie crust, cakes, cake mixes, yeast-raised doughs; gluten conditioner; reduces sticking in prod. of biscuits, cookies, sugar wafers, cones; *Usage level:* 0.5-2.0%; *Regulatory:* kosher approved; *Properties:* tan powd.; 1.2-1.5% oil content; *Storage:* store at 60-90 F; 1 yr shelf life in original, unopened container.

Alcolec® PG. [Am. Lecithin] Oil-free lecithin containing 97% phospholipids; CAS 8002-43-5; EINECS 232-307-2; food emulsifier; *Regulatory:* FDA 21CFR §184.1400, GRAS; kosher approval; *Properties:* sl. yel. waxy gran.; 1% max. moisture; *Storage:* store below 25 C; protect from light and moisture; 1 yr shelf life in original, unopened container.

Alcolec® S. [Am. Lecithin] Unbleached lecithin FCC; CAS 8002-43-5; EINECS 232-307-2; nonionic; w/o emulsifier, wetting and dispersing agent, stabilizer, release and lubricating agent, foam suppressant, solubilizer, choline source; *Regulatory:* FDA 21CFR §184.1400, GRAS; kosher approved; *Properties:* Gardner 17 liq.; acid no. 32 max.; 1% max. moisture; *Storage:* 1 yr shelf life in original, unopened containers.

Alcolec® SFG. [Am. Lecithin] An oil-free lecithin

containing a high inositol content; CAS 8002-43-5; EINECS 232-307-2; dough conditioner/volume improver, release aid, lubricant for food applics.; foam suppressant; *Regulatory:* FDA 21CFR §184.1400, GRAS; kosher approved; *Properties:* gran.; acid no. 32 max.; 1% max. moisture; *Storage:* store at 10-30 C; 1 yr shelf life in original unopened container.

Alcolec® XTRA-A. [Am. Lecithin] Premium lecithin containing 66% phospholipids; CAS 8002-43-5; EINECS 232-307-2; w/o emulsifier for food industry, esp. for use with oil or fat prods. where low fatty acid is important; *Regulatory:* FDA 21CFR §184.1400, GRAS; kosher approved; *Properties:* amber high visc. liq.; HLB 2.0; acid no. 25 max.; 0.8% max.moisture; *Storage:* 1 yr shelf life in original unopened container.

Alcolec® Z-3. [Am. Lecithin] Hydroxylated lecithin; CAS 8029-76-3; EINECS 232-440-6; nonionic; wetting agent, emulsifier, solubilizer, dispersant; improves dispersability of colors and flavors in aq. systems, improves wetting of fatty powds.; exc. instantizing props.; *Regulatory:* FDA 21CFR §172.814, GRAS; kosher approved; *Properties:* lt. amber liq.; sol. in most fat solv. except acetone; visc. 8000 cps; acid no. < 38; < 1.5% moisture, < 42% oil content; *Storage:* 1 yr shelf life in original unopened container.

Alco Whip® 1026. [A.E. Staley Mfg.] Specialty protein; aerating agent, primarily intended for bar mixes.

Aldo® DC. [Lonza] Propylene glycol dicaprylate/ dicaprate; nonionic; food emulsifier; *Properties:* Gardner 2 max. liq.; sol. in ethanol, min. and veg. oil; insol. in water; sp.gr. 0.92; HLB 2 ± 1; iodine no. 0.5 max.; sapon. no. 315-335; 100% conc.

Aldo® GLP FG. [Lonza] Glyceryl lactopalmitate; nonionic; food emulsifier; *Properties:* bead; HLB 2.5; 100% conc.; Unverified

Aldo® HMO FG. [Lonza] Glyceryl mono and dioleate (high mono); nonionic; food emulsifier; *Properties:* liq.; HLB 3.0; 100% conc.; Unverified

Aldo® HMS FG. [Lonza] Glyceryl stearate; CAS 31566-31-1; nonionic; emulsifier for baked goods, dairy prods., edible oils/shortenings, confectionery, peanut butter; improves volume, texture, fat absorption, dispersibility, aeration, chewability; emulsion stabilizer; prevents ice crystals; *Usage level:* 3-5% on shortening (cakes), 10-14% on shortening (doughnuts), 0.2-0.4% (dry coffee whiteners), 0.5-1.0% (liq. coffee whiteners), 0.1-0.2% (frozen desserts), 0.5-1.0% (whipped toppings), 0.4% (margarine), 0.4-0.8% (caramels); *Regulatory:* FDA 21CFR §184.1505 GRAS; *Properties:* wh. beads, mild odor, bland taste; m.p. 60-65 C; HLB 3.0; acid no. 2 max.; iodine no. 2 max.; 100% conc.; *Storage:* store in dry, cool area.

Aldo® HMS KFG. [Lonza] Glyceryl stearate, high mono; CAS 31566-31-1; food emulsifier, emulsion stabilizer, dispersant, softener; used in baked goods, dairy prods. (coffee whiteners, frozen desserts, whipped toppings), edible oils/shortenings,

confectionery, pet foods, peanut butter; *Usage level:* 3-5% on shortening (cakes), 0.1-0.2% (frozen desserts), 0.4% (margarine), 0.4-0.8% (caramels); *Regulatory:* FDA 21CFR §184.1505 (GRAS); kosher; *Properties:* wh. beads, bland odor and taste; m.p. 64 C; HLB 3; acid no. 2 max.; sapon. no. 170; 52% alpha monoglycerides; *Storage:* store in cool, dry area.

Aldo® LP FG. [Lonza] Glyceryl lactopalmitate; nonionic; emulsifier for cakes, whipped toppings, shortenings, prepared mixes; *Properties:* bead; HLB 2.5; 100% conc.

Aldo® LP KFG. [Lonza] Glyceryl-lacto ester of fatty acids; CAS 1335-49-5; alpha tending food emulsifier imparting aeration props., whippability, and foam stabilization; used for cakes, whipped toppings, frozen dairy systems; *Usage level:* 4-8% on shortening (cake), 6-12% (whipped toppings), 4-10% (frozen dairy systems); *Regulatory:* FDA 21CFR §172.852, kosher; *Properties:* ivory bead, mild odor; m.p. 40-45 C; HLB 2.5; acid no. 6 max.; 6-10% alpha monoglyceride; *Storage:* store in cool, dry area.

Aldo® MCT KFG. [Lonza] Caprylic/capric triglyceride; CAS 73398-61-5; EINECS 265-724-3; dietary supplement; dense source of calories; used to produce energy; lubricant for confectionery coatings; carrier for spray coatings; solvent, diluent, and flavor fixative; *Regulatory:* kosher; *Properties:* clear liq., neutral odor and taste; acid no. 0.10 max.; iodine no. 1.0 max.; sapon. no. 335-360; hyd. no. 10 max.; 0.1% max. moisture.

Aldo® ML. [Lonza] Glyceryl monodilaurate; nonionic; food emulsifier; *Properties:* liq.; HLB 5.2; 100% conc.; Unverified

Aldo® MLD FG. [Lonza] Glyceryl laurate SE; nonionic; o/w food emulsifier for vegetable oil; dispersant and solubilizer for flavors and spice oils; antifoam or defoamer in food processing systems; *Usage level:* 5-10% (veg. oil), 1-5% (flavors, spice oils); *Regulatory:* FDA 21CFR §172.863, 184.1505, GRAS; *Properties:* yel. liq. to soft solid, mild odor; m.p. 21-26 C; HLB 7; acid no. 5 max.; 30% mono alpha monoglycerides; *Storage:* store in cool, dry area.

Aldo® MOD. [Lonza] Glyceryl oleate SE; nonionic; emulsifier; avail. in animal, veg., food, and Kosher grades; *Properties:* yel. liq.; sol. in toluol, naphtha, min. and veg. oils, disp. in water; sp.gr. 0.95; m.p. < 0 C; HLB 5.0; sapon. no. 141-147; pH 8.5-9.5 (5% aq.); 100% conc.

Aldo® MOD FG. [Lonza] Glyceryl monodioleate, disp.; CAS 25496-72-4; nonionic; emulsifier and antifoam for foods; *Properties:* liq.; HLB 4.0; 100% conc.; Unverified

Aldo® MO FG. [Lonza] Glyceryl monooleate; CAS 25496-72-4; nonionic; emulsifier improving dispersibility of coffee whiteners; antifoam/defoamer for food processing systems; dispersant/solubilizer for flavors and spice oils; *Usage level:* 0.5% (coffee whiteners), 1-5% (flavors, spice oils); *Regulatory:* FDA 21CFR §184.1505, GRAS; *Properties:* liq.

11

with lt. haze, mild odor; HLB 3.0; acid no. 2 max.; sapon. no. 166-176; 42% min. alpha monoglyceride; *Storage:* store in cool, dry area.

Aldo® MS-20 FG. [Lonza] PEG-20 glyceryl stearate; nonionic; emulsifier, antifoam for foods; *Properties:* solid; HLB 13.0; 100% conc.; Unverified

Aldo® MSD FG. [Lonza] Glyceryl stearate SE (Glyceryl stearate and potassium stearate); nonionic; emulsifier for general use in foods; lubricant for confectionery cutting surfaces to prevent sticking; release agent for container surfaces; softener for dough systems; *Usage level:* 5-10% (lubricant); *Regulatory:* FDA 21CFR §172.863, 184.1505, GRAS; *Properties:* lt. tan beads, mild odor; m.p. 56-60 C; HLB 6.0; acid no. 3.5 max.; sapon. no. 135-150; 30% min. alpha monoglycerides; *Storage:* store in cool, dry area.

Aldo® MSD KFG. [Lonza] Glyceryl stearate SE (Glyceryl stearate and potassium stearate; nonionic; lubricant for confectionery cutting surfaces to prevent sticking; release agent on container surfaces; softener for dough systems; *Usage level:* 5-10% (lubricant); *Regulatory:* FDA 21CFR §172863, 184.1505 (GRAS); kosher; *Properties:* tan beads, bland odor; 100% thru 8 mesh, 5% max. thru 100 mesh; m.p. 58-60 C; HLB 6; acid no. 3 max.; sapon. no. 130-145; 30% min. alpha monoglycerides.

Aldo® MS FG. [Lonza] Glyceryl stearate; CAS 31566-31-1; nonionic; food emulsifier; improves antisticking, chewability, shelf life in chewing gum, caramels; prevents oil separation in fudges, peanut butter; aerating agent for whipped toppings; improves mouthfeel and texture in peanut butter; improves flow props.; *Usage level:* 0.5-1% (chewing gum), 0.5% (fudge), 2-4% (peanut butter), 1% (caramel); *Regulatory:* FDA 21CFR §184.1505, GRAS; *Properties:* wh. beads, mild odor, bland taste; 100% thru 8 mesh, 5% thru 100 mesh; m.p. 59 C; HLB 4.0; acid no. 3 max.; sapon. no. 160-165; 35% min. alpha monoglycerides; *Storage:* store in cool, dry area.

Aldo® MSLG FG. [Lonza] Glyceryl stearate, low glycerin; CAS 31566-31-1; nonionic; food emulsifier and emulsion stabilizer for baked goods, dairy prods., edible oils/shortenings, confectionery, pet foods; *Usage level:* 4-6% on shortening (cake), 0.1-0.2% (frozen desserts), 0.5% (margarine), 0.05-1% (dry potatoes); *Regulatory:* FDA 21CFR § 184.1505, GRAS; *Properties:* wh. beads, bland odor and taste; m.p. 58-62 C; HLB 3; acid no. 2 max.; sapon. no. 162; 40-45% alpha monoglycerides; *Storage:* store in cool, dry area.

Aldo® MS LG KFG. [Lonza] Glyceryl stearate, low glycerin; CAS 31566-31-1; nonionic; emulsifier for baked goods, dairy prods., edible oils/shortenings, confectionery; improves volume/texture in cakes; emulsion stabilizer in margarine; retards crystallization in starch jellies; improves dehydrated potatoes, moisture retention; *Usage level:* 4-6% on shortening (cake), 0.1-0.2% (frozen desserts), 0.5% (margarine), 0.25-0.6% (starch jellies); *Regulatory:* FDA 21 CFR §184.1505 (GRAS); kosher;

Properties: wh. beads, bland odor and taste; m.p. 60-65 C; HLB 3; acid no. 2 max.; iodine no. 5 max.; 40% min. alpha monoglycerides; *Storage:* store in cool, dry area.

Aldo® PGHMS. [Lonza] Propylene glycol stearate (high mono); CAS 1323-39-3; EINECS 215-354-3; nonionic; aerating, whipping agent in toppings, cake mixes, shortenings; *Properties:* bead; HLB 3.0; 100% conc.

Aldo® PGHMS KFG. [Lonza] Propylene glycol stearate, high mono; CAS 1323-39-3; EINECS 215-354-3; alpha tending food grade emulsifier, whipping agent; for cakes, cake mixes, whipped toppings, shortening; improves aeration and volume, increases moisture retention, promotes more stable foam structure; *Regulatory:* FDA 21CFR §172.856; kosher; *Properties:* off-wh. beads or flakes, bland odor and taste; m.p. 36-42 C; HLB 3; acid no. 2 max.; iodine no. 1 max.; 70-76% monoester; *Storage:* store in cool, dry area.

Aldo® PME. [Lonza] Propylene glycol stearate, glyceryl stearate; nonionic; emulsifier in cakes, shortenings, toppings; *Properties:* solid; HLB 3.0; 100% conc.; Unverified

Aldo® TC. [Lonza] Caprylic/capric triglyceride; nonionic; emulsifier for cakes, shortenings, toppings; *Properties:* lt. yel. clear liq., bland; HLB 1.0; acid no. 1 max.; iodine no. 1 max.; sapon. no. 330-360; pH 4.5 (5%); 100% conc.

Aldosperse® 30/70 FG. [Lonza] Mono- and diglycerides of edible fats/oils and ethoxylated mono- and diglycerides; used for bread, buns, rolls, yeast-raised doughnuts; *Usage level:* 4-8 oz/cwt flour (bread, doughnuts), 6-8 oz/cwt flour (buns, rolls); *Regulatory:* FDA 21CFR §172.834, 184.1505 (GRAS); *Properties:* straw-colored beads, bland odor and taste; 95% thru 20 mesh, 45-60% thru 40 mesh, 10-30% thru 60 mesh; HLB 6; acid no. 2 max.; sapon. no. 130-145; 36% min. alpha monoglycerides; *Storage:* store in tied polybags in dry area below 100 F.

Aldosperse® 30/70 KFG. [Lonza] Mono- and diglycerides of edible fats/oils and ethoxylated mono- and diglycerides; used for bread, buns, rolls, yeast-raised doughnuts; *Usage level:* 4-8 oz/cwt flour (bread, doughnuts), 6-8 oz/cwt flour (buns, rolls); *Regulatory:* FDA 21CFR §172.834, 184.1505 (GRAS); kosher; *Properties:* straw-colored beads, bland odor and taste; 100% thru 8 mesh, 5% thru 100 mesh; HLB 6; acid no. 2 max.; sapon. no. 130-145; pH 5.0-7.5 (5%); 36% min. alpha monoglycerides; *Storage:* store in tied polybags in dry area below 100 F.

Aldosperse® 40/60. [Lonza] 60% Glyceryl stearate and 40% PEG-20 glyceryl stearate; ice cream emulsifier, dough conditioner for food industry; *Properties:* Gardner 2 beadx; HLB 7 ± 1; m.p. 57 C; acid no. 2.

Aldosperse® 40/60 FG. [Lonza] 40% PEG-20 glyceryl stearate, 60% glyceryl stearate (high mono); nonionic; bakery and food emulsifier; dough strengthener, softener; for bread, buns, rolls, yeast-raised doughnuts; *Usage level:* 4-8 oz/cwt flour (bread, doughnuts), 6-8 oz/cwt flour (buns, rolls); *Regulatory:* FDA 21CFR 172.834, 184.1505 (GRAS); *Properties:* straw-colored beads, bland odor and taste; 100% thru 8 mesh, 5% thru 100 mesh; HLB 7.0; acid no. 2 max.; sapon. no. 120-135; 30% alpha monoglyceride; *Storage:* store in tied polybags in dry area below 100 F.

Aldosperse® 40/60 KFG. [Lonza] 60% Glyceryl stearate and 40% PEG-20 glyceryl stearate; nonionic; food emulsifier; baked goods strengthener; for bread, buns, rolls, yeast-raised doughnuts; *Usage level:* 4-8 oz/cwt flour (bread, doughnuts), 6-8 oz/cwt flour (buns, rolls); *Regulatory:* FDA 21CFR §172.834, 184.1505 (GRAS); kosher; *Properties:* cream-colored beads, bland odor and taste; 100% thru 8 mesh, 5% thru 100 mesh; m.p. 57 C; HLB 7; acid no. 2 max.; sapon. no. 120-135; 30% alpha monoglyceride; *Storage:* store in tied polybags in dry area below 100 F.

Aldosperse® 712 FG. [Lonza] Monodiglycerides and ethoxylated monodiglycerides blend; nonionic; food emulsifier; *Properties:* bead; 100% conc.; Unverified

Aldosperse® MO-50 FG. [Lonza] Mono- and diglycerides and 50% polysorbate 80; emulsifier for ice cream, frozen custard, yeast-defoamer formulations, edible fats/oils, shortening, whipped toppings; solubilizer; dispersant in pickles, vitamin-min. preps., gelatin desserts; defoamer for cottage cheese; wetting agent for colorants; *Regulatory:* FDA 21CFR 712.840, 172.515, 184.1505 GRAS; *Properties:* clear to sl. hazy semiliq.; HLB 9; acid no. 2 max.; iodine no. 45-55; sapon. no. 100-115; 15% alpha monoglyceride; *Storage:* store in cool, dry area.

Aldosperse® MS-20 FG. [Lonza] PEG-20 glyceryl stearate; CAS 68553-11-7; nonionic; food emulsifier; dough strengthener in yeast-raised baked goods; provides increased volume, uniform texture, improved moisture retention, crumb structure; emulsifier in whipped veg. oil toppings, frozen desserts, icings, icing mixes, nondairy creamer; *Usage level:* 2-4 oz/cwt flour (baked goods), 3-6 oz/cwt flour (cake), 0.2-0.45% (whipped toppings); *Regulatory:* FDA 21CFR §172.834; *Properties:* straw-colored liq.; m.p. 27 C; HLB 13.0; acid no. 2 max.; sapon. no. 68-75; hyd. no. 65-80; pH 6 (5%); 100% conc.; *Storage:* store with lid firmly in place in cool, dry area.

Aldosperse® MS-20 KFG. [Lonza] PEG-20 glyceryl stearate; CAS 68553-11-7; nonionic; food grade emulsifier; baked goods strengthener; increases volume, provides more uniform texture, improves moisture retention; emulsifier for whipped toppings, frozen desserts, icings, icing mixes; nondairy creamer in beverage coffee; *Usage level:* 2-4 oz/ cwt flour (baked goods), 3-6 oz/cwt flour (cake), 0.2-0.4% (whipped toppings); *Regulatory:* FDA 21CFR §172.834; kosher; *Properties:* straw-colored soft solid, mild odor, bland taste; m.p. 27 C; HLB 13; acid no. 2 max.; sapon. no. 65-75; hyd. no.

65-80; *Storage:* store with lid firmly in place in cool, dry area.

Aldosperse® O-20. [Lonza] 80% Glyceryl stearate, 20% polysorbate 80; ice cream emulsifier, dough conditioner for foods; *Properties:* straw beads; m.p. 55-65 C; HLB 5.2; acid no. 2; sapon. no. 140-150.

Aldosperse® O-20 FG. [Lonza] 80% Glyceryl stearate, 20% polysorbate 80; nonionic; emulsifier, solubilizer for frozen desserts; *Properties:* bead; HLB 5.0; 100% conc.

Aldosperse® O-20 KFG. [Lonza] 80% Glyceryl stearate, 20% polysorbate 80; nonionic; food grade emulsifier for ice cream, ice milk, and frozen desserts; provides exc. whipping and overrun props., max. dryness/texture/body; ideal for high temp. short time pasteurization; *Usage level:* 0.1-0.25% (ice cream), 0.15-0.25% (ice milk), 0.1-0.25% (frozen desserts); *Regulatory:* FDA 21CFR §172.840, 184.1505 (GRAS); kosher; *Properties:* lt. tan beads, mild odor, bland taste; 100% thru 8 mesh, 5% max. thru 100 mesh; m.p. 50-65 C; HLB 5; acid no. 2 max.; sapon. no. 140-155; *Storage:* store in cool, dry area below 35 C.

Aldosperse® TS-20. [Lonza] Glyceryl stearate (80%), polysorbate 65 (20%); ice cream emulsifier, dough conditioner for food industry; *Properties:* straw beads; m.p. 55-65 C; HLB 4.3; acid no. 2; sapon. no. 150-160.

Aldosperse® TS-20 FG. [Lonza] 80% Glyceryl stearate, 20% polysorbate 65; nonionic; food emulsifier for frozen desserts, ice cream, ice milk, frozen custard, soft-serve and other specialty prods.; *Regulatory:* FDA 21CFR §172.838, 182.4505; *Properties:* lt. tan beads, bland odor and taste; 20% max. 100 mesh; HLB 4.0; acid no. 2 max.; sapon. no. 150-160; hyd.no. 160-185; 100% conc.

Aldosperse® TS-20 KFG. [Lonza] Blend of mono- and diglycerides and polysorbate 65; food emulsifier for ice cream, frozen desserts, sherbets; provides exc. whipping and overrun props., dryness and body; *Usage level:* 0.1-0.25% (ice cream, frozen desserts), 0.25% (sherbets); *Regulatory:* FDA 21CFR §172.838, 184.1505 (GRAS); kosher; *Properties:* lt. tan beads, mild odor and taste; 100% thru 8 mesh, 5% max. thru 100 mesh; HLB 4; acid no. 2 max.; sapon. no. 150-160; hyd. no. 160-190; *Storage:* store in cool, dry area.

Aldosperse® TS-40. [Lonza] Glyceryl stearate (60%), polysorbate 65 (40%); ice cream emulsifier and dough conditioner for food industry; *Properties:* straw beads; m.p. 40-60 C; HLB 6.0; acid no. 2; sapon. no. 120-130.

Aldosperse® TS-40 FG. [Lonza] 60% Glyceryl stearate, 40% polysorbate 65; nonionic; frozen dessert emulsifier; *Properties:* bead; HLB 6.0; 100% conc.

Aldosperse® TS-40 KFG. [Lonza] 60% Glyceryl stearate, 40% Polysorbate 65; nonionic; food emulsifier for frozen desserts, ice cream, sherbet; provides max. whipping and overrun props., imparts dryness, body, and texture; *Usage level:* 0.1-0.25% (ice cream, frozen desserts), 0.25% (sherbets); *Regulatory:* FDA 21CFR §172.838, 184.1505

(GRAS); kosher; *Properties:* tan beads, mild odor; 100% thru 8 mesh, 5% max. thru 100 mesh; m.p. 55 C; HLB 6; acid no. 2 max.; sapon. no. 135-145; hyd. no. 140-155; *Storage:* store in cool, dry area.

Algel P. [Kimitsu Chem. Industries; Unipex] Algin deriv.; used for edible jelly; *Properties:* wh. or creamy powd.; 80 mesh pass; visc. gel; pH 6-8; < 15% water.

Alginade DC. [Kelco Int'l.] Alginate blend; thickener, gellant, stabilizer used for whipped cream, ice lollies.

Alginade MR. [Kelco Int'l.] Alginate blend; stabilizer improving resist. to heat shock used for ice cream, ice lollies, milk shakes.

Alginade MRE. [Kelco Int'l.] Alginate blend; stabilizer improving resist. to heat shock used for ice cream.

Alginade XK9. [Kelco Int'l.] Xanthan gum blend; CAS 11138-66-2; EINECS 234-394-2; stabilizer for ice cream, frozen desserts; provides processing stability and stable visc. during shelf life in UHT dairy prods. such as soups, sauces, milkshakes, instant drinks.

Alginic Acid FCC. [Meer] Alginic acid FCC; CAS 9005-32-7; EINECS 232-680-1; tablet binder for pharmaceutical and health food industries; holds tablet together during compression and disintegrates rapidly under acidic stomach conditions to allow active ingreds. to work faster; *Usage level:* 0.4-1.0%; *Properties:* wh. to off-wh. free-flowing powd.; ≥95% thru 80 mesh; visc. 5-50 cps (1%); pH 1.5-3.5 (1%).

Alitame. [Pfizer Food Science] A high-intensity sweetener with 200 times the potency of sucrose; *Regulatory:* awaiting FDA approval.

Alkamuls® PSTO-20. [Rhone-Poulenc Surf. & Spec.] Polysorbate 85; CAS 9005-70-3; nonionic; food emulsifier; *Properties:* amber liq., typ. odor; water disp.; dens. 1.0 g/ml; HLB 11; sapon. no. 80-95; 97% act.

Alkamuls® PSTS-20. [Rhone-Poulenc Surf. & Spec.] Polysorbate 65; CAS 9005-71-4; nonionic; wetting agent, emulsifier; *Properties:* tan solid; typ. odor; sol. in min. oils, min. spirits; water disp.; dens. 1.0 g/ml; HLB 10.5; sapon. no. 88-98; 97% act.; Discontinued

Alkamuls® S-6. [Rhone-Poulenc Surf. & Spec.] PEG-6 stearate; CAS 9004-99-3; nonionic; food emulsifier; *Properties:* wax; 99% conc.; formerly Rhodiasurf S-6.

Alkamuls® SMS. [Rhone-Poulenc Surf. & Spec.] Sorbitan stearate; CAS 1338-41-6; EINECS 215-664-9; nonionic; emulsifier, coupling agent; *Properties:* cream flakes; sol. (10%) in aromatic solv., perchloroethylene; dens. 1.0 g/ml; HLB 4.7; sapon. no 147-157; hyd. no. 235-260; 98.5% act.

Alkamuls® STS. [Rhone-Poulenc Surf. & Spec.] Sorbitan tristearate; CAS 26658-19-5; EINECS 247-891-4; nonionic; hydrophobic emulsifier; *Properties:* cream flakes; typ. odor; sol. (10%) in aromatic solv., perchloroethylene; insol. in water; dens. 1.0 g/ml; HLB 2.1; sapon. no. 176-188; hyd. no. 66-80; 100% act.; Discontinued

All Purpose Shortening 101-050. [ADM Packaged Oils] Partially hydrog. soybean and cottonseed oils; packaged veg. oil; *Properties:* wh. plasticized cube, bland flavor; m.p. 118 ± 3 F; iodine no. 77 ± 3.

Allura® Red AC. [Buffalo Color] FD&C Red no. 40; CAS 25956-17-6; food colorant; *Properties:* dk. red powd.; sol. 22.5% in water, 1.3% in 50% alcohol; m.w. 496.42.

Almond Filling Powd. [Bunge Foods] Produces almond flavored filling for sweet goods and danish pastries; *Usage level:* 16%; *Properties:* tan/cream gran. powd., almond aroma; pH 6.0; 12% moisture.

Aloe Con UP 10. [Florida Food Prods.] Aloe vera gel, ultra purified grade; used for beverages; *Properties:* Gardner 3 max. liq.; pH 4.0-5.0; 5% solids.

Aloe Con UP 40. [Florida Food Prods.] Aloe vera gel, ultra purified grade; used for beverages; *Properties:* Gardner 3 max. liq.; pH 4.0-5.0; 20% solids.

Aloe-Con UP-200. [Florida Food Prods.] Freeze-dried aloe vera gel, ultra purified grade; used for beverages; *Properties:* Gardner 3 max. powd.; pH 4.0-5.0; 100% solids.

Aloe Con WG 10. [Florida Food Prods.] Aloe vera whole gel; used for beverages; *Properties:* lt. tan, typ. aloe flavor; pH 4.0-5.0; 5 ± 1% solids.

Aloe Con WG 40. [Florida Food Prods.] Aloe vera whole gel; used in beverages; *Properties:* dk. tan to brn., typ. aloe flavor; sp.gr. 1.1 ± 0.05; pH 3.5-5.0; 20 ± 1% solids; *Storage:* store in container; protect from light; 6 mo shelf life.

Aloe Con WG 200. [Florida Food Prods.] Aloe vera whole gel; used for beverages; *Properties:* lt. tan, typ. aloe flavor; pH 4.0-5.0; 100% solids.

Aloe Con WLG 10. [Florida Food Prods.] Aloe vera whole leaf gel; used for beverages; *Properties:* Gardner 3 max. liq.; pH 4.0-5.0; 5% solids.

Aloe Con WLG 200. [Florida Food Prods.] Freeze-dried aloe vera whole leaf gel; used for beverages, health supplements; *Properties:* sl. tan free-flowing powd.; pH 3.5-5.5 (1:199); < 7% moisture.

Aloe Vera Gel 1X. [Terry Labs] Aloe vera gel, food grade; used in food and beverage applics. (digestion aid, health drinks, juice blends, health foods, sports/fitness drinks); *Properties:* yel.-grn. clear to sl. hazy liq., lt. veg. odor; sp.gr. 0.997-1.004; pH 3.5-5.0; 0.5% min. total solids; *Storage:* store @ R.T.; protect from oxidation; may darken with age.

Aloe Vera Gel 10X. [Terry Labs] Aloe vera gel, food grade; used in food and beverage applics. (digestion aid, health drinks, juice blends, health foods, sports/fitness drinks); *Properties:* lt. amber clear to sl. hazy liq., mod. veg. odor; sp.gr. 1.022-1.032; pH 3.5-5.0; 5% min. total solids; *Storage:* store @ R.T.; protect from oxidation; may darken with age.

Aloe Vera Gel 40X. [Terry Labs] Aloe vera gel, food grade; used in food and beverage applics. (digestion aid, health drinks, juice blends, health foods, sports/fitness drinks); *Properties:* amber clear to sl. hazy liq. with possible precipitate, strong veg. odor; sp.gr. 1.100-1.120; pH 3.5-5.0; 20% min. total solids; *Storage:* store @ R.T.; protect from oxidation;

may darken with age.

Aloe Vera Gel Thickened FG. [Terry Labs] Aloe vera gel and carrageenan; food grade aloe vera gel, visc. enhanced for internal or topical applics.; *Properties:* yel.-grn. clear to sl. hazy gelatinous liq.; pH 3.5-5.0; *Storage:* store @ R.T.; protect from heat and oxidation; may darken with age.

Aloe Vera Lipo-Quinone Extract™ (AVLQE) Food Grade. [Terry Labs] Aloe extract, extracted onto soybean oil base; used in internal/oral applics. (lip balms, lipsticks, oral ointments, soft gel encapsulation); *Properties:* yel. oil, veg. odor; sp.gr. 0.919-0.924; acid no. 1.1 max.; sapon. no. 185-200; < 10% moisture; *Storage:* store at R.T.; do not store above 65 C for > 24 h.

Alomine 042. [Vaessen-Schoemaker] Animal protein hydrolysate; nonfoaming protein for food applics. which require low gelling and low visc.; meat flavor; *Properties:* powd.; completely cold sol.; pH 6 (1%); < 5% moisture.

Alomine 223. [Vaessen-Schoemaker] Protein hydrolysate, taste enhancer, citric acid (antioxidant); phosphate-free brine additive, esp. for injected prods., such as cooked ham, poultry prods.; improves tenderness of final prod.; *Usage level:* 0.5% on final prod. wt.; *Properties:* free-flowing powd., lt. aromatic odor; readily sol. in water; bulk dens. 1.2 l/kg; *Storage:* store under cool, dry conditions.

Alomine-Mix. [Vaessen-Schoemaker] Phosphate-free curing compd. based on Alomine; curative for cooked hams, shoulders, poultry prods.

Alpha 500. [Paniplus] Dist. monoglyceride, polysorbate 60; emulsifier; *Properties:* dry powd.; Unverified

Alpha W6 HP 0.6. [Wacker-Chemie GmbH] Hydroxypropyl-α-cyclodextrin; complex hosting guest molecules; increases the sol. and bioavailability of other substances; masks flavor, odor, or coloration; stabilizes against light, oxidation, heat, and hydrolysis; turns liqs. or volatiles into stable solid powds.; *Properties:* m.w. 1184.

Alpha W6 M1.8. [Wacker-Chemie GmbH] Methyl-α-cyclodextrin; complex hosting guest molecules; increases the sol. and bioavailability of other substances; masks flavor, odor, or coloration; stabilizes against light, oxidation, heat, and hydrolysis; turns liqs. or volatiles into stable solid powds.; *Properties:* m.w. 1123.

Alpha W 6 Pharma Grade. [Wacker-Chemie GmbH] α-Cyclodextrin; CAS 7585-39-9; EINECS 231-493-2; complex hosting guest molecules; increases the sol. and bioavailability of other substances; masks flavor, odor, or coloration; stabilizes against light, oxidation, heat, and hydrolysis; turns liqs. or volatiles into stable solid powds.; *Properties:* wh. cryst. powd.; sol. 14.5 g/100 ml in water; m.w. 972; > 98% act.; *Toxicology:* LD50 (acute IV, rat) 500-750 mg/kg; eye irritant but not corrosive; nonirritating to skin.

Alphadim® 70K. [Am. Ingreds.] Mono- and diglycerides; food additive for puddings, coffee whiteners, ice cream, frozen desserts, whipped toppings, chip dips, and sour cream bases; *Properties:* wh. to cream colored powd.; m.p. 61-64 C; 72% min. mono.

Alphadim® 90AB. [Am. Ingreds.] High-purity, molecularly dist. monoglyceride prepared from fully hardened edible fats and glycerin; CAS 67701-33-1; nonionic; food additive improving processing, storage stability; stabilizes and disperses fat particles in coffee whiteners; as starch complexing and softening aid in breads; also for margarine, dehydrated potatoes, chewing gum, peanut butter; *Usage level:* 0.1-2.5%; *Regulatory:* FDA GRAS; *Properties:* wh. to cream fine bead; m.p. 65-71 C; iodine no. 3 max.; sapon. no. 150-165; 90% min. alpha monoester; *Storage:* store under cool, dry conditions.

Alphadim® 90NLK. [Am. Ingreds.] High-purity, molecularly dist. monoglyceride prepared from refined sunflower oil and glycerin with TBHQ and citric acid; nonionic; food additive providing a stable, finely dispersed emulsion in diet margarines; emulsifier system in cake and icing shortenings; *Usage level:* 0.3-1.0%; *Regulatory:* FDA GRAS; kosher; *Properties:* wh. to pale yel. fluid plastic; m.p. 37-48 C; iodine no. 110; sapon. no. 150-165; HLB 3.5; 90% min. alpha monoester; *Storage:* store at moderate temps.

Alphadim® 90PBK. [Am. Ingreds.] Mono- and diglycerides; food additive for margarine, coffee whiteners, dehydrated potatoes, peanut butter, chewing gum, pasta, snacks, and cereal prods.; *Properties:* wh. to cream colored bead; m.p. 64-68 C; iodine no. 5 max.; 90% min. mono.

Alphadim® 90SBK. [Am. Ingreds.] High-purity, molecularly dist. monoglyceride prepared from fully hardened soybean oil and glycerin; nonionic; provides a stable finely dispersed emulsion in margarines; stabilizes, disperses fat particles in coffee whiteners; starch complexing agent; prevents stickiness; pasta, snacks, cereals, dehydrated potatoes, confection coatings, chewing gum; *Regulatory:* FDA GRAS; kosher; *Properties:* wh. to cream fine bead; m.p. 70-75 C; iodine no. 3 max.; sapon. no. 150-165; HLB 3.5; 90% min. alpha monoester; *Storage:* store under cool, dry conditions.

Alphadim® 90SBK FG. [Am. Ingreds.] Mono- and diglycerides; food additive for pasta, snacks, and cereal prods.; *Properties:* wh. to cream colored bead; m.p. 70-75 C; iodine no. 3 max.; 90% min. mono.

Alphadim® 90VCK. [Am. Ingreds.] High-purity, molecularly dist. monoglyceride prepared from partially hardened soybean oil and glycerin with TBHQ and citric acid; nonionic; food additive providing a stable emulsion of finely dispersed water droplets for margarine and coffee whiteners; emulsifier for cake and icing shortenings; *Usage level:* 0.1-6.0%; *Regulatory:* FDA GRAS; kosher; *Properties:* ivory plastic; m.p. 58-62 C; iodine no. 54-65; sapon. no. 150-165; HLB 3.5; 90% min. alpha monoester; *Storage:* store under cool, dry conditions.

Alpine. [Karlshamns Food Ingreds.] Partially hydrog.

vegetable oil (soybean, palm) with vegetable mono- and diglycerides and Tween 60K; nontropical version also avail.; emulsified shortening; high volume type cream filling and icing; *Properties:* plastic; m.p. 118 F.

Alta®. [Quest Int'l.] Fermented specialty; provides shelf-life and flavor protection for minimally heat processed and non-heat processed foods.

Aluminum Stearate EA. [Witco/H-I-P] Aluminum distearate; CAS 300-92-5; EINECS 206-101-8; binder, emulsifier, anticaking agent for food applics.; *Properties:* wh. powd., 98% thru 200 mesh; sol. in turpentine, benzene, toluene, xylene, carbon tetrachloride, veg. and min. oil, oleic acid, waxes; insol. in water; sp.gr. 1.01; soften. pt. 160 C; Unverified

Amalty®. [Mitsubishi Int'l.; Towa Chem. Ind.] Maltitol; CAS 585-88-6; EINECS 209-567-0; mild sweetener with no aftertaste; does not cause tooth decay; heat-stable; humectant; for sugar-free confectionery, sauces, spices, processed foods, vitamin supplements, foods for diabetics; low hygroscopicity; *Properties:* wh. powd. or gran.; m.p. 140 C $\geq$ 93.5% act.

Amfal-46. [MLG Enterprises Ltd.] Ethoxylated mono- and diglycerides (40%), mono- and diglycerides (60%); dough strengthener/softener; protein/ starch complexing agents; improves absorption, shelf life, slicing qualities; *Usage level:* 1.25% max. on flour; *Properties:* free-flowing powd.; *Storage:* store up to 6 mos under cool, dry conditions; keep poly liner closed when not in use.

Amflex. [MLG Enterprises Ltd.] Fungal amylase from *Aspergillus oryzae* with cereal filler and other edible excipients; CAS 9000-92-4; EINECS 232-567-7; enzyme dough conditioner; converts damaged starch gran. in bread flour to fermentable sugars; provides more uniform gas prod., reduced proof time, greater volume, uniform crust color, better grain/texture, added softness in bread crumb; *Regulatory:* FDA GRAS; *Properties:* sl. off-wh. to beige tablet, char. odor; *Storage:* store cool and dry; 6 mos min. shelf life.

Amidan. [Grindsted Prods.; Grindsted Prods. Denmark] Cold-dispersible monoglycerides; nonionic; emulsifier, dough conditioner, starch complexing agent, crumb softener in bread, rolls, bread improvers, cakes, cereals; reduces stickiness in pasta prods.; aids extrusion of snacks; *Properties:* powd.; HLB 4.3; 100% conc.

Amidan 500 Series. [Grindsted Prods.] Cold-dispersible monoglycerides with enzymes and flow agents; food emulsifier.

Amidan 600 Series. [Grindsted Prods.] Cold-dispersible monoglycerides with enzymes and flow agents; food emulsifier.

Amidan ES. [Grindsted Prods.] Cold-dispersible monoglycerides from veg. source; food emulsifier; *Properties:* powd.; drop pt. 69 C; iodine no. 22 min.; 90% min. monoester.

Amidan ES Kosher. [Grindsted Prods.] Dist. monoglyceride from refined vegetable oil; emulsifying

and starch complexing agent; crumb softener for bread; amylose complexing agent in extruded snack foods, pasta, dehydrated potato prods.; starch complexing agent in cereals and starchy prods.; ideal for powd. mixes; *Usage level:* 0.3-1% on flour wt. (yeast-raised breads/rolls), 0.1-0.3% (extruded snack food), 0.1-0.3% (pasta), 0.1-0.5% (dehydrated potatoes); *Regulatory:* FDA 21CFR §184.1505, GRAS; FCC; kosher; *Properties:* powd.; drop pt. 69 C; iodine no. 22 min.; *Storage:* store in cool, dry area; storage above 21 C may cause caking.

Amidan SDM-T. [Grindsted Prods.] Cold-dispersible monoglycerides from veg. source; food emulsifier; avail. as kosher grade; *Properties:* powd.; drop pt. 66 C; iodine no. 35-45; 90% min. monoester.

Amidan XTR-A. [Grindsted Prods.] Cold-dispersible monoglycerides from animal source; food emulsifier; *Properties:* powd.; drop pt. 67 C; iodine no. 22 min.; 90% min. monoester.

Amisol™ 210-L. [Lucas Meyer] Lecithin with selected oil carriers; CAS 8002-43-5; EINECS 232-307-2; release agent for confections with ease of handling in cold weather and capacity for uniform spraying and thin films; *Regulatory:* FDA 21CFR §172, 182; *Properties:* liq.; acid no. 0.5-1.0; 0.2-0.5% moisture; *Storage:* 1 yr shelf life in original, unopened containers; store at 60-90 F.

Amisol™ 210 LP. [Lucas Meyer] Lecithin; CAS 8002-43-5; EINECS 232-307-2; release agent for spray or brush applic. to stainless steel belts for food applics.; *Regulatory:* FDA approved; *Properties:* amber liq.

Amisol™ 329. [Lucas Meyer] Lecithin; CAS 8002-43-5; EINECS 232-307-2; wetting agent, emulsifier, dispersant, and release agent for food applics.; Discontinued

Amisol™ 406-N. [Lucas Meyer] Lecithin, glycerin, propylene glycol, glyceryl oleate; coupling agent for water-sol. dyes in fatty systems; used for food, confections, food dyes; *Properties:* amber liq.

Amisol™ 683 A. [Lucas Meyer] Lecithin and selected fatty acids; CAS 8002-43-5; EINECS 232-307-2; lubricity, spreading and film-forming agent effective in low concs. as a release aid and lubricant; used for cheese release, confections; *Properties:* lt. gold, mild taste.

Amisol™ 697. [Lucas Meyer] Lecithin; CAS 8002-43-5; EINECS 232-307-2; reduces volatilization of oil-sol. flavors and enhances emulsification; used for confections, baked goods, flavorings incl. synthetic and imitation flavors; *Properties:* amber moderate visc. fluid, mild odor and flavor.

Amisol™ 785-15K. [Lucas Meyer] Lecithin, veg. oils, stearates, and antioxidants; release aid for use on belts and/or hot or cold slabs, esp. effective with nut brittles and hard candies; does not impart color to prod.; resist. to oxidation; *Properties:* milky noncongealing liq., odorless, tasteless; visc. 200 cps.

Amisol™ MS-10. [Lucas Meyer] Lecithin, polysorbate 21; emulsifier and wetting agent in foods; *Usage*

level: 1-3% (detergents, shampoos), 1-2% (shaving cream); *Properties:* lt. amber free-flowing liq., almost odorless; sol. in most fatty oils; readily water-disp.; 100% act.

Amisol™ MS-12 BA. [Lucas Meyer] Lecithin, polysorbate 20, polysorbate 60, and sorbitan laurate; emulsifier for preparing w/o emulsions for aq. food and nonfood systems; *Properties:* Gardner 14 max. color; disp. in water; visc. 8000 cps max.; acid no. 28 max.; 100% act.

Ammonium Alginate Type S. [Meer] Ammonium alginate FCC; CAS 9005-34-9; emulsifier for ice cream, sherbets, water ices; prevents freezer burn; *Usage level:* 0.4-1.0%; *Properties:* off-wh. to tan free-flowing powd.; ≥ 95% thru 80 mesh; sol. in cold water; visc. ≥ 500 cps (1%); pH 5.0-6.5.

Amoco Superla® DCO 55. [Amoco Lubricants] Highly refined food grade wh. min. oil; dust control oil for grains intended for human consumption; *Regulatory:* FDA §172.878, 178.3620, 573.680, approved up to 200 ppm; USDA approval H-1, 3H; *Properties:* colorless, odorless, tasteless; sp.gr. 0.850 (60/60 F); visc. 8.4 cSt (40 C); pour pt. -12 C; flash pt. 154 C.

Amoco Superla® DCO 75. [Amoco Lubricants] Highly refined food grade wh. min. oil; dust control oil for grains intended for human consumption; *Regulatory:* FDA §172.878, 178.3620, 573.680, approved up to 200 ppm; USDA approval H-1, 3H; *Properties:* colorless, odorless, tasteless; sp.gr. 0.853 (60/60 F); visc. 14 cSt (40 C); pour pt. -12 C; flash pt. 185 C.

Amoloid HV. [Kelco] Ammonium alginate; CAS 9005-34-9; food-grade gum used as gelling agent, thickener, emulsifier, film-forming agent, suspending agent, and stabilizer; *Regulatory:* FDA 21CFR §184.1133, GRAS; *Properties:* wh. to tan powd., sl. odor; water-sol.; bulk dens. ≈ 50 lb/ft³; pH ≈ 5.5 (1%); *Toxicology:* LD50 (oral, rat) > 5000 mg/kg; excessive dust inhalation may cause respiratory irritation; dry powd. may cause eye irritation; *Precaution:* not flamm., but powd. will burn if involved in a fire; spills may be slippery; incompat. with strong oxidizers, alkaline sol'ns.; *Storage:* store in cool, dry place.

Amoloid LV. [Kelco] Ammonium alginate; CAS 9005-34-9; food-grade gum used as gelling agent, thickener, emulsifier, film-forming agent, suspending agent, and stabilizer; *Regulatory:* FDA 21CFR §184.1133, GRAS; *Properties:* wh. to tan powd., sl. odor; water-sol.; bulk dens. ≈ 50 lb/ft³; pH ≈ 5.5 (1%); *Toxicology:* LD50 (oral, rat) > 5000 mg/kg; excessive dust inhalation may cause respiratory irritation; dry powd. may cause eye irritation; *Precaution:* not flamm., but powd. will burn if involved in a fire; spills may be slippery; incompat. with strong oxidizers, alkaline sol'ns.; *Storage:* store in cool, dry place.

Angrex Seasonings Code #02300. [MLG Enterprises Ltd.] Formulated seasoning blends for ketchup, mayonnaise, bologna sausage, pizza snack, etc.; *Usage level:* 0.12-1.5 g/kg.

Angrex Spices Oleoresins. [MLG Enterprises Ltd.] Natural oleoresins of aromatic herbs and spices constituted by the essential oils and sol. organic resins present in the herbs or spices; provides flavor uniformity, stability, freedom from bacteriological contamination; for use in homogenized foods, e.g., sauces, seasonings, canned meat, sausages; *Properties:* avail. in aniseed, capsicum, cinnamon, garlic, onion, ginger, mustard, rosemary, wh. pepper, blk. pepper, sage, thyme, etc.; sol. in.

Annatto ECA-1. [Hilton Davis] Purified annatto processed and emulsified with propylene glycol, polysorbate 80, and potassium hydroxide; natural colorant for beverages, candies, yogurts, and sherbets; color str. 2.5% bixin/norbixin blend; *Properties:* yel. orange to orange emulsion.

Annatto ECH-1. [Hilton Davis] Purified annatto processed and emulsified with propylene glycol, monoglycerides, and potassium hydroxide; high heat stable natural colorant for butters, margarine, cheeses, shortening, frosting, confectionery coatings; color str. 1.5% bixin/norbixin blend; *Properties:* yel. orange to orange emulsion.

Annatto ECPC. [Hilton Davis] Purified annatto processed and emulsified with propylene glycol, monoglycerides, and potassium hydroxides; natural colorant for baked goods, snack foods, salad dressings; color str. 2.5% bixin/norbixin blend; *Properties:* orange yel. to yel. orange emulsion.

Annatto OS #2894. [Crompton & Knowles/Ingred. Tech.] Annatto seed extract in suspension with propylene glycol, refined edible vegetable oil, lecithin, and potassium hydroxide (as processing aid); color additive for foods; *Regulatory:* FDA 21CFR §73.30, 73.1030, 73.2030; *Properties:* dk. reddish-brn. liq., bright yel. to yel.-orange when solubilized in fats/oils; sl. odor; sol. in fats and oils; sp.gr. 0.90 ± 0.10; 2.25% ± 0.15% bixin content; *Storage:* store in tight containers; avoid exposure to light and heat; 2 yrs shelf life.

Annatto OS #2922. [Crompton & Knowles/Ingred. Tech.] Annatto extract in suspension with propylene glycol, monoglycerides of refined vegetable oil, and potassium hydroxide (as processing aid); color additive for foods; *Regulatory:* FDA 21CFR §73.30, 73.1030, 73.2030; *Properties:* dk. reddish-brn. liq., bright yel. to yel.-orange when disp. in water or oils; sl. odor; misc. in oil and water; sp.gr. 0.95 ± 0.10; 1.3 ± 0.2% bixin content; *Storage:* store in tight containers; avoid exposure to light and heat; 2 yrs shelf life.

Annatto OS #2923. [Crompton & Knowles/Ingred. Tech.] Annatto extract in suspension with propylene glycol, monoglycerides of refined vegetable oil, and potassium hydroxide (as processing aid); color additive for foods; *Regulatory:* FDA 21CFR §73.30, 73.1030, 73.2030; *Properties:* dk. reddish-brn. liq., bright yel. to yel.-orange when disp. in water or oils; sl. odor; misc. in oil and water; sp.gr. 0.95 ± 0.10; 1.75 ± 0.25% bixin content; *Storage:* store in tight containers; avoid exposure to light and heat; 2 yrs shelf life.

Annatto OSL-1. [Hilton Davis] Annatto extract solubilized in veg. oil; natural colorant for butters, shortening, margarine, salad oils; color str. 0.2% bixin; *Properties:* lt. or butter yel. liq.; oil-sol.

Annatto OSS-1. [Hilton Davis] Annatto extract suspended in veg. oil; natural colorant for process cheeses, cake mixes, frostings, candy coatings, salad oil; color str. 3.8% bixin; *Properties:* yel. orange to orange suspension; oil-sol.

Annatto WSL-2. [Hilton Davis] Annatto extract in potassium hydroxide sol'n.; natural colorant for ice cream cones, cereals, cakes, drinks, and cheeses; color str. 2.7% norbixin; *Properties:* yel-orange to orange liq.; water-sol.

Annatto WSP-75. [Hilton Davis] Annatto extract with potassium carbonate and potassium hydroxide; natural colorant for dry mixes, soups, cakes, drinks, desserts, cereals; color str. 7.5% norbixin; *Properties:* yel. orange to deep orange powd.; water-sol.

Annatto WSP-150. [Hilton Davis] Purified annatto extract with potassium carbonate and potassium hydroxide; natural colorant for dry mixes, soups, cakes, drinks, desserts, cereals; color str. 15% norbixin; *Properties:* yel. orange to deep orange powd.; water-sol.

Annatto Extract P1003. [Phytone Ltd.] Aq. alkaline extract of annatto seeds, principle pigment norbixin; CAS 8015-67-6; food colorant; *Regulatory:* EEC E160b; JECFA compliance; *Properties:* deep orange mobile liq.; *Storage:* store @ 5-10 C in tightly closed containers in cool, dry area; protect from direct sunlight, freezing.

Annatto Liq. #3968, Acid Proof. [Crompton & Knowles/Ingred. Tech.] Annatto extract in suspension with propylene glycol, and potassium hydroxide (as processing aid); color additive for foods; *Regulatory:* FDA 21CFR §73.30, 73.1030, 73.2030; *Properties:* dk. reddish-brn. liq., bright yel. to yel.-orange when diluted with water; sl. odor; water-sol.; sp.gr. 0.95 ± 0.10; 3.25 ± 0.25% bixin content; *Storage:* store in tight containers; avoid exposure to light and heat; 2 yrs shelf life.

Annatto/Turmeric OSS-1. [Hilton Davis] Annatto extract and purified turmeric suspended in veg. oil; natural colorant for confectionery coatings, shortenings, frostings, cakes, pies, margarines, process cheeses, salad dressings; color str. 3.5% bixin, 3.5% curcumin; *Properties:* yel. suspension; oil-sol.

Annatto/Turmeric WML-1. [Hilton Davis] Purified annatto and turmeric emulsified in propylene glycol, polysorbate 80, and potassium hydroxide; natural colorant for baked goods, ice cream, frozen desserts, and candies; color str. 0.8% bixin and 3% curcumin; *Properties:* yel. liq.; water-misc.

Annatto/Turmeric WMP-1. [Hilton Davis] Purified annatto and turmeric emulsified in propylene glycol, polysorbate 80, and potassium hydroxide plated on silica gel; natural colorant for dry mixes, cereals, baked goods, ice cream, frozen desserts, candies; color str. 0.8% bixin, 3% curcumin; *Properties:* yel. powd.; water-misc.

Antelope Aerator. [Albright & Wilson Australia] Specially formulated chemical aerator based on sodium acid pyrophosphate; CAS 7758-16-9; EINECS 231-835-0; general purpose aerator for use by commercial pastry chefs when slow dough rate of reaction is required; *Regulatory:* FCC, Australian compliance; *Properties:* wh. free-flowing powd.; 0.1% +250 μm particle size, 0.8% +150 μm; 64.5% acidity; *Toxicology:* avoid inhaling dust, prolonged skin contact; *Storage:* store in cool, dry place.

Anterior Pituitary Substance. [Am. Labs] Vacuum-dried defatted glandular prod.; nutritive food additive; *Properties:* powd.

Anthocyanin Extract P2001. [Phytone Ltd.] Anthocyanin; extract of black grapes (*Vitis vinifera* food colorant; *Regulatory:* EEC E163; JECFA compliance; *Properties:* deep purple mobile liq.; *Storage:* store in tightly closed containers in cool, dry area; protect from direct sunlight.

Antispumin ZU. [Stockhausen] Ethoxylated/propoxylated fatty alcohols; nonionic; defoamer for beet sugar industry for flume water, diffusion, liming, and carbonation; *Properties:* yel. sl. cloudy oily visc. liq., typ. odor; forms unstable emulsion in water; sp.gr. 0.92; visc. 100 cp; b.p. 412 F; pour pt. -30 C; flash pt. 323 F; pH 5.5 (10%); 100% act.; *Toxicology:* skin irritant on prolonged contact.

APA. [Calgon Carbon] Activated carbon; CAS 64365-11-3; used for food and chemical decolorization, pharmaceutical purification, odor control; *Properties:* 12 x 40 sieve size.

APC. [Calgon Carbon] Activated carbon; CAS 64365-11-3; used for food and chemical decolorization; *Properties:* 12 x 40 sieve size.

Apo-Carotenal Suspension 20% in Veg. Oil No. 66417. [Roche] β-Apo-8′-carotenal suspension in partially hydrog. vegetable oils; colorant for processed cheese, salad dressing, whipped margarine, fats, oils, other fat-based foods; *Usage level:* 75 mg max. (foods); *Regulatory:* FDA 21CFR §73.90 (food use); *Properties:* reddish-brn. to dk. brn. homogeneous suspension, pourable fluid above 68 F, thick paste when refrigerated; 20% min. β-apo-8′-carotenal, 240,000 units provitamin A activity/g; *Storage:* store in cool place @ 46-59 F; protect from freezing; stir well before use.

Aqualon® 7H0F. [Hercules] Cellulose gum; CAS 9004-32-4; improves mouthfeel, body, and texture, controls ice crystal formation in sherbet, frozen breakfast drinks; thickener inhibiting syneresis, preventing shrinkage in pie fillings; thickener, suspending aid in reduced-calorie beverages; *Usage level:* 0.05-0.2% (beverages), 0.1-0.5% (sherbet), 0.1-0.5% (pie fillings), 0.1-0.3% (frozen breakfast drinks); *Regulatory:* FDA 21CFR §182.1745, GRAS; *Properties:* water-sol.; *Toxicology:* LD50 (oral, rat) 27 g/kg; sl. eye irritant; nonirritating to skin.

Aqualon® 7H0XF. [Hercules] Cellulose gum; CAS 9004-32-4; thickener for eggnog; *Usage level:* 0.1-0.3% (eggnog); *Regulatory:* FDA 21CFR §

182.1745, GRAS; *Properties:* water-sol.; *Toxicology:* LD50 (oral, rat) 27 g/kg; sl. eye irritant; nonirritating to skin.

Aqualon® 7H3SF. [Hercules] Cellulose gum; CAS 9004-32-4; thickener in barbecue sauces, pickle relish; improves mouthfeel, body, and texture, controls ice crystal formation in sherbet, frozen breakfast drinks; thickener inhibiting syneresis, preventing shrinkage in pie fillings; stabilizer in jams/jellies; *Usage level:* 0.05-0.1% (dietetic beverages), 0.1-0.5% (sherbet, pie fillings, barbecue sauces), 0.1-0.4% (jams), 0.1-0.7% (variegated syrup), 0.03-0.5% (instant hot cereal), 0.1-0.25% (sweet goods), 0.1-0.3% (breakfast drinks), 0.1-0.5% (meringues); *Regulatory:* FDA 21CFR §182.1745, GRAS; *Properties:* water-sol.; *Toxicology:* LD50 (oral, rat) 27 g/kg; sl. eye irritant; nonirritating to skin.

Aqualon® 7H3SXF. [Hercules] Cellulose gum; CAS 9004-32-4; controls texture and sugar crystal size in frostings and icings; inhibits syneresis in puddings; thickener, improves mouthfeel, body in reduced-calorie cocoa mixes; *Usage level:* 0.05-0.5% (frostings), 0.1-0.3% (puddings), 0.2-0.5% (cocoa mixes); *Regulatory:* FDA 21CFR § 182.1745, GRAS; *Properties:* water-sol.; *Toxicology:* LD50 (oral, rat) 27 g/kg; sl. eye irritant; nonirritating to skin.

Aqualon® 7H4F. [Hercules] Cellulose gum; CAS 9004-32-4; improves mouthfeel, body, and texture, controls ice crystal formation in ice cream; gravy thickener, processing aid in canned pet foods; *Usage level:* 0.1-0.25% (ice cream), 0.2-0.5% (canned pet food); *Regulatory:* FDA 21CFR §182.1745, GRAS; *Properties:* water-sol.; *Toxicology:* LD50 (oral, rat) 27 g/kg; sl. eye irritant; nonirritating to skin.

Aqualon® 7H4XF. [Hercules] Cellulose gum; CAS 9004-32-4; improves mouthfeel, body, and texture, controls ice crystal formation in soft serve; gravy thickener, extrusion aid in dry pet foods; *Usage level:* 0.1-0.5% (soft serve), 0.3-0.5% (pet food); *Regulatory:* FDA 21CFR §182.1745, GRAS; *Properties:* water-sol.; *Toxicology:* LD50 (oral, rat) 27 g/kg; sl. eye irritant; nonirritating to skin.

Aqualon® 7HC4F. [Hercules] Cellulose gum; CAS 9004-32-4; improves mouthfeel, body, and texture, controls ice crystal formation in ice cream, ice milk; *Usage level:* 0.1-0.25% (ice cream), 0.1-0.5% (ice milk); *Regulatory:* FDA 21CFR §182.1745, GRAS; *Properties:* water-sol.; *Toxicology:* LD50 (oral, rat) 27 g/kg; sl. eye irritant; nonirritating to skin.

Aqualon® 7HCF. [Hercules] Cellulose gum; CAS 9004-32-4; improves mouthfeel, body, and texture, controls ice crystal formation in ice cream, ice milk; *Usage level:* 0.1-0.25% (ice cream), 0.1-0.5% (ice milk); *Regulatory:* FDA 21CFR §182.1745, GRAS; *Properties:* water-sol.; *Toxicology:* LD50 (oral, rat) 27 g/kg; sl. eye irritant; nonirritating to skin.

Aqualon® 7HF. [Hercules] Cellulose gum; CAS 9004-32-4; improves mouthfeel, body, and texture, controls ice crystal formation in ice cream, soft serve;

inhibits syneresis in puddings; bodying agent in reduced-calorie spreads; suspending aid in syrups, fruit drinks; stabilizer; binder, oil/fat barrier; *Usage level:* 0.1-0.25% (ice cream), 0.1-0.5% (soft serve), 0.1-0.3% (puddings), 0.1-0.3% (spreads), 0.1-0.7% (variegated syrup), 0.1-0.3% (marshmallow topping), 0.1-0.4% (batters), 0.1-0.3% (cake dry mix), 0.1-0.25% (sweet goods), 0.1-0.4% (tortillas); *Regulatory:* FDA 21CFR §182.1745, GRAS; *Properties:* water-sol.; *Toxicology:* LD50 (oral, rat) 27 g/kg; sl. eye irritant; nonirritating to skin.

Aqualon® 7HXF. [Hercules] Cellulose gum; CAS 9004-32-4; improves mouthfeel, body, and texture, controls ice crystal formation in ice cream; improves mouthfeel and body in bar mixes, breakfast drinks, fruit drink mix; *Usage level:* 0.1-0.25% (ice cream), 0.1-0.3% (bar mixes), 0.2-0.8% (breakfast drinks, fruit drink mix); *Regulatory:* FDA 21CFR §182.1745, GRAS; *Properties:* water-sol.; *Toxicology:* LD50 (oral, rat) 27 g/kg; sl. eye irritant; nonirritating to skin.

Aqualon® 7LF. [Hercules] Cellulose gum; CAS 9004-32-4; thickener for table syrup; moisture retention, controls crystal size in icings; binder and lubricant for animal feed pellets; improves mouthfeel, body in bar mixes, hot chocolate mix; *Usage level:* 0.05-0.5% (icings), 0.1-1.0% (syrup), 0.1-0.3% (bar mixes), 0.1-0.4% (hot chocolate mix), 0.05-0.1% (animal feed); *Regulatory:* FDA 21CFR §182.1745, GRAS; *Properties:* water-sol.; *Toxicology:* LD50 (oral, rat) 27 g/kg; sl. eye irritant; nonirritating to skin.

Aqualon® 7LXF. [Hercules] Cellulose gum; CAS 9004-32-4; improves mouthfeel, body, flavor perception in breakfast drinks; *Usage level:* 0.2-0.8% (breakfast drinks); *Regulatory:* FDA 21CFR §182.1745, GRAS; *Properties:* water-sol.; *Toxicology:* LD50 (oral, rat) 27 g/kg; sl. eye irritant; nonirritating to skin.

Aqualon® 7M8SF. [Hercules] Cellulose gum; CAS 9004-32-4; improves moisture retention, texture, controls crystal size in icings; *Usage level:* 0.05-0.5% (icings); *Regulatory:* FDA 21CFR §182.1745, GRAS; *Properties:* water-sol.; *Toxicology:* LD50 (oral, rat) 27 g/kg; sl. eye irritant; nonirritating to skin.

Aqualon® 7MF. [Hercules] Cellulose gum; CAS 9004-32-4; thickener, stabilizer in marshmallow toppings; improves moisture retention, texture, controls crystal size in icings; protein stabilizer in acidified milk; suspending aid, improves mouthfeel, body in frozen breakfast drinks; *Usage level:* 0.05-0.5% (icings), 0.1-0.3% (marshmallow topping), 0.2-0.3% (acidified milk), 0.1-0.3% (breakfast drinks); *Regulatory:* FDA 21CFR §182.1745, GRAS; *Properties:* water-sol.; *Toxicology:* LD50 (oral, rat) 27 g/kg; sl. eye irritant; nonirritating to skin.

Aqualon® 9H4F. [Hercules] Cellulose gum; CAS 9004-32-4; thickener in chocolate toppings; *Usage level:* 0.2-0.4% (chocolate topping); *Regulatory:* FDA 21CFR §182.1745, GRAS; *Properties:* water-sol.; *Toxicology:* LD50 (oral, rat) 27 g/kg; sl. eye irritant; nonirritating to skin.

Aqualon® 9H4XF. [Hercules] Cellulose gum; CAS 9004-32-4; thickener, suspending aid providing rapid visc. build in salad dressing mixes; moisture retention, extrusion aid in semimoist pet foods; improves flavor in breakfast drinks, fruit drink mix; *Usage level:* 0.05-0.2% (salad dressing mix), 0.2-0.4% (pet food), 0.1-0.3% (dry cake mixes), 0.2-0.8% (breakfast drinks, fruit drink mix); *Regulatory:* FDA 21CFR §182.1745, GRAS; *Properties:* water-sol.; *Toxicology:* LD50 (oral, rat) 27 g/kg; sl. eye irritant; nonirritating to skin.

Aqualon® 9M31F. [Hercules] Cellulose gum; CAS 9004-32-4; thickener in syrups; improves mouth-feel and body, suspending aid in reduced-calorie beverages; *Usage level:* 0.05-0.2% (beverages), 0.1-1.0% (syrups); *Regulatory:* FDA 21CFR §182.1745, GRAS; *Properties:* water-sol.; *Toxicology:* LD50 (oral, rat) 27 g/kg; sl. eye irritant; nonirritating to skin.

Aqualon® 9M31XF. [Hercules] Cellulose gum; CAS 9004-32-4; thickener, suspending aid providing rapid visc. build in salad dressing mixes; improves mouthfeel in reduced-calorie cocoa mixes, fruit drink mix; binder, fat/oil barrier, suspending aid in batters for deep frying; *Usage level:* 0.02-0.5% (cocoa mixes), 0.2-0.8% (fruit drink mix), 0.05-0.2% (salad dressing mix), 0.1-0.4% (batters); *Regulatory:* FDA 21CFR §182.1745, GRAS; *Properties:* water-sol.; *Toxicology:* LD50 (oral, rat) 27 g/kg; sl. eye irritant; nonirritating to skin.

Aqualon® 9M8F. [Hercules] Cellulose gum; CAS 9004-32-4; thickener for table syrup; binder, fat and oil barrier, suspending aid, visc. control for batter for deep-fat frying; *Usage level:* 0.1-1.0% (syrup), 0.1-0.4% (batter); *Regulatory:* FDA 21CFR §182.1745, GRAS; *Properties:* water-sol.; *Toxicology:* LD50 (oral, rat) 27 g/kg; sl. eye irritant; nonirritating to skin.

Aqualon® 9M8XF. [Hercules] Cellulose gum; CAS 9004-32-4; thickener improving mouthfeel in re-duced-calorie cocoa mixes; *Usage level:* 0.2-0.5% (cocoa mixes); *Regulatory:* FDA 21CFR § 182.1745, GRAS; *Properties:* water-sol.; *Toxicology:* LD50 (oral, rat) 27 g/kg; sl. eye irritant; nonirritating to skin.

Aquasol CSL. [Rhone-Poulenc Food Ingreds.] Locust bean gum; CAS 9000-40-2; EINECS 232-541-5; food-grade natural vegetable gum; high visc., pH stable (3.5-9), freeze-thaw stable; requires heat to develop full visc.; synergistic with xanthan, carra-geenan; *Usage level:* 0.1-0.25% (ice cream/milk), 0.3-0.5% (cream cheese), 0.25-0.4% (fruit preps. for yogurt), 0.15-0.3% (pie fillings), 0.1-0.25% (sauces); *Properties:* fine gran.; fair dispersibility; visc. 3000 cps min. (1%).

Aratex. [Van Den Bergh Foods] Partially hydrog. veg. oil (cottonseed, soybean); EINECS 269-820-6; high performance fat; icing stabilizer; syrups; donut glazes; bakery dry mixes; structuring system for composition enhancement; *Regulatory:* kosher; *Properties:* Lovibond 3.0R max. flakes; m.p. 115-119 F.

Arbran™ Soy Fiber. [ADM Protein Spec.] Soy fiber; dietary fiber for snacks, breads, cereals; 70% total dietary fiber; *Properties:* tan color; 1% fat.

Archer Soybean Oil 86-070-0. [ADM Refined Oils] Soybean oil, refined, bleached, deodorized; CAS 8001-22-7; EINECS 232-274-4; veg. oil for mayon-naise, salad dressings, spreads, canned foods, sauces, bakery goods; *Properties:* clear brilliant oil, bland flavor; iodine no. 125 ± 4.

Arcon F. [ADM Protein Spec.] Soy protein conc.; CAS 68153-28-6; for protein supplement meat systems; 21% total dietary fiber; 290 calories/100 g; *Proper-ties:* free-flowing fine powd.; 69% min. protein, 9% max. moisture.

Arcon G, Fortified Arcon G. [ADM Protein Spec.] Soy protein conc.; CAS 68153-28-6; for protein supple-ment systems; 21% total dietary fiber; 290 calories/ 100 g; *Properties:* free-flowing grits; 69% min. pro-tein, 9% max. moisture.

Arcon S, Fortified Arcon S. [ADM Protein Spec.] Soy protein conc.; CAS 68153-28-6; emulsion stabi-lizer; protein supplement meat systems; 20% total dietary fiber; 290 calories/100 g; *Properties:* free-flowing flour; 70% min. protein, 8% max. moisture.

Arcon T, Fortified Arcon T. [ADM Protein Spec.] Textured soy protein conc.; CAS 68153-28-6; pro-tein for ground meats, fish, poultry; 21% total di-etary fiber; 290 calories/100 g; *Properties:* variety of textures, sizes, colors; 69% min. protein, 9% max. moisture.

Arcon VF, Fortified Arcon VF. [ADM Protein Spec.] Soy protein conc.; CAS 68153-28-6; for protein supplement meat systems; 21% total dietary fiber; 290 calories/100 g; *Properties:* free-flowing very fine powd.; 69% min. protein, 9% max. moisture.

Arctic Rye. [Brolite Prods.] Rye bread base specifi-cally designed for freezing; *Usage level:* 12 lb/100 lb flour.

Ardex D, D Dispersible. [ADM Protein Spec.] Isolated soy protein; CAS 68153-28-6; adhesive, emulsifier, emulsion stabilizer for meat systems, sauces, gra-vies, dairy blends; *Properties:* pH 6.8-7.3; 90% min. protein, 6.5% max. moisture.

Ardex D-HD. [ADM Protein Spec.] Isolated soy protein; CAS 68153-28-6; heavy dens. protein for high protein beverages and nutritional supple-ments; *Properties:* pH 6.1-6.4; 90% min. protein, 6.5% max. moisture.

Ardex DHV Dispersible. [ADM Protein Spec.] Iso-lated soy protein; CAS 68153-28-6; nondusting, salt tolerant protein for coarse and emulsion meat applics. (patties, loaves, sausages); *Properties:* highly sol.; pH 6.8-7.3 (1:10 aq. disp.); 90% min. protein, 6.5% max. moisture.

Ardex F, F Dispersible. [ADM Protein Spec.] Isolated soy protein; CAS 68153-28-6; nongelling protein for processed dairy foods, milk replacer, infant formulas; *Properties:* readily disp.; pH 6.8-7.3; 90% min. protein, 6.5% max. moisture.

Ardex FR. [ADM Protein Spec.] Isolated soy protein;

CAS 68153-28-6; very low visc. protein for infant formulas and similar prods.; *Properties:* pH 6.4-7.0; 90% min. protein, 6.0% max. moisture.

Ardex R. [ADM Protein Spec.] Isolated soy protein; CAS 68153-28-6; isoelectric pH protein for nutritional supplements, high protein foods, low sodium foods; *Properties:* pH 4.5-5.0; 90% min. protein, 6.5% max. moisture.

Arkady Yeast Food. [ADM Arkady] Provides nutrients for vigorous yeast growth and oxidants for greater protein strength; also avail. in double strength.

Arlac P. [ADM Protein Spec.] Potassium lactate; CAS 996-31-6; EINECS 213-631-3; effective antimicrobial agent to extend shelf life of meat, poultry, and seafood prods.; *Properties:* sl. saline taste; sp.gr. 1.310-1.340; dens. 12 lb/gal; pH 6.5-7.5; 58-62% lactate.

Arlac S. [ADM Protein Spec.] Sodium lactate; CAS 72-17-3; EINECS 200-772-0; effective antimicrobial agent to extend shelf life of meat, poultry, and seafood prods.; *Properties:* sl. saline taste; sp.gr. 1.310-1.340; dens. 12 lb/gal; pH 6.5-7.5; 58-62% lactate.

Arlacel® 165. [ICI Spec. Chem.; ICI Surf. Am.; ICI Atkemix; ICI Surf. Belgium] Glyceryl stearate, PEG-100 stearate; nonionic; self-emulsifying surfactant; *Properties:* ivory wh. beads; bland odor; disp. in water; HLB 11; flash pt. > 300 F; pour pt. 54 C; 100% act.

Arlacel® 186. [ICI Spec. Chem.; ICI Surf. Am.; ICI Atkemix; ICI Surf. Belgium] Glyceryl oleate, propylene glycol, 0.02% BHA and 0.01% citric acid as preservatives; nonionic; surfactant, omulsifier, *Properties:* pale yel. clear liq.; sol. in ethanol, IPA, cottonseed and min. oil; sp.gr. 1; visc. 150 cps; HLB 2.8; flash pt. > 300 F; 100% act.

Arlatone® 650. [ICI Spec. Chem.; ICI Surf. UK] Ethoxylated castor oil; vitamin solubilizer; *Properties:* pale yel. liq.; HLB 12.5.

Arlatone® 827. [ICI Spec. Chem.; ICI Surf. UK] Ethoxylated castor oil; nonionic; vitamin solubilizer; *Properties:* yel. liq.; HLB 11.9; 100% act.

Aromahop™. [Pfizer Food Science] Beta-acid hop oil fraction isolated from Pfico$_2$.Hop, hops extract; brewery additive providing consistent bitter-free source of hop aroma and flavor in beer brewed for light stability; for brewkettle addition; *Storage:* store at R.T. or under refrigeration.

Aromild® S. [Mitsubishi Int'l.; Kohjin Co. Ltd.] Soy sauce, yeast extract; natural flavoring created by fermentation and enzyme processing; for seasoning stocks for soups and dressing, processed meat, processed fish prods., canned foods, frozen foods, snacks, food flavoring; *Usage level:* 0.1-1.5% (seasoning stock), 5-20% (dry seasoning), 0.05-0.5% (vinegar, processed meat/fish), 0.1-0.5% (canned/frozen food, food flavoring, pet food), 0.1-1.5% (sauces); 36.7% protein, 5% moisture, 1% fat; *Storage:* keep cool and dry; 6 mos shelf life.

Arovas. [Vaessen-Schoemaker] Complete curing compd. with reduced phosphate content; curing agent for cooked hams, pork loins, bacon, poultry

prods.

Artodan CF 40. [Grindsted Prods.] Calcium stearoyl lactylate; CAS 5793-94-2; EINECS 227-335-7; emulsifier; *Regulatory:* EEC, FDA §172.844; ADI 0-20 mg/kg.

Artodan SP 55 Kosher. [Grindsted Prods.; Grindsted Prods. Denmark] Sodium stearyl-2-lactylate; CAS 25383-99-7; EINECS 246-929-7; anionic; food emulsifier, dough conditioner, starch complexing agent, bread improver, freeze/thaw emulsions; *Properties:* sm. beads; HLB 10.0; 100% act.; Discontinued

Asahi Aji®. [Asahi Chem. Industry] Monosodium glutamate; CAS 142-47-2; EINECS 205-538-1; food additive.

Ascorbic Acid USP/FCC, 100 Mesh. [Int'l. Sourcing] Ascorbic acid USP/FCC; CAS 50-81-7; EINECS 200-066-2; vitamin, antioxidant; *Properties:* translucent wh. to sl. yel. cryst. powd., pract. odorless, acidic taste; 95% min. thru 30 mesh, 50-65% thru 60 mesh; appreciable sol. in water; m.w. 176.13; sp.gr. 1.65; m.p. 375 F (dec.); pH 2.5-3.0; 99-100.5% assay; *Toxicology:* possible skin and eye irritant from powd. contact; dust may irritant respiratory tract; heated to decomp., emits acrid smoke and irritating fumes; *Storage:* store under cool, dry conditions free from metal contact; protect from light in well-closed containers.

Ascorbic Acid USP, FCC Fine Gran. No. 6045655. [Roche] Ascorbic acid USP, FCC; CAS 50-81-7; EINECS 200-066-2; source of vitamin C, antioxidant in foods; used where mesh range is critical in providing uniformity in granular foods, e.g., gelatin prods.; *Regulatory:* FDA GRAS; *Properties:* wh. or sl. yel. cryst. powd., pract. odorless, pleasantly tart taste; sol. 1 g/3 ml water, 30 ml alcohol; m.w. 176.13; bulk dens. 0.8-1.1 (tapped); m.p. 190 C; pH 1.9-2.4 (10% aq.); 99-100.5% assay; *Precaution:* may deteriorate on exposure to atmospheric moisture, oxidizes readily in aq. sol'n.; avoid contact with iron, copper, or nickel salts; *Storage:* store in tight, light-resist. containers, optimally @ ≤ 72 F; avoid exposure to moisture and excessive heat.

Ascorbic Acid USP, FCC Fine Powd. No. 6045652. [Roche] Ascorbic acid; CAS 50-81-7; EINECS 200-066-2; source of vitamin C, antioxidant in foods (instant beverages, cocoa prods., powd. foods, premixes); *Regulatory:* FDA GRAS; *Properties:* wh. or sl. yel. fine powd., pract. odorless, pleasantly tart taste; sol. 1 g/3 ml water, 30 ml alcohol; m.w. 176.13; bulk dens. 0.55-1.0 (tapped); m.p. 190 C; pH 1.9-2.4 (10% aq.); 99-100.5% assay; *Precaution:* may deteriorate on exposure to atmospheric moisture, oxidizes rapidly in aq. sol'n.; avoid contact with iron, copper, or nickel salts; *Storage:* store in tight, light-resist. containers, optimally @ ≤ 72 F; avoid exposure to moisture and excessive heat.

Ascorbic Acid USP, FCC Gran. No. 6045654. [Roche] Ascorbic acid USP, FCC; CAS 50-81-7; EINECS 200-066-2; source of vitamin C and antioxidant in foods; *Regulatory:* FDA GRAS; *Proper-*

ties: wh. or sl. yel. cryst. powd., pract. odorless, pleasantly tart taste; 100% thru 20 mesh, 20% max. thru 50 mesh; sol. 1 g/3 ml water, 30 ml alcohol; m.w. 176.13; bulk dens. 0.8-1.0 (tapped); m.p. 190 C; pH 1.9-2.4 (10% aq.); 99-100.5% assay; *Precaution:* may deteriorate on exposure to atmospheric moisture; oxidizes readily in aq. sol'n.; avoid contact with iron, copper, or nickel salts; *Storage:* store in tight, light-resist. containers, optimally @ ≤ 72 F; avoid exposure to moisture and excessive heat.

Ascorbic Acid USP, FCC Type S No. 6045660. [Roche] Ascorbic acid USP, FCC; CAS 50-81-7; EINECS 200-066-2; source of vitamin C, antioxidant for foods, beverages, esp. where the ascorbic acid is dissolved during processing or at the end of the process, and in most dry powd. applics.; *Regulatory:* FDA GRAS; *Properties:* wh. or sl. yel. cryst. powd., pract. odorless, pleasantly tart taste; sol. 1 g/3 ml water, 30 ml alcohol; m.w. 176.13; bulk dens. 0.8-1.2 (tapped); m.p. 190 C; pH 1.9-2.4 (10% aq.); 99-100.5% assay; *Precaution:* may deteriorate on exposure to atmospheric moisture, oxidizes readily in aq. sol'n.; avoid contact with iron, copper, or nickel salts; *Storage:* store in tight, light-resist. containers, optimally @ ≤ 72 F; avoid exposure to moisture and excessive heat.

Ascorbic Acid USP, FCC Ultra-Fine Powd No. 6045653. [Roche] Ascorbic acid USP, FCC; CAS 50-81-7; EINECS 200-066-2; source of vitamin C and antioxidant in foods (instant beverages, cocoa prods., powd. foods); for use where exceptionally fine, dense powd. is needed for uniformity; *Regulatory:* FDA GRAS; *Properties:* wh. or sl. yel. cryst. powd., pract. odorless, pleasantly tart taste; 100% thru 100 mesh, 95% min. thru 200 mesh; sol. 1 g/3 ml water, 30 ml alcohol; m.w. 176.13; bulk dens. 0.8-1.1 (tapped); m.p. 190 C; pH 1.9-2.4 (10% aq.); 99-100.5% assay; *Precaution:* may deteriorate on exposure to atmospheric moisture; oxidizes readily in aq. sol'n.; avoid contact with iron, coper, or nickel salts; *Storage:* store in tight, light-resist. containers, optimally @ ≤ 72 F; avoid exposure to moisture and excessive heat.

Ascorbo-120. [MLG Enterprises Ltd.] Coated ascorbic acid and other edible excipients; CAS 50-81-7; EINECS 200-066-2; dough conditioner; intermediate oxidizer providing machine tolerance and dough strength to bread flours; for high usage level applics.; esp. effective in whole grain prods.; *Properties:* tablets supplying 120 ppm ascorbic acid to each cwt of flour; *Storage:* 6 mos shelf life stored under cool, dry conditions.

Ascorbo-160-2 Powd. [MLG Enterprises Ltd.] Coated ascorbic acid, calcium sulfate, and tricalcium phosphate; dough conditioner; intermediate oxidizer providing machine tolerance and dough strength to dry mixes and bases; esp. for whole grain variety breads; *Usage level:* 0.5-3 oz/cwt flour; *Properties:* powd.; 1 oz supplies 60 ppm ascorbic acid/cwt of flour; *Storage:* 6 mos shelf life stored under cool, dry conditions; close poly liner when not in use.

Ascorbo-C Tablets. [MLG Enterprises Ltd.] Coated ascorbic acid and other edible excipients; CAS 50-81-7; EINECS 200-066-2; dough conditioner; intermediate oxidizer providing machine tolerance and dough strength to bread flours; used with bromate, azodicarbonamide, and iodates; *Usage level:* 200 ppm max. ascorbic acid/cwt flour; *Properties:* tablets supplying 30 ppm ascorbic acid to each cwt of flour.

Ascorbyl Palmitate NF, FCC No. 60412. [Roche] Ascorbyl palmitate NF, FCC; CAS 137-66-6; EINECS 205-305-4; fat-sol. form of ascorbic acid; antioxidant for foods; preservative for natural oils, oleates, fragrances, colors, vitamins, edible oils/waxes; *Usage level:* 0.02% on wt. of oil or fat (foods); *Regulatory:* FDA 21CFR §182.3149, GRAS; *Properties:* wh. or ylsh. wh. cryst. powd., sl. odor; very sl. sol. in water and veg. oils; sol. 1 g/4.5 ml alcohol; m.w. 414.54; apparent dens. 8.7 lb/ft^3; m.p. 107-117 C; 95-100.5% assay; *Storage:* store in tight containers in cool dry place @ 8-15 C; avoid exposure to light, moisture, excessive heat.

Asol. [Lucas Meyer] Lecithin fraction; CAS 8002-43-5; EINECS 232-307-2; nonionic; antispatter agent, release agent, emulsifier for foods (baking margarine, frying fats); *Properties:* liq.; 40-100% conc.

Atlas 70K. [ICI Surf. Am.] Polysorbate 60; CAS 9005-67-8; cake emulsifier; *Regulatory:* FDA 21CFR §172.836; *Properties:* powd.

Atlas 90K. [ICI Surf. Am.] Blend; doughnut improver; *Regulatory:* FDA 21CFR §172.856; *Properties:* powd.

Atlas 110K. [ICI Surf. Am.] Sorbitan stearate; CAS 1338-41-6; EINECS 215-664-9; cake emulsifier; *Regulatory:* FDA 21CFR §172.842; *Properties:* powd.

Atlas 800. [ICI Surf. Am.] Mono- and diglycerides; nonionic; food emulsifier; enhances aeration, volume, and texture in icings; *Usage level:* 4-6% in shortening (icings); *Regulatory:* FDA 21CFR §184.1505; Canada compliance; *Properties:* ivory wh. plastic solid; sol. above its m.p. in IPA, veg. oils, min. oil; m.p. 115-122 F; HLB 2.8; iodine no. 56-64; flash pt. > 300 F; 40% min. alpha monoglyceride.

Atlas 1400K. [ICI Surf. Am.] Mono- and diglycerides; food emulsifier; provides volume and texture in cakes; emulsion stabilizer and film-former in caramels; inhibits oil separation in peanut butter; antispatter aid in margarine; in chewing gum, ice cream, coffee whiteners; *Usage level:* 3-6% in shortening (cake), 0.5-1% (caramel), 0.5-4% on fat (peanut butter), 0.1-0.5% (ice cream), 0.5% (margarine); *Regulatory:* FDA 21CFR §184.1505; Canada compliance; kosher; *Properties:* ivory wh. powd. or flakes, bland odor and taste; sol. above its m.p. in veg., min. oils; m.p. 144 F; HLB 2.8; iodine no. < 5; flash pt. > 300 F; 40-44% alpha mono.

Atlas 1500. [ICI Surf. Am.] Glyceryl stearate; food emulsifier; emulsion stabilizer and film-former in caramels; retards starch crystallization in starch jellies; inhibits oil separation in peanut butter; in dehydrated potatoes; *Usage level:* 0.5-1% (caramel), 0.25-0.5% on starch (starch jellies), 0.5-4%

on fat (peanut butter), 0.5-1% (dehydrated potatoes), 0.25-0.5% on starch (starch jellies); *Regulatory:* FDA 21CFR §182.1324, GRAS, 184.1505; Canada compliance; *Properties:* ivory wh. flakes, bland odor and taste; sol. above its m.p. in IPA, veg., min. oils; m.p. 140 F; HLB 3.5; iodine no. < 5; flash pt. > 300 F; 52% min. alpha monoglyceride; *Toxicology:* nonirritating to skin, noncorrosive.

Atlas 2000. [ICI Surf. Am.] Mono- and diglycerides (40%); food surfactant; *Regulatory:* FDA 21CFR §182.1505; *Properties:* hydrated form.

Atlas 2200H. [ICI Surf. Am.] Hydrated blend of mono- and diglycerides and polysorbate 60; food emulsifier, conditioner/softener for yeast-raised baked goods, e.g., bread, rolls; provides aeration in cakes, esp. those containing veg. oil; *Usage level:* 1-1.2% on flour (bread), 1-2.75% on flour (cake); *Regulatory:* FDA 21CFR §184.1505, 172.360, 172.836; *Properties:* ivory wh. plastic solid; disp. in water; m.p. 138-142 F; HLB 7.2; iodine no. < 3; 19.5% alpha monoglyceride.

Atlas 3000. [ICI Surf. Am.] Glyceryl oleate and propylene glycol; nonionic; food emulsifier, humectant, release; antifoaming agent for starch jellies, sugar-protein syrup systems; provides dispersibility in coffee whiteners, flavors; also in bread, cocoa, ice cream, cottage cheese, infant formula, sausage casings, margarine; *Usage level:* 0.05-0.5% (starch jellies), 0.4% (coffee whiteners), 0.02-0.05% (defoaming); *Regulatory:* FDA 21CFR §184.1505; Canada compliance; *Properties:* lt. amber clear liq.; sol. in cottonseed oil, IPA; insol. in water; sp.gr. 0.96; visc. 130 cps; m.p. 70 F; HLB 2.8; iodine no. 68 ± 3; flash pt. > 300 F; 46% min. alpha monoglyceride.

Atlas 5000K. [ICI Surf. Am.] Mono- and diglycerides; food emulsifier; provides volume and texture in cakes; extends shelf life in doughnuts; enhances aeration, volume, and texture in icings; *Usage level:* 0.4-0.5% on flour (bread), 3-6% in shortening (cake), 10-14% on shortening (doughnuts), 4-6% in shortening (icings); *Regulatory:* FDA 21CFR §184.1505; kosher; Canada compliance; *Properties:* ivory wh. plastic solid; sol. above its m.p. in IPA, veg. oils, min. oil; m.p. 133 F; HLB 3.5; iodine no. 72 ± 3; flash pt. > 300 F; 52% min. alpha monoglcyeride.

Atlas 5520. [ICI Surf. Am.] Mono- and diglycerides, polysorbate 60 with BHA, citric acid; food emulsifier, conditioner/softener for yeast-raised baked goods, e.g., bread, rolls; pan-release agent; when sprayed on buns, functions as pan release and minimzes seed loss; extends shelf life in doughnuts; *Usage level:* 0.6-0.8% on flour (bread), 1.25% on flour (doughnuts); *Regulatory:* FDA 21CFR §184.1505, 172.360, 172.836; *Properties:* golden clear liq.; sol. in veg. oils; disp. in water; m.p. 45 F; HLB 8.1; iodine no. 41; flash pt. > 300 F; 29% alpha monoglyceride.

Atlas A. [ICI Surf. Am.] Polysorbate 60; CAS 9005-67-8; cake emulsifier; *Regulatory:* FDA 21CFR §172.836; *Properties:* hydrate.

Atlas B-60. [ICI Surf. Am.] Sodium stearoyl lactylate; CAS 25383-99-7; EINECS 246-929-7; cake emulsifier; *Regulatory:* FDA 21CFR §172.846; *Properties:* hyrate.

Atlas E. [ICI Surf. Am.] Polysorbate 80; CAS 9005-65-6; icing stabilizer; *Regulatory:* FDA 21CFR §172.840; *Properties:* hydrate.

Atlas G-986K. [ICI Surf. Am.] Mono- and diglycerides (42%); food surfactant; *Regulatory:* FDA 21CFR §184.1505; *Properties:* plastic.

Atlas MDA. [ICI Surf. Am.] Amylase blend; CAS 9000-92-4; EINECS 232-567-7; food surfactant; *Regulatory:* FDA 21CFR §184.1505, GRAS; *Properties:* bead.

Atlas P-44. [ICI Surf. Am.] Mono- and diglycerides, sodium stearoyl lactylate; food surfactant; *Regulatory:* FDA 21CFR §172.846; *Properties:* bead.

Atlas SSL. [ICI Surf. Am.] Sodium stearoyl lactylate; CAS 25383-99-7; EINECS 246-929-7; dough strengthener, emulsifier, processing aid, surfactant, texturizer; *Regulatory:* FDA 21CFR §172.846; *Properties:* bead.

Atmos® 150. [ICI Am.; ICI Atkemix; ICI Surf. UK; Witco/H-I-P] Glyceryl stearate; nonionic; food emulsifier for puddings, frozen desserts; emulsion stabilizer for icings; provides lubrication for taco shells; extrusion aid for pasta; also for coffee whiteners, whipped toppings; *Usage level:* 0.15-2.0% by wt. of resin; 4-6% in shortening (icings), 0.2-1% (puddings), 0.2-1% (tacos), 0.1-0.2% (frozen desserts); *Regulatory:* FDA 21CFR §182.1324 (GRAS), 184.1505; Canada compliance; *Properties:* ivory wh. powd., bland odor and taste; sol. above its m.p. in veg. oils, min. oil, IPA; insol. in water, cottonseed oil; m.p. 140 F; HLB 3.5; iodine no. ≤ 5; flash pt. > 300 F; 52% min. alpha monoglyceride; *Toxicology:* nonirritating to skin, noncorrosive.

Atmos® 150 K. [Witco/H-I-P] Mono- and diglycerides; food emulsifier; emulsion stabilizer and film-former in caramels; inhibits oil separation in peanut butter; also for coffee whiteners, whipped toppings; *Usage level:* 0.5-1% (caramel), 0.5-4% on fat (peanut butter); *Regulatory:* FDA 21CFR §184.1505; *Properties:* ivory wh. flakes; m.p. 149 F; HLB 3.5; iodine no. < 5; 52% min. alpha monoglyceride.

Atmos® 300. [ICI Am.; ICI Atkemix; ICI Surf. UK; Witco/H-I-P] Glyceryl oleate and propylene glycol; nonionic; food emulsifier, humectant, release; antifoaming agent for starch jellies, sugar-protein syrup systems; provides dispersibility in coffee whiteners, flavors; also in bread, cocoa, ice cream, cottage cheese, infant formula, sausage casings, margarine; *Usage level:* 0.05-0.5% (starch jellies), 0.4% (coffee whiteners), 0.02-0.05% (defoaming); *Regulatory:* FDA 21CFR §184.1505; Canada compliance; *Properties:* lt. amber clear liq.; sol. in cottonseed oil, IPA; insol. in water; sp.gr. 0.96; visc. 130 cps; m.p. 70 F; HLB 2.8; iodine no. 68 ± 3; flash pt. > 300 F; 46% min. alpha monoglyceride.

Atmos® 300 K. [ICI Atkemix; Witco/H-I-P] Mono- and diglycerides; food emulsifier; provides dispersibility in coffee whiteners, flavors; defoamer for sugar-protein syrup systems; *Usage level:* 0.4% (coffee whiteners), 0.02-0.05% (defoaming); *Regulatory:* FDA 21CFR §184.1505, kosher; *Properties:* lt. amber liq.; m.p. 71 F; HLB 2.8; iodine no. 68 ± 3; 46% min. alpha monoglyceride.

Atmos® 378 K. [Witco/H-I-P] Hydrated blend of mono- and diglycerides, polysorbate 60, and sodium stearoyl lactylate; emulsifier blend for cakes permitting use of vegetable oil in certain formulations; provides aeration and volume; *Usage level:* 1-2.75% on flour (cake); *Regulatory:* FDA 21CFR §184.1505, 172.360, 172.846; *Properties:* ivory wh. plastic solid; HLB 5.5; iodine no. 5; 19% alpha monoglyceride.

Atmos® 659 K. [ICI Am.; Witco/H-I-P] Propylene glycol mixed esters, mono- and diglycerides, and lecithin; emulsifier blend for cake formulations; provides enhanced aeration and volume; *Usage level:* 1-2.75% on flour (cake); *Regulatory:* FDA 21CFR §184.1505, 184.1400, 172.856; *Properties:* ivory wh. plastic solid; HLB 3.4; iodine no. 21; 22% alpha monoglyceride.

Atmos® 729. [ICI Am.; ICI Atkemix] Hydrated mono- and diglycerides, polysorbate 60 (11.3%) and sorbitan stearate (11.3%); food emulsifier; used in cakes; *Usage level:* 2-4% on flour (cake); *Properties:* creamy wh., soft plastic; water disp.; m.p. 138-142 F; HLB 6.5; iodine no. 1; flash pt. > 300 F; 45% act.

Atmos® 758. [ICI Am.; Witco/H-I-P] Hydrated blend of sorbitan stearate (22.5%), polysorbate 60 (13.5%), and mono- and diglycerides; food emulsifier designed for cake doughnuts; *Usage level:* 2.5% on flour (doughnuts); *Regulatory:* FDA 21CFR §184.1505, 172.360, 172.842; *Properties:* ivory wh. soft plastic; water-disp.; HLB 7.5; iodine no. < 3; flash pt. > 300 F; 45% act., 5% alpha monoglyceride.

Atmos® 758 K. [Witco/H-I-P] Hydrated blend of sorbitan stearate, polysorbate 60, and mono- and diglycerides; emulsifier blend for cake doughnuts; enhances aeration and volume; *Usage level:* 1-2.75% on flour (cake); *Regulatory:* FDA 21CFR §184.1505, 172.360, 172.842; *Properties:* ivory wh. plastic solid; HLB 7.5; iodine no. < 3; 5% alpha monoglyceride.

Atmos® 1069. [ICI Atkemix; Witco/H-I-P] Hydrated blend of mono- and diglycerides, polysorbate 60, and sodium stearoyl lactylate; emulsifier blend for cakes permitting use of vegetable oil in certain formulations; enhances aeration and volume; *Usage level:* 1-2.75% on flour (cake); *Regulatory:* FDA 21CFR §184.1505, 172.360, 172.846; *Properties:* ivory wh. plastic solid; disp. in water; m.p. 135-142 F; HLB 5.5; iodine no. < 4; 19% alpha monoglyceride.

Atmos® 2462. [ICI Am.] Hydrated mono- and diglycerides, polysorbate 60 (13%), sorbitan stearate (13%) and propylene glycol; food emulsifier; *Properties:* creamy wh. soft plastic; water disp.; HLB 7.4; flash pt. > 300 F; 39% act.; Unverified

Atmos® 7515 K. [Witco/H-I-P] Blend of mono- and diglycerides, polysorbate 60, and sodium stearoyl lactylate; food emulsifier for cakes; disperses easily in batter, producing cakes with fine, uniform grain, enhanced aeration and volume; *Usage level:* 1-2% on flour (cake); *Properties:* bead-shaped powd.

Atmul® 27 K. [Witco/H-I-P] Mono- and diglycerides; food emulsifier; *Regulatory:* FDA 21CFR § 184.1505; *Properties:* ivory wh. plastic solid; m.p. 92 F; HLB 2.8; iodine no. 70 ± 3; 25% alpha monoglyceride.

Atmul® 80. [ICI Am.; ICI Atkemix; Witco/H-I-P] Mono- and diglycerides; nonionic; food emulsifier; enhances aeration, volume, and texture in icings; *Usage level:* 4-6% in shortening (icings); *Regulatory:* FDA 21CFR §184.1505; Canada compliance; *Properties:* ivory wh. plastic solid; sol. above its m.p. in IPA, veg. oils, min. oil; m.p. 115-122 F; HLB 2.8; iodine no. 56-64; flash pt. > 300 F; 40% min. alpha monoglyceride.

Atmul® 82. [ICI Atkemix] Mono- and diglycerides; nonionic; food emulsifier; *Properties:* ivory wh. plastic solid; HLB 2.8; pour pt. 46 C.

Atmul® 84. [ICI Surf. Am.; ICI Atkemix; Witco/H-I-P] Glyceryl stearate; food emulsifier for puddings; improves volume and texture of cakes; emulsion stabilizer and film-former in caramels; inhibits oil separation in peanut butter; in chewing gum, starch jellies, dehydrated potatoes, coffee whiteners, whipped toppings; *Usage level:* 3-6% in shortening (cake), 0.5-1% (caramel), 0.5-4% on fat (peanut butter), 0.5-1% (dehydrated potatoes), 0.25-0.5% on starch (starch jellies); *Regulatory:* FDA 21CFR §184.1505; Canada compliance; *Properties:* ivory wh. beads or flakes, bland odor and taste; sol. above its m.p. in veg., min. oils; m.p. 140 F; HLB 2.8; iodine no. < 5; flash pt. > 300 F; 40% min. alpha monoglyceride.

Atmul® 84 K. [ICI Am.; ICI Atkemix; Witco/H-I-P] Mono- and diglycerides; food emulsifier; provides volume and texture in cakes; emulsion stabilizer and film-former in caramels; inhibits oil separation in peanut butter; antispatter aid in margarine; in chewing gum, ice cream, coffee whiteners; *Usage level:* 3-6% in shortening (cake), 0.5-1% (caramel), 0.5-4% on fat (peanut butter), 0.1-0.5% (ice cream), 0.5% (margarine); *Regulatory:* FDA 21CFR §184.1505; Canada compliance; kosher; *Properties:* ivory wh. powd. or flakes, bland odor and taste; sol. above its m.p. in veg., min. oils; m.p. 144 F; HLB 2.8; iodine no. < 5; flash pt. > 300 F; 40-44% alpha mono.

Atmul® 86 K. [ICI Am.; ICI Atkemix] Mono- and diglycerides; nonionic; food emulsifier; provides volume and texture in cake, icings; antispattering agent in margarine; *Usage level:* 3-6% in shortening (cake), 4-6% in shortening (icings), 0.5% (margarine); *Regulatory:* FDA 21CFR §184.1505; kosher; Canada compliance; *Properties:* ivory wh. plastic solid; sol. above its m.p. in IPA, veg. oils, min. oil; m.p. 117 F; HLB 2.8; iodine no. 74 ± 3; flash pt. > 300 F; 40% min. alpha monoglyceride.

Atmul® 122. [ICI Am.] Mono- and diglycerides; nonionic; food emulsifier; *Properties:* ivory wh. plastic solid; sol. above its m.p. in IPA, veg. oils, min. oil; m.p. 126 F; HLB 3.5; flash pt. > 300 F; Unverified

Atmul® 124. [ICI Surf. Am.; ICI Atkemix; Witco/H-I-P] Glyceryl stearate; food emulsifier; emulsion stabilizer and film-former in caramels; retards starch crystallization in starch jellies; inhibits oil separation in peanut butter; in dehydrated potatoes; *Usage level:* 0.5-1% (caramel), 0.25-0.5% on starch (starch jellies), 0.5-4% on fat (peanut butter), 0.5-1% (dehydrated potatoes), 0.25-0.5% on starch (starch jellies); *Regulatory:* FDA 21CFR § 182.1324, GRAS, 184.1505; Canada compliance; *Properties:* ivory wh. flakes, bland odor and taste; sol. above its m.p. in IPA, veg., min. oils; m.p. 140 F; HLB 3.5; iodine no. < 5; flash pt. > 300 F; 52% min. alpha monoglyceride; *Toxicology:* nonirritating to skin, noncorrosive.

Atmul® 400. [ICI Atkemix] Mono- and diglycerides with 2% sodium stearate as dispersant; surfactant for food industry; *Regulatory:* Canada compliance; *Properties:* wh. free-flowing powd.; disp. in water; m.p. 139-143 F; HLB 3.5; iodine no. < 2; 50-54% alpha mono; *Toxicology:* 0.4-0.5% on flour (bread).

Atmul® 500. [ICI Am.; ICI Atkemix; Witco/H-I-P] Mono- and diglycerides; food emulsifier; softener for bread; extends shelf life in doughnuts; enhances aeration, volume, and texture in icings; *Usage level:* 0.2-0.5% on flour (bread), 10-14% on shortening (doughnuts), 4-6% in shortening (icings); *Regulatory:* FDA 21CFR §184.1505; Canada compliance; *Properties:* ivory wh. plastic solid; sol. above its m.p. in IPA, veg. oils, min. oil; m.p. 132 F; HLB 3.5; iodine no. 48 ± 3; flash pt. > 300 F; 52% min. alpha monoglyceride.

Atmul® 600H. [ICI Atkemix] Hydrated mono- and diglycerides with 4% polysorbate 60 as stabilizer; surfactant for food industry; *Usage level:* 1-1.25% on flour (bread); *Regulatory:* Canada compliance; *Properties:* wh. creamy plastic; disp. in water; m.p. 135-140 F; HLB 3.5; iodine no. 9; 20% alpha mono.

Atmul® 601 K. [Witco/H-I-P] Mono- and diglycerides; food emulsifier; *Regulatory:* FDA 21CFR § 184.1505; *Properties:* ivory wh. plastic solid; m.p. 125 F; HLB 2.8; iodine no. 64 ± 3; 40% min. alpha monoglyceride.

Atmul® 651 K. [ICI Am.; ICI Atkemix; Witco/H-I-P] Mono- and diglycerides; food emulsifier; provides volume and texture in cakes; extends shelf life in doughnuts; enhances aeration, volume, and texture in icings; *Usage level:* 0.4-0.5% on flour (bread), 3-6% in shortening (cake), 10-14% on shortening (doughnuts), 4-6% in shortening (icings); *Regulatory:* FDA 21CFR §184.1505; kosher; Canada compliance; *Properties:* ivory wh. plastic solid; sol. above its m.p. in IPA, veg. oils, min. oil; m.p. 133 F; HLB 3.5; iodine no. 72 ± 3; flash pt. > 300 F; 52% min. alpha monoglcyeride.

Atmul® 695. [Witco/H-I-P] Mono- and diglycerides; food emulsifier for puddings; *Usage level:* 0.1-0.2% (puddings); *Regulatory:* FDA 21CFR §184.1505; *Properties:* amber semiliq.; m.p. 76 F; HLB 3.0; iodine no. 76 ± 3; 52% min. alpha monoglyceride.

Atmul® 695 K. [Witco/H-I-P] Mono- and diglycerides; food emulsifier; *Regulatory:* FDA 21CFR § 184.1505; *Properties:* amber semiliq.; m.p. 77 F; HLB 3.0; iodine no. 76 ± 3; 52% min. alpha monoglyceride.

Atmul® 700 H. [ICI Am.] Hydrated mono- and diglycerides with polysorbate 60 (1%); food emulsifier, softener for yeast-raised bakery prods.; *Properties:* creamy wh., soft plastic; water disp.; HLB 4.0; flash pt. > 300 F; Unverified

Atmul® 900. [ICI Atkemix] Distilled monoglyceride; food emulsifier; *Properties:* ivory wh. beads; HLB 4.5; pour pt. 70 C.

Atmul® 900K. [ICI Atkemix] Distilled monoglyceride; food emulsifier; *Regulatory:* kosher; *Properties:* ivory wh. beads; HLB 4.5; pour pt. 72 C.

Atmul® 918 K. [Witco/H-I-P] Mono- and diglycerides; food emulsifier; inhibits aeration, aids emulsion stability in icings; retards starch crystallization in starch jellies; provides dryness and overrun in ice cream; *Usage level:* 4-6% in shortening (icings), 0.25-0.5% on starch (starch jellies); *Regulatory:* FDA 21CFR §184.1505; *Properties:* ivory wh. powd.; m.p. 149 F; HLB 3.5; iodine no. < 5; 52% min. alpha monoglyceride.

Atmul® 1003 K. [Witco/H-I-P] Mono- and diglycerides; food emulsifier; *Regulatory:* FDA 21CFR §184.1505; *Properties:* ivory wh. plastic solid; m.p. 119 F; HLB 2.8; iodine no. 85 + 3; 40% min. alpha monoglyceride.

Atmul® 2622 K. [ICI Am.; Witco/H-I-P] Glyceryllactostearate; lipophilic food emulsifier useful for aerating props.; *Regulatory:* FDA 21CFR § 172.852, 172.860, 172,862; *Properties:* ivory flakes; HLB 3.7; hyd. no. 152-170; 6-8.5% alpha monoglyceride, 18.7-21.1% lactic acid.

Atmul® P-28. [ICI Surf. Am.] Mono- and diglycerides, sodium stearoyl-2-lactylate; food additive for bread and bakery prods.; *Usage level:* Canada: 0.81% max. on flour (bread); *Regulatory:* Canada compliance.

Atmul® P-36. [ICI Surf. Am.] Sodium stearoyl-2-lactylate, mono- and diglycerides; food additive for bread and bakery prods.; *Usage level:* Canada: 0.57% max. on flour (bread); *Regulatory:* Canada compliance.

Atmul® P-44. [ICI Surf. Am.] Sodium stearoyl-2-lactylate, mono- and diglycerides; food additive for bread and bakery prods.; *Usage level:* Canada: 0.55% max. on flour (bread); *Regulatory:* Canada compliance.

Atmul® P-50. [ICI Surf. Am.] Acetylated tartaric acid esters of mono- and diglycerides; food additive for bread and unstandardized foods; *Usage level:* Canada: 2% max. on flour (bread); *Regulatory:* Canada compliance.

Atmul® P-96. [ICI Surf. Am.] Mono- and diglycerides; food additive for bread and unstandardized foods; *Regulatory:* Canada compliance.

Atomite®. [ECC Int'l.] Calcium carbonate; CAS 1317-65-3; EINECS 207-439-9; *Regulatory:* NSF compliance; *Properties:* wh. powd., odorless; 3.0 μ mean particle size; fineness (Hegman) 6.0; negligible sol. in water; sp.gr. 2.71; bulk dens. 45 lb/ft³ (loose); surf. area 2.8 m²/g; oil absorp. 15; pH 9.5 (5% slurry); nonflamm.; 97.6% CaCO₃, 0.2% max. moisture; *Toxicology:* TLV/TWA 10 mg/m³, considered nuisance dust; *Precaution:* incompat. with acids.

Atsurf 456K. [ICI Surf. Am.] Polyglycerol ester; surfactant for food industry; *Regulatory:* FDA 21CFR §172.854; *Properties:* liq.

Atsurf 594. [ICI Surf. Am.; ICI Atkemix; ICI Surf. UK] Glyceryl oleate; food emulsifier; *Properties:* amber visc. liq.; HLB 2.8; pour pt. 19 C.

Atsurf 595. [ICI Surf. Am.; ICI Atkemix] Glyceryl oleate; surfactant; *Properties:* yel. clear liq.; HLB 2.8; pour pt. 11 C.

Atsurf 595K. [ICI Atkemix] Glyceryl oleate; surfactant; *Regulatory:* kosher; *Properties:* yel. clear liq.; HLB 2.8; pour pt. 11 C.

Atsurf 596. [ICI Surf. Am.; ICI Atkemix] Glyceryl oleate; surfactant; *Properties:* yel. clear liq.; visc. 130 cps; HLB 2.8.

Atsurf 596K. [ICI Surf. Am.] Glyceryl oleate; surfactant for food industry; *Regulatory:* FDA 21CFR §184.1505, GRAS; *Properties:* yel. clear liq.; visc. 130 cps; HLB 2.8.

Autolyzed Type G Torula Dried Yeast. [Lake States] Autolyzed yeast; flavor enhancer which brings out hearty savory meat flavors in high temp. processed foods, e.g., soups, stews, chili, sloppy joes, gravies, and sauces; *Properties:* pH 5.8 (5%); 40% min. protein; *Storage:* store in cool, dry conditions.

Autolyzed Type N Torula Dried Yeast. [Lake States] Autolyzed yeast; flavor potentiator in high temp. processed foods such as canned soups, sauces, gravies, stews, seafoods; can intensify top flavor notes associated with HVP; *Regulatory:* kosher; *Properties:* pH 5.8 (5%); 42% min. protein; *Storage:* store in cool, dry conditions.

Avgard™. [Rhone-Poulenc Food Ingreds.] Trisodium phosphate; food additive for meat, poultry, and seafood industries.

Avicel® CL-611. [FMC] Cellulose gel in colloidal disp.; CAS 9004-34-6; fat replacement system imparting creamy mouthfeel; stabilizer in low-fat emulsions; provides foam stability, controls ice crystal growth and syneresis, adds visc. and opacity, suspends particulates; for dressings, frozen desserts, whipped toppings; *Regulatory:* FDA GRAS.

Avicel® RC-501. [FMC] Cellulose gel in colloidal disp.; CAS 9004-34-6; fat replacement system imparting short creamy mouthfeel; stabilizer in low-fat emulsions; provides foam stability, controls ice crystal growth and syneresis, adds visc. and opacity, suspends particulates; *Regulatory:* FDA GRAS.

Avicel® RC-581. [FMC] Cellulose gel in colloidal disp.; CAS 9004-34-6; fat replacement system imparting short creamy mouthfeel; stabilizer in low-fat emulsions; provides foam stability, controls ice crystal growth and syneresis, adds visc. and opac-

ity, suspends particulates; for frozen desserts, whipped toppings; *Regulatory:* FDA GRAS.

Avicel® RC-591F. [FMC] Cellulose gel in colloidal disp.; CAS 9004-34-6; fat replacement system imparting creamy mouthfeel; stabilizer in low-fat emulsions; provides foam stability, controls ice crystal growth/syneresis, adds visc. and opacity, suspends particulates; for baked goods, sour cream, mayonnaise, frozen yogurt; *Regulatory:* FDA GRAS.

Avicel® RCN-10. [FMC] Cellulose gel coprocessed with 10% guar gum; fat replacement system; imparts creamy fat-like mouthfeel, simulates rheological props. of emulsion systems, increases solids; for dairy prods., frozen desserts, cream soup; *Regulatory:* FDA GRAS; *Properties:* spherical particles, 1-10 μ size.

Avicel® RCN-15. [FMC] Cellulose gel; dry co-processed mixt. of cellulose gel and guar gum (85:15); fat replacement system; simulates rheological props. of fat; provides water structuring, creamy mouthfeel and body, self-stabilization; absorbs moisture in low-solids systems; for baked goods, puddings, dressings, gravies, whipped toppings; *Regulatory:* FDA GRAS; *Properties:* spherical particle, 1-15 μ size.

Avicel® RCN-30. [FMC] Cellulose gel co-processed with maltodextrin and xanthan gum; fat replacement system; *Regulatory:* FDA GRAS; *Properties:* rod-shaped particles.

Avicel® WC-595. [FMC] Cellulose gel in colloidal disp.; CAS 9004-34-6; stabilizer, fat replacement system; provides creaminess, thickening, freeze/thaw stability, moisture retention, air cell uniformity to baked goods, dairy prods., dressings, sauces; *Regulatory:* FDA GRAS.

Axol® C 62. [Goldschmidt AG] Citric acid ester of glycerol mono/distearates; CAS 68990-05-6; anionic; food emulsifier, surfactant; *Regulatory:* EEC E472c, German food compliance; *Properties:* wh.-ivory powd., char. odor; disp. in water, warm essential oils; insol. in veg. oils; sp.gr. 0.900; m.p. 58-64 C; HLB 10.0 ± 1; iodine no. 3 max.; sapon. no. 215-265; flash pt. > 100 C; pH 6.0 (50 g/l water); 100% conc.; *Toxicology:* nonhazardous; avoid formation of dusts.

Axol® C 63. [Goldschmidt AG] Citric acid ester of glycerol mono/distearates; CAS 68990-05-6; anionic; food emulsifier; for cooked meat prods., sausage, etc.; prevents fat and gel separation, improves consistency of sausage mixes; *Usage level:* 3-5 g/1kg sausage mix; *Regulatory:* EEC E472c compliance; *Properties:* ivory powd., char. odor; misc. with water; sp.gr. 0.900; flash pt. > 100 C; pH 6.0 (50 g/l water); 100% conc.; *Toxicology:* nonhazardous; avoid formation of dusts.

Axol® E 41. [Goldschmidt AG] Acetic acid ester of mono- and diglycerides; CAS 68990-58-9; nonionic; food emulsifier; *Regulatory:* EEC E472a, FAO/WHO compliance; *Properties:* ivory slabs; misc. with fats and oils; insol. in water; m.p. 42-47 C; HLB 2-3; iodine no. 3 max.; sapon. no. 240-270;

hyd. no. 95-115; 100% conc.

Axol® E 61. [Goldschmidt AG] Acetylated hydrog. lard glyceride; CAS 8029-91-2; nonionic; food emulsifier, lubricant, solv., plasticizer and coating material for foodstuffs (meat prods., cheese, dried fruit, nuts); can be combined with hydrophilic emulsifiers to increase whipping props. in shortenings/toppings; Regulatory: EEC E472a, FAO/WHO compliance; Properties: ivory waxy slabs, faint char. odor; misc. with fats and oils; insol. in water; m.p. 34-38 C; HLB 2-3; iodine no. 3 max.; sapon. no. 310-340; hyd. no. 75-95; flash pt. > 100 C; 100% conc.; Toxicology: nonhazardous.

Axol® E 66. [Goldschmidt AG] Acetic acid ester of mono- and diglycerides; CAS 68990-55-6; nonionic; lubricant, solubilizer, solv. and plasticizer for foodstuffs; solubilizer for spice oils, essential oils, antioxidants; plasticizer for fats; Regulatory: EEC E472a compliance; Properties: pale yel. liq., faint char. odor; misc. with fats and oils; insol. in water; sp.gr. 0.900; HLB 2-3; iodine no. ≈ 30; sapon. no. 360-390; hyd. no. 20 max.; cloud pt. 10-15 C; flash pt. > 100 C; 100% conc.; Toxicology: nonhazardous.

Axol® L 61, L62. [Goldschmidt AG] Lactic acid fatty acid glyceride; nonionic; food emulsifier; Properties: solid; HLB 5; 100% conc.; Unverified

Axol® L 626. [Goldschmidt AG] Lactic acid ester of mono- and diglycerides; nonionic; food emulsifier; Properties: paste; HLB 5; 100% conc.; Unverified

AYE-2000. [Gist-brocades] Brewer's yeast extract; CAS 8013-01-2; imparts a subtle caramelized beef or sweet soy profile; useful to round flavors in savory systems; suggested for oriental dishes, rice and pasta dishes, finished entrees, sauces, and gravies.

AYE 2200. [Gist-brocades] Yeast extract; CAS 8013-01-2; char. by a sharp cheddar cheese like aftertaste combined with subtle bitter flavor notes; used in seasoning blends, cheese crackers, cheese spreads, beef soups, gravies, and sauce mixes.

AYE 2312. [Gist-brocades] Yeast extract; CAS 8013-01-2; contains more smokey flavor and bitterness with a hint of beer aroma; for use in cheese spreads; Properties: powd.

AYE Family. [Gist-brocades] Yeast extracts derived from spent brewer's yeast; CAS 8013-01-2; flavor for use where cheese-like or heavy beef after-taste is desirable in dough-based snacks, cheese soup, sauce, rice, or pasta dishes.

AYS 2311. [Gist-brocades] Yeast extract; CAS 8013-01-2; imparts a toasted cheddar cheese-like flavor with a strong after-taste similar to that found in common cheese cracker snacks; for use by bakers and manufacturers of dehydrated cheeses.

AYS 2350. [Gist-brocades] Autolyzed brewer's yeast extract; CAS 8013-01-2; imparts a baked or beer flavor in bread, rolls, and snack foods; esp. useful in frozen dough.

Aytex®-P Food Powd. Wheat Starch. [ADM Ogilvie] Wheat starch; CAS 9005-25-8; for food prods. incl. breakfast cereals, mixes, soups, sauces; Properties: brilliant wh. powd.; pH 6.0; 11% moisture, 0.4% protein.

B

B-45. [Brolite Prods.] A no-time dough accelerant and conditioner for all breads; *Properties:* powd.

B-50. [Brolite Prods.] Highly conc. no-time dough accelerant designed esp. for automatic lines and frozen dough systems; *Properties:* powd.

Bac-N-Fos®. [Rhone-Poulenc Food Ingreds.] Sodium hexametaphosphate FCC and sodium bicarbonate FCC; food additive for bacon processing; helps reduce formation of nitrosamines and increases yield of No. 1 slices; *Regulatory:* kosher; *Properties:* wh. powd., odorless; 7% max. on 60 mesh, 65% min. thru 100 mesh; pH 7.0-7.4 (1%); *Storage:* store cool and dry.

Bac-N-Fos® Formula 191. [Rhone-Poulenc Food Ingreds.] Sodium hexametaphosphate and sodium bicarbonate; food additive for bacon processing; reduces residual nitrite levels in bacon, stabilizes pickle sol'ns., produces improved slab texture, more appealing color and less spatter when frying.

Bagel Base. [MLG Enterprises Ltd.] Calcium sulfate, ascorbic acid, potassium bromate, L-cysteine hydrochloride; conc. dough conditioner system for bagel prod.; *Usage level:* 0.50-1.0%/cwt flour; *Properties:* free-flowing powd.; *Storage:* 6 mos shelf life stored under cool, dry conditions.

Bagel Buddy. [Brolite Prods.] Base for bagels; *Usage level:* 5 lb/100 lb flour.

Bakeall M/V All Purpose Shortening. [Bunge Foods] Meat fats and vegetable oil, antioxidant; shortening for general bakery use, cookies, biscuits, pie crusts, frying; *Properties:* 1.1 red color, bland flavor; m.p. 120 F; smoke pt. 450 F.

Baker's Ideal®. [Bunge Foods] Partially hydrog. soybean and cottonseed oils, water, salt, mono- and diglycerides, artifical flavor, Vitamin A; margarine for puff pastry; *Properties:* drop pt. 120 F; 80.2% fat, 17.2% moisture, 2.6% salt.

Bakers Nutrisoy. [ADM Arkady] Enzyme inactive defatted soy flour; CAS 68513-95-1; protein source for foods.

Bake-Well 52. [Am. Ingreds.] Soy oil, coconut oil, min. oil (25%), and lecithin; release aid for use on pans, molds, and conveyor belts to reduce sticking; *Properties:* amber liq.; iodine no. 50-60; smoke pt. 325 F.

Bake-Well 80/20. [Am. Ingreds.] Soy oil, white min. oil USP (20%), and lecithin (1.75-2.25%); release aid for use on pans, molds, and conveyor belts to reduce sticking; *Properties:* amber liq.; i.v. 100-110; smoke pt. 325 F.

Bake-Well Bun Release. [Am. Ingreds.] Mono- and diglycerides, polysorbate 60 (35-37%), water, and propylene glycol; bun pan release; *Properties:* amber liq.; i.v. 39-44; 24% min. mono.

Bake-Well Heavy Divider Oil. [Am. Ingreds.] Mineral oil USP; lubricant for use in dough dividers; *Properties:* clear liq.; sp.gr. 0.86-0.89; visc. 63-72 cps; flash pt. 445 F.

Bake-Well High Stability Pan Oil. [Am. Ingreds.] Soy oil, white min. oil (25%), and lecithin (0.5%); pan and mold release aid; *Properties:* amber liq.; i.v. 80-92; smoke pt. 330 F min.

Bake-Well K Machine Oil. [Am. Ingreds.] Mineral oil USP; lubricant for use in dough dividers; *Properties:* clear liq.; visc. 38-43 cps; flash pt. 400 F.

Banana Essence 1000 Fold Natural. [Confoco; Commodity Services Int'l.] Banana essence contg. acetaldehyde, ethanol, diacetyl, butyrate, butyl acetate, ethyl-2-methyl butyrate, 2-hexenal, isoamyl acetate, methyl amylketone, isoamyl alcohol, etc.; produces natural fresh banana effect comprising top and body notes of ripe banana fruit for water-based flavor systems; also in combination with other fruit extracts for tropical or exotic fruit effect; *Usage level:* 0.5-1% (fruit drinks, Italian ices), 1% (ice cream), 1-1.5% (sparkling waters), 1-2% (yoghurt); *Regulatory:* kosher certified; *Properties:* colorless clear liq., intense odor and taste of fresh bananas of the Cavendish variety; sol. in water, ethanol, propylene glycol; insol. in veg. oils; sp.gr. 1.0; m.p. 0 C; b.p. 99-100 C; flash pt. > 80 C; possible dizziness, light-headed feeling; *Storage:* > 12 mos stability stored below 7 C in sealed containers away from air and light; avoid freezing.

Barley*Complete® 25. [Zumbro; Garuda Int'l.] Hydrolyzed barley flour; controls body, provides mouthfeel, regulates osmolality; natural sweetener; helps lower serum cholesterol; for nutritional beverages, gravies, cereals, baked goods; *Properties:* lt. tan powd., bland odor/flavor; 100% thru 140 mesh; sol. in hot or cold water; 7% sol. fiber, 12% insol. fiber, < 8% moisture; *Storage:* store in cool, dry area.

Basic Natural™. [MLG Enterprises Ltd.] Calcium sulfate and vegetable powd.; patented natural dough conditioner, gluten modifier; produces doughs with superior machining qualities; gentle reducing action, more tolerance to mixing and dosage level; produces drier doughs; for breads, bun, English muffins, puff pastry; *Usage level:* 1-4 oz/ 100 lb flour (breads, buns), 1-6 oz/100 lb flour (English muffins), 1-5 oz/100 lb flour (puff pastry); *Properties:* free-flowing; *Storage:* 6 mos shelf life stored in cool, dry conditions; keep plastic bags sealed when not in use.

Basic Spice Mix. [Bunge Foods] Flour-type spice mix with all natural color; for breads, rolls, sweet goods, danish, donuts, cakes, and prods. where accenting of egg color is desirable; *Usage level:* 1-1.5%; *Properties:* yel. powd.; pH 6.2; 15% max. moisture.

Basmati Rice Extract. [Beeta; Commodity Services Int'l.] Basmati rice extract contg. natural heterocyclic materials, hexanol, propionaldehyde, acetaldehyde, 3-methyl-1-butanol, aliphatic esters; propylene glycol as carrier; flavor ingred. to build rounded natural baked and roasted effects in microwave food prods.; helps create roasted grain and cereal impressions in extruded snacks; *Regulatory:* IFRA, IOFI; *Properties:* pale yel.-orange clear to sl. cloudy liq.; sol. in water, 70% ethanol; sp.gr. 1.035-1.045; b.p. 180 C; flash pt. (TCC) > 300 F; ref. index 1.420-1.430.

Batter Up®. [A.E. Staley Mfg.] Food starch modified derived from dent corn; starch for batters and breading applics. for seafood (fillets, sticks, shrimp, scallops), poultry prods., onion rings; provides exc. adhesion; withstands freezing, deep fat frying, steam table conditions; *Regulatory:* FDA 21CFR §172.892; *Properties:* wh. powd.; pH 5.0-7.0; 10-12% moisture.

BBS. [Karlshamns Food Ingreds.] Partially hydrog. veg. oil (soybean and palm); nontropical version also avail.; specialty shortening for baking and frying where emulsifiers are not required; *Properties:* liq., plastic; m.p. 116 F.

B&C Caramel Powd. [MLG Enterprises Ltd.] Caramel color; CAS 8028-89-5; EINECS 232-435-9; natural colorant for dry mixes, soups, gravies, frozen foods, specialties; *Properties:* free-flowing powd.; 100% min. thru 40 mesh, 90% min. thru 100 mesh; pH 6.7-7.7 (1%); 4% max. moisture; *Storage:* unlimited shelf life under normal storage conditions.

Beakin LV1. [ADM Ross & Rowe Lecithin] Complexed lecithin; CAS 8002-43-5; EINECS 232-307-2; instantizer, spray oils, mold release; *Properties:* translucent fluid.

Beakin LV2. [ADM Ross & Rowe Lecithin] Complexed lecithin; CAS 8002-43-5; EINECS 232-307-2; for instantized mixes, pan/mold release; *Properties:* translucent fluid.

Beakin LV3. [ADM Ross & Rowe Lecithin] Complexed lecithin with vegetable oil; for processed cheese, instantized foods, dry mixes; *Properties:* translucent fluid.

Beakin LV4. [ADM Ross & Rowe Lecithin] Complexed lecithin with vegetable oil; dispersant, dust control, pan/mold release; *Properties:* translucent fluid.

Beakin LV30. [ADM Ross & Rowe Lecithin] Complexed lecithin with vegetable oil; instantizing

agent, drink mixes, pan/mold release; *Properties:* translucent fluid; water-disp.

Bealite™. [Kerry Ingreds./Beatreme] Food emulsifiers; *Properties:* free-flowing spray-dried prod.

Beanfeast. [Courtaulds plc] Proteinaceous substances used as food or as ingreds. for food.

BeaTrim™. [Kerry Ingreds./Beatreme] Formulated ingred. systems for low-fat applics.

Beef Extract #3041. [Ariake USA] Beef extract; flavorings for snacks, soup mixes, sauces, gravies; 25% moisture, 63% protein.

Beef Extract Paste #9225. [E.A. Miller] Beef extract; protein-rich, full-bodied beef flavor for use in snack foods, soups, gravies, and sauces; *Usage level:* 0.4-1.2% (soups, broths), 1.4-2% (gravies, sauces), 1-2% (snack foods); *Regulatory:* USDA approved; *Properties:* dk. brn. paste, no off odors or taste; pH 5.8 ± 0.4 (2%); 80 ± 2% total solids, 60 ± 2% protein, 5 ± 2% salt, 3 ± 2% fat; *Storage:* 12 mos shelf life; keep cool (< 70 F) and dry.

Beef Extract Powd. #9267. [E.A. Miller] Beef extract; protein-rich, full-bodied roast beef flavor for use in snack foods, soups, gravies, and sauces; *Usage level:* 0.4-1.2% (soups, broths), 1.4-2% (gravies, sauces), 1-2% (snack foods, sausages, reformed meat prods.); *Regulatory:* USDA approved; *Properties:* lt. brn. spray-dried powd.; pH 5.8 ± 0.4 (2%); 97 ± 2% total solids, 74 ± 2% protein, 5 ± 2% salt, 3 ± 2% fat; *Storage:* 12 mos shelf life; keep cool (< 70 F) and dry.

Beef Spicey No. 25748. [Universal Flavors] Mixt. of process flavorings and flavor enhancers; flavor char. of beef; for use in soups, sauces, beef prods.; *Usage level:* 0.4-0.8%; *Properties:* dk. brn. mobile liq.; dens. 1.30 ± 0.1; flash pt. > 65 c; ref. index 1.45 ± 0.02; *Toxicology:* nonhazardous; *Storage:* 6 mos storage life when stored in tightly sealed, well-filled containers in dark place, below 5 C.

Beer Extract. [Commodity Services Int'l.] Extract from beer brewed only from malt, water, and hops in accordance with German beer law; used for alcohol-reduced and alcohol-free beer, low-calorie beverages, and for improving malt beer and malty beverages; imparts an inimitable note to isotonic beverages; also as whiskey flavor; *Properties:* sol. 40 g/l in water; sol. 1:1 in propylene glycol, triacetin; *Storage:* 1 yr stability when stored in glass containers @ 7 C in absence of light and air.

Beetroot Conc. P3003. [Phytone Ltd.] Conc. juice of red beetroot (*Beta vulgaris*), contg. pigment betanin; CAS 89957-89-1; EINECS 289-610-8; food colorant; *Regulatory:* EEC E162; JECFA compliance; *Properties:* deep purple-red mobile liq.; *Storage:* store @ 5-10 C in tightly closed containers in cool, dry area; protect from direct sunlight.

Benecel® M 042. [Hercules] Methyl cellulose; CAS 9004-67-5; nonionic; thickener for cheese sauces; synergistic with starch, improving prod. cling; moisture retention, thermal gelation for meat patties, structured potato prods.; provides adhesion for microwavable foods; *Usage level:* 0.5-1.2% (cheese sauces), 0.05-0.15% (meat patties), 0.25-

0.75% (potato prods.), 0.1-2.5% (batters), 20-75% (microwave foods); *Regulatory:* FDA 21CFR §182.1480, GRAS; USDA 9CFR §318.7, 381.147.

Benecel® M 043. [Hercules] Methyl cellulose; CAS 9004-67-5; nonionic; thickener, suspending aid, binder; inhibits syneresis; prevents shrinkage in pie fillings; thermal gelation; for barbecue/cheese sauces, salad dressings; reduces oil absorp. during frying of onion rings; moisture retention in meat patties; *Usage level:* 0.3-0.6% (pie fillings), 0.2-1.0% (barbecue sauces), 0.5-1.2% (cheese sauces), 0.2-0.8% (salad dressings), 0.4-2.5% (onion rings), 0.05-0.15% (meat patties), 0.25-0.75% (potato prods.), 20-75% (microwave prods.), 0.1-0.75% (tortillas); *Regulatory:* FDA 21CFR §182.1480, GRAS; USDA 9CFR §318.7, 381.147.

Benecel® MP 643. [Hercules] Hydroxypropyl methylcellulose; CAS 9004-65-3; nonionic; thickener, suspending aid for salad dressings; controls batter visc., improves moisture retention, gas retention during baking, improving volume and texture of dry mix cakes; *Usage level:* 0.2-0.8% (salad dressings), 0.1-0.5% (cakes); *Regulatory:* FDA 21CFR §172.874.

Benecel® MP 824. [Hercules] Hydroxypropyl methylcellulose; CAS 9004-65-3; nonionic; binder; improves mouthfeel, body, texture; ice crystal control in frozen dairy desserts; thickener, suspending aid for salad dressings; reduces oil absorp. during frying; foam stabilizer for whipped toppings; thickener for low-solids syrup; *Usage level:* 0.1-0.5% (frozen dairy desserts); 0.2-0.8% (salad dressings), 0.5-3.0% (seafood prods.), 0.1-0.9% (whipped toppings), 0.2-0.8% (syrup); *Regulatory:* FDA 21CFR §172.874.

Benecel® MP 843. [Hercules] Hydroxypropyl methylcellulose; CAS 9004-65-3; nonionic; thickener, stabilizer, binder; improves texture, inhibits syneresis, entrains air, improves moisture/mouthfeel in mousse, frostings, cakes, doughnuts, diet jelly, syrup; ice crystal control in frozen dairy prods.; processing aid for tortillas; reduce; *Usage level:* 0.2-0.6% (mousse, doughnuts), 0.1-0.2% (frostings), 0.1-0.5% (frozen dairy desserts, dry mix cakes), 0.5-3.0% (seafood prods.), 0.2-0.5% (canned pet foods), 0.05-0.5% (tortillas), 0.25-0.75% (diet jelly), 0.2-0.8% (syrup); *Regulatory:* FDA 21CFR §172.874.

Benecel® MP 872. [Hercules] Hydroxypropyl methylcellulose; CAS 9004-65-3; nonionic; improves texture of structure potato prods.; thermal gelation reduces oil absorp.; *Usage level:* 0.25-0.75% (structured potato prods.); *Regulatory:* FDA 21CFR §172.874.

Benecel® MP 874. [Hercules] Hydroxypropyl methylcellulose; CAS 9004-65-3; nonionic; binder, thickener, inhibits syneresis, prevents shrinkage in pie fillings; thickener for barbecue sauces; thermal gelation reduces boil-over during baking; thickener, suspending aid for salad dressings; reduces oil absorp. in onion rings, tacos; *Usage level:* 0.3-0.6% (pie fillings), 0.2-1.0% (barbecue sauces), 0.2-

0.8% (salad dressings), 0.4-2.5% (onion rings), 0.2-0.5% (canned pet food), 0.1-2.5% (batters); *Regulatory:* FDA 21CFR §172.874.

Benecel® MP 943. [Hercules] Hydroxypropyl methylcellulose; CAS 9004-65-3; nonionic; thickener, suspending aid for salad dressings; *Usage level:* 0.2-0.8% (salad dressings); *Regulatory:* FDA 21CFR §172.874.

Benefiber®. [Sandoz Nutrition] Natural dietary vegetable fiber extracted from hydrolyzed guar gum; provides low visc., exc. clarity, low sweetness, physical, heat, and pH stability; resist. to heat, acid, salt, and digestive enzymes; exc. coating props.; improves foam stability; uniform ice crystals for smooth ice cream; *Properties:* water-sol.; visc. 10 cps (10%); pH 6.4-7.0; 80% min. dietary fiber.

Benol®. [Witco/Petroleum Spec.] Wh. min. oil NF; lubricant used in food industry; *Regulatory:* FDA 21CFR §172.878, §178.3620a; *Properties:* water wh., odorless, tasteless; sp.gr. 0.839-0.855; visc. 18-20 cSt (40 C); pour pt. -7 C; flash pt. 182 C.

Berol 374. [Berol Nobel AB] EO/PO block polymer; CAS 9003-11-6; nonionic; foam depressant/detergent for foodstuffs industry; *Properties:* wh. flakes; m.w. 2200; sol. in ethanol, xylene, trichloroethylene, wh. spirit; disp. water; sp.gr. 1.05; visc. 450 cps; cloud pt. 24-26 C (1% aq.); flash pt. > 100 C; pour pt. < -10 C; pH 5-7 (1% aq.); surf. tens. 40 dynes/cm (0.1%); Draves wetting 2 g/l (25 s); Ross-Miles foam 5 mm (initial, 0.05%, 50 C); 100% act.

Be Square® 175. [Petrolite] Microcryst. wax; CAS 63231-60-7; EINECS 264-038-1; plasticizer in chewing gum base; *Regulatory:* incl. FDA §172.230, 172.615, 175.105, 175.300, 176.170, 176.180, 176.200, 177.1200, 178.3710, 179.45; *Properties:* amber wax; dens. 0.93 g/cc; visc. 13 cps (99 C); m.p. 83 C; flash pt. 293 C.

Be Square® 185. [Petrolite] Hard microcryst. wax consisting of n-paraffinic, branched paraffinic, and naphthenic hydrocarbons; CAS 63231-60-7; EINECS 264-038-1; chewing gum base; *Regulatory:* incl. FDA §172.230, 172.615, 175.105, 175.300, 176.170, 176.180, 176.200, 177.1200, 178.3710, 179.45; *Properties:* amber wax; very low sol. in org. solvs.; sp.gr. 0.93; visc. 15 cps (99 C); m.p. 90.5 C.

Be Square® 195. [Petrolite] Hard microcryst. wax consisting of n-paraffinic, branched paraffinic, and naphthenic hydrocarbons; CAS 63231-60-7; EINECS 264-038-1; wax used in chewing gum; *Regulatory:* incl. FDA §172.230, 172.615, 175.105, 175.300, 176.170, 176.180, 176.200, 177.1200, 178.3710, 179.45; *Properties:* wh., amber wax; very low sol. in org. solvs.; sp.gr. 0.93; visc. 15.5 cps (99 C); m.p. 93 C.

Best One®. [Bunge Foods] Premier no-time dough system with conditioners, strengtheners, and emulsifiers; can be used in any yeast-raised dough from French bread to danish or donuts; *Usage level:* 1% based on flour.

Beta W 7. [Wacker-Chemie GmbH] β-Cyclodextrin; CAS 7585-39-9; EINECS 231-493-2; complex hosting guest molecules; increases the sol. and bioavailability of other substances; masks flavor, odor, or coloration; stabilizes against light, oxidation, heat, and hydrolysis; turns liqs. or volatiles into stable solid powds.; *Properties:* wh. cryst. powd.; sol. 1.85 g/100 ml in water; m.w. 1135.

Beta W7 HP 0.9. [Wacker-Chemie GmbH] Hydroxypropyl-β-cyclodextrin; complex hosting guest molecules; increases the sol. and bioavailability of other substances; masks flavor, odor, or coloration; stabilizes against light, oxidation, heat, and hydrolysis; turns liqs. or volatiles into stable solid powds.; *Properties:* m.w. 1507.

Beta W7 M1.8. [Wacker-Chemie GmbH] Methyl-β-cyclodextrin; complex hosting guest molecules; increases the sol. and bioavailability of other substances; masks flavor, odor, or coloration; stabilizes against light, oxidation, heat, and hydrolysis; turns liqs. or volatiles into stable solid powds.; *Properties:* m.w. 1311.

Beta W7 P. [Wacker-Chemie GmbH] β-Cyclodextrin polymer; complex hosting guest molecules; increases the sol. and bioavailability of other substances; masks flavor, odor, or coloration; stabilizes against light, oxidation, heat, and hydrolysis; turns liqs. or volatiles into stable solid powds.

Beta Carotene 1% CWS No. 65659. [Roche] β-Carotene with dextrin, gum acacia, partially hydrog. veg. oil, sucrose and antioxidants (sodium ascorbate, dl-α-tocopherol); colorant used in dry foods (beverage powds., cake mixes, instant pudding, icing mixes, dry salad dressings/sauces/gravies), aq. applics. (ice cream, prepared cakes, imitation dairy prods., egg substitutes, sauces/gravies); *Regulatory:* FDA 21CFR §73.95 (food use), 73.1095 (drug use), 73.2095 (cosmetic use), GRAS; *Properties:* orange fine powd.; 90% min. thru 80 mesh; disp. in water; 1% min. assay (β-carotene); *Storage:* store in tightly closed container in cool place protected from light and humidity; sensitive to air, heat.

24% Beta Carotene HS-E in Veg. Oil No. 65671. [Roche] β-Carotene FCC suspension in partially hydrogenated cottonseed oil and partially hydrogenated soybean oils, with 3.5% dl-α-tocopherol, 1.5% ascorbyl palmitate, and citric acid as preservatives; colorant and antioxidant developed for the popcorn industry; may be used in oil phase of other prods.; *Regulatory:* FDA 21CFR §73.95 (food use), 73.1095 (drug use), 73.2095 (cosmetic use), 182.1245, 182.5245 GRAS; *Properties:* brick-red suspension, pourable fluid above 68 F, thick paste at refrigerated temps.; particle size ≥ 90% < 10 μ; sol. in all veg. oils and fats; 22% min. assay (β-carotene); *Storage:* store in tightly closed container in cool, dry place (59-86 F); stir well before use.

24% Beta Carotene Semi-Solid Suspension No. 65642. [Roche] β-Carotene suspension with hydrog. coconut oil, 1% glyceryl stearate as stabilizer; colorant incorporated into oil phase of food prods. (shortening, pastry); esp. convenient for weighing out small amts.; *Regulatory:* FDA 21CFR

§73.95 (food use), 73.1095 (drug use), 73.2095 (cosmetic use), 182.5245 GRAS; *Properties:* brick-red homogeneous suspension, semisolid @ 70-90 F; particle size ≥ 90% < 10 μ; sol. in all veg. oils and fats; 24% min. assay (β-carotene); *Storage:* store in tightly closed container in cool, dry place.

30% Beta Carotene in Veg. Oil No. 65646. [Roche] β-Carotene in food-grade veg. oils (partially hydrog. cottonseed oil and partially hydrogenated soybean oils); colorant, antioxidant for incorporation into oil phase of food prods. (margarine, imitation dairy prods., shortening, pastry), and in soft gel capsules for pharmaceutical applics.; *Regulatory:* FDA 21CFR §73.95 (food use), 73.1095 (drug use), 73.2095 (cosmetic use), 182.5245 GRAS; *Properties:* terra-cotta red suspension, fresh char. odor; crystal size 90% ≤ 10 μ; sol. 0.1% in all veg. oils and fats @ R.T.; 500,000 I.U. Vitamin A activity/g; *Storage:* store in cool, dry place in tightly closed container; protect from freezing.

Beta Carotene Emulsion Beverage Type 3.6 No. 65392. [Roche] β-Carotene emulsion with orange oil, brominated veg. oil, aq. hydrolyzed protein base with preservatives (sorbic acid, sodium benzoate) and antioxidants (BHT, BHA); provides orange color and/or vitamin A activity to beverages incl. fruit juice blends, drinks, carbonated and noncarbonated beverages; *Usage level:* 150 ppm max. (beverages contg. no other source of BVO); *Regulatory:* FDA GRAS; *Properties:* orange emulsion, pourable fluid above 75 F, thick paste @ 45-55 F, pleasant orange oil odor; sp.gr. 1.052; 3.6% β-carotene, 60,000 IU provitamin A activity/g.

Beta Plus. [Van Den Bergh Foods] Partially hydrog. soybean oil with mono- and diglycerides, sodium stearoyl lactylate, ethoxylated mono- and diglycerides, TBHQ; high performance fat for continuous and conventional breads and other yeast-raised goods; *Regulatory:* kosher; *Properties:* Lovibond 5.0R max. fluid; 3.7-4.3% mono.

Beta-Tabs. [MLG Enterprises Ltd.] Sodium metabisulfite, starch, and other edible excipients; reducing agent, dough conditioner; produces rapid relaxation of pizza, puff pastry, English muffins, pie doughs, bread prods., flour tortillas; *Usage level:* 1-3 tablets/cwt of flour; *Properties:* tablets supplying 30 ppm sodium metabisulfite/cwt flour; *Storage:* 6 mos shelf life stored in cool, dry conditions.

Betricing. [Van Den Bergh Foods] Partially hydrog. veg. oil (soybean, cottonseed), mono- and diglycerides, < 0.9% polysorbate 60; multipurpose shortening for icings, fillings, yeast-raised goods; *Regulatory:* kosher; *Properties:* Lovibond 2.0R max. plastic; m.p. 114-120 F; 2.4-3.0% mono.

Betrkake. [Van Den Bergh Foods] Partially hydrog. veg. oil (soybean, cottonseed), mono- and diglycerides; high quality emulsified shortening for cakes, icings, and sweet doughs; *Regulatory:* kosher; *Properties:* Lovibond 2R max. plastic; m.p. 111-119 F; 3.1-3.6% mono.

BF-46 Icing Powd. [Bunge Foods] Boiling type icing stabilizer for translucent, firm icings with good stability; for sweet goods icings and glazes; *Usage level:* 3-4 lb/100 lb powd. sugar; *Properties:* wh. free-flowing powd., bland aroma/flavor; pH 9.1; 3.7% max. moisture.

BF-50 Icing Powd. [Bunge Foods] Boiling type icing stabilizer for transparent and firm icings, good stability; for sweet goods icings and glazes, wh. and chocolate; *Usage level:* 3-4 lb/100 lb powd. sugar; *Properties:* wh. free-flowing powd., bland aroma/flavor; pH 9.1; 3.7% max. moisture.

BFP 30. [Am. Ingreds.] Mono- and diglycerides; food emulsifier for flavor and color dispersions, snack, high protein food systems; *Properties:* liq.; iodine no. 90-93; 40% min. mono.

BFP 64. [Am. Ingreds.] Mono/diglyceride from hydrog. soybean oil and glycerin; food emulsifier, crumb softener; for bread, sweet goods, bakery mixes, icings, shortening, margarine, sour cream dips, fillings, tortillas; *Usage level:* 0.5-10%; *Regulatory:* FDA GRAS; *Properties:* votated ivory wh. plastic; c.m.p. 118-122 F; iodine no. 65-72; 42% min. alpha monoester; *Storage:* store at 65-95 F.

BFP 64A. [Am. Ingreds.] Mono/diglyceride from hydrog. soybean oil and glycerin, TBHQ; food ingred., emulsifier, crumb softener; for bread, sweet goods, bakery mixes, shortening, margarine; *Properties:* Lovibond 3R color; 42% monoester; Discontinued

BFP 64K. [Am. Ingreds.] Mono- and diglycerides from hydrog. soybean oil and glycerin, TBHQ, citric acid; food ingred., emulsifier, crumb softener; for bread, sweet goods, bakery mixes, shortening, margarine; *Regulatory:* kosher; *Properties:* Lovibond 3R plastic; c.m.p. 48-50 C; iodine no. 65-72; 42% min. alpha monoester.

BFP 65. [Am. Ingreds.] Mono- and diglycerides from hydrog. edible fats and glycerin with TBHQ, citric acid; nonionic; food additive, emulsifier for baked prods., mixes, shortenings, icings; *Usage level:* 0.375-10%; *Regulatory:* FDA GRAS; *Properties:* Lovibond 3R votated plastic; HLB 3.2; 100% conc.; *Storage:* store at 65-95 F; Discontinued

BFP 65A. [Am. Ingreds.] Mono- and diglycerides from animal lipid source; food emulsifier for bakery mixes, cakes, sweet goods, icings, margarines, sour cream dips, fillings, tortillas; *Regulatory:* FDA GRAS; *Properties:* votated ivory wh. plastic; c.m.p. 52-54 C; iodine no. 50-57; 54% min. alpha monoester.

BFP 65C. [Am. Ingreds.] Mono- and diglycerides; food emulsifier for bakery mixes, cakes, sweet goods, icings, margarines, sour cream dips, fillings, tortillas; *Properties:* votated ivory wh. plastic; c.m.p. 130 F; iodine no. 40-50; 52% min. mono.

BFP 65K. [Am. Ingreds.] Mono- and diglycerides from veg. lipid source; food emulsifier for bakery mixes, cakes, sweet goods, icings, margarines, sour cream dips, fillings, tortillas; *Regulatory:* FDA GRAS; kosher; *Properties:* votated ivory wh. plastic; c.m.p. 48-50 C; iodine no. 65-75; 52% min. alpha mono.

BFP 74. [Am. Ingreds.] Mono- and diglycerides from hydrog. soybean oil and glycerin with citric acid; nonionic; food emulsifier for coffee whiteners, whipped toppings, snack food, chewing gum, margarine, frozen desserts, jelly, gum, confectionery coatings, sour cream dips, caramel, nougats; *Usage level:* 0.25-10%; *Regulatory:* FDA GRAS; *Properties:* ivory wh. flake or bead; c.m.p. 140-145 F; HLB 3.2; iodine no. 3 max.; 100% conc., 42% min. alpha mono; *Storage:* store at 65-95 F.

BFP 74A. [Am. Ingreds.] Mono- and diglycerides from animal lipid source; food emulsifier; *Properties:* bead, flake; c.m.p. 60-63 C; iodine no. 3 max.; 42% min. alpha mono.

BFP 74K. [Am. Ingreds.] Mono- and diglycerides from veg. lipid source; food emulsifier; *Regulatory:* kosher; *Properties:* bead, flake; c.m.p. 60-63 C; iodine no. 3 max.; 42% min. alpha mono.

BFP 75. [Am. Ingreds.] Mono- and diglycerides from hydrog. soybean oil and glycerin with citric acid; nonionic; food emulsifier for coffee whiteners, whipped toppings, snack food, chewing gum, margarine, frozen dessert, jelly, confectionery coatings, sour cream dips, caramel, nougats; *Usage level:* 0.35-10%; *Regulatory:* FDA GRAS; *Properties:* wh. to cream flake or bead; c.m.p. 140-145 F; HLB 3.2; iodine no. 3 max.; 100% conc.; 52% min. alpha mono; *Storage:* store at 65-95 F.

BFP 75A. [Am. Ingreds.] Mono- and diglycerides from animal lipid source; food emulsifier; *Properties:* bead, flake; c.m.p. 60-63 C; iodine no. 3 max.; 52% min. alpha mono.

BFP 75K. [Am. Ingreds.] Mono- and diglycerides from veg. lipid source; food emulsifier; *Regulatory:* kosher; *Properties:* bead, flake; c.m.p. 60-63 C; iodine no. 3 max.; 52% min. alpha mono.

BFP 100. [Am. Ingreds.] Mono- and diglycerides and ethoxylated mono- and diglycerides; food emulsifier; dough strengthener and improver in yeast-raised doughs; *Properties:* wh. cream powd.; c.m.p. 130-140 F; 34% min. mono.

BFP 800 K. [Am. Ingreds.] Mono- and diglycerides; food additive for baked prods., bakery mixes, pasta, cereal, sauces, and gravies; *Properties:* wh. to cream colored powd.; m.p. 57-64 C; iodine no. 32-34; 90% min. mono.

BFP L Mono. [Am. Ingreds.] Mono- and diglycerides from veg. lipid source; food emulsifier for cakes, bread, sweet goods, shortening; sparging effect; *Properties:* votated ivory wh. plastic; c.m.p. 48-50 C; iodine no. 70-74; 26-28% min. alpha mono.

BFP White Sour. [Am. Ingreds.] Flour, lactic acid, monocalcium phosphate, phosphoric acid, salt, dry yeast, and acetic acid; sour for French bread, English muffins, snack crackers, and snacks; *Properties:* free-flowing powd.; pH 3.2-3.4; 6.8-7.4% protein.

Binasol™ 15. [A.E. Staley Mfg.] Food starch modified, derived from tapioca; pregelatinized starch, binder, moisture retention aid, stabilizer for neutral and highly acidic foods, restructured foods, dry mixes, dips, fruit fillings, relishes, fondants, pulpy textures; exc. freeze/thaw stability; *Regulatory:* FDA 21CFR

§172.892; kosher, Passover certified; *Properties:* wh. coarse powd.; 30% max. on 50 mesh, 28% max. thru 200 mesh; pH 6.0; 5% moisture.

Binasol™ 81. [A.E. Staley Mfg.] Food starch modified, derived from tapioca; pregelatinized starch, thickener, binder, moisture retention aid; imparts med. to high visc., stable to heat, acid, and freeze/thaw conditions; for pulp texture, relishes, sauces, tomato sauces, meat patties; *Regulatory:* FDA 21CFR §172.892; *Properties:* coarse powd.; 50% on 100 mesh, 30% thru 200 mesh; pH 6.0; 6% moisture.

Binasol™ 90. [A.E. Staley Mfg.] Food starch modified, derived from tapioca; pregelatinized starch, thickener for pulpy textures, relishes, sauces.

Bindox-HV-051. [Vaessen-Schoemaker] Sodium caseinate; gellant, emulsifier, water binder for meat prods.; provides homogeneous distribution of fat, prevents fat separation, improves consistency, and increases protein content (nutritional value); *Regulatory:* FAO/WHO compliance; *Properties:* visc. 10,000-20,000 cps (15%); pH 6.8-7.2 (2%); 88% min. protein.

Bindox-LV-050. [Vaessen-Schoemaker] Sodium caseinate; gellant, emulsifier, water binder for meat prods.; provides homogeneous distribution of fat, prevents fat separation, improves consistency, and increases protein content (nutritional value); *Regulatory:* FAO/WHO compliance; *Properties:* visc. 1200 cps (15%); pH 6.4-6.8 (2%); 88% min. protein.

Bioblend®. [ADM Ogilvie] Custom vitamin/mineral fortifications for food and related prods., incl. cereal, dietetic, milling, surimi, and structured prods.

Biodiastase. [Mitsubishi Int'l.; Amano Enzyme USA] Fungal amylase from *Aspergillus oryzae;* CAS 9000-92-4; EINECS 232-567-7; digestive enzyme; *Properties:* powd.

Biorion 450 Super. [Asahi Chem. Industry] Iron supplement; for food industry.

d-Biotin USP, FCC No. 63345. [Roche] d-Biotin USP, FCC; CAS 58-85-5; EINECS 200-399-3; component of enzyme systems involved in metabolism of fats and carbohydrates, in other biochemical processes; nutrient in food, and special dietary prods. incl. multivitamins in liq., tablet, capsule or powd. forms; *Properties:* pract. wh. cryst. powd.; 100% min. thru 80 mesh; sol. 1 g/5000 ml water, 1300 ml alcohol; more sol. in hot water, dil. alkali; insol. in other common org. solvs.; m.w. 244.31; m.p. 229-232 C; 97.5% min. assay; *Precaution:* oxidized by hydrogen peroxide or potassium permanganate; *Storage:* store @ 59-86 F.

Biozyme L. [Mitsubishi Int'l.; Amano Enzyme USA] Fungal amylase from *Aspergillus oryzae;* CAS 9000-92-4; EINECS 232-567-7; food enzyme; *Properties:* liq.

Biozyme M. [Mitsubishi Int'l.; Amano Enzyme USA] β-Amylase from malt; CAS 9000-92-4; EINECS 232-567-7; food enzyme; *Properties:* powd.

Biozyme S. [Mitsubishi Int'l.; Amano Enzyme USA] Fungal amylase from *Aspergillus oryzae;* CAS 9000-92-4; EINECS 232-567-7; enzyme for food

and feed industries; bromate replacer; for confectionery and maltose syrup mfg.; *Properties:* lt. ylsh. powd.; sol. in water; insol. in ethanol; *Toxicology:* nontoxic, nonpathogenic.

Bitrit-1™ (1% Biotin Trituration No. 65324). [Roche] Biotin FCC in dicalcium phosphate dihydrate carrier; component of enzyme systems involved in metabolism of fats and carbohydrates, in other biochemical processes; nutrient in pharmaceutical, food, and special dietary prods. incl. multivitamins in tablet, capsule or powd. forms; *Properties:* wh. free-flowing powd.; 98% min. thru 80 mesh; insol. in water, org. solvs.; bulk dens. 50 lb/ft³; 10 mg min. d-biotin/g; *Storage:* store @ 59-86 F.

Bittex. [MLG Enterprises Ltd.] Vegetal bitter; high conc. soluble extract providing natural bitter touch to soft drinks and aperitives; does not contain chinine derivs., free of metallic taste; *Usage level:* 0.2-0.5 ppm.

BK-102 Series. [Premier Malt Prods.] Extract of 60-65% malted barley and 35-40% corn with saccharifying enzymes; syrup with mellow, rich malt flavor; used in confections, baked goods, fermentation nutrient in breweries/distilleries, bread sticks, horse food; 55% the sweetness of sucrose; *Usage level:* 2-4% (candy sponge), 1-2.5% flour wt. (bread), 1-3% dough wt. (bagels), , 4-6% flour wt. (cookies), 2-3% flour wt. (bread sticks); 78.5-80.0% solids; 3-4% protein; 57-67% reducing sugars.

BK-305 Series. [Premier Malt Prods.] Extract of 65% corn and 35% malted barley with saccharifying enzymes; syrup with mild malt flavor and pleasantly sweet char.; used in ice cream prods., bakery goods, vinegar, snacks, cereal, cigarettoo, pretzel, bread sticks; 65% the sweetness of sucrose; 78.5-80.0% solids; 1.8-2.6% protein; 60-70% reducing sugars.

BK-PR2 Series. [Premier Malt Prods.] Extract of 80% corn and 20% malted barley with saccharifying enzymes; syrup with sweet but mild malt flavor; used in ice cream prods., bakery goods, nutrient for distilled vinegars, entomological, and fermentations, cereal, tobacco, baby foods, soybean milk, pretzel; 72% the sweetness of sucrose; *Usage level:* 1.5-2.5% flour wt. (saltines); 78.5-80.0% solids, 1.0-1.9% protein, 63-73% reducing sugars.

BL™. [Calgon Carbon] Activated carbon; CAS 64365-11-3; used for food and chemical decolorization; *Properties:* pulverized.

BL-60®. [Rhone-Poulenc Food Ingreds.] Sodium aluminum phosphate acidic with aluminum sulfate anhydrous; leavening agent for baking powds., cake/cookie/muffin mixes, cereals; more sensitive to heat than Levair, releases final acidity at a batter temp. of 120 F; used for cake mixes, esp. those containing highly emulsified shortenings; *Properties:* wh. free-flowing powd.; 1% max. on 60 mesh, 15% max. on 140 mesh; bulk dens. 50 lb/ft³; neutralizing value 100 min.; pH 2.7 (1% susp.).

Blandol®. [Witco/Petroleum Spec.] Wh. min. oil N.F.; lubricant used in food industry; *Regulatory:* FDA 21CFR §172.878, §178.3620a; *Properties:* water-wh., odorless, tasteless; sp.gr. 0.839-0.855; visc. 14-17 cSt (40 C); pour pt. -7 C; flash pt. 179 C.

Blend 424. [FMC] flavor protectant recommended for processing requiring rapid cure-color development, e.g., sausage and frankfurters; enhances prod. shelf life by inhibiting lipid oxidation; *Regulatory:* USDA, FDA GRAS; *Properties:* sol. 10 g/100 g water; pH 6.9 (1%).

Blendmax 322. [Central Soya] Enzyme-modified lecithin; CAS 8002-43-5; EINECS 232-307-2; amphoteric; o/w emulsifier with enhanced water dispersibility, wetting agent; for instantizing whole milk powds., emulsifying veg. and animal fats in milk replacer prods.; dough conditioner, antistaling agent in baking; *Properties:* amber liq.; oil-sol., water-disp.; visc. 8000 cP; HLB 8.0; acid no. 40; 100% conc.

Blendmax 322D. [Central Soya] Enzyme-modified soy lecithin; CAS 8002-43-5; EINECS 232-307-2; o/w emulsifier, wetting agent; enzyme modification enhances water dispersibility to instantize fatty powds. in cold water (milk powds., milk replacers); dough conditioner, antistaling agent in baked goods; source of phospholipids in aquaculture diets; *Properties:* amber fluid; oil-sol., water-disp.; visc. 10,000 cP max.; HLB 8; acid no. 42 max.; 1% max. moisture; *Storage:* store in closed containers @ 16-32 C; 1 yr shelf life in original, unopened container.

Blitz® Danish Conc. [Bunge Foods] Conc. formulated for prod. of fryable danish pastries; uses Blitz mixing method; delivers a highly machineable dough for high speed prod. of danish snack pastries, honey buns, other fried danish; *Usage level:* 33% based on flour.

1626 Blue Natural Liq. Colorant. [MLG Enterprises Ltd.] Indigotine sulfonic acid sol'n.; natural food colorant producing bright or light blue shades for confectionery, icicles, dried and candied fruit, and for obtaining green shades with Curcumex 1600; *Properties:* liq.; sol. in water; pH 1.4; 0 ± 0.1% color (as indigotin sulfonic acid); *Storage:* store in tightly closed containers away from light and heat; rotate stock every 8 mos.

Bohemian Rye Sour. [Brolite Prods.] Rich flavored rye sour designed to exhibit Bohemian rye bread chars.

Bone Gelatin Type B 200 Bloom. [Hormel] Gelatin; CAS 9000-70-8; EINECS 232-554-6; stabilizer, gellant, protein used in food applics; *Properties:* beige, weak bouillon-like odor; 30 mesh; visc. 52 ± 4 mps; pH 4.5-5.8; 12% max. moisture; *Storage:* stable for up to 1 yr when stored dry at ambient temps; keep containers tightly closed.

Bone Marrow Powd. [Am. Labs] Vacuum-dried defatted glandular prod; nutritive food additive; *Properties:* powd.

B.P. Pyro®. [Rhone-Poulenc Food Ingreds.] Sodium acid pyrophosphate, leavening, FCC; CAS 7758-16-9; EINECS 231-835-0; leavening agent for baking, cereals; relatively slow reaction rate; ideally suited for refrigerated doughs, bakers' high-

strength baking powds. and baking creams; *Properties:* wh. powd; 100% thru 60 mesh, 99% thru 200 mesh; sol. 13 g/100 g saturated sol'n; m.w. 221.94; bulk dens. 68 lb/ft³; pH 4.3 (1%); 95% min. act.

B.P. Pyro® Type K. [Rhone-Poulenc Food Ingreds.] Sodium acid pyrophosphate, leavening, FCC; CAS 7758-16-9; EINECS 231-835-0; leavening agent for baking powds., baking creams, prepared doughnut and cake mixes; *Properties:* wh. powd; 100% thru 60 mesh, 99% thru 200 mesh; sol. 13 g/100 g saturated sol'n; m.w. 221.94; bulk dens. 68 lb/ft³; pH 4.3 (1%).

Brain Substance. [Am. Labs] Vacuum-dried defatted glandular prod; nutritive food additive; *Properties:* powd.

Bread Pan Oil. [ADM Arkady] Mineral oil, soybean oil, and lecithin; quality release agent for foods.

Brem. [Am. Ingreds.] Hydrated sorbitan monostearate, polysorbate 60 (45%), propylene glycol, hydroxylated lecithin, potassium sorbate; cake emulsifier; *Properties:* cream plastic.

Brew Aid B. [ADM Arkady] Acidifer/oxidant blend providing rapid, even fermentation in brew system doughs; use in the dough while using Ferment Buffer in the brew.

Britesorb®. [PQ Corp.] Silica hydrogel-based composition; preservative, stabilizer for beer; *Properties:* wh. powd; water-insol.

Britesorb® A 100. [PQ Corp.] Silica hydrogel; preservative, stabilizer for beer; *Properties:* wh. powd; water-insol.

Britol®. [Witco/Petroleum Spec.] Wh. min. oil USP; lubricant, binder, carrier, moisture barrier, softener for food processing; *Regulatory:* FDA 21CFR §172.878, §178.3620a; *Properties:* water-wh., odorless, tasteless; sp.gr. 0.869-0.885; visc. 56-60 cSt (40 C); pour pt. -15 C; flash pt. 199 C.

Britol® 6NF. [Witco/Petroleum Spec.] White min. oil NF; white oil functioning as binder, carrier, defoamer, dispersant, extender, lubricant, moisture barrier, plasticizer, protective agent, release agent, and/or softener; for foods; *Properties:* sp.gr. 0.830-0.858; visc. 8.5-10.8 cst (40 C); pour pt. -24 C max; flash pt. 166 C min.

Britol® 7NF. [Witco/Petroleum Spec.] White min. oil NF; white oil functioning as binder, carrier, defoamer, dispersant, extender, lubricant, moisture barrier, plasticizer, protective agent, release agent, and/or softener; for foods; *Properties:* sp.gr. 0.840-0.858; visc. 10.8-13.6 cst (40 C); pour pt. -18 C max; flash pt. 171 C min.

Britol® 9NF. [Witco/Petroleum Spec.] White min. oil NF; white oil functioning as binder, carrier, defoamer, dispersant, extender, lubricant, moisture barrier, plasticizer, protective agent, release agent, and/or softener; for foods; *Properties:* sp.gr. 0.845-0.860; visc. 14.4-16.9 cst (40 C); pour pt. -18 C max; flash pt. 171 C min.

Britol® 20USP. [Witco/Petroleum Spec.] White min. oil USP; white oil functioning as binder, carrier, defoamer, dispersant, extender, lubricant, moisture barrier, plasticizer, protective agent, release agent, and/or softener; for foods; *Properties:* sp.gr. 0.858-0.870; visc. 37.9-40.1 cst (40 C); pour pt. -18 C max; flash pt. 193 C min.

Britol® 35USP. [Witco/Petroleum Spec.] White min. oil USP; white oil functioning as binder, carrier, defoamer, dispersant, extender, lubricant, moisture barrier, plasticizer, protective agent, release agent, and/or softener; for foods; *Properties:* sp.gr. 0.862-0.880; visc. 65.8-71.0 cst (40 C); pour pt. -15 C max; flash pt. 216 C min.

Britol® 50USP. [Witco/Petroleum Spec.] White min. oil USP; white oil functioning as binder, carrier, defoamer, dispersant, extender, lubricant, moisture barrier, plasticizer, protective agent, release agent, and/or softener; for foods; *Properties:* sp.gr. 0.870-0.890; visc. 91-102.4 cst (40 C); pour pt. -12 C max; flash pt. 249 C min.

Bro 3D 6451. [Brolite Prods.] Cultured, no-time dough accelerant that eliminates time-consuming sponge and dough methods while still retaining the fermentation flavor because it is cultured.

Bro Action 6467. [Brolite Prods.] Combination min. yeast food, dough accelerant, and natural cultured dough conditioner; dough improver system.

Bro-Eze. [Brolite Prods.] Highly conc. release agent to be used in place of dusting flour; increases number of bakes between glazes while reducing cripples; eliminates all pan oils.

Bro-Eze III. [Brolite Prods.] Release agent to be used in place of dusting flour; increases number of bakes between glazes while reducing cripples; eliminates all pan oils.

Brolite 1A. [Brolite Prods.] Premium cultured butter flavor.

Bromelain 1:10. [Solvay Enzymes] Mixt. of proteases, standardized with lactose; CAS 9014-01-1; EINECS 232-752-2; enzyme for hydrolysis of plant and animal proteins to peptides and amino acids; for tenderizer formulations for meat; in baking, chillproofing of beer, fermentation, fish/soya/egg processing, protein modification; eliminates protein haze in foods; *Properties:* tan to lt. brn. amorphous powd., free of offensive odor and taste; water-sol.

Bromelain 150 GDU. [Meer] Bromelain, enzyme derived from pineapple; CAS 37189-34-7; EINECS 253-387-5; proteolytic enzyme used as meat tenderizer, beer clarifier, in dough conditioners for baking, digestive aids; in detergents to remove blood stains, in cleaning sol'ns. for contact lenses; as antitussive agents in veterinary preps.

Bromelain 600 GDU. [Meer] Bromelain, enzyme derived from pineapple; CAS 37189-34-7; EINECS 253-387-5; proteolytic enzyme used as meat tenderizer, beer clarifier, in dough conditioners for baking, digestive aids; in detergents to remove blood stains, in cleaning sol'ns. for contact lenses; as antitussive agents in veterinary preps.

Bromelain 1200 GDU. [Meer] Bromelain, enzyme derived from pineapple; CAS 37189-34-7; EINECS 253-387-5; proteolytic enzyme used as meat tenderizer, beer clarifier, in dough conditioners for baking, digestive aids; in detergents to remove

blood stains, in cleaning sol'ns. for contact lenses; as antitussive agents in veterinary preps.

Bromelain 1500 GDU. [Meer] Bromelain, enzyme derived from pineapple; CAS 37189-34-7; EINECS 253-387-5; proteolytic enzyme used as meat tenderizer, beer clarifier, in dough conditioners for baking, digestive aids; in detergents to remove blood stains, in cleaning sol'ns. for contact lenses; as antitussive agents in veterinary preps.

Bromelain Conc. [Solvay Enzymes] Mixt. of proteases; CAS 9014-01-1; EINECS 232-752-2; enzyme for hydrolysis of plant and animal proteins to peptides and amino acids; for tenderizer formulations for meat; in baking; fish processing; eliminates protein haze in foods; *Properties:* tan to lt. brn. amorphous powd., free of offensive odor and taste; water-sol.

Bromette. [MLG Enterprises Ltd.] Potassium bromate and other edible excipients; CAS 7758-01-2; EINECS 231-829-8; dough conditioner; oxidizer for flour, improving dough strength, gas retention, bread quality; esp. effective with low protein or poor gluten quality flours; *Usage level:* 30 ppm; *Properties:* tablet supplying 60 ppm potassium bromate/ cwt flour; *Storage:* 6 mos shelf life stored under cool, dry conditions.

Bromitabs. [ADM Arkady] Each tablet provides 60 ppm potassium bromate; CAS 7758-01-2; EINECS 231-829-8; provides oxidation designed to increase loaf volume and improve crumb structure; *Properties:* tablets.

Bro Rye Sour. [Brolite Prods.] Natural, cultured rye sour; sour producing richly full-flavored rye breads.

Brosoft 6430. [Brolite Prods.] Mono- and diglycerides, lecithins; crumb softener and tenderizer for all baked goods; promotes max. absorption rates so that batches have increased yields; *Properties:* powd.

Bro White Sour. [Brolite Prods.] Natural sour flavor produced from a natural sour culture and a fermented sponge; the sponge is ground and dehydrated into a powd; adds zest to any baked prod. by adding natural flavor that is only produced from a sour culture.

Bubble Breaker® 3009-F. [Witco/H-I-P] Blend of org., nonsilicone compds; defoamer for wash water of sliced potatoes; Unverified

Buckeye C (Baker's Grade). [Karlshamns Food Ingreds.] Partially hydrog. vegetable oil (soybean, cottonseed), water, salt, nonfat milk, lecithin, mono- and diglycerides, sodium benzoate, artificial color/flavor, vitamin A palmitate; margarine/puff paste for baking and yeast-raised sweet goods, roll-in margarine; *Properties:* plastic; m.p. 109 F.

Building Blocks® Beef Extract. [Hormel] Beef extract; natural beef extract used to enhance beefy flavor notes in soups, gravies, and sauces; imparts a deep brn. color to the broth or gravy; *Properties:* deep brn. visc. grainy paste, beefy bouillon odor; 80% min. total solids; *Storage:* 1 yr stability when stored at ambient temps.

Building Blocks® Beef Extract B1. [Hormel] Juice

35

conc. from beef bones and meat (92%), salt (8%); shelf stable beef extract used to develop beefy flavor notes in soups, gravies, and sauces; *Properties:* deep brn. heavy paste, beefy peptide-like odor; sol. in water; pH 4.8-5.5; 68-74% total solids; *Storage:* 1 yr stability when stored at ambient temps.

Building Blocks® Dried Flavored Beef Extract B7. [Hormel] Beef extract (41%), hydrolyzed soy, corn, and wheat protein (40%), autolyzed yeast (19%); natural meat stock/broth/extract used in soups, gravies, and sauces; imparts natural beefy flavor and rich brn. color; *Properties:* dk. brn. free-flowing powd., beef bouillon odor; hygroscopic; sol. in water; pH 4.6-6.0; 93-97% total solids; *Storage:* 1 yr stability when stored at ambient temps; protect from moisture.

Building Blocks® Dried Flavored Beef Extract B8. [Hormel] Beef stock, broth, or extract (49%), autolyzed yeast extract (27%), salt (23%), flavors (1%); natural meat stock/broth/extract used to bring out beef flavor and color of soups, gravies, and sauces; can replace expensive imported beef extracts; *Properties:* deep dk. brn. spray-dried free-flowing powd., beefy bouillon-like odor; hygroscopic; sol. in water; pH 4.5-6.5; 94% min. total solids; *Storage:* 1 yr stability when stored at ambient temps; protect from moisture.

Building Blocks® Dried Flavored Beef Stock #6. [Hormel] Hydrolyzed vegetable (corn, wheat, soy) protein (46%), beef stock/broth (36%), MSG (18%); natural meat stock/broth used to impart a full, rich beefy flavor to soups, gravies, and sauces; inexpensive replacement for beef extract; *Properties:* dk. brn. free-flowing powd., beefy bouillon odor; hygroscopic; sol. in water; pH 4.8-5.5; 93-97% total solids; *Storage:* 1 yr stability when stored at ambient temps.

Building Blocks® Dried Flavored Beef Stock B4. [Hormel] Conc. beef stock/broth (93%), hydrolyzed soy protein (3%), hydrolyzed corn protein (2%), autolyzed yeast (2%), hydrolyzed wheat protein (1%); natural meat stock/broth used in soups, gravies, and sauces; provides natural beefy flavor and rich brn. color; *Properties:* dk. brn. heavy free-flowing powd., beef bouillon odor; hygroscopic; sol. in water; pH 5.0-7.0; 93-97% total solids; *Storage:* 1 yr stability when stored at ambient temps.

Building Blocks® Dried Flavored Meat Stock #6. [Hormel] Hydrolyzed vegetable (corn, wheat, soy) protein (46%), meat stock/broth (36%), MSG (18%); natural meat stock/broth used in soups, gravies, and sauces; imparts full rich beefy flavor; inexpensive replacement for beef extract; *Properties:* dk. brn. heavy free-flowing powd., meaty bouillon odor; hygroscopic; sol. in water; pH 4.8-5.5; 93-97% total solids; *Storage:* 1 yr stability when stored at ambient temps.

Building Blocks® Dried Ham Stock H5. [Hormel] Ham stock/broth (90%), hydrolyzed vegetable protein (10%); natural meat stock/broth used to reduce harsh flavors in soups, gravies, and sauces; pro-

vides natural smoky flavor; adds smoothing, enriching mouthfeel; *Properties:* tan to lt. brn. free-flowing particle, smoked ham odor; sol. in water; pH 5-7; 93-97% total solids; *Storage:* 1 yr stability when stored at ambient temps.

Building Blocks® Flavored Beef Extract B7. [Hormel] Beef extract (60.5%), hydrolyzed soy, corn, and wheat protein (27.5%), autolyzed yeast (12%); natural meat stock/broth/extract used to bring out beef flavor of soups, gravies, and sauces; can replace expensive imported beef extracts; *Properties:* deep brn. heavy paste, beefy bouillon-like odor; sol. in water; pH 4.8-5.5; 70-76% total solids; *Storage:* 1 yr stability when stored at ambient temps.

Building Blocks® Flavored Beef Extract B8. [Hormel] Beef stock, broth, or extract (66%), autolyzed yeast extract (18%), salt (15%), flavors (1%); natural meat stock/broth/extract used to bring out beef flavor and roast beef color of soups, gravies, and sauces; can replace expensive imported beef extracts; *Properties:* deep dk. brn. heavy paste, beefy bouillon-like odor; sol. in water; pH 4.5-6.5; 67-74% total solids; *Storage:* 1 yr stability when stored at ambient temps.

Building Blocks® Flavored Beef Extract B8 LS. [Hormel] Beef stock, broth, or extract (74.7%), autolyzed yeast extract (18.4%), salt (6.5%), flavors (0.4%); natural meat stock/broth/extract used to bring out beef flavor and roast beef color of soups, gravies, and sauces; can replace expensive imported beef extracts; *Properties:* deep dk. brn. heavy paste, beefy bouillon-like odor; sol. in water; pH 4.5-6.5; 67-74% total solids; *Storage:* 1 yr stability when stored at ambient temps.

Building Blocks® Flavored Beef Stock #6. [Hormel] Beef stock/broth (57%), hydrolyzed vegetable (corn, soy, wheat) protein (31%), MSG (12%); shelf stable natural flavored stock/broth used to enhance beefy flavor notes in soups, gravies, and sauces; imparts rich, full beefy flavor; inexpensive replacement for beef extract; *Properties:* deep brn. heavy paste, beefy bouillon-like odor; sol. in water; pH 4.8-5.5; 72-76% total solids; *Storage:* 1 yr stability when stored at ambient temps.

Building Blocks® Flavored Beef Stock B4. [Hormel] Beef stock/broth (87%), salt (5%), hydrolyzed soy protein (3%), autolyzed yeast (2%), hydrolyzed corn protein (2%), hydrolyzed wheat protein (1%); natural meat stock/broth used to bring out the beef flavor of soups, gravies, and sauces; adds a roast flavor while smoothing out harsh flavors; *Properties:* deep brn. heavy paste, beefy bouillon-like odor; sol. in water; pH 4.8-5.5; 68-74% total solids; *Storage:* 1 yr stability when stored at ambient temps.

Building Blocks® Flavored Meat Stock #6. [Hormel] Meat stock/broth (75%), hydrolyzed vegetable (corn, soy, wheat) protein (31%), MSG (12%); natural flavored stock/broth used to enhance beefy flavor notes in soups, gravies, and sauces; imparts rich, full beefy flavor; inexpensive replacement for beef extract; *Properties:* deep brn. heavy paste, beefy bouillon-like odor; sol. in water; pH 4.8-5.5; 72-76% total solids; *Storage:* 1 yr stability when stored at ambient temps.

Building Blocks® Frozen Roast Beef Juice. [Hormel] Roast beef juices (contg. salt, sodium phosphate, sugar, dextrose, caramel color, flavoring, and spice); natural meat stock/broth/extract; rich roast beef flavor for soups, gravies, sauces; imparts rich brn. color; *Properties:* dk. brn. frozen liq., roast beef odor; nearly sol. in water; pH 5-7; 8-13% total solids; *Storage:* 1 yr stability when kept frozen.

Building Blocks® Ham Stock H5. [Hormel] Ham stock/broth (90%), hydrolyzed vegetable protein (5%), salt (5%); natural meat stock/broth giving a full rich natural flavor to soups, gravies, and sauces; adds smoothing, enriching mouthfeel to foods; *Properties:* deep brn. heavy paste, smoked ham odor; sol. in water; pH 4.5-6.5; 68-74% total solids; *Storage:* 1 yr stability when stored at ambient temps.

Building Blocks® Pork Stock P1. [Hormel] Pork stock/broth (92%), salt (8%); natural meat stock/ broth used to enhance meaty flavor notes in soups, gravies, and sauces; smoothes out harsh flavors and improves mouthfeel while increasing body and flavor; *Properties:* deep brn. heavy paste, meaty bouillon-like odor; sol. in water; pH 4.8-5.5; 68-74%

total solids; *Storage:* 1 yr stability when stored at ambient temps.

Bunge All-Purpose Vegetable Shortening. [Bunge Foods] Partially hydrog. soybean and cottonseed oils; shortening for general bakery use, cookies, biscuits, pie crusts, frying; *Properties:* 1.1 red color, bland flavor; m.p. 120 F; smoke pt. 450 F.

Bunge Biscuit Flakes. [Bunge Foods] Partially hydrog. soybean oil; EINECS 232-410-2; shortening for biscuit mixes and other dry mixes; icing stabilizer, glazes; improves texture of confectionery centers; *Properties:* flakes, bland flavor; m.p. 117 F.

Bunge Cake and Icing Vegetable Shortening. [Bunge Foods] Partially hydrog. soybean and cottonseed oil, mono- and diglycerides; shortening for cakes and icings; *Properties:* 1.1 red color, bland flavor; 2.7% alpha-mono.

Bunge Heavy Duty Donut Frying Shortening. [Bunge Foods] Partially hydrogenated soybean oil; EINECS 232-410-2; solid shortening for donut frying, fried pies; *Properties:* solid, 0.8 red color, bland flavor; m.p. 103 F; smoke pt. 455 F.

Bunge Special Mix Shortening. [Bunge Foods] Partially hydrog. soybean oil, propylene glycol monoesters, mono- and diglycerides, lecithin; shortening for cake mixes, esp. lean cake formulations; *Properties:* solid, 3.0 red color, bland flavor; 5.8% mono.

C

C™. [Calgon Carbon] Activated carbon; CAS 64365-11-3; used for food and chemical decolorization; *Properties:* pulverized.

C-60 Icing Powd. [Bunge Foods] Boiling type icing stabilizer; makes an extremely white, opaque, firm icing with exc. stability; for wrapped sweet goods icings, wrapped falt and creme icings applics; *Usage level:* 10 lb/100 lb powd. sugar; *Properties:* wh. powd; pH 8.5; 5% max. moisture.

C-249 Icing Powd. [Bunge Foods] Boiling type icing stabilizer; makes a white firm icing of high stability; for sweet goods icings for wrapped flat and creme prods; *Usage level:* 10 lb/100 lb powd. sugar; *Properties:* wh. to off-wh. powd; pH 8.5; 3.2% max. moisture.

C-369. [Bunge Foods] Boiling type icing stabilizer for danish and sweet goods, icings and glazes, frozen danish, cake and fondant type icings; does not require addition of hard fat; semitransparent; *Usage level:* 10 lb/100 lb powd. sugar; *Properties:* wh. to off-wh. powd./flakes; pH 7.3.

Cab-O-Sil® HS-5. [Cabot] Fumed silica, undensed; CAS 112945-52-5; *Properties:* 0.02% 325 mesh residue; bulk dens. 2.5 lb/ft³; surf. area 325±25 m²/g; pH 3.7-4.3 (4% aq. slurry); > 99.8% assay; *Toxicology:* LD50 (oral, rat) > 5 g/kg; inert to mildly irritating to skin; inert to very mildly irritating to eyes; *Storage:* store in dry environment away from chemical vapors.

Cain's PDC (Pizza Dough Conditioner). [MLG Enterprises Ltd.] Calcium sulfate, salt, L-cysteine, vegetable powd., tricalcium phosphate, and fungal enzymes; patented dough conditioner/starch modifier for use where rapid reduction of flour proteins is required; provides rapid hydration, more extensible doughs with superior machining qualities; for pizza and tortilla doughs; *Usage level:* 4-12 oz/cwt flour; *Properties:* free-flowing powd; *Storage:* 6 mos shelf life under cool, dry conditions; keep container closed when not in use.

Cake Mix 96. [Van Den Bergh Foods] Partially hydrog. soybean oil, propylene glycol mono and diesters of fats and fatty acids, mono- and diglycerides, optional lecithin; high performance fat for single-stage household cake mixes; *Regulatory:* kosher; *Properties:* Lovibond 4.5R-40Y max. plastic; m.p. 108-115 F; 5.5-6.1% mono.

Cake Moist. [Bunge Foods] Stabilizer enhancing moisture retention in cakes and fruit cakes; *Usage level:* 10-14% based on flour; *Properties:* wh. powd., bland taste; pH 4.5.

Cake Pan Release. [ADM Arkady] Homogeneous blend of vegetable oils with fine corn flour; release agent for hot or cold pans in cake prod; will not clog greasing equip. or spray heads.

Cake Sta. [ADM Arkady] Stabilization system for use in layer cake batter, pancake and waffle batter; improves handling and increases moisture retention.

CAL®. [Calgon Carbon] Activated carbon; CAS 64365-11-3; used for food decolorization; *Properties:* 12 x 40 sieve size.

C.A.L. [Pfizer Food Science] Natural flavors.

Calcium Ascorbate FCC No. 60475. [Roche] Calcium ascorbate FCC; CAS 5743-27-1; preservative, antioxidant, sodium-free, acid-free source of vitamin C for foods, esp. dry preps; *Regulatory:* FDA GRAS; *Properties:* wh. to sl. yel. powd., pract. odorless ; freely sol. in water; sl. sol. in alcohol; insol. in ether; m.w. 426.35; pH 6.8-7.4 (10% aq.); ≥ 98% assay; *Storage:* store in tight light-resist. containers, optimally @ ≤ 72 F; avoid exposure to moisture, excessive heat.

Calcium Pantothenate USP, FCC Type SD No. 63924. [Roche] Calcium pantothenate USP FCC; CAS 137-08-6; EINECS 205-278-9; source of pantothenic acid to fortify dry foods and enrichment premixes, usually in combination with other vitamins; *Properties:* wh. spray-dried, free-flowing powd., odorless, sl. hygroscopic; freely sol. in water (1 g/3 ml water); sol. in glycerin; pract. insol. in alcohol, chloroform, ether; m.w. 476.54; 90-110% assay; *Precaution:* sl. sensitive to heat; *Storage:* store in dry place @ 59-96 F.

Calfskin Gelatin Type B 175 Bloom. [Hormel] Gelatin; CAS 9000-70-8; EINECS 232-554-6; used in dairy prods., bakery toppings, canned goods to produce a moderate set; *Regulatory:* kosher; *Properties:* beige free-flowing gran., weak bouillon-like odor; 40 mesh; requires agitation when water is added to reduce clumping; visc. 33±3 mps; pH 4.5-5.8; 12% max. moisture; *Storage:* stable for up to 1 yr when stored dry at ambient temps; keep containers tightly closed.

Calfskin Gelatin Type B 200 Bloom. [Hormel] Gelatin; CAS 9000-70-8; EINECS 232-554-6; used in canning, dairy prods., and bakery toppings to produce a moderate set and control moisture migration; adds protein to foods; *Properties:* beige free-flowing gran., weak bouillon-like odor; 40 mesh; visc. 36 ± 4 mps; pH 4.5-5.8; 12% max. moisture; *Storage:* stable for up to 1 yr when stored dry at ambient temps; keep containers tightly closed.

Calfskin Gelatin Type B 225 Bloom. [Hormel] Gelatin; CAS 9000-70-8; EINECS 232-554-6; used in canning, dairy prods., and bakery toppings to produce a moderate set and control moisture migration; adds protein to foods; *Properties:* beige free-flowing gran., weak bouillon-like odor; 40 mesh; visc. 39 ± 4 mps; pH 4.5-5.8; 12% max. moisture; *Storage:* stable for up to 1 yr when stored dry at ambient temps; keep containers tightly closed.

Calfskin Gelatin Type B 250 Bloom. [Hormel] Gelatin; CAS 9000-70-8; EINECS 232-554-6; used in canning, dairy prods., and bakery toppings to produce a moderate set and control moisture migration; adds protein to foods; *Properties:* beige free-flowing gran., weak bouillon-like odor; 40 mesh; visc. 45 ± 3 mps; pH 4.5-5.8; 12% max. moisture; *Storage:* stable for up to 1 yr when stored dry at ambient temps; keep containers tightly closed.

Calgon. [Calgon] Sodium hexametaphosphate; CAS 10124-56-8; EINECS 233-343-1; food ingred.

Calgon® T Powd. Food Grade. [Albright & Wilson Australia] Sodium hexametaphosphate; CAS 68915-31-1; EINECS 233-343-1; food additive in meat, fish, and cheese processing; *Regulatory:* FCC, Australian compliance; *Properties:* wh. free-flowing powd; 1.5% +850 µm particle size, 22% +250 µm; pH 7.0 (1%); 67% P_2O_5; *Toxicology:* avoid inhaling dust, prolonged skin contact; *Storage:* store in cool dry place to avoid caking; very hygroscopic.

Calgon® Type RB. [Calgon Carbon] Activated carbon; CAS 64365-11-3; pulverized form for purification applics. in food industries.

Calib. [MLG Enterprises Ltd.] Potassium bromate, calcium iodate, and other edible excipients; dough conditioner for continuous mix processing; provides balanced blend of fast and slow acting oxidizers to improve dough strength throughout baking process; *Usage level:* 1 tablet; 75 ppm max. (total bromates and iodates)/cwt flour; *Properties:* tablet supplying 45 ppm potassium bromate and 15 ppm calcium iodate/cwt flour; *Storage:* 6 mos shelf life stored under cool, dry conditions.

Caliment Dicalcium Phosphate Dihydrate. [Albright & Wilson UK] Dicalcium phosphate dihydrate food grade; CAS 7789-77-7; EINECS 231-826-1; source of calcium and phosphorus in foods for infants, invalids, and geriatric patients and in animal feeds; gel-forming agent with alginates in fruit pulps, desserts, and in pet food brawns; *Regulatory:* EEC, FCC compliance; *Properties:* wh. fine impalpable powd., odorless, tasteless; m.w. 172.09; 31.9-33.5% calcium as CaO; *Storage:* store in closed pkgs. in cool, dry place; protect from contamination.

Calsoft L-60. [Pilot] Sodium dodecylbenzene sulfonate; CAS 25155-30-0; EINECS 246-680-4; anionic; biodeg. emulsion stabilizer, wetting and foaming agent, detergent, emulsifier for washing fruits and vegetables; *Properties:* water-wh. pasty liq; odorless; dens. 8.7 lb/gal; pH 7.4; 60% solids.

Cameo Showcase. [ADM Arkady] stabilizer for retail-type icings and glazes; produces semitransparent icing; system is activated in hot tap water.

Cameo Velvet. [ADM Arkady] Agar-based; CAS 9002-18-0; EINECS 232-658-1; icing stabilizer for wh. glossy icing; esp. for foods exposed to high humidity or packed in poly bags.

Cameo Velvet WT. [ADM Arkady] Agar-based; CAS 9002-18-0; EINECS 232-658-1; icing stabilizer for semi-transparent icing; esp. for foods exposed to high humidity or packed in poly bags.

Canamulse 55. [Canada Packers/Edible Oils] Propylene glycol stearate; CAS 1323-39-3; EINECS 215-354-3; nonionic; food emulsifier for cakes, whipped toppings, starch complexing; *Properties:* flakes, beads; HLB 3.5; 100% conc.

Canamulse 70. [Canada Packers/Edible Oils] Propylene glycol stearate; CAS 1323-39-3; EINECS 215-354-3; nonionic; food emulsifier for emulsified shortenings for cakes, cake mixes, donut mixes, whipped toppings; *Properties:* flakes, beads; HLB 4.0; 100% conc.

Canamulse 90K. [Canada Packers/Edible Oils] Propylene glycol stearate, kosher; CAS 1323-39-3; EINECS 215-354-3; nonionic; food emulsifier for emulsified shortenings for cakes, cake mixes, donut mixes, whipped toppings; *Properties:* flakes; HLB 4.0; 100% conc.

Canamulse 100. [Canada Packers/Edible Oils] Mono- and diglycerides; CAS 68990-53-4; nonionic; food emulsifier for shortening, margarines, baked goods; *Properties:* plastic solid; HLB 2.8; 100% conc., 40-44% alpha monoglycerides.

Canamulse 110K. [Canada Packers/Edible Oils] Mono- and diglycerides; CAS 68990-53-4; nonionic; food emulsifier for shortening, margarines, baked goods; *Regulatory:* kosher; *Properties:* plastic solid; HLB 2.8; 100% conc.

Canamulse 150K. [Canada Packers/Edible Oils] Mono- and diglycerides; CAS 68990-53-4; nonionic; food emulsifier for margarines, peanut butter, icings, dairy substitutes; *Regulatory:* kosher; *Properties:* flakes/beads; HLB 3.0; 100% conc., 40-44% alpha monoglycerides.

Canamulse 155. [Canada Packers/Edible Oils] Mono- and diglycerides; CAS 68990-53-4; nonionic; food emulsifier for dairy substitutes, cake mixes; *Properties:* flakes/beads; HLB 3.0; 100% conc., 40-44% alpha monoglycerides.

Canasperse SBF. [Canada Packers/Food Ingreds.] Natural lecithin; CAS 8002-43-5; EINECS 232-307-2; amphoteric; food grade emulsifier, dispersant, wetting agent; avail. in kosher grade; *Properties:* fluid; HLB 4.0; 62% conc.

Canasperse UBF. [Canada Packers/Food Ingreds.] Unbleached natural lecithin; CAS 8002-43-5;

EINECS 232-307-2; amphoteric; food grade emulsifier, dispersant, wetting agent; avail. in kosher grade; *Properties:* fluid; HLB 4.0; 62% conc.

Canasperse UBF-LV. [Canada Packers/Food Ingreds.] Unbleached natural lecithin; CAS 8002-43-5; EINECS 232-307-2; amphoteric; lower visc., easier flow grade; *Properties:* fluid; HLB 4.0; 56% conc.

Canasperse WDF. [Canada Packers/Food Ingreds.] Modified natural lecithin; CAS 8002-43-5; EINECS 232-307-2; amphoteric; improved water dispersion; avail. in kosher grade; *Properties:* fluid; HLB 12.0; 52% conc.

Candex®. [Mendell] Dextrose with small amounts of higher glucose saccharides; CAS 50-99-7; EINECS 200-075-1; offers sweet, nongritty taste and is easily blended with flavors, lubricants, and other dry additives; exc. flow and compaction props; for use in chewable tablets, esp. those made by direct compression; *Properties:* wh. porous, spherical granules, sweet noncloying/nongritty taste; avg particle size 218 µ; 30% max. -100 mesh; very sol. in water; dens. (tapped) 0.77 g/ml; pH 3.5 min.

Cane Cal®. [Calgon Carbon] Activated carbon; CAS 64365-11-3; used for food decolorization; *Properties:* 12 x 40 sieve size.

Canola Spray Oil 81-599-0. [ADM Refined Oils] Canola oil; quick melting, high stability oil esp. for spraying applics; *Properties:* Lovibond 1.5R max. oil, bland flavor; m.p. 90 ± 2 F; iodine no. 80 ± 2.

Canola Vegetable All-Purpose Shortening 81-706-0. [ADM Refined Oils] Canola oil and cottonseed oil; shortening for baking, frying, and cooking use; *Properties:* Lovibond 2.0R max. oil, bland flavor; m.p. 112 ± 4 F; iodine no. 75 ± 4.

Canola Vegetable Frying Shortening 81-573-0. [ADM Refined Oils] Canola oil; high stability multipurpose frying shortening; *Properties:* Lovibond 1.5R max. oil, bland flavor; m.p. 90 ± 2 F; iodine no. 80 ± 4.

Canola Vegetable Shortening 81-577-0. [ADM Refined Oils] Canola oil; high stability oil for deep fat frying, nondairy fat for sour cream, dips, and confection; *Properties:* Lovibond 1.5R max. oil, bland flavor; m.p. 102 ± 2 F; iodine no. 70 ± 4.

Canthaxanthin 10% Type RVI No. 66523. [Roche] Canthaxanthin dispersed in a matrix of gelatin, vegetable oil with antioxidants (ascorbyl palmitate, dl-α-tocopherol); colorant imparting a purple-red or bluish-red color to surimi and other food prods; *Usage level:* 300 mg max./lb of solid or semisolid or pint of liq. food; *Properties:* red-violet powd; 78-80% min. thru 80 mesh; disp. in water; 10% min. assay (canthaxanthin); *Storage:* store in tightly closed container in cool, dry place, optimally below 46 F; avoid moisture.

Canthaxanthin Beadlets 10%. [Crompton & Knowles/Ingred. Tech.] Canthaxanthine in a matrix of gelatin, sucrose, food starch, vegetable oil with antioxidants, ascorbyl palmitate and dl-α-tocopherol; color additive for foods; *Regulatory:* FDA

21CFR §73.75, 73.1075; *Properties:* purplish red free-flowing beadlets, sl. odor; disp. readily in warm water producing tomato-red hue; 100% min. thru mesh 20; 25% max. thru mesh 80; 10% canthaxanthine; *Storage:* store in tight containers below 45 C; avoid exposure to heat, light, and moisture; 6 mos storage life.

CAO®-3. [PMC Specialties] 2,6-Ditert.-butyl-p-cresol (BHT); CAS 128-37-0; EINECS 204-881-4; antioxidant for food and feed prods., pkg., food contact uses; retards oxidative deterioration in animal, veg., min. oils, fats, greases; preserves/stabilizes flavor, color, freshness, nutritive value; *Regulatory:* FDA 21CFR §137.350, 166.110, 172.115, 172.615, 172.878, 172.880, 172.882, 172.886, 173.340, 174, 175, 176, 177, 178, 181.24, 182.3173, 582.3173, GRAS, USDA; kosher; *Properties:* wh. cryst., sl. odor; sol. (g/100 ml): 45-55 g aliphatic/aromatic solvs., 40-45 g animal fats, ketone, acetone, 25-30 g veg. oils; m.w. 220.36; sp.gr. 0.899 (80/4 C); bulk dens. 1.048; visc. 3.47 cst (80 C); f.p. 69.2 C min; m.p. 157 F; b.p. 260-262 C (760 mm); flash pt. (COC) 275 C; ref. index 1.4859; 99% min. purity; *Toxicology:* ACGIH TLV-TWA 10 mg/m³; may cause skin irritation, respiratory passage irritation, severe eye irritation/burns; *Storage:* stable when stored below 77 F under dry conditions in unopened, original containers; may yel. sl. on aging.

Capcithin™. [Lucas Meyer] Range of capsule grade fluid soy lecithin; CAS 8002-43-5; EINECS 232-307-2; specially formulated for encapsulation; for health food industry, soft gel capsules, tonics.

Capital Soya. [Karlshamns Food Ingreds.] Soybean oil; CAS 8001-22-7; EINECS 232-274-4; specialty oil for cooking, frying, and baking; *Properties:* liq.

Caplube 8350. [Karlshamns] Glyceryl monodioleate; CAS 25496-72-4; nonionic; o/w emulsifier for food applics; *Properties:* liq; HLB 2.7; 100% conc.

Capmul® EMG. [Karlshamns Food Ingreds.] PEG-20 glyceryl stearate; nonionic; food emulsifier; dough strengthener in yeast-raised baked goods; *Regulatory:* FDA 21CFR §172.834; *Properties:* liq., plastic; HLB 13.1; sapon. no. 65-75; hyd. no. 65-80; 100% conc.

Capmul® GMO. [Karlshamns; Karlshamns Food Ingreds.] Glyceryl oleate; CAS 111-03-5; nonionic; food emulsifier, dispersant, defoamer; *Regulatory:* FDA 21CFR §184.1323; *Properties:* liq; sol. in org. solvs. and oils; m.p. 25 C max; HLB 3.4; acid no. 3 max; iodine no. 75 max; sapon. no. 160-170; 40% min. mono.

Capmul® GMS. [Karlshamns] Glyceryl stearate; nonionic; food emulsifier used in margarine, yeast-raised baked goods; *Properties:* beads, flakes; m.p. 57-62 C; HLB 3.2; acid no. 3 max; iodine no. 5 max; sapon. no. 155-165; 100% conc.

Capmul® GMVS-K. [Karlshamns] Glyceryl mono shortening; nonionic; food emulsifier for shortenings for cakes and icings, margarine, whipped topping; *Properties:* plastic, solid; HLB 3.4; 100% conc.

Capmul® O. [Karlshamns] Sorbitan oleate; CAS 1338-43-8; EINECS 215-665-4; nonionic; food emulsifier and dispersant; *Properties:* Gardner 6 max. plastic, solid; HLB 4.3; acid no. 10 max; sapon. no. 147-157; hyd. no. 235-260; 100% conc; Discontinued

Capmul® PGME. [Karlshamns] Propylene glycol monoesters and monodiglycerides; nonionic; emulsifier for cake mixes, toppings; *Properties:* solid; HLB 2.2; 100% conc.

Capmul® POE-L. [Karlshamns] Polysorbate 20; CAS 9005-64-5; nonionic; food emulsifier and solubilizer for flavors; *Properties:* Gardner 3 max. liq; HLB 16.7; acid no. 2 max; sapon. no. 40-50; hyd. no. 96-108; 100% conc.

Capmul® POE-O. [Karlshamns] Polysorbate 80; CAS 9005-65-6; nonionic; food emulsifier for frozen desserts; solubilizer for oils into water systems; used for flavors, fragrances, vitamins; *Properties:* Gardner 6 max. liq; water-sol; HLB 15.0; acid no. 2 max; sapon. no. 45-55; hyd. no. 65-80; 100% conc; Discontinued

Capmul® POE-S. [Karlshamns] Polysorbate 60; CAS 9005-67-8; nonionic; food emulsifier; solubilizer for oils into water systems; *Properties:* Gardner 7 max. liq; water-sol; HLB 14.9; acid no. 2 max; sapon. no. 45-55; hyd. no. 81-96; 100% conc.

Capmul® S. [Karlshamns] Sorbitan stearate; CAS 1338-41-6; EINECS 215-664-9; nonionic; food emulsifier for chocolate and confectionery coatings, shortenings; *Properties:* Gardner 6 max. bead; HLB 4.7; acid no. 10 max; sapon. no. 147-157; hyd. no. 235-260; 100% conc.

Caprol® 2G4S. [Karlshamns] Polyglyceryl-2 tetrastearate; CAS 72347-89-8; nonionic; food emulcifler; *Regulatory:* FDA 21CFR §172.854; *Properties:* Gardner 8 max. powd; sol. in oils, waxes; HLB 2.5; acid no. 6 max; iodine no. 6 max; sapon. no. 165-185; 100% conc; Discontinued

Caprol® 3GO. [Karlshamns] Polyglyceryl-3 oleate; CAS 9007-48-1; nonionic; food emulsifier for frozen desserts, veg. dairy prods., diet spreads; *Regulatory:* FDA 21CFR §172.854; *Properties:* Gardner 7 max. liq; sol. in org. solvs. and oils; HLB 6.2; acid no. 6 max; iodine no. 78 max; sapon. no. 125-150; 100% conc.

Caprol® 3GS. [Karlshamns] Polyglyceryl-3 stearate; CAS 37349-34-1; EINECS 248-403-2; nonionic; food emulsifier, stabilizer and whipping agent used in frozen desserts and fat reduction; *Regulatory:* FDA 21CFR §172.854; *Properties:* Gardner 8 max. powd; HLB 6.2; acid no. 6 max; iodine no. 3 max; sapon. no. 120-135; 100% conc.

Caprol® 3GVS. [Karlshamns] Triglyceryl mono shortening; nonionic; food emulsifier for icings, shortenings; *Properties:* semisolid; HLB 6.0; 100% conc.

Caprol® 6G2O. [Karlshamns] Polyglyceryl-6 dioleate; CAS 76009-37-5; nonionic; food emulsifier for frozen desserts; *Regulatory:* FDA 21CFR §172.854; *Properties:* Gardner 10 max. liq; HLB 8.5; acid no. 6 max; iodine no. 75 max; sapon. no. 105-125; 100% conc.

Caprol® 6G2S. [Karlshamns] Polyglyceryl-6 distearate; CAS 34424-97-0; nonionic; food emulsifier for whipped toppings, frozen desserts, coffee whiteners; *Regulatory:* FDA 21CFR §172.854; *Properties:* Gardner 10 max. solid; HLB 8.5; acid no. 6 max; iodine no. 3 max; sapon. no. 105-125; 100% conc.

Caprol® 10G2O. [Karlshamns] Polyglyceryl-10 dioleate; CAS 33940-99-7; food emulsifier for frozen desserts; *Regulatory:* FDA 21CFR §172.854; *Properties:* Gardner 7 max. liq; sol. in alcohol; water-disp; HLB 10.0; acid no. 6 max; iodine no. 60 max; sapon. no. 100-120; 100% conc; Discontinued

Caprol® 10G4O. [Karlshamns] Polyglyceryl-10 tetraoleate; CAS 34424-98-1; EINECS 252-011-7; nonionic; food emulsifier; *Regulatory:* FDA 21CFR §172.854; *Properties:* Gardner 7 max. liq; HLB 6.2; acid no. 6 max; iodine no. 60 max; sapon. no. 125-150; 100% conc.

Caprol® 10G10O. [Karlshamns] Polyglyceryl-10 decaoleate; CAS 11094-60-3; EINECS 234-316-7; nonionic; food emulsifier; *Regulatory:* FDA 21CFR §172.854; *Properties:* Gardner 9 max. liq; sol. in oils and org. solvs; HLB 3.5; acid no. 8 max; iodine no. 85 max; sapon. no. 155-185; 100% conc.

Caprol® ET. [Karlshamns Food Ingreds.] Polyglyceryl mixed esters; nonionic; food emulsifier and crystal inhibitor in vegetable oils; *Regulatory:* FDA 21CFR §172.854(c); *Properties:* plastic; m.p. 37-39 C; HLB 2.5; acid no. 3 max; sapon. no. 190-197; 100% conc.

Caprol® JB. [Karlshamns] Decaglyceryl decastearate; CAS 39529-26-5; EINECS 254-495-5; nonionic; food emulsifier in confectionery coatings; *Properties:* flakes; HLB 2.5; 100% conc; Discontinued

Caprol® PGE860. [Karlshamns] Decaglyceryl mono-, dioleate; nonionic; food emulsifier and beverage cloud agent; *Regulatory:* FDA 21CFR §172.854; *Properties:* Gardner 10 max. liq; HLB 11.0; acid no. 6 max; iodine no. 60 max; sapon. no. 90-105; 100% conc; Discontinued

Cap-Shure® AS-125-50. [Balchem] Ascorbic acid encapsulated in edible hydrog. soybean oil coating; CAS 50-81-7; EINECS 200-066-2; nutritive additive contg. 40,000 IU/g vitamin A, 20,000 IU/g vitamin D; oxidation agent, dough strengthener/conditioner; *Usage level:* 200-400 ppm (doughs); *Regulatory:* FDA, kosher approved; *Properties:* wh. free-flowing powd; sp.gr. 0.915-0.925; m.p. 152-158 F; iodine no. 120-160; sapon. no. 178-190; 50% ascorbic acid, 50% hydrog. soybean oil.

Cap-Shure® AS-165-70. [Balchem] Ascorbic acid encapsulated with a continuous film of partially hydrog. soybean oil; nutritive encapsulant for nutritional prods; provides taste masking, enhanced flow props. temp./time release, and prevents reactivity with other ingreds; *Properties:* wh. to off-wh. free-flowing gran; 2% max. on 10 mesh; coating m.p. 152-158 F; 70% ascorbic acid, 30% hydrog. soybean oil; *Storage:* store in cool, dry, odor-free environment @ 10-32 C.

Cap-Shure® AS-165CR-70. [Balchem] Ascorbic acid encapsulate; CAS 50-81-7; EINECS 200-066-2; nutrient.

Cap-Shure® BC-140-70. [Balchem] Sodium bicarbonate encapsulated with partially hydrog. cottonseed oil; encapsulant for use in dry mix, refrigerated dough, frozen dough, and microwave prods; coating provides protection from pre-reaction and controls gas evolution during leavening; *Properties:* wh. to off-wh. free-flowing gran; 2% max. on 14 mesh; coating m.p. 141-147 F; 70% sodium bicarbonate, 30% oil; *Storage:* store in cool, dry, odor-free environment @ 10-32 C.

Cap-Shure® BC-900-85. [Balchem] Sodium bicarbonate encapsulated with mono- and diglycerides; encapsulant for refrigerated batters and frozen doughs; the coating provides delayed release, prevents pre-reaction in baked goods, and controls gas evolution during leavening; *Properties:* wh. to off-wh. free-flowing gran; 2% max. on 14 mesh; 85% sodium bicarbonate, 15% mono- and diglycerides; *Storage:* store in cool, dry, odor-free environment @ 10-32 C.

Cap-Shure® C-135-72. [Balchem] Citric acid encapsulated with continuous film of partially hydrog. vegetable oil; encapsulated acidulant for cured meat prods., (dry and semi-dry sausage, pepperoni, genoa, hard salami); *Regulatory:* USDA approved; *Properties:* wh. powd., tangy acid flavor; coating m.p. 135-144 F; 72% citric acid, 28% oil; *Storage:* store in cool, dry place @ 50-90 F.

Cap-Shure® C-140-72. [Balchem] Citric acid encapsulated in partially hydrog. cottonseed oil; acidulant encapsulant; coating provides delayed release in meat systems, baked goods, confections, dessert mixes, and microwave prods; *Properties:* wh. to off-wh. free-flowing gran; 2% max. on 10 mesh; coating m.p. 141-147 F; 72% citric acid, 28% oil; *Storage:* store in cool, dry, odor-free environment @ 10-32 C.

Cap-Shure® C-140-85. [Balchem] Citric acid encapsulated in partially hydrog. cottonseed oil; encapsulated acidulant; coating provides delayed release in meat systems, baked goods, confections, dessert mixes, and microwave prods; *Properties:* wh. to off-wh. free-flowing gran; 2% max. on 10 mesh; coating m.p. 141-147 F; 85% citric acid, 15% oil; *Storage:* store in cool, dry, odor-free environment @ 10-32 C.

Cap-Shure® C-140E-75. [Balchem] Citric acid encapsulated with low-temp. release coating; CAS 77-92-9; EINECS 201-069-1; encapsulated acidulant for meat and sausage prod; unique edible coating releases citric acid slowly in presence of ambient moisture; *Regulatory:* USDA approved; *Properties:* wh. to off-wh. free-flowing gran; coating m.p. 132-140 F; 75% citric acid, 25% coating; *Storage:* store in cool, dry place @ 50-90 F; one yr shelf life.

Cap-Shure® C-150-50. [Balchem] Citric acid encapsulated with continuous film of partially hydrog. soybean oil; encapsulated acidulant for frozen food applics., esp. meat applics. with below freezing temp. processing; *Properties:* wh. to off-wh. free-flowing gran; 2% max. on 10 mesh; coating m.p. 152-158 F; 50% citric acid, 50% oil; *Storage:* store in cool, dry, odor-free environment @ 10-32 C.

Cap-Shure® C-165-63. [Balchem] Citric acid encapsulated in partially hydrog. soybean oil; acidulant encapsulant; coating provides barrier protecting the acid from other ingreds. in candies, gums, baked goods, and dessert mixes; *Properties:* wh. to off-wh. free-flowing gran; 2% max. on 14 mesh; coating m.p. 152-158 F; 63% citric acid, 37% oil; *Storage:* store in cool, dry, odor-free environment @ 10-32 C.

Cap-Shure® C-165-85. [Balchem] Citric acid encapsulated with continuous film of partially hydrog. soybean oil; encapsulated acidulant for sanded, starch molded candies; eliminates moisture pickup and clumping, provides long term storage stability; also for dry dessert and beverage mixes, baked goods, microwave prods; barrier prevents pre-reaction; *Properties:* wh. to off-wh. free-flowing gran; 2% max. on 20 mesh; coating m.p. 152-158 F; 85% citric acid, 15% oil; *Storage:* store in cool, dry, odor-free environment @ 10-32 C.

Cap-Shure® C-900-85. [Balchem] Citric acid encapsulated in mono- and diglycerides; encapsulated acidulant; coating provides delayed release in meat systems, baked goods, confections, dessert mixes, and microwave prods; *Properties:* wh. to off-wh. free-flowing gran; 2% max. on 10 mesh; 85% citric acid, 15% mono- and diglycerides; *Storage:* store in cool, dry, odor-free environment @ 10-32 C.

Cap-Shure® CLF-165-70. [Balchem] Lemon flavored citric acid encapsulated with partially hydrog. soybean oil; encapsulated acidulant for confections, baked goods, dessert mixes, and microwave prods; coating provides delayed release; lemon oil enhances flavor; *Properties:* wh. to off-wh. free-flowing gran; 2% max. on 10 mesh; coating m.p. 150-156 F; 70% citric acid, 30% coating; *Storage:* store in cool, dry, odor-free environment @ 10-32 C.

Cap-Shure® F-125-85. [Balchem] Fumaric acid encapsulated with continuous film of partially hydrog. soybean oil; encapsulated acidulant for baked goods, dessert mixes, microwave systems; coating protects acid from other ingreds., reduces problems with hygroscopicity, flavor and color degradation, premature pH changes, and shelf life; *Properties:* wh. to off-wh. free-flowing gran; 2% max. on 20 mesh; coating m.p. 152-158 F; 85% fumaric acid, 15% oil; *Storage:* store in cool, dry, odor-free environment @ 10-32 C.

Cap-Shure® F-140-63. [Balchem] Fumaric acid encapsulated with continuous film of partially hydrog. cottonseed oil; encapsulated acidulant for baked goods, dessert mixes, starch-gel hard candies; coating provides barrier protecting acid from other ingreds., reduces problems with hygroscopicity, flavor/color degradation, premature pH changes, and shelf life; *Properties:* wh. to off-wh. free-flowing gran; 2% max. on 10 mesh; coating m.p. 141-147 F;

63% fumaric acid, 37% oil; *Storage:* store in cool, dry, odor-free environment @ 10-32 C.

Cap-Shure® FE-165-50. [Balchem] Specially processed iron encapsulated with partially hydrog. soybean oil; nutritive encapsulant for nutritional prods; provides taste masking, enhanced flow props. temp./time release, and prevents reactivity with other ingreds; *Properties:* blk. free-flowing gran; 2% max. on 20 mesh; coating m.p. 152-158 F; 50% Fe; 50% hydrog. soybean oil; *Storage:* store in cool, dry, odor-free environment @ 10-32 C.

Cap-Shure® FF-165-60. [Balchem] Ferrous fumarate USP/FCC encapsulated with partially hydrog. soybean oil; nutritive encapsulant for nutritional prods; provides taste masking, enhanced flow props. temp./time release, and prevents reactivity with other ingreds; *Properties:* maroon free-flowing gran; 2% max. on 20 mesh; coating m.p. 152-158 F; 60% ferrous fumarate, 40% hydrog. soybean oil; *Storage:* store in cool, dry, odor-free environment @ 10-32 C.

Cap-Shure® FS-165-60. [Balchem] Ferrous sulfate, dried encapsulated with partially hydrog. soybean oil; nutritive encapsulant for nutritional prods; provides taste masking, enhanced flow props. temp./ time release, and prevents reactivity with other ingreds; *Properties:* tan free-flowing gran; 2% max. on 20 mesh; coating m.p. 152-158 F; 60% ferrous sulfate dried, 40% hydrog. soybean oil; *Storage:* store in cool, dry, odor-free environment @ 10-32 C.

Cap-Shure® GDL-140-70. [Balchem] Glucono δ-lactone encapsulated with partially hydrog. cottonseed oil; as leavening agent in baking industry, as acidulant in meals; provides flavor to sour doughs without any harmful effects on gluten structure or yeast cells; *Properties:* wh. to off-wh. free-flowing gran; 2% max. on 20 mesh; coating m.p. 141-147 F; 70% glucono delta-lactone, 30% oil; *Storage:* store in cool, dry, odor-free environment @ 10-32 C.

Cap-Shure® KCL-140-50. [Balchem] Potassium chloride USP/FCC encapsulated with partially hydrogenated cottonseed oil; nutritive encapsulant for nutritional mixes and health foods; protective coating provides taste masking and delayed release; *Properties:* wh. to off-wh. free-flowing gran; 2% max. on 10 mesh; coating m.p. 141-147 F; 50% KCl; 50% hydrog. cottonseed oil; *Storage:* store in cool, dry, odor-free environment @ 10-32 C.

Cap-Shure® KCL-165-70. [Balchem] Potassium chloride USP/FCC encapsulated with partially hydrogenated soybean oil; nutritive encapsulant for nutritional mixes and health foods; protective coating provides taste masking and delayed release; *Properties:* wh. to off-wh. free-flowing gran; 2% max. on 10 mesh; coating m.p. 152-158 F; 70% KCl, 30% hydrog. soybean oil; *Storage:* store in cool, dry, odor-free environment @ 10-32 C.

Cap-Shure® LCL-135-50. [Balchem] Lactic acid on calcium lactate encapsulated with partially hydrog. palm oil; encapsulated acidulant for cured meat prods. (dry and semi-dry sausages, pepperoni,

genoa, hard salami); low pre-release leach; *Regulatory:* USDA approved; *Properties:* wh. powd; 2% max. on 10 mesh; coating m.p. 136-144 F; 30% act., 50% substrate, 50% oil; *Storage:* store in cool, dry place @ 10-32 C; one yr shelf life.

Cap-Shure® SC-135-63. [Balchem] Sodium chloride encapsulated with partially hydrog. palm oil; reduces water-binding effects improving meats and baking doughs, improves in-plant handling of meat, provides color retention, reduces oxidative rancidity; for meat patties, pork sausage, yeast-leavened baked goods, frozen pretzels; *Properties:* wh. to off-wh. free-flowing gran; 2% max. on 10 mesh; coating m.p. 136-144 F; 63% NaCl, 37% hydrog. palm oil; *Storage:* store in cool, dry, odor-free environment @ 10-32 C.

Cap-Shure® SC-140X-70. [Balchem] Sodium chloride encapsulated with partially hydrog. cottonseed and soybean oils; used in low fat and frozen beef patties to prevent color degradation and poor texture; prevents yeast inhibition in baked goods; improves flavor and texture in microwaveable prods; coating prevents cracking and improves performance in frozen prods; *Properties:* wh. free-flowing gran; 2% max. on 10 mesh; coating m.p. 136-144 F; 70% NaCl, 30% hydrog. oils; *Storage:* store in cool, dry, odor-free environment @ 10-32 C.

Cap-Shure® SC-165-85FT. [Balchem] Sodium chloride encapsulated with partially hydrog. soybean oil; coated salt which prevents yeast inhibition in baked goods; prevents color degradation and poor texture in meat systems, improves flavor and texture in microwaveable prods; *Properties:* wh. to off-wh. free-flowing gran; 2% max. on 10 mesh; coating m.p. 152-158 F; 85% NaCl, 15 hydrog. soybean oil; *Storage:* store in cool, dry, odor-free environment @ 10-32 C.

Cap-Shure® SC-165X-60FF. [Balchem] Sodium chloride encapsulated with partially hydrog. soybean and cottonseed oils; coated salt preventing yeast inhibition in baked goods; prevents color degradation and poor texture in meat systems; improves flavor and texture in microwaveable prods; coating prevents cracking and improves performance in frozen prods; *Properties:* wh. free-flowing gran; 2% max. on 10 mesh; coating m.p. 145-151 F; 60% NaCl, 40% oils; *Storage:* store in cool, dry, odor-free environment @ 10-32 C.

Capsulec 51-SB. [ADM Ross & Rowe Lecithin] Capsule-grade lecithin; CAS 8002-43-5; EINECS 232-307-2; for soft and hard gelatin encapsulation; *Properties:* translucent fluid.

Capsulec 51-UB. [ADM Ross & Rowe Lecithin] Capsule-grade lecithin; CAS 8002-43-5; EINECS 232-307-2; for soft and hard gelatin encapsulation; *Properties:* translucent fluid.

Capsulec 56-SB. [ADM Ross & Rowe Lecithin] Capsule-grade lecithin; CAS 8002-43-5; EINECS 232-307-2; for soft and hard gelatin encapsulation; *Properties:* translucent fluid.

Capsulec 56-UB. [ADM Ross & Rowe Lecithin] Capsule-grade lecithin; CAS 8002-43-5; EINECS 232-307-2; for soft and hard gelatin encapsulation; *Properties:* translucent fluid.

Capsulec 62-SB. [ADM Ross & Rowe Lecithin] Capsule-grade lecithin; CAS 8002-43-5; EINECS 232-307-2; for soft and hard gelatin encapsulation; *Properties:* translucent fluid.

Capsulec 62-UB. [ADM Ross & Rowe Lecithin] Capsule-grade lecithin; CAS 8002-43-5; EINECS 232-307-2; for soft and hard gelatin encapsulation; *Properties:* translucent fluid.

Captex® 200. [Karlshamns] Propylene glycol dicaprylate/dicaprate; CAS 68583-51-7; carrier, coupler, solv. for flavors, fragrance oil, sol. colorants, vitamins; *Properties:* sol. in alcohol, oils, hydrocarbons, ketones; visc. 9-13 mPa•s; acid no. 0.1 max; iodine no. 0.5 max; sapon. no. 315-335; cloud pt. -15 C; ref. index 1.4393.

Captex® 300. [Karlshamns; Karlshamns Food Ingreds.] Caprylic/capric triglyceride; CAS 65381-09-1; med. chain triglyceride for nutritional supplements; *Properties:* Gardner 2 max. liq., bland odor and flavor; sol. in alcohol, oils, hydrocarbons, ketones; visc. 24-30 mPa•s; acid no. 0.1 max; iodine no. 0.5; sapon. no. 335-350; cloud pt. < -5 C; ref. index 1.4481.

Captex® 350. [Karlshamns; Karlshamns Food Ingreds.] Caprylic/capric/lauric triglyceride; CAS 68991-68-4; med. chain triglyceride for nutritional supplements; *Properties:* Gardner 2 max. liq., bland odor and flavor; visc. 36-42 mPa•s; acid no. 0.1 max; iodine no. 1.52 max; sapon. no. 290-310; cloud pt. 0 C max; ref. index 1.4582.

Captex® 355. [Karlshamns; Karlshamns Food Ingreds.] Caprylic/capric triglyceride; CAS 65381-09-1; med. chain triglyceride for nutritional supplements; carrier for essential oils, flavors, and fragrances; *Properties:* Lovibond R1.0 max. clear liq; neutral odor; bland flavor; misc. with most org. solvs. incl. 95% ethanol; sp.gr. 0.92-0.96; visc. 26-32 mPa•s; acid no. 0.1 max; iodine no. 0.5 max; sapon. no. 325-345; cloud pt. < -5 C; ref. index 1.4486.

Captex® 800. [Karlshamns] Propylene glycol dioctanoate; CAS 7384-98-7; carrier for essential oils, flavors; vehicle for vitamins, nutritional prods; *Properties:* APHA 100 max. clear liq; neutral odor; bland flavor; misc. with most org. solvs. incl. 95% ethanol; visc. 9-13 mPa•s; acid no. 1.0 max; iodine no. 1.0 max; sapon. no. 315-335; cloud pt. < -20 C.

Captex® 810A. [Karlshamns] Caprylic/capric/linoleic triglyceride; CAS 67701-28-4; emollient, solv., fixative, and extender in nutritional applics; carrier for flavors and fragrances; *Properties:* Gardner 2 max. color; acid no. 0.1 max; iodine no. 25; sapon. no. 307-320.

Captex® 810B. [Karlshamns] Caprylic/capric/linoleic triglyceride; CAS 67701-28-4; emollient, solv., fixative, and extender in nutritional applics; carrier for flavors and fragrances; *Properties:* Gardner 2 max. color; visc. 30 cps; acid no. 0.1 max; iodine no. 55; sapon. no. 280-296; cloud pt. 10 C; Discontinued

Captex® 810C. [Karlshamns] Caprylic/capric/linoleic triglyceride; CAS 67701-28-4; emollient, solv., fixative, and extender in nutritional applics; carrier for flavors and fragrances; *Properties:* Gardner 2 max. color; acid no. 0.1 max; iodine no. 75; sapon. no. 257-275; Discontinued

Captex® 810D. [Karlshamns] Caprylic/capric/linoleic triglyceride; CAS 67701-28-4; emollient, solv., fixative, and extender in nutritional applics; carrier for flavors and fragrances; *Properties:* Gardner 2 max. color; acid no. 0.1 max; iodine no. 85; sapon. no. 235-253.

Captex® 910A. [Karlshamns] Caprylic/capric/oleic triglyceride; CAS 67701-28-4; emollient, solv., fixative, and extender in nutritional applics; carrier for flavors and fragrances; *Properties:* Gardner 2 max. color; acid no. 0.1 max; iodine no. 19; sapon. no. 304-318; Discontinued

Captex® 910B. [Karlshamns] Caprylic/capric/oleic triglyceride; CAS 67701-28-4; emollient, solv., fixative, and extender in nutritional applics; carrier for flavors and fragrances; *Properties:* Gardner 2 max. color; acid no. 0.1 max; iodine no. 35; sapon. no. 280-296; Discontinued

Captex® 910C. [Karlshamns] Caprylic/capric/oleic triglyceride; CAS 67701-28-4; emollient, solv., fixative, and extender in nutritional applics; carrier for flavors and fragrances; *Properties:* Gardner 2 max. color; acid no. 0.1 max; iodine no. 48; sapon. no. 260-275; Discontinued

Captex® 910D. [Karlshamns] Caprylic/capric/oleic triglyceride; CAS 67701-28-4; emollient, solv., fixative, and extender in nutritional applics; carrier for flavors and fragrances; *Properties:* Gardner 2 max. color; acid no. 0.1 max; iodine no. 60; sapon. no. 238-250; Discontinued

Captex® 8000. [Karlshamns] Tricaprylin; CAS 538-23-8; EINECS 208-686-5; carrier for essential oils, flavors; vehicle for vitamins, nutritional prods; *Properties:* APHA 150 max. clear liq; neutral odor; bland flavor; misc. with most org. solvs. incl. 95% ethanol; visc. 20-28 mPa•s; acid no. 1.0 max; iodine no. 1.0 max; sapon. no. 350-365; cloud pt. < -5 C; ref. index 1.4469.

Captex® 8227. [Karlshamns] Triundecanoin; CAS 13552-80-2; EINECS 236-935-8; emollient, solv., fixative, and extender in nutritional applics; carrier for flavors and fragrances; *Properties:* acid no. 3 max; sapon. no. 270-290; cloud pt. 21 C.

Caramel Color Double Strength. [Crompton & Knowles/Ingred. Tech.] Caramel; CAS 8028-89-5; EINECS 232-435-9; color additive for foods; *Regulatory:* FDA 21CFR §73.85, 73.1085, 73.2085; *Properties:* dk. brn. visc. liq., typ. odor; sol. in water; sp.gr. 1.2 ± 0.1; visc. 300 cps max; pH 2.75 ± 0.5; *Storage:* store in tight containers; avoid exposure to excessive heat, light, and moisture; 2 yrs storage life.

Caramel Color Single Strength. [Crompton & Knowles/Ingred. Tech.] Caramel, acid proof; CAS 8028-89-5; EINECS 232-435-9; color additive for foods; *Regulatory:* FDA 21CFR §73.85, 73.1085,

73.2085; *Properties:* dk. brn. visc. liq., typ. odor; sol. in water; sp.gr. 1.3 ± 0.1; visc. 400 cps max; pH 3.0 ± 0.3; *Storage:* store in tight containers; avoid exposure to excessive heat, light, and moisture; 2 yrs storage life.

Caramel Prep. with Fat No. 75955. [Universal Flavors] Glucose syrup, condensed skimmed sweetened milk, water, hydrogenized coco fat, butter, salt, carrageenan, flavor, disodium phosphate; nature identical flavor for ice cream; *Properties:* dens. 1.33 ± 0.01; visc. 25,000 ± 5000 cps; pH 5.9 ± 0.2; *Storage:* 3 mos min. storage; keep cool @ 4 C.

Carbital® 35. [ECC Int'l.] Calcium carbonate; CAS 1317-65-3; EINECS 207-439-9; *Properties:* wh. powd. or 72% solids slurry, odorless; 0.005% max. 325 mesh; fineness 35% 2 µ; negligible sol. in water; sp.gr. 2.71; dens. 15.32 lb/gal (72% solids); visc. 500 cps (72%); surf. area 3.5 m²/g; pH 9-10; nonflamm; *Toxicology:* TLV/TWA 10 mg/m³, considered nuisance dust; *Precaution:* incompat. with acids.

Carbital® 50. [ECC Int'l.] Calcium carbonate; CAS 1317-65-3; EINECS 207-439-9; *Properties:* wh. powd. or 75% solids slurry, odorless; 0.005% max. 325 mesh; fineness 55% 2 µ; negligible sol. in water; sp.gr. 2.71; dens. 15.87 lb/gal (75% solids); visc. 500 cps (75%); surf. area 5 m²/g; pH 9-10; nonflamm; *Toxicology:* TLV/TWA 10 mg/m³, considered nuisance dust; *Precaution:* incompat. with acids.

Carbital® 75. [ECC Int'l.] Calcium carbonate; CAS 1317-65-3; EINECS 207-439-9; *Properties:* wh. powd. or 75% solids slurry, odorless; 0.005% max. 325 mesh; fineness 75% 2 µ; negligible sol. in water; sp.gr. 2.71; dens. 15.87 lb/gal (75% solids); visc. 250 cps (75%); surf. area 8 m²/g; pH 9-10; nonflamm; *Toxicology:* TLV/TWA 10 mg/m³, considered nuisance dust; *Precaution:* incompat. with acids.

Carbowax® Sentry® PEG 300. [Union Carbide] PEG-6, FCC grade; CAS 25322-68-3; EINECS 220-045-1; diluent in color additive coatings for fresh citrus fruit; coating, binder, plasticizer, and lubricant in tablets; bodying agent, dispersant, defoamer for food processing; *Regulatory:* FDA approved; *Properties:* water-wh. visc. liq; sol. in water, alcohols, glycerin, glycols; m.w. 285-315; sp.gr. 1.1250; dens. 9.38 lb/gal; visc. 5.8 cSt (99 C); f.p. -15 to -8 C; hyd. no. 356-394; flash pt. (PMCC) > 180 C; ref. index 1.463; pH 4.5-7.5 (5% aq.); surf. tens. 44.5 dynes/cm.

Carbowax® Sentry® PEG 400. [Union Carbide] PEG-8, FCC grade; CAS 25322-68-3; EINECS 225-856-4; diluent in color additive coatings for fresh citrus fruit; coating, binder, plasticizer, and lubricant in tablets; bodying agent, dispersant, defoamer for food processing; *Regulatory:* FDA approved; *Properties:* water-wh. visc. liq; sol. in water, methanol, ethanol, acetone, trichloroethylene, Cellosolve®, Carbitol®, dibutyl phthalate, toluene; m.w. 380-420; sp.gr. 1.1254; dens. 9.39 lb/gal; visc. 7.3 cSt (99 C); f.p. 4-8 C; hyd. no. 267-295; flash pt.

(PMCC) > 180 C; ref. index 1.465; pH 4.5-7.5 (5% aq.); surf. tens. 44.5 dynes/cm.

Carbowax® Sentry® PEG 540 Blend. [Union Carbide] PEG-6 and PEG-32, FCC grade; diluent in color additive coatings for fresh citrus fruit; coating, binder, plasticizer, and lubricant in tablets; bodying agent, dispersant, defoamer for food processing; *Regulatory:* FDA approved; *Properties:* wh. soft waxy solid; sol. in methylene chloride, 73% in water, 50% in trichloroethylene, 48% in methanol; m.w. 500-600; sp.gr. 1.0930; dens. 9.17 lb/gal (55 C); visc. 15.1 cSt (99 C); m.p. 38-41 C; hyd. no. 187-224; flash pt. (PMCC) > 180 C; pH 4.5-7.5 (5% aq.).

Carbowax® Sentry® PEG 600. [Union Carbide] PEG-12, FCC grade; CAS 25322-68-3; EINECS 229-859-1; diluent in color additive coatings for fresh citrus fruit; coating, binder, plasticizer, and lubricant in tablets; bodying agent, dispersant, defoamer for food processing; *Regulatory:* FDA approved; *Properties:* water-wh. visc. liq; sol. in water, alcohols, glycols; m.w. 570-630; sp.gr. 1.1257; dens. 9.40 lb/gal; visc. 10.8 cSt (99 C); f.p. 20-25 C; hyd. no. 178-197; flash pt. (PMCC) > 180 C; ref. index 1.46; pH 4.5-7.5 (5% aq.); surf. tens. 44.5 dynes/cm.

Carbowax® Sentry® PEG 900. [Union Carbide] PEG-20, FCC grade; CAS 25322-68-3; diluent in color additive coatings for fresh citrus fruit; coating, binder, plasticizer, and lubricant in tablets; bodying agent, dispersant, defoamer for food processing; *Regulatory:* FDA approved; *Properties:* wh. soft waxy solid; sol. 86% in water; m.w. 855-945; sp.gr. 1.0927 (60 C); dens. 9.16 lb/gal (55 C); visc. 15.3 cSt (99 C); m.p. 32-36 C; hyd. no. 119-131; flash pt. (PMCC) > 180 C; pH 4.5-7.5 (5% aq.).

Carbowax® Sentry® PEG 1000. [Union Carbide] PEG-20, FCC grade; CAS 25322-68-3; diluent in color additive coatings for fresh citrus fruit; coating, binder, plasticizer, and lubricant in tablets; bodying agent, dispersant, defoamer for food processing; *Regulatory:* FDA approved; *Properties:* wh. soft waxy solid; sol. 80% in water; m.w. 950-1050; sp.gr. 1.0926 (60 C); dens. 9.16 lb/gal (55 C); visc. 17.2 cSt (99 C); m.p. 37-40 C; hyd. no. 107-118; flash pt. (PMCC) > 180 C; pH 4.5-7.5 (5% aq.).

Carbowax® Sentry® PEG 1450. [Union Carbide] PEG-32, FCC grade; CAS 25322-68-3; diluent in color additive coatings for fresh citrus fruit; coating, binder, plasticizer, and lubricant in tablets; bodying agent, dispersant, defoamer for food processing; *Regulatory:* FDA approved; *Properties:* wh. soft waxy solid or flake; sol. 72% in water; m.w. 1300-1600; sp.gr. 1.0919 (60 C); dens. 9.17 lb/gal (55 C); bulk dens. 30 lb/ft³ (flake); visc. 26.5 cSt (99 C); m.p. 43-46 C; hyd. no. 70-86; flash pt. (PMCC) > 180 C; pH 4.5-7.5 (5% aq.).

Carbowax® Sentry® PEG 3350. [Union Carbide] PEG-75, FCC grade; CAS 25322-68-3; diluent in color additive coatings for fresh citrus fruit; coating, binder, plasticizer, and lubricant in tablets; bodying agent, dispersant, defoamer for food processing; *Regulatory:* FDA approved; *Properties:* wh. hard waxy flake or powd; sol. 67% in water; m.w. 3000-3700; sp.gr. 1.0926 (60 C); dens. 8.94 lb/gal (80 C); bulk dens. 30 lb/ft³ (flake), 40 lb/ft³ (powd.); visc. 90.8 cSt (99 C); m.p. 54-58 C; hyd. no. 30-37; flash pt. (PMCC) > 180 C; pH 4.5-7.5 (5% aq.).

Carbowax® Sentry® PEG 4600. [Union Carbide] PEG-100, FCC grade; CAS 25322-68-3; diluent in color additive coatings for fresh citrus fruit; coating, binder, plasticizer, and lubricant in tablets; bodying agent, dispersant, defoamer for food processing; *Regulatory:* FDA approved; *Properties:* wh. hard waxy flake or powd; sol. 65% in water; m.w. 4400-4800; sp.gr. 1.0926 (60 C); dens. 8.95 lb/gal (80 C); bulk dens. 30 lb/ft³ (flake), 40 lb/ft³ (powd.); visc. 184 cSt (99 C); m.p. 57-61 C; hyd. no. 23-26; flash pt. PMCC) > 180 C; pH 4.5-7.5 (5% aq.).

Carbowax® Sentry® PEG 8000. [Union Carbide] PEG-150, FCC grade; CAS 25322-68-3; diluent in color additive coatings for fresh citrus fruit; coating, binder, plasticizer, and lubricant in tablets; bodying agent, dispersant, defoamer for food processing; *Regulatory:* FDA approved; *Properties:* wh. hard waxy flake or powd; sol. 63% in water; m.w. 7000-9000; sp.gr. 1.0845 (60 C); dens. 8.96 lb/gal (80 C); bulk dens. 30 lb/ft³ (flake), 40 lb/ft³ (powd.); visc. 822 cSt (99 C); m.p. 60-63 C; hyd. no. 13-16; flash pt. (PMCC) > 180 C; pH 4.5-7.5 (5% aq.).

Carbrea® Tabs. [MLG Enterprises Ltd.] Azodicarbonamide, carbamide, ascorbic acid, and other edible excipients; dough conditioner, yeast nutrient, fermentation aid for yeast-raised bakery prods; oxidation system provides dough str., gas retention; nitrogen source aids fermentation; *Usage level:* 1-2 tablets/cwt flour; *Regulatory:* FDA GRAS; *Properties:* beige tablets, odorless; disp. in water; *Storage:* 1 yr shelf life stored under cool, dry conditions.

Carcao. [Idea Srl; Commodity Services Int'l.] Carob pulp flour; nonionic; cocoa substitute with lower fat content and nearly total absence of theobromine; facilitates emulsifying action; enhances cocoa; for dietary prods., in food and confectionery sectors; *Properties:* avail. in dark, medium, and light with total sugar 20, 30, and 40% resp; pH 5-6; 2-3% moisture (dark), 3-5% moisture (medium, light).

Carex for Foods. [Nutritional Research Assoc.] Carotene from carrot oil with refined soybean oil and 50 ppm BHT (antioxidant); colorant imparting golden color and nutritional additive for foods incl. butter, cheese, oleomargarine, salad/cooking oils, salad dressing, mayonnaise, popcorn seasoning, breads, rolls, noodles, cakes, doughnuts, candies, frostings; *Regulatory:* FDA approved; *Properties:* reddish brn. to dk. orange liq., bland odor and flavor char. of carrots; sp.gr. 0.919; potency 5000 USP Vitamin A units; *Toxicology:* harmless although excessive intake of carotenes may cause temporary yellowing of skin.

Carmacid R. [MLG Enterprises Ltd.] Acid-stable ammonium carminate; natural colorant producing red shades for foods, alcoholic and nonalcoholic beverages; 7.5% carminic acid.

Carmacid Y. [MLG Enterprises Ltd.] Carminic acid sprayed with maltodextrin; CAS 1390-65-4; EINECS 215-724-4; natural colorant producing yellow-orange shades for foods, alcoholic and non-alcoholic beverages; *Properties:* powd; 50% carminic acid.

Carmine 1623. [MLG Enterprises Ltd.] Carminic acid aluminum-calcic lake; CAS 1390-65-4; EINECS 215-724-4; natural colorant producing berry red or violet shades after dilution; for milk prods., flavored yogurts, ice cream, confections, cakes, desserts, fondants, cracker fillings, dried/candied fruits, cherries, artificial fruits; *Usage level:* 0.05-0.2%; *Properties:* impalpable powd; 100% 80 mesh; sol. in alkaline sol'ns; insol. in water; 52 ± 1% (as carminic acid); *Storage:* store in tightly closed containers in fresh, dry room.

Carmine AS. [Crompton & Knowles/Ingred. Tech.] Carmine, acid stable; CAS 1390-65-4; EINECS 215-724-4; color additive for foods; exc. coloring props. in acidic environments; *Regulatory:* FDA 21CFR §73.100, 73.1100, 73.2087; *Properties:* sl. visc. liq., bright red color when diluted in water; sl. odor; highly sol. in water and alcoholic beverages; pH 3.2 ± 0.2; 34 ± 1.7% solids, 2.5% min. carminic acid; *Storage:* store in tight containers; avoid exposure to heat and light; 1 yr storage life.

Carmine Extract P4011. [Phytone Ltd.] Carmine extract; CAS 1390-65-4; EINECS 215-724-4; food colorant; *Regulatory:* EEC E120; JECFA compliance; *Properties:* deep red-purple mobile liq; *Storage:* store in tightly closed containers in cool, dry area; protect from direct sunlight.

Carmine FG. [MLG Enterprises Ltd.] Hydrated aluminum chelate of carminic acid, carmine lake-red powd; CAS 1390-65-4; EINECS 215-724-4; natural colorant producing red-blue shades for meats, ice cream, yogurt, dairy prods., soups, jellies, fruit fillings, beverages, confections, puddings; *Regulatory:* FDA 21CFR §73.100, 73.1100, 73.2087; EEC E120; *Properties:* sol. in alkaline media; insol. in water and alcohol; up to 60% carminic acid; *Storage:* keep in cool, dark place.

Carmine Nacarat 40. [MLG Enterprises Ltd.] Carminic acid aluminum-calcic lake; CAS 1390-65-4; EINECS 215-724-4; natural colorant producing berry red or violet shades after dilution; for milk prods., flavored yogurts, ice cream, confections, cakes, desserts, fondants, cracker fillings, dried/candied fruits, cherries, artificial fruits; *Usage level:* 0.05-0.2%; *Properties:* impalpable powd; 100% 80 mesh; sol. in alkaline sol'ns; insol. in water; 30.4 ± 0.7% (as carminic acid); *Storage:* store in tightly closed containers in fresh, dry room.

Carmine PG. [MLG Enterprises Ltd.] Hydrated aluminum chelate of carminic acid, carmine lake-red powd; CAS 1390-65-4; EINECS 215-724-4; natural food colorant producing purple shades; for meat prods., ice cream, yogurt, dairy prods., soups, jellies, fruit fillings, beverages, confections, puddings; *Regulatory:* FDA 21CFR §73.100, 73.1100, 73.2087; EEC E120; *Properties:* sol. in alkaline media; insol. in water and alcohol; up to 60%

carminic acid; *Storage:* keep in cool, dark place.

Carmine Powd. WS. [Crompton & Knowles/Ingred. Tech.] Carmine; CAS 1390-65-4; EINECS 215-724-4; color additive for foods; *Regulatory:* FDA 21CFR §73.100, 73.1100, 73.2087; *Properties:* free-flowing powd., bright red when diluted in water; sl. odor; 100% min. thru 100 mesh; highly sol. in water, alcoholic beverages; 50% min. carminic acid; *Storage:* store in tight containers; avoid exposure to heat, light, moisture; 6 mos storage life.

Carmine XY/UF. [MLG Enterprises Ltd.] Hydrated aluminum chelate of carminic acid, carmine laked powd; CAS 1390-65-4; EINECS 215-724-4; natural colorant producing red-yel. shades, with high tinting str; for meat prods., ice cream, yogurt, dairy prods., soups, jellies, fruit fillings, beverages, confections, puddings; *Regulatory:* FDA 21CFR §73.100, 73.1100, 73.2087; EEC E120; *Properties:* sol. in alkaline media; insol. in water and alcohol; up to 60% carminic acid; *Storage:* keep in cool, dark place.

Carmisol A. [MLG Enterprises Ltd.] Ammonium salt of hydrated aluminum chelate of carminic acid; CAS 1390-65-4; EINECS 215-724-4; natural colorant producing pink to magenta red shades for bakery mixes, confections, dairy prods; for sol'ns. or dry blends; *Regulatory:* FDA 21CRR §73.100, 73.1100, 73.2087; EEC E120; *Properties:* dk. red powd; sol. in water; 50% carminic acid; *Storage:* keep in cool, dark place.

Carmisol NA. [MLG Enterprises Ltd.] Sodium salt of hydrated aluminum chelate of carminic acid; CAS 1390-65-4; EINECS 215-724-4; natural colorant producing pink to magenta red shades for bakery mixes, confections, dairy prods; may be used in dry blends; *Regulatory:* FDA 21CFR §73.100, 73.1100, 73.2087; EEC E120; *Properties:* odorless; water-sol; 50% carminic acid; *Storage:* keep in cool, dark place.

Carnation®. [Witco/Petroleum Spec.] Wh. min. oil NF; lubricant for food processing; *Regulatory:* FDA 21CFR §172.878, §178.3620a; *Properties:* water-wh., odorless, tasteless; sp.gr. 0.837-0.853; visc. 11-14 cSt (40 C); pour pt. -7 C; flash pt. 177 C.

Carotenal Sol'n. #2 No. 66425. [Roche] β-Apo-8´-carotenal compded. with dl-α-tocopherol as stabilizer and fractionated triglycerides of coconut oil origin; colorant for use in process cheese and salad dressings; also incorporated into oil phase of other food prods., e.g., imitation dairy prods., pastry, whipped margarine; *Usage level:* 750 mg max./lb solid or semisolid food or pint of liq. food; *Regulatory:* FDA 21CFR §73.95 (food use); *Properties:* dk. red sol'n; sol. in veg. oils and fats; sp.gr. 0.93-0.96; dens. 27.5-28.4 g/fl oz; 2% β-apo-8´-carotenal, 24,000 IU provitamin A activity/g; *Storage:* store in tight container @ 65-85 F; avoid exposure to air and light; solidifies below 32 F.

Carotenal Sol'n. 4% No. 66424. [Roche] β-Apo-8´-carotenal compded. with acetylated monoglyceride and dl-α-tocopherol; colorant for incorporation into oil phase of any food prod. and for color plating of spice blends, breading mixes, etc; color value approximates that of paprika oleoresin; *Usage level:* 375 mg max./lb solid or semisolid food or pint of liq. food; *Regulatory:* FDA 21CFR §73.95 (food use); *Properties:* dk. red sol'n; sol. in veg. oils and fats; sp.gr. 0.96; visc. 240 cps; 4% β-apo-8´-carotenal, 48,000 units provitamin A activity/g; *Storage:* store in tight container @ 59-86 F; avoid exposure to air and light.

Carotenal Sol'n. #73 No. 66428. [Roche] 1.4% β-Apo-8´-carotenal and 0.6% β-carotene FCC compded. with dl-α-tocopherol (stabilizer) and fractionated coconut oil triglycerides; colorant for use in process cheese; also incorporated into oil phase of other food prods., e.g., imitation dairy prods., pastry, whipped margarine, nonstandardized salad dressings, french dressing; *Usage level:* 1.07 g max./lb or pint of food; *Properties:* dk. red sol'n; sol. in veg. oils and fats; sp.gr. 0.93-0.96; dens. 27.5-28.4 g/fl oz; 26,800 IU provitamin A activity/g; *Storage:* store in tight container @ 65-85 F; avoid exposure to air and light; solidifies below 32 F.

CarraFat™. [Carrageenan Co; Hormel] Carrageenan emulsion with water, salt, and flavoring; CAS 9000-07-1; EINECS 232-524-2; fat replacer in meat and poultry formulations esp. in ground or emulsified processed meat and poultry prods; low-calorie and cholesterol-free substitute for animal fat; heat stable; microwaveable; water binder; *Usage level:* 5-25% (ground beef), 10-30% (breakfast sausage); *Regulatory:* FDA approved; kosher certified, *Properties:* off-wh. glossy, smooth powd., fatty flavor; 8% max. moisture; *Storage:* stable for up to 1 yr when kept dry at ambient temps; keep container tightly closed.

Carralean™ CG-100. [Hormel] Carrageenan, pure refined kappa; CAS 9000-07-1; EINECS 232-524-2; stabilizer, gellant for water-based food systems, dairy applics., frozen desserts, chocolate milk prods., canned meats, gel dessert mixes, doughs, pasta, nondairy toppings; *Usage level:* 0.02% (frozen desserts), 0.03% (chocolate milk prods.), 0.2-0.5% (canned meats), 0.1% (doughs); *Properties:* lt. tan to wh; disp. in water or milk; moderately sol. in hot sugar sol'ns; visc. 60-85 cps; pH 7-8 (1.5%); 10% max. moisture; *Storage:* stable for up to 1 yr when stored dry at ambient temps; keep containers tightly closed.

Carralean™ CM-70. [Hormel] Carrageenan, refined kappa type; CAS 9000-07-1; EINECS 232-524-2; stabilizer, gellant for dairy and water-based food systems, frozen desserts, chocolate milk prods., cheese prods., pie fillings; *Usage level:* 0.02% (frozen desserts), 0.03% (chocolate milk prods.), 0.25-0.75% (cheese prods.), 0.3-0.5% (pies); *Properties:* lt. tan; disp. in water; visc. 25-75 cps; pH 7-9 (1.5%); 12% max. moisture; *Storage:* stable for up to 1 yr when stored dry at ambient temps; keep containers tightly closed.

Carralean™ CM-80. [Hormel] Carrageenan, high fiber kappa; CAS 9000-07-1; EINECS 232-524-2;

stabilizer, gellant for dairy and water-based food systems; *Properties:* lt. yel. to wh; disp. in water; visc. 25-75 cps; pH 7-9 (1.5%); 12% max. moisture; *Storage:* stable for up to 1 yr stored dry at ambient temps; keep containers tightly closed.

Carralean™ MB-60. [Hormel] Carrageenan, refined kappa, dextrose, KCL; CAS 9000-07-1; EINECS 232-524-2; gellant and stabilizer for meat and poultry water-based systems; *Usage level:* 0.3-1.5%; *Properties:* lt. yel; disp. in water; visc. 15-25 cps; pH 7-8 (1.5%); 12% max. moisture; *Storage:* stable for up to 1 yr when stored dry at ambient temps; keep containers tightly closed.

Carralean™ MB-93. [Hormel] Carrageenan, high fiber kappa; CAS 9000-07-1; EINECS 232-524-2; thickener, stabilizer for fish, meat, and poultry applics., esp. high moisture food systems; *Usage level:* 0.5-1%; *Properties:* lt. tan; disp. in water; visc. 20-40 cps; pH 7-10 (1.5%); 10% max. moisture; *Storage:* stable for 1 yr when kept dry at ambient temps; keep containers tightly closed.

Carralite™. [Carrageenan Co; Hormel] Fat flavored carrageenan; CAS 9000-07-1; EINECS 232-524-2; fat replacer, stabilizer, gellant, bodying agent for dairy, bakery, confectionery use, dietary and nutritional beverages, dessert mixes; *Usage level:* 0.8-1.6%; *Regulatory:* FDA approved; kosher certified; *Properties:* off-wh., fatty flavor; 95% thru 80 mesh; swells in cold water; completely sol. in hot water; visc. 40-50 cps; pH 7-9 (1.5%); 8% max. moisture; *Storage:* stable for up to 1 yr when stored dry at ambient temps; keep containers tightly closed.

CarraLizer™ CGB-10. [Carrageenan Co.] Carrageenan blend; CAS 9000-07-1; EINECS 232-524-2; ice cream stabilizer for standard formulations; prevents separation of fat and other solids, whey-off while unfrozen mix is in storage, formation of large ice crystals; improves prod. texture; *Usage level:* 0.08-0.15% (chocolate ice cream), 0.10-0.20% (vanilla and flavored ice cream), 0.20-0.30% (sugar-free ice cream); *Regulatory:* FDA approved; kosher certified.

CarraLizer™ CGB-20. [Carrageenan Co; Hormel] Carrageenan blend; CAS 9000-07-1; EINECS 232-524-2; stabilizer for dairy industry; controls migration of water which can form crystals in frozen desserts; *Usage level:* 0.02-0.03% of total mix (frozen desserts); *Regulatory:* FDA approved; kosher certified; *Properties:* cream to wh. free-flowing powd; visc. 380-420 cps; pH 7-9 (1.5%); 12% max. moisture; *Storage:* stable for up to 1 yr when stored dry at ambient temps; keep containers tightly closed.

CarraLizer™ CGB-40. [Carrageenan Co.] Carrageenan blend; CAS 9000-07-1; EINECS 232-524-2; ice cream stabilizer for premium formulations; prevents separation of fat and other solids, whey-off while unfrozen mix is in storage, formation of large ice crystals; improves prod. texture; *Usage level:* 0.08-0.15% (chocolate ice cream), 0.10-0.20% (vanilla and flavored ice cream), 0.20-0.30% (sugar-free ice cream); *Regulatory:* FDA approved; kosher certified.

CarraLizer™ CGB-50. [Carrageenan Co.] Carrageenan blend; CAS 9000-07-1; EINECS 232-524-2; ice cream stabilizer for premium formulations; prevents separation of fat and other solids, whey-off while unfrozen mix is in storage, formation of large ice crystals; improves prod. texture; *Usage level:* 0.08-0.15% (chocolate ice cream), 0.10-0.20% (vanilla and flavored ice cream), 0.20-0.30% (sugar-free ice cream); *Regulatory:* FDA approved; kosher certified.

Carsonol® SHS. [Lonza] Sodium 2-ethylhexyl sulfate; EINECS 204-812-8; anionic; low foaming detergent, wetting agent, penetrant, emulsifier used in caustic sol'ns. for peeling of fruits and vegetables; stable to high concs. of electrolytes; *Regulatory:* FDA approved; *Properties:* Gardner 3 clear liq; dens. 9.6 lb/gal; sp.gr. 1.15; visc. 35 cps; pH 10.3 (10%); 40% act.

Carudan 000 Range. [Grindsted Prods.] Locust bean gum; CAS 9000-40-2; EINECS 232-541-5; stabilizer, thickener for cream cheese, dairy desserts; extra high grade; *Regulatory:* EEC, FDA § 184.1343 (GRAS).

Carudan 100 Range. [Grindsted Prods.] Locust bean gum; CAS 9000-40-2; EINECS 232-541-5; stabilizer, thickener for ice cream, fruit preps., baked goods, pimento strips, dressings, and sauces; high grade; *Regulatory:* EEC, FDA §184.1343 (GRAS).

Carudan 200 Range. [Grindsted Prods.] Locust bean gum; CAS 9000-40-2; EINECS 232-541-5; stabilizer, thickener for ice cream, fruit preps., baked goods, pimento strips, dressings, and sauces; med. grade; *Regulatory:* EEC, FDA §184.1343 (GRAS).

Carudan 300 Range. [Grindsted Prods.] Locust bean gum; CAS 9000-40-2; EINECS 232-541-5; stabilizer, thickener for ice cream, fruit preps., baked goods, pimento strips, dressings, and sauces; std. grade; *Regulatory:* EEC, FDA §184.1343 (GRAS).

Carudan 400 Range. [Grindsted Prods.] Locust bean gum; CAS 9000-40-2; EINECS 232-541-5; stabilizer, thickener for canned pet food.

Carudan 700 Range. [Grindsted Prods.] Locust bean gum; CAS 9000-40-2; EINECS 232-541-5; stabilizer, thickener for dairy desserts, dressings, and sauces; purified grade; *Regulatory:* EEC, FDA §184.1343 (GRAS).

Castorwax® NF. [CasChem] Hydrog. castor oil NF; CAS 8001-78-3; EINECS 232-292-2; wax for food applics; *Regulatory:* FDA approval; *Properties:* flake.

Catalase L. [Solvay Enzymes] Bovine catalase; CAS 9001-05-2; EINECS 232-577-1; enzyme which catalyzes the decomposition of hydrogen peroxide to water and oxygen; for milk prod; *Properties:* liq.

Cavitron Cyclo-dextrin.™ [Am. Maize Prods.] Cyclodextrin; CAS 7585-39-9; EINECS 231-493-2; molecular encapsulation; protects act. ingreds. against oxidation and decomp., eliminates/reduces undesired taste/odor, and contamination, stabilizes food flavors and fragrances, enhances solubility; avail. as α, β, γ, derivs., and polymers; *Properties:* spherical beads; α: m.w. 973; sol. 12.7 g/100 ml in water; β: m.w. 1135; sol. 1.88 g/100 ml in water, > 41% in dimethyl sulfoxide, 28.3% dimethyl formamide, 7% ethylene glycol; γ: m.w. 1297; sol. 25.6 g/100 ml in water.

CBC #7 Shortening. [Karlshamns] Partially hydrog. soybean and cottonseed oils, polyglycerol esters of fatty acids, mono- and diglycerides, polysorbate 60, and lecithin; food ingred; *Properties:* drop pt. 50-52.5 C.

CC™-101. [ECC Int'l.] Calcium carbonate; CAS 471-34-1, 1317-65-3; EINECS 207-439-9; *Properties:* wh. powd., odorless; 93% thru 325 mesh; negligible sol. in water; sp.gr. 2.71; dens. 22.57 lb/gal; bulk dens. 70 lb/ft³ (loose); bulking value 0.044; ref. index 1.59; pH 9.5; nonflamm; *Toxicology:* TLV/TWA 10 mg/m³, considered nuisance dust; *Precaution:* incompat. with acids.

CC™-103. [ECC Int'l.] Calcium carbonate; CAS 471-34-1, 1317-65-3; EINECS 207-439-9; *Properties:* wh. powd., odorless; 80% thru 325 mesh; negligible sol. in water; sp.gr. 2.71; dens. 22.57 lb/gal; bulk dens. 72 lb/ft³ (loose); bulking value 0.044; ref. index 1.59; pH 9.5; nonflamm; *Toxicology:* TLV/TWA 10 mg/m³, considered nuisance dust; *Precaution:* incompat. with acids.

CC™-105. [ECC Int'l.] Calcium carbonate; CAS 471-34-1, 1317-65-3; EINECS 207-439-9; *Properties:* wh. powd., odorless; 65% thru 325 mesh; negligible sol. in water; sp.gr. 2.71; dens. 22.57 lb/gal; bulk dens. 80 lb/ft³ (loose); bulking value 0.044; ref. index 1.59; pH 9.5; nonflamm; *Toxicology:* TLV/TWA 10 mg/m³, considered nuisance dust; *Precaution:* incompat. with acids.

CC-603. [Stepan Food Ingreds.] Blend of carrageenan, guar gum and dextrose; food grade stabilizer for use in whipped toppings, fillings, coffee whiteners; Unverified

CDC-10A. [MLG Enterprises Ltd.] Wheat flour, diacetyl tartaric acid esters of mono- and diglycerides, mono- and diglycerides, ethoxylated monodiglycerides, soy flour, lecithin, ascorbic acid, potassium bromate, l-cysteine, fungal amylase; dough conditioning system for use in string line-produced hearth prods. where extensibility and crispy crust are important; *Usage level:* 0.75-1.25% on flour (no-time doughs); *Storage:* 6 mos shelf life stored in cool, dry conditions.

CDC-10A (NB). [MLG Enterprises Ltd.] Wheat flour, diacetyl tartaric acid esters of mono- and diglycerides, mono- and diglycerides, ethoxylated monodiglycerides, soy flour, calcium salts, lecithin, ascorbic acid, l-cysteine, fungal amylase, azodicarbonamide; bromate-free dough conditioning system for use in string line-produced hearth prods. where extensibility and crispy crust are important; *Usage level:* 0.75-1.25% on flour (no-time doughs); *Storage:* 6 mos shelf life stored in cool, dry conditions.

CDC-50. [MLG Enterprises Ltd.] Diacetyl tartaric acid esters of mono- and diglycerides, mono- and di-

glycerides; dough conditioner, strengthener, softener for yeast-raised bakery prods; improves volume, grain, keeping qualities of bread prods; produces superior crumb texture; *Usage level:* 0.5% on flour (bread, buns); *Regulatory:* FDA GRAS; *Properties:* sl. off-wh. free-flowing powd., sl. acrid odor, char. flavor; m.p. 60 C; acid no. 27-57; iodine no. 2.0 max; sapon. no. 325-350; pH 2.2-2.4 (10% aq. disp.); 100% act; *Storage:* 6 mos shelf life stored in cool, dry conditions; keep plastic bag sealed when not in use.

CDC-79. [MLG Enterprises Ltd.] Soya flour, wheat flour, lecithin, barley malt, ascorbic acid; dough conditioner for yeast-raised prods. where claims of natural or health food are made; dough strengthener; provides tolerance to mechanical abuse; improves volume, grain; vital wheat gluten replacement; *Usage level:* 1.5-2.25%/cwt flour (yeast-raised prods.); 1:1 replacement for vital wheat gluten; *Regulatory:* FDA GRAS; *Properties:* dry free-flowing; *Storage:* 6 mos shelf life stored in cool, dry conditions.

CDC-79DZ. [MLG Enterprises Ltd.] Soya flour, wheat flour, lecithin, barley malt, calcium peroxide, ascorbic acid; dough conditioner; *Properties:* cereal-colored free-flowing powd; 38% protein, < 1% calcium peroxide, 0.6% ascorbic acid; *Storage:* store under cool, dry conditions.

CDC-2001. [MLG Enterprises Ltd.] Wheat flour, dextrose, diacetyl tartaric acid esters of mono- and diglycerides, mono- and diglycerides, soy flour, lecithin, vegetable powd., ascorbic acid, potassium bromate, azodicarbonamide, carbamide, fungal amylase; dough conditioner; *Regulatory:* FDA GRAS; *Properties:* cream-colored free-flowing powd., char. odor; acid no. 20-25; sapon. no. 140-150; ester no. 114-120; pH 3.6-4.0 (10% disp.); 0.3 ± 2% potassium bromate; *Storage:* 6 mos min. shelf life stored under cool, dry conditions.

CDC-2002. [MLG Enterprises Ltd.] Wheat flour, dextrose, diacetyl tartaric acid esters of mono- and diglycerides, mono- and diglycerides, soy flour, lecithin, calcium phosphate, ascorbic acid, azodicarbonamide, carbamide, vegetable powd., fungal amylase; patented complete dough conditioner system for yeast-raised bakery prods; contains protein modifier, starch/protein complexer, yeast nutrient, and bromate-free oxidation system; ideal for frozen and no-time doughs incl. croissants; *Usage level:* 0.75-1.25%; *Properties:* cream-colored free-flowing powd., sl. acrid odor; acid no. 20-25; sapon. no. 140-150; pH 3.6-4.0 (10% disp.); 19.6% protein, 4-5% moisture; *Storage:* 6 mos min. shelf life stored under cool, dry conditions.

CDC-2500. [MLG Enterprises Ltd.] Wheat flour, diacetyl tartaric acid esters of mono- and diglycerides, mono- and diglycerides, dextrose, soy flour, lecithin, ascorbic acid, and fungal enzymes; dough conditioner for yeast-raised bakery prods; improves volume, dough tolerance and absorption, machining of doughs, grain, texture, keeping qualities; accelerates proof times; *Usage level:* 0.5-

0.75% on flour (sponge and dough systems), 0.75-1.25% on flour (straight or no-time doughs), 0.5-1.25% on flour (doughnut, other mixes); *Properties:* free-flowing powd; *Storage:* 6 mos min. shelf life stored under cool, dry conditions.

CDC-3000. [MLG Enterprises Ltd.] Wheat flour, dextrose, soy flour, diacetyl tartaric acid esters of mono- and diglycerides, mono- and diglycerides, lecithin, ascorbic acid, and fungal enzymes; dough conditioner for yeast-raised bakery prods., esp. straight or no-time doughs, crusty hearth-type prods. such as French bread, Kaiser rolls; produces thin crust prod. with improved volume; increases dough tolerance, absorption; *Usage level:* 1.5-2% on flour (sponge and dough systems), 2-3% on flour (straight or no-time doughs); *Properties:* free-flowing powd; *Storage:* 6 mos min. shelf life stored under cool, dry conditions.

C-D Stabilizer. [ADM Arkady] Special gum system; dough conditioner for cake doughnuts; improves texture, moistness, tenderness, and shelf life.

Cedepal® Range. [Grünau GmbH] Fat carrier compds; fat components in premixes and cake ready mixes and in yeast raised baked goods.

Cegemett® MZ 490. [Grünau GmbH] Mono- and diglycerides of edible fatty acids, esterified with citric acid; emulsifier for liver sausages and fat emulsions; prevents gelation and fat separation in mfg. of boiled and cooked sausages; *Usage level:* 0.5-1% (liver sausage); *Regulatory:* FDA § 172.832; *Properties:* yel.-wh. powd., neutral odor and taste; m.p. 60-64 C; acid no. 40-60; iodine no. < 2; sapon. no. 220-240; 20-30% monoglyceride; *Storage:* at least 1 yr storage life when stored in cool, dry place at temps. ≤ 30 C.

Cegepal® AF. [Grünau GmbH] Fat powd. from animal source; fat component for dessert and cake mixes, yeast-raised baked goods, confectionery, soups; features long shelf life, improved pourability and dispersibility; *Properties:* creamy-wh. spray-dried powd., neutral odor, neutral fatty taste; 80% on 0.2 mm sieve; dens. 200-300 g/l; m.p. 34-36 C; iodine no. 80-90; sapon. no. 180-200; pH 7-8 (10% aq.); 80% fat content, 20% skimmed milk; *Storage:* 1 yr shelf life stored under cool, dry conditions below 20 C.

Cegepal® AVF. [Grünau GmbH] Fat powd. from animal/vegetable source; fat component for dessert and cake mixes, yeast-raised baked goods, confectionery, soups; features long shelf life, improved pourability and dispersibility; *Properties:* creamy-wh. free-flowing powd., neutral odor, neutral fatty taste; 90% on 0.2 mm sieve; dens. 300-400 g/l; m.p. 34-36 C; iodine no. 40-50; sapon. no. 200-220; pH 7-8 (10% aq.); 80% fat content, 20% skimmed milk; *Storage:* 1 yr shelf life stored under cool, dry conditions below 20 C.

Cegepal® TG 126. [Grünau GmbH] Fat powd. from vegetable source, contg. 36% maltodextrin, 14% casein, 2% emulsifier; fat component for dessert and cake mixes, yeast-raised baked goods, confectionery, soups; esp. suitable for icing, cream, puddings, and desserts; features long shelf life, improved pourability and dispersibility; *Properties:* creamy-wh. free-flowing powd., neutral odor, neutral fatty taste; 80% on 0.2 mm sieve; dens. 300-400 g/l; m.p. 30-34 C; iodine no. < 2; sapon. no. 240-260; pH 7-8 (10% aq.); 48% fat content; *Storage:* 1 yr shelf life stored under cool, dry conditions below 20 C.

Cegepal® TG 809. [Grünau GmbH] Fat powd. from vegetable source, contg. 24% maltodextrin, 11% casein, 3% lecithin, 1% emulsifier; fat component for dessert and cake mixes, yeast-raised baked goods, confectionery, soups; esp. suitable for icing, cream, puddings, and desserts; features long shelf life, improved pourability and dispersibility; *Properties:* creamy-wh. free-flowing powd., neutral odor, neutral fatty taste; 80% on 0.2 mm sieve; dens. 300-400 g/l; m.p. 30-34 C; iodine no. 4-6; sapon. no. 220-240; pH 7-8 (10% aq.); 61% fat content; *Storage:* 1 yr shelf life stored under cool, dry conditions below 20 C.

Cegepal® VF 347. [Grünau GmbH] Fat powd. from vegetable source, contg. 14% glucose syrup, 7% casein, 1% emulsifier; fat component for dessert and cake mixes, yeast-raised baked goods, confectionery, soups; features long shelf life, improved pourability and dispersibility; *Properties:* creamy-wh. free-flowing powd., neutral odor, neutral fatty taste; 80% on 0.2 mm sieve; dens. 300-400 g/l; m.p. 30-34 C; iodine no. 60-70; sapon. no. 180-200; pH 7-8 (10% aq.); 78% fat content; *Storage:* 1 yr shelf life stored under cool, dry conditions below 20 C.

Cegepal® VF. [Grünau GmbH] Fat powd. from vegetable source; fat component for dessert and cake mixes, yeast-raised baked goods, confectionery, soups; features long shelf life, improved pourability and dispersibility; *Properties:* creamy-wh. free-flowing powd., neutral odor, neutral fatty taste; 90% on 0.2 mm sieve; dens. 250-350 g/l; m.p. 34-36 C; iodine no. 1-3; sapon. no. 225-245; pH 7-8 (10% aq.); 80% fat content, 20% skimmed milk; *Storage:* 1 yr shelf life stored under cool, dry conditions below 20 C.

Cegeprot® Range. [Grünau GmbH] Protein concs; food emulsifier, stabilizer, protein enrichment for meat and sausage industry.

Cegeskin® Range. [Grünau GmbH] Acetic acid esters of mono- and diglycerides of edible fatty acids; coating for shelf life and reduction of weight loss for raw and cooked sausages.

Cegesol® Range. [Grünau GmbH] Medium chain triglycerides; dust preventer for spice mixtures and other powd. blends in meat/sausage industry.

Cegesterin® Range. [Grünau GmbH] Monoglyceride hydrate dispersions; antistaling effect, formation of starch complexes in baked goods.

Celish® FD-100F. [Daicel USA] Dietary fiber; no-calorie functional food ingred., thickener, fat replacer, thixotrope; water binder for high-moisture, low-fat foods; stabilizer for emulsions and suspensions; for low-fat spreads, margarine, sauces, dressings, pasta, baked goods, hamburger steak;

Properties: insol. in water; 20% solids.

Cellogen HP-5HS. [Multi-Kem] Carboxymethylcellulose sodium; CAS 9004-32-4; moderate hydration thickening system for food use; resist. to heat, enzyme, and bacterial decomp; superb water-holding ability; *Properties:* visc. 2000-3000 cps (1%).

Cellogen HP-6HS9. [Multi-Kem] Carboxymethylcellulose sodium; CAS 9004-32-4; thickening system for food use with milk, salt, sugar sol'n. stability; resist. to heat, enzyme, and bacterial decomp; superb water-holding ability; *Properties:* visc. 3000-4000 cps (1%).

Cellogen HP-8A. [Multi-Kem] Carboxymethylcellulose sodium; CAS 9004-32-4; moisture retention system for food use; resist. to heat, enzyme, and bacterial decomp; superb water-holding ability; *Properties:* visc. 25-45 cps (2%).

Cellogen HP-12HS. [Multi-Kem] Carboxymethylcellulose sodium; CAS 9004-32-4; high efficiency thickening system for food use; resist. to heat, enzyme, and bacterial decomp; superb water-holding ability; *Properties:* visc. 6000-8000 cps (1%).

Cellogen HP-SB. [Multi-Kem] Carboxymethylcellulose sodium; CAS 9004-32-4; thickening system for food use with acid/milk sol'n. stability; minimizes turbidity; resist. to heat, enzyme, and bacterial decomp; superb water-holding ability; *Properties:* visc. 150-250 cps (2%).

Celluferm. [Finnsugar Bioprods.] Cellulase (fungal); CAS 9012-54-8; EINECS 232-734-4; enzyme for fruit and veg. processing; *Properties:* tan powd; Unverified

Cellulase 4000. [Solvay Enzymes] Fungal cellulase derived from *Aspergillus niger*, standardized with lactose; CAS 9012-54-8; EINECS 232-734-4; enzyme for brewing, fruit juices, essential oils and spices; *Usage level:* 0.1-1.0%; *Properties:* wh. to lt. tan powd., free of offensive odor and taste; water-sol.

Cellulase AC. [Solvay Enzymes] Fungal cellulase; CAS 9012-54-8; EINECS 232-734-4; enzyme for extraction of grains, vegetables, fruits, and other plant tissues; *Properties:* powd.

Cellulase AP. [Mitsubishi Int'l; Amano Enzyme USA] Cellulase from *Aspergillus niger*, CAS 9102-54-8; EINECS 232-734-4; food enzyme; *Properties:* powd.

Cellulase AP 3. [Mitsubishi Int'l; Amano Enzyme USA] Cellulase from *Aspergillus niger*, CAS 9102-54-8; EINECS 232-734-4; digestive enzyme; *Properties:* powd.

Cellulase L. [Mitsubishi Int'l; Amano Enzyme USA] Cellulase from *Aspergillus niger*, CAS 9102-54-8; EINECS 232-734-4; food enzyme; *Properties:* liq.

Cellulase TAP. [Mitsubishi Int'l; Amano Enzyme USA] Cellulase from *Trichoderma viride*; CAS 9102-54-8; EINECS 232-734-4; food enzyme; *Properties:* powd.

Cellulase TRL. [Solvay Enzymes] Cellulase; CAS 9012-54-8; EINECS 232-734-4; cellulose for conversion of cellulose to glucose; for brewing malt beverages, fruit juices (aids extract and clarification of citrus juice), essential oils and spices, animal feeds; *Properties:* liq.

Cellulase Tr Conc. [Solvay Enzymes] Cellulase derived from *Trichoderma reesei*; CAS 9012-54-8; EINECS 232-734-4; enzyme for animal feeds, brewing, fruit juices, essential oils and spices; *Usage level:* 0.1-1.0%; *Properties:* lt. tan to wh. powd., free of offensive odor and taste; water-sol.

Celynol F1. [Rhone-Poulenc Surf. & Spec.] Palm oil sucroglyceride; nonionic; emulsifier, coemulsifier, stabilizer, and dispersant for fats and oils in a range of nonstandardized baked goods; *Properties:* paste; 100% conc.

Celynol MSPO-11. [Rhone-Poulenc Surf. & Spec.] Palm oil sucroglyceride; nonionic; emulsifier, texture improver for food applics; improves stability of emulsion or dispersion; *Properties:* paste; HLB 6.0; 100% conc.

Celynol P1M. [Rhone-Poulenc Surf. & Spec.] Palm sucroglyceride on milk protein carrier; nonionic; emulsifier and texture improver for food applics; *Properties:* powd; 37% conc.

Centex®. [Central Soya] Textured soy flour; CAS 68513-95-1; protein for use in cooked and raw beef patties, pizza toppings, ground meat foods, and meatballs.

Centrobake® 100L. [Central Soya] Distilled soy monoglycerides and soy lecithin; crumb softener, dough conditioner for yeast-raised bakery goods; *Regulatory:* FDA 21CFR §184.1400, 184.1505, GRAS; kosher approved; *Properties:* cream-colored free-flowing powd; m.p. 152-158 F; 70-71% α-monoglycerides; 0.5% max. moisture; *Storage:* store in cool dry place (60-90 F).

Centrocap® 162SS. [Central Soya] Special grade lecithin, highly filtered; CAS 8002-43-5; EINECS 232-307-2; mixing/blending aid, visc. modifier, w/o emulsifier for food applics; designed for encapsulation where clarity and brilliance are required; *Regulatory:* FDA 21CFR §184.1400, GRAS; kosher approved; *Properties:* lt. amber fluid; oil-sol; visc. 3500 cP; HLB 4; acid no. 36 max; 0.3% moisture; *Storage:* store in closed containers @ 16-32 C; 1 yr shelf life in original, unopened containers.

Centrocap® 162US. [Central Soya] Special grade lecithin, highly filtered; CAS 8002-43-5; EINECS 232-307-2; mixing/blending aid, visc. modifier, w/o emulsifier for food applics; designed for encapsulation where clarity and brilliance are required; *Properties:* amber fluid; oil-sol.; visc. 3500 cP; acid no. 28; 0.3% moisture.

Centrocap® 273SS. [Central Soya] Special grade lecithin, highly filtered; CAS 8002-43-5; EINECS 232-307-2; mixing/blending aid, visc. modifier, w/o emulsifier for food applics.; designed for encapsulation where clarity and brilliance are required; *Regulatory:* FDA 21CFR §184.1400, GRAS; kosher approved; *Properties:* lt. amber low-visc. fluid; oil-sol.; visc. 1500 cP; acid no. 27; 0.3% moisture; *Storage:* store in closed container @ 16-32 C; 1 yr shelf life in original, unopened container.

Centrocap® 273US. [Central Soya] Special grade

lecithin, highly filtered; CAS 8002-43-5; EINECS 232-307-2; mixing/blending aid, visc. modifier, w/o emulsifier for food applics.; designed for encapsulation where clarity and brilliance are required; *Properties:* amber low-visc. fluid; oil-sol.; visc. 1500 cP; acid no. 27; 0.3% moisture.

Centrol® 2F SB. [Central Soya] Standard grade lecithin, single bleached; CAS 8002-43-5; EINECS 232-307-2; amphoteric; mixing/blending aid, visc. modifier, w/o emulsifier for food applics. incl. chocolate, margarine, shortenings; *Properties:* amber fluid; oil-sol.; visc. 6000 cP; acid no. 27; 0.3% moisture.

Centrol® 2F UB. [Central Soya] Standard grade lecithin, unbleached; CAS 8002-43-5; EINECS 232-307-2; amphoteric; mixing/blending aid, visc. modifier, w/o emulsifier for food applics. incl. chocolate, margarine, shortenings; *Properties:* amber fluid; oil-sol.; visc. 6000 cP; acid no. 27; 0.3% moisture.

Centrol® 3F SB. [Central Soya] Standard grade lecithin, single bleached; CAS 8002-43-5; EINECS 232-307-2; amphoteric; mixing/blending aid, visc. modifier, w/o emulsifier, dispersant for food applics. incl. chocolate, margarine, shortenings; more highly filtered than 2F type; *Regulatory:* FDA 21CFR §184.1400, GRAS; kosher approved; *Properties:* lt. amber med. visc. fluid; oil-sol.; visc. 12,000 cP max.; acid no. 32 max.; 0.8% max. moisture; *Storage:* store in closed container @ 16-32 C; 1 yr shelf life.

Centrol® 3F UB. [Central Soya] Standard grade lecithin, unbleached; CAS 8002-43-5; EINECS 232-307-2; amphoteric; mixing/blending aid, visc. modifier, w/o emulsifier for food applics. incl. chocolate, margarine, shortenings; more highly filtered than 2F type; *Properties:* amber fluid; oil-sol.; visc. 6000 cP; acid no. 27; 0.3% moisture.

Centrol® CA. [Central Soya] Special grade lecithin; CAS 8002-43-5; EINECS 232-307-2; amphoteric; o/w emulsifier and wetting agent for food applics., esp. infant formulas, meat gravies; emulsifies fats in infant formulas and other dairy replacer systems; synergistic with other hydrophilic emulsifiers; *Regulatory:* FDA 21CFR §184.1400; kosher; *Properties:* amber fluid; oil-sol., water-disp.; visc. 4000 cP; HLB 6.0; acid no. 20; 100% conc.; *Storage:* store in closed containers @ 16-32 C; 1 yr shelf life in original, unopened container.

Centrolene® A. [Central Soya] Hydroxylated soy lecithin; CAS 8029-76-3; EINECS 232-440-6; amphoteric; o/w emulsifier, release agent, wetting agent for food applics. (baked goods), cosmetic applics. incl. hair care; increased hydrophilic props.; solubilizer, dispersant aiding flavor/color incorporation; emulsion stabilizer in salad dressings; *Regulatory:* FDA 21CFR §172.814; kosher approved; *Properties:* lt. amber-yel. heavy-bodied fluid; oil-sol., water-disp.; HLB 9-10; acid no. 27; 1.2% moisture; *Storage:* store in closed containers @ 16-32 C; 1 yr shelf life in original, unopened container.

Centrolene® S. [Central Soya] Hydroxylated lecithin; CAS 8029-76-3; EINECS 232-440-6; amphoteric; o/w emulsifier and wetting agent for food applics., cosmetic applics. incl. hair care; solubilizer and dispersant aiding color/flavor incorporation; wetting agent for fatty powds.; *Regulatory:* FDA 21CFR §172.814; kosher approved; *Properties:* lt. amber-yel. heavy-bodied fluid; oil-sol., water-disp.; HLB 9-10; acid no. 27; 100% conc.; *Storage:* store in closed container @ 16-32 C; 1 yr shelf life in original, unopened container.

Centrolex® C. [Central Soya] Special grade lecithin; CAS 8002-43-5; EINECS 232-307-2; amphoteric; emulsifier, stabilizer, suspending agent for foods; controls fat in ground meat prods.; improves dough machinability in baked goods; when solubilized, instantizes dry powds.; *Properties:* yel. gran.; oil-sol., water-disp.; bulk dens. 0.44 g/cc; acid no. 27; 100% conc.

Centrolex® D. [Central Soya] Special grade lecithin; CAS 8002-43-5; EINECS 232-307-2; o/w emulsifier, visc. modifier, wetting agent for foods esp. instantizing applics. where it controls hydration rate of hydrophilic powds.; controls fat in ground meat prods.; improves dough machinability in baked goods; *Usage level:* 0.5-1.0%; *Properties:* yel. fine powd., low flavor; oil-sol., water-disp.; bulk dens. 0.46 g/cc; acid no. 27; 0.7% moisture.

Centrolex® F. [Central Soya] Special grade lecithin; CAS 8002-43-5; EINECS 232-307-2; o/w emulsifier, visc. modifier, wetting agent for foods (baked goods, cheese sauces, chocolate, frostings, granola bars, meat gravies, milk powds., nondairy creamers); reduces fat requirement; enhances smoothness, reduces tackiness in chewing gum; *Regulatory:* FDA 21CFR §184.1400, GRAS; kosher approved; *Properties:* lt. tan or yel. powd., bland odor and flavor; oil-sol., water-disp.; bulk dens. 0.45 g/cc; acid no. 36 max.; 1% max. moisture; *Storage:* store in dry, closed, original container below 25 C; 1 yr shelf life in original unopened container.

Centrolex® G. [Central Soya] Special grade lecithin; CAS 8002-43-5; EINECS 232-307-2; emulsifier for food applics.; *Properties:* yel. powd.; bulk dens. 0.46 g/cc; acid no. 27; 0.9% moisture.

Centrolex® P. [Central Soya] Special grade lecithin; CAS 8002-43-5; EINECS 232-307-2; o/w emulsifier, visc. modifier, wetting agent for foods; controls fat in ground meat prods.; improves dough machinability in baked goods; when solubilized, instantizes dry powds.; *Properties:* yel. gran., low flavor; oil-sol., water-disp.; bulk dens. 0.38 g/cc; acid no. 27; 0.7% moisture.

Centrolex® R. [Central Soya] Special grade lecithin; CAS 8002-43-5; EINECS 232-307-2; o/w emulsifier, visc. modifier, wetting agent for foods; controls fat in ground meat prods.; improves dough machinability in baked goods; when solubilized, instantizes dry powds.; *Regulatory:* FDA 21CFR §184.1400, GRAS; kosher approved; *Properties:* lt. tan or yel. gran., very bland odor and flavor; oil-sol, water-disp.; bulk dens. 0.38 g/cc; HLB 7; acid no. 36 max.; 1% max. moisture; *Storage:* 1 yr shelf life when stored in dry, closed, original containers below 25 C.

Centromix® CPS. [Central Soya] Lecithin, polysorbate 80; amphoteric; o/w emulsifier for food applics.; retards separation and stabilizes emulsion in pourable salad dressings; *Regulatory:* kosher; *Properties:* amber fluid; oil-sol., highly water-disp.; visc. 8500 cP; HLB 12; acid no. 23; 0.4% moisture; *Storage:* store in closed container @ 16-32 C; 1 yr shelf life in original, unopened container.

Centromix® E. [Central Soya] Lecithin, ethoxylated mono- and diglycerides, propylene glycol; amphoteric; o/w emulsifier, release and wetting agent for food applics., esp. instantizing, milk powds.; belt release in continuous cooking operations; *Regulatory:* kosher approved; *Properties:* amber fluid; oil-sol., highly water-disp.; visc. 6500 cP; HLB 12; acid no. 17; 0.3% moisture; *Storage:* store in closed container @ 16-32 C; 1 yr shelf life in original, unopened container.

Centrophase® 152. [Central Soya] Special grade lecithin; CAS 8002-43-5; EINECS 232-307-2; amphoteric; release and wetting agent; bland, sprayable food grade for spray release applics., oil-based aerosol spray release systems; *Regulatory:* FDA 21CFR §184.1400; kosher; *Properties:* amber low visc. fluid; oil-sol.; visc. 1200 cP; acid no. 22; 100% conc.; *Storage:* store in closed container at 16-32 C; mildly agitate before use.

Centrophase® C. [Central Soya] Special grade lecithin; CAS 8002-43-5; EINECS 232-307-2; amphoteric; w/o emulsifier, sprayable release and wetting agent for flavor-sensitive foods and drinks (caramel corn, chocolate, instantizing applics., milk powds.); *Regulatory:* FDA 21CFR §184.1400; kosher; *Properties:* amber low-visc. fluid, low flavor; oil-sol.; visc. 1000 cP; acid no. 27; 0.3% moisture; *Storage:* store in closed container @ 16-32 C; 1 yr shelf life in original, unopened container.

Centrophase® HR. [Central Soya] Lecithin; CAS 8002-43-5; EINECS 232-307-2; amphoteric; release and wetting agent, lubricant, and emulsifier; does not develop burned odor, color, or taste at high temps.; resists darkening at temps. of 350 F; *Regulatory:* FDA 21CFR §184.1400, GRAS; kosher approved; *Properties:* amber fluid; sol. in liq. veg. and min oil, aliphatic, aromatic and halogenated hydrocarbons, petroleum ether; sp.gr. 1.01; visc. 3000 cP max.; pour pt. < 5 F; acid no. 36 max.; 0.8% max. moisture; *Storage:* store in closed containers @ 16-32 C; 1 yr shelf life.

Centrophase® HR2B. [Central Soya] Special grade lecithin; CAS 8002-43-5; EINECS 232-307-2; amphoteric; patented, heat-resist. o/w emulsifier, release and wetting agent for food applics. esp. cheese sauces, nondairy creamers, shortening; will not develop burned odor, color, or taste on heating; *Regulatory:* FDA 21CFR §184.1400; kosher approved; *Properties:* lt. amber fluid; oil-sol., water-disp.; sp.gr. 1.02; visc. 5000 cP; acid no. 22;

pour pt. 10 F; 0.6% moisture; *Storage:* store in closed container @ 16-32 C; 1 yr shelf life in original, unopened container.

Centrophase® HR2U. [Central Soya] Special grade lecithin; CAS 8002-43-5; EINECS 232-307-2; amphoteric; heat-resist. o/w emulsifier, release and wetting agent for food applics. esp. cheese sauces, nondairy creamers, shortening; *Properties:* amber fluid; oil-sol., water-disp.; visc. 5000 cP; acid no. 22; 0.6% moisture.

Centrophase® HR4B. [Central Soya] Special grade lecithin; CAS 8002-43-5; EINECS 232-307-2; amphoteric; heat-resist. o/w emulsifier, release and wetting agent for food applics.; suitable for extended high heat cooking, e.g., in deep fat fryer or continuous cooking systems; will not develop burned odor, color, or taste; *Regulatory:* FDA 21CFR §184.1400; kosher approved; *Properties:* amber heavy-bodied sl. hazy fluid; oil-sol., water-disp.; sp.gr. 1.02; visc. 3300 cP; acid no. 24; pour pt. 10 F; 0.6% moisture; *Storage:* store in closed containers @ 16-32 C; 1 yr shelf life in original, unopened container.

Centrophase® HR4U. [Central Soya] Special grade lecithin; CAS 8002-43-5; EINECS 232-307-2; amphoteric; heat-resist. o/w emulsifier, release and wetting agent for food applics.; *Properties:* heavy-bodied fluid; oil-sol., water-disp.; visc. 3300 cP; acid no. 24; 0.6% moisture.

Centrophase® HR6B. [Central Soya] Special grade lecithin; CAS 8002-43-5; EINECS 232-307-2; amphoteric; heat-resist. o/w emulsifier, release and wetting agent for food applics.; offers max. heat resist. for deep fat frying and continuous cooking; *Properties:* amber fluid; oil-sol., water-disp.; visc. 2500 cP; acid no. 23; 0.8% moisture.

Centrophase® NV. [Central Soya] Special grade soy lecithin with added soybean oil; amphoteric; mixing/blending aid, sprayable release and wetting agent for flavor-sensitive food and drink applics., esp. cheese release; instantizing agent controlling hydration rate of hydrophilic powds.; *Regulatory:* kosher; *Properties:* lt. amber low-visc. fluid, low flavor; oil-sol.; visc. 215 cP; acid no. 21; 0.3% moisture; *Storage:* store in closed container @ 16-32 C; 1 yr shelf life in original, unopened container.

Centrophil® K. [Central Soya] Special grade lecithin; CAS 8002-43-5; EINECS 232-307-2; amphoteric; mixing/blending aid, visc. modifier for food applics. esp. chocolate; enhances smoothness, reduces tackiness in chewing gum; compat. with hard butters used in chocolate; *Regulatory:* FDA 21CFR §184.1400; kosher approved; *Properties:* amber solid plastic, low odor and flavor; acid no. 20; 0.4% moisture; *Storage:* store below 27 C; 1 yr shelf life in original, unopened container.

Centrophil® M. [Central Soya] Special grade soy lecithin containing coconut oil, cottonseed oil, and antioxidants (propylene glycol, BHA, citric acid); amphoteric; sprayable release and wetting agent for delicately flavored food systems, esp. cheese release; *Regulatory:* kosher; *Properties:* amber low-visc. fluid, low odor and flavor; oil-sol.; visc. 150 cP; acid no. 12; 0.3% moisture; *Storage:* store in closed container @ 16-32 C; 1 yr shelf life in original, unopened container.

Centrophil® W. [Central Soya] Special grade soy lecithin containing coconut oil, cottonseed oil, and antioxidants (propylene glycol, BHA, citric acid); amphoteric; sprayable release agent and wetting agent for delicately flavored food systems, esp. cheese release; *Regulatory:* kosher; *Properties:* amber low visc. fluid, low odor and flavor; oil-sol.; visc. 150 cP; acid no. 12; 0.3% moisture; *Storage:* store in closed container @ 16-32 C; 1 yr shelf life in original, unopened container.

Cetodan®. [Grindsted Prods.] Acetylated monoglycerides; nonionic; alpha-tending emulsifier, aerating agent for shortenings, toppings, cakes; edible coating; plasticizer for chewing gum base; *Regulatory:* EEC, FDA §172.828; *Properties:* solid, liq.; HLB 1.5; 100% conc.

Cetodan® 50-00A. [Grindsted Prods.; Grindsted Prods. Denmark] Monoglyceride acetic acid ester; food emulsifier; aerating and foam stabilizing agent in food prods.; *Properties:* block; m.p. 40 C; sapon. no. 280.

Cetodan® 50-00P Kosher. [Grindsted Prods.] Acetylated monoglyceride from hydrog. refined fats; food emulsifier for cake shortenings and fats for powd. toppings; edible coating agent for meat prods., candy, fruit and nuts; *Usage level:* 2-5% (cake shortenings), 5-10% (topping fats); *Regulatory:* EEC, FDA §172.828; *Properties:* block; m.p. 40 C; acid no. 2 max.; iodine no. 2 max.; sapon. no. 285; 0.5% acetylation; *Storage:* store under cool, dry conditions.

Cetodan® 70-00A. [Grindsted Prods.] Monoglyceride acetic acid ester; food emulsifier; aerating and foam stabilizing agent in food prods.; *Properties:* m.p. 35 C; sapon. no. 315; Unverified

Cetodan® 70-00P Kosher. [Grindsted Prods.] Acetylated monoglyceride from hydrog. refined fats; food emulsifier for cake shortenings and fats for powd. toppings; edible coating agent for meat prods., candy, fruit and nuts; *Usage level:* 2-5% (cake shortening), 5-10% (topping fats); *Regulatory:* EEC, FDA §172.828, kosher; *Properties:* ivory block; m.p. 35 C; acid no. 2 max.; iodine no. 2 max.; sapon. no. 315; 0.7% acetylation; *Storage:* store under cool, dry conditions.

Cetodan® 90-40. [Grindsted Prods.] Acetylated lard glyceride; CAS 8029-92-3; food emulsifier; aerating and foam stabilizing agent in food prods.; food-grade plasticizer for PVC; *Regulatory:* FDA 21CFR §172.828 (PVC); *Properties:* m.p. 10 C; sapon. no. 370; Discontinued

Cetodan® 90-50. [Grindsted Prods.] Acetylated monoglyceride from edible, partially hydrog. soybean oil with antioxidants (200 ppm max. BHT, 200 ppm max. citric acid); food emulsifier, plasticizer, lubricant, release agent for confectionery prods., chewing gum base, licorice; defoamer for bottling operations and in prod. of jams and jellies; release

agent for molding and slicing operations; avail. as kosher grade; *Usage level:* 6-10% (chewing gum base); *Regulatory:* EEC, FDA §172.828; *Properties:* liq.; acid no. 2 max.; iodine no. 45; sapon. no. 380; cloud pt. 5 C; 0.9% acetylation; *Storage:* store under cool, dry conditions; if crystallization occurs heat to 30 C before use.

CFI-10. [MLG Enterprises Ltd.] Mono- and diglycerides, whey, soya flour, ethoxylated mono- and diglycerides; dough conditioner; starch complexing agent for bakery prods. (bread, buns, rolls, sweet goods, doughnuts, tortillas); provides softness, improved crumb texture and resiliency, extends shelf life; *Usage level:* 4-6 oz/cwt of flour; *Properties:* sl. off-wh. to cream free-flowing powd., sl. dairy odor; pH 5.9 ± 0.1 (10% aq. disp.); 41% α-monodiglycerides, 13.2% protein, 1% max. moisture; *Storage:* 6 mos shelf life stored under cool, dry conditions.

CG Tase. [Mitsubishi Int'l.; Amano Enzyme USA] Cyclodextrin glucanotransferase from *Bacillus macerans*; food enzyme; *Properties:* liq.

Cheddar Harmony #274. [Gamay Flavors] Mild to med. cheddar flavor, Am. profile.

Ches® 500. [CasChem] Nonfat drymilk, xanthan gum, propylene glycol, alginate, glyceryl stearate, sodium glyceryl oleate phosphate; food grade stabilizer; unique ambient temp. emulsifier which yields stable, aesthetic o/w emulsions; *Properties:* powd.; 100% conc.

Chicken Extract #1083. [Ariake USA] Chicken extract, salt; flavorings for snacks, soup mixes, sauces, gravies; 46.9% moisture, 40% protein; *Storage:* keep in cool, dry place.

Choclin. [Van Den Bergh Foods] Vegetable oil (palm and shea); cocoa butter equivalent in confectioner's coatings, center fat, veg. dairy systems; *Regulatory:* kosher; *Properties:* solid; m.p. 98.8-93.2 F.

Chocolate Flavored Frosteen®. [Bunge Foods] Fully prepared chocolate icing for bakery prods., donuts, brownies, sweet rolls; *Properties:* chocolate flavor; pH 5.7-6.3.

Chocothin. [Lucas Meyer] Standardized lecithin fraction; CAS 8002-43-5; EINECS 232-307-2; amphoteric; visc. reducing agent for milk chocolates; *Properties:* liq.; 100% conc.

Chocotop™. [Lucas Meyer] Range of fractionated soya lecithins; CAS 8002-43-5; EINECS 232-307-2; amphoteric; emulsifer and visc. reducing agent for confections, esp. for chocolate mfg.; controls yield value; *Properties:* liq.; 100% conc.

Chy-Max®. [Pfizer Food Science] Standardized saline sol'n. of chymosin produced by fermentation, with $\leq$ 3% sodium benzoate preservative; dairy ingred. used to coagulate milk in the mfg. of cheese and other dairy prods.; props. identical to veal rennet chymnosin; *Regulatory:* kosher; *Properties:* clear liq.; sp.gr. 1.06-1.10 (15 C); pH 5.6-6.0; *Storage:* store refrigerated to maintain max. activity; avoid freezing.

Cirol. [Van Den Bergh Foods] Partially hydrog. veg. oil

(soybean, cottonseed); EINECS 269-820-6; domestic oil for nut roasting, margarines, frozen desserts, coffee whiteners; *Regulatory:* kosher; *Properties:* Lovibond 2.5R max. plastic; m.p. 91-95 F.

Cithrol GDO N/E. [Croda Chem. Ltd.] Glyceryl dioleate; CAS 25637-84-7; EINECS 247-144-2; nonionic; emulsifier, stabilizer, wetting agent, lubricant used in food applics.; *Properties:* liq.; HLB 2.0; 100% conc.

Cithrol GDO S/E. [Croda Chem. Ltd.] Glyceryl dioleate SE; anionic; emulsifier, stabilizer, wetting agent used in food applics.; *Properties:* liq.; HLB 2.9; 100% conc.

Cithrol GDS N/E. [Croda Chem. Ltd.] Glyceryl distearate; CAS 1323-83-7; EINECS 215-359-0; nonionic; emulsifier, stabilizer, wetting agent used in food applics.; *Properties:* solid; HLB 3.4; 100% conc.

Cithrol GDS S/E. [Croda Chem. Ltd.] Glyceryl distearate SE; anionic; emulsifier, stabilizer, wetting agent used in food applics.; *Properties:* solid; HLB 4.2; 100% conc.

Cithrol GML N/E. [Croda Chem. Ltd.] Glyceryl laurate; CAS 142-18-7; EINECS 205-526-6; nonionic; emulsifier, stabilizer, wetting agent used in food applics.; *Properties:* liq.; HLB 4.9; 100% conc.

Cithrol GML S/E. [Croda Chem. Ltd.] Glyceryl laurate SE; anionic; emulsifier, stabilizer, wetting agent used in food applics.; *Properties:* liq.; HLB 5.6; 100% conc.

Cithrol GMO N/E. [Croda Chem. Ltd.] Glyceryl oleate; CAS 111-03-5; nonionic; emulsifier, stabilizer, wetting agent used in food applics.; *Properties:* liq.; HLB 3.3; 100% conc.

Cithrol GMO S/E. [Croda Chem. Ltd.] Glyceryl oleate SE; anionic; emulsifier, stabilizer, wetting agent used in food applics.; *Properties:* liq.; HLB 4.1; 100% conc.

Cithrol GMR N/E. [Croda Chem. Ltd.] Glyceryl ricinoleate; CAS 141-08-2; EINECS 205-455-0; nonionic; emulsifier, stabilizer, wetting agent used in food applics.; *Properties:* liq.; HLB 2.7; 100% conc.

Cithrol GMR S/E. [Croda Chem. Ltd.] Glyceryl ricinoleate SE; anionic; emulsifier, stabilizer, wetting agent used in food applics.; *Properties:* liq.; HLB 3.6; 100% conc.

Cithrol GMS Acid Stable. [Croda Chem. Ltd.] Glyceryl stearate SE; CAS 31566-31-1; nonionic; emulsifier, stabilizer, wetting agent used in food applics.; *Properties:* solid; HLB 10.9; 100% conc.

Cithrol GMS N/E. [Croda Chem. Ltd.] Glyceryl stearate; CAS 31566-31-1; nonionic; emulsifier, stabilizer, wetting agent used in food applics.; *Properties:* solid; HLB 3.4; 100% conc.

Cithrol GMS S/E. [Croda Chem. Ltd.] Glyceryl stearate SE; CAS 31566-31-1; anionic; emulsifier, stabilizer, wetting agent used in food applics.; *Properties:* solid; HLB 4.4; 100% conc.

Cithrol PGML N/E. [Croda Chem. Ltd.] Propylene glycol laurate; CAS 142-55-2; EINECS 205-542-3; nonionic; emulsifier, stabilizer, wetting agent used in food applics.; *Properties:* liq.; HLB 2.7; 100% conc.

Cithrol PGML S/E. [Croda Chem. Ltd.] Propylene glycol laurate SE; anionic; emulsifier, stabilizer, wetting agent used in food applics.; *Properties:* liq.; HLB 3.6; 100% conc.

Cithrol PGMO N/E. [Croda Chem. Ltd.] Propylene glycol oleate; CAS 1330-80-9; EINECS 215-549-3; nonionic; emulsifier, stabilizer, wetting agent used in food applics.; *Properties:* liq.; HLB 3.1; 100% conc.

Cithrol PGMO S/E. [Croda Chem. Ltd.] Propylene glycol oleate SE; anionic; emulsifier, stabilizer, wetting agent used in food applics.; *Properties:* liq.; HLB 3.9; 100% conc.

Cithrol PGMR N/E. [Croda Chem. Ltd.] Propylene glycol ricinoleate; CAS 26402-31-3; EINECS 247-669-7; nonionic; emulsifier, stabilizer, wetting agent used in food applics.; *Properties:* liq.; HLB 2.7; 100% conc.

Cithrol PGMR S/E. [Croda Chem. Ltd.] Propylene glycol ricinoleate SE; anionic; emulsifier, stabilizer, wetting agent used in food applics.; *Properties:* liq.; HLB 3.6; 100% conc.

Cithrol PGMS N/E. [Croda Chem. Ltd.] Propylene glycol stearate; CAS 1323-39-3; EINECS 215-354-3; nonionic; emulsifier, stabilizer, wetting agent used in food applics.; *Properties:* solid; HLB 2.4; 100% conc.

Cithrol PGMS S/E. [Croda Chem. Ltd.] Propylene glycol stearate SE; anionic; emulsifier, stabilizer, wetting agent used in food applics.; *Properties:* solid; HLB 3.2; 100% conc.

Citraxine. [Vaessen-Schoemaker] Phosphate-free curing oompd. on citrate basis, curing compd. for traditional style cooked hams; improved natural flavor.

Citreatt Lemon 3123. [Florida Treatt] Deterpenated lemon oil; CAS 8008-56-8; flavoring; *Properties:* oil; sol. in 90% ethanol; dens. 0.901-0.909; flash pt. > 80 C; ref. index 1.482-1.486.

Citreatt Lemon 6122. [Florida Treatt] Partially deterpenated lemon oil; CAS 8008-56-8; flavoring; *Properties:* oil; sol. 1 in 5-10 in 90% ethanol; dens. 0.855-0.865; flash pt. > 51 C; ref. index 1.475-1.478.

Citreatt Lime 3135. [Florida Treatt] Deterpenated lime oil; CAS 8008-26-2; flavoring; *Properties:* oil; sol. in 90% ethanol; dens. 0.911-0.921; flash pt. > 68 C; ref. index 1.478-1.481.

Citreatt Lime 6134. [Florida Treatt] Partially deterpenated lime oil; CAS 8008-26-2; flavoring; *Properties:* oil; sol. in 90% ethanol; dens. 0.883-0.890; flash pt. > 54 C; ref. index 1.475-1.478.

Citreatt Orange 3111. [Florida Treatt] Deterpenated orange oil; CAS 8008-57-9; flavoring; *Properties:* oil; sol. 1 to 2-10 in 90% ethanol; dens. 0.873-0.883; flash pt. > 71 C; ref. index 1.460-1.466.

Citreatt Orange 6110. [Florida Treatt] Partially deterpenated orange oil; CAS 8008-57-9; flavoring; *Properties:* oil; sol. 1 to 8-10 in 90% ethanol; dens. 0.841-0.848; flash pt. > 49 C; ref. index 1.470-1.473.

Citric Acid Anhydrous USP/FCC. [PMC Specialties] Citric acid USP/FCC; CAS 77-92-9; EINECS 201-069-1; food additive; *Regulatory:* FDA compliance; *Properties:* colorless or wh. free-flowing gran. and powd., odorless, tart taste; sol. (g/100 ml): 181 g in water, 59.1 g in alcohol; m.w. 192.13; 99.5-100.5% assay; *Storage:* store in tightly sealed containers away from heat and humidity.

Citric Acid USP FCC Anhyd. Fine Gran. No. 69941. [Roche] Citric acid USP, FCC; CAS 77-92-9; EINECS 201-069-1; acidulant, flavor enhancer, sequestrant, dispersant in processed foods and beverages; *Regulatory:* GRAS; *Properties:* wh. cryst., pract. odorless, strong acid taste; 97% min. thru 30 mesh, 5% max. thru 100 mesh; sl. hygroscopic; very sol. in water (1 g/0.5 ml); freely sol. in alcohol (1 g/2 ml); sparingly sol. in ether (1 g/30 ml); m.w. 192.13; 99.5-100.5% assay; *Storage:* store in dry place; avoid excessive exposure to heat and humidity.

Citrid Acid USP FCC Anhyd. Gran. No. 69942. [Roche] Citric acid USP, FCC; CAS 77-92-9; EINECS 201-069-1; acidulant, flavor enhancer, sequestrant, dispersant in processed foods and beverages; *Regulatory:* FDA GRAS; *Properties:* wh. cryst., pract. odorless, strong acid taste; 98% min. thru 16 mesh, 10% max. thru 50 mesh; sl. hygroscopic; very sol. in water (1 g/0.5 ml); freely sol. in alcohol (1 g/2 ml); sparingly sol. in ether (1 g/30 ml); m.w. 192.13; 99.5-100.5% assay; *Storage:* store in dry place; avoid exposure to heat and humidity.

Citrocoat® A 1000 HP. [Jungbunzlauer Int'l. AG] Sat. vegetable fat-coated citric acid; CAS 77-92-9; EINECS 201-069-1; acidulant, pH control agent; coating prevents premature reactions and delays release of citric acid; for effervescent powd./tablets, confectionery, pudding mixes, meat prods.; *Regulatory:* EP compliance; *Properties:* wh. free-flowing gran., odorless to sl. fatty odor; partly sol. in water; bulk dens. 950 kg/m³; m.p. 58-60 C; pH 2 (1%); 90% acid, 10% fat; *Toxicology:* LD50 (oral, rat) 11,700 mg/kg; may cause eye or skin irritation; *Precaution:* avoid dust formation; incompat. with strong bases and oxidizing agents; *Storage:* store in original container tightly closed in dry, cool place.

Citrocoat® A 2000 HP. [Jungbunzlauer Int'l. AG] Sat. vegetable fat-coated citric acid; CAS 77-92-9; EINECS 201-069-1; acidulant, pH control agent; coating prevents premature reactions and delays release of citric acid; for effervescent powd./tablets, confectionery, pudding mixes, meat prods.; *Regulatory:* EP compliance; *Properties:* wh. free-flowing gran., odorless to sl. fatty odor; partly sol. in water; bulk dens. 950 kg/m³; m.p. 58-60 C; pH 2 (1%); 80% acid, 20% fat; *Toxicology:* LD50 (oral, rat) 11,700 mg/kg; may cause eye or skin irritation; *Precaution:* avoid dust formation; incompat. with strong bases and oxidizing agents; *Storage:* store in original container tightly closed in dry, cool place.

Citrocoat® A 4000 TP. [Jungbunzlauer Int'l. AG] Sat. vegetable fat-coated citric acid; CAS 77-92-9; EINECS 201-069-1; acidulant, pH control agent; coating prevents premature reactions and delays release of citric acid; for effervescent powd./tablets, confectionery, pudding mixes, meat prods.; *Regulatory:* EP compliance; *Properties:* wh. free-flowing gran., odorless to sl. fatty odor; partly sol. in water; bulk dens. 950 kg/m³; m.p. 58-60 C; pH 2 (1%); 60% acid, 40% fat; *Toxicology:* LD50 (oral, rat) 11,700 mg/kg; may cause eye or skin irritation; *Precaution:* avoid dust formation; incompat. with strong bases and oxidizing agents; *Storage:* store in original container tightly closed in dry, cool place.

Citrocoat® A 4000 TT. [Jungbunzlauer Int'l. AG] Sat. animal fat-coated citric acid; CAS 77-92-9; EINECS 201-069-1; acidulant, pH control agent; coating prevents premature reactions and delays release of citric acid; for effervescent powd./tablets, confectionery, pudding mixes, meat prods.; *Regulatory:* EP compliance; *Properties:* wh. free-flowing gran., odorless to sl. fatty odor; partly sol. in water; bulk dens. 950 kg/m³; m.p. 58-60 C; pH 2 (1%); 60% acid, 40% fat; *Toxicology:* LD50 (oral, rat) 11,700 mg/kg; may cause eye or skin irritation; *Precaution:* avoid dust formation; incompat. with strong bases and oxidizing agents; *Storage:* store in original container tightly closed in dry, cool place.

Citrostabil® NEU. [Jungbunzlauer Int'l. AG] Buffered, highly conc. citric acid aq. sol'n.; CAS 77-92-9; EINECS 201-069-1; acidulant, pH control agent; developed to allow continuous mfg. of cast hard caramels; provides clear, fresh, fruity taste of citric acid to candies without undesired inversion of the sucrose; also for fish prods.; *Properties:* colorless to lt. yel. clear liq., char. odor; completely misc. with water; dens. 1.3-1.4 g/cm³; visc. 50-80 mPa•s; b.p. 105 C; pH 2.0-2.7; readily biodeg.; ≈ 75% act.; *Toxicology:* may cause eye or skin irritation; *Storage:* store @ R.T. in original container; avoid freezing.

Citrostabil® S. [Jungbunzlauer Int'l. AG] Buffered, highly conc. citric acid aq. sol'n.; CAS 77-92-9; EINECS 201-069-1; acidulant, pH control agent; developed to allow continuous mfg. of cast hard caramels; provides clear, fresh, fruity taste of citric acid to candies without undesired inversion of the sucrose; also for fish prods.; *Properties:* colorless to lt. yel. clear liq., nearly odorless; completely misc. with water; dens. 1.3 kg/l; visc. 50-80 mPa•s; b.p. 105 C; pH 2.0-2.2 (10%); readily biodeg.; ≈ 75% act.; *Toxicology:* may cause eye or skin irritation; *Storage:* store @ R.T. in original container; avoid freezing.

CK-20L. [Mitsubishi Int'l.; Amano Enzyme USA] Pullulanase; CAS 9075-68-7; EINECS 232-983-9; food enzyme; *Properties:* liq.

Clam Powd. #60. [Ariake USA] Clam powd.; flavorings for snacks, soup mixes, sauces, gravies; 5% moisture, 10% protein.

Clarase® 5,000. [Solvay Enzymes] Fungal α-amylase from controlled fermentation of *Aspergillus oryzae* var.; FCC; CAS 9000-92-4; EINECS 232-567-7; enzyme for dextrinizing (liquefying), saccharifying (glucose and maltose liberating) starch; used for baking, brewing, cereal, chocolate syrup, jelly, juice, syrups, starch modification; reduces starch haze, facilitates juice filtration; *Properties:* wh. to lt. tan amorphous dry powd., free of offensive odor; readily water-sol.; *Storage:* activity loss < 10% in 1 yr stored in sealed containers under cool, dry conditions; 5 C storage extends life.

Clarase® 40,000. [Solvay Enzymes] Fungal α-amylase from controlled fermentation of *Aspergillus oryzae* var.; FCC; CAS 9000-92-4; EINECS 232-567-7; food-grade enzyme for dextrinizing (liquefying) and saccharifying (glucose and maltose liberating) starch; used for baking, brewing, cereal, chocolate syrup, jelly, juice, syrups, and starch modification; *Usage level:* 0.01-0.1%; *Properties:* wh. to lt. tan amorphous dry powd., free of offensive odor; readily water-sol.; *Storage:* activity loss < 10% in 1 yr stored in sealed containers under cool, dry conditions; 5 C storage extends life.

Clarase® Conc. [Solvay Enzymes] Fungal α-amylase from controlled fermentation of *Aspergillus oryzae* var.; FCC; CAS 9000-92-4; EINECS 232-567-7; food-grade enzyme for dextrinizing (liquefying) and saccharifying (glucose and maltose liberating) starch; used for baking, brewing, cereal, chocolate syrup, jelly, juice, syrups, and starch modification; *Properties:* wh. to lt. tan amorphous dry powd., free of offensive odor; readily water-sol.; *Storage:* activity loss < 10% in 1 yr stored in sealed containers under cool, dry conditions; 5 C storage extends life.

Clarase® L-40,000. [Solvay Enzymes] α-Amylase; CAS 9000-92-4; EINECS 232-567-7; food-grade enzyme for dextrinizing (liquefying) and saccharifying (glucose and maltose liberating) starch; used for baking, brewing, cereal, chocolate syrup, jelly, juice, syrups, and starch modification; *Properties:* amber to lt. brn. nonviscous liq., free of offensive odor; misc. with water; *Storage:* activity loss ≤ 10% in 3 mo when stored in sealed containers under cool, dry conditions.

Clarex® 5XL. [Solvay Enzymes] Pectinase derived from *Aspergillus niger* var.; CAS 9032-75-1; EINECS 232-885-6; food-grade enzyme; depectinization of fruit juices, grape processing, jams and jellies, berry processing and wine prod.; extraction of flavors and fragrances; *Usage level:* 0.1-1.0%; *Properties:* clear to amber brn. liq., free of offensive odor and taste; completely misc. with water; sp.gr. 1.05-1.15; *Storage:* activity loss ≤ 10% in 3 mo when stored in sealed containers at 25 C; 5 C storage extends life.

Clarex® L. [Solvay Enzymes] Pectinase derived from *Aspergillus niger* var.; CAS 9032-75-1; EINECS 232-885-6; food-grade enzyme; depectinization of fruit juices, grape processing, jams and jellies, berry processing and wine prod.; extraction of flavors and fragrances; *Usage level:* 0.1-0.05%; *Properties:* clear to amber brn. liq.; misc. with water; sp.gr. 1.05-1.15; *Storage:* activity loss ≤ 10% in 3 mo when stored in sealed containers at 25 C; 5 C

storage extends life.

Clarex® ML. [Solvay Enzymes] Pectinase, cellulase, and hemicellulase; food-grade enzyme for fruit processing (maceration of fruit to maximize fruit juice and solids extraction); *Usage level:* 25-75 ml/ton of fruit (maceration), 100-200 ml/ton of fruit (liquefaction); *Properties:* brn. clear liq.; completely misc. with water; sp.gr. 1.05-1.10; *Storage:* activity loss ≤ 10% in 6 mo stored in sealed containers under cool dry conditions; 5 C storage extends life.

Clarity. [Karlshamns Food Ingreds.] Partially hydrog. soybean oil; EINECS 232-410-2; specialty oil for salad dressing and light duty cooking; *Properties:* liq.

Clearate Special Extra. [W.A. Cleary] Lecithin; CAS 8002-43-5; EINECS 232-307-2; nonionic; dispersant in foods; *Properties:* liq.

Clearate WDF. [W.A. Cleary] Soya lecithin; CAS 8002-43-5; EINECS 232-307-2; nonionic; o/w emulsifier in the food industry (icings, cakes, instant cocoa); *Properties:* liq.; HLB 8.0; 60% conc.

Cleartic. [TIC Gums] water binder, processing aid which prevents syneresis, aids freeze/thaw props., adds sheen and gloss, prevents cracking; for pie fillings; *Usage level:* 1-2% (pie fillings).

Clinton #600 Dextrin. [MLG Enterprises Ltd.] Dextrin; CAS 9004-53-9; EINECS 232-675-4; carrier; *Properties:* wh. powd.; 45-55% water sol.

Clinton #655 Dextrin. [MLG Enterprises Ltd.] Dextrin; CAS 9004-53-9; EINECS 232-675-4; carrier; *Properties:* wh. powd.; 85-90% water sol.

Clinton #656 Dextrin. [MLG Enterprises Ltd.] Dextrin; CAS 9004-53-9; EINECS 232-675-4; carrier; *Properties:* wh. powd.; 85-90% water sol.

Clinton #700 Dextrin. [MLG Enterprises Ltd.] Dextrin; CAS 9004-53-9; EINECS 232-675-4; carrier; *Properties:* yel. powd.; 95% min. water sol.

Clinton #721 Dextrin. [MLG Enterprises Ltd.] Dextrin; CAS 9004-53-9; EINECS 232-675-4; carrier; *Properties:* yel. powd.; 95% min. water sol.

Clintose® A. [ADM Corn Processing; MLG Enterprises Ltd.] Dextrose; sweetener featuring mild sweetness, natural flavor enhancement, high fermentability, negative heat of sol'n., high osmotic pressure; for baking, confections, canning; *Properties:* wh. cryst., odorless, bland sweet taste; 99% min. thru #16 screen; 99.7% act., 8.5% moisture.

Clintose® F. [ADM Corn Processing] Dextrose; sweetener featuring mild sweetness, natural flavor enhancement, high fermentability, high osmotic pressure; for baking, confections, canning; *Properties:* wh. powd.; 98% thru 50 mesh, 60% thru 100 mesh; 99.7% act., 8.5% moisture.

Clintose® L. [ADM Corn Processing; MLG Enterprises Ltd.] Enzyme converted ion exchange refined dextrose syrup; sweetener with high fermentable sugars for use in baking, brewing, and fermentation processes; features mild sweetness, high osmotic pressure, natural flavor enhancement; *Properties:* water-wh. syrup, char. odor, sweet taste; dens. 11.01 lb/gal (130 F); visc. 35 cps (120 F); pH 4.5; DE 99.5; 99.5% dextrose, 29.5% moisture; *Storage:* store @ 130 F to avoid crystallization.

Clintose® VF. [ADM Corn Processing] Dextrose; sweetener featuring mild sweetness, natural flavor enhancement, high fermentability, high osmotic pressure; for baking, confections, canning; *Properties:* wh. powd.; 99% thru 100 mesh, 75% thru 200 mesh; 99.7% act., 8.5% moisture.

Clouding Agent Powd. [MLG Enterprises Ltd.] Powder emulsion which does not contain titanium dioxide; clouding agent for dehydrated drinks; *Usage level:* 0.5-0.7 ppm (drinks); 280 mg thiamine nitrate, 170 mg riboflavin, 2000 mg niacin, 4984 mg ferrous sulfate/$^1/_2$ oz; *Storage:* store in cool, dry place with kegs tightly closed.

CLSP 399. [Van Den Bergh Foods] Palm kernel oil with lecithin; lauric coating fat for pastel and chocolate flavored confectionery coatings; *Properties:* liq.; m.p. 86-92 F.

CLSP 499. [Van Den Bergh Foods] Palm kernel oil, lecithin; lauric coating fat for pastel and chocolate flavored confectionery coatings; *Regulatory:* kosher; *Properties:* solid; m.p. 31-33.8 C.

CLSP 555. [Van Den Bergh Foods] Hydrog. palm kernel oil, lecithin; lauric center fat or confectionery coating fat where a high degree of diluent fat is present; peanut butter fillings; *Regulatory:* kosher; *Properties:* solid; m.p. 33-55 C.

CLSP 874. [Van Den Bergh Foods] Partially hydrog. veg. oil (cottonseed, soybean); EINECS 269-820-6; nonlauric coating fat for no-tempering confectioner's coating, center fat; *Regulatory:* kosher; *Properties:* liq.; m.p. 93 F.

CMC Daicel 1150. [Daicel USA] Sodium carboxymethyl cellulose; CAS 9004-32-4; thickener, binder, stabilizer, excipient, protective colloid, suspending agent for sherbet; *Regulatory:* FDA 21CFR §182.1745, GRAS; *Properties:* odorless, tasteless; sol. in water; insol. in almost all org. solvs., oils, and fats; visc. 200-300 cps; *Storage:* hygroscopic-store unused portion in tightly closed containers.

CMC Daicel 1160. [Daicel USA] Sodium carboxymethyl cellulose; CAS 9004-32-4; thickener, binder, stabilizer, excipient, protective colloid, suspending agent for ice cream, instant cocoa, peanut butter, flour paste; *Regulatory:* FDA 21CFR §182.1745, GRAS; *Properties:* odorless, tasteless; sol. in water; insol. in almost all org. solvs., oils, and fats; visc. 300-500 cps; *Storage:* hygroscopic-store unused portion in tightly closed containers.

CMC Daicel 1220. [Daicel USA] Sodium carboxymethyl cellulose; CAS 9004-32-4; thickener, binder, stabilizer, excipient, protective colloid, suspending agent for jam, ketchup; improved acid-, salt-, heat-, and chemical-resist.; *Regulatory:* FDA 21CFR §182.1745, GRAS; *Properties:* odorless, tasteless; sol. in water; insol. in almost all org. solvs., oils, and fats; visc. 10-20 cps; *Storage:* hygroscopic-store unused portion in tightly closed containers.

CMC Daicel 1240. [Daicel USA] Sodium carboxymethyl cellulose; CAS 9004-32-4; thickener,

binder, stabilizer, excipient, protective colloid, suspending agent for milk drinks; improved acid-, salt-, heat-, and chemical-resist.; *Regulatory:* FDA 21CFR §182.1745, GRAS; *Properties:* odorless, tasteless; sol. in water; insol. in almost all org. solvs., oils, and fats; visc. 30-40 cps; *Storage:* hygroscopic-store unused portion in tightly closed containers.

CMC Daicel 1260. [Daicel USA] Sodium carboxymethyl cellulose; CAS 9004-32-4; thickener, binder, stabilizer, excipient, protective colloid, suspending agent for sauces, canned fruit, fish, and meat, juices, instant soup; improved acid-, salt-, heat-, and chemical-resist.; *Regulatory:* FDA 21CFR §182.1745, GRAS; *Properties:* odorless, tasteless; sol. in water; insol. in almost all org. solvs., oils, and fats; visc. 80-150 cps; *Storage:* hygroscopic-store unused portion in tightly closed containers.

CMC Daicel 2200. [Daicel USA] Sodium carboxymethyl cellulose; CAS 9004-32-4; thickener, binder, stabilizer, excipient, protective colloid, suspending agent for canned fruit, fish, and meat; improved acid-, salt-, heat-, and chemical-resist.; *Regulatory:* FDA 21CFR §182.1745, GRAS; *Properties:* odorless, tasteless; sol. in water; insol. in almost all org. solvs., oils, and fats; visc. 1500-2500 cps; *Storage:* hygroscopic-store unused portion in tightly closed containers.

CNC Antifoam 30-FG. [CNC Int'l. L.P.] Silicone emulsion; nonionic; antifoam/defoam avail. in food grade versions; *Properties:* wh. visc. emulsion; easily disp. with water; dens. 8.3 lb/gal; pH 4-5; 27-28% act. silicone.

CNP BRSHG40. [Calif. Natural Prods.] High glucose brown rice syrup; lightly filtered to make opaque syrup with buttery flavor notes; preferred sweetener for soy-based prods., granola, cereals, sauces, catsup; *Regulatory:* kosher certified; *Properties:* dens. 11.8 lb/gal (100 F); visc. 100 poise (100 F); DE 70; pH 6.0-6.6 (10%).

CNP BRSHG40CL. [Calif. Natural Prods.] Clarified high glucose brown rice syrup; patented syrup with further filtration to produce a clear syrup with low flavor profile; sweetener for soy-based prods., granola, cereals, sauces, catsup; *Regulatory:* kosher certified; *Properties:* dens. 11.8 lb/gal (100 F); visc. 100 poise (100 F); DE 70; pH 6.0-6.6 (5%).

CNP BRSHM. [Calif. Natural Prods.] High maltose brown rice syrup; lightly filtered to make opaque syrup with buttery flavor notes; sweetener for hard candies, cough drops, home brew beers, flavor carrier, extruded bars/confections; *Regulatory:* kosher certified; *Properties:* dens. 11.8 lb/gal (100 F); visc. 125 poise (100 F); DE 42; pH 6.0-6.6 (10%).

CNP BRSHMCL. [Calif. Natural Prods.] Clarified high maltose brown rice syrup; patented syrup with further filtration to produce a clear syrup with low flavor profile; sweetener for hard candies, cough drops, home brew beers, flavor carrier, extruded bars/confections; *Regulatory:* kosher certified; *Properties:* dens. 11.8 lb/gal (100 F); visc. 125

poise (100 F); DE 42; pH 6.0-6.6 (5%).

CNP BRSMC10. [Calif. Natural Prods.] Med. conversion brown rice syrup; lightly filtered to make opaque syrup with buttery flavor notes; sweetener, humectant for soy-based prods.; beverage sweetener, licorice, soft confections, caramel corn, granola, cereals, baked goods, cookies, table syrups, desserts; *Regulatory:* kosher certified; *Properties:* dens. 11.8 lb/gal (100 F); visc. 125 poise (100 F); DE 50; pH 6.0-6.6 (10%).

CNP BRSMC20. [Calif. Natural Prods.] Med. conversion brown rice syrup; lightly filtered to make opaque syrup with buttery flavor notes; sweetener, humectant for soy-based prods.; beverage sweetener, licorice, soft confections, caramel corn, granola, cereals, baked goods, cookies, table syrups, desserts; *Regulatory:* kosher certified; *Properties:* dens. 11.8 lb/gal (100 F); visc. 100 poise (100 F); DE 55; pH 6.0-6.6 (10%).

CNP BRSMC35. [Calif. Natural Prods.] Med. conversion brown rice syrup; lightly filtered to make opaque syrup with buttery flavor notes; sweetener, humectant for soy-based prods.; beverage sweetener, licorice, soft confections, caramel corn, granola, cereals, baked goods, cookies, table syrups, desserts; *Regulatory:* kosher certified; *Properties:* dens. 11.8 lb/gal (100 F); visc. 100 poise (100 F); DE 60; pH 6.0-6.6 (10%).

CNP BRSMC35CL. [Calif. Natural Prods.] Clarified med. conversion brown rice syrup; patented syrup with further filtration to produce a clear syrup with low flavor profile; sweetener, humectant for soy-based prods.; beverage sweetener, licorice, soft confections, caramel corn, granola, cereals, baked goods, cookies, table syrups; *Regulatory:* kosher certified; *Properties:* dens. 11.8 lb/gal (100 F); visc. 100 poise (100 F); DE 57; pH 6.0-6.6 (5%).

CNP BRSSHG40. [Calif. Natural Prods.] High glucose brown rice syrup solids; sweetener, humectant for instant soups, baking, soy-based prod. sweetening; *Regulatory:* kosher certified; *Properties:* lt. beige drum-dried powd., -30 mesh, caramel flavor; DE 70; pH 6.0-6.6 (5%).

CNP BRSSHG40CL. [Calif. Natural Prods.] Clarified high glucose brown rice syrup solids; patented syrup with further filtration to produce a clear syrup with low flavor profile; sweetener for instant soups, baking, soy-based prods.; *Regulatory:* kosher certified; *Properties:* lt. beige drum-dried powd., -30 mesh, caramel flavor; DE 70; pH 6.0-6.6 (5%).

CNP BRSSHM. [Calif. Natural Prods.] High maltose brown rice syrup solids; sweetener for hard candies, cough drops, home brew beers, flavor carrier, extruded bars/confections; *Regulatory:* kosher certified; *Properties:* off-wh. spray-dried powd., -100 mesh, bland taste or lt. beige drum-dried powd., -30 mesh, caramel flavor; DE 42; pH 6.0-6.6 (5%).

CNP BRSSHMCL. [Calif. Natural Prods.] Clarified high maltose brown rice syrup solids; patented syrup with further filtration to produce a clear syrup with low flavor profile; sweetener for hard candies, cough drops, home brew beers, flavor carrier, ex-

truded bars/confections; *Regulatory:* kosher certified; *Properties:* lt. beige drum-dried powd., -30 mesh, caramel flavor or off-wh. spray-dried powd., -100 mesh, bland taste; DE 42; pH 6.0-6.6 (5%).

CNP PPRSRDCL. [Calif. Natural Prods.] Clarified low conversion partially polished brown rice syrup; patented syrup with further filtration to produce a clear syrup with low flavor profile; preferred in high usage and nonfat prod. formulation; *Regulatory:* kosher certified; *Properties:* dens. 11.8 lb/gal (100 F); visc. 150 poise (100 F); DE 26; pH 6.0-6.6 (5%).

CNP PPRSSRD. [Calif. Natural Prods.] Low conversion partially polished brown rice syrup solids; carrier, sol. complex carbohydrate, bulking agent, low osmolality for dry beverage mixes for sports/oral rehydration, flavor carrier, home brew kits; *Regulatory:* kosher certified; *Properties:* off-wh. spray-dried powd., -100 mesh, bland taste; DE 26; pH 6.0-6.6 (5%).

CNP PPRSSRDCL. [Calif. Natural Prods.] Clarified low conversion partially polished brown rice syrup solids; patented syrup with further filtration to produce a clear syrup with low flavor profile; carrier, sol. complex carbohydrate, bulking agent, low osmolality for dry beverage mixes for sports/oral rehydration, flavor carrier, home brew kits; *Regulatory:* kosher certified; *Properties:* off-wh. spray-dried powd., -100 mesh, bland taste; DE 26; pH 6.0-6.6 (5%).

CNP Pregelled Rice Flour. [Calif. Natural Prods.] Pregelled rice flour; gellant providing instant thickening when added to aq. systems; for dry mixes for instant soups, gravies, sauces, and in instant baby food cereals; as in-process visc. control agents (gravies or sauces for frozen meals and sandwiches); *Regulatory:* kosher certified; *Properties:* powd., passes 30 mesh; < 5% moisture.

CNP RMDRD05. [Calif. Natural Prods.] Rice maltodextrin; visc. and body modifier for liq. systems; flavor carrier; hypoallergenic prods.; oral rehydration and sports beverage/carbohydrate loading prods.; control of osmolaity; bulking agent; cryst. inhibitor; *Regulatory:* kosher certified; *Properties:* bland taste; disp. readily into aq. sol'ns.; DE 5; pH 6.5-6.7 (10% in DI water); < 5% moisture.

CNP RMDRD18. [Calif. Natural Prods.] Rice maltodextrin; visc. and body modifier for liq. systems; flavor carrier; hypoallergenic prods.; oral rehydration and sports beverage/carbohydrate loading prods.; control of osmolaity; bulking agent; cryst. inhibitor; *Regulatory:* kosher certified; *Properties:* bland taste; disp. readily into aq. sol'ns.; DE 18; pH 6.5-6.7 (10% in DI water); < 5% moisture.

CNP RPCCG. [Calif. Natural Prods.] Rice protein conc.; patented prod. providing both conc. rice protein and dietary fiber; for baking applics. such as protein fortification of cookies, muffins, breads, animal feed; hypoallergenic protein prods.; *Regulatory:* kosher certified; *Properties:* coarse gran., -10 mesh; pH 6.0-6.6 (5% disp.); 53% protein, 7% dietary fiber; < 5% moisture.

CNP RPCXFG. [Calif. Natural Prods.] Rice protein

conc.; patented prod. providing both conc. rice protein and dietary fiber; for wide range of dry mixes or prepared foods, animal feed; hypoallergenic protein prods.; *Regulatory:* kosher certified; *Properties:* extra fine powd., > 90% -140 mesh; pH 6.0-6.6 (5% disp.); 53% protein, 7% dietary fiber; < 5% moisture.

CNP RSSHG40. [Calif. Natural Prods.] High glucose rice syrup solids; sweetener, humectant for instant soups, baking, soy-based prod. sweetening; *Regulatory:* kosher certified; *Properties:* lt. beige drum-dried powd., -30 mesh, caramel flavor; sol. in water; DE 70; pH 6.0-6.6 (5%).

CNP RSSHG40CL. [Calif. Natural Prods.] Clarified high glucose rice syrup solids; sweetener, humectant for instant soups, baking, soy-based prod. sweetening; *Regulatory:* kosher certified; *Properties:* lt. beige drum-dried powd., -30 mesh, caramel flavor; sol. in water; DE 70; pH 6.0-6.6 (5%).

CNP RSSHM. [Calif. Natural Prods.] High maltose rice syrup solids; drum-dried: sweetener, humectant for instant soups, baking, soy-based prod. sweetening; spray-dried: carrier, sol. complex carbohydrate, bulking agent, low osmolality for dry beverage mixes for sports/oral rehydration, flavor carrier, home brew kits; *Regulatory:* kosher certified; *Properties:* off-wh. spray-dried powd., -100 mesh, bland taste or lt. beige drum-dried powd., -30 mesh, caramel flavor; sol. in water; DE 42; pH 6.0-6.6 (5%).

CNP RSSHMCL. [Calif. Natural Prods.] Clarified high maltose rice syrup solids; drum-dried: sweetener, humectant for instant soups, baking, soy-based prod. sweetening; spray-dried: carrier, sol. complex carbohydrate, bulking agent, low osmolality for dry beverage mixes for sports/oral rehydration, flavor carrier, home brew kits; *Regulatory:* kosher certified; *Properties:* off-wh. spray-dried powd., -100 mesh, bland taste or lt. beige drum-dried powd., -30 mesh, caramel flavor; sol. in water; DE 42; pH 6.0-6.6 (5%).

CNP RSSRD. [Calif. Natural Prods.] Low conversion rice syrup solids; carrier, sol. complex carbohydrate, bulking agent, low osmolality for dry beverage mixes for sports/oral rehydration, flavor carrier, home brew kits; *Regulatory:* kosher certified; *Properties:* off-wh. spray-dried powd., -100 mesh, bland taste; sol. in water; DE 26; pH 6.0-6.6 (5%).

CNP RSSRDCL. [Calif. Natural Prods.] Clarified low conversion rice syrup solids; carrier, sol. complex carbohydrate, bulking agent, low osmolality for dry beverage mixes for sports/oral rehydration, flavor carrier, home brew kits; *Regulatory:* kosher certified; *Properties:* off-wh. spray-dried powd.; -100 mesh, bland taste; sol. in water; DE 26; pH 6.0-6.6 (5%).

CNP SRFCG. [Calif. Natural Prods.] Sweet rice flour; sweet white rice flour used for baking breads, waffles, pizza doughs, prepared foods (gravies, sauces, soups); *Regulatory:* kosher certified; *Properties:* coarse gran., 21-27% thru 325 mesh; 12% max. moisture.

CNP SRFFG. [Calif. Natural Prods.] Sweet rice flour; sweet white rice flour used for baking breads, waffles, pizza doughs, prepared foods (gravies, sauces, soups); *Regulatory:* kosher certified; *Properties:* fine gran., 33-42% thru 325 mesh; 12% max. moisture.

CNP SRFRG. [Calif. Natural Prods.] Sweet rice flour; sweet white rice flour used for baking breads, waffles, pizza doughs, prepared foods (gravies, sauces, soups); *Regulatory:* kosher certified; *Properties:* regular gran., 27-32% thru 325 mesh; 12% max. moisture.

CNP SRFXFG. [Calif. Natural Prods.] Sweet rice flour; sweet white rice flour used for baking breads, waffles, pizza doughs, prepared foods (gravies, sauces, soups); *Regulatory:* kosher certified; *Properties:* extra fine gran., 50-70% thru 325 mesh; 12% max. moisture.

CNP WRFCG. [Calif. Natural Prods.] Rice flour; white rice flour used for baking breads, waffles, pizza doughs, prepared foods (gravies, sauces, soups), extrusion (cereals, crisp snacks), baby foods; *Regulatory:* kosher certified; *Properties:* coarse gran., 21-27% thru 325 mesh; 12% max. moisture.

CNP WRFFG. [Calif. Natural Prods.] Rice flour; white rice flour used for baking breads, waffles, pizza doughs, prepared foods (gravies, sauces, soups), extrusion (cereals, crisp snacks), baby foods; *Regulatory:* kosher certified; *Properties:* fine gran., 33-42% thru 325 mesh; 12% max. moisture.

CNP WRFRG. [Calif. Natural Prods.] Rice flour; white rice flour used for baking breads, waffles, pizza doughs, prepared foods (gravies, sauces, soups), extrusion (cereals, crisp snacks), baby foods; *Regulatory:* kosher certified; *Properties:* regular gran., 27-32% thru 325 mesh; 12% max. moisture.

CNP WRFXFG. [Calif. Natural Prods.] Rice flour; white rice flour used for baking breads, waffles, pizza doughs, prepared foods (gravies, sauces, soups), extrusion (cereals, crisp snacks), baby foods; *Regulatory:* kosher certified; *Properties:* extra fine gran., 50-70% thru 325 mesh; 12% max. moisture.

CNP WRSHG40. [Calif. Natural Prods.] High glucose rice syrup; patented traditional syrup from wh. rice, lightly filtered to produce a translucent syrup with a light butter flavor note; *Regulatory:* kosher certified; *Properties:* dens. 11.8 lb/gal (100 F); visc. 100 poise (100 F); DE 70; pH 6.0-6.6 (5%).

CNP WRSHG40CL. [Calif. Natural Prods.] Clarified high glucose rice syrup; patented clear syrup with low flavor provile; used for applics. with high syrup usage or lightly flavored prods. such as vanilla-flavored desserts; *Regulatory:* kosher certified; *Properties:* dens. 11.8 lb/gal; visc. 100 poise (100 F); DE 70; pH 6.0-6.6 (5%).

CNP WRSHM. [Calif. Natural Prods.] High maltose rice syrup; patented traditional syrup from wh. rice, lightly filtered to produce a translucent syrup with a light butter flavor note; *Regulatory:* kosher certified; *Properties:* dens. 11.8 lb/gal (100 F); visc. 125 poise (100 F); DE 42; pH 6.0-6.6 (5%).

CNP WRSHMCL. [Calif. Natural Prods.] Clarified high

maltose rice syrup; patented clear syrup with low flavor profile; used for applics. with high syrup usage or lightly flavored prods. such as vanilla-flavored desserts; *Regulatory:* kosher certified; *Properties:* dens. 11.8 lb/gal; visc. 125 poise (100 F); DE 42; pH 6.0-6.6 (5%).

CNP WRSMC30. [Calif. Natural Prods.] Med. conversion rice syrup; patented traditional syrup from wh. rice, lightly filtered to produce a translucent syrup with a light butter flavor note; *Regulatory:* kosher certified; *Properties:* dens. 11.8 lb/gal (100 F); visc. 100 poise (100 F); DE 57; pH 6.0-6.6 (5%).

CNP WRSMC30CL. [Calif. Natural Prods.] Clarified med. conversion rice syrup; patented clear syrup with low flavor provile; used for applics. with high syrup usage or lightly flavored prods. such as vanilla-flavored desserts; *Regulatory:* kosher certified; *Properties:* dens. 11.8 lb/gal; visc. 100 poise (100 F); DE 57; pH 6.0-6.6 (5%).

CNP WRSRD. [Calif. Natural Prods.] Low conversion rice syrup; patented traditional syrup from wh. rice, lightly filtered to produce a translucent syrup with a light butter flavor note; *Regulatory:* kosher certified; *Properties:* dens. 11.8 lb/gal (100 F); visc. 150 poise (100 F); DE 26; pH 6.0-6.6 (5%).

CNP WRSRDCL. [Calif. Natural Prods.] Clarified low conversion rice syrup; patented clear syrup with low flavor profile; used for applics. with high syrup usage or lightly flavored prods. such as vanilla-flavored desserts; *Regulatory:* kosher; *Properties:* dens. 11.8 lb/gal; visc. 150 poise (100 F); DE 26%; pH 6.0-6.6 (5%).

Coated Ascorbic Acid 97.5% No. 60482. [Roche] L-Ascorbic acid USP, FCC coated with ethylcellulose; vitamin protected against reactive materials in dry food prods.; *Regulatory:* FDA GRAS; *Properties:* wh. to sl. off-wh. free-flowing fine gran. powd.; 100% thru 16 mesh, 35% max. thru 100 mesh; 97.5% min. assay; *Storage:* store in tight, light-resist. containers, optimally @ ≤ 72 F; avoid exposure to moisture and excessive heat.

Coatingum L. [MLG Enterprises Ltd.] Gum acacia; CAS 9000-01-5; EINECS 232-519-5; emulsifier, stabilizer; substitute for shellac for polishing and glazing of pan-coated sweets; *Properties:* visc. 45-80 cps (20%); pH 4.0-5.0 (20%); 99.9% min. purity.

Cobee 76. [Stepan Food Ingreds.] Refined, bleached, deodorized coconut oil; CAS 8001-31-8; EINECS 232-282-8; emollient, superfatting agent, clouding agent, used in food industry; *Properties:* Lovibond 20/2 soft solid; no odor or taste; m.p. 76-80 F; Unverified

Cobee 92. [Stepan Food Ingreds.] Refined, bleached, deodorized, hydrog. coconut oil; EINECS 284-283-8; emollient, superfatting agent, clouding agent, used in food industry; *Properties:* Lovibond 20/2 solid; no odor or taste; m.p. 98-114 F; Unverified

Cobee 110. [Stepan Food Ingreds.] Refined, bleached, deodorized, hydrog. coconut oil; EINECS 284-283-8; emollient, superfatting agent, clouding agent for food industry; *Properties:* Lovibond 20/2 solid; odorless; m.p. 110-114 F; Un-

verified

Coberine. [Van Den Bergh Foods] Vegetable oil (palm and shea); cocoa butter equivalent in confectioner's coatings, center fat, veg. dairy systems; *Regulatory:* kosher; *Properties:* solid; m.p. 91.4-96.8 F.

Cocoloid®. [Kelco] Alginate; used in chocolate milk, soft-serve frozen desserts, ice cream, ice milk, custard, variegated syrup, and fudge topping.

Coconad MT. [Kao/Edible Fat & Oil] Caprylic triglyceride; nonionic; food emulsifier; coffee whitener, flavor diluting agent; *Properties:* liq.; 100% conc.

Coconad PL. [Kao/Edible Fat & Oil] Caprylic/capric triglyceride; nonionic; food emulsifier; health and sports food; *Properties:* liq.; 100% conc.; Unverified

Coconad RK. [Kao/Edible Fat & Oil] Caprylic/capric triglyceride; nonionic; food emulsifier; coffee whitener, releasing agent for processing, flavor diluting agent; *Properties:* liq.; 100% conc.

Coconad SK. [Kao/Edible Fat & Oil] Capric triglyceride; nonionic; food emulsifier; health and sports food; *Properties:* liq.; 100% conc.; Unverified

Coconut Oil® 76. [Stepan Food Ingreds.] Coconut oil, refined, bleached, deodorized; CAS 8001-31-8; EINECS 232-282-8; oil with superior stability and resist. to oxidative rancidity; flavor solubilizer; base for ice cream coatings, chewy candies, cream fillings, icings, and frozen desserts; cracker sprays; clouding agent for beverages; *Regulatory:* FDA 21CFR §170.30 (GRAS); kosher grade avail.; *Properties:* sl. yel. soft solid, mild fatty odor, bland taste; iodine no. 8.4; sapon. no. 256; 0.04% moisture, *Toxicology:* low irritation potential; *Storage:* avoid prolonged storage above 90 F.

Coconut Oil® 92. [Stepan Food Ingreds.] Coconut oil, hydrog., refined, bleached, deodorized; CAS 8001-31-8; EINECS 232-282-8; oil with superior stability and resist. to oxidative rancidity; flavor solubilizer; base for ice cream coatings, chewy candies, cream fillings, icings, and frozen desserts; clouding agent for beverages; *Regulatory:* FDA 21CFR §170.30 (GRAS); kosher grade avail.; *Properties:* cream-colored solid, mild fatty odor, bland taste; iodine no. 1.3; sapon. no. 255; 0.04% moisture; *Toxicology:* low irritation potential; *Storage:* avoid prolonged storage above 90 F. Discontinued

Coco-Spred™. [MLG Enterprises Ltd.] Coconut oil, vegetable oil and partially hydrog. vegetable oil (soybean and/or cottonseed), calcium carbonate, lecithin, mono- and diglycerides, BHA (as antioxidant); processing aid, release lubricant; *Properties:* off-wh. paste, virtually odorless; negligible sol. in water; m.p. 100F; flash pt. 219 C.

Code 321. [Van Den Bergh Foods] Partially hydrog. soybean oil; EINECS 232-410-2; multipurpose shortening used for filler, snack frying, prepared foods, spray oil; *Regulatory:* kosher; *Properties:* Lovibond 1.5R max. plastic; m.p. 95-99 F; iodine no. 78 max.

Coffee Conc. [Universal Flavors] Coffee flavoring with 2.6% caffeine and potassium sorbate; natural flavor for ice cream, bakery prods.; *Usage level:* 2.5%

(ice cream), 3-5% (bakery prods.); *Properties:* liq.; water-sol.; dens. 1.35; *Storage:* 12 mos storage life in original sealed packaging in cool, dry area.

Cola H. [MLG Enterprises Ltd.] Precompounded base for prod. of cola drinks with more pronounced citric touch, requiring only addition of sugar, acidifier, and preservative.

Cola L. [MLG Enterprises Ltd.] Complex blend of natural extracts and essential oils; preformulated flavor for cola gasified soft drinks and syrups with more pronounced citric touch.

Cola No. 23443. [Universal Flavors] Cola with caramel, arabic gum, sodium benzoate, citric acid, and phosphoric acid; flavor for soft drinks; *Usage level:* 13 g/l (syrup); *Properties:* emulsion; water-sol.; dens. 1.19; *Storage:* 9 mos storage life in original sealed container in cool, dry area.

Cola Acid No. 23444. [Universal Flavors] Cola acid with phosphoric acid, citric acid, caffeine; flavor for soft drinks; *Usage level:* 3.56 g/l (cola syrup); *Properties:* liq.; water-sol.; dens. 1.620; *Precaution:* corrosive; *Storage:* 18 mos storage life in original sealed container in cool, dry area.

Cola Complex. [MLG Enterprises Ltd.] Precompounded base for prod. of traditional cola drinks, requiring only addition of sugar, acidifier, and preservative.

Cola Extra. [MLG Enterprises Ltd.] Complex blend of natural extracts and essential oils; preformulated traditional flavor for cola gasified soft drinks and syrups.

Colloid 451T. [TIC Gums] Texturizer, oil and shortening replacement, moisture retention aid for low-fat muffins; water binder, controls ice crystals in frozen dough; controls batter visc. in dry cake mixes; *Usage level:* 0.15-0.2% (low-fat muffins), 0.2-0.3% (frozen doughs), 0.1-0.15% (dry cake mixes).

Colloid 488T. [TIC Gums] Sodium alginate; CAS 9005-38-3; gellant, film former, emulsifier, suspending aid; reactive with milk; for foods (dairy prods., extruded foods, beverages, salad dressings, dessert gels, sauces, frozen desserts, bakery prods.); *Usage level:* 0.3-0.5% (pie fillings); *Properties:* sol. in cold water.

Colloid 515MT. [TIC Gums] Reduces tackiness, holds moisture, controls sugar crystals in icings; *Usage level:* 0.5-0.8% (icings).

Colloid 602. [TIC Gums] Propylene glycol alginate; CAS 9005-37-2; gellant, film former, emulsifier, suspending aid; reactive with milk; for foods (dairy prods., extruded foods, beverages, salad dressings, dessert gels, sauces, frozen desserts, bakery prods.); *Properties:* sol. in cold water.

Colloid 787. [TIC Gums] water binder, processing aid which prevents syneresis, aids freeze/thaw props., adds sheen and gloss, prevents cracking in pie fillings; water binder, controls ice crystals in frozen doughs; *Usage level:* 0.2-0.4% (pie fillings), 0.05-0.15% (frozen doughs).

Colloid 862. [TIC Gums] Texturizer for meringues; inhibits weeping; *Usage level:* 0.5-1.5% (meringues).

Colloid 886. [TIC Gums] Emulsifier, stabilizer for bakery prods.; *Usage level:* 0.6-0.7% (bakery emulsions).

Colloid 1023T. [TIC Gums] Water binder, controls ice crystals in frozen doughs; *Usage level:* 0.2-0.4% (frozen doughs).

Colloid XC7408. [TIC Gums] Moisture retention aid in baked goods; *Usage level:* 0.2-0.3% (baked goods).

Colloid Stick Tic. [TIC Gums] Used in a separate spray as film-former to which spices or sesame seeds adhere; *Usage level:* 5-10% (cracker adhesive).

Coloreze™. [MLG Enterprises Ltd.] Food color concs. made from FD&C dyes, lakes, and natural colors dispersed in FDA-approved ingreds. such as glycerin, propylene glycol, dextrose, water, veg. oil, etc.; color concs. for use in mfg. of candy prods., baked goods; *Regulatory:* kosher; *Properties:* paste and liq.

Colorin 102, 104, 202. [Sanyo Chem. Industries] Polyether polyol; antifoamer used in fermentation of glutamic acid, yeast; *Properties:* liq.; Unverified

Colorsorb. [Calgon Carbon] Activated carbon; CAS 64365-11-3; used for food decolorization; *Properties:* pulverized.

Complemix® 50. [Cytec Industries] Sulfonated ester; anionic; wetting agent, emulsifier, clouding agent used in the food and beverage industry; *Properties:* nearly colorless liq.; sp.gr. 0.93; 50% act.

Complemix® 100. [Cytec Industries] Sodium dioctyl sulfosuccinate NF; CAS 577-11-7; EINECS 209-406-4; anionic; surfactant, emulsifier, wetting or clouding agent used in the food and beverage industry; dispersant, solubilizer, modifier; *Regulatory:* TSCA, EEC compliance; *Properties:* wh. waxy solid, odor of octyl alcohol; sol. 1.5 g/100 g in water; sp.gr. 1.1; 100% act.; *Toxicology:* LD50 (oral, rat), 4.2 g/kg, (dermal, rabbit) > 1 g/kg; causes moderate eye and skin irritation.

Compritol 888 ATO. [Gattefosse; Gattefosse SA] Glyceryl behenate; EINECS 250-097-0; nonionic; food emulsifier and additive for tablet mfg.; *Properties:* solid; drop pt. ≈ 70 C; HLB 2.0; acid no. < 4; iodine no. < 3; sapon. no. 145-165; 100% conc.; *Toxicology:* LD0 (oral, rat) > 5 g/kg.

Concentrated Beef Stock #3021. [Ariake USA] Beef stock; flavorings for snacks, soup mixes, sauces, gravies; 32% moisture, 48% protein.

Concentrated Chicken Broth #1023. [Ariake USA] Chicken broth; flavorings for snacks, soup mixes, sauces, gravies; 54.2% moisture, 31.8% protein.

Concentrated Chicken Broth #1095. [Ariake USA] Chicken broth; flavorings for snacks, soup mixes, sauces, gravies; 34.7% moisture, 43.3% protein.

Concentrated Chicken Stock #1021. [Ariake USA] Chicken stock; flavorings for snacks, soup mixes, sauces, gravies; 43.2% moisture, 48.2% protein.

Concentrated Dariloid®. [Kelco] Alginate; dairy stabilizer used in puddings, pie fillings, custards, cheese dips, cheesecake, egg nog, milk shake, chiffon pie fillings, toppings, syrups, and baked cream fillings.

Concentrated Dariloid® KB. [Kelco] Propylene glycol alginate blend; CAS 9005-37-2; dairy stabilizer used for sour cream, cottage cheese dressings, cream cheese, egg nog, ice cream fruit, cottage cheese fruit, fruit beverages, water ice, sherbet and lowfat novelties; *Properties:* 28 mesh; milk-sol.; pH 5.4.

Concentrated Dariloid® XL. [Kelco] Alginate; dairy stabilizer used for egg nog and liq. dairy applics. requiring a low-visc. stabilizer.

Concentrated Pork Stock #2021. [Ariake USA] Pork stock; flavorings for snacks, soup mixes, sauces, gravies; 42.2% moisture, 43.1% protein.

Concentrated Turkey Stock #4021. [Ariake USA] Turkey stock; flavorings for snacks, soup mixes, sauces, gravies; 42% moisture, 40% protein.

Confectioners F & G. [A.E. Staley Mfg.] Food starch modified; thin-boiling starch for developing proper depositing visc. and finished textures in gum candies; *Regulatory:* FDA 21CFR §172.892; *Properties:* wh. powd.; pH 5.5; 11% moisture.

Confecto™ Lubes. [MLG Enterprises Ltd.] Edible vegetable oils and other ingreds.; confectionery lubricant, release, anti-stick agent for kettles, cooling slabs, belts, trays, mixers used in candy mfg.; *Properties:* off-wh. or pale yel. free-flowing oils with and without solids, bland taste.

Confecto No-Stick™. [MLG Enterprises Ltd.] Blend of partially hydrogenated vegetable oils and mono- and diglycerides; confectionery conditioning agent; reduces stickiness in candy; minimizes effects of humid and hot conditions; extends prod. shelf life; effective for hard candy, taffy, caramel, nougat, brittles, etc., *Usage level:* 1 oz/100 lb candy; *Regulatory:* kosher; *Properties:* off-wh. free-flowing beads, odorless, tasteless; m.p. 140-145 F.

Confecto™ Rubs. [MLG Enterprises Ltd.] Edible vegetable oils and other ingreds.; confectionery lubricant, release, anti-stick agent for kettles, cooling slabs, belts, trays, mixers used in candy mfg.; *Properties:* off-wh. or pale yel. hard plastics, bland taste.

Confecto™ Spreds. [MLG Enterprises Ltd.] Edible vegetable oils and other ingreds.; confectionery lubricant, release, anti-stick agent for kettles, cooling slabs, belts, trays, mixers used in candy mfg.; *Properties:* off-wh. or pale yel. plastics, bland taste.

Consista®. [A.E. Staley Mfg.] Food starch modified, derived from waxy maize; starch for prod. of cream-style corn, meat prods., and other low-acid foods requiring bland, high-visc. stabilizer/thickener; provides short, smooth textures, exc. clarity, resist. to syneresis, freeze/thaw cycles; *Usage level:* 1.5-5.0% (cream-style corn, gravies, sauces, soups, puddings); *Regulatory:* FDA 21CFR §172.892; *Properties:* wh. powd.; 90% thru 100 mesh; pH 5.5 (uncooked); 10-12% moisture.

Contraspum 210. [Zschimmer & Schwarz] Mixt. of fatty acids, paraffin oil, and nonionic emulsifier; nonionic; defoamer for yeast and alcohol prod.; *Properties:* clear yel. liq.; emulsifiable in water;

sp.gr. 0.92; visc. 100 mPa•s; 100% act.

Controx® AT. [Grünau GmbH] Tocopherol (40%), lecithin (40%), ascorbyl palmitate (10%), fatty acid mono- and diglycerides citrate (10%); antioxidant for fats in the food industry; esp. suitable for stabilization of animal fats (tallow, lard, butter fat, butter oil), vegetable oils (coconut, palm kernel, cocoa butter), syn. lipids, essential oils; *Usage level:* 100-500 ppm; *Regulatory:* EEC compliance; *Properties:* reddish-brn. visc. liq., typ. odor, mild taste; 28% tocopherol; *Storage:* 1 yr shelf life under cool, dry storage conditions in original containers, protected from light.

Controx® KS. [Grünau GmbH; Henkel KGaA/Cospha; Henkel Canada] Tocopherol, hydrog. tallow glycerides citrate; antioxidant for fats in the food industry; esp. suitable for stabilization of animal oils and fats, essential oils; *Usage level:* 0.02-0.1% on fat content in final prod.; *Regulatory:* EEC compliance; *Properties:* reddish-brn. visc. liq., typ. odor, mild taste; oil-sol.; 56% tocopherol; *Storage:* 1 yr shelf life under cool, dry storage in original container protected from light.

Controx® VP. [Grünau GmbH; Henkel/Cospha; Henkel Canada; Henkel KGaA/Cospha] Lecithin (40%), tocopherol (25%), ascorbyl palmitate (25%), hydrog. tallow glycerides citrate (10%); antioxidant for fats in the food industry; esp. suitable for stabilization of vegetable oils, frying fats, margarines, shortenings, release agents; *Usage level:* 0.02-0.1% on fat content in final prod.; *Regulatory:* EEC compliance; *Properties:* reddish-brn. paste, typ. odor, mild taste; oil-sol.; 17.5% tocopherol.

Cool 'N Fresh™. [Hormel] Line of drink mixes, instant and cooked puddings (flavored and unflavored), for salads or deli use.

Coral®. [Bunge Foods] Partially hydrog. soybean and cottonseed oils, water, salt, mono- and diglycerides, lecithin, artificial flavor and color, vitamin A; bakery magarine for danish, croissants, cookies; adds flavor to icings; general kitchen use; *Properties:* m.p. 102 F; 80.2% fat, 16.7% moisture, 3.0% salt.

Coral Star. [Ikeda] Coral sand deriv. containing high calcium and magnesium content; food supplement providing calcium and magnesium; *Properties:* grayish powd., faint char. odor; 90% min. thru 350 mesh; 34% min. Ca, 1.5% min Mg.

Corn Oil. [Van Den Bergh Foods] Cornoil, choice refined; CAS 8001-30-7; EINECS 232-281-2; general purpose oil for salad dressing, cooking, baking, and frying applics.; *Properties:* Lovibond 3R max. liq.

Corn Oil. [Bunge Foods] Corn oil, antioxidants; CAS 8001-30-7; EINECS 232-281-2; specialty oil for general frying, snack spray oil; *Properties:* 1.4 red clear bright oil, sl. corn flavor; smoke pt. 450 F.

Corn Salad Oil 104-250. [ADM Packaged Oils] Dewaxed corn oil, no additives; CAS 8001-30-7; EINECS 232-281-2; packaged veg. salad oil; *Properties:* clear bright oil, bland flavor; iodine no. 126 ± 4.

CornSweet® 42. [ADM Corn Processing; MLG Enterprises Ltd.] High fructose corn syrup; CAS 8029-43-4; EINECS 232-436-4; sweetener for use at 50-100% replacement for sucrose or invert sugar; clean, nonmasking sweetness for use in delicate flavored foods; *Properties:* water-wh. low-visc. syrup, sweet taste; dens. 11.23 lb/gal; visc. 225 cps (70 F); pH 3.5; 71% solids, 42% min. fructose, 29% moisture; *Storage:* store @ 80-90 F.

CornSweet® 55. [ADM Corn Processing; MLG Enterprises Ltd.] High fructose corn syrup; CAS 8029-43-4; EINECS 232-436-4; sweetener for carbonated drinks, still drinks, processed foods; *Properties:* water-wh. syrup, sweet taste; dens. 11.55 lb/gal; visc. 1200 cps (70 F); pH 3.5; 77% solids, 55% min. fructose, 23% moisture; *Storage:* store @ 80-85 F for ease in pumping and blending.

CornSweet® 95. [ADM Corn Processing; MLG Enterprises Ltd.] High fructose corn syrup; CAS 8029-43-4; EINECS 232-436-4; sweetener offering intense sweetness, sweetness synergism, flavor enhancement, humectancy, low water activity, f.p. depression, high osmotic pressure; for beverages, baked goods, frozen foods, cereals, dairy prods., reduced-calorie foods; *Properties:* water-wh. liq., very sweet taste; dens. 11.56 lb/gal; visc. 575 cps (80 F); pH 3.5; 77% dry solids, 95% fructose; *Storage:* store @ 65-75 F.

CronSweet® Crystalline Fructose. [ADM Corn Processing; MLG Enterprises Ltd.] Fructose FCC/USP; sweetener offering intense sweetness, sweetness synergism, flavor enhancement, humectancy, low water activity, f.p. depression, high osmotic pressure; for beverages, baked goods, frozen foods, cereals, dairy prods., reduced-calorie foods; *Properties:* wh. free-flowing cryst., odorless, clean very sweet taste; 1% max. on 16 mesh, 10% max. thru 100 mesh; bulk dens. 52 lb/ft^3; m.p. 103 C; 99.9% act.; *Storage:* store @ 25 C and 50% r.h.

Corn Syrup 36/43. [ADM Corn Processing] Corn syrup; CAS 8029-43-4; EINECS 232-436-4; nutritive carbohydrate sweetener for ice cream, hard candies, cough drops; *Properties:* dens. 11.84 lb/gal (100 F); visc. 1000 poise (80 F); pH 4.8 (1:1); DE 36; Baume 43 (100 F); 80.3% total solids, 19.7% moisture.

Corn Syrup 42/43. [ADM Corn Processing] Corn syrup; CAS 8029-43-4; EINECS 232-436-4; nutritive carbohydrate sweetener for candies, jams, jellies, fountain and table syrups; *Properties:* dens. 11.84 lb/gal (100 F); visc. 1000 poise (80 F); pH 4.8 (1:1); DE 42; Baume 43 (100 F); 80.7% total solids, 19.3% moisture.

Corn Syrup 42/44. [ADM Corn Processing] Corn syrup; CAS 8029-43-4; EINECS 232-436-4; nutritive carbohydrate sweetener for candies, jams, jellies, fountain and table syrups; *Properties:* dens. 11.95 lb/gal (100 F); visc. 4000 poise (80 F); pH 4.8 (1:1); DE 42; Baume 44 (100 F); 82.7% total solids, 17.3% moisture.

Corn Syrup 52/43. [ADM Corn Processing] Corn syrup; CAS 8029-43-4; EINECS 232-436-4; nutritive carbohydrate sweetener for marshmallows, gum drops, nougats; sl. sweeter than 42/43; *Properties:* dens. 11.84 lb/gal (100 F); visc. 550 poise (80 F); pH 4.8 (1:1); DE 52; Baume 43 (100 F); 81.2% total solids, 18.8% moisture.

Corn Syrup 62/43. [ADM Corn Processing] Corn syrup; CAS 8029-43-4; EINECS 232-436-4; nutritive carbohydrate sweetener for baked goods, nougats, fondants; sweeter than 42/43; *Properties:* dens. 11.84 lb/gal (100 F); visc. 400 poise (80 F); pH 4.8 (1:1); DE 62; Baume 43 (100 F); 82% total solids, 18% moisture.

Corn Syrup 62/44. [ADM Corn Processing] Corn syrup; CAS 8029-43-4; EINECS 232-436-4; nutritive carbohydrate sweetener for baked goods, nougats, fondants; *Properties:* dens. 11.95 lb/gal (100 F); visc. 1300 poise (80 F); pH 4.8 (1:1); DE 62; Baume 44 (100 F); 84% total solids, 16% moisture.

Corn Syrup 62/44-1. [ADM Corn Processing] Corn syrup; CAS 8029-43-4; EINECS 232-436-4; nutritive carbohydrate sweetener for baked goods, candies, canned fruits; *Properties:* dens. 11.95 lb/gal (100 F); visc. 1300 poise (80 F); pH 4.8 (1:1); DE 62; Baume 44 (100 F); 84% total solids, 16% moisture.

Corn Syrup 62/44-2. [ADM Corn Processing] Corn syrup; CAS 8029-43-4; EINECS 232-436-4; pH buffered nutritive carbohydrate sweetener for baked goods, candies, canned fruits; *Properties:* dens. 11.95 lb/gal (100 F); visc. 1300 poise (80 F); pH 5.2 (1:1); DE 62; Baume 44 (100 F); 84% total solids, 16% moisture.

Corn Syrup 97/71. [ADM Corn Processing] Corn syrup; CAS 8029-43-4; EINECS 232-436-4; nutritive carbohydrate sweetener with high sweetness and high fermentables; for fermentation, baked goods; *Properties:* dens. 11.07 lb/gal (100 F); visc. 0.36 poise (120 F); pH 4.5 (1:1); DE 97; 71% total solids, 29% moisture.

Cossacks. [Brolite Prods.] Rye bread mix for dark rye, deli-style bread; exc. for sandwiches.

Cottonseed Cooking Oil 82-060-0. [ADM Refined Oils] Cottonseed oil, refined, bleached, deodorized; CAS 8001-29-4; EINECS 232-280-7; veg. oil for frying; *Properties:* cloudy oil, pleasing flavor; iodine no. 100-115.

Cottonseed Oil. [Van Den Bergh Foods] Choice refined cottonseed oil; CAS 8001-29-4; EINECS 232-280-7; general purpose oil for cooking, baking, and frying applics.; *Properties:* Lovibond 4R max. liq.

Covi-Ox® T-70. [Henkel/Cospha; Henkel Canada] Tocopherol; natural antioxidant and blocking agent for food industry; *Properties:* brwnsh. red visc. liq..

Covipherol T-75. [Henkel/Cospha; Henkel Canada] Mixed d-tocopherols; antioxidant in food prods.; *Properties:* clear brnsh-red visc. oil.

Cozeen®. [Zumbro; Garuda Int'l.] Corn protein zein, special vegetable oils, glycerin; natural nutritious edible coating to extend shelf life of nutmeats, prevent stickiness of licorice pieces, coat nutritional supplements and pharmaceutical tablets; provides aesthetic and barrier props.; film-former, adhesive chars.; shellac alternate; *Properties:* sol. in alcohol; insol. in water; *Storage:* store @ R.T. (18-24 C); stir before use.

CP-2600. [Custom Ingreds.] Water, propionic acid, phosphoric acid, sorbic acid; corn tortilla preservative to extend shelf life of uncooked tortilla chips up to 21 days; inhibits growth of molds and yeast; *Properties:* clear apprance, sl. propionic odor; sol. in water; sp.gr. 1.085; *Storage:* store in clean, dry area @ 40-85 F for storage life of 9 mos.

CP-3650. [Custom Ingreds.] Water, propionic acid, phosphoric acid, dextrose; corn tortilla preservative which inhibits growth of molds and yeast; provides a corn tortilla with very clean flavor; *Properties:* clear liq., sl. propionic odor; sol. in water; *Storage:* store in clean, dry area @ 40-85 F for storage life of 9 mos.

CPG®. [Calgon Carbon] Activated carbon; CAS 64365-11-3; used for food decolorization; *Properties:* 12 x 40, 20 x 50 sieve sizes.

CPG® LF. [Calgon Carbon] Activated carbon; CAS 64365-11-3; used for food decolorization; *Properties:* 8 x 30, 12 x 40 sieve sizes.

Crab Extract #41. [Ariake USA] Crab extract; flavorings for snacks, soup mixes, sauces, gravies; 52% moisture, 6.25% protein.

Crackerase. [Mitsubishi Int'l.; Amano Enzyme USA] Blend of amylolytic and proteolytic enzymes derived from *B. subtilis;* enzyme improving bloom, even bake, and eliminating burnt ends to cookies and crackers; Regular for flour with 9-10.5% protein, Supcr for flour with > 11% protein content; *Usage level:* 15-45 g/450 kg of flour; *Regulatory:* FDA GRAS.

Creamtex. [Van Den Bergh Foods] Partially hydrog. veg. oil (soybean, cottonseed); EINECS 269-820-6; good quality bakery shortening for use where emulsifier is not required; *Regulatory:* kosher; *Properties:* Lovibond 2R max. plastic; m.p. 111-119 F.

Creamy Liquid Frying Shortening 102-050. [ADM Packaged Oils] Partially hydrog. soybean oil, TBHQ, dimethylpolysiloxane; packaged vegetable oil; *Properties:* opaque pourable liq., bland flavor; iodine no. 104 ± 3.

Cremelite®. [Bunge Foods] Highly conc. stabilizer giving exc. aeration to buttercremes, fillings, and toppings; *Usage level:* 1.5 lb/100 lb powd. sugar; *Properties:* wh. to off-wh. powd., vanilla flavor; 8% moisture.

Cremodan. [Grindsted Prods. Denmark] Emulsifier/ stabilizer blend; nonionic; emulsifier/stabilizer for ice cream and related prods.; *Properties:* powd.; 100% conc.

Cremol Plus™ 60 Vegetable. [Bunge Foods] Partially hydrog. soybean and cottonseed oils, mono- and diglycerides, polysorbate 60; shortening for creme fillers, cakes, icings; *Properties:* 1.5 red color, bland flavor; 2.5% alpha-mono.

Cremophor® NP 10. [BASF AG] Nonoxynol-10; CAS 9016-45-9; EINECS 248-294-1; nonionic; solubi-

lizer for essential oils and flavors; *Properties:* liq.; Discontinued

Cremophor® NP 14. [BASF AG] Nonoxynol-14; CAS 9016-45-9; nonionic; solubilizer for essential oils and flavors; *Properties:* liq.; Discontinued

Crestawhip. [Croda Food Prods. Ltd.] Emulsifier blend with propylene glycol, glycerin, sorbitol; emulsifier for improving sponges, Swiss roll, and cakes; also for fermented flour confectionery; *Usage level:* 1-2%; *Regulatory:* UK clearances; *Storage:* store under cool and dry conditions.

Crestawhip 2. [Croda Food Prods. Ltd.] Emulsifier blend with propylene glycol, sorbitol; emulsifier for improving sponges, Swiss roll, and cakes, esp. chocolage sponges and cakes; also for fermented flour confectionery; *Usage level:* 1-2%; *Regulatory:* UK clearances; *Storage:* store under cool and dry conditions.

Crestawhip Fl. [Croda Food Prods. Ltd.] Proprietary blend; food ingred. for icings and toppings.

Crestawhip H13. [Croda Food Prods. Ltd.] Emulsifier blend (fatty acid mono- and diglycerides, diacetylated tartaric acid esters of monoglycerides) with veg. oil and sodium propionate In aq. disp.; nonionic; emulsifier, improver for sponge cakes, bread, rolls, buns; provides finer, softer crumb structure, bigger volume, extended shelf life; *Usage level:* 1-2%; *Properties:* paste; 100% conc.

Crestawhip H13A. [Croda Food Prods. Ltd.] Emulsifier blend (fatty acid mono- and diglycerides, diacetylated tartaric acid esters of monoglcyerides) with veg. oil and acetic acid in aq. disp.; emulsifier for breads, rolls, buns, teacakes.

Crester KZ. [Croda Food Prods. Ltd.] Polyglycerol and edible fatty acid mixed esters; nonionic; food emulsifier; improves volume, texture, shelf life in cakes, flour confectionery; aids emulsion formation in coffee whiteners; gel stabilizer in sponges and Swiss rolls; improves whipping props. and foam stability in whipped desserts; *Usage level:* 2-4% basis shortening (cakes), 10-15% gel systems (sponges); *Properties:* solid; acid no. 3 max.; iodine no. 3 max.; sapon. no. 120-150; 100% conc.

Crester L. [Croda Food Prods. Ltd.] Polyglycerol and fatty acids mixed ester; nonionic; emulsifier for margarine; controlls crystallization; increases aerating and emulsifying props.; improves emulsion stablity of syn. cream; *Usage level:* 0.3-4%; *Properties:* paste; acid no. 5 max.; sapon. no. 130-140; 100% conc.

Crester MG. [Croda Food Prods. Ltd.] Mixed fatty acid monoglyceride and polyglyceryl ester blend; nonionic; margarine emulsifier; controlls crystallization; increases aerating and emulsifying props.; *Usage level:* 0.5-1.0%; *Properties:* buff colored solid block or pastilles; sapon. no. ≈ 160; hyd. no. 230-276; 100% conc.

Crester PR. [Croda Food Prods. Ltd.] Polyglycerol polyricinoleate; nonionic; visc. modifier in the food industry; emulsifier in bakery release spray; reduces yield value of chocolate or coating, enhancing flow chars.; *Usage level:* 0.2-0.5% (chocolate),

1-4% (pan grease); *Properties:* liq.; acid no. 6 max.; iodine no. 72-103; sapon. no. 170-180; 100% conc.; *Storage:* store under cool, dry conditions.

Crester RA. [Croda Food Prods. Ltd.] Glyceryl esters, polyglyceryl polyricinoleate, thermally oxidized soya bean oil, etc.; nonionic; emulsifier in bakery release spray; *Usage level:* 1-4%; *Properties:* liq.; acid no. 6 max.; iodine no. 100-120; sapon. no. 190-220; 100% conc.; *Storage:* store under cool, dry conditions.

Crester RB. [Croda Food Prods. Ltd.] Glyceryl esters, polyglyceryl polyricinoleate, thermally oxidized soya bean oil, etc.; nonionic; emulsifier in bakery release spray; *Usage level:* 1-4%; *Properties:* liq.; acid no. 6 max.; iodine no. 78-82; sapon. no. 158-175; 100% conc.; *Storage:* store under cool, dry conditions.

Crester RT. [Croda Food Prods. Ltd.] Polyglyceryl polyricinoleate; visc. modifier for chocolate, coatings, low fat spreads; pan release aid; *Properties:* pale amber liq., bland taste; acid no. 6 max.; iodine no. 72-103; sapon. no. 170-190.

Crill 1. [Croda Inc.; Croda Food Prods. Ltd.; Croda Surf. Ltd.] Sorbitan laurate; CAS 1338-39-2; nonionic; emulsifier, wetting agent for foods; antifoam for syrups, calf starters; *Usage level:* 0.1-0.2% (calf starters), 0.01-0.05% (syrups); *Regulatory:* UK clearance; *Properties:* pale yel. clear visc. liq.; sol. in ethanol, oleyl alcohol, min. oil; HLB 8.6; acid no. 4-7; sapon. no. 160-175; hyd. no. 330-358; 98% conc.; *Storage:* store under cool, dry conditions.

Crill 2. [Croda Inc.; Croda Surf. Ltd.] Sorbitan palmitate; CAS 26266-57-9; EINECS 247-568-8; nonionic; emulsifier, wetting agent for foods; antifoam for syrups, calf starters; *Properties:* pale tan hard waxy solid; partially sol. in propylene glycol, ethyl and oleyl alcohols, olive oil, oleic acid; m.p. 46 C; HLB 6.7; sapon. no. 140-150; 98% conc.

Crill 3. [Croda Inc.; Croda Food Prods. Ltd.; Croda Surf. Ltd.] Sorbitan stearate NF; CAS 1338-41-6; EINECS 215-664-9; nonionic; emulsifier for shortenings; controls crystallization and prevents fat bloom in chocolate and coatings; emulsion stabilizer in coffee whiteners, whipped desserts; *Usage level:* 0.5% (chocolate), 0.1-0.2% (coffee whiteners), 0.2-0.4% (shortenings, whipped desserts); *Regulatory:* UK clearance; *Properties:* pale tan hard waxy solid; low odor; partially sol. in oleyl alcohol, olive oil, oleic acid; insol. in water; m.p. 54 C; HLB 4.7; acid no. 5-10; sapon. no. 146-157; hyd. no. 235-260; 98% conc.; *Toxicology:* LD50 (oral, rat) > 31 g/kg; nonirritating to eyes; *Storage:* store under cool, dry conditions.

Crill 4. [Croda Inc.; Croda Food Prods. Ltd.; Croda Surf. Ltd.] Sorbitan oleate NF; CAS 1338-43-8; EINECS 215-665-4; nonionic; food process antifoam; emulsion stabilizer for calf starters, whipped desserts; aids wetting/defeathering in poultry; *Usage level:* 0.5-5%; 0.2-0.4% (calf starter, whipped dessert); 0.1-0.2% (cream), 175 ppm of scald water (poultry); *Properties:* amber visc. liq.; sol. in ethyl, isopropyl, and oleyl alcohols, min. oil, IPM, olive oil,

oleic acid; HLB 4.3; acid no. 5.5-7.5; sapon. no. 147-160; hyd. no. 193-209; 98% conc.; *Toxicology:* LD50 (oral, rat) > 40 g/kg; nonirritating to eyes.

Crill 35. [Croda Inc.; Croda Surf. Ltd.] Sorbitan tristearate; CAS 26658-19-5; EINECS 247-891-4; nonionic; emulsifier, lubricant for food applics.; *Properties:* pale tan hard waxy solid; partly sol. in oleyl alcohol, min. and olive oil, IPM, oleic acid; m.p. 48 C; HLB 2.1; sapon. no. 176-188; 98% conc.

Crill 41. [Croda Chem. Ltd.; Croda Food Prods. Ltd.] Sorbitan tristearate; CAS 26658-19-5; EINECS 247-891-4; food emulsifier; controls crystallization and prevents fat bloom in chocolate and confectioners coatings; *Usage level:* 1%; *Properties:* m.p. 53 C; HLB 2.1; acid no. 7 max.; sapon. no. 172-185; hyd. no. 60-80.

Crill 43. [Croda Inc.; Croda Surf. Ltd.] Sorbitan sesquioleate; CAS 8007-43-0; EINECS 232-360-1; nonionic; w/o emulsifier, wetting agent, pigment dispersant for food applics.; *Properties:* amber visc. liq.; sol. in oleyl alcohol, min. and olive oil, oleic acid; HLB 3.7; sapon. no. 149-160; 98% conc.

Crill 45. [Croda Inc.; Croda Surf. Ltd.] Sorbitan trioleate; CAS 26266-58-0; EINECS 247-569-3; nonionic; w/o emulsifier, wetting agent, pigment dispersant for food applics.; *Properties:* amber visc. liq.; sol. in oleyl alcohol, min. and olive oil, IPM, oleic acid; HLB 4.3; sapon. no. 172-186; 98% conc.

Crill 50. [Croda Surf. Ltd.] Sorbitan oleate, tech.; CAS 1338-43-8; EINECS 215-665-4; nonionic; emulsifier, wetting agent, dispersant for food applics.; *Properties:* liq.; 98% conc.

Crillet 1. [Croda Inc.; Croda Chem. Ltd.] Polysorbate 20; CAS 9005-64-5; nonionic; solubilizer, emulsifier, dispersant, wetting agent; often combined with a member of the Crill range in emulsification systems; for food applics.; *Properties:* clear, yel. clear liq.; low odor; sol. in water, ethyl and oleyl alcohol, oleic acid; HLB 16.7; sapon. no. 40-51; surf. tens. 38.5 dynes/cm (0.1%); 97% conc.

Crillet 2. [Croda Chem. Ltd.] PEG-20 sorbitan palmitate; CAS 9005-66-7; nonionic; emulsifier, solubilizer, wetting agent for food applics.; *Properties:* paste; HLB 15.6; 97% conc.

Crillet 3. [Croda Inc.; Croda Chem. Ltd.; Croda Food Prods. Ltd.] Polysorbate 60 NF; CAS 9005-67-8; nonionic; food emulsifier for icings; starch/protein complexing and aeration agent in cakes; emulsion stabilizer in calf starters, coffee whiteners, margarine; controlls crystallization; reduces palate cling in chocolate and coatings; prevents fat separation; *Usage level:* 0.5-5%; 0.2-0.4% (cake), 0.1-0.2% (calf starter), 0.5% (chocolate), 0.05-0.1% (coffee whitener); *Regulatory:* UK clearance; *Properties:* yel. liq. gels to soft solid on cooling; sol. in ethyl, isopropyl, and oleyl alcohol, oleic acid; partly sol. in water; HLB 14.9; acid no. 2 max.; sapon. no. 45-55; hyd. no. 81-96; pH 5-7 (5%); surf. tens. 42.5 dynes/cm (0.1%); 97% conc.; *Toxicology:* LD50 (oral, rat) > 38 g/kg; nonirritating to eyes; *Storage:* store under cool, dry conditions.

Crillet 4. [Croda Inc.; Croda Chem. Ltd.; Croda Food

Prods. Ltd.] Polysorbate 80 NF; CAS 9005-65-6; nonionic; food emulsifier; controls fat agglomeration in ice cream; solubilizer producing stable vitamin and essential oil emulsions; *Usage level:* 0.5-5%; 0.02-0.05% (ice cream), 5-15% of oil or vitamin; *Regulatory:* UK clearance; *Properties:* yel. amber clear liq.; faint char. odor; sol. in water, ethyl, isopropyl, and oleyl alcohols, oleic acid; HLB 15.0; acid no. 2 max.; sapon. no. 45-55; hyd. no. 65-80; pH 6-7 (5%); surf. tens. 42.5 dynes/cm (0.1%); 97% conc.; *Toxicology:* LD50 (oral, rat) > 38 g/kg; nonirritating to eyes; *Storage:* store under cool, dry conditions.

Crillet 11. [Croda Chem. Ltd.] PEG-4 sorbitan laurate; CAS 9005-64-5; nonionic; emulsifier, solubilizer, wetting agent for food applics.; *Properties:* liq.; 97% conc.

Crillet 31. [Croda Chem. Ltd.] Polysorbate 61; CAS 9005-67-8; nonionic; emulsifier, solubilizer, wetting agent for food applics.; *Properties:* solid; 97% conc.

Crillet 35. [Croda Chem. Ltd.] Polysorbate 65; CAS 9005-71-4; nonionic; emulsifier, solubilizer, wetting agent for food applics.; *Properties:* cream/buff waxy solid; sol. in ethyl and oleyl alcohols, oleic acid, trichlorethylene, partly sol. in water; HLB 10.5; sapon. no. 88-98; surf. tens. 42.5 dynes/cm (0.1%); 97% conc.

Crillet 41. [Croda Chem. Ltd.] PEG-5 sorbitan oleate; CAS 9005-65-6; nonionic; emulsifier, solubilizer, wetting agent for food applics.; *Properties:* liq.; HLB 10.0; 97% conc.

Crillet 45. [Croda Chem. Ltd.] Polysorbate 85; CAS 9005-70-3, nonionic, emulsifier, solubilizer, wetting agent for food applics.; *Properties:* clear amber visc. liq.; sol. in ethyl and oleyl alcohols, IPM, oleic acid, kerosene, trichlorethylene, butyl stearate; HLB 11.0; sapon. no. 82-95; surf. tens. 41 dynes/cm (0.1%); 97% conc.

Crodacreme. [Croda Food Prods. Ltd.] Blend of mono- and diglycerides of fatty acids, guar gum, sodium carboxymethyl cellulose, carrageenan, and glucose syrup solids; nonionic; ice cream emulsifier/stabilizer; *Regulatory:* UK clearance; *Properties:* solid; 100% conc.; *Toxicology:* ADI 150 mg/kg bodyweight; *Storage:* store under cool, dry conditions.

Crodaglaze. [Croda Food Prods. Ltd.] Shellac-based; enhances/protects gloss, prevents agglomeration, extends shelf life, and serves as a moisture barrier in chocolate and confectionery.

Crodamol GTC/C. [Croda Food Prods. Ltd.] Glyceryl tricaprylate/caprate; nonionic; bakery lubricant, release agent, glazing agent for sugar confectionery, flour confectionery; solv. carrier for flavors and fragrances; prevents excessive dehydration in dried fruit; hypoallergenic baby food formulations; *Usage level:* 0.25-1% (dried fruit, sugar confectionery); *Properties:* pale yel. liq.; acid no. 0.2 max.; iodine no. 1 max.; sapon. no. 325-345; 100% conc.; *Storage:* store under cool, dry conditions.

Crodascoop. [Croda Food Prods. Ltd.] Mono- and diglycerides of fatty acids, glycerin, guar gum, so-

dium carboxymethyl cellulose, and carrageenan; nonionic; emulsifier, stabilizer, freeze pt. depressant for soft ice cream; *Regulatory:* UK clearance; *Properties:* paste/gel; 100% conc.; *Storage:* store under cool, dry conditions.

Crodatem L. [Croda Food Prods. Ltd.] Diacetyl tartaric acid derivs. of fatty acid mono- and diglycerides; nonionic; antistaling agent, starch complexing agent in biscuits, cake mixes, gravy mix; toffee emulsifier; controls starch gelation in snack foods, pastas; improves dispersibility of coffee whiteners; emulsifies fat in meat prods.; *Properties:* amber visc. liq.; acid no. 110-130; iodine no. 70-80; sapon. no. 45-500; 100% conc.

Crodatem L50. [Croda Food Prods. Ltd.] Diacetyl tartaric acid derivs. of fatty acid mono- and diglycerides; nonionic; antistaling agent, starch complexing agent in biscuits, cake mixes, gravy mix; toffee emulsifier; controls starch gelation in snack foods, pastas; improves dispersibility of coffee whiteners; emulsifies fat in meat prods.; *Properties:* amber mobile liq.; acid no. 45-65; iodine no. 90-110; sapon. no. 315-346.

Crodatem T22. [Croda Food Prods. Ltd.] Diacetyl tartaric acid derivs. of fatty acid mono- and diglycerides with 10% tricalcium phosphate (anticaking agent); nonionic; antistaling agent, starch complexing agent in biscuits, cake mixes, gravy mix; toffee emulsifier; controls starch gelation in snack foods, pastas; improves dispersibility of coffee whiteners; emulsifies fat in meat prods.; *Properties:* off-wh. powd.; drop pt. 45 C; acid no. 110-130; iodine no. 2 max.; sapon. no. 450-500.

Crodatem T25. [Croda Food Prods. Ltd.] Diacetyl tartaric acid derivs. of fatty acid mono- and diglycerides with 10% tricalcium phosphate (anticaking agent); nonionic; antistaling agent, starch complexing agent in biscuits, cake mixes, gravy mix; toffee emulsifier; controls starch gelation in snack foods, pastas; improves dispersibility of coffee whiteners; emulsifies fat in meat prods.; *Properties:* off-wh. powd.; drop pt. 45 C; acid no. 110-130; iodine no. 2 max.; sapon. no. 460-510.

Crodroit CS. [Croda Colloids Ltd; O.C. Lugo] Partially hydrolyzed protein and mucopolysaccharides derived from bovine trachea incl. chondroitin sulfate; nutritive additive for dietary supplements, high-nutrition body-building formulations; useful in diets of patients with certain heart diseases; recognized in Japan as providing nutrition and energy to health food supplements; *Properties:* spray-dried powd.; pH 6 (1%); 4% moisture; 70% protein, 17.5% mucopolysaccharide.

Crodyne BY-19. [Croda Inc.] Pharmaceutical gelatin NF; CAS 9000-70-8; EINECS 232-554-6; humectant, thickener for food applics.; *Usage level:* 1-5%; *Properties:* buff cryst. powd., bland pleasant odor; water-sol.; m.w. 25,000; visc. 14-18 mps; pH 5.5-6.2 (10%); 85% act.; *Toxicology:* LD50 (oral, rat) > 5 g/kg; nonirritating to skin and eyes.

Crolactem. [Croda Food Prods. Ltd.] Glyceryl lactostearate; emulsion stabilizer in coffee whiteners; improves aerating and emulsifying props. in cake mixes, flour confectionery, shortening; stabilizes cake batter and extends shelf life; aids incorporation of fat into toffee mix and reduces stickiness; *Usage level:* 2-4% fat basis (cake), 2-4% batch wt. (shortening), 1-3% dry wt. (coffee whitener); *Properties:* m.p. 50 C; acid no. 3 max.; iodine no. 5 max.; sapon. no. 220-230; 20-30% monoglyceride, 1% max. moisture.

Crolactil CS2L. [Croda Surf. Ltd.] Calcium 2-stearoyl lactylate; CAS 5793-94-2; EINECS 227-335-7; anionic; ingredients in food prods. (bread, coffee whiteners, frozen dough); *Properties:* off-wh. powd.; HLB 5; sapon. no. 175-250.

Crolactil SISL. [Croda Surf. Ltd.] Sodium isostearoyl lactylate; CAS 66988-04-3; EINECS 266-533-8; anionic; ingredients in food prods; *Properties:* yel. liq.; HLB 6.5; sapon. no. 210-280.

Crolactil SS2L. [Croda Chem. Ltd.] Sodium stearoyl lactylate; CAS 25383-99-7; EINECS 246-929-7; ingredients in food prods. (bread, coffee whitener, frozen dough, snack foods); *Properties:* off-wh. powd.; HLB 6.5; sapon. no. 210-280.

Crystallization Inhibitor HL-13788. [Stepan Food Ingreds.] Hexaglyceryl ester of mixed fatty acids from coconut oil/tall oil, FCC; crystallization inhibitor; used in salad oils to increase cold stability; *Regulatory:* meets CFR specs; *Properties:* Gardner 8 solid; iodine no. 19; sapon. no. 189; 0.03% moisture.

Crystalsorb B-56. [Stepan Food Ingreds.] Hexaglycerol ester of mixed fatty acids (from coconut/tall/soy/palm/corn/canola oils)/sorbitan tristearate blend; emulsifier for food and beverage applics.; *Regulatory:* FDA 21CFR §172.854; avail. in kosher grade; *Properties:* cream-colored solid, mild fatty odor; iodine no. 11; sapon. no. 182; 0.5% moisture; *Storage:* avoid prolonged storage above 90 F. Custom prod.

CS-9 Icing Powd. [Bunge Foods] Boiling type icing stabilizer for transparent and pilable icings and glazes for sweet goods; wh. or chocolate; good stability; *Usage level:* 3-4 lb/100 lb powd. sugar; *Properties:* wh. free-flowing powd., bland aroma/flavor; pH 9.0; 6% max. moisture.

Culinox® 999 Chemical Grade Salt. [Morton Salt] Sodium chloride FCC; CAS 7647-14-5; EINECS 231-598-3; high purity food-grade salt; used for continuous churn butter, margarine, mayonnaise, salad dressings, cheddar, Swiss, or processed cheeses, peanut butter, baking, canning/pickling, confections, egg prods., meat/poultry; *Properties:* cubic crystals; 20-70 mesh range; bulk dens. 1.12-1.28 g/ml (loose); pH 6-8 (5% deionized water sol'n.); 99.95% min. NaCl.

Culinox® 999® Food Grade Salt. [Morton Salt] Sodium chloride FCC; CAS 7647-14-5; EINECS 231-598-3; high purity food-grade salt with consistent saltiness intensity and stringent stds. on visible insol. extraneous material; for comminuted meats, baking, canning/pickling, cheese, confections, egg. prods., meat/poultry, margarine, peanut but-

ter; *Properties:* cubic crystals; 44-64% retained on 50 mesh, 17-36% on 70 mesh; mean crystal size 340 μm; mean surf. area 84 cm²/g; bulk dens. 1.2-1.27 g/ml (loose); 99.97% NaCl.

Curacel. [Griffith Labs. Ltd.] Sodium acid pyrophosphate; CAS 7758-16-9; EINECS 231-835-0; functional food additive for sausages.

Curafos® 11-2. [Rhone-Poulenc Food Ingreds.] Sodium tripolyphosphate FCC, sodium polyphosphate glassy FCC; antioxidant providing better flavor and aroma; improves texture of final prod.; for meat, ham, pork, poultry, seafood; enhances/stabilizes cured meat color, inhibits color changes, off-flavors; increases retention of natural juices in meat prods.; *Regulatory:* kosher; *Properties:* wh. gran., odorless; 1% max. on 20 mesh, 30% max. thru 100 mesh; pH 9.1-9.8 (1%); *Storage:* store cool and dry.

Curafos® 22-4. [Rhone-Poulenc Food Ingreds.] Sodium tripolyphosphate FCC, sodium polyphosphate glassy FCC; food additive developed for use in pickle sol'ns. for cured beef prods., corned beef, pastrami, cured pork prods., bacon, pre-basted turkeys; provides more rapid cure color development, better natural juice retention, reduced shrinkage; *Regulatory:* kosher; *Properties:* wh. gran., odorless; 1% max. on 20 mesh, 30% max. thru 100 mesh; pH 8.8-9.2 (1%); *Storage:* store cool and dry.

Curafos® STP. [Rhone-Poulenc Food Ingreds.] Sodium tripolyphosphate FCC; CAS 7758-29-4; EINECS 231-694-5; food additive for meat processing, poultry, and seafood industries, processed foods, confections; *Regulatory:* kosher; *Properties:* wh. gran. or powd., odorless; 1% max. on 20 mesh, 20 max. thru 100 mesh (gran.), 1% max. on 40 mesh, 80% min. thru 100 mesh (powd.); m.w. 368; pH 9.5-10.3 (1%); 85% min. assay; *Storage:* store cool and dry.

Curavis® 150. [Rhone-Poulenc Food Ingreds.] Sodium acid pyrophosphate FCC; CAS 7758-16-9; EINECS 231-835-0; cure accelerator for rapid frank processing; *Regulatory:* kosher; *Properties:* wh. powd., odorless; 1% max. on 70 mesh, 95% min. thru 200 mesh; m.w. 222; pH 4-4.8 (1%); 95% min. assay; *Storage:* store cool and dry.

Curavis® 250. [Rhone-Poulenc Food Ingreds.] Sodium hexametaphosphate FCC, sodium tripolyphosphate FCC, sodium acid pyrophosphate FCC; food additive for use in cooked sausage items such as franks, bologna, vienna sausage, knockwurst; improves protein binding, stabilizes emulsion, produces good texture and mouthfeel, provides superior color development and stability; *Regulatory:* USDA, kosher approved; *Properties:* wh. gran., odorless; 5% max. on 20 mesh, 30% max. thru 100 mesh; pH 7.6-8.1 (1%); *Storage:* store cool and dry.

Curavis® 350. [Rhone-Poulenc Food Ingreds.] Sodium hexametaphosphate FCC, sodium tripolyphosphate FCC, sodium acid pyrophosphate FCC; food additive developed to allow faster cure color development in cooked sausage prods. for rapid processing techniques; for franks, bologna, Vienna sausage, knockwurst; *Regulatory:* USDA, kosher approved; *Properties:* wh. gran, odorless; 5% max. on 20 mesh, 30% max. thru 100 mesh; pH 7.0-7.5 (1%); *Storage:* store cool and dry.

Curcumex 1600, 1601. [MLG Enterprises Ltd.] Turmeric extract; CAS 84775-52-0; food color for flour-based foods (noodles, pasta, crackers), confectionery, candies, desserts, milk-based foods, ice cream, dried/candied fruits; lemon-yel. to intense yel. shade; substitute for tartrazine; *Properties:* liq.; water-disp.; 2.2 ± 0.2% and 1.1 ± 0.1% resp. (as curcumine); *Storage:* protect from air, light, and heat; rotate stock every 8 mos.

Curcumex Natural Powd. Colorant. [MLG Enterprises Ltd.] Turmeric extract; CAS 84775-52-0; natural colorant producing intensely lemon-yel. tartrazine-like shades in aq. sol'n.; for instant or dehydrated foods, beverages, fantasy desserts, custards, ice cream, sauces, soups, gelatins; *Usage level:* 0.06-0.2%; 0.3-1% (flour carriers, e.g.,

noodles, pasta, cake mixes); *Properties:* impalpable powd.; 100% 80 mesh; sol. in water; 0.3 ± 0.03% color (curcumin); *Storage:* keep in tightly closed containers, in fresh, dry rooms.

Curcumin Extract P8002. [Phytone Ltd.] Deodorized extract of turmeric (*Curcuma longa*) with permitted emulsifier; CAS 84775-52-0; food colorant; *Regulatory:* EEC E100; JECFA compliance; *Properties:* deep orange-yel. visc. liq.; misc. with water; *Storage:* store in tightly closed containers in cool, dry area; protect from exposure to direct sunlight.

Cyncal®. [Hilton Davis] Myristalkonium chloride; CAS 139-08-2; EINECS 205-352-0; cationic; antimicrobial, antistat, disinfectant, sanitizer for use in food, beverage processing; *Properties:* lt. yel. liq.; mild, pleasant odor; m.w. 359; sol. in water, lower alcohols, ketones, glycols; dens. 7.8 lb/gal; sp.gr. 0.94; 80% act.; Discontinued

D

Dairy-Lo™. [Pfizer Food Science] Whey protein conc.; proteinaceous fat replacer in frozen or cultured dairy prods. (ice cream, low-fat frozen desserts, yogurt, sour cream, processed cheese); stabilizes and controls water; provides enhanced creamy mouthfeel; final prods. don't shrink or become icy; *Regulatory:* FDA 21CFR §184.1979c, GRAS; *Properties:* cream wh. free-flowing powd., bland clean fresh odor; sol. in water; pH 6.0-6.5 (10%); 35% protein, 52% lactose, 3% butterfat; *Toxicology:* nontoxic; chronic ingestion may cause allergenic reactions; mild eye irritant; *Storage:* store in dry, well ventilated area below 80 F.

Dairytrim. [Rhone-Poulenc Food Ingreds.] Dairy ingred.; *Properties:* lt. creamy wh.; 1% max. on 40 mesh, 10% max. thru 200 mesh; bulk dens. ≥ 0.25 g/cc (tapped); visc. ≥ 6 cps (5%); pH 6.3-7.7; ≤ 8% moisture.

Danishine™. [MLG Enterprises Ltd.] Dextrose, vegetable colloids, guar gum, and buffer salts; danish pastry wash in powd. form; applied by brush or spraying; gloss stays bright and dry when applied to pastry hot from the oven; *Regulatory:* kosher.

Danish Snax® Conc. [Bunge Foods] Patented conc. formulated for prod. of fryable danish pastries; produces desserts which are crisp on the outside, fine, rich, and flaky on the inside; *Usage level:* 20% based on flour.

Danish Snax® Icing Powd. [Bunge Foods] Boiling type icing stabilizer; *Usage level:* 5 lb/100 lb powd. sugar; *Properties:* wh. powd.; pH 7.5.

Danpro DS. [Central Soya] Functional soya protein conc.; CAS 68153-28-6; high-quality nutritive source, binder, and emulsifier for use in cooked/raw meat patties, pizza topping, meatballs, coarse ground sausages, meat rolls, seafood, soup bases, gravies, protein drinks, dairy emulsions, breads/baking, cereals, baby food.

Danpro HV. [Central Soya] Functional soya protein conc.; CAS 68153-28-6; high-quality nutritive source, binder, and emulsifier for use in coarse ground sausages, paté, meat rolls, seafood, surimi, dairy emulsions, breads/baking, cereals, baby food.

Danpro S. [Central Soya] Functional soya protein conc.; CAS 68153-28-6; high-quality nutritive source, binder, and emulsifier for use in coarse ground sausages, saté, meat rolls, seafood,

breads/baking, cereals, and baby foods.

Danpro S-760. [Central Soya] Functional soya protein conc.; CAS 68153-28-6; high-quality nutritive source, binder, and emulsifier for use in cooked/raw meat patties, pizza topping, meatballs, paté, coarse ground sausages, meat rolls, seafood, dairy emulsions.

Dansk® Conc. [Bunge Foods] Sweet dough conc. formulated for super rich sweet goods and short, tender, flaky danish pastry of the highest quality; *Usage level:* 42% based on flour.

Dariloid®. [Kelco] Sodium alginate blend; dairy stabilizers and stabilizer/emulsifiers; thickener for food prep.; for chocolate milk, soft-serve frozen desserts, ice cream, ice milk, custard, variegated syrup, fudge toppings, puddings, pie filings, cheese dips, eggnog, whipping cream; *Properties:* lt. ivory gran.; 42 mesh; milk-sol. at pasteurization temps.; pH 10 (1%).

Dariloid® 100. [Kelco] Xanthan gum, guar gum, and locust bean gum; dairy stabilizer, thickener, emulsifier used for sherbert, water ice, cottage cheese dressings, and sour cream; provides good visc. control, ease of aeration, firm body, creamy texture, full flavor release; *Usage level:* 0.1-0.22% (frozen desserts); *Regulatory:* FDA 21CFR §172.695, 184.1339, 184.1343, GRAS; EEC E-415, E-412, E-410; kosher; *Properties:* cream-colored dry powd.; 40 mesh; visc. 2500 cps (1% aq.); pH 6.5 (1%).

Dariloid® 300. [Kelco] Guar gum, xanthan gum, and carrageenan; dairy stabilizer and thickener used for cottage cheese dressings, premium ice cream, lowfat frozen desserts, frozen yogurt, nonfat frozen dessert, milk shake mix, egg nog; binds and thickens water; provides exc. flavor release; *Usage level:* 0.2-0.3% (cottage cheese dressing), 0.12-0.16% (ice cream), 0.21-0.24% (frozen yogurt), 0.15-0.17% (milk shake mix), 0.18-0.30% (egg nog); 0.8% max. (cheese), 0.6% max. (milk prods.), 1% (dairy prod. analogs), 0.5% (other foods); *Regulatory:* FDA 21CFR §184.1339, 172.695, 182.7255, GRAS; EEC E412, E415, E407; *Properties:* 80 mesh; visc. 1000 cps (1%); pH 7.0.

Dariloid® 400. [Kelco] Xanthan gum; CAS 11138-66-2; EINECS 234-394-2; dairy stabilizer used for milk shake mix, soft-serve ice cream and ice milk, flavored milk drinks, whipping cream, egg nog; designed for UHT applics.; *Properties:* 42 mesh; visc.

2100 cps (1%); pH 6.7.

Dariloid® Q. [Kelco] Buffered standardized sodium alginate; CAS 9005-38-3; gellant, thickener for instant powd. mixes, sour cream-based chip dips, cheese dips, cheese sauce, pie filling, puddings, custards, eggnog, milk shake powd., chiffons, toppings, syrups, baked cream fillings, whipping cream; *Regulatory:* FDA 21CFR §184.7724, GRAS; kosher; *Properties:* lt. ivory fine gran. powd.; 100 mesh; milk-sol.; pH 10.2 (1%).

Dariloid® QH. [Kelco] Buffered standardized sodium alginate; CAS 9005-38-3; gelling agent, emulsifier and stabilizer for instant powd. mixes, puddings, cheesecake mix, whipped toppings, bakery fillings, instant egg nog, milk shake, and dessert dry mixes; *Regulatory:* FDA 21CFR §184.7724, GRAS; kosher; *Properties:* lt. ivory fine gran. powd.; 100 mesh; milk-sol.; pH 10.2 (1%).

Dariloid® XL. [Kelco Int'l.] Sodium alginate blend; gellant, suspending agent, thickener, stabilizer for milk-based systems (puddings, pie fillings, custards, cheese dips, cheesecake, eggnog, milk shake powd., chiffons, toppings, syrups, baked cream fillings, whipping cream); *Properties:* 42 mesh; milk-sol.; pH 10 (1%).

Dark Rye Sour. [Brolite Prods.] Sour for darkest of rye breads.

Dascolor Annatoo Enc. [Dasco Sales] Annatto; CAS 1393-63-1; EINECS 215-735-4; used as a colorant for margarine, sausage casings.

Datagel. [Croda Chem. Ltd.] Monoglyceride activated diacetyl tartaric acid ester; nonionic; food emulsifier; *Proportioa:* paoto, gel; 100% conc., Unverified

Datamuls® 42. [Goldschmidt AG] Fatty acid glyceride diacetyl tartaric acid ester; nonionic; emulsifier for yeast-raised bakery goods; *Properties:* pumpable; 100% conc.; Unverified

Datamuls® 43. [Goldschmidt AG] Diacetyl tartaric acid monoglyceride; nonionic; food emulsifier for yeast-raised bakery goods, esp. fatty baking agents; *Usage level:* 0.3-0.6% on flour (baked goods); *Regulatory:* EEC E472e compliance; *Properties:* ivory liq., vinegar-like odor; disp. in cold water, sol. in warm water; sp.gr. 0.900; acid no. 80-110; iodine no. 47-60; sapon. no. 420-470; flash pt. 106 C; pH 3.0 (50 g/l aq.); 100% conc.; *Toxicology:* nonhazardous; *Storage:* store cool and dry, below 20 C.

Datamuls® 4720. [Goldschmidt AG] Diacetyl tartaric acid monoglyceride; nonionic; food emulsifier for prod. of baking agents for yeast-raised doughs; *Usage level:* 0.3-0.6% on flour (baked goods); *Regulatory:* EEC E472e compliance; *Properties:* ivory powd., vinegar-like odor; disp. in cold water, sol. in warm water; sp.gr. 0.900; m.p. 54-60 C; acid no. 67-87; iodine no. 2 max.; sapon. no. 380-420; flash pt. > 100 C; pH 3.0 (50 g/l aq.); 80% conc.; *Toxicology:* nonhazardous; avoid formation of dusts; *Storage:* store cool and dry below 20 C.

Datamuls® 4820. [Goldschmidt AG] Fatty acid glyceride diacetyl tartaric acid ester; nonionic; food emulsifier for yeast-raised doughs and baking aids; for deep frozen doughs for rolls, brioche, hamburger rolls with optimal volume and very fine porosity; fermentation stability ≈ 15-25 min, longer proof-time; *Usage level:* 0.3-0.6% on flour (baked goods); *Regulatory:* EEC E472e compliance; *Properties:* ivory powd., vinegar-like odor; disp. in cold water; sol. in warm water; sp.gr. 0.900; m.p. 54-60 C; acid no. 82-116; iodine no. 2 max.; sapon. no. 460-510; flash pt. > 100 C; pH 3.0 (50 g/l aq.); 80% conc.; *Toxicology:* nonhazardous; avoid formation of dusts.

Datamuls® 4820 U. [Goldschmidt AG] Diacetyl tartaric acid ester fatty acid glyceride; nonionic; food emulsifier for yeast-raised doughs, esp. for rolls, baguette or white bread with high volume and optimal porosity; fermentation stability 10-15 min, longer proof-time; *Usage level:* 0.3-0.6% on flour (baked goods); *Regulatory:* EEC E472e compliance; *Properties:* ivory powd., vinegar-like odor; disp. in cold water; sol. in warm water; m.p. 54-60 C; acid no. 75-109; iodine no. 2 max.; sapon. no. 460-510; 80% conc.

Datem esters. [Croda Chem. Ltd.] Monoglyceride diacetyl tartaric acid ester; nonionic; emulsifier for use in yeast-raised bakery goods; *Properties:* solid; 100% conc.; Unverified

DCME. [Premier Malt Prods.] Extract of 65% corn and 35% malted barley; syrup with mild malt flavor and pleasantly sweet char.; used in ice cream prods., bakery goods, vinegar, snacks, cereal, cigarettes, pretzel, bread sticks; 65% the sweetness of sucrose; *Usage level:* 2-3% (novelty ice cream), 1-2.5% flour wt. (bread), 1-3% dough wt. (bagels), 2-4% dough mix (snacks), 6-14% (breakfast cereal), 2-4% (pretzels), 2-3% flour wt. (bread sticks); *Properties:* Lovibond 80-160 syrup; pH 5.1-5.8 (10%); 78.5-80.0% solids; 1.8-2.6% protein; 60-70% reducing sugars.

Deamizyme. [Mitsubishi Int'l.; Amano Enzyme USA] Deaminase from *Aspergillus*; food enzyme; *Properties:* powd.

Degressal® SD 22. [BASF AG] Fatty alcohol alkoxylate; foam suppressor for sugar and chemical industries.

Dehysan Z 4904. [Henkel/Organic Prods.] Mixt. of fatty alcohol polyglycol ethers and soaps; defoamer for sugar prod. (cane and beet); *Properties:* liq.; Unverified

Dehysan Z 7225. [Henkel/Organic Prods.] Mixt. of natural fats with hydrophilic fat derivs.; defoamer for beet sugar prod.; *Properties:* liq.; Unverified

Delios® C. [Grünau GmbH] Capric acid triglyceride; solv. for flavors and oil-sol. food additives; component for glazing and release agents; surf. treating agent for dried fruits, confectionery; fat malabsorption for dietary foodstuffs; fat for icings, fat powds.; ingred. for sports food; *Properties:* wh. cryst. mass, neutral odor, sl. to coconut taste; m.p. 30 C; acid no. 0.1 max.; iodine no. 0.5 max.; sapon. no. 300; hyd. no. 5 max.; *Storage:* store under cool, dry conditions in tightly closed containers below 25 C; 18 mos storage life in original pkg.

Delios® S. [Grünau GmbH] Caprylic/capric acid triglyceride; EINECS 265-724-3; solv. for flavors and oil-sol. food additives; very good oxidation stability; for dietetic prods., sport foods; glazing and release agent for dried fruit and confectionery; lubricant, release agent for baked goods; antidusting agent; visc. reducer; *Properties:* colorless to sl. ylsh. liq., neutral odor, sl. coconut taste; misc. with ethanol; visc. 27 mPa•s; cloud pt. 0 C max.; acid no. 0.1 max.; iodine no. 0.5 max.; sapon. no. 335-350; hyd. no. 5 max.; 70% caprylic acid, 30% capric acid; *Storage:* store in tightly closed containers @ 0-30 C; avoid contact with water; 18 mos storage life in original pkg.

Delios® V. [Grünau GmbH] Caprylic/capric acid triglyceride; EINECS 265-724-3; solv. for flavors and oil-sol. food additives; exc. oxidation stability; for dietetic prods., sport foods; glazing and release agent for dried fruit and confectionery; lubricant for cutting machines; antidusting agent; visc. reducer; *Properties:* colorless to sl. yel. liq., neutral odor, sl. to coconut taste; misc. with ethanol; visc. 30 mPa•s; cloud pt. 0 C max.; acid no. 0.1 max.; iodine no. 0.5 max.; sapon. no. 330-345; hyd. no. 5 max.; 60% caprylic acid, 40% capric acid; *Storage:* store in tightly closed containers @ 0-30 C; avoid contact with water; 18 mos storage life in original pkg.

Deli Rye Conc. [Bunge Foods] Special sour producing classical Eastern European type rye bread; just add flour, water, and yeast; *Usage level:* 10% based on flour.

Delta™ SD 7393. [A.E. Staley Mfg.] Food starch modified, derived from dent and waxy corn; cook-up starch for spoonable salad dressings; provides low pH stability, high shear tolerance, good stability to cold and contributes to body, texture, and flavor; *Regulatory:* FDA 21CFR §172.892; *Properties:* wh. powd.; pH 5.5; 11% moisture.

De-lux Y.R.D. Conc. [Bunge Foods] Standardized yeast raised donut conc. yielding donuts with light tender texture and very low fat absorp.; *Usage level:* 27.5% based on flour.

Deluxe Fudge. [Bunge Foods] Gives icings a rich, dark brn. appearance with good keeping and eating qualities; for chocolate fudge and creme icings, brownie batters; *Usage level:* 25-40 lb/100 lb. powd. sugar; *Properties:* dk. brn. semi-plastic base, cocoa aroma/flavor.

Delvocid. [Int'l. Bio-Synthetics] Natamycin; CAS 7681-93-8; EINECS 231-683-5; food grade mold and yeast inhibitor; *Properties:* powd.; Unverified

De-Mol® Molasses. [ADM Ogilvie] 70% Molasses; CAS 68476-78-8; sweetener for baked goods, desserts, gingerbread, mixes; *Properties:* dk. brn. powd./flake; 3.5% max. moisture.

Descote® Ascorbic Acid 60%. [Particle Dynamics] Encapsulated ascorbic acid USP; each gram of coated prod. contains 600 mg ascorbic acid in edible matrix; CAS 50-81-7; EINECS 200-066-2; encapsulated ingreds. for food prods., nutritional supplements, pharmaceuticals; provides taste/odor masking, prevents interaction of actives; for chewable tablets, nutritional powd. mixes, weight loss supplements, health bars; *Properties:* wh. to off-wh. relatively free-flowing material with some soft agglomerates, sl. char. odor, satisfactory taste; 95% min. thru 20 mesh, 20% max. thru 100 mesh; 58.8% min. assay; *Storage:* physically/chemically stable when stored in cool, dry area, preferably @ 59-86 F.

Descote® Citric Acid 50%. [Particle Dynamics] Encapsulated citric acid USP; each gram of coated prod. contains 500 mg of citric acid in an edible matrix; CAS 77-92-9; EINECS 201-069-1; designed for incorporation into tablets and other dry dosage forms where it masks the char. citric acid taste and protects actives with inert coating; *Properties:* wh. free-flowing material with some soft agglomerates, sl. char. odor; 98% min. thru 20 mesh, 30% max. thru 100 mesh; 49% min. assay; *Storage:* store in cool, dry area, preferably @ 59-86 F.

Descote® Sodium Ascorbate 50%. [Particle Dynamics] Encapsulated sodium ascorbate USP; each gram of coated prod. contains 500 mg sodium ascorbate in edible kosher matrix; CAS 134-03-2; EINECS 205-126-1; encapsulated ingreds. for food prods., nutritional supplements, pharmaceuticals; provides taste/odor masking, prevents interaction of actives; for chewable tablets, nutritional powd. mixes, weight loss supplements, health bars; *Properties:* lt. yel. to yel./tan free-flowing powd. with some soft agglomerates, sl. char. odor; 95% min. thru 20 mesh, 20% max. thru 200 mesh; 48.5% min. assay; *Storage:* 15 mos physical/chemical stability stored in cool, dry area, preferably @ 59-86 F.

Desiccated Beef Liver Granular Undefatted. [Am. Labs] Dried undefatted granular form processed from beef livers; nutritive food and pharmaceutical additive; *Properties:* gran.

Desiccated Beef Liver Granular Defatted. [Am. Labs] Dried defatted granular form processed from beef livers; nutritive food and pharmaceutical additive; *Properties:* gran.

Desiccated Beef Liver Powd. [Am. Labs] Dried, undefatted powd. processed from beef livers; nutritive food and pharmaceutical additive; *Properties:* powd.

Desiccated Beef Liver Powd. Defatted. [Am. Labs] Dried, defatted powd. processed from beef livers; nutritive food and pharmaceutical additive; *Properties:* powd.

Desiccated Hog Bile. [Am. Labs] Vacuum-dried prod. from fresh hog bile contg. hyodeoxycholic acid, sodium glycohyodeoxycholate, sodium taurohydeoxycholate; nutritive food and pharmaceutical additive; *Storage:* preserve in tight containers with moisture-proof liners, in cool, dry place.

Desiccated Ox Bile. [Am. Labs] Dried ox bile; nutritive food and pharmaceutical additive; *Properties:* powd.; pH 6.5-7.5; *Storage:* preserve in tight containers with moisture-proof liners, in cool, dry place.

Desiccated Pork Liver Powd. [Am. Labs] Dried, undefatted powd. processed from pork livers;

nutrittive food and pharmaceutical additive; *Properties:* powd.

Design C200. [Van Den Bergh Foods] Partially hydrog. canola oil; high stability oil with low solids and enhanced nutritional profile; provides natural stability to finished prod.; *Regulatory:* kosher; *Properties:* Lovibond 1.0R max. liq.; drop pt. 15 C max.

Design NH. [Van Den Bergh Foods] Canola oil; high stability oil low in saturated and high in mono-unsaturated fatty acids; provides for ingred. statement void of hydrog.; *Regulatory:* kosher; *Properties:* Lovibond 1.0R max. liq.; drop pt. 10 C max.

Destab™. [Particle Dynamics] Avail. in calcium carbonate, magnesium carbonate, magnesium oxide, calcium sulfates, ferrous fumarate, or oyster shell; direct compression ingreds. for tableting of pharmaceutical and nutritional supplements; provides source of minerals, functions as antacid, buffer in buffered aspirin, or filler/binder excipient; *Properties:* free-flowing gran.

Detergent CR. [Arol Chem. Prods.] Fatty ethanolamine condensate; detergent, wetting agent, emulsifier, thickener for food applics.; *Properties:* pale amber liq.; mild odor; readily sol. in water; sp.gr. 1.01; 100% act.

Dewaxed Sunflower Oil 83-070-0. [ADM Refined Oils] Sunflower oil, high linoleic, polyunsaturated; CAS 8001-21-6; EINECS 232-273-9; veg. oil for frying, cooking, baking, salad dressings; *Properties:* clear brilliant oil, pleasing flavor; iodine no. 130-140.

Dextranase L Amano. [Amano Enzyme USA; Unipex] Dextranase from *Chaetomium;* CAS 9025-70-1, EINECS 232-803-9; enzyme for sugar prod.; *Proportios:* liq.

Dextranase Novo 25 L. [Novo Nordisk] Fungal dextranase derived from *Penicillium lilacinum;* CAS 9025-70-1; EINECS 232-803-9; enzyme used in sugar industry to break down dextran in raw sugar juice; *Properties:* brn. liq.; slight smell typ. of fermentation prods.; water-sol.; dens. 1.25 g/ml; pH 5-7; Unverified

Diagum LBG. [Multi-Kem] Locust bean gum; CAS 9000-40-2; EINECS 232-541-5; thickener, suspending agent, water binder for food use; low cold sol'n. visc., post-cooked sol'n. visc. in 3000 cps range; min. bacteria count that meets or exceeds all food standards.

Diahope®-S60. [Mitsubishi Kasei] Activated carbon; CAS 64365-11-3; for sugar refining, water works, food additives purification.

Diaion® HP 10. [Mitsubishi Kasei] Syn. adsorbent for vitamins, enzymes, fatty acids, other bioactive substances; decolorization applics..

Diaion® WK10. [Mitsubishi Kasei] Weakly acidic cation exchange resin (methacrylic type) for refining of cane sugar.

Diamond D-21. [Van Den Bergh Foods] Partially hydrog. veg. oil (soybean, cottonseed), mono- and diglycerides, glycerol-lacto ester of fatty acids, lecithin; shortening for cakes; *Regulatory:* kosher; *Properties:* Lovibond 5.0R max. plastic; 4.7-5.3%

mono.

Diamond D 31. [Van Den Bergh Foods] Partially hydrog. soybean oil; EINECS 232-410-2; all-purpose deep fat frying filler fat for cookies, whipped toppings, mellorine; *Regulatory:* kosher; *Properties:* Lovibond 1.5R max. plastic; m.p. 102-106 F.

Diamond D 40. [Van Den Bergh Foods] Partially hydrog. veg. oil (soybean, cottonseed); EINECS 269-820-6; multipurpose, good quality, nonemulsified shortening with narrower plastic range; used for biscuits and crackers; *Regulatory:* kosher; *Properties:* Lovibond 1.5R max. plastic; m.p. 112-117 F.

Diamond D 42. [Van Den Bergh Foods] Partially hydrog. veg. oil (soybean, cottonseed); EINECS 269-820-6; multipurpose shortening for doughnut frying; *Regulatory:* kosher; *Properties:* Lovibond 2.0R max. plastic; m.p. 109-119 F.

Diamond D 75. [Van Den Bergh Foods] Partially hydrog. cottonseed oil; CAS 68334-00-9; EINECS 269-804-9; frying oil for snack foods; *Regulatory:* kosher; *Properties:* Lovibond 4.0R max. plastic; m.p. 100-104 F.

Diazyme® L-200. [Solvay Enzymes] Fungal glucoamylase FCC; CAS 9032-08-0; EINECS 232-877-2; food-grade enzyme for hydrolysis of starch dextrins to glucose; for glucose prod., alcohol prod., brewing and fermentation applics., vinegar prod., prod. of yeast for industrial and food purposes, in fruit juice processing to prevent starch haze; *Properties:* amber to lt. brn. nonviscous liq.; completely misc. with water; dens. 1.00-1.25 g/ml; *Storage:* activity loss ≤ 10% in 3 mo stored at R.T.; 5 C storage extends life.

Diet Imperial® Spread. [Van Den Bergh Foods] Water, liq. soybean oil, partially hydrog. soybean oil, salt, veg. mono- and diglycerides, soy lecithin, potassium sorbate, citric acid, and calcium disodium EDTA (preservatives), artificial flavor, beta carotene (colorant), vitamin A palmitate; diet margarine; *Properties:* lt. yel. plastic, butter-like flavor; 39-40% fat, 57.3-58.3% moisture, 2.4-2.6% salt.

Dimodan BP-T Kosher. [Grindsted Prods.] Palm oil monoglyceride with antioxidants (250 ppm max. tocopherol, 200 ppm max. citric acid, 200 ppm max. ascorbic acid); food emulsifier; *Regulatory:* EEC, FDA §184.1505 (GRAS), kosher; *Properties:* block; drop pt. 60 C; iodine no. 45; 90% min. monoester; *Storage:* 12 mo storage under cool, dry conditions (5 C preferred, below 20 C recommended).

Dimodan CP. [Grindsted Prods.; Grindsted Prods. Denmark] Cottonseed oil dist. monoglyceride, unsat.; CAS 8029-44-5; EINECS 232-438-5; nonionic; food emulsifier; *Regulatory:* EEC, FDA §184.1505 (GRAS); avail. as kosher grade; *Properties:* plastic; m.p. 52 C; iodine no. 80; 90% min. monoester.

Dimodan CP Kosher. [Grindsted Prods.] Distilled monoglyceride from refined cottonseed oil; CAS 8029-44-5; EINECS 232-438-5; emulsifier for diet table margarine, low-calorie spreads, icing shortenings, cake shortenings; *Usage level:* 0.1-0.3% (diet margarine), 0.3-0.6% (low-calorie spreads),

0.3-1% (icing shortening), 3-6% (cake shortenings); *Regulatory:* FDA 21CFR §184.1505, GRAS; FCC; *Properties:* plastic; m.p. 50 C; iodine no. 90; 90% min. monoester; *Storage:* store in cool, dry area, preferably refrigerated.

Dimodan LS Kosher. [Grindsted Prods.; Grindsted Prods. Denmark] Glyceryl linoleate with 200 ppm max. BHT, 200 ppm max. citric acid as antioxidants; CAS 2277-28-3; EINECS 218-901-4; nonionic; w/o food emulsifier for low-calorie spreads, icing shortenings, and cake shortenings; *Usage level:* 0.3-0.6% (low-calorie spreads), 0.3-1.0% (icing shortening), 3-6% (cake shortening); *Regulatory:* EEC, FDA §184.1505 (GRAS); *Properties:* soft plastic; m.p. 50 C; iodine no. 110; 90% min. monoester; *Storage:* store under cool, dry conditions; formerly Grindtek MOL 90.

Dimodan O Kosher. [Grindsted Prods.; Grindsted Prods. Denmark] Partially hydrogenated soybean oil glyceride with antioxidants (200 ppm max BHT, 200 ppm max. citric acid); nonionic; food emulsifier for margarine, icing shortenings, coffee whiteners; *Usage level:* 0.1-0.3% (margarine), 2-6% (cake shortening), 0.3-1% (icing shortening), 0.1-2% (coffee whiteners); *Regulatory:* EEC, FDA 21CFR §184.1505, GRAS, FCC, kosher; *Properties:* block; m.p. 57 C; iodine no. 60; 90% min. monoester; *Storage:* store under cool, dry conditions.

Dimodan P. [Grindsted Prods. Denmark] Lard dist. monoglyceride, sat.; CAS 61789-10-4; EINECS 263-032-6; nonionic; food emulsifier; *Properties:* m.p. 70 C; 90% min. monoester; Unverified

Dimodan PM. [Grindsted Prods.; Grindsted Prods. Denmark] Dist. monoglyceride from edible refined hydrog. lard or tallow; nonionic; emulsifier, starch complexing agent; for margarine, cake shortenings, confectionery coatings; softener for bread; peanut butter stabilizer; amylose complexing agent for dehydrated potatoes; *Usage level:* 0.1-0.2% (margarine), 2-6% (cake shortening), 0.5-1.5% (starchy foods), 0.1-0.5% (plastics); *Regulatory:* EEC E471, FDA 21CFR §184.1505, 184.4505, GRAS; *Properties:* beads; m.p. 70 C; iodine no. 2 max.; 90% min. monoester; *Storage:* store in cool, dry area.

Dimodan PM 300. [Grindsted Prods. Denmark] Glyceryl stearate; nonionic; food emulsiifer; *Properties:* fine powd.; m.p. 70 C; 90% min. monoester.

Dimodan PV. [Grindsted Prods.; Grindsted Prods. Denmark] Hydrog. soybean oil dist. monoglyceride, unsat.; nonionic; food emulsifier, starch complexing agent, antisticking agent; crumb softener for bread; aerating agent in cake mixes and frozen desserts; *Usage level:* 0.1-0.5% (plastics); *Regulatory:* EEC, FDA §184.1505 (GRAS), 184.4505 (GRAS); *Properties:* beads, powd.; m.p. 72 C; iodine no. 2 max.; 90% min. monoester.

Dimodan PV 300 Kosher. [Grindsted Prods.; Grindsted Prods. Denmark] Hydrogenated soybean oil dist. monoglyceride, unsat.; nonionic; food emulsifier, starch complexing agent, antisticking agent; crumb softener for bread; aerating agent in cake mixes and frozen desserts; amylose complexing agent for extruded snack foods, pasta, and dehydrated potato prods.; *Usage level:* 0.3-1% on flour (bread), 0.1-0.3% (pasta), 0.1-0.5% (dehydrated potatoes); *Regulatory:* EEC, FDA § 184.1505 (GRAS), kosher; *Properties:* fine powd.; m.p. 72 C; iodine no. 5 max.; 90% min. monoester; *Storage:* store in cool, dry area.

Dimodan PV Kosher. [Grindsted Prods.] Hydrog. veg. oil monoglyceride; starch complexer; emulsifier for margarine, cake shortening, confectionery coatings; crumb softener for bread; amylose complexing agent for starch and dehydrated potato prods.; aerating agent for cake mixes, frozen desserts; peanut butter stabilizer; *Usage level:* 0.1-0.2% (margarine), 2-6% (cake shortening), 0.3-1% on flour (bread), 1-2.5% (peanut butter); *Regulatory:* EEC, FDA §184.1505 (GRAS); *Properties:* beads; drop pt. 72 C; iodine no. 2 max.; 90% min. monoester; *Storage:* store under cool, dry conditions.

Dimodan PVP Kosher. [Grindsted Prods.; Grindsted Prods. Denmark] Hydrogenated palm oil dist. monoglyceride; CAS 97593-29-8; nonionic; emulsifier for margarine, cake shortening, confectionery coatings; crumb softener for bread; amylose complexer for dehydrated potato prods.; aerator for cake mixes, frozen desserts; peanut butter stabilizer providing creamy mouthfeel, surface sheen; *Usage level:* 0.1-0.2% (margarine), 2-6% (cake shortening), 1-2.5% (peanut butter), 0.1-0.5% (dried potato); *Regulatory:* EEC E471, FDA 21CFR §184.1505, GRAS, kosher; *Properties:* beads; m.p. 69 C; iodine no. 2 max.; 90% min. monoester; *Storage:* store in cool, dry area.

Dimodan S. [Grindsted Prods.] Lard dist. monoglyceride, unsat. with antioxidants (200 ppm max. BHT, 200 ppm max. citric acid); CAS 61789-10-4; EINECS 263-032-6; nonionic; food emulsifier for margarine, cake shortenings, icing shortenings, coffee whiteners; *Usage level:* 0.1-0.3% (margarine), 2-6% (cake shortening), 2-3% (icing shortening), 0.1-1% (coffee whitener); *Regulatory:* EEC, FDA 21CFR §184.1505, GRAS; FCC; *Properties:* block; m.p. 55 C; iodine no. 50; 90% min. monoester; *Storage:* store in cool, dry area.

Dimodan TH. [Grindsted Prods. Denmark] Tallow dist. monoglyceride, sat.; CAS 61789-13-7; EINECS 263-035-2; nonionic; food emulsifier; *Properties:* beads; m.p. 70 C; 90% min. monoester.

Dimul DDM K. [Witco/H-I-P] Monoglycerides; food emulsifier; softener for bread; *Usage level:* 0.2-0.5% on flour (bread); *Regulatory:* FDA 21CFR §184.1505; *Properties:* wh. flowable powd.; m.p. 153 F; HLB 4.3; iodine no. 22 ± 4; 90% min. alpha monoglyceride.

Dimul S. [Witco/H-I-P] Monoglycerides; food emulsifier for puddings; inhibits oil separation in peanut butter; improves palatability in dehydrated potatoes; extrusion aid in pasta; *Usage level:* 0.5-4% on fat (peanut butter), 0.5-1% (dehydrated potatoes), 0.2-1% (puddings), 0.1-0.5% (pasta); *Regulatory:*

FDA 21CFR §184.1505; *Properties:* wh. flowable powd.; m.p. 151 F; HLB 4.3; iodine no. < 4; 90% min. alpha monoglyceride.

Dimul S K. [Witco/H-I-P] Monoglycerides; food emulsifier; inhibits oil separation in peanut butter; aerating agent and emulsion stabilizer in whipped toppings; *Usage level:* 0.5-4% on fat (peanut butter); *Regulatory:* FDA 21CFR §184.1505; *Properties:* wh. flowable powd.; m.p. 151 F; HLB 4.3; iodine no. < 4; 90% min. alpha monoglyceride.

Dinner Rich. [Brolite Prods.] Bread base for rich dinner rolls; *Usage level:* 25 lb/50 lb flour; *Properties:* powd.

Dispersable Curcumex 2600, 2601. [MLG Enterprises Ltd.] Turmeric extract; CAS 84775-52-0; food color esp. suitable for noodle and pasta mfg. by machines; provides yel. to greenish-yel. shade; *Usage level:* 0.01-0.1%; *Properties:* liq.; waterdisp.; easier to disperse than Curcumex 1600 and 1601; 2.2 ± 0.2% and 1.1 ± 0.1% resp. (as curcumine); *Storage:* protect from light and heat; rotate stock every 3 mos.

Divergan® F, R. [BASF AG] Stabilizing and clarifying agents for drinks (beer, wine, clear fruit juices).

Divider Oil 90. [ADM Arkady] Mineral oil USP; release agent for food industry; type 90 visc.

Divider Oil 210. [ADM Arkady] Mineral oil USP; release agent for food industry; type 210 visc.

Dixie® Icing Base. [Bunge Foods] Nonboiling icing base for soft icings; baked, frozen and retail icings on cakes, danish pastries, sweet goods, creme fillings, white and chocolate icings; *Usage level:* 10-16 lb; *Properties:* wh. to off-wh. semi-plastic base, sweet vanilla flavor.

DK-Ester. [Grünau GmbH] Sugar esters; food emulsifier for baked goods.

DK-Ester. [Multi-Kem] Sucrose fatty acid esters; nonionic; surfactant; emulsifier and stabilizer for baked goods, baking mixes, dairy desserts/mixes, dairy prod. analogs, cookie mix texturizers; component for protective fruit coatings.

DK Ester F-10. [Dai-ichi Kogyo Seiyaku] Sucrose fatty acid ester; nonionic; food additive for stabilizing emulsion, dispersant, solubilizer; *Properties:* powd.; HLB 1.0; 100% conc.

DK Ester F-20. [Dai-ichi Kogyo Seiyaku] Sucrose fatty acid ester; nonionic; food additive for stabilizing emulsion, dispersant, solubilizer; *Properties:* powd.; HLB 2.0; 100% conc.

DK Ester F-50. [Dai-ichi Kogyo Seiyaku] Sucrose fatty acid ester; nonionic; food additive for stabilizing emulsion, dispersant, solubilizer; *Properties:* powd.; HLB 6.0; 100% conc.

DK Ester F 70. [Dai-ichi Kogyo Seiyaku] Sucrose fatty acid ester; nonionic; food additive for stabilizing emulsion, dispersant, solubilizer; *Properties:* powd.; HLB 8.0; 100% conc.

DK Ester F 90. [Dai-ichi Kogyo Seiyaku] Sucrose fatty acid ester; nonionic; food additive for stabilizing emulsion, dispersant, solubilizer; *Properties:* powd.; HLB 9.5; 100% conc.

DK Ester F-110. [Dai-ichi Kogyo Seiyaku] Sucrose

fatty acid ester; nonionic; food additive for stabilizing emulsion, dispersant, solubilizer; *Properties:* powd.; HLB 11.0; 100% conc.

DK Ester F-140. [Dai-ichi Kogyo Seiyaku] Sucrose fatty acid ester; nonionic; food additive for stabilizing emulsion, dispersant, solubilizer; *Properties:* powd.; HLB 13.0; 100% conc.

DK Ester F-160. [Dai-ichi Kogyo Seiyaku] Sucrose fatty acid ester; nonionic; food additive for stabilizing emulsion, dispersant, solubilizer; *Properties:* powd.; HLB 15.0; 100% conc.

DMCE. [Premier Malt Prods.] Extract of 60-65% malted barley and 35-40% corn; syrup with mellow, rich malt flavor; used in confections, ice cream prods., baked goods, vinegar, cereal, tobacco, soybean milk, pretzels, fermentation nutrient, bread sticks, horse food; 55% the sweetness of sucrose; *Usage level:* 1.3% (caramels), 1-5% (ice cream), 1-2.5% flour wt. (bread), 1-3% dough wt. (bagels), 1.5-2.5% flour wt. (saltines), 2-4% dough mix (snacks), 6-14% (breakfast cereal), 3-7% (soy bean milk), 2-4% (pretzels), 1-1.5% (chocolate syrup); *Properties:* Lovibond 100-250 syrup; pH 5.1-5.8 (10%); 78.5-80.0% solids; 3-4% protein; 57-67% reducing sugars.

DME Dry Malt Extract. [Premier Malt Prods.] Extract of 100% malted barley; CAS 8002-48-0; EINECS 232-310-9; syrup with mellow, rich malt flavor; used in confections, ice cream prods., baked goods, vinegar, cereal, tobacco, soybean milk, pretzels, fermentation nutrient, bread sticks, horse food; 40% the sweetness of sucrose; *Usage level:* 12-18% (malted milk balls), 1-5% (ice cream), 1-2.5% flour wt. (bread), 2-4% of dough mix wt. (snacks), 4-6% flour wt. (cookies), 6-14% (breakfast cereal), 3-4% (diet food/beverages), 1-1.5% (chocolate syrup); *Regulatory:* FDA GRAS; *Properties:* Lovibond 160-360 syrup; pH 5.1-5.8 (10%); 78.5-80.0% solids; 5-6% protein; 54-62% reducing sugars.

Do Crest 60. [Am. Ingreds.] Ethoxylated mono- and diglycerides and wheat flour; dough shortening for yeast-raised bakery prods.; *Properties:* wh. powd.; 4% max. moisture.

Do Crest Gold. [Am. Ingreds.] Soy flour, ethoxylated mono- and diglycerides; dough strengthener for yeast-raised prods.; *Properties:* cream powd.; 26% min. protein.

Do Crest Gold Plus. [Am. Ingreds.] Soy flour, ethoxylated mono- and diglycerides, and calcium peroxide (0.2-0.25%); dough improver and strengthener for yeast-raised bakery prods.; *Properties:* cream powd.; 26% min. protein.

Donut Frying Shortening 101-150. [ADM Packaged Oils] Partially hydrog. soybean oil; EINECS 232-410-2; packaged veg. oil; *Properties:* wh. plasticized cube, bland flavor; m.p. 102 ± 4 F; iodine no. 72 ± 4.

Donut Pyro®. [Rhone-Poulenc Food Ingreds.] Sodium acid pyrophosphate, leavening, FCC; CAS 7758-16-9; EINECS 231-835-0; leavening agent for baking, cereals, esp. doughnut mixes prepared in doughnut machines; *Properties:* wh. powd.;

100% thru 60 mesh, 99% thru 200 mesh; sol. 13 g/ 100 g saturated sol'n.; m.w. 221.94; bulk dens. 68 lb/ft³; pH 4.3 (1%); 95% min. act.

Donut SAPP. [FMC] Sodium acid pyrophosphate; CAS 7758-16-9; EINECS 231-835-0; leavening acid for cake doughnuts; fast acting grade; *Regulatory:* FDA GRAS; *Properties:* powd.; 98% thru 200 mesh; sol. 15 g/100 g water; m.w. 221.9; bulk dens. 65 lb/ft³; 98.3% assay.

Do-Pep Vital Wheat Gluten. [ADM Milling] Wheat gluten; CAS 8002-80-0; EINECS 232-317-7; baking additive for formulated foods; *Properties:* lt. tan powd.; 75-77% protein, 7.5% moisture.

Do Sure. [ADM Arkady] Dough conditioner with L-cysteine; CAS 52-90-4; EINECS 200-158-2; no-time dough conditioner providing rapid dough development.

Double Strength Acid Proof Caramel Colour. [MLG Enterprises Ltd.] Caramel color; CAS 8028-89-5; EINECS 232-435-9; natural colorant for general food and beverage use; prolonged stability and brilliance in acidulated carbonated beverages; also for nonacidulated drinks and processed foods; *Properties:* sp.gr. 1.266-1.277 (60 F); dens. 10.54-10.63 lb/gal (60 F); visc. 300 cps max. (68 F); pH 2.8-3.0; *Storage:* 2 yr min. shelf life.

Dow E300 NF. [Dow] PEG-6 NF; CAS 25322-68-3; EINECS 220-045-1; food applics.; *Properties:* clear visc. liq.; m.w. 300; sp.gr. 1.125; dens. 9.36 lb/gal; f.p. -10 C; visc. 69 cSt; flash pt. (PMCC) > 400 F; ref. index 1.463; sp. heat 0.508 cal/g/°C.

Dow E400 NF. [Dow] PEG-8 NF; CAS 25322-68-3; EINECS 225-856-4; food applics.; *Properties:* clear visc. liq.; m.w. 400; sp.gr. 1.125; dens. 9.36 lb/gal; f.p. 6 C; visc. 90 cSt; flash pt. (PMCC) > 450 F; ref. index 1.465; sp. heat 0.498 cal/g/°C.

Dow E600 NF. [Dow] PEG-12 NF; CAS 25322-68-3; EINECS 229-859-1; food applics.; *Properties:* clear visc. liq.; sol. in water, ethanol, cyclomethicone, sunscreens, lactic acid; m.w. 600; sp.gr. 1.126; dens. 9.37 lb/gal; f.p. 22 C; visc. 131 cSt; flash pt. (PMCC) > 450 F; ref. index 1.466; sp. heat 0.490 cal/g/°C.

Dow E900 NF. [Dow] PEG NF; CAS 25322-68-3; food applics.; *Properties:* wh. waxy solid; m.w. 900; sp.gr. 1.204; f.p. 34 C; visc. 100 cSt (100 F); flash pt. (PMCC) > 450 F.

Dow E1000 NF. [Dow] PEG-20 NF; CAS 25322-68-3; food applics.; *Properties:* wh. waxy solid; m.w. 1000; sp.gr. 1.214; f.p. 37 C; visc. 18 cSt (210 F); flash pt. (PMCC) > 450 F.

Dow E1450 NF. [Dow] PEG-32 NF; CAS 25322-68-3; food applics.; *Properties:* wh. waxy solid; m.w. 1450; sp.gr. 1.214; f.p. 44 C; visc. 29 cSt (210 F); flash pt. (PMCC) > 450 F.

Dow E3350 NF. [Dow] PEG-75 NF; CAS 25322-68-3; food applics.; *Properties:* wh. waxy solid, pract. odorless; sol. > 100 g/100 g in water; m.w. 3350; sp.gr. 1.224; visc. 93 cSt (210 F); f.p. 54 C; b.p. dec.; flash pt. (PMCC) > 232 C; pH 4.5-7.5 (5% aq.); *Toxicology:* single dose oral toxicity believed to be very low; may cause sl. transient eye irritation;

avoid prolonged/repeated contact with abraded skin; *Precaution:* dusts may present explosive hazard; incompat. with oxidizers, conc. min. acids.

Dow E4500 NF. [Dow] PEG-100 NF; CAS 25322-68-3; food applics.; *Properties:* wh. waxy solid, sl. polyether odor; colorless visc. liq. above 136 F; sol. > 100 g/100 g in water; m.w. 4500; sp.gr. 1.224; visc. 180 cSt (210 F); f.p. 58 C; b.p. dec.; flash pt. (PMCC) > 232 C; pH 4.5-7.5 (5% aq.); *Toxicology:* LD50 (oral, rat) > 50,000 mg/kg; extremely low oral toxicity; may cause sl. transient eye irritation; avoid prolonged/repeated exposure on abraded skin; contact with heated material may cause thermal burns; *Precaution:* supports combustion; do not breathe smoke; incompat. with oxidizers and strong acids.

Dow E8000 NF. [Dow] PEG-150 NF; CAS 25322-68-3; food applics.; *Properties:* wh. waxy solid, clear liq. above 65 C, pract. odorless; sol. > 100 g/100 g in water; m.w. 8000; sp.gr. 1.224; f.p. 60 C; visc. 800 cSt (210 F); flash pt. (PMCC) > 260 C; pH 4.5-7.5 (5% aq.); > 99% act.; *Toxicology:* LD50 (oral, rat) > 50 g/kg (low oral toxicity); may cause sl. transient temporary eye irritation; avoid prolonged/repeated exposure on abraded skin; *Precaution:* dusts may present explosive hazard; incompat. with oxidizers, conc. min. acids.

Dow EP530. [Dow] Poloxamer-181; CAS 53637-25-5; defoamer in scald baths for poultry defeathering and hog dehairing machines, for food-grade papers and coatings and sanitizing sol'ns. for food processing equip.; *Regulatory:* 21CFR §172.808, 176.200, 176.210, 178.1010; *Properties:* liq.; m.w. 2000; sp.gr. 1.017; dens. 8.46 lb/gal; visc. 321 cSt; pour pt. -32 C; flash pt. (PMCC) > 420 F; ref. index 1.452; surf. tens. 42.6 dynes/cm (100 ppm).

Dow Corning® 200 Fluid, Food Grade. [Dow Corning] Dimethyl silicone fluid; foam control agent for nonaq. systems, food pkg. and processing, meat, poultry and seafood processing, rendering; *Usage level:* 10 ppm; *Regulatory:* FDA, EPA, USDA, kosher approved; *Properties:* thin clear fluid; sol. in food-grade glycols; visc. 350 cSt; 100% act.

Dow Corning® 1920 Powdered Antifoam. [Dow Corning] Silicone dispersed on free-flowing powd.; nonionic; foam control agent for fermentation, beverages, preserves, meat, poultry, seafood, and rendering in food industry, ultrafiltration, waste and heating/cooling water treatment, biotechnology; *Usage level:* 50 ppm; *Regulatory:* FDA, USDA, kosher approved; *Properties:* wh. free-flowing powd.; sol. in food-grade glycols; 20% act.

Dow Corning® Antifoam 1500. [Dow Corning] Silicone compd.; foam control agent for foods, meat and poultry prod., fermentation; *Regulatory:* FDA, USDA, EPA, kosher approved; *Properties:* med. off-wh. liq.; dilutable in food-grade glycols; 100% act.

Dow Corning® Antifoam 1510-US. [Dow Corning] Silicone emulsion; nonionic; foam control agent for fermentation, beverages, preserves, meat, poultry, seafood, fruit and vegetable processing, waste water treatment, conditioners; *Regulatory:* FDA, EPA, USDA, kosher approved; *Properties:* med. wh. cream; 10% act.

Dow Corning® Antifoam 1520-US. [Dow Corning] Silicone emulsion; nonionic; foam control agent for food, fermentation, beverage mfg., meat/poultry/ seafood processing; *Usage level:* 10 ppm (foods); *Regulatory:* FDA 21CFR §173.340, 176.170, 176.180, EPA 40CFR §180.1001, USDA, kosher approved; *Properties:* milky-wh. thin cream; water-dilutable; sp.gr. 1.0; visc. 6000 cp; pH 4.0; 20% act.; *Toxicology:* may cause temporary eye discomfort; *Storage:* 12 mo shelf life when store @ 20-40 C.

Dow Corning® Antifoam A Compd., Food Grade. [Dow Corning] Simethicone; CAS 8050-81-5; antifoam for food processing, fermentation, edible oil mfg., rendering; *Regulatory:* FDA, USDA, EPA, kosher approved; *Properties:* med. off-wh., gray liq.; dilutable in aliphatic, aromatic or chlorinated solvs., food grade glycols; 100% act.

Dow Corning® Antifoam AF Emulsion. [Dow Corning] Aq. simethicone emulsion; CAS 8050-81-5; nonionic; foam control agent for beverages, preserves, fermentation, dairy prods., rendering in the food industry; *Usage level:* 30 ppm; *Regulatory:* FDA 21CFR §173.340, USDA, EPA 40CFR §180.1001, kosher; *Properties:* milky wh. thick cream; water-dilutable; sp.gr. 1.0; visc. 50,000 cp; pH 3.5; 30% act.; *Toxicology:* food grade; may cause temporary eye discomfort; *Storage:* 6 mo shelf life when stored @ 20-40 C.

Dow Corning® Antifoam C Emulsion. [Dow Corning] Aq. simethicone emulsion; CAS 8050-81-5; nonionic; food grade foam control agent for food processing, fruit juice processing, food sauce prep.; *Usage level:* 30 ppm; *Regulatory:* FDA, EPA, kosher, USDA approved; *Properties:* med. wh. cream; water dilutable; 30% act.

Dow Corning® Antifoam FG-10. [Dow Corning] Aq. silicone emulsion; nonionic; foam control agent for food industry, fermentation, beverages, preserves, meat, poultry, seafood, high sugar-content processes; *Usage level:* 100 ppm; *Regulatory:* FDA, EPA, USDA, kosher approved; *Properties:* thin wh. cream; 10% act.

Drakeol 5. [Penreco] Lt. min. oil NF; CAS 8042-47-5; mold release lubricant for foods, egg coatings, fruit/ veg. coatings; *Regulatory:* FDA 21CFR §172.878, 178.3620, 573.680; *Properties:* sp.gr. 0.831-0.842; dens. 6.89-7.00 lb/gal; visc. 7.6-8.7 cSt (40 C); pour pt. -9 C; flash pt. 154 C; ref. index 1.4600.

Drakeol 7. [Penreco] Lt. min. oil NF; CAS 8042-47-5; protective coating for foods, eggs, fruits, and vegetables; divider oil, mold release lubricant for food industry; *Regulatory:* FDA 21CFR §172.878, 178.3620, 573.680; *Properties:* sp.gr. 0.828-0.843; dens. 6.94-7.08 lb/gal; visc. 10.8-13.6 cSt (40 C); pour pt. -9 C; flash pt. 177 C; ref. index 1.4632.

Drakeol 9. [Penreco] Lt. min. oil NF; CAS 8042-47-5; divider oil, mold release lubricant for food industry, coating for fruits and vegetables; *Regulatory:* FDA 21CFR §172.878, 178.3620, 573.680; *Properties:*

sp.gr. 0.838-0.854; dens. 7.03-7.16 lb/gal; visc. 14.2-17.0 cSt (40 C); pour pt. 09 C; flash pt. 179 C; ref. index 1.4665.

Drakeol 10. [Penreco] Lt. min. oil NF; CAS 8042-47-5; food additive; *Regulatory:* FDA 21CFR §172.878, 178.3620, 573.680; *Properties:* sp.gr. 0.838-0.864; dens. 7.08-7.25 lb/gal; visc. 17.7-20.2 cSt (40 C); pour pt. -9 C; flash pt. 182 C; ref. index 1.4692.

Drakeol 10B. [Penreco] Lt. min. oil NF; CAS 8042-47-5; food additive; *Regulatory:* FDA 21CFR §172.878, 178.3620, 573.680; *Properties:* sp.gr. 0.867-0.878; visc. 17.7-20.2 cSt (40 C); pour pt. -40 C; flash pt. 160 C.

Drakeol 13. [Penreco] Lt. min. oil NF; CAS 8042-47-5; food grade lubes and greases, fruit/veg. coatings, divider oils; *Regulatory:* FDA 21CFR §172.878, 178.3620, 573.680; *Properties:* sp.gr. 0.848-0.867; dens. 7.11-7.27 lb/gal; visc. 24.2-26.3 cSt (40 C); pour pt. -9 C; flash pt. 185 C; ref. index 1.4726.

Drakeol 15. [Penreco] Lt. min. oil NF; CAS 8042-47-5; food additive; *Regulatory:* FDA 21CFR §172.878, 178.3620, 573.680; *Properties:* sp.gr. 0.850-0.873; dens. 7.13-7.30 lb/gal; visc. 28.1-30.3 cSt (40 C); pour pt. -12 C; flash pt. 188 C; ref. index 1.4740.

Drakeol 19. [Penreco] Min. oil USP; CAS 8042-47-5; ingred. in foods (lubes/greases, food pkg., divider oils); *Regulatory:* FDA 21CFR §172.878, 178.3620, 573.680; *Properties:* sp.gr. 0.852-0.876; dens. 7.14-7.31 lb/gal; visc. 34.0-37.3 cSt (40 C); pour pt. -12 C; flash pt. 188 C; ref. index 1.4725.

Drakeol 21. [Penreco] Min. oil USP; CAS 8042-47-5; ingred. in food lubes and greases; *Regulatory:* FDA 21CFR §172.878, 178.3620, 573.680; *Properties:* sp.gr. 0.853-0.876; dens. 7.15-7.32 lb/gal; visc. 38.4-41.5 cSt (40 C); pour pt. -12 C; flash pt. 193 C; ref. index 1.4733.

Drakeol 32. [Penreco] Min. oil USP; CAS 8042-47-5; food additive; *Regulatory:* FDA 21CFR §172.878, 178.3620, 573.680; *Properties:* sp.gr. 0.856-0.876; dens. 7.18-7.35 lb/gal; visc. 60.0-63.3 cSt (40 C); pour pt. -12 C; flash pt. 213 C; ref. index 1.4770.

Drakeol 34. [Penreco] Min. oil USP; CAS 8042-47-5; plasticizer/lube for bakery pan oils, food pkg., lubes/greases in food industry; *Regulatory:* FDA 21CFR §172.878, 178.3620, 573.680; *Properties:* sp.gr. 0.858-0.872; dens. 7.19-7.31 lb/gal; visc. 72.0-79.5 cSt (40 C); pour pt. -9 C; flash pt. 238 C; ref. index 1.4760.

Draketex 50. [Penreco] Lt. min. oil NF; CAS 8042-47-5; food additive; *Regulatory:* FDA 21CFR §172.878, 178.3620, 573.680; *Properties:* sp.gr. 0.817-0.832; dens. 6.86-6.96 lb/gal; visc. 6.5-7.8 cSt (40 C); pour pt. -9 C; flash pt. 152 C; ref. index 1.4570.

Dress All. [Van Den Bergh Foods] Partially hydrog. soybean oil, artificial flavor, TBHQ, artificial color; colored/flavored dressing oil for seafood, vegetables, frying, other prepared foods; *Regulatory:* kosher; *Properties:* Lovibond 4.0-6.0R fluid.

Dress'n® 300. [A.E. Staley Mfg.] Food starch modified, derived from dent and waxy corn; cook-up starch for semisolid or heavy cream type salad dressings; emulsion stabilizer with good acid tolerance, resist. to hear shear; produces heavy, visc. dressing with moderate set-back to a soft gel as it cools; *Regulatory:* FDA 21CFR §172.892; *Properties:* wh. powd.; pH 5.5; 11% moisture.

Dress'n® 400. [A.E. Staley Mfg.] Food starch modified, derived from dent and waxy corn; cook-up starch for semisolid or heavy cream type salad dressings; emulsion stabilizer with good acid tolerance, resist. to hear shear; produces visc. dressing with short, smooth texture and no set-back; *Regulatory:* FDA 21CFR §172.892; *Properties:* wh. powd.; pH 5.5; 11% moisture.

Drewlate 30. [Stepan Food Ingreds.] Acetylated tartrated monoglyceride of vegetable fat; food emulsifier; *Properties:* amber syrupy liq.; Unverified

Drewlene 10. [Stepan Food Ingreds.] Propylene glycol stearate; CAS 1323-39-3; EINECS 215-354-3; food emulsifier which increases whipping ability and aeration; *Properties:* yel. solid, flake, bead; m.p. 110 F; HLB 3.4; sapon. no. 170-190; Unverified

Drewmulse® 3-1-O. [Stepan Food Ingreds.] Triglyceryl oleate; CAS 9007-48-1; nonionic; emulsifier, solubilizer, dispersant for internal use; *Properties:* Gardner 8 liq.; sol. in IPA, peanut oil, min. oil, water disp.; HLB 7.0; sapon. no. 125-150; Unverified

Drewmulse® 3-1-S. [Stepan Food Ingreds.] Triglyceryl stearate; CAS 37349-34-1; EINECS 248-403-2; nonionic; emulsifier, solubilizer, dispersant for internal use; *Properties:* Gardner 10 liq.; disp. in water, IPA, peanut oil, min. oil, propylene glycol; HLB 7.0; sapon. no. 120-140; Unverified

Drewmulse® 3-1-SH. [Stepan Food Ingreds.] Triglyceryl monoshortening; emulsifier for margarine, icings, fillings, and whipped toppings; substitute for polysorbates; *Regulatory:* FDA 21CFR §172.854; avail. in kosher grade; *Properties:* amber visc. liq., sl. fatty odor; sapon. no. 120; 20% alpha monoester, 0.5% moisture; *Storage:* avoid prolonged storage above 90 F; Discontinued

Drewmulse® 6-2-S. [Stepan Food Ingreds.] Hexaglyceryl distearate; CAS 34424-97-0; nonionic; emulsifier, solubilizer, dispersant for internal use; *Properties:* Gardner 13 solid; sol. in IPA, peanut oil, min. oil; HLB 8.0; sapon. 105-125; Unverified

Drewmulse® 10-4-O. [Stepan Food Ingreds.] Decaglyceryl tetraoleate; CAS 34424-98-1; EINECS 252-011-7; nonionic; emulsifier, solubilizer, dispersant for internal use; *Properties:* Gardner 8 liq.; sol. in IPA, peanut oil, min. oil; water disp.; HLB 6.0; sapon. no. 125-145; Unverified

Drewmulse® 10-8-O. [Stepan Food Ingreds.] Decaglyceryl octaoleate; nonionic; emulsifier, solubilizer, dispersant for internal use; *Properties:* Gardner 8 liq.; sol. in peanut oil, min. oil; HLB 4.0; sapon. no. 155-175; Unverified

Drewmulse® 10-10-S. [Stepan Food Ingreds.] Deca-

glyceryl decastearate; CAS 39529-26-5; EINECS 254-495-5; nonionic; emulsifier, solubilizer, dispersant for internal use; *Properties:* Gardner 8 solid; sol. in peanut oil, min. oil; HLB 3.0; sapon. no. 160-180; Unverified

Drewmulse® 10K. [Stepan Food Ingreds.] Glyceryl mono-shortening from soya oil; CAS 1323-39-3; lipophilic emulsifier used in foods as dispersing aid, antistaling agent, antistick agent, stabilizer; for o/w emulsion systems; emulsifier for shortenings, icings, margarines, bakery mixes, bread and dairy mixes; *Regulatory:* FDA 21CFR §182.4505; kosher grade; *Properties:* wh. soft solid, sl. fatty odor; m.p. 46 C; HLB 2.8; iodine no. 70; sapon. no. 170; 40% alpha monoester; *Storage:* avoid prolonged storage above 90 F.

Drewmulse® 70. [Aquatec Quimica SA] Blend of glyceryl monoesters; nonionic; food emulsifier; *Properties:* liq.; 13.5-19% act.; Unverified

Drewmulse® 75. [Aquatec Quimica SA] Glyceryl cocoate; CAS 61789-05-7; EINECS 263-027-9; nonionic; food emulsifier; *Properties:* paste; 100% act.; Unverified

Drewmulse® 200. [Stepan Food Ingreds.] Glyceryl oleate from tallow and soya; nonionic; emulsifier for food industry as dispersing aid, antistaling agent, antistick agent; *Properties:* cream beads; m.p. 135-145; HLB 2.8; sapon. no. 154-170 (soya), 163-177 (tallow); 40% alpha mono; Unverified

Drewmulse® 200K. [Stepan Food Ingreds.; Stepan Europe] Glyceryl stearate; CAS 123-94-4; nonionic; emulsifier, emollient, antistat, stabilizer, visc. builder, opacifier for foods (breads, chewing gum base, frozen desserts, peanut butter, margarine, candies, toppings); *Regulatory:* FDA 21CFR §182.4505; kosher grade; *Properties:* ivory wh. flakes, sl. fatty odor; m.p. 60 C; iodine no. 2; sapon. no. 175; 42% monoester; *Storage:* avoid prolonged storage above 90 F.

Drewmulse® 365. [Aquatec Quimica SA] Glyceryl stearate (67%) and polysorbate 65 (33%); lipophilic emulsifier used in food industry as dispersing aid, antistaling agent, antistick agent; *Properties:* cream beads; m.p. 136 F; HLB 5.3; 25-30% alpha mono; Unverified

Drewmulse® 700K. [Stepan Food Ingreds.] Glyceryl stearate, polysorbate 80; food emulsifier; *Regulatory:* FDA 21CFR §182.4505; kosher grade; *Properties:* cream-colored beads, mild fatty odor; sapon. no. 151; 0.8% moisture; *Storage:* avoid prolonged storage above 90 F.

Drewmulse® 900K. [Stepan Food Ingreds.] Glyceryl stearate; CAS 123-94-4; nonionic; emulsifier for food industry as dispersing aid, antistaling agent, antistick agent (shortenings, icings, breads, chewing gum base, frozen desserts, peanut butter, candies, toppings); *Regulatory:* FDA 21CFR § 182.4505; kosher grade; *Properties:* off-wh. flakes; m.p. 63 C; HLB 3.2; iodine no. 2; sapon. 165; 50% alpha monoester; *Storage:* avoid prolonged storage above 90 F.

Drewmulse® 8731-S. [Stepan Food Ingreds.] Triglyceryl monoshortening; food emulsifier for margarine, icings, fillings, and whipped toppings; effective in both o/w and w/o emulsions; substitute for polysorbates; *Regulatory:* FDA 21CFR §172.854; avail. in kosher grade; *Properties:* amber visc. liq., sl. fatty odor; HLB 7.4; sapon. no. 110; 20% alpha monoester; *Storage:* avoid prolonged storage above 90 F; Discontinued

Drewmulse® D-4661. [Stepan Food Ingreds.] Mixed veg. oils, refined, deodorized; CAS 68938-35-2; defoamer, lubricant for food applics.; esp. well suited as a lubricant in downstream processing in the fermentation industry; *Properties:* yel. clear liq., bland sl. fatty odor; water-disp.; iodine no. 40; sapon. no. 190; 0.5% moisture; *Storage:* avoid prolonged storage above 90 F.

Drewmulse® M. [Aquatec Quimica SA] Blend of glyceryl monoesters; nonionic; food emulsifier; *Properties:* paste; 32% act.; Unverified

Drewmulse® PNO. [Aquatec Quimica SA] Blend of glyceryl monoesters; nonionic; food emulsifier; *Properties:* paste; 38-42% act.; Unverified

Drewmulse® POE-SML. [Stepan Food Ingreds.] Polysorbate 20; CAS 9005-64-5; nonionic; solubilizer for fragrances, flavors, vitamins, essential oils and germicides in aq. systems; *Properties:* liq., bland char. odor; HLB 15.1; acid no. 1; iodine no. 1; sapon. no. 47; hyd. no. 100; 2.5% moisture.

Drewmulse® POE-SMO. [Stepan Food Ingreds.] Polysorbate 80; CAS 9005-65-6; nonionic; emulsifier for whipped toppings, cake icings/fillings, shortenings, ice cream, frozen custard, ice milk, sherbets, solubilizer, dispersant for pickles, vitamin/mineral preps., essential oils, and flavors; yeast defoamer; *Regulatory:* FDA 21CFR §172.840; avail. in kosher grade; *Properties:* amber visc. liq., mild odor; HLB 15.0; iodine no. 25; sapon. no. 50; hyd. no. 70; 2% moisture; *Storage:* avoid prolonged storage above 90 F.

Drewmulse® SMO. [Stepan Food Ingreds.] Sorbitan oleate; CAS 1338-43-8; EINECS 215-665-4; nonionic; w/o emulsifier, thickener, solubilizer, dispersant, wetting agent, detergent for foods; *Properties:* liq.; HLB 4.7; iodine no. 69; sapon. no. 155; hyd. no. 196.8; 0.5% moisture.

Drewmulse® SMS. [Stepan Food Ingreds.] Sorbitan stearate; CAS 1338-41-6; EINECS 215-664-9; nonionic; w/o emulsifier, solubilizer, dispersant, wetting agent, detergent, visc. control agent for foods; flavor and fragrance solubilizer; vitamins; *Regulatory:* avail. in kosher grade (Drewsorb 60K); *Properties:* cream-colored beads, mild fatty odor; water-disp.; HLB 2.1; iodine no. 1; sapon. no. 150; hyd. no. 250; *Storage:* avoid prolonged storage above 90 F.

Drewplus® L-523. [Drew Ind. Div.] Blend of fatty oils, surfactants, and silica derivs.; defoamer for food/fermentation applics.; *Regulatory:* FDA compliance; *Properties:* off-wh. opaque liq.; disp. in water; sp.gr. 0.99; dens. 8.26 lb/gal; visc. 800 cps; pour pt. 10 C.

Drewplus® L-722. [Drew Ind. Div.] Silicone defoamer;

defoamer for food/fermentation applics.; *Regulatory:* FDA compliance; *Properties:* emulsifiable in water; sp.gr. 1.01; dens. 8.43 lb/gal.

Drewplus® L-768. [Drew Ind. Div.] Blend of silicone fluid, silica derivs. and surfactants; defoamer for food/fermentation applics.; *Regulatory:* FDA and EPA compliance; *Properties:* off-wh. opaque liq.; disp. in water; sp.gr. 1.02; dens. 8.51 lb/gal; visc. 1100 cps; pour pt. 5 C; flash pt. (COC) 93 C.

Drewplus® L-790. [Drew Ind. Div.] Blend of min. oils, silica derivs., and emulsifiers; defoamer for food/fermentation applics.; *Regulatory:* FDA and EPA compliance; *Properties:* off-wh. opaque liq.; disp. in water; sp.gr. 0.90; dens. 7.51 lb/gal; visc. 1400 cps; pour pt. 0 C; flash pt. > 149 C.

Drewplus® L-813. [Drew Ind. Div.] Blend of dimethylpolysiloxane, silica, and emulsifiers; nonionic; foam control agent for industrial food processing, veg. and fruit processing, sugar, instant coffee, fruit juices, dehydrating and evaporating systems; *Regulatory:* FDA compliance; *Properties:* off-wh. opaque liq.; dilutable in water; sp.gr. 1.00; visc. 400 cps; 10% silicone.

Drewplus® L-833. [Drew Ind. Div.] Blend of dimethylpolysiloxane, silica, emulsifiers; nonionic; defoamer for food processing operations at dosages as low as 2 ppm (fermentation, cheese whey, starch processing, fruit and veg. processing, dehydration, evaporation); effective over wide pH range and under high shear and high salt conditions; *Usage level:* 2-33 pp; *Regulatory:* FDA compliance; *Properties:* wh. opaque liq.; emulsifiable in water; sp.gr. 1.01; dens. 8.43 lb/gal; visc. 1200 cps; 30% silicone.

Drewplus® Y-250. [Drew Ind. Div.] Blend of min. oils, silica derivs., and esters; defoamer for food/fermentation applics.; *Properties:* off-wh. opaque liq.; disp. in water; sp.gr. 0.91; dens. 7.59 lb/gal; visc. 1800 cps.

Drewpol® 3-1-O. [Stepan; Stepan Food Ingreds.] Polyglyceryl-3 monooleate; CAS 9007-48-1; nonionic; food emulsifier; *Properties:* amber clear liq.; HLB 7.0; acid no. 5.0 max.; Unverified

Drewpol® 3-1-OK. [Stepan Food Ingreds.] Triglyceryl monooleate; CAS 9007-48-1; food emulsifier and additive for margarines, confectionery coatings, lowfat dietary prepared foods; solubilizer for essential oils, flavors, and colors; defoamer for processed foods; dispersant for high-solids preps.; *Regulatory:* FDA 21CFR §172.854; kosher grade; *Properties:* amber visc. liq., mild fatty odor; HLB 7.0; iodine no. 65; sapon. no. 145; 0.5% moisture.

Drewpol® 3-1-SHK. [Stepan Food Ingreds.] Triglyceryl monoshortening; nonionic; food emulsifier for margarine, icings, fillings, and whipped toppings; substitute for polysorbates; *Regulatory:* FDA 21CFR §172.854; kosher grade; *Properties:* amber visc. liq., mild fatty odor; HLB 7.4; sapon. no. 120; 20% alpha monoester, 0.5% moisture; *Storage:* avoid prolonged storage above 90 F.

Drewpol® 3-1-SK. [Stepan Food Ingreds.] Triglyceryl monostearate; CAS 37349-34-1; EINECS 248-

403-2; nonionic; food emulsifier for cake shortenings, nondairy creamers, whipped toppings, lowfat dietary prepared foods, coffee whiteners, confection coatings; *Regulatory:* FDA 21CFR §172.854; kosher grade; *Properties:* beads, mild fatty odor; HLB 7.0; iodine no. 1; sapon. no. 140; 0.5% moisture; *Storage:* avoid prolonged storage above 90 F.

Drewpol® 6-1-O. [Stepan; Stepan Canada] Polyglyceryl-6 oleate; CAS 9007-48-1; food emulsifier; *Properties:* amber clear liq.; HLB 8.5; acid no. 6 max.; Unverified

Drewpol® 6-1-OK. [Stepan Food Ingreds.] Hexaglyceryl oleate; CAS 9007-48-1; food emulsifier; *Properties:* Gardner 10 clear to hazy liq., mild typ. odor; HLB 8.5; acid no. 6 max.; iodine no. 60; sapon. no. 130; hyd. no. 275; 1% moisture.

Drewpol® 6-2-O. [Stepan Food Ingreds.] Polyglyceryl ester; nonionic; food emulsifier; *Properties:* liq.; HLB 9.6; 100% conc.; Unverified

Drewpol® 6-2-OK. [Stepan Food Ingreds.] Hexaglyceryl dioleate; CAS 76009-37-5; food emulsifier (confection coatings); polysorbate substitute; *Regulatory:* FDA 21CFR §172.854; kosher grade; *Properties:* HLB 9.6.

Drewpol® 6-2-SK. [Stepan Food Ingreds.] Hexaglyceryl distearate; CAS 61725-93-7; food emulsifier for cake shortenings, nondairy creamers, whipped toppings, lowfat dietary prepared foods, and coffee whiteners; *Regulatory:* FDA 21CFR §172.854; kosher grade; *Properties:* beads, mild fatty odor; HLB 4.0; iodine no. 1; sapon. no. 125; 18% monoester; *Storage:* avoid prolonged storage above 90 F.

Drewpol® 8-1-OK. [Stepan Food Ingreds.] Octaglyceryl monooleate; CAS 9007-48-1; nonionic; food emulsifier for vegetable dairy systems, lowfat dietary prepared foods, confection coatings; solubilizer for essential oils, flavors; suspending agent for food colors; defoamer; polysorbate replacement; *Regulatory:* FDA 21CFR §172.854; kosher grade; *Properties:* amber clear to hazy visc. liq., mild fatty odor; HLB 13.0; iodine no. 60; sapon. no. 130; 0.5% moisture; *Storage:* avoid prolonged storage above 90 F.

Drewpol® 10-1-CC. [Stepan Food Ingreds.; Stepan Europe] Decaglyceryl monocaprylate; nonionic; food emulsifier; *Properties:* amber clear to sl. hazy visc. liq.; HLB 14.5; acid no. 3; iodine no. 2; sapon. no. 100; hyd. no. 700; pH 7 (3% disp.); 100% conc.; Unverified

Drewpol® 10-1-CCK. [Stepan Food Ingreds.] Decaglyceryl monocaprylate; CAS 68937-16-6; nonionic; food emulsifier; *Regulatory:* kosher grade; *Properties:* amber clear to sl. hazy visc. liq.; HLB 14.5; acid no. 3; sapon. no. 100; hyd. no. 700; 100% conc.

Drewpol® 10-2-OK. [Stepan Food Ingreds.] Decaglyceryl dioleate; CAS 9009-48-1; food emulsifier for margarine; solubilizer for flavors and essential oils; dispersant for high-solids preps.; suspending agent for food colors; *Properties:* amber visc. liq.; iodine no. 55; sapon. no. 115; hyd. no. 315; 0.5% moisture; *Storage:* avoid prolonged storage above

90 F.

Drewpol® 10-4-O. [Stepan; Stepan Canada; Stepan Europe] Polyglyceryl-10 tetraoleate; CAS 34424-98-1; EINECS 252-011-7; nonionic; solubilizer for vitamins, flavors; *Properties:* amber clear liq.; HLB 6.0; acid no. 8.0 max.; 100% act.; Unverified

Drewpol® 10-4-OK. [Stepan Food Ingreds.] Decaglyceryl tetraoleate; CAS 34424-98-1; EINECS 252-011-7; food emulsifier for margarines; solubilizer for flavors and essential oils; suspending agent for food colors; dispersant for high-solids preps.; *Regulatory:* FDA 21CFR §172.854; kosher grade; *Properties:* amber visc. liq., mild fatty odor; HLB 6.0; acid no. 8.0 max.; sapon. no. 125 min.; 0.5% moisture; *Storage:* avoid prolonged storage above 90 F.

Drewpol® 10-6-OK. [Stepan Food Ingreds.] Decaglyceryl hexaoleate; food emulsifier for margarine; solubilizer for flavors and essential oils; dispersant for high-solids preps.; suspending agent for food colors; *Properties:* amber clear to hazy liq.; acid no. 2.1; iodine no. 69; sapon. no. 153; hyd. no. 105; 0.1% moisture.

Drewpol® 10-8-OK. [Stepan Food Ingreds.] Decaglyceryl octaoleate; food emulsifier for margarines; solubilizer for flavors and essential oils; suspending agent for food colors; dispersant for high-solids preps.; *Regulatory:* FDA 21CFR §172.854; kosher grade; *Properties:* amber visc. liq., mild fatty odor; iodine no. 75; sapon. no. 160; 0.5% moisture; *Storage:* avoid prolonged storage above 90 F.

Drewpol® 10-10-O. [Stepan; Stepan Canada; Stepan Europe] Polyglyceryl-10 decaoleate; CAS 11094-60-3; EINECS 234-316-7; nonionic; food emulsifier; *Properties:* amber clear liq.; HLB 3.0; acid no. 10.0 max.; Unverified

Drewpol® 10-10-OK. [Stepan Food Ingreds.] Decaglyceryl decaoleate; CAS 11094-60-3; EINECS 234-316-7; nonionic; food emulsifier for margarines; solubilizer for flavors and essential oils; suspending agent for food colors; dispersant for high-solids preps.; *Regulatory:* FDA 21CFR §172.854; kosher grade; *Properties:* amber visc. liq., mild fatty odor; HLB 3.0; acid no. 10.0 max.; iodine no. 80; sapon. no. 170; 0.5% moisture; *Storage:* avoid prolonged storage above 90 F.

Drewpone® 60K. [Stepan Food Ingreds.] Polysorbate 60; CAS 9005-67-8; nonionic; food emulsifier for whipped toppings, cake mixes, confectionery coatings, cake icings/fillings, baked goods, coffee whiteners, salad dressings; foaming agent; dough conditioner; *Regulatory:* FDA 21CFR §172.836; kosher grade; *Properties:* yel. clear to hazy visc. liq., mild odor; HLB 14.9; iodine no. 1; sapon. no. 50; hyd. no. 85; 2.5% moisture; *Storage:* avoid prolonged storage above 90 F.

Drewpone® 65K. [Stepan Food Ingreds.] Polysorbate 65; CAS 9005-71-4; nonionic; food emulsifier, foaming agent for ice cream, frozen custard, ice milk, sherbets, whipped toppings, coffee whiteners, cake icings and fillings; dough conditioner; *Regulatory:* FDA 21CFR §172.838; kosher grade; *Proper-*

ties: amber soft solid paste, mild odor; HLB 10.5; iodine no. 1; sapon no. 90; hyd. no. 46; *Storage:* avoid prolonged storage above 90 F.

Drewpone® 80K. [Stepan Food Ingreds.] Polysorbate 80; CAS 9005-65-6; nonionic; emulsifier, foaming agent for whipped toppings, cake icings/fillings, shortenings, ice cream, frozen custard, ice milk, sherbets; solubilizer, dispersant for pickles, vitamin/mineral preps., essential oils, flavors; yeast defoamer; dough conditioner; *Regulatory:* FDA 21CFR §172.840; kosher grade; *Properties:* amber visc. liq., mild odor; HLB 15; iodine no. 25; sapon. no. 50; hyd. no. 70; 2% moisture; *Storage:* avoid prolonged storage above 90 F.

Drewsorb® 60K. [Stepan Food Ingreds.] Sorbitan stearate; CAS 69005-67-8; EINECS 215-664-9; nonionic; food emulsifier for whipped toppings, cakes and cake mixes, confectionery coatings, cake icings/fillings, coffee whiteners; rehydration aid for dry yeast; defoaming agent for processed food; *Regulatory:* FDA 21CFR §172.842; kosher grade; *Properties:* cream-colored beads, mild fatty odor; water disp.; HLB 4.7; iodine no. 1; sapon. no. 150; hyd. no. 245; 100% conc.; *Storage:* avoid prolonged storage above 90 F.

Drewsorb® 65K. [Stepan Food Ingreds.] Sorbitan tristearate; nonionic; food emulsifier; *Properties:* cream-colored beads, mild odor; HLB 2.1; acid no. 15 max.; sapon. no. 176; hyd. no. 75; 1.5% max. moisture; *Storage:* avoid prolonged storage above 90 F.

Drewsorb® 80K. [Stepan Food Ingreds.] Sorbitan monooleate, EINECS 215-665-4; nonionic; food emulsifier; *Properties:* bead; HLB 4.3; 100% conc.; Unverified

Dricoid® 200. [Kelco] Guar gum, mono- and diglycerides, xanthan gum, carrageenan, citric acid blend; dairy stabilizer/emulsifier used for ice cream, premium ice cream, lowfat frozen dessert, frozen yogurt, ice milk, and novelties; provides firm to med. body, smooth creamy texture, exc. heat shock resist., uniform meltdown, enhanced flavor release; *Usage level:* 0.2-0.24% (ice cream), 0.26-0.38% (ice milk), 0.25-0.36% (frozen yogurt), 0.3-0.4% (lowfat frozen dessert); 1% max. (USA); *Regulatory:* FDA 21CFR §184.1339, 184.1505, 172.695, 182.7255, 182.6033, GRAS; EEC E412, E471, E415, E407, E330; *Properties:* 14 mesh; visc. 600 cps (1%); pH 7.0.

Dricoid® 280. [Kelco] Guar gum, mono- and diglycerides, xanthan gum, polysorbate 80, carrageenan; dairy stabilizer/emulsifier used for premium ice cream, ice milk, soft-serve ice cream/milk, lowfat frozen desserts, soft-serve yogurt, hard frozen yogurt, nonfat frozen desserts, dietetic ice cream, milk shake mixes; *Usage level:* 0.2-0.22% (ice cream), 0.24-0.36% (lowfat frozen dessert), 0.26-0.3% (soft-serve ice cream), 0.2-0.3% (frozen novelties), 0.35-0.5% (nonfat frozen dessert), 0.22-0.38% (artificially sweetened frozen dessert); *Properties:* 14 mesh; easily disp., hydrates readily in cold or hot systems; visc. 400 cps (1%); pH 7.0.

Dricoid® KB. [Kelco] Propylene glycol alginate; CAS 9005-37-2; dairy stabilizer/emulsifier used for ice cream, premium ice cream, novelties, and soft-serve yogurt; *Properties:* 14 mesh; milk-sol.; pH 8.5.

Dricoid® KBC. [Kelco] Propylene glycol alginate; CAS 9005-37-2; dairy stabilizer/emulsifier used for ice cream, premium ice cream, soft-serve ice cream, lowfat ice cream, ice milk, soft-serve and frozen yogurt, novelties, egg nog, and sour cream dips.

Dried Beef Seasoning #6540. [Ariake USA] Beef seasoning; flavorings for snacks, soup mixes, sauces, gravies; 5% moisture, 31% protein.

Dried Beef Stock #3024. [Ariake USA] Beef stock; flavorings for snacks, soup mixes, sauces, gravies; 5% moisture, 65.9% protein.

Dried Chicken Seasoning #1026. [Ariake USA] Chicken seasoning; flavorings for snacks, soup mixes, sauces, gravies; 2% moisture, 44.9% protein.

Dried Chicken Stock #120. [Ariake USA] Chicken stock, salt, lecithin; flavorings for snacks, soup mixes, sauces, gravies, specialty meat prods.; 5% moisture, 63% protein; *Storage:* keep in cool, dry place.

Dried Chicken Stock #1004. [Ariake USA] Chicken stock; flavorings for snacks, soup mixes, sauces, gravies; 5% moisture, 87.5% protein.

Dried Pork Stock #220. [Ariake USA] Pork stock; flavorings for snacks, soup mixes, sauces, gravies; 5% moisture, 87.3% protein.

Dried Pork Stock #2004. [Ariake USA] Pork stock; flavorings for snacks, soup mixes, sauces, gravies; 5% moisture, 93% protein.

Dried Turkey Stock #410. [Ariake USA] Turkey stock; flavorings for snacks, soup mixes, sauces, gravies; 5% moisture, 81% protein.

Dri-Mol® 604 Molasses. [ADM Ogilvie] 55% Molasses and other sweetener solids; CAS 68476-78-8; sweetener for baked goods, desserts, mixes; *Properties:* lt. brn. powd./flake; 4% max. moisture.

Dri-Mol® Molasses. [ADM Ogilvie] 55% Molasses; CAS 68476-78-8; sweetener for baked goods, desserts, barbeque sauces, mixes; *Properties:* dk. brn. powd./flake; 4% max. moisture.

DriRite Glaze Powd. [Bunge Foods] Boiling or nonboiling type icing stabilizer for donut glaze, pie glaze; good stability when boiled; *Usage level:* 1-2 lb/100 lb powd. sugar; *Properties:* off-wh. free-flowing powd., bland flavor; pH 10.

Drize-P Powd. [MLG Enterprises Ltd.] Calcium sulfate, soy flour, diammonium phosphate, calcium peroxide, tricalcium phosphate; dough conditioner for yeast-raised prods.; produces drier doughs with improved handling chars.; *Usage level:* 0.5 oz/cwt flour; *Properties:* powd.; *Storage:* 6 mos shelf life stored under cool, dry conditions.

Dry Beta Carotene Beadlets 10% CWS No. 65633. [Roche] β-Carotene compded. with sucrose, fish gelatin, food starch, peanut oil, with antioxidants (ascorbyl palmitate, dl-α-tocopherol); colorant for aq.-based foods and beverages; source of vitamin A in pharmaceutical hard shell capsules; *Usage level:* 2-5 ml/lb (baked goods), 5-10 ml/qt (lemon pudding), 2-6 ml/qt (ice cream), 2-4 ml/qt (orange drinks), 0.25-1 ml/qt (clam chowder), 1-10 ml/lb (candy cream filling, coatings); *Regulatory:* FDA 21CFR §73.95 (food use), 73.1095 (drug use), 73.2095 (cosmetic use), 182.5245 GRAS; *Properties:* orange-red to dk. red beadlets; 100% thru 20 mesh, 20% max. thru 100 mesh; readily disp. in water with stirring; 167,000 IU provitamin A activity/g; *Storage:* store in cool, dry place @ 46-59 F in tightly closed container; avoid moisture condensation.

Dry Beta Carotene Beadlets 10% No. 65661. [Roche] β-Carotene compded. with gelatin, sucrose, food starch, peanut oil with antioxidants (ascorbyl palmitate, dl-α-tocopherol; colorant for aq.-based foods or beverages; vitamin A source in pharmaceutical hard shell capsules; *Usage level:* 2-5 ml/lb (baked goods), 5-10 ml/qt (lemon pudding), 2-6 ml/qt (ice cream), 2-4 ml/qt (orange drinks), 0.25-1 ml/qt (clam chowder), 1-10 ml/lb (candy cream filling, coatings); *Regulatory:* FDA 21CFR §73.95 (food use), 73.1095 (drug use), 73.2095 (cosmetic use), 182.5245 GRAS; *Properties:* orange-red to dk. red beadlets; 100% thru 20 mesh, 20% max. thru 100 mesh; readily disp. in warm water; 167,000 IU provitamin A/g; *Storage:* store in cool, dry place @ 46-59 F in tightly closed containers; avoid moisture condensation.

Dry Beta Carotene Beadlets Yellow Type 2.4-S No. 653800100. [Roche] β-Carotene compded. with hydrog. coconut oil, gelatin, sucrose, modified food starch with antioxidants (BHT, BHA), preservatives (methyl paraben, propyl paraben, sodium benzoate, potassium sorbate); colorant for aq.-based foods and beverages; *Usage level:* 100-200 mg/lb (baked goods), 100-250 mg/qt (ice cream), 100 mg/qt (orange drinks), 12-50 mg/qt (clam chowder), 50-500 mg/lb (candy cream filling, coatings), 50-100 mg/qt (whipped dessert topping); *Regulatory:* FDA 21CFR §73.95 (food use), 73.1095 (drug use), 73.2095 (cosmetic use), 182.5245 GRAS; *Properties:* dk. orange-red beadlets; somewhat hygroscopic; 100% thru 20 mesh, 25% max. thru 80 mesh; rapidly disp. in water, fruit juices, aq. foods; 40,000 USP units provitamin A activity/g; *Storage:* store in tightly sealed containers in cool, dry place @ 46-59 F.

Dry Canthaxanthin 10% SD No. 66514. [Roche] 10% Canthaxanthin with gelatin, sucrose, vegetable oil and antioxidants (ascorbyl palmitate, dl-α-tocopherol); colorant producing tomato-red hues in pizza, barbecue sauce, Russian dressing, fruit punch, bloody Mary-type beverages; also produces range of hues in dressings, cakes, simulated meats, dry mixes; *Usage level:* 300 mg max./lb of solid or semisolid or pint of liq. food; *Regulatory:* FDA 21CFR §73.75 (food use), 73.1075 (drug use); *Properties:* purplish red free-flowing powd.; 95% min. thru 40 mesh; disp. in cold water; 10% min.

assay (canthaxanthin); *Storage:* store in tightly closed container below 46 F; avoid moisture.

Dry β-Carotene Beadlets. [Crompton & Knowles/ Ingred. Tech.] β-Carotene FCC compded. with gelatin, sucrose, starch, and oil, with antioxidants and preservatives; color additive for foods; *Regulatory:* FDA 21CFR §73.95, 73.1095, 73.2095; *Properties:* dk. red to dk. orange-red beadlets, sl. odor; readily disp. in hot water with moderate stirring; 100% min. 20 mesh, 90% min. thru 30 mesh; *Storage:* store in tight containers @ 46-59 F; avoid exposure to heat, light, moisture; 1 yr storage life.

Dry Phytonadione 1% SD No. 61748. [Roche] Vitamin K₁ USP compded. with dextrose, acacia, lactose, and 1% silicon dioxide; component of enzyme systems associated with blood-clotting mechanism; used in fortification of infant formulas; *Properties:* off-wh. to yel. free-flowing powd.; 98% min. thru 40 mesh, 50% min. thru 100 mesh; m.w. 450.68; 1% min. assay (phytonadione); *Storage:* store in cool, dry place protected from light; dec. on exposure to light.

Dry Vitamin A Palmitate Type 250-SD No. 65378. [Roche] Vitamin A palmitate USP FCC compded. with modified food starch, lactose, fractionated coconut oil, and BHT, sodium benzoate, sorbic acid (preservatives), and silicon dioxide; for vitamin enrichment premixes and baked prods., dehydrated potato flakes, dry milk, beverages; *Properties:* lt. yel. to tan fine powd.; 98% min. thru 50 mesh, 80% min. thru 100 mesh; 250,000 IU/g assay; *Toxicology:* vitamin A: sustained daily intakes exceeding 50,000 IU (adults), 20,000 IU (infants) may cause toxic effects (headache, vomiting, liver damage); US RDA 8000 IU (pregnant/lactating women); *Storage:* store in cool, dry place; avoid excessive heat and moisture; keep container tightly closed.

Dry Vitamin D₃ Beadlets Type 850 No. 652550401, 652550601. [Roche] Cholecalciferol USP, FCC in gelatin matrix with food-grade modified starch, peanut oil, sucrose, and antioxidants (BHA, BHT); used in foods, dry preps., multivitamin tablets/capsules; *Properties:* buff colored free-flowing spherical beadlets, sl. char. odor; 40 and 60 mesh resp.; 850,000 min. IU vitamin D/g; *Toxicology:* Vitamin D: potentially toxic esp. for young children; excessive ingestion may cause hypercalcemia, hypercalcuria; *Storage:* store in cool, dry place, optimally @ 46-59 F; keep tightly closed.

Dry Vitamin D₃ Type 100 CWS No. 65242. [Roche] Cholecalciferol compded. with edible fats finely dispersed in a starch-coated matrix of gelatin and sucrose; dl-α-tocopherol as antioxidant; used in food preps., and prods. which are reconstituted with liqs., esp. effervescent tablets; *Properties:* off-wh. to brnsh. fine gran. powd.; 100% thru 20 mesh, 16% max. thru 100 mesh; m.w. 384.64; 100,000 min. IU vitamin D/g; *Toxicology:* Vitamin D: potentially toxic esp. for young children; excessive ingestion may cause hypercalcemia, hypercalcuria; *Storage:* store in cool, dry place in tightly closed container; sensitive to air, heat, light, and humidity.

Dry Vitamin D₃ Type 100-SD No. 65216. [Roche] Cholecalciferol compded. with dicalcium phosphate, lactose, acacia, coconut oil, BHT, silicon dioxide, and preservatives (sodium benzoate, sorbic acid); food additive; *Properties:* wh. dry free-flowing powd.; 98% min. thru 60 mesh, 95% min. thru 100 mesh; m.w. 384.64; 100,000 min. IU vitamin D/g; *Toxicology:* Vitamin D: potentially toxic esp. for young children; excessive ingestion may cause hypercalcemia, hypercalcuria; *Storage:* store in cool, dry place in tightly closed container.

Dry Vitamin E Acetate 50% SD No. 65356. [Roche] dl-α-Tocopheryl acetate USP, FCC compded. with hydrolyzed protein, 3% silicon dioxide; vitamin E source for food prods., chewable or coated dietary supplements; *Regulatory:* FDA GRAS; *Properties:* wh. to off-wh. free-flowing powd., acceptable odor and taste; 95% min. thru 20 mesh; sl. hygroscopic; disp. readily in cold or warm water; 50% min. dl-α-tocopheryl acetate; *Storage:* store in dry place in tightly closed container; avoid excessive heat.

Dry Vitamin E Acetate 50% Type CWS/F No. 652530001. [Roche] dl-α-Tocopheryl acetate USP, FCC dispersed in a matrix of fish gelatin and maltodextrin; vitamin E source for dry food prods. which are reconstituted with liqs.; *Regulatory:* FDA GRAS; *Properties:* wh. to off-wh. spray-dried powd., acceptable odor and taste; 60% min. thru 80 mesh; sl. hygroscopic; disp. readily in cold or warm water; 50% min. dl-α-tocopheryl acetate; *Storage:* store in dry place in tightly closed container; avoid excessive heat.

Duodenal Substance. [Am. Labs] Vacuum-dried defatted glandular prod.; nutritive food additive; *Properties:* powd. or freeze-dried form.

Durasorb D. [Van Den Bergh Foods] Partially hydrog. veg. oil (soybean, cottonseed) with polyglycerol esters of fatty acids; high performance fat for peanut butter-filled chocolate, enrobed bars, granola snacks; structuring system for oil migration inhibition; *Properties:* Lovibond 4R max. flakes; drop pt. 52-55 C; iodine no. 4 max.

Duratex. [Van Den Bergh Foods] Partially hydrog. cottonseed oil NF; CAS 68334-00-9; EINECS 269-804-9; lubricant for tablets; release agent; candy dusting and coating for hydroscopic materials; *Regulatory:* kosher; *Properties:* Lovibond 5R max. powd.; m.p. 141-147 F; iodine no. 4 max.

Dur-Em® 114. [Van Den Bergh Foods] Mono- and diglcyerides with BHT and citric acid to protect flavor; nonionic; emulsifier for foods (icings, cakes, margarine, sweet doughs, whipped toppings, veg. dairy prods., reduced fat foods); *Regulatory:* FDA 21CFR §182.4505, GRAS; kosher; *Properties:* cream plastic, typ. odor/flavor; HLB 2.8; m.p. 43-49 C; iodine no. 65-75; flash pt. 300 F min.; 100% conc.; 40% min. alpha monoglyceride; *Storage:* store in cool, dry place away from odor-producing substances.

Dur-Em® 114K. [Van Den Bergh Foods] Mono- and diglycerides from hydrog. veg. oil with BHA, BHT and citric acid; food emulsifier in bakery, prepared mixes, convenience foods; *Properties:* cream plastic; m.p. 110-120 F; 100% act.

Dur-Em® 117. [Van Den Bergh Foods] Mono- and diglycerides with citric acid (preservative); nonionic; food emulsifier for margarine, bread, frozen desserts, veg. dairy prods., danish, candies, chewing gum, reduced fat foods; *Usage level:* 0.04-0.25% (frozen desserts); *Regulatory:* FDA 21CFR §182.4505, GRAS; kosher; *Properties:* wh. beads, flakes, typ. odor/flavor; m.p. 62-65 C; HLB 2.8; iodine no. 5 max.; flash pt. 300 F min.; 100% conc.; 40% min. alpha monoglyceride; *Storage:* store in cool, dry place away from odor-producing substances.

Dur-Em® 117K. [Van Den Bergh Foods] Mono- and diglycerides from hydrog. veg. oil with citric acid; nonionic; food emulsifier for bakery, prepared mixes, convenience foods; *Properties:* wh. bead; m.p. 145-150 F.

Dur-Em® 204. [Van Den Bergh Foods] Mono- and diglycerides with BHT and citric acid (preservatives); nonionic; food emulsifier for icings, cakes, margarine, sweet doughs, whipped toppings, veg. dairy prods., reduced fat foods; *Regulatory:* FDA 21CFR §182.4505, GRAS; kosher; *Properties:* cream plastic, typ. odor/flavor; m.p. 114-121 F; HLB 3.3; iodine no. 65-75; flash pt. 300 F min.; 100% conc., 52% min. mono; *Storage:* store in cool, dry place away from odor-producing substances.

Dur-Em® 204K. [Van Den Bergh Foods] Mono- and diglycerides from hydrog. veg. oil with BHA, BHT and citric acid; nonionic; food emulsifier; *Properties:* cream plastic; m.p. 120-130 F; 100% act.

Dur-Em® 207. [Van Den Bergh Foods] Mono- and diglycerides with citric acid to protect flavor; nonionic; food emulsifier for frozen desserts, whipped toppings, veg. dairy prods., candies, bread, chewing gum, sweet dough; peanut butter stabilizer; *Usage level:* 0.04-0.25% (frozen desserts); *Regulatory:* FDA 21CFR §182.4505, GRAS; kosher; *Properties:* wh. bead, flakes, typ. odor/flavor; m.p. 140-146 F; HLB 3.3; iodine no. 5 max.; flash pt. 300 F min.; 100% conc.; 52% min. alpha mono; *Storage:* store in cool, dry place away from odor-producing substances.

Dur-Em® 207-E. [Van Den Bergh Foods] Mono- and diglcyerides with citric acid to help protect flavor; nonionic; food emulsifier for use in continuous and conventional breads, sweet doughs; emulsifier hydrate for breads and nonfat cakes; crumb softener; also for pharmaceutical, cosmetic uses; *Regulatory:* FDA 21CFR §182.4505, 136.110; *Properties:* wh. free-flowing powd., bland odor/flavor; m.p. 140-146 F; HLB 4.2; iodine no. 5 max.; 100% act.; 50% min. alpha mono; *Storage:* store sealed in cool, dry area away from odor-producing substances; 90 days storage life @ 40-80 F.

Dur-Em® 207K. [Van Den Bergh Foods] Mono- and diglycerides from hydrog. veg. oil with citric acid; food emulsifier; *Properties:* wh. bead or plastic; m.p. 140-146 F; 100% act.

Dur-Em® 300. [Van Den Bergh Foods] Mono- and

diglycerides, propylene glycol; emulsifier, dispersant, solubilizer in flavor and color systems; wetting aid, antifoam agent for food applics.; *Regulatory:* FDA 21CFR §182.4505; *Properties:* clear liq.; HLB 3.0; iodine no. 63 min.; 45% min. mono.

Dur-Em® 300K. [Van Den Bergh Foods] Mono- and diglycerides, propylene glycol; emulsifier; dispersant, solubilizer in flavor and color systems; wetting agent in spray-dried foods; antifoam for high-sugar prods. (juices, jellies) and high protein systems; processing aid for yeast mfg.; *Regulatory:* FDA 21CFR §182.4505, 182.1666; kosher; *Properties:* yel. liq., may cloud at low temp., bland flavor; HLB 2.8; iodine no. 63 min.; flash pt. > 300 F; fire pt. > 300 F; 46% min. monoglyceride; *Storage:* store sealed in dry place away from odor-producing substances.

Dur-Em® GMO. [Van Den Bergh Foods] Glyceryl oleate; nonionic; solubilizer, dispersant, lubricant, wetting aid, penetrant for foods; *Properties:* HLB 2.8; Custom prod.

Durfax® 20. [Van Den Bergh Foods] Polysorbate 20; CAS 9005-64-5; nonionic; food emulsifier; *Properties:* HLB 16.5; sapon. no. 39-52; Custom prod.

Durfax® 60. [Van Den Bergh Foods] Polysorbate 60; CAS 9005-67-8; nonionic; food emulsifier for cakes, whipped toppings, coffee whiteners, icing, confectionery coating; *Usage level:* 0.46% max. (cake), 0.4% max. (whipped toppings, coffee whiteners), 0.46% max. (icings), 0.5% max. (confectionery coatings), 1% max. (shortenings); *Regulatory:* FDA 21CFR §172.836; kosher; *Properties:* soft gel; sol. in water; HLB 14.0; acid no. 2 max.; sapon. no. 45-55; hyd. no. 81-96; 100% conc.; *Storage:* store sealed in a cool, dry place away from odor-producing substances; 180 day storage life @ 40-95 F.

Durfax® 60K. [Van Den Bergh Foods] Polysorbate 60; CAS 9005-67-8; nonionic; food emulsifier for o/w emulsions; *Properties:* yel. visc. liq.; sol. in water, cottonseed oil; HLB 14.9; 100% act.; Custom prod.

Durfax® 65. [Van Den Bergh Foods] Polysorbate 65; CAS 9005-71-4; nonionic; food emulsifier for frozen desserts, cakes, icings, whipped toppings, coffee whiteners; foaming agent for nonalcoholic beverage mixes; pesticide dispersant; *Usage level:* 0.1% max. (frozen desserts), 0.32% max. (cakes, icings), 0.4% max. (whipped toppings, coffee whiteners); *Regulatory:* FDA 21CFR §172.838; kosher; *Properties:* cream plastic; water-disp.; HLB 10.5; acid no. 2 max.; sapon. no. 88-98; hyd. no. 44-60; 100% conc.; *Storage:* store sealed in a cool, dry place away from odor-producing substances; 180 day storage life @ 40-95 F.

Durfax® 65K. [Van Den Bergh Foods] Polysorbate 65; CAS 9005-71-4; nonionic; food emulsifier for o/w emulsions; *Properties:* tan solid; sol. in cottonseed oil, disp. in water; 100% act.; Custom prod.

Durfax® 80. [Van Den Bergh Foods] Polysorbate 80; CAS 9005-65-6; nonionic; food emulsifier for frozen desserts, whipped toppings, shortenings; solubilizer, dispersant in pickles, vitamins; yeast defoamer; personal care prods.; *Usage level:* 0.1%

max. (frozen desserts), 0.4% max. (whipped toppings), 1% max. (shortenings), 4 ppm max. (yeast); *Regulatory:* FDA 21CFR §172.840; kosher; *Properties:* yel. liq.; water-sol.; HLB 15.0; acid no. 3 max.; sapon. no. 45-55; hyd. no. 65-80; 100% conc.; *Storage:* store sealed in a cool, dry place away from odor-producing substances; 180 day storage life @ 40-95 F.

Durfax® 80K. [Van Den Bergh Foods] Polysorbate 80; CAS 9005-65-6; nonionic; food emulsifier for o/w emulsions; *Properties:* yel. liq.; sol. in water, cottonseed oil; HLB 15.9; 100% act.; Custom prod.

Durfax® EOM. [Van Den Bergh Foods] PEG-20 glyceryl stearate; nonionic; food emulsifier for cakes and icing; dough conditioner in yeast-raised bakery goods; bakery hydrates; *Properties:* plastic; HLB 13.5; sapon. no. 65-75; hyd. no. 65-80; 100% conc.; Custom prod.

Durfax® EOM K. [Van Den Bergh Foods] PEG-20 glyceryl stearate; nonionic; food emulsifier, dough conditioner; *Properties:* paste; m.p. 80-85 F; HLB 13.1; 100% act.; Custom prod.

Durkex 25BHA. [Van Den Bergh Foods] Partially hydrog. soybean oil with polyglycerol esters of fatty acids and BHA to help protect flavor; multipurpose shortening for salad oils; also avail. with BHT, TBHQ, lecithin, or no antioxidants; *Regulatory:* kosher; *Properties:* Lovibond 2.0R max. liq.

Durkex 100BHA. [Van Den Bergh Foods] Partially hydrog. soybean oil with BHA to protect flavor; higher stability oil used in frying (snacks, potato chips), coating/spraying (crackers, croutons), antidusting (spices, grains, dry mixes), moisture barrier (ready-to-freeze breads), color/flavor carrier, clouding agent, lubricant; confectionery; *Regulatory:* kosher; *Properties:* Lovibond 1.5R max. liq.; m.p. 67 F max.

Durkex 100BHT. [Van Den Bergh Foods] Partially hydrog. soybean oil with BHT to protect flavor; higher stability oil used in frying (snacks, potato chips), coating/spraying (crackers, croutons), antidusting (spices, grains, dry mixes), moisture barrier (ready-to-freeze breads), color/flavor carrier, clouding agent, lubricant; confectionery; *Regulatory:* kosher; *Properties:* Lovibond 1.5R max. liq.; m.p. 67 F max.

Durkex 100DS. [Van Den Bergh Foods] Partially hydrog. soybean oil, BHA, dimethyl silicone; oil used in frying to minimize foaming when high water content is present, i.e., vegetables; improves oxidative stability of the oil at frying temps.; *Regulatory:* kosher; *Properties:* Lovibond 1.5R max. liq.; m.p. 67 F max.

Durkex 100F. [Van Den Bergh Foods] Partially hydrog. soybean oil; EINECS 232-410-2; oil used in frying (snacks, potato chips), coating/spraying (crackers, croutons), antidusting (spices, grains, dry mixes), moisture barrier (ready-to-freeze breads), color/flavor carrier, clouding agent, lubricant, in confectionery, nondairy veg.; *Regulatory:* kosher; *Properties:* Lovibond 1.5R max. liq.; m.p. 67 F max.

Durkex 100L. [Van Den Bergh Foods] Partially hydrog. soybean oil with lecithin, BHT to protect flavor; lubricant/release agent for use on confectionery molds, cake pans, baking depositor conveyor belts; prevents sticking with no added flavor; *Regulatory:* kosher; *Properties:* Lovibond 3.0R max. liq.; m.p. 67 F max.

Durkex 100TBHQ. [Van Den Bergh Foods] Partially hydrog. soybean oil with TBHQ to protect flavor; higher stability oil used in frying (snacks, potato chips), coating/spraying (crackers, croutons), antidusting (spices, grains, dry mixes), moisture barrier (ready-to-freeze breads), color/flavor carrier, clouding agent, lubricant; confectionery; *Regulatory:* kosher; *Properties:* Lovibond 1.5R max. liq.; m.p. 67 F max.

Durkex 500. [Van Den Bergh Foods] Partially hydrog. veg. oil (cottonseed, soybean); EINECS 269-820-6; high stability oil for coating/spraying (dried fruit, crackers), frying/roasting (nuts), antidusting (gravies, soups), color/flavor carrier, lubricant, moisture barrier, min./coconut oil replacement, in confectionery, nondairy veg. prods.; *Properties:* Lovibond 5.0R max. liq.; drop pt. 21 C max.

Durkex 500S. [Van Den Bergh Foods] Partially hydrog. soybean oil; EINECS 232-410-2; high stability oil; for coating/spraying (dried fruit, crackers), frying/roasting (nuts), antidusting (gravies, soups), color/flavor carrier, lubricant, moisture barrier, min./coconut oil replacement; confectionery centers, nondairy veg. prods.; *Regulatory:* kosher; *Properties:* Lovibond 3.0R max. liq.; drop pt. 21 C max.

Durkex Durola. [Van Den Bergh Foods] Partially hydrog. canola oil; high stability oil for coating/spraying (crackers, snacks); food ingred. where enhanced nutritional profile is desired; *Regulatory:* kosher; *Properties:* Lovibond 1.5R max. liq.; drop pt. 23 C max.

Durkex Gold 77A. [Van Den Bergh Foods] Partially hydrog. soybean oil, artificial flavor, beta carotene; high stability oil used for coating/spraying (crackers, croutons, biscuits, bagels; used where butter color and flavor are desired; *Regulatory:* kosher; *Properties:* Lovibond 15.0R max. liq.; m.p. 67 F max.

Durkex Gold 77F. [Van Den Bergh Foods] Partially hydrog. soybean oil; EINECS 232-410-2; high stability oil used for coating/spraying applics. (crackers, croutons, biscuits, snacks); *Regulatory:* kosher; *Properties:* Lovibond 2.0R max. liq.; m.p. 67 F max.

Durkex Gold 77N. [Van Den Bergh Foods] Partially hydrog. soybean oil, natural flavor, annatto; high stability oil used for coating/spraying (crackers, croutons, biscuits, bagels); used where natural butter color and flavor are desired; *Regulatory:* kosher; *Properties:* Lovibond 8.0R max. liq.; m.p. 67 F max.

Durko. [Van Den Bergh Foods] Partially hydrog. veg. oil (soybean, cottonseed), mono- and diglycerides; shortening used in yeast-raised sweet goods; *Regulatory:* kosher; *Properties:* Lovibond 3.5R max. plastic; 9.5-10.5% mono.

Durkote Calcium Carbonate/Starch, Acacia Gum. [Van Den Bergh Foods] Encapsulated ingred. (calcium chloride substrate with starch/acacia gum coating), USP/FCC; highly conc., directly compressible form of calcium carbonate for use in direct compression tablets and food fortification; provides ease of tableting, improves flow, reduces dusting; *Properties:* wh. to off-wh. free-flowing powd.

Durkote Citric Acid/Hydrog. Veg. Oil. [Van Den Bergh Foods] Encapsulated ingred. (citric acid substrate with hydrog. veg. oil coating); acid used as leavening for prepared mixes, desserts, starch gel candies, confectionery tablets; produces uniform color and flavor replacing traditional fermentation in meat emulsions; *Properties:* wh. free-flowing gran.

Durkote Citric Acid/Maltodextrin, Acacia Gum. [Van Den Bergh Foods] Encapsulated ingred. (citric acid substrate with maltodextrin/acacia gum coating); acid used to adjust flavor and pH in dry mix puddings, beverages, desserts, and teas; prevents hygroscopicity and reactions with other formula components; *Properties:* off-wh. to tan free-flowing gran.; water-sol.

Durkote Ferrous Fumarate/Hydrog. Veg. Oil. [Van Den Bergh Foods] Encapsulated ingred. (ferrous fumarate substrate with hydrog. veg. oil coating); fortification used in general food applics. where taste-masking, time-release, and controlled reactivity are desired; also for vitamin/mineral tablets and capsules; contains no sugar, starch, or free-flow agents; *Properties:* maroon free-flowing gran.

Durkote Ferrous Sulfate/Hydrog. Veg. Oil. [Van Den Bergh Foods] Encapsulated ingred. (ferrous sulfate substrate with hydrog. veg. oil coating); used for iron fortification, improved flow, taste-masking, time-release props., and preventing reaction with other ingreds.; useful in formulating nutritional prods., vitamin/mineral tablets and capsules; *Properties:* tan free-flowing gran.

Durkote Fumaric Acid/Hydrog. Veg. Oil. [Van Den Bergh Foods] Encapsulated ingred. (fumaric acid substrate with hydrog. veg. oil coating); acid used in desserts, leavenings, prepared mixes, starch gel hard candies, and confectionery tablets; reduces problems relating to shelf life, hygroscopicity, color and flavor degradation, and premature pH change; *Properties:* wh. to off-wh. free-flowing gran.

Durkote Glucono-Delta-Lactone/Hydrog. Veg. Oil. [Van Den Bergh Foods] Encapsulated ingred. (glucono-δ-lactone substrate with hydrog. veg. oil coating); provides controlled release, masked taste, improved flow; alternative to natural flora or lactic acid starter culture fermentation in processed meats; also for desserts, fillings, prepared mixes for bakery and dairy, and in pet foods; *Properties:* wh. free-flowing gran.

Durkote Lactic Acid/Hydrog. Veg. Oil. [Van Den Bergh Foods] Encapsulated ingred. (lactic acid substrate with hydrog. veg. oil coating); provides controlled release, alternative to time-consuming and costly fermentation; also for dry mix food applics. where a stable, time-release, conc. form of lactic acid is desired; *Properties:* off-wh. to tan free-flowing gran.

Durkote Malic Acid/Maltodextrin. [Van Den Bergh Foods] Encapsulated ingred. (malic acid substrate with maltodextrin coating); acidification for dry mix beverages, prepared food mixes, and tableted prods.; prevents hygroscopicity, color and flavor degradation, reduces dusting, controls pH, and prevents reactions with incompat. ingreds.; *Properties:* wh. free-flowing gran.; water-sol.

Durkote Potassium Chloride/Hydrog. Veg. Oil. [Van Den Bergh Foods] Encapsulated ingred. (potassium chloride substrate with hydrog. veg. oil coating); used in nutritional prods. and tablets for fortification; and in meat prods. for binding effect where taste masking and reactivity control are desired; *Properties:* wh. free-flowing gran.

Durkote Sodium Aluminum Phosphate/Hydrog. Veg. Oil. [Van Den Bergh Foods] Encapsulated ingred. (sodium aluminum phosphate substrate with hydrog. veg. oil coating); heat-release bakery acid designed for use in frozen cake batters to restrict the release of gas during mixing; *Properties:* wh. free-flowing powd.

Durkote Sodium Bicarbonate/Hydrog. Veg. Oil. [Van Den Bergh Foods] Encapsulated ingred. (sodium bicarbonate substrate with hydrog. veg. oil coating); used in combination with food acids in dry mix baking and other chemically leavened prods. where it is desired to delay and control the reaction of the acid and the base; release occurs during baking cycle; *Properties:* wh. free-flowing powd.

Durkote Sodium Chloride/Hydrog. Veg. Oil. [Van Den Bergh Foods] Encapsulated ingred. (sodium chloride substrate with hydrog. veg. oil coating); used in yeast-containing mixes, bread or pastry doughs, pork sausages, meat patties, pretzel prods. where inhibition of yeast activity, rancidity, excessive salt binding, and water absorption need to be controlled; *Properties:* wh. free-flowing gran.

Durkote Sorbic Acid/Hydrog. Veg. Oil. [Van Den Bergh Foods] Encapsulated ingred. (sorbic acid substrate with hydrog. veg. oil coating); heat-release bakery acid designed for release during baking cycle; *Properties:* wh. to off-wh. free-flowing gran.

Durkote Vitamin B-1/Hydrog. Veg. Oil. [Van Den Bergh Foods] Encapsulated ingred. (vitamin B1 substrate with hydrog. veg. oil coating); used for fortified nutritional prods. and chewable tablets; provides taste masking, temp. and moisture protection, and a degree of time-release; *Properties:* off-wh. to tan free-flowing powd.

Durkote Vitamin C/Hydrog. Veg. Oil. [Van Den Bergh Foods] Encapsulated ingred. (vitamin C substrate with hydrog. veg. oil coating); used for food fortification and vitamin/mineral tablets and capsules; provides exc. stability, temp. and moisture protection, taste masking, protection from reactivity and degradation, and time release props.; *Properties:* tan or wh. to off-wh. free-flowing gran.

Durkote Vitamin C/Maltodextrin. [Van Den Bergh Foods] Encapsulated ingred. (vitamin C substrate with maltodexttrin coating); fortification for dry nutritional mixes; provides formulator with vitamin source protected from premature reactions, degradation and dusting, while prolonging shelf life and improving flow; compat. with pH sensitive sweeteners; *Properties:* wh. free-flowing gran.

Durlac® 100W. [Van Den Bergh Foods] Glyceryl lactopalmitate/stearate with citric acid to protect flavor; nonionic; alpha-tending emulsifier and aerating agent for prepared food mixes, whipped toppings, veg. dairy systems, cakes, cake shortenings; confectionery gloss enhancer; starch gelling agent in industrial processes; *Usage level:* 4-10% (cake shortening), 5-10% fat basis (whipped toppings), 3-5% (instant desserts), 1-3% fat basis (confectionery coatings); *Regulatory:* FDA 21CFR §172.852, EEC, DK, FAO/WHO clearance; *Properties:* Lovibond 10R max. flake, tart flavor; m.p. 46-54 C; HLB 2.4; acid no. 5 max.; iodine no. 5 max.; sapon. no. 245-260; flash pt. > 148 C; fire pt. > 148 C; 100% conc., 6-10% mono; *Storage:* store in cool, dry place away from odor-producing substances; reseal between use.

Durlac® 100WK. [Van Den Bergh Foods] Glyceryl lactoesters of fatty acids from hydrolyzed veg. oil; nonionic; food emulsifier; *Properties:* flake; m.p. 115-130 F; 100% act.

Dur-Lec® P. [Van Den Bergh Foods] Lecithin (phospholipids); CAS 8002-43-5; EINECS 232-307-2; for instant foods, wetting, chewing gum, egg replacement, ice cream; *Properties:* fine gran.; 97% conc.

Dur-Lec® UB. [Van Den Bergh Foods] Lecithin (phospholipids); CAS 8002-43-5; EINECS 232-307-2; emulsifier for baked goods, confections, margarine, pasta; *Properties:* liq.; HLB 4.0; 62% conc.

Durlite F. [Van Den Bergh Foods] Partially hydrog. vegetable oil (soybean, cottonseed); EINECS 269-820-6; multipurpose shortening for vegetable dairy, bakery, and processed foods; *Regulatory:* kosher; *Properties:* solid, flakes; m.p. 98-102 F.

Durlite Gold MBN II. [Van Den Bergh Foods] Partially hydrog. veg. oil (soybean, cottonseed), lecithin, TBHQ, beta carotene, natural flavors; shortening system for bakery items, replacement for cholesterol- and tropical oil-containing shortening systems; *Regulatory:* kosher.

Dur-Lo®. [Van Den Bergh Foods] Mono- and diglycerides with BHA and citric acid to help protect flavor; food emulsifier for reduced fat foods; fat replacement or reduction in sour dressings and other veg. dairy systems, bakery cake mixes, cookies; *Usage level:* 50% Dur-Lo and 50% water to replace fat in a formula; *Regulatory:* FDA 21CFR §182.4505, GRAS; kosher; *Properties:* ivory plastic; m.p. 111-122 F; HLB 2.1; iodine no. 66-70; 100% conc., 17-22% alpha mono; *Storage:* store sealed away from odor-producing substances; 6 mo storage life at 65-80 F.

Durola Select. [Van Den Bergh Foods] Partially hydrog. canola oil; multipurpose shortening for cookies, crackers, biscuits; filler fat for cream centers, whipped toppings; *Regulatory:* kosher; *Properties:* Lovibond 3.0R max. liq.; m.p. 101-105 F.

Duromel. [Van Den Bergh Foods] Partially hydrog. cottonseed oil; CAS 68334-00-9; EINECS 269-804-9; multipurpose shortening for veg. dairy, confectionery, bakery; *Regulatory:* kosher; *Properties:* solid; m.p. 101-105 F.

Duromel B108. [Van Den Bergh Foods] Partially hydrog. cottonseed oil; CAS 68334-00-9; EINECS 269-804-9; nonlauric coating fat for bakery coatings; *Regulatory:* kosher; *Properties:* liq.; m.p. 106-110 F.

Durpeg® 400MO. [Van Den Bergh Foods] PEG-8 oleate; CAS 9004-96-0; nonionic; food emulsiifer, processing aid in animal feeds; *Properties:* amber semisolid; HLB 11.1; sapon. no. 80-88; 100% act.

Durpro® 107. [Van Den Bergh Foods] Propylene glycol mono- and diesters of fats and fatty acids with citric acid; nonionic; emulsifier for cakes and cake mixes, whipped toppings, spray-dried foods; *Properties:* flake; HLB 2.2; m.p. 115-125 F; 100% conc.; 50% min. PGME, 10-15% mono; Custom prod.

Durtan® 60. [Van Den Bergh Foods] Sorbitan stearate; CAS 1338-41-6; EINECS 215-664-9; nonionic; food emulsifier, gloss enhancer for chocolate coatings, cakes, icing, whipped topping, coffee whiteners, confectionery coatings, beverage mix; dispersant for inorganics used in thermoplastics; *Usage level:* 0.61% max. (cake), 0.4% max. (whipped topplngs, coffee whiteners), 0.7% max. icings, 1% max. (confectionery coatings); *Regulatory:* FDA 21CFR §172.842; kosher; *Properties:* cream flakes; HLB 4.7; acid no. 5-10; sapon. no. 147-157; hyd. no. 235-260; 100% conc.; *Storage:* store sealed in a cool, dry place away from odor-producing substances; 180 day storage life @ 40-80 F.

Durtan® 60K. [Van Den Bergh Foods] Sorbitan stearate; CAS 1338-41-6; EINECS 215-664-9; nonionic; food emulsifier; *Properties:* cream beads; m.p. 121-127 F; HLB 4.7; 100% act.

Dycol™ 4000FC. [Rhone-Poulenc Food Ingreds.] Guar gum; CAS 9000-30-0; EINECS 232-536-8; premium food-grade natural gum which does not require heat to develop full visc.; moderate hydration, very good dispersibility, freeze-thaw stable, pH stable (4-9); *Usage level:* 0.2-0.4% (cake mixes), 0.1-0.3% (processed cheese), 0.1-0.25% (ice cream), 0.1-0.25% (instant soups/gravies), 0.15-0.25% (instant hot cereal), 0.2-0.35% (fresh tortillas), 0.1-0.3% (cocoa mix); *Properties:* coarse gran.; sol. in cold water; high visc.

Dycol™ 4500F. [Rhone-Poulenc Food Ingreds.] Guar gum; CAS 9000-30-0; EINECS 232-536-8; premium food-grade natural gum which does not require heat to develop full visc.; fast hydration, good dispersibility with proper agitation, freeze-thaw stable, pH stable (4-9); *Usage level:* 0.2-0.4% (cake mixes), 0.1-0.3% (processed cheese), 0.1-0.25% (ice cream), 0.1-0.25% (instant soups/gravies), 0.15-0.25% (instant hot cereal), 0.2-0.35% (fresh tortillas), 0.1-0.3% (cocoa mix); *Properties:* med. fine gran.; sol. in cold water; high visc.

Dycol™ HV400F. [Rhone-Poulenc Food Ingreds.] Guar gum; CAS 9000-30-0; EINECS 232-536-8; premium food-grade natural gum which does not require heat to develop full visc.; very fast hydration, fair dispersibility with proper agitation, freeze-thaw stable, pH stable (4-9); *Usage level:* 0.2-0.4% (cake mixes), 0.1-0.3% (processed cheese), 0.1-0.25% (ice cream), 0.1-0.25% (instant soups/gravies), 0.15-0.25% (instant hot cereal), 0.2-0.35% (fresh tortillas), 0.1-0.3% (cocoa mix); *Properties:* creamy wh. fine powd.; 95% min. thru 100 mesh, 80% min. thru 200 mesh; sol. in cold water; visc. 5000 cps (1%); pH ≈ 6; 11% moisture.

Dynacet® 212. [Hüls Am.; Hüls AG] Acetylated monoglycerides of edible fatty acids; nonionic; food emulsifier; fat base for prep. of dip coatings; coating agent for shortenings, toppings, meat and sausages; *Properties:* block; m.p. 35 C; acid no. 1 max.; iodine no. 3 max.; sapon. no. 350; 100% conc.

Dynacet® 278. [Hüls AG] Acetylated monoglyceride with plasticizer; dip coating for cooked and boiled sausages processed at 80-90 C; *Properties:* block; m.p. 35-45 C; acid no. 1; iodine no. 1 max.

Dynacet® 281. [Hüls AG] Acetylated monoglyceride with plasticizer; dip coating for cooked and boiled sausages processed at 120-140 C; *Properties:* block; m.p. 38-45 C; acid no. 1; iodine no. 1 max.

Dynacet® 282. [Hüls AG] Acetylated monoglyceride with plasticizer; dip coating for cooked and boiled sausages processed at 130-150 C; *Properties:* block; m.p. 45-50 C; acid no. 1; iodine no. 1 max.

Dynacet® 285. [Hüls AG] Acetoglyceride based on edible fats; lubricant and release agent in prod. of sausage casings; *Properties:* liq.; dens. 0.98 kg/dm³; visc. 45 mPa•s; acid no. 0.3 max.; iodine no. 2 max.; sapon. no. 385-415.

Dynasan® 110. [Hüls Am.; Hüls AG] Tricaprin; CAS 621-71-6; EINECS 210-702-0; for prep. of dietary foods, margarine, confectionery, milk prods., fruit diets; *Properties:* wh. block; m.p. 28-31 C; clear pt. 30 C; acid no. 0.2 max.; iodine no. 1 max.; sapon. no. 295-305.

Dynasan® 112. [Hüls Am.; Hüls AG] Trilaurin; CAS 538-24-9; EINECS 208-687-0; aux. in prod. of compressed sweets; *Properties:* wh. powd.; m.p. 43-47 C; clear pt. 45 C; acid no. 0.3 max.; iodine no. 1 max.; sapon. no. 257-266.

Dynasan® 114. [Hüls Am.; Hüls AG] Trimyristin; CAS 555-45-3; EINECS 209-099-7; binder, lubricant for tablets and compressed confectioneries; *Properties:* wh. powd./flakes; m.p. 55-58 C; clear pt. 57 C; acid no. 0.3 max.; iodine no. 1 max.; sapon. no. 230-238.

Dynasan® 116. [Hüls Am.; Hüls AG] Tripalmitin; CAS 555-44-2; EINECS 209-098-1; aux. in prod. of compressed sweets; *Properties:* wh. microcryst. powd.; sol. in ether and benzene; m.p. 61-65 C; clear pt. 63 C; acid no. 0.5 max.; iodine no. 1 max.; sapon. no. 205-210.

Dynasan® 118. [Hüls Am.; Hüls AG] Tristearin; CAS 555-43-1; EINECS 209-097-6; nonionic; crystallization accelerator in chocolate, compound coatings, and other fat prods.; *Properties:* wh. microcryst. powd./flakes; sol. in ether and benzene; m.p. 69-73 C; clear pt. 71 C; acid no. 0.5 max.; iodine no. 1 max.; sapon. no. 186-192.

Dynasan® 119. [Hüls AG] Mono- and diglycerides based on edible fats; lubricant and release aux. in prod. of compressed sweets; *Properties:* powd.; acid no. 1 max.; iodine no. 6 max.; sapon. no. 170-190.

Dynasan® 182. [Hüls AG] Tricaprin/skimmed milk powd.; fat for prod. of dietary foods containing MCT; *Properties:* powd.

Dynasan® P60. [Hüls Am.; Hüls AG] Hydrog. palm oil; consistency regulator for fats and fat preps.; coating fat for salt, spices, citric acid; stabilizer to prevent oil separation in nut paste; *Properties:* wh. to off-wh. powd.; m.p. 56-60 C; acid no. 0.3 max.; iodine no. 3 max.; sapon. no. 190-210.

E

Eastman Potassium Sorbate. [Eastman] Potassium sorbate NF, FCC; EINECS 246-376-1; antimicrobial preservative for foods incl. dairy prods., baked goods, fruit/berry prods., veg. prods., dry sausage casings, diet drinks, pet foods, mayonnaise, and salad dressings; *Usage level:* 0.2-0.3% (cheese), 0.025-0.075 (fruit drinks), 0.05-0.1% (cakes), 0.1% (margarine); *Regulatory:* FDA GRAS §182.3089, 182.3640, 21CFR 101.22(j); over 90 approvals; *Properties:* wh. free-flowing powd. and gran., mild char. odor; 2% max. on 40 mesh (powd.), 80% min. on 35 mesh (gran.); m.w. 150.22; sol.: 58.2% in water, 45.3% in 50% ethanol, 19% in propylene glycol; sp.gr. 1.36; flash pt. (COC) none; 98-101% assay; *Storage:* store under cool dry conditions; 12 mo shelf life; minimize exposure above 37.8 C.

Eastman Sorbic Acid. [Eastman] Sorbic acid NF, FCC; EINECS 203-768-7; antimicrobial preservative for foods incl. dairy prods., baked goods, fruit/ berry prods., veg. prods., dry sausage casings, diet drinks, pet foods, mayonnaise, and salad dressings; *Usage level:* 0.2-0.3% (cheese), 0.025-0.075 (fruit drinks), 0.05-0.1% (cakes), 0.1% (margarine); *Regulatory:* FDA GRAS §182.3089, 182.3640, 21CFR 101.22(j); over 90 approvals; *Properties:* wh. free-flowing powd. and dust-free powd., mild char. odor; 50% min. on 140 mesh (powd.), 50% min. 100 mesh (dust-free); m.w. 112.13; sol.: 12.9% in 100% ethanol, 11.5% in glac. acetic acid, 9.2% in acetone, 0.15% in water; m.p. 132-135 C; flash pt. (COC) 127 C; 99-101% assay; *Storage:* store under cool dry conditions; 12 mo shelf life; minimize exposure above 37.8 C.

EC-25®. [Van Den Bergh Foods] Propylene glycol mono- and diesters of fats and fatty acids, mono- and diglycerides, partially hydrog. soybean oil, lecithin, BHT, citric acid; nonionic; food emulsifier system for cakes, cake mixes; used with all-purpose shortenings for conventional cakes or with emulsified shortenings for very lean cakes, cookies, fat reduced foods; *Usage level:* 12-25% of fat (layer cake), 5-10% (dry mixes), 15-25% of fat (soft cookies); *Regulatory:* FDA 21CFR §172.856, 182.4505; kosher; *Properties:* lt. straw plastic; m.p. 90-100 F; HLB 2.6; 100% act., 34-38% propylene glycol monoester; *Storage:* store sealed in cool, dry place away from odor-producing substances; 6 mo storage life at 65-85 F.

Edible Beef Gelatin. [GMI Prods.] Type B or calfskin gelatin; CAS 9000-70-8; EINECS 232-554-6; biodeg. edible protein; *Regulatory:* GRAS; *Properties:* pale yel. vitreous brittle solid, nearly odorless and tasteless; sol. in warm water; swells due to hydration in cold water; sp.gr. 1.2; dens. 1.3-1.4; b.p. > 100 C (dec.); pH 5.0-7.5; flash pt. nonflamm.; 9-13% moisture; *Toxicology:* nuisance dust; contact with dust causes mild eye irritation; inhaling dust may cause respiratory irritation; *Storage:* store in airtight containers @ R.T.; avoid exposure to water or excessive heat.

Edible Gelatins. [Croda Colloids Ltd; O.C. Lugo] Gelatin, bloom gel strengths from 30-270; CAS 9000-70-8; EINECS 232-554-6; natural food used in cider fining, ice cream, jellies, licorice, canned meat, nougat, pies, soups, wine fining, yogurt, toffees, etc.; compat. with sugars, starches, gums, seasonings, colors, and flavorings used in food industry; *Properties:* colorless or pale yel. powd. or gran.; not sol. in cold water, but will absorb several times its wt. of water; pH 5.0-6.5; 13% max. moisture; *Toxicology:* nontoxic; nonirritant to skin and eyes; avoid prolonged exposure to finely milled powds.; *Precaution:* burns if ignited; dusts present potential explosive risk; *Storage:* store in sealed containers in cool, dry conditions away from odiferous materials; 5 yr max. storage life.

Edicol®. [Indian Gum Industries; Commodity Services Int'l.] Guar gum; CAS 9000-30-0; EINECS 232-536-8; nonionic; rheology control aid, viscosifier, thickener, processing aid, binder, protective colloid, dough improver in cheese, baked goods, beverages, canned foods, soups, salad dressing; prevents ice crystal formation; stabilizer in sour dairy prods.; *Properties:* wh. to cream-wh. powd., sl. char. grassy or beany odor, neutral taste; sol. in hot and cold water.

Edicol® ULV Series. [Indian Gum Industries; Commodity Services Int'l.] Ultra low visc. guar gum; CAS 9000-30-0; EINECS 232-536-8; nonionic; fat replacer, source of soluble fiber, protective colloid for food and animal feed industry, milk prods., instant powd. prods., dietary prods., diabetic foods, sour milk drinks, fruit concs., fruit juices, spray-dried flavors, vitamins, meat prods.; *Regulatory:* EEC, FAO/WHO, FCC compliance; *Properties:* cream-wh. powd.; > 90% thru 200 mesh; visc.:

various grades from 30-1100 cps; pH 5.5-6.5; 13% max. moisture; *Storage:* store in cool, dry place out of the sun; 1 yr shelf life.

Edifas. [Imperial Chem. Ind. plc] Vegetable derivs.; for use in food mfg. processes as binding, stabilizing, and emulsifying agents.

Edigel® 100 Powd. Wheat Starch. [ADM Ogilvie] Wheat starch; CAS 9005-25-8; filler binder for sausage extrusion and pelletized foods; *Properties:* off-wh. powd.; pH 3.8; 8% moisture, 3.5% protein.

Eez-Out™. [MLG Enterprises Ltd.] Vegetable oil-based; anti-sticking lubricants for processing of meat and poultry prods.; applied to molds, netting, racks, trays prior to cooking/smoking of processed meat and poultry prods.; *Regulatory:* USDA approved; *Properties:* pale yel. paste, thick or thin liq., bland taste.

Egg White Solids Type P-11. [Henningsen Foods] Dehydrated egg whites; CAS 9006-50-2; a nonwhipping egg white with higher than normal ovomucin content which enhances binding and toughening props.

Egg White Solids Type P-18. [Henningsen Foods] Dehydrated egg whites with sodium lauryl sulfate as whipping aid; a whipping egg white for the confection industry.

Egg White Solids Type P-18G. [Henningsen Foods] Dehydrated egg whites without additives; CAS 9006-50-2; a whipping egg white used in aerated confections or where clarity of sol'n. is important.

Egg White Solids Type P-19. [Henningsen Foods] Dehydrated egg whites; CAS 9006-50-2; similar to P-20 except it does not contain whipping additives; widely used for bakery and confectionery applics.

Egg White Solids Type P-20. [Henningsen Foods] Albumen; CAS 9006-50-2; 80% min. protein; extremely rapid foaming; suggested for large commercial bakeries; used in angel food cakes, meringues, toppings, chiffon pies, candy; *Properties:* powd., bland odor; 100% thru 80 mesh, 90% thru 100 mesh; pH 6.5-7.5; 8% max. moisture; *Storage:* store dry, no refrigeration needed.

Egg White Solids Type P-20 CMC. [Henningsen Foods] Dehydrated egg whites, cellulose gum; combines foaming props. of P-20 with exc. foam stability by addition of cellulose gum.

Egg White Solids Type P-21. [Henningsen Foods] Dehydrated egg whites; CAS 9006-50-2; exceptional flavor chars. for use alone or with yolk to prepare omelettes, custards, etc.

Egg White Solids Type P-25. [Henningsen Foods] Dehydrated egg whites; CAS 9006-50-2; nondusty modification of P-20 contg. a whipping additive.

Egg White Solids Type P-39. [Henningsen Foods] Dehydrated egg whites; CAS 9006-50-2; superior gelation and binding props.; pH similar to freshly broken liq. egg white.

Egg White Solids Type P-110. [Henningsen Foods] Dehydrated egg whites; CAS 9006-50-2; similar to P011 but contains all egg white proteins in natural proportion; used where binding and coagulation props. are essential.

Egg White Solids Type P-110 High Gel Strength. [Henningsen Foods] Dehydrated egg whites; CAS 9006-50-2; a neutral pH prod. with exc. gelation and binding ability.

Egg White Solids Type PF-1. [Henningsen Foods] Dehydrated egg whites; CAS 9006-50-2; high in ovomucin content; ideal for use as binder in pasta.

Egg Yolk Solids Type Y-1. [Henningsen Foods] Dried egg yolk with 40-60 ppm β-carotene color; 30% min. protein, 56% min. fat; for sweet doughs, cake mixes, baby foods, sauces, pie fillings, egg noodles, ice cream, eggnog, mayonnaise, salad dressings; *Properties:* powd.; 100% thru 16 mesh; pH 6.5 ± 0.3; 5% max. moisture; *Storage:* store @ 40-50 F for storage over 30 days.

Egg Yolk Solids Type Y-1-FF. [Henningsen Foods] Dehydrated egg yolk with sodium silicoaluminate; provides visc. and exc. stability to mayonnaise and salad dressings; contains free-flow agent; *Usage level:* 1 lb/1$^1/_4$ lb water to replace 2$^1/_4$ lb liq. or frozen yolk.

Egg Yolk Solids Type Y-2. [Henningsen Foods] Dehydrated egg yolk stabilized by removal of natural glucose; used in cake mixes, frozen doughs, and other foods where long shelf life is required.

Egg Yolk Solids Type Y-2-FF. [Henningsen Foods] Dehydrated egg yolk stabilized by removal of natural glucose contg. free-flow agent; used in cake mixes, frozen doughs, and other foods where long shelf life is required.

Eisai Natural Vitamin E Series. [Eisai; Unipex] Mixt. of α, β, γ, and δ tocopherols; antioxidant for foods; increases stability of animal fats, vegetable oils, processed food prods., liposoluble vitamins; prevents discoloration and browning of colorants; *Usage level:* 0.02-0.1% of fat and oil content; *Regulatory:* USA GRAS; *Properties:* oil, emulsion, or emulsified powd.

Elasdo +. [Paniplus] Mono- and diglycerides and ethoxylated mono- and diglycerides; food emulsifier; *Properties:* bead; HLB 7.7; Unverified

Elasdo Power 70. [ADM Arkady] Ethoxylated monoglyceride and mono- and diglycerides; dough conditioner/strengthener providing higher level of dough strength for variety breads; *Properties:* plastic; HLB 8.8; 100% conc.

Elvanol® 71-30. [DuPont] Fully hydrolyzed PVAL; CAS 9002-89-5; EINECS 209-183-3; food additive; *Properties:* wh. gran.; sol. in hot water, ethanol; insol. in cold water; sp.gr. 1.30; dens. 400-432 kg/m3; visc. 28-32 cps (4% aq.); sapon. no. 3-12; ref. index 1.54; pH 5.0-7.0; tens. str. 117 MPa; tens. elong. 10% (break) to 400%; hardness > 100; biodeg.

Emalex DISG-2. [Nihon Emulsion] Diglyceryl diisostearate; CAS 67938-21-0; EINECS 267-821-6; food emulsifier; *Regulatory:* JCID compliance; *Properties:* pale yel. liq.; HLB 2.

Emalex DISG-3. [Nihon Emulsion] Polyglyceryl-3 diisostearate; CAS 66082-42-6; food emulsifier; *Regulatory:* JCID compliance; *Properties:* pale yel. liq.; HLB 4.

Emalex DISG-5. [Nihon Emulsion] Polyglyceryl-5 diisostearate; food emulsifier; *Properties:* pale yel. liq.; HLB 6.

Emalex DSG-2. [Nihon Emulsion] Polyglyceryl-2 distearate; food emulsifier; *Properties:* cream-colored solid; HLB 2.

Emalex DSG-3. [Nihon Emulsion] Polyglyceryl-3 distearate; CAS 94423-19-5; food emulsifier; *Properties:* cream-colored solid; HLB 3.

Emalex DSG-5. [Nihon Emulsion] Polyglyceryl-5 distearate; food emulsifier; *Properties:* cream-colored solid; HLB 6.

Emalex GMS-10SE. [Nihon Emulsion] Glyceryl monostearate, SE; surfactant for food applics.; *Regulatory:* JSCI compliance; *Properties:* cream-colored wax; HLB 6.

Emalex GMS-15SE. [Nihon Emulsion] Glyceryl monostearate, SE; surfactant for food applics.; *Regulatory:* JSCI compliance; *Properties:* cream-colored soft wax; HLB 9.

Emalex GMS-20SE. [Nihon Emulsion] Glyceryl monostearate, SE; surfactant for food applics.; *Regulatory:* JSCI compliance; *Properties:* cream-colored wax; HLB 8.

Emalex GMS-25SE. [Nihon Emulsion] Glyceryl monostearate, SE; surfactant for food applics.; *Regulatory:* JSCI compliance; *Properties:* cream-colored soft wax; HLB 10.

Emalex GMS-45RT. [Nihon Emulsion] Glyceryl monostearate, SE; surfactant for food applics.; *Regulatory:* JSCI compliance; *Properties:* cream-colored wax; HLB 5.

Emalex GMS-50. [Nihon Emulsion] Glyceryl monostearate, SE; surfactant for food applics.; *Regulatory:* JSCI compliance; *Properties:* cream-colored wax; HLB 12.

Emalex GMS-55FD. [Nihon Emulsion] Glyceryl monostearate, SE; surfactant for food applics.; *Regulatory:* JSCI compliance; *Properties:* cream-colored wax; HLB 7.

Emalex GMS-195. [Nihon Emulsion] Glyceryl monostearate, SE; surfactant for food applics.; *Regulatory:* JSCI compliance; *Properties:* cream-colored wax; HLB 6.

Emalex GMS-A. [Nihon Emulsion] Glyceryl monostearate; surfactant for food applics.; *Regulatory:* JSCI compliance; *Properties:* cream-colored beads; HLB 3.

Emalex GMS-ASE. [Nihon Emulsion] Glyceryl monostearate, SE; surfactant for food applics.; *Regulatory:* JSCI compliance; *Properties:* cream-colored beads; HLB 7.

Emalex GMS-B. [Nihon Emulsion] Glyceryl monostearate; surfactant for food applics.; *Regulatory:* JSCI compliance; *Properties:* cream-colored beads; HLB 5.

Emalex GMS-P. [Nihon Emulsion] Glyceryl monopalmitate; CAS 26657-96-5; EINECS 247-887-2; surfactant for food applics.; *Regulatory:* JCID compliance; *Properties:* cream-colored beads; HLB 4.

Emalex GWIS-100. [Nihon Emulsion] Glyceryl monoisostearate; surfactant for food applics.; *Regulatory:* JCID compliance; *Properties:* pale yel. paste; HLB 5.

Emalex HC-50. [Nihon Emulsion] PEG-50 hydrogenated castor oil; CAS 61788-85-0; nonionic; solubilizer for oil-sol. vitamins; *Regulatory:* JSCI compliance; *Properties:* cream-colored soft wax; HLB 13.

Emalex HC-60. [Nihon Emulsion] PEG-60 hydrogenated castor oil; CAS 61788-85-0; nonionic; solubilizer for oil-sol. vitamins; *Regulatory:* JSCI compliance; *Properties:* cream-colored soft wax; HLB 14.

Emalex MSG-2. [Nihon Emulsion] Diglyceryl monostearate; CAS 12694-22-3; EINECS 235-777-7; food emulsifier; *Regulatory:* JCID compliance; *Properties:* cream-colored solid; HLB 7.

Emalex MSG-2MA. [Nihon Emulsion] Diglyceryl monostearate; CAS 12694-22-3; EINECS 235-777-7; food emulsifier; *Regulatory:* JCID compliance; *Properties:* cream-colored solid; HLB 5.

Emalex MSG-2MB. [Nihon Emulsion] Diglyceryl monostearate; CAS 12694-22-3; EINECS 235-777-7; food emulsifier; *Regulatory:* JCID compliance; *Properties:* cream-colored solid; HLB 6.

Emalex MSG-2ME. [Nihon Emulsion] Diglyceryl monostearate; CAS 12694-22-3; EINECS 235-777-7; food emulsifier; *Regulatory:* JCID compliance; *Properties:* cream-colored solid; HLB 7.

Emalex MSG-2ML. [Nihon Emulsion] Diglyceryl monostearate; CAS 12694-22-3; EINECS 235-777-7; food emulsifier; *Regulatory:* JCID compliance; *Properties:* cream-colored solid; HLB 8.

Emalex PGML. [Nihon Emulsion] Propylene glycol monolaurate; CAS 142-55-2; EINECS 205-542-3; nonionic; surfactant for food applics.; *Regulatory:* JCID compliance; *Properties:* pale yel. oil; HLB 4.

Emalex PGMS. [Nihon Emulsion] Propylene glycol monostearate; CAS 1323-39-3; EINECS 215-354-3; nonionic; surfactant for food applics.; *Regulatory:* JSCI compliance; *Properties:* wh. powd.; HLB 2.

Emalex PGO. [Nihon Emulsion] Propylene glycol monooleate; CAS 1330-80-9; EINECS 215-549-3; nonionic; surfactant for food applics.; *Regulatory:* JCID compliance; *Properties:* pale yel. oil; HLB 2.

Emalex PGS. [Nihon Emulsion] Propylene glycol monostearate; CAS 1323-39-3; EINECS 215-354-3; nonionic; surfactant for food applics.; *Regulatory:* JSCI compliance; *Properties:* wh. wax; HLB 2.

Emalex TISG-2. [Nihon Emulsion] Diglyceryl triisostearate; CAS 120486-24-0; food emulsifier; *Regulatory:* JCID compliance; *Properties:* pale yel. liq.; HLB 0.

Emalex VS-31. [Nihon Emulsion] Mixt. of nonionic surfactants; nonionic; solubilizer for oil-sol. vitamins; *Properties:* pale yel. soft paste; HLB 14.

Emargol® KL. [Witco/H-I-P] Sodium sulfoacetate of mono- and diglycerides; anionic; food processing defoamer and o/w emulsifier; dough modifier; *Properties:* paste; disp. in oil and water; Unverified

Emargol® L. [Witco/H-I-P] Sulfoacetate of fatty glycerides; anionic; food emulsifier; *Properties:* liq.; 100% conc.

EMB. [Bunge Foods] Blend of food acids, yeast nutrients, and conditioners; sour; English muffin base providing uniform fermentation, exc. flavor and porosity; can be used for many varieties of English muffins, e.g., high fiber oat bran, raisin, fruit and nut, corn; *Usage level:* 10% based on flour.

Embanox®. [Rhone-Poulenc Food Ingreds.] Antioxidant for use in processed foods, baking, cereal, and confections.

Embanox® 2. [Rhone-Poulenc Food Ingreds.] Veg. oil, BHT, and BHA.

Embanox® 3. [Rhone-Poulenc Food Ingreds.] Propylene glycol, BHA, propyl gallate, and citric acid.

Embanox® 4. [Rhone-Poulenc Food Ingreds.] Propylene glycol, BHA, and citric acid.

Embanox® 5. [Rhone-Poulenc Food Ingreds.] Veg. oil and BHA.

Embanox® 6. [Rhone-Poulenc Food Ingreds.] Propylene glycol, BHA, citric acid, and propyl gallate.

Embanox® BHA. [Rhone-Poulenc Food Ingreds.] BHA; food-grade antioxidant, preservative; *Properties:* wh. or sl. cream coarse cryst. powd., char. odor; m.w. 180.2; m.p. $\geq$ 50 C; $\geq$ 90% act.

Embanox® BHT. [Rhone-Poulenc Food Ingreds.] BHT; CAS 128-37-0; EINECS 204-881-4.

Embanox® TBHQ. [Rhone-Poulenc Food Ingreds.] t-Butyl hydroquinone; CAS 1948-33-0; EINECS 217-752-2; food-grade antioxidant; *Properties:* wh. to sl. cream cryst. powd., char. odor; m.w. 166.2; m.p. 126.5-128.5 C; 99-100.5% assay.

Emersol® 6313 NF. [Henkel/Emery] Low-titer oleic acid USP/NF; CAS 112-80-1; EINECS 204-007-1; food grade fatty acid; *Regulatory:* FDA 21CFR §172.860; EPA-exempt; *Properties:* lt. yel. clear liq., fatty acid odor; insol. in water; sp.gr. 0.891 (25/20 C); m.p. 14 C; b.p. 286 C (100 mm); acid no. 201-204; iodine no. 88-93; flash pt. (CC) 184-189 C; *Toxicology:* LD50 (oral, rat) > 21.5 ml/kg; mild eye and skin irritant; *Storage:* store in closed containers away from heat and open flames; avoid contact with strong oxidizers, alkalies.

Emersol® 6320. [Henkel/Emery] Stearic acid, double pressed; CAS 57-11-4; EINECS 200-313-4; food grade fatty acid; *Regulatory:* FDA 21CFR §172.860; EPA-exempt; *Properties:* wh. waxy solid, fatty acid odor; insol. in water; sp.gr. 0.85 (75/20 C); m.p. 52-57 C; b.p. 383 C; acid no. 205-210; iodine no. 3.5-5.0; flash pt. (COC) 185 C; *Toxicology:* LD50 (oral, rat) > 10 g/kg; mild eye and skin irritant; *Storage:* stored in closed containers away from het or open flames; avoid contact with strong oxidizers, alkalies.

Emersol® 6321 NF. [Henkel/Emery] Low-titer wh. oleic acid UPS/NF; CAS 112-80-1; EINECS 204-007-1; food grade fatty acid; *Regulatory:* FDA 21CFR §172.860; EPA-exempt; *Properties:* lt. clear liq., fatty acid odor; insol. in water; sp.gr. 0.891 (25/20 C); m.p. 14 C; b.p. 286 C (100 mm); acid no. 201-204; iodine no. 87-92; flash pt. (CC) 184-189 C; *Toxicology:* LD50 (oral, rat) > 21.5 ml/kg; mild eye, moderate skin irritant; *Storage:* store in closed containers away from heat and open flames; avoid

contact with strong oxidizers, alkalies.

Emersol® 6332 NF. [Henkel/Emery] Stearic acid, triple pressed USP/NF; CAS 57-11-4; EINECS 200-313-4; food grade fatty acid; *Regulatory:* FDA 21CFR §172.860; EPA-exempt; *Properties:* wh. waxy solid, fatty acid odor; insol. in water; sp.gr. 0.85 (75/20 C); m.p. 52-57 C; b.p. 383 C; acid no. 205-211; iodine no. 0.5 max.; flash pt. (COC) 185 C; *Toxicology:* LD50 (oral, rat) > 10 g/kg; mild eye and skin irritant; *Storage:* store in closed containers away from heat and open flames; avoid contact with strong oxidizers, alkalies.

Emersol® 6333 NF. [Henkel/Emery] Low-linoleic content oleic acid USP/NF; CAS 112-80-1; EINECS 204-007-1; food grade fatty acid; *Regulatory:* FDA 21CFR §172.860; EPA-exempt; *Properties:* lt. clear liq., fatty acid odor; insol. in water; sp.gr. 0.891 (25/20 C); m.p. 14 C; b.p. 286 C (100 mm); acid no. 200-204; iodine no. 86-91; flash pt. (CC) 184-189 C; *Toxicology:* LD50 (oral, rat) > 21.5 ml/kg; mild eye, moderate skin irritant; *Storage:* store in closed containers away from heat and open flames; avoid contact with strong oxidizers, alkalies.

Emersol® 6343. [Henkel/Emery] Palmitic acid; CAS 57-10-3; EINECS 200-312-9; food grade fatty acid; *Regulatory:* FDA 21CFR §172.860; *Properties:* wh. waxy solid, fatty acid odor; insol. in water; sp.gr. 0.85 (75/25 C); m.p. 58-61 C; b.p. 215 C (15 mm); flash pt. 185 C; *Toxicology:* LD50 (oral, rat) > 10 g/kg; nonirritating to skin and eyes; *Storage:* keep in closed containers; avoid heat and open flames, contact with strong oxidizers and alkalies.

Emersol® 6349. [Henkel/Emery] Stearic acid; CAS 57-11-4; EINECS 200-313-4; food grade fatty acid; *Regulatory:* FDA 21CFR §172.860; EPA-exempt; *Properties:* wh. waxy flakes, fatty acid odor; insol. in water; sp.gr. 0.85 (75/25 C); m.p. 59-60.5 C; b.p. 383 C (760 mm); acid no. 203-206; iodine no. 0.5 max.; flash pt. (COC) 185 C; *Toxicology:* LD50 (oral, rat) > 10 g/kg; mild eye irritant; *Storage:* keep in closed containrs; avoid heat and open flames, contact with strong oxidizers and alkalies.

Emersol® 6351. [Henkel/Emery] Stearic acid; CAS 57-11-4; EINECS 200-313-4; food grade fatty acid; *Regulatory:* FDA 21CFR §172.860; EPA-exempt; *Properties:* wh. waxy solid, fatty acid odor; insol. in water; sp.gr. 0.85 (75/25 C); m.p. 65-68 C; b.p. 383 C; acid no. 196-201; iodine no. 1.0 max.; flash pt. (COC) 185 C; *Toxicology:* LD50 (oral, rat) > 10 g/kg; mild eye and skin irritant; *Precaution:* flamm. dust; concs. as low as 0.017 oz/ft^3 in air can burn or explode if ignited in confined space; avoid contact with strong oxidizers and alkalies; *Storage:* store in closed containers away from heat and open flames.

Emersol® 6353. [Henkel/Emery] Stearic acid; CAS 57-11-4; EINECS 200-313-4; food grade fatty acid; *Regulatory:* FDA 21CFR §172.860; *Properties:* acid no. 195-199; iodine no. 1 max.

Emersol® 6357. [Henkel/Emery] Caprylic acid; CAS 124-07-2; EINECS 204-677-5; food grade fatty acid; *Regulatory:* FDA 21CFR §172.860; *Proper-*

ties: pract. water-wh. liq., penetrating short-chain fatty acid odor; sol. 0.25% in water (100 C); sp.gr. 0.910 (20/4 C); m.p. 16 C; b.p. 240 C; flash pt. (CC) 110 C; 99% act.; *Toxicology:* LD50 (oral, rat) 1.3-10.1 g/kg; severe eye irritant; corrosive; causes severe burns; avoid breathing vapor; *Precaution:* avoid contact with oxidizing materials and basic chems.; *Storage:* store in cool, dry, well-ventilated areas away from sunlight, heat, and flames; container hazardous when empty.

Emersol® 7021. [Henkel/Emery] Oleic acid; CAS 112-80-1; EINECS 204-007-1; food grade fatty acid; *Regulatory:* FDA 21CFR §172.860; EPA-exempt; kosher; *Properties:* acid no. 196-204; iodine no. 93-104.

Emersol® 7051. [Henkel/Emery; Henkel Canada] Stearic acid; CAS 57-11-4; EINECS 200-313-4; food grade fatty acid; *Regulatory:* kosher; *Properties:* wh. waxy solid, fatty acid odor; insol. in water; sp.gr. 0.85 (75/25 C); m.p. 65-68 C; b.p. 383 C; flash pt. (COC) 185 C; 88% act.; *Toxicology:* LD50 (oral, rat) > 10 g/kg; mild eye irritant; *Precaution:* avoid contact with strong oxidizers and alkalies; *Storage:* store in closed containers away from heat and open flames.

Emery® 912. [Henkel/Emery] CP/USP glycerin; CAS 56-81-5; EINECS 200-289-5; visc. modifier, flavor enhancer, solv., humectant, thickener, and solubilizer in food applics.; *Regulatory:* EPA-exempt; *Properties:* water-wh. visc. liq.; odorless; sol. in water; sp.gr. 1.2517; m.p. 18 C; b.p. 171 C; pour pt. 18 C; flash pt. (CC) 199 C; 96.0% min. glycerol; *Toxicology:* TLV:TWA 10 mg/m³ (mist), LD50 (oral, rat) 12.6 g/kg; mild eye irritant, moderate skin irritant; inhalation of mist may cause respiratory irritation; *Storage:* store in cool, dry area; avoid excessive heat, open flames; incompat. with strong acids and oxidizers.

Emery® 916. [Henkel/Emery] CP/USP glycerin; CAS 56-81-5; EINECS 200-289-5; visc. modifier, flavor enhancer, solv., humectant, thickener, and solubilizer in food applics.; *Regulatory:* EPA-exempt; *Properties:* water-wh. visc. liq.; odorless; sol. in water; sp.gr. 1.2607; m.p. 18 C; b.p. 171 C; pour pt. 18 C; flash pt. (CC) 199 C; 99.5% min. glycerol; *Toxicology:* TLV:TWA 10 mg/m³ (mist); LD50 (oral, rat) 12.6 g/kg; mild eye irritant, moderate skin irritant; inhalation of mist may cause respiratory irritation; *Storage:* store in cool, dry area; avoid excessive heat, open flames; incompat. with strong acids and oxidizers.

Emery® 917. [Henkel/Emery] CP/USP kosher glycerin; CAS 56-81-5; EINECS 200-289-5; visc. modifier, flavor enhancer, solv., humectant, thickener, and solubilizer in food applics.; *Properties:* water-wh. liq.; sol. in water; sp.gr. 1.2607; m.p. 18 C; b.p. 171 C; flash pt. (CC) 199 C; 99.7% min. glycerol; *Toxicology:* TLV:TWA 10 mg/m³ (mist); LD50 (oral, rat) 12.6 g/kg; mild eye irritant, moderate skin irritant; inhalation of mist may cause respiratory irritation; *Storage:* store in cool, dry area; avoid excessive heat, open flames; incompat. with strong

acids and oxidizers.

Emery® 918. [Henkel/Emery] CP/USP glycerin; CAS 56-81-5; EINECS 200-289-5; visc. modifier, flavor enhancer, solv., humectant, thickener, and solubilizer in food applics.; *Properties:* water-wh. liq.; sol. in water; sp.gr. 1.2612; m.p. 18 C; b.p. 171 C; flash pt. (CC) 199 C; 99.8% min. glycerol; *Toxicology:* TLV:TWA 10 mg/m³ (mist); LD50 (oral, rat) 12.6 g/kg; mild eye irritant, moderate skin irritant; inhalation of mist may cause respiratory irritation; *Storage:* store in cool, dry area; avoid excessive heat, open flames; incompat. with strong acids and oxidizers.

Emery® 6354. [Henkel/Emery] Lauric acid; CAS 143-07-7; EINECS 205-582-1; food grade fatty acid; *Properties:* wh. waxy solid, fatty acid odor; insol. in water; sp.gr. 0.89 (75/25 C); m.p. 43 C; b.p. 190 C (10 mm); flash pt. (COC) 160 C; *Toxicology:* LD50 (oral, rat) > 10 g/kg; skin and moderate eye irritant; avoid breathing vapor; *Precaution:* avoid contact with strong oxidizers and alkalies; *Storage:* keep in closed containers away from heat and open flames; container hazardous when emptied.

Emery® 6355. [Henkel/Emery] Myristic acid; CAS 544-63-8; EINECS 208-875-2; food grade fatty acid; *Properties:* wh. waxy solid, fatty acid odor; insol. in water; sp.gr. 0.89 (75/25 C); m.p. 53 C; b.p. 192 C (10 mm); flash pt. (COC) 207 C; 97% act.; *Toxicology:* LD50 (oral, rat) > 10 g/kg; nonirritating to skin and eyes; *Precaution:* avoid contact with strong oxidizers and alkalies; *Storage:* keep in closed containers away from heat and open flames,

Emery® 6358. [Henkel/Emery] Caprylic/capric acid; CAS 67762-36-1; food grade fatty acid; *Regulatory:* FDA 21CFR §172.860; *Properties:* pract. water-wh. liq., penetrating short-chain fatty acid odor; negligible sol. in water; sp.gr. 0.870 (75 C); b.p. 163 C (60 mm); flash pt. (COC) 270 F; 58% caprylic acid, 40% capric acid; *Toxicology:* LD50 (oral, rat) 1.3-10.1 g/kg; causes severe burns to skin and eyes; may be harmful absorbed thru skin; avoid breathing vapor; *Precaution:* avoid contact with oxidizing agents and basic chems.; *Storage:* store in cool, dry, well-ventilated areas away from sunlight, heat, and flames; container hazardous when empty.

Emery® 6359. [Henkel/Emery] Capric acid; CAS 334-48-5; EINECS 206-376-4; food grade fatty acid; *Properties:* pract. water-wh. liq., penetrating short-chain fatty acid odor; negligible sol. in water; sp.gr. 0.858 (75 C); b.p. 190 C (60 mm); flash pt. (COC) 270 F; 97% act.; *Toxicology:* LD50 (oral, rat) 3.3 g/kg; causes severe burns to eyes, severe skin irritation; avoid breathing vapor; *Precaution:* avoid contact with oxidizing agents and basic chems.; *Storage:* store in cool, dry, well-ventilated areas away from sunlight, heat, and flames; container hazardous when empty.

Emery® 6361. [Henkel/Emery] Coconut fatty acid; CAS 61788-47-4; food grade fatty acid; *Properties:* lt. color semisolid, fatty acid odor; insol. in water; sp.gr. 0.89 (75/25 C); m.p. 22-26 C; b.p. 160 C (6

mm); flash pt. (COC) 157 C; 48% lauric acid, 20% myristic acid, 9% palmitic acid, 8% oleic acid; *Toxicology:* LD50 (oral, rat) 18.5 g/kg; may cause skin and eye irritation; avoid breathing vapor; *Precaution:* avoid contact with strong oxidizers and alkalies; *Storage:* store in closed containers away from heat and open flames; container hazardous when empty.

Emery® 7021. [Henkel/Emery; Henkel Canada] Oleic acid; CAS 112-80-1; EINECS 204-007-1; food grade fatty acid; *Regulatory:* kosher; *Properties:* water-wh. liq.; insol. in water; sp.gr. 0.891 (25/20 C); m.p. 16 C; b.p. 286 C (100 mm); flash pt. (CC) 184-189 C; 82% act.; *Toxicology:* LD50 (oral, rat) > 21.5 ml/kg; mild eye and skin irritant; *Precaution:* avoid contact with strong oxidizers and alkalies; *Storage:* store in closed containers away from heat and open flames.

EMG 20. [Am. Ingreds.] Ethoxylated mono- and diglycerides, PEG (20) mono- and diglycerides; food emulsifier for icings, whipped toppings, frozen desserts; dough strengthener in yeast-raised doughs; *Properties:* plastic; sapon. no. 65-75; hyd. no. 65-80.

Emphos™ D70-30C. [Witco/H-I-P] Sodium glyceryl oleate phosphate; anionic; food processing surfactant; food-grade mold lubricant; *Properties:* liq.; oil-sol.

Emphos™ D70-31. [Witco/H-I-P] Mono- and diglycerides phosphate ester; anionic; w/o emulsifier, lubricant, release agent, and thickener in food processing; *Properties:* flake; oil-sol.; 100% conc.

Emphos™ F27-85. [Witco/H-I-P] Hydrog. veg. glycerides phosphate; CAS 85411-01-4; anionic; emulsifier, mold lubricant, release agent for food use; pigment dispersant in oil-based systems; moisture barrier; *Properties:* tan soft solid; insol. in water; sp.gr. 1.01; pour pt. 17 C; acid no. 40 (to pH 9.5); pH 6.9 (3% aq.).

Empilan® GMS LSE32. [Albright & Wilson UK] Glyceryl stearate SE; nonionic; food grade emulsifier; *Properties:* wh. wax-like powd.; m.w. 358 (of monoglyceride); m.p. 55-60 C; pH 7-9 (10% aq.); 32.5% min. monoglyceride.

Empilan® GMS LSE40. [Albright & Wilson UK] Glyceryl stearate SE; nonionic; emulsifier in the baking and food industry; *Properties:* wh. microbead powd.; dens. 0.5 g/cm³; m.p. 55-65 C; 40.0% min. monoglyceride.

Empilan® GMS LSE80. [Albright & Wilson UK] Glyceryl stearate SE; nonionic; emulsifier in the baking and food industry; *Properties:* wh. microbead powd.; dens. 0.5 g/cm³; m.p. 64 C; 80.0% min. monoglyceride.

Empilan® GMS MSE40. [Albright & Wilson UK] Glyceryl stearate SE; nonionic; emulsifier in the baking and food industry; *Properties:* wh. microbead powd.; dens. 0.5 g/cm³; m.p. 55-60 C; 36.0% min. monoglyceride.

Empilan® GMS NSE32. [Albright & Wilson UK] Glyceryl stearate; nonionic; food grade emulsifier; *Properties:* wh. wax-like powd.; m.p. 54-59 C; pH 6-8

(10% aq.); 32.5% monoglyceride.

Empilan® GMS NSE40. [Albright & Wilson UK] Glyceryl stearate SE; CAS 31566-31-1; nonionic; emulsifier in the baking and food industry; *Properties:* wh. microbead powd.; dens. 0.5 g/cm³; m.p. 55-60 C; pH 6-8 (10% aq.); 36-40% act.

Empilan® GMS NSE90. [Albright & Wilson UK] Glyceryl stearate SE; nonionic; emulsifier in the baking and food industry; *Properties:* wh. microbead powd.; dens. 0.5 g/cm³; m.p. 65 C; 90.0% min. monoglyceride.

Empilan® GMS SE40. [Albright & Wilson UK] Glyceryl stearate SE; CAS 31566-31-1; nonionic; emulsifier in the baking and food industry; *Properties:* wh. microbead powd.; dens. 0.5 g/cm³; m.p. 55-60 C; pH 7-9 (10% aq.); 36-40% act.; Discontinued

Empilan® GMS SE70. [Albright & Wilson UK] Glyceryl stearate SE; nonionic; emulsifier in the baking and food industry; *Properties:* wh. microbead powd.; dens. 0.5 g/cm³; m.p. 61 C; 70.0% min. monoglyceride.

Emplex. [Am. Ingreds.] Sodium stearoyl lactylate; CAS 25383-99-7; EINECS 246-929-7; anionic; starch and protein complexing agent for bakery prods.; emulsifier, conditioning agent, softener for processed foods; for dehydrated potatoes, cheese substitutes, nondairy coffee whiteners, puddings, snack dips, sauces, and gravies; *Usage level:* 0.2-0.6%, 4-8 oz/cwt; *Regulatory:* FDA §172.846; *Properties:* lt. tan powd., mild caramel odor; sol. in water and oil; acid no. 60-80; ester no. 150-190; 100% conc., 3.5-5.0% sodium; *Storage:* store under cool, dry conditions (< 90 F).

Empruv®. [Quest Int'l.] Fermented specialty; texturizer for bakery prods., puddings, soups, sauces.

Emrite® 6003. [Henkel/Emery] Glyceryl stearate, food grade; nonionic; lipophilic emulsifier used in food industry; *Properties:* bead; HLB 3.9; 100% conc.; Unverified

Emrite® 6008. [Henkel/Emery] Glyceryl oleate, food grade; nonionic; lipophilic emulsifier used in food industry; *Properties:* liq.; HLB 3.4; 100% conc.; Unverified

Emrite® 6105. [Henkel/Emery] SMS; nonionic; lipophilic emulsifier used in food industry; *Properties:* bead; HLB 5.2; 100% conc.; Unverified

Emrite® 6120. [Henkel/Emery] Polysorbate 80; CAS 9005-65-6; nonionic; food emulsifier; *Properties:* Gardner 6 max. liq.; sp.gr. 1.00-1.10; visc. 400 cSt; HLB 15.1; pour pt. -15 C; acid no. 2.0 max.; sapon. no. 45-55; ref. index 1.4700; flash pt. 605 F; 100% conc.; Unverified

Emrite® 6125. [Henkel/Emery] Polysorbate 60; CAS 9005-67-8; nonionic; food emulsifier; *Properties:* HLB 15.2; 100% conc.; Unverified

Emulbesto. [Lucas Meyer] Hydrophilic lecithin; CAS 8002-43-5; EINECS 232-307-2; nonionic; o/w emulsifier for improved feed utilization, calf milk replacement, piglet starters, fish feed; fat and protein enrichment; *Properties:* liq., powd.; 60-65% conc.

Emulcoll. [Lucas Meyer] Range of standardized guar gums; CAS 9000-30-0; EINECS 232-536-8; binding agent and stabilizer for foodstuffs; stabilizer for animal feeds, zero replacer, calf milk replacer, pet foods.

Emuldan. [Grindsted Prods.; Grindsted Prods. Denmark] Mono- and diglycerides; nonionic; food emulsifier, emulsion stabilizer for margarine, shortening, spreads; improves palatability and reduces stickiness in caramels, toffees; *Regulatory:* EEC, FDA §184.1505 (GRAS); *Properties:* bead, powd., plastic, liq.; HLB 2.8-4.1; 100% conc.

Emuldan HA 40. [Grindsted Prods.] Tallow or lard mono- and diglycerides; nonionic; food emulsifier and stabilizer; *Properties:* beads; m.p. 60 C; sapon. no. 160-180; 40% min. monoester.

Emuldan HA 52. [Grindsted Prods.] Tallow or lard mono- and diglycerides; nonionic; food emulsifier and stabilizer; *Properties:* beads; m.p. 62 C; sapon. no. 155-175; 52% min. monoester.

Emuldan HV 40 Kosher. [Grindsted Prods.] Hydrogenated veg. oil mono- and diglycerides; CAS 69028-36-0; nonionic; food emulsifier and stabilizer; *Regulatory:* EEC, FDA 21CFR §184.1505, GRAS, kosher; *Properties:* beads; m.p. 63 C; iodine no. 5 max.; sapon. no. 155-175; 40% min. monoester; *Storage:* store in cool, dry area.

Emuldan HV 52 Kosher. [Grindsted Prods.] Hydrogenated veg. oil mono- and diglycerides; CAS 69028-36-0; nonionic; food emulsifier and stabilizer; *Regulatory:* EEC, FDA §184.1505 (GRAS), kosher; *Properties:* beads; m.p. 64 C; iodine no. 5 max.; sapon. no. 100-180, 52% min. monoester; *Storage:* store in cool, dry area.

Emuldan HVF 52 K. [Grindsted Prods.] Veg. oil mono- and diglycerides; nonionic; food emulsifier and stabilizer; *Properties:* powd.; m.p. 65 C; sapon. no. 160-180; 52% min. monoester.

Emulfluid™. [Lucas Meyer] Range of hydrophilic lecithin; CAS 8002-43-5; EINECS 232-307-2; amphoteric; fat emulsifier, stabilizer for prod. of foodstuffs (baked goods, margarine, infant formulas, pasta, powd. mixes, release agents); *Properties:* liq.; water-disp.; 100% conc.

Emulfluid® A. [Lucas Meyer] Higher water dispersibility soy lecithin; CAS 8002-43-5; EINECS 232-307-2; emulsifier, dispersant, solubilizer for aq. systems; prepares o/w emulsions; stabilizes emulsions and dispersions in mfg. of foodstuffs; for baked goods, margarine, infant formulas, powd. mixes, release agents; *Usage level:* 0.2-0.5%; *Properties:* char. odor; water-disp.; acid no. 25 max.; pH 6 (1%); 1% moisture.

Emulfluid® AS. [Lucas Meyer] Lecithin; CAS 8002-43-5; EINECS 232-307-2; heat-stable emulsifier and dispersant for high heat systems, baking additive, coemulsifier; for baked goods, release agents; *Properties:* higher water dispersibility.

Emulfluid® E. [Lucas Meyer] Lecithin; CAS 8002-43-5; EINECS 232-307-2; emulsifier and dispersant for high fat systems, baking additive, coemulsifier for baked goods, margarine, powd. mixes; en-hanced water dispersibility and emulsifying; *Properties:* enhanced water dispersibility.

Emulgator 484. [BASF AG] POE glyceryl triricinoleate; nonionic; emulsifier and solubilizer for the feed industry; *Properties:* liq.; HLB 5.0; 100% conc.

Emulgel E-21. [MLG Enterprises Ltd.] Mono- and diglycerides, guar gum, locust bean gum, and xanthan gum; stabilizer/emulsifier for ice cream and ice desserts; inhibits ice crystal growth; provides firm body and creamy texture; *Usage level:* 0.24-0.48% (ice cream).

Emulgel S-32. [MLG Enterprises Ltd.] Guar gum, locust bean bum, and xanthan gum; stabilizer for water-based or low milk solids ice cream; inhibits ice crystal growth; provides firm body and creamy texture; *Usage level:* 0.18-0.25% (ice cream).

Emulgum™. [Lucas Meyer] Deoiled soya lecithin; CAS 8002-43-5; EINECS 232-307-2; softener, antistaling agent, moisture retention agent; aids freshness of chewing gums, foodstuffs; *Properties:* powd.

Emulgum BV. [MLG Enterprises Ltd.] Acacia gums purified and spray-dried; CAS 9000-01-5; EINECS 232-519-5; emulsifier, stabilizer; *Properties:* visc. 15-25 cps (10%); pH 4.0-5.0 (5%); 99.85% min. purity.

Emulpur™ N. [Lucas Meyer] Deoiled soy lecithin; CAS 8002-43-5; EINECS 232-307-2; emulsifier and dispersant esp. for flavor-sensitive systems; for crackers, cookies, baked goods, powd. mixes, frozen doughs, ice cream; *Properties:* powd..

Emulpur™ N P-1. [Lucas Meyer] Deoiled soybean lecithin fraction; CAS 8002-43-5; EINECS 232-307-2; amphoteric; natural emulsifier and release agent for the food industry, bakery goods; *Regulatory:* GRAS, FDA § 814.1400; *Properties:* lt. tan to med. yel. powd., bland to sl. nut-like flavor; 90% < 0.315 mm; sol. @ 60 C in fat; disp. in water; pH 5.7-6.5; 95% conc.; *Storage:* store below 25 C in sealed containers; sensitive to light; reclose drum immediately after opening.

Emulsifier D-1. [Stepan Food Ingreds.] Decaglyceryl hexaoleate; emulsifier for food applics.; *Properties:* amber clear to hazy liq.; acid no. 2; iodine no. 70; sapon. no. 152.5; hyd. no. 107; 0.1% moisture.

Emulsilac S. [Witco/H-I-P] Sodium stearoyl lactylate; CAS 25383-99-7; EINECS 246-929-7; food emulsifier for puddings; complexing agent with protein and starch, dough conditioner/strengthener for bread, rolls, taco shells; freeze/thaw stabilizer for frozen-fluid whipped toppings, coffee whiteners, frozen dough; *Usage level:* 0.2-0.5% on flour (bread), 0.1-0.2% (puddings), 0.1-0.3% (coffee whiteners, whipped toppings); *Regulatory:* FDA 21CFR §172.846; *Properties:* cream-colored powd.; m.p. 120 F; acid no. 60-80.

Emulsilac S K. [Witco/H-I-P] Sodium stearoyl lactylate; CAS 25383-99-7; EINECS 246-929-7; food emulsifier; complexing agent with protein and starch, dough conditioner/strengthener for bread and rolls; freeze/thaw stabilizer for frozen-fluid whipped toppings, coffee whiteners, frozen dough; Usage level: 0.2-0.5% on flour (bread), 0.1-0.3% (coffee whiteners); Regulatory: FDA 21CFR §172.846; Properties: cream-colored powd. or flakes; m.p. 120 F; acid no. 60-80.

Emulsi-Phos® 440. [Monsanto] Disodium phosphate and insol. sodium metaphosphate; food-grade; Regulatory: FCC compliance; Properties: wh. anhyd. powd., no foreign odor; 0.2% max. on 10 mesh.

Emulsi-Phos® 660. [Monsanto] Trisodium phosphate and insol. sodium metaphosphate; food-grade; Regulatory: FCC compliance; Properties: wh. anhyd. powd., no foreign odor; 0.2% max. on 10 mesh; Toxicology: causes eye burns, skin and respiratory tract irritation; avoid breathing dust; Storage: keep container closed; empty containers contain residue.

Emulsi-Phos® 990. [Monsanto] Trisodium phosphate and insol. sodium metaphosphate; food-grade; Regulatory: FCC compliance; Properties: wh. anhyd. powd., no foreign odor; 0.2% max. on 10 mesh; Toxicology: causes eye burns, skin and respiratory tract irritation; avoid breathing dust; Storage: keep container closed; empty containers contain residue.

Emulthin M-35. [Lucas Meyer] Lecithin; CAS 8002-43-5; EINECS 232-307-2; amphoteric; emulsifier, dispersant, wetting agents for all food uses esp. for baking industry; releasing agent for waffles; flour improver; Properties: spray-dried powd.; 50% conc.

Emulthin M-501. [Lucas Meyer] Lecithin; CAS 8002-43-5; EINECS 232-307-2; natural wheat flour improver; Properties: spray-dried powd.; 50% conc.; Unverified

Enrich®. [Quest Int'l.] Fermented specialty; all-natural fat replacement system and stabilizer for frozen desserts.

Enrichment R Tablets. [ADM Arkady] Thiamine, riboflavin, niacin,iron; supplies required amt. of thiamine, riboflavin, niacin, and iron to meet standard for enriched bread, buns, and rolls; Properties: tablets.

Enzyme Modified Beef #3083. [Ariake USA] Cooked beef; flavorings for snacks, soup mixes, sauces, gravies; 35% moisture, 56% protein.

Epamarine®. [Arista Industries] Marine lipid conc.; nutritive supplement; Properties: Gardner 5.5 color; sp.gr. 0.917-0.932; acid no. 2; iodine no. 180-205; sapon. no. 180-192.

Epicholin. [Lucas Meyer] Soya lecithin fractions with increased phosphatidylcholine content (25-60%); CAS 8002-43-5; EINECS 232-307-2; choline enrichment, carrier for dietetics; Discontinued

Epikuron™ 100 P, 100 G. [Lucas Meyer] Deoiled soy lecithin; CAS 8002-43-5; EINECS 232-307-2; nutritional supplement for health food industry, dietetic prods., tonics, tablets, powd. mixes; Properties: powd. and gran. resp.

Epikuron™ 100 X. [Lucas Meyer] Deoiled soy lecithin with selected components in custom extruded

granules; CAS 8002-43-5; EINECS 232-307-2; nutritional supplements, dietetic supplements and prods.; Properties: gran.

Epikuron™ 130 G. [Lucas Meyer] Phosphatidylcholine-enriched deoiled soy lecithin; CAS 8002-43-5; EINECS 232-307-2; for enhanced nutritional supplements, dietetic supplements, health food industry; Properties: gran.

Epikuron™ 130 P. [Lucas Meyer] Phosphatidylcholine enriched deoiled soy lecithin; CAS 8002-43-5; EINECS 232-307-2; choline enrichment, carrier for dietetics, tonics, tablets, nutritional supplements, powd. mixes; Properties: powd.

Epikuron™ 130 X. [Lucas Meyer] Phosphatidylcholine-enriched deoiled soy lecithin and selected components in custom extruded gran. form; CAS 8002-43-5; EINECS 232-307-2; for enhanced nutritional supplements, dietetic supplements and prods., health food industry; Properties: gran.

Epikuron™ 135 F. [Lucas Meyer] Phosphatidylcholine-enriched fluid soy lecithin; CAS 8002-43-5; EINECS 232-307-2; for health food industry, nutritonal supplements, soft gel capsules, granola prods., tonics; Properties: fluid.

Epikuron™ 145V. [Lucas Meyer] Phosphatidylcholine-enriched fraction of polar lipids; for health food industry, special nutritional applics., dietetic prods.

Equal®. [NutraSweet] Aspartame; CAS 22839-47-0; EINECS 245-261-3; low-calorie sweetener for food and hot/cold beverage prods. incl. yogurt, frozen desserts, soft drinks, puddings, breakfast food; limited use in cooking and baking; loss of sweetness may result on prolonged exposure to high temps.; 2 calories/tsp; Storage: indefinite shelf life if kept dry, away from heat and humidity.

Eribate®. [PMP Fermentation Prods.] Sodium erythorbate FCC; EINECS 228-973-9; antioxidant, preservative for fresh and cured meats, fish, beverages, fruit, vegetables; inhibits growth of microorganisms and formation of nitrosamines in cured meats; increases color stability, protects flavor, extends shelf life; Usage level: 87.5 oz/100 gal of pickle, 7/8 oz/100 lb (meat or meat byprods.), 10% sol'n. (applied to surfaces of cured cuts prior to pkg.); Regulatory: FDA GRAS; USDA; Properties: wh. or ylsh. wh. cryst. powd. or gran., odorless; readily water-sol.; m.w. 216.12; pH 5.5-8.0 (1 in 20 sol'n.); 98.0% min. assay; Toxicology: nontoxic; Precaution: stable in dry form, easily oxidized on exposure to air in aq. sol'n.; Storage: store in tight, light-resist. containers.

Eromenth. [Aromatics; Commodity Services Int'l.] Oil obtained from subspecies of Mentha spicata contg. no carvone or menthol; avail. as field mint oil or rectified oil; peppermint-like oil without cooling and irritating effects of menthol; flavoring for fresh bottled waters, herbal teas, liquors, salad dressings, toothpaste, mouthwash, wine coolers, savory flavors; Properties: colorless to pale yel. liq., aromatic odor of mint; sl. sol. in water; sp.gr. 0.894 g/cc; b.p. 209 C; flash pt. 67 C; ref. index 1.4552; biodeg.; Storage: store in cool, well ventilated area, void of ignition sources.

Ervol®. [Witco/Petroleum Spec.] Wh. min. oil NF; emulsifier, lubricant, emollient; for food contact; *Regulatory:* FDA 21CFR §172.878, §178.3620a; *Properties:* water-wh., odorless, tasteless; sp.gr. 0.849-0.865; visc. 24-26 cSt (40 C); pour pt. -7 C; flash pt. 185 C.

Essential Aloe™ GII-1X. [Garuda Int'l.] Aloe vera gel; stabilized prod. for food, beverage industries; avail. in natural and purified grades.

Essential Aloe™ GII-10X. [Garuda Int'l.] Aloe vera gel, ten-fold stabilized liq. conc.; stabilized prod. for food, beverage industries; avail. in natural and purified grades.

Essential Aloe™ GII-40X. [Garuda Int'l.] Aloe vera gel, forty-fold stabilized liq. conc.; stabilized prod. for food, beverage industries; avail. in natural and purified grades.

Essential Aloe™ GII-100X. [Garuda Int'l.] Aloe vera gel, hundred-fold powd. conc.; stabilized prod. for food, beverage industries; avail. in natural and purified grades; *Properties:* powd.

Essential Aloe™ GII-200X. [Garuda Int'l.] Aloe vera gel, two hundred-fold powd. conc.; stabilized prod. for food, beverage industries; avail. in natural and purified grades; *Properties:* powd.

Essiccum®. [Jungbunzlauer Int'l. AG] Dry soluble vinegar flavor (specially processed blend of sodium diacetate, citric acid, and lactose); unique dry prod. tasting like vinegar; mild acidulant; for flavoring soups, gravies, dressings, fish prods., spice preps., sour dough; avoids pungent, tart, astringent flavor nuances of the fruit acids; stabilizes color; antimicrobial; *Properties:* colorless to wh. cryst. powd., acetic acid odor; completely water-sol.; bulk dens. 900 g/l; pH 3.8 (1%); readily biodeg.; 1.7% max. water; *Toxicology:* LD50 (oral, rat) 4960 mg/kg (neutral); may cause eye or skin irritation; *Precaution:* incompat. with strong bases and oxidizing agents; *Storage:* store in original container tightly closed in dry, cool place.

Estasan GT 8-40 3578. [Unichema] Caprylic/capric triglyceride (40% C8 + 60% C10); CAS 65381-09-1; solv. and carrier for flavors, fragrances, food ingreds.; lubricant, mold release for food applics.; fat source in dietetic prods.; *Regulatory:* Japan approval; *Properties:* Lovibond 5Y/0.5R color, odorless, bland taste; visc. 29 mPa•s; acid no. 0.1 max.; iodine no. 2 max.; sapon. no. 300-315; hyd. no. 5 max.; cloud pt. 10 C.

Estasan GT 8-60 3575. [Unichema] Caprylic/capric triglyceride (60% C8 + 40% C10); CAS 65381-09-1; EINECS 265-724-3; solv. and carrier for flavors, fragrances, food ingreds.; lubricant, mold release for food applics.; fat source in dietetic prods.; *Regulatory:* Japan approval; *Properties:* Lovibond 5Y/0.5R color, odorless, bland taste; visc. 23 mPa•s; acid no. 0.1 max.; iodine no. 0.5 max.; sapon. no. 325-345; hyd. no. 5 max.; cloud pt. -8 C.

Estasan GT 8-60 3580. [Unichema] Caprylic/capric triglyceride (60% C8 + 40% C10); CAS 65381-09-1; EINECS 265-724-3; solv. and carrier for flavors,

fragrances, food ingreds.; lubricant, mold release for food applics.; fat source in dietetic prods.; *Regulatory:* kosher; *Properties:* Lovibond 5Y/0.5R color, odorless, bland taste; visc. 23 mPa•s; acid no. 0.1 max.; iodine no. 0.5 max.; sapon. no. 325-345; hyd. no. 5 max.; cloud pt. -8 C.

Estasan GT 8-65 3577. [Unichema] Caprylic/capric triglyceride (65% C8 + 35% C10); CAS 65381-09-1; EINECS 265-724-3; solv. and carrier for flavors, fragrances, food ingreds.; lubricant, mold release for food applics.; fat source in dietetic prods.; *Regulatory:* Japan approval; *Properties:* Lovibond 5Y/0.5R color, odorless, bland taste; visc. 23 mPa•s; acid no. 0.1 max.; iodine no. 0.5 max.; sapon. no. 325-360; hyd. no. 5 max.; cloud pt. -5 C.

Estasan GT 8-65 3581. [Unichema] Caprylic/capric triglyceride (65% C8 + 35% C10); CAS 65381-09-1; EINECS 265-724-3; solv. and carrier for flavors, fragrances, food ingreds.; lubricant, mold release for food applics.; fat source in dietetic prods.; *Regulatory:* kosher; *Properties:* Lovibond 5Y/0.5R color; visc. 23 mPa•s; acid no. 0.1 max.; iodine no. 0.5 max.; sapon. no. 335-360; hyd. no. 5 max.; cloud pt. -5 C.

Estric™. [Bunge Foods] Mono- and diglycerides; emulsifier for cakes, breads, rolls, and instant foods; *Properties:* 2.0 red color; iodine no. 5 max.; 52% mono content.

Evanstab® 12. [Evans Chemetics] Dilauryl thiodipropionate; CAS 123-28-4; EINECS 204-614-1; stabilizer for oils, fats, food applics.; *Properties:* wh. cryst. flakes or powd.; m.w. 514; f.p. 40.0 C min.; sol. in acetone, MEK, n-heptane, toluene, ethyl acetate, ethanol; acid no. 1.0 max.; 00.0% min. assay.

Evanstab® 18. [Evans Chemetics] Distearyl thiodipropionate; CAS 693-36-7; EINECS 211-750-5; antioxidant for food-pkg. materials and edible fats and oils; *Properties:* wh. cryst. flakes or powd.; m.w. 683; f.p. 64.0 C min.; sol. (g/100 g sol'n.) 11 g toluene, 2 g n-heptane; < 1 g ethyl acetate, 95% ethanol, MEK, water; 98.0% min. assay.

EverFresh®. [Pfizer Food Science] Blend of 4-hexylresorcinol in sodium chloride; antioxidant for food use; prevents blackspot in shrimp harvesting; *Properties:* off-wh. free-flowing powd., pungent odor; sol. 500 mg/l in water; *Toxicology:* may cause eye, skin or respiratory irritation; very lg. doses can cause vomiting, diarrhea, prostration; heated to decomp., emits toxic fumes of chloride/sodium oxide, carbon monoxide/dioxide; 4-Hexylresorcinol: LD50 (oral, rat) 550 mg/kg; *Storage:* store in cool, dry, ventilated area in tightly closed containers.

Eversoft Plus Kosher. [Am. Ingreds.] Wheat flour, calcium sulfate, salt, and enzyme; enzymatic softener for yeast-raised white bread, hamburger buns, sweet rolls, donuts, and sponge, snack, and foam cakes; *Properties:* cream-colored powd.; 8.9-11.9% moisture, 2.6-3.6% salt.

Exafine® 250. [Cosucra BV; Feinkost Ingred.] Pea outer fibers; dietary fiber with high water retention capacity; for bakery prods., pasta, snacks, cereals, dietary specialties, extruded prods.; *Properties:* beige powd., neutral taste; 90% < 250 µ; 90% min. dry matter; *Storage:* store under dry conditions.

Exafine® 500. [Cosucra BV; Feinkost Ingred.] Pea outer fibers; dietary fiber with high water retention capacity; for bakery prods., pasta, snacks, cereals, dietary specialties, extruded prods.; *Properties:* beige powd., neutral taste; 90% < 500 µ; 90% min. dry matter; *Storage:* store under dry conditions.

Exafine® 1000. [Cosucra BV; Feinkost Ingred.] Pea outer fibers; dietary fiber with high water retention capacity; for bakery prods., pasta, snacks, cereals, dietary specialties, extruded prods.; *Properties:* beige powd., neutral taste; 90% < 1 mm; 90% min. dry matter; *Storage:* store under dry conditions.

Exafine® 1000. [Cosucra BV; Feinkost Ingred.] Pea outer fibers; dietary fiber with high water retention capacity; for bakery prods., pasta, snacks, cereals, dietary specialties, extruded prods.; *Properties:* beige powd., neutral taste; 90% < 2 mm; 90% min. dry matter; *Storage:* store under dry conditions.

Excel 122. [Kao/Edible Fat & Oil] Mono- and diglycerides oleic and stearic acid base; nonionic; food emulsifier; *Properties:* plastic; HLB 3; 100% conc.

Excel 124. [Kao Corp. SA] Mono- and diglycerides stearic acid base; nonionic; food emulsifier; *Properties:* powd.; 100% conc.; Unverified

Excel 150. [Kao/Edible Fat & Oil] Mono- and diglycerides stearic acid base; nonionic; food emulsifier; *Properties:* powd.; HLB 3.2; 100% conc.

Excel 200. [Kao/Edible Fat & Oil] Monoglyceride blend, distilled; nonionic; food emulsifier; *Properties:* powd.; HLB 3.5; 100% conc.

Excel 300. [Kao/Edible Fat & Oil] Mono- and diglycerides/propylene glycol blend; nonionic; defoaming agent for beverages; *Properties:* liq.; HLB 2.8; 100% conc.

Excel 84. [Kao/Edible Fat & Oil] Mono- and diglycerides stearic acid base; nonionic; food emulsifier; *Properties:* powd.; HLB 2.8; 100% conc.

Excel O-95F. [Kao/Edible Fat & Oil] Monoglyceride oleic acid base, distilled; nonionic; food emulsifier; *Properties:* solid; HLB 3.5; 100% conc.

Excel O-95N. [Kao/Edible Fat & Oil] Monoglyceride, dist.; nonionic; food emulsifier; *Properties:* plastic; HLB 3.5; 100% conc.

Excel O-95R. [Kao/Edible Fat & Oil] Monoglyceride oleic acid base, dist.; nonionic; food emulsifier; *Properties:* solid; HLB 3.5; 100% conc.

Excel P-40S. [Kao/Edible Fat & Oil] Mono- and diglycerides of stearic acid; nonionic; food emulsifier; *Properties:* powd.; HLB 2.8; 100% conc.

Excel T-95. [Kao/Edible Fat & Oil] Monoglyceride, dist.; nonionic; food emulsifier; *Properties:* bead; HLB 3.8; 100% conc.

Excel VS-95. [Kao/Edible Fat & Oil] Monoglyceride, dist.; nonionic; food emulsifier; *Properties:* bead; HLB 3.8; 100% conc.

Extend® Yeast Raised Freezer System. [Bunge Foods] Blend of buffers, nutrients, emulsifiers, and conditioners; Conc. producing raw, frozen dough prods. with frozen shelf life to 24 wks., exc. volume and eating quality; for frozen dough hearth prods., kaiser rolls, sub rolls, French and Italian breads; *Usage level:* 4% based on flour.

Exter™ Family. [Gist-brocades] Hydrolyzed vegetable proteins; CAS 100209-45-8; flavor-specific profile incl. basic HVP flavors, general roasted/brown beef, chicken, pork, tomato, onion, cheese, mushroom, bacon, and ham; *Regulatory:* meets US and European regulatory standards; *Properties:* powd., liq., and paste forms.

Extlat®. [Asahi Chem. Industry] Hydrolyzed animal protein; for food industry.

Extractase L5X, P15X. [Finnsugar Bioprods.] Pectinase (fungal); CAS 9032-75-1; EINECS 232-885-6; broad act. enzyme for fruit processing; *Properties:* amber liq. and tan powd. resp.; Unverified

Extract of Hog Bile. [Am. Labs] Hog bile extract contg. hyodeoxycholic acid, sodium glycohyodeoxycholate, sodium taurohydeoxycholate; nutritive food and pharmaceutical additive; *Storage:* preserve in tight containers with moisture-proof liners, in cool, dry place.

Extract of Ox Bile NF XI. [Am. Labs] Ox bile extract contg. 45-50% sodium taurocholate, 25% sodium glycocholate; nutritive food and pharmaceutical additive; *Properties:* lt. color; sol. in water, alcohol; pH 6.5-7.5; *Storage:* preserve in tight containers with moisture-proof liners, in cool, dry place.

Extract of Whole Grapefruit. [Commodity Services Int'l.] Oil derived from entire grapefruit, both fruit and peel, of *Citrus paradisi;* contains aldehydes, alcohols, esters, nookatone, other sesquiterpenes, ketones, etc.; flavor ingred. where fresh, fruity grapefruit impact is desired; for alcoholic and non-alcoholic beverages, nectars, lemonade-type drinks, bitter beverages; emulsion base for cloudy beverages; *Properties:* yel. oil, intense odor and taste of fresh grapefruit; *Storage:* 6 mos stability stored at 7 C in absence of light and air in sealed containers.

Extract of Whole Orange. [Dierberger Oleos Essencias; Commodity Services Int'l.] Essential oil blend obtained from Brazilian sweet orange; flavor for reconsituted juices and drinks to impart natural fresh orange impression; in carbonated beverages, syrups, punches, liquors, and wine coolers; *Usage level:* 200 ml/100 l of finished beverage; *Regulatory:* GRAS (FEMA); *Properties:* amber-yel. clear liq., intense odor and taste of freshly cold-pressed oranges; sp.gr. 0.840-0.860; ref. index 1.455-1.475; 30-34% ethanol content.

Extram H. [ADM Arkady] Economical replacement of nonfat dry milk in bakery prods.; approved for use in Roman Meal prod.

Extramalt 10. [Sandoz Nutrition] 10% Nondiastatic dry malt extract with 90% corn syrup solids; natural food ingred. imparting flavor, sweetness, color, nutrition (protein, vitamins, minerals), humectant props. to bakery prods., cereals, confections, health and instant foods, pharmaceuticals, ice cream, and yogurt; *Properties:* lt. tan powd., sweet taste, mild malt flavor; pH 5.0-5.8 (10%); 10% act.;

Storage: 1 yr shelf life stored in cool, dry place (below 70 F) in sealed containers.

Extramalt 33. [Sandoz Nutrition] 34% Diastatic malted barley flour, 33% wheat flour, and 33% dextrose; natural food ingred. imparting flavor, sweetness, color, nutrition (protein, vitamins, minerals), humectant props., enzymatic activity to aid starch conversion, increase sugar prod., promote vigorous yeast action in dough; high quality baked goods; *Properties:* cream-colored powd., sl. sweet, mild malt flavor; pH 5-6 (10%); *Storage:* 1 yr shelf life stored in cool, dry place (below 70 F) in sealed containers.

Extramalt 35. [Sandoz Nutrition] 35% Nondiastatic dry malt extract with 65% corn syrup solids; natural food ingred. imparting flavor, sweetness, color, nutrition (protein, vitamins, minerals), humectant props. to bakery prods., cereals, confections, health and instant foods, pharmaceuticals, ice cream, and yogurt; *Properties:* lt. brn color, moderate malt flavor, pleasant sweet taste; pH 5-6 (10%); 35% act.; *Storage:* 1 yr shelf life stored in cool, dry place (below 70 F) in sealed containers.

Extramalt Dark. [Sandoz Nutrition] Nondiastatic dry malt extract; CAS 8002-48-0; EINECS 232-310-9; natural food ingred. imparting flavor, sweetness, color, nutrition (protein, vitamins, minerals), humectant props. to bakery prods., cereals, confections, health and instant foods, pharmaceuticals, ice cream, and yogurt; *Properties:* dk. brn. with deep, rich malt flavor and aroma; pH 5.8-7.4 (10%); 100% act.; *Storage:* 1 yr shelf life stored in cool, dry place (below 70 F) in sealed containers.

Extramalt Light. [Sandoz Nutrition] Nondiastatic dry malt and barley extract; natural food ingred. imparting flavor, sweetness, color, nutrition (protein, vitamins, minerals), humectant props. to bakery prods., cereals, confections, health and instant foods, pharmaceuticals, ice cream, and yogurt; *Properties:* golden powd., char. malt flavor and aroma; pH 5-6 (10%); 100% act.; *Storage:* 1 yr shelf life stored in cool, dry place (below 70 F) in sealed containers.

Eye Substance. [Am. Labs] Vacuum-dried defatted glandular prod.; nutritive food additive; *Properties:* powd.

F

Famodan MS Kosher. [Grindsted Prods.] Sorbitan stearate; CAS 1338-41-6; EINECS 215-664-9; nonionic; food emulsifier; fat crystal modifier and bloom retarder in cocoa butter substitutes, compd. coatings, imitation dairy systems; improves texture of frostings and icings; *Usage level:* 0.5-2% based on fat content; *Regulatory:* EEC, FDA 21CFR §172.842, FCC, kosher; ADI 0-25 mg/kg; *Properties:* coarse powd.; m.p. 57 C; acid no. 5-10; iodine no. 2 max.; sapon. no. 147-157; hyd. no. 235-260; 100% conc.; *Storage:* store in cool, dry area.

Famodan TS Kosher. [Grindsted Prods.] Sorbitan tristearate; fat crystal modifier preventing fat bloom in cocoa butter substitutes and compd. coatings; improves texture of frostings and icings; *Usage level:* 1-2.5% based on fat content; *Regulatory:* EEC, FDA 21CFR §170.30, GRAS, FCC, kosher; ADI 0-25 mg/kg; *Properties:* coarse beads; m.p. 57 C; acid no. 12-15; iodine no. 2 max.; sapon. no. 176-188; hyd. no. 66-80; *Storage:* store in cool, dry area.

Famous. [Karlshamns Food Ingreds.] Partially hydrog. soybean oil, stabilized; CAS 8016-70-4; EINECS 232-410-2; high stability oil used for frying, color/flavor carrier, antidusting applics.; *Properties:* liq.; iodine no. 88-93.

Fancol HON. [Fanning] Honey; CAS 8028-66-8; biological additive, flavoring; *Properties:* clear; pH 3.0-4.9; 80% min. solids.

Fancol Menthol. [Fanning] Menthol; CAS 89-78-1; EINECS 201-939-0; denaturant, flavoring agent, fragrance component for mouthwashes; *Properties:* m.p. 41-44 C.

Fancol OA-95. [Fanning] Oleyl alcohol; CAS 143-28-2; EINECS 205-597-3; nonionic; emulsion stabilizer, antifoam, detergent, release agent for food applics.; *Properties:* liq.; sol. in IPA, acetone, lt. min. oil, trichloroethylene, kerosene, VMP naphtha, benzene, turpentine; acid no. 0.05 max.; iodine no. 90-96; sapon no. 1 max.; hyd. no. 200-212; cloud pt. 5 C max.

Fargo®. [Quest Int'l.] Starter culture; for dairy prods.

Fatin. [Lucas Meyer] Fat enrichment for ready-mixed flours, bakery prods., dry mixes, soups, and sauces; *Properties:* spray-dried powds.; 80% conc.

Fat Replacer 785. [Vaessen-Schoemaker] Carbohydrates (maltodextrin, vegetable fiber), stabilizers (carrageenan, carboxymethylcellulose sodium), flavor; fat substitute for partial or full replacement of solid animal fats in processed meat prods. (burgers, cooked sausage, liver paté) without affecting taste, texture, or mouthfeel; also reduces cooking/frying losses; stable @ -18 to 160 C; 45-50% carbohydrates, 5% max. protein, 5% max. salt, 0.5% max. fat, 5% max. moisture.

FD-6. [MLG Enterprises Ltd.] Monocalcium phosphate, calcium sulfate, ammonium sulfate, potassium bromate, and enzymes; complete fast dough conditioning system for accelerated fermentation; for straight-dough breads, sweet goods, brew buns, prepared yeast-raised mixes, Kaiser rolls, etc.; *Usage level:* 2-6 oz/cwt flour; 4-6 oz/cwt (no-time straight dough); *Regulatory:* FDA approved.

FD-8. [MLG Enterprises Ltd.] Calcium sulfate, monocalcium phosphate, ammonium sulfate, potassium bromate, and enzymes; complete fast dough conditioning system for accelerated fermentation; for straight-dough breads, sweet goods, brew buns, prepared yeast-raised mixes, Kaiser rolls, etc.; *Usage level:* 2-5 oz/cwt flour; 4-5 oz/cwt (no-time straight dough); *Regulatory:* FDA approved.

Fermaloid Yeast Food. [ADM Arkady] Monocalcium phosphate; CAS 7758-23-8; acid-type yeast food providing acid salt to lower pH value in the dough.

Fermalpha. [Finnsugar Bioprods.] Fungal α-amylase; CAS 9000-92-4; EINECS 232-567-7; enzyme for food processing; used in maltose syrups and baking; optimum pH 5.0-6.0; optimum temp. 45-55 C; activity 30,000 and 52,500 SKG/g; *Properties:* tan powd.; Unverified

Fermcolase®. [Finnsugar Bioprods.] Catalase (fungal); CAS 9001-05-2; EINECS 232-577-1; enzyme for food processing, peroxide decomposition; wider pH, temp., and peroxide range than animal catalase; activity 1000 Baker units/mL; *Properties:* red-amber liq.; Unverified

Fermcozyme® 1307, BG, BGXX, CBB, CBBXX, M. [Finnsugar Bioprods.] Glucose oxidase (fungal); CAS 9001-37-0; EINECS 232-601-0; enzyme for food processing; antioxidant; maintains freshness by preventing oxygen or glucose deterioration; desugarization of egg prods.; optimum pH 4.5-7.5; special preparations avail. for use at pH down to 2.5; *Properties:* amber clear liqs.; Unverified

Fermentase®. [Premier Malt Prods.] Extract of 60-65% malted barley and 35-40% corn with saccharifying enzymes; syrup with mellow, rich malt flavor;

used in confections, ice cream prods., baked goods, vinegar, cereal, tobacco, soybean milk, pretzels, fermentation nutrient, bread sticks, horse food; 55% the sweetness of sucrose; 78.5-80.0% solids; 3-4% protein; 57-67% reducing sugars.

Ferment Buffer. [ADM Arkady] Blend of buffer agents and yeast food for controlled fermentation brew systems; produces even yeast growth for optimum prod. of fermentation flavors and yeast conditioning; for use in brew.

Fermented Soy Sauce Powd. [Nikken Foods Co. Ltd.] Naturally fermented soy sauce and salt spray-dried on a dextrin carrier; flavoring for soups, stews, gravies, seasonings, cocktail mixes, juices, dressings, vegetable spreads, snack foods, prepared foods; *Properties:* med. brn. free-flowing powd., mild typical fermented soy aroma; pH 4.9 (10%); 15.6% protein, 33% salt, 4% moisture; *Storage:* store in cool, dry, well-ventilated area in sealed containers; avoid storage at elevated temps.

Fermlipase. [Finnsugar Bioprods.] Lipase (pancreatic); CAS 9001-62-1; EINECS 232-619-9; enzyme for food processing; hydrolysis of yolk fat in egg albumin; digestive aid; additive to animal feed; low (reduced) proteolytic activity; activity 30X NF/mg; *Properties:* lt. tan powd.; Unverified

Fermvertase, 10X, XX. [Finnsugar Bioprods.] Invertase (yeast); enzyme for food processing; sucrose inversion for confections; optimum pH 4.0-5.5; optimum temp. 55-65 C; activity 3000, 30,000, and 6000 SU/mL resp.; *Properties:* clear liqs.; Unverified

Fibrex®. [Fibrex AD] Natural prod. from sugarbeets; dietary fiber enrichment for bakery use, meat prods.; provides calorie reduction, prolongs freshness in bread thru water-holding capacity; finer fractions as fat replacers; coarse fractions as texture improvers; gluten-free; *Usage level:* 1-3% on flour wt. (breads), 1-1.5% on meat quantity (meat loaf, sauce, pates, soups), 0.5-1% (hamburgers); *Properties:* avail. in various particle sizes, coarse small and large flakes; pH 4.5 ± 0.5; 73% dietary fiber of which 1/3 is soluble; *Storage:* store dry @ < 65% r.h.

Fibrim® 1020. [Protein Tech. Int'l.] Soy fiber; dietary fiber providing nutritional/economic benefits; bulking agent, water-binder; for dry-blended beverages, sauces, soups; *Properties:* low particle size; high water absorp., med. visc. and suspension; 66-77% dietary fiber.

Fibrim® 1250. [Protein Tech. Int'l.] Soy fiber; dietary fiber providing nutritional/economic benefits; bulking agent, water-binder; for flaked or extruded cereals; *Properties:* med. particle size; med. water absorp.; 66-77% dietary fiber.

Fibrim® 1255. [Protein Tech. Int'l.] Soy fiber; dietary fiber providing nutritional/economic benefits; bulking agent, water-binder; for food bars, pasta, extruded snacks, fresh/frozen baked goods; *Properties:* med. particle size; med. water absorp.; 66-77% dietary fiber.

Fibrim® 1450. [Protein Tech. Int'l.] Soy fiber; dietary fiber providing nutritional/economic benefits; bulking agent, water-binder; for food bars, pasta, fresh/frozen baked goods; *Properties:* high particle size; med.-low water absorp.; 66-77% dietary fiber.

Fibrim® 2000. [Protein Tech. Int'l.] Soy fiber; dietary fiber providing nutritional/economic benefits; bulking agent, water-binder; for liq. nutritional beverages; *Properties:* med.-low particle size; med. visc. and suspension; high water absorp.; 66-77% dietary fiber.

Finecol 8000. [Hormel] Hydrolyzed gelatin, sodium benzoate, and potassium sorbate; fining and clarification agent for apple juice, grape juice, beer, and wine; produces clear sparkling liq.; *Usage level:* 40-80 ppm solids/1000 gal; *Properties:* Gardner 9 max. syrupy liq., faint protein-like odor; dissolves readily into juice or liq.; pH 3.6-4.1; 35% min. total solids; *Storage:* stable for up to 1 yr; keep containers tightly closed.

Finmalt L. [Xyrofin UK] Maltitol syrup; CAS 585-88-6; EINECS 209-567-0; sweetener for foods and pharmaceuticals; *Properties:* colorless liq., odorless, sweet taste; pH 4-6; 75% dry substance; *Storage:* store below 30 C; avoid temp. fluctuations, water condensation.

FishGard® FP-55. [Rhone-Poulenc Food Ingreds.] Sodium polyphosphate, Disodium diphosphate, potassium sorbate, and sodium bicarbonate, all FCC grades; processing aid improving shelf life of unfrozen seafood; added to brine tanks to limit microbial growth on seafood and in the tank; for cod, flounder, haddock, other ice packed or refrigerated fillets, processed scallops, squid, shrimp, whole fish; *Regulatory:* kosher; *Properties:* wh. to sl. ylsh. powd., char. odor; 5% max. on 20 mesh, 65% min. thru 100 mesh; pH 6.2-6.9 (1%); *Storage:* store cool and dry.

Fish Liver Oil. [Arista Industries] Fish liver oil; nutritive additive; *Properties:* acid no. < 1.2; iodine no. 140-180; sapon. no. 155-175.

Flavor Enhancer Powd. [Nikken Foods Co. Ltd.] Flavor enhancement; *Properties:* lt. tan free-flowing powd., mild typical fermented soy aroma; *Storage:* store in cool, dry, well-ventilated area in sealed containers; avoid storage at elevated temps.

Flavorset® GP-2. [Hormel] Gelatin, Type A, 100 bloom; CAS 9000-70-8; EINECS 232-554-6; used when a soft set or low visc. or high protein content is required, e.g., consomme soups, high protein foods; *Properties:* ivory powd., weak boullion-like odor; 30 mesh; visc. 20 ± 3 mps; pH 4.5-5.8; 12% max. moisture; *Storage:* stable for up to 1 yr when stored dry at ambient temps.; keep containers tightly closed.

Flavorset® GP-3. [Hormel] Gelatin, Type A, 125 bloom; CAS 9000-70-8; EINECS 232-554-6; used when a soft set or low visc. is required, e.g., consomme soups, high protein foods; *Properties:* ivory powd., weak boullion-like odor; 30 mesh; visc. 22 ± 2 mps; pH 4.5-5.8; 12% max. moisture; *Storage:* stable for up to 1 yr when stored dry at ambient temps.; keep containers tightly closed.

Flavorset® GP-4. [Hormel] Gelatin, Type A, 150 bloom; CAS 9000-70-8; EINECS 232-554-6; used when a soft set or low visc. is required, e.g., consomme soups, soft candies; *Properties:* ivory powd., weak boullion-like odor; 30 mesh; visc. 24 ± 2 mps; pH 4.5-5.8; 12% max. moisture; *Storage:* stable for up to 1 yr when stored dry at ambient temps.; keep containers tightly closed.

Flavorset® GP-5. [Hormel] Gelatin, Type A, 175 bloom; CAS 9000-70-8; EINECS 232-554-6; used when a semifirm set and a noticeably increased visc. is desired, e.g., yogurt, ice cream, sour cream, bakery fillings, head cheese, aspics; *Properties:* ivory powd., weak boullion-like odor; 30 mesh; visc. 28 ± 2 mps; pH 4.5-5.8; 12% max. moisture; *Storage:* stable for up to 1 yr when stored dry at ambient temps.; keep containers tightly closed.

Flavorset® GP-6. [Hormel] Gelatin, Type A, 200 bloom; CAS 9000-70-8; EINECS 232-554-6; used when a semifirm set and a noticeably increased visc. is desired, e.g., yogurt, ice cream, sour cream, bakery fillings, head cheese, aspics; *Properties:* ivory powd., weak boullion-like odor; 30 mesh; visc. 32 ± 2 mps; pH 4.5-5.8; 12% max. moisture; *Storage:* stable for up to 1 yr when stored dry at ambient temps.; keep containers tightly closed.

Flavorset® GP-7. [Hormel] Gelatin, Type A, 225 bloom; CAS 9000-70-8; EINECS 232-554-6; used when a semifirm set and a noticeably increased visc. is desired, e.g., yogurt, ice cream, sour cream, bakery fillings, head cheese, aspics; *Properties:* ivory powd., weak boullion-like odor; 30 mesh; visc. 37 ± 2 mps; pH 4.5-5.8; 12% max. moisture; *Storage:* stable for up to 1 yr when stored dry at ambient temps.; keep containers tightly closed.

Flavorset® GP-8. [Hormel] Gelatin, Type A, 250 bloom; CAS 9000-70-8; EINECS 232-554-6; used when a semifirm set and a noticeably increased visc. is desired, e.g., yogurt, ice cream, sour cream, bakery fillings, head cheese, aspics; *Properties:* ivory powd., weak boullion-like odor; 30 mesh; visc. 42 ± 2 mps; pH 4.5-5.8; 12% max. moisture; *Storage:* stable for up to 1 yr when stored dry at ambient temps.; keep containers tightly closed.

Flavorset® GP-9. [Hormel] Gelatin, Type A, 275 bloom; CAS 9000-70-8; EINECS 232-554-6; used when a firm or brittle set and a significantly increased visc. is desired; suggested where the most effect with the least amount of protein is required; best film-forming capabilities; *Properties:* ivory powd., weak bouillon-like odor; 30 mesh; visc. 47 ± 4 mps; pH 4.5-5.8; 12% max. moisture; *Storage:* stable for up to 1 yr when stored dry at ambient temps.; keep containers tightly closed.

Flavorset® GP-10. [Hormel] Gelatin, Type A, 300 bloom; CAS 9000-70-8; EINECS 232-554-6; used when a firm or brittle set and a significantly increased visc. is desired; suggested where the most effect with the least amount of protein is required; best film-forming capabilities; *Properties:* ivory powd., weak bouillon-like odor; 30 mesh; visc. 54 ± 4 mps; pH 4.5-5.8; 12% max. moisture; *Storage:* stable for up to 1 yr when stored dry at ambient temps.; keep containers tightly closed.

Flav-R-Base® Primary Yeast Autolysate Extracts. [Rhone-Poulenc Food Ingreds.] Primary yeast autolysate extracts; used to enhance flavors of juices, gravies, and sauces, in chicken and veal prods., reduced-salt prods.; masks beany off-flavor in soy-extended meat prods.; ideal for seafood.

Flav-R-Keep®. [Rhone-Poulenc Food Ingreds.] Sodium tripolyphosphate FCC, lemon juice solids, rosemary extract; meat processing additive; *Regulatory:* kosher; *Properties:* off-wh. to sl. ylsh. gran., sl. spice aroma; pH 9.5-10.3 (1%); *Storage:* store cool and dry.

Flav-R-Keep® FP-51. [Rhone-Poulenc Food Ingreds.] Sodium tripolyphosphate, lemon juice solids, rosemary extract; natural antioxidant food additive reducing development of off-flavors in precooked frozen prods., meat and poultry; improves moisture retention during freezing, thawing, and cooking of precooked meat prods.; extends shelf life.

FloAm® 200. [Qualcepts Nutrients] Powdered cellulose, dextrose, enzymes; oxygen scavenger, anticaking and antimycotic agent reducing oxygen levels in packaged shredded cheese (mozzarella, muenster, Monterey jack); *Usage level:* 2% by wt. of cheese; *Properties:* wh. fibrous powd.; 5% max. 100 mesh, 20% max. 200 mesh; partly sol. in water; sp.gr. 1.5 g/cc; pH 5-7.5 (10% susp.); *Toxicology:* inhalation of enzyme dust may cause allergenic reactions on repeated exposure; skin/eye irritant on prolonged direct contact; *Storage:* store in cool, dry area; avoid exposure to high temps. and humidity.

FloAm® 200C. [Qualcepts Nutrients] Powdered cellulose, dextrose, annatto, enzymes; anticaking and antimycotic agent reducing oxygen levels in packaged shredded cheese (cheddar or Colby type cheeses); *Usage level:* 2% by wt. of cheese; *Properties:* cheese-colored fibrous powd.; 5% max. 100 mesh, 20% max. 200 mesh; partly sol. in water; sp.gr. 1.5 g/cc; pH 5-7.5 (10% susp.); *Toxicology:* inhalation of enzyme dust may cause allergenic reactions on repeated exposure; skin/eye irritant on prolonged direct contact; *Storage:* store in cool, dry area; avoid exposure to high temps. and humidity.

FloAm® 210 HC. [Qualcepts Nutrients] Powdered cellulose, dextrose, annatto, enzymes; anticaking and antimycotic agent reducing oxygen levels in packaged shredded cheese; *Usage level:* 2%; *Properties:* cheese-colored fibrous powd.; 5% max. 100 mesh, 20% max. 200 mesh; partly sol. in water; sp.gr. 1.5 g/cc; pH 5-7.5 (10% susp.); *Toxicology:* inhalation of enzyme dust may cause allergenic reactions on repeated exposure; skin/eye irritant on prolonged direct contact; *Storage:* store in cool, dry area; avoid exposure to high temps. and humidity.

FloAm® 221. [Qualcepts Nutrients] Powdered cellulose, potassium sorbate; anticaking and antimycotic agent reducing oxygen levels in packaged shredded cheese; *Properties:* powd.; 5% max. 100 mesh, 20% max. 200 mesh; pH 5-7.5 (10% susp.).

FloAm® 300. [Qualcepts Nutrients] Calcium silicate, dextrose, enzymes; anticaking agent, oxygen scavenging enzyme system which protects packaged prods. from mold by reducing oxygen content; for grated Parmesan, Roman cheeses; *Usage level:* 2% by wt. (packaged cheese).

FloAm® 400. [Qualcepts Nutrients] Sodium aluminum silicate, dextrose, enzymes; anticaking agent, oxygen scavenger which reduces free oxygen in pkg., reducing likelihood of mold; used for grated Parmesan or Romano cheeses; *Usage level:* 2% by wt. of cheese; *Properties:* wh. powd.

FloAm® 450. [Qualcepts Nutrients] Silicon dioxide, dextrose, enzymes; anticaking agent, oxygen scavenging enzyme system which protects packaged prods. from mold by reducing oxygen content; for grated cheeses; *Usage level:* 2% by wt. (packaged cheese).

FloAm® System. [Qualcepts Nutrients] Oxygen scavenging enzymes, anticaking agent (powd. cellulose, sodium aluminum silicate, microcryst. cellulose, etc.), dextrose (a processing aid); oxygen scavenging enzyme system which protects packaged prods. from mold by reducing oxygen content; produces no off flavors or odors; used for shredding or dicing of mozzarella, muenster, monterey jack, and other cheeses; *Usage level:* 2% max. (CFR); *Properties:* wh. powd.

Flo-Malt® Malt. [ADM Ogilvie] 65% Malt; for baked goods, mixes, candy, cereal, dairy prods.; *Properties:* lt. tan powd./flake; 8% max. moisture.

Fluid EEZ 1000. [Van Den Bergh Foods] Soybean oil with glycerol lacto esters of fatty acids, propylene glycol, mono- and diglycerides, TBHQ; fluid shortening for high quality cakes made without hydrogenated shortening; added emulsifiers for improved performance; *Regulatory:* kosher; *Properties:* Lovibond 3.0R fluid; 1.5% min. mono.

Fluid Flex. [Van Den Bergh Foods] Soybean oil, fatty acid glyceryl lactates, mono- and diglycerides, TBHQ; fluid shortening for high quality cakes made without hydrog. shortening; increases pliability and shelf life of soft tortillas; *Regulatory:* kosher; *Properties:* Lovibond 3.5R max. fluid; 2.5-3.5% mono.

Foamaster FGA. [Henkel/Organic Prods.] Nonsilicone emulsion; food grade defoamer; *Properties:* liq.; Unverified

Foamaster FLD. [Henkel/Emery] Silicone emulsion; food grade defoamer for aq. systems; *Properties:* wh. opaque liq.; pH 6.0; 15% act.; Discontinued

Foam Blast 5, 7. [Ross Chem.] Silicone antifoam compd.; food-grade antifoam for foam control in nonaq. systems; applics. incl. edible oil processing, meat processing, fat rendering; 100% act.

Foam Blast 10. [Ross Chem.] Silicone antifoam compd.; food-grade antifoam for foam control in nonaq. systems; applics. incl. edible oil processing, meat processing, fat rendering.

Foam Blast 10K. [Ross Chem.] Silicone compd.; defoamer for use in nonaq. foods; *Regulatory:* kosher; 100% act.

Foam Blast 100. [Ross Chem.] Silicone emulsion; defoamer for processing seafood, potatoes, freeze-dried beverages, syrup evaporations, and cooking of fruits and vegetables; 10% act.

Foam Blast 100 Kosher. [Ross Chem.] Silicone defoamer; nonionic; defoamer for starch/proteinaceous systems requiring acid or alkaline tolerance; for yeast fermentations, sugar beet processing, caustic potato peeling, veg. processing, soups, gravies, fruit juices, jellies/jams, freeze-dried coffee, clam processing; *Usage level:* 10-100 ppm; *Regulatory:* FDA §173.340, 100 ppm max., USDA, kosher compliance; exempt from labeling under 21CFR §101.100(a)(3)(ii)(c); *Properties:* wh. emulsion; disp. in water; dens. 8.50 lb/gal; visc. 1200-1600 cps; flash pt. (PMCC) none; pH 6.5-7.5; 10% act.

Foam Blast 102. [Ross Chem.] Silicone emulsion; nonionic; food-grade defoamer for starch and proteinaceous systems requiring an acid or alkaline tolerant defoamer; for hot and cold processes; for yeast fermentations, sugar beet processing, caustic potato peeling, starch, protein; *Usage level:* 0.01-0.1% based on foaming system; 10-100 ppm (foods); *Regulatory:* FDA 21CFR §173.340, 100 ppm max.; *Properties:* wh. emulsion; dens. 8.35 lb/gal; visc. 1500-2500 cps; flash pt. (PMCC) none; pH 7.2; 10% silicone.

Foam Blast 150. [Ross Chem.] Silicone emulsion; defoamer for processing seafood, potatoes, freeze-dried beverages, syrup evaporations, and cooking of fruits and vegetables; 30% act.

Foam Blast 150 Kosher. [Ross Chem.] Silicone emulsion; nonionic; defoamer for starch/proteinaceous systems requiring acid or alkaline tolerance; for yeast fermentations, sugar beet processing, caustic potato peeling, seafood/veg. processing, soups, gravies, fruit juices, jellies/jams, freeze-dried coffee; *Usage level:* 5-33 ppm; *Regulatory:* FDA §173.340, 33 ppm max., kosher compliance; *Properties:* wh. emulsion; disp. in water; dens. 8.30 lb/gal; visc. 2100 cps; flash pt. (PMCC) none; pH 7.0; 30% act.

Foam Blast SPD. [Ross Chem.] Nonsilicone defoamer; nonionic; defoamer for food processes, esp. extraction of protein from soybeans, potato and vegetable slicing; *Usage level:* 25-250 ppm, 1000 ppm max.; *Regulatory:* FDA 21CFR §173.340(a)(2); *Properties:* dk. amber opaque liq.; disp. in water; sp.gr. 0.950 g/ml; dens. 7.92 lb/gal; visc. 300 cps; pour pt. 45 F; flash pt. (PMCC) > 550 F; 100% organic; *Storage:* store above 40 F; if solidified, heat before use; if stratified, mix before use.

Foamkill® 8BA. [Crucible] Silicone compd.; defoamer for food applics.

Foamkill® 8G. [Crucible] Silicone compd.; defoamer for food applics. incl. general nonaq. systems, starch extractions and processing, anaerobic fermentations, vitamins; *Regulatory:* FDA §173.340; *Properties:* gray sl. hazy liq., bland odor; insol. in water; sp.gr. 1.020; b.p. > 300 F; flash pt. (TOC) > 300 F; 100% act.; *Toxicology:* may cause eye irritation, mild skin irritation on prolonged/repeated contact.

Foamkill® 30 Series. [Crucible] Org. and organosilicone conc.; defoamer for food applics.

Foamkill® 80J Series. [Crucible] Silicone compd.; defoamer for food applics.

Foamkill® 618 Series. [Crucible] Org. and organosilicone conc.; defoamer for food applics.

Foamkill® 634 Series. [Crucible] Org. and organosilicone conc.; defoamer for food applics.

Foamkill® 634B-HP. [Crucible] Org. and organosilicone conc.; defoamer for food applics.

Foamkill® 634C. [Crucible] Nonsilicone; nonionic; alkaline and acid-stable defoamer for food applics., starch and protein systems, yeast processing, soya slurries, sugar beet refining, caustic potato peeling, canning trade, pasteurizer defoaming, juice and wine making, fermentations; *Regulatory:* FDA §173.340, 173.315; *Properties:* pale straw med. visc. liq., mild fatty odor; disp. in water; dens. 7.4 lb/gal; b.p. > 400 F; flash pt. (TOC) 370 F; pH 7.4 (2%); 100% act.

Foamkill® 634D-HP. [Crucible] Org. and organosilicone conc.; defoamer for food applics.

Foamkill® 634F-HP. [Crucible] Org. and organosilicone conc.; defoamer for food applics.

Foamkill® 639J-F. [Crucible] Organo-silicone; nonionic; alkaline and acid-stable defoamer for food applics., warm starch and protein systems, soya slurries, sugar beet refining, caustic potato peeling, yeast fermentations, canning; *Usage level:* 333 ppm max.; *Regulatory:* FDA §172.340, 172.315; *Properties:* wh. semitransparent liq., bland odor; disp. in water; sp.gr. 0.895; dens. 7.45 lb/gal; b.p. 390 F; flash pt. (TOC) 160 F; pH 6.6 (2%); 100% act.; *Toxicology:* skin and eye irritant; may cause harmful effects on ingestion.

Foamkill® 644 Series. [Crucible] Org. and organosilicone conc.; defoamer for food applics.

Foamkill® 652H. [Crucible] Org. and organo-silicone conc.; defoamer for food applics.

Foamkill® 652-HF. [Crucible] Org. and organosilicone conc.; defoamer for food applics. and general nonaq. systems.

Foamkill® 663J. [Crucible] Organo-silicone; defoamer for food applics.; *Properties:* water-disp.

Foamkill® 684 Series. [Crucible] Org. and organosilicone conc.; defoamer for food applics.

Foamkill® 810F. [Crucible] Dimethicone emulsion; nonionic; defoamer for food applics., general aq. systems, egg washing, soft drink and wine making, vegetable processing, yeast processing, jam/jellies, starch sol'ns., protein processing; *Usage level:* 100 ppm max.; *Regulatory:* FDA §173.340, 176.210, 175.105, 175.320, 176.200; *Properties:* wh. pourable visc. liq., bland odor; disp. in water; sp.gr. 1.000; dens. 8.3 lb/gal; visc. 300 cps; b.p. 212 F; flash pt. (TOC) > 212 F; pH 7.0; 10% act.; *Toxicology:* may cause eye irritation, mild skin irritation on prolonged/repeated contact.

Foamkill® 830F. [Crucible] Dimethicone; nonionic; defoamer for food applics., soft drink mfg., fermen-

tation, vegetable washing, sugar beet processing, jams/jellies, starch and protein processing; *Usage level:* to 33 ppm; *Regulatory:* FDA §172.340, 173.340, 176.210, 175.105, 175.300, 176.200; *Properties:* wh. pourable visc. liq., bland odor; disp. in water; sp.gr. 0.993; dens. 8.3 lb/gal; visc. 3500 cps; b.p. 212 F; flash pt. (TOC) > 212 F; pH 7.0; 30% act.; *Toxicology:* may cause eye irritation, mild skin irritation on prolonged/repeated contact.

Foamkill® 836A. [Crucible] Silicone emulsions; defoamer for food applics.

Foamkill® 1001 Series. [Crucible] Org. and organosilicone conc.; defoamer for food applics.

Foamkill® GCP Series. [Crucible] Silicone fluid; defoamer for food applics.

Foamkill® MS Conc. [Crucible] Silicone emulsion; defoamer for food applics.

Foamkill® MSC Series. [Crucible] Org. and organosilicone conc.; defoamer for food applics.

Foamkill® MSF Conc. [Crucible] Organo-silicone; nonionic; conc. for formulating defoamers for wide variety of high and low temp. applics., esp. food applics. (fruit and veg. washing, egg washing, soft drinks, wine making, yeast processing, jam/jellies, sugar refining); *Usage level:* 66 ppm max.; *Regulatory:* FDA §173.340, 176.210, 175.105, 175.300, 176.200; *Properties:* wh. visc. liq., mild pleasant odor; readily disp. in water; sp.gr. 1.084; dens. 9.02 lb/gal; visc. 20,000 cps; b.p. 212 F; flash pt. > 212 F; pH 9.02 (1%); 15% dimethylpolysiloxane; *Toxicology:* may cause eye irritation, mild skin irritation on prolonged/repeated skin contact.

Foamkill® RP. [Crucible] Org. and organo-silicone conc.; defoamer for food applics.

Folic Acid 10% Trituration No. 69997. [Roche] Folic acid USP, FCC in calcium phosphate dibasic carrier; vitamin essential for forming certain body proteins and genetic materials for cell nucleus; for food and pharmaceutical formulations; *Properties:* lt. yel. to ylsh. orange powd.; 100% thru 80 mesh; insol. in water and org. solvs.; m.w. 441.40; 10% min. assay; *Storage:* store in tight, light-resist. containers; avoid exposure to moisture and excessive heat.

Folic Acid USP, FCC No. 20383. [Roche] Folic acid USP, FCC; CAS 59-30-3; EINECS 200-419-0; vitamin essential for forming certain body proteins and genetic materials for cell nucleus; for food and pharmaceutical formulations; *Properties:* yel. to ylsh. orange cryst. powd., odorless; readily sol. in dil. alkali; sol. in sol'ns. of hot dil. acids; very sl. sol. in water; insol. in alcohol, acetone, ether, chloroform; m.w. 441.40; 87% min. assay (as is); *Precaution:* destroyed by lt. and uv radiation; unstable in sol'ns. with pH < 5; *Storage:* store in tightly closed containers; protect from light.

Fond-Ice® Icing Base. [Bunge Foods] Nonboiling icing base for dry icings for danish pastries designed to be used with fondant; *Usage level:* 16 lb/100 lb fondant; *Properties:* wh. to off-wh. semiplastic base, vanilla flavor.

Fonoline® White. [Witco/Petroleum Spec.] Petrolatum USP; CAS 8027-32-5; EINECS 232-373-2; *Regulatory:* FDA 21CFR §172.880; *Properties:* wh., odorless; visc. 9-14 cSt (100 C); m.p. 53-58 C; pour pt. 20 F.

Fonoline® Yellow. [Witco/Petroleum Spec.] Petrolatum USP; CAS 8027-32-5; EINECS 232-373-2; *Regulatory:* FDA 21CFR §172.880; *Properties:* yel., odorless; visc. 9-14 cSt (100 C); m.p. 53-58 C; pour pt. 20 F; 99% solids.

42/43 Corn Syrup. [MLG Enterprises Ltd.] Acid converted corn syrup; CAS 8029-43-4; EINECS 232-436-4; sweetener used in candies, jams, jellies, fountain and table syrups where it imparts flavor, body chars., and moisture retention; *Properties:* water-wh. med. visc. liq., char. odor, sweetbland taste; sp.gr. 1.4201 (100/60 F); dens. 11.84 lb/gal (100 F); visc. 1000 poises (80 F); pH 4.8; DE 42; Baumé 43 (100 F); 80.7% total solids, 19.3% moisture.

FOS-6. [Griffith Labs. Ltd.] Sodium tripolyphosphate, sodium hexametaphosphate blend; functional food additive for meats.

400 Stabilizer®. [Am. Maize Prods.] Moderately inhibited waxy maize starch; versatile starch for neutral or acidic foods, bakery fillings/toppings, canned foods, soups, sauces, gravies, fruit filling, baby foods, aseptic/shelf-stable foods; for batch and continuous process systems; resist. to breakdown at high temps.; *Regulatory:* FDA 21CFR §172.892; *Properties:* waxy, bland flavor; 95% thru 200 mesh; pH 5.5; 10% moisture; *Storage:* store in dry, cool conditions, in odor-free area, ≤ 50% r.h. and ≤ 25 C.

4-Teen Base. [Brolite Prods.] Plastic bread base for rich, soft white bread or rolls; *Usage level:* 14 lb/100 lb flour.

4-T-5® Hearth Conc. [Bunge Foods] No-time system with conditioners, strengtheners, and emulsifiers; cost-effective conc. for high speed prod. of French or Italian bread, sub rolls, hoagie rolls; *Usage level:* 4-5% based on flour.

FP 940. [Protein Tech. Int'l.] Isolated soy protein; CAS 68153-28-6; enhanced functional protein for whipped toppings, salad dressing, fat powds.; *Properties:* high sol. and emulsification, med. visc., low gel str., med. water absorp.; 87% protein.

Franco Pani. [Brolite Prods.] Naturally flavored cultured sour bread base for old-fashioned, Italian Ciabatta style breads; *Usage level:* 15 lb/100 lb flour.

Freedom X-PGA. [TIC Gums] Blend of tragacanth, guar, and xanthan gums; replacement for propylene glycol alginate as stabilizer and emulsifier in salad dressings.

Freeze Dried Beef Liver Powd. [Am. Labs] Dried powd. processed from beef livers; nutritive food and pharmaceutical additive; *Properties:* powd.

Freeze Dried Beef Liver Powd. Defatted. [Am. Labs] Dried defatted powd. processed from beef livers; nutritive food and pharmaceutical additive; *Properties:* powd.; *Storage:* store in cool, dry area.

Freeze Dried Pork Liver Powd. [Am. Labs] Dried

powd. processed from pork livers; nutritive food and pharmaceutical additive; *Properties:* powd.

Freezer #4® Conc. [Bunge Foods] Conc. for prod. of frozen dough for yeast-raised donuts; yields good freezer shelf life (≈ 12 wks), exc. volume, tender crumb, delicate flavor, low fat absorp.; *Usage level:* 28% based on flour.

Freeze Thaw Stable Egg White Mix. [Bunge Foods] Stabilized egg white-based mix for meringes and toppings; freeze-thaw stable; *Properties:* creamy colored powd.

Freez-Gard® FP-15. [Rhone-Poulenc Food Ingreds.] Blend of sodium tripolyphosphate FCC, natural lemon juice solids, and rosemary extract; natural antioxidant which maintains quality shelf life in frozen fish and seafood prods., incl. fatty fish, tuna, shark, fillets, fish sticks, shrimp, scallops, and surimi; provides improved succulence and texture and processing efficiency; *Regulatory:* kosher; *Properties:* off-wh. to sl. ylsh. gran., sl. spice aroma; 5% max. on 20 mesh, 30% max. thru 100 mesh; pH 9.5-10.3 (1%); *Storage:* store cool and dry.

Freez-Gard® FP-19. [Rhone-Poulenc Food Ingreds.] Sodium tripolyphosphate FCC; CAS 7758-29-4; EINECS 231-694-5; functional food additive for seafood and processed foods; used in treatment of seafood prior to freezing; eliminating thaw drip loss; locks in natural moisture, flavor, and nutritonal elements; *Regulatory:* kosher; *Properties:* wh. crushed or powd., odorless; 1% max. on 20 mesh, 20% max. thru 100 mesh (crushed), 1% max. on 40 mesh, 80% min. thru 100 mesh (powd.); m w 368; pH 9.5-10.3 (1%); 85% min. assay; *Storage:* store cool and dry.

Freez-Gard® FP-88E. [Rhone-Poulenc Food Ingreds.] Sodium hexametaphosphate glassy FCC, sodium chloride FCC, and sodium erythorbate FCC; food additive for use with fatty or minced seafood prior to freezing; retards oxidative rancidity, reduces drip loss, reduces discoloration, locks in natural moisture and nutrients; *Regulatory:* kosher; *Properties:* wh. powd., odorless; 2% max. on 20 mesh, 30% min. thru 100 mesh; pH 6-7 (1%); *Storage:* store cool and dry.

Freezist® M. [A.E. Staley Mfg.] Food starch modified, derived from tapioca; starch with exc. freeze/thaw and cold temp. stability; forms smooth, creamy cooked preps. with resist. to heat and acid; thickener, stabilizer for canned puddings, frozen sauces, pies, sauces, gravies, soups, baby foods, fruit pie fillings; *Regulatory:* FDA 21CFR §172.892; *Properties:* powd.; 98% thru 60 mesh; pH 5.0 (50% uncooked sol'n.); 11.5% moisture.

Frescolat, Type ML. [Haarmann & Reimer GmbH] Menthyl lactate; CAS 59259-38-0; EINECS 261-678-3; mild cooling flavor in powd. drink mixes, chewing gums, sweets; *Properties:* colorless liq. or solid, faintly minty odor, pract. tasteless; sol. in ethanol (50 vol. %), diethyl phthalate, min. oil, 1,2-propanediol; m.w. 228.4; acid no. 1 max.; flash pt. > 100 C; pH 4-8; *Precaution:* avoid combustible substances.

Fri-Bind® 411. [A.E. Staley Mfg.] Food starch modified derived from dent corn; coating binder for foods that are batter coated and/or breaded; provides improved adhesion of batter and produces coatings that are crisp and golden-brown on frying; for fish sticks, meat patties, onion rings, shrimp, scallops, fillets, poultry parts; *Regulatory:* FDA 21CFR §172.892; *Properties:* Gardner 10 max. color; visc. 75 cp; pH 6 (uncooked slurry); 8.5% moisture.

Frigesa® D 890. [Grünau GmbH] Locust bean gum, carrageenan, potassium citrate; hydrocolloid compd., thickener, stabilizer for jelly desserts; *Regulatory:* EEC, FAO/WHO compliance; *Properties:* off-wh. to beige powd., neutral odor and taste; sol. in hot water, disp. in cold water; *Storage:* 1 yr storage in original pkg. under dry, cool conditions.

Frigesa® F. [Grünau GmbH] Hydrocolloid compd.; thickener and stabilizer for desserts, ice cream, dressings.

Frigesa® IC 178. [Grünau GmbH] Carboxymethylcellulose sodium, guar gum; hydrocolloid compd., thickener, stabilizer for ice cream and dessert prods.; *Regulatory:* FCC, FAO/WHO, EEC compliance; *Properties:* wh. to cream powd., typ. odor and taste without off-flavor; 90% < 0.1 mm, 100% < 0.2 mm sieve; visc. 3000 mPa•s (1% aq.); 12% max. water; *Storage:* 1 yr storage in original pkg. under dry, cool conditions.

Frigesa® IC 184. [Grünau GmbH] Pectin, guar gum; hydrocolloid compd., thickener, stabilizer for fruit-sherbet and water-ice, esp. powdery ice-mixes; *Usage level:* 0.2-0.4% on finished mix; *Regulatory:* FCC; FAO/WHO, EEC compliance; *Properties:* cream to beige powd., typ. odor and taste without off-flavor; 60% < 0.1 mm, 100% < 0.315 mm; visc. 50 mPa•s (1% aq.); 12% max. water; *Storage:* 1 yr storage in original pkg. under dry, cool conditions.

Frimulsion 10. [Hercules Food Ingreds.] Refined and standardized locust bean gum and guar gum; stabilizer and water-binding agent in food systems; good disp. properties; *Properties:* off-wh. powd., < 1% retained on 0.1-mm mesh; visc. 3000 ± 300 cps (1% aq.); pH 5-6; 13% max. moisture.

Frimulsion 6G. [Hercules Food Ingreds.] Pectin, dextrose, carrageenan, sucrose, potassium chloride, locust bean gum, potassium citrate; stabilizer and gelling agent for food systems; effective jellification at 0.4-1.0% by wt.; *Properties:* off-wh. powd., < 1% retained on 0.25-mm mesh; pH 6-7 (1%); 13% max. moisture.

Frimulsion Q8. [Hercules Food Ingreds.] Modified food starch, gelatin, locust bean gum, guar gum, pectin, and dextrose; stabilizer, mainly in heat-processed, low-fat skim milk cheese; *Properties:* particulate, < 1% retained on 0.25-mm mesh; pH 5-6; 13% max. moisture.

Frimulsion RA. [Hercules Food Ingreds.] Standardized blend of starch, gelatin, pectin, and dextrose; stabilizer in Swiss-style yogurt with act. cultures; added to milk before pasturization and incubation; *Properties:* particulate, < 1% retained on 25-mm mesh; pH 5.5-6.5; 13% max. moisture.

Frimulsion RF. [Hercules Food Ingreds.] Standardized blend of native starch and gelatin; stabilizer and texturizing agent for Swiss-style yogurt at levels of 1-2%; added to milk before pasturization and incubation; *Properties:* particulate, < 1% retained on 0.25-mm mesh; pH 5-6; 12% max. moisture.

Frimulsion X5. [Hercules Food Ingreds.] Standardized blend of refined guar gum and locust bean gum; stabilizer and thickener for food systems; *Properties:* off-wh. powd., < 1% retained on 0.18-mm mesh; visc. 3000 ± 300 cps; pH 5-6 (1% aq.); 13% max. moisture.

Fro-Dex® 24 Powd. [Am. Maize Prods.] Corn syrup solids; CAS 68131-37-3; bulking agent for soup, sauce and gravy mixes, spice blends, beverages mixes, coatings for nuts and snack foods, icing mixes, coffee whiteners; *Regulatory:* FDA 21CFR §184.1865, GRAS; *Properties:* powd., bland flavor with sl. sweetness; 100% thru 40 mesh; sol. in water; DE 26; bulk dens. 31 lb/ft³; pH 4.5; 5% moisture; *Storage:* store in dry, cool conditions, in odor-free area, ≤ 50% r.h. and ≤ 25 C.

Frost-N-Wrap™ Icing Powd. [Bunge Foods] Nonboiling type icing stabilizer; quick setting; optimum stability with skim milk solids @ 1:1 ratio; for wh. and chocolate icings for cupcakes, creme fillings; *Usage level:* 3-5 lb/100 lb powd. sugar; *Properties:* wh. to off-wh. powd., vanilla flavor; pH 8.5.

Frost-O-Loid Icing Powd. [Bunge Foods] Highly conc. boiling type icing stabilizer for clear icings and glazes for sweet goods; wh. and chocolate; *Usage level:* 10 oz/100 lb powd. sugar; *Properties:* wh. to off-wh. powd., bland flavor; pH 6.5-7.5; 8% moisture.

Fructodan. [Grindsted Prods. Denmark] Stabilizer blend; for use in ice lollies and fruit ice to improve water binding, texture, air distribution, etc.; *Properties:* powd.

Fructofin® C. [Xyrofin UK] Fructose FCC/USP/NF; food grade sweetener for foods and pharmaceutical applics.; stable to air and heat; *Properties:* wh. cryst. powd., pract. odorless, very sweet taste; very sol. in water (≈ 3.5 g/ml); m.w. 180.16; pH 4.5-7 (0.1 g/ml aq.); 98% min. act.; *Storage:* 1 yr stability in original sealed pkg. stored below 25 C and < 60% r.h.; hygroscopic.

Fruitfil® 1. [A.E. Staley Mfg.] Food starch modified, derived from tapioca; thickener, stabilizer for use in neutral and low-acid foods; provides bland flavor, soft tender gel, good clarity; for citrus-flavored puddings/fillings for pies and cakes, dry sauce, soup, and gravy mixes, canned foods, infant foods; *Regulatory:* FDA 21CFR §172.892; kosher, Passover certified; *Properties:* wh. free-flowing powd.; mixes readily in cold water; pH 5.7; 11% moisture.

FruitSource® Granular. [LSI] Prepared from grape juice conc. and whole rice syrup; patented nutritive sweetener 1¹/₂ times sweeter than sugar, fat replacer; for bakery, confectionery, dairy, athletic foods/beverages, icings, cereals, spice mixes, soups, health/diabetic/dietetic/infant foods, fat-free prods., chocolate, jams; *Regulatory:* GRAS, kosher; *Properties:* pale amber to off-wh. gran. powd., odorless, bland taste; avail. in 15 and 30 mesh; completely water-sol.; sp.gr. 0.8-0.9 g/cc; biodeg.; *Precaution:* incompat. with powerful oxidizers; *Storage:* 1 yr stability in unopened container; store below 70 F, 50% r.h.

FruitSource® Liquid Sweetener. [LSI] Prepared from grape juice conc. and whole rice syrup; patented nutritive sweetener $1^1/_2$ times sweeter than sugar, fat replacer; for bakery, confectionery, dairy, athletic foods/beverages, icings, cereals, spice mixes, soups, health/diabetic/dietetic/infant foods, fat-free prods., chocolate, jams; *Regulatory:* GRAS, kosher; *Properties:* amber to dk. brn. liq., odorless, bland taste; completely water-sol.; dens. 11.8 lb/gal; visc. 9295 cps; b.p. 220 F; biodeg.; *Precaution:* incompat. with powerful oxidizers; *Storage:* 1 yr stability in unopened container, 2 wks in opened container; store below 80 F; darkening may occur.

FruitSource® Liquid Sweetener Plus. [LSI] Prepared from grape juice conc. and whole rice syrup; patented nutritive sweetener $1^1/_2$ times sweeter than sugar, fat replacer; for bakery, confectionery, dairy, athletic foods/beverages, icings, cereals, spice mixes, soups, health/diabetic/dietetic/infant foods, fat-free prods., chocolate, jams; *Regulatory:* GRAS, kosher; *Properties:* amber liq., bland very sweet taste; dens. 11.8 lb/gal; visc. 4000 cps; *Storage:* 1 yr stability in unopened container; store below 80 F; darkening may occur.

Fru-T-Cake Conc. [Bunge Foods] Unique chemically leavened fryable snack pastry for frozen prods., *Usage level:* 40% based on flour.

Fungal Lactase 100,000. [Solvay Enzymes] Fungal lactase derived from *Aspergillus oryzae*; food-grade enzyme for hydrolyzing lactose in dairy prods. (milk, whey, cheese, yogurt); *Usage level:* 0.2%; *Properties:* lt. tan amorphous dry powd., free of offensive odor and taste; water-sol.; *Storage:* activity loss $\leq$ 10% in 1 yr stored in sealed containers under cool dry conditions; 5 C storage extends life.

Fungal Protease 31,000. [Solvay Enzymes] Protease derived from *Aspergillus oryzae var.*, maltodextrin diluent; CAS 9014-01-1; EINECS 232-752-2; enzyme for hydrolysis of peptide bonds; for baking (improves grain, texture, loaf volume), meat tenderizer formulations; hydrolyzes and modifies plant and animal protein under acid conditions; brewing, fermentation; *Properties:* lt. tan to wh. powd., free of offensive odor and taste; water-sol.; *Storage:* activity loss $\leq$ 10% in 1 yr stored in sealed containers under cool dry conditions; 5 C storage extends life.

Fungal Protease 60,000. [Solvay Enzymes] Protease derived from *Aspergillus oryzae var.*, maltodextrin diluent; CAS 9014-01-1; EINECS 232-752-2; enzyme for hydrolysis of peptide bonds; for baking (improves grain, texture, loaf volume), meat tenderizer formulations; hydrolyzes and modifies plant and animal protein under acid conditions; brewing, fermentation; *Usage level:* 0.01-0.1%; *Properties:* lt. tan to wh. powd., free of offensive odor and taste; water-sol.; *Storage:* activity loss $\leq$ 10% in 1 yr stored in sealed containers under cool dry conditions; 5 C storage extends life.

Fungal Protease 500,000. [Solvay Enzymes] Protease derived from *Aspergillus oryzae var.*, maltodextrin diluent; CAS 9014-01-1; EINECS 232-752-2; enzyme for hydrolysis of peptide bonds; for baking (improves grain, texture, loaf volume), meat tenderizer formulations; hydrolyzes and modifies plant and animal protein under acid conditions; brewing, fermentation; *Usage level:* 0.01-0.1%; *Properties:* lt. tan to wh. powd., free of offensive odor and taste; water-sol.; *Storage:* activity loss $\leq$ 10% in 1 yr stored in sealed containers under cool dry conditions; 5 C storage extends life.

Fungal Protease Conc. [Solvay Enzymes] Protease derived from *Aspergillus oryzae var.*; CAS 9014-01-1; EINECS 232-752-2; enzyme for hydrolysis of peptide bonds; for baking (improves grain, texture, loaf volume), meat tenderizer formulations; hydrolyzes and modifies plant and animal protein under acid conditions; brewing, fermentation; *Usage level:* 0.01-0.1%; *Properties:* lt. tan to wh. dry powd., free of offensive odor and taste; water-sol.; *Storage:* activity loss $\leq$ 10% in 1 yr stored in sealed containers under cool dry conditions; 5 C storage extends life.

G

G-695. [ICI Am.; ICI Atkemix] Mono- and diglycerides of edible fats or oils; food emulsifier; *Properties:* yel. amber visc. liq./paste, bland odor, taste; sol. in lower alcohols, veg. and min. oils; sp.gr. 0.96; visc. 528 cps; HLB 3; clear pt. 40 C; flash pt., pour pt. > 300 F; formerly Atlas G-695.

G-991. [ICI Am.] Mono- and diglycerides; food emulsifier; *Properties:* ivory wh., plastic solid; sol. above its m.p. in veg. oils and min. oil; m.p. 118 F; HLB 2.8; flash pt. > 300 F; formerly Atlas G-991; Unverified

G-2183. [ICI Am.] Propylene glycol palmitate; food emulsifier; *Properties:* ivory, waxy solid; sol. in veg. oils, common org. solv.; m.p. 87 F; HLB 3.2; formerly Atlas G-2183; Unverified

G-2185. [ICI Am.] Propylene glycol oleate; CAS 1330-80-9; EINECS 215-549-3; food emulsifier; *Properties:* yel. liq.; sol. in veg. oils, common org. solv.; HLB 3.2; formerly Atlas G-2185; Unverified

Gamma W8. [Wacker-Chemie GmbH] γ-Cyclodextrin; CAS 7585-39-9; EINECS 231-493-2; complex hosting guest molecules; increases the sol. and bioavailability of other substances; masks flavor, odor, or coloration; stabilizes against light, oxidation, heat, and hydrolysis; turns liqs. or volatiles into stable solid powds.; *Properties:* wh. cryst. powd.; sol. 23.2 g/100 ml in water; m.w. 1297; *Toxicology:* LD50 (acute IV, rat) > 3750 mg/kg; nonirritating to eye.

Gamma W8 HP0.6. [Wacker-Chemie GmbH] Hydroxypropyl-γ-cyclodextrin; complex hosting guest molecules; increases the sol. and bioavailability of other substances; masks flavor, odor, or coloration; stabilizes against light, oxidation, heat, and hydrolysis; turns liqs. or volatiles into stable solid powds.; *Properties:* m.w. 1580.

Gamma W8 M1.8. [Wacker-Chemie GmbH] Methyl-γ-cyclodextrin; complex hosting guest molecules; increases the sol. and bioavailability of other substances; masks flavor, odor, or coloration; stabilizes against light, oxidation, heat, and hydrolysis; turns liqs. or volatiles into stable solid powds.; *Properties:* m.w. 1499.

Gamma Oryzanol. [Ikeda] Oryzanol; CAS 11042-64-1; uv absorbent and antioxidant for foods, feeds; *Properties:* wh. or pale yel. cryst. powd., odorless; 98% min. assay; *Toxicology:* LD50 (oral, mice) > 10,000 mg/kg.

Garbefix 31. [Great Lakes] BHA (antioxidant), diso-dium EDTA (chelant), calcium phosphate (extender), glyceryl stearate (emulsifier), calcium carbonate (filler), silica (anticaking agent); antioxidant/stabilizer for animal feeds, vitamins, amino acids, meals based on meat/bone/fish, poultry feeds, lucerne powd., oil cake; inhibits rancidity, assures color stability, nutrition values, and flavor; *Usage level:* 100-200 ppm; *Regulatory:* EEC, FDA clearance for components; *Properties:* powd.

Gatodan 415. [Grindsted Prods.; Grindsted Prods. Denmark] Propylene glycol stearate and distilled monoglycerides; anionic/nonionic; food emulsifier for cake and sponge improvers; *Properties:* plastic powd.; HLB 3.0; 100% conc.; Discontinued

Gelatin XF. [Hormel] Gelatin USP/NF, Type A, 235 bloom; CAS 9000-70-8; EINECS 232-554-6; hydrates rapidly; in food applics. as moisture binder in cold water (i.e., meringue, fruit pies); *Properties:* ivory fine powd., bland bouillon-like odor; 100 mesh; dens. 0.7-0.8 g/cc; visc. 40 ± 8 mps; pH 4.5-5.8; 8% max. moisture; *Storage:* stable for at least 1 yr when stored dry at ambient temps.

Gelatinized Dura-Jel®. [A.E. Staley Mfg.] Food starch modified, derived from waxy maize; gelatinized starch, thickener, stabilizer; good freeze/thaw stability, clarity, blandness, resist. to syneresis; for acid and neutral systems, cake mixes, instant puddings, bakery fillings, soft, cookies, instant gravy/sauce mixes; *Regulatory:* FDA 21CFR §172.892; *Properties:* wh. powd., bland flavor; pH 5; 5% moisture.

Gelcarin® GP-359. [FMC] Fat replacement system for sour cream; provides creamy mouthfeel, soft gel structure, smoothness, and opacity.

Gelcarin® GP-379. [FMC] Fat replacement system for gravies providing creaminess, cling, hot visc., and mouthfeel; in whipped toppings, provides creamy texture, foam stability, syneresis control, and aids extrusion of proteins.

Gelcarin® GP-911. [FMC] Fat replacement system for nonfat pasteurized processed cheese; provides creamy mouthfeel, loaf structure, sliceability, smoothness, opacity; enhances flavor; reduces rubber texture.

Gelcarin® ME-621. [FMC] Fat replacement system for poultry patties, turkey sausage, poultry franks, turkey bacon; provides juiciness, tenderness, visible fat.

Gelcarin® PS-316. [FMC] Fat replacement system for nonfat pasteurized processed cheese; provides creamy mouthfeel, loaf structure, sliceability, smoothness, opacity; enhances flavor; reduces rubber texture.

Gelcarin® XP-1008. [FMC] Fat replacement system for nonfat pasteurized processed cheese; provides creamy mouthfeel, loaf structure, sliceability, smoothness, opacity; enhances flavor; reduces rubber texture.

Gelcarin® XP-8004. [FMC] Fat replacement system for frankfurters; provides juiciness, bite, and snap.

Gelex® Instant Starch. [Am. Maize Prods.] Highly stabilized moderately inhibited pregelatinized starch; starch featuring rapid hydration, high visc. without cooking, smooth creamy texture; provides moisture retention and extends shelf life in baked goods; exc. cold storage, freeze/thaw stability; in instant puddings, soup/sauce mixes, cheese sauces; *Regulatory:* FDA 21CFR §172.892; *Properties:* waxy; 90% thru 200 mesh; pH 5.5; 5% moisture; *Storage:* store in dry, cool conditions, in odor-free area, ≤ 50% r.h. and ≤ 25 C.

Gelite. [Mid-Am. Food Sales] Water substitute for fat-free and low-fat baking applics.; relatively low sweetness and flavor; low water activity similar to that of partially hydrog. fat.

Gelodan CC Range. [Grindsted Prods.] Carrageenan; CAS 9000-07-1; EINECS 232-524-2; gellant for meat applics.; reduces cooking loss, provides firm consistency to emulsified prods. such as patés, sausages, etc.; *Regulatory:* EEC, FDA §172.620.

Gelodan CW Range. [Grindsted Prods.] Carrageenan; CAS 9000-07-1; EINECS 232-524-2; gellant for water-based applics., low-sugar jams and jellies; *Regulatory:* EEC, FDA §172.620.

Gelodan CX Range. [Grindsted Prods.] Carrageenan; CAS 9000-07-1; EINECS 232-524-2; gellant for misc. food applics.; *Regulatory:* EEC, FDA §172.620.

Genu® 04CG or 04CB. [Hercules] Pectin, low-methoxyl; CAS 9000-69-5; EINECS 232-553-0; gellant for Turkish Delight-type jelly prods. in confectionery industry; *Properties:* pH 4.8-5.2 (2%); 35-45% conc.; Discontinued

Genu® 12CG. [Hercules] Pectin, low-methoxyl; CAS 9000-69-5; EINECS 232-553-0; gellant, thickener for fruit preparation used in fruit yogurt, bakery fillings; thickener, suspending agent for toppings and variegated syrups in bakery industry; *Usage level:* 0.8-1.4% (yogurt fruit, bakery fillings), 0.5-1.0% (toppings, sauces, syrups); *Regulatory:* FDA 21CFR §184.1588, GRAS; *Properties:* pH 3.8-4.4 (1%); 27-35% methoxylation; 65-73% free acid.

Genu® 18CG. [Hercules] Pectin, low-methoxyl; CAS 9000-69-5; EINECS 232-553-0; gellant/thickener for fruit preparation used in fruit yogurt, bakery fillings, fruit spreads with 45-68% sol. solids; gellant for canned fruit preparation used for fruit-flavored milk dessert; *Usage level:* 0.8-1.0% (yogurt fruit, bakery fillings); *Regulatory:* FDA 21CFR §

184.1588, GRAS; *Properties:* pH 3.2-4.0 (1%); 35-45% methoxylation; 55-65% free acid.

Genu® 18CG-YA. [Hercules] Pectin, low-methoxyl; CAS 9000-69-5; EINECS 232-553-0; protein stabilizer for cultured milk prods.; increases body; designed for cup set; *Usage level:* 0.1-0.2% (cultured milk prods.); *Regulatory:* FDA 21CFR §184.1588, GRAS.

Genu® 20AS. [Hercules] Pectin, low-methoxyl; CAS 9000-69-5; EINECS 232-553-0; gellant for jams and jellies, yogurt fruit, bakery fillings, fruit spreads with 45-68% sol. solids; thickener for toppings, ready-to-spread frostings, and variegated syrups in bakery industry; *Usage level:* 0.8-1.0% (yogurt fruit, bakery fillings), 0.2-0.6% (frostings); *Regulatory:* FDA 21CFR §184.1588, GRAS; *Properties:* pH 3.8-4.4 (1%); 33-40% methoxylation; 15-20% amidation.

Genu® 21AS or 21AB. [Hercules] Pectin, low-methoxyl; CAS 9000-69-5; EINECS 232-553-0; gellant for canned fruit preparation used for fruit-flavored milk dessert, and milk prods.; *Properties:* pH 3.8-4.4 (1%); 33-40% methoxylation; 15-20% amidation; Discontinued

Genu® 22CG. [Hercules] Pectin, low-methoxyl; CAS 9000-69-5; EINECS 232-553-0; gellant, thickener for yogurt fruit, bakery fillings, fruit spreads with 45-68% sol. solids; viscosity and bodying agent for ready-to-spread frostings; *Usage level:* 0.2-0.6% (frostings), 0.8-1.0% (yogurt fruit, bakery fillings); *Regulatory:* FDA 21CFR §184.1588, GRAS.

Genu® 102AS. [Hercules] Pectin, low-methoxyl; CAS 9000-69-5; EINECS 232-553-0; gellant, thickener for yogurt fruit, bakery fillings, fruit spreads with 45-68% sol. solids, firm gels in confectionery industry; *Usage level:* 0.8-1.0% (yogurt fruit, bakery fillings), 1.5-2.0% (firm jellies); *Regulatory:* FDA 21CFR §184.1588, GRAS.

Genu® 104AS. [Hercules] Pectin, low-methoxyl; CAS 9000-69-5; EINECS 232-553-0; gellant agent for jams and jellies, yogurt fruit, bakery fillings, fruit spreads with 5-25% sol. solids, milk prods.; thickener for toppings, ready-to-spread frostings, and variegated syrups in bakery industry; *Usage level:* 0.8-1.4% (yogurt fruit, bakery fillings), 0.2-0.6% (frostings); *Regulatory:* FDA 21CFR §184.1588, GRAS; *Properties:* pH 3.8-4.4 (1%); 27-33% methoxylation; 20-25% amidation.

Genu® 104AS-YA. [Hercules] Pectin, low-methoxyl; CAS 9000-69-5; EINECS 232-553-0; protein stabilizer for cultured milk prods.; increases body; designed for bulk set; *Usage level:* 0.1-0.2% (cultured milk prods.); *Regulatory:* FDA 21CFR §184.1588, GRAS.

Genu® AA Medium-Rapid Set, 150 Grade. [Hercules] Pectin, high-methoxyl; CAS 9000-69-5; EINECS 232-553-0; gellant for jams and preserves in jars; *Properties:* pH 3.4-4.2 (4%); 67-70% methoxylation; Discontinued

Genu® BA-KING. [Hercules] Pectin, high-methoxyl; CAS 9000-69-5; EINECS 232-553-0; gellant for heat-stable jams and jellies in fruit preserving and bakery industries; allows filling large containers @ elevated temps.; *Usage level:* 0.5-1.0% (fruit filling); *Regulatory:* FDA 21CFR §184.1588, GRAS; *Properties:* pH 2.7-3.2 (4%); 68-71% methoxylation.

Genu® BB Rapid Set. [Hercules] Pectin, high-methoxyl; CAS 9000-69-5; EINECS 232-553-0; gellant for jams and preserves in jars, jellies in bakery industry; stabilizer for pulp suspension and oil emulsion in citrus drink concs.; *Usage level:* 0.2-9.5% (jams, jellies in jars); *Regulatory:* FDA 21CFR §184.1588, GRAS; *Properties:* pH 3.4-4.2 (4%); 71-75% methoxylation.

Genu® DD Extra-Slow Set. [Hercules] Pectin, high-methoxyl; CAS 9000-69-5; EINECS 232-553-0; gellant for jams and preserves in lg. containers, e.g., drums; allows for lower filling temp.; *Usage level:* 0.3-0.7% (jams); *Regulatory:* FDA 21CFR §184.1588, GRAS; *Properties:* pH 3.4-4.2 (4%); 61-64% methoxylation.

Genu® DD Extra-Slow Set C, 150 Grade. [Hercules] Pectin, high-methoxyl; CAS 9000-69-5; EINECS 232-553-0; gellant for jellies; texturizer, and foam stabilizer for aerated confectionery prods.; e; *Properties:* pH 3.4-4.2 (4%); 60-63% methoxylation; Discontinued

Genu® DD Slow Set. [Hercules] Pectin, high-methoxyl; CAS 9000-69-5; EINECS 232-553-0; gellant agent for jams and preserves in lg. containers; *Usage level:* 0.3-0.7% (jellies in jars); *Regulatory:* FDA 21CFR §184.1588, GRAS; *Properties:* pH 3.4-4.2 (4%); 63-66% methoxylation.

Genu® JMJ. [Hercules] Pectin, high-methoxyl type; CAS 9000-69-5; EINECS 232-553-0; stabilizer for milk proteins in pasturized cultured milk drinks and milk/fruit juice drinks having a long shelf life; provides protective colloid effect; allows pasteurization of acidic milk prods. without curdling; *Usage level:* 0.4-0.7% (dairy drinks); *Regulatory:* FDA 21CFR §184.1588, GRAS; *Properties:* pH 2.7-3.2 (4%).

Genu® Type DJ. [Hercules] Pectin, high-methoxyl; CAS 9000-69-5; EINECS 232-553-0; gellant for jellies (jet cook applics.); *Usage level:* 1.0-2.5% (jellies); *Regulatory:* FDA 21CFR §184.1588, GRAS.

Genu® VIS. [Hercules] Pectin, high-methoxyl; CAS 9000-69-5; EINECS 232-553-0; bodying agent in diet soft drinks, fruit drinks; restores mouthfeel imparted by natural sweeteners; *Usage level:* 0.1-0.2% (diet soft drinks, fruit drinks); *Regulatory:* FDA 21CFR §184.1588, GRAS.

Genugel® CHP-2. [Hercules Food Ingreds.] Carrageenan; CAS 9000-07-1; EINECS 232-524-2; water binding gellant providing firmness, cohesiveness to cook-in-bag poultry prods., massage-tumbled; *Usage level:* 0.3-1.0% (process poultry prods.); *Regulatory:* FDA 21CFR §172.620; USDA 9CFR §318.7, 381.7.

Genugel® CHP-2 Fine Mesh. [Hercules Food Ingreds.] Carrageenan; CAS 9000-07-1; EINECS 232-524-2; water binding gellant providing firmness, cohesiveness to cook-in-bag poultry prods.,

injection-processed; *Usage level:* 0.3-1.0% (process poultry prods.); *Regulatory:* FDA 21CFR §172.620; USDA 9CFR §318.7, 381.7.

Genugel® CHP-200. [Hercules Food Ingreds.] Carrageenan; CAS 9000-07-1; EINECS 232-524-2; water binding gellant providing firmness, cohesiveness, suspension in brine to cook-in-bag poultry prods., injection-processed; *Usage level:* 0.3-1.0% (process poultry prods.); *Regulatory:* FDA 21CFR §172.620; USDA 9CFR §318.7, 381.7.

Genugel® CJ. [Hercules Food Ingreds.] Carrageenan; CAS 9000-07-1; EINECS 232-524-2; stabilizes foam, mouthfeel, and body in instant dairy-based whipped toppings; emulsion stiffener for cook-in-bag poultry prods.; eliminates leaks in hot-seal applics.; *Usage level:* 0.10-0.20% (whipped toppings); *Regulatory:* FDA 21CFR §172.620; USDA 9CFR §318.7, 381.7.

Genugel® LC-1. [Hercules Food Ingreds.] Carrageenan; CAS 9000-07-1; EINECS 232-524-2; gellant providing firm cohesive gelation for ultrahigh temp. water-gel desserts; *Usage level:* 0.8-1.1% (water-gel desserts); *Regulatory:* FDA 21CFR §172.620; USDA 9CFR §318.7, 381.7.

Genugel® LC-4. [Hercules Food Ingreds.] Carrageenan; CAS 9000-07-1; EINECS 232-524-2; gellant providing firm cohesive gelation for water-gel dessert powds.; *Usage level:* 0.6-1.0% (dessert powds.); *Regulatory:* FDA 21CFR §172.620; USDA 9CFR §318.7, 381.7.

Genugel® LC-5. [Hercules Food Ingreds.] Carrageenan; CAS 9000-07-1; EINECS 232-524-2; gellant providing soft cohesive gelation for ultrahigh temp. water-gel desserts; *Usage level:* 0.8-1.1% (water-gel desserts); *Regulatory:* FDA 21CFR §172.620; USDA 9CFR §318.7, 381.7.

Genugel® MB-51. [Hercules Food Ingreds.] Carrageenan; CAS 9000-07-1; EINECS 232-524-2; water binding gellant providing juiciness and cohesiveness in low-fat ground beef and sausage; *Usage level:* 0.36-0.45% (ground meat prods.); *Regulatory:* FDA 21CFR §172.620; USDA 9CFR §318.7, 381.7.

Genugel® MB-78F. [Hercules Food Ingreds.] Carrageenan; CAS 9000-07-1; EINECS 232-524-2; water binding gellant providing firm, cohesive gel in massage or injection cook-in-bag smoked hams; *Usage level:* 0.5-0.8% (ham); *Regulatory:* FDA 21CFR §172.620; USDA 9CFR §318.7, 381.7.

Genugel® UE. [Hercules Food Ingreds.] Carrageenan; CAS 9000-07-1; EINECS 232-524-2; gellant providing spreadable strong gel for sugarless or low-sugar fruit spreads; *Usage level:* 1.0-1.2% (fruit spreads); *Regulatory:* FDA 21CFR §172.620; USDA 9CFR §318.7, 381.7.

Genugel® UEU. [Hercules Food Ingreds.] Carrageenan; CAS 9000-07-1; EINECS 232-524-2; high functionality gellant providing spreadable gelation for sugarless or low-sugar fruit spreads; *Usage level:* 0.8-1.0% (fruit spreads); *Regulatory:* FDA 21CFR §172.620; USDA 9CFR §318.7, 381.7.

Genulacta® CP-100. [Hercules Food Ingreds.] Carrageenan; CAS 9000-07-1; EINECS 232-524-2; provides visc. and mouthfeel in instant puddings and flan; *Usage level:* 0.2-0.3% (instant puddings); *Regulatory:* FDA 21CFR §172.620; USDA 9CFR §318.7, 381.7.

Genulacta® CSM-2. [Hercules Food Ingreds.] Carrageenan; CAS 9000-07-1; EINECS 232-524-2; provides thickening and mouthfeel in instant aerated desserts, cocoa suspension in instant chocolate drinks; *Usage level:* 0.15-0.25% (aerated desserts), 0.05-0.10% (instant chocolate drink); *Regulatory:* FDA 21CFR §172.620; USDA 9CFR §318.7, 381.7.

Genulacta® K-100. [Hercules Food Ingreds.] Carrageenan; CAS 9000-07-1; EINECS 232-524-2; emulsion and fat stabilizer for dairy-based coffee whiteners, evaporated milk; prevents serum separation; provides cocoa suspension and mouthfeel in chocolate milk; *Usage level:* 0.1-0.2% (coffee whiteners), 0.005-0.010% (evaporated milk), 0.01-0.03% (chocolate milk); *Regulatory:* FDA 21CFR §172.620; USDA 9CFR §318.7, 381.7.

Genulacta® KM-1. [Hercules Food Ingreds.] Carrageenan; CAS 9000-07-1; EINECS 232-524-2; provides cocoa suspension and mouthfeel in chocolate milk; *Usage level:* 0.01-0.03% (chocolate milk); *Regulatory:* FDA 21CFR §172.620; USDA 9CFR §318.7, 381.7.

Genulacta® KM-5. [Hercules Food Ingreds.] Carrageenan; CAS 9000-07-1; EINECS 232-524-2; prevents whey separation, adds creaminess for frozen desserts (ice cream, ice milks, sherbet, soft-serve, milk shakes); controls large ice crystal growth; *Usage level:* 0.005-0.010% (frozen desserts); *Regulatory:* FDA 21CFR §172.620; USDA 9CFR §318.7, 381.7.

Genulacta® L-100. [Hercules Food Ingreds.] Carrageenan; CAS 9000-07-1; EINECS 232-524-2; prevents whey separation, adds creaminess for frozen desserts (ice cream, ice milks, sherbet, soft-serve, milk shakes); controls large ice crystal growth; *Usage level:* 0.01-0.02% (frozen desserts); *Regulatory:* FDA 21CFR §172.620; USDA 9CFR §318.7, 381.7.

Genulacta® LK-71. [Hercules Food Ingreds.] Carrageenan; CAS 9000-07-1; EINECS 232-524-2; provides cocoa suspension and mouthfeel in chocolate milk; *Usage level:* 0.02-0.03% (chocolate milk); *Regulatory:* FDA 21CFR §172.620; USDA 9CFR §318.7, 381.7.

Genulacta® LR-41. [Hercules Food Ingreds.] Carrageenan; CAS 9000-07-1; EINECS 232-524-2; gellant providing demoldable gel with creamy mouthfill (hot fill); for ready-to-eat dairy desserts; *Usage level:* 0.25-0.45% (dairy desserts); *Regulatory:* FDA 21CFR §172.620; USDA 9CFR §318.7, 381.7.

Genulacta® LR-60. [Hercules Food Ingreds.] Carrageenan; CAS 9000-07-1; EINECS 232-524-2; gellant providing flan-like relatively firm gel (hot fill); for ready-to-eat dairy desserts; *Usage level:* 0.10-0.25% (dairy desserts); *Regulatory:* FDA 21CFR

§172.620; USDA 9CFR §318.7, 381.7.

Genulacta® LRA-50. [Hercules Food Ingreds.] Carrageenan; CAS 9000-07-1; EINECS 232-524-2; gellant providing yogurt-like gel structure with creamy mouthfeel (hot fill) or creamy texture, rich body (cold fill); for yogurt desserts; *Usage level:* 0.20-1.0% (yogurt desserts); *Regulatory:* FDA 21CFR §172.620; USDA 9CFR §318.7, 381.7.

Genulacta® LRC-21. [Hercules Food Ingreds.] Carrageenan; CAS 9000-07-1; EINECS 232-524-2; gellant providing creamy sl. cohesive weak gel (hot fill) or creamy sl. weak gel (cold fill); for ready-to-eat dairy desserts/aseptic fill; *Usage level:* 0.25-0.50% (dairy desserts); *Regulatory:* FDA 21CFR §172.620; USDA 9CFR §318.7, 381.7.

Genulacta® LRC-30. [Hercules Food Ingreds.] Carrageenan; CAS 9000-07-1; EINECS 232-524-2; gellant providing flan-like creamy gel (hot fill) or creamy weak gel (cold fill); for ready-to-eat dairy desserts/aseptic fill; *Usage level:* 0.25-0.40% (dairy desserts); *Regulatory:* FDA 21CFR §172.620; USDA 9CFR §318.7, 381.7.

Genulacta® P-100. [Hercules Food Ingreds.] Carrageenan; CAS 9000-07-1; EINECS 232-524-2; gellant providing firm, brittle gel in cooked puddings; *Usage level:* 0.25-0.40% (cooked puddings); *Regulatory:* FDA 21CFR §172.620; USDA 9CFR §318.7, 381.7.

Genulacta® PL-93. [Hercules Food Ingreds.] Carrageenan; CAS 9000-07-1; EINECS 232-524-2; gellant providing firm, cohesive gel in cooked puddings; *Usage level:* 0.25-0.40% (cooked puddings); *Regulatory:* FDA 21CFR §172.620; USDA 9CFR §318.7, 381.7.

Genu® Pectins. [Hercules] Pectin; high-methoxyl and low-methoxyl purified natural hydrocolloid derived from citrus peels; CAS 9000-69-5; EINECS 232-553-0; gelling agent for jellies, jams, fruit preps.; visc. builder, bodying agent, suspending agent, protective colloid, and stabilizer for food systems, convenience foods, beverages, pharmaceutical, and cosmetic industries; *Regulatory:* FDA 21CFR §184.1588, GRAS; *Properties:* lt. cream to grayish powd.; no odor and flavor.

Genuvisco® CSW-2. [Hercules Food Ingreds.] Carrageenan; CAS 9000-07-1; EINECS 232-524-2; gellant providing visc., mouthfeel, and body to instant cold-water visc., nondairy prods.; *Usage level:* 0.1-0.3% (instant prods.); *Regulatory:* FDA 21CFR §172.620; USDA 9CFR §318.7, 381.7.

Genuvisco® J. [Hercules Food Ingreds.] Carrageenan; CAS 9000-07-1; EINECS 232-524-2; gellant providing very soft, cohesive gelation for water-gel dessert powds.; *Usage level:* 0.6-1.0% (dessert powds.); *Regulatory:* FDA 21CFR §172.620; USDA 9CFR §318.7, 381.7.

Genuvisco® MP-11. [Hercules Food Ingreds.] Carrageenan; CAS 9000-07-1; EINECS 232-524-2; water binding gellant providing juiciness and cohesiveness to low-fat ground beef and sausage prods.; *Usage level:* 0.45-0.5% (ground meat prods.); *Regulatory:* FDA 21CFR §172.620; USDA

9CFR §318.7, 381.7.

Geoldan CL Range. [Grindsted Prods. Denmark] Carrageenan; CAS 9000-07-1; EINECS 232-524-2; gellant for milk-based applics., e.g., in ice cream where it prevents whey separation, in milk gels where it improves consistency, and in chocolate milk where it prevents cocoa particles from precipitating.

GFS®. [Kelco] Xanthan gum, locust bean gum, guar gum; gum for use in food preparations; stabilizer, suspending agent, thickener; provides rheological control to water- and milk-based foods; for pasteurized processed cheese spread, Neufchatel cheese, cream cheese, cottage cheese dressings, sour cream; *Usage level:* USA: 2.7% max. (cheese), 1.7% (other foods); *Regulatory:* FDA 21CFR §172.695, 184.1343, 184.1339, GRAS; EEC E410, E412, E415; kosher; *Properties:* beige; 100% thru 60 mesh; partly sol. in water; visc. 1400 ± 500 cps (1%); pH 6.5 ± 1.1; 89 ± 3% solids.

Gistex®. [Gist-brocades] Baker's yeast extract; CAS 8013-01-2; developed to replace meat extract; imparts a savory stock type flavor and enhances other savory flavor notes; synergistically enhances hydrolyzed veg. protein used at 1 part Glistex to 2 parts HVP.

Gistex® Family. [Gist-brocades] Baker's yeast extracts; source of nondescript background/base flavors for savory flavor systems; to be used as a flavor base with beef, pork, chicken, fish, or vegetable flavor profiles in soups, sauces, side dishes, and finished entrees.

Gistex® LS. [Gist-brocades] Yeast extract; CAS 8013-01-2; low sodium version of Gistex®; imparts flavor and enhancement chars. of the Gistex® family.

Gistex® MR. [Gist-brocades] Yeast extract; CAS 8013-01-2; flavor enhancer; particularly efficient in lighter seasoned prods. like poultry, fish, and vegetables.

Gistex® X-II. [Gist-brocades] Yeast extract; CAS 8013-01-2; offers a clean flavor profile and exc. enhancing abilities; effective at maximizing the flavor of meat, fish, game, poultry, and vegetables.

Gistex® Xtra Powd. [Gist-brocades] Yeast extract; CAS 8013-01-2; imparts a clean savory taste; offers reduced sodium, instant solubility, and less dusting.

Glass H®. [FMC/Food Phosphates] Long chain sodium hexametaphosphate; EINECS 233-343-1; sequestrant, emulsifier, suspending agent, protein stabilizer for dairy prods. esp. process cheese, ice cream, and frozen desserts; protein dispersant in whey processing; water-binding agent for cured pork; *Usage level:* 2% (process cheese), 0.1-0.5% (whey), 0.1-0.2% (ice cream); *Regulatory:* FDA GRAS; *Properties:* powd., gran., plate; 99.9% thru 20 mesh (powd.), 100% thru 8 mesh (gran.); inifinitely sol. in water; m.w. 2204; pH 6.3 (1%).

Glazo™. [Bunge Foods] Clear glaze with stabilizers for donuts and sweet goods; *Properties:* sweet flavor; pH 5.5-6.0.

Gloria®. [Witco/Petroleum Spec.] Wh. mineral oil USP; emollient, lubricant for food industry; *Regulatory:* FDA 21CFR §172.878, §178.3620a; *Properties:* water-wh., odorless, tasteless; sp.gr. 0.859-0.880; visc. 39-42 cSt (40 C); pour pt. -12 C; flash pt. 193 C.

GLS. [Croda Food Prods. Ltd.] Glyceryl lactostearate; nonionic; used in the food industry; *Properties:* solid; 100% conc.

Glucanase GV. [Grindsted Prods. Denmark] β-Glucanase; enzyme for the brewing industry; added either to the mash or during fermentation or lagering to facilitate wort run-off, increase extract yield, improve clarity of the wort and beer, and optimize filtration.

Glucanex® L-300. [Solvay Enzymes] β-Glucanase; enzyme for hyrolysis of β-glucans; reduces visc. of barley, barley malt, rice, rye, and other cereal grains; for beer brewing; reduces wort visc., improves beer clarity and stability; *Properties:* liq.

Gluconal® CA A. [Akzo Chemie] Calcium gluconate anhydrous; CAS 299-28-5; EINECS 206-075-8; food grade mineral source for dietary supplements, fortified foods, animal feeds; *Properties:* wh. powd.; sol. 30 g/l water; m.w. 430.4; bulk dens. 250-350 kg/m³; pH 7.4 (1%); 95-100% act.; *Toxicology:* LD50 (oral, rat) > 5000 mg/kg.

Gluconal® CA M. [Akzo Chemie] Calcium gluconate monohydrate; CAS 299-28-5; EINECS 206-075-8; food grade mineral source for dietary supplements, fortified foods, animal feeds; *Properties:* wh. powd./gran.; sol. 40 g/l water; m.w. 448.4; bulk dens. 300-650 kg/m³; pH 7.5 (1%); 98.5-100% act.; *Toxicology:* LD50 (oral, rat) > 5000 mg/kg.

Gluconal® CA M B. [Akzo Chemie] Calcium borogluconate; CAS 5743-34-0; food grade mineral source for dietary supplements, fortified foods, animal feeds; *Properties:* wh. powd.; sol. 200 g/l water; m.w. 448.4 + 61.8; bulk dens. 550-650 kg/m³; pH 5.1 (1%); 82-89% act.; *Toxicology:* LD50 (oral, rat) > 2000 mg/kg.

Gluconal® CO. [Akzo Chemie] Cobalt gluconate; food grade mineral source for dietary supplements, fortified foods, animal feeds; *Properties:* pink powd.; sol. 200 g/l water; m.w. 449.3; bulk dens. 450-550 kg/m³; pH 6.5 (1%); 88-100% act.; *Toxicology:* LD50 (oral, rat) 1420 mg/kg.

Gluconal® CU. [Akzo Chemie] Copper gluconate; CAS 527-09-3; EINECS 208-408-2; food grade mineral source for dietary supplements, fortified foods, animal feeds; *Properties:* lt. blue powd.; sol. 500 g/l water; m.w. 453.8; bulk dens. 450-550 kg/m³; pH 4.6 (1%); 98-100% act.; *Toxicology:* LD50 (oral, rat) 1710 mg/kg.

Gluconal® FE. [Akzo Chemie] Ferrous gluconate; CAS 299-29-6; EINECS 206-076-3; food grade mineral source for dietary supplements, fortified foods, animal feeds; *Properties:* yel.-gray powd./gran.; sol. 100 g/l water; m.w. 446.1; bulk dens. 650-850 kg/m³; pH 4.5 (1%); 87.5-95% act.; *Toxicology:* LD50 (oral, rat) 4600 mg/kg.

Gluconal® K. [Akzo Chemie] Potassium gluconate;

CAS 299-27-4; EINECS 206-074-2; food grade mineral source for dietary supplements, fortified foods, animal feeds; *Properties:* wh. powd./gran.; sol. 1000 g/l water; m.w. 234.3; bulk dens. 500-650 kg/m³; pH 7.1 (1%); 95-100% act.; *Toxicology:* LD50 (oral, rat) 6060 mg/kg.

Gluconal® MG. [Akzo Chemie] Magnesium gluconate; CAS 3632-91-5; EINECS 222-848-2; food grade mineral source for dietary supplements, fortified foods, animal feeds; *Properties:* wh. powd./gran.; sol. 160 g/l water; m.w. 414.6; bulk dens. 500-750 kg/m³; pH 7.3 (1%); 86-99% act.; *Toxicology:* LD50 (oral, rat) 9100 mg/kg.

Gluconal® MN. [Akzo Chemie] Manganese gluconate; CAS 6485-39-8; EINECS 229-350-4; food grade mineral source for dietary supplements, fortified foods, animal feeds; *Properties:* off-wh. powd.; sol. 110 g/l water; m.w. 445.2; bulk dens. 700-800 kg/m³; pH 6.4 (1%); 90.5-100% act.; *Toxicology:* LD50 (oral, rat) 5850 mg/kg.

Gluconal® NA. [Akzo Chemie] Sodium gluconate; CAS 527-07-1; EINECS 208-407-7; food grade mineral source for dietary supplements, fortified foods and animal feed; *Properties:* wh. powd./gran.; sol. 600 g/l water; m.w. 218.1; bulk dens. 600-780 kg/m³; pH 6.9 (1%); 98-100% act.; *Toxicology:* LD50 (oral, rat) > 5000 mg/kg.

Gluconal® ZN. [Akzo Chemie] Zinc gluconate; CAS 4468-02-4; EINECS 224-736-9; food grade mineral source for dietary supplements, fortified foods, animal feeds; *Properties:* wh. powd./gran.; sol. 100 g/l water; m.w. 455.7; bulk dens. 600-800 kg/m³; pH 6.5 (1%); 85.5-100% act.; *Toxicology:* LD50 (oral, rat) > 5000 mg/kg.

Gluczyme. [Mitsubishi Int'l.; Amano Enzyme USA] Amyloglucosidase; CAS 9032-08-0; food enzyme; *Properties:* powd.

Glutaminase Amano. [Amano Enzyme USA; Unipex] Enzyme from *Bacillus*; enzyme for food applics.

Glutarom® Range. [Grünau GmbH] Amino acid salt mixt. on carrier; natural flavor enhancer for meat and sausage prod.

Glycerine (Pharmaceutical). [Asahi Denka Kogyo] Glycerin; CAS 56-81-5; EINECS 200-289-5; solvent, emollient, sweetener; *Properties:* clear liq.; sp.gr. > 1.2598; > 98% purity.

Glycomul® L. [Lonza] Sorbitan laurate; CAS 1338-39-2; nonionic; emulsifier for edible uses; *Properties:* amber liq.; sol. in methanol, ethanol, naphtha; sp.gr. 1.0; visc. 4500 cps; HLB 8.6; acid no. 5; sapon. no. 157-171; 100% conc.

Glycomul® MA. [Lonza] Sorbitan ester mixed fatty acids; nonionic; emulsifier for edible uses; *Properties:* dk. amber liq.; sol. in methanol, ethanol, naphtha, acetone, toluol, min. and veg. oil; sp.gr. 1.0; visc. 1000 cps; sapon. no. 139-156; 100% conc.; Unverified

Glycomul® S FG. [Lonza] Sorbitan stearate; CAS 1338-41-6; EINECS 215-664-9; nonionic; emulsifier for food uses; *Properties:* Gardner 5 beads; m.p. 53 C; HLB 5; acid no. 5.

Glycomul® S KFG. [Lonza] Sorbitan stearate; CAS 1338-41-6; EINECS 215-664-9; nonionic; emulsifier for food prods.; *Properties:* Gardner 5 beads; m.p. 53 C; HLB 5; acid no. 5.

Glycomul® TAO. [Lonza] Sorbitan ester mixed resin and fatty acids; nonionic; emulsifier for edible uses; *Properties:* amber liq.; sol. in methanol, ethanol, toluol, naphtha, min. and veg. oils, disp. in water; sp.gr. 1.02; visc. 1300 cps; HLB 4.3; sapon. no. 135-148; 100% conc.; Unverified

Glycomul® TS KFG. [Lonza] Sorbitan tristearate; CAS 26658-19-5; EINECS 247-891-4; nonionic; emulsifier for food uses; *Properties:* Gardner 2 beads; m.p. 55 C; HLB 2; acid no. 14.

Glycon® G 100. [Lonza] Glycerin; CAS 56-81-5; EINECS 200-289-5; humectant, bodying agent, moisture control agent for sugarless confections, controlled moisture foods; *Properties:* sp.gr. 1.2607; 99.5% act.

Glycon® G-300. [Lonza] Glycerin; CAS 56-81-5; EINECS 200-289-5; humectant, bodying agent, moisture control agent for sugarless confections, controlled moisture foods; *Properties:* sp.gr. 1.2517; 96% act.

Glycon® P-45. [Lonza] 45% Palmitic acid; CAS 57-10-3; EINECS 200-312-9; lubricant, defoamer, and component of other food additives; *Regulatory:* FDA 21CFR §172.860; *Properties:* acid no. 210; iodine no. 1.0; sapon. no. 211.

Glycon® S-65. [Lonza] Hydrogenated tallow fatty acid; lubricant, defoamer, and component of other food additives; *Regulatory:* FDA 21CFR §172.860; *Properties:* acid no. 204; iodine no. 0.0, sapon. no. 205.

Glycon® S-70. [Lonza] 70% Stearic acid; CAS 57-11-4; EINECS 200-313-4; lubricant, defoamer, and component of other food additives; *Regulatory:* FDA 21CFR §172.860; *Properties:* acid no. 203; iodine no. 0.8; sapon. no. 204.

Glycon® S-90. [Lonza] 90% Stearic acid; CAS 57-11-4; EINECS 200-313-4; lubricant, defoamer, and component of other food additives; *Regulatory:* FDA 21CFR §172.860; *Properties:* acid no. 199; iodine no. 0.7; sapon. no. 201.

Glycon® TP. [Lonza] Stearic acid, triple pressed; CAS 57-11-4; EINECS 200-313-4; lubricant, defoamer, and component of other food additives; *Regulatory:* FDA 21CFR §172.860; *Properties:* acid no. 210; iodine no. 0.8; sapon. no. 211.

Glycosperse® HTO-40. [Lonza] PEG-40 sorbitan hexatallate; food emulsifier; *Properties:* Gardner 7 liq.; HLB 10; acid no. 10.

Glycosperse® L-20. [Lonza] Polysorbate 20; CAS 9005-64-5; nonionic; emulsifier; flavor solubilizer and dispersant; *Properties:* yel. clear liq.; sol. in water, alcohol, acetone; sp.gr. 1.1; visc. 400 cps; HLB 16.7; acid no. 2 max.; sapon. no. 40-50; hyd. no. 96-108; pH 7 (5%); 3% max. moisture.

Glycosperse® O-5. [Lonza] Polysorbate 81; CAS 9005-65-6; nonionic; emulsifier; flavor solubilizer and dispersant; *Properties:* amber liq.; sol. in alcohol, ethyl acetate, min. oil; disp. in water; sp.gr. 1.0; visc. 450 cps; HLB 10.0; sapon. no. 95-105; 100%

conc.

Glycosperse® O-20. [Lonza] Polysorbate 80; CAS 9005-65-6; nonionic; emulsifier; flavor solubilizer and dispersant; *Properties:* yel. liq.; sol. in water, alcohol, ethyl acetate, toluol, veg. oil; sp.gr. 1.0; visc. 400 cps; HLB 15; sapon. no. 44-56; 100% conc.

Glycosperse® O-20 FG. [Lonza] Polysorbate 80 NF FCC; CAS 9005-65-6; nonionic; emulsifier for ice cream, frozen desserts; solubilizer and dispersant for shortenings, pickles, vitamin-min. preps., gelatin desserts; wetting agent in poultry defeathering scald, colorants; defoamer for cottage cheese; *Regulatory:* FDA 21CFR §172.515, 172.840; *Properties:* amber clear liq., mild odor; HLB 15.0; acid no. 1 max.; iodine no. 19-22; sapon. no. 45-55; hyd. no. 65-80; 3% max. moisture; *Storage:* store in cool, dry area.

Glycosperse® O-20 KFG. [Lonza] Polysorbate 80 NF FCC; CAS 9005-65-6; nonionic; emulsifier for ice cream, frozen desserts; solubilizer and dispersant for shortenings, pickles, vitamin-min. preps., gelatin desserts; wetting agent in poultry defeathering scald, colorants; defoamer for cottage cheese; *Regulatory:* FDA 21CFR §172.515, 172.840; kosher; *Properties:* amber clear liq., mild odor; HLB 15.0; acid no. 2 max.; sapon. no. 45-55; hyd. no. 65-80; 3% max. moisture; *Storage:* store in cool, dry area.

Glycosperse® O-20 Veg. [Lonza] Polysorbate 80, veg. grade; CAS 9005-65-6; nonionic; food emulsifier; flavor solubilizer and dispersant; *Properties:* yel. liq.; sol. in water, alcohol, ethyl acetate, toluol, veg. oil; sp.gr. 1.0; visc. 400 cps; sapon. no. 44-56; 100% conc.; Unverified

Glycosperse® O-20X. [Lonza] Polysorbate 80, anhyd.; CAS 9005-65-6; nonionic; food emulsifier; flavor solubilizer and dispersant; *Properties:* yel. liq.; sol. in water, alcohol, ethyl acetate, toluol, veg. oil; sp.gr. 1.0; visc. 600 cps; sapon. no. 44-56; 100% conc.; Unverified

Glycosperse® P-20. [Lonza] Polysorbate 40; CAS 9005-66-7; nonionic; food emulsifier; flavor solubilizer and dispersant; *Properties:* yel. liq.; sol. in water, methanol, ethanol, acetone, ethyl acetate; sp.gr. 1.0; visc. 550 cps; HLB 15.6; sapon. no. 40-53; 100% conc.

Glycosperse® S-20. [Lonza] Polysorbate 60; CAS 9005-67-8; nonionic; emulsifier; flavor solubilizer and dispersant; *Properties:* yel. liq.; sol. in water, ethyl acetate, toluol; sp.gr. 1.1; HLB 15.0; sapon. no. 44-56; 100% conc.

Glycosperse® S-20 FG. [Lonza] Polysorbate 60 NF FCC; CAS 9005-67-8; nonionic; emulsifier for whipped toppings, cakes, shortenings, nondairy creamers, protective coatings on raw fruits/vegs.; opacifier in confection coatings; foaming agent in beverages; dispersant in gelatin desserts; wetting agent for colors; *Regulatory:* FDA 21CFR §172.836; *Properties:* Gardner 7 max. soft solid; HLB 15.0; m.p. 28 C; acid no. 2; sapon. no. 45-55; hyd. no. 81-96; pH 7 (5%); 3% max. moisture;

Storage: store in cool, dry area.

Glycosperse® S-20 KFG. [Lonza] Polysorbate 60 NF FCC; CAS 9005-67-8; nonionic; emulsifier for whipped toppings, cakes, shortenings, nondairy creamers, protective coatings on raw fruits/vegs.; opacifier in confection coatings; foaming agent in beverages; dispersant in gelatin desserts; wetting agent for colors; *Regulatory:* FDA 21CFR §172.836; kosher; *Properties:* Gardner 7 max. soft solid; HLB 15.0; m.p. 28 C; acid no. 2 max.; sapon. no. 45-55; hyd. no. 81-96; pH 7 (5%); 3% max. moisture; *Storage:* store in cool, dry area.

Glycosperse® TO-20. [Lonza] Polysorbate 85; CAS 9005-70-3; nonionic; emulsifier; flavor solubilizer and dispersant; *Properties:* yel. liq., gels on standing; sol. in ethanol, methanol, ethyl acetate; water disp.; sp.gr. 1.0; visc. 300 cps; HLB 11.0; sapon. no. 82-95; 100% conc.

Glycosperse® TS-20. [Lonza] Polysorbate 65; CAS 9005-71-4; nonionic; emulsifier; flavor solubilizer and dispersant; *Properties:* tan waxy solid; sol. in ethanol, methanol, acetone, ethyl acetate, naphtha, min. and veg. oils; disp. water, toluol; sp.gr. 1.05; HLB 11.0; sapon. no. 88-98; 100% conc.

Glycosperse® TS-20 FG, TS-20 KFG. [Lonza] Polysorbate 65 FCC; CAS 9005-71-4; nonionic; emulsifier for ice cream, frozen desserts, cakes, whipped toppings, nondairy creamers, icings, fillings; *Regulatory:* FDA 21CFR §172.838; *Properties:* Gardner 7 max. soft solid, mild odor; m.p. 31 C; HLB 11.0; acid no. 2 max.; iodine no. 1 max.; sapon. no. 88-98 hyd. no. 44-60; 3% max. moisture; *Storage:* store in cool, dry area.

Glycosperse® TS-20 KFG. [Lonza] Polysorbate 65 FCC; CAS 9005-71-4; nonionic; emulsifier for ice cream, frozen desserts, cakes, whipped toppings, nondairy creamers, icings, fillings; *Regulatory:* FDA 21CFR §172.838; kosher; *Properties:* soft solid, bland odor; m.p. 31 C; HLB 11.0; acid no. 2 max.; iodine no. 1 max.; sapon. no. 88-98 hyd. no. 44-60; 3% max. moisture; *Storage:* store in cool, dry area.

GMO 0041. [Croda Food Prods. Ltd.] Glyceryl oleate; nonionic; for food industry; *Properties:* liq.; acid no. 3 max.; iodine no. 75-85; sapon. no. 160-170; 35% monoglyceride.

GMR-33. [Hefti Ltd.] Mixt. of glycerides of edible fatty acids; CAS 67701-32-0; nonionic; emulsifier, antifoamer for food industry; *Properties:* liq.; 100% conc.

GMS 52. [Am. Ingreds.] Hydrated monoglycerides; emulsifier for buns, sweet goods, tortillas, candy; *Properties:* wh. plastic; 25% mono, 16.8% diglyceride.

GMS 90. [Am. Ingreds.] Hydrated monoglycerides; emulsifier for breads, buns, sweet goods, tortillas, pizza; *Properties:* wh. plastic; i.v. 3 max.; 23% mono.

GMS 90 Dbl. Strength. [Am. Ingreds.] Hydrated monoglycerides; emulsifier for breads, buns, sweet goods, tortillas, pizza; *Properties:* wh. plastic; i.v. 3 max.; 45% mono.

GMS 300. [Croda Food Prods. Ltd.] Glyceryl stearate; improves fat dispersion and improves mouthfeel in toffees; *Usage level:* 0.2-0.5%; *Properties:* solid; m.p. 55-58 C; acid no. 2 max.; iodine no. 2 max.; sapon. no. 165-175; 30% min. monoglyceride.

GMS 305. [Croda Food Prods. Ltd.] Glyceryl stearate; starch complexing agent for bread; improves volume and extends shelf life in cakes; *Usage level:* 1-1.25% (bread), 4-6% (cake); *Properties:* solid; m.p. 57-59 C; acid no. 5 max.; iodine no. 2 max.; sapon. no. 152-160; 30% min. monoglyceride.

GMS 400. [Croda Food Prods. Ltd.] Glyceryl stearate; nonionic; prevents sticking and hardening in chewing gum; emulsion stabilizer for margarine; improves emulsification props. in shortenings increasing performance in cake and pastry prod.; *Usage level:* 0.5% (chewing gum), 0.25-0.5% (margarine), 6-8% (shortenings); *Regulatory:* UK clearance; *Properties:* powd.; m.p. 58-60 C; acid no. 3 max.; iodine no. 2 max.; sapon. no. 160-170; 45% min. monoglyceride; *Storage:* store under cool, dry conditions.

GMS 400V. [Croda Food Prods. Ltd.] Glyceryl stearate, veg. grade; prevents sticking and hardening in chewing gum; emulsion stabilizer for margarine; improves emulsification props. in shortenings increasing performance in cake and pastry prod.; *Usage level:* 0.5% (chewing gum), 0.25-0.5% (margarine), 6-8% (shortenings); *Properties:* solid; m.p. 57-60 C; acid no. 2 max.; iodine no. 2 max.; sapon. no. 160-170; 45% min. monoglyceride.

GMS 402. [Croda Food Prods. Ltd.] Glyceryl stearate SE; starch complexing agent for bread; extends shelf life; improves quality and volume in cakes; controls/improves emulsification in coffee whiteners, filled milk; stabilizer for mayonnaise; aids fat dispersion in soups; *Usage level:* 0.75-5% (bread, cake), 1-1.5% (coffee whitener), 0.5% (mayonnaise), 6-8% (shortenings); *Properties:* solid; m.p. 53-58 C; acid no. 3 max.; iodine no. 2 max.; sapon. no. 158-165; 42% min. monoglyceride.

GMS 402V. [Croda Food Prods. Ltd.] Glyceryl stearate SE, veg. grade; starch complexing agent for bread; extends shelf life; improves quality and volume in cakes; controls/improves emulsification in coffee whiteners, filled milk; stabilizer for mayonnaise; aids fat dispersion in soups; *Usage level:* 0.75-5% (bread, cake), 1-1.5% (coffee whitener), 0.5% (mayonnaise), 6-8% (shortenings); *Properties:* solid; m.w. 54-58 C; acid no. 2 max.; iodine no. 2 max.; sapon. no. 158-165; 42% min. monoglyceride.

GMS 600V. [Croda Food Prods. Ltd.] Glyceryl stearate; improves homogenization, emulsion stability in ice cream, margarine; improves fat dispersion, aids mouthfeel in toffee; *Usage level:* 0.34% (ice cream), 0.2-0.4% (margarine), 0.2-0.5% (toffee); *Properties:* iodine no. 2 max.; 58% min. monoglyceride.

GMS 602V. [Croda Food Prods. Ltd.] Glyceryl stearate SE; starch complexing agent for bread; extends shelf life; improves quality, volume, and shelf life in cakes, sponges; *Usage level:* 1-2% (bread), 2-3% (cakes), 1.75-2.5% (sponges); *Properties:* iodine no. 2 max.; 58% min. monoglyceride.

GMS 900. [Croda Food Prods. Ltd.] Glyceryl stearate; starch complexing agent controlling gelation and stickiness in dehydrated potatoes, pasta, snack foods; aids release from extruder; improves shelf life; *Usage level:* 0.5-0.75% (potato), 0.2-0.5% (pasta, snack foods); *Properties:* solid; m.p. 64-66 C; acid no. 3 max.; iodine no. 3 max.; sapon. no. 150-165; 90% min. monoglyceride.

GMS 902. [Croda Food Prods. Ltd.] Glyceryl stearate; starch complexing agent and shelf life extender in bread; improves volume, stability, and shelf life in cakes, sponges; improves homogenization and emulsion stability in ice cream; *Usage level:* 0.5-0.75% (bread), 0.5-1.5% (cake), 0.25% (ice cream), 1-1.5% (sponges); *Properties:* solid; m.p. 63-65 C; acid no. 3 max.; iodine no. 3 max.; sapon. no. 150-165; 80% min. monoglyceride.

GNL-3000. [Mitsubishi Int'l.; Amano Enzyme USA] Amyloglucosidase; CAS 9032-08-0; food enzyme; *Properties:* liq.

Golden Bake Sour V. [Bunge Foods] Blend of food acids; sour for prod. of San Francisco and California-style sour dough hearth breads, pan breads, and rolls with golden bake flavor; *Usage level:* 10% based on flour.

Golden Bake Sour Bread Base. [Bunge Foods] Specially formulated sour blend complete with fermentation accelerators, conditioning agents, and yeast nutrients; for prod. of San Francisco-type sour dough bread and rolls; controlled pH sour yielding uniform prod.; *Usage level:* 10% based on flour.

Golden Chef® Clear Liquid Frying Shortening 104-052 [ADM Packaged Oils] Soybean oil, TBHQ, dimethylpolysiloxane; packaged vegetable oil; *Properties:* clear bright liq., bland flavor; iodine no. 130 ± 4.

Golden Chef® Creamy Liquid Frying Shortening 104-065. [ADM Packaged Oils] Soybean oil, partially hydrog. soybean and cottonseed oils, TBHQ, dimethylpolysiloxane; packaged vegetable oil; *Properties:* opaque pourable liq., bland flavor; iodine no. 120 ± 4.

Golden Covo® Shortening. [Van Den Bergh Foods] Partially hydrog. soybean and cottonseed oils, citric acid (preservative), annatto and turmeric (colorants), TBHQ (to protect flavor); roll-in shortening for puffy pastry; creates outstanding lift and tender, flaky texture; *Properties:* lt. yel. plastic, bland flavor; drop pt. 123 F.

Golden Crescent. [Bunge Foods] Partially hydrog. soybean and cottonseed oils, milk, salt, monoglycerides, sodium benzoate, natural color; specialty hydrated shortening for danish, croissants, other sweet doughs; *Properties:* buttery flavor; m.p. 97 F; 80.8% fat, 16.4% moisture, 2.0% salt.

Golden Croissant/Danish™ Margarine. [Van Den Bergh Foods] Partially hydrog. soybean oil, liq. and partially hydrog. cottonseed oil, water, sugar, veg.

mono- and diglycerides, soy lecithin, potassium sorbate, citric acid, artifical flavor, annatto, turmeric, vitamin A palmitate; pastry shortening with superior functionality, exc. flavor; salt-free; roll-in for croissants, Danish; *Properties:* lt. yel. plastic, butter-like flavor; drop pt. 102 F; 80% min. fat.

Golden Crust® Conc. [Bunge Foods] Conc. producing premium hearth-baked goods incl. Vienna, French, and Italian breads, hard kaiser and hoagie rolls; *Usage level:* 7% based on flour.

Golden Flake™. [Bunge Foods] Partially hydrog. soybean and cottonseed oils, lecithin, artificial color, natural and artificial flavor; shortening, icing stabilizer, glazes, candy, refrigerated dough; *Properties:* flakes, bvutter-like flavor; m.p. 117 F.

Golden Gate Sour. [ADM Arkady] Authentic sour dough flavor contg. five flavor components.

Golden Gate Sour. [Brolite Prods.] Conc. acid-based sour; sour.

Goldex II. [ADM Arkady] Provides a rich yellow color in dry convenient-to-use form; readily disperses for even color distribution.

Gold 'N Flavor® 103-061. [ADM Packaged Oils] Partially hydrog. soybean oil with artificial butter flavoring and coloring, TBHQ, dimethylpolysiloxane; margarine; *Properties:* rich butter colored liq., buttery flavor; iodine no. 104 ± 3.

Gold Star 6MN Sherbet. [Grindsted Prods.] Emulsifier/hydrocolloid blend; stabilizer for sherbert, sorbet; creamy, fresh, smooth texture; *Usage level:* 0.35-0.5%.

Gold Star BMS 3200. [Grindsted Prods.] Emulsifier/hydrocolloid blend; stabilizer for buttermilk; good body and smooth mouthfeel; *Usage level:* 0.3-0.35%.

Gold Star CDS 3100. [Grindsted Prods.] Emulsifier/hydrocolloid blend; stabilizer for cottage cheese dressing; creamy texture; *Usage level:* 0.4%.

Gold Star DD Base. [Grindsted Prods.] Emulsifier/hydrocolloid blend; stabilizer for no-sugar-added frozen desserts; provides good body and texture with the right sweetness level; *Usage level:* 6.52%.

Gold Star HFY. [Grindsted Prods.] Emulsifier/hydrocolloid blend; stabilizer for frozen yogurt; creamy smooth texture, slow melt down; *Usage level:* 0.6-0.8%.

Gold Star LFB. [Grindsted Prods.] Emulsifier/hydrocolloid blend; stabilizer for no-sugar-added frozen desserts; *Usage level:* 9.22%.

Gold Star MSS 1800. [Grindsted Prods.] Emulsifier/hydrocolloid blend; stabilizer for milk shakes, esp. direct draw; *Usage level:* 0.25-0.45%.

Gold Star MSS 1803. [Grindsted Prods.] Emulsifier/hydrocolloid blend; stabilizer for milk shakes; *Usage level:* 0.45%.

Gold Star NFB. [Grindsted Prods.] Emulsifier/hydrocolloid blend; stabilizer for nonfat frozen desserts; for low solids prods., smooth body and texture; *Usage level:* 1-1.3%.

Gold Star SCS 3501. [Grindsted Prods.] Emulsifier/hydrocolloid blend; stabilizer for sour cream; very stiff; exc. for institutional type; *Usage level:* 0.9-

1.2%.

Gold Star SSS 1506. [Grindsted Prods.] Emulsifier/hydrocolloid blend; stabilizer for soft serve ice cream; inexpensive, smooth; *Usage level:* 0.4-0.45%.

Gold Star WCS 3400. [Grindsted Prods.] Emulsifier/hydrocolloid blend; stabilizer for whipping cream; good creamy texture, good whippability; *Usage level:* 0.1-0.2%.

Gold Star Eagle. [Grindsted Prods.] Emulsifier/hydrocolloid blend; stabilizer for premium ice cream and ice milk; *Usage level:* 0.35-0.45% (ice cream).

Gold Star Finch. [Grindsted Prods.] Emulsifier/hydrocolloid blend; stabilizer for sherbert, sorbet; good body, ice cream-like consistency; *Usage level:* 0.35-0.45%.

Gold Star Lark. [Grindsted Prods.] Emulsifier/hydrocolloid blend; stabilizer for ice cream and ice milk; good versatility, smooth, firm; *Usage level:* 0.35-0.5% (ice cream).

Gold Star Soft Serve. [Grindsted Prods.] Emulsifier/hydrocolloid blend; stabilizer for soft serve ice cream; creamy and firm body; *Usage level:* 0.35-0.45%.

Gold Star Sparrow. [Grindsted Prods.] Emulsifier/hydrocolloid blend; stabilizer for ice cream and ice milk; *Usage level:* 0.35-0.5% (ice cream).

Gold Star Swallow. [Grindsted Prods.] Emulsifier/hydrocolloid blend; stabilizer for frozen yogurt; extremely versatile; *Usage level:* 0.35-0.5%.

Gold Star Swan. [Grindsted Prods.] Emulsifier/hydrocolloid blend; stabilizer for nonfat frozen desserts; creamy, firm extrusion; *Usage level:* 0.7-1%.

GP-1200. [Custom Ingreds.] Calcium propionate, fumaric acid, corn starch, potassium sorbate; preservative for corn tortillas; provides unrefrigerated (hot rack) shelf life, exc. shelf life with a minimum of added preservative flavor and odors; *Usage level:* 1 lb/100 lb corn; *Storage:* store in clean, dry area @ 40-85 F; 6 mo storage life.

Granular Gum Arabic NF/FCC C-4010. [Meer] Gum arabic; CAS 9000-01-5; EINECS 232-519-5; protective colloid, stabilizer, thickener for confectionery, candies, jellies, glazes, chewing gum, flavor emulsions, citrus oil and beverage flavor emulsions; foam stabilizer in beer; *Properties:* colorless, odorless, tasteless; water-sol.

Granular Gum Ghatti #1. [Meer] Gum ghatti; stabilizer, binder, emulsifier forming o/w emulsions; used in beverage emulsions where it forms orange oil emulsions, in table syrup emulsions; *Properties:* water-sol.

Grape Skin Extract, 2X #3850. [Crompton & Knowles/Ingred. Tech.] Grape skin extract, dough strength; color additive for foods; *Regulatory:* FDA 21CFR §73.170; *Properties:* deep purple-red sl. visc. liq., sl. odor; sol. in water; sp.gr. 1.1 ± 0.1; pH 2.0 ± 0.1; *Storage:* store in tight containers @ 5-15 C; avoid exposure to heat, light, moisture; 6 mos storage life.

Grape Skin Extract, Double Strength. [Crompton & Knowles/Ingred. Tech.] Grape skin extract; color additive for foods; *Regulatory:* FDA 21CFR

§73.170; *Properties:* deep purple-red sl. visc. liq., sl. odor; sol. in water; sp.gr. 1.1±0.1; pH 2.3±0.1; *Storage:* store in tight containers @ 5-15 C; avoid exposure to heat, light, moisture; 6 mos storage life.

Grape Skin Extract, Single Strength. [Crompton & Knowles/Ingred. Tech.] Grape skin extract; color additive for foods; *Regulatory:* FDA 21CFR §73.170; *Properties:* deep purple-red sl. visc. liq., sl. odor; sol. in water; sp.gr. 1.1±0.1; pH 2.3±0.1; *Storage:* store in tight containers @ 5-15 C; avoid exposure to heat, light, moisture; 6 mos storage life.

Green Pea Fiber. [Garuda Int'l.] Pea fiber from cleaned and milled green pea hulls; highly absorptive fiber source for breads, snacks, dietary supplements; *Properties:* green powd., mild pea odor and flavor; 0.05% 25 mesh, 6% 35 mesh; bulk dens. 47 lb/ft^3; absorp. 220%; pH 6.5; 76% dietary fiber, 5% sol. fiber.

Green Pea Powd. [Garuda Int'l.] Pulverized, precooked green peas; flavoring agent for breads, crackers, soups without off-tastes; *Properties:* green fine powd., bland odor, mild pea flavor; 120 mesh; pH 6.7 (10% aq. disp.); 21% total dietary fiber, 3% sol. fiber, 47.2% carbohydrate, 23.5% protein, 4% moisture; *Storage:* 6 mos storage under cool, dry conditions.

Grindamyl. [Grindsted Prods.] Baking enzymes providing superior handling in dough, and larger volume, better crust color, finer crumb structure, better crumb softness, and extended shelf life in baked goods.

Grindtek MI 90. [Grindsted Prods.; Grindsted Prods. Denmark] Glyceryl laurate; CAS 142-18-7; EINECS 205-526-6; food additive; *Properties:* wh. powd.; sol. in ethanol, warm in propylene glycol, toluene, wh. spirit; m.p. 56 C; HLB 5.3; iodine no. 1 max.; sapon. no. 200-210; 90% min. monoester; *Storage:* 12 mo storage under cool, dry conditions; Discontinued

Grits 'N' Bran. [ADM Milling] Dietary fiber of popcorn; for cereals, snack foods; 12% moisture, 8% protein, 1% fat.

Guardan 100 Range. [Grindsted Prods. Denmark] Guar gum; CAS 9000-30-0; EINECS 232-536-8; thickener, texture modifier for wide range of food prods. incl. dressings, ice cream, cakes, pet food, beverages, and dairy prods.; controls syneresis; synergistic with xanthan gum; high grade; *Regulatory:* EEC, FDA §184.1339 (GRAS).

Guardan 600 Range. [Grindsted Prods. Denmark] Guar gum; CAS 9000-30-0; EINECS 232-536-8; thickener, texture modifier for wide range of food prods. incl. dressings, ice cream, cakes, pet food, beverages, and dairy prods.; controls syneresis; synergistic with xanthan gum; low-visc. grade; *Regulatory:* EEC, FDA §184.1339 (GRAS).

Guardan 700 Range. [Grindsted Prods. Denmark] Guar gum; CAS 9000-30-0; EINECS 232-536-8; thickener, texture modifier for wide range of food prods. incl. dressings, ice cream, cakes, pet food, beverages, and dairy prods.; controls syneresis; synergistic with xanthan gum; purified grade;

Regulatory: EEC, FDA §184.1339 (GRAS).

Guar Gum. [Hercules] Guar gum; CAS 9000-30-0; EINECS 232-536-8; thickener and stabilizer for food applics.; used to improve mouthfeel, retain moisture, retard crystal growth, or as binder or lubricant; *Properties:* powd.; visc. 1000-6000 cps.

Guar Gum HV. [MLG Enterprises Ltd.] Guar gum; CAS 9000-30-0; EINECS 232-536-8; emulsifier, stabilizer; *Properties:* beige powd.; 95% min. thru 200 mesh; visc. 4300-5500 cps (1%); pH 6.5-7.5 (1%); 14% max. moisture, 7% max. protein.

Gum Arabic NF/FCC Clean Amber Sorts. [Meer] Gum arabic; CAS 9000-01-5; EINECS 232-519-5; protective colloid, stabilizer, thickener for confectionery, candies, jellies, glazes, chewing gum, flavor emulsions, citrus oil and beverage flavor emulsions; foam stabilizer in beer; *Properties:* colorless, odorless, tasteless; water-sol.

Gum Arabic, Purified, Spray-Dried No. 1834. [MLG Enterprises Ltd.] Gum arabic; CAS 9000-01-5; EINECS 232-519-5; thickener improving consistency and palatability in soft drinks; replaces maltodextrins in powd. flavors; foam stabilizer in brewery; also in confectionery; *Properties:* wh. sl. ylsh. fine powd.; instantly water-sol.; visc. 28-35 cps (101%); pH 4.2-4.8 (10%); 5-6% moisture.

Gummi Gelatin P-5. [Hormel] Gelatin, Type A, 175 bloom; CAS 9000-70-8; EINECS 232-554-6; used in starch deposited confections where clarity, visc., and final elasticity are critical; offers slow flavor release and slow melt down; produces soft texture in gummy-type confections; wets easier with less agitation; *Usage level:* 6-7% of total formula (gummy bears); *Properties:* coarse mesh; visc. 26-30 mps; pH 4.5-5.5; 12% max. moisture; *Storage:* stable for up to 1 yr when stored dry at ambient temps.; keep containers tightly closed.

Gummi Gelatin P-6. [Hormel] Gelatin, Type A, 200 bloom; CAS 9000-70-8; EINECS 232-554-6; used in starch deposited confections where clarity, visc., and final elasticity are critical; offers slow flavor release and slow melt down; produces soft texture in gummy-type confections; wets easier with less agitation; *Usage level:* 6-7% of total formula (gummy bears); *Properties:* coarse mesh; visc. 30-34 mps; pH 4.5-5.5; 12% max. moisture; *Storage:* stable for up to 1 yr when stored dry at ambient temps.; keep containers tightly closed.

Gummi Gelatin P-7. [Hormel] Gelatin, Type A, 225 bloom; CAS 9000-70-8; EINECS 232-554-6; used in starch deposited confections where clarity, visc., and final elasticity are critical; wets easier with less agitation and dissolves with fewer entrapped air bubbles; *Usage level:* 7-8% of total formula (gummy bears); *Properties:* coarse mesh; visc. 32-36 mps; pH 4.5-5.8; 12% max. moisture; *Storage:* stable for up to 1 yr when stored dry at ambient temps.; keep containers tightly closed.

Gummi Gelatin P-8. [Hormel] Gelatin, Type A, 250 bloom; CAS 9000-70-8; EINECS 232-554-6; used in starch deposited confections where clarity, visc., and final elasticity are critical; rapid wetting chars.;

Usage level: 7-8% of total formula (gummy bears); *Properties:* coarse mesh; visc. 39-43 mps; pH 4.5-5.8; 11.5% max. moisture; *Storage:* stable for up to 1 yr when stored dry at ambient temps.; keep containers tightly closed.

GW. [Calgon Carbon] Activated carbon; CAS 64365-11-3; used for food decolorization; *Properties:* pulverized.

H

Halibut Liver Oil 60,000A/1,000D. [Arista Industries] Halibut liver oil; nutritive additive contg. 60,000 IU/g min. vitamin A, 1000 IU/g max. vitamin D; *Properties:* acid no. 2; iodine no. 112-150; sapon. no. 180 max.

Hamine. [Vaessen-Schoemaker] Mixt. of food grade phosphates; additive for prep. of cured meat and fish prods. subject to heat treatment, e.g., cooked hams, pork loins, poultry meats; buffering agent; reduces cooking losses, improves sliceability and color; *Properties:* completely sol. in cold water.

Hamine-Mix. [Vaessen-Schoemaker] Mixt. of food grade phosphates and functional additives; additive for prep. of cured meat and fish prods. subject to heat treatment, e.g., cooked hams, pork loins, poultry meats, beef rounds, bacon; buffering agent; *Usage level:* 10-30 g/kg final prod.

Hamine-ST. [Vaessen-Schoemaker] Special phosphate compd., curing agent for sterilized, canned meats, ox tongues, poultry meats.

Harina-Ase. [Mitsubishi Int'l.; Amano Enzyme USA] α-Amylase substitute for malt in baking prods.

HBF. [Croda Food Prods. Ltd.] Triglycerides of mixed edible fatty acids; nonionic; hardened bread fat used in snack food coatings; starch complexing agent in bread; *Properties:* powd.; m.p. 57-61 C; acid no. 2.5 max.; iodine no. 2 max.; sapon. no. 193-198; 2.5% max. monoglyceride.

HBI Icing Powd. [Bunge Foods] Boiling type icing stabilizer for glazes for honey buns, donuts, sweet goods, icings; avail. wh. or clear; clear recommended for chocolate; *Usage level:* 2-4 lb/100 lb powd. sugar; *Properties:* wh. powd.; pH 8.5.

Heart Substance. [Am. Labs] Vacuum-dried defatted glandular prod.; nutritive food additive; *Properties:* powd. or freeze-dried form.

Heavenly Danish Conc. [Bunge Foods] Conc. producing a light, flaky danish with above average volume, for American style sweet good pastries; also avail. in no-time dough formula; *Usage level:* 42% based on flour.

Heavy Duty Frying Shortening 101-100. [ADM Packaged Oils] Partially hydrog. soybean oil, dimethylpolysiloxane; packaged veg. oil; *Properties:* wh. plasticized cube, bland flavor; m.p. 102±4 F; iodine no. 72±4.

Heavy Duty Peanut Oil 104-210. [ADM Packaged Oils] Peanut oil, TBHQ, dimethylpolysiloxane; packaged veg. salad oil; *Properties:* clear bright oil, sl. nutty flavor; iodine no. 100 ± 10.

Hefti GMS-33. [Hefti Ltd.] Glyceryl stearate; nonionic; food emulsifier; *Properties:* flakes; HLB 3.5; 100% conc.

Hefti GMS-33-SES. [Hefti Ltd.] Glyceryl stearate SE; anionic; food emulsifier; *Properties:* flakes; HLB 5.0; 100% conc.

Hefti GMS-99. [Hefti Ltd.] Glyceryl stearate; nonionic; food emulsifier; *Properties:* powd.; HLB 5.0; 100% conc.

Hefti GMS-233. [Hefti Ltd.] Polyglyceryl stearate; nonionic; food emulsifier; *Properties:* flakes; HLB 7.5; 100% conc.

Hefti GMS-333. [Hefti Ltd.] Polyglyceryl-3 stearate; CAS 61790-95-2; EINECS 248-403-2; nonionic; food emulsifier; *Properties:* solid; HLB 7.5; 100% conc.

Hefti MO-55-F. [Hefti Ltd.] Polysorbate 80; CAS 61790-86-1; nonionic; food emulsifier; *Proportion:* liq.; HLB 15.0; 100% conc.

Hefti PMS-33. [Hefti Ltd.] Propylene glycol stearate; CAS 1323-39-3; EINECS 215-354-3; nonionic; food emulsifier; *Properties:* solid; HLB 3.5; 100% conc.

Hemi-Cellulase Amano 90. [Mitsubishi Int'l.; Amano Enzyme USA] Hemicellulase from *Aspergillus niger*; food enzyme; *Properties:* sl. ylsh. powd.; sol. in water; insol. in ethanol; *Toxicology:* nontoxic, nonpathogenic; enzyme dust may cause sensitization when inhaled; *Storage:* store in cool, dry place.

Hemicellulase B 1500. [Solvay Enzymes] Hemicellulase; enzyme for coffee (hydrolyzes coffee gums), flavors (extraction of essential oils and plant extracts); *Properties:* powd.

Hemicellulase CE 1500. [Solvay Enzymes] Hemicellulase; enzyme for coffee (hydrolyzes coffee gums), flavors (extraction of essential oils and plant extracts); *Properties:* powd.

Hemicellulase Conc. [Solvay Enzymes] Fungal hemicellulase; enzyme for coffee (hydrolyzes coffee gums), flavors (extraction of essential oils and plant extracts); *Properties:* powd.

Henneg-50. [Henningsen Foods] Prepared scrambled egg mix contg. whole egg, skim milk, and vegetable oil; add water for scrambled eggs or as food ingred.

Hentex-10. [Henningsen Foods] Whole egg and corn

syrup; used as replacement for frozen whole egg in cookies and cakes.

Hentex-20. [Henningsen Foods] Whole egg fortified with yolk and co-dried with corn syrup; used in cakes, soft cookies, doughnuts, or where extra richness is required.

Hentex-20A. [Henningsen Foods] Whole egg fortified with yolk and co-dried with corn syrup; modification of Hentex-20 for use where egg function is less critical.

Hentex-25. [Henningsen Foods] Whole egg fortified with yolk and co-dried with corn syrup; richer in yolk than Hentex-20 or -20A; for use in sweet doughs, pound cakes, or where good whipping chars. are required.

Hentex-30. [Henningsen Foods] Egg yolk, corn syrup, and dextrin blend; used in sweet doughs, pie bases, doughnuts, and ice cream.

Hentex-30A. [Henningsen Foods] Egg yolk, corn syrup, and dextrin blend modified with yolk and corn syrup; used in sweet doughs, pie bases, doughnuts, and ice cream, and where extra richness is required.

Hentex-35. [Henningsen Foods] Egg white, corn oil, and nonfat dry milk; whole egg substitute contg. no cholesterol; source of high quality protein.

Hentex-45. [Henningsen Foods] Whole egg prod. contg. 50% protein and reduced cholesterol; developed for making omelettes, quiches, scrambled egg dishes; nutritional labeling may read reduced cholesterol.

Hentex-70. [Henningsen Foods] Whole egg co-dried with sucrose; used for sponge cakes and other cakes where whipping performance is important.

Hentex-76. [Henningsen Foods] Whole egg, soy flour, corn syrup, salt, and lecithin blend; developed esp. for use in yeast doughs, cookies, and some cakes.

Hentex-81. [Henningsen Foods] Whole egg, sucrose, fortified with egg white; developed esp. as replacement for frozen and conc. sugared whole egg in mfg. of fine light sponge cakes and Swiss rolls.

Hentex-85. [Henningsen Foods] Egg yolk, whole egg, sucrose blend; used for applics. requiring extra richness of yolk as well as the structure building props. of whole egg.

Hentex-120. [Henningsen Foods] Whole egg, sucrose; recommended for cakes and cookies.

Hentex-505. [Henningsen Foods] Whole egg, yolk blend; offers structure building props. of whole egg with added richness of yolk; replacement for liq. or frozen whole egg in pasta formulations.

Hentex Type P-600. [Henningsen Foods] Albumen with sodium lauryl sulfate as whipping aid; high foaming instant egg white prod. for use in high protein drink mixes and other applics. where instant props. are required.

Hentex Type P-1100. [Henningsen Foods] Albumen with no additives; CAS 9006-50-2; instantized egg white prod. for use in high protein drink mixes and other applics. where instant props. are required.

Hentex Type P-1800. [Henningsen Foods] Albumen; CAS 9006-50-2; instantized egg white prod. for use in high protein drink mixes and other applics. where instant props. are required; produces a clear sol'n. when reconstituted.

Hentex Type P-2100. [Henningsen Foods] Albumen; CAS 9006-50-2; instantized form of Egg White Solids P-21 with exceptional flavor chars.

Hercules® AR 160. [Hercules] PEG-16 rosinate; nonionic; low foaming surfactant, detergent, emulsifier, wetting agent, suspending agent, dispersant for food related areas; *Regulatory:* FDA compliance; *Properties:* Gardner 14 soft wax; sol. in water, IPA, benzene, toluene, ester solvs.; visc. 250 cps (38 C); cloud pt. 68 C (2%); pH 8 (1%); surf. tens. 37.6 dynes/cm (0.1% aq.); 100% conc.

Hercules® Ester Gum 8BG. [Hercules] Purified glycerol ester of wood rosin, beverage grade; thermoplastic resin gum used in beverage industry as clouding agent; improves stability of citrus oils; *Properties:* USDA Rosin N max. flakes; sol. in aromatic and aliphatic hydrocarbons, terpenes, esters, ketones, and citrus and essential oils; dens. 1.08 kg/l; soften. pt. 90 C; acid no. 6.5.

Hercules® Ester Gum 8D. [Hercules] Deodorized glycerol ester of wood rosin, chewing gum grade; thermoplastic resin used as masticatory ingred. in chewing gums; *Properties:* USDA Rosin N max. flakes; sol. in aromatic and aliphatic hydrocarbons, esters, ketones, and CCl4; sp.gr. 1.08; dens. 1.07 kg/l; soften. pt. 90 C; acid no. 6.5.

Hercules® Ester Gum 8D-SP. [Hercules] Deodorized glycerol ester of tall oil rosin-chewing gum grade; hard thermoplastic resin used in chewing gums; *Properties:* USDA Rosin Scale: WG solid, flakes; low odor; sol.: see Hercules Ester Gum 8D; dens. 1.07 k/gl; R&B soften. pt. 80 C; acid no. 6.9.

Hercules® Ester Gum 10D. [Hercules] Glycerol ester of a partially dimerized rosin; thermoplastic resin used as a softener or plasticizer for elastomeric masticatory agents used in chewing gums; *Properties:* USDA Rosin M max. flakes; sol. in aromatic, aliphatic, and chlorinated hydrocarbons, esters, ketones; insol. in water; G-H visc. H; soften. pt. 116 C; acid no. 7.

Hercules® Locust Bean Gum FL 50-40. [Hercules Food Ingreds.] Locust bean gum; CAS 9000-40-2; EINECS 232-541-5; nonionic; food grade; *Regulatory:* FDA 21CFR §184.1343, GRAS; *Properties:* cream-colored; 25% max. on 100 mesh, 40% max. thru 200 mesh; visc. 2500-2900 cps (1%, 2 h); pH 5.5-6.5; 12% max. moisture; *Storage:* store in dry place away from heat and sunlight.

Hercules® Locust Bean Gum FL 50-50. [Hercules Food Ingreds.] Locust bean gum; CAS 9000-40-2; EINECS 232-541-5; nonionic; food grade; *Regulatory:* FDA 21CFR §184.1343, GRAS; *Properties:* cream-colored; 25% max. on 100 mesh, 60% max. thru 200 mesh; visc. 2400-2800 cps (1%, 2 h); pH 5.5-6.5; 12% max. moisture; *Storage:* store in dry place away from heat and sunlight.

Hetsorb O-5. [Heterene] Polysorbate 81; CAS 9005-65-6; food emulsifier; *Properties:* Gardner 8 max. liq.; sol. in IPA; disp. in water; HLB 10.0; acid no. 2

max.; sapon. no. 96-104; hyd. no. 134-150; 97% conc.

Hexaphos®. [FMC/Food Phosphates] Med. chain sodium hexametaphosphate; EINECS 233-343-1; sequestrant, emulsifier, suspending agent, protein stabilizer for dairy prods. esp. process cheese, ice cream, and frozen desserts; protein dispersant in whey processing; water-binding agent for cured pork; *Usage level:* 2% (process cheese), 0.1-0.5% (whey), 0.1-0.2% (ice cream); *Regulatory:* FDA GRAS; *Properties:* powd., gran., plate; 99.9% thru 20 mesh (powd.), 100% thru 8 mesh (gran.); infinitely sol. in water; m.w. 1388; bulk dens. 62 lb/ft³ (powd.), 79 lb/ft³ (gran.), 75 lb/ft³ (plate); pH 6.9 (1%).

HG-100. [Rhone-Poulenc Food Ingreds.] Locust bean gum; CAS 9000-40-2; EINECS 232-541-5; food-grade natural vegetable gum; high visc., pH stable (3.5-9), freeze-thaw stable; requires heat to develop full visc.; synergistic with xanthan, carrageenan; *Usage level:* 0.1-0.25% (ice cream/milk), 0.3-0.5% (cream cheese), 0.25-0.4% (fruit preps. for yogurt), 0.15-0.3% (pie fillings), 0.1-0.25% (sauces); *Properties:* coarse gran.; exc. dispersibility; visc. 3000 cps min. (1%).

HG-175. [Rhone-Poulenc Food Ingreds.] Locust bean gum; CAS 9000-40-2; EINECS 232-541-5; food-grade natural vegetable gum; high visc., pH stable (3.5-9), freeze-thaw stable; requires heat to develop full visc.; synergistic with xanthan, carrageenan; *Usage level:* 0.1-0.25% (ice cream/milk), 0.3-0.5% (cream cheese), 0.25-0.4% (fruit preps. for yogurt), 0.15-0.3% (pie fillings), 0.1-0.25% (sauces); *Properties:* fine gran.; good dispersibility; visc. 3000 cps min. (1%).

HG-200. [Rhone-Poulenc Food Ingreds.] Locust bean gum; CAS 9000-40-2; EINECS 232-541-5; food-grade natural vegetable gum; high visc., pH stable (3.5-9), freeze-thaw stable; requires heat to develop full visc.; synergistic with xanthan, carrageenan; *Usage level:* 0.1-0.25% (ice cream/milk), 0.3-0.5% (cream cheese), 0.25-0.4% (fruit preps. for yogurt), 0.15-0.3% (pie fillings), 0.1-0.25% (sauces); *Properties:* very fine gran.; fair dispersibility; visc. 2500 cps min. (1%).

Hickory Smoked Torula Dried Yeast. [Lake States] Torula dried yeast, natural smoke flavoring; CAS 68876-77-7; 45% min. protein; *Storage:* store in cool, dry conditions.

Hidelase. [Mitsubishi Int'l.; Amano Enzyme USA] Glucose oxidase from *Aspergillus*; CAS 9001-37-0; EINECS 232-601-0; food enzyme (U.S. market); replaces potassium bromate in yeast-raised bakery goods; improves mixing/proofing tolerance, provides high vol. yield in wheat doughs; replacement for ascorbic acid; removes glucose in egg white, oxygen in canned food; *Regulatory:* GRAS; *Properties:* powd.; *Toxicology:* enzyme dust may cause sensitization when inhaled; *Storage:* store in cool, dry area.

Hi Glo™. [Bunge Foods] Partially hydrog. soybean oil, natural and artificial butter flavor, Vitamin A, antioxi-

dants; butter-flavored liq. shortening for snacks, breads, and rolls; *Properties:* oily liq., 5.0 red color, buttery aroma and flavor; iodine no. 100.

Hi-Glo Wash. [Bunge Foods] Gives an instant, lasting shine to buns, soft rolls, specialty breads, danish, sweet goods, pies; *Usage level:* 20%; *Properties:* yel. to cream-colored paste, bland flavor; pH 6-7.

Hi-Jel™ S. [A.E. Staley Mfg.] Food starch modified, derived from tapioca; pregelatinized starch, thickener for instant puddings, dips, sauces, acidic systems requiring med. visc.; imparts smooth, creamy texture; *Regulatory:* FDA 21CFR §172.892; *Properties:* wh. powd., bland flavor; 0.5% on 60 mesh, 72% thru 200 mesh; pH 6; 6% moisture.

Hi-Lite™. [Bunge Foods] Partially hydrog. soybean oil, antioxidants, citric acid, dimethylpolysiloxane (antifoaming agent); specialty oil for frying croutons, snack spray oil, cereal, salad dressings; *Properties:* clear bright oil, 1.0 red color, bland flavor; iodine no. 95.

Hilton Davis Carmine. [Hilton Davis] Carmine; CAS 1390-65-4; EINECS 215-724-4; natural red colorant for food for obtaining hues from magenta to pink; at pink end of spectrum, exc. substitute for delisted FD&C Red #3 lake.

Hilton Davis Titanium Dioxide. [Hilton Davis] Anatase titanium dioxide; CAS 13463-67-7; EINECS 236-675-5; natural colorant, opacifier for food use; excellent for use with FD&C lakes in panning or tablet coating; rutile forms avail. on request; *Regulatory:* FDA 21CFR 73.575 (foods), 73.1575 (drugs), 73.2572 (cosmetics); *Properties:* wh.

Hi-Ratio Shortening. [Am. Ingreds.] Soy oil, ethoxylated mono- and diglycerides; high ratio shortening for yeast-raised prods. and bakery mixes; *Properties:* votated plastic; slip pt. 145-150 F; i.v. 32-40.

Hi-Tone™. [Bunge Foods] Partially hydrog. soybean oil, antioxidants, citric acid, dimethylpolysiloxane (antifoaming agent); specialty oil for frying croutons, snack spray oil; *Properties:* clear oil above 70 F, 1.0 red color, bland flavor; iodine no. 102.

H.L. 93. [Bunge Foods] Partially hydrog. canola oil, dimethylpolysiloxane (antifoaming agent); oil allowing formulators to eliminate cholesterol and reduce saturated fats; for general frying, snacks, crackers, breadings; *Properties:* oil; 0.7 red color.

H.L. 94. [Bunge Foods] Canola oil; CAS 8002-13-9; oil allowing formulators to eliminate cholesterol and reduce saturated fats; for salad oil, sauces, baking, light frying; *Properties:* oil; 0.7 red color.

H.L. PY™. [Bunge Foods] Partially hydrog. soybean and cottonseed oils, antioxidants, citric acid; functional replacement for lard in pie crust; allows formulators to eliminate cholesterol and reduce saturated fats; *Properties:* oil, bland flavor.

Hoagie Boy 9. [Brolite Prods.] Used for a hoagie and submarine sandwich buns; *Usage level:* 9 lb/100 lb flour.

Hodag CC-22. [Calgene] Propylene glycol dicaprylate/dicaprate; nonionic; surfactant for food industry; vehicle/diluent/carrier for vitamins, flavors, color, fragrance; *Regulatory:* FDA compliance as food additive; *Properties:* clear, almost colorless, tasteless, odorless liq.; sol. in ethanol, min. oil, acetone; sp.gr. 0.916; set pt. -20 C; iodine no. 0.5; sapon. no. 325; 100% conc.

Hodag CC-22-S. [Calgene] Propylene glycol dicaprylate/dicaprate; nonionic; surfactant for food industry; vehicle/diluent/carrier for vitamins, flavors, color, fragrance; *Regulatory:* FDA compliance as food additive; *Properties:* clear, almost colorless, tasteless, odorless liq.; sol. in ethanol, min. oil acetone; sp.gr. 0.919; set pt. -20 C; iodine no. 0.5; sapon. no. 325; 100% conc.

Hodag CC-33. [Calgene] Caprylic/capric triglyceride; CAS 65381-09-1; EINECS 265-724-3; nonionic; surfactant for food industry; vehicle/diluent/carrier for vitamins, flavors, color, fragrance; *Regulatory:* GRAS; *Properties:* clear, almost colorless, tasteless, odorless liq.; sol. in ethanol, min. oil, acetone; sp.gr. 0.945; set pt. -2 C; iodine no. 0.5; sapon. no. 340; 100% conc.

Hodag CC-33-F. [Calgene] Caprylic/capric triglyceride; CAS 65381-09-1; EINECS 265-724-3; nonionic; surfactant for food industry; vehicle/diluent/carrier for vitamins, flavors, color, fragrance; *Regulatory:* GRAS; *Properties:* clear, almost colorless, tasteless, odorless liq.; sol. in min. oil, acetone; sp.gr. 0.935; set pt. 5 C; iodine no. 8; sapon. no. 305; 100% conc.

Hodag CC-33-L. [Calgene] Caprylic/capric triglyceride; CAS 65381-09-1; EINECS 265-724-3; nonionic; surfactant for food industry; vehicle/diluent/carrier for vitamins, flavors, color, fragrance; *Regulatory:* GRAS; *Properties:* clear, almost colorless, tasteless, odorless liq.; sol. in min. oil, acetone; sp.gr. 0.938; set pt. 0 C; iodine no. 8; sapon. no. 308; 100% conc.

Hodag CC-33-S. [Calgene] Caprylic/capric triglyceride; CAS 65381-09-1; EINECS 265-724-3; nonionic; surfactant for food industry; vehicle/diluent/carrier for vitamins, flavors, color, fragrance; *Regulatory:* GRAS; *Properties:* clear, almost colorless, tasteless, odorless liq.; sol. in ethanol, min. oil, acetone; sp.gr. 0.945; set pt. -5 C; iodine no. 0.5; sapon. no. 347; 100% conc.

Hodag FD Series. [Calgene] Silicones; food defoamer; *Properties:* liq.; Unverified

Hodag GDO-V. [Calgene] food emulsifier; its use makes less shortening necessary in cakes, reduces fat level in sour cream and chip dips; *Usage level:* 2-5%; *Regulatory:* FDA §182.4505.

Hodag GML. [Calgene] Glyceryl laurate; CAS 142-18-7; EINECS 205-526-6; nonionic; emulsifier, opacifier, stabilizer for food industry; *Properties:* paste; HLB 3.0; 100% conc.

Hodag GMO. [Calgene] Glyceryl oleate; CAS 25496-72-4; nonionic; emulsifier, opacifier, stabilizer for food industry; improves overrun in frozen desserts; improves stability in pet foods, margarines; dough conditioner for tacos and tortillas; *Usage level:* 0.1-0.5%; *Regulatory:* FDA 21CFR §182.4505, Part 166; *Properties:* yel. opaque liq., bland odor; insol. in water; sp.gr. 0.95; b.p. > 200 C (760 mm); HLB 2.7; flash pt. (COC) > 350 F; 100% conc.; *Toxicology:* may cause eye irritation; *Precaution:* avoid strong oxidizing agents; *Storage:* store in well-ventilated areas @ 50-120 F; keep away from oxidizing agents, excessive heat, ignition sources.

Hodag GMO-D. [Calgene] Glyceryl oleate; anionic; emulsifier, opacifier, stabilizer for food industry; *Properties:* liq.; HLB 2.7; 100% conc.

Hodag GMR. [Calgene] Glyceryl ricinoleate; CAS 141-08-2; EINECS 205-455-0; nonionic; emulsifier, opacifier, stabilizer for food industry; *Properties:* amber liq.; nondispersible in water; sp.gr. 0.99; dens. 8.25 lb/gal; 100% conc.

Hodag GMR-D. [Calgene] Glyceryl ricinoleate; CAS 141-08-2; EINECS 205-455-0; anionic; emulsifier, opacifier, stabilizer for food industry; *Properties:* amber liq.; disp. in water; sp.gr. 1.00; dens. 8.25 lb/gal; 100% conc.

Hodag GMS. [Calgene] Glyceryl stearate; CAS 31566-31-1; nonionic; emulsifier, opacifier, stabilizer; improves chewing, reduces sticking for caramels, chewing gums; improves shelf life of starch jellies; emulsion stabilizer for canned puddings, coffee whiteners; holds moisture in margarine; inhibits oil separation; *Usage level:* 0.25-2.5%; *Regulatory:* FDA 21CFR §172.854, 182.4505, Part 136; *Properties:* wh. waxy solid, mild char. odor; insol. in water; sp.gr. 0.97; m.p. 58 C; b.p. > 200 C (760 mm); HLB 2.7; acid no. 5 max.; iodine no. 0.5; sapon. no. 160-175; flash pt. (COC > 175 C; pH 5.3 (5% in 25% IPA); 100% conc.; *Toxicology:* may cause eye irritation; *Precaution:* avoid strong oxidizing agents; dust is combustible; *Storage:* store in well-ventilated areas @ 50-120 F; keep away from oxidizing agents, heat, ignition sources.

Hodag GMSH. [Calgene] food emulsifier; improves tenderness and texture in cakes; improves volume in sweet goods, icings and fillings; provides better dispersion in dry coffee whiteners; provides faster whip for whipped toppings; emulsion stabilizer in margarine; *Usage level:* 4-20% basis shortening; *Regulatory:* FDA §182.4505.

Hodag GTO. [Calgene] Glyceryl trioleate; CAS 122-32-7; EINECS 204-534-7; nonionic; emulsifier, opacifier, stabilizer for foods; *Properties:* liq.; HLB 1.0; 100% conc.

Hodag PGL. [Calgene] Triglyceryl monolaurate; developmental; approved for food applics.; *Regulatory:* FDA 21CFR §172.854; avail. kosher; *Properties:* amber liq.; HLB 7.0; acid no. 6 max.; iodine no. 3 max.; sapon. no. 120-140.

Hodag PGL-101. [Calgene] Decaglyceryl monolaurate; CAS 34406-66-1; developmental; approved for food applics.; *Regulatory:* FDA 21CFR §172.854; avail. kosher; *Properties:* amber liq.; HLB 15.5; acid no. 4 max.; iodine no. 3 max.; sapon. no. 50-70.

Hodag PGML. [Calgene] Propylene glycol laurate; CAS 142-55-2; EINECS 205-542-3; nonionic; surfactant for foods; *Properties:* yel. liq.; disp. in water; sp.gr. 0.92; HLB 4.0; pour pt. -4 C; 100% conc.

Hodag PGMS. [Calgene] Propylene glycol stearate;

CAS 1323-39-3; EINECS 215-354-3; nonionic; surfactant for foods; *Properties:* waxy solid; nondisp. in water; sp.gr. 0.95; HLB 3.4; pour pt. 38 C; 100% conc.

Hodag PGO. [Calgene] Triglyceryl monooleate; CAS 9007-48-1; nonionic; approved for food applics.; *Regulatory:* FDA 21CFR §172.854; avail. kosher; *Properties:* amber liq.; HLB 7.0; acid no. 6 max.; iodine no. 75-90; sapon. no. 125-150; 100% conc.

Hodag PGO-61. [Calgene] Hexaglyceryl monooleate; CAS 9007-48-1; approved for food applics.; *Regulatory:* FDA 21CFR §172.854; avail. kosher; *Properties:* amber liq.; HLB 10.5; acid no. 7 max.; iodine no. 40 max.; sapon. no. 85-105.

Hodag PGO-62. [Calgene] Hexaglyceryl dioleate; CAS 76009-37-5; approved for food applics.; *Regulatory:* FDA 21CFR §172.854; avail. kosher; *Properties:* amber liq.; HLB 8.5; acid no. 8 max.; iodine no. 75 max.; sapon. no. 115-135.

Hodag PGO-101. [Calgene] Decaglyceryl monooleate; CAS 9007-48-1; approved for food applics.; *Regulatory:* FDA 21CFR §172.854; avail. kosher; *Properties:* amber liq.; HLB 13.5; acid no. 8 max.; iodine no. 25-35; sapon. no. 55-80.

Hodag PGO-102. [Calgene] Decaglyceryl dioleate; CAS 33940-99-7; solubilizer for flavor oils; *Regulatory:* FDA 21CFR §172.854; avail. kosher; *Properties:* amber liq., bland odor; disp. in water; sp.gr. 0.98; b.p. > 200 C (760 mm); HLB 10.5; acid no. 8 max.; iodine no. 45 max.; sapon. no. 85-105; flash pt. > 175 C; *Toxicology:* may cause eye irritation; *Precaution:* avoid strong oxidizing agents; combustion produces carbon dioxide, carbon monoxide, thick smoke; *Storage:* store in well-ventilated areas @ 50-120 F; keep away from excessive heat, ignition sources.

Hodag PGO-103. [Calgene] Decaglyceryl trioleate; CAS 102051-00-3; flavor carrier and clouding agent for beverages; *Usage level:* 40-80 ppm; *Regulatory:* FDA 21CFR §172.854; avail. kosher; *Properties:* amber liq., bland odor; disp. in water; sp.gr. 0.99; b.p. > 200 C (760 mm); HLB 8.5; acid no. 8 max.; iodine no. 65 max.; sapon. no. 110-125; flash pt. (COC) > 175 C; *Toxicology:* may cause eye irritation; *Precaution:* avoid strong oxidizing agents; combustion produces carbon dioxide, carbon monoxide, thick smoke; *Storage:* store in well-ventilated areas @ 50-120 F; keep away from excessive heat, ignition sources.

Hodag PGO-104. [Calgene] Decaglycerol tetraoleate; CAS 34424-98-1; EINECS 252-011-7; approved for food applics.; *Regulatory:* FDA 21CFR §172.854; avail. kosher; *Properties:* amber liq.; HLB 6.0; acid no. 8 max.; iodine no. 60 max.; sapon. no. 125-150; hyd. no. 210-250; formerly Hodag SVO-1047.

Hodag PGO-108. [Calgene] Decaglyceryl octaoleate; approved for food applics.; *Regulatory:* FDA 21CFR §172.854; avail. kosher; *Properties:* amber liq.; HLB 4.5; acid no. 10 max.; iodine no. 75 max.; sapon. no. 150-160.

Hodag PGO-1010. [Calgene] Decaglycerol deca-

oleate; CAS 11094-60-3; EINECS 234-316-7; nonionic; emulsifier for foods (coffee whiteners); aids aeration in frozen desserts; solubilizer for flavor oils; *Usage level:* 0.1-1.0%; *Regulatory:* FDA 21CFR §172.854; avail. kosher; *Properties:* amber liq., bland odor; sol. in flavoring oils, brominated veg. oils, ethanol, glycerin, propylene glycol; insol. in water; sp.gr. 0.93-0.97; b.p. > 200 C (760 mm); HLB 2.5; acid no. 15 max.; iodine no. 85 max.; sapon. no. 160-180; flash pt. (COC) > 175 C; pH 7-8; *Toxicology:* may cause eye irritation; *Precaution:* avoid strong oxidizing agents; *Storage:* store in well-ventilated areas @ 50-120 F; keep away from heat, ignition sources; formerly Hodag SVO-10107.

Hodag PGS. [Calgene] Triglyceryl stearate; CAS 37349-34-1; EINECS 248-403-2; nonionic; emulsifier for food processing; improves texture in cakes; minimizes syneresis in icings/filllings; dispersant for coffee whiteners; improves body in frozen dietetic desserts; improves whipped toppings, shelf life in canned puddings, margarine; *Regulatory:* FDA 21CFR §172.854; avail. kosher; *Properties:* tan solid, bland odor; < 5% sol. in water; sp.gr. 0.97; b.p. > 200 C (760 mm); HLB 7.0; acid no. 6 max.; iodine no. 3 max.; sapon. no. 120-140; flash pt. (COC) > 175 C; 100% conc.; *Toxicology:* may cause eye irritation; *Precaution:* avoid strong oxidizing agents; combustion produces carbon dioxide, carbon monoxide, thick smoke; *Storage:* store in well-ventilated areas @ 50-120 F; keep away from heat, ignition sources.

Hodag PGS-61. [Calgene] Hexaglyceryl monostearate; approved for food applics.; *Regulatory:* FDA 21CFR §172.854; avail. kosher; *Properties:* tan solid; HLB 12.0; acid no. 6 max.; iodine no. 3 max.; sapon. no. 70-95.

Hodag PGS-62. [Calgene] Hexaglyceryl distearate; provides better dispersion, increases whitening in dry coffee whiteners; improves body of frozen dietetic desserts, whipped toppings; *Usage level:* 0.1-4.0%; *Regulatory:* FDA 21CFR §172.854; avail. kosher; *Properties:* tan solid; HLB 8.5; acid no. 8 max.; iodine no. 3 max.; sapon. no. 105-125.

Hodag PGS-101. [Calgene] Decaglyceryl monostearate; CAS 79777-30-3; approved for food applics.; *Regulatory:* FDA 21CFR §172.854; avail. kosher; *Properties:* tan solid; HLB 14.5; acid no. 8 max.; iodine no. 3 max.; sapon. no. 55-80.

Hodag PGS-102. [Calgene] Decaglyceryl distearate; CAS 12764-60-2; approved for food applics.; *Regulatory:* FDA 21CFR §172.854; avail. kosher; *Properties:* tan paste; HLB 10.5; acid no. 8 max.; iodine no. 3 max.; sapon. no. 85-105.

Hodag PGS-103. [Calgene] Decaglyceryl tristearate; CAS 12709-64-7; approved for food applics.; *Regulatory:* FDA 21CFR §172.854; avail. kosher; *Properties:* tan solid; HLB 8.5; acid no. 8 max.; iodine no. 3 max.; sapon. no. 110-125.

Hodag PGS-104. [Calgene] Decaglyceryl tetrastearate; approved for food applics.; *Regulatory:* FDA 21CFR §172.854; avail. kosher; *Properties:* tan solid; HLB 6.0; acid no. 8 max.; iodine no. 3 max.;

sapon. no. 125-150.

Hodag PGS-108. [Calgene] Decaglyceryl octastearate; approved for food applics.; *Regulatory:* FDA 21CFR §172.854; avail. kosher; *Properties:* tan solid; HLB 4.5; acid no. 10 max.; iodine no. 3 max.; sapon. no. 150-160.

Hodag PGS-1010. [Calgene] Decaglyceryl decastearate; CAS 39529-26-5; EINECS 254-495-5; emulsifier for chocolate and confectionery coatings; *Regulatory:* FDA 21CFR §172.854; avail. kosher; *Properties:* tan solid; HLB 3.7; acid no. 15 max.; iodine no. 3 max.; sapon. no. 160-180.

Hodag PGSH. [Calgene] Triglyceryl monoshortening; emulsifier for icings, fillings; emulsion stabilizer for sour cream and chip dips, salad dressings; *Usage level:* 0.2-1.0%; *Regulatory:* FDA 21CFR § 172.854, 182.4505; *Properties:* tan semisolid; HLB 6.0; acid no. 8 max.; iodine no. 30-45; sapon. no. 110-125.

Hodag PGSH-61. [Calgene] Hexaglyceryl monoshortening; approved for food applics.; *Regulatory:* FDA 21CFR §172.854; avail. kosher; *Properties:* tan semisolid; HLB 12.0; acid no. 6 max.; iodine no. 30 max.; sapon. no. 65-90.

Hodag PGSH-62. [Calgene] Hexaglyceryl dishortening; approved for food applics.; *Regulatory:* FDA 21CFR §172.854; avail. kosher; *Properties:* tan semisolid; HLB 8.5; acid no. 8 max.; iodine no. 45-55; sapon. no. 90-120.

Hodag POE (20) GMS. [Calgene] PEG-20 glyceryl stearate; CAS 68153-76-4; emulsifier for cakes, icings, whipped toppings; dough conditioner for breads; dispersant for coffee whiteners; controls agglomeration in frozen desserts; *Usage level:* 0.1-0.4%; *Regulatory:* FDA 21CFR §172.834, Part 136; *Properties:* pale yel. semisolid, bland odor; sol. in ethanol, partly sol. in veg. oil, disp. in propylene glycol, water; sp.gr. 1.1 (35 C); b.p. > 200 C (760 mm); acid no. 2 max.; iodine no. 2 max.; sapon. no. 65-75; flash pt. (COC) > 300 F; *Toxicology:* may cause eye, digestive tract irritation; skin irritant on prolonged/repeated contact; *Precaution:* avoid strong oxidizing agents; combustion produces carbon monoxide, carbon dioxide; *Storage:* store in closed containers @ 50-120 F; keep away from oxidizing agents, excessive heat, ignition sources.

Hodag PSML-20. [Calgene] Polysorbate 20; CAS 9005-64-5; nonionic; food-grade emulsifier for bakery, dairy, confectionery, and convenience foods applics.; *Properties:* yel. liq.; sol. in water; sp.gr. 1.1; HLB 16.7; acid no. 2 max.; sapon. no. 40-50; hyd. no. 96-108; 100% conc.

Hodag PSML-80. [Calgene] PEG-80 sorbitan laurate; CAS 9005-64-5; nonionic; food-grade emulsifier for bakery, dairy, confectionery, and convenience foods applics.; *Properties:* yel. liq.; water-sol.; sp.gr. 1.1; HLB 19.4; acid no. 3 max.; sapon. no. 7-15; hyd. no. 25-40; 27-29% moisture.

Hodag PSMO-5. [Calgene] Polysorbate 81; CAS 9005-65-6; nonionic; food-grade emulsifier for bakery, dairy, confectionery, and convenience foods applics.; *Properties:* yel. liq. (may gel on standing);

disp. in water; sp.gr. 1.0; HLB 10.0; acid no. 2 max.; sapon. no. 95-105; hyd. no. 136-152.

Hodag PSMO-20. [Calgene] Polysorbate 80; CAS 9005-65-6; nonionic; emulsifier for food processing; improves dryness and melt down in frozen desserts; dispersant for flavor oils; aids flavor and color in pickles; *Usage level:* 0.05-0.2%; *Regulatory:* FDA 21CFR §172.836, 172.845, 172.854; *Properties:* yel. liq., bland odor; water-sol.; sp.gr. 1.1; b.p. > 350 F (760 mm); HLB 15.0; acid no. 2 max.; sapon. no. 45-55; hyd. no. 65-80; flash pt. (COC) > 150 C; 100% conc.; *Toxicology:* may cause eye irritation; *Precaution:* avoid strong oxidizing agents; *Storage:* store in well-ventilated areas @ 50-120 F; keep away from oxidizing agents, excessive heat, ignition sources.

Hodag PSMP-20. [Calgene] Polysorbate 40; CAS 9005-66-7; nonionic; emulsifier for food processing; *Properties:* yel. liq. (may gel on standing); sol. in water; sp.gr. 1.1; HLB 15.6; acid no. 2 max.; sapon. no. 43-49; hyd. no. 89-105; 100% conc.

Hodag PSMS-20. [Calgene] Polysorbate 60; CAS 9005-67-8; nonionic; emulsifier for food processing (chocolate, confectionery coatings, panned sugar); better aeration and volume in cakes, whipped toppings, fillings; dispersant in dry coffee whiteners; foaming agent in beverage mixes; stabilizer for salad dressings; *Usage level:* 0.05-1%; *Regulatory:* FDA 21CFR §172.836, 172.854; *Properties:* yel. liq. (may gel on standing), mild odor; water-sol.; sp.gr. 1.1; b.p. > 200 C (760 mm); HLB 14.9; acid no. 2 max.; sapon. no. 45-55; hyd. no. 81-96; flash pt. (COC) > 175 C; 100% conc.; *Toxicology:* may cause eye irritation; *Precaution:* avoid oxidizing agents; *Storage:* store in well-ventilated areas @ 50-120 F; keep away from oxidizing agents, excessive heat, ignition sources.

Hodag PSTO-20. [Calgene] Polysorbate 85; CAS 9005-70-3; nonionic; food-grade emulsifier for bakery, dairy, confectionery, and convenience foods applics.; *Properties:* yel. liq. (may gel on standing); disp. in water; sp.gr. 1.0; HLB 11.0; acid no. 2 max.; sapon. no. 82-95; hyd. no. 39-52.

Hodag PSTS-20. [Calgene] Polysorbate 65; CAS 9005-71-4; nonionic; emulsifier for food processing; dispersant for dry coffee whiteners; *Usage level:* 0.1-0.3%; *Properties:* yel. soft solid, bland odor; sol. in water; sp.gr. 1.0; b.p. > 200 C (760 mm); HLB 10.5; acid no. 2 max.; sapon. no. 88-98; hyd. no. 42-60; flash pt. (COC) > 175 C; 100% conc.; *Toxicology:* may cause eye irritation; *Precaution:* avoid strong oxidizing agents; combustion produces carbon dioxide, carbon monoxide, thick smoke; *Storage:* store in well-ventilated area @ 50-120 F; keep away from oxidizing agents, excessive heat, ignition sources.

Hodag S-35. [Calgene] Glycerol ester; nonionic; emulsifier for food processing; *Properties:* liq.; 100% conc.

Hodag SML. [Calgene] Sorbitan laurate; CAS 1338-39-2; nonionic; food-grade emulsifier; *Properties:* yel. liq.; water-disp.; oil-sol.; sp.gr. 1.0; HLB 8.6;

acid no. 7 max.; sapon. no. 158-170; hyd. no. 330-358; 100% conc.

Hodag SMO. [Calgene] Sorbitan oleate; CAS 1338-43-8; EINECS 215-665-4; nonionic; food-grade emulsifier; *Properties:* amber liq.; oil-sol.; sp.gr. 1.0; HLB 4.3; acid no. 7.5; sapon. no. 149-160; hyd. no. 193-209; 100% conc.

Hodag SMP. [Calgene] Sorbitan palmitate; CAS 26266-57-9; EINECS 247-568-8; nonionic; food-grade emulsifier; *Properties:* tan solid; oil-sol.; sp.gr. 1.0; HLB 6.7; acid no. 7.5 max.; sapon. no. 140-150; hyd. no. 275-305; 100% conc.

Hodag SMS. [Calgene] Sorbitan stearate; CAS 1338-41-6; EINECS 215-664-9; nonionic; emulsifier for food processing (chocolate and confectionery coatings); improves volume in icings and fillings; dispersant in dry coffee whiteners, flavor oils; *Regulatory:* FDA 21CFR §172.842, 172.845, 182.4504; *Properties:* cream flakes; insol. in water; sp.gr. 1.0; b.p. > 200 C (760 mm); HLB 4.7; acid no. 10 max.; sapon. no. 147-157; hyd. no. 235-260; flash pt. (COC) > 175 C; 100% conc.; *Toxicology:* may cause eye irritation; *Precaution:* avoid strong oxidizing agents; combustion produces carbon dioxide, carbon monoxide, thick smoke; *Storage:* store in well-ventilated areas @ 50-120 F; keep away from oxidizing agents, excessive heat, ignition sources.

Hodag STO. [Calgene] Sorbitan trioleate; CAS 26266-58-0; EINECS 247-569-3; nonionic; food-grade emulsifier; *Properties:* amber liq.; oil-sol.; sp.gr. 1.0; HLB 1.8; acid no. 13.5 max.; sapon. no. 171-185; hyd. no. 58-69; 100% conc.

Hodag STS. [Calgene] Sorbitan tristearate; CAS 26658-19-5; EINECS 247-891-4; nonionic; food-grade emulsifier; *Properties:* cream solid; oil-sol.; sp.gr. 1.0; HLB 2.1; acid no. 15 max.; sapon. no. 175-190; hyd. no. 65-80; 100% conc.

Hodag SVO-9. [Calgene] PEG-20 sorbitan oleate; CAS 9005-65-6; nonionic; food emulsifier; *Regulatory:* kosher; *Properties:* liq.; HLB 15.0; 100% conc.

Hodag SVO-629. [Calgene] Hexaglyceryl distearate; nonionic; food emulsifier; *Regulatory:* GRAS; *Properties:* waxy flake; HLB 4.0; acid no. 5 max.; iodine no. 3 max.; sapon. no. 100-130; hyd. no. 265-310; 100% conc.

Hodag SVS-18. [Calgene] PEG-20 sorbitan stearate; CAS 9005-67-8; nonionic; food emulsifier; *Regulatory:* kosher; *Properties:* liq.; HLB 14.9; 100% conc.

Hodag Antifoam CO-350. [Calgene] Silicone antifoam; antifoam for nonaq. foaming systems; for fermentation, foods; *Regulatory:* FDA approved; *Properties:* syrup-like fluid; sol. in amyl acetate, aromatic solvs., 2-ethylhexanol, kerosene; insol. in water, paraffin oil, glycols, ethanol, methanol, glycerin; sp.gr. 0.968; dens. 8.3 lb/gal; 100% act.

Hodag Antifoam F-1. [Calgene] Simethicone; CAS 8050-81-5; antifoam for aq. and nonaq. systems; for fermentation, food; *Regulatory:* FDA approved, to 10 ppm max. in final prod.; *Properties:* translucent syrup-like compd.; sol. in amyl acetate, aromatic solvs., 2-ethylhexanol, kerosene; insol. in

water, paraffin oil, glycols, ethanol, methanol, glycerin; sp.gr. 0.990; dens. 8.3 lb/gal; 100% act.

Hodag Antifoam F-2. [Calgene] Silicone antifoam; antifoam for aq. and nonaq. systems for fermentation, food applics.; *Regulatory:* FDA approved, 10 ppm max. in final prod.; *Properties:* translucent syrup-like compd.; sol. in amyl acetate, aromatic solvs., 2-ethylhexanol, kerosene; sl. disp. in water; insol. in paraffin oil, glycols, ethanol, methanol, glycerin; sp.gr. 1.000; dens. 8.4 lb/gal; 100% act.

Hodag Antifoam FD-62. [Calgene] Polymeric silicone emulsion; antifoam for aq. systems for fermentation, food applics.; *Regulatory:* FDA approved, 100 ppm max. in final prod.; *Properties:* wh. creamy emulsion; disp. in water; 1.000; dens. 8.4 lb/gal; 10% act.

Hodag Antifoam FD-82. [Calgene] Silicone emulsion; antifoam for aq. foaming problems; for fermentation, foods; *Regulatory:* FDA approved, 30 ppm max. in final prod.; *Properties:* wh. creamy emulsion; disp. in water; sp.gr. 0.970; dens. 8.1 lb/gal; 30% act.

Hoechst Wax KPS. [Hoechst Celanese; Hoechst AG] Glycol/butylene glycol montanate; wax for emulsions for citrus fruit coating; *Properties:* ylsh. flakes; insol. in water; dens. 1.00-1.02 g/cc (20 C); visc. $\approx$ 30 mm²/s (100 C); drop pt. 80-85 C; acid no. 30-40; sapon. no. 135-150.

Hoechst Wax KSL. [Hoechst Celanese; Hoechst AG] Glycol/butylene glycol montanate; wax for emulsions for citrus fruit coating.

Hoechst Wax PED 121. [Hoechst AG] Oxidized polyethylene; CAS 68441-17-8; wax for emulsions for citrus fruit coating.

Holland Fudge Standard and Extra Dark. [Bunge Foods] Fudge base that makes icings with a med. brn. color with mild chocolate flavor; for fudge and creme icings, brownie batters; *Usage level:* 25-40 lb/100 lb. powd. sugar; *Properties:* dk. brn. semiplastic base, cocoa aroma/flavor.

Home Style Bun. [Brolite Prods.] Bromate-free, no-time dough bread base for Kaiser rolls, French, Italian, Vienna, and other breads; *Usage level:* 6 lb/100 lb flour.

Homodan RD. [Grindsted Prods.; Grindsted Prods. Denmark] Special emulsifier blend from veg. sources; nonionic; emulsifier, emulsion stabilizer, plasticizer; improves frying props. and plasticity of margarine for flakier puff pastry; *Properties:* flakes; drop pt. 63 C; 30% min. monoester.

Homotex PS-200. [Kao/Edible Fat & Oil] Propylene glycol ester of fatty acids; nonionic; used for cake mix, margarine; *Properties:* bead; HLB 3.4; 100% conc.

Homotex PS-90. [Kao Corp. SA] Propylene glycol stearate; CAS 1323-39-3; EINECS 215-354-3; nonionic; cake mix, fluid shortening; *Properties:* powd.; HLB 3.4; 100% conc.; Unverified

Honi-Bake® 705 Honey. [ADM Ogilvie] 65% Honey; CAS 8028-66-8; sweetener for baked goods, mixes, sauces, snacks; *Properties:* yel. powd./ flake; 3.5% max. moisture.

Honi-Bake® Honey. [ADM Ogilvie] 65% Honey and other sweetener solids; CAS 8028-66-8; sweetener for baked goods, mixes, sauces, snacks; *Properties:* yel. powd./flake; 2.5% max. moisture.

HSC Aspartame. [Holland Sweetener N. Am.] Aspartame FCC, USP-NF; CAS 22839-47-0; EINECS 245-261-3; nutritive intense sweetener and flavor enhancer with sweetness ≈ 200 times that of sucrose; for drinks, table use, instant mixes, dairy prods., confectionery, pharmaceutical tablets, emulsions, sugar-free syrups; *Regulatory:* FDA 21CFR §172.804; JECFA compliance; *Properties:* wh. cryst. powd., odorless, clean sweet taste without bitter or metallic aftertaste; sparinlgy sol. in water; sl. sol. in alcohol; m.w. 294.31; pH 4.5-6.0 (0.8% aq.); 98-102% assay; *Storage:* 5 yr shelf life in tightly closed container with sealed inner bag under cool, dry conditions.

HT® Monocalcium Phosphate, Monohydrate (MCP) Spray Dried Coarse Granular. [Monsanto] Calcium phosphate; CAS 10031-30-8; EINECS 231-837-1; FCC grade; *Regulatory:* FCC compliance; *Properties:* wh. free-flowing gran.; 12% max. on 100 mesh, 85% min. on 200 mesh; m.w. 252.05.

HT® Monocalcium Phosphate, Monohydrate (MCP) Spray Dried Fines. [Monsanto] Calcium phosphate; CAS 10031-30-8; EINECS 231-837-1; FCC grade; *Regulatory:* FCC compliance; *Properties:* wh. free-flowing powd.; 30% max. on 200 mesh, 80% max. thru 325 mesh; m.w. 252.05.

HT® Monocalcium Phosphate, Monohydrate (MCP) Spray Dried Medium Granular. [Monsanto] Calcium phosphate; CAS 10031-30-8; EINECS 231-837-1; FCC grade; *Regulatory:* FCC compliance; *Properties:* wh. free-flowing gran.; 10% max. on 100 mesh, 35-70% thru 200 mesh; m.w. 252.05.

HT-Proteolytic® 200. [Solvay Enzymes] Bacterial protease; CAS 9014-01-1; EINECS 232-752-2; food-grade enzyme for hydrolysis of proteins over neutral and alkaline pH range; for baking (cracker/cookie gluten modification), brewing (malt supplementation), candy, fermentation, fish/soya processing, protein modification; *Usage level:* 0.01-0.1% on substrate; *Properties:* wh. to lt. tan amorphous dry powd., free of offensive odor; readily water-sol.; *Storage:* activity loss < 10% in 1 yr stored in sealed containers under cool, dry conditions; 5 C storage extends life.

HT-Proteolytic® Conc. [Solvay Enzymes] Bacterial protease; CAS 9014-01-1; EINECS 232-752-2; food-grade enzyme for hydrolysis of proteins over neutral and alkaline pH range; for baking (cracker/cookie gluten modification), brewing (malt supplementation), candy, fermentation, fish/soya processing, protein modification; *Properties:* wh. to lt. tan amorphous dry powd., free of offensive odor and taste; readily water-sol.; *Storage:* activity loss ≤ 10% in 1 yr stored in sealed containers under cool, dry conditions; 5 C storage extends life.

HT-Proteolytic® L-175. [Solvay Enzymes] Bacterial protease; CAS 9014-01-1; EINECS 232-752-2; food-grade enzyme for hydrolysis of proteins over neutral and alkaline pH range; for baking (cracker/cookie gluten modification), brewing (malt supplementation), candy, fermentation, fish/soya processing, protein modification; *Properties:* amber to lt. brn. nonvisc. liq., free of offensive odor and taste; completely misc. with water; *Storage:* activity loss < 10% in 3 mo stored in sealed containers under cool, dry conditions; 5 C storage extends life.

HVP 5-SD. [Hercules] Hydrolyzed veg. protein with salt and caramel color; flavor enhancer; beef-type flavor used in food applics.; *Properties:* tan powd.; pH 5.4; 97% total solids; 46% flavor solids.

HVP-A. [Hercules] Hydrolyzed veg. protein with salt and caramel color; flavor enhancer; smooth, beefy flavor useable in food applics. in highly seasoned prods.; *Properties:* tan powd.; pH 5.4; 97% total solids; 43.5% flavor solids.

HVP-LS. [Hercules] Hydrolyzed veg. protein, low sodium; CAS 100209-45-8; flavoring for sodium-restricted dietary applics. for meaty flavors in food; *Properties:* brn. powd.; pH 5.2; 97% total solids; 75% flavor solids.

HVP Replacer Powder. [Nikken Foods Co. Ltd.] Replacement for hydrolyzed vegetable protein; *Properties:* lt. tan free-flowing powd., mild typical fermented soy aroma; *Storage:* store in cool, dry, well-ventilated area in closed container.

HWDX Winterized Salad Oil. [Bunge Foods] Partially hydrog. soybean oil, antioxidants, citric acid, dimethylpolysiloxane, crystal inhibitor; specialty oil for salad dressings, general cooking, light duty frying; *Properties:* clear bright oil, 1.1 red color, bland flavor; iodine no. 110; smoke pt. 440 F.

Hyderase. [Mitsubishi Int'l.; Amano Enzyme USA] Glucose oxidase from *Aspergillus*; CAS 9001-37-0; EINECS 232-601-0; food enzyme (Europe and Japan markets); replaces potassium bromate in yeast-raised bakery goods; improves mixing/proofing tolerance, provides high vol. yield in wheat doughs; replacement for ascorbic acid; *Properties:* powd.

Hydex® 100 Coarse Powd. [Lonza] Sorbitol NF/FCC; CAS 50-70-4; EINECS 200-061-5; humectant, bodying agent, moisture control agent for sugarless confections, controlled moisture foods; *Properties:* wh. powd.; ≤ 5% +40 mesh, ≤ 10% -200 mesh; 100% act.

Hydex® 100 Coarse Powd. 35. [Lonza] Sorbitol NF/FCC; CAS 50-70-4; EINECS 200-061-5; humectant, bodying agent, moisture control agent for sugarless confections, controlled moisture foods; *Properties:* powd.; ≤ 5% +35 mesh, ≤ 10% -80 mesh; pH neutral.

Hydex® 100 Gran. 206. [Lonza] Sorbitol NF/FCC; CAS 50-70-4; EINECS 200-061-5; humectant, bodying agent, moisture control agent for sugarless confections, controlled moisture foods; *Properties:* wh. powd., 90% max. on 80 mesh screen; 100% act.

Hydex® 100 Powd. 60. [Lonza] Sorbitol NF/FCC; CAS 50-70-4; EINECS 200-061-5; humectant, bodying agent, moisture control agent for sugarless confections, controlled moisture foods; *Properties:* wh. powd.; ≤ 1% +60 mesh; 100% act.

Hydex® Tablet Grade. [Lonza] Sorbitol; CAS 50-70-4; EINECS 200-061-5; humectant, bodying agent, moisture control agent for sugarless confections, controlled moisture foods; *Properties:* wh. powd., 75% max. on 200 mesh screen; 100% act.

Hydradan D 42. [Grindsted Prods.] Monoglycerides disp. from edible fat source; food emulsifier; crumb softening agent for bread and other baked goods; 40% min. alpha monoester.

Hydradan S 21. [Grindsted Prods.] Monoglycerides disp. from edible fat source; food emulsifier; crumb softening agent for bread and other baked goods; 20% min. alpha monoester.

Hydrogenated Canola Oil 81-601-0. [ADM Refined Oils] Hydrog. canola oil; base for creamy liq. pourable shortenings; exc. for frying applics.; *Properties:* Lovibond 1.5R max. oil, bland flavor; m.p. 94 ± 2 F; iodine no. 92 ± 2.

Hydrol 92. [Van Den Bergh Foods] Partially hydrog. coconut oil; EINECS 284-283-8; high solids lauric system for veg. dairy prods., biscuits, crackers, frying; *Regulatory:* kosher; *Properties:* Lovibond 1.0R max. plastic; m.p. 92-96 F.

Hydrol 100. [Van Den Bergh Foods] Partially hydrog. coconut oil; EINECS 284-283-8; oil for whipped toppings, coffee whiteners, veg. dairy systems, biscuits, and crackers; *Regulatory:* kosher; *Properties:* Lovibond 1.0R max. plastic; m.p. 98-102 F.

Hydrol 110. [Van Den Bergh Foods] Partially hydrog. veg. oil (coconut, soybean, cottonseed); EINECS 209-020-0, high solids lauric system for high quality biscuits; also for crackers, fillers, cream centers, dressings; *Regulatory:* kosher; *Properties:* Lovibond 1.5R max. plastic; m.p. 108-112 F.

Hyfoama. [Quest Int'l.] Protein; whipping agents for dairy, confectionery, bakery, and beverage prods.

Hymo. [Karlshamns Food Ingreds.] Partially hydrog. vegetable oil (soybean, cottonseed) with vegetable mono- and diglycerides; emulsified shortening for cakes and icings; *Properties:* plastic; m.p. 120 F.

Hymono 1103. [Quest Int'l.] Dist. monoglycerides; nonionic; used in food industry; *Properties:* microbead; HLB 4.0; 100% conc.

Hymono 1123. [Quest Int'l.] Dist. monoglycerides; nonionic; used in food industry; *Properties:* microbead; HLB 4.0; 100% conc.

Hymono 1163. [Quest Int'l.] Dist. monoglycerides; nonionic; used in food industry; *Properties:* microbead; HLB 4.0; 100% conc.

Hymono 3203. [Quest Int'l.] Dist. monoglycerides; nonionic; used in food industry; *Properties:* microbead; HLB 4.0; 100% conc.

Hymono 4404. [Quest Int'l.] Dist. monoglycerides; nonionic; used in food industry; *Properties:* paste; HLB 4.0; 100% conc.

Hymono 7804. [Quest Int'l.] Dist. monoglycerides; nonionic; used in food industry; *Properties:* paste; HLB 4.0; 100% conc.

Hymono 8803. [Quest Int'l.] Dist. monoglycerides; nonionic; used in food industry; *Properties:*

microbead; HLB 4.0; 100% conc.

Hymono 8903. [Quest Int'l.] Dist. monoglycerides; nonionic; used in food industry; *Properties:* microbead; HLB 4.0; 100% conc.

Hyonic NP-60. [Henkel/Functional Prods.; Henkel/Organic Prods.; Henkel Canada] Nonoxynol-6; CAS 9016-45-9; nonionic; surfactant for food applics.; *Properties:* clear liq.; disp. in water; sol. in oil; dens. 1.04 g/ml; HLB 10.9; pour pt. -32 C; cloud pt. < 0 C (1% aq.); pH 7.0 (1% aq.); surf. tens. 30 dynes/cm (0.01%); > 99% act.

Hyonic NP-100. [Henkel/Functional Prods.; Henkel/Organic Prods.; Henkel Canada] Nonoxynol-10; CAS 9016-45-9; nonionic; surfactant, wetting agent, emulsifier for food applics.; *Properties:* clear liq.; sol. in water; dens. 1.07 g/ml; HLB 13.2; pour pt. 4 C; cloud pt. 68 C (1% aq.); pH 7.0 (1% aq.); surf. tens. 32 dynes/cm (0.01%); > 99% act.

Hyonic NP-110. [Henkel/Functional Prods.; Henkel Canada] Nonoxynol-11; CAS 9016-45-9; nonionic; surfactant, wetting agent, emulsifier for food applics.; *Properties:* clear liq.; sol. in water; dens. 1.07 g/ml; HLB 13.8; pour pt. 13 C; cloud pt. 72 C (1% aq.); pH 7.0 (1% aq.); surf. tens. 34 dynes/cm (0.01%); > 99% act.

Hyonic NP-120. [Henkel/Functional Prods.; Henkel/Organic Prods.] Nonoxynol-12; CAS 9016-45-9; nonionic; detergent, wetting agent, penetrant, co-emulsifier for food applics.; *Properties:* liq., semi-solid; sol. in water; dens. 1.07 g/ml; HLB 14.1; pour pt. 15 C; cloud pt. 91 C (1% aq.); pH 7.0 (1% aq.); surf. tens. 36 dynes/cm (0.01%); > 99% act.

Hyonic NP-407. [Henkel Canada] Nonoxynol-40; CAS 9016-45-9; nonionic; surfactant for food applics.; *Properties:* clear liq.; sol. in water; dens. 1.10 g/ml; HLB 17.6; pour pt. -6 C; cloud pt. 100 C (1% aq.); pH 7.0 (1% aq.); 70% act.

Hyonic NP-500. [Henkel/Organic Prods.] Nonoxynol-50; CAS 9016-45-9; nonionic; surfactant for food applics.; *Properties:* clear solid; sol. in water; dens. 1.08 g/ml; HLB 18; pour pt. 24 C; cloud pt. 100 C (1% aq.); pH 7.0 (1% aq.); > 99% act.; Unverified

Hypothalamus Substance. [Am. Labs] Vacuum-dried defatted glandular prod.; nutritive food additive; *Properties:* powd.

Hyprol. [Quest Int'l.] Proteins and peptides; for nutritional and sports foods and drinks.

Hystar® 3375. [Lonza] Hydrog. starch hydrolysate; CAS 68425-17-2; humectant, bodying agent, moisture control agent for sugarless confections, controlled moisture foods; *Regulatory:* kosher certification; *Properties:* water-wh. clear liq.; sp.gr. 1.32; visc. 1500 cps (40 C); pH neutral; 75% act.

Hystar® 4075. [Lonza] Hydrog. starch hydrolysate; CAS 68425-17-2; humectant, bodying agent, moisture control agent for sugarless confections, controlled moisture foods; *Regulatory:* kosher certification; *Properties:* water-wh. clear liq.; sp.gr. 1.33; visc. 1000 cps (40 C); pH neutral; 75% act.

Hystar® 5875. [Lonza] Hydrog. starch hydrolysate; CAS 68425-17-2; humectant, bodying agent, moisture control agent for sugarless confections, controlled moisture foods; *Properties:* water-wh. clear liq.; sp.gr. 1.35; visc. 500 cps (40 C); pH neutral; 75% act.

Hystar® 6075. [Lonza] Hydrog. starch hydrolysate; CAS 68425-17-2; humectant, bodying agent, moisture control agent for sugarless confections, controlled moisture foods; *Regulatory:* kosher certification; *Properties:* water-wh. clear liq., bland taste; sp.gr. 1.35; visc. 2000 cps (40 C); pH neutral; 75% act.

Hystar® 7570. [Lonza] Sorbitol; CAS 50-70-4; EINECS 200-061-5; humectant, bodying agent, moisture control agent for sugarless confections, controlled moisture foods; Discontinued

Hystar® HM-75. [Lonza] Hydrog. starch hydrolysate; CAS 68425-17-2; humectant, bodying agent, moisture control agent for sugarless confections, controlled moisture foods; *Regulatory:* kosher certification; *Properties:* water-wh. clear liq.; sp.gr. 1.48; visc. 500 cps (40 C); pH neutral; 75% act.

Hystrene® 5016 NF FG. [Witco/H-I-P] Stearic acid, triple pressed; CAS 57-11-4; EINECS 200-313-4; food grade acids used as lubricants, release agents, binders, and defoamers, and in components for producing other food grade additives; *Properties:* Lovibond 1.0Y-0.1R solid; solid pt. 54.5-56.5 C; acid no. 206-210; iodine no. 0.5 max.;

sapon. no. 206-211; 100% conc.

Hystrene® 7018 FG. [Witco/H-I-P] Stearic acid; CAS 57-11-4; EINECS 200-313-4; food grade acids used as lubricants, release agents, binders, and defoamers, and in components for producing other food grade additives; *Properties:* Lovibond 1.0Y-0.1R solid; solid. pt. 58.0-62.5 C; acid no. 200-205; iodine no. 0.8 max.; sapon. no. 200-206; 100% conc.

Hystrene® 8718 FG. [Witco/H-I-P] Stearic acid; CAS 57-11-4; EINECS 200-313-4; food grade acids used as lubricants, release agents, binders, and defoamers; intermediate for producing other food grade emulsifiers; *Properties:* solid; 100% conc.

Hystrene® 9718 NF FG. [Witco/H-I-P] Stearic acid NF (92%); CAS 57-11-4; EINECS 200-313-4; food grade acids used as lubricants, release agents, binders, and defoamers; intermediate for producing food grade emulsifiers; *Properties:* solid; solid. pt. 66.5-68.5 C; acid no. 196-201; iodine no. 0.8 max.; sapon no. 196-201; 100% conc.

Hytex. [Karlshamns Food Ingreds.] Partially hydrog. vegetable oil (soybean, palm) with vegetable mono- and diglycerides; nontropical version also avail.; emulsified shortening for yeast-raised sweet goods; *Properties:* plastic; m.p. 114 F.

I

Ice # 2. [Van Den Bergh Foods] Glyceryl stearate and polysorbate 80; nonionic; stabilizer and emulsifier for the frozen desserts providing body, overrun, and dryness; *Usage level:* 0.04-0.25% (frozen desserts); *Regulatory:* FDA 21CFR §182.4505, 172.840; *Properties:* ivory bead; HLB 5.2; m.p. 59-63 C; 32-38% alpha monoglyceride; *Storage:* store in cool, dry place; reseal drums between use; 3 mo life for optimum free-flowing props.

Ice #12. [Van Den Bergh Foods] 80/20 blend of mono- and diglycerides and polysorbate 65; nonionic; emulsifier used with stabilizers in frozen desserts; provides body, overrun, dryness; *Usage level:* 0.04-0.25% (frozen desserts); *Regulatory:* FDA 21CFR §182.4505, 172.838; *Properties:* ivory beads; m.p. 138-144 F; HLB 4.1; 100% conc., 32-36% mono; *Storage:* store in cool, dry place; reseal drums between use; 3 mo life for optimum free-flowing props.

Ice # 81. [Van Den Bergh Foods] Mono- and diglycerides and polyglyceryl esters of fatty acids; nonionic; ice cream stabilizer; *Usage level:* 0.04-0.25% (frozen desserts); *Regulatory:* FDA 21CFR §182.4505, 172.854, 135.110(c), 135.140(c); *Properties:* ivory bead; m.p. 135-145 F; HLB 4.9; 100% conc., 34-40% alpha monoglyceride; *Storage:* store in cool, dry place; reseal drums between use; 3 mo life for optimum free-flowing props.; Discontinued

Iconol NP-7. [BASF] Nonoxynol-7; CAS 9016-45-9; nonionic; food grade acids used as lubricants, release agents, binders, and defoamers; intermediate for producing food grade emulsifiers; *Properties:* APHA 100 max. liq.; water-sol.; m.w. 523; sp.gr. 1.05; visc. 300 cps; HLB 11.9; pour pt. 5 C; cloud pt. 22-27 C (1.0% aq.); surf. tens. 30 dynes/cm (0.1% aq.); 100% conc.; Discontinued

Idealgum 1A. [Idea Srl; Commodity Services Int'l.] Locust bean gum; CAS 9000-40-2; EINECS 232-541-5; nonionic; stabilizer, thickener, syneresis inhibitor; with K-carrageenan can enhance elasticity, strength, and water-retention of gels; for ice cream, desserts, sauces, canned meats, drinks, cheese, other milk derivs.; *Regulatory:* EEC compliance; *Properties:* powd.; 25-35% on 200 mesh, 10-20% on 140 mesh; water-sol.; visc. 2800±200 cps (1%); pH 5.5-6.0; 13% max. moisture.

Idealgum 1B. [Idea Srl; Commodity Services Int'l.]

Locust bean gum; CAS 9000-40-2; EINECS 232-541-5; nonionic; stabilizer, thickener, syneresis inhibitor; with K-carrageenan can enhance elasticity, strength, and water-retention of gels; for ice cream, desserts, sauces, canned meats, drinks, cheese, other milk derivs.; *Regulatory:* EEC compliance; *Properties:* powd.; 20-30% on 200 mesh, 30-40% on 140 mesh; water-sol.; visc. 2800±200 cps (1%); pH 5.5-6.0; 13% max. moisture.

Idealgum 1C. [Idea Srl; Commodity Services Int'l.] Locust bean gum; CAS 9000-40-2; EINECS 232-541-5; nonionic; stabilizer, thickener, syneresis inhibitor; with K-carrageenan can enhance elasticity, strength, and water-retention of gels; for ice cream, desserts, sauces, canned meats, drinks, cheese, other milk derivs.; *Regulatory:* EEC compliance; *Properties:* powd.; 20-30% on 200 mesh, 50-60% on 140 mesh; water-sol.; visc. 2900+200 cps (1%); pH 5.5-6.0; 13% max. moisture.

Idealgum 2A. [Idea Srl; Commodity Services Int'l.] Locust bean gum; CAS 9000-40-2; EINECS 232-541-5; nonionic; stabilizer, thickener, syneresis inhibitor; with K-carrageenan can enhance elasticity, strength, and water-retention of gels; for ice cream, desserts, sauces, canned meats, drinks, cheese, other milk derivs.; *Regulatory:* EEC compliance; *Properties:* powd.; 20-30% on 200 mesh, 10-20% on 140 mesh; water-sol.; visc. 2300±200 cps (1%); pH 5.5-6.0; 13% max. moisture.

Idealgum 2B. [Idea Srl; Commodity Services Int'l.] Locust bean gum; CAS 9000-40-2; EINECS 232-541-5; nonionic; stabilizer, thickener, syneresis inhibitor; with K-carrageenan can enhance elasticity, strength, and water-retention of gels; for ice cream, desserts, sauces, canned meats, drinks, cheese, other milk derivs.; *Regulatory:* EEC compliance; *Properties:* powd.; 20-30% on 200 mesh, 30-40% on 140 mesh; water-sol.; visc. 2400±200 cps (1%); pH 5.5-6.0; 13% max. moisture.

Idealgum 3F. [Idea Srl; Commodity Services Int'l.] Locust bean gum; CAS 9000-40-2; EINECS 232-541-5; nonionic; stabilizer, thickener, syneresis inhibitor; with K-carrageenan can enhance elasticity, strength, and water-retention of gels; for ice cream, desserts, sauces, canned meats, drinks, cheese, other milk derivs.; *Regulatory:* EEC compliance; *Properties:* powd.; 15-25% on 200 mesh, 45-55% on 140 mesh; water-sol.; visc. 2200±200 cps (1%);

pH 5.5-6.0; 13% max. moisture.

IDX. [MLG Enterprises Ltd.] Potassium bromate, calcium iodate, and other edible excipients; dough conditioner for conventional bread and bun prod.; slower sol. of calcium iodate allows for improved pan flow, closer grain structure with sufficient oxidation through proofing and baking stages; *Usage level:* 1-2 tablets/cwt flour; *Properties:* tablets; *Storage:* 6 mos shelf life under cool, dry conditions.

IDX-20 NB. [MLG Enterprises Ltd.] Calcium iodate and other edible excipients; CAS 7789-80-2; EINECS 232-191-3; dough conditioner, oxidation agent; *Regulatory:* FDA GRAS; *Properties:* wh. dust-free tablets, odorless; disp. in water; 0.908 g calcium iodate/tablet; *Storage:* 6 mos shelf life; store cool and dry.

I+G. [Ajinomoto; Unipex] Mixt. of disodium 5'-inosinate and disodium 5'-guanylate; flavor potentiator; *Properties:* colorless or wh. cryst. or wh. cryst. powd., odorless, char. taste; sol. in water; sparingly sol. in ethanol; pract. insol. in ether; pH 7.0-8.5 (1 in 20 sol'n.); 97-102% assay.

Imperial FD. [ADM Arkady] Composite dough conditioner improving frozen yeast-raised dough prods.; increased tolerance in freeze/thaw cycles.

Improved Paniplus M. [ADM Arkady] Calcium peroxide; CAS 1305-79-9; EINECS 215-139-4; dough conditioner; similar to Regular Paniplus with additional benefit of crumb whitening and additional amts. of calcium for nutritional fortification.

Imwitor® 175. [Hüls Am.; Hüls AG] Self-emulsifying mono- and diglycerides derived from edible oils; nonionic; emulsifier for baking industry, bread improvers, soups, sauces, dressings, sausages; *Properties:* powd.; m.p. 60 C; acid no. 5 max.; iodine no. 3 max.; sapon. no. 140-160; 50% monoglycerides.

Imwitor® 191. [Hüls Am.; Hüls AG] Glyceryl stearate; CAS 31566-31-1; nonionic; food emulsifier for bread improvers, soft caramels, potato prods., margarine, ice cream, pasta; *Properties:* ylsh. powd.; sol. in oils, molten fats, acetone, ether; m.p. 66-71 C; solid. pt. 63-68 C; HLB 4.4; acid no. 3 max.; iodine no. 3 max.; sapon. no. 155-170; 90% monoglycerides.

Imwitor® 325. [Hüls Am.; Hüls AG] Monodiglyceride derived from edible oils; nonionic; food emulsifier for bread improvers; *Properties:* powd.; m.p. 70 C; acid no. 3 max.; iodine no. 3 max.; sapon. no. 165-175; 60% monoglycerides.

Imwitor® 333. [Hüls Am.] Monodiglyceride lactic acid ester; nonionic; food emulsifier; *Properties:* solid; 100% conc.

Imwitor® 369. [Hüls Am.; Hüls AG] Monoglyceride citric ester; CAS 91744-38-6; nonionic; food emulsifier; *Properties:* solid; HLB 12.0; 100% conc.

Imwitor® 370. [Hüls Am.; Hüls AG] Glyceryl stearate citrate; CAS 91744-38-6; anionic; food emulsifier/stabilizer for boiled and cooked sausages, soups, sauces, dressings; *Properties:* ylsh. powd./flakes; m.p. 59-63 C; HLB 10-12; acid no. 10-25; iodine no. 3 max.; sapon. no. 240-260; 20% monoglycerides.

Imwitor® 371. [Hüls Am.; Hüls AG] Monoglyceride esterified with citric acid; nonionic; food emulsifier/stabilizer for boiled and cooked sausages, soups, sauces, dressings; *Properties:* powd./flakes; m.p. 60 C; acid no. 25-50; iodine no. 2; sapon. no. 265-285; 20% monoglycerides.

Imwitor® 372. [Hüls AG] Monoglyceride esterified with citric acid; food emulsifier/stabilizer for boiled and cooked sausages, soups, sauces, dressings; *Properties:* powd./flakes; m.p. 60 C; acid no. 25 max.; iodine no. 2; sapon. no. 240-260; 20% monoglycerides.

Imwitor® 373. [Hüls AG] Monodiglyceride/monoglyceride citric acid ester; food emulsifier/stabilizer for boiled and cooked sausages, soups, sauces, dressings; *Properties:* powd./flakes; m.p. 60 C; acid no. 20 max.; iodine no. 2; sapon. no. 170-185; 35% monoglycerides.

Imwitor® 375. [Hüls Am.; Hüls AG] Glyceryl citrate/lactate/linoleate/oleate; nonionic; food emulsifier/stabilizer for release agent emulsions; *Properties:* ylsh. highly visc. liq.; HLB 11.0; acid no. 5-15; iodine no. 90; sapon. no. 230-250; 20% monoglceyrides.

Imwitor® 440. [Hüls Am.; Hüls AG] Soya oil mono- and diglycerides; nonionic; emulsifier for food industry (bread improvers); *Properties:* pasty; HLB 3.7; acid no. 3 max.; iodine no. 120; sapon. no. 155-175; 45% monoglycerides.

Imwitor® 460. [Hüls Am.; Hüls AG] Soya oil mono- and diglycerides; nonionic; food emulsifier for bread improvers; *Properties:* pasty; acid no. 3 max.; iodine no. 110; sapon. no. 160-175; 60% monoglycerides.

Imwitor® 490. [Hüls Am.] Soya oil mono- and diglycerides; nonionic; food emulsifier; *Properties:* powd.; 100% conc.

Imwitor® 595. [Hüls Am.; Hüls AG] Monoglyceride based on edible fats; nonionic; food emulsifier for bread improvers, soft caramels, potato prods., margarine, ice cream, pasta; *Properties:* powd.; m.p. 70 C; HLB 4.4; acid no. 3 max.; iodine no. 3 max.; sapon. no. 155-170; 90% monoglycerides.

Imwitor® 845. [Hüls Am.] Lard or tallow mono- and diglycerides; nonionic; food emulsifier; *Properties:* paste, plastic; 100% conc.

Imwitor® 895. [Hüls Am.; Hüls AG] Mono/diglyceride based on edible fats; nonionic; food emulsifier for bakery goods, bread improvers; *Properties:* pasty; m.p. 55 C; acid no. 3 max.; iodine no. 45; sapon. no. 155-175; 90% monoglycerides.

Imwitor® 900. [Hüls Am.; Hüls AG] Glyceryl stearate; nonionic; food emulsifier for bread improvers, margarine, ice cream, sausages; *Properties:* ylsh. powd.; sol. in fats, oils, waxes; m.p. 56-61 C; HLB 3.0; acid no. 3 max.; iodine no. 3 max.; sapon. no. 160-175; 45% monoglycerides.

Imwitor® 910. [Hüls Am.; Hüls AG] Glyceryl caprate; CAS 26402-22-2; EINECS 247-667-6; nonionic; food emulsifier; *Properties:* wh. powd.; m.p. 38-41 C; HLB 3.0; acid no. 2 max.; iodine no. 1 max.; sapon. no. 240-260; 40% min. monoglycerides.

Imwitor® 928. [Hüls AG] Glyceryl cocoate; CAS

61789-05-7; EINECS 263-027-9; surfactant for nutritional fields; *Properties:* soft wh. substance; sol. in acetone, ether, water/ethanol; m.p. 33-37 C; acid no. 2 max.; iodine no. 3 max.; sapon. no. 200-220.

Imwitor® 940 K. [Hüls Am.] Glyceryl stearate/palmitate Ph. Eur.; nonionic; food emulsifier; *Properties:* Gardner 4 max. powd.; m.p. 53-57 C; solid. pt. 54-60; sapon. no. 165-178; 42-48% 1-monoglyceride.

Imwitor® 945. [Hüls Am.; Hüls AG] Glyceryl palmitate/stearate; nonionic; food emulsifier for bread improvers, margarine, ice cream, sausages; *Properties:* powd.; m.p. 60 C; HLB 3.8; acid no. 3 max.; iodine no. 3 max.; sapon. no. 160-175; 45% monoglycerides.

Imwitor® 960. [Hüls Am.; Hüls AG] Glyceryl stearate SE; anionic; food emulsifier for bread improvers, potato prods., sausages; *Properties:* ylsh. flakes; sol. in fats, oils, waxes; m.p. 56-61 C; HLB 12.0; acid no. 5 max.; iodine no. 3 max.; sapon. no. 150-180; 35% monoglycerides.

Imwitor® 988. [Hüls AG] Glyceryl caprylate; CAS 26402-26-6; EINECS 247-668-1; surfactant for nutritional fields; softener for chewing gum bases; *Properties:* almost colorless liq./semisolid; sol. in water/ethanol (25/75), acetone, ether, heptane; acid no. 3 max.; iodine no. 3 max.; sapon. no. 275-300; 50% monoglycerides.

Imwitor® 1330. [Hüls Am.; Hüls AG] Monoglyceride diacetylated tartaric acid ester; food emulsifier for bread improvers; *Properties:* flakes; m.p. 45 C; acid no. 100-120; iodine no. 2; sapon. no. 500-550; 100% conc.

Imwitor® 1339. [Hüls AG] Monoglyceride diacetyl tartaric acid ester with carrier; food emulsifier for bread improvers; *Properties:* powd.; acid no. 65-85; iodine no. 2; sapon. no. 380-420.

Imwitor® 1940. [Hüls AG] Lecithin/monoglyceride citric acid ester; food emulsifier, antispattering agent for margarine; *Properties:* powd.; m.p. 55 C; acid no. 20 max.; iodine no. 30; sapon. no. 175-185.

Imwitor® 2020. [Hüls Am.; Hüls AG] Monoglyceride diacetylated tartaric acid ester; CAS 91052-81-2; food emulsifier for bakery prods., bread improvers; *Properties:* pasty; acid no. 65-85; iodine no. 80; sapon. no. 420-455; 100% conc.

Imwitor® 2320. [Hüls Am.; Hüls AG] Monoglyceride diacetylated tartaric acid ester; food emulsifier for bread improvers; *Properties:* paste; acid no. 65-85; iodine no. 40; sapon. no. 420-455; 100% conc.

Industrene® 143. [Witco/H-I-P] Tallow acid; CAS 61790-37-2; EINECS 263-129-3; FG grades as lubricant, release agent, binder, defoamer in foods, intermediate for food emulsifiers; *Properties:* Gardner 5 paste; solid. pt. 39-43 C; acid no. 202-206; iodine no. 50-65; sapon. no. 202-207; 100% conc.

Industrene® 205. [Witco/H-I-P] Oleic acid; CAS 112-80-1; EINECS 204-007-1; FG grades as lubricant, release agent, binder, defoamer in foods, intermediate for food emulsifiers; *Properties:* wh. liq.; acid no. 195-204; iodine no. 85-95; sapon. no. 195-205;

100% conc.

Industrene® 225. [Witco/H-I-P] Soya acid, dist.; CAS 67701-08-0; EINECS 269-657-0; FG grades as lubricant, release agent, binder, defoamer in foods, intermediate for food emulsifiers; *Properties:* Gardner 3-4 liq.; solid. pt. 25 C max.; acid no. 195-201; iodine no. 135-145; sapon. no. 197-204; 100% conc.

Industrene® 226 FG. [Witco/H-I-P] Soya acid, dist.; EINECS 269-657-0; lubricant, release agent, binder, defoaming agent for foods; intermediate for food emulsifiers; *Properties:* liq.; Lovibond 25.0Y/2.5R; solid. pt. 26 C max.; acid no. 195-203; iodine no. 125-135; sapon. no. 195-204; 100% conc.

Industrene® 365. [Witco/H-I-P] Mixt. caprylic/capric acid; CAS 67762-36-1; FG grades as lubricant, release agent, binder, defoamer in foods, intermediate for food emulsifiers; *Properties:* acid no. 355-369; iodine no. 1 max.; sapon. no. 355-374; 100% conc.

Industrene® 4518. [Witco/H-I-P] Single pressed stearic acid; CAS 57-11-4; EINECS 200-313-4; FG grades as lubricant, release agent, binder, defoamer in foods, intermediate for food emulsifiers; *Properties:* Gardner 3 solid; acid no. 204-211; iodine no. 8-11; sapon. no. 204-212; 100% conc.

Industrene® 5016 NF FG. [Witco/H-I-P] Stearic acid NF; CAS 57-11-4; EINECS 200-313-4; lubricant, release agent, binder, defoamer in foods; intermediate for food-grade emulsifiers; *Properties:* acid no. 206-210; iodine no. 0.5 max.; sapon. no. 206-211.

Industrene® 7018 FG. [Witco/H-I-P] 70% Stearic acid; CAS 57-11-4; EINECS 200-313-4; lubricant, release agent, binder, defoamer for foods; intermediate for food-grade emulsifiers; *Properties:* Lovibond 2.0Y/0.2R solid; solid. pt. 58-62 C; acid no. 200-207; iodine no. 1 max.; sapon. no. 200-208; 100% conc.

Industrene® 8718 FG. [Witco/H-I-P] 87% Stearic acid; CAS 57-11-4; EINECS 200-313-4; lubricant, release agent, binder, defoamer for foods; intermediate for food-grade emulsifiers; *Properties:* Lovibond 2.0Y-0.2R; solid. pt. 64.5-67.5 C; acid no. 196-201; iodine no. 2 max.; sapon. no. 196-202; 92% conc.

Industrene® B. [Witco/H-I-P] Hydrog. stearic acid; FG grades as lubricant, release agent, binder, defoamer in foods; intermediate for food-grade emulsifiers; *Properties:* solid; acid no. 199-207; iodine no. 3 max.; sapon. no. 199-208; 100% conc.

Instant Remygel AX-P, AX-2-P. [Remy Industries SA] Waxy rice starch; pre-cooked version of Remygel AX and AX-2; binder, thickener, stabilizer, oil/fat substitute in food specialties, baby foods, sauces, soups, salad dressings, frozen foods, preserves, low-fat margarine/butter; *Properties:* wh. powd., neutral taste; sol. in cold water.

Instant Remyline AX-P. [Remy Industries SA] Waxy rice starch; pre-cooked version of Remyline AX; binder, thickener, stabilizer, oil/fat substitute in food specialties, baby foods, sauces, soups, salad dressings, frozen foods, preserves, low-fat margarine/butter; *Properties:* wh. powd., neutral taste; sol. in cold water.

Instant Tender-Jel® 419. [A.E. Staley Mfg.] Food starch modified, derived from waxy maize; pregelatinized starch providing stable visc. in acid and neutral food systems, clarity, sheen, short texture; for salad dressings, tomato prods., fruit mixes, pie fillings, instant soups, sauces, and gravies; *Regulatory:* FDA 21CFR §172.892; *Properties:* wh. powd.; 97% thru 100 mesh, 80% thru 200 mesh; pH 5; 3% moisture; *Storage:* store at ambient temps. and humidity; store under cool, dry conditions for sl. improved handling.

Instant Tender-Jel® 434. [A.E. Staley Mfg.] Food starch modified, derived from waxy maize; pregelatinized starch, high moisture capacity for high visc. sol'ns. and high moisture gels; used in extra moist cakes, doughnuts, soft cookies, fruit pie fillings, mixes, instant cream-style soups, instant gravies; *Regulatory:* FDA 21CFR §172.892; *Properties:* wh. powd.; 5% max. on 100 mesh, 75% min. thru 200 mesh; pH 4.5-6.0; 5.5% max. moisture; *Storage:* store at ambient temps. and humidity; store under cool, dry conditions for sl. improved handling.

Instant Tender-Jel® 479. [A.E. Staley Mfg.] Food starch modified, derived from waxy maize; precooked starch, water binding agent; produces med. visc. sol'ns. with exc. resist. to heat, refrigeration, acid, shear stresses; for high acid tomato-based sauces and fruit fillings, and for dry mixes; hydrates at slow rate; *Regulatory:* FDA 21CFR §172.892; *Properties:* powd.; 15% on 60 mesh, 65% on 100 mesh, 30% thru 100 mesh; pH 5.0-6.5; 5.5% moisture.

Instant Tender-Jel® 480. [A.E. Staley Mfg.] Food starch modified, derived from waxy maize; precooked starch producing med. to high visc. sol'ns. @ 1-5% usage levels; for sauces, gravies, dips, toppings, fillings, dry mixes; reduced lumping tendencies; resists heat, moderate stress; exc. freeze/thaw stability; for systems above pH 4; *Regulatory:* FDA 21CFR §172.892; *Properties:* powd.; 5% on 60 mesh, 65% on 100 mesh, 30% thru 100 mesh; pH 5.0-6.5; 5.5% moisture.

Instant Tender-Jel® C. [A.E. Staley Mfg.] Food starch modified, derived from waxy maize; pregelatinized starch, high moisture holding, forming high visc. sol'ns. and high moisture gels; for extra moist cakes, doughnuts, soft cookies, fruit pie fillings/mixes, instant cream-style soups, instant gravies; *Regulatory:* FDA 21CFR §172.892; *Properties:* wh. powd.; 5% max. on 100 mesh, 80% min. thru 200 mesh; pH 5-6; 2-5% moisture; *Storage:* store at ambient temps. and humidity; store under cool, dry conditions for sl. improved handling.

IPSO-C403. [Vaessen-Schoemaker] Soy protein conc. from high quality defatted soy beans; CAS 68153-28-6; protein source; *Properties:* powd.; 95% < 150 µ particle size; 70% min. protein.

IPSO-FC. [Vaessen-Schoemaker] Soy protein isolate; CAS 68153-28-6; low heat gelling protein source with good oil emulsification props.; for use in spray-dried processed prods., e.g., semidairy and nondairy fat concs., spray-dried milk replacers, etc.; *Properties:* cream-colored nondusting powd.; highly sol.; 92% protein, 5% moisture.

IPSO-MR Dispersible. [Vaessen-Schoemaker] Soy protein isolate; CAS 68153-28-6; protein source with strong gelling and emulsifying props. for comminuted meats, injection brines for additional binding and sliceability in cooked hams; also for cooked sausage, pork loins, poultry prods., patties, meat and fish balls, cream soups; *Properties:* cream-colored nondusting free-flowing powd., neutral taste and flavor; pH 7; 92.2% protein, 3.5% moisture.

IPSO-NGL. [Vaessen-Schoemaker] Soy protein isolate; CAS 68153-28-6; protein source for powd. drinks which are hydrated by consumer, protein/sports drinks, special dietary prods.; instant dispersibility, high protein digestibility; contains no soy carbohydrates, eliminating risk of flatulence; 91% protein.

Iron Bile Salts. [Am. Labs] Blend of ox bile with a source of ferric iron; nutritive food and pharmaceutical additive; *Properties:* brnsh.-yel. powd.; hygroscopic; pH 6.6.

Isoamyl Alcohol 95%. [CPS] Isoamyl alcohol; CAS 123-51-3; EINECS 204-633-5; used for synthetic flavors and fragrances; *Regulatory:* FDA 21CFR §172.515; *Properties:* APHA 50 color; m.w. 88.15; sp.gr. 0.811 (20/20 C); distillation range 126-132 C (760 mm); flash pt. 47 C; 95% purity, 0.5% water.

Isoamyl Alcohol 99%. [CPS] Isoamyl alcohol; CAS 123-51-3; EINECS 204-633-5; used for synthetic flavors and fragrances; *Regulatory:* FDA 21CFR §172.515; *Properties:* APHA 20 color; m.w. 88.15; sp.gr. 0.811 (20/20 C); distillation range 128-132 C (760 mm); flash pt. 47 C; 99% purity, 0.2% water.

Isobutanol HP. [CPS] Isobutyl alcohol; CAS 78-83-1; EINECS 201-148-0; used for synthetic flavors and fragrances; *Regulatory:* FDA 21CFR §172.515; *Properties:* APHA 20 color; m.w. 74.12; sp.gr. 0.803 (20/20 C); flash pt. 74 C; 96% purity, 0.5% water.

Isohop®. [Pfizer Food Science] Aq. sol'n. of potassium salts of isoalpha acids standardized at 30% w/v, from hop conc.; brewery additive providing economic, controlled bitterness to beer; for post-fermentation injection; *Storage:* store @ 2-8 C under refrigeration.

J

Jaguar® 1105. [Rhone-Poulenc Food Ingreds.] Functionally enhanced guar gum; CAS 9000-30-0; EINECS 232-536-8; performance grade with rapid hydration, pH stability (4-9), low bacterial counts; for dairy prods., beverages, bakery prods., frozen entrees, desserts, soups, and sauces; *Properties:* visc. 600 ± 100 cps (1%); 12% max. moisture.

Jaguar® 1110. [Rhone-Poulenc Food Ingreds.] Functionally enhanced guar gum; CAS 9000-30-0; EINECS 232-536-8; performance grade with rapid hydration, pH stability (4-9), low bacterial counts; for dairy prods., beverages, bakery prods., frozen entrees, desserts, soups, and sauces; *Properties:* visc. 1200 ± 150 cps (1%); 12% max. moisture.

Jaguar® 1120. [Rhone-Poulenc Food Ingreds.] Functionally enhanced guar gum; CAS 9000-30-0; EINECS 232-536-8; performance grade with rapid hydration, pH stability (4-9), low bacterial counts; for dairy prods., beverages, bakery prods., frozen entrees, desserts, soups, and sauces; *Properties:* visc. 2300 ± 300 cps (1%); 12% max. moisture.

Jaguar® 1140. [Rhone-Poulenc Food Ingreds.] Functionally enhanced guar gum; CAS 9000-30-0; EINECS 232-536-8; performance grade with rapid hydration, pH stability (4-9), low bacterial counts; for dairy prods., beverages, bakery prods., frozen entrees, desserts, soups, and sauces; *Properties:* visc. 4000 ± 250 cps (1%); 12% max. moisture.

Jaguar® 2209. [Rhone-Poulenc Food Ingreds.] Functionally enhanced guar gum; CAS 9000-30-0; EINECS 232-536-8; performance grade with rapid hydration, pH stability (4-9), low bacterial counts; for dairy prods., beverages, bakery prods., frozen entrees, desserts, soups, and sauces; *Properties:* visc. 400 ± 40 cps (1%); 12% max. moisture.

Jaguar® 2220. [Rhone-Poulenc Food Ingreds.] Functionally enhanced guar gum; CAS 9000-30-0; EINECS 232-536-8; performance grade with rapid hydration, pH stability (4-9), low bacterial counts; for dairy prods., beverages, bakery prods., frozen entrees, desserts, soups, and sauces; *Properties:* visc. 2700 ± 200 cps (1%); 12% max. moisture.

Jaguar® 2240. [Rhone-Poulenc Food Ingreds.] Functionally enhanced guar gum; CAS 9000-30-0; EINECS 232-536-8; performance grade with rapid hydration, pH stability (4-9), low bacterial counts; for dairy prods., beverages, bakery prods., frozen entrees, desserts, soups, and sauces; *Properties:* visc. 4000 ± 300 cps (1%); 12% max. moisture.

Jaguar® 6000. [Rhone-Poulenc Food Ingreds.] Functionally enhanced guar gum; CAS 9000-30-0; EINECS 232-536-8; performance grade providing superior visc., hydration rate, and synergy with xanthan gum; does not require heat to develop visc.; freeze-thaw stable; pH stable (4-9); for cake mixes, instant soups/gravies/sauces/beverages, pet foods; *Usage level:* 0.2-0.4% (cake mixes), 0.1-0.25% (instant soups/gravies/sauces), 0.1-0.3% (instant beverages), 0.25-0.6% (canned pet foods), 0.2-0.4% (gravy-style dog food); *Properties:* creamy wh. fine powd.; 100% min. thru 150 mesh; sol. in cold water; visc. 6000 cps (1%); pH 5.4-7.0; 8-12% moisture.

Jaguar® Guar Gum. [Rhone-Poulenc Food Ingreds.] Guar gum; CAS 9000-30-0; EINECS 232-536-8; hydrocolloid for food applics. (baking, cereal, dairy/cheese, processed foods, beverages).

Jerusalem Artichoke Flour (JAF). [Zumbro; Garuda Int'l.] Jerusalem artichoke flour; bulking agent; nutracetical ingred. stimulating friendly bifidobacteria and inhibiting undesirable bacteria associated with gastric upset, high blood pressure, elevated cholesterol levels; for baked goods, pasta, beverages, weight loss formulas; *Properties:* pale brn. flour, sl. sweet nutty flavor; 13.2% dietary fiber, 65.7% carbohydrates, 8.8% protein, 6% moisture; *Storage:* store in cool, dry area.

Jewel Oil™. [Bunge Foods] Winterized cottonseed oil; CAS 8001-29-4; EINECS 232-280-7; specialty oil for salad dressings, snack spray oil, mayonnaise; *Properties:* clear bright oil, 2.0 red color, bland flavor.

Jungbunzlauer GS 7097. [Jungbunzlauer Int'l. AG] Glucose syrup derived from corn starch; used in soft drinks, beer, other beverages, as feedstock in fermentation industry; *Properties:* pH 3.8-4.5; DE 96 min.; 94% min. glucose content.

Justfiber® CL-20-H. [Van Den Bergh Foods] Cellulose fiber; CAS 9004-34-6; EINECS 232-674-9; for use in frozen dairy novelties, sauces, dressings, diet drink mixes, batters, crackers, cookies, pasta; *Regulatory:* kosher; *Properties:* wh. powd.; avg. fiber length 20 μ; bulk dens. 2.8 cc/g; water absorp. 3.6-3.8 g H₂O/g.

Justfiber® CL-35-H. [Van Den Bergh Foods] Cellulose fiber; CAS 9004-34-6; EINECS 232-674-9;

anticlumping for grated cheese; calorie reduction/ fiber enrichment in breads, sauces; structures water, provides volume in cakes; prevents ice crystal growth in frozen novelties; water absorp. bulking agent in diet drinks; reduces fat absorp.; *Regulatory:* kosher; *Properties:* wh. powd.; avg. fiber length 35 μ; bulk dens. 3.0 cc/g; water absorp. 3.6-3.8 g H₂O/g.

Justfiber® CL-40-H. [Van Den Bergh Foods] Cellulose fiber; CAS 9004-34-6; EINECS 232-674-9; antidusting aid; binder in filling pastes; *Regulatory:* kosher; *Properties:* wh. powd.; avg. fiber length 40 μ; bulk dens. 3.2 cc/g; water absorp. 4.0-4.5 g H₂O/ g.

Justfiber® CL-100-H. [Van Den Bergh Foods] Cellulose fiber; CAS 9004-34-6; EINECS 232-674-9; to structure water in gums; to reduce fat and increase moisture retention in pancakes, waffles, muffins, cakes; to provide volume, moistness, reduce oil absorp. during frying in batters; provides noncaloric bulk, absorbs water in pet food; *Regulatory:* kosher; *Properties:* wh. powd.; avg. fiber length 100 μ; bulk dens. 7.8 cc/g; water absorp. 7.6-8.3 g H₂O/g.

Justfiber® CS-20-H. [Van Den Bergh Foods] Cottonseed fiber; used for very bland tasting systems; to absorb less moisture in crackers, cookies, pasta; for calorie reduction/fiber supplementation in frozen dairy novelties, sauces, dressings, diet drink mixes, batters; confectionery applics.; *Regulatory:* kosher; *Properties:* wh. powd.; avg. fiber length 20 μ; bulk dens. 2.8 cc/g; water absorp. 3.6-3.8 g H₂O/ g.

Justfiber® CS-35-H. [Van Den Bergh Foods] Cottonseed fiber; for very bland tasting systems; calorie reductiom/fiber enrichment in breads; to structure water and provide volume in cakes, muffins, mixes, brownies; to structure water and prevent ice crystal growth in frozen novelties; *Regulatory:* kosher; *Properties:* wh. powd.; avg. fiber length 35 μ; bulk dens. 3.1 cc/g; water absorp. 3.5-3.7 g H₂O/g.

K

K2B387. [Kelco] Xanthan gum/emulsifier blend; CAS 11138-66-2; EINECS 234-394-2; dairy stabilizer/ emulsifier for nonfat frozen dessert, nonfat frozen yogurt.

K8B243. [Kelco] Algin/emulsifier blend; dairy stabilizer/emulsifier for ice cream, premium ice cream, soft-serve ice cream, lowfat ice cream, ice milk, soft-serve and frozen yogurt, novelties, eggnog, sour cream dip.

K8B249. [Kelco] Xanthan gum/emulsifier blend; CAS 11138-66-2; EINECS 234-394-2; dairy stabilizer/ emulsifier for UHT/HTST whipping cream, half and half, table cream.

Kakebake® M/V Cake Icing Shortening. [Bunge Foods] Meat fats, vegetable oils with mono- and diglycerides, antioxidant; shortening for cakes and icings; *Properties:* 1.5 red color, bland flavor; 2.7% alpha-mono.

Kaokote. [Van Den Bergh Foods] Partially hydrog. veg. oil (cottonseed, soybean), sorbitan stearate, polysorbate 60; nonlauric coating fat for no-tempering low-liquor type coatings and centers; *Regulatory:* kosher; *Properties:* solid; m.p. 99-103 F.

Kaokote F. [Van Den Bergh Foods] Partially hydrog. veg. oil (cottonseed, soybean); EINECS 269-820-6; high performance fat for center fat, veg. dairy systems; *Regulatory:* kosher; *Properties:* solid; m.p. 99-103 F.

Kaola. [Van Den Bergh Foods] Partially hydrog. veg. oil (soybean, palm kernel); EINECS 269-820-6; high solids lauric oil for veg. dairy systems, candies, mellorines, ice cream bar coatings, nut roasting; *Regulatory:* kosher; *Properties:* Lovibond 2.5R max. plastic; m.p. 87-93 F.

Kaola D. [Van Den Bergh Foods] Partially hydrog. veg. oil (soybean, palm kernel); EINECS 269-820-6; high solids lauric system primarily for icing applics.; *Regulatory:* kosher; *Properties:* Lovibond 2.5R max. plastic; m.p. 99-104 F.

Kaomax 870. [Van Den Bergh Foods] Partially hydrog. veg. oil (cottonseed, soybean); EINECS 269-820-6; high performance fat for no-tempering confectioner's coating, center fat, veg. dairy systems; steep melting shortening, oil migration inhibitor; *Regulatory:* kosher; *Properties:* solid; m.p. 95-100 F.

Kaomax-S. [Van Den Bergh Foods] Partially hydrog. soybean oil; EINECS 232-410-2; nonlauric coating fat for confectioner's coatings; *Regulatory:* kosher; *Properties:* liq.; drop pt. 35-38 C.

Kaomel. [Van Den Bergh Foods] Partially hydrog. veg. oil (cottonseed, soybean); EINECS 269-820-6; high performance fat for no-tempering confectioner's coatings, center fat, veg. dairy systems; steep melting shortening, oil migration inhibitor; *Regulatory:* kosher; *Properties:* flake; m.p. 97-101 F.

Kaoprem-E. [Van Den Bergh Foods] Partially hydrog. soybean oil with sorbitan tristearate; nonlauric coating fat for no-tempering confectioner's coating; *Regulatory:* kosher; *Properties:* solid; drop pt. 37.9 C.

Kaorich. [Van Den Bergh Foods] Partially hydrog. veg. oil (cottonseed, soybean); EINECS 269-820-6; high performance fat for breading mixes and other dry mixes; structuring system for composition enhancement; *Regulatory:* kosher; *Properties:* Lovibond 2.5R max. beads; m.p. 124-130 F.

Kaorich Gold. [Van Den Bergh Foods] Partially hydrog. vegetable oil (soybean, cottonseed), artificial butter flavor, artificial color (β-carotene); shortening system for breading mixes and other dry mixes; *Regulatory:* kosher.

Karaya Gum #1 FCC. [Meer] Karaya gum; CAS 9000-36-6; stabilizer, water binder, emulsifier for water ices, sherbets, cheese spreads, meringue powds., meat processing; *Properties:* water-sol.

Kasal®. [Rhone-Poulenc Food Ingreds.] Sodium aluminum phosphate, basic FCC; CAS 7785-88-8; food additive for baking, cereals, dairy, and processed cheese; *Properties:* wh. powd., odorless; 0% on 200 mesh, 2% on 325 mesh; sl. sol.; bulk dens 42 lb/ft³; pH 9.3 (25% susp.).

Katch® Fish Phosphate. [Monsanto] Sodium tripolyphosphate and sodium hexametaphosphate; food grade blended esp. for fish processing; sequesterant inhibiting oxidative rancidity; reduces moisture during thawing and cooking; emulsifies fat and protein; improves solubility; *Usage level:* < 0.5%; *Regulatory:* FDA 21CFR §182.6810, 182.6760 GRAS resp.; 21CFR §100.161; FCC compliance; *Properties:* wh. anhyd. gran.; 15% max. on 20 mesh, 30% max. thru 100 mesh; *Toxicology:* may cause irritation to respiratory tract; avoid breathing dust; *Storage:* keep container closed.

Kaydol®. [Witco/Petroleum Spec.] Wh. min. oil USP;

emollient and lubricant in food processing; *Regulatory:* FDA 21CFR §172.878, §178.3620a; *Properties:* water-wh., odorless, tasteless; sp.gr. 0.869-0.885; visc. 64-70 cSt (40 C); pour pt. -23 C; flash pt. 216 C.

Kelacid®. [Kelco] Alginic acid; CAS 9005-32-7; EINECS 232-680-1; gelling agent, emulsifier and stabilizer in foods; *Regulatory:* FDA 21CFR §184.1011, GRAS; *Properties:* wh. fibrous particles, sl. odor; sol. in alkaline sol'n.; swells in water; bulk dens. ≈ 50 lb/ft³; pH 2.9 (1% aq.); surf. tens. 53 dynes/cm; 7% moisture; *Toxicology:* LD50 (oral, rat) > 5000 mg/kg; excessive dust inhalation may cause respiratory irritation; dry powd. may cause eye irritation; *Precaution:* not flamm., but powd. will burn if involved in fire; spills are slippery; incompat. with strong oxidizers; *Storage:* store in cool, dry place.

Kelco® HV. [Kelco] Low-calcium sodium alginates; CAS 9005-38-3; gelling agent, emulsifier and stabilizer in foods; *Properties:* cream fibrous particles; sp.gr. 1.64; dens. 43.38 lb/ft³; visc. 400 cps; ref. index 1.3342; pH 7.2; 9% moisture.

Kelco® LV. [Kelco] Low-calcium sodium alginates; CAS 9005-38-3; gelling agent, emulsifier and stabilizer in foods; *Properties:* cream fibrous particles; sp.gr. 1.64; dens. 43.38 lb/ft³; visc. 50 cps; ref. index 1.3342; pH 7.2; 9% moisture.

Kelcogel®. [Kelco] Purified gellan gum; CAS 71010-52-1; EINECS 275-117-5; anionic; gellant, texturizer, stabilizer, suspending agent, film-former for use in aspics, confections, dessert gels, icings, bakery fillings, dairy prods., frostings, glazes, jams/jellies, structured foods, lowfat foods, beverages; *Regulatory:* FDA 21CFR §172.665; EEC E418, JECFA, Japan, Canada approvals; *Properties:* cream to wh. dry free-flowing powd., sl. odor; 100% thru 28 mesh; water-sol.; bulk dens. ≈ 50 lb/ft³; pH 4.5-7.5; readily biodeg.; *Toxicology:* LD50 (oral, rat) > 5000 mg/kg; excessive dust inhalation may cause respiratory irritation; dry powd. may cause eye irritation; *Precaution:* not flamm., but powd. will burn if involved in a fire; spills are slippery when wet; incompat. with strong oxidizers; *Storage:* store in cool, dry place.

Kelcogel® BF. [Kelco] Gellan gum; CAS 71010-52-1; EINECS 275-117-5; gelling agent for bakery fillings (< 50% solids); *Regulatory:* FDA 21CFR §172.665; EEC E418, JECFA, Japan, Canada approvals; *Properties:* cream to wh. dry free-flowing powd., sl. odor; 100% thru 28 mesh; water-sol.; bulk dens. ≈ 50 lb/ft³; pH 7.0-9.0; readily biodeg.; *Toxicology:* LD50 (oral, rat) > 5000 mg/kg; excessive dust inhalation may cause respiratory irritation; dry powd. may cause eye irritation; *Precaution:* not flamm., but powd. will burn if involved in a fire; spills are slippery when wet; incompat. with strong oxidizers; *Storage:* store in cool, dry place.

Kelcogel® BF10. [Kelco] Gellan gum; CAS 71010-52-1; EINECS 275-117-5; gellant for bakery fillings (> 50% solids); *Regulatory:* FDA 21CFR §172.665; EEC E418, JECFA, Japan, Canada approvals;

Properties: wh. to tan powd., sl. odor; 42 mesh; water-sol.; bulk dens. ≈ 50 lb/ft³; pH 6.5-8.5; readily biodeg.; *Toxicology:* LD50 (oral, rat) > 5000 mg/kg; excessive dust inhalation may cause respiratory irritation; dry powd. may cause eye irritation; *Precaution:* not flamm., but powd. will burn if involved in a fire; spills are slippery when wet; incompat. with strong oxidizers; *Storage:* store in cool, dry place.

Kelcogel® CF. [Kelco] Gellan gum; CAS 71010-52-1; EINECS 275-117-5; gellant for confections, marshmallow, jelly candies, candy centers; *Regulatory:* FDA 21CFR §172.665; EEC E418, JECFA, Japan, Canada approvals; *Properties:* cream to wh. dry free-flowing powd., sl. odor; 100% thru 28 mesh; water-sol.; bulk dens. ≈ 50 lb/ft³; pH 5.5-7.5; readily biodeg.; *Toxicology:* LD50 (oral, rat) > 5000 mg/kg; excessive dust inhalation may cause respiratory irritation; dry powd. may cause eye irritation; *Precaution:* not flamm., but powd. will burn if involved in a fire; spills are slippery when wet; incompat. with strong oxidizers; *Storage:* store in cool, dry place.

Kelcogel® CF10. [Kelco] Gellan gum; CAS 71010-52-1; EINECS 275-117-5; gellant for gummi-type confections; *Regulatory:* FDA 21CFR §172.665; EEC E418, JECFA, Japan, Canada approvals; *Properties:* wh. to tan powd., sl. odor; 42 mesh; water-sol.; bulk dens. ≈ 50 lb/ft³; pH 5-7; readily biodeg.; *Toxicology:* LD50 (oral, rat) > 5000 mg/kg; excessive dust inhalation may cause respiratory irritation; dry powd. may cause eye irritation; *Precaution:* not flamm., but powd. will burn if involved in a fire; spills are slippery when wet; incompat. with strong oxidizers; *Storage:* store in cool, dry place.

Kelcogel® F. [Kelco] Gellan gum; CAS 71010-52-1; EINECS 275-117-5; gellant, film-former, stabilizer, suspension agent, texturizer for aspics, farinaceous foods, structured foods, beverages, dairy, sauces, batters, breadings, lowfat spreads, adhesion systems; *Regulatory:* FDA 21CFR §172.665; EEC E418, JECFA, Japan, Canada approvals; *Properties:* wh. to tan powd., sl. odor; 100 mesh; water-sol.; bulk dens. ≈ 50 lb/ft³; pH 4.5-7.5; readily biodeg.; *Toxicology:* LD50 (oral, rat) > 5000 mg/kg; excessive dust inhalation may cause respiratory irritation; dry powd. may cause eye irritation; *Precaution:* not flamm., but powd. will burn if involved in a fire; spills are slippery when wet; incompat. with strong oxidizers; *Storage:* store in cool, dry place.

Kelcogel® IF. [Kelco] Gellan gum; CAS 71010-52-1; EINECS 275-117-5; gelling agent, binder used for icings, frostings, and glazes; *Regulatory:* FDA 21CFR §172.665; EEC E418, JECFA, Japan, Canada approvals; *Properties:* cream to wh. dry free-flowing powd., sl. odor; 100% thru 28 mesh; water-sol.; bulk dens. ≈ 50 lb/ft³; pH 4.5-7.5; readily biodeg.; *Toxicology:* LD50 (oral, rat) > 5000 mg/kg; excessive dust inhalation may cause respiratory irritation; dry powd. may cause eye irritation; *Precaution:* not flamm., but powd. will burn if involved in a fire; spills are slippery when wet; incompat. with strong oxidizers; *Storage:* store in cool, dry place.

Kelcogel® JJ. [Kelco] Gellan gum; CAS 71010-52-1;

EINECS 275-117-5; gelling agent used for reduced-calorie jams and jellies, fruit spreads; *Regulatory:* FDA 21CFR §172.665; EEC E418, JECFA, Japan, Canada approvals; *Properties:* cream to wh. dry free-flowing powd., sl. odor; 100% thru 28 mesh; water-sol.; bulk dens. ≈ 50 lb/ft³; pH 7-9; readily biodeg.; *Toxicology:* LD50 (oral, rat) > 5000 mg/kg; excessive dust inhalation may cause respiratory irritation; dry powd. may cause eye irritation; *Precaution:* not flamm., but powd. will burn if involved in a fire; spills are slippery when wet; incompat. with strong oxidizers; *Storage:* store in cool, dry place.

Kelcogel® PD. [Kelco] Gellan gum; CAS 71010-52-1; EINECS 275-117-5; gellant for puddings; *Regulatory:* FDA 21CFR §172.665; EEC E418, JECFA, Japan, Canada approvals; *Properties:* cream to wh. dry free-flowing powd., sl. odor; 100% thru 28 mesh; water-sol.; bulk dens. ≈ 50 lb/ft³; readily biodeg.; *Toxicology:* LD50 (oral, rat) > 5000 mg/kg; excessive dust inhalation may cause respiratory irritation; dry powd. may cause eye irritation; *Precaution:* not flamm., but powd. will burn if involved in a fire; spills are slippery when wet; incompat. with strong oxidizers; *Storage:* store in cool, dry place.

Kelcoloid® D. [Kelco] Propylene glycol alginate; CAS 9005-37-2; gelling agent, emulsifier and stabilizer in foods; beer foam stabilizer; *Properties:* cream fibrous particles; sp.gr. 1.46; dens. 33.71 lb/ft³; visc. 170 cps; ref. index 1.3343; pH 4.4; surf. tens. 58 dynes/cm; 13% max. moisture.

Kelcoloid® DH. [Kelco] Propylene glycol alginate; CAS 9005-37-2; gelling agent, emulsifier and stabilizer in fruit and chocolate syrups, sauces, icings, frozen foods, salad dressings, relish, batters, barbeque sauces, food emulsions; *Regulatory:* FDA 21CFR §172.858, EEC E405 compliance; *Properties:* cream agglomerated, sl. odor; 840 µ particle size; sol. in water; sp.gr. 1.46; dens. 33.71 lb/ft³; visc. 400 cps (1%), 7000 cps (2%); ref. index 1.3343; pH 3.7; surf. tens. 58 dynes/cm; *Toxicology:* LD50 (oral, rat) > 5000 mg/kg; dry powd. may cause eye irritation; *Precaution:* incompat. with strong oxidizers.

Kelcoloid® DO. [Kelco] Propylene glycol alginate; CAS 9005-37-2; used as gelling agent, emulsifier and stabilizer in food applics.; *Properties:* cream agglomerated; sp.gr. 1.46; dens. 33.71 lb/ft³; visc. 25 cps; ref. index 1.3343; pH 4.3; surf. tens. 58 dynes/cm.

Kelcoloid® DSF. [Kelco] Propylene glycol alginate; CAS 9005-37-2; used as gelling agent, emulsifier and stabilizer in food applics., emulsions and low pH systems; *Properties:* cream agglomerated; sp.gr. 1.46; dens. 33.71 lb/ft³; visc. 20 cps; ref. index 1.3343; pH 4.0; surf. tens. 58 dynes/cm.

Kelcoloid® HVF. [Kelco] Propylene glycol alginate; CAS 9005-37-2; gelling agent, emulsifier and stabilizer in fruit and chocolate syrups, sauces, icings, frozen foods, salad dressings, relish, batters, barbeque sauces, food emulsions; *Regulatory:* FDA 21CFR §172.858, EEC E405 compliance;

Properties: cream fibrous particles, sl. odor; 175 µ particle size; sol. in water; sp.gr. 1.46; dens. 33.71 lb/ft³; visc. 400 cps (1%), 7000 cps (2%); ref. index 1.3343; pH 3.7; surf. tens. 58 dynes/cm; *Toxicology:* LD50 (oral, rat) > 5000 mg/kg; dry powd. may cause eye irritation; *Precaution:* incompat. with strong oxidizers.

Kelcoloid® LVF. [Kelco] Propylene glycol alginate; CAS 9005-37-2; gelling agent, emulsifier and stabilizer in fruit and chocolate syrups, sauces, icings, frozen foods, salad dressings, relish, batters, barbeque sauces, food emulsions; *Regulatory:* FDA 21CFR §172.858, EEC E405 compliance; *Properties:* cream fibrous particles; 175 µ particle size; sp.gr. 1.46; dens. 33.71 lb/ft³; visc. 120 cps (1%), 1200 cps (2%); ref. index 1.3343; pH 3.7; surf. tens. 58 dynes/cm; *Toxicology:* LD50 (oral, rat) > 5000 mg/kg; dry powd. may cause eye irritation; *Precaution:* incompat. with strong oxidizers.

Kelcoloid® O. [Kelco] Propylene glycol alginate; CAS 9005-37-2; gelling agent, emulsifier and stabilizer in foods; beer foam stabilizer; *Regulatory:* FDA 21CFR §172.858, EEC E405 compliance; *Properties:* cream fibrous particles; 175 µ particle size; sp.gr. 1.46; dens. 33.71 lb/ft³; visc. 25 cps (1%), 130 cps (2%); ref. index 1.3343; pH 4.0; surf. tens. 58 dynes/cm; *Toxicology:* LD50 (oral, rat) > 5000 mg/ kg; dry powd. may cause eye irritation; *Precaution:* incompat. with strong oxidizers.

Kelcoloid® S. [Kelco] Propylene glycol alginate; CAS 9005-37-2; gelling agent, emulsifier and stabilizer for fruit and chocolate syrups, sauces, icings, frozen foods, salad dressings, relish, batters, barbeque sauces, food emulsions; beer foam stabilizer; *Regulatory:* FDA 21CFR §172.858, EEC E405 compliance; *Properties:* cream fibrous particles; 175 µ particle size; sp.gr. 1.46; dens. 33.71 lb/ft³; visc. 20 cps (1%), 120 cps (2%); ref. index 1.3343; pH 4.1; surf. tens. 58 dynes/cm; *Toxicology:* LD50 (oral, rat) > 5000 mg/kg; dry powd. may cause eye irritation; *Precaution:* incompat. with strong oxidizers.

Kelcosol®. [Kelco] Algin; CAS 9005-38-3; gelling agent, suspending agent, emulsifier and stabilizer in foods (puddings, pie fillings, bakery fillings, chiffons, sauces, dessert gels); *Regulatory:* FDA 21CFR §184.1724, GRAS; EEC E401 compliance; *Properties:* cream fibrous particles, sl. odor; 80 mesh; water-sol.; sp.gr. 1.64; dens. 43.38 lb/ft³; visc. 1300 cps (1%), 15,000 cps (2%); pH 7; surf. tens. 70 dynes/cm; 9% moisture; *Toxicology:* LD50 (oral, rat) > 5000 mg/kg; dry powd. may cause eye irritation; *Precaution:* incompat. with strong oxidizers.

Kelgin® F. [Kelco] Algin, refined; CAS 9005-38-3; gelling agent, suspending agent, emulsifier and stabilizer in foods (batters, icings, bakery gels, sauces, gravies, frozen foods, syrups, fountain syrups, dry mixes, dietetic beverages); *Regulatory:* FDA 21CFR §184.1724, GRAS; EEC E401 compliance; *Properties:* ivory gran., sl. odor; 80 mesh; water-sol.; sp.gr. 1.59; dens. 54.62 lb/ft³; visc. 300

cps (1%), 4000 cps (2%); ref. index 1.3343; pH 7; surf. tens. 62 dynes/cm; 13% moisture; *Toxicology:* LD50 (oral, rat) > 5000 mg/kg; dry powd. may cause eye irritation; *Precaution:* incompat. with strong oxidizers.

Kelgin® HV. [Kelco] Algin; CAS 9005-38-3; gelling agent, suspending agent, emulsifier and stabilizer in foods (batters, icings, bakery gels, sauces, gravies, frozen foods, syrups, fountain syrups, dry mixes, dietetic beverages); *Regulatory:* FDA 21CFR §184.1724, GRAS; EEC E401 compliance; *Properties:* ivory gran., sl. odor; 30 mesh; water-sol.; sp.gr. 1.59; dens. 54.62 lb/ft³; visc. 800 cps (1%), 10,000 cps (2%); ref. index 1.3343; pH 7; surf. tens. 62 dynes/cm; 13% moisture; *Toxicology:* LD50 (oral, rat) > 5000 mg/kg; dry powd. may cause eye irritation; *Precaution:* incompat. with strong oxidizers.

Kelgin® LV. [Kelco] Algin; CAS 9005-38-3; gelling agent, suspending agent, emulsifier and stabilizer in foods (batters, icings, bakery gels, sauces, gravies, frozen foods, syrups, fountain syrups, dry mixes, dietetic beverages); *Regulatory:* FDA 21CFR §184.1724, GRAS; EEC E401 compliance; *Properties:* ivory gran., sl. odor; 42 mesh; water-sol.; sp.gr. 1.59; dens. 54.62 lb/ft³; visc. 60 cps (1%), 500 cps (2%); ref. index 1.3343; pH 7; surf. tens. 62 dynes/cm; 13% moisture; *Toxicology:* LD50 (oral, rat) > 5000 mg/kg; dry powd. may cause eye irritation; *Precaution:* incompat. with strong oxidizers.

Kelgin® MV. [Kelco] Algin; CAS 9005-38-3; gelling agent, suspending agent, emulsifier and stabilizer in foods (batters, icings, bakery gels, sauces, gravies, frozen foods, syrups, fountain syrups, dry mixes, dietetic beverages); *Regulatory:* FDA 21CFR §184.1724, GRAS; EEC E401 compliance; *Properties:* ivory gran., sl. odor; 30 mesh; water-sol.; sp.gr. 1.59; dens. 54.62 lb/ft³; visc. 400 cps (1%), 6000 cps (2%); ref. index 1.3343; pH 7; surf. tens. 62 dynes/cm; 13% moisture; *Toxicology:* LD50 (oral, rat) > 5000 mg/kg; dry powd. may cause eye irritation; *Precaution:* incompat. with strong oxidizers.

Kelgin® QL. [Kelco] Treated sodium alginate; CAS 9005-38-3; gelling agent, emulsifier and stabilizer in foods; *Properties:* ivory gran. particles; visc. 30 cps; pH neutral.

Kelgin® XL. [Kelco] Refined sodium alginate; CAS 9005-38-3; gelling agent, suspending agent, emulsifier and stabilizer in foods (batters, icings, bakery gels, sauces, gravies, frozen foods, syrups, fountain syrups, dry mixes, dietetic beverages); *Regulatory:* FDA 21CFR §184.1724, GRAS; EEC E401 compliance; *Properties:* ivory gran., sl. odor; 42 mesh; water-sol.; sp.gr. 1.59; dens. 54.62 lb/ft³; visc. 30 cps (1%), 160 cps (2%); ref. index 1.3343; pH 7; surf. tens. 62 dynes/cm; 13% moisture; *Toxicology:* LD50 (oral, rat) > 5000 mg/kg; dry powd. may cause eye irritation; *Precaution:* incompat. with strong oxidizers.

Kelgum®. [Kelco] Xanthan gum/locust bean gum blend; used for desserts, gelled confectioneries, gels.

Kel-Lite™ BK. [Kelco] Hydrocolloid/emulsifier blend; shortening-replacement blend for breads, biscuits, and tortillas, providing 15-40% calorie reduction in most bakery prods.; good organoleptic props.; improves moisture retention, enhances shelf life; *Usage level:* 0.2-0.5%; *Properties:* 20 mesh; visc. 170 cps (1%); pH 6.5.

Kel-Lite™ CM. [Kelco] Sodium stearoyl lactylate, xanthan gum, gum Arabic, dextrin, lecithin, and mono- and diglycerides; shortening-replacement blend for cakes, cookies, muffins, waffles, pancakes; *Properties:* 20 mesh; visc. 170 cps (1%); pH 6.5.

Kelmar®. [Kelco] Potassium alginate; CAS 9005-36-1; gellant, emulsifier, and stabilizer in foods; water holding capacity; used in dietetic and low-sodium foods, dry mixes; *Regulatory:* FDA 21CFR §184.1610, GRAS; *Properties:* cream gran., sl. odor; 100 mesh; water-sol.; bulk dens. ≈ 50 lb/ft³; visc. 270 cps (1%), 3200 cps (2%); pH 7.0; *Toxicology:* LD50 (oral, rat) > 5000 mg/kg; excessive dust inhalation may cause respiratory irritation; dry powd. may cause eye irritation; *Precaution:* not flamm., but powd. will burn if involved in a fire; spills may be slippery; incompat. with strong oxidizers.

Kelmar® Improved. [Kelco] Potassium alginate; CAS 9005-36-1; gellant, emulsifier, and stabilizer in foods (dietetic and low-sodium foods, dry mixes); used for water holding; *Regulatory:* FDA 21CFR §184.1610, GRAS; *Properties:* cream gran., sl. odor; water-sol.; bulk dens. ≈ 50 lb/ft³; visc. 400 cps (1%), 4500 cps (2%); pH 7.0; *Toxicology:* LD50 (oral, rat) > 5000 mg/kg; excessive dust inhalation may cause respiratory irritation; dry powd. may cause eye irritation; *Precaution:* not flamm., but powd. will burn if involved in a fire; spills may be slippery; incompat. with strong oxidizers.

Kelset®. [Kelco] Sodium alginate; CAS 9005-38-3; gelling agent, emulsifier and stabilizer in foods (bakery fillings, tomato sauces, glazes); *Regulatory:* FDA 21CFR §184.1724, GRAS; EEC E401 compliance; *Properties:* lt. ivory fibrous particles, sl. odor; 80 mesh; water-sol.; bulk dens. ≈ 50 lb/ft³; pH 7.0; *Toxicology:* LD50 (oral, rat) > 5000 mg/kg; dry powd. may cause eye irritation; *Precaution:* incompat. with strong oxidizers.

Keltone®. [Kelco] Algin; CAS 9005-38-3; gelling agent, emulsifier and stabilizer in foods (hot and cold water dessert gels); *Regulatory:* FDA 21CFR §184.1724, GRAS; EEC E401 compliance.

Keltone® HV. [Kelco] Sodium alginate; CAS 9005-38-3; gellant, thickener, stabilizer, suspending agent, film-former used in puddings, pie fillings, bakery fillings, chiffons, sauces, and dessert gels; *Regulatory:* FDA 21CFR §184.1724, GRAS; EEC E401 compliance; *Properties:* wh. to tan powd., sl. odor; 80 mesh; water-sol.; bulk dens. ≈ 50 lb/ft³; visc. 400 cps (1%), 3500 cps (2%); pH 7; *Toxicology:* LD50 (oral, rat) > 5000 mg/kg; dry powd. may cause eye irritation; *Precaution:*

incompat. with strong oxidizers.

Keltone® LV. [Kelco] Sodium alginate; CAS 9005-38-3; gellant, thickener, stabilizer, suspending agent, film-former used in puddings, pie fillings, bakery fillings, chiffons, sauces, and dessert gels; *Regulatory:* FDA 21CFR §184.1724, GRAS; EEC E401 compliance; *Properties:* wh. to tan powd., sl. odor; 150 mesh; water-sol.; bulk dens. ≈ 50 lb/ft³; visc. 50 cps (1%), 250 cps (2%); pH 7; *Toxicology:* LD50 (oral, rat) > 5000 mg/kg; dry powd. may cause eye irritation; *Precaution:* incompat. with strong oxidizers.

Keltose®. [Kelco] Calcium alginate and ammonium alginate; gellant, binder, emulsifier, and stabilizer in foods, icings, fruit fillings, meringues; *Regulatory:* FDA GRAS; *Properties:* ivory gran., sl. odor; 80 mesh; water-sol.; dens. ≈ 50 lb/ft³; visc. 250 cps; pH 7.0; readily biodeg.; *Toxicology:* LD50 (oral, rat) > 5000 mg/kg; dry powd. may cause eye irritation, skin irritation on prolonged contact; excessive dust inhalation may cause respiratory irritation; *Precaution:* not flamm., but powd. will burn if involved in a fire; spills may be slippery; incompat. with strong oxidizers, alkaline sol'ns.; *Storage:* store in cool, dry place.

Keltrol®. [Kelco] Food-grade xanthan gum; CAS 11138-66-2; EINECS 234-394-2; stabilizer for foods, bakery fillings, flavor emulsions, canned food, dry mixes, frozen foods, juice drinks, dressings, relishes, gravies, sauces, syrups, baked goods, batters, puddings and pie fillings; suspending agent for fruit pulp; *Regulatory:* FDA 21CFR §172.605; EEC E415 compliance; *Properties:* cream to wh. powd., sl. odor; ≥ 95% thru 80 mesh; sol. in hot and cold water; swells in glycerin and propylene glycol; sp.gr. 1.5; bulk dens. 52.2 lb/ft³; visc. 1200-1600 cps (1% gum in 1% KCl sol'n.); pH 7.0; surf. tens. 75 dynes/cm; readily biodeg.; 11% moisture; *Toxicology:* LD50 (oral, rat) > 5000 mg/kg; excessive dust inhalation may cause respiratory irritation; dry powd. may cause eye irritation; *Precaution:* not flamm., but powd. will burn if involved in a fire; spills are slippery; incompat. with strong oxidizers.

Keltrol® 1000. [Kelco] Xanthan gum; CAS 11138-66-2; EINECS 234-394-2; high microbiological purity grade thickener, stabilizer used for bakery fillings, flavor emulsions, canned foods, dry mixes, frozen foods, juice drinks, dressings, relishes, gravies, sauces, syrups, baked goods, batters, puddings, pie fillings; *Regulatory:* FDA 21CFR §172.695; EEC E415 compliance; *Properties:* wh. to tan powd., sl. odor; 175 μ particle size; water-sol.; bulk dens. ≈ 50 lb/ft³; visc. 1400 cps (1%); pH 7.0; readily biodeg.; *Toxicology:* LD50 (oral, rat) > 5000 mg/kg; excessive dust inhalation may cause respiratory irritation; dry powd. may cause eye irritation; *Precaution:* not flamm., but powd. will burn if involved in a fire; spills are slippery; incompat. with strong oxidizers.

Keltrol® BT. [Kelco] Xanthan gum; CAS 11138-66-2; EINECS 234-394-2; brine tolerant thickener, sus-

pending agent, stabilizer for bakery fillings, flavor emulsions, canned foods, dry mixes, frozen foods, juice drinks, dressings, relishes, gravies, sauces, syrups, baked goods, batters, puddings, pie fillings; *Regulatory:* FDA 21CFR §172.695; FCC, NF, EEC E415 compliance; kosher; *Properties:* cream to wh. gran. powd., sl. odor; ≥ 95% thru 80 mesh; sol. in hot or cold water; bulk dens. ≈ 50 lb/ft³; visc. 1200-1600 cps (1% gum in 1% KCl sol'n.); pH 6.1-8.1 (1% aq.); readily biodeg.; 6-14% moisture; *Toxicology:* LD50 (oral, rat) > 5000 mg/kg; excessive dust inhalation may cause respiratory irritation; dry powd. may cause eye irritation; *Precaution:* not flamm., but powd. will burn if involved in a fire; spills are slippery; incompat. with strong oxidizers.

Keltrol® CR. [Kelco] Food-grade xanthan gum; CAS 11138-66-2; EINECS 234-394-2; thickener, stabilizer, suspending agent for foods; *Regulatory:* FDA 21CFR §172.695; EEC E415 compliance; *Properties:* wh. to tan powd., sl. odor; water-sol.; bulk dens. ≈ 50 lb/ft³; pH neutral; readily biodeg.; *Toxicology:* LD50 (oral, rat) > 5000 mg/kg; excessive dust inhalation may cause respiratory irritation; dry powd. may cause eye irritation; *Precaution:* not flamm., but powd. will burn if involved in a fire; spills are slippery; incompat. with strong oxidizers.

Keltrol® F. [Kelco] Food-grade xanthan gum; CAS 11138-66-2; EINECS 234-394-2; stabilizer for foods, bakery fillings, flavor emulsions, canned foods, dry mixes, frozen foods, juice drinks, dressings, relishes, gravies, sauces, syrups, baked goods, batters, puddings, pie fillings; suspending agent for fruit pulp; *Regulatory:* FDA 21CFR §172.695; EEC E415 compliance; *Properties:* cream to wh. powd.; ≥ 92% thru 200 mesh; sol. in cold and hot water; swells in glycerin and propylene glycol; sp.gr. 1.5; bulk dens. 52.2 lb/ft³; visc. 1200-1600 cps (1% gum in 1% KCl sol'n.); pH 7.0; surf. tens. 75 dynes/cm; readily biodeg.; 11% moisture; *Toxicology:* LD50 (oral, rat) > 5000 mg/kg; excessive dust inhalation may cause respiratory irritation; dry powd. may cause eye irritation; *Precaution:* not flamm., but powd. will burn if involved in a fire; spills are slippery; incompat. with strong oxidizers.

Keltrol® GM. [Kelco] Xanthan gum; CAS 11138-66-2; EINECS 234-394-2; thickener, suspending agent, stabilizer for bakery fillings, flavor emulsions, canned foods, dry mixes, frozen foods, juice drinks, dressings, relishes, gravies, sauces, syrups, baked goods, batters, puddings, pie fillings; *Regulatory:* FDA 21CFR §172.695; FCC, NF, EEC E415 compliance; kosher; *Properties:* cream to wh. gran. powd., sl. odor; ≥ 95% thru 42 mesh, ≤ 25% thru 100 mesh; sol. in cold or hot water; bulk dens. ≈ 50 lb/ft³; visc. 1200-1600 cps (1% gum in 1% KCl sol'n.); pH 6.1-8.0 (1% aq.); readily biodeg.; *Toxicology:* LD50 (oral, rat) > 5000 mg/kg; excessive dust inhalation may cause respiratory irritation; dry powd. may cause eye irritation; *Precaution:* not flamm., but powd. will burn if involved in a fire; spills are slippery; incompat. with strong oxidizers.

Keltrol® RD. [Kelco] Xanthan gum; CAS 11138-66-2;

EINECS 234-394-2; thickener, stabilizer used for bakery fillings, flavor emulsions, canned foods, dry mixes, frozen foods, juice drinks, dressings, relishes, gravies, sauces, syrups, baked goods, batters, puddings, pie fillings; *Regulatory:* FDA 21CFR §172.695; EEC E415 compliance; *Properties:* wh. to tan powd., sl. odor; 990 µ particle size; watersol.; bulk dens. ≈ 50 lb/ft³; visc. 1400 cps (1%); pH 7.0; readily biodeg.; *Toxicology:* LD50 (oral, rat) > 5000 mg/kg; excessive dust inhalation may cause respiratory irritation; dry powd. may cause eye irritation; *Precaution:* not flamm., but powd. will burn if involved in a fire; spills are slippery; incompat. with strong oxidizers.

Keltrol® SF. [Kelco] Xanthan gum; CAS 11138-66-2; EINECS 234-394-2; thickener, suspending agent, stabilizer for bakery fillings, flavor emulsions, canned foods, dry mixes, frozen foods, juice drinks, dressings, relishes, gravies, sauces, syrups, baked goods, batters, puddings, pie fillings; *Regulatory:* FDA 21CFR §172.695; FCC, NF, EEC E415 compliance; kosher; *Properties:* wh. to cream powd., sl. odor; ≥ 100% thru 60 mesh, ≥ 95% thru 80 mesh; sol. in cold or hot water; bulk dens. ≈ 50 lb/ft³; visc. 800-1300 cps; pH 6.1-8.1; readily biodeg.; *Toxicology:* LD50 (oral, rat) > 5000 mg/kg; excessive dust inhalation may cause respiratory irritation; dry powd. may cause eye irritation; *Precaution:* not flamm., but powd. will burn if involved in a fire; spills are slippery; incompat. with strong oxidizers.

Keltrol® T. [Kelco] Xanthan gum; CAS 11138-66-2; EINECS 234-394-2; thickener, suspending agent, stabilizer for bakery fillings, flavor emulsions, canned foods, dry mixes, frozen foods, juice drinks, dressings, relishes, gravies, sauces, syrups, baked goods, batters, puddings, pie fillings; *Regulatory:* FDA 21CFR §172.695; FCC, NF, EEC E415 compliance; kosher; *Properties:* cream to wh. powd., sl. odor; 100% thru 60 mesh, ≥ 95% thru 80 mesh; sol. in hot or cold water; bulk dens. ≈ 50 lb/ft³; visc. 1200-1600 cps (1% gum in 1% KCl sol'n.); pH 7.0 ± 1.0; readily biodeg.; *Toxicology:* LD50 (oral, rat) > 5000 mg/kg; excessive dust inhalation may cause respiratory irritation; dry powd. may cause eye irritation; *Precaution:* not flamm., but powd. will burn if involved in a fire; spills are slippery; incompat. with strong oxidizers.

Keltrol® TF. [Kelco] Xanthan gum; CAS 11138-66-2; EINECS 234-394-2; thickener, suspending agent, stabilizer for bakery fillings, flavor emulsions, canned foods, dry mixes, frozen foods, juice drinks, dressings, relishes, gravies, sauces, syrups, baked goods, batters, puddings, pie fillings; *Regulatory:* FDA 21CFR §172.695; FCC, NF, EEC E415 compliance; kosher; *Properties:* cream to wh. powd., sl. odor; 100% thru 80 mesh, ≥ 92% thru 200 mesh; sol. in hot or cold water; bulk dens. ≈ 50 lb/ft³; visc. 1200-1600 cps (1% gum in 1% KCl sol'n.); pH 7.0 ± 1.0; readily biodeg.; *Toxicology:* LD50 (oral, rat) > 5000 mg/kg; excessive dust inhalation may cause respiratory irritation; dry powd. may cause eye irritation; *Precaution:* not flamm., but powd. will

burn if involved in a fire; spills are slippery; incompat. with strong oxidizers.

Kelvis®. [Kelco] Sodium alginate, refined; CAS 9005-38-3; gellant, suspending agent, emulsifier, and stabilizer in food batters; *Regulatory:* FDA 21CFR §184.1724, GRAS; EEC E401 compliance; *Properties:* ivory gran.; 150 mesh; water-sol.; sp.gr. 1.59; dens. 54.62 lb/ft³; visc. 760 cps (1%), 9000 cps (2%); ref. index 1.3343; pH 7; surf. tens. 62 dynes/cm; 13% moisture.

Kemamide® O. [Witco/H-I-P] Oleamide, tech.; CAS 301-02-0; EINECS 206-103-9; release agent migrating from food pkg.; *Regulatory:* FDA 21CFR §175.105, 175.300, 175.380, 175.390, 178.3910, 179.45, 181.28; *Properties:* Gardner 7 max. solid; m.p. 68-78 C; acid no. 5 max.; iodine no. 72-90; 100% conc.

Kemamide® S. [Witco/H-I-P] Stearamide; CAS 124-26-5; EINECS 204-693-2; release agent migrating from food pkg.; *Regulatory:* FDA 21CFR §175.105, 177.1210, 178.3860, 178.3910, 179.45, 181.28; *Properties:* Gardner 4 max. waxy solid, powd., and pellet; m.w. 278; sol. (g/100 g solv. @ 50 C) > 10 g in chloroform, 10 g in IPA, 4 g in MEK, 3 g in methyl alcohol, 2 g in toluene; dens. 0.809 g/ml (130 C); visc. 5.8 cP (130 C); m.p. 98-108 C; acid no. 4 max.; iodine no. 3 max.; flash pt. 246 C (COC); fire pt. 268 C (COC); 95% min. amide.

Kemamide® U. [Witco/H-I-P] Oleamide; CAS 301-02-0; EINECS 206-103-9; release agent migrating from food pkg.; *Regulatory:* FDA 21CFR §175.105, 175.300, 175.380, 175.390, 178.3910, 179.45, 181.28; *Properties:* Gardner 5 max. waxy solid, powd., and pellet; m.w. 275; sol. (g/100 g solv. @ 30 C) > 30 g in IPA, 28 g in methyl alcohol, 25 g in toluene, > 20 g in MEK; dens. 0.823 g/ml (130 C); visc. 5.5 cP (130 C); m.p. 68-78 C; acid no. 4 max.; iodine no. 72-90; flash pt. 245 C (COC); 95% min. amide.

Kemamide® W-39. [Witco/H-I-P] Ethylene distearamide; release agent migrating from food pkg.; *Regulatory:* FDA 21CFR §175.105, 175.300, 175.380, 175.390, 176.170, 177.1200, 177.2470, 177.2480, 181.28; *Properties:* Gardner 18 max. flakes; sol. (g/100 g solv. @ 70 C) 2.0 g toluene, 1.6 g dichloroethane; 1.4 g IPA, 0.9 g MEK; m.p. 140 C; acid no. 10 max.; flash pt. 299 C (COC); fire pt. 315 C (COC).

Kemamide® W-40. [Witco/H-I-P] Ethylene distearamide; release agent migrating from food pkg.; *Regulatory:* FDA 21CFR §175.105, 175.300, 175.380, 175.390, 176.170, 177.1200, 177.2470, 177.2480, 181.28; *Properties:* Gardner 3 max. powd. and flake; sol. (g/100 g solv. @ 70 C) 2.0 g toluene, 1.6 g dichloroethane; 1.4 g IPA, 0.9 g MEK; m.p. 140 C; acid no. 10 max.; flash pt. 299 C (COC); fire pt. 315 C (COC).

Kemamide® W-45. [Witco/H-I-P] Ethylene distearamide; release agent migrating from food pkg.; *Regulatory:* FDA 21CFR §175.105, 175.300, 175.380, 175.390, 176.170, 177.1200, 177.2470, 177.2480, 181.28; *Properties:* Gardner 3 max.

powd. and flake; sol. (g/100 g solv. @ 70 C) 2.0 g toluene, 1.6 g dichloroethane; 1.4 g IPA, 0.9 g MEK; m.p. 145 C; acid no. 10 max.; flash pt. 304 C (COC); fire pt. 322 C.

Kemester® 2050. [Witco/H-I-P] Methyl eicosenate; CAS 1120-28-1; EINECS 214-304-8; foam depressant and nutrient in fermentation; *Properties:* Gardner 14 max. color; acid no. 25 max.; iodine no. 90-110; sapon. no. 180-190.

Kemester® 4516. [Witco/H-I-P] Methyl stearate; foam depressant and nutrient in fermentation; *Properties:* Gardner 1 max. color; acid no. 3 max.; iodine no. 2.5 max.; sapon. no. 192-202.

Kena® FP-28. [Rhone-Poulenc Food Ingreds.] Sodium tripolyphosphate FCC, sodium hexametaphosphate FCC blend; food additive for poultry to be cooked, smoked, cured, or dehydrated, raw frozen items; provides oxidative stability, increases tenderness, provides better moisture retention, better texture; also for chicken/turkey rolls, barbecue chicken; *Regulatory:* kosher; *Properties:* wh. gran., odorless; 1% max. on 20 mesh, 30% max. thru 100 mesh; pH 9.1-9.8 (1%); *Storage:* store cool and dry.

Kena® FP-85. [Rhone-Poulenc Food Ingreds.] Sodium tripolyphosphate FCC, sodium hexametaphosphate FCC, and sodium acid pyrophosphate FCC; food additive for meat, poultry, and seafood industries; *Regulatory:* kosher; *Properties:* wh. gran., odorless; 5% max. on 20 mesh, 30% max. thru 100 mesh; pH 8.2-8.8 (1%); *Storage:* store cool and dry.

Kessco® Glycerol Distearate 300Г. [Stepan Food Ingreds.] Glyceryl distearate; CAS 1323-83-7; EINECS 215-359-0; food-grade emulsifier for pharmaceutical use; *Properties:* wh. to off-wh. waxy flake, typ. mild fatty odor; insol. in water; sol. in IPA, min. oil; partly sol. in peanut oil; m.p. 56-59 C; HLB 2.4; acid no. 5.0 max.; sapon. no. 182-188; flash pt. (COC) 450 F; 100% conc.

Kessco® GMC-8. [Stepan; Stepan Canada] Glyceryl caprylate/caprate; solubilizer and emulsifier for vitamins, flavors; *Properties:* lt. yel. liq.; HLB 8.3; acid no. 1.5 max.; Discontinued

Keycel®. [Protein Tech. Int'l.] Powd. cellulose; CAS 9004-34-6; EINECS 232-674-9; functional ingred. or noncaloric bulking agent for foods.

Kidney Substance. [Am. Labs] Vacuum-dried defatted glandular prod.; nutritive food additive; *Properties:* powd. or freeze-dried form.

Kimiloid HV. [Kimitsu Chem. Industries; Unipex] Propylene glycol alginate; CAS 9005-37-2; visc. promoter, shape-making agent, stabilizer for sauces; *Regulatory:* meets official std. of Food Additives Regs.; *Properties:* wh. or creamy powd.; 40 mesh pass; visc. 150 cp min.; pH 3.0-4.5; < 15% water.

Kimiloid MV. [Kimitsu Chem. Industries; Unipex] Propylene glycol alginate; CAS 9005-37-2; visc. promoter, shape-making agent, stabilizer for mayonnaise, dressings; *Regulatory:* meets official std. of Food Additives Regs.; *Properties:* wh. or creamy powd.; 40 mesh pass; visc. 100-150 cp; pH 3.0-4.5; < 15% water.

Kimiloid NLS-K. [Kimitsu Chem. Industries; Unipex] Propylene glycol alginate; CAS 9005-37-2; visc. promoter, shape-making agent, stabilizer for acidophilus beverage, sherbet, essences; *Regulatory:* meets official std. of Food Additives Regs.; *Properties:* wh. or creamy powd.; 40 mesh pass; visc. 30-60 cp; pH 3.0-4.5; < 15% water.

Kimitsu Acid. [Kimitsu Chem. Industries; Unipex] Alginic acid; CAS 9005-32-7; EINECS 232-680-1; used as tablet disintegrant and for edible uses; *Properties:* wh. powd., 80 mesh pass; pH 1.5-3.5; < 15% water.

Kimitsu Algin I-1. [Kimitsu Chem. Industries; Unipex] Algin; CAS 9005-38-3; visc. promoter, shape-making agent, stabilizer for edible uses esp. ice cream; *Regulatory:* meets Japanese official std.; *Properties:* wh. or creamy powd.; 80 mesh pass; visc. 90-130 cp; pH neutral; < 15% water.

Kimitsu Algin I-2. [Kimitsu Chem. Industries; Unipex] Algin; CAS 9005-38-3; visc. promoter, shape-making agent, stabilizer for edible uses esp. ice cream; *Regulatory:* meets Japanese official std.; *Properties:* wh. or creamy powd.; 80 mesh pass; visc. 220-280 cp; pH neutral; < 15% water.

Kimitsu Algin I-3. [Kimitsu Chem. Industries; Unipex] Algin; CAS 9005-38-3; visc. promoter, shape-making agent, stabilizer for edible uses (ice cream, instant soup, cakes, sausage); *Regulatory:* meets Japanese official std.; *Properties:* wh. or creamy powd.; 80 mesh pass; visc. 320-380 cp; pH neutral; < 15% water.

Kirnol® Range. [Grunau GmbH] Fatty acid mono- and diglycerides based or vegetable or animal sources; general purpose food emulsifiers for baked goods; < 60% mono content.

Klearol®. [Witco/Petroleum Spec.] Wh. min. oil NF; emollient in food processing; *Regulatory:* FDA 21CFR §172.878, §178.3620a; *Properties:* water-wh., odorless, tasteless; sp.gr. 0.822-0.833; visc. 7-10 cSt (40 C); pour pt. -7 C; flash pt. 138 C.

Klucel® F Grades. [Hercules] Premium grades hydroxypropylcellulose NF; grades designated FF also meet FCC specs.; CAS 9004-64-2; nonionic; surface active thickener, stabilizer, film-former, suspending agent, protective colloid for food applics.; *Regulatory:* FDA 21CFR §172.870; *Properties:* off-wh. powd., tasteless; 85% min. thru 30 mesh, 99% min. thru 20 mesh; water-sol.; dissolves easily in many polar org. solvs.; soften. pt. 100-150 C; pH 5.0-8.5; surf. tens. 43.6 dynes/cm (0.1%); 5% max. moisture; *Toxicology:* nontoxic orally; nonirritating to skin; may cause transient eye irritation.

K.L.X. [Van Den Bergh Foods] Partially hydrog. veg. oil (cottonseed, soybean); EINECS 269-820-6; high performance fat; icing stabilizer; adds solids to shortening; fat encapsulation; structuring system for composition enhancement; *Regulatory:* kosher; *Properties:* Lovibond 2.5R max. flakes; m.p. 124-130 F.

K.L.X. Flakes. [Van Den Bergh Foods] Partially hydrog. veg. oil; EINECS 269-820-6; icing stabilizer; *Properties:* m.p. 124-130 F.

KOB87. [Kelco] Xanthan gum/galactomannan blend; CAS 11138-66-2; EINECS 234-394-2; dairy stabilizer, bodying agent, thickener for pasteurized process cheese food, yogurt dry mix, novelty water and ice milk bars, salad dressing, gravies, pasteurized liq. egg preps., syrup, pudding dry mix, cookie filling, nonfat frozen desserts; *Usage level:* 0.1-0.3% (coldpack cheese), 0.21-0.28% (nonfat frozen desserts), 0.2-0.45% (salad dressing), 0.15-0.3% (dry pudding mix), 0.08-0.2% (imitation liq. creamer), 0.12-0.18% (novelty ice milk bars); *Properties:* 100 mesh; visc. 2700 cps (1%); pH 7.0.

KOB348. [Kelco] Xanthan gum/emulsifier blend; CAS 11138-66-2; EINECS 234-394-2; dairy stabilizer/ emulsifier for sherbet, water ice, extruded water ice.

KOB349. [Kelco] Xanthan gum; CAS 11138-66-2; EINECS 234-394-2; dairy stabilizer for sour cream and dips.

Kol Guard® 7373. [A.E. Staley Mfg.] Food starch modified, derived from waxy maize; stabilizer for neutral pH, frozen food systems; imparts freeze/ thaw stability, sheen, clarity, blandness; for frozen fruits, cream pies, boil-in-bag foods, pot pies, TV dinners, canned fillings, lemon puddings, refrigerated foods, baby foods; *Regulatory:* FDA 21CFR §172.892; *Properties:* wh. powd.; pH 5.6; 11% moisture.

Kol Guard® 7413. [A.E. Staley Mfg.] Food starch modified, derived from waxy maize; stabilizer for low-temp., low pH food systems; resists freeze/ thaw breakdown, maintains short paste texture, high sheen, clarity; for fresh/frozen fruit pies, meat prods., sauces, gravies, baby food; extends shelf life in cook-up puddings; *Regulatory:* FDA 21CFR §172.892; *Properties:* wh. powd.; pH 5.6; 11.5% moisture.

Konut. [Van Den Bergh Foods] Coconut oil; CAS 8001-31-8; EINECS 232-282-8; high solids lauric oil for ice cream, coatings, nut roasting, corn popping; *Regulatory:* kosher; *Properties:* Lovibond 1.0R max. plastic; m.p. 73-79 F.

Koster Keunen Candelilla. [Koster Keunen] Candelilla wax; CAS 8006-44-8; EINECS 232-347-0; gellant for candy and gum; *Regulatory:* FDA 21CFR §172.615, 175.105, 175.320, 176.180; *Properties:* wax; sol. hot in alcohol, benzene, petrleum ether; sp.gr. 0.9820-0.9930; m.p. 68.5-72.5 C; acid no. 11-19; iodine no. 19-44; sapon. no. 44-66; ref. index 1.4555; dielec. const. 2.50-2.63.

Koster Keunen Carnauba. [Koster Keunen] Carnauba; CAS 8015-86-9; EINECS 232-399-4; candies; *Regulatory:* FDA 21CFR §182.1978; *Properties:* wax; sol. (g/100 cc): 1.690 g chloroform, 0.610 g xylene, 0.518 g benzene, 0.440 g turpentine, 0.324 g acetone; sp.gr. 0.996-0.998; m.p. 82.5-86 C; acid no. 2-6; iodine no. 7-14; sapon. no. 78-88; flash pt. > 300 C; ref. index 1.463; dielec. const. 2.67-4.20.

Koster Keunen Paraffin Wax. [Koster Keunen] Paraffin; CAS 8002-74-2; EINECS 232-315-6; wax for protection of food, plants, fruits, cheese, and vegetables; *Regulatory:* FDA 21CFR §172.615, 175.250, 175.300; *Properties:* wax; sol. (g/100 cc): 40 g benzene, 9 g min. oil, 3 g dichloroethane, 0.4 g IPA; m.p. various grades from 118-165 F; ref. index 1.4219-1.4357.

Koster Keunen Synthetic Spermaceti. [Koster Keunen] Syn. spermaceti (cetyl palmitate and other esters); CAS 136097-97-7; wax for mfg. of sweetmeats, candies, confectionery; *Regulatory:* FDA approved; *Properties:* wax; sol. in boiling alcohol, chloroform, carbon disulfide, volatile oils; sp.gr. 0.940-0.946; visc. 6.7-7.4 (100 C); m.p. 45-49 C; acid no. 0-0.5; iodine no. < 3; sapon. no. 116-125; flash pt. > 240 C; ref. index 1.440 (60 C).

Kristel Gold. [Van Den Bergh Foods] Partially hydrog. veg. oil (palm kernel, soybean, cottonseed), lecithin, artificial color and flavor; high performance colored/flavored structuring system for composition enhancement; shortening system for Danish pastries, sweet rolls, coffee cakes; *Regulatory:* kosher; *Properties:* Lovibond 7-9R flakes, buttery flavor; m.p. 101-105 F.

Kristel Gold II. [Van Den Bergh Foods] Partially hydrog. veg. oil (soybean, cottonseed), lecithin, artificial color and flavor; high performance colored/ flavored structuring system for composition enhancement; shortening system for Danish pastries, sweet rolls, coffee cakes; higher melt pt. than Kristel Gold for high shear applics.; *Regulatory:* kosher; *Properties:* Lovibond 12-15R flakes, buttery flavor; m.p. 121-125 F.

Krunchy Krust 5. [Brolite Prods.] Plastic bread base for all hard crusted white breads; *Usage level:* 5 lb/ 100 lb flour.

Kureton 200. [Kao/Edible Fat & Oil] Dist. monoglyceride/food additives blend; defoaming agent for soybean curd; *Properties:* bead.

Kureton Power. [Kao/Edible Fat & Oil] Soybean curd; nonionic; special formulated prod. for the tofu industry; *Properties:* powd.; 100% conc.

L

Lactacel®. [Quest Int'l.] Starter culture; for meat prods.

Lactarin® MV-306. [FMC] Fat replacement system for cheese sauce; provides cling, creaminess, hot visc., and mouthfeel.

Lactarin® MV-308. [FMC] Fat replacement system for frozen shakes; prevents whey separation and improves mouthfeel.

Lactarin® MV-406. [FMC] Fat replacement system for buttermilk-type dry mix dressings; provides creaminess, thickening, body, cling, and mouthfeel.

Lactarin® PS-185X. [FMC] Fat replacement system for buttermilk-type dry mix dressings; provides creaminess, thickening, body, cling, and mouthfeel.

Lactase AIE. [Mitsubishi Int'l.; Amano Enzyme USA] Lactase from *Aspergillus oryzae*; digestive enzyme; *Properties:* powd.

Lactase F Amano. [Mitsubishi Int'l.; Amano Enzyme USA] Lactase from *Aspergillus oryzae*; food enzyme; *Properties:* powd.

Lacticol CFT. [Kelco Int'l.] Alginate blend; thickener, gellant used for savory dips, bakery cream fillings, cheesecake toppings, milk shakes, instant drinks, chilled desserts, instant puddings and mousses.

Lacticol F336. [Kelco Int'l.] Alginate blend; thickener, gellant used for milk-based systems, bakery cream fillings, cheesecake toppings, instant puddings; *Properties:* 60 mesh; milk-sol.; pH 9 (1%).

Lacticol F616. [Kelco Int'l.] Alginate blend; thickener, gellant used for bakery filling creams.

Lactitol MC. [Xyrofin UK] Lactitol monohydrate, food grade; CAS 81025-04-9; food grade sweetener for foods and pharmaceutical applics.; stable to air and heat; *Properties:* wh. cryst. powd., odorless, mild sweet taste; sol. in water (≈ 1.7 g/ml); m.w. 362.37; m.p. 95-105 C; pH 4.5-7 (0.1 g/ml aq.); 98% min. act.; *Storage:* 1 yr stability in original sealed pkg. stored below 25 C and < 65% r.h.; marginally hygroscopic.

Lactodan B 30. [Grindsted Prods. Denmark] Glyceryl stearate lactate; nonionic; food emulsifier, improves aeration and foam stabilization; *Properties:* solid, flakes; 100% act.

Lactodan LW. [Grindsted Prods.] Lactic acid ester of monoglycerides; food emulsifier; aerating agent, foam stabilizer in whipped toppings; enhances whippability of nondairy creams; improves aeration

and stability in cake batter in combination with Dimodans; avail. in kosher grade; *Regulatory:* EEC, FDA §172.852; *Properties:* pellets; drop pt. 50 C; sapon. no. 235-255; 13-16% lactic acid.

Lactodan P 22. [Grindsted Prods.] Lactylated monoglyceride; food emulsifier; aerating agent, foam stabilizer in whipped toppings; enhances whippability of nondairy creams; improves aeration and stability in cake batter in combination with Dimodans; avail. in kosher grade; *Regulatory:* EEC, FDA §172.852; *Properties:* sm. beads; drop pt. 45 C; sapon. no. 270-300; 20-35% lactic acid.

Lactodan P 22 Kosher. [Grindsted Prods.] Lactic acid ester of mono- and diglycerides from edible refined hydrog. vegetable fat; emulsifier for cakes, shortenings, whipped toppings for aeration; enhances gloss in confectionery coatings; *Usage level:* 3-8% (cake shortening), 8-12% (whipped topping powd.), 0.5-1% (liq. imitation whipping cream), 0.2-0.5% (UHT dairy creams), 0.5-2% (compd. coatings); *Regulatory:* FDA 21CFR §172.852; FCC; kosher certified; *Properties:* sm. beads; drop pt. ≈ 45 C; acid no. 4 max.; iodine no. 2 max.; sapon. no. 270-300; *Storage:* 12 mo storage life when kept cool and dry; storage above 20 C may cause caking.

Lactylate Hydrate. [Custom Ingreds.] Water, sodium stearoyl lactate, ethoxylated monoglycerides, polysorbate 60; dough conditioner for flour tortillas; fully hydrated to maximize functionality; *Usage level:* 0.75-1%/100 lb of flour; *Properties:* soft plastic; *Storage:* avoid extreme heat or cold; do not freeze.

Lamefrost® ES 216 G. [Grünau GmbH] Mono- and diglycerides of edible fatty acids, guar gum, sodium alginate, carrageenan; emulsifier/stabilizer for industrial ice cream; improves creaminess, body, and mouthfeel; *Usage level:* 0.55% (3% fat content ice cream), 0.35% (11% fat content ice cream); *Regulatory:* FAO/WHO, EEC compliance; *Properties:* creamy free-flowing powd., neutral odor and taste; 60-70% < 0.2 mm sieve, 90-100% 0.5 mm; bulk dens. 550-650 g/l; *Storage:* 1 yr shelf life stored in closed original pkg. under cool, dry conditions.

Lamefrost® ES 251 G. [Grünau GmbH] Mono- and diglycerides of edible fatty acids, locust bean gum, sodium alginate, carrageenan; emulsifier/stabilizer for industrial ice cream; improves creaminess, body, and mouthfeel; *Usage level:* 0.55% (3% fat content ice cream), 0.40% (11% fat content ice cream); *Regulatory:* FAO/WHO, EEC compliance; *Properties:* creamy free-flowing powd., neutral odor and taste; 60-70% < 0.2 mm sieve, 90-100% 0.5 mm; bulk dens. 550-650 g/l; *Storage:* 1 yr shelf life stored in closed original pkg. under cool, dry conditions.

Lamefrost® ES 315. [Grünau GmbH] Mono- and diglycerides of edible fatty acids, guar gum, carrageenan; emulsifier/stabilizer for industrial ice cream; improves creaminess, body, and mouthfeel; *Usage level:* 0.35% (3% fat content ice cream), 0.20% (11% fat content ice cream); *Regulatory:* FAO/WHO, EEC compliance; *Properties:* creamy free-flowing powd., neutral odor and taste; 60% < 0.1 mm sieve, 90% < 0.2 mm; bulk dens. 550-650 g/l; *Storage:* 1 yr shelf life stored in closed original pkg. under cool, dry conditions.

Lamefrost® ES 375. [Grünau GmbH] Mono- and diglycerides of edible fatty acids, guar gum, carboxymethylcellulose sodium, carrageenan; emulsifier/stabilizer for industrial ice cream; improves creaminess, body, and mouthfeel; *Usage level:* 0.45% (3% fat content ice cream), 0.30% (11% fat content ice cream); *Regulatory:* FAO/WHO, EEC compliance; *Properties:* creamy free-flowing powd., neutral odor and taste; 60-70% < 0.2 mm sieve, 90-100% 0.5 mm; bulk dens. 550-650 g/l; *Storage:* 1 yr shelf life stored in closed original pkg. under cool, dry conditions.

Lamefrost® ES 379. [Grünau GmbH] Mono- and diglycerides of edible fatty acids, guar gum, carboxymethylcellulose sodium, carrageenan; emulsifier/stabilizer for industrial ice cream; improves creaminess, body, and mouthfeel; *Usage level:* 0.45% (3% fat content ice cream), 0.30% (11% fat content ice cream); *Regulatory:* FAO/WHO, EEC compliance; *Properties:* creamy free-flowing powd., neutral odor and taste; 60-70% < 0.2 mm sieve, 90-100% 0.5 mm; bulk dens. 550-650 g/l; *Storage:* 1 yr shelf life stored in closed original pkg. under cool, dry conditions.

Lamefrost® ES 424. [Grünau GmbH] Mono- and diglycerides of edible fatty acids, guar gum, locust bean gum, carrageenan; emulsifier/stabilizer for industrial ice cream; improves creaminess, body, and mouthfeel; *Usage level:* 0.35% (3% fat content ice cream), 0.20% (11% fat content ice cream); *Regulatory:* FAO/WHO, EEC compliance; *Properties:* creamy free-flowing powd., neutral odor and taste; 60% < 0.1 mm sieve, 90% < 0.2 mm; bulk dens. 550-650 g/l; *Storage:* 1 yr shelf life stored in closed original pkg. under cool, dry conditions.

Lamegin® CSL. [Grünau GmbH] Calcium stearoyl lactylate; CAS 5793-94-2; EINECS 227-335-7; anionic; lipophilic emulsifier and antistaling agent for food industry; *Properties:* powd.; 100% conc.; Unverified

Lamegin® DW 8000-Range. [Grünau GmbH] Mono- and diglycerides of edible fatty acids esterified with monoacetyl- and diacetyl tartaric acid (80%), tricalcium phosphate (20%); emulsifier, baking ad-

ditive, dough conditioner; provides increased kneading and fermentation tolerance and increased volume; for yeast-raised wheaten baked goods; also for foodstuff systems requiring better emulsion or emulsion stability; *Properties:* wh. powd., sl. acetic acid odor, sourish taste; 70% 0.2 mm sieve, 100% 0.5 mm sieve; bulk dens. 550-650 g/l; m.p. 43-46 C; acid no. 95-105; sapon. no. 505-520; *Storage:* 1 yr storage in cool, dry place below 20 C.

Lamegin® DW 8000 HW. [Grünau GmbH; Henkel/ Functional Prods.] Diacetyl tartaric acid ester of mono- and diglycerides; dough conditioner for bread, rolls; *Properties:* powd.; 100% conc.

Lamegin® DW 8200 VA/HW. [Henkel/Functional Prods.] Diacetyl tartaric acid ester of mono- and diglycerides; dough conditioner for bread, rolls; *Properties:* powd.; 100% conc.

Lamegin® DW 9000 HW. [Henkel/Functional Prods.] Diacetyl tartaric acid ester of mono- and diglycerides; dough conditioner for bread, rolls; *Properties:* powd.; 100% conc.

Lamegin® DW Range. [Grünau GmbH] Mono- and diacetyl tartaric acid esters of fatty acids; emulsifiers for improvement of fermentation tolerance, volume and texture of yeast-raised baked goods.

Lamegin® DWF, DWH, DWP. [Grünau GmbH] Mono-diglyceride diacetyl tartaric acid ester; anionic; hydrophilic emulsifier, bread improver; *Properties:* liq., flakes, paste resp.; 100% conc.; Unverified

Lamegin® EE Range. [Grünau GmbH] Acetic acid esters of mono- and diglycerides of fatty acids; nonionic; coating agents, basic emulsifiers for aerating agents and whipped topping concs. for baked goods.

Lamegin® EE 50. [Grünau GmbH] Acetic acid ester of mono- and diglycerides of fatty acids; coating agents, basic emulsifiers for aerating agents and whipped topping concs. for baked goods; *Regulatory:* EEC E 472a, FDA 21CFR §172.828; *Properties:* wh. to cream-colored wax, neutral odor and taste; m.p. 40-46 C; acid no. 3 max.; iodine no. 2 max.; sapon. no. 250-290; *Storage:* 1 yr storage in original pkg. under dry, cool (R.T.) conditions.

Lamegin® EE 70. [Grünau GmbH] Acetic acid ester of mono- and diglycerides of fatty acids; coating agents, basic emulsifiers for aerating agents and whipped topping concs. for baked goods; *Regulatory:* EEC E 472a, FDA 21CFR §172.828; *Properties:* wh. to cream-colored wax-like plastic, neutral odor and taste; m.p. 36-40 C; acid no. 2 max.; iodine no. 2 max.; sapon. no. 300-320; hyd. no. 100-120; pH 5.0-6.5 (1:10 in methanol/water 1:1); *Storage:* 1 yr storage in original pkg. under dry, cool (R.T.) conditions.

Lamegin® EE 100. [Grünau GmbH] Acetic acid ester of mono- and diglycerides of fatty acids; coating agents, basic emulsifiers for aerating agents and whipped topping concs. for baked goods; *Regulatory:* EEC E 472a, FDA 21CFR §172.828; *Properties:* wh. to cream-colored wax, neutral odor and taste; m.p. 28-32 C; acid no. 2 max.; iodine no. 2

max.; sapon. no. 370-390; hyd. no. 10 max.; pH 5.0-6.5 (1:10 in methanol/water 1:1); *Storage:* 1 yr storage in original pkg. under dry, cool (R.T.) conditions.

Lamegin® GLP Range. [Grünau GmbH] Lactic acid esters of mono- and diglycerides of fatty acids; food emulsifiers for aerating agents and whipped topping concs. in baked goods.

Lamegin® GLP 10, 20. [Grünau GmbH] Hydrog. tallow glyceride lactate; CAS 68990-06-7; EINECS 273-576-6; nonionic; emulsifier, plasticizer for foods; *Properties:* solid; 100% conc.

Lamegin® NSL. [Grünau GmbH] Sodium stearoyl lactylate; CAS 25383-99-7; EINECS 246-929-7; anionic; food emulsifier; *Properties:* powd.; 100% conc.; Unverified

Lamegin® ZE 30, 60. [Grünau GmbH] Hydrog. tallow glyceride citrate; CAS 68990-59-0; EINECS 273-613-6; anionic; emulsifier for margarine and meat industry; *Properties:* solid; 100% conc.

Lamegin® ZE Range. [Grünau GmbH] Citric acid esters of mono- and diglycerides of fatty acids; emulsifiers for minced meat prods. and in mfg. of instant dry yeast for baking.

Lamemul® Range. [Grünau GmbH] Mono and diglycerides; food emulsifiers for baking additives, antistaling effect.

Lamemul® K 1000-Range. [Grünau GmbH] Mono and diglycerides of edible fatty acids (90%), tricalcium phosphate as anticaking agent (10%); raw material in mfg. of baking additives; improves crumb; delays staling; improves kneading and fermenting tolerance of dough; antistick agent for mfg. of noodles; improves sensory props. of potato based food prods.; *Regulatory:* EEC; *Properties:* wh. powd., neutral odor and taste; 80% on 0.2 mm sieve, 100% on 0.5 mm sieve; bulk dens. 550-650 g/l; m.p. 68-72 C; acid no. < 3; iodine no. < 2; sapon. no. 155-165; pH 4-5 (1:10 in methanol/water 1:1); 90-95% total monoglyceride; *Storage:* 1 yr shelf life under cool, dry conditions below 30 C.

Lamemul® K 2000-Range. [Grünau GmbH] Mono and diglycerides of edible fatty acids, microbeaded; raw material in mfg. of baking additives; improves crumb; delays staling; improves kneading and fermenting tolerance of dough; antistick agent for mfg. of noodles; improves sensory props. of potato-based food prods.; emulsifier in ice cream mfg.; *Regulatory:* EEC; *Properties:* wh. powd., neutral odor and taste; 60% on 0.2 mm sieve, 100% on 0.5 mm sieve; bulk dens. 550-650 g/l; m.p. 68-72 C; acid no. < 3; iodine no. < 2; sapon. no. 155-165; pH 4-5 (1:10 in methanol/water 1:1); 90-95% total monoglyceride; *Storage:* 1 yr shelf life under cool, dry conditions below 30 C.

Lamemul® K 3000-Range. [Grünau GmbH] Mono and diglycerides of edible fatty acids (60%), sodium caseinate (30%), maltodextrin (10%); raw material in mfg. of baking additives; improves crumb; delays staling; improves kneading and fermenting tolerance of dough; antistick agent for mfg. of noodles; improves sensory props. of potato-based food

prods.; emulsifier in ice cream mfg.; *Regulatory:* EEC; *Properties:* wh. powd., sl. odor and taste of milk powd.; 70% on 0.2 mm sieve, 80% on 0.5 mm sieve; disp. in cold water; bulk dens. 250-350 g/l; m.p. 68-72 C; acid no. < 3; iodine no. < 2; sapon. no. 155-165; pH 6.5-7.5 (1:10 in methanol/water 1:1); < 5% water; *Storage:* 1 yr shelf life under cool, dry conditions below 30 C.

Lamephos® Range. [Grünau GmbH] Mono- and diglcyerides esterifed with citric acid and phosphates on carrier; emulsifier for meat and sausage prod.; improves fat and jelly separation; *Properties:* spray-dried.

Lamequick® Range. [Grünau GmbH] Fat/emulsifier compds.; whipping base for toppings, desserts, other aerated foods; *Properties:* spray-dried.

Lamequick® AS 340. [Grünau GmbH] Hydrog. coconut and/or palm kernel oil, lactylated mono- and diglycerides, glucose syrup solids and/or maltodextrins, milk proteins; nonionic; whipped topping conc. for acidic dessert systems incl. aerated mousses, cake mousse or fillings, frozen desserts, soft-ice preps., or artisan ice cream; *Regulatory:* FCC, FAO/WHO, EEC compliance; *Properties:* cream to sl. yel. free-flowing powd., neutral milky taste; pH 6.5-7.5 (10% aq.); 40% fat/emulsifier content, 9% protein content; *Storage:* 1 yr shelf life stored under dry, cool conditions (16-24 C).

Lamequick® AS 342. [Grünau GmbH] Hydrog. coconut and/or palm kernel oil, lactylated mono- and diglycerides, glucose syrup and/or maltodextrins, milk proteins; whipped topping conc. for acidic dessert systems incl. aerated mousses, cake mousse or fillings, frozen desserts, soft-ice preps., or artisan ice cream; *Regulatory:* FCC, FAO/WHO, EEC compliance; *Properties:* cream to sl. yel. free-flowing powd., neutral milky taste; pH 6.5-7.5 (10% aq.); 40% fat/emulsifier content, 9% protein content; *Storage:* 1 yr shelf life stored under dry, cool conditions (16-24 C).

Lamequick® AS 370. [Grünau GmbH] Hydrog. coconut and/or palm kernel oil, lactylated mono- and diglycerides, glucose syrup and/or maltodextrins, milk proteins; whipped topping conc. for acidic dessert systems incl. aerated mousses, cake mousse or fillings, frozen desserts, soft-ice preps., or artisan ice cream; *Regulatory:* FCC, FAO/WHO, EEC compliance; *Properties:* cream to sl. yel. free-flowing powd., neutral milky taste; pH 6.5-7.5 (10% aq.); 50% fat/emulsifier content, 10.5% protein content; *Storage:* 1 yr shelf life stored under dry, cool conditions (16-24 C).

Lamequick® AS 400. [Grünau GmbH] Hydrog. coconut and/or palm kernel oil, acetylated mono- and diglycerides, glucose syrup and/or maltodextrins saccharose, milk protein gelatin; whipped topping conc. for acidic dessert systems incl. aerated mousses, cake mousse or fillings, frozen desserts, soft-ice preps., or artisan ice cream; *Regulatory:* FCC, FAO/WHO, EEC compliance; *Properties:* cream to off-wh. free-flowing powd., neutral milky taste; pH 6.5-7.5 (10% aq.); 55% fat/emulsifier content, 6% protein content; *Storage:* 1 yr shelf life stored under dry, cool conditions (16-24 C).

Lamequick® C. [Henkel/Functional Prods.] Emulsifier blend; nonionic; whipped topping conc. based on lactated glycerides; for desserts, dry powd. blends; *Properties:* powd.; 100% conc.

Lamequick® CE 1. [Grünau GmbH] Hydrog. coconut and/or palm kernel oil, acetylated mono- and diglycerides, corn syrup solids and/or maltodextrins sugar, caseinate protein; whipped topping conc. imparting high whippability, foam stiffness and stability, creaminess, and pleasant mouthfeel; for cream chantilly, mousses, frozen desserts, artisan ice cream, cake filling/decorating creams; *Regulatory:* FCC, FAO/WHO, EEC compliance; *Properties:* wh. to off-wh. free-flowing powd., neutral milky taste; pH 6.8-7.8 (10% aq.); 60% fat/emulsifier, 10% protein; *Storage:* 1 yr shelf life stored under dry, cool conditions (16-24 C).

Lamequick® CE 1/63. [Grünau GmbH] Hydrog. coconut and/or palm kernel oil, acetylated mono- and diglycerides, corn syrup solids and/or maltodextrins sugar, caseinate protein; whipped topping conc. imparting high whippability, foam stiffness and stability, creaminess, and pleasant mouthfeel; for cream chantilly, mousses, frozen desserts, artisan ice cream, cake filling/decorating creams; *Regulatory:* FCC, FAO/WHO, EEC compliance; *Properties:* wh. to off-wh. free-flowing powd., neutral milky taste; pH 6.8-7.8 (10% aq.); 60% fat/emulsifier, 10% protein; *Storage:* 1 yr shelf life stored under dry, cool conditions (16-24 C).

Lamequick® CE 1 SF. [Grünau GmbH] Hydrog. coconut and/or palm kernel oil, acetylated mono- and diglycerides, corn syrup solids and/or maltodextrins, caseinate protein; nonionic; whipped topping conc. imparting high whippability, foam stiffness and stability, creaminess, and pleasant mouthfeel; for cream chantilly, mousses, frozen desserts, artisan ice cream, cake filling/decorating creams; *Regulatory:* FCC, FAO/WHO, EEC compliance; *Properties:* wh. to off-wh. free-flowing powd., neutral milky taste; pH 6.8-7.8 (10% aq.); 60% fat/emulsifier, 10% protein; *Storage:* 1 yr shelf life stored under dry, cool conditions (16-24 C).

Lamequick® CE 5557. [Grünau GmbH] Hydrog. coconut and/or palm kernel oil, acetylated mono- and diglycerides, corn syrup solids and/or maltodextrins, caseinate protein; whipped topping conc. imparting high whippability, foam stiffness and stability, creaminess, and pleasant mouthfeel; for cream chantilly, mousses, frozen desserts, artisan ice cream, cake filling/decorating creams; *Regulatory:* FCC, FAO/WHO, EEC compliance; *Properties:* wh. to off-wh. free-flowing powd., neutral milky taste; pH 6.8-7.8 (10% aq.); 60% fat/emulsifier, 10% protein; *Storage:* 1 yr shelf life stored under dry, cool conditions (16-24 C).

Lamequick® ME 1. [Grünau GmbH] Hydrog. coconut and/or palm kernel oil, acetylated mono- and diglycerides, corn syrup solids and/or maltodextrins sugar, skim milk powd.; whipped topping conc. imparting high whippability, foam stiffness and stability, creaminess, and pleasant mouthfeel; for cream chantilly, mousses, frozen desserts, artisan ice cream, cake filling/decorating creams; *Regulatory:* FCC, FAO/WHO, EEC compliance; *Properties:* cream to off-wh. free-flowing powd., neutral milky taste; pH 6.8-7.8 (10% aq.); 60% fat/emulsifier, 9% protein; *Storage:* 1 yr shelf life stored under dry, cool conditions (16-24 C).

Lamequick® ME 1-30. [Grünau GmbH] Hydrog. coconut and/or palm kernel oil, acetylated mono- and diglycerides, corn syrup solids and/or maltodextrins sugar, skim milk powd.; whipped topping conc. imparting high whippability, foam stiffness and stability, creaminess, and pleasant mouthfeel; for cream chantilly, mousses, frozen desserts, artisan ice cream, cake filling/decorating creams; *Regulatory:* FCC, FAO/WHO, EEC compliance; *Properties:* cream to off-wh. free-flowing powd., neutral milky taste; pH 6.8-7.8 (10% aq.); 60% fat/emulsifier, 9% protein; *Storage:* 1 yr shelf life stored under dry, cool conditions (16-24 C).

Lamesorb® Range. [Grünau GmbH] Sorbitan esters of fatty acids/polysorbates; nonionic; food emulsifier for o/w and w/o emulsions for baking.

Lamesorb® SML. [Grünau GmbH] Sorbitan laurate; CAS 1338-39-2; emulsifier for whipped toppings, ice cream, salad dressing, sauces; dough improver in cakes, baked goods; crystallization retarder for confectionery coatings, shortenings; solubilizer for soft drinks; fruit/veg. coating; sugar/yeast defoamer; *Regulatory:* FDA, EEC compliance; *Properties:* yel. liq./paste, neutral odor, sl. bitter taste; HLB 8-10; acid no. 4-8; sapon. no. 155-170; hyd. no. 330-358; < 2% water; *Storage:* 1 yr storage life in protected drums below 30 C.

Lamesorb® SML-20. [Grünau GmbH] PEG-20 sorbitan laurate; CAS 9005-64-5; emulsifier for whipped toppings, ice cream, salad dressing, sauces; dough improver in cakes, baked goods; crystallization retarder for confectionery coatings, shortenings; solubilizer for soft drinks; fruit/veg. coating; sugar/yeast defoamer; *Regulatory:* FDA, EEC compliance; *Properties:* yel. visc. liq., neutral odor, sl. bitter taste; HLB 16-18; acid no. < 2; sapon. no. 40-50; hyd. no. 96-108; < 3% water; *Storage:* 1 yr storage life in protected drums below 30 C.

Lamesorb® SMO. [Grünau GmbH] Sorbitan oleate; CAS 1338-43-8; EINECS 215-665-4; emulsifier for whipped toppings, ice cream, salad dressing, sauces; dough improver in cakes, baked goods; crystallization retarder for confectionery coatings, shortenings; solubilizer for soft drinks; fruit/veg. coating; sugar/yeast defoamer; *Regulatory:* FDA, EEC compliance; *Properties:* yel. liq., neutral odor, sl. bitter taste; HLB 3-5; acid no. 5-8; sapon. no. 145-160; hyd. no. 193-210; < 2% water; *Storage:* 1 yr storage life in protected drums below 30 C.

Lamesorb® SMO-20. [Grünau GmbH] PEG-20 sorbitan oleate; CAS 9005-65-6; emulsifier for whipped toppings, ice cream, salad dressing, sauces; dough improver in cakes, baked goods; crystallization retarder for confectionery coatings, shortenings; solubilizer for soft drinks; fruit/veg. coating; sugar/yeast defoamer; *Regulatory:* FDA 21CFR §172.840, EEC compliance; *Properties:* yel. visc. liq., neutral odor, sl. bitter taste; HLB 14-16; acid no. < 2; sapon. no. 45-55; hyd. no. 65-80; < 3% water; *Storage:* 1 yr storage life in protected drums below 30 C.

Lamesorb® SMP. [Grünau GmbH] Sorbitan palmitate; CAS 26266-57-9; EINECS 247-568-8; emulsifier for whipped toppings, ice cream, salad dressing, sauces; dough improver in cakes, baked goods; crystallization retarder for confectionery coatings, shortenings; solubilizer for soft drinks; fruit/veg. coating; sugar/yeast defoamer; *Regulatory:* FDA, EEC compliance; *Properties:* wh. flakes, neutral odor and taste; HLB 6-8; acid no. 4-7; sapon. no. 140-150; hyd. no. 270-305; < 2% water; *Storage:* 1 yr storage life in protected drums below 30 C.

Lamesorb® SMP-20. [Grünau GmbH] PEG-20 sorbitan palmitate; CAS 9005-66-7; emulsifier for whipped toppings, ice cream, salad dressing, sauces; dough improver in cakes, baked goods; crystallization retarder for confectionery coatings, shortenings; solubilizer for soft drinks; fruit/veg. coating; sugar/yeast defoamer; *Regulatory:* FDA, EEC compliance; *Properties:* yel. gelatinous paste, neutral odor, sl. bitter taste; HLB 15-17; acid no. < 2; sapon. no. 41-52; hyd. no. 90-107; < 3% water; *Storage:* 1 yr storage life in protected drums below 30 C.

Lamesorb® 3MS. [Grünau GmbH] Sorbitan stearate; CAS 1338-41-6; EINECS 215-664-9; emulsifier for whipped topplngs, ice cream, salad dressing, sauces; dough improver in cakes, baked goods; crystallization retarder for confectionery coatings, shortenings; solubilizer for soft drinks; fruit/veg. coating; sugar/yeast defoamer; *Regulatory:* FDA 21CFR §172.842, EEC compliance; *Properties:* wh. flakes, neutral odor and taste; HLB 4-6; acid no. 5-10; sapon. no. 147-157; hyd. no. 235-260; < 2% water; *Storage:* 1 yr storage life in protected drums below 30 C.

Lamesorb® SMS-20. [Grünau GmbH] PEG-20 sorbitan stearate; CAS 9005-67-8; emulsifier for whipped toppings, ice cream, salad dressing, sauces; dough improver in cakes, baked goods; crystallization retarder for confectionery coatings, shortenings; solubilizer for soft drinks; fruit/veg. coating; sugar/yeast defoamer; *Regulatory:* FDA 21CFR §172.836, EEC compliance; *Properties:* yel. gelatinous paste, neutral odor, sl. bitter taste; HLB 14-16; acid no. < 2; sapon. no. 45-55; hyd. no. 81-96; < 3% water; *Storage:* 1 yr storage life in protected drums below 30 C.

Lamesorb® STS. [Grünau GmbH] Sorbitan tristearate; CAS 26658-19-5; EINECS 247-891-4; emulsifier for whipped toppings, ice cream, salad dressing, sauces; dough improver in cakes, baked goods; crystallization retarder for confectionery coatings, shortenings; solubilizer for soft drinks;

fruit/veg. coating; sugar/yeast defoamer; *Regulatory:* FDA, EEC compliance; *Properties:* yel. flakes, neutral odor and taste; HLB 1-3; acid no. 12-15; sapon. no. 176-188; hyd. no. 66-88; < 2% water; *Storage:* 1 yr storage life in protected drums below 30 C.

Lamesorb® STS-20. [Grünau GmbH] PEG-20 sorbitan tristearate; CAS 9005-71-4; emulsifier for whipped toppings, ice cream, salad dressing, sauces; dough improver in cakes, baked goods; crystallization retarder for confectionery coatings, shortenings; solubilizer for soft drinks; fruit/veg. coating; sugar/yeast defoamer; *Regulatory:* FDA 21CFR §172.838, EEC compliance; *Properties:* ivory wax-like solid, neutral odor, sl. bitter taste; HLB 10-12; acid no. < 2; sapon. no. 88-96; hyd. no. 44-60; < 3% water; *Storage:* 1 yr storage life in protected drums below 30 C.

Lamizell. [Kimitsu Chem. Industries; Unipex] Combination of sodium and calcium salts with carboxylic group of alginic acid; special viscosity, fluid curve, and appetizing ability for use in foods; *Properties:* wh. or creamy powd.; 80 mesh pass; visc. 100-4000 cp; pH 5-7; < 15% water.

Lanquell 206, 217. [Harcros UK] Polyglycol-based; low-toxicity antifoam for food industry; *Properties:* pale yel. liq.; faint odor; disp. in water; sp.gr. 0.984 and 1.005 resp.; visc. 180 cs and 355 cs resp.; flash pt. > 150 and > 200 F resp. (Abel CC); pour pt. -28 and -13 C resp.; pH 7 (1% aq.); 92 and 98% act.; Unverified

Lecidan [Grindsted Prods. Denmark] Blends of monoglycerides and special soya lecithin; nonionic; food emulsifier; *Properties:* pills; HLB 5; 100% act.

Lecigran™ 5750. [Riceland Foods] Deoiled lecithin; CAS 8002-43-5; EINECS 232-307-2; emulsifier, dry blending/instantizing agent for foods, dietary supplements; improves shortening, stabilizes dough structure in baked goods; release/dispersing aid in waffles; egg replacer/extender; emulsifies oils (baby food); *Usage level:* 0.2-0.5% (baked goods), 1-2% (waffles), 0.4-0.7% (egg replacer), 0.2-1% (baby food), 0.2-1% (instant powds.); *Regulatory:* FDA 21CFR §184.1400, GRAS; *Properties:* fine powd., superior odor and flavor profile; 97% min. phospholipid; *Storage:* hygroscopic; store in sealed containers below 25 C; protect from light and moisture.

Lecigran™ 6750. [Riceland Foods] Deoiled lecithin; CAS 8002-43-5; EINECS 232-307-2; emulsifier, blending agent for foods, dietary supplements; stabilizer, dispersant for color and flavor suspensions; for dairy prods., cocoa powd., high protein beverages, gravies, instant/prepared foods; *Usage level:* 0.2-0.5% (dairy prods., meat prods.), 0.2-1% (pudding mixes, coffee whiteners, infant formulas, icings), 0.5% (cocoa powd., egg replacers, ice cream); *Properties:* extra fine powd.; 97% min. phospholipid.

Lecigran™ A. [Riceland Foods] Deoiled lecithin with 1% flow agent to increase flowability; CAS 8002-

43-5; EINECS 232-307-2; emulsifier, blending agent for foods, dietary supplements; stabilizer, dispersant for color and flavor suspensions; for dairy prods., cocoa powd., high protein beverages, gravies, instant/prepared foods; *Usage level:* 0.1-0.5% flour basis (bakery mixes), 0.2-0.5% (meat prods.), 0.2-1% (icings, pudding mixes, coffee whiteners, infant formulas), 0.5% (ice cream, egg replacers), 0.75-1% (caramel); *Properties:* lt. tan to med. yel. fine gran., superior odor and flavor profiles; 97% min. phospholipid; *Storage:* hygroscopic; store in sealed containers below 25 C; protect from light and moisture.

Lecigran™ C. [Riceland Foods] Deoiled lecithin with ≤ 2% tricalcium phosphate to enhance flowability; emulsifier, blending agent for foods, dietary supplements; stabilizer, dispersant for color and flavor suspensions; for dairy prods., cocoa powd., high protein beverages, gravies, instant/prepared foods; *Usage level:* 0.1-0.5% flour basis (bakery mixes), 0.2-0.5% (meat prods.), 0.2-1% (icings, pudding mixes, coffee whiteners, infant formulas), 0.5% (ice cream, egg replacers), 0.75-1% (caramel); *Regulatory:* FDA 21CFR §184.1400, GRAS; *Properties:* lt. tan to med. yel. gran., superior odor and flavor profiles; 97% min. phospholipid; *Storage:* hygroscopic; store in sealed containers below 25 C; protect from light and moisture.

Lecigran™ F. [Riceland Foods] Deoiled lecithin; CAS 8002-43-5; EINECS 232-307-2; emulsifier, blending agent for foods, dietary supplements; stabilizer, dispersant for color and flavor suspensions; controls visc. in icings, filings; softener, plasticizer for chewing gum base; increases smoothness in ice cream; *Usage level:* 0.3-0.7% (coating compds.), 0.2-1% (icings, fillings), 0.7-2% (chewing gum base), 0.5% max. (ice cream), 2-4% of fat (pet food); *Regulatory:* FDA 21CFR §184.1400, GRAS; *Properties:* lt. tan to med. yel. fine gran., superior odor and flavor profiles; 97% min. phospholipid; *Storage:* hygroscopic; store in sealed containers below 25 C; protect from light and moisture.

Lecigran™ M. [Riceland Foods] Deoiled lecithin; CAS 8002-43-5; EINECS 232-307-2; emulsifier, blending agent for foods, high quality dietary supplements; stabilizer, dispersant for color and flavor suspensions; for dairy prods., cocoa powd., high protein beverages, gravies, instant/prepared foods; *Usage level:* 0.1-0.5% flour basis (bakery mixes), 0.2-0.5% (meat prods.), 0.2-1% (icings, pudding mixes, coffee whiteners, infant formulas), 0.5% (ice cream, egg replacers), 0.75-1% (caramel); *Regulatory:* FDA 21CFR §184.1400, GRAS; *Properties:* lt. tan to med. yel. free-flowing gran., superior odor and flavor profile; 97% min. phospholipid; *Storage:* hygroscopic; store in sealed containers below 25 C; protect from light and moisture.

Lecigran™ Super A. [Riceland Foods] Deoiled lecithin with calcium silicate as flow agent; emulsifier, blending agent for foods, dietary supplements; specially formulated for direct compression tableting; *Properties:* lt. tan to med. yel. free-flowing fine gran.; 94% min. phospholipid; *Storage:* hygroscopic; store in sealed containers below 25 C; protect from light and moisture.

Lecigran™ T. [Riceland Foods] Deoiled lecithin with tricalcium phosphate to enhance flowability; emulsifier, blending agent for foods, dietary supplements; stabilizer, dispersant for color and flavor suspensions; for dairy prods., cocoa powd., high protein beverages, gravies, instant/prepared foods; *Properties:* powd.; 97% min. phospholipid.

Lecimulthin. [Lucas Meyer] Lecithin; CAS 8002-43-5; EINECS 232-307-2; release agent, separating and emulsifying agent for wafers and dry bakery prods.; *Properties:* powd.; 50% conc.

Leciprime™. [Riceland Foods] Fluid soy lecithin; CAS 8002-43-5; EINECS 232-307-2; emulsifier, fat extender, release agent in baked goods; emulsifier, antispatter aid in margarine; visc. control in chocolate; flavor carrier in chewing gum; emulsifier/stabilizer in salad dressings; lubricant, extrusion aid; *Usage level:* 0.2-0.5% (margarine, chewing gum, salad dressing, batters/breadings), 0.2-1% (chocolate), 0.5-1% (baked goods); *Regulatory:* FDA 21CFR §184.1400, GRAS; *Properties:* fluid; disperses readily in oils and forms hydrates/emulsions in water; dens. 1.03 g/cc; visc. 7000 cps; acid no. 28; 62% acetone insolubles, 37.3% soybean oil; *Storage:* hygroscopic; store in sealed shipping container; may be heated to below 160 F to facilitate handling.

Lecipur™ 95 C. [Lucas Meyer] Deoiled soy lecithin; CAS 8002-43-5; EINECS 232-307-2; for health food industry, nutritional supplements, dietetic supplements; *Properties:* free-flowing gran.

Lecipur™ 95 R. [Lucas Meyer] Deoiled soy lecithin; CAS 8002-43-5; EINECS 232-307-2; for health food industry, nutritional supplements, dietetic supplements, tablets; *Properties:* gran.

Lecithin L 1000-Range. [Grünau GmbH] Native lecithin; CAS 8002-43-5; EINECS 232-307-2; food emulsifier and coemulsifier for baking additives.

Lecithin L 2000-Range. [Grünau GmbH] Lecithin (20%), wheat flour (80%); w/o emulsifier in aq. systems; used in powdery baking additives; improves dough, volume of baked goods, and fermentation tolerance; visc. and yield pt. reducer in chocolates; *Properties:* ylsh.-brn. powd., char. odor, neutral taste; bulk dens. 400-500 g/l; acid no. 20-30; iodine no. 50; *Storage:* 1 yr shelf life under cool, dark, dry storage conditions.

Lecithin L 3000-Range. [Grünau GmbH] Lecithin (40%), maltodextrin (40%), sodium caseinate (20%); w/o emulsifier in aq. systems; used in powdery baking additives; improves dough, volume of baked goods, and fermentation tolerance; visc. and yield pt. reducer in chocolates; *Properties:* ylsh. powd., neutral odor, sl. nutty taste; 70% on 0.2 mm sieve, 100% on 0.5 mm sieve; bulk dens. 200-300 g/l; acid no. 20-30; iodine no. 50; *Storage:* 1 yr shelf life under cool, dark, dry storage conditions.

Lecithin L 4000-Range. [Grünau GmbH] Lecithin on a carrier system; CAS 8002-43-5; EINECS 232-

307-2; food emulsifier and coemulsifier for baking additives.

Lecitreme™. [Kerry Ingreds./Beatreme] Food emulsifiers; *Properties:* free-flowing spray-dried prod.

Lemalta. [ADM Arkady] Conc. formula imparting a rich malt and citrus flavor in cakes and icings; *Properties:* liq.

Lem-N-Fil. [Brolite Prods.] Base for producing a high quality, gourmet lemon filling when mixed with water and boiled; *Properties:* powd.

Lem-O-Fos® 101. [Rhone-Poulenc Food Ingreds.] Sodium tripolyphosphate FCC with natural lemon juice conc.; antioxidant for meat, poultry, seafood; provides improved succulence, flavor, and processing for fresh beef/pork, chunked/formed meats, cooked meat prods., frozen patties, minced fish; improves color retention, sliceability; retards off-flavors; *Regulatory:* kosher; *Properties:* yel.-tan gran., lemon odor; 5% max. on 12 mesh, 30% max. thru 100 mesh; pH 8.4-9.0 (1%); *Storage:* store cool and dry.

Lem-O-Fos® 202. [Rhone-Poulenc Food Ingreds.] Blend of sodium phosphates and natural lemon juice conc.; food additive providing improved flavor, body, and shelf life to prefabricated steaks and meat patties and poultry prods.; improved antioxidant props. help retard formation of off-flavors; *Regulatory:* USDA approved.

Levair®. [Rhone-Poulenc Food Ingreds.] Sodium aluminum phosphate, acidic FCC; CAS 7785-88-8; slow-acting leavening agent for baking powds., prepared cake, cookie, muffin, and pancake mixes, cereals; exc. buttering action; produces baked goods with firm yet tender crumb texture; used with yeast in frozen doughs; *Properties:* wh. free-flowing powd.; < 1% on 60 mesh, 8% on 140 mesh; m.w. 949.9; bulk dens. 50 lb/ft³; neutralizing value 100; pH 2.7 (1% susp.); 95% min. assay.

Levn-Lite®. [Monsanto] Sodium aluminum phosphate with flow conditioners (calcium carbonate, precipitated amorphous silica); functional food additive; *Regulatory:* FCC; *Properties:* anhyd. wh. powd.; 12% max. on 140 mesh.

Lexein® 152D (Dry). [Inolex] Hydrolyzed animal protein; industrial fermenter; *Properties:* cream to tan fluffy powd., mild char. odor, proteinaceous note; pH 6.5-7.5 (10%); 8% max. moisture; *Storage:* store in tightly closed original containers below 32 C; Discontinued

Lexgard® M. [Inolex] Methylparaben USP; CAS 99-76-3; EINECS 202-785-7; preservative for foods, fruit butters, jellies, preserves; *Usage level:* 0.1-1.0%; 0.1% max. in foods; *Regulatory:* FDA 21CFR §150, 172.515, 184.1490, GRAS, Japan, Europe compliance; *Properties:* wh. fine powd.; sol. in ethanol, propylene glycol; sol. 0.25 g/100 g in water; m.p. 125-128 C; 100% conc.; *Toxicology:* nontoxic; *Storage:* store in tightly closed original containers below 38 C away from heat and moisture.

Lexgard® P. [Inolex] Propylparaben USP; CAS 94-13-3; EINECS 202-307-7; preservative for foods, fruit butters, jellies, preserves; *Usage level:* 0.02-1.0%; 0.1% max. in foods; *Regulatory:* FDA 21CFR 150, 172.515, 184.1670, GRAS, Japan, Europe compliance; *Properties:* wh. fine powd.; sol. in ethanol, propylene glycol; sol. 0.05 g/100 g in water; m.p. 95-98 C; 100% conc.; *Storage:* store in tightly closed original containers below 38 C away from heat and moisture.

Lexol® GT-855. [Inolex] Caprylic/capric triglyceride; CAS 65381-09-1; EINECS 265-724-3; solv. for flavor ingreds.; vehicle for vitamins; solubilizer; oxidative stability; *Properties:* APHA 100 max. clear liq., odorless; tasteless; alcohol-sol.; sp.gr. 0.943; visc. 27 cps; f.p. -19 C; acid no. 0.1 max.; iodine no. 1 max.; sapon. no. 325-355; flash pt. (PMCC) 224 C; ref. index 1.4479; 100% conc.

Lexol® GT-865. [Inolex] Caprylic/capric triglyceride; CAS 65381-09-1; EINECS 265-724-3; carrier for flavors; *Properties:* APHA 100 max. clear liq., odorless; sp.gr. 0.947; visc. 25 cps; f.p. -19 C; acid no. 0.1 max.; iodine no. 1 max.; sapon. no. 335-355; flash pt. (PMCC) 233 C; ref. index 1.4471; 100% conc.; Discontinued

Lexol® PG-800. [Inolex] Propylene glycol dioctanoate; CAS 56519-71-2; solv., carrier/vehicle for flavors, fragrance, vitamins; *Properties:* colorless clear liq., char. odor; sol. in alcohol, min. oil, acetone; sp.gr. 0.918; visc. 10 cps; f.p. -34 C; acid no. 1 max.; iodine no. 1 max.; sapon. no. 320-340; flash pt. (COC) 272 C; ref. index 1.4350; 100% conc.

Lexol® PG-855 [Inolex] Propylene glycol dicaprylate/dicaprate; nonionic; carrier/vehicle for flavors, fragrances; *Properties:* APHA 100 max. clear liq., odorless; sp.gr. 0.919; visc. 10 cps; f.p. -38 C; acid no. 0.1 max.; iodine no. 1 max.; sapon. no. 315-335; flash pt. (PMCC) 196 C; ref. index 1.4397; 100% act.

Lexol® PG-865. [Inolex] Propylene glycol dicaprylate/dicaprate; vehicle for flavors, fragrances, vitamins; solubilizer; *Properties:* APHA 100 max. clear liq., odorless; sol. in alcohol, min. and veg. oil, acetone; sp.gr. 0.922; visc. 10 cps; f.p. -38 C; acid no. 0.1 max.; iodine no. 1 max.; sapon. no. 315-335; flash pt. (PMCC) 195 C; ref. index 1.4391; 100% conc.

Lime/Lemon Essence 1574. [MLG Enterprises Ltd.] Blend of essential oils of lemon and lime; natural flavor to produce lemon/lime soft drinks.

Lipal 4-1S. [Aquatec Quimica SA] Polyglyceryl-4 stearate; CAS 37349-34-1; nonionic; emulsifier for food prods.; *Properties:* flakes; 100% conc.

Lipal 70. [Aquatec Quimica SA] Mono/diglycerides blend; nonionic; emulsifier for food prods.; *Properties:* liq.; 100% conc.

Lipal 75. [Aquatec Quimica SA] Sorbitan ester; nonionic; emulsifier for food prods.; *Properties:* liq.; 100% conc.

Lipal GMS. [Aquatec Quimica SA] Glyceryl stearate; CAS 31566-31-1; nonionic; food emulsifier; *Properties:* flakes; 100% conc.

Lipal M. [Aquatec Quimica SA] Polyglyceryl fatty

ester; nonionic; emulsifier for food prods.; *Properties:* paste; 100% conc.

Lipal PGMS. [Aquatec Quimica SA] Propylene glycol stearate; CAS 1323-39-3; EINECS 215-354-3; nonionic; food emulsifier; *Properties:* flakes; 100% conc.

Lipal PNO. [Aquatec Quimica SA] Glyceryl fatty ester; nonionic; emulsifier for food prods.; *Properties:* paste; 100% conc.

Lipase 8 Powd. [Am. Labs] Lipase; CAS 9001-62-1; EINECS 232-619-9; fat-splitting enzyme; for removal of yolk lipids from egg albumen; ingred. for vitamin-mineral mixes; min. 8 USP lipase units/mg; *Properties:* sl. yel. or cream-colored powd., sl. nonoffensive odor; *Storage:* preserve in tight containers in cool, dry place.

Lipase 16 Powd. [Am. Labs] Lipase; CAS 9001-62-1; EINECS 232-619-9; fat-splitting enzyme; for removal of yolk lipids from egg albumen; ingred. for vitamin-mineral mixes; min. 16 USP lipase units/mg; *Properties:* sl. yel. or cream-colored powd., sl. nonoffensive odor; *Storage:* preserve in tight containers in cool, dry place.

Lipase 24 Powd. [Am. Labs] Lipase; CAS 9001-62-1; EINECS 232-619-9; fat-splitting enzyme; for removal of yolk lipids from egg albumen; ingred. for vitamin-mineral mixes; min. 24 USP lipase units/mg; *Properties:* Sl. yel. or cream-colored powd., sl. nonoffensive odor; *Precaution:* preserve in tight containers in cool, dry place; *Storage:* preserve in tight containers in cool, dry place.

Lipase 30 Powd. [Am. Labs] Lipase; CAS 9001-62-1; EINECS 232-619-9; fat-splitting enzyme; for removal of yolk lipids from egg albumen; ingred. for vitamin-mineral mixes; min. 30 USP lipase units/mg; *Properties:* sl. yel. or cream-colored powd., sl. nonoffensive odor; *Storage:* preserve in tight containers in cool, dry place.

Lipase AK. [Mitsubishi Int'l.; Amano Enzyme USA] Lipase from *Pseudomonas*; CAS 9001-62-1; EINECS 232-619-9; food enzyme; *Properties:* powd.

Lipase AP. [Mitsubishi Int'l.; Amano Enzyme USA] Lipase from *Aspergillus niger*; CAS 9001-62-1; EINECS 232-619-9; food enzyme; *Properties:* powd.

Lipase AP 6. [Mitsubishi Int'l.; Amano Enzyme USA] Acid-stable lipase from *Aspergillus niger*; CAS 9001-62-1; EINECS 232-619-9; digestive enzyme; *Properties:* powd.

Lipase AY. [Mitsubishi Int'l.; Amano Enzyme USA] Lipase from *Candida*; CAS 9001-62-1; EINECS 232-619-9; food enzyme; *Properties:* powd.

Lipase FAP. [Mitsubishi Int'l.; Amano Enzyme USA] Lipase from *Rhizopus*; CAS 9001-62-1; EINECS 232-619-9; food enzyme; *Properties:* powd.

Lipase G. [Mitsubishi Int'l.; Amano Enzyme USA] Lipase from *Penicillium*; CAS 9001-62-1; EINECS 232-619-9; food enzyme; *Properties:* powd.

Lipase GC. [Mitsubishi Int'l.; Amano Enzyme USA] Lipase from *Geotricum candidum*; CAS 9001-62-1; EINECS 232-619-9; food enzyme; *Properties:* powd.

Lipase MAP. [Mitsubishi Int'l.; Amano Enzyme USA] Lipase from *Mucor*; CAS 9001-62-1; EINECS 232-619-9; food enzyme; *Properties:* powd.

Lipase N. [Mitsubishi Int'l.; Amano Enzyme USA] Lipase from *Rhizopus niveus*; CAS 9001-62-1; EINECS 232-619-9; food enzyme; *Properties:* powd.

Lipase PS. [Mitsubishi Int'l.; Amano Enzyme USA] Lipase from *Pseudomonas cepacia*; CAS 9001-62-1; EINECS 232-619-9; food enzyme; *Properties:* powd.

Lipo GMS. [Lipo] Glyceryl stearate; nonionic; emulsifier and defoamer for food industry; *Properties:* solid; HLB 3.7; 100% conc.; Unverified

Lipo GMS 410. [Lipo] Glyceryl stearate; nonionic; emulsifier for whipped toppings, ice cream, shortening, bread, rolls, etc.; defoamer for food processing; *Properties:* solid; HLB 3.7; 100% conc.

Lipo GMS 450. [Lipo] Glyceryl stearate; nonionic; food emulsifier, emulsion stabilizer for frozen desserts, dairy systems, whipped toppings, confections, chewing gum, baked goods, sweet doughs, and peanut butter; *Regulatory:* FDA 21CFR §182.4505; kosher grade avail.; *Properties:* wh. bead or flake; HLB 3.6 ± 1; acid no. 5 max.; sapon. no. 165-182; 100% act.

Lipo GMS 470. [Lipo] Glyceryl stearate SE; nonionic; food emulsifier, emulsion stabilizer for frozen desserts, dairy systems, whipped toppings, confections, chewing gum, baked goods, sweet doughs, and peanut butter; *Regulatory:* FDA 21CFR §182.4505; kosher grade avail.; *Properties:* wh. bead or flake; HLB 5.8 ± 1; acid no. 5 max.; sapon. no. 138-152; 100% act.

Lipo SS. [Lipo] Hydrog. veg. oil; CAS 68334-28-1; EINECS 269-820-6; emollient for food applics.; *Properties:* wh. to tan waxy flake; acid no. 0.1 max.; iodine no. 5 max.; sapon. no. 230-250.

Lipodan CDS Kosher. [Grindsted Prods.; Grindsted Prods. Denmark] Hydrog. cottonseed oil/dist. monoglycerides blend; nonionic; stabilizer for oils in peanut butter and fat-based systems; *Usage level:* 1-2%; *Regulatory:* FDA 21CFR §164.150, 184.1505, GRAS; FCC, kosher; *Properties:* beads; drop pt. 62 C; HLB 3.8; iodine no. 5 max.; 22% min. monoester; *Storage:* store in cool, dry area.

Lipodan CRE Kosher. [Grindsted Prods.] Hydrog. cottonseed/rapeseed oils and dist. monoglycerides blend; stabilizer for peanut butter and other oil-based systems; *Usage level:* 1-2%; *Regulatory:* FDA 21CFR §164.150, 184.1505, GRAS, FCC, kosher; *Properties:* beads; drop pt. 62 C; iodine no. 5 max.; *Storage:* store in cool, dry area.

Lipodan OM 30. [Grindsted Prods.] Food emulsifier; stabilizer for peanut butter and nut spreads; controls fat agglomeration in sour cream, dairy creams, nondairy creams; improves whippability; avail. in kosher grade; *Properties:* beads; drop pt. 60 C; iodine no. 25-30; 72% min. monoester.

Lipodan SET Kosher. [Grindsted Prods.] Hydrogenated veg. oils (rapeseed, cottonseed, soybean);

EINECS 269-820-6; stabilizer for peanut butter and other oil-based systems; *Usage level:* 1-2%; *Regulatory:* FDA 21CFR §164.150, FCC, kosher; *Properties:* beads; drop pt. 63 C; iodine no. 5 max.; *Storage:* store in cool, dry area.

Liponate GC. [Lipo] Caprylic/capric triglyceride; CAS 65381-09-1; EINECS 265-724-3; clouding agent, diluent for vitamins, solv. for oil-disp. materials, useful in flavor systems and dietary formulations; *Regulatory:* FDA 21CFR §170.30; kosher grade avail.; *Properties:* colorless liq., bland odor and taste; sol. in anhyd. alcohol, min. and veg. oils; insol. in water; acid no. 0.1 max.; sapon. no. 325-355.

Liponate PC. [Lipo] Propylene glycol dicaprylate/dicaprate; clouding agent, diluent for vitamins, solv. for oil-disp. materials, useful in flavor systems and dietary formulations; *Regulatory:* FDA 21CFR §172.860, 172.856; kosher grade avail.; *Properties:* colorless liq., bland odor and taste; sol. in min. and veg. oils, ethanol; water-insol.; acid no. 0.1 max.; sapon. no. 315-335.

Liponic Sorbitol Powd. [Lipo] Sorbitol; CAS 50-70-4; EINECS 200-061-5; humectant, stabilizes moisture, improves texture, intensifies colors, helps prevent crystallization; for aerated confections, processed meat; sweetener replacement in fruit jams for diabetics; carrier for liq. flavors; softener in chewing gum; *Regulatory:* FDA 21CFR §121.101; kosher grade avail.

Liponic Sorbitol Sol'n. 70% USP. [Lipo] Sorbitol; CAS 50-70-4; EINECS 200-061-5; humectant, stabilizes moisture, improves texture, intensifies colors, helps prevent crystallization; for aerated confections, processed meat; sweetener replacement in fruit jams for diabetics; carrier for liq. flavors; softener in chewing gum; *Regulatory:* FDA 21CFR §121.101; kosher grade avail.

Liposorb L-20. [Lipo] Polysorbate 20; CAS 9005-64-5; nonionic; o/w emulsifier, lubricant for foods (flavor systems, salad dressings, pickles, coffee whiteners, whipped toppings, frozen desserts, baked goods, icings, fillings, and confections); *Regulatory:* FDA 21CFR §172.515; kosher grade avail.; *Properties:* yel. liq.; HLB 16.7 ± 1; sapon. no. 40-50; hyd. no. 96-108; 100% act.

Liposorb O-20. [Lipo] Polysorbate 80; CAS 9005-65-6; nonionic; surfactant for food processing; flavor and color dispersant for pickles; defoamer for beet sugar, yeast processing; wetting agent for poultry defeathering; crystal control agent for salt; *Regulatory:* FDA 21CFR §172.840; kosher grade avail.; *Properties:* yel. liq.; HLB 15.0 ± 1; sapon. no. 45-55; hyd. no. 65-80; 100% act.

Liposorb S. [Lipo] Sorbitan stearate; CAS 1338-41-6; EINECS 215-664-9; nonionic; emulsifier, thickener, lubricant, antistat, all-purpose lipophilic surfactant used with POE Liposorb series for foods (coffee whiteners, whipped toppings, frozen desserts, baked goods, icings, fillings, and confectionery coatings),; *Regulatory:* FDA 21CFR §172.842; kosher grade avail.; *Properties:* cream beads or flakes; HLB 4.7 ± 1; sapon. no. 147-157; hyd. no. 235-260; 100% act.

Liposorb S-20. [Lipo] Polysorbate 60; CAS 9005-67-8; nonionic; food emulsifier, defoamer for flavor systems, salad dressings, pickles, coffee whiteners, whipped toppings, frozen desserts, baked goods, icings, fillings, and confections; *Regulatory:* FDA 21CFR §172.836; kosher grade avail.; *Properties:* yel. paste; HLB 14.9 ± 1; sapon. no. 45-55; hyd. no. 81-96; 100% act.

Liposorb TS. [Lipo] Sorbitan tristearate; CAS 26658-19-5; EINECS 247-891-4; nonionic; emulsifier, thickener for coffee whiteners, whipped toppings, frozen desserts, baked goods, icings, fillings, and confectionery coatings; *Regulatory:* FDA 21CFR §172.842; kosher grade avail.; *Properties:* cream flakes or beads; HLB 2.1 ± 1; sapon. no. 175-190; hyd. no. 65-80; 100% act.

Liposorb TS-20. [Lipo] Polysorbate 65; CAS 9005-71-4; nonionic; food emulsifier, defoamer for flavor systems, salad dressings, pickles, coffee whiteners, whipped toppings, frozen desserts, baked goods, icings, fillings, and confections; *Regulatory:* FDA 21CFR §172.838; kosher grade avail.; *Properties:* tan solid wax; HLB 10.5 ± 1; sapon. no. 88-98; hyd. no. 44-60; 100% act.

Lipovol A. [Lipo] Avocado oil; CAS 8024-32-6; EINECS 232-428-0; emollient for food applics.; *Properties:* yel. to green visc. oil, bland char. odor; sol. in oils; sp.gr. 0.908-0.925; acid no. 3 max.; iodine no. 65-95; sapon. no. 177-198; ref. index 1.460-1.470.

Lipovol ALM. [Lipo] Sweet almond oil; CAS 8007-69-0; emollient for food applics.; *Properties:* pale yel. clear oily liq., bland, odorless; sol. in min. oil, isopropyl esters, ether, chloroform, benzene, and solv. hexane; water-insol.; acid no. 2 max.; iodine no. 95-115; sapon. no. 185-200.

Lipovol HS. [Lipo] Hydrog. soybean oil; CAS 8016-70-4; EINECS 232-410-2; emollient, lubricant for food applics.; *Properties:* yel. oil; acid no. 0.5 max.; iodine no. 101-114; sapon. no. 186-197.

Lipovol P. [Lipo] Apricot kernel oil; CAS 72869-69-3; emollient for food applics.; *Properties:* straw oily liq., bland char. fatty odor; sol. in min. oil and isopropyl esters; insol. in water; acid no. 1 max.; iodine no. 90-115; sapon. no. 185-195.

Lipovol PAL. [Lipo] Palm oil; CAS 8002-75-3; EINECS 232-316-1; emollient, lubricant for food applics.; *Properties:* off-wh. paste; acid no. 1 max.; iodine no. 44-59; sapon. no. 195-205.

Lipovol SAF. [Lipo] Safflower oil; CAS 8001-23-8; EINECS 232-276-5; emollient for food applics.; *Properties:* yel. oil; acid no. 2 max.; iodine no. 135-155; sapon. no. 182-202.

Lipovol SES. [Lipo] Sesame oil; CAS 8008-74-0; EINECS 232-370-6; emollient, solv., and vehicle used in foods; *Properties:* yel. clear liq.; bland char. odor; sol. in isopropyl esters and min. oil; insol. in water; acid no. 0.2 max.; iodine no. 103-116; sapon. no. 188-195.

Lipovol SOY. [Lipo] Soybean oil; CAS 8001-22-7;

EINECS 232-274-4; emollient, lubricant for foods; *Properties:* yel. oil; acid no. 1 max.; iodine no. 120-145; sapon. no. 180-200.

Lipovol SUN. [Lipo] Sunflower seed oil; CAS 8001-21-6; EINECS 232-273-9; emollient; used in preparation of margarine; *Properties:* lt. yel. oily liq., bland, char. fatty odor; sol. in min. oil and isopropyl esters; insol. in water; acid no. 2 max.; iodine no. 120-140; sapon. no. 185-195; 0.05% max. moisture.

Lipovol WGO. [Lipo] Wheat germ oil; CAS 8006-95-9; emollient for food applics.; *Properties:* yel./brn. oil; char. fatty odor; sol. in oils; insol. in water; acid no. 5 max.; iodine no. 120-140; sapon. no. 175-195.

Liquid Canola Oil 81-071-0. [ADM Refined Oils] Canola oil, low sat. fat, refined, bleached, deodorized; oil for salad dressings, mayonnaise, frying, baking, cooking, processed foods; *Properties:* clear brilliant oil, bland flavor; iodine no. 116 ± 3.

Liquid Corn Oil 87-070-0. [ADM Refined Oils] High linoleic corn oil; CAS 8001-30-7; EINECS 232-281-2; refined veg. oil for salad dressings, frying, baking, cooking; *Properties:* clear brilliant oil, pleasing flavor; iodine no. 120 min.

Liquid Fish Gelatin Conc. [Croda Colloids Ltd; O.C. Lugo] Gelatin from fish sources; CAS 9000-70-8; EINECS 232-554-6; emulsion stabilizer, protective colloid, film-former, adhesive agent, flocculant; controls crystal growth; inferior gelling chars. to mammalian gelatin; *Properties:* pale amber clear visc. sol'n.; cold water sol. when dried as a film; visc. 60-85 mps; pH 4-6; 25% solids.

Liquid Vitamin D_3 No. 63643. [Roche] Cholecalciferol with edible corn oil; used in foods, oil dispersions, encapsulated prods., food fortification; *Properties:* yel. clear liq., sl. char. odor, bland fatty taste; misc. with edible oils/fats; sol. in ether, hydrocarbons, chlorinated hydrocarbons; sl. sol. in alcohol; insol. in water; m.w. 384.64; sp.gr. 0.9-0.95; visc. 50-90 cps; 100,000 min. IU vitamin D/g; *Toxicology:* Vitamin D: potentially toxic esp. for young children; excessive ingestion may cause hypercalcemia, hypercalcuria; *Storage:* store @ 70 F; refrigerate if stored for several mos, then hold @ 24 h before opening; close tightly.

Liquinat®. [Jungbunzlauer Int'l. AG] Citric acid; CAS 77-92-9; EINECS 201-069-1; acidulant and pH control agent for food industry for soft drinks, preserved fruit, baking materials, margarine, fats, oils, feeds; liq. form permits easier handling and processing; *Properties:* liq.

Lite Pack 350. [Custom Ingreds.] Modified food starch, mono- and diglycerides, cellulose, xanthan gum, guar gum, wheat gluten, lecithin, sucrose fatty acid esters, salt, L-cysteine; blend of emulsified, modified food starch, gums, fiber, and lecithin for use as shortening replacer for flour tortillas; can replace 50-100% of shortening in flour tortilla formula; *Properties:* free-flowing powd.

Litesse®. [Pfizer Food Science] Polydextrose FCC; CAS 68434-04-4; bulking agent, texturizer, formulation aid, humectant for foods; replaces functional-

ity lost when sugar is replaced by high-intensity sweeteners in lite prods., frozen desserts, confections, baked goods, chewing gum; enhances mouthfeel, lowers f.p.; *Regulatory:* FDA 21CFR §172.841; EEC, Japan, JECFA approvals; *Properties:* wh. to cream-colored powd., odorless, clean neutral taste; very water-sol.; sparingly to insol. in most org. solvs.; pH 2.5-3.5 (10% aq.); 90% min. assay, 1 calorie/g; *Toxicology:* LD50 (oral, mouse) 30 g/kg; may be mild eye irritant; sensitive individuals may experience laxative effect from excessive consumption; *Precaution:* incompat. with strong oxidizing agents, strong acids; *Storage:* store powd. prod. in well-sealed containers at or below R.T. at low humidity.

Litesse® II. [Pfizer Food Science] Polydextrose FCC; CAS 68434-04-4; bulking agent for foods; replaces functionality lost when sugar is replaced by high-intensity sweeteners in lite prods., frozen desserts, confections; provides smooth texture and meltdown, enhances mouthfeel, lowers f.p.; *Regulatory:* FDA 21CFR §172.841; *Properties:* wh. to lt. cream powd., odorless, bland taste; very sol. in water; sparingly to insol. in most org. solvs.; pH 2.5-3.5 (10%); 90% min. assay; *Toxicology:* LD50 (oral, mouse) 30 g/kg; may be mild eye irritant; sensitive individuals may experience laxative effect from excessive consumption; *Precaution:* incompat. with strong oxidizing agents, strong acids; *Storage:* store in tight containers @ or below R.T. under low humidity conditions.

Liver Conc. Paste. [Am. Labs] Dried liver extract; nutrittive food and pharmaceutical additive; *Properties:* lt. tan to dk. brn. powd. or paste, char. odor and taste; hygroscopic; partly sol. in water and alcohol; *Storage:* store in tight containers in cool, dry place.

Lobra. [Karlshamns] Partially hydrog. canola oil; *Properties:* m.p. 66-72 F; iodine no. 88-93; sapon. no. 178-195.

Lobra 70. [Karlshamns Food Ingreds.] Partially hydrog. canola oil; avail. with antioxidant on request; specialty oil for general frying, spray oil use; *Properties:* liq.; m.p. 70 F.

Lobra RBD. [Karlshamns Food Ingreds.] Canola oil; specialty oil for salad dressing, light duty cooking applics.; *Properties:* liq.

Locust Bean Gum Speckless Type D-200. [Meer] Locust bean gum; CAS 9000-40-2; EINECS 232-541-5; thickener, water binder, suspending agent, stabilizer for soft cheese and spreads, bread, cakes, biscuits, frozen pie fillings, sausages, stuffed meat prods., canned pet foods; freeze/thaw stabilizer in ice cream; *Properties:* wh. to cream-wh. powd., nearly odorless; 97% thru 100 mesh, $\geq$ 25% thru 200 mesh; visc. $\geq$ 2800 cps (1%); pH 5.0-6.4.

Lo-Dex® 10. [Am. Maize Prods.] Enzyme converted maltodextrin; CAS 9050-36-6; EINECS 232-940-4; bulk agent with exc. film-forming props.; spray drying/encapsulation aid for flavor oils; for beverage mixes, soup, sauce, and gravy mixes, icings, salad dressings, spice blends, specialty confections; *Regulatory:* FDA 21CFR §184.1444, GRAS;

Properties: waxy, very bland flavor; 100% thru 40 mesh; sol. in water; DE 11; bulk dens. 29 lb/ft³; pH 4.7; 5% moisture; *Storage:* store in dry, cool conditions, in odor-free area, ≤ 50% r.h. and ≤ 25 C.

Lonzest® SMO-20. [Lonza] Polysorbate 80; CAS 9005-65-6; nonionic; emulsifier, solubilizer and stabilizer used in foods; *Properties:* liq.; HLB 15.0; 100% conc.

Lonzest® SMP-20. [Lonza] Polysorbate 40; CAS 9005-66-7; nonionic; emulsifier, solubilizer and stabilizer used in foods; *Properties:* liq.; HLB 16.0; 100% conc.

Lonzest® SMS-20. [Lonza] Polysorbate 60; CAS 9005-67-8; nonionic; emulsifier, solubilizer and stabilizer used in foods; *Properties:* liq.; HLB 15.0; 100% conc.

Lonzest® STO-20. [Lonza] Polysorbate 85; CAS 9005-70-3; nonionic; emulsifier, solubilizer and stabilizer used in foods; *Properties:* liq.; HLB 11.0; 100% conc.

Lonzest® STS-20. [Lonza] Polysorbate 65; CAS 9005-71-4; nonionic; emulsifier, solubilizer and stabilizer used in foods; *Properties:* solid; HLB 11.0; 100% conc.

Lo-Temp® 452. [A.E. Staley Mfg.] Modified food starch derived from waxy corn; specialized starch for low temp. processing in neutral and moderately acidic food systems, esp. microwave foods; for sauces, gravies; improves moisture and texture in emulsified meats; good freeze/thaw stability; *Regulatory:* FDA 21CFR §172.892; *Properties:* wh. powd., good odor/flavor; 88% thru 60 mesh; pH 6.0-7.0 (5%), 5% moisture.

Lo-Temp® 588. [A.E. Staley Mfg.] Modified food starch derived from waxy corn; specialized starch for frozen pies, fruit fillings; acid-stable; *Regulatory:* FDA 21CFR §172.892.

Loving KL. [Kao Corp. SA] Sucrose ester and food additives; nonionic; cleaning compd. for the food industry; *Properties:* liq.; Unverified

Loving P. [Kao Corp. SA] Sucrose ester and food additives; nonionic; peeling agent for food industry; *Properties:* powd.; Unverified

Lowinox® BHA. [Lowi] BHA; antioxidant for foodstuffs.

Lowinox® WSP. [Great Lakes] CAS 77-62-3; antioxidant for food can adhesives; *Properties:* wh. to cream-colored powd.; 1.5 % max. > 250 µ; m.p. 132 C min.

L.S.B. Stabl. [Brolite Prods.] Conc. icing stabilizer designed only to be used in white icings.

Lucantin® CX. [BASF AG] Citranaxanthine; for pigmentation of egg yolks.

Lucantin® Red. [BASF AG] Canthaxanthine; CAS 514-78-3; EINECS 208-187-2; for pigmentation of egg yolks, skin and legs of broilers.

Lucantin® Yellow. [BASF AG] β-Apo-8´carotenal; CAS 1107-26-2; EINECS 214-171-6; for pigmentation of egg yolks, and skin and legs of broilers.

Lung Substance. [Am. Labs] Vacuum-dried defatted glandular prod.; nutritive food additive; *Properties:* powd.

Lutavit® Calpan. [BASF AG] Calcium D-pantothenate; CAS 137-08-6; EINECS 205-278-9; for food applics.; Discontinued

Lutavit® Niacin. [BASF AG] Nicotinic acid; CAS 59-67-6; EINECS 200-441-0; vitamin reinforcement for foods; Discontinued

Luwax ES 9668. [BASF] Isotactic polypropylene wax with high crystallinity; high hardness wax for coating of fruits; *Properties:* visc. 800-1200 mm²/s (200 C); m.p. 159-166 C; acid no. 0.

Luwax OA 5. [BASF] Oxidized polyethylene wax; CAS 68441-17-8; high hardness wax for coating of fruits; *Properties:* gran. or powd.; dens. 0.97 g/cc; visc. 410 mm²/s (120 C); m.p. 104 C; drop pt. 110 C; acid no. 16; penetration hardness < 1 dmm.

Luxor® 1517. [Champlain Industries] Hydrolyzed veg. protein with salt, MSG, and partially hydrog. veg. oil; premium flavor and flavor enhancer; mild chicken flavor, moderate mouthfeel; background flavor for poultry and seafood dishes; *Properties:* lt. yel. powd.; pH 4.0; 96% total solids; 53% flavor solids; 11% MSG.

Luxor® 1576. [Champlain Industries] Hydrolyzed veg. protein with salt, caramel color, disodium guanylate, and disodium inosinate; rich, meaty, beeflike flavor with sl. astringent note; for use in soups, meat prods., bouillon pastes, gravies, sauces; *Properties:* brn. powd.; pH 5.3; 97% total solids; 52% flavor solids.

Luxor® 1626. [Champlain Industries] Hydrolyzed veg. protein with salt, caramel color, corn syrup, disodium guanylate, and disodium inosinate; mild beef flavor for use in gravies, soups, stews; *Properties:* dk. brn. powd.; pH 5.2; 96% total solids; 55% flavor solids.

Luxor® 1639. [Champlain Industries] Hydrolyzed veg. protein with salt, partially hydrog. veg. oil, citric acid, disodium guanylate, disodium inosinate, and oleoresin celery; celery char.; high degree of flavor enhancement; used in soups and gravies for tangy and spicy flavor; *Properties:* lt. tan powd.; pH 6.1; 96% total solids; 37% flavor solids.

Luxor® 1658. [Champlain Industries] Hydrolyzed veg. protein with salt and partially hydrog. veg. oil; mild premium flavor, similar to Luxor 1517, but with no added MSG; *Properties:* lt. yel. powd.; pH 4.8; 96% total solids; 53% flavor solids.

Luxor® Century™ V. [Champlain Industries] Hydrolyzed vegetable protein; CAS 100209-45-8; pleasant mild meaty flavor with high MSG content; *Properties:* powd.

Luxor® CVP 5-SD. [Champlain Industries] Hydrolyzed vegetable protein; CAS 100209-45-8; strong beef broth flavor for all-purpose use; *Properties:* powd.

Luxor® CVP 1632. [Champlain Industries] Hydrolyzed vegetable protein; CAS 100209-45-8; beef type flavor with fermented char. for Oriental-type prods.; *Properties:* liq.

Luxor® CVP 1689. [Champlain Industries] Hydrolyzed vegetable protein; CAS 100209-45-8; mild flavor similar to CVP 5-SD but with lighter color and milder flavor; *Properties:* powd.

Luxor® CVP 1753. [Champlain Industries] Hydrolyzed vegetable protein; CAS 100209-45-8; mild flavor similar to HVP-A, but with lighter color and milder flavor; *Properties:* powd.

Luxor® CVP LS. [Champlain Industries] Hydrolyzed vegetable protein; CAS 100209-45-8; flavor with < 2% sodium for dietary foods; *Properties:* powd.

Luxor® E-40. [Champlain Industries] Hydrolyzed veg. protein with partially hydrog. soybean oil; flavor used as an MSG extender; beef background taste; *Properties:* lt. tan powd.; pH 5.6; 97% total solids; 54% flavor solids.

Luxor® E-50. [Champlain Industries] Hydrolyzed veg. protein with partially hydrog. soybean oil; blander flavor than E-40; MSG extender; contributes mild seafood char. similar to sweet clams; *Properties:* lt. tan powd.; pH 5.7; 97% total solids; 53% flavor solids.

Luxor® E-610. [Champlain Industries] Hydrolyzed vegetable protein; CAS 100209-45-8; beef base for table and cooking sauces; *Properties:* liq.

Luxor® EB-2. [Champlain Industries] Hydrolyzed veg. protein with MSG, caramel color, and partially hydrog. veg. oil; rich beef-like premium flavor; used in canned, frozen, and dehydrated foods requiring hearty meat flavor; *Properties:* tan powd.; pH 5.2; 96% total solids; 57% flavor solids; 12% MSG.

Luxor® EB-400. [Champlain Industries] Hydrolyzed veg. protein with MSG, partially hydrog. veg. oil, disodium guanylate, and disodium inosinate; rich beef premium flavor; light-colored, high in flavor enhancement; *Properties:* lt. tan powd.; pH 5.1; 96% total solids; 55% flavor solids; 12% MSG.

Luxor® FB-10. [Champlain Industries] Hydrolyzed vegetable protein; CAS 100209-45-8; flavor with well-balanced roast-beef flavor, high in aromatics; *Properties:* powd.

Luxor® FC-20. [Champlain Industries] Hydrolyzed vegetable protein; CAS 100209-45-8; flavor with well-balanced roast-chicken flavor, high in aromatics; *Properties:* powd.

Luxor® GR-100. [Champlain Industries] Hydrolyzed veg. protein; CAS 100209-45-8; flavor providing distinctive roasted char. to food applic.; *Properties:* dk. brn. powd.; pH 6.4; 97% total solids; 52% flavor solids.

Luxor® GR-150. [Champlain Industries] Hydrolyzed veg. protein; CAS 100209-45-8; flavor providing moderately browned flavor profile where browned notes are required; when less roasted char. is desired; *Properties:* golden brn. powd.; pH 5.6; 97% total solids; 54% flavor solids.

Luxor® GR-200. [Champlain Industries] Hydrolyzed veg. protein; CAS 100209-45-8; strongest browned flavor used when roasted note is desired; *Properties:* dk. brn. powd.; pH 5.5; 97% total solids; 53% flavor solids.

Luxor® HVP-A. [Champlain Industries] Hydrolyzed vegetable protein; CAS 100209-45-8; milder beef broth flavor; < 0.5% ammonium chloride; *Properties:* powd.

Luxor® KB-300. [Champlain Industries] Hydrolyzed vegetable protein with caramel color, partially hydrog. soybean oils; premium flavor with well-balanced beef-like taste and mild aroma; for soups, meat prods., bouillon pastes, gravies, sauces; *Properties:* brn. powd.; pH 5.2; 97% total solids; 56% flavor solids.

Luxor® KB-312. [Champlain Industries] Hydrolyzed vegetable protein; CAS 100209-45-8; premium flavor similar to KB-300 but without added caramel and with low ammonium chloride; suitable for highly spiced prods.; *Properties:* powd.

Luxor® KB-320. [Champlain Industries] Hydrolyzed veg. protein with caramel color, yeast, soy flour, and partially hydrog. soybean oil; premium flavor for use where meatiness and fully body are required, e.g., in thick soups and sauces; *Properties:* brn. powd.; pH 5.6; 97% total solids; 59% flavor solids.

Luxor® KB-330. [Champlain Industries] Hydrolyzed veg. protein with caramel color and partially hydrog. soybean oil; beef broth flavor for background of spice blends and food applic.; more economical version of Luxor KB-300; *Properties:* brn. powd.; pH 5.3; 96% total solids; 57% flavor solids.

Luxor® KB-350. [Champlain Industries] Hydrolyzed veg. protein with MSG, caramel color, and partially hydrog. soybean oil; premium beef-like flavor with high flavor enhancement; *Properties:* brn. powd.; pH 5.3; 97% total solids; 59% flavor solids; 12% MSG.

Luxor® KB-400. [Champlain Industries] Hydrolyzed veg. protein with partially hydrog. soybean oil; premium mild flavor with aroma of lightly roasted meat; high mouthfeel; for delicately flavored foods such as cream soups, enhancement of prods. based on pork, veal, poultry, or fish; *Properties:* tan powd.; pH 5.7; 97% total solids; 53% flavor solids.

Luxor® KB-500. [Champlain Industries] Hydrolyzed veg. protein with partially hydrog. soybean oil and caramel color; mild premium flavor with well-balanced chicken-like taste and mild aroma; *Properties:* lt. tan powd.; pH 4.8; 97% total solids; 53% flavor solids.

Luxor® KB-530. [Champlain Industries] Hydrolyzed veg. protein with partially hydrog. soybean oil; mild meaty flavor for middle-range meat types such as pork and veal; richer version of KB-500; *Properties:* lt. tan powd.; pH 5.1; 96% total solids; 56% flavor solids.

Luxor® KB-600. [Champlain Industries] Hydrolyzed veg. protein with partially hydrog. soybean oil; premium flavor with very mild seafood char.; *Properties:* tan powd.; pH 6.0; 97% total solids; 53% flavor solids.

Luxor® L-625. [Champlain Industries] Hydrolyzed vegetable protein; CAS 100209-45-8; mild flavor for pumping cures and mild meats; *Properties:* liq.

Luxor® MB-40. [Champlain Industries] Hydrolyzed veg. protein; CAS 100209-45-8; flavor used in petfood industry to improve palatability of animal foods; less roasted meat flavor than MB-110; *Prop-*

erties: brn. powd.; 97% total solids; 53% flavor solids.

Luxor® MB-110. [Champlain Industries] Hydrolyzed veg. protein; CAS 100209-45-8; flavor used in petfood industry to improve palatability of animal foods; *Properties:* brn. powd.; 97% total solids; 53% flavor solids.

Luxor® MB-120. [Champlain Industries] Hydrolyzed veg. protein; CAS 100209-45-8; flavor used in petfood industry to improve palatability of animal foods; milder version of MB-110; *Properties:* brn. powd.; 97% total solids; 53% flavor solids.

Luxor® No. 6 Sauce. [Champlain Industries] Hydrolyzed vegetable protein; CAS 100209-45-8; ready-to-use beef table sauce, primarily for Oriental-type dishes; *Properties:* liq.

Luxor® No. 700 Sauce. [Champlain Industries] Hydrolyzed vegetable protein; CAS 100209-45-8; beef type for use in conc. table and cooking sauces; *Properties:* liq.

Luxor® Prozate®. [Champlain Industries] Hydrolyzed vegetable protein; CAS 100209-45-8; mild flavor for use in canned foods; *Properties:* liq.

Luxor® R-100. [Champlain Industries] Hydrolyzed veg. protein with thiamine hydrochloride and partially hydrog. soybean oils; flavor providing stewed/cooked meaty aromatics to all types of meat or meat systems; *Properties:* lt. tan powd.; pH 5.0; 96% total solids; 54% flavor solids.

Luxor® Triple-H®. [Champlain Industries] Hydrolyzed vegetable protein; CAS 100209-45-8; mild flavor used in seasoning blends for meat prods. such as sausage, corned beef, other processed meats; *Properties:* powd.

Lymphatic Substance. [Am. Labs] Vacuum-dried defatted glandular prod.; nutritive food additive; *Properties:* powd. or freeze-dried form.

M

Macol® 5100. [PPG/Specialty Chem.] PPG-33 buteth-45; CAS 9038-95-3; nonionic; food processing; *Regulatory:* FDA compliance; *Properties:* APHA 100 max. clear visc. liq.; sol. in acetone, propylene glycol, oleic acid, castor oil, water, toluene, IPA; sp.gr. 1.050; dens. 8.75 lb/gal; visc. 1100 cst (100 F); acid no. 0-1; pour pt. -28 C; cloud pt. 55 C (1% aq.); flash pt. 430 F (COC); ref. index 1.462; pH 5-7; 100% conc.

Maco-O-Line 091. [Vaessen-Schoemaker] Carrageenan; CAS 9000-07-1; EINECS 232-524-2; binding, gelling agent, stabilizer, peeling agent, colorant for food industry; esp. suitable for injection brines where low visc. is required and where finished prod. should have firm/strong texture; also for massaging brines; *Usage level:* 0.1-0.3% on final prod. wt.; *Properties:* wh. free-flowing powd.; visc. < 300 cps (5%); pH 7-8 (1%); < 10% moisture.

Magathin. [Lucas Meyer] Lecithin deriv.; CAS 8002-43-5; EINECS 232-307-2; nonionic; natural fat emulsifier for toffees; *Properties:* liq.; 100% conc.; Unverified

Magna A. [Van Den Bergh Foods] Partially hydrog. veg. oil (cottonseed, soybean); EINECS 269-820-6; high melting, high performance fat for no-tempering coatings, centers, veg. dairy systems; shortening for bakery applics.; *Regulatory:* kosher; *Properties:* solid; m.p. 101-104 F.

Magna B. [Van Den Bergh Foods] Partially hydrog. veg. oil (cottonseed, soybean); EINECS 269-820-6; high melting, high performance fat for no-tempering coatings, centers, veg. dairy systems; shortening for bakery applics.; *Regulatory:* kosher; *Properties:* solid; m.p. 105-109 F.

Magna C. [Van Den Bergh Foods] Partially hydrog. veg. oil (cottonseed, soybean); EINECS 269-820-6; high melting, high performance fat for no-tempering coatings, centers, veg. dairy systems; shortening for bakery applics.; for use where high shear conditions exist; *Properties:* solid; m.p. 110-111 F.

Magnesium Hydroxide USP. [Morton Int'l.] Magnesium hydroxide; CAS 1309-42-8; EINECS 215-170-3; source of magnesium for vitamins, magnesium supplement in fortifiers, neutralizing agent; *Regulatory:* USP, FCC compliance; *Properties:* powd.; 99% thru 325 mesh; bulk dens. 0.2 g/ml; 98.4% act.

Magnesium Hydroxide USP DC. [Morton Int'l.] Magnesium hydroxide, dust-controlled form; CAS 1309-42-8; EINECS 215-170-3; source of magnesium for vitamins, magnesium supplement in fortifiers, neutralizing agent; directly compressible; *Regulatory:* USP, FCC compliance; *Properties:* micro-pellets; 99% thru 20 mesh; bulk dens. 0.4 g/ml; 98.5% act.

Magnesium Oxide USP 30 Light. [Morton Int'l.] Magnesium oxide; CAS 1309-48-4; EINECS 215-171-9; food additive; *Regulatory:* USP, FCC compliance; *Properties:* flat platelets; 3 μm avg. particle size; 99.5% min. thru 325 mesh; bulk dens. 0.1-0.2 g/ml; 98.4% act.

Magnesium Oxide USP 60 Light. [Morton Int'l.] Magnesium oxide; CAS 1309-48-4; EINECS 215-171-9; food additive; *Regulatory:* USP, FCC compliance; *Properties:* flat platelets; 3 μm avg. particle size; 99.5% min. thru 325 mesh; bulk dens. 0.1-0.2 g/ml; 98.4% act.

Magnesium Oxide USP 90 Light. [Morton Int'l.] Magnesium oxide; CAS 1309-48-4; EINECS 215-171-9; food additive; *Regulatory:* USP, FCC compliance; *Properties:* flat platelets; 3 μm avg. particle size; 99.5% min. thru 325 mesh; bulk dens. 0.1-0.2 g/ml; 98.4% act.

Magnesium Oxide USP Heavy. [Morton Int'l.] Magnesium oxide; CAS 1309-48-4; EINECS 215-171-9; food additive; *Regulatory:* USP, FCC compliance; *Properties:* flat platelets; 3 μm avg. particle size; 99.5% min. thru 325 mesh; bulk dens. 0.2-0.3 g/ml; 98.4% act.

Magnificent 7. [Brolite Prods.] Used for a richer, hearth-baked French, Vienna, and/or Kaiser rolls; *Usage level:* 7 lb/100 lb flour.

Maisine. [Gattefosse; Gattefosse SA] Corn glycerides; nonionic; food emulsifier for ice creams, sauces, bread, and rice cakes; *Properties:* liq.; HLB 3.0; 100% conc.

Maltoferm® 10001. [Mid-Am. Food Sales] Malted barley extract; CAS 8002-48-0; EINECS 232-310-9; provides rich natural malt flavor; nondiastatic; *Properties:* free-flowing powd.; hygroscopic.

Maltoferm® 10001 VDK. [Mid-Am. Food Sales] Malted barley extract; CAS 8002-48-0; EINECS 232-310-9; provides rich natural malt flavor and dark color; nondiastatic; *Properties:* free-flowing powd.; hygroscopic.

Maltoferm® A-6001. [Mid-Am. Food Sales] Malted

barley extract; CAS 8002-48-0; EINECS 232-310-9; provides soluble color; *Properties:* dk. red. to blk. free-flowing powd.; hygroscopic.

Maltoferm® MBF CR-40. [Mid-Am. Food Sales] Natural barley malt flour; provides flavor, color, enzymes, fermentable sugars, nutrients; nondiastatic; *Properties:* lt. to med. brn. powd., pleasant sweet caramel flavor.

Maltrin® M040. [Grain Processing] Maltodextrin; CAS 9050-36-6; EINECS 232-940-4; nonsweet, nutritive polymer useful for wet binding and anticaking, fat and oil replacement, spray-drying hygroscopic materials, pan coatings, soup and gravy mixes, nonbrowning carrier for drying sensitive prods., film-forming agent, adds sol'n. visc.; *Regulatory:* 9CFR §318.7(c)(4), 319.140, 319.143, 319.182, 319.180, 319.181, 319.281, 319.300, 319.301, 319.306, 319.312, 381.147(f)(4), 381.159, 381.160; 21CFR §184.1444, GRAS; *Properties:* wh. powd., bland flavor; can be dissolved up to 40% in water; bulk dens. 0.51 g/cc (packed); DE 5; pH 4-5; 6% max. moisture.

Maltrin® M050. [Grain Processing] Maltodextrin; CAS 9050-36-6; EINECS 232-940-4; nonsweet, nutritive film-forming polymer useful for adding visc., opacity, and mouthfeel to foods, beverages, pan coatings, soup and gravy mixes, as nonbrowning carrier for drying sensitive prods., as thickener for food systems; *Regulatory:* 9CFR §318.7(c)(4), 319.140, 319.143, 319.182, 319.180, 319.181, 319.281, 319.300, 319.301, 319.306, 319.312, 381.147(f)(4), 381.159, 381.160; 21CFR § 184.1444, GRAS; *Properties:* wh. powd.; can be dissolved up to 40% in water yielding opaque syrup; bulk dens. 0.50 g/cc (packed), DE 5; pH 4-5.

Maltrin® M100. [Grain Processing] Maltodextrin; CAS 9050-36-6; EINECS 232-940-4; nonsweet, nutritive polymer useful as carrier, filler, extender, and bulking agent for foods, dry mixes, spray-dried foods, frozen foods, confections, bakery goods, nondairy prods., snack foods; controls bloom and moisture in hard candy; *Regulatory:* 9CFR §318.7(c)(4), 319.140, 319.143, 319.182, 319.180, 319.181, 319.281, 319.300, 319.301, 319.306, 319.312, 381.147(f)(4), 381.159, 381.160; 21CFR §184.1444, GRAS; *Properties:* wh. powd., bland, low sweetness; disp. readily to produce clear sol'ns. at 30%; bulk dens. 0.56 g/cc (packed); DE 10; pH 4.0-4.7 (20%); 6% max. moisture.

Maltrin® M105. [Grain Processing] Maltodextrin; CAS 9050-36-6; EINECS 232-940-4; bulking and bodying agent for use in dry mixes, spray-dried foods, frozen foods, confections, baked goods, nondairy prods., snack foods; absorbs up to 13% moisture while remaining a free-flowing powd.; *Regulatory:* 9CFR §318.7(c)(4), 319.140, 319.143, 319.182, 319.180, 319.181, 319.281, 319.300, 319.301, 319.306, 319.312, 381.147(f)(4), 381.159, 381.160; 21CFR §184.1444, GRAS; *Properties:* wh. powd., bland odor, low sweetness; disperses easily to produce clear sol'ns. @ 30% solids; DE 9-12; pH 4.0-4.7 (20%); 6% max. moisture.

Maltrin® M150. [Grain Processing] Maltodextrin; CAS 9050-36-6; EINECS 232-940-4; very sl. sweet nutritive polymer useful for bulking, bodying, and binding properties; for cake and cookie mixes, confections, frozen desserts, glazes, dehydrated foods, flavored syrups, as carrier for seasonings, artificial sweeteners; *Regulatory:* 9CFR § 318.7(c)(4), 319.140, 319.143, 319.182, 319.180, 319.181, 319.281, 319.300, 319.301, 319.306, 319.312, 381.147(f)(4), 381.159, 381.160; 21CFR §184.1444, GRAS; *Properties:* wh. powd.; disp. to produce clear sol'n. at 50%; bulk dens. 0.61 g/cc (packed); DE 15; pH 4.0-4.7 (20%); 6% max. moisture.

Maltrin® M180. [Grain Processing] Maltodextrin; CAS 9050-36-6; EINECS 232-940-4; sl. sweet, nutritive polymer; bulking, bodying, and binding properties; for cake and cookie mixes, confections, frozen desserts, glazes, dehydrated foods, flavored syrups, as carrier for seasonings, artificial sweeteners, other powd. specialties; *Regulatory:* 9CFR §318.7(c)(4), 319.140, 319.143, 319.182, 319.180, 319.181, 319.281, 319.300, 319.301, 319.306, 319.312, 381.147(f)(4), 381.159, 381.160; 21CFR §184.1444, GRAS; *Properties:* wh. powd., bland; disp. easily to produce clear sol'ns. up to 60%; bulk dens. 0.63 g/cc (packed); DE 18; pH 4.0-4.7; 6% max. moisture.

Maltrin® M200. [Grain Processing] Corn syrup solids; CAS 68131-37-3; coating and binding agent; imparts smooth mouthfeel; for coffee whiteners, glazes, cake and cookie mixes, frozen desserts, fondants, frozen egg prods., meringue and icing powds., whipped prods., nut coatings, meat systems; *Regulatory:* FDA 21CFR §168.121, 184.1865, GRAS; *Properties:* wh. powd.; disp. rapidly in cold or hot water to produce clear sol'n. at 60%; bulk dens. 0.64 g/cc (packed); DE 20; pH 4.0-4.7 (20%); 6% max. moisture.

Maltrin® M205. [Grain Processing] Corn syrup solids; CAS 68131-37-3; sl. sweet carbohydrate for use in coffee whiteners, glazes, cake mixes, cookie mixes, frozen desserts, fondants, frozen egg prods., icings, whipped prods., nut coatings, where clarity, blandness, and rapid dispersibility are important; *Regulatory:* FDA 21CFR §168.121, 184.1865, GRAS; *Properties:* wh. free-flowing powd., bland; disperses easily to produce a clear sol'n. @ 60% solids; absorbs up to 10% moisture and remains free-flowing; DE 20-23; pH 4.0-4.7 (20%); 6% max. moisture.

Maltrin® M250. [Grain Processing] Corn syrup solids; CAS 68131-37-3; dried glucose syrup with good binding properties; imparts smooth mouthfeel; for coffee whiteners, imitation dairy prods., peanut butter, citrus powds., frozen desserts, dry mixes, beverages, dried eggs, as spray drying aid; *Regulatory:* FDA 21CFR §168.121, 184.1865, GRAS; *Properties:* wh. powd., sl. sweet, no undesirable aftertaste; disp. rapidly in hot or cold water to produce clear sol'n. at 70%; bulk dens. 0.67 g/cc (packed); DE 25; pH 4.5-5.5; 6% max. moisture.

Maltrin® M255. [Grain Processing] Corn syrup solids; CAS 68131-37-3; sl. sweet carbohydrate for use in coffee whiteners, imitation dairy prods., peanut butter, citrus powds., frozen desserts, dry mixes, beverages, dried eggs, as spray-drying aid; *Regulatory:* FDA 21CFR §168.121, 184.1865, GRAS; *Properties:* wh. free-flowing powd., sl. sweet, no undesirable aftertaste; disperses easily to produce clear sol'ns. to 70% concs.; DE 23-27; pH 4.5-5.2; 6% max. moisture.

Maltrin® M365. [Grain Processing] High maltose corn syrup solids; CAS 68131-37-3; dried glucose syrup with noticeable sweetness; imparts desirable mouthfeel; coffee whiteners, imitation dairy prods., meat prods., peanut butter, citrus powds., frozen desserts, dry mixes, beverages, dried eggs, as spray drying aid; *Regulatory:* FDA 21CFR §168.121, 184.1865, GRAS; *Properties:* very bland; disp. easily to produce clear sol'ns. up to 70%; bulk dens. 0.67 g/cc (packed); DE 36; pH 4.5-5.2; 6% max. moisture; Discontinued

Maltrin® M510. [Grain Processing] Agglomerated maltodextrin; CAS 9050-36-6; EINECS 232-940-4; flowable form of Maltrin M100; exhibits exc. dispersibility and dissolution; directly compressible binder and diluent for dry mixes; as carrier for artificial sweeteners, seasonings, and other powd. food specialties; *Regulatory:* 9CFR §318.7(c)(4), 319.140, 319.143, 319.182, 319.180, 319.181, 319.281, 319.300, 319.301, 319.306, 319.312, 381.147(f)(4), 381.159, 381.160; 21CFR § 184.1444, GRAS; *Properties:* wh. free-flowing fine gran., bland; disp. easily to produce clear sol'ns. to 30%; bulk dens. 0.56 g/cc (packed); DE 10; pH 4.0-4.7.

Maltrin® M520. [Grain Processing] Maltodextrin; CAS 9050-36-6; EINECS 232-940-4; carrier, filler for use in low-moisture food systems (cookie fillings, icings, frostings with smooth mouthfeel) and finely milled dry blends (powdered sugar, spices); *Regulatory:* 9CFR §318.7(c)(4), 319.140, 319.143, 319.182, 319.180, 319.181, 319.281, 319.300, 319.301, 319.306, 319.312, 381.147(f)(4), 381.159, 381.160; 21CFR §184.1444, GRAS; *Properties:* wh. fine powd., bland, low sweetness; 10% min. on 200 mesh; bulk dens. 34 lb/ft³ (packed); DE 9-12; pH 4.0-4.7 (20%); 6% max. moisture.

Maltrin® M700. [Grain Processing] Agglomerated maltodextrin NF; CAS 9050-36-6; EINECS 232-940-4; agglomerated form of Maltrin M100; bulking agent, carrier enhancing solubility and emulsification of absorbed components; exhibits exc. dissolution; dry mixes (esp. low calorie or high flavored mixes), carrying most nonaq. liqs. (flavor/veg. oils); *Regulatory:* 9CFR §318.7(c)(4), 319.140, 319.143, 319.182, 319.180, 319.181, 319.281, 319.300, 319.301, 319.306, 319.312, 381.147(f)(4), 381.159, 381.160; 21CFR §184.1444, GRAS; *Properties:* free-flowing gran.; disp. easily to clear sol'n. @ 30% solids, opaque sol'n. > 40% solids; bulk dens. 0.13 g/cc (packed); DE 9-12; pH 6.0-7.0

(20%); 6% max. moisture.

Maltrin® QD M440. [Grain Processing] Agglomerated maltodextrin; CAS 9050-36-6; EINECS 232-940-4; agglomerated form of Maltrin M040; exhibits exc. dispersibility and dissolution; *Regulatory:* 9CFR §318.7(c)(4), 319.140, 319.143, 319.182, 319.180, 319.181, 319.281, 319.300, 319.301, 319.306, 319.312, 381.147(f)(4), 381.159, 381.160; 21CFR §184.1444, GRAS; *Properties:* wh. free-flowing powd., bland; 90% min. on 200 mesh, 10% max. on 20 mesh; quickly dispersible at levels to 15% in water; bulk dens. 0.30 g/cc (packed); DE 4-7; pH 4.0-5.1 (20%); 6% max. moisture.

Maltrin® QD M500. [Grain Processing] Agglomerated maltodextrin; CAS 9050-36-6; EINECS 232-940-4; agglomerated form of Maltrin M100; exhibits exc. dispersibility and dissolution; directly compressible binder and good carrier; for food use; *Regulatory:* 9CFR §318.7(c)(4), 319.140, 319.143, 319.182, 319.180, 319.181, 319.281, 319.300, 319.301, 319.306, 319.312, 381.147(f)(4), 381.159, 381.160; 21CFR §184.1444, GRAS; *Properties:* wh. free-flowing gran., bland; 90% min. on 200 mesh, 10% min. on 20 mesh; quickly disp. in water to 30%; bulk dens. 0.34 g/cc (packed); DE 9-12; pH 4.0-5.1.

Maltrin® QD M550. [Grain Processing] Agglomerated maltodextrin; CAS 9050-36-6; EINECS 232-940-4; agglomerated form of Maltrin M150; exhibits exc. dispersibility and dissolution; binder, carrier; *Regulatory:* 9CFR §318.7(c)(4), 319.140, 319.143, 319.182, 319.180, 319.181, 319.281, 319.300, 319.301, 319.306, 319.312, 381.147(f)(4), 381.159, 381.160; 21CFR §184.1444, GRAS; *Properties:* wh. free-flowing gran.; 87% on 200 mesh, 4% on 20 mesh; quickly disp. up to 50% solids; bulk dens. 0.37 g/cc (packed); DE 13-17; pH 4.0-5.1; 6% max. moisture.

Maltrin® QD M580. [Grain Processing] Agglomerated maltodextrin; CAS 9050-36-6; EINECS 232-940-4; quickly disp. carbohydrate with high rate of sol'n., exc. particulate str.; *Regulatory:* 9CFR § 318.7(c)(4), 319.140, 319.143, 319.182, 319.180, 319.181, 319.281, 319.300, 319.301, 319.306, 319.312, 381.147(f)(4), 381.159, 381.160; 21CFR §184.1444, GRAS; *Properties:* wh. free-flowing gran., bland; 10% max. on 20 mesh, 85% min. on 200 mesh; quickly disp. in water to 60% conc.; absorbs up to 10-11% moisture remaining free-flowing; bulk dens. 0.40 g/cc (packed); DE 16.5-19.5; pH 4.0-5.1; 6% max. moisture.

Maltrin® QD M600. [Grain Processing] Agglomerated corn syrup solids; CAS 68131-37-3; agglomerated form of Maltrin M200; directly compressible binder and good carrier; *Regulatory:* FDA 21CFR §168.121, 184.1865, GRAS; *Properties:* wh. free-flowing gran., bland, sl. sweet; 87% on 200 mesh, 4% on 20 mesh; quickly disp. in water to 60% conc.; bulk dens. 0.40 g/cc (packed); DE 20-23; pH 4.0-5.1 (20%); 6% max. moisture.

Mammary Substance. [Am. Labs] Freeze-dried defatted glandular prod.; nutritive food additive; *Prop-*

erties: powd. or freeze-dried form.

Manucol DH. [Kelco Int'l.] Sodium alginate; CAS 9005-38-3; thickener, gellant, stabilizer used in bakery glazes, filling creams, cheesecake toppings, cheese spreads, processed cheeses, instant puddings, aspics; *Regulatory:* FDA 21CFR §184.1724, GRAS; EEC E401 compliance; *Properties:* wh. to tan powd., sl. odor; 250 μ particle size; water-sol.; bulk dens. ≈ 50 lb/ft³; visc. 65 mPa•s; pH neutral; *Toxicology:* LD50 (oral, rat) > 5000 mg/kg; dry powd. may cause eye irritation; *Precaution:* incompat. with strong oxidizers.

Manucol DM. [Kelco Int'l.] Sodium alginate; CAS 9005-38-3; gellant, suspending agent, thickener, stabilizer, film-former for ice cream, chilled desserts, softer reversible gels, bakery fillings; *Regulatory:* FDA 21CFR §184.1724, GRAS; EEC E401 compliance; *Properties:* wh. to tan powd., sl. odor; 250 μ particle size; water-sol.; bulk dens. ≈ 50 lb/ft³; visc. 250 cps (1%), 5500 cps (2%); pH 6.0; *Toxicology:* LD50 (oral, rat) > 5000 mg/kg; dry powd. may cause eye irritation; *Precaution:* incompat. with strong oxidizers.

Manucol DMF. [Kelco Int'l.] Sodium alginate; CAS 9005-38-3; gellant, suspending agent, thickener, stabilizer, film-former used for bakery filling creams, softer reversible gels, structured fish, potato prods.; *Regulatory:* FDA 21CFR §184.1724, GRAS; EEC E401 compliance; *Properties:* wh. to tan powd., sl. odor; 105 μ particle size; water-sol.; bulk dens. ≈ 50 lb/ft³; visc. 300 cps (1%), 7000 cps (2%); pH 6; *Toxicology:* LD50 (oral, rat) > 5000 mg/ kg; dry powd. may cause eye irritation; *Precaution:* incompat. with strong oxidizers.

Manucol JKT. [Kelco Int'l.] Specialty algin blend requiring addition of a food-grade acid or calcium salt; gellant, suspending agent, thickener, stabilizer, film-former for foods (structured glacé fruit, high solids fillings, bakery fillings, bakery jellies, dry mixes, instant puddings, instant mousses, egg prods.); *Properties:* 250 mesh; pH 7.

Manucol LB. [Kelco Int'l.] Algin; CAS 9005-38-3; gellant, suspending agent, thickener, stabilizer, film-former for foods (softer reversible gels, bakery fillings); *Regulatory:* FDA 21CFR §184.1724, GRAS; EEC E401 compliance; *Properties:* wh. to tan powd., sl. odor; 60 mesh; water-sol.; bulk dens. ≈ 50 lb/ft³; visc. 5 cps (1%), 10 cps (2%); pH 6; *Toxicology:* LD50 (oral, rat) > 5000 mg/kg; dry powd. may cause eye irritation; *Precaution:* incompat. with strong oxidizers.

Manucol SMF. [Kelco Int'l.] Alginate blend; thickener, gellant, stabilizer used for bakery filling creams, cheesecake toppings, chilled desserts, and instant puddings.

Manucol Ester B. [Kelco Int'l.] Propylene glycol alginate; CAS 9005-37-2; beer foam stabilizer; *Regulatory:* FDA 21CFR §172.858, EEC E405 compliance; *Properties:* wh. to tan powd., sl. odor; 250 μ particle size; water-sol.; bulk dens. ≈ 50 lb/ft³; visc. 20 cps (1%), 110 cps (2%); pH 4.0; *Toxicology:* LD50 (oral, rat) > 5000 mg/kg; dry powd. may cause

eye irritation; *Precaution:* incompat. with strong oxidizers.

Manucol Ester E/PL. [Kelco Int'l.] Propylene glycol alginate; CAS 9005-37-2; emulsifier, thickener, stabilizer used for pourable dressings, mustard/mint sauces; *Regulatory:* FDA 21CFR §172.858, EEC E405 compliance; *Properties:* wh. to tan powd., sl. odor; 250 μ particle size; water-sol.; bulk dens. ≈ 50 lb/ft³; visc. 300 cps (1%); pH 4.0; *Toxicology:* LD50 (oral, rat) > 5000 mg/kg; dry powd. may cause eye irritation; *Precaution:* incompat. with strong oxidizers.

Manucol Ester E/RK. [Kelco Int'l.] Propylene glycol alginate; CAS 9005-37-2; emulsifier, thickener, stabilizer used for fermented milks, fruit drinks, citrus concs.; *Regulatory:* FDA 21CFR §172.858, EEC E405 compliance; *Properties:* wh. to tan powd., sl. odor; 355 μ particle size; water-sol.; bulk dens. ≈ 50 lb/ft³; visc. 125 cps (1%), 1120 cps (2%); pH 4.0; *Toxicology:* LD50 (oral, rat) > 5000 mg/kg; dry powd. may cause eye irritation; *Precaution:* incompat. with strong oxidizers.

Manucol Ester M. [Kelco Int'l.] Propylene glycol alginate; CAS 9005-37-2; emulsifier, thickener, stabilizer used for confectionery, pourable dressings, mustard/mint sauces, meringues, yogurt/fruit preps., fruit drinks, citrus concs., topping syrups; *Regulatory:* FDA 21CFR §172.858, EEC E405 compliance; *Properties:* wh. to tan powd., sl. odor; 355 μ particle size; water-sol.; bulk dens. ≈ 50 lb/ft³; visc. 500 cps (1%); pH 4.0; *Toxicology:* LD50 (oral, rat) > 5000 mg/kg; dry powd. may cause eye irritation; *Precaution:* incompat. with strong oxidizers.

Manugel C. [Kelco Int'l.] Specialty algin blend requiring addition of a food-grade acid or calcium salt; gellant, suspending agent, thickener, stabilizer, film-former for foods (structured glacé fruit, high solids fillings, bakery fillings, bakery jellies, dry mixes); *Properties:* 60 mesh; pH 10.

Manugel DJX. [Kelco Int'l.] Sodium alginate; CAS 9005-38-3; gellant, suspending agent, thickener, stabilizer, film-former for foods (strong gels, pet foods, dessert gels, structured fish and meat, potato prods.); *Regulatory:* FDA 21CFR §184.1724, GRAS; EEC E401 compliance; *Properties:* wh. to tan powd., sl. odor; 65 μ particle size; water-sol.; bulk dens. ≈ 50 lb/ft³; visc. 200 cps (1%), 2200 cps (2%); pH 7; *Toxicology:* LD50 (oral, rat) > 5000 mg/ kg; dry powd. may cause eye irritation; *Precaution:* incompat. with strong oxidizers.

Manugel DMB. [Kelco Int'l.] Sodium alginate; CAS 9005-38-3; gellant, suspending agent, thickener, stabilizer, film-former for foods (glazes, instant water jelly, structured fish/meat/vegetables/anchovy, tomato prods., potato prods., aspics, coating batters, canned pet food); *Regulatory:* FDA 21CFR §184.1724, GRAS; EEC E401 compliance; *Properties:* wh. to tan powd., sl. odor; 105 μ particle size; water-sol.; bulk dens. ≈ 50 lb/ft³; visc. 300 cps (1%), 3000 cps (2%); pH 6.0; *Toxicology:* LD50 (oral, rat) > 5000 mg/kg; dry powd. may cause eye irritation; *Precaution:* incompat. with strong oxidizers.

Manugel GHB. [Kelco Int'l.] Sodium alginate; CAS 9005-38-3; gellant, suspending agent, thickener, stabilizer, film-former for foods (strong gels, pet foods, dessert gels, structured fish and meat); *Regulatory:* FDA 21CFR §184.1724, GRAS; EEC E401 compliance; *Properties:* wh. to tan powd., sl. odor; 250 μ particle size; water-sol.; bulk dens. ≈50 lb/ft³; visc. 75 cps (1%), 550 cps (2%); pH 6; *Toxicology:* LD50 (oral, rat) > 5000 mg/kg; dry powd. may cause eye irritation; *Precaution:* incompat. with strong oxidizers.

Manugel GMB. [Kelco Int'l.] Sodium alginate; CAS 9005-38-3; thickener, gellant, stabilizer used for glazes, instant water jelly, structured fish/meat/vegetables/anchovies/peppers, tomato prods., aspics, coating batters, canned pet food, aquaculture, pet food brawn; *Regulatory:* FDA 21CFR §184.1724, GRAS; EEC E401 compliance; *Properties:* wh. to tan powd., sl. odor; 250 μ particle size; water-sol.; bulk dens. ≈50 lb/ft³; visc. 185 cps (1%); pH neutral; *Toxicology:* LD50 (oral, rat) > 5000 mg/kg; dry powd. may cause eye irritation; *Precaution:* incompat. with strong oxidizers.

Manugel JKB. [Kelco Int'l.] Specialty algin blend requiring addition of a food-grade acid or calcium salt; gellant, suspending agent, thickener, stabilizer, film-former for foods (structured glacé fruit, high solids fillings, bakery fillings, bakery jellies, dry mixes); *Properties:* 60 mesh; pH 7.

Manugel L98. [Kelco Int'l.] Specialty algin blend requiring addition of a food-grade acid or calcium salt; gellant, suspending agent, thickener, stabilizer, film-former for foods (fruit fillings, glazes, instant puddings, instant mousses, instant water jelly); *Properties:* 150 mesh; pH 7.

Manugel PTB. [Kelco Int'l.] Alginate blend; thickener, gellant, stabilizer used for fruit fillings.

Manugel PTJ. [Kelco Int'l.] Specialty algin blend requiring addition of a food-grade acid or calcium salt; gellant, suspending agent, thickener, stabilizer, film-former for foods (structured glacé fruit, high solids fillings, bakery fillings, bakery jellies, dry mixes); *Properties:* 150 mesh; pH 7.

Manugel PTJA. [Kelco Int'l.] Alginate blend; thickener, gellant, stabilizer used for fruit fillings.

Mapron®. [Mitsubishi Kasei] Soybean milk; food additive.

Margarine Gelatin P-8. [Hormel] Gelatin, Type A, 240 bloom; CAS 9000-70-8; EINECS 232-554-6; used in extra light low-fat margarine prods.; wets easier with less agitation and dissolves with fewer entrapped air bubbles; *Properties:* coarse mesh; visc. 39-43 mps; pH 4.5-5.8; 11.5% max. moisture; *Storage:* stable for up to 1 yr when stored dry at ambient temps.; keep containers tightly closed.

Marinco CH. [Marine Magnesium] Magnesium carbonate USP, FCC; CAS 546-93-0; EINECS 208-915-9; antacid, alkaline buffer, and min. supplement used in foods; *Properties:* wh. free-flowing powd.; 100% thru 100 mesh, 97% min. thru 325 mesh; bulk dens. 10-14 lb/ft³ (loose); 40-43.5% MgO; *Storage:* store in dry environment in tightly sealed containers.

Marinco® CH-Granular. [Marine Magnesium] Magnesium carbonate USP, FCC; EINECS 208-915-9; antacid, alkaline buffer, mineral supplement; *Properties:* wh. free-flowing gran.; 100% min. thru 20 mesh, 20% max. on 40 mesh; bulk dens. 35 lb/ft³ min. (loose); 40-43.5% MgO; *Storage:* store in dry environment in tightly sealed containers.

Marinco® CL. [Marine Magnesium] Magnesium carbonate USP, FCC; EINECS 208-915-9; antacid, alkaline buffer, ingred. used in foods; decolorizer, deodorizer in various processes; *Properties:* wh. free-flowing powd.; 100% min. thru 100 mesh, 99% min. thru 325 mesh; bulk dens. 5-8 lb/ft³ (loose); 40-43.5% MgO; *Storage:* store in dry environment in tightly sealed containers.

Marinco H-USP. [Marine Magnesium] Magnesium hydroxide USP, FCC; CAS 1309-42-8; EINECS 215-170-3; antacid and alkaline buffer for milk of magnesia tablets and for making magnesium citrate; *Properties:* wh. free-flowing powd.; 100% min. thru 100 mesh, 99% min. thru 325 mesh; bulk dens. 25-33 lb/ft³; 95-100.5% MgOH; *Storage:* store in dry environment in tightly sealed containers.

Marinco OH. [Marine Magnesium] Magnesium oxide USP, FCC; CAS 1309-48-4; EINECS 215-171-9; antacid, alkaline buffer, and mineral supplement; *Properties:* wh. free-flowing powd.; 100% min. thru 100 mesh, 99.5% min. thru 325 mesh; bulk dens. 18-26 lb/ft³ (loose); 96-100.5% MgO; *Storage:* store in dry environment in tightly sealed containers.

Marinco OL. [Marine Magnesium] Magnesium oxide; CAS 1309-48-4; EINECS 215-171-9; antacid and alkaline buffer; *Regulatory:* FCC, USP compliance; *Properties:* wh. free-flowing powd.; 100% min thru 100 mesh, 99.5% min. thru 325 mesh; bulk dens. 4-8 lb/ft³; 96-100.5% MgO; *Storage:* store in dry environment in tightly sealed containers.

Marine Colloids™ Carrageenan. [FMC] Carrageenan; CAS 9000-07-1; EINECS 232-524-2; fat replacement systems, gellants; provides rheology, structure, and mouthfeel; nongelling carrageenan binds moisture; *Regulatory:* FDA 21CFR §172.620, 182.7255, GRAS; EEC compliance.

Marloid® CMS. [Kelco] Sodium alginate blend; stabilizer, gellant, thickener for milk-based systems (puddings, pie fillings, custards, cheese dips, cheesecake, eggnog, milk shake powd., chiffons, toppings, syrups, baked cream fillings, whipping cream); *Properties:* 28 mesh; milk-sol.; pH 10 (1%).

Masil® SF 5. [PPG/Specialty Chem.] Dimethicone; release aid, defoamer for nonaq. processes in foods industry; *Properties:* water-wh. oily, clear liq.; odorless, tasteless; sp.gr. 0.916; visc. 5 cSt; pour pt. -65 C; flash pt. (PMCC) 280 F; ref. index 1.3970.

Masil® SF 10. [PPG/Specialty Chem.] Dimethicone; release aid, defoamer for nonaq. processes in foods industry; *Properties:* sp.gr. 0.940; pour pt. -65 C; flash pt. (PMCC) 320 F; ref. index 1.3990.

Masil® SF 20. [PPG/Specialty Chem.] Dimethicone;

release aid, defoamer for nonaq. processes in foods industry; *Properties:* water-wh. oily, clear liq.; odorless, tasteless; sp.gr. 0.953; visc. 20 cSt; pour pt. -65 C; flash pt. (PMCC) 395 F; ref. index 1.4010.

Masil® SF 50. [PPG/Specialty Chem.] Dimethicone; release aid, defoamer for nonaq. processes in foods industry; *Properties:* water-wh. oily, clear liq.; odorless, tasteless; sp.gr. 0.963; visc. 50 cps; pour pt. -55 C; flash pt. (PMCC) 460 F; ref. index 1.4020.

Masil® SF 100. [PPG/Specialty Chem.] Dimethicone; release aid, defoamer for nonaq. processes in foods industry; *Properties:* water-wh. oily, clear liq.; odorless, tasteless; sp.gr. 0.968; visc. 100 cps; pour pt. -55 C; flash pt. (PMCC) 461 C; ref. index 1.4030.

Masil® SF 200. [PPG/Specialty Chem.] Dimethicone; release aid, defoamer for nonaq. processes in foods industry; *Properties:* water-wh. oily, clear liq.; odorless, tasteless; sp.gr. 0.972; visc. 200 cSt; pour pt. -50 C; flash pt. (PMCC) 460 F; ref. index 1.4031.

Masil® SF 350. [PPG/Specialty Chem.] Dimethicone; release aid, defoamer for nonaq. processes in foods industry; *Properties:* water-wh. oily, clear liq.; odorless, tasteless; sp.gr. 0.973; visc. 350 cSt; pour pt. -50 C; flash pt. (PMCC) 500 F; ref. index 1.4032.

Masil® SF 350 FG. [PPG/Specialty Chem.] Dimethicone; CAS 63148-62-9; release aid, defoamer for nonaq. processes in foods industry; *Properties:* colorless clear liq., bland odor; misc. in aliphatic/aromatic/halogenated solvs.; insol. in water, lower alcohols; sp.gr. 0.966-0.972; visc. 332.5-367.5 cst; b.p. > 300 F; pour pt. -50 C; flash pt. (PMCC) 500 F; ref. index 1.4025-1.4045; *Toxicology:* temporary eye irritant; nonirritating to skin; *Storage:* store in well-ventilated area below 120 F.

Masil® SF 500. [PPG/Specialty Chem.] Dimethicone; release aid, defoamer for nonaq. processes in foods industry; *Properties:* water-wh. oily, clear liq.; odorless, tasteless; sp.gr. 0.973; visc. 500 cSt; pour pt. -50 C; flash pt. (PMCC) 500 F; ref. index 1.4033.

Masil® SF 1000. [PPG/Specialty Chem.] Dimethicone; release aid, defoamer for nonaq. processes in foods industry; *Properties:* water-wh. oily, clear liq.; odorless, tasteless; sp.gr. 0.974; visc. 1000 cSt; pour pt. -50 C; flash pt. (PMCC) 500 F; ref. index 1.4035; CC flash pt. 260 C.

Masil® SF 5000. [PPG/Specialty Chem.] Dimethicone; release aid, defoamer for nonaq. processes in foods industry; *Properties:* water-wh. oily, clear liq.; odorless, tasteless; sp.gr. 0.975; visc. 5000 cSt; pour pt. -49 C; flash pt. (PMCC) 500 F; ref. index 1.4035.

Masil® SF 10,000. [PPG/Specialty Chem.] Dimethicone; release aid, defoamer for nonaq. processes in foods industry; *Properties:* water-wh. oily, clear liq.; odorless, tasteless; sp.gr. 0.975; visc. 10,000 cSt; pour pt. -47 C; flash pt. (PMCC) 500 F; ref. index 1.4035.

Masil® SF 12,500. [PPG/Specialty Chem.] Dimethicone; release aid, defoamer for nonaq. processes in foods industry; *Properties:* water-wh. oily, clear liq.; odorless, tasteless; sp.gr. 0.975; visc. 12,500 cSt; pour pt. -47 C; flash pt. (PMCC) 500 F; ref. index 1.4035.

Masil® SF 30,000. [PPG/Specialty Chem.] Dimethicone; release aid, defoamer for nonaq. processes in foods industry; *Properties:* water-wh. oily, clear liq.; odorless, tasteless; sp.gr. 0.976; visc. 30,000 cSt; pour pt. -46 C; flash pt. (PMCC) 500 F; ref. index 1.4035.

Masil® SF 60,000. [PPG/Specialty Chem.] Dimethicone; release aid, defoamer for nonaq. processes in foods industry; *Properties:* water-wh. oily, clear liq.; odorless, tasteless; sp.gr. 0.977; visc. 60,000 cSt; pour pt. -44 C; flash pt. (PMCC) 500 F; ref. index 1.4035.

Masil® SF 100,000. [PPG/Specialty Chem.] Dimethicone; release aid, defoamer for nonaq. processes in foods industry; *Properties:* water-wh. oily, clear liq.; odorless, tasteless; sp.gr. 0.978; visc. 100,000 cSt; pour pt. -40 C; flash pt. (PMCC) 500 F; ref. index 1.4035.

Masil® SF 300,000. [PPG/Specialty Chem.] Dimethicone; release aid, defoamer for nonaq. processes in foods industry; *Properties:* water-wh. oily, clear liq.; odorless, tasteless; sp.gr. 0.978; visc. 300,000 cSt; pour pt. -40 C; flash pt. (PMCC) 500 F; ref. index 1.4035.

Masil® SF 500,000. [PPG/Specialty Chem.] Dimethicone; release aid, defoamer for nonaq. processes in foods industry; *Properties:* sp.gr. 0.978; pour pt. -40 C; flash pt. (PMCC) 500 F; ref. index 1.4035.

Masil® SF 600,000. [PPG/Specialty Chem.] Dimethicone; release aid, defoamer for nonaq. processes in foods industry; *Properties:* water-wh. oily, clear liq.; odorless, tasteless; sp.gr. 0.979; visc. 600,000 cSt; pour pt. -34 C; flash pt. (PMCC) 500 F; ref. index 1.4035.

Masil® SF 1,000,000. [PPG/Specialty Chem.] Dimethicone; release aid, defoamer for nonaq. processes in foods industry; *Properties:* sp.gr. 0.979; pour pt. -25 C; flash pt. (PMCC) 500 F; ref. index 1.4035.

Massa Estarinum® AM. [Hüls Am.] Trilaurin; CAS 538-24-9; EINECS 208-687-0; raw material for food industry; *Properties:* wh. to ivory solid, natural odor; sol. in ether, sl. sol. in ethanol, pract. insol. in water; m.p. 33.5-35.5 C; solid. pt. 31-34 C; acid no. 0.5 max.; iodine no. 3 max.; sapon. no. 235-245; hyd. no. 3 max.

Maxarome® Family. [Gist-brocades] Yeast extract; CAS 8013-01-2; enhances natural savory flavor, saltiness, and mouthfeel; for use whenever cost reduction, flavor enhancement, sodium reduction, MSG replacement, HVP replacement or label statement improvements are desired.

Maxarome® MR. [Gist-brocades] Yeast extract; CAS 8013-01-2; low sodium extract imparting virtually no meaty or yeast char.; provides enhancement capabilities, clean flavor; suitable for vegetables, rice and pasta dishes.

Maxarome® Plus. [Gist-brocades] Yeast extract; CAS 8013-01-2; versatile flavor enhancer containing > 4% naturally occurring GMP and IMP flavor nucleotides; imparts a sweeter, savory flavor which is fully expressed when used in conjunction with other flavor systems.

Maxarome® Plus RS. [Gist-brocades] Yeast extract; CAS 8013-01-2; reduced sodium version of Maxarome® Plus containing > 6% naturally occurring GMP and IMP flavor nucleotides; usage levels generally 50% less than other Maxarome prods.

Maxavor™ MYE. [Gist-brocades] Autolyzed yeast extract system; CAS 8013-01-2; either imparts mushroom flavor to a prod. or enhances mushroom flavors already in the prod.

Maxavor™ RYE-A, RYE-AS. [Gist-brocades] Yeast extract; CAS 8013-01-2; designed to give a general roasted, brown flavor note to a savory system; RYE-A is a good starting ingred. when a pan-dripping type of flavor profile is desired; RYE-AS provides a more intesne darker roast beef flavor.

Maxavor™ RYE-B. [Gist-brocades] Yeast extract; CAS 8013-01-2; imparts a distinct roast beef note to a finished prod.

Maxavor™ RYE-C. [Gist-brocades] Yeast extract; CAS 8013-01-2; imparts a distinct baked/roasted chicken skin type note to a finished prod.

Maxavor™ RYE-CL. [Gist-brocades] Yeast extract; CAS 8013-01-2; imparts a cooked or boiled chicken flavor more subtle than the roasted notes of RYE-C.

Maxavor™ RYE-CR. [Gist-brocades] Yeast extract; CAS 8013-01-2; char. burnt flavor and appearance common to caramel coloring.

Maxavor™ RYE-D. [Gist-brocades] Yeast extract; CAS 8013-01-2; imparts a sweet roasted pork or poultry note to a finished prod.

Maxavor™ RYE-G. [Gist-brocades] Yeast extract; CAS 8013-01-2; imparts a roasted turkey or dark poultry note to a finished prod.

Maxavor™ RYE-PMR. [Gist-brocades] Yeast extract; CAS 8013-01-2; designed to replace meat extract and provide consistent taste and color; enhances savory prods. where a beef profile is desired.

Maxavor™ RYE-T. [Gist-brocades] Yeast extract; CAS 8013-01-2; imparts a unique taste sensation combining nuances of beef and tomato flavor; at high levels the beef notes dominate; compat. with pH 4 systems; *Properties:* reddish-brn.

Maxavor™ RYE Family. [Gist-brocades] Yeast extracts; CAS 8013-01-2; distinct savory flavors for combination with Gistex® base flavor notes and Maxarome® flavor enhancers for complete savory flavor system; avail. as general roasted/brown, beef, chicken, pork, tomato/beef, turkey and mushroom.

Maxi-Gel® 445. [A.E. Staley Mfg.] Food starch modified, derived from waxy maize; cook-up starch, thickener for neutral to sl. acidic foods, cream soups, cream/fruit pies, sauces, gravies, systems requiring high freeze/thaw, visc., and texture stability; *Regulatory:* FDA 21CFR §172.892; *Properties:* wh. free-flowing powd.; disp. in cold water; pH 5.3 (10% slurry); 11.5% moisture.

Maxi-Gel® 542. [A.E. Staley Mfg.] Food starch modi-

fied, derived from waxy maize; cook-up starch, thickener for puddings, frozen fruit pies.

Maxi-Gel® 7776. [A.E. Staley Mfg.] Food starch modified, derived from waxy corn; cook-up starch, stabilizer, thickener for highly acidic and asceptically processed foods; produces short, smooth pastes resist. to heat, shear, acid, freeze/thaw stresses; for canned fruit pie fillings, tomato-based sauces, gravies; *Regulatory:* FDA 21CFR §172.892; *Properties:* wh. powd.; 5% max. on 60 mesh; pH 4.5-6.0 (cooked); 10-13% moisture.

Maximaize® 7360. [A.E. Staley Mfg.] Food starch modified, derived from waxy maize; cook-up starch for low pH foods; provides sheen, clarity, blandness, mouthfeel, paste shortness, esp. in fruit and cream fillings for pies and pastries, retorted sauces, gravies; *Regulatory:* FDA 21CFR §172.892; *Properties:* wh. powd.; pH 5.7; 11% moisture.

Maximaize® 7367. [A.E. Staley Mfg.] Food starch modified, derived from waxy maize; cook-up starch with high processing tolerance; for salad dressings, high acid hot fill systems.

Maximaize® DJ. [A.E. Staley Mfg.] Food starch modified, derived from waxy maize; cook-up starch, thickener, stabilizer for neutral to acidic foods and foods subject to vigorous processing; provides clarity, sheen, blandness, stability, short texture; for soups, sauces, gravies, fruit pies, canned foods; *Regulatory:* FDA 21CFR §172.892; *Properties:* wh. powd., odorless, bland flavor; 90% min. thru 100 mesh; pH 4.5-6 (10% uncooked sol'n.); 10-13% moisture.

Maximaize® HV. [A.E. Staley Mfg.] Food starch modified, derived from waxy maize; cook-up starch, thickener, stabilizer for neutral systems and foods that are processed at high temps.; provides short smooth textures, exc. clarity, resist. to syneresis; for canned foods, cream-style corn, stew gravies, sauces, soups, pie fillings; *Regulatory:* FDA 21CFR §172.892; *Properties:* wh. powd., odorless, bland flavor; 90% min. thru 100 mesh; pH 4.5-6 (10% uncooked sol'n.); 10-13% moisture.

Mayberry Bread. [Brolite Prods.] Bread base for old-fashioned tasting white breads; *Usage level:* 20 lb/50 lb flour; *Properties:* powd.

Mayfair's English Muffin. [Brolite Prods.] Bread base for English muffins or English muffin bread; provides outstanding volume and porosity.

Mayodan. [Grindsted Prods. Denmark] Stabilizer blends; food emulsifier for mayonnaise and salad dressing; *Properties:* powd.; 100% conc.

Mayonat DF. [Feinkost Ingred.] Milk protein, modified starch, guar gum; emulsifier/stabilizer for dressings, toppings, decorations etc. for deep-freeze precooked meals; for dressings without egg yolk; *Usage level:* 8%; *Regulatory:* GRAS; *Properties:* off-wh. powd., neutral odor; 35-60% oil content; *Storage:* store under cool, dry conditions; 6 mos shelf life in original, unopened bag.

Mayonat PS. [Feinkost Ingred.] Whey protein, wheat flour, methylcellulose, xanthan gum, albumin, guar gum; emulsifier for mfg. of pasteurized salads and dressings; very good temp. stability; *Usage level:* 4% based on total recipe; *Regulatory:* FDA, USDA, GRAS; *Properties:* powd., neutral odor and flavor; disp. in cold water; 35-60% oil content; *Storage:* store under cool, dry conditions; 6 mos shelf life in original, unopened bag.

Mayonat V/100. [Feinkost Ingred.] Vegetable protein, starch, guar gum; emulsifier/stabilizer for cold prod. of strictly vegetable dressings without any animal components (milk or egg); suitable for vegetarian, kosher, cholesterol-free, lactose-free, gluten-free, dietary/health foods; *Usage level:* 2-5%; *Regulatory:* FAO/WHO; *Properties:* off-wh. powd., neutral odor; cold sol.; 35-60% oil content; *Storage:* store under cool, dry conditions; 6 mos shelf life in original, unopened bag.

Mazol® 80 MGK. [PPG/Specialty Chem.] PEG-20 glyceryl stearate; CAS 68153-76-4; nonionic; emulsifier for pan release aids, cake and cake mixes, icing, icing mixes, frozen desserts, coffee whiteners; dough conditioner for yeast-leavened baking prods.; *Usage level:* 0.5% (dough conditioner), 0.5% (cakes), 0.45% (whipped toppings), 0.5% (icings), 0.2% (frozen desserts), 0.4% (coffee whiteners), 0.5% (emulsifier in pan release agents); *Regulatory:* FDA 21CFR §172.834; kosher; *Properties:* pale yel. oily liq. or semigel, faint char. odor, mild taste; sol. in ethanol, water; sol. hot in min. oil; partly sol. in soybean oil, disp. in propylene glycol, veg. oil; sp.gr. 1.00 (70 C); m.p. 25-27 C; b.p. > 300 F; HLB 13.5; acid no. 2 max.; iodine no. 1; sapon. no. 65-75; hyd. no. 65-80; flash pt. (PMCC) > 350 F; 100% conc.; *Toxicology:* nonirritating; *Storage:* store in well-ventilated area below 120 F.

Mazol® 159. [PPG/Specialty Chem.] PEG-7 glyceryl cocoate; emulsifier for food prods.; *Properties:* amber liq.; sol. in water, min. oil; disp. in min. spirits, toluene, IPA; HLB 13.0; acid no. 5 max.; sapon. no. 82-98; flash pt. (PMCC) > 350 F.

Mazol® 300 K. [PPG/Specialty Chem.] Glyceryl oleate; CAS 111-03-5; nonionic; food emulsifier; coemulsifier for cake mixes; antifoam for sugar and protein processing; *Regulatory:* FDA 21CFR §184.1323, 184.1666, GRAS; kosher; *Properties:* amber liq.; sol. in min. oil, toluene, IPA, ethanol, veg. oil; disp. in min. spirits, propylene glycol; insol. in water; m.p. 20 C; HLB 3.8; acid no. 2 max.; iodine no. 80; sapon. no. 145-155; flash pt. (PMCC) > 350 F; 100% conc., 40% alpha mono.

Mazol® GMO. [PPG/Specialty Chem.] Glyceryl oleate; CAS 111-03-5; nonionic; dispersant for oil or solv. systems; antifoam for food processing; *Regulatory:* GRAS; *Properties:* yel. liq.; sol. in soybean oil, min. oil, toluene, IPA; disp. in ethanol, propylene glycol; HLB 3.8; acid no. 2 max.; sapon. no. 150-170; flash pt. (PMCC) > 350 F; 100% conc.

Mazol® GMO K. [PPG/Specialty Chem.] Glyceryl oleate; CAS 37220-82-9; w/o emulsifier for foods, bakery and prepared mixes, margarines, convenience foods, frozen desserts; antifoam for sugar and protein processing; coemulsifier for cake mixes; *Regulatory:* FDA 21CFR §182.4505,

184.1323, GRAS; kosher; *Properties:* tan paste, ester odor; sol. in min. oil, veg. oil, ethanol, toluene, IPA; disp. in min. spirits, propylene glycol; sp.gr. 0.94; m.p. 25 C; b.p. > 300 F; HLB 3.8; acid no. 2 max.; iodine no. 90; sapon. no. 160-170; flash pt. (PMCC) > 200 F; 100% act., 40% min. monoglycerides; *Toxicology:* nonirritating; *Storage:* store in well-ventilated area below 120 F.

Mazol® GMS. [PPG/Specialty Chem.] Glyceryl stearate; lubricant, emulsifier, plasticizer, and thickener for foods; *Properties:* tan flake; sol. in IPA; sol. hot in min. oil, toluene; HLB 3.9; acid no. 5 max.; sapon. no. 172; flash pt. (PMCC) > 350 F.

Mazol® GMS-90. [PPG/Specialty Chem.] Glyceryl stearate; emulsifier for food prods.; *Properties:* tan flake; sol. in IPA; sol. hot in min. oil, toluene; HLB 3.9; acid no. 2 max.

Mazol® GMS-K. [PPG/Specialty Chem.] Glyceryl stearate; CAS 11099-07-3; nonionic; w/o emulsifier for foods, bakery and prepared mixes, margarines, convenience foods, frozen desserts; antifoam for sugar and protein processing; *Regulatory:* FDA 21CFR §184.4505, GRAS; kosher; *Properties:* wh. to cream flakes, ester odor; sol. in IPA; sol. hot in min. oil, veg. oil, propylene glycol, toluene; disp. hot in ethanol; insol. in water; sp.gr. 0.908; m.p. 61 C; b.p. > 300 F; HLB 3.9; acid no. 3 max.; iodine no. 5 max.; sapon. no. 165-176; flash pt. (PMCC) > 350 F; 100% conc., 40% min. monoglyceride; *Toxicology:* nonirritating, noncomedogenic; *Storage:* store in well-ventilated area below 120 F.

Mazol® PGMS. [PPG/Specialty Chem.] Propylene glycol stearate; CAS 1323-39-3; EINECS 215-354-3; nonionic; coemulsifier for edible oil and shortenings, dispersing aid for nondairy creamers; *Properties:* wh.-cream solid; sol. in propylene glycol, disp. in ethanol; m.p. 39-46 C; HLB 3.4; acid no. 5 max.; sapon. no. 170-180; 100% conc.

Mazol® PGMSK. [PPG/Specialty Chem.] Propylene glycol stearate; CAS 1323-39-3; EINECS 215-354-3; nonionic; coemulsifier for edible oil and shortenings; dispersant for nondairy creamers; in defoamers for beet sugar and yeast processing; *Regulatory:* FDA 21CFR §172.856; kosher; *Properties:* tan solid; sol. in min. spirits, toluene, IPA, ethanol, propylene glycol; sol. hot in min. oil, veg. oil; disp. hot in water; m.p. 39-46 C; HLB 3.4; acid no. 3 max.; iodine no. 2; sapon. no. 170-190; flash pt. (PMCC) > 350 F; 100% conc., 70% alpha mono.

Mazol® PGO-31 K. [PPG/Specialty Chem.] Triglyceryl oleate; CAS 9007-48-1; nonionic; food emulsifier; solubilizer for flavors and essential oils; *Regulatory:* FDA 21CFR §172.854, kosher; *Properties:* Gardner 8 max. liq.; sol. in min. oil, veg. oil, propylene glycol, toluene, IPA; disp. in ethanol, min. spirits; insol. in water; HLB 6.2; acid no. 3 max.; iodine no. 78; sapon. no. 140-150; 100% conc.

Mazol® PGO-104. [PPG/Specialty Chem.] Decaglyceryl tetraoleate; CAS 34424-98-1; EINECS 252-011-7; food emulsifier; solubilizer and carrier for essential oils and flavors; *Regulatory:* FDA 21CFR §172.854; kosher; *Properties:* dk. liq.; sol. in IPA, min. oil, veg. oil, toluene; disp. in ethanol, propylene glycol, min. spirits; insol. in water; HLB 6.2; acid no. 8 max.; iodine no. 61; sapon. no. 125-145; flash pt. (PMCC) > 350 F.

Mazu® 10 P Mod 11. [PPG/Specialty Chem.] Organic defoamer; defoamer for yeast fermentation; *Properties:* liq.; sp.gr. 0.990; visc. 250 cps; flash pt. (PMCC) > 300 F.

Mazu® 22. [PPG/Specialty Chem.] Organic defoamer; defoamer for yeast fermentation; *Properties:* liq.; sp.gr. 1.015; visc. 520 cps; flash pt. (PMCC) > 350 F.

Mazu® 87-C. [PPG/Specialty Chem.] Organic defoamer; defoamer for alcohol fermentation; *Properties:* liq.; sp.gr. 0.920; visc. 55 cps; flash pt. (PMCC) > 300 F.

Mazu® 150. [PPG/Specialty Chem.] Organic defoamer; defoamer for alcohol fermentation; *Properties:* liq.; sp.gr. 1.010; visc. 400 cps; flash pt. (PMCC) > 350 F.

Mazu® 201. [PPG/Specialty Chem.] Organic defoamer; defoamer for protein and starch; *Properties:* liq.; sp.gr. 0.915; visc. 130 cps; flash pt. (PMCC) > 300 F.

Mazu® 201 A. [PPG/Specialty Chem.] Organic defoamer; defoamer for protein and starch; *Properties:* liq.; sp.gr. 0.950; visc. 1000 cps; flash pt. (PMCC) > 300 F.

Mazu® 201 B. [PPG/Specialty Chem.] Organic defoamer; defoamer for protein and starch, distillation; *Properties:* liq.; sp.gr. 0.956; visc. 700 cps; flash pt. (PMCC) > 300 F.

Mazu® 201 PM. [PPG/Specialty Chem.] Organic defoamer; defoamer for protein and starch; *Properties:* liq.; sp.gr. 0.966; visc. 400 cps; flash pt. (PMCC) > 300 F.

Mazu® 204. [PPG/Specialty Chem.] Organic defoamer; defoamer for yeast fermentation, distillation; *Properties:* liq.; sp.gr. 1.010; visc. 400 cps; flash pt. (PMCC) > 350 F.

Mazu® 208. [PPG/Specialty Chem.] Organic defoamer; defoamer for protein and starch; *Properties:* emulsion; sp.gr. 0.971; visc. 2000 cps; flash pt. (PMCC) none.

Mazu® 280. [PPG/Specialty Chem.] Organic defoamer; defoamer for alcohol fermentation; *Properties:* liq.; sp.gr. 1.002; visc. 300 cps; flash pt. (PMCC) > 350 F.

Mazu® 285. [PPG/Specialty Chem.] Organic defoamer; defoamer for yeast and vitamin fermentation; *Properties:* liq.; sp.gr. 1.002; visc. 400 cps; flash pt. (PMCC) > 350 F.

Mazu® 286. [PPG/Specialty Chem.] Organic defoamer; defoamer for alcohol fermentation; *Properties:* paste; sp.gr. 0.992; visc. 1500 cps; flash pt. (PMCC) > 350 F.

Mazu® 287. [PPG/Specialty Chem.] Organic defoamer; defoamer for alcohol fermentation; *Properties:* liq.; sp.gr. 1.001; visc. 600 cps; flash pt. (PMCC) > 350 F.

Mazu® 288. [PPG/Specialty Chem.] Organic defoamer; defoamer for alcohol fermentation; *Proper-*

ties: paste; sp.gr. 1.005; visc. 1700 cps; flash pt. (PMCC) > 350 F.

Mazu® 289. [PPG/Specialty Chem.] Organic defoamer; defoamer for distillation, fermentation; *Properties:* liq.; sp.gr. 1.010; visc. 500 cps; flash pt. (PMCC) > 350 F.

Mazu® 307. [PPG/Specialty Chem.] Organic defoamer; defoamer for hot processing, food applics.; *Properties:* emulsion; sp.gr. 0.99; visc. 1300 cps; flash pt. (PMCC) none.

Mazu® 309. [PPG/Specialty Chem.] Organic defoamer; defoamer for food processing; *Properties:* emulsion; sp.gr. 1.001; visc. 6200 cps; flash pt. (PMCC) none.

Mazu® 322 A. [PPG/Specialty Chem.] Organic defoamer; defoamer for alcohol fermentation; *Properties:* liq.; sp.gr. 0.912; visc. 2800 cps; flash pt. (PMCC) > 300 F.

Mazu® 330 Mod 2. [PPG/Specialty Chem.] Organic defoamer; defoamer for yeast fermentation; *Properties:* liq.; sp.gr. 0.892; visc. 110 cps; flash pt. (PMCC) > 300 F.

Mazu® 352. [PPG/Specialty Chem.] Organic defoamer; defoamer for yeast fermentation; *Properties:* liq.; sp.gr. 0.985; visc. 350 cps; flash pt. (PMCC) > 350 F.

Mazu® DF 100S. [PPG/Specialty Chem.] Silicone compd.; food-grade defoamer for fermentation, veg. oils; *Regulatory:* FDA compliance; *Properties:* sp.gr. 0.99; flash pt. (PMCC) > 350 F; 100% silicone.

Mazu® DF 110S. [PPG/Specialty Chem.] Silicone emulsion; food-grade defoamer for fermentation, brine systems, cheese processing, chewing gum, food processing, wine making, yeast processing; *Regulatory:* FDA compliance; *Properties:* sp.gr. 1.00; flash pt. (PMCC) none; pH 7.0; 10% silicone.

Mazu® DF 130SAV. [PPG/Specialty Chem.] Silicone; food-grade defoamer; *Properties:* sp.gr. 1.00; flash pt. (PMCC) none; pH 7.0; 30% silicone.

Mazu® DF 200S. [PPG/Specialty Chem.] Silicone compd.; food-grade defoamer for fermentation, veg. oils; *Regulatory:* FDA compliance; *Properties:* liq.; sp.gr. 0.99; dens. 8.3 lb/gal; visc. 2000 cSt; flash pt. (PMCC) > 350 F; 100% silicone.

Mazu® DF 200SP. [PPG/Specialty Chem.] Simethicone; CAS 8050-81-5; food grade defoamer for fermentation, veg. oils; *Regulatory:* FDA compliance; *Properties:* liq.; sp.gr. 0.99; dens. 8.3 lb/gal; visc. 1720 cSt; flash pt. (PMCC) > 350 F; 100% silicone.

Mazu® DF 210S. [PPG/Specialty Chem.] Silicone emulsion; nonionic; food grade defoamer for fermentation, brine systems, cheese processing, chewing gum, food processing, wine making, yeast processing; *Regulatory:* FDA compliance; *Properties:* liq.; water-disp.; sp.gr. 1.00; dens. 8.3 lb/gal; visc. 400 cS; flash pt. (PMCC) none; pH 7.0; 10% silicone.

Mazu® DF 220S. [PPG/Specialty Chem.] Silicone; food-grade defoamer; *Regulatory:* FDA compliance; *Properties:* sp.gr. 1.00; flash pt. (PMCC) none; pH 6.0; 20% silicone.

Mazu® DF 230SP. [PPG/Specialty Chem.] Simethicone emulsion; CAS 8050-81-5; defoamer for food processing, fermentation, wine making, yeast processing; *Regulatory:* FDA compliance; *Properties:* sp.gr. 1.00; flash pt. (PMCC) none; pH 4.5; 30% silicone.

MB 100. [FMC] Low-sodium flavor protectant and meat binder for corned beef, self-basting turkey, meat patties and nuggets, marinated meat and poultry; *Regulatory:* USDA, FDA GRAS; *Properties:* sol. in water; pH 8-9 (1%).

MB 300. [FMC] flavor protectant recommended for processing requiring rapid cure-color development, e.g., corned beef, chopped and formed meats; enhances prod. shelf life by inhibiting lipid oxidation; *Regulatory:* USDA, FDA GRAS; *Properties:* sol. in water; pH 7.2 (1%).

MC². [ADM Arkady] Protase; CAS 9014-01-1; EINECS 232-752-2; enzyme-type dough conditioner producing mellow, pliable dough; provides higher amylase activity than MLO.

M-C-Thin®. [Lucas Meyer] Soy lecithin, standardized; CAS 8002-43-5; EINECS 232-307-2; amphoteric; emulsifier and dispersant for general and instantizing applics.; used in confections, baked goods, margarine, fillings, icings, powd. mixes; *Properties:* liq., paste; 100% conc.

M-C-Thin® AF-1 Type DB. [Lucas Meyer] Lecithin, natural, double bleached; CAS 8002-43-5; EINECS 232-307-2; amphoteric; emulsifier, wetting agent and stabilizer used in the food industry; *Regulatory:* FDA 21CFR §184.1400, GRAS; *Properties:* Gardner 12 max. translucent to opaque liq.; visc. 10,000 cps; acid no. 32 max.; 100% conc.

M-C-Thin® AF-1 Type SB. [Lucas Meyer] Lecithin, single bleached; CAS 8002-43-5; EINECS 232-307-2; emulsifier, wetting agent and stabilizer used in the food industry; *Regulatory:* GRAS, FDA 21CFR §184.1400; *Properties:* Gardner 14 max. translucent to opaque liq.; visc. 10,000 cps; acid no. 32 max.; 1% max. moisture.

M-C-Thin® AF-1 Type UB. [Lucas Meyer] Lecithin, unbleached; CAS 8002-43-5; EINECS 232-307-2; emulsifier, wetting agent and stabilizer used in the food industry; *Regulatory:* GRAS, FDA 21CFR §184.1400; *Properties:* Gardner 16 max. translucent to opaque liq.; visc. 10,000 cps; acid no. 32 max.; 1% max. moisture.

M-C-Thin® ASOL. [Lucas Meyer] Range of specially processed soy lecithin and selected oil; CAS 8002-43-5; EINECS 232-307-2; instantizing agent esp. effective for cold water-sol. and other difficult applics.; used for powd. mixes.

M-C-Thin® ASOL 436. [Lucas Meyer] Lecithin and veg. oil; instantizing aid for mfg. of cold water-sol. food prods.; facilitates dispersion, improves taste, promotes wettability, reduces sedimentation and flotation; high oxidative stability; *Regulatory:* GRAS, FDA 21CFR 184.1400; *Properties:* Gardner 14 max. color; visc. 1.25 poise max.; acid no. 20 max.; 1% max. moisture.

M-C-Thin® FWD. [Lucas Meyer] Fluid compound of soy lecithin and food grade emulsifier; CAS 8002-43-5; EINECS 232-307-2; emulsifier and dispersant for general and baking applics.

M-C-Thin® FWD 425. [Lucas Meyer] Lecithin and ethoxylated mono- and diglycerides (polyglycerate 60); multipurpose food additive; *Regulatory:* GRAS, FDA 21CFR 184.1400, 172.834; *Properties:* Gardner 14 max. color; water-disp.; visc. 10,000 cps; acid no. 28 max.; 1% max. moisture.

M-C-Thin® HL. [Lucas Meyer] Range of hydroxylated fluid soy lecithins; CAS 8029-76-3; EINECS 232-440-6; emulsifier, dispersant for baked goods, confections, powd. mixes, chewing gums; esp. effective in o/w emulsions; *Properties:* higher water dispersibility.

Meatbind®-3000. [ADM Ogilvie] Wheat gluten; CAS 8002-80-0; EINECS 232-317-7; filler binder in meat prods.; *Properties:* off-wh. powd.; 13% moisture, 30% protein, 3.5% fat.

Medium 7. [Pfizer Food Science] Dry sweet whey, sodium caseinate, disodium phosphate, yeast extract; dairy culture media; *Properties:* cream wh. powd., odor of whey; sol. in water; pH 6.5-6.8 (11.5%); *Toxicology:* nontoxic; may cause eye irritation or conjunctivitis, allergenic reaction by skin contact or inhalation; *Precaution:* incompat. with strong oxidizers.

Medium 10. [Pfizer Food Science] Dry sweet whey, nonfat dry milk, disodium phosphate, yeast extract; dairy culture media; used to prepare culture media for growth of thermophilic lactic acid bacteria; *Properties:* cream wh. powd., odor of whey; pH 6.5-6.8 (11.5%); *Toxicology:* nontoxic; may cause conjunctivitis, allergenic reaction on skin contact or inhalation; *Precaution:* incompat. with oxidizers.

Medium 55. [Pfizer Food Science] Dry sweet whey, sodium citrate, nonfat dry milk, yeast extract; low phosphate phage-resist. dairy culture media; *Properties:* cream wh. powd., odor of whey; pH 6.2-6.6 (11.5%); *Toxicology:* nontoxic; may cause eye irritation or conjunctivitis, allergenic reaction on skin contact or inhalation; *Precaution:* incompat. with strong oxidizers.

Medium 700. [Pfizer Food Science] Dry sweet whey, nonfat dry milk, yeast extract, disodium phosphate; phage-resist. dairy culture media; used to prepare culture medium for growth of thermophilic lactic acid bacteria; *Properties:* cream-wh. powd., odor of whey; sol. in water; pH 6.5-6.8 (11.5%); *Toxicology:* nontoxic; may cause eye irritation or conjunctivitis, allergenic reaction on skin contact or inhalation; *Precaution:* incompat. with strong oxidizers.

Medium KL. [Pfizer Food Science] Nonfat dry milk, sol. growth factors derived from *Saccharomyces cerevisiae*, monosodium phosphate, and disodium phosphate; culture media; *Properties:* pH 6.2-6.7 (11.5%).

Medium VS. [Pfizer Food Science] Dry sweet whey, sodium caseinate, sodium citrate, yeast extract; low phosphate phage-resist. dairy culture media; *Properties:* cream wh. powd., odor of whey; sol. in water; pH 6.2-6.6 (11.5%); *Toxicology:* nontoxic; may cause eye irritation or conjunctivitis, allergenic reaction on skin contact or inhalation; *Precaution:* incompat. with strong oxidizers.

Mekon® White. [Petrolite] Hard microcryst. wax; CAS 63231-60-7; EINECS 264-038-1; chewing gum base; *Regulatory:* incl. FDA §172.230, 172.615, 175.105, 175.300, 176.170, 176.180, 176.200, 177.1200, 178.3710, 179.45; *Properties:* wh. wax; very low sol. in org. solvs.; dens. 0.78 g/cc (99 C); visc. 15 cps (99 C); m.p. 94 C.

Merecol® FA. [Meer] Tragacanth, guar, xanthan, and locust bean gums; stabilizer for pourable salad dressing, potato salad, bakery, confectionery, flavors, emulsions, carbonated and flavored beverages with pulp; provides economical emulsification, body, and stability; *Usage level:* 0.4-0.6% (salad dressing); *Regulatory:* FDA GRAS; kosher; *Properties:* cream-colored powd.; ≥ 98% thru 60 mesh; readily sol. in hot and cold water; visc. 1000-1700 cps; pH 6.2-6.6 (1%).

Merecol® FAL. [Meer] Guar gum, xanthan gum, gum arabic; emulsion stabilizer for institutional salad dressings; provides economical emulsification, body, and stability; *Usage level:* 0.4-0.6% (salad dressing); *Regulatory:* FDA GRAS; kosher; *Properties:* cream-colored powd.; ≥ 98% thru 60 mesh; readily sol. in hot and cold water; visc. 1500-2500 cps; pH 6.0-6.5 (1%).

Merecol® FT. [Meer] Water-sol. vegetable gums; stabilizer for oil-free salad dressing; achieves body and texture of a low-calorie dressing without the use of oil; *Regulatory:* FDA GRAS; kosher; *Properties:* cream-wh. fine powd., nearly odorless; sol.: must be heated to 90 C for complete hydration.

Merecol® G. [Meer] Natural gums, cellulose gums, and seaweed extracts; stabilizer for low-calorie syrups; *Regulatory:* FDA GRAS; kosher; *Properties:* off-wh. powd.; completely sol. in hot water.

Merecol® GL. [Meer] Locust bean gum, guar gum; stabilizer for prod. of cream cheese and flour doughs by tying up free water and minimizing syneresis; *Usage level:* 0.25% (flour doughs, cheeses); *Regulatory:* FDA GRAS; kosher; *Properties:* creamy wh. fine powd., nearly odorless; 98% thru 60 mesh; visc. ≥ 500 cps (1%); pH 5.7-6.4.

Merecol® GX. [Meer] Guar gum, xanthan gum; stabilizer for processing of dry soup, sauce, and beverage mixes; provides rapid hydration; stable over broad pH range while in aq. media; improves mouthfeel, body, and texture; *Usage level:* 0.1-0.2%; *Regulatory:* FDA GRAS; kosher; *Properties:* cream-colored fine powd., nearly odorless; ≥ 97% thru 80 mesh; readily sol. in hot and cold water; visc. ≥ 2400 cps (1%); pH 5.5-6.5.

Merecol® I. [Meer] Veg. gums and xanthan gum; CAS 11138-66-2; EINECS 234-394-2; stabilizer, gellant providing exc. emulsion/heat stability in milk- and water-based pudding and gelled foods, pie fillings, bakery jellies, imitation sour cream, tomato aspic, chip dips, baby foods; *Regulatory:* FDA GRAS;

kosher; *Properties:* off-wh. powd.; completely sol. in hot water.

Merecol® IC. [Meer] Guar gum, carrageenan, and psyllium husks, standardized with dextrose; stabilizer for ice cream, ice milk; aids prod. of homogeneous mixt. of fat, milk solids, sugar, and water which will result in small ice crystals, uniform air cells when frozen; provides extra dryness for soft-serve items; high in dietary fiber; *Usage level:* 0.35-0.60%; *Regulatory:* FDA GRAS; kosher; *Properties:* cream-colored dry powd., neutral flavor; readily disp., sol. above 60 C in water; visc. 2000 cps (2%).

Merecol® K. [Meer] Blend of water-sol. gums; general purpose food stabilizer formulated for its ability to hydrate in cold water and milk; adds creaminess and stability to instant puddings and mousse formulations; *Usage level:* 0.25-0.50%; *Regulatory:* FDA GRAS; kosher; *Properties:* lt. tan powd.; completely sol. in cold water and/or milk; visc. 300-500 cps (1%).

Merecol® LK. [Meer] Tragacanth, guar, carrageenan, standardized with potassium chloride; stabilizer formulated as a replacement for xanthan; offers superior flowability, improved mouthfeel; *Usage level:* 0.35-0.60%; *Regulatory:* FDA GRAS; kosher for Passover; *Properties:* cream-colored powd.; ≥ 98% thru 60 mesh; readily sol. in hot and cold water; visc. ≥ 1000 cps (1%); pH 6.2-6.6 (1%).

Merecol® MS. [Meer] Xanthan gum FCC standardized with veg. gums; CAS 11138-66-2; EINECS 234-394-2; stabilizer for food applics., esp. pourable salad dressings; *Regulatory:* FDA GRAS; kosher; *Properties:* cream-colored powd.; ≥ 97% thru 80 mesh; readily sol. in hot and cold water; visc. ≥ 1200 cps (1%); pH 6.2-6.6 (1%).

Merecol® R. [Meer] Guar gum, carrageenan, standardized with calcium sulfate; stabilizer aiding prod. of ricotta cheese by minimizing syneresis and improving body texture and mouthfeel; extra dry; *Usage level:* 0.2-0.3% (ricotta cheese); *Regulatory:* FDA GRAS; kosher; *Properties:* creamy wh. powd., odorless; 95% thru 80 mesh; visc. ≥ 500 cps (1%); pH 5.5-6.5.

Merecol® RB. [Meer] Guar gum, locust bean gum, and carrageenan; stabilizer aiding ricotta cheese prod. by tying up free water, minimizing syneresis, and improving mouthfeel, body, and texture; produces ricotta cheese with exc. sliceability, moisture retention, and extended shelf life; *Regulatory:* FDA GRAS; kosher; *Properties:* creamy wh. fine powd., nearly odorless; 98% thru 60 mesh; visc. ≥ 1000 cps (1%); pH 5.5-6.5.

Merecol® RCS. [Meer] Xanthan gum, locust bean gum, and guar gum; stabilizer aiding ricotta cheese prod. by tying up free water, minimizing syneresis, improving mouthfeel, body, texture; produces ricotta with exc. sliceability, moisture retention, extended shelf life; also for cottage cheese dressings, spreads; *Usage level:* 0.1-0.2% (ricotta cheese), 0.4-0.5% (spreads); *Regulatory:* FDA GRAS; kosher; *Properties:* creamy wh. fine powd., nearly odorless; 98% thru 60 mesh; visc. ≥ 1000 cps (1%); pH 5.5-6.5; *Storage:* store in cool, dry area in closed container.

Merecol® S. [Meer] Modified starch, carrageenan, vegetable gum, and dextrose; stabilizer for use as a gelatin substitute in buttermilk, yogurt, and sour cream dressings; provides desirable set and mouthfeel; *Regulatory:* FDA GRAS; kosher; *Properties:* creamy wh. powd., odorless; sol.: must be heated to 60 C for complete hydration; visc. ≥ 125 cps (2%); *Storage:* store in cool, dry area in closed container.

Merecol® SH. [Meer] Guar gum and psyllium husks standardized with dextrose; stabilizer for ice milk, sherbet, fruit ice, lollies, popsicles; assists in homogeneous mixt. of sugar and water which will result in small ice crystals and uniform air cells when frozen; high in dietary fiber; *Usage level:* 0.35-0.60%; *Regulatory:* FDA GRAS; kosher; *Properties:* cream-colored dry powd., neutral flavor; readily disp., sol. > 60 C in water; visc. ≥ 550 cps.

Merecol® Y. [Meer] Modified starch, carrageenan, veg. gum psyllium, and dextrose; stabilizer for use as a gelatin substitute in yogurt, sour cream, and salad dressings; provides desirable set and mouthfeel; high in dietary fiber; *Regulatory:* FDA GRAS; kosher; *Properties:* geige-tan powd., odorless; sol.: must be heated to 60 C for complete hydration; visc. ≥ 125 cps (2%).

Merezan® 8. [Meer] Xanthan gum FCC from *Xanthomonas campestris*; CAS 11138-66-2; EINECS 234-394-2; high m.w. polysaccharide for the food industry; *Properties:* cream-colored powd., ≥ 97% thru 80 mesh; readily sol. in hot and cold water; visc. ≥ 1200 cps (1%); pH 6.1-8.1 (1%); *Storage:* store in clean, cool, dry area; 1 yr min. shelf life.

Merezan® 20. [Meer] Xanthan gum FCC from *Xanthomonas campestris*; CAS 11138-66-2; EINECS 234-394-2; high m.w. polysaccharide for the food industry; *Properties:* cream-colored powd., ≥ 97% thru 200 mesh; readily sol. in hot and cold water; visc. ≥ 1200 cps (1%); pH 6.1-8.1; *Storage:* store in clean, cool, dry area; 1 yr min. shelf life.

Meringue Stabilizer. [Bunge Foods] Boiling-type stabilizer for meringue toppings, whipped toppings; *Usage level:* 2%; *Properties:* wh. to off-wh. powd., bland flavor; pH 6.7.

Metarin™. [Lucas Meyer] Range of specially processed fluid soy lecithins; CAS 8002-43-5; EINECS 232-307-2; instantizing and release agents for food release, powd. mixes; designed for ease of applic.

Metarin™ C. [Lucas Meyer] Lecithin; CAS 8002-43-5; EINECS 232-307-2; instantizing aid for caseinates; *Properties:* powd.

Metarin™ CP. [Lucas Meyer] Deoiled soy lecithin; CAS 8002-43-5; EINECS 232-307-2; instantizing agent with enhanced props.; esp. effective in flavor-sensitive systems, powd. mixes; *Properties:* powd.

Metarin™ DA 51. [Lucas Meyer] Lecithin; CAS 8002-43-5; EINECS 232-307-2; multipurpose food additive; sprayable at R.T.; can be dissolved in other fats and oils; *Regulatory:* GRAS, FDA 21CFR §184.1400; *Properties:* Gardner 14 max. color; visc. 1500 cps; acid no. 25 max.; 0.8% max. moisture.

Metarin™ F. [Lucas Meyer] Lecithin compd.; CAS 8002-43-5; EINECS 232-307-2; nonionic; hydrophilic instantizer for foods; cold sprayable; *Properties:* liq.; 100% conc.

Metarin™ P. [Lucas Meyer] Deoiled soy lecithin; CAS 8002-43-5; EINECS 232-307-2; nonionic; hydrophilic instantizer for powdery foods, milk powds., baby food; esp. effective in flavor-sensitive systems; *Properties:* powd.; sol. in oil and fat; 100% conc.

Methocel® A4C Premium. [Dow] Methylcellulose; CAS 9004-67-5; food gums used as thickener, stabilizer, emulsifier, adhesive, and gellant; contributes firm thermal gel structure to structured meat, veg., and poultry prods., meat, poultry, and seafood batters during frying; gravies; *Usage level:* 0.15% (structured meat), 0.2-0.5% (meat batter), 0.3-0.5% (gravy); *Regulatory:* FDA GRAS 21CFR §182.1480, 9CFR 318.7, 381.147, FCC, USP, EP, JP, BP; kosher; *Properties:* firm gel structure; water-sol.; visc. 400 mPa•s.

Methocel® A4M Premium. [Dow] Methylcellulose; CAS 9004-67-5; food gum, thickener, stabilizer, emulsifier, adhesive, and gellant; improves dough release, binding str. of tortillas, tacos; lubricity aid in extruded snack food; in pie fillings, structured potato/seafood/meat prods, meat/poultry battera, gravies; *Usage level:* 0.1-0.75% (taco, snack foods), 0.3-0.5% (pie fillings, structured potato prods.), 1-3% (seafood); *Regulatory:* FDA GRAS 21CFR §182.1480, 9CFR 318.7, 381.147, FCC, USP, EP, JP, BP; kosher; *Properties:* firm gel structure; water-sol.; visc. 4000 mPa•s.

Methocel® A15C Premium. [Dow] Methylcellulose; CAS 9004-67-5; food gums used as thickener, stabilizer, emulsifier, adhesive, and gellant; *Regulatory:* FDA GRAS 21CFR §182.1480, 9CFR 318.7, 381.147, FCC, USP, EP, JP, BP; kosher; *Properties:* visc. 1500 cps.

Methocel® A15LV Premium. [Dow] Methylcellulose; CAS 9004-67-5; food gums used as thickener, stabilizer, emulsifier, adhesive, and gellant; improves volume and moisture in microwave cakes; imparts body and fat-like texture to reduced fat cheese sauces; *Usage level:* 0.4-0.8% (cake), 1% (reduced-fat cheese sauce); *Regulatory:* FDA GRAS 21CFR §182.1480, 9CFR 318.7, 381.147, FCC, USP, EP, JP, BP; kosher; *Properties:* wh./off-wh. powd.; visc. 15 mPa•s.

Methocel® E3 Premium. [Dow] Hydroxypropyl methylcellulose; CAS 9004-65-3; visc. control agent, gellant, lather enhancer/stabilizer, film-former, dispersant, lubricant, binder, emulsion stabilizer, and suspending agent for foods; *Regulatory:* FDA 21CFR §172.874, USDA, FCC, USP, EP, JP, BP; kosher; *Properties:* wh./off-wh. powd.; visc. 3 mPa•s.

Methocel® E4M. [Dow] Hydroxypropyl methylcellulose; CAS 9004-65-3; food gums used as thickener, stabilizer, emulsifier, adhesive, and gellant; *Properties:* semifirm gel structure; water-sol.; visc. 4000 cps (2% aq.).

Methocel® E4M Premium. [Dow] Hydroxypropyl methylcellulose; CAS 9004-65-3; visc. control agent, gellant, lather enhancer/stabilizer, film-former, dispersant, lubricant, binder, emulsion stabilizer, suspending agent for foods; improves binding str., aids release in extruded seafood; improves mouthfeel in salad dressing; *Usage level:* 1-3% (seafood), 0.2-0.8% (dressings); *Regulatory:* FDA 21CFR §172.874, USDA, FCC, USP, EP, JP, BP; kosher; *Properties:* wh./off-wh. powd.; visc. 4000 mPa•s.

Methocel® E5 Food Grade. [Dow] Hydroxypropyl methylcellulose; CAS 9004-65-3; food gums used as thickener, stabilizer, emulsifier, adhesive, and gellant; *Regulatory:* FDA, FCC, USP, EP, JP, BP; kosher; *Properties:* visc. 5 cps.

Methocel® E6 Premium. [Dow] Hydroxypropyl methylcellulose; CAS 9004-65-3; visc. control agent, gellant, film-former, dispersant, lubricant, binder, emulsion stabilizer, and suspending agent for foods; *Regulatory:* FDA 21CFR §172.874, USDA, FCC, USP, EP, JP, BP; kosher; *Properties:* wh./off-wh. powd.; visc. 6 mPa•s.

Methocel® E15 Food Grade. [Dow] Hydroxypropyl methylcellulose; CAS 9004-65-3; food gums used as thickener, stabilizer, emulsifier, adhesive, and gellant; improves binding str. and reduces oil absorp. during frying in tacos, meat/poultry/seafood batters, french fry dips; *Usage level:* 0.1-0.75% (taco), 0.2-0.5% (meat batter); *Regulatory:* FDA, FCC, USP, EP, JP, BP; kosher; *Properties:* visc. 15 cps.

Methocel® E15LV Premium. [Dow] Hydroxypropyl methylcellulose; CAS 9004-65-3; visc. control agent, gellant, film-former, dispersant, lubricant, binder, emulsion stabilizer, and suspending agent for foods; *Regulatory:* FDA 21CFR §172.874, USDA, FCC, USP, EP, JP, BP; kosher; *Properties:* wh./off-wh. powd.; visc. 15 mPa•s.

Methocel® E50LV Premium. [Dow] Hydroxypropyl methylcellulose; CAS 9004-65-3; visc. control agent, gellant, film-former, lubricant, binder, emulsion stabilizer, and suspending agent for foods; controls body of whipped toppings; in edible moisture barrier coatings; emulsion stabilizer, dispersant in flavor oil emulsions; *Usage level:* 0.4% (whipped toppings), 30-90% (coatings), 10-16% (flavors); *Regulatory:* FDA 21CFR §172.874, USDA, FCC, USP, EP, JP, BP; kosher; *Properties:* visc. 50 mPa•s.

Methocel® F4M Premium. [Dow] Hydroxypropyl methylcellulose; CAS 9004-65-3; food gums used as thickener, stabilizer, emulsifier, adhesive, and gellant; provides moist texture in reduced-fat cakes; provides firm thermal gel structure to structured meat, veg., and poultry prods.; *Usage level:* 0.2-0.4% (cake), 0.8% (structured meat); *Regula-*

tory: FDA 21CFR §172.874, USDA, FCC, USP, EP, JP, BP; kosher; *Properties:* vics. 4000 cps.

Methocel® F50LV Premium. [Dow] Hydroxypropyl methylcellulose; CAS 9004-65-3; thickener, stabilizer, emulsifier, adhesive, gellant; improves moisture, volume, reduces surf. stickiness in muffins, cakes incl. microwave; improves pancake batters, meat/poultry/seafood batters during frying; also in whipped toppings, edible films; *Usage level:* 0.03-0.3% (muffins), 0.4% (cake), 0.5-1.0% (pancake), 0.2-0.5% (meat batter); *Regulatory:* FDA 21CFR §172.874, USDA, FCC, USP, EP, JP, BP; kosher; *Properties:* visc. 50 cps.

Methocel® K3 Premium. [Dow] Hydroxypropylmethylcellulose; CAS 9004-65-3; used for foods; improves freeze/thaw stability in bakery glazes; provides elasticity and improves adhesion of candy coatings; emulsion stabilizer, dispersant for flavor oil emulsions and encapsulation; *Usage level:* 0.1-10.2% (glaze), 0.5-12% (candy coatings), 10-16% (flavors); *Regulatory:* FDA 21CFR §172.874, USDA, FCC, USP, EP, JP, BP; kosher; *Properties:* wh./off-wh. powd.; visc. 3 mPa•s.

Methocel® K4M Premium. [Dow] Hydroxypropyl methylcellulose; CAS 9004-65-3; visc. control, gellant, film-former, dispersant, lubricant, binder, emulsion stabilizer, suspending agent; improves dough release in tortillas, binding in tacos; lubricity aid in extruded snacks; ice cream cones, icings, structured meats, desserts; *Usage level:* 1-3% basis flour (rice flour bread), 0.2-0.4% wet dough basis (tortilla), 0.1-0.75% (taco); *Regulatory:* FDA 21CFR §172.874, USDA, FCC, USP, EP, JP, BP; kosher; *Properties:* visc. 4000 mPa•s.

Methocel® K15M Premium. [Dow] Hydroxypropyl methylcellulose; CAS 9004-65-3; visc. control agent, gellant, film-former, dispersant, lubricant, binder, emulsion stabilizer, and suspending agent for foods; stabilizes and extends shelf life of salad dressings; also in gravies, whipped toppings, desserts, diet foods; *Usage level:* 0.3-0.5% (dressings, gravies), 0.1-0.5% (desserts), 0.3-0.8% (diet foods); *Regulatory:* FDA 21CFR §172.874, USDA, FCC, USP, EP, JP, BP; kosher; *Properties:* wh./off-wh. powd.; visc. 15,000 mPa•s.

Methocel® K35 Premium. [Dow] Hydroxypropyl methylcellulose; CAS 9004-65-3; food gums used as thickener, stabilizer, emulsifier, adhesive, and gellant; *Regulatory:* FDA 21CFR §172.874, USDA, FCC, USP, EP, JP, BP; kosher; *Properties:* visc. 35 cP.

Methocel® K100LV Premium. [Dow] Hydroxypropyl methylcellulose; CAS 9004-65-3; gellant, film-former, dispersant, lubricant, binder, emulsion stabilizer, suspending agent; improves moisture, volume in muffins, cakes; improves texture of butter cream icings; stabilizer for structured potato prods.; whipped toppings, mousses; *Usage level:* 0.03-0.25% (muffins), 0.15-0.8% (cake), 0.1-0.2% powd. sugar basis (icings), 0.3-0.5% (potato); *Regulatory:* FDA 21CFR §172.874, USDA, FCC, USP, EP, JP, BP; kosher; *Properties:* wh./off-wh.

powd.; visc. 100 mPa•s.

Methocel® K100M Premium. [Dow] Hydroxypropyl methylcellulose; CAS 9004-65-3; visc. control agent, gellant, film-former, dispersant, lubricant, binder, emulsion stabilizer, and suspending agent for foods; improves dough release in tortillas, binding in taco shells/chips, extruded onion rings; improves pie fillings; stabilizer; *Usage level:* 0.1-0.75% (taco), 0.3-0.5% (pie fillings, dressings), 0.5-2.5% (onion), 1% (glaze); *Regulatory:* FDA 21CFR §172.874, USDA, FCC, USP, EP, JP, BP; kosher; *Properties:* wh./off-wh. powd.; visc. 100,000 mPa•s.

Mexpectin HV 400 Range. [Grindsted Prods.] Pectin; CAS 9000-69-5; EINECS 232-553-0; gellant for fruit drinks; protein stabilizer for cultured milk prods.; *Regulatory:* EEC, FDA §184.1588 (GRAS).

Mexpectin LA 100 Range. [Grindsted Prods.] Amidated low-ester pectin; CAS 9000-69-5; EINECS 232-553-0; gellant and mouthfeel modifier for fruit-based prods, low-sugar jams/jellies with 50-60% solids, yogurt fruit; protein stabilizer for cultured milk prods.; *Regulatory:* EEC, FDA §184.1588 (GRAS).

Mexpectin LA 200 Range. [Grindsted Prods.] Amidated low-ester pectin; CAS 9000-69-5; EINECS 232-553-0; gellant for low-sugar fruit preps., esp. low-sugar jams and jellies with 45-55% solids, bakery fillings, yogurt and yogurt fruit; protein stabilizer for cultured milk prods.; *Regulatory:* EEC, FDA §184.1588 (GRAS).

Mexpectin LA 400 Range. [Grindsted Prods.] Amidated low-ester pectin; CAS 9000-69-5; EINECS 232-553-0; gellant for low-sugar fruit preps., esp. low-sugar jams and jellies with 30-45% solids, bakery jams, fruit glazes, yogurt, yogurt fruit; protein stabilizer for cultured milk prods.; *Regulatory:* EEC, FDA §184.1588 (GRAS).

Mexpectin LC 700 Range. [Grindsted Prods.] Low-ester pectin; CAS 9000-69-5; EINECS 232-553-0; thickener and mouthfeel modifier for fruit-based prods., high- and low-sugar jams/jellies, toppings, and ripples; protein stabilizer for cultured milk prods.; *Regulatory:* EEC, FDA §184.1588 (GRAS).

Mexpectin LC 800 Range. [Grindsted Prods.] Low-ester pectin; CAS 9000-69-5; EINECS 232-553-0; thickener and mouthfeel modifier for fruit-based prods., bakery jams, low-sugar jams/jellies, toppings and ripples; protein stabilizer for cultured milk prods.; *Regulatory:* EEC, FDA §184.1588 (GRAS).

Mexpectin LC 900 Range. [Grindsted Prods.] Low-ester pectin; CAS 9000-69-5; EINECS 232-553-0; thickener and mouthfeel modifier for fruit-based prods., bakery jams, low-sugar jams/jellies, yogurt fruit; protein stabilizer for cultured milk prods.; *Regulatory:* EEC, FDA §184.1588 (GRAS).

Mexpectin MRS 300 Range. [Grindsted Prods.] Pectin; CAS 9000-69-5; EINECS 232-553-0; gellant, protein stabilizer for jams, jellies, bakery fillings; protein stabilizer for cultured milk prods.; med. setting time; *Regulatory:* EEC, FDA §184.1588 (GRAS).

Mexpectin RS 400 Range. [Grindsted Prods.] Pectin; CAS 9000-69-5; EINECS 232-553-0; gellant, protein stabilizer for jams, jellies, cold-setting flan jellies, fruit drinks, carbonated soft drinks; protein stabilizer for cultured milk prods.; rapid setting time, high setting temp.; *Regulatory:* EEC, FDA §184.1588 (GRAS).

Mexpectin SS 200 Range. [Grindsted Prods.] Pectin; CAS 9000-69-5; EINECS 232-553-0; gellant, protein stabilizer for jams, jellies, fruit preps.; protein stabilizer for cultured milk prods.; slow setting time; *Regulatory:* EEC, FDA §184.1588 (GRAS).

Mexpectin XSS 100 Range. [Grindsted Prods.] Pectin; CAS 9000-69-5; EINECS 232-553-0; gellant and mouthfeel modifier for fruit-based prods., confectionery, fruit jellies, high-solids fruit fillings; protein stabilizer for cultured milk prods.; extra slow setting time, low setting temp.; *Regulatory:* EEC, FDA §184.1588 (GRAS).

MG-60. [Am. Ingreds.] Vital wheat gluten, polysorbate 60, partially hydrog. soy oil, sodium stearoyl lactylate; modified gluten for wh. bread, variety breads, hamburger and hot dog buns; *Properties:* honey-colored powd.; 5-8% moisture, 58% min. protein.

MG-100 K. [Am. Ingreds.] Vital wheat gluten and monoglyceride; modified gluten for wh. bread, variety breads, hamburger and hot dog buns; *Properties:* cream-colored fine powd.; 17% min. mono, 60% min. protein.

MG-F. [Am. Ingreds.] Vital wheat gluten, ethoxylated mono- and diglycerides; modified gluten for high-fiber bread prods.; *Properties:* free-flowing powd.; water-disp.; 6.5% moisture, 70.3% min. protein.

MG-S. [Am. Ingreds.] Vital wheat gluten, sodium stearoyl lactylate, calcium stearoyl-2-lactylate, lactylic stearate; modified gluten for standardized baked prods.; *Properties:* cream-colored fine powd.; 4.5% moisture, 56% protein.

Microcatalase®. [Solvay Enzymes] Bacterial catalase derived from *Micrococcus lysodeikticus*; CAS 9001-05-2; EINECS 232-577-1; enzyme which removes residual hydrogen peroxide after antimicrobial treatment of milk; *Properties:* liq.

Microduct®. [CasChem] Maltodextrin; CAS 9050-36-6; EINECS 232-940-4; food additive; *Properties:* powd.; 100% act.

Microlube System 999R. [Advanced Food Systems] Dry mix of stabilizers/emulsifiers which turn into emulsion when mixed with water and oil (water: Microlube: oil = 72:10:18); prevents dehydration and sticking of frozen or refrigerated precooked pasta, rice on microwave/conventional oven cooking.

MicroPlus™. [Pfizer Food Science] Proprietary emulsion; browning and crisping agent providing appealing brown surf. color to poultry, pot pies, and baked goods during microwaving and retaining crispness in pre-fried frozen foods in the microwave; colorant is released under microwave conditions; *Properties:* tan visc. liq., faint fatty odor; misc. with water; sp.gr. 0.88-1.00; visc. 2000-12,000 cP;

b.p. 110 C (760 mm); pH 3-4; *Toxicology:* nontoxic; may be mild eye irritant; *Precaution:* may be slip hazard when spilled; *Storage:* store refrigerated in closed container; do not freeze.

Micro-White® 10 Codex. [ECC Int'l.] Calcium carbonate; CAS 1317-65-3; EINECS 207-439-9; food-grade calcium supplement and extender meeting purity requirements for food and food contact applics. (flour, cake mixes, cereals, chewing gum, crackers); *Regulatory:* FCC, kosher; *Properties:* wh. powd., odorless, tasteless; 1.0 µm mean particle size; trace retained on 325 mesh; negligible sol. in water; sp.gr. 2.71; pH 9-10; 97% $CaCO_3$; *Toxicology:* considered nuisance dust; TLV-TWA 10 mg/m^3.

Micro-White® 25 Codex. [ECC Int'l.] Calcium carbonate; CAS 1317-65-3; EINECS 207-439-9; food-grade calcium supplement and extender meeting purity requirements for food and food contact applics. (flour, cake mixes, cereals, chewing gum, crackers); *Regulatory:* FCC, kosher; *Properties:* wh. powd., odorless, tasteless; 3 µm mean particle size; trace retained on 325 mesh; negligible sol. in water; sp.gr. 2.71; pH 9-10; 96.8% $CaCO_3$; *Toxicology:* considered nuisance dust; TLV-TWA 10 mg/m^3.

Micro-White® 50 Codex. [ECC Int'l.] Calcium carbonate; CAS 1317-65-3; EINECS 207-439-9; food-grade calcium supplement and extender meeting purity requirements for food and food contact applics. (flour, cake mixes, cereals, chewing gum, crackers); *Regulatory:* FCC, kosher; *Properties:* wh. powd., odorless, tasteless; 7 µm mean particle size; 0.35% retained on 325 mesh; negligible sol. in water; sp.gr. 2.71; pH 9-10; 95.5% $CaCO_3$; *Toxicology:* considered nuisance dust; TLV-TWA 10 mg/m^3.

Micro-White® 100 Codex. [ECC Int'l.] Calcium carbonate; CAS 1317-65-3; EINECS 207-439-9; food-grade calcium supplement and extender meeting purity requirements for food and food contact applics. (flour, cake mixes, cereals, chewing gum, crackers); *Regulatory:* FCC, kosher; *Properties:* wh. powd., odorless, tasteless; 17 µm mean particle size; 7% retained on 325 mesh; negligible sol. in water; sp.gr. 2.71; pH 9-10; 95% $CaCO_3$; *Toxicology:* considered nuisance dust; TLV-TWA 10 mg/m^3.

Miglyol® 808. [Hüls Am.; Hüls AG] Tricaprylin; CAS 538-23-8; EINECS 208-686-5; special oil for the prod. of dietary foods containing MCT; *Properties:* colorless to sl. yel. liq.; dens. 0.95 kg/dm^3; visc. 25-35 mPa•s; turbidity pt. -5 C; acid no. 0.2 max.; iodine no. 1 max.; sapon. no. 350-360.

Miglyol® 810. [Hüls Am.; Hüls AG] Caprylic/capric triglyceride; CAS 65381-09-1; EINECS 265-724-3; nonionic; lubricant, carrier/vehicle, solv. for dietary prods.; *Properties:* colorless liq., neutral odor; sol. in diethyl ether, petrol. ether, chloroform, IPA, toluene, alcohol, min. oil, acetone; sp.gr. 0.94-0.95; visc. 25-35 mPa•s; acid no. 0.1 max.; iodine no. 0.5 max.; sapon. no. 335-355; pH neutral; cloud pt. 0 C.

Miglyol® 810 N. [Hüls AG] Glyceryl tricaprylate/tricaprate; EINECS 265-724-3; special oil for the prod. of dietary foods containing MCT; *Properties:* liq.; dens. 0.95 kg/dm^3; visc. 25-35 mPa•s; turbidity pt. 0 C; acid no. 0.2 max.; iodine no. 1 max.; sapon. no. 335-355.

Miglyol® 812. [Hüls Am.; Hüls AG] Caprylic/capric triglyceride; CAS 65381-09-1; EINECS 265-724-3; nonionic; dispersant, lubricant, anticaking agent, carrier, solv., solubilizer, suspending agent for dietetic prods.; *Properties:* colorless liq.; sol. in alcohol, min. oil, acetone; visc. 25-35 mPa•s; acid no. 0.1 max.; iodine no. 0.5 max.; sapon. no. 325-345; cloud pt. 10 C; 100% conc.

Miglyol® 812 N. [Hüls AG] Glyceryl tricaprylate/tricaprate; EINECS 265-724-3; special oil for the prod. of dietary foods containing MCT; *Properties:* liq.; dens. 0.95 kg/dm^3; visc. 25-35 mPa•s; turbidity pt. 10 C; acid no. 0.2 max.; iodine no. 1 max.; sapon. no. 325-345.

Miglyol® 812 S. [Hüls AG] Glyceryl tricaprylate/tricaprate; EINECS 265-724-3; lubricant, release agent; oil for surf. treating sticky prods., e.g., sweets, liquorice, jelly babies, wine gums, confectionery, dried fruit; basic oil for bakery release agents; lubricant in prod. of sausage casings; carrier for flavors; *Properties:* liq.; dens. 0.95 kg/dm^3; visc. 25-35 mPa•s; turbidity pt. 10 C; acid no. 0.2 max.; iodine no. 1 max.; sapon. no. 325-345.

Miglyol® 820. [Hüls AG] Veg. oil triglycerides; lubricant and release agent for the surface treatment of prods. with a tendency to stickiness, e.g., sweets, liquorice, jelly babies, wine gums, confectionery, dried fruit; bakery release oil; *Properties:* liq.; dens. 0.93 kg/dm^3; visc. 60 mPa•s; turbidity pt. 15 C; acid no. 0.2 max.; iodine no. 30; sapon. no. 255-275.

Miglyol® 840. [Hüls Am.; Hüls AG] Propylene glycol dicaprylate/dicaprate; dispersant, lubricant, carrier, solv., solubilizer, suspending agent for dietetic prods.; *Properties:* colorless liq.; sol. in alcohol, min. oil, acetone; visc. 8-14 mPa•s; acid no. 0.1 max.; iodine no. 0.5 max.; sapon. no. 320-340; cloud pt. -30 C.

Miglyol® 850. [Hüls AG] Veg. oils triglyceride; lubricant, release agent; oil for the surf. treatment of prods. with a tendency to stickiness, e.g., sweets, liquorice, jelly babies, wine gums, confectionery, dried fruit; basic oil for bakery release agents; *Properties:* liq.; dens. 0.93 kg/dm^3; visc. 50 mPa•s (40 C); turbidity pt. 45 C; acid no. 1 max.; iodine no. 30; sapon. no. 260.

Miglyol® 8108. [Hüls AG] Glyceryl tricaprylate/tricaprate; EINECS 265-724-3; special oil for the prod. of dietary foods containing MCT; *Properties:* liq.; dens. 0.95 kg/dm^3; visc. 25-35 mPa•s; turbidity pt. 0 C; acid no. 0.2 max.; iodine no. 1 max.; sapon. no. 335-355.

Milk Calcium ND (Food Grade). [Garuda Int'l.] Natural calcium peptide source from cow's milk; used to prepare nutritional supplement tablets or capsules for prevention of osteoporosis and to provide usable source of mineral calcium; *Properties:* wh. to lt.

cream free-flowing powd., neutral odor, mineral-salt flavor; 7% max. moisture; *Storage:* store in cool, dry area; protect from moisture and strong odors; 1 yr min. shelf life.

Mira-Bake® 505. [A.E. Staley Mfg.] Modified food starch derived from dent corn; stabilizer; specialized starch providing unique two-stage thickening, particle suspension/protection during baking to fruit pie fillings, microwaveable foods, gravies, meat/vegetable entrees, stews; inhibits ice crystal formation in frozen foods; *Regulatory:* FDA 21CFR §172.892; *Properties:* wh. powd., good odor and flavor; 0.5% max. on 60 mesh, 95.5% min. thru 100 mesh; pH 5.0-6.5; 10% moisture.

Mira-Cap®. [A.E. Staley Mfg.] Food starch modified, derived from waxy maize; lipophilic starch for emulsification and encapsulation of oils incl. flavor oils in dry mixes (beverage powds., bakery and confectionery blends); emulsifier in salad dressings, imitation dairy prods., nondairy coffee whiteners; film-forming props.; *Regulatory:* FDA 21CFR §172.892; *Properties:* wh. to off-wh. noncaking powd., bland flavor; cold water-sol.; visc. 100 cp (80 F, 30% solids); pH 6.5; 6% max. moisture.

Mira-Cleer® 187. [A.E. Staley Mfg.] Food starch modified, derived from dent corn; high visc. thickener, stabilizer in low-acid or neutral foods, dry pudding mixes, ice cream toppings, bakery fillings, gravies, dairy-based sauces, stews; resists. thinning, congealing; contributes desirable mouthfeel, freeze/thaw stability; *Regulatory:* FDA 21CFR §172.892; *Properties:* wh. free-flowing powd.; readily mixes in cold water, pH 5.7 (10% slurry); 10% moisture.

Mira-Cleer® 340. [A.E. Staley Mfg.] Food starch modified, derived from dent corn; thickener for neutral to low pH foods, aseptic puddings, cheese sauces, tomato sauces, fruit fillings, canned retorted prods.; binder for cooked structured foods; resist. to congealing; provides stable visc., good shelf life under refrigeration; *Regulatory:* FDA 21CFR §172.892; *Properties:* wh. free-flowing powd.; disp. in cold water; pH 5.7 (10% slurry); 10% moisture.

Mira-Cleer® 516. [A.E. Staley Mfg.] Food starch modified, derived from dent corn; thickener for highly acidic foods, or foods processed at high retorting temps.; exc. temp. stability; for canned pet foods, condiment sauces, canned fruit fillings for pies; *Regulatory:* FDA 21CFR §172.892; *Properties:* wh. free-flowing powd.; disp. in cold water; pH 5.7 (10% slurry); 11.5% moisture.

Mira-Cleer® 516. [A.E. Staley Mfg.] Food starch modified, derived from dent corn; cook-up starch for highly acidic foods or foods processed at high retorting temps.; thickener for canned pet foods, condiment sauces, canned fruit fillings for pies, tarts, and pastries; *Regulatory:* FDA 21CFR §172.892; *Properties:* wh. powd.; disp. in cold water; pH 5.7 (10% slurry); 11.5% moisture.

Miracle Pie Dough Base. [ADM Arkady] Functional base for prod. of pie crust and other chemically leavened dough; provides immediate dough relaxation and prevents toughening/shrinkage by providing mixing tolerance.

Mira-Foam® 100. [A.E. Staley Mfg.] Soy albumen; CAS 68153-28-6; basic whipping agent for confections providing same aerating as egg albumen but with better fat tolerance and without the problem of egg beatdown; *Properties:* lt. colored free-flowing powd.; highly disp. in hot or cold water.

Mira-Foam® 100K. [A.E. Staley Mfg.] Soy albumen; CAS 68153-28-6; basic whipping agent for confections providing same aerating as egg albumen but with better fat tolerance and without the problem of egg beatdown; *Regulatory:* kosher; *Properties:* lt. colored free-flowing powd.; highly disp. in hot or cold water.

Mira-Foam® 120V. [A.E. Staley Mfg.] Soy albumen with stabilizers; CAS 68153-28-6; whipping agent; provides same aerating as egg albumen but with better fat tolerance and without the problem of egg beatdown; *Properties:* lt. colored free-flowing powd.; highly disp. in hot or cold water.

Mira-Foam® 130H. [A.E. Staley Mfg.] Soy albumen; CAS 68153-28-6; high performance whipping agent providing same aerating as egg albumen but with better fat tolerance and without the problem of egg beatdown; *Properties:* lt. colored free-flowing powd.; highly disp. in hot or cold water.

Mira-Gel® 463. [A.E. Staley Mfg.] Corn starch; CAS 9005-25-8; EINECS 232-679-6; cold water swelling gran. starch which forms gels @ R.T.; used for gel/structured applics., texture modification, high-moisture desserts, cold batters, cookie doughs, bakery fillings, low-moisture confections, icings, microwave preps.; *Properties:* wh. powd.; 0.1% max. on 30 mesh, 95% min. thru 100 mesh; pH 5.5; 5% moisture.

Miranol® J2M Conc. [Rhone-Poulenc Surf. & Spec.] Disodium capryloamphodiacetate; CAS 68608-64-0; amphoteric; emulsifier; food washing and peeling; *Properties:* clear liq.; pH 8.2-8.6; 48-50% solids.

Miranol® JB. [Rhone-Poulenc Surf. & Spec.] Disodium capryloamphodiacetate; CAS 68608-64-0; amphoteric; wetting agent used in caustic soda based cleaners used for food washing and peeling; *Properties:* liq.; 49% conc.; formerly Mirapon JB.

Mira-Quik® MGL. [A.E. Staley Mfg.] Food starch modified derived from potato; gelling agent, stabilizer for gelled food prods. such as confectionery jellies, bakery fillings, desserts; outstanding gel clarity; high amylose, gelatin replacement; *Regulatory:* FDA 21CFR §172.892; *Properties:* off-wh. powd., bland flavor; pH 5; 12% moisture.

Mira-Set® 285. [A.E. Staley Mfg.] Food starch modified derived from dent corn; thin-boiling confectioner's starch for prod. of jelly candies, extrusion of fruit pieces or fillings; *Regulatory:* FDA 21CFR §172.892; *Properties:* wh. powd.; 0.1% on 60 mesh, 99% thru 100 mesh; pH 5.5; 7% moisture.

Mira-Set® B. [A.E. Staley Mfg.] Dent corn starch; CAS 9005-25-8; EINECS 232-679-6; quick-setting confectioner's starch, gellant for producing starch jelly confections; *Properties:* wh. fine powd.; pH 5.0; 10% moisture.

Mira-Set® J. [A.E. Staley Mfg.] Food starch modified; quick-setting confectioner's starch used with fructose and HFCS sweeteners to produce high quality, tender-textured jujube and gelatin type chewy jellies and fruit snack confections; rapid gelling; reduces surface stickiness; *Regulatory:* FDA 21CFR §172.892; *Properties:* wh. fine powd.; pH 5.0; 10% moisture.

Mira-Sperse®. [A.E. Staley Mfg.] Food starch modified; instant starch for reduced-calorie, dry-mix formulations which lack carbohydrate loads; suspends flavors; for puddings, sauces, fillings, glazes, dressings, microwave applics.; disperses without aid of bulking agents; *Properties:* powd.; 2% on 100 mesh; hydrates in cold water @ 30-60 s delay; visc. 10,000 cps (10 min); pH 5-7; 4-8% moisture.

Mira-Sperse® 535. [A.E. Staley Mfg.] Food starch modified, derived from waxy maize; pregelatinized instant starch with exc. dispersibility, high visc., stability to acid and heat processing; for cold-processed or dry mix sauces, gravies, fillings; *Properties:* powd.; 10% mon 30 mesh, 30% mans. thru 120 mesh; bulk dens. 14-16 lb/ft³; pH 5.0-6.0; 8.5% max. moisture; *Storage:* store in cool, dry environment; avoid exposure to excessive heat and humidity.

Mira-Sperse® 623. [A.E. Staley Mfg.] Food starch modified, derived from waxy maize; cold water swelling starch with exc. flow props.; thickener, stabilizer in no-oil processed foods, low sugar systems, hot water dispersed dry mixes, acidic food systems, systems which undergo severe processing; *Regulatory:* FDA 21CFR §172.892; *Properties:* agglomerated powd., good odor/flavor; 20% on 30 mesh, 30% thru 120 mesh; bulk dens. 11-16 lb/ft³; pH 5-7; 6-8.5% moisture.

Mira-Sperse® 626. [A.E. Staley Mfg.] Food starch modified, derived from waxy maize; cold water swelling starch with exc. flow props.; thickener, stabilizer in no-oil processed foods, low sugar systems, hot water dispersed dry mixes, neutral to mildly acidic food systems, systems processed under mild conditions; *Regulatory:* FDA 21CFR §172.892; *Properties:* agglomerated powd., good odor/flavor; 20% on 30 mesh, 30% thru 120 mesh; bulk dens. 11-16 lb/ft³; pH 5-7; 6-8.5% moisture.

Mira-Thik® 468. [A.E. Staley Mfg.] Food starch modified, derived from dent corn starch; cold water hydrating starch, thickener, stabilizer for acidic foods; provides superior smoothness, syneresis control, freeze/thaw resist.; for dry instant foods, pourable/spoonable salad dressings, fruit pie fillings, pickle relishes, microwave foods; *Regulatory:* FDA 21CFR §172.892; *Properties:* wh. free-flowing powd., good odor/flavor; 0.5% on 60 mesh, 98% thru 100 mesh; bulk dens. 30 lb/ft³; pH 6.0; 5% moisture.

Mira-Thik® 469. [A.E. Staley Mfg.] Food starch

modified, derived from dent corn starch; cold water hydrating starch, thickener, stabilizer for neutral to mildly acidic foods; for dry instant foods, puddings, glazes, sauces, soups, gravies, pastry fillings, dairy/microwave prods.; improves texture, mouth-feel; exc. freeze/thaw stability; *Regulatory:* FDA 21CFR §172.892; *Properties:* wh. free-flowing powd., good odor/flavor; 0.5% on 60 mesh, 98% thru 100 mesh; bulk dens. 30 lb/ft³; pH 6.0; 5% moisture.

Mira-Thik® 603. [A.E. Staley Mfg.] Food starch modified, derived from waxy maize; instant starch, thickener, stabilizer for dry mix and processed foods, acidic food systems, systems which undergo severe processing, spoonable/pourable salad dressings, fruit pie fillings, acidic sauces, microwave applics.; exc. freeze/thaw stability; *Regulatory:* FDA 21CFR §172.892; *Properties:* wh. free-flowing powd., good odor/flavor; 0.5% on 60 mesh, 98% thru 100 mesh; bulk dens. 33 lb/ft³ (loose); pH 6.0; 7% moisture.

Mira-Thik® 606. [A.E. Staley Mfg.] Food starch modified, derived from waxy maize; instant starch, thickener, stabilizer for dry mix and processed foods, neutral to mildly acidic food systems, systems processed under mild conditions, gravies, sauces, bakery fillings, dairy prods.; exc. freeze/thaw stability; *Regulatory:* FDA 21CFR §172.892; *Properties:* wh. free-flowing powd., good odor/flavor; 0.5% on 60 mesh, 98% thru 100 mesh; bulk dens. 33 lb/ft³ (loose); pH 6.0; 7% moisture.

Mitas®. [Asahi Chem. Industry] Compound seasonings; for food industry.

Mix 41. [Vaessen-Schoemaker] Soy protein isolate, milk protein, and carbohydrates; food supplement for children and adults, esp. for sportsmen and for people who perform hard physical labor; *Properties:* pleasant taste and flavor; 41% protein.

Mix 63. [Vaessen-Schoemaker] Soy protein isolate and carbohydrates; CAS 68153-28-6; food supplement esp. suited for slimming purposes since it takes more energy to digest than it contains; low in fat, cholesterol- and lactose-free; daily consumption of 75-100 g supplies FAO/WHO-required amt. of protein; 63% protein.

Mix 70. [Vaessen-Schoemaker] Soy protein isolate, milk protein, and carbohydrates; food supplement similar to Mix 63, suited for slimming purposes; increased protein content permits lower daily intake.

MLO. [ADM Arkady] Protease-type; CAS 9014-01-1; EINECS 232-752-2; enzyme dough conditioner producing mellow, pliable dough, reducing mixing time by 15-25%; also avail. in double strength and tablet form; *Properties:* powd.

Moldban. [MLG Enterprises Ltd.] Water, propylene glycol, sodium benzoate, potassium sorbate, food grade enzymes, and citric acid; mold inhibitor, preservative for cakes, pies, pastries, pie fillings (custard and fruit), toppings, icings, jams, jellies, non-yeast raised baked goods; may also be used to wash down conveyors and other surfaces in contact with baked goods; *Usage level:* 4-7 oz/100 lb (cakes), 6-7 oz/100 lb (pie dough), 5-6 oz/1 gal water (washing conveyors); *Properties:* liq., odorless, tasteless; 20% sodium benzoate, 2% methyl paraben, 1% propyl paraben; *Storage:* 6 mos shelf life stored under cool, dry conditions.

Monolan® OM Range. [Harcros UK] EO/PO copolymers and esters; component for sugar beet and yeast defoamers.

Monolan® PK. [Harcros UK] EO/PO block polymer; nonionic; defoamer component for sugar beet and yeast prod.; *Properties:* pale yel. clear liq., mild odor; sp.gr. 1.031; visc. 500 cs; flash pt. (PMCC) > 150 C; 100% act.

Monolan® PL. [Harcros] High m.w. EO/PO copolymer; nonionic; defoamer component for sugar beet and yeast prod.; *Properties:* pale yel. clear liq., mild odor; sp.gr. 0.973; visc. 930 cs; flash pt. (PMCC) > 150 C; 100% act.

Monomax AH90 B. [Australian Bakels] Distilled monoglycerides; emulsifier/stabilizer for ice cream, margarine, shortening, peanut butter, confectionery, other foods; chewing gum plasticizer; starch complexing agent in pastas; *Properties:* wh. to off-wh. beads; sp.gr. 0.91 (80 C); m.p. 66-70 C; iodine no. 2 max.; sapon. no. 150-165; 90% min. monoester.

Monomax VH90 B. [Australian Bakels] Distilled monoglycerides; emulsifier/stabilizer for ice cream, margarine, shortening, peanut butter, confectionery, other foods; chewing gum plasticizer; starch complexing agent in pastas; *Properties:* wh. to off-wh. beads; sp.gr. 0.91 (80 C); m.p. 67-73 C; iodine no. 3 max.; sapon. no. 150-165; 90% min. monoester.

Monomuls® Range. [Grünau GmbH] Fatty acid mono- and diglycerides from vegetable or animal sources; general purpose food emulsifiers for baked goods; > 60% mono content.

Monomuls® 60-10. [Grünau GmbH] Pork lard mono- and diglycerides; nonionic; emulsifier for margarine, baking shortenings, mayonnaise, dressings, meat/milk prods., confectionery; stabilizer improving aeration in ice cream; starch complexing aid in baked goods, pasta, prolonging freshness; improves chewing gum and toffees; *Regulatory:* EEC E 471; *Properties:* cream-colored lardy solid, neutral fatty odor/taste; sol. in warm xylene, ethanol, IPA; sol. cloudy in warm IPM, propylene glycol; disp. in warm water; m.p. 40-44 C; acid no. < 3; iodine no. 40-50; sapon. no. 165-175; pH 6-7 (1:10 in methanol/water); 57-62% monoglyceride; *Storage:* 1 yr storage life.

Monomuls® 60-15. [Grünau GmbH] Hydrog. lard mono- and diglycerides; nonionic; emulsifier for margarine, baking shortenings, mayonnaise, dressings, meat/milk prods., confectionery; stabilizer improving aeration in ice cream; starch complexing aid in baked goods, pasta, prolonging freshness; improves chewing gum and toffees; *Regulatory:* EEC E 471; *Properties:* ylsh.-wh. powd., neutral odor and taste; sol. in warm ethanol, IPA,

xylene, soja oil; sol. cloudy in IPM; disp. in warm water; m.p. 60-64 C; acid no. < 3; iodine no. < 2; sapon. no. 165-175; pH 6-7 (1:10 in methanol/ water); 57-62% monoglycerides; *Storage:* 1 yr storage life.

Monomuls® 60-20. [Grünau GmbH] Beef tallow mono- and diglycerides; CAS 67701-27-3; nonionic; emulsifier for margarine, baking shortenings, mayonnaise, dressings, meat/milk prods., confectionery; stabilizer improving aeration in ice cream; starch complexing aid in baked goods, pasta, prolonging freshness; improves chewing gum and toffees; *Regulatory:* EEC E 471; *Properties:* cream-colored lardy solid, neutral fatty odor/taste; sol. in warm ethanol, IPA, IPM, xylene, paraffin oil; disp. in warm water; m.p. 48-52 C; acid no. < 3; iodine no. 32-42; sapon. no. 165-175; pH 6-7 (1:10 in methanol/water 1:1); 57-62% monoglycerides; *Storage:* 1 yr storage life.

Monomuls® 60-25. [Grünau GmbH] Hydrog. beef tallow glycerides; CAS 68308-54-3; EINECS 269-658-6; nonionic; emulsifier for margarine, baking shortenings, mayonnaise, dressings, meat/milk prods., confectionery; stabilizer improving aeration in ice cream; starch complexing aid in baked goods, pasta, prolonging freshness; improves chewing gum and toffees; *Regulatory:* EEC E 471; *Properties:* ylsh.-wh. powd., neutral odor and taste; sol. in warm ethanol, IPA, IPM, xylene, soja oil; disp. in warm water; m.p. 60-64 C; acid no. < 3; iodine no. < 2; sapon. no. 165-175; pH 6-7 (1:10 in methanol/ water 1:1); 57-62% monoglycerides; *Storage:* 1 yr storage life.

Monomuls® 60-25/2. [Grünau GmbH] Hydrog. beef tallow mono- and diglycerides with 2% sodium stearate; nonionic; self-emulsifying emulsifier for margarine, baking shortenings, mayonnaise, dressings, meat/milk prods., confectionery; stabilizer improving aeration in ice cream; starch complexing aid in baked goods, pasta, prolonging freshness; improves chewing gum; *Regulatory:* EEC E 471; *Properties:* cream-colored powd., neutral odor and taste; sol. in warm ethanol, IPA, xylene, paraffin oil, soja oil; disp. in warm water; m.p. 60-74 C; acid no. < 4; iodine no. < 2; sapon. no. 160-170; pH 8-9 (1:10 in methanol/water 1:1); 54-59% monoglycerides; *Storage:* 1 yr storage life.

Monomuls® 60-25/5. [Grünau GmbH] Hydrog. beef tallow mono- and diglycerides with 5% sodium stearate; self-emulsifying emulsifier for margarine, baking shortenings, mayonnaise, dressings, meat/milk prods., confectionery; stabilizer improving aeration in ice cream; starch complexing aid in baked goods, pasta, prolonging freshness; improves chewing gum; *Regulatory:* EEC E 471; *Properties:* cream-colored powd., neutral odor and taste; sol. warm in ethanol, IPA, IPM, xylene, paraffin oil, soja oil; disp. warm in water; m.p. 60-64 C; acid no. < 4; iodine no. < 2; sapon. no. 155-165; pH 8.5-9.5 (1:10 in methanol/water 1:1); 51-56% monoglycerides; *Storage:* 1 yr shelf life.

Monomuls® 60-30. [Grünau GmbH] Palm oil mono-

and diglycerides; anionic; emulsifier for margarine, baking shortenings, mayonnaise, dressings, meat/milk prods., confectionery; stabilizer improving aeration in ice cream; starch complexing aid in baked goods, pasta, prolonging freshness; improves chewing gum and toffees; *Regulatory:* EEC E 471; *Properties:* ylsh.-wh. lardy solid, sl. lardy odor/taste; sol. in warm ethanol, IPA, xylene, soya oil, cloudy in IPM; disp. in warm water; m.p. 38-42 C; acid no. < 3; iodine no. 40-50; sapon. no. 165-175; pH 6-7 (1:10 in methanol/water 1:1); 57-62% monoglycerides; *Storage:* 1 yr storage life.

Monomuls® 60-35. [Grünau GmbH] Hydrog. palm oil glycerides; nonionic; emulsifier for margarine, baking shortenings, mayonnaise, dressings, meat/milk prods., confectionery; stabilizer improving aeration in ice cream; starch complexing aid in baked goods, pasta, prolonging freshness; improves chewing gum and toffees; *Regulatory:* EEC E 471; *Properties:* ylsh.-wh. powd., neutral odor/taste; sol. in warm propylene glycol, ethanol, IPA, IPM, xylene; disp. in warm water; m.p. 58-62 C; acid no. < 3; iodine no. < 6; sapon. no. 165-175; pH 6-7 (1:10 in methanol/water 1:1); 57-62% monoglycerides; *Storage:* 1 yr storage life.

Monomuls® 60-40. [Grünau GmbH] Sunflower seed oil mono- and diglycerides; nonionic; emulsifier for margarine, baking shortenings, mayonnaise, dressings, meat/milk prods., confectionery; stabilizer improving aeration in ice cream; starch complexing aid in baked goods, pasta, prolonging freshness; improves chewing gum and toffees; *Regulatory:* EEC E 471; *Properties:* ylsh.-wh. lardy soft solid, fatty odor and taste; sol. in warm ethanol, IPA, xylene, soya oil, cloudy in IPM; disp. in warm water; m.p. 28-32 C; acid no. < 3; iodine no. 100-120; sapon. no. 165-175; pH 6-7 (1:10 in methanol/water 1:1); 57-62% monoglycerides; *Storage:* 1 yr storage life.

Monomuls® 60-45. [Grünau GmbH] Hydrog. soybean oil mono- and diglycerides; CAS 68201-48-9; nonionic; emulsifier for margarine, baking shortenings, mayonnaise, dressings, meat/milk prods., confectionery; stabilizer improving aeration in ice cream; starch complexing aid in baked goods, pasta, prolonging freshness; improves chewing gum and toffees; *Regulatory:* EEC E 471; *Properties:* ylsh.-wh. powd., neutral odor; sol. in warm ethanol, IPA, IPM, xylene; disp. in warm water; m.p. 62-66 C; acid no. < 3; iodine no. < 4; sapon. no. 165-175; pH 6-7 (1:10 in methanol/water 1:1); 57-62% monoglycerides; *Storage:* 1 yr storage life.

Monomuls® 90-10. [Grünau GmbH] Dist. pork lard glyceride; CAS 61789-10-4; EINECS 263-032-6; nonionic; emulsifier for margarine, baking shortenings, mayonnaise, dressings, meat/milk prods., confectionery; stabilizer improving aeration in ice cream; starch complexing aid in baked goods, pasta, prolonging freshness; improves chewing gum and toffees; *Regulatory:* EEC E 471; *Properties:* lt cream-colored lardy solid, neutral lardy odor and taste; sol. in warm propylene glycol, ethanol,

IPA, IPM, xylene, soja oil; insol. in water; m.p. 56-60 C; acid no. < 3; iodine no. 37-47; sapon. no. 155-165; pH 4-5 (1:10 in methanol/water 1:1); 90-95% monoglycerides; *Storage:* 1 yr storage life.

Monomuls® 90-15. [Grünau GmbH] Dist. hydrog. pork lard glyceride; CAS 8040-05-9; nonionic; emulsifier for margarine, baking shortenings, mayonnaise, dressings, meat/milk prods., confectionery; stabilizer improving aeration in ice cream; starch complexing aid in baked goods, pasta, prolonging freshness; improves chewing gum and toffees; *Regulatory:* EEC E 471; *Properties:* wh. powd., neutral odor/taste; sol. in warm ethanol, IPA, IPM, xylene; disp. in warm water; m.p. 69-73 C; acid no. < 3; iodine no. < 2; sapon. no. 155-165; pH 4-5 (1:10 in methanol/water 1:1); 90-95% monoglycerides; *Storage:* 1 yr storage life.

Monomuls® 90-20. [Grünau GmbH] Dist. beef tallow glyceride; CAS 61789-13-7; EINECS 263-035-2; nonionic; emulsifier for margarine, baking shortenings, mayonnaise, dressings, meat/milk prods., confectionery; stabilizer improving aeration in ice cream; starch complexing aid in baked goods, pasta, prolonging freshness; improves chewing gum and toffees; *Regulatory:* EEC E 471; *Properties:* lt. cream-colored lardy solid, neutral fatty odor/taste; sol. in warm propylene glycol, ethanol, IPA, IPM, xylene, soja oil; insol. in water; m.p. 57-61 C; acid no. < 3; iodine no. 30-40; sapon. no. 155-165; pH 4-5 (1:10 in methanol/water 1:1); 90-95% monoglycerides; *Storage:* 1 yr storage life.

Monomuls® 90-25. [Grünau GmbH] Dist. hydrog. beef tallow glyceride; CAS 61789-09-1; EINECS 263-031-0; nonionic; emulsifier for margarine, baking shortenings, mayonnaise, dressings, meat/milk prods., confectionery; stabilizer improving aeration in ice cream; starch complexing aid in baked goods, pasta, prolonging freshness; improves chewing gum and toffees; *Usage level:* 1-5%; *Regulatory:* EEC E 471; *Properties:* wh. powd., neutral odor; sol. in warm ethanol, IPA, IPM, xylene; disp. in warm water; m.p. 68-72 C; HLB 3.8; acid no. < 3; iodine no. < 2; sapon. no. 155-165; pH 4-5 (1:10 methanol/water 1:1); 90-95% monoglycerides; *Storage:* 1 yr storage life.

Monomuls® 90-25/2. [Grünau GmbH] Dist. hydrog. beef tallow glyceride with 2% sodium stearate; anionic; self-emulsifying emulsifier for margarine, baking shortenings, mayonnaise, dressings, meat/milk prods., confectionery; stabilizer improving aeration in ice cream; starch complexing aid in baked goods, pasta, prolonging freshness; improves chewing gum; *Regulatory:* EEC E 471; *Properties:* lt. cream-colored powd., neutral odor and taste; sol. warm in ethanol, IPA, xylene, paraffin oil; disp. warm in water; m.p. 66-70 C; acid no. < 4; iodine no. < 2; sapon. no. 150-160; pH 8-9 (1:10 in methanol/water 1:1); 87-92% monoglycerides; *Storage:* 1 yr storage life.

Monomuls® 90-25/5. [Grünau GmbH] Dist. hydrog. beef tallow glyceride, 5% sodium stearate; self-emulsifying emulsifier for margarine, baking shortenings, mayonnaise, dressings, meat/milk prods., confectionery; stabilizer improving aeration in ice cream; starch complexing aid in baked goods, pasta, prolonging freshness; improves chewing gum; *Regulatory:* EEC E 471; *Properties:* lt. cream-colored powd., neutral odor and taste; sol. warm in ethanol, IPA, xylene, paraffin oil; disp. warm in water; m.p. 66-70 C; acid no. < 4; iodine no. < 2; sapon. no. 145-155; pH 8.5-9.5 (1:10 in methanol/water 1:1); 80-85% monoglycerides; *Storage:* 1 yr shelf life.

Monomuls® 90-30. [Grünau GmbH] Dist. palm oil glyceride; nonionic; emulsifier for margarine, baking shortenings, mayonnaise, dressings, meat/milk prods., confectionery; stabilizer improving aeration in ice cream; starch complexing aid in baked goods, pasta, prolonging freshness; improves chewing gum and toffees; *Regulatory:* EEC E 471; *Properties:* lt. cream-colored lardy solid, fatty odor and taste; sol. in warm propylene glycol, ethanol, IPA, IPM, xylene, soja oil; insol. in water; m.p. 55-59 C; acid no. < 3; iodine no. 37-47; sapon. no. 155-165; pH 4-5 (1:10 in methanol/water 1:1); 90-95% monoglycerides; *Storage:* 1 yr storage life.

Monomuls® 90-35. [Grünau GmbH] Dist. hydrog. palm oil glyceride; CAS 67784-87-6; nonionic; emulsifier for margarine, baking shortenings, mayonnaise, dressings, meat/milk prods., confectionery; stabilizer improving aeration in ice cream; starch complexing aid in baked goods, pasta, prolonging freshness; improves chewing gum and toffees; *Regulatory:* EEC E 471; *Properties:* wh. powd., neutral odor and taste; sol. in warm propylene glycol, ethanol, IPA, IPM, xylene; disp. in warm water; m.p. 64-68 C; acid no. < 3; iodine no. < 6; sapon. no. 155-165; pH 4-5 (1:10 in methanol/water 1:1); 90-95% monoglycerides; *Storage:* 1 yr storage life.

Monomuls® 90-40. [Grünau GmbH] Dist. sunflower seed oil glyceride; nonionic; emulsifier for margarine, baking shortenings, mayonnaise, dressings, meat/milk prods., confectionery; stabilizer improving aeration in ice cream; starch complexing aid in baked goods, pasta, prolonging freshness; improves chewing gum and toffees; *Regulatory:* EEC E 471; *Properties:* wh. lardy soft solid, fatty odor and taste; sol. in warm propylene glycol, ethanol, IPA, IPM, xylene, soja oil; insol. in water; m.p. 29-33 C; acid no. < 3; iodine no. 100-120; sapon. no. 155-165; pH 4-5 (1:10 in methanol/water 1:1); 90-95% monoglycerides; *Storage:* 1 yr storage life.

Monomuls® 90-45. [Grünau GmbH] Dist. hydrog. soybean oil glyceride; nonionic; emulsifier for margarine, baking shortenings, mayonnaise, dressings, meat/milk prods., confectionery; stabilizer improving aeration in ice cream; starch complexing aid in baked goods, pasta, prolonging freshness; improves chewing gum and toffees; *Regulatory:* EEC E 471; *Properties:* wh. powd., neutral odor and taste; sol. in warm ethanol, IPA, IPM, xylene, paraffin oil; disp. in warm water; m.p. 70-74 C; acid no. < 3; iodine no. < 4; sapon. no. 155-165; pH 4-5 (1:10

in methanol/water 1:1); 90-95% monoglycerides; *Storage:* 1 yr storage life.

Monomuls® 90-O18. [Henkel/Cospha; Henkel KGaA/Cospha; Grünau GmbH] Glyceryl oleate; CAS 111-03-5; nonionic; w/o emulsifier, stabilizer, refatting agent, thickener for foods; *Properties:* almost wh. paste, sl. fatty odor; HLB 3.4; acid no. 3 max.; iodine no. 67-80; sapon. no. 150-160; 100% conc.

Morillol®. [BASF AG] 1-Octene-3-ol; mushroom-like, earthy, herbaceous fragrance and flavoring.

Morton® Flour Lite Salt™ Mixt. [Morton Salt] Sodium chloride, potassium chloride, with anticaking agents (tricalcium phosphate, magnesium carbonate); food-grade table salt blend; *Regulatory:* GRAS, exempt from label declaration under FDA 21CFR §101.100(a)(3) for tricalcium phosphate and magnesium carbonate; *Properties:* powd.; 11-28% retained 100 mesh, 22-35% retained 140 mesh, 8-21% retained 200 mesh; mean crystal size 130 μm; mean surf. area 310 cm²/g; bulk dens. 0.94-1.04 g/ml (loose); 49.23% NaCl, 49.23% KCl.

Morton® Flour Salt. [Morton Salt] Extra fine sodium chloride FCC with anticaking additive (1.5% tricalcium phosphate); food-grade salt ground to be extremely fine and free flowing; provides nutritionally avail. calcium from its anticaking additive; for food processing (cereal, flour, powd. cheese, dry mixes, dough conditioners, frostings, meat cures, tenderizers); *Regulatory:* tricalcium phosphate: GRAS, exempt from label declaration under 21CFR §101.100(a)(3); *Properties:* fine powd.; 25-37% retained on 140 mesh, 20-27% on 200 mesh; mean crystal size 100 μm; mean surf. area 400 cm²/g; bulk dens. 0.96-1.06 ml (loose); 98.5% NaCl.

Morton® H.G. Blending Prepared Salt. [Morton Salt] Sodium chloride FCC with anticaking agent (0.4% sodium silicoaluminate); food-grade salt; used for condiments; *Regulatory:* sodium silicoaluminate: GRAS, FDA 21CFR §182.2727, exempt from label declaration under 21CFR §101.100(a)(3); *Properties:* powd.; 33-89% retained 40 mesh, 12-49% retained 50 mesh (Hutchinson); mean crystal size 380 μm; mean surf. area 76 cm²/g; bulk dens. 1.15-1.23 g/ml (loose); 99.85% NaCl.

Morton® H.G. Blending Salt. [Morton Salt] Sodium chloride FCC; CAS 7647-14-5; EINECS 231-598-3; food-grade salt; *Regulatory:* sodium silicoaluminate: GRAS, FDA 21CFR §182.2727, exempt from label declaration under 21CFR § 101.100 (a)(3); *Properties:* gran.; 33-89% retained 40 mesh, 12-49% retained 50 mesh (Hutchinson); mean crystal size 380 μm; mean surf. area 76 cm²/g; bulk dens. 1.15-1.23 g/ml (loose); 99.85% NaCl.

Morton® Lite Salt® Mixt. [Morton Salt] Sodium chloride, potassium chloride, with free-flow agent (magnesium carbonate); food-grade table salt blend imparting similar degree of saltiness and seasoning enhancement as common salt while contributing only one-half the sodium; mild saline taste of NaCl masks disagreeable bitter aftertaste of KCl in most food preps.; *Regulatory:* magnesium carbonate: GRAS, exempt from label declaration under 21CFR §101.100(a)(3); *Properties:* powd.; 30-60% retained 50 mesh, 10-30% retained 70 mesh; mean crystal size 380 µm; mean surf. area 80 cm²/g; bulk dens. 1.09-1.15 g/ml (loose); 49.85% NaCl, 49.85% KCl.

Morton® Lite Salt® TFC Mixt. [Morton Salt] Sodium chloride, potassium chloride, with anticaking agent (sodium ferrocyanide); food-grade table salt blend; *Regulatory:* anticaking per 21CFR §172.490, exempt from food label declaration on food under 21CFR §101.100(a)(3); *Properties:* powd.; 30-60% retained 50 mesh, 10-30% retained 70 mesh; mean crystal size 380 µm; mean surf. area 80 cm²/g; bulk dens. 1.09-11.5 g/ml (loose); 49.95% NaCl, 49.95% KCl.

Morton® Star Flake® Dendritic ES Salt. [Morton Salt] Sodium chloride FCC with anticaking additive (0.5% sodium silicoaluminate), crystal modifying agent (sodium ferrocyanide); high purity food-grade salt; *Regulatory:* sodium silicoaluminate: GRAS, 21CFR §182.2727; sodium silicoaluminate and sodium ferrocyanide: exempt from label declaration under 21CFR 101.100(a)(3); *Properties:* powd.; 25-45% retained on 70 mesh, 16-38% on 100 mesh; mean crystal size 230 µm; mean surf. area 270 cm²/g; bulk dens. 0.85-0.93 g/ml (loose); 99.9% NaCl.

Morton® Star Flake® Dendritic Salt. [Morton Salt] Sodium chloride FCC with crystal modifying agent (sodium ferrocyanide); high purity food-grade salt with high specific surf. area, rapid dissolution rate/flowability, high liq. adsorptive capacity/caking resist., low apparent density; for comminuted meats, seasonings/flavors, cheese processing, oil-roasted snacks, etc.; *Regulatory:* sodium ferrocyanide: FDA 21CFR §172.490; *Properties:* porous star-shaped cryst. cubes; 25-45% retained on 70 mesh, 16-38% on 100 mesh; mean crystal size 230 µm; mean surf. area 270 cm²/g; bulk dens. 0.85-0.93 g/ml (loose); 99.9% NaCl.

Morton® Table Iodized Salt. [Morton Salt] Sodium chloride, sodium silicoaluminate (anticaking agent), dextrose (stabilizer for potassium iodide), potassium iodide (source of nutritional iodine); food-grade table salt serving as a source of dietary iodine; *Regulatory:* FCC, GRAS; *Properties:* powd.; 33-89% retained on 40 mesh, 12-49% retained on 50 mesh (Hutchinson); 380 µm crystal size (Rittman); bulk dens. 1.15-1.23 g/ml (loose); 99.3% NaCl.

Morton® TFC 999® Salt. [Morton Salt] Sodium chloride FCC with trace anticaking agent (sodium ferrocyanide); CAS 7647-14-5; EINECS 231-598-3; food-grade salt; used for continuous cheddar/colby cheese processing; *Regulatory:* sodium ferrocyanide: 21CFR §172.490; exempt from label declaration under 21CFR 101.100(a)(3); *Properties:* powd.; 44-64% retained on 50 mesh, 17-36% on 70 mesh; mean crystal size 340 µm; mean surf. area 84 cm²/g; bulk dens. 1.2-1.27 g/ml (loose); 99.97% NaCl.

Morton® TFC H.G. Blending Salt. [Morton Salt]

Sodium chloride FCC with anticaking agent (sodium ferrocyanide); food-grade salt; *Regulatory:* sodium ferrocyanide: FDA 21CFR §172.490, exempt from label declaration under 21CFR §101.100(a)(3); *Properties:* powd.; 33-89% retained 40 mesh, 12-49% retained 50 mesh (Hutchinson); mean crystal size 380 µm; mean surf. area 76 cm²/g; bulk dens. 1.15-1.23 g/ml (loose); 99.85% NaCl.

Morton® Top Flake Coarse Salt. [Morton Salt] Sodium chloride FCC with anticaking agent (sodium ferrocyanide); food-grade salt; used for vat process cheddar/colby cheese, Swiss cheese, soda crackers, condiments; *Regulatory:* sodium ferrocyanide: 21CFR §172.490, exempt from label declaration under 21CFR 101.100(a)(3); *Properties:* powd.; 19-57% retained on 20 mesh, 17-26% retinaed on 30 mesh; mean crystal size 940 µm; mean surf. area 70 cm²/g; bulk dens. 0.93-1.0 g/ml (loose); 99.82% NaCl.

Morton® Top Flake Extra Coarse Salt. [Morton Salt] Sodium chloride FCC with anticaking agent (sodium ferrocyanide); food-grade salt; *Regulatory:* sodium ferrocyanide: 21CFR §172.490, exempt from label declaration under 21CFR 101.100(a)(3); *Properties:* powd.; 16-39% retained on 16 mesh, 16-25% retained on 20 mesh, 10-20% retained on 30 mesh; mean crystal size 990 µm; mean surf. area 66 cm²/g; bulk dens. 0.9-1.0 g/ml (loose); 99.82% NaCl.

Morton® Top Flake Fine Salt. [Morton Salt] Sodium chloride FCC with anticaking agent (sodium ferrocyanide); food-grade salt; *Regulatory:* sodium ferrocyanide: 21CFR §172.490, exempt from label declaration under 21CFR 101.100(a)(3); *Properties:* powd.; 15-40% retained on 40 mesh, 32-50% on 50 mesh; mean crystal size 340 µm; mean surf. area 120 cm²/g; bulk dens. 1.07-1.12 g/ml (loose); 99.82% NaCl.

Morton® Top Flake Topping Salt. [Morton Salt] Sodium chloride FCC with anticaking agent (sodium ferrocyanide); food-grade salt; used for snack crackers; *Regulatory:* sodium ferrocyanide: 21CFR §172.490, exempt from label declaration under 21CFR 101.100(a)(3); *Properties:* powd.; 35-49% retained on 30 mesh, 34-50% retained on 40 mesh; mean crystal size 600 µm; mean surf. area 95 cm²/g; bulk dens. 1.0-1.1 g/ml (loose); 99.82% NaCl.

Mr. Chips™. [Bunge Foods] Partially hydrog. soybean and cottonseed oils, lecithin; shortening, icing stabilizer, glazes, candy, refrigerated dough; *Properties:* flakes, 2.0 red max. color, bland flavor; m.p. 119 F.

Muffin Kote. [ADM Arkady] Corn meal, corn flour, and a mold inhibitor; English muffin coating which uniformly disperses over muffin surface for exc. mold protection.

Muffit. [MLG Enterprises Ltd.] Calcium sulfate, citric acid, calcium citrate, sodium chloride, azodicarbonamide, and fungal enzymes; conc. dough conditioning system for English muffins; provides desirable porous grain and tart flavor; *Usage level:* 8

oz/cwt flour; *Properties:* free-flowing powd.; *Storage:* 6 mos shelf life under cool, dry conditions.

Multiwax® 180-M. [Witco/Petroleum Spec.] Microcryst. wax NF; CAS 63231-60-7; EINECS 264-038-1; chewing gum base; *Regulatory:* FDA §172.886, 178.3710; *Properties:* lt. yel.; misc. with petrol. prods., many essential oils, most animal and veg. fats, oils, and waxes; visc. 14.3-18.0 cSt (99 C); m.p. 82-88 C; flash pt. (COC) 277 C min.

Multiwax® ML-445. [Witco/Petroleum Spec.] Microcryst. wax NF; CAS 63231-60-7; EINECS 264-038-1; chewing gum base; *Regulatory:* FDA §172.886, 178.3710; *Properties:* lt. yel.; misc. with petrol. prods., many essential oils, most animal and veg. fats, oils, and waxes; visc. 14.3-18.0 cSt (99 C); m.p. 77-82 C; flash pt. (COC) 274 C min.

Multiwax® W-445. [Witco/Petroleum Spec.] Microcryst. wax NF; CAS 63231-60-7; EINECS 264-038-1; chewing gum base; *Regulatory:* FDA §172.886, 178.3710; *Properties:* wh. wax; misc. with petrol. prods., many essential oils, most animal and veg. fats, oils, and waxes; visc. 14.3-18.0 cSt (99 C); m.p.77-82 C; flash pt. (COC) 274 C min.

Multiwax® W-835. [Witco/Petroleum Spec.] Microcryst. wax; CAS 63231-60-7; EINECS 264-038-1; *Regulatory:* FDA §172.886, 178.3710; *Properties:* wh.; visc. 14.3-18.0 cSt (99 C); m.p. 74-79 C; flash pt. (COC) 246 C min.

Multiwax® X-145A. [Witco/Petroleum Spec.] Microcryst. wax NF; CAS 63231-60-7; EINECS 264-038-1; cheese coating; *Regulatory:* FDA §172.886, 178.3710; *Properties:* lt. yel.; misc. with petrol. prods., many essential oils, most animal and veg. fats, oils, and waxes; visc. 14.3-18.0 cSt (99 C); m.p. 66-71 C; flash pt. (COC) 260 C min.

Myrj® 45. [ICI Spec. Chem.; ICI Surf. Am.; ICI Surf. Belgium] PEG-8 stearate; CAS 9004-99-3; nonionic; food additive in bakery prods.; *Usage level:* Canada: 0.4% max. (baked goods); *Regulatory:* Canada compliance; *Properties:* wh. cream-colored soft waxy solid; sol. in alcohol, disp. in water; sp.gr. 1.0; HLB 11.1; pour pt. 28 C; sapon. no. 82-95; 100% conc.

Myrj® 52. [ICI Atkemix; ICI Spec. Chem.; ICI Surf. Am.; ICI Surf. Belgium] PEG-40 stearate NF; CAS 9004-99-3; nonionic; o/w emulsifier; antifoaming agent for processed foods, fruit jellies, preserves, sauces; *Usage level:* 0.02-0.05% (defoaming); *Regulatory:* FDA 21CFR §173.340; *Properties:* ivory waxy solid or flake; sol. in water, acetone, ether, alcohol; sp.gr. 1.1; HLB 16.9; pour pt. 38 C; sapon. no. 25-35; 100% conc.

Myrj® 52S. [ICI Spec. Chem.] PEG-40 stearate NF; CAS 9004-99-3; nonionic; defoaming agent; *Regulatory:* FDA 21CFR §173.340; *Properties:* wh. waxy granular solid; sol. in water, toluol, acetone, ether, Cellosolve, CCl₄, alcohol; sp.gr. 1.1; HLB 16.9; pour pt. 38 C; sapon. no. 25-35; 100% conc.

Myvacet® 5-07. [Eastman] Acetylated hydrog. cottonseed glyceride; nonionic; emulsifier; alpha crystalline stabilizer providing long-term aeration props. for dry whipped toppings; forms protective film on nuts, dried fruits, sausages, etc. providing moisture and oxygen barrier; plasticizer and softener for chewing gum; *Regulatory:* FDA 21CFR §172.828, EEC E472a; *Properties:* creamy wh. waxy solid; sol. in most common org. solvs.; sp.gr. 0.94 (80 C); m.p. 41-46 C; HLB 3.8-4.0; acid no. 3 max.; iodine no. 5 max.; sapon. no. 279-292; hyd. no. 133-152; 48.5-51.5% acetylation; *Precaution:* plasticizes/softens some common plastics; *Storage:* 24 mo shelf life.

Myvacet® 5-07K. [Eastman] Acetylated hydrog. veg. glyceride, dist.; nonionic; emulsifier, emollient; forms films with good moisture vapor barrier properties for nuts, dried fruits, sausages; *Properties:* waxy solid; sp.gr. 0.94 (80 C); m.p. 41-46 C; HLB 3.8-4.0; acid no. 3 max.; iodine no. 5 max.; sapon. no. 279-292; hyd. no. 133-152; 48.5-51.5% acetylation; Discontinued

Myvacet® 7-00. [Eastman] Acetylated hydrog. lard glyceride, dist.; CAS 8029-91-2; emulsifier, emollient; forms films with good moisture vapor barrier properties for nuts, dried fruits, sausages; *Properties:* waxy solid; sp.gr. 0.94 (80 C); m.p. 37-40 C; acid no. 3 max.; iodine no. 5 max.; sapon. no. 316-331; hyd. no. 80.5-95; 66.5-69.5% acetylation; Discontinued

Myvacet® 7-07. [Eastman] Acetylated hydrog. veg. oil monoglycerides; nonionic; emulsifier, emollient forming highly flexible films with good moisture-vapor barrier props. for foods; forms protective films on nuts, dried fruits, sausages, etc. as moisture and oxygen barriers; plasticizer/softener for chewing gums; *Regulatory:* FDA 21CFR §172.828, EEC E472a; *Properties:* creamy wh. waxy solid, sol. in most common org. solvs.; sp.gr. 0.94 (80 C); m.p. 37-40 C; HLB 3.8-4.0; acid no. 3 max.; iodine no. 5 max.; sapon. no. 316-331; hyd. no. 80.5-95.0; 66.5-69.5% acetylation; *Precaution:* plasticizes/softens some common plastics; *Storage:* 24 mo shelf life.

Myvacet® 7-07K. [Eastman] Acetylated hydrog. veg. glyceride; nonionic; forms thin transparent film to serve as oxygen moisture barrier for food pkg.; *Properties:* waxy solid; sp.gr. 0.94 (80 C); m.p. 37-40 C;HLB 3.8-4.0; acid no. 3 max.; iodine no. 5 max.; sapon. no. 316-331; hyd. no. 80.5-95; 66.5-69.5% acetylation; Discontinued

Myvacet® 9-08. [Eastman] Acetylated hydrog. coconut oil monoglycerides; nonionic; emulsifier, emollient, lubricant, and deaerator for foods; antidusting agent for powd. foods and mixes; lubricant and release agent for molding, slicing, stamping, and forming operations; plasticizer/softener for chewing gum; *Regulatory:* FDA 21CFR §172.828, EEC E472a; *Properties:* clear liq.; sol. in all common org. solvs., 80% w/w aq. ethanol, veg. and min. oils; sp.gr. 0.94 (80 C); m.p. -12 to -14 C; HLB 3.8-4.0; acid no. 3 max.; iodine no. 2 max.; sapon. no. 440-455; hyd. no. 20 max.; 96% min. acetylation; *Precaution:* plasticizes/softens some common plastics; *Storage:* 12 mo shelf life.

Myvacet® 9-08K. [Eastman] Acetylated hydrog. coconut oil glyceride, dist.; nonionic; emulsifier, emollient for food processing; *Properties:* liq.; sp.gr. 0.94 (80 C); m.p. 4-12 C; HLB 3.8-4.0; acid no. 3 max.; iodine no. 5 max.; sapon. no. 410-440; hyd. no. 0-20; 96% min. acetylation; Discontinued

Myvacet® 9-40. [Eastman] Acetylated lard glyceride, dist.; CAS 8029-92-3; emulsifier; food-grade lubricant and emollient; deaerator in some systems; *Properties:* liq.; sp.gr. 0.94 (80 C); m.p. 4-12 C; acid no. 3 max.; iodine no. 40-45; sapon. no. 375-385; hyd. no. 0-15; 96% min. acetylation; Discontinued

Myvacet® 9-45. [Eastman] Acetylated hydrog. soybean oil glycerides; emulsifier, lubricant, deaerating agent; antidusting agent for powd. foods and mixes; defoamer for bottling operations (jams, jellies, puddings); lubricant/release for molding, slicing, stamping, forming; plasticizer/softener for chewing gum; *Regulatory:* FDA 21CFR §172.828, EEC E472a; *Properties:* clear liq.; sol. in all common org. solvs., 80% w/w aq. ethanol, veg. and min. oils; sp.gr. 0.94 (80 C); visc. 44.5 cP; m.p. 4-12 C; HLB 3.8-4.0; acid no. 3 max.; iodine no. 43-53; sapon. no. 370-382; hyd. no. 0-15; 96% min. acetylation; *Precaution:* plasticizes/softens some common plastics; *Storage:* 12 mo shelf life.

Myvacet® 9-45K. [Eastman] Acetylated hydrog. soybean oil glyceride; nonionic; emulsifier, lubricant, solv. for icings and shortenings; *Properties:* liq.; sp.gr. 0.94 (80 C); m.p. 4-12 C; HLB 3.8-4.0; acid no. 3 max.; iodine no. 43-53; sapon. no. 370-382; hyd. no. 0-15; 96% min. acetylation; Discontinued

Myvaplex® 600. [Eastman] Glyceryl stearate from hydrog. soybean oil; starch complexing agent able to form insol. adducts with amylose; lubricant and processing aid for extruded foods; used in pasta, cereals, instant potatoes, snack foods, pet foods, ice cream, frozen desserts; *Regulatory:* FDA 21CFR §184.1324 GRAS, EEC E471; *Properties:* small beads; sp.gr. 0.92 (80 C); m.p. 69 C; acid no. 3 max.; iodine no. 5 max.; 90% min. monoester; *Storage:* 24 mo shelf life.

Myvaplex® 600K. [Eastman] Glyceryl stearate, food grade; nonionic; emulsifier in macaroni and cereal prods.; starch complexing agent, lubricant, processing aid for foods; *Properties:* wh. powd., char. odor; negligible sol. in water; sp.gr. 0.92 (80 C); m.p. 70 C; HLB 3.8-4.0; acid no. 3 max.; iodine no. 5 max.; flash pt. (COC) 258 C; 90% min. monoester; *Toxicology:* LD50 (oral, rat) > 5000 mg/kg, (dermal, rat) > 2000 mg/kg; nonirritating to skin and eyes; Discontinued

Myvaplex® 600P. [Eastman] Glyceryl monostearate (from hydrog. soybean oil); emulsifier for foods; starch complexing agent, lubricant, and processing aid for extruded foods; used for pasta, cereals, instant potatoes, snack foods, pet foods, ice cream, frozen desserts; *Usage level:* 1.0-1.5% (pasta); 0.5-1.0% (EPS); *Regulatory:* FDA 21CFR §182.1324, 182.4505; *Properties:* powd.; sp.gr. 0.92 (80 C); m.p. 69 C; congeal pt. 70 C; clear pt. 78 C; acid no. 3 max.; iodine no. 5 max.; 90% min. monoester; *Storage:* 24 mo shelf life.

Myvaplex® 600PK. [Eastman] Glyceryl monostear-

ate (from hydrog. soybean oil); nonionic; starch complexing agent, lubricant, processing aid for foods; *Properties:* powd.; sp.gr. 0.92 (80 C); m.p. 69 C; acid no. 3 max.; iodine no. 5 max.; 90% min. monoester; Discontinued

Myvatem® 06K. [Eastman] Diacetyl tartaric acid ester of dist. monoglycerides (from hydrog. soybean oil); emulsifier, dispersant for foods; *Regulatory:* FCC; *Properties:* solid; m.p. 47 C; Discontinued

Myvatem® 30. [Eastman] Diacetyl tartaric acid ester of dist. monoglycerides (from edible tallow); emulsifier, dispersant for foods; *Regulatory:* FCC; *Properties:* semisolid; m.p. 33 C; Discontinued

Myvatem® 35K. [Eastman] Diacetyl tartaric acid ester of dist. monoglycerides (from refined palm oil); emulsifier, dispersant for foods; *Regulatory:* FCC; *Properties:* semisolid; m.p. 26 C; Discontinued

Myvatem® 92K. [Eastman] Diacetyl tartaric acid ester of dist. monoglycerides (from refined sunflower oil); emulsifier, dispersant for foods; *Regulatory:* FCC; *Properties:* liq.; m.p. < 0 C; Discontinued

Myvatex® 3-50. [Eastman] Dist. monoglycerides and dist. propylene glycol monoesters, soybean oil source; emulsifier for foods; provides aeration in whipped toppings; *Regulatory:* FDA 21CFR §172.856; *Properties:* beads; sp.gr. 0.91 (80 C); m.p. 58 C; acid no. < 3; iodine no. 5 max.; 90% min. monoester; *Storage:* 24 mo shelf life.

Myvatex® 3-50K. [Eastman] Gyceryl stearate (dist.) and propylene glycol stearate (dist.); nonionic; food emulsifier; *Properties:* small beads; sp.gr. 0.91 (80 C); m.p. 58 C; acid no. < 3; iodine no. 5 max.; 90% max. monoester; Discontinued

Myvatex® 7-85. [Eastman] Dist. cottonseed oil monoglycerides; CAS 8029-44-5; EINECS 232-438-5; food emulsifier; provides emulsion stability, aeration, volume, extends shelf life in cakes and mixes, icings, shortenings, cream fillings; *Regulatory:* FDA 21CFR §184.1505; *Properties:* soft paste; sp.gr. 0.94 (80 C); m.p. 49 C; acid no. < 3; iodine no. 91-101; 63% min. monoester; *Storage:* 6 mo shelf life.

Myvatex® 7-85K. [Eastman] Cottonseed oil monoglycerides, dist.; CAS 8029-44-5; EINECS 232-438-5; food emulsifier; *Properties:* plastic; sp.gr. 0.94 (80 C); m.p. 49 C; acid no. < 3; iodine no. 91-101; 90% min. monoester; Discontinued

Myvatex® 8-06. [Eastman] Dist. hydrog. soybean oil monoglycerides; emulsifier, dispersant for coffee whiteners; imparts body, dryness to frozen desserts; emulsifies, suspends cocoa powd. in chocolate milk; *Properties:* beads; sp.gr. 0.93 (80 C); m.p. 67 C; acid no. < 3; iodine no. 24-30; 72% min. monoester; *Storage:* 12 mo shelf life.

Myvatex® 8-06K. [Eastman] Hydrog. soybean oil monoglyceride; nonionic; food emulsifier; *Properties:* beads; sp.gr. 0.93 (80 C); m.p. 67 C; acid no. < 3; iodine no. 24-30; 72% min. monoester; Discontinued

Myvatex® 8-16. [Eastman] Dist. hydrog. palm oil monoglycerides; CAS 67784-87-6; emulsifier, dispersant for coffee whiteners; imparts body, dryness to frozen desserts; emulsifies, suspends cocoa powd. in chocolate milk; provides aeration in whipped toppings; *Properties:* beads; m.p. 61 C; acid no. < 3; iodine no. 28 max.; 72% min. monoester; *Storage:* 12 mo shelf life.

Myvatex® 8-16K. [Eastman] Hydrog. palm oil monoglycerides, dist.; nonionic; food emulsifier; *Properties:* beads; m.p. 61 C; acid no. < 3; iodine no. 28; 72% min. monoester; Discontinued

Myvatex® 8-20. [Eastman] Dist. monoglycerides with 20% hydrog. soybean oil; nonionic; food emulsifier; *Properties:* small beads; sp.gr. 0.93 (80 C); m.p. 58 C; acid no. < 3; iodine no. 19-24; 72% min. monoester; Discontinued

Myvatex® 8-20E. [Eastman] Dist. monoglycerides with 20% hydrog. soybean oil; food emulsifier; *Properties:* beads or powd.; sp.gr. 0.93 (80 C); m.p. 57 C; acid no. < 3; iodine no. 24-30; 72% min. monoester; Discontinued

Myvatex® 25-07K. [Eastman] Hydrog. veg. oil glycerides, lecithin, water, propionic acid, sodium propionate; nonionic; emulsifier used in bread and sweet goods shortening; *Properties:* soft plastic; acid no. < 7.3; iodine no. 1 max.; 25% solids; Discontinued

Myvatex® 40-06S. [Eastman] Myvatex 3-50, lactylic stearate, potassium sorbate, and water; nonionic; food emulsifier; improves aeration, volume, texture, and extends shelf life in cakes; *Regulatory:* FDA 21CFR §172.848, 172.856, 182.1324, 184.1505; *Properties:* wh. creamy paste; HLB 3.8-4.0; acid no. < 13; iodine no. < 2; 25% solids; *Storage:* prod. must not be allowed to freeze; 6 mo shelf life.

Myvatex® 40-06S K. [Eastman] Dist. propylene glycol esters, dist. monoglycerides, lactylic esters of stearic acid, and water; nonionic; food emulsifier for cakes; *Properties:* soft plastic; HLB 3.8-4.0; 25% conc.; Discontinued

Myvatex® 90-10K. [Eastman] Hydrog. rapeseed oil, hydrog. cottonseed oil; nonionic; food emulsifier; stabilizer for peanut butter; *Usage level:* 1.3-17.5% (peanut butter), FDA 2.22% max.; *Properties:* beads; m.p. 61 C; acid no. < 6; iodine no. 4 max.; 0% monoglyceride; Discontinued

Myvatex® Do Control. [Eastman] Dist. succinylated monoglycerides and dist. monoglycerides from palm oil; food emulsifier, dough strengthener for yeast-raised baked goods; *Usage level:* 0.375-0.5% max. on flour wt. (baked goods); *Regulatory:* FDA 21CFR §172.830, 184.1505; *Properties:* powd.; sp.gr. 0.94 (80 C); m.p. 53 C; acid no. 64-74; iodine no. 4-7; 41% min. monoester; *Storage:* 9 mo shelf life.

Myvatex® Do Control K. [Eastman] Succinylated palm oil monoglycerides, palm oil monoglycerides, dist.; nonionic; dough strengthener for yeast-raised bakery goods; *Properties:* powd.; sp.gr. 0.94 (80 C); m.p. 53 C; HLB 3.8; acid no. > 74; iodine no. 4-7; 41% min. monoester; Discontinued

Myvatex® Liquid Lite® (K). [Eastman] Dist. propylene glycol monoetsers and dist. acetylated monoglycerides (from soybean and cottonseed oils); food emulsifier; *Properties:* beads; sp.gr. 0.97 (80

C); m.p. 44-48 C; acid no. < 3; iodine no. 3 max.; 90% min. monoester; Discontinued

Myvatex® Mighty Soft®. [Eastman] Dist. monoglyceride prepared from soybean oil; nonionic; self-hydrating dough strengthener, crumb softener, starch complexing agent, extrusion aid, bodying agent, extension aid for yeast-raised bakery goods, ice cream, frozen desserts, pasta, pet food; reduces stickiness, increases shelf life; *Usage level:* 0.25-0.5% on flour wt. (baked goods), 1% (cereals), 0.3% (snacks, pet foods), 0.25% (cake), 0.05-0.07% (ice cream), 1-1.75% (pasta); *Regulatory:* FDA 21CFR §184.1505; *Properties:* wh. powd., odorless; water-disp.; sp.gr. 0.94 (80 C); m.p. 67 C; acid no. 3 max.; iodine no. 19-36; 90% min. monoester; *Storage:* 12 mo shelf life.

Myvatex® Monoset®. [Eastman] Dist. monoglycerides, rapeseed and cottonseed oils; food emulsifier; stabilizes oil, improves mouth-feel and flavor in peanut butter; *Properties:* beads; m.p. 63 C; acid no. < 4; iodine no. 5 max.; 18% min. monoester; *Storage:* 24 mo shelf life.

Myvatex® Monoset® K. [Eastman] Rapeseed oil monoglycerides, cottonseed oil monoglycerides, dist.; food emulsifier; stabilizer for peanut butter; *Usage level:* 1.3-1.75% (peanut butter), FDA max. 2.77%; *Properties:* beads; m.p. 63 C; acid no. < 4; iodine no. 5 max.; 18% min. monoglyceride; Discontinued

Myvatex® MSPS. [Eastman] Soybean oil monoglycerides with 25% polysorbate 80; food emulsifier; imparts body, dryness to frozen desserts, *Regulatory:* FDA 21CFR §172.840; *Properties:* beads; m.p. 69 C; acid no. < 3; iodine no. 6 max.; 67.5% min. monoester; *Storage:* 12 mo shelf life.

Myvatex® Peanut Butter Stabilizer. [Eastman] Glyceryl stearate (dist.) and fully hydrog. palm oil; nonionic; peanut butter emulsifier and stabilizer; *Properties:* small beads; slt. odor; negligible sol. in water; sp.gr. 0.94; m.p. 60 C; flash pt. (COC) 230 C; sapon no. 176-180; 45% max. monoester content.

Myvatex® SSH. [Eastman] Dist. soybean oil monoglycerides, lecithin, water, and propionic acid; food emulsifier; improves crumb softness and extends shelf life of yeast-raised baked goods; *Usage level:* 0.5-1% of flour wt. (baked goods); *Regulatory:* FDA 21CFR §184.1505; *Properties:* soft plastic; acid no. < 7.3; iodine no. 6 min.; 45% min. monoester; *Storage:* prod. must not be allowed to freeze; 12 mo shelf life.

Myvatex® Super DO. [Eastman] Succinylated monoglycerides and dist. monoglycerides; nonionic; emulsifier for baked goods; *Properties:* powd.; 35-40% monoester content; Discontinued

Myvatex® Texture Lite®. [Eastman] Glyceryl stearate, propylene glycol stearate, sodium stearoyl lactylate, silicon dioxide; nonionic; food emulsifier; provides emulsion stability and aeration, extends shelf life in cakes, cake mixes, cake doughnuts; provides aeration and improves speed of whip in whipped toppings; *Regulatory:* FDA 21CFR §172.846, 172.856; *Properties:* powd.; sp.gr. 0.94

(80 C); m.p. 50-60 C; acid no. < 13; iodine no. 5 max.; 80% min. monoester; *Storage:* 12 mo shelf life.

Myvatex® Texture Lite® K. [Eastman] Soybean oil monoglycerides, propylene glycol monoesters, sodium stearoyl lactylate, silicon dioxide; nonionic; emulsifier for cakes, icings, cream fillings, whipped toppings, sauces; *Properties:* powd.; sp.gr. 0.94 (80 C); HLB 3.8-4.0; acid no. < 13; iodine no. 3 max.; 80% min. monoester; Discontinued

Myverol® 18-00. [Eastman] Dist. hydrog. lard or tallow glyceride; nonionic; emulsifier for baked goods, confectionery prods., dehydrated potatoes, etc.; *Usage level:* 0.4-0.6% (in PP); *Regulatory:* FDA 21CFR §182.1324, 182.4505; *Properties:* small beads; sp.gr. 0.91 (80 C); m.p. 68 C; acid no. 3 max.; iodine no. 5 max.; 90% min. monoester; Discontinued

Myverol® 18-04. [Eastman] Dist. hydrog. palm oil glyceride; CAS 67784-87-6; nonionic; emulsifier for baked goods, confectionery prods., dehydrated potatoes, etc.; improves texture, reduces stickiness in toffees; emulsifier for infant formulas; emulsion stabilizer, antispattering aid in margarine; improves mouth-feel in peanut butter; *Regulatory:* FDA 21CFR §184.1505 (GRAS), EEC E471; *Properties:* small beads; sp.gr. 0.94 (80 C); m.p. 66 C; HLB 3.8-4.0; acid no. 3 max.; iodine no. 5 max.; 90% min. monoester; *Storage:* 24 mo shelf life.

Myverol® 18-04K. [Eastman] Hydrog. palm oil glyceride; nonionic; food emulsifier, stabilizer for candy, infant formula, margarine, peanut butter; *Properties:* small bead; sp.gr. 0.94 (80 C); m.p. 66 C; HLB 4.0; acid no. 3 max.; iodine no. 5 max.; 90% min. monoglyceride; Discontinued

Myverol® 18-06. [Eastman] Hydrog. soy glyceride, dist.; CAS 61789-08-0; nonionic; food emulsifier; provides aeration, extends shelf life in cake mixes, whipped toppings; reduces stickiness in toffees; softener, plasticizer for chewing gum; emulsifier for infant formulas; improves gloss, crystallization in confectionery coatings; *Regulatory:* FDA 21CFR §184.1505 (GRAS), EEC E471; *Properties:* small beads; negligible sol. in water; sp.gr. 0.92 (80 C); m.p. 69 C; b.p. 460 C; HLB 3.8-4.0; acid no. 3 max.; iodine no. 5 max.; flash pt. (COC) 227 C; 90% min. monoester; *Toxicology:* LD50 (oral, rat) > 5 g/kg, (dermal, rat) > 2 g/kg; nonirritating to skin; sl. eye irritant; *Storage:* 24 mo shelf life.

Myverol® 18-06K. [Eastman] Hydrog. soy glyceride; nonionic; food emulsifier for candy, chewing gum base, cake mixes, infant formula, whipped toppings; *Properties:* wh. bead, char. odor; negligible sol. in water; sp.gr. 0.92 (80 C); m.p. 69 C; b.p. 460 C; HLB 4.0; acid no. 3 max.; iodine no. 5 max.; flash pt. (COC) 227 C; 90% min. monoester; *Toxicology:* LD50 (oral, rat) > 5 g/kg, (dermal, rat) > 2 g/kg; nonirritating to skin; sl. eye irritant; Discontinued

Myverol® 18-07. [Eastman] Hydrog. cottonseed glyceride, dist.; CAS 61789-07-9; nonionic; emulsifier for foods; improves texture and reduces stickiness in toffees; improves gloss and crystallization

in confectionery coatings; emulsion stabilizer, antispattering aid in margarine; aerator for whipped toppings; *Regulatory:* FDA 21CFR §184.1505 (GRAS), EEC E471; *Properties:* small beads; sp.gr. 0.92 (80 C); m.p. 68 C; HLB 3.8-4.0; acid no. 3 max.; iodine no. 5 max.; 90% min. monoester; *Storage:* 24 mo shelf life.

Myverol® 18-07K. [Eastman] Hydrog. cottonseed glyceride; CAS 61789-07-9; nonionic; food emulsifier, stabilizer for candy, chewing gum base, cake mixes, confectionery coatings, infant formula, whipped toppings, peanut butter; *Usage level:* 1.5-2.0% (peanut butter); *Properties:* small beads; sp.gr. 0.92 (80 C); m.p. 68 C; HLB 4.0; acid no. 3 max.; iodine no. 5 max.; 90% min. monoglyceride; Discontinued

Myverol® 18-30. [Eastman] Edible beef tallow glyceride, dist.; CAS 61789-13-7; EINECS 263-035-2; nonionic; emulsifier for baked goods, confectionery prods.; *Properties:* sp.gr. 0.92 (80 C); m.p. 60 C; acid no. 3 max.; iodine no. 27-40; 90% min. monoester; Discontinued

Myverol® 18-35. [Eastman] Palm oil glyceride, dist.; nonionic; emulsifier for baked goods, confectionery prods.; emulsifier, dispersant for coffee whiteners, infant formulas; improves gloss and crystallization in confectionery coatings; aerator for shortenings; *Regulatory:* FDA 21CFR §184.1505 (GRAS), EEC E471; *Properties:* plastic; sp.gr. 0.94 (80 C); m.p. 60 C; HLB 3.8-4.0; acid no. 3 max.; iodine no. 36-45; 90% min. monoester; *Storage:* 12 mo shelf life.

Myverol® 18-35K. [Eastman] Palm oil glyceride; nonionic; food emulsifer used in confectionery coatings, infant formulas, shortenings; *Properties:* plastic; sp.gr. 0.94 (80 C); m.p. 60 C; HLB 4.0; acid no. 3 max.; iodine no. 36-45; 90% min. monoester; Discontinued

Myverol® 18-40. [Eastman] Lard glyceride, dist.; CAS 61789-10-4; EINECS 263-032-6; nonionic; emulsifier for baked goods, confectionery prods.; *Properties:* plastic; sp.gr. 0.92 (80 C); m.p. 58 C; acid no. 3 max.; iodine no. 43-55; 90% min. monoester; Discontinued

Myverol® 18-50 [Eastman] Hydrog. veg. glyceride; CAS 61789-08-0; nonionic; emulsifier for foods; provides emulsion stability, aeration, extends shelf life in cake mixes, icings; emulsifier, dispersant for coffee whiteners; aerator for shortenings; *Regulatory:* FDA 21CFR §184.1505 (GRAS), EEC E471; *Properties:* plastic; sp.gr. 0.94 (80 C); m.p. 54 C; HLB 3.8-4.0; acid no. 3 max.; iodine no. 50-60; 90% min. monoester; *Storage:* 9 mo shelf life.

Myverol® 18-50K. [Eastman] Hydrog. veg. glyceride, dist.; CAS 61789-08-0; nonionic; food emulsifier used in icings, cream fillings, cake mixes, shortenings; *Properties:* plastic; sp.gr. 0.94 (80 C); m.p. 54 C; HLB 4.0; acid no. 3 max.; iodine no. 50-60; 90% min. monoester; Discontinued

Myverol® 18-85. [Eastman] Cottonseed glyceride, dist.; CAS 8029-44-5; EINECS 232-438-5; nonionic; emulsifier for baked goods, confectionery prods.; provides emulsification and aeration in icings, shortenings; *Regulatory:* FDA 21CFR §184.1505 (GRAS), EEC E471; *Properties:* soft plastic; sp.gr. 0.95 (80 C); m.p. 46 C; HLB 3.8-4.0; acid no. 3 max.; iodine no. 85-95; 90% min. monoester; *Storage:* 6 mo shelf life.

Myverol® 18-85K. [Eastman] Cottonseed glyceride; CAS 8029-44-5; EINECS 232-438-5; nonionic; food emulsifier for icings, cream fillings, shortenings; *Properties:* plastic; sp.gr. 0.95 (80 C); m.p. 46 C; HLB 4.0; acid no. 3 max.; iodine no. 85-95; 90% min. monoester; Discontinued

Myverol® 18-92. [Eastman] Sunflower seed oil glyceride; nonionic; emulsifier, dispersant for foods; antifoaming agent for food processing; emulsion stabilizer for diet spreads; emulsifier, aerator for icings; stabilizes oil, improves mouth-feel and flavor in peanut butter; *Regulatory:* FDA 21CFR §184.1505 (GRAS), EEC E471; *Properties:* soft plastic; sp.gr. 0.90 (80 C); m.p. 41 C; HLB 3.8-4.0; acid no. 3 max.; iodine no. 105-115; 90% min. monoester; *Storage:* 6 mo shelf life.

Myverol® 18-92K. [Eastman] Sunflower seed oil glyceride; nonionic; food emulsifier for icings, cream fillings, shortenings; *Properties:* semiplastic; sp.gr. 0.90 (80 C); m.p. 41 C; HLB 4.0; acid no. 3 max.; iodine no. 105-115; 100% conc.; Discontinued

Myverol® 18-98. [Eastman] Safflower glyceride, dist.; nonionic; emulsifier for diet margarines, icing and cream-fillings shortenings; *Properties:* soft plastic; sp.gr. 0.90 (80 C); 100% act.; Discontinued

Myverol® 18-99. [Eastman] Canola oil glyceride; nonionic; food emulsifier; antifoaming agent for food processing; emulsion stabilizer for diet spreads; emulsifier, aerator for icings; stabilizes oil, improves mouth-feel and flavor in peanut butter; *Regulatory:* FDA 21CFR §184.1505 (GRAS), EEC E471; *Properties:* semiplastic; sp.gr. 0.93 (80 C); m.p. 35 C; HLB 3.8-4.0; acid no. 3 max.; iodine no.

90-95; 90% min. monoester; *Storage:* 6 mo shelf life.

Myverol® 18-99K. [Eastman] Low-erucic rapeseed oil monoglyceride, dist.; nonionic; food emulsifier for icings, cream fillings, shortenings; *Properties:* semiplastic; sp.gr. 0.93 (80 C); m.p. 35 C; HLB 4.0; acid no. 3 max.; iodine no. 90-95; 90% min. monoester; Discontinued

Myverol® P-06. [Eastman] Propylene glycol hydrogenated soybean oil monostearate; CAS 1323-39-3; EINECS 215-354-3; aerating emulsifier for cakes and whipped toppings; provides emulsion stability, extends shelf life; aerator for shortenings; *Regulatory:* FDA 21CFR §172.856, EEC E477; *Properties:* beads; sp.gr. 0.89 (80 C); m.p. 45 C; acid no. 3 max.; iodine no. 5 max.; 90% min. monoester; *Storage:* 24 mo shelf life.

Myverol® P-06K. [Eastman] Dist. monoester from hydrog. soybean oil and propylene glycol; nonionic; food emulsifier, stabilizer, aerating agent; *Properties:* small beads; sp.gr. 0.89 (80 C); m.p. 45 C; acid no. 3 max.; iodine no. 5 max.; 90% min. monoester content; Discontinued

Myverol® SMG. [Eastman] Dist. succinylated monoglycerides from hydrog. palm oil or palm stearine; food emulsifier; dough strengthener for yeast-raised bakery prods.; provides good crumb softening and antistaling props. to baked goods; *Usage level:* 0.125-0.25% flour wt. (baked goods); *Regulatory:* FDA 21CFR 172.830; *Properties:* beads; sp.gr. 0.94 (80 C); m.p. 58 C; acid no. 70-120; iodine no. 3 max.; 12-20% monoester; 55% min. SMG; *Storage:* 12 mo shelf life.

Myverol® SMG® VK. [Eastman] Hydrog. palm oil or palm stearine succinylated monoglyceride; nonionic; emulsifier for shortenings, dough strengthener, softener for bread baking; *Properties:* beads; m.p. 58 C; acid no. 70-120; iodine no. 3 max.; 12-20% monoester; Discontinued

N

Nalco® 131. [Nalco] Glycol-type surfactants; nonionic; food grade antifoam for beet sugar and enzyme aq. operations; *Regulatory:* FDA compliance; *Properties:* straw liq.; dens. 8.2 lb/gal; flash pt. (PMCC) 188 F; 100% act.

Nalco® 1090. [Nalco] Silica-silicone-organic; food grade antifoam for food and beverage processing; *Regulatory:* FDA compliance; *Properties:* milky wh. appearance; dens. 8.3 lb/gal.

Narangrex Complex D. [MLG Enterprises Ltd.] Citric conc. contg. conc. natural juices, deterpenated essential oils, syn. colorants, acidifiers, thickeners, and preservatives; preformulated base for prod. of plain, nongasified soft drinks.

Narangrex Complex G. [MLG Enterprises Ltd.] Citric conc. contg. conc. natural juices, deterpenated essential oils, syn. colorants, acidifiers, thickeners, and preservatives; preformulated base for prod. of fizzy soft drinks.

Naringinase Amano. [Amano Enzyme USA; Unipex] Enzyme from *Penicillium;* enzyme for juice prod.

Nastar®. [Cosucra BV; Feinkost Ingred.] Native pea starch contg. 35% amylose and 65% amylopectin; gellant, viscosifier with exc. stability to high temps., to shearing, and to variations in pH; for instant soups and sauces, powd. puddings and desserts, canned foods, confectionery, meat prods., cereal prods., dehydrated prods., extruded prods.; *Properties:* wh. powd., neutral taste; 88% min. dry matter; *Storage:* store under dry conditions.

Nastar® Instant. [Cosucra BV; Feinkost Ingred.] Pregelatinized pea starch contg. 35% amylose, 65% amylopectin; gellant, viscosifier with exc. stability to high temps., to cold processes, and to variations in pH; suitable for cold processes; for dessert creams, dressings, chocolate mousses, instant soups; *Properties:* wh. powd., neutral taste; 93% min. dry matter; *Storage:* store under dry conditions.

Natamax™. [Pfizer Food Science] 50:50 Powd. blend of lactose and active ingred. natamycin; Antimycotic agent for inhibition of yeasts, molds, fungi in food prods., esp. cheese, meats; extends shelf life; esp. suitable for surface applics.; *Usage level:* 0.5% (dips for cheese), 4000 ppm (dips for meats); *Properties:* cryst.; low sol. in aq. sol'ns., most org. solvs.; more sol. at pH < 3 and > 9; *Toxicology:* may cause skin and eye irritation; *Precaution:* prevent exposure to chemical oxidants and sunlight; avoid contact with oxidizing agents, heavy metals.

Natural Arabic Type Gum Purified, Spray-Dried. [MLG Enterprises Ltd.] Gum arabic; CAS 9000-01-5; EINECS 232-519-5; thickener improving consistency and palatability in soft drinks, powd. flavors; foam stabilizer in brewery; also in confectionery (gum candies and drops, chewy candies, sugarless chewing gums); *Properties:* lt. brn. fine powd.; instantly water-sol.; visc. 51 cps (10%); pH 5.6 (10%); 4.54% moisture.

Natural Filling Powd. [Bunge Foods] Produces a base filling for dough (sweet goods, danish pastry); *Usage level:* 10%; *Properties:* tan/cream gran. powd., bland aroma; pH 6.0; 12% moisture.

Natural Flavor Enhancer No. 11.9743. [Bell Flavors & Fragrances] Natural flavor potentiator and enhancer; 1 lb replaces 4 lb of MSG; taste potentiator for all meats, snack food, cheese prods., dips, soups, sauces, salad dressings; resist. to microwave destruction; stable in freeze/thaw cycles; *Regulatory:* kosher, USDA approved.

Natural Flavor White Sour. [Am. Ingreds.] Wheat flour, potato flour, lactic acid, and soy flour; sour for French bread, English muffins, snack crackers, brown-and-serve prods., snacks; *Properties:* free-flowing powd.; pH 3.0-3.4; 11.5-13.5% protein.

Natural Glycerine USP 96%. [Dial] Glycerin USP; CAS 56-81-5; EINECS 200-289-5; used in food prods.; *Properties:* colorless clear liq., char. odor; sp.gr. 1.2553 (15/15 C); pH neutral; 96% min. act.

Natural Glycerine USP 99%. [Dial] Glycerin USP; CAS 56-81-5; EINECS 200-289-5; used in food prods.; *Properties:* colorless clear liq., char. odor; sp.gr. 1.2630 (15/15 C); pH neutral; 99% min. act.

Natural Glycerine USP 99.5%. [Dial] Glycerin USP; CAS 56-81-5; EINECS 200-289-5; used in food prods.; *Properties:* colorless clear liq., char. odor; sp.gr. 1.2643 (15/15 C); pH neutral; 99.5% min. act.

Natural Guarana Extract 420.09.2. [Industrias Reunidas Jaragua; Commodity Services Int'l.] Extract prepared from seeds of *Paulinia cupana* from Brazil; used in S. Am. to prepare soft drinks; *Usage level:* 50 ml/100 l of finished beverage; *Properties:* reddish-brn. liq.; sol. 0.2 vol. % in water; sp.gr. 0.9850-0.9950; ref. index 1.3580-1.3680; pH 6.0-6.8; 30-34% ethanol content.

Natural HVP Replacer. [Nikken Foods Co. Ltd.] Blend

of fermented soy sauce, yeast extract, and salt spray-dried on a dextrin carrier; natural replacement for hydrolyzed vegetable protein; replace 2.5 parts HVP with one part of prod.; 10% protein, 33.2% salt, 5% moisture.

Natural Lipolyzed Butter Oil 500. [Gamay Flavors] Butter flavor with cheese background; for use in veg. oils, lite margarine/butter, shortenings, candies, cookies, cakes, snack foods, toppings, icings, sauces, coffee whiteners.

Natural Liquid AP Carmine Colorant. [MLG Enterprises Ltd.] Stabilized carminic acid aluminumcalcic lake ammonia sol'n.; CAS 1390-65-4; EINECS 215-724-4; natural food colorant producing intense violet-red shades for acidic foods, beverages, desserts, gelatins, jams, jellies, fruit candy; *Usage level:* 0.05-0.2%; *Properties:* liq.; sol. in water and pH 3 acid sol'ns.; 3 ± 0.1% color (as carminic acid); *Storage:* store in tightly closed containers in fresh, dark rooms.

Natural Liquid Carmine Colorant (Type 100, 50, and Simple). [MLG Enterprises Ltd.] Carminic acid aluminum-calcic lake hydroalcoholic ammoniacal sol'n.; CAS 1390-65-4; EINECS 215-724-4; natural food colorant producing berry to strawberry red shades for flavored milk, yogurt, ice cream, candies, confectionery, fondants, cracker fillings, candied and dried fruits, cherries, artificial fruits; *Usage level:* 0.05-0.2%; *Properties:* liq.; water-sol.; 3.0 ± 0.1% (100), 1.5 ± 0.05% (50), 0.65 ± 0.05% (simple), as carminic acid; *Storage:* store in fresh and dark rooms.

Natural Pineapple Extract. [Commodity Services Int'l.] Extract derived from flesh of fresh ripened fruit of pineapple, *Ananas comosus* brown in Brazil, odor and flavoring for beverages; *Usage level:* 200 ml/100 l of finished beverage; *Properties:* dk. brn. liq.; sp.gr. 1.2900-1.2950 (20 C); ref. index 1.4455; pH 3.4-3.6; 30-34% ethanol content.

Natural Soluble Carmine Powd. [MLG Enterprises Ltd.] Solubilized, spray-dried carminic acid aluminum-calcic lake; CAS 1390-65-4; EINECS 215-724-4; natural food colorant producing red-violet shades for instant/dehydrated foods, soups, sauces, dressings, beverages, desserts, custards, cake mixes, ice cream, candy coatings; *Usage level:* 0.03-0.06%; *Properties:* impalpable powd.; 100% 80 mesh; water-sol.; 40 ± 1% (as carminic acid); *Storage:* keep in well closed containers, in fresh, dry room.

Natural Soluble Orange Powd. [MLG Enterprises Ltd.] Spray-dried annatto seed extract; CAS 8015-67-6; natural food colorant producing yel.-orange to peach-orange shades for instant/dehydrated foods, soups, sauces, dressings, beverages, desserts, custards, cake mixes, ice cream, candy coatings; *Usage level:* 0.03-0.06%; *Properties:* impalpable powd.; 100% 80 mesh; water-sol.; 6.3 ± 0.15% (as norbixin); *Storage:* keep in refrigerated, tightly closed containers, away from light; rotate stock every 4 mos.

Natural Soluble Powder AP Carmine Colorant.

[MLG Enterprises Ltd.] Carminic acid aluminumcalcic lake, solubilized, stabilized, and spray-dried; CAS 1390-65-4; EINECS 215-724-4; natural colorant producing red to violaceous shades, or stable cherry red @ pH 3; for instant or dehydrated prods. with acid reaction, e.g., beverages, gelatins, fantasy desserts, ice creams, sauces; *Usage level:* 0.025-0.075%; *Properties:* impalpable powd.; 100% 80 mesh; sol. in water and pH 3 acid sol'ns.; 15 ± 0.5% color (as carminic acid); *Storage:* keep in tightly closed containers, in fresh, dry rooms.

Natural Soluble Powder Green Colorant. [MLG Enterprises Ltd.] Solubilized and spray-dried blend of turmeric with indigotine blue; natural colorant producing apple to leaf green shades for instant or dehydrated prods., gelatins, beverages, fantasy desserts, ice cream, soups, sauces; *Usage level:* 0.1-0.3%; *Properties:* green impalpable powd.; 100% 80 mesh; water-sol.; *Storage:* keep in tightly closed containers in fresh, dry, dark rooms.

Natural Soluble Powder Q A/L 15 Yellow Colorant. [MLG Enterprises Ltd.] Turmeric and annatto solubilized and spray dried alcoholic extracts blend; natural colorant producing egg yel. shades for noodles, dry fresh pasta, instant foods, cake mixes, puddings, rolls, gelatins, fantasy desserts; β-carotene substitute; *Usage level:* 150 g/100 kg flour (pasta), 0.1-0.2% (rolls), 0.1-0.15% (desserts); *Properties:* yel. impalpable powd.; 100% 80 mesh; water-sol.; *Storage:* keep in tightly closed containers in fresh, dry, dark rooms.

Natural Starter Distillate Replacer 15X W.S. [Gamay Flavors] Butter top notes for use in process cheese, cottage cheese, toppings, icings, dessert mixes, baked goods, soups, snack mixes, margarines; *Properties:* water-sol.

Natural Yellow Colour Q-500, Q-1000, Q-2000. [MLG Enterprises Ltd.] Annatto seed hydroalcoholic extract; CAS 8015-67-6; natural food colorant producing pale yel. to orange shades for candies, confectionery, ice cream, milk-based desserts, cheese, sauces, dressings; *Usage level:* 0.05-0.2% (Q-500); 0.01-0.03% (Q-2000); *Properties:* liq.; sol. in water and ethyl alcohol; 0.25±0.025% (Q-500), 0.5 ± 0.05% (Q-1000), 1.0 ± 0.1% (Q-2000), as norbixin; *Storage:* protect from light and heat; rotate stock every 3 mos.

NatuReal™ Balance. [Alko Ltd.] Sl. hydrolyzed oat starch; fat replacer; strong cuttable texture for lowfat foods or fat shortenings, mouthfeel improver, natural food ingred. for salad dressings, mayonnaise, bakery prods., processed meat prods., dairy-based desserts; *Usage level:* 1-5% (functional purposes); *Regulatory:* kosher; *Properties:* wh. very fine powd., neutral odor/flavor; sol. in water; bulk dens. 0.55 kg/l (loose); pH 4-6 (10% slurry); 12% max. moisture; *Toxicology:* nontoxic; avoid inhalation of dust; *Precaution:* dust explosion is possible; *Storage:* store in cool, dry area away from strong odors.

NatuReal™ Essence. [Alko Ltd.] β-Glucan enriched (15%) oat fiber; shelf life extender in baked goods, mouthfeel improver/visc. enhancer in moist foods, water holding ingred. in meat prods., easy fiber for extrusion, fatty acid shield against oxidation; for cereals, snacks, dairy prods., health/nutritional foods; *Usage level:* 1-8% (functional purposes), up to 50% (in extrusion); *Regulatory:* kosher; US patent 5183677; *Properties:* lt. yel. to lt. brn. powd., neutral odor, bland flavor; 95% min. thru 1 mm sieve, 2% max. thru 0.125 mm sieve; insol. in water; bulk dens. 0.3 kg/l; pH 5 (10% slurry); 35% total dietary fiber, 25% protein, 25% starch, 11% fat; *Toxicology:* nontoxic; avoid inhalation of dust; *Precaution:* dust explosion is possible; *Storage:* > 10 mos shelf life stored in dry, cool place away from strong odors and uv light.

NatuReal™ Sensoral. [Alko Ltd.] Native waxy barley starch; thickener, stabilizer with slow meltdown; improves freeze/thaw stability, provides creamy mouthfeel; for low/no fat prods. (mayonnaise, salad dressing, frozen dessert, bakery prod., processed meat), replacement for modified starches in dairy desserts; *Usage level:* comparable with other native waxy starches; *Regulatory:* kosher; *Properties:* wh. free-flowing fine powd., neutral odor/flavor; insol. in water; bulk dens. 0.57 kg/l (loose); pH 406 (10% slurry); 12% moisture; *Toxicology:* nontoxic; avoid inhalation of dust; *Precaution:* dust explosion is possible; *Storage:* store in cool, dry area away from strong odors.

NatuReal™ Textural. [Alko Ltd.] Native oat starch; thickener, stabilizer for spoonable foods, low- and nofat foods (salad dressings, spreads, frozen desserts, bakery prods., processed meats), extruded prods.; replaces modified starch in dairy-based desserts, ketchup, jellied confectionery; *Usage level:* comparable with other native starches; *Regulatory:* kosher; *Properties:* wh. free-flowing fine powd., neutral odor/flavor; insol. in water; bulk dens. 0.55 kg/l (loose); pH 4-6 (10% slurry); 12% max. moisture; *Toxicology:* nontoxic; avoid inhalation of dust; *Precaution:* dust explosion is possible; *Storage:* store in cool, dry area away from strong odors.

NatureTones™. [Kerry Ingreds./Beatreme] Natural flavorings from beef and clam extracts, wine, vinegar, soy sauce, etc.

ND-201 Syrup. [Premier Malt Prods.] Extract of 100% malted barley; CAS 8002-48-0; EINECS 232-310-9; syrup with mellow, rich malt flavor; used in confections, ice cream prods., baked goods, vinegar, cereal, tobacco, soybean milk, pretzels, fermentation nutrient, bread sticks, horse food; 40% the sweetness of sucrose; *Usage level:* 1.3% (caramels), 1-5% (ice cream), 2-4% of dough mix wt. (snacks), 3-5% of dough wt. (fruit-filled cookies), 6-14% (breakfast cereal), 3-7% (soybean milk), 2-4% (pretzels); *Regulatory:* FDA GRAS; *Properties:* Lovibond 160-360 syrup; pH 5.1-5.8 (10%); 78.5-80.0% solids; 5-6% protein; 54-62% reducing sugars.

ND-201-C Syrups. [Premier Malt Prods.] Extract of 60-65% malted barley and 35-40% corn; syrup with mellow, rich malt flavor; used in confections, ice cream prods., baked goods, vinegar, cereal, tobacco, soybean milk, pretzels, fermentation nutrient, bread sticks, horse food; 55% the sweetness of sucrose; *Usage level:* 1.3% (caramels), 1-5% (ice cream), 1-2.5% flour wt. (bread), 1-3% dough wt. (bagels), 1.5-2.5% flour wt. (saltines), 2-4% dough mix (snacks), 6-14% (breakfast cereal), 3-7% (soy bean milk), 2-4% (pretzels), 1-1.5% (chocolate syrup); *Properties:* Lovibond 100-250 syrup; pH 5.1-5.8 (10%); 78.5-80.0% solids; 3-4% protein; 57-67% reducing sugars.

ND-305 Syrups. [Premier Malt Prods.] Extract of 65% corn and 35% malted barley; syrup with mild malt flavor and pleasantly sweet char.; used in ice cream prods., bakery goods, vinegar, snacks, cereal, cigarettes, pretzel, bread sticks; 65% the sweetness of sucrose; *Usage level:* 2-3% (novelty ice cream), 1-2.5% flour wt. (bread), 1-3% dough wt. (bagels), 2-4% dough mix (snacks), 6-14% (breakfast cereal), 2-4% (pretzels), 2-3% flour wt. (bread sticks); *Properties:* Lovibond 80-160 syrup; pH 5.1-5.8 (10%); 78.5-80.0% solids; 1.8-2.6% protein; 60-70% reducing sugars.

Neobee® 18. [Stepan Food Ingreds.; Stepan Europe] Hybrid safflower oil, refined, deodorized; CAS 8001-23-8; EINECS 232-276-5; highly stable oil for margarine, mayonnaise, salad dressings; solubilizer, solv., lubricant, nutritional fluid; *Regulatory:* FDA 21CFR §170.30, GRAS; avail. in kosher grade; *Properties:* sl. yel. liq., sl. fatty odor, bland taste; sol. in alcohol, min. oil, acetone; sp.gr. 0.915; visc. 70 cps; iodine no. 90; sapon. no. 190; surf. tens. 31.6 dynes/cm; *Toxicology:* nonirritating; *Storage:* avoid prolonged storage above 90 F.

Neobee® 20. [Stepan Food Ingreds.] Propylene glycol dicaprylate/dicaprate; carrier/extender for flavors, fragrances, and colors in foods; vehicle for vitamins; solubilizer, cosolv.; *Properties:* APHA 50 max. liq.; sol. in alcohol, min. oil, acetone; sp.gr. 0.920; visc. 9 cps; acid no. 0.10 max.; sapon. no. 315-335; surf. tens. 31.0 dynes/cm; Discontinued

Neobee® 62. [Stepan Food Ingreds.] Tristearin; CAS 555-43-1; EINECS 209-097-6; solubilizer, stabilizer used in food applics.; *Regulatory:* FDA 21CFR §170.30, GRAS; avail. in kosher grade; *Properties:* flakes; sapon. no. 189-195; Unverified

Neobee® 1053. [Stepan Food Ingreds.] Caprylic/capric triglyceride; CAS 65381-09-1; EINECS 265-724-3; solubilizer for flavors, vitamins; diluent for food colors, essential oils; clouding agent for beverages; *Regulatory:* FDA 21CFR §170.30, GRAS; avail. in kosher grade; *Properties:* sl. yel. liq., bland odor, tasteless; sol. in min. oil, acetone, alcohol; sp.gr. 0.930-0.960; iodine no. 0.5; sapon. no. 333; hyd. no. 5; 0.05% moisture; *Storage:* avoid prolonged storage above 90 F.

Neobee® 1054. [Stepan Food Ingreds.] Propylene glycol dicaprylate/dicaprate; cosolv., solubilizer; carrier for flavors and colors; *Regulatory:* FDA 21CFR §170.30, GRAS; avail. in kosher grade; *Properties:* sl. yel. liq., bland odor; sol. in alcohol, min. oil, acetone; sp.gr. 0.910-0.923; iodine no. 0.1; sapon. no. 325; hyd. no. 1; 0.05% moisture; *Storage:* avoid prolonged storage above 90 F.

Neobee® 1062. [Stepan Food Ingreds.] Coconut oil-derived triglycerides; cosolv., carrier for fat-sol. vitamins in foods; *Regulatory:* FDA 21CFR §170.30, GRAS; avail. in kosher grade; *Properties:* liq.; sol. in min. oil, acetone; sp.gr. 0.925-0.945; sapon. no. 295-315; Unverified

Neobee® C-10. [Stepan Food Ingreds.] Capric triglyceride; source of nutritional MCT; alternative to hydrog. fats for margarine, frostings, confections, bakery mixes; ideal for applics. where solid fat is desirable; *Properties:* sl. yel. solid, bland odor; m.p. 30-33 C; acid no. 0.1 max.; sapon. no. 285-315; hyd. no. 10 max.; 0.1% max. moisture; *Storage:* avoid prolonged storage above 90 F; Custom prod.

Neobee® M-5. [Stepan Food Ingreds.; Stepan Europe] Caprylic/capric triglyceride; CAS 65381-09-1; EINECS 265-724-3; diluent vehicle/carrier for vitamins, nutritional fluids; solubilizer, cosolv. for fragrance/flavoring, food colors; clouding agent for beverages; source of nutritional MCT; *Regulatory:* FDA 21CFR §170.30, GRAS; avail. in kosher grade; *Properties:* sl. yel. liq., bland odor, tasteless; sol. in alcohol, min. oil, acetone; sp.gr. 0.930-0.960; visc. 23 cps; acid no. 0.10 max.; iodine no. 1; sapon. no. 340; hyd. no. 5; surf. tens. 32.3 dynes/cm; 0.1% moisture; *Storage:* avoid prolonged storage above 90 F.

Neobee® M-20. [Stepan Food Ingreds.; Stepan Europe] Propylene glycol dicaprylate/dicaprate; CAS 00503-51-7, solubilizer, solv., lubricant; diluent for essential oils; *Regulatory:* FDA 21CFR§170.30, 172.856, avail. In kosher grade; *Properties:* pale yel. liq., bland odor; sol. in alcohol containing up to 20% water; iodine no. 1 max.; sapon. no. 326; hyd. no. 0.6; 0.04% moisture; *Storage:* avoid prolonged storage above 90 F.

Neobee® O. [Stepan Food Ingreds.] Caprylic/capric triglyceride; CAS 65381-09-1; EINECS 265-724-3; solubilizer, cosolv., carrier for fat-sol. vitamins, flavors; diluent for food colors, essential oils; clouding agent for beverages; source of nutritional MCT; *Regulatory:* FDA 21CFR §170.30, GRAS; avail. in kosher grade; *Properties:* sl. yel. liq., very sl. fatty odor; sol. in min. oil, acetone; sp.gr. 0.938; visc. 30 cps; iodine no. 4.2; sapon. no. 298; hyd. no. 5; surf. tens. 32 dynes/cm; 0.03% moisture; *Storage:* avoid prolonged storage above 90 F.

Neobee® RPO. [Stepan Food Ingreds.] Randomized palm olein; fat source in infant formulas; mfg. exclusively for Wyeth Labs; *Properties:* amber solid, free of rancid odor; iodine no. 55; sapon. no. 195; ref. index 1.459 (40 C); *Storage:* avoid prolonged storage above 90 F; Custom prod.

Neobee® SL-110. [Stepan Food Ingreds.] Interesterified MCT and soybean oil; nutritional supplements; specialty food ingreds.; *Properties:* yel. clear sl. visc. oil; iodine no. 23-33; sapon. no. 313-323; 0.1% max. moisture; *Storage:* avoid prolonged storage above 90 F.

Neobee® SL-120. [Stepan Food Ingreds.] Interesterified MCT and menhaden oil; nutritional supplements; specialty food ingreds.; *Properties:* yel. clear sl. visc. oil; iodine no. 36-44; sapon. no. 313-323; 0.1% max. moisture; *Storage:* avoid prolonged storage above 90 F.

Neobee® SL-130. [Stepan Food Ingreds.] Interesterified MCT and sunflower oils; nutritional supplements; specialty food ingreds.; *Properties:* yel. clear sl. visc. oil; iodine no. 49-59; sapon. no. 281-291; 0.1% max. moisture; *Storage:* avoid prolonged storage above 90 F.

Neobee® SL-140. [Stepan Food Ingreds.] Interesterified MCT, menhaden oil, soybean oil, and tributyrin; nutritional supplements; specialty food ingreds.; *Properties:* yel. clear sl. visc. oil; iodine no. 52-60; sapon. no. 340-350; 0.1% max. moisture; *Storage:* avoid prolonged storage above 90 F.

Neobee® SL-210. [Stepan Food Ingreds.] Interesterified coconut and soybean oil; nutritional supplements; specialty food ingreds.; *Properties:* yel. clear sl. visc. oil; iodine no. 30-40; sapon. no. 239-249; 0.1% max. moisture; *Storage:* avoid prolonged storage above 90 F.

Neobee® SL-220. [Stepan Food Ingreds.] Interesterified coconut, menhaden, and canola oils; nutritional supplements; specialty food ingreds.; *Properties:* yel. clear sl. visc. oil; iodine no. 86-94; sapon. no. 212-229; 0.1% max. moisture; *Storage:* avoid prolonged storage above 90 F.

Neobee® SL-230. [Stepan Food Ingreds.] Interesterified coconut, menhaden, and soybean oils, and tributyrin; nutritional supplements; specialty food ingreds.; *Properties:* yel. clear sl. visc. oil; iodine no. 54-64; sapon. no. 292-302; 0.1% max. moisture; *Storage:* avoid prolonged storage above 90 F.

Neobee® SL-310. [Stepan Food Ingreds.] Interesterified MCT and menhaden oil; CAS 8002-50-4; EINECS 232-311-4; nutritional supplements; specialty food ingreds.; *Properties:* yel. clear sl. visc. oil; iodine no. 139-147; sapon. no. 226-236; 0.1% max. moisture; *Storage:* avoid prolonged storage above 90 F.

Neobee® SL-410. [Stepan Food Ingreds.] Interesterified MCT, butter, and sunflower oil; nutritional supplements; specialty food ingreds.; *Properties:* yel. clear sl. visc. oil; iodine no. 33-40; sapon. no. 255-265; 0.1% max. moisture; *Storage:* avoid prolonged storage above 90 F.

Neo-Cebitate®. [Rhone-Poulenc Food Ingreds.] Sodium erythorbate; CAS 7378-23-6; EINECS 228-973-9; antioxidant for meat, poultry, seafood processing; cure accelerator providing attractive cured meat color with shorter processing times; reduces shrinkage in sausage prods.; *Regulatory:* kosher; *Properties:* wh. cryst. powd., odorless; 3% max. on 40 mesh, 25% max. thru 325 mesh; m.w. 216; pH 5.5-8.0 (5%); 98% min. assay; *Storage:* store cool and dry.

Neral. [BASF AG] Z-3,7-Dimethyl-2,6-octadiene-1-al; fresh, lemon-like, green, sl. lime-like fragrance and flavoring.

Nerolidol. [BASF AG] Z,E-3,7,11-Trimethyl-1,6,10-dodecatriene-3-ol; sweetly floral, green, woody, lilly-like fragrance and flavoring.

Neutral-lactase. [Pfizer Food Science] Food grade lactase conc. prepared from extract of *Candida pseudotropicalis*; dairy ingred. which hydrolyzes lactose in milk, whey, whey permeate, ice cream, cheese, conc. milk, milk powd., yogurt; provides improved texture, allows use of less sugar, flavoring, fruit; makes prods. suitable for lactose-intolerant people; *Storage:* store at refrigerated temps. (0-4 C.

New Fresh Conc. [ADM Arkady] Enzyme conditioner providing shelf life extension for 3-5 days longer than conventional emulsifiers.

Newlase. [Mitsubishi Int'l.; Amano Enzyme USA] Acid protease from *Phizopus niveus*; CAS 9014-01-1; EINECS 232-752-2; digestive enzyme; *Properties:* powd.

New York Roll® Base. [Bunge Foods] No-time dough system providing exc. volume, good color, thin crispy crust, and soft light crumb; for French bread, Italian bread; *Usage level:* 4-5% based on flour.

N'Hance CM. [MLG Enterprises Ltd.] Calcium salts, enzymes, azodicarbonamide, ascorbic acid, and calcium iodate; potassium bromate-free dough conditioner for use in continuous mix bread prod.; provides rapid recovery from abusive dough development in continuous mix systems; *Usage level:* 6 oz/cwt flour; *Properties:* free-flowing powd.; *Storage:* 6 mos shelf life under cool, dry conditions.

N'Hance (Phase I) Powd. [MLG Enterprises Ltd.] Soy flour, calcium salts, ammonium sulfate, salt, ascorbic acid, azodicarbonamide, and fungal enzymes; fermentation aid/dough conditioner for most yeast-raised bakery prods.; provides extensibility and gas-retaining props.; *Usage level:* 8 oz/cwt flour; *Properties:* free-flowing powd.; *Storage:* 6 mos shelf life under cool, dry conditions.

N'Hance (Phase II) Powd. [MLG Enterprises Ltd.] Calcium salts, calcium iodate, ascorbic acid, fungal enzymes; dough conditioner producing strengthening effect of potassium bromate; provides uniform oxidation, improves gas retention, bread quality; esp. effective with low protein or poor gluten quality flours; *Properties:* free-flowing powd.; *Storage:* 6 mos shelf life under cool, dry conditions.

N'Hance (Phase II) Tablets. [MLG Enterprises Ltd.] Calcium iodate, azodicarbonamide, fungal enzymes, and other edible excipients; dough conditioner producing strengthening effects of potassium bromate in yeast-raised bakery prods.; provide uniform oxidation of flour, improves gas retention, bread quality; esp. effective with low protein or poor gluten quality flours; *Usage level:* 1-2 tablets in conjunction with Phase I; *Properties:* tablets; *Storage:* 6 mos shelf life under cool, dry conditions.

N'Hance SD. [MLG Enterprises Ltd.] Soy flour, calcium salts, ascorbic acid, azodicarbonamide, calcium iodate, and fungal enzymes; dough conditioning system replacing potassium bromate; for use on dough side of liq. fermentation systems where no compounded yeast foods are used; also for straight doughs, sweet doughs, danish, frozen doughs; *Usage level:* 0.375-0.5%; *Properties:* free-flowing powd.; *Storage:* 6 mos shelf life under cool, dry conditions.

Niacet Calcium Acetate FCC. [Niacet] Calcium acetate FCC; CAS 62-54-4; EINECS 200-540-9; sequestrant, thickener, pH control agent for food industry; *Properties:* wh. powd.; sol. 26% in water, 2.5% in methanol, 0.1% in ethanol; m.w. 167.10; m.p. 445 C; pH 6.3-8.3 (10% aq.); 99% min. purity; *Toxicology:* relatively low toxicity; strong sol'ns. may cause some eye irritation; transient irritant on inhalation of dus.

Niacet Calcium Propionate FCC. [Niacet] Calcium propionate; CAS 4075-81-4; EINECS 223-795-8; antimicrobial for control of molds in food industry (baked goods, bread, rolls, pies, cakes, pie fillings, cheese), animal feeds; *Usage level:* 3-5 oz/cwt flour (wh. bread), 2-4 oz/cwt batter (cheese cake), 0.3% max. (cheese prods.); *Regulatory:* FDA 21CFR §184.1221 (GRAS); *Properties:* wh. powd. or gran.; sol. 28.5% in water, 2% in methanol; m.w. 186.22; m.p. 400 C dec.; pH 8-9.5 (10% aq.); 99% min. purity; *Toxicology:* food additives, low toxicity; minimal reaction on skin and eye contact; breathing concs. of dust will be irritating.

Niacet Sodium Acetate 60% FCC. [Niacet] Sodium acetate FCC; CAS 127-09-3; EINECS 204-823-8; flavoring agent and adjuvant; pH control aid; *Properties:* clear fine cryst.; sol. 73% in methanol, 58% in water, 12% in ethanol; m.w. 136.09; m.p. 58 C; pH 6.5-8.5 (10% aq.); 99% min. purity; *Toxicology:* low toxicity; strong sol'ns. will cause some eye irritation; only mild skin irritation on prolonged contact.

Niacet Sodium Acetate Anhyd. FCC. [Niacet] Sodium acetate anhyd. FCC; CAS 127-09-3; EINECS 204-823-8; flavoring agent and adjuvant; pH control aid; *Regulatory:* FDA 21CFR §184.1721 (GRAS); *Properties:* wh. powd.; sol. 26% in methanol, 3.2% in ethanol; sol. 58% in water @ 60 C; m.w. 82.03; m.p. 324 C; pH 7.5-9.2 (10% aq.); 99% min. purity; *Toxicology:* low toxicity; strong sol'ns. will cause some eye irritation; only mild skin irritation on prolonged contact.

Niacet Sodium Diacetate FCC. [Niacet] Sodium diacetate FCC; CAS 126-96-5; EINECS 204-814-9; antimicrobial agent, flavoring agent, adjuvant, and pH control agent; source of acetic acid and sodium acetate; *Regulatory:* FDA 21CFR § 184.1754 (GRAS); *Properties:* wh. cryst. powd.; sol. 40% i water, 26% in methanol, 5% in ethanol; m.w. 142.09; m.p. 328 C; pH 4.0-5.0 (10% aq.); 58-60% purity; 39-41% avail. acetic acid; *Toxicology:* moderately toxic; skin irritant, serious eye irritant on prolonged contact with dust or strong sol'ns.

Niacet Sodium Propionate FCC. [Niacet] Sodium propionate FCC; CAS 137-40-6; EINECS 205-290-4; antimicrobial for control of mold in food industry (baked goods, bread, rolls, pie crust and fillings, cake, cheese prods.), animal feeds; *Usage level:* 3-

5 oz/cwt flour (wh. bread), 2-4 oz/cwt batter (cheese cake), 0.3% max. (cheese prods.); *Regulatory:* FDA 21CFR §184.1784 (GRAS); *Properties:* wh. powd.; sol. 49% in water, 14% ln methanol, 1.4% in ethanol; m.w. 96.06; m.p. 285 C; pH 7.5-9.0 (10% aq.); 99% min. purity; *Toxicology:* food additive, low toxicity; minimal reaction on skin and eye contact; breathing concs. of dust will be irritating.

Niacin USP, FCC Fine Granular No. 69901. [Roche] Nicotinic acid USP-FCC; CAS 59-67-6; EINECS 200-441-0; vitamin, pellagra preventive for supplementing cereal grain foods; *Properties:* wh. fine gran. cryst. or cryst. powd., odorless; sol. 1 g/60 ml water; freely sol. in boiling alcohol, sol'ns. of alkali hydroxides and carbonates; pract. insol. in ether; m.w. 123.11; bulk dens. 0.56 g/ml (loose); m.p. 234-238 C; 99.5-101% assay; *Toxicology:* ingestion may produce vascular dilation; high level intakes may produce severe liver damage; *Storage:* store in tight, light-resist. containers; avoid exposure to moisture, excessive heat.

Niacin USP, FCC No. 69902. [Roche] Nicotinic acid USP-FCC; CAS 59-67-6; EINECS 200-441-0; vitamin, pellagra preventive added as supplement to cereal grain foods; *Properties:* wh. cryst. or cryst. powd., odorless or nearly odorless; sol. 1 g/60 ml water; freely sol. in boiling water and boiling alcohol, sol'ns. of alkali hydroxides and carbonates; pract. insol. in ether; m.w. 123.11; bulk dens. 0.36 g/cc; m.p. 234-238 C; 99.5-101% assay; *Toxicology:* ingestion may produce vascular dilation; high level intakes may produce severe liver damage; *Storage:* store in tight, light-resist. containers; avoid exposure to moisture, excessive heat.

Niacinamide Free Flowing No. 69914. [Roche] Nicotinamide USP, FCC with 0.5% silicon dioxide (anticaking agent); essential nutrient for multivitamin preps. and to fortify foods; *Properties:* wh. cryst. powd., nearly odorless, bitter taste; 100% thru 30 mesh; sol. 1 g/1 ml water, 1.5 ml alcohol, 10 ml glycerin; m.w. 122.13; m.p. 128-131 C; 98.5-101% assay; *Storage:* store in dry place @ 59-86 F.

Niacinamide USP, FCC No. 69905. [Roche] Niacinamide; CAS 98-92-0; EINECS 202-713-4; essential nutrient for solid and liq. multivitamin preps. and to fortify foods; *Properties:* wh. cryst. powd., pract. odorless, bitter taste; 90% min. thru 100 mesh; sol. 1 g/1 ml water, 1 g/1.5 ml alcohol, 1 g/10 ml glycerin; m.w. 122.13; m.p. 128-131 C; 98.5-101% assay; *Storage:* store in dry place @ 59-86 F in tightly closed container; optimum storage temp. 46-59 F.

Niacinamide USP, FCC Fine Granular No. 69916. [Roche] Niacinamide USP FCC; CAS 98-92-0; EINECS 202-713-4; essential nutient for multivitamin preps. and food fortification; *Properties:* wh. free-flowing cryst. powd., nearly odorless, bitter taste; 70% min. thru 40 mesh, 5% min. thru 100 mesh; sol. 1 g/1 ml water, 1.5 ml alcohol, 10 ml glycerin; m.w. 122.13; m.p. 128-131 C; 98.5-101% assay; *Storage:* store in dry place @ 59-86 F.

Nikkol DGMS. [Nikko Chem. Co. Ltd.] Polyglyceryl-2 stearate; CAS 12694-22-3; EINECS 235-777-7;

nonionic; w/o emulsifier for food applics.; *Properties:* pale yel. solid; HLB 5.0; 100% conc.

Nikkol MGO. [Nikko Chem. Co. Ltd.] Glyceryl oleate; nonionic; food emulsifier; *Properties:* pale yel. paste; HLB 3.0; 100% conc.

Nikkol MGS-150. [Nikko Chem. Co. Ltd.] Glyceryl stearate SE; food emulsifier; *Properties:* wh. solid.

Nikkol MGS-A. [Nikko Chem. Co. Ltd.] Glyceryl stearate; nonionic; emulsifier, stabilizer for foods; *Properties:* wh. flake; HLB 4.5; 100% conc.

Nikkol MGS-ASE. [Nikko Chem. Co. Ltd.] Glyceryl stearate SE; nonionic; emulsifier, stabilizer for foods; *Properties:* wh. flake; HLB 6.5; 100% conc.

Nikkol MGS-B. [Nikko Chem. Co. Ltd.] Glyceryl stearate; nonionic; lipophilic emulsifier, stabilizer for foods; *Properties:* wh. flake; HLB 5.0; 100% conc.

Nikkol MGS-BSE. [Nikko Chem. Co. Ltd.] Glyceryl stearate SE; nonionic; emulsifier, stabilizer for foods; *Properties:* flake; HLB 5.5; 100% conc.

Nikkol MGS-DEX. [Nikko Chem. Co. Ltd.] Glyceryl stearate SE; nonionic; lipophilic emulsifier, stabilizer for foods; *Properties:* wh. flake; HLB 6.5; 100% conc.

Nikkol MGS-F20. [Nikko Chem. Co. Ltd.] Glyceryl stearate; nonionic; emulsifier, stabilizer for foods; *Properties:* wh. flake; HLB 7.5; 100% conc.

Nikkol MGS-F40. [Nikko Chem. Co. Ltd.] Glyceryl stearate; nonionic; emulsifier, stabilizer for foods; *Properties:* wh. flake; HLB 4.0; 100% conc.

Nikkol MGS-F50. [Nikko Chem. Co. Ltd.] Glyceryl stearate; nonionic; emulsifier, stabilizer for foods; *Properties:* wh. flake; HLB 3.5; 100% conc.

Nikkol MGS-F60SE. [Nikko Chem. Co. Ltd.] Glyceryl stearate SE; food emulsifier; *Properties:* wh. flake.

Nikkol MGS-F75. [Nikko Chem. Co. Ltd.] Glyceryl stearate; nonionic; emulsifier, stabilizer for foods; *Properties:* wh. flake; HLB 0.5; 100% conc.

Nikkol MGS-TG. [Nikko Chem. Co. Ltd.] Glyceryl stearate; acid-stable emulsifier for foods; *Properties:* wh. flake.

Nikkol MGS-TGL. [Nikko Chem. Co. Ltd.] Glyceryl stearate; acid-stable emulsifier for foods; *Properties:* wh. flake.

Nikkol SL-10. [Nikko Chem. Co. Ltd.] Sorbitan laurate; nonionic; food emulsifier; *Properties:* pale yel. liq.; HLB 8.6; 100% conc.

Nikkol SO-10. [Nikko Chem. Co. Ltd.] Sorbitan oleate; CAS 1338-43-8; EINECS 215-665-4; nonionic; food emulsifier; *Properties:* yel. liq.; HLB 5.0; 100% conc.

Nikkol SO-15. [Nikko Chem. Co. Ltd.] Sorbitan sesquioleate; CAS 8007-43-0; EINECS 232-360-1; nonionic; w/o emulsifier for foods; *Properties:* yel. liq.; HLB 4.5; 100% conc.

Nikkol SP-10. [Nikko Chem. Co. Ltd.] Sorbitan palmitate; CAS 26266-57-9; EINECS 247-568-8; nonionic; food emulsifier; *Properties:* pale yel. flake; HLB 6.7; 100% conc.

Nikkol SS-15. [Nikko Chem. Co. Ltd.] Sorbitan sesquistearate; nonionic; lipophilic emulsifier for foods; *Properties:* wh. to pale yel. flake; HLB 4.2;

100% conc.

Nikkol TO-10. [Nikko Chem. Co. Ltd.] Polysorbate 80; CAS 9005-65-6; nonionic; hydrophilic emulsifier, solubilizer, and dispersant for foods; *Properties:* yel. liq.; HLB 15.0; 100% conc.

Nikkol TS-10. [Nikko Chem. Co. Ltd.] Polysorbate 60; CAS 9005-67-8; nonionic; emulsifier, solubilizer, and dispersant for o/w prods., foods; *Properties:* yel. visc. liq.; sol. in water, ethanol, ethyl acetate, toluene; HLB 14.9; sapon. no. 43-49; pH 5.7-7.7 (5%); 100% conc.

Nikkol Decaglyn 1-IS. [Nikko Chem. Co. Ltd.] Polyglyceryl-10 isostearate; CAS 133738-23-5; nonionic; emulsifier, solubilizer, dispersant for foods; *Properties:* pale yel. visc. liq.; HLB 12.5; 100% conc.

Nikkol Decaglyn 1-L. [Nikko Chem. Co. Ltd.] Polyglyceryl-10 laurate; CAS 34406-66-1; nonionic; emulsifier, solubilizer, dispersant for foods; *Properties:* pale yel. visc. liq.; HLB 17.0; 100% conc.

Nikkol Decaglyn 1-LN. [Nikko Chem. Co. Ltd.] Polyglyceryl-10 linoleate; nonionic; emulsifier, solubilizer, dispersant for foods; *Properties:* pale yel. visc. liq.; HLB 12.0; 100% conc.

Nikkol Decaglyn 1-M. [Nikko Chem. Co. Ltd.] Polyglyceryl-10 myristate; CAS 87390-32-7; nonionic; emulsifier, solubilizer, dispersant for foods; *Properties:* pale yel. visc. liq.; HLB 14.5; 100% conc.

Nikkol Decaglyn 1-O. [Nikko Chem. Co. Ltd.] Polyglyceryl-10 oleate; CAS 9007-48-1; nonionic; emulsifier, solubilizer, dispersant for foods; *Properties:* pale yel. visc. liq.; HLB 13.5; 100% conc.

Nikkol Decaglyn 1-S. [Nikko Chem. Co. Ltd.] Polyglyceryl-10 stearate; CAS 79777-30-3; nonionic; emulsifier, solubilizer, dispersant for foods; *Properties:* pale yel. plate; HLB 12.5; 100% conc.

Nikkol Decaglyn 2-O. [Nikko Chem. Co. Ltd.] Polyglyceryl-10 dioleate; CAS 33940-99-7; nonionic; o/w emulsifier for foods; *Properties:* pale yel. visc. liq.; HLB 10.0; 100% conc.

Nikkol Decaglyn 2-S. [Nikko Chem. Co. Ltd.] Polyglyceryl-10 distearate; CAS 12764-60-2; nonionic; o/w emulsifier for foods; *Properties:* pale yel. plate; HLB 12.0; 100% conc.

Nikkol Tetraglyn 1-O. [Nikko Chem. Co. Ltd.] Polyglyceryl-4 oleate; CAS 9007-48-1; nonionic; o/w emulsifier, anticrystallizing agent; food emulsifier; *Properties:* pale yel. visc. liq.; HLB 6.0; 100% conc.

Nikkol Tetraglyn 1-S. [Nikko Chem. Co. Ltd.] Polyglyceryl-4 stearate; CAS 37349-34-1; nonionic; hydrophilic emulsifier for foods; *Properties:* wh. flake; HLB 6.0; 100% conc.

Nikkol Tetraglyn 3-S. [Nikko Chem. Co. Ltd.] Polyglyceryl-4 tristearate; nonionic; o/w emulsifier for foods, anticrystallizing agent; *Properties:* wh. flake; 100% conc.

Nikkol Tetraglyn 5-O. [Nikko Chem. Co. Ltd.] Polyglyceryl-4 pentaoleate; nonionic; o/w emulsifier for foods, anticrystallizing agent; *Properties:* pale yel. liq.; 100% conc.

Nikkol Tetraglyn 5-S. [Nikko Chem. Co. Ltd.] Polyglyceryl-4 pentastearate; nonionic; o/w emulsifier

for foods; anticrystallizing agent for fats; *Properties:* wh. flake; 100% conc.

90/10 Blend. [FMC] Sodium tripolyphosphate, sodium hexametaphosphate; moisture binder, flavor protectant in cured meat, pastrami, ham, deli turkey breast; provides moisture retention and purge control in prep. of meat and fish for freezing, vacuum-pkg. prods.; fast dissolving; stable in conc. brine sol'n.; *Regulatory:* USDA, FDA GRAS; *Properties:* gran.; 100% turh 12 mesh; sol. 15 g/100 g water; bulk dens. 50 lb/ft³; pH 9.7 (1%).

Niox KI-29. [Pulcra SA] Ethoxylated castor oil; solubilizer for flavors, perfumes, vitamins, oils; *Properties:* water-sol.; Unverified

Nipabenzyl. [Nipa Labs] Benzylparaben; CAS 94-18-8; EINECS 202-311-9; preservative, bactericide, fungicide for foods; *Regulatory:* BP, NF, Euorpean pharmacopoeia, FCC compliance; *Properties:* wh. fine cryst. powd., odorless, tasteless; sol. (g/100 g solv.) 102 g acetone, 79 g methanol, 72 g ethanol, 60 g lanolin, 42 g ether; m.w. 228.25; m.p. 110-112 C; 99% assay; *Toxicology:* LD50 (oral, rat) > 5 g/kg; nonirritating to skin, sl. harmful by ingestion, sl. irritating to eyes.

Nipabutyl. [Nipa Labs] Butylparaben NF, BP, FCC, Eur. Ph.; CAS 94-26-8; EINECS 202-318-7; preservative, bactericide, fungicide for foods; *Properties:* wh. cryst. powd., odorless or very faint aromatic odor, tasteless; sol. (g/100 g solv.) 240 g acetone, 220 g methanol, 208 g ethanol, > 200 g IPA, 150 g ether, > 100 g lanolin; m.w. 194.23; m.p. 68-69 C; > 99% act.; *Toxicology:* LD50 (oral, mouse) > 5 g/kg; nonirritating to skin; sl. irritating to eyes.

Nipabutyl Potassium. [Nipa Labs] Potassium butylparaben; CAS 38566-94-8; preservative, bactericide, fungicide for foods; *Properties:* wh. fine hygroscopic powd.; sol. in cold water; m.w. 232.32; pH 9.5-10.5 (0.1% aq.); > 99% act.; *Toxicology:* nonirritating to skin; sl. irritating to eyes and nasal passages.

Nipabutyl Sodium. [Nipa Labs] Sodium butylparaben; CAS 36457-20-2; preservative, bactericide, fungicide for foods; *Properties:* wh. fine hygroscopic powd.; sol. in cold water; m.w. 216.21; pH 9.5-10.5 (0.1% aq.); > 99% act.; *Toxicology:* LD50 (oral, mouse) > 2 g/kg; pure material irritating to skin; sl. irritating to eyes and nasal passages.

Nipacide® OPP. [Nipa Labs] o-Phenyl phenol; CAS 90-43-7; EINECS 201-993-5; preservative for citrus fruit; *Usage level:* 10-15% (disinfectant concs.), 0.1-0.2% (cooling lubricants), 0.08-2% (adhesives), 0.15-1.3% (citrus dips); *Properties:* cream to wh. flakes, sl. phenolic odor; sol. (g/100 g): 800 g in methanol, 660 g acetone, 590 g ethanol, 490 g ether, 300 g propylene glycol, 275 g ethylene glycol, 270 g benzene; m.w. 170; bulk dens. 0.60-0.65; m.p. 56-58 C; b.p. 286 C; flash pt. (COC) 138 C; 99.5% min. purity; *Toxicology:* LD50 (rats) 2000 mg/kg; harmful by ingestion; corrosive to skin and eyes; *Storage:* unlimited shelf life if protected from light and moisture; keep containers tightly closed.

Nipacide® Potassium. [Nipa Labs] Preservative for

aq. foodstuffs; *Properties:* sol. in cold water.

Nipacide® Sodium. [Nipa Labs] Preservative for aq. foodstuffs; *Properties:* sol. in cold water.

Nipacombin PK. [Nipa Labs] Preservative for aq. foodstuffs; *Properties:* sol. in cold water.

Nipagin A. [Nipa Labs] Ethylparaben NF, BP, FCC, Eur.Ph.; CAS 120-47-8; EINECS 204-399-4; preservative, bactericide, fungicide for foods (beer, flavors, fruit juices, fish, pickles, sauces, soft drinks, yogurt); *Properties:* wh. fine cryst. powd., odorless or very faint aromatic odor, tasteless; sol. (g/100 g solv.) 84 g acetone, 81 g methanol, 70 g ethanol, 40 g ether, 30 g lanolin; m.w. 166.18; m.p. 115-117 C; > 99% act.; *Toxicology:* LD50 (oral, rat) > 8 g/kg; nonirritating to skin; sl. irritating to eyes.

Nipagin A Potassium. [Nipa Labs] Potassium ethylparaben; CAS 36457-19-9; preservative, bactericide, fungicide for foods (beer, flavors, fruit juices, fish, pickles, sauces, soft drinks, yogurt); *Properties:* wh. fine hygroscopic powd., tasteless; sol. in cold water; m.w. 204.27; pH 9.5-10.5 (0.1% aq.); > 99% act.; *Toxicology:* nonirritating to skin; sl. irritating to eyes and nasal passages.

Nipagin A Sodium. [Nipa Labs] Sodium ethylparaben; CAS 35285-68-8; EINECS 252-487-6; preservative, bactericide, fungicide for foods (beer, flavors, fruit juices, fish, pickles, sauces, soft drinks, yogurt); *Properties:* wh. fine hygroscopic powd., tasteless; sol. in cold water; m.w. 188.16; pH 9.5-10.5 (0.1% aq.); > 99% act.; *Toxicology:* LD50 (oral, mouse) 2.5 g/kg; pure material irritating to skin; sl. irritating to eyes and nasal passages.

Nipagin M. [Nipa Labs] Methylparaben NF, BP, FCC, Eur.Ph.; CAS 99-76-3; EINECS 202-785-7; preservative, bactericide, fungicide for foods (beer, flavors, fruit juices, fish, pickles, sauces, soft drinks, yogurt, soy sauce); *Regulatory:* GRAS; *Properties:* wh. fine cryst. powd., odorless or very faint aromatic odor, tasteless; sol. (g/100 g solv.): 64 g acetone, 60 g IPA, 58 g methanol, 48 g ethanol, 35 g propylene glycol, 23 g ether; m.w. 152.15; m.p. 125-128 C; > 99% act.; *Toxicology:* LD50 (oral, rat) > 8 g/kg; sl. irritating to eyes; nonirritating to skin.

Nipagin M Sodium. [Nipa Labs] Sodium methylparaben; CAS 5026-62-0; EINECS 225-714-1; preservative, bactericide, fungicide for foods (beer, flavors, fruit juices, fish, pickles, sauces, soft drinks, yogurt); *Properties:* wh. hygroscopic powd.; sol. in cold water; m.w. 174.1; pH 9.5-10.5 (0.1%); > 99% act.; *Toxicology:* LD50 (oral, mouse) 2.0 g/kg; nonirritating to skin; sl. irritating to eyes and nasal passages.

Nipaheptyl. [Nipa Labs] n-Heptyl p-hydroxybenzoate; preservative, bactericide, fungicide for foods; permitted for use in beer.

Nipanox® S-1. [Nipa Labs] Propyl gallate (20%), citric acid (10%) in propylene glycol (70%); antioxiant for veg. oil industry; *Properties:* clear to lt. amber liq., very sl. odor; sol. in animal and veg. fats.

Nipanox® Special. [Nipa Labs] BHA (13%), propyl gallate (13%), citric acid (4%) in propylene glycol (70%); antioxidant for foods; fat and oil stabilizer;

exc. carry-through props. for pastries, crackers, biscuits, potato chips; *Properties:* clear to lt. straw liq., sl. odor; sol. in animal and veg. fats, fatty acids, glycerides; insol. in water.

Nipasept Potassium. [Nipa Labs] Preservative for aq. foodstuffs; *Properties:* sol. in cold water.

Nipasept Sodium. [Nipa Labs] Sodium methylparaben, sodium propylparaben, and sodium ethylparaben; preservative for aq. foodstuffs; *Properties:* sol. in cold water.

Nipasol M. [Nipa Labs] Propylparaben NF, BP, FCC, Eur.Ph.; CAS 94-13-3; EINECS 202-307-7; preservative, bactericide, fungicide for foods; *Regulatory:* GRAS; *Properties:* wh. fine cryst. powd., odorless or very faint aromatic odor, tasteless; sol. (g/100 g solv.): 105 g acetone, 100 g in methanol, ethanol, 88 g IPA, 80 g lanolin, 50 g ether; m.w. 180.20; m.p. 95-98 C; > 99% act.; *Toxicology:* LD50 (oral, rat) > 8 g/kg (pract. nonharmful by ingestion); nonirritant to skin; sl. irritant to eyes.

Nipasol M Potassium. [Nipa Labs] Potassium propylparaben; CAS 84930-16-5; preservative, bactericide, fungicide for foods; *Properties:* wh. fine hygroscopic powd., odorless or very faint aromatic odor, tasteless; sol. in cold water; m.w. 218.29; pH 9.5-10.5 (0.1% aq.); > 99% act.; *Toxicology:* LD50 (oral, rat) > 8 g/kg (pract. nonharmful by ingestion); nonirritating to skin; sl. irritating to eyes.

Nipasol M Sodium. [Nipa Labs] Sodium propylparaben; CAS 35285-69-9; EINECS 252-488-1; preservative, bactericide, fungicide for foods; *Properties:* wh. hygroscopic powd.; sol. in cold water; m.w. 202.2; pH 9.5-10.5 (0.1%); > 99% act.; *Toxicology:* LD50 (oral, mouse) 3.7 g/kg (pract. nonharmful); nonirritating to skin; sl. irritating to eyes, nasal passages.

Nipastat. [Nipa Labs] Methylparaben (> 50%), butylparaben (> 20%), ethylparaben (<15%), and propylparaben (< 10%); preservative, bactericide, fungicide for foods, veg. extracts; *Usage level:* 0.05-0.3%; *Regulatory:* USA and Europe approvals; *Properties:* wh. fine cryst. powd., virtually odorless, tasteless; sol. 0.14% in water; m.p. 60-125 C; pH 7.0 (10% aq.); *Toxicology:* pract. nonharmful by ingestion; nonirritating to skin; sl. irritating to eyes.

Nissan Anon BF. [Nippon Oils & Fats] Dimethyl cocoalkyl betaine; amphoteric; germicide in foods; *Properties:* lt. yel. liq.; 25% min act.

Nissan Anon BL. [Nippon Oils & Fats] Dimethyl dodecyl betaine; amphoteric; antistatic agent, softener, germicide in foods; extraction aid in fermentation; *Properties:* lt. yel. liq.; pH 4.5-6.5; 30-33% min. act.; Unverified

Nissan Anon LG. [Nippon Oils & Fats] Alkyl di-(aminoethyl) glycine; amphoteric; antistatic agent, softener, germicide in foods; extraction aid for fermentation; *Properties:* lt. yel. liq.; pH 8.5-10.5; 30% act.; Unverified

Nissan Cation AB. [Nippon Oils & Fats] Octadecyl trimethyl ammonium chloride; cationic; germicide in foods; *Properties:* lt. yel. liq.; 23% min. act.

Nissan Nonion CP-08R. [Nippon Oils & Fats] Sorbitan caprylate; nonionic; food emulsifier; *Properties:* liq.; oil-sol.; HLB 7.3; 100% conc.

Nissan Nonion DN-202. [Nippon Oils & Fats] POE lauryl ether; nonionic; food emulsifier; *Properties:* liq.; oil-sol.; HLB 6.2; 100% conc.

Nissan Nonion DN-203. [Nippon Oils & Fats] POE lauryl ether; nonionic; food emulsifier; *Properties:* liq.; oil-sol.; HLB 7.9; 100% conc.

Nissan Nonion DN-209. [Nippon Oils & Fats] POE lauryl ether; nonionic; food emulsifier; *Properties:* liq.; oil-sol.; HLB 13.2; 100% conc.

Nissan Nonion LP-20R, LP-20RS. [Nippon Oils & Fats] Sorbitan laurate; CAS 1338-39-2; nonionic; food emulsifier; *Properties:* Gardner 5 max. oily liq.; sol. in methanol, ethanol, acetone, xylene, ethyl ether, kerosene, disp. in water; HLB 8.6; 100% conc.

Nissan Nonion LT-221. [Nippon Oils & Fats] Polysorbate 20; CAS 9005-64-5; nonionic; food emulsifier; *Properties:* Gardner 6 max. oily liq.; sol. in water, methanol, ethanol, acetone, xylene, ethyl ether, ethylene glycol, HLB 16.7; 100% conc.

Nissan Nonion MP-30R. [Nippon Oils & Fats] Sorbitan myristate; nonionic; food emulsifier; *Properties:* solid; oil-sol.; HLB 6.6; 100% conc.

Nissan Nonion OP-80R. [Nippon Oils & Fats] Sorbitan oleate; CAS 1338-43-8; EINECS 215-665-4; nonionic; food emulsifier; *Properties:* Gardner 9 max. oily liq.; oil-sol.; HLB 4.3; 100% conc.

Nissan Nonion OP-83RAT. [Nippon Oils & Fats] Sorbitan sesquioleate; CAS 8007-43-0; EINECS 232-360-1; nonionic; food emulsifier; *Properties:* Gardner 9 max. oily liq.; sol. in ethanol, acetone, xylene, ethyl ether, kerosene, methanol, warm in water; HLB 3.7; 100% conc.

Nissan Nonion OP-85R. [Nippon Oils & Fats] Sorbitan trioleate; CAS 26266-58-0; EINECS 247-569-3; nonionic; food emulsifier; *Properties:* Gardner 9 max. oily liq.; oil-sol.; HLB 1.8; 100% conc.

Nissan Nonion OT-221. [Nippon Oils & Fats] POE sorbitan monooleate; nonionic; food emulsifier; *Properties:* Gardner 6 max. oily liq.; sol. in water, ethanol, acetone, xylene, disp. in methanol; HLB 15.0; 100% conc.

Nissan Nonion PP-40R. [Nippon Oils & Fats] Sorbitan palmitate; CAS 26266-57-9; EINECS 247-568-8; nonionic; food emulsifier; *Properties:* Gardner 7 max. waxy solid; oil-sol.; HLB 6.7; 100% conc.

Nissan Nonion PT-221. [Nippon Oils & Fats] POE sorbitan monopalmitate; nonionic; food emulsifier; *Properties:* Gardner 8 max. oily liq.; sol. in water, methanol, ethanol, acetone, xylene, ethyl ether, ethylene glycol; HLB 15.3; 100% conc.

Nissan Nonion SP-60R. [Nippon Oils & Fats] Sorbitan stearate; CAS 1338-41-6; EINECS 215-664-9; nonionic; food emulsifier; *Properties:* Gardner 5 max. waxy solid; sol. in methanol, ethanol, xylene, kerosene, ethyl ether, disp. in warm water; HLB 4.7; 100% conc.

Nissan Nonion ST-221. [Nippon Oils & Fats] POE sorbitan monostearate; nonionic; food emulsifier; *Properties:* Gardner 5 max. oily liq.; sol. in water,

methanol, ethanol, acetone, xylene, ethyl ether, kerosene; HLB 14.9; 100% conc.

Nissan Panacete 810. [Nippon Oils & Fats] Med. chain triglyceride; nonionic; raw material for special foods; *Properties:* liq.; 100% conc.

No-GluAce™. [Mitsubishi Int'l.; Tokai Bussan] Hydrolyzed vegetable protein, dextrin, yeast extract; natural flavor enhancer; MSG-replacement at 1:1 ratio; for soups, sauces, cooked foods, cooked meats, boiled fish paste, frozen foods, snacks, health foods, childrens food; *Properties:* yel.-brn. powd., MSG-like flavor; pH 5.2; 36.7% protein, 5% moisture, 1% fat.

Non-Diastatic Malt Syrup #40600. [MLG Enterprises Ltd.] Liq. extract of corn and malted barley; CAS 8002-48-0; sweetener providing malt flavor; enriches color to a rich golden appearance; contributes to nutritional value; humectant increasing shelf life; enhances texture; for baked goods, baby foods, ice cream, confections, cereal; *Regulatory:* kosher; *Properties:* pH 4.3-5.5 (10%); 10-25% solids; *Storage:* store @ 40-90 F; 6 mos max. shelf life.

Norfox® Anionic 27. [Norman, Fox] Sodium 2-ethylhexyl sulfate; EINECS 204-812-8; anionic; wetting agent and peeling aid; *Properties:* liq.; 27% conc.

Norfox® GMS-FG. [Norman, Fox] Glyceryl stearate; nonionic; food grade emulsifier; *Properties:* flake; HLB 3.9; 100% conc.

Norfox® Sorbo S-60FG. [Norman, Fox] Sorbitan stearate; CAS 1338-41-6; EINECS 215-664-9; nonionic; food emulsifier in veg. and dairy prods.; *Properties:* flake; HLB 4.7; 100% conc.

Norfox® Sorbo S-80. [Norman, Fox] Sorbitan oleate; CAS 1338-43-8; EINECS 215-665-4; nonionic; hydrophobic emulsifier; *Properties:* liq.; HLB 4.3; 100% conc.

Norfox® Sorbo T-20. [Norman, Fox] Polysorbate 20; CAS 9005-64-5; nonionic; flavor and fragrance solubilizer; *Properties:* liq.; HLB 16.7; 100% act.

Norfox® Sorbo T-60. [Norman, Fox] PEG-20 sorbitan stearate; CAS 9005-67-8; nonionic; emulsifier for frozen desserts, salad dressings, cake mixes, icings; dough conditioner; *Properties:* liq.; HLB 20.0; 100% conc.

Norfox® Sorbo T-80. [Norman, Fox] Polysorbate 80; CAS 9005-65-6; nonionic; solubilizer for fat-sol. actives; emulsifier for shortening; whipped topping stabilizer; *Properties:* liq.; HLB 15.0; 100% act.

No Stick Emulsifier. [Custom Ingreds.] Mono and diglycerides, partially hydrog. soybean oil, water, lecithin, calcium propionate; developed to equilibrate moisture in corn tortillas; helps prevent sticking together of tortillas and aids moisture-holding props.; *Usage level:* 1-2 lb/100 lb masa; *Storage:* store in clean, dry area @ 40-85 F for storage life of 9 mos.

Nova-CPLN. [Champlain Industries] Primary yeast extract; CAS 8013-01-2; meat flavor, no added MSG; *Properties:* powd.; 97% solids, 44% protein, 26% salt.

Nova-Chef™ 4005. [Champlain Industries] Yeast

extract-based natural savory flavor; universal flavor enhancer with good mouthfeel and round, subtle, meaty taste; no added MSG; for snack foods, soups, gravies, processed meat systems.

Nova-Chef™ 4015. [Champlain Industries] Yeast extract-based natural savory flavor; good, full-bodied chicken flavor, pleasant mouthfeel; no added MSG; for snack foods, soups, gravies, processed meat systems.

Nova-Chef™ 4018. [Champlain Industries] Yeast extract-based natural savory flavor; light beefy flavor and aroma, good mouthfeel; no added MSG; for snack foods, soups, gravies, processed meat systems.

Nova-Chef™ 4021. [Champlain Industries] Yeast extract-based natural savory flavor; good vegetable flavor, good mouthfeel; no added MSG; for snack foods, soups, gravies, processed meat systems.

Nova-Chef™ 4022. [Champlain Industries] Yeast extract-based natural savory flavor; roast beef flavor; no added MSG; for soups, gravies, processed meat systems.

Nova-Chef™ 4025. [Champlain Industries] Yeast extract-based natural savory flavor; grilled chicken flavor; no added MSG; for soups, gravies, processed meat systems.

Nova-Chef™ 4026. [Champlain Industries] Yeast extract-based natural savory flavor; full-bodied, light chicken flavor; no added MSG; for soups, gravies, processed meat systems.

Nova-Chef™ 4028 [Champlain Industries] Yeast extract-based natural savory flavor; grilled beef flavor; no added MSG; for soups, gravies, processed meat systems.

Nova-Chef™ 4031. [Champlain Industries] Yeast extract-based natural savory flavor; light roasted chicken flavor; no added MSG; for soups, gravies, processed meat systems.

Nova-Chef™ 4032. [Champlain Industries] Yeast extract-based natural savory flavor; hearty, beefy flavor, good aroma, clear in sol'n.; no added MSG; for soups, gravies, processed meat systems.

Nova-Chef™ 4102. [Champlain Industries] Yeast extract-based natural savory flavor; prime rib-like beef flavor and aroma; contains MSG; for soups, gravies, processed meat systems.

Nova-Chef™ 4108. [Champlain Industries] Yeast extract-based natural savory flavor; mild meaty flavor with pleasant aftertaste; contains MSG; for soups, gravies, processed meat systems.

Nova-Chef™ 4110. [Champlain Industries] Yeast extract-based natural savory flavor; good roasted beef flavor and aroma, good mouthfeel; contains MSG; for soups, gravies, processed meat systems.

Nova-Chef™ 4114. [Champlain Industries] Yeast extract-based natural savory flavor; light chicken flavor with flavor enhancement; contains MSG; for soups, gravies, processed meat systems.

Nova-Chef™ 4120. [Champlain Industries] Yeast extract-based natural savory flavor; strong vegetable enhancer, good clarity; contains MSG; for snack foods, soups, gravies, processed meat systems.

Nova-Chef™ 4130. [Champlain Industries] Yeast extract-based natural savory flavor; strong roasted chicken flavor and aroma; contains MSG; for snack foods, soups, gravies, processed meat systems.

Nova-Flav™ 1000. [Champlain Industries] Primary yeast extract; CAS 8013-01-2; chicken flavor, no added MSG; *Properties:* powd.; 95% solids, 71% protein, 1% salt.

Nova-Flav™ 1001. [Champlain Industries] Primary yeast extract; CAS 8013-01-2; creamy chicken flavor, no added MSG; *Properties:* paste; 70% solids, 52% protein, 1% salt.

Nova-Flav™ 1002. [Champlain Industries] Primary yeast extract; CAS 8013-01-2; beef flavor, no added MSG; *Properties:* powd.; 95% solids, 40% protein, 1% salt.

Nova-Flav™ 1006. [Champlain Industries] Primary yeast extract; CAS 8013-01-2; yeast flavor, no added MSG; *Properties:* powd.; 95% solids, 35% protein, 0.5% salt.

Nova-Flav™ 1020. [Champlain Industries] Primary yeast extract; CAS 8013-01-2; grilled meat flavor, no added MSG; *Properties:* powd.; 95% solids, 71% protein, 1% salt.

Nova-Flav™ 1030. [Champlain Industries] Primary yeast extract; CAS 8013-01-2; charbroiled meat flavor, no added MSG; *Properties:* powd.; 95% solids, 71% protein, 1% salt.

Nova-Flav™ 5004. [Champlain Industries] Primary yeast extract; CAS 8013-01-2; beef flavor, no added MSG; *Properties:* powd.; 95% solids, 60% protein, 20% salt.

Nova-Flav™ 5006. [Champlain Industries] Primary yeast extract; CAS 8013-01-2; mushroom flavor, no added MSG; *Properties:* paste; 80% solids, 40% protein, 18% salt.

Nova-Flav™ 5009. [Champlain Industries] Primary yeast extract; CAS 8013-01-2; chicken flavor enhancer, no added MSG; *Properties:* powd.; 95% solids, 55% protein, 12% salt.

Nova-Flav™ 5010. [Champlain Industries] Primary yeast extract; CAS 8013-01-2; beef flavor, no added MSG; *Properties:* paste; 80% solids, 45% protein, 18% salt.

Nova-Flav™ 5030. [Champlain Industries] Primary yeast extract; CAS 8013-01-2; cheese flavor, no added MSG; *Properties:* powd.; 95% solids, 55% protein, 18% salt.

Nova-Flav™ 5100. [Champlain Industries] Primary yeast extract; CAS 8013-01-2; beef bouillon flavor, contains MSG; *Properties:* paste; 80% solids, 43% protein, 15% salt.

Nova-Flav™ 5101. [Champlain Industries] Primary yeast extract; CAS 8013-01-2; beef bouillon flavor, contains MSG; *Properties:* paste; 80% solids, 43% protein, 15% salt.

Nova-Flav™ 5102. [Champlain Industries] Primary yeast extract; CAS 8013-01-2; bouillon flavor, no added MSG; *Properties:* paste; 80% solids, 43% protein, 15% salt.

Nova-Flav™ 5103. [Champlain Industries] Primary yeast extract; CAS 8013-01-2; beef flavor, contains MSG; *Properties:* powd.; 95% solids, 51% protein, 19% salt.

Nova-Flav™ 5105. [Champlain Industries] Primary yeast extract; CAS 8013-01-2; beef flavor, contains MSG; *Properties:* paste; 80% solids, 43% protein, 15% salt.

Nova-Flav™ 7000. [Champlain Industries] Primary yeast extract; CAS 8013-01-2; chicken flavor, no added MSG; *Properties:* powd.; 95% solids, 41% protein, 38% salt.

Nova-Flav™ 7001. [Champlain Industries] Primary yeast extract; CAS 8013-01-2; smokey bacon flavor, no added MSG; *Properties:* powd.; 95% solids, 40% protein, 38% salt.

Nova-Flav™ 7003. [Champlain Industries] Primary yeast extract; CAS 8013-01-2; cheese flavor, no added MSG; *Properties:* powd.; 95% solids, 40% protein, 36% salt.

Nova-Flav™ 7004. [Champlain Industries] Primary yeast extract; CAS 8013-01-2; potato/earthy flavor, no added MSG; *Properties:* powd.; 95% solids, 41% protein, 38% salt.

Nova-Flav™ 7006. [Champlain Industries] Primary yeast extract; CAS 8013-01-2; bouillon flavor, no added MSG; *Properties:* powd.; 95% solids, 43% protein, 38% salt.

Nova-Flav™ 7007. [Champlain Industries] Primary yeast extract; CAS 8013-01-2; vegetable flavor, no added MSG; *Properties:* powd.; 95% solids, 40% protein, 38% salt.

Nova-Flav™ 7009. [Champlain Industries] Primary yeast extract; CAS 8013-01-2; chicken flavor, no added MSG; *Properties:* powd.; 95% solids, 42% protein, 36% salt.

Nova-Flav™ 7010. [Champlain Industries] Primary yeast extract; CAS 8013-01-2; roast beef flavor, no added MSG; *Properties:* powd.; 95% solids, 40% protein, 35% salt.

Nova-Flav™ 7102. [Champlain Industries] Primary yeast extract; CAS 8013-01-2; bouillon flavor, contains MSG; *Properties:* powd.; 95% solids, 37% protein, 38% salt.

Nova-Flav™ 7105. [Champlain Industries] Primary yeast extract; CAS 8013-01-2; chicken/bouillon flavor, contains MSG; *Properties:* powd.; 95% solids, 34% protein, 38% salt.

Nova-Flav™ 7107. [Champlain Industries] Primary yeast extract; CAS 8013-01-2; vegetable flavor, contains MSG; *Properties:* powd.; 95% solids, 34% protein, 38% salt.

Nova-Flav™ 7109. [Champlain Industries] Primary yeast extract; CAS 8013-01-2; chicken flavor, contains MSG; *Properties:* powd.; 95% solids, 37% protein, 36% salt.

Nova-Flav™ 8002. [Champlain Industries] Primary yeast extract; CAS 8013-01-2; poultry flavor, no added MSG; *Properties:* powd.; 95% solids, 40% protein, 21% salt.

Nova-Flav™ 8004. [Champlain Industries] Primary yeast extract; CAS 8013-01-2; poultry flavor, no added MSG; *Properties:* powd.; 95% solids, 44% protein, 14% salt.

Nova-Max 1. [Champlain Industries] Primary yeast extract; CAS 8013-01-2; brothy flavor, no added MSG; *Properties:* powd.; 96% solids, 27% protein, 37% salt.

Nova-Naturelle™ Anchovy Flavor WONF RC-92. [Champlain Industries] Natural seafood flavors contg. no added MSG; strong flavor and aroma; used to enhance snack foods, soups, gravies, processed meat systems.

Nova-Naturelle™ Anchovy Flavor WONF RC-93. [Champlain Industries] Natural seafood flavors contg. no added MSG; nice flavor and aroma with meaty notes; used to enhance snack foods, soups, gravies, processed meat systems.

Nova-Naturelle™ Bacon Type Flavor WONF RC-0027. [Champlain Industries] Natural flavor contg. added MSG; strong bacon type flavor and aroma; used to enhance snack foods, soups, gravies, processed meat systems.

Nova-Naturelle™ Bacon Type Flavor WONF RC-0038. [Champlain Industries] Natural flavor containing no added MSG or autolyzed yeast extract; strong bacon flavor and aroma; used to enhance snack foods, soups, gravies, processed meat systems.

Nova-Naturelle™ Bacon Type Flavor WONF RC-0056. [Champlain Industries] Kosher parve flavors containing no meat or animal prods.; contains autolyzed yeast extracts and natural flavors; strong bacon flavor; no MSG; used to enhance snack foods, soups, gravies, processed meat systems.

Nova-Naturelle™ Bacon Type Flavor WONF RC-112. [Champlain Industries] Natural flavor containing no added MSG or autolyzed yeast extract; strong pork flavor with a nice aftertaste; used to enhance snack foods, soups, gravies, processed meat systems.

Nova-Naturelle™ Beef Flavor WONF 9101. [Champlain Industries] Natural flavor contg. added MSG; hearty beefy taste, good aroma, clear in sol'n.; used to enhance snack foods, soups, gravies, processed meat systems.

Nova-Naturelle™ Beef Flavor WONF RC-0034. [Champlain Industries] Natural flavor containing no added MSG or autolyzed yeast extract; mild beef flavor and aroma; used to enhance snack foods, soups, gravies, processed meat systems.

Nova-Naturelle™ Beef Flavor WONF RC-0035. [Champlain Industries] Natural flavor containing no added MSG or autolyzed yeast extract; clean beef flavor and aroma, brothy type; used to enhance snack foods, soups, gravies, processed meat systems.

Nova-Naturelle™ Beef Flavor WONF RC-0044. [Champlain Industries] Natural flavor containing no added MSG or autolyzed yeast extract; dark beef flavor; used to enhance snack foods, soups, gravies, processed meat systems.

Nova-Naturelle™ Beef Flavor WONF RC-97. [Champlain Industries] Low-sodium flavor containing no added MSG; strong beefy flavor with bloody notes; used to enhance snack foods, soups, gravies, processed meat systems.

Nova-Naturelle™ Beef Flavor WONF RC-100. [Champlain Industries] Natural flavor containing no added MSG or autolyzed yeast extract; beefy flavor with bloody notes; used to enhance snack foods, soups, gravies, processed meat systems.

Nova-Naturelle™ Beef Flavor WONF RC-200. [Champlain Industries] Low-sodium flavor containing no added MSG; mild beef broth flavor and aroma; used to enhance snack foods, soups, gravies, processed meat systems.

Nova-Naturelle™ Beef Flavor WONF RC-300. [Champlain Industries] Flavor keys used as building blocks in formulations that require strong savory notes without addition of MSG; round beef flavor and aroma, nice broth flavor; used to enhance snack foods, soups, gravies, processed meat systems.

Nova-Naturelle™ Beef Stew Flavor WONF RC-0021. [Champlain Industries] Natural flavor contg. added MSG; roasted beef flavor and aroma, clear in sol'n.; used to enhance snack foods, soups, gravies, processed meat systems.

Nova-Naturelle™ Beef Teriyaki Flavor WONF RC-0046. [Champlain Industries] Natural flavor containing no added MSG or autolyzed yeast extract; mild teriyaki flavor; used to enhance snack foods, soups, gravies, processed meat systems.

Nova-Naturelle™ Beef Type Flavor WONF RC-0053. [Champlain Industries] Kosher parve flavors containing no meat or animal prods.; contains autolyzed yeast extracts and natural flavors; beefy taste, good aroma; used to enhance snack foods, soups, gravies, processed meat systems.

Nova-Naturelle™ Beef Type Flavor WONF RC-0058. [Champlain Industries] Kosher parve flavors containing no meat or animal prods.; contains autolyzed yeast extracts and natural flavors; nice beef flavor and odor, no added MSG; used to enhance snack foods, soups, gravies, processed meat systems.

Nova-Naturelle™ Breakfast Sausage Type Flavor WONF RC-0047. [Champlain Industries] Natural flavor containing no added MSG or autolyzed yeast extract; strong sausage flavor; used to enhance snack foods, soups, gravies, processed meat systems.

Nova-Naturelle™ Breakfast Sausage Type Flavor WONF RC-0062. [Champlain Industries] Kosher parve flavors containing no meat or animal prods.; contains autolyzed yeast extracts and natural flavors; nice mild sausage flavor; used to enhance snack foods, soups, gravies, processed meat systems.

Nova-Naturelle™ Chicken Flavor WONF RC-0028. [Champlain Industries] Natural flavor contg. added MSG; full-bodied chicken flavor and aroma; used to enhance snack foods, soups, gravies, processed meat systems.

Nova-Naturelle™ Chicken Flavor WONF RC-0031. [Champlain Industries] Natural flavor containing no added MSG or autolyzed yeast extract; strong chicken flavor and aroma; used to enhance snack foods, soups, gravies, processed meat systems.

Nova-Naturelle™ Chicken Flavor WONF RC-0040. [Champlain Industries] Natural flavor containing no added MSG or autolyzed yeast extract; good white meat flavor with fatty notes; used to enhance snack foods, soups, gravies, processed meat systems.

Nova-Naturelle™ Chicken Flavor WONF RC-107. [Champlain Industries] Natural flavor containing no added MSG or autolyzed yeast extract; white meat chicken flavor with brothy char.; used to enhance snack foods, soups, gravies, processed meat systems.

Nova-Naturelle™ Chicken Flavor WONF RC-108. [Champlain Industries] Natural flavor containing no added MSG or autolyzed yeast extract; strong chicken flavor with bloody notes; used to enhance snack foods, soups, gravies, processed meat systems.

Nova-Naturelle™ Chicken Flavor WONF RC-202. [Champlain Industries] Low-sodium flavor containing no added MSG; strong meaty flavor with bloody notes; used to enhance snack foods, soups, gravies, processed meat systems.

Nova-Naturelle™ Chicken Flavor WONF RC-203. [Champlain Industries] Low-sodium flavor containing no added MSG; white meat type flavor with brothy char.; used to enhance snack foods, soups, gravies, processed meat systems.

Nova-Naturelle™ Chicken Flavor WONF RC-305. [Champlain Industries] Flavor keys used as building blocks in formulations that require strong savory notes without addition of MSG; strong flavor dark meat with bloody notes; used to enhance snack foods, soups, gravies, processed meat systems.

Nova-Naturelle™ Chicken Flavor WONF RC-307. [Champlain Industries] Flavor keys used as building blocks in formulations that require strong savory notes without addition of MSG; white meat flavor with fatty notes, brothy char.; used to enhance snack foods, soups, gravies, processed meat systems.

Nova-Naturelle™ Clam Flavor WONF RC-95. [Champlain Industries] Natural seafood flavors contg. no added MSG; strong clam flavor and aroma, brothy type; used to enhance snack foods, soups, gravies, processed meat systems.

Nova-Naturelle™ Crab Flavor WONF RC-99. [Champlain Industries] Natural seafood flavors contg. no added MSG; nice mild crab flavor with a strong pleasant aroma; used to enhance snack foods, soups, gravies, processed meat systems.

Nova-Naturelle™ Dark Beef Flavor WONF RC-303. [Champlain Industries] Flavor keys used as building blocks in formulations that require strong savory notes without addition of MSG; strong dark meat flavor with light burn notes; used to enhance snack foods, soups, gravies, processed meat systems.

Nova-Naturelle™ Fish Flavor WONF RC-89. [Champlain Industries] Natural seafood flavors contg. no added MSG; strong fish flavor and aroma; used to enhance snack foods, soups, gravies, processed meat systems.

Nova-Naturelle™ Grilled Beef Flavor WONF RC-0022. [Champlain Industries] Natural flavor contg. added MSG; strong beef flavor with grilled flavor; used to enhance snack foods, soups, gravies, processed meat systems.

Nova-Naturelle™ Grilled Beef Flavor WONF RC-0042. [Champlain Industries] Natural flavor containing no added MSG or autolyzed yeast extract; nice grill or BBQ flavor; used to enhance snack foods, soups, gravies, processed meat systems.

Nova-Naturelle™ Grilled Beef Flavor WONF RC-301. [Champlain Industries] Flavor keys used as building blocks in formulations that require strong savory notes without addition of MSG; strong grill flavor and aroma with beefy background; used to enhance snack foods, soups, gravies, processed meat systems.

Nova-Naturelle™ Grilled Chicken Flavor WONF RC-0029. [Champlain Industries] Natural flavor contg. added MSG; chicken flavor accented with grilled notes; used to enhance snack foods, soups, gravies, processed meat systems.

Nova-Naturelle™ Grilled Chicken Flavor WONF RC-0049. [Champlain Industries] Natural flavor containing no added MSG or autolyzed yeast extract; chicken flavor accented with grilled notes; used to enhance snack foods, soups, gravies, processed meat systems.

Nova-Naturelle™ Grilled Chicken Flavor WONF RC-306. [Champlain Industries] Flavor keys used as building blocks in formulations that require strong savory notes without addition of MSG; BBQ chicken flavor and aroma with fatty notes; used to enhance snack foods, soups, gravies, processed meat systems.

Nova-Naturelle™ Grilled Meat Type Flavor WONF RC-0060. [Champlain Industries] Kosher parve flavors containing no meat or animal prods.; contains autolyzed yeast extracts and natural flavors; mild grilled flavor; used to enhance snack foods, soups, gravies, processed meat systems.

Nova-Naturelle™ Grilled Pork Type Flavor RC-0043. [Champlain Industries] Natural flavor containing no added MSG or autolyzed yeast extract; strong BBQ notes with pork flavor; used to enhance snack foods, soups, gravies, processed meat systems.

Nova-Naturelle™ Grilled Pork Type Flavor WONF RC-0020. [Champlain Industries] Natural flavor contg. added MSG; good grilled pork flavor, spare rib type; used to enhance snack foods, soups, gravies, processed meat systems.

Nova-Naturelle™ Hamburger Flavor WONF RC-0023. [Champlain Industries] Natural flavor contg. added MSG; strong hamburger flavor; used to enhance snack foods, soups, gravies, processed meat systems.

Nova-Naturelle™ Hamburger Flavor WONF RC-0030. [Champlain Industries] Natural flavor containing no added MSG or autolyzed yeast extract; strong hamburger flavor; used to enhance snack foods, soups, gravies, processed meat systems.

Nova-Naturelle™ Hamburger Type Flavor WONF RC-0052. [Champlain Industries] Kosher parve flavors containing no meat or animal prods.; contains autolyzed yeast extracts and natural flavors; strong hamburger flavor; used to enhance snack foods, soups, gravies, processed meat systems.

Nova-Naturelle™ Hamburger Type Flavor WONF RC-0057. [Champlain Industries] Kosher parve flavors containing no meat or animal prods.; contains autolyzed yeast extracts and natural flavors; good flavor and odor; no added MSG; used to enhance snack foods, soups, gravies, processed meat systems.

Nova-Naturelle™ Hamburger Type Flavor WONF RC-0063. [Champlain Industries] Kosher parve flavors containing no meat or animal prods.; contains autolyzed yeast extracts and natural flavors; strong grilled hamburger flavor; used to enhance snack foods, soups, gravies, processed meat systems.

Nova-Naturelle™ Italian Sausage Type Flavor WONF RC-0036. [Champlain Industries] Natural flavor containing no added MSG or autolyzed yeast extract; strong sausage flavor with nice spice background; used to enhance snack foods, soups, gravies, processed meat systems.

Nova-Naturelle™ Italian Sausage Type Flavor WONF RC-0061. [Champlain Industries] Kosher parve flavors containing no meat or animal prods.; contains autolyzed yeast extracts and natural flavors; strong sausage flavor profile; used to enhance snack foods, soups, gravies, processed meat systems.

Nova-Naturelle™ Liquid Bacon Flavor O.S. 62093. [Champlain Industries] Kosher parve flavors containing no meat or animal prods.; contains autolyzed yeast extracts and natural flavors; oil-based flavor; used to enhance snack foods, soups, gravies, processed meat systems.

Nova-Naturelle™ Lobster Flavor WONF RC-90. [Champlain Industries] Natural seafood flavors contg. no added MSG; nice lobster flavor with a strong aroma; used to enhance snack foods, soups, gravies, processed meat systems.

Nova-Naturelle™ Meat Type Flavor WONF RC-0059. [Champlain Industries] Kosher parve flavors containing no meat or animal prods.; contains autolyzed yeast extracts and natural flavors; nice mild flavor; can be used as pork or chicken flavor; used to enhance snack foods, soups, gravies, processed meat systems.

Nova-Naturelle™ Oyster Flavor WONF RC-96. [Champlain Industries] Natural seafood flavors contg. no added MSG; strong flavor and aroma, brothy type; used to enhance snack foods, soups, gravies, processed meat systems.

Nova-Naturelle™ Pork Fat Type Flavor WONF RC-0045. [Champlain Industries] Natural flavor containing no added MSG or autolyzed yeast extract;

mild pork fat taste; used to enhance snack foods, soups, gravies, processed meat systems.

Nova-Naturelle™ Pork Flavor WONF RC-304. [Champlain Industries] Flavor keys used as building blocks in formulations that require strong savory notes without addition of MSG; strong meaty flavor; may be used at low levels in poultry applics.; used to enhance snack foods, soups, gravies, processed meat systems.

Nova-Naturelle™ Pork Type Flavor WONF RC-0019. [Champlain Industries] Natural flavor contg. added MSG; strong pork chop taste; used to enhance snack foods, soups, gravies, processed meat systems.

Nova-Naturelle™ Pork Type Flavor WONF RC-0032. [Champlain Industries] Natural flavor containing no added MSG or autolyzed yeast extract; strong pork taste with spice background; used to enhance snack foods, soups, gravies, processed meat systems.

Nova-Naturelle™ Pork Type Flavor WONF RC-0033. [Champlain Industries] Natural flavor containing no added MSG or autolyzed yeast extract; good pork flavor, pork chop type; used to enhance snack foods, soups, gravies, processed meat systems.

Nova-Naturelle™ Pork Type Flavor WONF RC-0054. [Champlain Industries] Kosher parve flavors containing no meat or animal prods.; contains autolyzed yeast extracts and natural flavors; strong pork flavor with pork aroma; used to enhance snack foods, soups, gravies, processed meat systems.

Nova-Naturelle™ Pork Type Flavor WONF RC-106. [Champlain Industries] Natural flavor containing no added MSG or autolyzed yeast extract, strong pork flavor with a nice aftertaste; used to enhance snack foods, soups, gravies, processed meat systems.

Nova-Naturelle™ Pork Type Flavor WONF RC-204. [Champlain Industries] Low-sodium flavor containing no added MSG; nice mild flavor with a pork chop aroma; used to enhance snack foods, soups, gravies, processed meat systems.

Nova-Naturelle™ Roast Beef Flavor WONF 92592. [Champlain Industries] Natural flavor containing no added MSG or autolyzed yeast extract; strong roast flavor with pan drippings; used to enhance snack foods, soups, gravies, processed meat systems.

Nova-Naturelle™ Roast Beef Flavor WONF RC-0039. [Champlain Industries] Natural flavor containing no added MSG or autolyzed yeast extract; good roast meat flavor and aroma; used to enhance snack foods, soups, gravies, processed meat systems.

Nova-Naturelle™ Roast Beef Flavor WONF RC-0041. [Champlain Industries] Natural flavor containing no added MSG or autolyzed yeast extract; good roast flavor; used to enhance snack foods, soups, gravies, processed meat systems.

Nova-Naturelle™ Roast Beef Flavor WONF RC-201. [Champlain Industries] Low-sodium flavor containing no added MSG; strong roast notes with pan drippings; used to enhance snack foods, soups,

gravies, processed meat systems.

Nova-Naturelle™ Roast Beef Flavor WONF RC-302. [Champlain Industries] Flavor keys used as building blocks in formulations that require strong savory notes without addition of MSG; roast meaty flavor with fatty notes; used to enhance snack foods, soups, gravies, processed meat systems.

Nova-Naturelle™ Shrimp Flavor WONF RC-98. [Champlain Industries] Natural seafood flavors contg. no added MSG; strong shrimp flavor and aroma with pleasant aftertaste; used to enhance snack foods, soups, gravies, processed meat systems.

Nova-Naturelle™ Turkey Type Flavor WONF RC-0048. [Champlain Industries] Natural flavor containing no added MSG or autolyzed yeast extract; strong poultry notes; used to enhance snack foods, soups, gravies, processed meat systems.

Nova-Naturelle™ Turkey Type Flavor WONF RC-110. [Champlain Industries] Natural flavor containing no added MSG or autolyzed yeast extract; poultry flavor and aroma, white meat type; used to enhance snack foods, soups, gravies, processed meat systems.

Nova-Zyme™ MB-1. [Champlain Industries] Natural roasted savory flavor based on enzyme-modified vegetable proteins; mild roast beef flavor profile; *Properties:* spray-dried.

Nova-Zyme™ MB-0002. [Champlain Industries] Natural roasted savory flavor based on enzyme-modified vegetable proteins; roast beef flavor profile; *Properties:* spray-dried.

Nova-Zyme™ MB-0003. [Champlain Industries] Natural roasted savory flavor based on enzyme-modified vegetable proteins; roast meat flavor profile; *Properties:* vacuum-dried.

Nova-Zyme™ MB-0004. [Champlain Industries] Natural roasted savory flavor based on enzyme-modified vegetable proteins; au jus flavor profile; *Properties:* spray-dried.

Nova-Zyme™ MB-5. [Champlain Industries] Natural roasted savory flavor based on enzyme-modified vegetable proteins; rare roast beef flavor profile; *Properties:* vacuum-dried.

Nova-Zyme™ MB-6. [Champlain Industries] Natural roasted savory flavor based on enzyme-modified vegetable proteins; mild roast beef flavor profile; *Properties:* vacuum-dried.

Nova-Zyme™ MB-7. [Champlain Industries] Natural roasted savory flavor based on enzyme-modified vegetable proteins; grilled roast beef flavor profile; *Properties:* vacuum-dried.

Nova-Zyme™ MB-8. [Champlain Industries] Natural roasted savory flavor based on enzyme-modified vegetable proteins; roast beef flavor profile; *Properties:* vacuum-dried.

Nova-Zyme™ MB-9. [Champlain Industries] Natural roasted savory flavor based on enzyme-modified vegetable proteins; roast beef flavor profile; *Properties:* vacuum-dried.

Nova-Zyme™ MB-10. [Champlain Industries] Natural roasted savory flavor based on enzyme-modified vegetable proteins; roast chicken flavor profile; *Properties:* vacuum-dried.

Nova-Zyme™ MB-11. [Champlain Industries] Natural roasted savory flavor based on enzyme-modified vegetable proteins; roast chicken flavor profile; *Properties:* vacuum-dried.

Nu-Col™ 4227. [A.E. Staley Mfg.] Food starch modified, derived from dent corn starch; cold water swelling starch, thickener, suspending agent, stabilizer for dry instant mixes, convenience foods, instant soups, gravies, sauces, dip mixes, dessert toppings, meringues, cocoa/fruit powd. drinks, frozen milk shakes; enhances mouthfeel; *Usage level:* 0.85-2.85% (instant soups, mixes), 0.35-1.25% (marshmallow prods.), 0.85-3.05% (meringues), 0.35-0.55% (frozen desserts); *Regulatory:* FDA 21CFR §172.892; *Properties:* wh. powd., bland flavor; 99% thru 100 mesh, 85% thru 200 mesh; bulk dens. 36.7 lb/ft³ (loose); visc. 9500 cp (5%); pH 6.0 (20%); 5% moisture.

Nuodex S-1421 Food Grade. [Syn. Prods.] Calcium stearate FCC; CAS 1592-23-0; EINECS 216-472-8; tablet mold release, powder flow aid, direct food additive; *Properties:* wh. free-flowing powd.; 100% thru 325 mesh; apparent dens. 0.20 g/cc; m.p. 154 C; 2.5% moisture.

Nuodex S-1520 Food Grade. [Syn. Prods.] Calcium stearate FCC; CAS 1592-23-0; EINECS 216-472-8; tablet mold release, powder flow aid, direct food additive; *Properties:* wh. free-flowing powd.; 95% thru 200 mesh; apparent dens. 0.35 g/cc; m.p. 154 C; 2.5% moisture.

Nuodex Magnesium Stearate Food Grade. [Syn. Prods.] Magnesium stearate FCC; CAS 557-04-0; EINECS 209-150-3; tablet mold release, powder flow aid, direct food additive; *Properties:* wh. free-flowing powd.; 100% thru 325 mesh; apparent dens. 0.25 g/cc; m.p. 155 C; 1.0% moisture.

Nuruflakes. [Lucas Meyer] Soya; enrichment and improvement of protein, reduces caloric units for cereals, dry baked goods, confectionery; *Properties:* flakes.

Nurugran. [Lucas Meyer] Soya grits; protein enrichment for bread, dry baked goods, ready mixes.

Nurulat. [Lucas Meyer] Soya powd.; emulsifier with high protein solubility and dissolution; for foodstuffs; *Properties:* spray-dried powd.

Nurupan™. [Lucas Meyer] Fullfat toasted soy flour; CAS 68513-95-1; antioxidant, emulsifier extending shelf-life and freshness in dry baked goods, fancy pastry, confectionery; *Properties:* powd.

Nustar®. [A.E. Staley Mfg.] Food starch modified, derived from tapioca and potato; cook-up starch, thickener, stabilizer for fruit pie fillings, processed foods, frozen meat pies and gravies, soups, sauces, oriental foods; stable to heat, acid, freeze/thaw stress; maintains clarity and sheen, resists syneresis; *Regulatory:* FDA 21CFR §172.892; *Properties:* wh. powd.; 98% thru 100 mesh, 85% thru 200 mesh; pH 6.0; 13% moisture.

NutraSweet®. [NutraSweet] Aspartame; CAS 22839-47-0; EINECS 245-261-3; low-calorie sweetener for food and hot/cold beverage prods.; limited use in cooking and baking; loss of sweetness may result on prolonged exposure to high temps.; 2 calories/tsp; *Storage:* indefinite shelf life if kept dry, away from heat and humidity.

Nutricol® Konjac. [FMC] Konjac; fat replacement; provides barrier film and reduces oil uptake in fried foods, savory snacks; as visible fat in lean ground and emulsified meats, adds meltability, juiciness, tenderness, enhanced flavor; for baked goods, meat prods., soups; *Regulatory:* FDA GRAS.

Nutrifos® 088. [Monsanto] Sodium tripolyphosphate; CAS 7758-29-4; texturizer, retention agent in meat and dessert prods.; sequestrant inhibiting oxidative rancidity; reduces moisture loss during thawing/cooking; emulsifies fat and protein; in pet foods, prod. of food starch, as egg wash, boiler water additive; *Usage level:* 0.5% max. (processed meat and poultry, seafood); *Regulatory:* FDA 21CFR §182.6810, GRAS, 100.161, 582, 172.892, 173.310; USDA 9CFR §318.7, 318.147; *Properties:* anhyd. wh. free-flowing lt. dense gran.; 1% max. on 16 mesh, 30% max. thru 100 mesh; m.w. 367.9; pH 9.5-10.1 (1%); 90% min. assay; *Toxicology:* may cause irritation to respiratory tract; avoid breathing dust.

Nutrifos® B-75. [Monsanto] Sodium tripolyphosphate and sodium hexametaphosphate; functional food additive; sequestrant inhibiting oxidative rancidity; reduces moisture loss during thawing/cooking; emulsifies fat/protein; improves sol.; *Usage level:* 0.5% max. (seafood, processed meats and poultry); *Regulatory:* FDA 21CFR §182.6810, 182.6760 resp., GRAS, 100.161; USDA approved; *Properties:* anhyd. wh. gran.; 15% max. on 20 mesh, 30% max. thru 100 mesh; *Toxicology:* may cause irritation to respiratory tract; avoid breathing dust; *Storage:* keep container closed.

Nutrifos® B-90. [Monsanto] Sodium tripolyphosphate, sodium hexametaphosphate blend; functional food additive.

Nutrifos® H-30. [Monsanto] Polyphosphate blend; sequestrant inhibiting oxidative rancidity; reduces moisture loss during thawing/cooking; emulsifies fat/protein; improves emulsion stability; *Usage level:* 0.5% max. (processed meat and poultry, seafood); *Regulatory:* FDA 21CFR 100.161; GRAS; FCC, USDA approved; *Properties:* anhyd. wh. powd.; 59% min. assay (as P_2O_5); *Toxicology:* may cause irritation to respiratory tract; avoid breathing dust; *Storage:* keep container closed.

Nutrifos® L-50. [Monsanto] Sodium tripolyphosphate and tetrasodium pyrophosphate; sequestrant inhibiting oxidative rancidity; reduces moisture loss during thawing/cooking; emulsifies fat/protein; improves gelation props. in surimi; *Usage level:* 0.5% max. (processed meat and poultry, seafood); *Regulatory:* FDA 21CFR §182.6810, 182.6789 resp., GRAS, 100.161; USDA approved; *Properties:* anhyd. wh. powd.; 10% max. on 100 mesh, 65% min. thru 270 mesh; pH 9.8-10.5 (1%); 54% min. assay (as P_2O_5); *Toxicology:* may cause irritation to respiratory tract; avoid breathing dust; *Storage:* keep container closed.

Nutrifos® Powd. [Monsanto] Sodium tripolyphosphate, anhydrous; CAS 7758-29-4; sequestrant inhibiting oxidative rancidity; reduces moisture loss during thawing/cooking; emulsifies fat/protein; in pet foods, prod. of food starch, as boiler water additive for food contact, as egg washing compd.; *Usage level:* 0.5% max. (processed meat and poultry, seafood); *Regulatory:* FDA 21CFR §182.6810, GRAS, 100.161, 582, 172.892, 173.310; USDA 9CFR §318.7, 318.147; *Properties:* anhyd. wh. powd.; 10% max. on 100 mesh, 65% min. thru 270 mesh; pH 9.5-10.2 (1%); 90% min. assay; *Toxicology:* may cause irritation to respiratory tract; avoid breathing dust; *Storage:* keep container closed.

Nutrifos® SK. [Monsanto] Trisodium dipotassium tripolyphosphate; CAS 24315-83-1; sequestrant inhibiting oxidative rancidity; reduces moisture loss during thawing/cooking; emulsifies fat/protein; in pet foods; *Usage level:* 0.5% max. (processed meat and poultry, seafood); *Regulatory:* FDA GRAS; USDA approved; *Properties:* wh. free-flowing heavy dense gran.; 12% min on 20 mesh, 80% min. on 100 mesh; high sol.; bulk dens. 68-72 lb/ft³; pH 9.8-10.2 (1%); 92% min. assay.

Nutrilife®. [Grünau GmbH] Enzyme-protein compds. with primarily α-amylase activity; CAS 9000-90-2; enzymes with synergistic effects with emulsifiers for baking processes; improves mixing and fermentation tolerance, improves dough rheology; provides fine crumb structure; increases volume; delays staling; *Regulatory:* FAO/WHO compliance; *Properties:* powd.; bulk dens. 500-600 g/l; 5-8% moisture; *Storage:* 1 yr storage life if kept dry @ R.T.

Nutriloid® Agar. [TIC Gums] Agar; CAS 9002-18-0; EINECS 232-658-1; gellant for icings, sugar glazes, pie fillings, puddings, dessert gels, and some low-calorie formulations; provides soluble dietary fiber; 85% soluble fiber.

Nutriloid® Arabic. [TIC Gums] Gum arabic; CAS 9000-01-5; EINECS 232-519-5; provides soluble dietary fiber to powd. or liq. drinks, juices, milk shakes, soups, and sauces; helps emulsify, coat, and bind without increasing visc.; *Properties:* sol. in cold water; visc. 1 cps (1%), 10 cps (10%); 94.2% soluble fiber.

Nutriloid® Carrageenan. [TIC Gums] Carrageenan; CAS 9000-07-1; EINECS 232-524-2; gellant providing creamy consistency to soy milk prods., low-calorie jams and jellies, frozen dough to control ice crystallization, cocoa suspensions in hot/cold milk; provides soluble dietary fiber; 78% soluble fiber.

Nutriloid® Cellulose Gums. [TIC Gums] Cellulose gum; CAS 9004-32-4; provides wide range of water visc. from 10 to 4000 cps making them ideal for addition of fiber to specific formulations; *Properties:* visc. various grades 10-4000 cps (1%); 88% soluble fiber.

Nutriloid® Guar Special. [TIC Gums] Guar gum; CAS 9000-30-0; EINECS 232-536-8; thickener for soups, sauces, ice cream mixes, bakery batters; provides soluble dietary fiber; replacement for starch; low visc. grade allows higher usage levels, increasing fiber content; *Properties:* sol. in cold water; visc. 400 cps (1%), 2000 cps (2%); 91.8% soluble fiber.

Nutriloid® Guar Standard. [TIC Gums] Guar gum; CAS 9000-30-0; EINECS 232-536-8; thickener for soups, sauces, ice cream mixes, bakery batters; provides soluble dietary fiber; replacement for starch; *Properties:* sol. in cold water; visc. 3500 cps (1%), 20,000 cps (2%); 91.8% soluble fiber.

Nutriloid® Locust. [TIC Gums] Locust bean gum; CAS 9000-40-2; EINECS 232-541-5; thickener in low-calorie jams, jellies (with carrageenan), ice cream mixes, fruit stabilizers, pie fillings, cheese spreads, low pH sauces, retorted gravies; adds fiber, reduces syneresis of gels; synergistic with xanthan and carrageenan; *Properties:* visc. 3000 cps (1%), needs heat to hydrate; 90.3% soluble fiber.

Nutriloid® Tragacanth. [TIC Gums] Tragacanth gum; CAS 9000-65-1; EINECS 232-552-5; gum providing thickening and soluble dietary fiber to foods; produces creamy sol'ns.; 91.2% soluble fiber.

Nutriloid® Xanthan. [TIC Gums] Xanthan gum; CAS 11138-66-2; EINECS 234-394-2; thixotropic gum producing stable prods. and lower viscs.; stabilizes salad dressings, retains moisture in pickle relish, suspends spices in sauces, thickens gravies, provides body to fruit drinks; stable over wide pH and temp. ranges; *Properties:* visc. 1400 cps (1%); 85% soluble fiber.

Nutrimalt®. [Fleischmann-Kurth Malting] Malt; malt with high yield, high foaming chars.; nutrient for yeast; for beer prod., baking, cereal, candy; 4.2-5.2% moisture.

Nutrimalt® Range. [Grünau GmbH] Malt extract; CAS 8002-48-0; EINECS 232-310-9; taste and volume improver in baked good; *Properties:* powd.

Nutrisoft® 55. [Grünau GmbH] Mono- and diglycerides of edible fatty acids (pure vegetable origin); food emulsifier for baking additives, confectionery; crumb softener, antistaling agent for yeast-raised baked goods; improved sensory props.; *Regulatory:* FDA 21CFR 182.4505, GRAS; EEC E 471; *Properties:* wh. powd., neutral odor, neutral fatty taste; 50% on 0.2 mm sieve, 95% on 0.5 mm sieve; water-disp.; bulk dens. 450-550 g/l; m.p. 63-68 C; acid no. < 3; iodine no. 20-28; sapon. no. 155-165; > 90% monoglycerides; *Storage:* store in cool, dry place; 1 yr effective life stored below 20 C.

Nutrisoy 7B. [ADM Arkady] Lipoxidase enzyme active and defatted soy flour; dough conditioner, crumb whitener; the enzyme bleaches yel. pigments in flour to produce a brighter, whiter crumb.

Nutrisoy 7B Flour. [ADM Protein Spec.] Enzyme active flour; for bleaching wheat flour pigments in white bread; 18% total dietary fiber; 280 calories/100 g; *Properties:* flour; 53% protein, 9% max. moisture.

Nutrisoy 220T. [ADM Protein Spec.] Refatted soy flour; CAS 68513-95-1; partial or complete egg and/or shortening replacer; 16% total dietary fiber; 360 calories/100 g; *Properties:* free-flowing powd.; 46% min. protein, 15-17% fat, 8% max. moisture.

O

O.B. Stabilizer. [Brolite Prods.] Stabilizer for meringe toppings; prevents fillings from cracking during baking; also for bread wash, glazes, and icing.

n-Octenyl Succinic Anhydride. [Humphrey] Octenyl succinic anhydride; CAS 26680-54-6; starch modifier used as thickener, emulsifier and opacifier for food mixes; *Properties:* liq.; 100% conc.

Ointment Base No. 3. [Penreco] Wh. petrolatum USP; CAS 8027-32-5; EINECS 232-373-2; food additive; *Regulatory:* FDA 21CFR 172.880, 178.3700, 573.720; *Properties:* visc. 55-65 SUS (210 F); m.p. 118-125 F; congeal pt. 104-115 F.

Ointment Base No. 4. [Penreco] Wh. petrolatum USP; CAS 8027-32-5; EINECS 232-373-2; food additive; *Regulatory:* FDA 21CFR 172.880, 178.3700, 573.720; *Properties:* Lovibond 1.5Y color; visc. 60-70 SUS (210 F); m.p. 118-125 F; congeal pt. 109-119 F.

Ointment Base No. 6. [Penreco] Wh. petrolatum USP; CAS 8027-32-5; EINECS 232-373-2; food additive; *Regulatory:* FDA 21CFR 172.880, 178.3700, 573.720; *Properties:* Lovibond 1.5Y color; visc. 60-70 SUS (210 F); m.p. 122-133 F; congeal pt. 120-130 F.

OL®. [Calgon Carbon] Activated carbon; CAS 64365-11-3; used for food decolorization; *Properties:* 20 x 50 sieve size.

Oleoresin Carrot 10,000 CV. [Arista Industries] Carrot oleoresin, BHT (≤ 0.02%); *Properties:* orange, yel. in dil. form; readily sol. in oil; *Storage:* may be stored at 38 F, < 60% r.h. for 24 mos.

Olestra. [Procter & Gamble] Frying oil substitute; *Regulatory:* awaiting approval.

Olicine. [Gattefosse; Gattefosse SA] Peanut glycerides; nonionic; food emulsifier; *Properties:* liq.; HLB 3.0; 100% conc.

Onion Super. [Beeta; Commodity Services Int'l.] Conc. of cooked onions of *Allium cepa*; CAS 8002-72-0; flavor ingred. in meat prods., sausages, soups, soup mixes, sauces, gravies, seasonings, condiments, salad dressings; *Regulatory:* FDA 21CFR §182.20; GRAS (FEMA); EEC; IFRA; IOFI; *Properties:* brn. liq., strong pungent char. odor and taste of cooked onion; sol. in propylene glycol, 70% ethanol, water; sp.gr. 1.215 (32 C).

Optex. [Pfizer Food Science] A food ingred. that thickens vegetable oils, allowing processors to make table spreads without using hard fats; *Regu-*

latory: awaiting approval.

Optima 7B. [Van Den Bergh Foods] Partially hydrog. veg. oil (cottonseed, soybean); EINECS 269-820-6; center fat for bakeable fillings; butter fat replacement; *Regulatory:* kosher; *Properties:* solid; m.p. 81-85 F.

Optima 23B. [Van Den Bergh Foods] Partially hydrog. veg. oil (cottonseed, soybean); EINECS 269-820-6; center fat for confectionery and bakery applics.; *Regulatory:* kosher; *Properties:* solid; drop pt. 36 C.

Optima 77IC. [Van Den Bergh Foods] Partially hydrog. veg. oil (soybean, cottonseed); EINECS 269-820-6; high performance fat; replacement for 76° coconut oil; ice cream bar coatings; *Regulatory:* kosher; *Properties:* liq.; drop pt. 26.7 C.

Optima 871. [Van Den Bergh Foods] Partially hydrog. veg. oil (cottonseed, soybean); EINECS 269-820-6; soft center fat; *Properties:* liq.; drop pt. 15-20 C.

Optimase® APL-440. [Solvay Enzymes] Bacterial alkaline protease; CAS 9014 01 1; EINECS 232-752-2; food-grade enzyme for hydrolysis of proteins incl. hemoglobin, casein, egg yolk, soya, gelatin, fish proteins; used for baking (cracker/cookie gluten modification), candy, fermentation, fish/soya processing, protein modification; *Regulatory:* FDA 21CFR §184.1027; *Properties:* amber to lt. brn. liq., free of offensive odor; readily misc. with water; *Storage:* activity loss ≤ 10% in 6 mo stored in sealed containers under cool dry conditions; 5 C storage extends life.

1823 Orange Oil. [MLG Enterprises Ltd.] Highly conc. essence; flavor booster; pleasantly fresh and perfumed orange juice flavor for beverages; *Usage level:* 15-25 cm³/1000 l of beverage.

Orchic Substance. [Am. Labs] Vacuum-dried defatted glandular prod.; nutritive food additive; *Properties:* powd.

Orzol®. [Witco/Petroleum Spec.] Wh. min. oil USP; emollient, lubricant for foods; *Regulatory:* FDA 21CFR §172.878, 178.3620a; *Properties:* water-wh.; sp.gr. 0.869-0.885; visc. 61-64 cSt(40 C); pour pt. -20 C; flash pt. 202 C.

Ovarian Substance. [Am. Labs] Vacuum-dried defatted glandular prod.; nutritive food additive; *Properties:* powd. or freeze-dried form.

Ovazyme, XX. [Finnsugar Bioprods.] Glucose oxidase (fungal); CAS 9001-37-0; EINECS 232-601-0; enzyme for food processing; maintains freshness by preventing oxygen or glucose deterioration; desugarization of egg prods.; optimum pH 4.5-7.5; *Properties:* amber clear liq.; Unverified

Ovothin™. [Lucas Meyer] Range of native egg yolk phospholipids; emulsifying agent for flavor emulsions and oils; food supplement for dietetics; strengthening and build-up aid.

Ovothin™ 120. [Lucas Meyer] Natural mixt. of egg phospholipids and egg oil; specially formulated for unique emulsification and dietetic applics.

Ovothin™ 160. [Lucas Meyer] Natural mixt. of egg phospholipids and egg oil; emulsifier for health foods, dietetic prods.

Oxipur. [Vaessen-Schoemaker] Mixt. of saccharides of different degrees of polymerization; water binder for prep. of cooked meats, poultry prods., e.g., hams, cassler, tongues; improves color, color stability, taste, bacteriological stability, sliceability; better moisture retention; *Usage level:* 1-2% on final prod.; *Properties:* low sweetness profile; easily sol. in water; pH 5-6 (1%); DE 20%; 5% max. moisture.

Oxipur-Mix. [Vaessen-Schoemaker] Complete injection brine mix; provides moisture retention, color, and sliceability in cooked hams and shoulders, pork loins.

Oxitabs. [ADM Arkady] Each tablet provides 45 ppm potassium bromate and 15 ppm potassium iodate; provides oxidation designed to increase loaf volume and improve crumb structure; *Properties:* tablets.

Oxynex® K. [Rona; E. Merck] PEG-8 (62%), tocopherol (30%), ascorbyl palmitate (5%), ascorbic acid (1%), and citric acid (1%); antioxidant, stabilizer for fats and oils; esp. for protection of sat. and unsat. components of the oil phase and for inhibition of formation of free radicals; used for food, essential oils; *Usage level:* 0.05-0.2%; *Regulatory:* EEC, GRAS compliance; *Properties:* yel. to lt. brn. transparent liq., char. faint odor.

Oxynex® L. [Rona; E. Merck] Tocopherol (30%), ascorbyl palmitate (5%), ascorbic acid (1%), citric acid (1%), ethanol (53%), veg. oil (20%); antioxidant for stabilization of high-grade fats and fat-containing foodstuffs (e.g., shortenings), anti-adhesive emulsions, dried soups, instant cake mixes, essential oils; *Usage level:* 0.005-0.2%; *Properties:* yel. to reddish brn. transparent liq., char. odor; *Precaution:* highly flamm.

Oxynex® LM. [Rona; E. Merck] Tocopherol (25%), lecithin (25%), ascorbyl palmitate (20%), glyceryl stearate, glyceryl oleate, citric acid (2.5%); antioxidant esp. suitable for high-grade veg. fats and fat-containing foodstuffs (e.g., shortenings), anti-adhesive emulsions, mayonnaises, margarine, cake mixes, essential oils; *Usage level:* 0.005-0.1%; *Regulatory:* German, EEC, US GRAS compliances; *Properties:* lt. brn. waxy solid; odorless and tasteless in dilution; sol. in oils and fats.

P

P147 Caramel Color. [MLG Enterprises Ltd.] Caramel color; CAS 8028-89-5; EINECS 232-435-9; natural colorant; *Regulatory:* FDA 21CFR §73.85; *Properties:* sp.gr. 1.3810 (60 F); dens. 11.50 lb/gal (60 F); visc. 5000 cps (80 F); pH 4.6; *Storage:* 2 yr storage life.

Packers Powd. [Vaessen-Schoemaker] Nitrite curing salts; curative for cured meats.

Palmabeads® Type 500 No. 65332. [Roche] Vitamin A palmitate dispersed in gelatin matrix with sucrose, peanut oil, tricalcium phosphate, and BHA and BHT as antioxidants; for formulations with high moisture content, prods. incl. min. salts, food prods. not dispersed in cold water (e.g., compressed bars where high resist. to moisture and heat is required); *Properties:* yel. free-flowing spherical dry beadlets, bland typ. odor and taste; 95% max. thru 40 mesh, 15% max. thru 120 mesh; m.w. 524.9; *Toxicology:* sustained daily intakes of vitamin A exceeding 50,000 IU (adults), 20,000 IU (infants) may cause toxic effects (headache, vomiting, liver damage); US RDA 8000 IU (pregnant/lactating women); *Storage:* store in cool, dry place in tightly closed containers @ 46-59 F.

Palma-Sperse® Type 250-S No. 65322. [Roche] Vitamin A palmitate USP FCC compded. with gelatin, sorbitol, modified food starch, ascorbic acid, sodium citrate, sugar, caprylic/capric triglyceride, and BHT, BHA, dl-α-tocopherol (antioxidants); designed for dry mix and fluid milk prods.; used in instant breakfasts, fluid milk, nonfat dry milk, dry and fluid milk amplifiers; *Regulatory:* FDA GRAS; *Properties:* yel. to tan sperhical dry free-flowing beadlets, sl. char. odor; 90% min. thru 30 mesh, 25% max. thru 80 mesh; disp. readily in water and milk at wide temp. range (40-200 F); m.w. 524.9; 250,000 IU/g assay; *Toxicology:* sustained daily intakes of vitamin A exceeding 50,000 IU (adults), 20,000 IU (infants) may cause toxic effects (headache, vomiting, liver damage); US RDA 8000 IU (pregnant/lactating women); *Storage:* store in cool, dry place in tightly closed containers @ 46-59 F.

Palma-Sperse® Type 250A/50 D-S No. 65221. [Roche] Vitamin A palmitate USP FCC, vitamin D_3 USP FCC compded. with gelatin, sorbitol, modified food starch, peanut oil, sodium citrate, sugar, caprylic/capric triglyceride, and ascorbic acid, BHT, BHA, dl-α-tocopherol (antioxidants); designed for

dry mix and fluid milk prods.; used in instant breakfasts, fluid milk, nonfat dry milk, dry mix and fluid milk amplifiers; *Regulatory:* FDA GRAS; *Properties:* yel. spherical dry free-flowing beadlets, sl. typ. odor, satisfactory flavor; 90% min. thru 30 mesh, 25% max. thru 80 mesh; readily dispersible; 250,000 IU/g (vitamin A), 50,000 IU/g (vitamin D); *Toxicology:* vitamin A: sustained daily intakes exceeding 50,000 IU (adults), 20,000 IU (infants) may cause toxic effects (headache, vomiting, liver damage); US RDA 8000 IU (pregnant/lactating women); vitamin D: potentially toxic esp. in young children; *Storage:* store in cool, dry place.

Panaid. [ADM Arkady] Dough conditioner for no-time systems; reduces mix time, increases extensibility, improves prod. chars.

Panalite 40. [ADM Arkady] Mono- and diglycerides derived from animal or vegetable sources; food emulsifier; *Properties:* plastic or flakes; 40% min. alpha mono.

Panalite 40 HVK. [Paniplus] Mono- and diglycerides; emulsifier in food industry in w/o emulsions; *Properties:* flakes; HLB 2.8; Unverified

Panalite 40 SVK, 50 SA, 50 SVK. [Paniplus] Mono- and diglycerides; emulsifier in food industry in w/o emulsions; *Properties:* plastic; HLB 2.6, 3.5, and 3.5 resp.; Unverified

Panalite 50. [ADM Arkady] Mono- and diglycerides derived from animal or vegetable sources; food emulsifier; *Properties:* plastic or beads; 52% min. alpha mono.

Panalite 50 HVK. [Paniplus] Mono- and diglycerides; emulsifier in food industry in w/o emulsions; *Properties:* beads; HLB 3.5; Unverified

Panalite 90D. [ADM Arkady] Dist. mono- and diglycerides derived from vegetable sources; self-emulsifying food emulsifier; *Properties:* fine powd.; 90% min. alpha mono.

Panalite 100 K. [Paniplus] Mono- and diglycerides and polysorbate 80; emulsifier in food industry in w/o emulsions; *Properties:* beads; HLB 5.2; Unverified

Panalite EOM-K. [Paniplus] Ethoxylated mono- and diglycerides; emulsifier in food industry in w/o emulsions; *Properties:* plastic; HLB 13.1; Unverified

Panalite MP Hydrate. [Paniplus] Hydrated mono- and diglycerides, polysorbate 60; hydrated emulsifier system; *Properties:* plastic; HLB 6.3; 7% alpha mono; Unverified

Panalite MPH, MPS. [Paniplus] Hard and soft mono- and diglycerides and polysorbate 60; emulsifier in food industry in w/o emulsions; *Properties:* flakes and plastic resp.; HLB 6.0 and 8.1; Unverified

Panatex, Panatex HM. [Paniplus] Dist. monoglycerides; hydrated emulsifier system; *Properties:* HLB 3.5; 20 and 35% min. alpha mono resp.; Unverified

Pancreas Substance. [Am. Labs] Freeze-dried defatted glandular prod.; nutritive food additive; *Properties:* powd.

Pancreatic Lipase 250. [Solvay Enzymes] Lipase; CAS 9001-62-1; EINECS 232-619-9; enzyme for hydrolysis of triglycerides to glycerol and fatty acids; used for development of flavors; hydrolysis of egg yolk lipids; pet food improvement; fat modification; *Properties:* cream-colored amorphous dry powd., free from offensive odor; water-sol.; *Storage:* activity loss ≤ 10% in 1 yr stored in sealed containers under cool, dry conditions; 5 C storage extends life.

Pancreatin 3X USP Powd. [Am. Labs] Lipase, amylase, and protease enzymes; enzyme for food and pharmaceutical applics.; *Properties:* powd.

Pancreatin 4X USP Powd. [Am. Labs] Lipase, amylase, and protease enzymes; enzyme for food and pharmaceutical applics.; *Properties:* powd.

Pancreatin 5X USP Powd. [Am. Labs] Lipase, amylase, and protease enzymes; enzyme for food and pharmaceutical applics.; *Properties:* powd.

Pancreatin 6X USP Powd. [Am. Labs] Lipase, amylase, and protease enzymes; enzyme for food and pharmaceutical applics.; *Properties:* powd.

Pancreatin 8X USP Powd. [Am. Labs] Lipase, amylase, and protease enzymes; enzyme for food and pharmaceutical applics.; *Properties:* powd.

Pancreatin TA. [Mitsubishi Int'l.; Amano Enzyme USA] Pancreatin from procine pancreas; CAS 8049-47-6; digestive enzyme; *Properties:* powd.

Pancreatin USP Powd. [Am. Labs] Lipase, amylase, and protease enzymes; enzyme for food and pharmaceutical applics.; *Properties:* powd.

Pancrelipase USP. [Am. Labs] Lipase, amylase, and protease enzymes; enzyme for food and pharmaceutical applics.; *Properties:* powd.

Panicrust LC K. [ADM Arkady] Blend with L-cysteine; CAS 52-90-4; EINECS 200-158-2; strengthener/dough conditioner blend providing improved texture to pizza crust and yeast-raised doughs; also useful in no-time bread systems.

Paniplex CK. [ADM Arkady] Calcium stearoyl lactylate; CAS 5793-94-2; EINECS 227-335-7; conditioner, emulsifier improving dough tolerance, retarding crumb firming and improving prod. texture; for bakery foods, frozen dough, puddings, toppings, dehydrated potatoes, extruded prods.

Paniplex SK. [ADM Arkady] Sodium stearoyl lactylate; CAS 25383-99-7; EINECS 246-929-7; conditioner, emulsifier improving dough tolerance, retarding crumb firming and improving prod. texture; for bakery foods, frozen dough, puddings, toppings, dehydrated potatoes, extruded prods.

Panipower III. [ADM Arkady] Modified gluten prod.; vital wheat gluten coated with ethoxylated monoglyceride and calcium stearate; protein supplement/fiber prod. which is much more efficient than standard gluten; *Properties:* instantly disp. in water or doughs.

Paniset/Panistay. [ADM Arkady] Single-strength stabilizer for icings and glazes requiring long pot life without thickening; remains easily pumpable for extended periods, but dries rapidly after applic. to bakery foods.

Pano™. [MLG Enterprises Ltd.] Edible vegetable oil-based; anti-sticking lubricants (pan grease) for baked goods mfg.; provides release action without smoking, dripping, or running; *Properties:* off-wh. or pale yel. free-flowing oils (Lubes), semiplastic (Sprays), bland taste.

Panodan 120 Series. [Grindsted Prods.] Diacetyl tartaric esters of mono- and diglycerides, enzymes, and flow agents; food emulsifier, emulsion stabilizer; dough strengthener for baked goods; improves fat particle distribution in coffee whiteners.

Panodan 140 Series. [Grindsted Prods.] Diacetyl tartaric esters of mono- and diglycerides, enzymes, and flow agents; food emulsifier, emulsion stabilizer; dough strengthener for baked goods; improves fat particle distribution in coffee whiteners.

Panodan 150. [Grindsted Prods.] Diacetyl tartaric ester/mono- and diglycerides blend; food emulsifier, emulsion stabilizer; dough strengthener for baked goods; improves fat particle distribution in coffee whiteners; avail. in kosher grade; *Properties:* powd.; acid no. 47-57; sapon. no. 325-355.

Panodan 150 Kosher. [Grindsted Prods.] Blend of diacetyl tartaric acid ester of mono- and diglycerides and mono- and diglycerides from edible refined vegetable fats; dough conditioning agent for yeast-raised doughs; crumb softener for white bread; *Usage level:* 0.3-0.7% on flour in yeast-raised doughs; *Regulatory:* FDA 21CFR § 184.1101, 184.1505, GRAS; FCC, kosher; *Properties:* fine powd.; drop pt. 56 C; acid no. 47-57; iodine no. 2 max.; sapon. no. 325-355; *Storage:* store in dry area, below 20 C, with liner securely closed.

Panodan 205. [Grindsted Prods.] Diacetyl tartaric ester of monoglycerides with 5% flow agent; food emulsifier, emulsion stabilizer; dough strengthener for baked goods; improves fat particle distribution in coffee whiteners; avail. in kosher grade; *Properties:* fine powd.; acid no. 62-76; sapon. no. 380-425.

Panodan 235. [Grindsted Prods.] Monoglyceride diacetyl tartaric acid ester; anionic; emulsifier for bakery industry; *Properties:* wh. fine powd.; m.p. 45 C; sapon. no. 395-420.

Panodan FDP. [Grindsted Prods.] Diacetyl tartaric ester of monoglycerides; food emulsifier, emulsion stabilizer; dough strengthener for baked goods; improves fat particle distribution in coffee whiteners; *Regulatory:* EEC, FDA 21CFR §184.1101, GRAS; ADI 0-50 mg/kg; *Properties:* beads; acid no. 62-76; sapon. no. 380-425.

Panodan FDP Kosher. [Grindsted Prods.] Diacetyl

tartaric acid ester of mono- and diglycerides made from edible refined vegetable fat; dough conditioning agent for yeast-raised doughs giving more tolerance in mfg. and improved volume; emulsifier for liq. and powd. coffee whiteners, confectionery prods., vitamin dispersions, o/w emulsions; *Usage level:* 0.2-0.5% (yeast-raised doughs), 0.1-0.2% (liq. coffee whitener), 0.3-1% (powd. coffee whitener), 0.5-1% (confectionery coatings), 0.2-1% (o/w emulsions); *Regulatory:* FDA 21CFR §184.1101, GRAS; EEC E472(e), kosher; *Properties:* powd.; m.p. 45 C; acid no. 62-76; iodine no. 3 max.; sapon. no. 380-425; *Storage:* store in cool, dry area @ 10 C; storage above 20 C may cause caking.

Panodan SD. [Grindsted Prods.] Diacetyl tartaric ester of monoglycerides; food emulsifier, emulsion stabilizer; dough strengthener for baked goods; improves fat particle distribution in coffee whiteners; avail. in kosher grade; *Regulatory:* EEC, FDA §184.1101 (GRAS); ADI 0-50 mg/kg; *Properties:* amber semiliq.; m.p. 45 C; acid no. 62-76; sapon. no. 380-425.

Pan-O-Lite®. [Monsanto] Sodium aluminum phosphate with calcium phosphate monohydrate, calcium carbonate, precipitated amorphous silica; functional food additive; *Regulatory:* FCC compliance; *Properties:* anhyd. wh. powd.; 3% max. on 100 mesh, 12% max. on 140 mesh.

Pano-Lube™. [MLG Enterprises Ltd.] Vegetable oil, lecithin, TBHQ (as antioxidant); release lubricant for baked goods mfg.; *Properties:* off-wh. or pale yel. free-flowing oil, bland taste.

Panospray™ [MLG Enterprises Ltd.] Vegetable oil, partially hydrogenated vegetable oil, flour, lecithin, TBHQ (as antioxidant); release lubricant for baked goods mfg.; *Properties:* soft paste.

Panospray™ LS. [MLG Enterprises Ltd.] Vegetable oil, partially hydrogenated vegetable oil solids, lecithin, TBHQ (as antioxidant); release lubricant for baked goods mfg.; *Properties:* off-wh. or pale yel. thick liq., easily spayable, bland taste.

Panospray™ SQ. [MLG Enterprises Ltd.] Vegetable oil, partially hydrogenated vegetable oil, flour, lecithin, TBHQ (as antioxidant); release lubricant for baked goods mfg.; higher solids than Panospray for extra firmness; sprayable with high pressure equip.

Panospray™ WL. [MLG Enterprises Ltd.] Vegetable oil, partially hydrogenated vegetable oil, corn starch, lecithin, mono- and diglycerides, TBHQ (as antioxidant); release lubricant for baked goods mfg.; *Properties:* off-wh. or pale yel. thick liq., easily spayable, bland taste.

Pantex. [ADM Arkady] Hydrated dist. monoglycerides; emulsifier system, functional crumb softener and shelf life extender; also avail. in double strength; 20% min. alpha mono.

Papain 16,000. [Solvay Enzymes] Protease; CAS 9014-01-1; EINECS 232-752-2; enzyme for hydrolysis of proteins; for brewing (stabilizes and chillproofs beer); fermentation; fish/soya processing; meat tenderizers; protein modification; animal feed supplement; *Usage level:* 0.01-0.1%; *Proper-*

ties: tan to lt. brn. amorphous dry powd., free of offensive odors and taste; water-sol.; *Storage:* activity loss ≤ 10% in 1 yr stored in sealed containers under cool dry conditions; 5 C storage extends life.

Papain 30,000. [Solvay Enzymes] Protease; CAS 9014-01-1; EINECS 232-752-2; enzyme for hydrolysis of proteins; for brewing (stabilizes and chillproofs beer); fermentation; fish/soya processing; meat tenderizers; protein modification; animal feed supplement; *Usage level:* 0.01-0.1%; *Properties:* tan to lt. brn. amorphous dry powd., free of offensive odors and taste; water-sol.; *Storage:* activity loss ≤ 10% in 1 yr stored in sealed containers under cool dry conditions; 5 C storage extends life.

Papain A300. [Meer] Papain, enzyme derived from papaya; CAS 9001-73-4; EINECS 232-627-2; proteolytic enzyme used as meat tenderizer, beer clarifier, in dough conditioners for baking, digestive aids.

Papain A400. [Meer] Papain, enzyme derived from papaya; CAS 9001-73-4; EINECS 232-627-2; proteolytic enzyme used as meat tenderizer, beer clarifier, in dough conditioners for baking, digestive aids.

Papain AIE. [Mitsubishi Int'l.; Amano Enzyme USA] Papain from papaya; CAS 9001-73-4; EINECS 232-627-2; digestive enzyme; *Properties:* powd.

Papain Conc. [Solvay Enzymes] Protease; CAS 9014-01-1; EINECS 232-752-2; enzyme for hydrolysis of proteins; for brewing (stabilizes and chillproofs beer); fermentation; fish/soya processing; meat tenderizers; protein modification; animal feed supplement; *Usage level:* 0.01-0.1%; *Properties:* tan to lt. brn. amorphous dry powd., free of offensive odors and taste; water-sol.; *Storage:* activity loss ≤ 10% in 1 yr stored in sealed containers under cool dry conditions; 5 C storage extends life.

Papain M70. [Meer] Papain, enzyme derived from papaya; CAS 9001-73-4; EINECS 232-627-2; proteolytic enzyme used as meat tenderizer, beer clarifier, in dough conditioners for baking, digestive aids.

Papain P-100. [Finnsugar Bioprods.] Protease (papain); CAS 9014-01-1; EINECS 232-752-2; enzyme for food processing; meat tenderizer; baking; beer chillproofing; protein hydrolysates; *Properties:* tan powd.; Unverified

Papain S100. [Meer] Papain, enzyme derived from papaya; CAS 9001-73-4; EINECS 232-627-2; proteolytic enzyme used as meat tenderizer, beer clarifier, in dough conditioners for baking, digestive aids.

Paramount B. [Van Den Bergh Foods] Partially hydrog. palm kernel oil; lauric center fat for confectioner's coatings, veg. dairy systems, candy centers, icings; *Regulatory:* kosher; *Properties:* solid; m.p. 93-96 F.

Paramount C. [Van Den Bergh Foods] Partially hydrog. palm kernel oil, lecithin; lauric center fat for confectioner's coatings, veg. dairy systems, candy centers, icings; *Regulatory:* kosher; *Properties:* flake; m.p. 101-104 F.

Paramount H. [Van Den Bergh Foods] Partially hydrog. veg. oil (palm kernel, soybean, cottonseed); EINECS 269-820-6; lauric center fat for confectioner's coatings, veg. dairy systems, candy centers, icings; *Regulatory:* kosher; *Properties:* liq.; m.p. 107-109 F.

Paramount X. [Van Den Bergh Foods] Partially hydrog. veg. oil (palm kernel, soybean, cottonseed); EINECS 269-820-6; lauric center fat for confectioner's coatings, veg. dairy systems, candy centers, icings, doughnut glazes; icing stabilizer; *Regulatory:* kosher; *Properties:* liq.; m.p. 107-109 F.

Paramount XX. [Van Den Bergh Foods] Partially hydrog. veg. oil (palm kernel, soybean, cottonseed); EINECS 269-820-6; lauric center fat for confectioner's coatings, veg. dairy systems, candy centers, icings, doughnut glazes; icing stabilizer; *Regulatory:* kosher; *Properties:* flake; m.p. 117-119 F.

Parathyroid Substance. [Am. Labs] Vacuum-dried defatted glandular prod.; nutritive food additive; *Properties:* powd.

Parotid Substance. [Am. Labs] Vacuum-dried defatted glandular prod.; nutritive food additive; *Properties:* powd.

Partially Hydro Soybean Oil 86-505-0. [ADM Refined Oils] Partially hydrog. soybean oil; EINECS 232-410-2; oil with exc. oxidative stability; for breads, rolls, processed fried foods, packaged frying oil; *Properties:* Lovibond 1.5R max. oil, bland flavor; iodine no. 108 ± 4.

Pastry Wash Powd. [Bunge Foods] Wash for Danish pastry and hot cross buns; nonboiling; produces a stable, high gloss sugar wash; *Usage level:* 2.4%; *Properties:* wh. to off-wh. powd.; pH 9.3; 6.8% moisture.

Patco® 3. [Am. Ingreds.] Blend of Emplex sodium stearoyl lactylate and Verv® calcium stearoyl-2-lactylate; conditioner/softener; starch and protein complexing agent for use in yeast-leavened bakery prods., buns, rolls, bread, sweet goods; *Usage level:* 4-8 oz/cwt flour; *Regulatory:* FDA approved 8 oz/cwt flour max.; *Properties:* lt. tan powd., mild caramel odor; acid no. 55-83; ester no. 137-177; 100% act., 3.9-5.1% sodium/calcium; *Storage:* store under cool, dry conditions.

Patco 305 K. [Am. Ingreds.] Silicone-based defoamer; defoamer for beverages, preserves, institutional food prods., potato, waste water, fermentation systems; *Properties:* emulsion; 10% silicone.

Patco 306 K. [Am. Ingreds.] Silicone-based defoamer; defoamer for institutional food prods., fruits and vegetables, waste water, brine systems, condiment mfg., fermentation systems; *Properties:* emulsion; 20% silicone.

Patco 307 K. [Am. Ingreds.] Silicone-based defoamer; defoamer for institutional food prods., cheese whey, soya isolates, fruits and vegetables, condiment mfg., fermentation systems; *Properties:* emulsion; 30% silicone.

Patco 309 K. [Am. Ingreds.] Nonsilicone-based defoamer; defoamer for potato and starch processing; *Properties:* emulsion; 40% nonsilicone.

Patco 310 K. [Am. Ingreds.] Nonsilicone-based defoamer; defoamer for potato and starch processing; *Properties:* emulsion; 20% nonsilicone.

Patco 315 K. [Am. Ingreds.] Silicone-based defoamer; defoamer for beverages and preserves; *Properties:* emulsion; 10% silicone.

Patco 319 K. [Am. Ingreds.] Organic defoamer; defoamer for cheese whey, soya isolates, fermentation systems; 100% act.

Patco 333 K. [Am. Ingreds.] Nonsilicone defoamer; defoamer for institutional foods, puddings, jellos; 100% act.

Patco 337 K. [Am. Ingreds.] Veg. oil-based defoamer; defoamer for fermentation systems; *Properties:* emulsion; 100% act.

Patco 338 K. [Am. Ingreds.] Veg. oil-based defoamer; defoamer for edible and cooking oils; *Properties:* emulsion; 100% act.

Patco 501 K. [Am. Ingreds.] Silicone-based defoamer; defoamer for bottle washing, meat and poultry processing, fermentation systems; *Properties:* emulsion; 100% act.

Patco 502 K. [Am. Ingreds.] Silicone-based defoamer; defoamer for potato and starch processing; *Properties:* emulsion; 100% act.

Patco 555 K. [Am. Ingreds.] Silicone-based defoamer; defoamer for beverages, preserves, soya isolates, edible oils, cooking oils; *Properties:* emulsion; 100% act.

Patco 801 K. [Am. Ingreds.] Min. oil-based defoamer; defoamer for fruits, vegetables, fermentation alcohol and systems; *Properties:* emulsion; 100% act.

Patcote® 305. [Am. Ingreds.] 10% Filled silicone emulsion; defoamer for general food applics.; *Usage level:* 50-150 ppm; *Regulatory:* FDA §173.340; *Properties:* milky wh. emulsion; sp.gr. 1.002; dens. 8.35 lb/gal; pour pt. 30 F; flash pt. none; *Storage:* protect from freezing; avoid excessive agitation.

Patcote® 306. [Am. Ingreds.] 20% Filled silicone emulsion; defoamer for general food applics.; *Usage level:* 10-50 ppm; *Regulatory:* FDA §173.340; 50 ppm max. in ready-for-consumption food; not for use in milk; *Properties:* milky wh. emulsion; sp.gr. 1.004; dens. 8.36 lb/gal; pour pt. 1-3 C; *Storage:* protect from freezing; avoid excessive agitation.

Patcote® 307. [Am. Ingreds.] 30% Filled silicone emulsion; defoamer for food applics.; *Usage level:* 10-33 ppm; *Regulatory:* FDA §173.340; 33 ppm max. in ready-for-consumption food; not for use in milk; *Properties:* milky wh. emulsion; sp.gr. 1.008; dens. 8.40 lb/gal; pour pt. 1-3 C; *Storage:* protect from freezing; avoid excessive agitation.

Patcote® 308. [Am. Ingreds.] Silicone-containing; kosher grade; defoamer for use in alcohol fermentation; *Usage level:* 8-15 ppm; *Regulatory:* FDA §173.340, 33 ppm max.; *Properties:* milky wh. emulsion; sp.gr. 0.978; dens. 8.15 lb/gal; pour pt. 1-3 C; flash pt. 385 F; 100% act.; *Toxicology:* mild skin irritant on prolonged contact; do not take internally; *Storage:* keep away from open flame.

Patcote® 309. [Am. Ingreds.] Nonsilicone aq. emulsion; defoamer for food applics.; *Usage level:* 50-150 ppm; *Regulatory:* FDA §173.340; *Properties:* milky wh. emulsion; sp.gr. 0.9076; dens. 7.56 lb/gal; pour pt. 30 F; *Storage:* protect from freezing; avoid excessive agitation.

Patcote® 310. [Am. Ingreds.] Nonsilicone aq. emulsion; defoamer for general food applics., esp. starchy applics., e.g., potatoes; *Usage level:* 75-200 ppm; *Regulatory:* FDA §173.340; *Properties:* milky wh. emulsion; sp.gr. 0.9478; dens. 8.12 lb/gal; pour pt. 30 F; *Storage:* protect from freezing; avoid excessive agitation.

Patcote® 311. [Am. Ingreds.] 100% Filled silicone; defoamer for food processing; *Usage level:* 5-10 ppm; *Regulatory:* FDA §173.340; *Properties:* wh. semitransparent emulsion; sp.gr. 0.994; dens. 8.28 lb/gal; pour pt. 25 F; flash pt. > 300 F; *Toxicology:* eye or skin irritant on prolonged contact; do not take internally; *Storage:* may settle on aging; stir before use.

Patcote® 315. [Am. Ingreds.] 10% Silicone emulsion; defoamer for food applics.; *Usage level:* 50-100 ppm; *Regulatory:* FDA §173.340; *Properties:* milky blue wh. emulsion; sp.gr. 0.960; dens. 8.00 lb/gal; pour pt. 30 F; *Storage:* protect from freezing; avoid excessive agitation.

Patcote® 319. [Am. Ingreds.] Nonsilicone; defoamer for prod. of yeast; *Usage level:* 100-200 ppm; *Regulatory:* FDA §173.340; *Properties:* clear, sl. opalescent, mild aliphatic odor; sp.gr. 0.845; dens. 7.04 lb/gal; pour pt. < 0 F; flash pt. (PMCC) 150 F; 100% act.; *Precaution:* keep away from open flame.

Patcote® 323. [Am. Ingreds.] 15% filled silicone emulsion; defoamer for potato and starch processing; *Usage level:* 10-50 ppm; *Regulatory:* FDA §173.340, 65 ppm max.; *Properties:* milky wh. emulsion; infinitely dilutable with water; sp.gr. 1.000; dens. 8.34 lb/gal; pour pt. 30 F; *Storage:* store below 100 F; protect from freezing; avoid excessive agitation; Discontinued

Patcote® 333. [Am. Ingreds.] Nonsilicone; defoamer for wheat gluten prod. and wet milling of wheat and corn starch; *Usage level:* 100-150 ppm; *Regulatory:* FDA §173.340; *Properties:* off-wh. liq.; sp.gr. 0.917; dens. 7.64 lb/gal; pour pt. 15 F; flash pt. (PMCC) > 400 F; 100% active.

Patcote® 337. [Am. Ingreds.] Nonsilicone; defoamer for use in alcohol prod. via grain fermentation; *Usage level:* 100-150 ppm; *Regulatory:* FDA §173.340; *Properties:* amber liq.; sp.gr. 0.917; dens. 7.64 lb/gal; pour pt. 15 F; flash pt. (PMCC) > 400 F; 100% act.; Discontinued

Patcote® 337K. [Am. Ingreds.] Nonsilicone; defoamer for alcohol prod. via grain fermentation; *Usage level:* 100-150 ppm; *Regulatory:* FDA §173.340; kosher; *Properties:* amber liq.; sp.gr. 0.917; dens. 7.64 lb/gal; pour pt. 15 F; flash pt. (PMCC) > 400 F; 100% active.

Patcote® 338K. [Am. Ingreds.] Nonsilicone; defoamer for potato, cereal grain, and other food processing applics.; *Usage level:* 100 ppm; *Regulatory:* FDA §173.340; kosher; *Properties:* amber cloudy liq.; sp.gr. 0.955; dens. 7.96 lb/gal; pour pt. -10 F; flash pt. (PMCC) > 400 F; 100% active.

Patcote® 501K. [Am. Ingreds.] Silicone-containing; defoamer for use in alcohol fermentation, meat rendering, and other high temp. applics.; *Usage level:* 8-15 ppm; *Regulatory:* FDA §173.340, 33 ppm max.; kosher; *Properties:* milky wh. liq.; sp.gr. 0.952; dens. 7.92 lb/gal; pour pt. 1-3 C; flash pt. 385 F; 100% act.; *Toxicology:* mild skin irritant on prolonged contact; do not take internally; *Precaution:* keep away from open flame; *Storage:* may show sl. separation, stir before use.

Patcote® 502K. [Am. Ingreds.] Silicone-containing; defoamer for potato processing and alcohol fermentation; *Usage level:* 8-20 ppm; *Regulatory:* FDA §173.340, 33 ppm max.; kosher; *Properties:* milky wh. liq.; sp.gr. 0.952; dens. 7.92 lb/gal; pour pt. 1-3 C; flash pt. 385 F; 100% act.; *Toxicology:* mild skin irritant on prolonged contact; do not take internally; *Precaution:* keep away from open flame; *Storage:* may show sl. separation, stir before use.

Patcote® 555K. [Am. Ingreds.] 100% filled silicone; defoamer for food processing; *Usage level:* 5-10 ppm; *Regulatory:* FDA §173.340; kosher; *Properties:* wh. semitransparent; sp.gr. 0.994; dens. 8.28 lb/gal; pour pt. < -30 F; flash pt. > 300 F; *Toxicology:* eye or skin irritant on prolonged contact; do not take internally; *Storage:* may settle on aging, stir before use.

Patcote® 801K. [Am. Ingreds.] Defoamer for food processing; *Usage level:* 50-100 ppm; *Regulatory:* FDA §173.340; kosher; *Properties:* wh. opaque liq.; sp.gr. 0.911; dens. 7.59 lb/gal; pour pt. 15 F; flash pt. > 300 F; 100% act.; *Storage:* may show sl. separation on extended storage, stir before use.

Pationic® 900. [Am. Ingreds.] Dist. glycerol monoester derived from natural fats and oils; direct food substance affirmed as GRAS; *Usage level:* 0.2-0.5% (as mold release), 1.0-2.0 phr (PVC lube); *Regulatory:* FDA 21CFR §184.1324, GRAS; *Properties:* ivory wh. fine beads; m.p. 66 C; 96% monoglyceride.

Pationic® 901. [Am. Ingreds.] Dist. glyceryl stearate derived from fully hydrog. veg. oil; direct food substance affirmed as GRAS; *Usage level:* 0.2-0.5% (as mold release), 1.0-2.0 phr (PVC lube); *Regulatory:* FDA 21CFR §184.1324, GRAS; *Properties:* ivory wh. fine beads; m.p. 72 C; 96% monoglyceride.

Pationic® 902. [Am. Ingreds.] Dist. glyceryl stearate derived from fully hydrog. animal fat; direct food substance affirmed as GRAS; *Usage level:* 0.2-0.5% (as mold release), 1.0-2.0 phr (PVC lube); *Regulatory:* FDA 21CFR §184.1324, GRAS; *Properties:* ivory wh. fine beads; m.p. 68 C; 96% monoglyceride.

Pationic® 905. [Am. Ingreds.] Dist. glyceryl stearate; direct food substance affirmed as GRAS; *Usage level:* 0.2-0.5% (mold release in PP), 1-2 phr (internal lube in PVC); *Regulatory:* FDA 21CFR §184.1324, GRAS; *Properties:* ivory wh. fine beads; m.p. 69 C; 96% monoglyceride.

Pationic® 907. [Am. Ingreds.] Unsat. dist. glycerol ester derived from veg. oil; direct food substance affirmed as GRAS; *Usage level:* 1 phr (internal lube in rigid PVC); *Regulatory:* FDA 21CFR §184.1505, GRAS; *Properties:* ivory wh. liq. @ 40-45 C, soft paste at ambient temps.; 96% monoglyceride.

Pationic® 909. [Am. Ingreds.] Dist. glyceryl stearate derived from fully hydrog. veg. oil; direct food substance affirmed as GRAS; *Usage level:* 0.2-0.5% (as mold release), 1.0-2.0 phr (PVC lube); *Regulatory:* FDA 21CFR §184.1324, GRAS; *Properties:* ivory wh. fine beads; 98% thru 100 mesh; m.p. 72 C; 96% monoglyceride.

Pationic® 914. [Am. Ingreds.] Glyceryl mono/tristearate derived from fully hydrog. veg. oil; direct food substance affirmed as GRAS; *Regulatory:* FDA 21CFR §182.70, 184.1324, GRAS; *Properties:* ivory wh. very fine beads; m.p. 65 C; 65% monoglyceride.

Pationic® 919. [Am. Ingreds.] Glyceryl tristearate; CAS 555-43-1; EINECS 209-097-6; direct food additive; *Regulatory:* FDA 21CFR §182.70; *Properties:* ivory wh. fine beads; 98% thru 100 mesh; m.p. 65 C.

Pationic® 920. [Am. Ingreds.] Sodium stearoyl lactylate; CAS 25383-99-7; EINECS 246-929-7; direct food additive, dough strengthener, emulsifier, processing aid, surface-active agent, stabilizer, formulation aid, texturizer; *Regulatory:* FDA 21CFR §172.846; *Properties:* off-wh. powd.; 99% thru 20 mesh; m.p. 47-53 C; sapon. no. 210-235; 3.5-5.0% sodium.

Pationic® 925. [Am. Ingreds.] Calcium/sodium stearoyl lactylate; direct food additive; dough strengthener, emulsifier, processing aid, surface-active agent, stabilizer, formulation aid, texturizer, whipping agent, conditioning agent; *Regulatory:* FDA 21CFR §172.844, 172.846; *Properties:* off-wh. powd.; 99% thru 20 mesh; m.p. 45-55 C; sapon. no. 195-230.

Pationic® 930. [Am. Ingreds.] Calcium stearoyl-2-lactylate; CAS 5793-94-2; EINECS 227-335-7; direct food additive; dough conditioner, whipping agent, conditioning agent; *Regulatory:* FDA 21CFR §172.844; *Properties:* off-wh. powd.; 99% thru 20 mesh; m.p. 45-55 C; sapon. no. 195-230; 4.2-5.2% calcium.

Pationic® 940. [Am. Ingreds.] Calcium stearoyl-2-lactylate; CAS 5793-94-2; EINECS 227-335-7; direct food additive; dough conditioner, whipping agent, conditioning agent; *Regulatory:* FDA 21CFR §172.844; *Properties:* ivory free-flowing powd.; basic pH; 12% calcium.

Pationic® 1019. [Am. Ingreds.] Glyceryl tristearate; direct food additive; *Usage level:* 0.5% (lube in rigid PVC); *Regulatory:* FDA 21CFR §182.70; *Properties:* ivory wh. very fine flakes; m.p. 65 C.

Pationic® 1042. [Am. Ingreds.] Glyceryl stearate; avail. as kosher grade (1042K); direct food substance affirmed as GRAS; *Regulatory:* FDA 21CFR §184.1505, GRAS; *Properties:* ivory wh. beads or very fine flakes; m.p. 60-63 C; 43% monoglyceride.

Pationic® 1042K. [Am. Ingreds.] Glyceryl stearate, kosher; direct food substances affirmed as GRAS; *Regulatory:* FDA 21CFR §184.1505, GRAS; *Properties:* ivory wh. beads or very fine flakes; m.p. 62 C; 43% α-monoglyceride.

Pationic® 1052. [Am. Ingreds.] Glyceryl stearate; direct food substance affirmed as GRAS; *Regulatory:* FDA 21CFR §184.1505, GRAS; *Properties:* ivory wh. beads or very fine flakes; m.p. 62 C; 53% α-monoglyceride.

Pationic® 1052K. [Am. Ingreds.] Glyceryl stearate, kosher; direct food substances affirmed as GRAS; *Regulatory:* FDA 21CFR §184.1505, GRAS; *Properties:* ivory wh. very fine flakes; m.p. 62 C; 53% α-monoglyceride.

Pationic® 1061. [Am. Ingreds.] Glyceryl oleate; direct food substances affirmed as GRAS; *Regulatory:* FDA 21CFR §184.1555, GRAS; *Properties:* amber visc. liq. @ 19 C; m.p. 39-41 C; 44% α-monoglyceride.

Pationic® 1064. [Am. Ingreds.] Glyceryl oleate; CAS 111-03-5; direct food substance affirmed as GRAS; *Regulatory:* FDA 21CFR §184.1505, GRAS; *Properties:* ivory wh. semisolid, liq. above its m.p. 39-41 C; 44% α-monoglyceride.

Pationic® 1074. [Am. Ingreds.] Glyceryl oleate; CAS 111-03-5; direct food substance affirmed as GRAS; *Regulatory:* FDA 21CFR §184.1505, GRAS; *Properties:* ivory wh. semisolid, liq. above its m.p. 48-50 C; 43% α-monoglyceride.

Pationic® 1230. [Am. Ingreds.] Calcium lactate; CAS 814-80-2; EINECS 212-406-7; firming agent, flavor enhancer, flavoring agent/adjuvant, leavening agent, nutrient supplement, stabilizer, thickener; *Regulatory:* FDA 21CFR §184.1207, GRAS; *Properties:* wh. free-flowing powd.; pH 7.1 (1%); 13% calcium.

Pationic® 1240. [Am. Ingreds.] Calcium lactate; CAS 814-80-2; EINECS 212-406-7; firming agent, flavor enhancer, flavoring agent/adjuvant, leavening agent, nutrient supplement, stabilizer, thickener; *Regulatory:* FDA 21CFR §184.1207, GRAS; *Properties:* wh. free-flowing powd.; pH basic; 19.5% calcium.

Pationic® 1250. [Am. Ingreds.] Calcium lactate; CAS 814-80-2; EINECS 212-406-7; direct food additive; firming agent, flavor enhancer, flavoring agent/adjuvant, leavening agent, nutrient supplement, stabilizer, thickener; *Regulatory:* FDA 21CFR §172.844, 184.1207; *Properties:* wh. free-flowing powd.; pH basic; 15.5% calcium.

Patlac® NAL. [Am. Ingreds.; R.I.T.A.] Sodium lactate; CAS 72-17-3; EINECS 200-772-0; pH buffer, humectant, stabilizer, component of stratum corneum; for foods; *Properties:* clear liq.; sp.gr. 1.31-1.34; pH 8-9; 60% conc.; *Toxicology:* nontoxic if ingested; nonirritating to skin and eyes.

PAV. [Custom Ingreds.] Calcium propionate, modified food starch, citric acid, salt, mono- and diglycerides, and soybean oil; preservative/conditioner/antistaling agent for flour tortillas; helps produce a soft flexible tortilla with longer shelf life, better volume; *Usage level:* 1.5 lb/100 lb of flour; *Properties:* wh. free-flowing powd.; *Storage:* store in a dry area below 90°.

Paygel® 290.295 Pregelatinized Food Powd. Wheat Starch. [ADM Ogilvie] Wheat starch; CAS 9005-25-8; thickener, sausage binder for food prods. incl. soups, gravies, doughnuts, breading, compds., cereals, mixes; *Properties:* wh. powd.; pH 6.0; 7% moisture, 0.6% protein.

PB-3 Icing Powd. [Bunge Foods] Boiling type icing stabilizer for high stability icings or glazes for sweet goods, donuts; avail. wh. or clear; clear type recommended for chocolate; *Usage level:* 2 lb/100 lb powd. sugar; *Properties:* wh. free-flowing powd., bland flavor; pH 8.3.

PB-34 Icing Powd. [Bunge Foods] Boiling type icing stabilizer for extra white, high stability icings and glazes for sweet goods; *Usage level:* 3-4 lb/100 lb powd. sugar; *Properties:* wh. free-flowing powd., bland aroma/flavor; pH 9.1; 3.7% max. moisture.

PC-35. [Paniplus] Mono- and diglyceride hydrate; emulsifier used in food industry; *Properties:* plastic; HLB 3.2; Unverified

PD-23. [Witco/Petroleum Spec.] Min. oil; refined aliphatic solv. used as defoamer in froth flotation of vegetables; *Properties:* colorless clear liq., nearly odorless; sp.gr. 0.800 (60/60 C); visc. 35 SUS (100 F); b.p. 452 F; pour pt. -5 F; flash pt. (COC) 225 F; ref. index 1.442.

PD-25. [Witco/Petroleum Spec.] Petroluem distillate solv.; refined aliphatic solv. used as defoamer in froth flotation of vegetables; *Properties:* colorless clear liq., nearly odorless; sp.gr. 0.810 (60/60 C); visc. 38 SUS (100 F); b.p. 486 F; pour pt. 30 F; flash pt. (COC) 255 F; ref. index 1.446.

Peanut Oil. [Bunge Foods] Peanut oil, dimethylpolysiloxane (antifoam); specialty oil for seafood frying, oriental frying, snack spray oil; *Properties:* clear oil, 2.0 red color, sl. peanut flavor; smoke pt. 450 F.

Peanut Oil 85-060-0. [ADM Refined Oils] Peanut oil, refined, bleached; CAS 8002-03-7; EINECS 232-296-4; veg. oil for nut roasting, frying; *Properties:* clear oil, bland sl. nutty flavor; iodine no. 90-100.

Pea Protein Conc. [Garuda Int'l.] Pea protein conc.; functional/nutritious protein conc. with exc. binding props.; for meat extension, casein extenders, food supplements, baked goods, baby foods, protein hydrolysates, breakfast cereals, pet foods, dry mixes, snacks.

Pearex® L. [Solvay Enzymes] Fungal pectinase; CAS 9032-75-1; EINECS 232-885-6; enzyme used to prevent haze formation in pear and apple processing; clarification and stabilization of fruit juices; jams/jellies; vegetable processing; extraction of flavors and fragrances; increases solubility, reduces visc. in pectin; *Properties:* clear to amber brn. liq.; misc. with water; sp.gr. 1.05-1.15; *Storage:* activity loss ≤ 10% in 3 mo stored in sealed containers @ 25 C; 5 C storage extends life.

Pectinase AT. [Solvay Enzymes] Fungal pectinase; CAS 9032-75-1; EINECS 232-885-6; enzyme for

hydrolyzing and depolymerizing pectin in low pH fruits such as cranberries; improves juice clarity and color; *Usage level:* 0.07%; *Properties:* dk. amber to brn. liq., free of offensive odor and taste; misc. with water; dens. 1.10-1.20 g/ml; *Storage:* activity loss ≤ 10% in 6 mo in sealed containers under mild ambient conditions; 5 C storage extends life.

Pectinol® 59L. [Genencor] Pectinase; CAS 9032-75-1; EINECS 232-885-6; all purpose enzyme for juice, wine industry; *Properties:* liq.; Unverified

Pectinol® DL. [Genencor] Anthocyanase; enzyme for breakdown of anthocyanin pigments in wines, juices; *Properties:* liq.; Unverified

Pectinol® R10. [Genencor] Pectinase; CAS 9032-75-1; EINECS 232-885-6; enzyme for processing fruit juice, wine; *Properties:* powd.; Unverified

Pegol® 10R8. [Rhone-Poulenc Surf. & Spec.] EO/PO block copolymer; nonionic; surfactant for fermentation; *Properties:* HLB 18-23; pour pt. 46 C; cloud pt. 98 C (1% aq.); Discontinued

Pegol® 17R1. [Rhone-Poulenc Surf. & Spec.] EO/PO block copolymer; nonionic; surfactant for fermentation; *Properties:* liq.; HLB 6.0; pour pt. -27 C; cloud pt. 32 C (1% aq.); 100% conc.; Discontinued

Pektolase. [Grindsted Prods. Denmark] Pectinase; CAS 9032-75-1; EINECS 232-885-6; enzyme for fruit processing industry (apple juice, berry juice, wine); increases juice yield from mash-treated fruit, reduces pressing time; provides more rapid extraction of color and solids, better clarification, better storage stability

Penguin® M/V Specialty Icing Shortening. [Bunge Foods] Meat fats, vegetable oil, polyglycerol esters, antioxidants; shortening for highly aerated, well-structured icings, and creme fillers; *Properties:* 1.4 red color, bland flavor; 1% alpha mono.

Penguin® Vegetable Specialty Icing. [Bunge Foods] Partially hydrog. soybean and cottonseed oils, polyglycerol esters; shortening for highly aerated, well-structured icings, and creme fillers; *Properties:* 1.4 red color, bland flavor; 1% alpha mono.

Penreco 2251 Oil. [Penreco] Petroleum distillates; CAS 64742-14-9; high purity hydrocarbon processing solv., foam control agent for fruit/vegetable processing; *Regulatory:* FDA 21CFR §172.884, 178.3650, 573.740; *Properties:* sp.gr. 0.779-0.797 (60 F); dens. 6.56 lb/gal (60 F); visc. 30.5 SUS (100 F); b.p. 375 F min.; flash pt. (COC) 165 F; pour pt. -40 F; *Precaution:* combustible liq. (DOT).

Penreco 2257 Oil. [Penreco] High purity hydrocarbon solv.; CAS 64742-46-7; high purity solv., foam control agent for fruit and veg. processing; *Regulatory:* FDA 21CFR §172.884, 178.3650, 573.740; *Properties:* sp.gr. 0.793-0.806 (60 F); dens. 6.64 lb/gal (60 F); visc. 33.2 SUS (100 F); b.p. 430 F min.; flash pt. (COC) 220 F; pour pt. -10 F.

Penreco 2263 Oil. [Penreco] Petroleum distillates; CAS 64742-47-8; high purity hydrocarbon processing solv., foam control agent for fruit/vegetable processing; *Regulatory:* FDA 21CFR §172.884, 178.3650, 573.740; *Properties:* sp.gr. 0.779-0.797

(60 F); dens. 6.56 lb/gal (60 F); visc. 30.5 SUS (100 F); b.p. 375 F min.; flash pt. (COC) 165 F; pour pt. -40 F; *Precaution:* combustible liq. (DOT).

Penreco Amber. [Penreco] Petrolatum USP; CAS 8027-32-5; EINECS 232-373-2; fruit/vegetable coatings; animal feed supplements; *Regulatory:* FDA 21CFR 172.880, 178.3700, 573.720; *Properties:* visc. 68-82 SUS (210 F); m.p. 122-135 F; congeal pt. 123 F; solid. pt. 122 F.

Penreco Blond. [Penreco] Petrolatum USP; CAS 8027-32-5; EINECS 232-373-2; food additive; *Regulatory:* FDA 21CFR 172.880, 178.3700, 573.720; *Properties:* visc. 68-82 SUS (210 F); m.p. 122-135 F; congeal pt. 123 F; solid. pt. 122 F.

Penreco Cream. [Penreco] Wh. petrolatum USP; CAS 8027-32-5; EINECS 232-373-2; food additive; *Regulatory:* FDA 21CFR 172.880, 178.3700, 573.720; *Properties:* visc. 64-75 SUS (210 F); m.p. 122-135 F; congeal pt. 125 F; solid. pt. 122 F.

Penreco Lily. [Penreco] Wh. petrolatum USP; CAS 8027-32-5; EINECS 232-373-2; food additive; *Regulatory:* FDA 21CFR 172.880, 178.3700, 573.720; *Properties:* visc. 64-75 SUS (210 F); m.p. 122-135 F; congeal pt. 124 F; solid. pt. 123 F.

Penreco Regent. [Penreco] Wh. petrolatum USP; CAS 8027-32-5; EINECS 232-373-2; food additive; *Regulatory:* FDA 21CFR 172.880, 178.3700, 573.720; *Properties:* visc. 57-70 SUS (210 F); m.p. 118-130 F; congeal pt. 120 F; solid. pt. 119 F.

Penreco Royal. [Penreco] Petrolatum USP; CAS 8027-32-5; EINECS 232-373-2; food additive; *Regulatory:* FDA 21CFR 172.880, 178.3700, 573.720; *Properties:* visc. 57-70 SUS (210 F); m.p. 118-130 F; congeal pt. 118 F; solid. pt. 115 F.

Penreco Snow. [Penreco] Wh. petrolatum USP; CAS 8027-32-5; EINECS 232-373-2; emollient, base, solv., carrier for foods (animal feed supplement, fruit/veg. coatings); *Regulatory:* FDA 21CFR 172.880, 178.3700, 573.720; *Properties:* visc. 64-75 SUS (210 F); m.p. 122-135 F; congeal pt. 123 F; solid. pt. 121 F.

Penreco Super. [Penreco] Wh. petrolatum USP; CAS 8027-32-5; EINECS 232-373-2; food adiditve; *Regulatory:* FDA 21CFR 172.880, 178.3700, 573.720; *Properties:* visc. 60-75 SUS (210 F); m.p. 122-135 F; congeal pt. 125 F; solid. pt. 124 F.

Penreco Ultima. [Penreco] Wh. petrolatum USP; CAS 8027-32-5; EINECS 232-373-2; fruit/vegetable coatings; *Regulatory:* FDA 21CFR 172.880, 178.3700, 573.720; *Properties:* visc. 60-70 SUS (210 F); m.p. 130-140 F; congeal pt. 130 F; solid. pt. 128 F.

Pepsin 1:3000 NF XII Powd. [Am. Labs] Pepsin; CAS 9001-75-6; EINECS 232-629-3; proteolytic enzyme for food and pharmaceutical applics.; digests coagulated egg albumen; *Properties:* weak yel. to lt. brn. powd., nonoffensive char. odor, salty taste; freely sol. in water; pract. insol. in alcohol, chloroform, ether; pH 3-4 (2%); *Storage:* preserve in tight containers with moisture-proof liners, in cool, dry place.

Pepsin 1:10,000 Powd. or Gran. [Am. Labs] Pepsin; CAS 9001-75-6; EINECS 232-629-3; proteolytic enzyme for food and pharmaceutical applics.; digests coagulated egg albumen; *Properties:* weak yel. to lt. brn. gran. or powd., nonoffensive char. odor, salty taste; freely sol. in water; pract. insol. in alcohol, chloroform, ether; pH 3-4 (2%); *Storage:* preserve in tight containers with moisture-proof liners, in cool, dry place.

Pepsin 1:15,000 Powd. [Am. Labs] Pepsin; CAS 9001-75-6; EINECS 232-629-3; proteolytic enzyme for food and pharmaceutical applics.; digests coagulated egg albumen; *Properties:* weak yel. to lt. brn. powd., nonoffensive char. odor, salty taste; freely sol. in water; pract. insol. in alcohol, chloroform, ether; pH 3-4 (2%); *Storage:* preserve in tight containers with moisture-proof liners, in cool, dry place.

Perfection®. [Rhone-Poulenc Food Ingreds.] Sodium acid pyrophosphate, leavening, FCC; CAS 7758-16-9; EINECS 231-835-0; food-grade leavening acid for baking, cereals, doughnuts, cakes, prepared mixes, devil's food cake formulations; fast acting; often used with slower-acting sodium acid pyrophosphates to increase reaction rate; *Properties:* wh. powd.; 100% thru 60 mesh, 99% thru 200 mesh; sol. 13 g/100 g saturated sol'n.; m.w. 221.94; bulk dens. 68 lb/ft³; pH 4.3 (1%); 95% min. act.

Perfect Slice. [TIC Gums] water binder, processing aid which prevents syneresis, aids freeze/thaw props., adds sheen and gloss, prevents cracking; for pie fillings; *Usage level:* 2.25-3% (pie fillings).

Perflex. [Van Den Bergh Foods] Partially hydrog. veg. oil (soybean, cottonseed), propylene glycol mono- and diesters of fats and fatty acids, mono- and diglcyerides, lecithin, BHT; multipurpose shortening for cakes, nonstandard yeast-leavened prods.; *Regulatory:* kosher; *Properties:* Lovibond 4R max. plastic; m.p. 98-106 F; 4.6-5.0% mono.

Perlatum® 400. [IGI Petroleum Spec.] Wh. petrolatum USP; CAS 8027-32-5; EINECS 232-373-2; as dust control agent and lubricant in baking industry, animal feed processors, and food processors; *Properties:* wh.; visc. 50-60 SUS (210 F); m.p. 52-60 C.

Perlatum® 410. [IGI Petroleum Spec.] Wh. petrolatum USP; CAS 8027-32-5; EINECS 232-373-2; as dust control agent and lubricant in baking industry, animal feed processors, and food processors; *Properties:* wh.; visc. 50-60 SUS (210 F); m.p. 52-60 C.

Perlatum® 410 CG. [IGI Petroleum Spec.] Wh. petrolatum USP; CAS 8027-32-5; EINECS 232-373-2; as dust control agent and lubricant in baking industry, animal feed processors, and food processors; *Properties:* wh.; visc. 50-60 SUS (210 F); m.p. 52-60 C.

Perlatum® 415. [IGI Petroleum Spec.] Petrolatum USP; CAS 8027-32-5; EINECS 232-373-2; as dust control agent and lubricant in baking industry, animal feed processors, and food processors; *Properties:* amber or yel.; visc. 50-60 SUS (210 F); m.p. 52-60 C.

Perlatum® 415 CG. [IGI Petroleum Spec.] Petrolatum USP; CAS 8027-32-5; EINECS 232-373-2; as dust control agent and lubricant in baking industry, animal feed processors, and food processors; *Properties:* amber or yel.; visc. 50-60 SUS (210 F); m.p. 52-60 C.

Perlatum® 420. [IGI Petroleum Spec.] Wh. petrolatum USP; CAS 8027-32-5; EINECS 232-373-2; as dust control agent and lubricant in baking industry, animal feed processors, and food processors; *Properties:* wh.; visc. 50-60 SUS (210 F); m.p. 52-60 C.

Perlatum® 425. [IGI Petroleum Spec.] Petrolatum USP; CAS 8027-32-5; EINECS 232-373-2; as dust control agent and lubricant in baking industry, animal feed processors, and food processors; *Properties:* amber or yel.; visc. 50-60 SUS (210 F); m.p. 52-60 C.

Perlatum® 510. [IGI Petroleum Spec.] Wh. petrolatum USP; CAS 8027-32-5; EINECS 232-373-2; as dust control agent and lubricant in baking industry, animal feed processors, and food processors; *Properties:* wh.; visc. 60-80 SUS (210 F); m.p. 60-77 C.

Perma-Flo®. [A.E. Staley Mfg.] Food starch modified, derived from waxy maize; cook-up starch, thickener, stabilizer maintaining freeze-thaw/shelf-life stability, bland flavor, smooth uniform visc. in acidic and neutral foods, sauces, gravies, puddings, pie fillings, canned and frozen foods, baby foods, imitation dairy prods.; *Regulatory:* FDA 21CFR §172.892; *Properties:* wh. powd., bland flavor; 95% thru 100 mesh; pH 5.5 (uncooked); 12% moisture.

Perma Set/Perma Clear. [ADM Arkady] Multiple component stabilizers for best icing or glaze stability; ideal for frozen icings and glazes or R.T. prods. packed in air-tight pkgs.

Pescamine. [Vaessen-Schoemaker] Fish phosphates; reduces dripping losses in frozen fish prods.

Petrolite® C-1035. [Petrolite] Hard microcryst. wax consisting of n-paraffinic, branched paraffinic, and naphthenic hydrocarbons; CAS 63231-60-7; EINECS 264-038-1; chewing gum base, microcapsule for flavoring; *Regulatory:* incl. FDA §172.230, 172.615, 175.105, 175.300, 176.170, 176.180, 176.200, 177.1200, 178.3710, 179.45; *Properties:* color 0.5 (D1500) wax; very low sol. in org. solvs.; sp.gr. 0.93; visc. 15 cps (99 C); m.p. 94 C.

Petrolite® C-4040. [Petrolite] Syn. petroleum wax; hard, high melting food-grade wax; *Regulatory:* FDA 21CFR §172.888, 175.105, 175.125, 175.300, 175.320, 176.170, 176.180, 176.200, 176.210, 177.1200, 177.1210, 177.2600, 177.2800, 178.3720, 178.3850, 179.45; *Properties:* visc. 54 SUS (149 C); m.p. 104 C.

PF-80. [Hormel] Carrageenan; CAS 9000-07-1; EINECS 232-524-2; gellant designed for pet food applics.; *Usage level:* 0.5-1%; *Properties:* wh. to cream free-flowing powd.; disp. in water; visc. 20-50 cps; pH 9.5-10.5 (1.5%); *Storage:* stable for up to 1 yr when stored dry at ambient temps.; keep containers tightly closed.

Pfico₂Hop®. [Pfizer Food Science] Pure liq. CO_2 extract of hops contg. alpha acids, beta acids, and hop oils; CAS 8016-25-9; additive to brewkettle providing cleaner consistent flavor and aroma chars. and improved physical stability in beer; *Storage:* store at R.T. or under refrigeration.

PGMS 70. [Croda Food Prods. Ltd.] Propylene glycol stearate, veg. grade; CAS 1323-39-3; EINECS 215-354-3; nonionic; food emulsifier; improves volume, shelf life for cakes, creaming and emulsifying props. in shortenings, mouthfeel and whipping in whipped desserts, emulsion stability in whipped toppings; *Usage level:* 2-7% of shortening (cake), 2-7% (shortening), 2-5% (whipped desserts); *Regulatory:* UK clearance; *Properties:* pastille; m.p. 37-42 C; HLB 1.8; acid no. 2 max.; iodine no. 2 max.; 68% min. monoester; *Toxicology:* ADI 0-25 mg/kg bodyweight; *Storage:* store under dry, cool conditions.

Phenylacetaldehyde 50. [BASF] 2-Phenylacetaldehyde; CAS 122-78-1; EINECS 204-574-5; pungent, green, hyacinth-like fragrance and flavorin.

Phenylethyl Alcohol Extra. [BASF] 2-Phenylethanol; mildly foral, rose-like fragrance and flavoring with nuances of hyacinth and honey.

Phico₂hop®. [Pfizer Food Science] For brewery use.

Phosal® 50 SA. [Am. Lecithin; Rhone-Poulenc Rorer] Safflower oil and lecithin; solubilizer for lipophilic substances; phosphatidylcholine source for nutritional supplements, esp. as capsule filling mass; *Usage level:* 5-15% topical applic.; *Proportion:* honey yel. fluid; visc. 5000 mPa•s max.; 53 ± 3% phosphatidylcholine; *Storage:* store @ R.T.; warming reverses sedimentation; use soon after opening.

Phosal® 53 MCT. [Am. Lecithin; Rhone-Poulenc Rorer] Lecithin, caprylic/capric triglyceride, and alcohol (5% max.); solubilizer for lipophilic substances; phosphatidylcholine source for dietetics, esp. as filling mass for soft gelatin capsules; *Usage level:* 5-15% topical applic.; *Properties:* honey yel. fluid; visc. 5000 mPa•s max.; 56 ± 3% phosphatidylcholine; *Storage:* store at R.T. in closed containers; warming reverses sedimentation; use immediately after opening.

Phosal® 75 SA. [Am. Lecithin; Rhone-Poulenc Rorer] Lecithin, ethanol (≈ 9%), and safflower oil; solubilizer for lipophilic substances; phosphatidylcholine source for nutritional supplements, esp. as capsule filling mass; *Usage level:* 3-10% topical applic.; *Properties:* honey yel. fluid; visc. 5500 mPa•s max.; pH 6 ± 2; 75 ± 3% phosphatidylcholine; *Storage:* store @ R.T. in closed containers; warming reverses sedimentation; use soon after opening.

Phospholipon® 90/90 G. [Am. Lecithin; Rhone-Poulenc Rorer] Min. 90% soya 3-sn-phosphatidylcholine; CAS 97281-47-5; EINECS 232-307-2; phosphatidylcholine source for dietetics; *Usage level:* 1-3% topical use; *Properties:* yel. solid or gran.; acid no. 0.5 max.; 95% conc.; *Storage:* store @ R.T., closed under inert gas.

Phytonadione USP No. 61749. [Roche] Vitamin K₁ USP; component of enzyme systems associated with blood-clotting mechanism; used in foods and pharmaceuticals; *Properties:* yel. to amber clear very visc. liq., odorless to sl. odor; sol. in dehydrated alcohol, benzene, chloroform, ether, veg. oils; sl. sol. in alcohol; insol. in water; m.w. 450.68; ref. index 1.5230-1.5260; 1% min. assay (phytonadione); *Precaution:* destroyed by sol'ns. of alkali hydroxides and by reducing agents; dec. on exposure to sunlight; *Storage:* store in cool, dry place in tightly closed containers protected from light.

Pl-64 Icing Powd. [Bunge Foods] Boiling type icing stabilizer for wh. icings and glazes for sweet goods; whiter than BF-46; *Usage level:* 3-4 lb/100 lb powd. sugar; *Properties:* wh. free-flowing powd., bland aroma/flavor; pH 9.1; 3.7% max. moisture.

Piccolyte® C115. [Hercules] Polydipentene; CAS 9003-73-0; cleared under FDA regulations for use as masticatory agents in chewing gum compositions; *Properties:* Gardner 4 (50% in toluene) solid, flakes; sol. in aliphatic, aromatic, chlorinated hydrocarbons, min. oil, VM&P naphtha, turpentine, ether, amyl and butyl acetates, long-chain aliphatic alcohols; dens. 0.99 kg/l; melt visc. 204 C (1 poise); soften. pt. (R&B)115 C; flash pt. (COC) 235 C.

Pie Dough Culture 6450. [Brolite Prods.] Provides greater tolerances to mixing variation in pie doughs, improves crust color and flavor; produces finished crusts with dry, flaky char.

Pie Dough-N-Answer. [Brolite Prods.] Pie dough relaxer that produces tender, flaky crust while preventing doughs from shrinking.

Pineal Substance. [Am. Labs] Freeze-dried defatted glandular prod.; nutritive food additive; *Properties:* powd.

Pisane®. [Cosucra BV; Feinkost Ingred.] Pea protein isolate; functional and nutritional additive, water and fat binder, emulsifier, foaming agent for meat and fish prods., biscuits, pastry-making, desserts, prepared dishes, soups, sauces, dietary, health, and baby foods; *Properties:* cream-colored powd., neutral taste; < 150 μ; easily dispersible; pH 7.5 (10%); 94% min. dry matter; *Storage:* store under dry conditions for max. 1 yr.

Pita Dough Culture 6448. [Brolite Prods.] Natural flavored dough conditioner for pita breads; maintains natural flavor of the pita for days.

Pita Soft. [Brolite Prods.] Dough improver providing greater tolerance and keeping qualities in lean formulated baked goods such as pita breads, tortillas, soft pretzels, and bagels; keeps prods. soft for days.

Placental Substance. [Am. Labs] Vacuum-dried defatted glandular prod.; nutritive food additive; *Properties:* powd.

Plantex® Family. [Gist-brocades] Range of prods. based on natural raw materials manufactured under mild conditions where all hydrolysis is accomplished by enzymes and all processing is at low temps.; flavor profiles incl. basic flavors, general roasted/brown beef, chicken, pork, tomato, onion, cheese, mushroom, bacon, and ham; *Properties:* powd. and paste forms.

Pluracol® E300. [BASF] PEG-6; CAS 25322-68-3; EINECS 220-045-1; dispersant in food tablets and preparations; *Properties:* colorless clear liq.; m.w. 300; sol. in water and org. solvs. except aliphatic hydrocarbons; dens. 9.4 lb/gal; sp.gr. 1.12; visc. 5.9 cs (99 C); flash pt. (COC) 210 C; pour pt. -13 C; surf. tens. 62.9 dynes/cm (1%); pH 5.7 (5% aq.).

Pluracol® E400. [BASF] PEG-8; CAS 25322-68-3; EINECS 225-856-4; dispersant in food tablets and preparations; *Properties:* colorless clear liq.; sol. in water and org. solvs. except aliphatic hydrocarbons; dens. 9.4 lb/gal; sp.gr. 1.12; visc. 7.39 cs (210 F); flash pt. 460 F; surf. tens. 66.6 dynes/cm (1%); pH 6.2 (5% aq.).

Pluracol® E600. [BASF] PEG-12; CAS 25322-68-3; EINECS 229-859-1; dispersant in food tablets and preparations; *Properties:* colorless clear liq.; sol. in water and org. solvs. except aliphatic hydrocarbons; dens. 9.4 lb/gal; sp.gr. 1.12; visc. 10.83 cs (210 F); flash pt. 480 F; surf. tens. 65.2 dynes/cm (1%); pH 5.3 (5% aq.).

Pluracol® E1500. [BASF] PEG-6-32; dispersant in food tablets and preparations; *Properties:* wh. waxy solid; sol. in water and org. solvs. except aliphatic hydrocarbons; dens. 10.0 lb/gal; sp.gr. 1.20; m.p. 46.0-47.5 C; flash pt. > 490 F; surf. tens. 62.8 dynes/cm (1%); pH 6.7 (5% aq.); Unverified

Pluracol® E4000. [BASF] PEG-75; CAS 25322-68-3; dispersant in food tablets and preparations; *Properties:* wh. waxy solid; sol. in water and org. solvs. except aliphatic hydrocarbons; dens. 10.0 lb/gal; sp.gr. 1.20; m.p. 59.5 C; flash pt. > 490 F; surf. tens. 61.9 dynes/cm (1%); pH 6.7 (5% aq.).

Pluracol® W170. [BASF] PPG-5-buteth-7; CAS 74623-31-7; defoamer for food processing; *Properties:* APHA 50 max. visc. liq.; sol. in water, alcohols, ketones, esters, benzene, toluene, glycol ethers, chlorinated solvs.; dens. 8.58 lb/gal; sp.gr. 1.03; visc. 160-180 SUS (100 F); cloud pt. 73 C (1%); flash pt. (OC) 360 F; pour pt. -45 F; pH 5.5-7.0 (10% aq.); Discontinued

Pluracol® W260. [BASF] Polyalkoxylated polyether; defoamer for food processing; *Properties:* liq.; m.w. 1000; water-sol.; visc. 260 SUS (37.8 C); pour pt. -40 C; flash pt. (COC) 222 C; Discontinued

Pluracol® W660. [BASF] PPG-12-buteth-16; CAS 74623-31-7; defoamer for food processing; *Properties:* APHA 50 max. visc. liq.; sol. in water, alcohols, ketones, esters, benzene, toluene, glycol ethers, chlorinated solvs.; dens. 8.79 lb/gal; sp.gr. 1.055; visc. 660 SUS (100 F); cloud pt. 60.5 C (1%); flash pt. (OC) 440 F; pour pt. -34 C; pH 5.5-7.5 (10% aq.); Discontinued

Pluracol® W2000. [BASF] PPG-20-buteth-30; CAS 74623-31-7; defoamer for food processing; *Properties:* APHA 50 max. visc. liq.; sol. in water, alcohols, ketones, esters, benzene, toluene, glycol ethers, chlorinated solvs.; dens. 8.83 lb/gal; sp.gr. 1.06; visc. 2000 SUS (100 F); cloud pt. 57.0 C (1%); flash pt. (OC) 440 F; pour pt. -25 F; pH 5.5-7.5 (10% aq.);

Discontinued

Pluracol® W3520N. [BASF] PPG-28-buteth-35; CAS 9038-95-3; defoamer for food processing; *Properties:* APHA 40 max. visc. liq.; sol. in water, alcohols, ketones, esters, benzene, toluene, glycol ethers, chlorinated solvs.; dens. 8.83 lb/gal; sp.gr. 1.06; visc. 3520 SUS (100 F); cloud pt. 57 C (1%); flash pt. (OC) 437 F; pour pt. -20 F; pH 6.0-7.5 (10% aq.).

Pluracol® W3520N-RL. [BASF] PPG; defoamer for food processing; *Properties:* APHA 200 max. visc. liq.; sol. in water, alcohols, ketones, esters, benzene, toluene, glycol ethers, chlorinated solvs.; dens. 8.87 lb/gal; sp.gr. 1.065; visc. 1752-2500 SUS (100 F); cloud pt. 55.5 C (1%); flash pt. (OC) 440 F; pour pt. 28.4 F; pH 6.0-7.5 (10% aq.); Discontinued

Pluracol® W5100N. [BASF] PPG-33-buteth-45; CAS 74623-31-7; defoamer for food processing; *Properties:* APHA 40 max. visc. liq.; sol. in water, alcohols, ketones, esters, benzene, toluene, glycol ethers, chlorinated solvs.; dens. 8.83 lb/gal; sp.gr. 1.06; visc. 5100 SUS (100 F); cloud pt. 55 C (1%); flash pt. (OC) 437 F; pour pt. -20 F; pH 5.5-7.0 (10% aq.).

Pluracol® WD1400. [BASF] Polyalkoxylated polyether; defoamer for food processing; *Properties:* liq.; m.w. 2500; water-sol.; visc. 1400 SUS (37.8 C); pour pt. -20 C; flash pt. (COC) 255 C; Discontinued

Plurafac® LF 1300. [BASF AG] Alkoxylated straight chain alcohol; nonionic; defoamer for sugar industry; *Properties:* liq.; 100% conc.; Discontinued

Plurol Stearique WL 1009. [Gattefosse SA] Polyglyceryl-6 distearate; CAS 34424-97-0; nonionic; food emulsifier; *Regulatory:* FCC listed; *Properties:* Gardner < 10 waxy solid, faint odor; sol. in chloroform, methylene chloride; partly sol. in ethanol; disp. in water; HLB 9.0; drop pt. 48-53 C; acid no. < 5; iodine no. < 3; sapon. no. 120-140; pH 7.0-9.5 (10% aq.); 100% conc.; *Toxicology:* nonirritating to eyes.

Pluronic® PE 6100. [BASF AG] PO/EO block polymer; nonionic; defoamer for sugar industry; *Properties:* liq.; 100% conc.; Discontinued

Pluronic® PE 6200. [BASF AG] PO/EO block polymer; nonionic; defoamer for sugar industry; *Properties:* liq.; 100% conc.; Discontinued

Pluronic® PE 6400. [BASF AG] PO/EO block polymer; nonionic; defoamer for sugar industry; *Properties:* liq.; 100% conc.; Discontinued

Pluronic® PE 8100. [BASF AG] PO/EO block polymer; nonionic; defoamer for sugar industry; *Properties:* liq.; 100% conc.; Discontinued

Pluronic® RPE 2520. [BASF AG] EO/PO block polymer; nonionic; defoamer for sugar industry; *Properties:* liq.; 100% conc.; Discontinued

Polar-Gel® 5. [Am. Maize Prods.] Stabilized, moderately inhibited waxy maize starch; starch for fruit/bakery fillings, canned foods, barbecue/tomato sauce, aseptic/shelf-stable foods, baby foods, frozen entrees; ideal for high acid systems, batch and continuous processing; stable to high heat; good cold storage, freeze/thaw stability; *Regulatory:*

FDA 21CFR §172.892; *Properties:* waxy; 95% thru 200 mesh; pH 6.0; 10% moisture; *Storage:* store in dry, cool conditions, in odor-free area, ≤ 50% r.h. and ≤ 25 C.

Polar-Gel® 8. [Am. Maize Prods.] Stabilized, lightly inhibited waxy maize starch; starch for canned soups, sauces, canned/frozen oriental foods, baby foods, dry mixes, frozen entrees; high visc. in neutral systems; for batch processing; stable to prolonged heat; good cold storage, freeze/thaw stability; superior paste clarity; *Regulatory:* FDA 21CFR §172.892; *Properties:* waxy; 95% thru 200 mesh; pH 5.5; 10% moisture; *Storage:* store in dry, cool conditions, in odor-free area, ≤ 50% r.h. and ≤ 25 C.

Polar-Gel® 10. [Am. Maize Prods.] Stabilized, highly inhibited waxy maize starch; starch for use in fruit fillings, bakery fillings, toppings, syrups, salad dressings, aseptic/shelf-stable foods; resist. to high heat and acid conditions; high cold storage, freeze/thaw stability; exc. performance in continuous high shear systems; *Regulatory:* FDA 21CFR §172.892; *Properties:* waxy, bland flavor; 80% thru 200 mesh; pH 5.5; 10% moisture; *Storage:* store in dry, cool conditions, in odor-free area, ≤ 50% r.h. and ≤ 25 C.

Polar-Gel® 13. [Am. Maize Prods.] Stabilized, moderately inhibited waxy maize starch; starch for soups, sauces, gravies, bakery and fruit fillings, canned puddings, aseptic/shelf-stable foods; superior cold storage, freeze/thaw stability; stable to wide range of pH and temp.; for batch and continuous processing; exc. paste clarity; *Regulatory:* FDA 21CFR §172.892; *Properties:* waxy, bland flavor; 80% thru 200 mesh; pH 5.5; 10% moisture; *Storage:* store in dry, cool conditions, in odor-free area, ≤ 50% r.h. and ≤ 25 C.

Polar-Gel® 15. [Am. Maize Prods.] Stabilized, lightly inhibited waxy maize starch; starch for soups, sauces, gravies, frozen entrees, oriental foods, dry mixes, cream-style corn; high visc. in neutral systems; stable to prolonged heat; exc. cold storage and freeze/thaw stability, paste clarity; *Regulatory:* FDA 21CFR §172.892; *Properties:* waxy, bland flavor; 80% thru 200 mesh; pH 5.5; 10% moisture; *Storage:* store in dry, cool conditions, in odor-free area, ≤ 50% r.h. and ≤ 25 C.

Polar-Gel® 18. [Am. Maize Prods.] Stabilized, lightly inhibited waxy maize starch; starch for soups, sauces, gravies, frozen entrees, oriental foods, dry mixes, puddings, cream fillings; high visc. in neutral systems; stable under prolonged heat; exc. cold storage and freeze/thaw stability, paste clarity; *Regulatory:* FDA 21CFR §172.892; *Properties:* waxy, bland flavor; 80% thru 200 mesh; pH 5.5; 10% moisture; *Storage:* store in dry, cool conditions, in odor-free area, ≤ 50% r.h. and ≤ 25 C.

Polarin® Range. [Grünau GmbH] Glycerol and derivs.; solvs., humectants for food industry.

Poly E. [Am. Ingreds.] Polysorbate 60 (58-63%), soybean oil, mono- and diglycerides, propylene glycol; dough improver for yeast-raised prods. and bakery mixes; *Properties:* amber liq.; cloud pt. 59-63 F; 6-7% mono.

Polyaldo® 2O10 KFG. [Lonza] Polyglyceryl-10 dioleate; food emulsifier, surfactant; polysorbate 80 replacer; *Regulatory:* kosher; *Properties:* Gardner < 1 liq.; HLB 11.5; acid no. 2; sapon. no. 85.

Polyaldo® 2P10 KFG. [Lonza] Polyglyceryl-10 dipalmitate; nonionic; food emulsifier, surfactant, gelling agent; replacement for polysorbate 60; *Properties:* Gardner 1 soft waxy solid; m.p. 48 C; HLB 12.0; acid no. 2; sapon. no. 89; 100% conc.

Polyaldo® 2S6 KFG. [Lonza] Polyglyceryl-6 distearate; food grade emulsifier; aerating and whipping agent; *Regulatory:* kosher; *Properties:* Gardner < 1 beads; m.p. 57 C; HLB 6.3; acid no. 2; sapon. no. 136.

Polyaldo® 10-1-O KFG. [Lonza] Polyglyceryl-10 oleate; CAS 67784-82-1; emulsifier; *Regulatory:* FDA 21CFR §172.854, kosher; *Properties:* Hazy visc. liq., mild odor, bland taste; HLB 14; acid no. 2 max.; sapon. no. 70-80; hyd. no. 480-510.

Polyaldo® DGDO. [Lonza] Polyglyceryl-10 decaoleate; CAS 11094-60-3; EINECS 234-316-7; nonionic; emulsifier for baking, ice cream, general food use; *Properties:* liq.; HLB 3.0; 100% conc.

Polyaldo® DGDO KFG. [Lonza] Polyglyceryl-10 decaoleate; CAS 11094-60-3; EINECS 234-316-7; nonionic; dispersant, solubilizer in margarine; antifoamer and defoamer in sugar-protein syrups; inhibits crystallization in high quality salad oils for salad dressings and mayonnaise; *Usage level:* 0.02-0.05% (syrups), 0.1-1.0% (margarine), 0.1% (salad oils); *Regulatory:* FDA 21CFR §172.854; kosher; *Properties:* amber clear liq., mild odor; sol. in ethanol, min. and veg. oils, insol. in water; HLB 3 ± 1; acid no. 20 max.; iodine no. 80 max.; sapon. no. 155-185; hyd. no. 25-45; *Storage:* store in cool, dry area.

Polyaldo® HGDS. [Lonza] Polyglyceryl-6 distearate; nonionic; emulsifier for baking, ice cream, general food use; *Properties:* bead; HLB 7.0; 100% conc.

Polyaldo® HGDS KFG. [Lonza] Polyglyceryl-6 distearate; nonionic; food emulsifier for cake mixes; whipping agent providing overrun and foam stabilization props. to whipped creams; freeze/thaw stabilizer for frozen desserts; improves antisticking props. and promotes chewability in chewing gum; *Usage level:* 0.1-0.5% (frozen desserts); *Regulatory:* FDA 21CFR §172.854; kosher; *Properties:* tan beads, mild fatty odor, bland taste; sol. in ethanol; disp. in water; HLB 7 ± 1; m.p. 56 C; acid no. 6 max.; iodine no. 3 max.; sapon. no. 120-140; *Storage:* store in cool, dry area.

Polyaldo® TGMS. [Lonza] Polyglyceryl-3 stearate; CAS 37349-34-1; EINECS 248-403-2; nonionic; emulsifier for baking, ice cream, general food use; *Properties:* bead; HLB 7.0; 100% conc.

Polyaldo® TGMS KFG. [Lonza] Polyglyceryl-3 stearate; CAS 37349-34-1; EINECS 248-403-2; nonionic; food emulsifier; aerating and whipping agent in fat-based systems, cake, and cake icing formulations; permits substantial fat reduction in frozen

desserts; promotes hardness and improves gloss in hard butter confectionery coatings; *Regulatory:* FDA 21CFR §172.854; kosher; *Properties:* off-wh. beads, mild odor, bland taste; sol. in ethanol, min. and veg. oils; insol. in water; HLB 7 ± 1; m.p. 55 C; acid no. 8 max.; iodine no. 4 max.; sapon. no. 120-140; *Storage:* store in cool, dry area.

Polycon S60 K. [Witco/H-I-P] Sorbitan stearate; CAS 1338-41-6; EINECS 215-664-9; lipophilic emulsifier used where weaker water-binding props. and enhanced aeration are desired; crystallization promoter, surface film former, defoamer; improves dispersibility in flavors; for cakes, icings, confectionery, coffee whiteners, toppings; *Usage level:* 0.02-0.05% (defoaming), 0.5-2% (flavors); *Regulatory:* FDA 21CFR §172.842; *Properties:* cream-colored powd.; m.p. 130 F; HLB 4.7.

Polycon S80 K. [Witco/H-I-P] Sorbitan oleate; EINECS 215-665-4; lipophilic food emulsifier used in emulsions where weaker water-binding props. and enhanced aeration are desired; crystallization promoter, surface film former; *Regulatory:* FDA 21CFR §173.75; *Properties:* yel. liq.; m.p. 5 F; HLB 4.9.

Polycon T60 K. [Witco/H-I-P] Polysorbate 60; CAS 9005-67-8; hydrophilic emulsifier, surfactant for formulating o/w emulsions; emulsion stabilizer for salad dressing; defoamer; for cakes, icings, confectionery, beverage mixes, coffee whiteners, whipped toppings, flavors; *Usage level:* 1-3% (beverage), 0.3% (salad dressing), 0.1-0.4% (coffee whiteners), 0.02-0.05% (defoaming); *Regulatory:* FDA 21CFR §172.360; *Properties:* yel. liq.; m.p. 76 F; HLB 14.9.

Polycon T80 K. [Witco/H-I-P] Polysorbate 80; CAS 9005-65-6; hydrophilic food emulsifier, surfactant; used in pickle industry due to very low m.p.; defoaming agent for beet sugar prod.; provides dryness to ice cream; dispersant for flavors; *Usage level:* 0.02-0.1% (frozen desserts), 0.02-0.05% (defoaming); *Regulatory:* FDA 21CFR §172.840; *Properties:* yel. liq.; m.p. -8 F; HLB 15.0.

Polypro 2000® CF 45%. [Hormel] Cherry flavored hydrolyzed gelatin, citric acid, artificial flavor and color, sodium benzoate, potassium sorbate; used when a heavy flavored, high protein liq. is needed, e.g., in protein drinks for weight clinics and weight lifting groups; *Regulatory:* kosher; *Properties:* deep red liq., cherry/protein odor, wild cherry flavor; pH 3.8-4.0; 45-58% solids; *Storage:* stable for 1 yr when stored at ambient temps.

Polypro 2000® UF 45%. [Hormel] Unflavored hydrolyzed gelatin, citric acid, sodium benzoate, potassium sorbate; used when a sl. medicinal flavor is important, esp. for heavily fruit flavored high protein drinks; *Regulatory:* kosher; *Properties:* tan to beige liq., noticeable protein aroma, sharp flavor; pH 3.8-4.2; 45-58% solids; *Storage:* stable for up to 1 yr when stored at ambient temps.

Polypro 5000®. [Hormel] Hydrolyzed gelatin; food-grade protein with exc. film-forming, encapsulation, and moisturizing props.; for meal supplements where protein fortification is needed; replaces up to 25% egg white solids in food foam systems; *Regulatory:* kosher; *Properties:* lt. ivory free-flowing powd., bland char. odor; cold water-sol.; m.w. 5000; pH 5-6 (10 g/50 ml water); 93-97% solids; *Storage:* store at ambient temps. below 80 F for up to 1 yr.

Polypro 5000® CF or SF 45%. [Hormel] Hydrolyzed gelatin, citric acid, artificial color and flavor, sodium benzoate, potassium sorbate; cherry flavored (CF) or strawberry flavored (SF) protein for use in delicate fruit flavored high protein drinks; *Properties:* red liq., fruity sl. protein aroma; pH 3.8-4.2; 45-48% solids; *Storage:* stable for 1 yr stored at ambient temps.; protect from freezing.

Polypro 5000® UF 45%. [Hormel] Hydrolyzed gelatin, citric acid, sodium benzoate, potassium sorbate; clear bland protein for delicate fruit flavored high protein drinks or flavored soups; *Regulatory:* kosher; *Properties:* tan to beige liq., bland sl. protein aroma, mild flavor; pH 3.8-4.2; 45-48% solids; *Storage:* stable for 1 yr stored at ambient temps.

Polypro 15000® Food Grade. [Hormel] Hydrolyzed gelatin; food-grade protein with exc. film forming and encapsulation props.; in meal supplement prods. where protein fortification is needed; *Regulatory:* kosher; *Properties:* lt. ivory free-flowing powd., bland char. odor; cold water-sol.; m.w. 15,000; pH 5-6 (10 g/50 ml water); 93-97% solids; *Storage:* store at ambient temps. below 80 F for up to 1 yr.

Polypro 15000® Pharmaceutical Grade. [Hormel] Hydrolyzed gelatin; in meal supplement prods. where protein fortification is needed; *Regulatory:* kosher; *Properties:* lt. ivory free-flowing powd., bland char. odor; cold water-sol.; m.w. 15,000; pH 5-6 (10 g/50 ml water); 93-97% solids; *Storage:* store at ambient temps. below 80 F for up to 1 yr.

Polysar Butyl 101-3. [Miles/Polysar] Isobutylene/isoprene copolymer rubber; no antioxidant; CAS 9010-85-9; butyl rubber for use as chewing gum base; *Properties:* sp.gr. 0.92; Mooney visc. 44-60 (ML1+8, 125 C).

Polysar Butyl 402. [Miles/Polysar] Isobutylene/isoprene copolymer rubber; nonstaining stabilizer; CAS 9010-85-9; butyl rubber for specific food applics.; *Properties:* amber bales; sp.gr. 0.92; Mooney visc. 47 (ML1+8, 100 C).

Polysorbate 60. [Am. Ingreds.] Polysorbate 60; CAS 9005-67-8; emulsifier for icings, cakes, yeast-raised baked goods, toppings, sugar coatings; *Properties:* yel. visc. liq.; sapon. no. 45-55; hyd. no. 81-96.

Poly-Tergent® P-17A. [Olin] EO/PO block polymer; nonionic; emulsifier, foam control agent for egg washing, foods; *Properties:* APHA 50 max. clear liq.; mild odor; sol. in water, alcohols, glycol ethers, aromatic and aliphatic hydrocarbons, chlorinated solvs.; sp.gr. 1.01; dens. 8.4 lb/gal; flash pt. 241 C (COC); cloud pt. 28 C (1% aq.); pH 5.0-7.5 (1% aq.); surf. tens. 37 dynes/cm (0.1%); Draves wetting 35 s (0.25%); Ross-Miles foam 0 mm (0.1%, initial); 100% act.; *Toxicology:* LD50 (oral, rats) ≈ 2.5 g/kg;

nonirritating to skin and eyes.

Poly-Tergent® P-17B. [Olin] EO/PO block polymer; nonionic; emulsifier, foam control agent for egg washing, foods; *Properties:* APHA 50 max. clear liq.; mild odor; sol. in water, alcohols, glycol ethers, aromatic hydrocarbons, chlorinated solvs.; sp.gr. 1.04; dens. 8.6 lb/gal; flash pt. 246 C (COC); cloud pt. 31 (1% aq.); pH 5.0-7.5 (1% aq.); surf. tens. 40 dynes/cm (0.1%); Draves wetting > 180 s (0.1%); Ross-Miles foam 30 mm (0.1%, initial); 100% act.; *Toxicology:* toxicology: LD50 (oral, rats) ≈ 2.5 g/kg; nonirritating to skin and eyes.

Poly-Tergent® P-17BLF. [Olin] EO/PO block polymer; nonionic; emulsifier, foam control agent for egg washing, foods; *Properties:* APHA 50 clear liq., mild odor; sol. in water, alcohols, glycol ethers, aromatic and chlorinated hydrocarbons; sp.gr. 1.02; dens. 8.5 lb/gal; f.p. -10 C; cloud pt. 31 C (1% aq.); flash pt. (COC) 238 C; pH 4.5-7.5 (1% aq.); surf. tens. 40 dynes/cm (0.1%); Draves wetting > 300 s (0.1%); Ross-Miles foam 15 mm (0.1%, initial); 100% act.; *Toxicology:* LD50 (oral, rats) ≈ 2.5 g/kg; nonirritating to skin and eyes.

Poly-Tergent® P-17BX. [Olin] EO/PO block polymer; nonionic; emulsifier, foam control agent for egg washing, foods; *Properties:* APHA 50 clear liq., mild odor; sol. in water, alcohols, glycol ethers, aromatic and chlorinated solvs.; sp.gr. 1.02; dens. 8.5 lb/gal; f.p. -12 C; cloud pt. 32 C (1% aq.); flash pt. (COC) 232 C; pH 4.5-7.5 (1% aq.); surf. tens. 40 dynes/cm (0.1%); Draves wetting > 300 s (0.1%); Ross-Miles foam 35 mm (0.1%, initial); 100% act.; *Toxicology:* LD50 (oral, rats) ≈ 3 g/kg; nonirritating to skin; eye irritant.

Poly-Tergent® P-17D. [Olin] EO/PO block polymer; nonionic; emulsifier, foam control agent for egg washing, foods; *Properties:* APHA 50 clear liq., mild odor; sol. in water, alcohols, glycol ethers, aromatic and chlorinated solvs.; sp.gr. 1.05; dens. 8.7 lb/gal; f.p. 9 C; cloud pt. 59 C (1% aq.); flash pt. (COC) 238 C; pH 4.5-7.5 (1% aq.); surf. tens. 41 dynes/cm (0.1%); Draves wetting 265 s (0.1%); Ross-Miles foam 50 mm (0.1%, initial); 100% act.; *Toxicology:* LD50 (oral, rats) 1.85 g/kg; nonirritating to skin; irritating to eyes.

Poly-Tergent® P-32D. [Olin] EO/PO block polymer; nonionic; emulsifier, foam control agent for egg washing, foods; *Properties:* APHA 50 clear semi-solid, mild odor; sol. in water, alcohols, glycol ethers, aromatic hydrocarbons, chlorinated solvs.; sp.gr. 1.04; dens. 8.6 lb/gal; f.p. 15 C; cloud pt. 81 C (1% aq.); flash pt. (COC) 232 C; pH 4.5-7.5 (1% aq.); surf. tens. 34 dynes/cm (0.1%); Draves wetting 60 s (0.1%); Ross-Miles foam 48 mm (0.1%, initial); 100% act.; *Toxicology:* LD50 (oral, rats) ≈ 2.5 g/kg; nonirritating to skin and eyes.

Polywax® 500. [Petrolite] Polyethylene homopolymer; CAS 9002-88-4; EINECS 200-815-3; chewing gum base; *Regulatory:* incl. FDA §172.888, 175.105, 175.300, 176.170, 176.180, 176.200, 177.1200, 178.3720, 179.45; *Properties:* prilled; low sol. in org. solvs., esp. at R.T.; sol. in CCl₄, benzene, xylene, toluene, turpentine; m.w. 500; melt index > 5000 g/10 min; dens. 0.93 g/cc; visc. 6.6 cps (99 C); m.p. 88 C; soften. pt. (R&B) 88 C.

Polywax® 600. [Petrolite] Polyethylene homopolymer; CAS 9002-88-4; EINECS 200-815-3; syn. polymer with outstanding heat stability and resist. to chemical attack; food and food contact applics.; *Regulatory:* FDA 21CFR §172.888, 175.105, 175.125, 175.300, 175.320, 176.170, 176.180, 1176.200, 176.210, 177.1200, 177.1210, 177.2600, 177.2800, 178.3720, 178.3850, 179.45; *Properties:* visc. 64 SUS (149 C); m.p. 94 C.

Polywax® 655. [Petrolite] Polyethylene homopolymer; CAS 9002-88-4; EINECS 200-815-3; chewing gum base; *Regulatory:* incl. FDA §172.888, 175.105, 175.300, 176.170, 176.180, 176.200, 177.1200, 178.3720, 179.45; *Properties:* prilled; low sol. in org. solvs., esp. at R.T.; sol. in CCl₄, benzene, xylene, toluene, turpentine; m.w. 655; melt index > 5000 g/10 min; dens. 0.94 g/cc; visc. 5 cps (149 C); m.p. 99 C; soften. pt. (R&B) 99 C.

Pomelex Complex D. [MLG Enterprises Ltd.] Citric conc. contg. conc. natural juices, deterpenated essential oils, syn. colorants, acidifiers, thickeners, and preservatives; preformulated base for prod. of plain, non-gasified soft drinks.

Pomelex Complex G. [MLG Enterprises Ltd.] Citric conc. contg. conc. natural juices, deterpenated essential oils, syn. colorants, acidifiers, thickeners, and preservatives; preformulated base for prod. of fizzy soft drinks.

Pork Pancreas Substance. [Am. Labs] Freeze-dried defatted prod. derived from hog pancreas glands; nutritive food additive; *Properties:* powd.

Powdered Agar Agar NF M-100 (Gracilaria). [Meer] Agar agar; CAS 9002-18-0; EINECS 232-658-1; emulsifier, stabilizer, gellant for chiffon pies, meringues, pie fillings, icings, toppings, cookies, Neufchatel cream cheeses, yogurt, chocolate milk drinks, sherbets, jelly candies, sweets, poultry/meat canning, jams; antistaling agent for breads; *Properties:* water-sol.

Powdered Agar Agar NF MK-60. [Meer] Agar agar; CAS 9002-18-0; EINECS 232-658-1; emulsifier, stabilizer, gellant for chiffon pies, meringues, pie fillings, icings, toppings, cookies, Neufchatel cream cheeses, yogurt, chocolate milk drinks, sherbets, jelly candies, sweets, poultry/meat canning, jams; antistaling agent for breads; *Properties:* water-sol.

Powdered Agar Agar NF MK-80-B. [Meer] Agar agar; CAS 9002-18-0; EINECS 232-658-1; emulsifier, stabilizer, gellant for chiffon pies, meringues, pie fillings, icings, toppings, cookies, Neufchatel cream cheeses, yogurt, chocolate milk drinks, sherbets, jelly candies, sweets, poultry/meat canning, jams; antistaling agent for breads; *Properties:* water-sol.

Powdered Agar Agar NF MK-80 (Bacteriological). [Meer] Agar agar; CAS 9002-18-0; EINECS 232-658-1; emulsifier, stabilizer, gellant for chiffon pies, meringues, pie fillings, icings, toppings, cookies, Neufchatel cream cheeses, yogurt, chocolate milk drinks, sherbets, jelly candies, sweets, poultry/

meat canning, jams; antistaling agent for breads; *Properties:* water-sol.

Powdered Agar Agar NF S-100. [Meer] Agar agar; CAS 9002-18-0; EINECS 232-658-1; emulsifier, stabilizer, gellant for chiffon pies, meringues, pie fillings, icings, toppings, cookies, Neufchatel cream cheeses, yogurt, chocolate milk drinks, sherbets, jelly candies, sweets, poultry/meat canning, jams; antistaling agent for breads; *Properties:* water-sol.

Powdered Agar Agar NF S-100-B. [Meer] Agar agar; CAS 9002-18-0; EINECS 232-658-1; emulsifier, stabilizer, gellant for chiffon pies, meringues, pie fillings, icings, toppings, cookies, Neufchatel cream cheeses, yogurt, chocolate milk drinks, sherbets, jelly candies, sweets, poultry/meat canning, jams; antistaling agent for breads; *Properties:* water-sol.

Powdered Agar Agar NF S-150. [Meer] Agar agar NF; CAS 9002-18-0; EINECS 232-658-1; emulsifier, stabilizer, gellant for chiffon pies, meringues, pie fillings, icings, toppings, cookies, Neufchatel cream cheeses, yogurt, chocolate milk drinks, sherbets, jelly candies, sweets, poultry/meat canning, jams; antistaling for breads, cakes; *Usage level:* 0.05-0.8% (cream cheese), 0.5-2% (meat canning); *Properties:* wh. to grayish wh. fine powd.; 100% thru 80 mesh, ≥ 90% thru 150 mesh; water-sol.

Powdered Agar Agar NF S-150-B. [Meer] Agar agar; CAS 9002-18-0; EINECS 232-658-1; emulsifier, stabilizer, gellant for chiffon pies, meringues, pie fillings, icings, toppings, cookies, Neufchatel cream cheeses, yogurt, chocolate milk drinks, sherbets, jelly candies, sweets, poultry/meat canning, jams; antistaling agent for breads; *Properties:* water-sol.

Powdered Aloe Vera (1:200) Food Grade. [Tri-K Industries] Aloe vera gel; food grade; *Properties:* off-wh. to lt. beige powd., sl. vegetable odor to almost odorless, bland taste; pH 3.5-7.0 (0.5% aq.); 10% max. moisture; *Storage:* store in tightly sealed container in cool, dry area below 120 F, protected from direct sunlight; hygroscopic.

Powdered Caramel Color, Acid Proof. [Crompton & Knowles/Ingred. Tech.] Caramel powd.; CAS 8028-89-5; EINECS 232-435-9; color additive for foods; *Regulatory:* FDA 21CFR §73.85, 73.1085, 73.2085; *Properties:* brn. free-flowing powd., typ. odor; 100% min. thru 80 mesh; sol. in water; pH 5.0 ± 0.5 (1%); 4% max. moisture; *Storage:* store in tight containers; avoid exposure to excessive heat, light, and moisture; 2 yrs storage life.

Powdered Caramel Colour Non-Ammoniated-All Natural T-717. [MLG Enterprises Ltd.] Caramel color; CAS 8028-89-5; EINECS 232-435-9; natural colorant for dry food prods. incl. powd. desserts, dry cake mix, powd. gravy seasonings, powd. coffee or other beverages, dehydrated soups, bouillon powds., baked goods; *Regulatory:* FDA 21CFR §73.85; *Properties:* 1% max. on 30 mesh, 58% max. on 200 mesh; pH 3.3 ± 0.3 (50%); 3.5 ± 0.3% moisture.

Powdered Gum Arabic NF/FCC G-150. [Meer] Gum arabic; CAS 9000-01-5; EINECS 232-519-5; protective colloid, stabilizer, thickener for confectionery, candies, jellies, glazes, chewing gum, flavor emulsions, citrus oil and beverage flavor emulsions; foam stabilizer in beer; *Properties:* colorless, odorless, tasteless; water-sol.

Powdered Gum Arabic NF/FCC Superselect Type NB-4. [Meer] Gum arabic; CAS 9000-01-5; EINECS 232-519-5; protective colloid, stabilizer, thickener for confectionery, candies, jellies, glazes, chewing gum, flavor emulsions, citrus oil and beverage flavor emulsions; foam stabilizer in beer; *Properties:* colorless, odorless, tasteless; water-sol.

Powdered Gum Ghatti #1. [Meer] Gum ghatti; stabilizer, binder, emulsifier forming o/w emulsions; used in beverage emulsions where it forms orange oil emulsions, in table syrup emulsions; *Properties:* off-wh. to lt. amber gum; 99% thru 80 mesh, ≥ 90% thru 100 mesh; visc. ≥ 100 cps (5%).

Powdered Gum Ghatti #2. [Meer] Gum ghatti; stabilizer, binder, emulsifier forming o/w emulsions; used in beverage emulsions where it forms orange oil emulsions, in table syrup emulsions; *Properties:* water-sol.

Powdered Gum Guar NF Type 80 Mesh B/T. [Meer] Guar gum; CAS 9000-30-0; EINECS 232-536-8; thickener, binder, suspending agent, stabilizer, dietary fiber source for soft cheese, ice cream, fruit drinks, dietetic beverages, cocoa beverages, baked goods, frozen pie fillings, icings, salad dressings, relishes; binder/lubricant in sausages; *Properties:* water-sol.

Powdered Gum Guar Type 140 Mesh B/T. [Meer] Guar gum; CAS 9000-30-0; EINECS 232-536-8; thickener, binder, suspending agent, stabilizer, dietary fiber source for soft cheese, ice cream, fruit drinks, dietetic beverages, cocoa beverages, baked goods, frozen pie fillings, icings, salad dressings, relishes; binder/lubricant in sausages; *Properties:* water-sol.

Powdered Gum Guar Type ECM. [Meer] Guar gum; CAS 9000-30-0; EINECS 232-536-8; thickener, binder, suspending agent, stabilizer, dietary fiber source for soft cheese, ice cream, fruit drinks, dietetic beverages, cocoa beverages, baked goods, frozen pie fillings, icings, salad dressings, relishes; binder/lubricant in sausages; *Properties:* water-sol.

Powdered Gum Guar Type M. [Meer] Guar gum; CAS 9000-30-0; EINECS 232-536-8; thickener, binder, suspending agent, stabilizer, dietary fiber source for soft cheese, ice cream, fruit drinks, dietetic beverages, cocoa beverages, baked goods, frozen pie fillings, icings, salad dressings, relishes; binder/lubricant in sausages; *Properties:* water-sol.

Powdered Gum Guar Type MM FCC. [Meer] Guar gum; CAS 9000-30-0; EINECS 232-536-8; thickener, binder, suspending agent, stabilizer, dietary fiber source for soft cheese prods., ice cream, fruit drinks, dietetic/cocoa beverages, baked goods, frozen pie fillings, icings, salad dressings, relishes; binder/lubricant in sausages; *Properties:* wh. to cream-wh. fine powd., nearly odorless; 97% thru

140 mesh, ≥ 90% thru 200 mesh; visc. ≥ 3500 cps (1%); pH 5.0-6.4.

Powdered Gum Guar Type MM (HV). [Meer] Guar gum; CAS 9000-30-0; EINECS 232-536-8; thickener, binder, suspending agent, stabilizer, dietary fiber source for soft cheese, ice cream, fruit drinks, dietetic beverages, cocoa beverages, baked goods, frozen pie fillings, icings, salad dressings, relishes; binder/lubricant in sausages; *Properties:* water-sol.

Powdered Gum Guar Type MMM ¹/₂. [Meer] Guar gum; CAS 9000-30-0; EINECS 232-536-8; thickener, binder, suspending agent, stabilizer, dietary fiber source for soft cheese, ice cream, fruit drinks, dietetic beverages, cocoa beverages, baked goods, frozen pie fillings, icings, salad dressings, relishes; binder/lubricant in sausages; *Properties:* water-sol.

Powdered Gum Guar Type MMW. [Meer] Guar gum; CAS 9000-30-0; EINECS 232-536-8; thickener, binder, suspending agent, stabilizer, dietary fiber source for soft cheese, ice cream, fruit drinks, dietetic beverages, cocoa beverages, baked goods, frozen pie fillings, icings, salad dressings, relishes; binder/lubricant in sausages; *Properties:* water-sol.

Powdered Gum Karaya Superfine #1 FCC. [Meer] Karaya gum; CAS 9000-36-6; stabilizer, water binder, emulsifier for water ices, sherbets, cheese spreads, meringue powds., meat processing; *Properties:* water-sol.

Powdered Gum Karaya Superfine XXXX ГОО. [Meer] Karaya gum NF; CAS 9000-36-6; stabilizer, water binder, emulsifier for water ices, sherbets, cheese spreads, meringe powds., meat processing; *Usage level:* 0.2-0.4% (sherbets), < 0.8% (cheese spreads), 0.25-1% (meat); *Properties:* powd.; 100% thru 80 mesh, ≥ 60% thru 200 mesh; swells in water, insol. in alcohol; visc. ≥ 200 cps (1%); pH 4.4-4.8.

Powdered Gum Tragacanth T-150. [Importers Service; Commodity Services Int'l.] Gum tragacanth USP/NF/FCC; CAS 9000-65-1; EINECS 232-552-5; food additive; *Properties:* off-wh. free-flowing powd., bland mucilaginous odor and taste; 95% min. thru 140 mesh, 65% min. thru 200 mesh; visc. 1000 cps min. (1%); pH 5-6 (1%); 15% max. moisture.

Powdered Gum Tragacanth T-200. [Importers Service; Commodity Services Int'l.] Gum tragacanth USP/NF/FCC; CAS 9000-65-1; EINECS 232-552-5; food additive; *Properties:* lt. cream to lt. yel. free-flowing powd., bland mucilaginous odor and taste; 95% min. thru 140 mesh, 65% min. thru 200 mesh; visc. 700 cps min. (1%); pH 5-6 (1%); 15% max. moisture.

Powdered Gum Tragacanth T-300. [Importers Service; Commodity Services Int'l.] Gum tragacanth USP/NF/FCC; CAS 9000-65-1; EINECS 232-552-5; food additive; *Properties:* lt. cream to lt. yel. free-flowing powd., bland mucilaginous odor and taste; 95% min. thru 140 mesh, 65% min. thru 200 mesh;

visc. 490 cps min. (1%); pH 5-6 (1%); 15% max. moisture.

Powdered Gum Tragacanth T-400. [Importers Service; Commodity Services Int'l.] Gum tragacanth USP/NF/FCC; CAS 9000-65-1; EINECS 232-552-5; food additive; *Properties:* lt. cream to lt. yel. free-flowing powd., bland mucilaginous odor and taste; 95% min. thru 140 mesh, 65% min. thru 200 mesh; visc. 350 cps min. (1%); pH 5-6 (1%); 15% max. moisture.

Powdered Gum Tragacanth T-500. [Importers Service; Commodity Services Int'l.] Gum tragacanth USP/NF/FCC; CAS 9000-65-1; EINECS 232-552-5; food additive; *Properties:* lt. cream to lt. yel. free-flowing powd., bland mucilaginous odor and taste; 95% min. thru 140 mesh, 65% min. thru 200 mesh; visc. 280 cps min. (1%); pH 5-6 (1%); 15% max. moisture.

Powdered Locust Bean Gum Type D-200. [Meer] Locust bean gum FCC; CAS 9000-40-2; EINECS 232-541-5; thickener, water binder, suspending agent, and stabilizer for soft cheese, cheese spreads, ice cream, baked goods, cakes, frozen pie fillings, sausages, stuffed meat prods., canned pet foods; *Properties:* water-sol.

Powdered Locust Bean Gum Type D-300. [Meer] Locust bean gum FCC; CAS 9000-40-2; EINECS 232-541-5; thickener, water binder, suspending agent, and stabilizer for soft cheese, cheese spreads, ice cream, baked goods, cakes, frozen pie fillings, sausages, stuffed meat prods., canned pet foods; *Properties:* water-sol.

Powdered Locust Bean Gum Type P-100. [Meer] Locust bean gum FCC; CAS 9000-40-2; EINECS 232-541-5; thickener, water binder, suspending agent, and stabilizer for soft cheese, cheese spreads, ice cream, baked goods, cakes, frozen pie fillings, sausages, stuffed meat prods., canned pet foods; *Properties:* water-sol.

Powdered Locust Bean Gum Type PP-100. [Meer] Locust bean gum FCC; CAS 9000-40-2; EINECS 232-541-5; thickener, water binder, suspending agent, and stabilizer for soft cheese, cheese spreads, ice cream, baked goods, cakes, frozen pie fillings, sausages, stuffed meat prods., canned pet foods; *Properties:* water-sol.

Powdered Sta-Fudge. [Bunge Foods] Blend of stabilizers; fudge base for icings, yeast-raised donut fudge icing, cupcake icings; fast setting, very stable with chocolate flavor; *Usage level:* 22 lb/100 lb. powd. sugar; *Properties:* dk. brn. powd., cocoa flavor.

Powdered Tragacanth Gum Type A/10. [Meer] Tragacanth gum; CAS 9000-65-1; EINECS 232-552-5; thickener, water binder, suspending agent, emulsifier for salad dressings, bakery, flavor emulsions, fruit fillings, cream fillings, citrus beverages, barbecue and steak sauces; *Properties:* water-sol.

Powdered Tragacanth Gum Type E-1. [Meer] Tragacanth gum; CAS 9000-65-1; EINECS 232-552-5; thickener, water binder, suspending agent, emulsifier for salad dressings, bakery, flavor emulsions,

fruit fillings, cream fillings, citrus beverages, barbecue and steak sauces; *Properties:* water-sol.

Powdered Tragacanth Gum Type G-3. [Meer] Tragacanth gum; CAS 9000-65-1; EINECS 232-552-5; thickener, water binder, suspending agent, emulsifier for salad dressings, bakery, flavor emulsions, fruit fillings, cream fillings, citrus beverages, barbecue and steak sauces; *Properties:* water-sol.

Powdered Tragacanth Gum Type L. [Meer] Tragacanth gum; CAS 9000-65-1; EINECS 232-552-5; thickener, water binder, suspending agent, emulsifier for salad dressings, bakery, flavor emulsions, fruit fillings, cream fillings, citrus beverages, barbecue and steak sauces; *Properties:* water-sol.

Powdered Tragacanth Gum Type W. [Meer] Tragacanth gum NF; CAS 9000-65-1; EINECS 232-552-5; thickener, water binder, suspending agent, emulsifier for salad dressings, bakery, flavor emulsions, fruit fillings, cream fillings, citrus beverages, barbecue and steak sauces; *Properties:* wh. to cream-wh. fine powd.; 100% thru 80 mesh, ≥ 45% thru 200 mesh; visc. 420-520 cps (1%); pH 4.8-5.8.

Precision Base. [Sandoz Nutrition] Whey, partially hydrog. coconut oil, corn syrup solids, sodium caseinate, mono and diglceryides, dipotassium phosphate, artifical color, and artificial flavor; instantized blend of sweet dairy whey and other ingreds. for use as alternate ingred. for nonfat dry milk or whole milk; instant dissolving, free-flowing, noncaking, exc. whitening power; used for hot chocolate beverages, cream soups, cream sauces; *Properties:* powd.; 100% thru 20 mesh, < 25% thru 100 mesh; pH 6.4 ± 0.2 (10%); *Storage:* 1 yr shelf life stored in cool, dry place (below 70 F) in sealed containers.

Prefera® CSL. [Henkel/Cospha; Henkel Canada] Calcium stearoyl lactylate; CAS 5793-94-2; EINECS 227-335-7; anionic; dough conditioner for bread and rolls; *Properties:* powd.; 100% conc.

Prefera® NSL. [Henkel/Cospha; Henkel Canada] Sodium stearoyl lactylate; CAS 25383-99-7; EINECS 246-929-7; anionic; dough conditioner for bread and rolls; *Properties:* powd.; 100% conc.

Prefera® Range. [Grünau GmbH] Sodium or calcium stearyl-2-lactylates; food emulsifier for improvement of fermentation tolerance, volume and texture of yeast-raised baked goods; antistaling effect.

Prefera® SSL 6000 MB. [Grünau GmbH] Sodium stearoyl lactyl-2-lactate; CAS 25383-99-7; EINECS 246-929-7; food emulsifier for improvement of fermentation tolerance, volume and texture of yeast-raised baked goods; antistaling effect; *Regulatory:* EEC E 481; FDA 21CFR §172.846; *Properties:* cream-colored microbeads, char. odor and taste; 10% on 0.2 mm sieve, 65% on 0.5 mm sieve; bulk dens. 550-650 g/l; m.p. 42-50 C; acid no. 60-80; sapon. no. 210-240; pH 5-6 (1:10 in methanol/water 1:1); 30-34% lactic acid content; *Storage:* 1 yr shelf life stored under cool, dry conditions below 20 C.

Premium Fine Granular Gum Arabic. [Importers Service; Commodity Services Int'l.] Gum arabic;

CAS 9000-01-5; EINECS 232-519-5; food additive; *Properties:* wh. to off-wh. free-flowing fine gran., odorless, tasteless; 100% min. thru 40 mesh, 20% max. thru 140 mesh; visc. 200 cps max. (20%); pH 4.0-4.8 (20%); 15% max. moisture.

Premium Granular Gum Arabic. [Importers Service; Commodity Services Int'l.] Gum arabic; CAS 9000-01-5; EINECS 232-519-5; food additive; *Properties:* lt. amber coarse gran., odorless, tasteless; 90% min. thru 6 mesh, 20% max. thru 60 mesh; visc. 200 cps max. (20%); pH 4.0-4.8 (20%); 15% max. moisture.

Premium Powdered Gum Arabic. [Importers Service; Commodity Services Int'l.] Gum arabic; CAS 9000-01-5; EINECS 232-519-5; food additive; *Properties:* off-wh. to buff free-flowing powd., odorless, tasteless; 98% min. thru 140 mesh; pH 4-4.8 (20%); 15% max. moisture.

Premium Powdered Gum Ghatti G-1. [Importers Service; Commodity Services Int'l.] Gum ghatti; emulsifier for flavor oils; *Properties:* lt. brn. free-flowing powd., odorless; 99.9% min. thru 80 mesh, 98% min. thru 140 mesh; visc. 200 cps min. (5%); pH 4.5-5.2 (5%); 15% max. moisture.

Premium Powdered Gum Karaya No. 1. [Importers Service; Commodity Services Int'l.] Karaya gum; CAS 9000-36-6; food additive; *Properties:* buff to lt. tan free-flowing powd., faint acetic acid-like odor; 99% min. thru 80 mesh; visc. 300 cps min. (1%), 7000 cps min. (2%); pH 4.3-5.0 (1%); 19% max. moisture.

Premium Powdered Gum Karaya No. 1 Special. [Importers Service, Commodity Services Int'l.] Karaya gum; CAS 9000-36-6; food additive; *Properties:* off-wh. to buff free-flowing powd., faint acetic acid-like odor; 99% min. thru 80 mesh; visc. 400 cps min. (1%), 8000 cps min. (2%); pH 4.3-5.0 (1%); 20% max. moisture.

Premium Powdered Gum Karaya No. 2 Special HV. [Importers Service; Commodity Services Int'l.] Karaya gum; CAS 9000-36-6; food additive; *Properties:* tan to lt. brn. free-flowing powd., faint acetic acid-like odor; 99% min. thru 80 mesh; visc. 500 cps min. (1%), 8000 cps min. (2%); pH 4.3-5.0 (1%); 19% max. moisture.

Premium Powdered Gum Karaya No. 2. [Importers Service; Commodity Services Int'l.] Karaya gum; CAS 9000-36-6; food additive; *Properties:* buff to lt. brn. free-flowing powd., faint acetic acid-like odor; 99% min. thru 80 mesh; visc. 300 cps min. (1%), 6000 cps min. (2%); pH 4.3-5.0 (1%); 19% max. moisture.

Premium Powdered Gum Karaya No. 3. [Importers Service; Commodity Services Int'l.] Karaya gum; CAS 9000-36-6; food additive; *Properties:* tan to brn. free-flowing powd., faint acetic acid-like odor; 99% min. thru 80 mesh; visc. 200 cps min. (1%), 5000 cps min. (2%); pH 4.3-5.0 (1%); 19% max. moisture.

Premium Spray Dried Gum Arabic. [Importers Service; Commodity Services Int'l.] Gum arabic; CAS 9000-01-5; EINECS 232-519-5; food additive;

Properties: wh. to cream free-flowing powd., odorless, tasteless; 99% min. thru 60 mesh, 80% min. thru 140 mesh; visc. 150 cps max. (20%); pH 4.0-4.8 (20%); 12% max. moisture.

Premose® Syrup. [Premier Malt Prods.] Extract of 80% corn and 20% malted barley; syrup with sweet but mild malt flavor; used in ice cream prods., bakery goods, nutrient for distilled vinegars, entomological, and fermentations, cereal, tobacco, baby foods, soybean milk, pretzel; 72% the sweetness of sucrose; *Usage level:* 2-3% (novelty ice cream), 1-22.5% of flour wt. (bread), 1-3% dough wt. (bagels), 1.5-2.5% of flour wt. (saltines), 2-4% of dough mix wt. (snacks), 4-6% flour wt. (cookies), 6-14% (breakfast cereal), 3-4% (diet foods/beverages), 3-7% (soy bean milk); *Properties:* Lovibond 60-160 syrup; pH 5.1-5.8 (10%); 78.5-80.0% solids, 1.0-1.9% protein, 63-73% reducing sugars.

Prenol. [BASF AG] 3-Methyl-2-butene-1-ol; fragrance and flavoring (fresh, herbal, green, fruity, sl. lavender-like).

Pressure Cut Y.R.D. Conc. [Bunge Foods] Conc. delivering high quality donuts for automated pressure cut systems; *Usage level:* 22.5% based on flour.

Prevent-O. [Mitsubishi Int'l.; Amano Enzyme USA] Enzyme assisting in moisture retention and retarding staling; for fruitcakes and brownies.

Prime F-25. [Multi-Kem] Sodium alginate; CAS 9005-38-3; offers high rate of reactivity to form firm heat-stable gels in food ingred. applics.; *Properties:* tan to gray powd., sl. char. aroma; cold-water sol. forming a visc. sol'n.; visc. 25 cps (1%); pH 6-7; 15% max. moisture; *Storage:* store away from moisture, humidity, and high temps.

Prime F-40. [Multi-Kem] Sodium alginate; CAS 9005-38-3; offers high rate of reactivity to form firm heat-stable gels in food ingred. applics.; *Properties:* tan to gray powd., sl. char. aroma; cold-water sol. forming a visc. sol'n.; visc. 40 cps (1%); pH 6-7; 15% max. moisture; *Storage:* store away from moisture, humidity, and high temps.

Prime F-400. [Multi-Kem] Sodium alginate; CAS 9005-38-3; offers high rate of reactivity with calcium salts to form firm heat-stable gels in food ingred. applics.; *Properties:* tan to gray powd., sl. char. aroma; cold-water sol. forming a visc. sol'n.; visc. 400 cps (1%); pH 6-7; 15% max. moisture; *Storage:* store away from moisture, humidity, and high temps.

Prime F-600. [Multi-Kem] Sodium alginate; CAS 9005-38-3; offers high rate of reactivity with calcium salts to form firm heat-stable gels in food ingred. applics.; *Properties:* tan to gray powd., sl. char. aroma; cold-water sol. forming a visc. sol'n.; visc. 600+ cps (1%); pH 6-7; 15% max. moisture; *Storage:* store away from moisture, humidity, and high temps.

Prinza® 452. [Grünau GmbH] Guar gum; CAS 9000-30-0; EINECS 232-536-8; thickener and stabilizer for food industry esp. for desserts; *Regulatory:* FAO/WHO; EEC E 412; FDA 21CFR §184.1339,

GRAS; *Properties:* beige-wh. free-flowing powd., neutral odor/taste; 99% < 0.1 mm; disp. in cold water; bulk dens. 600 g/l; visc. 3800-4200 mPa•s (1% aq.); 75% min. galactomannan content; *Storage:* 1 yr shelf life stored in closed original pkg. under cool, dry conditions.

Prinza® 455. [Grünau GmbH] Guar gum; CAS 9000-30-0; EINECS 232-536-8; thickener and stabilizer for food industry esp. for desserts; *Regulatory:* FAO/WHO; EEC E 412; FDA 21CFR §184.1339, GRAS; *Properties:* beige-wh. free-flowing powd., neutral odor/taste; 99% < 0.1 mm; disp. in cold water; bulk dens. 600 g/l; visc. 4800-5200 mPa•s (1% aq.); 75% min. galactomannan content; *Storage:* 1 yr shelf life stored in closed original pkg. under cool, dry conditions.

Prinza® Range. [Grünau GmbH] Guar gum; CAS 9000-30-0; EINECS 232-536-8; thickener and stabilizer for food industry esp. for desserts.

Pristene® 180. [UOP] 70% Mixed tocopherols in veg. oil; food-grade antioxidant/stabilizer for edible fats and oils, natural foods, processed foods/meat/poultry, fresh and frozen sausage, salad dressings, dehydrated potatoes, seasonings, snacks, nuts, soup bases, pet foods; natural source of vitamin E; *Usage level:* 0.015-0.045% tocopherol on oil/fat content; *Regulatory:* FDA 21CFR §182.3890; USDA 9CFR §318.7; *Properties:* reddish brn. sl. visc. liq., mild sl. odor and taste; sol. in veg. oils, fats, essential oils, and ethanol; insol. in water; sp.gr. 0.92; ref. index 1.559; 70% act.; *Storage:* 6-8 mo shelf life at moderate temps.; purge with nitrogen after each opening.

Pristene® 181. [UOP] Mixed tocopherols (35%), propylene glycol (42%), ascorbyl palmitate (8%), vegetable oil (15%); food-grade antioxidant preventing oxidative rancidity; for edible fats and oils, natural foods, processed foods, meat, and poultry, fresh/frozen sausage, salad dressings, dehydrated potatoes, seasonings, snacks, nuts, soup bases, pet foods; *Properties:* amber liq.; oil-sol.; sp.gr. 0.9971; visc. 202 cS (38 C); flash pt. (PM) 225 F; ref. index 1.4770.

Pristene® 184. [UOP] Mixed tocopherols (28%), vegetable oils (72%); food-grade antioxidant/stabilizer for edible fats and oils, natural foods, processed foods/meat/poultry, fresh and frozen sausage, salad dressings, dehydrated potatoes, seasonings, nuts, soup bases, pet foods; *Usage level:* 0.015-0.045% of tocopherol on oil/fat content; *Regulatory:* FDA 21CFR §182.3890; USDA 9CFR §318.7; *Properties:* golden yel. sl. visc. liq., mild sl. odor and taste; sol. in veg. oils, fats, essential oils, and ethanol; insol. in water; sp.gr. 0.933; ref. index 1.487; *Storage:* 6-8 mo shelf life at moderate temps.; purge with nitrogen after each opening.

Pristene® 185. [UOP] Mixed tocopherols (50%), vegetable oils (50%); food-grade antioxidant, stabilizer for edible fats and oils, natural foods, processed foods/meat/poultry, fresh and frozen sausage, salad dressings, dehydrated potatoes, sea-

sonings, snacks, nuts, soup bases, pet foods; *Usage level:* 0.015-0.045% of tocopherols on oil/fat content; *Regulatory:* FDA 21CFR §182.3890; USDA 9CFR 318.7; *Properties:* reddish brn. sl. visc. liq., mild sl. odor and taste; sol. in veg. oils, animal fats, min. oils, essential oils, ethanol; insol. in water; sp.gr. 0.92; ref. index 1.4990; *Storage:* 6-8 mo shelf life at moderate temps.; purge with nitrogen after each opening.

Pristene® 186. [UOP] Mixed tocopherols (35%), vegetable oil (65%); food-grade antioxidant, stabilizer for edible fats and oils, natural foods, processed foods/meat/poultry, fresh and frozen sausage, salad dressings, dehydrated potatoes, seasonings, snacks, nuts, soup bases, pet foods; *Usage level:* 0.015-0.045% tocopherols on oil/fat content; *Regulatory:* FDA 21CFR §182.3890; USDA 9CFR 318.7; *Properties:* lt. amber sl. visc. liq., mild sl. odor and taste; sol. in veg. oils, fats, essential oils, ethanol; insol. in water; sp.gr. 0.9382; ref. index 1.4916; *Storage:* 6-8 mo shelf life at moderate temps.; purge with nitrogen after each opening.

Pristene® 189. [UOP] Mixed tocopherols (38%), propylene glycol (13%), citric acid (5%), mono/diglycerides (27%), vegetable oils (17%); food-grade antioxidant for edible fats and oils, natural foods, processed foods, meat, and poultry, fresh and frozen sausage, salad dressings, dehydrated potatoes, seasonings, snacks, nuts, soup bases, pet foods; *Properties:* reddish amber liq.; oil-sol.; sp.gr. 0.9864; visc. 441 cS (38 C); flash pt. (PM) 235 C; ref. index 1.4920.

Pristene® R20. [UOP] 20% Rosemary extract dispersed on maltodextrin; natural flavoring, stabilizer with antioxidant props. for edible fats and oils, natural foods, processed meat and poultry, fresh/frozen sausage, salad dressings, seasonings, snacks and nuts, soup bases, pet foods, and cosmetics; *Regulatory:* FDA 21CFR §182.20, GRAS; *Properties:* lt. tan to brn. free-flowing powd., char. rosemary odor but free of camphor; water-disp.; sp.gr. 0.65-0.75; b.p. > 177 C (of oleoresin); 20% act.; *Storage:* store in cool, dry location; 3 mo shelf life can be extended with refrigeration.

Pristene® RO. [UOP] Rosemary extract, veg. oil, mono- and diglycerides; natural flavor, stabilizer with antioxidant props. for edible fats and oils, natural foods, processed meat and poultry, fresh/frozen sausage, salad dressings, seasonings, snacks and nuts, soup bases, pet foods, and cosmetics; *Regulatory:* FDA 21CFR §182.20, GRAS; *Properties:* brnsh. free-flowing liq., char. rosemary odor and flavor, free of camphor; disp. in veg. oil, sl. disp. in water; sp.gr. 0.95-1.1; b.p. > 177 C; flash pt. (OC) 93 C; *Storage:* store in cool, dry location; 1 yr shelf life.

Pristene® RW. [UOP] Rosemary extract, veg. oil, mono-, di-, and triglycerides, and lecithin; natural flavor, stabilizer with antioxidant props. for edible fats and oils, natural foods, processed meat and poultry, fresh/frozen sausage, salad dressings, seasonings, snacks and nuts, soup bases, pet foods, and cosmetics; *Regulatory:* FDA 21CFR §182.20, GRAS; *Properties:* brnsh. liq., char. rosemary aroma and flavor; water-disp. sp.gr. 0.95-1.10; b.p. 149 C; flash pt. (OC) > 93.3 C; *Storage:* store in a cool, dry location; 1 yr shelf life.

Pristene® TR. [UOP] Mixed tocopherols (38%), vegetable oils (15%), rosemary extract (50%); food-grade antioxidant, stabilizer for edible fats and oils, natural foods, processed foods/meat/poultry, fresh and frozen sausage, salad dressings, dehydrated potatoes, food pkg. materials, seasonings, snacks, nuts, soup bases, pet foods, cosmetics; *Usage level:* 0.015-0.045% tocopherols on oil/fat content; *Regulatory:* FDA 21CFR §182.3890, 182.20, GRAS; USDA 9CFR §318.7; *Properties:* reddish-brn. sl. visc. liq., char. rosemary with mild veg. oil aroma and taste; oil-disp.; sp.gr. 0.9585; flash pt. > 400 F; ref. index 1.4960; *Storage:* 6-8 mo shelf life at moderate temps.; purge with nitrogen after each opening.

ProBenz. [DuCoa] Sodium benzoate FCC; CAS 532-32-1; EINECS 208-534-8; food preservative used in carbonated beverages, preserved fruit and its juices, syrups, jams, fillings, prepared salads and condiments; *Properties:* low-dust dense form.

Procon®. [Central Soya] Soy conc.; protein for use in raw beef patties, chicken or veal patties, ground meat foods, meatballs, and poultry or beef rolls; *Properties:* powd.

Procon® 2000. [Central Soya] Soya protein conc.; CAS 68153-28-6; protein with good water binding functionality but limited fat binding and emulsification props., for food industry; *Properties:* wh. fine grind powd.; formerly Promosoy 100.

Procon® 20/60. [Central Soya] Soy protein conc.; CAS 68153-28-6; protein for use in cooked or raw beef patties, ground meat foods, coarse ground sausages, and meatballs; good water binding agent but limited fat binding and emulsification props.; *Properties:* uncolored gran.; 97% min. thru #14 sieve, 10% max. thru #60 sieve; 70% min. protein, 10% max. moisture; *Storage:* store in cool, dry place below 26.5 C at ≤ 60% r.h.; formerly Promosoy® 20/60.

Procon® 20/60 Military. [Central Soya] Soya protein conc.; CAS 68153-28-6; protein with good water binding functionality but limited fat binding and emulsification props.; for food industry; *Properties:* lt. tan gran.; formerly Promosoy 20/60 Military.

Procon® 20/60-SL. [Central Soya] Soya protein conc.; CAS 68153-28-6; protein with good water binding functionality but limited fat binding and emulsification props.; for food industry; *Properties:* lt. tan gran., fortified; formerly Promosoy SL.

Proctin BUS. [Pronova Biopolymer] Sodium alginate; CAS 9005-38-3; heat-stable thickener, stabilizer, gellant for bakery marmalade and jam; *Properties:* wh. to tan powd., sl. odor; water-sol.; *Toxicology:* TLV 10 mg/m³ total dust; LD50 (oral, rat) > 5 g/kg (essentially nontoxic); may cause eye irritation; *Precaution:* incompat. with strong oxidizing agents; *Storage:* store in cool, dry place.

Pro Fam 646. [ADM Protein Spec.] Isolated soy protein; CAS 68153-28-6; protein for nutritional supplements, high protein drinks, injected meat prods.; *Properties:* spray-dried, bland; easily disp.; pH 6.3-6.7; 90% min. protein, 6.5% max. moisture.

Pro Fam 648. [ADM Protein Spec.] Isolated soy protein; CAS 68153-28-6; protein for injected meat prods.; *Properties:* spray-dried very fine powd., bland; readily disp.; pH 6.3-6.7; 90% min. protein, 6.5% max. moisture.

Pro Fam 780. [ADM Protein Spec.] Isolated soy protein; CAS 68153-28-6; very low visc. protein for nutritional supplements and dairy-free prods.; *Properties:* pH 6.9-7.3; 90% min. protein, 6.5% max. moisture.

Pro Fam 781. [ADM Protein Spec.] Isolated soy protein; CAS 68153-28-6; very low visc., low sodium protein for nutritional supplements and dairy-free prods.; *Properties:* pH 6.9-7.3; 90% min. protein, 6.5% max. moisture.

Pro Fam 955. [ADM Protein Spec.] Isolated soy protein; CAS 68153-28-6; protein for high protein tablets, extruded snack items, pasta, baked prods.; *Properties:* spray-dried, bland; low sol.; pH 6.0-6.4; 90% min. protein, 6.5% max. moisture.

Pro Fam 970. [ADM Protein Spec.] Isolated soy protein; CAS 68153-28-6; highly functional, salt tolerant protein for dairy-free prods., milk replacers, snack foods, sauces, spreads, canned specialties, special diet foods; *Properties:* spray-dried, bland; high sol.; pH 6.8-7.3; 90% min. protein, 6.5% max. moisture.

Pro Fam 972. [ADM Protein Spec.] Isolated soy protein; CAS 68153-28-6; highly functional, salt tolerant protein for dairy-free prods., milk replacers, snack foods, sauces, spreads, processed meats, emulsified meats, sausage-type prods.; *Properties:* dust-free spray-dried, bland; high sol.; pH 6.8-7.3; 90% min. protein, 6.5% max. moisture.

Pro Fam 974, 974 Fortified. [ADM Protein Spec.] Isolated soy protein; CAS 68153-28-6; highly functional, salt tolerant protein for dairy-free prods., milk replacers, snack foods, processed meats, emulsified meats, sausage-type prods.; *Properties:* dust-free spray-dried, bland; highly sol.; pH 6.8-7.3; 90% min. protein, 6.5% max. moisture.

Proflex. [MLG Enterprises Ltd.] Active enzyme deriv. of *Aspergillus oryzae* (high proportion of protease, lower amt. of α-amylase) with cereal filler, and other edible excipients; enzyme dough conditioner improving pliability, pan flow, softness, grain, and texture; *Regulatory:* FDA GRAS; *Properties:* sl. off-wh. to beige tablet, char. odor; *Storage:* 6 mos shelf life under cool, dry conditions.

Promax® 70. [Central Soya] Functional soya protein conc.; CAS 68153-28-6; protein with high emulsification and water/fat binding functionality for food industry; *Properties:* lt. cream-colored fine powd.; formerly Promine 70.

Promax® 70L. [Central Soya] Functional soya protein conc.; CAS 68153-28-6; protein with high emulsification and water/fat binding functionality for food industry; *Properties:* lt. cream-colored fine powd.; formerly Promine 70L.

Promax® 70LSL. [Central Soya] Functional soya protein conc.; CAS 68153-28-6; protein with high emulsification and water/fat binding functionality for food industry; *Properties:* lt. cream-colored fine powd.

Promax® Plus. [Central Soya] Functional soya protein conc.; CAS 68153-28-6; protein with med. emulsification and water/fat binding functionality for food industry; *Properties:* cream-colored fine powd.; formerly Promosoy Plus.

Promine® DS. [Central Soya] Functional soya protein conc.; CAS 68153-28-6; high-quality nutritive source, binder, and emulsifier for use in cooked/raw meat patties, pizza topping, meatballs, coarse ground sausages, meat rolls, seafood, soup bases, gravies, protein drinks, dairy emulsions, breads/baking, cereals, baby food; *Properties:* lt. cream-colored fine powd.; formerly Promax DS.

Promine® DS-SL. [Central Soya] Functional soya protein conc.; CAS 68153-28-6; protein with very high emulsification and water/fat binding functionality for food industry; *Properties:* lt. cream-colored fine powd.; formerly Promosoy 20/60.

Promine® HV. [Central Soya] Functional soya protein conc.; CAS 68153-28-6; nutritive source, fat emulsifier, emulsion stabilizer, water binder, highly digestible protein-rich food ingred. in coarse ground sausages, paté, meat rolls, seafood, surimi, dairy emulsions, breads/baking, cereal, baby food; *Properties:* lt. cream-colored fine powd.; formerly Promax® HV.

Promise Liq. Oil. [Van Den Bergh Foods] Partially hydrog. canola oil; high stability oil for frying, coating, or spray applics. for snack foods, prepared foods, and applics. where low solids is critical; high in monounsaturates and low in saturated fatty acids; *Regulatory:* kosher; *Properties:* Lovibond 1.5R max. liq.; drop pt. 15 C max.

Promodan SP. [Grindsted Prods.] Propylene glycol stearate; CAS 1323-39-3; EINECS 215-354-3; food emulsifier; *Properties:* m.p. 38-45 C; 90% min. monoester.

Promodan USV. [Grindsted Prods.] Propylene glycol ester; food emulsifier; aerating agent and foam stabilizer for whipped toppings; in combination with Dimodans as aerating agents for fine bakery goods and cake shortenings; *Regulatory:* EEC, FDA §172.856; *Properties:* sm. beads; drop pt. 41 C; 90% min. monoester.

Promodan USV Kosher. [Grindsted Prods.] Propylene glycol ester made from fully hydrog. vegetable oil with antioxidant (200 ppm max. citric acid in propylene glycol); emulsifier for cake mixes, cake shortenings, whipped toppings; *Usage level:* 3-6% based on fat (cake mixes), 3-6% (shoretning), 0.5-1% (prewhipped frozen toppings), 5-10% (powd. toppings), 0.5-1% (imitation whipped cream); *Regulatory:* FDA 21CFR §172.856; kosher; *Properties:* ivory beads; drop pt. 41 C; iodine no. 5 max.; 90% min. propylene glycol monoester; *Storage:*

store in cool, dry area; storage above 20 C may cause caking.

Pronal ST-1. [Toho Chem. Industry] POP ether; nonionic; defoaming agent for food and fermentation industry; *Properties:* liq.

Proplus®. [Protein Tech. Int'l.] Isolated soy protein with no cholesterol and virtually no fats or carbohydrates; CAS 68153-28-6; nutritional and functional protein additive performing like lean meat protein; 90% pure.

Propylene Glycol Alginate HV. [Meer] Propylene glycol alginate; CAS 9005-37-2; food grade emulsifier and stabilizer for thick, creamy, textured salad dressings, w/o emulsions; *Properties:* water-sol.

Propylene Glycol Alginate LV FCC. [Meer] Propylene glycol alginate FCC standardized with dextrose; emulsifier and stabilizer for salad dressings, meat and flavor sauces, milk shakes, fountain syrups, and fruit juices; *Usage level:* 0.4-1.0%; *Properties:* wh. to ylsh. powd., flavorless; ≥ 95% thru 80 mesh; visc. 1000-1500 cps (2%); pH 3.6-4.3.

Prostate Substance. [Am. Labs] Vacuum-dried defatted glandular prod.; nutritive food additive; *Properties:* powd. or freeze-dried form.

Protacream. [Pronova Biopolymer] Alginate; heat-stable thickener, stabilizer, gellant for bakery marmalade and jam; *Properties:* water-sol.

Protanal 686. [Pronova Biopolymer] Sodium alginate; CAS 9005-38-3; thickener, gellant for cake icings; anti-sticking, freeze-thaw stable; *Properties:* wh. to tan powd., sl. odor; water-sol.; *Toxicology:* TLV 10 mg/m³ total dust; LD50 (oral, rat) > 5 g/kg (essentially nontoxic); may cause eye irritation; *Precaution:* incompat. with strong oxidizing agents; *Storage:* store in cool, dry place.

Protanal HF 120 M. [Pronova Biopolymer] Sodium alginate; CAS 9005-38-3; stabilizer; provides good mouthfeel and texture, prevents syneresis; for low-fat margarines/spreads; *Properties:* wh. to tan powd., sl. odor; water-sol.; *Toxicology:* TLV 10 mg/m³ total dust; LD50 (oral, rat) > 5 g/kg (essentially nontoxic); may cause eye irritation; *Precaution:* incompat. with strong oxidizing agents; *Storage:* store in cool, dry place.

Protanal HFC 60. [Pronova Biopolymer] Sodium alginate; CAS 9005-38-3; thickener, stabilizer; imparts texture, mouthfeel, and flavor release to powd. soups and sauces; *Properties:* wh. to tan powd., sl. odor; water-sol.; *Toxicology:* TLV 10 mg/m³ total dust; LD50 (oral, rat) > 5 g/kg (essentially nontoxic); may cause eye irritation; *Precaution:* incompat. with strong oxidizing agents; *Storage:* store in cool, dry place.

Protanal KC 119. [Pronova Biopolymer] Sodium alginate; CAS 9005-38-3; thickener, stabilizer; provides good mouthfeel and texture, prevents syneresis; for tomato ketchup; *Properties:* wh. to tan powd., sl. odor; water-sol.; *Toxicology:* TLV 10 mg/m³ total dust; LD50 (oral, rat) > 5 g/kg (essentially nontoxic); may cause eye irritation; *Precaution:* incompat. with strong oxidizing agents; *Storage:* store in cool, dry place.

Protanal KP. [Pronova Biopolymer] Sodium alginate; CAS 9005-38-3; gellant; prevents syneresis, provides good mouthfeel in instant pudding/flan; *Properties:* wh. to tan powd., sl. odor; water-sol.; *Toxicology:* TLV 10 mg/m³ total dust; LD50 (oral, rat) > 5 g/kg (essentially nontoxic); may cause eye irritation; *Precaution:* incompat. with strong oxidizing agents; *Storage:* store in cool, dry place.

Protanal KPM. [Pronova Biopolymer] Sodium alginate; CAS 9005-38-3; gellant; prevents syneresis, provides good mouthfeel, soft texture in instant pudding/flan; *Properties:* wh. to tan powd., sl. odor; water-sol.; *Toxicology:* TLV 10 mg/m³ total dust; LD50 (oral, rat) > 5 g/kg (essentially nontoxic); may cause eye irritation; *Precaution:* incompat. with strong oxidizing agents; *Storage:* store in cool, dry place.

Protanal LF 5/60. [Pronova Biopolymer] Sodium alginate; CAS 9005-38-3; thickener, stabilizer; prevents syneresis and fruit separation; for yogurt mixes; *Properties:* wh. to tan powd., sl. odor; water-sol.; *Toxicology:* TLV 10 mg/m³ total dust; LD50 (oral, rat) > 5 g/kg (essentially nontoxic); may cause eye irritation; *Precaution:* incompat. with strong oxidizing agents; *Storage:* store in cool, dry place.

Protanal LF 20. [Pronova Biopolymer] Sodium alginate; CAS 9005-38-3; heat-resist. gellant for canned meat chunks for petfood; enables cold processing; cost effective; *Properties:* wh. to tan powd., sl. odor; water-sol.; *Toxicology:* TLV 10 mg/m³ total dust; LD50 (oral, rat) > 5 g/kg (essentially nontoxic); may cause eye irritation; *Precaution:* incompat. with strong oxidizing agents; *Storage:* store in cool, dry place.

Protanal LF 20/40. [Pronova Biopolymer] Sodium alginate; CAS 9005-38-3; heat-resist. gellant for canned meat chunks for petfood; enables cold processing; cost effective; *Properties:* wh. to tan powd., sl. odor; water-sol.; *Toxicology:* TLV 10 mg/m³ total dust; LD50 (oral, rat) > 5 g/kg (essentially nontoxic); may cause eye irritation; *Precaution:* incompat. with strong oxidizing agents; *Storage:* store in cool, dry place.

Protanal LF 60. [Pronova Biopolymer] Sodium alginate; CAS 9005-38-3; thickener, stabilizer, gellant; prevents syneresis; provides good mouthfeel; for reduced-sugar jam, yogurt mix, dessert jelly, egg liqueur, restructured fruit; forms thermoreversible gel with HM-pectin in jellies for bakery prods.; *Properties:* wh. to tan powd., sl. odor; water-sol.; *Toxicology:* TLV 10 mg/m³ total dust; LD50 (oral, rat) > 5 g/kg (essentially nontoxic); may cause eye irritation; *Precaution:* incompat. with strong oxidizing agents; *Storage:* store in cool, dry place.

Protanal LF 120 M. [Pronova Biopolymer] Sodium alginate; CAS 9005-38-3; stabilizer; provides good mouthfeel and texture, prevents syneresis; for low-fat margarines/spreads, mayonnaise, dressings; *Properties:* wh. to tan powd., sl. odor; water-sol.; *Toxicology:* TLV 10 mg/m³ total dust; LD50 (oral, rat) > 5 g/kg (essentially nontoxic); may cause eye irritation; *Precaution:* incompat. with strong oxidiz-

ing agents; *Storage:* store in cool, dry place.

Protanal LF 200. [Pronova Biopolymer] Sodium alginate; CAS 9005-38-3; thickener, stabilizer; imparts texture, mouthfeel, and flavor release to powd. soups and sauces; *Properties:* wh. to tan powd., sl. odor; water-sol.; *Toxicology:* TLV 10 mg/m³ total dust; LD50 (oral, rat) > 5 g/kg (essentially nontoxic); may cause eye irritation; *Precaution:* incompat. with strong oxidizing agents; *Storage:* store in cool, dry place.

Protanal LFS 40. [Pronova Biopolymer] Sodium alginate; CAS 9005-38-3; heat-resist. gellant for canned meat chunks for petfood; enables cold processing; cost effective; *Properties:* wh. to tan powd., sl. odor; water-sol.; *Toxicology:* TLV 10 mg/m³ total dust; LD50 (oral, rat) > 5 g/kg (essentially nontoxic); may cause eye irritation; *Precaution:* incompat. with strong oxidizing agents; *Storage:* store in cool, dry place.

Protanal SF 40. [Pronova Biopolymer] Sodium alginate; CAS 9005-38-3; heat-resist. gellant for canned meat chunks for petfood; enables cold processing; cost effective; *Properties:* wh. to tan powd., sl. odor; water-sol.; *Toxicology:* TLV 10 mg/m³ total dust; LD50 (oral, rat) > 5 g/kg (essentially nontoxic); may cause eye irritation; *Precaution:* incompat. with strong oxidizing agents; *Storage:* store in cool, dry place.

Protanal SF 60. [Pronova Biopolymer] Sodium alginate; CAS 9005-38-3; heat-stable gellant for pimento fillings for olives; *Properties:* wh. to tan powd., sl. odor; water-sol.; *Toxicology:* TLV 10 mg/m³ total dust; LD50 (oral, rat) > 5 g/kg (essentially nontoxic); may cause eye irritation; *Precaution:* incompat. with strong oxidizing agents; *Storage:* store in cool, dry place.

Protanal SF 120. [Pronova Biopolymer] Sodium alginate; CAS 9005-38-3; thickener, stabilizer; provides good mouthfeel and texture, prevents syneresis; for mayonnaise, dressings; encapsulant providing controlled yeast release in champagne; gellant for onion rings, restructured fish , meat, and noodles; heat-stable; *Properties:* wh. to tan powd., sl. odor; water-sol.; *Toxicology:* TLV 10 mg/m³ total dust; LD50 (oral, rat) > 5 g/kg (essentially nontoxic); may cause eye irritation; *Precaution:* incompat. with strong oxidizing agents; *Storage:* store in cool, dry place.

Protanal SF 120 M. [Pronova Biopolymer] Sodium alginate; CAS 9005-38-3; stabilizer providing controlled visc., even and slow meltdown, and prevents crystal formation in ice cream; *Properties:* wh. to tan powd., sl. odor; water-sol.; *Toxicology:* TLV 10 mg/m³ total dust; LD50 (oral, rat) > 5 g/kg (essentially nontoxic); may cause eye irritation; *Precaution:* incompat. with strong oxidizing agents; *Storage:* store in cool, dry place.

Protanal SP 5 H. [Pronova Biopolymer] Sodium alginate; CAS 9005-38-3; thickener, stabilizer; provides smooth texture, good mouthfeel and flavor release in toppings; prevents crystal formation in ice cream; *Properties:* wh. to tan powd., sl. odor;

water-sol.; *Toxicology:* TLV 10 mg/m³ total dust; LD50 (oral, rat) > 5 g/kg (essentially nontoxic); may cause eye irritation; *Precaution:* incompat. with strong oxidizing agents; *Storage:* store in cool, dry place.

Protanal VK 687. [Pronova Biopolymer] Sodium alginate; CAS 9005-38-3; heat-stable thickener, stabilizer, gellant providing med. texture; for filling creams and custards in bakery industry; *Properties:* wh. to tan powd., sl. odor; water-sol.; *Toxicology:* TLV 10 mg/m³ total dust; LD50 (oral, rat) > 5 g/ kg (essentially nontoxic); may cause eye irritation; *Precaution:* incompat. with strong oxidizing agents; *Storage:* store in cool, dry place.

Protanal VK 749. [Pronova Biopolymer] Sodium alginate; CAS 9005-38-3; heat-stable thickener, stabilizer, gellant providing very short texture; for filling creams and custards in bakery industry; *Properties:* wh. to tan powd., sl. odor; water-sol.; *Toxicology:* TLV 10 mg/m³ total dust; LD50 (oral, rat) > 5 g/kg (essentially nontoxic); may cause eye irritation; *Precaution:* incompat. with strong oxidizing agents; *Storage:* store in cool, dry place.

Protanal VK 805 IMP. [Pronova Biopolymer] Sodium alginate; CAS 9005-38-3; heat-stable thickener, stabilizer, gellant providing light creamy texture; for filling creams and custards in bakery industry; *Properties:* wh. to tan powd., sl. odor; water-sol.; *Toxicology:* TLV 10 mg/m³ total dust; LD50 (oral, rat) > 5 g/kg (essentially nontoxic); may cause eye irritation; *Precaution:* incompat. with strong oxidizing agents; *Storage:* store in cool, dry place.

Protanal VK 990. [Pronova Biopolymer] Sodium alginate; CAS 9005-38-3; heat-stable thickener, stabilizer, gellant providing light creamy texture; for filling creams and custards in bakery industry; *Properties:* wh. to tan powd., sl. odor; water-sol.; *Toxicology:* TLV 10 mg/m³ total dust; LD50 (oral, rat) > 5 g/kg (essentially nontoxic); may cause eye irritation; *Precaution:* incompat. with strong oxidizing agents; *Storage:* store in cool, dry place.

Protanal VK 998. [Pronova Biopolymer] Sodium alginate; CAS 9005-38-3; heat-stable thickener, stabilizer, gellant providing short texture; for filling creams and custards in bakery industry; *Properties:* wh. to tan powd., sl. odor; water-sol.; *Toxicology:* TLV 10 mg/m³ total dust; LD50 (oral, rat) > 5 g/ kg (essentially nontoxic); may cause eye irritation; *Precaution:* incompat. with strong oxidizing agents; *Storage:* store in cool, dry place.

Protanal VPM. [Pronova Biopolymer] Sodium alginate; CAS 9005-38-3; gellant; prevents syneresis, provides good mouthfeel in instant pudding/flan, mousse; *Properties:* wh. to tan powd., sl. odor; water-sol.; *Toxicology:* TLV 10 mg/m³ total dust; LD50 (oral, rat) > 5 g/kg (essentially nontoxic); may cause eye irritation; *Precaution:* incompat. with strong oxidizing agents; *Storage:* store in cool, dry place.

Protanal VSM. [Pronova Biopolymer] Sodium alginate; CAS 9005-38-3; thickener,stabilizer; freeze-thaw stable; prevents syneresis; provides smooth texture, good mouthfeel and flavor release; for whipped cream, vanilla sauce; *Properties:* wh. to tan powd., sl. odor; water-sol.; *Toxicology:* TLV 10 mg/m³ total dust; LD50 (oral, rat) > 5 g/kg (essentially nontoxic); may cause eye irritation; *Precaution:* incompat. with strong oxidizing agents; *Storage:* store in cool, dry place.

Protanal Ester Bl. [Pronova Biopolymer] Propylene glycol aginate; CAS 9005-37-2; stabilizer improving and maintaining foam levels in beer; *Properties:* wh. to tan powd., sl. odor; sol. in water; *Toxicology:* TLV 10 mg/m³ total dust; LD50 (oral, rat) > 5 g/kg (essentially nontoxic); may cause eye irritation; *Precaution:* incompat. with strong oxidizing agents; *Storage:* store in cool, dry place.

Protanal Ester CF. [Pronova Biopolymer] Propylene glycol aginate; CAS 9005-37-2; stabilizer, emulsifier providing good mouthfeel and texture in lactobacillus drinks; *Properties:* wh. to tan powd., sl. odor; sol. in water; *Toxicology:* TLV 10 mg/m³ total dust; LD50 (oral, rat) > 5 g/kg (essentially nontoxic); may cause eye irritation; *Precaution:* incompat. with strong oxidizing agents; *Storage:* store in cool, dry place.

Protanal Ester H. [Pronova Biopolymer] Propylene glycol aginate; CAS 9005-37-2; stabilizer for sorbet; controlls visc., prevents crystal formation; *Properties:* wh. to tan powd., sl. odor; sol. in water; *Toxicology:* TLV 10 mg/m³ total dust; LD50 (oral, rat) > 5 g/kg (essentially nontoxic); may cause eye irritation; *Precaution:* incompat. with strong oxidizing agents; *Storage:* store in cool, dry place.

Protanal Ester L. [Pronova Biopolymer] Propylene glycol aginate; CAS 9005-37-2; thickener, stabilizer, emulsifier providing good mouthfeel and texture, preventing syneresis; for mayonnaise/dressings; *Properties:* wh. to tan powd., sl. odor; sol. in water; *Toxicology:* TLV 10 mg/m³ total dust; LD50 (oral, rat) > 5 g/kg (essentially nontoxic); may cause eye irritation; *Precaution:* incompat. with strong oxidizing agents; *Storage:* store in cool, dry place.

Protanal Ester L-25 A/H. [Pronova Biopolymer] Propylene glycol aginate; CAS 9005-37-2; thickener, stabilizer, emulsifier providing good mouthfeel and texture, preventing syneresis and separation; for mayonnaise/dressings, fruit juice; *Properties:* wh. to tan powd., sl. odor; sol. in water; *Toxicology:* TLV 10 mg/m³ total dust; LD50 (oral, rat) > 5 g/kg (essentially nontoxic); may cause eye irritation; *Precaution:* incompat. with strong oxidizing agents; *Storage:* store in cool, dry place.

Protanal Ester PVH-A. [Pronova Biopolymer] Propylene glycol aginate; CAS 9005-37-2; thickener, stabilizer, emulsifier providing good mouthfeel and texture, preventing syneresis; for mayonnaise/ dressings; *Properties:* wh. to tan powd., sl. odor; sol. in water; *Toxicology:* TLV 10 mg/m³ total dust; LD50 (oral, rat) > 5 g/kg (essentially nontoxic); may cause eye irritation; *Precaution:* incompat. with strong oxidizing agents; *Storage:* store in cool, dry place.

Protanal Ester SD-H. [Pronova Biopolymer] Propylene glycol aginate; CAS 9005-37-2; thickener, stabilizer, emulsifier providing good mouthfeel and texture, preventing syneresis; for mayonnaise/dressings; *Properties:* wh. to tan powd., sl. odor; sol. in water; *Toxicology:* TLV 10 mg/m^3 total dust; LD50 (oral, rat) > 5 g/kg (essentially nontoxic); may cause eye irritation; *Precaution:* incompat. with strong oxidizing agents; *Storage:* store in cool, dry place.

Pro-Tein 1550. [A.E. Staley Mfg.] Soy peptone; protein fortification for foods and beverages; forms clear sol'ns. with no off-flavors or odors; *Properties:* bland; sol. in water over entire pH range.

Pro-Tein SF 1000. [A.E. Staley Mfg.] Soy peptone; protein for fermentation and medias, nutrient sources.

Pro-Tein SP 1000. [A.E. Staley Mfg.] Soy peptone; autoclave-stable protein for fermentation and medias, nutrient sources.

Prote-sorb SML. [Protex] Sorbitan laurate; CAS 1338-39-2; nonionic; food emulsifier; *Properties:* liq.; HLB 8.6; sapon. no. 162.

Prote-sorb SMO. [Protex] Sorbitan oleate; CAS 1338-43-8; EINECS 215-665-4; nonionic; food emulsifier; *Properties:* liq.; HLB 4.3; sapon. no. 153.

Prote-sorb SMP. [Protex] Sorbitan palmitate; CAS 26266-57-9; EINECS 247-568-8; nonionic; food emulsifier; *Properties:* solid; HLB 6.7; sapon. no. 155.

Prote-sorb SMS. [Protex] Sorbitan stearate; CAS 1338-41-6; EINECS 215-664-9; nonionic; food emulsifier; *Properties:* solid; HLB 4.7; sapon. no. 152.

Prote-sorb STO. [Protex] Sorbitan trioleate; CAS 26266-58-0, EINECS 247-569-3; nonionic; food emulsifier; *Properties:* liq.; HLB 1.8; sapon. no. 180.

Prote-sorb STS. [Protex] Sorbitan tristearate; CAS 26658-19-5; EINECS 247-891-4; food emulsifier; *Properties:* solid; HLB 2.1; sapon. no. 182.

Protex™ 20. [Mid-Am. Food Sales] High fiber rice bran; CAS 68553-81-1.

Protoferm. [Finnsugar Bioprods.] Protease (fungal); CAS 9014-01-1; EINECS 232-752-2; enzyme for food processing; meat tenderizer; protein hydrolysates; beer chillproofing; *Properties:* tan powd.; Unverified

Protol®. [Witco/Petroleum Spec.] Wh. min. oil USP; emollient, lubricant in food processing; *Regulatory:* FDA 21CFR §172.878, §178.3620a; *Properties:* water-wh., odorless, tasteless; sp.gr. 0.859-0.875; visc. 35-37 cSt (40 C); pour pt. -12 C; flash pt. 188 C.

Protopet® Alba. [Witco/Petroleum Spec.] Petrolatum USP; CAS 8027-32-5; EINECS 232-373-2; med. consistency and m.p. petrolatum functioning as carrier, lubricant, emollient, moisture barrier, protective agent, softener for food processing; *Regulatory:* FDA 21CFR §172.880; *Properties:* Lovibond 1.0Y color, odorless; visc. 10-16 cSt (100 C); m.p. 54-60 C.

Protopet® White 1S. [Witco/Petroleum Spec.] Petrolatum USP; CAS 8027-32-5; EINECS 232-373-2; med. consistency and m.p. petrolatum functioning as carrier, lubricant, emollient, moisture barrier, protective agent, softener for food processing; *Regulatory:* FDA 21CFR §172.880; *Properties:* Lovibond 1.5Y color, odorless; visc. 10-16 cSt (100 C); m.p. 54-60 C.

Protopet® White 2L. [Witco/Petroleum Spec.] Petrolatum USP; CAS 8027-32-5; EINECS 232-373-2; med. consistency and m.p. petrolatum functioning as carrier, lubricant, emollient, moisture barrier, protective agent, softener for food processing; *Regulatory:* FDA 21CFR §172.880; *Properties:* Lovibond 8Y0.6R color, odorless; visc. 10-16 cSt (100 C); m.p. 54-60 C.

Protopet® White 3C. [Witco/Petroleum Spec.] Petrolatum USP; CAS 8027-32-5; EINECS 232-373-2; med. consistency and m.p. petrolatum functioning as carrier, lubricant, emollient, moisture barrier, protective agent, softener for food processing; *Regulatory:* FDA 21CFR §172.880; *Properties:* Lovibond 25Y1.0R color, odorless; visc. 10-16 cSt (100 C); m.p. 54-60 C.

Protopet® Yellow 2A. [Witco/Petroleum Spec.] Petrolatum USP; CAS 8027-32-5; EINECS 232-373-2; med. consistency and m.p. petrolatum functioning as carrier, lubricant, emollient, moisture barrier, protective agent, softener for food processing; *Regulatory:* FDA 21CFR §172.880; *Properties:* Lovibond 30Y/2.5R color, odorless; visc. 10-16 cSt (100 C); m.p. 54-60 C.

Provim ESP®. [ADM Ogilvie] Vital wheat gluten; CAS 8002-80-0; EINECS 232-317-7; for food prods. incl. baked goods, cereals, pet foods, pasta, and cheese analog prods.; *Properties:* lt. tan powd.; 7% moisture, 75% protein, 1% fat.

Prox-onic EP 1090-1. [Protex] Difunctional block polymer ending in primary hydroxyl groups; nonionic; component for sugar beet and yeast defoamers; *Properties:* liq.; m.w. 2000; HLB 3.0; cloud pt. 24 C (1% aq.); 100% act.

Prox-onic EP 1090-2. [Protex] Difunctional block polymer ending in primary hydroxyl groups; nonionic; component for sugar beet and yeast defoamers; *Properties:* liq.; m.w. 2600; HLB 6.5; cloud pt. 28 C (1% aq.); 100% act.

Prox-onic EP 2080-1. [Protex] Difunctional block polymer ending in primary hydroxyl groups; nonionic; component for sugar beet and yeast defoamers; *Properties:* liq.; m.w. 2500; HLB 7.0; cloud pt. 30 C (1% aq.); 100% act.

Prox-onic EP 4060-1. [Protex] Difunctional block polymer ending in primary hydroxyl groups; nonionic; component for sugar beet and yeast defoamers; *Properties:* liq.; m.w. 3000; HLB 1.0; cloud pt. 16 C (1% aq.); 100% act.

Prozyme 6. [Mitsubishi Int'l.; Amano Enzyme USA] Semi-alkaline protease from *Aspergillus*; CAS 9014-01-1; EINECS 232-752-2; digestive enzyme; *Properties:* powd.

P/T 25. [Brolite Prods.] Pizza and tortilla conditioner resulting in shorter mixing times with increased extensibility; minimizes cracking of tortillas.

Purac®. [Purac Am.] Lactic acid; acidulant, preservative, flavoring for marinades, salad dressings, confectionery, beer, wine, beverages, dairy prods., desserts, baby food, jams, meat prods./flavors, pickles, baked goods, dry mixes; decontaminating agent in meat/poultry/fish; *Properties:* liq., bland acid taste.

Puracal® PP. [Purac Am.] Calcium lactate, pentahydrate; CAS 814-80-2; EINECS 212-406-7; calcium fortification for beverages, juices, powd. drinks providing good sol. and bioavailability; *Regulatory:* USA GRAS, FCC, USP, EP, DAB, JSFA compliance; *Properties:* wh. powd./chips; sol. 9 g/100 ml in water; m.w. 218 (anhyd.); pH 6.0-8.5 (10% aq.); 98% min. assay.

Purasal® P/USP 60. [Purac Am.] Potassium lactate; CAS 996-31-6; EINECS 213-631-3; antimicrobial; extends shelf life, enhances flavor, controls pathogens in cured and uncured meats, poultry prods., reduced fat prods., reduced sodium prods., convenience foods, pasta prods., dressings; improves water binding capacity, texture; *Usage level:* 2% max. (meat and poultry prods.); *Regulatory:* FDA GRAS, USDA 9CFR §319.180; *Properties:* APHA 100 max. liq., mildly saline taste; m.w. 128 (anhyd.); sp.gr. 1.32-1.35 g/ml (20 C); ref. index 1.415-1.422; pH 6.5-8.5; 58-62% assay.

Purasal® S/SP 60. [Purac Am.] Sodium lactate; CAS 72-17-3; EINECS 200-77-20; emulsifier, flavor enhancer, flavoring adjuvant, humectant, pH control agent, antimicrobial; extends shelf life, improves flavor of cooked beef, controls pathogens in cured and uncured meats, poultry prods., reduced fat prods.; *Usage level:* 2% max. (meat and poultry prods.); *Regulatory:* FDA GRAS, USDA 9CFR §319.180; JSFA compliance; *Properties:* APHA 200 max. liq., mildly saline taste; m.w. 112 (anhyd.); sp.gr. 1.32-1.34 g/ml (20 C); ref. index 1.422-1.425; pH 6.5-8.5; 59-61% assay.

Purasolv® ELECT. [Purac Am.] Ethyl lactate; CAS 97-64-3; food additive, flavor; *Usage level:* ADI not specified; *Regulatory:* FDA GRAS; FAO/WHO, JECFA approved; *Properties:* colorless clear liq., char. mild pleasant odor; misc. with water and most org. solvs.; m.w. 118; sp.gr. 1.030-1.038; b.p. 309 F; flash pt. (CC) 52 C; ref. index 1.410-1.420; readily biodeg.; 99% min. assay; *Toxicology:* LD50 (oral) > 2000 mg/kg; immediately degrades into lactic acid and ethanol; sl. erythema on percutaneous applic. of 2068 mg/kg/day; possible sl. temporary eye irritation; *Storage:* store in well-closed containers.

Purasolv® ELS. [Purac Am.] Ethyl lactate; CAS 97-64-3; food additive, flavor; *Usage level:* ADI not specified; *Regulatory:* FDA GRAS; FAO/WHO, JECFA approved; *Properties:* colorless clear liq., char. mild pleasant odor; misc. with water and most org. solvs.; m.w. 118; sp.gr. 1.030-1.038; b.p. 309 F; flash pt. (CC) 52 C; ref. index 1.410-1.420; readily biodeg.; 98% min. assay; *Toxicology:* LD50 (oral) > 2000 mg/kg; immediately degrades into lactic acid and ethanol; sl. erythema on percutane-

ous applic. of 2068 mg/kg/day; possible sl. temporary eye irritation; *Storage:* store in well-closed containers.

Pure-Bind® B910. [Grain Processing] Modified food starch FCC; CAS 65996-62-5; starch for batters and breadings; provides exc. adhesion to meat; does not interfere with delicate seasonings or color; *Regulatory:* FDA 21CFR §172.892; *Properties:* wh. powd., odorless, bland flavor; > 90% thru 325 mesh; pH 4.5-6.5; 8-11% moisture.

Pure-Bind® B923. [Grain Processing] Modified food starch FCC; CAS 65996-62-5; starch for batters and breadings; provides exc. adhesion to meat and vegetables; does not interfere with delicate seasonings or color; imparts crisp uniform coating to finished prod.; *Regulatory:* FDA 21CFR §172.892; *Properties:* wh. powd., odorless, bland flavor; 10% max. on 325 mesh; pH 5.0-6.5; 9-11% moisture.

Pureco® 76. [Karlshamns Food Ingreds.] Coconut oil; CAS 8001-31-8; EINECS 232-282-8; speicalty oil for pail coating, corn popping, nut roasting, mellorine, general purpose food applics.; *Properties:* Lovibond R1.5 max. liq.; m.p. 76 F; iodine no. 12 max.; sapon. no. 248-264.

Pureco® 92. [Karlshamns Food Ingreds.] Partially hydrog. coconut oil; EINECS 284-283-8; lauric fat for frying, vegetable dairy, biscuits, crackers; *Properties:* Lovibond R1.5 liq., plastic; m.p. 100-104 F; iodine no. 4 max.; sapon. no. 248-264.

Pureco® 100. [Karlshamns Food Ingreds.] Partially hydrog. coconut oil; EINECS 284-283-8; lauric fat for frying, vegetable dairy, biscuits, and crackers; *Properties:* liq. plastic; drop pt. 100 F.

Pureco® 110. [Karlshamns Food Ingreds.] Partially hydrog. coconut and palm oils; lauric fat for dressings, cream centers, biscuits, crackers; *Properties:* Lovibond R1.5 liq.; m.p. 112-115 F; iodine no. 4 max.; sapon. no. 246-262.

Pure-Dent® B700. [Grain Processing] Unmodified corn starch USP, NF; CAS 9005-25-8; EINECS 232-679-6; thickener for salad dressings, cooked puddings, bakery fillings, gravies; *Properties:* off-wh. powd., no odor, bland flavor; pH 5.5-6.5; 9-12.5% moisture.

Pure-Dent® B810. [Grain Processing] Corn starch FCC, NF; CAS 9005-25-8; EINECS 232-679-6; food additive complying with regs. for food starch-modified; high visc., exc. gel str.; *Regulatory:* FDA 21CFR §172.892; *Properties:* wh. powd., no odor, bland flavor; pH 4.5-7.0; 8-11% moisture.

Pure-Dent® B812. [Grain Processing] Corn starch USP; CAS 9005-25-8; EINECS 232-679-6; food additive complying with regs. for food starch-modified; *Regulatory:* FDA 21CFR §172.892; *Properties:* wh. powd., odorless, bland flavor; pH 6.0; 11% moisture.

Pure-Dent® B815. [Grain Processing] Corn starch NF; CAS 9005-25-8; EINECS 232-679-6; food additive complying with regs. for food starch-modified; *Regulatory:* FDA 21CFR §172.892; *Properties:* wh. powd., odorless, bland flavor; pH 4.5-7.0; 8-11% moisture.

Pure-Dent® B816. [Grain Processing] Topical corn starch USP; CAS 9005-25-8; EINECS 232-679-6; food additive complying with regs. for food starch-modified; *Regulatory:* FDA 21CFR §172.892; *Properties:* wh. powd., odorless, bland flavor; pH 6.0; 11% moisture.

Pure-Dent® B880. [Grain Processing] Corn starch NF; CAS 9005-25-8; EINECS 232-679-6; food additive complying with regs. for food starch-modified; *Regulatory:* FDA 21CFR §172.892; *Properties:* wh. powd., odorless, bland flavor; pH 6.0; 11% moisture.

Pure-Dent® B890. [Grain Processing] Food starch modified NF; CAS 53124-00-8; food additive complying with regs. for food starch-modified; *Regulatory:* FDA 21CFR §172.892; *Properties:* wh. powd., odorless, bland flavor; pH 6.0; 11% moisture.

Pure Food Powd. Starch 105-A. [ADM Corn Processing] Corn starch; CAS 9005-25-8; EINECS 232-679-6; for puddings, salad dressing, prepared mixes, gravies; *Properties:* wh. powd.; 98+% thru 200 mesh; pH 5.5; 10% moisture, 0.3% protein.

Pure Food Powd. Starch 131-C. [ADM Corn Processing] Corn starch; CAS 9005-25-8; EINECS 232-679-6; for brewing; *Properties:* wh. powd.; 98+% thru 200 mesh; pH 5.0; 10% moisture, 0.3% protein.

Pure Food Starch Bleached 142-A. [ADM Corn Processing] Corn starch; CAS 9005-25-8; EINECS 232-679-6; for bakery prods.; *Properties:* wh. (bleach) powd.; 98+% thru 200 mesh; pH 5.5; 10% moisture, 0.3% protein.

Pure-Gel® B990. [Grain Processing] Modified dent corn starch, CA3 9005-25-8, EINECS 232-679-6; starch for food systems requiring heat, shear, or acid stability; imparts exc. freeze/thaw stability; thickener for frozen foods, soups, sauces, gravies, fruit pie fillings, emulsified meats, ham prods.; *Regulatory:* FDA 21CFR §172.892; *Properties:* wh. powd., odorless, bland flavor; pH 6.0; 10.5% moisture.

Pure-Gel® B992. [Grain Processing] Modified dent food starch FCC; CAS 53124-00-8; starch for food systems requiring heat, shear, or acid stability; imparts exc. freeze/thaw stability; thickener for frozen foods, soups, sauces, gravies, fruit pie fillings requiring additional processing tolerance; *Regulatory:* FDA 21CFR §172.892; *Properties:* wh. powd., typ. odor, bland flavor; pH 6.0; 10.5% moisture.

Pure Malt Colorant A6000. [MLG Enterprises Ltd.] Nondiastatic extract of malted barley; CAS 8002-48-0; EINECS 232-310-9; natural colorant providing rich deep reddish-brn. to baked goods, breakfast cereals, cookies, snack goods, beverages, confections, crackers, spices, sauces; *Regulatory:* kosher; *Properties:* black liq.; pH 4.5-5.5 (10%); 79-80% solids; *Storage:* store in clean, dry area @ 40-65 F; avoid direct sunlight; 6 mos max. shelf life.

Pure Malt Colorant A6001. [MLG Enterprises Ltd.] Nondiastatic extract of malted barley; CAS 8002-48-0; EINECS 232-310-9; natural colorant providing rich deep reddish-brn. to baked goods, breakfast cereals, cookies, snack goods, beverages, confections, crackers, spices, sauces; *Regulatory:* kosher; *Properties:* dk. red to black free-flowing powd.; hygroscopic; pH 4.5-5.5 (10%); 4% moisture; *Storage:* store in clean, dry area @ 40-65 F; 6 mos max. shelf life.

Pure/Riviera Olive Oil NF. [Arista Industries] Olive oil; CAS 8001-25-0; EINECS 232-277-0; food ingred.; *Properties:* greenish-yel. bright and clear oily liq.; sp.gr. 0.910-0.915; solid. pt. 17-26 C; acid no. 0.5; iodine no. 79-88; sapon. no. 190-195.

Pure-Set® B950. [Grain Processing] Food starch modified FCC; CAS 65996-63-6; starch for confectionery use; provides gel str. props. for gum candies, jellies; binder for frankfurters and bologna; *Regulatory:* FDA 21CFR §172.892; *Properties:* wh. powd., sl. odor, bland flavor; pH 5.5-6.5; 9-12% moisture.

Pure-Set™ B965. [Grain Processing] Flash-dried acid-modified food starch FCC; CAS 65996-63-6; starch for confectionery use; provides gel str. and texture props. for gum candies, jellies; *Properties:* wh. powd., sl. odor, bland flavor; pH 5.5-6.5; 9-12% moisture.

Purex® All Purpose Salt. [Morton Salt] Sodium chloride FCC; CAS 7647-14-5; EINECS 231-598-3; food-grade salt; used for ham, bacon, corned beef prods.; *Properties:* unscreened granulated with cubic crystals; 25-57% retained on 50 mesh, 10-37% on 70 mesh; mean crystal size 410 μm; mean surf. area 70 cm²/g; bulk dens. 1.17-1.28 g/ml (loose); 99.82% NaCl.

PWA™. [Calgon Carbon] Activated carbon; CAS 64365-11-3; used for food decolorization; *Properties:* pulverized.

Py-ran®. [Monsanto] Monocalcium phosphate anhyd.; CAS 7758-23-8; functional food additive; buffer, dietary supplement, dough conditioner, firming agent, leavening agent, sequestrant, stabilizer in cereals, dough, preserves, fruit jelly, pkg. materials; *Regulatory:* FCC compliance; *Properties:* wh. powd.; 0.5% max. on 100 mesh, 15% max. on 200 mesh; 16.8-18.3% assay (as Ca).

Pyridoxine Hydrochloride USP, FCC Fine Powd. No. 60650. [Roche] Vitamin B₆; CAS 58-56-0; EINECS 200-386-2; dietary supplement for mfg. of dry and liq. pharmaceuticals and in vitamin fortification of foods; *Regulatory:* FDA 21CFR §182.5676, GRAS; *Properties:* wh. cryst. powd., odorless; 95% min. thru 100 mesh; freely sol. in water (1 g/5ml); sl. sol. in alcohol (1 g/100 ml); insol. in ether; m.w. 205.64; m.p. 202-206 C; pH 2-4 (10% aq.); 98% min. assay; *Toxicology:* ingestion of 500 mg or more daily for 6 mos may cause ataxia and severe neuropathy; *Precaution:* avoid exposure to excessive heat, alkali, light; *Storage:* store in tightly closed containers away from light, moisture, excessive heat @ optimum temp. < 72 F.

Q

Quaker™ Oatrim 5, 5Q. [Rhone-Poulenc Food Ingreds.] Phosphoric acid; CAS 7664-38-2; EINECS 231-633-2; used for baking, cereal, meat, poultry, seafood, dairy, cheese, processed foods, beverages, and confections; *Properties:* lt. creamy wh.; bulk dens. ≥ 0.3 g/cc; visc. ≥ 20 cps (5%); pH 6.3-7.7; DE ≤ 5.

Qual Flo™. [Qualcepts Nutrients] Powd. cellulose FCC; CAS 9004-34-6; EINECS 232-674-9; emulsifier, stabilizer, and flow agent for foodstuffs; designed for fast turnover foods that do not require prolonged shelf life, e.g., institutional size bags for pizzeria, restaurant, and retail trades; for mozzarella, provolone cheeses; *Properties:* wh. fibrous powd., odorless; 0% 300 μ residue, 30% max. 32 μ residue; 40 μ avg. fiber length, 18 μ avg. fiber thickness; partly sol. in water; sp.gr. 1.5 g/cc; apparent dens. 185-220 g/l; pH 5-7.5; 99.5% act.; *Toxicology:* inhalation of dust may cause allergenic reactions on repeated exposure; *Storage:* store in cool, dry area.

Qual Flo™ C. [Qualcepts Nutrients] Powd. cellulose FCC, annatto; emulsifier, stabilizer, and flow agent for foodstuffs; designed for fast turnover foods that do not require prolonged shelf life, e.g., institutional size bags for pizzeria, restaurant, and retail trades; for cheddar, Colby cheeses; *Properties:* yel.-orange fibrous powd., odorless; 0% 300 μ residue, 30% max. 32 μ residue; 40 μ avg. fiber length, 18 μ avg. fiber thickness; partly sol. in water; sp.gr. 1.5 g/cc; apparent dens. 185-220 g/l; pH 5-7.5; 99.5% act.; *Toxicology:* inhalation of dust may cause allergenic reactions on repeated exposure; *Storage:* store in cool, dry area.

Qual Guard™. [Qualcepts Nutrients] Oxygen scavenging enzyme system without anticaking agents; oxygen scavenging enzyme system which protects packaged prods. from mold by reducing oxygen content; used for tortillas, pasta, cold pack cheese, frozen entrees, etc.; *Usage level:* 0.25% (block cheese), 0.5% (shredded cheese); *Properties:* wh. powd.

Qual Guard™ 100. [Qualcepts Nutrients] Dextrose, enzymes; oxygen scavenging enzyme system which protects packaged prods. from mold by reducing oxygen content; used for block or shredded cheese, esp. large blocks or bulk containers; *Usage level:* 0.25% (block cheeses), 0.5% (shredded cheese); *Properties:* wh. powd.

Qual Guard™ WB. [Qualcepts Nutrients] Whey, enzymes; oxygen scavenging enzyme system which protects packaged prods. from mold by reducing oxygen content; designed for use in high moisture applics., e.g., cold pack cheese; *Usage level:* 3% by wt. of finished prod.; *Properties:* wh. powd.

Quattro. [Brolite Prods.] Plastic bread base for long French and Italian bread; provides dough which is very elastic and pliable and easily rolled into long loaves; *Usage level:* 4-5 lb/100 lb flour.

Querton 210Cl-50. [Berol Nobel AB] Didecyl dimonium chloride; CAS 7173-51-5; EINECS 270-331-5; bactericide, fungicide for food processing industry, breweries.

Querton 210Cl-80. [Berol Nobel AB] Didecyl dimonium chloride; CAS 7173-51-5; EINECS 270-331-5; bactericide, fungicide for food processing industry, breweries.

Quick Chocolate. [Brolite Prods.] An instant chocolate flavored cream filling.

Quickset® D-4. [Hormel] Gelatin, Type A, 150 bloom; CAS 9000-70-8; EINECS 232-554-6; used in sugar-sweetened gelatin desserts when a softer set yet higher protein is desired; *Properties:* ivory powd., weak bouillon-like odor; 50 mesh; visc. 24 ± 2 mps; pH 4.5-5.8; 12% max. moisture; *Storage:* stable for up to 1 yr when stored dry at ambient temps.; keep containers tightly closed.

Quickset® D-5. [Hormel] Gelatin, Type A, 175 bloom; CAS 9000-70-8; EINECS 232-554-6; used in sugar-sweetened gelatin desserts when a sl. softer set is desired; *Properties:* ivory powd., weak bouillon-like odor; 50 mesh; visc. 28 ± 2 mps; pH 4.5-5.8; 12% max. moisture; *Storage:* stable for up to 1 yr when stored dry at ambient temps.; keep containers tightly closed.

Quickset® D-6. [Hormel] Gelatin, Type A, 200 bloom; CAS 9000-70-8; EINECS 232-554-6; used in sugar-sweetened gelatin desserts and for production that requires rapid dissolving when clumping of gelatin can be controlled; *Properties:* ivory powd., weak bouillon-like odor; 50 mesh; visc. 32 ± 2 mps; pH 4.5-5.8; 12% max. moisture; *Storage:* stable for up to 1 yr when stored dry at ambient temps.; keep containers tightly closed.

Quickset® D-7. [Hormel] Gelatin, Type A, 225 bloom;

CAS 9000-70-8; EINECS 232-554-6; used in sugar-sweetened gelatin desserts and for production that requires rapid dissolving when clumping of gelatin can be controlled; *Properties:* ivory powd., weak bouillon-like odor; 50 mesh; visc. 37 ± 2 mps; pH 4.5-5.8; 12% max. moisture; *Storage:* stable for up to 1 yr when stored dry at ambient temps.; keep containers tightly closed.

Quickset® D-8. [Hormel] Gelatin, Type A, 250 bloom; CAS 9000-70-8; EINECS 232-554-6; used in sugar-sweetened gelatin desserts and for production that requires rapid dissolving when clumping of gelatin can be controlled; *Properties:* ivory powd., weak bouillon-like odor; 50 mesh; visc. 42 ± 2 mps; pH 4.5-5.8; 12% max. moisture; *Storage:* stable for up to 1 yr when stored dry at ambient temps.; keep containers tightly closed.

Quickset® D-9. [Hormel] Gelatin, Type A, 275 bloom; CAS 9000-70-8; EINECS 232-554-6; used in sugar-sweetened gelatin desserts and for production that requires rapid dissolving when clumping of gelatin can be controlled; *Properties:* ivory powd.,

weak bouillon-like odor; 50 mesh; visc. 47 ± 2 mps; pH 4.5-5.8; 12% max. moisture; *Storage:* stable for up to 1 yr when stored dry at ambient temps.; keep containers tightly closed.

Quickset® D-10. [Hormel] Gelatin, Type A, 300 bloom; CAS 9000-70-8; EINECS 232-554-6; used in sugar-sweetened gelatin desserts and for production that requires rapid dissolving when clumping of gelatin can be controlled; *Properties:* ivory powd., weak bouillon-like odor; 50 mesh; visc. 54 ± 4 mps; pH 4.5-5.8; 12% max. moisture; *Storage:* stable for up to 1 yr when stored dry at ambient temps.; keep containers tightly closed.

Quick Vanilla. [Brolite Prods.] An instant vanilla/custard flavored cream filling.

Quick Whip. [Brolite Prods.] Whipped topping base for various toppings and fillings; *Properties:* powd.

Quik Release. [ADM Arkady] Highly effective release agent for the confectionery industry.

Quso® WR55-FG. [Degussa] Precipitated silica; for food grade defoamers; *Regulatory:* FDA approved.

R

R.B.-4 Sour. [Bunge Foods] Specialized sour providing controlled fermentation, added volume, ease in scaling, light crispy crust, tender crumb, traditional rye flavor; for hearty rye breads, sandwich rye, rye rolls; *Usage level:* 10% based on flour.

RC®. [Calgon Carbon] Activated carbon; CAS 64365-11-3; used for food decolorization; *Properties:* pulverized.

Readi-Glaze Systems. [Advanced Food Systems] W/ o emulsions designed for microwave or convection cooking of frozen fish, shellfish, chicken, red meat, vegetable, other food prods. for retail and food service operations; reduces freeze burn, dehydration, and flavor deterioration; various flavors.

Readi-Ice Icing Powd. [Bunge Foods] Boiling or nonboiling icing stabilizer for extra white and creme icing; can be formulated for frozen baked goods; *Usage level:* 8-10 lb/100 lb powd. sugar; *Properties:* off-wh. free-flowing powd., bland flavor; pH 7.5.

Recodan. [Grindsted Prods. Denmark] Integrated emulsifiers and stabilizers; nonionic; emulsifier, heat stabilizer; improves fat disp., enhances palatability in sterilized or pasteurized, recombined, filled, imitation, chocolate, and flavored milk prods.; *Properties:* powd.; 100% conc.

Red Beet WSL-300. [Hilton Davis] Red beet extract with ascorbic and citric acids; natural colorant for puddings, yogurts, sherbets, ice creams, frozen and gelatin desserts, frostings, and candy; color str. 0.3% betanin; *Properties:* red to redsh blue liq; water-sol.

Red Beet WSL-400. [Hilton Davis] Red beet extract with ascorbic and citric acids; natural colorant for puddings, yogurts, sherbets, ice creams, frozen and gelatin desserts, frostings, and candy; color str. 0.4% betanin; *Properties:* red to redsh. blue liq; water-sol.

Red Beet WSP-300. [Hilton Davis] Red beet extract with ascorbic and citric acids dried with maltodextrin; natural colorant for dry mixes; color str. 0.3% betanin; *Properties:* red to redsh blue powd.; water-sol.

Redihop®. [Pfizer Food Science] Highly purified aq. sol'n. of the potassium salt of reduced isoalpha acids standardized @ 35% w/w, from hops extract; brewery additive which provides a smooth bitterness with less after-bitter and improved light stabil-ity for light or reduced-calorie beers; for post-fermentation injection; *Storage:* store at 20-22 C; may precipitate during storage.

Redisol® 78D. [A.E. Staley Mfg.] Food starch modified, derived from potato; pregelatinized starch, thickener producing med. flow, smooth, cream-like texture; film-former on dry roasted peanuts; in toppings, icings, cheese sauces, as meringue stabilizer; *Regulatory:* FDA 21CFR §172.892; *Properties:* wh. powd.; 5% on 60 mesh, 50% thru 200 mesh; pH 6; 6% moisture.

Redisol® 88. [A.E. Staley Mfg.] Food starch modified, derived from tapioca; pregelatinized starch, thickener, stabilizer providing heavy flow, smooth, creamy texture, sl. chewy tender mouthfeel; for instant foods, dry mixes, puddings, icings, pie fillings, chip dips, soup mixes, spreads; *Regulatory:* FDA 21CFR §172.892; *Properties:* powd.; 98% thru 100 mesh, 85% thru 200 mesh; pH 6.0; 6% moisture.

Redisol® 248. [A.E. Staley Mfg.] Food starch modified, derived from potato; pregelatinized starch, thickener, bulking agent providing smooth, visc. sol'ns., good acid resist.; for milk-gel instant puddings, cream soup mixes, sauces, gravies, snack dips, spreads; *Regulatory:* FDA 21CFR §172.892; kosher, Passover certified; *Properties:* powd.; 0.2% on 140 mesh, 92% thru 200 mesh; cold water sol.; pH 6.0; 7% moisture.

Redisol® 412. [A.E. Staley Mfg.] Food starch modified, derived from tapioca; pregelatinized starch, thickener, stabilizer for convenience foods, instant puddings, cake mixes, icings, nonbaked refrigerated desserts; produces smooth, soft, tender gels of short texture; *Regulatory:* FDA 21CFR §172.892; *Properties:* powd.; 98% thru 100 mesh, 85% thru 200 mesh; pH 6.0; 7% moisture.

Redi-Spread Pastry Fillings. [ADM Milling] Ready-to-use pastry filling for filled Danish pastry, fried yeast-raised sweet goods, cookies, cake filling.

Redi-Tex®. [A.E. Staley Mfg.] Food starch modified, derived from dent corn; cold water swelling starch providing pulpy texture in hot or cold processed food systems, apple sauce, tomato paste/sauce, apricot puree, sauces, gravies, dry mixes; extender for tomato pastes, juices, soups, stews, sauces; extends coconut in candies; *Regulatory:* FDA 21CFR §172.892; *Properties:* wh. flakes, bland flavor; 2% on 10 mesh, 97% on 100 mesh, 3% thru 100 mesh; visc. 34,000 cps; pH 6.0 (10%); 5-6% moisture.

Red Miso Powd. [Nikken Foods Co. Ltd.] Fermented soy beans and barley with salt; provides a salty, meaty flavor with bitter overtones; used for oriental sauces, soups, chili, spreads, and as replacement for MSG or HVP; most effective in beef or pork applics.; *Properties:* brn. freeze-dried free-flowing fine powd., mild typical fermented soy aroma; 4% moisture, 20.6% protein, 20% salt, 9% fat; *Storage:* store in cool, dry, well-ventilated area in sealed containers; avoid storage at elevated temps.

Red Soluble Powd. Natural Colorant. [MLG Enterprises Ltd.] Blend of solubilized carminic acid aluminum lake with norbixin; natural food colorant producing berry red shades for instant/dehydrated foods, soups, sauces, dressings, beverages, desserts, custards, cake mixes, ice cream, candy coatings; *Usage level:* 0.03-0.06%; *Properties:* impalpable powd.; 100% 80 mesh; water-sol.; $32 \pm 0.7\%$ (as carminic acid); *Storage:* keep in tightly closed containers away from light and in dry and preferably refrigerated room.

Reduce®-150. [Am. Ingreds.] Blend of sodium stearoyl lactylate, calcium sulfate, and sodium sulfite; dough conditioner for use in flour tortillas, pie crusts, and pizza shells; *Usage level:* 4-16 oz/ cwt flour; *Properties:* lt. tan free-flowing powd., mild caramel odor; acid no. 30-40; 11-15% calcium, 2.0-2.8% sodium; *Storage:* store under cool, dry conditions.

Regent® 12XX. [Rhone Poulenc Food Ingreds.] Monocalcium phosphate, monohydrate FCC; CAS 10031-30-0; EINECS 231-837-1; acidulant, leavening agent for baking, cereal, pancake mixes, cookie mixes, angel food cakes, beverages; very fast action; used with other acids in double-acting baking powds.; as bread improver in yeast-raised doughs; *Properties:* brilliant wh. free-flowing cryst.; 99% thru 100 mesh, 87% thru 200 mesh; bulk dens. 66 lb/ft³; neutralizing value 80; pH 3.7 (1%); 15.9-17% assay (Ca).

Regular Paniplus. [ADM Arkady] Calcium peroxide; CAS 1305-79-9; EINECS 215-139-4; oxidation-type dough conditioner which improves machinability and yield through increased absorption; improves grain, texture, and crumb color of baked foods; also avail. in double strength.

Remy BLM7-FG. [Remy Industries SA] Rice starch; neutral starch for confectionery molding, dusting of chewing gum, baking powd.; *Properties:* wh. finely milled powd., neutral taste.

Remy CX. [Remy Industries SA] Rice starch; neutral starch for confectionery, fancy foods; *Properties:* wh. oven-dried crystals.

Remy DR. [Remy Industries SA] Rice starch; milled neutral starch for foods, confectionery, chocolate filling, biscuits, extruded snacks, custards, frozen vegetables with cream sauce; *Properties:* wh. powd., neutral taste; 7% max. on 200 mesh; sol. in hot water; insol. in cold water; bulk dens. 0.40 g/

cm³; pH 6.0-7.5 (10%); 15% max. moisture, 0.55% max. protein, 0.1% max. fat; *Toxicology:* nonhazardous; *Precaution:* dust explosion is possible; *Storage:* 5 yr min. storage life when stored in unopened original packing in dry place free from odors, insects, rodents.

Remygel AX. [Remy Industries SA] Modified waxy rice starch; binder, thickener, stabilizer, oil/fat substitute in baby foods, sauces, soups, salad dressings, frozen foods, preserves, low-fat margarine/ butter; when cooked, forms a creamy gel which resists freezing, sterilization, acid environments; *Properties:* wh. powd., neutral taste; 7% max. on 200 mesh; sol. in hot water; insol. in cold water; bulk dens. 0.40 g/cm³; pH 6.0-7.5 (10%); 15% max. moisture, 0.55% max. protein, 0.1% max. fat; *Toxicology:* nonhazardous; *Precaution:* dust explosion is possible; *Storage:* 5 yr min. storage life in unopened original packing in dry place free from odors, insects, and rodents.

Remygel AX-2. [Remy Industries SA] Modified waxy rice starch; binder, thickener, stabilizer, oil/fat substitute in food specialties, baby foods, sauces, soups, salad dressings, frozen foods, preserves, low-fat margarine/butter; oven-stable, less sensitive to shear stress; suitable for fruit preps.; *Properties:* wh. powd., neutral taste; 7% max. on 200 mesh; sol. in hot water; insol. in cold water; bulk dens. 0.40 g/cm³; pH 6.0-7.5 (10%); 15% max. moisture, 0.55% max. protein, 0.1% max. fat; *Toxicology:* nonhazardous; *Precaution:* dust explosion is possible; *Storage:* 5 yr min. storage life in unopened original packing in dry place free from odors, insects, and rodents.

Remygel NBO. [Remy Industries SA] Rice starch; starch for preserves; max. level of visc. produced only during sterilization; *Properties:* wh. powd., neutral taste.

Remy Instant. [Remy Industries SA] Rice starch; pregelatinized starch for baby foods, food specialties, instant foods; *Properties:* wh. powd., neutral taste; cold water sol.

Remyline AX. [Remy Industries SA] Native waxy rice starch; binder, thickener, stabilizer, oil/fat substitute in food specialties, baby foods, sauces, soups, salad dressings, frozen foods, preserves, low-fat margarine/butter; when cooked, forms a smooth gel which resists freezing; *Properties:* wh. powd., neutral taste; 7% max. on 200 mesh; sol. in hot water; insol. in cold water; bulk dens. 0.40 g/cm³; pH 6.0-7.5 (10%); 15% max. moisture, 0.55% max. protein, 0.1% max. fat; *Toxicology:* nonhazardous; *Precaution:* dust explosion is possible; *Storage:* 5 yr min. storage life in unopened original packing in dry place free from odors, insects, and rodents.

Resinogum DD. [MLG Enterprises Ltd.] Gum resin damar purified and deodorized; emulsifier, stabilizer; *Properties:* yel. flakes; sp.gr. 1.04-1.08; m.p. 75-125 C.

Response®. [Central Soya] Textured soy conc.; protein for use in cooked and raw beef patties, chicken or veal patties, pizza toppings, ground meat foods, coarse ground sausages, meatballs, restructured meats, and seafood.

Rex® Vitamin Fortified Wheat Germ Oil. [Viobin] Wheat germ oil, fortified with vitamins A and D; nutritonally rich veg. oil contg. ≈ 6% lecithin-rich phosphatides, 4.7% unsaponifiable material rich in octacosonal, vitamin E, active tocopherols, sitosterols, and dihydrositosterol; *Properties:* oil.

Rezista®. [A.E. Staley Mfg.] Food starch modified, derived from waxy maize; cook-up starch, thickener, stabilizer for acidic foods which are held at high temps. for long periods; for tart fruit pie fillings, glazes, baker's jellies, yogurt fruit bases; provides good shelf life, freeze/thaw stability, resist. to syneresis; *Regulatory:* FDA 21CFR §172.892; *Properties:* wh. powd., bland flavor; pH 5.0 (uncooked); 12% moisture.

Rhodacal® DS-4. [Rhone-Poulenc Surf. & Spec.] Sodium dodecylbenzene sulfonate; CAS 25155-30-0; EINECS 246-680-4; anionic; surfactant for washing fruits and vegs.; *Regulatory:* FDA compliance; *Properties:* dk. brn. liq.; surf. tens. 32 dynes/cm (@ CMC); 23% act.; formerly Siponate® DS-4.

Rhodacal® LDS-10. [Rhone-Poulenc Surf. & Spec.] Sodium dodecylbenzene sulfonate; EINECS 246-680-4; anionic; emulsifier, surfactant for washing fruits and vegetables; *Properties:* flake; pH 7.5 (10%); surf. tens. 32 dynes/cm (@ CMC); 98% act.; formerly Siponate® LDS-10.

Rhodapon® BOS. [Rhone-Poulenc Surf. & Spec.] Sodium 2-ethylhexylsulfate; CAS 126-92-1; EINECS 204-812-8; anionic; surfactant for fruit and vegetable washing; *Regulatory:* FDA compliance; *Properties:* clear liq.; visc. 50 cps; cloud pt. < 10 C; pH 9.5-10.5 (10%); surf. tens. 33 dynes/cm (@ CMC); 39-40% act.; formerly Sipex® BOS.

Rhodiarome™. [Rhone-Poulenc Food Ingreds.] Ethyl vanillin; CAS 121-32-4; EINECS 204-464-7; flavor for baking, cereal, dairy, cheese, processed foods, beverages, confections.

Rhodigel®. [Rhone-Poulenc Food Ingreds.; R.T. Vanderbilt] Xanthan gum; CAS 11138-66-2; EINECS 234-394-2; emulsion stabilizer, suspending agent, thickener for foods; freeze-thaw and pH stable (3-9); synergistic with guar and locust bean gums; rapid visc. buildup, good dispersibility; *Usage level:* 0.15-0.5% (salad dressing), 0.05-0.25% (cake mixes), 0.1-0.3% (sauces), 0.1-0.25% (relish), 0.03-0.15% (fruit fillings), 0.05-0.15% (fruit beverages), 0.3-0.5% (dry instant soups); *Regulatory:* kosher; *Properties:* fine gran.; sol. in cold water.

Rhodigel® 200. [Rhone-Poulenc Food Ingreds.; R.T. Vanderbilt] Xanthan gum; CAS 11138-66-2; EINECS 234-394-2; emulsion stabilizer, suspending agent, thickener for foods; freeze-thaw and pH stable (3-9); synergistic with guar and locust bean gums; instant visc. buildup, fair dispersibility; *Usage level:* 0.15-0.5% (salad dressing), 0.05-0.25% (cake mixes), 0.1-0.3% (sauces), 0.1-0.25% (relish), 0.03-0.15% (fruit fillings), 0.05-0.15% (fruit beverages), 0.3-0.5% (dry instant soups); *Regulatory:* kosher; *Properties:* very fine gran.; sol. in cold water.

Rhodigel® EZ. [Rhone-Poulenc Food Ingreds.] Xanthan gum; CAS 11138-66-2; EINECS 234-394-2; emulsion stabilizer, suspending agent, thickener for foods; freeze-thaw and pH stable (3-9); synergistic with guar and locust bean gums; very fast visc. buildup, very good dispersibility; *Usage level:* 0.15-0.5% (salad dressing), 0.05-0.25% (cake mixes), 0.1-0.3% (sauces), 0.1-0.25% (relish), 0.03-0.15% (fruit fillings), 0.05-0.15% (fruit beverages), 0.3-0.5% (dry instant soups); *Properties:* coarse gran.; sol. in cold water.

Rhodigel® Granular. [Rhone-Poulenc Food Ingreds.] Xanthan gum; CAS 11138-66-2; EINECS 234-394-2; used for baking, cereal, meat, poultry, seafood, dairy, cheese, processed foods, beverages, confections.

Rhodigel® Supra. [Rhone-Poulenc Food Ingreds.] Xanthan gum; CAS 11138-66-2; EINECS 234-394-2; emulsion stabilizer, suspending agent, thickener for foods; freeze-thaw and pH stable (3-9); synergistic with guar and locust bean gums; slow visc. buildup, exc. dispersibility; *Usage level:* 0.15-0.5% (salad dressing), 0.05-0.25% (cake mixes), 0.1-0.3% (sauces), 0.1-0.25% (relish), 0.03-0.15% (fruit fillings), 0.05-0.15% (fruit beverages), 0.3-0.5% (dry instant soups); *Regulatory:* kosher; *Properties:* very coarse gran.; sol. in cold water.

Rhodigum OEH. [Rhone-Poulenc Food Ingreds.] Xanthan gum, guar gum, locust bean gum system; texturizing, stabilizing, and gelling agent for salsa, pickles, relish, salad dressings, sauces.

Rhodigum OEM. [Rhone-Poulenc Food Ingreds.] Xanthan gum, guar gum, locust bean gum system; texturizing, stabilizing, and gelling agent for dips, barbecue sauce, pickles, relish, whipped toppings, creams, custards, chocolate mousse.

Rhodigum WGH. [Rhone-Poulenc Food Ingreds.] Xanthan gum, guar gum, locust bean gum system; texturizing, stabilizing, and gelling agent for chocolate drinks, bakery icings, gelled broth, nonfat imitation cheese.

Rhodigum WGL. [Rhone-Poulenc Food Ingreds.] Xanthan gum, guar gum, locust bean gum system; texturizing, stabilizing, and gelling agent for salsa, creams, custards, canned puddings, fruit fillings, cheese spreads.

Rhodigum WGM. [Rhone-Poulenc Food Ingreds.] Xanthan gum, guar gum, locust bean gum system; texturizing, stabilizing, and gelling agent for dips, canned puddings, fruit fillings, cheese spreads.

Rhodigum WVH. [Rhone-Poulenc Food Ingreds.] Xanthan gum, guar gum, locust bean gum system; texturizing, stabilizing, and gelling agent for instant soups/sauces/beverages, hot cocoa mix, instant puddings, cake mixes, frozen drinks.

Rhodigum WVM. [Rhone-Poulenc Food Ingreds.] Xanthan gum, guar gum, locust bean gum system; texturizing, stabilizing, and gelling agent for barbecue sauce, sauces, instant beverages, hot cocoa mix, instant puddings, cake mixes, batters, breading.

Rhozyme® HP-150 Conc. [Genencor] Pentosanase-hexosanase; food grade enzyme for hydrolysis of veg. gums; *Properties:* powd.; Unverified

Rhozyme® P11. [Genencor] Protease; CAS 9014-01-1; EINECS 232-752-2; food grade enzyme for meat tenderization, removing flesh from bones; *Properties:* powd.; Unverified

Rhozyme® P41. [Genencor] Protease; CAS 9014-01-1; EINECS 232-752-2; food grade enzyme for baking operations; *Properties:* powd.; Unverified

Rhozyme® P53, P64. [Genencor] Protease; CAS 9014-01-1; EINECS 232-752-2; food grade enzyme for protein hydrolysis; *Properties:* powd. and liq. resp.; Unverified

Riboflavin USP, FCC No. 602940002. [Roche] Riboflavin USP, FCC; CAS 83-88-5; EINECS 201-507-1; source of vitamin B_2 in liq. and solid food and pharmaceutical preps.; *Properties:* orange-yel. cryst. powd., sl. odor, char. bitter taste; < 20 μ particle size; very sol. in dil. alkalies; very sl. sol. in water, alcohol, isotonic sodium chloride sol'n.; insol. in ether, chloroform; m.w. 376.37; m.p. 280 C; 98-102% assay; *Storage:* store in dry place @ 59-86 F in tight, light-resist. containers; avoid excessive heat.

Riboflavin-5´-Phosphate Sodium USP, FCC No. 60296. [Roche] Riboflavin-5´-phosphate sodium USP, FCC; CAS 130-40-5; EINECS 204-988-6; source of riboflavin in foods where it is essential to obtain potencies above the solubility of riboflavin; *Properties:* orange-yel. fine cryst., sl. odor, bitter taste; sol. 4.3 g/100 ml @ pH 3.8, 11.2 g/100 ml water @ pH 7, m.w. 514.37, pH 5-6.5 (1:100 aq.); 73-79% assay (riboflavin); *Precaution:* incompat. with sol'ns. contg. calcium or metallic salts which form insoluble phosphates; *Storage:* store in dry place @ 59-86 F in tight, light-resist. containers; avoid heat and prolonged exposure to light.

Rice Bran Oil. [Ikeda] Rice bran oil; EINECS 271-397-8; nutritive oil providing cholesterol control; for foods; *Properties:* acid no. 0.1 max.; Iodine no. 102-110; sapon. no. 185-195; cloud pt. -4 C max.

Rice Bran Oil SO. [Tsuno Rice Fine Chems.; Tri-K Industries] Rice bran oil; EINECS 271-397-8; suitable for food use; *Properties:* lt. yel. liq.; insol. in water; sp.gr. 0.916-0.922; b.p. 310 C; acid no. 0.1 max.; iodine no. 92-115; sapon. no. 180-195; cloud pt. -7 C max.; ref. index 1.470-1.473; trace moisture; *Toxicology:* nonhazardous; edible; *Storage:* store in cool area below 30 C.

Rice Complete® 3. [Zumbro; Garuda Int'l.] 10% Rice protein, 90% rice maltodextrin (3DE); mimics mouthfeel of fat; provides creamy texture in water; gels in cold water above 15%; for fat/oil replacement in salad dressings, baked goods, frostings, soups, gravies, dips; adhesive for granola, nut coatings; bodying agent for beverages; *Properties:* creamy wh. powd., bland odor, flavor; 100% thru 140 mesh; 10% rice protein, 8% max. moisture; *Storage:* store in cool, dry area.

Rice Complete® 10. [Zumbro; Garuda Int'l.] 10% Rice protein, 90% rice maltodextrin (10DE); mimics

mouthfeel of fat; provides creamy texture in water; gels in cold water above 15%; for fat/oil replacement in salad dressings, baked goods, frostings, soups, gravies, dips; adhesive for granola, nut coatings; bodying agent for beverages; *Properties:* creamy wh. powd., bland odor, flavor; 100% thru 140 mesh; 10% rice protein, 8% max. moisture; *Storage:* store in cool, dry area.

Rice Complete® 18. [Zumbro; Garuda Int'l.] 10% Rice protein, 90% rice maltodextrin (18DE); mimics mouthfeel of fat; provides creamy texture in water; gels in cold water above 15%; for fat/oil replacement in salad dressings, baked goods, frostings, soups, gravies, dips; adhesive for granola, nut coatings; bodying agent for beverages; *Properties:* creamy wh. powd., bland odor, flavor; 100% thru 140 mesh; 10% rice protein, 8% max. moisture; *Storage:* store in cool, dry area.

Rice Complete® 25. [Zumbro; Garuda Int'l.] 10% Rice protein, 90% rice syrup solids (25DE); mimics mouthfeel of fat; provides sweetness; used in baked goods, frostings, nutritional beverages; bodying agent in beverages; *Properties:* cream-colored powd., bland odor, flavor; 100% thru 140 mesh; 10% rice protein, 8% max. moisture; *Storage:* store in cool, dry area.

Rice-Pro® 35W. [Zumbro; Garuda Int'l.] 40% Rice protein, 60% 35DE rice syrup solids; very digestible protein, natural sweetener for baby foods, diet and nutritional beverages, desserts, snacks, confections, frostings, baked goods; low allergenicity; *Properties:* creamy wh. powd., bland odor and flavor; 100% thru 140 mesh; < 6% moisture; *Storage:* store in cool, dry area.

Rice-Trin® 10. [Zumbro; Garuda Int'l.] Rice maltodextrin (10DE); controls body, provides mouthfeel, regulates osmolality; natural carrier for flavors and minor ingreds.; provides bulk without excess sweetness in frostings, baked goods, soups, sauces, salad dressings, coffee whiteners, confections, snacks; *Properties:* creamy wh. powd., bland odor and flavor; 100% thru 140 mesh; sol. in cold or hot water; < 8% moisture; *Storage:* store in cool, dry area.

Rice-Trin® 18. [Zumbro; Garuda Int'l.] Rice maltodextrin (18DE); controls body, provides mouthfeel, regulates osmolality; natural carrier for flavors and minor ingreds.; provides bulk without excess sweetness in frostings, baked goods, soups, sauces, salad dressings, coffee whiteners, confections, snacks; *Properties:* creamy wh. powd., bland odor and flavor; 100% thru 140 mesh; sol. in cold or hot water; < 8% moisture; *Storage:* store in cool, dry area.

Rice-Trin® 25. [Zumbro; Garuda Int'l.] Rice syrup solids (25DE); controls body, provides mouthfeel, regulates osmolality; natural sweetener and humectant for nutritional beverages, confections, snacks, baked goods; *Properties:* creamy wh. powd., bland odor and flavor; 100% thru 140 mesh; sol. in cold or hot water; < 8% moisture; *Storage:* store in cool, dry area.

Rice-Trin® 35. [Zumbro; Garuda Int'l.] Rice syrup solids (35DE); controls body, provides mouthfeel, regulates osmolality; natural sweetener and humectant for nutritional beverages, confections, snacks, baked goods; *Properties:* creamy wh. powd., bland odor and flavor; 100% thru 140 mesh; sol. in cold or hot water; < 6% moisture; *Storage:* store in cool, dry area.

Rich-Pak®. [MLG Enterprises Ltd.] Thiamine hydrochloride, riboflavin, niacin, ferrous sulfate, and other edible excipients; nutrient supplement providing vitamins and iron per 100 lb flour to conform with FDA 21CFR §136.115A(1); *Usage level:* 1 tablet/ cwt flour; *Regulatory:* FDA GRAS; *Properties:* yel. tablet; disp. in R.T. water; 4.984 g ferrous sulfate, 2 g niacin, 0.29 g thiamine HCl, 0.17 g riboflavin/ tablet; *Storage:* 6 mos storage life under cool, dry conditions.

Rich-Pak® Powd. 160. [MLG Enterprises Ltd.] Calcium sulfate, ferrous sulfate, niacin, thiamine HCl, riboflavin, calcium phosphate; nutrient supplement providing vitamins and iron per 100 lb flour to conform with FDA 21CFR §136.115A(1); *Usage level:* $^1/_2$ oz/cwt flour; *Regulatory:* FDA GRAS; *Properties:* yel. free-flowing powd., char. odor; 280 mg thiamine HCl, 170 mg riboflavin, 2000 mg niacin, 4984 mg ferrous sulfate/$^1/_2$ oz; *Storage:* 6 mos storage life under cool, dry conditions.

Rich-Pak® Powd. 160-M. [MLG Enterprises Ltd.] Calcium sulfate, ferrous sulfate, niacin, thiamine mononitrate, riboflavin, calcium phosphate; nutrient supplement providing vitamins and iron per 100 lb flour to conform with FDA 21CFR §136.115A(1); *Usage level:* $^1/_2$ oz/cwt flour; *Regulatory:* FDA GRAS; *Properties:* yel. free-flowing powd., char. odor; 280 mg thiamine nitrate, 170 mg riboflavin, 2000 mg niacin, 4984 mg ferrous sulfate/$^1/_2$ oz; *Storage:* 6 mos storage life under cool, dry conditions.

Rocarna. [Vaessen-Schoemaker] Mixt. of reducing sugars, vitamins, organic and inorganic salts; complete additive for prod. of semi-dried sausage, salami, bacon; maturing/curing agent; used with Vascoferment starter cultures; regulates pH, improves color retention and staiblity; inhibits rancidity; *Usage level:* 10 g Rocarna with 10 g nitrite salt/ 1 kg sausage mix; 5% max. moisture.

Rocoat® Niacinamide 33$^1/_3$% No. 69907. [Roche] Niacinamide USP FCC coated with mono- and diglycerides of edible fatty acids and ≤ 1.5% silicon dioxide; incorporated into chewable multivitamin tablets and other dry dosage forms; provides taste-free and odorless form and protects the vitamin; *Regulatory:* FDA GRAS; *Properties:* wh. relatively free-flowing powd. with some soft agglomerates, sl. char. odor, satisfactory taste; 99% min. thru 20 mesh, 50% max. thru 200 mesh; 32.6-35.3% assay (niacinamide); *Storage:* store below 72 F.

Rocoat® Niacinamide 33$^1/_3$% Type S No. 69909. [Roche] Niacinamide USP FCC in inert coating of food-grade stearic acid with ≈ 1% silicon dioxide (flow agent); incorporated into chewable multivita-min tablets and other dry dosage forms; provides taste-free and odorless form and protects the vitamin; *Regulatory:* FDA GRAS; *Properties:* wh. relatively free-flowing powd. with some soft agglomerates, sl. char. odor, satisfactory taste; 99% min. thru 30 mesh, 65% min. thru 60 mesh; 32.6-35.3% assay (niacinamide); *Storage:* store below 72 F.

Rocoat® Pyridoxine Hydrochloride 33$^1/_3$% No. 60688. [Roche] Pyridoxine hydrochloride USP FCC in inert coating of mono- and diglycerides of edible fatty acids; incorporated into chewable multivitamin tablets and other dry dosage forms; provides taste-free and odorless form and protects the vitamin; *Regulatory:* FDA GRAS; *Properties:* wh. to off-wh. relatively free-flowing powd. with some soft agglomerates, odorless to sl. odor, satisfactory taste; 99% min. thru 20 mesh, 50% max. thru 200 mesh; 32.6-35.3% assay (pyridoxine HCl); *Storage:* store below 72 F.

Rocoat® Riboflavin 25% No. 60289. [Roche] Riboflavin USP-FCC in an inert coating of mono- and diglycerides of edible fatty acids and starch; nutrient incorporated into chewable multivitamin tablets and other dry dosage forms; provides taste-free and odor-free form of vitamin B_2, protects the vitamin, provides uniform distribution; *Regulatory:* FDA GRAS; *Properties:* orange relatively free-flowing powd. with some soft agglomerates, sl. char. odor, satisfactory taste; 99% min. thru 20 mesh, 50% max. thru 200 mesh; 24-5.28.8% assay; *Storage:* store below 72 F.

Rocoat® Riboflavin 33$^1/_3$ No. 60288. [Roche] Riboflavin USP-FCC and cornstarch in an inert coating of mono- and diglycerides of edible fatty acids; nutrient incorporated into chewable multivitamin tablets and other dry dosage forms; provides taste-free and odor-free form of vitamin B_2, protects the vitamin; *Regulatory:* FDA GRAS; *Properties:* orange relatively free-flowing powd. with some soft agglomerates, sl. char. odor, satisfactory taste; 99% min. thru 20 mesh, 50% max. thru 200 mesh; 32.6-35.3% assay; *Storage:* store below 72 F.

Rocoat® Thiamine Mononitrate 33$^1/_3$% No. 60188. [Roche] Thiamine nitrate USP FCC in inert coating of mono- and diglycerides of edible fatty acids; nutrient incorporated into chewable multivitamin tablets and other dry dosage forms; provides taste-free and odor-free form of vitamin B_1, protects the vitamin; *Regulatory:* FDA GRAS; *Properties:* wh. to off-wh. relatively free-flowing powd. with some soft agglomerates, sl. odor, satisfactory taste; 99% min. thru 30 mesh, 50% max. thru 200 mesh; 32.6-35.3% assay; *Storage:* store in a dry place @ 46-59 F.

Ross Candelilla Wax. [Frank B. Ross] Candelilla wax; CAS 8006-44-8; EINECS 232-347-0; chewing gum; *Regulatory:* FDA 21CFR §172.615, 175.105, 175.320, 176.180; *Properties:* grayish-green wax-like fat, aromatic odor; sp.gr. 0.977-0.982; m.p. 38-49 C; acid no. 5-24; iodine no. 2-10; sapon. no. 210-239; flash pt. 470 F min.; ref. index 1.4360.

Ross Carnauba Wax. [Frank B. Ross] Carnauba wax;

CAS 8015-86-9; EINECS 232-399-4; gum; *Regulatory:* FDA 21CFR §182.1978; *Properties:* flakes or powd.; sp.gr. 0.996-0.998; m.p. 181.4 F min.; acid no. 2-10; iodine no. 7-14; sapon. no. 78-88; flash pt. 570 F min.; ref. index 1.4540.

Ross Rice Bran Wax. [Frank B. Ross] Rice bran wax, refined grade; CAS 8016-60-2; EINECS 232-409-7; wax for use in confectionery, chewing gum, coatings for fruit and vegetables; *Regulatory:* FDA 21CFR §172.615, 172.890; *Properties:* tan to lt. brn. flake or powd.; m.p. 76-82 C; acid no. 5-15; iodine no. 10 max.; sapon. no. 70-105; flash pt. 520 F min.

Roxanthin® Red 10WS No. 66515. [Roche] 10% Canthaxanthin in matrix of gelatin, sucrose, food starch, veg. oil with antioxidants (dl-α-tocopherol, ascorbyl palmitate); colorant producing tomato-red hues in pizza, barbecue sauce, russian dressing, fruit punch, bloody mary-type beverages; also produces range of hues in dressings, cakes, simulated meats, dry mixes which will be reconstituted with water; *Usage level:* 300 mg max./lb of solid or semisolid or pint of liq. food; *Regulatory:* FDA 21CFR §73.75 (food use), 73.1075 (drug use); *Properties:* purplish red dry spherical free-flowing beadlet; 100% thru 20 mesh, 25% max. thru 80 mesh; disp. in warm water; m.w. 564.8; 10% assay (canthaxanthin); *Storage:* store in cool dry place in tightly closed containers, optimally below 46 F; avoid moisture.

Royal Baking Powd. [ADM Arkady] Sodium aluminum phosphate; CAS 7785-88-8; multiple-acting chemical leavener esp. for cakes having a high portion of corn syrup solids or other simple sugars in place of sucrose.

Royal Fudge Standard and Extra Dark. [Bunge Foods] Used for a batter or icing where a semi-sweet dark chocolate type is desired; for chocolate creme and fudge icings, brownie batters; *Usage level:* 25-40 lb/100 lb. powd. sugar; *Properties:* dk. brn. semi-plastic base, cocoa aroma/flavor.

Royal Set. [Brolite Prods.] Blend of hard fat flakes that prevents icings from breaking down in almost any condition.

Rudol®. [Witco/Petroleum Spec.] Lt. min. oil NF; white oil functioning as binder, carrier, defoamer, dispersant, extender, lubricant, moisture barrier, plasticizer, softener, protective agent for food; *Regulatory:* FDA 21CFR §172.878, §178.3620a; *Properties:* water-wh., odorless, tasteless; sp.gr. 0.852-0.870; visc. 28-30 cSt (40 C); pour pt. -7 C; flash pt. 188 C.

Rye Base #12. [Brolite Prods.] Rye bread base; *Usage level:* 12 lb/100 lb flour.

Rye-Niks. [Brolite Prods.] Full-flavored rye bread base naturally flavored with ground spices and cultured sours; *Usage level:* 12 lb/100 lb flour.

Rye Sour. [Bunge Foods] Sour producing uniform prod. with constant pH; reduces fermentation loss; increases shelf life; decreases molding tendencies; enhances flavor of bread; for variety of rye breads from dk. heavy types to light sandwich type; *Usage*

level: 5% based on flour.

Rye Sour #4. [Brolite Prods.] Conc. natural rye flavor for a distinctive rye flavor for high quality rye breads.

Ry-Fla-Vor. [ADM Arkady] Conc. rye bread flavor which imparts a pleasing aroma and tangy taste.

Ryoto Ester KA. [Mitsubishi Kasei Foods] Sugar ester, calcium carbonate, potassium carbonate, carbohydrates; quality improver for fish paste prods.; enhances water holding capacity; *Usage level:* 0.3-0.8% on surimi.

Ryoto Ester SP. [Mitsubishi Kasei Foods] Sugar ester, monoglyceride, sorbitol, propylene glycol; batter aerating agent for sponge and pound cake; *Usage level:* 6% on flour (sponge cake).

Ryoto Sugar Ester B-370. [Mitsubishi Kasei Foods] Sucrose tribehenate; biodeg. emulsifier with antibacterial, wetting, and dispersing effect for foods; anticaking agent for hygroscopic powd. foods (spices, powd. seasonings); antisticking for noodles; emulsifier for coffee whiteners, shortening; *Usage level:* 0.5-1.5% (spices), 0.1-0.5% (coffee whitener), 0.1-10% (shortening); *Properties:* powd.; partly sol. in propylene glycol, glycerin, liq. paraffin, soybean oil, cottonseed oil, water; m.p. 53-63 C; decomp. pt. 241 C; HLB 3.

Ryoto Sugar Ester ER-190. [Mitsubishi Kasei Foods] Sucrose erucate; emulsifier for creams, confections, chewing gum; demulsifier for whipping cream, ice cream; wetting agent, dispersant for colorants; fresh fruit coating; starch complexing, antistaling agent; dough conditioner; aerating agent; fat crystal control; *Properties:* HLB 1.

Ryoto Sugar Ester ER-290. [Mitsubishi Kasei Foods] Sucrose erucate; emulsifier for creams, confections, chewing gum; demulsifier for whipping cream, ice cream; wetting agent, dispersant for colorants; fresh fruit coating; starch complexing, antistaling agent; dough conditioner; aerating agent; fat crystal control; *Properties:* HLB 2.

Ryoto Sugar Ester L-195. [Mitsubishi Kasei Foods] Sucrose polylaurate; food emulsifier; antisticking for noodles; emulsifier for coffee whiteners, shortening; *Usage level:* 0.5-1% on flour (noodles), 0.1-0.5% (coffee whitener), 0.1-10% (shortening); *Properties:* HLB 1.

Ryoto Sugar Ester L-595. [Mitsubishi Kasei Foods] Sucrose dilaurate; emulsifier with antibacterial, wetting, and dispersing effect for foods; antisticking for noodles; emulsifier for coffee whitener, shortening; *Usage level:* 0.5-1% on flour (noodles), 0.1-0.5% (coffee whitener), 0.1-10% (shortening); *Properties:* pellet; sol. in water, ethanol; partly sol. in glycerin; HLB 5.

Ryoto Sugar Ester L-1570. [Mitsubishi Kasei Foods] Sucrose laurate; CAS 25339-99-5; EINECS 246-873-3; emulsifier with antibacterial, wetting, and dispersing effect for foods; antisticking for noodles; emulsifier for coffee whitener, shortening; *Usage level:* 0.5-1% on flour (noodles), 0.1-0.5% (coffee whitener), 0.1-10% (shortening); *Properties:* surf. tens. 31.7 dynes/cm (0.1% aq.); Ross-Miles foam 177 mm (0.25% aq., initial).

Ryoto Sugar Ester L-1695. [Mitsubishi Kasei Foods] Sucrose laurate; CAS 25339-99-5; EINECS 246-873-3; emulsifier with antibacterial, wetting, and dispersing effect for foods; antisticking for noodles; emulsifier for coffee whitener, shortening; *Usage level:* 0.5-1% on flour (noodles), 0.1-0.5% (coffee whitener), 0.1-10% (shortening); *Properties:* pellet; sol. in water, ethanol; partly sol. in glycerin; m.p. 35-47 C; decomp. pt. 235 C; HLB 16; surf. tens. 31.6 dynes/cm (0.1% aq.).

Ryoto Sugar Ester LN-195. [Mitsubishi Kasei Foods] Sucrose polylinoleate; antisticking for noodles; emulsifier for coffee whitener, shortening; *Usage level:* 0.5-1% on flour (noodles), 0.1-0.5% (coffee whitener), 0.1-10% (shortening).

Ryoto Sugar Ester LWA-1570. [Mitsubishi Kasei Foods] Sucrose laurate; CAS 25339-99-5; EINECS 246-873-3; nonionic; o/w and w/o emulsifier, softener, conditioner, and aerating agent in foods, vegetable/fruit scrubbing; *Usage level:* 0.5-1% on flour (noodles), 0.1-0.5% (coffee whitener), 0.1-10% (shortening); *Properties:* paste; HLB 15.0; 40% conc.

Ryoto Sugar Ester M-1695. [Mitsubishi Kasei Foods] Sucrose myristate; emulsifier with antibacterial, wetting, and dispersing effect for foods; prevents precipitation in canned drinks; antisticking for noodles; emulsifier for coffee whitener, shortening; *Usage level:* 0.03-0.3% (canned drinks), 0.5-1% on flour (noodles), 0.1-0.5% (coffee whitener), 0.1-10% (shortenin; *Properties:* pellet; sol. in water, propylene glycol, ethanol; partly sol. in glycerin; m.p. 27-40 C; decomp. pt. 243 C; HLB 16.

Ryoto Sugar Ester O-170. [Mitsubishi Kasei Foods] Sucrose polyoleate; antisticking for noodles; emulsifier for coffee whitener, shortening; *Usage level:* 0.5-1% on flour (noodles), 0.1-0.5% (coffee whitener), 0.1-10% (shortening); *Properties:* HLB 1.

Ryoto Sugar Ester O-1570. [Mitsubishi Kasei Foods] Sucrose oleate; emulsifier with antibacterial, wetting, and dispersing effect for foods, vegetable/fruit scrubbing; improves dryness in ice cream; fruit coating agent; antisticking for noodles; emulsifier for coffee whitener, shortening; *Usage level:* 0.1-0.3% (ice cream), 0.5-1% on flour (noodles), 0.1-0.5% (coffee whitener), 0.1-10% (shortening); *Properties:* pellet; sol. in water, ethanol; partly sol. in glycerin; m.p. 27-43 C; decomp. pt. 227 C; HLB 15; surf. tens. 34.5 dynes/cm (0.1% aq.); Ross-Miles foam 14 mm (0.25% aq., initial).

Ryoto Sugar Ester OWA-1570. [Mitsubishi Kasei Foods] Sucrose oleate; nonionic; emulsifier, conditioner, softener, detergent for foods, vegetable/fruit scrubbing; *Usage level:* 0.5-1% on flour (noodles), 0.1-0.5% (coffee whitener), 0.1-10% (shortening); *Properties:* paste; HLB 15.0; 40% conc.

Ryoto Sugar Ester P-170. [Mitsubishi Kasei Foods] Sucrose palmitate; CAS 26446-38-8; EINECS 247-706-7; emulsifier for creams, confections, chewing gum; demulsifier for whipping cream, ice cream; wetting agent, dispersant for colorants; fresh fruit coating; starch complexing, antistaling agent; dough conditioner; aerating agent; fat crystal control; *Properties:* HLB 1.

Ryoto Sugar Ester P-1570. [Mitsubishi Kasei Foods] Sucrose palmitate; CAS 26446-38-8; EINECS 247-706-7; nonionic; emulsifier, softener, detergent; dispersant for flavors, preservatives; fat crystal inhibitor, quality improver for fats, oils; dough strengthener; improves dryness in ice cream; fruit coating; antisticking for noodles; coffee whitener, shortening; *Usage level:* 0.3-1% on flour (bread), 0.1-0.3% (ice cream), 0.03-0.3% (canned drinks), 0.5-1% on flour (noodles); *Properties:* powd.; partly sol. in water, propylene glycol, glycerin, liq. paraffin, soybean and cottonseed oils; m.p. 47-54 C; decomp. pt. 237 C; HLB 15.0; surf. tens. 35.4 dynes/cm (0.1% aq.); Ross-Miles foam 15 mm (0.25% aq., initial); 100% conc.

Ryoto Sugar Ester P-1570S. [Mitsubishi Kasei Foods] Sucrose palmitate, high ignition residue; CAS 26446-38-8; EINECS 247-706-7; nonionic; emulsifier, conditioner, softener, detergent for foods; dough strengthener, antisticking for noodles; emulsifier for coffee whitener, shortening; *Usage level:* 0.3-1% on flour (bread), 0.5-1% on flour (noodles), 0.1-0.5% (coffee whitener), 0.1-10% (shortening); *Properties:* powd.; sol. in water; partly sol. in propylene glycol, glycerin, liq. paraffin, soybean and cottonseed oils; HLB 15.

Ryoto Sugar Ester P-1670. [Mitsubishi Kasei Foods] Sucrose palmitate; CAS 26446-38-8; EINECS 247-706-7; nonionic; emulsifier, softener, detergent; dispersant for flavors, preservatives; fat crystal inhibitor, quality improver for fats, oils; dough strengthener; improves dryness in ice cream; fruit coating; antisticking for noodles; coffee whitener, shortening; *Usage level:* 0.3-1% on flour (bread), 0.1-0.3% (ice cream), 0.03-0.3% (canned drinks), 0.01-0.002% (soy sauce); *Properties:* powd.; sol. in water; partly sol. in propylene glycol, glycerin, liq. paraffin, soybean and cottonseed oils; m.p. 40-48 C; decomp. pt. 235 C; HLB 16; surf. tens. 34.5 dynes/cm (0.1% aq.); Ross-Miles foam 24 mm (0.25% aq., initial).

Ryoto Sugar Ester S-070. [Mitsubishi Kasei Foods] Sucrose polystearate; nonionic; o/w and w/o emulsifier, softener, conditioner, and aerating agent in foods; antisticking for noodles; emulsifier for coffee whitener, shortening; *Usage level:* 0.5-1% on flour (noodles), 0.1-0.5% (coffee whitener), 0.1-10% (shortening); *Properties:* powd.; 100% conc.

Ryoto Sugar Ester S-170. [Mitsubishi Kasei Foods] Sucrose di, tristearate; nonionic; emulsifier, softener, detergent for foods; o/w emulsifier in caramels, coffee whitener, shortening, whipped toppings, ice cream, chocolate; anticaking agent in powd. foods (spices, seasonings); antifoam in soybean curd; antisticking for noodles; *Usage level:* 0.1-0.5% (caramel, margarine), 0.2-0.5% (chocolate), 0.5-1.5% (spices), 0.5-1% on flour (noodles); *Properties:* powd.; partly sol. in propylene glycol, glycerin, liq. paraffin, soybean and cottonseed oils; insol. in water; m.p. 51-61 C; decomp. pt. 260 C;

HLB 1.0; 100% conc.

Ryoto Sugar Ester S-170 Ac. [Mitsubishi Kasei] Sucrose tetrastearate triacetate; antisticking for noodles; emulsifier for coffee whitener, shortening; *Usage level:* 0.5-1% on flour (noodles), 0.1-0.5% (coffee whitener), 0.1-10% (shortening).

Ryoto Sugar Ester S-270. [Mitsubishi Kasei Foods] Sucrose di, tristearate; nonionic; emulsifier, softener, detergent for foods; o/w emulsifier in caramels, margarine, coffee whitener, shortening, chocolate, whipped toppings, ice cream; antifoam in soybean curd; antisticking for noodles; *Usage level:* 0.1-0.5% (caramel, margarine), 0.2-0.5% (chocolate), 0.5-1% on flour (noodles), 0.1-10% (shortening); *Properties:* powd.; partly sol. in propylene glycol, glycerin, liq. paraffin, soybean and cottonseed oils; insol. in water; m.p. 52-61 C; decomp. pt. 253 C; HLB 2.0; 100% conc.

Ryoto Sugar Ester S-370. [Mitsubishi Kasei Foods] Sucrose tristearate; nonionic; emulsifier, conditioner, softener, detergent for foods; o/w emulsifier in caramels, coffee whitener, shortening, chocolate, whipped toppings, ice cream; antifoam in soybean curd; antisticking for noodles; *Usage level:* 0.1-0.5% (caramel, soybean curd); 0.2-0.5% (chocolate), 0.5-1% on flour (noodles); *Properties:* powd.; partly sol. in propylene glycol, glycerin, liq. paraffin, soybean and cottonseed oils; insol. in water; m.p. 51-58 C; decomp. pt. 238 C; HLB 3; 100% conc.

Ryoto Sugar Ester S-370F. [Mitsubishi Kasei Foods] Sucrose tristearate; nonionic; emulsifier, conditioner, softener, detergent for tablet candy; Increases fluidity of raw materials and efficiency of filling in tableting machine, *Usage level:* 0.5-1.5% (tablet candy); *Properties:* fine powd.; HLB 3.

Ryoto Sugar Ester S-570. [Mitsubishi Kasei Foods] Sucrose distearate; CAS 27195-16-0; EINECS 248-317-5; nonionic; emulsifier, softener, detergent for foods; improves volume and crumb in sponge cake; w/o emulsifier in caramels, chewing gum, margarine, low-fat spreads, chocolate, sugar mfg., coffee whitener; antisticking for noodles; anticrystallization for jelly; *Usage level:* 0.3-1% (sponge cake), 0.1-0.5% (caramel), 0.2-0.5% (chocolate), 0.005-0.01% (sugar); *Properties:* powd.; partly sol. in propylene glycol, glycerin, liq. paraffin, soybean and cottonseed oils; insol. in water; m.p. 50-57 C; decomp. pt. 231 C; HLB 5.0; surf. tens. 38.1 dynes/cm (0.1% aq.); 100% conc.

Ryoto Sugar Ester S-770. [Mitsubishi Kasei Foods] Sucrose distearate; CAS 27195-16-0; EINECS 248-317-5; emulsifier, softener, detergent; improves volume, crumb in sponge cake; w/o emulsifier in caramels, chewing gum, coffee whitener, shortening, chocolate, ice cream; antisticking for noodles; anticrystallization for jelly; prevents syneresis in pudding; *Usage level:* 0.3-1% (sponge cake), 0.1-0.5% (caramel), 0.2-0.5% (chocolate), 0.1-0.3% (ice cream); *Properties:* powd.; partly sol. in propylene glycol, glycerin, liq. paraffin, soybean and cottonseed oils; water; m.p. 49-60 C; decomp.

pt. 233 C; HLB 7.0; surf. tens. 37.4 dynes/cm (0.1% aq.).

Ryoto Sugar Ester S-970. [Mitsubishi Kasei Foods] Sucrose mono, distearate; CAS 27195-16-0; EINECS 248-317-5; nonionic; emulsifier, softener, detergent for foods; improves volume, crumb in sponge cake; emulsifier in caramels, chewing gum, coffee whitener, shortening; antisticking for noodles; anticrystallization for jelly; prevents syneresis in pudding; ice cream; *Usage level:* 0.3-1% (sponge cake), 0.1-0.5% (caramel), 0.1-0.3% (ice cream), 0.1-0.5% (coffee whitener); *Properties:* powd.; partly sol. in propylene glycol, glycerin, liq. paraffin, soybean and cottonseed oils, water; m.p. 49-56 C; decomp. pt. 234 C; HLB 9.0; surf. tens. 35.8 dynes/cm (0.1% aq.); 100% conc.

Ryoto Sugar Ester S-1170. [Mitsubishi Kasei Foods] Sucrose stearate; CAS 25168-73-4; EINECS 246-705-9; nonionic; emulsifier, conditioner; solubilizer, dispersant for flavors, preservatives; dough strengthener; improves volume, crumb in sponge cake; antisticking for noodles; prevents syneresis in pudding; emulsifier for coffee whitener, shortening; in ice cream; *Usage level:* 0.3-1% on flour (bread, sponge cake), 0.1-0.3% (ice cream), 0.2-1% (frozen foods); *Properties:* powd.; partly sol. in propylene glycol, glycerin, liq. paraffin, soybean and cottonseed oils, water; m.p. 49-55 C; decomp. pt. 234 C; HLB 11; surf. tens. 34.8 dynes/cm (0.1% aq.); 100% conc.

Ryoto Sugar Ester S-1170S. [Mitsubishi Kasei Foods] Sucrose stearate, high ignition residue; CAS 25168-73-4; EINECS 246-705-9; nonionic; emulsifier, conditioner, detergent; solubilizer, dispersant for flavors, preservatives; dough strengthener; improves volume and crumb in sponge cake; antisticking for noodles; emulsifier for coffee whitener, shortening; in ice cream, frozen foods; *Usage level:* 0.3-1% on flour (bread, sponge cake), 0.1-0.3% (ice cream), 0.2-1% (frozen foods); *Properties:* powd.; sol. in water; partly sol. in propylene glycol, glycerin, liq. paraffinn, soybean and cottonseed oils; HLB 11.

Ryoto Sugar Ester S-1570. [Mitsubishi Kasei Foods] Sucrose stearate; CAS 25168-73-4; EINECS 246-705-9; nonionic; emulsifier, conditioner, detergent; solubilizer, dispersant for flavors, preservatives; dough strengthener; improves volume, crumb in sponge cake; fruit coating agent; antisticking for noodles; emulsifier for coffee whitener, shortening; in ice cream; *Usage level:* 0.3-1% on flour (bread, sponge cake), 0.1-0.3% (ice cream), 0.2-1% (frozen foods); *Properties:* powd.; sol. in water; partly sol. in propylene glycol, glycerin, liq. paraffin, soybean and cottonseed oils; m.p. 49-55 C; decomp. pt. 234 C; HLB 15.0; surf. tens. 34.7 dynes/cm (0.1% aq.); Ross-Miles foam 11 mm (0.25% aq., initial); 100% conc.

Ryoto Sugar Ester S-1670. [Mitsubishi Kasei Foods] Sucrose stearate; CAS 25168-73-4; EINECS 246-705-9; nonionic; emulsifier, conditioner; solubilizer, dispersant for flavors, preservatives; visc. reducer in chocolate; dough strengthener; improves volume, crumb in sponge cake; antisticking for noodles; emulsifier for coffee whitener, shortening; ice cream; *Usage level:* 0.3-1% on flour (bread, sponge cake), 0.1-0.3% (ice cream), 0.2-1% (frozen foods); *Properties:* powd.; sol. in water; partly sol. in propylene glycol, glycerin, liq. paraffin, soybean and cottonseed oils; m.p. 49-56 C; decomp. pt. 237 C; HLB 16; surf. tens. 34.7 dynes/cm (0.1% aq.); Ross-Miles foam 12 mm (0.25% aq., initial).

Ryoto Sugar Ester S-1670S. [Mitsubishi Kasei Foods] Sucrose stearate, high ignition residue; CAS 25168-73-4; EINECS 246-705-9; nonionic; emulsifier, conditioner, softener, detergent for foods; antisticking for noodles; emulsifier for coffee whitener, shortening; *Usage level:* 0.1-0.5% (coffee whitener), 0.1-10% (shortening); *Properties:* powd.; sol. in water; partly sol. in propylene glycol, glycerin, liq. paraffin, soybean and cottonseed oils; HLB 16.

Ry-So. [Am. Ingreds.] Rye flour, lactic acid, monocalcium phosphate, salt, phosphoric acid, acetic acid, and dry yeast; sour for rye breads and rye flour-based snacks; *Properties:* free-flowing powd.; pH 3.4-3.6; 7.5-9.5% protein.

S

SagaSalt. [Akzo Salt Europe] Low sodium content salt; for home and food industry; 41% NaCl.

SAIB-SG. [Eastman] Sucrose acetate isobutyrate; CAS 126-13-6; EINECS 204-771-6; for use in food drinks except in U.S.; *Regulatory:* kosher; *Properties:* Gardner 1 visc. liq.; sp.gr. 1.146; dens. 9.55 lb/gal; visc. 100,000 cP (30 C); flash pt. 260 C.

Saladizer® #228M. [TIC Gums] Gum system; alternative to xanthan/alginate blends in pourable salad dressings; provides creamy texture and exceptional cling; sensitive to salt which lowers visc. and delays hydration; tolerates vinegar.

Saladizer® #250P. [TIC Gums] Gum system with emulsifiers; polysorbate replacement for salad dressings; produces dressings which are creamier and lighter in color.

Saladizer® #250PM. [TIC Gums] Gum system; low-cost stabilizer for salad dressings which require only a 6-9 mo shelf life.

Saladizer® #251. [TIC Gums] Gum system; low-cost stabilizer for salad dressings which require only a 6-9 mo shelf life.

Santone® 3-1-S. [Van Den Bergh Foods] Polyglyceryl-3 stearate; CAS 37349-34-1; EINECS 248-403-2; nonionic; emulsifier; aeration of lipid systems; solubilizer; color/flavor dispersant in water; *Properties:* cream bead; m.p. 128-130 F; HLB 7.2; sapon. no. 122-139; 27-34% total mono- and diglycerides; Custom prod.

Santone® 3-1-SH. [Van Den Bergh Foods] Polyglyceryl-3 oleate; CAS 9007-48-1; nonionic; emulsifier and aerating agent used in food industry, cakes; replacement for polysorbates in icings and icing shortenings; *Regulatory:* FDA 21CFR §172.854; kosher; *Properties:* wh. plastic; m.p. 84-92 F; HLB 7.2; acid no. 8 max.; iodine no. 36-48; sapon no. 125-135; hyd. no. 345-390; 100% conc.; *Storage:* store sealed in cool, dry place away from odor-producing substances; 1 yr storage life at 40-95 F.

Santone® 3-1-S XTR. [Van Den Bergh Foods] Triglyceryl stearate; CAS 37349-34-1; EINECS 248-403-2; food emulsifier, aerator, and whipping agent for nonaq. lipid systems in bakery, dairy, and confectionery applics.; *Regulatory:* FDA 21CFR §172.854; kosher; *Properties:* wh. bead; m.p. 125-135 F; drop pt. 52-56 C; HLB 6.9; acid no. 8 max.; iodine no. 3 max.; sapon. no. 130-145; hyd. no. 300-350; *Storage:* store sealed in cool, dry place away

from odor-producing substances; 1 yr storage life at 40-80 F.

Santone® 8-1-O. [Van Den Bergh Foods] Polyglyceryl-8 oleate; CAS 9007-48-1; nonionic; food emulsifier, visc. reducer in protein systems; emulsion stabilizer in ice cream toppings; direct replacement for polysorbates (1:1); clouding agent for beverages; color/flavor dispersant; gloss enhancer; wetting agent for proteins, cocoa powd.; *Usage level:* 0.1-0.45% (whipped toppings), 0.1-0.4% (protein), 0.1-1% (beverages), 0.5-1% (ice cream toppings), 2% fat basis (compd. coatings); *Regulatory:* FDA 21CFR §172.854; kosher; *Properties:* amber liq.; visc. 16,000 cps (50 C); HLB 13.0; acid no. 5 max.; iodine no. 30-40; sapon. no. 75-85; hyd. no. 460-520; 100% conc.; *Storage:* store sealed in cool, dry place away from odor-producing substances; 1 yr storage life at 40-95 F; stir well.

Santone® 10-10-O. [Van Den Bergh Foods] Polyglycerol-10 decaoleate; CAS 11094-60-3; EINECS 234-316-7; nonionic; food emulsifier; solubilizer; emulsion stabilizer; dispersant aid in flavors and colors; antifoam in sugar-protein syrups; emulsifier, lubricant, stabilizer in chilled and frozen foods; *Usage level:* 0.5-1% (colors/flavors), 0.5-1% (emulsion stabilizer), 0.05% (antifoam); *Regulatory:* FDA 21CFR §172.854; kosher; *Properties:* Lovibond 4.0R max. liq.; visc. 88 cps (50 C); HLB 2.0; acid no. 6 max.; iodine no. 65-80; sapon. no. 165-180; hyd. no. 25-45; 100% conc.; *Storage:* store sealed in cool, dry place away from odor-producing substances; 1 yr storage life at 40-90 F; Custom prod.

Sapp #4. [Rhone-Poulenc Food Ingreds.] Sodium acid pyrophosphate, leavening, FCC; CAS 7758-16-9; EINECS 231-835-0; slowest action leavening agent for baking, cereals; esp. for use in canned refrigerated biscuit doughs; *Properties:* wh. powd.; 100% thru 60 mesh, 99% thru 200 mesh; sol. 13 g/100 g saturated sol'n.; m.w. 221.94; bulk dens. 68 lb/ft³; pH 4.3 (1%); 95% min. act.

SAPP 22. [FMC] Sodium acid pyrophosphate; CAS 7758-16-9; EINECS 231-835-0; double-acting leavening agent for baked goods; slowest grade with low initial reactivity; esp. suited to refrigerated dough; often combined with faster acting leavening acids for other baking applics.; *Regulatory:* FCC, kosher; *Properties:* powd.; 99.5% thru 200 mesh; sol. 15 g/100 g water; m.w. 221.9; bulk dens. 65 lb/

ft³; 96.5% assay.

SAPP 26. [FMC] Sodium acid pyrophosphate; CAS 7758-16-9; EINECS 231-835-0; leavening acid for dough; med. reactivity grade; *Regulatory:* FDA GRAS; *Properties:* powd.; 93% thru 200 mesh; sol. 15 g/100 g water; m.w. 221.9; bulk dens. 65 lb/ft³; 98.3% assay.

SAPP 28. [FMC] Sodium acid pyrophosphate; CAS 7758-16-9; EINECS 231-835-0; double-acting leavening agent for baked goods; slow-med. reactivity; general purpose leavening acid and major component of commercial baking powds., prepared mixes, bakery cakes; *Regulatory:* FCC, kosher; *Properties:* powd.; 93% thru 200 mesh; sol. 15 g/100 g water; m.w. 221.9; bulk dens. 65 lb/ft³; 98.5% assay.

SAPP 40. [FMC] Sodium acid pyrophosphate; CAS 7758-16-9; EINECS 231-835-0; double-acting leavening agent for baked goods; fastest grade in which 40% of avail. acidity released initially during mixing; used alone or in combination for machined cake donuts and prepared mixes; *Regulatory:* FCC, kosher.

SAS Baking Powd. [ADM Arkady] Multiple-acting chemical leavener for general use; similar to home-style baking powd.

Satina 44. [Van Den Bergh Foods] Partially hydrog. palm kernel oil, lecithin; lauric coating fat for butterscotch and chocolate flavored confectionery coatings; *Regulatory:* kosher; *Properties:* solid; drop pt. 34-36 C.

Satina 50. [Van Den Bergh Foods] Partially hydrog. palm kernel oil, lecithin; lauric coating fat for pastel confectioner's coatings, center fat; sharp m.p. chars.; nontempering version avail.; *Regulatory:* kosher; *Properties:* solid; drop pt. 33-35 C.

Satina 53NT. [Van Den Bergh Foods] Partially hydrog. veg. oil (palm kernel, palm) with mono- and diglycerides, 1.5% sorbitan stearate, 0.1% polysorbate 60, lecithin; lauric nontempering coating fat for pastel and chocolate flavored confections; *Regulatory:* kosher; *Properties:* solid; drop pt. 33-35 C.

Sea-Gard® FP-91. [Rhone-Poulenc Food Ingreds.] Sodium tripolyphosphate FCC; CAS 7758-29-4; food additive used in treatment of seafood prior to freezing; minimizes cellular deterioration and fluid loss, reduces thaw drip loss, locks in natural moisture, protein, vitamins, and other nutrients; *Regulatory:* kosher; *Properties:* wh. gran. or powd., odorless; 1% max. on 20 mesh, 20% max. thru 100 mesh (gran.), 1% max. on 40 mesh, 80% min. thru 100 mesh (powd.); m.w. 368; pH 9.5-10.3 (1%); *Storage:* store cool and dry.

SeaKem® CM-611. [FMC] Fat replacement system for fluid milk; provides mouthfeel and body; prevents whey separation.

SeaKem® GP-418. [FMC] Fat replacement system for nonfat pasteurized processed cheese; provides creamy mouthfeel, loaf structure, sliceability, smoothness, opacity; enhances flavor; reduces rubber texture.

SeaKem® IC-611. [FMC] Fat replacement system for hard-pack and soft-serve frozen desserts, frozen yogurt; provides creamy mouthfeel, ice crystal control, improved extrudability, and control of whey separation.

SeaKem® IC-624. [FMC] Fat replacement system for soft-serve frozen desserts; imparts creaminess, ice crystal control, improved extrudability, and prevents whey separation.

SeaKem® IC-632. [FMC] Fat replacement system for frozen yogurt; provides ice crystal control, creamy mouthfeel, foam stability, improved meltdown; prevents whey separation.

SeaKem® IC-912. [FMC] Fat replacement system for hard-pack frozen desserts, frozen yogurt; provides creamy mouthfeel, ice crystal control, foam stability, improved extrudability and meltdown, and control of whey separation.

Seal N Glaze Systems for Vegetables and Fruits. [Advanced Food Systems] Dry glaze system for frozen vegetables/fruits to be cooked in microwave or convection ovens; eliminates exudation of water on cooking; prevents water migration; enhances crispness; reduces freezer burn; becomes glossy glazes.

Seal N' Season Glazes and Seasonings. [Advanced Food Systems] Dry seasoning/sealing systems for poultry, meat, fish, shellfish; keeps seasonings on surf.; seals prods. own juice on cooking; produces glossy sheen, tender texture, improves cooked yield; various flavors (BBQ, garlic, fajita, teriyaki, etc.).

SeasonRite Marinade Systems. [Advanced Food Systems] Marinade systems for meat, poultry, fish, shellfish; sol. flavors are massaged into meat; improves cooked yield, tenderness, eating quality, freeze/thaw stability; various flavors.

Sekicel. [Asahi Chem. Industry] Cellulose starch dispersion; for food industry.

Selin® G deo. [Grünau GmbH] Triglyceride of natural fatty acids, predominantly oleic acid; special oil for food industry; release agent for baked goods; oil component for cutting machines; better oxidation stability, lower polymerization, and better cold stability than vegetable oils; *Properties:* yel. liq., neutral odor, neutral oily taste; acid no. 1 max.; iodine no. 85-95; sapon. no. 190-200; *Storage:* store cool and dry in tightly closed containers; protect from light; 6 mos storage life in original pkg.

Selin® O. [Grünau GmbH] Trioleate; basic oil for release agents, bread cutting oil.

Sentry Simethicone NF. [Union Carbide] Simethicone; CAS 8050-81-5; used in food-grade emulsifiers for antacid prods.; Discontinued

Serdas GLN. [Servo Delden BV] Surfactants in min. oil; defoamer for sugar industry; *Properties:* disp. in water; Unverified

Servil® O 38. [Grünau GmbH] Wax ester based on edible fatty acids and unbranched, straight chained fatty alcohols; special oil in release agents for baked goods; good oxidation stability, low resinification tendency at high temps.; *Usage level:* 10-20%; *Properties:* yel. liq., neutral odor, fatty taste; acid no. 1 max.; iodine no. 95; sapon. no. 110; turbidity pt. 15 C max.; *Storage:* store cool, dry in tightly closed containers (15-25 C); protect from light; 6 mo storage life in original pkg.

Servil® Range. [Grünau GmbH] Wax esters; release agents for confectionery and baked goods.

Servit® Range. [Grünau GmbH] Citric acid esters of mono- and diglycerides of fatty acids with other emulsifiers; food emulsifier for mfg. of instant dry yeast for baked goods.

Set-it Powd. [Bunge Foods] Boiling or nonboiling type icing stabilizer for donut glaze, pie glaze; good stability when boiled; *Usage level:* 1-2 lb/100 lb powd. sugar; *Properties:* wh. to off-wh. powd., bland flavor; 8% moisture.

Sett®. [Grünau GmbH] Hydrogenated triglycerides; visc. enhancer, hardstock for margarine, coating agent.

721-A™ Instant Starch. [Am. Maize Prods.] Moderately inhibited pregelatinized starch; starch featuring rapid hydration, high visc. without cooking; provides moisture retention and extends shelf life in baked goods; in fruit fillings, instant puddings, soup, sauce, and gravy mixes; resist. to heat, acid, and shear; good paste clarity; *Regulatory:* FDA 21CFR §172.892; *Properties:* waxy; 50% thru 200 mesh; pH 5.0; 5% moisture; *Storage:* store in dry, cool conditions, in odor-free area, ≤ 50% r.h. and ≤ 25 C.

SF18-350. [GE Silicones] Dimethicone fluid; lubricant, antifoam, mold release for food applics.; antifoam in fermentation, corn oil mfg., deep-fat frying; *Regulatory:* FDA compliance; *Properties:* water-wh. clear oily liq., tasteless, odorless; disp. in org. solvs.; sp.gr. 0.973; dens. 8.0 lb/gal; visc. 350 cs; pour pt. -58 F; flash pt. (PMCC) 204 C; ref. index 1.4030.

SGL®. [Calgon Carbon] Activated carbon; CAS 64365-11-3; used for food decolorization; *Properties:* 8 x 30 sieve size.

Shasta®. [Bunge Foods] Partially hydrogenated soybean oil; EINECS 232-410-2; shortening for non-dairy prods., mellorine, fillings, snack dips, coffee whiteners, cheese foods; *Properties:* solid, 1.3 red color, bland flavor; m.p. 98 F.

Shedd's Liquid Margarine. [Van Den Bergh Foods] Liq. soybean oil, partially hydrog. soybean oil, water, salt, veg. mono- and diglycerides, soy lecithin, sodium benzoate, citric acid, and calcium disodium EDTA (preservatives), β-carotene (colorant), vitamin A palmitate; pumpable margarine for wide variety of applics. where a full margarine flavor is required; pourable at R.T.; *Properties:* lt. yel. liq., mild butter-like flavor; 80% min. fat, 17.4-18.0% moisture, 1.9-2.1% salt.

Shedd's Margarine. [Van Den Bergh Foods] Liq. soybean oil, partially hydrog. soybean oil, salt, soy lecithin, citric acid and sodium benzoate (preservatives), whey, veg. mono- and diglycerides, artificial flavors, beta carotene (colorant), vitamin A palmitate; low m.p. table grade margarine for various bakery and processed food applics.; *Properties:* lt. yel. plastic, mild butter-like flavor; m.p. 92-98 F;

80% min. fat, 17.4-18% moisture, 1.8-2.2% salt.

Shedd's NP Margarine. [Van Den Bergh Foods] Liq. soybean oil, partially hydrog. soybean oil, water, salt, soy lecithin, veg. mono- and diglycerides, annatto and turmeric (colorants), natural flavor, vitamin A palmitate; nonpreserved margarine with natural color and flavor; table grade margarine for baking and various processed foods; *Properties:* lt. yel. plastic, butter-like flavor; m.p. 92-98 F; 80% min. fat, 17.5-18.1% moisture, 2% salt.

Shedd's Special 40 Butter Blend Margarine. [Van Den Bergh Foods] Butter, liq. soybean oil, partially hydrog. soybean oil, water, salt, soy lecithin, sodium benzoate and citric acid (preservatives), whey, veg. mono- and diglycerides, artificial flavor, β-carotene (colorant), vitamin A palmitate; blend of margarine and butter for use in baked goods, as a table spread, dressing or topping; *Properties:* lt. yel. plastic, butter-like flavor; m.p. 90-96 F; 80% min. fat, 16.4-18.4% moisture, 1.7-2.3% salt.

Shedd's Special 60 Butter Blend Margarine. [Van Den Bergh Foods] Butter, liq. soybean oil, partially hydrog. soybean oil, water, salt, soy lecithin, sodium benzoate and citric acid (preservatives), whey, veg. mono- and diglycerides, artificial flavor, β-carotene (colorant), vitamin A palmitate; blend of margarine and butter for use in baked goods, as a table spread, dressing or topping; *Properties:* lt. yel. plastic, butter-like flavor; 80% min. fat, 16.4-18.4% moisture, 1.7-2.3% salt.

Shedd's Wonder Shortening. [Van Den Bergh Foods] Partially hydrog. soybean and cottonseed oil, water, veg. mono- and diglycerides, salt, nonfat dry milk, TBHQ (flavor protectant), natural flavor; margarine for baked goods, esp. breads; wide plastic range for use over a variety of temps.; *Properties:* wh. plastic, milk butter-like flavor; m.p. 114-122 F; 90% min. fat, 7.2-7.8% moisture, 1.3-1.7% salt.

Sherbelizer®. [Kelco] Propylene glycol alginate blend; emulsifier, thickener, stabilizer used for sherbets, soft-serve mixes, dispensed milk shakes, ice cream, sour cream and dips; *Properties:* 14 mesh; milk-sol.; visc. 300 cps (2%); pH 7.0.

Shokusen SE. [Mitsubishi Kasei] Sugar esters, propylene glycol, ethanol, and sodium citrate; biodeg. detergent with minimal foaming for washing agrochemicals and bacteria from foods, vegetables, fruit.

Shrimp Extract #51. [Ariake USA] Shrimp extract; flavorings for snacks, soup mixes, sauces, gravies; 50% moisture, 18.75% protein.

Shrimp Powd. #50. [Ariake USA] Shrimp powd.; flavorings for snacks, soup mixes, sauces, gravies; 5% moisture, 33% protein.

Shur-Fil® 427. [A.E. Staley Mfg.] Food starch modified (waxy corn starch); processing aid during prep. and filling of canned, retorted foods, e.g., soups, stews, chop suey, chili, vegetables; cooks up at pre-canning temps. to a thick, stable visc. which holds food particles in uniform suspension for controlled dispensing; *Usage level:* 3.25-3.75%; *Regu-*

latory: FDA 21CFR §172.892.

Sicovit®. [BASF AG] Sol. colorants and pigments for coloring foodstuffs.

Silver Star BMS 3251. [Grindsted Prods.] Hydrocolloid blend; stabilizer for buttermilk; good body, clean flavor, smooth texture; *Usage level:* 0.35-0.4%.

Silver Star CDS 3151. [Grindsted Prods.] Hydrocolloid blend; stabilizer for cottage cheese dressing; creamy texture; *Usage level:* 0.35%.

Silver Star SCS 3001. [Grindsted Prods.] Hydrocolloid blend; stabilizer for sour cream; good body and texture; *Usage level:* 1.3-2.1%.

Silver Star WCS 3451. [Grindsted Prods.] Hydrocolloid blend; stabilizer for whipping cream; good foam stability; *Usage level:* 0.03-0.05%.

Silver Star YS 3302. [Grindsted Prods.] Hydrocolloid blend; stabilizer for yogurt; pudding-like texture; *Usage level:* 1%; *Regulatory:* kosher.

Single Strength Acid Proof Caramel Colour. [MLG Enterprises Ltd.] Caramel color; CAS 8028-89-5; EINECS 232-435-9; natural colorant for general food and beverage use incl. soft drinks, soups, sauces, other food prods., pharmaceuticals; *Properties:* sp.gr. 1.315-1.325 (60 F); dens. 10.95-11.03 lb/gal (60 F); visc. 400 cps max. (68 F); pH 2.7-3.0; *Storage:* 2 yr min. shelf life.

Sipernat® 22. [Degussa] Syn. amorphous precipitated silica; CAS 112926-00-8; adsorbent, anticaking and free-flow agents for foods (table salt, powd. sweetener); used as aid to convert liqs. into powds.; processing aid; *Usage level:* 0.1-0.5% (table salt), 0.5% (powd. sweetener); *Regulatory:* 21CFR §172.480, 173.340, 175.105, 175.300, 176.200, 176.210, 177.1200, 177.2600, 178.3210; *Properties:* wh. fluffy powd., particle size 18 nanometer; tapped dens. 260 g/l; surf. area 190 m²/g; pH 6.2 (5%); 98% SiO_2, 6% moisture.

Sipernat® 22S. [Degussa] Syn. amorphous precipitated silica; CAS 112926-00-8; free-flow/anticaking agent for foodstuffs (spices, guar gum, dried soup, powd. cheese); *Usage level:* 0.25-2%; 0.1-2% (spices), 1% (guar gum, powd. cheese), 1-2% (soup); *Regulatory:* 21CFR §172.480, 173.340, 175.105, 175.300, 176.200, 176.210, 177.1200, 177.2600, 178.3210; *Properties:* loose wh. powd.; tapped dens. 90 g/l; surf. area 190 m²/g; pH 6.2 (5%); 98% SiO_2, 5.5% moisture.

Sipernat® 50. [Degussa] Syn. amorphous precipitated silica; CAS 112926-00-8; carrier, free-flow agent, anticaking agent for food industries (spices, high fat seasoning, cake mixes); *Usage level:* 0.25-2%; 0.1-2% (spices), 1% (high fat seasoning), 1.5% (cake mixes); *Regulatory:* 21CFR §172.480, 173.340, 175.105, 175.300, 176.200, 176.210, 177.1200, 177.2600, 178.3210; *Properties:* loose wh. powd.; pH 6.0-7.4 (5%); 7% max. moisture; *Toxicology:* inhalation may cause respiratory tract irritation.

Sipernat® 50S. [Degussa] Syn. amorphous precipitated silica; CAS 112926-00-8; free-flow, anticaking and defoaming agent for food use (cocoa powd., dried soup, cake mix); *Usage level:* 0.25-2%; 0.5-2% (cocoa powd.), 0.5% (dried soup), 1.5% (cake mix); *Regulatory:* 21CFR §172.480, 173.340, 175.105, 176.200, 176.210, 175.300, 177.1200, 177.2600, 178.3210; *Properties:* wh. loose powd.; 7 μm avg. agglomerate size; tapped dens. 100 g/l; surf. area 480 m²/g; pH 6.7 (5%); 99% SiO_2, 5% moisture; *Toxicology:* inhalation may cause respiratory tract irritation.

Sipernat® 50S. [Degussa] Precipitated silica; CAS 112926-00-8; free-flow agent for food industry; *Regulatory:* FDA approved; *Properties:* loose wh. powd.; tapped dens. 100 g/l; surf. area 480 m²/g; pH 6.7 (5%).

Siverslice. [Courtaulds plc] Proteinaceous substances used as food or food ingreds.

62/43 Corn Syrup. [MLG Enterprises Ltd.] Acid-enzyme converted corn syrup; CAS 8029-43-4; EINECS 232-436-4; sweetener with high proportion of fermentable carbohydrates and increased sweetness; for baked goods (bread, rolls), nougats, fondants, etc.; *Properties:* water-wh. syrup, char. odor, sweet taste; sp.gr. 1.4201 (100/60 F); dens. 11.84 lb/gal (100 F); visc. 400 poises (80 F); pH 4.8 (1:1); DE 62; Baumé 43 (100 F); 82% total solids, 18% moisture.

Skipjack Liver Oil 30,000A/20,000D. [Arista Industries] Skipjack liver oil; nutritive additive contg. 30,000 IU/g vitamin A, 20,000 IU/g vitamin D; *Properties:* oil; sp.gr. 0.915-0.925; iodine no. 120-160; sapon. no. 178-190.

Skipjack Liver Oil 40,000A/20,000D. [Arista Industries] Skipjack liver oil; nutritive additive contg. 40,000 IU/g vitamin A, 20,000 IU/g vitamin D; *Properties:* oil; sp.gr. 0.915-0.925; iodine no. 120-160; sapon. no. 178-190.

S-Maz® 20. [PPG/Specialty Chem.] Sorbitan laurate; CAS 1338-39-2; nonionic; lubricant, process defoamer, opacifier, coemulsifier, solubilizer, dispersant, suspending agent, coupler; with T-Maz series used as o/w emulsifiers in food formulations; *Properties:* amber liq.; sol. in ethanol, naphtha, min. oils, toluene, veg. oil; water-disp.; sp.gr. 1.0; visc. 4500 cps; HLB 8.0; acid no. 7 max.; sapon. no. 158-170; hyd. no. 330-358; flash pt. (PMCC) > 350 F; 100% conc.

S-Maz® 60K. [PPG/Specialty Chem.] Sorbitan stearate; CAS 1338-41-6; EINECS 215-664-9; nonionic; food emulsifier for veg. and dairy prods., cakes; gloss aid in chocolate, nondairy creamers; synthetic flavoring; defoamer in food processing; protective coatings on raw fruits/vegetables; *Usage level:* 0.4% max. (whipped toppings), 0.61% (cakes), 0.7% (icings, fillings), 1% (confectionery coatings), GMP (fruit/veg. coating, flavor, defoamer); *Regulatory:* FDA 21CFR §172.515, 172.842, 173.340, 573.960; kosher; *Properties:* wh. to cream flakes, ester odor; sol. in ethanol; disp. in water; sp.gr. 0.954 (70 C); m.p. 52 C; b.p. > 300 F; HLB 4.7; acid no. 10 max.; sapon. no. 147-157; hyd. no. 235-260; flash pt. (PMCC) > 350 F; *Toxicology:* LD50 (oral, rat) > 32 g/kg; nonirritating;

Storage: store in well-ventilated area below 120 F.

S-Maz® 60KHS. [PPG/Specialty Chem.] Sorbitan stearate; CAS 1338-41-6; EINECS 215-664-9; nonionic; high melt grade food emulsifier; *Regulatory:* kosher; *Properties:* Gardner 3 flake; sol. in ethanol; insol. in water, min. oils; HLB 4.7; m.p. 55-60 C; acid no. 10 max.; sapon. no. 147-157; hyd. no. 235-260; flash pt. (PMCC) > 350 F.

S-Maz® 80K. [PPG/Specialty Chem.] Sorbitan oleate; CAS 1338-43-8; EINECS 215-665-4; nonionic; lubricant, process defoamer, opacifier, coemulsifier, solubilizer, dispersant, suspending agent, coupler; prepares exc. w/o emulsions; kosher grade; *Properties:* Gardner 6 liq.; sol. in min. oils, min. spirits, toluene, veg. oil; insol. in water; HLB 4.6; acid no. 8 max.; sapon. no. 149-160; hyd. no. 193-209; flash pt. (PMCC) > 350 F.

S-Maz® 85K. [PPG/Specialty Chem.] Sorbitan trioleate; CAS 26266-58-0; EINECS 247-569-3; nonionic; lubricant, process defoamer, opacifier, coemulsifier, solubilizer, dispersant, suspending agent, coupler; prepares exc. w/o emulsions; kosher grade; *Properties:* Gardner 6 liq.; sol. in min. oils, min. spirits, toluene, veg. oil; insol. in water; HLB 2.1; acid no. 14 max.; sapon. no. 172-186; hyd. no. 56-68; flash pt. (PMCC) > 350 F.

S-Maz® 95. [PPG/Specialty Chem.] Sorbitan tritallate; nonionic; lubricant, process defoamer, opacifier, coemulsifier; prepares w/o emulsions; together with T-Maz series, as o/w emulsifier for food formulations; *Properties:* amber liq.; sol. in toluol, naphtha, min. oil, veg. oil; misc. with ethanol, acetone; disp. in water; sp.gr. 0.9; visc. 200 cps; HLB 1.9; acid no. 15 max.; sapon. no. 168-186; hyd. no. 55-85; flash pt. (PMCC) 350 F.

Snac-Kote. [Van Den Bergh Foods] Partially hydrog. veg. oil (cottonseed, soybean), sorbitan stearate, polysorbate 60; coating fat for bakery coatings; *Regulatory:* kosher.

Snac-Kote XTR. [Van Den Bergh Foods] Partially hydrog. veg. oil (soybean, cottonseed); EINECS 269-820-6; nonlauric coating fat for bakery coatings, veg. dairy prods.; *Regulatory:* kosher; *Properties:* solid; drop pt. 39-41 C.

Snowflake® Sweet Dough & Danish Conc. [Bunge Foods] Conc. for prod. of frozen sweet goods and danish; *Usage level:* 33% based on flour.

Snow Fresh®. [Monsanto] Proprietary blend; food additive; *Usage level:* 0.5% max. (processed meat and poultry, seafood); *Regulatory:* FCC GRAS; *Properties:* wh. free-flowing powd.; pH 2.5 ± 0.3 (1%); 92% min. assay.

Snowite® Oat Fiber. [Canadian Harvest USA] Oat hull-based fiber; natural fiber for boosting dietary fiber levels in foods without masking delicate flavors; partial flour replacement for baked goods, cereals, beverages, snack foods, pasta, meat prods., sauces, supplements, low fat systems, pharmaceuticals; *Regulatory:* GRAS; USDA approval; *Properties:* avail. in granulations from med. to micro-fine, coarse avail. on request; creamy-wh. color, neutral odor, bland flavor; 90% total dietary fiber; pH 5.5-6.5; 5% max. moisture.

Soageena®. [Mitsubishi Int'l.] Carrageenan; CAS 9000-07-1; EINECS 232-524-2; thickener, gellant, suspending agent, foam stabilizer, binder, syneresis control agent, whey prevention; for dairy prods., desserts, bakery, canned meat/fish, frozen surimi; *Regulatory:* Japan, FAO/WHO 8.355, FDA 21CFR §172.620, 172.626, 182.7255, EEC E407 compliances.

Soageena® LX22. [Mitsubishi Int'l.] Carrageenan, milk gelation type; CAS 9000-07-1; EINECS 232-524-2; nongelling type for chocolate drinks; *Usage level:* 150-500 ppm (chocolate drinks); *Regulatory:* Japan, FAO/WHO 8.355, FDA 21CFR §172.620, 172.626, 182.7255, EEC E407 compliances; *Properties:* pH 6.5-9.5 (0.5%).

Soageena® ML300. [Mitsubishi Int'l.] Carrageenan, milk gelation type; CAS 9000-07-1; EINECS 232-524-2; nongelling type for chocolate drinks, coffee; *Usage level:* 150-500 ppm (chocolate drinks, coffee); *Regulatory:* Japan, FAO/WHO 8.355, FDA 21CFR §172.620, 172.626, 182.7255, EEC E407 compliances; *Properties:* pH 6.5-9.5 (0.5%).

Soageena® MM101. [Mitsubishi Int'l.] Carrageenan, milk gelation type; CAS 9000-07-1; EINECS 232-524-2; anti-whey off aid in ice cream; *Usage level:* 150-500 ppm (ice cream); *Regulatory:* Japan, FAO/WHO 8.355, FDA 21CFR §172.620, 172.626, 182.7255, EEC E407 compliances; *Properties:* pH 6.5-9.5 (0.5%).

Soageena® MM301. [Mitsubishi Int'l.] Carrageenan, milk gelation type; CAS 9000-07-1; EINECS 232-524-2; protein activity type for milk puddings; clarity agent for alcoholic beverages; *Usage level:* 0.1-0.2% (milk pudding), 100-200 ppm (alcoholic beverages); *Regulatory:* Japan, FAO/WHO 8.355, FDA 21CFR §172.620, 172.626, 182.7255, EEC E407 compliances; *Properties:* pH 6.5-9.5 (0.5%).

Soageena® MM330. [Mitsubishi Int'l.] Carrageenan, milk gelation type; CAS 9000-07-1; EINECS 232-524-2; anti-whey off aid in soft cream prods.; *Usage level:* 150-500 ppm (soft cream); *Regulatory:* Japan, FAO/WHO 8.355, FDA 21CFR §172.620, 172.626, 182.7255, EEC E407 compliances; *Properties:* pH 6.5-9.5 (0.5%).

Soageena® MM350. [Mitsubishi Int'l.] Carrageenan, milk gelation type; CAS 9000-07-1; EINECS 232-524-2; anticrystallization agent in ice cream; *Usage level:* 150-500 ppm (ice cream); *Regulatory:* Japan, FAO/WHO 8.355, FDA 21CFR §172.620, 172.626, 182.7255, EEC E407 compliances; *Properties:* pH 6.5-9.5 (0.5%).

Soageena® MM501. [Mitsubishi Int'l.] Carrageenan, milk gelation type; CAS 9000-07-1; EINECS 232-524-2; provides soft texturized gel in milk puddings; *Usage level:* 0.1-0.2% (milk pudding); *Regulatory:* Japan, FAO/WHO 8.355, FDA 21CFR §172.620, 172.626, 182.7255, EEC E407 compliances; *Properties:* pH 6.5-9.5 (0.5%).

Soageena® MV320. [Mitsubishi Int'l.] Carrageenan, visc. type; CAS 9000-07-1; EINECS 232-524-2; emulsion and suspension synersis control;

strengthens freeze-thaw stability; *Usage level:* 0.1-0.5%; *Regulatory:* Japan, FAO/WHO 8.355, FDA 21CFR §172.620, 172.626, 182.7255, EEC E407 compliances; *Properties:* pH 6.5-9.5 (0.5%).

Soageena® MW321. [Mitsubishi Int'l.] Carrageenan, kappa, water gelation type; CAS 9000-07-1; EINECS 232-524-2; provides elastic gel in jellies; *Usage level:* 0.25-0.4% (jelly); *Regulatory:* Japan, FAO/WHO 8.355, FDA 21CFR §172.620, 172.626, 182.7255, EEC E407 compliances; *Properties:* pH 6.5-9.5 (0.5%).

Soageena® MW351. [Mitsubishi Int'l.] Carrageenan, kappa, water gelation type; CAS 9000-07-1; EINECS 232-524-2; provides hard gel in jellies; *Usage level:* 0.25-0.4% (jelly); *Regulatory:* Japan, FAO/WHO 8.355, FDA 21CFR §172.620, 172.626, 182.7255, EEC E407 compliances; *Properties:* pH 6.5-9.5 (0.5%).

Soageena® MW371. [Mitsubishi Int'l.] Carrageenan, kappa, water gelation type; CAS 9000-07-1; EINECS 232-524-2; provides hard elastic gel in jellies; *Usage level:* 0.3-0.6% (jelly); *Regulatory:* Japan, FAO/WHO 8.355, FDA 21CFR §172.620, 172.626, 182.7255, EEC E407 compliances; *Properties:* pH 6.5-9.5 (0.5%).

Soageena® WX57. [Mitsubishi Int'l.] Carrageenan; CAS 9000-07-1; EINECS 232-524-2; meat binder; *Regulatory:* Japan, FAO/WHO 8.355, FDA 21CFR §172.620, 172.626, 182.7255, EEC E407 compliances; *Properties:* pH 6.5-9.5 (0.5%).

Soalocust®. [Mitsubishi Int'l.] Locust bean gum; CAS 9000-40-2; EINECS 232-541-5; nonionic; provides visc. (ice cream, sherbet, mayonnaise), water retention (sausage, ham, jelly, pudding), and dietary fiber to foods; *Regulatory:* Japan, USA, EEC, FAO/WHO approved; *Properties:* sol. in water.

Sobalg FD 000 Range. [Grindsted Prods.] Alginic acid; CAS 9005-32-7; EINECS 232-680-1; provides solubility and gelation control in baked goods, low-fat spreads, restructured fruits, vegetables, meat, and fish, fruit preps., cream fillings, and milk-based prods.; *Regulatory:* EEC, FDA §184.1011 (GRAS).

Sobalg FD 100 Range. [Grindsted Prods.] Sodium alginate; CAS 9005-38-3; food stabilizer, gellant, film-former for dairy prods., desserts, beverages, structured fruits and vegetables; avail. in various granulations and visc.; low calcium content; *Regulatory:* EEC E 401, FDA §184.1724, GRAS; *Properties:* off-wh. to ylsh. powd.; sol. in water; visc. < 50 to > 950 cP; pH 5.5-7.5 (1%); 90.8-106% assay; *Storage:* 12 mos storage under cool, dry conditions.

Sobalg FD 200 Range. [Grindsted Prods.] Potassium alginate; CAS 9005-36-1; gellant for foods (baked goods, low-fat spreads, restructured fruit/veg./meat/fish, fruit preps., cream fillings, milk-based prods.); avail. in various granulations and visc.; very low calcium content; *Regulatory:* EEC, FDA §184.1610 (GRAS); *Properties:* off-wh. to ylsh. powd.; sol. in water; visc. 300-650 cP; pH 5.5-7.5 (1%); 89.2-105.5% assay; *Storage:* 6 mo storage under cool, dry conditions.

Sobalg FD 300 Range. [Grindsted Prods.] Ammonium alginate; CAS 9005-34-9; gellant for foods (baked goods, low-fat spreads, restructured fruit/veg./meat/fish, fruit preps., cream fillings, milk-based prods.); avail. in various visc.; very low calcium content; *Regulatory:* EEC, FDA §184.1133 (GRAS); *Properties:* off-wh. to ylsh. powd.; sol. in water; visc. 125-850 cP; pH 5.5-7.5 (1%); 88.7-103.6% assay; *Storage:* 6 mo storage under cool, dry conditions.

Sobalg FD 460. [Grindsted Prods.] Calcium alginate; CAS 9005-35-0; gellant for foods (baked goods, low-fat spreads, restructured fruit/veg./meat/fish, fruit preps., cream fillings, milk-based prods.); *Regulatory:* EEC, FDA §184.1187 (GRAS); *Properties:* off-wh. to ylsh. powd.; 98% min. thru 100 mesh (French); insol. in water; 89.6-104.5% assay; *Storage:* 12 mo storage under cool, dry conditions.

Sobalg FD 900 Range. [Grindsted Prods.] Special alginate blend; provides solubility and gelation control in baked goods, low-fat spreads, restructured fruits, vegetables, meat, and fish, fruit preps., cream fillings, and milk-based prods.

Sodaphos®. [FMC/Food Phosphates] Short chain sodium hexametaphosphate; EINECS 233-343-1; emulsifier, sequestrant, suspending agent, protein stabilizer for dairy prods. esp. process cheese, ice cream, and frozen desserts; protein dispersant in whey processing; water binding agent for cured pork; *Usage level:* 2% (process cheese), 0.1-0.5% (whey), 0.1-0.2% (ice cream); *Regulatory:* FDA GRAS; *Properties:* powd., gran., plate; 99.9% thru 20 mesh (powd.), 100% thru 8 mesh (gran.); infinitely sol. in water; m.w. 664; bulk dens. 62 lb/ft³ (powd.), 79 lb/ft³ (gran.), 75 lb/ft³ (plate); pH 7.7 (1%).

Sodium Alginate HV NF/FCC. [Meer] Sodium alginate NF/FCC; CAS 9005-38-3; food grade emulsifier, stabilizer, suspending agent for ice cream, sherbets, soft drinks, bakery jellies, prevention of freezer burn; freeze/thaw stabilizer for lemon pie fillings; *Properties:* water-sol.

Sodium Alginate LV. [Meer] Sodium alginate; CAS 9005-38-3; food grade emulsifier, stabilizer, suspending agent for ice cream, sherbets, soft drinks, bakery jellies, prevention of freezer burn; freeze/thaw stabilizer for lemon pie fillings; *Properties:* water-sol.

Sodium Alginate LVC. [Meer] Sodium alginate; CAS 9005-38-3; emulsifier for ice cream, sherbets, water ices; suspending agent in soft drinks; freeze/thaw stabilizer; prevents freezer burn; gellant in puddings, fabrication of sausage casings; thickener in whipped cream, jellies, fillings, sauces, cheese, candy; *Usage level:* 0.4-1.0%; *Properties:* off-wh. to tan free-flowing powd.; ≥ 98% thru 30 mesh; visc. ≥ 50 cps (1%); pH 5.5-8.0.

Sodium Alginate MV NF/FCC. [Meer] Sodium alginate NF/FCC; CAS 9005-38-3; food grade emulsifier, stabilizer, suspending agent for ice cream, sherbets, soft drinks, bakery jellies, prevention of freezer burn; freeze/thaw stabilizer for lemon pie fillings; *Properties:* water-sol.

Sodium Ascorbate USP, FCC Fine Gran. No. 6047709. [Roche] Sodium ascorbate USP, FCC; CAS 134-03-2; EINECS 205-126-1; preservative, antioxidant, and vitamin C source in foods, esp. where neutral pH and less acid taste are important, and where free-flowing gran. prod. is required; *Regulatory:* FDA GRAS; *Properties:* wh. to ylsh. cryst. powd., pract. odorless, pleasantly saline taste with tart overtone; 100% thru 30 mesh, 15% max. thru 100 mesh; sol. 1 g/2 ml water; m.w. 198.11; bulk dens. 0.8-1.1 (tapped); pH 7-8 (10% aq.); 99-101% assay; *Precaution:* oxidizes readily in sol'n.; exposure to light or atmospheric moisture may darken prod.; avoid contact with iron, copper, or nickel salts; *Storage:* store in tight, light-resist. containers, optimally @ ≤ 72 F; avoid exposure to moisture and excessive heat.

Sodium Ascorbate USP, FCC Fine Powd. No. 6047708. [Roche] Sodium ascorbate USP, FCC; CAS 134-03-2; EINECS 205-126-1; preservative, antioxidant, and vitamin C source in foods, esp. where rapid sol., neutral pH, and less acid taste is important; *Regulatory:* FDA GRAS; *Properties:* wh. to ylsh. powd. or cryst., pract. odorless, pleasantly saline taste with tart overtone; 98% min. thru 80 mesh; sol. 1 g/2 ml water; m.w. 198.11; bulk dens. 0.6-1.1 (tapped); pH 7-8 (10% aq.); 99-101% assay; *Precaution:* oxidizes readily in sol'n.; exposure to light or atmospheric moisture may darken prod.; avoid contact with iron, copper, or nickel salts; *Storage:* store in tight, light-resist. containers, optimally @ ≤ 72 F; avoid exposure to moisture and excessive heat.

Sodium Ascorbate USP, FCC Type AG No. 6047710. [Roche] Sodium ascorbate USP, FCC; CAS 134-03-2; EINECS 205-126-1; preservative, antioxidant, and vitamin C source in foods, esp. where neutral pH and less acid taste are important; *Regulatory:* FDA GRAS; *Properties:* wh. to ylsh. cryst. powd., pract. odorless, pleasantly saline taste with tart overtone; 99.5% thru 40 mesh, 7% max. thru 140 mesh; sol. 1 g/2 ml water; m.w. 198.11; bulk dens. 0.6-0.8(tapped); pH 7-8 (10% aq.); 99-101% assay; *Precaution:* oxidizes readily in sol'n.; exposure to light or atmospheric moisture may darken prod.; avoid contact with iron, copper, or nickel salts; *Storage:* store in tight, light-resist. containers, optimally @ ≤72 F; avoid exposure to moisture and excessive heat.

Sodium Benzoate BP88. [Pentagon Chem. Ltd; Unipex] Sodium benzoate BP, EP; CAS 532-32-1; EINECS 208-534-8; antimicrobial preservative for beverages, soft drinks, fruit juices, preserves, jams, pie fillings, pickles, margarine, processed meat; antifermentation additive for wines, cider; *Usage level:* 0.01-0.20% (foods), 0.05-0.15% (wines); *Properties:* wh. gran. or flaky powd.; 99% min. assay.

Sodium Bicarbonate T.F.F. Treated Free Flowing. [Rhone-Poulenc Food Ingreds.] Sodium bicarbon-ate, tricalcium phosphate FCC; used for foods (self-rising four and corn meal); *Properties:* wh. free-flowing cryst., odorless; 2% min. on 80 mesh, 60% max. on 325 mesh; bulk dens. 60 lb/ft³; 99% min. assay.

Sodium Bicarbonate USP No. 1 Powd. [Rhone-Poulenc Food Ingreds.] Sodium bicarbonate USP; CAS 144-55-8; EINECS 205-633-8; used for foods (colors, conditioners, starches, candies); *Properties:* wh. free-flowing cryst., odorless; 2% min. on 80 mesh, 20% max. on 200 mesh; bulk dens. 60 lb/ft³; 99% min. assay.

Sodium Bicarbonate USP No. 2 Fine Gran. [Rhone-Poulenc Food Ingreds.] Sodium bicarbonate USP; CAS 144-55-8; EINECS 205-633-8; used for foods (baking powds., cakes, pancake, dry mixes); *Properties:* wh. free-flowing cryst., odorless; 2% min. on 80 mesh, 70% max. on 200 mesh; bulk dens. 65 lb/ft³; 99% min. assay.

Sodium Bicarbonate USP No. 5 Coarse Gran. [Rhone-Poulenc Food Ingreds.] Sodium bicarbon-ate USP; CAS 144-55-8; EINECS 205-633-8; used for foods (carbonated beverage mixes); *Properties:* wh. free-flowing cryst., odorless; 50% max. on 80 mesh, 90% max. 170 mesh; bulk dens. 62 lb/ft³; 99% min. assay.

Sodium Citrate USP, FCC Dihydrate Gran. No. 69976. [Roche] Sodium citrate dihydrate USP FCC; controls acidity in candies, jams, jellies, pre-serves, powd. beverages, gelatin desserts; emulsi-fier in processed cheese; stabilizer in dairy and veg.-based whipped creams; *Regulatory:* FDA GRAS; *Properties:* colorless cryst. or wh. cryst. powd., pract. odorless, saline taste; 3% max. on 16 mesh, 3% max. thru 140 mesh; sl. deliq. in moist air; sol. 65 g/100 ml water; insol. in alcohol; m.w. 294.10; 99-100.5% assay; *Storage:* 12 mos storage life in original, unopened containers at temps. below 25 C.

Sodium Citrate USP, FCC Dihydrate Fine Gran. No. 69975. [Roche] Sodium citrate USP, FCC; CAS 68-04-2; EINECS 200-675-3; pH adjustor, buffer-ing agent; controls acidity in candies, jams, jellies, preserves, powd. beverages, gelatin desserts; emulsifier in processed cheese; stabilizer in dairy and veg.-based whipped creams; *Regulatory:* FDA GRAS; *Properties:* colorless cryst. or wh. cryst. powd., pract. odorless, saline taste; 1% max. on 3 mesh, 10% max. thru 100 mesh; sl. deliq. in moist air; sol. 65 g/100 ml water; insol. in alcohol; m.w. 294.10; 99-100.5% assay; *Storage:* 12 mos storage life in original unopened pkgs. at temps. below 25 C.

Softenol® 3108. [Hüls Am.] C8-10 fatty acid triglycer-ide; lubricant for prod. of sausage casings and for machinery used in food and confectionery indus-tries; *Properties:* liq., neutral odor and taste; dens. 0.950 g/cm³; visc. 25-35 mPa•s; turbidity pt. 10 C; acid no. 0.2 max.; iodine no. 1 max.; sapon. no. 325-345.

Soft-N-Rich® Conc. [Bunge Foods] Sweet dough conc. formulated to deliver prods. with extended 14 day shelf life; yields light, tender, rich tasting sweet dough; for cinnamon rolls, raisin buns, coffee cake; *Usage level:* 33% based on flour.

Soft-Set®. [A.E. Staley Mfg.] Food starch modified, derived from dent corn starch; cold water swelling instant starch with low initial visc., forming soft gel in 60 min; stabilizer; exc. moisture binding props.; for frostings, salad dressings, instant desserts, cream fillings, microwave foods; improves whipping props.; *Regulatory:* FDA 21CFR §172.892; *Properties:* wh. fine powd.; 0.5% max. on 30 mesh, 95% min. thru 100 mesh; 70% cold water sol.; pH 5.5; 7.5% moisture.

Soft Touch. [ADM Arkady] Hydrated mono- and diglycerides, sodium stearoyl lactylate, and polysorbate 60; hydrated emulsifier system with softener/strengtheners for use in cake prods. and bread; *Properties:* HLB 4.8; 13.5% min. alpha mono.

Solid Invert Sugar. [MLG Enterprises Ltd.] Invert sugar; CAS 8013-17-0; sweetener used in baking and candy trades; *Properties:* solid; sp.gr. 1.369; pH 4.5-5.5; 90-95% invert on solids, 5-10% sucrose on solids.

Solka-Floc®. [Protein Tech. Int'l.] Powd. cellulose; CAS 9004-34-6; EINECS 232-674-9; functional ingred. or noncaloric bulking agent for food, pharmaceutical, and industrial applics.

Soluble Liver Powd. [Am. Labs] Enzymatic digest of liver; nutritional supplement for tablets and capsules; also for pharmaceutical and veterinary preps.; *Properties:* brn. powd., char. odor; hygroscopic; completely sol. in water; pH 4.8 (2%); *Storage:* store in cool, dry area.

Soluble Trachea CS 16 Substance. [Am. Labs] Glandular prod. contg. 16-17% chondroitin sulfate; nutritive food additive; *Properties:* off-wh. to lt. tan free-flowing powd.; hygroscopic; sol.: clear to hazy sol'n. (5% in water).

Soluble Trachea Substance. [Am. Labs] Glandular prod. contg. 8-10% chondroitin sulfate; nutritive food additive; *Properties:* off-wh. to lt. tan free-flowing powd.; hygroscopic; sol.: clear to hazy sol'n. (5% in water).

Sol-U-Tein EA. [Fanning] Albumen; CAS 9006-50-2; binder, coagulant, film-former for food applics.; *Properties:* yel. powd.; 100% thru 80 mesh; 90% thru 100 mesh; bland odor; pH 6.5-8.0; 75% ovalbumin, ovoconalbumin, ovomucoid, ovomucin, ovoglobulin, lysozyme, and avidin.

Sorbanox AL. [Witco/H-I-P] Sorbitan ester, ethoxylated; nonionic; detergent used in vegetable cleaning; *Properties:* liq.; 100% conc.; Unverified

Sorbax PML-20. [Chemax] Polysorbate 20; CAS 9005-64-5; nonionic; o/w emulsifier, solubilizer for flavors; *Properties:* liq.; water-sol.; sapon. no. 45; HLB 16.7; 100% conc.

Sorbax PMO-5. [Chemax] Polysorbate 81; CAS 9005-65-6; nonionic; o/w emulsifier, solubilizer for flavors; *Properties:* liq.; water-sol.; sapon. no. 100; HLB 10.0; 100% conc.

Sorbax PMO-20. [Chemax] Polysorbate 80; CAS 9005-65-6; nonionic; o/w emulsifier, solubilizer for flavors; *Properties:* liq.; water-sol.; sapon. no. 50; HLB 15.0; 100% conc.

Sorbax PMP-20. [Chemax] Polysorbate 40; CAS 9005-66-7; nonionic; o/w emulsifier, solubilizer for flavors; *Properties:* liq.; water-sol.; sapon. no. 46; HLB 15.6; 100% conc.

Sorbax PMS-20. [Chemax] Polysorbate 60; CAS 9005-67-8; nonionic; o/w emulsifier, solubilizer for flavors; *Properties:* soft paste; water-sol.; sapon. no. 50; HLB 14.9; 100% conc.

Sorbax PTO-20. [Chemax] Polysorbate 85; CAS 9005-70-3; nonionic; o/w emulsifier, solubilizer for flavors; *Properties:* liq., gel; water-sol.; sapon. no. 88; HLB 11.0; 100% conc.

Sorbax PTS-20. [Chemax] Polysorbate 65; CAS 9005-71-4; nonionic; o/w emulsifier, solubilizer for flavors; *Properties:* solid; water-sol.; sapon. no. 93; HLB 10.5; 100% conc.

Sorbestrin. [Pfizer Food Science] A frying oil substitute with 1 calorie of fat/gram; *Regulatory:* awaiting approval.

Sorbo®. [ICI Am.; ICI Atkemix] Sorbitol sol'n.; CAS 50-70-4; EINECS 200-061-5; bodying agent, humectant, flavor, anticaking agent, curing/pickling agent, drying agent, emulsifier, firming agent, formulation aid, lubricant/release, nutritive sweetener, sequestrant, stabilizer, thickener, texturizer; *Regulatory:* FDA 21CFR §184.1835, GRAS; kosher; *Properties:* water-wh. liq., essentially odorless, pleasantly sweet taste; sp.gr. > 1.285; visc. 110 cps; ref. index 1.458; 70% act.

Sorgen 30. [Dai-ichi Kogyo Seiyaku] Sorbitan sesquioleate; CAS 8007-43-0; EINECS 232-360-1; nonionic; antifoamer and emulsifier for foods; *Properties:* liq.; HLB 3.7; 100% conc.

Sorgen 40. [Dai-ichi Kogyo Seiyaku] Sorbitan oleate; CAS 1338-43-8; EINECS 215-665-4; nonionic; antifoamer and emulsifier for foods; *Properties:* liq.; HLB 4.3; 100% conc.

Sorgen 50. [Dai-ichi Kogyo Seiyaku] Sorbitan stearate; CAS 1338-41-6; EINECS 215-664-9; nonionic; antifoamer and emulsifier for foods; *Properties:* flake; HLB 4.7; 100% conc.

Sorgen 90. [Dai-ichi Kogyo Seiyaku] Sorbitan laurate; nonionic; antifoamer and emulsifier for foods; *Properties:* liq.; HLB 8.6; 100% conc.

Sorgen S-30-H. [Dai-ichi Kogyo Seiyaku] Sorbitan sesquioleate; CAS 8007-43-0; EINECS 232-360-1; nonionic; antifoamer and emulsifier for foods; *Properties:* liq.; HLB 3.7; 100% conc.

Sorgen S-40-H. [Dai-ichi Kogyo Seiyaku] Sorbitan oleate; CAS 1338-43-8; EINECS 215-665-4; nonionic; antifoamer and emulsifier for foods; *Properties:* liq.; HLB 4.3; 100% conc.

Sorgen TW20. [Dai-ichi Kogyo Seiyaku] POE-20 sorbitan laurate; CAS 9005-64-5; nonionic; emulsifier and antifoaming agent for foods; *Properties:* liq.; water-sol.; HLB 16.7; 100% conc.

Sorgen TW60. [Dai-ichi Kogyo Seiyaku] POE sorbitan stearate; nonionic; emulsifier and antifoaming agent for foods; *Properties:* paste; water-sol.; HLB 14.9; 100% conc.

Sorgen TW80. [Dai-ichi Kogyo Seiyaku] POE sorbitan monooleate; nonionic; emulsifier and antifoaming agent for foods; *Properties:* liq.; water-sol.; HLB 15.0; 100% conc.

Sour Base. [Brolite Prods.] Plastic bread base for San Francisco style sourdough bread; *Usage level:* 6 lb/ 100 lb flour.

Sour Base M. [ADM Arkady] Imparts a tangy, natural fermentation flavor to specialty prods., bread, and rolls.

Sour Dough Base. [Am. Ingreds.] Wheat flour, potato flour, buttermilk solids, monocalcium phosphate, whey powd., lactic acid, phosphoric acid, and acetic acid; sour for sour-dough breads and snacks; *Properties:* free-flowing powd.; pH 2.9-3.3; 10-12% protein.

Sovital™. [Lucas Meyer] Toasted edible soya bran; bulking agent providing dietary fiber to foodstuffs, baked goods.

Soyafluff® 200 W. [Central Soya] Soy flour; CAS 68513-95-1; functional soy flour for food industry; *Properties:* particle size 90% min. thru #200 sieve; 52% min. protein, 8.5% max. moisture; *Storage:* store in cool, dry place below 26.5 C at ≤ 60% r.h.

Soyamin 50 E. [Lucas Meyer] Defatted soya flour, enzyme active; CAS 68513-95-1; produces lighter crumb, extends shelf life and freshness, provides protein enrichment for bread improvers.

Soyamin 50 T. [Lucas Meyer] Defatted soya flour, toasted; CAS 68513-95-1; protein enrichment for foodstuffs.

Soyamin 70. [Lucas Meyer] Soya conc.; emulsifier, water and fat binding agent, protein enrichment for vegetable meals, dietetic foods, meat and sausage prods.

Soyamin 90. [Lucas Meyer] Soya isolate; emulsifier, water and fat binding agent, protein enrichment for baby food, dietetic food, vegetarian meals, meat and sausage prods.

Soyapan™. [Lucas Meyer] Fullfat soya flour, enzyme active; CAS 68513-95-1; emulsifying agent, flour improver; extends shelf life and freshness, provides lighter crumb; for white bread and toast, ready mixes, bread improvers, baking additives; *Properties:* powd.

Soyarich® 115 W. [Central Soya] Soy flour and lecithin (15%); functional soy flour for food industry; *Properties:* particle size 90% min. thru #100 sieve; 40% min. protein, 8.5% max. moisture; *Storage:* store in cool, dry place below 26.5 C at ≤ 60% r.h.

Soybean Oil. [Van Den Bergh Foods] Choice refined soybean oil; CAS 8001-22-7; EINECS 232-274-4; general purpose oil for cooking, baking, and frying applics.; *Properties:* Lovibond 1.5R max. liq.

Soybean Salad Oil 104-050. [ADM Packaged Oils] Soybean oil, TBHQ; packaged veg. salad oil; *Properties:* clear bright oil, bland flavor; iodine no. 130 ± 6.

Soy Flakes. [Am. Ingreds.] Partially hydrog. soybean oil; EINECS 232-410-2; shortening; *Properties:* 20R color flake; c.m.p. 65-70 C; iodine no. 3 max.

Soylec C-6. [ADM Protein Spec.] Premix prod. of lecithin and Bakers Nutrisoy; for bakery mixes (donuts, sweet goods, cakes, pastries); 16% total dietary fiber; 360 calories/100 g; 50% min. protein, 6-8% fat, 9% max. moisture.

Soylec C-15. [ADM Protein Spec.] Premix prod. of lecithin and Bakers Nutrisoy; for bakery mixes (donuts, sweet goods, cakes, pastries); 16% total dietary fiber; 360 calories/100 g; 46% min. protein, 15-17% fat, 8% max. moisture.

Soylec T-15. [ADM Protein Spec.] Premix prod. of lecithin and Toasted Nutrisoy Flour; partial nonfat milk replacer; 46% min. protein, 15-17% fat, 8% max. moisture.

Soyoco. [Lucas Meyer] Roasted full fat soya kernels; protein enrichment for cereals, ready mixes, confectionery.

Soy Oil. [Am. Ingreds.] Soy oil; CAS 8001-22-7; EINECS 232-274-4; lubricant/release agent for frying and salad oil applics.; *Properties:* yel. liq.; i.v. 128-136.

Soy/Peanut Liquid Frying Shortening 104-215. [ADM Packaged Oils] Partially hydrog. soybean oil, peanut oil, TBHQ, dimethylpolysiloxane; packaged vegetable oil; *Properties:* clear bright liq., bland flavor; iodine no. 109 ± 6.

Spa Gelatin. [Croda Colloids Ltd; O.C. Lugo] Gelatin; CAS 9000-70-8; EINECS 232-554-6; fining agent for wines; combines with tannin in wine to form a flocculent precipitate which settles as insol. sediment; *Properties:* lt. colored, low-visc., bland odor and taste; sol. in warm water.

Span® 20. [ICI Spec. Chem.; ICI Surf. Am.; ICI Surf. Belgium] Sorbitan laurate NF; CAS 1338-39-2; nonionic; food emulsifier, stabilizer, thickener; *Properties:* amber liq.; sol. (@ 1%) in IPA, perchloroethylene, xylene, cottonseed oil, min. oil; sol. (hazy) in propylene glycol; visc. 4250 cps; HLB 8.6; 100% act.

Span® 40. [ICI Spec. Chem.; ICI Surf. Am.; ICI Surf. Belgium] Sorbitan palmitate NF; CAS 26266-57-9; EINECS 247-568-8; nonionic; food emulsifier, stabilizer, thickener; *Properties:* tan solid; sol. (@ 1%) in IPA, xylene; sol. (hazy) in perchloroethylene; HLB 6.7; pour pt. 48 C; 100% act.

Span® 60. [ICI Atkemix] Sorbitan stearate NF; CAS 1338-41-6; EINECS 215-664-9; nonionic; food emulsifier; alone or in combination with Tweens to improve volume and texture in cakes, improve chocolate props., whipped toppings, confectionery coatings, coffee whiteners; flavor dispersant; defoaming for process foods; *Usage level:* 0.8% (milk chocolate), 0.8% (confectionery coatings), 0.02-0.05% (defoaming); *Regulatory:* FDA 21CFR §172.515, 172.842, 173.340; Canada compliance; *Properties:* tan beads; sol. (@1%): sol. in IPA; sol. (hazy) in perchloroethylene, xylene; insol. in water, cottonseed oil; HLB 4.7; pour pt. 127 F.

Span® 60K. [ICI Spec. Chem.; ICI Surf. Am.; ICI Surf. Belgium] Sorbitan stearate NF; CAS 1338-41-6; EINECS 215-664-9; nonionic; emulsifier, stabilizer, solubilizer, thickener; *Regulatory:* FDA 21CFR §172.842; *Properties:* pale cream beads; sol.

(@1%): sol. in IPA; sol. (hazy) in perchloroethylene, xylene; HLB 4.7; pour pt. 53 C; 100% act.

Span® 60 VS. [ICI Spec. Chem.] Sorbitan stearate; CAS 1338-41-6; EINECS 215-664-9; vegetable source used in foods; *Properties:* beads; HLB 4.7; 100% conc.

Span® 65. [ICI Spec. Chem.; ICI Surf. Am.; ICI Surf. Belgium] Sorbitan tristearate; CAS 26658-19-5; EINECS 247-891-4; nonionic; emulsifier, stabilizer, thickener; for margarine, shortening, confectionery coatings; *Usage level:* Canada: 1% max. in foods; *Regulatory:* Canada compliance; *Properties:* cream solid; sol. (@ 1%) in IPA, perchloroethylene, xylene; HLB 2.1; pour pt. 53 C; 100% act.

Span® 65K. [ICI Atkemix] Sorbitan tristearate; CAS 26658-19-5; EINECS 247-891-4; nonionic; emulsifier, stabilizer, thickener; for margarine, shortening, confectionery coatings; *Regulatory:* kosher; *Properties:* tan waxy bead; HLB 2.1; pour pt. 53 C.

Span® 80. [ICI Spec. Chem.; ICI Surf. Am.; ICI Surf. Belgium] Sorbitan oleate NF; CAS 1338-43-8; EINECS 215-665-4; nonionic; emulsifier in polymer dispersions used to clarify beet or cane sugar juice/liquor; *Usage level:* 7.5% max. in final polymer disp.; 0.70 ppm (sugar juice), 1.4 ppm (sugar liquor); *Regulatory:* FDA 21CFR 173.75; *Properties:* amber liq.; sol. (@ 1%) in IPA, perchloroethylene, xylene, cottonseed and min. oils; visc. 1000 cps; HLB 4.3; 100% act.

Span® 85. [ICI Spec. Chem.; ICI Surf. Am.; ICI Surf. Belgium] Sorbitan trioleate; CAS 26266-58-0; EINECS 247-569-3; nonionic; emulsifier, stabilizer, thickener; for sausage casings; *Usage level:* Canada: 0.35% of casing; *Regulatory:* Canada compliance; *Properties:* amber liq.; sol. (@ 1%) in IPA, perchloroethylene, xylene, cottonseed and min. oils; visc. 210 cps; HLB 1.8; 100% act.

Spark-L® HPG. [Solvay Enzymes] Fungal pectinase; CAS 9032-75-1; EINECS 232-885-6; enzyme for depectinization and pulp washing; for fruit juices, citrus peeling and processing, citrus oil emulsions, recovered solids in pulp wash liquor, lemon oil extraction; *Properties:* amber liq., free of offensive odor and taste; sp.gr. 1.05-1.15; *Storage:* activity loss ≤ 10% in 6 mo stored in sealed containers at ambient temps.; 5 C storage extends life.

Special C. [ADM Arkady] Modified acid yeast food providing one-half the acidifying action of regular acid yeast food.

Special Bro Soft #126 6433. [Brolite Prods.] Bread strenghtener and softener to increase volume and extend shelf life; *Properties:* powd.

Special Fat 42/44. [Hüls Am.] Hydrog. coconut oil; EINECS 284-283-8; raw material for food industry.

Special Fat 168T. [Hüls Am.; Hüls AG] Hydrog. tallow; CAS 8030-12-4; EINECS 232-442-7; raw material for food industry; hydrophobing agent for powd. preps.; coating fat for salt, spices, citric acid; *Properties:* powd., flakes; m.p. 56-60 C; acid no. 5 max.; iodine no. 3 max.; sapon. no. 190-210.

Special Liquid Invert Sugar. [MLG Enterprises Ltd.] Invert sugar; CAS 8013-17-0; sweetener used in

soft drinks, fruit drinks, and canning; *Properties:* liq.; pH 4.8 ± 0.5; 33.2-38.4% sucrose.

Special Oil 107. [Hüls AG] Glyceryl trienanthate; tracer oil for butter; release agent for conveyor belts in the prod. of sweets; *Properties:* liq.; dens. 0.96 kg/dm³; visc. 20 mPa•s; turbidity pt. -20 C; acid no. 0.3 max.; iodine no. 1 max.; sapon. no. 385-395.

Special Oil 1739. [Hüls AG] Triglyceride oil/lecithin; bakery release oil; *Properties:* liq.; dens. 0.95 kg/dm³; visc. 80 mPa•s; turbidity pt. -5 C; acid no. 3 max.; iodine no. 50; sapon. no. 220-250.

Spezyme BBA. [Genencor] β-Amylase (barley); CAS 9000-92-4; EINECS 232-567-7; enzyme for food processing in high maltose syrups, malt diastatic supplementation; *Properties:* liq.

Spezyme GA. [Genencor] Glucoamylase (fungal); CAS 9032-08-0; EINECS 232-877-2; enzyme for food processing in saccharification of glucose syrups, low-carbohydrate beer; *Properties:* amber clear liq.

Spezyme IGI. [Genencor] Immobilized glucose isomerase (Streptomyces); enzyme for food processing; continuous catalytic isomerization of glucose to fructose in deep bed reactors; nonswelling; noncompressible; bisulfite resistant; *Properties:* inert granulate.

Spice Mix 614. [Bunge Foods] Flour-type spice mix with all natural color; double str. as Basic Spice Mix; for breads, rolls, sweet goods, danish, donuts, cakes, and prods. where accenting of egg color is desirable; *Usage level:* 0.5-0.75%; *Properties:* yel. powd.; pH 6.2; 15% max. moisture.

Spice Mix 808. [Bunge Foods] Flour-type spice mix with all natural color; same as Basic Spice Mix without flavor; for breads, rolls, sweet goods, danish, donuts, cakes, and prods. where accenting of egg color is desirable; *Usage level:* 1-1.5%; *Properties:* yel. powd.; pH 6.2; 15% max. moisture.

Spleen Substance. [Am. Labs] Vacuum-dried defatted glandular prod.; nutritive food additive; *Properties:* powd. or freeze-dried form.

SPL Lipase 30. [Scientific Protein Labs] Lipolytic enzyme; fat-splitting enzyme used for treatment of pancreatic insufficiency and in food applics.; *Properties:* cream-colored amorphous powd., faint char. but not offensive odor; partly sol. in water; insol. in alcohol; ≤ 5% moisture; *Storage:* store in tight containers in cool, dry place.

SPL Lipase-CE. [Scientific Protein Labs] Lipolytic enzyme; fat-splitting enzyme used for treatment of pancreatic insufficiency and in food applics.; *Properties:* cream-colored amorphous powd., faint char. but not offensive odor; partly sol. in water; insol. in alcohol; ≤ 5% moisture; *Storage:* store in tight containers in cool, dry place; protect from high temps. and excessive moisture.

SPL Pancreatin 3X USP. [Scientific Protein Labs] Enzymes, principally proteases, amylase, lipase; used in food applics.; hydrolyzes fats to glycerol and fatty acids, changes protein into proteoses and derived substances, converts starch to dextrins and sugars; *Properties:* cream-colored amorphous powd., faint char. but not offensive odor; partly sol. in water; insol. in alcohol; *Storage:* store in tight containers in cool, dry place.

SPL Pancreatin 4X USP. [Scientific Protein Labs] Enzymes, principally proteases, amylase, lipase; used in food applics.; hydrolyzes fats to glycerol and fatty acids, changes protein into proteoses and derived substances, converts starch to dextrins and sugars; *Properties:* cream-colored amorphous powd., faint char. but not offensive odor; partly sol. in water; insol. in alcohol; *Storage:* store in tight containers in cool, dry place.

SPL Pancreatin 5X USP. [Scientific Protein Labs] Enzymes, principally proteases, amylase, lipase; used in food applics.; hydrolyzes fats to glycerol and fatty acids, changes protein into proteoses and derived substances, converts starch to dextrins and sugars; *Properties:* cream-colored amorphous powd., faint char. but not offensive odor; partly sol. in water; insol. in alcohol; *Storage:* store in tight containers in cool, dry place.

SPL Pancreatin 6X USP. [Scientific Protein Labs] Enzymes, principally proteases, amylase, lipase; used in food applics.; hydrolyzes fats to glycerol and fatty acids, changes protein into proteoses and derived substances, converts starch to dextrins and sugars; *Properties:* cream-colored amorphous powd., faint char. but not offensive odor; partly sol. in water; insol. in alcohol; *Storage:* store in tight containers in cool, dry place.

SPL Pancreatin 7X USP. [Scientific Protein Labs] Enzymes, principally proteases, amylase, lipase; used in food applics.; hydrolyzes fats to glycerol and fatty acids, changes protein into proteoses and derived substances, converts starch to dextrins and sugars; *Properties:* cream-colored amorphous powd., faint char. but not offensive odor; partly sol. in water; insol. in alcohol; *Storage:* store in tight containers in cool, dry place.

SPL Pancreatin 8X USP. [Scientific Protein Labs] Enzymes, principally proteases, amylase, lipase; used in food applics.; hydrolyzes fats to glycerol and fatty acids, changes protein into proteoses and derived substances, converts starch to dextrins and sugars; *Properties:* cream-colored amorphous powd., faint char. but not offensive odor; partly sol. in water; insol. in alcohol; *Storage:* store in tight containers in cool, dry place.

SPL Pancreatin USP. [Scientific Protein Labs] Pancreatin USP contg. enzymes, principally proteases, amylase, lipase; used in food applics.; hydrolyzes fats to glycerol and fatty acids, changes protein into proteoses and derived substances, converts starch to dextrins and sugars; *Properties:* cream-colored amorphous powd., faint char. but not offensive odor; partly sol. in water; insol. in alcohol; *Storage:* store in tight containers in cool, dry place.

SPL Pancrelipase USP. [Scientific Protein Labs] Enzymes, principally lipase, amylase, protease; used in food applics.; hydrolyzes fats to glycerol and fatty acids, changes protein into proteoses and derived substances, converts starch to dextrins

and sugars; *Properties:* cream-colored amorphous powd., faint char. but not offensive odor; partly sol. in water; insol. in alcohol; *Storage:* store in tight containers in cool, dry place.

SPL Undiluted Pancreatic Enzyme Conc. (PEC). [Scientific Protein Labs] Enzymes, principally proteases, amylase, lipase; used in food applics.; hydrolyzes fats to glycerol and fatty acids, changes protein into proteoses and derived substances, converts starch to dextrins and sugars; *Properties:* cream-colored amorphous powd., faint char. but not offensive odor; partly sol. in water; insol. in alcohol; *Storage:* store in tight containers in cool, dry place.

Spongolit® Range. [Grünau GmbH] Esterified glycerides; nonionic; aerating emulsifier for Madeira and sponge cakes, fine baked goods, flavors; *Properties:* powd.; 100% conc.

Sporban Regular. [MLG Enterprises Ltd.] Sodium benzoate, propylene glycol, methyl and propyl parabens, water, and citric acid; mold inhibitor, preservative; *Regulatory:* FDA GRAS; *Properties:* colorless clear oily liq., odorless, sharp astringent flavor; 20% sodium benzoate, 2% methyl paraben, 1% propyl paraben; *Storage:* 6 mos shelf life stored under cool, dry conditions.

Sporban Special. [MLG Enterprises Ltd.] Sodium benzoate, propylene glycol, methyl and propyl parabens, water, and citric acid; mold inhibitor; *Regulatory:* FDA GRAS; *Properties:* colorless clear oily liq., odorless, sharp astringent flavor; 10% sodium benzoate, 7% methyl paraben, 3% propyl paraben; *Storage:* 6 mos shelf life stored under cool, dry conditions.

Spray Dried Gum Arabic NF Type CSP. [Meer] Gum arabic NF from *Acacia senegal;* CAS 9000-01-5; EINECS 232-519-5; used in the food industry; in prep. of low-calorie dry food and beverage mixes; *Properties:* creamy wh. powd., 98% thru 120 mesh; sol. 1 g/2 ml of water yielding lt. amber sol'n.; pH 4.0-4.5.

Spray Dried Fish Gelatin. [Croda Colloids Ltd; O.C. Lugo] Gelatin from fish sources; CAS 9000-70-8; EINECS 232-554-6; emulsion stabilizer, protective colloid, film-former, adhesive agent, flocculant; controls crystal growth; inferior gelling chars. to mammalian gelatin; *Properties:* spray-dried powd.; dissolves in cold water; visc. 60-85 mps; pH 5.0-6.5 (10%); 7% max. moisture.

Spray Dried Fish Gelatin/Maltodextrin. [Croda Colloids Ltd; O.C. Lugo] 50% Gelatin from fish sources, 50% maltodextrin; emulsion stabilizer, protective colloid, film-former, adhesive agent, flocculant; controls crystal growth; inferior gelling chars. to mammalian gelatin; *Properties:* spray-dried powd.; dissolves readily in cold water; pH 5.0-6.5 (10%); 5% max. moisture.

Spray Dried Gum Arabic NF/FCC CM. [Meer] Gum arabic; CAS 9000-01-5; EINECS 232-519-5; protective colloid, stabilizer, thickener for confectionery, candies, jellies, glazes, chewing gum, flavor emulsions, citrus oil and beverage flavor emulsions; foam stabilizer in beer; *Properties:* colorless, odorless, tasteless; water-sol.

Spray Dried Gum Arabic NF/FCC CS (Low Bacteria). [Meer] Gum arabic; CAS 9000-01-5; EINECS 232-519-5; protective colloid, stabilizer, thickener for confectionery, candies, jellies, glazes, chewing gum, flavor emulsions, citrus oil and beverage flavor emulsions; foam stabilizer in beer; *Properties:* colorless, odorless, tasteless; water-sol.

Spray Dried Gum Arabic NF/FCC CS-R. [Meer] Gum arabic; CAS 9000-01-5; EINECS 232-519-5; protective colloid, stabilizer, thickener for confectionery, candies, jellies, glazes, chewing gum, flavor emulsions, citrus oil and beverage flavor emulsions; foam stabilizer in beer; *Properties:* colorless, odorless, tasteless; water-sol.

Spray Dried Hemoglobin. [Am. Labs] High protein prod. collected from edible whole beef blood; meat protein supplement for meat prods.

Spray Dried Hydrolysed Fish Gelatin. [Croda Colloids Ltd; O.C. Lugo] Gelatin from fish sources; CAS 9000-70-8; EINECS 232-554-6; source of high m.w. film-forming protein; emulsion stabilizer, protective colloid, film-former, adhesive agent, flocculant; controls crystal growth; useful where dietary requirements of some religions prohibit use of mammalian gelatin; *Properties:* spray-dried powd.; dissolves readily in cold water; visc. 30-45 mps (10%); pH 5.0-6.5 (10%); 5% max. moisture.

Spray Dried Nigerian Gum Arabic. [MLG Enterprises Ltd.] Gum arabic BP/EP; CAS 9000-01-5; EINECS 232-519-5; emulsifier, stabilizer; *Properties:* visc. 70-120 cps (25%); pH 4.0-4.8; 99.9% min. purity.

Spraygum C. [MLG Enterprises Ltd.] Acacia gums purified and spray-dried; CAS 9000-01-5; EINECS 232-519-5; emulsifier, stabilizer; *Properties:* visc. 60-110 cps (25%); pH 4.2-4.7 (25%); 99.85% min. purity.

Spraygum GD. [MLG Enterprises Ltd.] Acacia gums purified and spray-dried; CAS 9000-01-5; EINECS 232-519-5; emulsifier, stabilizer; *Properties:* visc. 75-110 cps (25%); pH 4.2-4.7 (25%); 99.85% min. purity.

SQ®-48. [ADM Ogilvie] Wheat gluten; CAS 8002-80-0; EINECS 232-317-7; filler binder in meat prods.; *Properties:* lt. tan powd.; 12% moisture, 47% protein, 1.5% fat.

Stabil® 9 High Calcium Blended Powd. [Monsanto] Sodium aluminum phosphate acidic, calcium phosphate anhydrous, calcium carbonate, and precipitated amorphous silica; high calcium functional food additive for leavening; *Regulatory:* FCC; *Properties:* anhyd. wh. powd.; 0.1% max. on 50 mesh, 10% min. on 325 mesh; 10.8-13.5% Ca.

Stabil® 9 Regular Blended Powd. [Monsanto] Sodium aluminum phosphate acidic, calcium phosphate anhyd., calcium carbonate, precipitated amorphous silica; food additive; *Usage level:* 0.5% max. (processed meat and poultry, seafood); *Regulatory:* FCC compliance; *Properties:* anhyd. wh. powd.; 20% max. on 200 mesh, 10% min. on 325 mesh; 92% min. assay.

Stabilized Cookie Blend. [Canadian Harvest USA] Blend of corn bran, oat fiber, and wheat bran; fiber ingred. in cookies, crackers, baked goods, cereals, snack foods; *Properties:* med. granulation, coarse and fine avail. by request; tan color, nutty toasted flavor; 75% total dietary fiber; 10% max. moisture.

Stabilized Corn Bran. [Canadian Harvest USA] Corn bran; fiber ingred. for baked goods, cereals, beverages, snack foods, supplements; boasts great taste and contains twice as much dietary fiber as wheat bran; *Properties:* avail. in med. and fine granulations, coarse avail. on request; lt. yel. color, bland flavor; 80% total dietary fiber; 10% max. moisture.

Stabilized Full-Fat Wheat Germ. [Canadian Harvest USA] Wheat germ; nutritive source with enhanced stability, texture, and shelf life; for baked goods, cereals, snack foods; *Properties:* coarse and med. granulations, fine avail. on request; golden tan color, sweet nutty flavor; 13% total dietary fiber; 10% max. moisture.

Stabilized Micro-Lite Corn Bran. [Canadian Harvest USA] Corn bran; natural fiber for high-fiber formulations, baked goods, cereals, beverages, snack foods, pasta, pharmaceuticals; contains higher total dietary fiber than Stabilized Corn Bran, is lighter in color and blander in taste; *Properties:* avail. in fine and ultrafine granulations, med. avail. on request; lt. yel. color, bland flavor; 82% total dietary fiber; 10% max. moisture.

Stabilized Oatex. [Canadian Harvest USA] Steel-cut oat groat, high-fiber fraction from oat hulls, and bran fraction of the oat endosperm; natural fiber providing higher total dietary fiber than whole bran, supplies a soluble fiber fraction contg. beta glucan, linked to cholesterol reduction in humans; for baked goods, cereals, snack foods, meat prods.; *Properties:* coarse granulation, med. and fine avail. on request; lt. tan color, grainlike flavor; 16% total dietary fiber; 10% max. moisture.

Stabilized Oat Fiber. [Canadian Harvest USA] Oat fiber; natural fiber for breads and baked goods; higher in total dietary fiber than wheat bran; contains 30% hemicellulose, linked to fecal regularity; exc. water absorp. props.; used in many health food formulations, cereals, snack foods, supplements; *Properties:* avail. in med. and fine granulations, coarse avail. on request; tan color, bland flavor; 80% total dietary fiber; 10% max. moisture.

Stabilized Red Wheat Bran. [Canadian Harvest USA] Wheat bran; natural fiber stabilized to protect against flavor degradation and to provide crispness and mouthfeel; for baked goods, cereals, snack foods, pasta, supplements, pharmaceuticals; *Properties:* avail. in coarse, med., and fine granulations; red brn. color, nutty flavor; 42% total dietary fiber; 10% max. moisture.

Stabilized White Wheat Bran. [Canadian Harvest USA] Wheat bran; for foods where a creamy colored wheat fiber is desired, e.g., baked goods, cereals, snack foods, pasta, supplements, pharmaceuticals; softer texture and milder flavor than Stabilized Red Wheat Bran; *Properties:* avail. in coarse, med., and fine granulations; tan color, nutty flavor; 38% total dietary fiber; 10% max. moisture.

Stabilizer C. [French's Ingreds.] Water, vinegar, mustard bran; thixotropic stabilizer, emulsifier, suspending agent for sauces, dressings, condiments, low-fat and fat-free applics.; stable to high acid, high salt, high heat; synergistic with other gums; provides opacity, cling.

Stabl Plus. [Brolite Prods.] Highly conc. icing stabilizer designed for wrapped icings; can be used in any icing.

Stabolic™ C. [Lucas Meyer] Lecithin with antioxidant (TBHQ); emulsifier, dispersant, and wetting agent for food use; *Properties:* water-disp.

Stadex® 9. [A.E. Staley Mfg.] Dextrin; CAS 9004-53-9; EINECS 232-675-4; dough improver in breads and rolls; adds crispness to batters for breading fish and poultry; binder in panning and extruding confections; binding spices, seasonings, colorings to food surfaces, e.g., dry roasted nuts; *Regulatory:* FDA 21CFR §184.1277 GRAS; *Properties:* wh. powd.; 5-15% water sol.; visc. 1550 cp; pH 4.5.

Stadex® 60K. [A.E. Staley Mfg.] Dextrin; CAS 9004-53-9; EINECS 232-675-4; dough improver in breads and rolls; adds crispness to batters for breading fish and poultry; binder in panning and extruding confections; binding spices, seasonings, colorings to food surfaces, e.g., dry roasted nuts; *Regulatory:* FDA 21CFR §184.1277 GRAS; *Properties:* wh. powd.; 55-75% water sol.; visc. 2000 cp; pH 4.5.

Stadex® 90. [A.E. Staley Mfg.] Dextrin and sodium bisulfite; dough improver in breads and rolls; adds crispness to batters for breading fish and poultry; binder in panning and extruding confections; binding spices, seasonings, colorings to food surfaces, e.g., dry roasted nuts; *Regulatory:* FDA 21CFR §184.1277 GRAS; *Properties:* wh. powd.; 85-100% water sol.; visc. 1700 cp; pH 4.5.

Stadex® 126. [A.E. Staley Mfg.] Dextrin; CAS 9004-53-9; EINECS 232-675-4; improves handling qualities of prods. such as frozen eggs; functional adjuvant in drink mixes; *Regulatory:* FDA 21CFR §184.1277 GRAS; *Properties:* pale yel. to buff powd.; 96% water sol.; pH 4.5.

Stadex® 128. [A.E. Staley Mfg.] Dextrin; CAS 9004-53-9; EINECS 232-675-4; improves handling qualities of prods. such as frozen eggs; functional adjuvant in drink mixes; *Regulatory:* FDA 21CFR §184.1277 GRAS; *Properties:* pale yel. to buff powd.; 98% water sol.; pH 4.5.

Staform P. [Meer] Gum ghatti; stabilizer, binder, emulsifier forming o/w emulsions; used in beverage emulsions where it forms orange oil emulsions, in table syrup emulsions; *Properties:* water-sol.

Staley® 7025. [A.E. Staley Mfg.] Unmodified corn starch; CAS 9005-25-8; EINECS 232-679-6; thick boiling pearl starch, thickener for food applics.; *Properties:* wh. fine gran.; bulk dens. 40 lb/ft³; pH 5.5; 11.5% moisture.

Staley® 7350 Waxy No. 1 Starch. [A.E. Staley Mfg.]

Waxy corn starch; CAS 9005-25-8; EINECS 232-679-6; thick boiling waxy starch for food applics.; greater thickening, lower gelatinization temp., and improved paste clarity and stability than regular unmodified corn starch; *Regulatory:* FDA 21CFR §182.1 GRAS; *Properties:* wh. powd.; bulk dens. 40 lb/ft³; pH 4.5-6.0; 10-13% moisture.

Staley® Dusting Starch. [A.E. Staley Mfg.] Food starch modified derived from dent corn; antisticking agent for packaged prods. such as marshmallows, chewing gums, candies; processing lubricants in handling of doughs; *Regulatory:* FDA 21CFR §172.892; *Properties:* wh. powd.; pH 4.5; 9% moisture.

Staley® Maltodextrin 3260. [A.E. Staley Mfg.] Liq. maltodextrin from waxy maize starch; CAS 9050-36-6; EINECS 232-940-4; carrier for flavored, spray-dried prods.; encapsulating and/or bulking agent; provides handling convenience, blandness, body and mouthfeel; *Regulatory:* FDA 21CFR §184.1444; *Properties:* nearly colorless visc. liq., bland odor and flavor; water-sol.; dens. 11.35 lb/gal; visc. 51 poises (100 F); pH 4.3; DE 18; 70% solids; *Precaution:* susceptible to microbial growth; *Storage:* store and load/unload @ 140 F min.

Staley® Moulding Starch. [A.E. Staley Mfg.] Unmodified corn starch lightly treated with oil; CAS 9005-25-8; EINECS 232-679-6; forms clean, sharp impressions for gum dancy depositing; *Properties:* wh. powd.; pH 5.0; 10% moisture.

Staley® Pure Food Powd. (PFP). [A.E. Staley Mfg.] Unmodified corn starch; CAS 9005-25-8; EINECS 232-679-6; thickener, bodying agent for bakery fillings, salad dressings, gravies, puddings, sauces, stews, pet foods; provides adhesive props. for breading mixes and texture for cakes, cookies, and snacks; *Properties:* wh. powd.; pH 4.0-6.5; 9-10% moisture.

Staley® Pure Food Powd. Starch Type I. [A.E. Staley Mfg.] Unmodified corn starch; CAS 9005-25-8; EINECS 232-679-6; thickener with clean flavor and exc. bodying props. for puddings, gravies, sauces, soups, stews, cream pie fillings, pet foods; *Properties:* wh. powd.; 0.1% max. on 100 mesh; pH 5.0-5.5; 9-11% moisture.

Staley® Pure Food Powd. Starch Type II. [A.E. Staley Mfg.] Unmodified corn starch; CAS 9005-25-8; EINECS 232-679-6; thickener with clean flavor and exc. bodying props. for puddings, gravies, sauces, soups, stews, cream pie fillings, pet foods; *Properties:* wh. powd.; 5% max. on 100 mesh; pH 6.4-7.4; 9% moisture.

Staley® Redried Starch A. [A.E. Staley Mfg.] Unmodified corn starch; CAS 9005-25-8; EINECS 232-679-6; starch for use where moisture control is critical, e.g., in packaged dry mixes; *Properties:* wh. powd.; pH 6.0; 5% moisture.

Staley® Redried Starch B. [A.E. Staley Mfg.] Unmodified corn starch; CAS 9005-25-8; EINECS 232-679-6; starch for use where moisture control is critical, e.g., in packaged dry mixes; *Properties:* wh. powd.; pH 6.0; 7.5% moisture.

Staley® Tapioca Dextrin 11. [A.E. Staley Mfg.] Dextrin; CAS 9004-53-9; EINECS 232-675-4; carrier for flavorings and colors in dry preps.; in liqs., produces bland, low visc. sol'ns.; provides bulking and water binding props.; for glazes, sauces, frozen foods; *Regulatory:* FDA 21CFR §184.1277 GRAS; *Properties:* off-wh. powd., bland flavor; 42% water-sol.; pH 4.5; 5% moisture.

Staley® Tapioca Dextrin 12. [A.E. Staley Mfg.] Dextrin; CAS 9004-53-9; EINECS 232-675-4; high sol. film-former, flavor/color carrier; reduction of fat pick-up.

Sta-Lite. [ADM Arkady] Agar and locust bean gum; single-strength icing stabilizer for semitransparent icings on all types of sweet goods.

Sta-Lite™ 100C. [A.E. Staley Mfg.] Polydextrose; CAS 68434-04-4; nonsweet low-calorie bulking agent providing body and texture in reduced- and low-calorie foods, e.g., baked goods and mixes, chewing gum, confections, frostings, candy, frozen dairy desserts, salad dressing, gelatins, puddings; humectant, texturizer; *Regulatory:* FDA 21CFR §172.841; *Properties:* wh. to tan coarse powd., odorless, bland taste; hygroscopic; 17% > 30 mesh, 44% > 60 mesh; water-sol.; bulk dens. 49 lb/ft³ (loose); m.p. 135-145 C; pH 2.5-3.5; 4% max. moisture; *Storage:* store in closed containers in cool, dry area.

Sta-Lite™ 100CN Neutralized. [A.E. Staley Mfg.] Polydextrose; CAS 68434-04-4; nonsweet low-calorie bulking agent providing body and texture in reduced- and low-calorie foods, e.g., baked goods and mixes, chewing gum, confections, frostings, candy, frozen dairy desserts, salad dressing, gelatins, puddings; humectant, texturizer; *Regulatory:* FDA 21CFR §172.841; *Properties:* wh. to tan coarse powd., odorless, bland taste; hygroscopic; 17% > 30 mesh, 44% > 60 mesh; water-sol.; bulk dens. 49 lb/ft³ (loose); m.p. 135-145 C; pH 5.0-6.0; 4% max. moisture; *Storage:* store in closed containers in cool, dry area.

Sta-Lite™ 100F. [A.E. Staley Mfg.] Polydextrose; CAS 68434-04-4; nonsweet low-calorie bulking agent providing body and texture in reduced- and low-calorie foods, e.g., baked goods and mixes, chewing gum, confections, frostings, candy, frozen dairy desserts, salad dressing, gelatins, puddings; humectant, texturizer; *Regulatory:* FDA 21CFR §172.841; *Properties:* wh. to tan fine powd., odorless, bland taste; hygroscopic; 0.2% > 30 mesh, 67% > 140 mesh; water-sol.; bulk dens. 35 lb/ft³ (loose); m.p. 135-145 C; pH 2.5-3.5; 4% max. moisture; *Storage:* store in closed containers in cool, dry area.

Sta-Lite™ 100FN Neutralized. [A.E. Staley Mfg.] Polydextrose; CAS 68434-04-4; nonsweet low-calorie bulking agent providing body and texture in reduced- and low-calorie foods, e.g., baked goods and mixes, chewing gum, confections, frostings, candy, frozen dairy desserts, salad dressing, gelatins, puddings; humectant, texturizer; *Regulatory:* FDA 21CFR §172.841; *Properties:* wh. to tan fine

powd., odorless, bland taste; hygroscopic; 0.2% > 30 mesh, 67% > 140 mesh; water-sol.; bulk dens. 35 lb/ft³ (loose); m.p. 135-145 C; pH 5.0-6.0; 4% max. moisture; *Storage:* store in closed containers in cool, dry area.

Stamere® CK FCC. [Meer] Carrageenan; CAS 9000-07-1; EINECS 232-524-2; thickener and gellant in instant prods.; used in lowfat meat prods. to provide texture and juiciness while reducing fat and increasing prod. yield; provides set, controls syneresis, and prevents separation in milk puddings, custards, and pies; *Properties:* lt. tan powd.; water-sol.; visc. > 250 cps (1%); pH 6-9.

Stamere® CKM FCC. [Meer] Carrageenan; CAS 9000-07-1; EINECS 232-524-2; stabilizer for instant puddings requiring no heat, ice cream, milk gel; controls moisture and fat content in meat prods.; also for water gel prods., gelatin-type desserts with sheen and mouthfeel of gelatin, low-calorie jams and jellies; *Usage level:* 0.5-0.7% (instant pudding), 0.4% (ice cream); *Properties:* lt. tan powd.; water-sol.; visc. > 500 cps (1%); pH 8.0-10.5.

Stamere® CK-S. [Meer] Carrageenan; CAS 9000-07-1; EINECS 232-524-2; suspending agent, stabilizer, thickener, gelling agent for food industry; *Properties:* water-sol.

Stamere® HT. [Meer] Fucelleran; used for water gel prods., gelatin-type desserts with sheen and mouthfeel of gelatin, low-calorie jams and jellies.

Stamere® HTMX. [Meer] Refined extract of the red seaweed, *Furcellaria fasticiata*; *Usage level:* 0.5-0.7% (instant pudding), 0.4% (ice cream); *Properties:* off-wh. to lt. tan powd., neutral odor and taste; ≥ 99% thru 60 mesh; pH 7.6-9.1.

Stamere® N-47. [Meer] Carrageenan; CAS 9000-07-1; EINECS 232-524-2; suspending agent for cocoa in chocolate milk drinks; *Usage level:* 280-350 ppm.

Stamere® N-55. [Meer] Carrageenan; CAS 9000-07-1; EINECS 232-524-2; suspending agent for cocoa in chocolate milk drinks; *Usage level:* 280-350 ppm.

Stamere® N-325. [Meer] Carrageenan; CAS 9000-07-1; EINECS 232-524-2; suspending agent, stabilizer, thickener, gelling agent for food industry; *Properties:* water-sol.

Stamere® N-350. [Meer] Carrageenan; CAS 9000-07-1; EINECS 232-524-2; suspending agent, stabilizer, thickener, gelling agent for food industry; *Properties:* water-sol.

Stamere® N-350 E FCC. [Meer] Carrageenan; CAS 9000-07-1; EINECS 232-524-2; used for water gel prods., gelatin-type desserts with sheen and mouthfeel of gelatin, low-calorie jams and jellies; *Properties:* off-wh. to lt. tan powd.; ≥ 97% thru 40 mesh; pH 8-10.

Stamere® N-350 S. [Meer] Carrageenan; CAS 9000-07-1; EINECS 232-524-2; suspending agent, stabilizer, thickener, gelling agent for food industry; *Properties:* water-sol.

Stamere® NI. [Meer] Carrageenan; CAS 9000-07-1; EINECS 232-524-2; suspending agent, stabilizer, thickener, gelling agent for food industry.

Stamere® NIC FCC. [Meer] Carrageenan; CAS 9000-07-1; EINECS 232-524-2; stabilizer for ice cream and related milk prods.; provides smooth, creamy texture with elimination of ice formation; produces slow, smooth melt down; *Usage level:* 0.3-0.12%; *Properties:* off-wh. to lt. tan powd.; ≥ 97% thru 60 mesh; pH 8-10.

Stamere® NK. [Meer] Carrageenan; CAS 9000-07-1; EINECS 232-524-2; used for milk gels such as puddings, custards, pies, etc.; provides set, controls syneresis, and prevents separation of other ingreds.

Sta-Mist® 365. [A.E. Staley Mfg.] Food starch modified, derived from waxy maize; lipophilic starch for fat binding retorted prods.

Sta-Mist® 454. [A.E. Staley Mfg.] Food starch modified, derived from waxy maize; lipophilic starch for fat binding instant retorted prods.

Sta-Mist® 7415. [A.E. Staley Mfg.] Food starch modified, derived from waxy maize; lipophilic starch producing an opaque emulsion when cooked and processed with certain oils; clouding agent in foods and beverages, esp. carbonated soft drinks; *Regulatory:* FDA 21CFR §172.892; *Properties:* wh. pearl powd.; pH 5.7; 11% moisture.

Standamul® 318. [Henkel/Organic Prods.] Caprylic/capric triglyceride; CAS 65381-09-1; EINECS 265-724-3; base, emollient, carrier used in food prods.; extender for flavors; *Properties:* clear visc. oily liq.; odorless; sol. in min. and castor oil, IPM, oleyl alcohol and anhyd. ethanol; sp.gr. 0.950; visc. 25 cps; HLB 12; cloud pt. -5 C max.; acid no. 0.5 max.; sapon. no. 340-350; 100% conc.; Unverified

Sta-Nut EE. [Van Den Bergh Foods] Partially hydrog. cottonseed oil, fatty acid mono- and diglycerides; high performance fat; peanut butter stabilizer; structuring system for oil migration inhibition; *Regulatory:* kosher; *Properties:* tan beads; m.p. 140-145 F; iodine no. 5 max.; 25-29% alpha mono.

Sta-Nut P. [Van Den Bergh Foods] Partially hydrog. palm oil with fatty acid mono- and diglycerides; high performance fat; peanut butter stabilizer; structuring system for oil migration inhibition; *Regulatory:* kosher; *Properties:* Lovibond 5R max. beads; m.p. 143-148 F; iodine no. 5 max.; 26-29% alpha mono.

Sta-Nut R. [Van Den Bergh Foods] Partially hydrog. veg. oils (rapeseed, cottonseed, soybean); EINECS 269-820-6; high performance fat; peanut butter stabilizer; structuring system for oil migration inhibition; *Regulatory:* kosher; *Properties:* Lovibond 2.5R max. beads; drop pt. 144-149 F; iodine no. 4 max.; 26-29% alpha mono.

Sta-O-Paque®. [A.E. Staley Mfg.] Food starch modified (waxy corn starch); produces smooth, opaque gravies and sauces that do not gel on cooling; retort and freeze-thaw stability allows use in canned meat and vegetable prods., frozen foods, pot pies; binder in meat emulsions; reduces unwanted surf. sheen on some prods.; *Usage level:* 4-5% (ready-to-eat preps.), 11% (concs.), 3.5% (meat emulsions); *Regulatory:* FDA 21CFR §172.892; *Properties:* wh. powd., odorless, bland flavor; 95% thru 60 mesh;

pH 5.5 (10% susp.); 10.5% moisture.

Star. [Procter & Gamble] Glycerin USP; CAS 56-81-5; EINECS 200-289-5; humectant in food prods.; *Properties:* APHA 10 max. color; sp.gr. 1.2517.

StarchPlus™ SPR. [Calif. Natural Prods.] Rice starch; short texture starch, gellant; easy mixing; as shortening or fat substitute in baked goods; retains most of rice protein while most of fat and fiber are removed; freeze/thaw stable in sucrose pudding systems; replaces up to 50% fat in baked goods; *Regulatory:* kosher certified; *Properties:* granulation 95% < 200 mesh; easily dispersed; bulk dens. 28 lb/ft³; pH 6.9 (10% disp.); < 5% moisture.

StarchPlus™ SPR-LP. [Calif. Natural Prods.] Low protein rice starch; offers short texture, low protein; for puddings, sauces; *Regulatory:* kosher certified; *Properties:* powd., granulation 95% < 140 mesh, low flavor; bulk dens. 28 lb/ft³; pH 6.9 (10% disp.); < 5% moisture.

StarchPlus™ SPW. [Calif. Natural Prods.] Waxy rice starch; long texture starch, easy mixing; for baked goods, as fat replacer in low or non-fat frozen desserts; retains most of rice protein while most of fat and fiber are removed; freeze/thaw stable; *Usage level:* 1-5%; *Regulatory:* kosher certified; *Properties:* powd., 95% < 200 mesh; easy to disperse; bulk dens. 28 lb/ft³; pH 6.3 (10% disp.); 98% amylopectin; < 5% moisture.

StarchPlus™ SPW-LP. [Calif. Natural Prods.] Low protein waxy rice starch; long texture gel with low flavor impact; fat replacer for frozen desserts; exc. freeze/thaw props.; *Usage level:* 1-5%; *Regulatory:* kosher certified; *Properties:* low flavor; bulk dens. 28 lb/ft³; pH 6.3 (10% disp.); < 5% moisture.

Starco™ 401. [A.E. Staley Mfg.] Food starch modified, derived from tapioca; pregelatinized starch, thickener, stabilizer for uncooked foods, instant puddings, dips, sauces, gravies, refrigerated desserts; *Regulatory:* FDA 21CFR §172.892; *Properties:* wh. powd.; 1% max. on 140 mesh, 90% min. thru 200 mesh; pH 5-6; 7% moisture.

Starco™ 447. [A.E. Staley Mfg.] Food starch modified, derived from tapioca; pregelatinized starch, thickener, stabilizer for instant foods, dry mixes, puddings, icings, cream fillings, dips, soup mixes, spreads; produces heavy-flow pastes with smooth, creamy texture; imparts sl. fudge-like delicate mouthfeel; *Regulatory:* FDA 21CFR §172.892; *Properties:* wh. powd.; 1% max. on 140 mesh, 90% min. thru 200 mesh; pH 5.5; 5.5% max. moisture.

Star-Dri® 1. [A.E. Staley Mfg.] Maltodextrin; CAS 9050-36-6; EINECS 232-940-4; flavor carrier, binder for bakery mixes; provides nutritive value, anticaking, binding, mouthfeel/body, and blandness; *Properties:* powd.; dens. 31 lb/ft³; DE 1.

Star-Dri® 5. [A.E. Staley Mfg.] Maltodextrin; CAS 9050-36-6; EINECS 232-940-4; flavor carrier for frostings, dairy desserts; provides solubility, nutritive value, anticaking, binding, mouthfeel/body, and blandness; *Properties:* powd.; dens. 32 lb/ft³; DE 5.

Star-Dri® 10. [A.E. Staley Mfg.] Maltodextrin; CAS

9050-36-6; EINECS 232-940-4; flavor carrier for dairy desserts, sugar fondants, batters; provides solubility, nutritive value, anticaking, binding, mouthfeel/body, and blandness; *Properties:* powd.; dens. 34 lb/ft³; DE 10.

Star-Dri® 15. [A.E. Staley Mfg.] Maltodextrin; CAS 9050-36-6; EINECS 232-940-4; flavor carrier for confections, beverage mixes, dairy desserts; provides solubility, nutritive value, anticaking, binding, mouthfeel/body, and blandness; *Properties:* powd.; dens. 35 lb/ft³; DE 15.

Star-Dri® 18. [A.E. Staley Mfg.] Maltodextrin; CAS 9050-36-6; EINECS 232-940-4; flavor carrier for dairy desserts; provides solubility, nutritive value, anticaking, binding, mouthfeel/body, and blandness; *Properties:* powd.; dens. 36 lb/ft³; DE 18.

Star-Dri® 20. [A.E. Staley Mfg.] Corn syrup solids; CAS 68131-37-3; maintains visc. and body as fat content is reduced in adult and infant nutritional formulas, salad dressings; *Properties:* powd.; m.w. 3736; dens. 38 lb/ft³; DE 20.

Star-Dri® 35F. [A.E. Staley Mfg.] Corn syrup solids; CAS 68131-37-3; maintains visc. and body as fat content is reduced in frostings, fondants, dry mixes, creamers; *Properties:* finely ground powd.; m.w. 1560; dens. 30 lb/ft³; DE 35.

Star-Dri® 35R. [A.E. Staley Mfg.] Corn syrup solids; CAS 68131-37-3; maintains visc. and body as fat content is reduced in ice cream, dry mixes; *Properties:* powd.; m.w. 1560; dens. 38 lb/ft³; DE 35.

Star-Dri® 42C. [A.E. Staley Mfg.] Corn syrup solids; CAS 68131-37-3; maintains visc. and body as fat content is reduced in meats, dry mixes, ice cream; *Properties:* coarsely ground powd.; m.w. 1120; dens. 38 lb/ft³; DE 44.

Star-Dri® 42F. [A.E. Staley Mfg.] Corn syrup solids; CAS 68131-37-3; maintains visc. and body as fat content is reduced in frostings, fondants, confections; *Properties:* finely ground powd.; m.w. 1120; dens. 30 lb/ft³; DE 44.

Star-Dri® 42R. [A.E. Staley Mfg.] Corn syrup solids; CAS 68131-37-3; maintains visc. and body as fat content is reduced in meats, dry mixes, ice cream; *Properties:* powd.; m.w. 1120; dens. 38 lb/ft³; DE 44.

Star-Dri® 42X. [A.E. Staley Mfg.] Corn syrup solids; CAS 68131-37-3; maintains visc. and body as fat content is reduced in icings, candied apples; *Properties:* very coarsely ground powd.; m.w. 1120; dens. 38 lb/ft³; DE 44.

Star-Dri® 55. [A.E. Staley Mfg.] Corn syrup solids; CAS 68131-37-3; humectant; provides chewy texture to confection centers, semi-moist baked goods; *Properties:* powd.; dens. 38 lb/ft³; DE 39.

Star-Dri® 1005A. [A.E. Staley Mfg.] Maltodextrin; CAS 9050-36-6; EINECS 232-940-4; dispersant, bulking agent for nutritional beverages, dry mixes, sauces, gravies, spice blends; carrier for oil-sol. flavors, emulsifiers, colors, fats; provides rapid solubility, nutritive value, anticaking, binding, mouthfeel/body, and blandness; *Properties:* agglomerated powd.; < 20% on 20 mesh, < 20% thru

200 mesh; dens. 5 lb/ft³; pH 4-6; DE 10; 6% max. moisture.

Star-Dri® 1015A. [A.E. Staley Mfg.] Maltodextrin; CAS 9050-36-6; EINECS 232-940-4; dispersant, bulking agent for nutritional beverages, dry mixes; provides solubility, nutritive value, anticaking, binding, mouthfeel/body, and blandness; *Properties:* agglomerated powd.; dens. 15 lb/ft³; DE 10.

Starplex® 90. [Am. Ingreds.] High-purity, molecularly distilled monoglyceride prepared from edible fats or oils and glycerin with TBHQ and citric acid; provides increased hydration and improved functionality of the monoglyceride; improves softness and extends shelf life in baked goods; emulsifies and stabilizes fat in sauces and gravies; starch complexing agent; also for pasta, cereal; *Usage level:* 0.2-1.0%; 2-6 oz/cwt flour; *Regulatory:* FDA GRAS; *Properties:* wh. to cream powd.; water-disp.; m.p. 63-69 C; acid no. 4.0 max.; iodine no. 25-30; sapon. no. 150-165; 90% min. alpha monoester; *Storage:* store below 90 F.

Starwax® 100. [Petrolite] Hard microcryst. wax consisting of n-paraffinic, branched paraffinic, and naphthenic hydrocarbons; CAS 63231-60-7; EINECS 264-038-1; chewing gum base; *Regulatory:* incl. FDA §172.230, 172.615, 175.105, 175.300, 176.170, 176.180, 176.200, 177.1200, 178.3710, 179.45; *Properties:* color 1.0 (D1500) wax; very low sol. in org. solvs.; sp.gr. 0.93; visc. 15 cps (99 C); m.p. 88 C.

Sta-Rx®. [A.E. Staley Mfg.] Corn starch NF meeting specs for food starch modified; CAS 9005-25-8; EINECS 232-679-6; filler, absorbent, diluent, disintegration agent in pharmaceutical industry for tablets and powds.; wet binder in tablet granulations; *Regulatory:* FDA 21CFR §172.892; *Properties:* wh. powd., free of objectionable odor, sl. char. taste; 0.5% max. on 80 mesh, 5% max. on 325 mesh; pH 64.5-7.0.0; 9.5-12.5% moisture.

Sta-Slim™ 142. [A.E. Staley Mfg.] Instant, pregelatinized modified food starch derived from potato; 4 calorie/g fat replacer for low-fat foods, salad dressings, baked goods; provides mouthfeel, appearance of higher fat prods.; aids perception of creaminess; *Regulatory:* FDA 21CFR §172.892; *Properties:* wh. powd., odorless, bland taste; pH 4.5; 4% moisture.

Sta-Slim™ 143. [A.E. Staley Mfg.] Food starch modified derived from potato starch; 4 calorie/g fat replacer for low-fat foods, salad dressings, baked goods, imitation cream cheese, soups, meat prods.; provides mouthfeel, appearance of higher fat prods.; aids perception of creaminess; *Regulatory:* FDA 21CFR §172.892; *Properties:* wh. powd., odorless, bland taste; pH 7; 11% moisture.

Sta-Slim™ 150. [A.E. Staley Mfg.] Instant, pregelatinized modified food starch derived from tapioca; 4 calorie/g fat replacer for low-fat foods, salad dressings, baked goods; provides mouthfeel, appearance of higher fat prods.; aids perception of creaminess; *Regulatory:* FDA 21CFR §172.892; *Properties:* wh. powd., odorless, bland taste; pH

4.5; 4% moisture.

Sta-Slim™ 151. [A.E. Staley Mfg.] Modified food starch derived from tapioca; 4 calorie/g fat replacer for low-fat foods, salad dressings, baked goods; provides mouthfeel, appearance of higher fat prods.; aids perception of creaminess; *Regulatory:* FDA 21CFR §172.892; *Properties:* wh. powd., odorless, bland taste; pH 7; 11% moisture.

Sta-Slim™ 171. [A.E. Staley Mfg.] Food starch modified derived from waxy corn starch; 4 calorie/g fat replacer for low-fat foods, meats, sauces, gravies; provides mouthfeel, appearance of higher fat prods.; aids perception of creaminess; *Regulatory:* FDA 21CFR §172.892; *Properties:* wh. powd.; pH 5.7; 11.5% moisture.

Staybelite® Ester 5. [Hercules] Glyceryl hydrog. rosinate; thermoplastic syn. resin; as softener/plasticizer for the masticatory agent in chewing gum bases; used in food processing operations; *Properties:* USDA Rosin solid, flakes; low odor; sol. in esters, ketones, higher alcohols, glycol ethers, and aliphatic, aromatic, and chlorinated hydrocarbons; dens. 1.06 kg/l; Hercules drop soften. pt. 81 C; acid no. 5.

STD-175. [Rhone-Poulenc Food Ingreds.] Locust bean gum; CAS 9000-40-2; EINECS 232-541-5; food-grade natural vegetable gum; high visc., pH stable (3.5-9), freeze-thaw stable; requires heat to develop full visc.; synergistic with xanthan, carrageenan; *Usage level:* 0.1-0.25% (ice cream/milk), 0.3-0.5% (cream cheese), 0.25-0.4% (fruit preps. for yogurt), 0.15-0.3% (pie fillings), 0.1-0.25% (sauces); *Properties:* fine gran.; good dispersibility; visc. 2500 cps min. (1%).

07 Stearine. [Van Den Bergh Foods] Partially hydrog. cottonseed oil; CAS 68334-00-9; EINECS 269-804-9; high performance fat for breads, shortenings, chewing gums, caramel coatings; structuring system for composition enhancement; *Regulatory:* kosher; *Properties:* Lovibond 4R max. bead, flake; m.p. 141-147 F; iodine no. 4 max.

17 Stearine. [Van Den Bergh Foods] Partially hydrog. soybean oil; EINECS 232-410-2; high performance fat for shortenings; structuring system for composition enhancement; *Regulatory:* kosher; *Properties:* Lovibond 3R max. flake; m.p. 152-158 F; iodine no. 4 max.

27 Stearine. [Van Den Bergh Foods] Partially hydrog. palm oil; high performance fat; peanut butter stabilizer; structuring system for oil migration inhibition; *Properties:* Lovibond 5R max. beads, flakes; m.p. 136-144 F; iodine no. 5 max.

37 Stearine. [Van Den Bergh Foods] Partially hydrog. canola oil; high performance fat for shortenings; structuring system for composition enhancement; *Regulatory:* kosher; *Properties:* Lovibond 3R max. bead; m.p. 150-160 F; iodine no. 4 max.

Stellar®. [A.E. Staley Mfg.] Food starch modified (corn starch); 1 calorie/g fat replacement for baked goods, frostings, fillings, dairy prods., salad dressings, cheese prods., table spreads, meat prods., confections; maintains texture, mouthfeel, visual appeal, and stability of full-fat foods; *Regulatory:* FDA 21CFR §172.892; *Properties:* wh. free-flowing powd., bland flavor; disperses/hydrates readily in water; bulk dens. 33 ± 5 lb/ft³ (loose); pH 3.0-5.0 (10% slurry); > 97% total carbohydrates, < 8% moisture; *Storage:* stable in sealed, unopened containers under normal storage conditions.

Sterling® Purified USP Salt. [Akzo Salt] Sodium chloride USP; CAS 7647-14-5; EINECS 231-598-3; high purity, food grade salt for use where highest quality is required, esp. in medical field as in the mfg. of saline sol'ns. for intravenous feeding, plasma separation, kidney dialysis, and other critical medical applics.; *Regulatory:* GRAS, FCC, kosher approvals; *Properties:* cryst.; 16% retained 40 USS mesh, 51%-50 USS mesh, 25%-70 USS mesh; sol. 36 g/100 cc in water @ 20 C; sp.gr. 2.165; bulk dens. 1.15-1.28 g/cc; b.p. 1465 C (760 mm); flash pt. none; *Toxicology:* LD50 (oral, rat) 3.75 g/kg; MLD I.V. 2.5 g/kg; *Storage:* store in dry, covered area at r.h. below 75% to retard caking.

Sterotex® C. [Karlshamns] Hydrog. soybean oil and carnauba wax; tablet lubricant for nutrition; *Regulatory:* kosher; *Properties:* wh. to lt. tan powd., odorless; insol. in water; sp.gr. 0.9 (100 F); m.p. 79-82 C; b.p. > 500 F; acid no. 2.5 max.; iodine no. 5 max.; sapon. no. 164-174; flash pt. (COC) > 500 F; *Toxicology:* veg. oil mists classified as nuisance particles; *Precaution:* oil-soaked materials may spontaneously combust.

Sterotex® HM NF. [Karlshamns] Hydrog. soybean oils; CAS 8016-70-4; EINECS 232-410-2; tablet lubricant for nutrition; *Regulatory:* kosher; *Properties:* fine wh. powd., yel. oil when melted; insol. in water; sp.gr. 0.9; m.p. 67-70 C; b.p. > 500 F; acid no. 0.4 max.; iodine no. 5 max.; sapon. no. 186-196; flash pt. (COC) > 550 F; *Toxicology:* veg. oil mists classified as nuisance particles; *Precaution:* oil-soaked materials may spontaneously combust; *Storage:* store in cool, dry place.

Sterotex® NF. [Karlshamns] Hydrog. veg. oil; CAS 68334-00-9; EINECS 269-820-6; lubricant for tableting and compaction in nutritional supplements; *Regulatory:* kosher; *Properties:* wh. powd. @ R.T., lt. yel. oil when melted; insol. in water; sp.gr. 0.9; m.p. 140-145 F; b.p. > 500 F; acid no. 0.4 max.; iodine no. 5 max.; sapon. no. 188-198; flash pt. (COC) > 640 F; *Toxicology:* veg. oil mists classified as nuisance particles; *Precaution:* oil-soaked materials may spontaneously combust.

Stir & Sperse®. [A.E. Staley Mfg.] Corn starch; CAS 9005-25-8; EINECS 232-679-6; instant starch providing controlled hydration in dry mixes; disperses easily in hot or cold water; for instant soups, barbecue/pizza/spaghetti sauces, baking/microwave applics.; provides cling, pulp-like texture (tomato sauces), stability to heat; *Properties:* wh. agglomerated powd.; 2% on 30 mesh, 10% thru 200 mesh; visc. 900 cps min.; pH 5.0-6.5; 5% moisture.

Stomach Substance. [Am. Labs] Vacuum-dried defatted glandular prod.; nutritive food additive; *Properties:* powd. or freeze-dried form.

Strawberry No. 25820. [Universal Flavors] Strawberry flavor with 66% propylene glycol and 10% glycerin; flavor for still drinks; *Usage level:* 1 g/kg (still drinks); *Properties:* pale reddish-brn. liq.; water-sol.; dens. 1.03; *Storage:* 12 mos min. storage life when stored in well-filled package in cool area.

Strawberry Solid Pack No. 75954. [Universal Flavors] Strawberries, granulated sugar, water, modified starch, carrageenan, citric acid, carrageenan (stabilizer); natural flavor for ice cream; *Usage level:* 150 g/kg (ice cream); *Properties:* dens. 1.19 ± 0.01; pH 3.4 ± 0.2; *Storage:* 3 mos storage @ 4 C depending on packaging.

Sucro Ester 7. [Gattefosse SA] Saccharose distearate; CAS 27195-16-0; nonionic; food emulsifier; o/w emulsifier wetting agent, crystallization inhibitor for veg. oils and fats; inhibitor of thermal denaturation of proteins; *Properties:* fine powd., faint odor; sol. in water @ 75 C, in ethanol @ 60 C; insol. in veg. and min. oils; HLB 7.0; acid no. < 5; sapon. no. 115-135; 100% conc.; *Toxicology:* LD50 (oral, rat) > 5 g/kg.

Sucro Ester 11. [Gattefosse SA] Saccharose mono/distearate; nonionic; food emulsifier; o/w emulsifier wetting agent, crystallization inhibitor for veg. oils and fats; inhibitor of thermal denaturation of proteins; *Properties:* fine powd., faint odor; sol. in water @ 75 C, in ethanol @ 60 C; insol. in min. and veg. oils; HLB 11.0; acid no. < 5; sapon. no. 110-130; 100% conc.; *Toxicology:* LD50 (oral, rat) > 5 g/kg.

Sucro Ester 15. [Gattefosse SA] Saccharose palmitate; CAS 26446-38-8; nonionic; food emulsifier; o/w emulsifier wetting agent, crystallization inhibitor for veg. oils and fats; inhibitor of thermal denaturation of proteins; *Properties:* fine powd., faint odor; sol. in water and ethanol @ 60 C; insol. in min. and veg. oils; HLB 15.0; acid no. < 5; sapon. no. 95-135; 100% conc.; *Toxicology:* LD50 (oral, rat) > 8 g/kg.

Sugarless Coretrate® Dessert. [Hormel] Dessert conc. in 17 fruit flavors without sugar; potassium sorbate added as preservative.

Sul-fon-ate AA-9. [Boliden Intertrade] Sodium dodecylbenzene sulfonate; CAS 25155-30-0; EINECS 246-680-4; anionic; wetting agent for food applics.; *Properties:* wh. crisp flake; odorless; surf. tens. 30.6 dynes/cm (86 F); pH 7-8 (1%); 90% act.; Unverified

Sul-fon-ate AA-10. [Boliden Intertrade] Sodium dodecylbenzene sulfonate; CAS 25155-30-0; EINECS 246-680-4; anionic; wetting agent for food applics.; *Properties:* wh. crisp flake; sol. in water; pH 6-8 (1%); surf. tens. 33.2 dynes/cm (0.05%, 86 F); 96% act.; *Toxicology:* toxicology: toxic orally; LD50 (male rat, oral) 1-5 g/kg; eye and skin irritant but nontoxic dermally.

Sul-fon-ate LA-10. [Boliden Intertrade] Sodium dodecylbenzene sulfonate; CAS 25155-30-0; EINECS 246-680-4; surfactant; *Properties:* wh. crisp flake; water-sol.; pH 6; Unverified

Sulfotex OA. [Henkel/Cospha; Henkel Canada] Sodium octyl sulfate; EINECS 204-812-8; anionic; foaming agent, wetting agent for food processing;

Properties: lt. amber; dens. 9.2 lb/gal; pH 8.0-10.0 (10%); 43-46% act.

Summit®. [Bunge Foods] Partially hydrog. soybean oil, butter, water, salt, lecithin, potassium sorbate, natural and artificial flavors, antioxidants, beta carotene (colorant); liq. alternative to butter; does not require refrigerated storage; for gravies, soups, basting for vegetables and seafood, wash for baked goods; *Properties:* liq., buttery flavor; iodine no. 90.

Summit® 25. [Bunge Foods] Partially hydrog. soybean and cottonseed oils, 25% butter, water, salt, mono- and diglycerides, antioxidant, artificial flavor, vitamin A; butter substitute for croissants, danish, cookies, icings; *Properties:* buttery flavor; 80.2% fat, 16.8% moisture, 2.6% salt.

Summit® 50. [Bunge Foods] Partially hydrog. soybean and cottonseed oils, 50% butter, water, salt, mono- and diglycerides, antioxidant, artificial flavor, vitamin A; butter substitute for croissants, danish, cookies, icings; *Properties:* buttery flavor; 80.2% fat, 16.8% moisture, 2.9% salt.

Sunmalt. [Mitsubishi Int'l.; Hayashibara] Maltose monohydrate, food grade; EINECS 200-716-5; sweetener, bulking agent; prolongs shelf life of baked goods; stabilizes and enhances flavors; microbial growth inhibitor on poultry skin; sugar crystallization inhibitor; oils/fats absorber; sweetness reducer; food processing improver; *Usage level:* 1.2-2.6% (drinks), 1-2% (bakery prods.), 10-20% (chocolate), 7-10% (chewing gum), 3-4% (ice cream), 20-25% (candy), 15-20% (jam), 30-100% (tablets); *Properties:* wh. cryst. powd., free of undesirable odor or taste; 100% thru 10 mesh; > 92% purity.

Sunny Safe. [Dai-ichi Kogyo Seiyaku] Sucrose fatty acid ester compd.; nonionic; cleansing agent for foods; *Properties:* liq.

Suparen®. [Procter & Gamble] Dairy ingred.

Superb® 1450 Soy Fiber. [ADM Protein Spec.] Soy fiber; dietary fiber for snacks, breads, cereals; 80% total dietary fiber; *Properties:* wh. color; 0.5% fat.

Superb® 2000 Soy Fiber. [ADM Protein Spec.] Soy fiber; dietary fiber for drink mixes; 80% total dietary fiber; *Properties:* wh. color; 0.5% fat.

Superb® Bakers Margarine 105-101. [ADM Packaged Oils] Partially hydrog. soybean and cottonseed oils, salt, cultured nonfat milk, mono- and diglycerides, lecithin, butter flavor, coloring, Vitamin A, sorbic acid, sodium benzoate; margarine; *Properties:* rich butter colored cube, buttery flavor; m.p. 100 ± 4 F; iodine no. 81 ± 3.

Superb® Bakers Shortening 105-200. [ADM Packaged Oils] Partially hydrog. soybean and cottonseed oils, water, mono- and diglycerides, salt, cultured milk solids; packaged vegetable oil; *Properties:* wh. plasticized cube, bland flavor; m.p. 117 ± 3 F; iodine no. 78 ± 2.

Superb® Canola Liquid Frying Shortening 104-455. [ADM Packaged Oils] Partially hydrog. canola oil, TBHQ, dimethylpolysiloxane; packaged vegetable oil; *Properties:* opaque pourable liq., bland flavor; iodine no. 97 ± 5.

Superb® Canola Salad Oil 104-450. [ADM Packaged Oils] Canola oil, TBHQ; packaged veg. salad oil; *Properties:* clear bright oil, bland flavor; iodine no. 118 ± 8.

Superb® Cookie Bake Shortening 101-057. [ADM Packaged Oils] Partially hydrog. soybean oil; EINECS 232-410-2; packaged veg. oil; *Properties:* wh. plasticized cube, bland flavor; m.p. 97 ± 2 F; iodine no. 79 ± 3.

Superb® Corn Liquid Frying Shortening 104-255. [ADM Packaged Oils] Partially hydrog. corn and cottonseed oils, TBHQ, dimethylpolysiloxane; packaged vegetable oil; *Properties:* opaque pourable liq., bland flavor; iodine no. 113 ± 3.

Superb® Fish and Chip Oil 102-060. [ADM Packaged Oils] Partially hydrog. soybean oil, peanut oil, corn oil, TBHQ, dimethylpolysiloxane; packaged vegetable oil; *Properties:* opaque pourable liq., bland flavor; iodine no. 106 ± 3.

Superb® Heavy Duty Frying Shortening 101-200. [ADM Packaged Oils] Partially hydrog. soybean and cottonseed oils, TBHQ, and dimethylpolysiloxane; packaged veg. oil; *Properties:* wh. plasticized cube, bland flavor; m.p. 104 ± 4 F; iodine no. 73 ± 2.

Superb® Icing Shortening 101-270. [ADM Packaged Oils] Partially hydrog. soybean and cottonseed oils, mono- and diglycerides, polysorbate 60; packaged vegetable oil; *Properties:* wh. plasticized cube, bland flavor; m.p. 126 ± 3 F; iodine no. 74 ± 2.

Superb® Liquid Frying Shortening 102-110. [ADM Packaged Oils] Partially hydrog. soybean oil, TBHQ, dimethylpolysiloxane; packaged vegetable oil; *Properties:* opaque pourable liq., bland flavor; iodine no. 94 ± 2.

Superb Oil Hydro Winterized Soybean Oil 86-091-0. [ADM Refined Oils] Partially hydrog. soybean oil; EINECS 232-410-2; veg. oil for salad dressings, cooking, baking, frying, processed foods; *Properties:* clear brilliant oil, bland flavor; iodine no. 115 ± 3.

Superb® Soybean Salad Oil 104-112. [ADM Packaged Oils] Partially hydrog. and winterized soybean oil, TBHQ, and dimethylpolysiloxane; packaged veg. salad oil; *Properties:* clear bright oil, bland flavor; iodine no. 115 ± 3.

Superb® Soy/Corn Frying Shortening 102-070. [ADM Packaged Oils] Partially hydrog. soybean and corn oil, TBHQ, dimethylpolysiloxane; packaged vegetable oil; *Properties:* opaque pourable liq., bland flavor; iodine no. 112 ± 2.

Super-Cel® Specialty Cake Shortening. [Bunge Foods] Soybean oil, partially hydrog. soybean oil, propylene glycol, monoesters, mono- and diglycerides, lecithin; shortening for highly emulsified cake formulas, frozen cakes, moist cakes with long shelf life; *Properties:* 2.2 red color, bland flavor; 3.5% mono.

Supercol® G2S. [Hercules] Guar gum; CAS 9000-30-0; EINECS 232-536-8; improves cling of gravies, sauces; good pour chars.; stabilizer, bodying agent for ice cream.

Supercol® GF. [Hercules] Guar gum; CAS 9000-30-0; EINECS 232-536-8; bodying agent; improves mouthfeel in beverages; improves suspension of meat/veg. chunks in soups; aids cling in gravies, sauces; prevents syneresis, improves shelf life in cheese spreads; thickener for dry mixes; frozen foods, frostings, jams, meats; *Properties:* rapid cold-water hydration.

Supercol® U. [Hercules] Guar gum; CAS 9000-30-0; EINECS 232-536-8; bodying agent; improves mouthfeel in beverages; prevents syneresis, improves shelf life in cheese spreads; thickener for dry mixes; frozen foods; *Properties:* rapid cold-water hydration.

Superfine Ground Mustard. [French's Ingreds.] Mustard; savory flavoring agent and enhancer; emulsifier, suspending agent for sauces, dressings, low-fat and fat-free applics.; stable to high heat, acid, salt; provides opacity and cling; *Properties:* disp. in water.

Super Flex. [Custom Ingreds.] High fructose corn syrup, water, sorbitol, salt; humectant and softener for flour and corn tortillas; *Properties:* clear liq.; sol. in water; sp.gr. 1.085; *Storage:* store in clean, dry area @ 40-85 F for storage life of 9 mos.

Super Fry. [Karlshamns Food Ingreds.] Partially hydrog. vegetable oil (soybean, palm); nontropical version also avail.; specialty shortening for general bakery, chip and snack frying; *Properties:* liq., plastic; m.p. 107 F; smoke pt. 450 F min.

Super Gel for Glazes and Icings. [Bunge Foods] Conc. gel stabilizer used with hot water for semi-clear icings or glazes; good stability when boiled; for glazes for donuts, sweet goods icings, clear and chocolate; *Usage level:* 20 lb/100 lb powd. sugar; *Properties:* off-wh. to amber gel, sl. acid flavor; pH 4.6.

Superior Mustard. [French's Ingreds.] Water, vinegar, ground mustard; savory flavoring agent and enhancer; emulsifier, suspending agent for sauces, dressings, low-fat and fat-free applics.; stable to high heat, acid, salt; provides opacity and cling.

Superla® No. 5. [Amoco Lubricants] Wh. min. oil USP; used in foods; *Regulatory:* FDA 21CFR §172.878, 178.3620, 5763.680, USDA approved; kosher; *Properties:* colorless, odorless, tasteless; sp.gr. 0.844; visc. 8.4 cSt (40 C); pour pt. -12 C; flash pt. 149 C; ref. index 1.4660.

Superla® No. 6. [Amoco Lubricants] Wh. min. oil USP; used in foods; *Regulatory:* FDA 21CFR §172.878, 178.3620, 5763.680, USDA approved; kosher; *Properties:* colorless, odorless, tasteless; sp.gr. 0.847; visc. 10.9 cSt (40 C); pour pt. -15 C; flash pt. 174 C; ref. index 1.4664.

Superla® No. 7. [Amoco Lubricants] Wh. min. oil USP; used in foods; *Regulatory:* FDA 21CFR §172.878, 178.3620, 5763.680, USDA approved; kosher; *Properties:* colorless, odorless, tasteless; sp.gr. 0.849; visc. 14.0 cSt (40 C); pour pt. -15 C; flash pt. 185 C; ref. index 1.4666.

Superla® No. 9. [Amoco Lubricants] Wh. min. oil USP; used in foods; *Regulatory:* FDA 21CFR §172.878, 178.3620, 5763.680, USDA approved; kosher; *Properties:* colorless, odorless, tasteless; sp.gr. 0.850; visc. 16.0 cSt (40 C); pour pt. -15 C; flash pt. 193 C; ref. index 1.4715.

Superla® No. 10. [Amoco Lubricants] Wh. min. oil USP; used in foods; *Regulatory:* FDA 21CFR §172.878, 178.3620, 5763.680, USDA approved; kosher; *Properties:* colorless, odorless, tasteless; sp.gr. 0.850; visc. 18.4 cSt (40 C); pour pt. -15 C; flash pt. 193 C; ref. index 1.4728.

Superla® No. 13. [Amoco Lubricants] Wh. min. oil USP; used in foods; *Regulatory:* FDA 21CFR §172.878, 178.3620, 5763.680, USDA approved; kosher; *Properties:* colorless, odorless, tasteless; sp.gr. 0.854; visc. 26.2 cSt (40 C); pour pt. -12 C; flash pt. 190 C; ref. index 1.4728.

Superla® No. 18. [Amoco Lubricants] Wh. min. oil USP; used in foods; *Regulatory:* FDA 21CFR §172.878, 178.3620, 5763.680, USDA approved; kosher; *Properties:* colorless, odorless, tasteless; sp.gr. 0.857; visc. 36.3 cSt (40 C); pour pt. -12 C; flash pt. 193 C; ref. index 1.4738.

Superla® No. 21. [Amoco Lubricants] Wh. min. oil USP; used in foods; *Regulatory:* FDA 21CFR §172.878, 178.3620, 5763.680, USDA approved; kosher; *Properties:* colorless, odorless, tasteless; sp.gr. 0.859; visc. 40.2 cSt (40 C); pour pt. -12 C; flash pt. 210 C; ref. index 1.4744.

Superla® No. 31. [Amoco Lubricants] Wh. min. oil USP; used in foods; *Regulatory:* FDA 21CFR §172.878, 178.3620, 5763.680, USDA approved; kosher; *Properties:* colorless, odorless, tasteless; sp.gr. 0.865; visc. 60.5 cSt (40 C); pour pt. -12 C; flash pt. 227 C; ref. index 1.4763.

Superla® No. 35. [Amoco Lubricants] Wh. min. oil USP; used in foods; *Regulatory:* FDA 21CFR §172.878, 178.3620, 5763.680, USDA approved; kosher; *Properties:* colorless, odorless, tasteless; sp.gr. 0.865; visc. 69.3 cSt (40 C); pour pt. -12 C; flash pt. 232 C; ref. index 1.4772.

Superol. [Procter & Gamble] Glycerin USP; CAS 56-81-5; EINECS 200-289-5; humectant in food prods.; *Properties:* APHA 10 max. color; sp.gr. 1.2612.

Super Salox. [Vaessen-Schoemaker] Color enhancing/retaining additives; protectant, stabilizer for natural meat color in minced meat, meat balls, fresh sausage, hams, corned beef, other processed/cooked meat prods.; *Usage level:* 0.5-2 g/kg end prod.; *Regulatory:* FAO/WHO, EEC compliance; 5% max. moisture.

Super Set. [Brolite Prods.] Highly conc. blend that prevents icings from breaking down in almost any condition.

Supersta Icing Powd. [Bunge Foods] Boiling type icing stabilizer; glaze for wrapped donuts and sweet goods; also avail. in clear and extra white form; *Usage level:* 4-5 lb/100 lb powd. sugar; *Properties:* wh. powd.; pH 7.5.

Super Timeflex. [MLG Enterprises Ltd.] Proteolytic enzyme derived from *Bacillus subtilis* with cereal derivs. and other edible excipients; bacterial enzyme tablet for crackers, cookies, and pizza prod.; provides reduction in mixing, accelerated fermentation, improved prod. quality; *Properties:* tablet; readily disintegrates in R.T. water.

Super Vita Rye. [Brolite Prods.] Blend of spices and natural rye sours that produces a unique rye flavor.

Super White Fonoline®. [Witco/Petroleum Spec.] Wh. petrolatum USP; CAS 8027-32-5; EINECS 232-373-2; low m.p. grade with superior snow white color, exhibiting elegant feel and texture; for food related applics.; *Regulatory:* FDA §172.880; *Properties:* visc. 55-65 SUS (210 F); m.p. 122-133 F.

Super White Protopet®. [Witco/Petroleum Spec.] Wh. petrolatum USP; CAS 8027-32-5; EINECS 232-373-2; high purity, med. consistency, med. m.p. grade for food applics. where extra whiteness is preferred; *Regulatory:* FDA 21CFR §172.880; *Properties:* Lovibond 1.0Y max. color; visc. 60-70 SUS (210 F); m.p. 130-140 F.

Suprarenal (Adrenal) Substance. [Am. Labs] Vacuum-dried defatted glandular prod.; nutritive food additive; *Properties:* powd. or freeze-dried form.

Suprarenal Cortex Substance. [Am. Labs] Vacuum-dried defatted glandular prod.; nutritive food additive; *Properties:* powd.

Supro® 425. [Protein Tech. Int'l.] Isolated soy protein; CAS 68153-28-6; high-quality protein, lactose- and cholesterol-free, low-fat; for cream-style sauces, condensed cream soup; *Properties:* high sol. and emulsification, med. visc., med.-low gel str., med. water absorp.; 87% protein.

Supro® 610. [Protein Tech. Int'l.] Isolated soy protein; CAS 68153-28-6; high-quality protein, lactose- and cholesterol-free, low-fat; for pasta; *Properties:* med. sol. and emulsification, high visc., med. gel str., high water absorp.; 88% protein.

Supro® 661. [Protein Tech. Int'l.] Isolated soy protein; CAS 68153-28-6; high-quality protein, lactose- and cholesterol-free, low-fat; for dry beverages, protein/wt. loss supplements, food bars, dry soups, meal replacements; *Properties:* low sol. and emulsification, high visc., low gel str., high water absorp.; 88% protein.

Supro® 670. [Protein Tech. Int'l.] Isolated soy protein; CAS 68153-28-6; high-quality protein, lactose- and cholesterol-free, low-fat; for dry beverages, protein/wt. loss supplements, food bars, dry soups, meal replacements; *Properties:* med. sol. and emulsification, med.-low visc. and water absorp.; 87% protein.

Supro® 710. [Protein Tech. Int'l.] Isolated soy protein; CAS 68153-28-6; high-quality protein, lactose- and cholesterol-free, low-fat; for processed cheese, nondairy prods. (coffee creamers, dips, frozen dessert, imitation yogurt); *Properties:* med. sol. and emulsification, med.-low visc. and water absorp.; 87% protein.

Supro® 760. [Protein Tech. Int'l.] Isolated soy protein; CAS 68153-28-6; high-quality protein, lactose/cholesterol-free, low-fat; for liq. beverages, nutritional beverages, meal replacements, soy milk fortification, yogurt, imitation sour cream and yogurt, frozen nondairy dessert, cream-style sauces, tofu, baked goods; *Properties:* high sol. and emulsification, high visc., gel str., water absorp.; 87% protein.

Supro Plus® 651. [Protein Tech. Int'l.] Isolated soy protein and calcium phosphate; protein with calcium ratio similar to cow's milk; for liq. beverages, nutritional beverages, meal replacements, soy milk fortification, yogurt, frozen nondairy desserts, imitation yogurt, tofu; *Properties:* high sol., emulsification, visc., gel str., and water absorp.; 81% protein.

Supro Plus® 675. [Protein Tech. Int'l.] Isolated soy protein and calcium phosphate; protein with calcium ratio similar to cow's milk; for dry beverages, protein/wt. loss supplements, meal replacements, fortified juice, frozen nondairy dessert, imitation yogurt, dry soups; *Properties:* med. sol. and emulsification, med.-low visc. and water absorp.; 81% protein.

Supro Plus® 2100. [Protein Tech. Int'l.] Isolated soy protein and calcium phosphate; protein with calcium ratio similar to cow's milk; for dry and liq. beverages, wt. loss supplements, meal replacements, fortified juice, processed cheese, ice cream, dry soups; *Properties:* high sol., med. emulsification, med.-low visc., low water absorp.; 57% protein.

Supro Plus® 3000. [Protein Tech. Int'l.] Isolated soy protein and calcium phosphate; protein with calcium ratio similar to cow's milk; for dry beverages, wt. loss supplements, meal replacements, ice cream; *Properties:* med. sol. and emulsification, med.-low visc. and water absorp.; 75% protein.

Sure-Curd™. [Pfizer Food Science] Fermentation-derived cheese milk-clotting enzyme produced from *Endothia parasitica*; triple-strength dairy ingred. providing comparable coagulation, cheese yield, and cheese quality to rennet extract; esp. for Swiss and Italian cheeses; *Regulatory:* FDA 21CFR §173.150, kosher; *Properties:* amber liq.

Surfactant® AR 150. [Hercules] Oxyethylene glycol ester of rosin; nonionic; hydrophilic emulsifier, low foaming detergent; used in food industry; *Properties:* liq. to wax; 100% conc.

Suspengel Elite. [Cimbar Perf. Minerals] Purified dry colloidal hydrated aluminum silicate; CAS 1327-36-2; EINECS 215-475-1; thixotropic thickener giving good sag control, leveling, and suspension props.; used for foods, beverages; *Usage level:* 0.25-8.0%; *Regulatory:* FDA GRAS §184.1155; kosher; *Properties:* lt. cream-colored powd.; odorless, tasteless; avg. particle size 0.5 µ; surf. area 1.2 m²/cc; visc. 10,000 cps (6% solids); bulk dens. 38 lb/ft³ (loose); oil absorp. 47; pH 9.4; 8.7% moisture.

Suspengel Micro. [Cimbar Perf. Minerals] Purified dry colloidal hydrated aluminum silicate; CAS 1327-36-2; EINECS 215-475-1; thixotropic thickener giving good sag control, leveling, and suspension props.; used for foods, beverages; *Usage level:* 0.25-8.0%; *Regulatory:* FDA GRAS §184.1155; kosher; *Prop-*

erties: lt. cream-colored powd.; odorless, tasteless; avg. particle size 1.0 µ; surf. area 0.9 m²/cc; visc. 10,000 cps (6% solids); bulk dens. 41 lb/ft³ (loose); oil absorp. 36; pH 9.4; 8.7% moisture.

Suspengel Ultra. [Cimbar Perf. Minerals] Purified dry colloidal hydrated aluminum silicate; CAS 1327-36-2; EINECS 215-475-1; thixotropic thickener giving good sag control, leveling, and suspension props.; used for foods, beverages; *Usage level:* 0.25-8.0%; *Regulatory:* FDA GRAS §184.1155; kosher; *Properties:* lt. cream-colored powd.; odorless, tasteless; avg. particle size 0.18 µ; surf. area 1.8 m²/cc; visc. 15,000 cps (6% solids); bulk dens. 29 lb/ft³ (loose); oil absorp. 52; pH 9.4; 8.7% moisture.

Sustane® 1-F. [UOP Food Antioxidants] BHA; CAS 25013-16-5; EINECS 246-563-8; preservative and antioxidant for foods, flavors, vitamins, oils, sausage, chewing gum base, shortening, lard, potatoes, and cereals; inhibits oxidation reaction of oils and fats in presence of air; retards rancidity, off-flavors caused by oxidation; *Regulatory:* FDA 21CFR §172.110, 172.115, 182.3169; USDA 9CFR §318.7, 381.147; *Properties:* wh. flakes; sol. > 30 g/100 g in soybean oil, cottonseed oil; > 25 g/100 g in acetone, ether; > 10 g/100 g in methanol; neg. sol. in water; m.w. 180.2; visc. 3.3 cS (99 C); m.p. 57 C; b.p. 270 F (@ 5 mm Hg); flash pt. (OC) 130 C; 98.5% min. conc.

Sustane® 3. [UOP Food Antioxidants] BHA (20%), propylene glycol (70%), propyl gallate (6%), citric acid (4%); preservative and antioxidant used in snack foods, spices, animal fats; *Regulatory:* FDA, USDA approved; *Properties:* pale yel. liq.; sp.gr. 1.066; dens. 8.7 lb/gal; visc. 40.64 cS (38 C); f.p. -23 C; flash pt. (CC) 99 C; ref. index. 1.4636.

Sustane® 4A. [UOP Food Antioxidants] BHA (30%), veg. oil (70%); food-grade preservative and antioxidant used in nuts, baked goods, and edible fats and oils; prevents oxidative rancidity; *Regulatory:* FDA, USDA approved; *Properties:* pale yel. liq.; sp.gr. 0.953 (20/4 C); dens. 7.9 lb/gal; visc. 43.49 cS (38 C); flash pt. (CC) 254 C; ref. index 1.490.

Sustane® 6. [UOP Food Antioxidants] BHT (22%), BHA (18%), veg. oil (60%); food-grade preservative and antioxidant used in edible fats, meat prods., vitamins, emulsifiers, and confections; prevents oxidative rancidity; *Properties:* pale yel. liq.; sp.gr. 0.928 (20/4 C); dens. 7.8 lb/gal; visc. 35.56 cS (38 C); f.p. -26 C; flash pt. (CC) 254 C; ref. index 1.490.

Sustane® 7. [UOP Food Antioxidants] BHA (28%), propyl gallate (12%), citric acid (6%), propylene glycol (34%), mono/di-glycerides (20%); food-grade antioxidant for animal fats; prevents oxidative rancidity; *Regulatory:* FDA, USDA approved; *Properties:* yel.-grn. liq.; sp.gr. 1.068 (20/4 C); dens. 8.9 lb/gal; visc. 118.3 cS (38 C); flash pt. (CC) > 37.8 C; ref. index 1.478.

Sustane® 7G. [UOP Food Antioxidants] BHA (28%), propyl gallate (12%), citric acid (6%), propylene glycol (54%); food-grade antioxidant; prevents oxidative rancidity; *Properties:* pale yel. liq.; sp.gr.

1.089 (20/4 C); ref. index 1.480.

Sustane® 8. [UOP Food Antioxidants] BHA (20%) citric acid (20%), propylene glycol (60%); preservative and antioxidant used in animal fats, pet foods; prevents oxidative rancidity; *Regulatory:* FDA, USDA approved; *Properties:* pale yel. liq.; sp.gr. 1.114 (20/4 C); dens. 9.1 lb/gal; visc. 39.12 cS (38 C); f.p. -17.8 C; b.p. 188 C (745 mm); flash pt. (CC) 99 C; ref. index 1.465.

Sustane® 11. [UOP Food Antioxidants] BHT (20%), vegetable oil (80%); food-grade antioxidant preventing oxidative rancidity in edible fats and oils, nuts; *Regulatory:* FDA, USDA approved; *Properties:* yel. to amber liq.; sp.gr. 0.923 (20/4 C); dens. 7.7 lb/gal; visc. 31.32 cS (38 C); flash pt. (CC) 254 C; ref. index 1.482.

Sustane® 15. [UOP Food Antioxidants] BHT (15%), citric acid (3%), mono/di-glycerides (54%), propylene glycol (16%), vegetable oil (12%); food-grade antioxidant preventing oxidative rancidity in edible fats and oils, nuts; *Properties:* pale yel. liq.; sp.gr. 0.9655 (20/4 C); visc. 65.9 cS (38 C); flash pt. (CC) 111.7 C; ref. index 1.472.

Sustane® 16. [UOP Food Antioxidants] BHT (10%), propyl gallate (10%), citric acid (3%), mono/di-glycerides (33%), vegetable oil (31%), propylene glycol (13%); food-grade antioxidant preventing oxidative rancidity; *Properties:* amber liq.; sp.gr. 0.99 (20/4 C); visc. 119.3 cS (38 C); flash pt. (CC) 116 C; ref. index 1.480.

Sustane® 18. [UOP Food Antioxidants] BHT (10%), TBHQ (10%), citric acid (3%), mono/di-glycerides (33%), vegetable oil (31%), propylene glycol (13%); food-grade antioxidant preventing oxidative rancidity; *Properties:* amber liq.; sp.gr. 0.99 (20/4 C); visc. 124.5 cS (38 C); flash pt. (CC) 116 C; ref. index 1.480.

Sustane® 20. [UOP Food Antioxidants] TBHQ (20%), citric acid (10%), propylene glycol (70%); food-grade preservative and antioxidant used for edible fats and oils; prevents oxidative rancidity; *Regulatory:* FDA, USDA approved; *Properties:* pale amber liq.; sp.gr. 1.086 (20/4 C); dens. 9.0 lb/gal; visc. 83.98 cS (38 C); flash pt. (CC) 109 C; ref. index 1.463.

Sustane® 20-3. [UOP Food Antioxidants] TBHQ (20%), citric acid (3%), propylene glycol (77%); food-grade preservative and antioxidant used for edible fats and oils; prevents oxidative rancidity; *Regulatory:* FDA, USDA approved; *Properties:* amber liq.; sp.gr. 1.060 (20/4 C); dens. 8.8 lb/gal; visc. 52.34 cS (38 C); f.p. -24 C; flash pt. (CC) 122 C; ref. index 1.458.

Sustane® 20A. [UOP Food Antioxidants] TBHQ (20%), citric acid (3%), veg. oil (30%), propylene glycol (15%), mono- and diglycerides (32%); food-grade antioxidant preventing oxidative rancidity in edible fats and oils, nuts, confections; *Regulatory:* FDA, USDA approved; *Properties:* amber liq.; sp.gr. 0.989 (20/4 C); dens. 8.1 lb/gal; visc. 159 cS (38 C); flash pt. (CC) 122 C; ref. index 1.482.

Sustane® 20B. [UOP Food Antioxidants] TBHQ (20%), citric acid (1%), propylene glycol (15%), mono- and diglycerides (32%), vegetable oil (32%); food-grade antioxidant preventing oxidative rancidity in edible fats and oils, nuts, confections; *Regulatory:* FDA, USDA approved; *Properties:* amber liq.; sp.gr. 0.987 (20/4 C); dens. 8.1 lb/gal; visc. 125.5 cS (38 C); ref. index 1.482.

Sustane® 21. [UOP Food Antioxidants] TBHQ (20%), propylene glycol (80%); food antioxidant; *Properties:* liq.

Sustane® 27. [UOP Food Antioxidants] BHA (28%), TBHQ (12%), citric acid (6%), emulsifier (20%), propylene glycol (34%); food antioxidant; *Properties:* liq.

Sustane® 31. [UOP Food Antioxidants] BHA (20%), TBHQ (6%), citric acid (4%), propylene glycol (70%); food-grade preservative and antioxidant used in dressing oils, nuts, confections, cereals, and flavors; *Regulatory:* FDA, USDA approved; *Properties:* amber liq.; sp.gr. 1.056 (20/4 C); dens. 8.7 lb/gal; visc. 42.04 cS (38 C); f.p. -29 C; flash pt. (CC) 113 C; ref. index 1.462.

Sustane® BHA. [UOP Food Antioxidants] BHA; CAS 25013-16-5; EINECS 246-563-8; antioxidant, stabilizer for fats, oils, other foods, flavors, vitamins, waxes, tallow, sausage, chewing gum base, shortening, lard, food pkg. materials, potatoes, and cereals; *Usage level:* 0.02% max.; *Regulatory:* FDA 21CFR §172.110, 172.115, 182.3169; USDA 9CFR §318.7, 381.147; *Properties:* wh. solid tablets; sol. in propylene glycol, ethanol, glyceryl oleate, soybean and cottonseed oils, acetone; insol. in water; m.w. 180.2; sp.gr. 1.020; visc. 3.3 cSt (100 C); m.p. 57 C; b.p. 270 C (760 mm); flash pt. (CC) 130 C; 98.5% min. purity; *Toxicology:* may cause irritation of skin, nostrils; *Precaution:* dusts may present an explosion hazard.

Sustane® BHT. [UOP Food Antioxidants] BHT; CAS 128-37-0; EINECS 204-881-4; preservative and antioxidant for foods, tallow, animal feeds; inhibits oxidation of fats and oils and retards rancidity and off-flavors caused by oxidation; *Usage level:* 0.02% max.; *Regulatory:* FDA 21CFR §172.115, 182.3173; USDA 9CFR §318.7, 381.147; *Properties:* wh. cryst.; sol. (g/100 solv.): 48 g in lard (50 C), 40 g in benzene, 30 g in min. oil, 28 g in linseed oils, 20 g in methanol, nil in water; m.w. 220.3; sp.gr. 1.01 (20/4 C); dens. 37.5 lb/ft³; visc. 3.5 cSt (80 C); m.p. 70 C; b.p. 265 C (760 mm Hg); flash pt. (CC) 118 C; ref. index 1.486; 100% act.; *Toxicology:* may cause irritation of skin, nostrils; *Precaution:* dusts may present an explosion hazard.

Sustane® CA. [UOP Food Antioxidants] Citric acid (30%), propylene glycol (70%); food-grade antioxidant for edible fats and oils; provides metal sequestering activity which results in synergism with antioxidants in prevention of oxidative rancidity; *Regulatory:* USDA approved; *Properties:* clear liq.; sp.gr. 1.158 (20/4 C); dens. 9.0 lb/gal; visc. 152.7 cSt (38 C); flash pt. (CC) 112 C; ref. index 1.450.

Sustane® HW-4. [UOP Food Antioxidants] BHA (20%), BHT (20%), veg. oil (60%); food-grade preservative and antioxidant used in edible fats, meat and sausage prods., vitamins, emulsifiers, and confections; prevents oxidative rancidity; *Regulatory:* FDA, USDA approved; *Properties:* pale yel. liq.; sp.gr. 0.946 (20/4 C); dens. 7.8 lb/gal; visc. 36.52 cS (38 C); f.p. -23 C; b.p. 258 C; flash pt. (CC) 132 C; ref. index 1.492.

Sustane® P. [UOP Food Antioxidants] BHA (25%), BHT (25%), ethyl alcohol (50%); food-grade preservative and antioxidant used in dehydrated vegetables; prevents oxidative rancidity; *Properties:* clear to pale yel. liq.; sp.gr. 0.884 (20/4 C); dens. 7.3 lb/gal; f.p. 12.8 C; flash pt. (CC) 16 C; ref. index 1.444.

Sustane® PA. [UOP Food Antioxidants] BHA (49.9%), ethyl alcohol (49.9%), citric acid (0.2%); food-grade preservative and antioxidant used in dehydrated vegetables, potatoes; prevents oxidative rancidity; *Regulatory:* FDA, USDA approved; *Properties:* pale yel. liq.; sp.gr. 0.909 (20/4 C); dens. 7.6 lb/gal; visc. 2.556 cS (38 C); f.p. -45.6 C; flash pt. (Seta) 18 C; ref. index 1.441.

Sustane® PB. [UOP Food Antioxidants] BHT (15%), ethanol (85%); food-grade antioxidant preventing oxidative rancidity in dehydrated potatoes; *Regulatory:* FDA, USDA approved; *Properties:* pale yel. liq.; sp.gr. 0.811 (20/4 C); dens. 7.6 lb/gal; visc. 2.6 cSt (38 C); flash pt. (PMCC) 12 C; ref. index 1.383.

Sustane® PG. [UOP Food Antioxidants] Propyl gallate; CAS 121-79-9; EINECS 204-498-2; preservative and antioxidant for fats and oils; synergistic with BHA; susceptible to discoloration-for use where color stability is not a problem and where the prod. will not contact iron or copper; *Regulatory:* FDA 21CFR §184.1660; USDA 9CFR §318.7, 381.147; *Properties:* off-wh. cryst. powd., sl. odor; sol. (g/100 g solv.): 170 g in methanol, 121 g in acetone, 103 g in ethanol, 83 g in ethyl ether, 67 g in propylene glycol; m.w. 212; sp.gr. 1.21 (20/4 C); m.p. 146-150 C; b.p. decomposes; flash pt. (CC) 187 C; 98% min. purity; *Toxicology:* may cause irritation of skin, nostrils; *Precaution:* dusts may present an explosion hazard.

Sustane® Q. [UOP Food Antioxidants] BHA (40%), citric acid (8%), propylene glycol (52%); food-grade preservative and antioxidant used in veg. oils, frozen foods, flavors, cereals, animal fats, pet foods; prevents oxidative rancidity; *Regulatory:* FDA, USDA approved; *Properties:* pale yel. liq.; sp.gr. 1.067 (20/4 C); dens. 8.8 lb/gal; visc. 57.09 cS (38 C); flash pt. (CC) 99 C; ref. index 1.477.

Sustane® S-1. [UOP Food Antioxidants] Propyl gallate (20%), citric acid (10%), propylene glycol (70%); food-grade antioxidant preventing oxidative rancidity in edible fats and oils; *Regulatory:* FDA, USDA approved; *Properties:* pale amber liq.; sp.gr. 1.116 (20/4 C); dens. 8.9 lb/gal; visc. 80.85 cS (38 C); f.p. -30 C; flash pt. (CC) 104 C; ref. index 1.465.

Sustane® T-6. [UOP Food Antioxidants] BHA (10%), BHT (10%), propyl gallate (6%), citric acid (6%), mono- and diglycerides (36%), vegetable oil (18%), propylene glycol (14%); food-grade antioxidant

preventing oxidative rancidity in spices, sausage prods., baked goods, animal fats; *Properties:* pale yel. liq.; sp.gr. 1.0019 (20/4 C); dens. 8.1 lb/gal; visc. 110.2 cSt (38 C); f.p. -23 C; flash pt. (CC) 240 C; ref. index 1.483.

Sustane® TBHQ. [UOP Food Antioxidants] TBHQ; CAS 1948-33-0; EINECS 217-752-2; food-grade preservative and antioxidant for foodstuffs and meat prods., esp. in preservation of shortenings and oils derived from cottonseed, soybean, safflower, etc.; color stable; as substitute for reactive antioxidants that form purple complexes; *Regulatory:* FDA 21CFR §172.185; USDA 9CFR §318.7, 381.147; *Properties:* off-wh. cryst. powd.; sol. (g/100 g): 100 g in methanol, 30 g in propylene glycol, 10 g in corn oil, 5 g in lard (50 C), nil in water; m.w. 166.2; dens. 27 lb/ft³; m.p. 126.5-128.5 C; b.p. 295 C (760 mm Hg); flash pt. (CC) 171 C; 99.0% min purity; *Toxicology:* may cause irritation of skin, nostrils; *Precaution:* dusts may present an explosion hazard.

Sustane® W-1. [UOP Food Antioxidants] BHA (10%), BHT (10%), TBHQ (6%), citric acid (6%), emulsifier (28%), vegetable oil (28%), propylene glycol (12%); food antioxidant; *Properties:* liq.

Sustane® W. [UOP Food Antioxidants] BHA, BHT, propyl gallate, citric acid, propylene glycol, veg. oil, glyceryl oleate, ratio 10:10:6:6:8:28:32; preservative and antioxidant used in spices, meat prods., baked goods, and snack foods; *Properties:* amber liq.; sp.gr. 0.978; dens. 8.1 lb/gal; visc. 102.4 cS (38 C); ref. index 1.484; Unverified

Swedex AR58P-15AC. [Australian Bakels] Diacetylated tartaric acid ester of mono- and diglycerides; emulsifier; *Properties:* cream-colored powd., acetic acid odor; sol. in warm water, alcohol, propylene glycol, glycerin, veg. oils; m.p. 45-50 C; acid no. ≈ 100; 85% emulsifier, 15% anticaking content.

Swedex SSL-5AC. [Australian Bakels] Sodium stearoyl-2-lactylate with 5% anticaking agent; CAS 25383-99-7; EINECS 246-929-7; emulsifier; *Properties:* cream-colored powd.; sol. in warm water, alcohol, propylene glycol, and veg. oils; m.p. 45-50 C; acid no. 55-75; 95% act.; *Storage:* store in cool, dry, dark conditions.

Sweet Almond Oil BP 73. [Anglia Spec. Oils] Sweet almond oil; CAS 8007-69-0; edible oil; *Properties:* iodine no. 93-106; sapon. no. 188-196; ref. index 1.4630-1.4660 (40 C).

#325 Sweet Dough Conc. [Bunge Foods] Conc. producing sweet doughs with very light flavor; provides stability, uniformity, machineability, and high quality; for sweet rolls, danish pastries; *Usage level:* 33% based on flour; *Properties:* avail. in flavored form.

Sweet Lime Oil 5 Fold. [Beeta; Commodity Services Int'l.] Oil obtained from Indian sweet limes, contg. linalyl acetate and other esters; CAS 8008-26-2; flavor ingred. in beverages, fruit flavors, liquors, and wine coolers; may be used in cologne-type fragrances to impart a fresh citrus lift; *Regulatory:* FDA 21CFR §182.20; GRAS (FEMA); EEC; IFRA;

IOFI; *Properties:* yel. clear liq.; sp.gr. 0.840-0.855; ref. index 1.455-1.475.

Sweet'n'Neat® 45. [ADM Ogilvie] 45% Honey; CAS 8028-66-8; sweetener for beverages, sauces, meats; *Properties:* yel. powd.; 4.0% max. moisture.

Sweet'n'Neat® 65 Molasses Powd. [ADM Ogilvie] 65% Molasses; CAS 68476-78-8; sweetener for beverages, sauces, mixes; *Properties:* tan powd.; 4% max. moisture.

Sweet'n'Neat® 2000. [ADM Ogilvie] 70% Honey; CAS 8028-66-8; sweetener for snacks, candy, baked goods, mixes, sauces, dairy, meats; *Properties:* yel. powd.; 3.5% max. moisture.

Sweet'n'Neat® 3000. [ADM Ogilvie] 70% Honey; CAS 8028-66-8; sweetener for snacks, candy, baked goods, mixes, sauces, dairy, meats; *Properties:* yel. flake; 3.5% max. moisture.

Sweet'n'Neat® 4000 Molasses Powd. [ADM Ogilvie] 75% Molasses; CAS 68476-78-8; sweetener for baked goods, barbeque sauces, mixes, candy, dairy; *Properties:* dk. brn. powd.; 4% max. moisture.

Sweet'n'Neat® 5000 Molasses Flake. [ADM Ogilvie] 75% Molasses; CAS 68476-78-8; sweetener for baked goods, barbeque sauces, mixes, candy, dairy; *Properties:* dk. brn. flake; 4% max. moisture.

Sweet'n'Neat Maple. [ADM Ogilvie] 62% Maple and other sweetener solids; sweetener for baked goods, mixes, cereal, snacks, candy, dairy prods.; *Properties:* lt. tan powd./flake; 4% max. moisture.

Sweet'n'Neat® Oil/Dry Roast. [ADM Ogilvie] Line of honey coatings used to mfg. dry and oil roasted honey-coated nuts, sunflower seeds, and baked or extruded snacks.

Sweet'n'Neat Raisin. [ADM Ogilvie] 70% Raisin; dry prod. for baked goods, mixes, cereal, snacks; *Properties:* lt. tan powd./flake; 4% max. moisture.

Sweet'n'Neat® Tack Blend. [ADM Ogilvie] Snack sweetener adhesive; *Properties:* wh. powd.; 5.0% max. moisture.

Swelite®. [Cosucra BV; Feinkost Ingred.] Pea inner fibers; dietary fiber with very high water retention capacity; enables development of lighter prods. while maintaining texture; for pastries, cereal prods., meat prods., desserts, dairy prods., soups, sauces, dietetic foods, extruded prods.; *Properties:* wh. powd., neutral taste; 95% < 400 μ (standard), 95% < 100 μ (micronized); 90% min. dry matter; *Storage:* store under dry conditions.

Swelite® Micro. [Cosucra BV; Feinkost Ingred.] Micronized pea inner fibers; dietary fiber, gellant, water retention aid; as substitute for high-calorie prods.; in meat/poultry prods., frozen foods, ready meals, soups, sauces, dairy prods., drinks, pastries, confectionery, dietetic foods, extruded prods.; *Properties:* wh. powd., neutral taste; 90% < 50 μ; 90% min. dry matter; *Storage:* store under dry conditions.

Syncal® CAS. [PMC Specialties] Calcium saccharin USP/FCC/BP; CAS 6485-34-3; EINECS 229-349-9; syn. sweetening agent for dietetic foods, beverages, snacks; for formulations where low sodium

content and improved taste are required; exc. free flow chars. and low degree of dust; *Regulatory:* FDA 21CFR §180.37; *Properties:* wh. fine free-flowing powd.; m.w. 404.45; 98% min. assay.

Syncal® GS. [PMC Specialties] Sodium saccharin USP/FCC/BP; CAS 128-44-9; EINECS 204-886-1; syn. sweetening agent for dietetic foods, beverages, snacks, tobacco; *Regulatory:* FDA 21CFR §180.37; *Properties:* wh. nondusting, free-flowing gran., uniformly sized particles, odorless; sol. 46% in water; m.w. 241.20; sp.gr. > 1; 98-101% assay; *Toxicology:* nuisance dust ACGIH 10 mg/m³; *Precaution:* incompat. with oxidizing agents; *Storage:* avoid excessive heat and humidity on storage to prevent caking.

Syncal® GSD. [PMC Specialties] Sodium saccharin USP/FCC/BP; CAS 128-44-9; EINECS 204-886-1; syn. sweetening agent for dietetic foods, beverages, snacks, tobacco; *Regulatory:* FDA 21CFR §180.37; *Properties:* wh. nondusting gran., mixed sized particles, odorless; sol. 46% in water; m.w. 241.20; sp.gr. > 1; 98-101% assay; *Toxicology:* nuisance dust ACGIH 10 mg/m³; *Precaution:* incompat. with oxidizing agents; *Storage:* avoid excessive heat and humidity on storage to prevent caking.

Syncal® S. [PMC Specialties] Sodium saccharin USP/FCC/BP; CAS 128-44-9; EINECS 204-886-1; syn. sweetening agent for dietetic foods, beverages, snacks; *Regulatory:* FDA 21CFR §180.37; *Properties:* wh. fine cryst. powd., odorless; m.w. 205.17; sp.gr. > 1; 98-101% assay; *Toxicology:* LD50 (oral, rat) 14,200 mg/kg; suspect carcinogen in rats on extreme exposure; nuisance dust; *Storage:* avoid excessive heat and humidity to prevent caking.

Syncal® SDI. [PMC Specialties] Saccharin insoluble USP/FCC/BP; CAS 81-07-2; EINECS 201-321-0; syn. sweetening agent for dietetic foods, beverages, snacks, tobacco; *Regulatory:* FDA 21CFR §180.37; *Properties:* wh. powd., odorless; sol. 0.03% in water; m.w. 183.18; sp.gr. > 1; m.p. 226-230 C; 98-101% assay; *Toxicology:* nuisance dust ACGIH 10 mg/m³; *Precaution:* incompat. with oxidizing agents; *Storage:* avoid excessive heat and humidity on storage to prevent caking.

Syncal® SDS. [PMC Specialties] Sodium saccharin USP/FCC/BP; CAS 128-44-9; EINECS 204-886-1; syn. sweetening agent for dietetic foods, beverages, snacks, tobacco; for applics. requiring free flow chars. and low degree of dust; *Regulatory:* FDA 21CFR §180.37; *Properties:* wh. fine free-flowing spray-dried powd.; very fine particle size; sol. 46% in water; m.w. 205.17; sp.gr. > 1; 98-101% assay; *Toxicology:* nuisance dust TLV 10 mg/m³; *Precaution:* incompat. with oxidizing agents; *Storage:* avoid excessive heat on storage to prevent caking.

Syncal® US. [PMC Specialties] Sodium saccharin USP/FCC/BP; CAS 128-44-9; EINECS 204-886-1; syn. sweetening agent for dietetic foods, beverages, snacks, tobacco; *Regulatory:* FDA 21CFR

§180.37; *Properties:* wh. nondusting granules, unsized particles, odorless; sol. 46% in water; m.w. 241.20; sp.gr. > 1; 98-101% assay; *Toxicology:* nuisance dust ACGIH 10 mg/m³; *Precaution:* incompat. with oxidizing agents; *Storage:* avoid excessive heat and humidity on storage to prevent caking.

Synperonic LF/RA43. [ICI Surf. Am.; ICI plc] Fatty alcohol alkoxylate; CAS 69227-21-0; nonionic; defoamer for washing of fruits and vegs.; *Properties:* colorless liq.; 100% conc.

Synpro® Calcium Stearate Food Grade. [Syn. Prods.] Calcium stearate; CAS 1592-23-0; EINECS 216-472-8; additive for food applics.; *Properties:* fineness 99% thru 200 mesh; soften. pt. 155 C; 2.5% moisture.

Synpro® Magnesium Stearate Food Grade. [Syn. Prods.] Magnesium stearate; CAS 557-04-0; EINECS 209-150-3; additive for food applics.; *Properties:* fineness 99% thru 200 mesh; soften. pt. 147 C; 3.0% moisture.

T

Tagat® I. [Goldschmidt; Goldschmidt AG] PEG-30 glyceryl isostearate; CAS 69468-44-6; nonionic; preparation of o/w emulsions; solubilizer for flavors, perfumes, vitamin oils; dispersant and antistat; *Properties:* pale yel. liq.; sol. in water; insol. in veg. and min. oils; HLB 15.6; acid no. 2 max.; iodine no. 4 max.; sapon. no. 25-40; hyd. no. 90-105; 100% conc.; Discontinued

Tagat® I2. [Goldschmidt; Goldschmidt AG] PEG-20 glyceryl isostearate; CAS 69468-44-6; nonionic; preparation of o/w emulsions; solubilizer for flavors, perfumes, vitamin oils; dispersant and antistat; *Properties:* pale yel. liq.; sol. in water; insol. in veg. and min. oils; HLB 14.2; acid no. 2 max.; iodine no. 4 max.; sapon. no. 45-65; hyd. no. 110-130; 100% conc.; Discontinued

Tagat® L. [Goldschmidt; Goldschmidt AG] PEG-30 glyceryl laurate; CAS 51248-32-9; nonionic; preparation of o/w emulsions; solubilizer for flavors, perfumes, vitamin oils; dispersant and antistat; *Properties:* pale yel. liq.; sol. in water; insol. in veg. and min. oils; HLB 17.0; acid no. 2 max.; iodine no. 4 max.; sapon. no. 35-55; hyd. no. 52-68; 100% conc.

Tagat® L2. [Goldschmidt; Goldschmidt AG] PEG-20 glyceryl laurate; CAS 51248-32-9; nonionic; preparation of o/w emulsions; solubilizer for flavors, perfumes, vitamin oils; dispersant and antistat; *Properties:* ivory liq.; sol. in water; insol. in veg. and min. oils; m.p. 60-80 C; HLB 15.7; acid no. 2 max.; iodine no. 4 max.; sapon. no. 50-70; hyd. no. 60-80; 100% conc.

Tagat® O. [Goldschmidt; Goldschmidt AG] PEG-30 glyceryl oleate; CAS 51192-09-7; nonionic; preparation of o/w emulsions; solubilizer for flavors, perfumes, vitamin oils; dispersant and antistat; *Properties:* yel. liq.; sol. in water; insol. in veg. and min. oils; HLB 16.4; acid no. 2 max.; iodine no. 15-19; sapon. no. 30-45; hyd. no. 50-65; 100% conc.

Tagat® O2. [Goldschmidt; Goldschmidt AG] PEG-20 glyceryl oleate; CAS 51192-09-7; nonionic; preparation of o/w emulsions; solubilizer for flavors, perfumes, vitamin oils; dispersant and antistat; *Properties:* yel. liq.; sol. in water; insol. in veg. and min. oils; m.p. 70-85 C; HLB 15.0; acid no. 2 max.; iodine no. 21-27; sapon. no. 40-55; hyd. no. 70-85; 100% conc.

Tagat® S. [Goldschmidt; Goldschmidt AG] PEG-30 glyceryl stearate; CAS 51158-08-8; nonionic; preparation of o/w emulsions; solubilizer for flavors, perfumes, vitamin oils; dispersant and antistat; *Properties:* ivory solid, partially liq.; sol. with sl. turbidity in water; insol. in veg. and min. oils; HLB 16.4; acid no. 2 max.; iodine no. 2 max.; sapon. no. 30-47; hyd. no. 53-70; 100% conc.

Tagat® S2. [Goldschmidt; Goldschmidt AG] PEG-20 glyceryl stearate; CAS 51158-08-8; nonionic; preparation of o/w emulsions; solubilizer for flavors, perfumes, vitamin oils; dispersant and antistat; *Properties:* ivory solid, partially liq.; sol. with sl. turbidity in water; insol. in veg. and min. oils; m.p. 65-85 c; HLB 15.0; acid no. 2 max.; iodine no. 2 max.; sapon. no. 40-60; hyd. no. 65-85; 100% conc.

Tagat® TO. [Goldschmidt; Goldschmidt AG] PEG-25 glyceryl trioleate; CAS 68958-64-5; nonionic; preparation of o/w emulsions; solubilizer for flavors, perfumes, vitamin oils; dispersant and antistat; *Properties:* amber liq.; sol. in veg. oils; disp. in water; insol. in min. oil; HLB 11.3; acid no. 12 max.; iodine no. 34-40; sapon. no. 75-90; hyd. no. 18-33; 100% conc.

Taka-Sweet®. [Solvay Enzymes] Glucose isomerase; enzyme for prod. of fructose syrups from glucose; immobilized.

Taka-Therm® L-340. [Solvay Enzymes] Bacterial α-amylase; CAS 9000-92-4; EINECS 232-567-7; enzyme for starch liquefaction; to produce high fructose corn syrup; in liquefaction of starch fermentation media; for brewing, candy, cane sugar (starch haze removal), cereal (reduces visc.), cocoa, distilling; *Properties:* golden to lt. brn. nonvisc. liq., free of offensive odor and taste; misc. with water; dens. 1.15-1.25 g/ml.

Tally® 100. [Van Den Bergh Foods] Mono- and diglycerides with 40% ethoxylated mono- and diglycerides; nonionic; dough strengthener, crumb softener, and antistaling agent for breads and sweet doughs; *Properties:* plastic; m.p. 118-125 F; HLB 7.7; 100% conc., 29-32% mono; Discontinued

Tally® 100 Plus. [Van Den Bergh Foods] Glyceryl stearate, PEG-20 glyceryl stearate, hydrog. soybean oil; nonionic; emulsifier, dough strengthener, crumb softener, and antistaling agent used in breads and other yeast-raised baked goods; *Regulatory:* FDA 21CFR §182.4505, 172.834; *Properties:* cream bead; HLB 6.1; m.p. 55-57 C; 100% conc., 32-37% mono.

Tandem 5K. [ICI Atkemix; Witco/H-I-P] Mono- and diglycerides, polysorbate 60 with BHA, citric acid; nonionic; food emulsifier, conditioner/softener for yeast-raised baked goods, e.g., bread, rolls; extends shelf life in doughnuts; enhances aeration, volume, and stability in icings; *Usage level:* 0.4-0.5% on flour (bread), 1.2fi% on flour (doughnuts), 5-7.5% in shortening (icings); *Regulatory:* FDA 21CFR §184.1505, 172.360; *Properties:* ivory wh. plastic solid; sol. in veg. oils; disp. in water; m.p. 128 F; HLB 8.1; iodine no. 30; flash pt. > 300 F; 31% alpha monoglyceride.

Tandem 8. [ICI Atkemix; Witco/H-I-P] Mono- and diglycerides, polysorbate 60 with BHA, citric acid; food emulsifier, conditioner/softener for yeast-raised baked goods, e.g., bread, rolls; extends shelf life in doughnuts; enhances aeration, volume, and stability in icings; *Usage level:* 0.4-0.5% on flour (bread), 1.25% on flour (doughnuts), 5-7.5% in shortening (icings); *Regulatory:* FDA 21CFR §184.1505, 172.360; *Properties:* ivory wh. plastic solid; HLB 8.1; iodine no. 22; pour pt. 52 C; 31% alpha monoglyceride.

Tandem 9. [Witco/H-I-P] Mono- and diglycerides and polysorbate 60 (25%); food emulsifier, conditioner/softener for yeast-raised baked goods, e.g., bread, rolls; *Usage level:* 0.4-0.5% on flour (bread); *Regulatory:* FDA 21CFR §182.70, 172, 360; *Properties:* ivory wh. flakes, bland odor; disp. in water; m.p. 135 F; HLB 6.7; iodine no. 10; flash pt. > 300 F; 34% alpha monoglyceride.

Tandem 11H. [ICI Atkemix] Hydrated mono- and diglycerides, polysorbate 00, surfactant for food industry; dough strengthener for breads; retards crumb firming; *Usage level:* 1-1.2% on flour (bread); *Properties:* wh. plasticized solid; disp. in water; m.p. 138-142 F; HLB 6.3; iodine no. 2.

Tandem 11H K. [Witco/H-I-P] Hydrated mono- and diglycerides, polysorbate 60 with sodium propionate and lactic acid; food emulsifier, conditioner/softener for yeast-raised baked goods, e.g., bread, rolls; provides aeration and volume in cakes; extends shelf life in doughnuts; *Usage level:* 1-1.2% on flour (bread), 1-2.75% on flour (cake), 2.5% on flour (doughnuts); *Properties:* ivory wh. soft plastic; disp. in water; HLB 6.3; iodine no. < 3; flash pt. > 300 F; pH 5; 50% conc., 19.5% alpha monoglyceride.

Tandem 22 H. [ICI Atkemix; Witco/H-I-P] Hydrated blend of mono- and diglycerides and polysorbate 60; food emulsifier, conditioner/softener for yeast-raised baked goods, e.g., bread, rolls; provides aeration in cakes, esp. those containing veg. oil; *Usage level:* 1-1.2% on flour (bread), 1-2.75% on flour (cake); *Regulatory:* FDA 21CFR §184.1505, 172.360; Canada compliance; *Properties:* ivory wh. plastic solid; disp. in water; m.p. 138-142 F; HLB 7.2; iodine no. < 3; 19.5% alpha monoglyceride.

Tandem 22 H K. [Witco/H-I-P] Hydrated blend of mono- and diglycerides and polysorbate 60; food emulsifier, conditioner/softener for yeast-raised baked goods, e.g., bread, rolls; *Usage level:* 1-1.2% on flour (bread); *Regulatory:* FDA 21CFR

§184.1505, 172.360; *Properties:* ivory wh. plastic solid; HLB 7.2; iodine no. < 3; 19.5% alpha monoglyceride.

Tandem 100 K. [Witco/H-I-P] Blend of mono- and diglycerides and polysorbate 80; food emulsifier for ice cream and frozen desserts; aerating agent; provides desirable drying and overrun props.; *Usage level:* 0.04-0.07% (ice cream); *Regulatory:* FDA 21CFR §184.1505, 172.840; *Properties:* ivory wh. powd.; HLB 5.2; iodine no. < 4; 32% alpha monoglyceride.

Tandem 552. [ICI Atkemix; Witco/H-I-P] Mono- and diglycerides, polysorbate 60 with BHA, citric acid; food emulsifier, conditioner/softener for yeast-raised baked goods, e.g., bread, rolls; pan-release agent; when sprayed on buns, functions as pan release and minimzes seed loss; extends shelf life in doughnuts; *Usage level:* 0.6-0.8% on flour (bread), 1.25% on flour (doughnuts); *Regulatory:* FDA 21CFR §184.1505, 172.360; *Properties:* golden clear liq.; sol. in veg. oils; disp. in water; m.p. 45 F; HLB 8.1; iodine no. 41; flash pt. > 300 F; 29% alpha monoglyceride.

Tandem 552 K. [ICI Atkemix; Witco/H-I-P] Blend of mono- and diglycerides and polysorbate 60; food emulsifier, conditioner/softener for yeast-raised baked goods, e.g., bread, rolls; extends shelf life in doughnuts; *Usage level:* 0.6-0.8% on flour (bread); *Regulatory:* FDA 21CFR §184.1505, 172.360; kosher; *Properties:* golden clear liq.; HLB 8.1; iodine no. 41; 29% alpha monoglyceride.

Tandem 930. [ICI Atkemix] Mono- and diglycerides, polysorbate 60, and sodium stearoyl-2-lactylate; surfactant for food industry; dough strenghthener for breads; retards crumb firming; *Usage level:* 0.6-0.8% on flour (breads); *Properties:* wh. free-flowing powd.; disp. in water; m.p. 138-142 F; HLB 6.0; iodine no. 1.

Tang. [MLG Enterprises Ltd.] Calcium citrate, citric acid, sodium chloride; highly conc. precipitated form of citric acid imparting distinctive tart flavor to rye, sour French, pizza crust, English muffins, pretzels, extruded dough prods.; replaces dry sours; *Usage level:* 3-8 oz/cwt flour; *Properties:* free-flowing powd.; *Storage:* store under cool, dry conditions; 6 mos shelf life.

Taterfos®. [Rhone-Poulenc Food Ingreds.] Sodium acid pyrophosphate, non-leavening; CAS 7558-16-9; EINECS 231-835-0; additive for processed foods; maintains natural white color of cooked potatoes; *Regulatory:* kosher; *Properties:* wh. powd., odorless; 1% max. on 70 mesh, 95% min. thru 200 mesh; sol. 13 g/100 g saturated sol'n.; m.w. 221.9; bulk dens. 68 lb/ft³; pH 4.0-4.8 (1%); 95% min. assay; *Storage:* store cool and dry.

TBHQ®. [Rhone-Poulenc Food Ingreds.] TBHQ; CAS 1948-33-0; EINECS 217-752-2; antioxidant for processed foods, baking, cereal, and confection.

Tegin® C-62 SE. [Goldschmidt AG] Hydrog. tallow glyceride citrate; CAS 68990-59-0; EINECS 273-613-6; nonionic; self-emulsifying hydrophilic emulsifier for foodstuffs; *Properties:* ivory powd.; disp.

warm in water, disp. in min. oil; insol. in veg. oil; m.p. 58-64 C; HLB 10.0; iodine no. 3 max.; sapon. no. 215-265; 100% conc.; Discontinued

Tegin® C-63 SE. [Goldschmidt] Hydrog. tallow glyceride citrate; CAS 68990-59-0; EINECS 273-613-6; emulsifier for foodstuffs; *Properties:* powd.; HLB 10.0; 100% conc.; Unverified

Tegin® E-61. [Goldschmidt] Acetylated hydrog. lard glyceride; CAS 8029-91-2; nonionic; food additive; *Properties:* solid; HLB 2.0-3.0; 100% conc.; Discontinued

Tegin® E-61 NSE. [Goldschmidt] Acetylated hydrog. lard glyceride; CAS 8029-91-2; nonionic; food additive; *Properties:* semisolid; HLB 2.0-3.0; 100% conc.; Unverified

Tegin® E-66. [Goldschmidt] Acetylated lard glyceride; CAS 8029-92-3; nonionic; food additive; *Properties:* liq.; HLB 2.0-3.0; 100% conc.; Discontinued

Tegin® E-66 NSE. [Goldschmidt] Acetylated lard glyceride; CAS 8029-92-3; nonionic; food additive; *Properties:* liq.; HLB 2.0-3.0; 100% conc.; Unverified

Tegin® L 61, L 62. [Goldschmidt] Hydrog. tallow glyceride lactate; CAS 68990-06-7; EINECS 273-576-6; nonionic; food additive; *Properties:* powd.; HLB 5.0 and 6.0 resp.; 100% conc.; Discontinued

Tegomuls® 19. [Goldschmidt AG] Lard mono- and diglycerides; nonionic; food emulsifier; for shortenings; *Properties:* semisolid; 100% conc.; Unverified

Tegomuls® 90. [Goldschmidt AG] Glyceryl stearate palmitate; nonionic; w/o fat emulsifier, oil retention aid, starch complexer; increases shelf-life in yeast-raised bakery goods; improves texture in potato foods, noodles; emulsion stabilizer in margarine; prevents oil separation in glazes; reduces stickiness in caramels; *Usage level:* 0.5% (baked goods), 0.5% (noodles), 0.5-1.0% (potato foods), 0.4% (starchy foods), 0.05-1.0% (margarine); *Regulatory:* EEC E471, FAO/WHO compliance; *Properties:* wh. free-flowing powd., faint char. odor; sol. in oils and fats above its m.p.; insol. in water; sp.gr. 0.900; m.p. 65-70 C; HLB 4.5; acid no. 3 max.; iodine no. 2 max.; sapon. no. 155-170; flash pt. (Abel-Pensky) > 200 C; 90% min. monoester; *Toxicology:* nonhazardous.

Tegomuls® 90 K. [Goldschmidt AG] Glyceryl stearate; texture improver for instant potato prods.; decreases stickiness; works against bursting of deep-frozen potato prods.; *Usage level:* 0.5-1.0% (potato prods.); *Regulatory:* EEC E471 compliance; *Properties:* wh. fine powd., faint char. odor; disp. in warm water forming homogeneous milky disps.; 80-85% monoester; *Toxicology:* nonhazardous.

Tegomuls® 90 S. [Goldschmidt AG] Glyceryl stearate, self-emulsifying; nonionic; o/w emulsifier; improves whipped volume, texture, foam stability, fat dispersion; reduces stickiness; starch complexing agent; for sponge mixes, fatty mixes, ice cream, sausages, yeast-raised goods, noodles, potato prods.; *Usage level:* 0.6% (sponge mixes), 0.5% (fatty mixes), 0.5% (yeast-raised goods), 0.5% (noodles), 0.5-1.0% (potato prods.); *Regulatory:*

EEC E471, FAO/WHO compliance; *Properties:* wh.-ivory powd., faint char. odor; dissolves in oils and fats above its m.p.; disp. in hot water; sp.gr. 0.900; m.p. 64-69 C; HLB 4.8; acid no. 3 max.; sapon. no. 140-155; flash pt. (Abel-Pensky) > 200 C; 80-85% monoester; *Toxicology:* nonhazardous.

Tegomuls® 4070. [Goldschmidt AG] Glyceryl monodistearate; nonionic; food emulsifier; for margarine, baking fat, shortening, caramels, pasta; *Properties:* powd.; 100% conc.; Unverified

Tegomuls® 4100. [Goldschmidt] Glyceryl stearate; nonionic; food emulsifier; water-binding agent for margarine; provides higher whipping capacity, finer cell structure, smooth creamy texture to ice cream; also for baking fat, shortenings, noddles, sweet goods; *Usage level:* 0.1-0.2% (margarine), 0.1-0.2% (ice cream, > 4% fat content), 0.3% (ice cream, < 4% fat content); *Regulatory:* EEC E471, FAO/WHO compliance; *Properties:* wh.-ivory powd., odorless; sol. in warm alcohol, toluene, acetic ester, liq. paraffin, animal/veg. oils and fats; pract. insol. in hot and cold water; sp.gr. 0.900; m.p. 58-63 C; HLB 3.8 ± 1; acid no. 2 max.; iodine no. 3 max.; sapon.no. 164-180; flash pt. (Abel-Pensky) > 200 C; 45% monoester; *Toxicology:* nonhazardous.

Tegomuls® 4101. [Goldschmidt] Glyceryl stearate; nonionic; food emulsifier; *Properties:* powd.; HLB 3.8; 100% conc.; Unverified

Tegomuls® 6070. [Goldschmidt AG] Glyceryl monodistearate; nonionic; food emulsifier; for margarine, baking fat, shortening, caramel, pasta; *Properties:* powd.; 100% conc.; Unverified

Tegomuls® 9050. [Goldschmidt AG] Glyceryl mono/distearate-palmitate; nonionic; food emulsifier; for margarine, baking fats, shortening, caramel, pasta; *Properties:* wh. powd., faint char. odor; insol. in water; sp.gr. 0.900; flash pt. (Abel-Pensky) > 200 C; 100% conc.; *Toxicology:* nonhazardous; avoid formation of dust.

Tegomuls® 9053 S. [Goldschmidt AG] Glyceryl stearate/palmitate with alkali soaps, self-emulsifying; o/w emulsifier; increases whipped volume, foam stability; starch complexing agent; improves texture; reduces stickiness; for sponge mixes, fatty mixes, ice cream, sausages, baked goods, noodles, potato prods.; *Usage level:* 0.6% (sponge mixes), 0.5% (fatty mixes, yeast-raised baked goods, noodles); 0.5-1.0% (potato prods.), 0.5% max. (ice cream); *Regulatory:* EEC E471, FAO/WHO compliance; *Properties:* wh.-ivory powd., faint char. odor; dissolves in oils and fats above its m.p.; disp. in hot water; sp.gr. 0.900; m.p. 64-69 C; acid no. 5 max.; sapon. no. 140-155; flash pt. > 200 C; pH 8.0 (100 g/l water); 80-85% monoester; *Toxicology:* nonhazardous; avoid formation of dust.

Tegomuls® 9102. [Goldschmidt AG] Glyceryl stearate with anticaking agent as carrier; nonionic; food emulsifier; prolongs freshness, improves fat dispersion for bakery goods; reduces stickiness in noodles; improves texture of potato foods; improves stability, consistency in ice cream; *Usage*

level: 0.5% (baked goods), 0.5% (noodles), 0.5-1.0% (potato foods), 0.5% max. (ice cream); *Regulatory:* EEC E471, FAO/WHO compliance; *Properties:* wh. powd., faint char. odor; insol. in water; sp.gr. 0.900; HLB 4.5; flash pt. (Abel-Pensky) > 200 C; 80% conc.; *Toxicology:* nonhazardous.

Tegomuls® 9102 S. [Goldschmidt AG] Glyceryl stearate, self-emulsifying, with anticaking agent as carrier; anionic; food emulsifier; prolongs freshness, improves fat dispersion for bakery goods; reduces stickiness in noodles; improves texture of potato foods; improves stability, consistency in ice cream; *Regulatory:* EEC E471, FAO/WHO compliance; *Properties:* wh. powd., faint char. odor; readily sol. in cold water; sp.gr. 0.900; HLB 4.8; flash pt. (Abel-Pensky) > 200 C; 80% conc.; *Toxicology:* nonhazardous.

Tegomuls® B. [Goldschmidt AG] Glyceryl monodistearate palmitate, self-emulsifying; anionic; food emulsifier; used in white bread to prolong freshness, improve structure; in baking cream to aid dispersion of fat in doughs and sponge mixes; in soft caramels for foam reduction and increase in volume; *Usage level:* 0.1-0.2% on flour (white bread), 0.1% (caramels); *Regulatory:* EEC E471, FAO/WHO compliance; *Properties:* wh.-ivory powd., faint char. odor; sol. in warm alcohol, toluene, acetic ester, liq. paraffin, animal/veg. fats and oils; disp. in wam water (60 C); bulk dens. 0.700; m.p. 57-63 C; HLB 4.5; acid no. 5 max.; iodine no. 4 max.; sapon. no. 150-168; flash pt. (Abel-Pensky) > 200 C; pH 8.0 (100 g/l water); 50% monoester; *Toxicology:* nonhazardous.

Tegomuls® M. [Goldschmidt AG] Glyceryl monodistearate, nonionic; food emulsifier; water-binding agent for margarine; also for noodles, sugary goods, ice cream, pasta; *Usage level:* 0.1-0.2% (margarine); *Regulatory:* EEC E471, FAO/WHO compliance; *Properties:* wh.-ivory powd., faint char. odor; sol. in warm alcohol, toluene, acetic ester, liq. paraffin, animal/veg. fats and oils; pract. insol. in water; sp.gr. 1.000; m.p. 58-63 C; HLB 3.2; acid no. 3 max.; iodine no. 4 max.; sapon. no. 160-175; flash pt. > 200 C; pH 7.0 (100 g/l aq. disp.); 55-60% monoester; *Toxicology:* nonhazardous.

Tegomuls® O. [Goldschmidt AG] Glyceryl monodioleate; CAS 25496-72-4; nonionic; defoamer in aq. sol'ns. in foodstuffs industry (e.g., milk and sugar industries); *Usage level:* 0.01-0.1% (as defoamer); *Regulatory:* EEC E471, FAO/WHO compliance; *Properties:* pale yel. semisolid; sol. in most org. solvs. (acetone, ethanol, ether, chloroform, acetic ester, wh. spirit, toluene), veg. oils, paraffin oil; pract. insol. in water; sp.gr. 0.900; HLB 3.3; acid no. 2 max.; iodine no. 70-76; sapon. no. 158-175; flash pt. > 200 C; pH 7.0 (100 g/l aq. disp.); 60% monoester; *Toxicology:* nonhazardous.

Tegomuls® O Spezial. [Goldschmidt] Glyceryl monodioleate; CAS 25496-72-4; nonionic; defoamer in food industry; *Properties:* semisolid; HLB 3.3; 100% conc.; Unverified

Tegomuls® P 411. [Goldschmidt AG] 1,2-Propylene glycol monodistearate; nonionic; food emulsifier; for whipped sponge mixtures, toppings, shortenings; *Properties:* solid; HLB 2.8; 100% conc.; Unverified

Tegomuls® SB. [Goldschmidt AG] Sunflower oil monoglyceride; nonionic; food emulsifier, stabilizer; for spiced oil emulsions, mayonnaise, margarine, liq. shortenings, baking aids, dressings; *Regulatory:* EEC E471, FAO/WHO compliance; *Properties:* ylsh. semisolid, fatty odor; sol. in ethanol, oils, fats; disp. in water; sp.gr. 0.850; HLB 3.3; acid no. 2 max.; sapon. no. 155-175; flash pt. (Abel-Pensky) > 200 C; pH 5.7 (100 g/l aq. disp.); 50-60% monoester; *Toxicology:* nonhazardous.

Tegomuls® SO. [Goldschmidt AG] Soybean oil mono- and diglycerides; nonionic; food emulsifier, stabilizer; for spiced oil emulsions, mayonnaise, margarine, liq. shortenings, baking aids, dressings; *Regulatory:* EEC E471, FAO/WHO compliance; *Properties:* ylsh. semisolid, fatty odor; sol. in ethanol, oils, fats; disp. in water; sp.gr. 0.850; HLB 3.3; acid no. 3 max.; sapon. no. 160-180; flash pt. (Abel-Pensky) ≈ 265 C; pH 5.7 (100 g/l aq. disp.); 50-60% monoester; *Toxicology:* nonhazardous.

Tem Cote® IE. [Bunge Foods] Partially hydrog. soybean and cottonseed oils; nontempering fat; lauric coating fat replacement; confectionery coating, candy centers, granola, fruit bars; *Properties:* oil, 1.5 red max. color, bland flavor; m.p. 104 F.

Tem Cote® II. [Bunge Foods] Partially hydrog. soybean and cottonseed oils with mono- and diglycerides; nontempering fat; lauric coating fat replacement with emulsifiers; chocolage and pastel confectioners coatings; *Properties:* oil, 1.5 red max. color, bland flavor; m.p. 102 F.

Tem Crunch™ 93. [Bunge Foods] Partially hydrog. soybean and cottonseed oils, antioxidants; nutritious alternative to tropical fats; for snacks, ready-to-eat cereals, nondairy creamers; *Properties:* oil, bland flavor; drop pt. 93 F.

Tem Crunch™ 110. [Bunge Foods] Partially hydrog. soybean oil; EINECS 232-410-2; nutritious alternative to tropical fats; for ready-to-eat cereals, coffee whiteners; coconut oil replacements; *Properties:* oil, 1.0 red max. color, bland flavor; m.p. 110 F.

Tem Freeze™. [Bunge Foods] Partially hydrog. soybean and corn oils; nutritious alternative to tropical fats; for ice cream coatings; *Properties:* oil, 1.2 red color, bland flavor; drop pt. 94 F.

Tem Plus® 95. [Bunge Foods] Partially hydrog. soybean oil; EINECS 232-410-2; nutritious alternative to tropical fat; for cookie fillers, frozen soft-serve desserts, imitation sour cream dips, coffee whiteners, whipped toppings; *Properties:* oil, 1.5 red max. color, bland flavor; m.p. 95 F.

Tem Top™. [Bunge Foods] Partially hydrog. soybean and cottonseed oils; nutritious alternative to tropical fats; for whipped toppings, coffee whiteners, coconut oil replacement, other nondairy applics.; *Properties:* oil, 1.0 red max. color, bland flavor; m.p. 102 F.

Tenase® 1200. [Solvay Enzymes] Bacterial α-amylase; CAS 9000-92-4; EINECS 232-567-7; enzyme for starch liquefaction; for baking (antistaling agent), brewing (starch haze removal), candy, cereal (reduces visc.), cocoa, distilling, fermentation, starch (prod. of low DE syrups); *Properties:* lt. tan to wh. impalpable dry powd., free of offensive odor and taste; water-sol.; *Storage:* activity loss ≤ 10% in 1 yr stored in sealed containers under cool dry conditions; 5 C storage extends life.

Tenase® L-340. [Solvay Enzymes] Bacterial α-amylase; CAS 9000-92-4; EINECS 232-567-7; enzyme for starch liquefaction; for baking (antistaling agent), brewing (starch haze removal), candy, cereal (reduces visc.), cocoa, distilling, fermentation, starch (prod. of low DE syrups); *Properties:* amber to brn. nonvisc. liq., free of offensive odor and taste; water-sol.; *Storage:* activity loss ≤ 10% in 1 yr stored in sealed containers under cool, dry conditions; 5 C storage extends life.

Tenase® L-1200. [Solvay Enzymes] Bacterial α-amylase; CAS 9000-92-4; EINECS 232-567-7; enzyme for starch liquefaction; for baking (antistaling agent), brewing (starch haze removal), candy, cereal (reduces visc.), cocoa, distilling, fermentation, starch (prod. of low DE syrups); *Properties:* amber to brn. nonvisc. liq., free of offensive odor and taste; water-sol.; *Storage:* activity loss ≤ 10% in 1 yr stored in sealed containers under cool, dry conditions; 5 C storage extends life.

Tenderbite SF(TB-SF). [Advanced Food Systems] Binding matrix that entraps moisture and natural juice inside seafood upon cooking; provides tender, juicy texture even in acidic sauces under microwave cooking; improves cooked yield, freeze/thaw stability; bland flavor without mealy/starchy texture.

Tenderfil® 8. [A.E. Staley Mfg.] Food starch modified, derived from tapioca; cook-up starch, thickener, stabilizer for canned and frozen foods, pie/cake fillings, sauces, gravies, soups, frozen dinners, baby foods; prevents syneresis during storage; high resist. to heat, freeze-thaw, and acid; *Regulatory:* FDA 21CFR §172.892; *Properties:* wh. powd.; 98% thru 100 mesh, 85% thru 200 mesh; pH 6.0; 13% moisture.

Tenderfil® 9. [A.E. Staley Mfg.] Food starch modified, derived from tapioca; cook-up starch, thickener, stabilizer for canned foods; good resist. to heat, acids; forms semisoft tender gels with creamy texture; resists syneresis; for baby foods, sauces, gravies, soups, dietetic foods, creme fillings, puddings, veg./fruit puree; *Regulatory:* FDA 21CFR §172.892; *Properties:* powd., 98% thru 100 mesh, 85% thru 200 mesh; pH 6.0; 12% moisture.

Tenderfil® 428. [A.E. Staley Mfg.] Food starch modified, derived from tapioca; cook-up starch for highly acidic food systems and those requiring visc. stability under high shear processing; provides good freeze/thaw stability in frozen preps.; for pie fillings, asceptic puddings, frozen fillings; *Regulatory:* FDA 21CFR §172.892; *Properties:* pH 5-6; 10-13% moisture.

Tenderfil® 473A. [A.E. Staley Mfg.] Food starch modified, derived from tapioca; cook-up starch, thickener for acidic foods; provides visc. stability at R.T. and under refrigeration; for fruit puddings, canned baby foods, bakers jellies, canned cream soups; *Regulatory:* FDA 21CFR §172.892; kosher, Passover certified; *Properties:* wh. free-flowing powd.; mixes readily in cold water; pH 5.7 (10% slurry); 11.5% moisture.

Tenderset® M-7. [Hormel] Gelatin, Type A, 225 bloom; CAS 9000-70-8; EINECS 232-554-6; used in marshmallows, frostings, toppings, and general bakery use; *Properties:* ivory powd., weak bouillon-like odor; 30 mesh; visc. 37 ± 2 mps; pH 4.5-5.8; 12% max. moisture; *Storage:* stable for up to 1 yr when stored dry at ambient temps.; keep containers tightly closed.

Tenderset® M-8. [Hormel] Gelatin, Type A, 250 bloom; CAS 9000-70-8; EINECS 232-554-6; used in marshmallows, frostings, toppings, and general bakery use; *Properties:* ivory powd., weak bouillon-like odor; 30 mesh; visc. 42 ± 2 mps; pH 4.5-5.8; 12% max. moisture; *Storage:* stable for up to 1 yr when stored dry at ambient temps.; keep containers tightly closed.

Tenderset® M-9. [Hormel] Gelatin, Type A, 275 bloom; CAS 9000-70-8; EINECS 232-554-6; used in marshmallows, frostings, toppings, and general bakery use; *Properties:* ivory powd., weak bouillon-like odor; 30 mesh; visc. 47 ± 2 mps; pH 4.5-5.8; 12% max. moisture; *Storage:* stable for up to 1 yr when stored dry at ambient temps.; keep containers tightly closed.

Tenderset® MWA-5. [Hormel] Gelatin, Type A, 175 bloom with 1% sodium hexametaphosphate; used in marshmallows, toppings, and frostings; addition of phosphates enhances whipping quality and set time; *Properties:* ivory powd., weak bouillon-like odor; 30 mesh; visc. 28 ± 2 mps; pH 4.5-5.8; 12% max. moisture; *Storage:* stable for up to 1 yr when stored dry at ambient temps.; keep containers tightly closed.

Tenderset® MWA-7. [Hormel] Gelatin, Type A, 225 bloom with 1% sodium hexametaphosphate; used in marshmallows, toppings, and frostings; addition of phosphates enhances whipping quality and set time; *Properties:* ivory powd., weak bouillon-like odor; 30 mesh; visc. 37 ± 2 mps; pH 4.5-5.8; 12% max. moisture; *Storage:* stable for up to 1 yr when stored dry at ambient temps.; keep containers tightly closed.

Tenderset® MWA-8. [Hormel] Gelatin, Type A, 250 bloom with 1% sodium hexametaphosphate; used in marshmallows, toppings, and frostings; addition of phosphates enhances whipping quality and set time; *Properties:* ivory powd., weak bouillon-like odor; 30 mesh; visc. 42 ± 2 mps; pH 4.5-5.8; 12% max. moisture; *Storage:* stable for up to 1 yr when stored dry at ambient temps.; keep containers tightly closed.

Tenite® [Eastman] 50% Tocopherols, 50% veg. oils; antioxidant for bakery prods./shortening, candies, chewing gum, citrus oils, cookies, crackers, edible/

inedible fats, fish oils/prods., flavors, frying oils, lard, min. oil, nuts, poultry prods., sausages, snack foods, spices, tallow; *Usage level:* 300 ppm (USDA); *Regulatory:* FDA 21CFR §161.175, 164.110, 165.175, 166.110, 182.3890; USDA 9CFR §318.7, 381.147; EEC listed; *Properties:* red to reddish-brn. sl. visc. liq., sl. mild odor, sl. mild taste; readily sol. in veg. oils, fats, essential oils, ethanol; insol. in water; Discontinued

Tenox® 2. [Eastman] 70% Propylene glycol, 20% BHA, 6% propyl gallate, 4% citric acid; food-grade antioxidant used for animal feeds, inedible fats, min. oil, pet foods; *Usage level:* 0.076% fat (FDA), 0.05% fat (USDA); *Properties:* colorless to lt. straw sol'n.; good sol. in fats and oils; sp.gr. 1.064; visc. 95 cP; *Storage:* 12 mo shelf life.

Tenox® 4. [Eastman] 60% Corn oil, 20% BHA, 20% BHT; food-grade antioxidant for baked goods, bakery mixes, shortening, candies, cereal, cookies, crackers, edible fats, lard, meat prods., nuts, pkg. materials, rice, sausages, breaded shrimp, spices, tallow, vitamins; exc. carry through in baked foods; *Usage level:* 0.05% fat (FDA), 0.05% fat (USDA); *Properties:* lt. straw sol'n.; exc. sol. in fats and oils; sp.gr. 0.942; visc. 61 cP; *Storage:* 12 mo shelf life.

Tenox® 4A. [Eastman] 30% BHA, 70% veg. oils; food-grade antioxidant; *Properties:* lt. amber liq.; Discontinued

Tenox® 4B. [Eastman] 80% Corn oil, 20% BHT; antioxidant for foods, baked goods and mixes, beverage mixes, candies, cereals, citrus oils, crackers, dessert mixes, edible fats, flavors, meat prods., min. oil, nuts, pkg. materials, rice, sausages, breaded shrimp, vitamins, yeast; *Usage level:* 0.1% fat (FDA), 0.05% fat (USDA); *Properties:* lt. amber sol'n.; exc. sol. in fats and oils; sp.gr. 0.951; visc. 69 cP; *Storage:* 12 mo shelf life.

Tenox® 5. [Eastman] 25% BHA, 25% BHT, 50% ethyl alcohol undenatured; food-grade antioxidant for instant potatoes; very good carry through in baked foods; *Usage level:* 0.04% fat (FDA), 0.04% fat (USDA); *Properties:* colorless to lt. straw sol'n.; poor sol. in fats and oils; sp.gr. 0.884; visc. 12 cP; *Storage:* 6 mo shelf life.

Tenox® 5B. [Eastman] 50% BHA, 50% ethyl alcohol undenatured; food-grade antioxidant; very good carry through in baked foods; *Usage level:* 0.04% fat (FDA), 0.02% fat (USDA); *Properties:* colorless to lt. straw sol'n.; poor sol. in fats and oils; sp.gr. 0.913; visc. 12 cP; *Storage:* 6 mo shelf life.

Tenox® 6. [Eastman] 28% Corn oil, 28% glyceryl oleate, 12% propylene glycol, 10% BHA, 10% BHT, 6% propyl gallate, 6% citric acid; antioxidant for food applics., baked goods and mixes, crackers, meat and poultry prods., sausages, spices, tallow, pet foods, animal feeds; very good carry through in baked foods; *Usage level:* 0.076% fat (FDA), 0.076% fat (USDA); *Properties:* golden brn. sol'n.; exc. sol. in fats and oils; sp.gr. 1.008; visc. 333 cP; *Storage:* 6 mo shelf life.

Tenox® 7. [Eastman] 34% Propylene glycol, 28% BHA, 20% glyceryl oleate, 12% propyl gallate, 6%

citric acid; food-grade antioxidant for baked goods and mixes, candies, carakcers, edible fats, margarine, meat prods., min. oil, poultry fat, sausages, snack foods, spices, tallow, pet foods, animal feeds; very good carry through in baked foods; *Usage level:* 0.05% fat (FDA), 0.035% fat (USDA); *Properties:* lt. brn. sol'n.; very good sol. in fats and oils; sp.gr. 1.081; visc. 291 cP; *Storage:* 12 mo shelf life.

Tenox® 8. [Eastman] 20% BHT, 80% veg. oils; food-grade antioxidant for baked prods., cereals, crackers, nuts, pkg. materials, rice, breaded shrimp, vitamins; good carry through in baked foods; *Usage level:* 0.1% fat (FDA), 0.05% fat (USDA); *Properties:* lt. straw to lt. amber sol'n.; exc. sol. in fats and oils; sp.gr. 0.925; visc. 49 cP; *Storage:* 12 mo shelf life.

Tenox® 20. [Eastman] 70% Propylene glycol, 20% t-butyl hydroquinone, 10% citric acid; food-grade antioxidant fish oils and prods., frying oils, potato chips, poultry prods., snack foods, veg. oils, inedible fats, pet foods, animal feeds; good carry through in baked foods; *Usage level:* 0.1% fat (FDA), 0.05% fat (USDA); *Properties:* lt. amber to golden brn. sol'n.; good sol. in fats and oils; sp.gr. 1.087; visc. 235 cP; *Storage:* 12 mos shelf life.

Tenox® 20A. [Eastman] 32% glyceryl oleate, 30% veg. oils, 20% t-Butylhydroquinone, 15% propylene glycol, 3% citric acid; food-grade antioxidant for candies, cookies, edible fats, fish oils and prods., frying oils, inedible fats, margarine, meat and poultry prods., nuts, potato chips, poultry fat, , sausages, snack foods, spices, tallow, veg. oil; *Usage level:* 0.1% fat (FDA), 0.05% fat (USDA); *Properties:* golden brn. sol'n.; exc. sol. in fats and oils; sp.gr. 0.998; visc. 369 cP; *Storage:* 6 mo shelf life.

Tenox® 20B. [Eastman] 77% Propylene glycol, 20% t-butyl hydroquinone, and 3% citric acid; antioxidant for foods, frying oils, potato chips, poultry prods., snack foods, veg. oils; good carry through in baked foods; *Usage level:* 0.1% fat (FDA), 0.05% fat (USDA); *Properties:* lt. amber to golden brn. sol'n.; good sol. in fats and oils; sp.gr. 1.047; visc. 120 cP; *Storage:* 12 mo shelf life; Discontinued

Tenox® 21. [Eastman] 32% Veg. oil, 32% glyceryl oleate, 20% t-butylhydroquinone, 15% propylene glycol, 1% citric acid; food-grade antioxidant for frying oils, margarine, snack foods, veg. oils; good carry through in baked foods; *Usage level:* 0.1% fat (FDA), 0.05% fat (USDA); *Properties:* golden brn. sol'n.; exc. sol. in fats and oils; sp.gr. 0.991; visc. 284 cP; *Storage:* 6 mo shelf life.

Tenox® 22. [Eastman] 70% Propylene glycol, 20% BHA, 6% t-butyl hydroquinone, and 4% citric acid; food-grade antioxidant for inedible fats, min. oil, pet foods, animal feeds; very good carry through in baked foods; *Usage level:* 0.076% fat (FDA), 0.05% fat (USDA); *Properties:* lt. amber to lt. brn. sol'n.; good sol. in fats and oils; sp.gr. 1.054; visc. 91 cP; *Storage:* 12 mo shelf life.

Tenox® 24. [Eastman] 20% BHA, 20% t-butylhydroquinone, 30% glyceryl oleate, 30% propylene glycol; food-grade antioxidant; *Properties:* lt. amber to lt. brn. liq.; Discontinued

Tenox® 25. [Eastman] 31% Corn oil, 31% glyceryl oleate, 15% propylene glycol, 10% BHT, 10% TBHQ, 3% citric acid; antioxidant for foods, bakery prods., mixes, shortening, candies, cookies, crackers, edible fats, lard, margarine, meat prods., sausages, snack foods, spices, tallow, animal feeds; exc. carry through in baked foods; *Usage level:* 0.10% fat (FDA), 0.10% fat (USDA); *Properties:* golden brn. sol'n.; exc. sol. in fats and oils; sp.gr. 0.938; visc. 190 cP; *Storage:* 12 mo shelf life.

Tenox® 26. [Eastman] 28% Corn oil, 28% glyceryl oleate, 12% propylene glycol, 10% BHA, 10% BHT, 6% t-butyl hydroquinone, and 6% citric acid; antioxidant for foods, bakery prods. and mixes, poultry prods., sausages, spices, pet foods, animal feeds; very good carry through in baked foods; *Usage level:* 0.076% fat (FDA), 0.076% fat (USDA); *Properties:* golden brn. sol'n.; exc. sol. in fats and oils; sp.gr. 0.997; visc. 235 cP; *Storage:* 612 mo shelf life mo shelf life.

Tenox® 27. [Eastman] 34% Propylene glycol, 28% BHA, 20% glyceryl oleate, 12% t-butylhydroquinone, 6% citric acid; food-grade antioxidant for bakery prods., mixes, shortening, candies, cookies, crackers, edible/inedible fats, lard, margarine, meat prods., min./veg. oils, nuts, poultry fat and prods., sausages, snack foods, spices, tallow, pet foods, animal feeds; *Usage level:* 0.05% fat (FDA), 0.035% fat (USDA); *Properties:* golden brn. sol'n.; very good sol. in fats and oils; sp.gr. 1.047; visc. 279 cP; *Storage:* 12 mo shelf life.

Tenox® A. [Eastman] 52% Propylene glycol, 40% BHA, 8% citric acid; food-grade antioxidant for beverage mixes, dessert mixes, rice, breaded shrimp, vitamins; exc. carry through in baked foods; *Usage level:* 0.05% fat (FDA), 0.025% fat (USDA); *Properties:* lt. straw sol'n.; good sol. in fats and oils; sp.gr. 1.071; visc. 144 cP; *Storage:* 12 mo shelf life.

Tenox® BHA. [Eastman] BHA; CAS 25013-16-5; antioxidant for foods, beverage/dessert mixes, chewing gum, flavorings, citrus oils, min. oil, nuts, vitamins, waxes, yeast; carry through effect in cooked foods, providing extended shelf life; stabilizer for fats, oils, vitamins; *Usage level:* 200 ppm (general), 1000 ppm (chewing gum base), 5000 ppm (flavors), 2-90 ppm (dry mixes); *Regulatory:* FDA 21CFR §161.175, 164.110, 165.175, 166.110, 172.110, 172.515, 182.3169; USDA 9CFR 318.7, 381.147; kosher; EEC listed; *Properties:* wh. waxy tablets or flakes, sl. odor; sol. in ethanol, diisobutyl adipate, propylene glycol, glyceryl oleate, soya oil, lard, yel. grease, and paraffin; insol. in water; m.w. 180.25; b.p. 264-270 C (733 mm); m.p. 48-63 C; *Storage:* 6 mo shelf life.

Tenox® BHT. [Eastman] BHT; CAS 128-37-0; EINECS 204-881-4; antioxidant for foods, chewing gum, edible fats, min. oil, nuts, , waxes; provides carry through effectiveness in cooked foods; stabilizes oils, fats, cereals; *Usage level:* 200 ppm (general), 1000 ppm (chewing gum base), 50 ppm (dry

potatoes), 100 ppm (USDA); *Regulatory:* FDA 21CFR §137.350, 161.175, 164.110, 165.175, 166.110, 172.115, 172.615, 182.3173; USDA 9CFR 318.7, 381.147; EEC listed; *Properties:* wh. gran. cryst., very sl. odor; sol. in paraffin, lard, yel. grease, diisobutyl adipate; insol. in water, glycerol, propylene glycol; m.w. 220; m.p. 69.7 C; b.p. 265 C (760 mm); *Storage:* 6 mo shelf life.

Tenox® GT-1. [Eastman] 50% Tocopherols, 50% veg. oils; natural antioxidant for food prods., baked goods/mixes/shortening, candies, chewing gum, cookies, crackers, edible/inedible fats, fish oils/prods., flavors, frying oils, lard, min. oil, nuts, snack foods, spices, tallow, pet/animal feeds; *Usage level:* 300 ppm (USDA); *Regulatory:* FDA 21CFR §161.175, 164.110, 165.175, 166.110, 182.3890; USDA 9CFR §318.7, 381.147; kosher; *Properties:* red to reddish-brn. sl. visc. liq., sl. mild odor, sl. mild taste; readily sol. in veg. oils, fats, essential oils, ethanol; insol. in water; *Storage:* 6-8 mo shelf life.

Tenox® GT-2. [Eastman] 70% Tocopherols, 30% veg. oils; natural antioxidant for food prods., baked goods/mixes/shortening, candies, chewing gum, cookies, crackers, edible/inedible fats, fish oils/prods., flavors, frying oils, lard, min. oil, nuts, snack foods, spices, tallow, pet/animal feeds; *Usage level:* 300 ppm (USDA); *Regulatory:* FDA 21CFR §161.175, 164.110, 165.175, 166.110, 182.3890; USDA 9CFR §318.7, 381.147; kosher; EEC listed; *Properties:* red to reddish-brn. sl. visc. liq., sl. mild odor, sl. mild taste; readily sol. in veg. oils, fats, essential oils, ethanol; insol. in water; *Storage:* 6-8 mo shelf life.

Tenox® PG. [Eastman] Propyl gallate; CAS 121-79-9; EINECS 204-498-2; antioxidant, stabilizer for veg. oils, poultry fat; improves storage life of butter oils, poultry fat, etc.; often in combination with citric acid, BHA, and/or BHT; suggested for use in veg. oils in countries where TBHQ not permitted; *Usage level:* 200 ppm (general), 1000 ppm (chewing gum base); *Regulatory:* FDA 21CFR §164.110, 165.175, 166.110, 172.615, 184.1660; USDA 9CFR 318.7, 381.147; kosher; EEC listed; *Properties:* wh. cryst. powd., very sl. odor; sol. in ethanol, propylene glycol, glycerol, < 1% in water; m.w. 212.20; m.p. 146-150 C; b.p. 148 C dec.; *Storage:* 6 mo shelf life.

Tenox® R. [Eastman] 60% propylene glycol, 20% BHA, 20% citric acid; food-grade antioxidant for inedible fats, pet foods, animal feeds; exc. carry through in baked foods; *Usage level:* 0.1% fat (FDA), 0.05% fat (USDA); *Properties:* lt. straw sol'n.; good sol. in fats and oils; sp.gr. 1.117; visc. 229 cP; *Storage:* 12 mo shelf life.

Tenox® S-1. [Eastman] 70% Propylene glycol, 20% propyl gallate, 10% citric acid; food-grade antioxidant for fish oils and prods., poultry fat, veg. oils, pet foods, animal feeds; fair carry through in baked foods; *Usage level:* 0.1% fat (FDA), 0.05% fat (USDA); *Properties:* lt. amber sol'n.; good sol. in fats and oils; sp.gr. 1.127; visc. 244 cP; *Storage:* 12 mo shelf life.

Tenox® TBHQ. [Eastman] t-Butyl hydroquinone; CAS 1948-33-0; EINECS 217-752-2; antioxidant, stabilizer used in edible fats, fish oils and prods., frying oils, margarine, nuts, poultry fat, veg. oils, waxes; used in combination with BHA and/or BHT in meat and poultry prods. to provide max. protection with USDA compliance; *Usage level:* 200 ppm (FDA); not permitted in combination with propyl gallate; *Regulatory:* FDA 21CFR §164, 110, 165.175, 166.110, 172.185; USDA 9CFR 318.7, 381.147; kosher; *Properties:* wh. to lt. tan cryst., very sl. odor; sol. in ethanol, ethyl acetate, propylene glycol, veg. oils; < 1% in water; m.w. 166.22; m.p. 126.5-128.5 C; b.p. 300 C (760 mm); *Storage:* 6 mo shelf life.

Tenox® WD-BHA. [Eastman] 25% BHA, 25% glyceryl oleate, 50% acacia gum; water-disp. food-grade antioxidant; *Properties:* wh. to lt. tan powd.; water-disp.; Discontinued

Terra Alba 114836. [Allied Custom Gypsum] Calcium sulfate dihydrate; Dough conditioner, firming agent, nutrient, dietary supplement in yeast-raised baked goods; carrier for bleaching agents; stabilizer, thickener, pH control agent, sequestrant; in brewing and fermentation processes; *Regulatory:* FDA 21CFR §184.1230, GRAS; *Properties:* Wh. fine free-flowing powd., odorless, tasteless; 95% thru 325 sieve; m.w. 172.17; sp.gr. 2.32; bulk dens. 45 lb/ft³ (loose); pH 7-7.5; 98% min. assay; *Toxicology:* Nuisance particulate; high exposure to dust may cause eye and respiratory system irritation.

Terra-Dry™ FD Aloe Vera Powd. Decolorized FDD. [Terry Labs] Freeze-dried aloe vera gel powd., decolorized, deodorized; used in foods/beverages (health/sports/fitness drinks, juice blends, health foods); *Properties:* wh. to lt. beige fine cryst. powd.; After reconstitution 1:99: sp.gr. 0.997-1.004; pH 3.5-5.0; 8% max. moisture; *Storage:* preserve after reconstitution; may darken with age.

Terra-Dry™ FD Aloe Vera Powd. Reg. FDR. [Terry Labs] Freeze-dried aloe vera gel powd.; used in foods/beverages (health/sports/fitness drinks, juice blends, health foods); *Properties:* beige to lt. brn. fine cryst. powd.; After reconstitution 1:99: sp.gr. 0.997-1.004; pH 3.5-5.0; 8% max. moisture; *Storage:* preserve after reconstitution; may darken with age.

Terra-Pulp™ FD Whole Aloe Vera Gel Fillet. [Terry Labs] Freeze-dried whole aloe vera gel fillet; used in health care, foods, beverage applics. (health drinks/foods, juice blends, sports/fitness drinks); *Properties:* off-wh. to beige flakes, mod. veg. odor; After rehydration 1:99: visc. gel with pulp fiber; pH 3.5-5.0; < 10% moisture; *Storage:* preserve after reconstitution; may darken with age.

Terra-Pure™ FD Non-Preserved Decolorized FDD. [Terry Labs] Freeze-dried aloe vera gel powd. 200X, decolorized, deodorized; used in foods/beverages (health/sports/fitness drinks, juice blends, health foods); *Properties:* wh. to lt. beige fine cryst. powd.; After reconstitution 1:199: sp.gr. 0.997-1.004; pH 3.5-5.0; 8% max. moisture; *Storage:* preserve after reconstitution; may darken with age.

Terra-Pure™ FD Non-Preserved Reg. FDR. [Terry Labs] Freeze-dried aloe vera gel powd. 200X; used in foods/beverages (health/sports/fitness drinks, juice blends, health foods); *Properties:* lt. cream to beige fine cryst. powd.; After reconstitution 1:199: sp.gr. 0.997-1.004; pH 3.5-5.0; 8% max. moisture; *Storage:* preserve after reconstitution; may darken with age.

Terra-Pure™ Non-Preserved Aloe Vera Powd. [Terry Labs] Aloe vera gel spray-dried without preservatives; additive to food or drink prods.; *Properties:* lt. cream to beige fine powd., mod. veg. odor; disperses rapidly in most aq. sol'ns.; After reconsitituting 1:199 with DI water: sp.gr. 0.997-1.004; pH 3.5-5.0; 12% max. moisture; *Storage:* highly hygroscopic; store in cool, dry place with desiccant; add preservatives on reconstitution.

Terra-Spray™ Spray Dried Aloe Vera Powd. Decolorized SDD. [Terry Labs] Aloe vera gel, decolorized, deodorized; used in foods/beverages (health/sports/fitness drinks, juice blends, health foods); *Properties:* wh. to lt. beige fine powd.; After reconstitution 1:99: sp.gr. 0.997-1.004; pH 3.5-5.0; 8% max. moisture; *Storage:* preserve after reconstitution; may darken with age.

Terra-Spray™ Spray Dried Aloe Vera Powd. Reg. SDR. [Terry Labs] Aloe vera gel; used in foods/beverages (health/sports/fitness drinks, juice blends, health foods); *Properties:* lt. cream to beige fine powd.; After reconstitution 1:99: sp.gr. 0.997-1.004; pH 3.5-5.0; 8% max. moisture; *Storage:* preserve after reconstitution; may darken with age.

Tetrahop®. [Pfizer Food Science] Highly purified aq. sol'n. of potassium salt of tetrahydroisoalpha acids standardized @ 10% w/v, from hops extract; brewery additive which produces light-stable beer with improved foam stand and cling and enhanced bitterness qualities and flavor stability; for post-fermentation injection; *Storage:* store at R.T.

Textratein. [Lucas Meyer] Textured soya protein; CAS 68153-28-6; protein enrichment for vegetarian meats, cereals, dry mixes.

TFC Purex® Salt. [Morton Salt] Sodium chloride FCC with anticaking agent (sodium ferrocyanide); food-grade salt; used for ham, bacon, corned beef prods.; *Regulatory:* sodium ferrocyanide: FDA 21CFR §172.490, exempt from label declaration under 21CFR §101.100(a)(3); *Properties:* powd.; 25-57% retained on 50 mesh, 10-37% on 70 mesh; mean crystal size 410 μm; mean surf. area 70 cm²/g; bulk dens. 1.17-1.28 g/ml (loose); 99.82% NaCl.

The Edge® C-1600. [Bunge Foods] Icing stabilizer; boiling or nonboiling; makes semiclear icings of pliable texture; for sweet goods, danish icing, fudge/creme icings, white and chocolate; can be formulated for frozen goods; *Usage level:* 8-10 lb/100 lb powd. sugar; *Properties:* off-wh. powd., bland flavor; pH 7.8; 5% max. moisture.

The Edge® Dryset® Icing Base. [Bunge Foods] Nonboiling for dry icings; baked, frozen and retail icings on cakes, danish pastries, sweet goods, creme fillings, white and chocolate icings; *Usage*

level: 10-16 lb; *Properties:* wh. to off-wh. semi-plastic base, vanilla flavor.

Thermolec 57. [ADM Ross & Rowe Lecithin] Modified soya lecithin; CAS 8002-43-5; EINECS 232-307-2; pan/mold release for baked goods; *Properties:* translucent fluid; 0.80% max. water.

Thermolec 68. [ADM Ross & Rowe Lecithin] Modified soya lecithin; CAS 8002-43-5; EINECS 232-307-2; dough conditioner, pan/mold release; *Properties:* opaque fluid; 0.80% max. water.

Thermolec 200. [ADM Ross & Rowe Lecithin] Modified soya lecithin; CAS 8002-43-5; EINECS 232-307-2; for infant formulas, coffee creamers, gravies, sauces; *Properties:* translucent fluid; 0.80% max. water.

Thermolec WFC. [ADM Ross & Rowe Lecithin] Hydroxylated soya lecithin; CAS 8029-76-3; EINECS 232-440-6; for infant formulas, coffee creamers, gravies, sauces, instant dry mixes; *Properties:* translucent fluid; 1% max. water.

Thiamine Hydrochloride USP, FCC Regular Type No. 601160. [Roche] Vitamin B_1 hydrochloride USP FCC; CAS 67-03-8; EINECS 200-641-8; nutrient, source of thiamine in pharmaceutical liqs. (polyvitamin drops), dry prods. (infant formulas, dry foods for reconstitution); not recommended for dry prods. with high moisture content; *Properties:* wh. fine cryst., char. sl. meat-like or yeast odor and taste; 90% thru 200 mesh; sol. 1 g/1 ml water, 1 g/100 ml alcohol; sol. in glycerin; insol. in ether, benzene; m.w. 337.27; pH 2.7-3.4 (1:100 aq.); 98-102% assay; *Precaution:* avoid heat, alkalies, prolonged light exposure; *Storage:* store in dry place in tightly closed container.

Thiamine Mononitrate USP, FCC Fine Powd. No. 601340. [Roche] Vitamin B_1 mononitrate USP FCC; CAS 532-43-4; EINECS 208-537-4; nutrient, thiamine source in dry mixts. that have appreciable moisture content, e.g., flour, certain infant cereals, tablets; not recommended for sol'ns.; *Regulatory:* FDA GRAS; *Properties:* wh. fine cryst. powd., char. sl. meat-like or yeasty odor and taste; 95% thru 200 mesh; sol. 1 g/35 ml water; sl. sol. in alcohol, chloroform; m.w. 327.36; pH 6.0-7.5 (2-100% aq.); 98-102% assay; *Precaution:* avoid heat, alkalies, prolonged light exposure; *Storage:* store in dry place in tight, light-resist. containers @ 72 F or below.

Thin-N-Thik® 99. [A.E. Staley Mfg.] Food starch modified, derived from waxy maize; cook-up starch providing controlled thin-to-thick visc. change to retorted and aseptically processed foods, acid and neutral systems; for puddings, sauces, gravies, fruit pie fillings, tomato-based prods., stews, meat prods.; *Regulatory:* FDA 21 CFR §172.892; *Properties:* wh. powd.; 90% thru 100 mesh; pH 5.5 (uncooked); 12% moisture.

Thixogum S. [MLG Enterprises Ltd.] Acacia gum, xanthan gum; emulsifier, stabilizer; *Properties:* off-wh. powd.; 95% thru 120 mesh; visc. 400-600 cps (1% aq.); 15% max. moisture.

Thixogum X. [MLG Enterprises Ltd.] Xanthan gum, guar gum, locust bean gum; emulsifier, stabilizer, binding agent for condimentary emulsions and sauces; provides good storage stability, cold emulsion processing, smooth/glazing texture, controllable visc.; allows regeneration without syneresis of deep-freezed sauces, gravies; *Usage level:* 0.8-1.2% (dressings, salad cream, mayonnaise).

Thymus Substance. [Am. Labs] Vacuum-dried defatted glandular prod.; nutritive food additive; *Properties:* powd. or freeze-dried form.

Ticaloid® 500. [TIC Gums] Agar replacer controlling syneresis in bakery glazes; *Usage level:* 0.3-0.7% (bakery glazes).

Ticaloid® 1024. [TIC Gums] Agar replacer controlling syneresis in bakery glazes; *Usage level:* 1-1.25% (bakery glazes).

Ticaloid® 1039. [TIC Gums] Texturizer for meringues; inhibits weeping; *Usage level:* 1.2-1.5% (meringues).

Ticaloid® 1070. [TIC Gums] water binder, processing aid which prevents syneresis, aids freeze/thaw props., adds sheen and gloss, prevents cracking; for pie fillings; *Usage level:* 1-2% (pie fillings).

Ticaloid® Lite. [TIC Gums] Improves moisture retention, antistaling props., reduces oil pickup in doughnuts; improves moisture retention, provides better pan release in cakes, baked goods; *Usage level:* 0.15-0.25% (doughnuts, cakes).

Ticaloid® No Fat 102B. [TIC Gums] Texturizer, oil and shortening replacement, moisture retention aid for low-fat muffins; reduces tackiness, holds moisture, controls sugar crystals in icings; cracker adhesive; agar replacement in bakery glazes; *Usage level:* 1-2% (low-fat muffins), 2-4% (icings), 2-4% (cracker adhesive), 1-2% (bakery glazes).

Ticaloid® S1-102. [TIC Gums] Hydrocolloid blend with emulsifiers and thickeners (tragacanth, gum Arabic, xanthan); oil replacement for fat-free salad dressings; provides mouthfeel of oil; *Properties:* visc. 1800-3200 cps; pH 4.5-6.5; 85% dietary fiber, 11% max. moisture.

Ticaloid® S2-102. [TIC Gums] Hydrocolloid blend with emulsifiers and thickeners (tragacanth, gum Arabic, starch, xanthan); oil replacement for fat-free salad dressings; provides mouthfeel of oil; *Properties:* visc. 1800-3200 cps; pH 4.5-6.5; 85% dietary fiber, 11% max. moisture.

Ticalose® 15. [TIC Gums] Cellulose gum; CAS 9004-32-4; anionic; thickener, moisture retention aid for carbonated beverages (gas entrapment), fruit drinks (mouthfeel enhancement), binder/lubricant in vitamins, ice crystallization control (icings); *Usage level:* 0.05-0.5% (icings), 0.1-1% (syrup), 0.1-0.3% (dry beverage mix), 0.05-0.1% (animal feed pellets); *Properties:* coarse (40 mesh), std. (80 mesh), or fine (120 mesh) particle size; visc. 10-15 cps (1%), 25-50 cps (2%).

Ticalose® 30. [TIC Gums] Cellulose gum; CAS 9004-32-4; anionic; thickener, moisture retention aid for carbonated beverages (gas entrapment), fruit drinks (mouthfeel enhancement), binder/lubricant in vitamins, ice crystallization control (icings); *Us-*

age level: 0.05-0.5% (icings), 0.1-1% (syrup), 0.1-0.3% (dry beverage mix), 0.05-0.1% (animal feed pellets); *Properties:* coarse (40 mesh), std. (80 mesh), or fine (120 mesh) particle size; visc. 20-30 cps (1%), 100-150 cps (2%).

Ticalose® 75. [TIC Gums] Cellulose gum; CAS 9004-32-4; anionic; med.-visc. grade for foods as viscosifier, texturizer, ice crystal control aid, protein stabilizer, pulp suspending agent, mouthfeel enhancer; for table syrups, icings, acidified milk, frozen juices; *Usage level:* 0.1-0.3% (marshmallow topping), 0.05-0.5% (icings), 0.2-0.3% (acidified milk), 0.1-0.3% (frozen breakfast drinks); *Properties:* coarse (40 mesh), std. (80 mesh), or fine (120 mesh) particle size; visc. 40-75 cps (1%), 400-600 cps (2%).

Ticalose® 100. [TIC Gums] Cellulose gum; CAS 9004-32-4; anionic; med.-visc. grade for foods as viscosifier, texturizer, ice crystal control aid, protein stabilizer, pulp suspending agent, mouthfeel enhancer; for table syrups, icings, acidified milk, frozen juices; *Properties:* coarse (40 mesh), std. (80 mesh), or fine (120 mesh) particle size; visc. 70-110 cps (1%), 800-1200 cps (2%).

Ticalose® 150 R. [TIC Gums] Cellulose gum, with high degree of substitution, providing better stability under high temp., low pH conditions, high salt concs., other high stress processes; CAS 9004-32-4; anionic; thickener, binder, suspending aid, mouthfeel enhancer, moisture retention aid; *Usage level:* 0.1-1% (table syrup), 0.1-0.4% (batter for deep frying), 0.2-0.8% (fruit drinks), 0.05-0.1% (dietetic beverages), 0.05-0.2% (salad dressing mix), 0.2-0.4% (semimoist pet foods), 0.2-0.5% (cocoa mixes); *Properties:* coarse (40 mesh), std. (80 mesh), or fine (120 mesh) particle size; visc. 50-150 cps (1%), 600-1000 cps (2%).

Ticalose® 700 R. [TIC Gums] Cellulose gum, with high degree of substitution, providing better stability under high temp., low pH conditions, high salt concs., other high stress processes; CAS 9004-32-4; anionic; thickener, binder, suspending aid, mouthfeel enhancer, moisture retention aid; *Properties:* coarse (40 mesh), std. (80 mesh), or fine (120 mesh) particle size; visc. 200-400 cps (1%).

Ticalose® 750. [TIC Gums] Cellulose gum; CAS 9004-32-4; anionic; high-visc. grade as viscosifier, ice crystal control aid, mouthfeel enhancer; for ice cream, sorbets, fried foods (controls fat absorp.), cake mixes (retains moisture), meringues (prevents syneresis), pizza/frozen dough, bar mixes, instant gravies; *Properties:* coarse (40 mesh), std. (80 mesh), or fine (120 mesh) particle size; visc. 550-750 cps (1%).

Ticalose® 2000 R. [TIC Gums] Cellulose gum, with high degree of substitution, providing better stability under high temp., low pH conditions, high salt concs., other high stress processes; CAS 9004-32-4; anionic; thickener, binder, suspending aid, mouthfeel enhancer, moisture retention aid; *Usage level:* 0.1-0.3% (eggnog), 0.05-0.1% (dietetic beverages), 0.1-0.3% (frozen breakfast drinks), 0.1-

0.5% (sherbet), 0.1-0.5% (pie fillings); *Properties:* coarse (40 mesh), std. (80 mesh), or fine (120 mesh) particle size; visc. 1000-1500 cps (1%).

Ticalose® 2500. [TIC Gums] Cellulose gum; CAS 9004-32-4; anionic; high-visc. grade as viscosifier, ice crystal control aid, mouthfeel enhancer; for ice cream, sorbets, fried foods (controls fat absorp.), cake mixes (retains moisture), meringues (prevents syneresis), pizza/frozen dough, bar mixes, instant gravies; *Usage level:* 0.1-0.7 (variegated syrup), 0.1-0.4% (batter for deep frying), 0.03-0.5% (instant hot cereals), 0.1-0.3% (cake mixes), 0.1-0.5% (soft serve desserts), 0.1-0.3% (reduced-calorie spreads), 0.1-0.4% (tortillas); *Properties:* coarse (40 mesh), std. (80 mesh), or fine (120 mesh) particle size; visc. 1500-2500 cps (1%).

Ticalose® 4000. [TIC Gums] Cellulose gum; CAS 9004-32-4; anionic; high-visc. grade as viscosifier, ice crystal control aid, mouthfeel enhancer; for ice cream, sorbets, fried foods (controls fat absorp.), cake mixes (retains moisture), meringues (prevents syneresis), pizza/frozen dough, bar mixes, instant gravies; *Usage level:* 0.1-0.25% (ice cream), 0.1-0.5% (ice milk), 0.1-0.5% (pie fillings), 0.1-0.5% (barbecue sauces, pickle relish); *Properties:* gran., 16 mesh particle size; visc. 2500-4000 cps (1%).

Ticalose® 4500. [TIC Gums] Cellulose gum; CAS 9004-32-4; anionic; high-visc. grade as viscosifier, ice crystal control aid, mouthfeel enhancer; for ice cream, sorbets, fried foods (controls fat absorp.), cake mixes (retains moisture), meringues (prevents syneresis), pizza/frozen dough, bar mixes, instant gravies; *Usage level:* 0.1-0.25% (ice cream), 0.1-0.5% (ice milk), 0.1-0.5% (pie fillings), 0.1-0.5% (barbecue sauces, pickle relish); *Properties:* coarse (40 mesh), std. (80 mesh), or fine (120 mesh) particle size; visc. 3000-4500 cps (1%).

Ticalose® 5000 R. [TIC Gums] Cellulose gum, with high degree of substitution, providing better stability under high temp., low pH conditions, high salt concs., other high stress processes; CAS 9004-32-4; anionic; thickener, binder, suspending aid, mouthfeel enhancer, moisture retention aid; *Usage level:* 0.2-0.8% (fruit-flavored drinks), 0.05-0.2% (salad dressing mix); *Properties:* coarse (40 mesh), std. (80 mesh), or fine (120 mesh) particle size; visc. 2500-3500 cps (1%).

Ticaxan® Regular. [TIC Gums] Xanthan gum; CAS 11138-66-2; EINECS 234-394-2; stabilizer, thickener, emulsifier, suspending agent, binder, moisture control aid in foods (salad dressings, sauces, fruit beverages, dry mixes, refrigerated dough, cake mixes, relish, pie fillings, bakery emulsions).

Ticolv. [TIC Gums] Guar gum; CAS 9000-30-0; EINECS 232-536-8; thickener, film former, binder featuring high visc., milk reactivity; for food (ice cream, dairy prods., instant soups, dressings, sauces, gravies, cakes, donuts, frozen foods, cheese, pet food); *Properties:* sol. in cold water.

TIC Pretested® Arabic FT Powd. [TIC Gums] Gum arabic FCC/NF; CAS 9000-01-5; EINECS 232-

519-5; emulsifier for beverage emulsions and flavor encapsulation; low visc. and high fiber for use in dry mixes and meal replacers; good adhesion and water binding chars.; said to be virtually noncaloric; *Regulatory:* FDA 21CFR §184.1330, GRAS; *Properties:* wh. to off-wh. spray-dried free-flowing powd., odorless, bland flavor; 80% min. thru 80 mesh; visc. 300 cps max.; pH 4-5; 85% dietary fiber, 11% max. moisture.

TIC Pretested® Arabic FT-1 USP. [TIC Gums] Gum arabic USP; CAS 9000-01-5; EINECS 232-519-5; carrier for spray-dried flavors; retains flavors, protects them from oxidation, provides clean taste; may be labeled natural; *Regulatory:* kosher; *Properties:* powd.; 85% min. dietary fiber, 15% max. moisture.

TIC Pretested® Arabic PH-FT. [TIC Gums] Gum arabic FCC/NF; CAS 9000-01-5; EINECS 232-519-5; emulsifier for beverage emulsions and flavor encapsulation; low visc. and high fiber for beverage mixes and meal replacers; exhibits good adhesion and water-binding qualities; *Regulatory:* FDA 21CFR §184.1330, GRAS; *Properties:* wh. to off-wh. free-flowing powd., odorless, bland flavor; 55% min. on 100 mesh; visc. 300 cps max.; pH 4-5; 85% min. dietary fiber, 15% max. moisture.

TIC Pretested® Aragum® 1067-T. [TIC Gums] Blend of gums and sugar solids; provides superior adhesion and low visc. for confectionery industry for snacks and seed cakes and to pet food industry for bird seed bells; *Regulatory:* FDA approved; *Properties:* off-wh. free-flowing powd., odorless, bland flavor; 20% max. on 80 mesh; sol. in cold water; visc. 75 cps max. (10%); pH 5-7; 15% max. moisture.

TIC Pretested® Aragum® 1070. [TIC Gums] Blend of gums and other natural solids; provides superior adhesion and low visc. to snack and baking industries for spice coating of nuts, seed or cereal cakes, crackers; prevents discoloration during roasting or baking; *Regulatory:* FDA approved; *Properties:* wh. free-flowing powd., odorless, bland flavor; 25% max. on 140 mesh; sol. in cold water; visc. 500-1500 cps (25%); pH 4-6; 10% max. moisture.

TIC Pretested® Aragum® 2000. [TIC Gums] Blend of starch and water-sol. gum; replaces gum arabic used in coatings on one-to-one basis; produces low visc. at high use levels; for foods; *Regulatory:* FDA approved; *Properties:* wh. free-flowing powd., odorless, bland flavor; 90% min. thru 80 mesh; sol. in cold water; visc. 500 cps max. (30%); pH 4-7; 10% max. moisture.

TIC Pretested® Aragum® 3000. [TIC Gums] Blend of starch and gum arabic; forms stable flavor emulsions with low visc.; replaces gum arabic on one-to-one basis; provides uniform cloud to finished beverage; *Regulatory:* FDA approved; *Properties:* wh. free-flowing powd., odorless, bland flavor; 90% min. thru 80 mesh; visc. 200 cps max. (30%); pH 4-7; 10% max. moisture.

TIC Pretested® Aragum® 9000 M. [TIC Gums] Blend of natural gums standardized with dextrin; replaces gum arabic in flavor emulsions on one-to-one basis; stable in presence of alcohol; can be stored refrigerated without setting; clouding agent; *Regulatory:* FDA approved; *Properties:* wh. free-flowing powd., bland odor, sl. sweet flavor; 20% min. on 80 mesh; visc. 600 cps max. (30%); pH 4-7; 12% max. moisture.

TIC Pretested® Aragum® 9500. [TIC Gums] Gum system contg. various species of acacia standardized with dextrose; replacement for modified starch; emulsifier yielding stable emulsions using weighted oils with ester gum and brominated vegetable oil; *Regulatory:* FDA approved; *Properties:* off-wh. free-flowing powd., nearly odorless, bland flavor; 25% max. on 80 mesh; visc. 80-120 cps (23%); pH 4-7; 10% max. moisture.

TIC Pretested® CMC 2500 S. [TIC Gums] Sodium carboxymethylcellulose; CAS 9004-32-4; thickener for foods and pharmaceuticals; controls texture and ice crystal growth in frozen dairy prods.; retains moisture in bakery and calorie applics.; synergistic with casein, other proteins; *Regulatory:* FDA GRAS, FCC; *Properties:* off-wh. to tan free-flowing powd., odorless, bland flavor; 25% min. on 140 mesh, 65% max. thru 140 mesh; sol. in cold or hot water; visc. 1400-2400 cps; pH 6.0-8.5; 99.5% min. purity, 8% max. moisture.

TIC Pretested® CMC PH-2500. [TIC Gums] Sodium carboxymethylcellulose USP/FCC; CAS 9004-32-4; pre-hydrated thickener; protein stabilizer; fat/oil barrier in batters; controls texture and ice crystal growth in frozen prods.; retains moisture in bakery prods.; synergistic with casein, most proteins; *Usage level:* 0.1-0.2% (ice cream/soft serve mixes, donut mixes), 0.25-0.4% (variegated syrup, pet food gravy), 0.125-0.2% (cake mixes), 0.075-0.175% (fresh tortilla); *Properties:* dust-free powd., odorless, tasteless; 60% min. on 100 mesh; sol. in hot or cold water; visc. 1000-2000 cps min.; pH 6.0-8.5; 99.5% min. purity.

TIC Pretested® Colloid 775 Powd. [TIC Gums] Specially processed λ-carrageenan with some kappa portions; CAS 9000-07-1; EINECS 232-524-2; thickener for cold milk, reactive with milk proteins; develops visc. @ low usage levels, soft gels @ higher levels; gels with heat; provides creamy texture, clean full flavor release; *Usage level:* 0.4-0.6% (milk shakes), 0.1-0.175% (fortified instant drinks), 0.4-.0.8% (aerated instant desserts), 0.2-0.25% (instant pudding), 0.5-0.7% (cheese spreads and sauces); *Regulatory:* kosher approved; *Properties:* 35% max. on 200 mesh; visc. 450 cps min. (1%); pH 8-9.5; 12% max. moisture.

TIC Pretested® Colloid 881 M. [TIC Gums] ι-Carrageenan; CAS 9000-07-1; EINECS 232-524-2; reactive in water and milk systems; yields pliable gels with superior organoleptic props.; suspends solids in liq. systems; combined with starches, used as partial fat replacer in minced and emulsified meats; hydrates fully above 70 C; *Regulatory:* FDA approved; *Properties:* lt. tan free-flowing powd., nearly odorless, bland flavor; 100% min. thru 40 mesh; pH 6-8; 15% max. moisture.

TIC Pretested® Gum Agar Agar 100 FCC/NF Powd. [TIC Gums] Agar agar gum; CAS 9002-18-0; EINECS 232-658-1; rigid gel which must boil to hydrate; stable under low pH conditions; tolerates large percentages of salts without changes in gel structure; for food applics.; *Usage level:* 0.15-0.25% (bakery icings), 0.25-0.45% (pie fillings), 0.1-0.2% (bakery glazes), 0.5-0.65% (low calorie jams/jellies), 1.1-1.25% (candy gels), 8-10% (dental impressions); *Regulatory:* kosher approved; *Properties:* med. mesh powd.; 90% thru 100 mesh; m.w. 5000-30,000; visc. forms gel on heating; gel pt. 88-103 F; pH 4-7; 20% max. moisture.

TIC Pretested® Gum Guar 8/22 FCC/NF Powd. [TIC Gums] Guar gum FCC/NF; CAS 9000-30-0; EINECS 232-536-8; provides high visc. and retains moisture in food applics.; quick hydration; freeze/thaw and pH stable (4-9); *Usage level:* 0.125-0.2% (instant soup), 0.125-0.25% (sauces, gravies, instant beverages), 0.25-0.35% (cake mixes), 0.1-0.4% (ice cream mixes), 0.3-0.45% (batters), 0.15-0.25% (fresh tortillas); *Regulatory:* kosher approved; *Properties:* 4% max. on 100 mesh, 75% min. thru 200 mesh; sol. in cold water; visc. 3000 cps min.; pH 4-7; 15% max. moisture.

TIC Pretested® Locust Bean POR/A. [TIC Gums] Locust bean gum; CAS 9000-40-2; EINECS 232-541-5; viscosifier, retains moisture for food applics.; needs heat to hydrate; freeze/thaw and pH stable (3.6-9); *Usage level:* 0.1-0.25% (cake mixes, dough), 0.1-0.3% (sauces), 0.1-0.4% (dairy stabilizer mix), 0.12-0.2% (fresh tortillas), 0.125-0.25% (pie fillings), 0.2-0.3% (cream cheese), 0.35-0.45% (fruit yogurt); *Regulatory:* kosher approved; *Properties:* 10% max. on 100 mesh; visc. 2700 cps min. (1%); pH 5-7; 15% max. moisture.

TIC Pretested® Pectin HM Rapid. [TIC Gums] Pectin; CAS 9000-69-5; EINECS 232-553-0; gellant for jams that are kept in small jars (< 1 kg); helps avoid fruit flotation; requires pH 3.3-3.5 and 63-67% sugar solids; 145-155% SAG; *Regulatory:* FDA 21CFR §184.1558, GRAS; *Properties:* ylsh. to tan free-flowing powd., nearly odorless, sl. sweet mucilaginous taste; 98% min. thru 60 mesh; dissolves in water forming opalescent colloid sol'n.; pract. insol. in alcohol; visc. 25-100 cps (1%); pH 3-4.2; 12% max. moisture.

TIC Pretested® Pectin HM Slow. [TIC Gums] Pectin; CAS 9000-69-5; EINECS 232-553-0; gellant for jams, jellies, preserves mfg. in large containers (> 1 kg); requires pH 3.3-3.5 and 63-67% sugar solids; 145-155% SAG; *Regulatory:* FDA 21CFR § 184.1558, GRAS; *Properties:* ylsh. to tan free-flowing powd., nearly odorless, bland mucilaginous taste; 98% min. thru 60 mesh; dissolves in water forming opalescent colloid sol'n.; pract. insol. in alcohol; visc. 25-100 cps (1%); pH 3-4; 12% max. moisture.

TIC Pretested® Saladizer 250. [TIC Gums] Blend of natural gums; emulsifier, viscosifier, stabilizer for salad dressings; hydrates slowly; heat and freeze/thaw stable, pH stable (3.2-9); *Usage level:* 0.15-0.35% (pourable dressings), 0.35-0.45% (spoonable dressings), 0.15-0.2% (relishes), 0.2-0.3% (fresh salad), 0.15-0.3% (barbeque sauce), 0.15-0.25% (mustard); *Properties:* 10% max. on 100 mesh; sol. in cold water; visc. 3800 cps min.; pH 4-7; 15% max. moisture.

TIC Pretested® Saladizer PH-250. [TIC Gums] Blend of natural gums; pre-hydrated salad dressing stabilizer; emulsifier, viscosifier; heat stable, freeze/thaw stable; pH stable (3.2-9); hydrates slowly; *Usage level:* 0.25-0.45% (pourable dressings), 0.45-0.55% (spoonable dressings), 0.25-0.3% (relishes), 0.3-0.4% (fresh salad), 0.25-0.4% (barbeque sauce), 0.25-0.35% (mustard); *Properties:* dust-free; 90% min on 140 mesh; sol. in cold water; visc. 3500-5000 cps; pH 6-7; 15% max. moisture.

TIC Pretested® Tragacanth 440. [TIC Gums] Gum tragacanth; CAS 9000-65-1; EINECS 232-552-5; emulsifier providing med. visc., creamy texture to foods; pH stable (3-9); slow to hydrate; exc. acid, heat, and salt tolerance; *Usage level:* 1.75-2.25% (candy lozenges), 0.6-1.2% (pourable dressings), 0.75-2% (pharmaceutical suspensions), 0.75-1.25% (car polish/wax emulsions), 1-1.5% (shampoo/conditioner), 1-1.5% (ceramic finish); *Properties:* 95% min. thru 140 mesh; sol. in cold water; m.w. ≈ 850,000; visc. 300 cps min. (1%); pH 4-7; 15% max. moisture; *Storage:* hygroscopic-store under cool, dry conditions.

TIC Pretested® Xanthan 200. [TIC Gums] Xanthan gum; CAS 11138-66-2; EINECS 234-394-2; stabilizer, thickener, processing aid featuring pseudoplasticity, heat and pH stability, high visc., good sol.; for food industry; *Regulatory:* FDA approved; *Properties:* cream-colored free-flowing powd., nearly odorless, typ. bland flavor; 95% min. thru 200 mesh, 70% min. thru 270 mesh; visc. 1000 cps; pH 6.5-7.5; 15% max. moisture.

TIC Pretested® Xanthan PH. [TIC Gums] Xanthan gum; CAS 11138-66-2; EINECS 234-394-2; pre-hydrated food additive, emulsifier, viscosifier for dressings, emulsions, cake mixes, fruit juices, frozen sauces; pH stable; *Usage level:* 0.2-0.35% (dressings), 0.3-0.5% (emulsions), 0.25-0.35% (cake mixes), 0.07-0.15% (fruit juices), 0.25-0.3% (frozen sauces); *Regulatory:* kosher approved; *Properties:* dust-free; 60% min on 100 mesh; sol. in cold water; visc. 1000 cps min.; pH 6-7; 15% max. moisture.

Timeflex. [MLG Enterprises Ltd.] Proteolytic enzyme derived from *Bacillus subtilis* with cereal derivs. and other edible excipients; bacterial enzyme tablet for crackers, cookies, and pizza prod.; provides reduction in mixing, accelerated fermentation, improved prod. quality; *Properties:* tablet; readily disintegrates in R.T. water.

Tinamul®. [Hüls AG] Monodiglyceride based on edible fats; emulsifier preventing oil separation in hazelnut and almond pastes; also for prep. of bakery goods, margarine, ice cream, release agents, sausages, mayonnaise, toffees, snacks; *Properties:* powd.; m.p. 60 C; acid no. 3 max.; iodine no. 3 max.; sapon. no. 160-170; 45% monoglycerides.

T-Maz® 20. [PPG/Specialty Chem.] Polysorbate 20; CAS 9005-64-5; nonionic; food emulsifier; *Regulatory:* kosher; *Properties:* yel. liq.; sol. in water, ethanol, acetone, toluene, veg. oil, propylene glycol; sp.gr. 1.1; visc. 400 cps; HLB 16.7; acid no. 2 max.; sapon. no. 40-50; hyd. no. 96-108; flash pt. (PMCC) > 350 F; 97% act.

T-Maz® 20K. [PPG/Specialty Chem.] Polysorbate 20; CAS 9005-64-5; flavor solubilizer; *Properties:* liq.; HLB 16.7; 100% conc.

T-Maz® 28. [PPG/Specialty Chem.] PEG-80 sorbitan laurate; CAS 9005-64-5; emulsifier and solubilizer of essential oils, wetting agent, visc. modifier, antistat, stabilizer and dispersant used in foods; *Properties:* pale yel. liq.; sol. in water; sp.gr. 1.0; visc. 1100 cps; HLB 19.2; acid no. 2 max.; sapon. no. 5-15; hyd. no. 25-40; flash pt. (PMCC) > 350 F; 30% max. water.

T-Maz® 60. [PPG/Specialty Chem.] Polysorbate 60; CAS 9005-67-8; emulsifier, solubilizer, wetting agent, antistat, stabilizer, dispersant, visc. modifier, suspending agent used in the food industry; *Properties:* yel. gel; sol. in water, min. spirits, toluol; sp.gr. 1.1; visc. 550 cps; HLB 14.9; pour pt. 23-25 C; sapon. no. 45-55; hyd. no. 81-96; flash pt. (PMCC) > 350 F; 97% min. act.

T-Maz® 60K. [PPG/Specialty Chem.] Polysorbate 60; CAS 9005-67-8; nonionic; food emulsifier for whipped toppings, shortenings, cake icings/fillings, confectionery coatings, dressings; dough conditioner for bakery goods; protective coatings on raw fruits/vegetables; *Usage level:* 0.4% (whipped toppings, coffee whiteners), 1% (shortening), 4.5% (nonalcoholic mixes), 0.5% (confectionery coatings), 0.5% (dough conditioner, gelatin desserts), 0.5-4.5% (colors), GMP (syn. flavor, defoamer); *Regulatory:* FDA 21CFR §172.515, 172.836, 173.340, 573.840; kosher; *Properties:* yel. to amber clear gel, bland odor; sol. in water, ethanol, min. spirits, toluene; disp. in propylene glycol; sp.gr. 1.08; m.p. 23-25 C; b.p. > 300 F; HLB 14.9; acid no. 2 max.; sapon. no. 45-55; hyd. no. 81-96; flash pt. (PMCC) > 350 F; 100% conc.; *Toxicology:* LD50 (oral, rat) 64 g/kg; nonirritating; *Storage:* store in well-ventilated area below 120 F.

T-Maz® 60KHS. [PPG/Specialty Chem.] Polysorbate 60; CAS 9005-67-8; emulsifier, solubilizer, wetting agent, antistat, stabilizer, dispersant, visc. modifier, suspending agent used for foods; high melt grade; *Regulatory:* kosher; *Properties:* Gardner 5 gel; sol. @ 5% in water, min. spirits, toluene; sol. @ 2% in ethanol, disp. in propylene glycol; m.p. 25-27 C; HLB 14.9; acid no. 2 max.; sapon. no. 45-55; hyd. no. 81-96; flash pt. (PMCC) > 350 F.

T-Maz® 65. [PPG/Specialty Chem.] Polysorbate 65; CAS 9005-71-4; nonionic; emulsifier, solubilizer, wetting agent, antistat, stabilizer, dispersant, visc. modifier, suspending agent used in the food indus-

try; *Properties:* sol. in ethanol, acetone, naphtha; disp. in water; sp.gr. 1.1; m.p. 30-32 C; HLB 10.5; sapon. no. 88-98; 97% min. act.

T-Maz® 65K. [PPG/Specialty Chem.] Polysorbate 65; CAS 9005-71-4; nonionic; emulsifier for frozen desserts, whipped toppings, cake icings and fillings, coffee whiteners; defoamer in food processing; *Usage level:* 0.1% (ice cream), 0.32% (cakes, icings, fillings), 0.4% (whipped toppings, coffee whiteners), GMP (defoamer); *Regulatory:* FDA 21CFR §172.838, 173.340; kosher; *Properties:* tan waxy paste, char. ester odor; sol. @ 2% in ethanol, veg. oil, disp. in water, insol. in min. oil, propylene glycol; m.p. 30-32 C; b.p. > 300 F; HLB 10.5; acid no. 2 max.; sapon. no. 88-98; hyd. no. 44-60; flash pt. (PMCC) > 350 F; 100% conc.; *Toxicology:* LD50 (oral, rat) > 39.8 g/kg; nonirritating, nonsensitizing; *Storage:* store in well-ventilated area below 120 F.

T-Maz® 80. [PPG/Specialty Chem.] Polysorbate 80; CAS 9005-65-6; nonionic; emulsifier, solubilizer, wetting agent, antistat, stabilizer, dispersant, visc. modifier, suspending agent used in the food industry; *Properties:* yel. liq.; sol. in water, ethanol, veg. oil, toluol; sp.gr. 1.0; visc. 400 cps; HLB 15.0; sapon. no. 45-55; hyd. no. 65-80; flash pt. (PMCC) > 350 F; 97% min. act.

T-Maz® 80K. [PPG/Specialty Chem.] Polysorbate 80; CAS 9005-65-6; nonionic; solubilizer and emulsifier for foods, vitamins, edible oils, ice cream, shortenings, pickles, whipped toppings, gelatin desserts; defoamer in cottage cheese, yeast; crystal modifier in NaCl; *Usage level:* 0.05% (pickle prods.), 0.1% (ice cream), 1% (shortenings), 4% (yeast defoamers), 175-475 mg/day (vitamins), 360 mg/day (dietary prods.), 80 ppm (defoamer in cottage cheese), 50 ppm (BBQ sauce colors); *Regulatory:* FDA 21CFR §172.840, 173.340, 573.860; kosher; *Properties:* yel. clear liq., bland odor; sol. in water, ethanol, veg. oil; disp. in propylene glycol, toluene; sp.gr. 1.09; b.p. > 300 F; HLB 15.0; acid no. 2 max.; sapon. no. 45-55; hyd. no. 65-80; flash pt. (PMCC) > 350 F; 100% conc.; *Toxicology:* LD50 (oral, rat) > 30 ml/kg (mild); nonirritating, noncomedogenic; *Storage:* store in well-ventilated area below 120 F.

T-Maz® 80KLM. [PPG/Specialty Chem.] Polysorbate 80, low melt pt.; CAS 9005-65-6; food emulsifier; *Regulatory:* kosher; *Properties:* Gardner 5 liq.; sol. @ 5% in water, veg. oil, @ 2% in ethanol; disp. in toluene, propylene glycol; HLB 15.0; acid no. 2 max.; sapon. no. 45-55; hyd. no. 65-80; flash pt. (PMCC) > 350 F.

T-Maz® 81. [PPG/Specialty Chem.] Polysorbate 81; CAS 9005-65-6; nonionic; emulsifier, solubilizer, wetting agent, antistat, stabilizer, dispersant, visc. modifier, suspending agent used in the food industry; *Properties:* Gardner 6 liq.; sol. in min. spirits; disp. in water, min. oils, toluene, veg. oils; HLB 10.0; sapon. no. 96-104; hyd. no. 134-150; flash pt. (PMCC) > 350 F; 100% conc.

T-Maz® 95. [PPG/Specialty Chem.] PEG-20 sorbitan tritallate; emulsifier, solubilizer, wetting agent, visc.

modifier, antistat, stabilizer, dispersant for foods; *Properties:* amber liq.; sol. in min. spirits; disp. in water, min. oils, toluene, veg. oils; sp.gr. 1.0; visc. 350 cps; HLB 11.0; sapon. no. 83-93; hyd. no. 39-52; flash pt. (PMCC) > 350 F; 3.0% max. water.

TMP-65. [FMC] Phosphate; moisture binder in cured meat; bacon processing aid for reducing nitrite residuals; *Properties:* powd.; 100% thru 20 mesh; infinitely sol. in water; bulk dens. 82 lb/ft³; pH 7.3-7.8 (1%); 98.3% assay.

Toffix® M. [Hüls AG] Nonlauric triglyceride; special fat containing an emulsifier for prep. of soft caramels, toffees, and chewing sweets; *Properties:* block; m.p. 35-37 C; acid no. 0.3 max.; iodine no. 70; sapon. no. 185-195.

Toffix® P. [Hüls AG] Palm kernel oil triglyceride; special fat containing an emulsifier for the prod. of soft caramels, toffees, and chewing sweets; *Properties:* block; m.p. 36-38 C; acid no. 0.3 max.; iodine no. 5 max.; sapon. no. 235-255.

Toffix® S. [Hüls AG] Soybean oil triglyceride; special fat containing an emulsifier for the prod. of soft caramels, toffees, and chewing sweets; *Properties:* block; m.p. 35-37 C; acid no. 0.3 max.; iodine no. 70; sapon. no. 185-195.

Tofupro-U. [Ikeda] Soya protein; CAS 68153-28-6; nutritive high-protein, low-calorie food without cholesterol; *Properties:* yel. to amber clear liq.; pH 3.8-6.2.

Tolerance. [ADM Arkady] Composite dough conditioner improving quality of buns and variety breads and reducing the need for vital wheat gluten.

Tomex®. [Quest Int'l.] Tomato enhancer for tomato prods.; enhances flavor and texture profiles.

Tona Enzyme 201. [Meer] Papain, enzyme derived from papaya; CAS 9001-73-4; EINECS 232-627-2; proteolytic enzyme used as meat tenderizer, beer clarifier, in dough conditioners for baking, digestive aids.

Tona Papain 14. [Meer] Papain, enzyme derived from papaya; CAS 9001-73-4; EINECS 232-627-2; proteolytic enzyme used as meat tenderizer, beer clarifier, in dough conditioners for baking, digestive aids.

Tona Papain 90L. [Meer] Papain, enzyme derived from papaya; CAS 9001-73-4; EINECS 232-627-2; proteolytic enzyme used as meat tenderizer, beer clarifier, in dough conditioners for baking, digestive aids.

Tona Papain 270L. [Meer] Papain, enzyme derived from papaya; CAS 9001-73-4; EINECS 232-627-2; proteolytic enzyme used as meat tenderizer, beer clarifier, in dough conditioners for baking, digestive aids.

Topcithin®. [Lucas Meyer] Soya lecithin, highly purified, with bacteriological and analytical controls; CAS 8002-43-5; EINECS 232-307-2; nonionic; food emulsifier, dispersant; improves emulsifying and flowing properties and physical appearance; extends shelf-life; low microbiological and heavy metal content; for infant formulas, powdered mixes, milk prods., chocolate; *Properties:* liq.; 100% conc.

Top Flite™ Icing Powd. [Bunge Foods] Boiling type icing stabilizer formulated with fats and gums for quick set for fresh/frozen distribution; for donut glaze, sweet goods icing; *Usage level:* 4-5 lb/100 lb powd. sugar; *Properties:* wh. coarse powd., sweet vanilla aroma/flavor; pH 6.5.

Top Mate Kosher. [Am. Ingreds.] Mono-diglycerides and ethoxylated mono- and diglycerides; food emulsifier for bakery mixes, bread, buns, and sweet goods; *Properties:* cream powd.; 56% min. mono.

Topmulgat. [Lucas Meyer] Natural stabilizer blend; nonionic; structure improver, consistency stabilizer used in soups, sauces, dressings, mayonnaise, frozen foods, deli prods.; *Properties:* powd.

Top Notch®. [Bunge Foods] Tallow, soybean oil, partially hydrog. tallow, water, salt, mono- and diglycerides, EDTA, artificial flavor, vitamin A; bakery magarine for roll-in applics., danish, croissants; 80.2% fat, 16.7% moisture, 3.0% salt.

Toppro. [Lucas Meyer] Hydrolyzed vegetable protein; CAS 100209-45-8; taste improvement and flavor enhancement for meat and fish processing, soups, sauces, and snacks.

Trachea Substance. [Am. Labs] Vacuum-dried defatted glandular prod.; nutritive food additive; *Properties:* powd.

Tragacanth Flake No. 27. [Meer] Tragacanth gum; CAS 9000-65-1; EINECS 232-552-5; thickener, water binder, suspending agent, emulsifier for salad dressings, bakery, flavor emulsions, fruit fillings, cream fillings, citrus beverages, barbecue and steak sauces; *Properties:* water-sol.

Tragacanth Gum Ribbon No. 1. [Meer] Tragacanth gum; CAS 9000-65-1; EINECS 232-552-5; thickener, water binder, suspending agent, emulsifier for salad dressings, bakery, flavor emulsions, fruit fillings, cream fillings, citrus beverages, barbecue and steak sauces; *Properties:* water-sol.

Trans-10, -10K. [Trans-Chemco] Silicone emulsion; -10K kosher grade; defoamer for aq. systems incl. food processing; *Properties:* disp. in water; 10% act.; Unverified

Trans-20, -20K. [Trans-Chemco] Silicone emulsion; -20K kosher grade; defoamer for aq. systems incl. food processing; *Properties:* water-disp.; 20% act.; Unverified

Trans-25, -25K. [Trans-Chemco] Silicone emulsion; -25K kosher grade; defoamer for aq. systems incl. food processing; *Properties:* water-disp.; 25% act.; Unverified

Trans-30, -30K. [Trans-Chemco] Silicone emulsion; -30K kosher grade; defoamer for aq. systems incl. food processing; *Properties:* water-disp.; 30% act.; Unverified

Trans-100. [Trans-Chemco] Silicone compd.; defoamer for food processing; *Properties:* sol. in aliphatic, aromatic, and chlorinated hydrocarbons; 100% act.; Unverified

Trans-175, -176. [Trans-Chemco] Nonsilicone; defoamer for mfg. of soya flour and isolate, egg washing; *Properties:* water-disp.; Unverified

Trans-177S, -177S K. [Trans-Chemco] Emulsion with 5% silicone; -177 S K kosher grade; defoamer for aq. systems incl. food processing; *Properties:* water-disp.; Unverified

Trans-179. [Trans-Chemco] Nonsilicone, water-based; defoamer for mfg. edible phosphates; *Properties:* water-sol.; Unverified

Trans-220. [Trans-Chemco] Nonsilicone, water-based; long-lasting antifoam for food processing; *Properties:* water-disp.; Unverified

Trans-224. [Trans-Chemco] Nonsilicone, water-based; aq. food processing defoamer; *Properties:* water-disp.; Unverified

Trans-225S. [Trans-Chemco] Silicone emulsion; aq. food processing antifoam; *Properties:* water-disp.; 12.5% act.; Unverified

Trans-1030 K. [Trans-Chemco] Emulsion with 3% silicone; defoamer for aq. systems incl. food processing; *Regulatory:* kosher; *Properties:* water-disp.; Unverified

Trans-134S K. [Trans-Chemco] Emulsion with 11% silicone; defoamer for aq. systems incl. food processing; *Regulatory:* kosher; *Properties:* water-disp.; Unverified

Trans-FG2. [Trans-Chemco] Nonsilicone, water-based; defoamer for veg. canning, food processing; *Properties:* water-disp.; Unverified

Trem-Tabs. [MLG Enterprises Ltd.] Active proteolytic enzymes derived from bromelain and other edible excipients; enzyme tablet for yeast-raised bakery prods., cookie and cracker doughs; produces fast, controlled mix reduction; provides doughs with superior machinability, improved grain, texture, crumb softness; *Usage level:* $1/2$-2 tablets/cwt flour; *Properties:* tablet; *Storage:* 6 mos storage life under cool, dry conditions.

Tri-Co® MV Yeast Raised Dough Shortening. [Bunge Foods] Meat fat, vegetable oil, mono- and diglycerides, antioxidants; shortening for all yeast-raised doughs; *Properties:* 1.8 red color, bland flavor; 8.7% alpha mono.

Tri-Co® Vegetable Yeast Raised Dough Shortening. [Bunge Foods] Partially hydrog. soybean and cottonseed oils, mono- and diglycerides; shortening for all yeast-raised doughs; *Properties:* 1.5 red color, bland flavor; 8.7% alpha mono.

TrimChoice™ 5. [A.E. Staley Mfg.] Hydrolyzed oat flour; fat replacer for low-fat foods, baked goods, salad dressings, confections, meats; provides visc., texture, appearance, flavor, and mouthfeel of full-fat foods; *Regulatory:* FDA 21CFR §182.1, GRAS; *Properties:* pH 5.5-6.5 (10%); DE 3.0-5.0; 4.5-5.5% beta-glucan, 4-8% moisture.

TrimChoice™ OC. [A.E. Staley Mfg.] Hydrolyzed oat and corn flour; fat replacer for low-fat foods, dairy prods., margarine, meats; provides visc., texture, appearance, flavor, and mouthfeel of full-fat foods; *Regulatory:* FDA 21CFR §182.1, GRAS; *Properties:* pH 5.5-6.5 (10%); DE 3.0-5.0; 2.0-3.0% beta-glucan, 4-8% moisture.

Triodan 20. [Grindsted Prods.] Polyglyceryl-3 oleate; CAS 9007-48-1; food emulsifier; *Properties:* visc.

liq.; sapon. no. 140-155; Discontinued

Triodan 55. [Grindsted Prods.] Polyglyceryl-3 stearate; CAS 37349-34-1; EINECS 248-403-2; food emulsifier improving cake batter performance, cake volume; *Regulatory:* EEC, FDA §172.854; ADI 0-25 mg/kg; *Properties:* sm. beads; drop pt. 57 C; iodine no. 2 max.; sapon. no. 130-145.

Triodan 55 Kosher. [Grindsted Prods.] Polyglycerol (mainly di-, tri-, and tetra-) ester of edible refined fatty acids; emulsifier for cakes, cake mixes, shortenings, confectionery prods.; *Usage level:* 2-4% (shortenings for cakes and cake mixes), 2-3% (confectionery prods.); *Regulatory:* FDA 21CFR §172.854; FCC; kosher; *Properties:* ivory sm. beads; drop pt. 57 C; iodine no. 2 max.; sapon. no. 130-145; *Storage:* store in cool, dry area.

Triodan 55/S6. [Grindsted Prods.] Fatty acid polyglycerol ester; food emulsifier; *Properties:* block; sapon. no. 110-130; Unverified

Triodan R 90. [Grindsted Prods.] Polyglcyerol polyricinoleate; food emulsifier improving cake batter performance, cake volume; *Regulatory:* EEC compliance; ADI 0-7.5 mg/kg; *Properties:* liq.; iodine no. 72-103; hyd. no. 80-100.

Triple-H. [Hercules] Hydrolyzed veg. protein with salt; flavoring used as background taste in chicken, veal, pork, and seafood applic.; *Properties:* very lt. tan powd.; pH 5.6; 97% total solids; 21% flavor solids.

Tristat. [Tri-K Industries] Sorbic acid NF/FCC; CAS 110-44-1; EINECS 203-768-7; antimycotic food preservative; for cake batters/fillings/toppings, carbonated beverages, cheese prods., dried/fresh fruits/juice, jams/jellies, margarine, pickles, refrigerated fresh salads, salad dressing, smoked fish, syrups; *Usage level:* 0.05-0.3% (cake batters), 0.02-0.05% (dried fruit), 0.1% max. (jams, margarine, salad dressings); *Regulatory:* FDA GRAS, FCC, EEC compliance.

Tristat K. [Tri-K Industries] Potassium sorbate NF/FCC; CAS 590-00-1; EINECS 246-376-1; food preservative; antimycotic effective; effective in acid media; for cake batters/fillings/toppings, carbonated beverages, cheese prods., dried/fresh fruits and juice, jams/jellies, margarine, pickles, salad dressing, smoked fish, syrups; *Usage level:* 0.05-0.3% (cake batters), 0.02-0.05% (dried fruit), 0.1% max. (jams, margarine, salad dressings); *Regulatory:* FDA GRAS, FCC, Japan, Europe approvals; *Properties:* wh. gran. and powd.; sol. 1 g/ml water; pH 8.5-9.8 (5%); 99% min. purity; *Storage:* store in closed containers below 100 F; heat and light may cause deterioration.

Tristat SDHA. [Tri-K Industries] Sodium dehydroacetate; CAS 4418-26-2; EINECS 224-580-1; preservative for beverages, butter and other food prods.; inhibits growth of molds, yeast, gram-positive and gram-negative bacteria; *Usage level:* 0.2-0.25% (butter, cheese, margarine), 0.3-0.5% (cosmetics); *Regulatory:* FCC approved; *Properties:* wh. to lt. yel. cryst. powd., char. odor; sol. in water; m.w. 208.15; bulk dens. 0.6; m.p. 109-112 C; 98% min. assay; *Toxicology:* nontoxic; *Storage:* store in dry, dark place.

Triton® X-45. [Union Carbide; Union Carbide Europe] Octoxynol-5; CAS 9002-93-1; nonionic; surfactant, emulsifier; *Regulatory:* FDA 21CFR §172.710, 175.105, 176.210, 178.3400, EPA compliance; *Properties:* clear liq.; sol. in oil; misc. with alcohol, glycols, ethers, ketones, aromatic hydrocarbons; m.w. 426; sp.gr. 1.040; dens. 8.7 lb/gal; visc. 290 cps; HLB 10.4; cloud pt. < 0 C (1%); flash pt. > 300 F (TOC); pour pt. -26 C; surf. tens. 28 dynes/cm (0.1%); Ross-Miles foam 16 mm (0.1%, initial, 120 F); 100% act.

Triton® X-100. [Union Carbide; Union Carbide Europe] Octoxynol-9 (9-10 EO); CAS 9002-93-1; nonionic; surfactant; *Regulatory:* FDA 21CFR §172.710, 175.105, 176.210, 178.3400, EPA compliance; *Properties:* clear liq.; sol. in water, toluene, xylene, trichlorethylene, ethylene glycol, alcohols; m.w. 628; sp.gr. 1.065; dens. 8.9 lb/gal; visc. 240 cps; HLB 13.5; cloud pt. 65 C (1% aq.); flash pt. > 300 F (TOC); pour pt. 45 F; pH 6 (5% aq.); surf. tens. 30 dynes/cm (1%); Ross-Miles foam 110 mm (0.1%, 120 F); 100% act.

Triton® X-102. [Union Carbide; Union Carbide Europe] Octoxynol-13 (12-13 EO); CAS 9002-93-1; nonionic; surfactant; *Regulatory:* FDA 21CFR §172.710, 175.105, 176.210, 178.3400, EPA compliance; *Properties:* lt. color clear liq.; sol. in water, many common org. solvs.; m.w. 756; sp.gr. 1.07; dens. 8.9 lb/gal; visc. 330 cps; HLB 14.6; cloud pt. 88 C (1% aq.); flash pt. > 300 F (TOC); pour pt. 60 F; surf. tens. 32 dynes/cm (1%); Ross-Miles foam 130 mm (0.1%, 120 F); 100% act.

Triton® X-114. [Union Carbide; Union Carbide Europe] Octoxynol-8 (7-8 EO); CAS 9002-93-1; nonionic; surfactant; *Regulatory:* FDA 21CFR §172.710, 175.105, 176.210, 178.3400, EPA compliance; *Properties:* clear liq.; sol. in water, many common solvs.; m.w. 536; sp.gr. 1.05-1.06; dens. 8.8 lb/gal; visc. 260 cps; HLB 12.4; cloud pt. 25 C (1% aq.); flash pt. > 300 F (TOC); pour pt. 15 F; pH 6 (5% aq.); surf. tens. 29 dynes/cm (1%); Ross-Miles foam 25 mm (0.1%, 50 C); 100% act.

Triton® X-120. [Union Carbide; Union Carbide Europe] Octoxynol-9 (9-10 EO); CAS 9002-93-1; nonionic; surfactant; *Regulatory:* FDA 21CFR §172.710, 175.105, 176.210, 178.3400; *Properties:* wh. solid; m.w. 628; 40% conc.

Triton® X-305-70%. [Union Carbide; Union Carbide Europe] Octoxynol-30; CAS 9002-93-1; nonionic; surfactant, emulsifier; *Regulatory:* FDA 21CFR §172.710, 175.105, 176.210, 178.3400, EPA compliance; *Properties:* APHA 150 liq.; sol. in inorg. salt sol'ns., aq. min. acids; misc. with water, alcohols, glycols, ethers, ketones; m.w. 1526; sp.gr. 1.095; dens. 9.1 lb/gal; visc. 470 cps; HLB 17.3; pour pt. 35 F; cloud pt. > 100 C (1%); flash pt. > 300 F (TOC); surf. tens. 37 dynes/cm (0.1%); Ross-Miles foam 150 mm (0.1%, initial, 120 F); 70% act.

Triton® X-405-70%. [Union Carbide; Union Carbide Europe] Octoxynol-40; CAS 9002-93-1; nonionic; surfactant, emulsifier; *Regulatory:* FDA 21CFR §172.710, 175.105, 176.210, 178.3400, EPA compliance; *Properties:* APHA 250 liq.; sol. in inorg. salt sol'ns., aq. min. acids; misc. with water, alcohols, glycols, ethers, ketones; m.w. 1966; sp.gr. 1.102; dens. 9.2 lb/gal; visc. 490 cps; HLB 17.9; cloud pt. > 100 C (1%); flash pt. > 212 F (TOC); pour pt. 25 F; surf. tens. 37 dynes/cm (0.1%); Ross-Miles foam 126 mm (0.1%, initial, 120 F); 70% act.

Triton® X-705-70%. [Union Carbide; Union Carbide Europe] Octoxynol-70; CAS 9002-93-1; nonionic; emulsifier; *Regulatory:* FDA 21CFR §172.710, 175.105, EPA compliance; *Properties:* APHA 500 liq.; sol. in inorg. salt sol'ns., aq. min. acids; misc. with water, alcohols, glycols, ethers, ketones; sp.gr. 1.1; dens. 9.2 lb/gal; visc. 505 cP; HLB 18.7; pour pt. 43 F; cloud pt. > 100 C (1% aq.); flash pt. (Seta CC) > 230 F; surf. tens. 39 dynes/cm (0.1%); Ross-Miles foam 95 mm (0.1%, initial, 120 F); 70% act.

Trophy®. [Bunge Foods] Partially hydrog. soybean and cottonseed oils, water, salt, mono- and diglycerides, lecithin, artificial flavor and color, vitamin A; bakery margarine for danish, croissants, cookies; adds flavors to icings; general kitchen use; *Properties:* m.p. 102 F; 80.2% fat, 14.6% moisture, 2.6% salt.

Trough Grease. [ADM Arkady] All-vegetable trough grease that simplifies dough handling by providing a clean release from dough troughs.

Trycol® 5888. [Henkel/Emery] Steareth-20; CAS 9005-00-9; nonionic; emulsifier for fruit coatings; *Regulatory:* EPA-exempt; *Properties:* Gardner 1 solid; sol. in water, xylene; HLB 15.3; m.p. 40 C; cloud pt. 91 C (5% saline); flash pt. 460 F

Trycol® OAL-23. [Henkel/Emery] Oleth-23; CAS 9004-98-2; nonionic; emulsifier for waxes used in coating citrus fruits; *Properties:* Gardner 2 solid; m.p. 40 C; HLB 15.9; cloud pt. > 100 C; 100% act.; Unverified

Trycol® SAL-20. [Henkel/Emery] Steareth-20; CAS 9005-00-9; nonionic; emulsifier for waxes used in coating citrus fruits; *Properties:* Gardner 1 solid; HLB 15.3; m.p. 40 C; flash pt. 560 F; cloud pt. 91 C; Unverified

Trypsin 1:75. [Am. Labs] Trypsin; CAS 9002-07-7; EINECS 232-650-8; proteolytic enzyme for food and pharmaceutical applics.; used for protein digestion, in tissue culture; *Properties:* cream-colored amorphous powd., char. nonoffensive odor; *Storage:* preserve in tight containers in cool, dry place.

Trypsin 1:80. [Am. Labs] Trypsin; CAS 9002-07-7; EINECS 232-650-8; proteolytic enzyme for food and pharmaceutical applics.; used for protein digestion, in tissue culture; *Properties:* cream-colored amorphous powd., char. nonoffensive odor; *Storage:* preserve in tight containers in cool, dry place.

Trypsin 1:150. [Am. Labs] Trypsin; CAS 9002-07-7; EINECS 232-650-8; proteolytic enzyme for food and pharmaceutical applics.; used for protein digestion, in tissue culture; *Properties:* cream-colored amorphous powd., char. nonoffensive odor; *Storage:* preserve in tight containers in cool, dry place.

Turmeric DP-300. [Hilton Davis] Purified turmeric extract in propylene glycol plated on silica gel; natural colorant for dry mixes, cereals, baked goods; color str. 30% curcumin; *Properties:* yel. dry powd.

Turmeric OSO-50. [Hilton Davis] Purified turmeric milled in veg. oils; natural colorant for confectionery coatings, shortenings, frostings, cakes, pies, margarine, salad dressing; color str. 5% curcumin; *Properties:* greenish-yel.; oil-sol.

Turmeric WMPE-80. [Hilton Davis] Purified turmeric in propylene glycol emulsion; natural colorant for candies, snacks, baked goods, beverages, and frozen desserts; color str. 8% curcumin; *Properties:* greenish-yel. emulsion; water-misc.

Turmeric WMPE-150. [Hilton Davis] Purified turmeric in propylene glycol emulsion; natural colorant for candies, snacks, baked goods, and frozen desserts; color str. 1.5% curcumin; *Properties:* greenish-yel. emulsion; water-misc.

Turmeric WMSE-80. [Hilton Davis] Purified turmeric in polysorbate 80 emulsion; natural colorant for candies, snacks, baked goods, and frozen desserts; color str. 8% curcumin; *Properties:* greenish-yel. emulsion; water-misc.

TVP, Fortified TVP. [ADM Protein Spec.] Textured vegetable protein; protein for ground meat in beef patties, sausage, vegetarian foods, meatloaf mix, etc.; 18% total dietary fiber; 280 calories/100 g; *Properties:* variety of textures, sizes, colors; 50-53% protein, 9% max. moisture.

Tween® 20. [ICI Atkemix; ICI Spec. Chem.; ICI Surf. Am.; ICI Surf. Belgium] Polysorbate 20 NF; CAS 9005-64-5; nonionic; solubilizer, emulsifier, dispersant for flavors, essential and vitamin oils; food surfactant; *Regulatory:* FDA 21CFR §172.515; *Properties:* pale yel. liq.; sol. in water, methanol, ethanol, IPA, propylene glycol, ethylene glycol, cottonseed oil; sp.gr. 1.1; visc. 400 cps; HLB 16.7; flash pt. > 300 F; sapon. no. 40-50; 100% act.

Tween® 20 SD. [ICI Spec. Chem.] POE sorbitan monolaurate; anionic; flavor emulsifier, solubilizer; *Properties:* liq.; 100% act.; Unverified

Tween® 60. [ICI Atkemix; ICI Surf. Am.; ICI Surf. UK] Polysorbate 60; CAS 9005-67-8; nonionic; solubilizer, emulsifier, dispersant for flavors, vitamin oils; foaming agent for nonalcoholic beverages; improves panned coatings; defoaming for process foods; emulsion stabilizer for salad dressing; in whipped toppings, cakes, icings, puddings; *Usage level:* 1-3% (beverages), 0.2% on coating (panned coatings), 0.02-0.05% (defoaming), 0.3% (salad dressing); *Regulatory:* FDA 21CFR §172.515, 172.836, 173.340; Canada compliance; *Properties:* yel. liq.; sol. in water, at low levels in cottonseed oil; visc. 600 cps; HLB 14.9.

Tween® 60K. [ICI Spec. Chem.; ICI Surf. Am.; ICI Atkemix; ICI Surf. Belgium] Polysorbate 60 NF; CAS 9005-67-8; nonionic; food emulsifier, dough strengthener for food industry; flavor emulsifier and dispersant; foaming agent for beverages; *Regulatory:* FDA 21CFR §172.836; *Properties:* pale yel. liq.; sol. in water, IPA, ethyl alcohol; sp.gr. 1.1; visc. 600 cps; HLB 14.9; flash pt. > 300 F; sapon. no. 45-55; 100% act.

Tween® 60 VS. [ICI Am.] Polysorbate 60; CAS 9005-67-8; nonionic; veg. source base stock; emulsifier for frozen desserts, salad dressings, cake mixes; dough conditioner for bakery; *Properties:* liq.; HLB 14.9; 100% conc.

Tween® 61. [ICI Spec. Chem.; ICI Surf. Am.; ICI Surf. Belgium] Polysorbate 61; CAS 9005-67-8; nonionic; emulsifier, solubilizer for perfume, flavor, vitamin oils; wetting agent, dispersant; *Properties:* ivory waxy solid; sol. in methanol, ethanol; disp. in water; sp.gr. 1.06; HLB 9.6; flash pt. > 300 F; pour pt. 100 F; sapon. no. 95-115; 100% act.

Tween® 65. [ICI Atkemix; ICI Surf. Am.; ICI Surf. UK] Polysorbate 65; CAS 9005-71-4; nonionic; solubilizer, emulsifier for flavors, essential and vitamin oils; provides dryness and overrun in ice cream; defoaming for process foods; increases whip in sherbet; whipped toppings; *Usage level:* 0.05-0.1% (ice cream, sherbet), 0.02-0.05% (defoaming); *Regulatory:* FDA 21CFR §172.838, 173.340; Canada compliance; *Properties:* tan solid; sol. at low levels in cottonseed oil; disp. in water; HLB 10.5; pour pt. 92 F.

Tween® 65K. [ICI Spec. Chem.; ICI Surf. Am.; ICI Atkemix; ICI Surf. Belgium] Polysorbate 65, kosher grade; CAS 9005-71-4; nonionic; emulsifier for cake mixes, icings, fillings, coffee whiteners, frozen desserts, whipped toppings; antifoam; *Regulatory:* kosher; *Properties:* pale yel. waxy solid; sol. in IPA, ethyl alcohol, veg. and min. oil; sp.gr. 1.05; HLB 10.5; flash pt. > 300 F; pour pt. 92 F; sapon. no. 88-98; 100% act.

Tween® 65 VS. [ICI Am.] Polysorbate 65; CAS 9005-71-4; nonionic; emulsifier for frozen desserts, whipped toppings, cake icings; veg. source base; *Properties:* solid; HLB 10.5; 100% conc.

Tween® 80. [ICI Atkemix; ICI Surf. Am.; ICI Surf. UK] Polysorbate 80; CAS 9005-65-6; nonionic; solubilizer, emulsifier for flavors, vitamin oils; stabilizes emulsions for dietary supplements; provides dryness in ice cream; disperses flavors and colors in pickles, yeast defoamers; antifoam for beet sugar prod.; crystal size control in salt; *Usage level:* 350 mg/daily dose max. (dietary supplements), 0.02-0.1% (ice cream), 0.05% (pickles), 0.02-0.05% (defoaming), 10 ppm (salt); *Regulatory:* FDA 21CFR §172.515, 172.840; Canada compliance; *Properties:* yel. brn. liq.; sol. in water, at low levels in cottonseed oil; visc. 460 cps; HLB 15.0.

Tween® 80K. [ICI Spec. Chem.; ICI Surf. Am.; ICI Atkemix; ICI Surf. Belgium] Polysorbate 80 NF; CAS 9005-65-6; nonionic; emulsifier, solubilizer for food, vitamins, oils; *Regulatory:* FDA 21CFR §172.840; *Properties:* yel. brn. liq.; sol. in water, IPA, ethanol; sp.gr. 1.08; visc. 425 cps; HLB 15.0; flash pt. > 300 F; sapon. no. 45-55; 100% act.

Tween® 80 VS. [ICI Am.] Polysorbate 80; CAS 9005-65-6; nonionic; veg. source base stock; solubilizer for fat-sol. vitamins, dill oil; emulsifier for edible shortenings and oils, whipped toppings; *Properties:* solid; HLB 15.0; 100% conc.

Tween® 81. [ICI Spec. Chem.; ICI Surf. Am.; ICI Surf. Belgium] Polysorbate 81; CAS 9005-65-6; nonionic; emulsifier, wetting agent, dispersant, solubilizer for perfume, flavor, vitamin oils; *Properties:* yel. brn. oily liq.; sol. in min. and corn oil, dioxane, Cellosolve, methanol, ethanol, ethyl acetate, aniline; disp. in water; sp.gr. 1; visc. 450 cps; HLB 10.0; flash pt. > 300 F; sapon. no. 96-104; 100% act.

Tween® 85. [ICI Spec. Chem.; ICI Surf. Am.; ICI Surf. Belgium] Polysorbate 85; CAS 9005-70-3; nonionic; emulsifier, wetting agent, dispersant, solubilizer for perfume, flavor, vitamin oils; *Properties:* yel. brn. liq.; sol. in veg. oil, Cellosolve, lower alcohols, aromatic solvs., ethyl acetate, min. oils and spirits, acetone, dioxane, CCl_4 and ethylene glycol; disp. in water; sp.gr. 1.0; visc. 300 cps; HLB 11.0; flash pt. > 300 F; sapon. no. 80-95; 100% act.

Tween®-MOS100K. [ICI Surf. Am.] Mono- and diglycerides, polysorbate 80; surfactant for food industry; *Regulatory:* FDA 21CFR §172.840; kosher; *Properties:* ivory wh. bead; HLB 5.3; pour pt. 58 C.

1216 Fast. [Bunge Foods] Specially blended no-time dough system; can be used to convert any yeast good to a no-time dough; *Usage level:* 0.75-1% based on flour.

Type B Dough Conditioner. [Custom Ingreds.] Modified food starch, vegetable gums, salt, partially hydrog. soybean oil; dough conditioner for use in frozen flour tortillas; helps retain moisture and prevents cracking after thawing; pre-frozen tortillas will remain moist and flexible; *Usage level:* 1 lb/100 lb flour; 5% max. moisture; *Storage:* store in clean, dry area @ 40-85 F for storage life of 9 mos.

Type B Torula Dried Yeast. [Lake States] Dried yeast USP; CAS 68876-77-7; nutritional yeast; 45% min. protein; *Storage:* store in cool, dry conditions.

U

Ultra-Bake® 3100, 3200. [A.E. Staley Mfg.] Specialty blend; improves moistness, eating quality, crumb softness, improves shelf life in bakery prods., e.g., cake doughnuts, layer cakes, brownies, specialty breads, pancakes, waffles, biscuits; can replace egg whites in some applics.

Ultra-Bake® 3100K, 3200K. [A.E. Staley Mfg.] Specialty blend; improves moistness, eating quality, crumb softness, improves shelf life in bakery prods., e.g., cake doughnuts, layer cakes, brownies, specialty breads, pancakes, waffles, biscuits; can replace egg whites in some applics.; *Regulatory:* kosher.

Ultra-Bake® NF. [A.E. Staley Mfg.] Specialty blend; improves moistness, eating quality, crumb softness, improves shelf life in nonfat bakery prods.; *Regulatory:* kosher.

Ultraflex®. [Petrolite] Microcryst. wax; CAS 63231-60-7; EINECS 264-038-1; chewing gum base; *Regulatory:* incl. FDA §172.230, 172.615, 175.105, 175.300, 176.170, 176.180, 176.200, 177.1200, 178.3710, 179.45; *Properties:* amber wax; also avail. in wh.; dens. 0.93 g/cc; visc. 13 cps (99 C); m.p. 69 C; flash pt. 293 C.

Ultra Flex. [ADM Arkady] Combination dough conditioner/crumb softener esp. useful in flour tortilla and pita bread prod.

Ultra-Freeze® 400. [A.E. Staley Mfg.] Protein blend; increases creaminess, smooths texture, provides fat mimicking props., standardizes quality, reduces heat shock in hard pack frozen prods. (nonfat and lowfat frozen desserts, frozen yogurts, novelties, ice cream, ice milk); *Regulatory:* kosher.

Ultra-Freeze® 400C. [A.E. Staley Mfg.] Protein blend; increases creaminess, smooths texture, provides fat mimicking props., standardizes quality, reduces heat shock in sugar-free hard pack frozen prods.; *Regulatory:* kosher.

Ultra-Freeze® 500. [A.E. Staley Mfg.] Protein blend; provides smooth texture and creamy mouthfeel to hard and soft frozen prods., lowfat and nonfat frozen desserts, yogurts, novelties, ice milk, frozen desserts; suitable for dry mixes incl. sugar-free versions.

Ultra Fresh. [ADM Arkady] Mono- and diglycerides; emulsifier system providing shelf life extension of 3-5 days longer than conventional emulsifiers; also avail. in double strength.

Unibix AP (Acid Proof). [MLG Enterprises Ltd.] Norbixin alkaline sol'n. (water, propylene glycol, polysorbate 80); natural colorant for ice cream, yogurt, beverages, candies, fruit juice drinks, and flavors; *Regulatory:* FDA 21CFR §73.30, 73.1030, 73.2030; EEC E160b; kosher incl. Passover use; *Properties:* dk. red sol'n.; sol. in water; stable down to pH 1; 4% norbixin; *Storage:* store in cook, dark place; avoid exposure to air, heat, and light.

Unibix CUS. [MLG Enterprises Ltd.] Encapsulated sodium norbixinate; natural colorant providing yellow tint for powd. mixes as flour, dry milk, soups, spice blends, end-prods. with near neutral pH; *Regulatory:* FDA 21CFR §73.30, 73.1030, 73.2030; EEC E160b; kosher incl. Passover use; *Properties:* orange powd.; water-sol.; 1% norbixin; *Storage:* store in closed container @ 40 F; avoid exposure to air, heat, and light.

Unibix ENC (Acid Proof). [MLG Enterprises Ltd.] Encapsulated sodium norbixinate; natural colorant providing yellow tint for powd. mixes as flour, dry milk, soups, spice blends, sausage caseins, end-prods. with near neutral pH; *Regulatory:* FDA 21CFR §73.30, 73.1030, 73.2030; EEC E160b; kosher incl. Passover use; *Properties:* orange powd.; water-sol.; *Storage:* store in closed container @ 40 F; avoid exposure to air, heat, and light.

Unibix W. [MLG Enterprises Ltd.] Annatto extract; CAS 8015-67-6; natural colorant providing yellow to orange shades in margarine, cheese, cereal, pasta, ice cream, ice cream cones, milk prods., soups, sausage casings, spice blends; *Regulatory:* FDA 21CFR §73.30, 73.1030, 73.2030; EEC E160b; kosher incl. Passover use; *Properties:* orange powd.; sol. in alkaline sol'n.s @ pH 10-11; 25-40% norbixin; *Storage:* store in closed container @ 40 F; avoid exposure to air, heat, and light.

Uniguar 20. [Rhone-Poulenc Food Ingreds.] Guar gum; CAS 9000-30-0; EINECS 232-536-8; food-grade natural gum which does not require heat to develop full visc.; used for cake mixes, processed cheese, ice cream, instant soups/gravies, canned pet foods, sauces, cocoa mix, fresh tortillas; very slow hydration, superior dispersibility; *Usage level:* 0.2-0.4% (cake mixes), 0.1-0.3% (processed cheese), 0.1-0.25% (ice cream), 0.1-0.25% (instant soups/gravies), 0.15-0.25% (instant hot cereal), 0.2-0.35% (fresh tortillas), 0.1-0.3% (cocoa mix);

Properties: very coarse gran.; sol. in cold water; med. visc.

Uniguar 40. [Rhone-Poulenc Food Ingreds.] Guar gum; CAS 9000-30-0; EINECS 232-536-8; food-grade natural gum which does not require heat to develop full visc.; slow hydration, exc. dispersibility, freeze-thaw stable, pH stable (4-9); *Usage level:* 0.2-0.4% (cake mixes), 0.1-0.3% (processed cheese), 0.1-0.25% (ice cream), 0.1-0.25% (instant soups/gravies), 0.15-0.25% (instant hot cereal), 0.2-0.35% (fresh tortillas), 0.1-0.3% (cocoa mix); *Properties:* med. coarse gran.; sol. in cold water; med. visc.

Uniguar 150. [Rhone-Poulenc Food Ingreds.] Guar gum; CAS 9000-30-0; EINECS 232-536-8; food-grade natural gum which does not require heat to develop full visc.; fast hydration, good dispersibility with proper agitation, freeze-thaw stable, pH stable (4-9); *Usage level:* 0.2-0.4% (cake mixes), 0.1-0.3% (processed cheese), 0.1-0.25% (ice cream), 0.1-0.25% (instant soups/gravies), 0.15-0.25% (instant hot cereal), 0.2-0.35% (fresh tortillas), 0.1-0.3% (cocoa mix); *Properties:* med. fine gran.; sol. in cold water; med. visc.

Uniguar 200. [Rhone-Poulenc Food Ingreds.] Guar gum; CAS 9000-30-0; EINECS 232-536-8; food-grade natural gum which does not require heat to develop full visc.; very fast hydration, fair dispersibility with proper agitation, freeze-thaw stable, pH stable (4-9); *Usage level:* 0.2-0.4% (cake mixes), 0.1-0.3% (processed cheese), 0.1-0.25% (ice cream), 0.1-0.25% (instant soups/gravies), 0.15-0.25% (instant hot cereal), 0.2-0.35% (fresh tortillas), 0.1-0.3% (cocoa mix); *Properties:* fine gran.; sol. in cold water; med. visc.

Uniplex 84. [Unitex] Acetyl tributyl citrate; CAS 77-90-7; EINECS 201-067-0; plasticizer for indirect and direct food contact applics.; *Regulatory:* FDA 21CFR §172.515, 175.105, 175.300, 175.320, 175.380, 175.390, 176.170, 176.180, 177.1200, 177.1210, 178.3910, 181.27; *Properties:* APHA 30 max. color, essentially odorless; insol. in water; m.w. 402.5; sp.gr. 1.045-1.055; dens. 8.74 lb/gal; visc. 33 cps; b.p. 173 C (1 mm Hg); pour pt. -75 C; flash pt. (COC) 204 C ref. index 1.441; 99% min. assay.

Unisweet 70. [Universal Preserv-A-Chem] Sorbitol; CAS 50-70-4; EINECS 200-061-5; food additive.

Unisweet Caramel. [Universal Preserv-A-Chem] Caramel; CAS 8028-89-5; EINECS 232-435-9; food coloring.

Unisweet EVAN. [Universal Preserv-A-Chem] Ethyl vanillin; CAS 121-32-4; EINECS 204-464-7; flavoring agent.

Unisweet L. [Universal Preserv-A-Chem] Lactose; CAS 63-42-3; EINECS 200-559-2; used in foods.

Unisweet Lactose. [Universal Preserv-A-Chem] Lactose; CAS 63-42-3; EINECS 200-599-2; used in foods.

Unisweet MAN. [Universal Preserv-A-Chem] Mannitol; CAS 69-65-8; EINECS 200-711-8; food additive.

Unisweet SAC. [Universal Preserv-A-Chem] Saccharin; CAS 81-07-2; EINECS 201-321-0; noncaloric sweetener.

Unisweet SOSAC. [Universal Preserv-A-Chem] Sodium saccharin; CAS 128-44-9; EINECS 204-886-1; artificial sweetener.

Uterus Substance. [Am. Labs] Vacuum-dried defatted glandular prod.; nutritive food additive; *Properties:* powd. or freeze-dried form.

V

V-90®. [Rhone-Poulenc Food Ingreds.] Coated mono-calcium phosphate anhyd. FCC; CAS 7758-23-8; delayed-action leavening agent for baking, cereal; used in cake mixes, self-rising flour and corn meal, pancake and waffle mixes, baking powd.; *Properties:* wh. gran. free-flowing powd.; 1% max. on 140 mesh, 90% min. thru 200 mesh; m.w. 234.1; bulk dens. 66 lb/ft³; neutralizing value 83; pH 4.2 (1%); 16.8-18.3% assay (Ca).

Vanade. [Am. Ingreds.] Sorbitan stearate and polysorbate 60; emulsifier blend used in the food industry; *Properties:* lt. tan soft, plastic solid; slight propionate odor; 32% sorbitan stearate.

Vanade Kosher. [Am. Ingreds.] Hydrated sorbitan stearate, polysorbate 60; emulsifier providing improved aeration in layer cakes; *Properties:* cream plastic; pH 4.9-5.2; 58-63% volatiles.

Vanall. [Am. Ingreds.] Hydrated sorbitan stearate, monoglycerides, polysorbate 60, and propylene glycol; emulsifier providing improved aeration in sponge and layer cakes; *Properties:* cream plastic; pH 4.5-5.0; 67-70% volatiles.

Vanall (K). [Am. Ingreds.] Sorbitan stearate, mono- and diglycerides, polysorbate 60 hydrated blend with propylene glycol, lactic acid, sodium propionate preservatives; emulsifier used in the food industry for cake formulations; not recommended for egg white foams, e.g., chiffon or angel cakes; *Usage level:* 0.25-4%; *Regulatory:* FDA compliance; *Properties:* wh. creamy, soft plastic paste; iodine no. 1.5 max.; sapon. no. 45-55; pH 4.5-5.0; 70% max. volatiles.

Vanease. [Am. Ingreds.] Hydrated polysorbate 80, glyceryl tristearate, glyceryl lactyl palmitate; emulsifier for cream icings and fillings; *Properties:* cream plastic; pH 4.4-4.9; 57-62% volatiles.

Vanease (K). [Am. Ingreds.] Hydrog. polysorbate 80, glyceryl lactyl palmitate, sodium carboxymethyl cellulose, sodium propionate, glyceryl tristearate, sodium benzoate, acetic acid; hydrated emulsifier used in food industry to increase aeration, smoothness and stability of icings and fillings; *Usage level:* 3-4%; *Properties:* wh. creamy soft plastic; pH 4.4-4.9; 38-43% solids, 57-62% volatiles; *Storage:* store at 50-95 F; protect from freezing.

Vanilla No. 25450. [Universal Flavors] Vanilla; CAS 8024-06-4; natural creamy vanilla flavor for ice cream, desserts; *Usage level:* 4 g/kg (ice cream,

desserts); *Properties:* brn. liq.; water-sol.; dens. 0.94; *Storage:* 12 mos min. storage life in original sealed packing; shake before use.

Vanillin Free Flow. [Rhone-Poulenc Food Ingreds.] Vanillin NF/FCC with 0.5% silicon dioxide; flavoring with better flowability for the confectionery industry; *Regulatory:* kosher; *Properties:* wh. fine cryst., char. vanilla-like odor; m.p. 81-83 C; flash pt. (TCC) > 153 C; 99.6% min. assay.

Vani-White. [Consolidated Flavor] Vanilla subsitute, artificially flavored; used as a 1:1 replacement for vanillin, 2:1 replacement for ethyl vanillin.

Vanlite. [Am. Ingreds.] Hydrated propylene glycol stearate, monoglycerides, and stearoyl lactylic acid; cake emulsifier; *Properties:* ivory wh. plastic; pH 3.3-4.3; 74-75% volatiles, 8-9.5% mono.

Vascan. [Vaessen-Schoemaker] Antioxidants, preservatives for cooked meats, sauces, salads, snack foods.

Vascomix 404. [Vaessen-Schoemaker] Spray-dried whey protein powd., cooking salt; improves fat and moisture binding chars., texture, and reduces frying losses in ground meat prods. like meat balls, meat patties; *Usage level:* 2.5-3%; *Properties:* powd.; 23.3% protein, 34.7% total sugars, 17.5% NaCl, 4% moisture; *Storage:* store cool and dry in unopened bags.

VCR. [Viobin] Defatted wheat germ, toasted with corn syrup solids; food ingred., nutrient as cocoa extender/replacer, flavor and/or color additive, tablet filler, spices; *Properties:* dk. rich brn. very fine powd., cocoa-like flavor.

Vegafoom. [Lucas Meyer] Soya protein; CAS 68153-28-6; whipping agent for foam sweets and desserts; *Properties:* spray-dried.

Vegetable All Purpose Shortening 86-713-0. [ADM Refined Oils] Soybean and cottonseed oils; shortening for baking, frying, and cooking; *Properties:* Lovibond 2.0R max. oil, bland flavor; m.p. 118 ± 4 F; iodine no. 74-79.

Vegetable Cake Shortening 101-250. [ADM Packaged Oils] Partially hydrog. soybean and cottonseed oils, mono- and diglycerides; packaged veg. oil; *Properties:* wh. plasticized cube, bland flavor; m.p. 117 ± 3 F; iodine no. 77 ± 3.

Vegetable Heavy-Duty Shortening 86-595-0. [ADM Refined Oils] Multipurpose high stability shortening for for frying applics.; *Properties:* Lovibond 1.5R max. oil, bland flavor; m.p. 109 ± 2 F; iodine no. 70 max.

Vegetable Shortening 86-569-0. [ADM Refined Oils] High stability shortening for deep fat frying, non-dairy fat for sour cream, dips, and confection; *Properties:* Lovibond 1.5R max. oil, bland flavor; m.p. 102 ± 2 F; iodine no. 74 max.

Veg-Juice. [Florida Food Prods.] Single strength liq. vegetable juices or extracts; incl. carrot, celery, corn, cucumber, garlic, mushroom, cantaloupe, aloe vera gel; avail. in refrigerated or frozen forms; for use in soups, sauces, marinades, food bases, salad dressings, beverages, snack foods, bakery prods., flavors, etc.

Veg-Juice Garlic. [Florida Food Prods.] Freshly extracted garlic juice with added salt; used for sauces, salad dressings, meat marinades, dips, appetizers, and side dishes; *Properties:* lt. amber clear appearance, strong garlic flavor; pH 4.5-5.5; 24-30% total solids; *Storage:* keep refrigerated for best flavor; 6 mo storage life.

Veg-X-Ten®. [Rhone-Poulenc Food Ingreds.] Citric acid FCC, lactose, and monocalcium phosphate FCC; vegetable processing additive; *Regulatory:* kosher; *Properties:* wh. powd., odorless; 1% max. on 20 mesh, 40% min. thru 100 mesh; pH 2.0-2.8 (1%); *Storage:* store cool and dry.

Veltol®. [Pfizer Food Science] Maltol; CAS 118-71-8; EINECS 204-271-8; flavor/fragrance enhancer/modifier for foods and beverages, esp. in lite foods; provides creaminess and sweetness enhancement of aspartame, allowing reduced usage levels; minimizes bitterness, acid taste; antioxidant; stabilizer in wines; *Usage level:* JECFA 1 mg/kg b.w./day ADI; 250 ppm (wine), 2-30 ppm (soft drinks), 10-150 ppm (soft drink powds.), 10-40 ppm (ice cream), 30-150 ppm (gelatin desserts), 75-250 ppm (baked goods), 100-200 ppm (fudge), 10-50 ppm (cocoa), 50-100 ppm (tomato beverages); *Regulatory:* FDA 21CFR §172.515, 27CFR §24.246; FCC compliance; FEMA GRAS; *Properties:* wh. cryst. powd., sweet caramellic aroma with lesser warm, fruity notes; sol. in alcohol; sol. 1 g/82 ml water, 21 ml alcohol, 80 ml glycerin, 28 ml propylene glycol; m.w. 126.11; m.p. 160-164 C; ignition temp. (dust) 740 C; 99-100.5% assay; *Toxicology:* LD50 (oral, rat) 2330 mg/kg; may cause mild eye irritation on prolonged contact; avoid formation of dust.

Veltol®-Plus. [Pfizer Food Science] Ethyl maltol; CAS 4940-11-8; flavor/fragrance enhancer/modifier for foods and beverages, esp. in lite foods; provides creaminess and sweetness enhancement of aspartame, allowing reduced usage levels; minimizes bitterness, acid taste; antioxidant; stabilizer in wines; *Usage level:* JECFA 2 mg/kg b.w./day ADI; 100 ppm (wine), 1-10 ppm (soft drinks), 5-50 ppm (soft drink powds.), 5-10 ppm (ice cream), 10-75 ppm (gelatin desserts), 25-150 ppm (baked goods), 40-50 ppm (fudge), 5-10 ppm (cocoa), 15-25 ppm (tomato beverages); *Regulatory:* FDA 21CFR §172.515, 27CFR §24.246; FCC compliance;

FEMA GRAS; *Properties:* wh. cryst. powd., char. odor, faint fruit-like odor on dilution; sol. in alcohol; sol. 1/g in 55 ml water, 10 ml alcohol, 17 ml propylene glycol, 5 ml chloroform; m.w. 140.14; m.p. 90 C; ignition temp. (dust) 700 C; 99-100.5% assay; *Toxicology:* LD50 (oral, rat) 1150 mg/kg; may cause mild eye irritation on prolonged contact; *Precaution:* avoid formation of heavy dust concs.-forms explosive mix; *Storage:* store in tight containers.

Veri-Lo® 100. [Pfizer Food Science] O/w emulsion of food-grade materials, with 0.13% potassium sorbate FCC to insure freshness; low-temp. fat extender in foods; provides opacity, fat texture, and mouthfeel; surrounds fat particles to make a little fat seem like more in the mouth; used for lite salad dressings, mayonnaise, and sandwich spreads; *Properties:* off-wh. to wh. pourable creamy smooth liq. emulsion; sp.gr. 0.9-1.0; dens. 0.85-0.97 g/ml; b.p. 110 C; pH 3.0-4.0; 35% solids, 33% fat, 3.1 calories/g; *Toxicology:* nontoxic; may be mild eye irritant; *Precaution:* may be slip hazard when spilled; *Storage:* requires refrigerated storage to achieve adequate shelf life; protect from freezing.

Veri-Lo® 200. [Pfizer Food Science] O/w emulsion of food-grade materials; fat extender in foods; proprietary emulsions that surround fat particles to make a little fat seem like more in the mouth; used for lite salad dressings, mayonnaise, and sandwich spreads; *Properties:* wh. to off-wh. pourable soft visc. paste or gel; dens. 0.9-1.0 g/ml; b.p. 110 C; pH 3-4; *Toxicology:* nontoxic; may be mild eye irritant; *Precaution:* may be slip hazard when spilled; *Storage:* store refrigerated; do not freeze.

Veri-Lo® 300. [Pfizer Food Science] Aq. emulsion of food-grade materials; formulation aid for reduced-fat prods.; *Properties:* wh. to off-wh. pourable visc. liq., essentially odorless; dens. 0.9-1.0 g/ml; b.p. 110 C; pH 4.0-4.7; *Toxicology:* nontoxic; may be mild eye irritant; *Precaution:* may be slip hazard if spilled; *Storage:* store in refrigerated cabinet in closed container.

Versa-Whip® 500, 510, 520. [A.E. Staley Mfg.] Modified soy protein; CAS 68153-28-6; high performance whipping agent for desserts, frozen desserts, confections, beverages, baked goods, marshmallows, icings; provides aeration, emulsification and/or texture modification; *Properties:* extremely bland flavor.

Versa-Whip® 520K, 600K, 620K. [A.E. Staley Mfg.] Modified soy protein; CAS 68153-28-6; high performance whipping agent for desserts, frozen desserts, confections, beverages, baked goods, marshmallows, icings; provides aeration, emulsification and/or texture modification; *Regulatory:* kosher; *Properties:* extremely bland flavor.

Versene CA. [Dow] Calcium-disodium EDTA; CAS 62-33-9; EINECS 200-529-9; chelating agent; high purity direct food additive preventing metal catalyzed oxidative breakdown; *Regulatory:* FCC, USP, and kosher compliance; *Properties:* wh. powd.; m.w. 410; dens. 40 lb/ft³; pH 6.5-7.5 (1%); 97-102%

act.

Versene NA. [Dow] Disodium EDTA dihydrate USP, FCC; CAS 139-33-3; EINECS 205-358-3; chelating agent; high purity direct food additive preventing metal catalyzed oxidative breakdown; *Regulatory:* FCC, USP, and Kosher compliance; *Properties:* wh. cryst.; m.w. 372; dens. 67 lb/ft³; pH 4.3-4.7 (1%); 99% act.

Verv®. [Am. Ingreds.] Calcium stearoyl-2-lactylate; CAS 5793-94-2; EINECS 227-335-7; anionic; starch and protein complexing agent, softener for use in yeast-leavened bakery prods.; conditioning agent in dehydrated potatoes; *Usage level:* 4-8 oz/cwt flour; 0.3-0.5%; *Regulatory:* FDA §172.844; *Properties:* cream powd., mild caramel odor, sl. tart taste; acid no. 50-86; ester no. 124-164; 100% conc., 4.2-5.2% calcium; *Storage:* store under cool dry conditions, below 90 F.

Victor®. [Bunge Foods] Partially hydrog. soybean and cottonseed oils, water, salt, mono- and diglycerides, EDTA, artificial flavor, vitamin A; bakery margarine for danish, croissants, cookies; *Properties:* m.p. 112 F; 80.2% fat, 16.8% moisture, 3.0% salt.

Victor Cream®. [Rhone-Poulenc Food Ingreds.] Sodium acid pyrophosphate, leavening, FCC; CAS 7758-16-9; EINECS 231-835-0; leavening agent for baking, cereals; used in mixes for institutional, bakery, and household apaplics.; *Properties:* wh. powd.; 100% thru 60 mesh, 99% thru 200 mesh; sol. 13 g/100 g saturated sol'n.; m.w. 221.94; bulk dens. 68 lb/ft³; pH 4.3 (1%).

Victory®. [Petrolite] Microcryst. wax; CAS 63231-60-7; EINECS 264-038-1; plastic wax offering high ductility, flexibility at very low temps.; as chewing gum base; *Regulatory:* incl. FDA §172.230, 172.615, 175.105, 175.300, 176.170, 176.180, 176.200, 177.1200, 178.3710, 179.45; *Properties:* wh., amber wax; dens. 0.93 g/cc; visc. 13 cps (99 C); flash pt. 293 C.

Vifcoll CCN-40, CCN-40 Powd. [Nikko Chem. Co. Ltd.] N-Cocoyl collagen peptide, sodium salt; food emulsifier; *Properties:* pale yel. liq. and wh. or pale yel. powd. resp.; m.w. 600; pH 5.5-7.5 (both, 3% aq. for powd.); 30% act. in water (CCN-40); Unverified

Viobin Defatted Wheat Germ #1. [Viobin] Defatted wheat germ; food ingred., nutrient, flavor for cereals, breads, muffins, biscuits, cakes, cookies, bagels, donuts, granola bars, crackers, toppings; *Properties:* lt. tan. gran. with dk. tan speckling, flavor of natural wheat germ.

Viobin Defatted Wheat Germ #2. [Viobin] Defatted wheat germ, toasted with corn syrup solids to obtain a rich med. brn. color; food ingred., nutrient, flavor for breads, muffins, cakes, cookies, bagels, donuts; *Properties:* brn. gran., toasty mocha-like flavor and aroma.

Viobin Defatted Wheat Germ #3. [Viobin] Defatted wheat germ; food ingred., nutrient, flavor for batter mixes, flour blends for breads, pastas, crackers, pretzels, etc., filler for meat prods., sauces, seasoning carrier, tablet filler; *Properties:* lt. tan flour, flavor of natural wheat germ.

Viobin Defatted Wheat Germ #5. [Viobin] Defatted wheat germ; food ingred., nutrient, flavor processed in particle size and color to simulate various gran. spices, e.g., spices, cinnamon; as extenders for spices for baking and cereal.

Viobin Defatted Wheat Germ #6. [Viobin] Defatted wheat germ; food ingred., nutrient, flavor for cereals, breads, muffins, biscuits, cakes, cookies, bagels, donuts, granola bars, crackers, toppings; *Properties:* toasted fine gran., flavor of natural wheat germ; 100% thru 60 mesh.

Viobin Defatted Wheat Germ #8. [Viobin] Defatted wheat germ, toasted; food ingred., nutrient, flavor for crackers, breads, muffins, biscuits, cakes, cookies, bagels, donuts, granola bars, toppings; *Properties:* med. tan gran. with dk. tan speckling, toasty mild lingering nut-like taste.

Viobin Defatted Wheat Germ #9. [Viobin] Defatted wheat germ, toasted; food ingred., nutrient, flavor; *Properties:* med. brn. flour, toasty mild lingering nut-like taste.

Viobin Defatted Wheat Germ #34. [Viobin] Defatted wheat germ, toasted and untoasted; food ingred., nutrient, flavor for cereals, breads, muffins, biscuits, cakes, cookies, bagels, donuts, granola bars, crackers, toppings; *Properties:* small, consistent flake-like granules, sl. toasty flavor of natural wheat grain with mild lingering nut-like taste.

Viobin Octacosanol. [Viobin] Octacosanol; CAS 557-61-9; high potency direct compression prod., suitable for tableting/capsules/powd. mixes.

Viobin Pork Liver Fat. [Viobin] Animal fat derived from high quality porcine livers for flavoring and palatability; *Properties:* dk. colored.

Viobin Wheat Germ Oil. [Viobin] Wheat germ oil; CAS 8006-95-9; natural source of vitamin E, nutritional supplement; substitute for heavily processed veg. oils in salad oil, batter mixes, cake mixes, etc.; *Properties:* oil.

Viscarin® GP-209. [FMC] Fat replacement system for cheese sauce; provides cling, creaminess, hot visc., and mouthfeel.

Viscarin® ME-389. [FMC] Fat replacement system for beef patties, pork sausage; provides lubricity, juiciness, tenderness, freeze/thaw stability.

Viscarin® SD-389. [FMC] Fat replacement system for spoonable mayonnaise providing creamy mouthfeel, body structure, rheology, and opacity; in gravies, provides creaminess, cling, hot visc. and mouthfeel.

Viscarin® XP-3160. [FMC] Fat replacement system for pork sausage; provides juiciness, tenderness.

Visco Compound™. [French's Ingreds.] Mustard bran; viscosifier, emulsifying and suspending agent for sauces, dressings, lowfat and fat-free applics., meat sauces, mustards, catsup; fat replacer; provides opacity for thick, rich appearance; imparts cling; high acid stability; tolerant to high heat; *Usage level:* 1-1.5% (salad dressings); *Properties:* disp. in water.

Vis*Quick™ GX 11. [Zumbro; Garuda Int'l.] Xanthan gum, guar gum; patented thickening agent for liq.

and dry mix food prods.; *Properties:* free-flowing powd.; < 25% above 40 mesh, < 10% below 200 mesh; sol. in cold water; bulk dens. 0.22-0.32 g/cc; 14% max. moisture; *Storage:* store in cool, dry area.

Vis*Quick™ GX 21. [Zumbro; Garuda Int'l.] Guar gum, xanthan gum; patented thickening agent for liq. and dry mix food prods.; *Properties:* free-flowing powd.; < 25% above 40 mesh, < 10% below 200 mesh; sol. in cold water; bulk dens. 0.22-0.32 g/cc; 14% max. moisture; *Storage:* store in cool, dry area.

Vita Choice. [Brolite Prods.] Dough accelerant and dough strengthener for use on automatic lines.

Vitalex A. [ADM Arkady] System of dough conditioners improving quality and replacing a portion or all of gluten in bread, buns, rolls, and variety breads; also avail. in double strength.

Vitalex P. [ADM Arkady] Synergistic blend of dough improvers incl. sodium stearoyl lactylate and DATEM esters; dough conditioner.

Vitamin A Palmitate Type 250-CWS No. 65312. [Roche] Vitamin A palmitate USP FCC compded. with acacia, sugar, modified food starch, and BHT, BHA, and dl-α-tocopherol (antioxidants); for use in dry food and pharmaceutical prods. that are to be reconstituted with liqs.; fortifies nonfat dry milk, mellorine, dehydrated foods, dry cereals, beverage powds.; not suitable where clarity of liqs. is essential or for tablet mfg.; *Properties:* lt. yel. spherical dry free-flowing beadlets, nearly odorless; 90% min. thru 30 mesh, 25% max. thru 80 mesh; disp. in cold water, fruit juices, milk, infant formulas; 250,000 IU/ g assay; *Toxicology:* vitamin A: sustained daily intakes exceeding 50,000 IU (adults), 20,000 IU (infants) may cause toxic effects (headache, vomiting, liver damage); US RDA 8000 IU (pregnant/ lactating women); *Storage:* store in cool, dry place.

Vitamin A Palmitate Type PIMO/BH No. 638280100. [Roche] Vitamin A palmitate, 1,000,000 IU/g, stabilized with BHA/BHT; CAS 79-81-2; EINECS 201-228-5; in aq. disps., oleaginous preps., capsulated prods. in food applics.; *Properties:* lt. yel. to amber clear liq., sl. char. odor, bland oily taste; misc. with edible oils and fats; sol. in ether, hydrocarbons, chlorinated hydrocarbons; sl. sol. in alcohol; insol. in water; m.w. 524.9; sp.gr. 0.90-0.95; 1,000,000 min. units vitamin A/g; *Toxicology:* Vitamin A: excessive daily intakes (> 50,000 IU in adults, 20,000 IU in infants) may cause toxic effects (headache, vomiting, liver damage); US RDA 8000 IU (pregnant/lactating women); *Storage:* store @ 70 F; refrigerate if stored for several mos, then hold @ R.T. for 24 h before use.

Vitamin A Palmitate USP, FCC Type P1.7 No. 262090000. [Roche] Vitamin A palmitate; CAS 79-81-2; EINECS 201-228-5; used when max. vitamin A conc. required in soft gelatin capsules, multivitamins, to fortify foods (margarine), liq. infant formulas; *Properties:* lt. yel. to amber clear liq., char. odor, bland oily taste; misc. with edible oils and fats; sol. in hydrocarbons and chlorinated hydrocarbons; sl.

sol. in alcohol; insol. in water; sp.gr.0.90-0.95; m.w. 524.9; 1,600,000 IU/g assay; *Toxicology:* vitamin A: sustained daily intakes exceeding 50,000 IU (adults), 20,000 IU (infants) may cause toxic effects (headache, vomiting, liver damage); US RDA 8000 IU (pregnant/lactating women); *Storage:* sealed under inert gas; store in cool place under refrigeration; contents may cryst.; warm to 40 C before use.

Vitamin A Palmitate USP, FCC Type P1.7/BHT No. 63693. [Roche] Vitamin A palmitate USP FCC, stabilized with BHT; used where max. vitamin A required for soft gelatin capsules, multivitamins, to fortify foods (margarine), liq. infant formulas; *Properties:* lt. yel. to amber clear liq., char. odor, bland oily taste; misc. with edible oils and fats; sol. in hydrocarbons, chlorinated hydrocarbons; sl. sol. in alcohol; insol. in water; m.w. 524.9; sp.gr. 0.90-0.95; 1,600,000 IU/g min. assay; *Toxicology:* vitamin A: sustained daily intakes exceeding 50,000 IU (adults), 20,000 IU (infants) may cause toxic effects (headache, vomiting, liver damage); US RDA 8000 IU (pregnant/lactating women); *Storage:* sealed under inert gas; store in cool place under refrigeration; contents may cryst.; warm to 40 C before use.

Vitamin A Palmitate USP, FCC Type P1.7/E No. 63699. [Roche] Vitamin A palmitate USP-FCC; CAS 79-81-2; EINECS 201-228-5; used when max. vitamin A conc. required in soft gelatin capsules, multivitamins, to fortify foods (margarine), liq. infant formulas; *Properties:* greenish-yel. to golden-yel. oily liq., char. odor; misc. with edible oils and fats; sol. in hydrocarbons, chlorinated hydrocarbons; sl. sol. in alcohol; insol. in water; m.w. 524.9; sp.gr. 0.90-0.95; 1,600,000 IU/g min.; *Toxicology:* sustained daily intakes exceeding 50,000 IU (adults), 20,000 IU (infants) may cause toxic effects (headache, vomiting, liver damage); US RDA 8000 IU (pregnant/lactating women); *Storage:* store in cool place under refrigeration; contents may crystallize; warm to 40 C to liquefy before use.

Vitamin B₁₂ 0.1% SD No. 65354. [Roche] Cyanocobalamin stabilized in a matrix of modified food starch with sodium citrate, citric acid, preservatives (sodium benzoate, sorbic acid), silicon dioxide; easy blending and distribution in foods and pharmaceuticals, premixes, dry prods., liq. mixts. or suspensions; *Properties:* pink fine powd.; sl. hygroscopic; 98% min. thru 40 mesh; disperses in cold or warm water; insol. in org. solvs.; *Storage:* store in cool dry place in tightly closed container, optimally @ 46-59 C; avoid excessive heat.

Vitamin B₁₂ 1% Trituration No. 69992. [Roche] Cyanocobalamin USP with carrier (calcium phosphate dibasic); used in dry formulations, pharmaceutical tablets/capsules and food premixes where water sol. is not critical and reducing agents or moisture are not present; *Properties:* lt. pink fine powd.; 99% min. thru 100 mesh; insol. in water and org. solvs.; m.w. 1355.38; *Storage:* store in cool dry place in tightly closed container, optimally @ 46-59 C.

Vitamin B₁₂ 1% Trituration No. 69993. [Roche]

Cyanocobalamin USP with mannitol carrier; source of vitamin B$_{12}$ in liq. and solid pharmaceuticals and food preps.; *Properties:* lt. pink fine powd.; 99.5% min. thru 100 mesh; sol. 1 g/10 ml water (clear red sol'n.); m.w. 1355.38; *Storage:* store in cool dry place in tightly closed container, optimally @ 46-59 C.

Vitamin B$_{12}$ 1.0% SD No. 65305. [Roche] Cyanocobalamin stabilized in a matrix of modified food starch with sodium citrate, citric acid, preservatives (sodium benzoate, sorbic acid), silicon dioxide; easy blending and distribution in foods and pharmaceuticals, premixes, dry prods., liq. mixts. or suspensions; *Properties:* pink fine powd.; sl. hygroscopic; 98% min. thru 40 mesh; disperses in cold or warm water; insol. in org. solvs.; *Storage:* store in cool dry place in tightly closed container, optimally @ 46-59 C; avoid excessive heat.

Vitamin E USP, FCC No. 60525. [Roche] dl-α-Tocopheryl acetate USP, FCC; CAS 1406-70-8; EINECS 231-710-0; vitamin E source for foods and pharmaceutical capsules and liqs., when precautions are taken to protect the formulation from oxygen; antioxidant in oil-based systems; *Regulatory:* FDA GRAS; *Properties:* yel. to amber clear visc. liq., pract. odorless; freely sol. in alcohol; misc. with ether, chloroform, acetone, veg. oils; insol. in water; m.w. 430.72; 96-102% assay; *Precaution:* unstable to air and light; degradation accelerated by ferric salts, cupric salts, silver salts, and rancid fats; *Storage:* store @ 59-86 F in tight, light-resist. containers.

Vitamin E USP, FCC No. 60526. [Roche] dl-α-Tocopheryl acetate USP, FCC; CAS 1406-70-8; EINECS 231-710-0; vitamin E source for foods and pharmaceuticals, tablets, capsules, and liqs.; *Regulatory:* FDA GRAS; *Properties:* yel. clear visc. oil, pract. odorless; may solidify at cold temps.; freely sol. in alcohol; misc. with ether, chloroform, acetone, veg. oils; insol. in water; m.w. 472.76; 96-102% assay; *Storage:* store @ 59-86 F in tight, light-resist. containers.

Vita Plus White Culture 6428. [Brolite Prods.] Natural dough conditioner and mineral yeast food; cultured conditioner that works at the mixer, during make-up, in the oven, and on the shelf; for all yeast-raised baked goods.

Vita-Rite A. [Consolidated Flavor] Vitamin A, oil-based; 1 ml provides 100,000 IU of vitamin A.

Vita-Rite A & D. [Consolidated Flavor] Vitamin A and vitamin D$_3$, oil-based; 1 ml provides 200,000 IU of vitamin A and 40,000 IU of vitamin D$_3$.

Vita-Rite A & D-2%. [Consolidated Flavor] Vitamin A and vitamin D$_3$, oil-based; 1 ml provides 140,000 IU of vitamin A and 40,000 IU of vitamin D$_3$.

Vita-Rite A & D H-W. [Consolidated Flavor] Vitamin A and vitamin D$_3$; high potency formulation; 1 ml provides 80,000 IU of vitamin A and 16,000 IU of vitamin D$_3$; *Properties:* water-sol.

Vita-Rite A & D-W. [Consolidated Flavor] Vitamin A and vitamin D$_3$; 1 ml provides 20,000 IU of vitamin A and 4000 IU of vitamin D$_3$; *Properties:* water-sol.

Vita-Rite A2. [Consolidated Flavor] Vitamin A, oil-based; 1 ml provides 200,000 IU of vitamin A.

Vita-Rite A-W. [Consolidated Flavor] Vitamin A; 1 ml provides 20,000 IU of vitamin A; *Properties:* water-sol.

Vita-Rite D3. [Consolidated Flavor] Vitamin D$_3$; 1 ml provides 200,000 IU of vitamin D$_3$; *Properties:* water-sol.

Vitiben®. [Cimbar Perf. Minerals] Natural bentonite; CAS 1302-78-9; EINECS 215-108-5; clarifying agent, heat stabilizer, and processing aid for wine and juice production; adsorbs and coagulates undesirable substances so they can be filtered out during fining or clarification; superior dispersibility; *Usage level:* 1-10 lb/1000 gal; *Regulatory:* FDA §184.1155; *Properties:* 83-87% thru 20 mesh, 29-33% thru 30 mesh; sp.gr. 2.8; bulk dens. 70 lb/ft^3; pH 9.6 (6% susp.); 63.59% SiO$_2$, 21.43% Al$_2$O$_3$; 7-9% moisture; *Storage:* store in tightly closed package.

Vitinc® Cholesterol NF XVII. [Vitamins, Inc.] Cholesterol NF; CAS 57-88-5; EINECS 200-353-2; emulsifying agent; *Properties:* m.p. 147-150.

Vitinc® Defatted Wheat Germ Nuggets. [Vitamins, Inc.] Defatted wheat germ; natural rich source of high quality protein, vitamins, and minerals; provides crunchy texture, mild nutty flavor, appealing bite; used in crackers, cookies, cakes, confections, nutriton snack foods, other bakery prods.; nut replacement.

Vitinc® Defatted Wheat Germ. [Vitamins, Inc.] Defatted wheat germ; provides rich dietary fiber, complex carbohydrates, high quality protein; improves mouthfeel and texture; low calories, no sodium or cholesterol; *Properties:* nuggets, gran., or flour.

Vitinc® Sesame Oil NF. [Vitamins, Inc.] Sesame oil; CAS 8008-74-0; EINECS 232-370-6; *Properties:* sp.gr. 0.916-0.921; iodine no. 103-116; sapon. no. 188-195; *Storage:* store in light-resist. containers; avoid exposure to heat.

Vitinc® dl-alpha Tocopheryl Acetate USP XXII. [Vitamins, Inc.] dl-α-Tocopheryl acetate USP; CAS 1406-70-8; EINECS 231-710-0; for food use.

Vitinc® Wheat Germ Oil. [Vitamins, Inc.] Wheat germ oil, cold processed; CAS 8006-95-9; source of vitamin E, polyunsaturated fatty acids, and omega-3 fatty acids, with zero cholesterol; *Properties:* sp.gr. 0.92; iodine no. 128; sapon. no. 186; ref. index 1.4753.

Vito®. [Bunge Foods] Meat fats, vegetable oil, water, salt, mono- and diglycerides, antioxidant, artificial flavor, Vitamin A; margarine for puff pastry; *Properties:* drop pt. 120 F; 80.2% fat, 18% moisture, 1.8% salt.

Vitrafos®. [Rhone-Poulenc Food Ingreds.] Sodium hexametaphosphate FCC; CAS 68915-31-1; EINECS 233-343-1; food additive for baking, cereals, meat, poultry, dairy/cheese, processed foods; in blends to process meats, poultry, seafood; tenderizing of vegetables; cheese emulsification; *Regulatory:* kosher; *Properties:* wh. amorphous prod. (crushed, powd., glass), odorless; 1% max.

on 20 mesh (crushed), 5% max. on 60 mesh (powd.); m.w. 1286; pH 6.7-7.2 (1%); 60-71% assay (as P_2O_5); *Storage:* store cool and dry.

Vitrafos® LC. [Rhone-Poulenc Food Ingreds.] Sodium polyphosphates, glassy, long chain, FCC; CAS 68915-31-1; EINECS 233-343-1; sequestrant and preservative adjunct for use in dairy, beverage, and miscellaneous foods; *Regulatory:* kosher; *Properties:* wh. free-flowing amorphous prod., crushed, powd., or glass, odorless; highly sol. in water; bulk dens. 75 lb/ft³ (loose, crushed or glass), 45-60 lb/ft³ (loose, powd.); pH 5.7-6.6 (1%); 68-71% assay (as P_2O_5); *Storage:* store cool and dry.

Vi-Vax. [Vitamins, Inc.] Vitinc® defatted wheat germ; natural source of high quality protein, 20% dietary fiber, and nutritionally significant amounts of certain vitamins and minerals; nutritional ingred. in baked prods., breakfast cereals, snack and dietetic foods; *Properties:* gran., flour, and extruded forms.

Vrest™. [Bunge Foods] Mono- and diglycerides, antioxidants; emulsifier for icing, cakes, mixes, margarine; *Properties:* 2.4 red color; iodine no. 75; 44% mono content.

Vrest Plus™. [Bunge Foods] Mono- and diglycerides, antioxidants; emulsifier for icing, cakes, mixes, margarine; *Properties:* 2.5 red color; iodine no. 75; 52% mono content.

V-V Vanilla. [ADM Arkady] Conc. artificial vanilla flavoring; *Properties:* dry powd.

Vykacet T/L. [Croda Food Prods. Ltd.] Acetylated mono- and diglycerides; nonionic; release agent in sugar confectionery; protective coating for dried fruit; glazing agent; *Properties:* liq.; 100% conc.

Vykamol 83G. [Croda Food Prods. Ltd.] Sorbitan stearate (Crill 3) and polysorbate 60 (Crillet 3); nonionic; emulsifier for cake mixes (improves hydration, volume, and shelf life), confectionery coatings (antibloom agent), icings/toppings (improves emulsification, volume), pastry fat; emulsion stabilizer in calf starters, syn. creams, whipped desserts; *Usage level:* 0.5-0.9% (cakes), 1% (chocolate), 0.5-0.6% (icings), 0.4-0.6% (cream), 0.2-0.4% (whipped dessert); *Regulatory:* UK clearance; *Properties:* solid; m.p. 50-55 C; HLB 6.2; acid no. 3-7; sapon. no. 130-135; 100% conc.; *Storage:* store under cool, dry conditions.

Vykasoid. [Croda Chem. Ltd.] Monoglyceride and veg. oil; nonionic; w/o emulsifier for dairy spreads; *Properties:* solid; 100% conc.; Unverified

W

W-300 FG. [Multi-Kem] Sodium alginate; CAS 9005-38-3; offers high rate of reactivity with calcium salts to form firm heat-stable gels in food ingred. applics.; *Properties:* tan to gray powd., sl. char. aroma; cold-water sol. forming a visc. sol'n.; visc. 300+ cps (1%); pH 6-7; 15% max. moisture; *Storage:* store away from moisture, humidity, and high temps.

Wacker Silicone Antifoam Emulsion SE 9. [Wacker-Chemie GmbH; Wacker Silicones] Simethicone; CAS 8050-81-5; antifoam, processing aid for foods, fermentation; *Usage level:* 0.02-0.5%; *Regulatory:* FDA §173.340, BGA II, VII; *Properties:* milky wh. med. visc. o/w emulsion; sp.gr. 1.0; pH 6-8; 15-16% solids in water.

Wakal® A 601. [Grünau GmbH] Calcium alginate, tetrasodium pyrophosphate, and disodium phosphate; thickener, stabilizer, gellant in food systems, esp. bakery creams, whipped desserts and toppings; produces bakery creams with freeze/thaw stability, heat stability, and sliceability; *Regulatory:* FAO/WHO, FCC, EEC compliance; *Properties:* wh.-ylsh. free-flowing powd., neutral odor and taste; sol. in cold water or milk; 99% < 120 μm; bulk dens. 650-850 g/l; visc. 200-250 mPa•s (1% aq.); pH 7.5-9.0 (1% aq.); *Storage:* store cool, dry in tightly closed containers (15-25 C); protect from light; 6 mo storage life in original pkg.

Wakal® A Range. [Grünau GmbH] Alginates; gelling agents for desserts and filling creams.

Wakal® J. [Grünau GmbH] Locust bean gum; CAS 9000-40-2; EINECS 232-541-5; nonionic; thickener, stabilizer for frozen desserts, soft drinks, salad dressings, condiments, meat/fish preserves, ready meals, puddings, cheese prods., baked goods; improves consistency; high water-binding capacity and swelling power; ice crystal inhibitor; *Regulatory:* FAO/WHO, EEC E410, FDA 21CFR §184.1343, GRAS; *Properties:* lt. beige free-flowing powd., neutral odor and taste; 75% on 0.1 mm sieve, 100% on 0.2 mm sieve; disp. in cold water; bulk dens. 500 g/l; visc. 200 mPa•s (1% aq., cold hydrated); 75% act.; *Storage:* 1 yr shelf life kept in closed original pkg. under cool, dry conditions.

Wakal® JG. [Grünau GmbH] Locust bean gum; CAS 9000-40-2; EINECS 232-541-5; thickener, stabilizer for frozen desserts, soft drinks, salad dressings, condiments, meat/fish preserves, ready meals, puddings, cheese prods., baked goods; improves consistency; high water-binding capacity and swelling power; ice crystal inhibitor; *Regulatory:* FAO/WHO, EEC E410, FDA 21CFR §184.1343, GRAS; *Properties:* lt. beige fine gran., neutral odor and taste; 70% on 0.1 mm sieve, 90% on 0.2 mm sieve; disp. in cold water; bulk dens. 550 g/l; visc. 200 mPa•s max. (1% aq., cold hydrated); 60% min. act.; *Storage:* 1 yr shelf life if kept in closed original pkg. under cool, dry conditions.

Wakal® K Range. [Grünau GmbH] Carrageenan; CAS 9000-07-1; EINECS 232-524-2; gelling agents for desserts and filling creams.

Wakal® K-e. [Grünau GmbH] Carrageenan; CAS 9000-07-1; EINECS 232-524-2; thickener, stabilizer for food systems, esp. desserts, whipped creams, mousses; improves mouthfeel; *Regulatory:* FAO/WHO, EEC E407, FDA 21CFR §172.620; *Properties:* beige-wh. powd., neutral odor and taste; 95% < 0.1 mm, 100% < 0.2 mm; sol. in cold water or milk; visc. 300-400 mPa•s (1% aq.); pH 7.0-9.5 (1% aq.); 12% max. water.

WaTox-20 P Powd. [MLG Enterprises Ltd.] Calcium sulfate and azodicarbonamide; fast-acting org. oxidizing agent for yeast-raised bakery prods., dry mixes and bases; provides exc. machine tolerance and dough str. in mixing/proof/oven stages; improves volume, grain, texture; *Usage level:* 5-25 ppm/cwt flour; *Properties:* free-flowing powd.; 1 oz supplies 20 ppm azodicarbonamide/cwt flour; *Storage:* 6 mos storage life under cool, dry conditions.

WaTox-20 Tablets. [MLG Enterprises Ltd.] Azodicarbonamide and other edible excipients; CAS 123-77-3; EINECS 204-650-8; organic oxidizer for use in yeast-raised bakery prods.; provides exc. machine tolerance and dough strength in mixing/proof/oven stages; improves volume, grain, texture; *Usage level:* 5-25 ppm/cwt flour; *Properties:* tablet supplying 20 ppm azodicarbonamide/cwt flour; *Storage:* 6 mos storage life under cool, dry conditions.

WaTox 60 Powd. [MLG Enterprises Ltd.] Calcium sulfate, azodicarbonamide, and calcium phosphate; dough conditioner; *Regulatory:* FDA GRAS; *Properties:* yel. free-flowing powd., odorless; *Storage:* 6 mos storage life under cool, dry conditions.

Wecobee® FS. [Stepan Food Ingreds.] Hydrog. veg. oil; CAS 68334-28-1; EINECS 269-820-6; cocoa butter replacement, emollient used for foods (salad dressing, snack dips, confection coatings); exc. mold release; *Regulatory:* FDA 21CFR §170.30, GRAS; avail. in kosher grade; *Properties:* wh. to off-wh. soft solid, bland odor; m.p. 39 C; iodine no. 3; sapon. no. 240; 0-1% moisture; *Storage:* avoid prolonged storage above 90 F.

Wecobee® FW. [Stepan Food Ingreds.] Hydrog. veg. oil; CAS 68334-28-1; EINECS 269-820-6; cocoa butter replacement, emollient used for foods (whipped toppings, salad dressing, snack dips, confection coatings); exc. mold release; *Regulatory:* FDA 21CFR §170.30, GRAS; avail. in kosher grade; *Properties:* wh. to off-wh. soft solid, bland odor; m.p. 38 C; iodine no. 3; sapon. no. 240; 0.1% moisture; Discontinued

Wecobee® HTR. [Stepan Food Ingreds.] Hydrog. veg. oil; CAS 68938-37-4; EINECS 269-820-6; cocoa butter replacement with exc. mold release chars.; *Properties:* wh. to off-wh. oil, bland odor; m.p. 36 C; iodine no. 1.2; 0.03% moisture; *Storage:* avoid prolonged storage above 90 F.

Wecobee® M. [Stepan Food Ingreds.; Stepan Europe] Hydrog. veg. oil; CAS 68938-37-4; EINECS 269-820-6; cocoa butter replacment with exc. mold release chars.; for food industry (coffee whiteners, whipped toppings, salad dressing, snack dips, confection coatings); *Regulatory:* FDA 21CFR §170.30, GRAS; avail. in kosher grade; *Properties:* wh. to off-wh. soft solid, bland odor; visc. 13.5 cps (150 F); m.p. 35 C; acid no. 0.20 max.; iodine no. 3; sapon. value 242; 0.1% moisture; *Storage:* avoid prolonged storage above 90 F.

Wecobee® R Mono. [Stepan Food Ingreds.] Hydrog. veg. oil; CAS 68334-28-1; EINECS 269-820-6; cocoa butter replacement with exc. mold release chars.; *Properties:* wh. to off-wh. oil, bland odor; m.p. 35 C; iodine no. 3; sapon. no. 189; 45% monoester; *Storage:* avoid prolonged storage above 90 F.

Wecobee® S. [Stepan Food Ingreds.; Stepan Europe] Hydrog. veg. oil; CAS 68334-28-1; EINECS 269-820-6; cocoa butter replacment with exc. mold release chars.; for food industry (coffee whiteners, whipped toppings, salad dressing, snack dips, confection coatings); *Regulatory:* FDA 21CFR §170.30, GRAS; avail. in kosher grade; *Properties:* ivory or sl. yel. flakes, bland odor; visc. 14.0 cps (150 F); m.p. 46 C; acid no. 0.20 max.; iodine no. 3; sapon. value 238; 0.1% moisture; *Storage:* avoid prolonged storage above 90 F.

Wecobee® SS. [Stepan Food Ingreds.] Hydrog. veg. oil; CAS 68334-28-1; EINECS 269-820-6; cocoa butter replacement, emollient used for foods (confection coatings); exc. mold release; *Regulatory:* FDA 21CFR §170.30, GRAS; avail. in kosher grade; *Properties:* ivory to sl. yel. flakes, bland odor; m.p. 47 C; iodine no. 3; sapon. no. 237; 0.1% max. moisture; *Storage:* avoid prolonged storage above 90 F.

Wecobee® W. [Stepan Food Ingreds.] Hydrog. veg. oil; CAS 68334-28-1; EINECS 269-820-6; cocoa butter replacement, emollient used for foods (cof-

fee whiteners, whipped toppings, salad dressing, snack dips), pharmaceuticals (suppositories), cosmetics (cream/ointment base); *Regulatory:* FDA 21CFR §170.30, GRAS; avail. in kosher grade; *Properties:* wh. to off-wh. soft solid, bland odor; m.p. 33-36 C; iodine no. 4 max.; sapon. no. 240-254; 0.1% max. moisture; Discontinued

Wecotop 1013. [Stepan Food Ingreds.] Soya and palm oil veg. fat; food emulsifier; *Properties:* Lovibond 20/2.0 solid; m.p. 120-125 F; Unverified

Wheat-Eze. [MLG Enterprises Ltd.] Active fungal enzyme deriv. of *Aspergillus oryzae* (protease) and other edible excipients; CAS 9014-01-1; EINECS 232-752-2; dough conditioner; reduces mixing requirements, improves machinability, gas retaining props., crust color, pan flow; for bread, buns, rolls, sweet goods; *Usage level:* $1/_2$-2 tablets/cwt flour; *Properties:* tablet; *Storage:* 6 mos storage life under cool, dry conditions.

Wheat Germ Oil. [Natural Oils Int'l.; Tri-K Industries] Wheat germ oil; CAS 8006-95-9; Unrefined grade for food applics. where natural vitamin E is required; Refined grade for vitamin E and natural antioxidant for formulations and other veg. oils; Expeller grade for gourmet food applics., natural vitamin E; *Properties:* golden-dk. reddish brn. clear oil, sl. nutty odor and taste; insol. in water; sp.gr. 0.93-0.94; iodine no. 115-130; sapon. no. 180-195; flash pt. (OC) 640 F; ref. index 1.469-1.478; 0.2% max. moisture; *Toxicology:* nonhazardous.

Whetpro®-75. [ADM Ogilvie] Vital wheat gluten; CAS 8002-80-0; EINECS 232-317-7; for food prods. incl. baked goods, cereals, pet foods, pasta, and cheese analog prods.; *Properties:* lt. tan powd.; 7% moisture, 75% protein, 1% fat.

Whetpro®-80. [ADM Ogilvie] Vital wheat gluten; CAS 8002-80-0; EINECS 232-317-7; for food prods. incl. baked goods, cereals, pet foods, pasta, and cheese analog prods.; *Properties:* lt. tan powd.; 7% moisture, 80% protein, 1% fat.

Whetstar® 7 Food Powd. Wheat Starch. [ADM Ogilvie] Wheat starch; CAS 9005-25-8; for sugar grinding applics.; *Properties:* wh. powd.; pH 4.0; 7% moisture, 0.4% protein.

White Frosteen®. [Bunge Foods] Fully prepared white icing for bakery prods., donuts, cake pastries; *Properties:* wh. color.

Whitened Pea Fiber. [Garuda Int'l.] Pea fiber from specially cleaned and processed pea hulls; highly absorptive fiber source for breads, snacks, dietary supplements; *Properties:* off-wh. powd., bland odor and flavor; 120 mesh; absorp. capacity 350%; 86% dietary fiber, 5% sol. fiber, 8% max. moisture.

White Sour. [Bunge Foods] Sour producing uniform prod. with constant pH; reduces fermentation loss; increases shelf life; decreases molding tendencies; enhances flavor of bread; for hearth prods., English muffins, French sourdough type breads; *Usage level:* 5% based on flour.

Whole Egg Solids Type W-1. [Henningsen Foods] Egg solids; 45% min. protein, 40% min. fat; used in cookies, layer cakes, pound cakes, pies, sweet goods, egg noodles, mayonnaise, salad dressings; *Usage level:* 1 lb/3 lb water to replace 4 lb liq. or frozen whole egg; *Properties:* powd.; 100% thru 16 mesh; pH 8.5 ± 0.5; 90% min. dry matter; *Storage:* store in cooler @ 40-50 F.

Whole Egg Solids Type W-1-FF. [Henningsen Foods] Dehydrated whole egg with sodium silicoaluminate; similar to W-1 with anticaking agent added to enhance free-flowing props. and aid in rehydration.

Whole Egg Solids Type W-2. [Henningsen Foods] Dehydrated whole egg stabilized by removing natural glucose; used in dry mixes which require long shelf life.

Whole Egg Solids Type W-2-FF. [Henningsen Foods] Dehydrated whole egg stabilized by removing natural glucose; similar to W-2 with free-flow agent.

Whole Egg Solids Type W-7. [Henningsen Foods] Dehydrated whole egg; exc. flavor chars. for use in scrambled eggs, custards, quiches, and other foods where flavor and texture are critical.

Whole Pituitary Substance. [Am. Labs] Vacuum-dried defatted glandular prod.; nutritive food additive; *Properties:* powd. or freeze-dried form.

Wickenol® 550. [CasChem] Maltodextrin; CAS 9050-36-6; EINECS 232-940-4; food-grade carrier for flavors; *Properties:* Gardner 2+ powd.; 100% act.

Winterized Clear Liquid Frying Shortening 104-112. [ADM Packaged Oils] Partially hydrog. and winterized soybean oil, TBHQ, dimethylpolysiloxane; packaged vegetable oil; *Properties:* clear bright liq., bland flavor; iodine no. 115 ± 3.

Witafrol® 7420. [Hüls Am.; Hüls AG] Caprylic/capric glycerides; emulsifier, solubilizer, dispersant, plasticizer, lubricant; antifoam for prod. of sweets, preserves, jams, potato prods., and flavorings; *Properties:* sl. ylsh. oil; sol. in water/ethanol (50/50), acetone; sol. cloudy in ether, heptane; dens. 1.02 kg/dm³; visc. 190 mPa•s; acid no. 2 max.; iodine no. 1 max.; sapon. no. 230-260; flash pt. > 180 C; 40-42% monoglycerides.

Witafrol® 7440. [Hüls AG] Mono- and diglycerides; foam preventer for food industry in prod. of sweets, preserves, jams, potato prods.; *Properties:* liq./paste; dens. 0.95 kg/dm³; acid no. 2 max.; sapon. no. 155-175; flash pt. > 180 C.

Witafrol® 7456. [Hüls AG] Fatty acid ester of polyalcohols; food antifoam for processing of dairy prods.; pronounced/spontaneous retardation effect on foam formation caused by proteins; *Properties:* liq.; dens. 1.04 kg/dm³; visc. 220 mPa•s; acid no. 2 max.; sapon. no. 45-60; flash pt. > 180 C.

Witafrol® 7480 N. [Hüls AG] Fatty acid ester of polyalcohols; defoamer and foam preventive for sugar beet processing, dairy industries, jams, fruit flavors/juices, seasonings, other food prods.; *Properties:* liq.; dens. 0.98 kg/dm³; visc. 80 mPa•s; acid no. 2 max.; sapon. no. 175-195; flash pt. > 180 C.

Witafrol® 7490. [Hüls AG] Fatty acid ester of polyalcohols; biodeg. antifoam agent for sugar beet processing; *Properties:* liq.; dens. 0.96 kg/dm³; visc. 70 mPa•s; acid no. 2 max.; sapon. no. 125-145; flash pt. > 200 C.

Witafrol® 7491. [Hüls AG] Fatty acid ester of polyalcohols; antifoam for sugar beet processing; *Properties:* liq.; dens. 0.97 kg/dm³; visc. 300 mPa•s; acid no. 2 max.; sapon. no. 40-70; flash pt. > 180 C.

Witafrol® 7497 N. [Hüls AG] Fatty acid ester of polyalcohols; foam preventer for sugar beet and potato processing, juice extraction/purification, external water circulation systems of sugar factories; biodeg.; *Properties:* liq.; dens. 0.93 kg/dm³; visc. 70 mPa•s; acid no. 2 max.; sapon. no. 90-110; flash pt. > 180 C.

Witarix® 125. [Hüls AG] Coconut/palm kernel oil triglyceride; special fat with exc. cooling effect and good summer stability; for prod. of high quality cool fillings, e.g., for ice cream confectionery; *Properties:* block; m.p. 28-30 C; acid no. 0.2 max.; iodine no. 2 max.; sapon. no. 250-270.

Witarix® 200. [Hüls AG] Coconut/palm kernel oil triglyceride; hard fat with well-balanced consistency and melting chars. for food applics.; *Properties:* block; m.p. 33-35 C; acid no. 0.2 max.; iodine no. 2 max.; sapon. no. 235-250.

Witarix® 210. [Hüls AG] Coconut/palm kernel oil triglyceride; low-melting filling fat with outstanding melting behavior; *Properties:* block; m.p. 30 C; acid no. 0.2 max.; iodine no. 10 max.; sapon. no. 235-250.

Witarix® 212. [Hüls Am.; Hüls AG] Hydrog. palm kernel oil; CAS 68990-82-9; hard fat for chocolate and confectionery prods. with soft consistency, good temp. stability; *Properties:* block; m.p. 36-38 C; acid no. 0.2 max.; iodine no. 5 max.; sapon. no. 240-255.

Witarix® 250. [Hüls Am.; Hüls AG] Cocoglycerides; soft fat and filling fat for ice cream, cool fillings, chocolate and confectionery prods. with soft consistency; *Properties:* block; m.p. 32-34 C; acid no. 0.2 max.; iodine no. 2 max.; sapon. no. 245-265.

Witarix® 251. [Hüls AG] Coconut oil triglyceride; soft fat and filling fat for prod. of cool fillings; esp. useful in combination with Witocan hard fats; *Properties:* liq./pasty; m.p. 25 C; acid no. 0.2 max.; iodine no. 11 max.; sapon. no. 245-265.

Witarix® 431. [Hüls AG] Palm oil triglyceride; oily/pasty fat for prod. of creamy fillings; *Properties:* liq./pasty; m.p. 20 C; acid no. 0.2 max.; iodine no. 60; sapon. no. 190-210.

Witarix® 440. [Hüls Am.] Hydrog. soybean oil; CAS 8016-70-4; EINECS 232-410-2; soft fat and filling fat for fillings, chocolate and confectionery prods. with soft consistency; *Properties:* block; m.p. 35-37 C; acid no. 0.2 max.; iodine no. 70; sapon. no. 190.

Witarix® 450. [Hüls Am.] Hydrog. peanut oil; CAS 68425-36-5; EINECS 270-350-9; fast solidifying filling fat for chocolate and confectionery prods. with firm consistency; *Properties:* block; m.p. 34-36 C; acid no. 0.2 max.; iodine no. 70; sapon. no. 180-200.

Witco® Aluminum Stearate EA Food Grade. [Witco/H-I-P] Aluminum distearate; CAS 300-92-5; EINECS 206-101-8; food applics.

Witconol™ 14F. [Witco/H-I-P] Polyglycerol fatty acid

ester; nonionic; w/o emulsifier for food processing; *Properties:* liq.; oil-sol.

Witconol™ 18F. [Witco/H-I-P] Polyglyceryl-4 stearate; CAS 37349-34-1; nonionic; w/o emulsifier for food processing; *Properties:* solid; oil-sol.; Unverified

Witconol™ GMOP. [Witco/H-I-P] Mono- and diglycerides; nonionic; o/w emulsifier for food processing; *Properties:* liq.; sol. in oil; Unverified

Witocan® 30 N. [Hüls AG] Coconut/palm kernel oil triglyceride; low melting hard fat for prod. of fillings with good melting chars. and rapid flavor release; *Properties:* block; m.p. 31-33 C; acid no. 0.1 max.; iodine no. 3 max.; sapon. no. 240-250.

Witocan® 42/44. [Hüls AG] Coconut/palm kernel oil triglyceride; high melting hard fat for prod. of food articles with good temp. stability; *Properties:* block, pellets; m.p. 42-44 C; acid no. 0.3 max.; iodine no. 2 max.; sapon. no. 220-230.

Witocan® H. [Hüls AG] Coconut/palm kernel oil triglyceride; hard fat for food applics., esp. for prod. of solid and hollow articles; *Properties:* block, pellets; m.p. 33-35 C; acid no. 0.1 max.; iodine no. 2 max.; sapon. no. 235-245.

Witocan® HS. [Hüls AG] Coconut/palm kernel oil triglyceride; hard fat with exc. consistency and melting chars.; esp. for thin coatings; *Properties:* block, pellets; m.p. 33-35 C; acid no. 0.1 max.; iodine no. 2 max.; sapon. no. 235-245.

Witocan® LM. [Hüls AG] Coconut/palm kernel oil triglyceride; low melting hard fat with outstanding stability; for prod. of high quality fillings and coatings; also for use with milk fats; *Properties:* block; m.p. 31-33 C; acid no. 0.1 max.; iodine no. 3 max.; sapon. no. 240-250.

Witocan® P. [Hüls AG] Coconut/palm kernel oil triglyceride; hard fat with pleasant melting props. in the mouth for a wide range of applics.; *Properties:* block, pellets; m.p. 34-36 C; acid no. 0.1 max.; iodine no. 2 max.; sapon. no. 235-245.

XYZ

Xpando. [Am. Ingreds.] Ethoxylated mono- and diglycerides, mono- and diglycerides; food emulsifier for icings; dough strengthener for yeast-raised bakery prods.; *Properties:* cream-colored plastic; c.m.p. 125-135 F; i.v. 14-18; 24-28% mono.

Xpando 70. [Am. Ingreds.] Ethoxylated mono- and diglycerides, mono- and diglycerides; dough strengthener for yeast-raised bakery prods. and mixes; *Properties:* votated cream-colored plastic; c.m.p. 110-120 F; i.v. 15-20; 16-18% mono.

Xpando Powd. [Am. Ingreds.] Ethoxylated mono- and diglycerides, mono- and diglycerides; emulsifier for yeast-raised bakery prods. and mixes; *Properties:* wh. beads; c.m.p. 128-135 F; i.v. 2 max.; 28-31% mono.

X-Pand'R®. [A.E. Staley Mfg.] Pregelatinized food starch modified (waxy corn starch); starch for extruded snack foods; forms pliable nonsticking doughs for cold processing; provides cohesiveness; expanding agent for snacks which are baked or fried; carrier for cereal flours, cheese/veg. powds., peanut butter, soy protein, etc.; *Regulatory:* FDA 21CFR §172.892; *Properties:* wh. powd., bland flavor; 95% thru 100 mesh, 75% thru 200 mesh; pH 6.0; 5% moisture.

Xylitol C. [Xyrofin UK] Xylitol FCC, USP, NF; CAS 87-99-0; EINECS 201-788-0; food grade sweetener for foods and pharmaceutical applics.; stable to air and heat; *Properties:* wh. cryst. powd., pract. odorless, very sweet cool taste; very sol. in water; sparingly sol. in ethanol; m.w. 152.15; m.p. 92-96 C; pH 5-7 (aq.); 98.5-101% assay; *Storage:* 1 yr stability in original sealed pkg. stored below 25 C and < 65% r.h.; marginally hygroscopic.

Yeast Lactase L-50,000. [Solvay Enzymes] Yeast lactase; enzyme for hydrolyzing lactose in dairy prods. (milk, ice cream, milk powd., whey, cheese, yogurt); *Properties:* lt. amber liq., free from offensive odors and taste; *Storage:* store at refrigeration temp.

Yelkin 1018. [ADM Ross & Rowe Lecithin] Hydroxylated soya lecithin; CAS 8029-76-3; EINECS 232-440-6; for baked goods, whipped/dry toppings, instant dry beverage mixes; *Properties:* turbid fluid; 1% max. water.

Yelkin DS. [ADM Ross & Rowe Lecithin] Standardized soya lecithin; CAS 8002-43-5; EINECS 232-307-2; for drink mixes, candies; *Properties:* translucent fluid.

Yelkin Gold. [ADM Ross & Rowe Lecithin] Purified soya lecithin; CAS 8002-43-5; EINECS 232-307-2; for instantized foods, liq. emulsions, infant formulas; *Properties:* transparent fluid.

Yelkin SS. [ADM Ross & Rowe Lecithin] Standardized soya lecithin; CAS 8002-43-5; EINECS 232-307-2; dough conditioner, pan lubricant; *Properties:* translucent fluid.

Yelkin T. [ADM Ross & Rowe Lecithin] Standardized soya lecithin; CAS 8002-43-5; EINECS 232-307-2; for candies, margarines, baked goods, icings, frostings; *Properties:* opaque plastic.

Yelkin TS. [ADM Ross & Rowe Lecithin] Standardized soya lecithin; CAS 8002-43-5; EINECS 232-307-2; for chocolate, frosting, baked goods, icings, margarines; *Properties:* translucent fluid.

Yelkinol F. [ADM Ross & Rowe Lecithin] Deoiled lecithin; CAS 8002-43-5; EINECS 232-307-2; dispersant, dietary supplement; for sauces, gravies, nondairy creamers, cake mixes, drink mixes; *Properties:* lt. gold fine gran.; 1% max. water.

Yelkinol G. [ADM Ross & Rowe Lecithin] Deoiled lecithin; CAS 8002-43-5; EINECS 232-307-2; dispersant, dietary supplement; for sauces, gravies, nondairy creamers, cake mixes, drink mixes; *Properties:* lt. gold gran.; 1% max. water.

Yelkinol P. [ADM Ross & Rowe Lecithin] Deoiled lecithin; CAS 8002-43-5; EINECS 232-307-2; dispersant, dietary supplement; for sauces, gravies, nondairy creamers, cake mixes, drink mixes; *Properties:* lt. gold powd.; 1% max. water.

Yellow Blend No. 3100 Powd. [ADM Ogilvie] Natural yel. coloring derived from vegetable pigments; food colorant for baked goods, dry mixes, sauces; alternative to FD&C Yel. No. 5; *Properties:* powd.; fat-sol.

Yellow Blend No. 3300 Powd. [ADM Ogilvie] Natural yel. coloring derived from vegetable pigments; food colorant for baked goods, dry mixes, sauces; alternative to FD&C Yel. No. 5; more intense yel. than Yellow Blend No. 3100; *Properties:* powd.; fat-sol.

Yellow Bun. [Brolite Prods.] Bread base for rich yellow breads, rolls, and buns.

Yellow Fat Soluble Liq. G-200. [MLG Enterprises Ltd.] Based on natural carotenoids (bixin and orelline) extracted from annatto seeds in an edible oil; natural food colorant producing deep yel. shades for fat-contg. foods, e.g., direct fats, hydrog. oils, biscuit fillings, mayonnaise-based dressings, cakes, melted cheeses; *Usage level:* 0.01-0.05%; *Properties:* visc. liq.; sol. in fats and oils; insol. in water; 1.3 ± 0.06% act. (as bixin); *Storage:* store in tightly closed containers away from light, heat; rotate stock every 4 mos; shake well before use.

Yellow Pea Fiber. [Garuda Int'l.] Pea fiber from cleaned and milled yel. pea hulls; highly absorptive fiber source for breads, snacks, dietary supplements; *Properties:* lt. beige powd., mild pea odor and flavor; 0.05% 25 mesh, 6% 35 mesh; bulk dens. 47 lb/ft³; absorp. 220%; pH 6.5; 76% dietary fiber, 5% sol. fiber.

Yellow Pea Powd. [Garuda Int'l.] Pulverized, precooked yel. peas; flavoring agent for breads, crackers, soups without off-tastes; *Properties:* yel. fine powd., bland odor, mild pea flavor; 120 mesh; pH 6.7 (10% aq. disp.); 21% total dietary fiber, 3% sol. fiber, 47.2% carbohydrate, 23.5% protein, 4% moisture; *Storage:* 6 mos storage under cool, dry conditions.

Yuccafoam™. [Bell Flavors & Fragrances] Yucca extract deriv.; natural food-grade foaming agent with exc. clarity for use in food and beverage industry; enhances carbonation; suggested for bar mixes, beer, cocktails, juices, wine coolers, schnapps, whipping cream; *Usage level:* 0.01-0.03% (carbonated beverages), 0.1-0.2% (bar mix), 0.005-0.01% (beer, cocktails), 0.005-0.002% (juices), 0.05-0.1% (wine coolers), 0.06-0.1% (schnapps), 0.2-0.25% (whipping cream); *Regulatory:* FDA approved; *Properties:* liq. and powd.; completely sol. in aq. sol'ns.

Zea Red. [MLG Enterprises Ltd.] Extract of the antocyanins of purple corn, a variety of *Zea mays*, stabilized with phosphoric acid, spray-dried; natural colorant in dry mixes, gelatins, soup powd., dehydrated foods, chewing gums, candies, beverages, yogurt, biscuits, chocolate fillings, cream cheese, frosting, fruit, wine, ice cream; *Properties:* red powd.; sol. in water.

Zonester® 85. [Arizona] Tall oil glycerides; thermoplastic resin ester used in chewing gums; *Regulatory:* FDA compliance; *Properties:* Gardner 7-8 solid or flakes; sp.gr. 1.06; R&B soften. pt. 82 C; CC flash pt. > 400 F; acid no. 5.

Part II
Chemical Additive Dictionary/
Cross-Reference

Chemical Dictionary/Cross Reference

Abies balsamea oleoresin. *See* Balsam Canada
ABS. *See* Acrylonitrile-butadiene-styrene
Absolute alcohol. *See* Alcohol

Acacia
CAS 9000-01-5; EINECS 232-519-5; EEC E414
Synonyms: Acacia gum; Sudan gum; Gum hashab; Kordofan gum; Gum arabic; Arabic gum
Definition: Dried gummy exudate from stems and branches of *Acacia farnesiana* or *A. senegal*
Properties: Sol. in water; insol. in alcohol; m.w. 240,000
Precaution: Combustible
Toxicology: LD50 (oral, rat) 18 g/kg; inh. or ingestion may produce hives, eczema, angiodema, asthma; allergic responses; people prone to allergies should avoid acacia; heated to decomp., emits acrid smoke
Uses: Emulsifier, flavoring agent, adjuvant, formulation aid, humectant, surface-finishing agent, stabilizer, thickener, firming agent, processing aid, texturizer, crystallization inhibitor; for confections, flavor syrups, dietary fiber; foam stabilizer
Usage level: Limitation 2% (beverages), 5.6% (chewing gum), 12.4% (confections), 1.3% (dairy prods.), 1.5% (fats, oils), 2.5% (gelatins, puddings), 46.5% (hard candy), 8.3% (nuts), 6% (frozen confections), 4% (snack foods), 85% (soft candy), 1% (other foods)
Regulatory: FDA 21CFR §169.179, 169.182, 184.1330, GRAS; Japan approved; Europe listed; UK approved; ADI not specified (JECFA)
Manuf./Distrib.: Ashland; Atomergic Chemetals; Bio-Botanica; Calaga Food Ingreds.; Chart; Commodity Services Int'l.; Cornelius; G Fiske; Florexco; Gumix Int'l.; Ikeda; Importers Service; Int'l. Ingreds.; Meer; Penta Mfg.; Quest; Rhone-Poulenc; Spice King; TIC Gums
Trade names: Coatingum L; Emulgum BV; Granular Gum Arabic NF/FCC C-4010; Gum Arabic NF/FCC Clean Amber Sorts; Gum Arabic, Purified, Spray-Dried No. 1834; Natural Arabic Type Gum Purified, Spray-Dried; Nutriloid® Arabic; Powdered Gum Arabic NF/FCC G-150; Powdered Gum Arabic NF/FCC Superselect Type NB-4; Premium Fine Granular Gum Arabic; Premium Granular Gum Arabic; Premium Powdered Gum Arabic; Premium Spray Dried Gum Arabic; Spray Dried Gum Arabic NF Type CSP; Spray Dried Gum Arabic NF/FCC CM; Spray Dried Gum Arabic NF/FCC CS (Low Bacteria); Spray Dried Gum Arabic NF/FCC CS-R; Spray Dried Nigerian Gum Arabic; Spraygum C; Spraygum GD; TIC Pretested® Arabic FT Powd.; TIC Pretested® Arabic FT-1 USP; TIC Pretested® Arabic PH-FT
Trade names containing: Cola No. 23443; Dry Phytonadione 1% SD No. 61748; Dry Vitamin D₃ Type 100-SD No. 65216; Durkote Calcium Carbonate/Starch, Acacia Gum; Durkote Citric Acid/Maltodextrin, Acacia Gum; Kel-Lite™ CM; Merecol® FAL; Tenox® WD-BHA; Thixogum S; Ticaloid® S1-102; Ticaloid® S2-102; TIC Pretested® Aragum® 3000; TIC Pretested® Aragum® 9500; Vitamin A Palmitate Type 250-CWS No. 65312

Acacia gum. *See* Acacia
Aceite de ricino. *See* Castor oil
Acesulfame K. *See* Acesulfame potassium

Acesulfame potassium
CAS 55589-62-3
Synonyms: Potassium 6-methyl-1,2,3-oxathiazine-4(3H)-1,2,2-dioxide; Potassium acesulfame; Acesulfame K; Sunnette
Definition: Potassium salt of 6-methyl-1,2,3-oxathiazine-4(3H)-one-2,2-dioxide
Empirical: $C_4H_4KNO_4S$
Properties: Wh. cryst. solid, odorless, sweet taste, very sl. bitter aftertaste; very sol. in water, DMF, DMSO; sol. in alcohol, glycerin-water; m.w. 201.24; about 200 times sweeter than sucrose; m.p. 250 C
Toxicology: Heated to decomp., emits toxic fumes of SO_x

Uses: Nonnutritive sweetening agent; used in sugar substitutes, chewing gum, dairy prods., confections, and dry bases for beverages, instant coffee/tea, gelatins, puddings
Usage level: ADI 9 mg/kg (WHO)
Regulatory: FDA 21CFR §172.800
Manuf./Distrib.: Dietary Foods; Hoechst AG; Hoechst UK; Sunette; Vrymer Commodities

Acetal
CAS 105-57-7; EINECS 203-310-6; FEMA 2002
Synonyms: Polyacetal; Acetaldehyde diethyl acetal; 1,1-Diethoxyethane; Diethyl acetal; Ethylidene diethyl ether
Empirical: $C_6H_{14}O_2$
Formula: $CH_3CH(OC_2H_5)_2$
Properties: Colorless volatile liq., fruity green flavor; sol. in heptane, ethyl acetate; misc. with alcohol, ether; m.w. 118.18; dens. 0.831; b.p. 103-104 C; flash pt. (CC) 36 C
Precaution: Highly flamm.; keep away from ignition sources; tends to polymerize on standing
Toxicology: LD50 (oral, rat) 4.57 g/kg; moderately toxic by ingestion, skin and eye irritant
Uses: Synthetic flavoring for nonalcoholic beverages, ice cream
Usage level: 7.3 ppm (nonalcoholic beverages), 52 ppm (ice cream, ices), 39 ppm (candy), 60-120 ppm (baked goods)
Regulatory: FDA §172.515; GRAS (FEMA)
Manuf./Distrib.: Aldrich

Acetal R. *See* Acetaldehyde phenethyl propyl acetal

Acetaldehyde
CAS 75-07-0; EINECS 200-836-8; FEMA 2003
Synonyms: Acetic aldehyde; Ethyl aldehyde; Ethanal
Classification: Aldehyde
Empirical: C_2H_4O
Formula: CH_3CHO
Properties: Colorless fuming liq., pungent fruity odor; misc. in water, alcohol, ether; m.w. 44.06; dens. 0.788 (16/4 C); m.p. -123.5 C; b.p. 20.8 C; flash pt. (CC) -38 C; ref. index 1.3316 (20 C)
Precaution: Flamm. liq. (DOT); can react violently with acid anhydrides, alcohols, ketones, phenols, NH_3, halogens, etc.; reaction with oxygen may lead to detonation; keep cold
Toxicology: LD50 (oral, rat) 1930 mg/kg; poison by intratracheal/IV routes; human systemic irritant by inh.; narcotic; human mutagenic data; experimental tumorigen, teratogen; skin and severe eye irritant; heated to decomp., emits acrid smoke and fumes
Uses: Synthetic flavoring; in nonalcoholic beverages, ice cream, ices, candy, baked goods
Usage level: 3.9 ppm (nonalcoholic beverages), 25 ppm (ice cream), 22 ppm (candy), 12 ppm (baked goods), 6.8 ppm (gelatins, puddings), 20-270 ppm (chewing gum)
Regulatory: FDA 21CFR §182.60, GRAS
Manuf./Distrib.: Aldrich; Eastman

Acetaldehyde benzyl β-methoxyethyl acetal. *See* Benzyl methoxyethyl acetal
Acetaldehyde diethyl acetal. *See* Acetal

Acetaldehyde phenethyl propyl acetal
FEMA 2004
Synonyms: Acetal R; Pepital; Propyl phenethyl acetal
Definition: From acetaldehyde with a mixture of propyl and β-phenyl ethyl alcohols
Empirical: $C_{16}H_{20}O_2$
Properties: Colorless, stable liquid; strong odor of green leaves; m.w. 208.30; flash pt. 95C
Uses: Synthetic flavoring agent
Usage level: 2.5 ppm (candy), 2.5 ppm (baked goods)
Regulatory: FDA 21CFR §172.515
Manuf./Distrib.: Aldrich

Acetanisole
CAS 100-06-1; FEMA 2005
Synonyms: p-Methoxyacetophenone; 4´-Methoxyacetophenone; p-Acetyl anisole; Navatone
Definition: From anisole and acetic acid in the presence of boron trifluoride
Empirical: $C_9H_{10}O_2$
Properties: Ylsh.-wh. cryst. @ R.T., floral, bitter; sl. sol. in water; sol. in most org. solvs.; bitter, unpleasant taste; m.w. 150.18; m.p. 38C; b.p. 152-154 C (26 mm)
Uses: Synthetic flavoring agent
Usage level: 2.3 ppm (nonalcoholic beverages), 2.5 ppm (ice cream, ices, etc.), 4.6 ppm (candy), 5.8 ppm

(baked goods), 840 ppm (chewing gum)
Regulatory: FDA 21CFR §172.515
Manuf./Distrib.: Aldrich

Acetate C-9. *See* n-Nonyl acetate
Acetate C-10. *See* Decyl acetate
Acetate C-11. *See* 10-Undecen-1-yl acetate
Acetate C-12. *See* Lauryl acetate
Acetate PA. *See* Allyl phenoxyacetate

Acetic acid
CAS 64-19-7; EINECS 200-580-7; FEMA 2006; EEC E260
Synonyms: Ethanoic acid; Vinegar acid; Methanecarboxylic acid; Ethyllic acid; Pyroligneus acid
Classification: Organic acid
Empirical: $C_2H_4O_2$
Formula: CH_3COOH
Properties: Clear colorless liq., pungent odor; misc. with water, alcohol, glycerol, ether; insol. in carbon
disulfide; m.w. 60.03; dens. 1.0492 (20/4 C); m.p. 16.63 C; b.p. 118 C (765 mm); visc. 1.22 cps (20 C); flash
pt. (OC) 43 C; ref. index 1.3715 (20 C)
Precaution: Combustible; moderate fire risk; DOT: corrosive, flamm. liq.
Toxicology: Pure acetic acid: moderately toxic by ingestion, inhalation; dilute approved FDA for food use;
strong irritant to skin and tissue; TLV 10 ppm in air; LD50 (oral, rat) 3310 mg/kg
Uses: Flavoring agent, flavor enhancer, acidifier, pH control agent, curing/pickling agent, solvent/vehicle,
boiler water additive, color diluent
Usage level: 39 ppm (nonalcoholic beverages), 32 ppm (ice cream), 52 ppm (candy), 38 ppm (baked goods),
15 ppm (pudding), 60 ppm (chewing gum), 5900 ppm (condiments); ADI no limit (EEC)
Regulatory: FDA 21CFR §73.85, 133, 172.814, 178.1010, 184.1005, GRAS; USDA 9CFR §318.7; Europe
listed; UK approved
Manuf./Distrib.: Albright & Wilson; Diamalt; Eastman; Ellis & Everard; Frutarom; Keith Harris; Hoechst AG;
Integrated Ingreds.; Penta; Pfaltz & Bauer; Siber Hegner; Tropic Agro; Van Waters & Rogers
Trade names containing: BFP White Sour; Ry-So; Sour Dough Base; Vanease (K)

Acetic acid, butyl ester. *See* n-Butyl acetate
Acetic acid, cinnamyl ester. *See* Cinnamyl acetate
Acetic acid citronellyl ester. *See* Citronellyl acetate
Acetic acid cyclohexyl ester. *See* Cyclohexyl acetate
Acetic acid esters of mono- and diglycerides of fatty acids. *See* Acetylated mono- and diglycerides of fatty
acids
Acetic acid, ethenyl ester, homopolymer. *See* Polyvinyl acetate (homopolymer)
Acetic acid, ethyl ester. *See* Ethyl acetate
Acetic acid methyl ester. *See* Methyl acetate
Acetic acid 1-methylethyl ester. *See* Isopropyl acetate
Acetic acid n-nonyl ester. *See* n-Nonyl acetate
Acetic acid pentyl ester. *See* Amyl acetate
Acetic acid 2-phenylethyl ester. *See* 2-Phenylethyl acetate
Acetic acid n-propyl ester. *See* Propyl acetate
Acetic acid, retinyl ester. *See* Retinyl acetate
Acetic acid sodium salt anhydrous. *See* Sodium acetate anhydrous
Acetic acid vinyl ester polymers. *See* Polyvinyl acetate (homopolymer)
Acetic aldehyde. *See* Acetaldehyde

Acetic anhydride
CAS 108-24-7; EINECS 203-564-8
Synonyms: Acetyl oxide; Acetic oxide
Empirical: $C_4H_6O_3$
Formula: $(CH_3CO)_2O$
Properties: Liq., strong acetic odor; slowly sol. in water; sol. in chloroform, ether; m.w. 102.09; dens. 1.08 (15/
4 C); m.p. -73 C; b.p. 138-140 C; flash pt. 49 C; ref. index 1.3904 (20 C)
Precaution: Readily combustible; fire hazard
Toxicology: LD50 (oral, rat) 1.78 g/kg; produces irritation and necrosis of tissues in liq. or vapor state
Uses: Esterifying agent for food starch
Regulatory: FDA, esterifier for food starch, in combination with adipic anhydride (0.12% max. adipic anhydride,
5% max. acetic anhydride)
Manuf./Distrib.: Eastman

Acetic ether. *See* Ethyl acetate
Acetic oxide. *See* Acetic anhydride
Acetin. *See* Triacetin

Acetisoeugenol
CAS 93-29-8; FEMA 2470
Synonyms: 4-Acetoxy-3-methoxy-1-propenylbenzene; Acetyl isoeugenol; Isoeugenol acetate; 2-Methoxy-4-propenylphenyl acetate
Empirical: $C_{12}H_{14}O_3$
Properties: Wh. cryst., clove odor; sol. in alcohol, chloroform, ether; insol. in water; m.w. 206.26; flash pt. 153 F
Precaution: Combustible liq.
Toxicology: LD50 (oral, rat) 3450 mg/kg; mod. toxic by ingestion; heated to decomp., emits acrid smoke and irritating fumes
Uses: Synthetic flavoring agent
Usage level: 0.44 ppm (nonalcoholic beverages), 2.1 ppm (ice cream, ices), 17 ppm (candy, baked goods), 100 ppm (chewing gum)
Regulatory: FDA 21CFR §172.515

Acetoacetic acid ethyl ester. *See* Ethylacetoacetate
Acetoacetone. *See* Acetylacetone
Acetocumene. *See* p-Isopropylacetophenone
Acetoin. *See* Acetyl methyl carbinol
β-Acetonaphthalene. *See* 2'-Acetonaphthone

2'-Acetonaphthone
CAS 93-08-3; EINECS 202-216-2; FEMA 2723
Synonyms: Orange crystals; Methyl 2-naphthyl ketone; Methyl β-naphthyl ketone; β-Acetonaphthalene; 2-Acetonaphthone; 2-Acetylnaphthalene
Empirical: $C_{12}H_{10}O$
Properties: Wh. or nearly wh. cryst. solid, orange blossom odor, strawberry-like flavor; sol. in most common org. solvs.; insol. in water; m.w. 170.21; m.p. 53 C; b.p. 301-303 C; flash pt. 168 C
Precaution: Combustible liq.
Toxicology: LD50 (oral, mouse) 599 mg/kg; mod. toxic by ingestion; skin irritant; heated to decomp., emits acrid smoke and fumes
Uses: Synthetic flavoring agent
Usage level: 0.50 ppm (nonalcoholic beverages), 0.75 ppm (ice cream, ices), 5.3 ppm (candy), 2.0 ppm (baked goods), 2.2-3.0 ppm (gelatins, puddings), 480-700 ppm (chewing gum)
Regulatory: FDA 21CFR §172.515; Japan approved as flavoring
Manuf./Distrib.: Aldrich

2-Acetonaphthone. *See* 2'-Acetonaphthone

Acetone
CAS 67-64-1; EINECS 200-662-2
Synonyms: Dimethylketone; 2-Propanone; β-Ketopropane; Pyroacetic ether
Classification: Aliphatic ketone
Empirical: C_3H_6O
Formula: CH_3COCH_3
Properties: Colorless volatile liq., sweetish odor, pungent sweetish taste; misc. with water, alcohol, ether, most oils; m.w. 58.09; dens. 0.792 (20/20 C); m.p. -94.3 C; b.p. 56.2 C; flash pt. (OC) 15 F
Precaution: DOT: flamm. liq.; dangerous fire risk; explosive limit in air 2.6-12.8%
Toxicology: LD50 (oral, mouse) 3000 mg/kg; TLV 750 ppm in air; narcotic in high conc.; moderately toxic by ingestion and inhalation; lg. doses may cause narcosis
Uses: Color diluent, extraction solvent for fruits, spice oleoresins, vegetables; processing aid
Regulatory: FDA 21CFR §73.1, 173.210, 30 ppm tolerance in spice oleoresins; Japan approved with restrictions
Manuf./Distrib.: Eastman

Acetone peroxides
Definition: Derived from a mixture of monomeric and linear dimeric acetone peroxides (mainly 2,2-hydroperoxypropane), with minor proportions of higher polymers, usually mixed with an edible carrier such as cornstarch
Properties: Liq. or absorbed on corn starch; trimeric form is cryst.; m.p. 97 C
Precaution: Flamm. by spontaneous chemical reaction; can react vigorously with reducing materials; trimeric form is shock-sensitive, static electricity-sensitive and may detonate

Toxicology: Severe skin and eye irritant
Uses: Bleaching agent, dough conditioner, maturing agent; used in bread, flour, rolls
Regulatory: FDA 21CFR §172.802

Acetonic acid. *See* Lactic acid

Acetophenone
CAS 98-86-2; EINECS 202-708-7; FEMA 2009
Synonyms: Ketone methyl phenyl; Methyl phenyl ketone; Acetylbenzene; 1-Phenylethanone; Hypnone
Empirical: C_8H_8O
Formula: $C_6H_5COCH_3$
Properties: Colorless liq., sweet pungent odor and taste; sol. in alcohol, chloroform, ether, fatty oils, glycerol; sl. sol. in water; m.w. 120.15; dens. 1.030 (20/20 C); m.p. 20.5 C; b.p. 201.7 C; flash pt. 82.2 C; ref. index 1.5339 (20 C)
Precaution: Combustible
Toxicology: Narcotic in high concs.; poison by intraperitoneal route; moderate toxicity by ingestion; skin and eye irritant; LD50 (oral, rat) 815 mg/kg
Uses: Synthetic flavoring agent
Usage level: 0.98 ppm (nonalcoholic beverage), 2.8 ppm (ice cream, ices), 3.6 ppm (candy), 5.6 ppm (baked goods), 7.0 ppm (gelatins, puddings)
Regulatory: FDA 21CFR §172.515; Japan approved for flavoring
Manuf./Distrib.: Aldrich; Penta Mfg.

4-Acetoxy-3-methoxy-1-propenylbenzene. *See* Acetisoeugenol
α-Acetoxytoluene. *See* Benzyl acetate

Acetylacetone
CAS 123-54-6; EINECS 204-634-0
Synonyms: 2,4-Pentanedione; Diacetylmethane; Acetoacetone
Empirical: $C_5H_8O_2$
Formula: $CH_3COCH_2COCH_3$
Properties: Colorless to sl. yel. liq., pleasant odor; misc. with alcohol, benzene, ether, chloroform, acetone, glac. acetic acid, propylene glycol; insol. in water; m.w. 100.13; dens. 0.952-0.962; m.p. -23.2 C; b.p. 139 C (746 mm); flash pt. (OC) 105 F; ref. in
Precaution: Flamm. liq. exposed to heat or flame; incompat. with oxidizing materials
Toxicology: LD50 (oral, rat) 1000 mg/kg; mod. toxic by ingestion, IP, inh. routes; skin and severe eye irritant
Uses: Synthetic flavoring agent
Regulatory: FDA 21CFR §172.515
Manuf./Distrib.: Aldrich; Penta Mfg.

N-Acetyl-L-2-amino-4-(methylthio) butyric acid. *See* N-Acetyl-L-methionine
p-Acetyl anisole. *See* Acetanisole
Acetylated hydrogenated coconut glyceride. *See* Acetylated hydrogenated coconut oil glyceride

Acetylated hydrogenated coconut glycerides
Synonyms: Acetylated hydrogenated coconut oil glycerides
Uses: Emulsifier, emollient, lubricant, and deaerator for foods; antidusting agent for powd. foods

Acetylated hydrogenated coconut oil glyceride
Synonyms: Acetylated hydrogenated coconut glyceride; Acetylated hydrogenated coconut oil monoglycerides
Uses: Emulsifier, emollient in food processing
Trade names: Myvacet® 9-08K

Acetylated hydrogenated coconut oil glycerides. *See* Acetylated hydrogenated coconut glycerides
Acetylated hydrogenated coconut oil monoglycerides. *See* Acetylated hydrogenated coconut oil glyceride

Acetylated hydrogenated cottonseed glyceride
Synonyms: Glycerides, cottonseed-oil, mono-, hydrogenated, acetates
Definition: Acetyl ester of the monoglyceride derived from hydrogenated cottonseed oil
Uses: Emulsifier, stabilizer, aerator for dry whipped toppings; protective film for foods; moisture and oxygen barrier; used in chewing gums as plasticizer and softener
Regulatory: FDA 21CFR §172.828, 175.230
Trade names: Myvacet® 5-07

Acetylated hydrogenated lard glyceride
CAS 8029-91-2
Synonyms: Glycerides, lard mono-, hydrogenated, acetates
Definition: Acetyl ester of the monoglyceride derived from hydrog. lard

Acetylated hydrogenated soybean oil glyceride

Uses: Food additive
Regulatory: FDA 21CFR §172.828
Trade names: Axol® E 61; Myvacet® 7-00; Tegin® E-61; Tegin® E-61 NSE

Acetylated hydrogenated soybean oil glyceride
Uses: Food additive
Regulatory: FDA 21CFR §172.828
Trade names: Myvacet® 9-45K
Trade names containing: Cetodan® 90-50

Acetylated hydrogenated soybean oil glycerides
Uses: Emulsifier, lubricant, deaerating agent; antidusting agent for powd. foods and mixes; defoamer for bottling operations (jams, jellies, puddings); lubricant/release for molding; plasticizer/softener for chewing gums

Acetylated hydrogenated tallow glyceride
CAS 68990-58-9; EINECS 273-612-0
Synonyms: Glycerides, tallow mono-, hydrogenated, acetates
Definition: Acetyl ester of hydrog. tallow glyceride
Uses: Food additive, edible coatings
Regulatory: FDA 21CFR §172.828, 175.230

Acetylated hydrogenated tallow glycerides
Synonyms: Glycerides, tallow mono-, di- and tri-, hydrogenated, acetates
Classification: Acetyl ester
Uses: Food additive
Regulatory: FDA 21CFR §172.828

Acetylated lard glyceride
CAS 8029-92-3
Synonyms: Glycerides, lard mono-, acetates
Classification: Acetyl ester
Uses: Food emulsifier
Regulatory: FDA 21CFR §172.828, 175.230
Trade names: Cetodan® 90-40; Myvacet® 9-40; Tegin® E-66; Tegin® E-66 NSE

Acetylated mono- and diglycerides of fatty acids
CAS 68990-55-6; 68990-58-9; EEC E472a
Synonyms: Acetic acid esters of mono- and diglycerides of fatty acids
Definition: Partial or complete esters of glycerin with a mixture of acetic acid and edible fat-forming fatty acids
Properties: White to pale yel. liq., bland taste; sol. in alcohol, acetone; insol. in water; HLB 2-3
Toxicology: Heated to decomp., emits acrid smoke and fumes
Uses: Coating agent, emulsifier, stabilizer, lubricant, solvent, texture modifying agent; used in baked goods, fruits, ice cream, meat products, peanut butter, pudding, shortening, whipped toppings
Usage level: ADI not specified (EEC)
Regulatory: FDA 21CFR §172.828; Europe listed; UK approved
Trade names: Axol® E 41; Axol® E 66; Cetodan® 50-00A; Cetodan® 50-00P Kosher; Cetodan® 70-00A; Cetodan® 70-00P Kosher; Dynacet® 212; Lamegin® EE 50; Lamegin® EE 70; Lamegin® EE 100; Vykacet T/L
Trade names containing: Carotenal Sol'n. 4% No. 66424; Lamequick® CE 1 SF

Acetylated tartaric acid esters of mono- and diglycerides of fatty acids
EEC E472e
Uses: Emulsifier, stabilizer
Usage level: ADI no limit (EEC)
Regulatory: Europe listed; UK approved
Trade names: Drewlate 30

Acetylbenzene. See Acetophenone

Acetyl butyryl
CAS 3848-24-6; EINECS 223-350-8; FEMA 2558
Synonyms: 2,3-Hexanedione; Methyl propyl diketone; Acetyl-n-butyryl
Empirical: $C_6H_{10}O_2$
Formula: $CH_3CH_2CH_2COCOCH_3$
Properties: Yel. oily liq., powerful creamy sweet buttery odor, butter cheese taste; sol. in alcohol, propylene glycol; sl. sol. in water; m.w. 114.15; dens. 0.934 (20/4 C); b.p. 128 C; flash pt. 83 F; ref. index 1.412 (20 C)

Precaution: Flammable
Toxicology: Irritant
Uses: Creamy, butter-like, sweet fragrance and flavoring (synthetic)
Usage level: 6.6 ppm (nonalcoholic beverages), 4.8 ppm (ice cream, ices), 7.3 ppm (candy), 6.6 ppm (baked goods)
Regulatory: FDA 21CFR §172.515
Manuf./Distrib.: Aldrich; BASF

Acetyl-n-butyryl. *See* Acetyl butyryl
Acetylcellulose. *See* Cellulose acetate
p-Acetyl cumol. *See* p-Isopropylacetophenone

3-Acetyl-2,5-dimethylfuran
CAS 10599-70-9; FEMA 3391
Synonyms: 2,5-Dimethyl-3-acetylfuran
Empirical: $C_8H_{10}O_2$
Properties: Yel. liq., strong roasted nut-like odor; sol. in alcohol, propylene glycol, fixed oils; sl. sol. in water; m.w. 138.16; dens. 1.027-1.048; ref. index 1.475-1.496
Toxicology: Heated to decomp., emits acrid smoke and irritating fumes
Uses: Flavoring agent
Manuf./Distrib.: Aldrich

Acetylene black. *See* Carbon black
Acetylene trichloride. *See* Trichloroethylene
Acetyl eugenol. *See* Eugenyl acetate
Acetyl formaldehyde. *See* Pyruvaldehyde
Acetylformic acid. *See* Pyruvic acid

N-Acetylglucosamine
CAS 7512-17-6; EINECS 231-368-2
Empirical: $C_8H_{15}NO_6$
Properties: m.w. 221.21; m.p. ≈ 215 (dec.)
Uses: Natural sweetener
Regulatory: Japan approved

Acetyl guaiacol. *See* Guaiacyl acetate
3-Acetyl-5-hydroxy-3-oxo-4-hexenoic acid δ-lactone. *See* Dehydroacetic acid
Acetyl isobutyryl. *See* 4-Methyl-2,3-pentanedione
Acetyl isoeugenol. *See* Acetisoeugenol
1,4-Acetyl-isopropyl benzol. *See* p-Isopropylacetophenone

N-Acetyl-L-methionine
CAS 65-82-7; EINECS 200-617-7
Synonyms: N-Acetyl-L-2-amino-4-(methylthio) butyric acid
Definition: Deriv. of the amino acid methionine; free or anhyd. form, or as sodium or potassium salts
Empirical: $C_7H_{13}NO_3S$
Formula: $CH_3SCH_2CH_2CH(NHCOCH_3)COOH$
Properties: Colorless or lustrous wh. cryst. or powd., odorless; sol. in water, alcohol, alkali, dil. min. acids; insol. in ether; m.w. 191.24; m.p. 104-107 C
Toxicology: Heated to decomp., emits toxic fumes of NO_x
Uses: Dietary supplement; source of L-methionine for use as nutrient; not for infant formulas
Usage level: Limitation 3.1% l- and dl-methionine (expressed as free amino acid) of total protein in food
Regulatory: FDA 21CFR §172.372

Acetyl methyl carbinol
CAS 513-86-0; EINECS 208-174-1; FEMA 2008
Synonyms: Acetoin; 2,3-Butanolone; γ-Hydroxy-β-oxobutane; 3-Hydroxy-2-butanone; Dimethylketol
Empirical: $C_4H_8O_2$
Formula: $CH_3COCH(OH)CH_3$
Properties: Colorless to sl. yel. liq. or cryst. solid, buttery odor; sol. in ethanol; sl. sol. in ether; misc. with water, alcohol, propylene glycol; insol. in veg. oil; m.w. 88.12; dens. 1.016; b.p. 147-148 C; m.p. 15 C; flash pt. 106 F; ref. index 1.417
Precaution: DOT: flamm. liq.
Toxicology: Moderate skin irritant; heated to decomp., emits acrid smoke and fumes
Uses: Aroma carrier; synthetic flavor; prep. of essences; buttery creamy flavor; for nonalc. beverages, ice cream, ices, candy, baked goods, gelatins, puddings, shortening, margarine, cottage cheese

Usage level: 7.4 ppm (nonalcoholic beverages), 3.3 ppm (ice cream, ices), 18 ppm (candy), 32 ppm (baked goods), 0.6-21 ppm (gelatins, puddings), 8.0 ppm (shortening), 0.80-50 ppm (margarine), 7.0 ppm (cheese)
Regulatory: FDA §182.60, GRAS
Manuf./Distrib.: Aldrich; Penta Mfg.

3-Acetyl-6-methyl-1,2-pyran-2,4(3H)-dione. *See* Dehydroacetic acid
3-Acetyl-6-methyl-2,4-pyrandione. *See* Dehydroacetic acid
2-Acetylnaphthalene. *See* 2´-Acetonaphthone
Acetyl nonanoyl. *See* 2,3-Undecadione
Acetyl nonyryl. *See* 2,3-Undecadione
Acetyl oxide. *See* Acetic anhydride
2-(Acetyloxy)-1,2,3-propanetricarboxylic acid, tributyl ester. *See* Acetyl tributyl citrate
2-(Acetyloxy)-1,2,3-propanetricarboxylic acid, triethyl ester. *See* Acetyl triethyl citrate
Acetyl pelargonyl. *See* 2,3-Undecadione
Acetyl pentanoyl. *See* 2,3-Heptanedione
Acetylpropionic acid. *See* Levulinic acid
β-Acetylpropionic acid. *See* Levulinic acid
Acetylpropionyl. *See* Pentane-2,3-dione
Acetyl pyrazine. *See* 2-Acetylpyrazine

2-Acetylpyrazine
CAS 22047-25-2; FEMA 3126
Synonyms: Acetyl pyrazine
Empirical: $C_6H_6N_2O$
Properties: Nutty odor; m.w. 122.13; m.p. 76-80 C
Toxicology: Irritant
Uses: Flavoring agent
Manuf./Distrib.: Aldrich

2-Acetylpyridine
CAS 1122-62-9; EINECS 214-355-6; FEMA 3251
Synonyms: Methyl 2-pyridyl ketone
Empirical: C_7H_7NO
Properties: m.w. 121.13; dens. 1.082; b.p. 76-79 C; flash pt. 164 F; ref. index 1.524
Toxicology: Irritant
Uses: Flavoring agent
Manuf./Distrib.: Aldrich; Penta Mfg.

3-Acetylpyridine
CAS 350-03-8; EINECS 206-496-7; FEMA 3424
Synonyms: Methyl 3-pyridyl ketone
Empirical: C_7H_7NO
Properties: Colorless to yel. liq., sweet nutty odor; sol. in acids, alcohol, ether, water; m.w. 121.14; dens. 1.102; m.p. 13-14 C; b.p. 220 C; flash pt. 302 F
Toxicology: Irritant
Uses: Flavoring agent
Manuf./Distrib.: Aldrich

2-Acetylpyrrole
CAS 1072-83-9; EINECS 214-016-2; FEMA 3202
Synonyms: Methyl 2-pyrrolyl ketone
Empirical: C_6H_7NO
Properties: Beige to yel. cryst., bread-like odor; sol. in acids, alcohol, ether, water @ 230 C; m.w. 109.12; m.p. 85-90 C; b.p. 220 C
Toxicology: Skin and eye irritant; heated to decomp., emits toxic fumes of NO_x
Uses: Flavoring agent
Manuf./Distrib.: Aldrich

Acetyl tributyl citrate
CAS 77-90-7; EINECS 201-067-0; FEMA 3080
Synonyms: 2-(Acetyloxy)-1,2,3-propanetricarboxylic acid, tributyl ester; Tributyl acetyl citrate
Classification: Aliphatic ester
Empirical: $C_{20}H_{34}O_8$
Formula: $CH_3COOC_3H_4(COOC_4H_9)_3$
Properties: Colorless sl. visc. liq., sweet herbaceous odor; sol. in alcohol; insol. in water; m.w. 402.49; dens.

1.14; b.p. > 300 C; flash pt. 204 C
Toxicology: Heated to decomp., emits acrid smoke and irritating fumes
Uses: Synthetic flavoring; plasticizer migrating from food pkg.
Regulatory: FDA 21CFR §172.515, 175.105, 175.300, 175.320, 178.3910, 181.22, 181.27
Manuf./Distrib.: Aldrich; Morflex; Pfizer Spec.
Trade names: Uniplex 84

Acetyl triethyl citrate
CAS 77-89-4; EINECS 201-066-5
Synonyms: 2-(Acetyloxy)-1,2,3-propanetricarboxylic acid, triethyl ester; Tricarballylic acid-β-acetoxytributyl ester; Triethyl acetylcitrate
Classification: Aliphatic ester
Empirical: $C_{14}H_{22}O_8$
Formula: $CH_3COOC_3H_4(COOC_2H_5)_3$
Properties: Colorless, odorless liq.; sl. sol. in water; m.w. 318.36; dens. 1.135 (25 C); flash pt. 187 C
Precaution: Combustible
Toxicology: Moderate toxicity by intraperitoneal route; mild toxicity by ingestion
Uses: Plasticizer migrating from food pkg.
Regulatory: FDA 21CFR §175.105, 175.300, 175.320, 178.3910, 181.22, 181.27
Manuf./Distrib.: Morflex

Acetyl valeryl. *See* 2,3-Heptanedione
Acetyl vanillin. *See* Vanillin acetate
Achillea. *See* Yarrow
Achillea millefolium extract. *See* Yarrow extract
Achillea millefolium oil. *See* Yarrow oil
Achilleic acid. *See* Aconitic acid
Achiote. *See* Annatto
Acid ammonium carbonate. *See* Ammonium bicarbonate
Acid Blue 9. *See* FD&C Blue No. 1
Acid calcium phosphate. *See* Calcium phosphate monobasic anhydrous
Acid calcium phosphate. *See* Calcium phosphate monobasic monohydrate
Acid Orange 137. *See* Orange B
Acid potassium tartrate. *See* Potassium acid tartrate
Acid quinine hydrochloride. *See* Quinine dihydrochloride
Acid Red 18. *See* Cochineal
Acid Red 27. *See* Amaranth
Acid Red 51. *See* FD&C Red No. 3
Acids, coconut. *See* Coconut acid
Acid sodium pyrophosphate. *See* Sodium acid pyrophosphate
Acid sodium sulfite. *See* Sodium bisulfite
Acids, soy. *See* Soy acid
Acids, tallow. *See* Tallow acid
Acid Yellow 23. *See* FD&C Yellow No. 5
Acid Yellow 23. *See* Tartrazine
Acimetion. *See* DL-Methionine

Aconitic acid
CAS 499-12-7; FEMA 2010
Synonyms: 1,2,3-Propenetricarboxylic acid; Achilleic acid; Citridic acid; 1-Propene-1,2,3-tricarboxylic acid; Equisetic acid
Definition: Occurs in leaves and tubers of *Aconitum napellus* and other *Ranunculaceae*
Empirical: $C_6H_6O_6$
Formula: $C_3H_3(COOH)_3$
Properties: Wh. or ylsh. cryst. solid; sol. in water, alcohol; sl. sol. in ether; m.w. 174.11; m.p. > 195 C; dec. 198-199 C
Uses: Flavoring agent, adjuvant
Usage level: Limitation 0.003% (baked goods), 0.002% (alcoholic beverages), 0.0015% (frozen dairy prods.), 0.0035% (soft candy), 0.0005% (other food)
Regulatory: FDA 21CFR §184.1007, GRAS

Acraldehyde. *See* Acrolein

Acrolein
CAS 107-02-8; EINECS 203-453-4

Acrylaldehyde

Synonyms: 2-Propenal; Acraldehyde; Acrylic aldehyde; Acrylaldehyde
Empirical: C_3H_4O
Properties: Liq., pungent odor; m.w. 56.06; dens. 0.8389 (20 C); m.p. -88 C; b.p. 52.5 C (760 mm); flash pt. (OC) -18 C
Precaution: Flamm.; unstable; polymerizes in light or presence of alkali or strong acid
Toxicology: LD50 (oral, rat) 0.046 g/kg; irritates skin and mucosa; vapors cause lacrimation; inh. may cause asthmatic reaction, pulmonary edema in lg. doses
Uses: Intermediate for synthetic glycerol

Acrylaldehyde. *See* Acrolein

Acrylamide/sodium acrylate copolymer
CAS 25085-02-3
Synonyms: 2-Propenamide, polymer with 2-propenoic acid, sodium salt; 2-Propenoic acid, sodium salt, polymer with 2-propenamide
Definition: Polymer of acrylamide and sodium acrylate monomers
Formula: $(C_3H_5NO \cdot C_3H_4O_2 \cdot Na)_x$
Uses: Boiler water additive for food contact
Regulatory: FDA 21CFR §173.310

Acrylates/acrylamide copolymer
Definition: Polymer of acrylamide and one or more monomers of acrylic acid, methacrylic acid or one of their simple esters
Toxicology: Heated to decomp., emits acrid smoke and irritating fumes
Uses: Flocculant in clarification of beet sugar juice/liquor, cane sugar juice/liquor, corn starch hydrolyzate; mineral scale control aid
Usage level: Limitation 10 ppm (cane or beet liquor for clarification), 2.5 ppm (cane or beet juice/liquor for min. scale control)
Regulatory: FDA 21CFR §173.5, 175.105, 176.110

Acrylic aldehyde. *See* Acrolein

Acrylonitrile-butadiene copolymer
Uses: Films
Regulatory: FDA 21CFR §181.32

Acrylonitrile-butadiene/PVC blend
Synonyms: NBR/PVC blend
Uses: Used only on paper/paperboard in contact with meats and lard; as extruded pipe in food pkg.
Regulatory: FDA 21CFR §181.32

Acrylonitrile-butadiene-styrene
CAS 9003-56-9
Synonyms: ABS; Acrylonitrile polymer with 1,3-butadiene
Definition: Thermoplastic resin grafted from the 3 monomers from which its name is derived
Properties: Dens. ≈ 1.04; tens. str. ≈ 6500 psi; flex. str. ≈ 10,000 psi
Precaution: Combustible
Uses: Films, coatings, rigid and semirigid containers for handling food prods. and for repeated-use articles contacting food
Regulatory: FDA 21CFR §181.32

Acrylonitrile polymer with 1,3-butadiene. *See* Acrylonitrile-butadiene-styrene

Acrylonitrile-styrene copolymer
CAS 9003-54-7
Synonyms: Acrylonitrile-styrene resin; Styrene-acrylonitrile copolymer; Polystyrene-acrylonitrile; 2-Propenenitrile polymer with ethenylbenzene
Empirical: $(C_8H_8 \cdot C_3H_3N)_x$
Toxicology: LD50 (oral, rat) 1800 mg/kg; mod. to highly toxic by ingestion; heated to decomp., emits toxic fumes of NO_x and CN^-
Uses: Films, coatings, rigid and semirigid containers; food pkg.
Regulatory: FDA 21CFR §181.32

Acrylonitrile-styrene resin. *See* Acrylonitrile-styrene copolymer
Activated charcoal. *See* Carbon, activated
Active carbon. *See* Carbon, activated

Acylase
CAS 9012-37-7; EINECS 232-732-3

Regulatory: Japan approved

Acyl lactylates
Uses: W/o emulsifiers, ingredients in food prods.

ADA. *See* Azodicarbonamide
Adermine hydrochloride. *See* Pyridoxine HCl
Adiantum capillus veneris extract. *See* Maidenhair fern extract
Adiantum pedatum extract. *See* Maidenhair fern extract

Adipic acid
CAS 124-04-9; EINECS 204-673-3; FEMA 2011; EEC E355
Synonyms: Dicarboxylic acid C_6; Hexanedioic acid; 1,4-Butanedicarboxylic acid
Classification: Organic dicarboxylic acid
Empirical: $C_6H_{10}O_4$
Formula: $HOOC(CH_2)_4COOH$
Properties: Wh. monoclinic prisms; very sol. in alcohol; sol. in acetone; m.w. 146.16; dens. 1.360 (25/4 C); m.p. 152 C; b.p. 337.5 C; flash pt. (CC) 385 F
Toxicology: LD50 (oral, mouse) 1900 mg/kg
Uses: Flavoring agent, leavening agent, neutralizer, acidulant, pH control agent for baked goods, baking powds., beverages, condiments, dairy prods., frozen desserts, edible oils, fats, gelatin, gravies, margarine, meat prods., oils, puddings, snack foods
Usage level: Limitation 0.05% (baked goods), 0.005% (nonalcoholic beverages), 5% (condiments), 0.45% (dairy prods.), 0.3% (fats/oil), 0.0004% (frozen dairy desserts), 0.55% (gelatin, puddings), 0.1% (gravies), 0.3% (meat prods.), 1.3% (snack foods), 0.02% (other)
Regulatory: FDA 21CFR §172.515, 184.1009, GRAS; GRAS (FEMA); USDA 9CFR §318.7; Japan approved; Europe listed (ADI 0-5 mg/kg, free acid basis); UK approved
Manuf./Distrib.: Aldrich; Asahi Chem Industry Co Ltd; Penta Mfg.

Agar
CAS 9002-18-0; EINECS 232-658-1; FEMA 2012; EEC E406
Synonyms: Agar-agar; Gelose; Bengal gelatin; Japan isinglass
Definition: A colloidal polygalactoside derived from *Gelidium* spp. or red algae; polysaccharide mixture of agarose and agaropectin
Properties: White to pale yel. color, either odorless or sl. char. odor, sol. in boiling water; insol. in cold water, org. solvs.
Toxicology: LD50 (oral, rat) 11 g/kg; mildly toxic by ingestion; heated to decomp., emits acrid smoke and fumes
Uses: Stabilizer, emulsifier, thickener, drying agent, flavoring agent, surface finisher, formulation aid, humectant; antistaling in baking, confections, meats, poultry; gellant in desserts, beverages; protective colloid in foods
Usage level: Limitation 0.8% (baked goods/mixes), 2% (confections, frostings), 1.2% (soft candy), 0.25% (other foods); ADI no limit (EEC)
Regulatory: FDA 21CFR §150.141, 150.161, 184.1115, GRAS; USDA 9CFR §318.7; GRAS (FEMA); Europe listed; UK approved
Manuf./Distrib.: Am. Roland; Atomergic Chemetals; Browne & Dureau Int'l.; Calaga Food Ingreds.; Chart; Commodity Services; Diamalt; G Fiske; Gumix Int'l.; A C Hatrick; Hercules Ltd.; Honeywill & Stein; Meer; MLG Enterprises; Quest; Spice King; TIC Gums
Trade names: Agar Agar NF Flake #1; Cameo Velvet; Cameo Velvet WT; Nutriloid® Agar; Powdered Agar Agar NF M-100 (Gracilaria); Powdered Agar Agar NF MK-60; Powdered Agar Agar NF MK-80-B; Powdered Agar Agar NF MK-80 (Bacteriological); Powdered Agar Agar NF S-100; Powdered Agar Agar NF S-100-B; Powdered Agar Agar NF S-150; Powdered Agar Agar NF S-150-B; TIC Pretested® Gum Agar Agar 100 FCC/NF Powd.
Trade names containing: Sta-Lite; Sta-Lite

Agar-agar. *See* Agar

Agarase
CAS 37288-57-6
Definition: From *Pseudomonas atlantica*
Regulatory: Japan approved

α-Alanine. *See* L-Alanine

DL-Alanine
CAS 302-72-7; EINECS 206-126-4
Synonyms: dl-2-Aminopropanoic acid
Definition: Racemic mixt. of D- and L-alanine

L-Alanine

Empirical: $C_3H_7NO_2$
Properties: Wh. cryst. powd., odorless, sweet taste; sol. in water; sl. sol. in alcohol; m.w. 89.09; m.p. 198 C
Toxicology: Heated to decomp., emits toxic fumes of NO_x
Uses: Dietary supplement, nutrient; flavor enhancer for sweeteners in pickling mixts., 1% max. of pickling spice
Regulatory: FDA 21CFR §172.320 (limitation 6.1%), 172.540; Japan approved
Manuf./Distrib.: Aldrich

L-Alanine
CAS 56-41-7; EINECS 200-273-8
Synonyms: 2-Aminopropionic acid; L-α-Aminopropionic acid; α-Alanine; L-α-Alanine; 1,2-Aminopropanoic acid
Classification: Amino acid
Empirical: $C_3H_7NO_2$
Formula: CH_3CHNH_2COOH
Properties: Wh. cryst. powd., odorless, sweet taste; sol. in water; sl. sol. in alcohol; insol. in ether; m.w. 89.09; dens. 1.401; dec. 297 C
Toxicology: Heated to decomp., emits toxic fumes of NO_x
Uses: Dietary supplement, nutrient
Regulatory: FDA 21CFR §172.320, limitation 6.1% by wt.; Japan approved
Manuf./Distrib.: Penta Mfg.

L-α-Alanine. *See* L-Alanine
Albacol. *See* Propyl alcohol

Albumen
CAS 9006-50-2
Synonyms: Dried egg white; Egg albumin
Definition: Dried whites of chicken eggs
Properties: Yel. amorphous lumps, scales, or powd.; swells in water, then dissolves gradually; decomp. in moist air
Toxicology: May cause allergic reactions in people allergic to milk or eggs
Uses: Protective colloid and emulsifier in baking; clarifying and refining wines and vinegars; in confectionery, food preps.
Regulatory: FDA 21CFR §160.145
Manuf./Distrib.: Am. Roland; Atomergic Chemetals; British Bakels; Dasco Sales; Farbest Brands; Food Additives & Ingreds.; Frigova Produce; Igreca; Industrial Proteins; Mitsubishi; Moore Fine Foods; Penta Mfg.; Spice King; Alfred L Wolff
Trade names: Egg White Solids Type P-11; Egg White Solids Type P-18G; Egg White Solids Type P-19; Egg White Solids Type P-20; Egg White Solids Type P-21; Egg White Solids Type P-25; Egg White Solids Type P-39; Egg White Solids Type P-110; Egg White Solids Type P-110 High Gel Strength; Egg White Solids Type PF-1; Hentex Type P-1100; Hentex Type P-1800; Hentex Type P-2100; Sol-U-Tein EA
Trade names containing: Egg White Solids Type P-18; Hentex-81; Hentex Type P-600

Albumin. *See also* Albumen and Albumin macro aggregates
Classification: Protein
Definition: Any of a large class of simple proteins that are usually characterized by their solubility in pure water, dilute salt solutions, half-saturated ammonium sulfate or sodium sulfate solutions that are heat coagulable.
Properties: Sol. in pure water and dilute salt sol'ns.
Trade names containing: Mayonat PS

Albumin macro aggregates
CAS 70536-17-3
Synonyms: Albumin
Toxicology: LD50 (IV, rat) 17 mg/kg; poison by IV route; heated to decomp., emits acrid smoke and irritating fumes
Uses: Binder, fining agent; used in imitation sausage, soups, stews, wine
Usage level: Limitation 1.5 gal sol'n./1000 gal wine (sol'n. contg. 2 lb albumin/gal of brine sol'n.)
Regulatory: BATF 27CFR §240.1051

Alcohol
CAS 64-17-5; EINECS 200-578-6
Synonyms: EtOH; Ethyl alcohol, undenatured; Ethanol, undenatured; Distilled spirits; Absolute alcohol
Definition: Undenatured ethyl alcohol
Empirical: C_2H_5OH
Formula: CH_3CH_2OH
Properties: Colorless limpid, volatile liq., vinous odor, pungent taste; misc. with water, methanol, ether,

chloroform, 95% acetone; m.w. 46.08; dens. 0.816 (15.5 C); b.p. 78.3 C; f.p. -117.3 C; flash pt. (CC) 12.7 C; ref. index 1.365 (15 C)

Precaution: Flamm. liq.; can react vigorously with oxidizers; reacts violently with many chemicals

Toxicology: Depressant drug; TLV 1000 ppm in air; moderately toxic by ingestion; experimental tumorigen, teratogen

Uses: Antimicrobial, extraction solv., vehicle; pizza crust, alcoholic beverages; processing aid (Japan)

Usage level: Limitation 2% (pizza crusts); ADI not specified (FAO/WHO)

Regulatory: FDA 21CFR §169.3, 169.175, 169.176, 169.177, 169.178, 169.180, 169.181, 172.340, 172.560, 175.105, 176.200, 176.210, 177.1440, 184.1293, GRAS; 27CFR §2.5, 2.12; Japan restricted

Manuf./Distrib.: ADM Ethanol Sales

Trade names containing: Adeka Menjust; Oxynex® L; Phosal® 53 MCT; Phosal® 75 SA; Sustane® P; Sustane® PA; Sustane® PB; Tenox® 5; Tenox® 5B

Alcohol C-3. *See* Propyl alcohol
Alcohol C-5. *See* n-Amyl alcohol
Alcohol C-6. *See* Hexyl alcohol
Alcohol C-7. *See* 3-Heptanol
Alcohol C-7. *See* Heptyl alcohol
Alcohol C-8. *See* Caprylic alcohol
Alcohol C-9. *See* Nonyl alcohol
Alcohol C-10. *See* n-Decyl alcohol
Alcohol C-11. *See* Undecyl alcohol
Alcohol C-12. *See* Lauryl alcohol
Aldehyde C-1. *See* Formaldehyde
Aldehyde C-3. *See* Propionaldehyde
Aldehyde C-5. *See* n-Valeraldehyde
Aldehyde C-6. *See* Hexanal
Aldehyde C-7. *See* Heptanal
Aldehyde C-8. *See* n-Octanal
Aldehyde C-9. *See* Nonanal
Aldehyde C-10. *See* Decanal
Aldehyde C-10 dimethyl acetal. *See* Decanal dimethyl acetal
Aldehyde C-11 undecyclic. *See* Undecanal
Aldehyde C-11 undecylenic. *See* 10-Undecenal
Aldehyde C-12. *See* Lauric aldehyde
Aldehyde C-12 MNA. *See* Methyl nonyl acetaldehyde
Aldehyde C-14. *See* Myristaldehyde
Aldehyde C-14 pure. *See* γ-Undecalactone
Aldehyde C-14 pure. *See* ς-Undecalactone
Aldehyde C-16. *See* Ethyl methylphenylglycidate
Aldehyde C-18. *See* γ-Nonalactone
Aleurites moluccana oil. *See* Kukui nut oil

Alfalfa
Definition: Herb and seed from *Medicago sativa*
Uses: Natural flavoring
Regulatory: FDA 21CFR §182.10, GRAS
Manuf./Distrib.: Am. Roland; Chart

Alfalfa extract
CAS 84082-36-0; EINECS 281-984-0
Synonyms: Medicago sativa extract; Lucerne extract; Purple medick extract
Definition: Extract of alfalfa, *Medicago sativa*
Uses: Natural flavoring agent
Regulatory: FDA 21CFR §182.20, GRAS
Manuf./Distrib.: Chart

Algae, brown
Synonyms: Brown algae
Definition: Seaweeds of *Analipus japonicus, Eisenia bicyclis, Hizikia fusiforme, Kjellmaniella gyrata, Laminaria angustata, L. claustonia, L. digitata, L. japonica, L. longicruris, Macrocystis pyrifera*, etc.
Toxicology: Heated to decomp., emits acrid smoke and irritating fumes
Uses: Flavor adjuvant, flavor enhancer for use in spices, seasonings, and flavorings
Regulatory: FDA 21CFR §184.1120, GRAS

Algae, red

Synonyms: Red algae
Definition: Seaweeds of *Gloiopeltis furcata, Porphyra crispata, P. deutata, P. perforata, P. suborbiculata, P. tenera, Rhodymenia palmata*
Toxicology: Heated to decomp., emits acrid smoke and irritating fumes
Uses: Flavor adjuvant, flavor enhancer for flavorings, seasonings, spices
Regulatory: FDA 21CFR §184.1121, GRAS

Alganet

Toxicology: Heated to decomp., emits acrid smoke and irritating fumes
Uses: Coloring agent for casings, rendered fat
Regulatory: USDA 9CFR §318.7

Algaroba. *See* Locust bean gum

Algin

CAS 9005-38-3; EEC E401
Synonyms: Sodium alginate; Sodium polymannuronate
Classification: Hydrophilic polysaccharide
Definition: Sodium salt of alginic acid
Empirical: $(C_6H_7O_6Na)_n$
Properties: Cream-colored powd., odorless, tasteless; sol. in water; insol. in alcohol, ether, chloroform; m.w. 198.11
Toxicology: LD50 (IV, rat) 1000 mg/kg; poison by IV and intraperitoneal routes; heated to decomp., emits toxic fumes of Na_2O
Uses: Emulsifier; firming agent; flavor enhancer/adjuvant; formulation aid; processing aid; surfactant; texturizer, stabilizer, thickener; for hard candy, confections, frostings, fruit juice, gelatins, sauces, toppings, ice cream; boiler water additive
Usage level: 0.1-2%; limitation 1% (condiments), 6% (pimento for stuffed olives), 0.3% (confections), 4% (gelatins, puddings), 10% (hard candy), 2% (processed fruits), 1% (other foods); ADI 0-50 mg/kg (EEC)
Regulatory: FDA 21CFR §133.133, 133.134, 133.162, 133.178. 133.179, 150.141, 150.161, 173.310, 184.1724, GRAS; Japan approved; Europe listed; UK approved
Manuf./Distrib.: Am. Roland; Multi-Kem
Trade names: Colloid 488T; Dariloid® Q; Dariloid® QH; Kelco® HV; Kelco® LV; Kelcosol®; Kelgin® F; Kelgin® HV; Kelgin® LV; Kelgin® MV; Kelgin® QL; Kelgin® XL; Kelset®; Keltone®; Keltone® HV; Keltone® LV; Kelvis®; Kimitsu Algin I-1; Kimitsu Algin I-2; Kimitsu Algin I-3; Manucol DH; Manucol DM; Manucol DMF; Manucol LB; Manugel DJX; Manugel DMB; Manugel GHB; Manugel GMB; Prime F-25; Prime F-40; Prime F-400; Prime F-600; Proctin BUS; Protanal 686; Protanal HF 120 M; Protanal HFC 60; Protanal KC 119; Protanal KP; Protanal KPM; Protanal LF 5/60; Protanal LF 20; Protanal LF 20/40; Protanal LF 60; Protanal LF 120 M; Protanal LF 200; Protanal LFS 40; Protanal SF 40; Protanal SF 60; Protanal SF 120; Protanal SF 120 M; Protanal SP 5 H; Protanal VK 687; Protanal VK 749; Protanal VK 805 IMP; Protanal VK 990; Protanal VK 998; Protanal VPM; Protanal VSM; Sobalg FD 100 Range; Sodium Alginate HV NF/FCC; Sodium Alginate LV; Sodium Alginate LVC; Sodium Alginate MV NF/FCC; W-300 FG
Trade names containing: Lamefrost® ES 216 G; Lamefrost® ES 251 G

Alginic acid

CAS 9005-32-7; EINECS 232-680-1; EEC E400
Synonyms: Norgine; Polymannuronic acid
Definition: Polysaccharide composed of β-d-mannuronic acid residues
Formula: $(C_6H_8O_6)_n$
Properties: White to yel. powd., tasteless; sol. in alkaline sol'ns.; very sl. sol. in water; insol. in org. solvs.; capable of absorbing 200-300 times its wt. of water; m.w. ≈ 240,000
Toxicology: LD50 (IP, rat) 1600 mg/kg; ; moderately toxic by intraperitoneal route; heated to decomp., emits acrid smoke and irritating fumes
Uses: Emulsifier, stabilizer, thickener, formulation aid for soups and soup mixes
Usage level: ADI 0-50 mg/kg (EEC)
Regulatory: FDA 21CFR §184.1011, GRAS; Japan approved; Europe listed; UK approved
Manuf./Distrib.: Kelco; Penta Mfg.; Protan Ltd.
Trade names: Alginic Acid FCC; Kelacid®; Kimitsu Acid; Sobalg FD 000 Range

Alginic acid, ammonium salt. *See* Ammonium alginate
Alginic acid, calcium salt. *See* Calcium alginate
Alginic acid, ester with 1,2-propanediol. *See* Propylene glycol alginate
Alginic acid, potassium salt. *See* Potassium alginate
Alkane C-4. *See* Butane

Alkane C-6. *See* Hexane

Alkanet extract
EINECS 286-469-4
Synonyms: Alkanna tinctoria extract; Spanish bugloss extract
Definition: Extract of the roots of *Alkanna tinctoria*
Uses: Coloring sausage casings, oleomargarine, shortenings, confectionery, wines
Regulatory: Cleared by MID; not listed as approved colorant for cosmetics under FDA 21CFR §73 and 74;
   Japan approved

Alkanna tinctoria extract. *See* Alkanet extract
Alligator pear oil. *See* Avocado oil
Allium cepa extract. *See* Onion extract
Allium sativum extract. *See* Garlic extract
Allium sativum oil. *See* Garlic oil
Allomaleic acid. *See* Fumaric acid

Allspice. *See also* Pimenta oil
Synonyms: Pimenta oil; Pimenta berries oil; Pimento oil
Definition: Distilled from the fruit of *Pimenta officinalis*
Properties: Yellow to red-yellow liq., odor and taste of allspice; dens. 1.018-1.048; ref. index 1.527-1.540
   (20 C)
Precaution: Combustible
Toxicology: Skin irritant
Uses: Natural flavoring agent; used in cakes, fruit pies, mincemeat, plum pudding, sauces, soups
Regulatory: FDA 21CFR §182.10, GRAS; Japan approved
Manuf./Distrib.: Chart

Allspice oil. *See* Pimenta oil
Allura Red. *See* FD&C Red No. 40
Allyl acetate. *See* 4-Pentenoic acid
Allylacetic acid. *See* 4-Pentenoic acid
Allyl 2-aminobenzoate. *See* Allyl anthranilate
Allyl o-aminobenzoate. *See* Allyl anthranilate
4-Allylanisole. *See* Estragole
p-Allylanisole. *See* Estragole

Allyl anthranilate
FEMA 2020
Synonyms: Allyl 2-aminobenzoate; Allyl o-aminobenzoate
Empirical: $C_{10}H_{11}NO_2$
Properties: m.w. 177.21
Uses: Synthetic flavoring agent
Usage level: 1.1 ppm (nonalcoholic beverages), 0.67 ppm (ice cream, ices, etc.), 2.0 ppm (candy), 0.02-1.0
   ppm (baked goods), 2.0 ppm (gelatins and puddings)
Regulatory: FDA 21CFR §172.515

5-Allyl-1,3-benzodioxole. *See* Safrol

Allyl butyrate
CAS 2051-78-7; FEMA 2021
Empirical: $C_7H_{12}O_2$
Properties: m.w. 128.17; dens. 0.902; b.p. 44-45 C; flash pt. 107 F; ref. index 1.4158
Toxicology: Irritant
Uses: Synthetic flavoring agent
Usage level: 1.2 ppm (nonalcoholic beverages), 0.50-1.0 ppm (ice cream, ices, etc.), 1.3 ppm (candy), 0.50-
   3.0 ppm (baked goods); 1.0 ppm (gelatins and puddings)
Regulatory: FDA 21CFR §172.515
Manuf./Distrib.: Aldrich

Allyl caproate
CAS 123-68-2; EINECS 204-642-4; FEMA 2032
Synonyms: 2-Propenyl hexanoate; Allyl capronate; Caproic acid allyl ester; Allyl hexanoate
Empirical: $C_9H_{16}O_2$
Properties: Colorless to lt. yel. liq., pineapple aroma; misc. with alcohol, ether; insol. in water; m.w. 156.23;
   dens. 0.887; b.p. 75-76 C; flash pt. 151 F; ref. index 1.4243
Toxicology: LD50 (oral, rat) 218 mg/kg; poison by ingestion and skin contact; skin irritant; heated to decomp.,

emits acrid smoke and irritating fumes
Uses: Pungent, fatty, fruity fragrance and flavoring; rum- and pineapple-like on dilution; for candy, puddings
Usage level: 7.0 ppm (nonalcoholic beverages), 11 ppm (ice cream, ices, etc.), 32 ppm (candy), 25 ppm (baked goods), 22 ppm (gelatins and puddings), 210 ppm (chewing gum)
Regulatory: FDA 21CFR §172.515; Japan approved
Manuf./Distrib.: Aldrich; Chr. Hansen's

Allyl capronate. *See* Allyl caproate
Allyl caprylate. *See* Allyl octanoate

Allyl cinnamate
CAS 1866-31-5; FEMA 2022
Synonyms: Allyl-3-phenylacrylate; Propenyl cinnamate; Vinyl carbinyl cinnamate
Empirical: $C_{12}H_{12}O_2$
Properties: Colorless to light yel. liq.; cherry odor; sol. in ether, alcohol; insol. in water; m.w. 188.24; dens. 1.052 (25/25 C); b.p. 150-152 C; flash pt. >230 F; ref. index 1.5661
Toxicology: LD50 (oral, rat) 1520 mg/kg; mod. toxic by ingestion; skin irritant; heated to decomp., emits acrid smoke and fumes
Uses: Synthetic flavoring agent
Usage level: 1.0 ppm (nonalcoholic beverages), 1.4 ppm (ice cream, ices, etc.), 1.8 ppm (candy), 2.6 ppm (baked goods)
Regulatory: FDA 21CFR §172.515
Manuf./Distrib.: Aldrich

Allyl cyclohexaneacetate
FEMA 2023
Empirical: $C_{11}H_{18}O_2$
Properties: Liq., intense fruital aroma; m.w. 182.26; b.p. 66 C; ref. index 1.4574
Uses: Synthetic flavoring agent
Usage level: 1.1 ppm (nonalcoholic beverages), 1.6 ppm (ice cream, ices, etc.), 3.5 ppm (candy), 4.0 ppm (baked goods)
Regulatory: FDA 21CFR §172.515

Allyl cyclohexanebutyrate
FEMA 2024
Empirical: $C_{13}H_{22}O_2$
Properties: Liq., pineapple odor; m.w. 210.31; b.p. 104 C; ref. index 1.4608
Uses: Synthetic flavoring agent
Usage level: 1.0 ppm (nonalcoholic beverages), 1.4 ppm (ice cream, ices, etc.), 3.3 ppm (candy), 3.8 ppm (baked goods)
Regulatory: FDA 21CFR §172.515

Allyl cyclohexanehexanoate
FEMA 2025
Synonyms: Allyl cyclohexylcaproate; Allyl cyclohexylcapronate
Empirical: $C_{15}H_{26}O_2$
Properties: m.w. 238.37
Uses: Synthetic flavoring agent
Usage level: 1.4 ppm (nonalcoholic beverages), 3.3 ppm (ice cream, ices, etc.), 8.0 ppm (candy), 8.5 ppm (baked goods)
Regulatory: FDA 21CFR §172.515

Allyl cyclohexanepropionate
CAS 2705-87-5; FEMA 2026
Synonyms: Allyl-3-cyclohexanepropionate; 3-Allylcyclohexyl propionate; Allyl hexahydrophenylpropionate
Empirical: $C_{12}H_{20}O_2$
Properties: Colorless liq.; pineapple odor; misc. in alcohol, chloroform, ether; insol. in glycerin, water; m.w. 196.32; dens. 0.945-0.950; b.p. 91 C; flash pt. 212 F; ref. index 1.457-1.463
Precaution: Combustible liq.
Toxicology: LD50 (oral, rat) 585 mg/kg; LD50 (oral, guinea pig) 380 mg/kg; poison by ingestion; heated to decomp., emits acrid smoke and fumes
Uses: Synthetic flavoring agent
Usage level: 3.7 ppm (nonalcoholic beverages), 3.1 ppm (ice cream, ices etc.), 13 ppm (candy), 7.1 ppm (baked goods), 7.7 ppm (gelatins and puddings), 30 ppm (chewing gum), 0.20 ppm (icings)
Regulatory: FDA 21CFR §172.515; Japan approved
Manuf./Distrib.: Aldrich

Allyl-3-cyclohexanepropionate. *See* Allyl cyclohexanepropionate

Allyl cyclohexanevalerate
FEMA 2027
Synonyms: Allyl cyclohexylpentanoate
Empirical: $C_{14}H_{24}O_2$
Properties: Liq., char. fruital aroma; m.w. 224.34; b.p. 119 C; ref. index 1.4605
Uses: Synthetic flavoring agent
Usage level: 1.2 ppm (nonalcoholic beverages), 2.3 ppm (ice cream, ices, etc.), 4.4 ppm (candy), 4.8 ppm (baked goods)
Regulatory: FDA 21CFR §172.515

Allyl cyclohexylcaproate. *See* Allyl cyclohexanehexanoate
Allyl cyclohexylcapronate. *See* Allyl cyclohexanehexanoate
Allyl cyclohexylpentanoate. *See* Allyl cyclohexanevalerate
3-Allylcyclohexyl propionate. *See* Allyl cyclohexanepropionate
4-Allyl-1,2-dimethoxybenzene. *See* Methyl eugenol
4-Allyl-1,2-dimethoxybenzene. *See* Methyl isoeugenol
Allyl trans-2,3-dimethylacrylate. *See* Allyl tiglate

Allyl disulfide
CAS 2179-57-9; EINECS 218-548-6; FEMA 2028
Synonyms: Diallyl disulfide
Empirical: $C_6H_{10}S_2$
Properties: Liq., char. garlic odor; sol. in most common org. solvs.; insol. in water; m.w. 146.28; dens. 1.008; b.p. 138-139 C; flash pt. 144 F; ref. index 1.541
Uses: Synthetic flavoring agent
Usage level: 6.5 ppm (condiments), 7.0 ppm (meats)
Regulatory: FDA 21CFR §172.515
Manuf./Distrib.: Aldrich

Allyl 2-ethylbutyrate
FEMA 2029
Empirical: $C_9H_{16}O_2$
Properties: Liq., ethereal aroma; m.w. 156.23; b.p. 105-107 O; ref. index 1.4240
Uses: Synthetic flavoring agent
Usage level: 0.50-1.0 ppm (nonalcoholic beverages), 2.0 ppm (candy), 1.0 ppm (gelatins and puddings)
Regulatory: FDA 21CFR §172.515

4-Allylguaiacol. *See* Eugenol
Allyl hendecenoate. *See* Allyl 10-undecenoate

Allyl heptanoate
CAS 142-19-8; FEMA 2031
Synonyms: Allyl heptoate; Allyl heptylate; 2-Propenyl heptanoate
Empirical: $C_{10}H_{18}O_2$
Properties: Colorless to pale yel. liq., fruity sweet pineapple odor; m.w. 170.28; dens. 0.880; flash pt. 154 F; ref. index 1.426
Precaution: Combustible liq.
Toxicology: LD50 (oral, rat) 500 mg/kg; LD50 (oral, mouse) 630 mg/kg; mod. toxic by ingestion, skin contact; skin irritant; heated to decomp., emits acrid smoke and fume
Uses: Synthetic flavoring agent
Usage level: 1.3 ppm (nonalcoholic beverages), 2.7 ppm (ice cream, ices), 6.4 ppm (candy, baked goods), 2.9 ppm (gelatins, puddings), 86 ppm (chewing gum)
Manuf./Distrib.: Aldrich

Allyl heptoate. *See* Allyl heptanoate
Allyl heptylate. *See* Allyl heptanoate
Allyl 2,4-hexadienoate. *See* Allyl sorbate
Allyl hexahydrophenylpropionate. *See* Allyl cyclohexanepropionate
Allyl hexanoate. *See* Allyl caproate
Allyl ionone. *See* Allyl α-ionone

Allyl α-ionone
CAS 79-78-7; FEMA 2033
Synonyms: 1-(2,6,6-Trimethyl-2-cyclohexene-1-yl)-1,6-heptadiene-3-one; Allyl ionone; Cetone V
Empirical: $C_{16}H_{24}O$

Allyl isosulfocyanate

>Properties: Colorless to yel. liq., fruity woody odor; sol. in alcohol; insol. in water; m.w. 232.40; dens. 0.928-
> 0.935; b.p. 102-104 C; flash pt. 212 F; ref. index 1.503-1.507
>Precaution: Combustible liq.
>Toxicology: Skin irritant; heated to decomp., emits acrid smoke and fumes
>Uses: Synthetic flavoring agent
>Usage level: 0.50 ppm (nonalcoholic beverages), 1.4 ppm (ice cream, ices, etc.), 2.6 ppm (candy), 3.1 ppm
> (baked goods), 1.0 ppm (gelatins and puddings), 2.0 ppm (jellies)
>Regulatory: FDA 21CFR §172.515

Allyl isosulfocyanate. See Allyl isothiocyanate

Allyl isothiocyanate

>CAS 57-06-7; EINECS 200-309-2; FEMA 2034
>Synonyms: Mustard oil; Isothiocyanic acid allyl ester; Allyl isosulfocyanate; 3-Isothiocyanato-1-propene
>Definition: Obtained by the distillation of Brassica species
>Empirical: C_4H_5NS
>Formula: $CH_2=CHCH_2NCS$
>Properties: Colorless to pale yel. liq., pungent irritable odor, sharp pungent mustard taste; misc. with alcohol,
> carbon disulfide, ether, most org. solvs.; sl. sol. in waterm.w. 99.16; dens. 1.013-1.016; m.p. -80 C; b.p.
> 150.7 C; flash pt. 115 F; ref. index 1.5
>Precaution: Volatile; combustible liq.
>Toxicology: LD50 (oral, rat) 339 mg/kg; poison by ingestion, skin contact; lachrymator; allergen; mutagenic;
> heated to decomp., emits highly toxic fumes
>Uses: Synthetic flavoring agent; used in baked goods, condiments, mustard oil, pickles, salad dressings,
> sauces
>Regulatory: FDA 21CFR §172.515; Japan approved as flavoring
>Manuf./Distrib.: Aldrich

Allyl isovalerate

>CAS 2835-39-4; FEMA 2045
>Synonyms: Isovaleric acid allyl ester; Allyl 3-methylbutyrate; 2-Propenyl isovalerate; 2-Propenyl 3-
> methylbutanoate
>Empirical: $C_8H_{14}O_2$
>Properties: Colorless to pale yel. liq., apple aroma; m.w. 142.20; b.p. 89-90 C; ref. index 1.4162
>Toxicology: LD50 (oral, rat) 230 mg/kg; poison by ingestion; mod. toxic by skin contact; experimental
> carcinogen, tumorigen; skin irritant; heated to decomp, emits acrid smoke and fumes
>Uses: Synthetic flavoring agent
>Usage level: 8.6 ppm (nonalcoholic beverages), 18 ppm (ice cream, ices etc.), 22 ppm (candy), 15-48 ppm
> (baked goods), 1.0 ppm (gelatins and puddings)
>Regulatory: FDA 21CFR §172.515

Allyl mercaptan

>CAS 870-23-5; EINECS 212-792-7; FEMA 2035
>Synonyms: 2-Propene-1-thiol; Allyl sulfhydrate; Allylthiol
>Empirical: C_3H_6S
>Formula: $CH_2:CHCH_2SH$
>Properties: White liq., strong garlic odor; m.w. 74.15; dens. 0.925; b.p. 68 C; flash pt. 14 F; ref. index 0.925 (23
> C)
>Precaution: Fire hazard
>Toxicology: Poison by inh., ingestion; strong irritant to skin, mucous membranes; heated to decomp. emits
> acrid fumes of SO_x
>Uses: Synthetic flavoring agent; used in baked goods, condiments
>Regulatory: FDA 21CFR §172.515
>Manuf./Distrib.: Aldrich

4-Allyl-2-methoxyphenol. See Eugenol
4-Allyl-2-methoxyphenyl acetate. See Eugenyl acetate
4-Allyl-2-methoxyphenyl benzoate. See Eugenyl benzoate
4-Allyl-2-methoxyphenyl formate. See Eugenyl formate
Allyl trans-2-methyl-2-butenoate. See Allyl tiglate
Allyl 3-methylbutyrate. See Allyl isovalerate
4-Allyl-1,2-methylenedioxybenzene. See Safrol

Allyl nonanoate

>CAS 7493-72-3; FEMA 2036
>Synonyms: Allyl pelargonate

Empirical: $C_{12}H_{22}O_2$
Properties: Mobil liq., pineapple odor; m.w. 198.31; b.p. 151 C; ref. index 1.4302
Uses: Synthetic flavoring agent
Usage level: 0.70 ppm (nonalcoholic beverages), 0.50-3.0 ppm (ice cream, ices, etc.), 5.0 ppm (candy), 3.0-5.0 ppm (baked goods), 1.0 ppm (meats)
Regulatory: FDA 21CFR §172.515
Manuf./Distrib.: Aldrich

Allyl octanoate
CAS 4230-97-1; FEMA 2037
Synonyms: Allyl caprylate; Octanoic acid allyl ester; Allyl octylate
Empirical: $C_{11}H_{20}O_2$
Properties: Colorless liq., fruity odor; sol. in alcohol, fixed oils; sl. sol. in propylene glycol; insol. in glycerin, water; m.w. 184.31; dens. 0.8550; b.p. 87-88 C; flash pt. 151 F; ref. index 1.425
Toxicology: LD50 (oral, rat) 570 mg/kg; mod. toxic by ingestion; skin irritant; heated to decomp., emits acrid smoke and fumes
Uses: Synthetic flavoring agent; used in beverages, candy, dessert gels, puddings
Regulatory: FDA 21CFR §172.515
Manuf./Distrib.: Aldrich

Allyl octylate. *See* Allyl octanoate
Allyl pelargonate. *See* Allyl nonanoate

Allyl phenoxyacetate
CAS 7493-74-5; FEMA 2038
Synonyms: Acetate PA
Empirical: $C_{11}H_{12}O_3$
Properties: Liq., honey and pineapple-like aroma; m.w. 192.22; b.p. 100-102 C; ref. index 1.5131 (25.5 C)
Toxicology: LD50 (oral, rat) 475 mg/kg; mod. toxic by ingestion; heated to decomp., emits acrid smoke and fumes
Uses: Synthetic flavoring agent; used in beverages, candy
Regulatory: FDA 21CFR §172.515

Allyl phenylacetate
CAS 1797-74-6; FEMA 2039
Synonyms: Allyl α-toluate; Benzeneacetic acid 2-propenyl ester
Empirical: $C_{11}H_{12}O_2$
Properties: Colorless to lt. yel. liq., fruit banana honey odor; m.w. 176.22; b.p. 89-93 C; ref. index 1.5122
Toxicology: LD50 (oral, rat) 650 mg/kg; mod. toxic by ingestion; human skin irritant; heated to decomp., emits acrid smoke and irritating fumes
Uses: Synthetic flavoring agent
Usage level: 0.06-3.0 ppm (nonalcoholic beverages), 8.0 ppm (ice cream, ices, etc.), 14 ppm (candy), 40 ppm (baked goods)
Regulatory: FDA 21CFR §172.515
Manuf./Distrib.: Aldrich

Allyl-3-phenylacrylate. *See* Allyl cinnamate

Allyl propionate
FEMA 2040
Empirical: $C_6H_{10}O_2$
Properties: Liq., apple odor; m.w. 114.15; b.p. 122-123 C; ref. index 1.4105 (20 C), 1.4142 (14 C)
Uses: Synthetic flavoring agent
Usage level: 0.06-3.0 ppm (nonalcoholic beverages), 16 ppm (ice cream, ices,etc.), 6.5 ppm (candy), 10 ppm (baked goods)
Regulatory: FDA 21CFR §172.515

Allylpyrocatechol methylene ether. *See* Safrol

Allyl sorbate
FEMA 2041
Synonyms: Allyl 2,4-hexadienoate
Empirical: $C_9H_{12}O_2$
Properties: m.w. 152.19
Uses: Synthetic flavoring agent
Usage level: 0.86 ppm (nonalcoholic beverages), 0.50 ppm (ice cream, ices etc.), 0.50 -5.0 ppm (candy), 1.0 ppm (baked goods), 2.0 ppm (gelatins and puddings)

Allyl sulfhydrate

Regulatory: FDA 21CFR §172.515

Allyl sulfhydrate. See Allyl mercaptan

Allyl sulfide
CAS 592-88-1; EINECS 209-775-1; FEMA 2042
Synonyms: Diallyl sulfide; Oil garlic; Thioallyl ether; 3,3′-Thiobis-1-propene
Empirical: $C_6H_{10}S$
Formula: $(CH_2:CHCH_2)_2S$
Properties: Colorless liq.; garlic odor; misc. with alcohol, ether, chloroform, CCl_4; insol. in water; m.w. 114.20; dens. 0.887 (20/4 C); ; m.p. -83 C; b.p. 138 C; flash pt. 23 C; ref. index 1.4877 (27 C)
Precaution: Flamm.
Uses: Synthetic flavoring agent
Usage level: 0.04 ppm (nonalcoholic beverages), 0.06 ppm (ice cream, ices, etc.), 0.07 ppm (candy), 0.05 ppm (baked goods), 13 ppm (condiments), 3.7 ppm (meats)
Regulatory: FDA 21CFR §172.515
Manuf./Distrib.: Aldrich

Allylthiol. See Allyl mercaptan

Allyl tiglate
CAS 7493-71-2; FEMA 2043
Synonyms: Allyl trans-2,3-dimethylacrylate; Allyl trans-2-methyl-2-butenoate
Empirical: $C_8H_{12}O_2$
Properties: Fruity odor; m.w. 140.18; dens. 0.926; flash pt. 140 F
Uses: Synthetic flavoring agent
Usage level: 0.28 ppm (nonalcoholic beverages), 0.50 ppm (ice cream, ices, etc.), 0.50-3.0 ppm (candy), 0.50-3.0 ppm (baked goods)
Regulatory: FDA 21CFR §172.515
Manuf./Distrib.: Aldrich

Allyl α-toluate. See Allyl phenylacetate

Allyl 10-undecenoate
FEMA 2044
Synonyms: Allyl hendecenoate; Allyl undecylenate; Allyl undecylenoate
Empirical: $C_{14}H_{24}O_2$
Properties: Liq., pineapple odor; misc. with most org. solvs.; insol. in water; m.w. 224.34; b.p. 180 C; ref. index 1.4448
Uses: Synthetic flavoring agent
Usage level: 0.25 ppm-1.0 ppm (nonalcoholic beverages), 0.50 ppm (ice cream, ices, etc.), 0.50 ppm (candy), 0.50 ppm (baked goods)
Regulatory: FDA 21CFR §172.515

Allyl undecylenate. See Allyl 10-undecenoate
Allyl undecylenoate. See Allyl 10-undecenoate
4-Allyl veratrole. See Methyl eugenol
Almond oil, bitter. See Bitter almond oil
Almond oil, sweet. See Sweet almond oil
Aloe barbadensis extract. See Aloe extract

Aloe extract
CAS 85507-69-3; EINECS 305-181-2
Synonyms: Aloe barbadensis extract; Barbados aloe extract; Curacao aloe extract
Definition: Extract of leaves of one or more species of Aloe
Properties: Bitter char. flavor
Uses: Natural flavoring for beverages; thickener/stabilizer
Usage level: 5-2000 ppm (nonalcoholic beverages), 130 ppm (alcoholic beverages)
Regulatory: FDA 21CFR §172.510; Japan approved
Manuf./Distrib.: Chart
Trade names containing: Aloe Vera Lipo-Quinone Extract™ (AVLQE) Food Grade

Aloe vera gel
Definition: Mucilage obtained from expression of juice from leaves of Aloe barbadensis
Uses: Used in beverages, vegetable drinks, health foods; juice enhancer
Regulatory: FDA 21CFR §172.510
Trade names: Activera™; Activera™ 1-1FA (Filtered); Activera™ 1-1 UA (Unfiltered); Activera™ 1-200 A; Aloe Con UP 10; Aloe Con UP 40; Aloe-Con UP-200; Aloe Con WG 10; Aloe Con WG 40; Aloe Con WG 200;

Aloe Con WLG 10; Aloe Con WLG 200; Aloe Vera Gel 1X; Aloe Vera Gel 10X; Aloe Vera Gel 40X; Essential Aloe™ GII-1X; Essential Aloe™ GII-10X; Essential Aloe™ GII-40X; Essential Aloe™ GII-100X; Essential Aloe™ GII-200X; Powdered Aloe Vera (1:200) Food Grade; Terra-Dry™ FD Aloe Vera Powd. Decolorized FDD; Terra-Dry™ FD Aloe Vera Powd. Reg. FDR; Terra-Pulp™ FD Whole Aloe Vera Gel Fillet; Terra-Pure™ FD Non-Preserved Decolorized FDD; Terra-Pure™ FD Non-Preserved Reg. FDR; Terra-Pure™ Non-Preserved Aloe Vera Powd; Terra-Spray™ Spray Dried Aloe Vera Powd. Decolorized SDD; Terra-Spray™ Spray Dried Aloe Vera Powd. Reg. SDR

Trade names containing: Aloe Vera Gel Thickened FG

Althea extract
CAS 97676-24-9; FEMA 2048
Synonyms: Althea officinalis extract; Marshmallow root extract
Definition: Extract of the roots of the marsh mallow, *Althea officinalis*
Uses: Natural flavoring for beverages
Usage level: 5.7-10 ppm (nonalcoholic beverages)
Regulatory: FDA 21CFR §172.510 (roots and flowers)
Manuf./Distrib.: Chart

Althea officinalis extract. *See* Althea extract
Alum. *See* Aluminum sulfate

Alumina
CAS 1344-28-1; EINECS 215-691-6
Synonyms: Aluminum oxide; Calcined alumina; Alumite
Classification: Inorganic compd.
Empirical: Al_2O_3
Properties: m.w. 101.96
Toxicology: Toxic by inhalation of dust; TLV:TWA 10 mg/m^3 (dust)
Uses: Food additive, dispersant
Regulatory: Exempt from certification
Manuf./Distrib.: Atomergic Chemetals

Alumina, hydrate
CAS 1333-84-2 (hydrate)
Synonyms: Alumina trihydrate; Aluminum hydroxide; Aluminum trihydroxide
Classification: Inorganic compd.
Empirical: $Al_2O_3 \cdot 3H_2O$
Properties: White powd., balls, or lumps; insol. in water, strong alkalis; m.w. 156.01; dens. 3.4-4.0; m.p. 2030 C
Uses: Food additive
Manuf./Distrib.: Atomergic Chemetals; Croxton & Garry Ltd

Aluminate, sodium. *See* Sodium aluminate
Alumina trihydrate. *See* Alumina, hydrate

Aluminum
CAS 7429-90-5; EINECS 231-072-3; EEC E173
Classification: Metallic element
Empirical: Al
Properties: Silvery wh. cryst. solid; m.w. 26.98; dens. 2.708; m.p. 660 C; b.p. 2450 C
Precaution: Flammable; explosive; no stable isotopes
Toxicology: TLV 10 mg/m^3 of air; (sol. salt) 2 mg/m^3 of air; (welding fumes) 5 mg/m^3 of air
Uses: Colorant for surface only; not permitted in certain foods
Regulatory: Japan approved

Aluminum ammonium sulfate. *See* Ammonium alum

Aluminum calcium silicate
CAS 1327-39-5; EEC E556
Synonyms: Calcium aluminum silicate
Toxicology: Nuisance dust
Uses: Anticaking agent in table salt, vanilla powd.
Regulatory: FDA 21CFR §182.2122 (2% max.), GRAS; Europe listed; UK approved

Aluminum caprylate
CAS 6028-57-5; EINECS 227-902-9
Synonyms: Aluminum octanoate
Definition: Aluminum salt of caprylic acid

Aluminum, dihydroxy (octadecanoato-o-)

Uses: Binder, emulsifier, anticaking agent
Regulatory: FDA 21CFR §172.863, 175.105, 175.300, 175.320, 176.170, 176.200, 176.210, 177.1200, 177.2260, 178.3910

Aluminum, dihydroxy (octadecanoato-o-). See Aluminum stearate

Aluminum distearate
CAS 300-92-5; EINECS 206-101-8
Definition: Aluminum salt of stearic acid
Empirical: $C_{36}H_{71}O_5Al$
Formula: $[CH_3(CH_2)_{16}COO]_2Al(OH)$
Properties: White powd.; insol. in water, alcohol, ether; forms gel w/ aliphatic and aromatic hydrocarbons; dens. 1.009; m.p. 145 C
Uses: Stabilizer migrating from food pkg.
Regulatory: FDA 21CFR §172.863, 173.340, 175.105, 175.300, 175.320, 176.170, 176.200, 176.210, 177.1200, 177.1460, 177.2260, 178.3910, 179.45, 181.22, 181.29
Trade names: Aluminum Stearate EA; Witco® Aluminum Stearate EA Food Grade

Aluminum hydrate. See Aluminum hydroxide

Aluminum hydroxide
CAS 21645-51-2; EINECS 244-492-7
Synonyms: Aluminum oxide trihydrate; Aluminum trihydrate; Aluminum hydrate; Hydrated alumina
Classification: Inorganic compd.
Empirical: AlH_3O_3
Formula: $Al(OH)_3$
Properties: White cryst. powd.; insol. in water; sol. in min. acids, caustic soda; m.w. 78.01; dens. 2.42; m.p. loses water @ 300 C
Toxicology: TLV:TWA 2 mg (Al)/m³; poison by intraperitoneal route
Uses: Migrating to food from paper/paperboard
Regulatory: FDA 21CFR §175.300, 177.1200, 177.2600, 182.90

Aluminum hydroxide. See Alumina, hydrate

Aluminum nicotinate
CAS 1976-28-9
Synonyms: Nicotinic acid aluminum salt; Tris (nicotinato) aluminum; 3-Pyridinecarboxylic acid aluminum salt
Empirical: $C_{18}H_{12}AlN_3O_6$
Properties: Solid; m.w. 393.30
Toxicology: Heated to decomp., emits toxic fumes of NO_x
Uses: Dietary supplement, nutrient; source of niacin in foods for special dietary use
Regulatory: FDA 21CFR §172.310

Aluminum octanoate. See Aluminum caprylate

Aluminum oleate
Synonyms: 9-Octadecenoic acid aluminum salt; Oleic acid aluminum salt
Empirical: $C_{54}H_{99}AlO_6$
Formula: $[CH_3(CH_2)_7CH=CH(CH_2)_7COO]_3Al$
Properties: Ylsh. visc. mass; sol. in alcohol, benzene, ether, oil turpentine; pract. insol. in water; m.w. 871.36
Uses: Migrating to foods from paper/paperboard
Regulatory: FDA 21CFR §182.90, GRAS

Aluminum oxide. See Alumina

Aluminum oxide trihydrate. See Aluminum hydroxide
Aluminum palmitate
Synonyms: Hexadecanoic acid aluminum salt; Palmitic acid aluminum salt
Empirical: $C_{48}H_{93}AlO_6$
Formula: $[CH_3(CH_2)_{14}COO]_3Al$
Properties: Wh. to yel. mass or powd.; pract. insol. in water or alcohol; dissolves in petrol. ether or oil turpentine when fresh; m.w. 793.25
Uses: Migrating to foods from paper/paperboard
Regulatory: FDA 21CFR §182.90, GRAS

Aluminum potassium sulfate. See Potassium alum dodecahydrate

Aluminum silicate
CAS 1327-36-2; 12141-46-7; EINECS 215-475-1

Synonyms: Pyrophyllite; CI 77004
Classification: Complex inorganic salt
Definition: Complex inorganic salt with 1 mole alumina, 1-3 moles silica
Formula: Al_2O_5Si
Properties: Varying proportions of Al_2O_3 and SiO_2; crystals or whiskers; high str.; m.w. 162.05
Uses: Thickener in beverages and foods
Regulatory: FDA 21CFR §175.300, 177.1200, 177.1460, 177.2600, 184.1155
Trade names: Suspengel Elite; Suspengel Micro; Suspengel Ultra

Aluminum sodium oxide. *See* Sodium aluminate
Aluminum sodium silicate. *See* Sodium silicoaluminate
Aluminum sodium sulfate. *See* Sodium alum

Aluminum starch octenyl succinate
CAS 9087-61-0
Synonyms: Starch, octenylbutanedioate, aluminum salt
Definition: Aluminum salt of the prod. of octenyl/succinic anhydride with starch
Uses: Food powds.
Regulatory: FDA 21CFR §172.892

Aluminum stearate
CAS 7047-84-9; EINECS 230-325-5
Synonyms: Aluminum, dihydroxy (octadecanoato-o-); Stearic acid aluminum dihydroxide salt
Definition: Aluminum salt of stearic acid
Empirical: $C_{18}H_{37}AlO_4$
Formula: $CH_3(CH_2)_{16}COOAl(OH)_2$
Properties: White powd.; insol. in water, alcohol, ether; sol. in alkali, petrol., turpentine oil; m.w. 344.48; dens. 1.070; m.p. 115 C
Toxicology: TLV:TWA 2 mg(Al)/m³; heated to decomp., emits acrid smoke and irritating fumes
Uses: Anticaking agent, binder, emulsifier; defoaming agent used in beet sugar, yeast processing; stabilizer migrating from food pkg.
Regulatory: FDA §121.1099, 172.863, 173.340, 181.29

Aluminum sulfate
CAS 10043-01-3; EINECS 233-135-0
Synonyms: Alum; Aluminum trisulfate; Cake alum
Classification: Inorganic salt
Empirical: $Al_2O_{12}S_3$
Formula: $Al_2(SO_4)_3 \cdot 14H_2O$
Properties: White powd., sweet taste; sol. in water; insol. in alcohol; m.w. 342.14 (anhyd.); dens. 2.71; b.p. dec. @ 770 C
Precaution: Forms sulfuric acid w/ water; decomposes to sulfur oxides at high temps.
Toxicology: Irritating to skin, eyes, respiratory tract; ingestion causes nausea, vomiting, abdominal pain; LD50 (oral, mousel) 6207 mg/kg; TLV-TWA 2 mg/m³(ACCGIH)
Uses: Food additive, firming agent, animal glue adjuvant; used in pickle relish, pickles, potatoes, pkg. materials
Regulatory: FDA 21CFR §172.892, 173.3120, 182.1125, GRAS
Manuf./Distrib.: Rhone-Poulenc
Trade names containing: BL-60®

Aluminum trihydrate. *See* Aluminum hydroxide
Aluminum trihydroxide. *See* Alumina, hydrate

Aluminum tristearate
CAS 637-12-7; EINECS 211-279-5
Definition: Aluminum salt of stearic acid
Empirical: $C_{54}H_{105}O_6 \cdot Al$
Formula: $[CH_3(CH_2)_{16}COO]_3Al$
Properties: Hard material; sol. in alcohol, benzene, oil turpentine, min. oils; pract. insol. in water; m.w. 877.35; m.p. 117-120 C
Uses: Stabilizer migrating from food pkg.
Regulatory: FDA 21CFR §172.863, 173.340, 175.105, 175.300, 175.320, 176.170, 176.200, 176.210, 177.1200, 177.1460, 177.260, 178.3910, 179.45, 181.22, 181.29

Aluminum trisulfate. *See* Aluminum sulfate
Alumite. *See* Alumina
Alum, potassium. *See* Potassium alum dodecahydrate

Amaranth
CAS 915-67-3; EEC E123
Synonyms: FD&C Red No. 2; Acid Red 27; Red Dye No. 2; Azorubin S; CI 16185
Classification: Azo dye
Empirical: $C_{20}H_{11}N_2Na_3O_{10}S_3$
Formula: $NaSO_3C_{10}H_5N=NC_{10}H_4(SO_3Na)_2OH$
Properties: Dk. red to purple powd.; sol. in water, glycerol, propylene glycol; very sl. sol. in alcohol; insol. in most
   org. solvs.; m.w. 604.48; dens. ≈ 1.50
Toxicology: Possible carcinogen
Uses: Food colorant
Regulatory: Banned by FDA for use in foods, drugs, and cosmetics; UK approved

Amber acid. *See* Succinic acid

Ambergris
EINECS 232-454-2
Synonyms: Physeter macrocephalus
Definition: Derived from *Phyester macrocephalus*
Uses: Natural flavoring agent, aroma
Usage level: 2.0 ppm (nonalcoholic beverages), 1.7 ppm (ice cream, ices, etc.), 9.7 ppm (candy), 0.10 ppm
   (baked goods)
Regulatory: FDA 21CFR §182.50, GRAS; Japan approved

Ambrette seed liq. *See* Ambrette seed oil

Ambrette seed oil
Synonyms: Ambrette seed liq.
Definition: Derived from *Hibiscus moschatus*
Properties: Oil; brandy-like odor; ref. index 1.4680 (20 C)
Toxicology: Heated to decomp., emits acrid smoke and fumes
Uses: Natural flavoring agent
Usage level: 0.30 ppm (nonalcoholic beverages), 0.30-0.50 ppm (ice cream, ices, etc.), 0.80 ppm (candy), 0.80
   ppm (baked goods)
Regulatory: FDA 21CFR §182.10, 182.20, GRAS; Japan approved
Manuf./Distrib.: Pierre Chauvet

Ambrettolide
Synonyms: ω-6-Hexadecenlactone; 16-Hydroxy-6-hexadecenoic acid, ω-lactone
Uses: Synthetic flavoring
Regulatory: FDA 21CFR §172.515

American dill seed oil. *See* Dill seed oil
Amide C-18. *See* Stearamide
Aminic acid. *See* Formic acid
Aminoacetic acid. *See* Glycine
(S)-α-Aminobenzenepropanoic acid. *See* L-Phenylalanine
2-Aminobenzoic acid methyl ester. *See* Methyl anthranilate
2-Amino-4-carbamoylbutanoic acid. *See* L-Glutamine
Aminocyclohexane. *See* Cyclohexylamine
2-Amino-2-deoxy-(1-4)-β-D-glucopyranan. *See* Chitosan
2-Aminoethanol. *See* Ethanolamine
2-Aminoethyl alcohol. *See* Ethanolamine
Aminoform. *See* Hexamethylene tetramine
L-2-Aminoglutaric acid. *See* L-Glutamic acid
L-2-Aminoglutaric acid hydrochloride. *See* L-Glutamic acid hydrochloride
1-1-Amino-4-guanidovaleric acid. *See* L-Arginine
L-1-Amino-4-guanidovaleric acid monohydrochloride. *See* L-Arginine monohydrochloride
Aminohexahydrobenzene. *See* Cyclohexylamine
α-Aminohydrocinnamic acid. *See* L-Phenylalanine
L-2-Amino-3-hydroxybutyric acid. *See* L-Threonine
α-Amino-β-hydroxybutyric acid. *See* L-Threonine
(S)-2-Amino-3-(4-hydroxyphenyl)propionic acid. *See* L-Tyrosine
(+)-2-Amino-3-hydroxypropionic acid. *See* DL-Serine
2-Amino-3-hydroxypropionic acid. *See* L-Serine
α-Amino-β-imidazolepropionic acid. *See* L-Histidine
l-α-Amino-3-indolepropionic acid. *See* L-Tryptophan
(±)-2-Amino-3-(3-indolyl)propionic acid. *See* DL-α-Tryptophan

α-**Aminoisocaproic acid.** *See* L-Leucine
α-**Aminoisovaleric acid.** *See* L-Valine
L-2-Amino-3-mercaptopropanoic acid. *See* L-Cysteine
L-2-Amino-3-mercaptopropanoic acid monohydrochloride monohydrate. *See* Cysteine hydrochloride
(+)-2-Amino-3-mercaptopropionic acid. *See* L-Cysteine
(±)-2-Amino-4-(methylmercapto)butyric acid. *See* DL-Methionine
(S)-2-Amino-4-(methylmercapto)butyric acid. *See* L-Methionine
2-Amino-3-methylpentanoic acid. *See* L-Isoleucine
3-[(4-Amino-2-methyl-5-pyrimidinyl)methyl]-4-(2-hydroxyethyl)-4-methylthiazolium nitrate (salt). *See* Thiamine nitrate
2-Amino-4-(methylthio)butyric acid. *See* DL-Methionine
2-Amino-4-(methylthio)butyric acid. *See* L-Methionine
2-Amino-3-methylvaleric acid. *See* L-Isoleucine
DL-2-Amino-3-methylvaleric acid. *See* DL-Isoleucine
DL-2-Amino-4-methylvaleric acid. *See* DL-Leucine
α-**Amino-β-methylvaleric acid.** *See* L-Isoleucine
α-**Amino-γ-methylvaleric acid.** *See* L-Leucine
2-Amino-4-methypentanoic acid. *See* L-Leucine
1-α-Amino-4 (or 5)-imidazolepropionic acid monohydrochloride. *See* Histidine hydrochloride monohydrate
2-Aminopentanedioic acid. *See* L-Glutamic acid
2-Aminopentanedioic acid hydrochloride. *See* L-Glutamic acid hydrochloride
DL-α-Amino-β-phenylpropionic acid. *See* DL-Phenylalanine
1,2-Aminopropanoic acid. *See* L-Alanine
dl-2-Aminopropanoic acid. *See* DL-Alanine
2-Aminopropionic acid. *See* L-Alanine
L-α-Aminopropionic acid. *See* L-Alanine
1-α-Aminosuccinamic acid. *See* L-Asparagine
2-Aminosuccinamic acid. *See* L-Asparagine
(-)-Aminosuccinic acid. *See* D-Aspartic acid
(+)-Aminosuccinic acid. *See* L-Aspartic acid
DL-Aminosuccinic acid. *See* DL-Aspartic acid
α-**Amino-β-thiolpropionic acid.** *See* L-Cysteine

Ammonia
  CAS 7664-41-7
  Empirical: NH₃
  Properties: Colorless gas or liquid; sharp, intensely irritating odor; easily liquefied by pressure; m.w. 18.02; b.p. -33.5 C; f.p. -77 C
  Precaution: Moderate fire risk
  Toxicology: Inh. of conc. fumes may be fatal
  Uses: Processing aid
  Regulatory: Japan approved

Ammonia solution, strong. *See* Ammonium hydroxide
Ammoniated glycyrrhizin. *See* Monoammonium glycyrrhizinate
Ammonia water. *See* Ammonium hydroxide

Ammonium alginate
  CAS 9005-34-9; EEC E403
  Synonyms: Ammonium polymannuronate; Alginic acid, ammonium salt
  Definition: Ammonium salt of alginic acid
  Formula: C₆H₇O₆ • NH₄
  Properties: Filamentous, grainy, granular, or powd.; colorless or sl. yel.; sl. smell or taste; slowly sol. in water forming visc. sol'n.; insol. in alcohol
  Toxicology: Heated to decomp., emits toxic fumes of NO_x
  Uses: Emulsifier, thickening agent, stabilizer, humectant, color diluent (gelatins, frostings, jams/jellies, sauces, toppings, ice cream); boiler water additive
  Usage level: Limitation 0.4% (confections), 0.5% (fats/oils), 0.5% (gelatins, puddings), 0.4% (gravies, sauces), 0.4% (jams, jellies), 0.5% (sweet sauces), 0.1% (other foods); ADI 0-50 mg/kg (EEC)
  Regulatory: FDA 21CFR §173.310, 184.1133, GRAS; GRAS (FEMA); USDA 9CFR 318.7; Europe listed; UK approved
  Manuf./Distrib.: Kelco Int'l.
  Trade names: Ammonium Alginate Type S; Amoloid HV; Amoloid LV; Sobalg FD 300 Range
  Trade names containing: Keltose®

Ammonium alum
CAS 7784-25-0; EINECS 232-055-3
Synonyms: Aluminum ammonium sulfate; Ammonium aluminum sulfate; Sulfuric acid, aluminum ammonium
  salt (2:1:1), dodecahydrate
Classification: Inorganic salt
Empirical: $Al \cdot H_3N \cdot 2H_2O_4S \cdot 12H_2O$
Formula: $AlNH_4(SO_4)_2 \cdot 12H_2O$
Properties: Wh. cryst. powd., strong sweet astringent taste
Toxicology: Irritating if inhaled or ingested; on decomp., emits toxic fumes of NO_x and SO_x
Uses: Buffer, neutralizing agent in baking powd.; clarifying agent, food additive; migrating to food from paper/
  paperboard
Regulatory: FDA 21CFR §178.3120, 182.90, 182.1127, GRAS; Japan approved except for miso

Ammonium aluminum sulfate. *See* Ammonium alum

Ammonium bicarbonate
CAS 1066-33-7; EINECS 213-911-5; EEC E503
Synonyms: Ammonium hydrogen carbonate; Acid ammonium carbonate; Carbonic acid, monoammonium salt
Classification: Inorganic salt
Empirical: $CH_2O_3 \cdot H_3N$
Formula: NH_4HCO_3
Properties: Colorless or wh. cryst., faint ammonia odor; sol. 17.4% in water (20 C); dec. by hot water; insol.
  in alcohol, acetone; m.w. 79.1; dens. 1.586; m.p. dec. 36-60 C
Precaution: Evolves irritating fumes on heating to 35 C
Toxicology: Poison by intravenous route; LD50 (IV, mouse) 245 mg/kg; heated to decomp., emits toxic fumes
  of NO_x and NH_3
Uses: Direct food additive, alkali, dough strengthener, leavening agent, pH control agent, texturizer for cookies,
  crackers, baking powd.
Regulatory: FDA 21CFR §163.110, 182.1135, 184.1135, GRAS; Japan approved; Europe listed; UK approved
Manuf./Distrib.: Browning
Trade names: ABC-Trieb®; ABC-Trieb®; ABC-Trieb®

Ammonium biphosphate. *See* Ammonium phosphate

Ammonium carbonate
CAS 506-87-6, 8000-73-5, 10361-29-2; EINECS 233-786-0; EEC E503
Synonyms: Carbonic acid, ammonium salt; Hartshorn
Definition: Mixt. of ammonium bicarbonate and ammonium carbamate
Empirical: $CH_2O_3 \cdot xH_3N$
Formula: $(NH_4)_2CO_3$
Properties: Wh. powd. or wh. or translucent hard mass, strong ammonia odor, sharp taste; slowly sol. in 4 parts
  water; dec. by hot water
Precaution: Decomp. on exposure to air; volatilizes at 60 C; keep tightly closed in cool place; incompat. with
  acids, acid salts, salts of iron and zinc, alkaloids, alum, tartar emetic
Toxicology: LD50 (IV, mouse) 96 mg/kg; poison by subcutaneous and IV routes; heated to decomp., emits toxic
  fumes of NO_x and NH_3
Uses: Direct food additive, buffer, neutralizing agent, food additive, leavening agent, pH control agent, yeast
  nutrient; for baked goods, baking powd., caramel, gelatins, wine
Regulatory: FDA 21CFR §184.1137, GRAS; BATF 27CFR 240.1051, limitation 0.2%; Japan approved; Europe
  listed; UK approved

Ammonium carrageenan
Definition: Ammonium salt of carrageenan
Uses: Emulsifier, stabilizer, thickener
Regulatory: FDA 21CFR §172.626

Ammonium caseinate
CAS 1336-21-6; 9005-42-9
Synonyms: Casein, ammonium salt
Uses: Dispersant, leveling agent, food additive
Regulatory: Granted prior sanction via clearance for optional use in food standards for frozen desserts

Ammonium chloride
CAS 12125-02-9; EINECS 235-186-4; EEC E510
Synonyms: Sal ammoniac; Salmiac; Ammonium muriate
Classification: Inorganic salt

Empirical: ClH₄N

Formula: NH₄Cl

Properties: Colorless to wh. cryst. or powd., cooling saline taste; hygroscopic; sol. 28.3% in water, glycerol; sol. in methanol, ethanol; m.w. 53.50; dens. 1.5274; m.p. 337.8 C; b.p. 520 C; pH 4.5-5.5 (1M water, 20 C)

Precaution: Incompat. with alkalies and their carbonates, lead and silver salts

Toxicology: LD50 (oral, rat) 1650 mg/kg; TLV (fume) 10 mg/m³ of air; poison by subcutaneous, IV and intramuscular routes; severe eye irritant; heated to decomp., emits very toxic fumes of NO$_x$, Cl⁻, and NH₃

Uses: Direct food additive, yeast food, dough conditioner/strengthener, flavor enhancer, leavening agent, processing aid, improving agent; in beer-brewing

Usage level: ADI no limit (EEC)

Regulatory: FDA 21CFR §184.1138, GRAS; Japan approved; Europe listed; UK approved

Manuf./Distrib.: Heico

Ammonium citrate

CAS 3012-65-5; EINECS 221-146-3

Synonyms: Ammonium citrate dibasic (DOT); Ammonium citrate secondary; Diammonium citrate; Diammonium hydrogen citrate; Citric acid diammonium salt

Empirical: C₆H₁₄N₂O₇

Properties: Gran. or cryst.; sol. in 1 part water; sl. sol. in alcohol; m.w. 226.19; dens. 1.48

Toxicology: Skin and eye irritant; heated to decomp., emits acrid smoke and irritating fumes

Uses: Stabilizer migrating from food pkg.

Regulatory: FDA 21CFR §181.29

Ammonium citrate dibasic. *See* Ammonium citrate

Ammonium citrate secondary. *See* Ammonium citrate

Ammonium dihydrogen orthophosphate. *See* Ammonium phosphate

Ammonium dihydrogen phosphate. *See* Ammonium phosphate

Ammonium ferric citrate. *See* Iron ammonium citrate

Ammonium furcelleran

Definition: Ammonium salt of furcelleran

Uses: Emulsifier, stabilizer, thickener

Regulatory: FDA 21CFR §172.660

Ammonium glutamate

CAS 7558-63-6

Synonyms: MAG; Monoammonium glutamate; Monoammonium L-glutamate

Empirical: C₅H₉NO₄ • H₃N

Properties: Wh. cryst. powd., odorless; sol. in water; insol. in common org. solvs.; m.w. 164.19

Toxicology: LD50 (IP, rat) 1000 mg/kg; mod. toxic by IP route; heated to decomp., emits toxic fumes of NO$_x$ and NH₃

Uses: Multipurpose food ingred.; flavor enhancer, salt substitute; used in meat, poultry

Regulatory: FDA 21CFR §182.1500, GRAS; USDA 9CFR §318.7, 381.147

Manuf./Distrib.: MLG Enterprises

Ammonium glycyrrhizinate. *See* Monoammonium glycyrrhizinate

Ammonium glycyrrhizinate pentahydrate. *See* Monoammonium glycyrrhizinate

Ammonium hydrogen carbonate. *See* Ammonium bicarbonate

Ammonium hydroxide

CAS 1336-21-6; EINECS 215-647-6; EEC E527

Synonyms: Ammonia solution, strong; Ammonia water; Aqua ammonium; Spirit of Hartshorn

Classification: Inorganic base

Empirical: H₅NO

Formula: NH₄OH

Properties: Clear colorless liq., very pungent odor; sol. in water; m.w. 35.06; dens. 0.90; m.p. -77 C

Precaution: DOT: Corrosive

Toxicology: LD50 (oral, rat) 350 mg/kg; poison by ingestion; inhalation irritant; severe eye irritant; liq. can inflict burns; heated to decomp., emits NH₃ and NO$_x$

Uses: Direct food additive, alkali, leavening agent, pH control agent, surface finishing agent for baked goods, caramel, cheese, processed fruits, puddings; boiler water additive; food coloring diluent/solvent; processing aid

Usage level: ADI no limit (EEC)

Regulatory: FDA 21CFR §163.110, 182.90, 184.1139, GRAS; Europe listed; UK approved

Ammonium isovalerate
FEMA 2054
Synonyms: Isovaleric acid ammonium salt
Empirical: $C_5H_{13}O_2N$
Properties: Deliq. cryst.; sol. in alcohol, water; m.w. 109
Uses: Synthetic flavoring agent; used in butter, nut, and cheese flavors
Usage level: 58 ppm (baked goods), 0.20 ppm (syrups)
Regulatory: FDA 21CFR §172.515

Ammonium monosulfide. *See* Ammonium sulfide
Ammonium muriate. *See* Ammonium chloride
Ammonium peroxydisulfate. *See* Ammonium persulfate

Ammonium persulfate
CAS 7727-54-0; EINECS 231-786-5
Synonyms: Ammonium peroxydisulfate; Peroxydisulfuric acid diammonium salt
Classification: Inorganic salt
Empirical: $H_3N \cdot \frac{1}{2}H_2O_8S_2$
Formula: $(NH_4)_2S_2O_8$
Properties: m.w. 228.20; dens. 1.982
Precaution: Oxidizer; corrosive
Uses: Food preservative; bleaching agent for food starch; flour treatment agent (Japan)
Regulatory: FDA 21CFR §172.892, 175.105, 176.170, 177.1200, 178.3520; Japan approved

Ammonium phosphate
CAS 7722-76-1; EINECS 231-764-5
Synonyms: MAP; Monoammonium phosphate; Primary ammonium phosphate; Ammonium dihydrogen orthophosphate; Ammonium phosphate monobasic; Ammonium biphosphate; Ammonium dihydrogen phosphate
Empirical: H_6NO_4P
Formula: $NH_4H_2PO_4$
Properties: Brilliant wh. cryst. or powd., odorless; mildly acidic in reaction; moderately sol. in water; sl. sol. in alcohol; pract. insol. in acetone; m.w. 115.04; dens. 1.803; m.p. 190 C
Uses: Direct food additive, mfg. of yeast, vinegar, yeast foods and bread improvers, food additive, buffer, pH control agent, dough strenghthener/conditioner, leavening agent for baked goods, baking powd., margarine, whipped toppings, frozen desserts
Regulatory: FDA 21CFR §184.1141a, GRAS; Japan approved
Manuf./Distrib.: Albright & Wilson Am.; Rhone-Poulenc Food Ingreds.
Trade names: Albrite Monoammonium Phosphate Food Grade

Ammonium phosphate, dibasic
CAS 7783-28-0; EINECS 231-987-8
Synonyms: DAP; Diammonium phosphate; Diammonium hydrogen phosphate; Diammonium hydrogen orthophosphate; Secondary ammonium phosphate
Classification: Inorganic salt
Empirical: $H_9N_2O_4P$
Formula: $(NH_4)_2HPO_4$
Properties: Wh. cryst. or powd., odorless, cooling salty taste; mildly alkaline in reaction; sol. 1 g/1.7 ml in water; pract. insol. in alcohol, acetone; m.w. 132.07; dens. 1.619; m.p. 155 C (dec.); pH ≈ 8
Precaution: Noncombustible; keep well closed
Toxicology: Low to moderate toxicity; heated to decomp., emits very toxic fumes of PO_x, NO_x, and NH_3
Uses: Direct food additive, nutrient in mfg. of yeast, vinegar, bread improvers, buffer, pH control agent, dough strengthener/conditioner, leavening agent, firming agent, processing aid; foods, wines, pharmaceuticals, purifying sugar; emulsifier (Japan)
Regulatory: FDA 21CFR §573.320, 184.1141b, GRAS; BATF 27CFR 240.1051, limitation 0.17% as yeast nutrient in wine prod., 0.8% in sparkling wines; Japan approved
Manuf./Distrib.: Albright & Wilson Am.; Rhone-Poulenc Food Ingreds.
Trade names: Albrite Diammonium Phosphate Food Grade

Ammonium phosphate monobasic. *See* Ammonium phosphate
Ammonium polymannuronate. *See* Ammonium alginate

Ammonium potassium hydrogen phosphate
Toxicology: Heated to decomp., emits acrid smoke and irritating fumes
Uses: Stabilizer migrating from food pkg.
Regulatory: FDA 21CFR §181.29

Ammonium saccharin
EINECS 228-971-8
Synonyms: 1,2-Benzisothiazolin-3-one 1,1-dioxide ammonium salt
Empirical: $C_7H_8N_2O_3S$
Properties: Wh. cryst. or cryst. powd., intense sweet taste; sol. in water; m.w. 200.21
Toxicology: Heated to decomp., emits toxic fumes of NO_x
Uses: Noncaloric sweetener for beverages, drink mixes, processed foods, chewable vitamin tablets, chewing gum, baked goods; as sugar substitute for cooking or table use
Usage level: Limitation 12 mg/fl oz (beverages, fruit juice drinks, beverage mixes), 30 mg/serving (processed foods)
Regulatory: FDA 21CFR §180.37; USDA 9CFR §318.7 (limitation 0.01%)

Ammonium sulfate
CAS 7783-20-2; EINECS 231-984-1
Synonyms: Sulfuric acid diammonium salt; Diammoniium sulfate; Mascagnite
Classification: Inorganic salt
Empirical: $H_8N_2O_4$ S
Formula: $(NH_4)_2SO_4$
Properties: Colorless or wh. cryst. or gran., odorless; sol. in water; insol. in alcohol, acetone; m.w. 132.16; dens. 1.77; m.p. > 280 C (dec.)
Toxicology: Moderately toxic by several routes; LD50 (oral, rat) 3000 mg/kg; heated to decomp., emits very toxic fumes of NO_x, NH_3, and SO_x
Uses: Food additive, dough strenghthener/conditioner, yeast food, firming agent, processing aid, fermentation aid for baked goods, gelatins, puddings
Usage level: limitation 0.15% (baked goods), 0.1% (gelatins, puddings)
Regulatory: FDA 21CFR §177.1200, 184.1143, GRAS; Japan approved
Manuf./Distrib.: Heico
Trade names containing: FD-6; FD-8; N'Hance (Phase I) Powd

Ammonium sulfide
CAS 12135-76-1; EINECS 235-223-4; FEMA 2053
Synonyms: Ammonium monosulfide
Empirical: H_8N_2S
Formula. $(NH_4)_2S$
Properties: Liq. @ R.T.; sol. in alcohol, ammonia, cold water; m.w. 68.15; dens. 0.997 (20/4 C); flash pt. 90 F
Precaution: Flamm.; corrosive
Toxicology: Strong irritant to skin and mucous membranes
Uses: Synthetic flavoring agent
Usage level: 5.0 ppm (baked good), 5.0 ppm (condiments)
Regulatory: FDA 21CFR §172.515
Manuf./Distrib.: Aldrich

Ammonium sulfite
CAS 10196-04-0; EINECS 233-484-9
Synonyms: Sulfurous acid, diammonium salt
Classification: Inorganic salt
Empirical: $H_8N_2O_3S$
Properties: Colorless cryst., acrid sulfurous taste; sol. in water; m.w. 116.14; dens. 1.41
Uses: Caramel production agent

Amorphous sodium polyphosphate. *See* Sodium hexametaphosphate
Amygdalin. *See* Bitter almond oil

Amyl acetate
CAS 628-63-7; EINECS 211-047-3
Synonyms: Amylacetic ester; Banana oil; Pear oil; Acetic acid pentyl ester; Pentyl acetate
Definition: Ester of amyl alcohol and acetic acid
Empirical: $C_7H_{12}O_2$
Formula: $CH_3COOC_5H_{11}$
Properties: Liq.; m.w. 130.19; dens. 0.876; m.p. -100; f.p. 75 F; b.p. 142 C; flash pt. 75 F
Precaution: Flamm.
Toxicology: Irritant
Uses: Flavoring agent
Regulatory: FDA 21CFR §175.105; GRAS (FEMA)

Amylacetic ester

Manuf./Distrib.: Aldrich; Penta Mfg.

Amylacetic ester. *See* Amyl acetate

n-Amyl alcohol
CAS 71-41-0; EINECS 200-752-1; FEMA 2056
Synonyms: 1-Pentanol; Alcohol C-5; Pentyl alcohol
Empirical: $C_5H_{12}O$
Formula: $CH_3(CH_2)_4OH$
Properties: Clear liq., mild char. odor; sol. in water; misc. with alcohol, ether; m.w. 88.15; dens. 0.812 (20/4 C); m.p. -79 C; b.p. 137-139 C; flash pt. (CC) 38 C; ref. index 1.409
Precaution: Flamm.
Toxicology: LD50 (oral, rat) 3030 mg/kg; irritating to eyes, respiratory tract; narcotic
Uses: Synthetic flavoring agent
Usage level: 18 ppm (nonalcoholic beverages), 15 ppm (ice cream, ices), 35 ppm (candy), 24 ppm (baked goods), 7.7-50 ppm (gelatins, puddings), 150-340 ppm (chewing gum),
Regulatory: FDA 21CFR §172.515; GRAS (FEMA)
Manuf./Distrib.: Aldrich; Ashland

sec-n-Amyl alcohol. *See* 3-Pentanol

t-Amyl alcohol
CAS 75-85-4; EINECS 136-135-1
Synonyms: 2-Methyl-2-butanol; t-Pentyl alcohol; t-Pentanol; Dimethyl ethyl carbinol; Amylene hydrate
Empirical: $C_5H_{12}O$
Formula: $CH_3CH_2C(CH_3)_2OH$
Properties: Volatile liq., char. odor, burning taste; sol. in 8 parts water; misc. with alcohol, ether, benzene, chloroform, glycerin, oils; m.w. 88.15; dens. 0.808 (20/4 C); m.p. -9 C; b.p. 100-103 C; ref. index 1.405
Precaution: Keep tightly closed; protect from light
Toxicology: LD50 (oral, rat) 1 g/kg; mod. irritating to human mucous membranes; narcotic in high concs.

Amylaldehyde. *See* n-Valeraldehyde

Amylase
CAS 9000-92-4; EINECS 232-567-7
Synonyms: Mylase 100; 1,4-D-Glucan glucanohydrolase
Classification: Enzyme
Definition: Starch-degrading enzyme
Properties: Off-white powd. or suspension
Uses: Conversion of starch to glucose sugar in syrups (esp. corn syrups), baking (to improve crumb softness and shelf life); brewing, distilling
Trade names: Amflex; Atlas MDA; Biodiastase; Biozyme L; Biozyme M; Biozyme S; Clarase® 5,000; Clarase® 40,000; Clarase® Conc; Clarase® L-40,000; Fermalpha; Spezyme BBA; Taka-Therm® L-340; Tenase® 1200; Tenase® L-340; Tenase® L-1200
Trade names containing: CDC-10A; CDC-10A (NB); CDC-2001; CDC-2002; Pancreatin 3X USP Powd; Pancreatin 4X USP Powd; Pancreatin 5X USP Powd; Pancreatin 6X USP Powd; Pancreatin 8X USP Powd; Pancreatin USP Powd; Pancrelipase USP; Proflex; SPL Pancreatin 3X USP; SPL Pancreatin 4X USP; SPL Pancreatin 5X USP; SPL Pancreatin 6X USP; SPL Pancreatin 7X USP; SPL Pancreatin 8X USP; SPL Pancreatin USP; SPL Pancrelipase USP; SPL Undiluted Pancreatic Enzyme Conc. (PEC)

α-Amylase
CAS 9000-90-2; EINECS 232-565-6
Uses: Natural enzyme for food use
Regulatory: FDA 21CFR §184.1027 (from *B. licheniformis*); GRAS for several other sources; UK, Japan approved

β-Amylase
CAS 9001-91-3; EINECS 232-566-1
Uses: Natural enzyme for food use
Regulatory: FDA GRAS; Japan approved

Amyl butyrate. *See also* Isoamyl butyrate
CAS 540-18-1; EINECS 208-739-2; FEMA 2059
Synonyms: Pentyl butyrate; n-Amyl butyrate; Butanoic acid pentyl ester
Empirical: $C_9H_{18}O_2$
Formula: $CH_3CH_2CH_2COOCH_2(CH_2)_3CH_3$
Properties: Colorless liq.; strong, penetrating apricot-like odor, sweet taste; very sol. in alcohol, ether; sol. 0.54 g/l in water (50 C); m.w. 158.24; dens. 0.8713 (15/4 C); m.p. -73.2 C; b.p.185-186 C; ref. index 1.4110

(20 C)
Toxicology: LD50 (oral, rat) 12,210 mg/kg
Uses: Synthetic flavoring agent for apricot, pineapple, pear, plum flavors
Usage level: 19 ppm (nonalcoholic beverages), 32 ppm (ice cream, ices, etc.), 76 ppm (candy), 43 ppm (baked
  goods), 0.50-1.4 ppm (gelatins and puddings), 760 ppm (chewing gum), 58 ppm (syrups)
Regulatory: FDA 21CFR §172.515
Manuf./Distrib.: Aldrich

n-Amyl butyrate. *See* Amyl butyrate
γ-N-Amylbutyrolactone. *See* γ-Nonalactone
Amyl caproate. *See* Amyl hexanoate
Amylcarbinol. *See* Hexyl alcohol
Amylcinnamaldehyde. *See* α-Amylcinnamaldehyde

α-Amylcinnamaldehyde
CAS 122-40-7; FEMA 2061
Synonyms: Amylcinnamaldeyhde; Jasminaldehyde; Amyl cinnamic aldehyde
Empirical: $C_{14}H_{18}O$
Properties: Yellow liq.; floral odor; sol. in most fixed oils; m.w. 202.30; dens. 0.970; b.p. 153-154 C (10 mm);
  flash pt. > 230 F; ref. index 1.5552 (20 C)
Toxicology: LD50 (oral, rat) 3730 mg/kg: mod. toxic by ingestion; mild skin irritant; heated to decomp., emits
  acrid smoke and fumes
Uses: Synthetic flavoring agent
Usage level: 1.3 ppm (nonalcoholic beverages), 1.5 ppm (ice cream, ices etc.), 4.0 ppm (candy), 4.5 ppm
  (baked goods), 0.03-0.05 ppm (gelatins and puddings), 15 ppm (chewing gum)
Regulatory: FDA 21CFR §172.515; Japan approved as flavoring
Manuf./Distrib.: Aldrich

α-Amylcinnamaldehyde dimethyl acetal
Uses: Synthetic flavoring agent
Regulatory: FDA 21CFR §172.515

Amyl cinnamate. *See* Isoamyl cinnamate
n-Amyl cinnamic alcohol. *See* α-Amylcinnamyl alcohol
Amyl cinnamic aldehyde. *See* α-Amylcinnamaldehyde

α-Amylcinnamyl acetate
Uses: Synthetic flavoring agent
Regulatory: FDA 21CFR §172.515

α-Amylcinnamyl alcohol
FEMA 2065
Synonyms: n-Amyl cinnamic alcohol; 2-Amyl-3-phenyl-2-propen-1-ol; 2-Benzylidene-heptanol
Empirical: $C_{14}H_{20}O$
Properties: Ylsh. liq.; m.w. 204.31; flash pt. > 100 C; ref. index 1.5330-1.5400 (20 C)
Uses: Synthetic flavoring agent
Usage level: 0.47 (nonalcoholic beverages), 1.5 ppm (ice cream, ices), 1.6 ppm (candy), 1.5 ppm (baked
  goods), 2.0 ppm (chewing gum)
Regulatory: FDA 21CFR §172.515

α-Amylcinnamyl formate
FEMA 2066
Synonyms: α-n-Amyl-β-phenylacryl formate; α-Pentylcinnamyl formate
Empirical: $C_{15}H_{20}O_2$
Properties: Colorless liq., herbaceous odor; sol. in alcohol; m.w. 232.33
Uses: Synthetic flavoring agent
Usage level: 0.17 ppm (nonalcoholic beverages), 0.93 ppm (ice cream, ices), 1.5 ppm (candy, baked goods),
  1.0 ppm (chewing gum)
Regulatory: FDA 21CFR §172.515

α-Amylcinnamyl isovalerate
FEMA 2067
Synonyms: α-n-Amyl-β-phenylacryl isovalerate; Floxin isovalerate; α-Pentylcinnamyl isovalerate
Empirical: $C_{19}H_{28}O_2$
Properties: Colorless liq., mild fruity odor, somewhat spicy flavor; sol. in alcohol; m.w. 288.43
Uses: Synthetic flavoring agent
Usage level: 0.36 ppm (nonalcoholic beverages), 1.2 ppm (ice cream, ices), 1.3 ppm (candy), 1.7 ppm (baked

goods), 1.0 ppm (chewing gum)
Regulatory: FDA 21CFR §172.515

Amylene hydrate. *See* t-Amyl alcohol
Amyl formate. *See* Isoamyl formate

Amyl furoate
CAS 1334-82-3; FEMA 2072
Manuf./Distrib.: Aldrich

Amyl heptanoate
FEMA 2073
Synonyms: Amyl heptoate; Amyl heptylate; Pentyl heptanoate
Empirical: $C_{12}H_{24}O_2$
Properties: Colorless liq., fruital odor; sol. in most organic solvents; m.w. 200.32; b.p. 245.4 C; ref. index 1.42627 (20 C)
Toxicology: Heated to decomp., emits acrid smoke and fumes
Uses: Synthetic flavoring agent
Usage level: 7.0 ppm (nonalcoholic beverages), 3.8 ppm (ice cream, ices, etc.), 7.5 ppm (candy), 3.0 ppm (baked goods), 3.5 ppm (gelatins and puddings), 53 ppm (chewing gum)
Regulatory: FDA 21CFR §172.515

Amyl heptoate. *See* Amyl heptanoate
Amyl heptylate. *See* Amyl heptanoate

Amyl hexanoate. *See also* Isoamyl hexanoate
FEMA 2074
Synonyms: Amyl caproate; Amyl hexylate; Pentyl hexanoate
Empirical: $C_{11}H_{22}O_2$
Properties: Colorless liq.; banana odor; m.w. 186.30; sp.gr. 0.8612; m.p. -47 C; b.p. 226 C; ref. index 1.4202 (25 C)
Uses: Synthetic flavoring agent
Usage level: 5.3 ppm (nonalcoholic beverages), 16 ppm (ice cream, ices, etc.), 22 ppm (candy), 8.3 ppm (baked goods), 0.30-3.7 ppm (gelatins and puddings), 110 ppm (chewing gum)
Regulatory: FDA 21CFR §172.515

Amyl hexylate. *See* Amyl hexanoate
Amyl isovalerate. *See* Isoamyl isovalerate
n-Amyl methyl ketone. *See* Methyl n-amyl ketone

Amyl octanoate
FEMA 2079
Synonyms: Isoamyl octanoate; Isoamyl caprylate; Pentyl octanoate
Empirical: $C_{13}H_{26}O_2$
Properties: Liq., orris odor; m.w. 214.35; sp.gr. 0.8562; m.p. -34 C; b.p. 260 C; ref. index 1.4262 (25 C)
Uses: Synthetic flavoring agent
Usage level: 5.0 ppm (nonalcoholic beverages), 3.5 ppm (ice cream, ices, etc.), 6.0 ppm (candy), 3.5 ppm (baked goods), 2.1 ppm (gelatins and puddings)
Regulatory: FDA 21CFR §172.515

Amyloglucosidase
CAS 9032-08-0
Definition: Derived from *Rhizopus niveus* with diatomaceous earth as carrier
Properties: Powd.
Toxicology: Heated to decomp., emits acrid smoke and irritating fumes
Uses: Enzyme, degrading agent; degrades gelatinized starch into sugars in prod. of distilled spirits and vinegar
Usage level: Limitation 0.1% (of gelatinized starch)
Regulatory: FDA 21CFR §173.110
Trade names: Gluczyme; GNL-3000

α-n-Amyl-β-phenylacryl formate. *See* α-Amylcinnamyl formate
α-n-Amyl-β-phenylacryl isovalerate. *See* α-Amylcinnamyl isovalerate
2-Amyl-3-phenyl-2-propen-1-ol. *See* α-Amylcinnamyl alcohol

Amyl propionate
FEMA 2082
Synonyms: Isoamyl propionate
Empirical: $C_8H_{16}O_2$

Properties: Colorless liq., fruity apricot-pineapple odor; sol. in alcohol, most fixed oils; insol. in glycol, water; m.w. 144.21; dens. 0.869-0.873; flash pt. 106 F; ref. index 1.405-1.409
Precaution: Combustible liq.
Toxicology: Heated to decomp., emits acrid smoke and fumes
Uses: Synthetic flavoring agent
Regulatory: FDA 21CFR §172.515; Japan approved as flavoring
Manuf./Distrib.: Aldrich

Amyl salicylate
CAS 87-20-7; EINECS 218-080-2; FEMA 2084
Synonyms: 2-Hydroxybenzoic acid, pentyl ester; Isoamyl salicylate; Isopentyl salicylate; Isoamyl 2-hydroxybenzoate; Isoamyl o-hydroxybenzoate
Definition: Ester of amyl alcohol and salicylic acid
Empirical: $C_{12}H_{16}O_3$
Properties: Wh. liq., orchid-like odor; sol. in alcohol, ether; insol. in water, glycerol; m.w. 208.26; dens. 1.053; b.p. 277-278 C; flash pt. > 230 F
Precaution: Combustible
Uses: Synthetic flavoring agent
Usage level: 1.4 ppm (nonalcoholic beverages), 2.9 ppm (ice cream, ices), 3.0 ppm (candy, baked goods)
Regulatory: FDA 21CFR §172.515; GRAS (FEMA)
Manuf./Distrib.: Aldrich

Amyl valerate. *See* Isoamyl isovalerate
Amyl-δ-valerolactone. *See* δ-Decalactone
Amyl vinyl carbinol. *See* 1-Octen-3-ol

Amyris oil, West Indian type
EINECS 291-076-6 (extract)
Synonyms: Sandalwood oil
Definition: Derived from *Amyris balsamifera* L.
Properties: Pale-yellow liq.; sandalwood odor; burning taste; ref. index 1.5035-1.5120 (20 C)
Toxicology: Heated to decomp., emits acrid smoke and fumes
Uses: Natural flavoring agent
Regulatory: FDA 21CFR §172.510

Ananas comosus extract. *See* Pineapple extract

Anethole
CAS 104-46-1; EINECS 203-205-5; FEMA 2086
Synonyms: 1-Methoxy-4-propenylbenzene; p-Propenylanisole; Anise camphor; p-Methoxy-β-methylstyrene
Classification: Substituted aromatic ether
Empirical: $C_{10}H_{12}O$
Properties: Leaves or lt. yel. liq. above 23 C, anise odor, sweet taste; very sl. sol. in water; misc. with abs. alcohol, ether, chloroform; m.w. 148.22; dens. 0.991 (20/20 C); m.p. 22.5 C; b.p. 235.3 C; flash pt. 198 F; ref. index 1.557-1.561
Precaution: Combustible
Toxicology: LD50 (oral, rat) 2090 mg/kg; poison by ingestion; experimental tumorigen; may cause human intolerance reaction; heated to decomp., emits acrid smoke and irritating fumes
Uses: Synthetic flavoring agent
Usage level: 11 ppm (nonalcoholic beverages), 1400 ppm (alcoholic beverages), 26 ppm (ice cream, ices), 340 ppm (candy), 150 ppm (baked goods), 1500 ppm (chewing gum),
Regulatory: FDA 21CFR §182.60, GRAS
Manuf./Distrib.: Acme-Hardesty

Anethum graveolens oil. *See* Dillweed oil
Aneurine hydrochloride. *See* Thiamine HCl
Aneurine mononitrate. *See* Thiamine nitrate

Angelica
Definition: Angelica archangelica
Properties: Bitter aromatic and tonic flavor
Uses: Natural flavoring agent
Regulatory: Japan approved
Manuf./Distrib.: Chart

Angelica archangelica extract. *See* Angelica extract

Angelica extract
CAS 84775-41-7; EINECS 283-871-1
Synonyms: Angelica archangelica extract; Archangelica officinalis; European angelica extract
Definition: Extract of roots of *Angelica archangelica*
Properties: Pale yel. to amber liq., pungent odor, bitter-sweet taste; sol. in fixed oils; sl. sol. in min. oil
Toxicology: Heated to decomp., emits acrid smoke and irritating fumes
Uses: Natural flavoring agent
Usage level: 49 ppm (nonalcoholic beverages), 46 ppm (ice cream, ices), 44 ppm (candy), 61 ppm (baked goods), 1-100 ppm (syrup)
Regulatory: FDA 21CFR §182.10, 182.20, GRAS
Manuf./Distrib.: Chart

Angelica lactone. *See* Pentadecalactone

Angelica root oil
Definition: Extracted from roots of *Angelica archangelica*
Properties: Pale yel. to amber liq., pungent odor, bittersweet taste; sol. in fixed oils; sl. sol. in min. oil; insol. in glycerin, propylene glycol, water; dens. 0.857-0.915 (15/15 C)
Precaution: Keep well closed; protect from light
Toxicology: Heated to decomp., emits acrid smoke and irritating fumes
Uses: Natural flavoring agent; mfg. of liqueurs
Usage level: 12 ppm (nonalcoholic beverages), 15 ppm (alcoholic beverages), 0.99 ppm (ice cream, ices), 0.86 ppm (candy)
Regulatory: FDA 21CFR §182.10, 182.20, GRAS
Manuf./Distrib.: Florida Treatt

Angelica seed oil
Definition: Extracted from seeds of *Angelica archangelica*
Properties: Lt. yel. liq., sweet taste; sol. in fixed oils; sl. sol. in min. oil; insol. in glycerin, propylene glycol
Toxicology: Heated to decomp., emits acrid smoke and irritating fumes
Uses: Natural flavoring agent
Usage level: 6.3 ppm (nonalcoholic beverages), 32 ppm (alcoholic beverages), 1.5 ppm (ice cream, ices), 1.9 ppm (candy), 2.2 ppm (baked goods), 5.0 ppm (gelatins, puddings)
Regulatory: FDA 21CFR §182.10, 182.20, GRAS
Manuf./Distrib.: Chart

Angola weed
Synonyms: Roccella fusiformis Ach.
Definition: Derived from *Roccella fuciformis* Ach.
Uses: Natural flavoring agent used in alcoholic beverages
Regulatory: FDA 21CFR §172.510

Angostura
EINECS 294-354-5 (extract)
Synonyms: Cusparia bark; Angostura bark; Carony bark
Definition: Bark of tree *Galipea officinalis*
Properties: Yellowish liq., aromatic unpleasant musty odor, burning bitter flavor
Uses: Natural flavoring agent
Usage level: 18 ppm (nonalcoholic beverages), 1700 ppm (alcoholic beverages)
Regulatory: FDA 21CFR §182.10, 182.20, GRAS; Japan approved
Manuf./Distrib.: Chart (extract)

Angostura bark. *See* Angostura
Anhydrite (natural form). *See* Calcium sulfate
1,4-Anhydro-D-glucitol, 6-hexadecanoate. *See* Sorbitan palmitate
Anhydrohexitol sesquioleate. *See* Sorbitan sesquioleate
Anhydrosorbitol monolaurate. *See* Sorbitan laurate
Anhydrosorbitol monooleate. *See* Sorbitan oleate
Anhydrosorbitol monostearate. *See* Sorbitan stearate
Anhydrosorbitol sesquioleate. *See* Sorbitan sesquioleate
Anhydrosorbitol sesquistearate. *See* Sorbitan sesquistearate
Anhydrosorbitol trioleate. *See* Sorbitan trioleate
Anhydrosorbitol tristearate. *See* Sorbitan tristearate
Anhydrous gypsum. *See* Calcium sulfate
Anhydrous lanolin. *See* Lanolin
Anisaldehyde. *See* p-Anisaldehyde

o-Anisaldehyde. *See* o-Methoxybenzaldehyde

p-Anisaldehyde
CAS 123-11-5; EINECS 204-602-6; FEMA 2670
Synonyms: 4-Methoxybenzaldehyde; Anisic aldehyde; Anisaldehyde; Aubépine; p-Methoxybenzaldehyde
Empirical: C$_8$H$_8$O$_2$
Properties: Colorless oil, hawthorn odor; sol. in propylene glycol; misc. in alcohol, ether, fixed oils; insol. in glycerin, water; m.w. 136.15; dens. 1.123 (20/4 C); m.p. 2.5 C; b.p. 247-248 C; flash pt. 121 C; ref. index 1.571-1.574
Precaution: Combustible; volatile in steam
Toxicology: LD50 (oral, rat) 1510 mg/kg; mod. toxic by ingestion; skin irritant; mutagenic data; heated to decomp., emits acrid smoke and irritating fumes
Uses: Sweetly floral, hawthorn-like fragrance and flavoring
Regulatory: FDA 21CFR §172.515; Japan approved for flavoring

Anise
Definition: Pimpinella anisum
Properties: Sweet, soft, mild flavor
Uses: Natural flavoring agent
Regulatory: Japan approved
Manuf./Distrib.: Chart

Anise alcohol. *See* Anisyl alcohol
Anise camphor. *See* Anethole
Aniseed extract. *See* Anise extract
Aniseed oil. *See* Anise oil

Anise extract
CAS 84775-42-8; 84650-59-9; EINECS 283-872-7
Synonyms: Aniseed extract; Pimpinella anisum extract
Definition: Extract of dried ripe fruit and seeds of *Pimpinella anisum*
Uses: Natural flavoring agent
Regulatory: FDA 21CFR §182.20, GRAS

Anise oil
OAO 0007-70-3
Synonyms: Aniseed oil; Pimpinella anisum oil; Star anise oil
Definition: Volatile oil derived from the dried ripe fruit and seeds of *Pimpinella anisum* or *Illicium verum*, contg. 80-90% anethole, methylchavicol, anisaldehyde
Properties: Colorless or pale yel. liq.; sol. in 3 vols. alcohol; freely sol. in chloroform, ether; sl. sol. in water; dens. 0.978-0.988 (25/25 C); ref. index 1.553-1.560
Precaution: Keep cool in well-closed containers; protect from light
Uses: Natural flavoring ingredient for bakery prods., beverages, candy, chewing gum, confections, ice cream, meat, soups, liqueurs
Usage level: 7.5 ppm (nonalcoholic beverages), 45 ppm (alcoholic beverages), 65 ppm (meats), 67 ppm (ice cream, ices), 500 ppm (candy), 120 ppm (baked goods), 3200 ppm (chewing gum)
Regulatory: FDA 21CFR §182.10, 182.20, GRAS; 27CFR §21.65, 21.151
Manuf./Distrib.: Chart; Florida Treatt

p-Anisic acid ethyl ester. *See* Ethyl-p-anisate
Anisic alcohol. *See* Anisyl alcohol
Anisic aldehyde. *See* p-Anisaldehyde
Anisic ketone. *See* 1-(p-Methoxyphenyl)-2-propanone

Anisole
CAS 100-66-3; EINECS 202-876-1; FEMA 2097
Synonyms: Methylphenyl ether; Methoxybenzene; Phenyl methylether
Empirical: C$_7$H$_8$O
Formula: CH$_3$OC$_6$H$_5$
Properties: Colorless liq.; aniselike odor; sol. in alcohol, ether; insol. in water; m.w.108.13; dens. 0.992; m.p. 37-38 C; b.p. 154 C; flash pt. 42 C; ref. index 1.515-1.518
Precaution: Combustible liq.
Toxicology: LD50 (oral, rat) 3700 mg/kg; mod. toxic by ingestion; skin irritant; heated to decomp., emits acrid fumes
Uses: Synthetic flavoring agent
Usage level: 9.0 ppm (nonalcoholic beverages), 16 ppm (ice cream, ices, etc.), 51 ppm (candy), 34 ppm (baked

goods)
Regulatory: FDA 21CFR §172.515
Manuf./Distrib.: Aldrich

Anisyl acetate
FEMA 2098
Synonyms: p-Methoxybenzyl acetate
Empirical: $C_{10}H_{12}O_3$
Properties: Colorless to sl. yellow liq.; floral, fruit-like odor; sweet taste; sol. in alcohol; insol. in water; m.w. 180.21; b.p. 270 C; ref. index 1.511-1.516 (20 C)
Precaution: Combustible liq.
Toxicology: Heated to decomp., emits acrid smoke and fumes
Uses: Synthetic flavoring agent
Usage level: 6.3 ppm (nonalcoholic beverages), 8.0 ppm (ice cream, ices, etc.), 15 ppm (candy), 12 ppm (baked goods), 11 ppm (gelatins and puddings), 30 ppm (chewing gum)
Regulatory: FDA 21CFR §172.515
Manuf./Distrib.: Aldrich

Anisyl acetone. *See* 4-p-Methoxyphenyl-2-butanone

Anisyl alcohol
CAS 105-13-5; EINECS 203-273-6; FEMA 2099
Synonyms: Anise alcohol; Anisic alcohol; 4-Methoxybenzenemethanol; p-Methoxybenzyl alcohol; 4-Methoxybenzyl alcohol
Empirical: $C_8H_{10}O_2$
Properties: Need. or colorless to sl. yellow liq.; floral odor; fruity (peach) taste; sol. in most fixed oils, alcohol, ether; pract. insol. in water; m.w. 138.17; dens. 1.113 (15/15 C); m.p. 25 C; b.p. 259 C; flash pt. 112 C; ref. index 1.543-1.545
Precaution: Combustible liq.
Toxicology: LD50 (oral, rat) 1200 mg/kg; moderately toxic by ingestion; skin irritant; heated to decomp., emits acrid smoke and irritating fumes
Uses: Mildly floral, sweet lilac-like fragrance and flavoring
Usage level: 7.4 ppm (nonalcoholic beverages), 8.0 ppm (ice cream, ices, etc.), 11 ppm (candy), 12 ppm (baked goods), 1.9 ppm (gelatins and puddings)
Regulatory: FDA 21CFR §172.515
Manuf./Distrib.: Aldrich

Anisyl butyrate
FEMA 2100
Synonyms: p- Methoxybenzyl butyrate
Empirical: $C_{12}H_{16}O_3$
Properties: Colorless liq.; weak plum-like odor; sol. in alcohol; insol. in water; m.w. 208.26; b.p. ≈ 270 C
Uses: Synthetic flavoring agent
Usage level: 3.1 ppm (nonalcoholic beverages), 5.7 ppm (ice cream, ices, etc.), 10 ppm (candy), 13 ppm (baked goods)
Regulatory: FDA 21CFR §172.515

Anisyl formate
CAS 122-91-8; FEMA 2101
Synonyms: p-Methoxybenzyl formate
Empirical: $C_9H_{10}O_3$
Properties: Colorless liq.; sweet, floral odor; strawberry taste; sol. in most organic solvents; insol. in water; m.w. 166.18; dens. 1.035; b.p. 220 C; flash pt. 100 C ref. index 1.5220-1.5240 (20 C)
Precaution: Combustible
Uses: Synthetic flavoring agent
Usage level: 3.2 ppm (nonalcoholic beverages), 3.9 ppm (ice cream, ices), 7.9 ppm (candy), 14 ppm (candy), 0.20 ppm (gelatins, puddings)
Regulatory: FDA 21CFR §172.515
Manuf./Distrib.: Aldrich

Anisylmethyl ketone. *See* 1-(p-Methoxyphenyl)-2-propanone

Anisyl phenylacetate
Synonyms: Anisyl α-toluate
Empirical: $C_{16}H_{16}O_3$
Properties: Colorless, oily liq.; honey-like odor; sol. in alcohol; m.w. 256.30; b.p. 370 C

Uses: Synthetic flavoring agent
Regulatory: FDA 21CFR §172.515

Anisyl propionate
CAS 7549-33-9; FEMA 2102
Synonyms: p-Methoxybenzyl propionate
Empirical: $C_{11}H_{14}O_3$
Properties: Herbaceous odor; fruity taste; m.w.194.23; dens. 1.070; b.p. 277 C; flash pt. >230 F; ref. index 1.5490 (20 C)
Uses: Synthetic flavoring agent
Usage level: 5.6 ppm (nonalcoholic beverages), 6.1 ppm (ice cream, ices, etc.), 16 ppm (candy), 20 ppm (baked goods), 0.25 ppm (gelatins and puddings)
Regulatory: FDA 21CFR §172.515
Manuf./Distrib.: Aldrich

Anisyl α-toluate. *See* Anisyl phenylacetate

Annatto
CAS 1393-63-1; EINECS 215-735-4; EEC E160b
Synonyms: CI 75120; Achiote; Annotta; Arnotta; Natural orange 4
Definition: Vegetable dye from seeds of *Bixa orellana* containing ethyl bixin
Formula: $C_{27}H_{34}O_4$
Properties: Sol. in alcohol, ether, oils
Toxicology: LD50 (intraperitoneal, mouse) 700 mg/kg; moderately toxic by intraperitoneal route; human systemic effect by skin contact; heated to decomp., emits acrid smoke and irritating fumes
Uses: As extract for coloring sausage casings, oleomargarine, cheese, beverages, cereal, ice cream, shortening, food-marking inks
Usage level: ADI 0-0.065 mg/kg (EEC, as bixin)
Regulatory: FDA 21CFR §73.30, 73.1030, 73.2030; USDA 9CFR §318.7, 381.147; Japan approved (water-sol.); Europe listed; UK approved
Manuf./Distrib.: Am. Roland; Chart; Consolidated Flavor; Crompton & Knowles; Haarmann & Reimer/Food Ingred.; MLG Enterprises; Quest Int'l.
Trade names: Dascolor Annatoo Enc
Trade names containing: Annatto ECA-1; Annatto ECH-1; Annatto ECPC; Annatto OS #2894; Annatto OS #2922; Annatto OS #2923; Annatto Liq. #3968, Acid Proof; Annatto/Turmeric WML-1; Annatto/Turmeric WMP-1; Durkex Gold 77N; FloAm® 200C; FloAm® 210 HC; Golden Covo® Shortening; Golden Croissant/ Danish™ Margarine; Qual Flo™ C; Shedd's NP Margarine

Annatto extract
CAS 8015-67-6; EINECS 289-561-2 (annatto tree extract)
Synonyms: Natural orange 4; CI 75120; Bixin
Definition: Carotenoid color contg. bixin (oil/fat extract) or norbixin (alkaline aq. extract)
Empirical: $C_{25}H_{30}O_4$ (bixin); $C_{24}H_{28}O_4$ (norbixin)
Properties: Bixin: brownish-red cryst.; m.w. 394.51; m.p. 198 C; Norbixin: m.w. 380.48
Toxicology: May cause asthma, rashes
Uses: Food colorant for butter, cheese, margarine, cooking oil, salad dressing, cereal, ice cream, sausage casings, bakery goods, spices
Usage level: 0.5-10 ppm (as pure color)
Regulatory: Japan restricted use
Manuf./Distrib.: Chr. Hansen's
Trade names: Annatto Extract P1003; Natural Soluble Orange Powd; Natural Yellow Colour Q-500, Q-1000, Q-2000; Unibix W
Trade names containing: Annatto OSL-1; Annatto OSS-1; Annatto WSL-2; Annatto WSP-75; Annatto WSP-150; Annatto/Turmeric OSS-1; Natural Soluble Powder Q A/L 15 Yellow Colorant

Annotta. *See* Annatto

Anoxomer
CAS 60837-57-2
Classification: Polymeric antioxidant
Definition: 1,4-Benzenediol, 2-(1,1-dimethylethyl) polymer with diethenylbenzene, 4-(1,1-dimethylethyl) phenol, 4-methoxyphenol, 4,4´-(1-methylethylidene)bis(phenol), 4-methylphenol
Toxicology: Heated to decomp., emits acrid smoke and fumes
Uses: Preservative, antioxidant
Usage level: 5000 ppm max. on fat and oil
Regulatory: FDA 21CFR §172.105

Anthemis nobilis extract. *See* Chamomile extract
Anthemis nobilis oil. *See* Chamomile oil, Roman

Anthocyanase
Uses: Natural enzyme
Regulatory: Japan approved
Trade names: Pectinol® DL

Anthocyanins
EEC E163
Definition: A flavenoid plant pigment consisting of glucose plus anthocyanidins
Properties: Water-sol.
Uses: Food colorant (violet, red, blue)
Trade names: Anthocyanin Extract P2001

Anthranilic acid cinnamyl ester. *See* Cinnamyl anthranilate
APM. *See* Aspartame
Apocarotenal. *See* β-Apo-8′-carotenal

β-Apo-8′-carotenal
CAS 1107-26-2; EINECS 214-171-6; EEC E160e
Synonyms: Food Orange 6; CI 40820; Apocarotenal
Classification: Carotenoid color
Empirical: $C_{30}H_{40}O$
Properties: Purplish-blk. cryst. fine powd.; sol. in chloroform; sl. sol. in acetone; insol. in water; m.w. 416.65; m.p. 136-140 C (dec.)
Precaution: Store in sealed containers under inert gas and preferably refrigeration
Toxicology: Heated to decomp., emits acrid smoke and irritating fumes
Uses: Food colorant with provitamin activity (1,200,000 IU vitamin A/g); for coloring juices, fruit drinks, soups, jams, jellies, gelatins; as vegetable oil sol'n. in fat-based foods incl. process cheese, margarine, salad dressings, fats, oils
Usage level: 1-20 ppm (as food colorant); ADI 0-5 mg/kg (EEC)
Regulatory: FDA 21CFR §73.90 (limitation 15 mg/lb of solid or /pint of liq. food); Europe listed; UK approved
Trade names: Lucantin® Yellow
Trade names containing: Apo-Carotenal Suspension 20% in Veg. Oil No. 66417; Carotenal Sol'n. #2 No. 66425; Carotenal Sol'n. 4% No. 66424; Carotenal Sol'n. #73 No. 66428

Apple acid. *See* N-Hydroxysuccinic acid

Apricot
Uses: Natural flavoring agent
Regulatory: Japan approved

Apricot kernel oil
CAS 72869-69-3
Synonyms: Persic oil; Prunus armeniaca
Definition: Oil expressed from kernels of *Prunus armeniaca*
Properties: Oily liq., nearly odorless; sp.gr. 0.910-0.923; acid no. 1 max.; iodine no. 90-115; sapon. no. 185-195; ref. index 1.4635-1.4655 (40 C)
Uses: Food ingred., flavoring
Regulatory: FDA 21CFR §182.40, GRAS
Trade names: Lipovol P

Aqua ammonium. *See* Ammonium hydroxide
Arabic gum. *See* Acacia

Arabinogalactan
CAS 9036-66-2; FEMA 3254
Synonyms: Larch gum; Polyarabinogalactan
Definition: Polysaccharide extracted from Western larch wood, having galactose and arabinose units in approx. ratio 6:1
Properties: Sol. in water; m.p. > 200 C (dec.)
Toxicology: Heated to decomp., emits acrid smoke and irritating fumes
Uses: Emulsifier, thickener, stabilizer, binder, bodying agent in essential oils, nonnutritive sweeteners, flavor bases, nonstandardized dressings, pudding mixes
Regulatory: FDA 21CFR §172.610; Japan approved
Manuf./Distrib.: Aldrich

L-Arabinose
 Uses: Natural sweetener
 Regulatory: Japan approved

d-Araboascorbic acid. *See* Erythorbic acid
Arachis oil. *See* Peanut oil
Arborvitae. *See* Cedar leaf oil
Archangelica officinalis. *See* Angelica extract
Argeol. *See* Santalol

L-Arginine
 CAS 74-79-3; EINECS 200-811-1
 Synonyms: 1-1-Amino-4-guanidovaleric acid
 Classification: Amino acid
 Empirical: $C_6H_{14}N_4O_2$
 Formula: $H_2NCNHNHCH_2CH_2CH_2CHNH_2COOH$
 Properties: Wh. cryst. powd.; sol. in water; sl. sol. in alcohol; insol. in ether; m.w. 174.20; strongly alkaline
 Toxicology: Heated to decomp., emits toxic fumes of NO_x
 Uses: Dietary supplement, nutrient, flavoring
 Regulatory: FDA 21CFR §172.320, limitation 6.6%; Japan approved
 Manuf./Distrib.: Am. Roland

L-Arginine L-glutamate
 CAS 147-81-9
 Empirical: $C_5H_{10}O_5$
 Properties: m.w. 150.13
 Uses: Dietary supplement (amino acid); flavorings
 Regulatory: Japan approved

L-Arginine monohydrochloride
 CAS 1119-34-2; EINECS 239-674-8
 Synonyms: L-1-Amino-4-guanidovaleric acid monohydrochloride
 Empirical: $C_6H_{15}ClN_4O_2$
 Formula: $C_6H_{14}N_4O_2 \cdot HCl$
 Properties: Wh. cryst. powd., odorless, sol. in water, sl. sol in hot alcohol; insol. in ether; m.w. 210.66; dec. 235 C
 Toxicology: LD50 (oral, rat) 12 g/kg, mildly toxic by ingestion; experimental teratogen; heated to decomp., emits toxic fumes of NO_x and HCl
 Uses: Dietary supplement, nutrient
 Manuf./Distrib.: Am. Roland

Arheol. *See* Sandalwood oil
Aritolochia extract. *See* Serpentaria extract

Arnica
 EINECS 273-579-2 (extract)
 Synonyms: Arnica montana; Arnica flowers; Mountain tobacco; Leopard's bane; Wolf's bane
 Definition: Plant material from dried flowerheads, roots, or rhizomes of *Arnica montana*
 Properties: Herbaceous sweet odor, sl. bitter flavor
 Uses: Natural flavoring used in liqueurs
 Regulatory: FDA 21CFR §172.510; Japan approved
 Manuf./Distrib.: Chart

Arnica extract
 CAS 8057-65-6; EINECS 273-579-2
 Synonyms: Arnica montana extract
 Definition: Extract of dried flowerheads of *Arnica montana*
 Uses: Natural flavoring agent for alcoholic beverages
 Regulatory: FDA §172.510
 Manuf./Distrib.: Chart

Arnica flowers. *See* Arnica
Arnica montana. *See* Arnica
Arnica montana extract. *See* Arnica extract
Arnotta. *See* Annatto

Artemisia
 CAS 8022-37-5; FEMA 3114

Artemisia absinthium oil

Synonyms: Wormwood; Artemisia spp.
Definition: Derived from *Artemesia absinthium* L
Toxicology: LD50 (oral, rat) 960 mg/kg; mod. toxic by ingestion; allergen; causes contact dermatitis; heated
to decomp., emits acrid smoke and fumes
Uses: Natural flavoring agent
Usage level: 360 ppm (nonalcoholic beverages), 5.0 ppm (alcoholic beverages)
Regulatory: FDA 21CFR §172.510; Japan approved

Artemisia absinthium oil. *See* Wormwood oil
Artemisia spp. *See* Artemisia

Artichoke

Definition: Leaves of *Cynara scolymus*
Properties: Bitter tonic flavor
Uses: Natural flavoring for alcoholic beverages only
Regulatory: FDA 21CFR §172.510; Japan approved
Manuf./Distrib.: Chart

Artichoke extract

CAS 84012-14-6; EINECS 281-659-3
Synonyms: Cynara scolymus extract
Definition: Extract of the leaves of *Cynara scolymus*
Uses: Natural flavoring
Regulatory: FDA 21CFR §172.510

Artificial almond oil. *See* Benzaldehyde
Artificial oil of ants. *See* Furfural

Asafetida

EINECS 232-522-1; FEMA 2107(gum); 2108(oil)
Synonyms: Devil's dung; Asafoetida; Food of the gods
Definition: Gum resin exudate from rhizome and roots from *Ferula asafoetida*
Properties: Strong garlic odor; sl. bitter acrid taste
Uses: Natural flavoring agent; used in nonalcoholic beverages, ice cream, ices, candy, baked goods, meats,
condiments, soups, Worcestershire sauce
Regulatory: FDA 21CFR §182.20, GRAS; Japan approved
Manuf./Distrib.: Florida Treatt

Asafetida extract

CAS 90028-70-9; EINECS 297-382-6; FEMA 2106
Synonyms: Ferula asafoetida extract
Definition: Brown, gum resin obtained from the roots of *Ferula fetida*, *F. asafoetida* or other species
Uses: Purified resin, tincture, and the oil as flavoring for sauces, meats, pickles, condiments, and candies
Regulatory: FDA 21CFR §182.20
Manuf./Distrib.: Chart

Asafoetida. *See* Asafetida
1-Ascorbic acid. *See* L-Ascorbic acid

L-Ascorbic acid

CAS 50-81-7; EINECS 200-066-2; FEMA 2109; EEC E300
Synonyms: 1-Ascorbic acid; Vitamin C; Cevitamic acid
Empirical: $C_6H_8O_6$
Properties: Wh. cryst.; sol. in water; sl. sol. in alcohol; insol. in ether, chloroform, benzene, petrol. ether, oils,
fats; m.w. 176.14; m.p. 192 C; flash pt. 99 C
Precaution: Combustible liq.
Toxicology: LD50 (IV, mouse) 518 mg/kg; moderately toxic; human blood systemic effects by IV route;
extremely high repeated doses may cause nausea, diarrhea, GI disturbances, flatus; emits acrid smoke
and irritating fumes when heated
Uses: Nutrient, dietary supplement, color fixing, flavoring, preservative, raising agent, oxidant, antioxidant,
abscission of citrus fruit in harvesting, reducing agent
Regulatory: FDA 21CFR §137.105, 137.155, 137.160, 137.165, 137.170, 137.175, 137.180, 137.185,
137.200, 137.205, 145.110, 145.115, 145.116, 145.135, 145.136, 145.170, 145.171, 146.113, 146.187,
150.141, 150.161, 155.200, 156.145, 161.175,; 182.3013, 182.3041, 182.5013, 182.8013, 240.1044,
GRAS; BATF 27CFR §240.1051; USDA 9CFR §318.7; Japan approved; Europe listed; UK approved
Manuf./Distrib.: ADM; Am. Roland; Browning; Gist-brocades Food Ingreds.; Int'l. Sourcing; Jungbunzlauer
Trade names: Ascorbic Acid USP/FCC, 100 Mesh; Ascorbic Acid USP, FCC Fine Gran. No. 6045655; Ascorbic

Acid USP, FCC Fine Powd. No. 6045652; Ascorbic Acid USP, FCC Gran. No. 6045654; Ascorbic Acid USP, FCC Type S No. 6045660; Ascorbic Acid USP, FCC Ultra-Fine Powd No. 6045653; Ascorbo-120; Ascorbo-C Tablets; Cap-Shure® AS-125-50; Cap-Shure® AS-165CR-70; Descote® Ascorbic Acid 60%

Trade names containing: Ascorbo-160-2 Powd; Bagel Base; Cap-Shure® AS-165-70; Carbrea® Tabs; CDC-10A; CDC-10A (NB); CDC-79; CDC-79DZ; CDC-2001; CDC-2002; CDC-2500; CDC-3000; Coated Ascorbic Acid 97.5% No. 60482; Durkote Vitamin C/Hydrog. Veg. Oil; Durkote Vitamin C/Maltodextrin; N'Hance CM; N'Hance (Phase I) Powd; N'Hance (Phase II) Powd; N'Hance SD; Oxynex® K; Oxynex® L; Palma-Sperse® Type 250-S No. 65322; Palma-Sperse® Type 250A/50 D-S No. 65221; Red Beet WSL-300; Red Beet WSL-400; Red Beet WSP-300

L-Ascorbic acid, 6-hexadecanoate. *See* Ascorbyl palmitate
L(+)-Ascorbic acid sodium salt. *See* Sodium ascorbate

Ascorbyl palmitate

CAS 137-66-6; EINECS 205-305-4; EEC E304
Synonyms: L-Ascorbic acid, 6-hexadecanoate; Palmitoyl L-ascorbic acid
Definition: Ester of ascorbic acid and palmitic acid
Empirical: $C_{22}H_{38}O_7$
Properties: White or yel.-wh. powd., citrus odor; sol. in alcohol, animal and veg. oils; sl. sol. in water; m.w. 414.54; m.p. 116-117 C
Toxicology: Heated to decomp., emits acrid smoke and irritating fumes
Uses: Antioxidant for fats and oils, color preservative, source of Vitamin C, stabilizer, emulsifier, sequestrant, synergist with α-tocopherol; for beverages, bread, lemon drinks, margarine, rolls, shortening
Regulatory: FDA 21CFR §166.110, 182.3149, GRAS; USDA 9CFR §318.7 (0.02% max. in margarine); Japan approved; Europe listed; UK approved
Trade names: Ascorbyl Palmitate NF, FCC No. 60412
Trade names containing: 24% Beta Carotene HS-E in Veg. Oil No. 65671; Canthaxanthin 10% Type RVI No. 66523; Canthaxanthin Beadlets 10%; Controx® AT; Controx® VP; Dry Beta Carotene Beadlets 10% CWS No. 65633; Dry Beta Carotene Beadlets 10% No. 65661; Dry Canthaxanthin 10% SD No. 66514; Oxynex® K; Oxynex® L; Oxynex® LM; Pristene® 181; Roxanthin® Red 10WS No. 66515

Ascorbyl stearate

Toxicology: Heated to decomp., emits acrid smoke and irritating fumes
Uses: Antioxidant for margarine; dietary supplement
Usage level: Limitation 0.02% (alone or in combination with other antioxidants in margarine)
Regulatory: USDA 9CFR §318.7; Japan approved

Asparagic acid. *See* D-Aspartic acid
Asparagic acid. *See* L-Aspartic acid

L-Asparagine

CAS 70-47-3 (anhyd.), 5794-13-8 (monohydrate); EINECS 200-735-9 (anhyd.), 200-735-9 (monohydrate)
Synonyms: 1-α-Aminosuccinamic acid; 2-Aminosuccinamic acid; (+)-Aspartic acid 4-amide; Aspartic acid β-monoamide
Classification: Amino acid
Definition: Commonly occurs as the monohydrate
Empirical: $C_4H_8N_2O_3$
Properties: Anhyd.: Wh. cryst. powd., sl. sweet taste; sol. in water; insol. in alcohol, ether; m.w. 132.12; m.p. 234 C; Monohydrate: Orthorhombic cryst.; sol. in acids, alkalies; pract. insol. in alcohols, ether, benzene; m.w. 150.14;
Toxicology: Heated to decomp., emits toxic fumes of NO_x
Uses: Nutrient, dietary supplement; flavoring
Regulatory: FDA 21CFR §172.320, 7% max. by wt.; Japan approved

Asparaginic acid. *See* D-Aspartic acid

Aspartame

CAS 22839-47-0; EINECS 245-261-3; EEC E951
Synonyms: APM; 1-Methyl N-L-α-aspartyl-L-phenylalanine; Aspartylphenylalanine methyl ester
Empirical: $C_{14}H_{18}N_2O_5$
Formula: $HOOCCH_2CH(NH_2)CONHCH(CH_2C_6H_5)COOCH_3$
Properties: Wh. cryst. powd. or colorless need., odorless, sweet taste, prolonged sweet aftertaste; sl. sol. in water, alcohol; m.w. 294.34; m.p. 246-248 C; 160 times sweeter than sucrose
Toxicology: Human systemic effects by ingestion (allergic dermatitis); possible link to neural problems; headaches; experimental reproductive effects; should not be used by individuals with PKU; heated to decomp., emits toxic fumes of NO_x

D-Aspartic acid

> *Uses:* Sweetening agent used in sugar substitutes for table use, breakfast cereals, chewing gum, beverages, chewable multivitamin food supplements, frozen desserts, dry mixes for gelatins, puddings, dairy prods.; flavor enhancer in chewing gum
> *Usage level:* ADI 40 mg/kg (WHO)
> *Regulatory:* FDA 21CFR §172.804; Japan, Canada approved
> *Manuf./Distrib.:* Ajinomoto; Browne & Dureau Int'l.; Calaga Food Ingreds.; Fruitsource; Holland Sweetener; NutraSweet AG; Quimdis; Sanofi; Scanchem; Sweeteners Plus; EH Worlee GmbH
> *Trade names:* Equal®; HSC Aspartame; NutraSweet®

D-Aspartic acid

> CAS 1783-96-6; EINECS 217-234-6
> *Synonyms:* Asparaginic acid; Asparagic acid; (-)-Aminosuccinic acid
> *Empirical:* $C_4H_7NO_4$
> *Formula:* $COOHCH_2CH(NH_2)COOH$
> *Properties:* m.w. 133.11; m.p. > 300 C
> *Uses:* Ingredient of aspartame
> *Manuf./Distrib.:* Am. Roland

DL-Aspartic acid

> CAS 617-45-8; EINECS 210-513-3
> *Synonyms:* DL-Aminosuccinic acid
> *Empirical:* $C_4H_7NO_4$
> *Properties:* Colorless or wh. cryst., odorless, acid taste; sl. sol in water; insol. in alcohol, ether; m.w. 133.10; m.p. > 300 C
> *Toxicology:* Heated to decomp., emits toxic fumes of NO_x
> *Uses:* Dietary supplement, nutrient

L-Aspartic acid

> CAS 56-84-8; EINECS 200-291-6
> *Synonyms:* (+)-Aminosuccinic acid; Asparagic acid
> *Classification:* Amino acid
> *Definition:* Commonly occurs in L-form
> *Empirical:* $C_4H_7NO_4$
> *Properties:* Colorless to wh. cryst., acid taste; sol. in acids, alkalies; sl. sol. in water; insol. in alcohol, ether; m.w. 133.11; dens. 1.661 (12.5 C); m.p. 270 C
> *Toxicology:* Possible brain damage; heated to decomp., emits toxic fumes of NO_x
> *Uses:* Nutrient, dietary supplement; flavoring
> *Regulatory:* FDA 21CFR §172.320, 7% max. by wt.; Japan approved

Aspartylphenylalanine methyl ester. *See* Aspartame

Attapulgite

> CAS 1337-76-4
> *Synonyms:* Fuller's earth; Palygorskite; Dioctrahedral smectite
> *Uses:* Dietary supplement
> *Regulatory:* FDA GRAS

Attar of rose. *See* Rose oil
Aubepine. *See* p-Anisaldehyde
Autolyzed yeast. *See* Yeast
Autolyzed yeast extract. *See* Bakers yeast extract

Avocado oil

> CAS 8024-32-6; EINECS 232-428-0
> *Synonyms:* Alligator pear oil
> *Definition:* Oil obtained by pressing dehydrated avocado pear Persea americana; consists principally of glycerides of fatty acids
> *Properties:* Yellowish-green to brownish-green oil, faint char. odor; sol. in min. oil, isopropyl esters, ethanol; insol. in water; dens. 0.908-0.925; iodine no. 84-95; sapon. no. 177-198; ref. index 1.460-1.470
> *Uses:* Food ingred.
> *Manuf./Distrib.:* Am. Roland; Arista Industries
> *Trade names:* Lipovol A

Axerophthol. *See* Retinol
Azobisformamide. *See* Azodicarbonamide

Azodicarbonamide

> CAS 123-77-3; EINECS 204-650-8; EEC E927

Synonyms: ADA; 1,1´-Azobisformamide; Azoformamide; Azobisformamide; Diazenedicarboxamide; Azodicarboxamide
Empirical: $C_2H_4N_4O_2$
Formula: $H_2NCON=NCONH_2$
Properties: Orange-red crystals; sol. in hot water; insol. in cold water, alcohol; m.w. 116.08; m.p. 225 C (dec.)
Precaution: Flamm.
Toxicology: Heated to decomp., emits toxic fumes of NO_x
Uses: Bleaching agent in cereal flour, maturing agent for flour; dough conditioner in breads
Usage level: 45 ppm (aging and bleaching ingred. in cereal flour), 5% (finished foamed polyethylene)
Regulatory: FDA 21CFR §172.806, 178.3010; Europe listed; UK approved
Manuf./Distrib.: Aldrich; Gist-Brocades Food Ingreds.
Trade names: ADA Tablets; WaTox-20 Tablets
Trade names containing: Adapt; Carbrea® Tabs; CDC-10A (NB); CDC-2001; CDC-2002; Muffit; N'Hance CM; N'Hance (Phase I) Powd.; N'Hance (Phase II) Tablets; N'Hance SD; WaTox-20 P Powd.

Azodicarboxamide. *See* Azodicarbonamide
Azoformamide. *See* Azodicarbonamide
Azole. *See* Pyrrole
Azorubin S. *See* Amaranth

Babassu oil
Synonyms: Oils, babassu
Definition: Oil obtained from the nuts of *Orbigyna oleifera*
Uses: Foods
Manuf./Distrib.: Karlshamns

Bakers yeast extract
Synonyms: Autolyzed yeast extract; Bakers yeast glycan
Definition: Comminuted, washed, pasteurized, and dried cell walls of the yeast *Saccharomyces cerevisiae*
Properties: Liq., paste, or powd.; sol. in water
Toxicology: Heated to decomp., emits acrid smoke and irritating fumes
Uses: Flavoring agent and adjuvant, emulsifier, nutrient supplement, stabilizer, thickener, yeast food, texturizer in salad dressings, frozen desserts, sour cream prods., cheese spreads, snack dips
Usage level: Limitation 5% (salad dressing), 1.5% (of hot carcass wt.)
Regulatory: FDA 21CFR §172.325, 172.896, 172.898, 184.1983, GRAS; BATF 27CFR §240.1051; USDA 9CFR §318.7, 318.147

Bakers yeast glycan. *See* Bakers yeast extract

Bakers yeast protein
Uses: Nutrient supplement
Regulatory: FDA 21CFR §172.325

Baking soda. *See* Sodium bicarbonate

Balm mint
Synonyms: Lemon balm
Definition: Leaves of balm mint
Properties: Citral odor, tonic-like flavor
Uses: Natural flavoring agent
Usage level: 2000 ppm (nonalcoholic beverages)
Regulatory: FDA 21CFR §182.10, GRAS

Balm mint extract
CAS 84082-61-1
Synonyms: Extract of balm mint; Lemon balm extract; Melissa officinalis extract
Definition: Extract of leaves and tops of *Melissa officinalis*
Uses: Natural flavoring agent
Regulatory: FDA 21CFR §182.20, GRAS

Balm mint oil
CAS 8014-71-9
Synonyms: Lemon balm; Melissa officinalis oil
Definition: Volatile oil obtained from leaves and tops of *Melissa officinalis*
Properties: Yel. to ylsh.-grn. liq.; sol. in alcohol; pract. insol. in water; dens. 0.89-0.925 (15/15 C)
Precaution: Keep well closed, cool; protect from light
Uses: Natural flavoring agent
Usage level: 8.5 ppm (nonalcoholic beverages), 1.7-15 ppm (ice cream, ices), 20 ppm (candy), 10-60 ppm

(baked goods)
Regulatory: FDA 21CFR §182.10, 182.20, GRAS

Balsam Canada
CAS 8007-47-4; EINECS 232-362-2
Synonyms: Abies balsamea oleoresin; Canadian balsam; Balsam of fir
Definition: Liq. oleoresin obtained from *Abies balsamea*
Toxicology: Heated to decomp., emits acrid smoke and irritating
Uses: Natural flavoring; chewing gum base (Japan)
Regulatory: FDA 21CFR §172.510; Japan approved

Balsam copaiba
CAS 8001-61-4; EINECS 232-288-0
Synonyms: Copaiba oil
Definition: Oil from steam distillation of *Copaifera* balsam
Properties: Colorless to yel. liq., char. aromatic odor, sl. bitter taste; very sol. in alcohol, ether, CS_2; sol. in fixed oils, min. oil; insol. in water; dens. 0.880-0.907; b.p. 250-275 C; ref. index 1.493-1.500 (20 C)
Precaution: Keep cool, well closed; protect from light
Toxicology: Heated to decomp., emits acrid smoke and irritating fumes
Uses: Natural flavoring agent; chewing gum base (Japan)
Regulatory: FDA 21CFR §172.510; Japan approved
Manuf./Distrib.: Florida Treatt

Balsam of fir. *See* Balsam Canada

Balsam Peru
CAS 8007-00-9; 8016-42-0; EINECS 232-352-8
Synonyms: Myroxylon pereirae oleoresin; Peruvian balsam
Definition: Oleoresin extracted from *Myroxylon pereirae*
Properties: Dk. brn. visc. liq., vanilla odor; sol. in fixed oils; sl. sol. in propylene glycol; insol. in glycerin
Precaution: Combustible when heated
Toxicology: Mild allergen; heated to decomp., emits acrid smoke and irritating fumes
Uses: Natural flavoring agent in food
Regulatory: FDA 21CFR §182.20, GRAS; Japan approved
Manuf./Distrib.: Florida Treatt

Balsam tolu
CAS 8011-89-0; EINECS 232-550-4
Synonyms: Resin tolu; Toluifera balsamam resin; Tolu resin
Definition: Resin derived from *Toluifera balsamam*
Uses: Natural flavoring
Regulatory: FDA 21CFR §172.510; Japan approved
Manuf./Distrib.: Chart (extract)

Banana oil. *See* Amyl acetate
Barbados aloe extract. *See* Aloe extract

Barley extract
CAS 94349-67-4; EINECS 286-476-2
Synonyms: Extract of barley
Definition: Extract of the cereal grass *Hordeum distichum* or *H. sativum*
Uses: Flavor and colorant in foods
Trade names containing: Extramalt Light

Barm. *See* Yeast

Basil
Definition: Ocimum basilicum L.
Properties: Warm, intense, spicy aroma; mint-like flavor
Uses: Natural flavoring agent
Regulatory: Japan approved
Manuf./Distrib.: Chart

Basil, bush
Definition: Derived from *Ocimum minimum*
Uses: Natural flavoring agent
Regulatory: FDA 21CFR §182.10, GRAS

Basil extract
CAS 84775-71-3
Synonyms: Common basil extract; Sweet basil extract; Ocimum basilicum extract
Definition: Extract of the leaves and flowers of *Ocimum basilicium*
Uses: Natural flavoring agent
Regulatory: FDA 21CFR §182.20, GRAS
Manuf./Distrib.: Chart

Basil oil
CAS 8015-73-4
Synonyms: Ocimum basilicum oil; Sweet basil oil
Definition: Volatile oil obtained from leaves of *Ocimum basilicum*, contg. methylchavicol, eucalyptol, linalool, estragol
Properties: Pale yel. to greenish liq., floral spicy odor; sol. in fixed oils, propylene glycol; misc. with ether, chloroform; insol. in glycerin, water; dens. 0.905-0.930 (20/20 C)
Precaution: Keep well closed; protect from light
Toxicology: LD50 (oral, rat) 1400 mg/kg; moderately toxic by ingestion; skin irritant; heated to decomp., emits acrid smoke and irritating fumes
Uses: Natural flavoring agent for foods
Usage level: 2 ppm (nonalcoholic beverages), 2.7 ppm (ice cream, ices), 4.2 ppm (baked goods), 0.01 ppm (gelatins, puddings), 15 ppm (condiments), 24 ppm (meats)
Regulatory: FDA 21CFR §182.10, 182.20, GRAS
Manuf./Distrib.: Acme-Hardesty; C.A.L.-Pfizer; Chart; Pierre Chauvet; Commodity Services; Florida Treatt

Battery acid. *See* Sulfuric acid
Bay leaf oil. *See* Bay oil
Bay leaf oil. *See* Laurel leaf oil

Bay oil
CAS 8006-78-8
Synonyms: Bay leaf oil; Myrcia oil
Definition: Volatile oil distilled from leaves of *Pimenta acris*, contg. 40-55% eugenol, myrcene, chavicol, etc.
Properties: Yel. to brnsh.-yel. liq., pleasant odor, sharpy spicy taste; very sol. in alcohol, CS_2, glac. acetic acid; insol. in water; dens. 0.962-0.990 (25/25 C); ref. index 1.500-1.520
Toxicology: LD50 (oral, rat) 1800 mg/kg; moderately toxic by ingestion; may cause human intolerance reaction; heated to decomp., emits acrid smoke
Uses: Natural flavoring agent for meats, stews, soups; mfg. of bay rum
Regulatory: FDA 21CFR §182.10, 182.20, GRAS; 27CFR §21.75, 21.151
Manuf./Distrib.: Chart; Florida Treatt

Beechwood creosote
Synonyms: Wood creosote
Definition: Mixt. of phenols (chiefly guaiacol, creosol, p-cresol)
Properties: Colorless or ylsh. oily liq., smoky odor; sol. in glycerol; misc. with alcohol, chloroform, ether; insol. in water; b.p. 200-220 C
Uses: Synthetic flavoring agent
Regulatory: FDA 21CFR §172.515

Beef extract
Uses: Food flavoring
Trade names: Beef Extract Paste #9225; Beef Extract Powd. #9267; Building Blocks® Beef Extract

Beef tallow. *See* Tallow

Beeswax
CAS 8006-40-4 (white), 8012-89-3 (yellow); EINECS 232-383-7; FEMA 2126; EEC E901
Synonyms: Cera alba
Definition: Purified wax from the honeycomb of the bee; commonly called white wax when bleached, yellow wax when not bleached
Properties: Brown or white (bleached) solid with faint odor; insol. in water; sl. sol. in alcohol; sol. in chloroform, ether, and oils; dens. 0.95; m.p. 62-65 C; sapon. no. 84
Precaution: Combustible when heated
Toxicology: Mild allergen; may cause human intolerance reaction
Uses: Food additive, flavoring agent, adjuvant, lubricant, candy glaze/polish, surface-finishing agent; chewing gum base; release agent
Usage level: 0.065% (chewing gum), 0.005% (confections, frostings), 0.04% (hard candy), 0.1% (soft candy),

Beet powder

      0.002% (other food)
      Regulatory: FDA 21CFR §184.1973, GRAS; Japan approved; Europe listed; UK approved
      Manuf./Distrib.: C.A.L.-Pfizer; Pierre Chauvet

Beet powder
      CAS 89957-89-1; EINECS 289-610-8; EEC E162
      Synonyms: Beetroot red; Dehydrated beets; Beets, dehydrated
      Definition: Color additive from edible beets, contg. red pigments, betacyanins (principally betanine, CAS 7659-95-2) and yel. pigments, betaxanthins, collectively known as betalains
      Properties: Dk. red powd.; readily dissolves in water
      Precaution: Degrades readily at temps. as low as 50 C, esp. on air/light exposure
      Uses: Colorant for foods with short shelf lives that do not require heat treatment, hard candies, yogurt, ice cream, salad dressing, frostings, cake mixes, meat substitutes, soft drinks, gelatin desserts
      Usage level: 0.1-1% (food colorant)
      Regulatory: FDA 21CFR §73.40; Europe listed; UK approved; Japan restricted
      Manuf./Distrib.: Am. Roland; Cham Foods (Israel); Chart; Crompton & Knowles; Quest Int'l.
      Trade names: Beetroot Conc. P3003

Beetroot red. *See* Beet powder
Beets, dehydrated. *See* Beet powder
Beet sugar. *See* Sucrose
Bengal gelatin. *See* Agar
Benne. *See* Sesame

Bentonite
      CAS 1302-78-9; EINECS 215-108-5; EEC E558
      Synonyms: Soap clay; Wilkinite; CI 77004
      Definition: Native hydrated colloidal aluminum silicate clay
      Formula: $Al_2O_3 \cdot 4SiO_2 \cdot nH_2O$
      Properties: Light to cream-colored impalpable powd.; forms colloidal suspension in water, thixotropic properties; insol. in water and org. solvs.
      Toxicology: LD50 (IV, rat) 35 mg/kg; poison by intravenous route causing blood clotting
      Uses: Direct food additive, colorant, pigment, processing aid, clarifying agent/stabilizer for wine; filtration aid; suspending agent; emulsifier
      Regulatory: FDA 21CFR §175.105, 175.300, 177.1460, 184.1155, GRAS; Japan restricted (0.5% max. residual); Europe listed; UK approved
      Trade names: Vitiben®

Benylate. *See* Benzyl benzoate
Benzalacetone. *See* Benzylidene acetone

Benzaldehyde
      CAS 100-52-7; EINECS 202-860-4; FEMA 2127
      Synonyms: Benzoic aldehyde; Artificial almond oil; Benzenecarbaldehyde
      Empirical: C_7H_6O
      Formula: C_6H_5CHO
      Properties: Colorless liq., bitter almond odor, burning taste; sl. sol. in water; misc. in alcohol, ether, oils; m.w. 106.13; sp.gr. 1.041; m.p. -26 C; b.p 179 C; flash pt. 148 F; ref. index 1.544
      Precaution: DOT: Combustible liq.; strong reducing agent; acts violently with oxidizers
      Toxicology: LD50 (oral, rat) 1300 mg/kg; poison by ingestion and intraperitoneal routes; allergen; feeble local anestethic; skin irritant; CNS depressant
      Uses: Synthetic flavoring agent
      Usage level: 36 ppm (nonalcoholic beverages), 50-60 ppm (alcoholic beverages), 42 ppm (ice cream, ices), 120 ppm (candy), 110 ppm (baked goods), 160 ppm (gelatins, puddings), 840 ppm (chewing gum),
      Regulatory: FDA 21CFR §182.60, GRAS; Japan approved as flavoring
      Manuf./Distrib.: Berje; Britannia Natural Prods.; H E Daniel; Dragoco Australia; Foote & Jenkins; Naturex; O'Laughlin Industries; Penta Mfg.; Quimdis

Benzaldehyde dimethyl acetal
      CAS 1125-88-8; EINECS 214-413-0; FEMA 2128
      Synonyms: α,α- Dimethoxytoluene
      Empirical: $C_9H_{12}O_2$
      Formula: $C_6H_5CH(OCH_3)_2$
      Properties: Liq.; floral odor; m.w. 152.19; dens. 1.014 (20/4 C); b.p. 198 C; flash pt. 55 C; ref. index 1.4950 (20 C)
      Precaution: Flamm.

Uses: Synthetic flavoring agent
Usage level: 26 ppm (nonalcoholic beverages), 60 ppm (alcoholic beverages), 22 ppm (ice cream, ices, etc.), 56 ppm (candy), 45 ppm (baked goods), 50 ppm (gelatins and puddings)
Regulatory: FDA 21CFR §172.515
Manuf./Distrib.: Aldrich

Benzaldehyde glyceryl acetal
CAS 1708-39-0; FEMA 2129
Synonyms: Benzal glyceryl acetal; 2-Phenyl-m-dioxan-5-ol
Empirical: $C_{10}H_{12}O_3$
Precaution: Combustible liq.
Toxicology: LD50 (oral, rat) 3150 mg/kg; LD50 (skin, rabbit) 5000 mg/kg; mod. toxic by ingestion and intraperitoneal routes; mildly toxic by skin contact; heated to decomp., emits acrid smoke and fumes
Uses: Synthetic flavoring agent
Usage level: 21 ppm (nonalcoholic beverages), 24 ppm (ice cream, ices, etc.), 110 ppm (candy), 73 ppm (baked goods), 100 ppm (gelatins and puddings), 840 ppm (chewing gum)
Regulatory: FDA 21CFR §172.515

Benzaldehyde propylene glycol acetal
FEMA 2130
Synonyms: 4-Methyl-2-phenyl-m-dioxolane
Empirical: $C_{10}H_{12}O_2$
Properties: m.w. 164.21
Uses: Synthetic flavoring agent
Usage level: 34 ppm (nonalcoholic beverages), 27 ppm (ice cream, ices, etc.), 110 ppm (candy), 96 ppm (baked goods), 50 ppm (gelatins and puddings)
Regulatory: FDA 21CFR §172.515

Benzal glyceryl acetal. *See* Benzaldehyde glyceryl acetal
1-Benzazine. *See* Quinoline
2-Benzazine. *See* Isoquinoline
1-Benzazole. *See* Indole
Benzeneacetaldehyde. *See* Phenylacetaldehyde
Benzeneacetic acid. *See* Phenylacetic acid
Benzeneacetic acid 2-propenyl ester. *See* Allyl phenylacetate
1-Benzene-azo-β-naphthylamine. *See* 1-(Phenylazo) 2 naphthylamine
Benzenecarbaldehyde. *See* Benzaldehyde
Benzenecarboxylic acid. *See* Benzoic acid
1,2-Benzenedicarboxylic acid, 2-butoxy-2-oxoethyl, butyl ester. *See* n-Butyl phthalyl-n-butyl glycolate
1,2-Benzenedicarboxylic acid dioctyl ester. *See* Dioctyl phthalate
Benzeneethanol. *See* Phenethyl alcohol
Benzenemethanol. *See* Benzyl alcohol
Benzenepropanal. *See* Hydrocinnamaldehyde
Benzenepropanoic acid. *See* Hydrocinnamic acid
3-Benzenepropanol. *See* Hydrocinnamic alcohol

Benzenethiol
CAS 108-98-5; EINECS 203-635-3; FEMA 3616
Synonyms: Thiophenol; Phenyl mercaptan
Empirical: C_6H_6S
Formula: C_6H_5SH
Properties: Liq., repulsive penetrating garlic-like odor; very sol. in alcohol; misc. with ether, benzene, CS_2; insol. in water; m.w. 110.04; dens. 1.078 (20/4 C); m.p. 70 C; b.p. 169.5 C; flash pt. 55 C; ref. index 1.5931 (14 C)
Precaution: Oxidizes in air
Toxicology: Toxic by inhalation, skin contact, ingestion; causes burns
Uses: Synthetic flavoring agent
Regulatory: FDA 21CFR §172.515
Manuf./Distrib.: Aldrich

Benzin. *See* Naphtha
1,2-Benzisothiazolin-3-one 1,1-dioxide ammonium salt. *See* Ammonium saccharin
2-Benzisothiazolin-3-one-1,1-dioxide sodium salt. *See* Saccharin sodium
1,2-Benzisothiazol-3(2H)-one, 1,1-dioxide, calcium salt. *See* Calcium saccharin
Benzoate of soda. *See* Sodium benzoate

Benzoate sodium. *See* Sodium benzoate
1,2-Benzodihydropyrone. *See* Dihydrocoumarin
1,3-Benzodioxole-5-carboxaldehyde. *See* Heliotropine
3,4-Benzodioxole-5-carboxaldehyde. *See* Heliotropine
1-[5-(1,3-Benzodioxol-5-l)-1-oxo-2,4-pentadienyl]piperidine. *See* Piperine

Benzoic acid
CAS 65-85-0; EINECS 200-618-2; FEMA 2131; EEC E210
Synonyms: Benzenecarboxylic acid; Phenylformic acid; Dracylic acid
Classification: Aromatic acid
Empirical: $C_7H_6O_2$
Properties: White scales, needles, crystals, benzoin odor; sol. in alcohol, ether, chloroform, benzene, carbon
disulfide; sl. sol. in water; m.w. 122.13; dens. 1.2659; m.p. 121.25 C; b.p. 249.2 C; flash pt. 121.1 C
Precaution: Combustible when exposed to heat or flame; reactive with oxidizing materials
Toxicology: LD50 (oral, rat) 2530 mg/kg; mod. toxic by ingestion, IP routes; poison by subcut. route; severe
eye/skin irritant; may cause human intolerance reaction, asthma, hyperactivity in children; heated to
decomp., emits acrid smoke and irritating fumes
Uses: Antimicrobial agent, flavoring agent, preservative
Usage level: 0.1% (preservative); 7.5 ppm (nonalcoholic beverages), 4.8 ppm (ice cream, ices), 8.9 ppm
(candy), 40 ppm (baked goods), 20-32 ppm (chewing gum), 250 ppm (icings); ADI 0-5 mg/kg (EEC)
Regulatory: FDA 21CFR §150.141, 150.161, 166.40, 166.110, 175.300, 184.1021, GRAS 0.1% max. in foods;
USA EPA registered; Japan 0.2% max.; Europe listed 0.5% max.; GRAS (FEMA); cleared by MID to retard
flavor reversion in oleomargarine at 0.1%; Japan approved with limitations; Europe listed; UK approved
Manuf./Distrib.: Aldrich; Ashland; Int'l. Sourcing; E. Merck; Penta Mfg.

Benzoic acid, benzyl ester. *See* Benzyl benzoate
Benzoic acid ethyl ester. *See* Ethyl benzoate
Benzoic acid, 4-hydroxy-, phenylmethyl ester. *See* Benzylparaben
Benzoic acid methyl ester. *See* Methyl benzoate
Benzoic acid potassium salt. *See* Potassium benzoate
Benzoic acid sodium salt. *See* Sodium benzoate
o-Benzoic acid sulfimide. *See* Saccharin
Benzoic aldehyde. *See* Benzaldehyde

Benzoin
CAS 119-53-9; EINECS 204-331-3; FEMA 2132
Synonyms: α-Hydroxybenzyl phenyl ketone; 2-Hydroxy-2-phenylacetophenone; α-Hydroxy-α-
phenylacetophenone; Bitter almond oil camphor
Empirical: $C_{14}H_{12}O_2$
Formula: $C_6H_5CH(OH)COC_6H_5$
Properties: Wh. or ylsh. cryst., sl. camphor odor; sol. in acetone, hot water; sl. sol. in water, ether; m.w. 212.22;
m.p. 137 C
Precaution: Combustible
Toxicology: Mutagenic data; heated to decomp., emits acrid smoke and irritating fumes
Uses: Synthetic flavoring agent, color diluent in fruit, vegetables
Usage level: 4.5 ppm (nonalcoholic beverages), 0.54 ppm (ice cream, ices), 2 ppm (candy), 1.4 ppm (baked
goods), 0.1 ppm (gelatins, puddings)
Regulatory: FDA 21CFR §73.1, no residue, 172.515; GRAS (FEMA); Japan approved
Manuf./Distrib.: C.A.L.-Pfizer

Benzoin extract
CAS 84696-18-4; 84929-79-3
Synonyms: Styrax benzoin extract; Styrax extract
Definition: Extract of resin obtained from *Styrax benzoin*
Uses: Natural flavoring agent
Usage level: 15 ppm (nonalcoholic beverages), 5.1 ppm (ice cream, ices), 8.7 ppm (candy), 20 ppm (baked
goods), 10 ppm (gelatins, puddings), 110 ppm (chewing gum)
Regulatory: FDA 21CFR §172.510
Manuf./Distrib.: Pierre Chauvet

Benzoin gum. *See* Gum benzoin
Benzoperoxide. *See* Benzoyl peroxide

Benzophenone
CAS 119-61-9; EINECS 204-337-6; FEMA 2134
Synonyms: Benzoylbenzene; Diphenyl ketone; Diphenylmethanone

Classification: Organic compd.
Empirical: $C_{13}H_{10}O$
Properties: Wh. rhombic cryst., persistent rose-like odor; sol. in fixed oils; sl. sol. in propylene glycol; m.w. 182.23; sp.gr. 1.0976 (α, 50/50 C), 1.108 (β, 23/40 C); m.p. 49 C (α), 26 C (β), 47 C (γ); b.p. 305 C
Precaution: Combustible when heated; incompat. with oxidizers
Toxicology: LD50 (oral, mouse) 2895 mg/kg; moderately toxic by ingestion and intraperitoneal routes; heated to decomp., emits acrid and irritating fumes
Uses: Synthetic flavoring agent
Usage level: 0.5 ppm (nonalcoholic beverages), 0.61 ppm (ice cream, ices), 1.7 ppm (candy), 2.4 ppm (baked goods)
Regulatory: FDA 21CFR §172.515
Manuf./Distrib.: Aldrich

Benzo[b]pyridine. *See* Quinoline
1H-Benzo[b]pyrrole. *See* Indole
2,3-Benzopyrrole. *See* Indole
1-Benzoxy-1-(2-methoxyethoxy)-ethane. *See* Benzyl methoxyethyl acetal
Benzoylacetic acid ethyl ester. *See* Ethyl benzoylacetate
Benzoylbenzene. *See* Benzophenone
Benzoyl eugenol. *See* Eugenyl benzoate

Benzoyl peroxide
CAS 94-36-0; EINECS 202-327-6
Synonyms: Dibenzoyl peroxide; Benzoperoxide; Benzoyl superoxide
Classification: Organic peroxide
Empirical: $C_{14}H_{10}O_4$
Formula: $[C_6H_5C(O)]_2O_2$
Properties: Colorless to wh. gran., crust. solid, faint odor of benzaldehyde, tasteless; sl. sol. in alcohols, veg. oils, water; sol. in benzene, chloroform, ether; m.w. 242.23; dens. 1.3340 (25 C); m.p. 103-105 C
Precaution: Flamm. oxidizing liq.; explosion hazard
Toxicology: LD50 (oral, rat) 7710 mg/kg; highly toxic; TLV 5 mg/m^3; poison by ingestion, intraperitoneal routes; allergen; eye irritant
Uses: Direct food additive, bleaching agent, oxidizing agent in bleaching oils, flours, and in cheese prod.; flour treatment agent
Regulatory: FDA 21CFR §184.1157, GRAS; Japan approved with limitations

Benzoyl superoxide. *See* Benzoyl peroxide
Benzylacetaldehyde. *See* Hydrocinnamaldehyde

Benzyl acetate
CAS 140-11-4; EINECS 205-399-7; FEMA 2135
Synonyms: Phenylmethyl acetate; α-Acetoxytoluene; Benzyl ethanoate
Definition: Ester of benzyl alcohol and acetic acid
Empirical: $C_9H_{10}O_2$
Formula: $CH_3COOCH_2C_6H_5$
Properties: Colorless liq., sweet floral fruity odor; sol. in alcohol, most fixed oils, propylene glycol; insol. in water, glycerin; m.w. 150.19; sp.gr. 1.06; m.p. -51.5 C; b.p. 213.5 C; flash pt. (CC) 216 F; ref. index 1.501
Precaution: Combustible liq.
Toxicology: LD50 (oral, rat) 2490 mg/kg; moderately toxic by ingestion and subcutaneous routes; poison by inhalation; antipsychotic; heated to decomp., emits irritating fumes
Uses: Synthetic flavoring agent
Usage level: 7.8 ppm (nonalcoholic beverages), 14 ppm (ice cream, ices), 34 ppm (candy), 22 ppm (baked goods), 23 ppm (gelatins, puddings), 760 ppm (chewing gum)
Regulatory: FDA 21CFR §172.515; GRAS (FEMA); Japan approved as flavoring
Manuf./Distrib.: Aldrich; Haarmann & Reimer; Chr. Hansen's; Penta Mfg.; Quest Int'l.

Benzylacetic acid. *See* Hydrocinnamic acid

Benzyl acetoacetate
CAS 5396-89-4; EINECS 226-416-4; FEMA 2136
Synonyms: Benzyl acetyl acetate; Benzyl β-ketobutyrate; Benzyl 3-oxobutanoate
Empirical: $C_{11}H_{12}O_3$
Formula: $CH_3COCH_2COOCH_2C_6H_5$
Properties: Oily liq.; sol. in alkali solutions at room temp.; m.w. 192.22; dens. 1.112; b.p. 156-159 C (10 mm); flash pt. >230 F
Toxicology: Irritating to skin, eyes

Benzyl acetyl acetate

Uses: Synthetic flavoring agent
Usage level: 2.7 ppm (nonalcoholic beverages), 6.0 ppm (ice cream, ices, etc.), 13 ppm (candy), 13 ppm (baked goods), 0.50-10 ppm (gelatins and puddings), 50 ppm (chewing gum)
Regulatory: FDA 21CFR §172.515
Manuf./Distrib.: Aldrich

Benzyl acetyl acetate. *See* Benzyl acetoacetate

Benzyl alcohol
CAS 100-51-6; EINECS 202-859-9; FEMA 2137
Synonyms: α-Hydroxytoluene; Phenylmethanol; Phenylcarbinol; Benzenemethanol
Classification: Aromatic alcohol
Empirical: $C_6H_5CH_2OH$
Formula: C_7H_8O
Properties: Water-wh. liq., faint aromatic odor, sharp burning taste; misc. with alcohol, chloroform, ether, water @ 206 C (dec.); m.w. 108.15; sp.gr. 1.042; m.p. -15.3 C; b.p. 205.7 C; flash pt. (CC) 213 F; ref. index 1.540
Precaution: Combustible liq.; dec. explosively at 180 C with sulfuric acid
Toxicology: LD50 (oral, rat) 1230 mg/kg; poison by ingestion, intraperitoneal, IV routes; mod. toxic by inh., skin contact; mod. skin and severe eye irritant; heated to decomp., emits acrid smoke and fumes
Uses: Synthetic flavoring agent
Usage level: 1-3%; 15 ppm (nonalcoholic beverages), 160 ppm (ice cream, ices), 47 ppm (candy), 220 ppm (baked goods), 21-45 ppm (gelatins, puddings), 1200 ppm (chewing gum)
Regulatory: FDA 21CFR §172.515; GRAS (FEMA); USA EPA registered; Japan approved as flavoring; Europe listed
Manuf./Distrib.: Aldrich; Ashland; Haarmann & Reimer; E. Merck; Penta Mfg.; Quest Int'l.

Benzyl benzoate
CAS 120-51-4; EINECS 204-402-9; FEMA 2138
Synonyms: Benylate; Phenylmethyl benzoate; Benzoic acid, benzyl ester
Definition: Ester of benzyl alcohol and benzoic acid
Empirical: $C_{14}H_{12}O_2$
Formula: $C_6H_5COOCH_2C_6H_5$
Properties: Colorless oily liq., sl. aromatic odor; misc. with alcohol, chloroform, ether; insol. in water, glycerin; m.w. 212.26; sp.gr. 1.116; m.p. 21 C; b.p. 324 C; flash pt. (CC) 298 F; ref. index 1.568
Precaution: Combustible liq.; reactive with oxidizing materials
Toxicology: LD50 (oral, rat) 500 mg/kg; mod. toxic by ingestion, skin contact; heated to decomp., emits acrid and irritating fumes and smoke
Uses: Synthetic flavoring agent
Usage level: 4.5 ppm (nonalcoholic beverages), 12 ppm (ice cream, ices), 39 ppm (candy), 33 ppm (baked goods), 280 ppm (chewing gum)
Regulatory: FDA 21CFR §172.515
Manuf./Distrib.: Aldrich; Haarmann & Reimer; Penta Mfg.

Benzyl butanoate. *See* Benzyl butyrate
Benzylbutyl alcohol. *See* α-Propylphenethyl alcohol

Benzyl butyl ether
FEMA 2139
Synonyms: Butyl benzyl ether
Empirical: $C_{11}H_{16}O$
Properties: Colorless liq., misc. with alcohol, ether; insol. in water; m.w.164.25; sp.gr. 0.931 (10 C); b.p. 220-221 C (744 mm)
Uses: Synthetic flavoring agent
Usage level: 0.50-2.0 ppm (nonalcoholic beverages), 3.5 ppm (ice cream, ices, etc.), 8.0 ppm (candy), 2.0-8.0 ppm (baked goods), 2.0 ppm (gelatins and puddings)
Regulatory: FDA 21CFR §172.515

Benzyl butyrate
CAS 103-37-7; FEMA 2140
Synonyms: Benzyl butanoate
Empirical: $C_{11}H_{14}O_2$
Properties: Colorless liq.; plum-like odor; sol. in alcohol; insol. in water; m.w. 178.23; dens. 1.009; b.p. 240 C; flash pt. 225 F; ref. index 1.492-1.496 (20 C)
Uses: Synthetic flavoring agent
Usage level: 4.5 ppm (nonalcoholic beverages); 6.9 ppm (ice cream, ices, etc.), 7.7 ppm (candy), 9.9 ppm (baked goods), 3.0 ppm (gelatins and puddings), 310 ppm (chewing gum)

Regulatory: FDA 21CFR §172.515
Manuf./Distrib.: Aldrich; Chr. Hansen's

Benzyl carbinol. *See* Phenethyl alcohol
Benzylcarbinyl acetate. *See* 2-Phenylethyl acetate
Benzyl carbinyl anthranilate. *See* Phenethyl anthranilate
Benzyl carbinyl cinnamate. *See* Phenethyl cinnamate
Benzyl carbinyl tiglate. *See* Phenethyl tiglate
Benzylcarbinyl-α-toluate. *See* Phenethyl phenylacetate

Benzyl cinnamate
CAS 103-41-3; EINECS 203-109-3; FEMA 2142
Synonyms: Cinnamein; Phenylmethyl 3-phenyl-2-propenoate; Benzyl 3-phenylpropenoate
Definition: Ester of benzyl alcohol and cinnamic acid
Empirical: $C_{16}H_{14}O_2$
Properties: Wh. cryst., aromatic odor; sol. in fixed oils; insol. in glycerin, propylene glycol; m.w. 238.30; m.p. 39 C; b.p. 350 C; flash pt. 100 C
Precaution: Combustible liq.
Toxicology: Mod. toxic by ingestion; mild allergen and skin irritant; heated to decomp., emits acrid smoke and irritating fumes
Uses: Mild, sweetly balsamic fragrance and flavoring; improves flavor of foodstuffs
Usage level: 1.4 ppm (nonalcoholic beverages), 2.5 ppm (ice cream, ices), 6.7 ppm (candy), 6.6 ppm (baked goods), 3-5 ppm (gelatins, puddings), 5.3-120 ppm (chewing gum)
Regulatory: FDA 21CFR §172.515; GRAS (FEMA)
Manuf./Distrib.: Aldrich

Benzyldiethyl [(2,6-xylylcarbomoyl)methyl]ammonium benzoate. *See* Denatonium benzoate NF
Benzyl dimethyl carbinol. *See* Dimethylbenzyl carbinol
Benzyl dimethyl carbinyl acetate. *See* Dimethylbenzyl carbinyl acetate
Benzyl dimethylcarbinyl butyrate. *See* α,α-Dimethylphenethyl butyrate
Benzyl dlmethylcarbinyl formate. *See* α,α-Dimethylphenethyl formate

Benzyl 2,3-dimethylcrotonate
Synonyms: Benzyl methyl tiglate
Uses: Synthetic flavoring agent
Regulatory: FDA 21CFR §172.515

Benzyldimethyldodecylammonium chloride. *See* Lauralkonium chloride

Benzyl dipropyl ketone
Synonyms: 3-Benzyl-4-heptanone
Uses: Synthetic flavoring agent
Regulatory: FDA 21CFR §172.515

Benzyl disulfide
CAS 150-60-7; EINECS 205-764-0; FEMA 3617
Synonyms: Dibenzyl disulfide; Di(phenylmethyl) disulfide; Bis(phenylmethyl) disulfide; α-(Benzyldithio)toluene
Empirical: $C_{14}H_{14}S_2$
Formula: $C_6H_5CH_2SSCH_2C_6H_5$
Properties: Pale-yellow leaflets; burnt caramel odor; sol. in hot alcohol, ether, benzene; pract. insol. in water; m.w. 246.40; m.p. 71-72 C; b.p. >270 C (with decomp.)
Toxicology: Irritating to respiratory system
Uses: Synthetic flavoring agent
Regulatory: FDA 21CFR §172.515
Manuf./Distrib.: Aldrich

α-(Benzyldithio)toluene. *See* Benzyl disulfide
Benzyl ethanoate. *See* Benzyl acetate

Benzyl ether
CAS 103-50-4; EINECS 203-118-2; FEMA 2371
Synonyms: Dibenzyl ether; 1,1′-[Oxybis(methylene)]bis[benzene]
Empirical: $C_{14}H_{14}O$
Formula: $(C_6H_5CH_2)_2O$
Properties: Colorless to pale yel. unstable liq.; misc. with ethanol, ether, chloroform, acetone; pract. insol. in water; m.w. 198.28; dens. 1.043 (20/4 C); m.p. 5 C; b.p. 298 C (dec.); flash pt. (CC) 275 F; ref. index 1.557
Precaution: Combustible exposed to heat or flame; reactive with oxidizing materials; mod. explosion hazard

347

by spontaneous chemical reaction
Toxicology: LD50 (oral, rat) 2500 mg/kg; mod. toxic by ingestion; vapors may be narcotic in high conc.; skin and eye irritant
Uses: Synthetic flavoring agent
Regulatory: FDA 21CFR §172.515
Manuf./Distrib.: Aldrich

Benzyl ethyl ether
FEMA 2144
Synonyms: Ethyl benzyl ether
Empirical: $C_9H_{12}O$
Properties: Oily liq.; pineapple odor; misc. with alcohol, ether; insol. in water; m.w. 136.19; b.p. 186 C; ref. index 1.4955 (20 C)
Precaution: Combustible
Uses: Synthetic flavoring agent
Usage level: 0.50-1.0 ppm (nonalcoholic beverages), 2.5 ppm (ice cream, ices, etc.), 7.5 ppm (candy), 7.5 ppm (baked goods)
Regulatory: FDA 21CFR §172.515

Benzyl formate
FEMA 2145
Synonyms: Formic acid benzyl ester
Empirical: $C_8H_8O_2$
Properties: Colorless liq.; floral-fruity odor; misc. with alcohols, ketones, oils; insol. in water; m.w. 136.15; b.p. 202-203 C; flash pt. 83 C; ref. index 1.5100-1.5120 (20 C)
Uses: Synthetic flavoring agent
Usage level: 2.4 ppm (nonalcoholic beverages), 8.0 ppm (ice cream, ices, etc.), 12 ppm (candy), 8.6 ppm (baked goods), 3.2 ppm (chewing gum)
Regulatory: FDA 21CFR §172.515

3-Benzyl-4-heptanone. *See* Benzyl dipropyl ketone
Benzylhexadecyldimethylammonium chloride. *See* Cetalkonium chloride
Benzyl hydrosulfide. *See* Benzyl mercaptan
Benzyl p-hydroxybenzoate. *See* Benzylparaben

Benzylidene acetone
CAS 122-57-6; EINECS 204-555-1; FEMA 2881
Synonyms: Benzalacetone; Methyl styryl ketone; trans-4-Phenyl-3-buten-2-one
Empirical: $C_{10}H_{10}O$
Formula: $C_6H_5CH{:}CHCOCH_3$
Properties: Colorless cryst., odor of coumarin; sol. in alcohol, ether, benzene, chloroform; insol. in water; f.w. 146.19; m.p. 39-42; f.p. 150 F; b.p. 260-262
Precaution: Combustible
Uses: Flavors
Manuf./Distrib.: Aldrich; Penta Mfg.

2-Benzylidene-heptanol. *See* α-Amylcinnamyl alcohol
2-Benzylidene hexanal. *See* α-Butylcinnamaldehyde
2-Benzylidene-octanal. *See* α-Hexylcinnamaldehyde
Benzyl isoamyl alcohol. *See* α-Isobutylphenethyl alcohol
Benzyl isobutyl ketone. *See* 4-Methyl-1-phenyl-2-pentanone

Benzyl isobutyrate
CAS 103-28-6; FEMA 2141
Synonyms: Isobutyric acid benzyl ester; Benzyl-2-methylpropionate
Empirical: $C_{11}H_{14}O_2$
Properties: Colorless liq., fruity floral jasmine odor; sol. in alcohol, fixed oils; sl. sol. in propylene glycol; insol. in glycerin; m.w. 178.25; dens. 1.001-1.005; b.p. 105-108 C (4 mm); flash pt. 100 C; ref. index 1.489-1.4920 (20 C)
Precaution: Combustible liq.
Toxicology: LD50 (oral, rat) 2850 mg/kg; mod. toxic by ingestion; heated to decomp, emits acrid smoke and fumes
Uses: Synthetic flavoring agent
Usage level: 5.2 ppm (nonalcoholic beverages), 12 ppm (ice cream, ices, etc.), 12 ppm (candy), 25 ppm (baked goods)
Regulatory: FDA 21CFR §172.515

Manuf./Distrib.: Hüls

Benzyl isoeugenol. *See* Isoeugenyl benzyl ether

Benzyl isovalerate
CAS 103-38-8; FEMA 2152
Synonyms: Benzyl 3-methyl butyrate
Empirical: $C_{12}H_{16}O_2$
Properties: Colorless liq.; apple, pineapple odor; sol. in alcohol, most fixed oils; insol. in water; m.w. 192.26; dens. 0.988; b.p. 245 C; flash pt. >230 F; ref. index 1.486-1.490
Uses: Synthetic flavoring agent
Usage level: 2.2 ppm (nonalcoholic beverages), 3.4 ppm (ice cream, ices, etc.), 16 ppm (candy), 9.4 ppm (baked goods), 56 ppm (gelatins and puddings), 200 ppm (chewing gum)
Regulatory: FDA 21CFR §172.515
Manuf./Distrib.: Aldrich

Benzyl β-ketobutyrate. *See* Benzyl acetoacetate
Benzyl ketone. *See* 1,3-Diphenyl-2-propanone

Benzyl mercaptan
CAS 100-53-8; EINECS 202-862-5; FEMA 2147
Synonyms: Benzyl hydrosulfide; α-Tolyl mercaptan; Thiobenzyl alcohol; α-Toluenethiol
Empirical: C_7H_8S
Formula: $C_6H_5CH_2SH$
Properties: Colorless liq.; leek-like odor; m.w. 124.19; dens. 1.058; b.p. 194-195 C; flash pt. 70 C; ref. index 1.576 (20 C)
Precaution: Oxidizes In alr to dibenzyl disulfide
Toxicology: May cause mild irritation to mucous membranes
Uses: Synthetic flavoring agent
Usage level: 0.15-0.25 ppm (nonalcoholic beverages), 0.15-0.50 ppm (ice cream, ices, etc.), 0.50-0.75 ppm (candy), 0.50-0.75 ppm (baked goods)
Regulatory: FDA 21CFR §172.515
Manuf./Distrib.: Aldrich

Benzyl methoxyethyl acetal
FEMA 2140
Synonyms: Acetaldehyde benzyl β-methoxyethyl acetal; 1-Benzoxy-1-(2-methoxyethoxy)-ethane
Empirical: $C_{12}H_{18}O_3$
Properties: Colorless liq.; fruital odor; m.w.210.27
Uses: Synthetic flavoring agent
Usage level: 0.50 ppm (nonalcoholic beverages), 1.0 ppm (ice cream, ices, etc.), 1.0 ppm (candy), 1.0 ppm (baked goods)
Regulatory: FDA 21CFR §172.515

Benzyl 3-methyl butyrate. *See* Benzyl isovalerate
Benzyl-2-methylpropionate. *See* Benzyl isobutyrate
Benzyl methyl tiglate. *See* Benzyl 2,3-dimethylcrotonate
Benzyl 3-oxobutanoate. *See* Benzyl acetoacetate

Benzylparaben
CAS 94-18-8; EINECS 202-311-9
Synonyms: Benzoic acid, 4-hydroxy-, phenylmethyl ester; Benzyl p-hydroxybenzoate; Phenylmethyl 4-hydroxybenzoate
Definition: Ester of benzyl alcohol and p-hydroxybenzoic acid
Empirical: $C_{14}H_{12}O_3$
Uses: Preservative, bactericide, fungicide for foods
Trade names: Nipabenzyl

Benzyl phenylacetate
FEMA 2149
Synonyms: Benzyl α-toluate
Empirical: $C_{15}H_{14}O_2$
Properties: Colorless liq.; floral odor; honey-like taste; misc. with alcohol, chloroform, ether; m.w. 226.28; dens. 1.097-1.099; b.p. 320 C; flash pt. >100 C; ref. index 1.553-1.558
Precaution: Combustible liq.
Toxicology: Heated to decomp., emits acrid smoke and fumes
Uses: Synthetic flavoring agent

Benzyl 3-phenylpropenoate

Usage level: 1.3 ppm (nonalcoholic beverages), 2.6 ppm (ice cream, ices, etc.), 6.6 ppm (candy), 4.3 ppm (baked goods), 5.0 ppm (toppings)
Regulatory: FDA 21CFR §172.515
Manuf./Distrib.: Chr. Hansen's

Benzyl 3-phenylpropenoate. *See* Benzyl cinnamate
Benzyl propanoate. *See* Benzyl propionate

Benzyl propionate
FEMA 2150
Synonyms: Benzyl propanoate
Empirical: $C_{10}H_{12}O_2$
Properties: Colorless liq.; floral-fruity odor; sol. in alcohol, most fixed oils; insol. in water; m.w. 164.20; dens. 1.036; b.p. 219-220 C; flash pt. 100 C; ref. index 1.496-1.500
Precaution: Combustible
Toxicology: Heated to decomp., emits acrid smoke and fumes
Uses: Synthetic flavoring agent
Usage level: 4.1 ppm (nonalcoholic beverages), 5.8 ppm (ice cream, ices, etc.), 19 ppm (candy), 17 ppm (baked goods), 19-150 ppm (chewing gum), 40 ppm (icings)
Regulatory: FDA 21CFR §172.515; Japan approved as flavoring

Benzylpropyl acetate. *See* Dimethylbenzyl carbinyl acetate
Benzyl-n-propyl carbinol. *See* α-Propylphenethyl alcohol

Benzylsalicylate
CAS 118-58-1; EINECS 204-262-9; FEMA 2151
Synonyms: Phenylmethyl 2-hydroxybenzoate
Definition: Ester of benzyl alcohol and salicylic acid
Empirical: $C_{14}H_{12}O_3$
Properties: Colorless visc. liq., pleasant odor; sol. in fixed oils; insol. in glycerin, propylene glycol; m.w. 228.26; sp.gr. 1.175 (20 C); b.p. 208 C (26 mm); ref. index 1.579
Precaution: Combustible when heated or exposed to flame; incompat. with oxidizing materials
Toxicology: LD50 (oral, rat) 2227 mg/kg; mod. toxic by ingestion; heated to decomp., emits acrid smoke and irritating fumes
Uses: Synthetic flavoring agent
Usage level: 1.4 ppm (nonalcoholic beverages), 0.89 ppm (ice cream, ices), 1.8 ppm (candy), 0.01-2.2 ppm (baked goods)
Regulatory: FDA 21CFR §172.515; GRAS (FEMA)

Benzyl α-toluate. *See* Benzyl phenylacetate
Bergamiol. *See* Linalyl acetate
Bergamol. *See* Linalyl acetate

Bergamot oil
CAS 8007-75-8; 85049-52-1; EINECS 289-612-9 (extract); FEMA 2153
Definition: Psoralen-free volatile oil obtained from the rind and fruit of *Citrus bergamia*, contg. 36-45% l-linalyl acetate, 6% l-linalool, and d-limonene, dipentene, bergaptene
Properties: Yel.-grn. liq., agreeable odor, bitter taste; sol. in fixed oils; misc. with alcohol, glac. acetic acid; insol. in glycerin, propylene glycol; dens. 0.875-0.880 (25/25 C); ref. index 1.464-1.467
Precaution: Keep cool, well closed; protect from light
Toxicology: LD50 (oral, rat) 11,520 mg/kg; mildly toxic by ingestion; mild skin irritant and allergen; heated to decomp., emits acrid smoke and irritating fumes
Uses: Natural flavoring agent
Usage level: 8.9 ppm (nonalcoholic beverages), 7.9 ppm (ice cream, ices), 27 ppm (candy), 29 ppm (baked goods), 5.3-90 ppm (gelatins, puddings), 43 ppm (chewing gum), 1-130 ppm (icings)
Regulatory: FDA 21CFR §182.20, GRAS; 27CFR §21.65, 21.151; Japan approved; Council of Europe listed
Manuf./Distrib.: Pierre Chauvet; Commodity Services; Florida Treatt

Betanine. *See also* Beet powd.
EEC E162
Definition: Coloring principal in beets
Properties: m.w. 568.5

Betula alba oil. *See* Birch oil

BHA
CAS 25013-16-5; EINECS 204-442-7, 246-563-8; FEMA 2183; EEC E320
Synonyms: Butylated hydroxyanisole; (1,1-Dimethylethyl)-4-methoxyphenol; 3-t-Butyl-4-hydroxyanisole

Definition: Mixture of isomers of tertiary butyl-substituted 4-methoxyphenols
Empirical: $C_{11}H_{16}O_2$
Properties: Wh. waxy solid, faint char. odor; insol. in water; sol. in petrol. ether, 50% or higher alcohol, propylene glycol, fats, oils; m.w. 180.27; m.p. 48-55 C; b.p. 264-270 C (733 mm)
Precaution: Combustible
Toxicology: LD50 (oral, mouse) 2000 mg/kg; suspected carcinogen; moderate toxicity by ingestion, intraperitoneal routes; may cause rashes, hyperactivity; heated to decomp., emits acrid and irritating fumes
Uses: Antioxidant, preservative for foods, etc.; in beet sugar and yeast defoamers
Usage level: 1000 ppm (active dry yeast), 2 ppm (beverages and desserts prepared from dry mixes), 32 ppm (dry diced glaceed fruits); ADI 0-0.5 mg/kg (EEC)
Regulatory: FDA 21CFR §166.110, 172.110, 172.515, 172.615, 173.340, 175.105, 175.125, 175.300, 175.380, 175.390, 176.170, 176.210, 177.1010, 177.1210, 177.1350, 178.3120, 178.3570, 179.45, 181.22, 181.24 (0.005% migrating from food pkg.), 182.3169 (0.02% max. of fat or oil), GRAS; GRAS (FEMA); USDA 9CFR 318.7, 381.147; Japan approved 0.2-1 g/kg; Europe approved; UK approved
Trade names: Embanox® BHA; Lowinox® BHA; Sustane® 1-F; Sustane® BHA; Tenox® BHA
Trade names containing: Arlacel® 186; Atlas 5520; Beta Carotene Emulsion Beverage Type 3.6 No. 65392; Centrophil® M; Centrophil® W; Coco-Spred™; Dry Beta Carotene Beadlets Yellow Type 2.4-S No. 653800100; Dry Vitamin D_3 Beadlets Type 850 No. 652550401, 652550601; Dur-Em® 114K; Dur-Em® 204K; Durkex 25BHA; Durkex 100BHA; Durkex 100DS; Dur-Lo®; Embanox® 2; Embanox® 3; Embanox® 4; Embanox® 5; Embanox® 6; Garbefix 31; Nipanox® Special; Palmabeads® Type 500 No. 65332; Palma-Sperse® Type 250-S No. 65322; Palma-Sperse® Type 250A/50 D-S No. 65221; Sustane® 3; Sustane® 4A; Sustane® 6; Sustane® 7; Sustane® 7G; Sustane® 8; Sustane® 27; Sustane® 31; Sustane® HW-4; Sustane® P; Sustane® PA; Sustane® Q; Sustane® T-6; Sustane® W-1; Sustane® W; Tandem 5K; Tandem 8; Tandem 552; Tenox® 2; Tenox® 4; Tenox® 4A; Tenox® 5; Tenox® 5B; Tenox® 6; Tenox® 7; Tenox® 22; Tenox® 24; Tenox® 26; Tenox® 27; Tenox® A; Tenox® R; Tenox® WD-BHA; Vitamin A Palmitate Type 250-CWS No. 65312

BHT

CAS 128-37-0; EINECS 204-881-4; FEMA 2184; EEC E321
Synonyms: DBPC; Butylated hydroxytoluene; 2,6-Di-t-butyl-p-cresol; 2,6-Bis (1,1-dimethylethyl)-4-methylphenol
Classification: Substituted toluene
Empirical: $C_{15}H_{24}O$
Properties: White cryst. solid; insol. in water; sol. in toluene, alcohols, MEK, acetone, Cellosolve, petrol. ether, benzene, most HC solvs.; m.w. 220.39; sp.gr. 1.048 (20/4 C); m.p. 68 C; b.p. 265 C, flash pt. (TOC) 260 F
Precaution: Combustible exposed to heat or flame; reactive with oxidizing materials
Toxicology: TLV: 10 mg/m³; LD50 (oral, rat) 890 mg/kg; moderately toxic by ingestion; poison by IP, IV routes; suspected carcinogen; human skin irritant; eye irritant; may cause rashes, hyperactivity; heated to decomp., emits acrid smoke and fumes
Uses: Antioxidant, preservative for foods, animal feed; prevents flavor deterioration, rancidity
Usage level: Limitation (total BHA and BHT): 50 ppm (dehydrated potato shreds), 50 ppm (dry breakfast cereals), 200 ppm (emulsion stabilizers for shortening), 10 ppm (potato granules); ADI 0-0.5 mg/kg (EEC)
Regulatory: FDA 21CFR §137.350, 166.110, 172.115, 172.615 (0.1% max.), 173.340 (0.1% of defoamer), 175.105, 175.125, 175.300, 175.380, 175.390, 176.170, 176.210, 177.1010, 177.1210, 177.1350, 177.2260, 177.2600, 178.3120, 178.3570, 179.45, 181.22, 181.24 (0.005% migrating from food pkg.), 182.3173 (0.02% max. of fat/oil), GRAS; USDA 9CFR §318.7 (0.003% in dry sausage, 0.01% in rendered animal fat, 0.02% in margarine), 381.147 (0.01% on fat in poultry); Japan 0.2-1 g/kg; Europe, UK approved
Manuf./Distrib.: Am. Roland
Trade names: CAO®-3; Embanox® BHT; Sustane® BHT; Tenox® BHT
Trade names containing: Beta Carotene Emulsion Beverage Type 3.6 No. 65392; Carex for Foods; Cetodan® 90-50; Dry Beta Carotene Beadlets Yellow Type 2.4-S No. 653800100; Dry Vitamin A Palmitate Type 250-SD No. 65378; Dry Vitamin D_3 Beadlets Type 850 No. 652550401, 652550601; Dry Vitamin D_3 Type 100-SD No. 65216; Dur-Em® 114; Dur-Em® 114K; Dur-Em® 204; Dur-Em® 204K; Durkex 100BHT; Durkex 100L; EC-25®; Embanox® 2; Oleoresin Carrot 10,000 CV; Palmabeads® Type 500 No. 65332; Palma-Sperse® Type 250-S No. 65322; Palma-Sperse® Type 250A/50 D-S No. 65221; Perflex; Sustane® 6; Sustane® 11; Sustane® 15; Sustane® 16; Sustane® 18; Sustane® HW-4; Sustane® P; Sustane® PB; Sustane® T-6; Sustane® W-1; Sustane® W; Tenox® 4; Tenox® 4B; Tenox® 5; Tenox® 6; Tenox® 8; Tenox® 25; Tenox® 26; Vitamin A Palmitate Type 250-CWS No. 65312; Vitamin A Palmitate USP, FCC Type P1.7/BHT No. 63693

Biacetyl. *See* Diacetyl
Bibenzene. *See* Biphenyl
Bicarbonate of soda. *See* Sodium bicarbonate

Biophyll. *See* Chlorophyll

d-Biotin
CAS 58-85-5; EINECS 200-399-3
Synonyms: [3aS-(3a-α,4b,6aα)]-Hexahydro-2-oxo-1H-thieno]3,4-d]imidazole-4-pentanoic acid; Vitamin H; cis-Hexahydro-2-oxo-1H-thieno (3,4)-imidazole-4-valeric acid; Coenzyme R
Classification: Organic compd.
Empirical: $C_{10}H_{16}N_2O_3S$
Properties: Wh. cryst. powd.; sl. sol. in water, alcohol; insol. in common org. solvs.; m.w. 244.31; m.p. 231-233 C
Toxicology: No human toxic symptoms reported on heavy dosage; heated to decomp., emits toxic fumes of NO_x, SO_x
Uses: Nutrient, dietary supplement for foods
Regulatory: FDA 21CFR §182.5159, 182.8159, GRAS
Manuf./Distrib.: Am. Roland
Trade names: d-Biotin USP, FCC No. 63345
Trade names containing: Bitrit-1™ (1% Biotin Trituration No. 65324)

Biphenyl
CAS 92-52-4; EINECS 202-163-5; FEMA 3129; EEC E230
Synonyms: PHPH; Bibenzene; Phenyl benzene; Diphenyl; Xenene; 1,1'-Biphenyl
Empirical: $C_{12}H_{10}$
Formula: $C_6H_5C_6H_5$
Properties: White scales; pleasant odor; sol. in alcohol, ether; insol. in water; m.w. 154.22; dens. 1.0; m.p. 70 C; b.p. 256 C; flash pt. 235 F (112.7 C)
Toxicology: Irritant; poison by intravenous route; moderately toxic by ingestion; TLV 0.2 ppm in air
Uses: Antimold (preservative) (0.07 g/kg in grapefruit, lemon, oranges); fungistat in packaging citrus fruits
Regulatory: Europe listed; UK approved; Japan approved
Manuf./Distrib.: Aldrich

1,1'-Biphenyl. *See* Biphenyl
(1,1'-Biphenyl)-2-ol. *See* o-Phenylphenol
(1,1'-Biphenyl)-2-ol, sodium salt. *See* Sodium o-phenylphenate

Birch oil
CAS 8001-88-5; 8027-43-8; EINECS 281-660-9 (extract)
Synonyms: Betula alba oil; Birch tar oil
Definition: Volatile oil obtained from *Betula alba*
Properties: Brn. liq., leather-like odor; sol. in fixed oils; insol. in glycerin, min. oil, propylene glycol; sp.gr. 0.886-0.950
Precaution: Combustible when exposed to heat or flame; reactive with oxidizing materials
Toxicology: Skin irritant; mod. irritating to eyes and mucous membranes; mild allergen
Uses: Natural flavoring agent in birch beer
Regulatory: GRAS
Manuf./Distrib.: Pierre Chauvet

Birch tar oil. *See also* Birch oil
Definition: Derived from *Betula pendula*
Properties: Clear, dark brown liq.; leather-like odor; sol. in most fixed oils; insol. glycerin, mineral oil, propylene glycol
Uses: Synthetic flavoring agent
Regulatory: FDA 21CFR §172.515

Birthwort extract. *See* Serpentaria extract

Bisbenthiamine
Uses: Dietary supplement (vitamin)
Regulatory: Japan approved

Bis(2-carboxyethyl) sulfide. *See* 3,3'-Thiodipropionic acid
N,N-Bis(carboxymethyl)glycine trisodium salt. *See* Trisodium NTA
1,4-Bis(3,4-dihydroxyphenyl)-2,3-dimethylbutane. *See* Nordihydroguaiaretic acid
2,6-Bis (1,1-dimethylethyl)-4-methylphenol. *See* BHT
Bis(dodecyloxycarbonylethyl) sulfide. *See* Dilauryl thiodipropionate
Bis(D-gluconato) copper. *See* Copper gluconate
N,N-Bis (2-hydroxyethyl) coco amides. *See* Cocamide DEA
1,7-Bis(4-hydroxy-3-methoxyphenyl)-1,6-heptadiene-3,5-dione. *See* Turmeric

Bis(2-methylpropyl) hexanedioate. *See* Diisobutyl adipate
Bisodium tartrate. *See* Sodium tartrate
Bis(phenylmethyl) disulfide. *See* Benzyl disulfide
Bissy nuts. *See* Cola

Bitter almond extract
EINECS 291-060-9
Synonyms: Prunus amygdalus amara extract
Definition: Extract of the seeds of *Prunus amygdalus amara*
Uses: Natural flavoring
Regulatory: FDA 21CFR §182.20, GRAS

Bitter almond oil
CAS 8013-76-1, 8015-75-6; EINECS 291-060-9 (extract); FEMA 2127
Synonyms: Amygdalin; Almond oil, bitter
Definition: The volatile essential oil distilled from ground kernels of bitter almonds or other sources of amygdalin (apricots, cherries, peaches)
Properties: Colorless liq., strong almond odor; sol. in fixed oils, propylene glycol; sl. sol. in water; insol. in glycerin; dens. 1.045-1.070 (15 C); b.p. 179 C; ref. index 1.5428-1.5439
Toxicology: LD50 (oral, rat) 960 mg/kg; human poison by ingestion; mod. toxic by skin contact; skin irritant; heated to decomp., emits toxic fumes of CN⁻
Uses: Natural flavoring agent for baked goods, liqueurs, beverages, cakes, chewing gum, confections, gelatin desserts, ice cream, maraschino cherries, pastries, puddings
Usage level: 80 ppm (nonalcoholic beverages), 130 ppm (alcoholic beverages), 66 ppm (ice cream, ices), 97 ppm (candy), 96 ppm (baked goods), 29 ppm (gelatins, puddings), 330 ppm (chewing gum), 340 ppm maraschino cherries
Regulatory: FDA 21CFR §182.20, GRAS; must be treated and redistilled to remove hydrocyanic acid; 27CFR §21.65, 21.151
Manuf./Distrib.: Aldrich

Bitter almond oil camphor. *See* Benzoin
Bitter ash. *See* Quassia
Bitter fennel oil. *See* Fennel oil

Bitter orange extract
Synonyms: Citrus aurantium fruit extract
Definition: Extract of fruit of *Citrus aurantium*
Uses: Natural flavoring
Regulatory: FDA 21CFR §182.20, GRAS

Bitter orange oil
CAS 68916-04-1
Definition: Volatile oil expressed from peel of *Citrus aurantium*, contg. about 90% d-limonene and citral, decyl aldehyde, methyl anthranilate, linalool, terpineol
Properties: Pale yel. to yel.-brn. liq., char. orange odor, bitter taste; sol. in fixed oils, min. oil; misc. with abs. alcohol, in 1 vol. glac. acetic acid; very sl. sol. in water; dens. 0.845-0.851; ref. index 1.470 (20 C)
Precaution: Keep cool, well closed; protect from light
Toxicology: Heated to decomp., emits acrid smoke and irritating fumes
Uses: Natural flavoring agent
Regulatory: FDA 21CFR §182.20, GRAS
Manuf./Distrib.: Pierre Chauvet; Florida Treatt

Bitter orange peel. *See* Curacao orange peel

Bitter orange peel extract
CAS 8028-48-6; EINECS 232-433-8
Synonyms: Citrus aurantium peel extract
Uses: Natural flavoring agent
Regulatory: FDA 21CFR §182.20, GRAS

Bitter root. *See* Gentian
Bitter salts. *See* Magnesium sulfate heptahydrate
Bitter wood. *See* Quassia
Bixin. *See* Annatto extract

Blackberry extract
CAS 84787-69-9
Synonyms: Dewberry extract; Rubus extract; Rubus villosus extract

Black caraway

Definition: Extract of *Rubus fruticocus* or *R. villosus*
Uses: Natural flavoring agent
Regulatory: FDA 21CFR §172.510

Black caraway

Synonyms: Black cumin
Definition: Derived from *Nigella sativa*
Uses: Natural flavoring agent
Regulatory: FDA 21CFR §182.10, GRAS

Black cumin. *See* Black caraway

Black currant

Definition: Buds and leaves from *Ribes nigrum*
Uses: Natural flavoring agent; colorant (Japan)
Regulatory: FDA 21CFR §172.510; Japan restricted use

Black currant extract

Definition: Extract of the fruit of *Ribes nigrum*
Uses: Natural flavoring agent
Regulatory: FDA 21CFR §172.510
Manuf./Distrib.: Am. Roland; Pierre Chauvet

Black currant oil

Definition: Oil produced from seed of the black currant
Properties: Lt. yel. clear oil
Uses: Food ingred.
Manuf./Distrib.: Arista Industries

Black pepper oil

CAS 8006-82-4
Definition: Derived from *Piper nigrum*
Toxicology: Mutagenic data; moderate skin irritant; heated to decomp., emits acrid smoke and fumes
Uses: Natural flavoring agent
Usage level: 2.7 ppm (nonalcoholic beverages), 0.10-20 ppm (ice cream, ices, etc.), 5.3 ppm (candy), 8.5 ppm (baked goods), 17 ppm (condiments), 40 ppm (meats)
Regulatory: FDA 21CFR §182.10, 182.20, GRAS
Manuf./Distrib.: Pierre Chauvet; Florida Treatt

Blackthorn berries

Synonyms: Sloe berries
Definition: *Prunus spinosa*
Uses: Natural flavoring agent
Usage level: 110 ppm (nonalcoholic beverages), 43,000 ppm (alcoholic beverages), 50-100 ppm (ice cream, ices), 40 ppm (candy), 45 ppm (baked goods)
Regulatory: FDA 21CFR §182.20, GRAS

Black wattle

EINECS 308-877-4 (extract)
Synonyms: Mimosa
Definition: Flowers from *Acacia decurrens*
Uses: Natural flavoring
Regulatory: FDA 21CFR §172.510

Bleached deodorized tallow. *See* Tallow
Bleached lard. *See* Lard
Bleached shellac. *See* Shellac
BLO. *See* Butyrolactone
Blood sugar. *See* D-Glucose anhyd.
Blue copperas. *See* Cupric sulfate (pentahydrate)
Blue stone. *See* Cupric sulfate (pentahydrate)
Blue vitriol. *See* Cupric sulfate (pentahydrate)

Bois de rose

FEMA 2156
Definition: Derived from *Aniba rosaeodora*
Properties: Colorless to pale yel. oily liq., flowery odor; sol. in alcohol; sl. sol. in glycerin
Toxicology: Heated to decomp., emits acrid smoke and fumes

Uses: Natural flavoring agent
Usage level: 0.65 ppm (nonalcoholic beverages), 2.6 ppm (ice cream, ices, etc.), 6.7 ppm (candy), 9.3 ppm (baked goods), 35 ppm (chewing gum)
Regulatory: FDA 21CFR §182.20, GRAS
Manuf./Distrib.: Chart

Boletic acid. *See* Fumaric acid
Bolus alba. *See* Kaolin
Boracic acid. *See* Boric acid

Borage seed oil
EINECS 281-661-4 (extract)
Definition: Oil obtained from seeds of *Borago officinalis*
Properties: Yel. golden clear oil, sl. fatty odor
Uses: Food ingred.
Manuf./Distrib.: Arista Industries

Borax. *See* Sodium borate

Boric acid
CAS 10043-35-3; EINECS 233-139-2; EEC E240
Synonyms: Boracic acid; Orthoboric acid
Classification: Inorganic acid
Empirical: BH_3O_3
Formula: H_3BO_3
Properties: Wh. or colorless cryst. powd.; m.w. 61.83; dens. 1.435
Precaution: Hygroscopic
Toxicology: Causes digestive upsets in man; irritant
Uses: Preservative; used in adhesives, sizes, and coatings for paper contacting food
Usage level: 0.01-1.0%
Regulatory: FDA 21CFR §175.105, 176.180, 181.22, 181.30; USA not restricted; Europe listed; permitted in Switzerland and Sweden as preservative in some processed seafoods
Manuf./Distrib.: Dragoco

2-Bornanol. *See* Borneol
2-Dornanone. *See* Camphor
Borneo camphor. *See* Borneol

Borneol
CAS 507-70-0; FEMA 2157
Synonyms: 2-Bornanol; d-Camphanol; Borneo camphor; Bornyl alcohol
Empirical: $C_{10}H_{18}O$
Properties: Colorless plates; pungent, camphor-like odor; burning taste; sol. in alcohol, chloroform, ether; insol. in water; m.w. 154.24; sp.gr. 1.011 (20 C); m.p. 208 C; b.p. 212 C; flash pt. 150 C
Precaution: Flamm.
Toxicology: May cause nausea, vomiting, mental confusio, dizziness, convulsions
Uses: Synthetic flavoring agent
Usage level: 0.25-1.4 ppm (nonalcoholic beverages), 1.4 ppm (ice cream, ices, etc.), 3.7 ppm (candy), 5.1 ppm (baked goods), 0.30 ppm (chewing gum), 0.30 ppm (syrups)
Regulatory: FDA 21CFR §172.515; Japan approved as flavoring
Manuf./Distrib.: Chr. Hansen's

Borneol isovalerate. *See* Bornyl isovalerate

Bornyl acetate
CAS 76-49-3; FEMA 2159
Synonyms: levo-Bornyl acetate; Bornyl acetic ether; Bornyl ethanoate
Empirical: $C_{12}H_{20}O_2$
Properties: Colorless liq., white crystalline solid; piney odor; fresh, burning taste; sol. in alcohol, most fixed oils; sl. sol. in water; insol. in propyl glycol; m.w. 196.29; m.p. 27.5 C; b.p. 225-226 C; flash pt. 89 C; ref. index 1.462-1.466
Precaution: Combustible liq.
Toxicology: Heated to decomp., emits acrid smoke and fumes
Uses: Synthetic flavoring agent
Usage level: 1.1 ppm (nonalcoholic beverages), 1.8 ppm (ice cream, ices, etc.), 1.9 ppm (candy), 1.4 ppm (baked goods), 70 ppm (gelatins and puddings), 0.30 ppm (chewing gum), 0.20 ppm (syrups)
Regulatory: FDA 21CFR §172.515

Manuf./Distrib.: Chr. Hansen's

levo-Bornyl acetate. *See* Bornyl acetate
Bornyl acetic ether. *See* Bornyl acetate
Bornyl alcohol. *See* Borneol
Bornyl ethanoate. *See* Bornyl acetate

Bornyl formate
FEMA 2161
Synonyms: Bornyl methanoate
Empirical: $C_{11}H_{18}O_2$
Properties: Colorless, oily liq.; m.w. 182.66; dens. 1.007-1.009; b.p. 106-108 C (21 mm); ref. index 1.4689
Precaution: Combustible
Uses: Synthetic flavoring agent
Usage level: 3.7 ppm (nonalcoholic beverages), 0.30-3.0 ppm (ice cream, ices, etc.), 0.80-2.0 ppm (candy, baked goods), 0.04 ppm (syrups)
Regulatory: FDA 21CFR §172.515

Bornyl isovalerate. *See also* Bornyl valerate
CAS 53022-14-3
Synonyms: (1R-endo)-3-Methylbutanoic acid 1,7,7-trimethylbicyclo[2.2.1]hept-2-yl ester; Borneol isovalerate; d-Bornyl isovalerate
Empirical: $C_{15}H_{26}O_2$
Formula: $(CH_3)_2CHCH_2COOC_{10}H_{17}$
Properties: Liq.; odor and taste of valerian and camphor; sol. in alcohol, ether; pract. insol. in water; m.w. 238.36; dens. 0.955; b.p. 255-260 C
Uses: Synthetic flavoring agent
Regulatory: FDA 21CFR §172.515

d-Bornyl isovalerate. *See* Bornyl isovalerate
Bornyl methanoate. *See* Bornyl formate
Bornyl 3-methylbutanoate. *See* Bornyl valerate
Bornylval. *See* Bornyl valerate

Bornyl valerate
FEMA 2165
Synonyms: Bornyl isovalerate; Bornyl 3-methylbutanoate; Bornylval
Empirical: $C_{15}H_{26}O_2$
Properties: Colorless oil; sol. in alcohol, ether; insol. in water; m.w. 238.37; dens. 0.951 (20 C); b.p. 255-260 C; ref. index 1.4605 (18 C)
Precaution: Combustible
Uses: Synthetic flavoring agent
Usage level: 0.06-1.0 ppm (nonalcoholic beverages, ice cream), 0.90-2.0 ppm (candy, baked goods), 1.2 ppm (syrups)
Regulatory: FDA 21CFR §172.515

Boronia flowers
EINECS 294-926-4 (extract)
Synonyms: Boronia megastigma flowers
Definition: Derived from *Boronia megastigma*
Uses: Natural flavoring
Regulatory: FDA 21CFR §172.510

Boronia megastigma flowers. *See* Boronia flowers
Boswellia carterii resin. *See* Olibanum
Bourbonal. *See* Ethyl vanillin

β-Bourbonene
Uses: Synthetic flavoring agent
Regulatory: FDA 21CFR §172.515

Bovine rennet. *See* Rennet
Brassica campestris oil. *See* Rapeseed oil
Brazil wax. *See* Carnauba

Brilliant Blue FCF. *See also* FD&C Blue No. 1
CAS 3844-45-9; EINECS 223-339-8; EEC E133
Synonyms: CI 42090

Classification: Triphenylmethane color
Empirical: $C_{37}H_{36}N_2O_9S_3 \cdot 2Na$
Properties: Greenish-blue powd., gran.; sol. (oz/gal): 52 oz propylene glycol, 36 oz glycerin, 25 oz dist. water, 2 oz 95% ethanol; sol. in ether, conc. sulfuric acid; m.w. 794.91
Toxicology: Experimental neoplastigen; mutagenic data; heated to decomp., emits very toxic fumes of NO_x, Na_2O, and SO_x
Uses: Food colorant for beverages, candy, confections, baked goods, ice cream, dessert powds.
Usage level: ADI 0-12.5 mg/kg (EEC)
Regulatory: FDA 21CFR §74.101, 74.1101, 74.2101, 82.101; Europe listed; UK approved; banned in France, Finland
Manuf./Distrib.: Hilton Davis

Bromelain
CAS 37189-34-7; EINECS 253-387-5
Synonyms: Bromelin
Definition: From pineapple stem
Toxicology: Heated to decomp., emits acrid smoke and fumes
Uses: Chillproofing of beer, natural enzyme, meat tenderizing, preparation of precooked cereals, processing aid, tissue softening agent; used in beer, bread, cereals, meat, poultry, wine
Regulatory: FDA GRAS; Canada, UK, Japan approved
Manuf./Distrib.: Am. Roland; Chart
Trade names: Bromelain 150 GDU; Bromelain 600 GDU; Bromelain 1200 GDU; Bromelain 1500 GDU

Bromelin. *See* Bromelain

Brominated soybean oil
CAS 68952-98-7
Synonyms: Brominated vegetable oil
Properties: Clear yel. viscous oily liq., bland odor; sol. in alcohol, chloroform, ether, hexane, fixed oils; insol. in water; dec. without flash at 480 F
Precaution: Avoid strong alkali, amines
Toxicology: Possible mild eye irritation; experimental reproductive effects; heated to decomp., emits toxic fumes of Br
Uses: Food additive in soft drinks for visc. adjustment
Usage level: 15 ppm max. (fruit flavored beverages)
Regulatory: FDA 21CFR §180.30
Trade names: Akwilox 133

Brominated vegetable oil. *See also* Brominated soybean oil
Properties: Pale yel. to dk. brn. visc. oily liq., bland or fruity odor, bland taste; sol. in alcohol, chloroform, ether, hexane, fixed oils; insol. in water
Toxicology: Experimental reproductive effects; heated to decomp., emits toxic fumes of Br
Uses: Flavoring agent, stabilizer for flavoring oils used in fruit-flavored beverages
Usage level: 15 ppm max. (finished beverage)
Regulatory: FDA 21CFR §180.30
Trade names containing: Beta Carotene Emulsion Beverage Type 3.6 No. 65392

Bromine
EINECS 231-778-1
Definition: Liq. nonmetallic element
Properties: Dk. red fuming liq., pungent odor; sol. in CCl_4, chloroform, methylene chloride, CS_2, ether, methanol, conc. HCl; sol. 3.38% in water (20 C); a.w. 159.83; sp.gr. 3.119 (20/4 C); dens. 25.9 lb/gal; f.p. -7.27 C; ref. index 1.6083 (20 C)
Precaution: Reactive and corrosive; nonflamm. but may ignite combustibles on contact
Toxicology: Severe skin, eye and mucous membrane irritant
Uses: Weighting agent for flavors and fragrances
Manuf./Distrib.: Great Lakes Chem.

Bronze powder. *See* Copper
Brown algae. *See* Algae, brown

Brown rice syrup
Uses: Sweetener for soy-based products
Trade names: CNP BRSHG40; CNP BRSHG40CL; CNP BRSHM; CNP BRSHMCL; CNP BRSMC10; CNP BRSMC20; CNP BRSMC35; CNP BRSMC35CL; CNP PPRSRDCL

B/S. *See* Butadiene-styrene copolymer

BT. *See* L-Carnitine

Buchu leaves
Definition: Derived from *Barosma betulina, B. crenulata,* or *B. serratifolia*
Properties: Strong sweet odor, fresh bitter flavor
Uses: Natural flavoring
Regulatory: FDA 21CFR §172.510
Manuf./Distrib.: C.A.L.-Pfizer; Chart; Florida Treatt

Buckbean extract
Uses: Natural flavoring
Regulatory: FDA 21CFR §172.510

Buckbean leaves
Definition: Derived from *Menyanthes trifoliata*
Properties: Bitter tonic flavor
Uses: Natural flavoring; used only in alcoholic beverages
Regulatory: FDA 21CFR §172.510

Burning bush. *See* Dittany
Burnt lime. *See* Calcium oxide
Burnt sugar. *See* Caramel
Burnt sugar coloring. *See* Caramel

Butadiene-styrene copolymer
CAS 9003-55-8
Synonyms: B/S; Styrene polymer with 1,3-butadiene; Butadiene-styrene resin; Butadiene-styrene rubber
Toxicology: Eye irritant; heated to decomp., emits acrid smoke and irritating fumes
Uses: Masticatory substance in chewing gum base; food containers; paper/paperboard used in food pkg.
Regulatory: FDA 21CFR §172.615, 181.30

Butadiene-styrene resin. *See* Butadiene-styrene copolymer
Butadiene-styrene rubber. *See* Butadiene-styrene copolymer

Butadiene-styrene 50/50 rubber
Uses: Masticatory substance in chewing gum base

Butadiene-styrene 75/25 rubber
Uses: Masticatory substance in chewing gum base

n-Butanal. *See* n-Butyraldehyde

Butane
CAS 106-97-8; EINECS 203-448-7
Synonyms: n-Butane; Alkane C-4
Classification: Hydrocarbon
Empirical: C_4H_{10}
Formula: $CH_3CH_2CH_2CH_3$
Properties: Colorless gas, faint disagreeable odor; easily liquefied under pressure @ R.T.; m.w. 58.12; dens. 0.599; b.p. -0.5 C; f.p. -138 C; flash pt. (CC) -76 F
Precaution: Flamm. gas; very dangerous fire hazard exposed to heat, flame, oxidizers; highly explosive; explosive limits 1.9-8.5%
Toxicology: TLV:TWA 800 ppm; mildly toxic by inh.; causes drowsiness; asphyxiant; narcotic in high concs.; heated to decomp., emits acrid smoke and fumes
Uses: Direct food additive, propellant, aerating agent, gas for foamed and sprayed food prods.
Regulatory: FDA 21CFR §173.350, 184.1165, GRAS

Butanecarboxylic acid. *See* n-Valeric acid
1,4-Butanedicarboxylic acid. *See* Adipic acid
1,4-Butanedioic acid. *See* Succinic acid
Butanedioic acid diethyl ester. *See* Diethyl succinate
Butanedioic anhydride. *See* Succinic anhydride
1,3-Butanediol. *See* Butylene glycol
2,3-Butanedione. *See* Diacetyl
n-Butane. *See* Butane
Butanoic acid. *See* n-Butyric acid
Butanoic acid-2-butoxy-1-methyl-2-oxoethyl ester. *See* Butyl butyryllactate
Butanoic acid methyl ester. *See* Methyl butyrate
Butanoic acid pentyl ester. *See* Amyl butyrate

Butanoic acid 1,2,3-propanetriyl ester. *See* Tributyrin
Butanoic acid propyl ester. *See* Propyl butyrate
1-Butanol. *See* Butyl alcohol

2-Butanol
CAS 15892-23-6; EINECS 240-029-8
Synonyms: s-Butyl alcohol; Methyl ethyl carbinol; Butylene hydrate; 2-Hydroxybutane
Empirical: $C_4H_{10}O$
Formula: $CH_3CH_2CH(OH)CH_3$
Properties: Liq.; m.w. 74.12; dens. 0.807 (20/4 C); b.p. 99-100 C; flash pt. 23 C; ref. index 1.397 (20 C)
Precaution: Flamm.
Uses: Synthetic flavoring agent
Regulatory: FDA 21CFR §172.515

4-Butanolide. *See* Butyrolactone
2,3-Butanolone. *See* Acetyl methyl carbinol
2-Butanone. *See* Methyl ethyl ketone

Butan-3-one-2-yl butyrate
FEMA 3332
Empirical: $C_8H_{14}O_3$
Properties: Wh. to sl. yel. liq., red berry odor; sol. in alcohol, propylene glycol, most oils; insol. in water; m.w.
158.19; dens. 0.972-0.992; flash pt. 179 F; ref. index 1.408-1.429
Toxicology: Heated to decomp., emits acrid smoke and irritating fumes
Uses: Flavoring agent
Regulatory: FDA GRAS
Manuf./Distrib.: Aldrich

2-Butenedioic acid. *See* Fumaric acid
trans-Butenedioic acid. *See* Fumaric acid
trans-2-Butenoic acid ethyl ester. *See* Ethyl crotonate
(2-Butenylidene) acetic acid. *See* Sorbic acid

Butoxyethanol
CAS 111-76-2; EINECS 203-905-0
Synonyms: 2-Butoxyethanol; Ethylene glycol monobutyl ether; Glycol butyl ether; Ethylene glycol butyl ether;
Butyl glycol
Classification: Ether alcohol
Empirical: $C_6H_{14}O_2$
Formula: $HOCH_2CH_2OC_4H_9$
Properties: Colorless liq.; mild pleasant odor; m.w. 118.20; dens. 0.9019 (20/20 C); m.p. -74.8 C; b.p. 171.2
C; flash pt. (COC) 160 F
Precaution: Flamm. liq. exposed to heat or flame; incompat. with oxidizers, heat, flame
Toxicology: TLV 25 ppm in air; LD50 (IP, rat) 20 mg/kg; poison by ing., skin contact, IP, IV routes; mod. toxic
by inh., subcut.; human systemic effects; experimental reproductive effects; skin/eye irritant; heated to
dec., emits acrid smoke, irritating fumes
Uses: Flume wash water additive for washing sugar beets prior to slicing
Usage level: Limitation 1 ppm (in wash water)
Regulatory: FDA 21CFR §173.315, 175.105, 176.210, 177.1650, 178.1010, 178.3297, 178.3570

2-Butoxyethanol. *See* Butoxyethanol

Butter acids
FEMA 2171
Properties: Waxy solid; sol. in oils; low melting, varible
Uses: Synthetic flavoring agent
Usage level: 2.0 ppm (nonalcoholic beverages), 3.0 ppm (ice cream, ices), 2800 ppm (candy), 8.3 ppm (baked
goods)
Regulatory: FDA 21CFR §172.515

Butter esters
FEMA 2172
Properties: Waxy solid, fatty oily waxy odor and flavor; sol. in alcohol; insol. in water
Uses: Synthetic flavoring agent
Usage level: 24 ppm (ice cream, ices), 78 ppm (candy), 86 ppm (baked goods), 2.0 ppm (toppings), 1200 ppm
(popcorn oil)
Regulatory: FDA 21CFR §172.515

Butyl acetate. *See* n-Butyl acetate

n-Butyl acetate
    CAS 123-86-4; EINECS 204-658-1; FEMA 2174
    Synonyms: Acetic acid, butyl ester; Butyl acetate
    Definition: Ester of butyl alcohol and acetic acid
    Empirical: $C_6H_{12}O_2$
    Formula: $CH_3COOC_4H_9$
    Properties: Colorless liq., fruity odor; sol. in alcohol, ether, hydrocarbons; sl. sol. in water; m.w. 116.18; dens.
        0.8826 (20/20 C); b.p. 126.3 C; f.p. -75 C; flash pt. (TOC) 36.6 C); ref. index 1.2951 (20 C)
    Precaution: Flamm. liq.; mod. explosive when exposed to flame
    Toxicology: LD50 (oral, rat) 14.13 g/kg; mildly toxic by inhalation, ingestion; moderately toxic by intraperitoneal
        route; skin irritant; severe eye irritant; mild allergen; TLV 150 ppm in air; heated to decomp., emits acrid
        and irritating fumes
    Uses: Synthetic flavoring agent
    Usage level: 11 ppm (nonalcoholic beverages), 16 ppm (ice cream, ices), 32 ppm (candy, baked goods), 13
        ppm (gelatins, puddings), 220 ppm (chewing gum)
    Regulatory: FDA 21CFR §172.515, 175.105, 175.320, 177.1200; GRAS (FEMA); Japan approved as flavoring
    Manuf./Distrib.: Aldrich; Eastman; Penta Mfg.

Butylacetic acid. *See* Caproic acid

Butyl acetoacetate
    CAS 591-60-6; FEMA 2176
    Synonyms: Butyl-β-ketobutyrate; Butyl-3-oxobutanoate
    Empirical: $C_8H_{14}O_3$
    Properties: Liq.; sol. in alcohol, ether; insol. in water; m.w. 158.20; dens. 0.9694; b.p. 213.9 C; flash pt. 185
        F; ref. index 1.4245
    Precaution: Combustible
    Uses: Synthetic flavoring agent
    Usage level: 4.2 ppm (nonalcoholic beverages), 7.3 ppm (ice cream, ices), 26 ppm (candy, baked goods)
    Regulatory: FDA 21CFR §172.515

Butyl alcohol
    CAS 71-36-3; EINECS 200-751-6; FEMA 2178
    Synonyms: n-Butyl alcohol; 1-Butanol; Propyl carbinol
    Classification: Aliphatic alcohol
    Empirical: C_4H10O
    Formula: $CH_3(CH_2)_2CH_2OH$
    Properties: Colorless liq., vinous odor; sol. in water; misc. with alcohol, ether; m.w. 74.14; dens. 0.8109 (20/
        20 C); m.p. -90 C; f.p. -89.0 C; b.p. 117.7 C; flash pt. 35 C; ref. index 1.3993 (20 C)
    Precaution: Flamm.; mod. explosive exposed to flame; incompat. with Al, oxidizing materials
    Toxicology: TLV:CL 50 ppm in air; LD50 (oral, rat) 790 mg/kg; poison by IV route; moderately toxic by skin
        contact, ingestion, subcutaneous, intraperitoneal routes; skin and severe eye irritant; heated to decomp.,
        emits acrid smoke and fumes
    Uses: Synthetic flavoring agent, color diluent for confectionery, tabletted food supplements, gum
    Usage level: 12 ppm (nonalcoholic beverages), 1 ppm (alcoholic beverages), 7 ppm (ice cream, ices), 34 ppm
        (candy), 32 ppm (baked goods), 4 ppm (cream)
    Regulatory: FDA 21CFR §73.1, 172.515, 172.560, 175.105, 175.320, 176.200, 177.1200, 177.1440,
        177.1650; 27CFR §21.99; GRAS (FEMA)
    Manuf./Distrib.: Aldrich; Eastman

n-Butyl alcohol. *See* Butyl alcohol
s-Butyl alcohol. *See* 2-Butanol
n-Butylaldehyde. *See* n-Butyraldehyde

Butylamine
    CAS 109-73-9; FEMA 3130
    Classification: Nitrogen compd.
    Empirical: $C_4H_{11}N$
    Formula: $CH_3(CH_2)_3NH_2$
    Properties: m.w. 73.14; dens. 0.740; m.p. -49 C; b.p. 78 C; flash pt. -14 C
    Uses: Fishy pungent flavor
    Manuf./Distrib.: Aldrich

Butyl-2-aminobenzoate. *See* Butyl anthranilate

Butyl-o-aminobenzoate. *See* Butyl anthranilate

Butyl anthranhilate
FEMA 2181
Synonyms: Butyl-2-aminobenzoate; Butyl-o-aminobenzoate
Empirical: C₁₁H₁₅NO₂
Properties: Liq. @ R.T.; m.w. 193.25; sp.gr. 1.060 (15.5 C); m.p. ≈0 C; b.p. 303 C; ref. index 1.5420 (20 C)
Uses: Synthetic flavoring agent
Usage level: 1.3 ppm (nonalcoholic beverages), 2.6 ppm (ice cream, ices), 9.0 ppm (candy), 6.7 ppm (baked goods)
Regulatory: FDA 21CFR §172.515
Manuf./Distrib.: Aldrich

Butylated hydroxyanisole. *See* BHA
Butylated hydroxymethyl phenol. *See* 4-Hydroxymethyl-2,6-di-t-butylphenol
Butylated hydroxytoluene. *See* BHT
Butyl benzyl ether. *See* Benzyl butyl ether
n-Butyl n-butanoate. *See* Butyl butyrate

Butyl butyrate
CAS 109-21-7; EINECS 203-656-8; FEMA 2186
Synonyms: n-Butyl n-butanoate; n-Butyl butyrate; n-Butyl n-butyrate
Empirical: C₈H₁₆O₂
Formula: CH₃CH₂CH₂COO(CH₂)₃CH₃
Properties: Colorless liq., pineapple odor; misc. with alcohol, ether, veg. oils; sl. sol. in propylene glycol, water; m.w. 144.24; dens. 0.67-0.871; b.p. 166 C; flash pt. (OC) 128 F; ref. index 1.405
Precaution: Combustible liq.; incompat. with oxidizers
Toxicology: LD50 (oral, rabbit) 9520 mg/kg, (IP, rat) 2300 mg/kg; mod. toxic by IP; mildly toxic by ingestion; mod. irritating to eyes, skin, mucous membranes; narcotic in high concs.; heated to decomp., emits acrid smoke and irritating fumes
Uses: Synthetic flavoring agent
Usage level: 8.6 ppm (nonalcoholic beverages), 22 ppm (ice cream, ices, baked goods), 24 ppm (candy), 14 ppm (gelatins, puddings), 150-1500 ppm (chewing gum)
Regulatory: FDA 21CFR §172.515; Japan approved as flavoring
Manuf./Distrib.: Aldrich

n-Butyl butyrate. *See* Butyl butyrate
n-Butyl n-butyrate. *See* Butyl butyrate
Butyl γ-butyrolactone. *See* Butyl levulinate
Butyl butyrol lactate. *See* Butyl butyryllactate

Butyl butyryllactate
CAS 7492-70-8; EINECS 231-326-3; FEMA 2190
Synonyms: Butanoic acid-2-butoxy-1-methyl-2-oxoethyl ester; Butyryllactic acid butyl ester; Butyl butyrol lactate
Empirical: C₁₁H₂₀O₄
Properties: Colorless liq., creamy buttery odor; sol. in propylene glycol; misc. with alcohol, fixed oils; insol. in water; m.w. 216.28; dens. 0.970; b.p. 90 C (2 mm); flash pt. > 100 C; ref. index 1.420
Precaution: Combustible liq.
Toxicology: Skin irritant; heated to decomp., emits acrid smoke and irritating fumes
Uses: Synthetic flavoring agent for baked goods, candy
Regulatory: FDA 21CFR §172.515
Manuf./Distrib.: Aldrich

Butyl caproate. *See* Butyl hexanoate
Butyl capronate. *See* Butyl hexanoate
Butyl carbobutoxymethyl phthalate. *See* n-Butyl phthalyl-n-butyl glycolate

α-Butylcinnamaldehyde
FEMA 2191
Synonyms: 2-Benzylidene hexanal; Butyl cinnamic aldehyde; α-Butyl-β-phenylacrolein
Empirical: C₁₃H₁₆O
Properties: Liq., lily-like odor; m.w. 188.27; sp.gr. 0.825 (15.5 C); b.p. 265 C
Uses: Synthetic flavoring agent
Usage level: 0.5-1.0 ppm (nonalcoholic beverages), 1.0-2.8 ppm (ice cream, ices), 2.0-8.0 ppm (candy, baked goods)

Regulatory: FDA 21CFR §172.515

Butyl cinnamate
FEMA 2192
Synonyms: Butyl β-phenyl acrylate; Butyl 3-phenyl propenoate
Empirical: $C_{13}H_{16}O_2$
Properties: Mobile liq., cocoa-like odor; sol. in 95% alcohol, chloroform, benzene, ether; insol. in water; m.w. 204.27
Uses: Synthetic flavoring agent
Usage level: 0.83 ppm (nonalcoholic beverages), 2.0 ppm (alcoholic beverages), 2.6 ppm (ice cream, ices), 1.0-15 ppm (candy, baked goods)
Regulatory: FDA 21CFR §172.515

Butyl cinnamic aldehyde. *See* α-Butylcinnamaldehyde

Butyl 2-decenoate
FEMA 2194
Synonyms: n-Butyl decylenate
Empirical: $C_{14}H_{26}O_2$
Properties: Colorless liq., peach/apricot odor, fruity green taste; sol. in alcohol; insol. in water; m.w. 226.36
Uses: Synthetic flavoring agent
Usage level: 8.0 ppm (nonalcoholic beverages), 1.5-22 ppm (candy), 30 ppm (baked goods), 2000 ppm (chewing gum)
Regulatory: FDA 21CFR §172.515

n-Butyl decylenate. *See* Butyl 2-decenoate

Butylene glycol
CAS 107-88-0; EINECS 203-529-7
Synonyms: 1,3-Butanediol; 1,3-Butylene glycol
Classification: Aliphatic diol
Empirical: $C_4H_{10}O_2$
Formula: $HOCH_2CH_2CHOHCH_3$
Properties: Visc. liq., sweet flavor with bitter aftertaste; m.w. 90.12; dens. 1.004-1.006 (20/20 C); b.p. 207.5 C; f.p. < -50 C; flash pt. 250 F
Precaution: Combustible when exposed to heat or flame; incompat. with oxidizing materials
Toxicology: LD50 (oral, rat) 23 g/kg; mildly toxic by ingestion and subcutaneous routes; eye irritant; heated to decomp., emits acrid smoke and irritating fumes
Uses: Solvent for natural and synthetic flavoring agents
Regulatory: FDA 21CFR §173.220, 184.1278 GRAS

1,3-Butylene glycol. *See* Butylene glycol
Butylene hydrate. *See* 2-Butanol
Butyl ethyl carbinol. *See* 3-Heptanol
Butyl ethylene. *See* 1-Hexene
Butyl ethyl ketone. *See* 3-Heptanone

Butyl ethyl malonate
CAS 32864-38-3; EINECS 251-269-8; FEMA 2195
Synonyms: Ethyl butyl malonate
Empirical: $C_9H_{16}O_4$
Formula: $(CH_3)_3CO_2CCH_2CO_2C_2H_5$
Properties: m.w. 188.22; dens. 0.994; b.p. 222 C; flash pt. 189 F
Uses: Synthetic flavoring agent
Usage level: 3.0 ppm (nonalcoholic beverages), 0.13 ppm (candy)
Regulatory: FDA 21CFR §172.515

Butyl formate
CAS 592-84-7; EINECS 209-772-5; FEMA 2196
Synonyms: Formic acid butyl ester
Empirical: $C_5H_{10}O_2$
Formula: $HCOO(CH_2)_3CH_3$
Properties: Colorless liq., plum-like odor; sl. sol. in water; misc. with alcohol, ether; m.w. 102.14; m.p. -90 C; b.p. 106.8 C; flash pt. 64 C; ref. index 1.3890-1.3891
Precaution: Flamm. liq.
Toxicology: Irritant
Uses: Synthetic flavoring agent

Usage level: 2.9 ppm (nonalcoholic beverages), 3.2 ppm (ice cream, ices), 11 ppm (candy), 9.1 ppm (baked goods), 5.0 ppm (gelatins, puddings)
Regulatory: FDA 21CFR §172.515
Manuf./Distrib.: Aldrich

Butyl glycol. *See* Butoxyethanol
Butyl 10-hendecenoate. *See* Butyl 10-undecenoate

Butyl heptanoate
FEMA 2199
Synonyms: Butyl heptoate; Butyl heptylate
Empirical: $C_{11}H_{22}O_2$
Properties: Colorless liq., sl. fruity odor; sol. in most org. solvs.; m.w. 186.30; sp.gr. 0.85553; m.p. -68.4 C; b.p. 226.2 C; ref. index 1.42280 (20 C)
Uses: Synthetic flavoring agent
Usage level: 0.5-1.0 ppm (nonalcoholic beverages), 2.0-10 ppm (ice cream, ices), 2.0-25 ppm (candy, baked goods)
Regulatory: FDA 21CFR §172.515
Manuf./Distrib.: Aldrich

Butyl heptoate. *See* Butyl heptanoate
Butyl heptylate. *See* Butyl heptanoate

Butyl hexanoate
CAS 626-82-4; FEMA 2201
Synonyms: Butyl caproate; Butyl capronate; Butyl hexylate
Empirical: $C_{10}H_{20}O_2$
Properties: Liq., pineapple-like odor; m.w. 172.27; sp.gr. 0.8623; m.p. -63 to -64 C; b.p. 208 C; ref. index 1.4153
Uses: Synthetic flavoring agent
Usage level: 1.7 ppm (nonalcoholic beverages), 3.9 ppm (ice cream, ices), 7.6 ppm (candy), 10 ppm (baked goods)
Regulatory: FDA 21CFR §172.515
Manuf./Distrib.: Aldrich

Butyl hexylate. *See* Butyl hexanoate

t-Butyl hydroquinone
CAS 1948-33-0; EINECS 217-752-2
Synonyms: TBHQ; Mono-t-butyl hydroquinone; 2-(1,1-Dimethylethyl)-1,4-benzenediol
Classification: Aromatic organic compd.
Empirical: $C_{10}H_{14}O_2$
Properties: Wh. to lt. tan cryst. solid; sol. in ethyl alcohol, ethyl acetate, acetone, ether; sl. sol. in water; m.w. 166.24; m.p. 126.5-128.5 C; flash pt. (COC) 171 C
Toxicology: LD50 (oral, rat) 700 mg/kg; poison by intraperitoneal route; moderately toxic by ingestion; irritant; heated to decomp., emits acrid smoke and irritating fumes
Uses: Antioxidant, food additive for beef patties, dry cereals, edible fats, margarine, dried meat, margarine, pizza toppings, potato chips, poultry, sausage, vegetable oils
Regulatory: FDA 21CFR 172.185 (limitation 0.02% of oil); USDA 9CFR §318.7 (limitation 0.003% in dry sausage, 0.006% with BHA/BHT, 0.01% in rendered animal fat, 0.02% with BHA and/or BHT, 0.02% in margarine), 381.147 (limitation 0.01% on fat in poultry)
Manuf./Distrib.: Aceto; Eastman; Penta Mfg.; UOP
Trade names: Embanox® TBHQ; Sustane® TBHQ; TBHQ®; Tenox® TBHQ
Trade names containing: Alphadim® 90NLK; Alphadim® 90VCK; Beta Plus; BFP 64A; BFP 64K; BFP 65; Creamy Liquid Frying Shortening 102-050; Dress All; Durkex 100TBHQ; Durlite Gold MBN II; Fluid Flex; Golden Chef® Clear Liquid Frying Shortening 104-052; Golden Chef® Creamy Liquid Frying Shortening 104-065; Golden Covo® Shortening; Gold 'N Flavor® 103-061; Heavy Duty Peanut Oil 104-210; Pano-Lube™; Panospray™; Panospray™ LS; Panospray™ SQ; Panospray™ WL; Shedd's Wonder Shortening; Soybean Salad Oil 104-050; Soy/Peanut Liquid Frying Shortening 104-215; Stabolic™ C; Starplex® 90; Superb® Canola Liquid Frying Shortening 104-455; Superb® Canola Salad Oil 104-450; Superb® Corn Liquid Frying Shortening 104-255; Superb® Fish and Chip Oil 102-060; Superb® Heavy Duty Frying Shortening 101-200; Superb® Liquid Frying Shortening 102-110; Superb® Soybean Salad Oil 104-112; Superb® Soy/Corn Frying Shortening 102-070; Sustane® 18; Sustane® 20; Sustane® 20-3; Sustane® 20A; Sustane® 20B; Sustane® 21; Sustane® 27; Sustane® 31; Sustane® W-1; Tenox® 20; Tenox® 20A; Tenox® 20B; Tenox® 21; Tenox® 22; Tenox® 24; Tenox® 25; Tenox® 26; Tenox® 27; Winterized Clear Liquid Frying Shortening 104-112

3-t-Butyl-4-hydroxyanisole. *See* BHA
n-Butyl p-hydroxybenzoate. *See* Butylparaben
Butyl p-hydroxybenzoate. *See* Butylparaben
n-Butyl-4-hydroxybenzoate potassium salt. *See* Potassium butyl paraben
4-Butyl-4-hydroxyoctanoic acid, γ-lactone. *See* 4,4-Dibutyl-γ-butyrolactone
Butyl 2-hydroxypropanoate. *See* Butyl lactate
Butyl α-hydroxypropionate. *See* Butyl lactate

Butyl isobutyrate
FEMA 2188
Synonyms: n-Butyl 2-methyl propanoate
Empirical: $C_8H_{16}O_2$
Properties: Colorless liq., fruity odor, pineapple taste; insol. in water; m.w. 144.21; b.p. 155-156 C; ref. index 1.4025
Precaution: Combustible liq.
Toxicology: Heated to decomp., emits acrid smoke and fumes
Uses: Synthetic flavoring agent
Usage level: 8.7 ppm (nonalcoholic beverages), 4.0-5.0 ppm (ice cream, ices), 19 ppm (candy), 39 ppm (baked goods), 2000 ppm (chewing gum)
Regulatory: FDA 21CFR §172.515
Manuf./Distrib.: Aldrich

n-Butyl isopentanoate. *See* n-Butyl isovalerate

n-Butyl isovalerate
CAS 109-19-3; FEMA 2218
Synonyms: Isovaleric acid butyl ester; n-Butyl isopentanoate; Butyl 3-methylbutyrate
Empirical: $C_9H_{18}O_2$
Properties: Colorless to pale yel. liq., fruity odor; misc. with alcohol, fixed oils; sl. sol. in propylene glycol; insol. in water; m.w. 158.27; dens. 0.851-0.857; b.p. 150 C; ref. index 1.407
Precaution: Flamm. when exposed to heat, flames, sparks, oxidizers
Toxicology: LD50 (oral, rat) 8200 mg/kg; mildly toxic by ingestion; skin irritant; heated to decomp, emits acrid smoke and fumes
Uses: Synthetic flavoring agent
Regulatory: FDA 21CFR §172.515
Manuf./Distrib.: Aldrich

Butyl-β-ketobutyrate. *See* Butyl acetoacetate

Butyl lactate
CAS 138-22-7; FEMA 2205
Synonyms: n-Butyl lactate; Butyl 2-hydroxypropanoate; Butyl α-hydroxypropionate
Classification: Ester
Empirical: $C_7H_{14}O_3$
Formula: $CH_3CHOHCOOC_4H_9$
Properties: Water-wh. liq., mild odor; misc. with many lacquer solvs., diluents, oils; sl. sol. in water; hydrolyzed in acids and alkalies; m.w. 146.19; dens. 0.974-0.984 (20/20 C); m.p. -43 C; flash pt. (TOC) 75.5 C
Precaution: Combustible
Toxicology: TLV 5 ppm in air
Uses: Synthetic flavoring agent
Usage level: 0.66 ppm (nonalcoholic beverages), 2.8 ppm (ice cream, ices), 6.5 ppm (candy), 7.7 ppm (baked goods)
Regulatory: FDA 21CFR §172.515; GRAS (FEMA)
Manuf./Distrib.: Aldrich; Penta Mfg.; Purac Biochem BV

n-Butyl lactate. *See* Butyl lactate

Butyl laurate
CAS 106-18-3; EINECS 203-370-3; FEMA 2206
Empirical: $C_{16}H_{32}O_2$
Formula: $CH_3(CH_2)_{10}COO(CH_2)_3CH_3$
Properties: Liq.; sol. in most org. solvs.; insol. in water; m.w. 256.43; dens. 0.860 (20/4 C); m.p. -7 C; b.p. 194 C (30 mm); ref. index 1.435 (20 C)
Uses: Synthetic flavoring agent
Usage level: 0.4-3.0 ppm (nonalcoholic beverages), 0.60 ppm (ice cream, ices), 17 ppm (candy), 1.0-40 ppm (baked goods)

Regulatory: FDA 21CFR §172.515
Manuf./Distrib.: Aldrich

Butyl levulinate
CAS 2052-15-5; FEMA 2207
Synonyms: Butyl γ-butyrolactone; Butyl 4-oxopentanoate
Empirical: C₉H₁₆O₃

Empirical: $C_9H_{16}O_3$
Properties: Liq., bitter taste; sol. in ether, alcohol, chloroform; sl. sol. in water; m.w. 172.23; dens. 0.974; b.p. 238 C; flash pt. 197 F; ref. index 1.4283 (20 C)
Uses: Synthetic flavoring agent
Usage level: 0.20-1.0 ppm (nonalcoholic beverages), 2.1 ppm (ice cream, ices), 4.6 ppm (candy, baked goods)
Regulatory: FDA 21CFR §172.515
Manuf./Distrib.: Aldrich

Butyl 3-methylbutyrate. *See* n-Butyl isovalerate
n-Butyl 2-methyl propanoate. *See* Butyl isobutyrate
Butyl octadecanoate. *See* Butyl stearate
n-Butyl octadecanoate. *See* Butyl stearate
Butyl-3-oxobutanoate. *See* Butyl acetoacetate
Butyl 4-oxopentanoate. *See* Butyl levulinate

Butylparaben
CAS 94-26-8; EINECS 202-318-7; FEMA 2203
Synonyms: Butyl p-hydroxybenzoate; 4-Hydroxybenzoic acid butyl ester; n-Butyl p-hydroxybenzoate
Definition: Ester of butyl alcohol and p-hydroxybenzoic acid
Empirical: $C_{11}H_{14}O_3$
Properties: Cryst. powd.; very sl. sol. in water, glycerin; freely sol. in acetone, alcohols, ether, chloroform, propylene glcyol; m.w. 194.22; m.p. 68-69 C
Precaution: Preserve in well-closed containers
Toxicology: Poison by intraperitoneal route; skin irritant
Uses: Preservative; flavoring
Usage level: 0.001-0.2%; 0.05-0.1% in mixtures
Regulatory: FDA 21CFR §172.515; EPA registered; Japan approved (0.012-1 g/kg as p-hydroxybenzoic acid); Europe listed
Manuf./Distrib.: Aldrich; Ashland; Int'l. Sourcing; E. Merck; Penta Mfg.
Trade names: Nipabutyl
Trade names containing: Nipastat

Butylparaben, potassium salt. *See* Potassium butyl paraben
Butylparaben, sodium salt. *See* Sodium butylparaben
n-Butyl pentanoate. *See* Butyl valerate

Butyl phenylacetate
CAS 122-43-0; FEMA 2209
Synonyms: Butyl α-toluate
Empirical: $C_{12}H_{16}O_2$
Properties: Colorless liq., honey-like odor; m.w. 192.26; dens. 0.991-0.994 (25/25 C); b.p. 260 C; flash pt. 74 C; ref. index 1.488-1.490 (20 C)
Precaution: Combustible
Uses: Synthetic flavoring agent
Usage level: 0.50 ppm (nonalcoholic beverages), 2.1 ppm (ice cream, ices), 4.5 ppm (candy), 4.6 ppm (baked goods), 5.0 ppm (gelatins, puddings)
Regulatory: FDA 21CFR §172.515
Manuf./Distrib.: Aldrich; Chr. Hansen's

α-Butyl-β-phenylacrolein. *See* α-Butylcinnamaldehyde
Butyl β-phenyl acrylate. *See* Butyl cinnamate
Butyl 3-phenyl propenoate. *See* Butyl cinnamate

p-t-Butylphenyl salicylate
CAS 87-18-3
Empirical: $C_{17}H_{18}O_3$
Properties: Off-wh. cryst., odorless; sol. in alcohol, ethyl acetate, toluene; insol. in water; m.w. 270.13; m.p. 62-64 C
Toxicology: Heated to decomp., emits acrid smoke and irritating fumes
Uses: Plasticizer migrating from food pkg.

n-Butyl phthalyl-n-butyl glycolate

Regulatory: FDA 21CFR §181.27

n-Butyl phthalyl-n-butyl glycolate
CAS 85-70-1; EINECS 201-624-8
Synonyms: 1,2-Benzenedicarboxylic acid, 2-butoxy-2-oxoethyl, butyl ester; Butyl carbobutoxymethyl phthalate; Dibutyl-o-carboxybenzoyloxyacetate; Dibutyl-o-(o-carboxybenzoyl) glycolate
Classification: Aromatic ester
Empirical: $C_{18}H_{24}O_6$
Properties: Colorless liq., odorless; insol. in water; m.w. 336.42; dens. 1.093-1.103 (25/25 C); b.p. 219 C (5 mm); flash pt. 390 F
Precaution: Combustible
Toxicology: LD50 (oral, rat) 7 g/kg; mildly toxic by IP; experimental teratogen, reproductive effects; mutagenic data; eye irritant; heated to decomp., emits acrid and irritating fumes
Uses: Plasticizer migrating from food pkg.
Regulatory: FDA 21CFR §175.105, 175.300, 175.320, 181.22, 181.27

Butyl propionate. *See* n-Butyl propionate

n-Butyl propionate
CAS 590-01-2; FEMA 2211
Synonyms: Propanoic acid butyl ester; Butyl propionate
Empirical: $C_7H_{14}O_2$
Formula: $CH_3CH_2COOC_4H_9$
Properties: Colorless liq., earthy faintly sweet odor, apricot-like taste; very sol. in alcohol, ether; very sl. sol. in water; m.w. 130.19; dens. 0.8754 (20/4 C); m.p. -89 C; f.p. 32 C; b.p. 145-146 C; ref. index 1.401 (20 C)
Precaution: Flamm.
Toxicology: Skin and eye irritant
Uses: Synthetic flavor ingredient
Usage level: 4 ppm (nonalcoholic beverages), 5.2 ppm (ice cream, ices), 25 ppm (candy), 27 ppm (baked goods)
Regulatory: FDA 21CFR §172.515; GRAS (FEMA)
Manuf./Distrib.: Aldrich; Penta Mfg.

Butyl rubber. *See* Isobutylene/isoprene copolymer

Butyl stearate
CAS 123-95-5; EINECS 204-666-5; FEMA 2214
Synonyms: Butyl octadecanoate; n-Butyl octadecanoate; Octadecanoic acid butyl ester
Definition: Ester of butyl alcohol and stearic acid
Empirical: $C_{22}H_{44}O_2$
Formula: $CH_3(CH_2)_{16}COO(CH_2)_3CH_3$
Properties: Crystals; sl. sol. in water; sol. in alcohol, ether; m.w. 340.60; dens. 0.86 (20/4 C); m.p. 17-22 C; b.p. 343 C; flash pt. (CC) 160 C; ref. index 1.4430 (20 C)
Toxicology: Heated to decomp., emits acrid smoke and irritating fumes
Uses: Synthetic flavoring agent; plasticizer for food pkg. materials
Usage level: 1 ppm (nonalcoholic beverages), 5 ppm (alcoholic beverages), 2 ppm (ice cream, ices), 190 ppm (candy), 340 ppm (baked goods), 330 ppm (chewing gum)
Regulatory: FDA 21CFR §172.515, 173.340, 175.105, 175.300, 175.320, 176.200, 176.210, 177.2600, 177.2800, 178.3910, 181.22, 181.27

Butyl sulfide
CAS 544-40-1; EINECS 208-870-5; FEMA 2215
Synonyms: Dibutyl sulfide; n-Butyl sulfide; Butylthiobutane; 1,1´-Thiobisbutane
Empirical: $C_8H_{18}S$
Formula: $CH_3(CH_2)_3S(CH_2)_3CH_3$
Properties: Liq.; very sol. in alcohol, ether; insol. in water; m.w. 146.30; dens. 0.839 (16/0 C); m.p. -80 C; b.p. 182 C; ref. index 1.4530 (20 C)
Uses: Synthetic flavoring agent
Usage level: 0.01-1.0 ppm (FDA)
Regulatory: FDA 21CFR §172.515
Manuf./Distrib.: Aldrich

n-Butyl sulfide. *See* Butyl sulfide
Butylthiobutane. *See* Butyl sulfide
Butyl α-toluate. *See* Butyl phenylacetate

Butyl 10-undecenoate
FEMA 2216
Synonyms: Butyl 10-hendecenoate; Butyl undecylenoate
Empirical: $C_{15}H_{28}O_2$
Properties: Liq.; m.w. 240.39; sp.gr. 0.8751 (20 C); b.p. 125-128 C (3 mm); ref. index 1.4426 (20 C)
Uses: Synthetic flavoring agent
Usage level: 0.90 ppm (nonalcoholic beverages), 5.0 ppm (alcoholic beverages, icings), 2.0 ppm (ice cream, ices), 6.6 ppm (candy), 7.8 ppm (baked goos), 0.40-60 ppm (chewing gum)
Regulatory: FDA 21CFR §172.515

Butyl undecylenoate. *See* Butyl 10-undecenoate

Butyl valerate
FEMA 2217
Synonyms: n-Butyl valerate; n-Butyl pentanoate; n-Butyl-n-valerianate
Empirical: $C_9H_{18}O_2$
Formula: Liq., fruity odor; sol. in propylene glycol; sl. sol. in water; m.w. 158.24; sp.gr. 0.8680 (20 C); b.p. 186.5 C; ref. index 1
Uses: Synthetic flavoring agent
Usage level: 3.0 ppm (nonalcoholic beverages), 2.6 ppm (ice cream, ices), 8.0 ppm (candy), 6.8 ppm (baked goods)
Regulatory: FDA 21CFR §172.515
Manuf./Distrib.: Aldrich

n-Butyl valerate. *See* Butyl valerate
n-Butyl-n-valerianate. *See* Butyl valerate

n-Butyraldehyde
CAS 123-72-8; EINECS 204-646-6; FEMA 2219
Synonyms: n-Butylaldehyde; Butyric aldehyde; n-Butanal
Classification: Aldehyde
Empirical: C_4H_8O
Formula: $CH_3CH_2CH_2CHO$
Properties: Colorless liq.; sol. in water; misc. with ether @ 75 C; m.w. 72.10; sp.gr. 0.800; m.p. -99 C; f.p. 12 F; b.p. 75 76 C; ref. index 1.3043
Precaution: Flamm. liq. (DOT); incompat. with oxidizng materials; reacts vigorously with chlorosulfonic acid
Toxioology: LD50 (oral, rat) 2490mg/kg; mod. toxic by ingestion, inh., skin contact, intraperitoneal, and subcutaneous routes; severe skin and eye irritant; heated to decomp., emits acrid smoke and fumes
Uses: Synthetic flavoring agent
Usage level: 0.71 ppm (nonalcoholic beverages), 0.5 ppm (alcoholic beverages), 4.8 ppm (ice cream, ices), 2.9 ppm (candy), 5.4 ppm (baked goods), 0.25 ppm (icings)
Regulatory: FDA 21CFR §172.515
Manuf./Distrib.: Aldrich; Penta Mfg.

Butyric acid isoamyl ester. *See* Isoamyl butyrate

n-Butyric acid
CAS 107-92-6; EINECS 203-532-3; FEMA 2221
Synonyms: Butanoic acid; Carboxylic acid C4; Ethylacetic acid; 1-Propanecarboxylic acid; Propylformic acid
Empirical: $C_4H_8O_2$
Formula: $CH_3CH_2CH_2COOH$
Properties: Colorless liq., strong rancid butter odor; m.w. 88.12; dens. 0.9590 (20/20 C); m.p. -7.9 C; f.p. -5.5 C; b.p. 163.5 C; flash pt. 161 F; ref. index 1.397
Precaution: Corrosive material; combustible liq.; may react with oxidizers
Toxicology: LD50 (oral, rat) 2940 mg/kg; mod. toxic by ingestion, skin contact, subcutaneous, intraperitoneal, and IV routes; severe skin and eye irritant; heated to decomp., emits acrid smoke and irritating fumes
Uses: Synthetic flavoring; pungent flavor, reminiscent of rancid butter
Usage level: 5.5 ppm (nonalcoholic beverages), 6.5 ppm (ice cream, ices), 82 ppm (candy), 32 ppm (baked goods), 0.19-45 ppm (gelatins, puddings), 60-270 ppm (chewing gum), 18 ppm (margarine)
Regulatory: FDA 21CFR §182.60, GRAS; Japan approved as flavoring
Manuf./Distrib.: Aldrich

Butyric aldehyde. *See* n-Butyraldehyde
Butyric ether. *See* Ethyl butyrate
Butyrin. *See* Tributyrin
Butyroin. *See* 5-Hydroxy-4-octanone

Butyrolactone
CAS 96-48-0; EINECS 202-509-5; FEMA 3291
Synonyms: BLO; Dihydro-2(3H)-furanone; γ-Butyrolactone; 4-Butanolide
Classification: Lactone
Empirical: $C_4H_6O_2$
Properties: Colorless oily liq., mild caramel odor; misc. with water; sol. in methanol, ethanol, acetone, ether, benzene; m.w. 86.09; dens. 1.120; m.p. -45 C; b.p. 204-205 C; flash pt. 98 C; ref. index 1.4348
Precaution: Combustible when exposed to heat or flame; reactive with oxidizing materials
Toxicology: LD50 (oral, rat) 1800 mg/kg; mod. toxic by ingestion, IV, intraperitoneal routes; suspected tumorigen; heated to decomp., emits acrid and irritating fumes
Uses: Flavoring agent for candy, soy milk
Regulatory: GRAS
Manuf./Distrib.: Aldrich; Spectrum Chem. Mfg.

γ-Butyrolactone. *See* Butyrolactone
Butyrone. *See* 4-Heptanone
Butyryllactic acid butyl ester. *See* Butyl butyryllactate
CA. *See* Cellulose acetate

Cacao
Synonyms: Theobroma cacao
Definition: From *Theobroma cacao*
Uses: Natural flavoring agent; used in liqueurs
Regulatory: FDA 21CFR §182.20, GRAS; Japan approved

Cacao butter. *See* Cocoa butter
C-8 acid. *See* Caprylic acid

Cade oil
Synonyms: Essence de Cadé; Kadeoel; Cadmium oil
Definition: Derived from *Juniperus oxycedrus* L.
Properties: Char. smoky acrid odor, bitter taste; sp.gr. 0.952-0.961 (25 C); ref. index 1.5110-1.5200 (25 C)
Uses: Natural flavoring agent
Regulatory: Japan approved

Cadinene
Classification: Sesquiterpene derived from essential oils
Empirical: $C_{15}H_{24}$
Properties: m.w. 204.34
Uses: Synthetic flavoring agent
Regulatory: FDA 21CFR §172.515

Cadmium oil. *See* Cade oil

Caffeine
CAS 58-08-2; EINECS 200-362-1; FEMA 2224
Synonyms: Theine; Coffeine; Guaramine; Methyltheobromine; 1,3,7-Trimethylxanthine
Classification: Heterocyclic organic compd.
Empirical: $C_8H_{10}N_4O_2$
Properties: Wh. flEECy mass, odorless, bitter taste; sol. in water, alcohol, chloroform, ether; m.w. 194.22; m.p. 238 C
Toxicology: Possible teratogen; diuretic effect; can cause heartburn, upset stomach, diarrhea; 200-500 mg can cause headache, tremors, nervousness, and irritability
Uses: Flavoring agent for beverages, stimulant, alkaloid; for cola and orange beverages
Usage level: 120 ppm (nonalcoholic beverages)
Regulatory: FDA 21CFR §165.175, 182.1180 (limitation 0.02%), GRAS
Manuf./Distrib.: Alfa; Am. Roland; Ashland; Bell Flavors & Fragrances; Berk; Browne & Dureau Int'l.; Dasco; Ellis & Everard; Food Ingred. Tech.; Jungbunzlauer; MLG Enterprises; Penta Mfg.; Pfizer; Quimdis; Robt. Bryce; Scanchem; Tesco; Van Waters & Rogers
Trade names containing: Coffee Conc.; Cola Acid No. 23444

Cajeputene. *See* dl-Limonene
Cajeputene. *See* l-Limonene
Cajeputi oil. *See* Cajeput oil

Cajeput oil
Synonyms: Cajeputi oil
Definition: Volatile oil from fresh leaves and twigs of *Melaleuca leucadendron*, contg. 50-60% eucalyptol and

l-pinene, terpineol, and aldehydes
Properties: Colorless of ylsh. liq., agreeable camphor odor, bitter aromatic taste; misc. with alcohol, chloroform, ether, CS_2; very sl. sol. in water; sol. in 1 vol. 80% alcohol; dens. 0.912-0.925; ref. index 1.4660-1.4710 (20 C)
Precaution: Keep well closed, cool, protected from light
Toxicology: LD50 (oral, rat) 3870 mg/kg
Uses: Natural flavoring
Regulatory: FDA 21CFR §172.510; Japan approved

Cajeputol. *See* Eucalyptol
Cake alum. *See* Aluminum sulfate

Calamus oil
EINECS 283-869-0 (extract)
Synonyms: Sweet flag oil
Definition: Volatile oil from rhizome of *Acorus calamus* contg. eugenol, asarone, stearopten
Properties: Yel. to ylsh.-brn. visc. liq., aromatic odor, bitter taste; very sl. sol. in water; misc. with alcohol; dens. 0.960-0.970 (20/20 C); sapon. no. 16-20; ref. index 1.507-1.515
Precaution: Keep cool, well closed; protect from light
Toxicology: LD50 (oral, rat) 777 mg/kg
Uses: Formerly as minor ingred. in bitter flavors such as vermouth and flavored wines
Regulatory: FDA 21CFR §189.110, prohibited from use in food; Japan approved (calamus)
Manuf./Distrib.: Acme-Hardesty

Calciferol. *See* Ergocalciferol
Calcined alumina. *See* Alumina
Calcined magnesia. *See* Magnesium oxide
Calciofon. *See* Calcium gluconate
Calciol. *See* Cholecalciferol

Calcium acetate
CAS 62-54-4; EINECS 200-540-9; EEC E263
Synonyms: Vinegar salts; Gray acetate; Lime acetate
Definition: Calcium salt of acetic acid
Empirical: $C_4H_6CaO_4 \cdot H_2O$
Formula: $(CH_3COO)_2Ca \cdot H_2O$
Properties: Wh. powd., sl. bitter taste; sol. in water; sl. sol. in alcohol; m.w. 158.18 + water
Precaution: Combustible; hygroscopic
Toxicology: LD50 (IV, mouse) 52 mg/kg; poison by IV route; heated to decomp., emits acrid smoke and fumes
Uses: Food additive, preservative, firming agent, pH control agent, processing aid, sequestrant, stabilizer, texturizer, thickener for baked goods, cake mixes, fillings, gelatins, syrups, sauces, toppings, pkg. materials
Usage level: Limitation 0.2% (baked goods), 0.02% (cheese), 0.2% (gelatins, puddings), 0.15% (sweet sauces, toppings), 0.001% (other foods); ADI not specified (EEC)
Regulatory: FDA 21CFR §175.300, 181.22, 181.29, 182.6197, 184.1185, GRAS; Europe listed; UK approved
Manuf./Distrib.: Lohmann; Niacet
Trade names: Niacet Calcium Acetate FCC

Calcium alginate
CAS 9005-35-0; EEC E404
Synonyms: Alginic acid, calcium salt
Definition: Calcium salt of alginic acid
Empirical: $[(C_6H_7O_6)_2Ca]_n$
Properties: White or cream-colored powd. or filaments, sl. odor and taste; insol. in water, acid; sol. in alkaline sol'n.; m.w. 195.16
Precaution: Flamm.
Toxicology: Heated to decomp., emits acrid smoke and irritating fumes
Uses: Emulsifier, stabilizer, thickener, gellant
Usage level: Limitation 0.002% (baked goods), 0.4% (alcoholic beverages), 0.4% (frostings), 0.6% (egg prods.), 0.5% (fats, oils), 0.25% (gelatins, puddings), 0.4% (gravies), 0.5% (jams, jellies), 0.5% (sweet sauces), 0.3% (other foods)
Regulatory: FDA 21CFR §184.1187, GRAS; GRAS (FEMA); Europe listed (ADI 0-25 mg/kg, as alginic acid); UK approved
Manuf./Distrib.: Pronova
Trade names: Sobalg FD 460

Calcium aluminum silicate

Trade names containing: Keltose®; Wakal® A 601

Calcium aluminum silicate. See Aluminum calcium silicate

Calcium ascorbate
CAS 5743-27-1; EEC E302
Empirical: $C_{12}H_{14}CaO_{12} \cdot 2H_2O$
Properties: Wh. to sl. yel. cryst. powd., odorless; sol. in water; sl. sol. in alcohol; insol. in ether; m.w. 426.35
Toxicology: Heated to decomp., emits acrid smoke and irritating fumes
Uses: Antioxidant, preservative in foods; vitamin; meat color preservative; used in meat prods., conc. dairy prods.
Regulatory: FDA 21CFR §182.3189, GRAS; Europe listed; UK approved
Trade names: Calcium Ascorbate FCC No. 60475

Calcium benzoate
CAS 2090-05-3; EINECS 218-235-4; EEC E213
Definition: Calcium salt of benzoic acid
Empirical: $C_{14}H_{10}O_4 \cdot 3H_2O$
Properties: Orthorhombic cryst. or powd.; sol. in water; m.w. 374.26; dens. 1.44
Precaution: Combustible exposed to heat or flame
Toxicology: May cause an intolerance reaction in some people; heated to decomp., emits acrid smoke and irritating fumes
Uses: Preservative for margarine
Usage level: Limitation 0.1% (alone), 0.2% (in combination with sorbic acid or its salts); ADI 0-5 mg/kg (EEC)
Regulatory: FDA 21CFR §166.110, 178.2010; USDA 9CFR §318.7; Europe listed

Calcium biphosphate. See Calcium phosphate monobasic anhydrous
Calcium biphosphate. See Calcium phosphate monobasic monohydrate
Calcium bis(dihydrogenphsophate) monohydrate. See Calcium phosphate monobasic monohydrate

Calcium borogluconate
CAS 5743-34-0
Synonyms: D-Gluconic acid cyclic 4,5-ester with boric acid calcium salt; Calcium diborogluconate
Empirical: $C_{12}H_{20}B_2C_9O_{16}$
Properties: Cryst.; freely sol. in water; m.w. 482.01
Uses: Mineral source
Trade names: Gluconal® CA M B

Calcium bromate
Empirical: $Ca(BrO_3)_2 \cdot H_2O$
Properties: Wh. cryst. powd.; sol. in water; m.w. 313.90; dens. 3.329
Toxicology: Nuisance dust
Uses: Dough conditioner, maturing agent

Calcium bromide
CAS 7789-41-5; EINECS 232-164-6
Empirical: $Br_2Ca \cdot 2H_2O$
Formula: $CaBr_2 \cdot 2H_2O$
Properties: Gran. or rhombic cryst., odorless, sharp saline taste; deliq.; very sol. in water, methanol, ethanol; sol. in acetone; pract. insol. in ether; m.w. 199.91 (anhyd.), 235.93 (dihydrate)
Precaution: Hygroscopic; keep well closed
Uses: Food preservative
Manuf./Distrib.: Atomergic Chemetals

Calcium carbonate. See also Limestone
CAS 471-34-1; 1317-65-3; EINECS 207-439-9; EEC E170
Synonyms: Carbonic acid calcium salt (1:1); CI 77220; Precipitated calcium carbonate; Precipitated chalk
Classification: Inorganic salt
Empirical: $CaCO_3$
Properties: White powd. or colorless crystals, odorless, tasteless; very sl. sol. in water; sol. in acids with CO_2; m.w. 100.09; dens. 2.7-2.95; m.p. 825 C (dec.)
Precaution: Ignites on contact with F_2; incompat. with acids, alum, ammonium salts
Toxicology: TLV 5 mg/m³ of air; LD50 (oral, rat) 6450 mg/kg; severe eye and moderate skin irritant
Uses: Nutrient, dietary supplement, dough conditioner, firming agent, yeast food, colorant, alkali; fortification of bread; release agent; stabilizer migrating from food pkg.
Usage level: ADI not specified; WHO limitation: 40 g/kg (cheese), 200 mg/kg (jams, jellies)
Regulatory: FDA 21CFR §73.1070, 137.105, 137.155, 137.160, 137.165, 137.170, 137.175, 137.180,

137.185, 137.350, 169.115, 175.300, 177.1460, 181.22, 181.29, 182.5191, 184.1191, 184.1409, GRAS; BATF 27CFR §240.1051, limitation 30 lb/1000 gal of wine; Japan approved (1-2%); Europe listed; UK approved

Manuf./Distrib.: Am. Roland

Trade names: Atomite®; Carbital® 35; Carbital® 50; Carbital® 75; CC™-101; CC™-103; CC™-105; Micro-White® 10 Codex; Micro-White® 25 Codex; Micro-White® 50 Codex; Micro-White® 100 Codex

Trade names containing: Coco-Spred™; Garbefix 31; Stabil® 9 High Calcium Blended Powd.; Stabil® 9 Regular Blended Powd.

Calcium carboxymethyl cellulose

Uses: Thickener; stabilizer; gellant (2% max.)
Regulatory: Japan approved; restricted

Calcium carrageenan

CAS 9049-05-2
Definition: Calcium salt of carrageenan
Uses: Emulsifier, stabilizer, thickener
Regulatory: FDA 21CFR §136.110, 136.115, 136.130, 136.160, 136.180, 139.121, 139.122, 150.141, 150.161, 172.626, 176.170

Calcium caseinate

Synonyms: Casein calcium salt
Definition: Calcium salt of casein
Properties: Wh. or sl. yel. powd., nearly odorless; insol. in cold water
Uses: Dietary supplement; in frozen desserts, nutritional drinks
Regulatory: FDA GRAS
Manuf./Distrib.: Blossom Farm Prods.; Excelpro

Calcium chloride

CAS 10043-52-4 (anhyd.), 10035-04-8 (dihydrate); EINECS 233-140-8; EEC E509
Classification: Inorganic salt
Empirical: $CaCl_2 \cdot 2H_2O$ (dihydrate)
Properties: White deliq. cryst., granules, lumps, or flakes; sol. in water and alcohol; m.w. 110.99 (anhyd.); dens. 2.150; b.p. > 1600 C; m.p. 782 C
Precaution: Keep well closed
Toxicology: LD50 (oral rat) 1000 mg/kg, (intraperitoneal, rat) 264 mg/kg; poison by intravenous, intramuscular, intraperitoneal, subcutaneous routes; moderately toxic by ingestion
Uses: Food sequestrant, firming agent, anticaking agent, antimicrobial, curing/pickling agent, flavor enhancer, humectant, nutrient supplement, pH control agent, processing aid, stabilizer, surfactant, texturizer, thickener, synergist; in beer malting
Usage level: Limitation 0.3% (baked goods, dairy prods.), 0.22% (nonalcoholic beverages), 0.2% (cheese, fruit juice), 0.32% (coffee), 0.4% (relishes), 0.2% (sauces), 0.1% (jams), 0.25% (meat prods.), 2% (plant protein prods.), 0.4% (processed veg.), 0.05% (other)
Regulatory: FDA 21CFR §184.1193, GRAS; USDA 9CFR §318.7, 381.147, 3% max.; Japan approved (1% max.); Europe listed (ADI not specified); UK approved; WHO limitation: 350-800 mg/kg (canned fruits/veg.), 200 mg/kg (preserves, processed cheese)
Manuf./Distrib.: Browning; Gist-brocades Food Ingreds.; Int'l. Sourcing; Lohmann

Calcium citrate

CAS 813-94-5 (tricalcium citrate anhyd.), 5785-44-4 (tricalcium citrate-4-hydrate); EINECS 212-391-7; EEC E333
Synonyms: Lime citrate; Tricalcium citrate; Calcium citrate tertiary; Dicalcium citrate
Empirical: $C_{12}H_{10}Ca_3O_{14} \cdot 4H_2O$
Formula: $Ca_3(C_6H_5O_7)_2 \cdot 4H_2O$
Properties: Wh. fine powd., odorless; sl. sol. in water; insol. in alcohol; m.w. 570.50
Toxicology: Heated to decomp., emits acrid smoke and irritating fumes
Uses: Dietary supplement, nutrient, sequestrant, buffer, antioxidant, and firming agent in foods; acidity regulator in jams, jellies, marmalades, soft drinks, and wines; raising agent; emulsifying salt
Usage level: ADI not specified (EEC); WHO limitation: 350 mg/kg (canned tomatoes)
Regulatory: FDA 21CFR §182.1195, 182.5195, 182.6195, 182.8195, GRAS; Japan approved (1% max. as calcium); Europe listed; UK approved
Manuf./Distrib.: Clofine; CRS; Int'l. Sourcing; Jungbunzlauer; Lohmann; Penta Mfg.; Qualcepts Nutrients; Quimdis
Trade names containing: Muffit; Tang

Calcium citrate tertiary. *See* Calcium citrate

Calcium cyclamate
CAS 139-06-0
Synonyms: Calcium cyclohexyl sulfamate
Empirical: $(C_6H_{11}NHSO_3)_2Ca \cdot 2H_2O$
Properties: Wh. cryst., odorless, sweet taste; freely sol. in water; insol. in alcohol, benzene, chloroform
Toxicology: Poison by ingestion; experimental tumorigen and neoplastigen; experimental reproductive effects; human mutagenic data; heated to decomp., emits acrid smoke and fume
Uses: Nonnutritive sweetener
Usage level: Prohibited from foods (U.S.)

Calcium cyclohexyl sulfamate. *See* Calcium cyclamate

Calcium diacetate
CAS 66905-25-7 (anhyd.); EINECS 226-516-5
Empirical: $C_6H_{10}CaO_6$
Uses: Sequestrant
Regulatory: FDA 21CFR §182.6197, GRAS

Calcium diborogluconate. *See* Calcium borogluconate
Calcium dihydrogen phosphate. *See* Calcium phosphate monobasic monohydrate

Calcium dihydrogen pyrophosphate
Uses: Dietary supplement (1% max. as calcium); emulsifier (1% max. as calcium); raising agent (1% max. as calcium)
Regulatory: Japan approved; restricted

Calcium N-(2,4-dihydroxy-3,3-dimethyl-1-oxobutyl-β-alanine. *See* Calcium pantothenate
Calcium dioxide. *See* Calcium peroxide
Calcium diphosphate. *See* Calcium pyrophosphate
Calcium disodium edetate. *See* Calcium disodium EDTA

Calcium disodium EDTA
CAS 62-33-9; EINECS 200-529-9; EEC E385
Synonyms: Calcium disodium ethylenediamine tetraacetic acid; Edetate calcium disodium; Calcium disodium edetate
Classification: Substituted diamine
Empirical: $C_{10}H_{12}CaN_2Na_2O_8 \cdot 2H_2O$
Formula: $CaNa_2C_{10}H_{12}N_2O_8 \cdot xHOH$
Properties: White cryst. powd., odorless, faint salt taste; sol. in water; insol. in org. solvs.; m.w. 410.30
Toxicology: Possible link to liver damage in test animals
Uses: Food preservative, sequestrant, chelating agent preventing discoloration, rancidity, off-odors; hog scald agent; antioxidant
Usage level: Limitation 220 ppm (pickled cabbage), 33 ppm (soft drinks), 25 ppm (alcoholic beverages), 200 ppm (egg prods.), 75 ppm (mayonnaise), 75 ppm (oleomargarine), 75 ppm (sauces); ADI 0-2.5 mg/kg (EEC)
Regulatory: FDA 21CFR §172.120; USDA 9CFR §318.7; Japan approved (0.035 g/kg max.); Europe listed; permitted in UK only in canned fish, shellfish
Manuf./Distrib.: Int'l. Sourcing
Trade names: Versene CA
Trade names containing: Diet Imperial® Spread; Shedd's Liquid Margarine

Calcium disodium ethylenediamine tetraacetic acid. *See* Calcium disodium EDTA

Calcium formate
CAS 544-17-2; EINECS 208-863-7, 208-376-3; EEC E238
Synonyms: Formic acid calcium salt
Empirical: $C_2H_2CaO_4$
Formula: $(HCOO)_2Ca$
Properties: Orthorhombic cryst. or cryst. powd., sl. acetic acid-like odor; sol. in water, pract. insol. in alcohol; m.w. 130.12
Uses: Preservative for food and silage
Manuf./Distrib.: Lohmann

Calcium fumarate
CAS 19855-56-2 (anhyd.); EINECS 243-376-3
Empirical: $C_4H_2CaO_4$
Uses: Dietary supplement
Regulatory: FDA 21CFR §172.350

Calcium furcelleran
Definition: Calcium salt of furcelleran
Uses: Emulsifier, stabilizer, thickener
Regulatory: FDA 21CFR §172.660

Calcium 4-(β-d-galactosido)-d-gluconate. *See* Calcium lactobionate

Calcium gluconate
CAS 299-28-5, 18016-24-5 (monohydrate); EINECS 206-075-8; EEC E578
Synonyms: D-Gluconic acid calcium salt; Calciofon; Glucal
Definition: Calcium salt of gluconic acid
Empirical: $C_{12}H_{22}O_{14} \cdot Ca$
Formula: $Ca[HOCH_2(CHOH)_4COO]_2$
Properties: White fluffy powd. or gran., odorless, pract. tasteless; sol. in hot water; insol. in alcohol, acetic acid, other org. solvs.; stable in air; m.w. 430.4; m.p. loses water @ 120 C
Toxicology: LD50 (IV, rat) 950 mg/kg; moderately toxic by subcutaneous, intraperitoneal, intravenous routes; heated to decomp., emits acrid smoke and fumes
Uses: Mineral source, food additive, buffer, firming agent, sequestrant, texturizer, stabilizer, thickener, formulation aid; anticaking agent in coffee powds.;
Usage level: Limitation 1.75% (baked goods), 0.4% (dairy prods.), 4.5% (gelatins, puddings), 0.01% (sugar substitutes); ADI 0-50 mg/kg (EEC, as total gluconic acid)
Regulatory: FDA 21CFR §184.1199, GRAS; Japan restricted (1% max. as calcium); Europe listed; UK approved; WHO limitation 350 ppm max. (fruits, veg.), 200 ppm (jams, jellies)
Manuf./Distrib.: Am. Roland; Int'l. Sourcing
Trade names: Gluconal® CA A; Gluconal® CA M

Calcium glycerinophosphate. *See* Calcium glycerophosphate

Calcium glycerophosphate
CAS 126-95-4, 27214-00-2 (hydrate); EINECS 248-328-5
Synonyms: Calcium glycerinophosphate; Calcium phosphoglycerate
Empirical: $C_3H_7CaO_6P$
Properties: Wh. fine powd., odorless, almost tasteless;sl. hygroscopic; sol. in water; insol. in alcohol; m.w. 210.14
Toxicology: Heated to decomp., emits toxic fumes of PO_x
Uses: Direct food additive, nutrient/dietary supplement, stabilizer; used in gelatins, puddings, fillings, baking powd., stabilizer migrating from food pkg.
Regulatory: FDA 21CFR §181.29, 182.5201, 184.1201, GRAS; Japan restricted (1% max. as calcium)
Manuf./Distrib.: Lohmann

Calcium hexametaphosphate
Toxicology: Nuisance dust
Uses: Sequestrant
Regulatory: FDA 21CFR §182.6203, GRAS

Calcium hydrate. *See* Calcium hydroxide
Calcium hydrogen orthophosphate. *See* Calcium phosphate dibasic
Calcium hydrogen phosphate anhydrous. *See* Calcium phosphate dibasic
Calcium hydrogen phosphate dihydrate. *See* Calcium phosphate dibasic dihydrate

Calcium hydroxide
CAS 1305-62-0; EINECS 215-137-3; EEC E526
Synonyms: Calcium hydrate; Hydrated lime; Slaked lime; Lime water
Classification: Inorganic base
Empirical: CaH_2O_2
Formula: $Ca(OH)_2$
Properties: Soft, white crystalline powd. with alkaline, sl. bitter taste; sl. sol. in water; sol. in glycerol, syrup, acid insol. in alcohols; m.w. 74.10; dens. 2.34; m.p. loses water at 580 C
Precaution: Violent reaction with maleic anhydride, nitroethane, nitromethane, nitroparaffins, etc.
Toxicology: TLV 5 mg/m³ in air; LD50 (oral, rat) 7.34 g/kg; mildly toxic by ingestion; severe eye irritant; skin, mucous membrane irritant; dust is industrial hazard
Uses: Food additive, buffer, neutralizing agent, firming agent, processing aid; in malting process in beer-making
Usage level: ADI no limit (EEC)
Regulatory: FDA 21CFR §135.110, 184.1205, GRAS; USDA 9CFR §318.7; Japan restricted (1% max. as calcium); Europe listed; UK approved

Calcium hydroxide phosphate

    Manuf./Distrib.: Pfizer

Calcium hydroxide phosphate. *See* Calcium phosphate tribasic
Calcium-L-2-hydroxypropionate. *See* Calcium lactate

Calcium iodate
    CAS 7789-80-2; EINECS 232-191-3
    Synonyms: Lautarite
    Empirical: $CaI_2O_6 \cdot H_2O$
    Formula: $Ca(IO_3)_2$
    Properties: Wh. powd., odorless or sl. odor; sl. sol. in water; insol. in alcohol; m.w. 407.90
    Precaution: Oxidizer
    Toxicology: Nuisance dust; irritant
    Uses: Dough strengthener/conditioner, maturing agent for bread
    Usage level: limitation 0.0075% of flour (bread)
    Regulatory: FDA 21CFR §184.1206, GRAS
    Manuf./Distrib.: Atomergic Chemetals
    Trade names: IDX-20 NB
    Trade names containing: Calib; IDX; N'Hance CM; N'Hance (Phase II) Powd.; N'Hance (Phase II) Tablets; N'Hance SD

Calcium lactate
    CAS 814-80-2 (anhyd.); EINECS 212-406-7; EEC E327
    Synonyms: 2-Hydroxypropanoic acid, calcium salt; Calcium-L-2-hydroxypropionate
    Empirical: $C_6H_{10}CaO_6 \cdot xH_2O$, x < 5
    Formula: $[CH_3CH(OH)COO]_2Ca \cdot xH_2O$
    Properties: Wh. to cream-colored cryst. powd. or granules containing up to 5 moles of water of crystallization, almost odorless; sol. in water; pract. insol. in alcohol; m.w. 218.22 (anhyd.)
    Toxicology: Heated to decomp., emits acrid smoke and irritating fumes
    Uses: Antioxidant, film-former, buffer, dough conditioner, yeast food, firming agent, flavor enhancer, flavoring agent, leavening agent, nutrient supplement, stabilizer, thickener; emulsifier in shortening; humectant in confectionery; yeast food
    Usage level: ADI no limit (EEC); WHO limitation: 350 ppm max. (canned fruits, veg., cheese)
    Regulatory: FDA 21CFR §184.1207, GRAS except for infant foods/formulas; USDA 9CFR §318.7 (0.6% max.); Japan approved (1% max. as calcium); Europe listed; UK approved
    Manuf./Distrib.: Am. Roland; Int'l. Sourcing; Jungbunzlauer; Lohmann; Qualcepts Nutrients
    Trade names: Pationic® 1230; Pationic® 1240; Pationic® 1250; Puracal® PP
    Trade names containing: Cap-Shure® LCL-135-50

Calcium lactobionate
    CAS 5001-51-4; EINECS 225-668-2
    Synonyms: Calcium 4-(β-d-galactosido)-d-gluconate
    Definition: Calcium salt of lactobionic acid
    Empirical: $C_{24}H_{42}CaO_{24}$
    Properties: Wh. powd.; sol. in water; insol. in alcohol, ether; m.w. 754.66; m.p. 120 C (dec.)
    Toxicology: Heated to decomp., emits acrid smoke and irritating fumes
    Uses: Firming agent in dry pudding mixes
    Regulatory: FDA 21CFR §172.720

Calcium lignosulfonate
    CAS 8061-52-7
    Synonyms: Lignosulfonic acid, calcium salt
    Definition: Calcium salt of polysulfonated lignin
    Toxicology: Heated to decomp., emits acrid smoke and irritating fumes
    Uses: Dispersant, stabilizer for pesticides for pre- or post-harvest applic. to bananas
    Regulatory: FDA 21CFR §172.715, 175.105, 176.170, 176.210

Calcium octadecanoate. *See* Calcium stearate

Calcium oleate
    CAS 142-17-6
    Synonyms: 9-Octadecenoic acid calcium salt; Oleic acid calcium salt
    Empirical: $C_{36}H_{66}CaO_4$
    Properties: Pale yel. transparent solid; sol. in benzene, chloroform; insol. in water, alcohol, acetone; m.w. 602.97; m.p. 83 C
    Toxicology: Heated to decomp., emits acrid smoke and irritating fumes

Uses: Stabilizer migrating from food pkg.
Regulatory: FDA 21CFR §181.29

Calcium orthophosphate. *See* Calcium phosphate tribasic

Calcium oxide
CAS 1305-78-8; EINECS 215-138-9; EEC E529
Synonyms: Lime; Quicklime; Calx; Burnt lime
Classification: Inorganic oxide
Empirical: CaO
Properties: White or gray cryst. or powd., odorless; sol. in acids, glycerol, sugar sol'n.; sol. in water forming
    $Ca(OH)_2$ and generating heat; pract. insol. in alcohol; m.w. 56.08; dens. 3.40; m.p. 2570 C; b.p. 2850 C
Precaution: Noncombustible; powd. may react explosively with water; mixts. with ethanol may ignite if heated
Toxicology: TLV 2 mg/m³; strong caustic; may cause severe irritation to skin, mucous membranes
Uses: Alkali, anticaking agent, nutrient, dietary supplement, dough conditioner, yeast food, hog/poultry scald
    agent, sugar processing
Usage level: ADI no limit (EEC)
Regulatory: FDA 21CFR §182.1210, 182.5210, 184.1210, GRAS; USDA 9CFR §318.7, 381.147; Europe
    listed; UK approved
Manuf./Distrib.: Pfizer

Calcium pantothenate
CAS 137-08-6 (D-); EINECS 205-278-9
Synonyms: Calcium N-(2,4-dihydroxy-3,3-dimethyl-1-oxobutyl-β-alanine; Vitamin B₅, calcium salt; Calcium d-
    pantothenate
Definition: Calcium salt of pantothenic acid
Empirical: $C_9H_{16}NO_5 \cdot \frac{1}{2}Ca$
Properties: Wh. powd., sl. hygroscopic, odorless, sweetish taste with sl. bitter aftertaste; stable in air; sol. in
    water, glycerol; insol. in alcohol, chloroform, ether; m.w. 490.63; m.p. 170-172 C; dec. 195-196 C
Toxicology: LD50 (oral, mouse) 10 g/kg, (IV, rat) 830 mg/kg; moderately toxic by intraperitoneal, subcutane-
    ous, and intravenous routes; mildly toxic by ingestion; heated to decomp., emits toxic fumes of NO_x
Uses: Direct food additive, nutrient, dietary supplement; may be used in infant formulas; only D isomer has
    vitamin activity, both D and DL used in food
Regulatory: FDA 21CFR §182.5212, 184.1212, GRAS; Japan approved (1% max. as calcium)
Manuf./Distrib.: Am. Roland; Daiichi Pharmaceutical; Unipex
Trade names: Calcium Pantothenate USP, FCC Type SD No. 63924; Lutavit® Calpan

Calcium d-pantothenate. *See* Calcium pantothenate

Calcium pantothenate calcium chloride double salt (d or dl).
CAS 6363-38-8
Definition: Calcium chloride double salt of dl- or d-calcium pantothenate
Empirical: $C_{19}H_{34}N_2O_{10} \cdot Ca_2Cl_2$
Properties: Wh. powd., odorless, bitter taste; sl. hygroscopic; sol. in water, glycerin; insol. in alcohol,
    chloroform, ether; m.w. 601.61
Toxicology: Mod. toxic by IP, subcut., IV routes; mildly toxic by ingestion; heated to decomp., emits toxic fumes
    of NO_x
Uses: Dietary supplement, nutrient, vitamin
Regulatory: FDA 21CFR §172.330, 181.5212, GRAS

Calcium peroxide
CAS 1305-79-9; EINECS 215-139-4
Synonyms: Calcium superoxide; Calcium dioxide
Empirical: CaO_2
Properties: Wh. or yellowish powd. or granular material, odorless, almost tasteless; decomp. in moist air; pract.
    insol. in water; dissolves in acids, forming hydrogen peroxide; m.w. 72.08; m.p. dec. @ 275 C
Precaution: Flamm. if hot and mixed with finely divided combustible material; oxidizer
Toxicology: Irritating in conc. form
Uses: Dough conditioner, oxidizing agent
Regulatory: GRAS
Manuf./Distrib.: FMC; Lohmann
Trade names: Improved Paniplus M; Regular Paniplus
Trade names containing: CDC-79DZ; Do Crest Gold Plus; Drize-P Powd.

Calcium phosphate dibasic
CAS 7757-93-9; EINECS 231-826-1; EEC E341b

Calcium phosphate dibasic dihydrate

Synonyms: DCP-0; Dicalcium phosphate; Dicalcium orthophosphate anhyd. (E341); Calcium hydrogen phosphate anhydrous; Calcium hydrogen orthophosphate; Phosphoric acid calcium salt (1:1)
Classification: Inorganic salt
Empirical: CaHPO₄
Properties: Wh. cryst. powd., odorless, tasteless; sol. in dilute HCl, nitric, and acetic acids; insol. in alcohol; sl. sol. in water; m.w. 136.07; dens. 2.306; loses water at 109 C
Toxicology: Skin and eye irritant; nuisance dust
Uses: Dietary supplement, dough conditioner, nutrient, stabilizer, firming agent, antioxidant synergist, yeast food for baked goods, cereal prods., dessert gels; stabilizer migrating from food pkg. materials
Regulatory: FDA 21CFR §181.29, 182.1217, 182.5217, 182.8217, GRAS; Japan approved (1% max. as calcium); Europe listed; UK approved
Manuf./Distrib.: Albrite & Wilson Am.; Browning; FMC; Rhone-Poulenc Food Ingreds.
Trade names: Albrite Dicalcium Phosphate Anhyd
Trade names containing: Folic Acid 10% Trituration No. 69997; Vitamin B₁₂ 1% Trituration No. 69992

Calcium phosphate dibasic dihydrate
CAS 7789-77-7; EINECS 231-826-1; EEC E540
Synonyms: DCP-2; Dicalcium phosphate dihydrate; Calcium hydrogen phosphate dihydrate; Dicalcium phosphate
Empirical: CaHPO₄ • 2H₂O
Properties: Monoclinic cryst.; loses water of cryst. slowly below 100 C; sol. in dil. HCl or HNO₃; sl. sol. in dil. acetic acid; pract. insol. in water, alcohol; m.w. 172.09; dens. 2.31
Toxicology: Irritant; heated to decomp., emits acrid smoke and irritating fumes
Uses: Dietary supplement, mineral supplement in cereals, foods, animal feeds; stabilizer migrating from food pkg.
Regulatory: FDA 21CFR §181.29
Manuf./Distrib.: Albright & Wilson Am.; Lohmann; Rhone-Poulenc Food Ingreds.
Trade names: Caliment Dicalcium Phosphate Dihydrate

Calcium phosphate monobasic. *See* Calcium phosphate monobasic anhydrous

Calcium phosphate monobasic anhydrous
CAS 7758-23-8; EEC E341a
Synonyms: MCP/A; Monocalcium phosphate anhydrous; Calcium tetrahydrogen diorthophosphate; Phosphoric acid calcium salt (2:1); Calcium phosphate monobasic; Acid calcium phosphate; Calcium biphosphate
Formula: Ca(H₂PO₄)₂
Properties: wh. powd.; m.w. 234.05
Uses: Leavening acid for self-rising flour, self-rising corn meal, pancake flour, prepared cake and muffin mixes
Usage level: ADI no limit (EEC)
Regulatory: Europe listed
Manuf./Distrib.: Albright & Wilson Am.; Browning; Rhone-Poulenc Food Ingreds.
Trade names: Fermaloid Yeast Food; Py-ran®; V-90®
Trade names containing: Veg-X-Ten®

Calcium phosphate monobasic monohydrate
CAS 10031-30-8; EINECS 231-837-1; EEC E341
Synonyms: MCP; Monocalcium phosphate monohydrate; Calcium dihydrogen phosphate; Acid calcium phosphate; Calcium bis(dihydrogenphsophate) monohydrate; Calcium biphosphate
Definition: Phosphoric acid calcium salt (2:1)
Empirical: CaH₄O₈P₂ • H₂O
Formula: CaH₄(PO₄)₂ • H₂O
Properties: Colorless pearly scales or powd., strong acid taste; deliq. in air; sol. in water and acids; m.w. 252.08; dens. 2.22 (18/4 C); m.p. loses water at 100 C, dec. 200 C
Precaution: Hygroscopic
Toxicology: Nuisance dust
Uses: Leavening acid in food prods., emulsifier, buffer, dough conditioner, firming agent, nutrient, dietary/ mineral supplement, yeast food, sequestrant, acidulant for foods, beverages; to control pH in malt; stabilizer migrating from food pkg.
Regulatory: FDA 21CFR §136.110, 136.115, 136.130, 136.160, 136.165, 136.180, 137.80, 137.165, 137.175, 137.270, 150.141, 150.161, 155.200, 175.300, 181.29, 182.1217, 182.5217, 182.6215, 182.8217, GRAS; Japan approved
Manuf./Distrib.: FMC; Rhone-Poulenc Food Ingreds.
Trade names: Ajax®; HT® Monocalcium Phosphate, Monohydrate (MCP) Spray Dried Coarse Granular; HT® Monocalcium Phosphate, Monohydrate (MCP) Spray Dried Fines; HT® Monocalcium Phosphate, Mono- hydrate (MCP) Spray Dried Medium Granular; Regent® 12XX

Calcium phosphate tertiary. *See* Calcium phosphate tribasic

Calcium phosphate tribasic
CAS 7758-87-4, 12167-74-7; EINECS 231-840-8; EEC E341c
Synonyms: TCP; Tricalcium phosphate; Tribasic calcium phosphate; Calcium hydroxide phosphate; Tricalcium orthophosphate; Calcium orthophosphate; Calcium phosphate tertiary; Precipitated calcium phosphate
Empirical: $Ca_3(PO_4)_2$
Formula: $3Ca_3(PO_4)_2 \cdot Ca(OH)_2$
Properties: Wh. cryst. powd.; odorless, tasteless; sol. in acid; insol. in water, alcohol, acetic acid; m.w. 1004.64; dens. 3.18; m.p. 1670 C; ref. index 1.63
Toxicology: Skin and eye irritant; nuisance dust
Uses: Anticaking agent, free-flow promoter for powds., dry beverage, table salt; buffer, mineral/dietary supplement, nutrient, fat rendering aid, stabilizer; yeast nutrient, chewing gum base, raising agent, emulsifier (Japan)
Usage level: ADI 0-70 mg/kg, total phosphorus intake (FAO/WHO)
Regulatory: FDA 21CFR §137.105, 137.155, 137.160, 137.165, 137.170, 137.175, 137.180, 137.185, 169.179, 169.182, 175.300, 181.29, 182.1217, 182.5217, 182.8217, GRAS; USDA 9CFR §318.7; Japan approved (1% max. as calcium), restricted; Europe listed; UK approved
Manuf./Distrib.: Albright & Wilson Am.; Browning; FMC; Lohmann; Rhone-Poulenc Food Ingreds.
Trade names containing: Lamemul® K 1000-Range

Calcium phosphoglycerate. *See* Calcium glycerophosphate

Calcium phytate
Synonyms: Hexacalcium phytate
Formula: $C_6H_6(CaPO_4)_6$
Properties: Wh. free-flowing powd.; sl. sol. in water
Uses: Sequestering agent used to remove excess metals from wine and vinegar

Calcium propionate
CAS 4075-81-4 (anhyd.); EINECS 223-795-8; EEC E282
Synonyms: Propionic acid, calcium salt; Propanoic acid, calcium salt
Definition: Calcium salt of propionic acid
Empirical: $C_6H_{10}CaO_4$
Formula: $Ca(OOCCH_2CH_3)_2$
Properties: White cryst. or cryst. powd., faint odor of propionic acid, sol. in water; sl. sol. in alcohol; m.w. 186.23; m.p. > 300 C
Precaution: Hygroscopic
Toxicology: Irritant; heated to decomp., emits acrid smoke and irritating fumes
Uses: Direct food additive, antimicrobial agent, preservative; used in cheese, confections, frostings, gelatins, puddings, fillings, jams, jellies; antimycotic migrating from food pkg.
Usage level: ADI no limit (EEC)
Regulatory: FDA 21CFR §133.123, 133.124, 133.173, 133.179, 136.110, 136.115, 136.130, 136.160, 136.180, 150.141, 150.161, 179.45, 181.22, 181.23, 184.1221, GRAS; USDA 9CFR §318.7, 0.32% max. on wt. of flour, 381.147, 0.3% max. on wt. of flour in fresh pie dough; Japan approved with limitations; Europe listed
Manuf./Distrib.: Browning; Gist-brocades Food Ingreds.; Lohmann; MLG Enterprises
Trade names: Niacet Calcium Propionate FCC
Trade names containing: GP-1200; No Stick Emulsifier; PAV

Calcium pyrophosphate
CAS 7790-76-3; EINECS 232-221-5
Synonyms: Calcium diphosphate; Diphosphoric acid, calcium salt (1:2)
Classification: Inorganic salt
Properties: Wh. fine powd.; sol. in dilute HCl, insol. in water; dens. 3.09; m.p. 1230 C
Toxicology: Nuisance dust
Uses: Buffer, neutralizing agent, nutrient, dietary supplement
Regulatory: FDA 21CFR §182.5223, 182.8223, GRAS

Calcium resinate
CAS 9007-13-0
Synonyms: Limed rosin
Formula: $Ca(C_{44}H_{62}O_4)_2$
Properties: Ylsh.-wh. powd., rosin odor; sol. in acid; insol. in water; m.w. 1349.5
Precaution: Flamm. solid; reactive with oxidizing materials

Calcium ricinoleate

Toxicology: Heated to decomp., emits acrid smoke and fumes
Uses: Color diluent for egg shells
Regulatory: FDA 21CFR §73.1

Calcium ricinoleate
Definition: Derived from castor oil
Formula: Ca[CH$_3$(CH$_2$)$_5$CHOHCH$_2$CHCH(CH$_2$)$_7$CO$_2$]$_2$
Properties: Wh. powd., sl. odor of fatty acids; sol. in alcohols, glycols, ether-alcohols; dens. 1.04; m.p. 85 C
Precaution: Combustible
Toxicology: Heated to decomp., emits acrid smoke and irritating fumes
Uses: Stabilizer migrating from food pkg.
Regulatory: FDA 21CFR §181.29

Calcium saccharin
CAS 6485-34-3; EINECS 229-349-9
Synonyms: 1,2-Benzisothiazol-3(2H)-one, 1,1-dioxide, calcium salt
Classification: Organic compd.
Empirical: C$_{14}$H$_8$CaN$_2$O$_6$S$_2$ • 3^1/$_2$H$_2$O
Properties: Wh. cryst. powd., faint aromatic odor; sol. in water; m.w. 467.48
Toxicology: Heated to decomp., emits toxic fumes of NO$_x$
Uses: Nonnutritive sweetener for beverages, drink mixes, processed foods, chewable vitamin tablets, chewing gum, baked goods; as sugar substitute for cooking or table use
Usage level: Limitation 12 mg/fluid oz (beverages), 30 mg/serving (processed foods)
Regulatory: FDA 21CFR §180.37; USDA 9CFR §318.7 (limitation 0.01% in bacon)
Manuf./Distrib.: Boots; Int'l. Sourcing
Trade names: Syncal® CAS

Calcium silicate
CAS 1344-95-2, 10101-39-0; EINECS 215-710-8; EEC E552
Synonyms: Silicic acid, calcium salt
Definition: Hydrous or anhydrous silicate with varying proportions of calcium oxide and silica
Empirical: Common forms: CaSiO$_3$, Ca$_2$SiO$_4$, Ca$_3$SiO$_5$
Properties: White or cream-colored powd.; pract. insol. in water; dens. 2.10; bulk dens. 15-16 lb/ft^3; absorp. power 600% (water); surf. area 95-175 m^2/g
Toxicology: Nuisance dust
Uses: Anticaking agent, filter aid for foods (baking powd., table salt); glazing, polishing, release agent (sweets); dusting agent (chewing gum); coating agent (rice); suspending agent; antacid (pharmacology)
Usage level: ADI not specified (FAO/WHO)
Regulatory: FDA 21CFR §172.410, 175.300, 177.1460, 182.2227, GRAS (limitation 2% in table salt, 5% in baking powd.); Europe listed; UK approved
Manuf./Distrib.: Crosfield; Degussa
Trade names containing: FloAm® 300; Lecigran™ Super A

Calcium sodium aluminosilicate hydrated. *See* Sodium calcium silicoaluminate, hydrated

Calcium/sodium stearoyl lactylate
Definition: Calcium/sodium salt derived from stearic and lactic acids
Uses: Dough strengthener, emulsifier, processing aid, surfactant, stabilizer, formulation aid, whipping agent, conditioning agent
Trade names: Pationic® 925

Calcium sorbate
EEC E203
Synonyms: 2,4-Hexadienoic acid calcium salt
Definition: Calcium salt of sorbic acid
Empirical: CaC$_{12}$H$_{14}$O$_4$
Properties: Solid; sl. sol. in water
Toxicology: Heated to decomp., emits acrid smoke and irritating fumes
Uses: Preservative, mold retardant; used in cheese, margarine, pkg. materials
Usage level: Limitation 0.1% (alone), 0.2% (in combination with its salts or benzoic acid or its salts); ADI 0-2.5 mg/kg (EEC)
Regulatory: FDA 21CFR §182.3225, GRAS; USDA 9CFR §318.7 (not allowed in cooked sausage); Europe listed

Calcium stearate
CAS 1592-23-0; EINECS 216-472-8; EEC E470

Synonyms: Calcium octadecanoate; Stearic acid calcium salt
Definition: Calcium salt of stearic acid
Empirical: $C_{18}H_{35}O_2 \cdot {}^1/_2Ca$
Formula: $Ca(C_{18}H_{35}O_2)_2$
Properties: White powd.; insol. in water; sl. sol. in hot alcohol, hot veg. and min. oils; m.w. 707.00; bulk dens. 20 lb/ft³; m.p. 149 C
Toxicology: Heated to decomp., emits acrid smoke and irritating fumes
Uses: Direct food additive, anticaking agent, binder, emulsifier, flavoring agent, adjuvant, lubricant, release agent, stabilizer, thickener; in beet sugar and yeast defoamers; stabilizer migrating from food pkg.
Usage level: ADI not specified (FAO/WHO)
Regulatory: FDA 21CFR §169.179, 169.182, 172.863, 173.340; must conform to FDA specs for fats or fatty acids derived from edible oils, 175.105, 175.300, 175.320, 176.170, 176.200, 176.210, 177.1200, 177.2260, 177.2410, 177.2600, 178.2010, 179.45, 181.22, 181.29, 184.1229, GRAS
Manuf./Distrib.: Lohmann; PPG Industries
Trade names: Nuodex S-1421 Food Grade; Nuodex S-1520 Food Grade; Synpro® Calcium Stearate Food Grade
Trade names containing: Panipower III

Calcium stearoyl lactylate
CAS 5793-94-2; EINECS 227-335-7; EEC E482
Synonyms: Calcium stearoyl-2-lactylate; Calcium stearyl-2-lactylate; Calcium stelate
Definition: Calcium salt of stearic acid ester of lactyl lactate
Empirical: $C_{24}H_{44}O_6 \cdot {}^1/_2Ca$
Properties: Cream-colored nonhygroscopic powd., caramel odor; sparingly sol. in water; m.w. 895.30; HLB 5-6; acid no. 50-86
Toxicology: When heated to decomp., emits acrid smoke and irritating fumes
Uses: Dough conditioner for bakery prods.; stabilizer; whipping agent; conditioning agent for dehydrated potatoes; emulsifier (Japan)
Usage level: Limitation 0.5 parts/100 parts flour (bakery prods.), .05% (egg. white), 0.3% (whipped vegetable topping), 0.5% (dehydrated potatoes); ADI 0-20 mg/kg (EEC)
Regulatory: FDA 21CFR §172.844; must conform to FDA specs for fats or fatty acids derived from edible oils; Japan approved with restrictions; Europe listed; UK approved
Trade names: Admul CSL 2007, CSL 2008; Artodan CF 40; Crolactil CS2L ; Lamegin® CSL; Paniplex CK; Pationic® 930; Pationic® 940; Prefera® CSL; Verv®
Trade names containing: MG-S; Patco® 3

Calcium stearoyl-2-lactylate. *See* Calcium stearoyl lactylate
Calcium stearyl-2-lactylate. *See* Calcium stearoyl lactylate
Calcium stelate. *See* Calcium stearoyl lactylate

Calcium sulfate
CAS 7778-18-9 (anhyd.), 10101-41-4 (dihydrate); EINECS 231-900-3; EEC E516
Synonyms: Anhydrite (natural form); Calcium sulfonate; Gypsum; Anhydrous gypsum; Plaster of Paris
Classification: Inorganic salt
Empirical: CaO_4S
Formula: $Ca \cdot H_2O_4S$
Properties: Wh. to sl. yel.-wh. powd. or crystals, odorless; sl. sol. in water; m.w. 136.14; dens. 2.964; m.p. 1450 C
Precaution: Reacts violently with aluminum when heated; mixts. with phosphorus ignite at high temps.
Toxicology: Irritant; heated to decomp., emits toxic fumes of SO_x
Uses: Nutrient/dietary supplement, yeast food, dough conditioner/strengthener, firming agent, sequestrant, anticaking, colorant, leavening agent, pH control agent, processing aid, stabilizer, texturizer, thickener, drying agent, synergist, flour treating
Usage level: Limitation 1.3% (baked goods), 3% (frostings), 0.5% (frozen dairy desserts), 0.4% (gelatins, puddings), 0.5% (grain prods., pasta), 0.35% (processed vegetables), 0.07% (other food); ADI not specified (EEC)
Regulatory: FDA 21CFR §133, 133.102, 133.106, 133.111, 133.141, 133.165, 133.181, 133.195, 137.105, 137.155, 137.160, 137.165, 137.170, 137.175, 137.180, 137.185, 150.141, 150.161, 155.200, 175.300, 177.1460, 184.1230, GRAS; BATF 27CFR §240.1051, limitation 16.69 lb/1000 gal; Japan approved with restrictions (1% max.); Europe listed; UK approved
Manuf./Distrib.: Am. Roland
Trade names: Terra Alba 114836
Trade names containing: Ascorbo-160-2 Powd.; Bagel Base; Basic Natural™; Cain's PDC (Pizza Dough Conditioner); Drize-P Powd.; Eversoft Plus Kosher; FD-6; FD-8; Merecol® R; Muffit; Reduce®-150; Rich-

Calcium sulfonate

Pak® Powd. 160; Rich-Pak® Powd. 160-M; WaTox-20 P Powd.; WaTox 60 Powd.

Calcium sulfonate. *See* Calcium sulfate
Calcium superoxide. *See* Calcium peroxide
Calcium tetrahydrogen diorthophosphate. *See* Calcium phosphate monobasic anhydrous
C-8 alcohols. *See* Caprylic alcohol

Calendula
Synonyms: Calendula officinalis; Pot marigold
Definition: Derived from flowers of *Calendula officinalis*
Uses: Natural flavoring agent
Regulatory: FDA 21CFR §182.10, GRAS
Manuf./Distrib.: Chart

Calendula extract
CAS 84776-23-8; EINECS 283-949-5
Synonyms: Calendula officinalis extract; Marigold extract; Extract of calendula
Definition: Extract of flowers of *Calendula officinalis*
Uses: Flavoring agent
Regulatory: FDA GRAS; not listed as approved colorant for cosmetics under FDA 21CFR §73 and 74
Manuf./Distrib.: Chart

Calendula officinalis. *See* Calendula
Calendula officinalis extract. *See* Calendula extract

Calumba root
Synonyms: Colombo
Definition: Derived from *Jateorhiza palmata*
Properties: Odorless, bitter tonic flavor
Uses: Natural flavoring; used in alcoholic beverages
Regulatory: FDA 21CFR §172.510

Calx. *See* Calcium oxide
Camelia oleifera extract. *See* Thea sinensis extract
Camellia sinensis extract. *See* Thea sinensis extract
d-Camphanol. *See* Borneol
2-Camphanone. *See* Camphor
2-Camphanyl acetate. *See* Isobornyl acetate

Camphene
CAS 79-92-5; FEMA 2229
Synonyms: 2,2-Dimethyl-3-methylenenorbornane; 3,3-Dimethyl-2-methylene norcamphane
Empirical: $C_{10}H_{16}$
Properties: Colorless cubic cryst., oily odor, terpene camphoraceous taste; sol. in alcohol; misc. with fixed oils; insol. in water; m.w. 136.26; dens. 0.842 (54/4 C); m.p. 50-51 C; b.p. 159 C; ref. index 1.452 (55 C)
Precaution: Combustible; yields flamm. vapors when heated; reacts with oxidizing materials
Toxicology: Mutagenic data; heated to decomp., emits acrid smoke and fumes
Uses: Synthetic flavoring agent
Usage level: 40-90 ppm (nonalcoholic beverages), 20 ppm (ice cream, ices), 160 ppm (candy), 27 ppm (baked goods)
Regulatory: FDA 21CFR §172.515
Manuf./Distrib.: Aldrich; Chr. Hansen's

Camphor
CAS 76-22-2, 464-49-3 (+); EINECS 207-355-2; FEMA 2230
Synonyms: 1,7,7-Trimethylbicyclo[2.2.1] heptan-2-one; Gum camphor; 2-Camphanone; 2-Bornanone
Definition: Ketone derived from wood of the camphor tree, *Cinnamomum camphora* or prepared synthetically
Empirical: $C_{10}H_{16}O$
Properties: Translucent cryst. mass, char. fragrant penetrating odor, sl. bitter and cooling taste; m.w. 152.24; dens. 0.992 (25/4 C); m.p. 179.7 C; b.p. 204 C; sublimes @ ambient temp. and pressure; flash pt. 150 F
Precaution: Combustible; keep tightly closed away from heat; incompat. with potassium permanganate
Toxicology: LD50 (IP, mouse) 3000 mg/kg; ingestion by humans may cause nausea, vomiting, vertigo, mental confusion, delirium, coma, respiratory failure, death
Uses: Flavoring
Regulatory: FDA 21CFR §172.510, 172.515, 175.105, 27CFR §21.65, 21.151; Japan approved
Manuf./Distrib.: Chart

Canadian balsam. *See* Balsam Canada

Cananga. *See* Cananga oil

Cananga oil
Synonyms: Cananga
Definition: Cananga odorata
Properties: Sl. woody floral odor, burning taste; sp.gr. 0.906-0.923 (20/20 C); ref. index 1.495-1.503 (20 C)
Toxicology: Heated to decomp., emits acrid smoke and fumes
Uses: Natural flavoring agent
Usage level: 7.0 ppm (nonalcoholic beverages), 1.0 ppm (ice cream, ices), 2.0 ppm (candy, baked goods)
Regulatory: FDA 21CFR §182.20, GRAS
Manuf./Distrib.: Florida Treatt

Candelilla wax
CAS 8006-44-8; EINECS 232-347-0
Definition: Wax from various *Euphorbiaceae* species
Properties: Yel.-brown to translucent solid; pract. insol. in water; sparingly sol. in alcohol; sol. in acetone, benzene, carbon disulfide, hot petrol. ether, gasoline, oils, turpentine, CCl_4; dens. 0.983; m.p. 67-68 C; ref. index 1.4555; sapon. no. 50-65
Precaution: Combustible
Toxicology: Heated to decomp., emits acrid smoke and irritating fumes
Uses: Direct food additive, lubricant, surface-finishing agent, protective coating for citrus; masticatory substance in chewing gum base; glazing agent; also used in hard candy
Regulatory: FDA 21CFR §175.105, 175.320, 176.180, 184.1976, GRAS; Japan approved
Manuf./Distrib.: Penta Mfg.
Trade names: Koster Keunen Candelilla; Ross Candelilla Wax

Candida guilliermondii.
Toxicology: Heated to decomp., emits acrid smoke and irritating fumes
Uses: Enzyme used in fermentation prod. of citric acid; production aid
Regulatory: FDA 21CFR §173.160

Candida lipolytica
Toxicology: Heated to decomp., emits acrid smoke and irritating fumes
Uses: Enzyme used in fermentation prod. of citric acid; production aid
Regulatory: FDA 21CFR §173.165

Candlenut oil. *See* Kukui nut oil
Cane sugar. *See* Sucrose

Canola oil
CAS 8002-13-9, 120962-03-0
Definition: Low-erucic rapeseed oil
Properties: Oil, odorless, bland flavor; iodine no. 110-120
Uses: Food ingred.
Manuf./Distrib.: Arista Industries; C&T Refinery; Calgene; Canada Packers; Cargill; Colombus Foods; De Choix Specialty; Karlshamns; Peerless Refining; Penta Mfg.; Spectrum Naturals; Tri-K Industries; Wensleydale Foods
Trade names: Canola Spray Oil 81-599-0; Canola Vegetable Frying Shortening 81-573-0; Canola Vegetable Shortening 81-577-0; Design NH; H.L. 94; Liquid Canola Oil 81-071-0; Lobra RBD
Trade names containing: Canola Vegetable All-Purpose Shortening 81-706-0; Neobee® SL-220; Superb® Canola Salad Oil 104-450

Canola oil glyceride
Definition: Monoglyceride derived from canola oil
Uses: Food emulsifier, antifoaming agent for food processing; emulsion stabilizer for diet spreads; aerator for icings; flavor improver; oil stabilizer
Trade names: Myverol® 18-99

Canola oil (low erucic acid rapeseed oil). *See* Rapeseed oil
Cantha. *See* Canthaxanthine
Canthaxanthin. *See* Canthaxanthine

Canthaxanthine
CAS 514-78-3; EINECS 208-187-2; EEC E161g
Synonyms: Cantha; Canthaxanthin; 4,4´-Diketo-β-carotene; β-Carotene-4,4´-dione; Food Orange 8; CI 40850
Definition: Carotenoid colorant
Empirical: $C_{40}H_{52}O_2$

Properties: Violet cryst. solid; sol. in chloroform and various oils; very sl. sol. in acetone; insol. in water; m.w. 564.86; m.p. 211-213 C (dec.)

Precaution: Sensitive to light and oxygen; store under inert gas at low temps.

Toxicology: Oral intake may cause loss of night vision; patients at risk for retinopathy should avoid excessive use; heated to decomp., emits acrid smoke and irritating fumes

Uses: Food additive, colorant; for pigmentation of egg yolks, broiler parts, tomato prods., salad dressings, fruit drinks, sausage prods., baked goods

Usage level: Limitation 30 mg/lb of solid or pint of liq. food, 4.41 mg/kg of complete food; 5-60 ppm (as pure color); ADI 0-25 mg/kg (EEC, 1974)

Regulatory: FDA 21CFR §73.75, 73.1075; Europe listed; UK approved

Manuf./Distrib.: Am. Roland

Trade names: Lucantin® Red

Trade names containing: Canthaxanthin 10% Type RVI No. 66523; Canthaxanthin Beadlets 10%; Dry Canthaxanthin 10% SD No. 66514; Roxanthin® Red 10WS No. 66515

Capers

Definition: Derived from *Capparis spinosa*

Properties: Sour astringent flavor

Uses: Natural flavoring agent; buds are often pickled and used in cooking

Regulatory: FDA 21CFR §182.10, GRAS; Japan approved

Capraldehyde. *See* Decanal

Capraldehyde dimethyl acetal. *See* Decanal dimethyl acetal

Capric acid

CAS 334-48-5; EINECS 206-376-4

Synonyms: n-Decanoic acid; Decoic acid; n-Capric acid; Carboxylic acid C_{10}

Classification: Fatty acid

Empirical: $C_{10}H_{20}O_2$

Formula: $CH_3(CH_2)_8COOH$

Properties: White crystals, unpleasant odor; sol. in org. solvs.; insol. in water; m.w. 172.27; dens. 0.8858 (40 C); m.p. 31.5 C; b.p. 270 C; ref. index 1.4288 (40 C)

Precaution: Combustible

Toxicology: LD50 (IV, mouse) 129 mg/kg; poison by IV route; mutagenic data; skin irritant; heated to decomp., emits acrid smoke and irritating fumes

Uses: Mfg. of food-grade additives, defoaming agent, lubricant

Regulatory: FDA 21CFR §172.860; GRAS (FEMA)

Manuf./Distrib.: Acme-Hardesty

Trade names: Emery® 6359

n-Capric acid. *See* Capric acid
Capric aldehyde. *See* Decanal

Capric triglyceride

Uses: Food emulsifier

Trade names: Coconad SK; Delios® C; Neobee® C-10

Caprinaldehyde. *See* Decanal

Caproic acid

CAS 142-62-1; EINECS 205-550-7; FEMA 2559

Synonyms: Carboxylic acid C-6; Hexanoic acid; Butylacetic acid; Pentylformic acid

Empirical: $C_6H_{12}O_2$

Formula: $CH_3(CH_2)_4COOH$

Properties: Colorless oily liq., odor of Limburger cheese; very sol. in ether, fixed oils; sl. sol. in water; m.w. 116.18; dens. 0.9295 (20/20 C); f.p. -3.4 C; b.p. 205 C; flash pt. (COC) 215 F; ref. index 1.415-1.418

Precaution: Combustible

Uses: Synthetic flavoring agent

Usage level: 1.8 ppm (nonalcoholic beverages), 4.3 ppm (ice cream, ices), 28 ppm (candy), 22 ppm (baked goods), 1.5 ppm (chewing gum), 450 ppm (condiments)

Regulatory: FDA 21CFR §172.515; GRAS (FEMA); Japan approved as flavoring

Manuf./Distrib.: Acme-Hardesty

Caproic acid allyl ester. *See* Allyl caproate
Caproic aldehyde. *See* Hexanal
γ-Caprolactone. *See* γ-Hexalactone
Capronaldehyde. *See* Hexanal

Capryl alcohol. *See* Caprylic alcohol

Caprylaldehyde. *See* n-Octanal

Caprylic acid
CAS 124-07-2; EINECS 204-677-5; FEMA 2799
Synonyms: n-Octanoic acid; Octanoic acid; Octoic acid; n-Octylic acid; C-8 acid
Classification: Fatty acid
Empirical: $C_8H_{16}O_2$
Formula: $CH_2(CH_2)_6COOH$
Properties: Colorless leaf or oily liq.; unpleasant odor; sl. sol. in water; sol. in alcohol, chloroform, ether, carbon disulfide, petrol. ether, glacial acetic acid; m.w. 144.21; dens. 0.91 (20/4 C); m.p. 16 C; b.p. 237.5 C; ref. index 1.4280
Toxicology: LD50 (oral, rat) 10,080 mg/kg; moderately toxic by intravenous route; mildly toxic by ingestion; skin irritant; yields irritating vapors which can cause coughing; heated to decomp., emits acrid smoke and irritating fumes
Uses: Flavoring agent, adjuvant, antimicrobial, mfg. of food-grade additives, defoamer, lubricant; antimicrobial preservative in cheese wraps
Usage level: Limitation 0.013% (baked goods), 0.04% (cheese), 0.005% (fats, oils, frozen dairy desserts, gelatins/puddings, meat prods., soft candy), 0.016% (snack foods), 0.001% (other food)
Regulatory: FDA 21CFR §172.210, 172.860, 173.340, 175.105, 175.320, 176.170, 176.200, 176.210, 177.1010, 177.1200, 177.2260, 177.2600, 177.2800, 178.3570, 178.3910, 184.1025, 186.1025,GRAS; GRAS as indirect additive; GRAS (FEMA)
Manuf./Distrib.: Acme-Hardesty
Trade names: Emersol® 6357

Caprylic acid, 1,2,3-propanetriyl ester. *See* Tricaprylin

Caprylic alcohol
CAS 111-87-5; EINECS 203-917-6; FEMA 2800
Synonyms: n-Octyl alcohol; Heptyl carbinol; 1-Octanol; n-Octanol; C-8 alcohols; Alcohol C-8; Capryl alcohol
Classification: Fatty alcohol
Empirical: $C_8H_{18}O$
Formula: $CH_3(CH_2)_6CH_2OH$
Properties: Colorless liq., fresh orange-rose odor, sl. herbaceous taste; sol. in water; misc. in alcohol, ether, and chloroform; m.w. 130.26; dens. 0.827; m.p. -16.7 C; b.p. 194.5 C; flash pt. 178 F; ref. index 1.429 (20 C)
Precaution: Combustible liq. when exposed to heat or flame; can react with oxidizers
Toxicology: LD50 (oral, mouse) 1790 mg/kg; poison by intravenous route; moderately toxic by ingestion; skin irritant
Uses: Synthetic flavoring agent, intermediate, solvent for beverages, candy, gelatin desserts, ice cream, pudding mixes
Usage level: Limitation 16 ppm max. residual in citric acid
Regulatory: FDA 21CFR §172.230, 172.515 (only for encapsulating lemon, dist. lime, orange, peppermint, and spearmint oils), 172.864, 173.280, 175.105, 175.300, 176.210, 177.1010, 177.1200, 177.2800, 178.3480; GRAS (FEMA)
Manuf./Distrib.: Aldrich; Penta Mfg.

Caprylic aldehyde. *See* n-Octanal

Caprylic/capric acid
CAS 67762-36-1
Uses: food grades as lubricant, release agent, binder, defoamer in foods, intermediate for food emulsifiers
Trade names: Emery® 6358; Industrene® 365

Caprylic/capric glycerides
CAS 26402-26-6
Definition: Mixture of mono, di and triglycerides of caprylic and capric acids
Uses: Antifoam for prod. of sweets, preserves, jams, potato prods. and flavorings
Trade names: Witafrol® 7420

Caprylic/capric/lauric triglyceride
CAS 68991-68-4
Definition: Mixed triester of glycerin with caprylic, capric and lauric acids
Uses: Emollient, solv., carrier, fixative, and extender for nutritional applics.
Trade names: Captex® 350

383

Caprylic/capric/linoleic triglyceride
CAS 67701-28-4
Definition: Mixed triester of glycerin with caprylic, capric, and linoleic acids
Uses: Emollient, solv., carrier, fixative, and extender for nutritional applics.; carrier for flavors and fragrances
Trade names: Captex® 810A; Captex® 810B; Captex® 810C; Captex® 810D

Caprylic/capric/oleic triglyceride
CAS 67701-28-4
Uses: Emollient, solvent, and extender in nutritional applics.; carrier for flavors and fragrances
Trade names: Captex® 910A; Captex® 910B; Captex® 910C; Captex® 910D

Caprylic/capric triglyceride
CAS 65381-09-1, 85409-09-2, 73398-61-5; EINECS 265-724-3
Synonyms: Octanoic/decanoic acid triglyceride
Definition: Mixed triester of glycerin and caprylic and capric acids
Uses: Emulsifier
Trade names: Aldo® MCT KFG; Aldo® TC; Captex® 300; Captex® 355; Coconad PL; Coconad RK; Delios®
  S; Delios® V; Estasan GT 8-40 3578; Estasan GT 8-60 3575; Estasan GT 8-60 3580; Estasan GT 8-65
  3577; Estasan GT 8-65 3581; Hodag CC-33; Hodag CC-33-F; Hodag CC-33-L; Hodag CC-33-S; Lexol®
  GT-855; Lexol® GT-865; Liponate GC; Miglyol® 810; Miglyol® 810 N; Miglyol® 812; Miglyol® 812 N;
  Miglyol® 812 S; Miglyol® 8108; Neobee® 1053; Neobee® M-5; Neobee® O; Standamul® 318
Trade names containing: Palma-Sperse® Type 250-S No. 65322; Palma-Sperse® Type 250A/50 D-S No.
  65221; Phosal® 53 MCT

Caprylic triglyceride
Uses: Food emulsifier; coffee whitener; flavor diluting agent
Trade names: Coconad MT

Capryloamphocarboxyglycinate. *See* Disodium capryloamphodiacetate
Capryloamphodiacetate. *See* Disodium capryloamphodiacetate
Capsicin. *See* Capsicum oleoresin

Capsicum
EINECS 297-599-6 (*C. annuum* extract)
Synonyms: Cayenne pepper; Red pepper
Definition: Plant material derived from dried ripe fruit of *Capsicum frutescens* or *C. annuum*
Uses: OTC drug active ingred.; natural flavoring agent
Usage level: 15-240 ppm (nonalcoholic beverages), 270 ppm (baked goods), 630 ppm (condiments), 310 ppm
  (meats), 11-59 ppm (pickles)
Regulatory: FDA 21CFR §182.10, GRAS; Japan approved

Capsicum extract
CAS 84625-29-6; EINECS 288-920-0
Synonyms: Capsicum frutescens extract
Definition: Extract from *Capsicum frutescens*
Uses: Natural flavoring agent
Usage level: 120 ppm (nonalcoholic beverages), 15 ppm (ice cream, ices), 12 ppm (candy), 12-14 ppm (baked
  goods), 50-100 ppm (condiments), 200 ppm (meats)
Regulatory: FDA §182.20, GRAS

Capsicum frutescens extract. *See* Capsicum extract

Capsicum oleoresin
CAS 8023-77-6
Synonyms: Capsicin; Oleoresin capsicum Africanus
Definition: Resinous extract of *Capsicum frutescens*
Uses: Natural flavoring
Regulatory: FDA 21CFR §182.20

Caramel
CAS 8028-89-5; EINECS 232-435-9; FEMA 2235; EEC E150
Synonyms: Burnt sugar coloring; Burnt sugar; Caramel color; Natural Brown 10
Definition: Conc. sol'n. obtained from heating sucrose or glucose sol'ns.
Properties: Dk. brn. to black liq. or solid, burnt sugar odor, pleasant bitter taste; sol. in water (colloidal); insol.
  in most org. solvs.; sp.gr. 1.25-1.38
Toxicology: Mutagenic data; may reduce wh. blood cells and destroy vitamin B_6; heated to decomp., emits acrid
  smoke and irritating fumes
Uses: Food additive, synthetic flavoring, food coloring for soft drinks (esp. root beer, colas), blended whiskey,

beer, baked goods, syrups, preserves, candies, pet food, canned meat prods.; processing aid

Usage level: 2200 ppm (nonalcoholic beverages), 590 ppm (ice cream, ices), 180 ppm (candy), 220 ppm (baked goods), 2100 ppm (meats), 2800 ppm (syrups); ADI 1000-10,000 mg/kg (FAC)

Regulatory: FDA 21CFR §73.85, 73.1085, 73.2085, 182.1235, GRAS; Japan restricted; Europe listed; UK approved

Manuf./Distrib.: Booths; Bradleys; Cairn Foods; Cerestar Int'l.; Crompton & Knowles; Dena AG; Dinoval; Ellis & Everard; Flavors of N. Am.; Foote & Jenkins; Fruitsource; Frutarom; Hershey Import; Kraft; Moore Fine Foods; Pacific Foods; Penta Mfg.; Universal Flavors

Trade names: Acid Proof Caramel Powd.; B&C Caramel Powd.; Caramel Color Double Strength; Caramel Color Single Strength; Double Strength Acid Proof Caramel Colour; P147 Caramel Color; Powdered Caramel Color, Acid Proof; Powdered Caramel Colour Non-Ammoniated-All Natural T-717; Single Strength Acid Proof Caramel Colour; Unisweet Caramel

Trade names containing: Cola No. 23443; HVP 5-SD; HVP-A; Luxor® 1576; Luxor® EB-2

Caramel color. *See* Caramel

Caraway oil
CAS 8000-42-8; EINECS 288-921-6 (extract)
Synonyms: Carum carvi oil
Definition: Volatile oil distilled from the dried ripe fruit of *Carum carvi*, contg. carvone, d-limonene
Properties: Colorless to pale yel. liq., caraway odor and taste; darkens and thickens with age; sol. in 8 vols. 80% alcohol; pract. insol. in water; dens. 0.900-0.910 (25/25 C); ref. index 1.485-1.497 (20 C)
Precaution: Keep cool, well closed; protect from light
Toxicology: LD50 (oral, rat) 3500 mg/kg; mod. toxic by ingestion, skin contact; skin irritant; mutagenic data; heated to decomp., emits acrid smoke and irritating fumes
Uses: Natural flavoring agent for cookies, candies; mfg. of liqueurs
Regulatory: FDA 21CFR §182.10, 182.20, GRAS; Japan approved
Manuf./Distrib.: Chart; Florida Treatt

Carbamide. *See* Urea
Carbamidic acid. *See* Urea
Carbinol. *See* Methyl alcohol

Carbohydrase
EINECS 232 575 0
Toxicology: Heated to decomp., emits acrid smoke and fumes
Uses: Production aid, enzyme

Carbohydrase-cellilase
Definition: Derived from *Aspergillus niger*
Toxicology: Heated to decomp., emits acrid smoke and irritating fumes
Uses: Tissue release agent for processing of clams and shrimp
Regulatory: FDA 21CFR §173.120

Carbohydrase from *Aspergillus niger*
Uses: Enzyme for removal of visceral mass in clam processing, removal of shell in shrimp processing
Regulatory: FDA 21CFR §173.120

Carbohydrase from *Rhizopus oryzae*
Toxicology: Heated to decomp., emits acrid smoke and irritating fumes
Uses: Enzyme used in prod. of dextrose from starch; production aid
Regulatory: FDA 21CFR §173.130; must be refrigerated from production to use

Carbohydrase-protease
Classification: Enzyme
Definition: Obtained from fermentation of *Bacillus licheniformis*
Properties: Brn. amorphous powd. or liq.; sol. in water; insol. i alcohol, chloroform, ether
Toxicology: Heated to decomp., emits acrid smoke and irritating fumes
Uses: Enzyme to hydrolyze proteins or carbohydrates; used in alcoholic beverages, candy, nutritive sweeteners, protein hydrolyzates, fish meal, starch syrups
Regulatory: FDA 21CFR §184.1027, GRAS

Carbon, activated
CAS 64365-11-3
Synonyms: Active carbon; Activated charcoal; Decolorizing carbon
Empirical: C
Properties: Black porous solid, coarse gran., or powd.; insol. in water, org. solvs.; m.w. 12.01; dens. 0.08-0.5
Precaution: Flamm. solid; combustible exposed to heat; dust is flamm. and explosive when exposed to heat,

flame, or oxides

Toxicology: Dust irritant, esp. to eyes and mucous membranes

Uses: Decolorizing agent, odor-removing agent, purification agent in food processing, taste-removing agent

Usage level: Limitation 0.9% (wine), 0.25% (pale dry sherry), 0.4% (red and blk. grape juice); ADI not specified (FAO/WHO)

Regulatory: FDA 21CFR §240.361, 240.365, 240.401, 240.405, 240.527, 240.527a, GRAS; BATF 27CFR §240.1051; USDA 9CFR §318.7

Manuf./Distrib.: Calgon Carbon

Trade names: ADP; APA; APC; BL™; C™; CAL®; Calgon® Type RB; Cane Cal®; Colorsorb; CPG®; CPG® LF; Diahope®-S60; GW; OL®; PWA™; RC®; SGL®

Carbonate magnesium. *See* Magnesium carbonate

Carbon black

CAS 1333-86-4; EINECS 215-609-9; EEC E153

Synonyms: Thermal black; Charcoal; Vegetable carbon; Channel black; Furnace black; Acetylene black; Lamp black; CI 77266

Definition: Finely divided particles of elemental carbon obtained by incomplete combustion of hydrocarbons (channel or impingement process)

Empirical: C

Properties: Insol. in water, ethanol, veg. oil

Toxicology: TLV 3.5 mg/m^3; low toxicity by ingestion, inhalation, and skin contact; nuisance dust in high concs.; suspected carcinogen

Uses: Purification agent in food processing; color additive

Regulatory: Banned by U.S. FDA; may be used in foods for the European community; UK approved

Manuf./Distrib.: Cabot; Degussa

Carbon dioxide

CAS 124-38-9; EINECS 204-696-9; EEC E290

Synonyms: Carbonic acid gas; Carbonic anhydride

Classification: Gas

Empirical: CO_2

Properties: Colorless gas, odorless; m.w. 44.01; m.p. subl. @ -78.5 C

Precaution: Noncombustible; various dusts explode in CO_2 atmospheres

Toxicology: TLV 5000 ppm; asphyxiant at > 10%; experimental teratogen; skin contact can cause burns

Uses: Direct food additive, preservative, aerating agent, cooling agent, freezant, leavening agent, pH control agent, processing aid, propellant, aerator; for carbonated beverages

Regulatory: FDA 21CFR §169.115, 169.140, 169.150, 184.120, GRAS, 193.45 (modified atm. for pest control); USDA 9CFR §318.7, 381.147; BATF 27CFR 240.1051; Japan approved; Europe listed; UK approved

Manuf./Distrib.: ADM Ethanol Sales

Carbonic acid ammonium salt. *See* Ammonium carbonate
Carbonic acid calcium salt (1:1). *See* Calcium carbonate
Carbonic acid dipotassium salt. *See* Potassium carbonate
Carbonic acid disodium salt. *See* Sodium carbonate
Carbonic acid gas. *See* Carbon dioxide
Carbonic acid monoammonium salt. *See* Ammonium bicarbonate
Carbonic acid monopotassium salt. *See* Potassium bicarbonate
Carbonic acid monosodium salt. *See* Sodium bicarbonate
Carbonic acid sodium salt (2:3). *See* Sodium sesquicarbonate
Carbonic anhydride. *See* Carbon dioxide
Carbonyldiamide. *See* Urea
Carbonyl iron. *See* Iron
Carboxylic acid C-4. *See* n-Butyric acid
Carboxylic acid C-5. *See* 2-Methylbutyric acid
Carboxylic acid C-5. *See* n-Valeric acid
Carboxylic acid C-6. *See* Caproic acid
Carboxylic acid C-6. *See* Diethylacetic acid
Carboxylic acid C-9. *See* Nonanoic acid
Carboxylic acid C-10. *See* Capric acid
Carboxylic acid C-18. *See* Stearic acid

Carboxymethylcellulose sodium

CAS 9004-32-4; EEC E466

Synonyms: CMC; Cellulose gum; Sodium carboxymethylcellulose; Sodium CMC

Definition: Sodium salt of the carboxylic acid R-O-CH$_2$COONa

Formula: [(C$_6$H$_7$O$_2$(OH)$_2$OCH$_2$COOH]$_n$

Properties: Colorless or white powd. or granules, odorless; water sol. depends on degree of substitution; insol. in org. liqs.; m.w. 21,000-500,000; visc. 25 cps (2% aq.); m.p. > 300 C

Toxicology: Nontoxic; LD50 (oral, rat) 27,000 mg/kg

Uses: Food stabilizer, binder, thickener, extender, boiler water additive; migrating from cotton in dry food pkg.

Usage level: ADI 0-25 mg/kg (FAO/WHO)

Regulatory: FDA 21CFR §133.134, 133.178, 133.179, 150.141, 150.161, 173.310, 175.105, 175.300, 182.70, 182.1745, GRAS; USDA 9CFR 318.7, limitation 1.5%, must be added dry; 9CFR 381.147; Japan restricted (2% max.); Europe listed; UK approved

Manuf./Distrib.: Am. Roland; Multi-Kem

Trade names: Aqualon® 7H0F; Aqualon® 7H0XF; Aqualon® 7H3SF; Aqualon® 7H3SXF; Aqualon® 7H4F; Aqualon® 7H4XF; Aqualon® 7HC4F; Aqualon® 7HCF; Aqualon® 7HF; Aqualon® 7HXF; Aqualon® 7LF; Aqualon® 7LXF; Aqualon® 7M8SF; Aqualon® 7MF; Aqualon® 9H4F; Aqualon® 9H4XF; Aqualon® 9M31F; Aqualon® 9M31XF; Aqualon® 9M8F; Aqualon® 9M8XF; Cellogen HP-5HS; Cellogen HP-6HS9; Cellogen HP-8A; Cellogen HP-12HS; Cellogen HP-SB; CMC Daicel 1150; CMC Daicel 1160; CMC Daicel 1220; CMC Daicel 1240; CMC Daicel 1260; CMC Daicel 2200; Nutriloid® Cellulose Gums; Ticalose® 15; Ticalose® 30; Ticalose® 75; Ticalose® 100; Ticalose® 150 R; Ticalose® 700 R; Ticalose® 750; Ticalose® 2000 R; Ticalose® 2500; Ticalose® 4000; Ticalose® 4500; Ticalose® 5000 R; TIC Pretested® CMC 2500 S; TIC Pretested® CMC PH-2500

Trade names containing: Crodacreme; Fat Replacer 785; Frigesa® IC 178; Lamefrost® ES 375; Lamefrost® ES 379; Vanease (K)

Carboxymethylmethylcellulose

Uses: Thickener, stabilizer, rheology control agent, film-former, suspending agent, water-retention aid, binder

Carboxyphenol. *See* Ethyl paraben

Cardamom oil

CAS 8000-66-6; EINECS 288-922-1 (extract)

Synonyms: Elettaria cardamomum oil; Cardamon oil

Definition: Volatile oil obtained from the dried ripe seeds of *Elettaria cardamomum*, contg. eucalyptol, sabinene, etc.

Properties: Colorless to pale yel. oily liq., aromatic penetrating odor of cardamom, pungent taste; sol. in ether; misc. with alcohol; insol. in water; dens. 0.917-0.947 (25/4 C); ref. index 1.4630-1.4660 (20 C)

Toxicology: Mutagenic data; heated to decomp., emits acrid smoke and fumes

Uses: Natural flavoring agent for curry sauces, confectionery, baked goods, liqueurs

Usage level: 1.9 ppm (nonalcoholic beverages), 10 ppm (alcoholic beverages), 1.3 ppm (ice cream, ices), 5.8 ppm (candy), 57 ppm (baked goods), 2.2 ppm (chewing gum), 10-16 ppm (pickles), 8 pp (condiments), 36 ppm (meats)

Regulatory: FDA 21CFR §182.10, 182.20, GRAS; Japan approved

Manuf./Distrib.: Chart; Pierre Chauvet; Florida Treatt

Cardamon oil. *See* Cardamom oil

Carmine

CAS 1390-65-4; EINECS 215-724-4; EEC E120

Synonyms: CI 75470; Carminic acid; Cochineal extract; Natural Red 4

Definition: Aluminum lake of the coloring agent, cochineal; cochineal is a natural pigment derived from the dried female insect *Coccus cacti*

Empirical: C$_{22}$H$_{20}$O$_{13}$

Properties: Bright red cryst., easily powdered; sol. in alkali, borax; insol. in dilute acids; sl. sol. in hot water; m.w. 492.39; decom. @ 250 C

Toxicology: Suspected of causing food intolerance; heated to decomp., emits acrid smoke and irritating fumes

Uses: Food colorant; useful for producing pink shades in retorted protein prods., candy, confections

Usage level: 25-1000 ppm (as food colorant); ADI 0-2.5 mg/kg (EEC)

Regulatory: FDA 21CFR §73.100, must be pasteurized to destroy *Salmonella*, 73.1100, 73.2087; GRAS (FEMA); Europe listed; UK approved

Manuf./Distrib.: Allchem Int'l.; Am. Fruit Processors; Am. Roland; Burlington Biomedical; Crompton & Knowles; Ellis & Everard; Food Additives & Ingreds.; Hilton Davis; MLG Enterprises; Penta Mfg.; Quest Int'l.; Sanofi UK; Universal Flavors; Warner-Jenkinson

Trade names: Carmacid Y; Carmine 1623; Carmine AS; Carmine Extract P4011; Carmine FG; Carmine Nacarat 40; Carmine PG; Carmine Powd. WS; Carmine XY/UF; Carmisol A; Carmisol NA; Hilton Davis Carmine; Natural Liquid AP Carmine Colorant; Natural Liquid Carmine Colorant (Type 100, 50, and Simple); Natural Soluble Carmine Powd.; Natural Soluble Powder AP Carmine Colorant

Carmine extract

Trade names containing: Red Soluble Powd. Natural Colorant

Carmine extract
CAS 1390-65-4; EINECS 215-724-4; EEC E120
Synonyms: CI 75470
Definition: Alum lake of carminic acid
Properties: Deep purple liq.
Uses: Food colorant
Regulatory: JEFCA compliance

Carminic acid. *See* Carmine

Carnauba
CAS 8015-86-9; EINECS 232-399-4; EEC E903
Synonyms: Brazil wax; Carnaubawax
Definition: Exudate from leaves of Brazilian wax palm tree *Copernicia prunifera*
Properties: Yel. greenish brown lumps, solid, characteristic odor; sol. in ether, boiling alcohol and alkalies; insol. in water; dens. 0.995 (15/15 C); m.p. 82-85.5 C; sapon. no. 78-89; ref. index 1.4500
Toxicology: Heated to decomp., emits acrid smoke and irritating fumes
Uses: Direct food additive, anticaking agent, candy glaze/polish, coating agent, formulation aid, lubricant, release agent, surface-finishing agent, chewing gum base; in baked goods, chewing gum, confections, fresh and processed fruits, gravies, soft candy
Regulatory: FDA 21CFR §175.320, 184.1978, GRAS; Japan approved; Europe listed (permitted only in chocolate prods.); UK approved
Manuf./Distrib.: Penta Mfg.; C.E. Roeper GmbH
Trade names: Koster Keunen Carnauba; Ross Carnauba Wax

Carnaubawax. *See* Carnauba

L-Carnitine
CAS 541-15-1; EINECS 208-768-0
Synonyms: Vitamin B$_7$; β-Hydroxy-γ-trimethylaminobutyrate; BT
Empirical: $C_7H_{15}NO_3$
Formula: $(CH_3)_3N^+CH_2CH(OH)CH_2COO^-$
Properties: Cryst.; sol. in water; insol. in acetone, ether; m.w. 161.20
Toxicology: No adverse human effects on excess intake
Uses: Nutritional additive for infant formulas

Carob bean extract. *See* Carob extract
Carob bean gum. *See* Locust bean gum

Carob extract
CAS 84961-45-5; EINECS 284-634-5
Synonyms: Ceratonia siliqua extract; Locust tree extract; Carob bean extract
Definition: Ceratonia siliqua
Properties: Sweet taste
Uses: Natural flavoring agent
Usage level: 66 ppm (nonalcoholic beverages), 93 ppm (ice cream, ices), 180 ppm (candy), 120 ppm (baked goods), 600 ppm (gelatins, puddings), 500-1000 ppm (icings, toppings)
Regulatory: FDA 21CFR §182.20, GRAS; Japan approved
Manuf./Distrib.: Chart

Carob flour. *See* Locust bean gum
Carony bark. *See* Angostura

Carotene
CAS 7235-40-7; EINECS 230-636-6; EEC E160a
Synonyms: β-Carotene; Provitamin A; Food Orange 5; Natural Yellow 26; CI 40800; CI 75130
Empirical: $C_{40}H_{56}$
Properties: Purple hexagonal prisms, red leaflets; sol. in carbon disulfide, benzene, chloroform; moderately sol. in ether, petrol. ether, oils; pract. insol. in water; m.w. 536.89; m.p. 178-179 C
Precaution: Sensitive to alkali, air, and light
Toxicology: Massive doses may cause yellowing of the skin; heated to decomp., emits acrid smoke and irritating fumes
Uses: Direct food additive, color additive, nutrient, dietary supplement; vitamin A precursor; for orange beverages, cheese, dairy prods., butter, ice cream, oleomargarine, puddings, fats and oils, processed fruits, fruit juices, infant formulas as vit. A
Usage level: 2-50 ppm (as food colorant)

Regulatory: FDA 21CFR §73.95, 73.1095, 73.2095, 166.110, 182.5245, 184.1245, GRAS; Japan restricted; Europe listed; UK approved

Manuf./Distrib.: Allchem Int'l.; Am. Roland; Atomergic Chemetals; Bronson & Jacobs; Cornelius; Dasco Sales; Ellis & Everard; Henkel; Hilton Davis; Hoffmann La Roche; Penta Mfg.; Phytone Ltd.; Produits Roche; Quest Int'l.; Quimdis; Spice King; Universal Flavors

Trade names containing: 24% Beta Carotene HS-E in Veg. Oil No. 65671; 24% Beta Carotene Semi-Solid Suspension No. 65642; 30% Beta Carotene in Veg. Oil No. 65646; Beta Carotene Emulsion Beverage Type 3.6 No. 65392; Carex for Foods; Carotenal Sol'n. #73 No. 66428; Diet Imperial® Spread; Dry Beta Carotene Beadlets 10% CWS No. 65633; Dry Beta Carotene Beadlets 10% No. 65661; Dry Beta Carotene Beadlets Yellow Type 2.4-S No. 653800100; Durkex Gold 77A; Durlite Gold MBN II; Kaorich Gold; Shedd's Liquid Margarine; Shedd's Margarine; Shedd's Special 40 Butter Blend Margarine; Shedd's Special 60 Butter Blend Margarine; Summit®

β-Carotene. *See* Carotene

Carotene cochineal
Toxicology: Heated to decomp., emits acrid smoke and irritating fumes
Uses: Coloring agent for casings, rendered fat
Regulatory: USDA 9CFR §318.7

β-Carotene-4,4´-dione. *See* Canthaxanthine
Carrageen. *See* Carrageenan

Carrageenan
CAS 9000-07-1; EINECS 232-524-2; EEC E407
Synonyms: Chondrus; Carrageen; Irish moss
Classification: Sulfated polysaccharide
Definition: Water extract obtained from various members of the *Gigartinaceae* or *Solieriaceae* familes of the red seaweed, *Rodophyceae*
Properties: Yel. white powd.; sol. in hot water, hot conc. NaCl sol'n.; insol. in oils and org. solvs.
Toxicology: Poison by intravenous route; experimental tumorigen; suspected carcinogen; linked to ulcers in colon, fetal damage in test animals; heated to decomp., emits acrid smoke and fumes
Uses: Emulsifier, binder, extender, stabilizer, thickener, gelling agent in food prods.; stabilizing aid in ice cream
Usage level: 1-5%; ADI 75 mg/kg
Regulatory: FDA 21CFR §172.020, 172.020, 172.020, limitation 5% polysorbate 80 in carrageenan and 500 ppm in final prod., 182.7255, GRAS; USDA 9CFR §318.7, 1.5% max. in restructured meat food prods., 381.147; ; Japan approved; JSCI, European listed, UK approved
Manuf./Distrib.: Browne & Dureau Int'l.; Browning; Calaga Food; Carrageenan Co.; Chart; Diamalt; G Fiske; FMC; Grindsted UK; A C Hatrick; Hercules; Honeywill & Stein; Hormel Foods; Marine Colloids; Meer; Mitsubishi; Quest Int'l.; Sanofi; Spice King; TIC Gums
Trade names: CarraFat™; Carralean™ CG-100; Carralean™ CM-70; Carralean™ CM-80; Carralean™ MB-60; Carralean™ MB-93; Carralite™; CarraLizer™ CGB-10; CarraLizer™ CGB-20; CarraLizer™ CGB-40; CarraLizer™ CGB-50; Gelodan CC Range; Gelodan CW Range; Gelodan CX Range; Genugel® CHP-2; Genugel® CHP-2 Fine Mesh; Genugel® CHP-200; Genugel® CJ; Genugel® LC-1; Genugel® LC-4; Genugel® LC-5; Genugel® MB-51; Genugel® MB-78F; Genugel® UE; Genugel® UEU; Genulacta® CP-100; Genulacta® CSM-2; Genulacta® K-100; Genulacta® KM-1; Genulacta® KM-5; Genulacta® L-100; Genulacta® LK-71; Genulacta® LR-41; Genulacta® LR-60; Genulacta® LRA-50; Genulacta® LRC-21; Genulacta® LRC-30; Genulacta® P-100; Genulacta® PL-93; Genuvisco® CSW-2; Genuvisco® J; Genuvisco® MP-11; Geoldan CL Range; Maco-O-Line 091; Marine Colloids™ Carrageenan; Nutriloid® Carrageenan; PF-80; Soageena®; Soageena® LX22; Soageena® ML300; Soageena® MM101; Soageena® MM301; Soageena® MM330; Soageena® MM350; Soageena® MM501; Soageena® MV320; Soageena® MW321; Soageena® MW351; Soageena® MW371; Soageena® WX57; Stamere® CK FCC; Stamere® CKM FCC; Stamere® CK-S; Stamere® N-47; Stamere® N-55; Stamere® N-325; Stamere® N-350; Stamere® N-350 E FCC; Stamere® N-350 S; Stamere® NI; Stamere® NIC FCC; Stamere® NK; TIC Pretested® Colloid 775 Powd.; TIC Pretested® Colloid 881 M; Wakal® K Range; Wakal® K-e
Trade names containing: Aloe Vera Gel Thickened FG; CC-603; Crodacreme; Dariloid® 300; Dricoid® 200; Dricoid® 280; Fat Replacer 785; Frigesa® D 890; Frimulsion 6G; Lamefrost® ES 216 G; Lamefrost® ES 251 G; Lamefrost® ES 315; Lamefrost® ES 375; Lamefrost® ES 379; Lamefrost® ES 424; Merecol® IC; Merecol® LK; Merecol® R; Merecol® RB; Merecol® S; Merecol® Y

Carrageenan extract
EEC E407
Synonyms: Chondrus crispus extract; Irish moss extract
Definition: Extract of *Chondrus crispus*

Carrageenan sodium salt

Properties: Ylsh. or tan to wh. powd., odorless, mucilaginous taste; sol. in water
Toxicology: Possible carcinogen
Uses: Stabilizer, thickener, gelling agent; used in milk prods., jams, jellies, soft drinks, soups
Regulatory: FDA 21CFR §182.7255, GRAS
Manuf./Distrib.: Carrageenan Co.; Cheil Foods (UK); Croda Colloids; Penta Mfg.; Siber Hegner

Carrageenan sodium salt. *See* Sodium carrageenan

Carrot extract
CAS 84929-61-3; EINECS 284-545-1
Synonyms: Daucus carota sativa extract
Definition: Extract of roots of *Daucus carota sativa*
Uses: Natural flavoring
Regulatory: FDA 21CFR §182.20, GRAS
Manuf./Distrib.: Chart

Carrot juice
Definition: Juice obtatined from carrots, *Daucus carota sativa*
Uses: Natural flavoring
Regulatory: FDA 21CFR §182.20
Manuf./Distrib.: AVOCO; Florida Food Prods.

Carrot oil
CAS 8015-88-1
Synonyms: Carrot seed oil
Definition: Oil obtained from the seeds of the carrot, *Daucus carota sativa*
Properties: Lt. yel. to amber liq., aromatic odor, sweet piquant flavor; sol. in fixed oils, min. oil; insol. in glycerin, propylene glycol; sp.gr. 0.866-0.940 (20 C); ref. index 1.4820-1.4910
Toxicology: Skin irritant; heated to decomp., emits acrid smoke and irritating fumes
Uses: Natural flavoring agent
Usage level: 3.1 ppm (nonalcoholic beverages), 5.5 ppm (ice cream, ices), 5.1 ppm (candy), 4.4 ppm (baked goods), 0.02 ppm (gelatins, puddings), 15 ppm (condiments), 1 ppm (soups)
Regulatory: FDA 21CFR §73.300, 182.10, 182.20, GRAS
Manuf./Distrib.: Am. Roland; C.A.L.-Pfizer; Pierre Chauvet

Carrot oleoresin. *See* Oleoresin carrot

Carrot seed extract
Synonyms: Daucus carota extract
Definition: Extract of the seeds of *Daucus carota sativa*
Uses: Flavoring agent in liqueurs
Regulatory: FDA GRAS

Carrot seed oil. *See* Carrot oil
Carthanus tinctorious oil. *See* Safflower oil
Carum carvi oil. *See* Caraway oil

Carvacrol
CAS 499-75-2; EINECS 207-889-6; FEMA 2245
Synonyms: 2-p-Cymenol; 2-Hydroxy-p-cymene; Isopropyl-o-cresol; Isothymol; 5-Isopropyl-2-methylphenol; 2-Methyl-5-(1-methylethyl)phenol
Empirical: $C_{10}H_{14}O$
Properties: Liq., thymol odor; freely sol. in alcohol, ether; pract. insol. in water; m.w. 150.22; dens. 0.976 (20/4 C); m.p. 0-2 C; b.p. 234-236 C; ref. index 1.523 (20 C)
Precaution: Combustible liq.; volatile with steam
Toxicology: Poison by ingestion and subcutaneous route; mod. toxic by skin contact; severe skin irritant; heated to decomp., emits acrid smoke and fumes
Uses: Synthetic flavoring agent
Usage level: 26 ppm (nonalcoholic beverages), 34 ppm (ice cream, ices), 92 ppm (candy), 120 ppm (baked goods), 37 ppm (condiments)
Regulatory: FDA 21CFR §172.515
Manuf./Distrib.: Aldrich

Carvacryl ethyl ether
FEMA 2246
Synonyms: 2-Ethoxy-p-cymene; Ethyl carvacrol; Ethyl carvacryl ether
Empirical: $C_{12}H_{18}O$
Properties: Oily light liq., carrot-like odor; m.w. 178.27; b.p. 235 C

Uses: Synthetic flavoring agent
Usage level: 2-13 ppm (nonalcoholic beverages), 10 ppm (ice cream, ices), 21 ppm (candy), 3.0-39 ppm (baked goods)
Regulatory: FDA 21CFR §172.515

Carvene. *See* l-Limonene

(+)-Carvene. *See* d-Limonene

Carveol
CAS 99-48-9; FEMA 2247
Synonyms: p-Mentha-6,8-dien-2-ol; 1-Methyl-4-isopropenyl-6-cyclohexen-2-ol
Empirical: $C_{10}H_{16}O$
Properties: m.w. 152.24; dens. 1.496; b.p. 226-227 C (15 mm); flash pt. 209 F
Toxicology: Irritant
Uses: Synthetic flavoring agent
Usage level: 1.5-13 ppm (nonalcoholic beverages), 3.0 ppm (ice cream, ices), 3.0-39 ppm (candy), 3.0-5.0 ppm (baked goods)
Regulatory: FDA 21CFR §172.515
Manuf./Distrib.: Aldrich

Carvol. *See* l-Carvone

4-Carvomenthenol
CAS 562-74-3; EINECS 209-235-5; FEMA 2248
Synonyms: (S)-1-Isopropyl-4-methyl-3-cyclohexen-1-ol; Origanol; 1-p-Menthen-4-ol; (S)-p-Menth-1-en-4-ol; 4-Terpinenol; (+)-Terpinen-4-ol
Empirical: $C_{10}H_{18}O$
Properties: m.w. 154.25; dens. 0.933 (20/4 C); b.p. 211-213 C; flash pt. 79 C; ref. index 1.479 (20 C)
Uses: Synthetic flavoring agent
Usage level: 1.0-21 ppm (nonalcoholic beverages), 1.0-84 ppm (ice cream, ices), 7.0-63 ppm (candy), 7.0 ppm (baked goods)
Regulatory: FDA 21CFR §172.515
Manuf./Distrib.: Aldrich

+-Carvone. *See* d-Carvone
(-)-Carvone. *See* l-Carvone

d-Carvone
CAS 2244-16-8; EINECS 218-827-2; FEMA 2249
Synonyms: d-p-Mentha-6,8,(9)-dien-2-one; d-1-Methyl-4-isopropenyl-6-cyclohexen-2-one; +-Carvone
Empirical: $C_{10}H_{14}O$
Properties: Colorless liq., caraway odor; sol. in propylene glycol, fixed oils; misc. in alcohol; insol. in glycerin; m.w. 150.24; dens. 0.956-0.960; ref. index 1.496-1.499
Toxicology: LD50 (skin, rabbit) 4 mg/kg; poison by ingestion and skin contact; skin irritant; heated to decomp., emits acrid smoke and irritating fumes
Uses: Synthetic flavoring agent in bakery prods., beverages, chewing gum, condiments, confections, ice cream
Regulatory: FDA 21CFR §182.60, GRAS
Manuf./Distrib.: Commodity Services; Florida Treatt

l-Carvone
CAS 99-49-0, 6485-40-1; EINECS 229-352-5; FEMA 2249
Synonyms: 1-1-Methyl-4-isopropenyl-6-cyclohexen-2-one; 1-6,8(9)-p-Menthadien-2-one; (-)-Carvone; Carvol
Empirical: $C_{10}H_{14}O$
Properties: Colorless to pale yel. liq., spearmint odor; sol. in propylene glycol, fixed oils; misc. in alcohol; insol. in glycerin; m.w. 150.22; dens. 0.956-0.960; ref. index 1.495-1.499
Toxicology: LD50 (oral, rat) 1640 mg/kg; mod. toxic by ingestion; heated to decomp., emits acrid smoke and irritating fumes
Uses: Synthetic flavoring agent used in baked goods, confectionery, ice cream, soft drinks
Regulatory: FDA 21CFR §172.515, 182.60, GRAS; GRAS (FEMA)
Manuf./Distrib.: Acme-Hardesty; Commodity Services Int'l.; Frutarom

cis-Carvone oxide
Synonyms: 1,6-Epoxy-p-menth-8-en-2-one
Uses: Synthetic flavoring agent
Regulatory: FDA 21CFR §172.515

Carvyl acetate
CAS 97-42-7; FEMA 2250
Empirical: $C_{12}H_{18}O_2$
Properties: Colorless liq., spearmint-like odor; m.w. 194.27; dens. 0.976; b.p. 115-116 C; flash pt. 208 F
Toxicology: Irritant
Uses: Synthetic flavoring agent
Usage level: 1.5-11 ppm (nonalcoholic beverages), 3.0-44 ppm (ice cream, ices), 20 ppm (candy, baked goods)
Regulatory: FDA 21CFR §172.515
Manuf./Distrib.: Aldrich

Carvyl propionate
CAS 97-45-0; FEMA 2251
Synonyms: l-Carvyl propionate; l-p-Mentha-6,8-dien-2-yl propionate
Empirical: $C_{13}H_{20}O_2$
Properties: Colorless liq., minty/fruity odor, sweet fruity minty taste; sol. in alcohol; insol. in water; m.w. 208.30; dens. 0.952; b.p. 239 C; flash pt. 226 F
Uses: Synthetic flavoring agent
Usage level: 1.0 ppm (nonalcoholic beverages), 2.0 ppm (ice cream, ices), 2.0-24 ppm (candy, baked goods)
Regulatory: FDA 21CFR §172.515
Manuf./Distrib.: Aldrich

l-Carvyl propionate. See Carvyl propionate
Caryophyllene. See β-Caryophyllene
Caryophyllene acetate. See Caryophyllene alcohol acetate

Caryophyllene alcohol
EINECS 260-364-3
Empirical: $C_{15}H_{26}O$
Properties: Wh. cryst. solid, moss-like spicy odor, minty earthy flavor; sol. in alcohol; insol. in water; m.w. 222.36; dens. 0.986 (17 C)
Uses: Synthetic flavoring agent
Regulatory: FDA 21CFR §172.515

Caryophyllene alcohol acetate
Synonyms: Caryophyllene acetate
Empirical: $C_{17}H_2iO_2$
Properties: Cryst. solid, mild fruity/woody odor; m.w. 264.41; dens. 1.003 (17 C); m.p. 40 C; b.p. 149-152 C (10 mm); ref. index 1.4919 (17 C)
Uses: Synthetic flavoring agent
Regulatory: FDA 21CFR §172.515

β-Caryophyllene
CAS 87-44-5; EINECS 201-746-1; FEMA 2252
Synonyms: (-)-trans-Caryophyllene; trans-(1R,9S)-8-Methylene-4,11,11-trimethylbicyclo[7.2.0]undec-4-ene; Caryophyllene
Definition: A mix of sesquiterpenes occurring in many essential oils
Empirical: $C_{15}H_{24}$
Properties: Colorless oil, terpene odor; sol. in alcohol; insol. in water; m.w. 204.36; dens. 0.902 (20/4 C); b.p. 262-264 C; flash pt. > 100 C; ref. index 1.429 (20 C)
Precaution: Combustible liq.
Toxicology: Skin irritant; heated to decomp., emits acrid smoke and fumes
Uses: Synthetic flavoring agent
Usage level: 14 ppm (nonalcoholic beverages), 2.0 ppm (ice cream, ices), 34 ppm (candy), 27 ppm (baked goods), 200 ppm (chewing gum), 50 ppm (condiments)
Regulatory: FDA 21CFR §172.515

β-Caryophyllene oxide
CAS 1139-30-6; EINECS 214-519-7
Synonyms: 4-12,12-Trimethyl-9-methylene-5-oxatricyclo[8.2.2.0.0.4.6]dodecane
Properties: m.w. 220.36; m.p. 45-59 C; flash pt. > 230 F
Toxicology: Irritant
Uses: Synthetic flavoring agent
Regulatory: FDA 21CFR §172.515

(-)-trans-Caryophyllene. See β-Caryophyllene

Caryophyllic acid. *See* Eugenol

Cascara extract
CAS 84650-55-5
Synonyms: Rhamnus purshiana extract
Definition: Extract of dried bark of *Rhamnus purshiana*
Uses: Natural flavoring
Regulatory: FDA 21CFR §172.510; Japan approved

Cascara sagrada. *See* Cascara sagraga

Cascara sagraga
EINECS 232-400-8 (extract); FEMA 2253
Synonyms: Cascara sagrada
Definition: Derived from *Rhamnus purshiana*
Properties: Bitter tonic flavor
Uses: Natural flavoring
Usage level: 100 ppm (nonalcoholic beverages), 50 ppm (ice cream, ices), 100 ppm (baked goods)
Regulatory: FDA 21CFR §172.510
Manuf./Distrib.: Chart

Cascarilla oil
EINECS 284-284-3 (extract)
Synonyms: Sweetwood bark oil
Definition: Volatile oil from bark of *Croton eluteria*
Properties: Yel. to greenish liq.; very sol. in alcohol, ether; dens. 0.890-0.925
Precaution: Keep cool, well closed; protect from light
Toxicology: Heated to decomp., emits acrid smoke and fumes
Uses: Natural flavoring agent
Usage level: 2.3 ppm (nonalcoholic beverages), 3.0 ppm (ice cream, ices), 8.7 ppm (candy), 13 ppm (baked goods), 50 ppm (condiments)
Regulatory: FDA 21CFR §182.20, GRAS; Japan approved
Manuf./Distrib.: Florida Treatt

Casein. *See also* Milk protein
CAS 0000 71 0; EINECS 232-555-1
Synonyms: Milk protein, casein
Definition: Mixt. of phosphoproteins obtained from cow's milk
Toxicology: Mild sensitive reactions in persons allergic to cow's milk
Uses: Dietary supplements; processing aid (Japan); frozen desserts; migrating to food from paper/paperboard
Regulatory: FDA 21CFR §166.110, 182.90, GRAS; Japan restricted
Manuf./Distrib.: Am. Roland; Blossom Farm Prods.

Casein ammonium salt. *See* Ammonium caseinate
Casein calcium salt. *See* Calcium caseinate
Casein sodium salt. *See* Sodium caseinate

Cassia gum
Uses: Natural thickener; stabilizer
Regulatory: Japan approved

Cassia oil. *See* Cinnamon oil

Cassie flowers
FEMA 2260
Definition: Flowers from *Acacia farnesiana*
Properties: Warm floral intense odor with a balsamic undertone
Uses: Natural flavoring
Usage level: 0.96 ppm (nonalcoholic beverages), 1.2 ppm (ice cream, ices), 4.1 ppm (candy, baked goods), 1.0 ppm (gelatins, puddings)
Regulatory: FDA 21CFR §172.510; Japan approved
Manuf./Distrib.: C.A.L.-Pfizer

Castoreum
EINECS 296-432-4 (Canada, extract); FEMA 2261 (extract), 2262 (liq.)
Definition: Derived from *Castor fiber* L. and *Castor canadensis*
Uses: Natural flavoring used in nonalcoholic beverages, ice cream, ices, candy, baked goods, chewing gum, toppings

Castor oil

Regulatory: FDA 21CFR §182.50, GRAS; Japan approved
Manuf./Distrib.: C.A.L.-Pfizer; Pierre Chauvet (absolute)

Castor oil
CAS 1323-38-2; 8001-79-4; EINECS 232-293-8; FEMA 2263
Synonyms: Ricinus oil; Aceite de ricino; Oil of Palma Christi; Tangantangan oil
Definition: Fixed oil obtained from seeds of *Ricinus communis*
Properties: Colorless to pale yel. viscous liq., characteristic odor; sol. in alcohol; misc. with glac. acetic acid, chloroform, ether; dens. 0.961; m.p. -12 C; b.p. 313 C; flash pt. (CC) 445 F; ref.index 1.478; sapon. no. 176-187
Precaution: Combustible when expose-d to heat; spontaneous heating may occur
Toxicology: Moderately toxic by ingestion; allergen; eye irritant; purgative, laxative in large doses
Uses: Antisticking agent, release agent for hard candy prod.; component of protective coatings for vitamin/ mineral tablets; drying oil; flavoring
Usage level: 500 ppm max. (finished foods), 500 ppm max. (hard candy)
Regulatory: FDA 21CFR §73.1, 172.510, 172.876, 175.300, 176.210, 177.2600, 177.2800, 178.3120, 178.3570, 178.3910, 181.22, 181.26, 181.28; GRAS (FEMA)
Manuf./Distrib.: Aldrich; Arista Industries; Ashland; CasChem
Trade names: AA USP

Castor oil, acetylated and hydrogenated. *See* Glyceryl (triacetoxystearate)

Castor oil, dehydrated
EINECS 264-705-7
Synonyms: DCO
Definition: Castor oil with 5% water removed
Uses: Drying oil as component of finished resin, migrating from food pkg.
Regulatory: FDA 21CFR §181.26

Castor oil, hydrogenated. *See* Hydrogenated castor oil
Castorwax. *See* Hydrogenated castor oil

Catalase
CAS 9001-05-2; EINECS 232-577-1
Classification: Oxidizing enzyme
Definition: Derived from bovine liver
Properties: m.w. ≈ 240,000
Toxicology: Heated to decomp., emits acrid smoke and irritating fumes
Uses: Enzyme; in food preservation; production aid in cheese; in decomposing residual hydrogen peroxide in cheese mfg., bleaching and oxidizing processes
Regulatory: FDA 21CFR §173.135; Canada, UK, Japan approved
Trade names: Catalase L; Fermcolase®; Microcatalase®

Catechu, black
Definition: From *Acacia catechu*
Properties: Bitter astringent flavor
Uses: Natural flavoring; used in nonalcoholic beverages, ice cream, ices, candy, baked goods
Regulatory: FDA 21CFR §172.510; Japan approved
Manuf./Distrib.: Chart (extract)

Catechu, pale
Synonyms: Gambir
Definition: From *Uncaria gambir*
Uses: Natural flavoring
Regulatory: FDA 21CFR §172.510

Caustic potash. *See* Potassium hydroxide
Caustic soda. *See* Sodium hydroxide
Cayenne pepper. *See* Capsicum

Cedar leaf oil
CAS 8007-20-3; FEMA 2267
Synonyms: Arborvitae
Definition: Volatile oil obtained by steam distillation from fresh leaves of *Thuja occidentalis*, not a true cedar
Properties: Ylsh. oil, strong camphoraceous odor reminiscent of sage; dens. 0.910-0.920; ref. index 1.4560-1.4590
Toxicology: LD50 (oral, rat) 830 mg/kg; mod. toxic by ingestion; skin irritant; ingestion of large quantities causes

hypertension, bradycardia, tachypnea, convulsions, death; heated to decomp., emits acrid smoke and fumes
Uses: Natural flavoring agent
Usage level: 0.01-0.5 ppm (nonalcoholic beverages), 16 ppm (alcoholic beverages), 0.01-1.0 ppm (ice cream, ices), 12 ppm (candy), 1.0-20 ppm (baked goods), 15 ppm (meats)
Regulatory: FDA 21CFR §172.510

Cedar, white
Definition: Leaves and twigs from *Thuja occidentalis*
Uses: Natural flavoring agent
Regulatory: FDA 21CFR §172.510; finished food must be thujone-free

Cedarwood camphor. *See* Cedarwood oil alcohols

Cedarwood oil alcohols
Synonyms: Cedarwood camphor; Cedrol; Cedrenol
Empirical: $C_{15}H_{26}O$
Properties: Wh. cryst., mild odor of cedarwood; sol. in benzyl benzoate; sl. sol. in glycol, min. oil; m.w. 222.36; dens. 0.98
Uses: Synthetic flavoring agent
Regulatory: FDA 21CFR §172.515

Cedarwood oil terpenes
EINECS 234-257-7
Synonyms: Cedrene
Empirical: $C_{15}H_{24}$
Properties: Sol. In alcohol; insol. in water; m.w. 204.34; ref. index 1.4983-1.4989
Uses: Synthetic flavoring agent
Regulatory: FDA 21CFR §172.515

Cedrene. *See* Cedarwood oil terpenes
Cedrenol. *See* Cedarwood oil alcohols
Cedrol. *See* Cedarwood oil alcohols
Cedro oil. *See* Lemon oil
Celery ketone. *See* 3-Methyl-5-propyl-2-cyclohexen-1-one
Celery oil. *See* Celery seed oil

Celery seed oil
EINECS 289-668-4 (extract); FEMA 2271
Synonyms: Celery oil
Definition: Volatile oil from celery seed, *Apium graveolens*
Properties: Colorless liq., celery odor; very sol. in alcohol; sl. sol. in water; dens. 0.870-0.895
Precaution: Keep cool, well closed; protect from light
Toxicology: Heated to decomp., emits acrid smoke and fumes
Uses: Natural flavoring agent for soft drinks
Usage level: 11 ppm (nonalcoholic beverages), 3.0-13 ppm (ice cream, ices), 13 ppm (candy), 12 ppm (baked goods), 28 ppm (chewing gum), 40 ppm (condiments), 40 ppm (meats), 10-35 ppm (pickles), 1.0 ppm (soups)
Regulatory: FDA 21CFR §182.10, 182.20, GRAS
Manuf./Distrib.: Acme-Hardesty; C.A.L.-Pfizer; Chart; Pierre Chauvet; Florida Treatt

Cellulase
CAS 9012-54-8; EINECS 232-734-4
Classification: Enzyme complex
Definition: Derived from *Aspergillus niger*
Properties: Off-white powd.; m.w. ≈ 31,000
Uses: Enzyme; digestive aid in medicine and brewing industry; aids bacteria in the hydrolysis of cellulose; aids in removal of visceral masses during clam processing and of shells in shrimp processing
Regulatory: FDA 21C FR §173.120, GRAS; UK, Japan approved
Trade names: Celluferm; Cellulase 4000; Cellulase AC; Cellulase AP; Cellulase AP 3; Cellulase L; Cellulase TAP; Cellulase TRL; Cellulase Tr Conc.
Trade names containing: Clarex® ML

Cellulose
CAS 9004-34-6; EINECS 232-674-9; EEC E460
Synonyms: Wood pulp, bleached; Cotton fiber; Cellulose powder; Cellulose gel
Definition: Natural polysaccharide derived from plant fibers

Properties: Colorless solid; insol. in water and org. solvs.; m.w. 160,000-560,000; dens. ≈ 1.5

Toxicology: Cannot be digested by humans; nuisance dust; heated to decomp., emits acrid smoke and irritating fumes

Uses: Stabilizer, emulsifier, thickener, dispersant, anticaking agent, binding agent, bulking agent, disintegrant, filter aid, texturizer in foods; beer clarifying aid

Usage level: ADI not specified

Regulatory: GRAS; Europe listed; use in baby foods not permitted in UK

Manuf./Distrib.: Allchem Int'l.; Am. Roland; Bioengineering AG; Courtaulds; Croxton & Garry; Dasco Sales; Dow Europe; FMC; A C Hatrick; Hercules BV; Honeywill & Stein; Mid-Am. Food Sales; Multi-Kem; Quimdis; Siber Hegner; Welding

Trade names: Justfiber® CL-20-H; Justfiber® CL-35-H; Justfiber® CL-40-H; Justfiber® CL-100-H; Keycel®; Qual Flo™; Solka-Floc®

Trade names containing: FloAm® 200; FloAm® 200C; FloAm® 210 HC; FloAm® 221; Lite Pack 350; Qual Flo™ C

Cellulose acetate

CAS 9004-35-7

Synonyms: CA; Cellulose acetate ester; Acetylcellulose

Classification: Cellulosics

Properties: Triacetate insol. in water, alcohol, ether, sol. in glacial acetic acid; tetraacetate insol. in water, alcohol, ether, glacial acetic acid, methanol; pentaacetate insol. in water, sol. in alcohol; m.w. ≈ 37,000; dens. 1.300

Uses: Migrating to food from paper/paperboard

Regulatory: FDA 21CFR §182.90, GRAS

Cellulose acetate ester. *See* Cellulose acetate
Cellulose ethyl ether. *See* Ethylcellulose
Cellulose gel. *See* Cellulose
Cellulose gel. *See* Microcrystalline cellulose
Cellulose gum. *See* Carboxymethylcellulose sodium
Cellulose 2-hydroxypropyl ether. *See* Hydroxypropylcellulose
Cellulose 2-hydroxypropyl methyl ether. *See* Hydroxypropyl methylcellulose
Cellulose methyl ether. *See* Methylcellulose
Cellulose, nitrate. *See* Nitrocellulose
Cellulose powder. *See* Cellulose
Cellulose tetranitrate. *See* Nitrocellulose

Centaury

Definition: Derived from *Centaurium umbellatum*

Properties: Bitter tonic

Uses: Natural flavoring for alcoholic beverages only

Regulatory: FDA 21CFR §172.510; Japan approved

Cera alba. *See* Beeswax
Ceratonia siliqua extract. *See* Carob extract

Cetalkonium chloride

CAS 122-18-9; EINECS 204-526-3

Synonyms: Cetyl dimethyl benzyl ammonium chloride; Benzylhexadecyldimethylammonium chloride; N-Hexadecyl-N,N-dimethylbenzenemethanaminium chloride

Classification: Quaternary ammonium salt

Empirical: $C_{25}H_{46}N \cdot Cl$

Formula: $C_6H_5CH_2N(CH_3)_2(C_{16}H_{33})Cl$

Properties: Colorless crystalline powd., odorless; sol. in water, alcohol, acetone, esters, propylene glycol, glycerol, CCl_4; m.w. 396.12; m.p. 58-60 C

Uses: Antimicrobial for use in cane-sugar mills

Regulatory: FDA 21CFR §173.320, 175.105, 178.1010

Cetone V. *See* Allyl α-ionone
Cetraria islandica extract. *See* Iceland moss extract

Cetyl alcohol

CAS 36653-82-4; EINECS 253-149-0; FEMA 2554

Synonyms: Palmityl alcohol; C16 linear primary alcohol; 1-Hexadecanol

Classification: Fatty alcohol

Empirical: $C_{16}H_{34}O$

Formula: $CH_3(CH_2)_{14}CH_2OH$

Properties: White waxy solid; partially sol. in alcohol and ether; insol. in water; m.w. 242.27; dens. 0.8176 (49.5 C); m.p. 49.3 C; b.p. 344 C; ref. index 1.4283

Precaution: Flamm. when exposed to heat or flame; can react with oxidizing materials

Toxicology: LD50 (oral, rat) 6400 mg/kg; mod. toxic by ingestion, intraperitoneal routes; eye and human skin irritant; heated to decomp., emits acrid smoke and fumes

Uses: Synthetic flavoring agent, color diluent, intermediate for confectionery, food supplements in tablet form, gum

Regulatory: FDA 21CFR §73.1, 73.1001, 172.515, 172.864, 175.105, 175.300, 176.200, 177.1010, 177.1200, 177.2800, 178.3480, 178.3910; GRAS (FEMA)

Manuf./Distrib.: Aarhus Oliefabrik A/S; Aldrich; Lipo; Lonza; Procter & Gamble

Cetyl dimethyl benzyl ammonium chloride. See Cetalkonium chloride

Cetyl esters
CAS 8002-23-1, 17661-50-6, 136097-97-7; EINECS 241-640-2
Synonyms: Synthetic spermaceti wax
Classification: Synthetic wax
Uses: Wax for mfg. of sweetmeats, candies, confectionery

Cetylic acid. See Palmitic acid
Cevitamic acid. See L-Ascorbic acid

C8-10 fatty acid triglyceride
Uses: Lubricant for prod. of sausage casings

CFC 113. See Trichlorotrifluoroethane

Chamomile
CAS 520-36-5
Synonyms: 2-(p-Hydroxyphenyl)-5,7-dihydroxychromone
Definition: Derived from Anthemis nobilis flower
Empirical: $C_{15}H_{10}O_5$
Properties: Blue visc. liq., strong pleasant aromatic odor, burning taste; becomes brnsh.-yel. on exposure to air and light; sol. in 6 vols. 70% alcohol; sl. sol. in water; m.w. 270.25; dens. 0.905-0.918 (15/15 C); sapon. no. 260-296; ref. index 1.442-1.448
Precaution: Keep cool, well closed; protect from light
Toxicology: Mild allergen; heated to dec., emits acrid smoke and irritating fumes
Uses: Aromatic bitter; natural flavoring
Regulatory: FDA 21CFR §182.10, GRAS; Japan approved
Manuf./Distrib.: Chart

Chamomile extract
CAS 84649-86-5; EINECS 283-467-5
Synonyms: Anthemis nobilis extract; English chamomile extract; Roman chamomile extract
Definition: Extract of flowers of Anthemis nobilis
Uses: Natural flavoring agent
Usage level: 13 ppm (nonalcoholic beverages), 9.3 ppm (ice cream, ices), 6.7 ppm (candy), 10 ppm (baked goods)
Regulatory: FDA 21CFR §182.20, GRAS

Chamomile extract, German or Hungarian. See Matricaria extract

Chamomile oil, Roman
CAS 8015-92-7
Synonyms: Anthemis nobilis oil; English chamomile oil; Roman chamomile oil
Definition: Volatile oil distilled from dried flower heads of Anthemis nobilis
Properties: Blue to brownish-yel. liq., strong aromatic odor; sol. in fixed oils, min. oil, propylene glycol; sol. in 6 vols. 70% alcohol; sl. sol. in water; insol. in glycerin; dens. 0.905-0.915 (15/15 C); sapon. no. 260-296; ref. index 1.442-1.448 (20 C)
Precaution: Combustible when heated
Toxicology: Mild allergen; skin irritant; heated to decomp., emits acrid smoke and irritating fumes
Uses: Natural flavoring agent
Usage level: 2.3 ppm (nonalcoholic beverages), 20 ppm (alcoholic beverages), 3.3 ppm (ice cream, ices), 4.3 ppm (candy, baked goods), 0.25 ppm (gelatins, puddings)
Regulatory: FDA 21CFR §182.10, 182.20, GRAS
Manuf./Distrib.: C.A.L.-Pfizer; Pierre Chauvet

Channel black. *See* Carbon black
Charcoal. *See* Carbon black
Chavicol methyl ether. *See* Estragole
Cherry kernel oil. *See* Cherry pit oil

Cherry-laurel oil
Definition: Volatile oil derived from leaves of *Prunus laurocerasus*, contg. HCN, benzaldehyde, etc.
Properties: Pale yel. liq., odor and taste similar to bitter almond oil; sol. in 2 vols 70% alcohol, in benzene, chloroform, ether; dens. 1.054-1.066 (20/20 C)
Precaution: Keep cool, well closed; protect from light
Toxicology: Very poisonous
Uses: Natural flavoring
Usage level: 75 ppm (baked goods), 77 ppm (maraschino cherries), 50-65 ppm (extracts)
Regulatory: FDA 21CFR §172.510; Japan approved

Cherry pit extract
FEMA 2278
Synonyms: Cherry pit oil extract
Uses: Natural flavoring agent
Usage level: 80-150 ppm (nonalcoholic beverages), 50-60 ppm (ice cream, ices)
Regulatory: FDA (cherry pits), not to exceed 25 ppm prussic acid; GRAS (FEMA)

Cherry pit oil
CAS 8022-29-5
Synonyms: Cherry kernel oil
Definition: Oil obtained from the kernels of cherries, *Prunus avium* or *P. cerasus*
Properties: Oil
Uses: Natural flavoring
Regulatory: FDA 21CFR §172.510; not to exceed 25 ppm prussic acid

Cherry pit oil extract. *See* Cherry pit extract

Cherry, wild, bark
FEMA 2276
Definition: Prunus serotina
Properties: Sweet tart cherry-like flavor
Uses: Natural flavoring agent
Usage level: 120 ppm (nonalcoholic beverages), 300-800 ppm (alcoholic beverages), 140 ppm (ice cream, ices), 200 ppm (candy), 76 ppm (baked goods), 3.5 ppm (gelatins, puddings), 30 ppm (syrup)
Regulatory: FDA 21CFR §182.20, GRAS
Manuf./Distrib.: Chart (extract)

Chervil
FEMA 2279
Definition: Derived from *Anthriscus cerefolium*
Uses: Natural flavoring agent
Usage level: 100 ppm (nonalcoholic beverages), 50 ppm (ice cream, ices), 150 ppm (baked goods), 60 ppm (condiments)
Regulatory: FDA 21CFR §182.10, 182.20, GRAS; Japan approved

Chestnut leaf extract
CAS 84695-99-8; EINECS 283-619-0
Definition: Extract of leaves of *Castanea vulgaris* or *C. dentata*
Uses: Natural flavoring
Regulatory: FDA 21CFR §172.510

Chicle
Definition: Obtained from the latex of the sapodilla tree native to Mexico and Central Am.
Properties: Thermoplastic gum-like substance; sol. in most org. solvs.; insol. in water
Toxicology: Ingestion should be avoided
Uses: Chewing gum base
Regulatory: Japan approved

Chicorium intybus extract. *See* Chicory extract

Chicory
Definition: Cichorium intybus L.
Properties: Bitter-tonic flavor

Uses: Natural flavoring agent
Regulatory: Japan approved

Chicory extract
CAS 68650-43-1; EINECS 272-045-6
Synonyms: Chicorium intybus extract
Definition: Extract of roots of *Chicorium intybus*
Properties: Bitter tonic flavor
Uses: Natural flavoring agent; colorant (Japan)
Usage level: 63 ppm (nonalcoholic beverages), 58 ppm (ice cream, ices), 57 ppm (candy), 100 ppm (baked goods)
Regulatory: FDA 21CFR §182.20, GRAS; Japan restricted

Chile saltpeter. *See* Sodium nitrate

Chilte
Definition: From *Cnidoscolus* species
Uses: Masticatory substance for chewing gum base
Regulatory: FDA 21CFR §172.615; Japan approved

China clay. *See* Kaolin
Chinawood oil. *See* Tung oil
Chinese bean oil. *See* Soybean oil
Chinese cinnamon oil. *See* Cinnamon oil
Chinese ginger. *See* Galanga
Chinese tea extract. *See* Thea sinensis extract
Chinese white. *See* Zinc oxide
Chinoline. *See* Quinoline

Chiquibul
Definition: From *Manikara zapotilla*
Uses: Masticatory substance for chewing gum base
Regulatory: FDA 21CFR §615

Chirata
EINECS 292-344-5 (extract)
Definition: Derived from *Swertia chirata*
Properties: Bitter tonic
Uses: Natural flavoring for alcoholic beverages only
Regulatory: FDA 21CFR §172.510

Chitin
CAS 1398-61-4; EINECS 215-744-3
Definition: Glucosamine polysaccharide
Empirical: $(C_8H_{13}NO_5)_n$
Properties: White, amorphous, semitransparent mass; sol. in conc. HCl; insol. in the common solvents
Uses: Natural thickener; stabilizer
Regulatory: Japan approved

Chitinase
CAS 9001-06-3; EINECS 232-578-7
Definition: From *Streptomyces griseus*
Properties: m.w. $\approx$ 56,000
Uses: Natural enzyme
Regulatory: Japan approved

Chitosamine. *See* Glucosamine

Chitosan
CAS 9012-76-4
Synonyms: 2-Amino-2-deoxy-(1-4)-β-D-glucopyranan
Definition: Deacylated derivative of chitin
Uses: Natural thickener; stabilizer
Regulatory: Japan approved

Chives
EINECS 289-695-1 (extract)
Definition: Derived from *Allium schoenoprasum*
Uses: Natural flavoring agent

Chloramizol

Regulatory: FDA 21CFR §182.10, GRAS

Chloramizol. *See* Imazalil

Chlorine
CAS 7782-50-5; EINECS 231-959-5; EEC E925
Formula: Cl_2
Properties: Greenish-yel. gas, liq. or rhombic crystal; sl. sol. in cold water; m.w. 70.90; sp.gr. (liq.) 1.47 (0 C,
3.65 atm); m.p. -101 C; b.p. -34.5 C
Precaution: Combines with moisture to liberate O_2 and form HCl; explodes on contact with many chemicals
Toxicology: TLV 1 ppm in air; poison by inhalation; human respiratory system effects; strong irritant to eyes,
mucous membranes; human mutagenic data
Uses: Food processing; bleaching agent; antimicrobial agent; preservative; oxidizing agent; improving agent
Regulatory: GRAS; Europe listed; UK approved
Manuf./Distrib.: Asahi Chem Industry Co Ltd

Chlorine dioxide
CAS 10049-04-4; EINECS 233-162-8; EEC E926
Synonyms: Chlorine peroxide
Empirical: ClO_2
Properties: Red to yel. gas, unpleasant odor; dec. in water; m.w. 67.46; dens. (liq.) 1.642; m.p. -59 C; b.p. 11
C; explodes when heated
Precaution: Strong oxidizing agent; reacts violently with organic materials
Toxicology: TLV 0.1 ppm in air; very irritating to skin and mucous membranes; may cause pulmonary edema
Uses: Bleaching agent for foods; oxidizing agent; flour treatment agent; purification of water; taste and odor
control; bactericide; antiseptic
Usage level: ADI 0-30 mg/kg (EEC)
Regulatory: Granted prior sanction via clearance for optional use as a bleaching agent for flour under food
standards §15; Japan approved; Europe listed; UK approved

Chlorine peroxide. *See* Chlorine dioxide
1-Chloro-2,2-dichloroethylene. *See* Trichloroethylene
Chlorofluorocarbon 113. *See* Trichlorotrifluoroethane

Chloromethylated aminated styrene-divinylbenzene resin
CAS 60177-39-1
Toxicology: Heated to decomp., emits toxic fumes of Cl⁻ and NO_2
Uses: Decolorizing and clarification agent for treatment of refinery sugar liquors and juices
Usage level: Limitation 500 ppm max. of sugar solids
Regulatory: FDA 21CFR §173.70

Chloropentafluoroethane
CAS 76-15-3
Empirical: C_2ClF_5
Properties: Colorless gas; sol. in alcohol, ether; insol. in water; m.w. 154.47; dens. 1.5678 (-42 C); m.p. -38
C; b.p. -39.3 C
Precaution: Nonflamm. gas
Toxicology: TLV:TWA 1000 ppm; mildly toxic by inhalation; heated to decomp., emits toxic fumes of F⁻ and Cl⁻
Uses: Aerating agent for foamed or sprayed food prods.
Regulatory: FDA 21CFR §173.345

Chlorophyll
CAS 1406-65-1; EINECS 215-800-7; EEC E140
Synonyms: Biophyll; Green chlorophyl; CI 75810
Properties: Dk. green sol'n.
Toxicology: LD50 (IV, mouse) 285 mg/kg; poison by IV and intraperitoneal routes; heated to decomp., emits
toxic fumes of NO_x
Uses: Food colorant for casings, rendered fat, oleomargarine, shortening
Usage level: ADI no limit (EEC)
Regulatory: Green chlorophyll cleared by MID for coloring sausage casings, oleomargarine, and shortening,
and in marking or branding inks for prods. in an amt. sufficient for the purpose; USDA 9CFR §318.7; Europe
listed; UK approved; may be mixed with approved synthetic dyes or harmless inert materials such as
common salt or sugar; Japan restricted
Manuf./Distrib.: Am. Roland; Bush Boake Allen Ltd.; Phytone Ltd.; Quest Int'l.

Cholalic acid. *See* Cholic acid

Cholecalciferol
CAS 67-97-0; EINECS 200-673-2
Synonyms: 5,7-Cholestadien-3-β-ol; 9,10-seco(5Z,7E)-5,7,10(19)-Cholestatrien-3-ol; 7-Dehydrocholesterol; Calciol; Vitamin D₃
Classification: Sterol; vitamin
Empirical: $C_{27}H_{44}O$
Properties: Colorless cryst.; unstable in light and air; sol. in acetone, alcohol, chloroform, fatty oils; insol. in water; m.w. 384.71; m.p. 84-88 C
Precaution: When heated to dec., emits acrid smoke and irritating fumes
Toxicology: LD50 (oral, rat) 42 mg/kg; poison by ingestion; experimental teratogen; heated to decomp., emits acrid smoke and irritating fumes
Uses: Dietary supplement, nutrient; may be used in infant formulas, margarine
Usage level: limitation 350 IU/100 g (breakfast cereals), 90 IU/100 g (grain prods., pasta), 42 IU/100 g (milk), 89 IU/100 g (milk prods.)
Regulatory: FDA 21CFR §166.110, 182.5953, 184.1950, GRAS; Japan approved
Manuf./Distrib.: Am. Roland
Trade names containing: Dry Vitamin D₃ Beadlets Type 850 No. 652550401, 652550601; Dry Vitamin D₃ Type 100 CWS No. 65242; Dry Vitamin D₃ Type 100-SD No. 65216; Liquid Vitamin D₃ No. 63643

5,7-Cholestadien-3-β-ol. *See* Cholecalciferol
9,10-seco(5Z,7E)-5,7,10(19)-Cholestatrien-3-ol. *See* Cholecalciferol
Cholesterin. *See* Cholesterol

Cholesterol
CAS 57-88-5; EINECS 200-353-2
Synonyms: Cholesterin
Classification: Common animal sterol
Empirical: $C_{27}H_{46}O$
Properties: White or faintly yellow pearly granules or crystals; almost odorless m.w. 386.73; dens. 1.067; m.p. 148.5 C; b.p. 360 C (with decomp.); sol. in benzene; sl. sol. in water
Uses: Natural emulsifier
Regulatory: Japan approved
Trade names: Vitinc® Cholesterol NF XVII

Cholic acid
CAS 81 26 1; EINECS 201-337-8
Synonyms: Cholalic acid; 3,7,12-Trihydroxycholanic acid
Formula: $C_{23}H_{49}O_3COOH$
Properties: Bitter taste with sweetish aftertaste; sol. in glacial acetic acid, acetone, alcohol; sl. sol. in chloroform; insol. in water, benzene; m.w. 408.58; m.p. 198 C
Uses: Emulsifying agent
Usage level: up to 0.1%
Regulatory: Japan approved

Cholic acid sodium salt. *See* Ox bile extract

Choline bitartrate
CAS 87-67-2
Synonyms: (2-Hydroxyethyl) trimethyl ammonium bitartrate
Empirical: $C_9H_{19}NO_7$
Formula: $(C_5H_{14}NO)C_4H_5O_6$
Properties: Wh. cryst. powd., acetic taste; sol. in water; sl. sol. in alcohol; insol. in ether, chloroform, benzene; m.w. 253.25; m.p. 151-153 C
Toxicology: Irritant; high dosages may cause salination, sweating, nausea, diarrhea, depression; heated to decomp., emits toxic fumes of NO_x
Uses: Dietary supplement, nutrient
Regulatory: FDA 21CFR §182.5250, 182.8250, GRAS
Manuf./Distrib.: Am. Roland

Choline chloride
CAS 67-48-1; EINECS 200-655-4
Synonyms: Choline hydrochloride; (2-Hydroxyethyl)trimethylammonium chloride
Empirical: $C_5H_{14}ONCl$
Formula: $(CH_3)_3N(Cl)CH_2CH_2OH$
Properties: Colorless to wh. hygroscopic cryst., sl. odor of trimethylamine; sol. in water and alcohol; m.w. 139.65

Choline hydrochloride

Toxicology: LD50 (oral, rat) 9 g/kg; poison by intraperitoneal and intravenous routes; moderately toxic by ingestion, subcutaneous routes; high dosages may cause adverse effects; suspected carcinogen; heated to decomp., emits toxic fumes of Cl⁻, NO_x, NH_3
Uses: Nutrient, dietary supplement, animal feed additive
Regulatory: FDA 21CFR §182.5252, 182.8252, GRAS
Manuf./Distrib.: Penta Mfg.

Choline hydrochloride. *See* Choline chloride

Choline phosphate
Uses: Fermentation regulator; taste improver (0.2 g/L in sake cmpd)
Regulatory: Japan approved

Chondrus. *See* Carrageenan
Chondrus crispus extract. *See* Carrageenan extract
Chymosin. *See* Rennet
CI 11380. *See* 1-(Phenylazo)-2-naphthylamine
CI 12156. *See* Citrus Red No. 2.
CI 14700. *See* FD&C Red No. 4
CI 15985. *See* FD&C Yellow No. 6
CI 16035. *See* FD&C Red No. 40
CI 16185. *See* Amaranth
CI 16255. *See* Cochineal
CI 19140. *See* FD&C Yellow No. 5
CI 19140. *See* Tartrazine
CI 19235. *See* Orange B
CI 40800. *See* Carotene
CI 40820. *See* β-Apo-8´-carotenal
CI 40850. *See* Canthaxanthine
CI 42053. *See* FD&C Green No. 3
CI 42090. *See* Brilliant Blue FCF
CI 42090. *See* FD&C Blue No. 1
CI 45430. *See* FD&C Red No. 3
CI 73015. *See* FD&C Blue No. 2
CI 75120. *See* Annatto
CI 75120. *See* Annatto extract
CI 75130. *See* Carotene
CI 75300. *See* Curcumin
CI 75300. *See* Turmeric
CI 75470. *See* Carmine
CI 75470. *See* Carmine extract
CI 75810. *See* Chlorophyll
CI 77004. *See* Aluminum silicate
CI 77004. *See* Bentonite
CI 77019. *See* Talc
CI 77220. *See* Calcium carbonate
CI 77266. *See* Carbon black
CI 77891. *See* Titanium dioxide
CI 77947. *See* Zinc oxide
Cincholepidine. *See* Lepidine

Cinchona extract
CAS 84929-25-9; 84776-28-3; EINECS 289-707-5, 283-952-1, 304-815-5, 289-708-0, 289-709-6
Synonyms: Peruvian bark extract
Definition: Extract of the bark of *Cinchona* spp.
Properties: Bitter tonic
Uses: Natural flavoring agent
Usage level: Red cinchona bark extract: 100 ppm (nonalcoholic beverages), 25 ppm (ice cream, ices), 20 ppm (baked goods), 60 ppm (condiments); Yellow cinchona bark extract: 10 ppm (nonalcoholic beverages), 1 ppm (candy)
Regulatory: FDA 21CFR §172.510; GRAS (FEMA)
Manuf./Distrib.: Chart

Cinene. *See* dl-Limonene
Cineol. *See* Eucalyptol

1,4-Cineole
CAS 470-67-7; FEMA 3658
Synonyms: 1,4-Epoxy-p-menthane
Empirical: C₁₀H₁₈O
Properties: Colorless liq., camphoraceous odor, spicy flavor; sol. in alcohol; insol. in water; m.w. 154.25; dens. 0.887; m.p. -46 C; b.p. 65 C (16 mm); flash pt. 118 F
Uses: Synthetic flavoring agent
Regulatory: FDA 21CFR §172.515
Manuf./Distrib.: Aldrich

Cinnamal
CAS 104-55-2, 14371-10-9; EINECS 203-213-9; FEMA 2286
Synonyms: Cinnamaldehyde; Phenylacrolein; Cinnamic aldehyde; 3-Phenyl-2-propenal
Classification: Aromatic aldehyde
Empirical: C₉H₈O
Formula: C₆H₅CH:CHCHO
Properties: Ylsh. oily liq., strong cinnamon odor; very sl. sol. in water; misc. with alcohol, ether, chloroform, fixed oils; m.w. 132.16; dens. 1.048-1.052; m.p. -7.5 C; b.p. 246 C (760 mm); flash pt. 248 F; ref. index 1.619-1.623 (20 C)
Precaution: Combustible liq.; may ignite after delay in contact with NaOH; volatile with steam
Toxicology: LD50 (oral, rat) 2220 mg/kg; poison by IV route; mod. toxic by ingestion and intraperitoneal routes; severe human skin irritant; heated to decomp., emits acrid smoke and fumes
Uses: Synthetic flavoring agent
Usage level: 9 ppm (nonalcoholic beverages), 7.7 ppm (ice cream, ices), 700 ppm (candy), 180 ppm (baked goods), 4900 ppm (chewing gum), 20 ppm (condiments), 60 ppm (meats)
Regulatory: FDA 21CFR §182.60, GRAS; 27CFR §21.65, 21.151; Japan approved as flavoring
Manuf./Distrib.: Florida Treatt

Cinnamaldehyde. *See* Cinnamal

Cinnamaldehyde ethylene glycol acetal
FEMA 2287
Synonyms: Cinnamic aldehyde ethylene glycol acetal; 2-Styryl-1,3-dioxolane; 2-Styryl-m-dioxolane
Definition: From cinnamic aldehyde and ethylene glycol
Empirical: C₁₁₁₂O₂
Properties: Colorless oily liq., cinnamon-like odor and flavor; sol. in alcohol; insol. in water; m.w. 176.22; b.p. 265 C
Uses: Synthetic flavoring agent
Usage level: 0.06-2.0 ppm (candy), 2.0 ppm (baked goods), 5.0 ppm (condiments)
Regulatory: FDA 21CFR §172.515

Cinnamein. *See* Benzyl cinnamate
Cinnamene. *See* Styrene

Cinnamic acid
CAS 140-10-3, 621-82-9; EINECS 205-398-1; FEMA 2288
Synonyms: β-Phenylacrylic acid; trans-3-Phenylacrylic acid; 3-Phenylpropenoic acid; Cinnamylic acid
Empirical: C₉H₈O₂
Formula: C₆H₅CH:CHCOOH
Properties: Wh. monoclinic cryst., honey floral odor; sol. in benzene, ether, acetone, glac. acetic acid, carbon disulfide, fixed oils; m.w. 148.17; dens. 1.2475 (4/4 C); m.p. 133 C; b.p. 300 C; flash pt. > 212 F
Precaution: Combustible liq.
Toxicology: LD50 (oral, rat) 2500 mg/kg; poison by IV and intraperitoneal routes; mod. toxic by ingestion; skin irritant; heated to decomp., emits acrid smoke and fumes
Uses: Synthetic flavoring agent
Usage level: 31 ppm (nonalcoholic beverages), 40 ppm (ice cream, ices), 30 ppm (candy), 36 ppm (baked goods), 10 ppm (chewing gum)
Regulatory: FDA 21CFR §172.515; GRAS (FEMA); Japan approved as flavoring
Manuf./Distrib.: Aceto; Aldrich; Penta Mfg.

Cinnamic acid cinnamyl ester. *See* Cinnamyl cinnamate
Cinnamic alcohol. *See* Cinnamyl alcohol
Cinnamic aldehyde. *See* Cinnamal
Cinnamic aldehyde ethylene glycol acetal. *See* Cinnamaldehyde ethylene glycol acetal
Cinnamol. *See* Styrene
Cinnamomum cassia oil. *See* Cinnamon oil

Cinnamon
EINECS 284-635-0 (extract)
Definition: Natural prod. derived from dried bark of *Cinnamomum cassia*
Uses: Natural flavoring agent
Usage level: 5.6 ppm (nonalcoholic beverages), 53 ppm (ice cream, ices), 10-4000 ppm (candy), 1900 ppm (baked goods), 78-450 ppm (apple butter), 110 ppm (condiments), 880 ppm (meats)
Regulatory: FDA 21CFR §182.10, GRAS; Japan approved
Manuf./Distrib.: C.A.L.-Pfizer; Chart

Cinnamon leaf oil
EINECS 283-479-0 (extract)
Synonyms: Cinnamon leaf oil, Ceylon; Cinnamon leaf oil, Seychelles
Definition: Oil obtained by steam distillation of leaves from *Cinnamomum zeylanicum*
Properties: Lt. to dk. brn. liq., spicy cinnamon, clove odor and taste; sol. in fixed oils, propylene glycol, min. oil; insol. in glycerin
Toxicology: Heated to decomp., emits acrid smoke and irritating fumes
Uses: Natural flavoring agent for baked goods, nonalcoholic beverages, chewing gum, condiments, ice cream, meat, pickles
Regulatory: FDA 21CFR §182.10, 182.20, GRAS
Manuf./Distrib.: Florida Treatt

Cinnamon leaf oil, Ceylon. *See* Cinnamon leaf oil
Cinnamon leaf oil, Seychelles. *See* Cinnamon leaf oil

Cinnamon oil
CAS 8007-80-5; EINECS 284-635-0 (extract)
Synonyms: Cassia oil; Cinnamomum cassia oil; Chinese cinnamon oil
Definition: Volatile oil distilled from the leaves and twigs of *Cinnamomum cassia*, contg. 80-90% cinnamaldehyde, plus cinnamyl acetate, eugenol
Properties: Ylsh. to brnsh. liq., cinnamon odor, spicy burning taste; darkens and thickens on exposure to air; sol. in fixed oils, propylene glycol; sl. sol. in water; insol. in glycerin, min. oil; dens. 1.045-1.063 (25/25 C); ref. index 1.6020-1.6060 (20 C)
Precaution: Keep cool, well closed; protect from light
Toxicology: LD50 (oral, rat) 2800 mg/kg; poison by skin contact; mod. toxic by ingestion, IP routes; human skin irritant; suspected weak carcinogen; mutagenic data; heated to decomp., emits acrid smoke and irritating fumes
Uses: Natural flavoring agent
Regulatory: FDA 21CFR §145.135, 145.140, 145.145, 145.180, 145.181, 182.10, 182.20, GRAS, 27CFR §21.151
Manuf./Distrib.: Berje; Formula One; Kato Worldwide; Meer; Penta Mfg.; SKW Chems.

Cinnamon oil, Ceylon
Definition: Volatile oil from bark of Ceylon cinnamon *Cinnamomum zeylanicum*, contg. 50-65% cinnamaldehyde, 4-8% eugenol, and phellandrene
Properties: Lt. yel. liq., char. odor; gradually becomes reddish; dens. 1.000-1.030; ref. index 1.565-1.582 (20 C)
Uses: Natural flavoring agent
Regulatory: FDA 21CFR §182.20, GRAS
Manuf./Distrib.: Berje; Cooke Aromatics; Diamalt; Dragoco Australia; Florida Treatt; Ikeda; Meer; O'Laughlin Industries; Penta Mfg.; Alfred L Wolff

Cinnamyl acetate
CAS 103-54-8; EINECS 203-121-9; FEMA 2293
Synonyms: Acetic acid, cinnamyl ester; 2-Phenyl-2-propen-1-ol acetate
Definition: Ester of cinnamyl alcohol and acetic acid
Empirical: $C_{11}H_{12}O_2$
Properties: Colorless liq., sweet floral odor; misc. with chloroform, ether, fixed oils; insol. in glycerin, water @ 264 C; m.w. 176.23; dens. 1.047-1.051; b.p. 265 C; flash pt. 244 F; ref. index 1.539-1.543
Precaution: Combustible liq.
Toxicology: LD5 (oral, rat) 3300 mg/kg; mod. toxic by ingestion and intraperitoneal routes; heated to decomp., emits acrid smoke and fumes
Uses: Synthetic flavoring agent
Usage level: 2.7 ppm (nonalcoholic beverages), 6.5 ppm (ice cream, ices), 16 ppm (candy), 11 ppm (baked goods), 8.7 ppm (chewing gum), 2 ppm (condiments)
Regulatory: FDA 21CFR §172.515; GRAS (FEMA); Japan approved as flavoring

Manuf./Distrib.: Aldrich

Cinnamyl alcohol
CAS 104-54-1; EINECS 203-212-3; FEMA 2294
Synonyms: 3-Phenyl-2-propen-1-ol; Cinnamic alcohol; γ-Phenylallyl alcohol; Styroen; Styryl carbinol
Classification: Organic compd.
Empirical: $C_9H_{10}O$
Formula: $C_6H_5CH:CHCH_2OH$
Properties: Needles or cryst. mass, hyacinth odor; sol. in water, glycerol, propylene glycol, alcohol, ether, other common org. solvs.; m.w. 134.19; dens. 1.0397 (35/35 C); m.p. 33 C; b.p. 250 C; flash pt. > 230 F; ref. index 1.58190
Toxicology: LD50 (oral, rat) 2000 mg/kg; mod. toxic by ingestion; skin irritant; heated to decomp., emits acrid smoke and fumes
Uses: Balsamic, sweetly floral fragrance and flavoring
Usage level: 8.8 ppm (nonalcoholic beverages), 5 ppm (alcoholic beverages), 8.7 ppm (ice cream, ices), 17 ppm (candy), 33 ppm (baked goods), 22 ppm (gelatins, puddings), 720 ppm (chewing gum)
Regulatory: FDA 21CFR §172.515; GRAS (FEMA); Japan approved as flavoring
Manuf./Distrib.: Aldrich

Cinnamyl-2-aminobenzoate. *See* Cinnamyl anthranilate

Cinnamyl anthranilate
CAS 87-29-6; FEMA 2295
Synonyms: Anthranilic acid cinnamyl ester; Cinnamyl-2-aminobenzoate; 3-Phenyl-2-propenylanthranilate
Empirical: $C_{16}H_{15}NO_2$
Properties: Reddish yel. powd., balsamic odor; sol. in alcohol, chloroform, ether; insol. in water; m.w. 253.32; dens. 1.180 (15.5 C); m.p. 60 C; b.p. 332 C; flash pt. > 100 C
Precaution: Combustible liq.
Toxicology: LD50 (oral, rat) 5000 mg/kg; experimental neoplastigen; heated to decomp., emits toxic fumes of NO_x
Uses: Synthetic flavoring agent for baked goods, beverages, candy
Regulatory: FDA 21CFR §172.515, 189.113; prohibited from direct addition or use in human food

Cinnamyl benzoate
Empirical: $C_{10}H_{14}O_2$
Properties: Wh. cryst. powd.; sol. in alcohol; insol. in water; m.w. 238.29; dens. 1.04; m.p. 31 C; b.p. 209 C (13 mm)
Uses: Synthetic flavoring agent
Regulatory: FDA 21CFR §172.515

Cinnamyl butyrate
CAS 103-61-7; FEMA 2296
Synonyms: Phenyl propenyl-n-butyrate
Empirical: $C_{13}H_{16}O_2$
Properties: Colorless to ylsh. liq., fruity sl. floral odor, honey-like taste; insol. in water; m.w. 204.27; dens. 1.010-1.015 (25/25 C); b.p. 300 C; flash pt. > 100 C
Uses: Synthetic flavoring agent
Usage level: 1.6 ppm (nonalcoholic beverages), 8.5 ppm (ice cream, ices), 7.6 ppm (candy), 11 ppm (baked goods), 1.2 ppm (gelatins, puddings)
Regulatory: FDA 21CFR §172.515
Manuf./Distrib.: Aldrich

Cinnamyl cinnamate
CAS 122-69-0
Synonyms: 3-Phenyl-2-propenoic acid 3-phenyl-2-propenyl ester; Cinnamic acid cinnamyl ester; Cinnyl cinnamate; Styracin
Empirical: $C_{18}H_{16}O_2$
Formula: $C_5H_5CH=CHCOOCH_2CH=CHC_6H_5$
Properties: sweet resinous odor; sol. in alcohol, benzene; m.w. 264.31; dens. 1.1565 (4 C); trans-trans: needles; pract. insol. in water; sol. 1 g/3 ml ether; m.p. 44 C
Uses: Synthetic flavoring agent
Usage level: 0.81 ppm (nonalcoholic beverages), 1.5 ppm (ice cream, ices), 10 ppm (candy), 7.0 ppm (baked goods)
Regulatory: FDA 21CFR §172.515

Cinnamyl formate
CAS 104-65-4; FEMA 2299

Cinnamylic acid

Empirical: $C_{10}H_{10}O_2$
Properties: Colorless to ylsh. liq., balsamic fruital-floral odor; sol. in most org. solvs.; insol. in water; m.w. 162.19; dens. 1.080; b.p. 250-254 C; flash pt. > 230 F; ref. index 1.5500-1.5560 (20 C)
Precaution: Combustible liq.
Toxicology: LD50 (oral, rat) 2900 mg/kg; mod. toxic by ingestion; heated to decomp., emits acrid smoke and fumes
Uses: Synthetic flavoring agent
Usage level: 1.3 ppm (nonalcoholic beverages), 9.1 ppm (ice cream, ices), 6.9 ppm (candy), 8.0 ppm (baked goods), 0.60 ppm (chewing gum)
Regulatory: FDA 21CFR §172.515
Manuf./Distrib.: Aldrich

Cinnamylic acid. *See* Cinnamic acid

Cinnamyl isobutyrate
CAS 103-59-3; FEMA 2297
Empirical: $C_{13}H_{16}O_2$
Properties: Colorless to ylsh. liq., sweet balsamic fruital odor; insol. in water; m.w. 204.27; dens. 1.01; b.p. 254 C; flash pt. > 230 F; ref. index 1.5230-1.5280 (20 C)
Uses: Synthetic flavoring agent
Usage level: 1.5 ppm (nonalcoholic beverages), 5.0 ppm (ice cream, ices), 7.7 ppm (candy), 8.5 ppm (baked goods), 0.02-1.2 ppm (gelatins, puddings), 140 ppm (chewing gum), 1.0 ppm (toppings)
Regulatory: FDA 21CFR §172.515
Manuf./Distrib.: Aldrich

Cinnamyl isovalerate
FEMA 2302
Definition: Obtained by esterification of cinnamic alcohol with isovaleric acid
Empirical: $C_{14}H_{18}O_2$
Properties: Colorless to ylsh. liq., rose-like odor, apple-like taste; insol. in water; m.w. 218.29; b.p. 313 C; flash pt. > 100 C
Precaution: Combustible liq.
Toxicology: Heated to decomp., emits acrid smoke and fumes
Uses: Synthetic flavoring agent
Usage level: 2.2 ppm (nonalcoholic beverages), 2.6 ppm (ice cream, ices), 4.1 ppm (candy), 3.6 ppm (baked goods), 11 ppm (gelatins, puddings), 19-30 ppm (chewing gum)
Regulatory: FDA 21CFR §172.515

Cinnamyl phenylacetate
FEMA 2300
Synonyms: Cinnamyl α-toluate
Empirical: $C_{17}H_{16}O_2$
Properties: Colorless liq., honey-like flavor; sol. in alcohol; insol. in water; m.w. 252.32; dens. 1.09; b.p. 333-335 C
Uses: Synthetic flavoring agent
Usage level: 2.7 ppm (nonalcoholic beverages), 0.25-2.0 ppm (ice cream, ices), 7.3 ppm (candy, baked goods)
Regulatory: FDA 21CFR §172.515

Cinnamyl propionate
FEMA 2301
Synonyms: γ-Phenylallyl propionate; 3-Phenyl-2-propenyl propanoate
Definition: Obtained by esterification of cinnamic alcohol with propionic acid
Empirical: $C_{12}H_{14}O_2$
Properties: Colorless to ylsh. liq., spicy fruital odor; insol. in water; m.w. 190.24; dens. 1.0370-1.0410 (15 C); b.p. 289 C; flash pt. > 100 C; ref. index 1.5180-1.5240 (20 C)
Precaution: Combustible liq.
Toxicology: Heated to decomp., emits acrid smoke and fumes
Uses: Synthetic flavoring agent
Usage level: 1.0 ppm (nonalcoholic beverages), 4.3 ppm (ice cream, ices), 7.5 ppm (candy), 8.8 ppm (baked goods), 2.4-4.0 ppm (gelatins, puddings), 20-53 ppm (chewing gum)
Regulatory: FDA 21CFR §172.515

Cinnamyl α-toluate. *See* Cinnamyl phenylacetate
Cinnyl cinnamate. *See* Cinnamyl cinnamate
Citral
CAS 5392-40-5; EINECS 226-394-6; FEMA 2303

Synonyms: 3,7-Dimethyl-2,6-octadienal; 2,6-Dimethyloctadien-2,6-al-8; Geranial; Neral
Classification: Aldehyde
Empirical: $C_{10}H_{16}O$
Formula: $(CH_3)_2C:CHCH_2CHC(CH_3)CHCHO$
Properties: Pale yel. mobile liq., strong lemon odor; sol. in glycerin, propylene glycol, min. oil, fixed oils, 95% alcohol; insol. in water; m.w. 152.24; dens. 0.891-0.987 (15 C); b.p. 220-225 C; flash pt. 198 F; ref. index 1.486-1.490
Precaution: Combustible liq.
Toxicology: LD50 (oral, rat) 4960 mg/kg; mod. toxic by intraperitoneal route; mildly toxic by ingestion; experimental reproductive effects; human skin irritant; heated to decomp., emits acrid smoke and irritating fumes
Uses: Fresh, lemon-like, green, sl. lime-like fragrance and flavoring; flavoring agent for baked goods, candy, ice cream; improves flavor of foodstuffs
Usage level: 9.2 ppm (nonalcoholic beverages), 23 ppm (ice cream, ices), 41 ppm (candy), 43 ppm (baked goods), 170 ppm (chewing gum)
Regulatory: FDA 21CFR §172.515, 182.60, GRAS; Japan approved as flavoring
Manuf./Distrib.: Florida Treatt

Citral diethyl acetal. *See* Citral diethyl acetate

Citral diethyl acetate
FEMA 2304
Synonyms: 3,7-Dimethyl-2,6-octadienal diethyl acetal; Citral diethyl acetal
Empirical: $C_{14}H_{26}O_2$
Properties: Colorless liq., mild green citrus odor; sol. in propylene glycol; insol. in water; m.w. 226.36; dens. 0.8745-0.8790 (15 C); b.p. 230 C; flash pt. 79 C; ref. index 1.4520-1.4545
Uses: Synthetic flavoring agent
Usage level: 0.03 ppm (nonalcoholic beverages), 0.13 ppm (candy), 110 ppm (condiments)
Regulatory: FDA 21CFR §172.515
Manuf./Distrib.: Aldrich

Citral dimethyl acetal
CAS 7549-37-3; FEMA 2305
Synonyms: 3,7-Dimethyl-2,6-octadienal dimethyl acetal
Empirical: $C_{12}H_{22}O_2$
Properties: Colorless to ylsh. liq., fresh lemon-like odor; m.w. 198.31; dens. 0.885; b.p. 198 C; flash pt. 180 F; ref. index 1.4560-1.4630 (20 C)
Uses: Synthetic flavoring agent
Usage level: 6.3 ppm (nonalcoholic beverages), 11 ppm (ice cream, ices), 60 ppm (candy, baked goods), 15 ppm (chewing gum)
Regulatory: FDA 21CFR §172.515
Manuf./Distrib.: Aldrich

Citral propylene glycol acetal
Definition: Obtained by condensation of citral with propylene glycol, using a catalyst
Empirical: $C_{13}H_{22}O_2$
Properties: Colorless oily liq., lemon-orange odor, mild lemon flavor; sol. in propylene glycol, alcohol, oils; sl. sol. in water; m.w. 210.32
Uses: Synthetic flavoring agent
Regulatory: FDA 21CFR §172.515

Citranaxanthine
Uses: For pigmentation of egg yolks, skin and legs of broilers
Trade names: Lucantin® CX

Citric acid
CAS 77-92-9 (anhyd.), 5949-29-1 (monohydrate), 526-95-4; EINECS 201-069-1; FEMA 2306; EEC E330
Synonyms: 2-Hydroxy-1,2,3-propanetricarboxylic acid; β-Hydroxytricarballylic acid
Classification: Organic acid
Empirical: $C_6H_8O_7$
Formula: $HOC(COOH)(CH_2COOH)_2$
Properties: Colorless translucent crystals or powd.; odorless, tart taste; very sol. in water and alcohol; sol. in ether; m.w. 192.43; dens. 1.542; m.p. 153 C
Precaution: Combustible; potentially explosive reaction with metal nitrates
Toxicology: LD50 (oral, rat) 6730 mg/kg; poison by IV route; mod. toxic by subcutaneous and intraperitoneal routes; mildly toxic by ingestion; severe eye, mod. skin irritant; some allergenic props.; heated to decomp.,

emits acrid smoke and fumes

Uses: Acidifier, flavoring extracts, confections, soft drinks; antioxidant in foods; sequestering agent; dispersant; curing accelerator

Usage level: 2500 ppm (nonalcoholic beverages), 1600 ppm (ice cream, ices), 4300 ppm (candy), 1200 ppm (baked goods), 3600 ppm (chewing gum); ADI no limit (EEC)

Regulatory: FDA 21CFR §131.111, 131.112, 131.136, 131.138, 131.144, 131.146, 133, 145.131, 145.145, 146.187, 150.141, 150.161, 155.130, 161.190, 166.40, 166.110, 169.115, 169.140, 169.150, 172.755, 173.160, 173.165, 173.280, 182.1033, 182.6033, GRAS; USDA 9CFR §318.7, 381.147; BATF 27CFR §240.1051, limitation 5.8 lb/1000 gal; Japan approved; Europe listed; UK approved

Manuf./Distrib.: ADM; Albright & Wilson; Am. Roland; Ashland; Browning; Consolidated Flavor; Ellis & Everard; Frutarom; Haarmann & Reimer; Hoffmann La Roche; Ikeda; Jungbunzlauer; Lohmann; Norman Fox; Penta Mfg.; Pfizer Food Science; Spice King

Trade names: Cap-Shure® C-140E-75; Citric Acid Anhydrous USP/FCC; Citric Acid USP FCC Anhyd. Fine Gran. No. 69941; Citrid Acid USP FCC Anhyd. Gran. No. 69942; Citrocoat® A 1000 HP; Citrocoat® A 2000 HP; Citrocoat® A 4000 TP; Citrocoat® A 4000 TT; Citrostabil® NEU; Citrostabil® S; Descote® Citric Acid 50%; Liquinat®

Trade names containing: Akopuff; Alphadim® 90NLK; Alphadim® 90VCK; Arlacel® 186; Atlas 5520; 24% Beta Carotene HS-E in Veg. Oil No. 65671; BFP 64K; BFP 65; BFP 74; BFP 75; Cap-Shure® C-165-85; Cap-Shure® CLF-165-70; Centrophil® M; Centrophil® W; Cetodan® 90-50; Cola No. 23443; Cola Acid No. 23444; Diet Imperial® Spread; Dricoid® 200; Dur-Em® 114; Dur-Em® 114K; Dur-Em® 117; Dur-Em® 117K; Dur-Em® 204; Dur-Em® 204K; Dur-Em® 207; Dur-Em® 207-E; Dur-Em® 207K; Durkote Citric Acid/ Hydrog. Veg. Oil; Durkote Citric Acid/Maltodextrin, Acacia Gum; Durlac® 100W; Dur-Lo®; Durpro® 107; EC-25®; Embanox® 3; Embanox® 4; Embanox® 6; Essiccum®; Golden Covo® Shortening; Golden Croissant/Danish™ Margarine; Hi-Lite™; Hi-Tone™; H.L. PY™; HWDX Winterized Salad Oil; Luxor® 1639; Moldban; Muffit; Nipanox® S-1; Nipanox® Special; Oxynex® K; Oxynex® L; Oxynex® LM; PAV; Polypro 2000® CF 45%; Polypro 2000® UF 45%; Polypro 5000® CF or SF 45%; Polypro 5000® UF 45%; Pristene® 189; Red Beet WSL-300; Red Beet WSL-400; Red Beet WSP-300; Shedd's Liquid Margarine; Shedd's Margarine; Shedd's Special 40 Butter Blend Margarine; Shedd's Special 60 Butter Blend Margarine; Sporban Regular; Sporban Special; Starplex® 90; Sustane® 3; Sustane® 7; Sustane® 7G; Sustane® 8; Sustane® 15; Sustane® 16; Sustane® 18; Sustane® 20; Sustane® 20-3; Sustane® 20A; Sustane® 20B; Sustane® 27; Sustane® 31; Sustane® CA; Sustane® PA; Sustane® Q; Sustane® S-1; Sustane® T-6; Sustane® W-1; Sustane® W; Tandem 5K; Tandem 8; Tandem 552; Tang; Tenox® 2; Tenox® 6; Tenox® 7; Tenox® 20; Tenox® 20A; Tenox® 20B; Tenox® 21; Tenox® 22; Tenox® 25; Tenox® 26; Tenox® 27; Tenox® A; Tenox® R; Tenox® S-1; Veg-X-Ten®; Vitamin B_{12} 0.1% SD No. 65354; Vitamin B_{12} 1.0% SD No. 65305

Citric acid diammonium salt. *See* Ammonium citrate

Citric acid esters of mono- and diglycerides of fatty acids
CAS 68990-05-6, 91744-38-6, 97593-31-2; EEC E472c
Synonyms: Mono- and diglycerides citrates; Citroglycerides
Properties: Sol. in hot water, lipids; HLB 10-12
Uses: Emulsifier; stabilizer; antispattering agent in margarine; improves baked goods; fat replacement in high-fat foods; synergist and solubilizer for antioxidants
Usage level: 200 ppm (oil, CFR); ADI no limit (EEC)
Regulatory: FDA 21CFR §172.832; Europe listed; UK approved
Trade names: Acidan; Axol® C 62; Axol® C 63; Cegemett® MZ 490; Imwitor® 369; Imwitor® 371; Servit® Range

Citric acid octadecyl ester. *See* Stearyl citrate
Citric acid tripotassium salt. *See* Potassium citrate
Citric acid trisodium salt. *See* Trisodium citrate
Citridic acid. *See* Aconitic acid
Citroglycerides. *See* Citric acid esters of mono- and diglycerides of fatty acids

Citronella
Definition: Cymbopogon nardus
Uses: Natural flavoring
Regulatory: FDA 21CFR §182.20, GRAS; Japan approved
Manuf./Distrib.: Florida Treatt

Citronellal
CAS 106-23-0; EINECS 203-376-6; FEMA 2307
Synonyms: 3,7-Dimethyl-6-octenal; Rhodinal
Empirical: $C_{10}H_{18}O$

Properties: Colorless to sl. yel. liq., strong lemon-citronnella-rose odor; sol. in alcohol, most oils; sl. sol. in propylene glycol; insol. in glycerin, water; m.w. 154.25; dens. 0.850-0.860; b.p. 83-85 C (11 mm); flash pt. 170 F; ref. index 1.446-1.456
Precaution: Combustible liq.
Toxicology: Heated to decomp., emits acrid smoke and irritating fumes
Uses: Synthetic flavoring; fresh, green, citrus-like fragrance and flavoring
Usage level: 4 ppm (nonalcoholic beverages), 1.3 ppm (ice cream, ices), 4.5 ppm (candy), 4.7 ppm (baked goods), 0.6 ppm (gelatins, puddings), 0.3 ppm (chewing gum)
Regulatory: FDA 21CFR §172.515; GRAS (FEMA); Japan approved as flavoring
Manuf./Distrib.: Florida Treatt

Citronellal hydrate. *See* Hydroxycitronellal
α-Citronellol. *See* Rhodinol

β-Citronellol
CAS 106-22-9; EINECS 203-375-0; FEMA 2309
Synonyms: 3,7-Dimethyl-6-octen-1-ol; (±)-β-Citronellol; d-Citronellol
Empirical: $C_{10}H_{20}O$
Formula: $(CH_3)_2C{:}CHCH_2CH_2CH(CH_3)CH_2CH_2OH$
Properties: Colorless oily liq., rose odor; sol. in fixed oils, propylene glycol; sl. sol. in water; insol. in glycerin @ 225 C; m.w. 156.30; dens. 0.850-0.860; b.p. 99 C (10 mm); flash pt. 215 F; ref. index 1.454-1.462
Precaution: Combustible liq.
Toxicology: LD50 (oral, rat) 3450 mg/kg; poison by IV route; mod. toxic by ingestion, skin contact; heated to decomp., emits acrid smoke and irritating fumes
Uses: Fresh floral, rose-like fragrance and flavoring
Usage level: 4.1 ppm (nonalcoholic beverages), 4.1 ppm (ice cream, ices), 16 ppm (candy), 18 ppm (baked goods), 5.8 ppm (gelatins, puddings), 29-52 ppm (chewing gum)
Regulatory: FDA 21CFR §172.515
Manuf./Distrib.: Int'l. Flavors & Fragrances; Penta Mfg.

(±)-β-Citronellol. *See* β-Citronellol
d-Citronellol. *See* β-Citronellol
l-Citronellol. *See* Rhodinol

Citronelloxyacetaldehyde
FEMA 2310
Empirical. $C_{12}H_{22}O_2$
Properties: Strong rose odor; m.w. 198.28; b.p. 130 C (12 mm)
Uses: Synthetic flavoring agent
Usage level: 0.005-1.0 ppm (nonalcoholic beverages), 1.4 ppm (ice cream, ices), 4.1 ppm (candy), 4.3 ppm (baked goods)
Regulatory: FDA 21CFR §172.515

Citronellyl acetate
CAS 150-84-5; EINECS 205-775-0; FEMA 2311
Synonyms: Acetic acid citronellyl ester; 3,7-Dimethyl-6-octen-1-ol acetate
Definition: Ester of citronellol and acetic acid
Empirical: $C_{12}H_{22}O_2$
Properties: Colorless liq., fruity odor; sol. in alcohol, fixed oils; insol. in glycerin, propylene glycol, water @ 229 C; m.w. 198.34; dens. 0.883-0.893; b.p. 240 C; flash pt. > 212 F; ref. index 1.440-1.450
Precaution: Combustible liq.
Toxicology: LD50 (oral, rat) 6800 mg/kg; mildly toxic by ingestion; human skin irritant; heated to decomp., emits acrid smoke and irritating fumes
Uses: Synthetic flavoring agent
Usage level: 3.4 ppm (nonalcoholic beverages), 4.2 ppm (ice cream, ices), 19 ppm (candy), 32 ppm (baked goods), 63-100 ppm (chewing gum)
Regulatory: FDA 21CFR §172.515, 182.60, GRAS; GRAS (FEMA); Japan approved as flavoring
Manuf./Distrib.: Chr. Hansen's

α-Citronellyl acetate. *See* Rhodinyl acetate

Citronellyl butyrate. *See also* Rhodinyl butyrate
CAS 141-16-2; FEMA 2312
Synonyms: 3,7-Dimethyl-6-octen-1-yl butyrate
Empirical: $C_{14}H_{26}O_2$
Properties: Colorless liq., rose-like odor, sweet plum-like taste; m.w. 226.36; dens. 0.880-0.886; b.p. 134-135

Citronellyl formate

C (12 mm); flash pt. > 230 F; ref. index 1.444-1.448
Precaution: Combustible liq.
Toxicology: Heated to decomp., emits acrid smoke and fumes
Uses: Synthetic flavoring agent
Usage level: 3.8 ppm (nonalcoholic beverages), 11 ppm (ice cream, ices, baked goods), 13 ppm (candy), 3.1-4.2 ppm (gelatins, puddings), 2.3 ppm (chewing gum)
Regulatory: FDA 21CFR §172.515
Manuf./Distrib.: Aldrich

Citronellyl formate
CAS 105-85-1; FEMA 2314
Synonyms: 3,7-Dimethyl-6-octen-1-yl formate
Empirical: $C_{11}H_{20}O_2$
Properties: Colorless oily liq., fruity rose-like odor, sweet fruity taste; insol. in water; m.w. 184.28; dens. 0.897; b.p. 235 C; flash pt. 198 F; ref. index 1.4430-1.4490
Precaution: Combustible liq.
Toxicology: LD50 (oral, rat) 8400 mg/kg; mildly toxic by ingestion; human skin irritant; heated to decomp., emits acrid smoke and fumes
Uses: Synthetic flavoring agent
Usage level: 14 ppm (nonalcoholic beverages), 13 ppm (ice cream, ices), 19 ppm (candy), 32 ppm (baked goods), 63-100 ppm (chewing gum)
Regulatory: FDA 21CFR §172.515; Japan approved as flavoring
Manuf./Distrib.: Aldrich

Citronellyl isobutyrate
FEMA 2313
Synonyms: 3,7-Dimethyl-6-octen-1-yl isobutyrate
Definition: Obtained by direct esterification of citronellol with isobutyric acid via azeotropic conditions or using isobutyric anhydride
Properties: Colorless liq., sweet fruity rose-like odor, apricot-like taste; m.w. 226.36; dens. 0.8760-0.8830; b.p. 249 C; flash pt. 100 C; ref. index 1.4400-1.4480
Precaution: Combustible liq.
Toxicology: Heated to decomp., emits acrid smoke and fumes
Uses: Synthetic flavoring agent
Usage level: 2.3 ppm (nonalcoholic beverages), 1.7 ppm (ice cream, ices), 8.2 ppm (candy), 12 ppm (baked goods), 3.1 ppm (gelatins, puddings)
Regulatory: FDA 21CFR §172.515

Citronellyl phenylacetate
FEMA 2315
Synonyms: Citronellyl α-toluate; 3,7-Dimethyl-6-octen-1-yl phenylacetate
Empirical: $C_{18}H_{26}O_2$
Properties: Honey rose-like odor; m.w. 274.41; dens. 0.992 (15.5 C); b.p. 342 C; ref. index 1.5100
Uses: Synthetic flavoring agent
Usage level: 1.3 ppm (nonalcoholic beverages), 0.95 ppm (ice cream, ices), 2.4 ppm (candy), 17 ppm (baked goods)
Regulatory: FDA 21CFR §172.515

Citronellyl propionate. *See also* Rhodinyl propionate
FEMA 2316
Synonyms: 3,7-Dimethyl-6-octen-1-yl propionate
Definition: Obtained by direct esterification of citronellol with propionic cid under azeotropic conditions or using propionic anhydride
Empirical: $C_{13}H_{24}O_2$
Properties: Colorless liq., rose-like odor, bittersweet plum-like taste; m.w. 212.33; dens. 0.8810-0.8840; b.p. 242 C; flash pt. > 100 C; ref. index 1.4430-1.4490
Precaution: Combustible liq.
Toxicology: Heated to decomp., emits acrid smoke and fumes
Uses: Synthetic flavoring agent
Usage level: 3.1 ppm (nonalcoholic beverages), 9.0 ppm (ice cream, ices), 18 ppm (candy), 19 ppm (baked goods), 0.80-15 ppm (chewing gum)
Regulatory: FDA 21CFR §172.515
Manuf./Distrib.: Chr. Hansen's

Citronellyl α-toluate. *See* Citronellyl phenylacetate

Citronellyl valerate
FEMA 2317
Synonyms: 3,7-Dimethyl-6-octen-1-yl valerate
Empirical: $C_{15}H_{28}O_2$
Properties: Liq., rose herb honey-like odor; m.w. 240.39; dens. 0.890; b.p. 237 C; ref. index 1.4435
Uses: Synthetic flavoring agent
Usage level: 1.0 ppm (nonalcoholic beverages), 2.5 ppm (ice cream, ices), 3.0 ppm (candy), 7.7 ppm (baked goods)
Regulatory: FDA 21CFR §172.515
Manuf./Distrib.: Aldrich

Citrus aurantium fruit extract. *See* Bitter orange extract
Citrus aurantium peel extract. *See* Bitter orange peel extract

Citrus extract
EINECS 304-454-3
Synonyms: Citrus spp.
Uses: Natural flavoring agent
Regulatory: FDA 21CFR §182.20, GRAS
Manuf./Distrib.: Caminiti Foti & Co. Srl

Citrus limon extract. *See* Lemon extract
Citrus limon juice extract. *See* Lemon juice extract
Citrus limon oil. *See* Lemon oil
Citrus paradisi peel oil. *See* Grapefruit oil
Citrus pectin. *See* Pectin

Citrus Red No. 2
CAS 6358-53-8
Synonyms: Solvent Red 80; CI 12156
Classification: Monoazo color
Empirical: $C_{18}H_{16}N_2O_3$
Properties: Mod. sol. in alcohol; sl. sol. in water; m.w. 308.36; m.p. 156 C
Toxicology: Experimental carcinogen; mutagenic data; heated to decomp., emits toxic fumes of NO_x
Uses: Colorant for oranges
Regulatory: FDA 21CFR §74.302 (limitation 2 ppm)

Citrus sinensis oil. *See* Orange oil
Citrus spp. *See* Citrus extract

Civet
EINECS 272-826-1; FEMA 2319
Synonyms: Zibeth; Zibet; Zibetum
Definition: Derived from civet cats, *Viverra civetta* and *V. zibetha*
Uses: Natural flavoring agent
Usage level: 1.0 ppm (nonalcoholic beverages), 3.0 ppm (ice cream, ices), 3.7 ppm (candy), 2.8 ppm (baked goods), 0.10 ppm (gelatins, puddings), 2.2 ppm (chewing gum)
Regulatory: FDA 21CFR §182.50, GRAS; Japan approved
Manuf./Distrib.: Pierre Chauvet (absolute)

Clary oil
CAS 8016-63-5
Synonyms: Clary sage oil; Muscatel oil
Definition: Oil from steam distillation of flowering tops and leaves of *Salvia sclarea*
Properties: Pale yel. liq., herbaceous odor; sol. in fixed oils, min. oil; insol. in glycerin, propylene glycol
Toxicology: Heated to decomp., emits acrid smoke and irritating fumes
Uses: Natural flavoring agent
Usage level: 1.8 ppm (nonalcoholic beverages), 100 ppm (alcoholic beverages), 3.9 ppm (ice cream, ices), 5.3 ppm (candy), 13 ppm (baked goods), 20 ppm (condiments)
Regulatory: FDA 21CFR §182.10, 182.20, GRAS; Japan approved
Manuf./Distrib.: C.A.L.-Pfizer; Pierre Chauvet; Commodity Services; Florida Treatt

Clary sage oil. *See* Clary oil
C18 linear alcohol. *See* Stearyl alcohol
C6 linear alpha olefin. *See* 1-Hexene
C12 linear primary alcohol. *See* Lauryl alcohol
C16 linear primary alcohol. *See* Cetyl alcohol

Clove
Definition: Eugenia caryophyllata
Properties: Spicy, clove-like aroma; warm flavor
Uses: Natural flavoring agent
Regulatory: Japan approved

Clove extract
EINECS 284-638-7
Synonyms: Eugenia caryophyllus extract
Uses: Natural flavoring agent and adjuvant; antioxidant (Japan)
Regulatory: FDA 21CFR §184.1257, GRAS; Japan approved
Manuf./Distrib.: C.A.L.-Pfizer; Chart

Clove leaf oil
CAS 8015-97-2
Definition: Volatile oil obtained by steam distillation of leaves of Eugenia caryophyllus, contg. mostly eugenol
Toxicology: LD50 (oral, rat) 1370 mg/ kg; mod. toxic by ingestion; severe skin irritant; heated to decomp., emits
    acrid smoke and fumes
Uses: Natural flavoring agent
Regulatory: FDA 21CFR §184.1257, GRAS
Manuf./Distrib.: Chart; Florida Treatt

Clove oil
CAS 8000-34-8; FEMA 2323 (bud), 2325 (leaf), 2328 (stem)
Synonyms: Eugenia caryophyllus oil
Definition: Volatile oil distilled from the dried flower buds of Eugenia caryophyllus, contg. 82-87% eugenol, 10%
    acetyleugenol, etc.
Properties: Colorless to pale yel. liq.; becomes darker and thicker with age; very sol. in strong alcohol, ether,
    glac. acetic acid; insol. in water; dens. 1.036-1.060; b.p. 250 C; ref. index 1.527-1.538 (20 C)
Precaution: Keep cool, well closed; protect from light
Uses: Natural flavoring agent and adjuvant; in confectionery
Regulatory: FDA 21CFR §184.1257, GRAS; 27CFR §21.65, 21.151; Council of Europe listed
Manuf./Distrib.: Chart; Pierre Chauvet; Florida Treatt

Clover
FEMA 2326
Definition: Derived from Trifolium spp.
Properties: Intense unpleasant odor
Uses: Natural flavoring agent
Usage level: 2.0 ppm (nonalcoholic beverages), 3.0 ppm (ice cream, ices), 20 ppm (candy), 9.0 ppm (baked
    goods)
Regulatory: FDA 21CFR §182.10, 182.20, GRAS; Japan approved

Clover blossom extract
CAS 85085-25-2; EINECS 285-356-7
Synonyms: Trifolium extract; Trifolium pratense extract
Definition: Extract of the flowers of Trifolium pratense
Uses: Natural flavoring
Regulatory: FDA 21CFR §182.20, GRAS

CMC. See Carboxymethylcellulose sodium
Coal tar naphtha. See Naphtha

Cobalt
CAS 7440-48-4; EINECS 231-158-0
Classification: Metallic element
Empirical: Co
Properties: Steel gray, shining, hard, somewhat malleable metal; ferromagnetic; m.w. 58.93; dens. 8.9; m.p.
    1493 C; b.p. 3100 C
Uses: Natural processing aid
Regulatory: Japan approved; not permitted in certain foods

Cobalt caprylate
Toxicology: Heated to decomp., emits acrid smoke and irritating fumes
Uses: Drier migrating from food pkg.
Regulatory: FDA 21CFR §181.25

Cobalt gluconate
Empirical: $C_{12}H_{22}O_{14}Co$
Properties: m.w. 449.3
Uses: Mineral source
Manuf./Distrib.: Am. Roland
Trade names: Gluconal® CO

Cobalt linoleate
Synonyms: Cobaltous linoleate
Definition: Obtained by boiling a cobalt salt and sodium linoleate
Formula: $Co(C_{18}H_{31}O_2)_2$
Properties: Brown amorphous powd.; sol. in alcohol, ether, acids; insol. in water
Precaution: Combustible
Toxicology: Heated to decomp., emits acrid smoke and irritating fumes
Uses: Drier migrating from food pkg.
Regulatory: FDA 21CFR §181.25

Cobalt naphthenate
CAS 61789-51-3
Synonyms: Naphthenic acid cobalt salt; Cobaltous naphthenate
Definition: Obtained by treating cobaltous hydroxide with naphthenic acid
Properties: Brn. amorphous powd. or bluish-red solid; sol. in oil, alcohol, ether; insol. in water; dens. 0.95; flash pt. 120 F; autoignition temp. 529 F; contains 6% Co
Precaution: Flamm. when exposed to heat or flame
Toxicology: TWA 0.1 mg(Co)/m³ (fume, dust); LD50 (oral, rat) 3900 mg/kg; mod. toxic by ingestion; heated to decomp., emits acrid smoke and irritating fumes
Uses: Drier when migrating from food pkg.
Regulatory: FDA 21CFR §181.25

Cobaltous linoleate. *See* Cobalt linoleate
Cobaltous naphthenate. *See* Cobalt naphthenate
Cobaltous sulfate. *See* Cobalt sulfate (ous)
Cobalt (II) sulfate (1:1). *See* Cobalt sulfate (ous)

Cobalt sulfate (ous)
CAS 10124-43-3
Synonyms: Cobaltous sulfate; Cobalt (II) sulfate (1:1)
Definition: Obtained by action of sulfuric acid on cobaltous oxide
Empirical: $O_4S \cdot Co$
Formula: $CoSO_4$
Properties: Red to lavender dimorphic, orthorhombic cryst.; dissolves slowly in boiling water; m.w. 154.99; dens. 3.71; stable to 708 C
Toxicology: TLV:TWA 0.05 mg(Co)/m³; LD50 (oral, rat) 424 mg/kg; poison by IV, IP routes; mod. toxic by ingestion; heated to decomp., emits toxic fumes of SO_x
Uses: Boiler water additive; foam stabilizer; fermented malt beverages (prohibited)
Regulatory: FDA 21CFR §173.310 (catalyst in boiler water), 189.120 (prohibited from direct addition or use in human food)
Manuf./Distrib.: Atomergic Chemetals

Cobalt tallate
Definition: Cobalt deriv. of refined tall oil, of varying composition
Precaution: Combustible
Toxicology: Heated to decomp., emits acrid smoke and irritating fumes
Uses: Drier migrating from food pkg.
Regulatory: FDA 21CFR §181.25

Cocamide DEA
CAS 8051-30-7; 61791-31-9; 68603-42-9; EINECS 263-163-9
Synonyms: Coconut diethanolamide; Cocoyl diethanolamide; N,N-Bis (2-hydroxyethyl) coco amides
Definition: Ethanolamides of coconut acid
Formula: $RCO-N(CH_2CH_2OH)_2$, RCO represents the coconut acid radical
Uses: Delinting of cottonseed for prod. of cottonseed oil; byprods. for use in animal feed
Regulatory: FDA 21CFR §172.710, 173.322 (0.2% max.), 175.105, 176.210, 177.1200, 177.2260, 177.2800

Coccus. *See* Cochineal
Cochin. *See* Lemongrass oil East Indian

Cochineal
CAS 2611-82-7; EINECS 215-680-6; EEC E120
Synonyms: Coccus; Acid Red 18; CI 16255
Classification: Monoazo color
Definition: Red coloring matter consistg. of the dried bodies of the female insects of *Coccus cacti*. The coloring principle is carminic acid, $C_{22}H_{20}O_{13}$
Empirical: $C_{20}H_{14}N_2O_{10}S_3 \cdot 3Na$
Properties: m.w. 604.48
Uses: Food colorant
Usage level: ADI 0-2.5 mg/kg (EEC)
Regulatory: Europe listed; Japan restricted (extract)

Cochineal extract. *See* Carmine

Cocoa butter
CAS 8002-31-1
Synonyms: Theobroma oil; Cacao butter
Definition: Yellowish white solid obtained from roasted seeds of *Theobroma cacao*
Properties: Ylsh.-wh. solid, chocolate-like odor and taste; sol. in ether, chloroform; sl. sol. in alcohol; insol. in water; m.p. 30-35 C; ref. index 1.4537-1.4585 (40 C)
Precaution: Combustible
Uses: Chewing gum base
Regulatory: FDA 21CFR §182.20, GRAS
Manuf./Distrib.: Karlshamns
Trade names containing: Alcolec® C-150

Cocoa butter substitute
Definition: Mixture of triglycerides derived primarily from palm oils
Properties: White, waxy, odorless solid; m.p. 33.8-35.5 C
Uses: Coating agent, formulation aid, texturizer

Cocoa extract
EINECS 283-480-6
Synonyms: Theobroma cacao extract
Definition: Extract of *Theobroma cacao*
Uses: Natural flavoring
Regulatory: FDA 21CFR §182.20

Coco fatty acid. *See* Coconut acid

Cocoglycerides
Synonyms: Glycerides, coconut, mono-, di-, and tri-
Definition: Mixture of mono, di and triglycerides derived from coconut oil
Uses: Fat for ice cream, chocolate, and confectionery prods.
Trade names: Witarix® 250

Coconut acid
CAS 61788-47-4; 67701-05-7, 68937-85-9; EINECS 262-978-7
Synonyms: Coco fatty acid; Coconut oil acids; Coconut fatty acids; Acids, coconut
Definition: Mixtures of fatty acids
Uses: Food grade fatty acid
Regulatory: FDA 21CFR §175.105, 175.320, 176.200, 176.210, 177.1010, 177.2260, 177.2600, 177.2800, 178.3570, 178.3910
Trade names: Emery® 6361

Coconut aldehyde. *See* γ-Nonalactone
Coconut butter. *See* Coconut oil
Coconut diethanolamide. *See* Cocamide DEA
Coconut fatty acids. *See* Coconut acid

Coconut oil
CAS 8001-31-8; EINECS 232-282-8
Synonyms: Copra oil; Coconut butter; Coconut palm oil
Definition: Fixed oil obtained from kernels of seeds of *Cocos nucifera*
Properties: Wh. fatty solid or liq., sweet nutty taste; very sol. in chloroform, ether, CS_2; pract. insol. in water; dens. 0.903 (0/4 C); m.p. 21-27 C; acid no. < 6; iodine no. 8-9.5; sapon. no. 255-258; ref. index 1.4485-1.4495; surf. tens. 33.4 dynes/cm
Precaution: Flamm. solid when exposed to heat or flame; may spontaneously heat and ignite if stored wet and

hot
Uses: Edible fat, coating agent, emulsifier, release agent, formulation aid, texturizer in baked goods, candy, desserts, margarine; migrating to foods from cotton in dry food pkg.
Regulatory: FDA 21CFR §175.105, 175.300, 176.200, 176.210, 177.2800, 182.70; GRAS
Manuf./Distrib.: Aarhus; Arista Industries; British Arkady; Calgene; Croda Singapore; Edw. Gittens; Karlshamns; Peerless Refining; Penta Mfg.; Spectrum Naturals
Trade names: Cobee 76; Coconut Oil® 76; Coconut Oil® 92; Konut; Pureco® 76
Trade names containing: Bake-Well 52; Centrophil® M; Centrophil® W; Coco-Spred™; Dry Vitamin A Palmitate Type 250-SD No. 65378; Dry Vitamin D_3 Type 100-SD No. 65216; Neobee® SL-210; Neobee® SL-220; Neobee® SL-230

Coconut oil acids. *See* Coconut acid
Coconut oil, hydrogenated. *See* Hydrogenated coconut oil

Coconut oil triglycerides
Uses: Soft fat and filling for prod. of cooling fillings
Trade names containing: Carotenal Sol'n. #2 No. 66425; Carotenal Sol'n. #73 No. 66428

Coconut palm oil. *See* Coconut oil
Cocoyl diethanolamide. *See* Cocamide DEA

Cod liver oil
CAS 8001-69-2; EINECS 232-289-6
Synonyms: Morrhua oil
Definition: Fixed oil expressed from fresh livers of *Gadus morrhua* and other species of codfish, contg. vitamins A and D, omega 3 fatty acid
Properties: Clear amber liq., sl. fishy odor and taste; sol. in ether, chloroform, ethyl acetate, carbon disulfide, petroleum ether; sl. sol. in alcohol; dens. 0.918-0.927; iodine no. 145-180; sapon. no. 180-192; ref. index 1.4705-1.4745
Precaution: Combustible
Uses: Dietary supplement
Regulatory: FDA 21CFR §175.105, 176.200, 176.210, 177.2800; GRAS for use in dietary supplements
Manuf./Distrib.: Am. Roland; Arista Industries

Coenzyme R. *See* d-Biotin

Coffee
Definition: Berries of the evergreen plant *Coffea arabica*
Uses: Food flavoring
Trade names containing: Coffee Conc.

Coffee bean extract
CAS 8001-67-0; 84650-00-0; EINECS 283-481-1
Definition: Extract of beans of *Coffea arabica*
Uses: Natural flavoring agent; antioxidant (Japan)
Regulatory: FDA 21CFR §182.20, GRAS; Japan approved
Manuf./Distrib.: Chart; Commodity Services

Coffeine. *See* Caffeine

Cognac oil, green or white
FEMA 2331 (green), 2332 (white)
Synonyms: Wine yeast oil; Ethyl oenanthate
Properties: Char. cognac aroma with a fruital note
Toxicology: Heated to decomp., emits acrid smoke and fumes
Uses: Natural flavoring agent
Regulatory: FDA 21CFR §182.50, GRAS
Manuf./Distrib.: C.A.L.-Pfizer; Pierre Chauvet; Florida Treatt

Cognac oil, synthetic. *See* Ethyl heptanoate

Cola
Synonyms: Kola; Kola nut; Kola seeds; Guru; Bissy nuts; Soudan coffee
Definition: Dried cotyledons of *Cola nitida*
Uses: Flavoring in soft drinks
Trade names containing: Cola No. 23443; Cola Acid No. 23444

Cola nut
EINECS 289-720-6 (extract)

Synonyms: Kola nut
Definition: Cola acuminata and other species
Uses: Natural flavoring agent
Regulatory: FDA 21CFR §182.20, GRAS; Japan approved

C12 alpha olefin. *See* Dodecene-1
Collodion cotton. *See* Nitrocellulose
Colloidal silicon dioxide. *See* Silica
Cologel. *See* Methylcellulose
Colombo. *See* Calumba root
Colophony. *See* Rosin
Columbian spirits. *See* Methyl alcohol
Colza oil. *See* Rapeseed oil
Commiphora extract. *See* Myrrh extract
Common basil extract. *See* Basil extract
Common salt. *See* Sodium chloride
Common speedwell extract. *See* Veronica extract
Confectioner's sugar. *See* Sucrose
Copaiba oil. *See* Balsam copaiba

Copal resin
Properties: Insol. in oils, water
Uses: Natural chewing gum base
Regulatory: Japan approved

Copper
CAS 7440-50-8; EINECS 231-159-6
Synonyms: Bronze powder; Copper bronze; Gold bronze
Formula: Cu
Properties: Reddish metal; at.wt. 63.54; dens. 8.96; m.p. 1083 C; b.p. 2595 C
Toxicology: TLV 0.2 mg/m^3 (fume), 1 mg/m^3 (dusts and mists); poison to humans by ingestion
Uses: Dietary supplement; herbicide for potable water; processing aid (Japan)
Regulatory: FDA 21CFR §193.90, herbicides residue tolerance 1 ppm in potable water; GRAS for use in dietary supplements; Japan restricted

Copperas. *See* Ferrous sulfate
Copperas. *See* Ferrous sulfate heptahydrate
Copper bronze. *See* Copper

Copper gluconate
CAS 527-09-3; EINECS 208-408-2
Synonyms: Bis(D-gluconato) copper; Cupric gluconate
Definition: Copper salt of gluconic acid
Empirical: $C_{12}H_{22}O_{14}Cu$
Formula: $[HOCOO^-CHOHCHCHOHCHOHCH_2OH]_2Cu^{++}$
Properties: Lt. blue powd., odorless; sol. in water; sl. sol. in alcohol; insol. in acetone, ether; m.w. 453.8
Toxicology: Heated to decomp., emits acrid smoke and irritating fumes
Uses: Direct food additive, mineral source, nutrient, dietary supplement, synergist
Usage level: up to 0.005%
Regulatory: FDA 21CFR §182.5260, 184.1260, GRAS; Japan approved (0.6 mg/L as copper in milk)
Manuf./Distrib.: Am. Roland
Trade names: Gluconal® CU

Copper (I) iodide. *See* Copper iodide (ous)

Copper iodide (ous)
CAS 7681-65-4; EINECS 231-674-6
Synonyms: Cuprous iodide; Copper (I) iodide
Empirical: CuI
Properties: Wh. cryst. powd.; sol. in ammonia and KI sol'ns.; insol. in water; dens. 5.653; m.p. 606 C; b.p. 1290 C
Toxicology: Heated to decomp., emits toxic fumes of I⁻
Uses: Source of dietary iodine in table salt, feed additive
Usage level: Limitation 0.01% (table salt)
Regulatory: FDA 21CFR §184.1265, GRAS
Manuf./Distrib.: Atomergic Chemetals

Copper sulfate. *See* Cupric sulfate (anhydrous)
Copper sulfate (ic). *See* Cupric sulfate (pentahydrate)
Copper sulfate pentahydrate. *See* Cupric sulfate (pentahydrate)
Copra oil. *See* Coconut oil
Cordycepic acid. *See* D-Mannitol

Coriander oil
CAS 8008-52-4; EINECS 283-880-0 (extract); FEMA 2334
Definition: Volatile oil from steam distillation of ripe fruit of *Coriandrum sativum*, contg. d-linalool and its acetate
Properties: Colorless to pale yel. liq., char. odor and taste; very sol. in chloroform, ether, glac. acetic acid; sol. in stronger alcohol; pract. insol. in water; dens. 0.863-0.875 (25/25 C); ref. index 1.4620-1.4720 (20 C)
Precaution: Keep cool, well closed; protect from light
Toxicology: LD50 (oral, rat) 4130 mg/kg; mod. toxic by ingestion; mutagenic data; skin irritant; heated to decomp., emits acrid smoke and fumes
Uses: Natural flavoring agent for curry powd., meat, sausages, alcoholic beverages; preservative
Usage level: 3.1 ppm (nonalcoholic beverages), 10-30 ppm (alcoholic beverages), 4.5 ppm (ice cream, ices), 9.3 ppm (baked goods), 7.4 ppm (chewing gum), 12 ppm (condiments), 47 ppm (meats), 8.8 ppm (candy)
Regulatory: FDA 21CFR §182.10, 182.20, GRAS; GRAS (FEMA); Japan approved; Council of Europe listed
Manuf./Distrib.: Chart; Pierre Chauvet; Commodity Services; Florida Treatt

Cork, oak
Definition: From *Querrcus suber* or *Q. occidentalis*
Uses: Natural flavoring for alcoholic beverages only
Regulatory: FDA 21CFR §172.510

Corn bran
Uses: Source of dietary fiber
Trade names: Stabilized Corn Bran; Stabilized Micro-Lite Corn Bran
Trade names containing: Stabilized Cookie Blend

Corn gluten
CAS 66071-96-3
Synonyms: Corn gluten meal
Toxicology: Heated to decomp., emits acrid smoke and irritating fumes
Uses: Direct food additive, nutrient supplement
Regulatory: FDA 21CFR §184.1321, GRAS

Corn gluten meal. *See* Corn gluten

Corn glycerides
Definition: Mixture of mono, di, and triglycerides derived from corn oil
Uses: Food emulsifier
Trade names: Maisine

Corn meal
CAS 66071-96-3; EINECS 266-116-0
Synonyms: Glutens, corn
Definition: Coarse flour from milling of *Zea mays*
Uses: Nutrient supplement, texturizer
Regulatory: FDA 21CFR §137.250, 137.275, 184.1321, GRAS
Trade names containing: Muffin Kote

Cornmint oil, partially dementholized. *See* Mentha arvensis oil

Corn oil
CAS 8001-30-7; EINECS 232-281-2
Synonyms: Maize oil; Zea mays oil
Definition: Refined fixed oil obtained from wet milling of corn, *Zea mays*
Properties: Pale yel. liq., faint char. odor and taste; insol. in water; sol. in ether, chloroform, amyl acetate, benzene, CS_2; sl. sol. in alcohol; dens. 0.914-0.921; m.p. -10 C; acid no. 2-6; iodine no. 109-133; flash pt. 321 C; ref. index 1.470-1.474
Precaution: Combustible liq. when exposed to heat or flame; dangerous spontaneous heating may occur
Toxicology: Nontoxic; human skin irritant; experimental teratogen; may be an allergen
Uses: Used in preparing foodstuffs, coating agent, emulsifier, formulation aid, texturizer, as salad and cooking oil, in bakery prods., margarine, dietary supplement
Regulatory: FDA 21CFR §175.105, 175.300, 176.200, 176.210, GRAS
Manuf./Distrib.: ADM; Arista Industries; British Arkady; Calgene; Cargill; Cerestar Int'l.; Colombus Foods; Spectrum Naturals; A E Staley Mfg.; Vandemoortele Professiona; Wensleydale Foods

Corn oil, hydrogenated

Trade names: Corn Oil; Corn Oil; Corn Salad Oil 104-250; Liquid Corn Oil 87-070-0
Trade names containing: Hentex-35; Liquid Vitamin D₃ No. 63643; Superb® Fish and Chip Oil 102-060; Tenox® 4; Tenox® 4B; Tenox® 6; Tenox® 25; Tenox® 26

Corn oil, hydrogenated. *See* Hydrogenated corn oil

Corn silk
FEMA 2335
Definition: Fresh styles and stigmas of *Zea mays*
Toxicology: Heated to decomp., emits acrid smoke and irritating fumes
Uses: Flavoring agent
Usage level: Limitation 30 ppm (baked goods/mixes), 20 ppm (nonalcoholic beverages), 10 ppm (frozen dairy desserts), 20 ppm (soft candy), 4 ppm (other food)
Regulatory: FDA 21CFR §184.1262
Manuf./Distrib.: Chart

Corn silk extract
Synonyms: Zea mays extract; Stigmata maydis extract
Definition: Extract of the stigmas of *Zea mays*
Toxicology: Heated to decomp., emits acrid smoke and irritating fumes
Uses: Flavoring agent
Usage level: Limitation 30 ppm (baked goods/mixes), 20 ppm (nonalcoholic beverages), 10 ppm (frozen dairy desserts), 20 ppm (soft candy), 4 ppm (other food)
Regulatory: FDA 21CFR §184.1262
Manuf./Distrib.: Chart

Corn starch
CAS 9005-25-8; EINECS 232-679-6
Synonyms: Starch, corn
Definition: Obtained from grains of *Zea mays*; carbohydrate polymer consisting primarily of amylose and amylopectin
Formula: White powd.
Uses: Source of glucose; filler in baking powder; thickening agent in food prods.; dietary supplement; migrating to foods from cotton in dry food pkg., paper/paperboard
Regulatory: FDA 21CFR §175.105, 178.3520, 182.70, 182.90; GRAS for use in dietary supplements
Manuf./Distrib.: ADM; Am. Maize Prods.; Cerestar UK; Grain Processing; Nat'l. Starch & Chem.; A.E. Staley Mfg.
Trade names: Mira-Gel® 463; Mira-Set® B; Pure-Dent® B700; Pure-Dent® B810; Pure-Dent® B812; Pure-Dent® B815; Pure-Dent® B816; Pure-Dent® B880; Pure Food Powd. Starch 105-A; Pure Food Powd. Starch 131-C; Pure Food Starch Bleached 142-A; Pure-Gel® B990; Staley® 7025; Staley® 7350 Waxy No. 1 Starch; Staley® Moulding Starch; Staley® Pure Food Powd. (PFP); Staley® Pure Food Powd. Starch Type I; Staley® Pure Food Powd. Starch Type II; Staley® Redried Starch A; Staley® Redried Starch B; Sta-Rx®; Stir & Sperse®
Trade names containing: GP-1200; Panospray™ WL

Corn sugar. *See* D-Glucose anhyd.
Corn sugar gum. *See* Xanthan gum
Corn sugar syrup. *See* Corn syrup

Corn syrup
CAS 8029-43-4; EINECS 232-436-4
Synonyms: Glucose syrup; Corn sugar syrup
Definition: Mixture of D-glucose, maltose, and maltodextrins; obtained by partial hydrolysis of corn starch
Properties: Aq. syrup
Uses: Food industry as nutritive carbohydrate sweetener, thickener, bodying agent in soft drinks, beer, other beverages, dietary supplement; feedstock in fermentation industry
Regulatory: FDA 21CFR §182.1866, 184.1865, GRAS; cleared by MID to flavor sausage, hamburger, meat loaf, luncheon meat, chopped or pressed ham; for use alone at 2% or in combination with corn syrup solids or glucose syrup, with combination totaling 2% on dry basis
Manuf./Distrib.: ADM; Am. Maize Prods.; Gist-Brocades Food Ingreds.; Jungbunzlauer; A.E. Staley Mfg.
Trade names: CornSweet® 42; CornSweet® 55; CornSweet® 95; Corn Syrup 36/43; Corn Syrup 42/43; Corn Syrup 42/44; Corn Syrup 52/43; Corn Syrup 62/43; Corn Syrup 62/44; Corn Syrup 62/44-1; Corn Syrup 62/44-2; Corn Syrup 97/71; 42/43 Corn Syrup; 62/43 Corn Syrup
Trade names containing: Hentex-10; Hentex-20; Hentex-20A; Hentex-25; Hentex-30; Hentex-30A; Hentex-76; Luxor® 1626; Super Flex

Corn syrup, hydrogenated. *See* Hydrogenated starch hydrolysate

Corn syrup solids
CAS 68131-37-3
Synonyms: Glucose syrup solids
Properties: Clear visc. syrup
Uses: Flavoring agent for sausage, hamburger, meat loaf, luncheon meat, chopped or pressed ham
Usage level: May be used alone at a level of 2%
Trade names: Fro-Dex® 24 Powd.; Maltrin® M200; Maltrin® M205; Maltrin® M250; Maltrin® M255; Maltrin® M365; Maltrin® QD M600; Star-Dri® 20; Star-Dri® 35F; Star-Dri® 35R; Star-Dri® 42C; Star-Dri® 42F; Star-Dri® 42R; Star-Dri® 42X; Star-Dri® 55
Trade names containing: Extramalt 10; Extramalt 35; Lamequick® CE 1 SF; Precision Base; VCR; Viobin Defatted Wheat Germ #2

Cosmetic talc. *See* Talc

Costmary
EINECS 286-479-9 (extract)
Definition: From *Chrysanthemum balsamita*
Properties: Pleasant clary sage, tansy, or mint-like odor
Uses: Natural flavoring; for use in alcoholic beverages
Regulatory: FDA 21CFR §172.510

Costus root oil
FEMA 2336
Definition: From *Saussurea lappa*
Toxicology: Heated to decomp., emits acrid smoke and fumes
Uses: Natural flavoring agent
Usage level: 0.08 ppm (nonalcoholic beverages), 0.90 ppm (ice cream, ices), 1.9 ppm (candy), 1.2 ppm (baked goods), 0.10 ppm (gelatins, puddings)
Regulatory: FDA 21CFR §172.510
Manuf./Distrib.: Acme-Hardesty; Pierre Chauvet

Cotton fiber. *See* Cellulose

Cottonseed fiber
Uses: Source of dietary fiber
Trade names: Justfiber® CS-20-H; Justfiber® CS-35-H

Cottonseed flour, partially defatted, cooked
EINECS 269-668-0 (cottonseed flour)
Toxicology: Heated to decomp., emits acrid smoke and fumes
Uses: Processing aid, colorant; used in hard candy

Cottonseed glyceride
CAS 8029-44-5; EINECS 232-438-5
Synonyms: Cottonseed oil monoglyceride; Glyceryl mono cottonseed oil; Glycerides, cottonseed oil, mono-
Definition: Monoglyceride derived from cottonseed oil
Uses: Emulsifier
Trade names: Dimodan CP; Dimodan CP Kosher; Myvatex® 7-85; Myvatex® 7-85K; Myverol® 18-85; Myverol® 18-85K
Trade names containing: Myvatex® Monoset® K

Cottonseed oil
CAS 8001-29-4; EINECS 232-280-7
Synonyms: Deodorized winterized cottonseed oil
Definition: Oil from seeds of *Gossypium hirsutum*
Properties: Pale yel. oily liq., nearly odorless; sol. in ether, benzene, chloroform; sl. sol. in alcohol; dens. 0.915-0.921; f.p. 0-5 C; solid. pt. 31-33 C; iodine no. 109-120; sapon. no. 190-198; flash pt. (CC) 486 F
Precaution: Combustible liq. when exposed to heat of flame; may be dangerous hazard due to spontaneous heating
Toxicology: Experimental tumorigen and teratogen; allergen
Uses: Coating agent, emulsifier, lubricant, formulation aid, texturizer for use in cooking oil, margarine, salad oil, shortening
Regulatory: FDA 21CFR §175.105, 175.300, 176.200, 176.210, 177.2800, GRAS
Manuf./Distrib.: Arista Industries; British Arkady; Calgene; Cargill; Colombus Foods; Good Food; Karlshamns; Nat'l. Cottonseed; Penta Mfg.
Trade names: Cottonseed Cooking Oil 82-060-0; Cottonseed Oil; Jewel Oil™

Trade names containing: Canola Vegetable All-Purpose Shortening 81-706-0; Centrophil® M; Centrophil® W; Vegetable All Purpose Shortening 86-713-0

Cottonseed oil monoglyceride. *See* Cottonseed glyceride

Coumarone-indene resin
Classification: Thermosetting resin
Properties: Soft and sticky at R.T.; hardens on heating to solid; soften. pt. 126 C; ref. index 1.63-1.64
Toxicology: Heated to decomp., emits acrid smoke and irritating fumes
Uses: Protective coating on grapefruit, lemons, limes, oranges, tangelos, tangerines; in chewing gum
Usage level: Limitation 200 ppm on fresh wt. basis (citrus coating)
Regulatory: FDA 21CFR §172.215
Manuf./Distrib.: Allchem Industries

C11 primary alcohol. *See* Undecyl alcohol
Cream of tartar. *See* Potassium acid tartrate
Creeping thyme. *See* Thyme, wild
Creosol. *See* 2-Methoxy-4-methylphenol
4-Cresol. *See* p-Cresol

p-Cresol
CAS 106-44-5; EINECS 203-398-6; FEMA 2337
Synonyms: 4-Methylphenol; p-Cresylic acid; 4-Cresol
Empirical: C_7H_8O
Formula: $CH_3C_6H_4OH$
Properties: Crystalline mass, phenolic odor; sol. in alcohol, ether, chloroform, hot water; m.w. 108.14; dens. 1.0341 (20/4 C); m.p. 32-35 C; b.p. 202 C; flash pt. (CC) 86 C; ref. index 1.5395
Toxicology: LD50 (oral, rat) 1.8 g/kg
Uses: Synthetic food flavors
Usage level: 0.67 ppm (nonalcoholic beverages), 0.01-1 ppm (ice cream, ices), 0.01-2 ppm (candy), 0.01-2 ppm (baked goods)
Regulatory: FDA 21CFR §172.515; GRAS (FEMA)
Manuf./Distrib.: Allchem Industries; Penta Mfg.; PMC Specialties; Spectrum Chem. Mfg.

o-Cresyl acetate
Synonyms: o-Tolyl acetate
Empirical: $C_9H_{10}O_2$
Formula: $CH_3COOC_6H_4CH_3$
Properties: Liq.; sol. in hot water; b.p. 208 C
Precaution: Combustible
Uses: Synthetic flavoring agent
Regulatory: FDA 21CFR §172.515

p-Cresyl acetate
CAS 140-39-6; FEMA 3073
Synonyms: p-Tolyl acetate
Empirical: $C_9H_{10}O_2$
Formula: $CH_3CO_2C_6H_4CH_3$
Properties: Colorless liq., anise sweet fragrant odor; m.w. 150.18; dens. 1.047; b.p. 210-211 C; flash pt. 90 C; ref. index 1.5010
Toxicology: Irritant
Uses: Synthetic flavoring agent
Regulatory: FDA 21CFR §172.515
Manuf./Distrib.: Aldrich

p-Cresyl dodecanoate. *See* p-Tolyl laurate
p-Cresylic acid. *See* p-Cresol
p-Cresyl isobutyrate. *See* p-Tolyl isobutyrate
p-Cresyl laurate. *See* p-Tolyl laurate
o-Cresyl methyl ether. *See* o-Methylanisole
p-Cresyl methyl ether. *See* p-Methylanisole
Crispmint oil. *See* Spearmint oil
Crocus. *See* Saffron

Crown gum
Definition: From *Manikara zapotilla* and *M. chicle*
Uses: Natural chewing gum base

Regulatory: FDA 21CFR §172.615; Japan approved

CSP. *See* Cupric sulfate (pentahydrate)

Cubeb oil
  Definition: Volatile oil from unripe fruit of *Piper cubeba*, contg. dipentene, cadinene, cubeb camphor
  Properties: Colorless, pale grn. or ylsh. liq.; sol. in 10 vols alcohol; misc. with abs. alcohol, chloroform; insol. in water; dens. 0.905-0.925 (25/25 C); ref. index 1.4800-1.5020 (20 C)
  Precaution: Keep cool, well closed; protect from light
  Toxicology: Heated to decomp., emits acrid smoke and fumes
  Uses: Natural flavoring agent
  Usage level: 2.4 ppm (nonalcoholic beverages), 0.25 ppm (ice cream, ices), 1.8 ppm (candy), 4.6 ppm (baked goods), 33 ppm (condiments), 25-30 ppm (meats)
  Regulatory: FDA 21CFR §172.510

Cubic niter. *See* Sodium nitrate
Cucumber alcohol. *See* 2,6-Nonadien-1-ol
Cumaldehyde. *See* Cuminaldehyde
o-Cumaric aldehyde methyl ether. *See* o-Methoxycinnamaldehyde
Cuminal. *See* Cuminaldehyde

Cuminaldehyde
  CAS 122-03-2; EINECS 204-516-9; FEMA 2341
  Synonyms: Cuminal; Cumaldehyde; p-Cuminic aldehyde; p-Isopropyl benzaldehyde; 4-Isopropyl-benzaldehyde; 4-(1-Methylethyl)benzaldehyde
  Definition: Constituent of eucalyptus, myrrh, cassia, cumin, and other essential oils or prepared synthetically
  Empirical: $C_{10}H_{12}O$
  Formula: $(CH_3)_2CHC_6H_4CHO$
  Properties: Colorless to ylsh. oily liq., strong persistent odor, acrid burning taste; sol. in alcohol, ether; pract. insol. in water; m.w. 148.21; dens. 0.978 (20 C); b.p. 235-236 C (760 mm); ref. index 1.5301 (20 C)
  Toxicology: LD50 (oral, rat) 1390 mg/kg
  Uses: Synthetic flavoring agent
  Regulatory: FDA 21CFR §172.515
  Manuf./Distrib.: Aldrich

Cumin extract
  CAS 84775-51-9; EINECS 283-881-6
  Synonyms: Cuminum cyminum extract
  Definition: Extract of dried seeds of *Cuminum cyminum*
  Uses: Natural flavoring
  Regulatory: FDA 21CFR §182.20, GRAS
  Manuf./Distrib.: Commodity Services Int'l.

Cuminic acetaldehyde. *See* p-Isopropylphenylacetaldehyde

Cuminic alcohol
  CAS 536-60-7; FEMA 2933
  Synonyms: p-Isopropylbenzyl alcohol; p-Cymen-7-ol; Cuminyl alcohol
  Empirical: $C_{10}H_{14}O$
  Properties: Floral odor; m.w. 150.22; dens. 0.982; b.p. 135-136 C (26 mm); flash pt. > 230 F
  Uses: Synthetic flavoring
  Regulatory: FDA 21CFR §172.515
  Manuf./Distrib.: Aldrich

p-Cuminic aldehyde. *See* Cuminaldehyde

Cumin oil
  CAS 8014-13-9
  Definition: Volatile oil from fruit of *Cuminum cyminum*, contg. 30-40% cuminaldehyde, p-cymene, β-pinene, dipentene
  Properties: Colorless to yel. liq.; sol. in 10 vols 80% alcohol; very sol. in chloroform, ether; pract. insol. in water; dens. 0.900-0.935 (25/25 C); ref. index 1.4950-1.5090 (20 C)
  Precaution: Keep cool, well closed; protect from light
  Toxicology: LD50 (oral, rat) 2500 mg/kg; mod. toxic by ingestion and skin contact; skin irritant; mutagenic data; heated to decomp., emits acrid smoke and fumes
  Uses: Natural flavoring agent for curry powders
  Regulatory: FDA 21CFR §182.10, 182.20, GRAS
  Manuf./Distrib.: Pierre Chauvet; Florida Treatt

Cuminum cyminum extract. *See* Cumin extract
Cuminyl acetaldehyde. *See* 3-(p-Isopropylphenyl)-propionaldehyde
Cuminyl alcohol. *See* Cuminic alcohol
Cupric gluconate. *See* Copper gluconate

Cupric sulfate (anhydrous)
CAS 7758-98-7; EINECS 231-847-6
Synonyms: Copper sulfate
Empirical: CuSO$_4$
Properties: Blue cryst. or cryst. gran. or powd., nauseous metallic taste; m.w. 159.60
Toxicology: LD50 (oral, rat) 300 mg/kg; human poison by ingestion; experimental tumorigen; human systemic effects by ingestion: gastritis, diarrhea, nausea, vomiting, hemolysis; mutagenic data; reacts violently with hydroxylamine, magnesium; heated, emits toxic fumes of SO$_x$
Uses: Nutrient supplement, processing aid; used in brandy, distilling spirits, infant formula, wine
Regulatory: Japan approved (0.6 mg/L as copper in milk)

Cupric sulfate (pentahydrate)
CAS 7758-99-8 (pentahydrate); EINECS 231-847-6
Synonyms: CSP; Copper sulfate pentahydrate; Copper sulfate (ic); Blue vitriol; Blue stone; Blue copperas
Classification: Inorganic salt
Empirical: CuO$_4$S • 5H$_2$O (pentahydrate)
Formula: CuSO$_4$ • 5H$_2$O (pentahydrate)
Properties: Blue cryst. or cryst. gran. or powd., nauseous metallic taste; very sol. in water; sol. in methanol, glycerin; sl. sol. in ethanol; m.w. 249.70 (pentahydrate); dens. 2.286 (15.6/4 C)
Toxicology: TLV:TWA 1 mg (Cu)/m^3; LD50 (oral, rat) 960 mg/kg; human poison by unspecified routes; moderately toxic by ingestion; heated to dec., emits toxic fumes of SO$_x$
Uses: Direct food additive, nutrient supplement, processing aid; may be used in infant formulas
Regulatory: FDA 21CFR §184.1261, GRAS
Manuf./Distrib.: Allchem Industries

Cuprous iodide. *See* Copper iodide (ous)
Curacao aloe extract. *See* Aloe extract

Curacao orange peel
Synonyms: Bitter orange peel
Definition: Citrus aurantium
Properties: Char. orange odor, bitter flavor
Uses: Natural flavoring agent
Usage level: 67 ppm (nonalcoholic beverages), 4.0 ppm (alcoholic beverages), 71 ppm (ice cream, ices), 150 ppm (candy), 110 ppm (baked goods), 300 ppm (gelatins, puddings), 500 ppm (chewing gum)
Regulatory: FDA 21CFR §182.20, GRAS

Curcuma domestica extract. *See* Turmeric extract
Curcuma longa extract. *See* Turmeric extract

Curcumin. *See also* Turmeric
CAS 8024-37-1; EEC E100
Synonyms: Turmeric yellow; CI 75300
Definition: The coloring principle from curcuma
Properties: Orange-yel. needles; sol. in water, ether, alcohol; m.p. 183 C
Toxicology: LD50 (IP, mouse) 1500 mg/kg; mutagenic data; skin irritant; mod. toxic by intraperitoneal route; heated to decomp., emits acrid smoke and fumes
Uses: Color additive
Usage level: ADI 0.01 mg/kg (EEC)
Regulatory: Europe listed; UK approved

Curled mint oil. *See* Spearmint oil
Cusparia bark. *See* Angostura
Cyamopsis gum. *See* Guar gum

Cyanocobalamin
CAS 68-19-9; EINECS 200-680-0
Synonyms: Vitamin B$_{12}$; Cyanocon(III)alamin; α-(5,6-Dimethylbenzimidazolyl)cyanocobamide
Classification: Organic compd.
Definition: Produced commercially from cultures of *Streptomyces griseus*
Empirical: C$_{63}$H$_{88}$CoN$_{14}$O$_{14}$P
Properties: Dark red cryst. or powd., very hygroscopic, odorless and tasteless; sl. sol. in water; sol. in alcohol;

insol. in acetone, ether; m.w. 1355.55

Toxicology: No hazard to humans from excessive ingestion in foods; poison by subcutaneous route, moderately toxic by intraperitoneal route; experimental teratogen, reproductive effects; heated to decomp., emits very toxic fumes of PO_x, NO_x

Uses: Direct food additive, nutrient, dietary supplement, nutrient, infant formulas, animal feed supplements

Regulatory: FDA 21CFR §182.5945, 184.1945, GRAS; Japan approved

Manuf./Distrib.: Am. Roland

Trade names containing: Vitamin B_{12} 0.1% SD No. 65354; Vitamin B_{12} 1% Trituration No. 69992; Vitamin B_{12} 1% Trituration No. 69993; Vitamin B_{12} 1.0% SD No. 65305

Cyanocon(III)alamin. *See* Cyanocobalamin
Cyanourotriamine. *See* Melamine
Cyanuramide. *See* Melamine
Cyanurotriamide. *See* Melamine
Cyclamal. *See* Cyclamen aldehyde

Cyclamen aldehyde
CAS 103-95-7; EINECS 203-161-7; FEMA 2743
Synonyms: Cyclamal; 2-Methyl-3-(p-isopropylphenyl)-propionaldehyde; α-Methyl-p-isopropyl-hydrocinnamaldehyde; p-Isopropyl-α-methylhydrocinnamic aldehyde; p-Isopropyl-α-methylphenyl-propyl aldehyde
Empirical: $C_{13}H_{18}O$
Properties: Colorless liq., strong floral odor; sol. in fixed oils, 1 vol. of 80% alcohol; insol. in propylene glycol, glycerin; sp.gr. 0.946-0.952; ref. index 1.503-1.508
Toxicology: LD50 (oral, rat) 3810 mg/kg; mod. toxic by ingestion; human skin irritant; heated to decomp., emits acrid smoke and irritating fumes
Uses: Synthetic flavoring agent
Regulatory: FDA 21CFR §172.515, GRAS; GRAS (FEMA)

Cyclamic acid
CAS 100-88-9
Synonyms: Cyclohexanesulfamic acid
Properties: Wh. cryst. solid, odorless, sweet-sour taste; sol. in water, alcohol; insol. in oils; m.p. 170 C
Toxicology: Suspected carcinogen
Uses: Nonnutritive sweetener; acidulant

α-Cyclocitrylideneacetone. *See* α-Ionone
β-Cyclocitrylideneacetone. *See* β-Ionone
α-Cyclocitrylidene butanone. *See* α-Isomethylionone

Cyclodextrin
CAS 7585-39-9, 10016-20-3; EINECS 231-493-2, 233-007-4
Synonyms: α-Cyclodextrin; Cyclomaltohexaose
Properties: m.w. 972.86; m.p. 278 (dec)
Uses: Natural processing aid
Regulatory: Japan approved; not permitted in certain foods
Trade names: Alpha W 6 Pharma Grade; Beta W 7; Beta W7 P; Cavitron Cyclo-dextrin.™; Gamma W8

α-Cyclodextrin. *See* Cyclodextrin

Cyclodextrin glucanotransferase
Uses: Natural enzyme
Regulatory: Japan approved
Trade names: CG Tase

Cyclo-1,13-ethylenedioxytridecane-1,13-dione. *See* Ethylene brassylate
Cyclohexanamine. *See* Cyclohexylamine

Cyclohexane
CAS 110-82-7; EINECS 203-806-2
Synonyms: Hexahydrobenzene; Hexamethylene; Hexanaphthene
Formula: C_6H_{12}
Properties: Colorless mobile liq., pungent odor; insol. in water; sol. in alcohol, acetone, benzene; m.w. 84.16; dens. 0.779 (20/4 C); m.p. 6.5 C; b.p. 80.7 C; f.p. 6.3 C; flash pt. (CC) -18.3 C; ref. index 1.4264; flamm.
Precaution: Flamm. liq.; dangerous fire hazard exposed to heat or flame; reactive with oxidizers; mod. explosion hazard as vapor exposed to flame; explosive mixed hot with liq. dinitrogen tetraoxide
Toxicology: TLV 300 ppm in air; LD50 (oral, rat) 29,820 mg/kg; poison by IV; mod. toxic by ingestion, inhalation, skin contact; systemic and skin irritant; high concs. may act as narcotic; mutagenic data; heated to

Cyclohexaneacetic acid

decomp., emits acrid smoke, irritating fumes
Uses: Color diluent
Regulatory: FDA 21CFR 73

Cyclohexaneacetic acid
CAS 5292-21-7; EINECS 226-132-0; FEMA 2347
Synonyms: Cyclohexylacetic acid
Definition: Obtained by catalytic reduction of phenylacetic acid
Empirical: $C_8H_{14}O_2$
Properties: Cryst. solid; sol. in most org. solvs.; sl. sol. in water; m.w. 142.19; dens. 1.007; m.p. 27-33 C; b.p. 244-246 C; flash pt. > 230 F; ref. index 1.4558 (30 C)
Uses: Synthetic flavoring agent
Usage level: 1.0 ppm (nonalcoholic beverages), 2.0 ppm (ice cream, ices, candy, baked goods)
Regulatory: FDA 21CFR §172.515
Manuf./Distrib.: Aldrich

Cyclohexaneethyl acetate. *See* Cyclohexylethyl acetate
Cyclohexane ethyl propionate. *See* Ethyl cyclohexanepropionate
cis-1,2,3,5-trans-4,6-Cyclohexanehexol. *See* Inositol
Cyclohexanesulfamic acid. *See* Cyclamic acid

Cyclohexyl acetate
CAS 622-45-7; EINECS 210-736-6; FEMA 2349
Synonyms: Acetic acid cyclohexyl ester
Empirical: $C_8H_{14}O_2$
Properties: Oily liq.; misc. with alcohol, ether; insol. in water; m.w. 142.20; dens. 0.966; b.p. 172-173 C; flash pt. 136 F; ref. index 1.4400-1.4410
Precaution: Flammable when exposed to heat or flame
Toxicology: LD50 (oral, rat) 6730 mg/kg; mod. toxic by subcutaneous route; mildly toxic by ingestion, skin contact; human systemic effects by inhalation: conjunctiva irritation and unspecified respiratory changes; systemic irritant to humans
Uses: Synthetic flavoring agent
Usage level: 20 ppm (nonalcoholic beverages), 15 ppm (ice cream, ices), 100 ppm (candy), 110 ppm (baked goods)
Regulatory: FDA 21CFR §172.515; Japan approved as flavoring
Manuf./Distrib.: Aldrich

Cyclohexylacetic acid. *See* Cyclohexaneacetic acid

Cyclohexylamine
CAS 108-91-8; EINECS 203-629-0
Synonyms: Aminocyclohexane; Aminohexahydrobenzene; Hexahydrobenzenamine; Cyclohexanamine; Hexahydroaniline
Empirical: $C_6H_{13}N$
Properties: Liq., strong fishy odor; m.w. 99.20; dens. 0.865; m.p. -17.7 C; b.p. 134.5 C; flash pt. 69.8 F
Precaution: Flamm. liq.; dangerous fire hazard exposed to heat, flame, or oxidizers; corrosive; heated to dec., emits toxic fumes of NO_x
Toxicology: TLV:TWA 10 ppm (skin); LD50 (oral, rat) 156 mg/kg; poison by ingestion, skin contact, IP; mod. toxic by subcut.; experimental teratogen, reproductive effects; severe human skin irritant; can cause dermatitis; human mutagenic data
Uses: Boiler water additive for food contact except milk prods.
Regulatory: FDA 21CFR §173.310 (10 ppm max. in steam)
Manuf./Distrib.: PMC Specialties

Cyclohexyl 2-aminobenzoate. *See* Cyclohexyl anthranilate
Cyclohexyl o-aminobenzoate. *See* Cyclohexyl anthranilate

Cyclohexyl anthranilate
FEMA 2350
Synonyms: Cyclohexyl 2-aminobenzoate; Cyclohexyl o-aminobenzoate
Definition: Obtained from isatoic anhydride and cyclohexanol
Empirical: $C_{13}H_{17}NO_2$
Properties: Pale yel. liq., orange blossom odor, grape-like taste; sol. in alcohol; insol. in water; m.w. 219.28; dens. 1.01; b.p. 318 C
Uses: Synthetic flavoring agent
Usage level: 10 ppm (nonalcoholic beverages), 3.7 ppm (ice cream, ices, candy), 2.0-10 ppm (baked goods),

1.0 ppm (gelatins, puddings)
Regulatory: FDA 21CFR §172.515

Cyclohexyl butyrate
CAS 1551-44-6; FEMA 2351
Definition: Obtained from esterification of cyclohexanol with isobutyric acid
Empirical: $C_{10}H_{18}O_2$
Properties: Colorless liq., fresh floral odor, intense sweet taste; sol. in alcohol; almost insol. in water; m.w. 170.25; dens. 0.957; b.p. 212 C; flash pt. 173 F; ref. index 1.4490
Uses: Synthetic flavoring agent
Usage level: 3.9 ppm (nonalcoholic beverages), 5.7 ppm (ice cream, ices), 9.2 ppm (candy), 28 ppm (baked goods), 0.54 ppm (gelatins, puddings)
Regulatory: FDA 21CFR §172.515; Japan approved as flavoring
Manuf./Distrib.: Aldrich

Cyclohexyl cinnamate
FEMA 2352
Synonyms: Cyclohexyl β-phenylacrylate; Cyclohexyl 3-phenylpropenoate
Definition: Obtained from cyclohexanol and cinnamic acid
Empirical: $C_{15}H_{18}O_2$
Properties: Colorless visc. liq., solidifies when cold; sol. in alcohol; insol. in water; m.w. 230.31; b.p. 195 C (12 mm); m.p. 28 C
Uses: Synthetic flavoring agent
Usage level: 2.0 ppm (nonalcoholic beverages), 1-5 ppm (ice cream, ices), 4-10 ppm (candy), 4-20 ppm (baked goods)
Regulatory: FDA 21CFR §172.515

Cyclohexylethyl acetate
CAS 21722-83-8; FEMA 2348
Synonyms: 2-Cyclohexylethyl acetate; Cyclohexaneethyl acetate; Hexahydrophenethyl acetate
Definition: Obtained from the corresponding alcohol by acetylation with sodium acetate in acetic acid sol'n.
Empirical: $C_{10}H_{18}O_2$
Properties: Liq., sweet fruity odor; m.w. 170.25; dens. 0.950; b.p. 104 C (15 mm); flash pt. 178 F
Uses: Synthetic flavoring; fruity, floral, green, raspberry, apple fragrance and flavoring
Usage level: 2.0 ppm (nonalcoholic beverages), 3.0 ppm (ice cream, ices), 6.0 ppm (candy), 20 ppm (baked goods)
Regulatory: FDA 21CFR §172.515
Manuf./Distrib.: Aldrich; BASF

2-Cyclohexylethyl acetate. *See* Cyclohexylethyl acetate

Cyclohexyl formate
FEMA 2353
Empirical: $C_7H_{12}O_2$
Properties: Lt. colorless liq., cherry-like odor; sol. in alcohol; insol. in water; m.w. 128.17; b.p. 162-163 C; ref. index 1.4417 (24 C)
Uses: Synthetic flavoring agent
Usage level: 11 ppm (nonalcoholic beverages), 2.8 ppm (ice cream, ices), 8.0 ppm (candy), 7-10 ppm (baked goods)
Regulatory: FDA 21CFR §172.515

Cyclohexyl isovalerate
CAS 7774-44-9; FEMA 2355
Definition: Synthesized from cyclohexane and isovaleric acid in the presence of perchloric acid
Empirical: $C_{11}H_{20}O_2$
Properties: Liq., apple-banana odor; m.w. 184.28; dens. 0.925; b.p. 223 C; flash pt. 192 F; ref. index 1.4410
Uses: Synthetic flavoring agent
Usage level: 13 ppm (nonalcoholic beverages), 7.0-25 ppm (ice cream, ices), 9.3 ppm (candy), 1.7-60 ppm (baked goods)
Regulatory: FDA 21CFR §172.515
Manuf./Distrib.: Aldrich

Cyclohexyl β-phenylacrylate. *See* Cyclohexyl cinnamate
Cyclohexyl 3-phenylpropenoate. *See* Cyclohexyl cinnamate

Cyclohexyl propionate
CAS 6222-35-1; FEMA 2354

Cyclomaltohexaose

Empirical: C₉H₁₆O₂

Empirical: $C_9H_{16}O_2$
Formula: Liq., apple-banana odor; m.w. 156.22; dens. 0.954; b.p. 72-73 C (10 mm); flash pt. 154 F; ref. index 1.4430
Uses: Synthetic flavoring agent
Usage level: 24 ppm (nonalcoholic beverages), 2.7 ppm (ice cream, ices), 2.0-3.0 ppm (candy), 3.0 ppm (baked goods), 5.0 ppm (gelatins, puddings)
Regulatory: FDA 21CFR §172.515
Manuf./Distrib.: Aldrich

Cyclomaltohexaose. *See* Cyclodextrin
Cyclooctafluorobutane. *See* Octafluorocyclobutane
Cydonia oblonga seed. *See* Quince seed
Cymbopogon citratus extract. *See* Lemongrass extract
p-Cymen-7-carboxaldehyde. *See* p-Isopropylphenylacetaldehyde

p-Cymene
CAS 99-87-6; EINECS 202-796-7; FEMA 2356
Synonyms: Cymol; 4-Isopropyl-1-methylbenzene; 1-Isopropyl-4-methylbenzene; p-Isopropyltoluene; 4-Isopropyltoluene
Definition: Obtained chiefly from the wash water of sulfite paper
Empirical: $C_{10}H_{14}$
Formula: $(CH_3)_2CHC_6H_4CH_3$
Properties: Colorless to pale yel. liq., odorless; sol. in alcohol, ether, acetone, benzene; m.w. 134.24; dens. 0.853; m.p. -68 C; b.p. 176 C; flash pt. (CC) 117 F; ref. index 1.489
Precaution: Flamm. or combustible liq.; sl. explosion hazard in vapor form
Toxicology: LD50 (oral, rat) 4750 mg/kg; mildly toxic by ingestion; human CNS effects at low doses; mutagenic data; skin irritant; heated to decomp., emits acrid smoke and fumes
Uses: Synthetic flavoring agent
Usage level: 3.3 ppm (nonalcoholic beverages), 5.3 ppm (ice cream, ices), 7 ppm (candy), 250 ppm (chewing gum), 10-130 ppm (condiments)
Regulatory: FDA 21CFR §172.515; GRAS (FEMA)
Manuf./Distrib.: Aldrich

2-p-Cymenol. *See* Carvacrol
3-p-Cymenol. *See* Thymol
p-Cymen-7-ol. *See* Cuminic alcohol
Cymol. *See* p-Cymene
p-Cymyl propanal. *See* 3-(p-Isopropylphenyl)-propionaldehyde
Cynara scolymus extract. *See* Artichoke extract

L-Cysteine
CAS 52-90-4; EINECS 200-158-2; FEMA 3263
Synonyms: (+)-2-Amino-3-mercaptopropionic acid; L-2-Amino-3-mercaptopropanoic acid; α-Amino-β-thiolpropionic acid; l-Cysteine; 3,3´-Dithiobis(2-aminopropanoic acid)
Classification: A nonessential amino acid
Empirical: $C_3H_7NO_2S$
Formula: $HSCH_2CH(NH_2)COOH$
Properties: m.w. 121.16; m.p. 220 C (dec.)
Toxicology: Irritant
Uses: Reducing agent; supplies up to 0.009 part of total L-cysteine/100 parts flour in dough as dough strengthener; used in yeast-leavened baked goods and mixes; nutrient; dietary supplement
Regulatory: FDA 21CFR §172.320 (2.3% max. by wt.), 184.1271, GRAS
Manuf./Distrib.: Aldrich; Am. Roland; Int'l. Sourcing
Trade names containing: Cain's PDC (Pizza Dough Conditioner); Do Sure; Lite Pack 350; Panicrust LC K

l-Cysteine. *See* L-Cysteine
Cysteine chlorhydrate. *See* Cysteine hydrochloride
Cysteine disulfide. *See* L-Cystine
Cysteine HCl. *See* Cysteine hydrochloride

Cysteine hydrochloride
CAS 52-89-1; EINECS 200-157-7; EEC E920
Synonyms: L-Cysteine hydrochloride anhydrous; L-2-Amino-3-mercaptopropanoic acid monohydrochloride monohydrate; Cysteine chlorhydrate; Cysteine HCl; L-Cystene monohydrochloride
Empirical: $C_3H_7NO_2S \cdot HCl \cdot H_2O$
Formula: $HSCH_2CH(NH_2)COOH \cdot HCl \cdot H_2O$

Properties: Wh. cryst. powd., char. acetic taste; sol. in water, alcohol; m.p. 175 C (dec.)
Toxicology: LD50 (IP, mouse) 1250 mg/kg; mod. toxic by intraperitoneal, IV routes; mutagenic data; heated to decomp., emits very toxic fumes of NO_x, SO_x, and Cl^-
Uses: Dietary supplement, dough strengthener, flavor, nutrient for yeast-leavened baked goods, baking mixes; supplies up to 0.009 parts total L-cysteine/100 parts of flour in dough
Regulatory: FDA 21CFR §172.320, limitation 2.3%, 184.1272, GRAS; Japan approved; Europe listed; UK approved
Manuf./Distrib.: Am. Roland

L-Cysteine hydrochloride anhydrous. *See* Cysteine hydrochloride
L-Cystene monohydrochloride. *See* Cysteine hydrochloride

L-Cystine
CAS 56-89-3; EINECS 200-296-3
Synonyms: Di(α-amino-β-thiolpropionic acid); (-)-3,3′-Dithio-bis(2-aminopropionic acid); β,β'-Dithiobisalanine; Cysteine disulfide
Classification: Amino acid
Empirical: $C_6H_{12}N_2O_4S_2$
Formula: $HOOCCH(NH_2)CH_2SSCH_2CH(NH_2)COOH$
Properties: Colorless to wh. hexagonal tablets; sl. sol. in water, alcohol; m.w. 240.30; dec. 260-261 C
Toxicology: Experimental reproductive effects; heated to decomp., emits toxic fumes of PO_x and SO_x
Uses: Biochemical and nutrition research, nutrient and dietary supplement, dough strengthener; flavoring (Japan)
Regulatory: FDA 21CFR §172.320, 2.3% max.; 184.1271; GRAS 0.009 parts/100 parts flour max.; Japan approved
Manuf./Distrib.: Am. Roland

Dakins sol'n. *See* Sodium hypochlorite
Dalmatian sage oil. *See* Sage oil, Dalmatian type

Damiana
Definition: Leaves of *Turnera diffusa*
Properties: Bitter tonic, aromatic
Uses: Natural flavoring; used mainly in pharmacology and sometimes in liqueurs
Regulatory: FDA 21CFR §172.510
Manuf./Distrib.: Chart

Dammar
CAS 9000-16-2; EINECS 232-528-4
Synonyms: Gum Dammar
Uses: Emulsifier, stabilizer

Dandelion
Definition: *Taraxacum officinale* and *T. laevigatum*
Properties: Bitter tonic, aromatic
Uses: Natural flavoring
Usage level: 10 ppm (nonalcoholic beverages), 2.5-20 ppm (ice cream, ices), 8.0-40 ppm (candy), 27 ppm (baked goods)
Regulatory: FDA 21CFR §182.20, GRAS; Japan approved
Manuf./Distrib.: Chart

Dandelion extract
CAS 84775-55-3; EINECS 273-624-6; FEMA 2357
Synonyms: Taraxacum extract; Taraxacum officinale extract; Dandelion fluid extract
Definition: Extract of rhizomes and roots of *Taraxacum officinale*
Properties: Bitter tonic, aromatic
Uses: Natural flavoring agent
Usage level: 3.5 ppm (nonalcoholic beverages), 6 ppm (ice cream, ices), 2-8 ppm (candy), 53 ppm (baked goods)
Regulatory: FDA 21CFR §182.20, GRAS
Manuf./Distrib.: Chart

Dandelion fluid extract. *See* Dandelion extract

Dandelion root
FEMA 2358
Synonyms: Taraxacum
Definition: Dried rhizome and roots of *Taraxacum officinale*

Properties: Bitter tonic, aromatic
Uses: Natural flavoring agent
Usage level: 10 ppm (nonalcoholic beverages), 2.5-20 ppm (ice cream, ices), 8-40 ppm (candy), 27 ppm (baked goods)
Regulatory: FDA 21CFR §182.20, GRAS

Danish agar. *See* Furcelleran
DAP. *See* Ammonium phosphate, dibasic
DATEM. *See* Diacetyl tartaric acid esters of mono- and diglycerides
Daucus carota extract. *See* Carrot seed extract
Daucus carota sativa extract. *See* Carrot extract

Davana
EINECS 295-155-6 (extract); FEMA 2359
Definition: From *Artemisia pallens*
Properties: Strong penetrating green odor
Uses: Natural flavoring
Usage level: 3.0 ppm (nonalcoholic beverages), 6.5 ppm (ice cream, ices), 8.0 ppm (candy), 11 ppm (baked goods), 5.0 ppm (chewing gum)
Regulatory: FDA 21CFR §172.510
Manuf./Distrib.: C.A.L.-Pfizer; Florida Treatt

DBNPA. *See* 2,2-Dibromo-3-nitrilopropionamide
DBPC. *See* BHT
DBS. *See* Dibutyl sebacate
1,2-DCE. *See* Ethylene dichloride
DCM. *See* Methylene chloride
DCO. *See* Castor oil, dehydrated
DCP-0. *See* Calcium phosphate dibasic
DCP-2. *See* Calcium phosphate dibasic dihydrate
DEAE. *See* Diethylaminoethanol
DEAE-cellulose. *See* Diethylaminoethyl cellulose

Deaminase
Uses: Food enzyme
Trade names: Deamizyme

Debricin. *See* Ficin

trans-trans-2,4-Decadienal
CAS 25152-84-5; EINECS 246-668-9; FEMA 3135
Empirical: $C_{10}H_{16}O$
Properties: Powerful fatty citrus odor; m.w. 152.24; dens. 0.871; b.p. 114-116 C (10 mm); flash pt. 214 F
Precaution: Combustible liq.
Toxicology: Heated to decomp., emits acrid smoke and fumes
Uses: Synthetic flavoring agent
Manuf./Distrib.: Aldrich

Decaglycerin dioleate. *See* Polyglyceryl-10 dioleate
Decaglycerin distearate. *See* Polyglyceryl-10 distearate
Decaglycerin monoisostearate. *See* Polyglyceryl-10 isostearate
Decaglycerin monolaurate. *See* Polyglyceryl-10 laurate
Decaglycerin monolinoleate. *See* Polyglyceryl-10 linoleate
Decaglycerin monomyristate. *See* Polyglyceryl-10 myristate
Decaglycerin monooleate. *See* Polyglyceryl-10 oleate
Decaglycerin monostearate. *See* Polyglyceryl-10 stearate
Decaglycerin pentaoleate. *See* Polyglyceryl-10 pentaoleate
Decaglycerin trioleate. *See* Polyglyceryl-10 trioleate
Decaglycerin tristearate. *See* Polyglyceryl-10 tristearate
Decaglycerol decaoleate. *See* Polyglyceryl-10 decaoleate
Decaglycerol decastearate. *See* Polyglyceryl-10 decastearate
Decaglycerol dioleate. *See* Polyglyceryl-10 dioleate
Decaglycerol octaoleate. *See* Polyglyceryl-10 octaoleate
Decaglycerol tetraoleate. *See* Polyglyceryl-10 tetraoleate
Decaglyceryl decaoleate. *See* Polyglyceryl-10 decaoleate
Decaglyceryl decastearate. *See* Polyglyceryl-10 decastearate

Decaglyceryl dioleate. *See* Polyglyceryl-10 dioleate
Decaglyceryl dipalmitate. *See* Polyglyceryl-10 dipalmitate
Decaglyceryl distearate. *See* Polyglyceryl-10 distearate
Decaglyceryl hexaoleate. *See* Polyglyceryl-10 hexaoleate
Decaglyceryl monoisostearate. *See* Polyglyceryl-10 isostearate
Decaglyceryl monolaurate. *See* Polyglyceryl-10 laurate
Decaglyceryl monooleate. *See* Polyglyceryl-10 oleate
Decaglyceryl monostearate. *See* Polyglyceryl-10 stearate
Decaglyceryl octaoleate. *See* Polyglyceryl-10 octaoleate
Decaglyceryl octastearate. *See* Polyglyceryl-10 octastearate
Decaglyceryl tetraoleate. *See* Polyglyceryl-10 tetraoleate
Decaglyceryl tetrastearate. *See* Polyglyceryl-10 tetrastearate
Decaglyceryl trioleate. *See* Polyglyceryl-10 trioleate
Decaglyceryl tristearate. *See* Polyglyceryl-10 tristearate

δ-Decalactone
CAS 705-86-2; EINECS 211-889-1; FEMA 2361
Synonyms: Amyl-δ-valerolactone; Decanolide-1,5
Classification: Heterocyclic compd.
Empirical: $C_{10}H_{18}O_2$
Properties: Colorless liq., coconut fruity odor, butterlike on dilution; very sol. in alcohol, propylene glycol; insol. in water; m.w. 170.28; m.p. -27 C; b.p. 117-120 C (0.02 mm); flash pt. > 230 F; ref. index 1.456-1.459
Toxicology: Skin and eye irritant; heated to decomp., emits acrid smoke and irritating fumes
Uses: Synthetic flavoring agent for oleomargarine
Usage level: Limitation 10 ppm (oleomargarine)
Regulatory: FDA 21CFR §172.515
Manuf./Distrib.: Aldrich

γ-Decalactone
CAS 706-14-9; FEMA 2360
Synonyms: 4-Hydroxydecanoic acid, γ-lactone
Empirical: $C_{10}H_{18}O_2$
Properties: Colorless liq., fruity peach odor; sl. sol. in water; m.w. 170.25 dens. 0.952; b.p. 281 C; flash pt. > 200 F
Uses: Synthetic flavoring agent
Usage level: 2.0 ppm (nonalcoholic beverages), 4.5 ppm (ice cream, ices), 5.7 ppm (candy), 7.1 ppm (baked goods), 0.08-8.0 ppm (gelatins, puddings)
Regulatory: FDA 21CFR §172.515
Manuf./Distrib.: Acme-Hardesty; Aldrich

Decanal
CAS 112-31-2; EINECS 203-957-4; FEMA 2362
Synonyms: Aldehyde C-10; Capraldehyde; Capric aldehyde; Caprinaldehyde; 1-Decyl aldehyde; n-Decyl aldehyde; Decylic aldehyde
Empirical: $C_{10}H_{20}O$
Formula: $CH_3(CH_2)_8CH_3$
Properties: Colorless to lt. yel. liq., floral fatty odor; sol. in 80% alcohol, fixed oils, volatile oils, min. oils; insol. in water, glycerin; m.w. 156.30; dens. 0.830 (15/4 C); m.p. 17-18 c; b.p. 208 C; flash pt. 185 F; ref. index 1.4260-1.4300
Precaution: Combustible liq.
Toxicology: LD50 (oral, rat) 3730 mg/kg; mod. toxic by ingestion; mildly toxic by skin contact; severe skin irritant; heated to decomp., emits acrid smoke and irritating fumes
Uses: Wax-like, orange-peel fragrance and flavoring; fresh citrus-like on dilution
Usage level: 2.3 ppm (nonalcoholic beverages), 4.1 ppm (ice cream, ices), 5.7 ppm (candy), 6.6 ppm (baked goods), 3 ppm (gelatins, puddings), 0.6 ppm (chewing gum)
Regulatory: FDA 21CFR §182.60, GRAS; Japan approved as flavoring
Manuf./Distrib.: Aldrich; BASF; Florida Treatt

Decanal dimethyl acetal
FEMA 2363
Synonyms: Aldehyde C-10 dimethyl acetal; Capraldehyde dimethyl acetal; 1,1-Dimethoxy decane
Definition: Synthesized from decanal and methyl alcohol
Empirical: $C_{12}H_{26}O_2$
Properties: Colorless liq., citrus-floral odor, brandy cognac flavor; sol. in alcohol; insol. in water; m.w. 202.34; dens. 0.830 (15.5 C); b.p. 218 C; ref. index 1.4244 (24 C)

Decanedioic acid, dibutyl ester

Uses: Synthetic flavoring agent
Usage level: 1.0-2.0 ppm (nonalcoholic beverages), 8.0 ppm (alcoholic beverages), candy, baked goods), 2.0 ppm (ice cream, ices), 3.0 ppm (gelatins, puddings)
Regulatory: FDA 21CFR §172.515

Decanedioic acid, dibutyl ester. See Dibutyl sebacate

n-Decanoic acid. See Capric acid
Decanoic acid ethyl ester. See Ethyl decanoate
Decanoic acid, 1-methyl-1,2-ethanediyl ester mixed with 1-methyl-1,2-ethanediyl dioctanoate. See Propylene glycol dicaprylate/dicaprate
Decanoic acid, monoester with 1,2,3-propanetriol. See Glyceryl caprate
1-Decanol. See n-Decyl alcohol
Decanolide-1,5. See δ-Decalactone
Decanyl acetate. See Decyl acetate
2-Decenal. See trans-2-Decenal

cis-4-Decen-1-al
FEMA 3264
Empirical: $C_{10}H_{18}O$
Properties: Colorless to sl. yel. liq., orange-like fatty odor; sol. in alcohol; insol. in water; m.w. 154.28; dens. 0.847-0.848; ref. index 1.442-1.444
Toxicology: Heated to decomp., emits acrid smoke and irritating fumes
Uses: Synthetic flavoring agent
Regulatory: GRAS

trans-2-Decenal
CAS 3913-71-1; EINECS 223-472-1; FEMA 2366
Synonyms: 2-Decenal
Empirical: $C_{10}H_{18}O$
Properties: m.w. 154.25; dens. 0.841; b.p. 78-80 C (3 mm); flash pt. 205 F
Toxicology: LD50 (oral, rat) 5000 mg/kg; mod. toxic by skin contact; mildly toxic by ingestion; severe skin irritant; heated to decomp., emits acrid smoke and fumes
Uses: Synthetic flavoring agent
Regulatory: FDA 21CFR §172.515
Manuf./Distrib.: Aldrich

Decene-1
CAS 872-05-9; EINECS 212-819-2
Synonyms: Linear C10 alpha olefin; Decylene
Empirical: $C_{10}H_{20}$
Formula: $H_2C:CH(CH_2)_7CH_3$
Properties: Colorless liq., mild hydrocarbon odor; sol. in alcohol; sl. sol. in water; m.w. 140.27; dens. 0.741 (20/4 C); f.p. -66.3 C; b.p. 166-171 C; flash pt. (Seta) 114 F; ref. index 1.421 (20 C)
Precaution: Combustible
Toxicology: Irritating to skin and eyes; low acute inhalation toxicity; sl. toxic by ingestion
Uses: Intermediate for flavors

3-Decen-2-one
EINECS 234-059-0
Synonyms: Heptylidene acetone; Oenanthylidene acetone
Empirical: $C_{10}H_{18}O$
Properties: Needles, fruital-floral jasmine-like odor; sol. in alcohol, perfume oils; insol. in water; m.w. 154.25; m.p. 16-17 C; b.p. 125-126 C (12 mm)
Uses: Synthetic flavoring agent
Regulatory: FDA 21CFR §172.515

Decoic acid. See Capric acid
Decolorizing carbon. See Carbon, activated

Decyl acetate
CAS 112-17-4; EINECS 203-942-2; FEMA 2367
Synonyms: Acetate C-10; Decanyl acetate
Definition: Synthesized by direct acetylation of n-decanol with acetic acid
Empirical: $C_{12}H_{24}O_2$
Formula: $CH_3(CH_2)_9OOCCH_3$
Properties: Liq., floral orange-rose odor; sol. in 80% alcohol, ether, benzene; insol. in water; m.w. 200.32; dens.

0.862-0.866; b.p. 187-190 C; ref. index 1.4250-1.4300
Precaution: Combustible
Uses: Synthetic flavoring agent
Usage level: 3.4 ppm (nonalcoholic beverages), 2.7 ppm (ice cream, ices), 6.1 ppm (candy), 10 ppm (baked goods), 1.2 ppm (gelatins, puddings), 12 ppm (chewing gum)
Regulatory: FDA 21CFR §172.515
Manuf./Distrib.: Aldrich

3-Decylacrolein. *See* 2-Tridecenal

n-Decyl alcohol
CAS 112-30-1; 68526-85-2; EINECS 203-956-9; FEMA 2365
Synonyms: Alcohol C-10; Noncarbinol; Nonylcarbinol; 1-Decanol; Decylic alcohol
Classification: Fatty alcohol
Empirical: $C_{10}H_{22}O$
Formula: $CH_3(CH_2)_8CH_2OH$
Properties: Mod. visc. liq., sweet odor; sol. in alcohol, ether; insol. in water; m.w. 158.32; dens. 0.8297 (20/4 C); m.p. 7 C; b.p. 232.9 C; flash pt. (OC) 180 F; ref. index 1.43587
Precaution: Flamm. when exposed to heat or flame
Toxicology: LD50 (oral, rat) 4720 mg/kg; moderately toxic by skin contact; irritating to eyes, skin, respiratory system; heated to decomp., emits acrid smoke and irritating fumes
Uses: Synthetic flavoring agent, intermediate
Usage level: 2.1 ppm (nonalcoholic beverages), 4.6 ppm (ice cream, ices), 5.2 ppm (candy), 5.2 ppm (baked goods), 3 ppm (chewing gum)
Regulatory: FDA 21CFR §172.515, 172.864, 175.300, 176.170, 178.3480, 178.3910; Japan approved as flavoring
Manuf./Distrib.: Aldrich; Penta Mfg.

1-Decyl aldehyde. *See* Decanal
n-Decyl aldehyde. *See* Decanal
Decylbenzene sodium sulfonate. *See* Sodium decylbenzene sulfonate
Decylbenzenesulfonic acid, sodium salt. *See* Sodium decylbenzene sulfonate

Decyl butyrate
CAS 5454-09-1; FEMA 2368
Empirical: $C_{14}H_{28}O_2$
Properties: Oily liq., apricot-like odor; m.w. 228.37, dens. 0.802, b.p. 134-135 C (8 mm), flash pt. > 230 F
Uses: Synthetic flavoring agent
Usage level: 0.18 ppm (nonalcoholic beverages), 1.4 ppm (ice cream, ices), 5.9 ppm (candy), 7.5 ppm (baked goods)
Regulatory: FDA 21CFR §172.515
Manuf./Distrib.: Aldrich

N-Decyl-N,N-dimethyl-1-decanaminium chloride. *See* Didecyldimonium chloride
Decylene. *See* Decene-1
Decylic alcohol. *See* n-Decyl alcohol
Decylic aldehyde. *See* Decanal
Decyl octyl alcohol. *See* Stearyl alcohol

Decyl propionate
CAS 5454-19-3; FEMA 2369
Empirical: $C_{13}H_{26}O_2$
Properties: Liq., ethereal rum fruity odor; m.w. 214.35; dens. 0.864; b.p. 123-134 C (8 mm); flash pt. 225 F; ref. index 1.4291
Uses: Synthetic flavoring agent
Usage level: 0.81 ppm (nonalcoholic beverages), 1.4 ppm (ice cream, ices), 5.9 ppm (candy), 7.5 ppm (baked goods)
Regulatory: FDA 21CFR §172.515
Manuf./Distrib.: Aldrich

DEHP. *See* Dioctyl phthalate
Dehydrated beets. *See* Beet powder

Dehydroacetic acid
CAS 520-45-6; 771-03-9; EINECS 208-293-9, 212-227-4
Synonyms: DHA; DHS; 3-Acetyl-5-hydroxy-3-oxo-4-hexenoic acid δ-lactone; Methylacetopyronone; 3-Acetyl-6-methyl-1,2-pyran-2,4(3H)-dione; 3-Acetyl-6-methyl-2,4-pyrandione

Classification: Cyclic ketone
Empirical: $C_9H_8O_4$
Properties: Wh. cryst. or cryst. powd.; mod. sol. in water and org. solvs.; m.w. 168.16; m.p. 109-111 C; b.p. 269 C
Precaution: Combustible when exposed to heat or flame
Toxicology: LD50 (oral, rat) 500 mg/kg; poison by ingestion and IV routes; mod. toxic by intraperitoneal route; experimental tumorigen; heated to decomp., emits acrid smoke and irritating fumes
Uses: Preservative for cut or peeled squash
Regulatory: FDA 21CFR §172.130, 65 ppm max. residue in or on prepared squash, GRAS; 175.105

7-Dehydrocholesterol. *See* Cholecalciferol
Delphinic acid. *See* Isovaleric acid

Denatonium benzoate NF
CAS 3734-33-6; EINECS 223-095-2
Synonyms: N-[2-[(2,6-Dimethylphenyl)amino]-2-oxoethyl]-N,N-diethylbenzenemathanaminium benzoate; Benzyldiethyl [(2,6-xylylcarbomoyl)methyl]ammonium benzoate; Lignocaine benzyl benzoate
Classification: Organic compd.
Empirical: $C_{21}H_{29}N_2O \cdot C_7H_5O_2$
Properties: Crystals; extremely bitter; sol. in water, alcohol; sparilngly sol. in acetone; pract. insol. in ether; m.w. 446.59; m.p. 174-176 C
Uses: Pharmaceutical aid (alcohol denaturant, flavor); added to toxic substances as a deterrent to ingestion
Regulatory: FDA 27CFR §21.151
Manuf./Distrib.: Atomergic Chemetals

Deodorized winterized cottonseed oil. *See* Cottonseed oil
6-Deoxy-L-mannose. *See* Rhamnose
DEP. *See* Diethyl phthalate
Devil's dung. *See* Asafetida
Devitalized wheat gluten. *See* Wheat gluten
Dewberry extract. *See* Blackberry extract
Dexpanthenol. *See* D-Panthenol
Dexpanthenol. *See* DL-Panthenol

Dextran
CAS 9004-54-0; EINECS 232-677-5
Synonyms: Macrose
Definition: Polymers of glucose with chain-like structures and m.w. to 200,000
Formula: $(C_6H_{10}O_5)_n$
Precaution: Combustible
Uses: Food additive; constituent of food contact surfaces; in soft center confections; as partial substitute for barley malt; thickener/stabilizer (Japan)
Regulatory: FDA 21CFR §186.1275, GRAS as indirect additive; Japan approved
Manuf./Distrib.: Spectrum Chem. Mfg.

Dextranase
CAS 9025-70-1; EINECS 232-803-9
Definition: From *Penicillium lilacinum*
Uses: Natural enzyme
Regulatory: UK, Japan approved
Trade names: Dextranase L Amano; Dextranase Novo 25 L

Dextrin
CAS 9004-53-9; EINECS 232-675-4
Synonyms: Dextrine; Starch gum; Tapioca; Vegetable gum
Definition: Gum produced by incomplete hydrolysis of starch
Empirical: $(C_6H_{10}O_5)_n \cdot xH_2O$
Properties: Yel. or wh. powd. or gran.; sol. in water; insol. in alcohol, ether; colloidal in props.
Toxicology: Mildly toxic by IV route; heated to decomp., emits acrid smoke and irritating fumes
Uses: Direct food additive, binder, colloidal stabilizer, extender, formulation aid, processing aid, surface-finishing agent, thickener for pasta, pizza dough, confections, snacks
Usage level: ADI 0-70 mg/kg (JECFA)
Regulatory: FDA 21CFR §184.1277, 186.1275, GRAS; USDA 9CFR §318.7, 381.147
Manuf./Distrib.: Am. Maize Prods.; Avebe; Cerestar Int'l.; Fruitsource; Grain Processing; Nat'l. Starch & Chem.; Roquette UK
Trade names: Clinton #600 Dextrin; Clinton #655 Dextrin; Clinton #656 Dextrin; Clinton #700 Dextrin; Clinton

#721 Dextrin; Stadex® 9; Stadex® 60K; Stadex® 126; Stadex® 128; Staley® Tapioca Dextrin 11; Staley® Tapioca Dextrin 12
Trade names containing: Fermented Soy Sauce Powd.; Hentex-30; Hentex-30A; Kel-Lite™ CM; Natural HVP Replacer; Stadex® 90

Dextrine. *See* Dextrin
Dextronic acid. *See* D-Gluconic acid
Dextrose. *See* D-Glucose anhyd.
DHA. *See* Dehydroacetic acid
DHS. *See* Dehydroacetic acid

Diacetyl
CAS 431-03-8; EINECS 207-069-8; FEMA 2370
Synonyms: 2,3-Butanedione; Biacetyl; 2,3-Diketobutane; Dimethyl diketone; Dimethylglyoxal
Empirical: $C_4H_6O_2$
Formula: $CH_3COCOCH_3$
Properties: Yel. to greenish-yel. liq., strong rancid butter odor; sol. in glycerin, water; misc. with alcohol, fixed oils, propylene glycol; m.w. 86.10; dens. 0.9904 (15/15 C); b.p. 88 C; flash pt. 80 F; ref. index 1.393-1.397
Precaution: Flamm. liq.; dangerous fire hazard when exposed to heat or flame
Toxicology: LD50 (oral, guinea pig) 990 mg/kg; poison by intraperitoneal route; mod. toxic by ingestion; skin irritant; human mutagenic data; heated to decomp., emits acrid smoke and fumes
Uses: Direct food additive; aroma carrier in food products; pungent, sweet, butter-like flavoring agent and adjuvant
Usage level: 2.5 ppm (nonalcoholic beverages), 5.9 ppm (ice cream, ices), 21 ppm (candy), 44 ppm (baked goods), 19 ppm (gelatins, puddings), 35 ppm (chewing gum), 11 ppm (shortening)
Regulatory: FDA 21CFR §184.1278, GRAS; USDA 9CFR §318.7
Manuf./Distrib.: Aldrich; BASF

Diacetylated tartaric acid ester glycerides. *See* Diacetyl tartaric acid esters of mono- and diglycerides
Diacetylmethane. *See* Acetylacetone

Diacetyl tartaric acid esters of mono- and diglycerides
EEC E472e
Synonyms: DATEM; Diacetylated tartaric acid ester glycerides
Toxicology: Heated to decomp., emits acrid smoke and irritating fumes
Uses: Emulsifier, flavoring agent for baked goods, mixes, beverages, coffee whiteners, confections, dairy prods., frostings, margarine, oils
Regulatory: FDA 21CFR §182.4101, 184.1101, GRAS; USDA 9CFR §318.7, 0.5% in margarine; 381.147; ADI 0-50 mg kg⁻¹ body wt.
Regulatory: FDA 21CFR §182.4101, 184.1101, GRAS; USDA 9CFR §318.7, 0.5% in margarine; 381.147; ADI 0-50 mg kg^{-1} body wt.
Trade names: Admul Datem; Datagel; Datamuls® 42; Datamuls® 43; Datamuls® 4720; Datamuls® 4820; Datamuls® 4820 U; Datem esters; Imwitor® 1330; Imwitor® 2020; Imwitor® 2320; Lamegin® DW 8000 HW; Lamegin® DW 8200 VA/HW; Lamegin® DW 9000 HW; Lamegin® DWF, DWH, DWP; Panodan 235; Swedex AR58P-15AC
Trade names containing: CDC-10A; CDC-50; CDC-2001; CDC-2002; CDC-2500; CDC-3000; Crestawhip H13; Lamegin® DW 8000-Range; Vitalex P

Dialkyl dimethyl ammonium chloride
Synonyms: Quaternium 31; Dimethyl dialkyl ammonium chloride
Toxicology: Heated to decomp., emits acrid smoke and irritating fumes
Uses: Decolorizing agent in clarification of refinery sugar liquors
Usage level: Limitation 700 ppm of sugar solids
Regulatory: FDA 21CFR §173.400

Diallyl disulfide. *See* Allyl disulfide
Diallyl sulfide. *See* Allyl sulfide
Diamide. *See* Hydrazine
Diamine. *See* Hydrazine
α, ε-Diaminocaproic acid. *See* L-Lysine
1,2-Diaminoethane. *See* Ethylenediamine
2,6-Diaminohexanoic acid. *See* L-Lysine
2,6-Diaminohexanoic acid hydrochloride. *See* L-Lysine hydrochloride
Di(α-amino-β-thiolpropionic acid). *See* L-Cystine
Diammoniium sulfate. *See* Ammonium sulfate
Diammonium citrate. *See* Ammonium citrate
Diammonium hydrogen citrate. *See* Ammonium citrate
Diammonium hydrogen orthophosphate. *See* Ammonium phosphate, dibasic

Diammonium hydrogen phosphate. *See* Ammonium phosphate, dibasic
Diammonium phosphate. *See* Ammonium phosphate, dibasic

Diatomaceous earth
    CAS 7631-86-9, 68855-54-9; EINECS 231-545-4
    Synonyms: Kieselguhr; Diatomite; Diatomaceous silica; Infusorial earth
    Definition: Mineral material consisting chiefly of the siliceous frustules and fragments of various species of
    diatoms
    Properties: Soft bulky solid; insol. in acids except HF; sol. in strong alkalies; dens. 1.9-2.35; bulk dens. 8-15
    lb/ft³; oil absorp. 135-185%; 88% silica; noncombustible
    Toxicology: TLV:TWA 10 mg/m³ (dust); poison by inhalation and ingestion; dust may cause fibrosis of the lungs
    Uses: Fat refining aid for rendered fats, filter aid, insecticide; migrating to foods from paper/paperboard;
    processing aid (Japan)
    Regulatory: FDA 21CFR §73.1, 133.146, 160.105, 160.185, 172.230, 172.480, 173.340, 175.300, 177.1460,
    177.2410, 182.90, 193.135, 561.145, 573.940; USDA 9CFR §318.7; Japan restricted (0.5% max. residual)

Diatomaceous silica. *See* Diatomaceous earth
Diatomite. *See* Diatomaceous earth
Diazenedicarboxamide. *See* Azodicarbonamide
DIBA. *See* Diisobutyl adipate
Dibasic potassium phosphate. *See* Potassium phosphate dibasic
Dibasic sodium phosphate. *See* Sodium phosphate dibasic
Dibasic sodium phosphate duohydrate. *See* Disodium phosphate, dihydrate
Dibenzoyl peroxide. *See* Benzoyl peroxide

Dibenzoyl thiamine
    Uses: Dietary supplement (vitamin)
    Regulatory: Japan approved

Dibenzoyl thiamine hydrochloride
    Uses: Dietary supplement (vitamin)
    Regulatory: Japan approved

Dibenzyl disulfide. *See* Benzyl disulfide
Dibenzyl ether. *See* Benzyl ether
Dibenzyl ketone. *See* 1,3-Diphenyl-2-propanone
Dibromocyanoacetamide. *See* 2,2-Dibromo-3-nitrilopropionamide
α,α-Dibromo-α-cyanoacetamide. *See* 2,2-Dibromo-3-nitrilopropionamide

2,2-Dibromo-3-nitrilopropionamide
    CAS 10222-01-2
    Synonyms: DBNPA; α,α-Dibromo-α-cyanoacetamide; Dibromocyanoacetamide
    Empirical: $C_3H_2Br_2N_2O$
    Properties: m.w. 241.89
    Toxicology: LD50 (oral, mammal) 118 mg/kg, (IV, mouse) 10 mg/kg; poison by ingestion and IV routes; severe
    skin and eye irritant; heated to decomp., emits very toxic fumes of Br and NO_x
    Uses: Antimicrobial for use in cane-sugar and beet-sugar mills
    Regulatory: FDA 21CFR §173.320 (limitation 2-10 ppm)

Dibutyl butyrolactone. *See* 4,4-Dibutyl-γ-butyrolactone

4,4-Dibutyl-γ-butyrolactone
    Synonyms: Dibutyl butyrolactone; 4-Butyl-4-hydroxyoctanoic acid, γ-lactone; 4,4-Dibutyl-4-hydroxybutyric
    acid, γ-lactone
    Definition: Synthesized from butyl pentanol and methyl acrylate using a catalyst
    Empirical: $C_{12}H_{22}O_2$
    Properties: Colorless oily liq., oily coconut-butter odor, coconut-like flavor; sol. in alcohol; insol. in water; m.w.
    198.31
    Uses: Synthetic flavoring agent
    Usage level: 2.8-3.5 ppm (ice cream, ices), 4.4-15 ppm (candy), 15 ppm (baked goods)
    Regulatory: FDA 21CFR §172.515

Dibutyl-o-(o-carboxybenzoyl) glycolate. *See* n-Butyl phthalyl-n-butyl glycolate
Dibutyl-o-carboxybenzoyloxyacetate. *See* n-Butyl phthalyl-n-butyl glycolate
2,6-Di-t-butyl-p-cresol. *See* BHT
Dibutyl decanedioate. *See* Dibutyl sebacate
4,4-Dibutyl-4-hydroxybutyric acid, γ-lactone. *See* 4,4-Dibutyl-γ-butyrolactone

Dibutyl sebacate
CAS 109-43-3; EINECS 203-672-5; FEMA 2373
Synonyms: DBS; Dibutyl decanedioate; Di-n-butyl sebacate; Decanedioic acid, dibutyl ester
Definition: Diester of butyl alcohol and sebacic acid
Empirical: $C_{18}H_{34}O_4$
Formula: $C_4H_9OCO(CH_2)_8OCOC_4H_9$
Properties: Clear colorless odorless liq.; insol. in water; m.w. 314.47; dens. 0.936 (20/20 C); b.p. 349 C (760 mm); f.p. -11 C; flash pt. 350 F; ref. index 1.4433 (15 C); stable, nonvolatile
Precaution: Combustible when exposed to heat or flame; can react with oxidizing materials
Toxicology: LD50 (oral, rat) 16 g/kg; mildly toxic by ingestion; experimental reproductive effects; heated to decomp., emits acrid smoke and fumes
Uses: Synthetic flavoring agent; plasticizer migrating from food pkg.
Usage level: 1-5 ppm (nonalcoholic beverages), 2-5 ppm (ice cream, ices), 15 ppm (candy), 15 ppm (baked goods)
Regulatory: FDA 21CFR §172.515, 175.105, 175.300, 175.320, 176.170, 177.2600, 178.3910, 181.22, 181.27

Di-n-butyl sebacate. *See* Dibutyl sebacate
Dibutyl sulfide. *See* Butyl sulfide
Dicalcium citrate. *See* Calcium citrate
Dicalcium orthophosphate anhyd. *See* Calcium phosphate dibasic
Dicalcium phosphate. *See* Calcium phosphate dibasic
Dicalcium phosphate. *See* Calcium phosphate dibasic dihydrate
Dicalcium phosphate dihydrate. *See* Calcium phosphate dibasic dihydrate
Dicarboxylic acid C-6. *See* Adipic acid
1,1-Dichloro-2-chloroethylene. *See* Trichloroethylene
1,6-Dichloro-1,6-dideoxy-β-D-fructofuranosyl-4-chloro-4-deoxy-α-D-galactopyranoside. *See* Sucralose

Dichlorodifluoromethane
CAS 75-71-8
Synonyms: Difluorodichloromethane; Fluorocarbon-12; Food Freezant 12
Empirical: CCl_2F_2
Properties: Colorless gas, odorless; noncorrosive; insol. in water; sol. in most org. solvs.; m.w. 120.9; b.p. -29.8 C; f.p. -158 C
Precaution: Nonflamm. gas, can react violently with Al
Toxicology: TLV 1000 ppm in air; human systemic effects; narcotic in high concs.; heated to decomp., emits highly toxic fumes of phosgene, Cl⁻, and F⁻
Uses: Direct-contact freezing agent for foods
Usage level: Estimated ADI 0-1.5 mg/kg (JECFA)
Regulatory: FDA 21CFR §173.355

1,2-Dichloroethane. *See* Ethylene dichloride
Dichloroethylene. *See* Ethylene dichloride
Dichloromethane. *See* Methylene chloride
Didecyl dimethyl ammonium chloride. *See* Didecyldimonium chloride

Didecyldimonium chloride
CAS 7173-51-5; EINECS 230-525-2
Synonyms: Didecyl dimethyl ammonium chloride; N-Decyl-N,N-dimethyl-1-decanaminium chloride; Dimethyl didecyl ammonium chloride
Classification: Quaternary ammonium chloride
Empirical: $C_{22}H_{48}ClN$
Formula: $[CH_3(CH_2)_9N(CH_3)_2(CH_2)_9CH_3]^+Cl^-$
Properties: Sol. in acetone; extremely sol. in benzene; m.w. 362.08
Toxicology: Causes eye damage and skin irritation; harmful if swallowed
Uses: Bactericide, fungicide for food processing industry, breweries
Regulatory: FDA 21CFR §172.712, 177.1200
Trade names: Querton 210Cl-50; Querton 210Cl-80

Didodecyl 3,3´-thiodipropionate. *See* Dilauryl thiodipropionate
8,8-Diethoxy-2,6-dimethyl-octanol-2. *See* Hydroxycitronellal diethyl acetal
1,1-Diethoxyethane. *See* Acetal
Diethyl acetal. *See* Acetal

Diethylacetic acid
CAS 88-09-5; EINECS 201-796-4; FEMA 2429
Synonyms: 2-Ethyl butanoic acid; Carboxylic acid C-6; α-Ethylbutyric acid; 3-Pentanecarboxylic acid;

2-Ethylbutyric acid
Empirical: $C_6H_{12}O_2$
Formula: $(C_2H_5)_2CHCOOH$
Properties: Colorless volatile liq., rancid odor; sol. in alcohol, ether; sl. sol. in water; m.w. 116.18; dens. 0.917; m.p. -15 C; b.p. 194-195 C; flash pt. (CC) 78 F
Precaution: Flamm. liq.
Toxicology: LD50 (oral, rat) 2200 mg/kg; mod. toxic by ingestion, skin contact; irritant to skin, mucous membranes; severe eye irritant; narcotic in high concs.; heated to decomp., emits acrid smoke and irritating fumes
Uses: Synthetic flavoring agent
Regulatory: FDA 21CFR §172.515
Manuf./Distrib.: Aldrich

Diethylaminoethanol
CAS 100-37-8; EINECS 202-845-2
Synonyms: DEAE; Diethylethanolamine; 2-Hydroxytriethylamine; β-Diethylaminoethyl alcohol
Empirical: $C_6H_{15}NO$
Formula: $(C_2H_5)_2NCH_2CH_2OH$
Properties: Colorless hygroscopic liq. base having props. of amines and alcohols; sol. in water, alcohol, ether, benzene; m.w. 117.19; dens. 0.88-0.89 (20/20 C); b.p. 161 C; flash pt. (OC) 140 F; f.p. -70 C; ref. index 1.4389
Precaution: Combustible; moderate fire risk; reactive with oxidizing materials
Toxicology: TLV 10 ppm in air; LD50 (oral, rat) 1300 mg/kg; poison by IP, IV routes; mod. toxic by ingestion, skin contact, subcut. routes; human systemic effects; skin, severe eye irritant; corrosive; heated to decomp., emits toxic fumes of NO_x
Uses: Boiler water additive for food contact except milk prods.
Regulatory: FDA 21CFR §173.310 (15 ppm max. in steam)

β-Diethylaminoethyl alcohol. *See* Diethylaminoethanol

Diethylaminoethyl cellulose
CAS 9013-34-7
Synonyms: DEAE-cellulose
Uses: Fixing agent in immobilization of glucose isomerase enzyme preps. for mfg. of high fructose corn syrup
Regulatory: FDA 21CFR §173.357

Diethyl 1,2-benzenedicarboxylate. *See* Diethyl phthalate
Diethyl butanedioate. *See* Diethyl succinate
Diethyl carbinol. *See* 3-Pentanol
Diethyl decanedioate. *See* Diethyl sebacate
Diethyl 2,3-dihydroxybutanedioate. *See* Diethyl tartrate
Diethyl 2,3-dihydroxysuccinate. *See* Diethyl tartrate
Diethyleneimide oxide. *See* Morpholine
Diethylene oximide. *See* Morpholine
Diethylethanolamine. *See* Diethylaminoethanol

Diethyl fumarate
CAS 623-92-6; EINECS 210-819-7
Empirical: $C_8H_{12}O_4$
Formula: $C_2H_5O_2CCH=CHCO_2C_2H_5$
Properties: m.w. 172.18; dens. 1.052; m.p. 1-2 C; b.p. 218-219 C; flash pt. 91 C
Uses: Inhibits fungal growth in tomato juice
Usage level: 0.05%

Di(2-ethylhexyl) phthalate. *See* Dioctyl phthalate
Diethylhydroxysuccinate. *See* Diethyl malate

Diethyl ketone
CAS 96-22-0; EINECS 202-490-3
Synonyms: 3-Pentanone; Dimethylacetone; Propione; Methacetone
Empirical: $C_5H_{10}O$
Formula: $CH_3CH_2COCH_2CH_3$
Properties: Liq., acetone odor; misc. with alcohol, ether; sol. in 25 parts water; m.w. 86.14; dens. 0.814; m.p. -42 C; b.p. 99-103 C; flash pt. 55 F; ref. index 1.392
Toxicology: LD50 (oral, rat) 2.1 g/kg
Uses: Flavorings
Manuf./Distrib.: Aldrich; Hüls AG

Diethyl malate
CAS 626-11-9; FEMA 2374
Synonyms: Diethylhydroxysuccinate; Ethyl malate
Empirical: $C_8H_{14}O_5$
Properties: Liq., fruital odor; m.w. 190.20; dens. 1.128; b.p. 122-124 C (12 mm); flash pt. 185 F
Uses: Synthetic flavoring agent
Usage level: 5.5 ppm (nonalcoholic beverages), 6.5 ppm (ice cream, ices), 18 ppm (candy), 44 ppm (baked goods), 1.5 ppm (gelatins, puddings)
Regulatory: FDA 21CFR §172.515
Manuf./Distrib.: Aldrich

Diethyl malonate
CAS 105-53-3; EINECS 203-305-9; FEMA 2375
Synonyms: Ethyl malonate; Malonic ester; Propanedioic acid diethyl ester
Empirical: $C_7H_{12}O_4$
Formula: $CH_2(COOC_2H_5)_2$
Properties: Liq., sl. aromatic pleasant odor; misc. with alcohol, ether; sol. 1 g/50 ml in water; m.w. 160.17; dens. 1.055 (20/4 C); m.p. -50 C; b.p. 94-95 C (11 mm); flash pt. 80 C; ref. index 1.413 (20 C)
Precaution: Combustible
Uses: Synthetic flavoring agent
Usage level: 5.6 ppm (nonalcoholic beverages), 17 ppm (ice cream, ices), 20 ppm (candy, gelatin, puddings), 19 ppm (baked goods)
Regulatory: FDA 21CFR §172.515
Manuf./Distrib.: Aldrich

Diethyl 1,8-octanedicarboxylate. *See* Diethyl sebacate

Diethyl phthalate
CAS 84-66-2; EINECS 201-550-6
Synonyms: DEP; Ethyl phthalate; Phthalic acid, diethyl ester; Diethyl 1,2-benzenedicarboxylate
Definition: Aromatic diester of ethyl alcohol and phthalic acid
Empirical: $C_{12}H_{14}O_4$
Formula: $C_6H_4(CO_2C_2H_5)_2$
Properties: Water-white liq., odorless, bitter taste; misc. with alcohols, ketones, esters, aromatic hydrocarbons; insol. in water; m.w. 222.24; dens. 1.120 (25/25 C); f.p. -40.5 C; b.p. 298 C; flash pt. (OC) 325 F; visc. 31.3 cs (0 C); stable
Precaution: Combustible when exposed to heat or flame; heated to decomp., emits acrid smoke and irritating fumes
Toxicology: TLV 5 mg/m³ of air; LD50 (oral, rat) 8600 mg/kg; poison by IV route; mod. toxic by ingestion, subcut., IP routes; human systemic effects; strong irritant to eyes and mucous membranes; narcotic in high concs.; experimental reproductive effects
Uses: Plasticizer migrating from food pkg.
Regulatory: FDA 21CFR §175.105, 175.300, 175.320, 178.3910, 181.22, 181.27, 212.177; 27CFR §21.105
Manuf./Distrib.: Eastman; Morflex; Penta Mfg.

Diethyl sebacate
CAS 110-40-7; EINECS 203-764-5; FEMA 2376
Synonyms: Diethyl decanedioate; Ethyl sebacate; Diethyl 1,8-octanedicarboxylate
Empirical: $C_{14}H_{26}O_4$
Properties: Colorless to sl. yel. liq., faint fruity odor; misc. with alcohol, ether, other org. solvs., fixed oils; insol. in water; m.w. 258.40; dens. 0.960-0.965; m.p. 1-3 C; b.p. 312 C; flash pt. > 230 F; ref. index 1.435
Toxicology: LD50 (oral, rat) 14,470 mg/kg; mildly toxic by ingestion; skin irritant; heated to decomp., emits acrid smoke and irritating fumes
Uses: Synthetic flavoring agent
Usage level: 4.1 ppm (nonalcoholic beverages), 9.1 ppm (ice cream, ices), 21 ppm (candy), 41 ppm (baked goods), 3.2-19 ppm (gelatins, puddings), 2.7-450 ppm (chewing gum)
Regulatory: FDA 21CFR §172.515
Manuf./Distrib.: Aldrich

Diethyl succinate
CAS 123-25-1; EINECS 204-612-0; FEMA 2377
Synonyms: Succinic acid diethyl ester; Ethyl succinate; Butanedioic acid diethyl ester; Diethyl butanedioate
Empirical: $C_8H_{14}O_4$
Formula: $C_2H_5OCOCH_2CH_2COOC_2H_5$
Properties: Colorless mobile liq., pleasant odor; sol. in alcohol, fixed oils, water; m.w. 174.22; dens. 1.039 (20/

Diethyl sulfide-2,2′-dicarboxylic acid

4 C); m.p. -20 C; b.p. 97-99 C (10 mm); flash pt. 230 F; ref. index 1.420
Precaution: Combustible liq.
Toxicology: LD50 (oral, rat) 8530 mg/kg; mildly toxic by ingestion; skin and eye irritant; heated to decomp., emits acrid smoke and irritating fumes
Uses: Synthetic flavoring agent
Usage level: 7.3 ppm (nonalcoholic beverages), 11 ppm (ice cream, ices), 38 ppm (candy), 45 ppm (baked goods)
Regulatory: FDA 21CFR §172.515
Manuf./Distrib.: Aldrich;

Diethyl sulfide-2,2′-dicarboxylic acid. *See* 3,3′-Thiodipropionic acid

Diethyl tartrate
CAS 87-91-2; EINECS 201-783-3; FEMA 2378
Synonyms: Diethyl 2,3-dihydroxybutanedioate; (R)-2,3-Dihydroxybutanedioic acid diethyl ester; Ethyl tartrate; (+)-Diethyl L-tartrate; Diethyl 2,3-dihydroxysuccinate
Empirical: $C_8H_{14}O_6$
Formula: $C_2H_5OOCCH(OH)CH(OH)COOC_2H_5$
Properties: Colorless thick oily liq.; sl. sol. in water; misc. with alcohol, ether; m.w. 206.19; dens. 1.204 (20/4 C); m.p. 17 C; b.p. 280 C; flash pt. 93 C; ref. index 1.4476 (20 C)
Uses: Synthetic flavoring agent
Usage level: 50 ppm (nonalcoholic beverages), 200 ppm (ice cream, ices, candy, baked goods)
Regulatory: FDA 21CFR §172.515
Manuf./Distrib.: Aldrich

(+)-Diethyl L-tartrate. *See* Diethyl tartrate

2,5-Diethyltetrahydrofuran
Empirical: $C_8H_{16}O$
Properties: Colorless liq., sweet herbaceous caramellic odor; sol. in alcohol, propylene glycol; sl. sol. in water; m.w. 128.21; b.p. 116 C
Uses: Synthetic flavoring agent; used in reconstructing mint flavors
Regulatory: FDA 21CFR §172.515

Difluorodichloromethane. *See* Dichlorodifluoromethane
Diglycerol tetrastearate. *See* Polyglyceryl-2 tetrastearate
Diglyceryl diisostearate. *See* Polyglyceryl-2 diisostearate
Diglyceryl monooleate. *See* Polyglyceryl-2 oleate
Diglyceryl monostearate. *See* Polyglyceryl-2 stearate
Diglyceryl sesquioleate. *See* Polyglyceryl-2 sesquioleate
Diglyceryl tetrastearate. *See* Polyglyceryl-2 tetrastearate
Diglyceryl triisostearate. *See* Polyglyceryl-2 triisostearate
Dihydroanethole. *See* p-Propyl anisole

Dihydrocarveol
CAS 17699-09-1; FEMA 2379
Synonyms: 8-p-Menthen-2-ol; 6-Methyl-3-isopropenylcyclohexanol; Tuberyl alcohol
Definition: Synthesized by reducing carvone and separating the resulting isomers
Empirical: $C_{10}H_{18}O$
Properties: Almost colorless straw-colored liq., floral woody odor, sweet somewhat spicy flavor; sol. in alcohol; m.w. 154.24; dens. 0.924; b.p. 224-225 C; flash pt. 197 F
Precaution: Combustible liq.
Toxicology: Moderate skin and eye irritant; heated to decomp., emits acrid smoke and fumes
Uses: Synthetic flavoring agent
Usage level: 84 ppm (nonalcoholic beverages), 500 ppm (alcoholic beverages), 300 ppm (ice cream, ices), 10-250 ppm (candy, baked goods)
Regulatory: FDA 21CFR §172.515
Manuf./Distrib.: Aldrich

d-Dihydrocarvone
CAS 7764-50-3; FEMA 3565
Synonyms: d-2-Methyl-5-(1-methylethenyl)-cyclohexanone
Empirical: $C_{10}H_{16}O$
Properties: Herbaceous spearmint odor; m.w. 152.24; dens. 0.926; b.p. 87-88 C (6 mm); flash pt. 178 F
Toxicology: Mod. toxic by subcutaneous route; heated to decomp., emits acrid smoke and fumes
Uses: Synthetic flavoring agent

Regulatory: FDA 21CFR §172.515
Manuf./Distrib.: Aldrich

Dihydrocarvyl acetate
CAS 20777-49-5; EINECS 244-029-9; FEMA 2380
Synonyms: 8-p-Menthen-2-yl acetate; 6-Methyl-3-isopropenyl cyclohexyl acetate; p-Menth-8-(9)-en-2-yl acetate
Definition: Acetylation of dihydrocarveol
Empirical: $C_{12}H_{20}O_2$
Properties: Colorless liq., sweet floral rose-like odor, sl. minty flavor; sol. in aclohol; sl. sol. in water; m.w. 196.29; dens. 0.96; b.p. 232-234 C; flash pt. 194 F
Uses: Synthetic flavoring agent
Usage level: 2.0-5.0 ppm (nonalcoholic beverages), 20 ppm (ice cream, ices), 22 ppm (candy, baked goods), 10 ppm (condiments)
Regulatory: FDA 21CFR §172.515
Manuf./Distrib.: Aldrich

Dihydrocinnamaldehyde. *See* Hydrocinnamaldehyde
Dihydrocitronellol. *See* 3,7-Dimethyl-1-octanol

Dihydrocoumarin
CAS 119-84-6; EINECS 204-354-9; FEMA 2381
Synonyms: Hydrocoumarin; 3,4-Dihydrocoumarin; 1,2-Benzodihydropyrone; Melilotin
Classification: Heterocyclic compd.
Empirical: $C_9H_8O_2$
Properties: Colorless to pale yel. liq., coconut odor; m.w. 148.17; dens. 1.186; m.p. 24-25 C; b.p. 272 C; flash pt. 266 F; ref. index 1.555
Precaution: Combustible liq.
Toxicology: LD50 (oral, rat) 1460 mg/kg; poison by intraperitoneal route; mod. toxic by ingestion; skin irritant; heated to decomp., emits acrid smoke and fumes
Uses: Synthetic flavoring agent
Usage level: 7.8 ppm (nonalcoholic beverages), 21 ppm (ice cream, ices), 44 ppm (candy), 28 ppm (baked goods), 10 ppm (gelatins, puddings), 78 ppm (chewing gum)
Regulatory: GRAS; GRAS (FEMA)
Manuf./Distrib.: Aldrich

3,4-Dihydrocoumarin. *See* Dihydrocoumarin
1,2-Dihydro-6-ethoxy-2,2,4-trimethylquinoline. *See* 6-Ethoxy-1,2-dihydro-2,2,4-trimethylquinoline
Dihydro-2,5-furandione. *See* Succinic anhydride
Dihydro-2(3H)-furanone. *See* Butyrolactone

Dihydro-β-ionone
Synonyms: 4-(2,6,6-Trimethyl cyclohexene-1-yl)-butane-2-one
Uses: Floral, woody fragrance and flavoring

4,5-Dihydro-5-methyl-2(3H)-furanone. *See* γ-Valerolactone
2-(1,3-Dihydro-3-oxo-5-sulfo-2H-indol-2-ylidene)-2,3-dihydro-3-oxo-1H-indole-5-sulfonic acid disodium salt. *See* FD&C Blue No. 2

Dihydrosafrole
Synonyms: 3,4-Methylenedioxy-propylbenzene
Empirical: $C_{10}H_{12}O_2$
Properties: Colorless to ylsh. liq., sassafras-like odor; m.w. 164.21; dens. 1.063-1.070; b.p. 112 C (15 mm); ref. index 1.5170-1.5200
Regulatory: Use in foods not permitted in U.S.

2,3-Dihydrosuccinic acid. *See* L-Tartaric acid
2 3-Dihydroxy-1,2-benzothiazolin-3-one-1,1-dioxide. *See* Saccharin
L-2,3-Dihydroxybutanedioic acid. *See* L-Tartaric acid
(R)-2,3-Dihydroxybutanedioic acid diethyl ester. *See* Diethyl tartrate
2,3-Dihydroxybutanedioic acid, monopotassium monosodium salt. *See* Potassium sodium tartrate anhyd.
2,4-Dihydroxy-N-(3-hydroxypropyl)-3,3-dimethylbutanamide. *See* D-Panthenol
2,4-Dihydroxy-N-(3-hydroxypropyl)-3,3-dimethylbutanamide. *See* DL-Panthenol
1,2-Dihydroxypropane. *See* Propylene glycol
2,3-Dihydroxypropyl docosanoate. *See* Glyceryl behenate
2,3-Dihydroxypropyl octadecanoate. *See* Glyceryl stearate
d-α,β-Dihydroxysuccinic acid. *See* L-Tartaric acid

3´,6´-Dihydroxy-2´,4´,5´,7´-tetraiodospiro[isobenzofuran-1(3H),9´-(9H)xanthen]-3-one disodium salt

3´,6´-Dihydroxy-2´,4´,5´,7´-tetraiodospiro[isobenzofuran-1(3H),9´-(9H)xanthen]-3-one disodium salt. *See* FD&C Red No. 3

Diisobutyl adipate
CAS 141-04-8; EINECS 205-450-3
Synonyms: DIBA; Bis(2-methylpropyl) hexanedioate; Diisobutyl hexanedioate; Hexanedioic acid diisobutyl ester
Definition: Diester of isobutyl alcohol and adipic acid
Empirical: $C_{14}H_{26}O_4$
Formula: $[C_2H_4COOCH_2CH(CH_3)_2]_2$
Properties: Colorless liq., odorless; sol. in most org. solvs.; insol. in water; m.w. 258.40; dens. 0.950 (25 C); b.p. 278-280 C; f.p. -20 C
Precaution: Combustible
Toxicology: LD50 (IP, rat) 5950 mg/kg; mod. toxic by IP route; mildly toxic by ingestion; experimental teratogen, reproductive effects; heated to decomp., emits acrid smoke and irritating fumes
Uses: Plasticizer migrating from food pkg.
Regulatory: FDA 21CFR §175.105, 181.22, 181.27
Manuf./Distrib.: Aceto

Diisobutyl hexanedioate. *See* Diisobutyl adipate

Diisooctyl phthalate
CAS 27554-26-3
Synonyms: DIOP; Isooctyl phthalate
Classification: Isomeric esters
Empirical: $C_{24}H_{38}O_4$
Formula: $(C_8H_{17}COO)_2C_6H_4$
Properties: Nearly colorless visc. liq., mild odor; insol. in water; m.w. 390.62; dens. 0.980-0.983 (20/20 C); m.p. -50 C; b.p. 370 C; flash pt. 450 F
Precaution: Combustible
Toxicology: LD50 (oral, rat) 22 g/kg; mod. toxic by ingestion; mildly toxic by skin contact; skin irritant; heated to decomp, emits acrid smoke and irritating fumes
Uses: Plasticizer migrating from food pkg. (for foods of high water content only)
Regulatory: FDA 21CFR §181.27

2,3-Diketobutane. *See* Diacetyl
4,4´-Diketo-β-carotene. *See* Canthaxanthine
2,5-Diketotetrahydrofuran. *See* Succinic anhydride

Dilauryl thiodipropionate
CAS 123-28-4; EINECS 204-614-1
Synonyms: Didodecyl 3,3´-thiodipropionate; Thiobis(dodecyl propionate); Thiodipropionic acid dilauryl ester; Bis(dodecyloxycarbonylethyl) sulfide
Classification: Diester
Definition: Diester of lauryl alcohol and 3,3´-thiodipropionic acid
Empirical: $C_{30}H_{58}O_4S$
Formula: $(C_{12}H_{25}OOCCH_2CH_2)_2S$
Properties: White flakes, sweetish odor; insol. in water; sol. in benzene, toluene, acetone, ether, chloroform; sl. sol. in alcohols, ethyl acetate; m.w. 514.94; dens. 0.975; m.p. 40 C; b.p. 240 C (1 mm); acid no. < 1
Toxicology: LD50 (oral, rat) > 10.3 g/kg; eye irritant; heated to decomp., emits toxic fumes
Uses: Antioxidant, preservative for vegetable oils, animal fats, fatty foods, pkg. materials
Regulatory: FDA 21CFR §175.300, 181.22, 181.24 (0.005% migrating from food pkg.), 182.3280 (0.02% max. fat/oil)
Trade names: Evanstab® 12

Dill
Uses: Natural flavoring agent
Regulatory: Japan approved

Dill oil. *See* Dillweed oil

Dill seed oil
FEMA 2383
Synonyms: American dill seed oil; European dill seed oil
Definition: Volatile oil from dried ripe fruit of *Anethum graveolens*, contg. 50% carvone d-limonene, phellandrene, other terpenes
Properties: Colorless or pale yel. liq., char. odor; sol. in 1 vol. 90% alcohol; insol. in water; dens. 0.900-0.915

(15/15 C); ref. index 1.481-1.492 (20 C)
Precaution: Keep cool, well closed; protect from light
Uses: Natural flavoring agent and adjuvant
Regulatory: FDA 21CFR §184.1282, GRAS

Dill seed oil, Indian type
EINECS 283-912-3 (extract)
Synonyms: Indian dill seed oil
Definition: From *Anethum sowa*
Properties: Caraway-like odor and flavor
Uses: Natural flavoring agent and adjuvant
Usage level: 400 ppm (baked goods), 200 ppm (condiments), 3.3-100 ppm (meats)
Regulatory: FDA 21CFR §172.510, 184.1282, GRAS

Dillweed oil
CAS 8006-75-5
Synonyms: Dill oil; Anethum graveolens oil
Toxicology: LD50 (oral, rat) 4040 mg/kg; mildly toxic by ingestion; skin irritant; mutagenic data; heated to decomp., emits acrid smoke and fumes
Uses: Natural flavoring agent and adjuvant for seasoning blends, pickles
Regulatory: FDA 21CFR §184.1282, GRAS; GRAS (FEMA); Europe listed, no restrictions
Manuf./Distrib.: Florida Treatt

Dimagnesium orthophosphate. *See* Magnesium phosphate dibasic
Dimagnesium phosphate. *See* Magnesium phosphate dibasic

Dimethicone
CAS 9006-65-9; 9016-00-6; 63148-62-9; 68037-74-1 (branched); EEC E900
Synonyms: PDMS; Dimethylpolysiloxane; Dimethyl silicone; Polydimethylsiloxane
Definition: Silicone oil consisting of dimethylsiloxane polymers
Empirical: $(C_2H_6OSi)_xC_4H_{12}Si$
Formula: $(CH_3)_3SiO[Si(CH_3)_2O]_nSi(CH_3)_3$
Properties: Colorless visc. oil; sol. in hydrocarbon solvs.; misc. in chloroform, ether; insol. in water; m.w. 340-250,000; dens. 0.96-0.97; visc. 300-1050 cst; ref. index 1.400-1.404
Toxicology: Suspected neoplastigen; heated to dec., emits acrid smoke and irritating fumes
Uses: Defoamer/antifoaming agent for food applics., fruit juices, wine-making, sugar refining; hog/poultry scald agent; release agent migrating from food pkg., chewing gum base; anticaking agent; In brewing, fermentation
Usage level: Limitation 10 ppm (food), zero (milk), 110 ppm (dry gelatin dessert mixes), 250 ppm (salt); ADI 0-1.5 mg/kg (EEC)
Regulatory: FDA 21CFR §145.180, 145.181, 146.185, 173.340, 175.105, 175.300, 176.170, 176.200, 176.210, 177.2260, 177.2600, 177.2800, 178.3570, 178.3910, 181.22, 181.28; USDA 9CFR §318.7, 381.147; Europe listed; UK approved
Trade names: AF 10 FG; AF 30 FG; AF 9020; Dow Corning® 200 Fluid, Food Grade; Foamkill® 810F; Foamkill® 830F; Masil® SF 5; Masil® SF 10; Masil® SF 20; Masil® SF 50; Masil® SF 100; Masil® SF 200; Masil® SF 350; Masil® SF 350 FG; Masil® SF 500; Masil® SF 1000; Masil® SF 5000; Masil® SF 10,000; Masil® SF 12,500; Masil® SF 30,000; Masil® SF 60,000; Masil® SF 100,000; Masil® SF 300,000; Masil® SF 500,000; Masil® SF 600,000; Masil® SF 1,000,000; SF18-350
Trade names containing: Creamy Liquid Frying Shortening 102-050; Durkex 100DS; Golden Chef® Clear Liquid Frying Shortening 104-052; Golden Chef® Creamy Liquid Frying Shortening 104-065; Gold 'N Flavor® 103-061; Heavy Duty Frying Shortening 101-100; Heavy Duty Peanut Oil 104-210; Hi-Lite™; Hi-Tone™; H.L. 93; HWDX Winterized Salad Oil; Peanut Oil; Soy/Peanut Liquid Frying Shortening 104-215; Superb® Canola Liquid Frying Shortening 104-455; Superb® Corn Liquid Frying Shortening 104-255; Superb® Fish and Chip Oil 102-060; Superb® Heavy Duty Frying Shortening 101-200; Superb® Liquid Frying Shortening 102-110; Superb® Soybean Salad Oil 104-112; Superb® Soy/Corn Frying Shortening 102-070; Winterized Clear Liquid Frying Shortening 104-112

3,4-Dimethoxybenzaldehyde. *See* Veratraldehyde
1,3-Dimethoxybenzene. *See* m-Dimethoxybenzene
1,4-Dimethoxybenzene. *See* p-Dimethoxybenzene

m-Dimethoxybenzene
CAS 151-10-0; EINECS 205-783-4; FEMA 2385
Synonyms: 1,3-Dimethoxybenzene; Dimethylresorcinol; Resorcinol dimethyl ether
Definition: Synthesized from resorcinol by methylation using dimethyl sulfate and alkali
Empirical: $C_8H_{10}O_2$

Properties: Liq., acrid fruity odor reminiscent of nerolin; sol. in alcohol, ether, benzene; sl. sol. in water; m.w. 138.17; dens. 1.067 (20/4 C); b.p. 85-87 C (10 mm); flash pt. 190 F; ref. index 1.525 (20 C)
Toxicology: Skin irritant
Uses: Synthetic flavoring agent
Usage level: 3.0 ppm (nonalcoholic beverages), 5.0 ppm (ice cream, ices, candy), 8.0 ppm (baked goods)
Regulatory: FDA 21CFR §172.515
Manuf./Distrib.: Aldrich

p-Dimethoxybenzene
CAS 150-78-7; EINECS 205-771-9; FEMA 2386
Synonyms: Dimethyl hydroquinone; Hydroquinone dimethyl ether; 1,4-Dimethoxybenzene
Definition: Obtained by methylation of hydroquinone using dimethyl sulfate and alkali
Empirical: $C_8H_{10}O_2$
Formula: $C_6H_4(OCH_3)_2$
Properties: Low melting solid; m.w. 138.17; dens. 1.036 (66 C); m.p. 54-56 C; b.p. 109 C (20 mm); flash pt. 125 C
Uses: Synthetic flavoring agent
Usage level: 8.1 ppm (nonalcoholic beverages), 5.0 ppm (ice cream, ices), 4.7 ppm (candy), 5.8 ppm (baked goods)
Regulatory: FDA 21CFR §172.515
Manuf./Distrib.: Aldrich

3,4-Dimethoxybenzenecarbonal. See Veratraldehyde
1,1-Dimethoxy decane. See Decanal dimethyl acetal
8,8-Dimethoxy-2,6-dimethyl-octanol-2. See Hydroxycitronellal dimethyl acetyl
(2,2-Dimethoxyethyl)-benzene. See Phenylacetaldehyde dimethyl acetal
1,1-Dimethoxy heptane. See Heptanal dimethyl acetal
1,1,-Dimethoxy-2-phenylethane. See Phenylacetaldehyde dimethyl acetal
1,1-Dimethoxy-2-phenylpropane. See 2-Phenylpropionaldehyde dimethylacetal
1,2-Dimethoxy-4-(2-propenyl)benzene. See Methyl eugenol
α,α- Dimethoxytoluene. See Benzaldehyde dimethyl acetal
Dimethylacetic acid. See Isobutyric acid
Dimethylacetone. See Diethyl ketone

2,4-Dimethylacetophenone
CAS 89-74-7; EINECS 201-935-9; FEMA 2387
Synonyms: Methyl 2,4-dimethylphenyl ketone
Definition: Obtained by condensation of acetyl chloride and m-xylene in presence of aluminum or ferric chloride
Empirical: $C_{10}H_{12}O$
Properties: Colorless to ylsh. oil liq., floral sweet odor; m.w. 148.21; dens. 0.998 (20/4 C); b.p. 120 C (10 mm); flash pt. > 100 C; ref. index 1.543 (20 C)
Precaution: Combustible
Uses: Synthetic flavoring agent
Usage level: 0.78 ppm (nonalcoholic beverages), 1.0 ppm (alcoholic beverages), 0.77 ppm (ice cream, ices), 3.9 ppm (candy), 2.7 ppm (baked goods)
Regulatory: FDA 21CFR §172.515
Manuf./Distrib.: Aldrich

2,5-Dimethyl-3-acetylfuran. See 3-Acetyl-2,5-dimethylfuran
1,1-Dimethylallyl alcohol. See 2-Methyl-3-buten-2-ol
3,3-Dimethylallyl alcohol. See 3-Methyl-2-buten-1-ol

Dimethylamine/epichlorohydrin copolymer
CAS 25988-97-0
Synonyms: Epichlorohydrin-dimethylamine copolymer
Properties: visc. 175 cps (50% aq.)
Toxicology: Heated to decomp., emits acrid smoke and irritating fumes
Uses: Decolorizing agent and/or flocculant in clarification of refinery sugar liquors and juices
Usage level: 150 ppm max. of sugar solids
Regulatory: FDA 21CFR §173.60

Dimethyl anthranilate
CAS 85-91-6; EINECS 201-642-6; FEMA 2718
Synonyms: MMA; Methyl methylaminobenzoate; N-Methyl methyl anthranilate; 2-Methylamino methyl benzoate
Empirical: $C_9H_{11}NO_2$

Properties: Pale yel. liq., grape-like odor; sol. in fixed oils; sl. sol. in propylene glycol; insol. in water, glycerin; m.w. 165.21; dens. 1.126-1.132; flash pt. 196 F; ref. index 1.578-1.581
Precaution: Combustible
Toxicology: Poison by IV route; mod. toxic by ingestion; heated to decomp., emits toxic fumes of NO_x
Uses: Synthetic flavoring for baked goods, confectionery, ice cream, soft drinks
Regulatory: FDA 21CFR §172.515; Japan approved as flavoring
Manuf./Distrib.: Florida Treatt

α-(5,6-Dimethylbenzimidazolyl)cyanocobamide. *See* Cyanocobalamin

Dimethylbenzyl carbinol
CAS 100-86-7; FEMA 2393
Synonyms: α,α-Dimethylphenethyl alcohol; 1,1-Dimethyl-2-phenylethanol; Benzyl dimethyl carbinol
Definition: Synthesized from acetone and benzyl magnesium chloride
Empirical: $C_{10}H_{14}O$
Properties: Wh. cryst. solid, fresh floral odor, bitter taste; m.w. 150.22; m.p. 24 C; b.p. 108 C (11 mm); flash pt. 92 C
Precaution: Combustible liq.
Toxicology: LD50 (oral, rat) 1280 mg/kg; mod. toxic by ingestion; heated to decomp., emits acrid smoke and fumes
Uses: Synthetic flavoring agent
Usage level: 3.3 ppm (nonalcoholic beverages), 3.2 ppm (ice cream, ices, jellies), 4.0 ppm (candy), 4.9 ppm (baked goods), 0.01 ppm (gelatins, puddings)
Regulatory: FDA 21CFR §172.515

Dimethylbenzyl carbinyl acetate
FEMA 2392
Synonyms: α,α-Dimethylphenethyl acetate; 1,1-Dimethyl-2-phenyl acetate; Benzylpropyl acetate; Benzyl dimethyl carbinyl acetate
Definition: Obtained by acetylation of dimethylbenzyl carbinol
Empirical: $C_{12}H_{16}O_2$
Properties: Water-wh. liq., floral fruity odor; sol. in alcohol (4 parts/70%); m.w. 192.26; sp.gr. 0.995-0.999 (supercooled); m.p. 29.5 C; flash pt. > 100 C; ref. index 1.4900-1.4940 (supercooled)
Precaution: Combustible liq.
Toxicology: Heated to decomp., emits acrid smoke and fumes
Uses: Synthetic flavoring agent
Usage level: 2.8 ppm (nonalcoholic beverages), 8.0 ppm (ice cream, ices), 22 ppm (candy), 19 ppm (baked goods), 2.9 ppm (chewing gum)
Regulatory: FDA 21CFR §172.515

Dimethylbenzylcarbinyl butyrate. *See* α,α-Dimethylphenethyl butyrate
Dimethylbenzylcarbinyl formate. *See* α,α-Dimethylphenethyl formate
4,5-Dimethyl-2-benzyl-1,3-dioxolan. *See* Phenylacetaldehyde 2,3-butylene glycol acetal

α,α-Dimethylbenzyl isobutyrate
FEMA 2388
Synonyms: Phenyldimethyl carbinyl isobutyrate; 2-Phenylpropan-2-yl isobutyrate
Definition: Obtained by esterification of dimethylphenyl carbinol with isobutyric acid
Empirical: $C_{13}H_{18}O_2$
Properties: Colorless liq., sweet fruity apricot/peach/plum odor, plum-like flavor; sol. in alcohol; insol. in water; m.w. 206.28
Uses: Synthetic flavoring agent
Usage level: 5.0 ppm (nonalcoholic beverages), 40 ppm (ice cream, ices), 30 ppm (candy), 20 ppm (baked goods)
Regulatory: FDA 21CFR §172.515

Dimethyl butanedioate. *See* Dimethyl succinate
4,4´-(2,3-Dimethyl-1,4-butanediyl)bis[1,2-benzenediol]. *See* Nordihydroguairetic acid
Dimethyl carbinol. *See* Isopropyl alcohol
Dimethyl dialkyl ammonium chloride. *See* Dialkyl dimethyl ammonium chloride
2,3-Dimethyl-1,4-diazine. *See* 2,3-Dimethylpyrazine

Dimethyl dicarbonate
CAS 4525-33-1; EINECS 224-859-8
Synonyms: DMDC; Dimethyl pyrocarbonate; Pyrocarbonic acid dimethyl ester
Empirical: $C_4H_6O_5$

Dimethyl didecyl ammonium chloride

Formula: (CH₃OCO)₂O
Properties: m.w. 134.09; dens. 1.254 (20/4 C); m.p. 15-17 C; b.p. 45-46 C (5 mm); ref. index 1.393
Toxicology: Heated to decomp., emits acrid smoke and fumes
Uses: Preservative for wine, dealcoholized and low alcohol wine; yeast inhibitor; flavoring agent
Usage level: 200 ppm max. (wine)
Regulatory: FDA 21CFR §172.133

Dimethyl didecyl ammonium chloride. *See* Didecyldimonium chloride
Dimethyl diketone. *See* Diacetyl
Dimethyl disulfide. *See* Methyl disulfide
Dimethyldithiocarbamic acid sodium salt. *See* Sodium dimethyldithiocarbamate
N,N-Dimethyl-N-dodecylbenzenemethanaminium chloride. *See* Lauralkonium chloride

Dimethyl dodecyl betaine
Uses: Germicide
Trade names: Nissan Anon BL

Dimethylenediamine. *See* Ethylenediamine
2-(1,1-Dimethylethyl)-1,4-benzenediol. *See* t-Butyl hydroquinone
Dimethyl ethyl carbinol. *See* t-Amyl alcohol
(1,1-Dimethylethyl)-4-methoxyphenol. *See* BHA

Dimethyl fumarate
CAS 624-49-7
Formula: CH₃O₂CCH=CHCO₂CH₃
Properties: m.w. 144.13; b.p. 102-103 C
Uses: Inhibits fungal growth in tomato juice
Usage level: 0.05%

Dimethylglyoxal. *See* Diacetyl

2,6-Dimethyl-5-heptenal
CAS 106-72-9; FEMA 2389
Empirical: C₉H₁₆O
Properties: Pale yel. liq., melon odor; m.w. 140.23; dens. 0.852-0.858; b.p. 80 C (19 mm); flash pt. 144 F; ref. index 1.443-1.448
Toxicology: Skin and eye irritant; heated to decomp., emits acrid smoke and irritating fumes
Uses: Synthetic flavoring agent
Usage level: 2.8 ppm (nonalcoholic beverages), 1.7 ppm (ice cream, ices), 8.4 ppm (candy), 19 ppm (baked goods), 0.02-10 ppm (gelatins, puddings), 0.80 ppm (chewing gum)
Regulatory: FDA 21CFR §172.515
Manuf./Distrib.: Aldrich

Dimethyl hexynediol
CAS 142-30-3
Synonyms: 2,5-Dimethyl-3-hexyne-2,5-diol
Definition: Di-tertiary acetylenic diol
Empirical: C₈H₁₄O₂
Formula: (CH₃)₂COHCCCOH(CH₃)₂
Properties: Wh. cryst.; sol. in water; sl. sol. in benzene, CCl₄, naphtha; very sol. in acetone, alcohol, ethyl acetate; dens. 0.949 (20/20 C); m.p. 94-95 C; b.p. 205-206 C
Uses: Intermediate in synthesis of flavors

2,5-Dimethyl-3-hexyne-2,5-diol. *See* Dimethyl hexynediol
Dimethyl hydroquinone. *See* p-Dimethoxybenzene
1,4-Dimethyl-7-(α-hydroxyisopropyl)-δ9,10-octahydroazulene acetate. *See* Guaiol acetate
3,7-Dimethyl-7-hydroxy-octane-1-al. *See* Hydroxycitronellal
5,6-Dimethyl-8-isopropenyl bicyclo[4.4.0]-dec-1-en-3-one. *See* Nootkatone
1,4-Dimethyl-7-isopropenyl-δ9,10-octahydroazulene. *See* Guaiene
Dimethylketol. *See* Acetyl methyl carbinol
Dimethylketone. *See* Acetone
Dimethylmethane. *See* Propane
6,6-Dimethyl-2-methylenebicyclo[3.1.1]heptane. *See* β-Pinene
2,2-Dimethyl-3-methylenenorbornane. *See* Camphene
3,3-Dimethyl-2-methylene norcamphane. *See* Camphene
N,N-Dimethyl-N-octadecylbenzenemethanaminium chloride. *See* Stearalkonium chloride
Dimethyloctadecylbenzyl ammonium chloride. *See* Stearalkonium chloride

2,6-Dimethyloctadien-2,6-al-8. *See* Citral
3,7-Dimethyl-2,6-octadienal. *See* Citral
3,7-Dimethyl-2,6-octadienal diethyl acetal. *See* Citral diethyl acetate
3,7-Dimethyl-2,6-octadienal dimethyl acetal. *See* Citral dimethyl acetal
2,6-Dimethyl-2,6-octadien-8-ol. *See* Nerol
2,6-Dimethyl-2,7-octadien-6-ol. *See* Linalool
3,7-Dimethyl-1,6-octadien-3-ol. *See* Linalool
3,7-Dimethyl-1,6-octadien-3-ol acetate. *See* Linalyl acetate
trans-3,7-Dimethyl-2,6-octadien-1-ol acetate. *See* Geranyl acetate
cis-3,7-Dimethyl-2,6-octadien-1-ol. *See* Nerol
trans-3,7-Dimethyl-2,6-octadien-1-ol. *See* Geraniol
3,7-Dimethyl-1,6-octadien-3-yl acetate. *See* Linalyl acetate
cis-3,7-Dimethyl-2,6-octadien-1-yl-acetate. *See* Neryl acetate
trans-3,7-Dimethyl-2,6-octadien-1-yl acetoacetate. *See* Geranyl acetoacetate
3,7-Dimethyl-1,6-octadien-3-yl anthranilate. *See* Linalyl anthranilate
3,7-Dimethyl-1,6-octadien-3-yl benzoate. *See* Linalyl benzoate
3,7-Dimethyl-2,6-octadien-1-yl benzoate. *See* Geranyl benzoate
3,7-Dimethyl-2,6-octadien-1-yl butyrate. *See* Geranyl butyrate
3,7-Dimethyl-1,6-octadien-3-yl formate. *See* Linalyl formate
3,7-Dimethyl-2,6-octadien-1-yl formate. *See* Geranyl formate
3,7-Dimethyl-2,6-octadien-3-yl isobutyrate. *See* Linalyl isobutyrate
3,7-Dimethyl-2,6-octadien-1-yl phenylacetate. *See* Geranyl phenylacetate
3,7-Dimethyl-2,6-octadien-1-yl propionate. *See* Geranyl propionate
3,7-Dimethyl-2,6-octadien-3-yl propionate. *See* Linalyl propionate

2,6-Dimethyl octanal
　　FEMA 2390
　　Synonyms: 2,6-Dimethyl octanoic aldehyde; Isoaldehyde C-10; Isodecylaldehyde
　　Empirical: $C_{10}H_{20}O$
　　Properties: Colorless liq., sweet fruity odor, somewhat green flavor; sol. in alcohol; insol. in water; m.w. 156.26
　　Uses: Synthetic flavoring agent
　　Usage level: 0.44 ppm (nonalcoholic beverages), 3.2 ppm (ice cream, ices), 1.9 ppm (candy, baked goods)
　　Regulatory: FDA 21CFR §172.515

3,7-Dimethyloctane-1,7-diol. *See* Hydroxycitronellol
3,7-Dimethyl-1,7-octanediol. *See* Hydroxycitronellol
3,7-Dimethyloctane-3-ol. *See* Tetrahydrolinalool
2,6-Dimethyl octanoic aldehyde. *See* 2,6-Dimethyl octanal
Dimethyl octanol. *See* 3,7-Dimethyl-1-octanol

3,7-Dimethyl-1-octanol
　　CAS 106-21-8; EINECS 203-374-5; FEMA 2391
　　Synonyms: Dimethyl octanol; Tetrahydrogeraniol; Dihydrocitronellol
　　Definition: Usually prepared by hydrogenation of geraniol, citronellol, or citronellal
　　Empirical: $C_{10}H_{22}O$
　　Formula: $(CH_3)_2CH(CH_2)_3CH(CH_3)CH_2CH_2OH$
　　Properties: Colorless liq., sweet rosy odor, bitter taste; sol. in min. oil; insol. in glycerol; m.w. 158.29; dens. 0.828 (20/4 C); b.p. 98-99 C (9 mm); flash pt. 97 C; ref. index 1.435 (20 C)
　　Precaution: Combustible
　　Toxicology: LD50 (skin, rabbit) 2400 mg/kg; mod. toxic by skin contact; skin irritant; heated to decomp., emits acrid smoke and fumes
　　Uses: Synthetic flavoring agent
　　Usage level: 4.3 ppm (nonalcoholic beverages), 2.0-44 ppm (ice cream, ices), 15 ppm (candy), 19 ppm (baked goods), 2.9 ppm (chewing gum)
　　Regulatory: FDA 21CFR §172.515
　　Manuf./Distrib.: Aldrich

3,7-Dimethyl-3-octanol. *See* Tetrahydrolinalool
3,7-Dimethyl-1,3,6-octatriene. *See* Ocimene
3,7-Dimethyl-6-octenal. *See* Citronellal
3,7-Dimethyl-6-octen-1-ol. *See* β-Citronellol
S-(-)-3,7-Dimethyl-7-octen-1-ol. *See* Rhodinol
3,7-Dimethyl-6-octen-1-ol acetate. *See* Citronellyl acetate
3,7-Dimethyl-7-octen-1-ol acetate. *See* Rhodinyl acetate
3,7-Dimethyl-6-octen-1-yl butyrate. *See* Citronellyl butyrate

3,7-Dimethyl-6-octen-1-yl formate. *See* Citronellyl formate
3,7-Dimethyl-6-octen-1-yl isobutyrate. *See* Citronellyl isobutyrate
3,7-Dimethyl-6-octen-1-yl phenylacetate. *See* Citronellyl phenylacetate
3,7-Dimethyl-6-octen-1-yl propionate. *See* Citronellyl propionate
3,7-Dimethyl-6-octen-1-yl valerate. *See* Citronellyl valerate
α,α-Dimethylphenethyl acetate. *See* Dimethylbenzyl carbinyl acetate
α,α-Dimethylphenethyl alcohol. *See* Dimethylbenzyl carbinol

α,α-Dimethylphenethyl butyrate
FEMA 2394
Synonyms: Benzyl dimethylcarbinyl butyrate; Dimethylbenzylcarbinyl butyrate; DMBC butyrate
Definition: Obtained by esterification of dimethyl benzyl carbinol with n-butyric acid
Empirical: $C_{14}H_{20}O_2$
Properties: Colorless liq., plum-prune odor, apricot/peach/plum-like taste; sol. in alcohol; insol. in water; m.w. 220.31
Precaution: Combustible liq.
Toxicology: Heated to decomp., emits acrid smoke and fumes
Uses: Synthetic flavoring agent
Usage level: 10 ppm (nonalcoholic beverages), 20 ppm (ice cream, ices, candy, gelatin, puddings)
Regulatory: FDA 21CFR §172.515

α,α-Dimethylphenethyl formate
Synonyms: Benzyl dimethylcarbinyl formate; Dimethylbenzylcarbinyl formate
Definition: Synthesized from dimethyl benzyl carbinol and formic acid using acetic anhydride
Empirical: $C_{11}H_{14}O_2$
Properties: Colorless liq., lily-jasmine odor, spicy taste; sol. in alcohol; almost insol. in water; m.w. 178.23
Uses: Synthetic flavoring agent
Usage level: 2.0 ppm (nonalcoholic beverages), 10 ppm (ice cream, ices, candy)
Regulatory: FDA 21CFR §172.515

1,1-Dimethyl-2-phenyl acetate. *See* Dimethylbenzyl carbinyl acetate
N-[2-[(2,6-Dimethylphenyl)amino]-2-oxoethyl]-N,N-diethylbenzenemathanaminium benzoate. *See* Denatonium benzoate NF
1,1-Dimethyl-2-phenylethanol. *See* Dimethylbenzyl carbinol

Dimethylphenylethylcarbinol
Synonyms: 4-Phenyl-2-methyl-butanol-2
Uses: Green, floral, mildly herbal, hyacinth-like fragrance and flavoring

Dimethylphenylethyl carbinyl acetate. *See* 2-Methyl-4-phenyl-2-butyl acetate
Dimethylphenylethyl carbinyl isobutyrate. *See* 2-Methyl-4-phenyl-2-butyl isobutyrate
Dimethylpolysiloxane. *See* Dimethicone

2,3-Dimethylpyrazine
CAS 5910-89-4; FEMA 3271
Synonyms: 2,3-Dimethyl-1,4-diazine
Empirical: $C_6H_8N_2$
Properties: Colorless liq.,nutty cocoa odor; misc. with water, org. solvs.; m.w. 108.16; dens. 1.000-1.022 (20 C); b.p. 182 C; flash pt. (OC) 147 F; ref. index 1.506-1.509
Precaution: Combustible liq.
Toxicology: LD50 (oral, rat) 613 mg/kg; mod toxic by ingestion and intraperitoneal routes; heated to decomp., emits toxic fumes of NO_x
Uses: Synthetic flavoring agent
Regulatory: GRAS
Manuf./Distrib.: Aldrich

2,5-Dimethylpyrazine
CAS 123-32-0; EINECS 204-618-3; FEMA 3272
Empirical: $C_6H_8N_2$
Properties: m.w. 108.14; dens. 0.990; b.p. 155 C; flash pt. 147 F
Precaution: Combustible liq.
Toxicology: LD50 (oral, rat) 1020 mg/kg; irritant; mod. toxic by ingestion and IP routes; mutagenic data; heated to decomp., emits toxic fumes of NO_x
Uses: Synthetic flavoring agent
Manuf./Distrib.: Aldrich

2,6-Dimethylpyrazine
CAS 108-50-9; FEMA 3273
Empirical: $C_6H_8N_2$
Properties: m.w. 108.14; m.p. 35-40 C; b.p. 154 C; flash pt. 127 F
Toxicology: LD50 (IP, mouse) 1080 mg/kg; mod. toxic by IP route; heated to decomp., emits toxic fumes of NO_x
Uses: Synthetic flavoring agent
Manuf./Distrib.: Aldrich

Dimethyl pyrocarbonate. *See* Dimethyl dicarbonate

2,5-Dimethylpyrrole
CAS 625-84-3; FEMA 7071
Properties: m.w. 95.15; dens. 0.935; b.p. 165 C (740 mm); flash pt. 130 F
Toxicology: Irritant; heated to decomp., emits toxic fumes of NO_x
Uses: Flavoring agent

Dimethylresorcinol. *See* m-Dimethoxybenzene
6,7-Dimethyl-9-d-ribitylisoalloxazine. *See* Riboflavin
7,8-Dimethyl-10-(d-ribityl) isoalloxazine. *See* Riboflavin
Dimethyl silicone. *See* Dimethicone

Dimethyl succinate
CAS 106-65-0; EINECS 203-419-9; FEMA 2396
Synonyms: Dimethyl butanedioate; Methyl succinate
Empirical: $C_6H_{10}O_4$
Formula: $CH_3OCOCH_2CH_2COOCH_3$
Properties: Colorless liq., ethereal winey odor; m.w. 146.14; dens. 1.119 (20/4 C); m.p. 16-19 C; b.p. 190-193 C; flash pt. 90 C; ref. index 1.419
Uses: Synthetic flavoring agent
Usage level: 1.0-100 ppm (nonalcoholic beverages), 5.0 ppm (ice cream, ices, chewing gum), 15 ppm (candy, baked goods)
Regulatory: FDA 21CFR §172.515
Manuf./Distrib.: Aldrich

Dimethyl sulfide. *See* Methyl sulfide
3-[(2,4-Dimethyl-5-sulfophenyl)azo]-4-hydroxy-1-naphthalenesulfonic acid disodium salt. *See* FD&C Red No. 4
N,N-Dimethyl-N-tetradecylbenzenemethanaminium chloride. *See* Myristalkonium chloride
4,4´-(2,3-Dimethyltetramethylene)dipyrocatechol. *See* Nordihydroguairetic acid
3,7-Dimethyl-9-(2,6,6-trimethyl-1-cyclohexen-1-yl)-2,4,6,8-nonatretraen-1-ol. *See* Retinol
6,10-Dimethyl-5,9-undecadien-2-one. *See* Geranyl acetone
6,10-Dimethyl-9-undecen-2-one. *See* Tetrahydro-pseudo-ionone
Dinitrogen monoxide. *See* Nitrous oxide
Dinkum oil. *See* Eucalyptus oil
3,3´-Dioctadecyl thiodipropionate. *See* Distearyl thiodipropionate
Dioctrahedral smectite. *See* Attapulgite

Dioctyl phthalate
CAS 117-81-7; EINECS 204-211-0
Synonyms: DOP; DEHP; Di(2-ethylhexyl) phthalate; Di-s-octyl phthalate; 1,2-Benzenedicarboxylic acid dioctyl ester
Definition: Diester of 2-ethylhexyl alcohol and phthalic acid
Empirical: $C_{24}H_{38}O_4$
Formula: $C_6H_4[COOCH_2CH(C_2H_5)C_4H_9]_2$
Properties: Lt.-colored liq., odorless; insol. in water; misc. with min. oil; m.w. 390.62; dens. 0.9861 (20/20 C); b.p. 231 C (5 mm); flash pt. 218 C; ref. index 1.4836
Precaution: Combustible
Toxicology: TLV 5 mg/m^3; STEL 10 mg/m^3; LD50 (oral, rat) 30,600 mg/kg; poison by IV; mildly toxic by ingestion; skin and severe eye irritant; suspected human carcinogen; experimental teratogen; affects human GI tract; heated to decomp., emits acrid smoke
Uses: Plasticizer migrating from food pkg. (for foods of high water content only)
Regulatory: FDA 21CFR §175.105, 175.300, 175.310, 715.380, 175.390, 176.170, 176.210, 176.1210, 177.1010, 177.1200, 177.1210, 177.1400, 177.2600, 178.3120, 178.3910, 181.22, 181.27
Manuf./Distrib.: Eastman

Di-s-octyl phthalate. *See* Dioctyl phthalate

Dioctyl sodium sulfosuccinate
CAS 577-11-7; 1369-66-3; EINECS 209-406-4
Synonyms: DSS; Sodium dioctyl sulfosuccinate; Sodium di(2-ethylhexyl) sulfosuccinate; Docusate sodium
Definition: Sodium salt of the diester of 2-ethylhexyl alcohol and sulfosuccinic acid
Empirical: $C_{20}H_{38}O_7S \cdot Na$
Formula: $C_8H_{17}OOCCH_2CH(SO_3Na)COOC_8H_{17}$
Properties: White wax-like solid, octyl alcohol odor; slowly sol. in water; freely sol. in alcohol, glycerol, CCl_4, acetone, xylene; m.w. 445.63; m.p. 173-179 C
Toxicology: LD50 (oral, rat) 1900 mg/kg; moderately toxic by ingestion, intraperitoneal routes; poison by intravenous route; skin, severe eye irritant; heated to decomp., emits toxic fumes of SO_x and Na_2O
Uses: Emulsifier, processing aid in sugar industry, wetting agent, hog/poultry scald agent
Usage level: Limitation 9 ppm (finished food), 75 ppm (cocoa beverage), 15 ppm (gelatin dessert), 10 ppm (beverage/fruit drinks), 25 ppm (molasses), 0.5% (gums)
Regulatory: FDA 21CFR §73.1, with cocoa, 131.130, 131.132, 133.124, 133.133, 133.134, 133.162, 133.178, 133.179, 163.114, 163.117, 169.115, 169.150, 172.520, 172.808, 172.810, 175.105, 175.300, 175.320, 176.170, 176.210, 177.1200, 177.2800, 178.1010, 178.3400; USDA 9 CFR §318.7, 381.147
Manuf./Distrib.: Am. Cyanamid
Trade names: Complemix® 100

DIOP. *See* Diisooctyl phthalate
1,4-Dioxacycloheptadecane-5,17-dione. *See* Ethylene brassylate
1,1-Dioxide-1,2-benzisothiazol-3(2H)-one sodium salt. *See* Sodium saccharin
Dioxymethyleneprotocatechuic aldehyde. *See* Heliotropine
Dipentene. *See* dl-Limonene
Dipentene. *See* l-Limonene
Diphenyl. *See* Biphenyl
Diphenyl-2-ethylhexyl phosphate. *See* Diphenyl octyl phosphate
Diphenyl ketone. *See* Benzophenone
Diphenylmethanone. *See* Benzophenone
Di(phenylmethyl) disulfide. *See* Benzyl disulfide

Diphenyl octyl phosphate
CAS 1241-94-7
Synonyms: Diphenyl-2-ethylhexyl phosphate; 2-Ethylhexyl diphenyl ester phosphoric acid
Empirical: $C_{20}H_{27}O_4P$
Properties: m.w. 362.44
Toxicology: Poison by IV route; heated to decomp., emits toxic fumes of PO_x
Uses: Plasticizer migrating from food pkg.
Regulatory: FDA 21CFR §181.27

1,3-Diphenyl-2-propanone
CAS 102-04-5; EINECS 203-000-0; FEMA 2397
Synonyms: Benzyl ketone; Dibenzyl ketone
Empirical: $C_{15}H_{14}O$
Formula: $(C_6H_5CH_2)_2CO$
Properties: Low-melting solid, bitter almond odor; sol. in ether; insol. in water; m.w. 210.28; m.p. 32-34 C; b.p. 330 C; flash pt. > 230 F
Uses: Synthetic flavoring agent
Usage level: 1.7 ppm (nonalcoholic beverages), 4.5 ppm (ice cream, ices), 9.5 ppm (candy), 13 ppm (baked goods)
Regulatory: FDA 21CFR §172.515
Manuf./Distrib.: Aldrich

Diphosphoric acid, calcium salt (1:2). *See* Calcium pyrophosphate
Diphosphoric acid disodium salt. *See* Sodium acid pyrophosphate
Diphosphoric acid tetrapotassium salt. *See* Tetrapotassium pyrophosphate
Diphosphoric acid tetrasodium salt. *See* Tetrasodium pyrophosphate
Dipotassium carbonate. *See* Potassium carbonate

Dipotassium citrate
Toxicology: Heated to decomp., emits acrid smoke and irritating fumes
Uses: Stabilizer migrating from food pkg.
Regulatory: FDA 21CFR §181.29

Dipotassium dichloride. *See* Potassium chloride
Dipotassium disulfite. *See* Potassium metabisulfite

Dipotassium hydrogen orthophosphate. *See* Potassium phosphate dibasic
Dipotassium hydrogen phosphate. *See* Potassium phosphate dibasic
Dipotassium orthophosphate. *See* Potassium phosphate dibasic
Dipotassium persulfate. *See* Potassium persulfate
Dipotassium phosphate. *See* Potassium phosphate dibasic
Dipotassium L-(+)-tartrate. *See* Potassium acid tartrate
Dipropionyl. *See* 3,4-Hexanedione
Dipropyl disulfide. *See* Propyl disulfide
Dipropyl ketone. *See* 4-Heptanone

Dismutase superoxide
CAS 9054-89-1; EINECS 232-943-0
Properties: m.w. ≈ 32,000
Uses: Natural enzyme
Regulatory: Japan approved

Disodium capryloamphodiacetate
CAS 7702-01-4; 68608-64-0; EINECS 231-721-0; 271-792-5
Synonyms: 1H-Imidazolium, 1-[2-(carboxymethoxy)ethyl]-1-(carboxymethyl)-2-heptyl-4,5-dihydro-, hydroxide, disodium salt; Capryloamphocarboxyglycinate; Capryloamphodiacetate
Classification: Amphoteric organic compd.
Empirical: $C_{16}H_{31}N_2O_6 \cdot 2Na$
Uses: Emulsifier, food washing and peeling
Trade names: Miranol® J2M Conc.; Miranol® JB

Disodium carbonate. *See* Sodium carbonate

Disodium citrate
CAS 144-33-2; EEC E331b
Synonyms: Disodium hydrogen citrate
Empirical: $C_6H_6O_7 \cdot 2Na$
Properties: Wh. cryst. or gran. powd., odorless; sol. in water; insol. in alcohol; m.w. 236.10; m.p. loses water @ 150 C; b.p. dec. @ red heat
Toxicology: LD50 (IP, rat) 1724 mg/kg; poison by IV route; mod. toxic by IP and subcut. routes; heated to decomp., emits acrid smoke, toxic fumes of Na_2O
Uses: Buffer, nutrient for cultured buttermilk, sequestrant, curing accelerator, antioxidant, antioxidant synergist, emulsifier; stabilizer migrating from food pkg.; in poultry, cured meat prods., coffee whiteners, margarine, evaporated milk, pork
Usage level: Limitation 10% sol'n. (cured cuts of meat), 250 ppm (citric acid and/or sodium citrate on fresh pork cuts)
Regulatory: FDA 21CFR §181.29, 182.1751, 182.6751, GRAS; USDA 9CFR §318.7, 381.147; Europe listed

Disodium cyanodithioimidocarbamate
Uses: Used in mfg. of paper/paperboard contacting food
Regulatory: FDA 21CFR §181.30

Disodium cyanodithioimidocarbonate
Uses: Antimicrobial for use in cane-sugar and beet-sugar mills
Regulatory: FDA 21CFR §173.320

Disodium dihydrogen diphosphate. *See* Sodium acid pyrophosphate
Disodium dihydrogen pyrophosphate. *See* Sodium acid pyrophosphate
Disodium diphosphate. *See* Sodium acid pyrophosphate
Disodium dithionite. *See* Sodium hydrosulfite
Disodium edetate. *See* Disodium EDTA

Disodium EDTA
CAS 139-33-3; EINECS 205-358-3
Synonyms: Disodium edetate; Ethylenediaminetetraacetic acid, disodium salt; Edetate disodium
Classification: Substituted diamine
Empirical: $C_{10}H_{16}N_2O_8 \cdot 2Na$
Properties: White crystalline powd.; freely sol. in water; m.w. 336.24; m.p. 252 (dec.)
Toxicology: LD50 (oral, rat) 2 g/kg; poison by intravenous route; moderately toxic by ingestion; experimental teratogen, reproductive effects; mutagenic data; heated to decomp., emits toxic fumes of NO_x and Na_2O
Uses: Food preservative, stabilizer, chelating and sequestering agent; promotes color retention in canned peas; hog scald agent; sequestrant in nonnutritive sweeteners; antioxidant (Japan)
Usage level: 0.1-0.5%; to 145 ppm (canned black eye peas), to 165 ppm (canned kidney beans, as

449

preservative), to 500 ppm (canned strawberry pie filling), to 75 ppm (dressings, mayonnaise), to 100 ppm (sandwich spread), to 150 ppm (aq. multivitamins)
Regulatory: FDA §155.200, 169.115, 169.140, 169.150, 172.135, 175.105, 176.150, 176.170, 177.1200, 177.2800, 178.3570, 178.3910, 573.360; USDA 9CFR §318.7; Japan approved (0.25 g/kg max. as calcium disodium EDTA)
Manuf./Distrib.: Int'l. Sourcing
Trade names: Versene NA
Trade names containing: Garbefix 31

Disodium ethylene bisdithiocarbamate. *See* Nabam
Disodium ethylene-1,2-bisdithiocarbamate. *See* Nabam

Disodium glycyrrhizinate
Uses: Sweetener
Regulatory: Japan approved

Disodium guanylate
CAS 5550-12-9; EEC E627
Synonyms: GMP; Sodium guanylate; Guanosine 5′-disodium phosphate; Disodium 5′-guanylate; Guanylic acid sodium salt; Sodium guanosine-5′-monophosphate
Empirical: $C_{10}H_{14}N_5O_8P \cdot 2Na$
Formula: $Na_2C_{10}H_{12}N_5O_8P \cdot 2HOH$
Properties: Colorless to wh. cryst., char. taste; sol. in cold water, very sol. in hot water; sl. sol. in alcohol; insol. in ether; m.w. 409.24
Toxicology: LD50 (oral, mouse) 15 g/kg; moderately toxic by ingestion, intraperitoneal, subcutaneous, intravenous routes; mildly toxic by ingestion; heated to decomp., emits toxic fumes of PO_x, NO_x, Na_2O
Uses: Flavor potentiator in canned foods, poultry, sauces, snack foods, soups
Regulatory: FDA 21CFR §172.530; USDA 9CFR §318.7, 381.147; Japan approved; Europe listed; UK approved
Trade names containing: I+G; Luxor® 1576; Luxor® 1626; Luxor® 1639; Luxor® EB-400

Disodium 5′-guanylate. *See* Disodium guanylate
Disodium hydrogen citrate. *See* Disodium citrate
Disodium hydrogen phosphate. *See* Sodium phosphate dibasic
Disodium hydrogen phosphate dihydrate. *See* Disodium phosphate, dihydrate
Disodium IMP. *See* Disodium inosinate
Disodium indigo-5,5-disulfonate. *See* FD&C Blue No. 2

Disodium inosinate
CAS 4691-65-0; EEC E631
Synonyms: IMP; Disodium IMP; Sodium inosinate; Sodium 5-inosinate; Disodium 5′-inosinate; Inosine 5′-disodium phosphate
Definition: A 5′-nucleotide derived from seaweed or dried fish
Empirical: $C_{10}H_{13}N_4O_8P \cdot 2Na$
Properties: Colorless to wh. cryst., char. taste; sol. in water; sl. sol. in alcohol; insol. in ether; m.w. 394.22
Toxicology: LD50 (oral, rat) 15,900 mg/kg; moderately toxic by several routes; experimental teratogen; mutagenic data; heated to decomp., emits toxic fumes of PO_x, NO_x, Na_2O
Uses: Flavor potentiator in foods (hams, cured meat, poultry, sausage)
Regulatory: FDA 21CFR §172.535, must contain ≤ 150 ppm sol. barium; USDA 9CFR §318.7, 381.147; UK, Japan approved
Trade names containing: I+G; Luxor® 1626; Luxor® 1639

Disodium 5′-inosinate. *See* Disodium inosinate
Disodium monohydrogen orthophosphate. *See* Sodium phosphate dibasic
Disodium monohydrogen orthophosphate dihydrate. *See* Disodium phosphate, dihydrate
Disodium orthophosphate. *See* Sodium phosphate dibasic
Disodium phosphate. *See* Sodium phosphate dibasic

Disodium phosphate, dihydrate
CAS 7758-79-4, 10028-24-7; EINECS 231-448-7
Synonyms: DSP-2; Sodium phosphate dibasic dihydrate; Dibasic sodium phosphate duohydrate; Disodium hydrogen phosphate dihydrate; Disodium monohydrogen orthophosphate dihydrate
Classification: Sodium phosphate
Formula: $Na_2HPO_4 \cdot 2H_2O$
Properties: Wh. gran., essentially odorless; sol. 15 g/100 g water; m.w. 178.00; pH 9.1 (1%)
Uses: Emulsifier for process cheese; buffer, texturizer, nutritional mineral supplement
Manuf./Distrib.: Albright & Wilson Am.; FMC; Rhone-Poulenc Food Ingreds.

Disodium pyrophosphate. *See* Sodium acid pyrophosphate
Disodium pyrosulfite. *See* Sodium metabisulfite

Disodium succinate. *See also* Sodium succinate
FEMA 3277
Uses: Acidity regulator; food acid; flavoring
Regulatory: Japan approved
Manuf./Distrib.: Aldrich

Disodium sulfate. *See* Sodium sulfate
Disodium tartrate. *See* Sodium tartrate
Disodium L-(+)-tartrate. *See* Sodium tartrate

Distearyl citrate
Toxicology: Heated to decomp., emits acrid smoke and irritating fumes
Uses: Flavor preservative, plasticizer, sequestrant; used for oleomargarine, pkg. materials
Regulatory: FDA 21CFR §181.27, 182.6851 (limitation 0.15% as sequestrant)

Distearyl thiodipropionate
CAS 693-36-7; EINECS 211-750-5
Synonyms: 3,3´-Thiobispropanoic acid, dioctadecyl ester; 3,3´-Dioctadecyl thiodipropionate; Thiodipropionic
acid, distearyl ester
Definition: Diester of stearyl alcohol and 3,3´-thiodipropionic acid
Empirical: $C_{42}H_{82}O_4S$
Formula: $(C_{18}H_{37}OOCCH_2CH_2)_2S$
Properties: White flakes; insol. in water; sol. in benzene, toluene, chloroform, and olefin polymers; m.w. 683;
m.p. 58-62 C; b.p. 360 C (dec.)
Toxicology: Nonhazardous; heated to decomp., emits toxic fumes of SO_x
Uses: Antioxidant
Regulatory: FDA 21CFR §175.105, 175.300, 181.22, 181.24 (0.005% migrating from food pkg.)
Trade names: Evanstab® 18

Distilled spirits. *See* Alcohol
Disulfurous acid dipotassium salt. *See* Potassium metabisulfite
Disulfurous acid disodium salt. *See* Sodium metabisulfite
Dithane A-40. *See* Nabam
β,β´-Dithiobisalanine. *See* L-Cystine
3,3´-Dithiobis(2-aminopropanoic acid). *See* L-Cysteine
(-)-3,3´-Dithio-bis(2-aminopropionic acid). *See* L-Cystine

Dittany
EINECS 289-766-7 (extract)
Synonyms: Fraxinella; Burning bush; Gas plant
Definition: Roots of *Dictamnus albus*
Properties: Tonic, aromatic flavor
Uses: Natural flavoring; used in alcoholic beverages
Regulatory: FDA 21CFR §172.510

Dittany of Crete
Synonyms: Spanish hops
Definition: From *Origanum dictamnus*
Properties: Intense pleasant odor, bitter aromatic flavor
Uses: Natural flavoring
Regulatory: 25 ppm (nonalcoholic beverages), 8.8 ppm (baked goods)

Divinylbenzene copolymer
Toxicology: Heated to decomp., emits acrid smoke and irritating fumes
Uses: Removal of organic substances from aq. foods
Regulatory: FDA 21CFR §173.65

Divinylenimine. *See* Pyrrole
DKP. *See* Potassium phosphate dibasic
DMBC butyrate. *See* α,α-Dimethylphenethyl butyrate
DMDC. *See* Dimethyl dicarbonate
Docusate sodium. *See* Dioctyl sodium sulfosuccinate

δ-Dodecalactone
CAS 713-95-1; FEMA 2401

γ-Dodecalactone

Synonyms: 5-Hydroxydodecanoic acid, δ-lactone
Definition: Obtained by lactonization of 5-hydroxydodecanoic acid
Empirical: $C_{12}H_{22}O_2$
Properties: Colorless to very pale straw-yel. visc. liq., fresh-fruit oily odor; sol. in alcohol; insol. in water; m.w. 198.31; dens. 0.942; m.p. -12 C; b.p. 140-141 C (1 mm); flash pt. > 230 F
Uses: Synthetic flavoring agent
Usage level: 0.06 ppm (baked goods, gelatins, puddings), 10 ppm (toppings)
Regulatory: FDA 21CFR §172.515
Manuf./Distrib.: Aldrich

γ-Dodecalactone

CAS 2305-05-7; FEMA 2400
Synonyms: 4-Hydroxydodecanoic acid γ-lactone; Dodecanolide-1,4
Empirical: $C_{12}H_{22}O_2$
Properties: Colorless oily liq., fatty peachy odor, butter peach-like flavor; sol. in alcohol; insol. in water; m.w. 198.31; dens. 0.936; m.p. 17-18 C; b.p. 258 C
Uses: Synthetic flavoring agent
Usage level: 3.3 ppm (nonalcoholic beverages), 4.3 ppm (ice cream, ices), 13 ppm (candy), 11 ppm (baked goods), 0.15 ppm (gelatins, puddings), 0.01 ppm (jellies)
Regulatory: FDA 21CFR §172.515
Manuf./Distrib.: Acme-Hardesty; Aldrich

Dodecanal. See Lauric aldehyde
Dodecanoic acid. See Lauric acid
n-Dodecanoic acid. See Lauric acid
Dodecanoic acid, 2,3-dihydroxypropyl ester. See Glyceryl laurate
Dodecanoic acid ethyl ester. See Ethyl laurate
Dodecanoic acid, 2-hydroxypropyl ester. See Propylene glycol laurate
Dodecanoic acid methyl ester. See Methyl laurate
Dodecanoic acid, monoester with 1,2-propanediol. See Propylene glycol laurate
Dodecanoic acid, monoester with 1,2,3-propanetriol. See Glyceryl laurate
Dodecanoic acid, 1,2,3-propanetriyl ester. See Trilaurin
Dodecanoic acid sodium salt. See Sodium laurate
1-Dodecanol. See Lauryl alcohol
Dodecanolide-1,4. See γ-Dodecalactone
Dodecanyl acetate. See Lauryl acetate

2-Dodecenal

EINECS 225-402-5; FEMA 2402
Synonyms: n-Dodecen-2-ol; 3-Nonyl acrolein; trans-2-Dodecen-1-al
Empirical: $C_{12}H_{22}O$
Properties: Colorless oily liq., orange-like odor, mandarin taste; sol. in alcohol; insol. in water; m.w. 182.31; b.p. 272 C
Toxicology: Heated to decomp., emits acrid smoke and fumes
Uses: Synthetic flavoring agent
Usage level: 2.9 ppm (nonalcoholic beverages), 3.1 ppm (ice cream, ices), 2.8 ppm (candy), 2.8 ppm (baked goods)
Regulatory: FDA 21CFR §172.515

trans-2-Dodecen-1-al. See 2-Dodecenal

Dodecene-1

CAS 112-41-4; 6842-15-5; EINECS 203-968-4
Synonyms: C12 alpha olefin; 1-Dodecene; α-Dodecylene; Tetrapropylene
Empirical: $C_{12}H_{24}$
Formula: $H_2C:CH(CH_2)_9CH_3$
Properties: Colorless liq.; insol. in water; sol. in alcohol, acetone, ether, petrol., coal tar solvs.; m.w. 168.32; dens. 0.764; m.p. -31.5 C; b.p. 213-215 C; flash pt. (Seta) 168 F; ref. index 1.430
Toxicology: Irritating to eyes and skin; low acute inhalation toxicity; sl. toxic by ingestion; narcotic in high concs.
Uses: Flavors
Manuf./Distrib.: Aldrich

1-Dodecene. See Dodecene-1
n-Dodecen-2-ol. See 2-Dodecenal
Dodecoic acid. See Lauric acid
Dodecyl acetate. See Lauryl acetate

Dodecyl alcohol. *See* Lauryl alcohol
n-Dodecyl aldehyde. *See* Lauric aldehyde
N-Dodecyl-*ar*-ethyl-N,N-dimethylbenzenemethanaminium chloride. *See* Quaternium-14
Dodecylbenzene sodium sulfonate. *See* Sodium dodecylbenzenesulfonate
Dodecylbenzenesulfonic acid sodium salt. *See* Sodium dodecylbenzenesulfonate
Dodecyl dimethyl benzyl ammonium chloride. *See* Lauralkonium chloride
Dodecyl dimethyl ethylbenzyl ammonium chloride. *See* Quaternium-14
α-Dodecylene. *See* Dodecene-1

Dodecyl gallate
　　CAS 1166-52-5; EINECS 214-620-6; EEC E312
　　Synonyms: 3,4,5-Trihydroxybenzoic acid, dodecyl ester; Dodecyl-3,4,5-trihydroxybenzoate; Lauryl gallate
　　Definition: Ester of gallic acid
　　Empirical: $C_{19}H_{30}O_5$
　　Properties: m.w. 338.49; m.p. 96-97 C
　　Toxicology: LD50 (oral, mouse) 1600 mg/kg; mod. toxic by ingestion; may cause intolerance and liver damage;
　　　　can irritate intestines; heated to decomp., emits acrid smoke and irritating fumes
　　Uses: Antioxidant; used in cream cheese, fats, margarine, oils, instant mashed potatoes
　　Usage level: Limitation 0.02% (alone or in combination with other antioxidants in margarine); ADI 0-0.5 mg/
　　　　kg (EEC, total gallates)
　　Regulatory: FDA 21CFR §166.110; USDA 9CFR §318.7; Europe listed; UK approved

Dodecyl-3,4,5-trihydroxybenzoate. *See* Dodecyl gallate
DOP. *See* Dioctyl phthalate
Dracylic acid. *See* Benzoic acid
Dried egg white. *See* Albumen

Dried egg yolk
　　Definition: Obtained by dehydration of chicken egg yolks
　　Uses: Protein source
　　Manuf./Distrib.: Enthoven BV; Henningsen
　　Trade names: Egg Yolk Solids Type Y-1; Egg Yolk Solids Type Y-2; Egg Yolk Solids Type Y-2-FF
　　Trade names containing: Egg Yolk Solids Type Y-1-FF; Hentex-20; Hentex-20A; Hentex-25; Hentex-30;
　　　　Hentex-30A; Hentex-85; Hentex-505

Dried whey. *See* Whey, dry
Dry whey. *See* Whey, dry
DSE. *See* Nabam
DSP-2. *See* Disodium phosphate, dihydrate
DSP-O. *See* Sodium phosphate dibasic
DSPP. *See* Sodium acid pyrophosphate
DSS. *See* Dioctyl sodium sulfosuccinate
Earthnut oil. *See* Peanut oil
East Indian geranium. *See* Palmarosa oil
East Indian lemongrass oil. *See* Lemongrass oil East Indian
East Indian nutmeg oil. *See* Nutmeg oil
East Indian sandalwood. *See* Sandalwood
East Indian sandalwood oil. *See* Sandalwood oil
EC. *See* Ethylcellulose
Edathamil. *See* Edetic acid
Edetate calcium disodium. *See* Calcium disodium EDTA
Edetate disodium. *See* Disodium EDTA
Edetate sodium. *See* Tetrasodium EDTA

Edetic acid
　　CAS 60-00-4; EINECS 200-449-4
　　Synonyms: EDTA; N,N´-1,2-Ethanediylbis[N-(carboxymethyl) glycine]; Ethylene diamine tetraacetic acid;
　　　　Edathamil
　　Classification: Substituted diamine
　　Empirical: $C_{10}H_{16}N_2O_8$
　　Formula: $(HOOCCH_2)_2NCH_2CH_2N(CH_2COOH)_2$
　　Properties: Colorless crystals; sl. sol. in water; insol. in common org. solvs.; m.w. 292.28; dec. 240 C
　　Toxicology: Irritant; poison by intraperitoneal route; mutagenic data
　　Uses: Antioxidant in foods
　　Regulatory: FDA 21CFR §175.105, 176.170

EDTA. *See* Edetic acid
EDTA Na₄. *See* Tetrasodium EDTA
EEA. *See* Ethylacetoacetate
Egg albumin. *See* Albumen

Egg powder
Synonyms: Egg solids; Whole dried egg
Definition: Powder obtained from the dried whole chicken egg
Properties: Sol. in water
Uses: Protein source
Regulatory: FDA 21CFR §42.30
Manuf./Distrib.: Cham Foods; Clofine; Cornelius; Croda Bakery; Dairy Crest; Dohler UK; EPI Bretagne; Farbest Brands; G Fiske; Freeman Foods; Henningsen Foods; Hormel Foods; Mitsubishi; Moore Fine Foods; Nat'l. Egg Prods.; Nordmann Rassmann; Sanovo Foods; Spice King
Trade names: Hentex-45
Trade names containing: Hentex-10; Hentex-20; Hentex-20A; Hentex-25; Hentex-35; Hentex-70; Hentex-76; Hentex-81; Hentex-85; Hentex-120; Hentex-505

Egg solids. *See* Egg powder

Egg yolk extract
Definition: Extract of egg yolk
Properties: Yel. semisolid mass; dens. 0.95; m.p. 22 C
Uses: Baking, dairy prods.
Manuf./Distrib.: Enthoven BV; Henningsen

Eglantine. *See* Isobutyl benzoate
Eicosanoic acid methyl ester. *See* Methyl eicosenate
Elainic acid. *See* Oleic acid

Elder flowers
EINECS 294-458-0 (*S. canadensis* extract), 283-259-4 (*S. nigra* extract)
Synonyms: Sambucus; Sweet elder
Definition: From *Sambucus canadensis* and *S. nigra*
Properties: Aromatic bitter tonic
Uses: Natural flavoring agent
Usage level: 340 ppm (nonalcoholic beverages), 25 ppm (alcoholic beverages), 1.0 ppm (ice cream, ices, candy, baked goods)
Regulatory: FDA 21CFR §182.10, 182.20, GRAS
Manuf./Distrib.: C.A.L.-Pfizer; Chart

Elder tree leaves
Definition: From *Sambucus nigra*
Uses: Natural flavoring
Regulatory: FDA 21CFR §172.510

Elecampane
Synonyms: Inula; Scabwort
Definition: Rhizome and roots of *Inula helenium*
Properties: Bitter aromatic flavor
Uses: Natural flavoring for alcoholic beverages only
Regulatory: FDA 21CFR §172.510; Japan approved

Elecampane extract
CAS 84012-20-4; EINECS 281-666-1
Synonyms: Inula helenium extract
Definition: Extract of *Inula helenium*
Uses: Natural flavoring
Regulatory: FDA 21CFR §172.510

Electrolyte acid. *See* Sulfuric acid
Elettaria cardamomum oil. *See* Cardamom oil
EMQ. *See* 6-Ethoxy-1,2-dihydro-2,2,4-trimethylquinoline
Emulsin. *See* β-Glucosidase
Enanthal. *See* Heptanal
Enanthaldehyde. *See* Heptanal
Enanthic alcohol. *See* Heptyl alcohol
Enanthyl alcohol. *See* Heptyl alcohol

English chamomile extract. *See* Chamomile extract
English chamomile oil. *See* Chamomile oil, Roman

English oak extract
 Synonyms: Quercus robur extract
 Definition: Extract of the bark of *Quercus robur*
 Uses: Natural flavoring
 Regulatory: FDA 21CFR §172.510

Enocianina. *See* Grape skin extract
Enocyanin. *See* Grape skin extract
Enzactin. *See* Triacetin

Enzyme-modified milkfat
 Properties: Light- to medium-tan liquid, paste, or powder; strong fatty acid odor and flavor
 Uses: Flavoring agent

Epichlorohydrin-dimethylamine copolymer. *See* Dimethylamine/epichlorohydrin copolymer

Epoxidized soybean oil
 CAS 8013-07-8; EINECS 232-391-0
 Synonyms: Soybean oil, epoxidized
 Definition: Modified oil obtained from soybean oil by epoxidation
 Properties: Clear pale yel. liq., low odor; sol. < 0.1% in water; sp.gr. 0.99; dec. 550 F; m.p. 25 F; iodine no. 6 max.; flash pt. (CC) 430 F
 Precaution: Avoid oxidizing agents, strong acids, bases and amines
 Toxicology: Nonhazardous; LD50 (oral, rat) 30 gm/kg; no eye or skin irritation (rabbit); heated to decomp., emits acrid smoke and irritating fumes
 Uses: Plasticizer migrating from food pkg.
 Regulatory: FDA 21CFR §175.105, 177.1650, 178.3910, 181.22, 181.27

1,4-Epoxy-p-menthane. *See* 1,4-Cineole
1,8-Epoxy-p-menthane. *See* Eucalyptol
1,2-Epoxy-p-menth-4-(8)-en-3-one. *See* Piperitenone oxide
1,6-Epoxy-p-menth-8-en-2-one. *See* cis-Carvone oxide
Epsom salts. *See* Magnesium sulfate heptahydrate
Equisetic acid. *See* Aconitic acid
Ercalciol. *See* Ergocalciferol

Ergocalciferol
 CAS 50-14-6; EINECS 200-014-9
 Synonyms: Calciferol; 9,10-seco(5Z,7E,22E)-5,7,10(19),22-Ergostatetraen-3-ol; Ergosterol, activated; Viosterol; Ercalciol; Vitamin D_2
 Definition: Derived from ergosterol by irradiation with UV light
 Empirical: $C_{28}H_{44}O$
 Properties: Wh. cryst., odorless; sol. in alcohol, chloroform, ether, fatty oils; insol. in water; m.w. 396.72; m.p. 115-118 C
 Precaution: Light-sensitive
 Toxicology: Poison by ingestion, intraperitoneal, IV, and intramuscular routes; experimental teratogen, reproductive effects; human systemic effects by ingestion (anorexia, nausea, etc.); heated to decomp., emits acrid smoke and irritating fumes
 Uses: Dietary supplement, nutrient for cereals, grain prods., pastas, margarine, milk and milk prods., infant formula
 Usage level: Limitation 350 IU/100 g (breakfast cereal), 90 IU/100 g (grain prods., pasta), 42 IU/100 g (milk), 89 IU/100 g (milk prods.)
 Regulatory: FDA 21CFR §166.110, 182.5950, 182.5953, 184.1950, GRAS; Japan approved
 Manuf./Distrib.: Am. Roland

9,10-seco(5Z,7E,22E)-5,7,10(19),22-Ergostatetraen-3-ol. *See* Ergocalciferol
Ergosterol, activated. *See* Ergocalciferol
Eriodictyon californicum extract. *See* Yerba santa extract
Eriodictyon glutinosum extract. *See* Yerba santa extract

Erythorbic acid
 CAS 89-65-6; EINECS 201-928-0
 Synonyms: D-Erythro-hex-2-enonic acid, γ-lactone; Isoascorbic acid; d-Araboascorbic acid
 Definition: Isomer of ascorbic acid
 Empirical: $C_6H_8O_6$

D-Erythro-hex-2-enonic acid, γ-lactone

Properties: Wh. or sl. yel. cryst. or powd.; sol. in water, alcohol; sl. sol. in glycerin; m.p. 164-171 C (dec.)
Toxicology: Heated to decomp., emits acrid smoke and irritating fumes
Uses: Antioxidant , preservative (food, brewing); enhances curing action of nitrites on meat, stabilizes color
   and flavor in meat, meat prods., and fruit
Regulatory: FDA 21CFR §101.33, 145.110, 175.105, 182.3041, GRAS; GRAS (FEMA); USDA 9CFR §318.7,
   381.147; Japan restricted for purpose of antioxidation
Manuf./Distrib.: Pfizer Food Science

D-Erythro-hex-2-enonic acid, γ-lactone. *See* Erythorbic acid
D-Erythro-hex-2-enonic acid, γ-lactone, monosodium salt. *See* Sodium erythorbate
Erythrosine. *See* FD&C Red No. 3
Erythrosine bluish. *See* FD&C Red No. 3
Esdragol. *See* Estragole
Essence de Cadé. *See* Cade oil
Essence of rose. *See* Rose oil

Esterase
   CAS 9013-79-0
   Definition: From hog liver
   Uses: Natural enzyme
   Regulatory: Japan approved

Esterase-lipase
   Definition: Derived from *Mucor miehei*
   Toxicology: Heated to decomp., emits acrid smoke and irritating fumes
   Uses: Enzyme used as flavor enhancer in cheeses, fats and oils
   Regulatory: FDA 21CFR §173.140

Ester gum
   Definition: Ester of natural resins (especially rosin) and polyhydric alcohols (principally glycerol)
   Properties: Hard, semisynthetic resin; flash pt. 375 F (190 C)
   Precaution: Combustible
   Uses: Chewing gum base
   Regulatory: Japan approved

Estragole
   CAS 140-67-0; FEMA 2411
   Synonyms: p-Allylanisole; 4-Allylanisole; Esdragol; 1-Methoxy-4-(2-propenyl) benzene; Chavicol methyl ether
   Definition: Main constituent of tarragon oil derived from *Artemisia dracunculus*
   Empirical: $C_{10}H_{12}O$
   Properties: Liq.; sol. in alcohol, chloroform; forms azeotropic mixts. with water; m.w. 148.20; dens. 0.9645 (21/
      4 C); b.p. 216 C (764 mm); flash pt. 178 F; ref. index 1.5230 (17.5 C)
   Toxicology: LD50 (oral, rat) 1820 mg/kg
   Uses: Synthetic flavoring agent; in foods and liqueurs
   Regulatory: FDA 21CFR §172.515
   Manuf./Distrib.: Aldrich

Estragon. *See* Tarragon
Ethanal. *See* Acetaldehyde
1,2-Ethanediamine. *See* Ethylenediamine
1,2-Ethanedicarboxylic acid. *See* Succinic acid
N,N´-1,2-Ethanediylbis[N-(carboxymethyl) glycine]. *See* Edetic acid
N,N´-1,2-Ethanediylbisoctadecanamide. *See* Ethylene distearamide
Ethanoic acid. *See* Acetic acid

Ethanolamine
   CAS 141-43-5; EINECS 205-483-3
   Synonyms: MEA; 2-Aminoethanol; 2-Aminoethyl alcohol; Monoethanolamine; Glycinol; 2-Hydroxyethylamine
   Classification: Monoamine
   Empirical: C_2H_7NO
   Formula: $NH_2CH_2CH_2OH$
   Properties: Colorless liq., ammoniacal odor; hygroscopic; misc. with water, alcohol; sol. in chloroform; sl. sol.
      in benzene; m.w. 61.10; dens. 1.012; m.p. 10.5 C; b.p. 170 C; flash pt. 93 C
   Precaution: Corrosive; flamm. exposed to heat or flame; powerful reactive base
   Toxicology: TLV:TWA 3 ppm; LD50 (oral, rat) 214 mg/kg, (skin, rat) 1500 mg/kg; poison by IP route; mod. toxic
      by ingestion, skin contact, subcut., IV routes; corrosive irritant to eyes, skin, mucous membranes; heated

to decomp., emits toxic fumes of NO$_x$

Uses: Flume water wash additive for sugar beets; paring (antibrowning) agent; used on peeled fruit and vegetables; animal glue adjuvant for pkg. materials

Usage level: Limitation 0.3 ppm (in wash water)

Regulatory: FDA 21CFR §173.315, 175.105, 176.210, 176.300, 178.3120; not permitted for use in foods intended for babies and young infants in UK

Ethanol, undenatured. *See* Alcohol
Ethene, homopolymer. *See* Polyethylene
Ethene, homopolymer, oxidized. *See* Polyethylene, oxidized
Ethenol homopolymer. *See* Polyvinyl alcohol
Ethenyl acetate, homopolymer. *See* Polyvinyl acetate (homopolymer)
Ethenylbenzene. *See* Styrene
1-Ethenyl-2-pyrrolidinone homopolymer. *See* PVP
Ethinyl trichloride. *See* Trichloroethylene
Ethocel. *See* Ethylcellulose
Ethone. *See* 1-(p-Methoxyphenyl)-1-penten-3-one
Ethovan. *See* Ethyl vanillin
6-Ethoxy-m-anol. *See* Propenylguaethol
4-Ethoxybenzaldehyde. *See* p-Ethoxybenzaldehyde

p-Ethoxybenzaldehyde
CAS 10031-82-0; EINECS 233-093-3; FEMA 2413
Synonyms: 4-Ethoxybenzaldehyde
Definition: Obtained by ethylation of p-hydroxybenzaldehyde using aluminum chloride catalyst
Empirical: C$_9$H$_{10}$O$_2$
Properties: Red-brn., sweet floral odor and taste; sol. in alcohol; pract. insol. in water; m.w. 150.18; dens. 1.081 (20/4 C); m.p. 13-16 C; b.p. 255 C; flash pt. 75 C; ref. index 1.559 (20 C)
Toxicology: Skin irritant
Uses: Synthetic flavoring agent
Usage level: 0.06-0.08 ppm (nonalcoholic beverages), 0.36-0.50 ppm (ice cream, ices), 1.0 ppm (candy, baked goods)
Regulatory: FDA 21CFR §172.515
Manuf./Distrib.: Aldrich

Ethoxycarbonylethylene. *See* Ethyl acrylate
2-Ethoxy-p-cymene. *See* Carvacryl ethyl ether

6-Ethoxy-1,2-dihydro-2,2,4-trimethylquinoline
CAS 91-53-2; EINECS 202-075-7
Synonyms: EMQ; Ethoxyquin; 1,2-Dihydro-6-ethoxy-2,2,4-trimethylquinoline; Santoquine
Empirical: C$_{14}$H$_{19}$NO
Properties: Yel. liq.; m.w. 217.34; dens. 1.029-1.031 (25 C); m.p. $\approx$ 0 C; b.p. 125 C (2 mm); ref. index 1.569-1.572 (25 C)
Precaution: Combustible when exposed to heat or flame; can react with oxidizing materials
Toxicology: LD50 (oral, rat) 800 mg/kg; moderately toxic by ingestion; poison by intraperitoneal route; mutagenic data; heated to decomp., emits toxic fumes of NO$_x$
Uses: Post-harvest preservation additive for food (human or animal); antioxidant for apples and pears; scald inhibitor; stabilizer, preservative; antioxidant for preservation of color in prod. of chili powd., paprika, ground chili
Regulatory: FDA §172.140, limitation 100 ppm in chili powd., 100 ppm in paprika, 5 ppm in uncooked meat fat, 3 ppm in uncooked poultry fat, 0.5 ppm in eggs, zero tolerance in milk

3-Ethoxy-4-hydroxybenzaldehyde. *See* Ethyl vanillin
1-Ethoxy-2-hydroxy-4-propenylbenzene. *See* Propenylguaethol

Ethoxylated mono- and diglycerides
Synonyms: Polyoxyethylene monoglycerides
Uses: Emulsifier in pan release agents in yeast-raised prods., cakes, whipped vegetable toppings, icings, frozen desserts, nondairy creamers
Usage level: 0.2-0.5% (flour, dairy, etc.)
Regulatory: FDA 21CFR §172.834
Trade names: EMG 20; Panalite EOM-K
Trade names containing: Aldosperse® 30/70 FG; Aldosperse® 30/70 KFG; Aldosperse® 712 FG; Amfal-46; Beta Plus; BFP 100; CDC-10A; CDC-10A (NB); Centromix® E; CFI-10; Do Crest 60; Do Crest Gold; Do Crest Gold Plus; Elasdo +; Elasdo Power 70; Hi-Ratio Shortening; Lactylate Hydrate; MG-F; Panipower

III; Tally® 100; Top Mate Kosher; Xpando; Xpando 70; Xpando Powd.

2-Ethoxy-5-propenyl anisole. *See* Isoeugenyl ethyl ether

Ethoxyquin. *See* 6-Ethoxy-1,2-dihydro-2,2,4-trimethylquinoline

Ethyl 2-acetal-3-phenylpropionate
CAS 620-79-1; FEMA 2416
Synonyms: Ethyl α-acetylhydroxycinnamate; Ethyl benzylacetoacetate; Ethyl-3-oxo-2-benzylbutanoate
Definition: Obtained by reacting benzyl chloride over hot sodium acetoacetate
Properties: Colorless liq., balsamic fruity jasmine odor; misc. with alcohol, ether; insol. in water; m.w. 220.27; dens. 1.036; b.p. 276 C; flash pt. > 230 F
Uses: Synthetic flavoring agent
Usage level: 0.10-5.0 ppm (nonalcoholic beverages), 2.0 ppm (ice cream, ices), 7.0 ppm (candy)
Regulatory: FDA 21CFR §172.515
Manuf./Distrib.: Aldrich

Ethyl acetate
CAS 141-78-6; EINECS 205-500-4; FEMA 2414
Synonyms: Acetic ether; Acetic acid, ethyl ester; Vinegar naphtha
Definition: Ester of ethyl alcohol and acetic acid
Empirical: $C_4H_8O_2$
Formula: $CH_3COOC_2H_5$
Properties: Colorless liq., fragrant; sol. in chloroform, alcohol, ether; sl. sol. in water; m.w. 88.12; dens. 0.902 (20/4 C); bulk dens. 0.8945 g/ml (25 C); b.p. 77 C; f.p. -83.6 C; flash pt. -4.4 C; ref. index 1.3723
Precaution: Flamm.; very dangerous fire hazard exposed to heat or flame; can react vigorously with oxidizers
Toxicology: TLV 400 ppm in air; LD50 (oral, rat) 5620 mg/kg; poison by inhalation; mildly toxic by ingestion; irritant to eyes, skin, mucous membranes; mutagenic data; mildly narcotic; heated to decomp., emits acrid smoke and irritating fumes
Uses: Color diluent, flavoring agent, solvent used in the decaffeination of coffee and tea
Usage level: 67 ppm (nonalcoholic beverages), 50-65 ppm (alcoholic beverages), 99 ppm (ice cream, ices), 170 ppm (candy), 170 ppm (baked goods), 200 ppm (gelatins, puddings), 1400 ppm (chewing gum); ADI 0-25 mg/kg (FAO/WHO)
Regulatory: FDA 21CFR § 73.1, 172.560, 173.228, 175.320, 177.1200, 182.60, GRAS; 27CFR §21.106; Japan approved with restrictions
Manuf./Distrib.: Aldrich

Ethylacetic acid. *See* n-Butyric acid

Ethylacetoacetate
CAS 141-97-9; EINECS 205-516-1; FEMA 2415
Synonyms: EEA; 3-Oxobutanoic acid ethyl ester; Acetoacetic acid ethyl ester; Ethyl 3-oxobutanoate
Empirical: $C_6H_{10}O_3$
Formula: $CH_3COCH_2COOC_2H_5$
Properties: Colorless liq., fruity odor; sol. in ≈ 35 parts water; misc. with common org. solvs.; m.w. 130.14; dens. 1.0213 (25/4 C); m.p. -45 C; b.p. 180.8 C (760 mm); flash pt. (CC) 184 F; ref. index 1.4180-1.4195
Precaution: Combustible liq. when exposed to heat or flame; can react with oxidizing materials
Toxicology: LD50 (oral, rat) 3.98 g/kg; mod. toxic by ingestion; mod. irritating to skin, mucous membranes, eyes; heated to decomp., emits acrid smoke and irritating fumes
Uses: Synthetic flavoring agent
Usage level: 17 ppm (nonalcoholic beverages), 24 ppm (ice cream, ices), 110 ppm (candy), 120 ppm (baked goods), 93 ppm (gelatins, puddings), 530 ppm (chewing gum)
Regulatory: FDA 21CFR §172.515; GRAS (FEMA); Japan approved as flavoring
Manuf./Distrib.: Aldrich

Ethyl acetone. *See* Methyl propyl ketone

Ethyl α-acetylhydroxycinnamate. *See* Ethyl 2-acetal-3-phenylpropionate

Ethyl 2-acetyl-3-phenylpropionate
FEMA 2416
Synonyms: Ethylbenzyl acetoacetate
Uses: Synthetic flavoring agent
Regulatory: FDA 21CFR §172.515
Manuf./Distrib.: Aldrich

Ethyl aconitate, mixed esters
FEMA 2417
Synonyms: Ethyl-2-carboxyglutaconate; Ethyl 1-propene-1,2,3-tricarboxylate
Properties: Colorless oily liq., sweet fruity winey odor and flavor; sol. in alcohol; sl. sol. in water; dens. 1.0961;

b.p. 260 C; ref. index 1.45771 (14.5 C)
Uses: Synthetic flavoring agent
Usage level: 3.6 ppm (nonalcoholic beverages), 12 ppm (ice cream, ices), 55 ppm (candy), 66 ppm (baked goods), 2.5 ppm (gelatins, puddings)
Regulatory: FDA 21CFR §172.515

Ethyl acrylate
CAS 140-88-5; EINECS 205-438-8; FEMA 2418
Synonyms: Ethyl propenoate; Ethoxycarbonylethylene
Definition: Esterification of acrylic acid
Empirical: $C_5H_8O_2$
Properties: Liq., penetrating and persistent odor; sl. sol. in water; m.w. 100.12; sp.gr. 0.918; m.p. -71 to -75 C; f.p. 60 F; b.p. 99-100 C; ref. index 1.4068
Precaution: Flamm. liq.; very dangerous fire hazard exposed to heat/flame; can react vigorously with oxidizers
Toxicology: LD50 (oral, rat) 800 mg/kg; poison by ingestion and inh.; mod. toxic by skin contact and intraperitoneal routes; suspected human carcinogen; skin and eye irritant; heated to decomp., emits acrid smoke and irritating fumes
Uses: Synthetic flavoring agent
Usage level: 0.13-0.26 ppm (nonalcoholic beverages), 0.06-1 ppm (ice cream, ices), 1.1 ppm (candy), 1.1 ppm (baked goods), 0.1 ppm (chewing gum)
Regulatory: FDA 21CFR §172.515
Manuf./Distrib.: Aldrich

Ethyl alcohol, undenatured. *See* Alcohol
Ethyl aldehyde. *See* Acetaldehyde
Ethyl 2-aminobenzoate. *See* Ethyl anthranilate
Ethyl o-aminobenzoate. *See* Ethyl anthranilate
Ethyl amyl ketone. *See* 3-Octanone

Ethyl-p-anisate
CAS 94-30-4; FEMA 2420
Synonyms: p-Anisic acid ethyl ester; Ethyl-4-methoxybenzoate; Ethyl-p-methoxybenzoate
Definition: Obtained by esterification of anisic acid with ethanol in presence of an acid catalyst
Empirical: $C_{10}H_{12}O_3$
Properties: Colorless liq., fruity anise odor; sol. in alcohol, ether; sl. sol. in water; m.w. 180.21; dens. 1.103; m.p. 7-8 C; b.p. 269-270 C; flash pt. > 100 C; ref. index 1.522-1.526
Precaution: Combustible liq.
Toxicology: LD50 (oral, rat) 2040 mg/kg; mod. toxic by ingestion; heated to decomp., emits acrid smoke and irritating fumes
Uses: Synthetic flavoring agent
Usage level: 2.6 ppm (nonalcoholic beverages), 0.96 ppm (ice cream, ices), 8.8 ppm (candy), 7.2 ppm (baked goods)
Regulatory: FDA 21CFR §172.515
Manuf./Distrib.: Aldrich

Ethyl anthranilate
CAS 87-25-2; EINECS 201-735-1; FEMA 2421
Synonyms: Ethyl o-aminobenzoate; Ethyl 2-aminobenzoate
Definition: Obtained by esterification of anthranilic acid with ethanol in presence of acid catalysts
Empirical: $C_9H_{11}NO_2$
Formula: $NH_2C_6H_4COOC_2H_5$
Properties: m.w. 165.19; dens. 1.118 (20/4 C); m.p. 13-15 C; b.p. 264-268 C; ref. index 1.564 (20 C)
Precaution: Combustible liq.
Toxicology: LD50 (oral, rat) 3750 mg/kg; mod. toxic by ingestion; skin irritant; heated to decomp., emits toxic fumes of NO_x
Uses: Synthetic flavoring agent
Usage level: 5.9 ppm (nonalcoholic beverages), 7.6 ppm (ice cream, ices), 19 ppm (candy), 23 ppm (baked goods), 14 ppm (gelatins, puddings), 79 ppm (chewing gum)
Regulatory: FDA 21CFR §172.515
Manuf./Distrib.: Aldrich

Ethyl benzeneacetate. *See* Ethyl phenylacetate
Ethyl benzenecarboxylate. *See* Ethyl benzoate

Ethyl benzoate
CAS 93-89-0; EINECS 202-284-3; FEMA 2422

Ethyl benzoylacetate

Synonyms: Benzoic acid ethyl ester; Ethyl benzenecarboxylate
Empirical: $C_9H_{10}O_2$
Formula: $C_6H_5COOC_2H_5$
Properties: Colorless liq., aromatic odor; misc. with alcohol, chloroform, ether, petroleum ether; pract. insol. in water; m.w. 150.18; dens. 1.046 (20/4 C); m.p. -34 C; b.p. 211-214 C; flash pt. 184 F; ref. index 1.505 (20 C)
Precaution: Combustible liq.
Toxicology: LD50 (oral, rat) 2100 mg/kg; mod. toxic by ingestion; mildly toxic by skin contact; skin and eye irritant; vapors cause cough; heated to decomp., emits acrid smoke and fumes
Uses: Synthetic flavoring agent
Usage level: 2.8 ppm (nonalcoholic beverages), 0.50 ppm (alcoholic beverages), 2.8 ppm (ice cream, ices), 9.0 ppm (candy), 10 ppm (baked goods), 0.06 ppm (gelatins, puddings), 59 ppm (chewing gum)
Regulatory: FDA 21CFR §172.515
Manuf./Distrib.: Aldrich

Ethyl benzoylacetate
CAS 94-02-0; EINECS 202-295-3; FEMA 2423
Synonyms: Benzoylacetic acid ethyl ester; β-Oxobenzenepropanoic acid ethyl ester; Ethyl β-keto-β-phenylpropionate; Ethyl 3-phenyl-3-oxopropanoate
Definition: Obtained by condensation of ethyl benzoate with ethyl acetate using sodium ethoxide
Empirical: $C_{11}H_{12}O_3$
Formula: $C_6H_5COCH_2COOC_2H_5$
Properties: Liq., pleasant odor; becomes yel. on exposure to air and light; misc. with alcohol, ether; insol. in water; m.w. 192.21; dens. 1.122 (15 C); b.p. 265-270 C (dec.); flash pt. 140 C
Precaution: Volatile with steam; keep protected from air and light
Uses: Synthetic flavoring agent
Usage level: 0.70 ppm (nonalcoholic beverages), 5.0 ppm (ice cream, ices), 10 ppm (candy, baked goods)
Regulatory: FDA 21CFR §172.515
Manuf./Distrib.: Aldrich

Ethyl benzylacetoacetate. See Ethyl 2-acetal-3-phenylpropionate
Ethylbenzyl acetoacetate. See Ethyl 2-acetyl-3-phenylpropionate

α-Ethylbenzyl butyrate
FEMA 2424
Synonyms: Ethyl phenyl carbinyl butyrate; α-Phenylpropyl butyrate
Empirical: $C_{13}H_{18}O_2$
Properties: Liq., floral fruity odor, sweet plum-like taste; m.w. 206.28; dens. 0.9875-0.9905 (15 C); b.p. 282 C; flash pt. 118 C; ref. index 1.4875-1.4895
Uses: Synthetic flavoring agent
Usage level: 0.13-1.0 ppm (nonalcoholic beverages), 0.12-0.20 ppm (ice cream, ices), 1.0 ppm (candy), 0.14 ppm (baked goods)
Regulatory: FDA 21CFR §172.515

Ethyl benzyl ether. See Benzyl ethyl ether
Ethyl brassylate. See Ethylene brassylate
Ethyl butanoate. See Ethyl butyrate
2-Ethyl butanoic acid. See Diethylacetic acid
Ethyl trans-2-butenoate. See Ethyl crotonate
Ethyl butylacetate. See Ethyl caproate

2-Ethylbutyl acetate
CAS 10031-87-5; FEMA 2425
Definition: Obtained by reacting 2-ethylbutanol with acetic anhydride in the presence of sulfuric acid
Empirical: $C_8H_{16}O_2$
Properties: Liq.; m.w. 144.21; dens. 0.876; b.p. 160-163 C; flash pt. 52 C; ref. index 1.4109
Precaution: Moderate fire risk
Uses: Synthetic flavoring agent
Usage level: 5.0 ppm (nonalcoholic beverages), 2.0 ppm (ice cream, ices), 0.03-7.0 ppm (candy)
Regulatory: FDA 21CFR §172.515

2-Ethyl-3-butylacrolein. See 2-Ethyl-2-heptenal
Ethyl butyl carbinol. See 3-Heptanol
Ethyl butyl ketone. See 3-Heptanone
Ethyl-n-butyl ketone. See 3-Heptanone
Ethyl butyl malonate. See Butyl ethyl malonate

2-Ethylbutyraldehyde
 CAS 97-96-1; EINECS 202-623-5; FEMA 2426
 Empirical: $C_6H_{12}O$
 Properties: Colorless liq., pungent odor; sol. in alcohol, ether; sl. sol. in water; m.w. 100.16; dens. 0.811; m.p. -89 C; b.p. 117 C; flash pt. 70 F; ref. index 1.40398
 Precaution: Flamm. liq.
 Toxicology: Irritant
 Uses: Synthetic flavoring agent
 Usage level: 10 ppm (nonalcoholic beverages), 40 ppm (ice cream, ices), 0.12-25 ppm (candy), 0.20-20 ppm (baked goods)
 Regulatory: FDA 21CFR §172.515
 Manuf./Distrib.: Aldrich

Ethyl butyrate
 CAS 105-54-4; EINECS 203-306-4; FEMA 2427
 Synonyms: Ethyl butanoate; Ethyl n-butyrate; Butyric ether
 Definition: Obtained by esterification of n-butyric acid with ethyl alcohol in presence of Twitchell's reagent or MgCl_2
 Empirical: $C_6H_{12}O_2$
 Formula: $CH_3CH_2CH_2COOC_2H_5$
 Properties: Colorless liq., banana-pineapple odor; sol. in water, fixed oils, propylene glycol; misc. with alcohol, ether; insol. in glycerin @ 121 C; m.w. 116.18; dens. 0.874; m.p. -100.8 C; b.p. 121.6 C; flash pt. (CC) 78 F; ref. index 1.391
 Precaution: Flamm. liq.; can react vigorously with oxidizing materials
 Toxicology: LD50 (oral, rat) 13 g/kg; mildly toxic by ingestion; skin irritant; heated to decomp., emits acrid smoke and irritating fumes
 Uses: Ethereal, strawberry/pineapple-like fragrance and flavoring (synthetic); flavor improver for foodstuffs; mfg. of rum
 Usage level: 28 ppm (nonalcoholic beverages), 44 ppm (ice cream, ices), 98 ppm (candy), 93 ppm (baked goods), 54 ppm (gelatins, puddings), 1400 ppm (chewing gum)
 Regulatory: FDA 21CFR §182.60, GRAS; Japan approved as flavoring
 Manuf./Distrib.: BASF

Ethyl n-butyrate. *See* Ethyl butyrate
2-Ethylbutyric acid. *See* Diethylacetic acid
γ-Ethylbutyric acid. *See* Diethylacetic acid
Ethyl butyrolactone. *See* γ-Hexalactone
Ethyl caprate. *See* Ethyl decanoate

Ethyl caproate
 CAS 123-66-0; EINECS 204-640-3; FEMA 2439
 Synonyms: Ethyl hexanoate; Ethyl capronate; Ethyl hexylate; Ethyl butylacetate; Hexanoic acid ethyl ester
 Definition: Obtained by esterification of caproic acid with ethyl alcohol in presence of conc. H_2SO_4 or HCl
 Empirical: $C_8H_{16}O_2$
 Formula: $CH_3(CH_2)_4COOC_2H_5$
 Properties: Colorless to ylsh. liq., mild wine odor; sol. in fixed oils; sl. sol. in propylene glycol; misc. with alcohol, ether; insol. in water, glycerin; m.w. 144.24; dens. 0.867-0.871; b.p. 163 C; flash pt. (OC) 130 F; ref. index 1.406-1.409
 Precaution: Flamm. or combustible liq.; can react with oxidizing materials
 Toxicology: Skin irritant; heated to decomp., emits acrid smoke and irritating fumes
 Uses: Fruity, sl. fermented, apple-like, pineapple-like fragrance and flavoring, synthetic; flavor improver for foodstuffs; mfg. of artificial fruit flavors
 Usage level: 7.0 ppm (nonalcoholic beverages), 18 ppm (ice cream, ices), 12 ppm (candy, baked goods), 10 ppm (gelatins, puddings), 32 ppm (chewing gum), 1.3 ppm (jellies)
 Regulatory: FDA 21CFR §172.515; Japan approved as flavoring
 Manuf./Distrib.: Aldrich; BASF

Ethyl capronate. *See* Ethyl caproate
Ethyl caprylate. *See* Ethyl octanoate
Ethyl-2-carboxyglutaconate. *See* Ethyl aconitate, mixed esters
Ethyl carvacrol. *See* Carvacryl ethyl ether
Ethyl carvacryl ether. *See* Carvacryl ethyl ether

Ethylcellulose
 CAS 9004-57-3; EEC E462

Ethyl cinnamate

Synonyms: EC; Cellulose, ethyl ether; Ethocel
Definition: Ethyl ether of cellulose
Properties: White granular thermoplastic solid; sol. in most org. liqs.; insol. in water, glycerol; dens. 1.07-1.18; ref. index 1.47
Toxicology: Heated to decomp., emits acrid smoke and irritating fumes
Uses: Food additive; protective coating and tablet binder in pharmaceutical vitamin/mineral preps., color diluent, flavor fixative; bulking agent; migrating to food from paper/paperboard
Usage level: ADI 0-25 mg/kg (FAO/WHO)
Regulatory: FDA 21CFR §73.1, 172.868, 175.300, 182.90, 573.420, GRAS
Manuf./Distrib.: FMC
Trade names containing: Coated Ascorbic Acid 97.5% No. 60482

Ethyl cinnamate

CAS 103-36-6; EINECS 203-104-6; FEMA 2430
Synonyms: Ethyl-trans-cinnamate; Ethyl-β-phenylacrylate; Ethyl-3-phenylpropenoate
Definition: Obtained by heating cinnamic acid, alcohol, and sulfuric acid to 100 C in presence of aluminum sulfate
Empirical: $C_{11}H_{12}O_2$
Properties: Nearly colorless oily liq., faint cinnamon odor; misc. with alcohol, ether, fixed oils; insol. in glycerin, water; m.w. 176.23; dens. 1.049 (20/4 C); m.p. 9 C; b.p. 271 C; flash pt. > 212 F; ref. index 1.558-1.561
Precaution: Combustible liq.
Toxicology: LD50 (oral, rat) 4000 mg/kg; mod. toxic by ingestion; heated to decomp., emits acrid smoke and irritating fumes
Uses: Synthetic flavoring agent
Usage level: 4.1 ppm (nonalcoholic beverages), 8.8 ppm (ice cream, ices), 9.5 ppm (candy), 12 ppm (baked goods), 2.4 ppm (gelatins, puddings), 11-40 ppm (chewing gum)
Regulatory: FDA 21CFR §172.515; GRAS (FEMA); Japan approved as flavoring
Manuf./Distrib.: Aldrich

Ethyl-trans-cinnamate. See Ethyl cinnamate
Ethyl citrate. See Triethyl citrate

Ethyl crotonate

CAS 623-70-1; EINECS 210-808-7; FEMA 3486
Synonyms: trans-2-Butenoic acid ethyl ester; Ethyl trans-2-butenoate; Ethyl β-methylacrylate
Definition: Obtained by esterification of crotonic acid with ethyl alcohol in presence of conc. H_2O_4
Empirical: $C_6H_{10}O_2$
Formula: $CH_3CH:CHCOOC_2H_5$
Properties: Water-wh. solid or liq., char. pungent persistent odor; sol. in alcohol, ether; insol. in water; m.w. 114.15; dens. 0.916 (20/4 C); b.p. 134-137 C; flash pt. 2 C; ref. index 1.425 (20 C)
Precaution: Highly flamm.; keep away from ignition sources
Toxicology: Irritating to eyes, skin, respiratory system
Uses: Synthetic flavoring agent
Regulatory: FDA 21CFR §172.515

Ethyl cyclohexanepropionate

CAS 10094-36-7; FEMA 2431
Synonyms: Cyclohexane ethyl propionate; Ethyl 3-cyclohexylpropanoate; Hexahydro phenylethyl propionate
Definition: Obtained by esterification of ethyl cyclohexanol with propionic acid or anhydride
Empirical: $C_{11}H_{20}O_2$
Properties: Colorless oily liq., fruity sweet pineapple-like odor; sol. in alcohol; insol. in water; m.w. 184.28; dens. 0.940; b.p. 91-94 C (8 mm); flash pt. 122 F; ref. index 1.4480
Uses: Synthetic flavoring agent
Usage level: 9 ppm (nonalcoholic beverages), 0.03-30 ppm (candy), 24 ppm (baked goods)
Regulatory: FDA 21CFR §172.515
Manuf./Distrib.: Aldrich

Ethyl 3-cyclohexylpropanoate. See Ethyl cyclohexanepropionate

Ethyl decanoate

CAS 110-38-3; EINECS 203-761-9; FEMA 2432
Synonyms: Ethyl caprate; Decanoic acid ethyl ester; Ethyl decylate
Definition: Obtained by esterification of decanoic acid and ethyl alcohol in presence of HCl or H_2SO_4
Empirical: $C_{12}H_{24}O_2$
Formula: $CH_3(CH_2)_8COOC_2H_5$
Properties: Colorless liq.; misc. with alcohol, chloroform, ether; insol. in water; m.w. 200.32; dens. 0.862 (20

C); m.p. -20 C; b.p. 243-245 C; flash pt. 216 F; ref. index 1.425 (20 C)
Precaution: Combustible liq.
Toxicology: Skin irritant; reacts with oxidizing materials; heated to decomp., emits acrid smoke and fumes
Uses: Synthetic flavoring agent
Usage level: 2.1 ppm (nonalcoholic beverages), 3.0-10 ppm (alcoholic beverages), 4.5 ppm (ice cream, ices), 8.3 ppm (candy), 23 ppm (baked goods), 5.3 ppm (gelatins, puddings)
Regulatory: FDA 21CFR §172.515; Japan approved as flavoring
Manuf./Distrib.: Aldrich

Ethyl decylate. *See* Ethyl decanoate
Ethyl trans-2,3-dimethyl acrylate. *See* Ethyl tiglate

2-Ethyl-3,5(6)-dimethylpyrazine
CAS 55031-15-7; FEMA 3149
Empirical: $C_8H_{12}N_2$
Properties: m.w. 136.20; dens. 0.965; b.p. 180-181 C; flash pt. 157 F
Precaution: Combustible liq.
Toxicology: Heated to decomp., emits toxic fumes of NO_x
Uses: Synthetic flavoring agent
Manuf./Distrib.: Aldrich

Ethyl dodecanoate. *See* Ethyl laurate
Ethyl dodecylate. *See* Ethyl laurate
Ethyl enanthate. *See* Ethyl heptanoate
Ethylenebis (dithiocarbamate), disodium salt. *See* Nabam
N,N′-Ethylene bisstearamide. *See* Ethylene distearamide

Ethylene brassylate
CAS 105-95-3; EINECS 203-347-8
Synonyms: 1,4-Dioxacycloheptadecane-5,17-dione; Cyclo-1,13-ethylenedioxytridecane-1,13-dione; Ethylene undecane dicarboxylate; Ethyl brassylate
Classification: Cyclic ester
Definition: Obtained by esterification of brassylic acid
Empirical: $C_{15}H_{26}O_4$
Properties: Wh. to lt. yel. liq., sweet odor, sol. in alcohol, most org. solvs.; insol. in water; m.w. 270.37; dens. 1.05; b.p. 332 C; ref. index 1.4690-1.4730
Uses: Synthetic flavoring agent
Regulatory: FDA 21CFR §172.515

Ethylene chloride. *See* Ethylene dichloride

Ethylenediamine
CAS 107-15-3; EINECS 203-468-6
Synonyms: 1,2-Diaminoethane; 1,2-Ethanediamine; Dimethylenediamine
Empirical: $C_2H_8N_2$
Formula: $NH_2CH_2CH_2NH_2$
Properties: Colorless volatile liq., ammonia-like odor; hygroscopic; m.w. 60.12; dens. 0.8994 (20/4 C); m.p. 8.5 C; b.p. 117.2 C; flash pt. (CC) 110 F; ref. index 1.4565
Precaution: Flamm. exposed to heat, flame, oxidizers; can react violently with acetic acid, acetic anhydride, acrylic acid, epichlorohydrin, many others; corrosive
Toxicology: TLV:TWA 10 ppm; LD50 (oral, rat) 500 mg/kg; human irritant poison by inh.; mod. toxic by ingestion, skin contact; corrosive; severe skin and eye irritant; allergen, sensitizer; mutagenic data; heated to decomp., emits toxic fumes of NO_x and NH_3
Uses: Antimicrobial for use in cane-sugar and beet-sugar mills; flume wash water additive; in paper for food pkg.; animal glue adjuvant; animal drug
Usage level: Limitation 0.1 ppm (in wash water), zero tolerance (milk)
Regulatory: FDA 21CFR §173.315, 173.320 (1 ppm max.), 178.3120, 556.270, 181.30

Ethylene diamine tetraacetic acid. *See* Edetic acid
Ethylenediaminetetraacetic acid, disodium salt. *See* Disodium EDTA
Ethylene diamine tetraacetic acid, sodium salt. *See* Tetrasodium EDTA
trans-1,2-Ethylenedicarboxylic acid. *See* Fumaric acid

Ethylene dichloride
CAS 107-06-2; EINECS 203-458-1
Synonyms: 1,2-DCE; Dichloroethylene; Dutch oil; 1,2-Dichloroethane; Ethylene chloride
Classification: Halogenated aliphatic hydrocarbon

Ethylene distearamide

Empirical: $C_2H_4Cl_2$
Formula: Cl • CH_2CH_2 • Cl
Properties: Colorless oily liq., chloroform-like odor, sweet taste; misc. with most common solvs.; sl. sol. in water; m.w. 98.96; dens. 1.2554 (20/4 C); b.p. 83.5 C; f.p. -35.5 C; flash pt. 56 C; ref. index 1.445
Precaution: Fire hazard exposed to heat, flame, or oxidizers; violent reaction with Al, NH_3
Toxicology: TLV:TWA 10 ppm; LD50 (oral, rat) 670 mg/kg; human poison by ingestion; mod. toxic by inh., intraperitoneal route; experimental carcinogen; strong narcotic; skin/severe eye irritant; heated to dec., emits highly toxic fumes of Cl., phosgene
Uses: Extract solvent for foods, flume wash water additive, pesticide
Usage level: Limitation 30 ppm (spice oleoresins), 30 ppm (chlorinated solv. residues), 0.1 ppm (wash water)
Regulatory: FDA 21CFR §172.560, 172.710, 173.230, 173.315, 175.105, 573.440
Manuf./Distrib.: Albright & Wilson Am.; Ashland; PPG Industries

Ethylene distearamide
CAS 110-30-5, 68955-45-3; EINECS 203-755-6, 273-277-0
Synonyms: N,N´-Ethylene bisstearamide; N,N´-1,2-Ethanediylbisoctadecanamide
Classification: Diamide
Empirical: $C_{39}H_{76}O_2$
Formula: $CCH_3(CH_2)_{16}CONH(CH_2)_2NHCO(CH_2)_{16}CH_3$
Uses: Release agent migrating from food pkg.
Trade names: Kemamide® W-39; Kemamide® W-40; Kemamide® W-45

Ethylene glycol butyl ether. *See* Butoxyethanol
Ethylene glycol methyl ether. *See* Methoxyethanol
Ethylene glycol monobutyl ether. *See* Butoxyethanol
Ethylene homopolymer. *See* Polyethylene

Ethylene oxide polymer
Classification: Polymer
Properties: visc. 1500 cps min. (1% aq.)
Toxicology: Heated to decomp., emits acrid smoke and irritating fumes
Uses: Foam stabilizer in fermented malt beverages
Usage level: Limitation 300 ppm (fermented malt beverages)
Regulatory: FDA 21CFR §172.770

Ethylenesuccinic acid. *See* Succinic acid
Ethylene undecane dicarboxylate. *See* Ethylene brassylate

2-Ethyl fenchol
CAS 18368-91-7; EINECS 242-243-7; FEMA 3491
Empirical: $C_{12}H_{22}O$
Properties: Earthy sharp camphoraceous odor; m.w. 182.31; dens. 0.956; b.p. 105 C (15 mm); flash pt. 192 F
Toxicology: Heated to decomp., emits acrid smoke and fumes
Uses: Synthetic flavoring agent
Manuf./Distrib.: Aldrich

Ethyl formate
CAS 109-94-4; EINECS 203-721-0; FEMA 2434
Synonyms: Ethyl formic ester; Ethyl methanoate; Formic ether
Definition: Ester of ethyl alcohol and formic acid
Empirical: $C_3H_6O_2$
Formula: $HCOOC_2H_5$
Properties: Colorless liq., sharp rum-like odor; sol. in fixed oils, propylene glycol, water (dec.); sl. sol. in min. oil; m.w. 74.09; dens. 0.9236 (20/20 C); m.p. -79 C; b.p. 54 C; flash pt. (CC) -4 F; ref. index 1.359
Precaution: Flamm. liq.; dangerous fire and explosion hazard exposed to heat, flame, oxidizers
Toxicology: LD50 (oral, rat) 1850 mg/kg; mod. toxic by ingestion, subcutaneous routes; mildly toxic by skin contact and inh.; experimental tumorigen; skin and eye irritant; heated to decomp., emits acrid smoke and irritating fumes
Uses: Synthetic flavoring agent and adjuvant; ethereal, fruity fragrance and flavoring, rum-like on dilution
Usage level: limitation 0.05% (baked goods), 0.04% (chewing gum, candy), 0.02% (frozen dairy desserts), 0.03% (gelatins, puddings), 0.1% (other foods)
Regulatory: FDA 21CFR §172.515, 184.1295, GRAS; 193.210, insecticide residue tolerance of 250 ppm in raisins and currants
Manuf./Distrib.: Aldrich

Ethylformic acid. *See* Propionic acid
Ethyl formic ester. *See* Ethyl formate

Ethyl fumarate
CAS 2459-05-4; EINECS 219-544-7
Synonyms: Monoethyl fumarate
Empirical: C₆H₈O₄
Formula: C₂H₅OCOCH:CHCOOH
Properties: m.w. 144.13; m.p. 67-69 C
Uses: Inhibits fungal growth in tomato juice
Usage level: 0.2%

2-Ethylfuran
CAS 3208-16-0; FEMA 3673
Definition: Obtained by dehydration of furyl methyl carbinol followed by reduction
Empirical: C₆H₈O
Properties: Colorless liq., powerful sweet burnt odor, coffee-like flavor; sol. in alcohol; almost insol. in water; m.w. 96.13; dens. 0.912; b.p. 92-93 C; flash pt. 28 F; ref. index 1.4390
Precaution: Flamm. liq.
Uses: Synthetic flavoring agent
Regulatory: FDA 21CFR §172.515
Manuf./Distrib.: Aldrich

Ethyl-2-furanpropionate
FEMA 2435
Synonyms: Ethyl furfuralacetate; Ethyl-3-(2-furyl)-propanoate; Ethyl furylpropionate
Empirical: C₉H₁₂O₃
Properties: Low-melting solid; turns yel. on exposure to air; fruity odor; m.w. 168.19; m.p. 24.5 C; b.p. 260 C; ref. index 1.54876
Uses: Synthetic flavoring agent
Usage level: 1.6 ppm (nonalcoholic beverages), 1.6 ppm (ice cream, ices), 5.6 ppm (candy), 7.5 ppm (baked goods)
Regulatory: FDA 21CFR §172.515

Ethyl furfuralacetate. *See* Ethyl 2-furanpropionate
Ethyl-3-(2-furyl)-propanoate. *See* Ethyl-2-furanpropionate
Ethyl furylpropionate. *See* Ethyl-2-furanpropionate

4-Ethylguaiacol
CAS 2785-89-9; FEMA 2436
Synonyms: 4-Ethyl-2-methoxyphenol
Empirical: C₉H₁₂O₂
Properties: Oily liq., smoky bacon-like odor; m.w. 152.19; dens. 1.063; m.p. 15 C; b.p. 234-236 C; flash pt. 226 F
Uses: Synthetic flavoring agent
Usage level: 0.05 ppm (nonalcoholic beverages), 1.1 ppm (ice cream, ices), 0.23 ppm (gelatins, puddings)
Regulatory: FDA 21CFR §172.515
Manuf./Distrib.: Aldrich

Ethyl hendecanoate. *See* Ethyl undecanoate
Ethyl 10-hendecenoate. *See* Ethyl 10-undecenoate

Ethyl heptanoate
CAS 106-30-9; EINECS 203-382-9; FEMA 2437
Synonyms: Ethyl heptoate; Ethyl n-heptoate; Heptanoic acid ethyl ester; Ethyl enanthate; Cognac oil, synthetic; Oil of grapes; Ethyl oenanthate
Definition: Ester of heptoic acid
Empirical: C₉H₁₈O₂
Formula: CH₃(CH₂)₅COOC₂H₅
Properties: Liq., fruity wine-like odor and taste with burning aftertaste; misc. with alcohol, ether, chloroform; insol. in water; m.w. 158.24; dens. 0.868 (20/4 C); m.p. -66.3 C; b.p. 186-188 C; flash pt. 74 C; ref. index 1.413 (20 C)
Precaution: Combustible liq.
Toxicology: LD50 (oral, rat) > 34,640 mg/kg; heated to decomp., emits acrid smoke and fumes
Uses: Synthetic flavoring agent for mfg. of liqueurs; in formulation of raspberry, gooseberry, grape, cherry, apricot, currant, bourbon, and other artificial essences

465

Regulatory: FDA 21CFR §172.515; Japan approved as flavoring
Manuf./Distrib.: Aldrich

2-Ethyl-2-heptenal
FEMA 2438
Synonyms: 2-Ethyl-3-butylacrolein
Properties: m.w. 140.23
Uses: Synthetic flavoring agent
Usage level: 0.40 ppm (nonalcoholic beverages), 0.03-2.0 ppm (candy)
Regulatory: FDA 21CFR §172.515

Ethyl heptoate. See Ethyl heptanoate
Ethyl n-heptoate. See Ethyl heptanoate
Ethyl-2,4-hexadienoate. See Ethyl sorbate
Ethyl hexanoate. See Ethyl caproate
2-Ethylhexanoic acid, 1-methyl-1,2-ethanediyl ester. See Propylene glycol dioctanoate
Ethyl hexylate. See Ethyl caproate
2-Ethylhexyl diphenyl ester phosphoric acid. See Diphenyl octyl phosphate
Ethyl hydrocinnamate. See Ethyl-3-phenylpropionate
Ethyl 4-hydroxybenzoate. See Ethyl paraben
Ethyl-o-hydroxybenzoate. See Ethyl salicylate
Ethyl p-hydroxybenzoate. See Ethyl paraben
Ethyl-4-hydroxybenzoate potassium salt. See Potassium ethylparaben
Ethyl-S(-)-2-hydroxypropionate. See Ethyl lactate
Ethyl α-hydroxy propionate. See Ethyl lactate
2-Ethyl-3-hydroxy-4H-pyran-4-one. See Ethyl maltol
Ethylidene diethyl ether. See Acetal
Ethylidenelactic acid. See Lactic acid

Ethyl isobutyrate
CAS 97-62-1; EINECS 202-595-4; FEMA 2428
Synonyms: 2-Methylpropanoic acid ethyl ester
Empirical: $C_6H_{12}O_2$
Formula: $(CH_3)_2CHCOOC_2H_5$
Properties: Liq., aromatic fruity odor; misc. with alcohol, ether; sl. sol. in water; m.w. 116.16; dens. 0.867 (20/4 C); m.p. -88 C; b.p. 107-110 C; flash pt. 20 C; ref. index 1.388
Precaution: Flammable liq.
Toxicology: LD50 (IP, mouse) 800 mg/kg; mod. toxic by Ip route; skin irritant; reacts with oxidizing materials; heated to decomp., emits acrid smoke and fumes
Uses: Synthetic flavoring agent; mfg. of flavoring compds. and essences
Usage level: 10 ppm (nonalcoholic beverages), 25 ppm (ice cream, ices), 73 ppm (candy), 200 ppm (baked goods), 6.0 ppm (gelatins, puddings), 1.5 ppm (toppings)
Regulatory: FDA 21CFR §172.515
Manuf./Distrib.: Aldrich

Ethyl isoeugenyl. See Isoeugenyl ethyl ether

Ethyl isovalerate
CAS 108-64-5; EINECS 203-602-3; FEMA 2463
Synonyms: Isovaleric acid ethyl ester; 3-Methylbutanoic acid ethyl ester; Ethyl 3-methylbutyrate; Ethyl β-methylbutyrate
Definition: Obtained by esterification of isovaleric acid with ethyl alcohol in presence of conc. H_2SO_4
Empirical: $C_7H_{14}O_2$
Formula: $(CH_3)_2CHCH_2COOC_2H_5$
Properties: Colorless oily liq., apple odor; sol. in propylene glycol; sl. sol. in water @ 135 C; misc. with alcohol, fixed oils, benzene, ether; m.w. 130.21; dens. 0.868 (20/20 C); b.p. 135 C; m.p. -99 C; flash pt. 77 F; ref. index 1.395-1.399
Precaution: Flamm. liq. when exposed to heat, flame, or sparks
Toxicology: LD50 (oral, rabbit) 7031 mg/kg; mod. toxic by intraperitoneal route; mildly toxic by ingestion; skin irritant; heated to decomp., emits acrid smoke and fumes
Uses: Etheral, wine-like, apple-like fragrance and flavoring, synthetic; flavor improver for foodstuffs
Usage level: 4.9 ppm (nonalcoholic beverages), 7.5 ppm (ice cream, ices), 29 ppm (candy), 27 ppm (baked goods), 5.0 ppm (gelatins, puddings), 80-430 ppm (chewing gum), 1.0 ppm (condiments)
Regulatory: FDA 21CFR §172.515; Japan approved as flavoring
Manuf./Distrib.: Aldrich; BASF

Ethyl β-keto-β-phenylpropionate. *See* Ethyl benzoylacetate
Ethyl α-ketopropionate. *See* Ethyl pyruvate
Ethyl γ-ketovalerate. *See* Ethyl levulinate

Ethyl lactate
CAS 97-64-3; FEMA 2440
Synonyms: Lactic acid ethyl ester; Ethyl α-hydroxy propionate; Ethyl-S(-)-2-hydroxypropionate
Definition: Ethyl ester of lactic acid
Empirical: $C_5H_{10}O_3$
Formula: $CH_3CHOHCOOC_2H_5$
Properties: Colorless liq., mild odor; misc. with water, alcohol, ketones, esters, hydrocarbons, oil; m.w. 118.13; dens. 1.020-1.036 (20/20 C); m.p. -26 C; b.p. 154 C; flash pt. 46.1 C; ref. index 1.410-1.420
Precaution: Flamm. or combustible; can react with oxidizers; sl. explosion hazard in vapor form exposed to flame
Toxicology: LD50 (oral, mouse) 2500 mg/kg; moderately toxic by ingestion, intraperitoneal, subcutaneous, intravenous routes; heated to decomp., emits acrid smoke and irritating fumes
Uses: Synthetic flavoring agent
Usage level: 5.4 ppm (nonalcoholic beverages), 1000 ppm (alcoholic beverages), 17 ppm (ice cream, ices), 28 ppm (candy), 71 ppm (baked goods), 8.3 ppm (gelatins, puddings), 580-3100 ppm (chewing gum), 35 ppm (syrups)
Regulatory: FDA 21CFR §172.515
Manuf./Distrib.: Aldrich; Jungbunzlauer
Trade names: Purasolv® ELECT; Purasolv® ELS

Ethyl laurate
CAS 106-33-2; EINECS 203-386-0; FEMA 2441
Synonyms: Dodecanoic acid ethyl ester; Ethyl dodecanoate; Ethyl dodecylate
Definition: Synthesized from lauroyl chloride and ethyl alcohol in presence of Mg in ether sol'n.
Empirical: $C_{14}H_{28}O_2$
Properties: Colorless oily liq., fruity-floral odor; misc. with alcohol, chloroform, ether; insol. in water; m.w. 228.37; dens. 0.858; b.p. 272-273 C; flash pt. > 212 F; ref. Index 1.430
Precaution: Combustible liq.
Toxicology: Heated to decomp., emits acrid smoke and irritating fumes
Uses: Synthetic flavoring agent
Usage level: 1.7 ppm (nonalcoholic beverages), 3 ppm (alcoholic beverages), 3.7 ppm (ice cream, ices), 17 ppm (candy), 17 ppm (baked goods), 4.4 ppm (gelatins, puddings), 39 ppm (chewing gum)
Regulatory: FDA 21CFR §172.515; GRAS (FEMA)
Manuf./Distrib.: Aldrich

Ethyl levulinate
CAS 539-88-8; EINECS 208-728-2; FEMA 2442
Synonyms: Ethyl γ-ketovalerate; Ethyl-4-oxopentanoate; 4-Oxopentanoic acid ethyl ester
Definition: Ester of levulinic acid and ethyl alcohol
Empirical: $C_7H_{12}O_3$
Formula: $CH_3COCH_2CH_2COOC_2H_5$
Properties: Liq.; freely sol. in water; misc. with alcohol; m.w. 144.17; dens. 1.012 (20/4 C); b.p. 203-205 C; flash pt. 94 C; ref. index 1.423 (20 C)
Precaution: Combustible
Uses: Synthetic flavoring agent
Usage level: 5.8 ppm (nonalcoholic beverages), 11 ppm (ice cream, ices), 12 ppm (candy, baked goods)
Regulatory: FDA 21CFR §172.515
Manuf./Distrib.: Aldrich

Ethyllic acid. *See* Acetic acid
Ethyl malate. *See* Diethyl malate
Ethyl malonate. *See* Diethyl malonate

Ethyl maltol
CAS 4940-11-8; FEMA 3487; EEC E637
Synonyms: 3-Hydroxy-2-ethyl-4-pyrone; 2-Ethyl-3-hydroxy-4H-pyran-4-one
Empirical: $C_7H_8O_3$
Properties: Caramel sweet odor; m.w. 140.14; m.p. 88-92 C
Toxicology: Mod. toxic by ingestion, subcutaneous routes; mutagenic data; heated to decomp., emits acrid smoke and fumes
Uses: Synthetic flavoring agent imparting sweete taste, flavor enhancer, processing aid

Ethyl methacrylate

Usage level: ADI 0-2 mg/kg (EEC)
Regulatory: FDA 21CFR §172.515; Europe listed; UK approved
Manuf./Distrib.: Aldrich
Trade names: Veltol®-Plus

Ethyl methacrylate
CAS 97-63-2; EINECS 202-597-5
Synonyms: Ethyl 2-methyl-2-propenoate; Ethyl-α-methyl acrylate
Definition: Ester of ethyl alcohol and methacrylic acid
Empirical: $C_6H_{10}O_2$
Formula: $H_2C:CCH_3COOC_2H_5$
Properties: Colorless liq.; insol. in water; m.w. 114.16; dens. 0.911; b.p. 119 C; f.p. -75 C; flash pt. (OC) 21.1 C; ref. index 1.4116
Precaution: Explosive
Toxicology: Skin irritant; moderately toxic by ingestion, intraperitoneal routes; mildly toxic by inhalation

Ethyl methanoate. *See* Ethyl formate
Ethyl-4-methoxybenzoate. *See* Ethyl-p-anisate
Ethyl-p-methoxybenzoate. *See* Ethyl-p-anisate
4-Ethyl-2-methoxyphenol. *See* 4-Ethylguaiacol
Ethyl-α-methyl acrylate. *See* Ethyl methacrylate
Ethyl β-methylacrylate. *See* Ethyl crotonate
Ethyl trans-2-methyl-2-butenoate. *See* Ethyl tiglate

Ethyl 2-methylbutyrate
CAS 7452-79-1; EINECS 231-225-4; FEMA 2443
Empirical: $C_7H_{14}O_2$
Formula: $CH_3CH_2CH(CH_3)COOC_2H_5$
Properties: Powerful green fruity pungent odor; m.w. 130.19; dens. 0.868 (20/4 C); b.p. 130-133 C; ref. index 1.397 (20 C)
Precaution: Combustible liq.
Toxicology: Heated to decomp., emits acrid smoke and fumes
Uses: Synthetic flavoring agent
Usage level: 0.50 ppm (nonalcoholic beverages), 3.0 ppm (ice cream, ices), 5.0 ppm (candy)
Regulatory: FDA 21CFR §172.515
Manuf./Distrib.: Aldrich

Ethyl 3-methylbutyrate. *See* Ethyl isovalerate
Ethyl β-methylbutyrate. *See* Ethyl isovalerate
Ethylmethylcellulose. *See* Methyl ethyl cellulose
Ethyl methyl ketone. *See* Methyl ethyl ketone

Ethyl methylphenylglycidate
CAS 77-83-8; FEMA 2444
Synonyms: Aldehyde C-16; Strawberry aldehyde; 3-Methyl-3-phenyl glycidic acid ethyl ester
Empirical: $C_{12}H_{14}O_3$
Properties: Colorless to pale yel. liq., fruity strawberry-like odor; sol. in most fixed oils; m.w. 206.24; dens. 1.104-1.123; ref. index 1.509-1.511
Precaution: Combustible liq.
Toxicology: Mildly toxic by ingestion; heated to decomp., emits acrid smoke and fumes
Uses: Synthetic flavoring agent
Regulatory: FDA 21CFR §182.60, GRAS

Ethyl 2-methyl-2-propenoate. *See* Ethyl methacrylate

2-Ethyl-3-methylpyrazine
CAS 15707-23-0; FEMA 3155
Empirical: $C_7H_{10}N_2$
Properties: Colorless to sl. yel. liq., strong raw potato odor; sol. in water; m.w. 122.17; dens. 0.985; b.p. 57 C (10 mm); flash pt. 138 F; ref. index 1.502-1.505
Toxicology: Irritant; heated to decomp., emits acrid smoke and fumes
Uses: Synthetic flavoring agent
Manuf./Distrib.: Aldrich

Ethyl myristate
CAS 124-06-1; EINECS 204-675-4; FEMA 2445
Synonyms: Ethyl tetradecanoate; Tetradecanoic acid ethyl ester

Definition: Ester of ethyl alcohol and myristic acid
Empirical: $C_{16}H_{23}O_2$
Formula: $CH_3(CH_2)_{12}COOC_2H_5$
Properties: Colorless to pale yel. liq., waxy odor; insol. in water; sol. in alcohol; sl. sol. in ether; m.w. 256.43; dens. 0.856; m.p. 12 C; f.p. > 110 C; b.p. 295 C; flash pt. > 212 F; ref. index 1.434
Precaution: Combustible liq.
Toxicology: Heated to decomp., emits acrid smoke and irritating fumes
Uses: Synthetic flavoring agent
Usage level: 6.7 ppm (nonalcoholic beverages), 30 ppm (alcoholic beverages), 8 ppm (ice cream, ices), 10 ppm (candy), 14 ppm (baked goods)
Regulatory: FDA 21CFR §172.515
Manuf./Distrib.: Aldrich

Ethyl nitrite
CAS 109-95-5; FEMA 2446
Synonyms: Nitrous ether; Sweet spirit of nitre
Empirical: $C_2H_5O_2N$
Properties: Ylsh. volatile liq., ether-like odor; sol. in alcohol; sl. sol. in water; m.w. 75.04; dens. 0.90; b.p. 16.4 C; flash pt. -31 F
Precaution: Flamm.; explodes
Uses: Synthetic flavoring
Usage level: 3.0 ppm (nonalcoholic beverages), 4.5 ppm (ice cream, ices), 0.10-8.0 ppm (candy), 0.10 ppm (baked goods), 3.9 ppm (chewing gum), 52 ppm (syrups), 13 ppm (icings)
Regulatory: FDA 21CFR §172.515

Ethyl nonanoate. *See* Ethyl pelargonate

Ethyl 2-nonynoate
FEMA 2448
Synonyms: Ethyl octyne carbonate
Empirical: $C_{11}H_{18}O_2$
Properties: Oily liq., green violet-like odor; m.w. 182.26; dens. 0.9032 (25 C); b.p. 227 C; ref. index 1.4527
Uses: Synthetic flavoring
Usage level: 0.56 ppm (nonalcoholic beverages), 0.55 ppm (ice cream, ices), 0.52 ppm (candy), 1.2 ppm (baked goods)
Regulatory: FDA 21CFR §172.515

Ethyl 9-octadecenoate. *See* Ethyl oleate

Ethyl octanoate
CAS 106-32-1; EINECS 203-385-5; FEMA 2449
Synonyms: Ethyl caprylate; Ethyl octoate; Ethyl octylate; Octanoic acid ethyl ester
Definition: Ester of cparylic acid and ethyl alcohol
Empirical: $C_{10}H_{20}O_2$
Formula: $CH_3(CH_2)_6COOC_2H_5$
Properties: Colorless very mobile liq., pleasant pineapple odor; misc. with alcohol, ether; insol. in water; m.w. 172.27; dens. 0.867 (20/4 C); m.p. -47 C; b.p. 207-209; flash pt. 167 F; ref. index 1.418 (20 C)
Precaution: Combustible liq.
Toxicology: LD50 (oral, rat) 25,960 mg/kg; mildly toxic by ingestion; skin irritant; heated to decomp., emits acrid smoke and fumes
Uses: Synthetic flavoring agent; mfg. of fruit ethers
Usage level: 4.1 ppm (nonalcoholic beverages), 2.4 ppm (ice cream, ices), 9.0 ppm (candy), 11 ppm (baked goods), 0.10-2.7 ppm (gelatins, puddings), 4.0-60 ppm (chewing gum)
Regulatory: FDA 21CFR §172.515; Japan approved as flavoring
Manuf./Distrib.: Aldrich

Ethyl octoate. *See* Ethyl octanoate
Ethyl octylate. *See* Ethyl octanoate
Ethyl octyne carbonate. *See* Ethyl 2-nonynoate
Ethyl oenanthate. *See* Cognac oil, green or white
Ethyl oenanthate. *See* Ethyl heptanoate

Ethyl oleate
CAS 111-62-6, 85049-36-1; EINECS 203-889-5, 285-206-0; FEMA 2450
Synonyms: 9-Octadecenoic acid ethyl ester; Ethyl 9-octadecenoate
Definition: Ester of ethyl alcohol and oleic acid

Ethyl-3-oxo-2-benzylbutanoate

Empirical: $C_{20}H_{38}O_2$
Formula: $CH_3(CH_2)_7CH:CH(CH_2)_7COOC_2H_5$
Properties: Ylsh. oily liq.; misc. with alcohol, ether; insol. in water; m.w. 310.52; dens. 0.870 (20/4 C); m.p. -32 C; b.p. 216-218 (15 mm); flash pt. 175 C; ref. index 1.451
Precaution: Combustible
Uses: Synthetic flavoring agent
Usage level: 0.1 ppm (nonalcoholic beverages), 0.1 ppm (ice cream, ices), 0.1-40 ppm (candy), 0.1-55 ppm (baked goods), 0.1 ppm (gelatins, puddings)
Regulatory: FDA 21CFR §172.515; GRAS (FEMA)

Ethyl-3-oxo-2-benzylbutanoate. *See* Ethyl 2-acetal-3-phenylpropionate
Ethyl 3-oxobutanoate. *See* Ethylacetoacetate
Ethyl-4-oxopentanoate. *See* Ethyl levulinate
Ethyl 2-oxopropanoate. *See* Ethyl pyruvate

Ethyl oxyhydrate
FEMA 2996
Synonyms: Rum ether
Toxicology: Heated to decomp., emits acrid smoke and fumes
Uses: Synthetic flavoring agent
Regulatory: FDA 21CFR §172.515

Ethyl paraben
CAS 120-47-8; EINECS 204-399-4; EEC E214
Synonyms: Ethyl 4-hydroxybenzoate; 4-Hydroxybenzoic acid ethyl ester; Ethyl p-hydroxybenzoate; Carboxyphenol
Empirical: $C_9H_{10}O_3$
Formula: $HOC_6H_4CO_2C_2H_5$
Properties: Ivory to wh. powd.; sol. in alcohol, ether; m.w. 166.18; m.p. 114-118 C; b.p. 297-298 C
Toxicology: Irritant; local anesthetic effect
Uses: Antimicrobial; preservative (0.012-1 g/kg as p-hydroxybenzoic acid)
Regulatory: Europe listed; UK permitted; Japan approved
Manuf./Distrib.: Int'l. Sourcing
Trade names: Nipagin A
Trade names containing: Nipastat

Ethylparaben, potassium salt. *See* Potassium ethylparaben
Ethylparaben, sodium salt. *See* Sodium ethylparaben

Ethyl pelargonate
CAS 123-29-5; EINECS 204-615-7; FEMA 2447
Synonyms: Ethyl nonanoate; Nonanoic acid ethyl ester; Wine ether
Definition: Ester of ethyl alcohol and pelargonic acid
Empirical: $C_{11}H_{22}O_2$
Formula: $CH_3(CH_2)_7COOCH_2CH_3$
Properties: Colorless liq., fruity odor; insol. in water; sol. in alcohol, ether; m.w. 186.33; dens. 0.866 (18/4 C); b.p. ≈ 220 C; f.p. -44 C; flash pt. 185 F; ref. index 1.4220 (20 C)
Precaution: Combustible liq.
Toxicology: LD50 (oral, rat) > 43,000 mg/kg; mildly toxic by ingestion; skin irritant; heated to decomp., emits acrid smoke and irritating fumes
Uses: Synthetic flavoring agent for alcoholic beverages
Usage level: 3.9 ppm (nonalcoholic beverages), 20 ppm (alcoholic beverages), 4 ppm (ice cream, ices), 14 ppm (candy), 15 ppm (baked goods), 580 ppm (chewing gum), 39 ppm (icings)
Regulatory: FDA 21C FR §172.515; GRAS (FEMA)
Manuf./Distrib.: Aldrich

Ethyl pentyl carbinol. *See* 3-Octanol
Ethyl pentyl ketone. *See* 3-Octanone
Ethyl phenacetate. *See* Ethyl phenylacetate

Ethyl phenylacetate
CAS 101-97-3; EINECS 202-993-8; FEMA 2452
Synonyms: Ethyl benzeneacetate; Ethyl phenacetate; Ethyl-2-phenylethanoate; Ethyl-α-toluate
Definition: Ester of ethyl alcohol and phenylacetic acid
Empirical: $C_{10}H_{12}O_2$
Properties: Colorless liq., sweet honey-like odor; sol. in fixed oils; insol. in glycerin, propylene glycol, water;

m.w. 164.22; dens. 1.033 (20 C); b.p. 227 C; flash pt. > 100 C; ref. index 1.496-1.500
Precaution: Combustible liq.
Toxicology: LD50 (oral, rat) 3300 mg/kg; mod. toxic by ingestion; heated to decomp., emits acrid smoke and
  irritating fumes
Uses: Synthetic flavoring agent
Usage level: 2.4 ppm (nonalcoholic beverages), 5.2 ppm (ice cream, ices), 8.1 ppm (candy), 6 ppm (baked
  goods), 24 ppm (syrups)
Regulatory: FDA 21CFR §172.515; GRAS (FEMA); Japan approved as flavoring
Manuf./Distrib.: Aldrich

Ethyl-β-phenylacrylate. *See* Ethyl cinnamate
Ethyl phenylbutyrate. *See* Ethyl-4-phenylbutyrate

Ethyl-4-phenylbutyrate
Synonyms: Ethyl phenylbutyrate; Ethyl-γ-phenylbutyrate
Definition: Ester of ethanol and γ-phenylbutyric acid
Empirical: $C_{12}H_{16}O_2$
Properties: Colorless somewhat oily liq., plum-like odor, plum-prune taste; sol. in alcohol; insol. in water; m.w.
  192.26
Uses: Synthetic flavoring
Usage level: 0.06-1.0 ppm (nonalcoholic beverages), 0.06 ppm (ice cream, ices)
Regulatory: FDA 21CFR §172.515

Ethyl-γ-phenylbutyrate. *See* Ethyl-4-phenylbutyrate
Ethyl phenyl carbinol. *See* 1-Phenyl-1-propanol
Ethyl phenyl carbinyl butyrate. *See* α-Ethylbenzyl butyrate
Ethyl 3-phenyl-2,3-epoxypropionate. *See* Ethyl phenylglycidate
Ethyl-2-phenylethanoate. *See* Ethyl phenylacetate

Ethyl phenylglycidate
CAS 121-39-1; FEMA 2454
Synonyms: Ethyl 3-phenylglycidate; Ethyl 3-phenyl-2,3-epoxypropionate
Empirical: $C_{11}H_{12}O_3$
Properties: Colorless to pale yel. liq., fruity odor; m.w. 192.22; flash pt. > 100 C; ref. index 1.5190-1.5230
Toxicology: Irritant; mutagenic data; mod. toxic by ingestion; heated to decomp., emits acrid smoke and fumes
Uses: Synthetic flavoring agent
Usage level: 4.6 ppm (nonalcoholic beverages), 12 ppm (ice cream, ices), 18 ppm (candy), 20 ppm (baked
  goods), 10-70 ppm (gelatins, puddings)
Regulatory: FDA 21CFR §172.515
Manuf./Distrib.: Aldrich

Ethyl 3-phenylglycidate. *See* Ethyl phenylglycidate
Ethyl 3-phenyl-3-oxopropanoate. *See* Ethyl benzoylacetate
Ethyl-3-phenylpropenoate. *See* Ethyl cinnamate

Ethyl-3-phenylpropionate
CAS 2021-28-5; EINECS 217-966-6; FEMA 2455
Synonyms: Ethyl hydrocinnamate
Empirical: $C_{11}H_{14}O_2$
Formula: $C_6H_5CH_2CH_2COOC_2H_5$
Properties: Colorless liq., floral odor; sol. in most org. solvs.; insol. in water; m.w. 178.23; dens. 1.013 (20/4
  C); b.p. 247-249 C; flash pt. > 98 C; ref. index 1.494 (20 C)
Uses: Synthetic flavoring
Usage level: 1.8 ppm (nonalcoholic beverages), 1.0 ppm (ice cream, ices), 2.5 ppm (candy), 0.50-30 ppm
  (baked goods)
Regulatory: FDA 21CFR §172.515
Manuf./Distrib.: Aldrich

Ethyl phthalate. *See* Diethyl phthalate

Ethylphthalyl ethyl glycolate
Empirical: $C_{14}H_{16}O_6$
Formula: $C_2H_5OCOC_6H_4COOCH_2COOC_2H_5$
Properties: Liq., sl. odor; misc. with org. solvs.; dens. 1.180 (25 C); b.p. 190 C (5 mm); ref. index 1.498
Precaution: Combustible
Toxicology: Heated to decomp., emits acrid smoke and irritating fumes
Uses: Plasticizer migrating from food pkg.

Ethyl 1-propene-1,2,3-tricarboxylate

Regulatory: FDA 21CFR §181.27

Ethyl 1-propene-1,2,3-tricarboxylate. *See* Ethyl aconitate, mixed esters
Ethyl propenoate. *See* Ethyl acrylate

Ethyl propionate
CAS 105-37-3; EINECS 203-291-4; FEMA 2456
Synonyms: Propanoic acid ethyl ester
Definition: Ester of propionic acid and ethyl alcohol
Empirical: $C_5H_{10}O_2$
Formula: $CH_3CH_2COOC_2H_5$
Properties: Colorless liq., fruity odor; sol. in ≈ 60 parts water; misc. with alcohol, ether; m.w. 102.14; dens. 0.890 (20/4 C); m.p. -73 C; b.p. 96-99 C; flash pt. (CC) 12 C; ref. index 1.384 (20 C)
Precaution: Highly flamm. liq.
Toxicology: LD50 (oral, rabbit) 3500 mg/kg; mod. toxic by ingestion and Ip routes; skin irritant; reacts with oxidizing materials; heated to decomp., emits acrid smoke and fumes
Uses: Synthetic flavoring
Usage level: 7.7 ppm (nonalcoholic beverages), 29 ppm (ice cream, ices), 78 ppm (candy), 110 ppm (baked goods), 10-15 ppm (gelatins, puddings), 1100 ppm (chewing gum)
Regulatory: FDA 21CFR §172.515; Japan approved as flavoring
Manuf./Distrib.: Aldrich

Ethyl pyruvate
CAS 617-35-6; EINECS 210-511-2; FEMA 2457
Synonyms: Ethyl α-ketopropionate; Ethyl 2-oxopropanoate
Definition: Ester of pyruvic acid and absolute ethyl alcohol
Empirical: $C_5H_8O_3$
Formula: $CH_3COCOOC_2H_5$
Properties: Liq., vegetable caramel odor; sl. sol. in water; misc. with alcohol, ether; m.w. 116.12; dens. 1.047 (20/4 C); m.p. -50 c; b.p. 148-150 C; flash pt. 114 F; ref. index 1.405
Uses: Synthetic flavoring agent
Usage level: 50 ppm (nonalcoholic beverages), 20-150 ppm (ice cream, ices), 35 ppm (candy), 40 ppm (baked goods)
Regulatory: FDA 21CFR §172.515; GRAS (FEMA)
Manuf./Distrib.: Aldrich

Ethyl salicylate
CAS 118-61-6; EINECS 204-265-5; FEMA 2458
Synonyms: Ethyl-o-hydroxybenzoate; Salicylic ethyl ester; Salicylic ether
Definition: Ester of salicylic acid and ethyl alcohol
Empirical: $C_9H_{10}O_3$
Formula: $HO \cdot C_6H_4 \cdot CO_2 \cdot C_2H_5$
Properties: Colorless liq., wintergreen odor; sol. in alcohol, ether, acetic acid, fixed oils; sl. sol. in water, glycerin; m.w. 166.18; dens. 1.127; m.p. 1.3 C; b.p. 233-234 C; flash pt. 225 F; ref. index 1.520
Precaution: Combustible
Toxicology: LD50 (oral, rat) 1320 mg/kg; mod. toxic by ingestion, subcutaneous routes; skin irritant; heated to decomp., emits acrid smoke and irritating fumes
Uses: Synthetic flavoring agent
Usage level: 2.8 ppm (nonalcoholic beverages), 11 ppm (ice cream, ices), 10 ppm (candy), 16 ppm (baked goods), 0.04 ppm (gelatins, puddings), 16 ppm (chewing gum)
Regulatory: FDA 21CFR §172.515
Manuf./Distrib.: Aldrich

Ethyl sebacate. *See* Diethyl sebacate

Ethyl sorbate
CAS 2396-84-1; FEMA 2459
Synonyms: Ethyl-2,4-hexadienoate
Definition: Ester of sorbyl chloride and ethyl alcohol
Empirical: $C_8H_{12}O_2$
Properties: Warm fruity ethereal odor; m.w. 140.18; dens. 0.956; b.p. 81 C (15 mm); flash pt. 157 F; ref. index 1.502
Uses: Synthetic flavoring
Usage level: 5.5 ppm (nonalcoholic beverages), 14 ppm (ice cream, ices), 15 ppm (candy), 18 ppm (baked goods)
Regulatory: FDA 21CFR §172.515

Manuf./Distrib.: Aldrich

Ethyl succinate. *See* Diethyl succinate
Ethyl tartrate. *See* Diethyl tartrate
Ethyl tetradecanoate. *See* Ethyl myristate

Ethyl tiglate
CAS 5837-78-5; FEMA 2460
Synonyms: Ethyl trans-2,3-dimethyl acrylate; Ethyl trans-2-methyl-2-butenoate
Definition: Ester of tiglic acid and ethyl alcohol
Empirical: $C_7H_{12}O_2$
Properties: Liq., fruity caramel odor; sol. in most org. solvs.; m.w. 128.17; dens. 0.923; b.p. 154-156 C; flash pt. 112 F; ref. index 1.4347 (16.8 C)
Uses: Synthetic flavoring agent
Usage level: 5.3 ppm (nonalcoholic beverages), 6.0 ppm (ice cream, ices), 20 ppm (candy), 6.5 ppm (baked goods)
Regulatory: FDA 21CFR §172.515
Manuf./Distrib.: Aldrich

Ethyl-α-toluate. *See* Ethyl phenylacetate

Ethyl undecanoate
CAS 627-90-7; FEMA 3492
Synonyms: Ethyl hendecanoate; Ethyl undecylate
Empirical: $C_{13}H_{26}O_2$
Properties: Liq., cognac coconut odor; insol. in water; m.w. 214.35; dens. 0.859; b.p. 255 C; flash pt. > 230 F; ref. index 1.9325
Uses: Synthetic flavoring agent
Regulatory: FDA 21CFR §172.515
Manuf./Distrib.: Aldrich

Ethyl 10-undecenoate
CAS 692-86-4; FEMA 2461
Synonyms: Ethyl 10-hendecenoate; Ethyl undecylenoate
Empirical: $C_{13}H_{24}O_2$
Properties: Colorless to pale yel. liq., wine-like odor; m.w. 212.34; dens. 0.879; b.p. 258-259 C; flash pt. > 230 F; ref. index 1.4382
Uses: Synthetic flavoring agent
Usage level: 1.7 ppm (nonalcoholic beverages), 5.0 ppm (alcoholic beverages), 8.7 ppm (ice cream, ices), 10 ppm (candy), 11 ppm (baked goods)
Regulatory: FDA 21CFR §172.515
Manuf./Distrib.: Aldrich

Ethyl undecylate. *See* Ethyl undecanoate
Ethyl undecylenoate. *See* Ethyl 10-undecenoate

Ethyl valerate
CAS 539-82-2; EINECS 208-726-1; FEMA 2462
Synonyms: Ethyl n-valerate
Empirical: $C_7H_{14}O_2$
Formula: $CH_3(CH_2)_3COOC_2H_5$
Properties: Liq.; misc. with alcohol; insol. in water; m.w. 130.19; dens. 0.877 (20/4 C); b.p. 142-146 C; flash pt. 34 C; ref. index 1.372-1.400 (20 C)
Precaution: Flamm.
Toxicology: Irritant
Uses: Synthetic flavoring agent
Usage level: 4.2 ppm (nonalcoholic beverages), 4.4 ppm (ice cream, ices), 15 ppm (candy), 8.3 ppm (baked goods), 5.5 ppm (gelatins, puddings), 260 ppm (chewing gum)
Regulatory: FDA 21CFR §172.515
Manuf./Distrib.: Aldrich

Ethyl n-valerate. *See* Ethyl valerate

Ethyl vanillin
CAS 121-32-4; EINECS 204-464-7; FEMA 2464
Synonyms: 3-Ethoxy-4-hydroxybenzaldehyde; Bourbonal; Ethovan; Vanillal
Classification: Substituted phenolic
Empirical: $C_9H_{10}O_3$

Ethyl vinyl carbinol

Properties: Fine, wh. cryst., vanilla odor; sol in alc., chloroform, and ether; sl. sol. in water; m.w. 166.19; m.p. 76.5 C; b.p. 285 C; flash pt. > 212 F
Precaution: Flamm. liq.
Toxicology: Mod. toxic by ingestion, intraperitoneal, subcutaneous, and IV routes; human skin irritant; heated to decomp., emits acrid smoke and irritating fumes
Uses: Synthetic flavoring agent; migrating to foods from paper/paperboard
Usage level: 100 ppm (alcoholic beverages), 47 ppm (ice cream, ices), 65 ppm (candy), 63 ppm (baked goods), 74 ppm (gelatins, puddings), 110 ppm (chewing gum), 140-200 ppm (icings, toppings), 250 ppm (chocolate), 28000 ppm (imitation vanilla extracts)
Regulatory: FDA 21CFR §163.111, 163.112, 163.113, 163.114, 163.117, 163.123, 163.130, 163.135, 163.140, 163.145, 163.150, 163.153, 163.155, 182.60, 182.90; GRAS; Japan approved as flavoring
Manuf./Distrib.: Am. Fruit Processors; Berje; Calaga Food Ingreds.; Chart; H E Daniel; Dragoco Australia; Flavors of N. Am.; Foote & Jenkins; Ikeda; Penta Mfg.; Quimdis; Rhone-Poulenc Food Ingreds.; Soda Aromatic USA; R C Treatt; Van Waters & Rogers
Trade names: Rhodiarome™; Unisweet EVAN

Ethyl vinyl carbinol. *See* 1-Penten-3-ol
EtOH. *See* Alcohol

Eucalyptol
CAS 470-82-6; EINECS 207-431-5; FEMA 2465
Synonyms: Cineol; Cajeputol; 1,8-Epoxy-p-menthane; 1,3,3-Trimethyl-2-oxabicyclo[2.2.2]octane
Definition: Constituent of oil of eucalyptus
Empirical: $C_{10}H_{18}O$
Properties: Colorless liq., camphor-like odor, spicy cooling taste; misc. with alcohol, chloroform, ether, glac. acetic acid, oils; pract. insol. in water; m.w. 154.25; dens. 0.924 (20/4 C); m.p. > 1.5 C; b.p. 175-179 C; flash pt. 49 C; ref. index 1.458 (20 C)
Precaution: Combustible liq.
Toxicology: LD50 (oral, rat) 2480 mg/kg; poison by subcutaneous and intramuscular routes; mod. toxic by ingestion; experimental reproductive effects; heated to decomp., emits acrid smoke and fumes
Uses: Synthetic flavoring for bakery prods., beverages, chewing gum, confections, ice cream
Usage level: 0.13 ppm (nonalcoholic beverages), 0.5 ppm (ice cream, ices), 15 ppm (candy), 0.5-4 ppm (baked goods), 190 ppm (chewing gum)
Regulatory: FDA 21CFR §172.515; 27CFR §21.65, 21.151; GRAS (FEMA); Japan approved as flavoring
Manuf./Distrib.: Aldrich; Florida Treatt

Eucalyptus extract
CAS 84625-32-1; EINECS 283-406-2 (E. globulus extract)
Synonyms: Eucalyptus globulus extract
Definition: Extract of fresh leaves of *Eucalyptus globulus* and other species
Properties: Tonic, astringent
Uses: Natural flavoring agent; antioxidant (Japan)
Regulatory: FDA 21CFR §172.510; Japan approved

Eucalyptus globulus extract. *See* Eucalyptus extract

Eucalyptus oil
CAS 8000-48-4; FEMA 2466
Synonyms: Dinkum oil
Definition: Volatile oil obtained from leaves of *Eucalyptus globulus* or other species, contg. 70-80% eucalyptol
Properties: Colorless to pale yel. liq., char. camphoraceous odor, pungent spicy cooling taste; sol. in 5 vols 70% alcohol; misc. with abs. alcohol, oils, fats; pract. insol. in water; dens. 0.905-0.925; m.p. -15.4 C
Precaution: Keep cool, well closed; protect from light
Toxicology: LD50 (oral, rat) 2480 mg/kg; human poison and human systemic effects by ingestion; skin and eye irritant; heated to decomp., emits acrid smoke and irritating fumes
Uses: Natural flavoring agent
Usage level: 1.7 ppm (nonalcoholic beverages), 1 ppm (alcoholic beverages), 0.5-50 ppm (ice cream, ices), 130 ppm (candy), 76 ppm (baked goods)
Regulatory: FDA 21CFR §172.510, 27CFR §21.65, 21.151; Japan approved; Europe listed, no restrictions
Manuf./Distrib.: Chart; Florida Treatt

Eugenia caryophyllus extract. *See* Clove extract
Eugenia caryophyllus oil. *See* Clove oil
Eugenic acid. *See* Eugenol

Eugenol
CAS 97-53-0; EINECS 202-589-1; FEMA 2467

Synonyms: 4-Allyl-2-methoxyphenol; 2-Methoxy-4-(2-propenyl) phenol; Caryophyllic acid; Eugenic acid; 4-Allylguaiacol
Classification: Substituted phenol
Formula: $C_{10}H_{14}O_2$
Properties: Colorless or ylsh. liq., pungent clove odor; sol. in alcohol, chloroform, ether, volatile oils; very sl. sol. in water; m.w. 164.22; dens. 1.064-1.070; m.p. 10.3 C; b.p. 253.5 C; flash pt. 219 F; ref. index 1.540
Precaution: Combustible liq.
Toxicology: LD50 (oral, rat) 1930 mg/kg; mod. toxic by ingestion, intraperitoneal, and subcutaneous routes; experimental carcinogen, tumorigen; human mutagenic data; human skin irritant; heated to decomp., emits acrid smoke and irritating fumes
Uses: Synthetic flavoring agent and adjuvant; mfg. of vanillin
Usage level: 1.4 ppm (nonalcoholic beverages), 3.1 ppm (ice cream, ices), 32 ppm (candy), 33 ppm (baked goods), 0.6 ppm (gelatins, puddings), 500 ppm (chewing gum), 9.6-100 ppm (condiments), 40-2000 ppm (meats)
Regulatory: FDA 21CFR §177.2800, 184.1257, GRAS; 27CFR §21.65, 21.151; Japan approved as flavoring
Manuf./Distrib.: Aldrich; Florida Treatt

Eugenol acetate. *See* Eugenyl acetate
Eugenol benzoate. *See* Eugenyl benzoate
Eugenol formate. *See* Eugenyl formate

Eugenyl acetate
CAS 93-28-7; FEMA 2469
Synonyms: 4-Allyl-2-methoxyphenyl acetate; Eugenol acetate; Acetyl eugenol
Empirical: $C_{12}H_{14}O_3$
Properties: Semisolid mass, clove oil-like odor, burning aromatic flavor; m.w. 206.24; m.p. 29-30 C; b.p. 281-282 C (752 mm); ref. index 1.52069
Precaution: Combustible liq.
Toxicology: LD50 (oral, rat) 1670 mg/kg; mod. toxic by ingestion; skin irritant; heated to decomp., emits acrid smoke and fumes
Uses: Synthetic flavoring agent
Usage level: 0.43 ppm (nonalcoholic beverages), 33 ppm (ice cream, ices), 20 ppm (candy), 10 ppm (baked goods), 25-100 ppm (chewing gum), 2.0 ppm (condiments)
Regulatory: FDA 21CFR §172.515

Eugenyl benzoate
FEMA 2471
Synonyms: Benzoyl eugenol; Eugenol benzoate; 4-Allyl-2-methoxyphenyl benzoate
Empirical: $C_{17}H_{16}O_3$
Properties: Colorless cryst. solid, balsamic odor; sol. in alcohol, ether; insol. in water; m.w. 268.32; m.p. 69-70 C; b.p. 360 C
Uses: Synthetic flavoring agent
Usage level: 0.03-0.13 ppm (nonalcoholic beverages), 0.25-2.0 ppm (ice cream, ices), 0.25-10 ppm (candy, baked goods)
Regulatory: FDA 21CFR §172.515

Eugenyl formate
FEMA 2473
Synonyms: 4-Allyl-2-methoxyphenyl formate; Eugenol formate
Empirical: $C_{11}H_{12}O_3$
Properties: Orris-like odor; m.w. 192.22; dens. 1.120; b.p. 270 C; flash pt. 102 C; ref. index 1.5240-1.5265
Uses: Synthetic flavoring agent
Usage level: 0.20 ppm (condiments)
Regulatory: FDA 21CFR §172.515

Eugenyl methyl ether. *See* Methyl eugenol
European angelica extract. *See* Angelica extract
European dill seed oil. *See* Dill seed oil
European pennyroyal. *See* Pennyroyal extract
European pennyroyal oil. *See* Pennyroyal oil

Everlasting extract
CAS 90045-56-0
Definition: Extract derived from flowering plant *Helichyrsum italicum* and other species
Uses: Natural flavoring
Regulatory: FDA 21CFR §182.20, GRAS

Evernia prunastri extract. *See* Oakmoss extract
Ext. D&C Yellow No. 9. *See* 1-(Phenylazo)-2-naphthylamine
Extract of balm mint. *See* Balm mint extract
Extract of barley. *See* Barley extract
Extract of calendula. *See* Calendula extract
Extract of rosemary. *See* Rosemary extract
Extract of yeast. *See* Yeast extract

Farnesol
CAS 4602-84-0; EINECS 225-004-1; FEMA 2478
Synonyms: 3,7,11-Trimethyl-2,6,10-dodecatrien-1-ol; Farnesyl alcohol
Classification: Organic compd.
Empirical: $C_{15}H_{26}O$
Properties: Lt. yel. liq., mild oily odor; insol. in water; dens. 0.8871 (20/4 C); b.p. 111 C; ref. index 1.487-1.492
Precaution: Combustible
Toxicology: LD50 (oral, rat) 6000 mg/kg; mod. toxic by intraperitoneal route; mildly toxic by ingestion; mutagenic data; heated to decomp., emits acrid smoke and irritating fumes
Uses: Synthetic flavoring agent
Usage level: 0.76 ppm (nonalcoholic beverages), 0.4 ppm (ice cream, ices), 1.4 ppm (candy), 1.7 ppm (baked goods), 0.1 ppm (gelatins, puddings)
Regulatory: FDA 21CFR §172.515; GRAS (FEMA)

Farnesyl alcohol. *See* Farnesol
Fast Green FCF. *See* FD&C Green No. 3

Fatty acids
Toxicology: Heated to decomp., emits acrid smoke and fumes
Uses: Flavorings
Regulatory: Japan approved as flavoring

Fatty acids, soya. *See* Soy acid
Fatty acids, tallow. *See* Tallow acid

FD&C Blue No. 1
CAS 3844-45-9; EINECS 223-339-8; EEC E133
Synonyms: Acid Blue 9; CI 42090; Brilliant Blue FCF
Classification: Triphenylmethane color
Empirical: $C_{37}H_{36}N_2O_9S_3$ • 2Na
Properties: Greenish-blue powd., gran.; sol. (oz/gal): 52 oz propylene glycol, 36 oz glycerin, 25 oz dist. water, 2 oz 95% ethanol; sol. in ether, conc. sulfuric acid; m.w. 794.91
Toxicology: Experimental neoplastigen; mutagenic data; heated to decomp., emits very toxic fumes of NO_x, Na_2O, and SO_x
Uses: Food colorant for beverages, candy, confections, baked goods, ice cream, dessert powds.
Usage level: 16 mg/kg max. daily intake; ADI 0-12.5 mg/kg (EEC)
Regulatory: FDA 21CFR §74.101, 74.1101, 74.2101, 82.101; banned in France, Finland
Manuf./Distrib.: Crompton & Knowles; Hilton Davis; Tricon Colors

FD&C Blue No. 2
CAS 860-22-0; EINECS 212-728-8; EEC E132
Synonyms: 2-(1,3-Dihydro-3-oxo-5-sulfo-2H-indol-2-ylidene)-2,3-dihydro-3-oxo-1H-indole-5-sulfonic acid di-sodium salt; Indigotine; Indigo carmine; Disodium indigo-5,5-disulfonate; Food Blue 1; CI 73015
Classification: Indigoid color
Definition: Disodium salt of 5,5´-indigotin disulfonic acid
Empirical: $C_{16}H_8N_2O_8S_2$ • 2Na
Properties: Blue-brn. to red-brn. powd.; sol. in water, conc. sulfuric acid; sl. sol. in alcohol; m.w. 466.36
Precaution: Sensitive to light, oxidizing agents
Toxicology: LD50 (oral, rat) 2 g/kg; poison by IV route mod. toxic by ingestion, subcut. routes; experimental neoplastigen; mutagenic data; may cause brain tumors, asthma, rashes, hyperactivity; heated to decomp., emits very toxic fumes of SO_x, NO_x, Na_2O
Uses: Food colorant for candy, confections, beverages, dessert powds., pet foods; reagent for testing milk
Usage level: 7.8 mg/kg max. daily intake; ADI 0.5 mg/kg (EEC)
Regulatory: FDA 21CFR §74.102, 74.1102, 81.1; Europe listed; banned in Norway
Manuf./Distrib.: Crompton & Knowles; Hilton Davis; Tricon Colors

FD&C Green No. 3
CAS 2353-45-9; EINECS 219-091-5

Synonyms: Fast Green FCF; Food Green 3; CI 42053
Classification: Triphenylmethane color
Empirical: $C_{37}H_{34}O_{10}N_2S_3Na_2$
Properties: Red to brn.-violet powd.; sol. in water, conc. sulfuric acid, ethanol; m.w. 808.86
Toxicology: LD50 (oral, rat) > 2 g/kg; experimental neoplastigen; may cause bladder tumors; mutagenic data; heated to decomp., emits very toxic fumes of NO_x and SO_x
Uses: Food colorant for beverages, candy, confections, maraschino cherries
Usage level: 4.3 mg/kg max. daily intake
Regulatory: FDA 21CFR §74.203, 74.1203, 74.2203, 82.203; banned in EEC
Manuf./Distrib.: Crompton & Knowles; Tricon Colors

FD&C Red No. 2. *See* Amaranth

FD&C Red No. 3
CAS 16423-68-0; EINECS 240-474-8; EEC E127
Synonyms: 3′,6′-Dihydroxy-2′,4′,5′,7′-tetraiodospiro[isobenzofuran-1(3H),9′-(9H)xanthen]-3-one disodium salt; Erythrosine; Erythrosine bluish; Acid Red 51; Tetraiodofluorescein sodium salt; CI 45430; Food Red 14
Classification: Xanthene color
Empirical: $C_{20}H_6O_5I_4Na_2$
Properties: Bluish pink to brn. powd., gran.; sol. (oz/gal): 30 oz glycerin, 28 oz propylene glycol, 12 oz dist. water, 2 oz 95% ethanol; m.w. 879.86
Toxicology: LD50 (oral, rat) 1840 mg/kg; poison by IV route; mod. toxic by ingestion; human mutagenic data; may cause thyroid tumors, chromosomal damage, asthma, rashes, hyperactivity; heated to decomp., emits very toxic fumes of Na_2O and I.
Uses: Food colorant for candy, confections, dessert powds., maraschino cherries, baked goods, sausage, cereals, beverages, etc.
Usage level: 24 mg/kg max. daily intake; ADI 0-1.25 mg/kg (EEC)
Regulatory: FDA 21CFR §74.303, 74.1303, 81.1; Europe listed
Manuf./Distrib.: Crompton & Knowles; Hilton Davis; Tricon Colors

FD&C Red No. 4
CAS 4548-53-2; EINECS 224-909-9
Synonyms: 3-[(2,4-Dimethyl-5-sulfophenyl)azo]-4-hydroxy-1-naphthalenesulfonic acid disodium salt; Ponceau SX; Food Red 1; CI 14700
Classification: Monoazo color
Empirical: $C_{18}H_{14}N_2O_7S_2Na_2$
Properties: m.w. 480.42
Uses: Food colorant for maraschino cherries
Regulatory: FDA 21CFR §74.1304, 174.2304, 82.304
Manuf./Distrib.: Crompton & Knowles

FD&C Red No. 40
CAS 25956-17-6; EEC 129
Synonyms: 6-Hydroxy-5-[(2-methoxy-5-methyl-4-sulfophenyl)azo]-2-naphthalenesulfonic acid disodium salt; Food Red 17; NT red; Allura Red; CI 16035
Classification: Monoazo color
Empirical: $C_{18}H_{14}N_2O_8S_2Na_2$
Properties: Ylsh.-red powd., gran.; sol. (oz/gal): 26 oz dist. water, 4 oz glycerin, 2 oz propylene glycol; m.w. 496.42
Toxicology: Experimental reproductive effects; may cause lymph tumors; heated to decomp., emits very toxic fumes of NO_x and SO_x
Uses: Food colorant for beverages, pet foods, dessert powds., candy, confections, and cereals
Usage level: 100 mg/kg max. daily intake; ADI 0-0.7 mg/kg (EEC)
Regulatory: FDA 21CFR §74.340, 74.1340, 74.2340; banned in EEC, Japan, Norway, Sweden, Finland, Austria
Manuf./Distrib.: Crompton & Knowles; Hilton Davis; Tricon Colors
Trade names: Allura® Red AC

FD&C Yellow No. 3. *See* 1-(Phenylazo)-2-naphthylamine

FD&C Yellow No. 5. *See also* Tartrazine
CAS 1934-21-0; EINECS 217-699-5; EEC E102
Synonyms: Trisodium-3-carboxy-5-hydroxy-1-p-sulfophenyl-4-p-sulfophenylazopyrazole; Tartrazine; Acid Yellow 23; Food Yellow 4; CI 19140
Classification: Pyrazole color

FD&C Yellow No. 6

Empirical: $C_{16}H_9N_4O_9S_2Na_3$

Properties: Lemon yel. powd., gran.; greenish-yel. in sol'n.; sol. (oz/gal): 28 oz glycerin, 17 oz dist. water, 12 oz propylene glycol; m.w. 534.36

Precaution: Heated to dec., emits very toxic fumes of NO_x, SO_x, and Na_2O

Toxicology: LD50 (oral, mouse) 12,750 mg/kg; mildly toxic by ingestion; experimental teratogen, reproductive effects; human mutagenic data; may cause allergies, asthma, rashes, hyperactivity, thyroid tumors, lyphocytic lymphomas

Uses: Food colorant for beverages, baked goods, butter, cheese, candy, confections, dessert powds., cereals, ice cream, pet foods

Usage level: 43 mg/kg max. daily intake; ADI 0-7.5 mg/kg (EEC)

Regulatory: FDA 21CFR §74.705, 74.1705, 74.2705, 82.705; Europe listed; banned in Norway, Austria

Manuf./Distrib.: Crompton & Knowles; Hilton Davis; Tricon Colors

FD&C Yellow No. 6

CAS 2783-94-0; EINECS 220-491-7; EEC E110

Synonyms: 6-Hydroxy-5-[(4-sulfophenyl)azo]-2-naphthalenesulfonic acid disodium salt; Sunset Yellow; Food Yellow 3; CI 15985

Classification: Monoazo color

Definition: Disodium salt of 1-p-sulfophenylazo-2-naphthol-6-sulfonic acid

Empirical: $C_{16}H_{10}N_2O_7S_2Na_2$

Properties: Reddish-yel. powd., gran.; sol. (oz/gal): 23 oz dist. water, 14 oz glycerin, 2 oz propylene glycol; sol. in conc. sulfuric acid; sl. sol. in abs. alcohol; m.w 452.36

Toxicology: LD50 (IP, rat) 4600 mg/kg; mod. toxic by IP route; may cause allergies, kidney tumors, chromosomal damage; heated to decomp., emits very toxic fumes of NO_x and SO_x

Uses: Food colorant for beverages, sausages, candy, confections, dessert powds., cereals, baked goods, ice cream, pet foods

Usage level: 37 mg/kg max. daily intake

Regulatory: FDA 21CFR §74.706, 74.1706, 74.2706, 81.1, 82.706, 201.20; banned in Norway, Sweden

Manuf./Distrib.: Crompton & Knowles; Hilton Davis; Tricon Colors

2-Fenchanol. See Fenchyl alcohol
d-2-Fenchanone. See d-Fenchone
(+)-Fenchol. See Fenchyl alcohol
Fenchone. See d-Fenchone

d-Fenchone

FEMA 2479

Synonyms: d-1,3,3-Trimethyl-2-norbornanone; d-2-Fenchanone; Fenchone

Definition: Isolated from cedarleaf oil

Empirical: $C_{10}H_{16}O$

Properties: Colorless oily liq., camphor-like odor; sol. in alcohol, ether; insol. in water; m.w. 152.23; b.p. 193 C

Precaution: Combustible

Uses: Synthetic flavoring agent

Usage level: 0.13-0.80 ppm (nonalcoholic beverages), 5.0 ppm (alcoholic beverages), 0.25 ppm (ice cream, ices, baked goods), 0.25-30 ppm (candy)

Regulatory: FDA 21CFR §172.515

Manuf./Distrib.: Chr. Hansen's

Fenchyl alcohol

CAS 1632-73-1; EINECS 216-639-5; FEMA 2480

Synonyms: (1S)-1,3,3-Trimethylbicyclo[2.2.1]heptan-2-ol; (+)-Fenchol; 2-Fenchanol; 1,3,3-Trimethyl-2-norbornanol

Empirical: $C_{10}H_{18}O$

Properties: Camphor-like odor; sol. in alcohol; sl. sol. in water; m.w. 154.25; m.p. 40-43 C; b.p. 201-202 C; flash pt. 165 F; ref. index 1.473

Uses: Synthetic flavoring agent

Usage level: 1.8 ppm (nonalcoholic beverages), 0.25 ppm (ice cream, ices), 4.7 ppm (candy), 0.25 ppm (baked goods)

Regulatory: FDA 21CFR §172.515

Manuf./Distrib.: Aldrich; Chr. Hansen's

Fennel

Uses: Natural flavoring agent

Regulatory: Japan approved

Manuf./Distrib.: Chart

Fennel extract

CAS 84625-39-8; 85085-33-2; EINECS 283-414-6; FEMA 2481
Synonyms: Sweet fennel extract; Foeniculum extract; Foeniculum vulgare extract
Definition: Extract of the fruit of *Foeniculum vulgare*
Properties: Warm camphoraceous odor, bitter sl. burning flavor
Uses: Botanical; natural flavoring agent
Usage level: 800 ppm (nonalcoholic beverages), 300-6500 ppm (baked goods), 50 ppm (condiments), 2400 ppm (meats)
Regulatory: FDA 21CFR §182.20, GRAS
Manuf./Distrib.: Chart

Fennel oil

CAS 8006-84-6; EINECS 283-414-6 (extract)
Synonyms: Bitter fennel oil
Definition: Volatile oil from steam distillation of *Foeniculum vulgare*, contg. 50-60% anethole, 20% fenchone, pinene, limonene, dipentene, phellandrene
Properties: Colorless to pale yel. liq., fennel odor and taste; sol. in 1 vol 90% alcohol; very sol. in chloroform, ether; sl. sol. in water; dens. 0.953-0.973 (25/25 C); ref. index 1.5280-1.5380 (20 C)
Precaution: Keep cool, well closed; protect from light
Toxicology: LD50 (oral, rat) 3120 mg/kg; mod. toxic by ingestion; mutagenic data; severe skin irritant; heated to decomp., emits acrid smoke and irritating fumes
Uses: Natural flavoring agent
Usage level: 3.9 ppm (nonalcoholic beverages), 10-20 ppm (alcoholic beverages), 0.38 ppm (ice cream, ices), 22 ppm (candy), 19 ppm (baked goods), 0.1-10 ppm (gelatins, puddings), 2 ppm (condiments), 40-100 ppm (meats)
Regulatory: FDA 21CFR §182.10, 182.20, GRAS
Manuf./Distrib.: Chart

Fenugreek

EINECS 283-415-1 (extract)
Definition: From *Trigonella foenum-graecum*
Properties: Intensely sweet spicy protein-like aroma
Uses: Natural flavoring agent
Regulatory: FDA 21CFR §182.10, 182.20, GRAS, Japan approved
Manuf./Distrib.: C.A.L.-Pfizer; Chart; Pierre Chauvet (concrete, absolute)

Fermented soy sauce. *See* Soy sauce
Ferric ammonium citrate. *See* Iron ammonium citrate
Ferric ammonium citrate, green. *See* Iron ammonium citrate

Ferric chloride

CAS 7705-08-0 (anhyd.), 10025-77-1 (hexahydrate); EINECS 231-729-4
Synonyms: Iron chlorides; Iron trichloride; Iron (III) chloride; Ferric trichloride
Empirical: $FeCl_3$ (anhyd.), $FeCl_3 \cdot 6H_2O$ (hexahydrate)
Properties: Anhyd.: Black-brown solid; sol. in water, alcohol, glycerol, methanol, ether, acetone; readily absorbs water in air to form hexahydrate; m.w. 162.21; dens. 2.898; m.p. 292 C; b.p. 319 C; Hexahydrate: brnsh.-yel. cryst.; m.p. 37 C; dens. 1.82
Precaution: Corrosive; catalyzes potentially explosive polymerization of ethylene oxide, chlorine + monomers; violent reaction with allyl chloride; keep containers well closed
Toxicology: TLV:TWA 1 mg (Fe)/m³; LD50 (oral, rat) 1872 mg/kg; poison by IV; mod. toxic by ingestion; strong irritant to skin and tissue; heated to decomp., emits highly toxic fumes of HCl
Uses: Flavoring agent; dietary supplement (Japan)
Regulatory: FDA 21CFR §184.1297, GRAS; Japan approved
Manuf./Distrib.: Penta Mfg.

Ferric choline citrate

CAS 1336-80-7
Synonyms: Iron (III) choline citrate; Iron choline citrate complex; Ferrocholinate
Empirical: $C_6H_{10}FeO_{10} \cdot C_5H_{14}NO$
Properties: Greenish-brn., reddish-brn. or brn. amorphous solid with glistening surface on fracture; sol. in water, acids, alkalies; m.w. 402.21
Toxicology: LD50 (oral, mouse) 5500 mg/kg, (IP, mouse) 151 mg/kg, (IV, mouse) 210 mg/kg; poison by IV, IP routes; mildly toxic by ingestion; heated to decomp., emits toxic fumes of NO_x
Uses: Nutrient, dietary supplement; used as source of iron in foods for special dietary use
Regulatory: FDA 21CFR §172.370

Ferric citrate
CAS 3522-50-7 (anhyd.), 2338-05-8 (monohydrate); EINECS 219-045-4
Synonyms: Iron (III) citrate
Empirical: $C_6H_5FeO_7$ (anhyd.); $C_6H_5FeO_7 \cdot H_2O$ (monohydrate)
Properties: Garnet-red transparent scales or pale brn. powd., odorless, sl. ferruginous taste; slowly but completely sol. in cold water, readily sol. in hot water; pract. insol. in alcohol; m.w. 244.95 (anhyd.), 262.97 (monohydrate)
Toxicology: Heated to decomp., emits acrid smoke and irritating fumes
Uses: Nutrient supplement; may be used in infant formulas
Regulatory: FDA 21CFR §184.1298, GRAS; Japan approved

Ferric orthophosphate. *See* Ferric phosphate

Ferric oxide
CAS 1309-37-1 (anhyd.); EINECS 215-168-2
Synonyms: Ferric oxide red; Iron (III) oxide; Red iron trioxide; Ferrosoferric oxide
Empirical: Fe_2O_3
Properties: Red-brn. to blk. cryst.; sol. in acids; insol. in water, alcohol, ether; m.w. 159.69; dens. 5.240; m.p. 1538 C (dec.)
Precaution: Catalyzes the potentially explosive polymerization of ethylene oxide
Toxicology: TLV:TWA 5 mg (Fe)/m^3 (vapor, dust); LD50 (IP, rat) 5500 mg/kg; poison by subcutaneous route; suspected human carcinogen; experimental tumorigen
Uses: Color additive; used in cat and dog food; constituent of paperboard for pkg. materials
Regulatory: FDA 21CFR §73.200 (limitation 0.25%), 186.1300, 186.1374, GRAS as indirect food additive

Ferric oxide red. *See* Ferric oxide

Ferric phosphate
CAS 10045-86-0 (anhyd.), 14940-41-1 (tetrahydrate); EINECS 233-149-7, 239-018-0 (tetrahydrate)
Synonyms: Ferric orthophosphate; Ferric (III) phosphate; Ferriphosphate; Iron phosphate
Empirical: $FePO_4 \cdot xH_2O$, x = 1-4
Properties: Ylsh.-wh. to buff-colored powd.; sol. in HCl; slowly sol. in HNO_3; pract. insol. in water; m.w. 150.83; dens. 2.87 (dihydrate)
Toxicology: Heated to decomp., emits toxic fumes of PO_x
Uses: Nutrient, dietary supplement; bread enrichment; may be used in infant formulas; feed supplement
Regulatory: FDA 21CFR §182.5301, 184.1301, GRAS

Ferric (III) phosphate. *See* Ferric phosphate

Ferric pyrophosphate
CAS 10058-44-3 (anhyd.); EINECS 233-190-0
Synonyms: Iron (III) pyrophosphate
Empirical: $Fe_4(P_{207})_3 \cdot xH_2O$
Properties: Tan or ylsh. wh. powd.; sol. in min. acids; pract. insol. in water, acetic acid; m.w. 745.25 (anhyd.)
Toxicology: Heated to decomp., emits toxic fumes of PO_x
Uses: Nutrient, dietary supplement; source of nutritional iron; may be used in infant formulas
Regulatory: FDA 21CFR §182.5304, 184.1304, GRAS; Japan approved
Manuf./Distrib.: Lohmann

Ferric sesquisulfate. *See* Ferric sulfate

Ferric sodium pyrophosphate
CAS 35725-46-3 (anhyd.); EINECS 252-700-2
Synonyms: Iron (III) sodium pyrophosphate
Formula: $Na_8Fe_4(P_2O_7)_5 \cdot xH_2O$
Properties: m.w. 1277.00 (anhyd.)
Uses: Dietary supplement
Regulatory: FDA 21CFR §182.5306, GRAS
Manuf./Distrib.: Lohmann

Ferric sulfate
CAS 10028-22-5 (anhyd.); EINECS 233-072-9
Synonyms: Iron (III) sulfate; Iron sulfate (2:3); Ferrisulfate; Ferric trisulfate; Iron persulfate; Iron tersulfate; Ferric sesquisulfate
Empirical: $Fe_2O_{12}S_3$
Formula: $Fe(SO_4)_3$
Properties: Grayish-wh. to yel. powd. or cryst.; slowly sol. in water; sparingly sol. in alcohol; pract. insol. in acetone, ethyl acetate; m.w. 399.88; dens. 3.097 (18 C)

Precaution: Keep well closed and protected from light
Toxicology: TLV:TWA 1 mg/m³; STEL 2 mg/m³; heated to decomp., emits toxic fumes of SOₓ and Fe⁻
Uses: Flavoring agent
Regulatory: FDA 21CFR §184.1307, GRAS

Ferric trichloride. *See* Ferric chloride
Ferric trisulfate. *See* Ferric sulfate
Ferriphosphate. *See* Ferric phosphate
Ferrisulfate. *See* Ferric sulfate
Ferrocholinate. *See* Ferric choline citrate
Ferrosoferric oxide. *See* Ferric oxide

Ferrous ascorbate
CAS 14536-17-5
Synonyms: Iron (II) ascorbate
Properties: Blue-violet solid
Toxicology: Nuisance dust
Uses: Nutrient supplement; may be used in infant formulas
Regulatory: FDA 21CFR §184.1307a, GRAS

Ferrous carbonate
CAS 563-71-3
Empirical: CFeO₃
Formula: FeCO₃
Properties: Wh. solid, odorless; m.w. 115.86
Toxicology: Nuisance dust
Uses: Nutrient supplement; may be used in infant formulas
Regulatory: FDA 21CFR §184.1307b, GRAS

Ferrous citrate
CAS 23383-11-1 (anhyd.); EINECS 245-625-1
Synonyms: Iron (II) citrate
Empirical: C₆H₆FeO₇
Properties: Sl. colored powd. or wh. cryst.; sol. in water; insol. in alcohol; m.w. 245.96
Toxicology: Heated to decomp., emits acrid smoke and irritating fumes
Uses: Nutrient supplement; may be used in infant formulas
Regulatory: FDA 21CFR §184.1307c, GRAS

Ferrous fumarate
CAS 141-01-5; EINECS 205-447-7
Synonyms: Iron (II) fumarate
Definition: Contg. 31.3% min. total iron, ≤ 2% ferric iron
Empirical: C₄H₂O₄ • Fe
Formula: FeC₄H₂O₄
Properties: Reddish-orange to reddish-brn. gran. powd., odorless, almost tasteless; sol. 0.14 g/100 ml water, < 0.01 g/100 ml alcohol; m.w. 169.91; dens. 2.435;
Toxicology: LD50 (oral, rat) 3850 mg/kg; poison by intraperitoneal route; mod. toxic by ingestion and subcutaneous routes; heated to decomp., emits acrid smoke and irritating fumes; ACGIH TLV: TWA 1 mg/(Fe)/m³
Uses: Dietary supplement, nutrient, iron source; used in cereals, frozen waffles; may be used in infant formulas
Regulatory: FDA 21CFR §172.350, 184.1307d, GRAS
Manuf./Distrib.: Am. Roland; Int'l. Sourcing; Lohmann
Trade names containing: Cap-Shure® FF-165-60; Durkote Ferrous Fumarate/Hydrog. Veg. Oil

Ferrous gluconate
CAS 299-29-6 (anhyd.), 6047-12-7; EINECS 206-076-3
Synonyms: Iron (II) gluconate; Niconate
Empirical: C₁₂H₂₂O₁₄Fe • 2H₂O
Formula: Fe(C₆H₁₁O₇)₂•2H₂O
Properties: Yellowish-gray or pale greenish-yel. fine powd. or gran., sl. odor of caramel; sol. in water, glycerol; insol. in alcohol; m.w. 446.1 (anhyd.)
Precaution: Combustible
Toxicology: LD50 (oral, rat) 2237 mg/kg; poison by intraperitoneal and intravenous routes; moderately toxic by ingestion; experimental tumorigen and teratogen; heated to decomp., emits acrid smoke and irritating fumes; ACGIH TLV:TWA 1 mg(Fe)/m³
Uses: Mineral source, coloring, flavoring, nutrient, dietary supplement; improves and stabilizes coloring of

Ferrous lactate

> black olives; may be used in infant formulas
> *Regulatory:* FDA 21CFR §73.160, 182.5308, 182.8308, 184.1308, GRAS; Japan approved (0.15 g/kg max. as iron)
> *Manuf./Distrib.:* Am. Roland; Lohmann
> *Trade names:* Gluconal® FE

Ferrous lactate

> CAS 5905-52-2; EINECS 227-608-0
> *Synonyms:* Iron (II) lactate; Lactic acid iron (2+) salt (2:1)
> *Empirical:* $C_6H_{10}O_6$ • Fe (anhyd.); $C_6H_{10}FeO_6$ • $3H_2O$ (trihydrate)
> *Properties:* Anhyd.: m.w. 233.99; Trihydrate: Greenish-wh. powd., sl. char. odor, mild sweet ferruginous taste; sol. in water, alkali citrates; almost insol. in alcohol; m.w. 287.97
> *Precaution:* Combustible; keep well closed and protected from light
> *Toxicology:* TLV:TWA 1 mg(Fe)/m³; LD50 (oral, mouse) 147 mg/kg, (IV, rabbits) 287 mg/kg; poison by ingestion; experimental tumorigen; heated to decomp, emits acrid smoke and irritating fumes
> *Uses:* Nutrient, dietary supplement (mineral); may be used in infant formulas
> *Regulatory:* FDA 21CFR §182.5311, 182.8311, 184.1311, GRAS; Japan approved
> *Manuf./Distrib.:* Am. Roland

Ferrous pyrophosphate

> *Uses:* Dietary supplement (mineral)
> *Regulatory:* Japan approved

Ferrous sulfate

> CAS 7720-78-7
> *Synonyms:* Iron sulfate (ous); Iron (II) sulfate (1:1); Iron vitriol; Copperas; Green vitriol; Sal chalybis
> *Empirical:* O_4S • Fe (anhyd.), FeO_4S•$7H_2O$ (heptahydrate)
> *Properties:* Grayish wh. to buff powd.; slowly sol. in water; insol. in alcohol; m.w. 151.91; dens. 1.89; m.p. 64 C
> *Toxicology:* TLV:TWA 1 mg(Fe)/m³; LD50 (oral, rat) 319 mg/kg; human poison by ingestion; human systemic effects; experimental tumorigen, reproductive effects; mutagenic data; heated to decomp., emits toxic fumes of SO_x
> *Uses:* Dietary supplement, supplement
> *Regulatory:* FDA 21CFR §182.5315, 182.8315, GRAS; Japan approved
> *Manuf./Distrib.:* Int'l. Sourcing
> *Trade names containing:* Durkote Ferrous Sulfate/Hydrog. Veg. Oil; Rich-Pak®; Rich-Pak® Powd. 160; Rich-Pak® Powd. 160-M

Ferrous sulfate (dried)

> CAS 7720-78-7 (anhyd.); EINECS 231-753-5
> *Definition:* Consists primarily of ferrous sulfate monohydrate (CAS 17375-41-6) with varying amts. of ferrous sulfate tetrahydrate (CAS 20908-72-9)
> *Empirical:* $FeSO_4$
> *Properties:* Grayish-wh. to buff-colored powd.; sol. in water; loses water at 300 C; dec. at higher temps.; m.w. 151.91 (anhyd.)
> *Toxicology:* Human G.I. disturbances
> *Uses:* Nutrient supplement; may be used in infant formulas; feed supplement
> *Regulatory:* FDA 21CFR §184.1315, GRAS; Japan approved
> *Manuf./Distrib.:* Lohmann
> *Trade names containing:* Cap-Shure® FS-165-60

Ferrous sulfate heptahydrate

> CAS 7782-63-0; EINECS 231-753-5
> *Synonyms:* Iron (II) sulfate heptahydrate; Ferrous sulfate-7-hydrate; Copperas; Green vitriol; Iron vitriol
> *Empirical:* O_4S • Fe • $7H_2O$
> *Formula:* $FeSO_4$ • $7H_2O$
> *Properties:* Pale bluish-gr. cryst. or gran., odorless; sol. in water; pract. insol. in alcohol; m.w. 278.05; dens. 2.99-3.08
> *Precaution:* Incompat. with alkalies, sol. carbonates, Au and Ag salts, lime water, K and Na tartrate, tannin, etc.
> *Toxicology:* TLV:TWA 1 mg(Fe)/m³; LD50 (oral, mouse) 1.52 g/kg; poison by IV, IP, subcut. routes; mod. toxic by ingestion; human G.I. disturbances; mutagenic data; heated to decomp., emits toxic fumes of SO_x
> *Uses:* Clarifying agent, dietary/nutrient supplement, processing aid, stabilizer; for baking mixes, cereals, pasta prods., wine; may be used in infant formulas; feed supplement
> *Usage level:* Limitation 3 oz/1000 gal (wine)
> *Regulatory:* FDA 21CFR §182.5315, 182.8315, 184.1315, GRAS; BATF 27CFR §240.1051

Manuf./Distrib.: Lohmann

Ferrous sulfate-7-hydrate. *See* Ferrous sulfate heptahydrate
Ferula asafoetida extract. *See* Asafetida extract

Ficin
CAS 9001-33-6; EINECS 232-599-1
Synonyms: Debricin; Ficus protease; Ficus proteinase
Definition: Proteolytic enzyme in the crude latex of the fig tree *Ficus*
Properties: Wh. powd.; very sol. in water
Toxicology: LD50 (oral, rat) 10 g/kg; poison by inhalation and intravenous routes; mildly toxic by ingestion; heated to decomp., emits toxic fumes
Uses: Enzyme for chillproofing of beer, meat tenderizing, prep. of precooked cereals; processing aid; tenderizing agent; tissue softening agent
Regulatory: USDA 9CFR §318.7, 381.147; BATF 27CFR §240.1051, GRAS; Canada approved
Manuf./Distrib.: Am. Roland

Ficus protease. *See* Ficin
Ficus proteinase. *See* Ficin

Fir needle oil
EINECS 289-870-2 (extract)
Synonyms: Silver pine oil; Silver fir oil
Definition: Volatile oil from needles and young twigs of *Abies alba*, contg. l-pinene, l-limonene, l-bornyl acetate
Properties: Colorless clear liq., balsamic odor, terebinthinate taste; sol. in 5 vols 90% alcohol, in ether; insol. in water; dens. 0.869-0.875 (15/15 C)
Precaution: Keep cool, well closed; protect from light
Uses: Natural flavoring agent

Fir needle oil, Siberian type
CAS 8021-29-2; EINECS 294-351-9 (extract); FEMA 2905
Synonyms: Pine needle oil
Definition: Oil derived from needles and twigs of *Abies sibirica*
Properties: Colorless to pale yel. liq., aromatic odor, pungent taste; sol. in equal vol. 90% alcohol; dens. 0.905-0.925 (15/15 C); ref. index 1.4660-1.476 (20 C)
Precaution: Keep cool, well closed; protect from light
Toxicology: LD50 (oral, rat) 10200 mg/kg; mildly toxic by ingestion; skin irritant; heated to decomp., emits acrid omoke and fumes
Uses: Natural flavoring agent
Usage level: 1.5 ppm (nonalcoholic beverages), 0.62 ppm (ice cream, ices), 5.2 ppm (candy), 2.7 ppm (baked goods)
Regulatory: FDA 21CFR §172.510

Fir-wood oil. *See* Pine needle oil, Scotch type
Fischer-Tropsch wax. *See* Synthetic wax
Fischer-Tropsch wax, oxidized. *See* Synthetic wax

Fish protein isolate
Definition: Consists principally of dried fish protein prepared from edible portions of fish
Uses: Food supplement
Regulatory: FDA 21CFR §172.340
Manuf./Distrib.: Am. Roland

Flavaxin. *See* Riboflavin

Flavoxanthin
CAS 512-29-8; EEC E161a
Definition: Carotenoid pigment
Empirical: $C_{40}H_{56}$
Properties: Golden-yel. prisms; freely sol. in chloroform, benzene, acetone; less sol. in methanol, ethanol; almost insol. in petrol. ether; m.w. 584.85; m.p. 184 C
Uses: Colorant
Regulatory: UK approved

Flaxseed oil. *See* Linseed oil
Flour, soy. *See* Soy flour
Flowers of zinc. *See* Zinc oxide
Floxin isovalerate. *See* α-Amylcinnamyl isovalerate

Fluorocarbon-12. *See* Dichlorodifluoromethane
Fluorocarbon 113. *See* Trichlorotrifluoroethane
Foeniculum extract. *See* Fennel extract
Foeniculum vulgare extract. *See* Fennel extract
Folacin. *See* Folicacid
Folate. *See* Folicacid

Folicacid
  CAS 59-30-3; EINECS 200-419-0
  Synonyms: Folacin; Folate; Pteroylglutamic acid; Vitamin Bc; Vitamin M
  Empirical: $C_{19}H_{19}N_7O_6$
  Properties: Orange-yel. need. or platelets, odorless; sol. in dil. alkali hydroxide, carbonate sol'ns.; sl. sol. in water; insol. in lipid solvs., acetone, alcohol, ether; m.w. 441.45
  Toxicology: LD50 (IV, rat) 500 mg/kg; poison by intraperitoneal and IV routes; experimental teratogen, reproductive effects; mutagenic data; heated to decomp., emits toxic fumes of NO_x
  Uses: Nutrient, dietary/vitamin supplement
  Usage level: Limitation 0.4 mg daily intake, 0.1 mg (infants), 0.3 mg (children under 4 yrs), 0.8 mg (pregnant or lactating women)
  Regulatory: FDA 21CFR §172.345; Japan approved
  Manuf./Distrib.: Am. Roland; Unipex
  Trade names: Folic Acid USP, FCC No. 20383
  Trade names containing: Folic Acid 10% Trituration No. 69997

Food Blue 1. *See* FD&C Blue No. 2
Food Freezant 12. *See* Dichlorodifluoromethane
Food of the gods. *See* Asafetida
Food Green 3. *See* FD&C Green No. 3
Food Orange 5. *See* Carotene
Food Orange 6. *See* β-Apo-8´-carotenal
Food Orange 8. *See* Canthaxanthine
Food Red 1. *See* FD&C Red No. 4
Food Red 14. *See* FD&C Red No. 3
Food Red 17. *See* FD&C Red No. 40

Food starch modified
  CAS 53124-00-8, 65996-62-5, 65996-63-6
  Synonyms: Modified food starch
  Properties: Wh. powd., odorless, tasteless; insol. in water, alcohol, ether, chloroform
  Toxicology: Heated to decomp., emits acrid smoke and irritating fumes
  Uses: Thickener, colloidal stabilizer, binder for foods
  Regulatory: FDA 21CFR §172.892, 178.3520
  Manuf./Distrib.: ADM; Avebe BV; Cargill; Cerestar Int'l.; Grain Processing; Humphrey; Nat'l. Starch & Chem.; Norba; Penwest Foods; J L Priestley; Roquette UK; Sandoz Nutriuton; A E Staley Mfg.; Zumbro
  Trade names: Batter Up®; Binasol™ 15; Binasol™ 81; Binasol™ 90; Confectioners F & G; Consista®; Delta™ SD 7393; Dress'n® 300; Dress'n® 400; 400 Stabilizer®; Freezist® M; Fri-Bind® 411; Fruitfil® 1; Gelatinized Dura-Jel®; Gelex® Instant Starch; Hi-Jel™ S; Instant Tender-Jel® 419; Instant Tender-Jel® 434; Instant Tender-Jel® 479; Instant Tender-Jel® 480; Instant Tender-Jel® C; Kol Guard® 7373; Kol Guard® 7413; Lo-Temp® 452; Lo-Temp® 588; Maxi-Gel® 445; Maxi-Gel® 542; Maxi-Gel® 7776; Maximaize® 7360; Maximaize® 7367; Maximaize® DJ; Maximaize® HV; Mira-Bake® 505; Mira-Cap®; Mira-Cleer® 187; Mira-Cleer® 340; Mira-Cleer® 516; Mira-Cleer® 516; Mira-Quik® MGL; Mira-Set® 285; Mira-Set® J; Mira-Sperse®; Mira-Sperse® 535; Mira-Sperse® 623; Mira-Sperse® 626; Mira-Thik® 468; Mira-Thik® 469; Mira-Thik® 603; Mira-Thik® 606; Nu-Col™ 4227; Nustar®; Perma-Flo®; Polar-Gel® 5; Polar-Gel® 8; Polar-Gel® 10; Polar-Gel® 13; Polar-Gel® 15; Polar-Gel® 18; Pure-Bind® B910; Pure-Bind® B923; Pure-Dent® B890; Pure-Gel® B992; Pure-Set® B950; Pure-Set™ B965; Redisol® 78D; Redisol® 88; Redisol® 248; Redisol® 412; Redi-Tex®; Rezista®; 721-A™ Instant Starch; Shur-Fil® 427; Soft-Set®; Staley® Dusting Starch; Sta-Mist® 365; Sta-Mist® 454; Sta-Mist® 7415; Sta-O-Paque®; Starco™ 447; Starco™ 401; Sta-Slim™ 142; Sta-Slim™ 143; Sta-Slim™ 150; Sta-Slim™ 151; Sta-Slim™ 171; Stellar®; Tenderfil® 8; Tenderfil® 9; Tenderfil® 428; Tenderfil® 473A; Thin-N-Thik® 99; X-Pand'R®
  Trade names containing: Canthaxanthin Beadlets 10%; Dry Beta Carotene Beadlets 10% CWS No. 65633; Dry Beta Carotene Beadlets 10% No. 65661; Dry Beta Carotene Beadlets Yellow Type 2.4-S No. 653800100; Dry Vitamin A Palmitate Type 250-SD No. 65378; Dry Vitamin D_3 Beadlets Type 850 No. 652550401, 652550601; Frimulsion Q8; Lite Pack 350; Mayonat DF; Palma-Sperse® Type 250-S No. 65322; Palma-Sperse® Type 250A/50 D-S No. 65221; PAV; Roxanthin® Red 10WS No. 66515; Type B Dough Conditioner; Vitamin A Palmitate Type 250-CWS No. 65312; Vitamin B_{12} 0.1% SD No. 65354;

Vitamin B$_{12}$ 1.0% SD No. 65305

Food Yellow 3. *See* FD&C Yellow No. 6
Food Yellow 4. *See* FD&C Yellow No. 5
Food Yellow 10. *See* 1-(Phenylazo)-2-naphthylamine

Formaldehyde
CAS 50-00-0; EINECS 200-001-8
Synonyms: Oxymethylene; Formalin; Formic aldehyde; Aldehyde C-1; Methanal
Empirical: CH$_2$O
Formula: HCHO
Properties: Gas, strong pungent odor; avail. commercially as aq. sol'ns. (37-50% in methanol); sol. in water, alcohol; m.w. 30.03; dens. 1.083; b.p. -19 C; f.p. -118 C
Precaution: Combustible liq. when exposed to heat, or flame; mod. fire risk; explosive limits 7-73%; can react vigorously with oxidizers
Toxicology: TLV 1 ppm in air; LD50 (oral, rat) 800 mg/kg; human poison by ingestion; toxic by inhalation; human systemic effects, skin/eye irritant, mutagenic data; strongly irritating; suspected human carcinogen; heated to decomp., emits acrid smoke and fumes
Uses: Preservative in defoaming agents containing dimethicone, defoamers for beet sugar and yeast processing; animal glue adjuvant
Regulatory: FDA 21CFR §173.340 (1% of dimethicone content), 175.105, 175.210, 176.170, 1876.180, 176.200, 176.210, 177.2410, 177.2800, 178.3120, 573.460
Manuf./Distrib.: Hoechst Celanese; Monsanto

Formalin. *See* Formaldehyde

Formic acid
CAS 64-18-6; EINECS 200-579-1; FEMA 2487; EEC E236
Synonyms: Hydrogen carboxylic acid; Methanoic acid; Aminic acid
Classification: Organic acid
Empirical: CH$_2$O$_2$
Formula: HCOOH
Properties: Colorless fuming liq., penetrating odor; sol. in water, alcohol, ether; m.w. 46.03; dens. 1.22 (20/4 C); m.p. 8.3 C; b.p. 100.8 C; flash pt. (OC) 69 C; ref. index 1.3714
Precaution: Flamm.; corrosive material; can react vigorously with oxidizers, explosive with furfuryl alcohol
Toxicology: LD50 (oral, rat) 1100 mg/kg; poison by IV route; mod. toxic by ingestion, intraperitoneal routes; skin and severe eye irritant; TLV 5 ppm in air; heated to decomp., emits acrid smoke and irritating fumes
Uses: Synthetic flavoring adjuvant, antibacterial preservative; constituent of paper/paperboard used for food pkg.
Usage level: 1 ppm (nonalcoholic beverages), 5 ppm (ice cream, ices), 5-18 ppm (candy), 5-6.1 ppm (baked goods); ADI 0-3 mg/kg (EEC)
Regulatory: FDA 21CFR §172.515, 186.1316, GRAS as indirect additive, 573.480; GRAS (FEMA); Europe listed; prohibited in UK
Manuf./Distrib.: Aldrich; Hoechst Celanese

Formic acid benzyl ester. *See* Benzyl formate
Formic acid butyl ester. *See* Butyl formate
Formic acid calcium salt. *See* Calcium formate
Formic acid propyl ester. *See* Propyl formate
Formic acid sodium salt. *See* Sodium formate
Formic aldehyde. *See* Formaldehyde
Formic ether. *See* Ethyl formate
4-Formyl-2-methoxyphenyl acetate. *See* Vanillin acetate
Fossil wax. *See* Ozokerite
Frambinone. *See* 4-(p-Hydroxyphenyl)-2-butanone
Frankincense. *See* Olibanum
Fraxinella. *See* Dittany
French chalk. *See* Talc
Freon C-318. *See* Octafluorocyclobutane

β-D-Fructofuranosyl-α-D-glucopyranoside. *See* Sucrose

Fructose
CAS 57-48-7 (D-), 7660-25-5; EINECS 200-333-3 (D-)
Synonyms: Levulose; D-Fructose; Fruit sugar; Laevosan
Definition: Sugar occurring in fruit and honey

D-Fructose

Empirical: $C_6H_{12}O_6$

Properties: Wh. hygroscopic cryst. or cryst. powd., odorless, sweet taste; sol. in methanol, ethanol, water, pyridine; m.w. 180.18; dens. 1.6; m.p. 103-105 C

Toxicology: Experimental tumorigen; heated to decomp., emits acrid smoke and fumes

Uses: Nutritive sweetener, processing aid, formulation aid for foods; flavor enhancer in fruits and berries; prevents sandiness in ice cream; used in detetic and diabetic foods, cereals, canned fruits, cakes, powd. juices, frozen prods.

Regulatory: GRAS

Manuf./Distrib.: Advanced Sweeteners; Am. Roland; British Arkady; Cargill; Cerestar Int'l.; Corn Prods.; Dasco Sales; Farbest Brands; Fruitsource; Maruzen Fine Chems.; MLG Enterprises; Penta Mfg.; Siber Hegner; A.E. Staley Mfg.; Xyrofin UK

Trade names: CronSweet® Crystalline Fructose; Fructofin® C

D-Fructose. *See* Fructose

Fruit sugar. *See* Fructose

Fuller's earth. *See* Attapulgite

Fumaric acid

CAS 110-17-8; EINECS 203-743-0; FEMA 2488; EEC E297

Synonyms: Allomaleic acid; Boletic acid; trans-1,2-Ethylenedicarboxylic acid; 2-Butenedioic acid; trans-Butenedioic acid; Lichenic acid

Classification: Dicarboxylic acid

Empirical: $C_4H_4O_4$

Formula: HOOCCH:CHCOOH

Properties: Wh. cryst., odorless; sol. in water, ether; very sl. sol. in chloroform; m.w. 116.08; dens. 1.635 (20/4 C); m.p. 287 C; b.p. 290 C

Precaution: Combustible when exposed to heat or flame; can react vigorously with oxidizers

Toxicology: LD50 (oral, rat) 10,700 mg/kg; poison by intraperitoneal route; mildly toxic by ingestion and skin contact; skin and eye irritant; heated to decomp., emits acrid smoke and irritating fumes

Uses: Food additive, acidulant, flavoring agent, leavening agent, dietary supplement, antioxidant

Usage level: 50 ppm (nonalcoholic beverages), 1300 ppm (baked goods), 3600 ppm (gelatins, puddings); ADI 0-6 mg/kg (EEC)

Regulatory: FDA 21CFR §172.350; USDA 9CFR §318.7, 381.147; BATF 27CFR §240.1051, limitation 25 lb/1000 gal wine; GRAS (FEMA); Japan approved; Europe listed; UK approved

Manuf./Distrib.: AB Tech.; Am. Roland; Ashland; Bartek; Browne & Dureau Int'l.; Browning; Cheil Foods UK; Cornelius; Forum Chems.; Haarmann & Reimer; Int'l. Sourcing; Jungbunzlauer; W Kündig; Penta Mfg.; Pfizer; Quimdis; Tesco; Thymly Prods.; Unipex; Welltep Int'l.

Trade names containing: Cap-Shure® F-125-85; Cap-Shure® F-140-63; Durkote Fumaric Acid/Hydrog. Veg. Oil; Durkote Glucono-Delta-Lactone/Hydrog. Veg. Oil; GP-1200

Fumed silica. *See* Silica

2-Furaldehyde. *See* Furfural

2-Furancarboxaldehyde. *See* Furfural

Furan-2-carboxylic acid. *See* 2-Furoic acid

2-Furanmethanol. *See* Furfuryl alcohol

Furcelleran

CAS 9000-21-9; EINECS 232-531-0; EEC E408

Synonyms: Furcelleran gum; Danish agar

Definition: Refined hydrocolloid obtained by aq. extraction of *Furcellaria fastigiata* of the class of red seaweed, *Rodophyceae*

Properties: Wh. powd., odorless; sol. in warm water

Toxicology: LD50 (oral, rat) 5000 mg/kg; mod. toxic by ingestion; heated to decomp., emits acrid smoke and fumes

Uses: Emulsifier, stabilizer, thickener for flans, gelled meat prods., jams, jellies, milk puddings

Regulatory: FDA 21CFR §172.655; Japan approved

Furcelleran gum. *See* Furcelleran

Furfural

CAS 98-01-1; EINECS 202-627-7; FEMA 2489

Synonyms: 2-Furaldehyde; 2-Furancarboxaldehyde; Artificial oil of ants

Classification: Cyclic aldehyde

Empirical: $C_5H_4O_2$

Formula: C_4H_3OCHO

Properties: Colorless liq. (pure); reddish-brown (on exposure to air and light); almond-like odor; sol. in alcohol, ether, benzene, 8.3% in water; m.w. 96.08; dens. 1.1598 (20/4 C); f.p. -36.5 C; b.p. 161.7 C; ref. index

1.5260 (20 C); flash pt. 60 C
Precaution: Flamm. or combustible
Toxicology: LD50 (oral, rat) 65 mg/kg; poison by ingestion, intraperitoneal, subcutaneous routes; moderately toxic by inhalation; TLV 2 ppm in air; irritates mucous membranes, acts on CNS
Uses: Synthetic flavoring agent
Usage level: 4 ppm (nonalcoholic beverages), 10 ppm (alcoholic beverages), 13 ppm (ice cream, ices), 12 ppm (candy), 17 ppm (baked goods), 0.8 ppm (gelatins, puddings), 45 ppm (chewing gum), 30 ppm (syrups)
Regulatory: FDA 21CFR §175.105; GRAS (FEMA); Japan approved as flavoring
Manuf./Distrib.: Aldrich

Furfuralcohol. *See* Furfuryl alcohol

Furfural mercaptan
CAS 98-20-2; EINECS 202-628-2; FEMA 2493
Empirical: C_5H_6OS
Properties: Fish oily smokey strong coffee-like odor; m.w. 114.17; b.p. 155 C
Uses: Flavoring
Manuf./Distrib.: Aldrich

Furfuryl alcohol
CAS 98-00-0; EINECS 202-626-1; FEMA 2491
Synonyms: 2-Furanmethanol; 2-Furylcarbinol; Furfuralcohol
Empirical: $C_5H_6O_2$
Formula: $C_4H_3OCH_2OH$
Properties: Colorless mobile liq., brn.-dk. red (air/lt. exposed), low odor, cooked sugar taste; sol. in alc., chloroform, benzene; misc. with water but unstable; m.w. 98.10; dens. 1.1285 (20/4 C); m.p. -29 C; b.p. 170 C; flash pt. (OC) 75 C; ref. index 1.485
Toxicology: TLV 10 ppm in air; TLV:TWA 2 ppm (skin); LD50 (rat, oral) 275 mg/kg; poisonous
Uses: Synthetic flavoring
Regulatory: GRAS (FEMA)
Manuf./Distrib.: Aldrich

Furnace black. *See* Carbon black

2-Furoic acid
CAS 88-14-2; EINECS 201-803-0
Synonyms: Furan-2-carboxylic acid; Pyromucic acid
Empirical: $C_5H_4O_3$
Properties: Monoclinic prisms; sol. 1 g/26 ml water (15 C); sol. in alcohol, ether; m.w. 112.09; m.p. 128-130 C; b.p. 230-232 C (760 mm)
Uses: Furoates for flavoring and perfume

Furyl acetone. *See* (2-Furyl)-2-propanone
2-Furylcarbinol. *See* Furfuryl alcohol

(2-Furyl)-2-propanone
Synonyms: Furyl acetone
Uses: Synthetic flavoring
Regulatory: FDA 21CFR §172.515

Fusel oil refined
CAS 8013-75-0; EINECS 232-395-2; FEMA 2497
Definition: Mixed amyl alcohols
Properties: Water-wh. liq. disagreeable odor; sol. in water, alcohol, ether; flash pt. 123 F
Precaution: Combustible, flammable liq.
Toxicology: Mutagenic data; suspected of containing carcinogens; reacts with oxidizing materials; heated to decomp., emits acrid smoke and fumes
Uses: Synthetic flavoring agent
Usage level: 21 ppm (nonalcoholic beverages), 2.5 ppm (alcoholic beverages), 4.1 ppm (ice cream, ices), 30 ppm (candy), 34 ppm (baked goods), 4.0 ppm (gelatins, puddings), 290 ppm (chewing gum)
Regulatory: FDA 21CFR §172.515

GA₃. *See* Gibberellic acid
4-O-β-D-Galactopyranosyl-D-glucitol. *See* Lactitol monohydrate
4-O-β-Galactopyranosyl D-glucose. *See* Lactose

α-Galactosidase
EINECS 232-792-0

Galanga

Definition: Derived from Mortierella vinaceae
Toxicology: Heated to decomp, emits acrid smoke and irritating fumes
Uses: Enzyme used in prod. of sugar from sugar beets; production aid
Regulatory: FDA 21CFR §173.145; Japan approved

Galanga

EINECS 291-068-2 (extract)
Synonyms: Galangal; Chinese ginger
Definition: From Alpinia officinarum
Properties: Bitter aromatic taste
Uses: Natural flavoring agent
Regulatory: FDA 21CFR §182.10, 182.20, GRAS; Japan approved

Galangal. See Galanga

Gallic acid

CAS 149-91-7; EINECS 205-749-9
Synonyms: 3,4,5-Trihydroxybenzoic acid
Empirical: $C_7H_6O_5$
Formula: $C_6H_2(OH)_3CO_2H$
Properties: Need.; sol. (1 g/ml): 87 ml water, 6 ml alcohol, 100 ml ether, 10 ml glycerin, 5 ml acetone; pract.
    insol. in benzene, chloroform, petroleum ether; m.w. 170.12; dens. 1.694; m.p. 258-265 C (dec.)
Precaution: Protect from light
Toxicology: LD50 (oral, rabbit) 5 g/kg
Uses: Food additive; natural antioxidant (Japan)
Regulatory: Japan approved
Manuf./Distrib.: Penta Mfg.

Gallic acid propyl ester. See Propyl gallate
Gallotannic acid. See Tannic acid
Gallotannin. See Tannic acid
Gambir. See Catechu, pale

Garlic extract

CAS 8008-99-9; EINECS 232-371-1
Synonyms: Allium sativum extract
Definition: Extract derived from Allium sativum
Properties: Pungent acrid aromatic garlic odor
Uses: Natural flavoring agent and adjuvant; processing aid (Japan)
Usage level: 0.01-0.3 ppm (beverages), 40 ppm (ice cream, ices), 2.9 ppm (candy), 6 ppm (baked goods), 12
    ppm (chewing gum), 16 ppm (condiments)
Regulatory: FDA 21CFR §184.1317, GRAS; Japan restricted
Manuf./Distrib.: Chart; Commodity Services

Garlic juice

Definition: Juice from the garlic bulb Albium sativum
Uses: As a spice and seasoning in food
Trade names: Veg-Juice Garlic

Garlic oil

EINECS 232-371-1 (extract); FEMA 2503
Synonyms: Allium sativum oil
Definition: Volatile oil from bulb or entire plant Allium sativum
Properties: Yel. liq., strong garlic odor; dens. 1.046-1.057 (15/15 C)
Uses: Natural flavoring agent for seasoning blends
Regulatory: GRAS (FEMA); Europe listed, no restrictions
Manuf./Distrib.: Florida Treatt

Gas plant. See Dittany
GDL. See Gluconolactone

Gelatin

CAS 9000-70-8; EINECS 232-554-6
Synonyms: Gelatine
Definition: Obtained from partial hydrolysis of collagen derived from animal skin, connective tissues, and bones
Properties: Flake or powd., tasteless, odorless; sol. in warm water, glycerol, acetic acid; insol. in org. solvs.;
    amphoteric
Toxicology: LD50 (rat, oral) 5 g/kg

Uses: Clarifying agent; protective colloid in ice cream; stabilizer, thickener, gelling agent, texturizer in food; dietary supplements; chewing gum base; migrating to foods from cotton in dry food pkg.
Usage level: ADI not specified (FAO/WHO)
Regulatory: FDA 21CFR §133.133, 133.134, 133.162, 133.178.133.179, 182.70; GRAS; Japan approved
Manuf./Distrib.: Alfa; Browning; Cheil Foods UK; Chemcolloids Ltd.; Croda Food; Dena AG; DynaGel; Ellis & Everard; Foodtech; Gelatine Prods.; Hormel Foods; N I Ibrahim; Lucas Meyer; Penta Mfg.; Sanofi; Siber Hegner; Spice King; Vyse Gelatine; World Trade Service
Trade names: Bone Gelatin Type B 200 Bloom; Calfskin Gelatin Type B 175 Bloom; Calfskin Gelatin Type B 200 Bloom; Calfskin Gelatin Type B 225 Bloom; Calfskin Gelatin Type B 250 Bloom; Crodyne BY-19; Edible Beef Gelatin; Edible Gelatins; Flavorset® GP-2; Flavorset® GP-3; Flavorset® GP-4; Flavorset® GP-5; Flavorset® GP-6; Flavorset® GP-7; Flavorset® GP-8; Flavorset® GP-9; Flavorset® GP-10; Gelatin XF; Gummi Gelatin P-5; Gummi Gelatin P-6; Gummi Gelatin P-7; Gummi Gelatin P-8; Liquid Fish Gelatin Conc.; Margarine Gelatin P-8; Quickset® D-4; Quickset® D-5; Quickset® D-6; Quickset® D-7; Quickset® D-8; Quickset® D-9; Quickset® D-10; Spa Gelatin; Spray Dried Fish Gelatin; Spray Dried Hydrolysed Fish Gelatin; Tenderset® M-7; Tenderset® M-8; Tenderset® M-9
Trade names containing: Canthaxanthin 10% Type RVI No. 66523; Canthaxanthin Beadlets 10%; Dry Beta Carotene Beadlets 10% CWS No. 65633; Dry Beta Carotene Beadlets 10% No. 65661; Dry Beta Carotene Beadlets Yellow Type 2.4-S No. 653800100; Dry Canthaxanthin 10% SD No. 66514; Dry Vitamin D_3 Beadlets Type 850 No. 652550401, 652550601; Dry Vitamin D_3 Type 100 CWS No. 65242; Dry Vitamin E Acetate 50% Type CWS/F No. 652530001; Frimulsion Q8; Frimulsion RA; Frimulsion RF; Palma-Sperse® Type 250-S No. 65322; Palma-Sperse® Type 250A/50 D-S No. 65221; Roxanthin® Red 10WS No. 66515; Spray Dried Fish Gelatin/Maltodextrin; Tenderset® MWA-5; Tenderset® MWA-7; Tenderset® MWA-8

Gelatine. *See* Gelatin

Gellan gum
CAS 71010-52-1; EINECS 275-117-5; EEC E418
Synonyms: Gum gellan
Definition: High m.w. heteropolysaccharide gum produced by pure-culture fermentation of a carbohydrate with *Pseudomonas elodea*
Uses: Stabilizer, thickener, gelling agent for use in foods, pet foods
Regulatory: FDA 21CFR §172.665; Japan, JECFA, Canada approvals
Manuf./Distrib.: Kelco Int'l
Trade names: Kelcogel®; Kelcogel® BF; Kelcogel® BF10; Kelcogel® CF; Kelcogel® CF10; Kelcogel® F; Kelcogel® IF; Kelcogel® JJ; Kelcogel® PD

Gelose. *See* Agar

Gentian
Synonyms: Bitter root; Gentiana lutea
Definition: Dried rhizome and roots of *Gentiana lutea*
Uses: Natural flavoring
Regulatory: FDA 21CFR §172.510
Manuf./Distrib.: Chart

Gentiana lutea. *See* Gentian
Gentiana lutea extract. *See* Gentian extract

Gentian extract
CAS 97676-22-7; EINECS 289-890-1, 289-891-7, 289-892-2, 277-139-0, etc.
Synonyms: Gentiana lutea extract; Gentian root extract
Definition: Extract of rhizomes and roots of various species of *Gentiana*
Uses: Natural flavoring; antioxidant, bittering agent (Japan)
Regulatory: FDA 21CFR §172.510; not listed as approved colorant for cosmetics under FDA 21CFR §73 and 74; Japan approved
Manuf./Distrib.: Chart

Gentian root extract. *See* Gentian extract
Geranial. *See* Citral

Geraniol
CAS 106-24-1; EINECS 203-377-1; FEMA 2507
Synonyms: trans-3,7-Dimethyl-2,6-octadien-1-ol; Geranyl alcohol; Guaniol
Classification: A terpene alcohol
Empirical: $C_{10}H_{18}O$
Properties: Colorless to pale yel. oily liq., pleasant geranium odor; sol. in fixed oils, propylene glycol; sl. sol.

in water; insol. in glycerin; m.w. 154.28; dens. 0.870-0.890 (15 C); m.p. 15 C; b.p. 230 C; flash pt. 214 F; ref. index 1.469-1.478
Precaution: Combustible
Toxicology: LD50 (oral, rat) 3600 mg/kg; poison by IV route; mod. toxic by ingestion and intramuscular routes; heated to decomp., emits acrid smoke and irritating fumes
Uses: Synthetic flavoring agent
Usage level: 2.1 ppm (nonalcoholic beverages), 3.3 ppm (ice cream, ices), 10 ppm (candy), 11 ppm (baked goods), 2 ppm (gelatins, puddings), 0.8-2.9 ppm (chewing gum), 1 ppm (toppings)
Regulatory: FDA 21CFR §182.60, GRAS; Japan approved as flavoring
Manuf./Distrib.: Aldrich; Florida Treatt

Geraniol acetate. *See* Geranyl acetate
Geraniol butyrate. *See* Geranyl butyrate

Geranium extract
EINECS 296-192-0, 283-491-6 (G. maculatum ext.), 283-492-1, 307-497-6
Definition: Extract of various species of *Geranium*
Uses: Natural flavoring
Regulatory: FDA 21CFR §182.20

Geranium oil
CAS 8000-46-2; FEMA 2508
Synonyms: Pelargonium oil; Rose geranium oil Algerian
Definition: Volatile oil from steam distillation of leaves from *Pelargonium graveolens* or *Geranium maculatum*
Properties: Yel. liq., rose and geraniol odor; sol. in fixed oils, min. oil; insol. in glycerin; dens. 0.886-0.898; ref. index 1.454-1.472 (20 C)
Precaution: Keep cool, well closed; protect from light
Toxicology: Skin irritant; heated to decomp., emits acrid smoke and irritating fumes
Uses: Natural flavoring agent
Usage level: 1.6 ppm (nonalcoholic beverages), 2.8 ppm (ice cream, ices), 6.9 ppm (candy), 8.1 ppm (baked goods), 1.1-2 ppm (gelatins, puddings), 210 ppm (chewing gum), 5.2 ppm (jellies)
Regulatory: FDA 21CFR §182.10, 182.20, GRAS
Manuf./Distrib.: Pierre Chauvet; Commodity Services; Florida Treatt

Geranyl acetate
CAS 105-87-3, 16409-44-2; EINECS 203-341-5, 240-458-0; FEMA 2509
Synonyms: Geraniol acetate; trans-3,7-Dimethyl-2,6-octadien-1-ol acetate
Empirical: $C_{12}H_{20}O_2$
Formula: $(CH_3)_2C{:}CHCH_2CH_2C(CH_3){:}CHCH_2OCOCH_3$
Properties: Colorless clear liq., lavender oil, sweet taste; sol. in alcohol, fixed oils, ether; sl. sol. in propylene glycol; insol. in water, glycerin; m.w. 196.32; dens. 0.907-0.918 (15 C); b.p. 128-129 C; flash pt. 219 F; ref. index 1.458-1.464
Precaution: Combustible
Toxicology: LD50 (oral, rat) 6330 mg/kg; mildly toxic by ingestion; heated to decomp., emits acrid smoke and irritating fumes
Uses: Synthetic flavoring agent
Usage level: 1.6 ppm (nonalcoholic beverages), 6.5 ppm (ice cream, ices), 15 ppm (candy), 17 ppm (baked goods), 6.8-7.5 ppm (gelatins, puddings), 0.3-1.2 ppm (chewing gum), 1 ppm (syrups)
Regulatory: FDA 21CFR §182.60, GRAS; Japan approved as flavoring
Manuf./Distrib.: Aldrich; Firmenich; Int'l. Flavors & Fragrances; Penta Mfg.

Geranyl acetoacetate
Synonyms: trans-3,7-Dimethyl-2,6-octadien-1-yl acetoacetate
Empirical: $C_{14}H_{22}O_3$
Properties: Liq.; m.w. 238.33; dens. 0.9625; ref. index 1.4670
Uses: Synthetic flavoring agent
Usage level: 0.50 ppm (nonalcoholic beverages), 1.0 ppm (ice cream, ices), 1.0-3.0 ppm (candy), 1.0-10 ppm (baked goods)
Regulatory: FDA 21CFR §172.515

Geranyl acetone
CAS 3796-70-1; EINECS 223-269-8; FEMA 3542
Synonyms: 6,10-Dimethyl-5,9-undecadien-2-one
Empirical: $C_{13}H_{22}O$
Properties: m.w. 194.32; dens. 0.869 (20/4 C); b.p. 254-258 C; flash pt. 65 C; ref. index 1.467 (20 C)
Uses: Synthetic flavoring agent

Regulatory: FDA 21CFR §172.515
Manuf./Distrib.: Aldrich

Geranyl alcohol. *See* Geraniol

Geranyl benzoate
FEMA 2511
Synonyms: 3,7-Dimethyl-2,6-octadien-1-yl benzoate
Empirical: $C_{17}H_{22}O_2$
Properties: Ylsh. oily liq., ylang-ylang-like odor; m.w. 258.37; b.p. 198-200 C (15 mm); flash pt. > 100 C
Uses: Synthetic flavoring agent
Usage level: 0.10-0.13 ppm (nonalcoholic beverages), 0.16-0.25 ppm (ice cream, ices), 0.50 ppm (baked goods, gelatins, puddings)
Regulatory: FDA 21CFR §172.515

Geranyl butyrate
FEMA 2512
Synonyms: 3,7-Dimethyl-2,6-octadien-1-yl butyrate; Geraniol butyrate
Empirical: $C_{14}H_{24}O_2$
Properties: Liq., char. fragrant odor; sol. in alcohol, ether; almost insol. in water; m.w. 224.34; dens. 0.901 (17/4 C); b.p. 152 C (18 mm); flash pt. 93 C; ref. index 1.455 (20 C)
Precaution: Combustible
Uses: Synthetic flavoring
Usage level: 1.6 ppm (nonalcoholic beverages), 2.8 ppm (ice cream, ices), 10 ppm (candy), 10 ppm (baked goods), 5.3 ppm (gelatins, puddings), 0.30-1.5 ppm (chewing gum)
Regulatory: FDA 21CFR §172.515
Manuf./Distrib.: Chr. Hansen's

Geranyl caproate. *See* Geranyl hexanoate

Geranyl formate
CAS 105-86-2
Synonyms: 3,7-Dimethyl-2,6-octadien-1-yl formate
Definition: Ester of geraniol and formic acid
Empirical: $C_{11}H_{18}O_2$
Properties: Colorless to sl. yel. liq., rose/green rose leaf odor; pract. insol. in alcohol, ether; m.w. 102.20, dens. 0.927 (20/4 C); b.p. 113-114 C (15 mm); ref. index 1.4580-1.4660
Precaution: Combustible
Toxicology: Nontoxic
Uses: Synthetic flavoring
Usage level: 1.9 ppm (nonalcoholic beverages), 1.6 ppm (ice cream, ices), 7.5 ppm (candy), 4.1 ppm (baked goods), 3.4 ppm (gelatins, puddings), 0.80 ppm (chewing gum)
Regulatory: FDA 21CFR §172.515; Japan approved as flavoring
Manuf./Distrib.: Chr. Hansen's

Geranyl hexanoate
FEMA 2515
Synonyms: Geranyl caproate; Geranyl hexylate
Empirical: $C_{16}H_{28}O_2$
Properties: Rose-geranium odor; m.w. 252.40; dens. 0.890 (15.5 C); b.p. 240 C; ref. index 1.4500
Uses: Synthetic flavoring
Usage level: 1.3 ppm (nonalcoholic beverages), 0.90 ppm (ice cream, ices), 3.2 ppm (candy), 2.0 ppm (baked goods)
Regulatory: FDA 21CFR §172.515

Geranyl hexylate. *See* Geranyl hexanoate

Geranyl isobutyrate
FEMA 2513
Empirical: $C_{14}H_{24}O_2$
Properties: Liq., lt. rose odor, sweet apricot-like taste; sol. in most org. solvs.; insol. in water; m.w. 224.34; dens. 0.8997 (15 C); ref. index 1.4576
Uses: Synthetic flavoring
Usage level: 1.0 ppm (nonalcoholic beverages), 0.80 ppm (ice cream, ices), 5.0 ppm (candy), 4.9 ppm (baked goods), 0.60 ppm (gelatins, puddings), 15 ppm (chewing gum)
Regulatory: FDA 21CFR §172.515
Manuf./Distrib.: Chr. Hansen's

Geranyl isovalerate
FEMA 2518
Properties: Liq., rose odor, sweet apple taste; m.w. 238.37; ref. index 1.4538
Uses: Synthetic flavoring agent
Usage level: 4.2 ppm (nonalcoholic beverages), 11 ppm (ice cream, ices), 10 ppm (candy), 6.8 ppm (baked goods)
Regulatory: FDA 21CFR §172.515

Geranyl linalool
Synonyms: 3,3,11,15-Tetramethyl-1,6,10,14-hexadecatetraene-3-ol
Uses: Green, floral, sweet, woody fragrance and flavoring

Geranyl phenylacetate
FEMA 2516
Synonyms: 3,7-Dimethyl-2,6-octadien-1-yl phenylacetate; Geranyl α-toluate
Empirical: $C_{18}H_{24}O_2$
Properties: Ylsh liq., honey- and rose-like odor; clearly sol. in alcohol (4 parts/90%); m.w. 272.39; sp.gr. 0.971-0.978; flash pt. 99 C; ref. index 1.5070-1.5110
Uses: Synthetic flavoring agent
Usage level: 1.1 ppm (nonalcoholic beverages), 3.1 ppm (ice cream, ices), 6.7 ppm (candy), 4.7 ppm (baked goods), 11 ppm (chewing gum)
Regulatory: FDA 21CFR §172.515

Geranyl propionate
FEMA 2517
Synonyms: 3,7-Dimethyl-2,6-octadien-1-yl propionate
Definition: Ester of geraniol and propionic acid
Empirical: $C_{13}H_{22}O_2$
Properties: Colorless liq., fruity flowery odor, bitter taste; m.w. 210.31; b.p. 253 C; flash pt. 99 C; ref. index 1.4570-1.4650
Uses: Synthetic flavoring agent
Usage level: 1.5 ppm (nonalcoholic beverages), 1.3 ppm (ice cream, ices), 3.7 ppm (candy), 4.9 ppm (baked goods), 3.0 ppm (gelatins, puddings), 30-70 ppm (chewing gum)
Regulatory: FDA 21CFR §172.515
Manuf./Distrib.: Chr. Hansen's

Geranyl α-toluate. *See* Geranyl phenylacetate
German chamomile oil. *See* Matricaria oil

Germander extract
CAS 90131-55-8; EINECS 290-367-5
Synonyms: Teucrium scorodonia extract
Definition: Extract of herb, *Teucrium scorodonia*
Uses: Natural flavoring agent
Regulatory: FDA 21CFR §172.510; Japan approved (germander)

Gibberellic acid
CAS 77-06-5; EINECS 201-001-0
Synonyms: GA₃; 2,4a,7-Trihydroxy-1-methyl-8-methylenegibb-3-ene-1,10-carboxylic acid 1-4-lactone; Gibberellin
Empirical: $C_{19}H_{22}O_6$
Properties: Wh. cryst. or cryst. powd.; sol. in methanol, ethanol, acetone, aq. sol'ns. of sodium bicarbonate and sodium acetate; sl. sol. in water, ether; mod. sol. in ethyl acetate; m.w. 346.41; m.p. 233-235 C
Toxicology: LD50 (oral, rat) 6300 mg/kg; mildly toxic by ingestion; experimental tumorigen; mutagenic data; heated to decomp., emits acrid smoke and irritating fumes
Uses: Malting of barley with improved enzymatic characteristics; enzyme activator
Regulatory: FDA 21CFR §172.725, limitation 2 ppm (treated barley malt), 0.5 ppm (finished beverage)
Manuf./Distrib.: Atomergic Chemetals

Gibberellin. *See* Gibberellic acid

Ginger oil
CAS 8007-08-7; FEMA 2522
Synonyms: Zingiber officinale oil
Definition: Volatile oil obtained from dried rhizomes of *Zingiber officinale*, contg. l-zingiberene, d-camphene, phellandrene, borneol, cineol, citral
Properties: Yel. visc. liq., ginger odor; sol. in fixed oils, min. oil, alcohol; pract. insol. in water; insol. in glycerin,

propylene glycol; dens. 0.875-0.885 (15/15 C); ref. index 1.4880-1.4950 (20 C)

Toxicology: Skin irritant; mutagenic data; heated to decomp., emits acrid smoke and irritating fumes

Uses: Natural flavoring agent for ginger beverages, liqueurs, spice blends for bakery/confectionery

Usage level: 17 ppm (nonalcoholic beverages), 20 ppm (ice cream, ices), 14 ppm (candy), 47 ppm (baked goods), 13 ppm (condiments), 12 ppm (meats)

Regulatory: FDA 21CFR §182.10, 182.20, GRAS; GRAS (FEMA); Japan approved (ginger); Europe listed, no restrictions

Manuf./Distrib.: Chart; Pierre Chauvet; Commodity Services; Florida Treatt

Gingilli oil. *See* Sesame oil

Glauber's salt. *See* Sodium sulfate

Glucal. *See* Calcium gluconate

Glucanase

CAS 9074-99-1 (β- from *Aspergillus niger*), 9012-54-8 (from *Bacillus subtilis*); EINECS 232-980-2, 232-734-4 resp.

Uses: Enzyme for beer filtration

Regulatory: FDA GRAS; Canada, Japan approved

Trade names: Glucanase GV; Glucanex® L-300

1,4-D-Glucan glucanohydrolase. *See* Amylase

Glucase. *See* α-Glucosidase

D-Glucitol. *See* Sorbitol

Glucoamylase

CAS 9032-08-0; EINECS 232-877-2

Definition: Amyloglucosidase from *Aspergillus niger*

Properties: Powd.; m.w. ≈ 97,000

Uses: Enzyme; hydrolysis of starch dextrins to glucose; food processing, low-carbohydrate beer

Regulatory: FDA GRAS; Canada, Japan approved

Trade names: Diazyme® L-200; Spezyme GA

D-Gluconic acid

CAS 526-95-4; EINECS 208-401-4

Synonyms: Glyconic acid; Glycogenic acid; Pentahydroxycaproic acid; Dextronic acid; Maltonic acid

Empirical: $C_6H_{12}O_7$

Formula: $CH_2OH(CHOH)_4COOH$

Properties: Pure: oryotalo, mild acid taste; sol. in water; sl. sol. in alcohol, insol. in most org. solvs.; m.w. 196.16; m.p. 131 C; Commercial 50% aq. sol'n.: lt. amber, faint odor of vinegar; dens. 1.24 (25/4 C)

Uses: Dietary supplement; acidity regulator, acidulant (Japan)

Regulatory: FDA GRAS; Japan approved

Manuf./Distrib.: PMP Fermentation Prods.

D-Gluconic acid calcium salt. *See* Calcium gluconate

D-Gluconic acid cyclic 4,5-ester with boric acid calcium salt. *See* Calcium borogluconate

D-Gluconic acid δ-lactone. *See* Gluconolactone

D-Gluconic acid magnesium salt. *See* Magnesium gluconate

D-Gluconic acid monosodium salt. *See* Sodium gluconate

D-Gluconic acid potassium salt. *See* Potassium D-gluconate

Gluconic acid sodium salt. *See* Sodium gluconate

Gluconolactone

CAS 90-80-2; EINECS 202-016-5; EEC E575

Synonyms: GDL; D-Gluconic acid δ-lactone; Glucono δ-lactone; D-Glucono-1,5-lactone

Empirical: $C_6H_{10}O_6$

Properties: Wh. cryst. powd., pract. odorless; sol. in water; sl. sol. in alcohol; insol. in ether, chloroform; m.w. 178.14; m.p. 153 C (dec.)

Precaution: Avoid dust formation

Toxicology: May cause eye irritation; heated to decomp., emits acrid smoke and irritating fumes

Uses: Direct food additive, acidifier, binder, curing/pickling agent, leavening agent, pH control agent, sequestrant, flavor; used in cured meats, frankfurters, genoa salami, sausages, dessert mixes, processed cheese, fish prods., spice preps.

Usage level: ADI 0-50 mg/kg (EEC, as total gluconic acid)

Regulatory: FDA 21CFR §131.144, 133.129, 184.1318, GRAS; USDA 9CFR §318.7 (limitation 8 oz/100 lb meat, 6 oz/100 lb genoa salami); Japan approved; Europe listed; UK approved

Manuf./Distrib.: ADM; Am. Roland; Int'l. Sourcing; Jungbunzlauer; PMP Fermentation Prods.

Trade names containing: Cap-Shure® GDL-140-70

Glucono δ-lactone. *See* Gluconolactone
D-Glucono-1,5-lactone. *See* Gluconolactone
α-D-Glucopyranose. *See* D-Glucose anhyd.
α-D-Glucopyranoside, β-D-fructofuranosyl, dioctadecanoate. *See* Sucrose distearate
α-D-Glucopyranoside, β-D-fructofuranosyl, monododecanoate. *See* Sucrose laurate
α-D-Glucopyranoside, β-D-fructofuranoysl, monooctadecanoate. *See* Sucrose stearate
4-O-β-D-Glucopyranosyl-D-glucitol. *See* Maltitol
r-O-α-D-Glucopyranosyl-D-glucose. *See* Maltose
4-O-α-Glucopyranosyl-D-sorbitol. *See* Maltitol

Glucosamine
CAS 3416-24-8
Synonyms: Chitosamine
Empirical: $C_6H_{13}NO_5$
Properties: Colorless needles; sol. in water; sl. sol. in methanol, ethanol; insol. in ether, chloroform; m.w.
  179.20
Uses: Natural thickener; stabilizer
Regulatory: Japan approved

D-Glucose anhyd.
CAS 50-99-7, 492-62-6; EINECS 200-075-1, 207-757-8
Synonyms: Dextrose; Grape sugar; Blood sugar; Corn sugar; α-D-Glucopyranose
Definition: Sugar obtained from the hydrolysis of starch
Empirical: $C_6H_{12}O_6$
Properties: Colorless cryst. or wh. gran. powd., odorless, sweet taste; sol. in water; sl. sol. in alcohol; m.w.
  180.18; dens. 1.544; m.p.146 C
Precaution: Potentially explosive reaction with potassium nitrate + sodium peroxide on heating
Toxicology: LD50 (oral, rat) 25,800 mg/kg; mildly toxic by ingestion; experimental reproductive effects;
  mutagenic data; large doses can cause diabetes; heated to decomp., emits acrid smoke and irritating
  fumes; mixts. with alkali release CO on heating
Uses: Dietary supplement, flavoring agent, nutritive sweetener for confectionery, foods, medicine, brewing,
  baking, canning
Regulatory: FDA 21CFR §133.124, 133.178. 133.179, 145, 145.134, 145.180, 145.181, 146, 155.170,
  155.194, 155.200, 163.123, 163.150, 163.153, 168.110, 168.111, 169.175, 169.176, 169.177, 169.178,
  169.179, 169.180, 169.181,169.182, 184.1857, GRAS; USDA 9CFR §318.7, 381.147
Manuf./Distrib.: ADM; Am. Roland; Avebe; Bestoval; Browning; Cargill; Cerestar Int'l.; Corn Prods.; Dasco
  Sales; Farbest Brands; Fruitsource; Penta Mfg.; Penwest Foods; Ragus Sugars; Roquette UK; Sefcol;
  Siber Hegner; A.E. Staley; Sweeteners Plus; Van Waters & Rogers
Trade names: Candex®; Clintose® A; Clintose® F; Clintose® L; Clintose® VF
Trade names containing: CC-603; Danishine™; Frimulsion 6G; Frimulsion Q8; Qual Guard™ 100

Glucose isomerase
Toxicology: Heated to decomp, emits acrid smoke and irritating fumes
Uses: Direct food additive, enzyme which converts glucose to fructose; used in prod. of high fructose corn syrup
Regulatory: FDA 21CFR §184.1372, GRAS; UK, Japan approved
Trade names: Spezyme IGI; Taka-Sweet®

D-Glucose monohydrate
CAS 5996-10-1; EINECS 200-075-1
Empirical: $C_6H_{12}O_6 \cdot H_2O$
Properties: Cryst.; m.w. 198.17; m.p. 83 C
Toxicology: LD50 (IV, rabbit) 35 g/kg
Uses: Sweetening agent
Regulatory: FDA 21CFR §184.1857, GRAS

Glucose oxidase
CAS 9001-37-0; EINECS 232-601-0
Synonyms: Oxidase, glucose
Definition: Enzyme which catalyzes the oxidation of glucose to gluconic acid; derived from *Aspergillus niger*
Properties: Amorphous powd. or crystal; sol. in water; m.w. ≈ 186,000
Toxicology: Poison by subcutaneous, intravenous, intraperitoneal routes
Uses: Enzyme, food preservative, stabilizer for Vitamins C and B$_{12}$; for soft drinks, liq. whole egg, egg white,
  liq. egg yolk
Regulatory: FDA GRAS; Canada, UK, Japan approved
Trade names: Fermcozyme® 1307, BG, BGXX, CBB, CBBXX, M; Hidelase; Hyderase; Ovazyme, XX

Glucose pentaacetate. *See* α-D-Glucose pentaacetate

α-D-Glucose pentaacetate
CAS 604-68-2; EINECS 210-073-2; FEMA 2524
Synonyms: Glucose pentaacetate; 1,2,3,4,6-Penta-O-acetyl-α-D-glucopyranose
Empirical: $C_{16}H_{22}O_{11}$
Properties: m.w. 390.35; m.p. 111-113 C
Uses: Synthetic flavoring agent
Regulatory: FDA 21CFR §172.515
Manuf./Distrib.: Aldrich

Glucose syrup. *See* Corn syrup
Glucose syrup solids. *See* Corn syrup solids

α-Glucosidase
Synonyms: Maltase; Glucase
Uses: Natural enzyme; hydrolyzes maltose to glucose
Regulatory: Japan approved

β-Glucosidase
Synonyms: Emulsin
Uses: Natural enzyme
Regulatory: Japan approved

L-Glutamic acid
CAS 56-86-0; EINECS 200-293-7; FEMA 3285; EEC E620
Synonyms: α-Glutamic acid; L-2-Aminoglutaric acid; 2-Aminopentanedioic acid
Classification: Amino acid
Empirical: $C_5H_9NO_4$
Properties: Wh. cryst. or cryst. powd.; sl. sol. in water; m.w. 147.15; dens. 1.538 (20/4 C); m.p. 224-225 C
Toxicology: Human systemic effects by ingestion and IV route (headache, vomiting); heated to decomp., emits toxic fumes of NO_x
Uses: Flavor enhancer, nutrient, dietary supplement, salt substitute
Usage level: ADI 0-120 mg/kg (EEC)
Regulatory: FDA 21CFR §172.320 (12.4% max.), 182.1045, GRAS; Japan approved; Europe listed; UK approved
Manuf./Distrib.: Aldrich; Penta Mfg

α-Glutamic acid. *See* L-Glutamic acid
L-Glutamic acid 5-amide. *See* L-Glutamine

L-Glutamic acid hydrochloride
CAS 138-15-8; EINECS 205-315-9
Synonyms: L-2-Aminoglutaric acid hydrochloride; 2-Aminopentanedioic acid hydrochloride
Empirical: $C_5H_9NO_4 \cdot HCl$
Formula: $HOOCCH_2CH_2CH(NH_2)COOH \cdot HCl$
Properties: Orthorombic bisphenoidal plates; m.w. 183.60; dec. 214 C
Uses: Multipurpose food ingred.
Regulatory: FDA 21CFR §182.1047, GRAS

L-Glutamic acid L-lysine salt. *See* L-Lysine L-glutamate
Glutamic acid monosodium salt. *See* MSG

L-Glutamine
CAS 56-85-9; EINECS 200-292-1; FEMA 3684
Synonyms: L-Glutamic acid 5-amide; 2-Amino-4-carbamoylbutanoic acid
Classification: Amino acid
Empirical: $C_5H_{10}N_2O_3$
Properties: Needles; sol. in water; pract. insol. in methanol, ethanol, ether, benzene, acetone, ethyl acetate, chloroform; m.w. 146.17; dec. 185-186 C
Toxicology: LD50 (oral, rat) 7500 mg/kg; mildly toxic by ingestion; experimental reproductive effects; heated to decomp., emits toxic fumes of NO_x
Uses: Nutrient, dietary supplement, flavoring, feed additive
Regulatory: FDA 21CFR §172.320, limitation 12.4%; Japan approved
Manuf./Distrib.: Aldrich; Degussa; Penta Mfg.

Glutaral
CAS 111-30-8; EINECS 203-856-5

Glutaraldehyde

Synonyms: Glutaraldehyde; Glutaric dialdehyde; Pentanedial
Classification: Dialdehyde
Formula: OHC(CH₂)₃CHO
Properties: Liq.; sol. in water, alcohol; m.w. 100.12; dens. 0.72; b.p. 188 C (dec.); f.p. -14 C; flash pt. none
Precaution: Corrosive
Toxicology: Irritant; TLV (ceiling) 0.1 ppm in air
Uses: Antimicrobial for use in beet-sugar mills; fixing agent in immobilization of enzyme preps.
Usage level: 0.02-0.2% (50%)
Regulatory: FDA 21CFR §173.320 (250 ppm max.), 173.357, 175.105, 176.170, 176.180; EPA reg. 10352-39; Japan MITI; Europe provisional list 0.1% max.
Manuf./Distrib.: Allchem Industries

Glutaraldehyde. *See* Glutaral
Glutaric dialdehyde. *See* Glutaral
Gluten. *See* Wheat gluten
Glutens, corn. *See* Corn meal
Glyccyrrhiza. *See* Licorice
Glycerides, coconut, mono-, di-, and tri-. *See* Cocoglycerides
Glycerides, coconut oil mono-. *See* Glyceryl cocoate
Glycerides, cottonseed oil, hydrogeanted. *See* Hydrogenated cottonseed glyceride
Glycerides, cottonseed oil, mono-. *See* Cottonseed glyceride
Glycerides, cottonseed-oil, mono-, hydrogenated, acetates. *See* Acetylated hydrogenated cottonseed glyceride
Glycerides, hydrogenated lard mono-. *See* Hydrogenated lard glyceride
Glycerides, hydrogenated tallow mono-. *See* Hydrogenated tallow glyceride
Glycerides, hydrogenated vegetable mono-. *See* Hydrogenated vegetable glyceride
Glycerides, lard mono-. *See* Lard glyceride
Glycerides, lard mono-, acetates. *See* Acetylated lard glyceride
Glycerides, lard mono-, di- and tri-, hydrogenated. *See* Hydrogenated lard glycerides
Glycerides, lard mono-, hydrogenated, acetates. *See* Acetylated hydrogenated lard glyceride
Glycerides, palm oil mono-, di- and tri-. *See* Palm glycerides
Glycerides, palm oil mono-, di- and tri, hydrogenated. *See* Hydrogenated palm glycerides
Glycerides, palm oil mono-, hydrogenated. *See* Hydrogenated palm glyceride
Glycerides, peanut oil, mono-, di-, and tri-. *See* Peanut glycerides
Glycerides, safflower mono-. *See* Safflower glyceride
Glycerides, soybean oil, hydrogenated, mono. *See* Hydrogenated soy glyceride
Glycerides, sunflower seed mono-. *See* Sunflower seed oil glyceride
Glycerides, sunflower seed mono-, di- and tri-. *See* Sunflower seed oil glycerides
Glycerides, tallow mono-. *See* Tallow glyceride
Glycerides, tallow mono-, di- and tri-. *See* Tallow glycerides
Glycerides, tallow mono-, di- and tri-, hydrogenated. *See* Hydrogenated tallow glycerides
Glycerides, tallow mono-, di- and tri-, hydrogenated, acetates. *See* Acetylated hydrogenated tallow glycerides
Glycerides, tallow mono-, hydrogenated, acetates. *See* Acetylated hydrogenated tallow glyceride
Glycerides, tallow mono-, hydrogenated, citrates. *See* Hydrogenated tallow glyceride citrate
Glycerides, tallow mono-, hydrogenated, lactates. *See* Hydrogenated tallow glyceride lactate
Glycerides, vegetable mono-, di- and tri, hydrogenated. *See* Hydrogenated vegetable glycerides
Glycerides, vegetable mono-, hydrogenated. *See* Hydrogenated vegetable glyceride

Glycerin
CAS 56-81-5; EINECS 200-289-5; EEC E422
Synonyms: Glycerol; Glycyl alcohol; 1,2,3-Propanetriol; Glycerine
Classification: Polyhydric alcohol
Empirical: C₃H₈O₃
Formula: HOCH₂COHHCH₂OH
Properties: Clear colorless syrupy liq., odorless, sweet taste; hygroscopic; sol. in water, alcohol; insol. in ether, benzene, chloroform; m.w. 92.09; dens. 1.26201 (25/25 C); m.p. 17.8 C; b.p. 290 C (dec.); flash pt. (OC) 176 C; ref. index 1.4730 (25 C)
Precaution: Combustible liq. exposed to heat, flame, strong oxidizers; highly explosive with hydrogen peroxide
Toxicology: LD50 (oral, ratl) > 20 ml/kg, (IV, rat) 4.4 ml/kg; poison by subcutaneous route; mildly toxic by ingestion; human systemic effects by ingestion; human mutagenic data; skin and eye irritant; nuisance dust; heated to dec., emits acrid smoke
Uses: Solvent, humectant, plasticizer, bodying agent for baked goods, candy, marshmallows, low-fat prods., pkg. materials; migrating to food from paper/paperboard; chewing gum plasticizer (Japan)
Usage level: ADI not specified (FAO/WHO)

Regulatory: FDA 21CFR §169.175, 169.176, 169.177, 169.178, 169.180, 169.181, 172.866, 175.300, 178.3500, 182.90, 182.1320, GRAS; Japan approved; Europe listed; UK approved

Manuf./Distrib.: Acme-Hardesty; Ashland; Browne & Dureau Int'l.; Croda; Crompton & Knowles; Dasco Foods; Ellis & Everard; Fina; Gr.nau; K&K Greeff; Henkel; Int'l. Sourcing; Norman Fox; Penta Mfg.; Procter & Gamble; Siber Hegner; Thew Arnott; Van Waters & Rogers

Trade names: Emery® 912; Emery® 916; Emery® 917; Emery® 918; Glycerine (Pharmaceutical); Glycon® G 100; Glycon® G-300; Natural Glycerine USP 96%; Natural Glycerine USP 99%; Natural Glycerine USP 99.5%; Star; Superol

Trade names containing: Amisol™ 406-N

Glycerine. *See* Glycerin
Glycerite. *See* Tannic acid
Glycerol. *See* Glycerin
Glycerol mono coconut oil. *See* Glyceryl cocoate
Glycerol shortening. *See* Glyceryl mono shortening
Glycerol tricaprinate. *See* Tricaprin
Glycerol trilaurate. *See* Trilaurin

Glyceryl behenate
CAS 6916-74-1, 30233-64-8; EINECS 250-097-0
Synonyms: 2,3-Dihydroxypropyl docosanoate; Glyceryl monobehenate
Definition: Monoester of glycerin and behenic acid
Properties: acid no. 4 max.; iodine no. 3 max.; sapon. no. 145-165
Toxicology: Heated to decomp., emits acrid smoke and irritating fumes
Uses: Direct food additive, formulation aid; excipient for tablets
Regulatory: FDA 21CFR §184.1328, GRAS
Trade names: Compritol 888 ATO

Glyceryl caprate
CAS 26402-22-2; EINECS 247-667-6
Synonyms: Glyceryl monocaprate; Decanoic acid, monoester with 1,2,3-propanetriol
Definition: Monoester of glycerin and capric acid
Empirical: $C_{13}H_{26}O_4$
Formula: $CH_3(CH_2)_8COOCH_2COHHCH_2OH$
Uses: Emulsifier
Regulatory: FDA 21CFR §176.180, 176.210, 177.2800
Trade names: Imwitor® 910

Glyceryl caprylate
CAS 26402-26-6; EINECS 247-668-1
Synonyms: Glyceryl monocaprylate; Octanoic acid, monoester with 1,2,3-propanetriol; Monooctanoin
Definition: Monoester of glycerin and caprylic acid
Empirical: $C_{11}H_{22}O_4$
Formula: $CH_3(CH_2)_6COOCH_2COHHCH_2OH$
Properties: Crystals; m.p. 39.5-40.5 C
Uses: Softener for chewing gum bases, surfactant
Regulatory: FDA 21CFR §176.210, 177.2800
Trade names: Imwitor® 988

Glyceryl caprylate/caprate
Definition: Mixture of monoglycerides of caprylic and capric acids
Uses: Solubilizer and emulsifier for vitamins and flavors
Regulatory: FDA 21CFR §176.210, 177.2800
Trade names: Kessco® GMC-8

Glyceryl citrate/lactate/linoleate/oleate
Definition: Ester of glycerin and a blend of citric, lactic, linoleic and oleic acids
Uses: Food emulsifier/stabilizer
Trade names: Imwitor® 375

Glyceryl cocoate
CAS 61789-05-7; EINECS 263-027-9
Synonyms: Glycerides, coconut oil mono-; Glycerol mono coconut oil; Glyceryl coconate
Definition: Monoester of glycerin and coconut fatty acids
Formula: $RCO—OCH_2COHHCH_2OH$, RCO- represents the fatty acids derived from coconut oil
Uses: Emulsifier

Glyceryl coconate

Regulatory: FDA 21CFR §175.105, 176.210, 177.2800
Trade names: Drewmulse® 75; Imwitor® 928

Glyceryl coconate. See Glyceryl cocoate

Glyceryl cottonseed oil
Uses: Emulsifier, dispersant, antistaling agent, antistick agent

Glyceryl dioleate
CAS 25637-84-7; EINECS 247-144-2
Synonyms: 9-Octadecenoic acid, diester with 1,2,3-propanetriol
Definition: Diester of glycerin and oleic acid
Empirical: $C_{39}H_{72}O_5$
Uses: Emulsifier, stabilizer, wetting agent, lubricant for food applics.
Regulatory: FDA 21CFR §175.105, 176.210, 177.2800
Trade names: Cithrol GDO N/E

Glyceryl dioleate SE
Uses: Emulsifier, coemulsifier, stabilizer, wetting agent, lubricant, and antistat

Glyceryl distearate
CAS 1323-83-7; EINECS 215-359-0; EEC E471
Synonyms: Octadecanoic acid, diester with 1,2,3-propanetriol
Definition: Diester of glycerin and stearic acid
Empirical: $C_{39}H_{76}O_5$
Uses: Emulsifier, coemulsifier, stabilizer, wetting agent, lubricant, and antistat
Regulatory: FDA 21CFR §175.105, 176.210, 177.2800; Europe listed
Trade names: Cithrol GDS N/E; Kessco® Glycerol Distearate 386F

Glyceryl distearate SE
Uses: Emulsifier, coemulsifier, stabilizer, wetting agent, lubricant, and antistat

Glyceryl di/tribehenate
Uses: Food emulsifier and additive for tablet mfg.

Glyceryl ester of tall oil rosin
Properties: Pale amber resin; sol. in acetone, benzene; insol. in water
Toxicology: Heated to decomp, emits acrid smoke and irritating fumes
Uses: Masticatory substance in chewing gum base
Regulatory: FDA 21CFR §172.615

Glyceryl hydrogenated rosinate
Synonyms: Rosin, hydrogenated, glycerol ester
Definition: Monoester of glycerin and hydrogenated mixed long chain acids derived from rosin
Uses: Softener/plasticizer for the masticatory agent in chewing gum bases
Trade names: Staybelite® Ester 5

Glyceryl isostearate
CAS 32057-14-0; 66085-00-5; 61332-02-3; EINECS 262-710-9, 266-124-4
Synonyms: Glyceryl monoisostearate; Isooctadecanoic acid, monoester with 1,2,3-propanetriol
Definition: Monoester of glycerin and isostearic acid
Empirical: $C_{21}H_{42}O_4$
Properties: Pale yel. paste
Uses: Surface-active agent
Trade names: Emalex GWIS-100

Glyceryl lactoesters
Uses: Food emulsifiers
Trade names: Durlac® 100WK

Glyceryl lactooleate
Toxicology: Heated to decomp, emits acrid smoke and irritating fumes
Uses: Emulsifier for rendered animal fats
Regulatory: USDA 9CFR §318.7, 381.147

Glyceryl lactopalmitate. See Glyceryl palmitate lactate

Glyceryl lactopalmitate/stearate
Uses: Emulsifying and aerating agent, starch gelling agent
Trade names containing: Durlac® 100W

Glyceryl lactyl palmitate. *See* Glyceryl palmitate lactate

Glyceryl laurate
CAS 142-18-7; EINECS 205-526-6
Synonyms: Glyceryl monolaurate; Dodecanoic acid, monoester with 1,2,3-propanetriol; Dodecanoic acid, 2,3-dihydroxypropyl ester
Definition: Monoester of glycerin and lauric acid
Empirical: $C_{15}H_{30}O_4$
Formula: $CH_3(CH_2)_{10}COOCH_2COHHCH_2OH$
Properties: Cream-colored paste, faint odor; disp. in water; sol. in methanol, ethanol, toluene, naphtha, min. oil; dens. 0.98; m.p. 23-27 C; pH 8-8.6
Precaution: Combustible
Uses: Emulsifier, dispersant
Usage level: 0.1-1.0%
Regulatory: FDA 21CFR §175.105, 176.210, 177.2800, GRAS; Japan approved; Europe listed
Manuf./Distrib.: Grindsted; Lonza
Trade names: Cithrol GML N/E; Grindtek ML 90; Hodag GML

Glyceryl laurate SE
CAS 27215-38-9
Definition: Self-emulsifying grade of glyceryl laurate containing some sodium and/or potassium laurate
Uses: Emulsifier, coemulsifier, stabilizer, wetting agent, lubricant, and antistat
Trade names: Aldo® MLD FG; Cithrol GML S/E

Glyceryl linoleate
CAS 2277-28-3; EINECS 218-901-4
Synonyms: Monolinolein; 9,12-Octadecadienoic acid, 2,3-dihydroxypropyl ester; 9,12-Octadecadienoic acid, monoester with 1,2,3-propanetriol
Definition: Monoester of glycerin and linoleic acid
Empirical: $C_{21}H_{38}O_4$
Uses: w/o food emulsifier for low-cal spreads, icings, cake shortenings
Trade names: Dimodan LS Kosher

Glyceryl monobehenate. *See* Glyceryl behenate
Glyceryl monocaprate. *See* Glyceryl caprate
Glyceryl monocaprylate. *See* Glyceryl caprylate
Glyceryl mono cottonseed oil. *See* Cottonseed glyceride

Glyceryl mono/dilaurate
Uses: Emulsifier
Trade names: Aldo® ML

Glyceryl mono/dioleate
CAS 25496-72-4
Properties: Yel. oil or soft solid; dens. 0.95; m.p. 14-19 C
Precaution: Combustible
Uses: Emulsifier, antifoam, flavoring
Trade names: Aldo® HMO FG; Aldo® MOD FG; Caplube 8350; Tegomuls® O; Tegomuls® O Spezial

Glyceryl mono/distearate
Uses: Emulsifier
Trade names: Tegomuls® 4070; Tegomuls® 6070; Tegomuls® M

Glyceryl mono/distearate-palmitate
Uses: Food emulsifier for margarine, baking fats, shortening, caramel, pasta
Trade names: Tegomuls® 9050; Tegomuls® B

Glyceryl monoisostearate. *See* Glyceryl isostearate
Glyceryl monolaurate. *See* Glyceryl laurate
Glyceryl monooleate. *See* Glyceryl oleate
Glyceryl monoricinoleate. *See* Glyceryl ricinoleate
Glyceryl monorosinate. *See* Glyceryl rosinate

Glyceryl mono shortening
Synonyms: Glyceryl shortening; Glycerol shortening
Uses: Food emulsifier for shortenings, cakes, icings, bread, dairy mixes; dispersing aid, antistaling agent, antistick agent; stabilizer for o/w emulsion systems
Trade names: Capmul® GMVS-K; Drewmulse® 10K

Glyceryl mono soya oil. *See* Soy glyceride
Glyceryl monosoyate. *See* Soy glyceride
Glyceryl monostearate. *See* Glyceryl stearate
Glyceryl monostearate SE. *See* Glyceryl stearate SE
Glyceryl monotristearate. *See* Tristearin

Glyceryl oleate
CAS 111-03-5; 25496-72-4, 37220-82-9; EINECS 203-827-7; 253-407-2
Synonyms: Glyceryl monooleate; Monoolein; 9-Octadecenoic acid, monoester with 1,2,3-propanetriol
Definition: Monoester of glycerin and oleic acid
Empirical: $C_{21}H_{40}O_4$
Formula: $CH_3(CH_2)_7CH=CH(CH_2)_7COOCH_2CCH_2OHHOH$
Toxicology: Heated to decomp., emits acrid smoke and irritating fumes
Uses: Direct food additive, flavoring agent, adjuvant, solv., vehicle, defoamer, dispersant, emulsifier, plasticizer for food use (coffee whiteners, baking mixes, beverages, chewing gum, meat prods., pkg. materials, veg. oil)
Regulatory: FDA 21CFR §175.105, 175.300, 176.210, 177.2800, 181.22, 181.27, 182.4505, 184.1323, GRAS
Manuf./Distrib.: Calgene; Grindsted; Karlshamns; Patco
Trade names: Aldo® MO FG; Atsurf 594; Atsurf 595; Atsurf 595K; Atsurf 596; Atsurf 596K; Capmul® GMO; Cithrol GMO N/E; Drewmulse® 200; Dur-Em® GMO; Emrite® 6008; GMO 0041; Hodag GMO; Hodag GMO-D; Mazol® 300 K; Mazol® GMO; Mazol® GMO K; Monomuls® 90-O18; Nikkol MGO; Pationic® 1061; Pationic® 1064; Pationic® 1074
Trade names containing: Amisol™ 406-N; Arlacel® 186; Atlas 3000; Atmos® 300; Oxynex® LM; Sustane® W; Tenox® 6; Tenox® 7; Tenox® 20A; Tenox® 21; Tenox® 24; Tenox® 25; Tenox® 26; Tenox® 27; Tenox® WD-BHA

Glyceryl oleate SE
Definition: Self-emulsifying grade of glyceryl oleate that contains some sodium and/or potassium oleate
Uses: Emulsifier, coemulsifier, stabilizer, wetting agent, lubricant, and antistat
Trade names: Aldo® MOD; Cithrol GMO S/E

Glyceryl palmitate
CAS 26657-96-5; EINECS 247-887-2
Uses: Food emulsifier
Trade names: Emalex GMS-P

Glyceryl palmitate lactate
Synonyms: Glyceryl lactopalmitate; Glyceryl lactyl palmitate
Definition: Lactic acid ester of glyceryl palmitate
Empirical: $C_{22}H_{42}O_6$
Toxicology: Heated to decomp., emits acrid smoke and irritating fumes
Uses: Food emulsifier (cakes, rendered animal fats, whipped toppings)
Regulatory: FDA 21CFR §172.852; USDA 9CFR §318.7, 381.147
Trade names: Aldo® GLP FG; Aldo® LP FG

Glyceryl palmitate stearate
Synonyms: Glyceryl stearate palmitate
Classification: Triester
Definition: Monoester of glycerin and a blend of palmitic and stearic acids
Empirical: $C_{63}H_{116}O_{12}$
Uses: Emulsifier
Trade names: Imwitor® 940 K; Imwitor® 945; Tegomuls® 9053 S

Glyceryl ricinoleate
CAS 141-08-2; EINECS 205-455-0
Synonyms: 12-Hydroxy-9-octadecenoic acid, monoester with 1,2,3-propanetriol; Monoricinolein; Glyceryl monoricinoleate
Definition: Monoester of glycerin and ricinoleic acid
Formula: $C_3H_5(OOCC_{16}H_{32}OH)_3$
Uses: Emulsifying agent, stabilizer, wetting agent
Regulatory: FDA 21CFR §175.105, 176.170, 176.210, 178.3130
Trade names: Cithrol GMR N/E; Hodag GMR; Hodag GMR-D

Glyceryl ricinoleate SE
Synonyms: Glyceryl triricinoleate SE
Uses: Emulsifier, coemulsifier, stabilizer, wetting agent, lubricant, and antistat
Trade names: Cithrol GMR S/E

Glyceryl rosinate
Synonyms: Glyceryl monorosinate; Rosin, glyceryl ester
Definition: Monoester of glycerin and mixed long chain acids derived from rosin
Properties: drop soften. pt. 88-96 C
Toxicology: Heated to decomp., emits acrid smoke and irritating fumes
Uses: Clouding agent in beverages; adjusts density of citrus oils in beverages; in chewing gums
Regulatory: FDA 21CFR §172.615, 172.735, limitation 100 ppm in finished beverage, 175.105, 175.300, 178.3120, 178.3800, 178.3870
Trade names: Hercules® Ester Gum 8BG; Hercules® Ester Gum 8D; Hercules® Ester Gum 8D-SP; Hercules® Ester Gum 10D

Glyceryl shortening. *See* Glyceryl mono shortening

Glyceryl soyate
Uses: Emulsifier

Glyceryl stearate
CAS 123-94-4; 11099-07-3; 31566-31-1; 61789-08-0; 85666-92-8; 85251-77-0; EINECS 250-705-4; 234-325-6; 204-664-4; 286-490-9; EEC E471
Synonyms: Monostearin; 1,2,3-Propanetriol octadecanoate; Glyceryl monostearate; 2,3-Dihydroxypropyl octadecanoate
Definition: Monoester of glycerin and stearic acid
Empirical: $C_{21}H_{42}O_4$
Formula: $CH_3(CH_2)_{16}COOCH_2COHHCH_2OH$
Properties: Wh. to cream flakes; insol. in water, ethanol, glycerin, propylene glycol; disp. in min. oil; m.p. 56-59 C; sapon. no. 162-175
Precaution: Combustible
Toxicology: LD50 (IP, mouse) 200 mg/kg; poison by IP route; heated to decomp., emits acrid smoke and irritating fumes
Uses: Food additive; coating agent, emulsifier, lubricant, solvent, texture modifier, thickener; used in baked goods, cake shortening, desserts, fruits, ice cream, nuts, peanut butter, puddings, shortening, whipped toppings
Regulatory: FDA 21CFR §139.110, 139.115, 139.117, 139.120, 139.121, 139.122, 139.125, 139.135, 139.138, 139.140, 139.150, 139.155, 139.160, 139.165, 139.180, 175.105, 175.210, 175.300, 176.200, 176.210, 177.2800, 184.1324, GRAS; Europe listed
Manuf./Distrib.: AB Tech.; Alfa; Calgene; Croda; Durkee; Eastman; Fina; Grindsted; Gr̦nau; A C Hatriok; Henkel; Honeywill & Stein; Hüls; Nordmann Rassmann; Norman Fox; Penta Mfg.; PPG; Quest; Riken Vitamin; Siber Hegner; Stepan; Unipex; Peter Whiting
Trade names: Aldo® HMS FG; Aldo® HMS KFG; Aldo® MS FG; Aldo® MSLG FG; Aldo® MS LG KFG; Atlas 1500; Atmos® 150; Atmul® 84; Atmul® 124; Capmul® GMS; Cithrol GMS N/E; Dimodan PM 300; Drewmulse® 200K; Drewmulse® 900K; Emalex GMS-A; Emalex GMS-B; Empilan® GMS NSE32; Empilan® GMS NSE40; Emrite® 6003; GMS 300; GMS 305; GMS 400; GMS 400V; GMS 600V; GMS 900; GMS 902; Hefti GMS-33; Hefti GMS-99; Hodag GMS; Imwitor® 191; Imwitor® 900; Lipal GMS; Lipo GMS; Lipo GMS 410; Lipo GMS 450; Mazol® GMS; Mazol® GMS-90; Mazol® GMS-K; Myvaplex® 600; Myvaplex® 600K; Myvaplex® 600P; Myvaplex® 600PK; Nikkol MGS-A; Nikkol MGS-B; Nikkol MGS-F20; Nikkol MGS-F40; Nikkol MGS-F50; Nikkol MGS-F75; Nikkol MGS-TG; Nikkol MGS-TGL; Norfox® GMS-FG; Pationic® 901; Pationic® 902; Pationic® 905; Pationic® 909; Pationic® 1042; Pationic® 1042K; Pationic® 1052; Pationic® 1052K; Tegomuls® 90 K; Tegomuls® 90 S; Tegomuls® 4100; Tegomuls® 4101; Tegomuls® 9102; Tegomuls® 9102 S
Trade names containing: Aldo® MSD FG; Aldo® MSD KFG; Aldo® PME; Aldosperse® 40/60; Aldosperse® 40/60 FG; Aldosperse® 40/60 KFG; Aldosperse® O-20; Aldosperse® O-20 FG; Aldosperse® O-20 KFG; Aldosperse® TS-20; Aldosperse® TS-20 FG; Aldosperse® TS-40; Aldosperse® TS-40 FG; Aldosperse® TS-40 KFG; Arlacel® 165; 24% Beta Carotene Semi-Solid Suspension No. 65642; Ches® 500; Drewmulse® 365; Drewmulse® 700K; Garbefix 31; Ice # 2; Myvatex® 3-50K; Myvatex® 40-06S; Myvatex® Peanut Butter Stabilizer; Myvatex® Texture Lite®; Oxynex® LM; Tally® 100 Plus

Glyceryl stearate SE
CAS 31566-31-1; 11099-07-3; 85666-92-8; 977053-96-5
Synonyms: GMS-SE; Glyceryl monostearate SE
Definition: Self-emulsifying grade of glyceryl stearate containing some sodium and/or potassium stearate
Properties: Wh. to cream flakes; sol. in oleyl alcohol; partly sol. in water, veg. oil, ethanol, propylene glycol; m.p. 57-59 C; sapon. no. 150-160
Uses: Thickener, emulsifier, stabilizer, wetting agent for margarine, shortening, baking, and food prods., flavoring
Trade names: Cithrol GMS Acid Stable; Cithrol GMS S/E; Emalex GMS-10SE; Emalex GMS-15SE; Emalex

GMS-20SE; Emalex GMS-25SE; Emalex GMS-45RT; Emalex GMS-50; Emalex GMS-55FD; Emalex GMS-195; Emalex GMS-ASE; Empilan® GMS LSE32; Empilan® GMS LSE40; Empilan® GMS LSE80; Empilan® GMS MSE40; Empilan® GMS NSE90; Empilan® GMS SE40; Empilan® GMS SE70; GMS 402; GMS 402V; GMS 602V; Hefti GMS-33-SES; Imwitor® 960; Lipo GMS 470; Nikkol MGS-150; Nikkol MGS-ASE; Nikkol MGS-BSE; Nikkol MGS-DEX; Nikkol MGS-F50SE

Glyceryl stearate citrate
CAS 39175-72-9, 91744-38-6
Synonyms: 2-Hydroxy-1,2,3-propanetricarboxylic acid, monoester with 1,2,3-propanetriol monooctadecanoate
Definition: Citric acid ester of glyceryl stearate
Empirical: $C_{27}H_{48}O_{10}$
Uses: Emulsifier
Trade names: Imwitor® 370

Glyceryl stearate lactate
Definition: Lactic acid ester of glyceryl stearate
Empirical: $C_{24}H_{46}O_6$
Toxicology: Heated to decomp., emits acrid smoke and irritating fumes
Uses: Emulsifier for cake mixes, chocolate coatings, rendered animal fats, shortening, whipped vegetable toppings
Regulatory: USDA 9CFR §318.7, 381.147
Trade names: GLS; Lactodan B 30

Glyceryl stearate palmitate. *See* Glyceryl palmitate stearate
Glyceryl triacetate. *See* Triacetin

Glyceryl (triacetoxystearate)
CAS 139-43-5; EINECS 295-625-0
Synonyms: Castor oil, acetylated and hydrogenated; Glyceryl tri(12-acetoxystearate)
Formula: $C_3H_5(OOCC_{17}H_{34}OCOCH_3)_3$
Properties: Clear pale yel. oily liq., mild odor; sol. in most org. solvs.; insol. in water; dens. 0.967 (25/25 C)
Precaution: Combustible
Toxicology: Heated to decomp, emits acrid smoke and irritating fumes
Uses: Polymer adjuvant for food pkg.
Regulatory: FDA 21CFR §178.3505, limitation with $CaCO_3$ 1% of total mixt.

Glyceryl tri(12-acetoxystearate). *See* Glyceryl (triacetoxystearate)
Glyceryl tributyrate. *See* Tributyrin
Glyceryl tricaprate. *See* Tricaprin

Glyceryl tricaprate/caprylate
Uses: Food additive

Glyceryl tricaprylate. *See* Tricaprylin

Glyceryl tricaprylate/caprate
Uses: Bakery lubricant, release agent, glazing agent for sugar confectionery, flour confectionery; solv. carrier for flavors and fragrances
Trade names: Crodamol GTC/C

Glyceryl tridodecanoate. *See* Trilaurin

Glyceryl trienanthate
Uses: Tracer oil for butter; release agent for processing of sweets
Trade names: Special Oil 107

Glyceryl tri(2-ethylhexanoate). *See* Trioctanoin

Glyceryl triheptanoate
Uses: Food additive for food and feed industries

Glyceryl trilaurate. *See* Trilaurin
Glyceryl trimyristate. *See* Trimyristin
Glyceryl trioctanoate. *See* Trioctanoin
Glyceryl trioleate. *See* Triolein
Glyceryl tripalmitate. *See* Tripalmitin

Glyceryl tripropanoate
FEMA 3286

Synonyms: Tripropionin
Manuf./Distrib.: Aldrich

Glyceryl triricinoleate SE. *See* Glyceryl ricinoleate SE
Glyceryl tristearate. *See* Tristearin

Glycine
CAS 56-40-6; EINECS 200-272-2; FEMA 3287
Synonyms: Aminoacetic acid; Glycocoll
Classification: Amino acid
Empirical: $C_2H_5NO_2$
Formula: H_2NCH_2COOH
Properties: Wh. cryst., odorless, sweet taste; sol. in water; insol. in alcohol, ether; m.w. 75.08; dens. 1.1607; m.p. 232-236 C (dec.)
Toxicology: LD50 (oral, rat) 7930 mg/kg; moderately toxic by IV route; mildly toxic by ingestion; heated to decomp., emits toxic fumes of NO_x
Uses: Buffering agent, chicken-feed additive, flavor enhancer, sweetener, stabilizer, nutrient, dietary supplement, reduces bitter taste of saccharin, retards rancidity in animal and vegetable fats; in beverages, rendered fats
Regulatory: FDA 21CFR §170.50, GRAS for animal feed (582.5049), 172.320, limitation 3.5%, 172.812, 0.2% in finished beverage; USDA 9CFR §318.7, 0.01% in rendered animal fat; Japan approved
Manuf./Distrib.: Aldrich; Allchem Industries; Degussa

Glycinol. *See* **Ethanolamine**
Glycocoll. *See* Glycine
Glycogenic acid. *See* D-Gluconic acid

Glycol/butylene glycol montanate
Uses: Wax for emulsions for citrus fruit coating
Trade names: Hoechst Wax KPS; Hoechst Wax KSL

Glycol butyl ether. *See* Butoxyethanol
Glycolic acid phenyl ether. *See* Phenoxyacetic acid
Glyconic acid. *See* D-Gluconic acid
Glycyl alcohol. *See* Glycerin
Glycyrrhiza extract. *See* Licorice root extract
Glycyrrhiza glabra extract. *See* Licorice extract
Glyoxaline-5-alanine. *See* L-Histidine
Glyoxaline-5-alanine monohydrochloride. *See* Histidine hydrochloride monohydrate
GMP. *See* Disodium guanylate
GMS-SE. *See* Glyceryl stearate SE

Gold
CAS 7440-57-5; EINECS 231-165-9; EEC E175
Classification: Metallic element
Empirical: Au
Properties: Yel. soft metal; at.wt. 196.9665
Uses: Colorant
Regulatory: UK approved

Gold bronze. *See* Copper
Graham's salt. *See* Sodium hexametaphosphate
Graham's salt. *See* Sodium metaphosphate
Granulated sugar. *See* Sucrose

Grape color extract
Toxicology: Heated to decomp, emits acrid smoke and irritating fumes
Uses: Color additive for nonbeverage foods
Regulatory: FDA 21CFR §73.169
Manuf./Distrib.: Quest Int'l.

Grapefruit oil
CAS 8016-20-4; FEMA 2530
Synonyms: Shaddock oil; Citrus paradisi peel oil
Definition: Volatile oil from the fresh peel of *Citrus paradisi*
Properties: Yel. liq.; sol. in fixed oils, min. oil; sl. sol. in propylene glycol; insol. in glycerin
Toxicology: Experimental tumorigen; mutagenic data; skin irritant; heated to decomp., emits acrid smoke and irritating fumes

Grape juice

> *Uses:* Natural flavoring agent
> *Usage level:* 160 ppm (nonalcoholic beverages), 180 ppm (ice cream), 630 ppm (candy), 370 ppm (baked goods), 250 ppm (gelatins, puddings), 1500 ppm (chewing gum), 400 ppm (toppings)
> *Regulatory:* FDA 21CFR §182.20, GRAS; GRAS (FEMA); Europe listed, no restrictions
> *Manuf./Distrib.:* Commodity Services; Florida Treatt

Grape juice
> *Definition:* Liq. expressed from fresh grapes

Grape skin extract
> EEC E163
> *Synonyms:* Enocianina; Enocyanin
> *Definition:* From aq. extraction of fresh deseeded marc remaining after grapes are pressed to produce wine or juice; contains anthocyanins, tartaric acid, tannins, sugars, minerals
> *Properties:* Deep purple; sol. in water
> *Uses:* Color additive for still and carbonated beverages, ales, alcoholic beverages
> *Regulatory:* FDA 21CFR §73.170; Japan restricted
> *Manuf./Distrib.:* Am. Roland
> *Trade names:* Grape Skin Extract, 2X #3850; Grape Skin Extract, Double Strength; Grape Skin Extract, Single Strength

Grape sugar. *See* D-Glucose anhyd.
Gray acetate. *See* Calcium acetate
Green chlorophyl. *See* Chlorophyll
Green vitriol. *See* Ferrous sulfate
Green vitriol. *See* Ferrous sulfate heptahydrate
Groundnut oil. *See* Peanut oil

Guaiac extract
> CAS 84650-13-5; FEMA 2533
> *Synonyms:* Guaiacum officinalis extract; Guaiacum
> *Definition:* Extract of wood of the guaiacum tree, *Guaiacum officinalis*
> *Properties:* Pleasant rose-like odor
> *Uses:* Natural flavoring agent
> *Usage level:* 760 ppm (nonalcoholic beverages), 4.0 ppm (ice cream, ices), 8.0 ppm (candy), 70 ppm (baked goods)
> *Regulatory:* FDA 21CFR §172.510

Guaiac gum
> CAS 9000-29-7
> *Synonyms:* Resin guaiac; Gum guaiac; Guaiacum
> *Definition:* Resin from wood of *Guaiacum officinale*
> *Properties:* Brn. or greenish-brn. lumps.; sol. in alcohol, chloroform, ether, creosote, alkalies; sl. sol. in benzene, carbon disulfide; insol. in water; m.p. 85-90 C
> *Toxicology:* LD50 (oral, rat) > 5000 mg/kg
> *Uses:* Antioxidant when migrating from food pkg.; gelling agent in cheeses and low-sugar jams; chewing gum base (Japan)
> *Regulatory:* FDA 21CFR §181.24 (0.005% migrating from food pkg.); Japan approved (1.0 g/kg max.)
> *Manuf./Distrib.:* Chart

Guaiacol
> CAS 90-05-1; EINECS 201-964-7; FEMA 2532
> *Synonyms:* Methylcatechol; Pyrocatechol methyl ether; o-Methoxyphenol; 2-Methoxyphenol; o-Hydroxyanisole
> *Definition:* Obtained from hardwood tar
> *Empirical:* $C_7H_8O_2$
> *Formula:* $OHC_6H_4OCH_3$
> *Properties:* Wh. to sl. yel. cryst. or colorless to ylsh. liq., char. odor; sol. (1 g/ml): 60-70 ml water, 1 ml glycerin; misc. with alcohol, chloroform, ether, oils, glac. acetic acid; m.w. 124.14; dens. 1.129 (cryst.), 1.112 (liq.); m.p. 26-29 C; b.p. 204-206 C
> *Precaution:* Protect from light
> *Toxicology:* LD50 (oral, rat) 725 mg/kg
> *Uses:* Synthetic flavor
> *Usage level:* 0.95 ppm (nonalcoholic beverages), 0.52 ppm (ice cream), 0.96 ppm (candy), 0.75 ppm (baked goods)
> *Regulatory:* FDA 21CFR §172.515

Manuf./Distrib.: Aldrich; Penta Mfg.

Guaiacol phenylacetate. *See* Guaiacyl phenylacetate
Guaiacum. *See* Guaiac extract
Guaiacum. *See* Guaiac gum
Guaiacum officinalis extract. *See* Guaiac extract

Guaiacyl acetate
Synonyms: Acetyl guaiacol; o-Methoxyphenyl acetate
Empirical: $C_9H_{10}O_3$
Properties: Colorless liq.; misc. with alcohol, ether; insol. in water; m.w. 166.18; b.p. 235-240 C
Uses: Synthetic flavoring agent
Regulatory: FDA 21CFR §172.515

Guaiacyl phenylacetate
CAS 4112-89-4; FEMA 2535
Synonyms: Guaiacol phenylacetate; o-Methoxyphenyl phenylacetate
Empirical: $C_{15}H_{14}O_3$
Properties: Amber visc. liq., woody herbaceous odor, spicy flavor; sol. in alcohol; insol. in water; m.w. 242.28
Uses: Synthetic flavoring
Usage level: 0.38 ppm (nonalcoholic beverages), 1.0 ppm (ice cream, ices, toppings), 2.2 ppm (candy), 3.2 ppm (baked goods)
Regulatory: FDA 21CFR §172.515
Manuf./Distrib.: Aldrich

Guaiene
Synonyms: 1,4-Dimethyl-7-isopropenyl-δ9,10-octahydroazulene
Empirical: $C_{15}H_{24}$
Properties: Greenish-yel. mobile liq., sl. delicate woody odor; sol. in alcohol
Uses: Synthetic flavoring agent
Regulatory: FDA 21CFR §172.515

Guaiol acetate
Synonyms: 1,4-Dimethyl-7-(α-hydroxyisopropyl)-δ9,10-octahydroazulene acetate
Empirical: $C_{17}H_{28}O_2$
Properties: Yish. liq.; m.w. 264.41; flash pt. > 100 C; ref. index 1.487-1.495
Uses: Synthetic flavoring agent
Regulatory: FDA 21CFR §172.515

Guaniol. *See* Geraniol
Guanosine 5´-disodium phosphate. *See* Disodium guanylate
Guanylic acid sodium salt. *See* Disodium guanylate
Guaramine. *See* Caffeine
Guar flour. *See* Guar gum

Guar gum
CAS 9000-30-0; EINECS 232-536-8; FEMA 2537; EEC E412
Synonyms: Guar flour; Jaguar gum; Gum cyamopsis; Cyamopsis gum
Definition: Natural material derived from the ground endosperms of *Cyamopsis tetragonolobus*
Properties: Yellowish-white free-flowing powd.; aq. sol'ns. tasteless, odorless; sol. in hot or cold water; insol. in oil, greases, hydrocarbons, ketones, esters; m.w. ≈ 220,000
Toxicology: Mildly toxic by ingestion; heated to decomp., emits acrid smoke and irritating fumes
Uses: Stabilizer, thickener, emulsifier, firming agent, formulation aid, suspending agent; dietary bulking agent; helps diabetics control blood sugar levels
Usage level: Limitation 0.35% (baked goods), 1.2% (cereal), 0.8% (cheese), 1% (dairy prods.), 2% (fats, oils), 1.2% (gravies), 1% (jams), 0.6% (milk prods.), 2% (processed vegetables), 0.8% (soups), 1% (sweet sauces), 0.5% (other foods)
Regulatory: FDA 21CFR §133.124, 133.133, 133.134, 133.162, 133.178, 133.179, 150.141, 150.161, 184.1339, GRAS; GRAS (FEMA); Japan approved; Europe listed; UK approved; ADI not specified (WHO)
Manuf./Distrib.: Ashland; Atomergic Chemetals; Bio-Botanica; Chart; Cornelius; Courtaulds; Diamalt; Ellis & Everard; Goorden; Grindsted; Gumix Int'l.; A C Hatrick; Hercules Food; Lucas Meyer; Meer; Multi-Kem; Quest; Rhone-Poulenc; Sanofi; Spice King; TIC Gums
Trade names: Dycol™ 4000FC; Dycol™ 4500F; Dycol™ HV400F; Edicol®; Edicol® ULV Series; Emulcoll; Guardan 100 Range; Guardan 600 Range; Guardan 700 Range; Guar Gum; Guar Gum HV; Jaguar® 1105; Jaguar® 1110; Jaguar® 1120; Jaguar® 1140; Jaguar® 2209; Jaguar® 2220; Jaguar® 2240; Jaguar® 6000; Jaguar® Guar Gum; Nutriloid® Guar Special; Nutriloid® Guar Standard; Powdered Gum Guar NF

Type 80 Mesh B/T; Powdered Gum Guar Type 140 Mesh B/T; Powdered Gum Guar Type ECM; Powdered Gum Guar Type M; Powdered Gum Guar Type MM FCC; Powdered Gum Guar Type MM (HV); Powdered Gum Guar Type MMM $^1/_2$; Powdered Gum Guar Type MMW; Prinza® 452; Prinza® 455; Prinza® Range; Supercol® G2S; Supercol® GF; Supercol® U; Ticolv; TIC Pretested® Gum Guar 8/22 FCC/NF Powd.; Uniguar 20; Uniguar 40; Uniguar 150; Uniguar 200

Trade names containing: Avicel® RCN-10; Avicel® RCN-15; CC-603; Crodacreme; Danishine™; Dariloid® 100; Dariloid® 300; Dricoid® 200; Dricoid® 280; Emulgel E-21; Emulgel S-32; Freedom X-PGA; Frigesa® IC 178; Frigesa® IC 184; Frimulsion 10; Frimulsion Q8; Frimulsion X5; GFS®; Lamefrost® ES 216 G; Lamefrost® ES 315; Lamefrost® ES 375; Lamefrost® ES 379; Lamefrost® ES 424; Lite Pack 350; Mayonat DF; Mayonat PS; Mayonat V/100; Merecol® FA; Merecol® FAL; Merecol® GL; Merecol® GX; Merecol® IC; Merecol® LK; Merecol® R; Merecol® RB; Merecol® RCS; Merecol® SH; Rhodigum OEH; Rhodigum OEM; Rhodigum WGH; Rhodigum WGL; Rhodigum WGM; Rhodigum WVH; Rhodigum WVM; Thixogum X; Vis*Quick™ GX 11; Vis*Quick™ GX 21

Guatemala lemongrass oil. *See* Lemongrass oil West Indian

Guava
Definition: Psidium spp.
Uses: Natural flavoring agent
Regulatory: FDA 21CFR §182.20, GRAS
Manuf./Distrib.: Commodity Services

Gum arabic. *See* Acacia

Gum benzoin
CAS 9000-05-9; EINECS 232-523-7
Synonyms: Benzoin gum; Gum sumatra; Siam benzoin; Sumatra benzoin
Definition: Balsamic resin obtained from various *Styrax* species
Uses: Natural flavoring agent; chewing gum base (Japan)
Regulatory: FDA 21CFR §73.1, 172.510; Japan approved

Gum camphor. *See* Camphor
Gum cyamopsis. *See* Guar gum
Gum Dammar. *See* Dammar
Gum dragon. *See* Tragacanth gum
Gum gellan. *See* Gellan gum

Gum ghatti
Synonyms: Indian gum
Definition: Exudate from wounds in the bark of *Anogeissus latifolia*
Uses: Emulsifier for use in flavor oils, beverages, other foods; thickener/stabilizer (Japan)
Usage level: Limitation 0.2% (nonalcoholic beverages), 0.1% (other food)
Regulatory: FDA 21CFR §184.1333, GRAS; Japan approved
Manuf./Distrib.: Importers Service
Trade names: Granular Gum Ghatti #1; Powdered Gum Ghatti #1; Powdered Gum Ghatti #2; Premium Powdered Gum Ghatti G-1; Staform P

Gum guaiac. *See* Guaiac gum
Gum hashab. *See* Acacia
Gum olibanum. *See* Olibanum
Gum rosin. *See* Rosin
Gum sandarac. *See* Sandarac gum
Gum sumatra. *See* Gum benzoin
Gum-tara. *See* Tara gum
Gum tragacanth. *See* Tragacanth gum
Guncotton. *See* Nitrocellulose
Guru. *See* Cola

Gutta percha
Synonyms: trans-Polyisoprene
Definition: Geometric isomer of natural rubber
Properties: insol. in water; m.p.100 C
Uses: Natural chewing gum base
Regulatory: Japan approved

Gypsum. *See* Calcium sulfate
Hard paraffin. *See* Paraffin
Hartshorn. *See* Ammonium carbonate

Hazelnut oil
Definition: Oil obtained from the nuts of various species of the hazelnut tree, genus *Corylus*
Properties: Amber yel. oil; dens. 0.911-0.917; iodine no. 85-100; sapon. no. 189-196; ref. index 1.4615-1.4725
Uses: Food ingred.
Manuf./Distrib.: Arista Industries

Heavy or light mineral oil. *See* Mineral oil
Hedeoma oil. *See* Pennyroyal oil, American
Helioptropyl isobutyrate. *See* Piperonyl isobutyrate
Heliotropin. *See* Heliotropine

Heliotropine
CAS 120-57-0; EINECS 204-409-7; FEMA 2911
Synonyms: Heliotropin; Piperonal; Piperonyl aldehyde; Dioxymethyleneprotocatechuic aldehyde; 3,4-Benzodioxole-5-carboxaldehyde; 1,3-Benzodioxole-5-carboxaldehyde; 3,4-Methylene dihydroxy-benzaldehyde
Empirical: $C_8H_6O_3$
Properties: Colorless lustrous cryst., floral heliotrope odor; very sol. in alcohol, ether; sol. in propylene glycol, fixed oils; insol. in water, glycerin; m.w. 150.14; m.p. 37 C; b.p. 263 C; flash pt. > 230 F
Precaution: Combustible when exposed to heat, flame; reactive with oxidizers; keep in cool place; photosensitive—protect from light
Toxicology: LD50 (oral, rat) 2700 mg/kg; mod. toxic by ingestion and intraperitoneal routes; can cause CNS depression; skin irritant
Uses: Synthetic flavoring agent; for cherry and vanilla flavors
Usage level: 6 ppm (nonalcoholic beverages), 7 ppm (ice cream), 7.4 ppm (candy), 18 ppm (baked goods), 5.8 ppm (gelatins, puddings), 36 ppm (chewing gum)
Regulatory: FDA 21CFR §182.60, GRAS
Manuf./Distrib.: Aldrich; Chr. Hansen's

Heliotropyl acetate. *See* Piperonyl acetate

Helium
CAS 7440-59-7; EINECS 231-168-5
Empirical: He
Properties: Colorless inert gas, odorless, flavorless; m.w. 4.0; b.p. -269 C
Precaution: Nonflamm.
Uses: Direct food additive, processing aid, propellant; for foamed and sprayed food prods.
Regulatory: FDA 21CFR §184.1355, GRAS

Hemicellulase
Uses: Food enyzme; hydrolyzes coffee gums; used in the extraction of essential oils and plant extracts
Trade names: Hemi-Cellulase Amano 90; Hemicellulase B 1500; Hemicellulase CE 1500; Hemicellulase Conc.
Trade names containing: Clarex® ML

Hendecanal. *See* Undecanal
Hendecanoic alcohol. *See* Undecyl alcohol
1-Hendecanol. *See* Undecyl alcohol
Hendecenal. *See* 10-Undecenal
Hendecen-9-al. *See* 9-Undecenal
10-Hendecenyl acetate. *See* 10-Undecen-1-yl acetate
Hendecyl alcohol. *See* Undecyl alcohol

γ-Heptalactone
CAS 105-21-5; FEMA 2539
Synonyms: 4-Hydroxyheptanoic acid, γ-lactone
Empirical: $C_7H_{12}O_2$
Properties: Colorless sl. oily liq., sweet nut-like caramel odor; sol. in alcohol; insol. in water; m.w. 128.17
Uses: Synthetic flavoring agent
Usage level: 18 ppm (nonalcoholic beverages), 40 ppm (ice cream, ices), 28 ppm (candy), 26 ppm (baked goods)
Regulatory: FDA 21CFR §172.515
Manuf./Distrib.: Aldrich

Heptaldehyde. *See* Heptanal
1-Heptaldehyde. *See* Heptanal

Heptanal
CAS 111-71-7; EINECS 203-898-4; FEMA 2540

1-Heptanal

Synonyms: Aldehyde C-7; 1-Heptanal; Enanthal; Oenanthol; Heptaldehyde; 1-Heptaldehyde; Heptylaldehyde; Enanthaldehyde; Oenanthaldehyde
Empirical: $C_7H_{14}O$
Formula: $CH_3(CH_2)_5CHO$
Properties: Liq., penetrating fruity odor; misc. with alcohol, ether; sl. sol. in water; m.w. 114.19; dens. 0.80902 (30/4 C); visc. 0.977 cp (15 C); m.p. -43.3 C; b.p. 152.8 C (760 mm); flash pt. 95 F; ref. index 1.42571 (20 C)
Precaution: Flamm.
Uses: Synthetic flavoring agent
Usage level: 4.9 ppm (nonalcoholic beverages), 4.0 ppm (alcoholic beverages), 1.2 ppm (ice cream, ices), 2.0 ppm (candy), 2.6 ppm (baked goods)
Regulatory: FDA 21CFR §172.515
Manuf./Distrib.: Aldrich

1-Heptanal. See Heptanal

Heptanal dimethyl acetal
FEMA 2541
Synonyms: 1,1-Dimethoxy heptane
Empirical: $C_9H_{20}O_2$
Properties: Oily liq., walnut cognac odor; m.w. 160.26; dens. 0.849 (20/20 C); b.p. 164-165 C; ref. index 1.4130
Uses: Synthetic flavoring agent
Usage level: 0.10-0.13 ppm (nonalcoholic beverages), 0.25 ppm (ice cream, ices, candy, baked goods), 1.0 ppm (condiments)
Regulatory: FDA 21CFR §172.515

Heptanal 1,2-glyceryl acetal
FEMA 2542
Empirical: $C_{10}H_{20}O_3$
Properties: Colorless visc. liq., fungus-like sweet odor, mushroom taste; sol. in alcohol; sl. sol. in water; m.w. 188.27
Uses: Synthetic flavoring
Usage level: 5.0 ppm (nonalcoholic beverages), 10 ppm (ice cream, ices, candy, baked goods), 100 ppm (condiments)
Regulatory: FDA 21CFR §172.515

2,3-Heptanedione
FEMA 2543
Synonyms: Acetyl valeryl; Acetyl pentanoyl; Valeryl acetyl
Empirical: $C_7H_{12}O_2$
Properties: Ylsh. liq., cheesy pungent odor, sweet butter taste; sol. in alcohol; sl. sol. in water; m.w. 128.17
Uses: Synthetic flavoring agent
Usage level: 0.96 ppm (nonalcoholic beverages), 3.1 ppm (ice cream, ices), 8.2 ppm (candy), 7.9 ppm (baked goods), 1.7 ppm (chewing gum)
Regulatory: FDA 21CFR §172.515

Heptanoic acid ethyl ester. See Ethyl heptanoate
Heptanoic acid methyl ester. See Methyl heptanoate
1-Heptanol. See Heptyl alcohol

3-Heptanol
CAS 589-82-2; EINECS 209-661-1; FEMA 3547
Synonyms: Alcohol C-7; Butyl ethyl carbinol; Ethyl butyl carbinol
Empirical: $C_7H_{16}O$
Formula: $CH_3(CH_2)_3CH(OH)CH_2CH_3$
Properties: Colorless oily liq., herbaceous odor, pungent sl. bitter taste; sol. in alcohol; insol. in water; m.w. 116.20; dens. 0.820 (20/4 C); b.p. 155-157 C; flash pt. 60 C; ref. index 1.422 (20 C)
Toxicology: Skin and eye irritant
Uses: Synthetic flavoring agent
Regulatory: FDA 21CFR §172.515
Manuf./Distrib.: Aldrich

2-Heptanone. See Methyl n-amyl ketone

3-Heptanone
CAS 106-35-4; EINECS 203-388-1; FEMA 2545
Synonyms: Ethyl butyl ketone; Butyl ethyl ketone; Ethyl-n-butyl ketone

Empirical: $C_7H_{14}O$
Formula: $CH_3(CH_2)_3COC_2H_5$
Properties: Colorless liq., powerful green fatty fruity odor, melon banana flavor; insol. in water; m.w. 114.19; dens. 0.818 (20/4 C); m.p. -39 C; b.p. 145-148 C; flash pt. 38 C; ref. index 1.409 (20 C)
Precaution: Flamm.
Toxicology: Eye irritant
Uses: Synthetic flavoring agent
Usage level: 0.13-2.0 ppm (nonalcoholic beverages), 0.25-170 ppm (ice cream, ices), 67 ppm (candy), 0.25-130 ppm (baked goods)
Regulatory: FDA 21CFR §172.515
Manuf./Distrib.: Hüls

4-Heptanone
CAS 123-19-3; EINECS 204-608-9
Synonyms: Dipropyl ketone; Butyrone
Empirical: $C_7H_{14}O$
Formula: $(CH_3CH_2CH_2)_2CO$
Properties: Colorless liq., penetrating odor, burning taste; misc. with alcohol, ether; insol. in water; m.w. 114.19; dens. 0.814 (20/4 C); m.p. -32.6 C; b.p. 142-144 C; flash pt. 49 C; ref. index 1.4073 (22 C)
Precaution: Flamm.
Uses: Synthetic flavoring agent
Usage level: 7.8 ppm (nonalcoholic beverages), 11 ppm (ice cream, ices), 19 ppm (candy), 27 ppm (baked goods), 0.60-8.0 ppm (gelatins, puddings)
Regulatory: FDA 21CFR §172.515
Manuf./Distrib.: Hüls

3,6,9,12,15,18,21-Heptaoxatricosane-1,23-diol. *See* PEG-8
cis-4-Heptenal. *See* cis-4-Hepten-1-al

cis-4-Hepten-1-al
CAS 6728-31-0; FEMA 3289
Synonyms: cis-4-Heptenal
Empirical: $C_7H_{12}O$
Properties: m.w. 112.17; dens. 0.856; flash pt. 105 F
Uses: Synthetic flavoring
Regulatory: FDA 21CFR §172.515
Manuf./Distrib.: Aldrich

Heptyl acetate
CAS 112-06-1; FEMA 2547
Empirical: $C_9H_{18}O_2$
Properties: Colorless liq., pear-like odor, apricot-like taste; sol. in alcohol, ether; insol. in water; m.w. 150.18; dens. 0.87505 (15 C); m.p. -50 C; b.p. 192-193 C; flash pt. 154 F; ref. index 1.4150
Uses: Synthetic flavoring agent
Usage level: 4.1 ppm (nonalcoholic beverages), 3.3 ppm (ice cream, ices), 4.9 ppm (candy), 4.8 ppm (baked goods)
Regulatory: FDA 21CFR §172.515
Manuf./Distrib.: Aldrich

Heptyl alcohol
CAS 111-70-6; EINECS 203-897-9; FEMA 2548
Synonyms: 1-Heptanol; Hydroxy heptane; Alcohol C-7; Enanthyl alcohol; Enanthic alcohol
Empirical: $C_7H_{16}O$
Formula: $CH_3(CH_2)_6OH$
Properties: Colorless liq., citrus odor; sl. sol. in water @ 175 C; misc. with alcohol, fixed oils, ether; m.w. 116.23; dens. 0.824 (20/4 C); m.p. -34.6 C; b.p. 175.8 C; flash pt. 70 C; ref. index 1.423-.1427
Precaution: Combustible liq.; reactive with oxidizing materials
Toxicology: LD50 (oral, rat) 500 mg/kg; mod. toxic by ingestion, skin contact; mildly toxic by inh.; heated to decomp., emits acrid smoke and fumes
Uses: Synthetic flavoring agent
Usage level: 0.9 ppm (nonalcoholic beverages), 1-5 ppm (ice cream), 3 ppm (candy), 3 ppm (baked goods)
Regulatory: FDA 21CFR §172.515
Manuf./Distrib.: Aldrich

Heptylaldehyde. *See* Heptanal
n-Heptyl-n-butanoate. *See* Heptyl butyrate

Heptyl butyrate
CAS 5870-93-9; FEMA 2549
Synonyms: n-Heptyl-n-butanoate; n-Heptyl-n-butyrate
Empirical: $C_{11}H_{22}O_2$
Properties: Colorless liq., fruity camomile-like odor, sweet green tea-like taste; sol. in alcohol; almost insol. in water; m.w. 186.30; m.p. -58 C; b.p. 225-226 C; flash pt. 195 F; ref. index 1.4231
Uses: Synthetic flavoring agent
Usage level: 0.66 ppm (nonalcoholic beverages), 0.74 ppm (ice cream, ices), 2.7 ppm (candy), 2.4 ppm (baked goods)
Regulatory: FDA 21CFR §172.515
Manuf./Distrib.: Aldrich

n-Heptyl-n-butyrate. *See* Heptyl butyrate
Heptyl caprylate. *See* Heptyl octanoate
Heptyl carbinol. *See* Caprylic alcohol

Heptyl cinnamate
FEMA 2551
Synonyms: n-Heptyl cinnamate; Heptyl-β-phenylacrylate; Heptyl-3-phenyl propenoate
Definition: Ester of n-heptanol with cinnamic acid
Empirical: $C_{16}H_{22}O_2$
Properties: Colorless to pale straw-yel. liq., green leafy odor; sol. in alcohol; insol. in water; m.w. 246.36
Uses: Synthetic flavoring agent
Usage level: 3.3 ppm (nonalcoholic beverages), 1.0-2.0 ppm (ice cream, ices), 1.0-6.0 ppm (candy), 1.0 ppm (baked goods), 270 ppm (chewing gum)
Regulatory: FDA 21CFR §172.515

n-Heptyl cinnamate. *See* Heptyl cinnamate
5-Heptyldihydro-2(3H)-furanone. *See* ç-Undecalactone
n-Heptyl dimethylacetate. *See* Heptyl isobutyrate

Heptyl formate
CAS 112-23-2; FEMA 2552
Empirical: $C_8H_{16}O_2$
Properties: Colorless liq., fruity floral odor, plum-like taste; sol. in ether; insol. in water
Uses: Synthetic flavoring agent
Regulatory: FDA 21CFR §172.515
Manuf./Distrib.: Aldrich

n-Heptyl p-hydroxybenzoate. *See* Heptyl paraben
Heptylidene acetone. *See* 3-Decen-2-one
n-Heptyl isobutanoate. *See* Heptyl isobutyrate

Heptyl isobutyrate
FEMA 2550
Synonyms: n-Heptyl dimethylacetate; n-Heptyl isobutanoate
Definition: Ester of n-heptanol and isobutyric acid
Empirical: $C_{11}H_{22}O_2$
Properties: Colorless liq., woody odor; sol. in most org. solvs.; insol. in water; m.w. 186.30; b.p. 98 C (10 mm); ref. index 1.4190
Uses: Synthetic flavoring agent
Usage level: 1.2 ppm (nonalcoholic beverages), 0.82 ppm (ice cream, ices), 2.6 ppm (candy), 3.0 ppm (baked goods)
Regulatory: FDA 21CFR §172.515

Heptyl methyl ketone. *See* 2-Nonanone

Heptyl octanoate
Synonyms: n-Heptyl octanoate; Heptyl caprylate; Heptyl octylate
Definition: Ester of n-heptanol and n-octanoic acid
Empirical: $C_{15}H_{30}O_2$
Properties: Colorless oily liq., oily green odor and flavor; insol. in water; m.w. 242.41; dens. 0.8520 (30 C); m.p. -10 C; b.p. 291 C; ref. index 1.4340
Uses: Synthetic flavoring agent
Usage level: 1.0 ppm (nonalcoholic beverages)
Regulatory: FDA 21CFR §172.515

n-Heptyl octanoate. *See* Heptyl octanoate

Heptyl octylate. *See* Heptyl octanoate

Heptyl paraben
    Synonyms: n-Heptyl p-hydroxybenzoate
    Empirical: $C_{14}H_{20}O_3$
    Properties: Colorless sm. cryst. or wh. cryst. powd., odorless or faint char. odor, sl. burning taste; sol. in alcohol, ether; sl. sol. in water; m.w. 236.31; m.p. 48-51 C
    Toxicology: Heated to decomp., emits acrid smoke and irritating fumes
    Uses: Antioxidant, preservative, antimicrobial agent for beer, fermented malt beverages, noncarbonated soft drinks and fruit drinks
    Regulatory: FDA 21CFR §172.145, 12 ppm max. in fermented malt beverages, 20 ppm max. in soft/fruit drinks; BATF 27CFR §240.1051, 12 ppm max. in wine

Heptyl-β-phenylacrylate. *See* Heptyl cinnamate
Heptyl-3-phenyl propenoate. *See* Heptyl cinnamate

Hesperetin
    Uses: Natural antioxidant
    Regulatory: Japan approved

Hexacalcium phytate. *See* Calcium phytate
Hexadecanoic acid. *See* Palmitic acid
Hexadecanoic acid aluminum salt. *See* Aluminum palmitate
Hexadecanoic acid, 1,2,3-propanetriyl ester. *See* Tripalmitin
Hexadecanoic acid sodium salt. *See* Sodium palmitate
1-Hexadecanol. *See* Cetyl alcohol
ω-6-Hexadecenlactone. *See* Ambrettolide
N-Hexadecyl-N,N-dimethylbenzenemethanaminium chloride. *See* Cetalkonium chloride
Hexadecylic acid. *See* Palmitic acid
Hexadienic acid. *See* Sorbic acid
2,4-Hexadienoic acid. *See* Sorbic acid
2,4-Hexadienoic acid calcium salt. *See* Calcium sorbate
2,4-Hexadienoic acid potassium salt. *See* Potassium sorbate
Hexaethylene glycol. *See* PEG-6
Hexaglycerin monooleate. *See* Polyglyceryl-6 oleate
Hexaglycerol dioleate. *See* Polyglyceryl-6 dioleate
Hexaglycerol distearate. *See* Polyglyceryl-6 distearate
Hexaglyceryl dioleate. *See* Polyglyceryl-6 dioleate
Hexaglyceryl distearate. *See* Polyglyceryl-6 distearate
Hexaglyceryl hexaoleate. *See* Polyglyceryl-6 hexaoleate
Hexaglyceryl oleate. *See* Polyglyceryl-6 oleate
Hexahydroaniline. *See* Cyclohexylamine
Hexahydrobenzenamine. *See* Cyclohexylamine
Hexahydrobenzene. *See* Cyclohexane
[3aS-(3a-α,4b,6aα)]-Hexahydro-2-oxo-1H-thieno]3,4-d]imidazole-4-pentanoic acid. *See* d-Biotin
cis-Hexahydro-2-oxo-1H-thieno (3,4)-imidazole-4-valeric acid. *See* d-Biotin
Hexahydrophenethyl acetate. *See* Cyclohexylethyl acetate
Hexahydro phenylethyl propionate. *See* Ethyl cyclohexanepropionate
Hexahydropyridine. *See* Piperidine
Hexahydrothymol. *See* Menthol
Hexahydroxy cyclohexane. *See* Inositol

γ-Hexalactone
    CAS 695-06-7; FEMA 2556
    Synonyms: 4-Hydroxyhexanoic acid γ-lactone; Ethyl butyrolactone; Tonkalide; γ-Caprolactone
    Empirical: $C_6H_{10}O_2$
    Properties: Colorless liq., herbaceous sweet odor, sweet coumarin-caramel taste; sol. in alcohol, propylene glycol; sl. sol. in water; m.w. 114.15; dens. 1.023; m.p. -18 C; b.p. 220 C; flash pt. 209 F
    Uses: Synthetic flavoring agent
    Usage level: 7.0 ppm (nonalcoholic beverages), 0.07-84 ppm (ice cream, ices), 21 ppm (candy, baked goods)
    Regulatory: FDA 21CFR §172.515
    Manuf./Distrib.: Aldrich

Hexaldehyde. *See* Hexanal
Hexamethylene. *See* Cyclohexane

Hexamethylene tetramine
CAS 100-97-0; EINECS 202-905-8; EEC E239
Synonyms: HMTA; Methenamine; Aminoform; Urotropine; Hexamine
Empirical: $C_6H_{12}N_4$
Formula: $(CH_2)_6N_4$
Properties: White crystalline powd. or colorless lustrous crystals, pract. odorless; sol. in water, alcohol, chloroform; insol. in ether; m.w. 140.22; dens. 1.27 (25 C)
Precaution: Flamm.
Toxicology: Skin irritant; poison by subcutaneous route; moderately toxic by ingestion, intraperitoneal routes
Uses: Antimicrobial preservative; setting agent for proteins incl. casein used in food pkg. paper/paperboard
Usage level: ADI 0-0.15 mg/kg (EEC)
Regulatory: FDA 21CFR §181.30; Europe listed; UK approved
Manuf./Distrib.: Allchem Industries

Hexamine. *See* Hexamethylene tetramine

Hexanal
CAS 66-25-1; EINECS 200-624-5; FEMA 2557
Synonyms: Aldehyde C-6; Caproic aldehyde; Capronaldehyde; Hexaldehyde; n-Hexanal
Empirical: $C_6H_{12}O$
Formula: $CH_3(CH_2)_4CHO$
Properties: Colorless liq., powerful fatty-green odor; sol. in alcohol, fixed oils, propylene glycol; very sl. sol. in water; m.w. 100.18; dens. 0.808-0.812; m.p. -56.3 C; b.p. 128.7 C; flash pt. (OC) 90 F; ref. index 1.402-1.407
Precaution: Flamm.; dangerous fire hazard exposed to heat or flame; can react vigorously with oxidizing materials
Toxicology: LD50 (oral, rat) 4890 mg/kg; mildly toxic by ingestion, inh.; skin and eye irritant; heated to decomp., emits acrid smoke and fumes
Uses: Green fatty penetrating fragrance and flavoring; apple-like on dilution
Usage level: 1.3 ppm (nonalcoholic beverages), 2.8 ppm (ice cream, ices), 3.6 ppm (candy), 4.2 ppm (baked goods), 2.0-2.5 ppm (gelatins, puddings), 3.0 ppm (chewing gum)
Regulatory: FDA 21CFR §172.515
Manuf./Distrib.: Aldrich; BASF; Hüls

n-Hexanal. *See* Hexanal
Hexanaphthene. *See* Cyclohexane

Hexane
CAS 110-54-3; 64742-49-0; EINECS 203-777-6
Synonyms: n-Hexane; Alkane C-6
Classification: Aliphatic compd.
Empirical: C_6H_{14}
Formula: $CH_3(CH_2)_4CH_3$
Properties: Colorless volatile liq., faint odor; sol. in alcohol, acetone, ether; insol. in water; m.w. 86.20; dens. 0.65937 (20/4 C); m.p. -95 C; b.p. 68.742 C; flash pt. -22.7 C; ref. index 1.37486 (20 C)
Precaution: Flamm.; very dangerous fire/explosion hazard exposed to heat or flame; can react vigorously with oxidizers
Toxicology: TLV 50 ppm in air; LD50 (oral, rat) 28.710 mg/kg; sl. toxic by ingestion, inh.; human systemic effects by inh.; mutagenic data; eye irritant; irritating to respiratory tract; narcotic in high concs.; heated to decomp., emits acrid smoke and fumes
Uses: Extraction solvent for hops extract, spice oleoresins; processing aid (Japan)
Usage level: Limitation 25 ppm (spice oleoresins), 2.2% (hops extract)
Regulatory: FDA 21CFR §173.270; Japan approved with restrictions
Manuf./Distrib.: Ashland

Hexanedioic acid. *See* Adipic acid
Hexanedioic acid diisobutyl ester. *See* Diisobutyl adipate
2,3-Hexanedione. *See* Acetyl butyryl

3,4-Hexanedione
CAS 4437-51-8; EINECS 224-651-7; FEMA 3168
Synonyms: Dipropionyl
Empirical: $C_6H_{10}O_2$
Formula: $CH_2CH_2COCOCH_2CH_3$
Properties: m.w. 114.15; dens. 0.946 (20/4 C); b.p. 123-125 C; flash pt. 27 C; ref. index 1.411 (20 C)
Precaution: Flamm.

Uses: Pungent, butter-like fragrance and flavoring
Manuf./Distrib.: Aldrich; BASF

1,2,3,4,5,6-Hexanehexol. *See* D-Mannitol
1,2,3,4,5,6-Hexanehexol. *See* Sorbitol
n-Hexane. *See* Hexane
Hexanoic acid. *See* Caproic acid
Hexanoic acid ethyl ester. *See* Ethyl caproate
1-Hexanol. *See* Hexyl alcohol
n-Hexanol. *See* Hexyl alcohol
Hexazane. *See* Piperidine

2-Hexenal
CAS 6728-26-3; EINECS 229-778-1; FEMA 2560
Synonyms: α-β-Hexylenaldehyde; trans-2-Hexenal; β-Propyl acrolein
Empirical: $C_6H_{10}O$
Properties: Oily liq., char. green leafy odor; sol. in most org. solvs.; m.w. 98.15; dens. 0.844; b.p.47-48 C (17 mm); flash pt. 101 F; ref. index 1.446 (20 C)
Precaution: Flamm.
Toxicology: Skin irritant
Uses: Synthetic flavoring agent
Usage level: 3.1 ppm (nonalcoholic beverages), 0.70 ppm (ice cream, ices), 15 ppm (candy), 16 ppm (baked goods)
Regulatory: FDA 21CFR §172.515
Manuf./Distrib.: Aldrich

trans-2-Hexenal. *See* 2-Hexenal

1-Hexene
CAS 592-41-6; EINECS 209-753-1
Synonyms: C6 linear alpha olefin; Hexylene; Butyl ethylene
Empirical: C_6H_{12}
Formula: $CH_3CH_2CH_2CH_2CH:CH_2$
Properties: Colorless liq., mild hydrocarbon odor; sol. in alcohol; insol. in water; m.w. 84.16; dens. 0.678; f.p. -139.8 C; b.p. 62-63 C; flash pt. (Seta) 11 F; ref. index 1.3870 (20 C)
Precaution: Highly flamm.; dangerous fire risk
Toxicology: Irritating to eyes and skin; sl. toxic by ingestion; inhalation may produce CNS depression
Uses: Intermediate for flavors

2-Hexene-1-yl acetate. *See* trans-2-Hexenyl acetate

2-Hexenol
CAS 928-95-0; EINECS 213-191-2; FEMA 2562
Synonyms: 2-Hexen-1-ol; α,β-Hexenol; trans-2-Hexen-1-ol; Leaf alcohol; γ-Propylallyl alcohol
Empirical: $C_6H_{12}O$
Properties: Colorless liq., powerful fruity green wine-like odor, sweet fruity flavor; sol. in alcohol, propylene glycol; sl. sol. in water; m.w. 100.16; dens. 0.843 (20/4 C); b.p. 158-160 C; flash pt. 64 C; ref. index 1.438 (20 C)
Toxicology: Irritant
Uses: Synthetic flavoring agent
Usage level: 1.0 ppm (nonalcoholic beverages), 0.63 ppm (ice cream, ices), 3.8 ppm (candy), 4.1 ppm (baked goods)
Regulatory: FDA 21CFR §172.515
Manuf./Distrib.: Aldrich

2-Hexen-1-ol. *See* 2-Hexenol

3-Hexenol
CAS 928-96-1; EINECS 213-192-8; FEMA 2563
Synonyms: cis-3-Hexen-1-ol; 3-Hexen-1-ol; Leaf alcohol; β-γ-Hexenol
Classification: Organic compd.
Empirical: $C_6H_{12}O$
Formula: $CH_3CH_2CH=CHCH_2CH_2OH$
Properties: Colorless liq., strong grassy-green odor; sol. in alcohol, propylene glycol, fixed oils; very sl. sol. in water; m.w. 100.18; dens. 0.846-0.850; b.p. 156-157 C; flash pt. 44 C; ref. index 1.43-1.441
Precaution: Flamm.
Toxicology: LD50 (oral, rat) 4700 mg/kg, (IP, rat) 600 mg/kg; poison by intraperitoneal route; mildly toxic by

ingestion; heated to decomp., emits acrid smoke and fumes
Uses: Synthetic flavoring agent
Usage level: 1.0 ppm (nonalcoholic beverages), 3.7 ppm (ice cream, ices), 5.0 ppm (candy, baked goods)
Regulatory: FDA 21CFR §172.515
Manuf./Distrib.: Aldrich

3-Hexen-1-ol. *See* 3-Hexenol
α,β-Hexenol. *See* 2-Hexenol
β-γ-Hexenol. *See* 3-Hexenol
cis-3-Hexen-1-ol. *See* 3-Hexenol
trans-2-Hexen-1-ol. *See* 2-Hexenol
2-Hexen-1-yl acetate. *See* trans-2-Hexenyl acetate

trans-2-Hexenyl acetate
CAS 2497-18-9; FEMA 2564
Synonyms: 2-Hexen-1-yl acetate; 2-Hexene-1-yl acetate
Empirical: $C_8H_{14}O_2$
Properties: Liq., pleasant fruity odor and taste; m.w. 142.19; dens. 0.898; b.p. 165-166 C; flash pt. 58 C; ref. index 1.4270
Toxicology: Irritant
Uses: Synthetic flavoring agent
Usage level: 0.28 ppm (nonalcoholic beverages), 0.40 ppm (ice cream, ices), 1.7 ppm (candy, baked goods)
Regulatory: FDA 21CFR §172.515

β-γ-Hexenyl isopentanoate. *See* cis-3-Hexenyl isovalerate
3-Hexenyl isovalerate. *See* cis-3-Hexenyl isovalerate

cis-3-Hexenyl isovalerate
Synonyms: 3-Hexenyl isovalerate; β-γ-Hexenyl isopentanoate
Definition: Ester of cis-3-hexenol and isovaleric acid
Empirical: $C_{11}H_{20}O_2$
Properties: Colorless liq., powerful sweet green odor of apple, butter apple-like taste; sol. in alcohol, propylene glycol; insol. in water; m.w. 184.28; dens. 0.89; b.p. 199 C
Uses: Synthetic flavoring agent
Regulatory: FDA 21CFR §172.515

3-Hexenyl-2-methylbutyrate. *See* cis-3-Hexenyl 2-methylbutyrate

cis-3-Hexenyl 2-methylbutyrate
FEMA 3497
Synonyms: 3-Hexenyl-2-methylbutyrate; cis-3-Hexenyl-α-methylbutyrate
Empirical: $C_{11}H_{20}O_2$
Properties: Colorless liq., warm fruity apple-like odor, sweet apple-like taste; sol. in alcohol; insol. in water; m.w. 184.28
Uses: Synthetic flavoring agent
Regulatory: FDA 21CFR §172.515
Manuf./Distrib.: Aldrich

cis-3-Hexenyl-α-methylbutyrate. *See* cis-3-Hexenyl 2-methylbutyrate

3-Hexenyl phenylacetate
Synonyms: cis-3-Hexenyl phenylacetate
Uses: Synthetic flavoring agent
Regulatory: FDA 21CFR §172.515

cis-3-Hexenyl phenylacetate. *See* 3-Hexenyl phenylacetate
Hexone. *See* Methyl isobutyl ketone

Hexyl acetate
CAS 142-92-7; EINECS 205-572-7; FEMA 2565
Empirical: $C_8H_{16}O_2$
Formula: $CH_3COO(CH_2)_5CH_3$
Properties: Colorless oily liq., pleasant fruity odor, bittersweet taste; sol. in alcohol, ether; insol. in water; m.w. 144.22; dens. 0.873 (20/4 C); m.p. -81 C; b.p. 167-169 C; flash pt. 41 C; ref. index 1.409 (20 C)
Precaution: Flamm.
Uses: Synthetic flavoring agent
Usage level: 4.6 ppm (nonalcoholic beverages, ice cream, ices), 36 ppm (candy), baked goods), 3.0 ppm (chewing gum)

Regulatory: FDA 21CFR §172.515
Manuf./Distrib.: Aldrich

2-Hexyl-4-acetoxytetrahydrofuran
FEMA 2566
Synonyms: 2-Hexyl-tetrahydrofuran-4-yl acetate
Empirical: $C_{12}H_{22}O_3$
Properties: Colorless liq., sweet floral-fruity odor, peach-apricot taste; sol. in alcohol; sl. sol. in water; m.w. 214.31
Uses: Synthetic flavoring agent
Usage level: 1.0 ppm (nonalcoholic beverages), 3.0 ppm (ice cream, ices, candy, baked goods)
Regulatory: FDA 21CFR §172.515

Hexyl alcohol
CAS 111-27-3; 68526-79-4; EINECS 203-852-3; FEMA 2567
Synonyms: 1-Hexanol; n-Hexanol; Alcohol C-6; Pentylcarbinol; Amylcarbinol
Classification: Aliphatic alcohol
Empirical: $C_6H_{14}O$
Formula: $CH_3(CH_2)_4CH_2OH$
Properties: Colorless liq., fruity odor, aromatic flavor; sol. in alcohol and ether; sl. sol. in water; m.w. 102.20; dens. 0.8186; f.p. -51.6 C; b.p. 157.2 C; flash pt. (TOC) 65 C; ref. index 1.1469 (25 C)
Precaution: Flamm. or combustible liq.; reactive with oxidizing materials
Toxicology: LD50 (rat, oral) 4.59 g/kg; poison by intravenous route; moderately toxic by ingestion, skin contact; skin and severe eye irritant;
Uses: Synthetic flavoring agent, intermediate in mfg. of food additives
Usage level: 6.6 ppm (nonalcoholic beverages), 26 ppm (ice cream, ices), 18 ppm (baked goods), 0.22-0.28 ppm (gelatins, puddings)
Regulatory: FDA 21CFR §172.515, 172.864, 178.3480; GRAS (FEMA)
Manuf./Distrib.: Aldrich

Hexyl-2-butenoate
CAS 19089-92-0; FEMA 3354
Synonyms: Hexyl trans-2-butenoate
Empirical: $C_{10}H_{18}O_2$
Properties: Fruity pineapple odor; m.w. 170.25; dens. 0.885; flash pt. 193 F
Uses: Flavoring agent
Manuf./Distrib.: Aldrich

Hexyl trans-2-butenoate. *See* Hexyl-2-butenoate

Hexyl butyrate
CAS 2639-63-6; FEMA 2568
Empirical: $C_{10}H_{20}O_2$
Properties: Liq., apricot-like odor, pineapple-like taste; m.w. 172.27; dens. 0.851; m.p. -78 C; b.p. 205 C
Uses: Synthetic flavoring agent
Usage level: 2.6 ppm (nonalcoholic beverages), 2.1 ppm (ice cream, ices), 7.8 ppm (candy), 8.6 ppm (baked goods)
Regulatory: FDA 21CFR §172.515
Manuf./Distrib.: Aldrich

Hexyl caproate. *See* Hexyl hexanoate
Hexyl capronate. *See* Hexyl hexanoate
Hexyl caprylate. *See* Hexyl octanoate

α-Hexylcinnamaldehyde
CAS 101-86-0; FEMA 2569
Synonyms: 2-Benzylidene-octanal; α-n-Hexyl cinnamic aldehyde
Empirical: $C_{15}H_{20}O$
Properties: Pale yel. liq., jasmine-like odor; m.w. 216.33; dens. 0.95; m.p. 4 C; b.p. 174-176 C (15 mm); flash pt. > 230 f; ref. index 1.5480-1.5520
Toxicology: Irritant
Uses: Synthetic flavoring agent
Usage level: 0.80 ppm (nonalcoholic beverages), 2.6 ppm (ice cream, ices), 6.5 ppm (candy), 2.4 ppm (baked goods), 0.05 ppm (gelatins, puddings)
Regulatory: FDA 21CFR §172.515
Manuf./Distrib.: Aldrich

α-n-Hexyl cinnamic aldehyde. *See* α-Hexylcinnamaldehyde
α-β-Hexylenaldehyde. *See* 2-Hexenal
Hexylene. *See* 1-Hexene
Hexylene glycol diacetate. *See* 1,3-Nonanediol acetate, mixed esters

Hexyl formate
CAS 629-33-4; FEMA 2570
Empirical: $C_7H_{14}O_2$
Properties: Colorless liq., green fruity odor, sweet taste; sl. sol. in water; misc. with alcohol, ether; m.w. 130.19; dens. 0.879; m.p. -63 C; b.p. 155-156 C; flash pt. 118 F; ref. index 1.4071
Uses: Synthetic flavoring agent
Usage level: 12 ppm (nonalcoholic beverages), 45 ppm (ice cream, ices), 39 ppm (candy), 52 ppm (baked goods)
Regulatory: FDA 21CFR §172.515
Manuf./Distrib.: Aldrich

Hexyl hexanoate
CAS 6378-65-0; FEMA 2572
Synonyms: Hexyl caproate; Hexyl capronate; Hexyl hexylate
Empirical: $C_{12}H_{24}O_2$
Properties: Oily liq., herbaceous odor; m.w. 201.33; dens. 0.863; m.p. -55 C; b.p. 244-246 C; flash pt. 211 F; ref. index 1.4070-1.4090
Uses: Synthetic flavoring agent
Usage level: 2.5-3.0 ppm (nonalcoholic beverages), 2.5 ppm (ice cream, ices), 3.6-10 ppm (candy), 10 ppm (baked goods)
Regulatory: FDA 21CFR §172.515
Manuf./Distrib.: Aldrich

Hexyl hexylate. *See* Hexyl hexanoate

2-Hexylidene cyclopentanone
Uses: Synthetic flavoring
Regulatory: FDA 21CFR §172.515

n-Hexyl isopentanoate. *See* Hexyl isovalerate

Hexyl isovalerate
FEMA 3500
Synonyms: n-Hexyl isopentanoate
Definition: Ester of n-hexanol and isovaleric acid
Empirical: $C_{11}H_{22}O_2$
Properties: Colorless liq., pungent fruity odor; sol. in alcohol, fixed oils; insol. in water; m.w. 186.30; dens. 0.853; flash pt. 215 C; ref. index 1.417
Toxicology: Heated to decomp, emits acrid smoke and irritating fumes
Uses: Synthetic flavoring agent
Regulatory: FDA 21CFR §172.515

Hexyl 2-methylbutyrate
Synonyms: 2-Methylbutanoic acid n-hexyl ester
Definition: Ester of n-hexanol and 2-methylbutanoic acid
Empirical: $C_{11}H_{22}O_2$
Properties: Colorless liq., strong green fruity odor, unripe strawberry taste; sol. in alcohol; insol. in water; m.w. 186.30
Uses: Synthetic flavoring agent
Regulatory: FDA 21CFR §172.515

Hexyl methyl carbinol. *See* 2-Octanol
Hexyl methyl ketone. *See* Methyl hexyl ketone

Hexyl octanoate
FEMA 2575
Synonyms: Hexyl caprylate; n-Hexyl-n-octanoate; n-Hexyl octylate
Definition: Ester of n-hexanol and caproic acid
Empirical: $C_{14}H_{28}O_2$
Properties: Liq., sl. fruity odor, sweet green fruity taste; sol. in alcohol; insol. in water; m.w. 228.37; dens. 0.87; m.p. -31 C; b.p. 277 C
Uses: Synthetic flavoring
Usage level: 1.0 ppm (nonalcoholic beverages), 0.70 ppm (gelatins, puddings)

Regulatory: FDA 21CFR §172.515

n-Hexyl-n-octanoate. *See* Hexyl octanoate
n-Hexyl octylate. *See* Hexyl octanoate

Hexyl phenylacetate
Synonyms: n-Hexyl phenylacetate
Uses: Synthetic flavoring
Regulatory: FDA 21CFR §172.515

n-Hexyl phenylacetate. *See* Hexyl phenylacetate
n-Hexyl propanoate. *See* Hexyl propionate

Hexyl propionate
CAS 2445-76-3; FEMA 2576
Synonyms: n-Hexyl propanoate
Definition: Ester of n-hexanol and propionic acid
Empirical: $C_9H_{18}O_2$
Properties: Liq., earthy acrid odor, sweet metallic-fruity taste; sol. in alcohol, propylene glycol; insol. in water; m.w. 158.24; dens. 0.871; b.p. 180 C; flash pt. 149 F; ref. index 1.4105
Uses: Synthetic flavoring
Usage level: 5.7 ppm (nonalcoholic beverages), 23 ppm (ice cream, ices), 21 ppm (candy), 22 ppm (baked goods)
Manuf./Distrib.: Aldrich

4-Hexylresorcinol
CAS 136-77-6
Trade names containing: EverFresh®

2-Hexyl-tetrahydrofuran-4-yl acetate. *See* 2-Hexyl-4-acetoxytetrahydrofuran

Hickory bark
FEMA 2577
Definition: Carya spp.
Uses: Natural flavoring agent
Usage level: 21-40 ppm (nonalcoholic beverages), 70 ppm (alcoholic beverages), 0.01-25 ppm (ice cream, iooo), 18 ppm (candy, baked goods), 05 ppm (condiments)
Regulatory: FDA 21CFR §182.20, GRAS

Histidine. *See* L-Histidine

L-Histidine
CAS 71-00-1; EINECS 200-745-3
Synonyms: Histidine; α-Amino-β-imidazolepropionic acid; Glyoxaline-5-alanine
Empirical: $C_6H_9N_3O_2$
Formula: $HOOCCH(NH_2)CH_2C_3H_3N_2$
Properties: Wh. need., plates, or cryst. powd., sl. bitter taste; sol. in water; very sl. sol. in alcohol; insol. in ether; m.p. 285-287 C (dec.)
Toxicology: Experimental reproductive effects; human mutagenic data; heated to decomp., emits toxic fumes of NO_x
Uses: Nutrient, dietary supplement, feed additive; flavoring (Japan)
Regulatory: FDA 21CFR §172.310, limitation 2.4%, 172.320; Japan approved
Manuf./Distrib.: Degussa; Penta Mfg.

Histidine hydrochloride monohydrate
CAS 6341-24-8 (D-), 5934-29-2 (L-); EINECS 228-733-3, 211-438-9 resp.
Synonyms: Histidine monohydrochloride; 1-α-Amino-4 (or 5)-imidazolepropionic acid monohydrochloride; Glyoxaline-5-alanine monohydrochloride
Empirical: $C_6H_9N_3O_2$ • ClH • H_2O
Properties: Wh. need., plates, or cryst. powd., sl. bitter taste; sol. in water; insol. in alcohol, ether; m.w. 209.63; decomp. 250 C
Toxicology: Heated to decomp., emits toxic fumes of NO_x
Uses: Nutrient, dietary supplement; flavoring
Regulatory: FDA 21CFR §172.310, limitation 2.4%; Japan approved

Histidine monohydrochloride. *See* Histidine hydrochloride monohydrate
HMTA. *See* Hexamethylene tetramine
Hoarhound. *See* Horehound
Homo-cuminic aldehyde. *See* p-Isopropylphenylacetaldehyde

Honey
CAS 8028-66-8
Definition: Saccharic secretion gathered by honey bees, *Apis mellifera*
Uses: Sweetener for foods and beverages
Trade names: Fancol HON; Honi-Bake® 705 Honey; Sweet'n'Neat® 45; Sweet'n'Neat® 2000; Sweet'n'Neat® 3000
Trade names containing: Honi-Bake® Honey

Hops extract
CAS 8016-25-9
Synonyms: Humulus lupulus extract
Definition: Derived from *Humulus lupulus*
Properties: Bitter tonic, aromatic flavor
Toxicology: Heated to decomp., emits acrid smoke and irritating fumes
Uses: Natural flavoring agent for beer making
Usage level: 160 ppm (nonalcoholic beverages)
Regulatory: FDA 21CFR §172.560, 182.20, GRAS; Japan approved
Manuf./Distrib.: C.A.L.-Pfizer; Chart
Trade names: Pfico$_2$.Hop®

Hops oil
CAS 8007-04-3
Definition: Volatile oil derived from *Humulus lupulus* or *H. americanus*, containing lupulin, alkaloids, valerianic acid
Properties: Lt. yel. to brnsh. liq., aromatic odor; sol. in fixed oils, min. oil; sl. sol. in alcohol; insol. in glycerin, propylene glycol, water; dens. 0.855-0.880 (15/15 C); ref. index 1.1470-1.494 (20 C)
Precaution: Keep cool, well closed; protect from light
Toxicology: Heated to decomp., emits acrid smoke and irritating fumes
Uses: Natural flavoring agent, beer making, medicine; aromatic bitter
Usage level: 1.7 ppm (nonalcoholic beverages), 1.7 ppm (ice cream), 2.5 ppm (candy), 2.9 ppm (baked goods), 2.2 ppm (chewing gum), 20-35 ppm (condiments)
Regulatory: FDA 21CFR §182.20, GRAS; Japan approved

Horehound
FEMA 2581
Synonyms: Hoarhound; Horehound extract
Definition: From *Marrubium vulgare*
Properties: Bitter tonic, balsamic
Uses: Natural flavoring agent
Usage level: 8.7 ppm (nonalcoholic beverages), 2.0 ppm (ice cream, ices), 680 ppm (candy), 2.0 ppm (baked goods)
Regulatory: FDA 21CFR §182.10, 182.20, GRAS
Manuf./Distrib.: Chart

Horehound extract. *See* Horehound

Horsemint
FEMA 2582
Synonyms: Monarda; Wild bergamot
Definition: Monarda punctata
Properties: Thymol-like odor, harsh burning aromatic flavor; dens. 0.923-0.933
Uses: Natural flavoring agent
Usage level: 600 ppm (nonalcoholic beverages)
Regulatory: FDA 21CFR §182.20, GRAS

Horseradish
Definition: From *Armoracia lapathifolia*
Properties: Odor similar to mustard seed, sharp burning pungent aromatic flavor
Uses: Natural flavoring agent; used in cooking to prepare sharp piquant sauces
Regulatory: FDA 21CFR §182.10, GRAS

Humulus lupulus extract. *See* Hops extract
Hungarian chamomile oil. *See* Matricaria oil
Hyacinthin. *See* Phenylacetaldehyde
Hydrated alumina. *See* Aluminum hydroxide
Hydrated lime. *See* Calcium hydroxide

Hydratropalcohol. *See* 2-Phenyl propanol-1
Hydratropaldehyde. *See* 2-Phenylpropanal
Hydratropaldehyde dimethyl acetal. *See* 2-Phenylpropionaldehyde dimethylacetal
Hydratropic alcohol. *See* 2-Phenyl propanol-1
Hydratropic aldehyde. *See* 2-Phenylpropanal
Hydratropic aldehyde dimethyl acetal. *See* 2-Phenylpropionaldehyde dimethylacetal

Hydrazine
CAS 302-01-2
Synonyms: Hydrazine base; Hydrazine anhydrous; Diamide; Diamine
Empirical: H_4N_2
Formula: H_2NNH_2
Properties: Colorless oily fuming liq. or wh. cryst.; hygroscopic; misc. with water, alcohol; insol. in chloroform, ether; m.w. 32.06; dens. 1.004 (25/4 C); m.p. 2.0 C; b.p. 113.5 C; flash pt. 52 C; explodes during distilling with traces of air
Precaution: Flamm. liq., corrosive; very dangerous fire hazard exposed to heat, flame, or oxidizers; severe explosive hazard exposed to heat, flame, or by chemical reaction; potentially explosive with many chems.
Toxicology: TLV 0.1 ppm (skin); LD50 (oral, rat) 60 mg/kg; poison by ing., skin contact, IP, IV; mod. toxic by inh.; experimental carcinogen, teratogen; human mutagenic data; corrosive to eyes, skin, mucous membranes; heated to dec., emits highly toxic fumes
Uses: Boiler water additive for food contact
Regulatory: FDA 21CFR §173.310 (zero tolerance in steam)

Hydrazine anhydrous. *See* Hydrazine
Hydrazine base. *See* Hydrazine

Hydrocarbon solvent
CAS 64742-46-7
Uses: Solvent, foam control agent in fruit and veg. processing
Trade names: Penreco 2257 Oil

Hydrochloric acid
CAS 7647-01-0; EINECS 231-595-7; EEC E507
Synonyms: Muriatic acid
Classification: Inorganic acid
Empirical: ClH
Formula: HCl
Properties: Colorless fuming gas or liq., pungent odor; misc. with water, alcohol; m.w. 36.46; dens. 1.639 g/L (gas, 0 C), 1.194 (liq., -26 C); m.p. -114.3 C; b.p. -84.8 C
Precaution: Nonflamm. gas; explosive reaction with many chems.; potentially dangerous reaction with sulfuric acid releases HCl gas; strongly corrosive
Toxicology: TLV: CL 5 ppm; human poison; mildly toxic to humans by inh.; corrosive irritant to skin, eyes, mucous membranes; mutagenic data; experimental teratogen; 35 ppm causes throat irritation on short exposure; heated to decomp., emits toxic fumes of Cl⁻
Uses: Acid, buffer and neutralizing agent; auxiliary for inversion of sucrose, hydrolysis of starch and protein; processing aid; in beer malting
Usage level: ADI no limit (EEC)
Regulatory: FDA 21CFR §131.144, 131.129, 160.105, 160.185, 172, 172.560, 182.1057, GRAS; Japan restricted; Europe listed; UK approved
Manuf./Distrib.: Asahi Chem Industry Co Ltd; Hüls AG; PPG Industries

Hydrocinnamaldehyde
CAS 104-53-0; EINECS 203-211-8; FEMA 2887
Synonyms: Benzenepropanal; Benzylacetaldehyde; Dihydrocinnamaldehyde; Hydrocinnamic aldehyde; 3-Phenylpropanal; 3-Phenylpropionaldehyde
Empirical: $C_9H_{10}O$
Formula: $C_6H_5CH_2CH_2CHO$
Properties: Colorless to sl. yel. liq., strong floral hyacinth odor; misc. with alcohol, ether; insol. in water; m.w. 134.19; dens. 1.010-1.020; b.p. 221-224 C; flash pt. 203 F; ref. index 1.520-1.532
Precaution: Combustible liq.
Toxicology: LD50 (IV, mouse) 56 mg/kg; poison by IV route; human skin irritant; eye irritant; heated to decomp, emits acrid smoke and irritating fumes
Uses: Synthetic flavoring agent
Regulatory: FDA 21CFR §172.515

Hydrocinnamic acetate. *See* Hydrocinnamyl acetate

Hydrocinnamic acid
CAS 501-52-0; EINECS 207-924-5; FEMA 2889
Synonyms: 3-Phenylpropionic acid; Benzenepropanoic acid; Benzylacetic acid
Empirical: $C_9H_{10}O_2$
Formula: $C_6H_5CH_2CH_2COOH$
Properties: Wh. cryst. powd., sweet rosy odor; sol. in 170 parts cold water; sol. in alcohol, benzene, chloroform, ether, glac. acetic acid, petroleum ether, CS_2; m.w. 150.18; m.p. 45-48 C; b.p. 280 C; flash pt. > 230 F
Uses: Synthetic flavoring agent
Regulatory: FDA 21CFR §172.515; GRAS (FEMA)
Manuf./Distrib.: Aldrich

Hydrocinnamic alcohol
CAS 122-97-4; EINECS 204-587-6; FEMA 2885
Synonyms: 3-Phenylpropanol; Hydrocinnamyl alcohol; 3-Phenylpropyl alcohol; 3-Benzenepropanol
Empirical: $C_9H_{12}O$
Formula: $C_6H_5(CH_2)_3OH$
Properties: Colorless sl. visc. liq., sweet hyacinth-mignonette odor; sol. in fixed oils, propylene glycol; insol. in glycerin; m.w. 136.21; dens. 0.998-1.002; b.p. 119-121 C (12 mm); flash pt. 120 C; ref. index 1.524-1.528
Precaution: Combustible liq.
Toxicology: LD50 (oral, rat) 2300 mg/kg; mod. toxic by ingestion; mildly toxic by skin contact; skin irritant; heated to decomp., emits toxic fumes
Uses: Synthetic flavoring agent; mild, floral/balsamic, sl. hyacinth-like fragrance and flavoring
Regulatory: FDA 21CFR §172.515
Manuf./Distrib.: BASF; Hüls AG

Hydrocinnamic aldehyde. *See* Hydrocinnamaldehyde
Hydrocinnamic isobutyrate. *See* 3-Phenylpropyl isobutyrate

Hydrocinnamyl acetate
CAS 122-72-5; FEMA 2890
Synonyms: Hydrocinnamic acetate; Phenylpropyl acetate; 3-Phenyl-1-propanol acetate; 3-Phenylpropyl acetate
Empirical: $C_{11}H_{14}O_2$
Properties: Colorless liq., spicy floral odor; sol. in alcohol; insol. in water; m.w. 178.25; dens. 1.012; flash pt. > 212 F; ref. index 1.494
Precaution: Combustible liq.
Toxicology: LD50 (oral, rat) 4700 mg/kg; mildly toxic by ingestion; heated to decomp, emits acrid smoke and fumes
Uses: Synthetic flavoring agent
Regulatory: FDA 21CFR §172.515
Manuf./Distrib.: Hüls AG

Hydrocinnamyl alcohol. *See* Hydrocinnamic alcohol
Hydrocoumarin. *See* Dihydrocoumarin

Hydrogenated canola oil
Uses: Base for liq. or creamy shortenings, for frying applics.
Trade names: Akorex C; Design C200; Durkex Durola; Durola Select; Hydrogenated Canola Oil 81-601-0; Lobra; Lobra 70; Promise Liq. Oil; 37 Stearine
Trade names containing: H.L. 93; Superb® Canola Liquid Frying Shortening 104-455

Hydrogenated castor oil
CAS 8001-78-3; EINECS 232-292-2
Synonyms: Opalwax; Castorwax; Castor oil, hydrogenated
Definition: End prod. of controlled hydrogenation of castor oil
Properties: Hard white wax; very insol. in water and in the more common org. solvs.; m.w. ≈ 932; m.p. 86-88 C
Uses: Wax for food applics.
Regulatory: FDA 21CFR §175.105, 175.300, 176.170, 176.210, 177.1200, 177.1210, 177.2420, 177.2800, 178.3280
Trade names: Castorwax® NF

Hydrogenated coconut oil
EINECS 284-283-8
Synonyms: Coconut oil, hydrogenated
Definition: End prod. of controlled hydrogenation of coconut oil

Uses: Clouding agent
Regulatory: FDA 21CFR §175.105, 176.210, 177.2800
Trade names: Cobee 92; Cobee 110; Hydrol 92; Hydrol 100; Pureco® 92; Pureco® 100; Special Fat 42/44
Trade names containing: Akodel 95; Akodel 102; Akodel 108; Akodel 112; Akodel 118; 24% Beta Carotene Semi-Solid Suspension No. 65642; Dry Beta Carotene Beadlets Yellow Type 2.4-S No. 653800100; Precision Base; Pureco® 110

Hydrogenated corn oil
EINECS 271-200-5
Synonyms: Corn oil, hydrogenated
Uses: Shortenings
Trade names containing: Superb® Corn Liquid Frying Shortening 104-255; Superb® Soy/Corn Frying Shortening 102-070; Tem Freeze™

Hydrogenated corn syrup. *See* Hydrogenated starch hydrolysate

Hydrogenated cottonseed glyceride
CAS 61789-07-9
Synonyms: Glycerides, cottonseed oil, hydrogeanted
Definition: End prod. of controlled hydrogenation of cottonseed glyceride
Uses: Emulsifier
Trade names: Myverol® 18-07; Myverol® 18-07K

Hydrogenated cottonseed oil
CAS 68334-00-9; EINECS 269-804-9
Definition: End prod. of controlled hydrogenation of cottonseed oil
Properties: Pale yel.; dens. 0.915-0.921; flash pt. 486 F
Precaution: Combustible
Uses: Dietary supplement, food additive
Regulatory: FDA 21CFR §175.105, 176.210, 177.2800
Trade names: Akoleno D; Akolizer C; Diamond D 75; Duratex; Duromel; Duromel B108; 07 Stearine
Trade names containing: Akocote 102; Akocote 106; Akocote 109; Akocote 112; Akoleno SC; Akolizer PKC; Akolizer RSC; Akolizer SC; Akopol R; Akopuff; Akorex B; Akorine 2A; Akorine 9F; All Purpose Shortening 101-050; Baker's Ideal®; 24% Beta Carotene HS-E in Veg. Oil No. 65671; 30% Beta Carotene in Veg. Oil No. 65646; Buckeye C (Baker's Grade); Bunge All-Purpose Vegetable Shortening; Bunge Cake and Icing Vegetable Shortening; Cap-Shure® BC-140-70; Cap-Shure® C-140-72; Cap-Shure® C-140-85; Cap-Shure® F-140-63; Cap-Shure® GDL-140-70; Cap-Shure® SC 140X 70; Cap-Shure® SC-105X-00FF, CBC #7 Shortening; Coral®; Cremol Plus™ 60 Vegetable; Golden Chef® Creamy Liquid Frying Shortening 104-065; Golden Covo® Shortening; Golden Crescent; Golden Croissant/Danish™ Margarine; Golden Flake™; H.L. PY™; Hymo; Lipodan CDS Kosher; Mr. Chips™; Myvatex® 90-10K; Penguin® Vegetable Specialty Icing; Shedd's Wonder Shortening; Sta-Nut EE; Summit® 25; Summit® 50; Superb® Bakers Margarine 105-101; Superb® Bakers Shortening 105-200; Superb® Corn Liquid Frying Shortening 104-255; Superb® Heavy Duty Frying Shortening 101-200; Superb® Icing Shortening 101-270; Tem Cote® IE; Tem Cote® II; Tem Crunch™ 93; Tem Top™; Tri-Co® Vegetable Yeast Raised Dough Shortening; Trophy®; Vegetable Cake Shortening 101-250; Victor®

Hydrogenated isomaltulose. *See* Isomalt

Hydrogenated lard glyceride
CAS 8040-05-9
Synonyms: Glycerides, hydrogenated lard mono-
Definition: End prod. of controlled hydrogenation of lard glyceride
Uses: Emulsifier, stabilizer, dispersant
Regulatory: FDA 21CFR §175.105, 176.210
Trade names: Monomuls® 90-15

Hydrogenated lard glycerides
Synonyms: Hydrogenated lard mono-, di- and tri- glycerides; Glycerides, lard mono-, di- and tri-, hydrogenated
Definition: End prod. of controlled hydrogenation of lard glycerides
Uses: Emulsifier, stabilizer, dispersant
Regulatory: FDA 21CFR §175.105, 176.210
Trade names: Monomuls® 60-15

Hydrogenated lard mono-, di- and tri- glycerides. *See* Hydrogenated lard glycerides

Hydrogenated menhaden oil
CAS 93572-53-3, 68002-72-2
Synonyms: Menhaden oil, hydrogenated

Hydrogenated palatinose

Definition: End prod. of controlled hydrogenation of menhaden oil
Properties: Wh. opaque solid, odorless; iodine no. 10 max.; sapon. no. 180-200
Uses: Edible fats or oils
Regulatory: FDA 21CFR §175.105, 176.170, 176.210, 177.2800, 184.1472, 186.1551

Hydrogenated palatinose. See Isomalt

Hydrogenated palm glyceride
CAS 67784-87-6, 97593-29-8
Synonyms: Palm oil glyceride, hydrogenated; Glycerides, palm oil mono-, hydrogenated
Definition: End prod. of controlled hydrogenation of palm glyceride
Uses: Emulsifier, stabilizer, dispersant
Regulatory: FDA 21CFR §176.210, 177.2800
Trade names: Dimodan PVP Kosher; Monomuls® 90-35; Myvatex® 8-16; Myvatex® 8-16K; Myverol® 18-04; Myverol® 18-04K

Hydrogenated palm glycerides
Synonyms: Hydrogenated palm mono-, di- and tri-glycerides; Glycerides, palm oil mono-, di- and tri, hydrogenated
Definition: End prod. of controlled hydrogenation of palm oil glycerides
Uses: Emulsifier, stabilizer, dispersant
Trade names: Monomuls® 60-35

Hydrogenated palm kernel oil
CAS 68990-82-9
Synonyms: Oils, palm kernel, hydrogenated; Palm kernel oil, hydrogenated
Definition: End prod. of controlled hydrogenation of palm kernel oil
Uses: Hard fat for chocolate and confectionery prods.
Trade names: Akorine 4; Akowesco 2; Akowesco 5; Akowesco 7; Akowesco 45 AC; Paramount B; Witarix® 212
Trade names containing: Akodel 95; Akodel 102; Akodel 108; Akodel 112; Akodel 118; Akolizer PKC; Akorine 2A; Akorine 3; Akorine 9F; Akowesco 90; CLSP 555; Paramount C; Satina 44; Satina 50

Hydrogenated palm mono-, di- and tri-glycerides. See Hydrogenated palm glycerides

Hydrogenated palm oil
CAS 8033-29-2; 68514-74-9
Synonyms: Oils, palm, hydrogenated; Palm oil, hydrogenated
Definition: End prod. of controlled hydrogenation of palm oil
Uses: Consistency regulator for fat and fat preps.; coating fat; stabilizer to prevent oil separation in nut paste; crystallization enhancer
Regulatory: FDA 21CFR §175.105, 176.210, 177.2800
Trade names: Akolizer P; Dynasan® P60; 27 Stearine
Trade names containing: Akodel 102; Akodel 108; Akodel 112; Akodel 118; Alpine; BBS; Cap-Shure® LCL-135-50; Cap-Shure® SC-135-63; Hytex; Myvatex® Peanut Butter Stabilizer; Pureco® 110; Sta-Nut P; Super Fry

Hydrogenated peanut oil
CAS 68425-36-5; EINECS 270-350-9
Synonyms: Oils, peanut, hydrogenated
Definition: End prod. of controlled hydrogenation of peanut oil
Uses: Fast solidifying filling fat for chocolate and confectionery prods.
Regulatory: FDA 21CFR §175.105, 176.210, 177.2800
Trade names: Witarix® 450

Hydrogenated rapeseed oil
Definition: End-prod. of controlled hydrogenation of rapeseed oil
Properties: acid no. 6 max.; iodine no. 4 max.
Toxicology: Heated to decomp., emits acrid smoke and irritating fumes
Uses: Emulsifier, stabilizer, thickener in peanut butter
Usage level: Limitation 2% max. (peanut butter)
Regulatory: FDA 21CFR §184.1555, GRAS
Trade names containing: Akolizer RSC; Myvatex® 90-10K

Hydrogenated soybean glyceride. See Hydrogenated soy glyceride

Hydrogenated soybean glycerides
CAS 68201-48-9
Synonyms: Hydrogenated soybean oil mono-, di- and tri- glycerides; Hydrogenated soybean oil glycerides

Definition: End prod. of controlled hydrogenation of a mixture of mono, di and triglycerides derived from soybean oil
Uses: Emulsifier, stabilizer, dispersant
Regulatory: FDA 21CFR §176.210, 177.2800
Trade names: Monomuls® 60-45

Hydrogenated soybean oil
CAS 8016-70-4, 68002-71-1; EINECS 232-410-2
Synonyms: Soybean oil hydrogenated
Definition: End prod. of controlled hydrogenation of soybean oil
Uses: Migrating from cotton in dry food pkg.
Regulatory: FDA 21CFR §175.105, 176.210, 177.2800, 182.70, 182.170
Trade names: Akofame; Akofil A; Akofil N; Akoleno S; Akolizer S; Akorex; Bunge Biscuit Flakes; Bunge Heavy Duty Donut Frying Shortening; Clarity; Code 321; Diamond D 31; Donut Frying Shortening 101-150; Durkex 100F; Durkex 500S; Durkex Gold 77F; Famous; Kaomax-S; Lipovol HS; Partially Hydro Soybean Oil 86-505-0; Shasta®; Soy Flakes; 17 Stearine; Sterotex® HM NF; Superb® Cookie Bake Shortening 101-057; Superb Oil Hydro Winterized Soybean Oil 86-091-0; Tem Crunch™ 110; Tem Plus® 95; Witarix® 440
Trade names containing: Akocote 102; Akocote 106; Akocote 109; Akocote 112; Akoleno SC; Akolizer RSC; Akolizer SC; Akopol E-1; Akopol R; Akopuff; Akorex B; Akorine 3; All Purpose Shortening 101-050; Alpine; Baker's Ideal®; BBS; 24% Beta Carotene HS-E in Veg. Oil No. 65671; 30% Beta Carotene in Veg. Oil No. 65646; Beta Plus; Buckeye C (Baker's Grade); Bunge All-Purpose Vegetable Shortening; Bunge Cake and Icing Vegetable Shortening; Bunge Special Mix Shortening; Cake Mix 96; Cap-Shure® AS-165-70; Cap-Shure® C-150-50; Cap-Shure® C-165-63; Cap-Shure® C-165-85; Cap-Shure® CLF-165-70; Cap-Shure® F-125-85; Cap-Shure® FE-165-50; Cap-Shure® FF-165-60; Cap-Shure® FS-165-60; Cap-Shure® KCL-165-70; Cap-Shure® SC-140X-70; Cap-Shure® SC-165-85FT; Cap-Shure® SC-165X-60FF; CBC #7 Shortening; Coral®; Creamy Liquid Frying Shortening 102-050; Cremol Plus™ 60 Vegetable; Diet Imperial® Spread; Dress All; Durkex 25BHA; Durkex 100BHA; Durkex 100BHT; Durkex 100DS; Durkex 100L; Durkex 100TBHQ; Durkex Gold 77A; Durkex Gold 77N; Golden Chef® Creamy Liquid Frying Shortening 104-065; Golden Covo® Shortening; Golden Crescent; Golden Croissant/Danish™ Margarine; Golden Flake™; Gold 'N Flavor® 103-061; Heavy Duty Frying Shortening 101-100; Hi Glo™; Hi-Lite™; Hi-Tone™; H.L. PY™; HWDX Winterized Salad Oil; Hymo; Hytex; Kaoprem-E; Luxor® E-40; Luxor® E-50; Luxor® KB-300; Luxor® KB-320; Luxor® KB-330; Luxor® KB-350; Luxor® KB-400; Luxor® KB-500; Luxor® KB-530; Luxor® KB-600; Luxor® R-100; Mr. Chips™; No Stick Emulsifier; Penguin® Vegetable Specialty Icing; Shedd's Liquid Margarine; Shedd's Margarine; Shedd's NP Margarine; Shedd's Special 40 Butter Blend Margarine; Shedd's Special 60 Butter Blend Margarine; Shedd's Wonder Shortening; Soy/Peanut Liquid Frying Shortening 104-215; Sterotex® C; Summit®; Summit® 25; Summit® 50; Superb® Bakers Margarine 105-101; Superb® Bakers Shortening 105-200; Superb® Fish and Chip Oil 102-060; Superb® Heavy Duty Frying Shortening 101-200; Superb® Icing Shortening 101-270; Superb® Liquid Frying Shortening 102-110; Superb® Soybean Salad Oil 104-112; Superb® Soy/Corn Frying Shortening 102-070; Super-Cel® Specialty Cake Shortening; Super Fry; Tally® 100 Plus; Tem Cote® IE; Tem Cote® II; Tem Crunch™ 93; Tem Freeze™; Tem Top™; Tri-Co® Vegetable Yeast Raised Dough Shortening; Trophy®; Type B Dough Conditioner; Vegetable Cake Shortening 101-250; Victor®; Winterized Clear Liquid Frying Shortening 104-112

Hydrogenated soybean oil glycerides. *See* Hydrogenated soybean glycerides
Hydrogenated soybean oil mono-, di- and tri- glycerides. *See* Hydrogenated soybean glycerides
Hydrogenated soybean oil monoglyceride. *See* Hydrogenated soy glyceride

Hydrogenated soy glyceride
CAS 61789-08-0; 68002-71-1
Synonyms: Hydrogenated soybean oil monoglyceride; Glycerides, soybean oil, hydrogenated, mono; Hydrogenated soybean glyceride
Definition: End prod. of controlled hydrogenation of soybean monoglycerides
Uses: Emulsifier, stabilizer, dispersant
Regulatory: FDA 21CFR §176.210, 177.2800
Trade names: Dimodan O Kosher; Dimodan PV; Dimodan PV 300 Kosher; Monomuls® 90-45; Myvatex® 8-06; Myvatex® 8-06K; Myverol® 18-06; Myverol® 18-06K

Hydrogenated sperm oil
Toxicology: Heated to decomp, emits acrid smoke and irritating fumes
Uses: Component of release agent or lubricant in bakery pans
Regulatory: FDA 21CFR §173.275

Hydrogenated starch hydrolysate
CAS 68425-17-2

Hydrogenated stearic acid

Synonyms: Hydrogenated corn syrup; Corn syrup, hydrogenated
Definition: End prod. of controlled hydrogenation of corn syrup
Uses: Tobacco, teat dip, pet food
Trade names: Hystar® 3375; Hystar® 4075; Hystar® 5875; Hystar® 6075; Hystar® HM-75

Hydrogenated stearic acid
Uses: Lubricant, release agent, binder, defoamer in foods, intermediate for food emulsifiers
Trade names: Industrene® B

Hydrogenated tallow
CAS 8030-12-4; EINECS 232-442-7
Synonyms: Tallow, hydrogenated
Definition: End prod. of controlled hydrogenation of tallow
Uses: Defoamer for beet sugar and yeast processing; migrating from cotton in dry food pkg.
Regulatory: FDA 21CFR §173.340, 175.105, 176.170, 176.210, 177.2800, 182.70
Trade names: Special Fat 168T
Trade names containing: Top Notch®

Hydrogenated tallow alcohol
Definition: End prod. of controlled hydrogenation of tallow alcohol
Uses: Defoamer for beet sugar and yeast processing
Regulatory: FDA 21CFR §173.340, 175.105, 176.170, 176.210, 177.2800

Hydrogenated tallow glyceride
CAS 61789-09-1; EINECS 263-031-0
Synonyms: Hydrogenated tallow monoglyceride; Glycerides, hydrogenated tallow mono-
Definition: Monoglyceride of hydrogenated tallow
Uses: Emulsifier, stabilizer, dispersant
Regulatory: FDA 21CFR §176.210, 177.2800
Trade names: Monomuls® 90-25
Trade names containing: Monomuls® 90-25/2; Monomuls® 90-25/5

Hydrogenated tallow glyceride citrate
CAS 68990-59-0; EINECS 273-613-6
Synonyms: Glycerides, tallow mono-, hydrogenated, citrates
Definition: Citric acid ester of hydrogenated tallow glyceride
Uses: Emulsifier for margarine, meats
Trade names: Lamegin® ZE 30, 60; Tegin® C-62 SE; Tegin® C-63 SE
Trade names containing: Controx® KS; Controx® VP

Hydrogenated tallow glyceride lactate
CAS 68990-06-7; EINECS 273-576-6
Synonyms: Glycerides, tallow mono-, hydrogenated, lactates
Definition: Lactic acid ester of hydrogenated tallow glyceride
Uses: Emulsifier, food additive
Trade names: Lamegin® GLP 10, 20; Tegin® L 61, L 62

Hydrogenated tallow glycerides
CAS 68308-54-3, 67701-27-3; EINECS 269-658-6
Synonyms: Hydrogenated tallow mono-, di- and tri- glycerides; Glycerides, tallow mono-, di- and tri-, hydrogenated
Definition: Mixture of mono, di and triglycerides of hydrogenated tallow acid
Uses: Emulsifier, stabilizer, dispersant
Regulatory: FDA 21CFR §176.210, 177.2800
Trade names: Monomuls® 60-25
Trade names containing: Monomuls® 60-25/2; Monomuls® 60-25/5

Hydrogenated tallow mono-, di- and tri- glycerides. See Hydrogenated tallow glycerides
Hydrogenated tallow monoglyceride. See Hydrogenated tallow glyceride

Hydrogenated vegetable glyceride
CAS 61789-08-0
Synonyms: Glycerides, hydrogenated vegetable mono-; Glycerides, vegetable mono-, hydrogenated
Definition: End prod. of controlled hydrogenation of vegetable monoglyceride
Uses: Emulsifier
Regulatory: FDA 21CFR §176.210
Trade names: Myverol® 18-50; Myverol® 18-50K

Hydrogenated vegetable glycerides
CAS 69028-36-0
Synonyms: Hydrogenated vegetable oil mono-, di-, and tri-glycerides; Glycerides, vegetable mono-, di-, and tri, hydrogenated
Definition: Mixture of hydrogenated mono-, di-, and tri-glycerides of vegetable oil
Uses: Food emulsifier for icings, cream fillings; emulsion stabilizer; extends shelf-life in cake mixes, icings; dispersant for coffee whiteners; aerator for shortenings
Regulatory: FDA 21CFR §176.210
Trade names: Emuldan HV 40 Kosher; Emuldan HV 52 Kosher
Trade names containing: Myvatex® 25-07K

Hydrogenated vegetable glycerides phosphate
CAS 85411-01-4; 25212-19-5
Definition: Complex mixture of esters of phosphoric acid and mono- and di-glycerides derived from hydrog. vegetable oil
Uses: Mold lubricant for food use
Trade names: Emphos™ F27-85

Hydrogenated vegetable oil
CAS 68334-00-9; 68334-28-1, 68938-37-4; EINECS 269-820-6
Synonyms: Vegetable oil, hydrogenated
Definition: End prod. of controlled hydrogenation of vegetable oil
Uses: Food ingred.; emulsifier for food processing; binder, lubricant in pharmaceutical tableting
Regulatory: FDA 21CFR §175.105, 176.210
Manuf./Distrib.: Arista Industries; Karlshamns; Lipo; A.E. Staley Mfg.
Trade names: Aratex; Cirol; CLSP 874; Creamtex; Diamond D 40; Diamond D 42; Durkex 500; Durlite F; Hydrol 110; Kaokote F; Kaola; Kaola D; Kaomax 870; Kaomel; Kaorich; K.L.X; K.L.X. Flakes; Lipo SS; Lipodan SET Kosher; Magna A; Magna B; Magna C; Optima 7B; Optima 23B; Optima 77IC; Optima 871; Paramount H; Paramount X; Paramount XX; Snac-Kote XTR; Sta-Nut R; Sterotex® NF; Wecobee® FS; Wecobee® FW; Wecobee® HTR; Wecobee® M; Wecobee® R Mono; Wecobee® S; Wecobee® SS; Wecobee® W
Trade names containing: Apo-Carotenal Suspension 20% in Veg. Oil No. 66417; Betricing; Betrkake; Cap-Shure® C-135-72; Coco-Spred™; Confecto No-Stick™; Diamond D-21; Durko; Durkote Citric Acid/ Hydrog. Veg. Oil; Durkote Ferrous Fumarate/Hydrog. Veg. Oil; Durkote Ferrous Sulfate/Hydrog. Veg. Oil; Durkote Fumaric Acid/Hydrog. Veg. Oil; Durkote Glucono-Delta-Lactone/Hydrog. Veg. Oil; Durkote Lactic Acid/Hydrog. Veg. Oil; Durkote Potassium Chloride/Hydrog. Veg. Oil; Durkote Sodium Aluminum Phosphate/Hydrog. Veg. Oil; Durkote Sodium Bicarbonate/Hydrog. Veg. Oil; Durkote Sodium Chloride/Hydrog. Veg. Oil; Durkote Sorbic Acid/Hydrog. Veg. Oil; Durkote Vitamin B-1/Hydrog. Veg. Oil; Durkote Vitamin C/ Hydrog. Veg. Oil; Durlite Gold MBN II; Kaokote; Kaorich Gold; Kristel Gold; Kristel Gold II; Lamequick® AS 340; Lamequick® CE 1 SF; Luxor® 1517; Luxor® 1639; Luxor® 1658; Luxor® EB-2; Luxor® EB-400; Panospray™; Panospray™ LS; Panospray™ SQ; Panospray™ WL; Perflex; Satina 53NT; Snac-Kote

Hydrogenated vegetable oil mono-, di-, and tri-glycerides. *See* Hydrogenated vegetable glycerides
Hydrogen carboxylic acid. *See* Formic acid
Hydrogen dioxide. *See* Hydrogen peroxide

Hydrogen peroxide
CAS 7722-84-1; EINECS 231-765-0
Synonyms: Hydrogen dioxide
Classification: Inorganic oxide
Empirical: H_2O_2
Formula: HOOH
Properties: Colorless visc. liq., cryst. solid at low temp., bitter taste; sol. in ether, alcohol; misc. with water; dec. by many org. solvs.; m.w. 34.02; dens. (liq.) 1.450 g/cc (20 C); f.p. -0.41 C; b.p. 150.2 C
Precaution: Dangerous fire hazard by chem. reaction with flamm. materials; explosion hazard; strong oxidizer
Toxicology: LD50 (oral, mouse) 2 g/kg; mod. toxic by inh., ingestion, skin contact; corrosive irritant to skin, eyes, mucous membranes; tumorigenic; human mutagenic data; TLV: 1 ppm in air
Uses: Bleaching agent, oxidizing and reducing agent, starch modifier, preservative, antimicrobial agent; used in cheese whey, corn syrup, dried eggs, emulsifiers, milk, pkg. materials, starch, tea, wine, wine vinegar
Usage level: Limitation 0.05% (milk), 0.04% (whey), 0.15% (starch), 0.15% (corn syrup), 0.05% (annatto colored cheese whey), 1.25% (emulsifiers contg. fatty acid esters)
Regulatory: FDA 21CFR §133.133, 160.105, 160.145, 160.185, 172.814, 172.892, 175.105, 178.1005, 178.1005 (35% sol'n. max.), 178.1010, 184.1366, GRAS; BATF 27CFR §240.1051 (3 ppm max. in wine), 240.1051a (200 ppm max. in distilling materials); Japan restricted
Manuf./Distrib.: Browning

Hydrogen sulfate. *See* Sulfuric acid

Hydrolyzed barley flour
Uses: Natural sweetener; controls body, regulates osmalality
Trade names: Barley*Complete® 25

Hydrolyzed casein. *See* Hydrolyzed milk protein

Hydrolyzed gelatin
Uses: Nutrient, protective colloid in ice cream, meal supplements, desserts, jellies; film former; moisturizer
Trade names: Polypro 5000®; Polypro 15000® Food Grade; Polypro 15000® Pharmaceutical Grade
Trade names containing: Polypro 2000® CF 45%; Polypro 2000® UF 45%; Polypro 5000® CF or SF 45%; Polypro 5000® UF 45%

Hydrolyzed milk protein
CAS 65072-00-6; 92797-39-2; EINECS 265-363-1
Synonyms: Proteins, milk, hydrolysate; Hydrolyzed casein
Definition: Hydrolysate of milk protein derived by acid, enzyme or other method of hydrolysis
Properties: Sol. in water
Toxicology: Heated to decomp., emits acrid smoke and irritating fumes
Uses: Flavoring agent for bologna, salami, sauces, stuffing
Regulatory: USDA 9CFR §318.7

Hydrolyzed oat flour
Uses: Fat replacer for low-fat foods, salad dressings, confections, meats; thickener, texturizer
Trade names: TrimChoice™ 5
Trade names containing: TrimChoice™ OC

Hydrolyzed protein
CAS 73049-73-7
Uses: Nutritive supplement
Trade names containing: Beta Carotene Emulsion Beverage Type 3.6 No. 65392; Dry Vitamin E Acetate 50% SD No. 65356

Hydrolyzed vegetable protein
CAS 100209-45-8
Synonyms: Vegetable protein hydrolysate; Proteins, vegetable, hydrolysate
Definition: Hydrolysate of vegetable protein derived by acid, enzyme or other method of hydrolysis
Properties: Sol. n water
Toxicology: Heated to decomp., emits acrid smoke and irritating fumes
Uses: Flavor enhancer for bologna, salami, sauces, stuffing, savory packet mixes, sauces mixes, snack foods
Regulatory: USDA 9CFR §318.7
Manuf./Distrib.: Am. Roland
Trade names: Exter™ Family; HVP-LS; Luxor® Century™ V; Luxor® CVP 5-SD; Luxor® CVP 1632; Luxor® CVP 1689; Luxor® CVP 1753; Luxor® CVP LS; Luxor® E-610; Luxor® FB-10; Luxor® FC-20; Luxor® GR-100; Luxor® GR-150; Luxor® GR-200; Luxor® HVP-A; Luxor® KB-312; Luxor® L-625; Luxor® MB-40; Luxor® MB-110; Luxor® MB-120; Luxor® No. 6 Sauce; Luxor® No. 700 Sauce; Luxor® Prozate®; Luxor® Triple-H®; Toppro
Trade names containing: HVP 5-SD; HVP-A; Luxor® 1517; Luxor® 1576; Luxor® 1626; Luxor® 1639; Luxor® 1658; Luxor® E-40; Luxor® E-50; Luxor® EB-2; Luxor® EB-400; Luxor® KB-300; Luxor® KB-320; Luxor® KB-330; Luxor® KB-350; Luxor® KB-400; Luxor® KB-500; Luxor® KB-530; Luxor® KB-600; Luxor® R-100; Triple-H

Hydroquinone dimethyl ether. *See* p-Dimethoxybenzene
Hydrous magnesium silicate. *See* Talc
4-Hydroxy-m-anisaldehyde. *See* Vanillin
o-Hydroxyanisole. *See* Guaiacol
2-Hydroxybenzaldehyde. *See* Salicylaldehyde
o-Hydroxybenzaldehyde. *See* Salicylaldehyde
2-Hydroxybenzoic acid. *See* Salicylic acid
4-Hydroxybenzoic acid butyl ester. *See* Butylparaben
4-Hydroxybenzoic acid ethyl ester. *See* Ethyl paraben
4-Hydroxybenzoic acid, ethyl ester, sodium salt. *See* Sodium ethylparaben
4-Hydroxybenzoic acid, methyl ester. *See* Methylparaben
4-Hydroxybenzoic acid, methyl ester, sodium salt. *See* Sodium methylparaben
o-Hydroxybenzoic acid. *See* Salicylic acid
2-Hydroxybenzoic acid, pentyl ester. *See* Amyl salicylate

4-Hydroxybenzoic acid, propyl ester. *See* Propylparaben
4-Hydroxybenzoic acid, propyl ester, sodium salt. *See* Sodium propylparaben
p-Hydroxybenzyl acetone. *See* 4-(p-Hydroxyphenyl)-2-butanone
α-Hydroxybenzyl phenyl ketone. *See* Benzoin
2-Hydroxybiphenyl. *See* o-Phenylphenol
2-Hydroxybutane. *See* 2-Butanol
Hydroxybutanedioic acid. *See* N-Hydroxysuccinic acid
3-Hydroxy-2-butanone. *See* Acetyl methyl carbinol

Hydroxycitronellal
CAS 107-75-5; EINECS 203-518-7; FEMA 2583
Synonyms: 3,7-Dimethyl-7-hydroxy-octane-1-al; Citronellal hydrate; 7-Hydroxy-3,7-dimethyloctanol; Lilyl aldehyde
Empirical: $C_{10}H_{20}O_2$
Formula: $CH_3CH_3OHCCH_2CH_2CH_2CH_3CHCH_2CHO$
Properties: Colorless liq., sweet floral, lily odor; sol. in fixed oils, propylene glycol; insol. in glycerin; m.w. 172.30; dens. 0.918-0.923; b.p. 94-96 C (1 mm); flash pt. > 212 F; ref. index 1.447-1.450
Precaution: Combustible liq.
Toxicology: Skin irritant; heated to decomp., emits acrid smoke and irritating fumes
Uses: Synthetic flavoring; fresh, floral, green, typical lily-of-the-valley type fragrance and flavoring
Usage level: 3.5 ppm (nonalcoholic beverages), 13 ppm (ice cream), 9.4 ppm (candy), 10 ppm (baked goods), 0.3 ppm (gelatins, puddings), 16 ppm (chewing gum)
Regulatory: FDA 21CFR §172.515; GRAS (FEMA); Japan approved as flavoring
Manuf./Distrib.: BASF

Hydroxycitronellal diethyl acetal
FEMA 2584
Synonyms: 8,8-Diethoxy-2,6-dimethyl-octanol-2
Empirical: $C_{14}H_{30}O_3$
Properties: Colorless sl. oily liq., delicate green-floral taste; sol. in alcohol; insol. in water; m.w. 246.39; b.p. 260 C
Uses: Synthetic flavoring agent
Usage level: 2.7 ppm (nonalcoholic beverages), 0.50-1.0 ppm (ice cream, ices), 7.3 ppm (candy), 2.2 ppm (baked goods)
Regulatory: FDA 21CFR §172.515

Hydroxycitronellal dimethylacetal. *See* Hydroxycitronellal dimethyl acetyl

Hydroxycitronellal dimethyl acetyl
FEMA 2585
Synonyms: 7-Hydroxy-3,7-dimethyl octanal:acetal; Hydroxycitronellal dimethylacetal; 8,8-Dimethoxy-2,6-dimethyl-octanol-2
Empirical: $C_{12}H_{26}O_3$
Properties: Colorless liq., lt. green flowery odor; m.w. 218.34; flash pt. > 100 C; ref. index 1.4410-1.4440
Uses: Synthetic flavoring agent
Usage level: 10 ppm (nonalcoholic beverages), 0.50 ppm (ice cream, ices), 24 ppm (candy), 0.50-20 ppm (baked goods)
Regulatory: FDA 21CFR §172.515; Japan approved as flavoring

Hydroxycitronellol
FEMA 2586
Synonyms: 3,7-Dimethyloctane-1,7-diol; 3,7-Dimethyl-1,7-octanediol
Empirical: $C_{10}H_{22}O_2$
Properties: Colorless visc. liq., rosy grape odor; sl. sol. in toluene, benzene; m.w. 174.29; dens. 0.935; b.p. 156 C (15 mm); flash pt. > 100 C; ref. index 1.4550-1.4600
Uses: Synthetic flavoring; floral, green, begonia-like fragrance and flavoring
Usage level: 2.0 ppm (nonalcoholic beverages), 1.6 ppm (ice cream, ices), 3.6 ppm (candy), 3.5 ppm (baked goods), 0.30 ppm (gelatins, puddings, chewing gum)
Manuf./Distrib.: BASF

2-Hydroxy-p-cymene. *See* Carvacrol
3-Hydroxy-p-cymene. *See* Thymol
4-Hydroxydecanoic acid, γ-lactone. *See* γ-Decalactone
3-Hydroxy-4,5-dihydroxymethyl-2-methylpyridine HCl. *See* Pyridoxine HCl

4-Hydroxy-2,5-dimethyl-3(2H)furanone
CAS 3658-77-3; FEMA 3174

7-Hydroxy-3,7-dimethyl octanal:acetal. *See* Hydroxycitronellal dimethyl acetyl
7-Hydroxy-3,7-dimethyloctanol. *See* Hydroxycitronellal
4-Hydroxydodecanoic acid γ-lactone. *See* γ-Dodecalactone
5-Hydroxydodecanoic acid, δ-lactone. *See* δ-Dodecalactone
1-Hydroxyethane 1-carboxylic acid. *See* Lactic acid
1-Hydroxy-1,2-ethanedicarboxylic acid. *See* N-Hydroxysuccinic acid
2-Hydroxyethylamine. *See* Ethanolamine

1-(2-Hydroxyethyl)-1-(4-chlorobutyl)-2-alkyl (C6-17) imidazolinium chloride
   Uses: Used in mfg. of paper/paperboard contacting food
   Regulatory: FDA 21CFR §181.30

3-(1-Hydroxyethylidene)-6-methyl-2H-pyran-2,4(3H)-dione sodium salt. *See* Sodium dehydroacetate
3-Hydroxy-2-ethyl-4-pyrone. *See* Ethyl maltol
(2-Hydroxyethyl) trimethyl ammonium bitartrate. *See* Choline bitartrate
(2-Hydroxyethyl)trimethylammonium chloride. *See* Choline chloride
Hydroxy heptane. *See* Heptyl alcohol
4-Hydroxyheptanoic acid, γ-lactone. *See* γ-Heptalactone
16-Hydroxy-6-hexadecenoic acid, omega-lactone. *See* Ambrettolide
4-Hydroxyhexanoic acid γ-lactone. *See* γ-Hexalactone

Hydroxylated lecithin
   CAS 8029-76-3; EINECS 232-440-6
   Synonyms: Lecithin, hydroxylated
   Definition: Prod. obtained by the controlled hydroxylation of lecithin
   Properties: Lt. yel. liq. to paste, char. odor; mod. sol. in water
   Toxicology: Heated to decomp., emits acrid smoke and irritating fumes
   Uses: Emulsifier, clouding agent, defoamer in bakery prods., beet sugar, dry mix beverages, margarine, yeast
   Usage level: ADI not specified (FAO/WHO)
   Regulatory: FDA 21CFR §136.110, 136.115, 136.130, 136.160, 136.165, 136.180, 172.814, 173.340, 176.170, 176.200
   Trade names: Alcolec® Z-3; Centrolene® A; Centrolene® S; M-C-Thin® HL; Thermolec WFC; Yelkin 1018
   Trade names containing: Alcolec® HWS; Brem

4-Hydroxy-3-methoxybenzaldehyde. *See* Vanillin
N-(4-Hydroxy-3-methoxybenzyl)nonanamide. *See* Pelargonyl vanillylamide
6-Hydroxy-5-[(2-methoxy-5-methyl-4-sulfophenyl)azo]-2-naphthalenesulfonic acid disodium salt. *See* FD&C Red No. 40
4-(4-Hydroxy-3-methoxyphenyl)-2-butanone. *See* Zingerone
(4-Hydroxy-3-methoxyphenyl)ethyl methyl ketone. *See* Zingerone
1-Hydroxy-2-methoxy-4-propenylbenzene. *See* Isoeugenol
Hydroxy methyl anethol. *See* Propenylguaethol
2-Hydroxy-5-methylanisole. *See* 2-Methoxy-4-methylphenol
3-Hydroxy-3-methyl-1-butene. *See* 2-Methyl-3-buten-2-ol
2-Hydroxy-3-methyl-2-cyclopenten-1-one. *See* Methyl cyclopentenolone

4-Hydroxymethyl-2,6-di-t-butylphenol
   Synonyms: Butylated hydroxymethyl phenol
   Properties: solid. pt. 140-141 C
   Uses: Food preservative, antioxidant
   Usage level: 0.02% max. of oil or fat content of food
   Regulatory: FDA 21CFR §172.150

3-Hydroxy-2-methyl-4H-pyran-4-one. *See* Maltol
5-Hydroxy-6-methyl-3,4-pyridinedimethanol hydrochloride. *See* Pyridoxine HCl
3-Hydroxy-2-methyl-4-pyrone. *See* Maltol
4-Hydroxynonanoic acid, γ-lactone. *See* γ-Nonalactone
1-Hydroxy-3-nonanone acetate. *See* 3-Nonanon-1-yl acetate
12-Hydroxy-9-octadecenoic acid, monoester with 1,2-propanediol. *See* Propylene glycol ricinoleate
12-Hydroxy-9-octadecenoic acid, monoester with 1,2,3-propanetriol. *See* Glyceryl ricinoleate
4-Hydroxyoctanoic acid, γ-lactone. *See* γ-Octalactone

5-Hydroxy-4-octanone
   FEMA 2587
   Synonyms: Butyroin; 5-Octanol-4-one
   Empirical: $C_8H_{16}O_2$

Properties: Ylsh. liq., buttery nut-like odor, sweet buttery oily taste; m.w. 144.21; dens. 0.9231; b.p. 182 C; ref. index 1.4290
Uses: Synthetic flavoring agent
Usage level: 0.5-5.0 ppm (nonalcoholic beverages), 1.0-20 ppm (ice cream, ices), 10 ppm (candy), 7.8 ppm (baked goods)
Regulatory: FDA 21CFR §172.515

α-Hydroxy-omega-hydroxy poly(oxy-1,2-ethanediyl). *See* Polyethylene glycol
γ-Hydroxy-β-oxobutane. *See* Acetyl methyl carbinol
4-Hydroxypentanoic acid lactone. *See* γ-Valerolactone
2-Hydroxy-2-phenylacetophenone. *See* Benzoin
α-Hydroxy-α-phenylacetophenone. *See* Benzoin
3-(4-Hydroxyphenyl)alanine. *See* L-Tyrosine
l-β-(p-Hydroxyphenyl) alanine. *See* L-Tyrosine

4-(p-Hydroxyphenyl)-2-butanone
CAS 5471-51-2; FEMA 2588
Synonyms: p-Hydroxybenzyl acetone; Frambinone; Raspberry ketone; Oxyphenalon
Empirical: $C_{10}H_{12}O_2$
Properties: Wh. cryst. solid, raspberry odor; sol. in alcohol, ether; m.w. 164.22; m.p. 81-86 C; flash pt. > 212 F
Precaution: Combustible liq.
Toxicology: LD50 (oral, rat) 1320 mg/kg; poison by IP route; mod. toxic by ingestion; heated to decomp., emits acrid smoke and irritating fumes
Uses: Synthetic flavoring agent
Usage level: 16 ppm (nonalcoholic beverages), 34 ppm (ice cream, ices), 44 ppm (candy), 54 ppm (baked goods), 5.0-50 ppm (gelatins, puddings), 40-320 ppm (chewing gum)
Regulatory: FDA 21CFR §172.515
Manuf./Distrib.: Aldrich; Chr. Hansen's

2-(p-Hydroxyphenyl)-5,7-dihydroxychromone. *See* Chamomile
2-Hydroxy-1,2,3-propanetricarboxylic acid. *See* Citric acid
2-Hydroxy-1,2,3-propanetricarboxylic acid, monoester with 1,2,3-propanetriol monooctadecanoate. *See* Glyceryl stearate citrate
2-Hydroxy-1,2,3-propanetricarboxylic acid, monooctadecyl ester. *See* Stearyl citrate
2-Hydroxy-1,2,3-propanetricarboxylic acid, triethyl ester. *See* Triethyl citrate
2-Hydroxy-1,2,3-propanetricarboxylic acid, trioctadecyl ester. *See* Tristearyl citrate
2-Hydroxypropanoic acid. *See* Lactic acid
2-Hydroxypropanoic acid, calcium salt. *See* Calcium lactate
2-Hydroxypropanoic acid monosodium salt. *See* Sodium lactate
2-Hydroxypropionic acid. *See* Lactic acid
α-Hydroxypropionic acid. *See* Lactic acid
Hydroxypropyl alginate. *See* Propylene glycol alginate

Hydroxypropylcellulose
CAS 9004-64-2; EEC E463
Synonyms: Cellulose, 2-hydroxypropyl ether; Oxypropylated cellulose
Definition: Propylene glycol ether of cellulose
Properties: White powd.; sol. in water, methanol, ethanol, other org. solvs.; insol. in water > 37.7 C; thermoplastic; can be extruded and molded; softens at 130 C
Precaution: Combustible
Toxicology: LD50 (oral, rat) 10,200 mg/kg; sl. toxic by ingestion; heated to decomp., emits acrid smoke and fumes
Uses: Emulsifier, film-former, protective colloid, stabilizer, suspending agent, thickener, food additive; used in glazes, oils, vitamin tablets, whipped toppings
Usage level: ADI 0-25 mg/kg (FAO/WHO)
Regulatory: FDA 21CFR §172.870, 177.1200; Europe listed; UK approved

Hydroxypropyl-α-cyclodextrin
CAS 99241-24-4
Properties: m.p. 245 C
Uses: Increases solubility and bioavailability of other substances; masks flavor, odor, or coloration

Hydroxypropyl-β-cyclodextrin
CAS 94035-02-6
Properties: m.w. 1500 (avg.)

Hydroxypropyl-γ-cyclodextrin

Uses: Increases solubility and bioavailability of other substances; masks flavor, odor, or coloration

Hydroxypropyl-γ-cyclodextrin
CAS 99241-25-5
Properties: m.p. 250 C
Uses: Increases solubility and bioavailability of other substances; masks flavor, odor, or coloration

Hydroxypropyl methylcellulose
CAS 9004-65-3; EEC E464
Synonyms: MHPC; Methyl hydroxypropyl cellulose; Cellulose 2-hydroxypropyl methyl ether; Hypromellose
Definition: Propylene glycol ether of methyl cellulose
Properties: White powd.; swells in water to produce a clear to opalescent visc. colloidal sol'n.; nonionic; insol. in anhyd. alcohol, ether, chloroform; sol. in most polar solvs.
Precaution: Combustible
Toxicology: LD50 (intraperitoneal, rat) 5200 mg/kg; mildly toxic by intraperitoneal route; heated to decomp., emits acrid smoke and fumes
Uses: Thickener, gellant, stabilizer, emulsifier, film-former, protective colloid, suspending agent in food prods. (bakery goods, ice cream, breading, dressing, salad dressings, sauce mixes); fat barrier
Regulatory: FDA 21CFR §172.874, 175.105, 175.300; Europe listed
Manuf./Distrib.: Aqualon; S Black; Cornelius; Courtaulds; Croxton & Garry; Hercules; Nordmann Rassmann; Penta; SPCI
Trade names: Benecel® MP 643; Benecel® MP 824; Benecel® MP 843; Benecel® MP 872; Benecel® MP 874; Benecel® MP 943; Methocel® E3 Premium; Methocel® E4M; Methocel® E4M Premium; Methocel® E5 Food Grade; Methocel® E6 Premium; Methocel® E15 Food Grade; Methocel® E15LV Premium; Methocel® E50LV Premium; Methocel® F4M Premium; Methocel® F50LV Premium; Methocel® K3 Premium; Methocel® K4M Premium; Methocel® K15M Premium; Methocel® K35 Premium; Methocel® K100LV Premium; Methocel® K100M Premium

N-Hydroxysuccinic acid
CAS 6915-15-7; 97-67-6 (L); 617-48-1 (DL); EINECS 202-601-5; FEMA 2655; EEC E296
Synonyms: Malic acid; Apple acid; 1-Hydroxy-1,2-ethanedicarboxylic acid; Hydroxybutanedioic acid
Empirical: $C_4H_6O_5$
Formula: COOHCH$_2$CH(OH)COOH
Properties: Wh. or colorless cryst., acid taste; dl, l, and d isomeric forms; very sol. in water, alcohol; sl. sol. in ether; m.w. 134.09; dens. 1.595 (20/40 C, d or l), 1.601 (dl); m.p. 100 C (d or l), 128 C (dl); b.p. 140 C (dec., d or l), 150 C (dl)
Toxicology: Mod. toxic by ingestion; skin and severe eye irritant; heated to decomp., emits acrid smoke and irritating fumes
Uses: Food acidulant, flavoring, adjuvant, flavor enhancer, pH control agent, synergist for antioxidants; used in dry mix beverages, candy, chewing gum, fats, fruits, gelatins, jams/jellies, lard, nonalcoholic beverages, puddings, shortening, wine
Usage level: Limitation 3.4% (nonalcoholic beverages), 3% (chewing gum), 0.8% (gelatins, puddings), 6.9% (hard candy), 2.6% (jams), 3.5% (processed fruits, fruit juice), 3% (soft candy), 0.7% (other foods)
Regulatory: FDA 21CFR §146.113, 150, 150.161, 169.115, 169.140, 169.150, 184.1069, GRAS; USDA 9CFR §318.7, 0.01% max.; BATF 27CFR §240.1051, GRAS; not for use in baby foods; Japan approved; Europe listed; UK approved
Manuf./Distrib.: AB Tech.; Ashland; Bartek; Browning; Cornelius; Ellis & Everard; Haarmann & Reimer; Honeywill & Stein; Int'l. Sourcing; Jungbunzlauer; Penta Mfg.; Quimdis; Van Waters & Rogers

6-Hydroxy-5-[(4-sulfophenyl)azo]-2-naphthalenesulfonic acid disodium salt. *See* FD&C Yellow No. 6
α-Hydroxytoluene. *See* Benzyl alcohol
β-Hydroxytricarballylic acid. *See* Citric acid
2-Hydroxytriethylamine. *See* Diethylaminoethanol
β-Hydroxy-γ-trimethylaminobutyrate. *See* L-Carnitine
5-Hydroxyundecanoic acid γ-lactone. *See* γ-Undecalactone
4-Hydroxyvaleric acid lactone. *See* γ-Valerolactone
Hypnone. *See* Acetophenone
Hypromellose. *See* Hydroxypropyl methylcellulose

Hyssop
Definition: From *Hyssopus officinalis*
Properties: Warm aromatic camphor-like odor, warm sweet sl. burning flavor
Uses: Natural flavoring agent
Usage level: 600 ppm (bitters)
Regulatory: FDA 21CFR §182.10, 182.20, GRAS; Japan approved

Manuf./Distrib.: Chart; Pierre Chauvet (oil)

Iceland moss extract
CAS 84776-25-0
Synonyms: Cetraria islandica extract; Lichen islandicus extract
Definition: Extract of *Cetraria islandicus*
Uses: Natural flavoring
Regulatory: FDA 21CFR §172.510; Japan approved (Iceland moss)

IIR. *See* Isobutylene/isoprene copolymer
Ilex paraguariensis extract. *See* Mate extract

Imazalil
CAS 35554-44-0
Synonyms: Chloramizol
Empirical: $C_{14}H_{14}Cl_2N_2O$
Properties: m.w. 297.20
Uses: Antimold (preservative) (0.002-0.005 g/kg residual)
Regulatory: Japan approved

1H-Imidazolium, 1-[2-(carboxymethoxy)ethyl]-1-(carboxymethyl)-2-heptyl-4,5-dihydro-, hydroxide, diso-dium salt. *See* Disodium capryloamphodiacetate
Imidole. *See* Pyrrole
IMP. *See* Disodium inosinate
Inactive limonene. *See* dl-Limonene
Incense. *See* Olibanum
Indian dill seed oil. *See* Dill seed oil, Indian type
Indian gum. *See* Gum ghatti
Indian tragacanth. *See* Karaya gum
India tragacanth. *See* Karaya gum
Indigo carmine. *See* FD&C Blue No. 2
Indigotine. *See* FD&C Blue No. 2

Indole
CAS 120-72-9; EINECS 204-420-7; FEMA 2593
Synonyms: 1H-Benzo[b]pyrrole; 2,3-Benzopyrrole; 1-Benzazole
Empirical: C_8H_7N
Properties: Leaflets; sol. in hot water, hot alcohol, ether, benzene; m.w. 117.15; m.p. 52-53 C; b.p. 253 C (762 mm); flash pt. 121 C
Precaution: Photosensitive; volatile with steam
Toxicology: LD50 (oral, rat) 1 g/kg
Uses: Synthetic flavoring agent
Usage level: 0.26 ppm (nonalcoholic beverages), 0.28 ppm (ice cream, ices), 0.50 ppm (candy), 0.58 ppm (baked goods), 0.02-0.40 ppm (gelatins, puddings)
Regulatory: FDA 21CFR §172.515; Japan approved as flavoring
Manuf./Distrib.: Aldrich

Indole-3-alanine. *See* L-Tryptophan
Industrial talc. *See* Talc
Infusorial earth. *See* Diatomaceous earth
Inosine 5´-disodium phosphate. *See* Disodium inosinate

Inositol
CAS 87-89-8; EINECS 201-781-2
Synonyms: Hexahydroxy cyclohexane; cis-1,2,3,5-trans-4,6-Cyclohexanehexol; i-Inositol; meso-Inositol; myo-Inositol
Classification: Cyclic polyol
Definition: Constituent of body tissue
Empirical: $C_6H_{12}O_6$
Formula: $C_6H_6(OH)_6$ • 2HOH
Properties: White cryst., odorless, sweet taste; sol. in water; insol. in abs. alcohol and ether; m.w. 180.16; dens. 1.524; ; m.p. 215-227 C
Toxicology: No toxic effects with high dosages; heated to decomp., emits acrid smoke and irritating fumes
Uses: Direct food additive, nutrient, dietary supplement; may be used in infant formulas
Regulatory: FDA 21CFR §182.5370, 184.1370, GRAS; Japan approved
Manuf./Distrib.: Am. Roland

i-Inositol. *See* Inositol
meso-Inositol. *See* Inositol
myo-Inositol. *See* Inositol
Inula. *See* Elecampane
Inula helenium extract. *See* Elecampane extract

Invertase
Synonyms: Sucrase; Invertin
Definition: Enzyme derived from *Saccharomyces*
Properties: White powder; sol. in water
Uses: Natural enzyme; used in production of invert sugar for syrups and candy; catalyzes conversion of sucrose to glucose and levulose during fermentation of sugar
Regulatory: FDA GRAS; Canada, Japan approved
Trade names: Fermvertase, 10X, XX

Invertin. *See* Invertase

Invert sugar
CAS 8013-17-0
Synonyms: Invert sugar syrup
Definition: Aq. sol'n. of inverted/partly inverted refined/partly refined sucrose; mixt. of approx. 50% dextrose and 50% fructose
Properties: Colorless sol'n., odorless, sweet flavor
Uses: Sweetening agent; in food prods., confectionery, brewing; humectant to hold moisture and prevent drying out
Regulatory: FDA 21CFR §184.1859, GRAS
Manuf./Distrib.: MLG Enterprises
Trade names: Solid Invert Sugar; Special Liquid Invert Sugar

Invert sugar syrup. *See* Invert sugar

Iodine
CAS 7553-56-2; EINECS 231-442-4
Classification: Nonmetallic halogen element
Empirical: I_2
Properties: Bluish-black scales or plates, metallic luster, char. odor, sharp acrid taste; sol. (g/100 g): 14.09 benzene, 16.47 CS_2, 21.43 ethanol; sol. in chloroform, glac. acetic acid; at.wt. 126.9045; dens. 4.98; m.p. 113.6 C; b.p. 185.24 C
Precaution: Incompat. with alkaloids, starch, tannins
Toxicology: Human ingestion of large quantities causes abdominal pain, nausea, vomiting, diarrhea; 2-4 g have been fatal; intensely irritating to eyes, skin, mucous membranes
Uses: Dietary supplement
Manuf./Distrib.: Atomergic Chemetals

α-Ionone
CAS 127-41-3; EINECS 204-841-6; FEMA 2594
Synonyms: 4-(2,6,6-Trimethyl-2-cyclohexene-1-yl)-3-butene-2-one; α-Cyclocitrylideneacetone
Empirical: $C_{13}H_{20}O$
Properties: Colorless oil, woody violet odor; sol. in alcohol, fixed oils, propylene glycol; sl. sol. in water; misc. with ether; insol. in glycerin; m.w. 192.33; dens. 0.930; b.p. 136 C; flash pt. 118 C; ref. index 1.497-1.502
Toxicology: LD50 (oral, rat) 4590 mg/kg; mildly toxic by ingestion; heated to decomp., emits acrid smoke and fumes
Uses: Synthetic flavoring; sweetly floral, balsamic, violet-like, woody fragrance and flavoring
Regulatory: FDA 21CFR §172.515
Manuf./Distrib.: BASF

β-Ionone
CAS 79-77-6, 14901-07-6; EINECS 201-224-3; FEMA 2595
Synonyms: 4-(2,6,6-Trimethyl-1-cyclohexene-1-yl)-3-butene-2-one; β-Cyclocitrylideneacetone
Empirical: $C_{13}H_{20}O$
Properties: Colorless oil, woody odor; sol. in alcohol, fixed oils, propylene glycol; sl. sol. in water; misc. with ether; insol. in glycerin; m.w. 192.33; dens. 0.944; b.p. 140 C; flash pt. > 234 F; ref. index 1.517-1.522
Precaution: Combustible liq.
Toxicology: LD50 (oral, rat) 4590 mg/kg; mildly toxic by ingestion; heated to decomp., emits acrid smoke and fumes
Uses: Synthetic flavoring; woody, dry, floral fragrance and flavoring
Regulatory: FDA 21CFR §172.515

Manuf./Distrib.: BASF

IPA. *See* Isopropyl alcohol
Iris florentina extract. *See* Orris root extract
Irish moss. *See* Carrageenan
Irish moss extract. *See* Carrageenan extract

Iron
CAS 7439-89-6; EINECS 231-096-4
Synonyms: Carbonyl iron
Classification: Metallic element
Formula: Fe
Properties: Silver white malleable metal; at.wt. 55.847; dens. 7.87 (20 C); m.p. 1536 C; b.p. 3000 C; highly reactive chemically; strong reducing agent
Precaution: Ultrafine powd. is potentially explosive
Toxicology: Poison by intraperitoneal route; potentially toxic by all forms and routes
Uses: Nutrient, dietary supplement used in baked goods, cereal prods., flour, pasta; processing aid (Japan)
Regulatory: FDA 21CFR §182.5375, 182.8375, 184.1375, GRAS; Japan restricted
Trade names containing: Cap-Shure® FE-165-50; Enrichment R Tablets

Iron ammonium citrate
CAS 1185-57-5 (green), 1185-57-6 (brown), 1333-00-2, 1332-98-5; EINECS 214-686-6; EEC E381
Synonyms: Ferric ammonium citrate; Ammonium ferric citrate; Iron (III) ammonium citrate; Ferric ammonium citrate, green
Classification: A complex salt
Properties: Transparent grn. scales, gran., powd., or cryst., ammoniacal odor, mild iron-metallic taste; sol. in water; insol. in alcohol; deliq.
Toxicology: Heated to decomp., emits acrid smoke and irritating fumes
Uses: Anticaking agent, dietary supplement, nutrient; may be used in infant formulas
Regulatory: FDA 21CFR §172.430, 25 ppm max. in finished salt, 184.1296, GRAS; Japan approved; Europe listed; UK approved
Manuf./Distrib.: Lohmann

Iron (III) ammonium citrate. *See* Iron ammonium citrate
Iron (II) ascorbate. *See* Ferrous ascorbate

Iron caprylate
Toxicology: Heated to decomp., emits acrid smoke and irritating fumes
Uses: Drier migrating from food pkg.
Regulatory: FDA 21CFR §181.25

Iron (III) chloride. *See* Ferric chloride
Iron chlorides. *See* Ferric chloride
Iron (III) choline citrate. *See* Ferric choline citrate
Iron choline citrate complex. *See* Ferric choline citrate
Iron (II) citrate. *See* Ferrous citrate
Iron (III) citrate. *See* Ferric citrate

α-Irone
CAS 79-69-6; EINECS 201-219-6; FEMA 2597
Synonyms: 4-(2,5,6,6-Tetramethyl-2-cyclohexene-1-yl)-3-buten-2-one; 6-Methylionone; 6-Methyl-α-ionone
Definition: The fragrant principle of violets
Empirical: $C_{14}H_{22}O$
Properties: Visc. liq., char. odor of violets; m.w. 206.32; dens. 0.934 (20/4 C); b.p. 146 C; ref. index 1.492 (20 C)
Uses: Synthetic flavoring agent
Usage level: 1.2 ppm (nonalcoholic beverages), 2.3 ppm (ice cream, ices), 4.1 ppm (candy), 5.4 ppm (baked goods), 1.4 ppm (chewing gum)
Regulatory: FDA 21CFR §172.515

Iron (II) fumarate. *See* Ferrous fumarate
Iron (II) gluconate. *See* Ferrous gluconate
Iron (II) lactate. *See* Ferrous lactate

Iron linoleate
Toxicology: Heated to decomp., emits acrid smoke and irritating fumes
Uses: Drier migrating from food pkg.
Regulatory: FDA 21CFR §181.25

Iron naphthenate
Toxicology: Heated to decomp., emits acrid smoke and irritating fumes
Uses: Drier migrating from food pkg.
Regulatory: FDA 21CFR §181.25

Iron (III) oxide. *See* Ferric oxide
Iron persulfate. *See* Ferric sulfate
Iron phosphate. *See* Ferric phosphate
Iron (III) pyrophosphate. *See* Ferric pyrophosphate

Iron sequioxide
Uses: Color for banana
Regulatory: Japan approved; restricted

Iron (III) sodium pyrophosphate. *See* Ferric sodium pyrophosphate
Iron (II) sulfate (1:1). *See* Ferrous sulfate
Iron (III) sulfate. *See* Ferric sulfate
Iron sulfate (2:3). *See* Ferric sulfate
Iron (II) sulfate heptahydrate. *See* Ferrous sulfate heptahydrate
Iron sulfate (ous). *See* Ferrous sulfate

Iron tallate
Toxicology: Heated to decomp, emits acrid smoke and irritating fumes
Uses: Drier migrating from food pkg.
Regulatory: FDA 21CFR §181.25

Iron tersulfate. *See* Ferric sulfate
Iron trichloride. *See* Ferric chloride
Iron vitriol. *See* Ferrous sulfate
Iron vitriol. *See* Ferrous sulfate heptahydrate
Isoaldehyde C-10. *See* 2,6-Dimethyl octanal

Isoamyl acetate
CAS 123-92-2; EINECS 204-662-3; FEMA 2055
Synonyms: 3-Methyl-1-butanol acetate; β-Methylbutyl acetate; Isopentyl alcohol acetate; Isopentyl acetate
Definition: Ester of isoamyl alcohol and acetic acid
Empirical: $C_7H_{14}O_2$
Formula: $CH_3COOCH_2CH_2CH(CH_3)_2$
Properties: Colorless liq., banana-like odor; sl. sol. in water; misc. with alcohol, ether, ethyl acetate, fixed oils; insol. in glycerin, propylene glycol; m.w. 130.21; dens. 0.876; b.p. 142 C; flash pt. 77 F; ref. index 1.400
Precaution: Highly flamm.; exposed to heat or flame, can react vigorously with reducing materials
Toxicology: TLV 100 ppm; LD50 (oral, rat) 16,600 mg/kg; mildly toxic by ingestion, inh., subcutaneous routes; heated to decomp., emits acrid smoke and fumes
Uses: Synthetic flavoring; fresh fruity fragrance and flavoring, pear- or banana-like on dilution
Usage level: 28 ppm (nonalcoholic beverages), 56 ppm (ice cream), 190 ppm (candy), 120 ppm (baked goods), 100 ppm (gelatins, puddings), 2700 ppm (chewing gum)
Regulatory: FDA 21CFR §172.515; Japan approved as flavoring
Manuf./Distrib.: Aldrich; Chr. Hansen's; Penta Mfg.

Isoamyl acetoacetate
Synonyms: Isoamyl β-ketobutyrate; Isoamyl 3-oxobutanoate
Empirical: $C_9H_{16}O_3$
Properties: Colorless liq., sweet winey odor, green-apple flavor; sol. in alcohol; insol. in water; m.w. 172.23; dens. 0.954 (10 C); b.p. 222-224 C
Uses: Synthetic flavoring
Regulatory: FDA 21CFR §172.515

Isoamyl alcohol
CAS 123-51-3; EINECS 204-633-5; FEMA 2057
Synonyms: 3-Methylbutanol; 3-Methyl-1-butanol; Isopentyl alcohol; Isobutyl carbinol
Empirical: $C_5H_{12}O$
Formula: $(CH_3)_2CHCH_2CH_2OH$
Properties: Colorless liq., pungent taste, disagreeable odor; sl. sol. in water; misc. with alcohol, ether; m.w. 88.15; dens. 0.813 (15/4 C); b.p. 132 C; f.p. -117.2 C; flash pt. (CC) 42.7 C; ref. index 1.407 (20 C)
Precaution: Combustible; explosive limits in air 1.2-9%; vapor is toxic and irritant
Toxicology: TLV 100 ppm in air
Uses: Synthetic flavoring agent

Usage level: 17 ppm (nonalcoholic beverages), 100 ppm (alcoholic beverages), 7.6 ppm (ice cream), 52 ppm (candy), 24 ppm (baked goods), 46 ppm (gelatins, puddings), 300 ppm (chewing gum)
Regulatory: FDA 21CFR §172.515
Manuf./Distrib.: Aldrich; Chr. Hansen's
Trade names: Isoamyl Alcohol 95%; Isoamyl Alcohol 99%

Isoamylase
Uses: Natural enzyme
Regulatory: Japan approved

Isoamyl benzoate
CAS 94-46-2; FEMA 2058
Synonyms: Isopentyl benzoate
Empirical: $C_{12}H_{16}O_2$
Properties: Colorless liq., fruity sl. pungent odor; insol. in water; m.w. 192.26; b.p. 261-262 C; flash pt. > 100 C; ref. index 1.492-1.495
Precaution: Combustible
Uses: Synthetic flavoring agent
Usage level: 3.0 ppm (nonalcoholic beverages), 2.5 ppm (ice cream, ices), 3.5 ppm (candy), 7.4 ppm (baked goods), 4.6 ppm (gelatins, puddings), 200 ppm (chewing gum)
Regulatory: FDA 21CFR §172.515
Manuf./Distrib.: Chr. Hansen's

Isoamyl butyrate
CAS 106-27-4; FEMA 2060
Synonyms: Butyric acid isoamyl ester; Amyl butyrate; Isopentyl butyrate
Definition: Ester of isoamyl alcohol and butyric acid
Empirical: $C_9H_{18}O_2$
Properties: Colorless liq., fruity odor; sol. in alcohol, fixed oils; insol. in glycerin, propylene glycol, water; m.w. 158.24; dens. 0.860; flash pt. 149 F; ref. index 1.409-1.414
Precaution: Combustible liq.
Toxicology: Heated to decomp., emits acrid smoke and irritating fumes
Uses: Synthetic flavoring; sweet, fruity apricot, banana, pineapple fragrance and flavoring; used in baked goods, dessert gels, puddings
Regulatory: FDA 21CFR §172.515; Japan approved as flavoring
Manuf./Distrib.: BASF

Isoamyl caproate. *See* Isoamyl hexanoate
Isoamyl caprylate. *See* Amyl octanoate

Isoamyl cinnamate
FEMA 2063
Synonyms: Amyl cinnamate; Isoamyl β-phenylacrylate; Isoamyl 3-phenylpropenoate; Isoamyl 3-phenyl propenate; Isopentyl cinnamate
Definition: Ester of cinnamic acid and commercial isoamyl alcohols
Empirical: $C_{14}H_{18}O_2$
Properties: Colorless to pale yel. liq., balsamic odor; sol. in most fixed oils; sl. sol. in alcohol; m.w. 218.29; ref. index 1.535-1.539
Precaution: Combustible liq.
Toxicology: Heated to decomp., emits acrid smoke and fumes
Uses: Synthetic flavoring agent
Usage level: 3.1 ppm (nonalcoholic beverages), 4.2 ppm (ice cream, ices), 13 ppm (candy, baked goods)
Regulatory: FDA 21CFR §172.515

Isoamyl dodecanoate. *See* Isoamyl laurate
Isoamyl dodecylate. *See* Isoamyl laurate

Isoamyl formate
CAS 110-45-2; FEMA 2069
Synonyms: Amyl formate; Isopentyl formate
Empirical: $C_6H_{12}O_2$
Properties: Colorless liq., plum-like odor; sol. in alcohol; sl. sol. in water; misc. with ether; m.w. 116.16; dens. 0.859; b.p. 123-124 C; flash pt. 86 F; ref. index 1.3960-1.40
Precaution: Combustible
Toxicology: Strong irritant
Uses: Synthetic flavoring agent

Isoamyl 2-furanbutyrate

Usage level: 8.4 ppm (nonalcoholic beverages), 14 ppm (ice cream, ices), 22 ppm (candy), 16 ppm (baked goods), 2.0-28 ppm (gelatins, puddings), 250 ppm (chewing gum)
Regulatory: FDA 21CFR §172.515; Japan approved as flavoring
Manuf./Distrib.: Aldrich

Isoamyl 2-furanbutyrate
FEMA 2070
Synonyms: α-Isoamyl furfurylpropionate; Isopentyl-2-furanbutyrate
Empirical: $C_{13}H_{20}O_3$
Properties: Pale yel. liq., sweet buttery fruity odor, caramel-like flavor; sol. in alcohol; insol. in water; m.w. 224.30
Uses: Synthetic flavoring
Usage level: 0.03-5.0 ppm (nonalcoholic beverages), 2.8 ppm (ice cream, ices), 6.0 ppm (candy), 0.50-8.0 ppm (baked goods), 5.0 ppm (gelatins, puddings)
Regulatory: FDA 21CFR §172.515

Isoamyl 2-furanpropionate
FEMA 2071
Synonyms: Isoamyl furfurhydracrylate; α-Isoamyl furfurylacetate
Empirical: $C_{12}H_{18}O_3$
Properties: Pale yel. liq., sl. floral odor; sol. in alcohol; insol. in water; m.w. 210.27
Uses: Synthetic flavoring
Usage level: 0.02-0.33 ppm (nonalcoholic beverages), 0.33-0.65 ppm (ice cream, ices), 1.6-3.6 ppm (candy, baked goods)
Regulatory: FDA 21CFR §172.515

Isoamyl furfurhydracrylate. *See* Isoamyl 2-furanpropionate
α-Isoamyl furfurylacetate. *See* Isoamyl 2-furanpropionate
α-Isoamyl furfurylpropionate. *See* Isoamyl 2-furanbutyrate

Isoamyl hexanoate
CAS 2198-61-0; FEMA 2075
Synonyms: Amyl hexanoate; Isoamyl caproate; Pentyl hexanoate
Definition: Ester of caproic acid and isomeric amyl alcohols
Empirical: $C_{11}H_{22}O_2$
Properties: Colorless liq., fruity odor; insol. in water; m.w. 186.30; dens. 0.860; b.p. 94-96 C (10 mm); flash pt. 88 C
Uses: Synthetic flavoring
Usage level: 7.8 ppm (nonalcoholic beverages), 14 ppm (ice cream, ices), 17 ppm (candy), 15 ppm (baked goods), 3.7 ppm (gelatins, puddings)
Regulatory: FDA 21CFR §172.515
Manuf./Distrib.: Aldrich

Isoamyl 2-hydroxybenzoate. *See* Amyl salicylate
Isoamyl o-hydroxybenzoate. *See* Amyl salicylate

Isoamyl isobutyrate
Empirical: $C_9H_{18}O_2$
Properties: Liq.; m.w. 158.24; dens. 0.8627; b.p. 170 C
Uses: Synthetic flavoring
Regulatory: FDA 21CFR §172.515

Isoamyl isovalerate
CAS 659-70-1; FEMA 2085
Synonyms: Amyl valerate; Amyl isovalerate; Isopentyl isovalerate
Empirical: $C_{10}H_{20}O_2$
Properties: Liq., fruity odor, apple-like flavor; sol. in most org. solvs.; insol. in water; m.w. 172.27; dens. 0.8583; b.p. 193 C; flash pt. 72 C; ref. index 1.4125-1.4135
Uses: Synthetic flavoring
Usage level: 8.5 ppm (nonalcoholic beverages), 14 ppm (ice cream, ices), 33 ppm (candy), 41 ppm (baked goods), 1.0-61 ppm (gelatins, puddings), 390 ppm (chewing gum), 10 ppm (jellies)
Regulatory: FDA 21CFR §172.515; Japan approved as flavoring
Manuf./Distrib.: Aldrich

Isoamyl β-ketobutyrate. *See* Isoamyl acetoacetate
Isoamyl α-ketopropionate. *See* Isoamyl pyruvate

Isoamyl laurate
CAS 6309-51-9; FEMA 2077
Synonyms: Isoamyl dodecanoate; Isoamyl dodecylate; Isopentyl laurate
Empirical: $C_{17}H_{34}O_2$
Properties: Colorless oily liq., fatty odor and flavor; sol. in alcohol; insol. in water; m.w. 270.46
Uses: Synthetic flavoring
Usage level: 0.04-3.0 ppm (nonalcoholic beverages), 0.16-6.0 ppm (ice cream, ices), 0.50-6.0 ppm (candy, baked goods)
Regulatory: FDA 21CFR §172.515
Manuf./Distrib.: Aldrich

Isoamyl-2-methylbutyrate
Synonyms: Isopentyl-2-methylbutyrate
Uses: Synthetic flavoring
Regulatory: FDA 21CFR §172.515

Isoamyl nonanoate
CAS 7779-70-6; FEMA 2078
Synonyms: Isoamyl nonylate; Isoamyl pelargonate; Isopentyl nonanoate; Nonate
Empirical: $C_{14}H_{28}O_2$
Properties: Colorless oily liq., nutty oily apricot-like odor, fruity winey cognac-rum flavor; sol. in alcohol; insol. in water; m.w. 228.37; dens. 0.86; b.p. 260-265 C
Uses: Synthetic flavoring
Usage level: 1.5 ppm (nonalcoholic beverages), 3.3 ppm (ice cream, ices), 3.0 ppm (candy), 4.0 ppm (baked goods)
Regulatory: FDA 21CFR §172.515
Manuf./Distrib.: Aldrich

Isoamyl nonylate. *See* Isoamyl nonanoate
Isoamyl octanoate. *See* Amyl octanoate
Isoamyl 3-oxobutanoate. *See* Isoamyl acetoacetate
Isoamyl pelargonate. *See* Isoamyl nonanoate

Isoamyl phenylacetate
FEMA 2081
Synonyms: Isoamyl α-toluate; Isopentyl phenylacetate
Definition: Ester of phenylacetic acid and isoamyl alcohol
Empirical: $C_{13}H_{18}O_2$
Properties: Liq., cocoa-like odor; m.w. 206.28; dens. 0.982; b.p. 265-266 C (723 mm); ref. index 1.4850-1.4870
Uses: Synthetic flavoring
Usage level: 5.0 ppm (nonalcoholic beverages), 16 ppm (ice cream, ices), 12 ppm (candy), 14 ppm (baked goods), 0.15-3.4 ppm (gelatins, puddings), 0.25-0.80 ppm (toppings)
Regulatory: FDA 21CFR §172.515; Japan approved as flavoring
Manuf./Distrib.: Chr. Hansen's

Isoamyl β-phenylacrylate. *See* Isoamyl cinnamate
Isoamyl 3-phenyl propenate. *See* Isoamyl cinnamate
Isoamyl 3-phenylpropenoate. *See* Isoamyl cinnamate
Isoamyl propionate. *See* Amyl propionate

Isoamyl pyruvate
FEMA 2083
Synonyms: Isoamyl α-ketopropionate; Isopentyl pyruvate
Empirical: $C_8H_{14}O_3$
Properties: m.w. 158.20; dens. 0.978 (17 C); b.p. 185 C
Uses: Synthetic flavoring
Usage level: 4.7 ppm (nonalcoholic beverages), 8.1 ppm (ice cream, ices), 9.2 ppm (candy), 12 ppm (baked goods)
Regulatory: FDA 21CFR §172.515

Isoamyl salicylate. *See* Amyl salicylate
Isoamyl α-toluate. *See* Isoamyl phenylacetate
Isoascorbic acid. *See* Erythorbic acid

Isoborneol
FEMA 2158
Synonyms: Isobornyl alcohol; Isocamphol

Isobornyl acetate

Empirical: $C_{10}H_{18}O$
Properties: Wh. cryst. solid, piney camphoraceous odor; m.w. 154.24; m.p. 212-213 C
Uses: Synthetic flavoring
Usage level: 6.2 ppm (nonalcoholic beverages), 23 ppm (ice cream, ices), 11 ppm (candy), 8.3 ppm (baked goods), 0.80 ppm (chewing gum)
Regulatory: FDA 21CFR §172.515

Isobornyl acetate
FEMA 2160
Synonyms: 2-Camphanyl acetate
Empirical: $C_{12}H_{20}O_2$
Properties: Colorless clear liq., camphor-like odor, fresh burning taste; sol. in most org. solvs.; m.w. 196.29; b.p. 102-103 C (12 mm); ref. index 1.4620-1.4650
Uses: Synthetic flavoring
Usage level: 9.6 ppm (nonalcoholic beverages), 12 ppm (ice cream, ices), 3.9 ppm (candy), 9.5 ppm (baked goods), 70 ppm (gelatins, puddings)
Regulatory: FDA 21CFR §172.515

Isobornyl alcohol. *See* Isoborneol

Isobornyl formate
FEMA 2162
Empirical: $C_{11}H_{18}O_2$
Properties: Liq., aromatic pine needles odor; sol. in most org. solvs.; insol. in water; m.w. 182.26; dens. 1.000 (18 C); b.p. 110 C (20 mm); ref. index 1.4717
Uses: Synthetic flavoring
Usage level: 0.06-1.0 ppm (nonalcoholic beverages), 0.03-1.0 ppm (ice cream, ices), 0.74 ppm (candy), 0.80 ppm (baked goods)
Regulatory: FDA 21CFR §172.515

Isobornyl isovalerate
FEMA 2166
Empirical: $C_{15}H_{26}O_2$
Properties: Colorless liq., warm herbaceous camphoraceous odor; sol. in alcohol; insol. in water; m.w. 238.37
Uses: Synthetic flavoring
Usage level: 0.60-1.0 ppm (nonalcoholic beverages), 0.30-1.0 ppm (ice cream, ices), 0.90 ppm (candy), 0.80-2.0 ppm (baked goods)
Regulatory: FDA 21CFR §172.515

Isobornyl propionate
FEMA 2163
Empirical: $C_{13}H_{22}O_2$
Properties: Colorless oily liq.; sol. in alcohol; insol. in water; m.w. 210.31; b.p. 254 C; ref. index 1.4640
Uses: Synthetic flavoring
Usage level: 0.01-1.0 ppm (nonalcoholic beverages), 0.80-1.0 ppm (ice cream, ices), 1.2 ppm (candy), 1.8 ppm (baked goods)
Regulatory: FDA 21CFR §172.515

Isobutanal. *See* 2-Methylpropanal

Isobutane
CAS 75-28-5; EINECS 200-857-2
Synonyms: 2-Methylpropane
Classification: Hydrocarbon gas
Empirical: C_4H_{10}
Formula: $CH(CH_3)_3$
Properties: Colorless gas, odorless; easily liquefied under pressure at R.T.
Precaution: Flamm. gas
Toxicology: Asphyxiant; narcotic at high concs.
Uses: Direct food additive, propellant, aerating agent, gas for foamed and sprayed food prods.
Regulatory: FDA 21CFR §184.1165

Isobutanol. *See* Isobutyl alcohol

Isobutyl acetate
CAS 110-19-0; EINECS 203-745-1; FEMA 2175
Synonyms: 2-Methylpropyl acetate; β-Methylpropyl ethanoate
Definition: Ester of isobutyl alcohol and acetic acid

Empirical: $C_6H_{12}O_2$
Formula: $CH_3COOCH_2CH(CH_3)_2$
Properties: Colorless liq., fruit-like odor; very sol. in alcohol, fixed oils, propylene glycol; sl. sol. in water; m.w. 116.18; dens. 0.8685 (15 C); m.p. -98.9 C; b.p. 118 C; flash pt. (CC) 18 C; ref. index 1.389
Precaution: Highly flamm.; very dangerous fire and mod. explosion hazard on exposure to heat, flame, oxidizers
Toxicology: TLV 150 ppm; LD50 (oral, rat) 13,400 mg/kg; mildly toxic by ingestion and inh.; skin and eye irritant; heated to decomp., emits acrid smoke and fumes
Uses: Synthetic flavoring agent
Usage level: 11 ppm (nonalcoholic beverages), 16 ppm (ice cream), 36 ppm (candy), 35 ppm (baked goods), 170 ppm (gelatins, puddings), 860 ppm (chewing gum), 5.5 ppm (icings)
Regulatory: FDA 21CFR §172.515
Manuf./Distrib.: Aldrich; Eastman

Isobutyl acetoacetate
FEMA 2177
Synonyms: Isobutyl β-ketobutyrate; Isobutyl-3-oxobutanoate
Empirical: $C_8H_{14}O_3$
Properties: Colorless liq., brandy-like odor, sweet sl. fruity flavor; sol. in alcohol; insol. in water; m.w. 158.20; dens. 0.9697; b.p. 84.5 C (11 mm); ref. index 1.4219
Uses: Synthetic flavoring
Usage level: 4.0 ppm (nonalcoholic beverages), 7.0 ppm (ice cream, ices), 25 ppm (candy, baked goods)
Regulatory: FDA 21CFR §172.515

Isobutyl alcohol
CAS 78-83-1; EINECS 201-148-0; FEMA 2179
Synonyms: Isobutanol; Isopropylcarbinol; 2-Methyl-1-propanol; 2-Methylpropanol
Classification: Alcohol
Empirical: $C_4H_{10}O$
Formula: $(CH_3)_2CHCH_2OH$
Properties: Colorless liq., sweet odor; partly sol. in water; sol. in alcohol, ether; m.w. 74.12; dens. 0.806 (15 C); m.p. -108 C; b.p. 106-109 C; f.p. -108 C; flash pt. (TCC) 29 C; ref. index 1.396
Precaution: Flamm.; dangerous fire hazard with heat, flame; mod. explosive as vapor with heat, flame, oxidizers
Toxicology: TLV 50 ppm in air; LD50 (oral, rat) 2460 mg/kg; poison by IV, intraperitoneal route; mod. toxic by ingestion, skin contact; experimental carcinogen, tumorigen; severe skin/eye irritant; mutagenic data; heated to decomp., emits acrid smoke and fumes
Uses: Synthetic flavoring agent
Usage level: 12 ppm (nonalcoholic beverages), 1 ppm (alcoholic beverages), 7 ppm (ice cream), 34 ppm (candy), 32 ppm (baked goods), 4 ppm (cream)
Regulatory: FDA 21CFR §172.515
Manuf./Distrib.: Aldrich; Eastman
Trade names: Isobutanol HP

Isobutyl aldehyde. *See* 2-Methylpropanal
Isobutyl 2-aminobenzoate. *See* Isobutyl anthranilate
Isobutyl o-aminobenzoate. *See* Isobutyl anthranilate

Isobutyl angelate
CAS 7779-81-9; FEMA 2180
Synonyms: Isobutyl cis-2-methyl-2-butenoate; Isobutyl cis-α-methylcrotonate
Definition: Ester of isobutyl alcohol and angelic acid
Empirical: $C_9H_{16}O_2$
Properties: Colorless liq., camomile-like odor; sol. in alcohol; insol. in water; m.w. 156.23; dens. 0.877; b.p. 176-177 C; flash pt. 140 F
Uses: Synthetic flavoring
Usage level: 1.5 ppm (nonalcoholic beverages, ice cream, ices), 5.0 ppm (candy), 3.0-100 ppm (icings)
Regulatory: FDA 21CFR §172.515
Manuf./Distrib.: Aldrich

Isobutyl anthranilate
FEMA 2182
Synonyms: Isobutyl 2-aminobenzoate; Isobutyl o-aminobenzoate
Empirical: $C_{11}H_{15}O_2N$
Properties: Colorless liq., faint orange-flowers odor; m.w. 193.25; b.p. 169-170 C (13.5 mm)

Isobutyl benzoate

Uses: Synthetic flavoring
Usage level: 2.0 ppm (nonalcoholic beverages), 4.0 ppm (ice cream, ices), 12 ppm (candy, baked goods), 5.0-1700 ppm (chewing gum)
Regulatory: FDA 21CFR §172.515

Isobutyl benzoate
FEMA 2185
Synonyms: Eglantine
Empirical: $C_{11}H_{14}O_2$
Properties: Colorless liq., floral-leafy odor; misc. with alcohol, ether; insol. in water; m.w. 178.23; dens. 1.002; b.p. 237 C; flash pt. 96 C; ref. index 1.493-1.496
Uses: Synthetic flavoring
Usage level: 2.0-9.0 ppm (nonalcoholic beverages), 7.9 ppm (ice cream, ices), 12 ppm (candy), 10-23 ppm (baked goods)
Regulatory: FDA 21CFR §172.515

Isobutyl benzyl carbinol. See α-Isobutylphenethyl alcohol

Isobutyl-2-butenoate
CAS 589-66-2; FEMA 3432
Synonyms: Isobutyl trans-2-butenoate
Empirical: $C_8H_{14}O_2$
Properties: Fruity odor; m.w. 142.19; dens. 0.890; b.p. 171 C; flash pt. 131 F
Manuf./Distrib.: Aldrich

Isobutyl trans-2-butenoate. See Isobutyl-2-butenoate

Isobutyl butyrate
CAS 539-90-2; FEMA 2187
Synonyms: Isobutyl n-butyrate; 2-Methyl propanyl butyrate
Definition: Ester of butyric acid and isobutyl alcohol
Empirical: $C_8H_{16}O_2$
Formula: $CH_3CH_2CH_2COOCH_2CH(CH_3)_2$
Properties: Liq., fruity odor, sweet rum-like flavor; sl. sol. in water; misc. with alcohol, ether; m.w. 144.21; dens. 0.866; b.p. 157 C; flash pt. 114 F; ref. index 1.4035 (20 C)
Uses: Synthetic flavoring agent
Usage level: 8.3 ppm (nonalcoholic beverages), 2.0 ppm (alcoholic beverages), 16 ppm (ice cream, ices), 25 ppm (candy), 24 ppm (baked goods), 14 ppm (gelatins, puddings), 2000 ppm (chewing gum)
Regulatory: FDA 21CFR §172.515
Manuf./Distrib.: Aldrich

Isobutyl n-butyrate. See Isobutyl butyrate
Isobutyl caproate. See Isobutyl hexanoate
Isobutyl capronate. See Isobutyl hexanoate
Isobutyl carbinol. See Isoamyl alcohol

Isobutyl cinnamate
CAS 122-67-8; FEMA 2193
Synonyms: Isobutyl-3-phenylpropenoate; Isobutyl-β-phenylacrylate
Empirical: $C_{13}H_{16}O_2$
Properties: Colorless liq., sweet fruity balsamic odor, sweet taste; m.w. 204.27; dens. 1.003; b.p. 287 C; flash pt. > 230 F; ref. index 1.541
Precaution: Combustible
Uses: Synthetic flavoring
Usage level: 1.3 ppm (nonalcoholic beverages), 2.0 ppm (alcoholic beverages), 3.4 ppm (ice cream, ices), 5.4 ppm (candy, baked goods)
Regulatory: FDA 21CFR §172.515
Manuf./Distrib.: Aldrich

Isobutylene/isoprene copolymer
CAS 9010-85-9
Synonyms: IIR; Butyl rubber; 3-Methyl-1,3-butadiene polymer with 2-methyl-1-propene
Definition: Copolymer of isobutylene and isoprene monomers
Formula: $(C_5H_8 \cdot C_4H_8)_x$
Toxicology: Heated to decomp., emits acrid smoke and irritating fumes
Uses: Masticatory substance in chewing gum base
Regulatory: FDA 21CFR §172.615, 175.105, 177.1210, 177.2600

Trade names: Polysar Butyl 101-3; Polysar Butyl 402

Isobutyl formate
CAS 542-55-2; FEMA 2197
Synonyms: Tetryl formate
Empirical: $C_5H_{10}O_2$
Formula: $HCOOCH_2CH(CH_3)_2$
Properties: Liq., fruity ether-like odor, rum-like taste; sol. in 100 parts water; misc. with alcohol, ether; m.w. 102.13; dens. 0.885 (20/4 C); m.p. -95 C; b.p. 98-99 C; flash pt. 50 F; ref. index 1.3858 (20 C)
Precaution: Flamm. liq.
Uses: Synthetic flavoring
Usage level: 2.2 ppm (nonalcoholic beverages), 7.1 ppm (ice cream, ices), 19 ppm (candy), 8.2 ppm (baked goods), 5.0 ppm (gelatins, puddings)
Regulatory: FDA 21CFR §172.515
Manuf./Distrib.: Aldrich

Isobutyl 2-furanpropionate
FEMA 2198
Synonyms: Isobutyl furfurylacetate; Isobutyl furylpropionate
Definition: Ester of 2-furanpropionic acid and isobutanol
Empirical: $C_{11}H_{16}O_3$
Properties: Colorless to pale straw-yel. liq., fruity winey brandy-like odor; sol. in alcohol; insol. in water; m.w. 196.25
Uses: Synthetic flavoring
Usage level: 8.1 ppm (nonalcoholic beverages), 14 ppm (ice cream, ices), 17 ppm (candy), 21 ppm (baked goods), 4.0-30 ppm (gelatins, puddings), 12 ppm (chewing gum), 20 ppm (icings)
Regulatory: FDA 21CFR §172.515

Isobutyl furfurylacetate. *See* Isobutyl 2-furanpropionate
Isobutyl furylpropionate. *See* Isobutyl 2-furanpropionate

Isobutyl heptanoate
FEMA 2200
Synonyms: Isobutyl heptoate; Isobutyl heptylate
Definition: Ester of heptanoic acid and isobutyl alcohol
Empirical: $C_{11}H_{22}O_2$
Properties: Colorless liq., green odor; sol. in most org. solvs.; m.w. 186.30; dens. 0.8593, b.p. 209 C
Uses: Synthetic flavoring
Usage level: 0.50-1.5 ppm (nonalcoholic beverages), 2.4-10 ppm (ice cream, ices), 7.0-25 ppm (candy, baked goods)
Regulatory: FDA 21CFR §172.515

Isobutyl heptoate. *See* Isobutyl heptanoate
Isobutyl heptylate. *See* Isobutyl heptanoate

Isobutyl hexanoate
CAS 105-79-3; FEMA 2202
Synonyms: Isobutyl caproate; Isobutyl capronate; Isobutyl hexylate
Empirical: $C_{10}H_{20}O_2$
Properties: Colorless liq., fruity apple-like odor; m.w. 172.27; dens. 0.856; flash pt. 169 F; ref. index 1.412-1.416
Uses: Synthetic flavoring agent
Usage level: 5.4 ppm (nonalcoholic beverages), 3.9 ppm (ice cream, ices), 8.1 ppm (candy), 8.3 ppm (baked goods), 2.0 ppm (chewing gum)
Regulatory: FDA 21CFR §172.515
Manuf./Distrib.: Aldrich

Isobutyl hexylate. *See* Isobutyl hexanoate
Isobutyl o-hydroxybenzoate. *See* Isobutyl salicylate

Isobutyl p-hydroxybenzoate
Uses: Preservative (0.012-1 g/kg as p-hydroxybenzoic acid)
Regulatory: Japan approved

Isobutyl isobutyrate
CAS 97-85-8; EINECS 202-612-5; FEMA 2189
Synonyms: 2-Methylpropanoic acid 2-methylpropyl ester
Empirical: $C_8H_{16}O_2$
Formula: $(CH_3)_2CHCOOCH_2CH(CH_3)_2$

Isobutyl isovalerate

> *Properties:* Liq., fruity odor; misc. with alcohol; insol. in water; m.w. 144.22; dens. 0.854 (20/4 C); m.p. -81 C; b.p. 149-151 C; flash pt. 40 C; ref. index 1.399 (20 C)
> *Precaution:* Flamm.
> *Toxicology:* Irritant
> *Uses:* Synthetic flavoring
> *Usage level:* 7.5 ppm (nonalcoholic beverages), 2.0 ppm (alcoholic beverages), 7.4 ppm (ice cream, ices), 16 ppm (candy), 17 ppm (baked goods), 3.3-10 ppm (gelatins, puddings)
> *Regulatory:* FDA 21CFR §172.515
> *Manuf./Distrib.:* Aldrich

Isobutyl isovalerate. *See* Isobutyl valerate
Isobutyl β-ketobutyrate. *See* Isobutyl acetoacetate

2-Isobutyl-3-methoxy-pyrazine

> CAS 24683-00-9; FEMA 3132
> *Empirical:* $C_9H_{14}N_2O$
> *Properties:* m.w. 166.22; dens. 0.990; flash pt. 80 C
> *Uses:* Powerful earthy flavoring
> *Manuf./Distrib.:* Aldrich

Isobutyl cis-2-methyl-2-butenoate. *See* Isobutyl angelate
Isobutyl cis-α-methylcrotonate. *See* Isobutyl angelate
Isobutyl methyl ketone. *See* Methyl isobutyl ketone
Isobutyl-3-oxobutanoate. *See* Isobutyl acetoacetate

α-Isobutylphenethyl alcohol

> FEMA 2208
> *Synonyms:* Isobutyl benzyl carbinol; 4-Methyl-1-phenyl-2-pentanol; Benzyl isoamyl alcohol
> *Empirical:* $C_{12}H_{18}O$
> *Properties:* Colorless oily liq., green floral herbaceous odor, buttery oily caramellic flavor; sol. in alcohol; insol. in water; m.w. 178.27; dens. 0.96; b.p. 250 C
> *Uses:* Synthetic flavoring
> *Usage level:* 1.0-10 ppm (nonalcoholic beverages), 50 ppm (alcoholic beverages), 38 ppm (ice cream, ices), 54 ppm (candy), 15-50 ppm (baked goods)
> *Regulatory:* FDA 21CFR §172.515

Isobutyl phenylacetate

> FEMA 2210
> *Synonyms:* Isobutyl α-toluate
> *Definition:* Ester of phenylacetic acid and isobutyl alcohol
> *Empirical:* $C_{12}H_{16}O_2$
> *Properties:* Colorless liq., sweet musk-like fragrance, sweet honey-like flavor; sol. in most fixed oils; insol. in glycerol; m.w. 192.26; dens. 0.984-0.988; b.p. 253 C; flash pt. 116 C; ref. index 1.4860-1.4880
> *Uses:* Synthetic flavoring
> *Usage level:* 2.8 ppm (nonalcoholic beverages, ice cream, ices), 5.5 ppm (candy), 5.0 ppm (baked goods, gelatins, puddings), 3.0 ppm (maraschino cherries)
> *Regulatory:* FDA 21CFR §172.515; Japan approved as flavoring
> *Manuf./Distrib.:* Chr. Hansen's

Isobutyl-β-phenylacrylate. *See* Isobutyl cinnamate
Isobutyl-3-phenylpropenoate. *See* Isobutyl cinnamate

Isobutyl propionate

> CAS 540-42-1; FEMA 2212
> *Empirical:* $C_7H_{14}O_2$
> *Formula:* $CH_3CH_2COOCH_2CH(CH_3)_2$
> *Properties:* Liq., agreeable ethereal odor; misc. with alcohol; insol. in water; m.w. 130.18; dens. 0.888 (0/4 C); m.p. -71 C; b.p. 137 C; ref. index 1.3975 (20 C)
> *Precaution:* Combustible
> *Uses:* Synthetic flavoring; mfg. of fruit essences
> *Usage level:* 5.4 ppm (nonalcoholic beverages), 4.2 ppm (ice cream, ices), 25 ppm (candy), 35 ppm (baked goods)
> *Regulatory:* FDA 21CFR §172.515
> *Manuf./Distrib.:* Aldrich

Isobutyl salicylate

> CAS 87-19-4; FEMA 2213
> *Synonyms:* Isobutyl o-hydroxybenzoate

542

Definition: Ester of isobutyl alcohol and salicylic acid
Empirical: $C_{11}H_{14}O_3$
Properties: Colorless liq., orchid wintergreen-like odor, bitter taste; sol. in alcohol, min. oil; insol. in water; m.w. 194.23; dens. 1.064-1.065; b.p. 259 C; flash pt. 121 C; ref. index 1.5070-1.5100
Precaution: Combustible
Uses: Synthetic flavoring
Usage level: 3.5 ppm (nonalcoholic beverages), 1.8 ppm (ice cream, ices), 2.6 ppm (candy), 5.0 ppm (baked goods)
Regulatory: FDA 21CFR §172.515
Manuf./Distrib.: Chr. Hansen's

2-Isobutylthiazole
CAS 18640-74-9; FEMA 3134
Empirical: $C_7H_{11}NS$
Properties: m.w. 141.24; dens. 0.995; b.p. 180 C; flash pt. 136 F
Uses: Synthetic flavoring
Regulatory: FDA 21CFR §172.515
Manuf./Distrib.: Aldrich

Isobutyl α-toluate. *See* Isobutyl phenylacetate

Isobutyl valerate
Synonyms: Isobutyl isovalerate
Empirical: $C_9H_{18}O_2$
Formula: $(CH_3)_2CHCH_2COOCH_2CH(CH_3)_2$
Properties: Liq., ethereal odor; misc. with alcohol, ether; insol. in water; dens. 0.853 (20 C); b.p. 170-172 C; ref. index 1.4064 (20 C)
Uses: Flavoring and mfg. fruit essences

Isobutyraldehyde. *See* 2-Methylpropanal

Isobutyric acid
CAS 79-31-2; EINECS 201-195-7; FEMA 2222
Synonyms: Dimethylacetic acid; 2-Methylpropanoic acid; Isopropylformic acid
Classification: Organic acid
Empirical: $C_4H_8O_2$
Formula: $(CH_3)_2CHCOOH$
Properties: Colorless liq., pungent odor of rancid butter; sol. in 6 parts of water; misc. with alcohol, ether, chloroform; m.w. 88.11; dens. 0.946-0.950 (20/20 C); b.p. 154.4 C (760 mm); f.p. -47 C; flash pt. (TOC) 76.6 C; ref. index 1.393 (20 C)
Precaution: Flamm., corrosive; reactive with oxidizing materials
Toxicology: LD50 (oral, rat) 280 mg/kg; poison by ingestion; mod. toxic by skin contact; corrosive irritant to tissue; heated to decomp., emits acrid smoke and fumes
Uses: Synthetic flavoring agent
Usage level: 4.1 ppm (nonalcoholic beverages), 12 ppm (ice cream), 41 ppm (candy), 38 ppm (baked goods), 470 ppm (chewing gum), 30 ppm (margarine)
Regulatory: FDA 21CFR §172.515
Manuf./Distrib.: Aldrich; Eastman

Isobutyric acid benzyl ester. *See* Benzyl isobutyrate
Isocamphol. *See* Isoborneol

Isocetyl alcohol
CAS 36311-34-9; EINECS 252-964-9
Synonyms: Isohexadecanol; Isohexadecyl alcohol; Isopalmityl alcohol
Definition: Mixture of branched chain C16 aliphatic alcohols
Empirical: $C_{16}H_{34}O$
Formula: $C_8H_{17}CHC_6H_{13}CH_2OH$
Properties: Liq.; sol. in min. oil, 95% ethanol, IPM, oleyl alcohol, castor oil, cyclomethicone; insol. in water, glycerin, propylene glycol; m.w. 242; sp.gr. 0.830-0.840; flash pt. (COC) 156 C
Toxicology: LD50 (rat, oral) > 50 g/kg; mildly irritating to skin and eyes
Uses: Carrier and extender for flavor and fragrance oils; solubilizer

Isodecylaldehyde. *See* 2,6-Dimethyl octanal
Isodulcit. *See* Rhamnose

Isoeugenol
CAS 97-54-1; EINECS 202-589-1; FEMA 2468

Isoeugenol acetate

Synonyms: 1-Hydroxy-2-methoxy-4-propenylbenzene; 2-Methoxy-4-propenylphenol; 4-Propenylguaiacol
Empirical: $C_{10}H_{12}O_2$
Properties: Pale yel. oil, carnation odor; sol. in fixed oils, propylene glycol; very sl. sol. in water; misc. with alcohol, ether; insol. in glycerin; m.w. 164.22; dens. 1.079-1.085; m.p. -10 C; b.p. 266 C; flash pt. > 230 F; ref. index 1.572-1.577
Precaution: Combustible
Toxicology: LD50 (oral, rat) 1560 mg/kg; mod. toxic by ingestion; human mutagenic data; heated to decomp., emits acrid smoke and fumes
Uses: Synthetic flavoring agent
Usage level: 3.7 ppm (nonalcoholic beverage), 3.8 ppm (ice cream), 5 ppm (candy), 11 ppm (baked goods), 0.3-1000 ppm (chewing gum), 1 ppm (condiments)
Regulatory: FDA 21CFR §172.515; Japan approved as flavoring
Manuf./Distrib.: Aldrich

Isoeugenol acetate. *See* Acetisoeugenol
trans-Isoeugenol benzyl ether. *See* Isoeugenyl benzyl ether

Isoeugenyl benzyl ether
CAS 92666-21-2; FEMA 3698
Synonyms: Benzyl isoeugenol; trans-Isoeugenol benzyl ether
Empirical: $C_{17}H_{18}O_2$
Properties: Wh. to ivory-colored cryst. powd., rose-carnation odor; m.w. 254.33; b.p. 282 C; m.p. 59-63 C
Uses: Synthetic flavoring
Regulatory: FDA 21CFR §172.515
Manuf./Distrib.: Aldrich

Isoeugenyl ethyl ether
FEMA 2472
Synonyms: Ethyl isoeugenyl; 2-Ethoxy-5-propenyl anisole
Empirical: $C_{12}H_{16}O_2$
Properties: Cryst. solid; sol. in ether, benzene, alcohol; m.w. 192.26; m.p. 63-64 C; b.p. 245 C
Uses: Synthetic flavoring
Usage level: 7.8 ppm (nonalcoholic beverages), 0.50 ppm (ice cream, ices), 17 ppm (candy), 1.0-3.5 ppm (baked goods)
Regulatory: FDA 21CFR §172.515

Isoeugenyl formate
FEMA 2474
Empirical: $C_{11}H_{12}O_3$
Properties: Colorless to pale straw-yel. visc. liq., orris-like green sweet woody odor, warm spicy flavor; sol. in alcohol; insol. in water; m.w. 192.22; b.p. 282 C; ref. index 1.5660
Uses: Synthetic flavoring
Usage level: 0.20 ppm (condiments)
Regulatory: FDA 21CFR §172.515

Isoeugenyl methyl ether. *See* Methyl isoeugenol

Isoeugenyl phenylacetate
CAS 120-24-1; FEMA 2477
Empirical: $C_{18}H_{18}O_3$
Properties: Ylsh. visc. liq.; insol. in water; m.w 282.34; dens. 1.119; flash pt. > 230 F; ref. index 1.575-1.577
Uses: Synthetic flavoring
Usage level: 0.05 ppm (nonalcoholic beverages), 0.20 ppm (ice cream, ices), 3.0 ppm (candy), 2.0-3.0 ppm (baked goods)
Regulatory: FDA 21CFR §172.515
Manuf./Distrib.: Aldrich

Isohexadecanol. *See* Isocetyl alcohol
Isohexadecyl alcohol. *See* Isocetyl alcohol

Isojasmone
Definition: Mixt. of 2-hexylidenecyclopentanone and 2-hexyl-2-cyclopenten-1-one
Empirical: $C_{11}H_{18}O$
Properties: Yel. to yel.-brn. liq., jasmine-like odor; m.w. 166.26; dens. 0.911; m.p. 144 C (10 mm); flash pt. > 100 C; ref. index 1.4750-1.4800
Uses: Synthetic flavoring
Regulatory: FDA 21CFR §172.515

Isolated soy protein. *See* Soy protein

DL-Isoleucine
CAS 443-79-8; EINECS 207-139-8
Synonyms: DL-2-Amino-3-methylvaleric acid
Empirical: $C_6H_{13}NO_2$
Properties: Wh. cryst. powd., odorless, sl. bitter taste; sol. in water; insol. in alcohol, ether; m.w. 131.17; m.p. 292 C (with dec.)
Uses: Nutrient; dietary supplement

L-Isoleucine
CAS 73-32-5; EINECS 200-798-2
Synonyms: 2-Amino-3-methylpentanoic acid; α-Amino-β-methylvaleric acid; 2-Amino-3-methylvaleric acid
Classification: A natural protein amino acid
Empirical: $C_6H_{13}NO_2$
Formula: $CH_3CH_2CH(CH_3)CH(NH_2)COOH$
Properties: Wh. cryst. powd., bitter taste; sl. sol. in water; pract. insol. in alcohol; insol. in ether; m.w. 131.17; m.p. 283-284 C (dec.)
Toxicology: LD50 (intraperitoneal, rat) 6822 mg/kg; mildly toxic by intraperitoneal route; heated to decomp., emits toxic fumes of NO_x
Uses: Nutrient, dietary supplement; flavorings
Regulatory: FDA 21CFR §172.320, 6.6% max.; Japan approved

Isomalt
Synonyms: Hydrogenated isomaltulose; Hydrogenated palatinose
Toxicology: 50% metabolized in humans; breaks down to form sorbitol, mannitol, and glucose; short-term animal studies show increase in bilirubin levels
Uses: Nutritive sweetener; sugar substitute in confections, chewing gum, soft drinks, desserts
Usage level: ADI not specified (WHO)

α-Isomethyllonone
FEMA 2714
Synonyms: 4-(2,6,6-Trimethyl-2-cyclohexen-1-yl)-3-methyl-3-buten-2-one; Methyl γ-ionone; α-Cyclo-citrylidene butanone
Empirical: $C_{14}H_{22}O$
Properties: Ylsh. liq., violet and orris-like odor; m.w. 206.30; dens. 0.9304; b.p. 93 C (31 mm); ref. index 1.5000-1.5020
Uses: Synthetic flavoring
Usage level: 0.97 ppm (nonalcoholic beverages), 0.98 ppm (ice cream, ices), 4.9 ppm (candy), 4.3 ppm (baked goods), 0.05 ppm (gelatins, puddings), 0.80 ppm (chewing gum)
Regulatory: FDA 21CFR §172.515

Isooctadecanoic acid, diester with diglycerol. *See* Polyglyceryl-2 diisostearate
Isooctadecanoic acid, monoester with 1,2,3-propanetriol. *See* Glyceryl isostearate
Isooctadecanoic acid, triester with diglycerol. *See* Polyglyceryl-2 triisostearate
Isooctyl phthalate. *See* Diisooctyl phthalate
Isopalmityl alcohol. *See* Isocetyl alcohol

Isoparaffinic petroleum hydrocarbons, synthetic
Toxicology: Heated to decomp, emits acrid smoke and irritating fumes
Uses: Coating agent, float; used for eggs, fruit, pickles, vegetables, vinegar, wine; froth-flotation cleaning; insecticide formulations component
Regulatory: FDA 21CFR §172.882, 561.365

Isopentanoic acid. *See* Isovaleric acid
Isopentyl acetate. *See* Isoamyl acetate
Isopentyl alcohol. *See* Isoamyl alcohol
Isopentyl alcohol acetate. *See* Isoamyl acetate
Isopentyl benzoate. *See* Isoamyl benzoate
Isopentyl butyrate. *See* Isoamyl butyrate
Isopentyl cinnamate. *See* Isoamyl cinnamate
Isopentyl formate. *See* Isoamyl formate
Isopentyl-2-furanbutyrate. *See* Isoamyl 2-furanbutyrate
Isopentyl isovalerate. *See* Isoamyl isovalerate
Isopentyl laurate. *See* Isoamyl laurate
Isopentyl-2-methylbutyrate. *See* Isoamyl-2-methylbutyrate

Isopentyl nonanoate. *See* Isoamyl nonanoate
Isopentyl phenylacetate. *See* Isoamyl phenylacetate
Isopentyl pyruvate. *See* Isoamyl pyruvate
Isopentyl salicylate. *See* Amyl salicylate
Isopropanol. *See* Isopropyl alcohol
Isopropenyl carbinyl-n-butyrate. *See* 2-Methylallyl butyrate
4-Isopropenyl-1-cyclohexene-1-carboxaldehyde. *See* Perillaldehyde
(R)-4-Isopropenyl-1-methyl-1-cyclohexene. *See* d-Limonene

Isopropyl acetate
CAS 108-21-4; EINECS 203-561-1; FEMA 2926
Synonyms: Acetic acid 1-methylethyl ester; 2-Propyl acetate; 1-Methylethyl acetate
Empirical: $C_5H_{10}O_2$
Formula: $CH_3COOCH(CH_3)2$
Properties: Colorless aromatic liq., fruity odor; sl. sol. in water; misc. with alcohol, ether, fixed oils; m.w. 102.15; dens. 0.874 (20/20 C); m.p. 073 C; f.p. -69.3 C; b.p. 88.4 C; flash pt. 40 F; ref. index 1.377
Precaution: Highly flamm.; dangerous fire hazard with heat, flame, oxidizers; mod. explosive with heat or flame
Toxicology: TLV 250 ppm; LD50 (oral, rat) 3000 mg/kg; mod. toxic by ingestion; mildly toxic by inh.; human systemic effects on inh.; narcotic in high conc.; chronic exposure can cause liver damage
Uses: Synthetic flavoring agent
Usage level: 16 ppm (nonalcoholic beverages), 17 ppm (ice cream, ices), 58 ppm (candy), 75 ppm (baked goods)
Regulatory: FDA 21CFR §172.515, 175.105, 177.1200
Manuf./Distrib.: Aldrich

Isopropylacetic acid. *See* Isovaleric acid
Isopropylacetone. *See* Methyl isobutyl ketone

p-Isopropylacetophenone
FEMA 2927
Synonyms: Acetocumene; 1,4-Acetyl-isopropyl benzol; p-Acetyl cumol
Empirical: $C_{11}H_{14}O$
Properties: Colorless liq.; sol. in alcohol; insol. in water; m.w. 162.23; dens. 0.975; b.p. 252-254 C
Uses: Synthetic flavoring
Usage level: 0.08 ppm (nonalcoholic beverages), 0.10 ppm (ice cream, ices), 0.50 ppm (candy), 1.0 ppm (baked goods), 5.0 ppm (pickles)
Regulatory: FDA 21CFR §172.515

Isopropyl alcohol
CAS 67-63-0; EINECS 200-661-7; FEMA 2929
Synonyms: IPA; Isopropanol; 2-Propanol; Dimethyl carbinol
Classification: Aliphatic alcohol
Empirical: C_3H_8O
Formula: $(CH_3)_2CHOH$
Properties: Colorless liq., pleasant odor, sl. bitter taste; sol. in water, alcohol, ether, chloroform; m.w. 60.11; dens. 0.7863 (20/20 C); f.p. -86 C; b.p. 82.4 C (760 mm); flash pt. (TOC) 11.7 C; ref. index 1.3756 (20 C)
Precaution: Flamm.; very dangerous fire hazard with heat, flame, oxidizers; reacts with air to form dangerous peroxides; heated to decomp., emits acrid smoke and fumes
Toxicology: TLV:TWA 400 ppm; STEL 500 ppm; LD50 (oral, rat) 5045 mg/kg; poison by ingestion, subcutaneous routes; human systemic effects by ingestion/inhalation (headache, nausea, vomiting, narcosis); 100 ml can be fatal; experimental reproductive effects
Uses: Synthetic flavoring agent, color diluent, extraction agent, defoamer; used in beet sugar, confectionery, tabletted food supplements, gum, hops extract, lemon oil, spice oleoresins, yeast
Usage level: 25 ppm (nonalcoholic beverages), 10-75 ppm (candy), 75 ppm (baked goods); ADI not specified (FAO/WHO)
Regulatory: FDA 21CFR §73.1 (no residue), 73.1001, 172.515, 172.560, 172.712, 173.240 (limitation 50 ppm in spice oleoresins, 6 ppm in lemon oil, 2% in hops extract), 173.340; 175.105, 176.200, 176.210, 177.1200, 177.2800, 178.1010, 178.3910; 27CFR §21.112; use in bread is permitted in Ireland and Japan
Manuf./Distrib.: Aldrich; Eastman

4-Isopropylbenzaldehyde. *See* Cuminaldehyde
p-Isopropyl benzaldehyde. *See* Cuminaldehyde

Isopropyl benzoate
FEMA 2932
Empirical: $C_{10}H_{12}O_2$

Properties: Liq.; sol. in alcohol, ether; insol. in water; m.w. 164.21; dens. 1.0263 (4 C); b.p. 218-219 C
Uses: Synthetic flavoring
Usage level: 0.50 ppm (nonalcoholic beverages), 1.0 ppm (ice cream, ices, candy, baked goods)
Regulatory: FDA 21CFR §172.515

p-Isopropylbenzyl alcohol. *See* Cuminic alcohol

Isopropyl butyrate
CAS 638-11-9; FEMA 2935
Empirical: $C_7H_{14}O_2$
Properties: Colorless liq., fruity odor; m.w. 130.19; dens. 0.859; b.p. 130-131 C; flash pt. 86 F; ref. index 1.3936
Uses: Synthetic flavoring
Usage level: 9.7 ppm (nonalcoholic beverages), 21 ppm (ice cream, ices), 39 ppm (candy, baked goods)
Regulatory: FDA 21CFR §172.515
Manuf./Distrib.: Aldrich

Isopropyl caproate. *See* Isopropyl hexanoate
Isopropyl capronate. *See* Isopropyl hexanoate
Isopropylcarbinol. *See* Isobutyl alcohol

Isopropyl cinnamate
FEMA 2939
Synonyms: Isopropyl β-phenylacrylate; Isopropyl 3-phenylpropenoate
Definition: Ester of isopropanol and cinnamic acid
Empirical: $C_{12}H_{14}O_2$
Properties: Colorless visc. liq., balsamic sweet dry amber-like odor, fresh fruity flavor; sol. in alcohol; insol. in water; m.w. 190.24; dens. 1.03; b.p. 268-270 C
Uses: Synthetic flavoring
Usage level: 0.52 ppm (nonalcoholic beverages), 0.75 ppm (ice cream, ices), 1.3 ppm (candy), 2.3 ppm (baked goods)
Regulatory: FDA 21CFR §172.515

Isopropyl citrate
Synonyms: Monoisopropyl citrate
Toxicology: Heated to decomp., emits acrid smoke and irritating fumes
Uses: Sequestrant; synergist for antioxidants; used in poultry fats, lard, dried meat, oleomargarine, sausage, shortening; plasticizer migrating from food pkg.; antioxidant (Japan)
Usage level: ADI 0-14 mg/kg (FAO/WHO)
Regulatory: FDA 21CFR §181.27, 182.6386 (0.2% max.), 182.6511, GRAS; USDA 9CFR §318.7 (limitation 0.02%), 381.147 (0.01% in poultry fats); Japan approved (0.1 g/kg max.)

6-Isopropyl-m-cresol. *See* Thymol
Isopropyl-o-cresol. *See* Carvacrol

Isopropyl formate
FEMA 2944
Empirical: $C_4H_8O_2$
Properties: Colorless liq., fruity ether-like odor, plum-like taste; sl. sol. in water; misc. with alcohol, ether; m.w. 88.10; dens. 0.8774; b.p. 67-68 C; ref. index 1.3678
Uses: Synthetic flavoring
Usage level: 18-25 ppm (nonalcoholic beverages, ice cream, ices), 55-100 ppm (candy), 60-100 ppm (baked goods)
Regulatory: FDA 21CFR §172.515

Isopropylformic acid. *See* Isobutyric acid
Isopropyl hex. *See* Isopropyl hexanoate

Isopropyl hexanoate
FEMA 2950
Synonyms: Isopropyl capronate; Isopropyl caproate; Isopropyl hex
Definition: Ester of hexanoic acid and isopropyl alcohol
Empirical: $C_9H_{18}O_2$
Properties: Colorless liq., pineapple-like odor, fresh sweet berry-like taste; sol. in alcohol; insol. in water; m.w. 158.24; dens. 0.8570; b.p. 176 C
Uses: Synthetic flavoring
Usage level: 0.50 ppm (nonalcoholic beverages), 5.5-10 ppm (ice cream, ices), 20-40 ppm (candy, baked goods)
Regulatory: FDA 21CFR §172.515

p-Isopropylhydrocinnamaldehyde. *See* 3-(p-Isopropylphenyl)-propionaldehyde

Isopropyl p-hydroxybenzoate
Uses: Preservative (0.012-1 g/kg as p-hydroxybenzoic acid)
Regulatory: Japan approved

(R)-2-Isopropylidene-5-methylcyclohexanone. *See* Pulegone

Isopropyl isobutyrate
FEMA 2937
Empirical: $C_7H_{14}O_2$
Properties: Liq., intense fruity ether-like odor; sol. in most org. solvs.; insol. in water; m.w. 130.18; dens. 0.8687 (0 C); b.p. 121 C
Uses: Synthetic flavoring
Usage level: 12-25 ppm (nonalcoholic beverages), 18-25 ppm (ice cream, ices), 58-100 ppm (candy), 60-100 ppm (baked goods)
Regulatory: FDA 21CFR §172.515

Isopropyl isovalerate
FEMA 2961
Empirical: $C_8H_{16}O_2$
Properties: Liq., ether-like odor, sweet apple-like taste; sol. in most org. solvs.; insol. in water; m.w. 144.21; b.p. 68-70 C (55 mm); ref. index 1.3960
Uses: Synthetic flavoring
Usage level: 3.4 ppm (nonalcoholic beverages, ice cream, ices), 11 ppm (candy, baked goods)
Regulatory: FDA 21CFR §172.515

1-Isopropyl-4-methylbenzene. *See* p-Cymene
4-Isopropyl-1-methylbenzene. *See* p-Cymene
1-Isopropyl-4-methyl-1,3-cyclohexadiene. *See* α-Terpinene
1-Isopropyl-4-methyl-1,4-cyclohexadiene. *See* γ-Terpinene
5-Isopropyl-2-methyl-1,3-cyclohexadiene. *See* α-Phellandrene
2-Isopropyl-5-methylcyclohexanol. *See* Menthol
2-Isopropyl-5-methylcyclohexanol. *See* d-Neomenthol
(S)-1-Isopropyl-4-methyl-3-cyclohexen-1-ol. *See* 4-Carvomenthenol
p-Isopropyl-α-methylhydrocinnamic aldehyde. *See* Cyclamen aldehyde
2-Isopropyl-5-methylphenol. *See* Thymol
5-Isopropyl-2-methylphenol. *See* Carvacrol
p-Isopropyl-α-methylphenylpropyl aldehyde. *See* Cyclamen aldehyde
α-Isopropyl phenylacetaldehyde. *See* 3-Methyl-2-phenylbutyraldehyde

p-Isopropylphenylacetaldehyde
FEMA 2954
Synonyms: p-Cymen-7-carboxaldehyde; Cuminic acetaldehyde; Homo-cuminic aldehyde
Empirical: $C_{11}H_{14}O$
Properties: Colorless liq., char. bark odor, citrus bittersweet fruity flavor; m.w. 162.23; dens. 0.0955; b.p. 230 C; ref. index 1.5200
Uses: Synthetic flavor
Usage level: 0.10 ppm (nonalcoholic beverages), 0.50 ppm (ice cream, ices, candy)
Regulatory: FDA 21CFR §172.515

Isopropyl phenylacetate
FEMA 2956
Synonyms: Isopropyl α-toluate
Empirical: $C_{11}H_{14}O_2$
Properties: Liq., fragrant rose-like scent, honey-like flavor; m.w. 178.23; dens. 1.0096; b.p. 253 C
Uses: Synthetic flavoring
Usage level: 0.20 ppm (nonalcoholic beverages), 1.8 ppm (ice cream, ices), 0.50-8.0 ppm (candy), 3.0-8.0 ppm (baked goods)
Regulatory: FDA 21CFR §172.515

Isopropyl β-phenylacrylate. *See* Isopropyl cinnamate
Isopropyl 3-phenylpropenoate. *See* Isopropyl cinnamate

3-(p-Isopropylphenyl)-propionaldehyde
FEMA 2957
Synonyms: p-Isopropylhydrocinnamaldehyde; Cuminyl acetaldehyde; p-Cymyl propanal
Empirical: $C_{12}H_{16}O$

Properties: Colorless visc. liq., powerful sweet green floral odor, sweet green fruity flavor; sol. in alcohol; insol. in water; m.w. 176.26
Uses: Synthetic flavoring
Usage level: 0.80 ppm (nonalcoholic beverages, ice cream, ices), 1.3 ppm (candy), 3.0 ppm (baked goods), 5.0 ppm (chewing gum)
Regulatory: FDA 21CFR §172.515

Isopropyl propionate
FEMA 2959
Empirical: $C_6H_{12}O_2$
Properties: Liq., plum-like taste; misc. with alcohol; m.w. 116.16; dens. 0.8660; b.p. 108-110 C; ref. index 1.3872
Uses: Synthetic flavoring
Usage level: 9.7 ppm (nonalcoholic beverages), 5.0-50 ppm (ice cream, ices), 40-50 ppm (candy), 30-50 ppm (baked goods)
Regulatory: FDA 21CFR §172.515

Isopropyl α-toluate. *See* Isopropyl phenylacetate
4-Isopropyltoluene. *See* p-Cymene
p-Isopropyltoluene. *See* p-Cymene

Isopulegol
CAS 7786-67-6; FEMA 2962
Synonyms: p-Menth-4-en-3-ol; p-Menth-8-en-3-ol; 1-Methyl-4-isopropenyl cyclohexan-3-ol
Classification: Terpene deriv.
Empirical: $C_{10}H_{18}O$
Properties: Colorless liq., mint-like odor; m.w. 154.24; dens. 0.904-0.911; ref. index 1.471-1.474
Precaution: Combustible
Uses: Synthetic flavoring
Usage level: 7.4 ppm (nonalcoholic beverages), 29 ppm (ice cream, ices), 23 ppm (candy, baked goods)
Regulatory: FDA 21CFR §172.515

Isopulegone
FEMA 2964
Synonyms: p-Menth-8-en-3-one; 1-Methyl-4-isopropenyl cyclohexan-3-one
Empirical: $C_{10}H_{18}O$
Properties: Colorless liq.; dens. 0.92177, b.p. 101-102 C (17 mm); ref. index 1.46787
Uses: Synthetic flavoring
Usage level: 4.0 ppm (nonalcoholic beverages), 12 ppm (ice cream, ices), 16 ppm (candy, baked goods)
Regulatory: FDA 21CFR §172.515

Isopulegyl acetate
CAS 89-49-6; 57576-09-7; FEMA 2965
Empirical: $C_{12}H_{20}O_2$
Properties: Colorless liq., sweet mint-like odor; m.w. 196.29; dens. 0.932-0.936; b.p. 104-105 C (10 mm); flash pt. 87 C; ref. index 1.4572
Uses: Synthetic flavoring
Usage level: 5.8 ppm (nonalcoholic beverages), 22 ppm (ice cream, ices), 19 ppm (candy, baked goods)
Regulatory: FDA 21CFR §172.515
Manuf./Distrib.: Aldrich

Isoquinoline
CAS 119-65-3; EINECS 204-341-8; FEMA 2978
Synonyms: 2-Benzazine
Empirical: C_9H_7N
Properties: Colorless plates or liq.; sol. in most org. solvs.; sl. sol. in water; m.w. 129.16; dens. 1.09 (20/4 C); m.p. 23-25 C; b.p. 243 C; flash pt. 102 C; ref. index 1.615
Precaution: Combustible
Toxicology: Skin and eye irritant
Uses: Synthetic flavoring
Usage level: 0.25 ppm (nonalcoholic beverages, ice cream, ices), 1.0 ppm (candy), 0.004-1.0 ppm (baked goods)
Regulatory: FDA 21CFR §172.515
Manuf./Distrib.: Aldrich

Isosafroeugenol. *See* Propenylguaethol

Isosvalerianic acid. *See* Isovaleric acid
3-Isothiocyanato-1-propene. *See* Allyl isothiocyanate
Isothiocyanic acid allyl ester. *See* Allyl isothiocyanate
Isothymol. *See* Carvacrol
Isourea. *See* Urea
Isovaleral. *See* Isovaleraldehyde

Isovaleraldehyde
CAS 590-86-3; EINECS 209-691-5; FEMA 2692
Synonyms: 3-Methylbutyraldehyde; 3-Methylbutanal; Isovaleral; Isovaleric aldehyde
Empirical: $C_5H_{10}O$
Formula: $(CH_3)_2CHCH_2CHO$
Properties: Colorless liq., pungent apple-like odor; sparingly sol. in water; misc. with alcohol, ether; m.w. 86.14; dens. 0.797 (20/4 C); m.p. -51 C; b.p. 91-93 C; flash pt. -5 C; ref. index 1.388 (20 C)
Precaution: Highly flamm.
Toxicology: Irritating to eyes, respiratory tract
Uses: Synthetic flavoring agent
Manuf./Distrib.: Aldrich

Isovaleric acid
CAS 503-74-2; EINECS 207-975-3; FEMA 3102
Synonyms: 3-Methylbutyric acid; 3-Methylbutanoic acid; Isosvalerianic acid; Delphinic acid; Isopentanoic acid; Isopropylacetic acid
Empirical: $C_5H_{10}O_2$
Formula: $(CH_3)_2CHCH_2COOH$
Properties: Colorless liq., disagreeable rancid cheese odor, acid taste; sol. in water @ 16 C; misc. with alcohol, chloroform, ether; m.w. 102.15; dens. 0.931 (20/4 C); solid. pt. -37 C; m.p. -34.5 C; b.p. 175-177 C; ref. index 1.403
Precaution: Corrosive material; keep tightly closed
Toxicology: LD50 (oral, rat) 2000 mg/kg; poison by skin contact; mod. toxic by ingestion, IV routes; corrosive skin and eye irritant; heated to decomp.,emits acrid smoke and fumes
Uses: Intense synthetic flavoring reminiscent of decomposing cheese
Usage level: 1.2 ppm (nonalcoholic beverages), 14 ppm (ice cream, ices), 12 ppm (candy), 5.5 ppm (baked goods), 2.4 ppm (cheese)
Regulatory: FDA 21CFR §172.515
Manuf./Distrib.: BASF

Isovaleric acid allyl ester. *See* Allyl isovalerate
Isovaleric acid ammonium salt. *See* Ammonium isovalerate
Isovaleric acid butyl ester. *See* n-Butyl isovalerate
Isovaleric acid ethyl ester. *See* Ethyl isovalerate
Isovaleric aldehyde. *See* Isovaleraldehyde

Itaconic acid (polymerized)
CAS 97-65-4; EINECS 202-599-6
Synonyms: Methylenesuccinic acid; Propylenedicarboxylic acid
Empirical: $C_5H_6O_4$
Formula: $HOOCCH_2C(:CH_2)COOH$
Properties: Wh. cryst., char. odor; hygroscopic; sol. (1 g/ml): 12 ml water, 5 ml alcohol; very sl. sol. in benzene, chloroform, ether, CS_2, petroleum ether; m.w. 130.10; dens. 1.63; m.p. 165-167 C (dec.)
Precaution: Keep well closed
Uses: Natural food acid (Japan); used in mfg. of paper/paperboard contacting food
Regulatory: FDA 21CFR §181.30; Japan approved (itaconic acid)

Jaguar gum. *See* Guar gum
Japanese tea extract. *See* Thea sinensis extract
Japan isinglass. *See* Agar
Japan tallow. *See* Japan wax

Japan wax
CAS 8001-39-6, 67701-27-3
Synonyms: Rhus succedanea wax; Japan tallow; Sumac wax
Definition: Fat expressed from the mesocarp of the fruit of *Rhus succedanea*, contg. 10-15% palmitin, stearin, olein, 1% japanic acid
Properties: Pale yel. solid, greasy feel, rancid odor and taste; sol. in benzene, CS_2, ether, hot alcohol, alkalies; insol. in water; dens. 0.97-0.98; m.p. 53.5-55 C; acid no. 22-23; iodine no. 10-15; sapon. no. 217-237

Precaution: Combustible
Uses: Migrating to foods from cotton in dry food pkg.
Regulatory: FDA 21CFR §73.1, 175.105, 175.350, 176.170, 182.70

Jasminaldehyde. *See* α-Amylcinnamaldehyde

Jasmine extract
CAS 90045-94-6, 84776-64-7; EINECS 289-960-1, 283-993-5
Synonyms: White jasmine extract; Jasminum grandiflorum extract; Jasminum officinale extract
Definition: Extract of the leaves and flowers of *Jasminum officinale* or *J. grandiflorum*
Uses: Natural flavoring agent
Regulatory: FDA 21CFR §182.20, GRAS; Japan approved
Manuf./Distrib.: Pierre Chauvet

Jasminum grandiflorum extract. *See* Jasmine extract
Jasminum officinale extract. *See* Jasmine extract

cis-Jasmone
CAS 488-10-8; FEMA 3196
Synonyms: 3-Methyl-2-(2-pentenyl)-2-cyclopenten-1-one
Classification: Ketone
Empirical: $C_{11}H_{16}O$
Properties: Odor of jasmine; m.w. 164.25; dens. 0.940; b.p. 134-135 C (12 mm); flash pt. 225 F
Uses: Synthetic flavoring
Regulatory: FDA 21CFR §172.515
Manuf./Distrib.: Aldrich

Jerusalem artichoke flour
Uses: Bulking agent, nutritive additive
Trade names: Jerusalem Arthichoke Flour (JAF)

Jojoba oil
CAS 61789-91-1
Synonyms: Oils, jojoba
Definition: Oil from the seeds of the Jojoba desert shrub (*Simmondsia chinensis*)
Properties: Colorless waxy liq., odorless
Uses: Food ingred.
Manuf./Distrib.: Am. Roland; Arista Industries

Jojoba wax
Uses: Natural chewing gum base
Regulatory: Japan approved

Juniper berry oil. *See* Juniper oil

Juniper extract
CAS 84603-69-0
Synonyms: Juniperus communis extract
Definition: Extract of ripe fruits of *Juniperus communis*
Uses: Natural flavoring agent
Usage level: 53 ppm (nonalcoholic beverages), 5 ppm (ice cream), 5 ppm (candy), 5 ppm (baked goods)
Regulatory: FDA 21CFR §182.20, GRAS
Manuf./Distrib.: Chart

Juniper oil
CAS 8012-91-7, 73049-62-4
Synonyms: Juniper berry oil
Definition: Volatile oil obtained from the berries of *Juniperus communis*, contg. pinene, cadinene, camphene, terpineol, juniper camphor
Properties: Colorless to pale grnsh.-yel. liq., aromatic bitter taste; sol. in fixed oils, min. oil; pract. insol. in water; insol. in glycerin, propylene glycol; dens. 0.854-0.879 (25/25 C); ref. index 1.4780-1.4840 (20 C)
Precaution: Keep cool, well closed; protect from light
Toxicology: LD50 (oral, rat) 6280 mg/kg; mildly toxic by ingestion; human skin and systemic irritant; allergen; taken internally may cause severe kidney irritation; heated to decomp., emits acrid smoke and fumes
Uses: Natural flavoring agent; mfg. liqueurs
Usage level: 32 ppm (nonalcoholic beverages), 95 ppm (alcoholic beverages), 1.9 ppm (ice cream), 4.3 ppm (candy), 11 ppm (baked goods), 0.01 ppm (gelatins, puddings), 0.1 ppm (chewing gum), 20 ppm (meats)
Regulatory: FDA 21CFR §182.20, GRAS; Japan approved

Manuf./Distrib.: Acme-Hardesty; Chart; Pierre Chauvet; Florida Treatt

Juniperus communis extract. See Juniper extract
Kadaya gum. See Karaya gum
Kadeoel. See Cade oil
Kalinite. See Potassium alum dodecahydrate

Kaolin
CAS 1332-58-7; EINECS 296-473-8; EEC E559
Synonyms: Bolus alba; China clay
Definition: Native hydrated aluminum silicate
Formula: $\approx Al_2O_3 \cdot 2SiO_2 \cdot 2H_2O$
Properties: White to yel. or grayish fine powd., earthy taste; insol. in water, dilute acids, alkali hydroxides; dens. 1.8-2.6
Toxicology: Nuisance dust
Uses: Anticaking agent, clarifying agent for wine; paper mfg. aid for food pkg.; processing aid (Japan)
Usage level: ADI no limit (EEC)
Regulatory: FDA 21CFR §182.2727, 182.2729, 186.1256, GRAS as indirect additive; BATF 27CFR §240.1051; Japan restricted (0.5% max. residual); Europe listed; UK approved
Manuf./Distrib.: ECC Int'l.

Karaya gum
CAS 9000-36-6; EINECS 232-539-4; EEC E416
Synonyms: Sterculia gum; Sterculia urens gum; India tragacanth; Indian tragacanth; Kadaya gum
Definition: A hydrophilic polysaccharide from trunks of the genus Sterculia
Properties: Wh. fine powd., sl. acetic acid odor; insol. in alcohol; swells in water to a gel
Toxicology: Very mildly toxic by ingestion; mild allergen; may cause intolerance; laxative effect, may reduce nutrient intake
Uses: Stabilizer, thickener, emulsifier, formulation aid; used in baked goods, soft candy, frozen dairy desserts, milk prods., toppings
Usage level: Limitation 0.3% (frozen dairy desserts), 0.02% (milk prods.), 0.9% (soft candy), 0.002% (other foods); ADI 0-20 mg/kg (EEC)
Regulatory: FDA 21CFR §133.133, 133.134, 133.162, 133.178, 133.179, 150.141, 150.161, 184.1349, GRAS; Japan approved; Europe listed; UK approved
Manuf./Distrib.: Aarhus Olie; Agrisales; Ashland; Bio-Botanica; Arthur Branwell; Bronson & Jacobs; Browne & Dureau Int'l.; Colloids Naturels; Florexco; Gum Tech.; Importers Service; Mar-Gel; Meer; Penta Mfg.; Red Carnation; Thew Arnott; TIC Gums; Alfred L Wolff
Trade names: Karaya Gum #1 FCC; Karaya Gum #1 FCC; Powdered Gum Karaya Superfine #1 FCC; Powdered Gum Karaya Superfine XXXX FCC; Premium Powdered Gum Karaya No. 1; Premium Powdered Gum Karaya No. 1 Special; Premium Powdered Gum Karaya No. 2 Special HV; Premium Powdered Gum Karaya No. 2; Premium Powdered Gum Karaya No. 3

Katchung oil. See Peanut oil
Katemfe. See Thaumatin

Kauri gum
Uses: Natural chewing gum base
Regulatory: Japan approved

Kelp
Definition: Dehydrated seaweed from Macrocystis pyriferae, Laminaria digitata, L. saccharina, L. cloustoni
Properties: Dk. grn. to olive-brn. color, salty char. taste
Toxicology: Heated to decomp., emits acrid smoke and irritating fumes
Uses: Dietary supplement, source of iodine, chewing gum base ingred.
Usage level: Limitation of total iodine 45 µg (infants), 105 µg (children under 4), 225 µg (adults), 300 µg (pregnant and lactating women)
Regulatory: FDA 21CFR §172.365, 184.1120, 184.1121
Manuf./Distrib.: Am. Roland; Chart

Kelp extract
Synonyms: Macrocystis pyriferae extract
Definition: Extract of kelp from Macrocystis pyriferae or other species
Uses: Natural thickener/stabilizer
Regulatory: Japan approved

α-Ketoglutaric acid
CAS 328-50-7

Synonyms: 2-Oxoglutaric acid; α-Oxoglutaric acid; 2-Oxopentanedioic acid
Empirical: $C_6H_6O_5$
Properties: Sol. in water, alcohol; m.w. 146.10; m.p. 113.5 C
Toxicology: Irritant
Uses: Natural food acid
Regulatory: Japan approved

Ketone methyl phenyl. *See* Acetophenone
β-Ketopropane. *See* Acetone
α-Ketopropionic acid. *See* Pyruvic acid
γ-Ketovaleric acid. *See* Levulinic acid
Kieselguhr. *See* Diatomaceous earth
Kola. *See* Cola
Kola nut. *See* Cola
Kola nut. *See* Cola nut
Kola seeds. *See* Cola
Konjac. *See* Konjac flour

Konjac flour
Synonyms: Konjac; Konnyaku; Konjac gum
Manuf./Distrib.: Am. Roland; Atomergic Chemetals
Trade names: Nutricol® Konjac

Konjac gum. *See* Konjac flour
Konnyaku. *See* Konjac flour
Kordofan gum. *See* Acacia
KTPP. *See* Potassium tripolyphosphate

Kukui nut oil
CAS 8015-80-3
Synonyms: Aleurites moluccana oil; Candlenut oil
Definition: Oil expressed from the nuts of the kukui tree, *Aleurites moluccana*
Properties: Lt. yel. oil; sp.gr. 0.926; m.p. -12 C; solid. pt. -22 C; acid no. 0.55; iodine no. 184; sapon. no. 192-193
Uses: Food ingred.
Manuf./Distrib.: Arista Industries

Labdanum
CAS 8016-26-0; EINECS 289-711-7 (extract)
Synonyms: Labdanum oil
Definition: Resinous exudation of *Cistus labdaniferus*
Uses: Natural flavoring
Regulatory: FDA 21CFR §172.510
Manuf./Distrib.: Pierre Chauvet; Florida Treatt

Labdanum oil. *See* Labdanum
Lacolin. *See* Sodium lactate

Lactase
Uses: Natural enzyme; catalyzes the production of glucose and galactose from lactose
Regulatory: FDA GRAS; Canada, Japan approved
Trade names: Fungal Lactase 100,000; Lactase AIE; Lactase F Amano; Yeast Lactase L-50,000

Lactase from *Kluyveromyces lactis*.
Definition: Enzyme prep. prepared from yeast *Kluyveromyces lactis*, contg. B-galactoside galactohydrase
Toxicology: Heated to decomp, emits acrid smoke and irritating fumes
Uses: Enzyme converting lactose to glucose and galactose; used in milk to produce lactase-treated milk
Regulatory: FDA 21CFR §184.1388, GRAS

Lactic acid
CAS 50-21-5, 598-82-3 (DL), 79-33-4 (L), 10326-41-7 (D); EINECS 200-018-0, 209-954-4 (DL), 201-296-2 (L), 233-713-2 (D); EEC E270
Synonyms: 2-Hydroxypropanoic acid; 1-Hydroxyethane 1-carboxylic acid; 2-Hydroxypropionic acid; α-Hydroxypropionic acid; Milk acid; Acetonic acid; Ethylidenelactic acid
Classification: Organic acid
Empirical: $C_3H_6O_3$
Formula: $CH_3CHOHCOOH$
Properties: Colorless to yellowish cryst. or syrupy liq.; misc. with water, alcohol, glycerol, furfural; m.w. 90.09;

dens. 1.249; m.p. 18 C; b.p. 122 C (15 mm)

Precaution: Mixts. with nitric acid + hydrofluoric acid may react vigorously

Toxicology: LD50 (oral, rat) 3730 mg/kg; mod. toxic by ingestion, rectal routes; mutagenic data; severe skin and eye irritant; heated to decomp., emits acrid smoke and irritating fumes

Uses: Acidulant in beverages, antimicrobial agent, preservative, curing/pickling agent, flavoring agent, adjuvant, flavor enhancer, pH control agent, solvent, vehicle for cheese, confectionery, cultured dairy prods., olives, poultry, wine; raising agent

Usage level: ADI no limit (EEC)

Regulatory: FDA 21CFR §131.144, 133, 150.141, 150.161, 172.814, 184.1061, GRAS; USDA 9CFR §318.7, 381.147; BATF 27CFR §240.1051, GRAS; not for use in infant foods; Japan approved; Europe listed; UK approved

Manuf./Distrib.: AB Tech.; ADM; Ashland; Dinoval; Ellis & Everard; Chr Hansen's Lab; Honeywill & Stein; Int'l. Sourcing; Jungbunzlauer; Lohmann; Mitsubishi; Penta Mfg.; Pointing; Purac Am.; Siber Hegner; Tesco; Todd's; Van Waters & Rogers

Trade names: Purac®

Trade names containing: BFP White Sour; Cap-Shure® LCL-135-50; Durkote Lactic Acid/Hydrog. Veg. Oil; Natural Flavor White Sour; Ry-So; Sour Dough Base; Tandem 11H K; Vanall (K)

Lactic acid esters of mono- and diglycerides of fatty acids

EEC E472b

Synonyms: Lactoglycerides

Properties: Disp. in hot water; HLB 3-4

Uses: Emulsifier, stabilizer

Usage level: ADI no limit (EEC)

Regulatory: FDA 21CFR §172.852; Europe listed; UK approved

Trade names: Admul GLP; Admul GLS; Axol® L 626; Imwitor® 333

Trade names containing: Diamond D-21; Lamequick® AS 340

Lactic acid ethyl ester. *See* Ethyl lactate

Lactic acid iron (2+) salt (2:1). *See* Ferrous lactate

Lactic acid menthyl ester. *See* Menthyl lactate

Lactitol monohydrate

CAS 81025-04-9

Synonyms: 4-O-β-D-Galactopyranosyl-D-glucitol

Classification: Disaccharide sugar alcohol

Empirical: $C_{12}H_{24}O_{11} \cdot H_2O$

Properties: m.w. 362.37; m.p. 95-98 C

Toxicology: Large doses may cause diarrhea

Uses: Nutritive sweetener

Usage level: ADI not specified (WHO)

Manuf./Distrib.: Am. Xyrofin

Trade names: Lactitol MC

Lactoflavin. *See* Riboflavin

Lactoglycerides. *See* Lactic acid esters of mono- and diglycerides of fatty acids

Lactose

CAS 63-42-3; EINECS 200-559-2

Synonyms: 4-O-β-Galactopyranosyl D-glucose; Milk sugar; Saccharum lactis

Classification: Disaccharide

Empirical: $C_{12}H_{22}O_{11}$

Formula: $C_6H_7O(OH)_4OC_6H_7O(OH)_4$

Properties: White hard cryst. mass or white powd., odorless to sl. char. odor, mildly sweet taste, odorless; sol. in water, alcohol, ether; sl. sol. in alcohol; m.w. 342.34; dens. 1.525 (20 C); m.p. dec. 203.5 C; b.p. dec.; stable in air

Toxicology: Moderately toxic by intravenous route

Uses: Dietary supplements, nutritive sweetener, formulation aid, processing aid, humectant, texturizer; infant foods, baking, confectionery, margarine and butter mfg.

Regulatory: FDA 21CFR §133.124, 133.178, 133.179, 168.122, 169.179, 169.182, GRAS

Manuf./Distrib.: Allchem Int'l.; Am. Dairy Prods.; Arla; Calaga Food Ingreds.; DMV Int'l.; G Fiske; K&K Greeff; Land O'Lakes; New Zealand Milk Prods.; Penta Mfg.; Quest USA; Siber Hegner; Tesco; Unilait France; Van Waters & Rogers; Westin

Trade names: Unisweet L; Unisweet Lactose

Trade names containing: Dry Phytonadione 1% SD No. 61748; Dry Vitamin A Palmitate Type 250-SD No.

65378; Dry Vitamin D₃ Type 100-SD No. 65216; Essiccum®; Natamax™; Veg-X-Ten®

Lacto-serum. *See* Whey
Lactylated fatty acid esters of glycerol and propane-1,2-diol. *See* Lactylated fatty acid esters of glycerol and propylene glycol

Lactylated fatty acid esters of glycerol and propylene glycol
EEC E478
Synonyms: Lactylated fatty acid esters of glycerol and propane-1,2-diol
Properties: Soft to hard waxy solid; disp. in hot water; mod. sol. in hot IPA, benzene, chloroform, soybean oil
Toxicology: Heated to decomp, emits acrid smoke and irritating fumes
Uses: Emulsifier, plasticizer, stabilizer, surface-active agent, whipping agent, surfactant; used in cake mixes, icings, toppings, coffee whiteners
Regulatory: FDA 21CFR §172.850; Europe listed; UK approved

Lactylic esters of fatty acids
Properties: Hard waxy solid to liq.; disp. in hot water; sol. in org. solvs., vegetable oil
Toxicology: Heated to decomp, emits acrid smoke and irritating fumes
Uses: Emulsifier, plasticizer, surface-active agent; used in baked goods, frozen desserts, dehydrated fruits and fruit juices, coffee whiteners, pudding mixes, liq. shortening, dehydrated vegetables and vegetable juices
Regulatory: FDA 21CFR §172.848, 172.860b, 172.862

Lactylic stearate
Uses: Used in baking
Trade names containing: MG-S; Myvatex® 40-06S

Laevosan. *See* Fructose
Lamp black. *See* Carbon black

Lanolin
CAS 8006-54-0 (anhyd.), 8020-84-6 (hyd.); EINECS 232-348-6
Synonyms: Anhydrous lanolin; Wool wax; Wool fat
Definition: Deriv. of unctuous fatty sebaceous secretion of sheep consistg. of complex mixt. of esters of high m.w. aliphatic, steroid, or triterpenoid alcohol and fatty acids
Properties: Yel.-wh. semisolid; sol. in chloroform, ether; insol. in water
Toxicology: Heated to decomp., emits acrid smoke and irritating fumes
Uses: Masticatory substance in chewing gum base; glazing agent
Regulatory: FDA 21CFR §172.615, 175.300, 176.170, 176.210, 177.1200, 177.2800, 178.3910; Japan approved
Manuf./Distrib.: Croda

Larch gum. *See* Arabinogalactan

Lard
CAS 71789-99-9; EINECS 263-100-5
Synonyms: Bleached lard; Unhydrogenated lard
Definition: Purified fat from abdomen of the hog
Properties: Whitish fat, faint odor, bland taste; sol. in ether, chloroform; insol. in water; m.p. 36-42 C
Precaution: Combustible
Toxicology: Heated to decomp., emits acrid smoke and irritating fumes
Uses: Coating agent, emulsifier, formulation aid, texturizer, chewing gum base; migrating to foods from cotton in dry food pkg.
Regulatory: FDA 21CFR §182.70, GRAS

Lard glyceride
CAS 61789-10-4, 97593-29-8; EINECS 263-032-6
Synonyms: Lard monoglyceride; Glycerides, lard mono-
Definition: Monoglyceride derived from lard
Uses: Emulsifier
Trade names: Dimodan P; Dimodan S; Monomuls® 90-10; Myverol® 18-40

Lard glycerides
Synonyms: Lard mono, di- and tri-glycerides
Definition: Mixture of mono, di and triglycerides derived from lard
Uses: Emulsifier, stabilizer, dispersant
Trade names: Monomuls® 60-10; Tegomuls® 19

Lard mono-, di- and tri-glycerides. *See* Lard glycerides
Lard monoglyceride. *See* Lard glyceride

Lard oil
Properties: Colorless or ylsh. liq., peculiar odor, bland taste; sol. in benzene, ether, chloroform; sl. sol. in alcohol; dens. 0.915; m.p. -2 C; flash pt. 420 F; ref. index 1.470
Precaution: Combustible
Uses: Migrating to foods from cotton in dry food pkg.
Regulatory: FDA 21CFR §182.70

Larixinic acid. *See* Maltol
Laughing gas. *See* Nitrous oxide

Lauralkonium chloride
CAS 139-07-1; EINECS 205-351-5
Synonyms: Lauryl dimethyl benzyl ammonium chloride; Benzyldimethyldodecylammonium chloride; N,N-Dimethyl-N-dodecylbenzenemethanaminium chloride; Dodecyl dimethyl benzyl ammonium chloride
Classification: Quaternary ammonium salt
Empirical: $C_{21}H_{38}N \cdot Cl$
Properties: m.w. 340.05
Toxicology: Skin and eye irritant; heated to decomp., emits very toxic fumes of NO_x, NH_3, and Cl.
Uses: Antimicrobial for use in cane-sugar mills
Regulatory: FDA 21CFR §172.165 (limitation 0.25-1.0 ppm), 173.320 (0.05 ± 0.005 ppm), 175.105

Laurel berries
Definition: Laurus nobilis
Uses: Natural flavoring agent
Regulatory: FDA 21CFR §182.20, GRAS

Laurel leaf oil
Synonyms: Bay leaf oil; Sweet bay oil; Laurel oil
Definition: Volatile oil from leaves of *Laurus nobilis*
Properties: Pale yel. to greenish liq.; sol. in alcohol; insol. in water; dens. 0.92-0.93 (15/15 C)
Precaution: Keep cool, well closed; protect from light
Uses: Natural flavoring agent
Regulatory: FDA 21CFR §182.10, 182.20, GRAS
Manuf./Distrib.: Pierre Chauvet

Laurel oil. *See* Laurel leaf oil

Lauric acid
CAS 143-07-7; EINECS 205-582-1; FEMA 2614
Synonyms: n-Dodecanoic acid; Dodecanoic acid; Dodecoic acid
Classification: Fatty acid
Empirical: $C_{12}H_{24}O_2$
Formula: $CH_3(CH_2)_{10}COOH$
Properties: Colorless needles; insol. in water; sol. in benzene and ether; m.w. 200.36; dens. 0.833; m.p. 44 C; b.p. 225 (100 mm); ref. index 1.4323 (45 C)
Precaution: Combustible when exposed to heat or flame; reactive with oxidizing materials
Toxicology: LD50 (oral, rat) 12 g/kg; poison by intravenous route; mildly toxic by ingestion; mutagenic data; heated to decomp., emits acrid smoke and irritating fumes
Uses: Component in mfg. of other food-grade additives; defoaming agent, lubricant; used in coconut oil, vegetable fats
Regulatory: FDA 21CFR §172.210, 172.860, 173.340, 175.105, 175.320, 176.170, 176.200, 176.210, 177.1010, 177.1200, 177.2260, 177.2600, 177.2800, 178.3570, 178.3910; GRAS (FEMA)
Manuf./Distrib.: Acme-Hardesty
Trade names: Emery® 6354

Lauric acid sodium salt. *See* Sodium laurate
Lauric acid triglyceride. *See* Trilaurin

Lauric aldehyde
CAS 112-54-9; EINECS 203-983-6; FEMA 2615
Synonyms: Aldehyde C-12; Dodecanal; n-Decyl aldehyde; Lauryl aldehyde; Laurinaldehyde
Empirical: $C_{12}H_{24}O$
Formula: $CH_3(CH_2)_{10}CHO$
Properties: Colorless to yel. liq., char. fatty odor; m.w. 184.32; dens. 0.830 (20/4 C); b.p. 237 C; flash pt. 101 C; ref. index 1.435 (20 C)
Toxicology: Skin irritant
Uses: Synthetic flavoring

Usage level: 0.93 ppm (nonalcoholic beverages), 1.5 ppm (ice cream, ices), 2.4 ppm (candy), 2.8 ppm (baked goods), 0.10 ppm (gelatins, puddings), 0.20-110 ppm (chewing gum)
Regulatory: FDA 21CFR §172.515
Manuf./Distrib.: Aldrich

Laurinaldehyde. *See* Lauric aldehyde

Lauryl acetate
CAS 112-66-3; FEMA 2616
Synonyms: Acetate C-12; Dodecanyl acetate; Dodecyl acetate
Empirical: $C_{14}H_{28}O_2$
Properties: Colorless liq., citrus-rose odor; sol. in most org. solvs.; m.w. 228.38; dens. 0.865; b.p. 150 C 915 mm); flash pt. > 230 F
Uses: Synthetic flavoring
Usage level: 2.3 ppm (nonalcoholic beverages), 1.7 ppm (ice cream, ices), 4.6 ppm (candy), 5.6 ppm (baked goods)
Regulatory: FDA 21CFR §172.515
Manuf./Distrib.: Aldrich

Lauryl alcohol
CAS 112-53-8; 68526-86-3; EINECS 203-982-0; FEMA 2617
Synonyms: 1-Dodecanol; C-12 linear primary alcohol; Alcohol C-12; Dodecyl alcohol
Classification: Fatty alcohol
Empirical: $C_{12}H_{26}O$
Formula: $CH_3(CH_2)_{10}CH_2OH$
Properties: Colorless leaflets, liq. above 21 C, floral odor; insol. in water; sol. in alcohol, ether; m.w. 186.33; dens. 0.8309 (24/4 C); m.p. 24 C; b.p. 259 C (760 mm); flash pt. (CC) > 212 F; ref. index 1.440-1.444
Precaution: Combustible; reactive with oxidizing materials
Toxicology: LD50 (oral, rat) 12,800 mg/kg; moderately toxic by intraperitoneal route; mildly toxic by ingestion; severe human skin irritant; heated to decomp., emits acrid smoke and irritating fumes
Uses: Synthetic flavoring agent, intermediate
Usage level: 2 ppm (nonalcoholic beverages), 1 ppm (ice cream), 2.8 ppm (candy), 1.7 ppm (baked goods), 16-27 ppm (chewing gum), 7 ppm (syrups)
Regulatory: FDA 21CFR §172.515, 172.864, 175.105, 175.300, 177.1010, 177.1200, 177.2800, 178.3480, 178.3910
Manuf./Distrib.: Aldrich

Lauryl aldehyde. *See* Lauric aldehyde
Lauryl dimethyl benzyl ammonium chloride. *See* Lauralkonium chloride
Lauryl gallate. *See* Dodecyl gallate
Lautarite. *See* Calcium iodate

Lavandin oil
CAS 8022-15-9; FEMA 2618
Definition: Oil from steam distillation of flowering stalks of *Lavandula hybrida reverchon, L. abrialis, L. officinalis,* or *L. latifolia*
Properties: Yel. liq., lavender camphoraceous odor; sol. in fixed oils, propylene glycol, min. oil; insol. in glycerin; dens. 0.885; ref. index 1.460 (20 C)
Toxicology: Skin irritant; heated to decomp., emits acrid smoke and irritating fumes
Uses: Natural flavoring agent
Usage level: 5.5 ppm (nonalcoholic beverages), 12 ppm (ice cream), 18 ppm (candy), 18 ppm (baked goods), 0.3 ppm (chewing gum)
Regulatory: FDA 21CFR §182.20, GRAS
Manuf./Distrib.: C.A.L.-Pfizer; Pierre Chauvet; Florida Treatt

Lavender flowers oil. *See* Lavender oil

Lavender oil
CAS 8000-28-0; FEMA 2622
Synonyms: Lavendula officinalis oil; Lavender flowers oil
Definition: Volatile oil obtained from flowers of *Lavendula officinalis,* contg. linalyl acetate, linalool, pinene, limonene, geraniol, etc.
Properties: Colorless to yel. liq.; sol. in 4 vols 70% alcohol; misc. with abs. alcohol, CS_2; sl. sol. in water; dens. 0.875-0.888; ref. index 1.459-1.470 (20 C)
Precaution: Keep cool, well closed; protect from light
Toxicology: LD50 (oral, rat) 9040 mg/kg; mildly toxic by ingestion; skin irritant; heated to decomp., emits acrid

smoke and irritating fumes
Uses: Natural flavoring agent
Usage level: 2.9 ppm (nonalcoholic beverages), 7.8 ppm (ice cream), 5.5 ppm (candy), 8.3 ppm (baked goods), 220 ppm (chewing gum)
Regulatory: FDA 21CFR §182.10, 182.20, GRAS; 27CFR §21.65, 21.151; Japan approved
Manuf./Distrib.: C.A.L.-Pfizer; Chart; Pierre Chauvet; Florida Treatt

Lavendula officinalis oil. *See* Lavender oil
Leaf alcohol. *See* 2-Hexenol
Leaf alcohol. *See* 3-Hexenol

Lecithin
CAS 8002-43-5; 97281-47-5; EINECS 232-307-2; EEC E322
Definition: Mixture of the diglycerides of stearic, palmitic and oleic acids linked to the choline ester of phosphoric acid; found in plants and animals
Formula: $C_8H_{17}O_5NRR'$, R and R´ are fatty acid groups
Properties: Nearly white to yel. or brown waxy mass or thick fluid; insol. but swells in water and salt sol'ns.; sol. in chloroform, ether, petrol. ether, min. oils, fatty acids; sol. in 12 parts abs. alcohol; dens. 1.0305 (24/4 C); sapon. no. 196
Toxicology: Heated to decomp., emits acrid smoke and irritating fumes
Uses: Antioxidant, emulsifier, stabilizer, release agent; used in oleomargarine, shortening, baked goods, beverage powds., cocoa powd., meat prods., poultry
Regulatory: FDA 21CFR §133.169, 133.173, 133.179, 136.110, 136.115, 136.130, 137.160, 136.165, 136.180, 163.123, 163.130, 163.135, 163.140, 163.145, 163.150, 163.155, 166.40, 166.110, 169.115, 169.140, 169.150, 175.300, 184.1400, GRAS; USDA 9CFR §318.7, 0.5% max. in oleomargarine, 381.147; Japan approved; Europe listed; UK approved
Manuf./Distrib.: ADM Ingreds.; Am. Lecithin; Am. Roland; D F Anstead; Canada Packers; Central Soya; W A Cleary; K&K Greeff; Grünau; A C Hatrick; Lucas Meyer GmbH; Penta Mfg.; Quest UK; Rhone-Poulenc; Riken Vitamin; Siber Hegner; Stern-France; Vamo Mills; Westin
Trade names: Actiflo® 68 SB; Actiflo® 68 UB; Actiflo® 70 SB; Actiflo® 70 UB; Alcolec® 140; Alcolec® 495; Alcolec® BS; Alcolec® Extra A; Alcolec® F-100; Alcolec® FF-100; Alcolec® Granules; Alcolec® PG; Alcolec® S; Alcolec® SFG; Alcolec® XTRA-A; Amisol™ 210-L; Amisol™ 210 LP; Amisol™ 329; Amisol™ 697; Asol; Beakin LV1; Beakin LV2; Blendmax 322; Blendmax 322D; Canasperse SBF; Canasperse UBF; Canasperse UBF-LV; Canasperse WDF; Capcithin™; Capsulec 51-SB; Capsulec 51-UB; Capsulec 56-SB; Capsulec 56-UB; Capsulec 62-SB; Capsulec 62-UB; Centrocap® 162SS; Centrocap® 162US; Centrocap® 273SS; Centrocap® 273US; Centrol® 2F SB; Centrol® 2F UB; Centrol® 3F SB; Centrol® 3F UB; Centrol® CA; Centrolex® C; Centrolex® D; Centrolex® F; Centrolex® G; Centrolex® P; Centrolex® R; Centrophase® 152; Centrophase® C; Centrophase® HR; Centrophase® HR2B; Centrophase® HR2U; Centrophase® HR4B; Centrophase® HR4U; Centrophase® HR6B; Centrophil® K; Chocothin; Chocotop™; Clearate Special Extra; Clearate WDF; Dur-Lec® P; Dur-Lec® UB; Emulbesto; Emulfluid™; Emulfluid® A; Emulfluid® AS; Emulfluid® E; Emulgum™; Emulpur™ N; Emulpur™ N P-1; Emulthin M-35; Emulthin M-501; Epicholin; Epikuron™ 100 P, 100 G; Epikuron™ 100 X; Epikuron™ 130 G; Epikuron™ 130 P; Epikuron™ 130 X; Epikuron™ 135 F; Lecigran™ 5750; Lecigran™ 6750; Lecigran™ A; Lecigran™ F; Lecigran™ M; Lecimulthin; Leciprime™; Lecipur™ 95 C; Lecipur™ 95 R; Lecithin L 1000-Range; Lecithin L 4000-Range; Magathin; M-C-Thin®; M-C-Thin® AF-1 Type DB; M-C-Thin® AF-1 Type SB; M-C-Thin® AF-1 Type UB; M-C-Thin® ASOL; M-C-Thin® FWD; Metarin™; Metarin™ C; Metarin™ CP; Metarin™ DA 51; Metarin™ F; Metarin™ P; Phospholipon® 90/90 G; Thermolec 57; Thermolec 68; Thermolec 200; Topcithin®; Yelkin DS; Yelkin Gold; Yelkin SS; Yelkin T; Yelkin TS; Yelkinol F; Yelkinol G; Yelkinol P
Trade names containing: Akolizer PKC; Akopuff; Akowesco 90; Alcolec® 30-A; Alcolec® C-150; Alcolec® EYR SM; Amisol™ 406-N; Amisol™ 683 A; Amisol™ 785-15K; Amisol™ MS-10; Amisol™ MS-12 BA; Annatto OS #2894; Atmos® 659 K; Bake-Well 52; Bake-Well 80/20; Bake-Well High Stability Pan Oil; Beakin LV3; Beakin LV4; Beakin LV30; Bread Pan Oil; Brosoft 6430; Buckeye C (Baker's Grade); Bunge Special Mix Shortening; Cake Mix 96; CBC #7 Shortening; CDC-10A; CDC-10A (NB); CDC-79; CDC-79DZ; CDC-2001; CDC-2002; CDC-2500; CDC-3000; Centrobake® 100L; Centromix® CPS; Centromix® E; Centrophase® NV; Centrophil® M; Centrophil® W; CLSP 399; CLSP 499; CLSP 555; Coco-Spred™; Controx® AT; Controx® VP; Coral®; Diamond D-21; Diet Imperial® Spread; Durkex 100L; Durlite Gold MBN II; EC-25®; Golden Croissant/Danish™ Margarine; Golden Flake™; Hentex-76; Kel-Lite™ CM; Kristel Gold; Kristel Gold II; Lecidan; Lecigran™ C; Lecigran™ Super A; Lecigran™ T; Lecithin L 2000-Range; Lecithin L 3000-Range; Lite Pack 350; M-C-Thin® ASOL 436; M-C-Thin® FWD 425; Mr. Chips™; Myvatex® 25-07K; Myvatex® SSH; No Stick Emulsifier; Oxynex® LM; Pano-Lube™; Panospray™; Panospray™ LS; Panospray™ SQ; Panospray™ WL; Paramount C; Perflex; Phosal® 50 SA; Phosal® 53 MCT; Phosal® 75 SA; Pristene® RW; Satina 44; Satina 50; Satina 53NT; Shedd's Liquid Margarine; Shedd's Margarine; Shedd's NP Margarine; Shedd's Special 40 Butter Blend Margarine; Shedd's Special

60 Butter Blend Margarine; Soyarich® 115 W; Soylec C-6; Soylec C-15; Soylec T-15; Stabolic™ C; Summit®; Superb® Bakers Margarine 105-101; Super-Cel® Specialty Cake Shortening; Trophy®

Lecithin, hydroxylated. *See* Hydroxylated lecithin
Lemon balm. *See* Balm mint
Lemon balm. *See* Balm mint oil
Lemon balm extract. *See* Balm mint extract

Lemon extract
CAS 8008-56-8; 84929-31-7
Synonyms: Citrus limon extract
Definition: Extract of the lemon, *Citrus limon*
Properties: Char. lemon-leaf odor, sour bitter taste
Uses: Natural flavoring agent
Usage level: 1000 ppm (nonalcoholic beverages), 540-4000 ppm (ice cream), 400-12,000 ppm (candy), 8900 ppm (baked goods), 10,000 ppm (icings)
Regulatory: FDA 21CFR §182.20, GRAS; Japan approved (lemon)

Lemongrass extract
CAS 89998-14-1; EINECS 289-752-0
Synonyms: Cymbopogon citratus extract
Definition: Extract of *Cymbopogon citratus*
Uses: Natural flavoring agent
Regulatory: FDA 21CFR §182.20, GRAS; Japan approved (lemongrass)
Manuf./Distrib.: Chart

Lemongrass oil East Indian
FEMA 2624
Synonyms: Cochin; East Indian lemongrass oil
Definition: Oil from steam distillation of grasses of *Cymbopogon flexosus* and *Andropogon nardus*, contg. citral
Properties: Dk. yel. to brn.-red liq., heavy lemon odor; sol. in min. oil, propylene glycol, alcohol; insol. in water, glycerin; dens. 0.894-0.902; ref. index 1.483
Toxicology: LD50 (oral, rat) 5600 mg/kg; mildly toxic by ingestion; skin irritant; heated to decomp, emits acrid smoke and irritating fumes
Uses: Natural flavoring agent for baked goods, nonalcoholic beverages, chewing gum, confections, gelatin desserts, ice cream, puddings
Regulatory: FDA 21CFR §182.20, GRAS; GRAS (FEMA); Europe listed, no restrictions
Manuf./Distrib.: Chart; Florida Treatt

Lemongrass oil West Indian
CAS 8007-02-1; EINECS 289-752-0 (extract); FEMA 2624
Synonyms: West Indian lemongrass oil; Guatemala lemongrass oil; Madagascar lemongrass oil
Definition: Volatile oil from steam distillation of fresh *Cymbopogon citratus* grasses
Properties: Lt. yel. to brn. liq., lt. lemon odor; sol. in min. oil, propylene glycol; insol. in water; dens. 0.869-0.894; ref. index 1.483
Precaution: Keep cool, well closed; protect from light
Toxicology: Skin irritant; heated to decomp., emits acrid smoke and irritating fumes
Uses: Natural flavoring agent
Usage level: 4.4 ppm (nonalcoholic beverages), 9.2 ppm (ice cream), 38 ppm (candy), 38 ppm (baked goods), 290 ppm (gelatins, puddings), 220 ppm (chewing gum)
Regulatory: FDA 21CFR §182.20, GRAS; Europe listed, no restrictions
Manuf./Distrib.: Florida Treatt

Lemon juice
CAS 68916-88-1
Definition: Liq. expressed from the fresh pulp of *Citrus limon*
Uses: Natural flavoring agent
Regulatory: FDA 21CFR §182.20, GRAS
Manuf./Distrib.: Vicente Trapani SA
Trade names containing: Lem-O-Fos® 101

Lemon juice extract
Synonyms: Citrus limon juice extract
Uses: Natural flavoring agent

Lemon oil
CAS 8008-56-8; FEMA 2625

Lemon oil, distilled

Synonyms: Citrus limon oil; Cedro oil
Definition: Volatile oil expressed from the fresh peel of fruit of *Citrus limon*, contg. limonene, terpinene, phellandrene, pinene
Properties: Pale yel. to greenish-yel. liq., lemon peel odor and taste; misc. with dehydrated alcohol, CS_2, glac. acetic acid; sl. sol. in water; dens. 0.849-0.855 (25/25 C); ref. index 1.4742-1.4755 (20 C)
Precaution: Keep cool, well closed; protect from light; do not use if terebinthine odor can be detected
Toxicology: LD50 (oral, rat) 2840 mg/kg; mod. toxic by ingestion; experimental tumorigen; skin irritant; heated to decomp., emits acrid smoke and irritating fumes
Uses: Natural flavoring agent for pastry, foods, beverages, liqueurs, other flavors
Usage level: 230 ppm (nonalcoholic beverages), 380 ppm (ice cream), 1100 ppm (candy), 580 ppm (baked goods), 340 ppm (gelatins, puddings), 1900 ppm (chewing gum), 140 ppm (breakfast cereals), 25-40 ppm (meats), 10-80 ppm (condiments), 65-600 ppm (icings)
Regulatory: FDA 21CFR §146.114, 146.120, 146.121, 146.126, 161.190, 182.20, GRAS; GRAS (FEMA); Europe listed, no restrictions
Manuf./Distrib.: Commodity Services; Florida Treatt
Trade names: Citreatt Lemon 3123; Citreatt Lemon 6122
Trade names containing: Cap-Shure® CLF-165-70

Lemon oil, distilled

Definition: Oil from distillation of fresh peel of *Citrus limon*
Properties: Pale yel. liq., fresh lemon peel odor and taste; misc. with dehydrated alcohol, glac. acetic acid; dens. 0.842; ref. index 1.470 (20 C)
Toxicology: Skin irritant; heated to decomp., emits acrid smoke and irritating fumes
Uses: Natural flavoring agent
Regulatory: FDA 21CFR §182.20, GRAS
Manuf./Distrib.: Caminitti Foti & Co. Srl; Vicente Trapani

Lemon peel extract

Definition: Extract of rinds of *Citrus limon*
Uses: Natural flavoring agent
Usage level: 190 ppm (nonalcoholic beverages), 420 ppm (ice cream), 480 ppm (candy), 480 ppm (baked goods)
Regulatory: FDA 21CFR §182.20, GRAS
Manuf./Distrib.: Chart

Lemon verbena extract

Synonyms: Lippia citriodora extract
Definition: Extract of flowering ends of *Lippia citriodora*
Uses: Natural flavoring
Regulatory: FDA 21CFR §172.510

Leopard's bane. *See* Arnica

Lepidine

CAS 491-35-0; EINECS 207-734-2; FEMA 2744
Synonyms: 4-Methylquinoline; Cincholepidine; γ-Methylquinoline
Empirical: $C_{10}H_9N$
Properties: Colorless oily liq., quinoline odor; turns reddish-brn. in light; sl. sol. in water; misc. with alcohol, benzene, ether; m.w. 143.19; dens. 1.083 (20/4 C); b.p. 260-263 C; congeal pt. 0 C; ref. index 1.620 (20 C)
Precaution: Photosensitive; protect from light
Uses: Synthetic flavoring
Usage level: 0.22 ppm (nonalcoholic beverages), 1.4 ppm (ice cream, ices), 1.8 ppm (candy, baked goods)
Regulatory: FDA 21CFR §172.515
Manuf./Distrib.: Aldrich

DL-Leucine

Synonyms: DL-2-Amino-4-methylvaleric acid
Empirical: $C_6H_{13}NO_2$
Properties: Wh. sm. cryst. or cryst. powd., odorless, sl. bitter taste; sol. in water; sl. sol. in alcohol; insol. in ether; m.w. 131.17; m.p. 290 C (dec.)
Uses: Nutrient, dietary supplement

L-Leucine

CAS 61-90-5; EINECS 200-522-0
Synonyms: α-Amino-γ-methylvaleric acid; α-Aminoisocaproic acid; 2-Amino-4-methypentanoic acid;

4-Methylnorvaline
Classification: Essential amino acid
Empirical: $C_6H_{13}NO_2$
Properties: Wh. cryst.; sol. in water, dil. HCl; sl. sol. in alcohol; insol. in ether; m.w. 131.20; dens. 1.291 (18 C); m.p. 295 C; subl. @ 145-148 C
Toxicology: LD50 (intraperitoneal, rat) 5379 mg/kg; mod. toxic by subcutaneous route; experimental teratogen, reproductive effects; heated to decomp., emits toxic fumes of NO_x
Uses: Nutrient and dietary supplement ; flavorings (Japan)
Regulatory: FDA 21CFR §172.320, 8.8% max.; Japan approved
Manuf./Distrib.: Degussa; Nippon Rikagakuyakuhin

Levisticum officinale oil. *See* Lovage oil
Levulic acid. *See* Levulinic acid

Levulinic acid
CAS 123-76-2; EINECS 204-649-2; FEMA 2627
Synonyms: γ-Ketovaleric acid; 4-Oxo-n-valeric acid; Acetylpropionic acid; β-Acetylpropionic acid; 4-Oxopentanoic acid; Levulic acid
Empirical: $C_5H_8O_3$
Formula: $CH_3COCH_2CH_2COOH$
Properties: Yel. plates or leaflets; freely sol. in water, alcohol, ether; insol. in aliphatic hydrocarbons; m.w. 116.12; dens. 1.1447; m.p. 33-35 C; b.p. 245-246 C; flash pt. ≈ 98 C; ref. index 1.442 (16 C)
Precaution: Protect from light
Uses: Synthetic flavoring agent
Usage level: 14 ppm (nonalcoholic beverages), 14 ppm (ice cream), 53 ppm (candy), 53 ppm (baked goods), 4 ppm (gelatins, puddings)
Regulatory: FDA 21CFR §172.515
Manuf./Distrib.: Aldrich; Penta Mfg.

Levulose. *See* Fructose
Lichenic acid. *See* Fumaric acid
Lichen islandicus extract. *See* Iceland moss extract

Licorice
Synonyms: Glycyrrhiza; Licorice root; Liquorice
Definition: Dried rhizome and roots of *Glycyrrhiza glabra*
Uses: Natural flavoring agent, flavor enhancer, surface-active agent
Usage level: Limitation 0.05% (baked goods), 0.1% (alcoholic beverages), 0.15% (nonalcoholic beverages), 1.1% (chewing gum), 16% (hard candy), 0.15% (herbs, seasonings, plant protein prods.), 3.1% (soft candy), 0.5% (vitamin/min. supplements), 0.1% (other foods)
Regulatory: FDA 21CFR §184.1408, GRAS; not permitted as nonnutritive sweetener in sugar substitutes; Japan approved
Manuf./Distrib.: Biddle Sawyer; Bio-Botanica; Arthur Branwell; Chart; Cornelius; MacAndrews & Forbes; Dr Madis Labs; Maruzen Fine Chems.; Meer; MLG Enterprises; Penta Mfg.; Rhone-Poulenc; Roeper; W Ruitenberg; E H Worlee GmbH

Licorice extract
CAS 97676-23-8; 84775-66-6
Synonyms: Glycyrrhiza glabra extract
Definition: Extract of *Glycyrrhiza glabra*
Uses: Natural flavoring agent, flavor enhancer, surface-active agent
Usage level: Limitation 0.05% (baked goods), 0.1% (alcoholic beverages), 0.15% (nonalcoholic beverages), 1.1% (chewing gum), 16% (hard candy), 0.15% (herbs, seasonings, plant protein prods.), 3.1% (soft candy), 0.5% (vitamin/min. supplements), 0.1% (other foods)
Regulatory: FDA 21CFR §184.1408; Japan approved
Manuf./Distrib.: Chart

Licorice root. *See* Licorice

Licorice root extract
CAS 8008-94-4
Synonyms: Glycyrrhiza extract
Toxicology: LD50 (oral, rat) 14,200 mg/kg; mod. toxic by IP, subcut. routes; mildly toxic by ingestion; mutagenic data; heated to decomp., emits acrid smoke and irritating fumes
Uses: Natural flavoring agent, flavor enhancer, surface-active agent; used in bacon, baked goods, beverages, candy, chewing gum, herbs, ice cream, seasonings, syrups, dietary supplements

Light petroleum hydrocarbons, odorless

Usage level: Limitation (as glycyrrhizin) 0.05% (baked goods), 0.1% (alcoholic beverages), 0.15% (nonalcoholic beverages), 16% (hard candy), 0.15% (herbs), 3.1% (soft candy), 0.5% (dietary supplements), 0.1% (other foods); not permitted in sugar substitutes
Regulatory: FDA 21CFR §184.1408, GRAS
Manuf./Distrib.: Chart

Light petroleum hydrocarbons, odorless

Properties: Liq., faint odor; b.p. 300-650 C
Toxicology: Heated to decomp., emits acrid smoke and irritating fumes
Uses: Coating agent, defoamer, float, froth-flotation cleaning, insecticide formulations component; used for beet sugar, eggs, fruit, pickles, vegetables, vinegar, wine
Regulatory: FDA 21CFR §172.884

Lignocaine benzyl benzoate. *See* Denatonium benzoate NF
Lignosulfonic acid, calcium salt. *See* Calcium lignosulfonate
Lignosulfonic acid, sodium salt. *See* Sodium lignosulfonate
Lilyl aldehyde. *See* Hydroxycitronellal
Lime. *See* Calcium oxide
Lime acetate. *See* Calcium acetate
Lime citrate. *See* Calcium citrate
Limed rosin. *See* Calcium resinate

Lime oil

CAS 8008-26-2; FEMA 2631
Definition: Oil from *Citrus aurantifolia*
Properties: Intensely fresh citrus aroma, astringent sweet-sour flavor
Uses: Natural flavoring agent for cola and other soft drinks
Usage level: 130 ppm (nonalcoholic beverages), 160 ppm (ice cream, ices), 680 ppm (candy), 370 ppm (baked goods), 200 ppm (gelatins, puddings), 3100 ppm (chewing gum), 20 ppm (condiments)
Regulatory: FDA 21CFR §182.20, GRAS; GRAS (FEMA); Japan approved (lime); Europe listed, no restrictions
Manuf./Distrib.: Chart; Pierre Chauvet; Commodity Services; Florida Treatt
Trade names: Citreatt Lime 3135; Citreatt Lime 6134

Lime-tree extract. *See* Linden extract
Lime water. *See* Calcium hydroxide
(±)-Limonene. *See* dl-Limonene

d-Limonene

CAS 5989-27-5; EINECS 227-813-5; FEMA 2633
Synonyms: d-p-Mentha-1,8-diene; R(+)-Limonene; (+)-Carvene; (R)-4-Isopropenyl-1-methyl-1-cyclohexene
Classification: Terpene
Empirical: $C_{10}H_{16}$
Properties: Colorless liq., citrus odor; insol. in water; misc. with alcohol and ether; m.w. 136.26; dens. 0.8411 (20 C); b.p. 176-176.4 C; ref. index 1.471
Toxicology: LD50 (oral, rat) 4400 mg/kg; poison by IV route; mod. toxic by intraperitoneal route; mildly toxic by ingestion; skin irritant; experimental tumorigen, reproductive effects; heated to decomp., emits acrid smoke and irritating fumes
Uses: Synthetic flavoring agent
Usage level: 31 ppm (nonalcoholic beverages), 68 ppm (ice cream), 49 ppm (candy), 120 pm (baked goods), 48-400 ppm (gelatins, puddings), 2300 ppm (chewing gum)
Regulatory: FDA 21CFR §182.60, GRAS
Manuf./Distrib.: Aldrich; Allchem Industries; Int'l. Flavors & Fragrances; Penta Mfg.

dl-Limonene

CAS 138-86-3; EINECS 205-341-0
Synonyms: Dipentene (INCI); Cinene; Cajeputene; Inactive limonene; (±)-Limonene
Empirical: $C_{10}H_{16}$
Properties: Colorless liq., pleasant lemon-like odor; misc. with alcohol; pract. insol. in water; m.w. 136.23; dens. 0.847 (15.5/15.5 C); m.p. -96.9 C; b.p. 175-176 C (763 mm); ref. index 1.4744
Precaution: Flamm.
Toxicology: Skin irritant
Uses: Synthetic flavoring agent
Regulatory: FDA 21CFR §175.105, 177.2600, 182.60, GRAS
Manuf./Distrib.: Penta Mfg.

l-Limonene

Synonyms: l-p-Mentha-1,8-diene; Dipentene; Carvene; Cajeputene

Empirical: $C_{10}H_{16}$
Properties: Liq.; misc. with alcohol; insol. in water; m.w. 136.23; dens. 0.8407 (20.5/4 C); b.p. 175.5-176.5 C
  (763 mm); ref. index 1.474 (21 C)
Toxicology: Skin irritant, sensitizer
Uses: Synthetic flavoring agent
Regulatory: FDA 21CFR §182.60, GRAS

R(+)-Limonene. *See* d-Limonene

Linaloe wood oil
FEMA 2634
Definition: Volatile oil distilled from a Mexican wood, *Bersera delpechiana* and other species, contg. linalool,
  geraniol, methylheptenone
Properties: Colorless to ylsh liq., pleasant odor; sol. in 2 vols 70% alcohol; sol. in ether, chloroform; sl. sol. in
  water; dens. 0.875-0.890 (15/15 C); ref. index 1.4638 (20 C)
Precaution: Keep cool, well closed; protect from light
Uses: Natural flavoring
Usage level: 4.3 ppm (nonalcoholic beverages), 1.0 ppm (alcoholic beverages), 3.8 ppm (ice cream, ices), 16
  ppm (candy), 15 ppm (baked goods)
Regulatory: FDA 21CFR §172.510

Linalol. *See* Linalool

Linalool
CAS 78-70-6; EINECS 201-134-4; FEMA 2635
Synonyms: 2,6-Dimethyl-2,7-octadien-6-ol; 3,7-Dimethyl-1,6-octadien-3-ol; Linalyl alcohol; Linalol
Classification: Terpene
Empirical: $C_{10}H_{18}O$
Properties: Colorless liq., odor similar to Bergamot oil, French lavender; sol. in alcohol, ether, fixed oils,
  propylene glycol; insol. in glycerin; m.w. 154.28; dens. 0.858-0.868; b.p. 195-199 C; flash pt. 172 F; ref.
  index 1.461
Toxicology: LD50 (oral, rat) 2790 mg/kg; mod. toxic by ingestion, mildly toxic by skin contact; skin irritant;
  heated to decomp., emits acrid smoke and irritating fumes
Uses: Floral, herbaceous, woody, rosewood-like synthetic fragrance and flavoring
Usage level: 2 ppm (nonalcoholic beverages), 3.6 ppm (ice cream), 0.4 ppm (candy), 9.0 ppm (baked goods),
  2.3 ppm (gelatins, puddings), 0.8-90 ppm (chewing gum), 40 ppm (condiments)
Regulatory: FDA 21CFR §182.60, GRAS; Japan approved as flavoring
Manuf./Distrib.: BASF; Florida Treatt

Linalool oxide
Synonyms: cis- and trans-2-Vinyl-2-methyl-5-(1′-hydroxy-1′-methylethyl) tetrahydrofuran
Empirical: $C_{10}H_{17}O_2$
Properties: Liq.; ref. index 1.4523
Uses: Synthetic flavoring
Regulatory: FDA 21CFR §172.515

Linalyl acetate
CAS 115-95-7; EINECS 204-116-4; FEMA 2636
Synonyms: 3,7-Dimethyl-1,6-octadien-3-ol acetate; 3,7-Dimethyl-1,6-octadien-3-yl acetate; Bergamiol;
  Bergamol
Definition: Ester of linalool and acetic acid
Empirical: $C_{12}H_{20}O_2$
Formula: $CH_3COOC_{10}H_{17}$
Properties: Colorless clear oily liq., bergamot odor; sol. in alcohol, ether, diethyl phthalate, benzyl benzoate,
  min. oil, fixed oils; sl. sol. in propylene glycol; insol. water; m.w. 196.32; dens. 0.898-0.914; b.p. 108-110
  C; flash pt. 185 F; ref. index 1.4500
Precaution: Combustible
Toxicology: LD50 (oral, rat) 14,550 mg/kg; mildly toxic by ingestion; heated to decomp., emits acrid smoke and
  irritating fumes
Uses: Fresh, floral synthetic fragrance and flavoring, reminiscent of bergamot, petitgrain
Usage level: 1.9 ppm (nonalcoholic beverages), 3.8 ppm (ice cream), 11 ppm (candy), 8.9 ppm (baked goods),
  3.8 ppm (gelatins, puddings), 13 ppm (chewing gum)
Regulatory: FDA 21CFR §182.60, GRAS; Japan approved as flavoring
Manuf./Distrib.: Aldrich; BASF; Florida Treatt

Linalyl alcohol. *See* Linalool

Linalyl 2-aminobenzoate. *See* Linalyl anthranilate

Linalyl anthranilate
FEMA 2637
Synonyms: 3,7-Dimethyl-1,6-octadien-3-yl anthranilate; Linalyl 2-aminobenzoate
Empirical: $C_{17}H_{23}NO_2$
Properties: Liq., sweet orange-like flavor; sol. in alcohol; insol. in water; m.w. 273.38; b.p. 350 C; ref. index 1.4970
Uses: Synthetic flavoring
Usage level: 1.8 ppm (nonalcoholic beverages), 0.72 ppm (ice cream, ices), 4.7 ppm (candy), 0.20-8.0 ppm (baked goods)
Regulatory: FDA 21CFR §172.515

Linalyl benzoate
FEMA 2638
Synonyms: 3,7-Dimethyl-1,6-octadien-3-yl benzoate
Empirical: $C_{17}H_{22}O_2$
Properties: Yel. or ylsh.-brn. liq.; m.w. 258.37; b.p. 263 C; flash pt. 98 C; ref. index 1.505-1.520
Uses: Synthetic flavoring
Usage level: 0.31 ppm (nonalcoholic beverages), 0.42 ppm (ice cream, ices), 1.2 ppm (candy), 1.6 ppm (baked goods), 0.28 ppm (gelatins, puddings)
Regulatory: FDA 21CFR §172.515

Linalyl butyrate
CAS 78-36-4; FEMA 2639
Synonyms: Linalyl-n-butyrate
Empirical: $C_{14}H_{24}O2$
Properties: Colorless or ylsh. liq., fruity banana odor; m.w. 224.34; dens. 0.8890-0.8903; b.p. 232 C; flash pt. > 100 C; ref. index 1.4510-1.4560 (20 C)
Uses: Synthetic flavoring
Usage level: 1.2 ppm (nonalcoholic beverages), 4.3 ppm (ice cream, ices), 2.2 ppm (candy), 13 ppm (baked goods), 0.09 ppm (gelatins, puddings)
Regulatory: FDA 21CFR §172.515
Manuf./Distrib.: Aldrich

Linalyl-n-butyrate. *See* Linalyl butyrate

Linalyl cinnamate
FEMA 2641
Synonyms: Linalyl 3-phenylpropenoate
Empirical: $C_{19}H_{24}O_2$
Properties: Colorless liq., soft floral odor, fruity flavor; sol. in alcohol; insol. in water; m.w. 284.40
Uses: Synthetic flavoring
Usage level: 0.57 ppm (nonalcoholic beverages), 0.59 ppm (ice cream, ices), 2.0 ppm (candy), 2.1 ppm (baked goods)
Regulatory: FDA 21CFR §172.515

Linalyl formate
CAS 115-99-1; FEMA 2642
Synonyms: 3,7-Dimethyl-1,6-octadien-3-yl formate
Uses: Synthetic flavoring
Regulatory: FDA 21CFR §172.515
Manuf./Distrib.: Aldrich

Linalyl hexanoate
Uses: Synthetic flavoring
Regulatory: FDA 21CFR §172.515

Linalyl isobutyrate
CAS 78-35-3; FEMA 2640
Synonyms: 3,7-Dimethyl-2,6-octadien-3-yl isobutyrate
Uses: Synthetic flavoring
Regulatory: FDA 21CFR §172.515

Linalyl isovalerate
Uses: Synthetic flavoring
Regulatory: FDA 21CFR §172.515

Linalyl octanoate
Uses: Synthetic flavoring
Regulatory: FDA 21CFR §172.515

Linalyl 3-phenylpropenoate. *See* Linalyl cinnamate

Linalyl propionate
Synonyms: 3,7-Dimethyl-2,6-octadien-3-yl propionate
Uses: Synthetic flavoring
Regulatory: FDA 21CFR §172.515

Linden extract
CAS 84929-52-2
Synonyms: Lime-tree extract; Tilia cordata extract
Definition: Extract of the flowers of the linden tree, *Tilia cordata* or *T. europa*
Uses: Natural flavoring
Regulatory: FDA 21CFR §172.510; Japan approved

Linden flowers
Definition: From *Tilia* spp.
Properties: Soothing aromatic flavor
Uses: Natural flavoring agent
Regulatory: FDA 21CFR §182.10, 182.20, GRAS
Manuf./Distrib.: Chart

Linear C10 alpha olefin. *See* Decene-1

Linoleamide
CAS 3999-01-7; EINECS 223-644-6
Synonyms: Linoleic acid amide; 9,12-Octadecadienamide
Definition: Aliphatic amide of linoleic acid
Empirical: $C_{18}H_{33}NO$
Toxicology: Heated to decomp., emits acrid smoke and irritating fumes
Uses: Release agent migrating from food pkg.
Regulatory: FDA 21CFR §175.105, 175.300, 178.3910, 179.45, 181.22, 181.28

Linoleic acid
CAS 60-33-3; EINECS 200-470-9
Synonyms: 9,12-Octadecadienoic acid; (Z,Z)-9,12-Octadecadienoic acid; Linolic acid
Classification: Unsaturated fatty acid
Empirical: $C_{18}H_{32}O_2$
Formula: $CH_3(CH_2)_4=CHCH_2CH=CH(CH_2)_7COOH$
Properties: Colorless to pale yel. oil; freely sol. in ether; sol. in chloroform, abs. alcohol; misc. with dimethylformamide, fat solvs., oils; insol. in water; m.w. 280.44; dens. 0.9007 (22/4 C); m.p. -12 C; b.p. 230 C (16 mm); ref. index 1.4699 (20 C)
Precaution: Combustible; easily oxidized by air
Toxicology: Human skin irritant; ingestion can cause nausea and vomiting; heated to decomp., emits acrid smoke and irritating fumes
Uses: Dietary supplement, nutrient, flavoring agent, adjuvant; suitable for infant formulas
Regulatory: FDA 21CFR §175.105, 182.5065, 184.1065, GRAS
Manuf./Distrib.: CasChem; Hercules

Linoleic acid amide. *See* Linoleamide
Linolic acid. *See* Linoleic acid

Linseed oil
CAS 8001-26-1; EINECS 232-278-6
Synonyms: Oils, linseed; Flaxseed oil
Definition: Expressed oil from the dried ripe seed of *Linum usitatissimum*
Properties: Golden-yel., amber, or brown drying oil, peculiar odor, bland taste; sol. in ether, chloroform, carbon disulfide, turpentine; sl. sol. in alcohol; dens. 0.921-0.936; m.p. -19 C; b.p. 343 C; flash pt. 222 C
Precaution: Combustible liq. exposed to heat or flame; can react with oxidizers; subject to spontaneous heating; violent reaction with Cl_2
Toxicology: Allergen and skin irritant to humans
Uses: Drying oil as component of finished resin, migrating from food pkg.
Regulatory: FDA 21CFR §175.105, 175.300, 176.200, 176.210, 181.22, 181.26
Manuf./Distrib.: Penta Mfg.

Lipase
CAS 9001-62-1; EINECS 232-619-9
Classification: Esterase
Definition: Digestive enzyme
Uses: Enzyme for food use; hydrolyzes fat to glycerol and fatty acid
Regulatory: FDA GRAS; Canada approved
Trade names: Fermlipase; Lipase 8 Powd.; Lipase 16 Powd.; Lipase 24 Powd.; Lipase 30 Powd.; Lipase AK; Lipase AP; Lipase AP 6; Lipase AY; Lipase FAP; Lipase G; Lipase GC; Lipase MAP; Lipase N; Lipase PS; Pancreatic Lipase 250
Trade names containing: Pancreatin 3X USP Powd.; Pancreatin 4X USP Powd.; Pancreatin 5X USP Powd.; Pancreatin 6X USP Powd.; Pancreatin 8X USP Powd.; Pancreatin USP Powd.; Pancrelipase USP; SPL Pancreatin 3X USP; SPL Pancreatin 4X USP; SPL Pancreatin 5X USP; SPL Pancreatin 6X USP; SPL Pancreatin 7X USP; SPL Pancreatin 8X USP; SPL Pancreatin USP; SPL Pancrelipase USP; SPL Undiluted Pancreatic Enzyme Conc. (PEC)

Lipoxidase
CAS 9029-60-1; EINECS 232-853-1
Definition: Enzyme from soybean which catalyzes the addition of oxygen to the double bonds of unsaturated fatty acids of plant origin
Uses: Enzyme; bleaches yel. pigments in flour; used to whiten bread
Trade names containing: Nutrisoy 7B

Lippia citriodora extract. *See* Lemon verbena extract
Liquid paraffin. *See* Mineral oil
Liquid petrolatum. *See* Mineral oil
Liquid rosin. *See* Tall oil
Liquid smoke. *See* Pyroligneous acid extract
Liquorice. *See* Licorice

Locust bean gum
CAS 9000-40-2; EINECS 232-541-5; EEC E410
Synonyms: Carob flour; Carob bean gum; St. John's bread; Algaroba
Classification: Polysaccharide plant mucilage
Definition: Ground seed of the ripe fruit of St. John's Bread (*Ceratonia siliqua*)
Properties: Wh. powd., odorless, tasteless; swells in cold water; insol. in org. solvs.; visc. increases when heated; m.w. [ae] 310,000
Precaution: Combustible
Toxicology: LD50 (oral, rat) 13 g/kg; mildly toxic by ingestion; heated to decomp., emits acrid smoke and irritating fumes
Uses: Stabilizer, thickener, emulsifier for baked goods, beverages, candy, cheese, fillings, gelatins, ice cream, jams/jellies, pies, puddings, soups
Usage level: Limitation 0.15% (baked goods), 0.25% (beverages), 0.8% (cheese), 0.75% (gelatins, puddings), 0.75% (jams/jellies), 0.5% (other foods)
Regulatory: FDA 21CFR §133.133, 133.134, 133.162, 133.178. 133.179, 150.141, 150.161, 182.20, 184.1343, 240.1051, GRAS; Japan approved; Europe listed; UK approved; ADI not specified (JECFA)
Manuf./Distrib.: Agrisales; Ashland; Bio-Botanica; Calaga Food Ingreds.; Chart; Diamalt; Grindsted; Hercules Ltd.; Lucas Meyer; Meer; Multi-Kem; Quest UK; Rhone-Poulenc Food Ingreds.; Sanofi; Thew Arnott; TIC Gums; Valmar; E H Worlee GmbH; Wykefold
Trade names: Aquasol CSL; Carudan 000 Range; Carudan 100 Range; Carudan 200 Range; Carudan 300 Range; Carudan 400 Range; Carudan 700 Range; Diagum LBG; Hercules® Locust Bean Gum FL 50-40; Hercules® Locust Bean Gum FL 50-50; HG-100; HG-175; HG-200; Idealgum 1A; Idealgum 1B; Idealgum 1C; Idealgum 2A; Idealgum 2B; Idealgum 3F; Locust Bean Gum Speckless Type D-200; Nutriloid® Locust; Powdered Locust Bean Gum Type D-200; Powdered Locust Bean Gum Type D-300; Powdered Locust Bean Gum Type P-100; Powdered Locust Bean Gum Type PP-100; Soalocust®; STD-175; TIC Pretested® Locust Bean POR/A; Wakal® J; Wakal® JG
Trade names containing: Dariloid® 100; Emulgel E-21; Emulgel S-32; Frigesa® D 890; Frimulsion 10; Frimulsion 6G; Frimulsion Q8; Frimulsion X5; GFS®; Kelgum®; Lamefrost® ES 251 G; Lamefrost® ES 424; Merecol® FA; Merecol® GL; Merecol® RB; Merecol® RCS; Rhodigum OEH; Rhodigum OEM; Rhodigum WGH; Rhodigum WGL; Rhodigum WGM; Rhodigum WVH; Rhodigum WVM; Sta-Lite; Sta-Lite; Thixogum X

Locust tree extract. *See* Carob extract

Lovage oil
CAS 8016-31-7

Synonyms: Levisticum officinale oil
Definition: Volatile oil from steam distillation of fresh root of *Levisticum officinale*
Properties: Yel.-green-brn. liq., strong odor and taste; sol. in fixed oils; sl. sol. in min. oil; insol. in glycerin, propylene glycol; dens. 1.034-1.057; ref. index 1.536-1.554 (20 C)
Toxicology: LD50 (oral, mouse) 3400 mg/kg; mod. toxic by ingestion; skin irritant; heated to decomp., emits acrid smoke and irritating fumes
Uses: Natural flavoring agent
Usage level: 1.3 ppm (nonalcoholic beverages), 0.6 ppm (ice cream), 0.83 ppm (candy), 2.4 ppm (baked goods), 3.7 ppm (condiments), 10 ppm (icings), 6.8 ppm (syrups)
Regulatory: FDA 21CFR §172.510
Manuf./Distrib.: C.A.L.-Pfizer; Pierre Chauvet

Lucerne extract. *See* Alfalfa extract

Lupulin
Definition: Humulus lupulus
Uses: Natural flavoring agent
Regulatory: FDA 21CFR §182.20, GRAS

Lutein. *See* Xanthophyll
Lye. *See* Potassium hydroxide
Lye. *See* Sodium hydroxide

L-Lysine
CAS 56-87-1; EINECS 200-294-2
Synonyms: α, ε-Diaminocaproic acid; 2,6-Diaminohexanoic acid
Classification: Amino acid
Empirical: $C_6H_{14}N_2O_2$
Properties: Need. or hexagonal plates; very sol. in water; very sl. sol. in alcohol; pract. insol. in ether; m.w. 146.19; m.p. 215 C (dec.)
Uses: Dietary and nutritional additive for food enrichment ; flavorings
Regulatory: FDA 21CFR §172.320; Japan approved
Manuf./Distrib.: Degussa

L-Lysine L-glutamate
CAS 5408-52-6
Synonyms: L-Glutamic acid L-lysine salt
Empirical: $C_{11}H_{23}N_3O_6$
Formula: $H_2N(CH_2)_4CH(NH_2)COOH \cdot HOOC(CH_2)_2CH(NH_2)COOH$
Properties: m.w. 293.32
Uses: Flavor and nutritive additive to foods
Regulatory: Japan approved

L-Lysine hydrochloride
CAS 657-27-2; EINECS 211-519-9
Synonyms: 2,6-Diaminohexanoic acid hydrochloride; L-Lysine monohydrochloride
Empirical: $C_6H_{14}N_2O_2 \cdot HCl$
Formula: $NH_2(CH_2)_4CH(NH_2)COOH \cdot HCl$
Properties: Wh. powd., odorless; sol. in water; insol. in alcohol, ether; m.w. 182.65; m.p. 265-270 C (with decomp.)
Toxicology: LD50 (oral, rat) 10 g/kg; mildly toxic by ingestion; heated to decomp., emits very toxic fumes of HCl and NO_x
Uses: Dietary supplement, nutrient; flavorings
Regulatory: FDA 21CFR §172.320, limitation 6.4% as free amino acid; Japan approved
Manuf./Distrib.: Am. Roland

L-Lysine monohydrochloride. *See* L-Lysine hydrochloride

Macadamia nut oil
Definition: Oil obtained from the nuts of *Macadamia ternifolia*
Properties: Colorless to lt. yel. clear oil, bland to sl. nutty flavor; sp.gr. 0.910-0.929; iodine no. 70-80; sapon. no. 190-200; ref. index 1.460-1.479
Uses: Food ingred.
Manuf./Distrib.: Arista Industries

Mace
FEMA 2653 (oil)
Definition: From *Myristica fragrans*

Macrocystis pyriferae extract

Uses: Natural flavoring agent
Regulatory: FDA 21CFR §182.10, 182.20, GRAS; FEMA GRAS (oil); Europe listed (oil, < 1 to 15 ppm safrole levels)
Manuf./Distrib.: Chart; Pierre Chauvet

Macrocystis pyriferae extract. *See* Kelp extract
Macrogol 300. *See* PEG-6
Macrogol 600. *See* PEG-12
Macrogol 1000. *See* PEG-20
Macrogol 1540. *See* PEG-32
Macrogol 6000. *See* PEG-150
Macrose. *See* Dextran
Madagascar lemongrass oil. *See* Lemongrass oil West Indian
MAG. *See* Ammonium glutamate
Magnesia. *See* Magnesium oxide
Magnesia magma. *See* Magnesium hydroxide
Magnesia usta. *See* Magnesium oxide
Magnesite. *See* Magnesium carbonate
Magnesite. *See* Magnesium carbonate hydroxide
Magnesium (II) carbonate (1:1). *See* Magnesium carbonate

Magnesium carbonate
CAS 546-93-0, 29409-82-0; EINECS 208-915-9; EEC E504
Synonyms: Magnesium (II) carbonate (1:1); Magnesite; Magnesium carbonate precipitated; Carbonate magnesium
Definition: Basic dehydrated magnesium carbonate or a normal hydrated magnesium carbonate
Empirical: $CO_3 \cdot Mg$
Properties: Light bulky wh. powd., odorless; sol. in acids; insol. in alcohol, water; m.w. 84.32; dens. 3.04; dec. 350 C; ref. index 1.52
Precaution: Noncombustible; incompat. with formaldehyde
Toxicology: TLV:TWA 10 mg/m³; heated to decomp., emits acrid smoke and irritating fumes
Uses: Alkali, acidity regulator, anticaking agent, carrier, color retention aid, antibleaching agent, drying agent; used in dry mixes, table salt
Usage level: ADI not specified (EEC)
Regulatory: FDA 21CFR §133.102, 133.106, 133.111, 133.141, 133.165, 133.181, 133.183, 133.195, 137.105, 137.155 137.160, 137.165, 137.170, 137.175, 137.180, 137.185, 163.110, 177.2600, 184.1425, GRAS; Japan approved (0.5% max.); Europe listed; UK approved
Trade names: Marinco CH; Marinco® CH-Granular; Marinco® CL
Trade names containing: Morton® Flour Lite Salt™ Mixt; Morton® Lite Salt® Mixt

Magnesium carbonate basic. *See* Magnesium carbonate hydroxide

Magnesium carbonate hydroxide
CAS 12125-28-9, 39409-82-0, 56378-72-4 (pentahydrate); EINECS 235-192-7
Synonyms: Magnesite; Magnesium carbonate basic; Magnesium, tetrakis[carbonato(2-)] dihydroxypenta-; Magnesium hydroxidecarbonate
Classification: Inorganic basic carbonate
Formula: $(MgCO_3)_4 \cdot Mg(OH)_2 \cdot 5H_2O$
Properties: Wh. powd., odorless; sol. in about 3300 parts CO_2.free water; sol. in dil. acids with effervescence; insol. in alcohol; m.w. 485.69
Uses: Anticaking and free-flow agent, flour treating agent, lubricant, release agent, nutrient supplement, pH control agent, processing aid, synergist
Regulatory: FDA 21CFR §184.1425, GRAS
Manuf./Distrib.: Lohmann

Magnesium carbonate precipitated. *See* Magnesium carbonate

Magnesium chloride
CAS 7786-30-3; EINECS 232-094-6
Synonyms: Magnesium chloride anhydrous
Classification: Inorganic salt
Empirical: Cl_2Mg
Formula: $MgCl_2$
Properties: Wh. to opaque gray gran. or flakes, deliq.; sol. in water evolving heat, alcohol; m.w. 95.21; dens. 2.325; m.p. 708 C; b.p. 1412 C
Precaution: Causes steel to rust very rapidly in humid environments

Toxicology: LD50 (oral, rat) 2800 mg/kg; poison by intraperitoneal and IV routes; mod. toxic by ingestion, subcutaneous routes; human mutagenic data; heated to decomp., emits toxic fumes of Cl⁻

Uses: Color retention aid, firming agent, flavoring agent, tissue softening agent; used in raw meat, poultry; processing aid, yeast nutrient, tofu coagulant (Japan)

Usage level: ADI not specified

Regulatory: FDA 21CFR §172.560, 177.1650, 182.5446, 184.1426, GRAS; USDA 9CFR §318.7, 381.147, limitation ≤ 3% of 0.8 molar sol'n.; Japan approved

Magnesium chloride anhydrous. *See* Magnesium chloride

Magnesium chloride hexahydrate
CAS 7791-18-6; EINECS 232-094-6
Formula: MgCl₂ • 6H₂O
Properties: Colorless cryst., deliq.; m.w. 203.31; dens. 1.56; m.p. 118 C (dec.)
Precaution: Keep well closed
Toxicology: LD50 (oral, rat) 8.1 g/kg
Uses: Flavoring agent and adjuvant, nutrient supplement; may be used in infant formulas
Regulatory: FDA 21CFR §184.1426, GRAS

Magnesium fumarate
CAS 7704-71-4 (anhyd.); EINECS 231-724-7
Uses: Dietary supplement
Regulatory: FDA 21CFR §172.350

Magnesium gluconate
CAS 3632-91-5 (anhyd.); EINECS 222-848-2
Synonyms: Magnesium D-gluconate; D-Gluconic acid magnesium salt
Empirical: C₁₂H₂₂O₁₄Mg
Formula: (C₆H₁₁O₇)₂Mg
Properties: Wh. powd. or fine need., odorless; sol. in water; m.w. 414.6; flash pt. > 100 C
Precaution: Combustible
Uses: Mineral source, dietary supplement
Manuf./Distrib.: Am. Roland
Trade names: Gluconal® MG

Magnesium D-gluconate. *See* Magnesium gluconate
Magnesium glycerinophosphate. *See* Magnesium glycerophosphate

Magnesium glycerophosphate
CAS 927-20-8 (anhyd.); EINECS 213-149-3
Synonyms: Magnesium glycerinophosphate
Empirical: C₃H₇MgO₆P
Properties: Colorless powd.; sol. in water; insol. in alcohol
Toxicology: Heated to decomp., emits acrid smoke and irritating fumes
Uses: Plasticizer, stabilizer migrating from food pkg.
Regulatory: FDA 21CFR §181.27, 181.29

Magnesium hydrate. *See* Magnesium hydroxide
Magnesium hydrogen metasilicate. *See* Talc
Magnesium hydrogen phosphate. *See* Magnesium phosphate dibasic

Magnesium hydroxide
CAS 1309-42-8 (anhyd.); EINECS 215-170-3; EEC E528
Synonyms: Magnesium hydrate; Milk of magnesia; Magnesia magma
Classification: Inorganic base
Empirical: H₂MgO₂
Formula: Mg(OH)₂
Properties: White amorphous powd., odorless; sol. in sol'n. of ammonium salts and dilute acids; almost insol. in water and alcohol; m.w. 58.33; dens. 2.36; m.p. 350 C (dec.)
Precaution: Noncombustible; incompat. with maleic anhdyride
Toxicology: Variable toxicity
Uses: Alkali, nutrient supplement, color retention aid, drying agent, pH control agent, processing aid; used in frozen desserts
Usage level: ADI not specified (EEC)
Regulatory: FDA 21CFR §184.1428, GRAS; Europe listed; UK approved
Manuf./Distrib.: Croxton & Garry Ltd; Lohmann
Trade names: Magnesium Hydroxide USP; Magnesium Hydroxide USP DC; Marinco H-USP

Magnesium hydroxidecarbonate. *See* Magnesium carbonate hydroxide
Magnesium octadecanoate. *See* Magnesium stearate

Magnesium oxide
CAS 1309-48-4; EINECS 215-171-9; EEC E530
Synonyms: Magnesia; Periclase; Calcined magnesia; Magnesia usta
Classification: Inorganic oxide
Empirical: MgO
Properties: White powd., odorless; sl. sol. in water; sol. in acids, ammonium salt sol'ns.; m.w. 40.31; dens. 0.36; m.p. 2800 C; b.p. 3600 C
Precaution: Noncombustible; violent reaction or ignition with interhalogens; incandescent reaction with phopshorus pentachloride
Toxicology: Toxic by inhalation of fume; experimental tumorigen; TLV (as magnesium) 10 mg/m³ (fume)
Uses: Alkali, anticaking and free-flow agent, firming agent, lubricant, release agent, neutralizing agent, nutrient supplement, pH control agent; may be used in infant formulas; processing aid (Japan)
Usage level: ADI not specified (FAO/WHO)
Regulatory: FDA 21CFR §163.110, 175.300, 177.1460, 177.2400, 177.2600, 182.5431, 184.1431, GRAS; Japan restricted; Europe listed
Manuf./Distrib.: Lohmann
Trade names: Magnesium Oxide USP 30 Light; Magnesium Oxide USP 60 Light; Magnesium Oxide USP 90 Light; Magnesium Oxide USP Heavy; Marinco OH; Marinco OL

Magnesium phosphate dibasic
CAS 7757-86-0, 7782-75-4 (trihydrate); EINECS 231-823-5 (trihydrate)
Synonyms: Dimagnesium orthophosphate; Dimagnesium phosphate; Magnesium phosphate secondary; Magnesium hydrogen phosphate
Empirical: HMgO₄P (anhyd.), MgHPO₄ • 3H₂O (trihydrate)
Properties: Anhyd.: m.w. 120.29; Trihydrate: Wh. cryst. powd.; sol. in dil. acid; sl. sol. in water; insol. in alcohol; m.w. 174.33; dens. 2.13; dec. 550-650 C
Precaution: Nonflamm.
Toxicology: Nuisance dust; heated to decomp., emits acrid smoke and irritating fumes
Uses: Dietary supplement, nutrient, pH control agent; plasticizer, stabilizer migrating from food pkg.
Regulatory: FDA 21CFR §181.27, 181.29, 182.5434, 184.1434, GRAS
Manuf./Distrib.: Lohmann; Spectrum Chem. Mfg.

Magnesium phosphate neutral. *See* Magnesium phosphate, tribasic
Magnesium phosphate secondary. *See* Magnesium phosphate dibasic
Magnesium phosphate tertiary. *See* Magnesium phosphate, tribasic

Magnesium phosphate, tribasic
CAS 7757-87-1 (anhyd.), 10233-87-1 (pentahydrate); EINECS 231-824-0
Synonyms: Magnesium phosphate neutral; Phosphoric acid magnesium salt (2:3); Trimagnesium phosphate; Tertiary magnesium phosphate; Magnesium phosphate tertiary
Classification: Inorganic salt
Empirical: Mg₃O₈P₂ (anhyd.)
Formula: Mg₃(PO₄)₂ • xH₂O, x = 4, 5, or 8
Properties: Wh. soft, bulky powd., odorless, tasteless; sol. in acids; insol. in water; m.w. 262.86 (anhyd.)
Precaution: Nonflamm.
Toxicology: Nuisance dust
Uses: Nutrient/dietary supplement, pH control agent; may be used in infant formulas
Regulatory: FDA 21CFR §175.300, 181.22, 181.29, 182.5434, 184.1434, GRAS
Manuf./Distrib.: Lohmann

Magnesium phosphide
CAS 12057-74-8
Empirical: Mg₃P₂
Properties: m.w. 134.87
Precaution: Flamm. when exposed to heat, flame, or oxidizers; reacts with water to evolve flamm. phosphine gas; ignites when heated in chlorine, bromine, or iodine vapors; incandescent reaction with nitric acid
Toxicology: Poison; mod. toxic by inh.; heated to decomp., emits toxic fumes of POₓ and phosphine
Uses: Fumigant for animal feed, processed foods
Regulatory: FDA 21CFR §193.255 (residue tolerance 0.01 phosphine in processed foods), 561.268 (0.1 ppm in animal feeds)

Magnesium silicate
CAS 1343-88-0; EINECS 215-681-1; EEC E553a

Synonyms: Silicic acid, magnesium salt (1:1)
Classification: Inorganic salt of variable composition
Empirical: $MgO \cdot SiO_2 \cdot xH_2O$, $3MgSiO_3 \cdot 5HOH$ (variable)
Properties: Fine white powd.; insol. in water, alcohol
Precaution: Noncombustible
Toxicology: Toxic by inhalation; use in foods restricted to 2%
Uses: Anticaking agent for table salt; adsorbent; release agent in some European countries for bread, baked goods, and confectionery; glazing, polishing, release agent (sweets); dusting agent (chewing gum); coating agent (rice)
Usage level: ADI not specified (FAO/WHO)
Regulatory: FDA 21CFR §169.179, 169.182, 182.2437 (2% max.), GRAS; Europe listed; UK approved

Magnesium silicate hydrate

CAS 1343-90-4
Empirical: $Mg_2O_8Si_3 \cdot H_2O$
Properties: Wh. fine powd., odorless, tasteless; insol. in water, alcohol
Toxicology: Human skin irritant
Uses: Anticaking agent, filter aid in table salt
Regulatory: FDA 21CFR §182.2437, GRAS, limitation 2% in table salt

Magnesium stearate

CAS 557-04-0; EINECS 209-150-3; EEC E572
Synonyms: Magnesium octadecanoate; Octadecanoic acid, magnesium salt
Definition: Magnesium salt of stearic acid
Empirical: $C_{36}H_{70}MgO_4$
Formula: $[CH_3(CH_2)_{16}COO]_2Mg$
Properties: Soft white powd., tasteless, odorless; insol. in water, alcohol; dec. by dilute acids; m.w. 591.27; dens. 1.028; m.p. 88.5 C (pure)
Precaution: Nonflamm.
Toxicology: Heated to decomp., emits acrid smoke and toxic fumes
Uses: Anticaking agent, binder, emulsifier, lubricant, release agent, nutrient supplement, processing aid, stabilizer, filter aid, defoamer; used in candy, mint, gum; stabilizer migrating from food pkg.; release agent for bread in some European countries
Usage level: ADI not specified (FAO/WHO)
Regulatory: FDA 21CFR §172.863, 173.340; 175.105, 175.300, 175.320, 176.170, 176.200, 176.210, 177.1200, 177.2260, 178.3910, 179.45, 181.22, 181.29, 184.1440, GRAS; must conform to FDA specs for salts of fats or fatty acids derived from edible oils; Europe listed; UK approved
Trade names: Nuodex Magnesium Stearate Food Grade; Synpro® Magnesium Stearate Food Grade

Magnesium sulfate anhyd.

CAS 7487-88-9; EINECS 231-298-2; EEC E518
Synonyms: Sulfuric acid magnesium salt (1:1)
Classification: Inorganic salt
Empirical: $O_4S \cdot Mg$
Formula: $MgSO_4$
Properties: Colorless crystals, odorless, saline bitter taste; sol. in water; slowly sol. in glycerin; sl. sol. in alcohol; m.w. 120.37; dens. 2.65; dec. at 1124 C
Precaution: Noncombustible; potentially explosive when heated with ethoxyethynyl alcohols
Toxicology: Moderately toxic by ingestion, intraperitoneal, subcutaneous routes; experimental teratogen; heated to decomp., emits toxic fumes of SO_x
Uses: Nutrient, dietary supplement, flavor enhancer, processing aid; fermentation aid, yeast nutrient, tofu coagulant (Japan)
Regulatory: FDA 21CFR §182.5443, 184.1443, GRAS; Japan approved; Europe listed; UK approved
Manuf./Distrib.: Heico

Magnesium sulfate heptahydrate

CAS 10034-99-8; EINECS 231-298-8; EEC E518
Synonyms: Bitter salts; Epsom salts
Empirical: $MgO_4S \cdot 7H_2O$
Formula: $MgSO_4 \cdot 7H_2O$
Properties: Efflorescent cryst. or powd., bitter saline cooling taste; sol. 71 g/100 ml in water (20 C); sl. sol. in alcohol; m.w. 246.48; dens. 1.670; pH 6-7
Precaution: Keep well closed
Uses: Flavor enhancer, nutrient/dietary supplement, processing aid, firming agent; in beer-making
Regulatory: FDA 21CFR §184.1443, GRAS; Europe listed; UK approved

Magnesium, tetrakis[carbonato(2-)] dihydroxypenta-

Manuf./Distrib.: Lohmann

Magnesium, tetrakis[carbonato(2-)] dihydroxypenta-. *See* Magnesium carbonate hydroxide

Maidenhair fern extract
CAS 84649-72-9; EINECS 283-457-0 (*A. capillus-veneris*)
Synonyms: Adiantum capillus veneris extract; Adiantum pedatum extract
Definition: Extract of leaves of the fern, *Adiantum capillus-veneris* or *A. pedatum*
Uses: Natural flavoring
Regulatory: FDA 21CFR §172.510; Japan approved
Manuf./Distrib.: Chart

Maize oil. *See* Corn oil
MAK. *See* Methyl n-amyl ketone
Malic acid. *See* N-Hydroxysuccinic acid
Malonic ester. *See* Diethyl malonate

Malt
Properties: Yel. or amber-colored grains of barley which have been partially germinated; char. odor and taste; sol. in water, ethanol; insol. in veg. oil
Uses: Colorant producing brown shades; brewing, malted milk, similar food prods., extract of malt (with 10% glycerol)
Regulatory: Japan approved
Manuf./Distrib.: ADM; Calgene; Cargill; Edw. Gittens; Chr Hansen's Lab; Int'l. Grain Prods.; Malt Prods.; Mid-Am. Food Sales; North Western; Pure Food Ingreds.; Pure Malt Prods.
Trade names: Flo-Malt® Malt; Nutrimalt®
Trade names containing: CDC-79; CDC-79DZ

Maltase. *See* α-Glucosidase

Malt extract
CAS 8002-48-0; EINECS 232-310-9
Synonyms: Maltine; Malt syrup
Definition: Dark syrup obtained by evaporating an aq. extract of partially germinated and dried barley seeds; derived from *Hordeum vulgare*; contains dextrin, maltose, a little glucose, and an amylolytic enzyme
Properties: Lt. brown, visc. liq., sweet; sol. in cold water; dens. 1.35-1.43
Toxicology: Heated to decomp., emits acrid smoke and irritating fumes
Uses: Nutritive sweetener, adjuvant, flavoring agent, color, enzyme, humectant, stabilizer, thickener, texturizer; used in cured meat, poultry, vanilla and chocolate flavors
Regulatory: FDA 21CFR §133.178, 184.1445, GRAS; USDA 9CFR §318.7, limitation 2.5% in cured meats, 381.147
Manuf./Distrib.: Beck Flavors; Calgene; Chart; Diamalt; Enco Prods.; Folexco; Grünau; Goorden; Henkel; Malt Prods.; Mid-Am. Food Sales; MLG Enterprises; J W Pike; Pure Malt Prods.; Regency Mowbray; Sandoz Nutrition; Wander
Trade names: DME Dry Malt Extract; Extramalt Dark; Maltoferm® 10001; Maltoferm® 10001 VDK; Maltoferm® A-6001; ND-201 Syrup; Non-Diastatic Malt Syrup #40600; Nutrimalt® Range; Pure Malt Colorant A6000; Pure Malt Colorant A6001
Trade names containing: BK-102 Series; BK-305 Series; BK-PR2 Series; DCME; DMCE; Extramalt 10; Extramalt 35; Extramalt Light; Fermentase®; ND-201-C Syrups; ND-305 Syrups; Premose® Syrup

Maltine. *See* Malt extract

Maltitol
CAS 585-88-6; EINECS 209-567-0
Synonyms: 4-O-α-Glucopyranosyl-D-sorbitol; 4-O-β-D-Glucopyranosyl-D-glucitol
Empirical: $C_{12}H_{24}O_{11}$
Properties: Liq., cryst.; easily sol. in water; m.w. 344.32; very stable at different pH conditions and temps.
Toxicology: LD50 (oral) > 24 g/kg; low acute toxicity; nonmutagenic; nonteratogenic
Uses: Nutritive sweetener; 60% of sucrose (liq.), 90% of sucrose (crystalline)
Trade names: Amalty®; Finmalt L

Maltobiose. *See* Maltose

Maltodextrin
CAS 9050-36-6; EINECS 232-940-4
Classification: Saccharide
Definition: Saccharide material obtained by hydrolysis of starch
Empirical: $(C_6H_{10}O_5)_n$

Properties: Wh. powd. or sol'n.
Toxicology: Heated to decomp., emits acrid smoke and irritating fumes
Uses: Nonsweet nutritive polymer useful as carrier, bulking agent, bodying agent, crystallization inhibitor, texturizer in candy, crackers, puddings
Regulatory: FDA 21CFR §184.144, GRAS
Manuf./Distrib.: Avebe UK; Cerestar Int'l.; Clofine; Fruitsource; Grain Processing; Kingfood Australia; Nat'l. Starch & Chem. UK; Roquette UK; Sweeteners Plus; Westin; Zumbro
Trade names: Lo-Dex® 10; Maltrin® M040; Maltrin® M050; Maltrin® M100; Maltrin® M105; Maltrin® M150; Maltrin® M180; Maltrin® M510; Maltrin® M520; Maltrin® M700; Maltrin® QD M440; Maltrin® QD M500; Maltrin® QD M550; Maltrin® QD M580; Microduct®; Staley® Maltodextrin 3260; Star-Dri® 1; Star-Dri® 5; Star-Dri® 10; Star-Dri® 15; Star-Dri® 18; Star-Dri® 1005A; Star-Dri® 1015A; Wickenol® 550
Trade names containing: Alcolec® EYR SM; Avicel® RCN-30; Dry Vitamin E Acetate 50% Type CWS/F No. 652530001; Durkote Citric Acid/Maltodextrin, Acacia Gum; Durkote Malic Acid/Maltodextrin; Durkote Vitamin C/Maltodextrin; Fat Replacer 785; Lamemul® K 3000-Range; Lamequick® AS 340; Lecithin L 3000-Range; Pristene® R20; Red Beet WSP-300; Spray Dried Fish Gelatin/Maltodextrin

Maltol
CAS 118-71-8; EINECS 204-271-8; FEMA 2656; EEC E636
Synonyms: 3-Hydroxy-2-methyl-4-pyrone; 2-Methyl pyromeconic acid; 3-Hydroxy-2-methyl-4H-pyran-4-one; Larixinic acid
Empirical: $C_6H_6O_3$
Properties: Cryst., fragrant caramel-like odor; sol. 1 g/85 ml water; freely sol. in hot water, chloroform; sol. in alcohol; sparingly sol. in benzene, ether, petroleum ether; m.w. 126.11; m.p. 160-162 C; begins to sublime @ 93 C; pH 5.3 (0.5% aq.)
Precaution: Volatile with steam
Toxicology: Skin irritant
Uses: Synthetic flavoring; imparts freshly baked odor and flavor to bread and cakes
Usage level: ADI 0-1 mg/kg (EEC)
Regulatory: FDA 21CFR §172.515; Japan approved as flavoring; Europe listed; UK approved
Manuf./Distrib.: Aldrich
Trade names: Veltol®

Maltonic acid. *See* D-Gluconic acid

Maltose
CAS 69-79-4, 6363-53-7 (monohydrate); EINECS 200-716-5
Synonyms: Malt sugar; Maltobiose; r-O-α-D-Glucopyranosyl D glucose
Classification: Malt sugar, an isomer of cellobiose
Empirical: $C_{12}H_{22}O_{11} \cdot H_2O$
Properties: Anhyd.: m.w. 342.31; Monohydrate: Cryst.; hygroscopic; sol. in water, sl. sol. in alcohol; pract. insol. in ether; m.w. 360.32; m.p. 102-103 C; about one-third as sweet as sucrose
Precaution: Combustible
Uses: Nutrient, sweetener, fermentable intermediate in brewing
Manuf./Distrib.: Alfa; Cargill; Cerestar Int'l.; Fruitsource; K&K Greeff; Mid-Am. Food Sales; Mitsubishi; North Western; Penta Mfg.; Ragus Sugars
Trade names: Sunmalt

Malt sugar. *See* Maltose
Malt syrup. *See* Malt extract
Mandarin oil. *See* Mandarin orange oil

Mandarin orange oil
CAS 8008-31-9; 84696-35-5; FEMA 2657
Synonyms: Mandarin oil
Definition: Oil expressed from the peel of *Citrus reticulata*
Properties: Clear orange to brn.-orange liq., orange odor; sol. in fixed oils, min. oil; sl. sol. in propylene glycol; insol. in glycerin; dens. 0.846
Toxicology: Heated to decomp., emits acrid smoke and irritating fumes
Uses: Natural flavoring agent for soft drinks, confectionery, other natural fruit flavors
Usage level: 62 ppm (nonalcoholic beverages), 160 ppm (ice cream), 350 ppm (candy), 190 ppm (baked goods), 30 ppm (gelatins, puddings), 83 ppm (chewing gum)
Regulatory: FDA 21CFR §182.20, GRAS; FEMA GRAS; Europe listed, no restrictions
Manuf./Distrib.: Commodity Services; Florida Treatt

Manganese caprylate
Toxicology: Heated to decomp., emits acrid smoke and irritating fumes

Uses: Drier migrating from food pkg.
Regulatory: FDA 21CFR §181.25

Manganese chloride tetrahydrate
CAS 13446-34-9; EINECS 231-869-6
Synonyms: Manganese dichloride; Manganous chloride; Manganese (II) chloride tetrahydrate
Formula: $MnCl_2 \cdot 4H_2O$
Properties: Pink translucent cryst., sl. deliq.; sol. in 0.7 part water, sol. in alcohol; insol. in ether; m.w. 197.91; dens. 2.01; m.p. 58 C; pH 5.5 (0.2 molar aq.)
Precaution: Hygroscopic; keep well closed
Toxicology: Irritant
Uses: Nutrient, dietary supplement; may be used in infant formulas
Regulatory: FDA 21CFR §182.5446, 184.1446, GRAS
Manuf./Distrib.: Aldrich

Manganese (II) chloride tetrahydrate. *See* Manganese chloride tetrahydrate

Manganese citrate
CAS 10024-66-5
Synonyms: Manganous citrate
Formula: $Mn_3(C_6H_5O_7)_2$
Properties: Pale orange or pinkish wh. powd.; sol. in water in presence of sodium citrate; m.w. 543.02
Precaution: Combustible
Toxicology: Heated to decomp., emits acrid smoke and irritating fumes
Uses: Nutrient, dietary supplement used in baked goods, nonalcoholic beverages, dairy prods., fish prods., meat prods., milk prods., poultry prods., infant formulas; feed additive
Regulatory: FDA 21CFR §182.5449, 184.1449, GRAS

Manganese dichloride. *See* Manganese chloride tetrahydrate

Manganese gluconate
CAS 6485-39-8 (anhyd.); EINECS 229-350-4
Synonyms: Manganese (II) gluconate
Empirical: $C_{12}H_{22}O_{14}Mn \cdot 2H_2O$
Formula: $Mn(C_6H_{11}O_7)_2 \cdot 2H_2O$
Properties: Lt. pinkish powd. or gran.; sol. in water; insol. in alc. and benzene; m.w. 481.27
Precaution: Combustible
Toxicology: Heated to decomp., emits toxic fumes of manganese
Uses: Dietary supplement, nutrient used in baked goods, nonalcoholic beverages, dairy prods., fish/meat/ poultry prods., milk prods., infant formulas
Regulatory: FDA 21CFR §182.5452, 184.1452, GRAS
Manuf./Distrib.: Lohmann; Spectrum Chem. Mfg.
Trade names: Gluconal® MN

Manganese (II) gluconate. *See* Manganese gluconate

Manganese glycerophosphate
CAS 1320-46-3; EINECS 215-301-4
Synonyms: Manganese (II) glycerophosphate
Empirical: $C_3H_7MnO_6P \cdot xH_2O$
Properties: Wh. or pinkish powd., odorless, nearly tasteless; sol. in citric acid sol'n.; sl. sol. in water; insol. in alcohol; m.w. 225.00
Toxicology: Heated to decomp., emits toxic fumes of manganese
Uses: Dietary supplement, nutrient
Regulatory: FDA 21CFR §182.5455, 182.8455, GRAS

Manganese (II) glycerophosphate. *See* Manganese glycerophosphate

Manganese hypophosphite
CAS 10043-84-2
Empirical: $H_4MnO_4P_2$
Formula: $Mn(PH_2O_2)_2 \cdot xH_2O$
Properties: Monohydrate: Pink gran. or cryst. powd., odorless, tasteless; sol. in water; insol. in alcohol; m.w. 184.91
Precaution: Heated, spontaneously evolves flamm. phosphine
Toxicology: Heated to decomp., emits toxic fumes of manganese
Uses: Nutrient, dietary supplement
Regulatory: FDA 21CFR §182.5458, 182.8458, GRAS

Manganese linoleate
Formula: $Mn(C_{18}H_{31}O_2)_2$
Properties: Dk. brn. plaster-like mass; sol. in linseed oil
Precaution: Combustible
Toxicology: Heated to decomp., emits acrid smoke and irritating fumes
Uses: Drier migrating from food pkg.
Regulatory: FDA 21CFR §181.25

Manganese naphthenate
Properties: Hard brn. resinous mass; sol. in min. spirits; m.p. 130-140 C
Precaution: Combustible; sol'n. is flamm.
Toxicology: Heated to decomp., emits acrid smoke and irritating fumes
Uses: Drier migrating from food pkg.
Regulatory: FDA 21CFR §181.25

Manganese oxide
CAS 1344-43-0
Synonyms: Manganous oxide
Uses: Dietary supplement
Regulatory: FDA 21CFR §182.5464, GRAS

Manganese (II) sulfate (1:1). See Manganese sulfate (ous)

Manganese sulfate (ous)
CAS 7785-87-7 (anhyd.), 10034-96-5 (monohydrate); EINECS 232-089-9
Synonyms: Manganous sulfate; Manganese (II) sulfate (1:1)
Empirical: $MnSO_4$ (anhyd.), $MnSO_4 \cdot H_2O$ (monohydrate)
Properties: Anhyd.: Pink gran. powd., odorless; very sol. in water; insol. in alcohol; m.w. 151.00; dens. 3.25; m.p. 700 C; b.p. 850 C (dec.); Monohydrate: Pale red, sl. efflorescent cryst.; sol. in 1 part water; insol. in alcohol; m.w. 169.00
Toxicology: TLV 5 mg(Mn)/m³; LD50 (intraperitoneal, mouse) 332 mg/kg; poison by intraperitoneal route; experimental neoplastigen; mutagenic data; heated to decomp., emits toxic fumes of SO_x and manganese
Uses: Nutrient, dietary supplement used in baked goods, nonalcoholic beverages, dairy/fish/meat/poultry/milk prods., infant formulas
Regulatory: FDA 21CFR §182.5461, 184.1461, GRAS

Manganese tallate
Definition: Manganese salts of tall oil fatty acids
Precaution: Combustible
Toxicology: Heated to decomp., emits acrid smoke and irritating fumes
Uses: Drier migrating from food pkg.
Regulatory: FDA 21CFR §181.25

Manganous chloride. See Manganese chloride tetrahydrate
Manganous citrate. See Manganese citrate
Manganous oxide. See Manganese oxide
Manganous sulfate. See Manganese sulfate (ous)
Manna sugar. See D-Mannitol
Mannite. See D-Mannitol

D-Mannitol
CAS 69-65-8; EINECS 200-711-8; EEC E421
Synonyms: 1,2,3,4,5,6-Hexanehexol; Mannite; Cordycepic acid; Manna sugar; Mannose sugar
Classification: Hexahydric alcohol
Empirical: $C_6H_{14}O_6$
Properties: Orthorhombic need., sweetish taste; hygroscopic; sol. 1 g/5.5 ml water, 1 g/18 ml glycerol; 1 g/83 ml alcohol; sol. in pyridine, aniline, aq. sol'ns. of alkalies; m.w. 182.18; dens. 1.52 (20 C); m.p. 165-167 C; b.p. 290-295 C (3.5 mm)
Toxicology: Excess consumption may have a laxative effect; may cause diarrhea and flatulence
Uses: Dietary supplement, texturizer, humectant, flavoring agent, lubricant, release agent, nutritive sweetener, bulking agent, anticaking and free-flow agent, stabilizer, thickener, formulation/processing aid, firming agent; inert base for tablets
Usage level: Limitation 98% (pressed mints), 5% (hard candy), 31% (chewing gum), 40% (soft candy), 8% (confections, frostngs), 15% (jams, jellies), < 2.5% (other foods); ADI not specified (WHO)
Regulatory: FDA 21CFR §180.25; GRAS; Japan approved; Europe listed; UK approved
Manuf./Distrib.: Allchem Int'l.; Am. Roland; Atomergic Chemetals; Cerestar Int'l.; Dasco Sales; Food Additives

& Ingreds.; Fruitsource; ICI Atkemix; Int'l. Sourcing; E Merck; Penta Mfg.; Roquette UK; Treasure Island Food; Van Waters & Rogers; Peter Whiting
Trade names: Unisweet MAN
Trade names containing: Vitamin B$_{12}$ 1% Trituration No. 69993

L-Mannomethylose. *See* Rhamnose
Mannose sugar. *See* D-Mannitol
MAP. *See* Ammonium phosphate
Marigold extract. *See* Calendula extract
Marigold extract. *See* Tagetes extract

Marjoram oil
Definition: Oil from steam distillation of herb *Marjorana hortensis*
Properties: Yel. to grn.-yel. liq., spicy odor; sol. in fixed oils, min. oil; partly sol. in propylene glycol; insol. in glycerin
Toxicology: Heated to decomp., emits acrid smoke and irritating fumes
Uses: Natural flavoring agent for fish, meat, sauces, soups
Regulatory: FDA 21CFR §182.10, 182.20, GRAS; Japan approved
Manuf./Distrib.: Chart; Pierre Chauvet

Marjoram oil, Spanish
CAS 8015-01-8
Synonyms: Spanish marjoram oil
Definition: Oil from steam distillation of flowering plant material from the shrub *Thymus mastichina*; main constituent is cineole
Properties: Pale yel. liq.; sol. in fixed oils; insol. in glycerin, propylene glycol, min. oil; dens. 0.904-0.920; ref. index 1.463 (20 C)
Toxicology: Skin irritant; heated to decomp., emits acrid smoke and irritating fumes
Uses: Natural flavoring agent
Regulatory: FDA 21CFR §182.10, GRAS

Marshmallow root extract. *See* Althea extract
Mascagnite. *See* Ammonium sulfate

Mate extract
CAS 84082-59-7
Synonyms: Ilex paraguariensis extract; St. Bartholomew's tea; Paraguay tea extract; Yerba mate extract
Definition: Extract of leaves of *Ilex paraguariensis*
Properties: Rich herbaceous green odor, somewhat bitter refreshing taste
Uses: Natural flavoring agent
Regulatory: FDA 21CFR §182.20, GRAS
Manuf./Distrib.: C.A.L.-Pfizer; Chart

Matricaria chamomilla extract. *See* Matricaria extract

Matricaria extract
Synonyms: Chamomile extract, German or Hungarian; Matricaria chamomilla extract; Wild chamomile extract
Definition: Extract of flowerheads of *Matricaria chamomilla*
Uses: Natural flavoring agent
Regulatory: FDA 21CFR §182.20, GRAS

Matricaria oil
CAS 8002-66-2
Synonyms: Hungarian chamomile oil; German chamomile oil; Wild chamomile oil
Definition: Volatile oil distilled from dried flower heads of *Matricaria chamomilla*
Properties: Blue-ylsh.-brn. liq., butter-like consistency when cooled, strong odor, bitter aromatic taste; very sol. in alcohol; sol. in fixed oils, propylene glycol; insol. in min. oil, glycerin; dens. 0.905-0.915 (15/15 C); sapon. no. 45
Precaution: Keep cool, well closed; protect from light
Toxicology: Mild allergen, skin irritant; heated to decomp., emits acrid and irritating fumes
Uses: Natural flavoring agent
Regulatory: FDA 21CFR §182.10, 182.20, GRAS

MC. *See* Methylcellulose
MCP. *See* Calcium phosphate monobasic monohydrate
MCP/A. *See* Calcium phosphate monobasic anhydrous
MEA. *See* Ethanolamine
Medicago sativa extract. *See* Alfalfa extract

MEK. *See* Methyl ethyl ketone

Melamine
CAS 108-78-1; EINECS 203-615-4
Synonyms: Cyanuramide; sym-Triaminotriazine; 1,3,5-Triazine-2,4,6-triamine; Cyanurotriamide; Cyanourotriamine; 2,4,6-Triamino-s-triazine; 2,4,6-Triamino-sym-triazine
Empirical: $C_3H_6N_6$
Properties: Colorless monoclinic prisms; sl. sol. in water; very sl. sol. in hot alcohol; insol. in ether; m.w. 126.15; dens. 1.573 (250 C); m.p. < 250 C; flash pt. > 300 C; b.p. sublimes
Toxicology: LD50 (oral, rat) 3161 mg/kg; mod. toxic by ingestion, IP routes; eye, skin, and mucous membrane irritant; causes dermatitis in humans; experimental carcinogen, tumorigen; mutagenic data; heated to decomp., emits toxic fumes of NO_x and CN⁻
Uses: Mfg. of paper and paperboard for food pkg.
Regulatory: FDA 21CFR §181.30

Melamine/formaldehyde resin
CAS 9003-08-1
Synonyms: Melamine resin; 1,3,5-Triazine,2,4,6-triamine, polymer with formaldehyde
Classification: Amino resin
Definition: Reaction prod. of melamine and formaldehyde
Formula: $(CH_3H_6N_6 \cdot CH_2O)_x$
Properties: Syrup or insol. powd.; sol. in water
Uses: Used in paper/paperboard for food pkg.
Regulatory: FDA 21CFR §175.105, 175.300, 175.320, 177.1200, 177.1460, 177.2260, 177.2470, 181.22, 181.30
Manuf./Distrib.: Monsanto

Melamine resin. *See* Melamine/formaldehyde resin
Melilotin. *See* Dihydrocoumarin
Melissa officinalis extract. *See* Balm mint extract
Melissa officinalis oil. *See* Balm mint oil

Menhaden oil
CAS 8002-50-4; EINECS 232-311-4
Synonyms: Oils, menhaden; Pogy oil; Mossbunker oil
Definition: Oil obtained from the small North Atlantic fish, *Brevoortia tyrannus*
Properties: Yellowish-brown or reddish-brown oil, characteristic fishy odor and taste; sol. in ether, benzene, petrol. ether, naphtha, kerosene, CS_2; dens. 0.925-0.933; m.p. 38.5-47.2 C; ref. index 1.480 (20 C); sapon. no. 191-200
Precaution: Combustible
Uses: Hydrogenated fats for cooking
Regulatory: FDA 21CFR §175.300, 176.200, 176.210, 177.2800
Trade names containing: Neobee® SL-220; Neobee® SL-230; Neobee® SL-310

Menhaden oil, hydrogenated. *See* Hydrogenated menhaden oil

Mentha arvensis oil
Synonyms: Cornmint oil, partially dementholized; Wild pennyroyal
Definition: Oil from *Mentha arvensis*
Properties: Colorless to yel. liq., minty odor; sol. in fixed oils, min. oil, propylene glycol; insol. in glycerin; dens. 0.888-0.908; ref. index 1.458 (20 C)
Toxicology: Experimental reproductive effects; heated to decomp., emits acrid smoke and irritating fumes
Uses: Natural flavoring agent giving economic peppermint flavor; often blended with peppermint oil
Regulatory: Europe listed
Manuf./Distrib.: Acme-Hardesty; Florida Treatt

p-Mentha-1,8-dien-7-al. *See* Perillaldehyde
d-p-Mentha-1,8-diene. *See* d-Limonene
l-p-Mentha-1,8-diene. *See* l-Limonene
p-Mentha-1,3-diene. *See* α-Terpinene
p-Mentha-1,4-diene. *See* γ-Terpinene
p-Mentha-1,5-diene. *See* α-Phellandrene
Menthadienol. *See* p-Mentha-1,8(10)-dien-9-ol
p-Mentha-1,8-dien-7-ol. *See* Perillyl alcohol

p-Mentha-1,8(10)-dien-9-ol
Synonyms: Menthadienol

Uses: Synthetic flavoring
Regulatory: FDA 21CFR §172.515

p-Mentha-6,8-dien-2-ol. *See* Carveol
1-6,8(9)-p-Menthadien-2-one. *See* l-Carvone
d-p-Mentha-6,8,(9)-dien-2-one. *See* d-Carvone
p-Mentha-1,4(8)-dien-3-one. *See* Piperitenone
Menthadienyl acetate. *See* p-Mentha-1,8(10)-dien-9-yl acetate

p-Mentha-1,8(10)-dien-9-yl acetate
Synonyms: Menthadienyl acetate
Uses: Synthetic flavoring
Regulatory: FDA 21CFR §172.515

l-p-Mentha-6,8-dien-2-yl propionate. *See* Carvyl propionate
3-p-Menthanol. *See* Menthol
p-Menthan-3-ol. *See* Menthol
l-p-Menthan-3-one. *See* l-Menthone
dl-p-Menthan-3-yl acetate. *See* dl-Menthyl acetate
l-p-Menthan-3-yl acetate. *See* l-Menthyl acetate
Mentha piperita extract. *See* Peppermint extract
Mentha piperita leaves. *See* Peppermint leaves
Mentha piperita oil. *See* Peppermint oil
Mentha pulegium extract. *See* Pennyroyal extract
Mentha spicata oil. *See* Spearmint oil

Mentha-8-thol-3-one
CAS 38462-22-5; FEMA 3177
Empirical: $C_{10}H_{18}OS$
Properties: m.w. 186.32; dens. 1.00; flash pt. 227 F
Uses: Black currant-like flavoring
Manuf./Distrib.: Aldrich

p-Menth-1,4(8)-diene. *See* Terpinolene
p-Menthen-1-en-8-ol. *See* α-Terpineol
1-p-Menthen-4-ol. *See* 4-Carvomenthenol
8-p-Menthen-2-ol. *See* Dihydrocarveol
(S)-p-Menth-1-en-4-ol. *See* 4-Carvomenthenol
p-Menth-1-en-8-ol. *See* α-Terpineol
p-Menth-1-en-8-ol. *See* Terpineol anhydrous

p-Menth-3-en-1-ol
Synonyms: 1-Terpinenol
Empirical: $C_{10}H_{18}O$
Properties: Colorless oily liq., dry woody somewhat musty odor; sol. in alcohol; sl. sol. in water; m.w. 154.24; dens. 0.9210; b.p. 210 C; ref. index 1.4778
Uses: Synthetic flavoring
Regulatory: FDA 21CFR §172.515

p-Menth-4-en-3-ol. *See* Isopulegol
p-Menth-8-en-3-ol. *See* Isopulegol
R-(+)-p-Menth-4(8)-en-3-one. *See* Pulegone
p-Menth-1-en-3-one. *See* d-Piperitone
p-Menth-4(8)-en-3-one. *See* Pulegone
p-Menth-8-en-3-one. *See* Isopulegone
Menthen-1-yl-8 acetate. *See* Terpinyl acetate
1-p-Menthen-9-yl acetate. *See* p-Menth-1-en-9-yl acetate
8-p-Menthen-2-yl acetate. *See* Dihydrocarvyl acetate

p-Menth-1-en-9-yl acetate
Synonyms: 1-p-Menthen-9-yl acetate
Uses: Synthetic flavoring
Regulatory: FDA 21CFR §172.515

p-Menth-8-(9)-en-2-yl acetate. *See* Dihydrocarvyl acetate
Menthen-1-yl-8 propionate. *See* Terpinyl propionate

Menthol
CAS 89-78-1; EINECS 201-939-0; FEMA 2665

Synonyms: Hexahydrothymol; 2-Isopropyl-5-methylcyclohexanol; 3-p-Menthanol; p-Menthan-3-ol; 5-Methyl-2-(1-methylethyl) cyclohexanol; Racemic menthol
Classification: Diterpene
Empirical: $C_{10}H_{20}O$
Formula: $CH_3C_6H_9(C_3H_7)OH$
Properties: Wh. cryst., cooling odor and taste; very sol. in alcohol, light petrol. solv., glacial acetic acid; sl. sol. in water; m.w. 156.26; dens. 0.89; m.p. 42.5 C; b.p. 215 C; flash pt. 200 F; ref. index 1.461
Precaution: Combustible; incompat. with phenol, β-naphthol, others
Toxicology: LD50 (oral, rat) 3180 mg/kg; poison by IV route; mod. toxic by ingestion, IP routes; severe eye irritant; irritant to mucous membranes on inh.; may cause human intolerance reaction; heated to decomp., emits acrid smoke and irritating fumes
Uses: Synthetic flavoring agent
Usage level: 35 ppm (nonalcoholic beverages), 68 ppm (ice cream), 400 ppm (candy), 130 ppm (baked goods), 1100 ppm (chewing gum)
Regulatory: FDA 21CFR §172.515, 182.20, GRAS; 27CFR §21.65, 21.151; Japan approved as flavoring
Manuf./Distrib.: Acme-Hardesty; Am. Fruit Processors; Berje; Berk; Biddle Sawyer; Charabot; Chart; Commodity Services Int'l.; Dasco Sales; Diamalt; Forrester Wood; Frutarom UK; Haarmann & Reimer; Ikeda; Penta Mfg.; Polarome; Quimdis; Robt. Bryce; R C Treatt
Trade names: Fancol Menthol

l-Menthone
CAS 14073-97-3; FEMA 2667
Synonyms: l-p-Menthan-3-one
Uses: Synthetic flavoring agent
Regulatory: FDA 21CFR §172.515
Manuf./Distrib.: Aldrich

Menthyl acetate. *See* p-Menth-3-yl acetate

dl-Menthyl acetate
CAS 16409-45-3; FEMA 2668
Synonyms: dl-p-Menthan-3-yl acetate
Manuf./Distrib.: Aldrich

l Menthyl acetate
Synonyms: l-p-Menthan-3-yl acetate
Uooo: Flavoring agent
Regulatory: Japan approved as flavoring agent
Manuf./Distrib.: Chr. Hansen's

p-Menth-3-yl acetate
Synonyms: Menthyl acetate
Uses: Synthetic flavoring
Regulatory: FDA 21CFR §172.515

Menthyl isovalerate
CAS 16409-46-4; FEMA 2669
Synonyms: p-Menth-3-yl isovalerate
Uses: Synthetic flavoring
Regulatory: FDA 21CFR §172.515
Manuf./Distrib.: Aldrich

p-Menth-3-yl isovalerate. *See* Menthyl isovalerate

Menthyl lactate
CAS 59259-38-0; EINECS 261-678-3
Synonyms: Lactic acid menthyl ester
Uses: Mild cooling flavor in powd. drink mixes, chewing gums, sweets
Trade names: Frescolat, Type ML

Mercaptan C-1. *See* Methyl mercaptan
Mercaptan C-3. *See* Propyl mercaptan

2-Mercaptopropionic acid
CAS 79-42-5; FEMA 3180
Empirical: $C_3H_6O_2S$
Properties: Colorless to pale-yellow liq.; roasted, meaty odor; miscible in water, alcohol, ether, and acetone; m.w. 106.16; b.p. 117 C

Uses: Flavoring agent
Manuf./Distrib.: Aldrich

2-Mercaptothiophene. See 2-Thienyl mercaptan
Metaphosphoric acid trisodium salt. See Sodium trimetaphosphate
Methacetone. See Diethyl ketone

Methacrylic acid-divinylbenzene copolymer
Toxicology: Heated to decomp., emits acrid smoke and irritating fumes
Uses: Carrier of vitamin B_{12} in foods for special dietary use
Regulatory: FDA 21CFR §172.775

Methallyl butyrate. See 2-Methylallyl butyrate
Methanal. See Formaldehyde
Methanecarboxylic acid. See Acetic acid
Methane dichloride. See Methylene chloride
Methanethiol. See Methyl mercaptan
Methanoic acid. See Formic acid
Methanol. See Methyl alcohol
Methenamine. See Hexamethylene tetramine
Methional. See 3-Methylthiopropionaldehyde
(±)-Methionine. See DL-Methionine

DL-Methionine
CAS 59-51-8; EINECS 200-432-1
Synonyms: 2-Amino-4-(methylthio)butyric acid; (±)-2-Amino-4-(methylmercapto)butyric acid; Acimetion; (±)-Methionine
Classification: Amino acid
Empirical: $C_5H_{11}NO_2S$
Formula: $CH_3SCH_2CH_2CH(NH_2)COOH$
Properties: Wh. cryst. platelets or powd., char. odor; sol. in water, dil. acids and alkalies; very sl. sol. in alcohol; ins. in ether; m.w. 149.21; m.p. approx. 280 C (dec.)
Toxicology: Mod. toxic by ingestion and other routes; experimental reproductive effects; heated to decomp., emits toxic fumes of SO_x and NO_x
Uses: Dietary supplement, nutrient, vegetable oil enrichment; feed additive; flavoring
Regulatory: FDA 21CFR §172.320, 3.1% max., not for infant foods; Japan approved
Manuf./Distrib.: Am. Roland

L-Methionine
CAS 63-68-3; EINECS 200-562-9
Synonyms: 2-Amino-4-(methylthio)butyric acid; (S)-2-Amino-4-(methylmercapto)butyric acid
Classification: Amino acid
Empirical: $C_5H_{11}NO_2S$
Formula: $CH_3SCH_2CH_2CH(NH_2)COOH$
Properties: Wh. cryst. powd. or platelets, sl. char. odor; sol. in water, dil. acids, alkalies; insol. in abs. alcohol, alcohol, benzene, acetone, ether; m.w. 149.21; dens. 1.340; m.p. 281 C (dec.)
Toxicology: LD50 (oral, rat) 36 g/kg; mildly toxic by ingestion, intraperitoneal routes; human mutagenic data; experimental teratogen, reproductive effects; heated to decomp., emits very toxic fumes of NO_x and SO_x
Uses: Nutrient, dietary supplement, vegetable oil enrichment, feed additive; flavoring
Regulatory: FDA 21CFR §172.320, limitation 3.1%; Japan approved

Methocel. See Methylcellulose
4´-Methoxyacetophenone. See Acetanisole
p-Methoxyacetophenone. See Acetanisole
2-Methoxybenzaldehyde. See o-Methoxybenzaldehyde
4-Methoxybenzaldehyde. See p-Anisaldehyde

o-Methoxybenzaldehyde
CAS 135-02-4; EINECS 205-171-7
Synonyms: o-Anisaldehyde; Methyl salicylaldehyde; 2-Methoxybenzaldehyde; Salicylaldehyde methyl ether
Empirical: $C_8H_8O_2$
Properties: Colorless or cream-colored cryst., faint sweet floral odor, spice-like flavor; m.w. 136.15; dens. 1.1326; m.p. 35-37 C; b.p. 243-244 C; ref. index 1.560
Uses: Synthetic flavoring
Regulatory: FDA 21CFR §172.515
Manuf./Distrib.: Aldrich

p-Methoxybenzaldehyde. *See* p-Anisaldehyde
Methoxybenzene. *See* Anisole
4-Methoxybenzenemethanol. *See* Anisyl alcohol
p-Methoxybenzyl acetate. *See* Anisyl acetate
4-Methoxybenzyl alcohol. *See* Anisyl alcohol
p-Methoxybenzyl alcohol. *See* Anisyl alcohol
p- Methoxybenzyl butyrate. *See* Anisyl butyrate
p-Methoxybenzyl formate. *See* Anisyl formate
p-Methoxybenzyl propionate. *See* Anisyl propionate
6´-Methoxycinchonan-9-ol. *See* Quinine
6´-Methoxycinchonan-9-ol dihydrochloride. *See* Quinine dihydrochloride
6´-Methoxycinchonan-9-ol monohydrochloride. *See* Quinine hydrochloride

o-Methoxycinnamaldehyde
 CAS 1504-74-1; FEMA 3181
 Synonyms: o-Cumaric aldehyde methyl ether
 Empirical: $C_{10}H_{10}O_2$
 Properties: Pale yel. cryst. flakes, spicy-floral odor; sol. in alcohol, ether, chloroform; sl. sol. in water; m.w.
 162.18; m.p. 45-46 C; b.p. 295 C
 Uses: Synthetic flavoring
 Regulatory: FDA 21CFR §172.515
 Manuf./Distrib.: Aldrich

2-Methoxy-p-cresol. *See* 2-Methoxy-4-methylphenol

Methoxyethanol
 CAS 109-86-4; EINECS 203-713-7
 Synonyms: Ethylene glycol methyl ether; 2-Methoxyethanol; Methyl Cellosolve®
 Classification: Aliphatic ether alcohol
 Empirical: $C_3H_8O_2$
 Formula: $CH_3OCH_2CH_2OH$
 Properties: Colorless liq., mild agreeable odor.; misc. with water, alcohol, ether, benzene, glycerol, acetone,
 dimethylformamide; m.w. 76.10; dens. 0.964 (20/4 C); f.p. -86.5 C; b.p. 123-124 C (760 mm); flash pt. (OC)
 115 F; ref. index 1.4028 (20 C)
 Precaution: Flamm. on exposure to heat and flame; mod. explosion hazard; can react with oxidizers to form
 explosive peroxides
 Toxicology: TLV:TWA 5 ppm (skin); LD50 (oral, rat) 2460 mg/kg; mod. toxic to humans by ingestion; human
 systemic effects; experimental teratogen, reproductive effects; mutagenic data; skin and eye irritant;
 heated to dec., emits acrid smoke, irritating fumes
 Uses: Binder, color diluent, extender; used in confectionery, tabletted food supplements, gum, poultry
 Usage level: Limitation 0.15% (poultry)
 Regulatory: FDA 21CFR §73.1 (no residue), 175.105; USDA 9CFR §381.147
 Manuf./Distrib.: Ashland

2-Methoxyethanol. *See* Methoxyethanol
1-Methoxy-4-methylbenzene. *See* p-Methylanisole

2-Methoxy-4-methylphenol
 CAS 93-51-6; EINECS 202-252-9; FEMA 2671
 Synonyms: 4-Methylguaiacol; Creosol; 2-Methoxy-p-cresol; 2-Hydroxy-5-methylanisole
 Empirical: $C_8H_{10}O_2$
 Properties: Colorless to ylsh. liq., vanilla-like odor; sol. in alcohol, ether, benzene; m.w. 138.17; dens. 1.092
 (20/4 C); m.p. 5 C; b.p. 221-222 C; flash pt. 99 C; ref. index 1.537 (20 C)
 Toxicology: Irritating to eyes, skin, respiratory system
 Uses: Synthetic flavoring
 Usage level: 10-21 ppm (nonalcoholic beverages), 0.02 ppm (alcoholic beverages), 0.05 ppm (ice cream,
 ices), 0.77 ppm (candy), 1.0 ppm (baked goods)
 Regulatory: FDA 21CFR §172.515
 Manuf./Distrib.: Aldrich

2-Methoxy-3(5)-methylpyrazine
 CAS 2847-30-5; FEMA 3183
 Manuf./Distrib.: Aldrich

p-Methoxy-β-methylstyrene. *See* Anethole
2-Methoxyphenol. *See* Guaiacol

o-Methoxyphenol. *See* Guaiacol
o-Methoxyphenyl acetate. *See* Guaiacyl acetate

4-p-Methoxyphenyl-2-butanone
CAS 104-20-1; FEMA 2672
Synonyms: Anisyl acetone
Uses: Synthetic flavoring
Regulatory: FDA 21CFR §172.515
Manuf./Distrib.: Aldrich

1-(4-Methoxyphenyl)-4-methyl-1-penten-3-one
Synonyms: Methoxystyryl isopropyl ketone
Uses: Synthetic flavoring
Regulatory: FDA 21CFR §172.515

1-(p-Methoxyphenyl)-1-penten-3-one
Synonyms: α-Methylanisylidene acetone; Ethone
Uses: Synthetic flavoring
Regulatory: FDA 21CFR §172.515

o-Methoxyphenyl phenylacetate. *See* Guaiacyl phenylacetate

1-(p-Methoxyphenyl)-2-propanone
FEMA 2674
Synonyms: Anisylmethyl ketone; Anisic ketone
Uses: Synthetic flavoring
Regulatory: FDA 21CFR §172.515
Manuf./Distrib.: Aldrich

1,2-Methoxy-4-propenylbenzene. *See* Methyl isoeugenol
1-Methoxy-4-propenylbenzene. *See* Anethole
1-Methoxy-4-(2-propenyl) benzene. *See* Estragole
2-Methoxy-4-propenylphenol. *See* Isoeugenol
2-Methoxy-4-(2-propenyl) phenol. *See* Eugenol
2-Methoxy-4-propenylphenyl acetate. *See* Acetisoeugenol
1-Methoxy-4-n-propylbenzene. *See* p-Propyl anisole

2-Methoxypyrazine
CAS 3149-28-8; FEMA 3302
Manuf./Distrib.: Aldrich

Methoxystyryl isopropyl ketone. *See* 1-(4-Methoxyphenyl)-4-methyl-1-penten-3-one
p-Methoxytoluene. *See* p-Methylanisole

2-Methoxy-4-vinylphenol
Synonyms: p-Vinylguaiacol
Uses: Synthetic flavoring
Regulatory: FDA 21CFR §172.515

Methylacetaldehyde. *See* Propionaldehyde

Methyl acetate
CAS 79-20-9; EINECS 201-185-2; FEMA 2676
Synonyms: Acetic acid methyl ester
Definition: Ester of methyl alcohol and acetic acid
Empirical: $C_3H_6O_2$
Formula: CH_3COOCH_3
Properties: Colorless volatile liq., fragrant odor, sl. bitter flavor; misc. with common hydrocarbon solvs.; sol. in water; m.w. 74.09; dens. 0.92438; f.p. -98.05 C; b.p. 54.05 C; flash pt. (CC) -10 C; ref. index 1.3614 (20 C)
Precaution: Flamm.; dangerous fire and explosion risk; explosive limits 3-16% in air
Toxicology: Irritant to respiratory tract; narcotic in high concs.; TLV 200 ppm in air
Uses: Synthetic flavoring agent
Usage level: 28 ppm (nonalcoholic beverages), 0.2 ppm (alcoholic beverages), 29 ppm (ice cream), 11 ppm (candy), 14 ppm (baked goods), 0.1 ppm (gelatins, puddings)
Regulatory: FDA 21CFR §172.515, 175.105; GRAS (FEMA)
Manuf./Distrib.: Aldrich; Hoechst Celanese; Penta Mfg.

Methylacetic acid. *See* Propionic acid

Methyl acetone. *See* Methyl ethyl ketone

4´-Methyl acetophenone
CAS 122-00-9; EINECS 204-514-8; FEMA 2677
Synonyms: Methyl p-tolyl ketone; p-Methylacetophenone
Empirical: $C_9H_{10}O$
Properties: Colorless liq., fruity-floral odor, strawberry-like flavor; insol. in water; m.w. 134.18; dens. 1.004 (20/4 C); m.p. -23 C; b.p. 220-223 C; flash pt. 82 C; ref. index 1.534 (20 C)
Uses: Synthetic flavoring
Usage level: 1.1 ppm (nonalcoholic beverages), 1.6 ppm (ice cream, ices), 5.2 ppm (candy), 4.9 ppm (baked goods), 870 ppm (chewing gum), 5.8 ppm (condiments), 8.0 ppm (maraschino cherries)
Regulatory: FDA 21CFR §172.515; Japan approved as flavoring
Manuf./Distrib.: Aldrich

p-Methylacetophenone. *See* 4´-Methyl acetophenone
Methylacetopyronone. *See* Dehydroacetic acid
Methyl 12-acetoxyoleate. *See* Methyl acetyl ricinoleate

Methyl acetyl ricinoleate
CAS 140-03-4
Synonyms: Methyl 12-acetoxyoleate
Properties: Sol. in most organic solvents; insol. in water; pale-yellow liquid; mild odor; m.w. 354.59
Precaution: Combustible
Uses: Chewing gum base
Regulatory: Japan approved

Methyl acrylate polymer
CAS 96-33-3; EINECS 202-500-6
Empirical: $C_4H_6O_2$
Properties: Transparent elastic substance, pract. odorless; little adhesive power; resists the usual solvs.; m.w. 86.09; dens. 0.956; m.p. -75 C; b.p. 80 C; flash pt. 44 F
Precaution: Flamm. liq.
Regulatory: FDA 21CFR §181.30
Manuf./Distrib.: Hoechst Celanese

Methyl alcohol
CAS 67-56-1; EINECS 200-659-6
Synonyms: Methanol; Wood alcohol; Wood naphtha; Wood spirit; Carbinol; Columbian spirits; Methyl hydroxide; Methylol
Empirical: CH_4O
Formula: CH_3OH
Properties: Clear colorless liq., alcoholic odor (pure), pungent odor (crude); highly polar; misc. with water, alcohol, ether, benzene, ketones; m.w. 32.05; dens. 0.7924; m.p. -97.8 C; b.p. 64.5 C (760 mm); flash pt.(OC) 54 F; ref. index 1.3292 (20 C)
Precaution: Flamm.; dangerous fire risk; explosive limits 6.0-36.5% vol. in air; reacts vigorously with oxidizers; heated to decomp., emits acrid smoke and irritating fumes
Toxicology: LD50 (oral, rat) 5628 mg/kg; toxic (causes blindness); poisonous by ingestion, inhalation, or percutaneous absorption; experimental teratogen, reproductive effects; eye and skin irritant; narcotic; usual fatal dose 100-250 ml; TLV 200 ppm in air
Uses: Extraction solv.; used for hops extract, spice oleoresins
Usage level: Limitation 50 ppm (spice oleoresins), 2.2% (hops extract)
Regulatory: FDA 21CFR §172.560, 173.250, 173.385, 175.105, 176.180, 176.200, 176.210, 177.2420, 177.2460, 27CFR §21.115
Manuf./Distrib.: Albright & Wilson; Eastman; Hoechst Celanese

2-Methylallyl butyrate
FEMA 2678
Synonyms: 2-Methyl-2-propenyl butyrate; Methallyl butyrate; Isopropenyl carbinyl-n-butyrate
Definition: Ester of β-methylallyl alcohol and butyric acid
Empirical: $C_8H_{14}O_2$
Properties: Colorless liq., fruity ethereal odor; sol. in alcohol; insol. in water; m.w. 142.19; b.p. 168 C
Uses: Synthetic flavoring
Usage level: 0.20 ppm (nonalcoholic beverages, baked goods)
Regulatory: FDA 21CFR §172.515

Methyl 2-aminobenzoate. *See* Methyl anthranilate

Methyl-o-aminobenzoate. *See* Methyl anthranilate
2-Methylamino methyl benzoate. *See* Dimethyl anthranilate

Methyl n-amyl ketone
 CAS 110-43-0; EINECS 203-767-1; FEMA 2544
 Synonyms: MAK; 2-Heptanone; n-Amyl methyl ketone; Methyl pentyl ketone
 Empirical: $C_7H_{14}O$
 Formula: $CH_3CH_2CH_2CH_2CH_2COCH_3$
 Properties: Water-white liq., penetrating fruity odor; almost insol. in water; misc. with org. solvs.; m.w. 114.18; dens. 0.8166 (20/20 C); b.p. 150.6 C; flash pt. 49 C; ref. index 1.4110 (20 C)
 Precaution: Combustible; moderate fire risk; reactive with oxidizing materials
 Toxicology: LD50 (oral, rat) 1670 mg/kg; mod. toxic by ingestion; mildly toxic by inhalation, skin contact; skin irritant; narcotic in high concs.; TLV 50 ppm in air; heated to decomp., emits acrid smoke and fumes
 Uses: Synthetic flavoring agent
 Regulatory: FDA 21CFR §172.515
 Manuf./Distrib.: Aldrich; Ashland; Eastman

Methyl anisate
 Uses: Synthetic flavoring
 Regulatory: FDA 21CFR §172.515

4-Methylanisole. *See* p-Methylanisole

o-Methylanisole
 Synonyms: o-Cresyl methyl ether
 Uses: Synthetic flavoring
 Regulatory: FDA 21CFR §172.515

p-Methylanisole
 CAS 104-93-8; EINECS 203-253-7; FEMA 2681
 Synonyms: p-Cresyl methyl ether; 4-Methylanisole; Methyl p-cresol; 1-Methoxy-4-methylbenzene; p-Methoxytoluene; p-Tolyl methyl ether
 Empirical: $C_8H_{10}O$
 Properties: Colorless liq., pungent ylang-ylang odor; sol. in fixed oils; insol. in glycerin, propylene glycol; m.w. 122.18; dens. 0.996-0.970; flash pt. 144 F; ref. index 1.510-1.513
 Precaution: Flamm.
 Toxicology: LD50 (oral, rat) 1920 mg/kg; mod. toxic by ingestion; skin irritant; heated to decomp., emits acrid smoke and irritating fumes
 Uses: Pungent, sweet, ylang-ylang, cresol-like synthetic fragrance and flavoring
 Usage level: 2.7 ppm (nonalcoholic beverages, ice cream, ices), 4.8 ppm (candy), 7.6 ppm (baked goods), 0.50-4.0 ppm (gelatins, puddings), 2.0 ppm (condiments), 8.0 ppm (syrups)
 Regulatory: FDA 21CFR §172.515
 Manuf./Distrib.: BASF

α-**Methylanisylidene acetone.** *See* 1-(p-Methoxyphenyl)-1-penten-3-one

Methyl anthranilate
 CAS 134-20-3; EINECS 205-132-4; FEMA 2682
 Synonyms: Methyl-o-aminobenzoate; Methyl 2-aminobenzoate; Neroli oil, artificial; 2-Aminobenzoic acid methyl ester
 Definition: Ester of methyl alcohol and 2-aminobenzoic acid
 Empirical: $C_8H_9NO_2$
 Formula: $H_2NC_6H_4CO_2CH_3$
 Properties: Cryst. or pale yel. liq., bluish fluorescence, grape-like odor; sol. in fixed oils, propylene glycol; sl. sol. in water; insol. in glycerol; m.w. 151.18; dens. 1.167-1.175 (15 C); m.p. 23.8 C; b.p. 258-261 C; flash pt. 123 C; ref. index 1.583 (20 C)
 Toxicology: LD50 (oral, rat) 2910 mg/kg; poison by intravenous route; mod. toxic by ingestion; skin irritant; experimental tumorigen; heated to decomp., emits toxic fumes of NO_x
 Uses: Synthetic flavoring agent
 Usage level: 16 ppm (nonalcoholic beverages), 0.2 ppm (alcoholic beverages), 21 ppm (ice cream), 56 ppm (candy), 20 ppm (baked goods), 23 ppm (gelatins, puddings), 2200 ppm (chewing gum)
 Regulatory: FDA 21CFR §182.60, GRAS; Japan approved as flavoring
 Manuf./Distrib.: Aldrich; Bell Flavors & Fragrances; Haarmann & Reimer; PMC Specialties

Methyl arachidate. *See* Methyl eicosenate
1-Methyl N-L-α-aspartyl-L-phenylalanine. *See* Aspartame
4-Methylbenzaldehyde. *See* p-Tolyl aldehyde

p-Methyl benzaldehyde. *See* p-Tolyl aldehyde
Methyl benzaldehydes, mixed o-, m-, p-. *See* Tolualdehydes, mixed o-, m-, p-
Methyl benzenecarboxylate. *See* Methyl benzoate

Methyl benzoate
    CAS 93-58-3; EINECS 202-259-7; FEMA 2683
    Synonyms: Benzoic acid methyl ester; Niobe oil; Oil niobe; Methyl benzenecarboxylate
    Classification: Ester
    Empirical: $C_8H_8O_2$
    Formula: $C_6H_5COOCH_3$
    Properties: Colorless liq., fragrant odor; sol. in alcohol, fixed oils, propylene glycol, water @ 30 C; misc. with alcohol, ether; insol. in glycerin; m.w. 136.15; dens. 1.082-1.088; m.p. -12.5 C; b.p. 199.6 C; flash pt. 181 F; ref. index 1.515
    Precaution: Combustible; reactive with oxidizing materials
    Toxicology: LD50 (oral, rat) 1350 mg/kg; mod. toxic by ingestion; mildly toxic by skin contact; skin and eye irritant; heated to decomp., emits acrid smoke and irritating fumes
    Uses: Synthetic flavoring agent
    Usage level: 2.2 ppm (nonalcoholic beverages), 4.5 ppm (ice cream), 8.4 ppm (candy), 9.9 ppm (baked goods), 61 ppm (chewing gum)
    Regulatory: FDA 21CFR §172.515
    Manuf./Distrib.: Hüls AG

α-Methylbenzyl acetate
    CAS 93-92-5; FEMA 2684
    Synonyms: Styralyl acetate; 1-Phenylethyl acetate; α-Phenylethyl acetate; Methyl phenylcarbinyl acetate
    Empirical: $C_{10}H_{12}O_2$
    Properties: Colorless liq., gardenia odor; sol. in fixed oils, glycerin; insol. in water; m.w. 164.20; dens. 1.023; b.p. 94-95 C (12 mm); flash pt. 196 F; ref. index 1.493-1.497
    Precaution: Combustible liq.
    Toxicology: Heated to decomp., emits acrid smoke and irritating fumes
    Uses: Synthetic flavoring
    Regulatory: FDA 21CFR §172.515
    Manuf./Distrib.: Aldrich; Hüls AG

Methylbenzyl acetate, mixed o-, m-, p-
    CAS 93-92-5; FEMA 2684
    Synonyms: Tolyl acetate
    Empirical: $C_{10}H_{12}O_2$
    Properties: Colorless liq.; sol. in benzyl benzoate, min. oil; sl. sol. in propylene glycol; insol. in glycerin; m.w. 164.20; ref. index 1.5015-1.5040
    Uses: Synthetic flavoring
    Regulatory: FDA 21CFR §172.515
    Manuf./Distrib.: Aldrich

p-Methylbenzylacetone. *See* 4-(p-Tolyl)-2-butanone

α-Methylbenzyl alcohol
    CAS 98-85-1; FEMA 2685
    Synonyms: Methyl phenylcarbinol; α-Phenethyl alcohol; Styralyl alcohol
    Uses: Synthetic flavoring
    Regulatory: FDA 21CFR §172.515
    Manuf./Distrib.: Aldrich

α-Methylbenzyl butyrate
    CAS 3460-44-4; FEMA 2686
    Synonyms: Styralyl butyrate
    Uses: Synthetic flavoring
    Regulatory: FDA 21CFR §172.515
    Manuf./Distrib.: Aldrich

α-Methylbenzyl formate
    Synonyms: Styralyl formate
    Uses: Synthetic flavoring
    Regulatory: FDA 21CFR §172.515

Methylbenzyl isobutyrate
    CAS 7775-39-5

α-**Methylbenzyl propionate**

    Synonyms: Styralyl isobutyrate; 1-Phenylethyl 2-methylpropionate
    Uses: Synthetic flavoring
    Regulatory: FDA 21CFR §172.515
    Manuf./Distrib.: Hüls AG

α-**Methylbenzyl propionate**
    CAS 120-45-6; FEMA 2689
    Synonyms: Styralyl propionate; 1-Phenylethyl propionate
    Uses: Synthetic flavoring
    Regulatory: FDA 21CFR §172.515
    Manuf./Distrib.: Hüls AG

3-Methyl-1,3-butadiene polymer with 2-methyl-1-propene. *See* Isobutylene/isoprene copolymer
2-Methylbutanal-1. *See* 2-Methylbutyraldehyde
3-Methylbutanal. *See* Isovaleraldehyde
3-Methylbutanoic acid. *See* Isovaleric acid
3-Methylbutanoic acid ethyl ester. *See* Ethyl isovalerate
2-Methylbutanoic acid n-hexyl ester. *See* Hexyl 2-methylbutyrate
3-Methylbutanoic acid methyl ester. *See* Methyl isovalerate
(1R-endo)-3-Methylbutanoic acid 1,7,7-trimethylbicyclo[2.2.1]hept-2-yl ester. *See* Bornyl isovalerate
2-Methyl-2-butanol. *See* t-Amyl alcohol
3-Methylbutanol. *See* Isoamyl alcohol
3-Methyl-1-butanol. *See* Isoamyl alcohol
3-Methyl-1-butanol acetate. *See* Isoamyl acetate

2-Methyl-3-buten-2-ol
    CAS 115-18-4; EINECS 204-068-4
    Synonyms: 1,1-Dimethylallyl alcohol; 3-Hydroxy-3-methyl-1-butene
    Empirical: $C_5H_{10}O$
    Formula: $CH_2:CHC(CH_3)_2OH$
    Properties: Herbaceous earthy oily odor; m.w. 86.14; dens. 0.824 (20/4 C); b.p. 96-99 C; flash pt. 10 C; ref.
        index 1.418 (20 C)
    Precaution: Highly flamm.
    Uses: Synthetic flavoring
    Regulatory: FDA 21CFR §172.515
    Manuf./Distrib.: Aldrich

3-Methyl-2-buten-1-ol
    CAS 556-82-1; EINECS 209-141-4; FEMA 3647
    Synonyms: Prenol; 3,3-Dimethylallyl alcohol
    Empirical: $C_5H_{10}O$
    Formula: $(CH_3)_2C:CHCH_2OH$
    Properties: m.w. 86.14; dens. 0.861 (20/4 C); b.p. 143-144 C; flash pt. 110 F; ref. index 1.443
    Precaution: Flamm.
    Toxicology: Harmful if swallowed; irritating to skin
    Uses: Fresh, herbal, green, fruity, sl. lavender-like flavor and fragrance
    Manuf./Distrib.: Aldrich

β-**Methylbutyl acetate.** *See* Isoamyl acetate

2-Methylbutyl isovalerate
    CAS 2445-77-4; FEMA 3506
    Synonyms: 2-Methylbutyl-3-methylbutanoate
    Uses: Synthetic flavoring
    Regulatory: FDA 21CFR §172.515
    Manuf./Distrib.: Aldrich

2-Methylbutyl-3-methylbutanoate. *See* 2-Methylbutyl isovalerate

Methyl p-t-butylphenylacetate
    Uses: Synthetic flavoring
    Regulatory: FDA 21CFR §172.515

2-Methylbutyraldehyde
    CAS 96-17-3; EINECS 202-485-6; FEMA 2691
    Synonyms: Methyl ethyl acetaldehyde; 2-Methylbutanal-1
    Empirical: $C_5H_{10}O$
    Formula: $C_2H_5CH(CH_3)CHO$

Properties: Colorless liq.; sol. in alcohol; sl. sol. in water; m.w. 86.13; dens. 0.82 (20/4 C); b.p. 90-92 C; flash pt. 40 F
Precaution: Highly flamm.
Uses: Synthetic flavoring
Usage level: 1.5-2.0 ppm (nonalcoholic beverages), 2.0-8.0 ppm (ice cream, ices), 6.6 ppm (candy), 5.7 ppm (baked goods)
Regulatory: FDA 21CFR §172.515
Manuf./Distrib.: Aldrich

3-Methylbutyraldehyde. *See* Isovaleraldehyde

Methyl butyrate
CAS 623-42-7; EINECS 210-792-1; FEMA 2693
Synonyms: Butanoic acid methyl ester
Empirical: $C_5H_{10}O_2$
Formula: $CH_3CH_2CH_2COOCH_3$
Properties: Colorless liq.; sol. in about 60 parts water; misc. with alcohol, ether; m.w. 102.14; dens. 0.898 (20/4 C); m.p. -95 C; b.p. 99-102 C; flash pt. 14 C; ref. index 1.388 (20 C)
Precaution: Highly flamm.
Uses: Synthetic flavoring; rum and fruit essences
Usage level: 17 ppm (nonalcoholic beverages), 31 ppm (ice cream, ices), 86 ppm (candy), 48-200 ppm (baked goods)
Regulatory: FDA 21CFR §172.515
Manuf./Distrib.: Aldrich

2-Methylbutyric acid
CAS 600-07-7; EINECS 209-982-7; FEMA 2695
Synonyms: Carboxylic acid C_5
Empirical: $C_5H_{10}O_2$
Formula: $CH_3CH_2CH(CH_3)COOH$
Properties: m.w. 102.14; dens. 0.934 (20/4 C); b.p. 173-176 C; flash pt. 83 C; ref. index 1.405
Toxicology: Causes burns
Uses: Synthetic flavoring for Roquefort cheese; fruity on dilution
Regulatory: FDA 21CFR §172.515
Manuf./Distrib.: Aldrich

3-Methylbutyric acid. *See* Isovaleric acid
γ-Methyl-γ-butyrolactone. *See* γ-Valerolactone

Methyl caproate
CAS 106-70-7; EINECS 203-425-1; FEMA 2708
Synonyms: Methyl hexanoate
Definition: Ester of methyl alcohol and caproic acid
Empirical: $C_7H_{14}O_2$
Formula: $CH_3(CH_2)_4COOCH_3$
Properties: m.w. 130.19; dens. 0.884 (20/4 C); b.p. 150-151 C; flash pt. 43 C; ref. index 1.405
Precaution: Flamm.
Uses: Synthetic flavoring agent
Regulatory: FDA 21CFR §172.515
Manuf./Distrib.: Aldrich

Methyl caprylate
CAS 111-11-5; EINECS 203-835-0; FEMA 2728
Synonyms: Methyl octanoate
Definition: Ester of methyl alcohol and caprylic acid
Empirical: $C_9H_{18}O_2$
Formula: $CH_3(CH_2)_6COOCH_3$
Properties: m.w. 158.24; dens. 0.875 (20/4 C); b.p. 193-194 C; flash pt. 69 C; ref. index 1.417 (20 C)
Uses: Synthetic flavoring agent
Regulatory: FDA 21CFR §172.225, 172.515, 176.200, 176.210, 177.2800
Manuf./Distrib.: Aldrich

Methyl caprylate/caprate
CAS 67762-39-4
Definition: Mixture of esters of methyl alcohol and caprylic and capric acids
Uses: Coatings

Methylcatechol

Regulatory: FDA 21CFR §172.225, 176.200, 176.210, 177.2800

Methylcatechol. *See* Guaiacol

Methyl Cellosolve®. *See* Methoxyethanol

Methylcellulose
CAS 9004-67-5; EEC E461
Synonyms: MC; Cellulose methyl ether; Cologel; Methocel
Definition: Methyl ether of cellulose
Properties: Grayish-white fibrous powd., odorless, tasteless; aq. suspension swells in water to visc. colloidal sol'n.; sol. in cold water, glacial acetic acid, some org. solvs.; insol. in alcohol, ether, chloroform, warm water; m.w. 86,000-115,000
Precaution: Combustible
Toxicology: Heated to decomp., emits acrid smoke and irritating fumes
Uses: Thickener, stabilizer, emulsifier, bodying agent, bulking agent, binder, film-former; used in baked goods, fruit pie fillings; extends and stabilizes meat and vegetable patties
Usage level: ADI 0-25 mg/kg (FAO/WHO)
Regulatory: FDA 21CFR §150.141, 150.161, 175.105, 175.210, 175.300, 176.200, 182.1480, GRAS; USDA 9CFR §318.7, limitation 0.15% in meat and vegetable prods.; Japan restricted (2% max.); Europe listed; UK approved
Manuf./Distrib.: Am. Roland; Aqualon; Courtaulds; Croxton & Garry; Hercules; N I Ibrahim; Nordmann Rassmann; Penta Mfg.; Roeper; Van Waters & Rogers
Trade names: Benecel® M 042; Benecel® M 043; Methocel® A4C Premium; Methocel® A4M Premium; Methocel® A15C Premium; Methocel® A15LV Premium
Trade names containing: Mayonat PS

α-Methylcinnamaldehyde
CAS 101-39-3; FEMA 2697
Uses: Synthetic flavoring
Regulatory: FDA 21CFR §172.515
Manuf./Distrib.: Aldrich

p-Methylcinnamaldehyde
Uses: Synthetic flavoring
Regulatory: FDA 21CFR §172.515

Methyl cinnamate
CAS 103-26-4; EINECS 203-093-8; FEMA 2698
Synonyms: Methyl-3-phenyl propenoate; Methyl cinnamylate
Empirical: $C_{10}H_{10}O_2$
Formula: $C_6H_5CH{:}CHCOOCH_3$
Properties: Wh. to sl. yel. cryst., fruity odor; very sol. in alcohol, ether; sol. in fixed oils, glycerin, propylene glycol; insol. in water; m.w. 162.20; dens. 1.042 (36/0 C); m.p. 33.4 C; b.p. 263 C; flash pt. > 212 F
Precaution: Combustible
Toxicology: LD50 (oral, rat) 2610 mg/kg; mod. toxic by ingestion; heated to decomp., emits acrid smoke and irritating fumes
Uses: Balsamic, fruity fragrance, flavoring, strawberry-like on dilution; improves flavor of foodstuffs
Regulatory: FDA 21CFR §172.515; Japan approved as flavoring
Manuf./Distrib.: Aldrich

Methyl cinnamylate. *See* Methyl cinnamate

Methyl cocoate
CAS 61788-59-8; EINECS 262-988-1
Definition: Ester of methyl alcohol and coconut fatty acids
Formula: RCO—OCH_3, RCO⁻ represents the fatty acids derived from coconut oil
Uses: Coatings
Regulatory: FDA 21CFR §172.225, 175.105, 176.200, 176.210, 177.2260, 177.2800, 178.3910

6-Methylcoumarin
FEMA 2699
Manuf./Distrib.: Aldrich

Methyl p-cresol. *See* p-Methylanisole

Methyl-γ-cyclodextrin
Definition: Complex hosting guest molecule
Uses: Increases solubility and bioavailability of other substances; masks odors, flavors, and coloration

Trade names: Alpha W6 M1.8; Beta W7 M1.8; Gamma W8 M1.8

2-Methyl-1,3-cyclohexadiene
Uses: Synthetic flavoring
Regulatory: FDA 21CFR §172.515

3-Methylcyclopentane-1,2-dione. *See* Methyl cyclopentenolone
3-Methyl-1,2-cyclopentanedione. *See* Methyl cyclopentenolone

Methyl cyclopentenolone
CAS 765-70-8; EINECS 212-154-8; FEMA 2700
Synonyms: 3-Methylcyclopentane-1,2-dione; 3-Methyl-1,2-cyclopentanedione; 2-Hydroxy-3-methyl-2-cyclopenten-1-one
Empirical: $C_6H_8O_2$
Properties: m.w. 112.13; m.p. 105-107 C
Uses: Synthetic flavoring
Regulatory: FDA 21CFR §172.515
Manuf./Distrib.: Chart

Methyl decyne carbonate. *See* Methyl 2-undecynoate
Methyl dihydroabietate. *See* Methyl ester of rosin, partially hydrogenated
Methyl dimethylacetate. *See* Methyl isobutyrate
Methyl 2,4-dimethylphenyl ketone. *See* 2,4-Dimethylacetophenone

Methyl disulfide
CAS 624-92-0; EINECS 210-871-0; FEMA 3536
Synonyms: Dimethyl disulfide
Empirical: $C_2H_6S_2$
Formula: CH_3SSCH_3
Properties: Pale yel. liq., onion odor; sol. in alcohol; sl. sol. in water; m.w. 94.20; dens. 1.062 (20/4 C); b.p. 108-110 C; flash pt. 15 C; ref. index 1.527 (20 C)
Precaution: Highly flamm.
Uses: Synthetic flavoring
Regulatory: FDA 21CFR §172.515
Manuf./Distrib.: Aldrich

Methyl dodecanoate. *See* Methyl laurate

Methyl eicosenate
CAS 1120-28-1; EINECS 214-304-8
Synonyms: Methyl arachidate; Eicosanoic acid methyl ester
Classification: Ester
Empirical: $C_{21}H_{42}O_2$
Formula: $CH_3(CH_2)_{18}COOCH_3$
Properties: m.w. 326.57
Uses: Foam depressant and nutrient in fermentation
Trade names: Kemester® 2050

Methyl enanthate. *See* Methyl heptanoate

Methylene chloride
CAS 75-09-2; EINECS 200-838-9
Synonyms: DCM; Dichloromethane; Methylene dichloride; Methane dichloride
Classification: Halogenated organic compd.
Empirical: CH_2Cl_2
Properties: Colorless volatile liq., penetrating ether-like odor; sol. in alcohol, ether; sl. sol. in water; m.w. 84.93; dens. 1.335 (15/4 C); f.p. -97 C; b.p. 40.1 C; ref. index 1.4244 (20 C)
Precaution: Nonflamm.; explosive as vapor exposed to heat or flame; contact with hot surfaces cause decomp., yielding toxic fumes; heated to decomp., emits highly toxic fumes of phosgene and Cl⁻
Toxicology: LD50 (oral, rat) 2136 mg/kg; poison by IV route; mod. toxic by ingestion, subcutaneous, IP routes; experimental carcinogen; human systemic effects; eye and severe skin irritant; human mutagenic data; narcotic in high concs.; ACGIH TLV:TWA 50 ppm
Uses: Extraction solvent, color diluent; used in decaffeinated coffee, fruits, hops extract, spice oleoresins, vegetables
Usage level: Limitation 30 ppm (spice oleoresins), 2.2% (hops extract), 10 ppm (decaf coffee)
Regulatory: FDA 21CFR §73.1 (no residue), 173.255
Manuf./Distrib.: Ashland; ICI Specialties

Methylene dichloride. *See* Methylene chloride
3,4-Methylene dihydroxybenzaldehyde. *See* Heliotropine
3,4-Methylenedioxybenzyl acetate. *See* Piperonyl acetate
3,4-Methylenedioxybenzyl isobutyrate. *See* Piperonyl isobutyrate
3,4-Methylenedioxy-propylbenzene. *See* Dihydrosafrole
Methylenesuccinic acid. *See* Itaconic acid (polymerized)
trans-(1R,9S)-8-Methylene-4,11,11-trimethylbicyclo[7.2.0]undec-4-ene. *See* β-Caryophyllene

Methyl ester of rosin, partially hydrogenated
  Synonyms: Methyl dihydroabietate
  Uses: Synthetic flavoring
  Regulatory: FDA 21CFR §172.515

Methyl ethyl acetaldehyde. *See* 2-Methylbutyraldehyde
1-Methylethyl acetate. *See* Isopropyl acetate
4-(1-Methylethyl)benzaldehyde. *See* Cuminaldehyde
Methyl ethyl carbinol. *See* 2-Butanol

Methyl ethyl cellulose
  EEC E465
  Synonyms: Ethylmethylcellulose
  Formula: $[C_6H_{(10-x-y)}O_5(CH_3)_x(C_2H_5)_y]_n$, x = no. of methyl groups, y = no. of ethyl groups
  Properties: Wh. fibrous solid or powd.; disp. in water; visc. 2-60 cps (2.5 g/100 ml aq.)
  Toxicology: Heated to decomp., emits acrid smoke and irritating fumes
  Uses: Food additive, emulsifier, stabilizer, thickener, suspending agent, gelling agent, aerating agent, foaming
    agent, foam stabilizer; used in meringues, whipped toppings
  Usage level: ADI 0-25 mg/kg (EEC)
  Regulatory: FDA 21CFR §172.872; Europe listed; UK approved
  Manuf./Distrib.: Courtaulds Fine Chems.; Dutch Protein; N I Ibrahim

Methyl ethyl ketone
  CAS 78-93-3; EINECS 201-159-0; FEMA 2170
  Synonyms: MEK; Ethyl methyl ketone; 2-Butanone; 2-Oxobutane; Methyl acetone
  Classification: Aliphatic ketone
  Empirical: C_4H_8O
  Formula: $CH_3COCH_2CH_3$
  Properties: Colorless liq., acetone-like odor; sol. in 4 parts water, benzene, alcohol, ether; misc. with oils; m.w.
    72.10; dens. 0.8255 (0/4 C); m.p. -86 C; b.p. 79.6 C; flash pt. (TOC) 24 F; visc. 0.40 cp (25 C); ref. index
    1.3814 (15 C)
  Precaution: Flammable, dangerous fire risk; explosive limits in air 2-10%
  Toxicology: LD50 (oral, rat) 2737 mg/kg; mod. toxic by ingestion, skin contact, IP routes; toxic by inhalation;
    experimental teratogen, reproductive effects; strong irritant; affects CNS; TLV 200 ppm in air; heated to
    decomp., emits acrid smoke and fumes
  Uses: Synthetic flavoring agent
  Regulatory: FDA 21CFR §172.515, 175.105, 175.320, 177.1200
  Manuf./Distrib.: Aldrich; Hoechst Celanese

Methyl eugenol
  CAS 93-15-2; EINECS 202-223-0; FEMA 2475
  Synonyms: 4-Allyl-1,2-dimethoxybenzene; 1,2-Dimethoxy-4-(2-propenyl)benzene; 4-Allyl veratrole; Eugenyl
    methyl ether
  Classification: Aromatic compd.
  Empirical: $C_{11}H_{14}O_2$
  Formula: $(CH_3O)_2C_6H_3CH_2CH:CH_2$
  Properties: Colorless to pale yel. liq., clove, carnation odor; sol. in fixed oils; insol. in glycerin, propylene glycol;
    m.w. 178.25; dens. 1.032-1.036; b.p. 128-130 C (10 mm); flash pt. 110 C; ref. index 1.534 (20 C)
  Precaution: Combustible liq.
  Toxicology: LD50 (oral, rat) 1179 mg/kg; poison by IV route; mod. toxic by ingestion, IP routes; skin irritant;
    mutagenic data; heated to decomp., emits acrid smoke and irritating fumes
  Uses: Synthetic flavoring agent
  Regulatory: FDA 21CFR §172.515
  Manuf./Distrib.: Aldrich; Penta Mfg.

Methyl fumarate
  Uses: Inhibits fungal growth in tomato juice
  Usage level: 0.2%

Methyl glucoside-coconut oil ester
Uses: Surfactant in molasses prod.
Regulatory: FDA 21CFR §172.816

Methyl glycol. See Propylene glycol
Methylglyoxal. See Pyruvaldehyde
4-Methylguaiacol. See 2-Methoxy-4-methylphenol

6-Methyl-3,5-heptadien-2-one
Uses: Synthetic flavoring
Regulatory: FDA 21CFR §172.515

Methyl heptanoate
CAS 106-73-0; EINECS 203-428-8; FEMA 2705
Synonyms: Methyl enanthate; Heptanoic acid methyl ester
Empirical: $C_8H_{16}O_2$
Formula: $CH_3(CH_2)_5COOCH_3$
Properties: Liq.; m.w. 144.22; dens. 0.881 (20/4 C); m.p. -55.8 C; b.p. 173.8 C; flash pt. 55 C; ref. index 1.412 (20 C)
Precaution: Flamm.
Uses: Synthetic flavoring
Regulatory: FDA 21CFR §172.515
Manuf./Distrib.: Aldrich

2-Methylheptanoic acid
CAS 1188-02-9; FEMA 2706
Uses: Synthetic flavoring
Regulatory: FDA 21CFR §172.515
Manuf./Distrib.: Aldrich

6-Methyl-5-heptene-2-one. See Methyl heptenone

Methyl-5-hepten-2-ol
CAS 110-93-0; FEMA 2707
Uses: Synthetic flavoring
Regulatory: FDA 21CFR §172.515
Manuf./Distrib.: Aldrich

Methyl heptenone
CAS 110-93-0, 409-02-9; EINECS 203-816-7; FEMA 2707
Synonyms: 6-Methyl-5-heptene-2-one
Empirical: $C_8H_{14}O$
Formula: $(CH_3)_2C:CHCH_2CH_2COCH_3$
Properties: Sl. yel. liq., citrus-lemongrass odor; misc. with alcohol, ether chloroform; insol. in water; m.w. 126.22; dens. 0.846-0.851; m.p. -67 C; b.p. 173-174 C; flash pt. 55 C; ref. index 1.438-1.442
Precaution: Combustible
Toxicology: LD50 (oral, rat) 3500 mg/kg; mod. toxic by ingestion; skin irritant; heated to decomp., emits acrid smoke and irritating fumes
Uses: Fresh, green, sl. pungent, fruity fragrance and flavoring
Regulatory: FDA 21CFR §172.515

Methyl heptine carbonate. See Methyl 2-octynoate
Methyl heptyl ketone. See 2-Nonanone
Methyl heptyne carbonate. See Methyl 2-octynoate
Methyl hexanoate. See Methyl caproate

Methyl 2-hexanoate
CAS 106-70-7; FEMA 2708
Uses: Synthetic flavoring
Regulatory: FDA 21CFR §172.515
Manuf./Distrib.: Aldrich

Methyl hexyl acetaldehyde. See 2-Methyloctanal
Methyl hexyl carbinol. See 2-Octanol

Methyl hexyl ketone
CAS 111-13-7; EINECS 203-837-1
Synonyms: 2-Octanone; Hexyl methyl ketone
Empirical: $C_8H_{16}O$

Formula: $CH_3(CH_2)_5COCH_3$
Properties: Liq., apple odor, camphor taste; misc. with alcohol, ether; insol. in water; m.w. 128.22; dens. 0.818 (20/4 C); m.p. -16 C; b.p. 170-172 C; flash pt. 56 C; ref. index 1.416 (20 C)
Precaution: Flamm.
Uses: Synthetic flavoring
Regulatory: FDA 21CFR §172.515

p-Methylhydratropic aldehyde. *See* 2-(p-Tolyl) propionaldehyde
Methyl hydrocinnamate. *See* Methyl 3-phenylpropionate
Methyl hydroxide. *See* Methyl alcohol
Methyl 2-hydroxybenzoate. *See* Methyl salicylate
Methyl 4-hydroxybenzoate. *See* Methylparaben
Methyl p-hydroxybenzoate. *See* Methylparaben
Methyl hydroxypropyl cellulose. *See* Hydroxypropyl methylcellulose
3-Methylindole. *See* Skatole
β-Methylindole. *See* Skatole
6-Methylionone. *See* α-Irone
6-Methyl-α-ionone. *See* α-Irone

Methyl α-ionone
Synonyms: 5-(2,6,6-Trimethyl-2-cyclohexen-1-yl)-4-penten-3-one
Uses: Synthetic flavoring
Regulatory: FDA 21CFR §172.515

Methyl β-ionone
Synonyms: 5-(2,6,6-Trimethyl-1-cyclohexen-1-yl)-4-penten-3-one
Uses: Synthetic flavoring
Regulatory: FDA 21CFR §172.515

Methyl δ-ionone
Synonyms: 5-(2,6,6-Trimethyl-3-cyclohexen-1-yl)-4-penten-3-one
Uses: Synthetic flavoring
Regulatory: FDA 21CFR §172.515

Methyl γ-ionone. *See* α-Isomethylionone

Methyl isobutyl ketone
CAS 108-10-1; EINECS 203-550-1; FEMA 2731
Synonyms: MIBK; 4-Methyl-2-pentanone; Hexone; Isopropylacetone; Isobutyl methyl ketone
Classification: Aliphatic ketone
Empirical: $C_6H_{12}O$
Formula: $CH_3COCH_2CH(CH_3)_2$
Properties: Colorless stable liq., pleasant odor; sl. sol. in water; misc. with most org. solvs.; m.w. 100.18; dens. 0.8042 (20/20 C); f.p. -85 C; b.p. 115.8 C; ref. index 1.396
Precaution: Flamm.; dangerous fire risk; may form explosive peroxides on exposure to air; can react vigorously with reducing materials; explosive limits 1.4-7.5% in air
Toxicology: LD50 (oral, rat) 2080 mg/kg; poison by IP route; mod. toxic by ingestion; mildly toxic by inh.; very irritating to skin, eyes, mucous membranes; narcotic in high conc.; TLV 50 ppm in air
Uses: Denaturant for alcohol, flavoring agent
Regulatory: FDA 21CFR §172.515
Manuf./Distrib.: Aldrich; Eastman

Methyl isobutyrate
CAS 547-63-7; EINECS 208-929-5; FEMA 2694
Synonyms: 2-Methylpropanoic acid methyl ester; Methyl dimethylacetate
Definition: Ester of methanol and isobutyric acid
Empirical: $C_5H_{10}O_2$
Formula: $(CH_3)_2CHCOOCH_3$
Properties: Colorless mobile liq.; sl. sol. in water; misc. with alcohol, ether; m.w. 102.14; dens. 0.891 (20/4 C); m.p. -84 to -85 C; b.p. 91-93 C; flash pt. 12 C; ref. index 1.384 (20 C)
Precaution: Highly flamm.
Uses: Synthetic flavoring
Usage level: 22 ppm (nonalcoholic beverages), 38 ppm (ice cream, ices), 48-200 ppm (candy, baked goods)
Regulatory: FDA 21CFR §172.515
Manuf./Distrib.: Aldrich

Methyl isoeugenol
CAS 93-16-3; FEMA 2476

Synonyms: 4-Allyl-1,2-dimethoxy benzene; Propenyl guaiacol; Isoeugenyl methyl ether; 1,2-Methoxy-4-propenylbenzene; 4-Propenylveratrole
Empirical: $C_{11}H_{14}O_2$
Properties: Colorless to pale yel. liq., clove-carnation odor, burning bitter taste; m.w. 178.23; dens. 1.050; b.p. 262-264 C; flash pt. > 230 F; ref. index 1.5650-15.690 (20 C)
Uses: Synthetic flavoring
Usage level: 4.0 ppm (nonalcoholic beverages), 7.7 ppm (ice cream, ices), 13 ppm (candy), 18 ppm (baked goods), 0.10 ppm (gelatins, puddings), 110 ppm (chewing gum)
Regulatory: FDA 21CFR §172.515
Manuf./Distrib.: Aldrich

1-Methyl-4-isopropenyl cyclohexan-3-ol. *See* Isopulegol
6-Methyl-3-isopropenylcyclohexanol. *See* Dihydrocarveol
1-Methyl-4-isopropenyl cyclohexan-3-one. *See* Isopulegone
1-Methyl-4-isopropenyl-6-cyclohexen-2-ol. *See* Carveol
1-1-Methyl-4-isopropenyl-6-cyclohexen-2-one. *See* l-Carvone
d-1-Methyl-4-isopropenyl-6-cyclohexen-2-one. *See* d-Carvone
6-Methyl-3-isopropenyl cyclohexyl acetate. *See* Dihydrocarvyl acetate
1-Methyl-4-isopropyl-1-cyclohexen-8-ol. *See* α-Terpineol
α-Methyl-p-isopropylhydrocinnamaldehyde. *See* Cyclamen aldehyde
1-Methyl-4-isopropylidene-3-cyclohexanone. *See* Pulegone
2-Methyl-3-(p-isopropylphenyl)-propionaldehyde. *See* Cyclamen aldehyde

Methyl isovalerate
CAS 556-24-1; EINECS 209-117-3
Synonyms: 3-Methylbutanoic acid methyl ester
Empirical: $C_6H_{12}O_2$
Formula: $(CH_3)_2CHCH_2COOCH_3$
Properties: Liq., valerian odor; sl. sol. in water; misc. with alcohol, ether; m.w. 116.16; dens. 0.880 (20/4 C); b.p. 115-117 C; flash pt. 16 C; ref. index 1.393 (20 C)
Precaution: Highly flamm.
Uses: Synthetic flavoring
Regulatory: FDA 21CFR §172.515

Methyl laurate
CAS 111-82-0; 67762-40-7; EINECS 203-911-3; FEMA 2715
Synonyms: Methyl dodecanoate; Dodecanoic acid methyl ester
Definition: Ester of methyl alcohol and lauric acid
Empirical: $C_{13}H_{26}O_2$
Formula: $CH_3(CH_2)_{10}COOCH_3$
Properties: Water-white liq., fatty floral odor; insol. in water; m.w. 214.35; dens. 0.8702 (20/4 C); m.p. 4.8 C; b.p. 262 C (766 mm); flash pt. > 230 F; ref. index 1.4320
Precaution: Combustible; noncorrosive
Uses: Synthetic flavoring agent
Usage level: 0.5-5 ppm (nonalcoholic beverages), 0.5-5 ppm (ice cream), 0.02-0.5 ppm (candy), 1 ppm (baked goods)
Regulatory: FDA 21CFR §172.225, 172.515, 176.200, 176.210, 177.2260, 177.2800
Manuf./Distrib.: Aldrich; Procter & Gamble

Methyl linoleate
CAS 112-63-0; EINECS 203-993-0
Synonyms: 9,12-Octadecadienoic acid methyl ester; Methyl cis,cis-9,12-octadecadienoate
Definition: Ester of methyl alcohol and linoleic acid
Empirical: $C_{19}H_{34}O_2$
Formula: $CH_3(CH_2)_4CH:CHCH_2CH:CH(CH_2)_7COOCH_3$
Properties: Colorless oil; misc. with dimethylformamide, fat solvs., oils; m.w. 294.48; dens. 0.887 (20/4 C); m.p. -35 C; b.p. 207-208 C (11 mm); iodine no. 172.4; ref. index 1.466
Precaution: Combustible
Uses: Coatings; used in vitamin industry
Regulatory: FDA 21CFR §172.225

Methyl mercaptan
CAS 74-93-1; EINECS 200-822-1; FEMA 2716
Synonyms: Methanethiol; Mercaptan C_1
Empirical: CH_4S

3-(Methylmercapto)propionaldehyde

Formula: CH_3SH
Properties: Objectionable rotting-cabbage odor; m.w. 48.11; flash pt. -18 C
Precaution: Extremely flamm.
Uses: Synthetic flavoring agent

3-(Methylmercapto)propionaldehyde. *See* 3-Methylthiopropionaldehyde

Methyl o-methoxybenzoate
Uses: Synthetic flavoring
Regulatory: FDA 21CFR §172.515

Methyl methylaminobenzoate. *See* Dimethyl anthranilate
N-Methyl methyl anthranilate. *See* Dimethyl anthranilate
Methyl 2-methylbutanoate. *See* Methyl 2-methylbutyrate

Methyl 2-methylbutyrate
CAS 868-57-5; EINECS 212-778-0; FEMA 2719
Synonyms: Methyl 2-methylbutanoate
Empirical: $C_6H_{12}O_2$
Formula: $CH_3CH_2CH(CH_3)COOCH_3$
Properties: m.w. 116.16; dens. 0.885 (20/4 C); b.p. 113-115 C; flash pt. 18 C; ref. index 1.394 (20 C)
Precaution: Highly flamm.
Uses: Synthetic flavoring
Regulatory: FDA 21CFR §172.515
Manuf./Distrib.: Aldrich

7-Methyl-3-methylene-1,6-octadiene. *See* Myrcene
d-2-Methyl-5-(1-methylethenyl)-cyclohexanone. *See* d-Dihydrocarvone
1-Methyl-4-(1-methylethyl)-1,3-cyclohexadiene. *See* α-Terpinene
2-Methyl-5-(1-methylethyl)-1,3-cyclohexadiene. *See* α-Phellandrene
5-Methyl-2-(1-methylethyl) cyclohexanol. *See* Menthol
5-Methyl-2-(1-methylethylidene)cyclohexanone. *See* Pulegone
2-Methyl-5-(1-methylethyl)phenol. *See* Carvacrol
5-Methyl-2-(1-methylethyl) phenol. *See* Thymol
Methyl 3-(methylmercapto)propionate. *See* Methyl 3-methylthiopropionate

Methyl 2-methylthiopropionate
Uses: Synthetic flavoring
Regulatory: FDA 21CFR §172.515

Methyl 3-methylthiopropionate
CAS 13532-18-8; EINECS 236-883-6; FEMA 2720
Synonyms: Methyl 3-(methylmercapto)propionate
Empirical: $C_5H_{10}O_2S$
Formula: $CH_3SCH_2CH_2COOCH_3$
Properties: Colorless to pale-yellow liq.; onion-like odor; m.w. 134.19; dens. 1.073 (20/4 C); b.p. 184-189 C; flash p t. 72 C; ref. index 1.465 (20 C)
Uses: Flavoring agent
Manuf./Distrib.: Aldrich

Methyl 4-methylvalerate
CAS 2412-80-8; FEMA 2721
Uses: Synthetic flavoring
Regulatory: FDA 21CFR §172.515
Manuf./Distrib.: Aldrich

Methyl myristate
CAS 124-10-7; EINECS 204-680-1; FEMA 2722
Synonyms: Methyl tetradecanoate; Tetradecanoic acid, methyl ester
Definition: Ester of methyl alcohol and myristic acid
Empirical: $C_{15}H_{30}O_2$
Formula: $CH_3(CH_2)_{12}COOCH_3$
Properties: Colorless liq., honey and orris-like odor; insol. in water; m.w. 242.40; dens. 0.866 (20/4 C); m.p. 17.8 C; b.p. 186.8 C (30 mm); flash pt. > 112 C; ref. index 1.438 (20 C)
Precaution: Combustible
Uses: Synthetic flavoring agent
Usage level: 0.25-0.5 ppm (nonalcoholic beverages), 0.25-5 ppm (ice cream), 2.4 ppm (candy), 0.3-2 ppm (baked goods), 0.24 ppm (gelatins, puddings)

Regulatory: FDA 21CFR §172.225, 172.515, 176.200, 176.210, 177.2260, 177.2800
Manuf./Distrib.: Aldrich

Methyl namate. *See* Sodium dimethyldithiocarbamate
Methyl 2-naphthyl ketone. *See* 2′-Acetonaphthone
Methyl β-naphthyl ketone. *See* 2′-Acetonaphthone

Methyl nonanoate
CAS 1731-84-6; EINECS 217-052-7
Synonyms: Methyl pelargonate
Empirical: $C_{10}H_{20}O_2$
Formula: $CH_3(CH_2)_7COOCH_3$
Properties: m.w. 172.27; dens. 0.873 (20/4 C); b.p. 91-92 C (11 mm); flash pt. 87 C; ref. index 1.422 (20 C)
Uses: Synthetic flavoring
Regulatory: FDA 21CFR §172.515

Methyl 2-nonenoate
CAS 111-79-5; FEMA 2725
Uses: Synthetic flavoring
Regulatory: FDA 21CFR §172.515
Manuf./Distrib.: Aldrich

Methyl nonyl acetaldehyde
CAS 110-41-8; FEMA 2749
Synonyms: Aldehyde C-12 MNA; 2-Methyl undecanal; Methyl n-nonyl acetaldehyde
Empirical: $C_{12}H_{24}O$
Properties: Colorless to sl. yel. liq., fatty odor; sol. in alcohol, fixed oils, propylene glycol; insol. in glycerin; m.w. 184.32; dens. 0.822-0.830; ref. index 1.431
Toxicology: Heated to decomp., emits acrid smoke and irritating fumes
Uses: Amber-like, dry, fresh, sl. fruity fragrance and flavoring
Regulatory: FDA 21CFR §172.515

Methyl n-nonyl acetaldehyde. *See* Methyl nonyl acetaldehyde
Methyl nonyl ketone. *See* 2-Undecanone

Methyl 2-nonynoate
FEMA 2726
Synonyms: Methyloctyne carbonato
Uses: Synthetic flavoring
Regulatory: FDA 21CFR §172.515
Manuf./Distrib.: Aldrich

4-Methylnorvaline. *See* L-Leucine
Methyl cis,cis-9,12-octadecadienoate. *See* Methyl linoleate
Methyl octadecanoate. *See* Methyl stearate

2-Methyloctanal
Synonyms: Methyl hexyl acetaldehyde
Uses: Synthetic flavoring
Regulatory: FDA 21CFR §172.515

Methyl octanoate. *See* Methyl caprylate
Methyl octine carbonate. *See* Methyl 2-octynoate
Methyloctyne carbonate. *See* Methyl 2-nonynoate

Methyl 2-octynoate
CAS 111-12-6; EINECS 203-836-6; FEMA 2729
Synonyms: Methyl heptine carbonate; Methyl heptyne carbonate; Methyl octine carbonate
Empirical: $C_9H_{14}O_2$
Formula: $CH_3(CH_2)_4C \equiv CCOOCH_3$
Properties: m.w. 154.20; dens. 0.924 (20/4 C); b.p. 215-217 C; flash pt. 89 C; ref. index 1.447 (20 C)
Uses: Synthetic flavoring
Regulatory: FDA 21CFR §172.515
Manuf./Distrib.: Aldrich

Methylol. *See* Methyl alcohol

Methylparaben
CAS 99-76-3; EINECS 202-785-7; FEMA 2710; EEC E218

Methylparaben, sodium salt

Synonyms: Methyl 4-hydroxybenzoate; 4-Hydroxybenzoic acid, methyl ester; Methyl p-hydroxybenzoate
Definition: Ester of methyl alcohol and p-hydroxybenzoic acid
Empirical: $C_8H_8O_3$
Formula: $CH_3OOCC_6H_4OH$
Properties: Colorless crystals or wh. cryst. powd., odorless or faint char. odor, sl. burning taste; sol. in alcohol, ether; sl. sol. in water, benzene, CCl_4; m.w. 152.14; m.p. 125-128 C; b.p. 270-280 C (dec.)
Toxicology: LD50 (oral, dog) 3000 mg/kg; mod. toxic by ingestion, subcutaneous, and IP routes; mutagenic data; may cause asthma, rashes, hyperactivity; heated to decomp., emits acrid smoke and fumes
Uses: Antimicrobial agent, preservative, flavoring agent; for baked goods, beverages, food colors, milk, wine; antimycotic migrating from food pkg.
Usage level: 0.1-1.0%; use in foods restricted to 0.1%; ADI 0-10 mg/kg (EEC)
Regulatory: FDA 21CFR §150.141, 150.161, 172.515, 181.22, 181.23, 184.1490, GRAS, limitation 0.1%, 556.390, zero limitation in milk; USA CIR approved, EPA reg.; Japan listed; Europe listed; UK approved
Manuf./Distrib.: Aldrich; Int'l. Sourcing
Trade names: Lexgard® M; Nipagin M
Trade names containing: Dry Beta Carotene Beadlets Yellow Type 2.4-S No. 653800100; Nipastat; Sporban Regular; Sporban Special

Methylparaben, sodium salt. *See* Sodium methylparaben
Methyl pelargonate. *See* Methyl nonanoate

4-Methyl-2,3-pentanedione
FEMA 2730
Synonyms: Acetyl isobutyryl
Empirical: $C_6H_{10}O_2$
Properties: Yel. oil, char. pungent odor; m.w. 114.15; dens. 0.9215 (11 C); m.p. -2.4 C; b.p. 116 C
Uses: Synthetic flavoring
Usage level: 7.6 ppm (nonalcoholic beverages), 5.6 ppm (ice cream, ices), 6.2 ppm (candy), 8.3 ppm (baked goods), 1.2-18 ppm (gelatins, puddings)
Regulatory: FDA 21CFR §172.515

2-Methylpentanoic acid
CAS 97-61-0; FEMA 2754
Synonyms: 2-Methylvaleric acid
Uses: Penetrating char. synthetic flavoring
Regulatory: FDA 21CFR §172.515
Manuf./Distrib.: Aldrich

4-Methylpentanoic acid
CAS 646-07-1; FEMA 3463
Manuf./Distrib.: Aldrich

4-Methyl-2-pentanone. *See* Methyl isobutyl ketone

2-Methyl-2-pentenoic acid
CAS 16957-70-3; FEMA 3195
Manuf./Distrib.: Aldrich

3-Methyl-2-(2-pentenyl)-2-cyclopenten-1-one. *See* cis-Jasmone
Methyl pentyl ketone. *See* Methyl n-amyl ketone
β-Methylphenethyl alcohol. *See* 2-Phenyl propanol-1
4-Methylphenol. *See* p-Cresol
p-Methyl phenylacetaldehyde. *See* p-Tolylacetaldehyde

Methyl phenylacetate
CAS 101-41-7; EINECS 202-940-9; FEMA 2733
Empirical: $C_9H_{10}O_2$
Formula: $C_6H_5CH_2COOCH_3$
Properties: m.w. 150.18; dens. 1.066 (20/4 C); b.p. 218-221 C; flash pt. 96 C; ref. index 1.507 (20 C)
Toxicology: Skin and eye irritant
Uses: Synthetic flavoring
Regulatory: FDA 21CFR §172.515
Manuf./Distrib.: Aldrich; Chr. Hansen's

3-Methyl-4-phenyl-3-buten-2-one
CAS 1901-26-4; FEMA 2734
Uses: Synthetic flavoring
Regulatory: FDA 21CFR §172.515

2-Methyl-4-phenyl-2-butyl acetate
Synonyms: Dimethylphenylethyl carbinyl acetate
Uses: Synthetic flavoring
Regulatory: FDA 21CFR §172.515

2-Methyl-4-phenyl-2-butyl isobutyrate
Synonyms: Dimethylphenylethyl carbinyl isobutyrate
Uses: Synthetic flavoring
Regulatory: FDA 21CFR §172.515

3-Methyl-2-phenylbutyraldehyde
Synonyms: α-Isopropyl phenylacetaldehyde
Uses: Synthetic flavoring
Regulatory: FDA 21CFR §172.515

Methyl 4-phenylbutyrate
Uses: Synthetic flavoring
Regulatory: FDA 21CFR §172.515

Methyl phenylcarbinol. *See* α-Methylbenzyl alcohol
Methyl phenylcarbinyl acetate. *See* α-Methylbenzyl acetate
4-Methyl-2-phenyl-m-dioxolane. *See* Benzaldehyde propylene glycol acetal
Methylphenyl ether. *See* Anisole
3-Methyl-3-phenyl glycidic acid ethyl ester. *See* Ethyl methylphenylglycidate
Methyl phenyl ketone. *See* Acetophenone
4-Methyl-1-phenyl-2-pentanol. *See* α-Isobutylphenethyl alcohol

4-Methyl-1-phenyl-2-pentanone
FEMA 2740
Synonyms: Benzyl isobutyl ketone
Uses: Synthetic flavoring
Regulatory: FDA 21CFR §172.515
Manuf./Distrib.: Aldrich

Methyl-3-phenyl propenoate. *See* Methyl cinnamate

Methyl 3-phenylpropionate
CAS 103-25-3; FEMA 2741
Synonyms: Methyl hydrocinnamate
Uses: Synthetic flavoring
Regulatory: FDA 21CFR §172.515
Manuf./Distrib.: Aldrich

2-Methylpropanal
CAS 78-84-2; EINECS 201-149-6; FEMA 2220
Synonyms: Isobutyraldehyde; 2-Methylpropionaldehyde; Isobutanal; Isobutyl aldehyde
Empirical: C_4H_8O
Formula: $(CH_3)_2CHCHO$
Properties: Colorless liq., pungent odor; sol. 11 5/100 ml water; misc. with alcohol, ether, benzene, carbon disulfide, acetone, toluene, chloroform; m.w. 72.12; dens. 0.7938 (20/4 C); m.p. -65 C; b.p. 64 C (760 mm); flash pt. (CC) -40 F; ref. index 1.374
Precaution: Flamm.; dangerous fire hazard exposed to heat, flames, oxidizers; can react vigorously with reducing materials; explosive limits 1.6-10.6%
Toxicology: LD50 (oral, rat) 2810 mg/kg; mod. toxic by ingestion; mildly toxic by skin contact, inh.; severe skin and eye irritant; heated to decomp., emits acrid smoke and fumes
Uses: Synthetic flavoring agent
Usage level: 0.3 ppm (nonalcoholic beverages), 5 ppm (alcoholic beverages), 0.25-0.5 ppm (ice cream), 0.67 ppm (candy), 0.5-1 ppm (baked goods)
Regulatory: FDA 21CFR §172.515
Manuf./Distrib.: Aldrich

2-Methylpropane. *See* Isobutane
Methyl propanoate. *See* Methyl propionate
2-Methylpropanoic acid. *See* Isobutyric acid
2-Methylpropanoic acid ethyl ester. *See* Ethyl isobutyrate
2-Methylpropanoic acid methyl ester. *See* Methyl isobutyrate
2-Methylpropanoic acid 2-methylpropyl ester. *See* Isobutyl isobutyrate
2-Methylpropanol. *See* Isobutyl alcohol

2-Methyl-1-propanol. *See* Isobutyl alcohol
2-Methyl propanyl butyrate. *See* Isobutyl butyrate
2-Methyl-1-propene, homopolymer. *See* Polyisobutene
2-Methyl-2-propenyl butyrate. *See* 2-Methylallyl butyrate
2-Methylpropionaldehyde. *See* 2-Methylpropanal

Methyl propionate
　　CAS 554-12-1; EINECS 209-060-4; FEMA 2742
　　Synonyms: Propanoic acid methyl ester; Methyl propanoate
　　Empirical: $C_4H_8O_2$
　　Formula: $CH_3CH_2COOCH_3$
　　Properties: Colorless liq., fruity rum-like odor; sol. in 16 parts water; misc. with alcohol, ether; m.w. 88.11; dens.
　　　　0.915 (20/4 C); m.p. -87 C; b.p. 78-80 C; flash pt. -2 C; ref. index 1.377 (20 C)
　　Precaution: Highly flamm.
　　Uses: Synthetic flavoring
　　Usage level: 20 ppm (nonalcoholic beverages), 29 ppm (ice cream, ices), 96 ppm (candy), 130 ppm (baked
　　　　goods)
　　Regulatory: FDA 21CFR §172.515
　　Manuf./Distrib.: Aldrich

2-Methylpropyl acetate. *See* Isobutyl acetate

3-Methyl-5-propyl-2-cyclohexen-1-one
　　Synonyms: Celery ketone
　　Empirical: $C_{10}H_{16}O$
　　Properties: Pale yel. to colorless liq., warm spicy woody odor; sol. in ether, alcohol, oils; insol. in water; m.w.
　　　　152.23; dens. 0.9267; b.p. 242-244 C
　　Uses: Synthetic flavoring
　　Regulatory: FDA 21CFR §172.515

Methyl propyl diketone. *See* Acetyl butyryl
β-Methylpropyl ethanoate. *See* Isobutyl acetate

Methyl propyl ketone
　　CAS 107-87-9; EINECS 203-528-1; FEMA 2842
　　Synonyms: MPK; 2-Pentanone; Ethyl acetone
　　Empirical: $C_5H_{10}O$
　　Formula: $CH_3COCH_2CH_2CH_3$
　　Properties: Water-wh. liq., fruity ethereal odor; sl. sol. in water; misc. with alcohol, ether; m.w. 86.14; dens.
　　　　0.801-0.806; m.p. -78 C; b.p. 216 F; flash pt. 45 F
　　Precaution: Highly flamm.; very dangerous fire hazard exposed to heat or flame; reacts vigorously with
　　　　oxidizers; explosive limits 1.5-8.2%
　　Toxicology: ACGIH TLV:TWA 200 ppm; LD50 (oral, rat) 3730 mg/kg; mod. toxic by ingestion, IP; mildly toxic
　　　　by skin contact, inh.; human systemic effects; skin irritant; mutagenic data; heated to decomp., emits acrid
　　　　smoke and irritating fumes
　　Uses: Synthetic flavoring agent
　　Regulatory: FDA 21CFR §172.515
　　Manuf./Distrib.: Aldrich

Methylprotocatechualdehyde. *See* Vanillin

2-Methylpyrazine
　　CAS 109-08-0; FEMA 3309
　　Empirical: $C_5H_6N_2$
　　Properties: m.w. 94.12; dens. 1.020; m.p. -29 C; b.p. 135 C (761 mm); flash pt. 122 F
　　Uses: Synthetic flavoring agent
　　Manuf./Distrib.: Aldrich

Methyl 2-pyridyl ketone. *See* 2-Acetylpyridine
Methyl 3-pyridyl ketone. *See* 3-Acetylpyridine
2-Methyl pyromeconic acid. *See* Maltol
Methyl 2-pyrrolyl ketone. *See* 2-Acetylpyrrole
4-Methylquinoline. *See* Lepidine
γ-Methylquinoline. *See* Lepidine
Methyl salicylaldehyde. *See* o-Methoxybenzaldehyde

Methyl salicylate
　　CAS 119-36-8; EINECS 204-317-7; FEMA 2745

Synonyms: Methyl 2-hydroxybenzoate; Sweet birch oil; Oil of wintergreen
Definition: Ester of methyl alcohol and salicylic acid
Empirical: $C_8H_8O_3$
Formula: $C_6H_4OHCOOCH_3$
Properties: Yel. to red liq., wintergreen odor; sol. in ether, glacial acetic acid; sl. sol. in water; m.w. 152.14; dens. 1.180-1.185 (25/25 C); m.p. -8.6 C; b.p. 222.2 C; flash pt. (CC) 101 C; ref. index 1.535-1.538
Precaution: Combustible when exposed to heat or flame; reactive with oxidizing materials; heated to decomp., emits acrid smoke and irritating fumes
Toxicology: LD50 (oral, rat) 887 mg/kg; human poison by ingestion; mod. toxic experimentally by ingestion, IP, IV, subcutaneous routes; experimental teratogen, reproductive effects; human systemic effects; severe skin and eye irritant
Uses: Synthetic flavoring agent in foods, beverages, baked goods, candy, chewing gum
Usage level: 59 ppm (nonalcoholic beverages), 27 ppm (ice cream), 840 ppm (candy), 54 ppm (baked goods), 8400 ppm (chewing gum), 200 ppm (syrups)
Regulatory: FDA 21CFR §175.105, 177.1010; 27CFR §21.65, 21.151; GRAS (FEMA); Japan approved as flavoring
Manuf./Distrib.: Aldrich

Methyl stearate
CAS 112-61-8; 85586-21-6; EINECS 203-990-4; 287-824-6
Synonyms: Methyl octadecanoate; Octadecanoic acid, methyl ester
Definition: Ester of methyl alcohol and stearic acid
Empirical: $C_{19}H_{38}O_2$
Formula: $CH_3(CH_2)_{16}COOCH_3$
Properties: White crystals; insol. in water; sol. in ether, alcohol; m.w. 298.57; m.p. 37.8 C; b.p. 234.5 C (30 mm); flash pt. 307 F
Precaution: Combustible
Uses: Foam depressant and nutrient in fermentation
Regulatory: FDA 21CFR §172.225, 176.200, 176.210, 177.2260, 177.2800, 178.3910
Manuf./Distrib.: Penta Mfg.
Trade names: Kemester® 4516

Methyl styryl carbinol. *See* 4-Phenyl-3-buten-2-ol
Methyl styryl ketone. *See* Benzylidene acetone
Methyl succinate. *See* Dimethyl succinate

Methyl sulfide
CAS 75-18-3; FEMA 2746
Synonyms: Dimethyl sulfide; Thiobismethane
Empirical: C_2H_6S
Formula: $(CH_3)_2S$
Properties: Liq., disagreeable cabbage-like odor; sol. in alcohol, ether; insol. in water; m.w. 62.13; dens. 0.847 (20/4 C); m.p. -83 C; b.p. 36-37 C; flash pt. -25 C; ref. index 1.435 (20 C)
Precaution: Flamm.
Uses: Synthetic flavoring
Usage level: 1.1 ppm (nonalcoholic beverages), 0.30 ppm (ice cream, ices), 1.4 ppm (candy), 1.6 ppm (baked goods), 0.13 ppm (gelatins, puddings), 0.50 ppm (syrups)
Regulatory: FDA 21CFR §172.515
Manuf./Distrib.: Aldrich

Methyl tetradecanoate. *See* Methyl myristate
Methyltheobromine. *See* Caffeine

5-Methyl-2-thiophene-carboxyaldehyde
CAS 13679-70-4; EINECS 237-178-6; FEMA 3209
Classification: Thiophene
Empirical: C_6H_6OS
Properties: m.w. 126.18; dens. 1.170; b.p. 114 C (25 mm); flash pt. 87 C
Uses: Nutty meaty flavoring
Manuf./Distrib.: Aldrich

3-Methylthiopropanol. *See* 3-Methylthiopropionaldehyde

3-Methylthiopropionaldehyde
CAS 3268-49-3; EINECS 221-882-5; FEMA 2747
Synonyms: Methional; 3-(Methylmercapto)propionaldehyde; 3-Methylthiopropanol

Methyl p-tolyl ketone

Empirical: C_4H_8OS
Formula: $CH_3SCH_2CH_2CHO$
Properties: Pale yel. mobile liq., powerful onion/meat-like odor; sol. in alcohol; insol. in water; m.w. 104.17; dens. 1.052 (20/4 C); b.p. 165-166 C; flash pt. 60 C; ref. index 1.486 (20 C)
Toxicology: Irritating to skin, eyes, respiratory system
Uses: Synthetic flavoring
Usage level: 0.35 ppm (nonalcoholic beverages), 0.01-1.0 ppm (ice cream, ices, candy), 0.66 ppm (baked goods), 0.62 ppm (condiments), 1.9 ppm (meats)
Regulatory: FDA 21CFR §172.515
Manuf./Distrib.: Aldrich

Methyl p-tolyl ketone. *See* 4′-Methyl acetophenone

2-Methyl-3-tolylpropionaldehyde, mixed o-, m-, p-
Uses: Synthetic flavoring
Regulatory: FDA 21CFR §172.515

2-Methyl undecanal. *See* Methyl nonyl acetaldehyde

Methyl 9-undecenoate
Uses: Synthetic flavoring
Regulatory: FDA 21CFR §172.515

Methyl 2-undecynoate
Synonyms: Methyl decyne carbonate
Uses: Synthetic flavoring
Regulatory: FDA 21CFR §172.515

Methyl valerate
CAS 624-24-8; EINECS 210-838-0; FEMA 2752
Synonyms: Methyl n-valerate
Empirical: $C_6H_{12}O_2$
Formula: $CH_3(CH_2)_3COOCH_3$
Properties: m.w. 116.16; dens. 0.889 (20/4 C); b.p. 126-128 C; flash pt. 27 C; ref. index 1.397 (20 C)
Precaution: Flamm.
Uses: Synthetic flavoring
Regulatory: FDA 21CFR §172.515
Manuf./Distrib.: Aldrich

Methyl n-valerate. *See* Methyl valerate
2-Methylvaleric acid. *See* 2-Methylpentanoic acid
MHPC. *See* Hydroxypropyl methylcellulose
MIBK. *See* Methyl isobutyl ketone

Microcrystalline cellulose. *See also* Cellulose
CAS 9004-34-6; EEC E460
Synonyms: Cellulose gel
Definition: Isolated, colloidal crystalline portion of cellulose fibers
Toxicology: LD50 (oral, rat) > 5 g/kg, no significant hazard; irritant by inhalation (dust); may be damaging to lungs
Uses: Binder, release agent, lubricant, nonnutritive bulking agent, anticaking agent, dietary fiber, emulsion/heat stabilizer, carrier, dispersant for foods, tablets
Usage level: ADI not specified (FAO/WHO)
Regulatory: FDA GRAS; Europe listed; UK approved
Manuf./Distrib.: Asahi Chem.
Trade names: Avicel® CL-611; Avicel® RC-501; Avicel® RC-581; Avicel® RC-591F; Avicel® WC-595
Trade names containing: Avicel® RCN-10; Avicel® RCN-15; Avicel® RCN-30

Microcrystalline wax. *See also* Petroleum wax
CAS 63231-60-7; 64742-42-3; EINECS 264-038-1; EEC E907
Synonyms: Petroleum wax, microcrystalline; Waxes, microcrystalline
Definition: Wax derived from petroleum and char. by fineness of crystals; consists of high m.w. saturated aliphatic hydrocarbons
Toxicology: May be carcinogenic
Uses: Chewing gum ingred.; polishing and release agent; stiffening agent; tablet coating
Regulatory: Europe listed; UK approved for restricted use
Trade names: Be Square® 175; Be Square® 185; Be Square® 195; Mekon® White; Multiwax® 180-M; Multiwax® ML-445; Multiwax® W-445; Multiwax® W-835; Multiwax® X-145A; Petrolite® C-1035;

Starwax® 100; Ultraflex®; Victory®

Microparticulated protein prod.
Definition: Protein source prepared from egg whites and/or milk protein
Uses: Thickener, texturizer used in frozen deserts; may not be used to replace milk fat required in standardized frozen desserts
Regulatory: FDA 21CFR §184.1498, GRAS; label must include source of protein

Milfoil. *See* Yarrow
Milfoil extract. *See* Yarrow extract
Milk acid. *See* Lactic acid

Milk-clotting enzyme from *Bacillus cereus*
Toxicology: Heated to decomp., emits acrid smoke and irritating fumes
Uses: Milk clotting enzyme used in cheese prod.
Regulatory: FDA 21CFR §173.150

Milk-clotting enzyme from *Endothia parasitica*
Toxicology: Heated to decomp., emits acrid smoke and irritating fumes
Uses: Milk clotting enzyme used in cheese prod.
Regulatory: FDA 21CFR §173.150; Canada approved

Milk-clotting enzyme from *Mucor miehei*
Toxicology: Heated to decomp., emits acrid smoke and irritating fumes
Uses: Milk clotting enzyme used in cheese prod.
Regulatory: FDA 21CFR §173.150; Canada approved

Milk-clotting enzyme from *Mucor pusillus*
Toxicology: Heated to decomp., emits acrid smoke and irritating fumes
Uses: Milk clotting enzyme used in cheese prod.
Regulatory: FDA 21CFR §173.150; Canada approved

Milk of magnesia. *See* Magnesium hydroxide
Milk, nonfat dry. *See* Nonfat dry milk

Milk protein
CAS 0000-71-0; EINECS 202-555-1
Synonyms: Casein
Definition: Mixture of proteins obtained from cow's milk
Properties: Light-yel. powd.
Uses: Cheesemaking, dietetic preps., foods and feeds
Manuf./Distrib.: Adams Food Ingreds.; Am. Casein; Am. Dairy Prods.; Byrton Dairy Prods.; Dena AG; Dutch Protein; EPI Bretagne; G Fiske; K&K Greeff; Kerry Foods; MD Foods; Meggle; New Zealand Milk Prods.; Sanofi France; Unilait France; Westin
Trade names containing: Lamequick® AS 340; Mayonat DF; Mix 70

Milk protein, casein. *See* Casein
Milk sugar. *See* Lactose
Mimosa. *See* Black wattle

Mineral oil
CAS 8012-95-1; 8020-83-5, 8042-47-5; EINECS 232-384-2; 232-455-8
Synonyms: Heavy or light mineral oil; Paraffin oil; Liquid paraffin; Liquid petrolatum
Definition: Liq. mixture of hydrocarbons obtained from petroleum
Properties: Colorless oily liq., tasteless, odorless; insol. in water, alcohol; sol. in benzene, chloroform, ether, petrol. ether, oils; dens. 0.83-0.86 (light), 0.875-0.905 (heavy); flash pt. (OC) 444 F; surf. tension < 35 dynes/cm
Precaution: Combustible
Toxicology: Eye irritant; human carcinogen and teratogen by inhalation; heated to decomp., emits acrid smoke and fumes
Uses: Binder, defoamer, lubricant, release agent, fermentation aid, protective coating; used in bakery prods., beet sugar, confections, fruit, grain, frozen meat, pickles, potatoes, sorbic acid, vegetables, vinegar, wine, yeast
Usage level: Limitation 0.6% (tablets contg. spices, nutrients, food for dietary use), 0.15% (baked goods), 0.02% (dried fruits/vegetables), 0.095% (frozen meat), 0.3% (molding starch), 0.15% (yeast), 0.25% (sorbic acid), 0.2% (confectionery), 0.02% (grain)
Regulatory: FDA 21CFR §172.878, 173.340 (limitation 0.008% in wash water for sliced potatoes, 150 ppm in yeast), 175.105, 175.210, 175.230, 175.300, 176.170, 176.200, 176.210, 177.1200, 177.2260, 177.2600,

177.2800, 178.3570, 178.3620, 178.3740, 178.3910, 179.45, 573.680; ADI not specified (FAO/WHO)
Manuf./Distrib.: Penreco
Trade names: Amoco Superla® DCO 55; Amoco Superla® DCO 75; Bake-Well Heavy Divider Oil; Bake-Well K Machine Oil; Benol®; Blandol®; Britol®; Britol® 6NF; Britol® 7NF; Britol® 9NF; Britol® 20USP; Britol® 35USP; Britol® 50USP; Carnation®; Divider Oil 90; Divider Oil 210; Drakeol 5; Drakeol 7; Drakeol 9; Drakeol 10; Drakeol 10B; Drakeol 13; Drakeol 15; Drakeol 19; Drakeol 21; Drakeol 32; Drakeol 34; Draketex 50; Ervol®; Gloria®; Kaydol®; Klearol®; Orzol®; PD-23; Protol®; Rudol®; Superla® No. 5; Superla® No. 6; Superla® No. 7; Superla® No. 9; Superla® No. 10; Superla® No. 13; Superla® No. 18; Superla® No. 21; Superla® No. 31; Superla® No. 35
Trade names containing: Bake-Well 52; Bake-Well 80/20; Bake-Well High Stability Pan Oil; Bread Pan Oil

Mineral wax. *See* Ozokerite

Miraculin
Classification: Glycoprotein
Definition: Taste-modifying protein from the African plant *Richardella dulcifa*
Properties: m.w. ≈ 44,000
Uses: Sweetening agent in sour or tart foods

MKP. *See* Potassium phosphate
MMA. *See* Dimethyl anthranilate
Modified food starch. *See* Food starch modified

Molasses
CAS 68476-78-8
Definition: Residue after sucrose has been removed from the mother liquor in sugar manufacture
Properties: Thick liq.
Uses: Sweetener
Trade names: De-Mol® Molasses; Dri-Mol® Molasses; Sweet'n'Neat® 65 Molasses Powd.; Sweet'n'Neat® 4000 Molasses Powd.; Sweet'n'Neat® 5000 Molasses Flake
Trade names containing: Dri-Mol® 604 Molasses

Molasses extract
Definition: Saccarum officinarum
Properties: Usually sweet, also can taste burnt
Uses: Natural flavoring agent; tobacco flavoring, baking
Regulatory: FDA 21CFR §182.20, GRAS

Monarda. *See* Horsemint

Monellin
Classification: Protein
Definition: Two polypeptide chains derived from the noncultivated African plant *Dioscoreophyllum cumminsii*
Properties: Sol. in water; m.w. ≈ 10,700
Uses: Low-calorie sweetener; 3000 times sweetness of sucrose on wt. basis

Monoammonium glutamate. *See* Ammonium glutamate
Monoammonium L-glutamate. *See* Ammonium glutamate

Monoammonium glycyrrhizinate
CAS 1407-03-0
Synonyms: Ammonium glycyrrhizinate, pentahydrate; Ammonium glycyrrhizinate; Ammoniated glycyrrhizin
Definition: Obtained by extraction from ammoniated glycyrrhizin, derived from roots of *Glycyrrhiza glabra*
Empirical: $C_{42}H_{65}NO_{165}H_2O$
Properties: White powder; sweet taste; sol. in ammonia; insol. in glacial acetic acid; m.w. 839.91
Toxicology: Heated to decomp., emits acrid smoke and irritating fumes
Uses: Flavor enhancer, flavoring agent, surface-active agent
Usage level: Limitation (as glycyrrhizin) 0.05% (baked goods), 0.1% (alcoholic beverages), 0.15% (nonalcoholic beverages), 1.1% (chewing gum), 16% (hard candy), 0.15% (herbs, seasonings), 3.1% (soft candy), 0.5% (dietary supplements), 0.1% (other foods)
Regulatory: FDA 21CFR §184.1408, GRAS; not permitted for use as nonnutritive sweetener in sugar substitutes
Manuf./Distrib.: Am. Roland

Monoammonium phosphate. *See* Ammonium phosphate
Monobasic sodium phosphate. *See* Sodium phosphate
Mono-t-butyl hydroquinone. *See* t-Butyl hydroquinone
Monocalcium phosphate anhydrous. *See* Calcium phosphate monobasic anhydrous

Monocalcium phosphate monohydrate. *See* Calcium phosphate monobasic monohydrate

Mono- and diglycerides citrates. *See* Citric acid esters of mono- and diglycerides of fatty acids

Mono- and diglycerides of fatty acids
CAS 67701-32-0; 67701-33-1; 68990-53-4; EEC E471
Properties: Yel. liqs. to ivory plastics to hard solids, bland odor and taste; sol. in alcohol, ethyl acetate, chloroform, other chlorinated hydrocarbons; insol. in water
Toxicology: Heated to decomp., emits acrid smoke and irritating fumes
Uses: Dough strengthener, emulsifier, flavoring agent, formulation aid, lubricant, release agent, softener, solvent, stabilizer, surface-active agent, surface-finishing agent, texturizer, thickener, vehicle
Usage level: 0.5% (oleomargarine); ADI not specified (EEC)
Regulatory: FDA 21CFR §172.863, 182.4505, 184.1505, GRAS; USDA 9CFR §318.7, 381.147; Europe listed; UK approved
Manuf./Distrib.: Browning
Trade names: Admul Emulsponge; Admul MG 4103; Admul MG 4123; Admul MG 4143; Admul MG 4163; Admul MG 4203; Admul MG 4223; Admul MG 4304; Admul MG 4404; Admul MG 4904; Admul MG 6404; Admul MG 6504; Alphadim® 70K; Alphadim® 90AB; Alphadim® 90PBK; Alphadim® 90SBK; Alphadim® 90SBK FG; Amidan; Atlas 800; Atlas 1400K; Atlas 2000; Atlas 5000K; Atlas G-986K; Atmos® 300 K; Atmul® 27 K; Atmul® 80; Atmul® 82; Atmul® 84 K; Atmul® 86 K; Atmul® 122; Atmul® 500; Atmul® 601 K; Atmul® 651 K; Atmul® 695; Atmul® 695 K; Atmul® 918 K; Atmul® 1003 K; Atmul® P-96; BFP 30; BFP 64; BFP 65A; BFP 65C; BFP 65K; BFP 74A; BFP 74K; BFP 75A; BFP 75K; BFP 800 K; BFP L Mono; Canamulse 100; Canamulse 110K; Canamulse 150K; Canamulse 155; Cegesterin® Range; Dimul DDM K; Emuldan; Estric™; Excel VS-95; G-695; G-991; Hymono 1103; Hymono 1123; Hymono 1163; Hymono 3203; Hymono 4404; Hymono 7804; Hymono 8803; Hymono 8903; Imwitor® 175; Imwitor® 325; Imwitor® 595; Imwitor® 895; Kirnol® Range; Kureton 200; Lamemul® Range; Lamemul® K 2000-Range; Lipal 70; Monomax AH90 B; Monomax VH90 B; Myvatex® 8-20; Myvatex® Mighty Soft®; Nutrisoft® 55; Panalite 40; Panalite 40 HVK; Panalite 40 SVK, 50 SA, 50 SVK; Panalite 50; Panalite 50 HVK; Panalite 90D; Panatex, Panatex HM; PC-35; Tinamul®; Ultra Fresh; Witconol™ GMOP
Trade names containing: Akocote 106; Aldosperse® 30/70 FG; Aldosperse® 30/70 KFG; Aldosperse® 712 FG; Aldosperse® MO-50 FG; Aldosperse® TS-20 KFG; Alpha 500; Alphadim® 90NLK; Alphadim® 90VCK; Alpine; Amfal-46; Annatto ECH-1; Annatto ECPC; Atlas 2200H; Atlas 5520; Atlas P-44; Atmos® 378 K; Atmos® 659 K; Atmos® 729; Atmos® 758; Atmos® 758 K; Atmos® 1069; Atmos® 2462; Atmos® 7515 K; Atmul® 100; Atmul® 600H; Atmul® 700 H; Atmul® P-28; Atmul® P-36; Atmul® P-44; Baker's Ideal®; Bake-Well Bun Release; Beta Plus; Betricing; Betrkake; BFP 64A; BFP 64K; BFP 65; BFP 74; BFP 76; BFP 100; Brocoft 6430; Buckeye C (Baker's Grade); Bunge Cake and Icing Vegetable Shortening; Bunge Special Mix Shortening; Cake Mix 96; Cap-Shure® BC-900-85; Cap-Shure® C-900-85; CBC #7 Shortening; CDC-10A; CDC-50; CDC-2001; CDC-2002; CDC-2500; CDC-3000; CFI-10; Coco-Spred™; Confecto No-Stick™; Coral®; Cremol Plus™ 60 Vegetable; Crestawhip H13; Crodacreme; Diamond D-21; Dricoid® 200; Dricoid® 280; Dur-Em® 114; Dur-Em® 114K; Dur-Em® 117; Dur-Em® 117K; Dur-Em® 204; Dur-Em® 204K; Dur-Em® 207; Dur-Em® 207-E; Dur-Em® 207K; Dur-Em® 300; Dur-Em® 300K; Durko; Dur-Lo®; EC-25®; Elasdo +; Elasdo Power 70; Emulgel E-21; Fluid Flex; Gatodan 415; Golden Crescent; Hymo; Hytex; Ice #12; Ice # 81; Kakebake® M/V Cake Icing Shortening; Kel-Lite™ CM; Lamefrost® ES 216 G; Lamefrost® ES 251 G; Lamefrost® ES 315; Lamefrost® ES 375; Lamefrost® ES 379; Lamefrost® ES 424; Lamemul® K 1000-Range; Lamemul® K 3000-Range; Lecidan; Lipodan CDS Kosher; Lite Pack 350; Myvatex® 40-06S K; Myvatex® Super DO; No Stick Emulsifier; Panalite 100 K; Panalite MP Hydrate; Panalite MPH, MPS; Panospray™ WL; PAV; Perflex; Poly E; Precision Base; Pristene® RO; Pristene® RW; Rocoat® Niacinamide 33$^1/_3$% No. 69907; Rocoat® Pyridoxine Hydrochloride 33$^1/_3$% No. 60688; Rocoat® Riboflavin 25% No. 60289; Rocoat® Riboflavin 33$^1/_3$ No. 60288; Rocoat® Thiamine Mononitrate 33$^1/_3$% No. 60188; Satina 53NT; Soft Touch; Sta-Nut EE; Sta-Nut P; Starplex® 90; Summit® 25; Summit® 50; Superb® Bakers Margarine 105-101; Superb® Bakers Shortening 105-200; Superb® Icing Shortening 101-270; Super-Cel® Specialty Cake Shortening; Sustane® 7; Sustane® 15; Sustane® 16; Sustane® 18; Sustane® 20A; Sustane® 20B; Sustane® T-6; Tally® 100; Tandem 5K; Tandem 8; Tandem 9; Tandem 11H; Tandem 11H K; Tandem 22 H; Tandem 22 H K; Tandem 100 K; Tandem 552; Tandem 552 K; Tandem 930; Tem Cote® II; Top Mate Kosher; Top Notch®; Tri-Co® Vegetable Yeast Raised Dough Shortening; Trophy®; Tween®-MOS100K; Vanall (K); Vegetable Cake Shortening 101-250; Victor®; Vrest™; Vrest Plus™; Vykasoid; Xpando; Xpando 70; Xpando Powd.

Mono- and diglycerides, sodium phosphate derivs.
Toxicology: Heated to decomp., emits acrid smoke and irritating fumes
Uses: Emulsifier, lubricant, release agent, surface-active agent; used in soft candy, dairy prods.
Usage level: ADI not specified
Regulatory: FDA 21CFR §184.1521, GRAS

Monoethanolamine. *See* Ethanolamine

Monoethyl fumarate. *See* Ethyl fumarate

Monoglyceride citrate
Definition: Mixt. of glyceryl monooleate and its citric acid monoester
Properties: Wh. to ivory-colored soft waxy solid, bland odor/taste; sol. in most common fat solvs., alcohol; insol. in water
Uses: Synergist and solubilizer for antioxidants added to fats and oils
Usage level: ADI 0-100 ppm (FAO/WHO)
Regulatory: FDA limitation 200 ppm (cured meats); WHO limitation 100 ppm (fats, oils, margarine)

Monoisopropyl citrate. *See* Isopropyl citrate
Monolinolein. *See* Glyceryl linoleate
Monooctanoin. *See* Glyceryl caprylate
Monoolein. *See* Glyceryl oleate
Monopotassium carbonate. *See* Potassium bicarbonate
Monopotassium glutamate. *See* Potassium glutamate
Monopotassium l-glutamate. *See* Potassium glutamate
Monopotassium orthophosphate. *See* Potassium phosphate
Monopotassium D(-)-pentahydroxy capronate. *See* Potassium D-gluconate
Monopotassium phosphate. *See* Potassium phosphate
Monopotassium L-(+)-tartrate. *See* Potassium acid tartrate
Monoricinolein. *See* Glyceryl ricinoleate
Monosodium ascorbate. *See* Sodium ascorbate
Monosodium carbonate. *See* Sodium bicarbonate
Monosodium citrate anhydrous. *See* Sodium citrate
Monosodium dihydrogen orthophosphate. *See* Sodium phosphate
Monosodium dihydrogen phosphate. *See* Sodium phosphate
Monosodium fumarate. *See* Sodium fumarate
Monosodium gluconate. *See* Sodium gluconate
Monosodium glutamate. *See* MSG
Monosodium L-glutamate monohydrate. *See* MSG
Monosodium-2-hydroxypropane-1,2,3-tricarboxylate. *See* Sodium citrate
Monosodium D(-)-pentahydroxy capronate. *See* Sodium gluconate
Monostearin. *See* Glyceryl stearate

Morpholine
CAS 110-91-8; EINECS 203-815-1
Synonyms: Tetrahydro-1,4-oxazine; 1-Oxa-4-azacyclohexane; Tetrahydro-2H-1,4-oxazine; Diethylene oximide; Diethyleneimide oxide
Classification: Heterocyclic organic compd.
Empirical: C_4H_9NO
Formula: C_4H_8ONH
Properties: Colorless clear hygroscopic liq., amine-like odor; misc. with water, acetone, benzene, ether, castor oil, alcohol; m.w. 87.14; dens. 1.002 (20/20 C); b.p. 128.9 C; f.p. -4.9 C; flash pt. (OC) 37.7 C; autoignition temp. 590 F; ref. index 1.4540 (20 C)
Precaution: Flamm.; dangerous fire hazard exposed to flame, heat or oxidizers; reactive with oxidizers; explosive with nitromethane
Toxicology: TLV:TWA 20 ppm; LD50 (oral, rat) 1050 mg/kg; mod. toxic by ing., inh., skin contact, IP; corrosive irritant to skin, eyes, mucous membranes; experimental neoplastigen; mutagenic data; kidney damage; heated to dec., emits highly toxic fumes of NO_x
Uses: Used as salt of one or more fatty acids as a component of protective coatings for fresh fruits and vegetables; boiler water additive for food contact except milk prods.
Regulatory: FDA 21CFR §172.235, 173.310 (10 ppm max. in steam), 175.105, 176.210, 178.3300
Manuf./Distrib.: PMC Specialties

Morrhua oil. *See* Cod liver oil
Mossbunker oil. *See* Menhaden oil
Mother of thyme extract. *See* Wild thyme extract
Mountain tobacco. *See* Arnica
MPG. *See* Potassium glutamate
MPK. *See* Methyl propyl ketone

MSG
CAS 142-47-2; EINECS 205-538-1; EEC E621
Synonyms: Sodium glutamate; Monosodium glutamate; Sodium hydrogen L-glutamate; Glutamic acid

monosodium salt; Monosodium L-glutamate monohydrate
Definition: Monosodium salt of L-form of glutamic acid
Empirical: $C_5H_9NO_4$ • Na
Formula: $HOOCCH_2CH_2CHNH_2COONa$
Properties: Wh. cryst. or powd., sl. peptone-like odor, meal-like taste; very sol. in water; sl. sol. in alcohol; m.w. 170.14
Toxicology: LD50 (oral, rat) 17 g/kg; mod. toxic by IV; mildly toxic by ingestion, etc.; experimental teratogen, reproductive effects; human systemic effects by ingestion and IV; Chinese Restaurant Syndrome; heated to dec., emits toxic fumes of NO_x and Na_2O
Uses: Flavor enhancer for meat, poultry, sauces, soups; dietary supplement (Japan)
Usage level: ADI 0-120 mg/kg (FAO/WHO)
Regulatory: FDA 21CFR §145.131, 155.120, 155.130, 155.170, 155.200, 158.170, 161.190, 169.115, 169.140, 169.150, 172.320, 182.1, GRAS; USDA 9CFR §318.7, 381.147; not permitted for use in baby foods in UK
Regulatory: Japan approved; Europe listed; UK approved
Manuf./Distrib.: Able Prods.; ADM; Ajinomoto; Ashland; Browning; Calaga Food Ingreds.; Cornelius; Croxton & Garry; Penta Mfg.; Scanchem; Tesco; Todd's; Van Waters & Rogers; E H Worlee GmbH
Trade names: Asahi Aji®
Trade names containing: Luxor® 1517; Luxor® EB-2; Luxor® EB-400; Luxor® KB-350

MSP. *See* Sodium phosphate

Mullein extract
CAS 90064-13-4; 84012-25-9; 84650-17-9
Synonyms: Verbascum thapsis extract
Definition: Extract of common mullein, *Verbascum thapsis*
Uses: Natural flavoring
Regulatory: FDA 21CFR §172.510

Muriatic acid. *See* Hydrochloric acid
Muscatel oil. *See* Clary oil

Musk
Synonyms: Tonquin musk
Definition: Derived from musk deer, *Moschus moschiferus*
Properties: Potent penetrating musk odor with strong animal note
Uses: Natural flavoring agent
Usage level: 0.67 ppm (nonalcoholic beverages), 0.62 ppm (ice cream, ices), 2.0 ppm (candy), 2.7 ppm (baked goods), 3.0 ppm (syrups)
Regulatory: FDA 21CFR §182.50, GRAS; Japan approved

Mustard, black
Definition: From *Brassica nigra*
Uses: Natural flavoring agent
Regulatory: FDA 21CFR §182.10, 182.20, GRAS

Mustard oil. *See* Allyl isothiocyanate

Mustard oil, expressed
Definition: Fixed oil expressed from mustard seeds of *Brassica alba* and *B. nigra*, contg. chiefly the glycerides of oleic and other fatty acids
Properties: Straw-colored to brnsh. liq.; sl. sol. in alcohol; misc. with chloroform, ether, petrol. ether; insol. in water; dens. 0.914-0.916 (15/15 C); solid. pt. -8 to -16 C; iodine no. 92-97; sapon. no. 170-174; ref. index 1.4655-1.4670 (40 C)
Uses: Mfg. oleomargarine; salad oil

Mutton tallow. *See* Tallow
Mylase 100. *See* Amylase

Myrcene
CAS 123-35-3; FEMA 2762
Synonyms: 7-Methyl-3-methylene-1,6-octadiene
Empirical: $C_{10}H_{16}$
Properties: Oily liq., pleasant odor; sol. in alcohol, chloroform, ether; insol. in water; m.w. 136.23; dens. 0.791; b.p. 167 C; flash pt. 103 F; ref. index 1.4650
Precaution: Combustible
Uses: Synthetic flavoring
Usage level: 4.4 ppm (nonalcoholic beverages), 6.4 ppm (ice cream, ices), 0.50-13 ppm (candy), 4.9 ppm

(baked goods)
Regulatory: FDA 21CFR §172.515
Manuf./Distrib.: Aldrich

Myrcia oil. *See* Bay oil

Myristaldehyde
CAS 124-25-4; EINECS 204-692-7; FEMA 2763
Synonyms: Tetradecanal; Tetradecyl aldehyde; Aldehyde C-14; Myristic aldehyde
Empirical: $C_{14}H_{28}O$
Formula: $CH_3(CH_2)_{12}CHO$
Properties: Colorless to sl. yel. liq., strong fatty orris-like odor; insol. in water; m.w. 212.38; dens. 0.825-0.835; m.p. 23 C; b.p. 260 C; ref. index 1.43.80-1.4450
Uses: Synthetic flavoring
Usage level: 2.7 ppm (nonalcoholic beverages), 0.06-8.0 ppm (ice cream, ices), 1.9 ppm (candy), 0.08-24 ppm (baked goods), 0.15 ppm (gelatins, puddings)
Regulatory: FDA 21CFR §172.515

Myristalkonium chloride
CAS 139-08-2; EINECS 205-352-0
Synonyms: N,N-Dimethyl-N-tetradecylbenzenemethanaminium chloride; Myristyl dimethyl benzyl ammonium chloride; Tetradecyl dimethyl benzyl ammonium chloride
Classification: Quaternary ammonium salt
Empirical: $C_{23}H_{42}N•Cl$
Properties: m.w. 368.11
Toxicology: Skin and eye irritant; heated to decomp., emits very toxic fumes of NO_x, NH_3, and Cl^-
Uses: Antimicrobial for use in cane-sugar mills, sanitizer for food and beverage processing
Regulatory: FDA 21CFR §172.165 (limitation 3-12 ppm), 173.320 (limitation 0.6 ppm on wt. of raw sugarcane or raw beets), 175.105, 178.1010
Trade names: Cyncal®

Myristic acid
CAS 544-63-8; EINECS 208-875-2; FEMA 2764
Synonyms: Tetradecanoic acid; l-Tridecanecarboxylic acid; n-Tetradecoic acid
Classification: Organic acid
Empirical: $C_{14}H_{28}O_2$
Formula: $CH_3(CH_2)_{12}COOH$
Properties: Oily wh. cryst. solid; sol. in alcohol, ether, water; m.w. 228.36; dens. 0.8739 (80 C); m.p. 54.5 C; b.p. 326.2 C; flash pt. > 230 F
Precaution: Combustible
Toxicology: LD50 (IV, mouse) 43 mg/kg; poison by intravenous route; human skin irritant; mutagenic data; heated to decomp., emits acrid smoke and irritating fumes
Uses: Component of food-grade additives; defoaming agent, lubricant
Regulatory: FDA 21CFR §172.210, 172.860, 173.340, 175.105, 175.320, 176.170, 176.200, 176.210, 177.1010, 177.1200, 177.2260, 177.2600, 177.2800, 178.3570, 178.3910; GRAS (FEMA)
Manuf./Distrib.: Acme-Hardesty
Trade names: Emery® 6355

Myristic acid sodium salt. *See* Sodium myristate
Myristic aldehyde. *See* Myristaldehyde
Myristica oil. *See* Nutmeg oil
Myristin. *See* Trimyristin

Myristyl alcohol
CAS 112-72-1
Synonyms: 1-Tetradecanol
Empirical: $C_{14}H_{30}O$
Properties: Colorless to wh. waxy solid flakes; waxy odor; sol. in ether; sl. sol. in alcohol; insol. in water; m.w. 214.38; dens. 0.8355 (20/20 C); m.p. 38 C; b.p. 167 C; flash pt. 285 F
Precaution: Combustible
Uses: Synthetic flavoring agent

Myristyl alcohol mixed isomers
CAS 27196-00-5
Synonyms: Tetradecanol mixed isomers; Tetradecyl alcohol mixed isomers
Empirical: $C_{14}H_{30}O$

Properties: m.w. 214.44
Precaution: Combustible exposed to heat or flame; reactive with oxidizing materials
Toxicology: LD50 (oral, rat) 33 g/kg; mildly toxic by ingestion, skin contact; heated to decomp., emits acrid smoke and irritating fumes
Uses: Food additive
Regulatory: FDA 21CFR §172.864

Myristyl dimethyl benzyl ammonium chloride. *See* Myristalkonium chloride
Myroxylon pereirae oleoresin. *See* Balsam Peru

Myrrh extract
CAS 100084-96-6; EINECS 284-510-0
Synonyms: Commiphora extract
Definition: Extract of *Commiphora* spp.
Uses: Natural flavoring
Regulatory: FDA 21CFR §172.510; Japan approved
Manuf./Distrib.: Chart

Myrrh oil
Uses: Flavoring
Manuf./Distrib.: Pierre Chauvet

Myrtle extract
CAS 84082-67-7
Synonyms: Myrtus communis extract
Uses: Natural flavoring
Regulatory: FDA 21CFR §172.510; Japan approved
Manuf./Distrib.: Chart

Myrtus communis extract. *See* Myrtle extract

Nabam
CAS 142-59-6
Synonyms: DSE; Disodium ethylene bisdithiocarbamate; Ethylenebis (dithiocarbamate), disodium salt; Disodium ethylene-1,2-bisdithiocarbamate; Dithane A-40
Empirical: $O_4I_6N_2O_4 \cdot 2Na$
Formula: $NaSSCNHCH_2CH_2NHCSSNa$
Properties: Colorless crystals; sol. in water; m.w. 256.34
Toxicology: LD50 (oral, rat) 395 mg/kg; poison by ingestion; moderately toxic by IP route; skin irritant; experimental teratogen, reproductive effects; mutagenic data; heated to decomp., emits toxic fumes of Na_2O, NO_x, SO_x
Uses: Antimicrobial for use in cane-sugar and beet-sugar mills
Usage level: Limitation 3 ppm (raw sugarcane or raw beets)
Regulatory: FDA 21CFR §173.320

NaMBT. *See* Sodium 2-mercaptobenzothiazole

Naphtha
CAS 8030-30-6
Synonyms: Coal tar naphtha; Benzin; Petroleum naphtha; VM&P naphtha; Petroleum benzin; Petroleum ether; Petroleum spirit
Definition: Petroleum distillate
Properties: Dark straw-colored to colorless liq.; sol. in benzene, toluene, xylene; dens. 0.862-0.892; b.p. 149-216 C; flash pt. (CC) 107 F
Precaution: Flamm. when exposed to heat or flame; sl. explosion hazard; can react with oxidizing materials; keep containers tightly closed
Toxicology: Mildly toxic by inhalation; human poison and systemic effects by IV route; common air contaminant
Uses: Color diluent, protective coating solvent; used in egg shells, fresh fruits and vegetables
Regulatory: FDA 21CFR §73.1, 172.250
Manuf./Distrib.: Ashland

Naphthenic acid cobalt salt. *See* Cobalt naphthenate
Natacyn. *See* Natamycin

Natamycin
CAS 7681-93-8; EINECS 231-683-5
Synonyms: Pimaricin; Tennecetin; Natacyn
Definition: Antibiotic produced by a strain of *Streptomyces chattanoogensis*

Empirical: $C_{33}H_{47}NO_{13}$
Properties: Crystals; pract. insol. in higher alcohols, ether, esters, aromatic, aliphatic, or chlorinated hydrocarbons, ketones, oils; m.w. 665.75; dec. 280-300 C; sensitive to light
Toxicology: LD50 (oral, male rat) 2.73 g/kg; poison by IV, intramuscular, subcutaneous, IP routes; mod. toxic by ingestion; heated to decomp., emits toxic fumes of NO_x
Uses: Fungicide, mold inhibitor for cheese and wine use
Regulatory: FDA 21CFR §172.155 (limitation 200-300 ppm applied to cut cheese)
Trade names: Delvocid
Trade names containing: Natamax™

Natural Brown 10. *See* Caramel
Natural Orange 4. *See* Annatto
Natural Orange 4. *See* Annatto extract
Natural Red 4. *See* Carmine
Natural Yellow 3. *See* Turmeric
Natural Yellow 26. *See* Carotene
Navatone. *See* Acetanisole
NBR/PVC blend. *See* Acrylonitrile-butadiene/PVC blend
NDGA. *See* Nordihydroguairetic acid
(+)-Neomenthol. *See* d-Neomenthol

d-Neomenthol
CAS 2216-52-6; EINECS 218-691-4; FEMA 2666
Synonyms: 2-Isopropyl-5-methylcyclohexanol; (+)-Neomenthol; d-β-Pulegomenthol
Empirical: $C_{10}H_{20}O$
Properties: Liq., menthol-like odor; sol. in alcohol, acetone; insol. in water; m.w. 156.27; dens. 0.899 (20/4 C); m.p. -22 C; b.p. 209-210 C; flash pt. 83 C; ref. index 1.461 (20 C)
Uses: Synthetic flavoring
Usage level: 10 ppm (nonalcoholic beverages), 31 ppm (ice cream, ices), 50 ppm (candy), 48 ppm (baked goods)
Regulatory: FDA 21CFR §172.515
Manuf./Distrib.: Aldrich

Neral. *See* Citral

Nerol
CAS 106-25-2; EINECS 203-378-7
Synonyms: cis-3,7-Dimethyl-2,6-octadien-1-ol; 2,6-Dimethyl-2,6-octadien-8-ol
Definition: The cis-isomer of geraniol; found in many essential oils
Empirical: $C_{10}H_{18}O$
Formula: $(CH_3)_2C:CHCH_2CH_2C(CH_3):CHCH_2OH$
Properties: Liq., sweet rose odor, bitter flavor; sol. in abs. alcohol; m.w. 154.25; dens. 0.8813 (15 C); b.p. 224-225 (745 mm), 125 C (25 mm); ref. index 1.4730-1.4780 (20 C)
Precaution: Combustible
Uses: Synthetic flavoring
Usage level: 1.4 ppm (nonalcoholic beverages), 3.9 ppm (ice cream, ices), 16 ppm (candy), 19 ppm (baked goods), 1.0-1.3 ppm (gelatins, puddings), 0.80 ppm (chewing gum)
Regulatory: FDA 21CFR §172.515

Nerolidol. *See* Trimethyldodecatrieneol
Neroli oil. *See* Orange flower oil
Neroli oil, artificial. *See* Methyl anthranilate

Neryl acetate
CAS 141-12-8; EINECS 205-459-2; FEMA 2773
Synonyms: cis-3,7-Dimethyl-2,6-octadien-1-yl-acetate
Definition: Ester of nerol and acetic acid
Empirical: $C_{12}H_{20}O_2$
Formula: $(CH_3)_2C:CHCH_2CH_2C(CH_3):CHCH_2OCOCH_3$
Properties: Colorless to sl. yel. oily liq., sweet floral orange-blossom and rose-like odor, honey-like flavor; m.w. 196.29; dens. 0.912 (20/4 C); b.p. 234-236 C; flash pt. 210 F; ref. index 1.460 (20 C)
Uses: Synthetic flavoring
Usage level: 1.3 ppm (nonalcoholic beverages), 1.6 ppm (ice cream, ices), 5.1 ppm (candy), 15 ppm (baked goods)
Regulatory: FDA 21CFR §172.515
Manuf./Distrib.: Aldrich

Neryl butyrate
FEMA 2774
Uses: Synthetic flavoring
Regulatory: FDA 21CFR §172.515
Manuf./Distrib.: Aldrich

Neryl formate
CAS 2142-94-1; FEMA 2776
Uses: Synthetic flavoring
Regulatory: FDA 21CFR §172.515

Neryl isobutyrate
CAS 2345-24-6; FEMA 2775
Uses: Synthetic flavoring
Regulatory: FDA 21CFR §172.515
Manuf./Distrib.: Aldrich

Neryl isovalerate
FEMA 2778
Uses: Synthetic flavoring
Regulatory: FDA 21CFR §172.515
Manuf./Distrib.: Aldrich

Neryl propionate
FEMA 2777
Definition: Ester of nerol and propionic acid
Empirical: $C_{13}H_{22}O_2$
Properties: Colorless oily liq., ether-like sweet intense fruity odor, plum-like taste; sol. in alcohol; sl. sol. in water; m.w. 210.31; b.p. 233 C; ref. index 1.4550
Uses: Synthetic flavoring
Usage level: 6.3 ppm (nonalcoholic beverages), 23 ppm (ice cream, ices), 21 ppm (candy, baked goods)
Regulatory: FDA 21CFR §172.515

Niacin. *See* Nicotinic acid

Niacinamide
CAS 98-92-0; EINECS 202-713-4
Synonyms: Nicotinamide; 3-Pyridinecarboxamide; Nicotinic acid amide
Classification: Heterocyclic aromatic amide
Empirical: $C_6H_6N_2O$
Formula: $C_5H_4NCONH_2$
Properties: Colorless needles or wh. cryst. powd., odorless, bitter taste; sol. in water, ethanol, glycerin; m.w. 122.14; dens. 1.40; m.p. 129 C
Toxicology: Moderately toxic by ingestion, intravenous, intraperitoneal and subcutaneous routes; mutagenic data; heated to decomp., emits toxic fumes of NO_x
Uses: Direct food additive, nutrient, dietary supplement to prevent pellagra; may be used in infant formulas
Regulatory: FDA 21CFR §182.5535, 184.1535, GRAS
Manuf./Distrib.: Am. Roland
Trade names: Niacinamide USP, FCC No. 69905; Niacinamide USP, FCC Fine Granular No. 69916
Trade names containing: Rocoat® Niacinamide 33$^1/_3$% No. 69907; Rocoat® Niacinamide 33$^1/_3$% Type S No. 69909

Niacinamide ascorbate
Synonyms: Nicotinamide-ascorbic acid complex
Definition: Complex of ascorbic acid and niacinamide
Properties: Lemon-yel. powd., odorless; sol. in water, alcohol; sl. sol. in glycerol; insol. in benzene; m.p. 141-145 C
Uses: Nutrient, dietary supplement; used as a source of ascorbic acid and nicotinamide in multivitamin preps.
Regulatory: FDA 21CFR §172.315

Nickel
CAS 7440-02-0; EINECS 231-111-4
Synonyms: Nickel catalysts
Classification: Metallic element
Empirical: Ni
Properties: Malleable silvery metal; high ductility and malleability; insol. in water; at.wt. 58.69; dens. 8.9; m.p. 1452 C; b.p. 2900 C; corrosion resistant

Precaution: Powders can ignite spontaneously in air; incompat. with oxidants

Toxicology: TLV (metal) 1 mg/m^3 of air; poison by ingestion, IV, IP, subcut. routes; experimental carcinogen, neoplastigen, tumorigen, teratogen, reproductive effects; mutagenic data; hypersensitivity can cause dermatitis, pulmonary sathma, conjunctivitis

Uses: Direct food additive, catalyst; used in hydrogenation of fats and oils; processing aid (Japan)

Regulatory: FDA 21CFR §184.1537, GRAS; USDA 9CFR §318 (must be eliminated during processing); Japan approved

Manuf./Distrib.: Atomergic Chemetals

Nickel catalysts. *See* Nickel
Niconate. *See* Ferrous gluconate
Nicotinamide. *See* Niacinamide
Nicotinamide-ascorbic acid complex. *See* Niacinamide ascorbate

Nicotinic acid

CAS 59-67-6; EINECS 200-441-0; EEC E375

Synonyms: Niacin; Vitamin B$_3$; 3-Picolinic acid; Pyridine-3-carboxylic acid

Classification: Heterocyclic aromatic compd.

Empirical: C$_6$H$_5$NO$_2$

Properties: Colorless needles or wh. cryst. powd., sl. odor, sour taste; nonhygroscopic; sol. in water, alcohol; insol. in most lipid solvs., ether; m.w. 123.12; dens. 1.473; m.p. 236 C, subl. above m.p.

Toxicology: LD50 (oral, rat) 7000 mg/kg; poison by intraperitoneal route; moderately toxic by ingestion, intravenous and subcutaneous routes; megadoses may cause itching, nausea, headaches; experimental carcinogen; heated to decomp., emits toxic fumes of NO$_x$

Uses: Direct food additive, nutrient, dietary supplement; anti-pellagra vitamin; color protectant; may be used in infant formulas; color retention agent, dietary supplement (Japan)

Regulatory: FDA 21CFR §135.115, 137, 139, 182.5530, 184.1530, GRAS; Japan restricted; Europe listed; UK approved

Manuf./Distrib.: Am. Roland

Trade names: Lutavit® Niacin; Niacin USP, FCC Fine Granular No. 69901; Niacin USP, FCC No. 69902

Nicotinic acid aluminum salt. *See* Aluminum nicotinate
Nicotinic acid amide. *See* Niacinamide
Niobe oil. *See* Methyl benzoate

Nisin

CAS 1414-45-5; EINECS 215-807-5; EEC E234

Definition: Derived from pure culture fermentation of *Streptococcus lactis* (*Lactococcus lactis*)

Empirical: C$_{143}$H$_{230}$N$_{42}$O$_{37}$S$_7$

Properties: Cryst.; sol. in dil. acids; m.w. 3354.25

Toxicology: Nontoxic; LD50 (oral, mice) 6950 mg/kg; heated to decomp., emits acrid smoke and irritating fumes

Uses: Preservative, antimicrobial agent to inhibit *Clostridium botulinum* spores and toxin formation in pasteurized cheese spreads, canned fruits and vegetables

Usage level: ADI 2.9 mg/day (FDA), 33,000 units/kg (EEC)

Regulatory: FDA 21CFR §184.1538, GRAS; limitation 250 ppm nisin in finished prod.; Europe listed; UK approved

Niter. *See* Potassium nitrate
Nitre. *See* Potassium nitrate
Nitre cake. *See* Sodium bisulfate solid
Nitrilotriacetic acid sodium salt. *See* Trisodium NTA
2,2',2''-Nitrilotris(ethanol). *See* Triethanolamine

Nitrites

Definition: Salts of nitrous acid

Properties: Wh. to pale yel. cryst., hygroscopic; sol. in water, liq. ammonia

Precaution: Generally powerful oxidizers; fire and explosion hazards variable; violent reaction may occur on contact with readily oxidized materials; organic nitrites can decompose violently

Toxicology: Ingestion of large amts. can cause nausea, vomiting, cyanosis, collapse, coma; small doses can cause blood pressure decrease, rapid pulse, headache; may form carcinogenic nitrosamines in the body

Uses: Curing agent for meat, bacon, poultry prods.

Usage level: Limitation 200 ppm of nitrite (calculated as sodium nitrate) in finished prod.

Regulatory: FDA 21CFR §170.60, 172.170, 172.175; USDA 9CFR §318.7

Nitrocellulose

CAS 9004-70-0

Synonyms: Cellulose, nitrate; Cellulose tetranitrate; Pyroxylin; Nitrocotton; Collodion cotton; Guncotton
Classification: Cellulose deriv.
Empirical: $C_{12}H_{16}O_{18}N_4$
Formula: $C_{12}H_{16}(ONO_2)_4O_6$
Properties: Colorless liq. or wh. amorphous solid; sol. in acetone; insol. in ether-alcohol mixt.; m.w. 504.3; dens. 1.66; flash pt. 55 F
Precaution: Flamm. solid; highly dangerous exposed to heat, flame, strong oxidizers; ignites easily; explodes
Uses: Used in paper/paperboard for food pkg.
Regulatory: FDA 21CFR §175.105, 175.300, 176.170, 177.1200, 181.22, 181.30
Manuf./Distrib.: Allchem Industries; Asahi Chem Industry Co Ltd; Hercules

Nitrocotton. *See* Nitrocellulose

Nitrogen
CAS 7727-37-9; EINECS 231-783-9
Classification: Gaseous element
Empirical: N_2
Properties: Colorless gas, odorless, flavorless; sl. sol. in alcohol; sparingly sol. in water; at. wt. 14.0067; m.w. 28.01; m.p. -210 C; b.p. -195.79 C
Precaution: Combustible
Toxicology: Asphyxiant in high concs.
Uses: Direct food additive, propellant, aerating agent, processing aid, and gas; liq. nitrogen in food-freezing processes
Regulatory: FDA 21CFR §169.115, 169.140, 169.150, 184.1540, GRAS; Japan approved
Manuf./Distrib.: Aldrich

Nitrogen monoxide. *See* Nitrous oxide
Nitrogen oxide. *See* Nitrous oxide

2-Nitropropane
CAS 79-46-9; EINECS 201-209-1
Synonyms: sec-Nitropropane
Classification: Nitroparaffin
Empirical: $C_3H_7NO_2$
Formula: $CH_3CH(NO_2)CH_3$
Properties: Colorless liq.; sl. sol. in water; misc. with many org. solvs.; m.w. 89.09; dens. 0.992 (20/20 C); m.p. -93 C; b.p. 119-122 C; flash pt. (TOC) 75 F
Precaution: Flamm. when exposed to heat, open flame, oxidizers; may explode on heating
Toxicology: TLV:CL 25 ppm in air; LD50 (oral, rat) 725 mg/kg; poison by ingestion, inhalation, and intraperitoneal routes (nausea, diarrhea, anorexia); suspected carcinogen; mutagenic
Uses: Solv. for fractionating edible fats and oils
Manuf./Distrib.: Ashland

sec-Nitropropane. *See* 2-Nitropropane
Nitrous acid potassium salt. *See* Potassium nitrite
Nitrous acid sodium salt. *See* Sodium nitrite
Nitrous ether. *See* Ethyl nitrite

Nitrous oxide
CAS 10024-97-2; EINECS 233-032-0
Synonyms: Nitrogen monoxide; Nitrogen oxide; Dinitrogen monoxide; Laughing gas
Classification: Gas
Empirical: N_2O
Properties: Colorless gas, sl. sweet odor; m.w. 44.01; m.p. -91 C; b.p. -88 C
Precaution: Does not burn but will support combustion; oxidizer
Toxicology: Asphyxiant at high concs.
Uses: Direct food additive, propellant, aerating agent, gas; used in dairy prods.
Regulatory: FDA 21CFR §184.1545, GRAS
Manuf./Distrib.: Aldrich

trans,trans-2,4-Nonadienal
CAS 5910-87-2; FEMA 3212
Empirical: $C_9H_{14}O$
Properties: m.w. 138.21
Uses: Synthetic flavoring
Manuf./Distrib.: Aldrich

Nonadienol. *See* 2,6-Nonadien-1-ol

2,6-Nonadien-1-ol
CAS 28069-72-9; FEMA 2780
Synonyms: Cucumber alcohol; Nonadienol; Violet leaf alcohol
Empirical: $C_9H_{16}O$
Properties: Colorless oily liq.; sol. in alcohol, propylene glycol; sl. sol. in water; m.w. 140.23; dens. 0.87; b.p. 196 C
Uses: Synthetic flavoring
Usage level: 0.01 ppm (nonalcoholic beverages, alcoholic beverages, baked goods), 0.05 ppm (ice cream, ices), 0.05-0.50 ppm (candy)
Regulatory: FDA 21CFR §172.515
Manuf./Distrib.: Aldrich

γ-Nonalactone
CAS 104-61-0; EINECS 203-219-1; FEMA 2781
Synonyms: Coconut aldehyde; Aldehyde C-18; γ-N-Amylbutyrolactone; 4-Hydroxynonanoic acid, γ-lactone
Classification: Heterocyclic compd.
Empirical: $C_9H_{16}O_2$
Properties: Colorless to sl. yel. liq., coconut odor; sol. in alcohol, fixed oils, propylene glycol; insol. in water; m.w. 156.25; dens. 0.958-0.966; b.p. 243 C; flash pt. > 212 F; ref. index 1.446-1.450
Precaution: Combustible liq.
Toxicology: LD50 (oral, rat) 6600 mg/kg; mod. toxic by ingestion; skin irritant; heated to decomp., emits acrid smoke and irritating fumes
Uses: Synthetic flavoring agent for baked goods, candy, gelatin, ice cream, puddings
Regulatory: FDA 21CFR §172.515; Japan approved as flavoring
Manuf./Distrib.: Aldrich

Nonalol. *See* Nonyl alcohol

Nonanal
CAS 124-19-6; EINECS 204-688-5; FEMA 2782
Synonyms: Aldehyde C-9; Nonanoic aldehyde; Pelargonic aldehyde; Pelargonaldehyde
Empirical: $C_9H_{18}O$
Formula: $CH_3(CH_2)_7CHO$
Properties: Colorless to lt. yel. liq., strong fatty odor; sol. in alcohol; insol. in water; m.w. 142.24; dens. 0.823 (20/4 C); b.p. 79-81 C (12 mm); flash pt. 63 C; ref. index 1.425 (20 C)
Precaution: Combustible
Toxicology: Skin and eye irritant
Uses: Synthetic flavoring
Usage level: 1.3 ppm (nonalcoholic beverages, ice cream, ices), 4.1 ppm (candy), 2.3 ppm (baked goods), 6.0 ppm (gelatins, puddings), 0.20-38 ppm (chewing gum)
Regulatory: FDA 21CFR §172.515
Manuf./Distrib.: Aldrich

Nonanediol-1,3-acetate. *See* 1,3-Nonanediol acetate, mixed esters

1,3-Nonanediol acetate, mixed esters
FEMA 2783
Synonyms: Hexylene glycol diacetate; Nonanediol-1,3-acetate
Empirical: $C_{13}H_{24}O_4$
Properties: Colorless to sl. yel. liq., floral odor; sl. sol. in water; m.w. 244.34; flash pt. > 100 C; ref. index 1.4410-1.4450
Uses: Synthetic flavoring
Usage level: 0.30-1.0 ppm (nonalcoholic beverages), 0.50-1.0 ppm (ice cream, ices), 1.5-6.0 ppm (candy), 1.5-4.0 ppm (baked goods)
Regulatory: FDA 21CFR §172.515

Nonanoic acid
CAS 112-05-0; EINECS 203-931-2; FEMA 2784
Synonyms: Pelargonic acid; Carboxylic acid C_9; Nonylic acid; Nonoic acid
Empirical: $C_9H_{18}O_2$
Formula: $CH_3(CH_2)_7COOH$
Properties: Colorless oily liq., char. fatty odor; cryst. when cooled; sol. in alcohol, chloroform, ether; pract. insol. in water; m.w. 158.24; dens. 0.907 (20/4 C); m.p. 10-12 C; b.p. 252-253 C (756 mm); flash pt. 129 C; ref. index 1.433 (20 C)
Precaution: Corrosive

Toxicology: LD50 (IV, mouse) 224 ± 4.6 mg/kg; strong irritant to skin and eyes
Uses: Synthetic flavoring
Usage level: 1.8 ppm (nonalcoholic beverages), 7.8 ppm (ice cream, ices), 6.6 ppm (candy), 13 ppm (baked goods), 10 ppm (shortening)
Regulatory: FDA 21CFR §172.515, 173.315
Manuf./Distrib.: Aldrich

Nonanoic acid ethyl ester. *See* Ethyl pelargonate
Nonanoic aldehyde. *See* Nonanal
1-Nonanol. *See* Nonyl alcohol
Nonanol acetate. *See* n-Nonyl acetate

2-Nonanone
CAS 821-55-6; EINECS 212-480-0; FEMA 2785
Synonyms: Methyl heptyl ketone; Heptyl methyl ketone
Empirical: $C_9H_{18}O$
Formula: $CH_3(CH_2)_6COCH_3$
Properties: Colorless oily liq., char. rue odor, rose tea-like flavor; sol. in alcohol; insol. in water; m.w. 142.24; dens. 0.82 (20/4 C); m.p. -21 C; b.p. 72-74 C (10 mm); flash pt. 68 C; ref. index 1.421 (20 C)
Uses: Synthetic flavoring
Usage level: 0.55 ppm (nonalcoholic beverages), 0.10-1.0 ppm (ice cream, ices), 0.40-4.0 ppm (candy, baked goods)
Regulatory: FDA 21CFR §172.515
Manuf./Distrib.: Hüls AG

3-Nonanon-1-yl acetate
Synonyms: 1-Hydroxy-3-nonanone acetate
Uses: Synthetic flavoring
Regulatory: FDA 21CFR §172.515

Nonate. *See* Isoamyl nonanoate
Noncarbinol. *See* n-Decyl alcohol

trans-2-Nonenal
FEMA 3213
Empirical: $C_9H_{16}O$
Properties: Wh. to sl. ylsh. liq., fatty violet odor; sol. in alcohol; insol. in water; m.w. 140.22; dens. 0.850-0.870; ref. index 1.457-1.460
Uses: Synthetic flavoring agent
Manuf./Distrib.: Aldrich

cis-6-Nonen-1-ol
FEMA 3465
Empirical: $C_9H_{18}O$
Properties: Wh. to sl. yel. liq., powerful melon-like odor; insol. in water; m.w. 142.23; dens. 0.850-0.870; ref. index 1.448-1.450
Uses: Synthetic flavoring agent

trans-2-Nonen-1-ol
FEMA 3379
Empirical: $C_9H_{18}O$
Properties: Wh. liq., fatty violet odor; insol. in water; m.w. 142.23; dens. 0.830-0.850; ref. index 1.444-1.448
Uses: Synthetic flavoring agent

Nonfat dry milk
Synonyms: Milk, nonfat dry; Powdered skim milk; Nonfat milk
Definition: Solid residue from dehydration of defatted cow's milk
Uses: Food ingredient; used in processed meat prods.
Regulatory: FDA 21CFR §131.125
Manuf./Distrib.: Blossom Farm Prods.; Browning
Trade names containing: Ches® 500; Hentex-35; Medium 10; Medium 55; Medium 700; Medium KL; Shedd's Wonder Shortening

Nonfat milk. *See* Nonfat dry milk
Nonoic acid. *See* Nonanoic acid

Nonoxynol-6
CAS 9016-45-9 (generic); 26027-38-3 (generic); 37205-87-1 (generic); 27177-01-1; 27177-05-5

Nonoxynol-7

Synonyms: PEG-6 nonyl phenyl ether; POE (6) nonyl phenyl ether; PEG 300 nonyl phenyl ether
Classification: Ethoxylated alkyl phenol
Empirical: $C_{27}H_{48}O_7$
Formula: $C_9H_{19}C_6H_4(OCH_2CH_2)_nOH$, avg. n = 6
Properties: Yel to almost colorless liq.
Toxicology: Moderately toxic by ingestion, skin contact; severe eye and mild skin irritant in humans; heated to dec., emits acrid smoke and fumes
Uses: Surfactant for food applics.
Regulatory: FDA 21CFR §175.105, 176.180, 176.210, 178.3400
Trade names: Hyonic NP-60

Nonoxynol-7

CAS 9016-45-9 (generic); 26027-38-3 (generic); 27177-05-5; 37205-87-1 (generic); EINECS 248-292-0
Synonyms: PEG-7 nonyl phenyl ether; POE (7) nonyl phenyl ether
Classification: Ethoxylated alkyl phenol
Empirical: $C_{29}H_{52}O_8$
Formula: $C_9H_{19}C_6H_4(OCH_2CH_2)_nOH$, avg. n = 7
Properties: Yel to almost colorless liq.
Toxicology: Moderately toxic by ingestion, skin contact; severe eye and mild skin irritant in humans; heated to dec., emits acrid smoke and fumes
Uses: Intermediate for prod. of food-grade emulsifiers, lubricants, release agents, binders
Regulatory: FDA 21CFR §175.105, 176.180, 176.210, 178.3400
Trade names: Iconol NP-7

Nonoxynol-10

CAS 9016-45-9 (generic); 26027-38-3 (generic); 27177-08-8; 37205-87-1 (generic); 27942-26-3; EINECS 248-294-1
Synonyms: PEG-10 nonyl phenyl ether; POE (10) nonyl phenyl ether; PEG 500 nonyl phenyl ether
Classification: Ethoxylated alkyl phenol
Empirical: $C_{35}H_{64}O_{11}$
Formula: $C_9H_{19}C_6H_4(OCH_2CH_2)_nOH$, avg. n = 10
Properties: Yel to almost colorless liq.
Toxicology: Moderately toxic by ingestion, skin contact; severe eye and mild skin irritant in humans; heated to dec., emits acrid smoke and fumes
Uses: Solubilizer for essential oils, surfactant, wetting agent, emulsifier for food applics.
Regulatory: FDA 21CFR §175.105, 176.180, 176.210, 178.3400
Trade names: Cremophor® NP 10; Hyonic NP-100

Nonoxynol-11

CAS 9016-45-9 (generic); 26027-38-3 (generic); 37205-87-1 (generic)
Synonyms: PEG-11 nonyl phenyl ether; POE (11) nonyl phenyl ether
Classification: Ethoxylated alkyl phenol
Formula: $C_9H_{19}C_6H_4(OCH_2CH_2)_nOH$, avg. n = 11
Properties: Yel to almost colorless liq.
Toxicology: Moderately toxic by ingestion, skin contact; severe eye and mild skin irritant in humans; heated to dec., emits acrid smoke and fumes
Uses: Surfactant, wetting agent, emulsifier for food applics.
Regulatory: FDA 21CFR §175.105, 176.180, 176.210, 178.3400
Trade names: Hyonic NP-110

Nonoxynol-12

CAS 9016-45-9 (generic); 26027-38-3 (generic); 37205-87-1 (generic)
Synonyms: PEG-12 nonyl phenyl ether; POE (12) nonyl phenyl ether; PEG 600 nonyl phenyl ether
Classification: Ethoxylated alkyl phenol
Formula: $C_9H_{19}C_6H_4(OCH_2CH_2)_nOH$, avg. n = 12
Properties: Yel to almost colorless liq.
Toxicology: Moderately toxic by ingestion, skin contact; severe eye and mild skin irritant in humans; heated to dec., emits acrid smoke and fumes
Uses: Detergent, wetting agent, coemulsifier for food applics.
Regulatory: FDA 21CFR §175.105, 176.180, 176.210, 178.3400
Trade names: Hyonic NP-120

Nonoxynol-14

CAS 9016-45-9 (generic); 26027-38-3 (generic); 37205-87-1 (generic)
Synonyms: PEG-14 nonyl phenyl ether; POE (14) nonyl phenyl ether

Classification: Ethoxylated alkyl phenol
Formula: $C_9H_{19}C_6H_4(OCH_2CH_2)_nOH$, avg. n = 14
Properties: Yel to almost colorless liq.
Toxicology: Moderately toxic by ingestion, skin contact; severe eye and mild skin irritant in humans; heated to dec., emits acrid smoke and fumes
Uses: Solubilizer for essential oils and flavors
Regulatory: FDA 21CFR §175.105, 176.180, 176.210, 178.3400
Trade names: Cremophor® NP 14

Nonoxynol-40

CAS 9016-45-9 (generic); 26027-38-3 (generic); 37205-87-1 (generic)
Synonyms: PEG-40 nonyl phenyl ether; POE (40) nonyl phenyl ether; PEG 2000 nonyl phenyl ether
Classification: Ethoxylated alkyl phenol
Formula: $C_9H_{19}C_6H_4(OCH_2CH_2)_nOH$, avg. n = 40
Properties: Pale yel. to off-white pastes or waxes
Toxicology: Moderately toxic by ingestion, skin contact; severe eye and mild skin irritant in humans; heated to dec., emits acrid smoke and fumes
Uses: Surfactant for food applics.
Regulatory: FDA 21CFR §175.105, 176.180, 178.3400
Trade names: Hyonic NP-407

Nonoxynol-50

CAS 9016-45-9 (generic); 26027-38-3 (generic); 37205-87-1 (generic)
Synonyms: PEG-50 nonyl phenyl ether; POE (50) nonyl phenyl ether
Classification: Ethoxylated alkyl phenol
Formula: $C_9H_{19}C_6H_4(OCH_2CH_2)_nOH$, avg. n = 50
Properties: Pale yel. to off-white pastes or waxes
Toxicology: Moderately toxic by ingestion, skin contact; severe eye and mild skin irritant in humans; heated to dec., emits acrid smoke and fumes
Uses: Surfactant for food applics.
Regulatory: FDA 21CFR §176.180, 178.3400
Trade names: Hyonic NP-500

Nonyl acetate. *See* n-Nonyl acetate

n-Nonyl acetate

CAS 143-13-5; EINECS 205-585-8; FEMA 2788
Synonyms: Acetic acid n-nonyl ester; Nonyl acetate; n-Nonyl ethanoate; Pelargonyl acetate; Nonanol acetate; Acetate C-9
Definition: Ester of nonyl alcohol and acetic acid
Empirical: $C_{11}H_{22}O_2$
Formula: $CH_3COO(CH_2)_8CH_3$
Properties: Colorless liq., pungent odor, suggestive of mushrooms, gardenia when dil.; sol. in abs. alcohol, ether; insol. in water; m.w. 186.29; dens. 0.864; b.p. 208-212 C; flash pt. > 153 F; ref. index 1.422
Precaution: Combustible liq.
Toxicology: Heated to decomp., emits acrid smoke and irritating fumes
Uses: Synthetic flavoring agent
Usage level: 0.81 ppm (nonalcoholic beverages), 0.81 ppm (ice cream), 1.9 ppm (candy), 3.1 ppm (baked goods)
Regulatory: FDA 21CFR §172.515; GRAS (FEMA)

3-Nonyl acrolein. *See* 2-Dodecenal

Nonyl alcohol

CAS 143-08-8; EINECS 205-583-7; FEMA 2789
Synonyms: 1-Nonanol; Nonalol; Alcohol C-9; n-Nonyl alcohol
Empirical: $C_9H_{20}O$
Formula: $CH_3(CH_2)_8OH$
Properties: Colorless to ylsh. liq., citronella oil odor; misc. with alcohol, ether; pract. insol. in water; m.w. 144.26; dens. 0.8279 (20/4 C); m.p. -6 to -4 C; b.p. 210-213 C; flash pt. 98 C; ref. index 1.4338 (20 C)
Uses: Synthetic flavoring; mfg. of artificial lemon oil
Regulatory: FDA 21CFR §172.515
Manuf./Distrib.: Aldrich

n-Nonyl alcohol. *See* Nonyl alcohol
Nonyl caprylate. *See* Nonyl octanoate

Nonylcarbinol. *See* n-Decyl alcohol
n-Nonyl ethanoate. *See* n-Nonyl acetate
Nonylic acid. *See* Nonanoic acid

Nonyl isovalerate
Uses: Synthetic flavoring
Regulatory: FDA 21CFR §172.515

Nonyl octanoate
FEMA 2790
Synonyms: Nonyl caprylate; n-Nonyl octoate; Nonyl octylate
Definition: Ester of n-nonanol and n-octanoic acid
Empirical: $C_{17}H_{34}O_2$
Properties: Colorless oily liq., sweet rose odor; sol. in alcohol; insol. in water; m.w. 270.46; dens. 0.86; b.p. 315 C
Uses: Synthetic flavoring
Usage level: 2.0 ppm (nonalcoholic beverages), 0.06 ppm (baked goods)
Regulatory: FDA 21CFR §172.515

n-Nonyl octoate. *See* Nonyl octanoate
Nonyl octylate. *See* Nonyl octanoate

Nootkatone
Synonyms: 5,6-Dimethyl-8-isopropenyl bicyclo[4.4.0]-dec-1-en-3-one
Uses: Synthetic flavoring
Regulatory: FDA 21CFR §172.515
Manuf./Distrib.: Acme-Hardesty

Nopinene. *See* β-Pinene

Nordihydroguairetic acid
CAS 500-38-9; EINECS 207-903-0
Synonyms: NDGA; 4,4´-(2,3-Dimethyl-1,4-butanediyl)bis[1,2-benzenediol]; 4,4´-(2,3-Dimethyltetramethyl-ene)dipyrocatechol; 1,4-Bis(3,4-dihydroxyphenyl)-2,3-dimethylbutane
Classification: Organic compd.
Empirical: $C_{18}H_{22}O_4$
Properties: Cryst.; sol. in ethanol, methanol, ether, acetone, glycerin, propylene glycol; sl. sol. in hot water, chloroform; sol. in dil. alkalies; m.w. 302.36; m.p. 184-185 C
Toxicology: Irritant; harmful if swallowed
Uses: Antioxidant when migrating from food pkg.
Regulatory: FDA 21CFR §175.300, 181.22, 181.24 (0.005% migrating from food pkg.), 189.165; Japan approved (0.1 g/kg max.)
Manuf./Distrib.: Aldrich

Norgine. *See* Alginic acid
NT red. *See* FD&C Red No. 40

Nutmeg oil
CAS 8008-45-5; FEMA 2793
Synonyms: Myristica oil; East Indian nutmeg oil
Definition: Oil extracted from kernel of *Myristica fragrans*; consists of α- and β-pinene, camphene, myristicin, dipentene, sabanene
Properties: Colorless to pale yel. liq., nutmeg odor and taste; very sol. in hot alcohol, chloroform, ether; sol. in fixed oils, min. oil; sl. sol. in cold alcohol; insol. in glycerin, propylene glycol, water; dens. 0.880-0.910; ref. index 1.474-1.488
Precaution: Keep cool, well closed; protect from light
Toxicology: LD50 (oral, rat) 2620 mg/kg; highly toxic—as little as 5 g can cause nausea, vomiting, and death; experimental reproductive effects; mutagenic data; skin irritant; heated to decomp., emits acrid smoke and irritating fumes
Uses: Natural flavoring agent for cakes, eggnog, fruit, puddings, cola flavors, meat seasonings
Usage level: 14 ppm (nonalcoholic beverages), 13 ppm (ice cream), 19 ppm (candy), 75 ppm (baked goods), 1.2-640 ppm (chewing gum), 21 ppm (condiments), 150 ppm (meat), 2-30 ppm (icings), 16 ppm (syrups)
Regulatory: FDA 21CFR §182.10, 182.20, GRAS; GRAS (FEMA); Japan approved; Europe listed (< 1 to 15 ppm safrole)
Manuf./Distrib.: Chart; Pierre Chauvet; Commodity Services; Florida Treatt

Nutmeg oil, expressed
CAS 8007-12-3
Definition: Oil from steam distillation of dried arillode of ripe seed of *Myristica fragrans*

Properties: Colorless to pale yel. liq., nutmeg odor and taste; sol. in fixed oils, min. oil; very sol. in hot alcohol, chloroform, ether; E. Indian: dens. 0.880-0.930; ref. index 1.474-1.488; W. Indian: dens. 0.854-0.880; ref. index 1.469-1.480

Toxicology: LD50 (oral, rat) 3640 mg/kg; mod. toxic by ingestion; skin irritant; human ingestion causes symptoms similar to volatile oil of nutmeg; heated to decomp., emits acrid smoke and irritating fumes

Uses: Natural flavoring agent in bread, cakes, chocolate pudding, fruit salad

Regulatory: FDA 21CFR §182.10, 182.20, GRAS

Oak bark extract

Definition: Extract of the bark of oak trees, *Quercus* species

Uses: Natural flavoring

Regulatory: FDA 21CFR §172.510

Oakmoss extract

CAS 90028-68-5; EINECS 289-861-3

Synonyms: Evernia prunastri extract

Definition: Extract of *Evernia prunastri*

Uses: Natural flavoring

Regulatory: FDA 21CFR §172.510

Manuf./Distrib.: Pierre Chauvet

Oat fiber

Uses: Dietary fiber supplement for cereals, snacks, dairy prods.; humectant for meat prods.; partial flour replacement for baked goods, cereals, snacks

Trade names containing: Stabilized Cookie Blend

Ocimene

Synonyms: trans-β-Ocimene; 3,7-Dimethyl-1,3,6-octatriene

Definition: Derived from leaves of *Ocimum basilicum*, etc.

Empirical: $C_{10}H_{16}$

Properties: Oil; sol. in alcohol, chloroform, ether, glac. acetic acid; pract. insol. in water; m.w. 136.23; dens. 0.799 (20/4 C); b.p. 81 C (30 mm); ref. index 1.4893 (20 C)

Precaution: Combustible

Uses: Synthetic flavoring

Regulatory: FDA 21CFR §172.515

trans-β-Ocimene. *See* Ocimene
Ocimum basilicum extract. *See* Basil extract
Ocimum basilicum oil. *See* Basil oil

Octacosanol

CAS 557-61-9

Synonyms: 1-Octacosanol; n-Octacosanol; Octacosyl alcohol

Definition: Constituent of vegetable waxes

Empirical: $C_{28}H_{58}O$

Properties: Sol. in carbon disulfide, other fat solvs., oils; insol. in water; m.w. 410.74; m.p. 83.4 C

Uses: Used in high potency direct compression prods. for tabletting, capsules, powd. mixes

Trade names: Viobin Octacosanol

1-Octacosanol. *See* Octacosanol
n-Octacosanol. *See* Octacosanol
Octacosyl alcohol. *See* Octacosanol
9,12-Octadecadienamide. *See* Linoleamide
9,12-Octadecadienoic acid. *See* Linoleic acid
(Z,Z)-9,12-Octadecadienoic acid. *See* Linoleic acid
9,12-Octadecadienoic acid, 2,3-dihydroxypropyl ester. *See* Glyceryl linoleate
9,12-Octadecadienoic acid methyl ester. *See* Methyl linoleate
9,12-Octadecadienoic acid, monoester with 1,2,3-propanetriol. *See* Glyceryl linoleate
Octadecanamide. *See* Stearamide
1-Octadecanamine. *See* Stearamine
n-Octadecanoic acid. *See* Stearic acid
Octadecanoic acid butyl ester. *See* Butyl stearate
Octadecanoic acid, 2-(1-carboxyethoxy)-1-methyl-2-oxoethyl ester, sodium salt. *See* Sodium stearoyl lactylate
Octadecanoic acid, decaester with decaglycerol. *See* Polyglyceryl-10 decastearate
Octadecanoic acid, diester with 1,2,3-propanetriol. *See* Glyceryl distearate

Octadecanoic acid, magnesium salt. *See* Magnesium stearate
Octadecanoic acid, methyl ester. *See* Methyl stearate
Octadecanoic acid, monoester with decaglycerol. *See* Polyglyceryl-10 stearate
Octadecanoic acid, monoester with octaglycerol. *See* Polyglyceryl-8 stearate
Octadecanoic acid, monoester with 1,2-propanediol. *See* Propylene glycol stearate
Octadecanoic acid, monoester with tetraglycerol. *See* Polyglyceryl-4 stearate
Octadecanoic acid, potassium salt. *See* Potassium stearate
Octadecanoic acid, 1,2,3-propanetriyl ester. *See* Tristearin
Octadecanoic acid sodium salt. *See* Sodium stearate
Octadecanoic acid, tetraester with diglycerol. *See* Polyglyceryl-2 tetrastearate
Octadecanoic acid, zinc salt. *See* Zinc stearate
1-Octadecanol. *See* Stearyl alcohol
n-Octadecanol. *See* Stearyl alcohol
9-Octadecenamide. *See* Oleamide
9-Octadecenoic acid. *See* Oleic acid
cis-9-Octadecenoic acid. *See* Oleic acid
9-Octadecenoic acid aluminum salt. *See* Aluminum oleate
9-Octadecenoic acid calcium salt. *See* Calcium oleate
9-Octadecenoic acid, diester with 1,2,3-propanetriol. *See* Glyceryl dioleate
9-Octadecenoic acid, diester with triglycerol. *See* Polyglyceryl-3 dioleate
9-Octadecenoic acid ethyl ester. *See* Ethyl oleate
9-Octadecenoic acid, monoester with octaglycerol. *See* Polyglyceryl-8 oleate
9-Octadecenoic acid, monoester with 1,2-propanediol. *See* Propylene glycol oleate
9-Octadecenoic acid, monoester with 1,2,3-propanetriol. *See* Glyceryl oleate
9-Octadecenoic acid, 1,2,3-propanetriyl ester. *See* Triolein
9-Octadecenoic acid sodium salt. *See* Sodium oleate
9-Octadecenoic acid, tetraester with decaglycerol. *See* Polyglyceryl-10 tetraoleate
9-Octadecen-1-ol. *See* Oleyl alcohol
cis-9-Octadecen-1-ol. *See* Oleyl alcohol
Octadecyl alcohol. *See* Stearyl alcohol
Octadecylamine. *See* Stearamine
Octadecyl citrate. *See* Stearyl citrate
Octadecyl dimethyl benzyl ammonium chloride. *See* Stearalkonium chloride

Octafluorocyclobutane
CAS 115-25-3
Synonyms: Cyclooctafluorobutane; Propellant C-318; Freon C-318; Perfluorocyclobutane
Empirical: C_4F_8
Properties: Colorless gas, odorless; m.w. 200.03; dens. 1.513 @ -70 F (liq.); m.p. -41.4 C; b.p. -6.04 C
Toxicology: Mildly toxic by ingestion, inh.; can cause sl. transient effects at high concs.; mutagenic data; heated to decomp., emits highly toxic fumes of F⁻
Uses: Propellant and aerating agent in foamed or sprayed food prods.
Regulatory: FDA 21CFR §173.360

Octaglyceryl stearate. *See* Polyglyceryl-8 stearate

γ-Octalactone
CAS 104-50-7; FEMA 2796
Synonyms: 4-Hydroxyoctanoic acid, γ-lactone; n-Octalactone
Empirical: $C_8H_{14}O_2$
Properties: Sl. yel. liq., strong fruity odor, sweet taste; m.w. 142.20; dens. 0.975; b.p. 234 C; flash pt. > 230 F
Uses: Synthetic flavoring
Usage level: 4.8 ppm (nonalcoholic beverages), 16 ppm (ice cream, ices, candy), 17 ppm (baked goods), 15 ppm (gelatins, puddings), 57 ppm (syrups)
Regulatory: FDA 21CFR §172.515
Manuf./Distrib.: Aldrich

n-Octalactone. *See* γ-Octalactone

n-Octanal
CAS 124-13-0; EINECS 204-683-8; FEMA 2797
Synonyms: Aldehyde C-8; Caprylic aldehyde; Caprylaldehyde; 1-Octanal; n-Octyl aldehyde
Empirical: $C_8H_{16}O$
Formula: $CH_3(CH_2)_6CHO$
Properties: Colorless to lt. yel. liq., fatty-orange odor; sol. in alcohol, fixed oils, propylene glycol; insol. in

glycerin; m.w. 128.24; dens. 0.821 (20/4 C); b.p. 163.4 C; flash pt. (CC) 125 F; ref. index 1.417-1.425
Precaution: Combustible exposed to heat or flame; can react with oxidizing materials
Toxicology: LD50 (oral, rat) 5630 mg/kg, (skin, rabbit) 6350 mg/kg; mildly toxic by ingestion and skin contact; skin and eye irritant
Uses: Fatty penetrating fragrance and flavoring; sweet and orange-like on dilution; synthetic
Regulatory: FDA 21CFR §172.515; Japan approved as flavoring
Manuf./Distrib.: Florida Treatt

1-Octanal. *See* n-Octanal

Octanal dimethyl acetal
Uses: Synthetic flavoring
Regulatory: FDA 21CFR §172.515

Octanoic acid. *See* Caprylic acid
n-Octanoic acid. *See* Caprylic acid
Octanoic acid allyl ester. *See* Allyl octanoate
Octanoic acid ethyl ester. *See* Ethyl octanoate
Octanoic acid, monoester with 1,2,3-propanetriol. *See* Glyceryl caprylate
Octanoic acid, 1,3-propanediyl ester. *See* Propylene glycol dioctanoate
Octanoic acid, 1,2,3-propanetriol ester. *See* Trioctanoin
Octanoic/decanoic acid triglyceride. *See* Caprylic/capric triglyceride
1-Octanol. *See* Caprylic alcohol

2-Octanol
CAS 4128-31-8; EINECS 223-938-4; FEMA 2801
Synonyms: (±)-2-Octanol; Hexyl methyl carbinol; Methyl hexyl carbinol; Secondary caprylic alcohol
Empirical: $C_8H_{18}O$
Formula: $CH_3(CH_2)_5CH(OH)CH_3$
Properties: m.w. 130.23; dens. 0.819 (20/4 C); b.p. 178.5 C; flash pt. 76 C; ref. index 1.426 (20 C)
Uses: Synthetic flavoring
Regulatory: FDA 21CFR §172.515
Manuf./Distrib.: Aldrich

(±)-2-Octanol. *See* 2-Octanol

3-Octanol
CAS 589-98-0; EINECS 200-667-4; FEMA 3581
Synonyms: Ethyl pentyl carbinol
Empirical: $C_8H_{18}O$
Formula: $CH_3(CH_2)_4CH(OH)CH_2CH_3$
Properties: m.w. 130.23; dens. 0.822 (20/4 C); b.p. 177-179 C; flash pt. 68 C; ref. index 1.427 (20 C)
Uses: Synthetic flavoring
Regulatory: FDA 21CFR §172.515
Manuf./Distrib.: Acme-Hardesty

n-Octanol. *See* Caprylic alcohol
5-Octanol-4-one. *See* 5-Hydroxy-4-octanone
2-Octanone. *See* Methyl hexyl ketone

3-Octanone
CAS 106-68-3; EINECS 203-423-0
Synonyms: Ethyl amyl ketone; Ethyl pentyl ketone; n-Pentyl ethyl ketone
Empirical: $C_8H_{16}O$
Formula: $CH_3(CH_2)_4COC_2H_5$
Properties: m.w. 128.22; dens. 0.821 (20/4 C); b.p. 166-169 C; flash pt. 51 C; ref index 1.415 (20 C)
Precaution: Flamm.
Uses: Synthetic flavoring
Regulatory: FDA 21CFR §172.515
Manuf./Distrib.: Acme-Hardesty; Hüls AG

3-Octanon-1-ol
Uses: Synthetic flavoring
Regulatory: FDA 21CFR §172.515

trans-2-Octen-1-al
CAS 2363-89-5; FEMA 3215
Manuf./Distrib.: Aldrich

1-Octene-3-ol. *See* 1-Octen-3-ol

1-Octen-3-ol
CAS 3391-86-4; EINECS 222-226-0; FEMA 2805
Synonyms: Amyl vinyl carbinol; Pentyl vinyl carbinol; 1-Octene-3-ol
Empirical: $C_8H_{16}O$
Formula: $CH_3(CH2)_4CH(OH)CH:CH_2$
Properties: m.w. 128.22; dens. 0.837 (20/4 C); b.p. 173-177 C; flash pt. 87 C; ref. index 1.437 (20 C)
Toxicology: Skin and eye irritant
Uses: Synthetic mushroom-like flavoring
Regulatory: FDA 21CFR §172.515
Manuf./Distrib.: Aldrich

cis-3-Octen-1-ol
FEMA 3467
Empirical: $C_8H_{16}O$
Properties: Wh. to ylsh. liq., musty mushroom odor; insol. in water; m.w. 128.22; dens. 0.830-0.850; ref. index 1.440
Toxicology: Heated to decomp., emits acrid smoke and irritating fumes
Uses: Mushroom-like, earthy, herbaceous fragrance and flavoring
Regulatory: FDA 21CFR §172.515

1-Octen-3-yl acetate
Uses: Synthetic flavoring
Regulatory: FDA 21CFR §172.515

1-Octen-3-yl butyrate

Octenyl succinic anhydride
CAS 26680-54-6
Synonyms: OSA
Empirical: $C_{12}H_{18}O_3$
Formula: $CH_3CH_2CH_2CH_2CH_2CH=CHCH_2CHCH_2C_2O_3$
Properties: m.w. 210; dens. 1.0 (25 C); b.p. 168 C (10 mm); flash pt. (COC) 185 C
Uses: Starch modifier, thickener, emulsifier, and opacifier for food mixes
Trade names: n-Octenyl Succinic Anhydride

Octoic acid. *See* Caprylic acid

Octoxynol-5
CAS 9002-93-1 (generic); 9036-19-5 (generic); 9004-87-9 (generic); 2315-64-2; 27176-99-4
Synonyms: PEG-5 octyl phenyl ether; POE (5) octyl phenyl ether; 14-(Octylphenoxy)-3,6,9,12-tetraoxatetradecan-1-ol
Classification: Ethoxylated alkyl phenol
Empirical: $C_{24}H_{42}O_6$
Formula: $C_8H_{17}C_6H_4(OCH_2CH_2)_nOH$, avg. n = 5
Uses: Surfactant, emulsifier
Regulatory: FDA 21CFR §172.710, 175.105, 176.180, 176.210, 178.3400
Trade names: Triton® X-45

Octoxynol-8
CAS 9004-87-9 (generic); 9036-19-5 (generic); 9002-93-1 (generic)
Synonyms: PEG-8 octyl phenyl ether; PEG 400 octyl phenyl ether; POE (8) octyl phenyl ether
Classification: Ethoxylated alkyl phenol
Empirical: $C_{30}H_{54}O_9$
Formula: $C_8H_{17}C_6H_4(OCH_2CH_2)_nOH$, avg. n = 8
Uses: Surfactant
Regulatory: FDA 21CFR §172.710, 175.105, 176.180, 176.210, 178.3400
Trade names: Triton® X-114

Octoxynol-9
CAS 9002-93-1 (generic); 9004-87-9 (generic); 9010-43-9; 9036-19-5 (generic); 42173-90-0
Synonyms: PEG-9 octyl phenyl ether; POE (9) octyl phenyl ether; PEG 450 octyl phenyl ether
Classification: Ethoxylated alkyl phenol
Empirical: $C_{32}H_{58}O_{10}$
Formula: $C_8H_{17}C_6H_4(OCH_2CH_2)_nOH$, avg. n = 9
Uses: Surfactant
Regulatory: FDA 21CFR §175.105, 176.180, 176.210, 178.3400

Trade names: Triton® X-100; Triton® X-120

Octoxynol-13
CAS 9002-93-1 (generic); 9004-87-9 (generic); 9036-19-5 (generic)
Synonyms: PEG-13 octyl phenyl ether; POE (13) octyl phenyl ether
Classification: Ethoxylated alkyl phenol
Formula: $C_8H_{17}C_6H_4(OCH_2CH_2)_nOH$, avg. n = 13
Uses: Surfactant
Regulatory: FDA 21CFR §172.710, 175.105, 176.180, 176.210, 178.3400
Trade names: Triton® X-102

Octoxynol-30
CAS 9004-87-9 (generic); 9036-19-5 (generic); 9002-93-1 (generic)
Synonyms: PEG-30 octyl phenyl ether; POE (30) octyl phenyl ether
Classification: Ethoxylated alkyl phenol
Formula: $C_8H_{17}C_6H_4(OCH_2CH_2)_nOH$, avg. n = 30
Uses: Surfactant, emulsifier
Regulatory: FDA 21CFR §172.710, 175.105, 176.180, 178.3400
Trade names: Triton® X-305-70%

Octoxynol-40
CAS 9002-93-1 (generic); 9004-87-9 (generic); 9036-19-5 (generic);
Synonyms: PEG-40 octyl phenyl ether; POE (40) octyl phenyl ether
Classification: Ethoxylated alkyl phenol
Formula: $C_8H_{17}C_6H_4(OCH_2CH_2)_nOH$, avg. n = 40
Uses: Surfactant, emulsifier
Regulatory: FDA 21CFR §172.710, 175.105, 176.180, 178.3400
Trade names: Triton® X-405-70%

Octoxynol-70
CAS 9004-87-9 (generic); 9036-19-5 (generic); 9002-93-1 (generic)
Synonyms: PEG-70 octyl phenyl ether; POE (70) octyl phenyl ether
Classification: Ethoxylated alkyl phenol
Formula: $C_8H_{17}C_6H_4(OCH_2CH_2)_nOH$, avg. n = 70
Uses: Emulsifier
Regulatory: FDA 21CFR §172.710, 176.180
Trade names: Triton® X-705-70%

Octyl acetate
CAS 112-14-1; FEMA 2806
Formula: $CH_3(CO_2(CH_2)_7CH_3$
Properties: m.w. 172.27; dens. 0.868; b.p. 211 C; flash pt. 86 C; ref. index 1.4180 (20 C)
Uses: Synthetic flavoring
Regulatory: FDA 21CFR §172.515

3-Octyl acetate
Uses: Synthetic flavoring
Regulatory: FDA 21CFR §172.515
Manuf./Distrib.: Aldrich

n-Octyl alcohol. *See* Caprylic alcohol
n-Octyl aldehyde. *See* n-Octanal

Octyl butyrate
CAS 110-39-4; FEMA 2807
Uses: Synthetic flavoring
Regulatory: FDA 21CFR §172.515
Manuf./Distrib.: Aldrich

n-Octyldecyl alcohol. *See* Stearyl alcohol

Octyl formate
CAS 112-32-3; FEMA 2809
Uses: Synthetic flavoring
Regulatory: FDA 21CFR §172.515
Manuf./Distrib.: Aldrich

Octyl gallate
CAS 1034-01-1; EINECS 213-853-0

Octyl heptanoate

> *Empirical:* $C_{15}H_{22}O_5$
> *Formula:* 3,4,5-(HO)$_3$C$_6$H$_2$CO$_2$(CH$_2$)$_7$CH$_3$
> *Properties:* More oil-sol. and less water-sol. than propyl gallate; m.w. 282.34; m.p. 101-104 C
> *Uses:* Antioxidant used in food fats with less discoloration than propyl gallate

Octyl heptanoate
> *Uses:* Synthetic flavoring
> *Regulatory:* FDA 21CFR §172.515

n-Octylic acid. *See* Caprylic acid

Octyl isobutyrate
> CAS 109-15-9; FEMA 2808
> *Synonyms:* Octyl 2-methylpropanoate
> *Uses:* Synthetic flavoring
> *Regulatory:* FDA 21CFR §172.515
> *Manuf./Distrib.:* Aldrich

Octyl isovalerate
> CAS 7786-58-5; FEMA 2814
> *Uses:* Synthetic flavoring
> *Regulatory:* FDA 21CFR §172.515
> *Manuf./Distrib.:* Aldrich

Octyl 2-methylpropanoate. *See* Octyl isobutyrate

Octyl octanoate
> *Uses:* Synthetic flavoring
> *Regulatory:* FDA 21CFR §172.515

14-(Octylphenoxy)-3,6,9,12-tetraoxatetradecan-1-ol. *See* Octoxynol-5

Octyl phenylacetate
> *Uses:* Synthetic flavoring
> *Regulatory:* FDA 21CFR §172.515

Octyl propionate
> FEMA 2813
> *Definition:* Ester of n-octanol and propionic acid
> *Empirical:* $C_{11}H_{22}O_2$
> *Properties:* Colorless liq.; waxy odor; sol. in alcohol, propylene glycol; insol. in water; m.w. 186.30; b.p. 228 C; ref. index 1.4225 (20 C)
> *Uses:* Synthetic flavoring agent
> *Usage level:* 0.84 ppm (nonalcoholic beverages), 0.57 ppm (ice cream, ices), 3.6 ppm (candy), 2.0-4.0 ppm (baked goods)
> *Regulatory:* FDA 21CFR §172.515

Octyl sulfate sodium salt. *See* Sodium octyl sulfate
Oenanthaldehyde. *See* Heptanal
Oenanthol. *See* Heptanal
Oenanthylidene acetone. *See* 3-Decen-2-one
Oil garlic. *See* Allyl sulfide
Oil of grapes. *See* Ethyl heptanoate
Oil niobe. *See* Methyl benzoate
Oil of Palma Christi. *See* Castor oil
Oils, babassu. *See* Babassu oil
Oils, jojoba. *See* Jojoba oil
Oils, linseed. *See* Linseed oil
Oils, menhaden. *See* Menhaden oil
Oils, palm. *See* Palm oil
Oils, palm, hydrogenated. *See* Hydrogenated palm oil
Oils, palm kernel, hydrogenated. *See* Hydrogenated palm kernel oil
Oils, peanut, hydrogenated. *See* Hydrogenated peanut oil
Oils, rice bran. *See* Rice bran oil
Oils, vegetable. *See* Vegetable oil
Oil of vitriol. *See* Sulfuric acid
Oil of wintergreen. *See* Methyl salicylate
Olea europaea oil. *See* Olive oil

Oleamide

CAS 301-02-0; EINECS 206-103-9
Synonyms: 9-Octadecenamide; Oleyl amide; Oleic acid amide
Classification: Aliphatic amide
Empirical: $C_{18}H_{35}NO$
Formula: $CH_3(CH_2)_7CH:CH(CH_2)_7CONH_2$
Properties: Ivory-colored powd.; dens. 0.94; m.p. 72 C
Precaution: Combustible
Toxicology: Heated to decomp., emits acrid smoke and irritating fumes
Uses: Release agent migrating from food pkg.
Regulatory: FDA 21CFR §175.105, 175.300, 178.3860, 178.3910, 179.45, 181.22, 181.28
Manuf./Distrib.: Croda Universal
Trade names: Kemamide® O; Kemamide® U

Oleic acid

CAS 112-80-1; EINECS 204-007-1; FEMA 2815
Synonyms: cis-9-Octadecenoic acid; Red oil; Elainic acid; 9-Octadecenoic acid
Classification: Unsaturated fatty acid
Empirical: $C_{18}H_{34}O_2$
Formula: $CH_3(CH_2)_7CH:CH(CH_2)_7COOH$
Properties: Colorless liq., odorless; insol. in water; sol. in alcohol, ether, benzene, chloroform, fixed and volatile oils; m.w. 282.47; dens. 0.895 (25/25 C); m.p. 6 C; b.p. 286 C (100 mm); flash pt. 100 C; ref. index 1.463 (18 C)
Precaution: Combustible when exposed to heat or flame; incompat. with Al and perchloric acid
Toxicology: LD50 (oral, rat) 74 g/kg; poison by intravenous route; mildly toxic by ingestion; experimental tumorigen; irritant to skin, mucous membranes; heated to decomp., emits acrid smoke and irritating fumes
Uses: Lubricant, food-grade additives, defoaming agent, binder, coatings, in dietary supplements; flume wash water additive; used in beet sugar, citrus fruit, yeast; migrating to foods from paper/paperboard, from cotton in dry food pkg.
Regulatory: FDA 21CFR §172.210, 172.860, 172.862, 173.315 (0.1 ppm max. in wash water), 173.340, 175.105, 175.320, 176.170, 176.200, 176.210, 177.1010, 177.1200, 177.2260, 177.2600, 177.2800, 178.3570, 178.3910, 182.70, 182.90; GRAS (FEMA)
Manuf./Distrib.: Aldrich; Hercules; Unichema
Trade names: Emersol® 6313 NF; Emersol® 6321 NF; Emersol® 6333 NF; Emersol® 7021; Emery® 7021; Industrene® 205

Oleic acid aluminum salt. *See* Aluminum oleate
Oleic acid amide. *See* Oleamide
Oleic acid calcium salt. *See* Calcium oleate
Oleic acid potassium salt. *See* Potassium oleate
Olein. *See* Triolein

Oleoresin angelica seed

Definition: Obtained by solv. extraction of dried seed of *Angelica archangelica*
Properties: Dk. brn. or grn. liq.
Uses: Flavoring

Oleoresin anise

Definition: Obtained by solv. extraction of the dried ripe fruit of *Pimpinella anisum* or *Illicium verum*
Properties: Dk. brn. or grn. liq.
Uses: Flavoring

Oleoresin basil

Definition: Obtained by solv. extraction of the dried plant of *Ocimum basilicum*
Properties: Dk. brn. or grn. semisolid
Uses: Flavoring

Oleoresin black pepper

Definition: Obtained by solv. extraction of dried fruit of *Piper nigrium*
Properties: Dk. grn., olive grn., or olive drab extract
Uses: Flavoring

Oleoresin capsicum

Definition: Obtained by solv. extraction of dried pods of *Capsicum frutescens* or *C. annum*
Properties: Clear red to dk. red somewhat visc. liq., char. odor and flavor; sol. in most fixed oils; partly sol. in alcohol

Uses: Flavoring

Oleoresin capsicum Africanus. *See* Capsicum oleoresin

Oleoresin caraway
Definition: Obtained by solv. extraction of the dried seeds of *Carum carvi*
Properties: grn. yel. to brn. liq.
Uses: Flavoring
Manuf./Distrib.: Commodity Services

Oleoresin cardamom
Definition: Obtained by solv. extraction of dried seeds of *Elettaria cardamomum Maton*
Properties: Dk. brn. or grn. liq.
Uses: Flavoring

Oleoresin carrot
Synonyms: Carrot oleoresin
Uses: Flavoring

Oleoresin celary
Uses: Flavoring
Trade names containing: Luxor® 1639

Oleoresin celery
Definition: Obtained by solv. extraction of dried ripe seed of *Apium graveolens*
Properties: Dk. grn. somewhat visc. nonhomogeneous liq., char. celery odor and flavor; sol. in most fixed oils, partly sol. in alcohol
Uses: Flavoring
Manuf./Distrib.: Commodity Services

Oleoresin coriander
Definition: Obtained by solv. extraction of the dried seeds of *Coriandrum sativum*
Properties: Brn. yel. to grn. liq.
Uses: Flavoring
Manuf./Distrib.: Commodity Services; Chr. Hansen's

Oleoresin cubeb
Definition: Obtained by solv. extraction of the dried fruit of *Piper cubeba*
Properties: Grn. or grn.-brn. liq.
Uses: Flavoring

Oleoresin cumin
Definition: Obtained by solv. extraction of the dried seeds of *Cuminum cyminum*
Properties: Brn. to yel. grn. liq.
Uses: Flavoring
Manuf./Distrib.: Commodity Services; Chr. Hansen's

Oleoresin dillseed
Definition: Obtained by solv. extraction of the dried seeds of *Anthenum graveolens*
Properties: Brn. or grn. liq.
Uses: Flavoring

Oleoresin fennel
Definition: Obtained by solv. extraction of the dried fruit of *Foeniculum vulgare*
Properties: Brn. grn. liq.
Uses: Flavoring
Manuf./Distrib.: Chr. Hansen's

Oleoresin ginger
Definition: Obtained by solv. extraction of dried rhizomes of *Zingiber officinalis*
Properties: Dk. brn. visc. to highly visc. liq., char. ginger odor and flavor; sol. in alcohol
Uses: Flavoring
Manuf./Distrib.: Commodity Services

Oleoresin laurel leaf
Definition: Obtained by solv. extraction of the dried leaves of *Laurus nobilis*
Properties: Dk. brn. or grn. semisolid
Uses: Flavoring

Oleoresin marjoram
Definition: Obtained by solv. extraction of the dried herb of *Majorama hortensis*

Properties: Dk. grn. to brn. visc. liq. or semisolid
Uses: Flavoring

Oleoresin origanum
Definition: Obtained by solv. extraction of the dried flowering herb *Origanum vulgare*
Properties: Dk. brn. grn. semisolid
Uses: Flavoring

Oleoresin paprika
Definition: Obtained by solv. extract of pods of *Capsicum annum*
Properties: Deep red to deep purplish red somewhat visc. liq., char. odor and flavor; sol. in most fixed oils; partly sol. in alcohol
Uses: Flavoring
Manuf./Distrib.: Chr. Hansen's

Oleoresin parsley leaf
FEMA 2837
Definition: Obtained by solv. extraction of the dried herb *Petroselinum crispum*
Properties: Brn. to grn. liq.
Uses: Flavoring

Oleoresin parsley seed
Definition: Obtained by solv. extraction of the dried seeds of *Petroselinum crispum*
Properties: Deep grn. semivisc. liq.
Uses: Flavoring

Oleoresin pimenta berries
Definition: Obtained by solv. extraction of dried fruit of *Pimenta officinalis*
Properties: Brn. grn. to dk. grn. liq.
Uses: Flavoring

Oleoresin thyme
Definition: Obtained by solv. extraction of the dried flowering plant *Thymus vulgaris*
Properties: Dk. brn. to grn. visc. semisolid
Uses: Flavoring
Manuf./Distrib.: Chr. Hansen's

Oleoresin turmeric
Definition: Obtained by solv. extraction of the dried rhizomes of *Curcuma longa*
Properties: Yel. orange to red brn. visc. liq., char. odor and flavor
Uses: Flavoring

Oleth-23
CAS 9004-98-2 (generic)
Synonyms: PEG-23 oleyl ether; POE (23) oleyl ether
Definition: PEG ether of oleyl alcohol
Formula: $CH_3(CH_2)_7CH=CH(CH_2)_7CH_2(OCH_2CH_2)_nOH$, avg. n = 23
Uses: Emulsifier for waxes used in coating fruit
Regulatory: FDA 21CFR §176.180, 176.200, 177.2800
Trade names: Trycol® OAL-23

Oleyl alcohol
CAS 143-28-2; EINECS 205-597-3
Synonyms: 9-Octadecen-1-ol; cis-9-Octadecen-1-ol
Classification: Unsaturated fatty alcohol
Empirical: $C_{18}H_{36}O$
Formula: $CH_3(CH_2)_7CH=CH(CH_2)_8OH$
Properties: Pale yel. oily visc. liq.; insol. in water; sol. in alcohol, ether; m.w. 268.49; dens. 0.84; m.p. 13-19 C; b.p. 207 C (13 mm); flash pt. > 110 C; ref. index 1.4582 (27.5 C)
Precaution: Gives off acrid fumes when heated
Toxicology: Irritant
Uses: Emulsion stabilizer, antifoam, detergent, release agent for food applics.
Regulatory: FDA 21CFR §176.170, 176.210, 177.1010, 177.1210, 177.2800, 178.3910
Trade names: Fancol OA-95

Oleyl amide. *See* Oleamide

Olibanum
CAS 8050-07-5; EINECS 232-474-1

Olibanum oil

Synonyms: Boswellia carterii resin; Frankincense; Gum olibanum; Resin olibanum; Incense
Definition: Gum resin obtained from Boswellia carterii; contains 3-8% volatile oil (pinene, dipentene, etc.), 60%
  resins, 20% gum (polysaccharide fraction), 6-8% bassorin
Toxicology: Skin irritant; heated to decomp., emits acrid smoke and irritating fumes
Uses: Natural flavoring agent; chewing gum base (Japan)
Regulatory: FDA 21CFR §172.510; Japan approved
Manuf./Distrib.: C.A.L.-Pfizer

Olibanum oil

EINECS 289-620-2 (extract); FEMA 2816
Definition: Derived from Boswellia carterii
Uses: Natural flavoring agent
Usage level: 0.60 ppm (nonalcoholic beverages), 1.2 ppm (ice cream, ices), 3.3 ppm (candy), 3.7 ppm (baked
  goods)
Manuf./Distrib.: Chart; Pierre Chauvet

Olive oil

CAS 8001-25-0; EINECS 232-277-0
Synonyms: Olea europaea oil
Definition: Fixed oil obtained from the ripe fruit of Olea europaea
Properties: Yel. liq., pleasant odor; sl. sol. in alcohol; misc. with ether, chloroform, carbon disulfide; dens. 0.909-
  0.915 (25/25 C); m.p. -6 C; flash pt. (CC) 437 F; ref. index 1.466-1.468 (25 C); sapon. no. 187-196
Precaution: Photosensitive
Toxicology: Human skin irritant; heated to dec., emits acrid smoke and fumes
Uses: Salad dressing and other foods
Regulatory: FDA 21CFR §175.105, 176.200, 176.210, GRAS; Japan approved (olive)
Manuf./Distrib.: Arista Industries; Croda; Penta Mfg.
Trade names: Pure/Riviera Olive Oil NF

Onion extract

Synonyms: Allium cepa extract
Definition: Extract of the bulbs of onion, Allium cepa
Uses: Natural flavoring agent
Regulatory: FDA 21CFR §182.20, GRAS
Manuf./Distrib.: Chart; Commodity Services

Onion oil

Definition: Derived from Allium cepa
Properties: Strong pungent lasting odor, char. onion flavor
Uses: Natural flavoring
Manuf./Distrib.: Florida Treatt

Opalwax. See Hydrogenated castor oil
Optal. See Propyl alcohol

Orange B

CAS 15139-76-1
Synonyms: 1-(4-Sulfophenyl)-3-ethylcarboxy-4-(4-sulfonaphthylazo)-5-hydroxypyrazole; CI 19235; Acid Or-
  ange 137
Classification: Pyrazolone color
Empirical: $C_{22}H_{16}N_4O_9S_2Na_2$
Properties: Dull orange cryst.; m.w. 590.49
Toxicology: Heated to decomp., emits toxic fumes of SO_x
Uses: Food colorant for frankfurters, sausages
Usage level: Limitation 150 ppm (frankfurters, sausages)
Regulatory: FDA 21CFR §74.250

Orange crystals. See 2´-Acetonaphthone

Orange flower extract

Definition: Extract of the flowers of Citrus sinensis
Uses: Natural flavoring agent
Regulatory: FDA 21CFR §182.20, GRAS; Japan approved (orange flower)
Manuf./Distrib.: C.A.L.-Pfizer; Pierre Chauvet

Orange flower oil

CAS 8016-38-4
Synonyms: Neroli oil

Definition: Volatile oil obtained from flowers of the orange tree, *Citrus sinensis*
Properties: Ylsh. fluorescent liq., very intense pleasant odor; becomes brn. on exposure to light; sol. in 1.5-2 vols 80% alcohol; sl. sol. in water; dens. 0.86-0.88 (25/25 C); ref. index 1.475 (20 C)
Precaution: Keep cool, well closed; protect from light
Uses: Natural flavoring agent
Regulatory: FDA 21CFR §182.20, GRAS

Orange leaf
Definition: Citrus sinensis
Uses: Natural flavoring agent
Regulatory: FDA 21CFR §182.20, GRAS

Orange oil
CAS 8008-57-9; FEMA 2821
Synonyms: Citrus sinensis oil; Orange oil, coldpressed; Sweet orange oil
Definition: Volatile oil obtained by expression from the fresh peel of the ripe fruit *Citrus sinensis*
Properties: Yel. to deep orange liq., char. orange odor and taste; sol. in 2 vols 90% alcohol, 1 vol glac. acetic acid; sl. sol. in water; misc. with abs. alcohol, carbon disulfide; dens. 0.842-0.846; ref. index 1.472 (20 C)
Precaution: Keep well closed, cool, protected from light
Toxicology: Experimental neoplastigen; skin irritant; heated to decomp., emits acrid smoke and irritating fumes
Uses: Natural flavoring agent
Usage level: 130 ppm (nonalcoholic beverages), 140 ppm (ice cream), 690 ppm (candy), 440 ppm (baked goods), 45-500 ppm (gelatins, puddings), 930 ppm (chewing gum)
Regulatory: FDA 21CFR §182.20, GRAS; FEMA GRAS; Europe listed, no restrictions
Manuf./Distrib.: Caminiti Foti & Co. Srl; Commodity Services; Florida Treatt; Chr. Hansen's
Trade names: Citreatt Orange 3111; Citreatt Orange 6110
Trade names containing: Beta Carotene Emulsion Beverage Type 3.6 No. 65392

Orange oil, coldpressed. *See* Orange oil

Orange oil, distilled
Definition: Oil from steam distillation of fresh peel of *Citrus sinensis*
Properties: Colorless to pale yel. liq., fresh orange peel odor; sol. in fixed oils, min. oil, alcohol; insol. in glycerin, propylene glycol
Toxicology: Heated to decomp., emits acrid smoke and irritating fumes
Uses: Natural flavoring agent
Regulatory: FDA 21CFR §182.20, GRAS
Manuf./Distrib.: Florida Treatt; Chr. Hansen's

Orange peel extract
Synonyms: Sweet orange peel extract
Definition: Extract of rinds of oranges, *Citrus sinensis*
Uses: Natural flavoring agent
Usage level: 99 ppm (nonalcoholic beverages), 170 ppm (ice cream), 320-330 ppm (candy), 320 ppm (baked goods)
Regulatory: FDA 21CFR §182.20, GRAS
Manuf./Distrib.: Chart

Oregano
Definition: From *Lippia* spp.
Uses: Natural flavoring agent; processing aid (Japan)
Regulatory: FDA 21CFR §182.10, GRAS; Japan approved (oregano extract)

Organosiloxane. *See* Silicone
Origanol. *See* 4-Carvomenthenol
Origanum majorana oil. *See* Sweet marjoram oil

Origanum oil
Definition: Origanum spp.
Properties: Fresh herbaceous odor, warm burning flavor
Uses: Natural flavoring agent
Regulatory: FDA 21CFR §182.20, GRAS
Manuf./Distrib.: Chr. Hansen's

Origanum vulgare extract. *See* Wild marjoram extract
γ-Orizanol. *See* Oryzanol

Orris root extract
CAS 90045-89-9

Orthoboric acid

Synonyms: Iris florentina extract
Definition: Extract of *Iris florentina* or *I. pallida*
Properties: Violet-like odor with a fruity undertone
Uses: Natural flavoring agent
Regulatory: FDA 21CFR §172.510; Japan approved (orris)
Manuf./Distrib.: C.A.L.-Pfizer; Chart; Pierre Chauvet

Orthoboric acid. *See* Boric acid
Orthophosphoric acid. *See* Phosphoric acid

Oryzanol
CAS 11042-64-1
Synonyms: γ-Orizanol
Definition: Ester of ferulic acid and a terpene alcohol; derived from rice bran oil
Empirical: $C_{40}H_{58}O_3$
Uses: UV absorber, antioxidant for food applics.
Trade names: Gamma Oryzanol

OSA. *See* Octenyl succinic anhydride
Otto of rose. *See* Rose oil
1-Oxa-4-azacyclohexane. *See* Morpholine
1-Oxa-2-cyclohexadecanone. *See* Pentadecalactone
Oxacyclohexadecan-2-one. *See* Pentadecalactone

Ox bile extract
CAS 361-09-1, 8008-63-7; EINECS 206-643-5
Synonyms: Purified oxgall; Sodium cholate; Cholic acid sodium salt
Definition: Purified portion of the bile of an ox
Empirical: $C_{24}H_{39}NaO_5$
Properties: Ylsh.-green soft solid, partly sweet/partly bitter disagreeable taste; m.w. 430.57
Toxicology: LD50 (IV, mouse) 200 mg/kg; poison by IV route; heated to decomp., emits toxic fumes of Na_2O
Uses: Surfactant
Usage level: Limitation 0.002% (cheese)
Regulatory: FDA 21CFR §184.1560, GRAS

Oxidase, glucose. *See* Glucose oxidase
Oxidized polyethylene. *See* Polyethylene, oxidized
β-Oxobenzenepropanoic acid ethyl ester. *See* Ethyl benzoylacetate
2-Oxobutane. *See* Methyl ethyl ketone
3-Oxobutanoic acid ethyl ester. *See* Ethylacetoacetate
3-Oxo-2,3-dihydro-1,2-benzisothiazole-1,1-dioxide. *See* Saccharin
2-Oxoglutaric acid. *See* α-Ketoglutaric acid
α-Oxoglutaric acid. *See* α-Ketoglutaric acid
2-Oxopentanedioic acid. *See* α-Ketoglutaric acid
4-Oxopentanoic acid. *See* Levulinic acid
4-Oxopentanoic acid ethyl ester. *See* Ethyl levulinate
2-Oxopropanal. *See* Pyruvaldehyde
2-Oxopropionic acid. *See* Pyruvic acid
4-Oxo-n-valeric acid. *See* Levulinic acid
2,2´-[Oxybis(2,1-ethanediyloxy)]bisethanol. *See* PEG-4
1,1´-[Oxybis(methylene)]bis[benzene]. *See* Benzyl ether
Oxymethylene. *See* Formaldehyde
Oxyphenalon. *See* 4-(p-Hydroxyphenyl)-2-butanone
Oxypropylated cellulose. *See* Hydroxypropylcellulose

Oxystearin
Definition: Mixt. of the glycerides of partially oxidized stearic and other fatty acids
Properties: Tan to lt. brn. waxy solid, bland taste; sol. in ether, hexane, chloroform; acid no. 15 max.; iodine no. 15 max.; sapon.no. 225-240; hyd. no. 30-45; ref. index 1.465
Toxicology: Heated to decomp., emits acrid smoke and irritating fumes
Uses: Crystallization inhibitor, release agent in salad and cooking oils; sequestrant; defoamer for beet sugar and yeast processing
Usage level: ADI -25 mg/kg (FAO/WHO)
Regulatory: FDA 21CFR §172.818 (limitation 0.125% of oils; must conform to FDA specs for fats or fatty acids derived from edible oils), 173.340

Ozocerite. *See* Ozokerite

Ozokerite
CAS 8021-55-4
Synonyms: Ozocerite; Mineral wax; Fossil wax
Classification: Hydrocarbon wax
Definition: Hydrocarbon wax derived from mineral or petroleum sources
Properties: Yel.-brown to black or green translucent (pure); sol. in lt. petrol. hydrocarbons, benzene, turpentine, kerosene, ether, carbon disulfide; sl. sol. in alcohol; insol. in water; dens. 0.85-0.95; m.p. 55-110 C (usually 70 C)
Precaution: Combustible
Uses: Natural chewing gum base
Regulatory: Japan approved

Ozone
CAS 10028-15-6
Synonyms: Triatomic oxygen
Empirical: O_3
Properties: Unstable colorless gas or dk. blue liq., pungent char. odor; m.w. 48; dens. 2.144 g/L (gas), 1.614 g/mL (liq., -195.4 C); m.p. -193 C; b.p. -111.9 C
Precaution: Powerful highly reactive oxidizing agent; severe explosion hazard in liq. form when shocked, exposed to heat or flame, or in conc. form by reaction with powerful reducing agents; incompat. with rubber
Toxicology: TLV:TWA CL 0.1 ppm; human poison by inh., systemic effects; experimental neoplastigen, tumorigen, teratogen, reproductive effects; human mutagenic data; skin, eye, upper respiratory, mucous membrane irritant
Uses: Antimicrobial for treating bottled water; processing aid (Japan)
Usage level: Limitation 0.4 mg/L residual of bottled water
Regulatory: FDA 21CFR §184.1563, GRAS; Japan approved

PAA. *See* Phenylacetaldehyde

Palmarosa oil
EINECS 283-461-2 (extract); FEMA 2831
Synonyms: East Indian geranium
Definition: Cymbopogon martini
Properties: Sweet rose-like odor with herbaceous undertone
Uses: Natural flavoring agent
Usage level: 4.2 ppm (nonalcoholic beverages), 1.7 ppm (ice cream, ices), 12 ppm (candy), 13 ppm (baked goods)
Regulatory: FDA 21CFR §182.20, GRAS
Manuf./Distrib.: Commodity Services; Florida Treatt

Palm butter. *See* Palm oil

Palm glyceride
Synonyms: Palm oil glyceride
Definition: Monoglyceride derived from palm oil
Uses: Emulsifier, stabilizer, dispersant
Trade names: Myverol® 18-35; Myverol® 18-35K

Palm glycerides
CAS 129521-59-1
Synonyms: Glycerides, palm oil mono-, di- and tri-
Definition: Mixt. of mono-, di-, and triglycerides derived from palm oil
Uses: Emulsifier; dispersant for coffee whiteners

Palm grease. *See* Palm oil

Palmitamide
Synonyms: Palmitic acid amide
Toxicology: Heated to decomp., emits acrid smoke and irritating fumes
Uses: Release agent migrating from food pkg.
Regulatory: FDA 21CFR §181.28

Palmitic acid
CAS 57-10-3; EINECS 200-312-9; FEMA 2832
Synonyms: Hexadecanoic acid; Cetylic acid; Hexadecylic acid
Classification: Saturated fatty acid

Palmitic acid aluminum salt

Empirical: $C_{16}H_{32}O_2$
Formula: $CH_3(CH_2)_{14}COOH$
Properties: White cryst. scales, sl. char. odor/taste; insol. in water; sl. sol. in cold alcohol, petrol. ether; sol. in hot alcohol, ether, propyl alcohol, chloroform; m.w. 256.42; dens. 0.853 (62/4 C); m.p. 63-64 C; b.p. 215 C (15 mm); ref. index 1.4273 (80 C)
Precaution: Combustible
Toxicology: LD50 (IV, mouse) 57 mg/kg; acute poison by intravenous route; experimental neoplastigen; human skin irritant; heated to decomp., emits acrid smoke and fumes
Uses: Mfg. of food-grade additives, defoaming agent, lubricant
Regulatory: FDA 21CFR §172.210, 172.860, 173.340, 175.105, 175.320, 176.170, 176.200, 176.210, 177.1010, 177.1200, 177.1200, 177.2260, 177.2600, 177.2800, 178.3570, 178.3910; must conform to FDA specs for fats or fatty acids derived from edible oils; GRAS (FEMA)
Manuf./Distrib.: Acme-Hardesty
Trade names: Emersol® 6343; Glycon® P-45

Palmitic acid aluminum salt. *See* Aluminum palmitate
Palmitic acid amide. *See* Palmitamide
Palmitic acid sodium salt. *See* Sodium palmitate
Palmitin. *See* Tripalmitin
Palmitoyl L-ascorbic acid. *See* Ascorbyl palmitate

1-Palmitoyl-2-oleoyl-3-stearin
Classification: Triglyceride
Definition: Cocoa butter substitute primarily from palm oil
Properties: Clear color, free from rancid odor and taste
Uses: Used in confections, frostings, soft candy coatings, sweet sauces and toppings
Regulatory: FDA 21CFR §184.1259, GRAS

Palmityl alcohol. *See* Cetyl alcohol

Palm kernel oil
CAS 8023-79-8; EINECS 232-425-4
Definition: Oil obtained from seeds of *Elaeis guineensis*
Properties: Fatty solid, char. sweet nutty flavor
Toxicology: Heated to decomp., emits acrid smoke and irritating fumes
Uses: Coating agent, emulsifier, formulation aid, texturizer; used in confections, margarine
Regulatory: FDA 21CFR §175.105, 176.200, 176.210, GRAS
Manuf./Distrib.: Alba Int'l.; Karlshamns; Penta Mfg.
Trade names: Akowesco 1
Trade names containing: CLSP 399; CLSP 499

Palm kernel oil, hydrogenated. *See* Hydrogenated palm kernel oil

Palm oil
CAS 8002-75-3; EINECS 232-316-1
Synonyms: Oils, palm; Palm butter; Palm grease
Definition: Natural oil obtained from pulp of the fruit of *Elaeis guineensis*
Properties: Yel.-brown buttery, edible solid at R.T.; sol. in alcohol, ether, chloroform, carbon disulfide; dens. 0.952; m.p. 26-30 C; iodine no. 15; sapon. no. 247
Precaution: Combustible
Toxicology: Heated to decomp., emits acrid smoke and irritating fumes
Uses: Coating agent, emulsifier, lubricant, formulation aid, texturizer, food shortening, margarine
Regulatory: FDA 21CFR §175.105, 175.300, 176.200, 176.210, 177.2800, GRAS
Manuf./Distrib.: Aarhus Olie; Booths; British Arkady; Calgene; Canada Packers; Cargill; Croda Singapore; Enco Prods.; Karlshamns; Peerless Refining; Penta Mfg.; J W Pike; Vamo-Fuji Specialities; Wensleydale Foods; Wynmouth Lehr
Trade names: Lipovol PAL

Palm oil glyceride. *See* Palm glyceride
Palm oil glyceride, hydrogenated. *See* Hydrogenated palm glyceride
Palm oil, hydrogenated. *See* Hydrogenated palm oil

Palm oil sucroglyceride
Uses: Emulsifier, coemulsifier, texture improver, stabilizer, and dispersant for fats, oils, and baked goods
Trade names: Celynol F1; Celynol MSPO-11; Celynol P1M

Palygorskite. *See* Attapulgite
Panama bark. *See* Quillaja

Panama wood extract. *See* Quillaja extract

Pancreatin
CAS 8049-47-6
Properties: Cream colored, amorphous powder; sol. in water; insol. in alcohol
Uses: Natural enzyme; nutrient; for egg white, instant cereals, starch
Regulatory: Canada, Japan approved
Trade names: Pancreatin TA

Pansy
Definition: From *Viola tricolor*
Uses: Natural flavoring used in alcoholic beverages
Regulatory: FDA 21CFR §172.510

Pansy extract
CAS 84012-42-0
Synonyms: Viola tricolor extract
Definition: Extract obtained from *Viola tricolor*
Uses: Natural flavoring
Regulatory: FDA 21CFR §172.510

D-Panthenol
CAS 81-13-0; EINECS 201-327-3
Synonyms: Dexpanthenol; Pantothenol; 2,4-Dihydroxy-N-(3-hydroxypropyl)-3,3-dimethylbutanamide; Pantothenyl alcohol; Provitamin B_5
Classification: Alcohol
Empirical: $C_9H_{19}NO_4$
Formula: $HOCH_2C(CH_3)_2CH(OH)CONH(CH_2)_2CH_2OH$
Properties: Visc. liq., sl. bitter taste; hygroscopic; freely sol. in water, alcohol, methanol, ether; sl. sol. in glycerin; m.w. 205.29; dens. 1.2 (20/20 C); b.p. 118-120 C; easily dec. on distillation; ref. index 1.500 (20 C); pH 9.5
Toxicology: LD50 (IP, mouse) 9 g/kg; mod. toxic by IV route; heated to decomp., emits toxic fumes of NO_x
Uses: Dietary supplement, nutrient
Regulatory: FDA 21CFR §182.5580, GRAS
Manuf./Distrib.: Hoffmann-LaRoche

DL-Panthenol
CAS 16485-10-2
Synonyms: 2,4-Dihydroxy-N-(3-hydroxypropyl)-3,3-dimethylbutanamide; Dexpanthenol; DL-Pantothenyl; Racemic pantothenyl alcohol; Pantothenyl alcohol; Provitamin B_5
Classification: Alcohol
Empirical: $C_9H_{19}NO_4$
Formula: $HOCH_2C(CH_3)_2CH(OH)CONH(CH_2)_2CH_2OH$
Properties: Visc. liq.; sol. in water and alcohol; m.w. 205.25; ref. index 1.497 (20 C)
Uses: Vitamin, nutrient, dietary supplement
Manuf./Distrib.: Hoffmann-LaRoche

D-Pantothenamide
CAS 7757-97-3
Uses: Dietary supplement, source of pantothenic acid in foods for special dietary use
Regulatory: FDA 21CFR §172.335

Pantothenol. *See* D-Panthenol
DL-Pantothenyl. *See* DL-Panthenol
Pantothenyl alcohol. *See* D-Panthenol
Pantothenyl alcohol. *See* DL-Panthenol

Papain
CAS 9001-73-4; EINECS 232-627-2
Synonyms: Vegetable pepsin; Papayotin
Definition: Proteolytic enzyme derived from latex of the green fruit and leaves of *Carica papaya*
Properties: Wh. to gray powd.; sl. hygroscopic; sol. in water, glycerin; insol. in other common org. solvs.; m.w. ≈ 21,000
Toxicology: Poison by intraperitoneal route; moderately toxic by ingestion; allergen; experimental teratogenic and reproductive effects; heated to decomp., emits toxic fumes of NO_x
Uses: Direct food additive, enzyme, processing aid, texturizer; meat tenderizer, tissue softening agent; chillproofing, antihazing agent for beer

Papayotin

> *Regulatory:* FDA 21CFR §184.1585, GRAS; USDA 9CFR §318.7, 381.147; BATF 27CFR §240.1051, GRAS; Canada, UK, Japan approved
> *Manuf./Distrib.:* Am. Roland; Chart
> *Trade names:* Papain A300; Papain A400; Papain AIE; Papain M70; Papain S100; Tona Enzyme 201; Tona Papain 14; Tona Papain 90L; Tona Papain 270L

Papayotin. *See* Papain

Paprika
EEC E160c
Definition: Deep red, sweet, pungent powd. obtained from the ground dried pod of mild capsicum, *Capsicum annum*
Toxicology: Heated to decomp., emits acrid smoke and irritating fumes
Uses: Natural flavoring agent; colorant producing orange to bright red shades and flavoring for chorizo sausage
Usage level: 0.2-100 ppm (as food colorant)
Regulatory: FDA 21CFR §73.340, 182.10, 182.20, GRAS; USDA 9CFR §318.7; Japan approved, restricted as color; Europe listed (extract), UK approved
Manuf./Distrib.: C.A.L.-Pfizer; Chart; Phytone Ltd.; Quest Int'l.

Paraffin
CAS 8002-74-2; EINECS 232-315-6
Synonyms: Paraffin wax; Hard paraffin; Petroleum wax, crystalline
Classification: Hydrocarbon
Definition: Solid mixture of hydrocarbons obtained from petroleum; characterized by relatively large crystals
Empirical: C_nH_{2n+2}
Properties: Colorless to white solid, odorless; insol. in water, alcohol; sol. in benzene, chloroform, ether, carbon disulfide, oils; misc. with fats; dens. ≈ 0.9; m.p. 50-57 C; flash pt. 340 F
Precaution: Dangerous fire hazard
Toxicology: Anesthetic effect; ACGIH TLV:TWA 2 mg/m³ (fume); experimental tumorigens by implantation; many paraffin waxes contain carcinogens
Uses: Masticatory substance in chewing gum base, adhesive component, coatings; glazing agent (Japan)
Regulatory: FDA 21CFR §133.181, 172.615, 173.3210, 175.105, 175.210, 175.250, 175.300, 175.320, 176.170, 176.200, 177.1200, 177.2420, 177.2600, 177.2800, 178.3710, 178.3800, 178.3910, 179.45; Canada, Japan approved
Manuf./Distrib.: Koster Keunen
Trade names: Koster Keunen Paraffin Wax

Paraffin oil. *See* Mineral oil
Paraffin wax. *See* Paraffin
Paraguay tea extract. *See* Mate extract

Parsley
FEMA 2835
Definition: Petroselinum crispum
Properties: Warm herbaceous fresh odor, warm spicy aromatic bitter taste
Uses: Natural flavoring agent
Usage level: 3000 ppm (nonalcoholic beverages), 850 ppm (baked goods), 2700 ppm (condiments), 1000 ppm (meats), 200-500 ppm (soups)
Regulatory: FDA 21CFR §182.10, 182.20, GRAS; Japan approved
Manuf./Distrib.: Chart

Parsley extract
CAS 84012-33-9
Synonyms: Petroselinum sativum extract
Uses: Natural flavoring agent
Regulatory: FDA 21CFR §182.20, GRAS
Manuf./Distrib.: Chart

Parsley seed oil
CAS 8000-68-8; FEMA 2836
Synonyms: Petroselinum sativum seed oil; Petroselinum hortense seed oil
Definition: Volatile oil obtained from seeds of *Petroselinum sativum*
Properties: Colorless or yel. visc. liq.; sol. in 8 vols 80% alcohol; sol. in ether; very sl. sol. in water; dens. 1.040-1.100 (15/15 C); ref. index 1.510-1.519 (20 C)
Uses: Natural flavoring agent for meat sauces, seasonings, spice blends, pickles
Regulatory: FDA 21CFR §172.510

Manuf./Distrib.: Florida Treatt

Passiflora incarnata extract. *See* Passionflower extract
Passiflora quadrangularis extract. *See* Passionflower extract

Passionflower extract
CAS 84012-31-7
Synonyms: Passiflora incarnata extract; Passiflora quadrangularis extract
Definition: Extract of various species of *Passiflora incarnata*
Uses: Natural flavoring agent
Regulatory: FDA 21CFR §172.510

Patchouli
Properties: Yellowish to brownish
Uses: Natural flavoring agent
Regulatory: Japan approved
Manuf./Distrib.: Chart; Pierre Chauvet

PDMS. *See* Dimethicone
Peach aldehyde. *See* γ-Undecalactone
Peach aldehyde. *See* ςs]-Undecalactone

Peach kernel oil
CAS 8023-98-1
Synonyms: Persic oil
Definition: Oil expressed from kernels of the peach, *Prunus persica*
Uses: Natural flavoring agent
Regulatory: FDA 21CFR §182.40, GRAS

Peach leaves
Definition: Leaves from *Prunus persica*
Uses: Natural flavoring for alcoholic beverages only
Regulatory: FDA 21CFR §172.510; not to exceed 25 ppm prussic acid in the flavor

Peanut glycerides
Synonyms: Glycerides, peanut oil, mono-, di-, and tri-
Definition: Mixture of mono, di, and triglycerides derived from peanut oil
Uses: Food emulsifier
Trade names: Olicine

Peanut oil
CAS 8002-03-7; EINECS 232-296-4
Synonyms: Arachis oil; Groundnut oil; Katchung oil; Earthnut oil; Pecan shell powder
Classification: Fixed oil
Definition: Refined fixed oil obtained from seed kernels of one or more cultivated varieties of *Arachis hypogaea*
Properties: Pale yel. liq., nutty odor, bland taste; sol. in benzene, alcohol, ether, chloroform; insol. in alkalies; dens. 0.916-0.922; solid. pt. -5 C; flash pt. 540 F; ref. index 1.466-1.470 (25 C)
Precaution: Combustible exposed to heat or flame; can react with oxidizing materials
Toxicology: Experimental tumorigen; human skin irritant, mild allergen; mutagenic data; heated to decomp., emits acrid smoke and irritating fumes
Uses: Edible oil, dietary supplement; substitute for olive oil; coating agent, emulsifier, formulation aid, texturizer; in mfg. of margarine, salad oil; migrating from cotton in dry food pkg.
Regulatory: FDA 21CFR §175.105, 176.200, 176.210, 177.2800, 182.70; GRAS
Manuf./Distrib.: Arista Industries; S Black; Bunge Foods; Calgene; Canada Packers; Cargill; H E Daniel; Durkee; Flavors of N. Am.; Edw. Gittens; MLG Enterprises; Peerless Refining; Penta Mfg.; SKW Chemicals; Vamo-Fuji Specialities
Trade names: Peanut Oil 85-060-0
Trade names containing: Dry Beta Carotene Beadlets 10% CWS No. 65633; Dry Beta Carotene Beadlets 10% No. 65661; Dry Vitamin D₃ Beadlets Type 850 No. 652550401, 652550601; Heavy Duty Peanut Oil 104-210; Palmabeads® Type 500 No. 65332; Palma-Sperse® Type 250A/50 D-S No. 65221; Peanut Oil; Soy/Peanut Liquid Frying Shortening 104-215; Superb® Fish and Chip Oil 102-060

Pea protein concentrate
Uses: Nutritional supplement with fat and water binding props.; emulsifier, foaming agent for meat and fish
Trade names: Pea Protein Conc.

Pearl ash. *See* Potassium carbonate
Pear oil. *See* Amyl acetate

Pecan shell powder. *See* Peanut oil

Pectin

CAS 9000-69-5; EINECS 232-553-0; EEC E440a
Synonyms: Citrus pectin
Classification: Polysaccharide
Definition: Purified carbohydrate prod. obtained from the dilute acid extract of the inner portion of the rind of citrus fruits or from apple pomace
Properties: White powd. or syrupy conc., pract. odorless; sol. in water; insol. in alcohol, org. solvs.; m.w. 30,000-100,000
Toxicology: Heated to decomp., emits acrid smoke and irritating fumes
Uses: Direct food additive, emulsifier, gelling agent, stabilizer, thickener for beverages, jams, jellies
Regulatory: FDA 21CFR §135.140, 145, 150, 173.385, 184.1588, GRAS; Japan approved; Europe listed; UK approved
Manuf./Distrib.: Am. Roland; Atomergic Chemetals; Copenhagen Pectin; Ellis & Everard; G Fiske; Fruitsource; Grindsted; A C Hatrick; Herbstreith & Fox; Hercules; Ikeda; Lucas Meyer; Penta Mfg.; Pomosin GmbH; Rit-Chem; Sanofi; Spice King; TIC Gums
Trade names: Genu® 04CG or 04CB; Genu® 12CG; Genu® 18CG; Genu® 18CG-YA; Genu® 20AS; Genu® 21AS or 21AB; Genu® 22CG; Genu® 102AS; Genu® 104AS; Genu® 104AS-YA; Genu® AA Medium-Rapid Set, 150 Grade; Genu® BA-KING; Genu® BB Rapid Set; Genu® DD Extra-Slow Set; Genu® DD Extra-Slow Set C, 150 Grade; Genu® DD Slow Set; Genu® JMJ; Genu® Type DJ; Genu® VIS; Genu® Pectins; Mexpectin HV 400 Range; Mexpectin LA 100 Range; Mexpectin LA 200 Range; Mexpectin LA 400 Range; Mexpectin LC 700 Range; Mexpectin LC 800 Range; Mexpectin LC 900 Range; Mexpectin MRS 300 Range; Mexpectin RS 400 Range; Mexpectin SS 200 Range; Mexpectin XSS 100 Range; TIC Pretested® Pectin HM Rapid; TIC Pretested® Pectin HM Slow
Trade names containing: Frigesa® IC 184; Frimulsion 6G; Frimulsion Q8; Frimulsion RA

Pectinase

CAS 9032-75-1; EINECS 232-885-6
Definition: Enzyme derived from *Aspergillus niger* or *Rhizopus oryzae*
Uses: Enzyme for wine, cider, fruit juice, natural flavor/color extracts, citrus fruit skins for jams, vegetable stock for soup mfg.
Regulatory: FDA 173.130, GRAS; Canada approved
Trade names: Clarex® 5XL; Clarex® L; Extractase L5X, P15X; Pearex® L; Pectinase AT; Pectinol® 59L; Pectinol® R10; Pektolase; Spark-L® HPG
Trade names containing: Clarex® ML

PEG. *See* Polyethylene glycol

PEG-4

CAS 25322-68-3 (generic); 112-60-7; EINECS 203-989-9
Synonyms: PEG 200; POE (4); 2,2´-[Oxybis(2,1-ethanediyloxy)]bisethanol
Definition: Polymer of ethylene oxide
Empirical: $C_8H_{18}O_5$
Formula: $H(OCH_2CH_2)_nOH$, avg. n = 4
Properties: Visc. liq., sl. char. odor; hygroscopic; m.w. 190-210; dens. 1.127 (25/25 C); visc. 4.3 cSt (210 F); supercools on freezing
Precaution: Solvent action on some plastics
Toxicology: LD50 (oral, rat) 28,900 mg/kg; mildly toxic by ingestion; heated to decomp., emits acrid smoke and irritating fumes
Uses: Coating, binder, plasticizer, lubricant for food tablets; adjuvant to improve flavor, bodying agent in nonnutritive sweeteners; adjuvant in dispersing vitamins/minerals; coating on sodium nitrite to inhibit hygroscopic props.; defoamer; food pkg.
Regulatory: FDA 21CFR §73.1, 172.210, 172.770, 172.820, 173.310, 173.340, 175.105, 175.300, 178.3750

PEG-6

CAS 25322-68-3 (generic); 2615-15-8; EINECS 220-045-1
Synonyms: PEG 300; Hexaethylene glycol; Macrogol 300
Definition: Polymer of ethylene oxide
Empirical: $C_{12}H_{26}O_7$
Formula: $H(OCH_2CH_2)_nOH$, avg. n = 6
Toxicology: LD50 (oral, rat) 27,500 mg/kg; mildly toxic by ingestion; heated to decomp., emits acrid smoke and irritating fumes
Uses: Coating, binder, plasticizer, lubricant for food tablets; adjuvant to improve flavor, bodying agent in nonnutritive sweeteners; adjuvant in dispersing vitamins/minerals; coating on sodium nitrite to inhibit

hygroscopic props.; defoamer; food pkg.

Usage level: Limitation zero tolerance (milk)

Regulatory: FDA 21CFR §172.210, 172.770, 172.820, 173.310, 173.340, 175.105, 175.300, 178.3750, 178.3910

Trade names: Carbowax® Sentry® PEG 300; Dow E300 NF; Pluracol® E300

Trade names containing: Carbowax® Sentry® PEG 540 Blend; Pluracol® E1500

PEG-8

CAS 25322-68-3 (generic); 5117-19-1; EINECS 225-856-4

Synonyms: PEG 400; POE (8); 3,6,9,12,15,18,21-Heptaoxatricosane-1,23-diol

Definition: Polymer of ethylene oxide

Empirical: $C_{16}H_{34}O_9$

Formula: $H(OCH_2CH_2)_nOH$, avg. n = 8

Properties: Visc. liq., sl. char. odor; sl. hygroscopic; m.w. 380-420; dens. 1.128 (25/25 C); m.p. 4-8 C; visc. 7.3 cSt (210 F)

Toxicology: LD50 (rat, oral) 30 ml/kg; low toxicity by ingestion, IV, IP routes; heated to decomp., emits acrid smoke and irritating fumes

Uses: Coating, binder, plasticizer, lubricant for food tablets; adjuvant to improve flavor, bodying agent in nonnutritive sweeteners; adjuvant in dispersing vitamins/minerals; coating on sodium nitrite to inhibit hygroscopic props.; defoamer; food pkg.

Usage level: Limitation zero tolerance (milk)

Regulatory: FDA 21CFR §172.210, 172.770, 172.820, 173.310, 173.340, 175.105, 175.300, 178.3750, 178.3910, 181.22, 181.30

Trade names: Carbowax® Sentry® PEG 400; Dow E400 NF; Pluracol® E400

Trade names containing: Oxynex® K

PEG-9

CAS 25322-68-3 (generic); 3386-18-3; EINECS 222-206-1

Synonyms: PEG 450; POE (9)

Definition: Polymer of ethylene oxide

Empirical: $C_{18}H_{38}O_{10}$

Formula: $H(OCH_2CH_2)_nOH$, avg. n = 9

Uses: Coating, binder, plasticizer, lubricant for food tablets; adjuvant to improve flavor, bodying agent in nonnutritive sweeteners; adjuvant in dispersing vitamins/minerals; coating on sodium nitrite to inhibit hygroscopic props.; defoamer; food pkg.

Regulatory: FDA 21CFR §172.210, 172.770, 172.820, 173.310, 173.340, 175.105, 175.300, 178.3750, 178.3910

PEG-12

CAS 25322-68-3 (generic); 6790-09-6; EINECS 229-859-1

Synonyms: PEG 600; POE (12); Macrogol 600

Definition: Polymer of ethylene oxide

Empirical: $C_{24}H_{50}O_{13}$

Formula: $H(OCH_2CH_2)_nOH$, avg. n = 12

Properties: Visc. liq., char. odor; sl. hygroscopic; m.w. 570-630; dens. 1.128 (25/25 C); m.p. 20-25 C; visc. 10.5 cSt (210 F)

Toxicology: LD50 (oral, rat) 38,100 mg/kg; low toxicity by ingestion; eye irritant; heated to decomp., emits acrid smoke and irritating fumes

Uses: Coating, binder, plasticizer, lubricant for food tablets; adjuvant to improve flavor, bodying agent in nonnutritive sweeteners; adjuvant in dispersing vitamins/minerals; coating on sodium nitrite to inhibit hygroscopic props.; defoamer; food pkg.

Usage level: Limitation zero tolerance (milk)

Regulatory: FDA 21CFR §172.210, 172.770, 172.820, 173.310, 173.340, 175.105, 175.300, 178.3750, 178.3910

Trade names: Carbowax® Sentry® PEG 600; Dow E600 NF; Pluracol® E600

PEG-14

CAS 25322-68-3 (generic)

Synonyms: POE (14)

Definition: Polymer of ethylene oxide

Empirical: $C_{28}H_{58}O_{15}$

Formula: $H(OCH_2CH_2)_nOH$, avg. n = 14

Uses: Coating, binder, plasticizer, lubricant for food tablets; adjuvant to improve flavor, bodying agent in nonnutritive sweeteners; adjuvant in dispersing vitamins/minerals; coating on sodium nitrite to inhibit hygroscopic props.; defoamer; food pkg.

Regulatory: FDA 21CFR §172.210, 172.770, 172.820, 173.310, 173.340, 175.105, 175.300, 178.3750, 178.3910

PEG-16

CAS 25322-68-3 (generic)
Synonyms: PEG 800; POE (16)
Definition: Polymer of ethylene oxide
Empirical: $C_{32}H_{66}O_{17}$
Formula: $H(OCH_2CH_2)_nOH$, avg. n = 16
Uses: Coating, binder, plasticizer, lubricant for food tablets; adjuvant to improve flavor, bodying agent in nonnutritive sweeteners; adjuvant in dispersing vitamins/minerals; coating on sodium nitrite to inhibit hygroscopic props.; defoamer; food pkg.
Regulatory: FDA 21CFR §172.210, 172.770, 172.820, 173.310, 173.340, 175.105, 175.300, 178.3750, 178.3910

PEG-20

CAS 25322-68-3 (generic);
Synonyms: PEG 1000; Macrogol 1000; POE (20)
Definition: Polymer of ethylene oxide
Empirical: $C_{40}H_{82}O_{21}$
Formula: $H(OCH_2CH_2)_nOH$, avg. n = 20
Toxicology: LD50 (oral, rat) 42 g/kg; mod. toxic by IP, IV routes; mildly toxic by ingestion; experimental tumorigen; heated to decomp., emits acrid smoke and irritating fumes
Uses: Coating, binder, plasticizer, lubricant for food tablets; adjuvant to improve flavor, bodying agent in nonnutritive sweeteners; adjuvant in dispersing vitamins/minerals; coating on sodium nitrite to inhibit hygroscopic props.; defoamer; food pkg.
Usage level: Limitation zero tolerance (milk)
Regulatory: FDA 21CFR §172.210, 172.770, 172.820, 173.310, 173.340, 175.105, 175.300, 178.3750, 178.3910
Trade names: Carbowax® Sentry® PEG 900; Carbowax® Sentry® PEG 1000; Dow E1000 NF

PEG-32

CAS 25322-68-3 (generic)
Synonyms: PEG 1540; Macrogol 1540; POE (32)
Definition: Polymer of ethylene oxide
Empirical: $C_{64}H_{130}O_{33}$
Formula: $H(OCH_2CH_2)_nOH$, avg. n = 32
Properties: White powd.
Toxicology: LD50 (oral, rat) 44,200 mg/kg; mildly toxic by ingestion; human skin irritant; heated to decomp., emits acrid smoke and irritating fumes
Uses: Coating, binder, plasticizer, lubricant for food tablets; adjuvant to improve flavor, bodying agent in nonnutritive sweeteners; adjuvant in dispersing vitamins/minerals; coating on sodium nitrite to inhibit hygroscopic props.; defoamer; food pkg.
Usage level: Limitation zero tolerance (milk)
Regulatory: FDA 21CFR §172.210, 172.770, 172.820, 173.310, 173.340, 175.105, 175.300, 178.3750, 178.3910
Trade names: Carbowax® Sentry® PEG 1450; Dow E1450 NF
Trade names containing: Carbowax® Sentry® PEG 540 Blend; Pluracol® E1500

PEG-40

CAS 25322-68-3 (generic)
Synonyms: PEG 2000; POE (40)
Definition: Polymer of ethylene oxide
Empirical: $C_{80}H_{162}O_{41}$
Formula: $H(OCH_2CH_2)_nOH$, avg. n = 40
Uses: Coating, binder, plasticizer, lubricant for food tablets; adjuvant to improve flavor, bodying agent in nonnutritive sweeteners; adjuvant in dispersing vitamins/minerals; coating on sodium nitrite to inhibit hygroscopic props.; defoamer; food pkg.
Regulatory: FDA 21CFR §172.210, 172.770, 172.820, 173.310, 173.340, 175.105, 175.300, 178.3750, 178.3910

PEG-75

CAS 25322-68-3 (generic)
Synonyms: PEG 4000; POE (75)
Definition: Polymer of ethylene oxide

Empirical: $C_{150}H_{302}O_{76}$
Formula: $H(OCH_2CH_2)_nOH$, avg. n = 75
Properties: White powd. or creamy-white flakes; m.w. 3000-3700; dens. 1.212 (25/25 C); m.p. 54-58 C; visc. 76-110 cSt (210 F)
Toxicology: LD50 (oral, rat) 50 g/kg; mildly toxic by ingestion; skin irritant; heated to decomp., emits acrid smoke and irritating fumes
Uses: Coating, binder, plasticizer, lubricant for food tablets; adjuvant to improve flavor, bodying agent in nonnutritive sweeteners; adjuvant in dispersing vitamins/minerals; coating on sodium nitrite to inhibit hygroscopic props.; defoamer; food pkg.
Usage level: Limitation zero tolerance (milk)
Regulatory: FDA 21CFR §172.210, 172.770, 172.820, 173.310, 173.340, 175.105, 175.300, 178.3750, 178.3910
Trade names: Carbowax® Sentry® PEG 3350; Dow E3350 NF; Pluracol® E4000

PEG-100
CAS 25322-68-3 (generic)
Synonyms: PEG (100); POE (100)
Definition: Polymer of ethylene oxide
Empirical: $C_{200}H_{402}O_{101}$
Formula: $H(OCH_2CH_2)_nOH$, avg. n = 100
Uses: Coating, binder, plasticizer, lubricant for food tablets; adjuvant to improve flavor, bodying agent in nonnutritive sweeteners; adjuvant in dispersing vitamins/minerals; coating on sodium nitrite to inhibit hygroscopic props.; defoamer; food pkg.
Regulatory: FDA 21CFR §172.210, 172.770, 172.820, 173.310, 173.340, 175.105, 175.300, 178.3750, 178.3910
Trade names: Carbowax® Sentry® PEG 4600; Dow E4500 NF

PEG (100). *See* PEG-100

PEG-150
CAS 25322-68-3 (generic)
Synonyms: PEG 6000; Macrogol 6000; POE (150)
Definition: Polymer of ethylene oxide
Formula: $H(OCH_2CH_2)_nOH$, avg. n = 150
Properties: Powd. or creamy-white flakes; water-sol.; m.w. 7000-9000; dens. 1.21 (25/25 C); m.p. 56-63 C; visc. 470-900 cSt (210 F); flash pt. > 887 F
Precaution: Combustible exposed to heat or flame
Toxicology: LD50 (rat, oral) > 50 g/kg; mildly toxic by ingestion; mutagenic data; skin irritant; heated to decomp., emits acrid smoke and irritating fumes
Uses: Coating, binder, plasticizer, lubricant for food tablets; adjuvant to improve flavor, bodying agent in nonnutritive sweeteners; adjuvant in dispersing vitamins/minerals; coating on sodium nitrite to inhibit hygroscopic props.; defoamer; food pkg.
Usage level: Limitation zero tolerance (milk)
Regulatory: FDA 21CFR §172.210, 172.770, 172.820, 173.310, 173.340, 175.300, 177.2420, 178.3750, 178.3910
Trade names: Carbowax® Sentry® PEG 8000; Dow E8000 NF

PEG-200
CAS 25322-68-3 (generic)
Synonyms: PEG 9000; POE (200)
Definition: Polymer of ethylene oxide
Formula: $H(OCH_2CH_2)_nOH$, avg. n = 200
Uses: Coating, binder, plasticizer, lubricant for food tablets; adjuvant to improve flavor, bodying agent in nonnutritive sweeteners; adjuvant in dispersing vitamins/minerals; coating on sodium nitrite to inhibit hygroscopic props.; defoamer; food pkg.
Regulatory: FDA 21CFR §172.210, 172.770, 172.820, 173.310, 173.340, 175.300, 178.3750, 178.3910

PEG 200. *See* PEG-4
PEG 300. *See* PEG-6

PEG-350
CAS 25322-68-3 (generic)
Synonyms: PEG 20000; POE (350)
Definition: Polymer of ethylene oxide
Formula: $H(OCH_2CH_2)_nOH$, avg. n = 350
Uses: Food and food pkg.

Regulatory: FDA 21CFR §172.770, 173.310, 175.300, 178.3910

PEG 400. *See* PEG-8
PEG 450. *See* PEG-9
PEG 600. *See* PEG-12
PEG 800. *See* PEG-16
PEG 1000. *See* PEG-20
PEG 1540. *See* PEG-32
PEG 2000. *See* PEG-40
PEG 4000. *See* PEG-75
PEG 6000. *See* PEG-150
PEG 9000. *See* PEG-200
PEG 20000. *See* PEG-350

PEG-6 dilaurate
CAS 9005-02-1 (generic)
Synonyms: POE (6) dilaurate; PEG 300 dilaurate
Definition: PEG diester of lauric acid
Formula: $CH_3(CH_2)_{10}CO(OCH_2CH_2)_nOCO(CH_2)_{10}CH_3$, avg. n = 6
Uses: Surfactant, solubilizer, thickener, emollient, opacifier, spreading agent, wetting agent, dispersant
Regulatory: FDA 21CFR §175.105, 175.300, 176.210

PEG-20 dilaurate
CAS 9005-02-1 (generic)
Synonyms: POE (20) dilaurate; PEG 1000 dilaurate
Definition: PEG diester of lauric acid
Empirical: $C_{64}H_{126}O_{23}$
Formula: $CH_3(CH_2)_{10}CO(OCH_2CH_2)_nOCO(CH_2)_{10}CH_3$, avg. n = 20
Uses: Dispersant, emulsifier, wetting agent, cosolvent, solubilizer, thickener, emollient, opacifier, spreading agent
Regulatory: FDA 21CFR §175.300, 176.210, 177.2260, 177.2800

PEG-32 dilaurate
CAS 9005-02-1 (generic)
Synonyms: POE (32) dilaurate; PEG 1540 dilaurate
Definition: PEG diester of lauric acid
Empirical: $C_{88}H_{174}O_{35}$
Formula: $CH_3(CH_2)_{10}CO(OCH_2CH_2)_nOCO(CH_2)_{10}CH_3$, avg. n = 32
Uses: Surfactant, solubilizer, thickener, emollient, opacifier, spreading agent, wetting agent, dispersant
Regulatory: FDA 21CFR §175.300, 176.210, 177.2260, 177.2800

PEG-75 dilaurate
CAS 9005-02-1 (generic)
Synonyms: POE (75) dilaurate; PEG 4000 dilaurate
Definition: PEG diester of lauric acid
Formula: $CH_3(CH_2)_{10}CO(OCH_2CH_2)_nOCO(CH_2)_{10}CH_3$, avg. n = 75
Uses: Surfactant, solubilizer, thickener, emollient, opacifier, spreading agent, wetting agent, dispersant
Regulatory: FDA 21CFR §175.300

PEG-150 dilaurate
CAS 9005-02-1 (generic)
Synonyms: POE (150) dilaurate; PEG 6000 dilaurate
Definition: PEG diester of lauric acid
Formula: $CH_3(CH_2)_{10}CO(OCH_2CH_2)_nOCO(CH_2)_{10}CH_3$, avg. n = 150
Uses: Surfactant, solubilizer, thickener, emollient, opacifier, spreading agent, wetting agent, dispersant
Regulatory: FDA 21CFR §175.300

PEG 300 dilaurate. *See* PEG-6 dilaurate
PEG 1000 dilaurate. *See* PEG-20 dilaurate
PEG 1540 dilaurate. *See* PEG-32 dilaurate
PEG 4000 dilaurate. *See* PEG-75 dilaurate
PEG 6000 dilaurate. *See* PEG-150 dilaurate

PEG-6 dioleate
CAS 9005-07-6 (generic); 52688-97-0 (generic)
Synonyms: POE (6) dioleate; PEG 300 dioleate
Definition: PEG diester of oleic acid

Uses: Surfactant, solubilizer, thickener, emollient, opacifier, spreading agent, wetting agent, dispersant; defoamer for beet sugar and yeast processing
Regulatory: FDA 21CFR §175.105, 175.300, 176.210

PEG-8 dioleate
CAS 9005-07-6 (generic); 52688-97-0 (generic)
Synonyms: POE (8) dioleate; PEG 400 dioleate
Definition: PEG diester of oleic acid
Uses: Defoamer for beet sugar and yeast processing
Regulatory: FDA 21CFR §173.340, 175.105, 175.300, 176.170, 176.200, 176.210, 177.1210, 177.2260, 177.2800

PEG-12 dioleate
CAS 9005-07-6 (generic); 52688-97-0 (generic); 85736-49-8; EINECS 288-459-5
Synonyms: POE (12) dioleate; PEG 600 dioleate
Definition: PEG diester of oleic acid
Uses: Defoamer for beet sugar and yeast processing
Regulatory: FDA 21CFR §173.340, 175.105, 175.300, 176.200, 176.210, 177.2260, 177.2800

PEG-20 dioleate
CAS 9005-07-6 (generic); 52688-97-0 (generic)
Synonyms: POE (20) dioleate; PEG 1000 dioleate
Definition: PEG diester of oleic acid
Uses: Surfactant, solubilizer, thickener, emollient, opacifier, spreading agent, wetting agent, dispersant
Regulatory: FDA 21CFR §175.300, 176.210, 177.2260, 177.2800

PEG-32 dioleate
CAS 9005-07-6 (generic); 52688-97-0 (generic)
Synonyms: POE (32) dioleate; PEG 1540 dioleate
Definition: PEG diester of oleic acid
Uses: Surfactant, solubilizer, thickener, emollient, opacifier, spreading agent, wetting agent, dispersant
Regulatory: FDA 21CFR §175.300, 176.210, 177.2260, 177.2800

PEG-75 dioleate
CAS 9005-07-6 (generic); 52688-97-0 (generic)
Synonyms: POE (75) dioleate; PEG 4000 dioleate
Definition: PEG diester of oleic acid
Uses: Surfactant, solubilizer, thickener, emollient, opacifier, spreading agent, wetting agent, dispersant
Regulatory: FDA 21CFR §175.300

PEG-150 dioleate
CAS 9005-07-6 (generic); 52688-97-0 (generic)
Synonyms: POE (150) dioleate; PEG 6000 dioleate
Definition: PEG diester of oleic acid
Uses: Surfactant, solubilizer, thickener, emollient, opacifier, spreading agent, wetting agent, dispersant
Regulatory: FDA 21CFR §175.300

PEG 300 dioleate. *See* PEG-6 dioleate
PEG 400 dioleate. *See* PEG-8 dioleate
PEG 600 dioleate. *See* PEG-12 dioleate
PEG 1000 dioleate. *See* PEG-20 dioleate
PEG 1540 dioleate. *See* PEG-32 dioleate
PEG 4000 dioleate. *See* PEG-75 dioleate
PEG 6000 dioleate. *See* PEG-150 dioleate

PEG-6 distearate
CAS 9005-08-7 (generic); 52668-97-0
Synonyms: POE (6) distearate; PEG 300 distearate
Definition: PEG diester of stearic acid
Formula: $CH_3(CH_2)_{16}CO(OCH_2CH_2)_nOCO(CH_2)_{16}CH_3$, avg. n = 6
Uses: Surfactant, solubilizer, thickener, emollient, opacifier, spreading agent, wetting agent, dispersant
Regulatory: FDA 21CFR §175.105, 175.300, 176.210

PEG-20 distearate
CAS 9005-08-7 (generic); 52668-97-0
Synonyms: POE (20) distearate; PEG 1000 distearate
Definition: PEG diester of stearic acid
Formula: $CH_3(CH_2)_{16}CO(OCH_2CH_2)_nOCO(CH_2)_{16}CH_3$, avg. n = 20

Uses: Surfactant, solubilizer, thickener, emollient, opacifier, spreading agent, wetting agent, dispersant
Regulatory: FDA 21CFR §175.300, 176.210, 177.2260, 177.2800

PEG-32 distearate
CAS 9005-08-7 (generic); 52668-97-0
Synonyms: POE (32) distearate; PEG 1540 distearate
Definition: PEG diester of stearic acid
Formula: $CH_3(CH_2)_{16}CO(OCH_2CH_2)_nOCO(CH_2)_{16}CH_3$, avg. n = 32
Uses: Surfactant, solubilizer, thickener, emollient, opacifier, spreading agent, wetting agent, dispersant
Regulatory: FDA 21CFR §175.300, 176.210, 177.2260, 177.2800

PEG-75 distearate
CAS 9005-08-7 (generic); 52668-97-0
Synonyms: POE (75) distearate; PEG 4000 distearate
Definition: PEG diester of stearic acid
Formula: $CH_3(CH_2)_{16}CO(OCH_2CH_2)_nOCO(CH_2)_{16}CH_3$, avg. n = 75
Uses: Surfactant, solubilizer, thickener, emollient, opacifier, spreading agent, wetting agent, dispersant
Regulatory: FDA 21CFR §175.300

PEG 300 distearate. *See* PEG-6 distearate
PEG 1000 distearate. *See* PEG-20 distearate
PEG 1540 distearate. *See* PEG-32 distearate
PEG 4000 distearate. *See* PEG-75 distearate

PEG-7 glyceryl cocoate
CAS 66105-29-1; 68201-46-7 (generic)
Synonyms: POE (7) glyceryl monococoate; PEG (7) glyceryl monococoate
Definition: PEG ether of glyceryl cocoate
Formula: $RCO—OCH_2COHHCH_2(OCH_2CH_2)_nOH$, RCO- rep. fatty acids from coconut oil, avg. n=7
Uses: Emulsifier for food prods.
Regulatory: FDA 21CFR §175.300
Trade names: Mazol® 159

PEG-20 glyceryl isostearate
CAS 69468-44-6
Synonyms: PEG 1000 glyceryl isostearate; POE (20) glyceryl isostearate
Definition: PEG ether of glyceryl isostearate
Formula: $C_{17}H_{35}COOCH_2CHOHCH_2(OCH_2CH_2)_nOH$, avg. n = 20
Uses: Solubilizer for flavors, vitamin oils
Trade names: Tagat® I2

PEG-30 glyceryl isostearate
CAS 69468-44-6
Synonyms: POE (30) glyceryl isostearate
Definition: PEG ether of glyceryl isostearate
Formula: $C_{17}H_{35}COOCH_2CHOHCH_2(OCH_2CH_2)_nOH$, avg. n = 30
Uses: Solubilizer for flavors, vitamin oils
Trade names: Tagat® I

PEG 1000 glyceryl isostearate. *See* PEG-20 glyceryl isostearate

PEG-20 glyceryl laurate
CAS 59070-56-3 (generic); 51248-32-9
Synonyms: POE (20) glyceryl monolaurate; PEG 1000 glyceryl monolaurate
Definition: PEG ether of glyceryl laurate
Formula: $CH_3(CH_2)_{10}COOCH_2COHHCH_2(OCH_2CH_2)_nOH$, avg. n = 20
Uses: Solubilizer for flavors, vitamin oils
Regulatory: FDA 21CFR §175.300, 176.210, 177.2800
Trade names: Tagat® L2

PEG-30 glyceryl laurate
CAS 59070-56-3 (generic); 51248-32-9
Synonyms: POE (30) glyceryl laurate
Definition: PEG ether of glyceryl laurate
Formula: $CH_3(CH_2)_{10}COOCH_2COHHCH_2(OCH_2CH_2)_nOH$, avg. n = 30
Uses: Solubilizer for flavors, vitamin oils
Regulatory: FDA 21CFR §175.300, 177.2800
Trade names: Tagat® L

PEG (7) glyceryl monococoate. *See* PEG-7 glyceryl cocoate
PEG 1000 glyceryl monolaurate. *See* PEG-20 glyceryl laurate
PEG 1000 glyceryl monooleate. *See* PEG-20 glyceryl oleate
PEG 1000 glyceryl monostearate. *See* PEG-20 glyceryl stearate

PEG-20 glyceryl oleate
CAS 68889-49-6 (generic); 51192-09-7
Synonyms: POE (20) glyceryl oleate; PEG 1000 glyceryl monooleate
Definition: PEG ether of glyceryl oleate
Formula: $CH_3[CH(CH_2)_7]_2COOCH_2CHOHCH_2(OCH_2CH_2)_nOH$, avg. n = 20
Uses: Solubilizer for flavors, vitamin oils
Regulatory: FDA 21CFR §175.300, 176.210, 177.2800
Trade names: Tagat® O2

PEG-30 glyceryl oleate
CAS 68889-49-6 (generic); 51192-09-7
Synonyms: POE (30) glyceryl oleate
Definition: PEG ether of glyceryl oleate
Formula: $CH_3[CH(CH_2)_7]_2COOCH_2CHOHCH_2(OCH_2CH_2)_nOH$, avg. n = 30
Uses: Solubilizer for flavors, vitamin oils
Regulatory: FDA 21CFR §175.300, 177.2800
Trade names: Tagat® O

PEG-15 glyceryl ricinoleate
CAS 51142-51-9 (generic); 39310-72-0
Synonyms: POE (15) glyceryl monoricinoleate
Definition: PEG ether of glyceryl ricinoleate
Uses: Solubilizer for flavors, vitamin oils
Regulatory: FDA 21CFR §175.300, 176.210, 177.2800

PEG-20 glyceryl stearate
CAS 68153-76-4, 68553-11-7; 51158-08-8
Synonyms: PEG 1000 glyceryl monostearate; POE (20) glyceryl monostearate
Definition: PEG ether of glyceryl stearate
Formula: $CH_3(CH_2)_{16}COOCH_2COHHCH_2(OCH_2CH_2)_nOH$, avg. n = 20
Uses: Emulsifier, dough conditioner
Trade names: Aldo® MS-20 FG; Aldosperse® MS-20 FG; Aldosperse® MS-20 KFG; Capmul® EMG; Durfax®
    EOM; Durfax® EOM K; Hodag POE (20) GMS; Mazol® 80 MGK; Tagat® S2
Trade names containing: Aldosperse® 40/60; Aldosperse® 40/60 FG; Aldosperse® 40/60 KFG; Tally® 100
    Plus

PEG-30 glyceryl stearate
CAS 51158-08-8
Synonyms: POE (30) glyceryl monostearate
Definition: PEG ether of glyceryl stearate
Formula: $CH_3(CH_2)_{16}COOCH_2COHHCH_2(OCH_2CH_2)_nOH$, avg. n = 30
Uses: Solubilizer for flavors, vitamin oils
Regulatory: FDA 21CFR §175.300, 177.2800
Trade names: Tagat® S

PEG-25 glyceryl trioleate
CAS 68958-64-5
Synonyms: POE (25) glyceryl trioleate
Definition: Triester of oleic acid and a PEG ether of glycerin
Uses: Solubilizer for flavors, vitamin oils
Regulatory: FDA 21CFR §175.300, 177.2800
Trade names: Tagat® TO

PEG-50 hydrogenated castor oil
CAS 61788-85-0 (generic)
Synonyms: POE (50) hydrogenated castor oil
Definition: PEG deriv. of hydrogenated castor oil with avg. 50 moles of ethylene oxide
Uses: Solubilizer for oil-sol. vitamins
Regulatory: FDA 21CFR §177.2800
Trade names: Emalex HC-50

PEG-60 hydrogenated castor oil
CAS 61788-85-0 (generic)
Synonyms: POE (60) hydrogenated castor oil
Definition: PEG deriv. of hydrogenated castor oil with avg. 60 moles of ethylene oxide
Uses: Solubilizer for oil-sol. vitamins
Regulatory: FDA 21CFR §177.2800
Trade names: Emalex HC-60

PEG-32 laurate
CAS 9004-81-3 (generic)
Synonyms: POE (32) monolaurate; PEG 1540 monolaurate
Definition: PEG ester of lauric acid
Formula: $CH_3(CH_2)_{10}CO(OCH_2CH_2)_nOH$, avg. n = 32
Uses: Surfactant, solubilizer, thickener, emollient, opacifier, spreading agent, wetting agent, dispersant
Regulatory: FDA 21CFR §175.300, 176.210, 177.2260, 177.2800, 178.3910

PEG 1540 monolaurate. *See* PEG-32 laurate
PEG 300 monooleate. *See* PEG-6 oleate
PEG 400 monooleate. *See* PEG-8 oleate
PEG 1540 monooleate. *See* PEG-32 oleate
PEG 4000 monooleate. *See* PEG-75 oleate
PEG (100) monostearate. *See* PEG-100 stearate
PEG 300 monostearate. *See* PEG-6 stearate
PEG 400 monostearate. *See* PEG-8 stearate
PEG 2000 monostearate. *See* PEG-40 stearate
PEG 4000 monostearate. *See* PEG-75 stearate
PEG-6 nonyl phenyl ether. *See* Nonoxynol-6
PEG-7 nonyl phenyl ether. *See* Nonoxynol-7
PEG-10 nonyl phenyl ether. *See* Nonoxynol-10
PEG-11 nonyl phenyl ether. *See* Nonoxynol-11
PEG-12 nonyl phenyl ether. *See* Nonoxynol-12
PEG-14 nonyl phenyl ether. *See* Nonoxynol-14
PEG-40 nonyl phenyl ether. *See* Nonoxynol-40
PEG-50 nonyl phenyl ether. *See* Nonoxynol-50
PEG 300 nonyl phenyl ether. *See* Nonoxynol-6
PEG 500 nonyl phenyl ether. *See* Nonoxynol-10
PEG 600 nonyl phenyl ether. *See* Nonoxynol-12
PEG 2000 nonyl phenyl ether. *See* Nonoxynol-40
PEG-5 octyl phenyl ether. *See* Octoxynol-5
PEG-8 octyl phenyl ether. *See* Octoxynol-8
PEG-9 octyl phenyl ether. *See* Octoxynol-9
PEG-13 octyl phenyl ether. *See* Octoxynol-13
PEG-30 octyl phenyl ether. *See* Octoxynol-30
PEG-40 octyl phenyl ether. *See* Octoxynol-40
PEG-70 octyl phenyl ether. *See* Octoxynol-70
PEG 400 octyl phenyl ether. *See* Octoxynol-8
PEG 450 octyl phenyl ether. *See* Octoxynol-9

PEG-6 oleate
CAS 9004-96-0 (generic); 60344-26-5
Synonyms: POE (6) monooleate; PEG 300 monooleate
Definition: PEG ester of oleic acid
Empirical: $C_{30}H_{58}O_8$
Formula: $CH_3(CH_2)_7CHCH(CH_2)_7CO(OCH_2CH_2)_nOH$, avg. n = 6
Uses: Emulsifier, lubricant, chemical intermediate, antifoam, dispersant
Regulatory: FDA 21CFR §175.105, 175.300, 176.210
Trade names: Acconon 300-MO

PEG-8 oleate
CAS 9004-96-0 (generic)
Synonyms: POE (8) monooleate; PEG 400 monooleate
Definition: PEG ester of oleic acid
Formula: $CH_3(CH_2)_7CHCH(CH_2)_7CO(OCH_2CH_2)_nOH$, avg. n = 8
Uses: Emulsifier, dispersant, lubricant, chemical intermediate, solubilizer, visc. control agent
Regulatory: FDA 21CFR §175.105, 175.300, 176.170, 176.200, 177.1200, 177.1210, 177.2260, 177.2800

Trade names: Durpeg® 400MO

PEG-32 oleate
CAS 9004-96-0 (generic)
Synonyms: POE (32) monooleate; PEG 1540 monooleate
Definition: PEG ester of oleic acid
Formula: $CH_3(CH_2)_7CHCH(CH_2)_7CO(OCH_2CH_2)_nOH$, avg. n = 32
Uses: Surfactant, solubilizer, thickener, emollient, opacifier, spreading agent, wetting agent, dispersant
Regulatory: FDA 21CFR §175.300, 176.200, 177.2261, 177.2800

PEG-75 oleate
CAS 9004-96-0 (generic)
Synonyms: POE (75) monooleate; PEG 4000 monooleate
Definition: PEG ester of oleic acid
Formula: $CH_3(CH_2)_7CHCH(CH_2)_7CO(OCH_2CH_2)_nOH$, avg. n = 75
Uses: Surfactant, solubilizer, thickener, emollient, opacifier, spreading agent, wetting agent, dispersant
Regulatory: FDA 21CFR §175.300, 176.200

PEG-23 oleyl ether. *See* Oleth-23

PEG-12 ricinoleate
CAS 9004-97-1 (generic)
Synonyms: POE (12) monoricinoleate
Definition: PEG ester of ricinoleic acid
Formula: $CHCH_2COHH(CH_2)_5CH_3CH(CH_2)_7CO(OCH_2CH_2)_nOH$, avg. n = 12
Uses: Defoamer for beet sugar and yeast processing
Regulatory: FDA 21CFR §173.340

PEG-40 sorbitan diisostearate
Synonyms: POE (40) sorbitan diisostearate
Definition: Ethoxylated sorbitan diester of isostearic acid with avg. 40 moles ethylene oxide
Uses: Solubilizer for flavors in mouthwashes

PEG-40 sorbitan hexatallate
Uses: Food emulsifier
Trade names: Glycosperse® HTO-40

PEG-4 sorbitan laurate. *See* Polysorbate 21
PEG-20 sorbitan laurate. *See* Polysorbate 20

PEG-80 sorbitan laurate
CAS 9005-64-5 (generic)
Synonyms: POE (80) sorbitan monolaurate
Definition: Ethoxylated sorbitan monoester of lauric acid with avg. 80 moles ethylene oxide
Uses: Emulsifier for bakery, dairy, confectionery, convenience food applics.; solubilizer for essential oils; dispersant
Regulatory: FDA 21CFR §175.300
Trade names: Hodag PSML-80; T-Maz® 28

PEG-5 sorbitan oleate. *See* Polysorbate 81
PEG-20 sorbitan oleate. *See* Polysorbate 80
PEG-4 sorbitan stearate. *See* Polysorbate 61
PEG-20 sorbitan stearate. *See* Polysorbate 60
PEG-20 sorbitan trioleate. *See* Polysorbate 85
PEG-20 sorbitan tristearate. *See* Polysorbate 65

PEG-20 sorbitan tritallate
Definition: Triester of tall oil acid and a PEG ether of sorbitol, avg. 20 moles ethylene oxide
Uses: Emulsifier and solubilizer of essential oils, wetting agent, visc. modifier, antistat, stabilizer and dispersant
Trade names: T-Maz® 95

PEG-6 stearate
CAS 9004-99-3 (generic); 10108-28-8
Synonyms: POE (6) stearate; PEG 300 monostearate
Definition: PEG ester of stearic acid
Empirical: $C_{30}H_{60}O_8$
Formula: $CH_3(CH_2)_{16}CO(OCH_2CH_2)_nOH$, avg. n = 6
Toxicology: Poison by intravenous, intraperitoneal route; mildly toxic by ingestion
Uses: Food emulsifier

Regulatory: FDA 21CFR §175.105, 175.300, 176.210
Trade names: Alkamuls® S-6

PEG-8 stearate
CAS 9004-99-3 (generic); 70802-40-3; EEC E430
Synonyms: POE (8) stearate; PEG 400 monostearate
Definition: PEG ester of stearic acid
Empirical: $C_{34}H_{68}O_{10}$
Formula: $CH_3(CH_2)_{16}CO(OCH_2CH_2)_nOH$, avg. n = 8
Toxicology: Poison by intravenous, intraperitoneal route; mildly toxic by ingestion
Uses: Emulsifier, stabilizer
Usage level: ADI 0-25 mg/kg body wt.
Regulatory: Europe listed; UK approved
Trade names: Myrj® 45

PEG-32 stearate
CAS 9004-99-3 (generic)
Synonyms: PEG 1540 stearate
Definition: PEG ester of stearic acid
Formula: $CH_3(CH_2)_{16}CO(OCH_2CH_2)_nOH$, avg. n = 32
Toxicology: Poison by intravenous, intraperitoneal route; mildly toxic by ingestion
Uses: Surfactant, solubilizer, thickener, emollient, opacifier, spreading agent, wetting agent, dispersant
Regulatory: FDA 21CFR §175.300, 176.210, 177.2260, 177.2800

PEG-40 stearate
CAS 9004-99-3 (generic); 31791-00-2; EEC E431
Synonyms: POE (40) stearate; PEG 2000 monostearate
Definition: PEG ester of stearic acid
Formula: $CH_3(CH_2)_{16}CO(OCH_2CH_2)_nOH$, avg. n = 40
Toxicology: Poison by intravenous, intraperitoneal route; mildly toxic by ingestion
Uses: Defoamer; emulsifier; freshness additive in bread
Usage level: ADI 0-25 mg/kg (EEC)
Regulatory: FDA 21CFR §173.340, 175.105, 175.300, 176.200, 176.210, 177.2260, 177.2800; Europe listed; UK approved
Trade names: Myrj® 52; Myrj® 52S
Trade names containing: AF 72; AF 75

PEG-75 stearate
CAS 9004-99-3 (generic)
Synonyms: POE (75) stearate; PEG 4000 monostearate
Definition: PEG ester of stearic acid
Formula: $CH_3(CH_2)_{16}CO(OCH_2CH_2)_nOH$, avg. n = 75
Toxicology: Poison by intravenous, intraperitoneal route; mildly toxic by ingestion
Uses: Surfactant, solubilizer, thickener, emollient, opacifier, spreading agent, wetting agent, dispersant
Regulatory: FDA 21CFR §175.300

PEG-100 stearate
CAS 9004-99-3 (generic)
Synonyms: POE (100) stearate; PEG (100) monostearate
Definition: PEG ester of stearic acid
Formula: $CH_3(CH_2)_{16}CO(OCH_2CH_2)_nOH$, avg. n = 100
Toxicology: Poison by intravenous, intraperitoneal route; mildly toxic by ingestion
Uses: Surfactant
Regulatory: FDA 21CFR §175.300, 176.210
Trade names containing: Arlacel® 165

PEG 1540 stearate. *See* PEG-32 stearate
PEG-20 stearyl ether. *See* Steareth-20
PEG 1000 stearyl ether. *See* Steareth-20
Pelargonaldehyde. *See* Nonanal
Pelargonic acid. *See* Nonanoic acid
Pelargonic aldehyde. *See* Nonanal
Pelargonium oil. *See* Geranium oil
Pelargonyl acetate. *See* n-Nonyl acetate

Pelargonyl vanillylamide
Synonyms: N-(4-Hydroxy-3-methoxybenzyl)nonanamide

Uses: Synthetic flavoring agent
Regulatory: FDA 21CFR §172.515

Pennyroyal extract
CAS 90064-00-9
Synonyms: Mentha pulegium extract; European pennyroyal
Definition: Extract of the flowering herb, *Mentha pulegium*
Uses: Natural flavoring agent
Regulatory: FDA 21CFR §172.510

Pennyroyal oil
FEMA 2839
Synonyms: European pennyroyal oil
Definition: Volatile oil from steam distillation of *Mentha pulegium*, contg. 85% pulegone
Properties: Yel. to greenish-yel. liq., aromatic mint odor, aromatic taste; sol. in fixed oils, propylene glycol, min. oil; insol. in glycerin; dens. 0.960 (15/15 C); ref. index 1.475-1.496 (20 C)
Toxicology: LD50 (oral, rat) 400 mg/kg; experimental poison by ingestion; skin irritant; heated to decomp., emits acrid smoke and irritating fumes
Uses: Natural flavoring agent
Usage level: 1.5-5 ppm (nonalcoholic beverages), 3.7 ppm (ice cream), 14 ppm (candy), 20-24 ppm (baked goods)
Regulatory: FDA 21CFR §172.10
Manuf./Distrib.: Chr. Hansen's

Pennyroyal oil, American
Synonyms: Hedeoma oil
Definition: Volatile oil from leaves and flowering tops of *Hedeoma pulegioides*, contg. chiefly pulegone
Properties: Pale yel. liq., aromatic odor; sol. in 3 vols 70% alcohol; very sol. in chloroform, ether; sl. sol. in water; dens. 0.920-0.935 (25/25 C); ref. index 1.482 (20 C)
Precaution: Keep cool, well closed; protect from light
Uses: Natural flavoring
Regulatory: FDA 21CFR §172.510

1,2,3,4,6-Penta-O-acetyl-α-D-glucopyranose. *See* α-D-Glucose pentaacetate
Pentachlorophenate sodium. *See* Sodium pentachlorophenate

Pentadecalactone
CAS 106-02-5; EINECS 203-354-6; FEMA 2840
Synonyms: Oxacyclohexadecan-2-one; ω-Pentadecalactone; Pentadecanolide; Angelica lactone; 1-Oxa-2-cyclohexadecanone
Definition: Lactone of 15-hydroxypentadecanoic acid
Empirical: $C_{15}H_{28}O_2$
Properties: Persistent musk-like odor; sol. in alcohol; insol. in water; m.w. 240.39; dens. 0.9447 (33 C); m.p. 34-36 C; b.p. 137 C (2 mm); flash pt. 62 C; ref. index 1.4669 (33 C)
Uses: Synthetic flavoring agent
Usage level: 0.27 ppm (nonalcoholic beverages), 0.50 ppm (alcoholic beverages), 0.68 ppm (ice cream, ices), 1.4 ppm (candy), 1.5 ppm (baked goods), 0.10 ppm (gelatins, puddings)
Regulatory: FDA 21CFR §172.515
Manuf./Distrib.: Aldrich

ω-Pentadecalactone. *See* Pentadecalactone
Pentadecanolide. *See* Pentadecalactone
Pentaerythritol ester of rosin. *See* Pentaerythrityl rosinate
Pentaerythritol rosinate. *See* Pentaerythrityl rosinate

Pentaerythrityl rosinate
CAS 8050-26-8; EINECS 232-479-9
Synonyms: Pentaerythritol rosinate; Pentaerythritol ester of rosin; Rosin pentaerythritol ester
Definition: Ester of rosin acids with the polyol, pentaerythritol
Properties: Amber hard solid; sol. in acetone, benzene; insol. in water
Toxicology: Heated to decomp., emits acrid smoke and irritating fumes
Uses: Masticatory substance in chewing gum base
Regulatory: FDA 21CFR §172.615, 175.105, 175.300, 176.170, 176.210, 176.2600, 178.3120, 178.3800, 178.3870

Pentahydroxycaproic acid. *See* D-Gluconic acid
1,2,3,4,5-Pentahydroxypentane. *See* Xylitol

γ-Pentalactone. *See* γ-Valerolactone
n-Pentanal. *See* n-Valeraldehyde
3-Pentanecarboxylic acid. *See* Diethylacetic acid
Pentanedial. *See* Glutaral

Pentane-2,3-dione
CAS 600-14-6; EINECS 209-984-8; FEMA 2841
Synonyms: 2,3-Pentanedione; Acetylpropionyl
Empirical: $C_5H_8O_2$
Formula: $CH_3CH_2COCOCH_3$
Properties: Yel. liq., penetrating buttery odor on dilution; sl. sol. in water; m.w. 100.12; dens. 0.959 (20/4 C); m.p. -52 c; b.p. 110-115 C; flash pt. 19 C; ref. index 1.404 (20 C)
Precaution: Highly flamm.; keep away from ignition sources
Uses: Synthetic pungent, oily, butter-like fragrance and flavoring
Usage level: 0.60 ppm (nonalcoholic beverages), 3.3 ppm (ice cream, ices), 5.9 ppm (candy), 9.6 ppm (baked goods), 0.28 ppm (gelatins, puddings), 0.30 ppm (toppings)
Regulatory: FDA 21CFR §172.515
Manuf./Distrib.: BASF

2,3-Pentanedione. *See* Pentane-2,3-dione
2,4-Pentanedione. *See* Acetylacetone
n-Pentanoic acid. *See* n-Valeric acid
1-Pentanol. *See* n-Amyl alcohol

3-Pentanol
CAS 584-02-1; EINECS 209-526-7
Synonyms: Diethyl carbinol; sec-n-Amyl alcohol
Empirical: $C_5H_{12}O$
Formula: $(C_2H_5)_2CHOH$
Properties: Liq., char. odor; sl. sol. in water; sol. in alcohol, ether; m.w. 88.15; dens. 0.819 (20/4 C); b.p. 114-116 C; flash pt. 34 C; ref. index 1.410
Precaution: Flamm.
Toxicology: LD50 (oral, rat) 1.87 g/kg; irritant to eyes, nose, and throat
Uses: Solvent; used in pharmaceuticals

t-Pentanol. *See* t-Amyl alcohol
2-Pentanone. *See* Methyl propyl ketone
3-Pentanone. *See* Diethyl ketone
Pentapotassium triphosphate. *See* Potassium tripolyphosphate
Pentapotassium tripolyphosphate. *See* Potassium tripolyphosphate

Pentasodium triphosphate
CAS 7758-29-4, 13573-18-7; EINECS 231-694-5; EEC E450b
Synonyms: STPP; Sodium tripolyphosphate; Pentasodium tripolyphosphate; Sodium triphosphate; Triphosphoric acid pentasodium salt
Classification: Inorganic salt
Empirical: $O_{10}P_3 \cdot 5Na$
Formula: $Na_5P_3O_{10}$
Properties: Wh. gran. or powd., sl. hygroscopic; sol. 20 g/100 ml water; m.w. 367.91
Toxicology: LD50 (oral, rat) 6.5 g/kg; poison by IV route; mod. toxic by ingestion, subcutaneous, IP routes; moderately irritating to skin, mucous membranes; ingestion can cause violent purging; heated to decomp., emits toxic fumes of PO_x and Na_2O
Uses: Boiler water additive; sequestrant; cooked out juices retention agent; hog/poultry scald agent; texturizer; used in angel food cake mix, meat/poultry prods., sausage, desserts, gelling juices, canned ham, meringues, canned peas; migrating from paper
Usage level: Limitation 5% (pickle for meat prods.), 0.5% (meat prods.), 0.5% (total poultry prod.); ADI 0-70 mg/kg (EEC, total phosphates)
Regulatory: FDA 21CFR §172.892, 173.310, 182.70, 182.90, 182.1810, 182.6810, GRAS; USDA 9CFR §318.7, 381.147; Europe listed; UK approved
Manuf./Distrib.: Albright & Wilson Am.; Browning; FMC; Rhone-Poulenc Food Ingreds.
Trade names: Albrite STPP-F; Albrite STPP-FC; Curafos® STP; Freez-Gard® FP-19
Trade names containing: Albriphos™ Blend 75-25; Albriphos™ Blend 90-10; Albriphos™ Blend 928; Curafos® 11-2; Curafos® 22-4; Curavis® 250; Curavis® 350; Flav-R-Keep®; FOS-6; Freez-Gard® FP-15; Kena® FP-28; Kena® FP-85

Pentasodium tripolyphosphate. *See* Pentasodium triphosphate

4-Pentenoic acid
   CAS 591-80-0; EINECS 209-732-7; FEMA 2843
   Synonyms: Allylacetic acid; 3-Vinylpropionic acid; Allyl acetate
   Empirical: $C_5H_8O_2$
   Formula: CH_2:$CHCH_2CH_2COOH$
   Properties: Colorless liq., cheesy odor; sol. in alcohol, ether; sl. sol. in water; m.w. 100.12; dens. 0.978 (20/
      4 C); m.p. -22 c; b.p. 187-189 C; flash pt. 193 F; ref. index 1.429 (20 C)
   Toxicology: Corrosive; irritating to eyes, respiratory system
   Uses: Synthetic flavoring agent
   Usage level: 1.0 ppm (nonalcoholic beverages), 2.0 ppm (ice cream, ices, margarine), 5.0 ppm (candy, baked
      goods)
   Regulatory: FDA 21CFR §172.515
   Manuf./Distrib.: Aldrich

1-Penten-3-ol
   CAS 616-25-1; EINECS 210-472-1; FEMA 3584
   Synonyms: Ethyl vinyl carbinol
   Empirical: $C_5H_{10}O$
   Formula: $CH_3CH_2CH(OH)CH$:CH_2
   Properties: Liq., butter mild green odor; sl. sol. in water; misc. with alcohol, ether; m.w. 86.14; dens. 0.838 (20/
      4 C); b.p. 114-116 C; flash pt. 28 C; ref. index 1.425 (20 C)
   Precaution: Flamm.
   Uses: Synthetic flavoring
   Regulatory: FDA 21CFR §172.515
   Manuf./Distrib.: Aldrich

Pentosanase-hexosanase
   Uses: Food grade enzyme for hydrolysis of veg. gums
   Trade names: Rhozyme® HP-150 Conc.

Pentyl acetate. *See* Amyl acetate
Pentyl alcohol. *See* n-Amyl alcohol
t-Pentyl alcohol. *See* t-Amyl alcohol
Pentyl butyrate. *See* Amyl butyrate
Pentylcarbinol. *See* Hexyl alcohol
α-Pentylcinnamyl formate. *See* α-Amylcinnamyl formate
α-Pentylcinnamyl isovalerate. *See* α-Amylcinnamyl isovalerate
n-Pentyl ethyl ketone. *See* 3-Octanone
Pentylformic acid. *See* Caproic acid
Pentyl heptanoate. *See* Amyl heptanoate
Pentyl hexanoate. *See* Amyl hexanoate
Pentyl hexanoate. *See* Isoamyl hexanoate
Pentyl octanoate. *See* Amyl octanoate
Pentyl vinyl carbinol. *See* 1-Octen-3-ol
Pepital. *See* Acetaldehyde phenethyl propyl acetal

Peppermint extract
   CAS 84082-70-2
   Synonyms: Mentha piperita extract
   Definition: Extract of leaves of *Mentha piperita*
   Uses: Natural flavoring agent
   Regulatory: FDA 21CFR §182.20, GRAS
   Manuf./Distrib.: Chart; Pierre Chauvet

Peppermint leaves
   Synonyms: Mentha piperita leaves
   Definition: Dried leaves and tops of *Mentha piperita*
   Uses: Natural flavoring agent; used to prepare the beverage peppermint tea and the infusion
   Regulatory: FDA 21CFR §182.10, GRAS; Japan approved (peppermint)
   Manuf./Distrib.: Chr. Hansen's

Peppermint oil
   CAS 8006-90-4; FEMA 2848
   Synonyms: Mentha piperita oil
   Definition: Volatile oil from steam distillation of *Mentha piperita*
   Properties: Colorless to pale yel. liq., strong penetrating peppermint odor, pungent taste; sol. in 4 vols 90%

Pepsin

alcohol; very sl. sol. in water; dens. 0.896-0.908 (25/25 C); ref. index 1.460-1.471 (20 C)
Toxicology: LD50 (oral, rat) 2426 mg/kg; mod. toxic by ingestion and IP routes; allergen; mutagenic data; heated to decomp., emits acrid smoke and irritating fumes
Uses: Natural flavoring agent; mfg. of liqueurs
Usage level: 99 ppm (nonalcoholic beverages), 240 ppm (alcoholic beverages), 110 ppm (ice cream), 1200 ppm (candy), 300 ppm (baked goods), 75-200 ppm (gelatins, puddings), 8300 ppm (chewing gum), 8 ppm (meats), 5-54 ppm (icings), 650 ppm (toppings)
Regulatory: FDA 21CFR §182.10, 182.20, GRAS; 27CFR §21.65, 21.151; FEMA GRAS; Europe listed (pulegone levels: 25 ppm in food to 350 ppm in mint confectionery)
Manuf./Distrib.: Acme-Hardesty; Chart; Commodity Services; Florida Treatt

Pepsin
CAS 9001-75-6; EINECS 232-629-3
Synonyms: Pepsinum
Definition: A digestive enzyme of gastric juice which hydrolyzes certain linkages of proteins to produce peptones
Properties: Wh. or ylsh. wh. powd. or lustrous transparent or translucent scales, odorless; sol. in water; insol. in alcohol, chloroform, ether; m.w. ≈ 36,000
Uses: Enzyme; subsititute for rennet in cheesemaking
Regulatory: FDA GRAS; Canada, Japan approved
Manuf./Distrib.: G Fiske & Co Ltd
Trade names: Pepsin 1:3000 NF XII Powd.; Pepsin 1:10,000 Powd. or Gran; Pepsin 1:15,000 Powd.

Pepsinum. *See* Pepsin
Perfluorocyclobutane. *See* Octafluorocyclobutane

Perfluorohexane
CAS 355-42-0; EINECS 206-585-0
Synonyms: Tetradecafluorohexane
Empirical: C_6F_{14}
Formula: $CF_3(CF_2)_4CF_3$
Properties: m.w. 338.04; dens. 1.684 (20/4 C); b.p. 54-58 C; ref. index 1.2520
Uses: Used with 99% 1,1,2-trichloro-1,2,2-trifluoroethane to quick-cool or crust-freeze chickens sealed in intact bags
Regulatory: FDA 21CFR §173.342

Periclase. *See* Magnesium oxide

Perillaldehyde
CAS 18031-40-8
Synonyms: 4-Isopropenyl-1-cyclohexene-1-carboxaldehyde; p-Mentha-1,8-dien-7-al
Empirical: $C_{10}H_{14}O$
Properties: Green oily fatty cherry odor; m.w. 150.22; dens. 0.965; b.p. 104-105 C (10 mm); flash pt. 204 F; ref. index 1.5072
Toxicology: Irritant
Uses: Synthetic flavoring
Regulatory: FDA 21CFR §172.515; Japan approved as flavoring
Manuf./Distrib.: Aldrich

Perillyl acetate
Uses: Synthetic flavoring
Regulatory: FDA 21CFR §172.515

Perillyl alcohol
CAS 536-59-4; EINECS 208-639-9; FEMA 2664
Synonyms: p-Mentha-1,8-dien-7-ol; (S)-(-)-Perillyl alcohol
Empirical: $C_{10}H_{16}O$
Properties: Green pungent fatty odor; m.w. 152.24; dens. 0.960; b.p. 119-121 C (11 mm); flash pt. > 230 F; ref. index 1.5010
Toxicology: Irritant
Uses: Synthetic flavoring
Regulatory: FDA 21CFR §172.515
Manuf./Distrib.: Aldrich

(S)-(-)-Perillyl alcohol. *See* Perillyl alcohol
Peroxydisulfuric acid diammonium salt. *See* Ammonium persulfate
Peroxydisulfuric acid dipotassium salt. *See* Potassium persulfate

Persic oil. *See* Apricot kernel oil
Persic oil. *See* Peach kernel oil
Peruvian balsam. *See* Balsam Peru
Peruvian bark extract. *See* Cinchona extract
Peruviol. *See* Trimethyldodecatrieneol

Petitgrain oil
 Definition: Volatile oil from leaves, twigs, and unripe fruit of *Citrus vulgaris*
 Properties: Yel. liq.; sol. in 2 vols 80% alcohol; sl. sol. in water; dens. 0.887-0.900 (15/15 C); ref. index 1.4623
 (20 C)
 Precaution: Keep cool, well closed; protect from light
 Uses: Natural flavoring agent
 Regulatory: FDA 21CFR §182.20, GRAS; Japan approved
 Manuf./Distrib.: Pierre Chauvet; Commodity Services; Florida Treatt

Petrolatum
 CAS 8009-03-8 (NF); 8027-32-5 (USP); EINECS 232-373-2
 Synonyms: Petroleum jelly; Petrolatum amber; Petrolatum white
 Classification: Petroleum hydrocarbons
 Definition: Semisolid mixture of hydrocarbons obtained from petroleum
 Properties: Yellowish to lt. amber or white semisolid, unctuous mass; pract. odorless and tasteless; sol. in
 benzene, chloroform, ether, petrol. ether, oils; pract. insol. in water; dens. 0.820-0.865 (60/25 C); m.p. 38-
 54 C; ref. index 1.460-1.474 (60 C)
 Toxicology: Heated to decomp., emits acrid smoke and irritating fumes
 Uses: Defoaming agent in beet sugar and yeast; protective coating on raw fruits and vegetables; release agent/
 lubricant in bakery prods., confectionery, dehydrated fruits and vegetables, egg white solids; polishing
 agent; sealing agent, lubricant
 Usage level: Limitation 0.15% (with wh. min. oil, bakery prods.), 0.2% (confections), 0.02% (dehydrated fruits/
 vegetables), 0.1% (egg white solids)
 Regulatory: FDA 21CFR §172.880, 172.884, 173.340, 175.105, 175.125, 175.176, 175.300, 176.170,
 176.200, 176.210, 177.2600, 177.2800, 1787.3570, 178.3700, 178.3910, 573.720
 Trade names: Fonoline® White; Fonoline® Yellow; Ointment Base No. 3; Ointment Base No. 4; Ointment Base
 No. 6; Penreco Amber; Penreco Blond; Penreco Cream; Penreco Lily; Penreco Regent; Penreco Royal;
 Penreco Snow; Penreco Super; Penreco Ultima; Perlatum® 400; Perlatum® 410; Perlatum® 410 CG;
 Perlatum® 415; Perlatum® 415 CG; Perlatum® 420; Perlatum® 425; Perlatum® 510; Protopet® Alba;
 Protopet® White 1S; Protopet® White 2L; Protopet® White 3C; Protopet® Yellow 2A; Super White
 Fonoline®; Super White Protopet®

Petrolatum amber. *See* Petrolatum
Petrolatum white. *See* Petrolatum
Petroleum benzin. *See* Naphtha

Petroleum distillates
 CAS 8002-05-9; 64742-14-9; 64742-47-8; EINECS 232-298-5
 Classification: Petroleum hydrocarbons
 Definition: Mixture of volatile hydrocarbons obtained from petroleum
 Uses: Hydrocarbon processing solvent, foam control agent for fruit and vegetable processing
 Trade names: Penreco 2251 Oil; Penreco 2263 Oil

Petroleum ether. *See* Naphtha
Petroleum jelly. *See* Petrolatum
Petroleum naphtha. *See* Naphtha
Petroleum spirit. *See* Naphtha

Petroleum wax
 CAS 8002-74-2
 Synonyms: Microcrystalline wax; Petroleum wax, synthetic; Refined petroleum wax
 Classification: Petroleum hydrocarbon
 Definition: Hydrocarbon derived from petroleum
 Properties: Translucent wax, odorless, tasteless; very sl. sol. in org. solvs.; insol. in water; m.p. 48-93 C
 Toxicology: Heated to decomp., emits acrid smoke and irritating fumes
 Uses: Masticatory substance in chewing gum base; protective coating on cheese, raw fruits and vegetables;
 defoaming agent for beet sugar and yeast processing; component of microcapsules for spice-flavoring
 substances
 Regulatory: FDA 21CFR §172.230, 172.615, 172.886 (1050 ppm max. of poly(alkylacrylate) as an antioxidant), 172.888, 173.340, 178.3710, 178.3720

Petroleum wax, crystalline. *See* Paraffin
Petroleum wax, microcrystalline. *See* Microcrystalline wax
Petroleum wax, synthetic. *See* Petroleum wax
Petroselinum hortense seed oil. *See* Parsley seed oil
Petroselinum sativum extract. *See* Parsley extract
Petroselinum sativum seed oil. *See* Parsley seed oil

α-Phellandrene
>CAS 99-83-2; FEMA 2856
>*Synonyms:* p-Mentha-1,5-diene; 2-Methyl-5-(1-methylethyl)-1,3-cyclohexadiene; 5-Isopropyl-2-methyl-1,3-cyclohexadiene
>*Empirical:* $C_{10}H_{16}$
>*Properties:* Colorless mobile oil, minty herbaceous odor; sol. in ether; insol. in water; m.w. 136.23; dens. 0.850; flash pt. 117 F
>*Toxicology:* Ingestion can cause vomiting, diarrhea; can be irritating to, and absorbed through, skin
>*Uses:* Synthetic flavoring agent
>*Usage level:* 10 ppm (nonalcoholic beverages), 28 ppm (ice cream, ices), 130 ppm (candy), 41 ppm (baked goods)
>*Regulatory:* FDA 21CFR §172.515
>*Manuf./Distrib.:* Aldrich

Phenacetaldehyde dimethyl acetal. *See* Phenylacetaldehyde dimethyl acetal
β-Phenethanol. *See* Phenethyl alcohol
2-Phenethyl acetate. *See* 2-Phenylethyl acetate
β-Phenethyl acetate. *See* 2-Phenylethyl acetate

Phenethyl alcohol
>CAS 60-12-8; EINECS 200-456-2; FEMA 2858
>*Synonyms:* Benzeneethanol; Benzyl carbinol; 2-Phenylethanol; β-Phenethanol; Phenylethyl alcohol; β-Phenylethyl alcohol
>*Classification:* Aromatic alcohol
>*Empirical:* $C_8H_{10}O$
>*Formula:* $C_6H_5CH_2CH_2OH$
>*Properties:* Colorless liq., floral rose odor; misc. with alcohol, ether; sol. in fixed oils, glycerin, propylene glycol; m.w. 122.18; dens. 1.0245 (15 C); m.p. -27 C; b.p. 220 C; flash pt. 102 C; ref. index 1.532 (20 C)
>*Precaution:* Combustible when exposed to heat or flame; reactive with oxidizing materials
>*Toxicology:* LD50 (oral, rat) 1790 mg/kg; poison by ingestion, IP routes; mod. toxic by skin contact; skin and eye irritant; experimental teratogenic effects; severe CNS injury; heated to decomp., emits acrid smoke and irritating fumes
>*Uses:* Mildly floral, rose-like fragrance and flavoring with nuances of hyacinth and honey
>*Usage level:* 1% max.; 1.5 ppm (nonalcoholic beverages), 8.3 ppm (ice cream), 12 ppm (candy), 16 ppm (baked goods), 0.15 ppm (gelatins, puddings), 21-80 ppm (chewing gum)
>*Regulatory:* FDA 21CFR §172.515, CIR approved, EPA reg.; JSCI listed
>*Manuf./Distrib.:* Aldrich; Chr. Hansen's

α-Phenethyl alcohol. *See* α-Methylbenzyl alcohol

Phenethyl anthranilate
>CAS 133-18-6; FEMA 2859
>*Synonyms:* Benzyl carbinyl anthranilate; β-Phenylethyl-o-aminobenzoate
>*Definition:* Ester of anthranilic acid and phenylethyl alcohol
>*Empirical:* $C_{15}H_{15}NO_2$
>*Properties:* Wh. colorless cryst. mass; insol. in water; m.w. 241.29; m.p. 39-44 C; b.p. 226 C; flash pt. > 230 F
>*Uses:* Synthetic flavoring agent
>*Usage level:* 1.4 ppm (nonalcoholic beverages), 1.9 ppm (ice cream, ices), 6.2 5.9 ppm (candy), 5.8 ppm (baked goods)
>*Regulatory:* FDA 21CFR §172.515
>*Manuf./Distrib.:* Aldrich

Phenethyl benzoate
>CAS 94-47-3; FEMA 2860
>*Synonyms:* 2-Phenylethyl benzoate
>*Empirical:* $C_{15}H_{14}O_2$
>*Properties:* Colorless to ylsh. oily liq., rose honey-like odor; insol. in water; m.w. 226.28; flash pt. > 100 C; ref. index 1.558-1.562
>*Uses:* Synthetic flavoring agent

Usage level: 1.0 ppm (nonalcoholic beverages, ice cream, ices), 2.0 ppm (candy), 4.0 ppm (baked goods), 3.8 ppm (chewing gum)
Regulatory: FDA 21CFR §172.515
Manuf./Distrib.: Aldrich

β-Phenethyl-n-butanoate. *See* Phenethyl butyrate

Phenethyl butyrate
CAS 103-52-6; FEMA 2861
Synonyms: β-Phenethyl-n-butanoate
Definition: Ester of phenylethyl alcohol and n-butyric acid
Empirical: $C_{12}H_{16}O_2$
Properties: Colorless liq., rose-like odor, sweet honey-like taste; insol. in water; m.w. 192.26; dens. 0.994; b.p. 260 C; flash pt. > 230 F; ref. index 1.488-1.492
Uses: Synthetic flavoring agent
Usage level: 3.2 ppm (nonalcoholic beverages), 8.9 ppm (ice cream, ices), 13 ppm (candy, baked goods)
Regulatory: FDA 21CFR §172.515
Manuf./Distrib.: Aldrich

Phenethyl cinnamate
FEMA 2863
Synonyms: Benzyl carbinyl cinnamate; Phenylethyl cinnamate
Empirical: $C_{17}H_{16}O_2$
Properties: Wh. cryst. solid, sweet balsamic odor; sol. in hot alcohol; m.w. 252.32; m.p. 58 C; b.p. > 300 C
Uses: Synthetic flavoring agent
Usage level: 1.7 ppm (nonalcoholic beverages), 0.80 ppm (ice cream, ices), 3.2 ppm (candy), 3.1 ppm (baked goods), 0.10 ppm (gelatins, puddings)
Regulatory: FDA 21CFR §172.515

Phenethyl 3,3-dimethylacrylate. *See* Phenethyl senecioate

Phenethyl formate
FEMA 2864
Uses: Synthetic flavoring agent
Regulatory: FDA 21CFR §172.515

Phenethyl isobutyrate
CAS 103-48-0; FEMA 2862
Uses: Synthetic flavoring agent
Regulatory: FDA 21CFR §172.515
Manuf./Distrib.: Aldrich

Phenethyl isovalerate
CAS 140-26-1; FEMA 2871
Uses: Synthetic flavoring agent
Regulatory: FDA 21CFR §172.515
Manuf./Distrib.: Aldrich

2-Phenethyl 2-methylbutyrate
CAS 24817-51-4; FEMA 3632
Uses: Synthetic flavoring
Regulatory: FDA 21CFR §172.515
Manuf./Distrib.: Aldrich

Phenethyl phenylacetate
CAS 102-20-5; FEMA 2866
Synonyms: Benzylcarbinyl-α-toluate; 2-Phenylethyl phenylacetate; β-Phenylethyl phenylacetate; 2-Phenylethyl-α-toluate
Empirical: $C_{16}H_{16}O_2$
Properties: Colorless to sl. yel. liq. above 26 C, rosy hyacinth odor; sol. in alcohol; insol. in water; m.w. 240.32; dens. 1.079-1.082; flash pt. > 212 F
Precaution: Combustible
Toxicology: LD50 (oral, mouse) 3190 mg/kg; mod. toxic by ingestion; heated to decomp. emits acrid smoke and irritating fumes
Uses: Synthetic flavoring agent
Regulatory: FDA 21CFR §172.515
Manuf./Distrib.: Chr. Hansen's

Phenethyl propionate
Uses: Synthetic flavoring
Regulatory: FDA 21CFR §172.515

Phenethyl salicylate
FEMA 2868
Empirical: $C_{15}H_{14}O_3$
Properties: Wh. cryst., balsamic odor; sol. in alcohol; insol. in water; m.w. 242.27; solid. pt. 41 C; flash pt. > 212 F
Precaution: Combustible
Toxicology: Heated to decomp., emits acrid smoke and irritating fumes
Uses: Synthetic flavoring agent
Regulatory: FDA 21CFR §172.515

Phenethyl senecioate
Synonyms: Phenethyl 3,3-dimethylacrylate
Uses: Synthetic flavoring
Regulatory: FDA 21CFR §172.515

Phenethyl tiglate
CAS 55719-85-2; FEMA 2870
Synonyms: Benzyl carbinyl tiglate; Phenylethyl tiglate
Empirical: $C_{13}H_{16}O_2$
Properties: Colorless liq., warm rose-like odor, sweet winey taste; sol. in alcohol; insol. in water; m.w. 204.27; dens. 1.018; b.p. 259 C; flash pt. > 230 F
Uses: Synthetic flavoring
Usage level: 0.80-0.90 ppm (nonalcoholic beverages), 4.3 ppm (ice cream, ices), 10 ppm (candy, baked goods)
Regulatory: FDA 21CFR §172.515
Manuf./Distrib.: Aldrich

Phenoxyacetic acid
CAS 122-59-8; EINECS 204-556-7; FEMA 2872
Synonyms: Phenylium; o-Phenylglycolic acid; Phenoxyethanoic acid; Glycolic acid phenyl ether
Empirical: $C_8H_8O_3$
Formula: $C_6H_5OCH2COOH$
Properties: Cryst. solid, sour odor, honey-like taste; readily sol. in alcohol, ether, benzene, CS_2, glac. acetic acid; sl. sol. in water; m.w. 152.14; m.p. 95 C; b.p 285 C (some decomp.)
Toxicology: Mild irritant
Uses: Synthetic flavoring agent
Usage level: 0.37 ppm (nonalcoholic beverages), 1 ppm (ice cream), 2.2 ppm (candy), 2.2 ppm (baked goods)
Regulatory: FDA 21CFR §172.515; GRAS (FEMA)
Manuf./Distrib.: Aldrich; Penta Mfg.

Phenoxyethanoic acid. *See* Phenoxyacetic acid

Phenoxyethyl isobutyrate
FEMA 2873
Synonyms: 2-Phenoxyethyl 2-methylpropionate; 2-Phenoxyethyl isobutyrate
Empirical: $C_{12}H_{16}O_3$
Properties: Colorless liq., honey roselike odor, sweet peach-like taste; misc. in alcohol, chloroform, ether; insol. in water; m.w. 208.26; dens. 1.044; b.p. 265 C; flash pt. > 212 F; ref. index 1.492
Precaution: Combustible
Toxicology: Heated to decomp., emits acrid smoke and irritating fumes
Uses: Synthetic flavoring agent
Usage level: 0.90-5.0 ppm (nonalcoholic beverages), 5.0-30 ppm (ice cream, ices), 15-30 ppm (candy, baked goods)
Regulatory: FDA 21CFR §172.515
Manuf./Distrib.: Hüls AG

2-Phenoxyethyl isobutyrate. *See* Phenoxyethyl isobutyrate
2-Phenoxyethyl 2-methylpropionate. *See* Phenoxyethyl isobutyrate

Phenylacetaldehyde
CAS 122-78-1; EINECS 204-574-5; FEMA 2874
Synonyms: PAA; Benzeneacetaldehyde; Hyacinthin; α-Tolualdehyde; α-Toluic aldehyde; Phenylacetic aldehyde; Phenylethanal

Empirical: C$_8$H$_8$O
Formula: C$_5$H$_5$CH$_2$CHO
Properties: Colorless oily liq., becomes more visc. on standing, hyacinth, lilac odor; sol. in alcohol, ether, propylene glycol; sl. sol. in water; m.w. 120.16; dens. 1.0123-1.030; m.p. 33-34 C; b.p. 78 C (10 mm); flash pt. 68 C; ref. index 1.525-1.545
Precaution: Combustible liq.
Toxicology: LD50 (oral, rat) 1550 mg/kg; mod. toxic by ingestion; human skin irritant; heated to decomp., emits acrid smoke and irritating fumes
Uses: Synthetic pungent, green, hyacinth-like fragrance and flavoring; for bakery prods., nonalcoholic beverages, chewing gum, confections, gelatin desserts, ice cream, maraschino cherries, puddings
Regulatory: FDA 21CFR §172.515
Manuf./Distrib.: BASF
Trade names: Phenylacetaldehyde 50

Phenylacetaldehyde 2,3-butylene glycol acetal

FEMA 2875
Synonyms: 4,5-Dimethyl-2-benzyl-1,3-dioxolan
Empirical: C$_{12}$H$_{16}$O$_2$
Properties: Colorless visc. liq., earthy fragrance, fruity flavor; sol. in alcohol; insol. in water; m.w. 192.26
Uses: Synthetic flavoring
Usage level: 4.0 ppm (candy)
Regulatory: FDA 21CFR §172.515

Phenylacetaldehyde diisobutyl acetal

FEMA 3384
Definition: Synthetic acetal aromatic
Properties: Sweet floral green taste
Uses: Flavoring

Phenylacetaldehyde dimethyl acetal

CAS 101-48-4; EINECS 202-945-6; FEMA 2876
Synonyms: (2,2-Dimethoxyethyl)-benzene; α-Tolyl aldehyde dimethyl acetal; 1,1,-Dimethoxy-2-phenylethane; Phenacetaldehyde dimethyl acetal
Empirical: C$_{10}$H$_{14}$O$_2$
Formula: C$_6$H$_5$CH$_2$CH(OCH$_3$)$_2$
Properties: Colorless liq., strong odor; sol. in fixed oils, propylene glycol; insol. in glycerin; m.w. 166.24; dens. 1.000-1.006; b.p. 95-98 C (10 mm); flash pt. 89 C; ref. index 1.493
Precaution: Combustible
Toxicology: LD50 (oral, rat) 3500 mg/kg; mod. toxic by ingestion; heated to decomp., emits acrid smoke and irritating fumes
Uses: Synthetic flavoring agent
Regulatory: FDA 21CFR §172.515
Manuf./Distrib.: Aldrich

Phenylacetaldehyde glyceryl acetal

FEMA 2877
Definition: Mixt. of 60% 5-hydroxymethyl-2-benzyl-1,3-dioxolan and 40% 5-hydroxy-2-benzyl-1,3-dioxan
Empirical: C$_{11}$H$_{14}$O$_3$
Properties: Colorless visc. liq., faint sweet rosy odor, sweet green flavor; insol. in water; m.w. 194.23; dens. 1.1650-1.1680 (15 C); flash pt. 95 C; ref. index 1.5315-1.5345
Uses: Synthetic flavoring
Usage level: 5.0 ppm (nonalcoholic beverages), 20 ppm (ice cream, ices), 0.06-20 ppm (candy)
Regulatory: FDA 21CFR §172.515

Phenylacetic acid

CAS 103-82-2; EINECS 203-148-6; FEMA 2878
Synonyms: α-Toluic acid; Benzeneacetic acid; α-Tolylic acid
Empirical: C$_8$H$_8$O$_2$
Formula: C$_6$H$_5$CH$_2$COOH
Properties: Wh. crystals, disagreeable geranium odor; sol. in alcohol, ether, hot water; m.w. 136.16; dens. 1.0809; m.p. 77-78 C; b.p. 265.5 C; flash pt. > 212 F
Precaution: Combustible
Toxicology: LD50 (oral, rat) 2250 mg/kg; mod. toxic by ingestion, subcutaneous, IP routes; experimental teratogen; heated to decomp., emits acrid smoke and irritating fumes
Uses: Synthetic flavoring agent

Usage level: 1.8 ppm (nonalcoholic beverages), 0.1 ppm (alcoholic beverages), 5.3 ppm (ice cream), 5.9 ppm (candy), 12 ppm (baked goods), 27 ppm (gelatins, puddings), 5.4-11 ppm (chewing gum), 0.1 ppm (syrups)
Regulatory: FDA 21CFR §172.515
Manuf./Distrib.: Aldrich

Phenylacetic aldehyde. *See* Phenylacetaldehyde
Phenylacrolein. *See* Cinnamal
β-Phenylacrylic acid. *See* Cinnamic acid
trans-3-Phenylacrylic acid. *See* Cinnamic acid

DL-Phenylalanine
CAS 150-30-1; EINECS 205-756-7
Synonyms: DL-α-Amino-β-phenylpropionic acid
Classification: Amino acid
Empirical: $C_9H_{11}NO_2$
Properties: Wh. cryst. platelets, odorless; sol. in water; sl. sol. in alcohol; m.w. 165.19
Uses: Nutrient; dietary supplement

L-Phenylalanine
CAS 63-91-2; EINECS 200-568-1
Synonyms: (S)-α-Aminobenzenepropanoic acid; α-Aminohydrocinnamic acid; β-Phenylalanine
Classification: Amino acid
Empirical: $C_9H_{11}NO_2$
Properties: Wh. cryst. or cryst. powd., sl. odor, bitter taste; sol. in water; very sl. sol. in alcohol, ether; m.w. 165.21; m.p. 275-283 C (dec.)
Toxicology: LD50 (IP, rat) 5287 mg/kg; mildly toxic by IP route; experimental reproductive effects; hazard to PKU individuals; heated to decomp., emits toxic fumes of NO_x
Uses: Nutrient, dietary supplement; flavorings
Regulatory: FDA 21CFR §172.320, 5.8% max., GRAS; Japan approved
Manuf./Distrib.: Am. Roland

β-Phenylalanine. *See* L-Phenylalanine
γ-Phenylallyl alcohol. *See* Cinnamyl alcohol
γ-Phenylallyl propionate. *See* Cinnamyl propionate

1-(Phenylazo)-2-naphthylamine
CAS 85-84-7
Synonyms: 1-Benzene-azo-β-naphthylamine; FD&C Yellow No. 3; CI 11380; Food Yellow 10; Solvent Yellow 5; Ext. D&C Yellow No. 9
Empirical: $C_{16}H_{13}N_3$
Properties: m.w. 247.32
Toxicology: Mod. toxic by ingestion, subcutaneous routes; experimental tumorigen; mutagenic data; heated to decomp., emits toxic fumes of NO_x
Uses: Color additive
Regulatory: FDA 21CFR §81.10

Phenyl benzene. *See* Biphenyl

4-Phenyl-2-butanol
Synonyms: Phenylethyl methyl carbinol
Uses: Synthetic flavoring
Regulatory: FDA 21CFR §172.515

4-Phenyl-3-buten-2-ol
Synonyms: Methyl styryl carbinol
Uses: Synthetic flavoring
Regulatory: FDA 21CFR §172.515

4-Phenyl-3-buten-2-one
Uses: Synthetic flavoring
Regulatory: FDA 21CFR §172.515

trans-4-Phenyl-3-buten-2-one. *See* Benzylidene acetone

4-Phenyl-2-butyl acetate
Synonyms: Phenylethyl methyl carbinyl acetate
Uses: Synthetic flavoring
Regulatory: FDA 21CFR §172.515

Phenylcarbinol. *See* Benzyl alcohol
Phenyldimethyl carbinyl isobutyrate. *See* α,α-Dimethylbenzyl isobutyrate
2-Phenyl-m-dioxan-5-ol. *See* Benzaldehyde glyceryl acetal
Phenylethanal. *See* Phenylacetaldehyde
2-Phenylethanol. *See* Phenethyl alcohol
1-Phenylethanone. *See* Acetophenone
1-Phenylethyl acetate. *See* α-Methylbenzyl acetate

2-Phenylethyl acetate
   CAS 103-45-7; EINECS 203-113-5; FEMA 2857
   Synonyms: 2-Phenethyl acetate; β-Phenethyl acetate; Acetic acid 2-phenylethyl ester; Benzylcarbinyl acetate
   Definition: Ester of phenethyl alcohol and acetic acid
   Empirical: $C_{10}H_{12}O_2$
   Formula: $CH_3COOCH_2CH_2C_6H_5$
   Properties: Colorless liq., sweet rosy honey odor; sol. in alcohol, fixed oils, propylene glycol; insol. in water, glycerin; m.w. 164.21; dens. 1.033 (20/4 C); m.p. 164 C; b.p. 232-234 C; flash pt. ≈ 105 C; ref. index 1.498 (20 C)
   Precaution: Combustible exposed to heat or flame; can react vigorously with oxidizing agents
   Toxicology: LD50 (oral, rat) 3670 mg/kg; mod. toxic by ingestion; mildly toxic by skin contact; heated to decomp., emits acrid smoke and irritating fumes
   Uses: Synthetic flavoring agent
   Usage level: 1.4 ppm (nonalcoholic beverages), 2.2 ppm (ice cream), 4.2 ppm (candy), 5.6 ppm (baked goods)
   Regulatory: FDA 21CFR §172.515; GRAS (FEMA); Japan approved as flavoring
   Manuf./Distrib.: Aldrich

α-Phenylethyl acetate. *See* α-Methylbenzyl acetate
Phenylethyl alcohol. *See* Phenethyl alcohol
β-Phenylethyl alcohol. *See* Phenethyl alcohol
β-Phenylethyl-o-aminobenzoate. *See* Phenethyl anthranilate
2-Phenylethyl benzoate. *See* Phenethyl benzoate
Phenylethyl carbinol. *See* 1-Phenyl-1-propanol
Phenylethyl cinnamate. *See* Phenethyl cinnamate
Phenylethylene. *See* Styrene
Phenylethyl methyl carbinol. *See* 4-Phenyl-2-butanol
Phenylethyl methyl carbinyl acetate. *See* 4-Phenyl-2-butyl acetate

2-Phenylethyl methyl ether
   Uses: Penetrating floral, spicy, green fragrance and flavoring, honey-like on dilution

Phenylethyl methyl ethyl carbinol. *See* 1-Phenyl-3-methyl-3-pentanol
1-Phenylethyl 2-methylpropionate. *See* Methylbenzyl isobutyrate
2-Phenylethyl phenylacetate. *See* Phenethyl phenylacetate
β-Phenylethyl phenylacetate. *See* Phenethyl phenylacetate
1-Phenylethyl propionate. *See* α-Methylbenzyl propionate
Phenylethyl tiglate. *See* Phenethyl tiglate
2-Phenylethyl-α-toluate. *See* Phenethyl phenylacetate
Phenylformic acid. *See* Benzoic acid
o-Phenylglycolic acid. *See* Phenoxyacetic acid
Phenylium. *See* Phenoxyacetic acid
Phenyl mercaptan. *See* Benzenethiol
Phenylmethanol. *See* Benzyl alcohol
Phenylmethyl acetate. *See* Benzyl acetate
Phenylmethyl benzoate. *See* Benzyl benzoate
4-Phenyl-2-methyl-butanol-2. *See* Dimethylphenylethylcarbinol
Phenyl methylether. *See* Anisole
Phenylmethyl 2-hydroxybenzoate. *See* Benzylsalicylate
Phenylmethyl 4-hydroxybenzoate. *See* Benzylparaben

1-Phenyl-3-methyl-3-pentanol
   Synonyms: Phenylethyl methyl ethyl carbinol
   Uses: Synthetic flavoring
   Regulatory: FDA 21CFR §172.515

Phenylmethyl 3-phenyl-2-propenoate. *See* Benzyl cinnamate
1-Phenyl-2-pentanol. *See* α-Propylphenethyl alcohol
2-Phenylphenol. *See* o-Phenylphenol

o-Phenylphenol
  CAS 90-43-7; EINECS 201-993-5; EEC E231
  Synonyms: (1,1'-Biphenyl)-2-ol; 2-Hydroxybiphenyl; 2-Phenylphenol; o-Xenol
  Classification: Substituted aromatic compd.
  Empirical: $C_{12}H_{10}O$
  Properties: Nearly white or lt. buff crystals, mild char. odor; sol. in alcohol, sodium hydroxide sol'n., most org.
    solvs.; insol. in water; m.w. 170.22; dens. 1.217 (25/25 C); m.p. 56-58 C; b.p. 280-284 C
  Toxicology: LD50 (rat, oral) 2.48 g/kg; poison by intraperitoneal route; moderately toxic by ingestion, others;
    irritating to eyes, skin
  Uses: Antibacterial preservative
  Usage level: 0.05-0.5%
  Regulatory: USA EPA reg.; FDA 21CFR §175.105, 176.210, 177.2600; Japan approved; Europe listed 0.2%
    max. as phenol; UK approved
  Manuf./Distrib.: Ashland; Nipa Labs
  Trade names: Nipacide® OPP

o-Phenylphenol sodium salt. *See* Sodium o-phenylphenate

2-Phenylpropanal
  CAS 93-53-8, 34713-70-7; FEMA 2886
  Synonyms: Hydratropaldehyde; Hydratropic aldehyde; 2-Phenylpropionaldehyde
  Empirical: $C_9H_{10}O$
  Formula: $C_6H_5CH(CH_3)CHO$
  Properties: Colorless to pale yel. liq., floral odor; m.w. 134.18; dens. 1.002 (20/4 C); b.p. 205-206 C; flash pt.
    169 F; ref. index 1.518
  Toxicology: Skin irritant
  Uses: Synthetic pungent green, earthy, hyacinth-like fragrance and flavoring
  Usage level: 0.61 ppm (nonalcoholic beverages), 0.30 ppm (ice cream, ices), 0.85 ppm (candy, baked goods)
  Regulatory: FDA 21CFR §172.515
  Manuf./Distrib.: Aldrich; BASF; Hüls AG

3-Phenylpropanal. *See* Hydrocinnamaldehyde
2-Phenylpropanal dimethyl acetal. *See* 2-Phenylpropionaldehyde dimethylacetal

1-Phenyl-1-propanol
  CAS 93-54-9; EINECS 202-256-0; FEMA 2884
  Synonyms: Phenylethyl carbinol; Ethyl phenyl carbinol
  Empirical: $C_9H_{12}O$
  Formula: $C_2H_5CH(C_6H_5)OH$
  Properties: Colorless oily liq., floral fragrance; sol. in alcohol; m.w. 136.19; dens. 0.994; b.p. 103 C (14 mm);
    flash pt. 90 C; ref. index 1.5200
  Uses: Synthetic flavoring
  Usage level: 0.50 ppm (nonalcoholic beverages, ice cream, ices), 1.5 ppm (candy, baked goods)
  Regulatory: FDA 21CFR §172.515

2-Phenyl propanol-1
  CAS 1123-85-9; EINECS 214-379-7
  Synonyms: Hydratropalcohol; β-Methylphenethyl alcohol; Hydratropic alcohol; 2-Phenyl-1-propanol
  Empirical: $C_9H_{12}O$
  Formula: $C_6H_5CH(CH_3)CH_2OH$
  Properties: m.w. 136.20; dens. 1.003 (20/4 C); b.p. 224-225 C; flash pt. 108 C; ref. index 1.527
  Uses: Sweetly foral, lilac, hyacinth fragrance and flavoring
  Regulatory: FDA 21CFR §172.515
  Manuf./Distrib.: BASF; Hüls AG

2-Phenyl-1-propanol. *See* 2-Phenyl propanol-1
3-Phenylpropanol. *See* Hydrocinnamic alcohol
3-Phenyl-1-propanol acetate. *See* Hydrocinnamyl acetate
2-Phenylpropan-2-yl isobutyrate. *See* α,α-Dimethylbenzyl isobutyrate
3-Phenyl-2-propenal. *See* Cinnamal
3-Phenylpropenoic acid. *See* Cinnamic acid
3-Phenyl-2-propenoic acid 3-phenyl-2-propenyl ester. *See* Cinnamyl cinnamate
3-Phenyl-2-propen-1-ol. *See* Cinnamyl alcohol
2-Phenyl-2-propen-1-ol acetate. *See* Cinnamyl acetate
3-Phenyl-2-propenylanthranilate. *See* Cinnamyl anthranilate
Phenyl propenyl-n-butyrate. *See* Cinnamyl butyrate

3-Phenyl-2-propenyl propanoate. *See* Cinnamyl propionate
2-Phenylpropionaldehyde. *See* 2-Phenylpropanal
3-Phenylpropionaldehyde. *See* Hydrocinnamaldehyde

2-Phenylpropionaldehyde dimethylacetal
CAS 90-87-9; FEMA 2888
Synonyms: Hydratropaldehyde dimethyl acetal; 1,1-Dimethoxy-2-phenylpropane; Hydratropic aldehyde dimethyl acetal; 2-Phenylpropanal dimethyl acetal
Empirical: $C_{11}H_{16}O_2$
Properties: Colorless to sl. yel. liq., mushroom odor; sol. in alcohol, ether; insol. in water; m.w. 180.25; dens. 0.989-0.994; b.p. 240-241 C; flash pt. 92 C; ref. index 1.492-1.497
Toxicology: Heated to decomp., emits acrid smoke and irritating fumes
Uses: Synthetic green, spicy, earthy, fruity, wild mushroom-like fragrance and flavoring
Usage level: 0.26 ppm (nonalcoholic beverages), 0.51 ppm (ice cream, ices), 1.5 ppm (candy), 3.1 ppm (baked goods), 5.0 ppm (chewing gum, condiments)
Regulatory: FDA 21CFR §172.515
Manuf./Distrib.: BASF; Hüls AG

3-Phenylpropionic acid. *See* Hydrocinnamic acid
Phenylpropyl acetate. *See* Hydrocinnamyl acetate
3-Phenylpropyl acetate. *See* Hydrocinnamyl acetate
3-Phenylpropyl alcohol. *See* Hydrocinnamic alcohol

2-Phenylpropyl butyrate
FEMA 2891
Uses: Synthetic flavoring
Regulatory: FDA 21CFR §172.515
Manuf./Distrib.: Aldrich

α-Phenylpropyl butyrate. *See* α-Ethylbenzyl butyrate

3-Phenylpropyl cinnamate
Uses: Synthetic flavoring
Regulatory: FDA 21CFR §172.515

3-Phenylpropyl formate
Uses: Synthetic flavoring
Regulatory: FDA 21CFR §172.515

3-Phenylpropyl hexanoate
Uses: Synthetic flavoring
Regulatory: FDA 21CFR §172.515

2-Phenylpropyl isobutyrate
FEMA 2892
Uses: Synthetic flavoring
Regulatory: FDA 21CFR §172.515
Manuf./Distrib.: Aldrich

3-Phenylpropyl isobutyrate
CAS 103-58-2
Synonyms: 3-Phenylpropyl 2-methylpropionate; Hydrocinnamic isobutyrate
Uses: Synthetic flavoring
Regulatory: FDA 21CFR §172.515
Manuf./Distrib.: Hüls AG

3-Phenylpropyl isovalerate
FEMA 2899
Uses: Synthetic flavoring
Regulatory: FDA 21CFR §172.515
Manuf./Distrib.: Aldrich

3-Phenylpropyl 2-methylpropionate. *See* 3-Phenylpropyl isobutyrate

3-Phenylpropyl propionate
Uses: Synthetic flavoring
Regulatory: FDA 21CFR §172.515

2-(3-Phenylpropyl)-tetrahydrofuran
Uses: Synthetic flavoring

Regulatory: FDA 21CFR §172.515

Phosphatidylcholine
CAS 97281-47-5
Definition: Purified grade of lecithin containing no less than 95% of the phospholipid
Uses: Emulsifying dispersing, wetting, penetrating agent, antioxidant in margarine, mayonnaise, chocolate candies, baked goods

Phosphinic acid sodium salt. *See* Sodium hypophosphite

Phosphoric acid
CAS 7664-38-2; EINECS 231-633-2; EEC E338
Synonyms: Orthophosphoric acid
Classification: Inorganic acid
Empirical: H_3O_4P
Formula: H_3PO_4
Properties: Colorless liq. or rhombic crystals, odorless; sol. in water, alcohol; m.w. 98.00; dens. 1.70 (20/4 C); m.p. 42.4 C; b.p. 158 C
Precaution: Corrosive; mixts. with nitromethane are explosive
Toxicology: LD50 (oral, rat) 1530 mg/kg; mod. toxic by ingestion and skin contact; conc. sol'ns. irritating to skin, mucous membranes; corrosive irritant to eyes; TLV:TWA 1 mg/m^3 of air; heated to decomp., emits toxic fumes of PO_x
Uses: Acidulant in cola beverages; acid flavoring agent in jams and jellies; antioxidant, sequestrant; in sugar refining; prod. of caramel; in bread doughs, cake flour; yeast nutrient; synergist for antioxidants
Usage level: ADI 0-70 mg/kg (EEC)
Regulatory: FDA 21CFR §131.144, 133, 175.300,177.2260, 178.3520, 182.1073, GRAS; USDA 9CFR §318.7, 381.147 (0.01% max. in lard, shortening, poultry fat)
Regulatory: Japan approved; Europe listed; UK approved
Manuf./Distrib.: Albright & Wilson Am.; Browning; FMC; Rhone-Poulenc Food Ingreds.
Trade names: Albrite Phosphoric Acid 85% Food Grade; Quaker™ Oatrim 5, 5Q
Trade names containing: BFP White Sour; Cola No. 23443; Cola Acid No. 23444; CP-2600; CP-3650; Ry-So; Sour Dough Base

Phosphoric acid calcium salt (1:1). *See* Calcium phosphate dibasic
Phosphoric acid calcium salt (2:1). *See* Calcium phosphate monobasic anhydrous
Phosphoric acid dipotassium salt. *See* Potassium phosphate dibasic
Phosphoric acid magnesium salt (2:3). *See* Magnesium phosphate, tribasic
Phosphoric acid sodium aluminum salt, basic. *See* Sodium aluminum phosphate, basic
Phosphoric acid trisodium salt. *See* Sodium phosphate tribasic
PHPH. *See* Biphenyl
Phthalic acid, diethyl ester. *See* Diethyl phthalate
Phylloquinone. *See* Vitamin K_1
Physeter macrocephalus. *See* Ambergris
Phytodione. *See* Vitamin K_1
Phytonadione. *See* Vitamin K_1
3-Picolinic acid. *See* Nicotinic acid
Pigment White 4. *See* Zinc oxide
Pigment White 6. *See* Titanium dioxide
Pigment White 26. *See* Talc
Pimaricin. *See* Natamycin
Pimenta berries oil. *See* Allspice

Pimenta leaf oil
Definition: Pimenta officinalis
Uses: Natural flavoring agent
Regulatory: FDA 21CFR §182.20, GRAS
Manuf./Distrib.: Florida Treatt

Pimenta officinalis extract. *See* Pimento extract

Pimenta oil. *See also* Allspice
Synonyms: Pimento oil; Allspice oil
Definition: Volatile oil from fruit of *Pimenta officinalis*
Properties: Colorless, yel. or reddish liq., allspice odor and taste; becomes darker with age; sol. in glac. acetic acid, 1 vol 90% alcohol; very sl. sol. in water; dens. 1.018-1.048 (25/25 C)
Precaution: Keep cool, well closed; protect from light

Uses: Natural flavoring agent
Regulatory: FDA 21CFR §182.20, GRAS
Manuf./Distrib.: Chart; Pierre Chauvet; Florida Treatt

Pimento extract

EINECS 284-540-4
Synonyms: Pimenta officinalis extract; Allspice extract
Definition: Extract of the fruit of *Pimenta officinalis*
Uses: Natural flavoring; antioxidant (Japan)
Regulatory: FDA 21CFR §182.20; Japan approved

Pimento oil. *See* Allspice
Pimento oil. *See* Pimenta oil
Pimpinella anisum extract. *See* Anise extract
Pimpinella anisum oil. *See* Anise oil

Pineapple extract

Synonyms: Ananas comosus extract
Definition: Extract of the fruit of *Ananas comosus*

Pineapple juice

Definition: Liq. obtained from the fruit of the pineapple, *Ananas comosus*

2-Pinene. *See* α-Pinene
2(10)-Pinene. *See* β-Pinene

α-Pinene

CAS 80-56-8; 7785-26-4 (-), 7785-70-8 (+); EINECS 232-077-3 (-), 232-087-8 (+); FEMA 2902
Synonyms: 2-Pinene; 2,6,6-Trimethylbicyclo(3.1.1)-2-hept-2-ene
Classification: Terpene hydrocarbon
Empirical: $C_{10}H_{16}$
Properties: Colorless liq., turpentine odor; insol. in water; sol. in alcohol, chloroform, ether, glacial acetic acid; m.w. 136.26; dens. 0.8592 (20/4 C); m.p. -55 C; b.p. 155 C; flash pt. 91 F; ref. index 1.464-1.468
Precaution: Flammable; dangerous fire hazard exposed to heat, flame, oxidizers; explodes on contact with nitrosyl perchlorate
Toxicology. LD50 (oral, rat) 3700mg/kg; deadly poison by inh., mod. toxic by ingestion, eye, mucous membrane, and severe skin irritant; toxic effects similar to turpentine
Uses. Synthetic flavoring agent
Usage level: 16-54 ppm (nonalcoholic beverages), 64 ppm (ice cream), 48 ppm (candy), 160 ppm (baked goods), 2.6-150 ppm (condiments)
Regulatory: FDA 21CFR §172.515
Manuf./Distrib.: Aldrich

β-Pinene

CAS 127-91-3; FEMA 2903
Synonyms: 6,6-Dimethyl-2-methylenebicyclo[3.1.1]heptane; Nopinene; 2(10)-Pinene; Pseudopinene
Classification: Terpene hydrocarbon
Empirical: $C_{10}H_{16}$
Properties: Colorless liq., terpene odor; insol. in water; sol. in alcohol, chloroform, ether; m.w. 136.24; dens. 0.859; m.p. -61 C; b.p. 165-167 C; soften. pt. 112-118 C; flash pt. 32 C
Precaution: Flamm.; fire risk
Toxicology: LD50 (oral, rat) 4700 mg/kg; mildly toxic by ingestion; skin irritant; heated to decomp., emits acrid smoke and irritating fumes
Uses: Synthetic flavoring agent; moisture barrier on soft gelatin capsules and on powders of ascorbic acid or its salts
Usage level: 0.05-16 ppm (nonalcoholic beverages), 64 ppm (ice cream), 48-600 ppm (candy), 48-600 ppm (baked goods), 0.07% max. (gelatin capsules), 7% max. (ascorbic acid powds.)
Regulatory: FDA 21CFR §172.280, 172.515
Manuf./Distrib.: Aldrich

Pine needle oil. *See* Fir needle oil, Siberian type

Pine needle oil, dwarf

FEMA 2904
Definition: Oil from needles and twigs from *Pinus montana*, contg. l-pinene, l-phellandrene, sylvestrene, dipentene, cadinene, bornyl acetate
Properties: Colorless to pale yel. liq., pleasant odor, bitter taste; very sol. in chloroform, ether; sol. in 5-8 vols 90% alcohol; insol. in water; dens. 0.853-0.869 (25/25 C); ref. index 1.4750-1.4800 (20 C)

Pine needle oil, Scotch type

Precaution: Keep cool, well closed; protect from light
Uses: Natural flavoring
Usage level: 0.39 ppm (nonalcoholic beverages), 0.63 ppm (ice cream, ices), 1.9 ppm (candy, baked goods)
Regulatory: FDA 21CFR §172.510

Pine needle oil, Scotch type
FEMA 2906
Synonyms: Scotch fir oil; Fir-wood oil
Definition: Volatile oil from needles and twigs from *Pinus sylvestris*, contg. dipentene, pinene, sylvestrene, cadinene, bornyl acetate
Properties: Ylsh. liq.; sol. in 10 vols 90% alcohol; insol. in water; dens. 0.884-0.886 (15/15 C)
Precaution: Keep cool, well closed; protect from light
Uses: Natural flavoring
Usage level: 6.0 ppm (nonalcoholic beverages), 3.0 ppm (candy), 2.0 ppm (baked goods)
Regulatory: FDA 21CFR §172.510

2-Pinen-4-ol. *See* Verbenol
2(10)-Pinen-3-ol. *See* Pinocarveol

Pine oil
CAS 8002-09-3
Synonyms: Yarmor
Definition: Volatile oil obtained from distillation of the species *Pinus*
Properties: Colorless to pale yel. liq., penetrating odor; insol. in water; sol. in org. solvs.; dens. 0.86; b.p. 200-220 C; flash pt. (CC) 172 F
Toxicology: Moderately toxic by ingestion; mildly toxic by skin contact; irritating to skin, mucous membranes; large doses may cause CNS depression
Uses: Natural flavoring agent
Regulatory: FDA 21CFR §172.510, 175.105, 176.180, 176.200, 176.210, 177.2800; 27CFR §21.65, 21.151
Manuf./Distrib.: Allchem Industries; Chr. Hansen's; Hercules; Penta Mfg.

Pine tar oil
FEMA 2907
Definition: Volatile oil from steam distillation of pine tar
Uses: Synthetic flavoring
Usage level: 2.0 ppm (ice cream, ices), 10 ppm (candy)
Regulatory: FDA 21CFR §172.515

Pine, white, bark
Definition: Bark from *Pinus strobus*
Uses: Natural flavoring used in alcoholic beverages
Regulatory: FDA 21CFR §172.510

Pinocarveol
Synonyms: 2(10)-Pinen-3-ol
Uses: Synthetic flavoring
Regulatory: FDA 21CFR §172.515

Piperidine
CAS 110-89-4; EINECS 203-813-0; FEMA 2908
Synonyms: Hexahydropyridine; Hexazane
Empirical: $C_5H_{11}N$
Properties: Liq., char. heavy sweet animal-like odor; hygroscopic; sol. in alcohol, benzene, chloroform; misc. with water; m.w. 85.15; dens. 0.862 (20/4 C); m.p. -11 to -9 C; b.p. 104-106 C; flash pt. 16 C; ref. index 1.453 (20 C)
Precaution: Highly flamm.
Toxicology: LD50 (oral, rat) 0.52 ml/kg; toxic by inhalation and skin contact; causes burns
Uses: Synthetic flavoring agent
Usage level: 3.0 ppm (nonalcoholic beverages), 5.0 ppm (candy), 0.05-5.0 ppm (baked goods), 0.05 ppm (condiments, meats, soups)
Regulatory: FDA 21CFR §172.515
Manuf./Distrib.: Aldrich

Piperine
CAS 94-62-2; EINECS 202-348-0; FEMA 2909
Synonyms: 1-[5-(1,3-Benzodioxol-5-l)-1-oxo-2,4-pentadienyl]piperidine; 1-Piperoylpiperidine
Empirical: $C_{17}H_{19}NO_3$

Properties: Prisms, tasteless with burning aftertaste; sol. in benzene, acetic acid; 1 g sol. in 15 ml alcohol, 36 ml ether; pract. insol. in water; m.w. 285.33; m.p. 130 C
Uses: Synthetic flavoring; imparts pungent taste to brandy
Usage level: 0.01 ppm (nonalcoholic beverages)
Regulatory: FDA 21CFR §172.515
Manuf./Distrib.: Aldrich

Piperitenone
Synonyms: p-Mentha-1,4(8)-dien-3-one
Uses: Synthetic flavoring
Regulatory: FDA 21CFR §172.515

Piperitenone oxide
Synonyms: 1,2-Epoxy-p-menth-4-(8)-en-3-one
Uses: Synthetic flavoring
Regulatory: FDA 21CFR §172.515

d-Piperitone
FEMA 2910
Synonyms: p-Menth-1-en-3-one
Empirical: $C_{10}H_{16}O$
Properties: Colorless liq., camphor-like odor, sharp minty flavor; sol. in alcohol; insol. in water; m.w. 152.23
Uses: Synthetic flavoring
Usage level: 1.0-11 ppm (nonalcoholic beverages), 18 ppm (ice cream, ices, candy, baked goods)
Regulatory: FDA 21CFR §172.515

Piperonal. *See* Heliotropine

Piperonyl acetate
CAS 326-61-4; FEMA 2912
Synonyms: Heliotropyl acetate; 3,4-Methylenedioxybenzyl acetate
Empirical: $C_{10}H_{10}O_4$
Properties: Colorless to lt. yel. liq., cherry/strawberry/heliotrope odor; sol. in alcohol; almost insol. in water; m.w. 194.19; dens. 1.227; b.p. 150-151 C (10 mm); flash pt. > 230 F
Toxicology: LD50 (oral, rat) 2100 mg/kg; mod. toxic by ingestion; skin irritant; heated to decomp., emits acrid smoke and irritating fumes
Uses: Synthetic flavoring agent
Usage level: 27-50 ppm (nonalcoholic beverages), 80-110 ppm (ice cream, ices), 70-80 ppm (candy), 55-80 ppm (baked goods)
Regulatory: FDA 21CFR §172.515
Manuf./Distrib.: Aldrich

Piperonyl aldehyde. *See* Heliotropine

Piperonyl isobutyrate
CAS 5461-08-5; FEMA 2913
Synonyms: Helioptropyl isobutyrate; 3,4-Methylenedioxybenzyl isobutyrate
Empirical: $C_{12}H_{14}O_4$
Properties: Colorless oily liq., fruity berry odor; sol. in alcohol; insol. in water; m.w. 222.24; dens. 1.154; b.p. 91-92 C (0.005 mm); flash pt. > 230 F
Uses: Synthetic flavoring
Usage level: 0.05-1.0 ppm (nonalcoholic beverages), 0.05 ppm (ice cream, ices), 0.05-3.5 ppm (candy), 0.10-3.5 ppm (baked goods)
Regulatory: FDA 21CFR §172.515
Manuf./Distrib.: Aldrich

1-Piperoylpiperidine. *See* Piperine
Plaster of Paris. *See* Calcium sulfate
Platy talc. *See* Talc

Pliofilm
CAS 9006-00-2
Synonyms: Rubber hydrochloride
Empirical: $(C_3H_5Cl)_n$
Toxicology: Experimental tumorigen; heated to decomp., emits toxic fumes of Cl⁻
Uses: Mfg. of paper/paperboard for food contact
Regulatory: FDA 21CFR §181.30

POE (4). *See* PEG-4
POE (8). *See* PEG-8
POE (9). *See* PEG-9
POE (12). *See* PEG-12
POE (14). *See* PEG-14
POE (16). *See* PEG-16
POE (20). *See* PEG-20
POE (32). *See* PEG-32
POE (40). *See* PEG-40
POE (75). *See* PEG-75
POE (100). *See* PEG-100
POE (150). *See* PEG-150
POE (200). *See* PEG-200
POE (350). *See* PEG-350
POE (6) dilaurate. *See* PEG-6 dilaurate
POE (20) dilaurate. *See* PEG-20 dilaurate
POE (32) dilaurate. *See* PEG-32 dilaurate
POE (75) dilaurate. *See* PEG-75 dilaurate
POE (150) dilaurate. *See* PEG-150 dilaurate
POE (6) dioleate. *See* PEG-6 dioleate
POE (8) dioleate. *See* PEG-8 dioleate
POE (12) dioleate. *See* PEG-12 dioleate
POE (20) dioleate. *See* PEG-20 dioleate
POE (32) dioleate. *See* PEG-32 dioleate
POE (75) dioleate. *See* PEG-75 dioleate
POE (150) dioleate. *See* PEG-150 dioleate
POE (6) distearate. *See* PEG-6 distearate
POE (20) distearate. *See* PEG-20 distearate
POE (32) distearate. *See* PEG-32 distearate
POE (75) distearate. *See* PEG-75 distearate
POE (20) glyceryl isostearate. *See* PEG-20 glyceryl isostearate
POE (30) glyceryl isostearate. *See* PEG-30 glyceryl isostearate
POE (30) glyceryl laurate. *See* PEG-30 glyceryl laurate
POE (7) glyceryl monococoate. *See* PEG-7 glyceryl cocoate
POE (20) glyceryl monolaurate. *See* PEG-20 glyceryl laurate
POE (15) glyceryl monoricinoleate. *See* PEG-15 glyceryl ricinoleate
POE (20) glyceryl monostearate. *See* PEG-20 glyceryl stearate
POE (30) glyceryl monostearate. *See* PEG-30 glyceryl stearate
POE (20) glyceryl oleate. *See* PEG-20 glyceryl oleate
POE (30) glyceryl oleate. *See* PEG-30 glyceryl oleate
POE (25) glyceryl trioleate. *See* PEG-25 glyceryl trioleate
POE (50) hydrogenated castor oil. *See* PEG-50 hydrogenated castor oil
POE (60) hydrogenated castor oil. *See* PEG-60 hydrogenated castor oil
POE (32) monolaurate. *See* PEG-32 laurate
POE (6) monooleate. *See* PEG-6 oleate
POE (8) monooleate. *See* PEG-8 oleate
POE (32) monooleate. *See* PEG-32 oleate
POE (75) monooleate. *See* PEG-75 oleate
POE (12) monoricinoleate. *See* PEG-12 ricinoleate
POE (6) nonyl phenyl ether. *See* Nonoxynol-6
POE (7) nonyl phenyl ether. *See* Nonoxynol-7
POE (10) nonyl phenyl ether. *See* Nonoxynol-10
POE (11) nonyl phenyl ether. *See* Nonoxynol-11
POE (12) nonyl phenyl ether. *See* Nonoxynol-12
POE (14) nonyl phenyl ether. *See* Nonoxynol-14
POE (40) nonyl phenyl ether. *See* Nonoxynol-40
POE (50) nonyl phenyl ether. *See* Nonoxynol-50
POE (5) octyl phenyl ether. *See* Octoxynol-5
POE (8) octyl phenyl ether. *See* Octoxynol-8
POE (9) octyl phenyl ether. *See* Octoxynol-9
POE (13) octyl phenyl ether. *See* Octoxynol-13
POE (30) octyl phenyl ether. *See* Octoxynol-30

POE (40) octyl phenyl ether. *See* Octoxynol-40
POE (70) octyl phenyl ether. *See* Octoxynol-70
POE (23) oleyl ether. *See* Oleth-23
POE (7) POP (5) monobutyl ether. *See* PPG-5-buteth-7
POE (16) POP (12) monobutyl ether. *See* PPG-12-buteth-16
POE (30) POP (20) monobutyl ether. *See* PPG-20-buteth-30
POE (35) POP (28) monobutyl ether. *See* PPG-28-buteth-35
POE (45) POP (33) monobutyl ether. *See* PPG-33-buteth-45
POE (40) sorbitan diisostearate. *See* PEG-40 sorbitan diisostearate
POE (4) sorbitan monolaurate. *See* Polysorbate 21
POE (20) sorbitan monolaurate. *See* Polysorbate 20
POE (80) sorbitan monolaurate. *See* PEG-80 sorbitan laurate
POE (5) sorbitan monooleate. *See* Polysorbate 81
POE (20) sorbitan monooleate. *See* Polysorbate 80
POE (20) sorbitan monopalmitate. *See* Polysorbate 40
POE (4) sorbitan monostearate. *See* Polysorbate 61
POE (20) sorbitan monostearate. *See* Polysorbate 60
POE (20) sorbitan trioleate. *See* Polysorbate 85
POE (20) sorbitan tristearate. *See* Polysorbate 65
POE (6) stearate. *See* PEG-6 stearate
POE (8) stearate. *See* PEG-8 stearate
POE (40) stearate. *See* PEG-40 stearate
POE (75) stearate. *See* PEG-75 stearate
POE (100) stearate. *See* PEG-100 stearate
POE (20) stearyl ether. *See* Steareth-20
Pogy oil. *See* Menhaden oil

Poloxamer 105
CAS 9003-11-6 (generic)
Classification: Polyoxyethylene, polyoxypropylene block polymer
Formula: $HO(CH_2CH_2O)_x(CCH_3HCH_2O)_y(CH_2CH_2O)_zH$, avg. x=11, y=16, z=11
Toxicology: Moderately toxic by ingestion, intraperitoneal route
Uses: Solubilizer, stabilizer in flavor concs.; processing aid, wetting agent; surfactant, defoaming agent in scald baths for poultry defeathering, hog dehairing; dough conditioner
Regulatory: FDA 21CFR §172.808, 173.340, 175.105, 176.180, 176.200, 176.210, 177.1200

Poloxamer 108
CAS 9003-11-6 (generic)
Classification: Polyoxyethylene, polyoxypropylene block polymer
Formula: $HO(CH_2CH_2O)_x(CCH_3HCH_2O)_y(CH_2CH_2O)_zH$, avg. x=46, y=16, z=46
Toxicology: Moderately toxic by ingestion, intraperitoneal route
Uses: Solubilizer, stabilizer in flavor concs.; processing aid, wetting agent; surfactant, defoaming agent in scald baths for poultry defeathering, hog dehairing; dough conditioner
Regulatory: FDA 21CFR §172.808, 173.340, 175.105, 176.180, 176.200, 176.210, 177.1200, 177.1210

Poloxamer 122
CAS 9003-11-6 (generic)
Classification: Polyoxyethylene, polyoxypropylene block polymer
Formula: $HO(CH_2CH_2O)_x(CCH_3HCH_2O)_y(CH_2CH_2O)_zH$, avg. x=5, y=21, z=5
Toxicology: Moderately toxic by ingestion, intraperitoneal route
Uses: Solubilizer, stabilizer in flavor concs.; processing aid, wetting agent; surfactant, defoaming agent in scald baths for poultry defeathering, hog dehairing; dough conditioner
Regulatory: FDA 21CFR §172.808, 173.340, 175.105, 176.180, 176.200, 176.210, 177.1200

Poloxamer 123
CAS 9003-11-6 (generic)
Classification: Polyoxyethylene, polyoxypropylene block polymer
Formula: $HO(CH_2CH_2O)_x(CCH_3HCH_2O)_y(CH_2CH_2O)_zH$, avg. x=7, y=21, z=7
Toxicology: Moderately toxic by ingestion, intraperitoneal route
Uses: Solubilizer, stabilizer in flavor concs.; processing aid, wetting agent; surfactant, defoaming agent in scald baths for poultry defeathering, hog dehairing; dough conditioner
Regulatory: FDA 21CFR §172.808, 173.340, 175.105, 176.180, 176.200, 176.210, 177.1200

Poloxamer 124
CAS 9003-11-6 (generic)

Classification: Polyoxyethylene, polyoxypropylene block polymer
Formula: $HO(CH_2CH_2O)_x(CCH_3HCH_2O)_y(CH_2CH_2O)_zH$, avg. x=11, y=21, z=11
Toxicology: Moderately toxic by ingestion, intraperitoneal route
Uses: Solubilizer, stabilizer in flavor concs.; processing aid, wetting agent; surfactant, defoaming agent in scald baths for poultry defeathering, hog dehairing; dough conditioner
Regulatory: FDA 21CFR §172.808, 173.340, 175.105, 176.180, 176.200, 176.210, 177.1200

Poloxamer 181

CAS 9003-11-6 (generic); 53637-25-5
Classification: Polyoxyethylene, polyoxypropylene block polymer
Formula: $HO(CH_2CH_2O)_x(CCH_3HCH_2O)_y(CH_2CH_2O)_zH$, avg. x=3, y=30, z=3
Toxicology: Moderately toxic by ingestion, intraperitoneal route
Uses: Solubilizer, stabilizer in flavor concs.; processing aid, wetting agent; surfactant, defoaming agent in scald baths for poultry defeathering, hog dehairing; dough conditioner
Regulatory: FDA 21CFR §172.808, 173.340, 175.105, 176.180, 176.200, 176.210, 177.1200
Trade names: Dow EP530

Poloxamer 182

CAS 9003-11-6 (generic)
Classification: Polyoxyethylene, polyoxypropylene block polymer
Formula: $HO(CH_2CH_2O)_x(CCH_3HCH_2O)_y(CH_2CH_2O)_zH$, avg. x=8, y=30, z=8
Toxicology: Moderately toxic by ingestion, intraperitoneal route
Uses: Solubilizer, stabilizer in flavor concs.; processing aid, wetting agent; surfactant, defoaming agent in scald baths for poultry defeathering, hog dehairing; dough conditioner
Regulatory: FDA 21CFR §172.808, 173.340, 175.105, 176.180, 176.200, 176.210, 177.1200

Poloxamer 183

CAS 9003-11-6 (generic)
Classification: Polyoxyethylene, polyoxypropylene block polymer
Formula: $HO(CH_2CH_2O)_x(CCH_3HCH_2O)_y(CH_2CH_2O)_zH$, avg. x=10, y=30, z=10
Toxicology: Moderately toxic by ingestion, intraperitoneal route
Uses: Solubilizer, stabilizer in flavor concs.; processing aid, wetting agent; surfactant, defoaming agent in scald baths for poultry defeathering, hog dehairing; dough conditioner
Regulatory: FDA 21CFR §172.808, 173.340, 175.105, 176.180, 176.200, 176.210, 177.1200

Poloxamer 184

CAS 9003-11-6 (generic)
Classification: Polyoxyethylene, polyoxypropylene block polymer
Formula: $HO(CH_2CH_2O)_x(CCH_3HCH_2O)_y(CH_2CH_2O)_zH$, avg. x=13, y=30, z=13
Toxicology: Moderately toxic by ingestion, intraperitoneal route
Uses: Solubilizer, stabilizer in flavor concs.; processing aid, wetting agent; surfactant, defoaming agent in scald baths for poultry defeathering, hog dehairing; dough conditioner
Regulatory: FDA 21CFR §172.808, 173.340, 175.105, 176.180, 176.200, 176.210, 177.1200, 177.1210

Poloxamer 185

CAS 9003-11-6 (generic)
Classification: Polyoxyethylene, polyoxypropylene block polymer
Formula: $HO(CH_2CH_2O)_x(CCH_3HCH_2O)_y(CH_2CH_2O)_zH$, avg. x=19, y=30, z=19
Toxicology: Moderately toxic by ingestion, intraperitoneal route
Uses: Solubilizer, stabilizer in flavor concs.; processing aid, wetting agent; surfactant, defoaming agent in scald baths for poultry defeathering, hog dehairing; dough conditioner
Regulatory: FDA 21CFR §172.808, 173.340, 175.105, 176.180, 176.200, 176.210, 177.1200, 177.1210

Poloxamer 188

CAS 9003-11-6 (generic)
Classification: Polyoxyethylene, polyoxypropylene block polymer
Formula: $HO(CH_2CH_2O)_x(CCH_3HCH_2O)_y(CH_2CH_2O)_zH$, avg. x=75, y=30, z=75
Properties: Flakeable solid; m.p. 50 C min.; cloud pt. > 100 C (10% aq.)
Toxicology: Moderately toxic by ingestion, intraperitoneal route
Uses: Solubilizer, stabilizer in flavor concs.; processing aid, wetting agent; surfactant, defoaming agent in scald baths for poultry defeathering, hog dehairing; dough conditioner
Regulatory: FDA 21CFR §172.808, 173.340, 175.105, 176.180, 176.200, 176.210, 177.1200, 177.1210

Poloxamer 212

CAS 9003-11-6 (generic)
Classification: Polyoxyethylene, polyoxypropylene block polymer

Formula: $HO(CH_2CH_2O)_x(CCH_3HCH_2O)_y(CH_2CH_2O)_zH$, avg. x=8, y=35, z=8
Toxicology: Moderately toxic by ingestion, intraperitoneal route
Uses: Solubilizer, stabilizer in flavor concs.; processing aid, wetting agent; surfactant, defoaming agent in scald baths for poultry defeathering, hog dehairing; dough conditioner
Regulatory: FDA 21CFR §172.808, 173.340, 175.105, 176.180, 176.200, 176.210, 177.1200, 177.1210

Poloxamer 215
CAS 9003-11-6 (generic)
Classification: Polyoxyethylene, polyoxypropylene block polymer
Formula: $HO(CH_2CH_2O)_x(CCH_3HCH_2O)_y(CH_2CH_2O)_zH$, avg. x=24, y=35, z=24
Toxicology: Moderately toxic by ingestion, intraperitoneal route
Uses: Solubilizer, stabilizer in flavor concs.; processing aid, wetting agent; surfactant, defoaming agent in scald baths for poultry defeathering, hog dehairing; dough conditioner
Regulatory: FDA 21CFR §172.808, 173.340, 175.105, 176.180, 176.200, 176.210, 177.1200, 177.1210

Poloxamer 217
CAS 9003-11-6 (generic)
Classification: Polyoxyethylene, polyoxypropylene block polymer
Formula: $HO(CH_2CH_2O)_x(CCH_3HCH_2O)_y(CH_2CH_2O)_zH$, avg. x=52, y=35, z=52
Toxicology: Moderately toxic by ingestion, intraperitoneal route
Uses: Solubilizer, stabilizer in flavor concs.; processing aid, wetting agent; surfactant, defoaming agent in scald baths for poultry defeathering, hog dehairing; dough conditioner
Regulatory: FDA 21CFR §172.808, 173.340, 175.105, 176.180, 176.200, 176.210, 177.1200, 177.1210

Poloxamer 231
CAS 9003-11-6 (generic)
Classification: Polyoxyethylene, polyoxypropylene block polymer
Formula: $HO(CH_2CH_2O)_x(CCH_3HCH_2O)_y(CH_2CH_2O)_zH$, avg. x=6, y=39, z=6
Toxicology: Moderately toxic by ingestion, intraperitoneal route
Uses: Solubilizer, stabilizer in flavor concs.; processing aid, wetting agent; surfactant, defoaming agent in scald baths for poultry defeathering, hog dehairing; dough conditioner
Regulatory: FDA 21CFR §172.808, 173.340, 175.105, 176.180, 176.200, 176.210, 177.1200, 177.1210

Poloxamer 234
CAS 0003-11-6 (generic)
Classification: Polyoxyethylene, polyoxypropylene block polymer
Formula: $HO(CH_2CH_2O)_x(CCH_3HCH_2O)_y(CH_2CH_2O)_zH$, avg. x=22, y=39, z=22
Toxicology: Moderately toxic by ingestion, intraperitoneal route
Uses: Solubilizer, stabilizer in flavor concs.; processing aid, wetting agent; surfactant, defoaming agent in scald baths for poultry defeathering, hog dehairing; dough conditioner
Regulatory: FDA 21CFR §172.808, 173.340, 175.105, 176.180, 176.200, 176.210, 177.1200, 177.1210

Poloxamer 235
CAS 9003-11-6 (generic)
Classification: Polyoxyethylene, polyoxypropylene block polymer
Formula: $HO(CH_2CH_2O)_x(CCH_3HCH_2O)_y(CH_2CH_2O)_zH$, avg. x=27, y=39, z=27
Toxicology: Moderately toxic by ingestion, intraperitoneal route
Uses: Solubilizer, stabilizer in flavor concs.; processing aid, wetting agent; surfactant, defoaming agent in scald baths for poultry defeathering, hog dehairing; dough conditioner
Regulatory: FDA 21CFR §172.808, 173.340, 175.105, 176.180, 176.200, 1876.210, 177.1200, 177.1210

Poloxamer 237
CAS 9003-11-6 (generic)
Classification: Polyoxyethylene, polyoxypropylene block polymer
Formula: $HO(CH_2CH_2O)_x(CCH_3HCH_2O)_y(CH_2CH_2O)_zH$, avg. x=62, y=39, z=62
Toxicology: Moderately toxic by ingestion, intraperitoneal route
Uses: Solubilizer, stabilizer in flavor concs.; processing aid, wetting agent; surfactant, defoaming agent in scald baths for poultry defeathering, hog dehairing; dough conditioner
Regulatory: FDA 21CFR §172.808, 173.340, 175.105, 176.180, 176.200, 176.210, 177.1200, 177.1210

Poloxamer 238
CAS 9003-11-6 (generic)
Classification: Polyoxyethylene, polyoxypropylene block polymer
Formula: $HO(CH_2CH_2O)_x(CCH_3HCH_2O)_y(CH_2CH_2O)_zH$, avg. x=97, y=39, z=97
Toxicology: Moderately toxic by ingestion, intraperitoneal route
Uses: Solubilizer, stabilizer in flavor concs.; processing aid, wetting agent; surfactant, defoaming agent in scald

baths for poultry defeathering, hog dehairing; dough conditioner
Regulatory: FDA 21CFR §172.808, 173.340, 175.105, 176.180, 176.200, 176.210, 177.1200, 177.1210

Poloxamer 282
CAS 9003-11-6 (generic)
Classification: Polyoxyethylene, polyoxypropylene block polymer
Formula: $HO(CH_2CH_2O)_x(CCH_3HCH_2O)_y(CH_2CH_2O)_zH$, avg. x=10, y=47, z=10
Toxicology: Moderately toxic by ingestion, intraperitoneal route
Uses: Solubilizer, stabilizer in flavor concs.; processing aid, wetting agent; surfactant, defoaming agent in scald
baths for poultry defeathering, hog dehairing; dough conditioner
Regulatory: FDA 21CFR §172.808, 173.340, 175.105, 176.180, 176.200, 176.210, 177.1200, 177.1210

Poloxamer 284
CAS 9003-11-6 (generic)
Classification: Polyoxyethylene, polyoxypropylene block polymer
Formula: $HO(CH_2CH_2O)_x(CCH_3HCH_2O)_y(CH_2CH_2O)_zH$, avg. x=21, y=47, z=21
Toxicology: Moderately toxic by ingestion, intraperitoneal route
Uses: Solubilizer, stabilizer in flavor concs.; processing aid, wetting agent; surfactant, defoaming agent in scald
baths for poultry defeathering, hog dehairing; dough conditioner
Regulatory: FDA 21CFR §172.808, 173.340, 175.105, 176.180, 176.200, 176.210, 177.1200, 177.1210

Poloxamer 288
CAS 9003-11-6 (generic)
Classification: Polyoxyethylene, polyoxypropylene block polymer
Formula: $HO(CH_2CH_2O)_x(CCH_3HCH_2O)_y(CH_2CH_2O)_zH$, avg. x=122, y=47, z=122
Toxicology: Moderately toxic by ingestion, intraperitoneal route
Uses: Solubilizer, stabilizer in flavor concs.; processing aid, wetting agent; surfactant, defoaming agent in scald
baths for poultry defeathering, hog dehairing; dough conditioner
Regulatory: FDA 21CFR §172.808, 173.340, 175.105, 176.180, 176.200, 176.210, 177.1200, 177.1210

Poloxamer 331
CAS 9003-11-6 (generic)
Classification: Polyoxyethylene, polyoxypropylene block polymer
Formula: $HO(CH_2CH_2O)_x(CCH_3HCH_2O)_y(CH_2CH_2O)_zH$, avg. x=7, y=54, z=7
Properties: Colorless liq.; sol. in alcohol; very sl. sol. in water; m.w. 3800; dens. 1.018 (25/25 C); visc. 756 cp;
cloud pt. 11 C (10% aq.)
Toxicology: Moderately toxic by ingestion, intraperitoneal route; heated to decomp., emits acrid smoke and
irritating fumes
Uses: Solubilizer and stabilizer in flavor concentrates; dough conditioner, foam control agent, surfactant,
poultry scald agent
Usage level: Limitation 0.05% (scald baths for poultry), 5 g/hog (in dehairing machines), 0.5% on wt. of flour
(dough)
Regulatory: FDA 21CFR §172.808, 173.340, 175.105, 176.180, 176.200, 176.210, 177.1200, 177.1210;
9CFR §381.147

Poloxamer 333
CAS 9003-11-6 (generic)
Classification: Polyoxyethylene, polyoxypropylene block polymer
Formula: $HO(CH_2CH_2O)_x(CCH_3HCH_2O)_y(CH_2CH_2O)_zH$, avg. x=20, y=54, z=20
Toxicology: Moderately toxic by ingestion, intraperitoneal route
Uses: Solubilizer, stabilizer in flavor concs.; processing aid, wetting agent; surfactant, defoaming agent in scald
baths for poultry defeathering, hog dehairing; dough conditioner
Regulatory: FDA 21CFR §172.808, 173.340, 175.105, 176.180, 176.200, 176.210, 177.1200, 177.1210

Poloxamer 334
CAS 9003-11-6 (generic)
Classification: Polyoxyethylene, polyoxypropylene block polymer
Formula: $HO(CH_2CH_2O)_x(CCH_3HCH_2O)_y(CH_2CH_2O)_zH$, avg. x=31, y=54, z=31
Toxicology: Moderately toxic by ingestion, intraperitoneal route
Uses: Solubilizer, stabilizer in flavor concs.; processing aid, wetting agent; surfactant, defoaming agent in scald
baths for poultry defeathering, hog dehairing; dough conditioner
Regulatory: FDA 21CFR §172.808, 173.340, 175.105, 176.180, 176.200, 176.210, 177.1200, 177.1210

Poloxamer 335
CAS 9003-11-6 (generic)
Classification: Polyoxyethylene, polyoxypropylene block polymer

Formula: $HO(CH_2CH_2O)_x(CCH_3HCH_2O)_y(CH_2CH_2O)_zH$, avg. x=38, y=54, z=38
Toxicology: Moderately toxic by ingestion, intraperitoneal route
Uses: Solubilizer, stabilizer in flavor concs.; processing aid, wetting agent; surfactant, defoaming agent in scald baths for poultry defeathering, hog dehairing; dough conditioner
Regulatory: FDA 21CFR §172.808, 173.340, 175.105, 176.180, 176.200, 176.210, 177.1200, 177.1210

Poloxamer 338
CAS 9003-11-6 (generic)
Classification: Polyoxyethylene, polyoxypropylene block polymer
Formula: $HO(CH_2CH_2O)_x(CCH_3HCH_2O)_y(CH_2CH_2O)_zH$, avg. x=128, y=54, z=128
Toxicology: Moderately toxic by ingestion, intraperitoneal route
Uses: Solubilizer, stabilizer in flavor concs.; processing aid, wetting agent; surfactant, defoaming agent in scald baths for poultry defeathering, hog dehairing; dough conditioner
Regulatory: FDA 21CFR §172.808, 173.340, 175.105, 176.180, 176.200, 176.210, 177.1200, 177.1210

Poloxamer 401
CAS 9003-11-6 (generic)
Classification: Polyoxyethylene, polyoxypropylene block polymer
Formula: $HO(CH_2CH_2O)_x(CCH_3HCH_2O)_y(CH_2CH_2O)_zH$, avg. x=6, y=67, z=6
Toxicology: Moderately toxic by ingestion, intraperitoneal route
Uses: Solubilizer, stabilizer in flavor concs.; processing aid, wetting agent; surfactant, defoaming agent in scald baths for poultry defeathering, hog dehairing; dough conditioner
Regulatory: FDA 21CFR §172.808, 173.340, 175.105, 176.180, 176.200, 176.210, 177.1200, 177.1210

Poloxamer 402
CAS 9003-11-6 (generic)
Classification: Polyoxyethylene, polyoxypropylene block polymer
Formula: $HO(CH_2CH_2O)_x(CCH_3HCH_2O)_y(CH_2CH_2O)_zH$, avg. x=13, y=67, z=13
Toxicology: Moderately toxic by ingestion, intraperitoneal route
Uses: Solubilizer, stabilizer in flavor concs.; processing aid, wetting agent; surfactant, defoaming agent in scald baths for poultry defeathering, hog dehairing; dough conditioner
Regulatory: FDA 21CFR §172.808, 173.340, 175.105, 176.180, 176.200, 176.210, 177.1200, 177.1210

Poloxamer 403
CAS 9003-11-6 (generic)
Classification: Polyoxyethylene, polyoxypropylene block polymer
Formula: $HO(CH_2CH_2O)_x(CCH_3HCH_2O)_y(CH_2CH_2O)_zH$, avg. x=21, y=67, z=21
Toxicology: Moderately toxic by ingestion, intraperitoneal route
Uses: Solubilizer, stabilizer in flavor concs.; processing aid, wetting agent; surfactant, defoaming agent in scald baths for poultry defeathering, hog dehairing; dough conditioner
Regulatory: FDA 21CFR §172.808, 173.340, 175.105, 176.180, 176.200, 176.210, 177.1200, 177.1210

Poloxamer 407
CAS 9003-11-6 (generic)
Classification: Polyoxyethylene, polyoxypropylene block polymer
Formula: $HO(CH_2CH_2O)_x(CCH_3HCH_2O)_y(CH_2CH_2O)_zH$, avg. x=98, y=67, z=98
Toxicology: Moderately toxic by ingestion, intraperitoneal route
Uses: Solubilizer, stabilizer in flavor concs.; processing aid, wetting agent; surfactant, defoaming agent in scald baths for poultry defeathering, hog dehairing; dough conditioner
Regulatory: FDA 21CFR §172.808, 173.340, 175.105, 176.180, 176.200, 176.210, 177.1200, 177.1210

Polyacetal. *See* Acetal

Polyacrylamide
CAS 9003-05-8
Synonyms: 2-Propenamide, homopolymer
Definition: Polyamide of acrylic monomers
Empirical: $(C_3H_5NO)_x$
Formula: $[CH_2CHCONH_2]_x$
Properties: Wh. solid; water-sol. high polymer; m.w. 10,000-18,000,000
Toxicology: LD50 (mouse, IP) 170 mg/kg (monomer); heated to decomp., emits acrid smoke and irritating fumes
Uses: Film-former in imprinting of soft-shell gelatin capsules; flocculant in clarification of beet or cane sugar juice (5 ppm max.); washing/lye peeling of fruits and vegetables
Regulatory: FDA 21CFR §172.255, 173.10, 173.315 (10 ppm in wash water), 175.105, 176.180
Trade names: Accofloc® A100 PWG; Accofloc® A110 PWG; Accofloc® A120 PWG; Accofloc® A130 PWG;

Polyacrylamide resins, modified

Accofloc® N100 PWG

Polyacrylamide resins, modified
Definition: Produced by copolymerization of acrylamide and not more than 5-mole % of β-methacrylyloxyethyl trimethylammonium methyl sulfate
Toxicology: Heated to decomp., emits acrid smoke and irritating fumes
Uses: Flocculant used in sugar liquor or juice
Regulatory: FDA 21CFR §173.10 (limitation 5 ppm)

Polyacrylic acid, sodium salt. *See* Sodium polyacrylate
Polyarabinogalactan. *See* Arabinogalactan

Polydextrose
CAS 68434-04-4
Definition: Random polymer formed from condensation of D-glucose
Properties: Off-wh. to lt. tan solid or clear straw-colored liq. (sol'n.); sol. in water
Toxicology: Heated to decomp., emits acrid smoke and irritating fumes
Uses: Bulking agent, formulation aid, humectant, and texturizer used in baked goods, baking mixes, chewing gum, confections, frostings, salad dressings, frozen dairy desserts, gelatins, puddings, fillings, hard and soft candy
Regulatory: FDA 21CFR §172.841
Manuf./Distrib.: Agric. & Chem.; Amylum; A Arnaud; Fibrisol Service; Food Ingred. Services; Pfizer; Siber Hegner; A E Staley Mfg.; Treasure Island Food; Tunnel Refineries
Trade names: Litesse®; Litesse® II; Sta-Lite™ 100C; Sta-Lite™ 100CN Neutralized; Sta-Lite™ 100F; Sta-Lite™ 100CN Neutralized

Polydimethylsiloxane. *See* Dimethicone

Polydipentene
CAS 9003-73-0
Definition: Prod. formed by polymerization of terpene hydrocarbons
Uses: Masticatory agents in chewing gums
Trade names: Piccolyte® C115

Polyether glycol. *See* Polyethylene glycol

Polyethylene
CAS 9002-88-4; EINECS 200-815-3
Synonyms: Ethene, homopolymer; Ethylene homopolymer
Definition: Polymer of ethylene monomers
Empirical: $(C_2H_4)_x$
Formula: $[CH_2CH_2]_x$
Properties: Wh. translucent partially cryst./partially amorphous plastic solid, odorless; sol. in hot benzene; insol. in water; m.w. 1500-100,000; dens. 0.92 (20/4 C); m.p. 85-110 C
Precaution: Combustible; store in well closed containers; reacts violently with F_2
Toxicology: Suspected carcinogen and tumorigen by implants; heated to decomp., emits acrid smoke and irritating fumes
Uses: Masticatory substance in chewing gum base; protective coating on fruits and vegetables
Regulatory: FDA 21CFR §172.260, 172.615 (m.w. 2000-21,000), 173.20, 175.105, 175.300, 176.180, 176.200, 176.210, 177.1200, 177.1520, 177.2600, 178.3570, 178.3850
Manuf./Distrib.: Asahi Chem Industry Co Ltd; Eastman
Trade names: Polywax® 500; Polywax® 600; Polywax® 655

Polyethylene glycol. *See also* PEG
CAS 25322-68-3; EINECS 203-473-3
Synonyms: PEG; α-Hydroxy-omega-hydroxy poly(oxy-1,2-ethanediyl); Polyglycol; Polyether glycol
Definition: Condensation polymers of ethylene glycol
Formula: $H(OC_2H_4)_nOH$
Properties: Clear liq. or wh. solid; sol. in org. solvs., aromatic hydrocarbons; dens. 1.110-1.140 (20 C); m.p. 4-10 C; flash pt. 471 C
Precaution: Combustible liq.
Toxicology: LD50 (oral, rat) 33,750 mg/kg; sl. toxic by ingestion; skin and eye irritant; heated to decomp., emits acrid smoke and irritating fumes
Uses: Binding agent, boiler water additive, coating agent, dispersant, flavoring adjuvant, lubricant, plasticizing agent; used in carbonated beverages, citrus fruit, nonnutritive sweeteners, tablets, vitamin or mineral preps.
Usage level: Limitation zero tolerance (milk)

Regulatory: FDA 21CFR §172.210, 172.820, 173.340
Manuf./Distrib.: Calgene

Polyethylene, oxidized
CAS 68441-17-8
Synonyms: Oxidized polyethylene; Ethene, homopolymer, oxidized
Definition: Reaction prod. of polyethylene and oxygen
Properties: dens. 0.930
Uses: Protective coating for avocados, bananas, beets, coconuts, eggplant, garlic, grapefruit, lemons, limes, mango, onions, oranges, papaya, peas (in pods), pineapple, pumpksin, rutabaga, watermelon, nuts in shells, etc.
Regulatory: FDA 21CFR §172.260, 175.105, 175.125, 176.170, 176.200, 176.210, 177.1200, 177.1620, 177.2800
Trade names: Hoechst Wax PED 121; Luwax OA 5

Polyethylenimine reaction prod. with 1,2-dichloroethane
CAS 68130-97-2
Definition: Reaction prod. of homopolymerization of ethyleneimine in aq. HCl @ 100 C and of crosslinking with 1,2-dichloroethane
Properties: m.w. 50,000-70,000
Uses: Fixing agent in immobilization of glucose isomerase enzyme preps. for mfg. of high fructose corn syrup
Regulatory: FDA 21CFR §173.357

Polyglycerate 60
Trade names containing: M-C-Thin® FWD 425

Polyglycerol esters of fatty acids
EEC E475
Synonyms: Polyglyceryl esters of fatty acids
Properties: Yel. to amber oily visc. liqs., lt. tan to brn. soft solids, tan to brn. waxy solids; sol. in org. solvs., oils; disp. in water; HLB 5-13
Toxicology: Heated to decomp., emits acrid smoke and irritating fumes
Uses: Cloud inhibitor, stabilizer, emulsifier in cake mixes, confections, rendered animal fats, margarine, salad and vegetable oils, whipped toppings
Usage level: ADI 0-25 mg/kg (EEC)
Regulatory: FDA 21CFR §172.854, 172.860b, 172.862; USDA 9CFR §318.7 (0.5% in margarine); Europe listed; UK approved

Polyglyceryl-10 caprylate
CAS 68937-16-6
Trade names: Drewpol® 10-1-CC; Drewpol® 10-1-CCK

Polyglyceryl-10 decaoleate
CAS 11094-60-3; EINECS 234-316-7
Synonyms: Decaglycerol decaoleate; Decaglyceryl decaoleate
Definition: Decaester of oleic acid and a glycerin polymer containing an avg. 10 glycerin units
Empirical: $C_{210}H_{382}O_{31}$
Uses: Food emulsifier
Regulatory: FDA 21CFR §172.854
Trade names: Caprol® 10G10O; Drewpol® 10-10-O; Drewpol® 10-10-OK; Hodag PGO-1010; Polyaldo® DGDO; Polyaldo® DGDO KFG; Santone® 10-10-O

Polyglyceryl-10 decastearate
CAS 39529-26-5; EINECS 254-495-5
Synonyms: Decaglycerol decastearate; Decaglyceryl decastearate; Octadecanoic acid, decaester with decaglycerol
Definition: Decaester of stearic acid and a glycerin polymer containing an avg. 10 glycerin units
Empirical: $C_{210}H_{402}O_{31}$
Uses: Food emulsifier
Regulatory: FDA 21CFR §172.854
Trade names: Caprol® JB; Drewmulse® 10-10-S; Hodag PGS-1010

Polyglyceryl-2 diisostearate
CAS 67938-21-0; EINECS 267-821-6
Synonyms: Diglyceryl diisostearate; Isooctadecanoic acid, diester with diglycerol
Definition: Diester of isostearic acid and a dimer of glycerin
Empirical: $C_{42}H_{82}O_7$

Polyglyceryl-3 diisostearate

    Uses: Food emulsifier
    Trade names: Emalex DISG-2

Polyglyceryl-3 diisostearate
    CAS 66082-42-6; 85404-84-8
    Synonyms: Triglyceryl diisostearate
    Definition: Diester of isostearic acid and a glycerin polymer with avg. 3 glycerin units
    Uses: Food emulsifier
    Trade names: Emalex DISG-3

Polyglyceryl-5 diisostearate
    Uses: Food emulsifier
    Trade names: Emalex DISG-5

Polyglyceryl-3 dioleate
    CAS 79665-94-4
    Synonyms: Triglyceryl dioleate; 9-Octadecenoic acid, diester with triglycerol
    Definition: Diester of oleic acid and a glycerin polymer containing an avg. of 3 glycerin units
    Empirical: $C_{45}H_{84}O_9$
    Uses: Food emulsifier
    Regulatory: FDA 21CFR §172.854

Polyglyceryl-6 dioleate
    CAS 76009-37-5
    Synonyms: Hexaglycerol dioleate; Hexaglyceryl dioleate
    Definition: Diester of oleic acid and a glycerin polymer containing an avg. of 6 glycerin units
    Empirical: $C_{54}H_{102}O_{15}$
    Uses: Food emulsifier
    Regulatory: FDA 21CFR §172.854
    Trade names: Caprol® 6G2O; Drewpol® 6-2-OK; Hodag PGO-62

Polyglyceryl-10 dioleate
    CAS 9009-48-1, 33940-99-7
    Synonyms: Decaglycerol dioleate; Decaglycerin dioleate; Decaglyceryl dioleate
    Definition: Diester of oleic acid and a glycerin polymer containing an avg. 10 glycerin units
    Empirical: $C_{66}H_{126}O_{23}$
    Uses: Food emulsifier
    Trade names: Caprol® 10G2O; Drewpol® 10-2-OK; Hodag PGO-102; Nikkol Decaglyn 2-O; Polyaldo® 2O10
      KFG

Polyglyceryl-10 dipalmitate
    Synonyms: Decaglyceryl dipalmitate
    Uses: Food emulsifier and surfactant
    Trade names: Polyaldo® 2P10 KFG

Polyglyceryl-6 dishortening
    Trade names: Hodag PGSH-62

Polyglyceryl-2 distearate
    Uses: Food emulsifier
    Trade names: Emalex DSG-2

Polyglyceryl-3 distearate
    CAS 94423-19-5
    Synonyms: Triglyceryl distearate
    Definition: Diester of stearic acid and a glycerin polymer with avg. 3 glycerin units
    Empirical: $C_{45}H_{88}O_9$
    Uses: Food emulsifier
    Trade names: Emalex DSG-3

Polyglyceryl-5 distearate
    Uses: Food emulsifier
    Trade names: Emalex DSG-5

Polyglyceryl-6 distearate
    CAS 34424-97-0, 61725-93-7
    Synonyms: Hexaglycerol distearate; Hexaglyceryl distearate
    Definition: Diester of stearic acid and a glycerin polymer containing an avg. 6 glycerin units
    Empirical: $C_{54}H_{106}O_{15}$

Uses: Food emulsifier
Regulatory: FDA 21CFR §172.854
Trade names: Caprol® 6G2S; Drewmulse® 6-2-S; Drewpol® 6-2-SK; Hodag PGS-62; Hodag SVO-629; Plurol Stearique WL 1009; Polyaldo® 2S6 KFG; Polyaldo® HGDS; Polyaldo® HGDS KFG

Polyglyceryl-10 distearate
CAS 12764-60-2
Synonyms: Decaglycerin distearate; Decaglyceryl distearate
Definition: Diester of stearic acid and a glycerin polymer containing an avg. 10 glycerin units
Empirical: $C_{66}H_{130}O_{23}$
Uses: Food emulsifier
Regulatory: FDA 21CFR §172.854
Trade names: Hodag PGS-102; Nikkol Decaglyn 2-S

Polyglyceryl esters of fatty acids. *See* Polyglycerol esters of fatty acids

Polyglyceryl-6 hexaoleate
CAS 95482-05-6
Synonyms: Hexaglyceryl hexaoleate
Definition: Hexaester of oleic acid and a glycerin polymer containing an avg. 6 glycerin units
Empirical: $C_{126}H_{230}O_{19}$
Uses: Food emulsifier
Regulatory: FDA 21CFR §172.854

Polyglyceryl-10 hexaoleate
Synonyms: Decaglyceryl hexaoleate
Definition: Hexaester of oleic acid and a glycerin polymer containing an avg. 10 glycerin units
Uses: Emulsifier for margarine, solubilizer for flavor and essential oils, dispersant for high-solids preps., suspending agent for food colors
Trade names: Drewpol® 10-6-OK; Emulsifier D-1

Polyglyceryl-10 isostearate
CAS 133738-23-5
Synonyms: Decaglycerin monoisostearate; Decaglyceryl monoisostearate
Definition: Ester of isostearic acid and a glycerin polymer containing an avg. 10 glycerin units
Empirical: $C_{48}H_{96}O_{22}$
Uses: Food emulsifier
Trade names: Nikkol Decaglyn 1-IS

Polyglyceryl-3 laurate
Synonyms: Triglyceryl monolaurate
Trade names: Hodag PGL

Polyglyceryl-10 laurate
CAS 34406-66-1
Synonyms: Decaglycerin monolaurate; Decaglyceryl monolaurate
Definition: Ester of lauric acid and a glycerin polymer containing an avg. 10 glycerin units
Empirical: $C_{42}H_{84}O_{22}$
Uses: Food emulsifier
Regulatory: FDA 21CFR §172.854
Trade names: Hodag PGL-101; Nikkol Decaglyn 1-L

Polyglyceryl-10 linoleate
Synonyms: Decaglycerin monolinoleate
Definition: Ester of linoleic acid and a glycerin polymer containing an avg. 10 glycerin units
Uses: Food emulsifier
Trade names: Nikkol Decaglyn 1-LN

Polyglyceryl-10 myristate
CAS 87390-32-7
Synonyms: Decaglycerin monomyristate
Definition: Ester of myristic acid and a glycerin polymer containing an avg. 10 glycerin units
Empirical: $C_{44}H_{88}O_{22}$
Uses: Food emulsifier
Regulatory: FDA 21CFR §172.854
Trade names: Nikkol Decaglyn 1-M

Polyglyceryl-10 octaoleate
Synonyms: Decaglycerol octaoleate; Decaglyceryl octaoleate
Definition: Octaester of oleic acid and a glycerin polymer containing an avg. 10 glycerin units
Uses: Emulsifier, solubilizer, dispersant for food applics.
Regulatory: FDA 21CFR §172.854
Trade names: Drewmulse® 10-8-O; Drewpol® 10-8-OK; Hodag PGO-108

Polyglyceryl-10 octastearate
Synonyms: Decaglyceryl octastearate
Uses: Food applics.
Regulatory: FDA 21CFR §172.854
Trade names: Hodag PGS-108

Polyglyceryl-2 oleate
CAS 9007-48-1 (generic); 49553-76-6
Synonyms: Diglyceryl monooleate
Definition: Ester of oleic acid and a dimer of glycerin
Empirical: $C_{24}H_{46}O_6$
Uses: Food emulsifier
Regulatory: FDA 21CFR §172.854

Polyglyceryl-3 oleate
CAS 9007-48-1 (generic); 33940-98-6
Synonyms: Triglyceryl oleate
Definition: Ester of oleic acid and a glycerin polymer containing an avg. 3 glycerin units
Empirical: $C_{27}H_{52}O_8$
Uses: Food emulsifier
Regulatory: FDA 21CFR §172.854
Trade names: Caprol® 3GO; Drewmulse® 3-1-O; Drewpol® 3-1-O; Drewpol® 3-1-OK; Hodag PGO; Mazol® PGO-31 K; Santone® 3-1-SH; Triodan 20

Polyglyceryl-4 oleate
CAS 9007-48-1 (generic); 71012-10-7
Synonyms: Tetraglyceryl monooleate
Definition: Ester of oleic acid and a glycerin polymer containing an avg. 4 glycerin units
Empirical: $C_{30}H_{58}O_{10}$
Uses: Food emulsifier
Regulatory: FDA 21CFR §172.854
Trade names: Nikkol Tetraglyn 1-O

Polyglyceryl-6 oleate
CAS 9007-48-1 (generic); 79665-92-2
Synonyms: Hexaglycerin monooleate; Hexaglyceryl oleate
Definition: Ester of oleic acid and a glycerin polymer containing an avg. 6 glycerin units
Empirical: $C_{36}H_{70}O_{14}$
Uses: Food emulsifier
Regulatory: FDA 21CFR §172.854
Trade names: Drewpol® 6-1-O; Drewpol® 6-1-OK; Hodag PGO-61

Polyglyceryl-8 oleate
CAS 9007-48-1 (generic); 75719-56-1
Synonyms: 9-Octadecenoic acid, monoester with octaglycerol
Definition: Ester of oleic acid and a glycerin polymer containing an avg. 8 glycerin units
Empirical: $C_{42}H_{82}O_{18}$
Uses: Food emulsifier; beverage clouding agent
Regulatory: FDA 21CFR §172.854
Trade names: Drewpol® 8-1-OK; Santone® 8-1-O

Polyglyceryl-10 oleate
CAS 9007-48-1 (generic); 67784-82-1; 79665-93-3
Synonyms: Decaglycerin monooleate; Decaglyceryl monooleate
Definition: Ester of oleic acid and a glycerin polymer containing an avg. 10 glycerin units
Empirical: $C_{48}H_{94}O_{22}$
Uses: Food emulsifier
Regulatory: FDA 21CFR §172.854
Trade names: Hodag PGO-101; Nikkol Decaglyn 1-O; Polyaldo® 10-1-O KFG

Polyglyceryl-4 pentaoleate
Synonyms: Tetraglyceryl pentaoleate
Definition: Pentaester of oleic acid and a glycerin polymer containing an avg. 4 glycerin units
Uses: O/w emulsifier, anticrystallizing agent for foods
Trade names: Nikkol Tetraglyn 5-O

Polyglyceryl-10 pentaoleate
CAS 86637-84-5
Synonyms: Decaglycerin pentaoleate
Definition: Pentaester of oleic acid and a glycerin polymer containing an avg. 10 glycerin units
Empirical: $C_{120}H_{222}O_{26}$
Uses: Food emulsifier
Regulatory: FDA 21CFR §172.854

Polyglyceryl-4 pentastearate
Synonyms: Tetraglyceryl pentastearate
Definition: Pentaester of stearic acid and a glycerin polymer containing an avg. 4 glycerin units
Uses: O/w emulsifier, anticrystallizing agent for foods
Trade names: Nikkol Tetraglyn 5-S

Polyglyceryl polyricinoleate
EEC E476
Uses: Emulsifier, stabilizer; with lecithin, improves fluidity of chocolate for coating
Usage level: ADI 0-75 mg/kg body wt. (EEC)
Regulatory: Europe listed; UK approved
Trade names: Admul WOL 1403; Crester PR; Crester RT; Triodan R 90
Trade names containing: Crester RA; Crester RB

Polyglyceryl-2 sesquioleate
Synonyms: Diglyceryl sesquioleate
Definition: Mixture of mono and diesters of oleic acid and a dimer of glycerin
Uses: Food emulsifier
Regulatory: FDA 21CFR §172.854

Polyglyceryl-2 stearate
CAS 12694-22-3; EINECS 235-777-7
Synonyms: Diglyceryl monostearate
Definition: Ester of stearic acid and a dimer of glycerin
Empirical: $C_{24}H_{48}O_6$
Uses: W/o emulsifier for food applics.
Trade names: Emalex MSG-2; Emalex MSG-2MA; Emalex MSG-2MB; Emalex MSG-2ME; Emalex MSG-2ML; Nikkol DGMS

Polyglyceryl-3 stearate
CAS 37349-34-1 (generic); 27321-72-8; 26855-43-6; 61790-95-2; EINECS 248-403-2
Synonyms: Triglyceryl stearate
Definition: Ester of stearic acid and a glycerin polymer containing an avg. 3 glycerin units
Uses: Food emulsifier, stabilizer, whipping agent
Regulatory: FDA 21CFR §172.854
Trade names: Caprol® 3GS; Drewmulse® 3-1-S; Drewpol® 3-1-SK; Hefti GMS-333; Hodag PGS; Polyaldo® TGMS; Polyaldo® TGMS KFG; Santone® 3-1-S; Santone® 3-1-S XTR; Triodan 55

Polyglyceryl-4 stearate
CAS 37349-34-1 (generic); 68004-11-5; 26855-44-7
Synonyms: Tetraglyceryl monostearate; Octadecanoic acid, monoester with tetraglycerol
Definition: Ester of stearic acid and a glycerin polymer containing an avg. 4 glycerin units
Empirical: $C_{30}H_{60}O_{10}$
Uses: Emulsifier
Regulatory: FDA 21CFR §172.854
Trade names: Lipal 4-1S; Nikkol Tetraglyn 1-S; Witconol™ 18F

Polyglyceryl-8 stearate
CAS 37349-34-1 (generic); 75719-57-2
Synonyms: Octadecanoic acid, monoester with octaglycerol; Octaglyceryl stearate
Definition: Ester of stearic acid and a glycerin polymer containing an avg. 8 glycerin units
Empirical: $C_{42}H_{84}O_{18}$
Uses: Food emulsifier

Polyglyceryl-10 stearate

Regulatory: FDA 21CFR §172.854

Polyglyceryl-10 stearate
CAS 79777-30-3
Synonyms: Decaglycerin monostearate; Decaglyceryl monostearate; Octadecanoic acid, monoester with decaglycerol
Definition: Ester of stearic acid and a glycerin polymer containing an avg. 10 glycerin units
Empirical: $C_{48}H_{96}O_{22}$
Uses: Food emulsifier, dispersant
Regulatory: FDA 21CFR §172.854
Trade names: Hodag PGS-101; Nikkol Decaglyn 1-S

Polyglyceryl-10 tetraoleate
CAS 34424-98-1; EINECS 252-011-7
Synonyms: Decaglycerol tetraoleate; Decaglyceryl tetraoleate; 9-Octadecenoic acid, tetraester with decaglycerol
Definition: Tetraester of oleic acid and a glycerin polymer containing an avg. 10 glycerin units
Empirical: $C_{102}H_{190}O_{25}$
Uses: Food emulsifier
Regulatory: FDA 21CFR §172.854
Trade names: Caprol® 10G4O; Drewmulse® 10-4-O; Drewpol® 10-4-O; Drewpol® 10-4-OK; Hodag PGO-104; Mazol® PGO-104

Polyglyceryl-2 tetrastearate
CAS 72347-89-8
Synonyms: Diglycerol tetrastearate; Diglyceryl tetrastearate; Octadecanoic acid, tetraester with diglycerol
Definition: Tetraester of stearic acid and a dimer of glycerin
Empirical: $C_{78}H_{150}O_9$
Uses: Food emulsifier
Regulatory: FDA 21CFR §172.854
Trade names: Caprol® 2G4S

Polyglyceryl-10 tetrastearate
Synonyms: Decaglyceryl tetrastearate
Trade names: Hodag PGS-104

Polyglyceryl-2 triisostearate
CAS 120486-24-0
Synonyms: Diglyceryl triisostearate; Isooctadecanoic acid, triester with diglycerol
Definition: Triester of isostearic acid and a dimer of glycerin
Empirical: $C_{60}H_{116}O_8$
Uses: Food emulsifier
Trade names: Emalex TISG-2

Polyglyceryl-10 trioleate
CAS 102051-00-3
Synonyms: Decaglycerin trioleate; Decaglyceryl trioleate
Definition: Triester of oleic acid and a glycerin polymer containing an avg. 10 glycerin units
Uses: Flavor carrier and clouding agent for beverages
Trade names: Hodag PGO-103

Polyglyceryl-4 tristearate
Synonyms: Tetraglyceryl tristearate
Definition: Triester of stearic acid and a glycerin polymer containing an avg. 4 glycerin units
Uses: O/w emulsifier, anticrystallizing agent for foods
Trade names: Nikkol Tetraglyn 3-S

Polyglyceryl-10 tristearate
CAS 12709-64-7
Synonyms: Decaglycerin tristearate; Decaglyceryl tristearate
Definition: Triester of stearic acid and a glycerin polymer containing an avg. 10 glycerin units
Uses: Food emulsifier
Trade names: Hodag PGS-103

Polyglycol. *See* Polyethylene glycol

Polyisobutene
CAS 9003-27-4; 9003-29-6

Synonyms: Polyisobutylene; 2-Methyl-1-propene, homopolymer
Definition: Homopolymer of isobutylene
Empirical: $(C_4H_8)_x$
Formula: $[CH_2C(CH_3)HCH_2]_x$
Uses: Masticatory substance for chewing gum base
Regulatory: FDA 21CFR §172.615 (min. m.w. 37,000), 175.105, 175.125, 175.300, 176.180, 177.1200, 177.1210, 177.1420, 178.3570, 178.3740, 178.3910; Japan approved

Polyisobutylene. *See* Polyisobutene
trans-Polyisoprene. *See* Gutta percha

Polylimonene
Uses: Synthetic flavoring
Regulatory: FDA 21CFR §172.515

Polymaleic acid
CAS 26099-09-2
Toxicology: Heated to decomp., emits acrid smoke and irritating fumes
Uses: Production aid; controls mineral scale in processing of beet or cane sugar juice and liquor; boiler water additive for food contact
Regulatory: FDA 21CFR §173.45 (0.4 ppm max. as acid), 173.310 (1 ppm max. as acid)

Polymaleic acid sodium salt
CAS 30915-61-8, 70247-90-4
Toxicology: Heated to decomp., emits acrid smoke and irritating fumes
Uses: Production aid; controls mineral scale in processing of beet or cane sugar juice and liquor; boiler water additive for food contact
Regulatory: FDA 21CFR §173.45 (0.4 ppm max., as acid), 173.310 (1 ppm max. as acid)

Polymannuronic acid. *See* Alginic acid
Polynoxylin. *See* Urea-formaldehyde resin
Polyoxyethylene monoglycerides. *See* Ethoxylated mono and diglycerides
Polyoxymethylene urea. *See* Urea-formaldehyde resin

Polypropylene glycol. *See also* PPG
CAS 25322-69-4; EINECS 200-338-0
Empirical: $(C_3H_6O_2)_n$
Formula: $HO(C_3H_6O)_nH$
Properties: Colorless clear liq.; sol. in water, aliphatic ketones, alcohol; insol. in ether, aliphatic hydrocarbons; m.w. 400-2000; dens. 1.001-1.007; flash p t. > 390 F
Precaution: Combustible exposed to heat or flame; reactive with oxidizers
Toxicology: LD50 (oral, rat) 4190 mg/kg; mildly toxic by ingstion; skin and eye irritant; heated to decomp., emits acrid smoke and irritating fumes
Uses: Defoaming agent in beet sugar and yeast processing, beverages, candy, shredded coconut, icings
Regulatory: FDA 21CFR §173.340
Manuf./Distrib.: Ashland; Calgene

Polysorbate 20
CAS 9005-64-5 (generic); EEC E432
Synonyms: POE (20) sorbitan monolaurate; PEG-20 sorbitan laurate; Sorbimacrogol laurate 300
Definition: Mixture of laurate esters of sorbitol and sorbitol anhydrides, with ≈ 20 moles ethylene oxide
Properties: Lemon to amber liq., char. odor, bitter taste; sol. in water, alcohol, ethyl acetate, methanol, dioxane; insol. in min. oil, min. spirits
Toxicology: LD50 (oral, rat) 37 g/kg; mod. toxic by intraperitoneal, intravenous routes; mildly toxic by ingestion; human skin irritant; heated to decomp., emits acrid smoke and irritating fumes
Uses: Emulsifier, stabilizer, flavoring agent, dispersant
Usage level: ADI 0-25 mg/kg (EEC)
Regulatory: FDA 21CFR §172.515, 175.105, 175.300, 178.3400; Europe listed
Trade names: Capmul® POE-L; Crillet 1; Drewmulse® POE-SML; Durfax® 20; Glycosperse® L-20; Hodag PSML-20; Lamesorb® SML-20; Liposorb L-20; Nissan Nonion LT-221; Norfox® Sorbo T-20; Sorbax PML-20; Sorgen TW20; T-Maz® 20; T-Maz® 20K; Tween® 20
Trade names containing: Amisol™ MS-12 BA

Polysorbate 21
CAS 9005-64-5 (generic)
Synonyms: POE (4) sorbitan monolaurate; PEG-4 sorbitan laurate
Definition: Mixture of laurate esters of sorbitol and sorbitol anhydrides, with ≈ 4 moles ethylene oxide

Polysorbate 40

Polysorbate 40

Toxicology: Moderately toxic by intraperitoneal, intravenous routes; mildly toxic by ingestion; skin irritant
Uses: Food emulsifier
Regulatory: FDA 21CFR §175.300
Trade names: Crillet 11
Trade names containing: Amisol™ MS-10

Polysorbate 40

CAS 9005-66-7; EEC E434
Synonyms: POE (20) sorbitan monopalmitate; Sorbimacrogol palmitate 300; Sorbitan, monohexadecanoate, poly(oxy-1,2-ethaneidyl) derivs.
Definition: Mixture of palmitate esters of sorbitol and sorbitol anhydrides, with ≈ 20 moles of ethylene oxide
Toxicology: Moderately toxic by intravenous route
Uses: Emulsifier, dough improver, crystallization retarder, solubilizer, stabilizer, flavor dispersant, wetting agent; veg./fruit coating; yeast/sugar defoamer
Usage level: ADI 0-25 mg/kg (EEC)
Regulatory: Europe listed; UK approved
Trade names: Crillet 2; Glycosperse® P-20; Hodag PSMP-20; Lamesorb® SMP-20; Lonzest® SMP-20; Sorbax PMP-20

Polysorbate 60

CAS 9005-67-8 (generic); EEC E435
Synonyms: POE (20) sorbitan monostearate; PEG-20 sorbitan stearate; Sorbimacrogol stearate 300
Definition: Mixture of stearate esters of sorbitol and sorbitol anhydrides, with ≈ 20 moles ethylene oxide
Empirical: $C_{64}H_{126}O_{26}$
Properties: Lemon to orange oily liq., faint odor, bitter taste; sol. in water, aniline, ethyl acetate, toluene; m.w. 1311.90; acid no. 2 max.; sapon. no. 45-55; hyd. no. 81-96
Toxicology: LD50 (IV, rat) 1220 mg/kg; moderately toxic by intravenous route; experimental tumorigen, reproductive effects; heated to decomp., emits acrid smoke and irritating fumes
Uses: Emulsifier (whipped toppings, cakes, confection coatings, dressings, shortenings/edible oils, coffee whiteners); opacifier, protective coating; foamer; dough conditioner; dispersant; wetting agent; stabilizer, defoamer, flavor; poultry scald agent
Usage level: Limitation 0.4% (whipped edible oil topping), 0.46% (cake), 0.5% (confectionery coating), 0.3% (dressing), 1% (shortening), 0.4% (milk), 4.5% (nonalcoholic beverage mix), 0.5% (baked goods, gelatin desserts); ADI 0-25 mg/kg (EEC)
Regulatory: FDA 21CFR §73.1001, 172.515, 172.836, 172.878, 172.886, 173.340, 175.105, 175.300, 178.3400; USDA CFR9 §318.7, 381.147 (limitation 1% max., 1% total combined with polysorbate 80, 0.0175% in scald water); Europe listed; UK approved
Manuf./Distrib.: Am. Ingreds.
Trade names: Atlas 70K; Atlas A; Capmul® POE-S; Crillet 3; Drewpone® 60K; Durfax® 60; Durfax® 60K; Emrite® 6125; Glycosperse® S-20; Glycosperse® S-20 FG; Glycosperse® S-20 KFG; Hodag PSMS-20; Hodag SVS-18; Lamesorb® SMS-20; Liposorb S-20; Lonzest® SMS-20; Nikkol TS-10; Norfox® Sorbo T-60; Polycon T60 K; Polysorbate 60; Sorbax PMS-20; T-Maz® 60; T-Maz® 60K; T-Maz® 60KHS; Tween® 60; Tween® 60K; Tween® 60 VS
Trade names containing: Alpha 500; Alpine; Amisol™ MS-12 BA; Atlas 2200H; Atlas 5520; Atmos® 378 K; Atmos® 729; Atmos® 758 K; Atmos® 1069; Atmos® 2462; Atmos® 7515 K; Atmul® 600H; Atmul® 700 H; Bake-Well Bun Release; Betricing; Brem; CBC #7 Shortening; Cremol Plus™ 60 Vegetable; Kaokote; Lactylate Hydrate; MG-60; Panalite MP Hydrate; Panalite MPH, MPS; Poly E; Satina 53NT; Snac-Kote; Soft Touch; Superb® Icing Shortening 101-270; Tandem 5K; Tandem 8; Tandem 9; Tandem 11H; Tandem 11H K; Tandem 22 H; Tandem 22 H K; Tandem 552; Tandem 552 K; Tandem 930; Vanade; Vanade Kosher; Vanall; Vanall (K); Vykamol 83G

Polysorbate 61

CAS 9005-67-8 (generic)
Synonyms: POE (4) sorbitan monostearate; PEG-4 sorbitan stearate
Definition: Mixture of stearate esters of sorbitol and sorbitol anhydrides, with ≈ 4 moles ethylene oxide
Toxicology: Moderately toxic by intravenous route
Uses: Emulsifier, solubilizer for flavors, vitamin oils
Regulatory: FDA 21CFR §175.300
Trade names: Crillet 31; Tween® 61

Polysorbate 65

CAS 9005-71-4; EEC E436
Synonyms: POE (20) sorbitan tristearate; PEG-20 sorbitan tristearate; Sorbimacrogol tristearate 300
Definition: Mixture of stearate esters of sorbitol and sorbitol anhydrides, with ≈ 20 moles ethylene oxide
Properties: Tan waxy solid, faint odor, bitter taste; sol. in min. and veg. oils, min. spirits, acetone, ether, dioxane,

alcohol, methanol; disp. in water, CCl₄; acid no. 2 max.; sapon. no. 88-98; hyd. no. 44-60

Toxicology: Heated to decomp., emits acrid smoke and irritating fumes

Uses: Emulsifier in ice cream, frozen desserts, cakes, cake mixes/icings/fillings, whipped toppings, coffee whiteners; stabilizer; defoamer

Usage level: Limitation 0.1% (ice cream), 0.32% (cakes), 0.4% (whipped topping); ADI 0-25 mg/kg (EEC)

Regulatory: FDA 21CFR §73.1001, 172.838, 173.340, 175.300, 178.3400; Europe listed; UK approved

Trade names: Alkamuls® PSTS-20; Crillet 35; Drewpone® 65K; Durfax® 65; Durfax® 65K; Glycosperse® TS-20; Glycosperse® TS-20 FG, TS-20 KFG; Glycosperse® TS-20 KFG; Hodag PSTS-20; Lamesorb® STS-20; Liposorb TS-20; Lonzest® STS-20; Sorbax PTS-20; T-Maz® 65; T-Maz® 65K; Tween® 65; Tween® 65K; Tween® 65 VS

Trade names containing: Aldosperse® TS-20; Aldosperse® TS-20 FG; Aldosperse® TS-20 KFG; Aldosperse® TS-40; Aldosperse® TS-40 FG; Aldosperse® TS-40 KFG; Drewmulse® 365; Ice #12

Polysorbate 80

CAS 9005-65-6 (generic); 37200-49-0; 61790-86-1; EEC E433

Synonyms: POE (20) sorbitan monooleate; PEG-20 sorbitan oleate; Sorbimacrogol oleate 300

Definition: Mixture of oleate esters of sorbitol and sorbitol anhydrides, with ≈ 20 moles ethylene oxide

Properties: Amber visc. liq.; very sol. in water; sol. in alcohol, cottonseed oil, corn oil, ethyl acetate, methanol, toluene; insol. in min. oil; dens. 1.06-1.10; visc. 270-430 cSt; acid no. 2 max.; sapon. no. 45-55; hyd. no. 65-80; pH 5-7 (5% aq.)

Toxicology: LD50 (oral, mouse) 25 g/kg; mod. toxic by intravenous route; mildly toxic by ingestion; eye irritant; experimental tumorigen, reproductive effects; mutagenic data; heated to decomp., emits acrid smoke and irritating fumes

Uses: Emulsifier in ice cream, edible fat/oils; yeast defoamer; solubilizer/dispersant in pickles, vitamin-mineral preps., gelatin dessert; surfactant in prod. of coarse cryst. NaCl; stabilizer; wetting agent in poultry scalds; color diluent; flavoring

Usage level: Limitation 0.1% (ice cream), 4% (yeast defoamers), 500 ppm (pickles), 300-475 mg/day (vitamin/mineral preps.); ADI 0-25 mg/kg (EEC)

Regulatory: FDA 21CFR §73.1, 73.1001, 172.515, 172.840, 173.340, 175.105, 175.300, 178.3400; USDA 9CFR §318.7, 381.147 (limitation 1% alone, 1% total combined with polysorbate 60); Europe listed; UK approved

Trade names: Atlas E; Capmul® POE-O; Crillet 4; Drewmulse® POE-SMO; Drewpone® 80K; Durfax® 80; Durfax® 80K; Emrite® 6120; Glycosperse® O-20; Glycosperse® O-20 FG; Glycosperse® O-20 KFG; Glycosperse® O-20 Veg; Glycosperse® O-20X; Hefti MO-55-F; Hodag PSMO-20; Hodag SVO-9; Lamesorb® SMO-20; Liposorb O-20; Lonzest® SMO-20; Nikkol TO-10; Norfox® Sorbo T-80; Polycon T80 K; Sorbax PMO-20; T-Maz® 80; T-Maz® 80K; T-Maz® 80KLM; Tween® 80; Tween® 80K; Tween® 80 VS

Trade names containing: Aldosperse® MO-50 FG; Aldosperse® O-20; Aldosperse® O-20 FG; Aldosperse® O-20 KFG; Annatto ECA-1; Annatto/Turmeric WML-1; Annatto/Turmeric WMP-1; Centromix® CPS; Drewmulse® 700K; Dricoid® 280; Ice # 2; Myvatex® MSPS; Panalite 100 K; Tandem 100 K; Turmeric WMSE-80; Tween®-MOS100K; Unibix AP (Acid Proof); Vanease; Vanease (K)

Polysorbate 81

CAS 9005-65-5 (generic)

Synonyms: POE (5) sorbitan monooleate; PEG-5 sorbitan oleate

Definition: Mixture of oleate esters of sorbitol and sorbitol anhydrides, with ≈ 5 moles ethylene oxide

Uses: Emulsifier for bakery, confectionery, convenience foods; solubilizer for flavors, vitamin oils; visc. modifier, suspending agent for food industry

Regulatory: FDA 21CFR §175.300

Trade names: Crillet 41; Glycosperse® O-5; Hetsorb O-5; Hodag PSMO-5; Sorbax PMO-5; T-Maz® 81; Tween® 81

Polysorbate 85

CAS 9005-70-3

Synonyms: POE (20) sorbitan trioleate; PEG-20 sorbitan trioleate; Sorbimacrogol trioleate 300

Definition: Mixture of oleate esters of sorbitol and sorbitol anhydrides, with ≈ 20 moles ethylene oxide

Toxicology: Skin irritant

Uses: Solubilizer for flavors, vitamin oils; o/w emulsifier for bakery, dairy, confectionery, convenience food applics.; wetting agent, dispersant, stabilizer

Regulatory: FDA 21CFR §175.300, 178.3400

Trade names: Alkamuls® PSTO-20; Crillet 45; Glycosperse® TO-20; Hodag PSTO-20; Lonzest® STO-20; Sorbax PTO-20; Tween® 85

Polystyrene-acrylonitrile. *See* Acrylonitrile-styrene copolymer

Polyvinyl acetate (homopolymer)

CAS 9003-20-7

Polyvinyl alcohol

    Synonyms: PVAc; Acetic acid, ethenyl ester, homopolymer; Acetic acid vinyl ester polymers; Ethenyl acetate, homopolymer
    Classification: Homopolymer
    Definition: Homopolymer of vinyl acetate
    Empirical: $(C_4H_6O_2)_x$
    Formula: $[CH_2CHOOCOCH_3]_x$
    Properties: Water-wh. clear solid resin; sol. in benzene, acetone; insol. in water
    Toxicology: Heated to decomp., emits acrid smoke and irritating fumes
    Uses: Color diluent, masticatory substance in chewing gum base; in paper/paperboard for food pkg.; coating material (Japan)
    Regulatory: FDA 21CFR §73.1, 172.615 (m.w. 2000 min.), 175.105, 175.300, 175.320, 176.170, 176.180, 177.1200, 177.2800, 181.22, 181.30; Japan approved
    Manuf./Distrib.: Nat'l. Starch & Chem.

Polyvinyl alcohol

    CAS 9002-89-5 (super and fully hydrolyzed); EINECS 209-183-3
    Synonyms: PVA; PVAL; Ethenol homopolymer; PVOH; Vinyl alcohol polymer
    Classification: Polymer
    Empirical: $(C_2H_4O)_x$
    Formula: $[CH_2CHOH]_x$
    Properties: White to cream amorphous powd.; sol. in water; insol. in petrol. solvs.; avg. m.w. 120,000; dens. 1.329; softens at 200 C with dec.; flash pt. (OC) 175 F; ref. index 1.49-1.53
    Precaution: Flamm. exposed to heat or flame; reactive with oxidizers; dust exposed to flame presents sl. explosion hazard
    Toxicology: Experimental carcinogen and tumorigen; heated to decomp., emits acrid smoke and irritating fumes
    Uses: Color diluent for egg shells; in mfg. of paper/paperboard for food pkg. (for fatty foods only)
    Regulatory: FDA 21CFR §73.1, 175.105, 175.300, 175.320, 176.170, 176.180, 177.1200, 177.1670, 177.2260, 177.2800, 178.3910, 181.22, 181.30
    Trade names: Elvanol® 71-30

Poly(n-vinylbutyrolactam). *See* Poly(1-vinyl-2-pyrrolidinone) homopolymer

Polyvinylpolypyrrolidone

    CAS 9003-39-8
    Synonyms: PVPP
    Definition: Insoluble homopolymer of purified vinylpyrrolidone
    Properties: Wh. powd., faint bland odor; hygroscopic; insol. in water
    Toxicology: Heated to decomp., emits acrid smoke and irritating fumes
    Uses: Clarifier and stabilizer for beverages, wine, and vinegar; bulking agent and stabilizer in low-calorie foods
    Regulatory: FDA 21CFR §173.50 (must be removed by filtration); BATF 27CFR §240.1051 (limitation 6 lb/1000 gal in wine)

Poly(1-vinyl-2-pyrrolidinone) homopolymer. *See* PVP
Polyvinylpyrrolidone. *See* PVP
Ponceau SX. *See* FD&C Red No. 4
POP (5) POE (6) monobutyl ether. *See* PPG-5-buteth-7
POP (12) POE (16) monobutyl ether. *See* PPG-12-buteth-16
POP (20) POE (30) monobutyl ether. *See* PPG-20-buteth-30
POP (28) POE (35) monobutyl ether. *See* PPG-28-buteth-35
POP (33) POE (45) monobutyl ether. *See* PPG-33-buteth-45

Poppy seed

    Definition: Papayer somniferum
    Uses: Natural flavoring agent
    Regulatory: FDA 21CFR §182.10, GRAS; Japan approved

Potash. *See* Potassium carbonate
Potash alum. *See* Potassium alum dodecahydrate
Potash lye. *See* Potassium hydroxide
Potassa. *See* Potassium hydroxide
Potassium acesulfame. *See* Acesulfame potassium

Potassium acetate

    CAS 127-08-2; EINECS 204-822-2; EEC E261
    Empirical: $C_2H_3KO_2$

Formula: CH₃COOK

Properties: Colorless lustrous cryst. or wh. cryst. powd. or flakes; deliq.; sol. in water, alcohol; insol. in ether; m.w. 98.14; dens. 1.57; m.p. 292 C

Precaution: Keep tightly closed

Toxicology: LD50 (oral, rat) 3.25 g/kg

Uses: Synthetic flavoring; preservative; buffer; neutralizer

Regulatory: FDA 21CFR §172.515; GRAS (FEMA); Europe listed

Manuf./Distrib.: Heico; Honeywil & Stein Ltd; Lohmann; Niacet

Potassium acid tartrate

CAS 868-14-4; EINECS 212-769-1; EEC E336

Synonyms: Potassium bitartrate; Monopotassium L-(+)-tartrate; Potassium hydrogen tartrate; Cream of tartar; L-Tartaric acid monopotassium salt; Acid potassium tartrate; Dipotassium L-(+)-tartrate

Definition: Salt of L(+)-tartaric acid

Empirical: $C_4H_5KO_6$

Formula: $KHC_4H_4O_6$

Properties: Colorless or sl. opaque cryst. or wh. cryst. powd., pleasant acid taste; sol. in water, sl. sol. in alcohol; m.w. 188.18; dens. 1.984

Toxicology: Heated to decomp., emits acrid smoke and irritating fumes

Uses: Direct food additive, anticaking, antimicrobial, formulation aid, humectant, leavening agent, pH control agent, processing aid, stabilizer, thickener, surface-active agent; in baking powder, baked goods, confections, gelatins, puddings, candy, jams

Usage level: ADI 0-30 mg/kg (EEC, as tartaric acid)

Regulatory: FDA 21CFR §184.1077, GRAS; USDA 9CFR §318.7; BATF 27CFR §240.1051 (limitation 25 lb/1000 gal grape wine); Europe listed; UK approved; Japan approved

Manuf./Distrib.: AB Tech.; Ashland; British Pepper & Spice; Browning; Calaga Food Ingreds.; Dinoval; Ellis & Everard; Int'l. Sourcing; Jungbunzlauer; Lohmann; New England Spice; Pacific Foods; Penta Mfg.; Siber Hegner; Spice King; Tesco; Van Waters & Rogers

Potassium alginate

CAS 9005-36-1; EEC E402

Synonyms: Potassium polymannuronate; Alginic acid, potassium salt

Definition: Potassium salt of alginic acid

Empirical: $(O_6 I_7 O_6 K)_x$

Properties: Wh. gran., odorless, tasteless; sol. in water; insol. in alcohol, chloroform, ether; m.w. 214.22

Toxicology: Heated to decomp., emits acrid smoke and irritating fumes

Uses: Gellant, thickener, emulsifier, stabilizer; used for water holding in foods

Usage level: Limitation 0.1% (confections, frostings), 0.7% (gelatins, puddings), 0.25% (processed fruits, fruit juices), 0.01% (other food); ADI 0-25 mg/kg (EEC, as alginic acid)

Regulatory: FDA 21CFR §184.1610, GRAS; Europe listed; UK approved

Manuf./Distrib.: Atomergic Chemetals; Kelco Int'l.

Trade names: Kelmar®; Kelmar® Improved; Sobalg FD 200 Range

Potassium alum dodecahydrate

CAS 7784-24-9 (dodecahydrate), 10043-67-1 (sesquihydrate); EINECS 233-141-3

Synonyms: Aluminum potassium sulfate; Potash alum; Alum, potassium; Potassium aluminum sulfate; Kalinite

Classification: Inorganic salt

Formula: $KAl(SO_4)_2 \cdot 12H_2O$ (dodecahydrate)

Properties: Dodecahydrate: Transparent cryst. or wh. cryst. powd., odorless, sweetish astringent taste; sol. glycerin; sol. 1 g/7.2 ml water; insol. in alcohol; insol. in alcohol; m.w. 474.38; dens. 1.725; m.p. 92.5 C

Toxicology: Nuisance dust

Uses: Multipurose food additive; buffer, firming agent for pickles, neutralizing agent; clarifying sugar, hardening gelatin, baking powds., purifying water; migrating to food from paper/paperboard

Regulatory: FDA 21CFR §133.102, 133.106, 133.111, 133.141, 133.165, 133.181, 133.183, 133.195, 137.105, 137.155, 137.160, 137.165, 137.170, 137.175, 137.180, 137. 185. 178.3120, 182.90, 182.1129, GRAS; Japan approved except for miso

Potassium aluminum sulfate. *See* Potassium alum dodecahydrate

Potassium benzoate

CAS 582-25-2 (anhyd.); EINECS 209-481-3; EEC E212

Synonyms: Benzoic acid potassium salt

Empirical: $C_7H_5KO_2$

Formula: C_6H_5COOK

Potassium bicarbonate

Properties: m.w.160.22; m.p. >300C
Uses: Preservative for margarine, oleomargarine, wine
Usage level: ADI 0-5 mg/kg (EEC)
Regulatory: USDA 9CFR §318.7 (limitation 0.1%), BATF 27CFR §240.1051 (limitation 0.1% in wine); Europe listed
Manuf./Distrib.: Lohmann; Pfizer Food Science

Potassium bicarbonate
CAS 298-14-6; EINECS 206-059-0
Synonyms: Carbonic acid monopotassium salt; Potassium hydrogen carbonate; Monopotassium carbonate
Classification: Inorganic salt
Empirical: $CH_2O_3 \cdot K$
Formula: $KHCO_3$
Properties: Colorless prisms or wh. gran. powd., odorless; sol. in water; insol. in alcohol; m.w. 100.12
Toxicology: Nuisance dust
Uses: Direct food additive, alkali, leavening agent, formulation aid, nutrient supplement, pH control agent, processing aid; in dietary drinks, dried milk prods., baked goods, margarine
Regulatory: FDA 21CFR §163.110, 184.1613, GRAS; USDA 9CFR §318.7; BATF 27CFR §240.1051
Manuf./Distrib.: Browning; Church & Dwight; Hays; Hüls AG; Lohmann; Penta Mfg.; Van Waters & Rogers

Potassium biphosphate. *See* Potassium phosphate

Potassium bisulfite
Uses: Preservative; not for use in meats, sources of vitamin B_1, raw fruits and vegetables, or fresh potatoes
Regulatory: FDA 21CFR §182.3616 (not for use in meats), GRAS

Potassium bitartrate. *See* Potassium acid tartrate

Potassium bromate
CAS 7758-01-2; EINECS 231-829-8; EEC E924
Empirical: $BrKO_3$
Formula: $KBrO_3$
Properties: Wh. cryst. or gran.; sol. in 12.5 parts water; almost insol. in alcohol; m.w. 167.01; dens. 3.27; m.p. 350 C, dec. 370 C
Precaution: Dangerous fire risk in contact with organic materials
Toxicology: May cause mild intestinal irritation; ingestion may cause vomiting, diarrhea, renal injury
Uses: Food additive, dough conditioner, maturing agent, bleaching agent, improving agent, malting of barley for prod. of fermented malt beverages or distilled spirits; processing aid (Japan)
Usage level: ADI 0-75 mg/kg (EEC)
Regulatory: FDA 21CFR §136.110, 136.115, 136.130, 136.160, 136.180, 137.155, 137.205, 172.730, 75 ppm max.; Japan restricted (0.03 g/kg max. of flour as bromic acid); Europe listed; UK approved
Manuf./Distrib.: Allchem Industries; Gist-Brocades Food Ingreds.
Trade names: Bromette; Bromitabs
Trade names containing: Adapt; Bagel Base; Calib; CDC-10A; CDC-2001; FD-6; FD-8; IDX; Oxitabs

Potassium bromide
CAS 7758-02-3; EINECS 231-830-3
Empirical: BrK
Formula: KBr
Properties: Colorless cubic cryst.; sl. hygroscopic; m.w. 119.01; dens. 2.75; m.p. 730 C; b.p. 1380 C
Precaution: Violent reaction with BrF_3
Toxicology: Large doses can cause CNS depression; prolonged inh. can cause skin eruptions; mutagenic data; heated to decomp., emits toxic fumes of K_2O and Br
Uses: Washing water agent for washing/lye peeling of fruits and vegetables
Regulatory: FDA 21CFR §173.315

Potassium t-butylate
Properties: powd.
Uses: Catalyst for transesterification and isomerization of fats and oils
Manuf./Distrib.: Hüls AG

Potassium butyl paraben
CAS 38566-94-8
Synonyms: Butylparaben, potassium salt; n-Butyl-4-hydroxybenzoate potassium salt
Definition: Potassium salt of butylparaben
Empirical: $C_{11}H_{14}O \cdot K$
Uses: Preservative, bactericide, fungicide for foods

Trade names: Nipabutyl Potassium

Potassium carbonate
CAS 584-08-7; EINECS 209-529-3; EEC E501
Synonyms: Carbonic acid dipotassium salt; Dipotassium carbonate; Pearl ash; Potash
Classification: Inorganic salt
Empirical: $CO_3 \cdot 2K$
Formula: K_2CO_3
Properties: White deliq. gran., translucent powd.; hygroscopic; alkaline reaction; sol. in water; insol. in alcohol; m.w. 138.20; dens. 2.428 (19 C); m.p. 891 C
Precaution: Noncombustible; incompatible with KCO, magnesium
Toxicology: LD50 (oral, rat) 1870 mg/kg; poison by ingestion; solutions irritating to tissue; strong caustic; heated to decomp., emits toxic fumes of K_2O
Uses: Direct food additive, flavoring agent, adjuvant, nutrient supplement, pH control agent, alkali, processing aid; in margarine, soups; auxiliary for processing cocoa beans, fermentation of tobacco, drying of fruit; boiler water additive; yeast nutrient
Usage level: ADI no limit (EEC)
Regulatory: FDA 21CFR §163.110, 172.560, 173.310, 184.1619, GRAS; USDA 9CFR §318.7; BATF 27CFR §240.1051; Japan approved; Europe listed; UK approved
Manuf./Distrib.: Browning; Church & Dwight; Hays; Hüls AG; Lohmann; Penta Mfg.; Van Waters & Rogers; Peter Whiting
Trade names containing: Annatto WSP-75; Annatto WSP-150

Potassium carrageenan
Definition: Potassium salt of carrageenan
Uses: Emulsifier, stabilizer, thickener
Regulatory: FDA 21CFR §172.626, 176.170

Potassium caseinate
Properties: off-wh. solid
Uses: More milk-like flavor than other caseinates; for dairy prods., margarine
Manuf./Distrib.: Excelpro

Potassium chloride
CAS 7447 40 7; EINECS 231 211 8; EEC E500
Synonyms: Dipotassium dichloride; Potassium monochloride
Classification: Inorganic salt
Empirical: ClK
Formula: KCl
Properties: Colorless to white crystals or powd., odorless, saline taste at low concs.; sol. in water, glycerin; sl. sol. in alcohol; insol. in abs. alcohol, ether, acetone; m.w. 74.55; dens. 1.987; m.p. 773 C (sublimes 1500 C); pH 7
Precaution: Explosive reaction with BrF_3
Toxicology: LD50 (oral, rat) 2600 mg/kg; human poison by ingestion; moderately toxic by subcutaneous route; eye irritant; human systemic effects; mutagenic data; heated to decomp., emits toxic fumes of K_2O and Cl^-
Uses: Direct food additive, flavor enhancer, flavoring agent, nutrient, dietary supplement, pH control agent, stabilizer, thickener, gelling agent, salt substitute, yeast food, tissue softening agent; in beer malting; may be used in infant formulas
Regulatory: FDA 21CFR §150.141, 150.161, 166.110, 182.5622, 184.1622, GRAS, 201.306; USDA 9CFR §318.7, 381.147 (limitation ≤ 3% of 2 molar sol'n.); Japan approved; Europe listed; UK approved
Manuf./Distrib.: Browning
Trade names containing: Cap-Shure® KCL-140-50; Cap-Shure® KCL-165-70; Durkote Potassium Chloride/ Hydrog. Veg. Oil; Frimulsion 6G; Merecol® LK; Morton® Flour Lite Salt™ Mixt; Morton® Lite Salt® Mixt; Morton® Lite Salt® TFC Mixt

Potassium citrate
CAS 866-84-2 (anhyd.), 6100-05-6 (monohydrate); EINECS 212-755-5, 231-905-0 (monohydrate); EEC E332
Synonyms: Citric acid tripotassium salt; Tripotassium citrate monohydrate; Potassium citrate tertiary
Empirical: $C_6H_5O_7 \cdot 3K$ (anhyd.), $C_6H_5K_3O_7 \cdot H_2O$ (monohydrate)
Properties: Colorless or white crystals or powd., odorless; deliq.; sol. in water, glycerol; insol. in alcohol; m.w. 306.41 (anhyd.), 324.42 (monohydrate); dens. 1.98; dec. 230 C; pH 8.5
Toxicology: LD50 (IV, dog) 167 mg/kg; poison by intravenous routes; heated to decomp., emits toxic fumes of K_2O, acrid smoke and irritating fumes
Uses: Miscellaneous and general purpose food additive, antioxidant, buffer, pH control agent, sequestrant; emulsifying salt; stabilizer migrating from food pkg.; for soft drinks, meat prods., processed cheese;

Potassium citrate tertiary

flavoring (Japan)
Usage level: ADI no limit (EEC)
Regulatory: FDA 21CFR §181.29, 182.1625, 182.6625, GRAS; USDA 9CFR §318.7; BATF 27CFR §240.1051 (limitation 25 lb/1000 gal wine); Japan approved; Europe listed; UK approved
Manuf./Distrib.: ADM; Browning; Jungbunzlauer; Lohmann
Trade names containing: Frigesa® D 890; Frimulsion 6G

Potassium citrate tertiary. *See* Potassium citrate
Potassium dihydrogen orthophosphate. *See* Potassium phosphate
Potassium dihydrogen phosphate. *See* Potassium phosphate
Potassium disulfite. *See* Potassium metabisulfite

Potassium ethylate
Properties: powd./liq.
Uses: Catalyst for transesterification and isomerization of fats and oils
Manuf./Distrib.: Hüls AG

Potassium ethylparaben
CAS 36457-19-9
Synonyms: Ethylparaben, potassium salt; Ethyl-4-hydroxybenzoate potassium salt
Definition: Potassium salt of ethylparaben
Empirical: $C_9H_{10}O_3 \cdot K$
Uses: Preservative, fungicide, bactericide for foods (beers, flavors, fruit juices, fish, pickles, sauces)
Trade names: Nipagin A Potassium

Potassium fumarate
CAS 4151-35-3 (anhyd.); EINECS 223-979-8
Empirical: $C_4H_2K_2O_4$
Uses: Dietary supplement
Regulatory: FDA 21CFR §172.350

Potassium furcelleran
Definition: Potassium salt of furcelleran
Uses: Emulsifier, stabilizer, thickener
Regulatory: FDA 21CFR §172.660

Potassium gibberellate
Definition: Potassium salt of gibberellic acid
Empirical: $C_{19}H_{21}KO_6$
Properties: Wh. to sl. off-wh. cryst. powd., odorless; deliq.; sol. in water, alcohol, acetone; m.w. 384.47
Toxicology: Heated to decomp., emits acrid smoke and irritating fumes
Uses: Food additive, enzyme activator; for use in the malting of barley, for prod. of fermented malt beverages or distilled spirits
Regulatory: FDA 21CFR §172.725, 2 ppm max. (malted barley), 0.5ppm (finished beverage)

Potassium D-gluconate
CAS 299-27-4; EINECS 206-074-2; EEC E577
Synonyms: D-Gluconic acid potassium salt; Monopotassium D(-)-pentahydroxy capronate
Empirical: $C_6H_{11}O_7K$
Properties: Wh. or ylsh. fine powd., odorless, salty taste; sol. in water, glycerin; insol in alc. and benzene; m.w. 234.3; m.p. 180 C (dec.)
Precaution: Avoid dust formation
Toxicology: LD50 (oral, rat) 10,380 mg/kg; mod. toxic by IP route; mildly toxic by ingestion; heated to decomp., emits toxic fumes of K_2O
Uses: Mineral source, nutrient, dietary supplement, sequestrant, denuding agent; used in dry mix beverages and desserts, cake mixes, tripe, dietetic foods
Usage level: ADI 0-50 mg/kg (EEC, as total gluconic acid)
Regulatory: USDA 9CFR §318.7; Europe listed; UK approved
Manuf./Distrib.: Am. Roland; Jungbunzlauer; Lohmann
Trade names: Gluconal® K

Potassium glutamate
CAS 19473-49-5 (monohydrate); EINECS 363-737-7 (monohydrate); EEC E622
Synonyms: MPG; Monopotassium glutamate; Potassium L-glutamate; Potassium hydrogen L-glutamate; Monopotassium l-glutamate; Potassium glutaminate
Empirical: $C_5H_8NO_4 \cdot K$, $C_5H_8KNO_4 \cdot H_2O$ (monohydrate)
Formula: $KOOCCH_2CH_2CH(NH_2)COOH \cdot H_2O$ (monohydrate)

Properties: Wh. free-flowing cryst. powd., pract. odorless; hygroscopic; freely sol. in water; sl. sol. in alcohol; m.w. 185.24 (anhyd.), 203.24 (monohydrate)

Toxicology: LD50 (oral, mouse) 4500 mg/kg; mildly toxic by ingestion; human systemic effects; heated to decomp., emits toxic fumes of K_2O and NO_x

Uses: Multipurpose food ingred.; flavor enhancer, salt substitute; used for meat

Regulatory: FDA 21CFR §182.1516, GRAS; Japan approved; Europe listed; UK approved

Potassium L-glutamate. *See* Potassium glutamate
Potassium glutaminate. *See* Potassium glutamate

Potassium glycerophosphate
CAS 1319-69-3 (anhyd.); EINECS 215-291-1
Empirical: $C_3H_7K_2O_6P \cdot 3H_2O$
Properties: Pale yel. syrupy liq.; sol. in water; m.w. 302.20
Toxicology: Heated to decomp., emits acrid smoke and irritating fumes
Uses: Dietary supplement, nutrient
Regulatory: FDA 21CFR §182.5628, 182.8628, GRAS
Manuf./Distrib.: Lohmann

Potassium 2,4-hexadienoate. *See* Potassium sorbate
Potassium hydrate. *See* Potassium hydroxide
Potassium hydrogen carbonate. *See* Potassium bicarbonate
Potassium hydrogen L-glutamate. *See* Potassium glutamate
Potassium hydrogen tartrate. *See* Potassium acid tartrate

Potassium hydroxide
CAS 1310-58-3; EINECS 215-181-3; EEC E525
Synonyms: Caustic potash; Potassium hydrate; Lye; Potash lye; Potassa
Classification: Inorganic base
Empirical: HKO
Formula: KOH
Properties: Wh. flakes, lumps or pellets, highly deliq.; sol. in water, alcohol, glycerol; sl. sol. in ether; m.w. 56.11; dens. 2.044; m.p. 405 C; b.p. 1320 C
Precaution: Corrosive
Toxicology: LD50 (oral, rat) 365 mg/kg; toxic by ingestion, inh.; strong caustic; eye and severe skin irritant; mutagenic data; TLV 2 mg/m³ of air (ACGIH); heated to dec., emits toxic fumes of K_2O; above 84 C reacts with reducing sugars to form CO
Uses: Direct food additive, formulation aid, pH control agent, alkali, processing aid, stabilizer, thickener, oxidizer; poultry scald agent; used in baked goods, cocoa, dairy prods., frozen vegetables, ice cream, soft drinks, black olives, poultry
Regulatory: FDA 21CFR §163.110, 175.210, 184.1631, GRAS; USDA 9CFR §381.147; Europe listed; UK approved
Trade names containing: Annatto ECA-1; Annatto ECH-1; Annatto ECPC; Annatto OS #2894; Annatto OS #2922; Annatto OS #2923; Annatto WSL-2; Annatto WSP-75; Annatto WSP-150; Annatto Liq. #3968, Acid Proof; Annatto/Turmeric WML-1; Annatto/Turmeric WMP-1

Potassium-L-2-hydroxypropionate. *See* Potassium lactate

Potassium iodate
CAS 7758-05-6; EINECS 231-831-9
Empirical: IKO_3
Formula: KIO_3
Properties: Colorless cryst. or wh. cryst. powd.; sol. in water; insol. in alcohol; m.w. 214.00; dens. 3.89; m.p. 560 C
Precaution: Violent reaction with organic matter
Toxicology: LD50 (IP, mouse) 136 mg/kg; poison by ingestion, IP routes; heated to decomp., emits very toxic fumes of I⁻ and K_2O
Uses: Dough strengthener/conditioner, maturing agent, feed additive; used for bread mfg.
Usage level: Limitation 0.0075% on wt. of flour (bread)
Regulatory: FDA 21CFR §184.1635, GRAS
Manuf./Distrib.: Atomergic Chemetals; Spectrum Chem. Mfg.
Trade names containing: Oxitabs

Potassium iodide
CAS 7681-11-0; EINECS 231-659-4
Classification: Inorganic salt

Potassium Kurrol's salt

Definition: Potassium salt of hydriodic acid
Formula: KI
Properties: White cryst. granules or powd., strong bitter, saline taste; sl. hygroscopic; sol. in water, alcohol, acetone, glycerol; m.w. 166.02; dens. 3.123; b.p. 1420 C; m.p. 723 C
Precaution: Moisture sensitive; explosive reaction with charcoal + ozone; incompat. with oxidants, BrF_3, FClO, metallic salts
Toxicology: Irritant; poison by intravenous route; moderately toxic by ingestion and intraperitoneal routes; human teratogenic effects; experimental reproductive effects; mutagenic data; heated to decomp., emits very toxic fumes of K_2O and I⁻
Uses: Nutrient, dietary supplement, source of iodine in table salt; animal feed additive
Usage level: Limitation 225 μg daily intake, 45 μg (infants), 105 μg (children under 4), 225 μg (adults), 300 μg (pregnant or lactating women), 0.01% (table salt)
Regulatory: FDA 21CFR §172.375, 178.1010, 184.1634, GRAS
Manuf./Distrib.: Atomergic Chemetals
Trade names containing: Morton® Table Iodized Salt

Potassium Kurrol's salt. *See* Potassium polymetaphosphate

Potassium lactate
CAS 996-31-6; EINECS 213-631-3; EEC E326
Synonyms: Potassium-L-2-hydroxypropionate; Propanoic acid, 2-hydroxy-, monopotassium salt
Definition: Potassium salt of lactic acid
Empirical: $C_3H_5KO_3$
Formula: $CH_3CHOHCOOK$
Properties: Wh. solid, odorless; hygroscopic; m.w. 128.17
Toxicology: Heated to decomp., emits acrid smoke and irritating fumes
Uses: Direct food additive, flavor enhancer, flavoring agent, adjuvant, humectant, pH control agent, antioxidant synergist; used in confectionery, jams, jellies, marmalades, margarine
Usage level: ADI no limit (EEC)
Regulatory: FDA 21CFR §184.1639, GRAS; not authorized for infant formulas; Europe listed; UK approved
Manuf./Distrib.: Lohmann
Trade names: Arlac P; Purasal® P/USP 60

Potassium metabisulfite
CAS 16731-55-8; EINECS 240-795-3; EEC E224
Synonyms: Disulfurous acid dipotassium salt; Potassium pyrosulfite; Potassium disulfite; Dipotassium disulfite
Classification: Inorganic salt
Empirical: $K_2O_5S_2$
Formula: $K_2S_2O_5$
Properties: White granules or powd., pungent sharp odor; sol. in water, insol. in alcohol; m.w. 222.32; dens. 2.3; m.p. > 300 C; dec. at 150-190 C; oxidizes in air and moisture to sulfate
Precaution: Moisture sensitive; keep dry and well closed
Toxicology: Irritant; experimental tumorigen, reproductive effects; heated to decomp., emits toxic fumes of SO_x and K_2O
Uses: Food preservative, antioxidant, sterilizer; brewing, wine making; not for use in meats, sources of vitamin B_1, raw fruits and vegetables, or fresh potatoes
Usage level: ADI 0-0.7 mg/kg (EEC)
Regulatory: FDA 21CFR §182.3637, GRAS; BATF 27CFR §240.1051; Europe listed
Manuf./Distrib.: Allchem Industries

Potassium metaphosphate. *See* Potassium polymetaphosphate

Potassium methylate
Properties: powd./liq.
Uses: Catalyst for transesterification and isomerization of fats and oils
Manuf./Distrib.: Hüls AG

Potassium N-methyldithiocarbamate
Uses: Antimicrobial for use in cane-sugar and beet-sugar mills; in paper for food pkg.
Regulatory: FDA 21CFR §173.320, 181.30

Potassium 6-methyl-1,2,3-oxathiazine-4(3H)-1,2,2-dioxide. *See* Acesulfame potassium
Potassium monochloride. *See* Potassium chloride

Potassium nitrate
CAS 7757-79-1; EINECS 231-818-8; EEC E252
Synonyms: Niter; Nitre; Saltpeter

Empirical: KNO₃

Properties: Transparent colorless or wh. cryst. or cryst. powd., odorless, cooling pungent salty taste; sol. in glycerin, water; mod. sol. in alcohol; m.w. 101.11; dens. 2.109 (16 C); m.p. 334 C; b.p. dec. @ 400 C

Precaution: Oxidizer

Toxicology: LD50 (oral, rat) 3750 mg/kg; poison by IV route; mod. toxic by ingestion; experimental reproductive effects; mutagenic data; can produce nitrosamines linked to cancer; can reduce blood oxygen levels; heated to dec., emits very toxic fumes NO$_x$, K₂O

Uses: Preservative; antimicrobial agent; source of nitrite in prod. of cured meats and poultry; curing agent in processing of cod roe (200 ppm max. of finished roe); color fixative

Usage level: Limitation 200 ppm (cod roe), 7 lb/100 gal (pickles), 3.5 oz/100 lb (meat), 2.75 oz/100 lb (chopped meat)

Regulatory: FDA 21CFR §172.160, 181.33; USDA 9CFR §318.7, 381.147; Japan approved with limitations; Europe listed

Manuf./Distrib.: Lohmann; Whiting, Peter Ltd

Potassium nitrite

CAS 7758-09-0; EINECS 231-832-4; EEC E249

Synonyms: Nitrous acid potassium salt; Potassium nitrite (1:1)

Empirical: NO₂ • K

Formula: KNO₂

Properties: Wh. or sl. ylsh. prisms or sticks, deliq.; very sol. in water; sl. sol. in alcohol; m.w. 85.11; dens. 1.915; m.p. 387 C; b.p. dec.

Precaution: Flamm. exposed to heat or flame; powerful oxidizer; sl. explosion hazard exposed to heat

Toxicology: LD50 (oral, rabbit) 200 mg/kg; poison by ingestion; experimental reproductive effects; mutagenic data; linked to increased incidence of cancer; on decomp., emits toxic fumes of K₂O

Uses: Color fixative, antimicrobial agent, preservative in the curing of red meats, bacon, and poultry prods.

Usage level: Limitation 2 lb/100 gal (pickle), 1 oz/100 lb (meat), 0.25 oz/100 lb (chopped meat); 200 ppm max. nitrite calculated as sodium nitrate in finished prod.; ADI 0-0.2 mg/kg (EEC)

Regulatory: FDA 21CFR §181.34; USDA 9CFR §318.7, 381.147; Europe listed

Manuf./Distrib.: Aldrich

Potassium nitrite (1:1). *See* Potassium nitrite

Potassium 9-octadecenoate. *See* Potassium oleate

Potassium oleate

CAS 143-18-0; EINECS 205-590-5

Synonyms: Potassium 9-octadecenoate; Oleic acid potassium salt

Definition: Potassium salt of oleic acid

Empirical: C₁₈H₃₃O₂ • K

Formula: CH₃(CH₂)₇CH=CH(CH₂)₇COOK

Properties: Ylsh. or brownish soft mass; sol. in water, alcohol; m.w. 320.56

Toxicology: Eye irritant; heated to decomp., emits toxic fumes of K₂O

Uses: Anticaking agent, binder, emulsifier; stabilizer migrating from food pkg.

Regulatory: FDA 21CFR §172.863, 175.105, 175.300, 176.170, 176.200, 176.210, 177.1200, 177.2260, 177.2600, 177.2800, 178.3910, 181.22, 181.29

Potassium pentachlorphenate

Uses: Slime control agent used in mfg. of paper/paperboard contacting food

Regulatory: FDA 21CFR §181.30

Potassium peroxydisulfate. *See* Potassium persulfate

Potassium persulfate

CAS 7727-21-1; EINECS 231-781-8

Synonyms: Peroxydisulfuric acid dipotassium salt; Potassium peroxydisulfate; Dipotassium persulfate

Empirical: K₂O₈S₂

Formula: K₂S₂O₈

Properties: Colorless or wh. cryst., odorless; sol. in 50 parts water; insol. in alcohol; m.w. 270.33; dens. 2.477; m.p. 100 C (dec.)

Precaution: Powerful oxidizer; flamm. when exposed to heat or by chemical reaction; reactive with reducing materials; liberates oxygen above 100 C (dry), 50 C (sol'n.); keep well closed in cool place

Toxicology: TLV:TWA 5 mg(S₂O₈)/m³; mod. toxic; irritant; allergen; heated to decomp., emits highly toxic fumes of SO$_x$ and K₂O

Uses: Defoaming agent, dispersing adjuvant, poultry scald agent

Regulatory: FDA 21CFR §172.210, 175.105, 175.210, 176.170, 177.1210, 177.2600; USDA 9CFR §381.147

Manuf./Distrib.: Allchem Industries; FMC

Potassium phosphate

CAS 7778-77-0; EINECS 231-913-4; EEC E340a

Synonyms: MKP; Potassium phosphate monobasic; Monopotassium phosphate; Potassium phosphate primary; Potassium dihydrogen orthophosphate; Potassium dihydrogen phosphate; Monopotassium orthophosphate; Potassium biphosphate

Classification: Inorganic salt

Empirical: KH_2PO_4

Formula: $H_3O_4P \cdot K$

Properties: Colorless cryst.; acid in reaction; sol. in water; insol. in alcohol; m.w. 136.09; dens. 2.338; m.p. 253 C

Toxicology: Nuisance dust

Uses: Buffer, sequestrant, yeast food; cooked out juices retention aid; baking powder, prod. of caramel; meat treating; nutrient in antibiotic and yeast prod.; antioxidant

Usage level: Limitation 5% (pickle for meat prods.), 0.5% (meat prods.), 0.5% (total poultry prod.)

Regulatory: FDA 21CFR §160.110, 175.105; USDA 9CFR §318.7, 381.147; Japan approved; Europe listed; UK approved

Manuf./Distrib.: Albright & Wilson Am.; Browning; FMC; Lohmann

Potassium phosphate dibasic

CAS 7758-11-4; EINECS 231-834-5; EEC E340b

Synonyms: DKP; Dipotassium phosphate; Phosphoric acid dipotassium salt; Dibasic potassium phosphate; Dipotassium hydrogen orthophosphate; Dipotassium hydrogen phosphate; Dipotassium orthophosphate

Classification: Inorganic salt

Empirical: K_2HPO_4

Properties: Colorless or wh. cryst. or powd.; deliq., hygroscopic; sol. in water; insol. in alcohol; m.w. 174.18

Toxicology: Nuisance dust

Uses: Buffer, acidity regulator, alkaline agent, raising agent, sequestrant, yeast food; cooked out juices retention aid; emulsifier in nondairy creamers; flavorings; humectant; nutrient in antibiotic prod., pharmaceuticals

Usage level: Limitation 5% (pickle for meat prods.), 0.5% (meat prods.), 0.5% (total poultry)

Regulatory: FDA 21CFR §133.169, 133.173, 133.179, 175.105, 182.6285, GRAS; USDA 9CFR §318.7, 381.147; Europe listed; UK approved; Japan approved

Manuf./Distrib.: Albright & Wilson Am.; Browning; FMC; Rhone-Poulenc Food Ingreds.

Potassium phosphate monobasic. *See* Potassium phosphate

Potassium phosphate primary. *See* Potassium phosphate

Potassium phosphate tribasic

CAS 7778-53-2; EINECS 231-907-1; EEC E340c

Synonyms: TKP; Tripotassium phosphate; Tripotassium orthophosphate

Formula: K_3PO_4

Properties: Wh. gran. powd., hygroscopic, deliq.; sol. in water; insol. in alcohol; m.w. 212.28; dens. 2.564 (17 C, anhyd.); m.p. 1340 C (anhyd.)

Toxicology: Nuisance dust

Uses: Emulsifier; alkaline buffer; sequestrant; antioxidant synergist

Regulatory: GRAS; Europe listed; UK approved

Manuf./Distrib.: FMC

Potassium polymannuronate. *See* Potassium alginate

Potassium polymetaphosphate

Synonyms: Potassium metaphosphate; Potassium Kurrol's salt

Definition: Straight-chain polyphosphate

Empirical: $(KPO_3)_x$

Properties: Wh. powd., odorless; insol. in water

Uses: Binding agent, fat emulsifier, moisture-retaining agent, alkaline agent, raising agent, emulsifier for cheese food (Japan)

Regulatory: Japan approved

Potassium propylparaben

CAS 84930-16-5

Synonyms: Propylparaben, potassium salt; n-Propyl-4-hydroxybenzoate potassium salt

Definition: Potassium salt of propylparaben

Empirical: $C_{10}H_{12}/_3 \cdot K$

Uses: Preservative, fungicide, bactericide for foods

Trade names: Nipasol M Potassium

Potassium pyrophosphate. *See* Tetrapotassium pyrophosphate
Potassium pyrophosphate, normal. *See* Tetrapotassium pyrophosphate
Potassium pyrosulfite. *See* Potassium metabisulfite
Potassium sodium L-(+)-tartrate. *See* Potassium sodium tartrate anhyd.

Potassium sodium tartrate anhyd.
CAS 304-59-6; EINECS 206-156-8; EEC E337
Synonyms: Rochelle salt; Sodium potassium tartrate; Potassium sodium L-(+)-tartrate; Seignette salt; 2,3-Dihydroxybutanedioic acid, monopotassium monosodium salt
Classification: Organic salt
Definition: Sodium potassium salt of L-tartaric acid
Empirical: $C_4H_4KNaO_6$
Properties: Sol. in water; m.w. 210.16
Uses: Food additive, emulsifier, buffer, pH control agent, sequestrant, emulsifying salt, stabilizer; synergist for antioxidants; used in cheeses, jams, jellies
Usage level: ADI 0-30 mg/kg (EEC)
Regulatory: FDA 21CFR §133.169, 133.173, 133.179, 150.141, 150.161, 184.1804; Europe listed; UK approved
Manuf./Distrib.: Browning; Int'l. Sourcing; Jungbunzlauer

Potassium sodium tartrate tetrahydrate
CAS 6381-59-5; EINECS 205-698-2, 206-156-8; EEC E337
Synonyms: Sodium potassium tartrate; Rochelle salt; Seignette salt
Empirical: $C_4H_4KNaO_6 \cdot 4H_2O$
Formula: $KOCOCH(OH)CH(OH)COONa \cdot 4H_2O$
Properties: Colorless cryst. or wh. cryst. powd., cooling saline taste; sol. in 0.9 part water; almost insol. in alcohol; m.w. 282.23; dens. 1.79; m.p. 70-80 C; loses 3 H_2O @ 100 C, becomes anhyd. @ 130-140 C, dec. @ 220 C; pH 7-8
Precaution: Incompat. with acids, calcium or lead salts, magnesium sulfate, silver nitrate
Toxicology: Heated to decomp., emits acrid smoke and irritating fumes
Uses: Direct food additive, buffer, emulsifier, pH control agent, sequestrant; used in cheeses, jams, jellies
Regulatory: FDA 21CFR §184.1804, GRAS; USDA 9CFR §318.7
Manuf./Distrib.: Lohmann; Pfizer Spec. Chem.

Potassium sorbate
OAO 590-00-1, 24004-01-5; EINECS 240-070-1, EEC E202
Synonyms: 2,4-Hexadienoic acid potassium salt; Sorbic acid potassium salt; Potassium 2,4-hexadienoate
Empirical: $C_6H_7O_2 \cdot K$
Formula: $CH_3CH=CHCH=CHCOOK$
Properties: Wh. cryst., cryst. powd., pellets; sol. 58.2% in water @ 20 C, 6.5% in alcohol @ 20 C; m.w. 150.23; dens. 1.363 (25/20 C); m.p. 270 C (dec.)
Toxicology: LD50 (oral, rat) 4920 mg/kg; moderately toxic by intraperitoneal route; mildly toxic by ingestion; mutagenic data; heated to decomp., emits toxic fumes of K_2O
Uses: Mold retardant, preservative in baked goods, beverages, cakes, cheese, fish, fruit juice, margarine, pickled goods, salad dressings, fresh salad, wine; migrating to foods from paper/paperboard
Usage level: 0.05-0.5%; limitation 0.1% (alone), 0.2% (with its salts), 300 mg/100 gal (wine); ADI 0-2.5 mg/kg (EEC)
Regulatory: FDA 21CFR §133, 150.141. 150.161, 166.110, 182.90, 182.3640, GRAS; USDA 9CFR §318.7; BATF 27CFR §240.1051; GRAS (FEMA), CIR approved; Japan approved with limitations; JSCI approved 0.5% max.; Europe listed 0.8% max.
Manuf./Distrib.: Browning; Consolidated Flavor; Daicel; Gist-brocades Food Ingreds.; Int'l. Sourcing; Jungbunzlauer; Ueno
Trade names: Eastman Potassium Sorbate; Tristat K
Trade names containing: Brem; Coffee Conc.; Diet Imperial® Spread; Dry Beta Carotene Beadlets Yellow Type 2.4-S No. 653800100; FishGard® FP-55; FloAm® 221; Golden Croissant/Danish™ Margarine; GP-1200; Moldban; Myvatex® 40-06S; Polypro 2000® CF 45%; Polypro 2000® UF 45%; Polypro 5000® CF or SF 45%; Polypro 5000® UF 45%; Summit®

Potassium stearate
CAS 593-29-3; EINECS 209-786-1
Synonyms: Octadecanoic acid, potassium salt; Stearic acid, potassium salt
Definition: Potassium salt of stearic acid
Empirical: $C_{18}H_{35}O_2 \cdot K$
Formula: $CH_3(CH_2)_{16}COOK$
Properties: White powd., sl. fatty odor; slowly sol. in cold water; readily sol. in hot water, alcohol; m.w. 322.57

Potassium sulfate

Toxicology: Heated to decomp., emits acrid smoke and irritating fumes
Uses: Masticatory substance in chewing gum base; anticaking agent, binder, emulsifier; defoamer in beet sugar and yeast processing; stabilizer migrating from food pkg.
Regulatory: FDA 21CFR §172.615, 172.863, 173.340, 175.105, 175.300, 176.170, 176.200, 176.210, 177.1200, 177.2260, 177.2600, 177.2800, 178.3910, 179.45, 181.22, 181.29
Trade names containing: Aldo® MSD FG; Aldo® MSD KFG

Potassium sulfate

CAS 7778-80-5; EINECS 231-915-5; EEC E515
Synonyms: Sulfuric acid, dipotassium salt; Potassium sulfate (2:1)
Classification: Inorganic salt
Empirical: K_2O_4S
Formula: K_2SO_4
Properties: Colorless to wh. cryst. or cryst. powd., bitter saline taste; sol. in water; insol. in alcohol; m.w. 174.26; dens. 2.66; m.p. 1067 C
Toxicology: LD50 (oral, rat) 6600 mg/kg; mod. toxic to humans by ingestion; ingesting large doses causes severe GI effects; heated to decomp., emits toxic fumes of K_2O and SO_x
Uses: Miscellaneous and general purpose food additive, flavoring agent, adjuvant; water corrective; salt substitute for dietetic use; prod. of caramel
Usage level: Limitation 0.015% (nonalcoholic beverages)
Regulatory: FDA 21CFR §184.1643, GRAS; Europe listed; UK approved
Manuf./Distrib.: Heico

Potassium sulfate (2:1). *See* Potassium sulfate

Potassium sulfite

CAS 10117-38-1; EINECS 233-321-1
Synonyms: Sulfurous acid dipotassium salt; Sulfurous acid potassium salt
Classification: Inorganic salt
Empirical: $O_3S \cdot 2K$
Formula: K_2SO_3
Properties: Wh. cryst. or gran. powd., odorless; sol. in water; sl. sol. in alcohol; m.w. 158.26
Toxicology: Heated to decomp., emits toxic fumes of K_2O and SO_x
Uses: Preservative; antioxidant; prod. of caramel
Regulatory: GRAS

Potassium trichlorophenate

Empirical: $Cl_3C_6H_2OH$
Properties: dens. 1.3; f.p. -9 C
Uses: Slime control agent used in mfg. of paper/paperboard contacting food
Regulatory: FDA 21CFR §181.30

Potassium triphosphate. *See* Potassium tripolyphosphate

Potassium tripolyphosphate

CAS 13845-36-8; EINECS 237-574-9; EEC E450b
Synonyms: KTPP; Pentapotassium tripolyphosphate; Potassium triphosphate; Pentapotassium triphosphate; Triphosphoric acid pentapotassium salt
Classification: Inorganic salt
Empirical: $K_5O_{10}P_3$
Formula: $K_5P_3O_{10}$
Properties: Wh. cryst. solid; hygroscopic; sol. in water; m.w. 448.4; dens. 2.54; m.p. 620-640 C
Toxicology: Nuisance dust
Uses: Emulsifier, suspending agent, texturizer, buffer, sequestrant, stabilizer; in chewing gum, meat and poultry prods.; moisture binder in meat; boiler water additive for food contact
Usage level: Limitation 5% (pickle for meat prods.), 0.5% (meat prods.), 0.5% (total poultry); ADI 0-70 mg/kg (EEC, total phosphates)
Regulatory: FDA 21CFR §173.310, 175.105, 182.1810, GRAS; USDA 9CFR §318.7 (with limitation), 381.147 (limitation 0.5% of total poultry prod.); Europe listed
Manuf./Distrib.: Albright & Wilson Am.; Browning; FMC

Potato starch

CAS 9005-25-8
Definition: Natural substance obtained from potatoes, *Solanum tuberosum*, contg. amylose and amylopectin
Uses: Migrating from cotton in dry food pkg.
Regulatory: FDA 21CFR §175.105, 178.3520, 182.70

Pot marigold. *See* Calendula
Povidone. *See* PVP
Powdered skim milk. *See* Nonfat dry milk

PPG-5-buteth-7
    CAS 9038-95-3 (generic); 9065-63-8 (generic); 74623-31-7
    Synonyms: POE (7) POP (5) monobutyl ether; POP (5) POE (6) monobutyl ether
    Definition: Polyoxypropylene, polyoxyethylene ether of butyl alcohol
    Empirical: $(C_7H_{14}O_2 \cdot C_6H_{12}O_2)_x$
    Formula: $C_4H_9(OCH_3CHCH_2)_x(OCH_2CH_2)_yOH$, avg. x = 5, avg. y = 7
    Uses: Defoamer; boiler water additive
    Regulatory: FDA 21CFR §173.310, 175.105, 176.210, 178.3570
    Trade names: Pluracol® W170

PPG-12-buteth-16
    CAS 9038-95-3 (generic); 9065-63-8 (generic); 74623-31-7
    Synonyms: POE (16) POP (12) monobutyl ether; POP (12) POE (16) monobutyl ether
    Definition: Polyoxypropylene, polyoxyethylene ether of butyl alcohol
    Empirical: $(C_7H_{14}O_2 \cdot C_6H_{12}O_2)_x$
    Formula: $C_4H_9(OCH_3CHCH_2)_x(OCH_2CH_2)_yOH$, avg. x = 12, avg. y = 16
    Uses: Defoamer for food processing; boiler water additive
    Regulatory: FDA 21CFR §173.310, 175.105, 176.210, 178.3570
    Trade names: Pluracol® W660

PPG-20-buteth-30
    CAS 9038-95-3 (generic); 9065-63-8 (generic)
    Synonyms: POE (30) POP (20) monobutyl ether; POP (20) POE (30) monobutyl ether
    Definition: Polyoxypropylene, polyoxyethylene ether of butyl alcohol
    Empirical: $(C_7H_{14}O_2 \cdot C_6H_{12}O_2)_x$
    Formula: $C_4H_9(OCH_3CHCH_2)_x(OCH_2CH_2)_yOH$, avg. x = 20, avg. y = 30
    Uses: Defoamer for food processing
    Regulatory: FDA 21CFR §173.310, 175.105, 176.210, 178.3570
    Trade names: Pluracol® W2000

PPG-20-buteth-35
    CAS 9038-95-3 (generic); 9065-63-8 (generic)
    Synonyms: POE (35) POP (28) monobutyl ether; POP (28) POE (35) monobutyl ether
    Definition: Polyoxypropylene, polyoxyethylene ether of butyl alcohol
    Empirical: $(C_7H_{14}O_2 \cdot C_6H_{12}O_2)_x$
    Formula: $C_4H_9(OCH_3CHCH_2)_x(OCH_2CH_2)_yOH$, avg. x = 28, avg. y = 35
    Uses: Defoamer for food processing
    Regulatory: FDA 21CFR §173.310, 175.105, 176.210, 178.3570
    Trade names: Pluracol® W3520N

PPG-33-buteth-45
    CAS 9038-95-3 (generic); 9065-63-8 (generic)
    Synonyms: POE (45) POP (33) monobutyl ether; POP (33) POE (45) monobutyl ether
    Definition: Polyoxypropylene, polyoxyethylene ether of butyl alcohol
    Empirical: $(C_7H_{14}O_2 \cdot C_6H_{12}O_2)_x$
    Formula: $C_4H_9(OCH_3CHCH_2)_x(OCH_2CH_2)_yOH$, avg. x = 33, avg. y = 45
    Uses: Defoamer for food processing; boiler water additive
    Regulatory: FDA 21CFR §173.310, 173.340, 175.105, 176.210, 178.3570
    Trade names: Macol® 5100; Pluracol® W5100N

Precipitated calcium carbonate. *See* Calcium carbonate
Precipitated calcium phosphate. *See* Calcium phosphate tribasic
Precipitated chalk. *See* Calcium carbonate
Precipitated silica. *See* Silica, hydrated
Prenol. *See* 3-Methyl-2-buten-1-ol
Primary ammonium phosphate. *See* Ammonium phosphate
Proline. *See* L-Proline

L-Proline
    CAS 147-85-3; EINECS 205-702-2
    Synonyms: Proline; 2-Pyrrolidine carboxylic acid
    Classification: Amino acid

Propanal

Empirical: $C_5H_9NO_2$
Formula: $(CH_2)_3NHCHCOOH$
Properties: Colorless or wh. cryst. or cryst. powd.; sol. in alcohol, water; insol. in ether; m.w. 115.13; m.p. 220-222 C with dec.
Toxicology: Nonirritating
Uses: Dietary supplement, nutrient; flavoring
Regulatory: FDA 21CFR §172.320 (4.2% max.); Japan approved
Manuf./Distrib.: Nippon Rikagakuyakuhin; Penta Mfg.

Propanal. *See* Propionaldehyde

Propane
CAS 74-98-6; EINECS 200-827-9
Synonyms: Dimethylmethane; Propyl hydride
Classification: Hydrocarbon
Empirical: C_3H_8
Formula: $CH_3CH_2CH_3$
Properties: Colorless gas, natural gas odor; easily liquefied under pressure at R.T.; noncorrosive; sol. in ether, alcohol; sl. sol. in water; m.w. 44.09; dens. 0.513 (0 C, as liq.), 1.56 (0 C, as vapor); m.p. -188 C; b.p. -42.5 C; f.p. -189.9 C; flash pt. -156
Precaution: Flamm.; autoignit. temp. 467 C; dangerous fire risk; explosive limits in air 2.4-9.5%; reactive with oxidizers
Toxicology: Asphyxiant; narcotic in high concs.; TWA 1000 ppm; heated to decomp., emits acrid smoke and irritating fumes
Uses: Direct food additive, propellant, aerating agent, gas; for foamed and sprayed food prods.
Regulatory: FDA 21CFR §173.350, 184.1655, GRAS
Manuf./Distrib.: Aldrich

1-Propanecarboxylic acid. *See* n-Butyric acid
Propanedioic acid diethyl ester. *See* Diethyl malonate
1,2-Propanediol. *See* Propylene glycol
Propane-1,2-diol. *See* Propylene glycol
Propane-1,2-diol alginate. *See* Propylene glycol alginate
Propane-1,2-diol esters of fatty acids. *See* Propylene glycol esters of fatty acids
1-Propanethiol. *See* Propyl mercaptan
1,2,3-Propanetriol. *See* Glycerin
1,2,3-Propanetriol octadecanoate. *See* Glyceryl stearate
1,2,3-Propanetriol triacetate. *See* Triacetin
1,2,3-Propanetriol tridodecanoate. *See* Trilaurin
1,2,3-Propanetriol trioctadecanoate. *See* Tristearin
1,2,3-Propanetriol trioctanoate. *See* Tricaprylin
1,2,3-Propanetriol tritetradecanoate. *See* Trimyristin
Propanoic acid. *See* Propionic acid
Propanoic acid butyl ester. *See* n-Butyl propionate
Propanoic acid, calcium salt. *See* Calcium propionate
Propanoic acid ethyl ester. *See* Ethyl propionate
Propanoic acid, 2-hydroxy-, monopotassium salt. *See* Potassium lactate
Propanoic acid methyl ester. *See* Methyl propionate
Propanoic acid propyl ester. *See* Propyl propionate
Propanoic acid sodium salt. *See* Sodium propionate
1-Propanol. *See* Propyl alcohol
2-Propanol. *See* Isopropyl alcohol
1,2,3-Propanol tridecanoate. *See* Tricaprin
2-Propanone. *See* Acetone
Propellant C-318. *See* Octafluorocyclobutane
2-Propenal. *See* Acrolein
2-Propenamide, homopolymer. *See* Polyacrylamide
2-Propenamide, polymer with 2-propenoic acid, sodium salt. *See* Acrylamide/sodium acrylate copolymer
2-Propenenitrile polymer with ethenylbenzene. *See* Acrylonitrile-styrene copolymer
2-Propene-1-thiol. *See* Allyl mercaptan
1-Propene-1,2,3-tricarboxylic acid. *See* Aconitic acid
1,2,3-Propenetricarboxylic acid. *See* Aconitic acid
2-Propenoic acid, 2-methyl-, homopolymer, sodium salt. *See* Sodium polymethacrylate
2-Propenoic acid, sodium salt, polymer with 2-propenamide. *See* Acrylamide/sodium acrylate copolymer

2-Propenylacrylic acid. *See* Sorbic acid
p-Propenylanisole. *See* Anethole
Propenyl cinnamate. *See* Allyl cinnamate

Propenylguaethol
CAS 94-86-0; FEMA 2922
Synonyms: Isosafroeugenol; Hydroxy methyl anethol; 6-Ethoxy-m-anol; 1-Ethoxy-2-hydroxy-4-propenylbenzene
Empirical: $C_{11}H_{14}O_2$
Properties: Wh. cryst. powd., vanilla odor; sol. in fixed oils; insol. in water; m.w. 178.25; m.p. 85-86 C; flash pt. > 212 F
Precaution: Combustible liq.
Toxicology: LD50 (oral, rat) 2400 mg/kg; mod. toxic by ingestion; heated to decomp, emits acrid smoke and fumes
Uses: Synthetic flavoring agent
Usage level: 5.9 ppm (nonalcoholic beverages), 6.3 ppm (ice cream, ices), 20 ppm (candy, baked goods), 2.5 ppm (gelatins, puddings)
Regulatory: FDA 21CFR §172.515
Manuf./Distrib.: Chart

Propenyl guaiacol. *See* Methyl isoeugenol
4-Propenylguaiacol. *See* Isoeugenol
2-Propenyl heptanoate. *See* Allyl heptanoate
2-Propenyl hexanoate. *See* Allyl caproate
2-Propenyl isovalerate. *See* Allyl isovalerate
2-Propenyl 3-methylbutanoate. *See* Allyl isovalerate
2-Propenyl phenylacetate. *See* Allyl phenylacetate
4-Propenylveratrole. *See* Methyl isoeugenol

Proplonaldehyde
CAS 123-38-6; EINECS 204-623-0; FEMA 2923
Synonyms: Aldehyde C-3; Propanal; Propionic aldehyde; Methylacetaldehyde; Propylaldehyde
Empirical: C_3H_6O
Formula: Cl I₃Cl I₂Cl IO
Properties: Colorless mobile liq., suffocating odor; sol. in 5 vols water; misc. with alcohol, ether; m.w. 58.08; dens. 0.807 (20/4 C); m.p. -81 C; b.p. 47-49 C; flash pt. -40 C; ref. index 1.362 (20 C)
Precaution: Highly flamm.; dangerous fire hazard exposed to heat or flame; reacts vigorously with oxidizers; explosive limits 2.9-17%
Toxicology: LD50 (oral, rat) 1.4 g/kg; mod. toxic by skin contact, ingestion, subcut. routes; mildly toxic by inh.; skin, severe eye irritant; heated to decomp., emits acrid smoke and irritating fumes
Uses: Synthetic flavoring agent
Usage level: 3.9 ppm (nonalcoholic beverages), 12 ppm (ice cream, ices), 11 ppm (candy), 13 ppm (baked goods)
Regulatory: FDA 21CFR §172.515
Manuf./Distrib.: Aldrich

Propione. *See* Diethyl ketone

Propionic acid
CAS 79-09-4; EINECS 201-176-3; EEC E280
Synonyms: Methylacetic acid; Propanoic acid; Ethylformic acid
Classification: Acid
Empirical: $C_3H_6O_2$
Formula: C_2H_5COOH
Properties: Oily liq., sl. pungent rancid odor; misc. in water, alcohol, ether, chloroform; m.w. 74.09; dens. 0.998 (15/4 C); visc. 1.020 cp (15 C); m.p. -21.5 C; b.p. 141. C (760 mm); flash pt. (OC) 58 C; ref. index 1.3862; surf. tension 27.21 dynes/cm (15 C)
Precaution: Corrosive; highly flamm. exposed to heat, flame, oxidizers; incompat. with $CaCl_2$; dec. at high temp.
Toxicology: TLV:TWA 10 ppm; LD50 (oral, rat) 3500 mg/kg; poison by intraperitoneal route; mod. toxic by ingestion, skin contact, IV route; irritant to eye, skin; heated to decomp., emits acrid smoke and irritating fumes
Uses: Direct food additive, antimicrobial agent, mold inhibitior, preservative, flavoring agent
Usage level: 1% (foods); limitation 0.32% (flour in wh. bread/rolls), 0.38% (whole wheat), 0.3% (cheese prods.)
Regulatory: FDA 21CFR §172.515, 184.1081, GRAS; GRAS (FEMA), EPA reg.; Japan restricted to flavoring use, limitation with sorbic acid 3 g/kg total; Europe listed; UK approved

Propionic acid, calcium salt

Manuf./Distrib.: Eastman; Hays; Honeywill & Stein; Lohmann; Penta Mfg.; Tournay; Unipex; Van Waters & Rogers
Trade names containing: CP-2600; CP-3650; Myvatex® 25-07K; Myvatex® SSH

Propionic acid, calcium salt. See Calcium propionate
Propionic acid sodium salt. See Sodium propionate
Propionic aldehyde. See Propionaldehyde

Propyl acetate
CAS 109-60-4; EINECS 203-686-1; FEMA 2925
Synonyms: Acetic acid n-propyl ester; n-Propyl acetate
Definition: Ester of propyl alcohol and acetic acid
Empirical: $C_5H_{10}O_2$
Formula: $CH_3COOCH_2CH_2CH_3$
Properties: Liq., pear-like odor; misc. with alcohol, ether; m.w. 102.14; dens. 0.887 (20/20 C); m.p. -92 C; b.p. 99-102 C; flash pt. (CC) 14 C; ref. index 1.384 (20 C)
Precaution: Highly flamm.
Toxicology: LD50 (oral, rat) 9370 mg/kg; may be irritating to skin, mucous membranes; narcotic in high concs.
Uses: Synthetic flavoring agent
Usage level: 4 ppm (nonalcoholic beverages), 16 ppm (ice cream), 12 ppm (candy), 14 ppm (baked goods)
Regulatory: FDA 21CFR §172.515, 177.1200; GRAS (FEMA)
Manuf./Distrib.: Aldrich

2-Propyl acetate. See Isopropyl acetate
n-Propyl acetate. See Propyl acetate
Propylacetic acid. See n-Valeric acid
β-Propyl acrolein. See 2-Hexenal

Propyl alcohol
CAS 71-23-8; EINECS 200-746-9; FEMA 2928
Synonyms: 1-Propanol; Propylic alcohol; Albacol; Optal; n-Propyl alcohol; Alcohol C-3
Classification: Aliphatic alcohol
Empirical: C_3H_8O
Formula: $CH_3CH_2CH_2OH$
Properties: Liq., alcoholic and sl. stupefying odor; misc. with water, alcohol, ether; m.w. 60.10; dens. 0.804 (20/4 C); m.p. -127 C; b.p. 97-98 C; flash pt. 23 C; ref. index 1.385 (20 C)
Precaution: Highly flamm.
Toxicology: LD50 (oral, rat) 1.87 g/kg; mildly irritating to eyes, mucous membranes; depressant action
Uses: Synthetic flavoring agent
Usage level: 0.50-5.0 ppm (nonalcoholic beverages), 0.50 ppm (ice cream, ices, candy), 0.65 ppm (baked goods)
Regulatory: FDA 21CFR §172.515, 175.105, 177.1200, 573.880
Manuf./Distrib.: Aldrich

n-Propyl alcohol. See Propyl alcohol
Propylaldehyde. See Propionaldehyde
γ-Propylallyl alcohol. See 2-Hexenol

p-Propyl anisole
FEMA 2930
Synonyms: Dihydroanethole; 1-Methoxy-4-n-propylbenzene
Empirical: $C_{10}H_{14}O$
Properties: Colorless to pale yel. oily liq., anise-type odor; m.w. 150.22; dens. 0.94718; b.p. 212-213 C; ref. index 1.5025-1.5055
Uses: Synthetic flavoring agent
Usage level: 4.3 ppm (nonalcoholic beverages), 9.9 ppm (ice cream, ices), 64 ppm (candy), 67 ppm (baked goods)
Regulatory: FDA 21CFR §172.515

Propyl benzoate
CAS 2315-68-6; FEMA 2931
Synonyms: n-Propyl benzoate
Empirical: $C_{10}H_{12}O_2$
Properties: Colorless oily liq., balsamic nutty odor, sweet fruity nut-like taste; sol. in alcohol; insol. in water; m.w. 164.21; dens. 1.026; b.p. 230-231 C; m.p. -51 to -52 C; flash pt. 98 C; ref. index 1.5100
Uses: Synthetic flavoring

Usage level: 11 ppm (nonalcoholic beverages), 44 ppm (ice cream, ices), 33 ppm (candy, baked goods)
Regulatory: FDA 21CFR §172.515

n-Propyl benzoate. *See* Propyl benzoate

Propyl butyrate
CAS 105-66-8; EINECS 203-320-0; FEMA 2934
Synonyms: Butanoic acid propyl ester
Empirical: $C_7H_{14}O_2$
Formula: $CH_3CH_2CH_2COOCH_2CH_2CH_3$
Properties: Colorless liq.; sl. sol. in water; misc. with alcohol, ether; m.w. 130.19; dens. 0.87 (20/4 C); m.p. -95 C; b.p. 142-144 C; flash pt. 37 C; ref. index 1.400 (20 C)
Precaution: Flamm.
Toxicology: LD50 (oral, rat) 15,000 mg/kg
Uses: Synthetic flavoring
Usage level: 6.8 ppm (nonalcoholic beverages), 4.6 ppm (ice cream, ices), 24 ppm (candy), 16 ppm (baked goods)
Regulatory: FDA 21CFR §172.515
Manuf./Distrib.: Aldrich

Propyl carbinol. *See* Butyl alcohol

Propyl cinnamate
Uses: Synthetic flavoring
Regulatory: FDA 21CFR §172.515

Propyl disulfide
CAS 629-19-6; EINECS 211-079-8; FEMA 3228
Synonyms: Dipropyl disulfide
Empirical: $C_6H_{14}S_2$
Formula: $CH_3CH_2CH_2SSCH_2CH_2CH_3$
Properties: m.w. 150.31; dens. 0.957 (20/4 C); b.p. 195 C; flash pt. 64 C; ref. index 1.498 (20 C)
Uses: Synthetic flavoring
Regulatory: FDA 21CFR §172.515
Manuf./Distrib.: Aldrich

Propylenedicarboxylic acid. *See* Itaconic acid (polymerized)

Propylene glycol
CAS 57-55-6; EINECS 200-338-0
Synonyms: 1,2-Propanediol; Propane-1,2-diol; 1,2-Dihydroxypropane; Methyl glycol
Classification: Aliphatic alcohol
Empirical: $C_3H_8O_2$
Formula: $CH_3CHOHCH_2OH$
Properties: Colorless visc. liq., odorless; hygroscopic; sol. in essential oils; misc. with water, acetone, chloroform; m.w. 76.11; dens. 1.0362 (25/25 C); b.p. 188.2 C; flash pt. (OC) 210 F
Precaution: Combustible exposed to heat or flame; reactive with oxidizers; explosive limits 2.6-12.6%
Toxicology: LD50 (oral, rat) 25 ml/kg; eye/human skin irritant; sl. toxic by ingestion, IP, IV, subcutaneous routes; human systemic effects; experimental teratogenic, reproductive effects; mutagenic data; heated to decomp., emits acrid smoke and irritating fumes
Uses: Anticaking agent, antioxidant, dough strengthener, emulsifier, flavor agent, formulation aid, humectant, processing aid, solvent, vehicle, stabilizer, thickener, surface-active agent, texturizer, wetting agent, clarifying agent, hog/poultry scald
Usage level: Limitation 5% (alcoholic beverages), 24% (confections), 2.5% (frozen dairy prods.), 97% (seasonings), 5% (nuts), 2% (other food), 40 ppm (wine); ADI 25 mg/kg (FAO/WHO)
Regulatory: FDA 21CFR §169.175, 169.176, 169.177, 169.178, 169.180, 169.181, 175.300, 177.2600, 178.3300, 184.1666, 582.4666, GRAS; USDA 9CFR §318.7, 381.147; BATF 27CFR §240.1051; EPA reg., approved for some drugs; Japan approved with limitations; Europe listed
Manuf./Distrib.: Ashland; Crompton & Knowles; Ellis & Everard; K&K Greeff; Hays; Honeywill & Stein; Meer; Norman Fox; Penta Mfg.; Todd's; R C Treatt; Van Waters & Rogers
Trade names: Adeka Propylene Glycol (P)
Trade names containing: Adeka Menjust; Amisol™ 406-N; Annatto ECA-1; Annatto ECH-1; Annatto ECPC; Annatto OS #2894; Annatto OS #2922; Annatto OS #2923; Annatto Liq. #3968, Acid Proof; Annatto/Turmeric WML-1; Annatto/Turmeric WMP-1; Arlacel® 186; Atlas 3000; Atmos® 300; Atmos® 2462; Bake-Well Bun Release; Brem; Centromix® E; Centrophil® M; Centrophil® W; Dur-Em® 300; Dur-Em® 300K; Embanox® 3; Embanox® 4; Embanox® 6; Moldban; Nipanox® S-1; Nipanox® Special; Poly E; Pristene®

Propylene glycol alginate

181; Pristene® 189; Sporban Regular; Sporban Special; Super-Cel® Specialty Cake Shortening; Sustane® 3; Sustane® 7; Sustane® 7G; Sustane® 8; Sustane® 15; Sustane® 16; Sustane® 18; Sustane® 20; Sustane® 20-3; Sustane® 20A; Sustane® 20B; Sustane® 21; Sustane® 27; Sustane® 31; Sustane® CA; Sustane® Q; Sustane® S-1; Sustane® T-6; Sustane® W-1; Sustane® W; Tenox® 2; Tenox® 6; Tenox® 7; Tenox® 20; Tenox® 20A; Tenox® 20B; Tenox® 21; Tenox® 22; Tenox® 24; Tenox® 25; Tenox® 26; Tenox® 27; Tenox® A; Tenox® R; Tenox® S-1; Turmeric DP-300; Turmeric WMPE-80; Turmeric WMPE-150; Unibix AP (Acid Proof); Vanall; Vanall (K)

Propylene glycol alginate
CAS 9005-37-2; EEC E405
Synonyms: Hydroxypropyl alginate; Alginic acid, ester with 1,2-propanediol; Propane-1,2-diol alginate
Definition: Mixture of propylene glycol esters of alginic acid
Empirical: $(C_9H_{14}O_7)_8$
Properties: White powd., odorless; sol. in water, dilute organic acids; m.w. 1873.6
Toxicology: LD50 (oral, rat) 7200 mg/kg; mildly toxic by ingestion; heated to decomp., emits acrid smoke and irritating fumes
Uses: Emulsifier, flavoring adjuvant, formulation aid, stabilizer, surfactant, thickener, solvent, defoamer; used in frozen dairy desserts, baked goods, cheeses, fats/oils, gelatins, puddings, gravies, jams, jellies, condiments, seasonings/flavors
Usage level: Limitation 0.5% (frozen dairy desserts), 0.5% (baked goods), 0.9% (cheese), 1.1% (fats/oils), 0.6% (gelatins, puddings), 0.5% (gravies), 0.4% (jams/jellies), 0.6% (condiments), 1.7% (seasonings), 0.3% (other foods); ADI 0-25 mg/kg (EEC)
Regulatory: FDA 21CFR §133.133, 133.134, 133.162, 133.178. 133.179, 172.210, 172.820, 172.858, 173.340, 176.170, GRAS; Japan approved (1% max.); Europe listed; UK approved
Manuf./Distrib.: Kelco Int'l.; Meer; Pronova
Trade names: Colloid 602; Concentrated Dariloid® KB; Dricoid® KB; Dricoid® KBC; Kelcoloid® D; Kelcoloid® DH; Kelcoloid® DO; Kelcoloid® DSF; Kelcoloid® HVF; Kelcoloid® LVF; Kelcoloid® O; Kelcoloid® S; Kimiloid HV; Kimiloid MV; Kimiloid NLS-K; Manucol Ester B; Manucol Ester E/PL; Manucol Ester E/RK; Manucol Ester M; Propylene Glycol Alginate HV; Protanal Ester BI; Protanal Ester CF; Protanal Ester H; Protanal Ester L; Protanal Ester L-25 A/H; Protanal Ester PVH-A; Protanal Ester SD-H
Trade names containing: Ches® 500; Propylene Glycol Alginate LV FCC

Propylene glycol dicaprylate/dicaprate
CAS 9062-04-8; 58748-27-9; 68583-51-7; 68988-72-7
Synonyms: Decanoic acid, 1-methyl-1,2-ethanediyl ester mixed with 1-methyl-1,2-ethanediyl dioctanoate
Definition: Mixture of the propylene glycol diesters of caprylic and capric acids
Uses: Food additive, emulsifier, stabilizer, defoaming agent; solv. for flavors, vitamins
Regulatory: FDA 21CFR §172.856, 173.340, 175.300, 176.170, 176.210, 177.2800
Trade names: Aldo® DC; Captex® 200; Hodag CC-22; Hodag CC-22-S; Lexol® PG-855; Lexol® PG-865; Liponate PC; Miglyol® 840; Neobee® 20; Neobee® 1054; Neobee® M-20

Propylene glycol dioctanoate
CAS 7384-98-7; 56519-71-2
Synonyms: 2-Ethylhexanoic acid, 1-methyl-1,2-ethanediyl ester; Octanoic acid, 1,3-propanediyl ester
Definition: Diester of propylene glycol and 2-ethylhexanoic acid
Empirical: $C_{19}H_{36}O_4$
Formula: $CH_3(CH_2)_3CHCH_3CH_2COOCH_2CHCH_3OCOCHCH_2CH_3(CH_2)_3CH_3$
Properties: Sol. in alcohol, min. oil
Uses: Carrier for essential oils, flavors; vehicle for vitamins, nutritional prods.
Trade names: Captex® 800; Lexol® PG-800

Propylene glycol esters of fatty acids
EEC E477
Synonyms: Propane-1,2-diol esters of fatty acids
Properties: Clear liq. or wh. to yel. beads or flakes, bland odor and taste; sol. in alcohol, ethyl acetate, chloroform; insol. in water
Toxicology: Heated to decomp., emits acrid smoke and irritating fumes
Uses: Emulsifier, stabilizer for beet sugar, cake batter/icing/shortening, margarine, oils, rendered poultry fat, whipped toppings, yeast
Usage level: Limitation 2% (margarine)
Regulatory: FDA 21CFR §172.856, 172.860, 172.862, 173.340; USDA 9CFR §318.7, 381.147; Europe, UK approved; ADI 0-25 mg kg⁻¹ body wt. (EEC)
Trade names: Homotex PS-200
Trade names containing: Atmos® 659 K; Bunge Special Mix Shortening; Cake Mix 96; Durpro® 107; EC-25®; Myvatex® 40-06S K; Myvatex® Texture Lite® K

Propylene glycol laurate
CAS 142-55-2; 27194-74-7; EINECS 205-542-3
Synonyms: Dodecanoic acid, 2-hydroxypropyl ester; Dodecanoic acid, monoester with 1,2-propanediol; Propylene glycol monolaurate
Definition: Ester of propylene glycol and lauric acid
Empirical: $C_{15}H_{30}O_3$
Formula: $CH_3(CH_2)_{10}COOCH_2CHCH_3OH$
Uses: Food additive, emulsifier, stabilizer, defoamer
Regulatory: FDA 21CFR §172.856, 173.340, 175.105, 175.300, 176.170, 176.210, 177.2800
Trade names: Cithrol PGML N/E; Emalex PGML; Hodag PGML

Propylene glycol laurate SE
Uses: Emulsifier, stabilizer
Trade names: Cithrol PGML S/E

Propylene glycol monodistearate
Definition: Mixt. of propylene glycol mono- and diesters of stearic and palmitic acids
Properties: Wh. beads or flakes, bland odor and taste; insol. in water; sol. in alcohol, ethyl acetate, chloroform, other chlorinated hydrocarbons
Uses: Food emulsifier, stabilizer for whipped sponge mixts., toppings, shortenings
Trade names: Tegomuls® P 411

Propylene glycol monolaurate. *See* Propylene glycol laurate
Propylene glycol monoricinoleate. *See* Propylene glycol ricinoleate
Propylene glycol monostearate. *See* Propylene glycol stearate

Propylene glycol oleate
CAS 1330-80-9; EINECS 215-549-3
Synonyms: 9-Octadecenoic acid, monoester with 1,2-propanediol
Definition: Ester of propylene glycol and oleic acid
Empirical: $C_{21}H_{40}O_3$
Formula: $CH_3(CH_2)_7CH=CH(CH_2)_7COOCH_2CHCH_3OH$
Uses: Food additive, emulsifier, stabilizer, defoamer
Regulatory: FDA 21CFR §172.856, 173.340, 175.300, 176.210, 177.2800
Trade names: Cithrol PGMO N/E, Emalex PGO, G-2185

Propylene glycol oleate SE
Definition: Self-emulsifying grade of propylene glycol oleate that contains some sodium and/or potassium oleate
Uses: Emulsifier, stabilizer
Trade names: Cithrol PGMO S/E

Propylene glycol palmitate
Definition: Ester of propylene glycol and palmitic acid
Uses: Emulsifier
Trade names: Admul PGMP; G-2183

Propylene glycol ricinoleate
CAS 26402-31-3; EINECS 247-669-7
Synonyms: 12-Hydroxy-9-octadecenoic acid, monoester with 1,2-propanediol; Propylene glycol monoricinoleate
Definition: Ester of propylene glycol and ricinoleic acid
Empirical: $C_{21}H_{40}O_4$
Uses: Emulsifier, stabilizer
Trade names: Cithrol PGMR N/E

Propylene glycol ricinoleate SE
Uses: Emulsifier, stabilizer
Trade names: Cithrol PGMR S/E

Propylene glycol stearate
CAS 1323-39-3; EINECS 215-354-3
Synonyms: Propylene glycol monostearate; Octadecanoic acid, monoester with 1,2-propanediol
Definition: Ester of propylene glycol and stearic acid
Empirical: $C_{21}H_{42}O_3$
Formula: $CH_3(CH_2)_{16}COOCH_2CHCH_3OH$
Properties: Wh. to cream flakes, bland typ. odor; sol. in min. oil, IPM, oleyl alcohol; insol. in water, glycerin, propylene glycol; m.p. 35-38 C; sapon. no. 181-191

Propylene glycol stearate SE

Toxicology: Poison by intraperitoneal route
Uses: Food additive, emulsifier, stabilizer, defoamer
Usage level: Limitation 2% (margarine)
Regulatory: FDA 21CFR §172.856, 172.860, 172.862, 173.340, 175.105, 175.300, 176.170, 176.210, 177.2800; USDA 9CFR §318.7, 381.147; GRAS (FEMA)
Trade names: Admul PGMS; Aldo® PGHMS; Aldo® PGHMS KFG; Canamulse 55; Canamulse 70; Canamulse 90K; Cithrol PGMS N/E; Drewlene 10; Emalex PGMS; Emalex PGS; Hefti PMS-33; Hodag PGMS; Homotex PS-90; Lipal PGMS; Mazol® PGMS; Mazol® PGMSK; Myverol® P-06; PGMS 70; Promodan SP
Trade names containing: Aldo® PME; Gatodan 415; Myvatex® 3-50K; Myvatex® 40-06S; Myvatex® Texture Lite®; Vanlite

Propylene glycol stearate SE

Definition: Self-emulsifying grade of propylene glycol stearate containing some sodium and/or potassium stearate
Uses: Emulsifier, stabilizer
Trade names: Cithrol PGMS S/E

Propyl formate

CAS 110-74-7; EINECS 203-798-0; FEMA 2943
Synonyms: Formic acid propyl ester; n-Propyl formate; n-Propyl methanoate
Empirical: $C_4H_8O_2$
Formula: $HCOOC_3H_7$
Properties: Colorless liq., pleasant odor, bittersweet flavor; sol. in 45 parts water; misc. with alcohol, ether; m.w. 88.10; dens. 0.901 (20 C); m.p. -93 C; b.p. 81-82 C; flash pt. (CC) -3 C; ref. index 1.3771 (20 C)
Toxicology: LD50 (oral, rat) 3980 mg/kg
Uses: Synthetic flavoring
Usage level: 20 ppm (nonalcoholic beverages), 57 ppm (ice cream, ices), 65 ppm (candy), 85 ppm (baked goods)
Regulatory: FDA 21CFR §172.515
Manuf./Distrib.: Aldrich

n-Propyl formate. *See* Propyl formate
Propylformic acid. *See* n-Butyric acid

Propyl 2-furanacrylate

FEMA 2945
Synonyms: Propyl β-furylacrylate; Propyl-3-furylpropenoate
Definition: Ester of n-propanol and furanacrylic acid
Empirical: $C_{10}H_{12}O_3$
Properties: Colorless liq., lt. strawberry apple-/pear-like odor; sol. in alcohol; insol. in water; m.w. 180.21; dens. 1.0744; b.p. 236 C; ref. index 1.5229
Uses: Synthetic flavoring
Usage level: 3.0 ppm (nonalcoholic beverages),0.03 ppm (candy)
Regulatory: FDA 21CFR §172.515

Propyl β-furylacrylate. *See* Propyl 2-furanacrylate
Propyl-3-furylpropenoate. *See* Propyl 2-furanacrylate

Propyl gallate

CAS 121-79-9; EINECS 204-498-2; EEC E310
Synonyms: 3,4,5-Trihydroxybenzoic acid, n-propyl ester; n-Propyl 3,4,5-trihydroxybenzoate; Gallic acid propyl ester
Classification: Aromatic ester
Definition: Aromatic ester of propyl alcohol and gallic acid
Empirical: $C_{10}H_{12}O_5$
Formula: $(HO)_3C_6H_2COOCH_2CH_2CH_3$
Properties: Fine ivory powd. or crystals, odorless; insol. in water; sol. in alcohol, oils; m.w. 212.22; m.p. 147-149 C; b.p. dec. > 148 C
Precaution: Combustible exposed to heat or flame; reactive with oxidizers
Toxicology: LD50 (oral, rat) 3.8 g/kg; poison by ingestion and intraperitoneal route; experimental tumorigen, reproductive effects; mutagenic data; heated to decomp., emits acrid smoke and irritating fumes
Uses: Food and feed antioxidant, flavor and pkg. material; synergistic with BHA and BHT
Usage level: Limitation 0.003% (dry sausage), 0.01% (rendered animal fat), 0.02% (margarine), 0.01% (on fat content in poultry); ADI 0-0.5 mg/kg (EEC, total gallates)
Regulatory: FDA 21CFR §172.615, 175.125, 175.300, 181.22, 181.24 (0.005% migrating from food pkg.), 184.1660 (0.02% max. of fat or oil), GRAS; USDA 9CFR §318.7, 381.147; Japan approved (0.1 g/kg max.);

Europe listed; UK approved
Manuf./Distrib.: Aceto; Eastman; Nipa Labs; UOP
Trade names: Sustane® PG; Tenox® PG
Trade names containing: Embanox® 3; Embanox® 6; Nipanox® S-1; Nipanox® Special; Sustane® 3; Sustane® 7; Sustane® 7G; Sustane® 16; Sustane® S-1; Sustane® T-6; Sustane® W; Tenox® 2; Tenox® 6; Tenox® 7; Tenox® S-1

Propyl heptanoate
FEMA 2948
Uses: Synthetic flavoring
Regulatory: FDA 21CFR §172.515
Manuf./Distrib.: Aldrich

Propyl hexanoate
FEMA 2949
Uses: Synthetic flavoring
Regulatory: FDA 21CFR §172.515
Manuf./Distrib.: Aldrich

Propyl hydride. *See* Propane
Propyl 4-hydroxybenzoate. *See* Propylparaben
Propyl p-hydroxybenzoate. *See* Propylparaben
n-Propyl-4-hydroxybenzoate potassium salt. *See* Potassium propylparaben
Propyl-4-hydroxybenzoate, sodium salt. *See* Sodium propylparaben
Propylic alcohol. *See* Propyl alcohol

3-Propylidenephthalide
FEMA 2952
Uses: Synthetic flavoring
Regulatory: FDA 21CFR §172.515
Manuf./Distrib.: Aldrich

Propyl isobutyrate
Uses: Synthetic flavoring
Regulatory: FDA 21CFR §172.515

Propyl isovalerate
FEMA 2960
Synonyms: n-Propyl isovalerate; n-Propyl-β-methylbutyrate
Empirical: $C_8H_{16}O_2$
Properties: Colorless mobile liq., fruity odor, bittersweet flavor; sol. in alcohol; insol. in water; m.w. 144.21; dens. 0.8617; b.p. 156-157 C; ref. index 1.4031
Uses: Synthetic flavoring
Usage level: 5.0 ppm (nonalcoholic beverages), 16 ppm (ice cream, ices), 17 ppm (candy), 20 ppm (baked goods)
Regulatory: FDA 21CFR §172.515

n-Propyl isovalerate. *See* Propyl isovalerate

Propyl mercaptan
CAS 107-03-9; EINECS 203-455-5; FEMA 3521
Synonyms: 1-Propanethiol; Mercaptan C_3
Empirical: C_3H_8S
Formula: $CH_3CH_2CH_2SH$
Properties: Colorless mobile liq., cabbage-like odor; m.w. 76.16; dens. 0.840 (20/4 C); m.p. -113 C; b.p. 67-68 C; flash pt. -20 C; ref. index 1.438 (20 C)
Precaution: Highly flamm.
Uses: Synthetic flavoring
Regulatory: FDA 21CFR §172.515
Manuf./Distrib.: Aldrich

n-Propyl methanoate. *See* Propyl formate
n-Propyl-β-methylbutyrate. *See* Propyl isovalerate

Propylparaben
CAS 94-13-3; EINECS 202-307-7; EEC E216
Synonyms: Propyl p-hydroxybenzoate; Propyl 4-hydroxybenzoate; 4-Hydroxybenzoic acid, propyl ester; Propyl parahydroxybenzoate

Propylparaben, potassium salt

Classification: Organic ester
Definition: Ester of n-propyl alcohol and p-hydroxybenzoic acid
Empirical: $C_{10}H_{12}O_3$
Properties: Colorless crystals or white powd.; sl. sol. in boiling water; sol. in alcohol, ether, acetone; m.w. 180.22; m.p. 95-98 C
Toxicology: Poison by intraperitoneal route; moderately toxic by subcutaneous route; mildly toxic by ingestion; may cause asthma, rashes, hyperactivity
Uses: Synthetic flavoring, food preservative, antimicrobial agent, mold control in sausage casings; antimycotic migrating from food pkg.
Usage level: 0.02-1.0%; 0.1% max. in food; ADI 0-10 mg/kg (EEC)
Regulatory: FDA 21CFR §150.141, 150.161, 172.515, 181.22, 181.23, 184.1670; USA CIR approved, EPA reg.; Japan listed; Europe listed; UK approved
Manuf./Distrib.: Aceto; Allchem Industries; Ashland; Int'l. Sourcing; Kraft; E. Merck; Nipa Labs; Penta Mfg.
Trade names: Lexgard® P; Nipasol M
Trade names containing: Dry Beta Carotene Beadlets Yellow Type 2.4-S No. 653800100; Nipastat; Sporban Regular; Sporban Special

Propylparaben, potassium salt. *See* Potassium propylparaben
Propylparaben, sodium salt. *See* Sodium propylparaben
Propyl parahydroxybenzoate. *See* Propylparaben
Propyl phenethyl acetal. *See* Acetaldehyde phenethyl propyl acetal

α-Propylphenethyl alcohol

FEMA 2953
Synonyms: Benzylbutyl alcohol; Benzyl-n-propyl carbinol; 1-Phenyl-2-pentanol
Empirical: $C_{11}H_{16}O$
Properties: Colorless oily liq., mild green sweet odor; sol. in alcohol; almost insol. in water; m.w. 164.25; dens. 0.98; b.p. 247 C
Uses: Synthetic flavoring
Usage level: 1.0 ppm (nonalcoholic beverages), 5.0 ppm (ice cream, ices), 5.0 ppm (candy, gelatins, puddings)
Regulatory: FDA 21CFR §172.515

Propyl phenylacetate

CAS 4606-15-9; FEMA 2955
Synonyms: n-Propyl-α-toluate
Definition: Ester of n-propanol and phenylacetic acid
Empirical: $C_{11}H_{14}O_2$
Properties: Colorless liq., honey-like apricot-rose odor, sweet honey-like taste; sol. in alcohol; almost insol. in water; m.w. 178.23; dens. 0.990 (15.5 C); b.p. 253 C; ref. index 1.4955
Uses: Synthetic flavoring
Usage level: 0.30-1.0 ppm (nonalcoholic beverages), 0.30-1.5 ppm (ice cream, ices), 2.7 ppm (candy), 1.0-5.0 ppm (baked goods)
Regulatory: FDA 21CFR §172.515
Manuf./Distrib.: Aldrich

Propyl propionate

CAS 106-36-5; EINECS 203-389-7; FEMA 2958
Synonyms: Propanoic acid propyl ester; n-Propyl propionate
Empirical: $C_6H_{12}O_2$
Formula: $CH_3CH_2COOCH_2CH_2CH_3$
Properties: Liq., complex fruity odor; sol. in 200 parts water; misc. with alcohol, ether; m.w. 116.16; dens. 0.881 (20/4 C); m.p. -76 C; b.p. 120-122 C; flash pt. 22 C; ref. index 1.3935 (20 C)
Precaution: Flamm.
Uses: Synthetic flavoring
Usage level: 6.0 ppm (nonalcoholic beverages), 12 ppm (ice cream, ices), 25 ppm (candy, baked goods)
Regulatory: FDA 21CFR §172.515
Manuf./Distrib.: Aldrich

n-Propyl propionate. *See* Propyl propionate
n-Propyl-α-toluate. *See* Propyl phenylacetate
n-Propyl 3,4,5-trihydroxybenzoate. *See* Propyl gallate

Protease

CAS 9014-01-1; EINECS 232-752-2
Classification: Enzyme
Properties: Sol. in water; m.w. $\approx$ 27,000

Toxicology: Irritant
Uses: Enzyme which breaks down protein; used in food processing, protein hydrolysis
Regulatory: Canada, Japan approved
Manuf./Distrib.: PMP Fermentation Prods.; Solvay Enzymes
Trade names: AFP 2000; Bromelain 1:10; Bromelain Conc.; Fungal Protease 31,000; Fungal Protease 60,000; Fungal Protease 500,000; Fungal Protease Conc.; HT-Proteolytic® 200; HT-Proteolytic® Conc.; HT-Proteolytic® L-175; MC²; MLO; Newlase; Optimase® APL-440; Papain 16,000; Papain 30,000; Papain Conc.; Papain P-100; Protoferm; Prozyme 6; Rhozyme® P11; Rhozyme® P41; Rhozyme® P53, P64; Wheat-Eze
Trade names containing: Pancreatin 3X USP Powd.; Pancreatin 4X USP Powd.; Pancreatin 5X USP Powd.; Pancreatin 6X USP Powd.; Pancreatin 8X USP Powd.; Pancreatin USP Powd.; Pancrelipase USP; Proflex; SPL Pancreatin 3X USP; SPL Pancreatin 4X USP; SPL Pancreatin 5X USP; SPL Pancreatin 6X USP; SPL Pancreatin 7X USP; SPL Pancreatin 8X USP; SPL Pancreatin USP; SPL Pancrelipase USP; SPL Undiluted Pancreatic Enzyme Conc. (PEC)

Proteins, milk, hydrolysate. *See* Hydrolyzed milk protein
Proteins, vegetable, hydrolysate. *See* Hydrolyzed vegetable protein
Protocatechualdehyde dimethyl ether. *See* Veratraldehyde
Protovanol. *See* Vanilla
Provitamin A. *See* Carotene
Provitamin B₅. *See* D-Panthenol
Provitamin B₅. *See* DL-Panthenol
Prunus amygdalus amara extract. *See* Bitter almond extract
Prunus armeniaca. *See* Apricot kernel oil
Prunus serotina extract. *See* Wild cherry bark extract
Pseudopinene. *See* β-Pinene

Psyllium
Synonyms: Psyllium gum; Psyllium husks
Definition: Vegetable mucilate preparation
Properties: Dark-redsh. brn., odorless, almost tasteless
Uses: Stabilizer for ice cream, ice milk, sherbet

Psyllium gum. *See* Psyllium
Psyllium husks. *See* Psyllium
Pteroylglutamic acid. *See* Folic acid
d-β-Pulegomenthol. *See* d-Neomenthol

Pulegone
CAS 89-82-7; EINECS 201-943-2; FEMA 2963
Synonyms: p-Menth-4(8)-en-3-one; R-(+)-p-menth-4(8)-en-3-one; 1-Methyl-4-isopropylidene-3-cyclohexanone; (R)-2-Isopropylidene-5-methylcyclohexanone;; 5-Methyl-2-(1-methylethylidene)cyclohexanone
Classification: Ketone
Empirical: $C_{10}H_{16}O$
Properties: Oil, pleasant odor of peppermint/camphor; misc. with alcohol, ether, chloroform; pract. insol. in water; m.w. 152.24; dens. 0.936 (20/4 C); b.p. 223-224 C; flash pt. 55 C; ref. index 1.487 (20 C)
Precaution: Flamm.
Uses: Synthetic flavoring
Usage level: 5.0-8.0 ppm (nonalcoholic beverages), 5.0-32 ppm (ice cream, ices), 17 ppm (candy), 24-25 ppm (baked goods); EEC limitations: 25 ppm (food), 100 ppm (beverages), 250 ppm (mint beverages), 350 ppm (mint confectionery)
Regulatory: FDA 21CFR §172.515

Pullulanase
CAS 9075-68-7; EINECS 232-983-9
Properties: Wh. powd.
Uses: Food enzyme in prod. of maltose
Trade names: CK-20L

Pumpkin seed oil
EINECS 289-741-0 (extract)
Definition: Oil expressed from seeds of the pumpkin *Cucurbita pepo*
Properties: Oil; sp.gr. 0.90-0.92; acid no. 1-6; iodine no. 110-125; sapon. no. 189-199
Manuf./Distrib.: Arista Industries

Purified oxgall. *See* Ox bile extract
Purple medick extract. *See* Alfalfa extract
PVA. *See* Polyvinyl alcohol
PVAc. *See* Polyvinyl acetate (homopolymer)
PVAL. *See* Polyvinyl alcohol
PVOH. *See* Polyvinyl alcohol

PVP
CAS 9003-39-8; EINECS 201-800-4
Synonyms: Polyvinylpyrrolidone; Poly (n-vinylbutyrolactam); Povidone; 1-Ethenyl-2-pyrrolidinone homopolymer
Classification: Linear polymer
Definition: Polymer of 1-vinyl-2-pyrrolidone monomers
Empirical: $(C_6H_9NO)_x$
Properties: Wh. free-flowing amorphous powd.; sol. in water, chlorinated hydrocarbons, alcohol, amines, nitroparaffins, lower m.w. fatty acids; m.w. ≈ 10,000, ≈ 24,000, ≈ 40,000 (food use), ≈ 160,000, ≈ 360,000 (beer); dens. 1.23-1.29; flash pt. > 215 C
Toxicology: LD50 (IP, mouse) 12 g/kg; mildly toxic by IP and IV routes; heated to decomp., emits toxic fumes of NO_x
Uses: Clarifying agent for beer, wine, vinegar; tableting adjuvant in flavor concs.; stabilizer, bodying agent, dispersant in tableted nonnutritive sweeteners, vitamin/mineral concs. in liq. or tablet form; color diluent; also for confectionery, fruit, gum
Usage level: Limitation 10 ppm residual (beer), 40 ppm residual (vinegar), 60 ppm residual (wine); ADI 0-25 mg/kg (JECFA)
Regulatory: FDA 21CFR §73.1, 73.1001, 172.210, 173.50, 173.55, 175.105, 175.300, 176.170, 176.180, 176.210; BATF 27CFR §240.1051
Manuf./Distrib.: Allchem Industries

PVPP. *See* Polyvinylpolypyrrolidone

Pyridine
CAS 110-86-1; EINECS 203-809-9; FEMA 2966
Empirical: C_5H_5N
Properties: Colorless liq., char. disagreeable odor, sharp taste; misc. with water, alcohol, ether, petroleum ether, oils, other org. liqs.; m.w. 79.10; dens. 0.98272 (20/4 C); m.p. -41.6 C; b.p. 115-116 C; flash pt. (CC) 20 C; ref. index 1.510 (20 C)
Precaution: Highly flamm.; volatile with steam
Toxicology: LD50 (oral, rat) 1.58 g/kg; may cause human CNS depression, irritation of skin and respiratory tract; large does may produce GI disturbances, kidney and liver damage
Uses: Synthetic flavoring
Usage level: 1.0 ppm (nonalcoholic beverages), 0.02-0.12 ppm (ice cream, ices), 0.40 ppm (candy, baked goods)
Regulatory: FDA 21CFR §172.515
Manuf./Distrib.: Aldrich

3-Pyridinecarboxamide. *See* Niacinamide
Pyridine-3-carboxylic acid. *See* Nicotinic acid
3-Pyridinecarboxylic acid aluminum salt. *See* Aluminum nicotinate

Pyridoxine HCl
CAS 58-56-0; EINECS 200-386-2
Synonyms: 5-Hydroxy-6-methyl-3,4-pyridinedimethanol hydrochloride; 3-Hydroxy-4,5-dihydroxymethyl-2-methylpyridine HCl; Vitamin B_6 hydrochloride; Adermine hydrochloride; Pyridoxine hydrochloride; Pyridoxol hydrochloride
Classification: Substituted aromatic compd.
Empirical: $C_8H_{11}NO_3$ • HCl
Properties: Colorless to wh. platelets or cryst. powd., odorless; sol. in water, alcohol, acetone, propylene glycol; sl. sol. in other org. solvs.; insol. in ether, chloroform; m.w. 205.66; m.p. 204-206 C (dec.); pH 2.0-3.5 (5% aq.)
Toxicology: LD50 (oral, rat) 4000 mg/kg; poison by IV route; mod. toxic by ingestion; prolonged high doses may cause ataxia; human reproductive effects; experimental teratogen; human mutagenic data; heated to decomp., emits very toxic fumes of NO_x and HCl
Uses: Direct food additive, nutrient, dietary supplement; used in baked goods, nonalcoholic beverages, breakfast cereals, dairy prods., meat prods., milk prods., plant protein prods., snack foods, infant formula
Regulatory: FDA 21CFR §182.5676, 184.1676, GRAS; Japan approved

Manuf./Distrib.: Am. Roland; Daiichi Pharmaceutical; Unipex
Trade names: Pyridoxine Hydrochloride USP, FCC Fine Powd. No. 60650
Trade names containing: Rocoat® Pyridoxine Hydrochloride 33¹/₃% No. 60688

Pyridoxine hydrochloride. *See* Pyridoxine HCl
Pyridoxol hydrochloride. *See* Pyridoxine HCl
Pyroacetic ether. *See* Acetone
Pyrocarbonic acid dimethyl ester. *See* Dimethyl dicarbonate
Pyrocatechol methyl ether. *See* Guaiacol

Pyroligneous acid extract
FEMA 2968
Synonyms: Liquid smoke; Pyroligneous vinegar
Properties: Yel. to red liq., smoke odor; dens. 1.018-1.030
Uses: Synthetic flavoring
Usage level: 20 ppm (nonalcoholic beverages), 50-200 ppm (baked goods), 100-300 ppm (meats)
Regulatory: FDA 21CFR §172.515

Pyroligneous vinegar. *See* Pyroligneous acid extract
Pyroligneus acid. *See* Acetic acid
Pyromucic acid. *See* 2-Furoic acid
Pyrophyllite. *See* Aluminum silicate
Pyroracemic acid. *See* Pyruvic acid
Pyroxylin. *See* Nitrocellulose

Pyrrole
CAS 109-97-7; EINECS 203-724-7; FEMA 3386
Synonyms: Azole; Imidole; Divinylenimine
Definition: Constituent of coal tar and bone oil
Empirical: C_4H_5N
Properties: Yel.-brn.; m.w. 67.09; dens. 0.966 (20/4 C); b.p. 129-131 C; flash pt. 36 C
Precaution: Flamm.; photosensitive
Manuf./Distrib.: Aldrich

2-Pyrrolidine carboxylic acid. *See* L-Proline

Pyruvaldehyde
CAS 78-98-8; EINECS 201-164-8; FEMA 2969
Synonyms: Methylglyoxal; 2-Oxopropanal; Pyruvic aldehyde; Acetyl formaldehyde
Empirical: $C_3H_4O_2$
Formula: CH_3COCHO
Properties: Yel. mobile liq., pungent stinging odor, caramellic sweet odor; hygroscopic; sol. in alcohol, ether; m.w. 72.06; dens. 1.178; b.p. 72 C; ref. index 1.4002 (17.5 C)
Toxicology: Irritant
Uses: Synthetic flavoring
Usage level: 1.0 ppm (nonalcoholic beverages, ice cream, ices), 0.03-5.0 ppm (candy, baked goods)
Regulatory: FDA 21CFR §172.515
Manuf./Distrib.: Aldrich

Pyruvic acid
CAS 127-17-3; EINECS 204-824-3; FEMA 2970
Synonyms: 2-Oxopropionic acid; Pyroracemic acid; α-Ketopropionic acid; Acetylformic acid
Empirical: $C_3H_4O_3$
Formula: $CH_3COCOOH$
Properties: Liq., acetic acid odor; misc. with water, alcohol, ether; m.w. 88.06; dens. 1.265 (20/4 C); m.p. 11.8 C; b.p. 165 C (760 mm, dec.); flash pt. 82 C; ref. index 1.4138 (20 C)
Precaution: Polymerizes and dec. on standing unless pure and kept in airtight container
Toxicology: Causes burns
Uses: Synthetic flavoring
Usage level: 0.25 ppm (nonalcoholic beverages), 0.25-20 ppm (ice cream, ices), 27 ppm (candy), 30 ppm (baked goods), 110 ppm (chewing gum)
Regulatory: FDA 21CFR §172.515
Manuf./Distrib.: Aldrich

Pyruvic aldehyde. *See* Pyruvaldehyde

Quassia
Synonyms: Bitter wood; Bitter ash

Definition: Wood of *Picrasma excelsa* or *Quassia amara*, contg. bitter principle quassin
Properties: Ylsh-wh. to bright yel. chips or fibrous coarse grains, sl. odor, very bitter taste
Uses: Natural flavoring
Regulatory: FDA 21CFR §172.510; Japan approved
Manuf./Distrib.: Chart

Quaternium-14
CAS 27479-28-3; EINECS 248-486-5
Synonyms: Dodecyl dimethyl ethylbenzyl ammonium chloride; N-Dodecyl-*ar*-ethyl-N,N-dimethylbenzene-methanaminium chloride
Classification: Quaternary ammonium salt
Empirical: $C_{23}H_{42}N \cdot Cl$
Uses: Antimicrobial for use in cane-sugar mills
Regulatory: FDA 21CFR §172.165, 173.320

Quaternium 31. *See* Dialkyl dimethyl ammonium chloride
Quercus robur extract. *See* English oak extract
Quicklime. *See* Calcium oxide

Quillaja
CAS 68990-67-0; EINECS 273-620-4; FEMA 2973
Synonyms: Quillaja bark; Soapbark; Panama bark
Definition: Dried bark of *Quillaja saponaria* contg. sapotoxin, tannin, and quillaja
Properties: Bittersweet aromatic flavor
Uses: Natural flavoring producing foam in carbonated beverages, bar mixes, beer, juices, wine coolers, schnapps, barley drinks, root beer, and other prods. needing exaggerated foam
Usage level: 95 ppm (nonalcoholic beverages), 0.12 ppm (ice cream, ices), 18 ppm (candy), 6.8 ppm (syrups)
Regulatory: FDA 21CFR §172.510; Japan approved
Manuf./Distrib.: Chart

Quillaja bark. *See* Quillaja

Quillaja extract
Synonyms: Panama wood extract; Quillaja saponaria extract
Definition: Extract of bark of *Quillaja saponaria*
Uses: Natural flavoring
Regulatory: FDA 21CFR §172.510; Japan approved
Manuf./Distrib.: Chart

Quillaja saponaria extract. *See* Quillaja extract

Quince seed
Synonyms: Cydonia oblonga seed
Definition: Dried seeds of *Cydonia oblonga*
Uses: Natural flavoring
Regulatory: FDA 21CFR §182.40; Japan approved

Quinine
CAS 130-95-0; EINECS 205-003-2
Synonyms: 6´-Methoxycinchonan-9-ol
Definition: Alkaloid from bark of *Cinchona officinalis*
Empirical: $C_{20}H_{24}N_2O_2$
Properties: Orthorhombic need., odorless, intense bitter taste; sol. 1 g/1900 ml water, 0.8 ml alcohol, 80 ml benzene, 1.2 ml chloroform, 20 ml glycerol, 1900 ml of 10% ammonia water; almost insol. in petroleum ether; m.w. 324.43; m.p. 175 C (dec.)
Precaution: Photosensitive
Toxicology: Irritant
Uses: Flavoring in carbonated beverages
Usage level: up to 83 ppm (hydrochloride or sulfate of quinine)
Regulatory: FDA 21CFR §172.575

Quinine bimuriate. *See* Quinine dihydrochloride
Quinine bisulfate. *See* Quinine sulfate
Quinine chloride. *See* Quinine hydrochloride
Quinine dichloride. *See* Quinine dihydrochloride

Quinine dihydrochloride
CAS 60-93-5

Synonyms: Acid quinine hydrochloride; 6´-Methoxycinchonan-9-ol dihydrochloride; Quinine bimuriate; Quinine dichloride

Empirical: $C_{20}H_{24}N_2O_2 \cdot 2ClH$

Properties: Wh. need. or cryst. powd., odorless, very bitter taste; sol. in water, alcohol, glycerin; sl. sol. in chloroform; very sl. sol. in ether; m.w. 397.38

Precaution: Protect from light

Toxicology: LD50 (oral, rat) 1392, mg/kg, (IV, rat) 78 mg/kg; poison by IV and subcutaneous routes; mod. toxic by ingestion; mutagenic data; heated to decomp., emits very toxic fumes of NO_x and HCl

Uses: Flavoring agent for carbonated beverages

Regulatory: FDA 21CFR §172.575 (limitation 83 ppm)

Quinine hydrochloride

CAS 6119-47-7; EINECS 231-437-7; FEMA 2976

Synonyms: Quinine chloride; Quinine muriate; Quinine monohydrochloride; 6´-Methoxycinchonan-9-ol monohydrochloride

Empirical: $C_{20}H_{25}ClN_2O_2$

Formula: $C_{20}H_{24}N_2O_2 \cdot HCl \cdot 2H_2O$ (dihydrate)

Properties: Anhyd.: m.w. 360.88; Dihydrate: Silky need., bitter taste; effloresces on expsoure to warm air; sol. 1 g/16 ml water, 1 ml alcohol, 7 ml glycerol, 1 ml chloroform; pH 6-07 (1% aq.)

Precaution: Photosensitive; protect from light

Uses: Food additive, flavoring agent for carbonated beverages

Usage level: 110 ppm (nonalcoholic beverages)

Regulatory: FDA 21CFR §172.575

Quinine hydrogen sulfate. *See* Quinine sulfate
Quinine monohydrochloride. *See* Quinine hydrochloride
Quinine muriate. *See* Quinine hydrochloride

Quinine sulfate

CAS 804-63-7; FEMA 2977

Synonyms: Quinine bisulfate; Quinine hydrogen sulfate

Empirical: $C_{20}H_{24}N_2O_2 \cdot O_4S$

Properties: Wh. fine needle-like cryst., odorless, very bitter taste; sol. in water, alcohol; sl. sol. in chloroform; m.w. 420.52

Precaution: Keep well closed; protect from light; incompat. with ammonia, alkalies, limewater, tannic acid, iodine, iodides, acetates, citrates, tartrates, benzoates, salicylates

Toxicology: Human poison by ingestion; human systemic effects; experimental reproductive effects; mutagenic data; heated to decomp., emits very toxic fumes of SO_x and NO_x

Uses: Flavoring agent for carbonated beverages, quinine water, tonic water

Regulatory: FDA 21CFR §172.575 (limitation 83 ppm)

Quinoline

CAS 91-22-5; EINECS 202-051-6; FEMA 3470

Synonyms: Chinoline; 1-Benzazine; Benzo[b]pyridine

Empirical: C_9H_7N

Properties: Liq., penetrating odor; hygroscopic; difficultly sol. in cold water; misc. with aclohol, ether, CS_2; m.w. 129.16; dens. 1.093 (20/4 C); m.p. -17 to -13 C; b.p. 108-110 C (11 mm); flash pt. 92 C; ref. index 1.625 (20 C)

Precaution: Photosensitive; protect from light and moisture

Toxicology: LD50 (oral, rat) 460 mg/kg

Uses: Synthetic flavoring agent

Regulatory: GRAS (FEMA)

Manuf./Distrib.: Aldrich

Racemic menthol. *See* Menthol
Racemic pantothenyl alcohol. *See* DL-Panthenol

Rapeseed oil

CAS 8002-13-9; EINECS 232-299-0

Synonyms: Brassica campestris oil; Colza oil; Canola oil (low erucic acid rapeseed oil)

Definition: Vegetable oil expressed from seeds of *Brassica campestris*

Properties: Brn. viscous liq., yel. when refined; sol. in chloroform, ether, CS_2; dens. 0.913-0.916; m.p. 17-22 C; solidifies at 0 C; flash pt. 325 F; iodine no. 97-105; sapon. no. 170-177 C; ref. index 1.4720-1.4752

Precaution: Subject to spontaneous heating

Uses: Edible fats and oils; margarine; not for use in infant formulas; antioxidant (Japan)

Regulatory: FDA 21CFR §175.105, 176.210, 177.1200, 177.2800, 184.1555; Japan approved (extract)

Rapeseed oil glyceride

    Manuf./Distrib.: Arista Industries; Penta Mfg.

Rapeseed oil glyceride
    Synonyms: Rapeseed oil monoglyceride
    Uses: Food emulsifier, stabilizer for peanut butter
    Trade names: Myverol® 18-99K
    Trade names containing: Myvatex® Monoset® K

Rapeseed oil monoglyceride. *See* Rapeseed oil glyceride
Raspberry ketone. *See* 4-(p-Hydroxyphenyl)-2-butanone
Red algae. *See* Algae, red

Red beet extract
    Uses: Natural colorant used in yogurt, sherbet, ice cream, frozen and gelatin desserts, frostings, and candy
    Trade names containing: Red Beet WSL-300; Red Beet WSL-400; Red Beet WSP-300

Red Dye No. 2. *See* Amaranth
Red iron trioxide. *See* Ferric oxide
Red oil. *See* Oleic acid
Red pepper. *See* Capsicum
Red sandalwood. *See* Red saunders

Red saunders
    Synonyms: Red sandalwood
    Definition: From *Pterocarpus san alinus*
    Uses: Natural flavoring for alcoholic beverages only; color (Japan)
    Regulatory: FDA 21CFR §172.510; Japan restricted as color
    Manuf./Distrib.: Chart (extract)

Reduced lactose whey. *See* Whey, reduced lactose
Refined bleached shellac. *See* Shellac, bleached, wax-free
Refined petroleum wax. *See* Petroleum wax

Rennet
    CAS 9001-98-3
    Synonyms: Bovine rennet; Rennin; Chymosin
    Definition: Digestive enzyme from calf stomach; contains the enzyme rennin
    Properties: Ylsh.-wh. powd., peculiar odor, sl. salty taste; sl. sol. in water
    Toxicology: Heated to decomp., emits acrid smoke and irritating fumes
    Uses: Binder, enzyme, extender, processing aid, stabilizer, thickener; used to coagulate milk in cheesemaking
    Regulatory: FDA 21CFR §184.1685, GRAS; USDA 9CFR §318.7 (rennet treated calcium reduced dried milk and calcium lactate, limitation 3.5% in sausages), 381.147; Canada, Japan approved
    Manuf./Distrib.: Aplin & Barrett; Gist-brocades; Chr Hansen's Lab; N I Ibrahim; Miles; New Zealand Milk Prods.; Pfizer; Sanofi; Texel; Trustin Foods

Rennet/chymosin prep.
    Definition: Rennet: animal derived extract containing enzyme rennin (chymosin) (CAS 9001-98-3) and chymosin prep.: clear sol'n. contg. enzyme chymosin, fermentation derived from *E. coli* or *Kluyveromyces marxianus*
    Properties: Rennet: clear amber to dk. brn. liq. or wh. to tan powd.; Chymosin prep.: clear sol'n.
    Uses: Direct food additive, milk-clotting enzyme, processing aid, stabilizer, thickener; used in cheese, frozen dairy desserts, gelatins, puddings, fillings, milk prods.
    Regulatory: FDA 21CFR §184.1685, GRAS

Rennin. *See* Rennet
Resin guaiac. *See* Guaiac gum
Resin olibanum. *See* Olibanum
Resin tolu. *See* Balsam tolu
Resorcinol dimethyl ether. *See* m-Dimethoxybenzene

Retinol
    CAS 68-26-8; 11103-57-4; EINECS 200-683-7; 234-328-2
    Synonyms: 3,7-Dimethyl-9-(2,6,6-trimethyl-1-cyclohexen-1-yl)-2,4,6,8-nonatretraen-1-ol; Vitamin A; Vitamin A alcohol; all-trans-Retinol; Axerophthol
    Classification: Organic compd.
    Empirical: $C_{20}H_{30}O$
    Properties: Yel. prisms or cryst., nearly odorless or mild fishy odor; sol. in abs. alcohol, methanol, chloroform, ether, fats, oils; pract. insol. in water, glycerin; m.w. 286.46; m.p. 54-58 C

Toxicology: LD50 (oral, rat) 2000 mg/kg; mod. toxic by ingestion; human teratogenic effects; experimental teratogen, reproductive effects; human mutagenic data; heated to decomp., emits acrid smoke and irritating fumes
Uses: Dietary supplement (vitamin)
Regulatory: FDA 21CFR §131, 133, 135.130, 166.40, 166.110, 182.5930, 184.1930, GRAS; Japan approved
Manuf./Distrib.: Am. Roland
Trade names: Vita-Rite A; Vita-Rite A2; Vita-Rite A-W
Trade names containing: Baker's Ideal®; Coral®; Hi Glo™; Rex® Vitamin Fortified Wheat Germ Oil; Summit® 25; Summit® 50; Superb® Bakers Margarine 105-101; Top Notch®; Trophy®; Victor®; Vita-Rite A & D-2%; Vita-Rite A & D-W

Retinol, hexadecanoate. *See* Retinyl palmitate
all-trans-Retinol. *See* Retinol

Retinyl acetate
CAS 127-47-9; EINECS 204-844-2
Synonyms: Acetic acid, retinyl ester; Vitamin A acetate; Vitamin A alcohol acetate
Definition: Ester of retinol and acetic acid
Empirical: $C_{22}H_{32}O_2$
Properties: Pale yel. prismatic cryst.; m.w. 328.54; m.p. 57-58 C
Toxicology: LD50 (oral, mouse, 10 day) 4100mg/kg; mod. toxic by ingestion; experimental neoplastigen, teratogen, reproductive effects; mutagenic data; heated to decomp., emits acrid smoke and irritating fumes
Uses: Direct food additive, nutrient, dietary supplement; may be used in infant formulas
Regulatory: FDA 21CFR §182.5933, 184.1930, GRAS

Retinyl palmltate
CAS 79-81-2; EINECS 201-228-5
Synonyms: Retinol, hexadecanoate; Vitamin A palmitate
Definition: Ester of retinol and palmitic acid
Empirical: $C_{36}H_{60}O_2$
Properties: Amorphous or cryst.; m.w. 524.88; m.p. 28-29 C
Toxicology: LD50 (oral, rat, 10 day) 7910 mg/kg; mildly toxic by ingestion; heated to dec., emits acrid smoke and irritating fumes
Uses: Direct food additive, nutrient, dietary supplement; may be used in infant formulas
Regulatory: FDA 21CFR §182.5936, 184.1930, GRAS
Trade names: Vitamin A Palmitate Type PIMO/BH No. 638280100; Vitamin A Palmitate USP, FCC Type P1.7 No. 262090000; Vitamin A Palmitate USP, FCC Type P1.7/E No. 63699
Trade names containing: Buckeye C (Baker's Grade); Dry Vitamin A Palmitate Type 250-SD No. 65378; Golden Croissant/Danish™ Margarine; Palmabeads® Type 500 No. 65332; Palma-Sperse® Type 250-S No. 65322; Palma-Sperse® Type 250A/50 D-S No. 65221; Shedd's Special 40 Butter Blend Margarine; Shedd's Special 60 Butter Blend Margarine; Vitamin A Palmitate Type 250-CWS No. 65312; Vitamin A Palmitate USP, FCC Type P1.7/BHT No. 63693

Rhamnose
CAS 3615-41-6
Synonyms: 6-Deoxy-L-mannose; L-Rhamnose; L-Mannomethylose; Isodulcit
Empirical: $C_6H_{12}O_5$
Properties: Wh. cryst.; sol. in water, methanol; m.w. 164.16; m.p. 82-92 C
Uses: Synthetic sweetener
Regulatory: Japan approved
Manuf./Distrib.: Penta Mfg.

L-Rhamnose. *See* Rhamnose
Rhamnus purshiana extract. *See* Cascara extract

Rhamsan gum
Uses: Natural thickener; stabilizer
Regulatory: Japan approved

Rheum palmatum root extract. *See* Rhubarb extract
Rheum rhabarbarum root extract. *See* Rhubarb extract
Rhodinal. *See* Citronellal

Rhodinol
CAS 6812-78-8; FEMA 2980
Synonyms: S-(-)-3,7-Dimethyl-7-octen-1-ol; α-Citronellol; l-Citronellol
Empirical: $C_{10}H_{20}O$

Rhodinol acetate

Properties: Oily liq., rose odor; misc. with alcohol, ether; very sl. sol. in water; m.w. 156.26; dens. 0.8549 (20/4 C); b.p. 114-115 C (12 mm); ref. index 1.4556 (20 C)
Uses: Synthetic flavoring agent
Regulatory: FDA 21CFR §172.515
Manuf./Distrib.: Commodity Services; Florida Treatt

Rhodinol acetate. *See* Rhodinyl acetate

Rhodinyl acetate
CAS 141-11-7; FEMA 2981
Synonyms: α-Citronellyl acetate; 3,7-Dimethyl-7-octen-1-ol acetate; Rhodinol acetate
Definition: Mixt. of acetates of geraniol and l-citronellol, found in geranium oil
Empirical: $C_{12}H_{22}O_2$
Properties: Colorless to sl. yel. liq., fresh rose odor; sol. in alcohol, fixed oils; insol. in glycerin, propylene glycol, water; m.w. 198.34; dens. 0.895-0.908; b.p. 237 C; ref. index 1.450-1.458
Toxicology: Skin irritant; heated to decomp., emits acrid smoke and irritating fumes
Uses: Synthetic flavoring agent
Usage level: 2.8 ppm (nonalcoholic beverages), 1.4 ppm (ice cream, ices), 9.4 ppm (candy), 18 ppm (baked goods)
Regulatory: FDA 21CFR §172.515

Rhodinyl butyrate
FEMA 2982
Synonyms: Citronellyl butyrate
Empirical: $C_{14}H_{26}O_2$
Properties: Colorless to ylsh. or greenish liq., fruity sweet odor; m.w. 226.36; b.p. 137 C (13 mm); flash pt. > 100 C; ref. index 1.451-1.455
Uses: Synthetic flavoring
Usage level: 0.94 ppm (nonalcoholic beverages), 1.1 ppm (ice cream, ices, chewing gum), 3.0 ppm (candy), 9.7 ppm (baked goods)
Regulatory: FDA 21CFR §172.515

Rhodinyl formate
FEMA 2984
Empirical: $C_{11}H_{20}O_2$
Uses: Synthetic flavoring agent
Regulatory: FDA 21CFR §172.515

Rhodinyl isobutyrate
Uses: Synthetic flavoring
Regulatory: FDA 21CFR §172.515

Rhodinyl isovalerate
Uses: Synthetic flavoring
Regulatory: FDA 21CFR §172.515

Rhodinyl phenylacetate
Uses: Synthetic flavoring
Regulatory: FDA 21CFR §172.515

Rhodinyl propionate
FEMA 2986
Synonyms: Citronellyl propionate
Properties: Colorless oily liq., sweet odor/flavor; sol. in alcohol; almost insol. in water; m.w. 212.26; b.p. 255 C; flash pt. 100 C; ref. index 1.4570
Uses: Synthetic flavoring
Usage level: 1.8 ppm (nonalcoholic beverages), 2.4 ppm (ice cream, ices), 4.9 ppm (candy), 5.8 ppm (baked goods)
Regulatory: FDA 21CFR §172.515

Rhubarb extract
CAS 90106-27-7, 8016-55-5; EINECS 290-249-3
Synonyms: Rheum palmatum root extract; Rheum rhabarbarum root extract
Definition: Extract of the stalks or roots of *Rheum palmatum* or *R. rhabarbarum*
Uses: Natural flavoring agent
Regulatory: FDA 21CFR §172.510
Manuf./Distrib.: Chart

Rhus succedanea wax. *See* Japan wax

Riboflavin
CAS 83-88-5; EINECS 201-507-1; EEC E101b
Synonyms: Vitamin B$_2$; 6,7-Dimethyl-9-d-ribitylisoalloxazine; 7,8-Dimethyl-10-(d-ribityl) isoalloxazine; Flavaxin; Lactoflavin; Vitamin G
Classification: Organic compd.
Empirical: C$_{17}$H$_{20}$N$_4$O$_6$
Properties: Orange to yel. cryst., sl. odor, bitter taste; sl. sol. in water, alcohol; insol. in ether, chloroform; m.w. 376.41; m.p. 282 C (dec.)
Toxicology: LD50 (IP, rat) 560 mg/kg; poison by intravenous route; moderately toxic by intraperitoneal and subcutaneous routes; mutagenic adata; heated to decomp., emits fumes of NO$_x$
Uses: Direct food additive, dietary supplement, nutrient, color additive; principal growth-promoting factor of Vitamin B$_2$ complex (functions as flavor protein in tissue respiration); may be used in infant formulas
Usage level: ADI 0-0.5 mg/kg (EEC)
Regulatory: FDA 21CFR §73.450, 136.115, 137, 139, 182.5695, 184.1695, GRAS; Europe listed; UK, Japan approved
Manuf./Distrib.: ADM; Am. Roland; Browne & Dureau Int'l.; Crompton & Knowles; CSR Food Ingreds.; Food Additives & Ingreds.; Forum; Hoffmann La Roche; LaMonde; E Merck; Penta Mfg.; Produits Roche; Roche; Scanchem; SCI; Takeda Europe; Warner-Jenkinson Netherlands
Trade names: Riboflavin USP, FCC No. 602940002
Trade names containing: Enrichment R Tablets; Rich-Pak®; Rich-Pak® Powd. 160; Rich-Pak® Powd. 160-M; Rocoat® Riboflavin 25% No. 60289; Rocoat® Riboflavin 33$^1/_3$ No. 60288

Riboflavin 5'-monophosphate sodium salt dihydrate. *See* Riboflavin-5'-phosphate sodium
Riboflavin 5'-phosphate ester monosodium salt. *See* Riboflavin-5'-phosphate sodium

Riboflavin-5'-phosphate sodium
CAS 130-40-5; EINECS 204-988-6; EEC E101a
Synonyms: Riboflavin 5'-monophosphate sodium salt dihydrate; Vitamin B$_2$ phosphate sodium; Riboflavin 5'-phosphate ester monosodium salt
Formula: C$_{17}$H$_{20}$N$_4$O$_9$PNa • 2H$_2$O
Properties: Yel. to orange-yel. cryst. powd., sl. odor; hygroscopic; sol. in water; m.w. 514.36
Precaution: Decomposed by light when in sol'n.
Toxicology: Heated to decomp., emits toxic fumes of NO$_x$ and NaO$_2$
Uses: Direct food additive, nutrient, dietary supplement; for milk prods.; may be used in infant formulas; colorant
Usage level: ADI 0-0.5 mg/kg (EEC)
Regulatory: FDA 21CFR §182.5697, 184.1697, GRAS; Europe listed; UK, Japan approved
Trade names: Riboflavin-5'-Phosphate Sodium USP, FCC No. 60296

Riboflavin tetrabutyrate
Uses: Color; Dietary supplement (vitamin)
Regulatory: Japan approved

Rice bran
CAS 68553-81-1
Uses: Dietary fiber supplement
Trade names: Protex™ 20

Rice bran extract
Definition: Extract of rice bran
Uses: Natural antioxidant (Japan)
Regulatory: Japan approved (rice bran oil extract)

Rice bran oil
CAS 68553-81-1; 84696-37-7; EINECS 271-397-8
Synonyms: Oils, rice bran; Rice oil
Definition: Oil expressed from rice bran
Properties: Golden yel. oil; misc. with hexane and other fat solvs.; dens. 0.916-0.921; cloud pt. < -7 C; acid no. < 0.1; iodine no. 92-115; sapon. no. 181-189; ref. index 1.470-1.473;
Uses: Edible oil, hydrogenated shortenings
Regulatory: FDA 21CFR §175.105, 176.200, 176.210, 177.2260, 177.2800
Manuf./Distrib.: Arista Industries
Trade names: Rice Bran Oil; Rice Bran Oil SO

Rice bran wax
CAS 8016-60-2; EINECS 232-409-7

Rice maltodextrin

Synonyms: Waxes, rice bran
Definition: Wax obtained from rice bran, the broken hulls of rice grains, Oryza sativa
Properties: Tan to brn. hard wax; sol. in chloroform, benzene; insol. in water; m.p. 75-80 C; iodine no. 20 max.; sapon. no. 75-120
Toxicology: Heated to decomp., emits acrid smoke and irritating fumes
Uses: Masticatory substance, plasticizer in chewing gum base; coating agent for candy, fresh fruits and vegetables; release agent; glazing agent (Japan)
Usage level: Limitation 50 ppm (candy coating), 50 ppm (fresh fruits and vegetables), 2.5% (chewing gum)
Regulatory: FDA 21CFR §172.615, 172.890, 178.3860 (limitation 1% in polymer); Japan approved
Trade names: Ross Rice Bran Wax

Rice maltodextrin

Uses: Fat replacement providing creamy texture, bodying agent, gelling agent; used in granola and nut coatings
Trade names: CNP RMDRD05; CNP RMDRD18; Rice-Trin® 10; Rice-Trin® 18
Trade names containing: Rice Complete® 3; Rice Complete® 10; Rice Complete® 18

Rice oil. See Rice bran oil

Rice protein

Trade names containing: Rice Complete® 3; Rice Complete® 10; Rice Complete® 18; Rice Complete® 25; Rice-Pro® 35W

Rice starch

CAS 9005-25-8
Uses: Gellant, fat substitute
Trade names: Instant Remygel AX-P, AX-2-P; Instant Remyline AX-P; Remy BLM7-FG; Remy CX; Remy DR; Remygel AX-2; Remygel AX; Remygel NBO; Remy Instant; Remyline AX; StarchPlus™ SPR; StarchPlus™ SPR-LP; StarchPlus™ SPW; StarchPlus™ SPW-LP

Rice syrup

Uses: Sweetener for soy-based prods.
Trade names: CNP WRSHG40; CNP WRSHG40CL; CNP WRSHM; CNP WRSHMCL; CNP WRSMC30; CNP WRSMC30CL; CNP WRSRD; CNP WRSRDCL

Ricinus oil. See Castor oil
Roccella fusiformis Ach. See Angola weed
Rochelle salt. See Potassium sodium tartrate anhyd.
Rochelle salt. See Potassium sodium tartrate tetrahydrate
Rock salt. See Sodium chloride
Roman chamomile extract. See Chamomile extract
Roman chamomile oil. See Chamomile oil, Roman

Rose buds and flowers

Definition: Rosa spp.
Uses: Natural flavoring agent
Regulatory: FDA 21CFR §182.20, GRAS; Japan approved (rose)
Manuf./Distrib.: Chart

Rose fruit. See Rose hips

Rose geranium oil

Definition: Pelargonium graveolens
Uses: Natural flavoring agent
Regulatory: FDA 21CFR §182.20, GRAS

Rose geranium oil Algerian. See Geranium oil

Rose hips

Synonyms: Rose fruit
Definition: The fruit of the rose, Rosa spp.
Uses: Natural flavoring agent; rich source of vitamin C
Regulatory: FDA 21CFR §182.20, GRAS
Manuf./Distrib.: Am. Roland; Chart

Rose leaves

Definition: Rosa spp.
Uses: Natural flavoring agent
Regulatory: FDA 21CFR §182.20, GRAS

Rosemary extract
CAS 84604-14-8
Synonyms: Extract of rosemary; Rosmarinum officinalis extract
Definition: Oleoresin extracted from rosemary leaves, *Rosmarinus officinalis* L.
Uses: Flavoring
Regulatory: FDA 21CFR §182.20
Trade names containing: Flav-R-Keep®; Flav-R-Keep® FP-51; Freez-Gard® FP-15; Pristene® R20; Pristene® RO; Pristene® RW; Pristene® TR

Rosemary oil
CAS 8000-25-7; FEMA 2992
Definition: Volatile oil obtained from flowering tops of *Rosmarinus officinalis*
Properties: Colorless to pale yel. liq., rosemary odor, camphoraceous taste; sol. in 10 vols 80% alcohol; pract. insol. in water; dens. 0.894-0.912 (25/25 C); ref. index 1.464-1.476 (20 C)
Precaution: Keep cool, well closed; protect from light
Toxicology: LD50 (oral, rat) 5000 mg/kg; mildly toxic by ingestion; skin irritant; heated to decomp., emits acrid smoke and irritating fumes
Uses: Natural flavoring agent
Usage level: 3.6 ppm (nonalcoholic beverages), 0.5-4 ppm (ice cream), 7.5 ppm (candy), 6.3 ppm (baked goods), 2.9 ppm (condiments), 40 ppm (meats)
Regulatory: FDA 21CFR §182.10, 182.20, GRAS; 27CFR 21.151; FEMA GRAS; Japan approved (rosemary); Europe listed, no restrictions
Manuf./Distrib.: Chart; Pierre Chauvet; Florida Treatt

Rose oil
CAS 8007-01-0; 84603-93-0; FEMA 2989
Synonyms: Otto of rose; Essence of rose; Attar of rose
Definition: Volatile oil obtained from the flowers of *Rosa spp.*, contg. geraniol, citronellol
Properties: Colorless to pale yel. visc. liq., rose odor and taste; congeals @ 18-22 C to a translucent cryst. mass; sol. in fatty oils, chloroform; sparingly sol. in alcohol; very sl. sol. in water; dens. 0.848-0.863 (30/15 C); ref. index 1.457-1.463 (30 C)
Precaution: Keep cool, well closed; protect from light
Uses: Natural flavoring agent for baked goods, nonalcoholic beverages, condiments, confections, ice cream, meat
Regulatory: FDA 21CFR §182.20, GRAS; FEMA GRAS; Europe listed, no restrictions
Manuf./Distrib.: C.A.L.-Pfizer; Pierre Chauvet; Florida Treatt

Rose water
Definition: Aq. sol'n. of the odoriferous principles of flowers or *Rosa centifloia*
Uses: Natural flavoring agent
Regulatory: FDA 21CFR §182.20, GRAS

Rosin
CAS 8050-09-7; 8052-10-6; EINECS 232-475-7
Synonyms: Colophony; Gum rosin; Rosin gum
Definition: Residue from distilling off the volatile oil from the oleoresin obtained from *Pinus palustris* and other species of *Pinaceae*
Properties: Pale yel. to amber translucent, sl. turpentine odor and taste; insol. in water; sol. in alcohol, benzene, ether, glacial acetic acid, oils, carbon disulfide; dens. 1.07-1.09; m.p. 100-150 C; flash pt. 187 C
Precaution: Combustible
Uses: Coating on fresh citrus fruit; natural flavoring; masticatory substance in chewing gum base
Regulatory: FDA 21CFR §73.1, 172.210, 172.510, 172.615, 175.105, 175.125, 175.300, 176.170, 176.200, 176.210, 177.1200, 177.1210, 177.2600, 178.3120, 178.3800, 178.3870; Japan approved
Manuf./Distrib.: Meer

Rosin, glyceryl ester. *See* Glyceryl rosinate
Rosin gum. *See* Rosin
Rosin, hydrogenated, glycerol ester. *See* Glyceryl hydrogenated rosinate
Rosin pentaerythritol ester. *See* Pentaerythrityl rosinate
Rosmarinum officinalis extract. *See* Rosemary extract
Rubber hydrochloride. *See* Pliofilm
Rubus extract. *See* Blackberry extract
Rubus villosus extract. *See* Blackberry extract

Rue
Definition: Herb of *Ruta montana, R. graveolens, R. bracteosa,* or *R. calepensis*

Properties: Yel. to amber liq., fatty odor; sol. in fixed oils, min. oil; insol. in glycerin, propylene glycol
Toxicology: Heated to decomp., emits acrid smoke and irritating fumes
Uses: Natural flavoring agent; for baked goods, soft candy, frozen dairy desserts and mixes
Usage level: 6.0 ppm (baked goods)
Regulatory: FDA 21CFR §184.1698 (limitation 2 ppm), GRAS

Rue oil

CAS 8014-29-7
Synonyms: Ruta graveolens oil
Definition: Volatile oil distilled from herb *Ruta graveolens* or other species, contg. about 90% methyl nonyl ketone, methyl anthranilate
Properties: Pale yel. to amber liq., char. sharp fatty odor; sol. in fixed oils, min. oil, 3 vols 70% alcohol; pract. insol. in water; insol. in glycerin, propylene glycol; dens. 0.832-0.845 (15/15 C); solid. pt. 8-10 C; ref. index 1.430-1.440 (20 C)
Precaution: Keep cool, well closed; protect from light
Toxicology: Frequent dermal contact produces erythema, vesication; ingestion of large quantities causes epigastric pain, nausea, vomiting, confusion, convulsions, death; may cause abortion; heated to decomp., emits acrid smoke and irritating fumes
Uses: Natural flavoring agent and adjuvant
Usage level: Limitation 10 ppm (frozen dairy desserts), 10 ppm (soft candy), 10 ppm (baked goods), 4 ppm (other food)
Regulatory: FDA 21CFR §184.1699, GRAS
Manuf./Distrib.: Chr. Hansen's

Rum ether. *See* Ethyl oxyhydrate
Ruta graveolens oil. *See* Rue oil

Saccharin

CAS 81-07-2; EINECS 201-321-0; 220-120-9
Synonyms: Saccharin insoluble; o-Benzoic acid sulfimide; 3-Oxo-2,3-dihydro-1,2-benzisothiazole-1,1-dioxide; 2 3-Dihydroxy-1,2-benzothiazolin-3-one-1,1-dioxide; o-Sulfobenzimide; Saccharin acid form
Classification: Organic compd.
Empirical: $C_7H_5NO_3S$
Properties: Wh. cryst., bitter metallic aftertaste; sol. in alcohol, benzene, amyl acetate, ethyl acetate; sl. sol. in water, ether; m.w. 183.18; dens. 0.828; m.p. 229 C
Toxicology: Has been linked to bladder cancer in test animals; not considered human carcinogen by recent findings
Uses: Noncaloric sweetener for beverages, drink mixes, processed foods, chewable vitamin tablets, chewing gum, baked goods; as sugar substitute for cooking or table use; 200-700 times the sweetness of sucrose
Usage level: Limitation 12 mg/fl oz (beverages), 30 mg/serving (processed foods), 20 mg/tsp of sugar sweetening equivalency (sugar substitutes); ADI 5.0 mg/kg (FAO/WHO)
Regulatory: FDA 21CFR §145.116, 145.126, 145.131, 145.136, 145.171, 145.181, 150.141, 150.161, 180.37; Japan approved (0.05 g/kg max. in chewing gum)
Manuf./Distrib.: Atomergic Chemetals; Boots; Cornelius; Dasco Sales; Dinoval; Ellis & Everard; Fruitsource; K&K Greeff; Hays; Hilton Davis; Int'l. Sourcing; Jungbunzlauer; Maruzen Fine Chems.; Penta Mfg.; PMC Specialties; Quimdis; Rit-Chem; Scanchem; Spice King
Trade names: Syncal® SDI; Unisweet SAC

Saccharin acid form. *See* Saccharin
Saccharine soluble. *See* Saccharin sodium
Saccharin insoluble. *See* Saccharin

Saccharin sodium

CAS 128-44-9; EINECS 204-886-1
Synonyms: Saccharine soluble; Sodium saccharide; Sodium saccharin; 1,1-Dioxide-1,2-benzisothiazol-3(2H)-one sodium salt; Sodium 2,3-dihydro-1,2-benzisothiazolin-3-one-1,1-dioxide; Sodium o-benzosulfimide; Sodium benzosulfimide
Classification: Organic compd.
Definition: Sodium salt of saccharin
Empirical: $C_7H_4NO_3S \cdot Na$
Properties: Wh. cryst. or cryst. powd., odorless or faint aromatic odor, very sweet taste; very sol. in water, sl. sol. in alcohol; m.w. 205.17
Toxicology: LD50 (oral, rat) 14,200 mg/kg; mod. toxic by ingestion and IP routes; experimental carcinogen, neoplastigen, tumorigen, teratogen, reproductive effects; human mutagenic data; heated to decomp., emits very toxic fumes of SO_x, Na_2O, and NO_x

Uses: Artificial nonnutritive sweetener for bacon, bakery prods., beverage mixes, chewing gum, dessert s, fruit juice drinks, jam, relishes, chewable vitamin tablets; as sugar substitute for cooking or table use
Usage level: Limitation 12 mg/fl oz (beverages), 12 mg/fl oz (fruit juice drinks), 12 mg/fl oz (beverage mixes), 30 mg/serving (processed foods), 0.01% (bacon)
Regulatory: FDA 21CFR §145.126, 145.131, 145.136, 145.171, 145.181, 150.141, 150.161, 180.37, GRAS; USDA 9CFR §318.7; Japan approved (0.1-2 g/kg residual)
Manuf./Distrib.: Boots; Int'l. Sourcing; Jungbunzlauer

Saccharose. *See* Sucrose
Saccharose distearate. *See* Sucrose distearate
Saccharose mono/distearate. *See* Sucrose polystearate
Saccharose palmitate. *See* Sucrose palmitate
Saccharum lactis. *See* Lactose

Safflower glyceride
Synonyms: Glycerides, safflower mono-; Safflower oil monoglyceride
Definition: Monoglyceride derived from refined safflower oil
Uses: Emulsifier for diet margarines, icings
Trade names: Myverol® 18-98

Safflower oil
CAS 8001-23-8; EINECS 232-276-5
Synonyms: Carthanus tinctorious oil
Definition: Oily liq. obtained from seeds of *Carthanus tinctorius* consisting principally of triglycerides of linoleic acid
Properties: Lt. ycl. oil; sol. in oil and fat solvs.; dens. 0.9211-0.9215 (25/25 C); iodine no. 135-150; sapon. no. 188-194; ref. index 1.472-1.475
Toxicology: Human skin irritant; ingestion in large volumes produces vomiting; heated to decomp., emits acrid smoke and irritating fumes
Uses: Coating agent, emulsifier, formulation aid, texturizer
Regulatory: FDA 21CFR §175.105, 175.300, 176.200, 176.210, GRAS; Japan approved (safflower)
Manuf./Distrib.: Arista Industries; Croda; Lipo
Trade names: Lipovol SAF; Neobee® 18
Trade names containing: Phosal® 50 SA; Phosal® 75 SA

Safflower oil monoglyceride. *See* Safflower glyceride

Saffron
FEMA 2998
Synonyms: Crocus
Definition: Dried stigmata of *Crocus sativus*, contg. glycoside picrocrocin, coloring principles crocin and crocetin
Empirical: $C_{44}H_{64}O_{26}$ (crocin), $C_{20}H_{24}O_4$ (crocetin)
Properties: Reddish-brn. or golden yel. odiferous powd., sl. bitter taste; Crocin: yel.-orange; sol. in hot water; sl. sol. in abs. alcohol, glycerin, propylene glycol; m.w. 1008.97; m.p. 186 C (dec.); Crocetin: brick-red rhomb.; m.w. 328.41; m.p. 285 C (dec.)
Toxicology: Heated to decomp., emits acrid smoke and irritating fumes
Uses: Natural food colorant, flavoring agent; used in alcoholic drinks, nonalcoholic beverages, baked goods, confectionery, margarine, sausage casings
Usage level: 1-260 ppm (as food colorant)
Regulatory: FDA 21CFR §73.500, 182.10, 182.20, GRAS; USDA 9CFR §318.7; Japan approved, restricted as color
Manuf./Distrib.: British Bakels; Calaga Food Ingreds.; Chart; Ikeda; Lebermuth; New England Spice; Norba; Scanchem; Todd's; Whole Herb Co.; World of Spice; E H Worlee GmbH

Safrol
CAS 94-59-7; EINECS 202-345-4
Synonyms: 5-Allyl-1,3-benzodioxole; Safrole; 4-Allyl-1,2-methylenedioxybenzene; Allylpyrocatechol methyl-ene ether
Empirical: $C_{10}H_{10}O_2$
Properties: Colorless liq. or cryst., sassafras odor; very sol. in alcohol; misc. with chloroform, ether; insol. in water; m.w. 162.20; dens. 1.0960 (20 C); m.p. 11 C; b.p. 234.5 C; flash pt. 100 C; ref. index 1.5360-1.5385
Precaution: Combustible exposed to heat or flame
Toxicology: LD50 (oral, rat) 1950 mg/kg; poison by IP, IV routes; mod. toxic by ingestion, subcutaneous routes; experimental carcinogen, neoplastigen, reproductive effects; human mutagenic data; skin irritant; heated to dec., emits acrid smoke, irritating fumes

Uses: Synthetic flavoring agent; prohibited in foods (U.S.)
Usage level: EEC limitations: 1 ppm (food), 2 ppm (alcoholic beverages < 25%), 5 ppm (alcoholic beverages > 25%, 15 ppm in foods contg. mace and nutmeg)
Regulatory: FDA 21CFR §189.180; prohibited from direct addition or use in human food

Safrole. *See* Safrol

Sage
Synonyms: Salvia
Definition: Salvia officinalis
Properties: Warm spicy odor, flavor
Uses: Natural flavoring agent
Usage level: 300 ppm (nonalcoholic beverages), 170 ppm (baked goods), 1500 ppm (meats)
Regulatory: FDA 21CFR §182.10, 182.20, GRAS; Japan approved
Manuf./Distrib.: C.A.L.-Pfizer; Chart

Sage extract
CAS 84082-79-1
Definition: Extract of the leaves of the herbal plant, *Salvia officinalis* of the mint family
Uses: Natural flavoring agent; antioxidant (Japan)
Regulatory: FDA 21CFR §182.20; Japan approved
Manuf./Distrib.: Chart

Sage oil
CAS 8022-56-8, 84776-73-8
Definition: Essential oil obtained from steam distillation of *Salvia lavandulaefolia* or *S. hisspanorium*, contg. essential oil thujone, α-pinene, cineol, borneol, and d-camphor
Properties: Colorless to yel. oil; sol. in fixed oils, glycerin, min. oil, propylene glycol; dens. 0.909-0.932; ref. index 1.468 (20 C)
Toxicology: LD50 (oral, rat) 2600 mg/kg; mod. toxic by ingestion; heated to decomp., emits acrid smoke and irritating fumes
Uses: Natural flavoring agent for fish, pork, poultry, seasonings, soups
Regulatory: FDA 21CFR §182.20, GRAS
Manuf./Distrib.: Chart; Florida Treatt

Sage oil, Dalmatian type
CAS 8022-56-8, 84776-73-8
Synonyms: Dalmatian sage oil; Salvia oil
Definition: Oil obtained from steam distillation of *Salvia officinalis* of the mint family
Properties: Yel. liq., thujone odor and taste; sol. in fixed oils, min. oil; sl. sol. in propylene glycol; insol. in glycerin; dens. 0.903-0.925; ref. index 1.457 (20 C)
Toxicology: LD50 (oral, rat) 2600 mg/kg; mod. toxic by ingestion; mutagenic data; human skin irritant; heated to decomp., emits acrid smoke and irritating fumes
Uses: Natural flavoring agent for fish, pork, poultry, seasonings, soups
Regulatory: FDA 21CFR §182.10, 182.20, GRAS
Manuf./Distrib.: Florida Treatt

SAIB. *See* Sucrose acetate isobutyrate
Sal ammoniac. *See* Ammonium chloride
Sal chalybis. *See* Ferrous sulfate
Salicylal. *See* Salicylaldehyde

Salicylaldehyde
CAS 90-02-8; EINECS 201-961-0; FEMA 3004
Synonyms: Salicylal; Salicylic aldehyde; o-Hydroxybenzaldehyde; 2-Hydroxybenzaldehyde
Empirical: $C_7H_6O_2$
Formula: C_6H_4OHCHO
Properties: Colorless oily liq., bitter almond-like odor, burning taste; sl. sol. in water; sol. in alcohol, ether; m.w. 122.12; dens. 1.166 (20/4 C); m.p. 1-5 C; b.p. 79-80 C (11 mm); flash pt. 77 C; ref. index 1.573 (20 C)
Precaution: Photosensitive
Toxicology: Skin irritant
Uses: Synthetic flavoring agent
Usage level: 0.55 ppm (nonalcoholic beverages), 5 ppm (alcoholic beverages), 1.1 ppm (ice cream), 1.8 ppm (candy), 6.3 ppm (baked goods), 11-18 ppm (chewing gum), 2 ppm (condiments)
Regulatory: FDA 21CFR §172.515
Manuf./Distrib.: Aldrich; Penta Mfg.

Salicylaldehyde methyl ether. *See* o-Methoxybenzaldehyde

Salicylic acid
CAS 69-72-7; EINECS 200-712-3
Synonyms: 2-Hydroxybenzoic acid; o-Hydroxybenzoic acid
Classification: Aromatic acid
Empirical: $C_7H_6O_3$
Formula: HOC_6H_4COOH
Properties: Crystals or cryst. powd., acrid taste; sol. in water, alcohol, ether; m.w. 138.12; dens. 1.443 (20/4 C); m.p. 157-159 C; b.p. 211 C (20 mm) sublimes at 76 C
Precaution: Protect from light
Toxicology: LD50 (oral, rat) 891 mg/kg, (IV, mouse) 500 mg/kg; poison by ingestion, IV, IP routes; mod. toxic by subcutaneous route; skin and severe eye irritant; experimental teratogen, reproductive effects; mutagenic data; heated to decomp., emits acrid smoke
Uses: Preservative for foods; animal drug, fungicide; prohibited in milk and wine
Usage level: Limitation zero tolerance in milk
Regulatory: FDA 21CFR §175.105, 175.300, 177.2600, 556.590
Manuf./Distrib.: Allchem Industries; Hilton Davis; PMC Specialties

Salicylic aldehyde. *See* Salicylaldehyde
Salicylic ether. *See* Ethyl salicylate
Salicylic ethyl ester. *See* Ethyl salicylate
Salmiac. *See* Ammonium chloride
SALP. *See* Sodium aluminum phosphate acidic
Salt. *See* Sodium chloride
Sal tartar. *See* Sodium tartrate
Salt cake. *See* Sodium sulfate
Saltpeter. *See* Potassium nitrate
Salvia. *See* Sage
Salvia oil. *See* Sage oil, Dalmatian type
Sambucus. *See* Elder flowers

Sambucus oil
CAS 68916-55-2
Definition: Volatile oil obtained from *Sambucus nigra* and other species of *Sambucus*
Uses: Natural flavoring agent
Regulatory: FDA 21CFR §172.510, 182.20

Sandalwood
Synonyms: Sandalwood resin; White sandalwood; Yellow sandalwood; East Indian sandalwood; Santalum album
Definition: Plant material obtained from *Santalum album*
Properties: Strong persistent warm woody odor
Uses: Natural flavoring agent
Regulatory: FDA 21CFR §172.510; Japan approved
Manuf./Distrib.: Chart

Sandalwood extract
CAS 84787-70-2
Definition: Extract of the tree *Santalum album*
Uses: Natural flavoring agent
Regulatory: FDA 21CFR §172.510

Sandalwood oil. *See also* Amyris oil, West Indian type
CAS 8006-87-9; FEMA 3005
Synonyms: Santalum album oil; Santal oil; Arheol; East Indian sandalwood oil
Definition: Volatile oil obtained from heartwood of *Santalum album*
Properties: Pale yel. liq., char. sandalwood odor and taste; sol. in 5 vols 70% alcohol; very sl. sol. in water; dens. 0.965-0.980 (25/25 C); ref. index 1.500-1.510 (20 C)
Precaution: Keep cool, well closed; protect from light
Uses: Natural flavoring agent
Usage level: 2.4 ppm (nonalcoholic beverages), 7.5 ppm (ice cream), 7.7 ppm (candy), 6.6 ppm (baked goods), 47 ppm (chewing gum)
Regulatory: FDA 21CFR §172.510
Manuf./Distrib.: Chart; Pierre Chauvet; Florida Treatt

Sandalwood resin. *See* Sandalwood

Sandarac gum
CAS 9000-57-1; EINECS 232-547-8
Synonyms: Gum sandarac
Definition: Resin from *Callitris quadrivalvis*, contg. approx. 80% pimaric acid, 10% callitrolic acid
Properties: Turpentine-like fresh resinous odor
Uses: Natural flavoring agent
Regulatory: FDA 21CFR §172.510, 175.105, 175.300

Santal oil. *See* Sandalwood oil

Santalol
CAS 115-71-9 (α), 77-42-9 (β); FEMA 3006
Synonyms: Argeol; d-α-Santalol; α-Santalol; β-Santalol
Empirical: $C_{15}H_{24}O$
Properties: Liq., sweet sandalwood odor; m.w. 220.34
Uses: Synthetic flavoring agent
Regulatory: FDA 21CFR §172.515

α-Santalol, acetate. *See* Santalyl acetate
β-Santalol, acetate. *See* Santalyl acetate
d-α-Santalol. *See* Santalol
Santalum album. *See* Sandalwood
Santalum album oil. *See* Sandalwood oil

Santalyl acetate
FEMA 3007
Synonyms: α-Santalol, acetate; β-Santalol, acetate
Properties: Colorless to ylsh. liq., sandalwood-like odor; sol. in most common org. solvs.; m.w. 262.40; b.p. 20.8 C (3 mm); ref. index 1.4894-1.4901
Uses: Synthetic flavoring agent
Usage level: 0.53 ppm (nonalcoholic beverages), 0.78 ppm (ice cream, ices), 2.0 ppm (candy, baked goods), 2.3 ppm (chewing gum)
Regulatory: FDA 21CFR §172.515

Santalyl phenylacetate
FEMA 3008
Empirical: $C_{23}H_{30}O_2$
Properties: Colorless liq., sandalwood-like odor; sl. sol. in alcohol; m.w. 338.49
Uses: Synthetic flavoring
Usage level: 1.0 ppm (nonalcoholic beverages), 0.95 ppm (ice cream, ices), 2.0 ppm (candy, baked goods)
Regulatory: FDA 21CFR §172.515

Santoquine. *See* 6-Ethoxy-1,2-dihydro-2,2,4-trimethylquinoline
SAP. *See* Sodium aluminum phosphate acidic
SAPP. *See* Sodium acid pyrophosphate

Sarsaparilla extract
FEMA 3009
Synonyms: Smilax aristolochiaefolia extract
Definition: Extract of the roots of *Smilax aristolochiaefolia*
Uses: Natural flavoring agent
Usage level: 190 ppm (nonalcoholic beverages), 130 ppm (ice cream, ices), 1000 ppm (candy), 2000 ppm (baked goods)
Regulatory: FDA 21CFR §172.510; Japan approved (sarsaparilla)

Sassafras albidum extract. *See* Sassafras extract
Sassafras albidum oil. *See* Sassafras oil

Sassafras extract
Synonyms: Sassafras albidum extract
Definition: Extract of bark and roots of *Sassafras albidum*
Properties: Spicy aromatic odor, flavor
Uses: Natural flavoring
Usage level: 290 ppm (nonalcoholic beverages), 10 ppm (ice cream, ices), 100 ppm (candy), 50 ppm (baked goods)
Regulatory: FDA 21CFR §172.580; must be safrole-free; Japan approved (sassafras)

Manuf./Distrib.: Chr. Hansen's

Sassafras oil
CAS 8006-80-2
Synonyms: Sassafras albidum oil
Definition: Volatile oil obtained from root of *Sassafras albidum*, contg. about 80% safrol, plus eugenol, pinene, phellandrene, sesquiterpene, d-camphor
Properties: Ylsh.-reddish volatile oil, pungent aromatic odor and taste of sassafras; sol. in alcohol, ether, chloroform, glac. acetic acid, CS_2, 2 vols 90% alcohol; very sl. sol. in water; dens. 1.065-1.077 (25/25 C); ref. index 1.5250-1.5350 (20 C)
Precaution: Keep cool, well closed; protect from light
Toxicology: Experimental neoplastigen; skin irritant; heated to decomp., emits acrid smoke and irritating fumes
Uses: Natural flavoring agent for baked goods, nonalcoholic beverages (root beer), confections, gelatin desserts, puddings
Regulatory: FDA 21CFR §172.510, 172.580
Manuf./Distrib.: Florida Treatt

Satureia hortensis extract. *See* Savory extract

Savory
Synonyms: Summer savory
Definition: A plant, *Satureia hortensis*, with strongly flavored leaves
Uses: Natural flavoring agent
Usage level: 800-850 ppm (baked goods), 200 ppm (condiments), 1100 ppm (meats)
Regulatory: FDA 21CFR §182.10, 182.20, GRAS
Manuf./Distrib.: Chr. Hansen's

Savory extract
CAS 84775-98-4
Synonyms: Satureia hortensis extract
Definition: Extract of *Satureia hortensis*
Uses: Natural flavoring agent
Regulatory: FDA 21CFR §182.20, GRAS
Manuf./Distrib.: Chr. Hansen's

Scabwort. *See* Elecampane
Scotch fir oil. *See* Pine needle oil, Scotch type
SDDC. *See* Sodium dimethyldithiocarbamate
SDS. *See* Sodium lauryl sulfate
Secondary ammonium phosphate. *See* Ammonium phosphate, dibasic
Secondary caprylic alcohol. *See* 2-Octanol
Seignette salt. *See* Potassium sodium tartrate anhyd.
Seignette salt. *See* Potassium sodium tartrate tetrahydrate

Senna extract
CAS 85085-71-8
Definition: Extract of leaves of *Cassia obovata*
Uses: Natural flavoring
Regulatory: FDA 21CFR §172.510
Manuf./Distrib.: Chart

DL-Serine
CAS 302-84-1; EINECS 206-130-6
Synonyms: (+-2-Amino-3-hydroxypropionic acid
Empirical: $C_3H_7NO_3$
Formula: $HOCH_2CH(NH_2)COOH$
Properties: Wh. cryst. or cryst. powd.; sol. in water; insol. in alcohol, ether; m.w. 105.09; m.p. 246 C (dec.)
Toxicology: Heated to decomp., emits toxic fumes of NO_x
Uses: Dietary supplement, nutrient; feed additive
Regulatory: FDA 21CFR §172.310 (limitation 8.4%)
Manuf./Distrib.: Degussa

L-Serine
CAS 56-45-1; EINECS 200-274-3
Synonyms: 2-Amino-3-hydroxypropionic acid
Classification: Amino acid
Empirical: $C_3H_7NO_3$

Serpentaria

Formula: HOCH$_2$CH(NH$_2$)COOH
Properties: Wh. cryst. or cryst. powd., odorless, sweet taste; sol. in water; insol. in alcohol, ether; m.w. 105.10; m.p. 228 C (dec.)
Uses: Dietary supplement, nutrient, feed additive; flavoring
Regulatory: FDA 21CFR §172.320, 8.4% max.; Japan approved
Manuf./Distrib.: Degussa; Nippon Rikagakuyakuhin

Serpentaria

EINECS 289-574-3 (extract)
Synonyms: Virginia snakeroot
Definition: From *Aristolochia serpentaria*
Uses: Natural flavoring for alcoholic beverages only
Regulatory: FDA 21CFR §172.510

Serpentaria extract

CAS 84775-44-0; EINECS 283-873-2
Synonyms: Birthwort extract; Aritolochia extract
Definition: Extract of rhizomes of *Aristolochia clematitis*
Uses: Natural flavoring
Regulatory: FDA 21CFR §172.510

Sesame

Synonyms: Benne
Definition: Sesamum indicum
Uses: Natural flavoring agent for bread (seeds)
Regulatory: FDA 21CFR §182.10, GRAS; Japan approved
Manuf./Distrib.: Chart

Sesame oil

CAS 8008-74-0; EINECS 232-370-6
Synonyms: Gingilli oil
Definition: Oil obtained from the seeds of *Sesamum indicum*
Properties: Bland yellowish vegetable oil; dens. 0.9; iodine no. 103-116; sapon. no. 188-195; flash pt. 491 F; ref. index 1.4575-1.4598 (60 C)
Toxicology: Poison by intravenous route; human skin irritant
Uses: Natural flavorant for baking and confections
Regulatory: FDA 21CFR §175.105, 175.300, 176.200, 176.210
Manuf./Distrib.: Arista Industries
Trade names: Lipovol SES; Vitinc® Sesame Oil NF

Shaddock oil. *See* Grapefruit oil

Shark liver oil

CAS 68990-63-6; EINECS 273-616-2
Definition: Oil expressed from fresh livers of sharks and other *Elasmobranchii* species
Properties: Yel.-red-brown liq., strong odor; sol. in ether, chloroform, benzene, carbon disulfide; dens. 0.917-0.928; iodine no. 125-155; sapon. no. 170-187; ref. index 1.4784 (20 C); sapon. no. 140-146
Uses: Nutritive additive
Regulatory: FDA CFR §175.105, 176.210, 177.2800
Manuf./Distrib.: Am. Roland; Arista Industries

Shellac

CAS 9000-59-3; EINECS 232-549-9; EEC E904
Synonyms: White shellac; Bleached shellac
Definition: Resinous secretion of the insect *Laccifer (Tachardia) lacca*
Properties: Off-wh. amorphous gran. solid; sol. in alcohol; sl. sol. in acetone, ether; insol. in water
Toxicology: Heated to decomp., emits acrid smoke and irritating fumes
Uses: Coating agent, color diluent, surface-finishing agent, glazing/polishing agent; used in confectionery, food supplements in tablet form, gum; chewing gum base
Usage level: 0.4% max. (EEC)
Regulatory: FDA 21CFR §73.1, 175.105, 175.300, 175.380, 175.390, 182.99; 27CFR §21.126, 212.61, 212.90; 40CFR §180.1001; Japan approved; Europe listed; UK approved
Manuf./Distrib.: Bradshaw-Praeger; Classic Flavors; Colony Import & Export; H E Daniel; Deutsche Nichimen; Ikeda; Mantrose-Haeuser; Roeper; Thew Arnott; Alfred L Wolff; E H Worlee GmbH

Shellac, bleached, wax-free

Synonyms: Refined bleached shellac

Silicic acid

Uses: Coating agent; surface-finishing agent; glaze

Shellac wax
Definition: Waxy fraction of bleached shellac obtained by physical means

SHMP. See Sodium hexametaphosphate
Siam benzoin. See Gum benzoin

Silica
CAS 7631-86-9; 112945-52-5; EEC E551
Synonyms: Silicon dioxide, fumed; Colloidal silicon dioxide; Silicon dioxide; Fumed silica; Silicic anhydride
Classification: Inorganic oxide
Definition: Occurs in nature as agate, amethyst, chalcedony, cristobalite, flint, quartz, sand, tridymite
Empirical: O_2Si
Formula: SiO_2
Properties: Transparent crystals or amorphous powd.; pract. insol. in water or acids except hydrofluoric; m.w.
  60.09; dens. 2.2 (amorphous), 2.65 (quartz, 0 C); lowest coeff. of heat expansion; melts to a glass
Toxicology: LD50 (oral, rat) 3160 mg/kg; poison by intraperitoneal, intravenous, intratracheal routes;
  moderately toxic by ingestion; prolonged inhalation of dust can cause silicosis
Uses: Anticaking agent, carrier, thickener, conditioner, color diluent; chillproofing agent in malt beverages;
  clarifying agent in wine-making; thickener/stabilizer in soft drinks; antifoam in animal fats, vegetable oils;
  migrating from paper
Usage level: Limitation 2% (of ink solids), 4% (in dry tocopherol-contg. bacon curing agent); ADI not specified
  (EEC)
Regulatory: FDA 21CFR §73.1, 172.230, 172.480 (limitation 2%), 173.340, 175.105, 175.300, 176.200,
  176.210, 177.1200, 177.1460, 177.2420, 177.2600, 182.90, 182.1711, GRAS; USDA 9CFR §318.7
Regulatory: Japan approved (2% max. as anticaking), other restrictions; Europe listed; UK approved
Manuf./Distrib.: Cabot Carbon Ltd; Degussa
Trade names: Aerosil® 200; Aerosil® 380; Cab-O-Sil® HS-5
Trade names containing: AF 72; AF 75; Annatto/Turmeric WMP-1; Dry Phytonadione 1% SD No. 61748; Dry
  Vitamin A Palmitate Type 250-SD No. 65378; Dry Vitamin D₃ Type 100-SD No. 65216; Dry Vitamin E
  Acetate 50% SD No. 65356; FloAm® 450; Garbefix 31; Myvatex® Texture Lite®; Myvatex® Texture Lite®
  K; Rocoat® Niacinamide 33¹/₃% No. 69907; Rocoat® Niacinamide 33¹/₃% Type S No. 69909; Stabil® 9
  High Calcium Blended Powd.; Stabil® 9 Regular Blended Powd.; Turmeric DP-300; Vanillin Free Flow;
  Vitamin B₁₂ 0.1% SD No. 65354; Vitamin B₁₂ 1.0% SD No. 65305

Silica aerogel
CAS 7631-86-9
Synonyms: Silica amorphous hydrated; Silica gel; Silicic acid
Definition: Fine powd. microcellular silica foam, 89.5% min. silica content
Empirical: O_2Si
Properties: m.w. 60.09
Toxicology: TLV:TWA 10 mg/m³; pure unaltered form considered nontoxic; some deposits are fibrogenic
Uses: Clarifying agent for wine; component for antifoaming agents
Usage level: Limitation 20 lb/colloidal silicon dioxide @ 30% conc./1000 gal wine (silicon dioxide to be removed
  by filtration)
Regulatory: FDA 21CFR §182.1711, GRAS; BATF 27CFR §240.1051

Silica amorphous hydrated. See Silica aerogel
Silica gel. See Silica aerogel
Silica gel. See Silica, hydrated

Silica, hydrated
CAS 1343-98-2 (silicic acid); 112926-00-8; EINECS 215-683-2
Synonyms: Silicic acid; Silica gel; Precipitated silica
Classification: Inorganic oxide
Definition: Occurs in nature as opal
Formula: $SiO_2 \cdot xH_2O$, x varies with method of precipitation and extent of drying
Properties: White amorphous powd.; insol. in water or acids except hydrofluoric
Toxicology: Eye irritant; poison by intravenous route; TLV:TWA 10 mg/m³ (total dust)
Uses: Adsorbent, anticaking/free-flow agent for table salt and powd. sweeteners
Regulatory: FDA 21CFR §73.1, 160.105, 160.185, 172.480, 173.340, 175.105, 175.300, 176.170, 176.180,
  176.200, 176.210, 177.1200, 177.2420, 177.2600, 182.90, 573.940
Trade names: Quso® WR55-FG; Sipernat® 22; Sipernat® 22S; Sipernat® 50; Sipernat® 50S; Sipernat® 50S

Silicic acid. See Silica aerogel

717

Silicic acid. *See* Silica, hydrated
Silicic acid, calcium salt. *See* Calcium silicate
Silicic acid, disodium salt. *See* Sodium metasilicate
Silicic acid, magnesium salt (1:1). *See* Magnesium silicate
Silicic acid, sodium salt. *See* Sodium silicate
Silicic anhydride. *See* Silica
Silicon dioxide. *See* Silica
Silicon dioxide, fumed. *See* Silica

Silicone

Synonyms: Organosiloxane
Classification: Siloxane polymers
Properties: Liq., semisolid, or solid; cis. 1->1,000,000 cs; water repellant; exc. ; sol. in most organic solventsdielectric prop.
Precaution: Unhalogenated types combustible
Uses: Defoamer in food processing
Trade names: AF 70; AF 100 FG; AF 8805 FG; AF 8810 FG; AF 8820 FG; AF 8830 FG; AF 9000; Hodag FD Series; Patcote® 308; Patcote® 311; Patcote® 323; Patcote® 555K

Silicone emulsions and compounds. *See also* Dimethicone, Cyclomethicone

Classification: Organosiloxane
Uses: Defoamer in food processing
Trade names: CNC Antifoam 30-FG; Dow Corning® Antifoam 1510-US; Dow Corning® Antifoam 1520-US; Dow Corning® Antifoam FG-10; Drewplus® L-813; Drewplus® L-833; Foamaster FLD; Foam Blast 100; Foam Blast 102; Foam Blast 150; Foamkill® MS Conc.; Hodag Antifoam FD-62; Hodag Antifoam FD-82; Mazu® DF 110S; Mazu® DF 210S; Patcote® 305; Patcote® 306; Patcote® 307; Patcote® 315; Trans-10, -10K; Trans-20, -20K; Trans-25, -25K; Trans-30, -30K; Trans-177S, -177S K; Trans-225S; Trans-1030 K; Trans-134S K

Silver

CAS 7440-22-4; EINECS 231-131-3; EEC E174
Synonyms: Silver, colloidal
Classification: Metallic element
Empirical: Ag
Properties: Soft ductile malleable lustrous wh. metal; sol. in fused alkali hydroxides in presence of air, in fused alkali peroxides; not attacked by water or atmospheric oxygen; inert to most acids; at.wt. 107.868; dens. 10.50 (15 C); m.p. 961.93 C; b.p. 2212 C
Toxicology: TLV:TWA 0.1mg/m^3 (metal); nontoxic but prolonged absorption of compds. can cause grayish discoloration of skin (argyria); avoid inhalation of dust
Manuf./Distrib.: Degussa

Silver, colloidal. *See* Silver
Silver fir oil. *See* Fir needle oil
Silver pine oil. *See* Fir needle oil

Simethicone

CAS 8050-81-5; EEC E900
Definition: Mixture of dimethicone with an avg. chain length of 200-350 dimethylsiloxane units and silica gel
Formula: $(CH_3)_3SiO[SI(CH_3)_2O]_nSi(CH_3)_3$, n = 200-350
Uses: Foam control agent for foods; water repellent; chewing gum base; anticaking agent; in brewing industry, fermentation
Usage level: ADI 0-1.5 mg/kg (EEC)
Regulatory: Europe listed; UK approved
Trade names: Dow Corning® Antifoam A Compd., Food Grade; Dow Corning® Antifoam AF Emulsion; Dow Corning® Antifoam C Emulsion; Hodag Antifoam F-1; Mazu® DF 200SP; Mazu® DF 230SP; Sentry Simethicone NF; Wacker Silicone Antifoam Emulsion SE 9

Skatole

CAS 83-34-1; EINECS 201-471-7; FEMA 3019
Synonyms: 3-Methylindole; β-Methylindole
Empirical: C_9H_9N
Properties: Wh. scales or powd., fecal odor; m.w. 131.18; sol. 1 g/10 ml methanol; m.p. 95-97 C; b.p. 265-266 C; flash pt. 132 C
Uses: Synthetic flavoring
Regulatory: FDA 21CFR §172.515
Manuf./Distrib.: Aldrich

Slaked lime. *See* Calcium hydroxide
Sloe berries. *See* Blackthorn berries
Smilax aristolochiaefolia extract. *See* Sarsaparilla extract
SMO. *See* Sorbitan oleate
SMS. *See* Sorbitan stearate
Soapbark. *See* Quillaja
Soap clay. *See* Bentonite
Soda alum. *See* Sodium alum
Soda ash. *See* Sodium carbonate
Soda calcined. *See* Sodium carbonate
Soda lye. *See* Sodium hydroxide
Sodamide. *See* Sodium amide
Soda niter. *See* Sodium nitrate
Sodium acetate. *See* Sodium acetate anhydrous
Sodium acetate acidic. *See* Sodium diacetate

Sodium acetate anhydrous
CAS 127-09-3; EINECS 204-823-8; EEC E262
Synonyms: Sodium acetate; Acetic acid sodium salt anhydrous
Empirical: $C_2H_3NaO_2$
Formula: CH_3COONa
Properties: Wh. gran. powd.; sol. in water, alcohol; m.w. 82.04; dens. 1.45; m.p. 58 C
Toxicology: LD50 (oral, rat) 3530 mg/kg; poison by IV route; mod. toxic by ingestion; skin and eye irritant; heated to decomp., emits toxic fumes of Na_2O
Uses: Food additive, flavoring agent, adjuvant, pH control agent, buffer, meat preservation; boiler water additive for food contact; migrating from cotton in dry food pkg.
Usage level: Limitation 0.007% (breakfast cereals), 0.5% (fats, oils), 0.6% (grain prods., pasta, snack food), 0.15% (hard candy), 0.12% (jams, jellies, meat prods.), 0.2% (soft candy), 0.05% (soups, sweet sauces); ADI no limit (EEC)
Regulatory: FDA 21CFR §173.310, 182.70, 184.1721, GRAS; Japan approved; Europe listed; UK approved
Manuf./Distrib.: Int'l. Sourcing; Lohmann
Trade names: Niacet Sodium Acetate 60% FCC; Niacet Sodium Acetate Anhyd. FCC

Sodium acetate trihydrate
CAS 6131-90-4; EINECS 204-823-8
Empirical: $C_2H_3NaO_2 \cdot 3H_2O$
Formula: $CH_3COONa \cdot 3H_2O$
Properties: m.w. 136.08
Uses: Flavoring agent, adjuvant, pH control agent
Usage level: Limitation 0.007% (breakfast cereals), 0.5% (fats, oils), 0.6% (grain prods., pasta, snack food), 0.15% (hard candy), 0.12% (jams, jellies, meat prods.), 0.2% (soft candy), 0.05% (soups, sweet sauces)
Regulatory: FDA 21CFR §184.1721, GRAS
Manuf./Distrib.: Lohmann

Sodium acid acetate. *See* Sodium diacetate
Sodium acid phosphate. *See* Sodium phosphate

Sodium acid pyrophosphate
CAS 7758-16-9; EINECS 231-835-0; EEC E450a
Synonyms: SAPP; DSPP; Disodium dihydrogen pyrophosphate; Diphosphoric acid disodium salt; Acid sodium pyrophosphate; Disodium dihydrogen diphosphate; Disodium pyrophosphate; Disodium diphosphate
Classification: Inorganic salt
Empirical: $Na_2H_2P_2O_7$
Properties: Wh. cryst. powd.; sol. in water; m.w. 221.94; dens. 1.862; m.p. 220 C (dec.)
Toxicology: LD50 (oral, mouse) 2650 mg/kg; poison by IV route; mod. toxic by ingestion and subcutaneous routes; irritant to skin, eyes, mucous membranes; heated to decomp., emits toxic fumes of PO_x and Na_2O
Uses: Sequestrant; leavening acid; peptizing agent in cheese/meat prods., frozen desserts; buffer; suspending agent; flavor preservative; crystal formation preventive; meat treating; emulsifier; hog/poultry scald agent; color improver
Usage level: ADI 0-70 mg/kg (EEC, total phosphates)
Regulatory: FDA 21CFR §133.169, 133.173, 133.179, 137.180, 161.190, 182.1087, 182.6787, GRAS; USDA 9CFR §318.7, 381.147; Europe listed; Japan approved
Manuf./Distrib.: Albright & Wilson Am.; Browning; FMC; Rhone-Poulenc Food Ingreds.
Trade names: Aerophos P; Albrite SAPP Food Grade; Antelope Aerator; B.P. Pyro®; B.P. Pyro® Type K; Curacel; Curavis® 150; Donut Pyro®; Donut SAPP; Perfection®; Sapp #4; SAPP 22; SAPP 26; SAPP 28;

SAPP 40; Taterfos®; Victor Cream®
Trade names containing: Albriphos™ Blend 928; Curavis® 250; Curavis® 350; Kena® FP-85

Sodium acid sulfate. *See* Sodium bisulfate
Sodium acid sulfate solid. *See* Sodium bisulfate solid
Sodium acid sulfite. *See* Sodium bisulfite
Sodium alginate. *See* Algin

Sodium alum
CAS 10102-71-3; EINECS 233-277-3
Synonyms: Aluminum sodium sulfate; Soda alum; Sodium aluminum sulfate; Sulfuric acid, aluminum sodium
salt (2:1:1)
Classification: Inorganic salt
Empirical: $AlNaO_8S_2$ (anhyd.), $NaAl(SO_4)_2$ • $12H_2O$ (dodecahydrate)
Properties: Anhyd.: Sol. in alcohol; sl. sol. in water; m.w. 242.10; Dodecahydrate: Colorless cryst. or wh. gran.
or powd.; sol. in water; pract. insol. in alcohol; m.w. 458.29; dens. 1.675; m.p. 61 C
Toxicology: Weak sensitizer; local contact may cause dermatitis; irritant; heated to decomp., emits toxic fumes
of SO_x and Na_2O
Uses: Multipurpose food additive; buffer, firming agent, neutralizing agent; for baked goods, pickles; migrating
to food from paper/paperboard
Regulatory: FDA 21CFR §175.105, 178.3120, 182.90, 182.1131, GRAS

Sodium aluminate
CAS 1302-42-7, 11138-49-1; EINECS 215-100-1
Synonyms: Aluminum sodium oxide; Sodium aluminum oxide; Aluminate, sodium; Sodium polyaluminate
Classification: Inorganic compd.
Empirical: AlO_2 • Na
Formula: $AlNaO_2$
Properties: Wh. powd.; hygroscopic; sol. in water; insol. in alcohol; m.w. 82.0; m.p. 1650 C
Precaution: Corrosive material
Toxicology: TLV:TWA 2 mg(Al)/m³; mod. irritant to skin, eyes, mucous membranes; corrosive substance;
heated to decomp., emits toxic fumes of Na_2O
Uses: Boiler water additive for food contact; migrating to food from paper/paperboard
Regulatory: FDA 21CFR §173.310, 182.90

Sodium aluminosilicate. *See* Sodium silicoaluminate
Sodium aluminum oxide. *See* Sodium aluminate
Sodium aluminum phosphate. *See* Sodium aluminum phosphate acidic

Sodium aluminum phosphate acidic
CAS 7785-88-8; EEC E541
Synonyms: SALP; SAP; Sodium aluminum phosphate
Formula: $NaAl_3H_{14}(PO_4)_8$ • $4H_2O$
Properties: Wh. powd., odorless; sol. in HCl; insol. in water; m.w. 949.88
Toxicology: Nuisance dust
Uses: Emulsifier, aerator, acidulant, leavening agent for cakes, pancake mixes
Usage level: ADI 0-6 mg/kg (EEC, as sodium aluminum phosphates)
Regulatory: FDA 21CFR §182.1781, GRAS; Europe listed; UK approved
Manuf./Distrib.: Monsanto; Rhone-Poulenc Food Ingreds.
Trade names: AeroLite LP; Levair®; Royal Baking Powd.
Trade names containing: actif•8®; actif•8® Hi-Calcium; BL-60®; Stabil® 9 Regular Blended Powd.

Sodium aluminum phosphate basic
CAS 7785-88-8; EEC E541
Synonyms: Phosphoric acid sodium aluminum salt, basic
Definition: Mixt. of alkaline sodium aluminum phosphate and dibasic sodium phosphate
Properties: Wh. powd., odorless; very sl. sol. in water
Uses: Emulsifier for processed cheese (U.S.)
Usage level: ADI 0-6 mg/kg (EEC, as sodium aluminum phosphates)
Regulatory: Europe listed; UK approved
Manuf./Distrib.: Rhone-Poulenc Food Ingreds.
Trade names: Kasal®

Sodium aluminum silicate. *See* Sodium silicoaluminate
Sodium aluminum sulfate. *See* Sodium alum

Sodium amide
CAS 7782-92-5; EINECS 231-971-0

Synonyms: Sodamide
Empirical: H$_2$NNa
Formula: NaNH$_2$
Properties: Wh. cryst. powd.; hygroscopic; m.w. 39.02; m.p. 210 C; b.p. 400 C
Precaution: Flamm. by chemical reaction; ignites or explodes with heat or grinding; explosive reaction with moisture, halocarbons oxidants, sodium nitrite, air; reacts with water to produce heat and toxic fumes
Toxicology: Intense irritant to tissue, skin, eyes; heated to decomp., emits highly toxic fumes of NH$_3$ and Na$_2$O
Uses: Catalyst for transesterification of rendered animal fats
Regulatory: USDA 9CFR §318.7 (must be eliminated during processing)

Sodium ascorbate

CAS 134-03-2; EINECS 205-126-1; EEC E301
Synonyms: L(+)-Ascorbic acid sodium salt; Vitamin C sodium salt; Monosodium ascorbate
Empirical: C$_6$H$_8$NaO$_6$
Properties: Wh. to yel. cryst., odorless; sol. in water; very sl. sol. in alcohol; insol. in chloroform, ether; m.w. 199.13; dec. 218 C
Precaution: Combustible
Toxicology: Human mutagenic data; heated to decomp., emits toxic fumes of Na$_2$O
Uses: Antioxidant, preservative, color preservative, dietary supplement, nutrient in food products; also in curing meat
Usage level: Limitation 87.5 oz/100 gal (pickle), $^7/_8$ oz/100 lb (meat), 500 ppm (singly or in combination with ascorbic acid, erythorbic acid)
Regulatory: FDA 21CFR §182.3731, GRAS; USDA 9CFR §318.7; Japan approved; Europe listed; UK approved
Manuf./Distrib.: Browning; Spice King
Trade names: Descote® Sodium Ascorbate 50%; Sodium Ascorbate USP, FCC Fine Gran. No. 6047709; Sodium Ascorbate USP, FCC Fine Powd. No. 6047708; Sodium Ascorbate USP, FCC Type AG No. 6047710

Sodium aspartate

CAS 17090-93-6, 3792-50-5; EINECS 241-155-6
Synonyms: Aspartic acid sodium salt
Definition: Sodium salt of aspartic acid
Empirical: C$_4$H$_7$NO$_4$ • Na
Uses: Dietary and nutritional additive; flavorings
Regulatory: FDA 21CFR §172.320; Japan approved

Sodium benzoate

CAS 532-32-1; EINECS 208-534-8; EEC E211
Synonyms: Benzoic acid sodium salt; Benzoate of soda; Benzoate sodium
Definition: Sodium salt of benzoic acid
Empirical: C$_7$H$_5$O$_2$ • Na
Formula: C$_6$H$_5$COONa
Properties: White gran. or cryst. powd., odorless, sweetish astringent taste; sol. in water, alcohol; m.w. 144.11; pH ≈ 8
Precaution: Combustible when exposed to heat or flame; incompat. with acids, ferric salts; heated to decomp., emits toxic fumes of Na$_2$O
Toxicology: LD50 (oral, rat) 4.07 g/kg; poison by subcutaneous, IV routes; mod. toxic by ingestion, IP routes; may cause human intolerance reaction, asthma, rashes, hyperactivity; experimental teratogen, reproductive effects; mutagenic data
Uses: Preservative, antimocrobial agent, flavoring agent, adjuvant; antimycotic migrating from food pkg.
Usage level: 0.5% max.; 0.1% max. in food; 0.1% (in distilling materials); ADI 0-5 mg/kg (EEC)
Regulatory: FDA 21CFR §146.152, 146.154, 150.141, 150.161, 166.40, 166.110, 181.22, 181.23, 184.1733; GRAS; USDA 9CFR §318.7; BATF 27CFR §240.1051; EPA reg.; GRAS (FEMA); Japan approved with limitations; Europe listed 0.5% as acid
Manuf./Distrib.: Ashland; Browning; Consolidated Flavor; Int'l. Sourcing; Jungbunzlauer; Lohmann; Mallinckrodt; MLG Enterprises; Pfizer
Trade names: ProBenz; Sodium Benzoate BP88
Trade names containing: Beta Carotene Emulsion Beverage Type 3.6 No. 65392; Buckeye C (Baker's Grade); Cola No. 23443; Dry Beta Carotene Beadlets Yellow Type 2.4-S No. 653800100; Dry Vitamin A Palmitate Type 250-SD No. 65378; Dry Vitamin D$_3$ Type 100-SD No. 65216; Golden Crescent; Moldban; Polypro 2000® CF 45%; Polypro 2000® UF 45%; Polypro 5000® CF or SF 45%; Polypro 5000® UF 45%; Shedd's Liquid Margarine; Shedd's Margarine; Shedd's Special 40 Butter Blend Margarine; Shedd's Special 60 Butter Blend Margarine; Sporban Regular; Sporban Special; Superb® Bakers Margarine 105-101;

Sodium benzosulfimide

Vanease (K); Vitamin B$_{12}$ 0.1% SD No. 65354; Vitamin B$_{12}$ 1.0% SD No. 65305

Sodium benzosulfimide. *See* Sodium saccharin
Sodium o-benzosulfimide. *See* Saccharin sodium

Sodium bicarbonate
CAS 144-55-8; EINECS 205-633-8; EEC E500
Synonyms: Baking soda; Sodium hydrogen carbonate; Bicarbonate of soda; Carbonic acid monosodium salt; Monosodium carbonate
Classification: Inorganic salt
Empirical: CHNaO$_3$
Formula: NaHCO$_3$
Properties: White powd. or cryst. lumps, sl. alkaline taste; sol. in water; insol. in alcohol; m.w. 84.01; dens. 2.159
Toxicology: Nuisance dust
Uses: Alkali, leavening agent, aerator, diluent, pH control agent, poultry scald agent; mfg. of effervescent salts and beverages, artificial mineral water
Usage level: ADI no limit (EEC)
Regulatory: FDA 21CFR §137.180, 137.270, 163.110, 173.385, 182.1736, 184.1736, GRAS; USDA 9CFR §318.7, 381.147; Japan approved; Europe listed
Manuf./Distrib.: Browning; FMC; Rhone-Poulenc Food Ingreds.
Trade names: Sodium Bicarbonate USP No. 1 Powd.; Sodium Bicarbonate USP No. 2 Fine Gran; Sodium Bicarbonate USP No. 5 Coarse Gran
Trade names containing: Bac-N-Fos®; Bac-N-Fos® Formula 191; Cap-Shure® BC-140-70; Cap-Shure® BC-900-85; Cap-Shure® C-135-72; Cap-Shure® C-140-72; Cap-Shure® C-140-85; Cap-Shure® C-150-50; Cap-Shure® C-165-63; Cap-Shure® C-900-85; Durkote Sodium Bicarbonate/Hydrog. Veg. Oil; FishGard® FP-55; Sodium Bicarbonate T.F.F. Treated Free Flowing

Sodium biphenyl-2-yl oxide. *See* Sodium o-phenylphenate
Sodium biphosphate. *See* Sodium phosphate

Sodium bisulfate
CAS 7681-38-1; EINECS 231-665-7
Synonyms: Sulfuric acid monosodium salt; Sodium acid sulfate; Sodium hydrogen sulfate; Sodium pyrosulfate
Empirical: HNaO$_4$S
Formula: NaHSO$_4$
Properties: Colorless cryst. or wh. fused lumps; hygroscopic; sol. in water; m.w. 120.07
Toxicology: Causes burns; irritating to respiratory system
Uses: Acidulant
Regulatory: FDA 21CFR §175.105

Sodium bisulfate fused. *See* Sodium bisulfate solid

Sodium bisulfate solid
CAS 7681-38-1
Synonyms: Nitre cake; Sodium acid sulfate solid; Sodium bisulfate fused; Sodium pyrosulfate
Empirical: HO$_4$S • Na
Properties: Wh. cryst. or gran.; hygroscopic; sol. in 2 parts water, 1 part boiling water; dec. by alcohol; m.w. 120.06; dens. 2.435 (13 C); m.p. > 315 C (dec.)
Precaution: Corrosive; reacts with moisture to form sulfuric acid; incompat. with calcium hypochlorite; keep well closed
Toxicology: Corrosive irritant to skin, eyes, mucous membrane; mutagenic data; heated to decomp., emits toxic fumes of SO$_x$ and Na$_2$O
Uses: Acidulant, preservative; for fresh and dried fruits, lemon juice, meat, fresh vegetables
Regulatory: FDA 21CFR §182.3739, GRAS (not for use in meats, vitamin B$_1$ sources, raw fruits and vegetables)

Sodium bisulfite. *See also* Sodium metabisulfite
CAS 7631-90-5; EINECS 231-548-0; EEC E222
Synonyms: Sodium acid sulfite; Acid sodium sulfite; Sodium bisulfite (1:1); Sulfurous acid monosodium salt; Sodium hydrogen sulfite; Sodium sulhydrate
Classification: Inorganic salt
Empirical: HO$_3$S • Na
Formula: NaHSO$_3$
Properties: Wh. cryst. powd., SO$_2$ odor, diasgreeable taste; sol. in water; sl. sol. in alcohol; m.w. 104.06; dens. 1.48; m.p. 315 C
Precaution: Corrosive

Toxicology: LD50 (oral, rat) 2000 mg/kg; poison by intravenous and intraperitoneal routes; mod. toxic by ingestion; corrosive irritant to skin, eyes, mucous membranes; mutagenic data; allergen; heated to decomp., emits toxic fumes of SO_x and Na_2O
Uses: Preservative; not for use on meats, sources of vitamin B_1, raw fruits and vegetables, or fresh potatoes
Usage level: ADI 0-0.7 mg/kg (EEC)
Regulatory: FDA 21CFR §161.173, 173.310, 182.3739, GRAS; Europe listed
Manuf./Distrib.: Hoechst Celanese; Penreco
Trade names containing: Stadex® 90

Sodium bisulfite (1:1). *See* Sodium bisulfite

Sodium borate
CAS 1303-96-4 (decahydrate), 1330-43-4 (anhyd.); EINECS 215-540-4
Synonyms: Sodium tetraborate; Sodium pyroborate; Borax
Classification: Inorganic salt
Formula: $Na_2B_4O_7 \cdot 10H_2O$
Properties: Wh. cyrstals or powd., odorless; hygroscopic; sol. in water, glycerol; m.w. 381.37; dens. 1.730
Toxicology: Irritant
Uses: Used in adhesives, sizes, and coatings in paper contacting food
Regulatory: FDA 21CFR §175.105, 175.210, 176.180, 177.2800, 181.22, 181.30
Manuf./Distrib.: Aldrich

Sodium n-butyl-4-hydroxybenzoate. *See* Sodium butylparaben
Sodium butyl-p-hydroxybenzoate. *See* Sodium butylparaben

Sodium butylparaben
CAS 36457-20-2
Synonyms: Butylparaben, sodium salt; Sodium n-butyl-4-hydroxybenzoate; Sodium butyl-p-hydroxybenzoate
Definition: Sodium salt of butylparaben
Empirical: $C_{11}H_{14}O_3 \cdot Na$
Uses: Preservative, bactericide, fungicide for foods
Trade names: Nipabutyl Sodium

Sodium calcium aluminosilicate hydrated. *See* Sodium calcium silicoaluminate, hydrated

Sodium calcium caseinate
Uses: Used in nutritional drinks
Manuf./Distrib.: Excelpro

Sodium calcium silicoaluminate, hydrated
Synonyms: Calcium sodium aluminosilicate hydrated; Sodium calcium aluminosilicate hydrated
Toxicology: Nuisance dust
Uses: Anticaking agent, general purpose food additive; for tea, cofee, chocolate drinks, dried fruit
Regulatory: FDA 21CFR §182.2729 (2% max.), GRAS; UK permitted

Sodium carbonate
CAS 497-19-8 (anhyd.), 5968-11-6 (monohydrate), 6132-02-1 (decahydrate); EINECS 207-838-8; EEC E500
Synonyms: Soda ash; Soda calcined; Carbonic acid disodium salt; Disodium carbonate
Classification: Inorganic salt
Empirical: $CO_3 \cdot 2Na$
Formula: Na_2CO_3
Properties: Wh. small cryst. or cryst. powd., alkaline taste, odorless; hygroscopic; sol. in water, glycerol; insol. in alcohol; m.w. 105.99; dens. 2.509 (0 C); m.p. 109 C (loses water 851 C)
Toxicology: LD50 (oral, rat) 4090 mg/kg; skin and eye irritant; poison by intraperitoneal route; moderately toxic by inhalation and subcutaneous routes; mildly toxic by ingestion; experimental reproductive effects; heated to decomp., emits toxic fumes of Na_2O
Uses: Antioxidant, curing/pickling agent, flavoring agent/adjuvant, pH control agent, alkali, processing aid, sequestrant, fat rendering agent; boiler water additive; hog scald agent; in beer malting
Usage level: ADI no limit (EEC)
Regulatory: FDA 21CFR §173.310, 184.1742, GRAS; USDA 9CFR §318.7, 381.147; Japan approved; Europe listed
Manuf./Distrib.: Lohmann

Sodium carboxymethylcellulose. *See* Carboxymethylcellulose sodium

Sodium carrageenan
CAS 9061-82-9; 60616-95-7
Synonyms: Carrageenan, sodium salt; Sodium carrageenate

Sodium carrageenate

Definition: Sodium salt of carrageenan
Uses: Emulsifier, stabilizer, thickener
Regulatory: FDA 21CFR §136.110, 136.115, 136.130, 136.160, 136.180, 139.121, 139.122, 150.141, 150.161, 172.626, 176.170

Sodium carrageenate. *See* Sodium carrageenan

Sodium caseinate
CAS 9005-46-3; 9004-36-3
Synonyms: Casein sodium salt
Definition: Sodium salt of casein
Properties: White coarse powd., odorless, tasteless; insol. in water, alcohol
Toxicology: Experimental tumorigen; heated to decomp., emits toxic fumes of Na_2O
Uses: Food additive; binder and extender in sausage, soups, etc.; emulsifier, stabilizer; clarifying agent; processing aid (Japan)
Regulatory: FDA 21CFR §135.110, 135.140, 166.110, 182.1748, GRAS; USDA 9CFR §318.7, 381.147; BATF 27CFR §240.1051, GRAS; Japan approved
Manuf./Distrib.: Blossom Farm Prods.; Excelpro
Trade names: Bindox-HV-051; Bindox-LV-050
Trade names containing: Lamemul® K 3000-Range; Lecithin L 3000-Range; Medium 7; Medium VS; Precision Base

Sodium chloride
CAS 7647-14-5; EINECS 231-598-3
Synonyms: Rock salt; Salt; Table salt; Common salt
Classification: Inorganic salt
Definition: Occurs in nature as the mineral halite
Empirical: ClNa
Formula: NaCl
Properties: Colorless transparent crystals, or white crystalline powd.; sol. 1 g/2.8 ml water; sol. in glycerin; very sl. sol. in alcohol; m.w. 58.45; dens. 2.17; m.p. 804 C; f.p. -20.5 C (23% aq.); pH 6.7-7.3
Precaution: Heated to decomp., emits toxic fumes of Cl⁻ and Na_2O
Toxicology: LD50 (oral, rat) 3.75 g/kg; mod. toxic by ingestion, IV, subcutaneous routes; experimental teratogen; human systemic effects (blood pressure increase); terminates human pregnancy by intraplacental route; human mutagenic data; skin/eye irritant
Uses: Preservative; flavoring agent and intensifier; seasoning for foods; chilling media; curing agent; dough conditioner; nutrient; for baked goods, butter, cheese, nuts, poultry; migrating to food from paper/paperboard, from cotton in dry food pkg.
Usage level: Limitation 700 lb/10,000 gal water (poultry chilling)
Regulatory: FDA 21CFR §131.111, 131.112, 131.136,131.138, 131.144, 131.146, 131.162, 131.170, 131.185, 131.187, 133.123, 133.124, 133.169, 133.173, 133.179, 133.187, 133.188, 133.189, 133.190, 133.195, 136.110, 136.115, 136.130, 136.160, 136.180, 145.110, 145.130, 155, 156, 158, 161.170, 161.173, 161.190, 163.111, 163.112, 163.113, 163.114, 163.117, 163.123, 163.130, 163.135, 163.140, 163.145, 163.150, 163.155, 166.110, 169.115, 169.140, 169.150, 182.1, 182.70, 182.90, GRAS; 9CFR §381.147
Manuf./Distrib.: Akzo Salt; Heico; Morton Salt
Trade names: Culinox® 999 Chemical Grade Salt; Culinox® 999® Food Grade Salt; Morton® H.G. Blending Salt; Morton® TFC 999® Salt; Purex® All Purpose Salt; Sterling® Purified USP Salt
Trade names containing: Cap-Shure® SC-135-63; Cap-Shure® SC-140X-70; Cap-Shure® SC-165-85FT; Cap-Shure® SC-165X-60FF; Diet Imperial® Spread; Durkote Calcium Carbonate/Starch, Acacia Gum; Durkote Sodium Chloride/Hydrog. Veg. Oil; EverFresh®; Freez-Gard® FP-88E; Morton® Flour Lite Salt™ Mixt; Morton® Flour Salt; Morton® H.G. Blending Prepared Salt; Morton® Lite Salt® Mixt; Morton® Lite Salt® TFC Mixt; Morton® Star Flake® Dendritic ES Salt; Morton® Star Flake® Dendritic Salt; Morton® Table Iodized Salt; Morton® TFC H.G. Blending Salt; Morton® Top Flake Coarse Salt; Morton® Top Flake Extra Coarse Salt; Morton® Top Flake Fine Salt; Morton® Top Flake Topping Salt; Shedd's Liquid Margarine; Shedd's Margarine; Shedd's NP Margarine; Shedd's Special 40 Butter Blend Margarine; Shedd's Special 60 Butter Blend Margarine; Shedd's Wonder Shortening; Tang; TFC Purex® Salt

Sodium cholate. *See* Ox bile extract

Sodium citrate
CAS 68-04-2; 18996-35-5; EINECS 242-734-6; 200-675-3; EEC E331
Synonyms: Monosodium citrate anhydrous; Monosodium-2-hydroxypropane-1,2,3-tricarboxylate; Sodium dihydrogen citrate; Sodium citrate primary
Empirical: $C_6H_7NaO_7$

Properties: Wh. cryst. powd. or gran., pract. odorless, saline sl. acidic taste; sol. in water; pract. insol. in ethanol; m.p. 200 C; dec. > 310 C; biodeg.
Precaution: Combustible; avoid dust formation
Toxicology: Heated to decomp., emits acrid smoke and irritating fumes
Uses: Sequestrant, antioxidant, buffer, nutrient for cultured buttermilk; curing accelerator; soft drinks, frozen desserts, meat prods., cheeses, poultry; stabilizer migrating from food pkg.
Usage level: ADI no limit (EEC)
Regulatory: FDA GRAS; Europe listed; UK approved
Manuf./Distrib.: ADM; Ashland; Browning; Cargill; Cheil Foods; Haarmann & Reimer; Hays; N I Ibrahim; Int'l. Sourcing; Jungbunzlauer; Lohmann; MLG Enterprises; Penta Mfg.; Produits Roche; Quimdis; Roche; SAPA; Tesco; Van Waters & Rogers
Trade names: Sodium Citrate USP, FCC Dihydrate Gran. No. 69976; Sodium Citrate USP, FCC Dihydrate Fine Gran. No. 69975
Trade names containing: Medium 55; Medium VS; Palma-Sperse® Type 250-S No. 65322; Palma-Sperse® Type 250A/50 D-S No. 65221; Vitamin B_{12} 0.1% SD No. 65354; Vitamin B_{12} 1.0% SD No. 65305

Sodium citrate primary. *See* Sodium citrate
Sodium citrate tertiary. *See* Trisodium citrate
Sodium CMC. *See* Carboxymethylcellulose sodium

Sodium cyclamate
Synonyms: Sodium cyclohexylsulfamate; Sucrosa; Sucaryl sodium
Empirical: $C_6H_{12}NNaO_3S$
Properties: Pleasantly sweet; relatively stable in sol'n. and during heating
Toxicology: LD50 (oral, rat) 15.25 g/kg, (oral, mouse) 17.0 g/kg
Uses: Nonnutritive sweetener; 30-60 times the sweetness of sucrose
Usage level: ADI (WHO) 11 mg/kg body wt.

Sodium cyclohexylsulfamate. *See* Sodium cyclamate
Sodium decylbenzenesulfonamide. *See* Sodium decylbenzene sulfonate

Sodium decylbenzene sulfonate
CAS 1322-98-1; EINECS 215-347-5
Synonyms: Decylbenzenesulfonic acid, sodium salt; Decylbenzene sodium sulfonate; Sodium decylbenzenesulfonamide
Classification: Substituted aromatic compd.
Empirical: $C_{16}H_{25}O_3S \cdot Na$
Properties: m.w. 320.46
Toxicology: LD50 (oral,mouse) 2000 mg/kg; poison by intravenous route; moderately toxic by ingestion; severe eye irritant; heated to decomp., emits toxic fumes of SO_x
Uses: Defoaming agent, dispersing adjuvant; used on fresh citrus fruit
Regulatory: FDA 21CFR §172.210

Sodium dehydroacetate
CAS 4418-26-2; EINECS 224-580-1
Synonyms: 3-(1-Hydroxyethylidene)-6-methyl-2H-pyran-2,4(3H)-dione sodium salt; Sodium dehydroacetic acid
Classification: Heterocyclic compd.
Empirical: $C_8H_7O_4 \cdot Na$
Properties: White powd., odorless, sl. char. taste; sol. in water, propylene glycol, glycerin; insol. in most org. solvs.; heat stable to 120 C; m.w. 190.14; m.p. 109-111 C
Toxicology: LD50 (oral, mouse) 1175 mg/kg; poison by IV route; mod. toxic by ingestion; experimental teratogen, reproductive effects; mutagenic data; heated to decomp., emits toxic fumes of Na_2O
Uses: Preservative in food
Usage level: 0.2-0.5%; 65 ppm max. (squash)
Regulatory: FDA 21CFR §172.130, 175.105; CIR approved; Japan approved (0.5 g/kg as dehydroacetic acid); Europe listed 0.6% max. (acid)
Manuf./Distrib.: Int'l. Sourcing
Trade names: Tristat SDHA

Sodium dehydroacetic acid. *See* Sodium dehydroacetate

Sodium diacetate
CAS 126-96-5 (anhyd.); EINECS 204-814-9; EEC E262
Synonyms: Sodium acid acetate; Sodium hydrogen diacetate; Sodium acetate acidic
Empirical: $C_4H_7O_4Na \cdot xH_2O$

Sodium di(2-ethylhexyl) sulfosuccinate

Formula: $CH_3COONa \cdot CH_3COOH$

Properties: Wh. cryst. powd., odor of acetic acid; sol. in water, liberating 42.25% avail. acetic acid; m.w. 142.09 + water; dec. above 150 C

Precaution: Combustible

Toxicology: LD50 (oral, rat) 4960 mg/kg (neutrasl); may cause eye or skin irritation; heated to decomp., emits acrid smoke and irritating fumes

Uses: Antimicrobial agent, flavoring agent, adjuvant, pH control agent, mold inhibitor, preservative, sequestrant; for baking materials, meat prods., spice preps.

Usage level: Limitation 0.4% (baked goods), 0.1% (fats, oils, meat prods., soft candy), 0.25% (gravies, sauces), 0.05% (snack foods, soups); ADI 0-15 mg/kg (EEC)

Regulatory: FDA 21CFR §184.1754, GRAS; Europe listed

Manuf./Distrib.: Jungbunzlauer; Lohmann

Trade names: Niacet Sodium Diacetate FCC

Trade names containing: Essiccum®

Sodium di(2-ethylhexyl) sulfosuccinate. *See* Dioctyl sodium sulfosuccinate
Sodium 2,3-dihydro-1,2-benzisothiazolin-3-one-1,1-dioxide. *See* Saccharin sodium
Sodium dihydrogen citrate. *See* Sodium citrate

Sodium dimethyldithiocarbamate

CAS 128-04-1 (hydrate); 148-18-5; EINECS 204-876-7 (hydrate)

Synonyms: SDDC; Sodium N,N-dimethyl dithiocarbamate; Dimethyldithiocarbamic acid sodium salt; Methyl namate

Empirical: $C_3H_6NS_2 \cdot Na$ (anhyd.), $C_3H_6NNaS_2 \cdot aq.$ (hydrate)

Formula: $(CH_3)_2NCS_2Na$

Properties: Crystals; m.w. 143.19 + aq.; dens. 1.1 (20/20 C); m.p. 95 C

Toxicology: LD50 (oral, rat) 1000 mg/kg; moderately toxic by ingestion, intraperitoneal and subcutaneous routes; mutagenic data; heated to decomp., emits toxic fumes of NO_x, SO_x, and Na_2O

Uses: Antimicrobial for use in cane-sugar and beet-sugar mills

Regulatory: FDA 21CFR §173.320 (3 ppm max.)

Sodium N,N-dimethyl dithiocarbamate. *See* Sodium dimethyldithiocarbamate
Sodium dioctyl sulfosuccinate. *See* Dioctyl sodium sulfosuccinate

Sodium dodecylbenzenesulfonate

CAS 25155-30-0; 68081-81-2; 85117-50-6; EINECS 246-680-4

Synonyms: Sodium lauryl benzene sulfonate; Dodecylbenzenesulfonic acid sodium salt; Dodecylbenzene sodium sulfonate

Classification: Substituted aromatic compd.

Empirical: $C_{18}H_{29}O_3S \cdot Na$

Properties: White to lt. yel. flakes, granules, or powd.; m.w. 348.52

Precaution: Combustible

Toxicology: LD50 (oral, rat) 1260 mg/kg, (oral, mouse) 2 g/kg, (IV, mouse) 105 mg/kg; poison by intravenous route; moderately toxic by ingestion; skin and severe eye irritant; heated to decomp., emits toxic fumes of Na_2O

Uses: Washing/lye peeling of fruits and vegetables; denuding agent; poultry scald agent; washing water agent; animal glue adjuvant

Usage level: Limitation 0.2% (in wash water)

Regulatory: FDA 21CFR §173.315, 175.105, 175.300, 175.320, 176.210, 177.1010, 177.1200, 177.1630, 177.2600, 177.2800, 178.3120, 178.3130, 178.3400; USDA 9CFR §318.7, 381.147

Trade names: Calsoft L-60; Rhodacal® DS-4; Rhodacal® LDS-10; Sul-fon-ate AA-9; Sul-fon-ate AA-10; Sul-fon-ate LA-10

Sodium dodecyl sulfate. *See* Sodium lauryl sulfate

Sodium erythorbate

CAS 6381-77-7, 7378-23-6; EINECS 228-973-9

Synonyms: D-Erythro-hex-2-enonic acid, γ-lactone, monosodium salt; Sodium isoascorbate

Definition: Sodium salt of erythorbic acid

Empirical: $C_6H_7NaO_6 \cdot H_2O$

Properties: White free-flowing cryst. powd., odorless; sol. in water; m.w. 216.12

Uses: Antioxidant, preservative, curing accelerator; for baked goods, beef, beverages, cured meat prods., fresh pork, poultry

Usage level: Limitation 87.5 oz/100 gal (pickle), $^7/_8$ oz/100 lb (meat)

Regulatory: USDA 9CFR §318.7, 381.147; cleared by MID to accelerate color fixing in cured pork and beef cuts; Japan restricted for purpose of antioxidation

Manuf./Distrib.: Browning; Int'l. Sourcing; Pfizer Food Science
Trade names: Eribate®; Neo-Cebitate®
Trade names containing: Freez-Gard® FP-88E

Sodium ethylate
Properties: powd.
Uses: Catalyst for transesterification and isomerization of fats and oils
Manuf./Distrib.: Hüls AG

Sodium 2-ethylhexyl sulfate. *See* Sodium octyl sulfate

Sodium ethylparaben
CAS 35285-68-8; EINECS 252-487-6
Synonyms: 4-Hydroxybenzoic acid, ethyl ester, sodium salt; Ethylparaben, sodium salt
Definition: Sodium salt of ethylparaben
Empirical: $C_9H_{10}O_3 \cdot Na$
Toxicology: May cause asthma, rashes, hyperactivity
Uses: Preservative, bactericide, fungicide for beer, flavors, fruit juices, fish, pickles, sauces, soft drinks, yogurt
Trade names: Nipagin A Sodium
Trade names containing: Nipasept Sodium

Sodium ferric pyrophosphate
CAS 10045-87-1
Synonyms: Sodium iron pyrophosphate
Formula: $Na_8Fe_4(P_2O_7)_5 \cdot xH_2O$
Properties: Wh. to tan powd., odorless; sol. in HCl; insol. in water; m.w. 1277.02 (anhyd.)
Uses: Nutritional additive; dietary supplement
Regulatory: FDA 21CFR §182.5306, 182.8306

Sodium ferrocyanide
CAS 13601-19-9; EINECS 237-081-9; EEC E535
Synonyms: Sodium ferrocyanide decahydrate; Sodium hexacyanoferrate (II); Yellow prussiate of soda
Empirical: $C_6FeN_6Na_4 \cdot 10H_2O$
Formula: $Na_4Fe(CN)_6 \cdot 10H_2O$
Properties: Yel. cryst. or cryst. powd.; sol. in water; pract. insol. in most org. solvs.; m.w. 484.06; dens. 1.458; dec. 435 C
Precaution: Do not mix with hot or conc. acids; protect from sunlight to avoid generation of hydrogen cyanide
Toxicology: Heated to decomp., emits toxic fumes of CN
Uses: Anticaking agent for salt, processing aid, crystal modifier; also in wine
Usage level: ADI 0-0.025 mg/kg (EEC)
Regulatory: FDA 21CFR §172.490 (13 ppm max.), GRAS; BATF 27CFR §240.1051 (1 ppm max. residue in finished wine); Europe listed; UK approved
Manuf./Distrib.: Atomergic Chemetals; Degussa Ltd; Rit-Chem
Trade names containing: Morton® Lite Salt® TFC Mixt; Morton® Star Flake® Dendritic ES Salt; Morton® Star Flake® Dendritic Salt; Morton® TFC H.G. Blending Salt; Morton® Top Flake Coarse Salt; Morton® Top Flake Extra Coarse Salt; Morton® Top Flake Fine Salt; Morton® Top Flake Topping Salt; TFC Purex® Salt

Sodium ferrocyanide decahydrate. *See* Sodium ferrocyanide

Sodium formate
CAS 141-53-7; EINECS 205-488-0; EEC E237
Synonyms: Formic acid sodium salt
Empirical: $CHNaO_2$
Formula: HCOONa
Properties: Wh. deliq. gran. or cryst. powd., sl. formic acid odor; sol. in glycerol; sl. sol. in alcohol; m.w. 68.01; dens. 1.92; m.p. 253 C
Uses: Preservative
Regulatory: FDA 21CFR §186.1756, GRAS as indirect food additive; Europe listed
Manuf./Distrib.: Heico; Hoechst Celanese; Lohmann

Sodium fumarate
CAS 5873-57-4; EINECS 227-535-4
Synonyms: Sodium hydrogen fumarate; Sodium fumarate acidic; Monosodium fumarate
Empirical: $C_4H_3NaO_4$
Properties: m.w. 138.05
Uses: Dietary supplement; acidity regulator, food acid, raising agent, flavorings (Japan)
Regulatory: FDA 21CFR §172.350; Japan approved

Sodium fumarate acidic

Sodium fumarate acidic. *See* Sodium fumarate

Sodium furcelleran
  Definition: Sodium salt of furcelleran
  Uses: Emulsifier, stabilizer, thickener
  Regulatory: FDA 21CFR §172.660

Sodium glucoheptonate
  Empirical: $C_7H_{13}O_8Na$
  Formula: $HOCH_2(CHOH)_5COONa$
  Properties: Lt. tan crystalline powd.
  Uses: Boiler water additive for food contact
  Regulatory: FDA 21CFR §173.315 (limitation < 1 ppm cyanide)

Sodium gluconate
  CAS 527-07-1; EINECS 208-407-7; EEC E576
  Synonyms: D-Gluconic acid monosodium salt; Monosodium D(-)-pentahydroxy capronate; Gluconic acid sodium salt; Monosodium gluconate; Sodium d-gluconate
  Definition: Sodium salt of gluconic acid
  Empirical: $C_6H_{12}NaO_7$
  Formula: $HOCH_2CHOHCHOHCOHHCHOHCOONa$
  Properties: Wh. to ylsh. cryst. powd., pleasant odor; sol. 59 g/100 ml water; sl. sol. in alcohol; insol. in ether; m.w. 219.17
  Precaution: Avoid dust formation
  Toxicology: Low toxicity by IV route; heated to decomp., emits acrid smoke and irritating fumes
  Uses: Nutrient, dietary supplement, sequestering agent
  Usage level: ADI 0-50 mg/kg (EEC, as total gluconic acid)
  Regulatory: FDA 21CFR §182.6757, GRAS; Europe listed; UK approved
  Manuf./Distrib.: ADM Ingreds.; Ashland; Hays; Jungbunzlauer; Pfaltz & Bauer; PMP Fermentation Prods.
  Trade names: Gluconal® NA

Sodium d-gluconate. *See* Sodium gluconate
Sodium glutamate. *See* MSG

Sodium glyceryl oleate phosphate
  Definition: Sodium salt of a complex mixture of phosphate esters of glyceryl monooleate
  Uses: Aerosol formulation and food processing surfactant; food-grade mold lubricant
  Trade names: Emphos™ D70-30C
  Trade names containing: Ches® 500

Sodium guanosine-5′-monophosphate. *See* Disodium guanylate
Sodium guanylate. *See* Disodium guanylate
Sodium hexacyanoferrate (II). *See* Sodium ferrocyanide
Sodium hexadecanoate. *See* Sodium palmitate

Sodium hexametaphosphate
  CAS 10124-56-8, 68915-31-1; EINECS 233-343-1
  Synonyms: SHMP; Sodium polyphosphates glassy; Sodium tetrapolyphosphate; Graham's salt; Amorphous sodium polyphosphate; Sodium phosphate glass; Sodium metaphosphate
  Classification: Inorganic salt
  Empirical: $H_6O_{18}P_6 \cdot 6Na$
  Formula: $(NaPO_3)_6$
  Properties: White powd. or flakes; sol. in water; m.w. 101.98
  Toxicology: LD50 (oral, mouse) 7250 mg/kg, (IP, mouse) 870 mg/kg; poison by intravenous route; moderately toxic by intraperitoneal, subcutaneous routes; mildly toxic by ingestion; heated to decomp., emits toxic fumes of PO_x and Na_2O
  Uses: Emulsifier, sequestrant, texturizer, dietary supplement; water binding agent in food prods.; boiler water additive; hog/poultry scald agent; migrating to food from paper/paperboard
  Regulatory: FDA 21CFR §173.310, 182.90, 182.6760, 182.6769, GRAS; USDA 9CFR §318.7, 381.147
  Manuf./Distrib.: Albright & Wilson Am.; Browning
  Trade names: Calgon; Calgon® T Powd. Food Grade; Glass H®; Hexaphos®; Sodaphos®; Vitrafos®; Vitrafos® LC
  Trade names containing: Albriphos™ Blend 75-25; Albriphos™ Blend 90-10; Albriphos™ Blend 928; Bac-N-Fos®; Bac-N-Fos® Formula 191; Curafos® 11-2; Curafos® 22-4; Curavis® 250; Curavis® 350; FishGard® FP-55; FOS-6; Freez-Gard® FP-88E; Katch® Fish Phosphate; Kena® FP-28; Kena® FP-85; 90/10 Blend; Nutrifos® B-75; Nutrifos® B-90; Tenderset® MWA-5; Tenderset® MWA-7; Tenderset® MWA-8

Sodium hydrate. *See* Sodium hydroxide
Sodium hydrogen carbonate. *See* Sodium bicarbonate
Sodium hydrogen diacetate. *See* Sodium diacetate
Sodium hydrogen fumarate. *See* Sodium fumarate
Sodium hydrogen L-glutamate. *See* MSG
Sodium hydrogen sulfate. *See* Sodium bisulfate
Sodium hydrogen sulfite. *See* Sodium bisulfite

Sodium hydrosulfite
CAS 7775-14-6; 16721-80-5; EINECS 231-890-0
Synonyms: Disodium dithionite; Sodium hyposulfite
Classification: Inorganic salt
Empirical: $Na_2O_4S_2$
Formula: NaO_2SSO_2Na
Properties: Large transparent cryst., bitter taste; sol. in water; insol. in alcohol, conc. HCl; dens. 2.189; dec. 267 C; m.p. 55C
Precaution: Combustible
Uses: Migrating to food from paper/paperboard; bleaching agent (Japan)
Regulatory: FDA 21CFR §177.2800, 182.90; Japan approved (0.03-5 g/kg residual as sulfur dioxide), not permitted in certain foods
Manuf./Distrib.: Hoechst Celanese

Sodium hydroxide
CAS 1310-73-2; EINECS 215-185-5; EEC E524
Synonyms: Caustic soda; Sodium hydrate; White caustic; Soda lye; Lye
Classification: Inorganic base
Empirical: HNaO
Formula: NaOH
Properties: White deliq. solid beads or pellets; absorbs water and CO_2 from air; sol. in water, alcohol, glycerin; m.w. 40; dens. 2.12 (20/4 C); m.p. 318 C; b.p. 1390 C
Precaution: Strong base; may ignite or react violently with many org. compds.; dangerous material to handle
Toxicology: TLV:Cl 2 mg/m³ of air; LD50 (IP, mouse) 40 mg/kg; poison by IP route; mod. toxic by ingestion; mutagenic data; corrosive irritant to eye, skin, mucous membrane; mists and dusts cause small burns; heated to decomp., emits toxic fumes of Na_2O
Uses: Direct food additive, pH control agent, alkali; solvent for food colorings; cooked out juices retention agent; hog/poultry scald agent; processing aid; modifier for food starch; boiler water additive, cleaning dairies and breweries; processing (Japan
Usage level: ADI no limit (EEC)
Regulatory: FDA 21CFR §163.110, 172.560, 172.814, 172.892, 173.310, 184.1763, GRAS; USDA 9CFR §318.7, 381.147; BATF 27CFR §21.101, 240.1051a; Japan restricted; Europe listed; UK approved
Manuf./Distrib.: Asahi Chem Industry Co Ltd; Hüls AG

Sodium-L-2-hydroxypropionate. *See* Sodium lactate

Sodium hypochlorite
CAS 7681-52-9
Synonyms: Dakins sol'n.
Formula: NaOCl • 5HOH
Properties: Pale greenish, disagreeable sweet odor; sol. in cold water; dec. by hot water; m.p. 18 C
Precaution: Anhyd.: highly explosive, sensitive to heat or friction; forms explosive prods. with amines; unstable in air; fire risk; exposure limits: none
Toxicology: Toxic by ingestion; strong irritant to tissue; eye irritant; human mutagenic data; heated to decomp., emits toxic fumes of Na_2O and Cl⁻
Uses: Washing/lye peeling of fruits and vegetables; bleaching of starch; disinfection and deodorization in food, beverage, and tobacco plants
Regulatory: FDA 21CFR §173.315; Japan approved as bleaching agent, restricted as sterilizing agent
Manuf./Distrib.: Hüls AG

Sodium hypophosphite
CAS 7681-53-0
Synonyms: Phosphinic acid sodium salt; Sodium phosphinate
Formula: NaH_2PO_2
Properties: Wh. gran. powd. or colorless pearly cryst. plates, odorless, bittersweet saline taste, deliq.; sol. in water, alcohol, glycerin; m.w. 87.97
Precaution: Flamm. when exposed to heat or flame; aq. sol'ns. may explode on evaporation; potentially

explosive with oxidants; with heat, evolves phosphine
Toxicology: LD50 (IP, mouse) 1584 mg/kg; poison by subcutaneous route; mod. toxic by IP route; heated to decomp., emits toxic fumes of PO_x and Na_2O
Uses: Direct food additive, antioxidant, preservative, emulsifier, stabilizer; used in cod liver oil emulsions
Regulatory: FDA 21CFR §184.1764, GRAS

Sodium hyposulfite. *See* Sodium hydrosulfite
Sodium hyposulfite. *See* Sodium thiosulfate anhyd.
Sodium hyposulfite. *See* Sodium thiosulfate pentahydrate
Sodium inosinate. *See* Disodium inosinate
Sodium 5-inosinate. *See* Disodium inosinate
Sodium iron pyrophosphate. *See* Sodium ferric pyrophosphate
Sodium isoascorbate. *See* Sodium erythorbate

Sodium isostearoyl lactylate
CAS 66988-04-3; EINECS 266-533-8
Definition: Sodium salt of the isostearic acid ester of lactyl lactate
Empirical: $C_{24}H_{44}O_6 \cdot Na$
Formula: $C_{17}H_{35}COOCHCH_3COOCHCH_3COONa$
Trade names: Crolactil SISL

Sodium lactate
CAS 72-17-3; EINECS 200-772-0; EEC E325
Synonyms: 2-Hydroxypropanoic acid monosodium salt; Lacolin; Sodium-L-2-hydroxypropionate
Definition: Sodium salt of lactic acid
Empirical: $C_3H_5O_3 \cdot Na$
Formula: $CH_3CHOHCOONa$
Properties: Colorless or yellowish syrupy liq., odorless; sl. salt taste; very hygroscopic; misc. in water, alcohol; m.w. 112.07; m.p. 17 C; dec. 140 C
Precaution: Combustible
Toxicology: LD50 (IP, rat) 2000 mg/kg; moderately toxic by intraperitoneal; eye irritant; heated to decomp., emits toxic fumes of Na_2O
Uses: Food additive, emulsifier, flavor enhancer, flavoring agent/adjuvant, humectant, pH control agent, cooked out juices retention aid, corrosion preventative, denuding agent, hog scald agent, lye peeling agent, washing agent, glycerol substitute
Usage level: Limitation 5% (pickle for meat prods.), 0.5% (meat prods.); ADI no limit (EEC)
Regulatory: FDA 21CFR §184.1768, GRAS (not for infant formulas); USDA 9CFR §318.7; Japan approved; Europe listed; UK approved
Manuf./Distrib.: Lohmann; Patco
Trade names: Arlac S; Patlac® NAL; Purasal® S/SP 60

Sodium laurate
CAS 629-25-4; EINECS 211-082-4
Synonyms: Dodecanoic acid sodium salt; Lauric acid sodium salt
Definition: Sodium salt of lauric acid
Empirical: $C_{12}H_{23}O_2 \cdot Na$
Formula: $CH_3(CH_2)_{10}COONa$
Properties: m.w. 222.31
Uses: Binder, emulsifier, anticaking agent
Regulatory: FDA 21CFR §172.863, 175.105, 175.320, 176.170, 176.200, 176.210, 177.1200, 177.2260, 177.2600, 177.2800, 178.3910

Sodium lauryl benzene sulfonate. *See* Sodium dodecylbenzenesulfonate

Sodium lauryl sulfate
CAS 151-21-3; 68585-47-7; 68955-19-1; EINECS 205-788-1
Synonyms: SDS; Sulfuric acid monododecyl ester sodium salt; Sodium dodecyl sulfate
Definition: Sodium salt of lauryl sulfate
Empirical: $C_{12}H_{26}O_4S \cdot Na$
Formula: $CH_3(CH_2)_{10}CH_2OSO_3Na$
Properties: White to cream crystals, flakes, or powd., faint odor; sol. in water; m.w. 289.43
Precaution: Heated to decomp., emits toxic fumes of SO_x and Na_2O
Toxicology: LD50 (oral, rat) 1288 mg/kg; poison by IV, IP routes; mod. toxic by ingestion; experimental teratogen, reproductive effects; human skin irritant; experimental eye, severe skin irritant; mild allergen; mutagenic data

Uses: Surface-active agent in fumaric acid-acidulated dry beverage bases, fruit juice drinks; wetting agent in vegetable oils and animal fats; emulsifier in or with egg whites; whipping agent for marshmallows; hog/poultry scald agent

Usage level: Limitation 1000 ppm (egg white solids), 125 ppm (liq. or frozen egg whites), 0.5% (marshmallows), 10 ppm (vegetable oils, animal fats), 25 ppm (fruit juice drink)

Regulatory: FDA 21CFR §172.210, 172.822, 175.105, 175.300, 175.320, 176.170, 176.210, 177.1200, 177.1210, 177.1630, 177.2600, 177.2800, 178.1010, 178.3400; USDA 9CFR §318.7, 381.147

Manuf./Distrib.: Albright & Wilson Ltd

Trade names containing: Egg White Solids Type P-18; Hentex Type P-600

Sodium lignosulfonate
CAS 8061-51-6
Synonyms: Sodium polignate; Lignosulfonic acid, sodium salt
Definition: Sodium salt of polysulfonated lignin, a dark brown polymeric material from wood
Properties: Tan free-flowing powd.
Precaution: Combustible
Uses: Boiler water additive for food contact
Regulatory: FDA 21CFR §173.310, 175.105, 176.170, 176.210, 177.1210

Sodium MBT. *See* Sodium 2-mercaptobenzothiazole

Sodium 2-mercaptobenzothiazole
CAS 2492-26-4
Synonyms: NaMBT; Sodium MBT
Empirical: $C_7H_4NS_2 \cdot Na$
Properties: m.w. 189.23
Toxicology: Moderately toxic by ingestion
Uses: Slimicide used in paper/paperboard contacting food
Regulatory: FDA 21CFR §181.30

Sodium metabisulfite
CAS 7681-57-4; EINECS 231-673-0; EEC E223
Synonyms: Disulfurous acid disodium salt; Sodium pyrosulfite; Disodium pyrosulfite; Sodium bisulfite
Classification: Inorganic salt
Empirical: $O_5S_2 \cdot 2Na$
Formula: $Na_2S_2O_5$
Properties: Colorless cryst. or wh. to ylsh. powd., SO_2 odor, sol. in water, sl. sol. in alcohol, m.w. 190.10
Toxicology: TWA 5 mg/m³; LD50 (IV, rat) 115 mg/kg; poison by intravenous route; moderately toxic by parenteral route; experimental reproductive effects; mutagenic data; heated to decomp., emits toxic fumes of SO_x and Na_2O
Uses: Preservative, antioxidant; flavoring in cherries and shrimp; boiler water additive for food contact; bleaching agent; not for use on meats, sources of vitamin B_1, raw fruits and vegetables, or fresh potatoes
Usage level: ADI 0-0.7 mg/kg (EEC)
Regulatory: FDA 21CFR §173.310, 177.1200, 182.3766, GRAS; Europe listed
Manuf./Distrib.: Browning
Trade names containing: Beta-Tabs

Sodium metaphosphate. *See also* Sodium hexametaphosphate
CAS 10361-03-2
Synonyms: Graham's salt; Sodium polyphosphates glassy; Sodium tetrapolyphosphate
Empirical: $O_3P \cdot Na$
Formula: $(NaPO_3)_n$, n= 3-10 (cyclic) or larger (polymers)
Properties: Amorphous wh. solids; very sol. in water; m.w. 101.96
Uses: Sequestrant, emulsifier, texturizer, cooked out juices retention agent, whey processing; binding agent, alkaline agent, raising agent (Japan)
Usage level: Limitation 5% (in pickle for meat prods.), 0.5% (meat prods.), 0.5% (total poultry prod.)
Regulatory: FDA 21CFR §182.6760, GRAS; USDA 9CFR §318.7, 381.147; Japan approved
Trade names containing: Emulsi-Phos® 440; Emulsi-Phos® 660; Emulsi-Phos® 990

Sodium metaphosphate insoluble
CAS 68915-31-1
Formula: $(NaPO_3)_n$, n = 1000-5000
Properties: m.w. 102,000-510,000

Sodium metasilicate
CAS 6834-92-0; EINECS 229-912-9

Sodium metasilicate anhydrous

Synonyms: Silicic acid, disodium salt; Sodium metasilicate anhydrous
Classification: Inorganic salt
Empirical: Na$_2$O$_3$Si
Formula: Na$_2$SiO$_3$ (anhyd.), Na$_2$SiO$_3$ • 5H$_2$O (pentahydrate), Na$_2$SiO$_3$ • 9H$_2$O (nonahydrate)
Properties: Anhyd.: White gran., dustless; sol. in water; insol. in alcohol, acids, salt sol'ns.; m.w. 122.07 (anhyd.); dens. 2.614; m.p. 1089 C; ref. index 1.520 (glass, 25 C)
Precaution: Strongly alkaline; caustic material
Toxicology: LD50 (oral, rat) 1280 mg/kg; poison by ingestion, intraperitoneal routes; severe eye, skin, mucous membrane irritant; experimental reproductive effects; ingestion causes GI tract upset; heated to decomp., emits toxic fumes of Na$_2$O
Uses: Direct food additive, processing aid; boiler water additive for food contact; washing/lye peeling of fruits, vegetables, nuts; denuding agent in tripe; hog scald agent in removing hair; corrosion preventative in canned and bottled water; saponifier
Regulatory: FDA 21CFR §173.310, 184.1769a, GRAS; USDA 9CFR §318.7
Manuf./Distrib.: Rhone-Poulenc

Sodium metasilicate anhydrous. *See* Sodium metasilicate
Sodium methane napthalene sulfonate. *See* Sodium methylnaphthalenesulfonate
Sodium methoxide. *See* Sodium methylate

Sodium methylate
CAS 124-41-4
Synonyms: Sodium methoxide
Empirical: CH$_3$ONa
Properties: Wh. powd./liq.; sol. in fats, esters, alcohols; m.p. > 127 C
Precaution: Flamm.
Toxicology: Caustic
Uses: Catalyst for transesterification and isomerization of fats and oils; processing aid (Japan)
Regulatory: Japan restricted
Manuf./Distrib.: Hüls AG

Sodium methylnaphthalenesulfonate
CAS 26264-58-4; EINECS 247-561-6
Synonyms: Sodium methane napthalene sulfonate; Sodium polynapthalene methane sulfonate
Definition: Mixture of mono- and dimethyl substituted naphthalene sulfonates
Properties: m.w. 245-260
Uses: Crystallization agent for sodium carbonate for use in potable water systems; anticaking agent in sodium nitrite used in cured fish/meats; fruit/vegetable lye peeling and washing
Regulatory: FDA 21CFR §172.824, 173.315 (0.2% in wash water), 175.105, 176.170, 176.180, 176.210

Sodium methylparaben
CAS 5026-62-0; EINECS 225-714-1
Synonyms: 4-Hydroxybenzoic acid, methyl ester, sodium salt; Methylparaben, sodium salt
Definition: Sodium salt of methylparaben
Empirical: C$_8$H$_8$O$_3$•Na
Uses: Preservative, bactericide, fungicide for foods
Trade names: Nipagin M Sodium
Trade names containing: Nipasept Sodium

Sodium methyl sulfate
CAS 512-42-5, 123333-89-1 (hydrate)
Empirical: CH$_3$NaO$_4$S
Formula: NaCH$_3$SO$_4$; CH$_3$OSO$_3$Na • xH$_2$O (hydrate)
Properties: m.w. 134.08; Monohydrate: wh. cryst.; hygroscopic; sol. in water, alcohol, methanol
Precaution: Hygroscopic; keep well closed
Uses: May be present in pectin at 0.1% max.
Regulatory: FDA 21CFR §173.385
Manuf./Distrib.: Aldrich

Sodium myristate
CAS 822-12-8; EINECS 212-487-9
Synonyms: Tetradecanoic acid sodium salt; Myristic acid sodium salt
Definition: Sodium salt of myristic acid
Empirical: C$_{14}$H$_{27}$O$_2$ • Na
Formula: CH$_3$(CH$_2$)$_{12}$COONa
Properties: m.w. 250.36

Uses: Binder, emulsifier, anticaking agent
Regulatory: FDA 21CFR §172.863, 175.105, 175.320, 176.170, 176.200, 176.210, 177.1200, 177.2260, 177.2600, 177.2800, 178.3910

Sodium nitrate

CAS 7631-99-4; EINECS 231-554-3; EEC E251
Synonyms: Soda niter; Cubic niter; Chile saltpeter
Empirical: $NNaO_3$
Formula: $NaNO_3$
Properties: Colorless transparent crystals, odorless; saline, sl. bitter taste; sol. in water, glycerol; sl. sol. in alcohol; m.w. 85.01; dens. 2.267; m.p. 308 C; dec. @ 380 C
Precaution: Strong oxidizer; ignites with heat or friction; explodes @ 537 C
Toxicology: LD50 (oral, rabbit) 2680 mg/kg; poison by intavenous route; moderately toxic by ingestion; can produce nitrosamines associated with cancers; can reduce blood oxygen levels; human mutagenic data; heated to decomp., emits toxic fumes of NO_x and Na_2O
Uses: Antimicrobial agent, preservative; source of nitrite, color fixative in cured meats, fish, poultry; boiler water additive; curing salt
Usage level: Limitation 500 ppm (smoked fish, meat curing), 200 ppm (smoked chub), 7 lb/100 gal (pickle), 3.5 oz/100 lb (meat), 2.75 oz/100 lb (chopped meat); ADI 0-5 mg/kg (EEC)
Regulatory: FDA 21CFR §171.170, 172.170, 172.177, 173.310, 181.33; USDA 9CFR §318.7, 381.147; Japan approved with limitations; Europe listed; UK approved
Manuf./Distrib.: Lohmann; Spice King

Sodium nitrite

CAS 7632-00-0; EINECS 231-555-9; EEC E250
Synonyms: Nitrous acid sodium salt
Empirical: $NNaO_2$
Formula: $NaNO_2$
Properties: Sl. ylsh. or wh. cryst. or powd.; deliq. in air; sol. in water, sl. sol. in alcohol; m.w. 69.00; dens. 2.168; m.p. 271 C; b.p. dec. @ 320 C
Precaution: Strong oxidizing agent; flamm.; ignites by friction in contact with organic matter; heated to decomp., emits toxic fumes of NO_x and Na_2O
Toxicology: LD50 (oral, rat) 85 mg/kg; human poison by ingestion; experimental neoplastigen, tumorigen, teratogen; human systemic effects; ; human mutagenic data; eye irritant; can produce nitrosamines associated with cancers; can reduce blood oxygen levels
Uses: Antimicrobial agent, color fixative, preservative in cured meats, meat prods., smoked fish
Usage level: Limitation 10 ppm (smoked cured tunafish), 200 ppm (smoked cooked salmon, shad, meat curing), 2 lb/100 gal (pickle), 1 oz/100 lb (meat), 0.25 oz/100 lb (chopped meat); ADI 0-0.2 mg/kg (EEC)
Regulatory: FDA 21CFR §172.175, 172.177, 181.34; USDA 9CFR §318.79, 381.147; Japan approved (0.005-0.07 g/kg); Europe listed; UK approved
Manuf./Distrib.: Browning

Sodium octadecanoate. *See* Sodium stearate
Sodium 9-octadecenoate. *See* Sodium oleate

Sodium octyl sulfate

CAS 126-92-1; 142-31-4; EINECS 204-812-8
Synonyms: Sodium 2-ethylhexyl sulfate; Sulfuric acid, mono (2-ethylhexyl) ester sodium salt; Octyl sulfate sodium salt
Definition: Sodium salt of 2-ethylhexyl sulfate
Empirical: $C_8H_{18}O_4S$ • Na
Formula: $CH_3(CH_2)_3CHCH_3CH_2CH_2OSO_3Na$
Properties: m.w. 233.31; m.p. 195 C
Toxicology: LD50 (oral, mouse) 1550 mg/kg; poison by intraperitoneal route; moderately toxic by ingestion; skin contact; skin and eye irritant; heated to decomp., emits very toxic fumes of SO_x and Na_2O
Uses: Washing/lye peeling of fruits and vegetables; poultry scald agent; washing water agent
Usage level: Limitation 0.2% (in wash water)
Regulatory: FDA 21CFR §173.315, 175.105, 176.170; USDA 9CFR §381.147
Manuf./Distrib.: Aldrich
Trade names: Carsonol® SHS; Norfox® Anionic 27; Rhodapon® BOS; Sulfotex OA

Sodium oleate

CAS 143-19-1; EINECS 205-591-0
Synonyms: Sodium 9-octadecenoate; 9-Octadecenoic acid sodium salt
Definition: Sodium salt of oleic acid

Sodium palmitate

Empirical: $C_{18}H_{34}O_2 \cdot Na$
Formula: $CH_3(CH_2)_7CH=CH(CH_2)_7COONa$
Properties: White powd., sl. tallow-like odor; sol. in ≈ 10 parts water, ≈ 20 parts alcohol; m.w. 304.50; m.p. 232-235 C
Precaution: Combustible when exposed to heat or flame
Toxicology: LD50 (IV, mouse) 152 mg/kg; poison by intravenous route; heated to decomp., emits acrid toxic fumes of Na_2O
Uses: Stabilizer, anticaking agent, binder, emulsifier, paper mfg. aid; coating material (Japan)
Regulatory: FDA 21CFR §172.863, 175.105, 175.300, 175.320, 176.170, 176.200, 176.210, 177.1200, 177.2260, 177.2600, 177.2800, 178.3910, 181.29, 186.1770, GRAS as indirect food additive; Japan approved

Sodium palmitate
CAS 408-35-5; EINECS 206-988-1
Synonyms: Hexadecanoic acid sodium salt; Sodium hexadecanoate; Palmitic acid sodium salt; Sodium pentadecanecarboxylate
Definition: Sodium salt of palmitic acid
Empirical: $C_{16}H_{31}O_2 \cdot Na$
Formula: $CH_3(CH_2)_{14}COONa$
Properties: Wh. to yel. powd.; m.w. 278.47
Precaution: Combustible
Toxicology: Heated to decomp., emits acrid smoke and irritating fumes
Uses: Binder, emulsifier, stabilizer, anticaking agent; constituent of paperboard for food pkg.
Regulatory: FDA 21CFR §172.863, 175.105, 175.320, 176.170, 176.200, 176.210, 177.1200, 177.2600, 177.2800, 178.3910, 186.1771, GRAS

Sodium pantothenate
CAS 867-81-2
Toxicology: Nuisance dust
Uses: Dietary supplement, nutrient
Regulatory: FDA 21CFR §182.5772, GRAS; Japan approved

Sodium PCP. *See* Sodium pentachlorophenate

Sodium pentachlorophenate
CAS 131-52-2
Synonyms: Pentachlorophenate sodium; Sodium PCP
Empirical: $C_6Cl_5O \cdot Na$
Properties: Tan powd.; sol. in water, ethanol, acetone; insol. in benzene; m.w. 288.30
Toxicology: LD50 (oral, rat) 126 mg/kg, (inh., rat) 11,700 μg/kg; poison by ingestion, inh., skin contact, IV, IP, subcutaneous routes; experimental reproductive effects; mutagenic data; heated to decomp., emits toxic fumes of Cl^- and Na_2O
Uses: Preservative, slime control agent used in paper/paperboard for food pkg.
Regulatory: FDA 21CFR §178.3120, 178.3900 (for use as preservative only), 181.30

Sodium pentadecanecarboxylate. *See* Sodium palmitate
Sodium peroxydisulfate. *See* Sodium persulfate

Sodium persulfate
CAS 7775-27-1; EINECS 231-892-1
Synonyms: Sodium peroxydisulfate
Empirical: $Na_2O_8S_2$
Formula: $Na_2S_2O_8$
Properties: Wh. cryst. powd.; sol. in water; dec. by alcohol; m.w. 238.10; dens. 2.400
Precaution: Powerful oxidizer; can cause fires
Toxicology: LD50 (IP, mouse) 226 mg/kg; poison by IP, IV routes; heated to decomp., emits toxic fumes of SO_x and Na_2O
Uses: Denuding agent for tripe
Regulatory: USDA 9CFR §318.7
Manuf./Distrib.: Degussa; FMC

Sodium o-phenylphenate
CAS 132-27-4; EINECS 205-055-6; EEC E232
Synonyms: o-Phenylphenol sodium salt; Sodium o-phenylphenol; Sodium biphenyl-2-yl oxide; (1,1´-Biphenyl)-2-ol, sodium salt; Sodium o-phenylphenolate
Definition: Sodium salt of o-phenylphenol

Empirical: C$_{12}$H$_9$O • Na
Formula: C$_6$H$_4$(C$_6$H$_5$)ONa • 4HOH
Properties: White flakes; sol./100 g solvent: 122 g water, 156 g acetone, 138 g methanol, 28 g propylene glycol; pract. insol. in petrol. fractions, pine oil; m.w. 192.20
Toxicology: Moderately toxic by ingestion; human skin irritant
Uses: Preservative, antimicrobial, mold inhibitor for apples, etc.
Usage level: 0.05-0.5%; ADI 0-0.02 mg/kg (EEC)
Regulatory: FDA 21CFR §175.105, 175.300, 176.170, 176.210, 177.1210, 178.3120; USA not restricted; Europe listed; Japan approved (0.01 g/kg residual)
Manuf./Distrib.: Ashland

Sodium o-phenylphenolate. *See* Sodium o-phenylphenate
Sodium o-phenylphenol. *See* Sodium o-phenylphenate

Sodium phosphate
CAS 7558-80-7 (anhyd.), 13472-35-0 (dihydrate); EINECS 231-449-2; EEC E339
Synonyms: MSP; Sodium phosphate monobasic; Sodium acid phosphate; Monobasic sodium phosphate; Monosodium dihydrogen phosphate; Monosodium dihydrogen orthophosphate; Sodium biphosphate; Sodium phosphate primary
Classification: Inorganic salt
Empirical: H$_2$NaO$_4$P
Formula: NaH$_2$PO$_4$
Properties: Wh. cryst. powd., gran., odorless; hygroscopic; sol. 87 g/100 g water; insol. in alcohol; m.w. 119.98
Toxicology: LD50 (oral, rat) 8290 mg/kg; poison by intramuscular route; mildly toxic by ingestion; eye irritant; heated to decomp., emits toxic fumes of PO$_x$ and Na$_2$O
Uses: Nutrient, dietary supplement, buffer, emulsifier, sequestrant, preservative; acid ingred. in effervescent powds. and laxatives; prod. of caramel; boiler water additive; poultry scald agent
Usage level: Limitation 5% (in pickling meat prods.), 0.5% (meat prods.), 0.5% (total poultry prod.)
Regulatory: FDA 21CFR §133, 150, 160, 163.123, 163.130, 163.135, 163.140, 163.145, 163.150, 163.153, 163.155, 172.892, 173.310, 182.1778, 182.5778, 182.6085, 182.6778, 182.8778, GRAS; USDA 9CFR §318.7, 381.147
Regulatory: Japan approved
Manuf./Distrib.: Albright & Wilson Am.; Browning; FMC; Lohmann; Rhone-Poulenc Food Ingreds.
Trade names: Albrite MSP Food Grade
Trade names containing: Medium KL

Sodium phosphate dibasic
CAS 7558-79-4 (anhyd.); 7782-85-6; 10039-32-4 (dodecahydrate); 10140-65-5; EINECS 231-448-7; EEC E339b
Synonyms: DSP-O; Disodium phosphate; Dibasic sodium phosphate; Disodium hydrogen phosphate; Disodium monohydrogen orthophosphate; Disodium orthophosphate
Classification: Inorganic salt
Definition: Phosphoric acid, disodium salt
Empirical: HO$_4$P • 2Na
Formula: Na$_2$HPO$_4$
Properties: Colorless translucent cryst. or wh. powd., saline taste; sol. in water; very sol. in alcohol; m.w. 141.97; dens. 1.5235; m.p. 35 C, loses water at 92.5 C
Toxicology: LD50 (oral, rat) 17 g/kg; skin and eye irritant; poison by intravenous route, moderately toxic by intraperitoneal, subcutaneous route; mildly toxic by ingestion; when heated to decomp., emits toxic fumes of PO$_x$ and Na$_2$O
Uses: Buffer, dietary supplement, emulsifier, hog/poultry scald agent, nutrient, sequestrant, stabilizer, texturizer in foods (evaporated milk, process cheese, cereals, meat prods.); water treatment in food processing; stabilizer migrating from food pkg.
Regulatory: FDA 21CFR §133.169, 133.173, 133.179, 135.110, 137.305, 139.110, 139.115, 139.117, 139.135, 150.141, 150.161, 173.310, 175.210, 175.300, 181.22, 181.29, 182.1778, 182.5778, 182.6290, 182.6778, 182.8778, 182.8890, GRAS; USDA 9CFR 318.7, 381.147
Regulatory: Japan approved
Manuf./Distrib.: Albright & Wilson Am.; FMC; Lohmann; Monsanto; Rhone-Poulenc Food Ingreds.

Sodium phosphate dibasic dihydrate. *See* Disodium phosphate, dihydrate
Sodium phosphate glass. *See* Sodium hexametaphosphate
Sodium phosphate monobasic. *See* Sodium phosphate
Sodium phosphate primary. *See* Sodium phosphate
Sodium phosphate tertiary dodecahydrate. *See* Sodium phosphate tribasic dodecahydrate

Sodium phosphate tribasic
CAS 7601-54-9; EINECS 231-509-8
Synonyms: TSP-O; Trisodium phosphate; Trisodium orthophosphate; Phosphoric acid trisodium salt
Classification: Inorganic salt
Empirical: Na_3PO_4
Formula: $H_3O_4P \cdot 3Na$
Properties: Wh. cryst., cryst. powd., gran., odorless; sol. 14 g/100 g water; insol. in alcohol; m.w. 163.97
Precaution: Strong caustic material
Toxicology: Moderately toxic by intravenous route; mutagenic data; heated to decomp., emits toxic fumes of Na_2O and PO_x
Uses: Dietary supplement, nutrient; sequestrant; buffer, pH control agent in food systems; emulsifier and alkalizer for process cheese; boiler water additive; denuding agent; fat rendering aid; hog/poultry scald agent
Regulatory: FDA 21CFR §133.169, 133.173, 133.179, 150.141, 150.161, 173.310, 182.1778, 182.5778, 182.6778, 182.8778, GRAS; USDA 9CFR §318.7, 381.147; Japan approved
Manuf./Distrib.: Browning; FMC

Sodium phosphate tribasic dodecahydrate
CAS 10101-89-0 (dodecahydrate); EINECS 231-509-8
Synonyms: TSP-12; Trisodium phosphate dodecahydrate; Tribasic sodium phosphate dodecahydrate; Sodium phosphate tertiary dodecahydrate
Classification: Sodium phosphate
Empirical: $Na_3PO_4 \cdot 12H_2O$
Properties: m.w. 380.14
Uses: Emulsifier for processed cheese; buffer for processed foods
Manuf./Distrib.: Albright & Wilson Am.; Lohmann; Rhone-Poulenc Food Ingreds.

Sodium phosphinate. *See* Sodium hypophosphite

Sodium phosphoaluminate
Properties: Wh. powd.
Uses: Migrating to foods from paper/paperboard
Regulatory: FDA 21CFR §182.90, GRAS

Sodium polignate. *See* Sodium lignosulfonate

Sodium polyacrylate
CAS 9003-04-7
Synonyms: Polyacrylic acid, sodium salt
Empirical: $(C_3H_4O_2)_x \cdot xNa$
Properties: m.w. 2000-2300
Toxicology: Eye irritant; heated to decomp., emits toxic fumes of Na_2O
Uses: Controls mineral scale during beet or cane sugar processing; boiler water additive; stabilizer and thickener in defoaming agents containing dimethicone
Usage level: 3.6 ppm max. of raw juice
Regulatory: FDA 21CFR §173.73, 173.310, 173.340; Japan approved (0.2% max.)
Manuf./Distrib.: Allchem Industries

Sodium polyacrylate-acrylamide resin
Definition: Produced by polymerization and subsequent hydrolysis of acrylonitrile in a sodium silicate-sodium hydroxide aq. sol'n.
Uses: Controls organic and mineral scale in beet sugar juice/liquor or cane sugar juice/liquor
Regulatory: FDA 21CFR §173.5, limitation 2.5 ppm

Sodium polyaluminate. *See* Sodium aluminate
Sodium polymannuronate. *See* Algin

Sodium polymethacrylate
CAS 25086-62-8; 54193-36-1
Synonyms: 2-Propenoic acid, 2-methyl-, homopolymer, sodium salt
Classification: Polymer
Empirical: $(C_4H_6O_2)_x \cdot xNa$
Formula: $[CH_3CHCHCOONa]_x$
Uses: Boiler water additive for food contact
Regulatory: FDA 21CFR §173.310, 175.105

Sodium polynapthalene methane sulfonate. *See* Sodium methylnaphthalenesulfonate
Sodium polyphosphates glassy. *See* Sodium hexametaphosphate

Sodium polyphosphates glassy. *See* Sodium metaphosphate
Sodium potassium tartrate. *See* Potassium sodium tartrate anhyd.
Sodium potassium tartrate. *See* Potassium sodium tartrate tetrahydrate

Sodium propionate
CAS 137-40-6; EINECS 205-290-4; EEC E281
Synonyms: Propanoic acid sodium salt; Propionic acid sodium salt
Definition: Sodium salt of propionic acid
Empirical: $C_3H_5O_2 \cdot Na$
Formula: CH_3CH_2COONa
Properties: Transparent cryst. or gran., nearly odorless, deliq. in moist air; very sol. in water; sl. sol. in alcohol; m.w. 96.07; m.p. 287-289 C
Precaution: Combustible
Toxicology: LD50 (skin, rabbit) 1640 mg/kg; moderately toxic by skin contact and subcutaneous route
Uses: Direct food additive, antimicrobial agent, preservative, mold and rope inhibitor, flavoring agent; used in baked goods, nonalcoholic beverages, cheese, confections, frostings, gelatins, puddings, frostings, jams, jellies, meat prods., soft candy
Usage level: Limitation 0.32% alone or with calcium propionate on wt. of flour; ADI no limit (EEC)
Regulatory: FDA 21CFR §133.123, 133.124, 133.169, 133.173, 133.179, 150.141, 150.161, 179.45, 181.22, 181.23, 184.1784, GRAS; USDA 9CFR §318.7, 381.147; Japan approved with limitations; Europe listed
Manuf./Distrib.: Browning; Gist-brocades Food Ingreds.; Lohmann; MLG Enterprises
Trade names: Niacet Sodium Propionate FCC
Trade names containing: Crestawhip H13; Myvatex® 25-07K; Tandem 11H K; Vanall (K); Vanease (K)

Sodium propylparaben
CAS 35285-69-9; EINECS 252-488-1
Synonyms: 4-Hydroxybenzoic acid, propyl ester, sodium salt; Propyl-4-hydroxybenzoate, sodium salt; Propylparaben, sodium salt
Definition: Sodium salt of propylparaben
Empirical: $C_{10}H_{12}O_3 \cdot Na$
Toxicology: May cause asthma, rashes, hyperactivity
Uses: Preservative, bactericide, fungicide for foods
Trade names: Nipasol M Sodium
Trade names containing: Nipasept Sodium

Sodium pyroborate. *See* Sodium borate
Sodium pyrophosphate. *See* Tetrasodium pyrophosphate
Sodium pyrosulfate. *See* Sodium bisulfate
Sodium pyrosulfate. *See* Sodium bisulfate solid
Sodium pyrosulfite. *See* Sodium metabisulfite
Sodium saccharide. *See* Saccharin sodium
Sodium saccharin. *See* Saccharin sodium

Sodium sesquicarbonate
CAS 533-96-0; EINECS 208-580-9; EEC E500
Synonyms: Carbonic acid, sodium salt (2:3)
Classification: Inorganic salt
Empirical: $CH_2O_3 \cdot {}^3/_2Na \cdot H_2O$
Formula: $Na_2CO_3 \cdot NaHCO_3 \cdot 2H_2O$
Properties: White needle-shaped crystals; sol. in water; m.w. 226.04; dens. 2.112; m.p. dec.
Precaution: Noncombustible
Toxicology: Irritant to tissue; nuisance dust
Uses: Direct food additive; pH control agent, alkali, neutralizer in dairy prods.; used to control lactic acid prior to pasteurization and churning of cream into butter; poultry scald agent
Regulatory: FDA 21CFR §184.1792, GRAS; USDA 9CFR §381.147; Europe listed; UK approved

Sodium silicate
CAS 1344-09-8; EINECS 215-687-4
Synonyms: Silicic acid, sodium salt; Water glass; Soluble glass
Definition: Sodium salt of silicic acid
Empirical: Na_2SiO_3 or $Na_6Si_2O_7$ or $Na_2Si_3O_7 \cdot$ aq.
Properties: Colorless to white or grayish-white crystalline lumps; very sl. sol. in cold water; m.w. 242.23 + water; dens. 1.390 (20C)
Toxicology: Corrosive; irritating and caustic to skin, mucous membranes; if swallowed, causes vomiting and diarrhea

Sodium silicoaluminate

Uses: Boiler water additive for food contact; migrating to foods from paper/paperboard, from cotton in dry food pkg.
Regulatory: FDA 21CFR §173.310, 177.1200, 182.70, 182.90
Manuf./Distrib.: Crosfield

Sodium silicoaluminate
CAS 1344-00-9; EINECS 215-684-8; EEC E554
Synonyms: Sodium aluminosilicate; Sodium aluminum silicate; Aluminum sodium silicate
Definition: Series of hydrated sodium aluminum silicates
Empirical: $Na_2O : Al_2O_3 : SiO_2$ with mole ratio $\approx$ 1:1:13.2
Properties: Fine white amorphous powd. or beads, odorless and tasteless; insol. in water, alcohol, org. solvs.; partly sol. in strong acids and alkali hydroxides @ 80-100 C; pH 6.5-10.5 (20% slurry)
Precaution: Noncombustible
Toxicology: Irritant to skin, eyes, mucous membranes; heated to decomp., emits toxic fumes of Na_2O
Uses: Anticaking agent in cake mixes, dry mixes, nondairy creamers, salt, powdered sugar
Usage level: ADI no limit (EEC)
Regulatory: FDA 21CFR §133.146, 160.105, 160.185, 182.2727 (2% max.), 582.2727, GRAS; Europe listed; UK approved
Trade names containing: Egg Yolk Solids Type Y-1-FF; FloAm® 400; Morton® H.G. Blending Prepared Salt; Morton® Star Flake® Dendritic ES Salt; Morton® Table Iodized Salt

Sodium sorbate
CAS 7757-81-5; EEC E201
Synonyms: Sorbic acid sodium salt
Empirical: $C_6H_7O_2$ • Na
Properties: m.w. 134.12
Toxicology: LD50 (oral, rat) 7160 mg/kg; mod. toxic by intraperitoneal route; mildly toxic by ingestion; mutagenic data; heated to decomp., emits toxic fumes of Na_2O
Uses: Preservative for baked goods, cheese, margarine; migrating to foods from paper/paperboard
Usage level: Limitation 0.1% (alone) or 0.2% (in combination with its salts or benzoic acid or salts, of finished prod.)
Regulatory: FDA 21CFR §182.90, 182.3795, GRAS; USDA 9CFR §318.7; Europe listed

Sodium stearate
CAS 822-16-2; EINECS 212-490-5
Synonyms: Sodium octadecanoate; Octadecanoic acid sodium salt; Stearic acid sodium salt
Definition: Sodium salt of stearic acid
Empirical: $C_{18}H_{36}O_2$ • Na
Formula: $CH_3(CH_2)_{16}COONa$
Properties: White powd., fatty odor; sol. in hot water and hot alcohol; slowly sol. in cold water and cold alcohol; insol. in many org. solvs.; m.w. 306.52
Toxicology: Poison by intravenous and other routes; heated to decomp., emits toxic fumes of Na_2O
Uses: Anticaking agent, binder, emulsifier, masticatory substance in chewing gum base; stabilizer migrating from food pkg.
Regulatory: FDA 21CFR §172.615, 172.863, 175.105, 175.300, 175.320, 176.170, 176.200, 176.210, 177.1200, 177.2260, 177.2600, 177.2800, 178.3910, 179.45, 181.22, 181.29
Trade names containing: Atmul® 400; Monomuls® 60-25/2; Monomuls® 60-25/5; Monomuls® 90-25/2; Monomuls® 90-25/5

Sodium stearoyl fumarate. *See* Sodium stearyl fumarate

Sodium stearoyl lactylate
CAS 25383-99-7; EINECS 246-929-7; EEC E481
Synonyms: Octadecanoic acid, 2-(1-carboxyethoxy)-1-methyl-2-oxoethyl ester, sodium salt; Sodium stearyl-2-lactylate
Definition: Sodium salt of the stearic acid ester of lactyl lactate
Empirical: $C_{24}H_{44}O_6$•Na
Formula: $CH_3(CH_2)_{16}COOCHCH_3COOCHCH_3COONa$
Properties: White or cream-colored powd., caramel odor; sol. in hot oil or fat; disp. in warm water; m.p. 46-52 C; HLB 10-12
Toxicology: Heated to decomp., emits acrid smoke and irritating fumes
Uses: Emulsifier, stabilizer, dough strengthener/conditioner, formulation/processing aid, texturizer, whipping agent in baked prods., desserts, mixes, icings, puddings, dehydrated potatoes; complexing agent for starches and proteins; surface-active agent
Usage level: Limitation 0.5 part/100 part flour (baked goods), 0.2% (icings, puddings, fillings), 0.3% (milk

substitute), 0.5% (dehydrated potatoes), 0.2% (snack dips), 0.2% (cheese substitute), 0.25% (sauces, gravies); ADI 0-20 mg/kg (EEC)

Regulatory: FDA 21CFR §172.846, 177.1200; Europe listed; UK approved

Trade names: Admul SSL 2003; Admul SSL 2004; Artodan SP 55 Kosher; Atlas B-60; Atlas SSL; Crolactil SS2L; Emplex; Emulsilac S; Emulsilac S K; Lamegin® NSL; Paniplex SK; Pationic® 920; Prefera® NSL; Prefera® SSL 6000 MB; Swedex SSL-5AC

Trade names containing: Atlas P-44; Atmos® 378 K; Atmos® 1069; Atmos® 7515 K; Atmul® P-28; Atmul® P-36; Atmul® P-44; Beta Plus; Kel-Lite™ CM; Lactylate Hydrate; MG-60; MG-S; Myvatex® Texture Lite®; Myvatex® Texture Lite® K; Patco® 3; Reduce®-150; Soft Touch; Tandem 930; Vitalex P

Sodium stearyl fumarate

CAS 4070-80-8

Synonyms: Sodium stearoyl fumarate

Empirical: $C_{22}H_{39}NaO_4$

Properties: Wh. fine powd.; sol. in methanol; insol. in water; m.w. 390.54

Toxicology: Heated to decomp., emits toxic fumes of Na_2O

Uses: Dough conditioner in yeast-raised baked goods; conditioning agent in dehydrated potatoes, processed cereals, starch- or flour-thickened foods; stabilizer in non-yeast-leavened baked goods

Usage level: Limitation 0.5% on flour (baked goods), 1% (dehydrated potatoes), 1% on flour (non-yeast-raised baked goods), 1% (dry cereals), 0.2% (starch thickened foods)

Regulatory: FDA 21CFR §172.826

Sodium stearyl-2-lactylate. *See* Sodium stearoyl lactylate
Sodium subsulfite. *See* Sodium thiosulfate anhyd.

Sodium succinate

FEMA 3277

Synonyms: Succinic acid sodium salt; Disodium succinate; Soduxin

Empirical: $C_4H_4Na_2O_4$

Properties: Hexahydrate gran. or cryst. powd.; sol. in 5 parts water; insol. in alcohol; m.w. 162.05

Toxicology: LD50 (IV, mouse) 4.5 g/kg

Uses: Acidulant, flavoring agent, protective coating for fruits and vegetables

Manuf./Distrib.: Aldrich

Sodium sulfate

CAS 7757-82-6 (anhyd.), 7727-73-3 (decahydrate); EINECS 231-820-9; EEC E514

Synonyms: Sodium sulfate anhydrous; Sodium sulfate (2:1); Glauber's salt; Salt cake; Disodium sulfate

Classification: Inorganic salt

Empirical: Na_2O_4S

Formula: Na_2SO_4

Properties: Wh. cryst. or powd., odorless, bitter saline taste; sol. in water, glycerol; insol. in alcohol; m.w. 142.04; dens. 2.671; m.p. 888 C

Toxicology: LD50 (oral, mouse) 5989 mg/kg; moderately toxic by intravenous route; mildly toxic by ingestion; experimental teratogen, reproductive effects; heated to decomp., emits toxic fumes of SO_x and Na_2O

Uses: Food additive; agent in caramel prod.; boiler water additive; hog/poultry scald agent; constituent of paper/paperboard used in food pkg., cotton in dry food pkg.; processing aid; diluent; in malt process for beer-making

Regulatory: FDA 21CFR §172.615, 173.310, 177.1200, 186.1797, GRAS as indirect food additive; USDA 9CFR §318.7, 381.147; Japan approved; Europe listed; UK approved

Sodium sulfate (2:1). *See* Sodium sulfate
Sodium sulfate anhydrous. *See* Sodium sulfate

Sodium sulfite

CAS 7757-83-7; EINECS 231-821-4; EEC E221

Synonyms: Sulfurous acid sodium salt (1:2); Sulfurous acid disodium salt; Sodium sulfite (2:1); Sodium sulfite anhydrous

Classification: Inorganic salt

Empirical: $O_3S \cdot 2Na$

Formula: Na_2SO_3

Properties: White powd. or hexagonal crystals, odorless, salty sulfurous taste; sol. in 3.2 parts water; sol. in glycerol; pract. insol. in alcohol; m.w. 126.04; dens. 2.633 (15.4 C); b.p. dec.; pH ≈ 9

Precaution: Reducing agent

Toxicology: LD50 (IV, rat) 115 mg/kg; poison by intravenous, subcutaneous routes; mod. toxic by ingestion, intraperitoneal routes; human mutagenic data; may provoke asthma; destroys vitamin B_1; heated to decomp., emits very toxic fumes of Na_2O and SO_x

Sodium sulfite (2:1)

Uses: Food preservative and antioxidant; boiler water additive; not for use on meats, sources of vitamin B₁, raw fruits and vegetables, or fresh potatoes; bleaching agent (Japan)
Usage level: ADI 0-0.7 mg/kg (EEC)
Regulatory: FDA 21CFR §172.615, 173.310, 177.1200, 182.3798, GRAS; Japan approved (0.03-5 g/kg max. residual as sulfur dioxide); Europe listed
Manuf./Distrib.: Browning; General Chem.
Trade names containing: Reduce®-150

Sodium sulfite (2:1). *See* Sodium sulfite
Sodium sulfite anhydrous. *See* Sodium sulfite

Sodium sulfoacetate derivs. of mono- and diglycerides
Toxicology: Heated to decomp., emits acrid smoke and irritating fumes
Uses: Flavoring agent for meat, poultry
Regulatory: USDA 9CFR §318.7, 381.147 (limitation 0.5%)

Sodium sulhydrate. *See* Sodium bisulfite

Sodium tartrate
CAS 868-18-8 (anhyd.); EINECS 212-773-3; EEC E335
Synonyms: Sal tartar; Disodium tartrate; Disodium L-(+)-tartrate; Bisodium tartrate; L-Tartaric acid disodium salt
Definition: Disodium salt of L(+)-tartaric acid
Empirical: $C_4H_4Na_2O_6$
Properties: Colorless transparent cryst., odorless; sol. in water; m.w. 194.05
Toxicology: Mod. toxic by ingestion; heated to decomp., emits acrid smoke and irritating fumes
Uses: Direct food additive, sequestrant, stabilizer, antioxidant, antioxidant synergist, pH control agent, acidity regulator, emulsifier; used in cheeses, fats and oils, jams and jellies
Usage level: ADI 0-30 mg/kg (EEC)
Regulatory: FDA 21CFR §184.1801, GRAS; USDA 9CFR §318.7; Europe listed; UK approved
Manuf./Distrib.: Lohmann

Sodium tetraborate. *See* Sodium borate
Sodium tetrapolyphosphate. *See* Sodium hexametaphosphate
Sodium tetrapolyphosphate. *See* Sodium metaphosphate

Sodium thiosulfate anhyd.
CAS 7772-98-7; EINECS 231-867-5
Synonyms: Sodium hyposulfite; Sodium subsulfite; Thiosulfuric acid disodium salt
Classification: Inorganic salt
Empirical: $Na_2O_3S_2$
Formula: $Na_2S_2O_3$
Properties: Colorless cryst. or cryst. powd.; sol. in water; pract. insol. in alcohol; m.w. 158.11; dens. 1.667
Precaution: Incompat. with metal nitrates, sodium nitrite
Toxicology: Irritant; mod. toxic by subcutaneous route; heated to decomp., emits very toxic fumes of Na_2O and SO_x
Uses: Antioxidant, formulation aid, reducing agent, sequestrant
Usage level: Limitation 0.0005% (alcoholic beverages), 0.1% (table salt)
Regulatory: FDA 21CFR §184.1807, GRAS

Sodium thiosulfate pentahydrate
CAS 10102-17-7; EINECS 231-867-5
Synonyms: Sodium hyposulfite
Empirical: $Na_2O_3S_2 \cdot 5H_2O$
Formula: $Na_2S_2O_3 \cdot 5H_2O$
Properties: Cryst. or gran., odorless; sl. deliq. in moist air; sol. in water; insol. in alcohol; m.w. 248.18; dens. 1.69; m.p. 48 C; pH 6.5-8.0
Precaution: Incompat. with iodine, acids, lead, mercury, and silver salts
Toxicology: LD50 (IV, rat) > 2.5 g/kg
Uses: Food additive, sequestrant, antioxidant, formulation aid, reducing agent, emulsifier, pH control agent; used in beverages, table salt, cheeses, jams, jellies
Usage level: Limitation 0.00005% (alcoholic beverages), 0.1% (table salt)
Regulatory: FDA 21CFR §184.1807, GRAS

Sodium-2,4,5-trichlorophenate
Empirical: $C_6H_2Cl_3ONa \cdot 1.5HOH$
Properties: Buff to lt. brown flakes; sol. in water, methanol, acetone

Uses: Slime control agent in mfg. of paper/paperboard for food pkg.
Regulatory: FDA 21CFR §181.30

Sodium trimetaphosphate
CAS 7785-84-4; EINECS 232-088-3
Synonyms: Trimetaphosphate sodium; Metaphosphoric acid trisodium salt
Classification: Inorganic salt
Empirical: $O(9P_3 \cdot 3Na$
Formula: $(NaPO_3)_3$
Properties: Wh. cryst. or wh. cryst. powd.; sol. in water; m.w. 305.88
Toxicology: LD50 (IP, rat) 3650 mg/kg; poison by IV route; mod. toxic by intraperitoneal route; heated to decomp., emits toxic fumes of PO_x and Na_2O
Uses: Starch modifying agent
Regulatory: GRAS

Sodium triphosphate. *See* Pentasodium triphosphate
Sodium tripolyphosphate. *See* Pentasodium triphosphate
Soduxin. *See* Sodium succinate
Soluble glass. *See* Sodium silicate
Solvent Red 80. *See* Citrus Red No. 2
Solvent Yellow 5. *See* 1-(Phenylazo)-2-naphthylamine

Sorbic acid
CAS 110-44-1; 22500-92-1; EINECS 203-768-7; EEC E200
Synonyms: 2,4-Hexadienoic acid; Hexadienic acid; (2-Butenylidene) acetic acid; 2-Propenylacrylic acid
Classification: Organic acid
Empirical: $C_6H_8O_2$
Formula: $CH_3CH=CHCH=CHCOOH$
Properties: Colorless needles or wh. powd., char. odor; sol. in hot water; very sol. in alcohol, ether; m.w. 112.14; m.p. 134.5 C; b.p. 228 C (dec.); flash pt. (OC) 260 F
Precaution: Combustible exposed to heat or flame; reactive with oxidizers
Toxicology: LD50 (oral, rat) 7360 mg/kg; mod. toxic by IP, subcutaneous routes; mildly toxic by ingestion; severe human skin irritant; experimental tumorigen, reproductive effects; mutagenic data; heated to decomp., emits acrid smoke, irritating fumes
Uses: Preservative; acidulant; flavoring; antimycotic when migrating from food pkg.
Usage level: 0.05-0.5%; limitation 0.1% (alone), 0.2% (with salts or benzoic acid or salts), 300 mg/1000 gal (wine); ADI 0-25 mg/kg (EEC)
Regulatory: FDA 21CFR §133, 146.115, 146.152, 146.154, 150.141, 150.161, 166.110, 181.22, 181.23, 182.3089, GRAS; USDA 9CFR §318.7; BATF 27CFR §240.1051; USA CIR approved, EPA reg.; JSCI approved 0.5% max.; Europe listed 0.6% max.
Regulatory: Japan approved with limitations; Europe listed; UK approved
Manuf./Distrib.: Ashland; Daicel; Dinoval; Eastman; Ellis & Everard; Hoechst AG; Honeywill & Stein; Int'l. Sourcing; Jungbunzlauer; E Merck; Penta Mfg.; Quimdis; Robt. Bryce; Sanofi; Siber Hegner; Tesco; Ueno; Unipex; Van Waters & Rogers
Trade names: Eastman Sorbic Acid; Tristat
Trade names containing: Beta Carotene Emulsion Beverage Type 3.6 No. 65392; CP-2600; Dry Vitamin A Palmitate Type 250-SD No. 65378; Dry Vitamin D_3 Type 100-SD No. 65216; Durkote Sorbic Acid/Hydrog. Veg. Oil; Superb® Bakers Margarine 105-101; Vitamin B_{12} 0.1% SD No. 65354; Vitamin B_{12} 1.0% SD No. 65305

Sorbic acid potassium salt. *See* Potassium sorbate
Sorbic acid sodium salt. *See* Sodium sorbate
Sorbimacrogol laurate 300. *See* Polysorbate 20
Sorbimacrogol oleate 300. *See* Polysorbate 80
Sorbimacrogol palmitate 300. *See* Polysorbate 40
Sorbimacrogol stearate 300. *See* Polysorbate 60
Sorbimacrogol trioleate 300. *See* Polysorbate 85
Sorbimacrogol tristearate 300. *See* Polysorbate 65

Sorbitan caprylate
Uses: Food emulsifier
Trade names: Nissan Nonion CP-08R

Sorbitan laurate
CAS 1338-39-2; 5959-89-7; EINECS 215-663-3; 227-729-9; EEC E493
Synonyms: Sorbitan monolaurate; Anhydrosorbitol monolaurate; Sorbitan monododecanoate

Sorbitan monododecanoate

Definition: Monoester of lauric acid and hexitol anhydrides derived from sorbitol
Empirical: $C_{18}H_{34}O_6$
Toxicology: Experimental neoplastigen
Uses: Food emulsifier, solubilizer, crystallization retarder, dough improver, antifoam, stabilizer
Usage level: ADI 0-25 mg/kg (EEC, total sorbitan esters)
Regulatory: FDA 21CFR §175.320, 178.3400; Europe listed; UK approved
Trade names: Ablunol S-20; Crill 1; Glycomul® L; Hodag SML; Lamesorb® SML; Nikkol SL-10; Nissan Nonion LP-20R, LP-20RS; Prote-sorb SML; S-Maz® 20; Sorgen 90; Span® 20
Trade names containing: Amisol™ MS-12 BA

Sorbitan monododecanoate. *See* Sorbitan laurate
Sorbitan, monohexadecanoate, poly(oxy-1,2-ethaneidyl) derivs. *See* Polysorbate 40
Sorbitan monolaurate. *See* Sorbitan laurate
Sorbitan monooctadecanoate. *See* Sorbitan stearate
Sorbitan mono-9-octadecenoate. *See* Sorbitan oleate
Sorbitan monooleate. *See* Sorbitan oleate
Sorbitan monopalmitate. *See* Sorbitan palmitate
Sorbitan monostearate. *See* Sorbitan stearate

Sorbitan myristate

Definition: Monoester of myristic acid and hexitol anhydrides derived from sorbitol
Uses: Food emulsifier
Trade names: Nissan Nonion MP-30R

Sorbitan, 9-octadecenoate (2:3). *See* Sorbitan sesquioleate

Sorbitan oleate

CAS 1338-43-8; 5938-38-5; EINECS 215-665-4; EEC E494
Synonyms: SMO; Sorbitan monooleate; Sorbitan mono-9-octadecenoate; Anhydrosorbitol monooleate
Definition: Monoester of oleic acid and hexitol anhydrides derived from sorbitol
Empirical: $C_{24}H_{44}O_6$
Properties: sapon. no. 145-160; hyd. no. 193-210
Toxicology: Heated to decomp., emits acrid smoke and irritating fumes
Uses: Emulsifier in polymer dispersions used in clarification of beet or cane sugar juice or liquor; stabilizer
Usage level: 0.7 ppm max. (sugar juice), 1.4 ppm max. (sugar liquor); ADI 0-25 mg/kg (EEC, total sorbitan esters)
Regulatory: FDA 21CFR §73.1001, 173.75, 175.105, 175.320, 178.3400; Europe listed; UK approved
Trade names: Ablunol S-80; Capmul® O; Crill 4; Crill 50; Drewmulse® SMO; Drewsorb® 80K; Hodag SMO; Lamesorb® SMO; Nikkol SO-10; Nissan Nonion OP-80R; Norfox® Sorbo S-80; Polycon S80 K; Prote-sorb SMO; S-Maz® 80K; Sorgen 40; Sorgen S-40-H; Span® 80

Sorbitan palmitate

CAS 26266-57-9; EINECS 247-568-8; EEC E495
Synonyms: Sorbitan monopalmitate; 1,4-Anhydro-D-glucitol, 6-hexadecanoate
Empirical: $C_{22}H_{42}O_6$
Uses: Food emulsifier; stabilizer
Usage level: ADI 0-25 mg/kg (EEC, total sorbitan esters)
Regulatory: FDA 21CFR §175.320, 178.3400; Europe listed; UK approved
Trade names: Ablunol S-40; Crill 2; Hodag SMP; Lamesorb® SMP; Nikkol SP-10; Nissan Nonion PP-40R; Prote-sorb SMP; Span® 40

Sorbitan sesquioleate

CAS 8007-43-0; EINECS 232-360-1
Synonyms: Anhydrosorbitol sesquioleate; Anhydrohexitol sesquioleate; Sorbitan, 9-octadecenoate (2:3)
Definition: Mixture of mono and diesters of oleic acid and hexitol anhydrides derived from sorbitol
Uses: W/o emulsifier, wetting agent, pigment dispersant for food applics.
Regulatory: FDA 21CFR §175.320
Trade names: Crill 43; Nikkol SO-15; Nissan Nonion OP-83RAT; Sorgen 30; Sorgen S-30-H

Sorbitan sesquistearate

Synonyms: Anhydrosorbitol sesquistearate
Definition: Mixture of mono and diesters of stearic acid and hexitol anhydrides derived from sorbitol
Uses: Food emulsifier
Regulatory: FDA 21CFR §175.320
Trade names: Nikkol SS-15

Sorbitan stearate

CAS 1338-41-6, 69005-67-8; EINECS 215-664-9; FEMA 3028; EEC E491

Synonyms: SMS; Sorbitan monostearate; Sorbitan monooctadecanoate; Anhydrosorbitol monostearate
Definition: Monoester of stearic acid and hexitol anhydrides derived from sorbitol
Empirical: $C_{24}H_{46}O_6$
Properties: m.w. 430.70; acid no. 5-10; sapon. no. 147-157; hyd. no. 235-260
Toxicology: LD50 (oral, rat) 31 g/kg; very mildly toxic by ingestion; experimental reproductive effects; heated to decomp., emits acrid smoke and irritating fumes
Uses: Emulsifier for whipped toppings, cakes, cake mixes/icings/fillings, confectionery coatings, coffee whitener; rehydration agent in prod. of active dry yeast; protective coatings on fruits and vegetables; stabilizer, defoaming agent, flavoring
Usage level: Limitation 0.4% (whipped toppings), 1% (confectionery coatings), 0.4% (milk substitute), 0.7% (cake icings), 1% (active dry yeast); ADI 0-25 mg/kg (EEC, total sorbitan esters)
Regulatory: FDA 21CFR §73.1001, 163.123, 163.130, 163.135, 163.140, 163.145, 163.150, 163.153, 163.155, 172.515, 172.842, 173.340, 175.105, 175.320, 178.3400, 573.960; Europe listed; UK approved
Manuf./Distrib.: Aldrich
Trade names: Ablunol S-60; Alkamuls® SMS; Atlas 110K; Capmul® S; Crill 3; Drewmulse® SMS; Drewsorb® 60K; Durtan® 60; Durtan® 60K; Famodan MS Kosher; Glycomul® S FG; Glycomul® S KFG; Hodag SMS; Lamesorb® SMS; Liposorb S; Nissan Nonion SP-60R; Norfox® Sorbo S-60FG; Polycon S60 K; Prote-sorb SMS; S-Maz® 60K; S-Maz® 60KHS; Sorgen 50; Span® 60; Span® 60K; Span® 60 VS
Trade names containing: AF 72; AF 75; Atmos® 729; Atmos® 758 K; Atmos® 2462; Brem; Kaokote; Satina 53NT; Snac-Kote; Vanade; Vanade Kosher; Vanall; Vanall (K); Vykamol 83G

Sorbitan trioctadecanoate. *See* Sorbitan tristearate
Sorbitan tri-9-octadecenoate. *See* Sorbitan trioleate

Sorbitan trioleate
CAS 26266-58-0; 85186-88-5; EINECS 247-569-3; 286-074-7
Synonyms: STO; Anhydrosorbitol trioleate; Sorbitan tri-9-octadecenoate
Definition: Triester of oleic acid and hexitol anhydrides derived from sorbitol
Empirical: $C_{60}H_{108}O_8$
Uses: Food emulsifier
Regulatory: FDA 21CFR §175.320, 178.3400
Trade names: Ablunol S-85; Crill 45; Hodag STO; Nissan Nonion OP-85R; Prote-sorb STO; S-Maz® 85K; Span® 85

Sorbitan tristearate
CAS 26658-19-5; 72869-62-6; EINECS 247-891-4; 276-951-2; EEC E492
Synonyms: STS; Anhydrosorbitol tristearate; Sorbitan trioctadecanoate
Definition: Triester of stearic acid and hexitol anhydrides derived from sorbitol
Empirical: $C_{60}H_{114}O_8$
Uses: Food emulsifier; stabilizer
Usage level: ADI 0-25 mg/kg (EEC, total sorbitan esters)
Regulatory: FDA 21CFR §175.320, 178.3400; Europe listed; UK approved
Manuf./Distrib.: ICI Spec.
Trade names: Alkamuls® STS; Crill 35; Crill 41; Drewsorb® 65K; Famodan TS Kosher; Glycomul® TS KFG; Hodag STS; Lamesorb® STS; Liposorb TS; Prote-sorb STS; Span® 65; Span® 65K
Trade names containing: Akopol E-1; Akorine 9F; Kaoprem-E

Sorbitan tritallate
Uses: O/w emulsifier, dispersant for food applics.
Trade names: S-Maz® 95

D-Sorbite. *See* Sorbitol

Sorbitol
CAS 50-70-4; EINECS 200-061-5; EEC E420
Synonyms: D-Glucitol; D-Sorbitol; D-Sorbite; 1,2,3,4,5,6-Hexanehexol
Classification: Hexahydric alcohol
Empirical: $C_6H_{14}O_6$
Formula: CH$_2$OHHCOHHOCHHCOHHCOHCH$_2$OH
Properties: White cryst. powd., odorless, sweet taste; sol. in water; sol. in hot alcohol, methanol, IPA, DMF, acetic acid, phenol, acetamide sol'ns.; m.w. 182.20; dens. 1.47 (-5 C); m.p. 93-97.5 C; b.p. 105 C; pH ≈ 7.0
Toxicology: LD50 (oral, rat) 17.5 mg/kg; mildly toxic by ingestion; excess consumption may have laxative effect
Uses: Anticaking/free-flow, curing/pickling, drying, firming, emulsifier, flavoring, adjuvant, formulation aid, humectant, lubricant, release, nutritive sweetener, sequestrant, stabilizer, thickener, surface-finisher, dietary supplement, texturizing agent

D-Sorbitol

    Usage level: Limitation 99% (hard candy), 75% (chewing gum), 98% (soft candy), 30% (jams, jellies), 30% (baked goods), 17% (frozen dairy desserts), 12% (other foods); up to 7% (nutrient, dietary supplement); ADI not specified (FAO/WHO)
    Regulatory: FDA 21CFR §175.300, 182.90, 184.1835, GRAS; Japan approved; Europe listed; UK approved
    Manuf./Distrib.: ADM; Am. Roland; Am. Xyrofin; Ashland; Browning; Calaga Food Ingreds.; Cerestar Int'l.; Cheil Foods; Ellis & Everard; Fruitsource; ICI Atkemix; Int'l. Sourcing; E Merck; Penta Mfg.; Pfizer Food Science; Regency Mowbray; Roquette UK; Treasure Island
    Trade names: Hydex® 100 Coarse Powd.; Hydex® 100 Coarse Powd. 35; Hydex® 100 Gran. 206; Hydex® 100 Powd. 60; Hydex® Tablet Grade; Hystar® 7570; Liponic Sorbitol Powd.; Liponic Sorbitol Sol'n. 70% USP; Sorbo®; Unisweet 70
    Trade names containing: Palma-Sperse® Type 250-S No. 65322; Palma-Sperse® Type 250A/50 D-S No. 65221; Super Flex

D-Sorbitol. *See* Sorbitol
Soudan coffee. *See* Cola
Soya bean oil. *See* Soybean oil
Soya bran. *See* Soy bran

Soy acid
    CAS 68308-53-2; 67701-08-0; EINECS 269-657-0
    Synonyms: Acids, soy; Fatty acids, soya
    Definition: Mixture of fatty acids derived from soybean oil
    Uses: Lubricant, release agent, binder, defoaming agent and intermediate for food additives
    Regulatory: FDA 21CFR §175.105, 177.2800, 178.3570
    Trade names: Industrene® 225; Industrene® 226 FG

Soya flour. *See* Soy flour
Soya oil. *See* Soybean oil
Soya oil monoglyceride. *See* Soy glyceride

Soybean milk
    Uses: Nutritive food additive
    Trade names: Mapron®

Soybean oil
    CAS 8001-22-7; EINECS 232-274-4
    Synonyms: Soya oil; Soya bean oil; Chinese bean oil
    Classification: Fixed oil
    Definition: Oil obtained from soybeans by extraction or expression; consists of triglycerides of oleic, linoleic, linolenic, and saturated acids
    Properties: Pale yel. oil, sl. char. odor and taste; sol. in alcohol, ether, chloroform, carbon disulfide; dens. 0.924-0.929; m.p. 22-31 C; flash pt. 540 F; ref. index 1.471-1.475 (25 C); visc. 50.09 cps (25 C); sapon. no. 189-195
    Precaution: Combustible
    Toxicology: Heated to decomp., emits acrid smoke and irritating fumes
    Uses: Dietary supplement, food ingred.; coating agent, emulsifier, formulation aid, texturizer, lubricant; used in cooking oil, margarine, salad oil, shortening; cattle feeds
    Regulatory: FDA 21CFR §175.105, 175.300, 176.200, 176.210, 177.2800, GRAS
    Manuf./Distrib.: Aarhus Olie; Amcan France; Am. Ingreds.; Am. Roland; Arista Industries; S Black; Calgene; Canada Packers; Cargill; Central Soya; Karlshamns; Peerless Refining; Penta Mfg.; Riceland Foods; Soya Mainz; Spectrum Naturals; Vamo-Fuji Specialities
    Trade names: Archer Soybean Oil 86-070-0; Capital Soya; Lipovol SOY; Soybean Oil; Soy Oil
    Trade names containing: Aloe Vera Lipo-Quinone Extract™ (AVLQE) Food Grade; Bake-Well 52; Bake-Well 80/20; Bake-Well High Stability Pan Oil; Bread Pan Oil; Carex for Foods; Centrophase® NV; Crester RA; Crester RB; Diet Imperial® Spread; EC-25®; Fluid Flex; Golden Chef® Clear Liquid Frying Shortening 104-052; Golden Chef® Creamy Liquid Frying Shortening 104-065; Hi-Ratio Shortening; MG-60; Neobee® SL-210; Neobee® SL-230; PAV; Poly E; Shedd's Liquid Margarine; Shedd's Margarine; Shedd's NP Margarine; Shedd's Special 40 Butter Blend Margarine; Shedd's Special 60 Butter Blend Margarine; Soybean Salad Oil 104-050; Super-Cel® Specialty Cake Shortening; Top Notch®; Vegetable All Purpose Shortening 86-713-0

Soybean oil, epoxidized. *See* Epoxidized soybean oil
Soybean oil glyceride. *See* Soy glyceride
Soybean oil hydrogenated. *See* Hydrogenated soybean oil

Soy bran
    Synonyms: Soya bran

Uses: Bulking agent providing dietary fiber to foodstuffs
Trade names: Sovital™

Soy fiber
Uses: Dietary fiber supplement
Trade names: Arbran™ Soy Fiber; Fibrim® 1020; Fibrim® 1250; Fibrim® 1255; Fibrim® 1450; Fibrim® 2000; Superb® 1450 Soy Fiber; Superb® 2000 Soy Fiber

Soy flour
CAS 68513-95-1
Synonyms: Flour, soy; Soya flour
Definition: Powd. prepared from fine grinding of soybean, *Glycine max.*
Uses: High protein food prods., esp. textured proteins and meat analogs
Trade names: Bakers Nutrisoy; Centex®; Nurupan™; Nutrisoy 220T; Soyafluff® 200 W; Soyamin 50 E; Soyamin 50 T; Soyapan™
Trade names containing: CDC-10A (NB); CDC-2001; CDC-2002; CDC-2500; CDC-3000; CFI-10; Do Crest Gold; Do Crest Gold Plus; Drize-P Powd.; Hentex-76; Luxor® KB-320; Natural Flavor White Sour; N'Hance (Phase I) Powd.; N'Hance SD; Nutrisoy 7B; Soyarich® 115 W; Soylec C-6; Soylec C-15; Soylec T-15

Soy glyceride
Synonyms: Glyceryl mono soya oil; Glyceryl monosoyate; Soybean oil glyceride; Soya oil monoglyceride
Uses: Food additive, emulsion stabilizer
Trade names: Imwitor® 460; Imwitor® 490; Tegomuls® SO
Trade names containing: Centrobake® 100L; Myvatex® MSPS; Myvatex® SSH; Myvatex® Texture Lite® K

Soy protein
CAS 68153-28-6
Synonyms: Soy protein isolated; Isolated soy protein
Definition: Protein obtained from the soybean, *Glycine max*
Toxicology: Heated to decomp., emits acrid smoke and irritating fumes
Uses: Binder, emulsifier, extender, stabilizer; used in meatballs, poultry, sausage, snack foods, whipped toppings, frozen spaghetti; migrating to food from paper/paperboard
Usage level: Limitation 2% (sausage)
Regulatory: FDA 21CFR §182.90; GRAS; USDA 9CFR §318.7, 319, 381.147
Manuf./Distrib.: Aarhus Olie; ADM Ingreds., Am. Roland, Ashland, Avebe BA, Bush Boake Allen; Cargill; Central Soya; Cheil Foods; Farbest Brands; KWR; Lucas Meyer; Nasoya Foods; PMS Foods; Protein Tech.; Quest; Seah Int'l; Stern-Lecithin; UFL Foods; Vaessen-Schoemaker
Trade names: Arcon F; Arcon G, Fortified Arcon G; Arcon S, Fortified Arcon S; Arcon T, Fortified Arcon T; Arcon VF, Fortified Arcon VF; Ardex D, D Dispersible; Ardex D-HD; Ardex DHV Dispersible; Ardex F, F Dispersible; Ardex FR; Ardex R; Danpro DS; Danpro HV; Danpro S; Danpro S-760; FP 940; IPSO-C403; IPSO-FC; IPSO-MR Dispersible; IPSO-NGL; Mira-Foam® 100; Mira-Foam® 100K; Mira-Foam® 120V; Mira-Foam® 130H; Procon® 2000; Procon® 20/60; Procon® 20/60 Military; Procon® 20/60-SL; Pro Fam 646; Pro Fam 648; Pro Fam 780; Pro Fam 781; Pro Fam 955; Pro Fam 970; Pro Fam 972; Pro Fam 974, 974 Fortified; Promax® 70; Promax® 70L; Promax® 70LSL; Promax® Plus; Promine® DS; Promine® DS-SL; Promine® HV; Proplus®; Supro® 425; Supro® 610; Supro® 661; Supro® 670; Supro® 710; Supro® 760; Textratein; Tofupro-U; Vegafoom; Versa-Whip® 500, 510, 520; Versa-Whip® 520K, 600K, 620K
Trade names containing: Mix 41; Mix 63; Mix 70; Supro Plus® 651; Supro Plus® 675; Supro Plus® 2100; Supro Plus® 3000

Soy protein isolated. *See* Soy protein

Soy sauce
Synonyms: Fermented soy sauce
Definition: Hydrolysis prod. of soybeans; a mixt. of amino acids, peptides, polypeptides, peptones, simple proteins, purines, carbohydrates, and lesser org. compds. suspended in an 18% sol'n. of sodium chloride
Uses: Natural flavoring
Trade names containing: Aromild® S; Natural HVP Replacer

Spanish bugloss extract. *See* Alkanet extract
Spanish hops. *See* Dittany of Crete
Spanish marjoram oil. *See* Marjoram oil, Spanish
Spanish thyme oil. *See* Thyme oil red

Spearmint oil
CAS 8008-79-5; FEMA 3032
Synonyms: Mentha spicata oil; Crispmint oil; Curled mint oil
Definition: Volatile oil obtained from the dried tops and leaves of *Mentha spicata*, contg. chiefly carvone

Speedwell extract

Properties: Colorless or greenish-yel. liq., spearmint odor and taste; sol. in equal vol 80% alcohol; very sl. sol. in water; dens. 0.917-0.934 (25/25 C); ref. index 1.4820-1.4900 (20 C)

Precaution: Keep cool, well closed; protect from light

Toxicology: LD50 (oral, rat) 5 g/kg; mildly toxic by ingestion; mutagenic data; skin irritant; allergen; heated to decomp., emits acrid smoke and irritating fumes

Uses: Natural flavoring agent for chewing gum, oral hygiene flavors, etc.

Usage level: 100 ppm (nonalcoholic beverages), 100 ppm (alcoholic beverages), 81 ppm (ice cream), 830 ppm (candy), 270 ppm (baked goods), 75 ppm (gelatins, puddings), 6200 ppm (chewing gum), 72-1900 ppm (jellies)

Regulatory: FDA 21CFR §182.10, 182.20, GRAS; 27CFR §21.65, 12.128, 21.151; FEMA GRAS; Japan approved (spearmint); Europe listed, no restrictions

Manuf./Distrib.: Acme-Hardesty; Florida Treatt

Speedwell extract. *See* Veronica extract

Spike lavender oil

FEMA 3033

Synonyms: Spike oil

Definition: Volatile oil from leaves and tops of *Lavandula spica* or *L. latifolia*, contg. about 35% eucalyptol, and camphor, linalool, borneol, terpineol, dd-camphene, sesquiterpene

Properties: Colorless or pale yel. liq., lavender and eucalyptol odor; sol. in 3 vols 70% alcohol; pract. insol. in water; dens. 0.900-0.920 (15/15 C); ref. index 1.462-1.469 (20 C)

Precaution: Keep cool, well closed; protect from light

Uses: Natural flavoring agent

Regulatory: FDA 21CFR §182.20, GRAS

Spike oil. *See* Spike lavender oil

Spirit of Hartshorn. *See* Ammonium hydroxide

Spruce

Definition: Needles and twigs of *Picea glauca* or *P. mariana*

Uses: Natural flavoring

Regulatory: FDA 21CFR §172.510

Stannous chloride anhyd.

CAS 7772-99-8; EINECS 231-868-0

Synonyms: Tin (II) chloride anhydrous; Tin crystals; Tin (II) chloride (1:2); Tin salt; Tin dichloride; Tin protochloride

Definition: Chloride salt of metallic tin

Empirical: Cl_2Sn

Formula: $SnCl_2$

Properties: Colorless orthorhombic cryst. mass or flakes, fatty appearance; sol. in water, ethanol, acetone, ether, methyl acetate, MEK, isobutyl alcohol; pract. insol. in min. spirits, xylene; m.w. 189.60; dens. 3.95

Precaution: Potentially explosive reaction with metal nitrates; violent reactions with hydrogen peroxide, ethylene oxide, nitrates, K, Na

Toxicology: TLV:TWA 2 mg(Sn)/m³; LD50 (oral, rat) 700 mg/kg, (IP, mouse) 66 mg/kg; poison by ingestion, IP, IV, subcutaneous routes; experimental reproductive effects; human mutagenic data; heated to decomp., emits toxic fumes of Cl⁻

Uses: Food preservative, reducing agent, antioxidant, color retention aid for asparagus packed in glass

Usage level: Limitation 0.0015% (in food, calculated as tin), 20 ppm (asparagus packed in glass)

Regulatory: FDA 21CFR §155.200, 172.180, 175.300, 177.2600, 184.1845, GRAS

Stannous chloride dihydrate

CAS 10025-69-1; EINECS 231-868-0

Synonyms: Tin (II) chloride dihydrate

Empirical: $Cl_2Sn \cdot 2H_2O$

Formula: $SnCl_2 \cdot 2H_2O$

Properties: Cryst.; sol. in less than own wt. in water; sol. in alcohol, ethyl acetate, glac. acetic acid, sodium hydroxide sol'n.; very sol. in dil. or conc. HCl; m.w. 225.63; dens. 2.71; m.p. 37-38 C; dec. on strong heating

Precaution: Keep tightly closed in cool place

Uses: Antioxidant

Usage level: Limitation 0.0015% in food

Regulatory: FDA 21CFR §184.1845, GRAS

Stannous stearate

Synonyms: Tin stearate

Toxicology: Heated to decomp., emits acrid smoke and irritating fumes
Uses: Stabilizer migrating from food pkg.
Regulatory: FDA 21CFR §181.29 (50 ppm max. in finished food)

Star anise

FEMA 2096 (oil)
Definition: Derived from *Illicium verum*
Uses: Natural flavoring agent in alcoholic drinks, confectionery, oral hygiene
Usage level: 2-30 ppm (nonalcoholic beverages), 96-5000 ppm (condiments), 1200 ppm (meats), 1-4 ppm (ice cream, ices, etc.), 3-4 ppm (candy), 490 ppm (baked goods)
Regulatory: FDA 21CFR §182.10, GRAS; FEMA GRAS (oil); Japan approved; Europe listed, no restrictions (oil)

Star anise oil. *See* Anise oil

Starch

CAS 9005-84-9; EINECS 232-686-4
Classification: Carbohydrate polymer
Definition: Complex polysaccharide composed of units of glucose consisting of about one quarter amylose and three quarters amylopectin
Empirical: $(C_6H_{10}O_5)_n$
Properties: White amorphous powd. or gran., tasteless; gel formation in hot water; m.w. 162.14_n
Uses: Gellant for foods, chelating and sequestering agent, filler in baking powder; migrating to food from paper/paperboard
Regulatory: FDA 21CFR §182.90, GRAS
Trade names containing: Beta-Tabs; Durkote Calcium Carbonate/Starch, Acacia Gum; Frimulsion RA; Frimulsion RF; Mayonat V/100; Merecol® S; Merecol® Y; Rocoat® Riboflavin 25% No. 60289; Ticaloid® S2-102; TIC Pretested® Aragum® 3000

Starch, corn. *See* Corn starch
Starch gum. *See* Dextrin
Starch, octenylbutanedioate, aluminum salt. *See* Aluminum starch octenyl succinate
St. Bartholomew's tea. *See* Mate extract

Stearalkonium chloride

CAS 122-19-0, EINECS 204-527-9
Synonyms: Stearyl dimethyl benzyl ammonium chloride; Dimethyloctadecylbenzyl ammonium chloride; Octadecyl dimethyl benzyl ammonium chloride; N,N-Dimethyl N octodooylbonzonomethanaminium chloride
Classification: Quaternary ammonium salt
Empirical: $C_{27}H_{50}N \cdot Cl$
Properties: m.w. 424.23
Toxicology: LD50 (oral, rat) 125 mg/kg; poison by intraperitoneal route; moderately toxic by ingestion; human skin and severe eye irritant; heated to decomp., emits very toxic fumes of NO_x, NH_3, and Cl⁻
Uses: Antimicrobial for use in cane-sugar mills
Regulatory: FDA 21CFR §172.165 (limitation 1.5-6 ppm), 173.320 (limitation 0.05 ppm raw sugarcane or raw beets), 175.105

Stearamide

CAS 124-26-5; EINECS 204-693-2
Synonyms: Octadecanamide; Stearic acid amide; Amide C-18
Classification: Aliphatic amide
Empirical: $C_{18}H_{37}NO$
Formula: $CH_3(CH_2)_{16}CONH_2$
Properties: Colorless leaflets; sl. sol. in alcohol, ether; insol. in water; m.w. 283.56; m.p. 98-102 C; b.p. 250-251 C (12 mm)
Toxicology: Experimental tumorigen
Uses: Release agent migrating from food pkg.
Regulatory: FDA 21CFR §175.105, 177.1210, 178.3860, 178.3910, 179.45, 181.22, 181.28
Manuf./Distrib.: Aldrich
Trade names: Kemamide® S

Stearamine

CAS 124-30-1; EINECS 204-695-3
Synonyms: Stearylamine; Octadecylamine; 1-Octadecanamine
Empirical: $C_{18}H_{39}N$

Formula: $CH_3(CH_2)_{16}CH_2NH_2$
Properties: m.w. 269.58
Toxicology: LD50 (IP, mouse) 250 mg/kg; poison by intraperitoneal route; skin irritant; heated to decomp., emits toxic fumes of NO_x
Uses: Boiler water additive for food contact except milk prods.
Regulatory: FDA 21CFR §173.310 (3 ppm max. in steam)
Manuf./Distrib.: Aldrich

Steareth-20

CAS 9005-00-9 (generic)
Synonyms: PEG-20 stearyl ether; POE (20) stearyl ether; PEG 1000 stearyl ether
Definition: PEG ether of stearyl alcohol
Empirical: $C_{58}H_{118}O_{21}$
Formula: $CH_3(CH_2)_{16}CH_2(OCH_2CH_2)_nOH$, avg. n = 20
Uses: Emulsifier for wax used in fruit coatings
Regulatory: FDA 21CFR §177.2800
Trade names: Trycol® 5888; Trycol® SAL-20

Stearic acid

CAS 57-11-4; EINECS 200-313-4; EEC E570
Synonyms: n-Octadecanoic acid; Carboxylic acid C_{18}
Classification: Fatty acid
Empirical: $C_{18}H_{36}O_2$
Formula: $CH_3(CH_2)_{16}COOH_9$
Properties: White to ylsh.-wh. amorphous solid, tallow-like odor and taste; very sl. sol. in water; sol. in alcohol, ether, acetone, CCl_4; m.w. 284.47; dens. 0.847 (70 C); m.p. 69.3 C; b.p. 383 C; flash pt. (CC) 385 F; ref. index 1.4299 (80 C)
Precaution: Combustible when exposed to heat or flame; heats spontaneously
Toxicology: LD50 (IV, rat) 21.5 ± 1.8 mg/kg; poison by intravenous route; experimental tumorigen; human skin irritant; heated to decomp., emits acrid smoke and irritating fumes
Uses: Direct food additive, flavoring agent, adjuvant; component in mfg. of other food-grade additives; lubricant; defoamer; anticaking agent; masticatory substance in chewing gum base; clarifying agent (wine)
Regulatory: FDA 21CFR §172.210, 172.615, 172.860, 175.105, 175.300, 175.320, 176.170, 176.200, 176.210, 177.1010, 177.1200, 177.2260, 177.2600, 177.2800, 178.3570, 178.3910, 184.1090, GRAS; GRAS (FEMA); Europe listed
Manuf./Distrib.: Acme-Hardesty
Trade names: Emersol® 6320; Emersol® 6332 NF; Emersol® 6349; Emersol® 6351; Emersol® 6353; Emersol® 7051; Glycon® S-70; Glycon® S-90; Glycon® TP; Hystrene® 5016 NF FG; Hystrene® 7018 FG; Hystrene® 8718 FG; Hystrene® 9718 NF FG; Industrene® 4518; Industrene® 5016 NF FG; Industrene® 7018 FG; Industrene® 8718 FG
Trade names containing: Rocoat® Niacinamide 33$^1/_3$% Type S No. 69909

Stearic acid aluminum dihydroxide salt. *See* Aluminum stearate
Stearic acid amide. *See* Stearamide
Stearic acid calcium salt. *See* Calcium stearate
Stearic acid, potassium salt. *See* Potassium stearate
Stearic acid sodium salt. *See* Sodium stearate
Stearin. *See* Tristearin
Stearoyl lactylic acid. *See* Stearyl-2-lactylic acid
Stearoyl propylene glycol hydrogen succinate. *See* Succistearin

Stearyl alcohol

CAS 112-92-5; EINECS 204-017-6
Synonyms: n-Octadecanol; Octadecyl alcohol; 1-Octadecanol; n-Octyldecyl alcohol; C18 linear alcohol; Decyl octyl alcohol
Classification: Fatty alcohol
Empirical: $C_{18}H_{38}O$
Formula: $CH_3(CH_2)_{16}CH_2OH$
Properties: Unctuous white flakes or gran., faint odor, bland taste; sol. in alcohol, acetone, ether; insol. in water; m.w. 270.56; dens. 0.8124 (59/4 C); m.p. 59 C; b.p. 210.5 C (15 mm)
Precaution: Flamm. when exposed to heat or flame; can react with oxidizers
Toxicology: LD50 (oral, rat) 20 g/kg; mildly toxic by ingestion; experimental neoplastigen; heated to decomp., emits acrid smoke and irritating fumes
Uses: Food additive, intermediate in mfg. of food additives
Regulatory: FDA 21CFR §172.755, 172.864, 175.105, 175.300, 176.200, 176.210, 177.1010, 177.1200,

177.2800, 178.3480, 178.3910
Manuf./Distrib.: Aarhus Oliefabrik A/S; Croda; Procter & Gamble

Stearylamine. *See* Stearamine

Stearyl citrate
CAS 1337-33-3; EINECS 215-654-4
Synonyms: 2-Hydroxy-1,2,3-propanetricarboxylic acid, monooctadecyl ester; Octadecyl citrate; Citric acid, octadecyl ester
Definition: Ester of stearyl alcohol and citric acid
Empirical: $C_{24}H_{44}O_7$
Formula: $HOCH_2COOHCCOOHCH_2COOCH_2(CH_2)_{16}CH_3$
Toxicology: Heated to decomp., emits acrid smoke and irritating fumes
Uses: Sequestrant, flavor preservative, plasticizer; used in oleomargarine, pkg. materials
Regulatory: FDA 21CFR §166.40, 166.110, 175.300, 178.3910, 181.22, 181.27, 182.6851 (0.15% max. as sequestrant), GRAS

Stearyl dimethyl benzyl ammonium chloride. *See* Stearalkonium chloride

Stearyl glyceridyl citrate
Synonyms: Stearyl monoglyceridyl citrate
Properties: Off-wh. to tan soft waxy solid, pract. tasteless; sol. in chloroform, ethylene glycol; insol. in water; acid no. 40-52; sapon. no. 215-255
Toxicology: Heated to decomp., emits acrid smoke and irritating fumes
Uses: Emulsion stabilizer in or with shortenings containing emulsifiers; in margarine, oleomargarine
Usage level: Limitation 0.5% (margarine)
Regulatory: FDA 21CFR §172.755, 172.860; USDA 9 CFR §318.7

Stearyl-2-lactylic acid
Synonyms: Stearoyl lactylic acid
Toxicology: Heated to decomp., emits acrid smoke and irritating fumes
Uses: Emulsifier for shortening
Usage level: Limitation 3% (shortening for cake icings/fillings)
Regulatory: USDA 9CFR §318.7

Stearyl monoglyceridyl citrate. *See* Stearyl glyceridyl citrate
Sterculia gum. *See* Karaya gum
Sterculia urens gum. *See* Karaya gum

Stevioside
Synonyms: Steviosin
Definition: Plant extract from leaves of rebaudium
Empirical: $C_{38}H_{50}O_{18}$
Properties: Bitter aftertaste; m.w. 804.90; relatively stable in sol'n. and on heating
Uses: Nonnutritive sweetener in soft drinks, candy, chewing gum; 100-300 times the sweetness of sucrose
Regulatory: Not approved in North America or Europe as a food additive; widely used in Japan
Manuf./Distrib.: Am. Roland

Steviosin. *See* Stevioside
Stigmata maydis extract. *See* Corn silk extract
St. John's bread. *See* Locust bean gum
STO. *See* Sorbitan trioleate

Storax
CAS 8023-62-9
Synonyms: Styrax; Sweet oriental gum
Definition: Balsam obtained from trunk of *Liquidambar orientalis*
Properties: Semiliq. to semisolid, char. odor and taste; sol. in 1 part warm alcohol, ether, acetone, CS_2; insol. in water
Uses: Natural flavoring
Regulatory: FDA 21CFR §172.510

STPP. *See* Pentasodium triphosphate
Strawberry aldehyde. *See* Ethyl methylphenylglycidate
STS. *See* Sorbitan tristearate
Styracin. *See* Cinnamyl cinnamate
Styralyl acetate. *See* α-Methylbenzyl acetate
Styralyl alcohol. *See* α-Methylbenzyl alcohol

Styralyl butyrate. *See* α-Methylbenzyl butyrate
Styralyl formate. *See* α-Methylbenzyl formate
Styralyl isobutyrate. *See* Methylbenzyl isobutyrate
Styralyl propionate. *See* α-Methylbenzyl propionate
Styrax. *See* Storax
Styrax benzoin extract. *See* Benzoin extract
Styrax extract. *See* Benzoin extract

Styrene
CAS 100-42-5; EINECS 202-851-5
Synonyms: Phenylethylene; Ethenylbenzene; Styrol; Styrolene; Vinylbenzene; Cinnamene; Cinnamol
Empirical: C_8H_8
Formula: $C_6H_5CH{:}CH_2$
Properties: Colorless to ylsh. oily liq., penetrating odor; sol. in alcohol, ether, methanol, acetone, CS_2; sparingly sol. in water; m.w. 104.15; dens. 0.906 (20/4 C); m.p. -30.6 C; b.p. 145-146 C; flash pt. (CC) 31 C; ref. index 1.546 (20 C)
Precaution: Flamm.; slowly undergoes polymerization and oxidation on exposure to light and air, yielding peroxides, etc.
Toxicology: LD50 (IV, mouse) 90 ± 5.2 mg/kg, (IP, mouse) 660 ± 44.3 mg/kg; may be irritating to eyes, mucous membranes; narcotic in high concs.
Uses: Synthetic flavoring
Regulatory: FDA 21CFR §172.515
Manuf./Distrib.: Aldrich

Styrene-acrylonitrile copolymer. *See* Acrylonitrile-styrene copolymer
Styrene polymer with 1,3-butadiene. *See* Butadiene-styrene copolymer
Styroen. *See* Cinnamyl alcohol
Styrol. *See* Styrene
Styryl carbinol. *See* Cinnamyl alcohol
2-Styryl-1,3-dioxolane. *See* Cinnamaldehyde ethylene glycol acetal
2-Styryl-m-dioxolane. *See* Cinnamaldehyde ethylene glycol acetal
Sucaryl sodium. *See* Sodium cyclamate

Succinic acid
CAS 110-15-6; EINECS 203-740-4; EEC E363
Synonyms: 1,4-Butanedioic acid; Amber acid; 1,2-Ethanedicarboxylic acid; Ethylenesuccinic acid
Classification: Dicarboxylic acid
Empirical: $C_4H_6O_4$
Formula: $HOOCCH_2CH_2COOH$
Properties: Colorless monoclinic prisms, odorless, sour acid taste; very sol. in alcohol, ether, acetone, glycerin; sol. in water; m.w. 118.09; dens. 1.552; m.p. 185 C; b.p. 235 C
Precaution: Combustible
Toxicology: LD50 (oral, rat) 2260 mg/kg; moderately toxic by subcutaneous route; severe eye irritant; heated to decomp., emits acrid smoke and irritating fumes
Uses: Flavor enhancer, pH control agent, sequestrant, buffer, neutralizing agent, miscellaneous and general purpose food chemical; used in beverages, condiments, meat prods., relishes, hot sausages
Usage level: Limitation 0.084% (condiments, relishes), 0.0061% (meat prods.)
Regulatory: FDA 21CFR §131.144, 184.1091, GRAS; Japan approved; Europe listed; UK approved
Manuf./Distrib.: Aldrich

Succinic acid anhydride. *See* Succinic anhydride
Succinic acid diethyl ester. *See* Diethyl succinate
Succinic acid sodium salt. *See* Sodium succinate

Succinic anhydride
CAS 108-30-5; EINECS 203-570-0
Synonyms: Dihydro-2,5-furandione; Butanedioic anhydride; 2,5-Diketotetrahydrofuran; Succinic acid anhydride; Succinyl oxide
Empirical: $C_4H_4O_3$
Properties: Orthorhombic prisms; sol. in chloforom, CCl_4, alcohol; very sl. sol. in ether, water; m.w. 100.08; dens. 1.503; m.p. 119-120 C; b.p. 261 C (760 mm); sublimes @ 115 C and 5 mm pressure
Toxicology: Irritating to eyes, respiratory system
Uses: Starch modifier for foods; buffer, neutralizing agent; dehydrating agent for removal of moisture from foods; imparts stability to dry mixes; aids controlled release of carbon dioxide during leavening
Usage level: up to 4% (modifier for food starch)

Regulatory: FDA

Succinylated monoglycerides
Properties: HLB 5-7
Uses: Emulsifier in shortenings; dough conditioner
Usage level: 3% (shortening), 0.5% (flour)
Regulatory: FDA 21CFR §172.830

Succinylated palm oil glyceride
Synonyms: Succinylated palm oil monoglyceride
Uses: Dough strengthener for yeast-based bakery goods
Trade names containing: Myvatex® Do Control K

Succinylated palm oil monoglyceride. *See* Succinylated palm oil glyceride
Succinyl oxide. *See* Succinic anhydride

Succistearin
Synonyms: Stearoyl propylene glycol hydrogen succinate
Properties: acid no. 50-150; hyd. no. 15-50
Toxicology: Heated to decomp., emits acrid smoke and irritating fumes
Uses: Emulsifier in or with shortenings and edible oils; used in cakes, cake mixes, fillings, icings, pastries, toppings
Regulatory: FDA 21CFR §172.765

Sucralose
CAS 56038-13-2
Synonyms: 1,6-Dichloro-1,6-dideoxy-β-D-fructofuranosyl-4-chloro-4-deoxy-α-D-galactopyranoside; 4,1′,6′-Trichlorogalactosucrose
Empirical: $C_{12}H_{19}Cl_3O_8$
Properties: White, crystalline powder; odorless; sweet taste; sol. in water, methanol, alcohol; sl. sol. in ethyl acetate
Uses: Nonnutritive sweetener, flavor enhancer; 600 times sweeter than sucrose; for beverages, processed foods

Sucrase. *See* Invertase
Sucrosa. *See* Sodium cyclamate

Sucrose
CAS 57-50-1; EINECS 200-334-9
Synonyms: β-D-Fructofuranosyl-α-D-glucopyranoside; Saccharose; Granulated sugar; Table sugar; Sugar; Beet sugar; Cane sugar; Confectioner's sugar
Classification: Disaccharide
Definition: Molecule of glucose linked to one of fructose
Empirical: $C_{12}H_{22}O_{11}$
Properties: Wh. crystals or powd., sweet taste; sol. in water, alcohol; insol. in ether; m.w. 342.30; dens. 1.587 (25/4 C); m.p. 160-186 C (dec.); ref. index 1.34783 (10%, 20 C)
Precaution: Vigorous reaction with nitric acid or sulfuric acid
Toxicology: TLV:TWA 10 mg/m³; LD50 (oral, rat) 29,700 mg/kg; mildly toxic by ingestion; experimental teratogen, reproductive effects; mutagenic data; heated to decomp., emits acrid smoke and irritating fumes
Uses: Sweetening agent in food; flavor, preservative, antioxidant; hog scald agent
Regulatory: FDA 21CFR §184.1854, GRAS; USDA 9CFR §318.7, 381.147; cleared by MID to flavor sausage, ham, misc. meat prods.
Trade names containing: Canthaxanthin Beadlets 10%; Dry Beta Carotene Beadlets 10% CWS No. 65633; Dry Beta Carotene Beadlets 10% No. 65661; Dry Beta Carotene Beadlets Yellow Type 2.4-S No. 653800100; Dry Canthaxanthin 10% SD No. 66514; Dry Vitamin D_3 Beadlets Type 850 No. 652550401, 652550601; Dry Vitamin D_3 Type 100 CWS No. 65242; Frimulsion 6G; Hentex-70; Hentex-81; Hentex-85; Hentex-120; Palmabeads® Type 500 No. 65332; Roxanthin® Red 10WS No. 66515

Sucrose acetate isobutyrate
CAS 126-13-6; EINECS 204-771-6
Synonyms: SAIB
Classification: Sucrose derivative
Definition: Mixed ester of sucrose and acetic and isobutyric acids
Empirical: $C_{40}H_{62}O_{19}$
Formula: $(CH_3COO)_2C_{12}H_{14}O_3[OOCCH(CH_3)_2]_6$
Properties: Clear semisolid or sol'n.; sp.gr. 1.146; flash pt. (COC) 260 C
Precaution: Combustible

Sucrose dilaurate

Uses: For use in food drinks except in the U.S.
Regulatory: FDA 21CFR §175.105
Trade names: SAIB-SG

Sucrose dilaurate
Definition: Diester of lauric acid and sucrose
Uses: Emulsifier with antibacterial, wetting, and dispersing props. for foods; antisticking for noodles; emulsifier for coffee whitener and shortening
Trade names: Ryoto Sugar Ester L-595

Sucrose distearate
CAS 27195-16-0; EINECS 248-317-5
Synonyms: α-D-Glucopyranoside, β-D-fructofuranosyl, dioctadecanoate; Saccharose distearate
Definition: Mixture of sucrose esters of stearic acid; consists mainly of the diester
Empirical: $C_{48}H_{90}O_{13}$
Uses: Emulsifier, softener, detergent for foods; improves volume and crumb in sponge cake; w/o emulsifier in caramel, chewing gum, margarine, low-fat spreads, chocolate, sugar mfg.; anticrystallization agent in jellies; antisticking in noodles
Trade names: Ryoto Sugar Ester S-570; Ryoto Sugar Ester S-770; Ryoto Sugar Ester S-970

Sucrose erucate
Uses: Emulsifier for creams and confections; demulsifier for whipping cream, ice cream; wetting agent and dispersant for colorants; coating for fresh fruits; starch complexing, dough and antistaling agent; aerating agent; fat crystallization control
Trade names: Ryoto Sugar Ester ER-190; Ryoto Sugar Ester ER-290

Sucrose esters. See Sucrose fatty acid esters

Sucrose fatty acid esters
EEC E473
Synonyms: Sucrose esters
Toxicology: Heated to decomp., emits acrid smoke and irritating fumes
Uses: Emulsifier, stabilizer, protective coating, texturizer; for apples, baked goods, bananas, dairy prods., frozen dairy desserts, whipped milk prods.
Regulatory: FDA 21CFR §172.859; ADI 0-2.5 mg kg body wt.; Europe listed; UK approved; FAO/WHO 8.446
Manuf./Distrib.: Multi-Kem

Sucrose laurate
CAS 25339-99-5; EINECS 246-873-3
Synonyms: α-D-Glucopyranoside, β-D-fructofuranosyl, monododecanoate
Definition: Mixture of sucrose esters of lauric acid; consists mainly of monoester
Empirical: $C_{24}H_{44}O_{12}$
Uses: Emulsifier with antibacterial, wetting, and dispersing effect for foods; antisticking for noodles; emulsifier for coffee whiteners, shortening
Trade names: Ryoto Sugar Ester L-1570; Ryoto Sugar Ester L-1695; Ryoto Sugar Ester LWA-1570

Sucrose mono/distearate. See Sucrose polystearate

Sucrose myristate
Definition: Monoester of myristic acid and sucrose
Empirical: $C_{26}H_{48}O_{12}$
Uses: Emulsifier with antibacterial, wetting, and dispersing effect for foods; prevents precipitation in canned drinks
Trade names: Ryoto Sugar Ester M-1695

Sucrose octaacetate
CAS 126-14-7; EINECS 204-772-1; FEMA 3038
Synonyms: D-(+)-Sucrose octaacetate
Empirical: $C_{28}H_{38}O_{19}$
Properties: Need., intensely bitter; hygroscopic; sol. in 1100 parts water, 1.1 parts acetal, 0.7 parts glac. acetic acid, 11 parts alcohol, 22 parts CCl_4; m.w. 678.60; m.p. 82-85 C; dec. above 285 C; b.p. 260 C (1 mm); ref. index 1.4660
Precaution: Combustible
Uses: Synthetic flavoring
Regulatory: FDA 21CFR §172.515
Manuf./Distrib.: Aldrich

D-(+)-Sucrose octaacetate. See Sucrose octaacetate

Sucrose oleate
Definition: Monoester of oleic acid and sucrose
Uses: Emulsifier, softener, conditioner, texturizer, and aerating agent in foods (baked goods, baking mixes, frozen dairy desserts); component of protective coatings on fruit
Regulatory: FDA 21CFR §172.859
Trade names: Ryoto Sugar Ester O-1570; Ryoto Sugar Ester OWA-1570

Sucrose palmitate
CAS 26446-38-8; EINECS 247-706-7
Synonyms: Saccharose palmitate
Definition: Monoester of palmitic acid and sucrose
Uses: Emulsifier, softener, conditioner, texturizer, and aerating agent in foods (baked goods, baking mixes, frozen dairy desserts); component of protective coatings on fruit
Regulatory: FDA 21CFR §172.859
Trade names: Ryoto Sugar Ester P-170; Ryoto Sugar Ester P-1570; Ryoto Sugar Ester P-1570S; Ryoto Sugar Ester P-1670

Sucrose polylaurate
Definition: Mixture of esters of lauric acid and sucrose
Uses: Food emulsifier; antisticking for noodles; emulsifier for coffee whiteners, shortenings
Trade names: Ryoto Sugar Ester L-195

Sucrose polylinoleate
Definition: Mixture of esters of linoleic acid and sucrose
Uses: Food emulsifier; antisticking for noodles; emulsifier for coffee whiteners, shortenings
Trade names: Ryoto Sugar Ester LN-195

Sucrose polyoleate
Definition: Mixture of esters of oleic acid and sucrose
Uses: Antisticking for noodles; emulsifier for coffee whiteners, shortening
Trade names: Ryoto Sugar Ester O-170

Sucrose polystearate
Synonyms: Sucrose mono/distearate; Saccharose mono/distearate
Definition: Mixture of esters of stearic acid and sucrose
Uses: O/w and w/o emulsifier, softener, conditioner, and aerating agent in foods
Trade names: Ryoto Sugar Ester S-070; Ryoto Sugar Ester S-170; Ryoto Sugar Ester S-270

Sucrose stearate
CAS 25168-73-4; EINECS 246-705-9
Synonyms: α-D-Glucopyranoside, β-D-fructofuranosyl, monooctadecanoate
Definition: Monoester of stearic acid and sucrose
Empirical: $C_{30}H_{56}O_{12}$
Uses: Dispersant, solubilizer, conditioner for flavors, preservatives; dough strengthener; improves volume, crumb in sponge cake; prevents syneresis in pudding; antisticking in noodles; emulsifier for coffee whitener, shortening, ice cream; visc. reducer
Trade names: Ryoto Sugar Ester S-1170; Ryoto Sugar Ester S-1170S; Ryoto Sugar Ester S-1570; Ryoto Sugar Ester S-1670; Ryoto Sugar Ester S-1670S

Sucrose tetrastearate triacetate
Definition: Mixt. of esters of stearic acid, acetic acid, and sucrose
Uses: Antisticking for noodles; emulsifier for coffee whiteners, shortening
Trade names: Ryoto Sugar Ester S-170 Ac

Sucrose tribehenate
Definition: Triester of behenic acid and sucrose
Uses: Emulsifier with antibacterial, wetting, and dispersing effect for foods; anticaking agent for hygroscopic powd. foods, e..g, spices, seasonings; antisticking for noodles; emulsifier for coffee whiteners, shortening
Trade names: Ryoto Sugar Ester B-370

Sucrose tristearate
Definition: Triester of stearic acid and sucrose
Uses: Emulsifier, softener, conditioner, texturizer, and aerating agent in foods (baked goods, baking mixes, frozen dairy desserts); component of protective coatings on fruit
Regulatory: FDA 21CFR §172.859
Trade names: Ryoto Sugar Ester S-370; Ryoto Sugar Ester S-370F

Sudan gum. *See* Acacia

Sugar. *See* Sucrose

Sulfated butyl oleate
> *Uses:* Used to dehydrate grapes in raisin prod.
> *Usage level:* 2% max. aq. emulsion for dehydrating grapes, 100 ppm residue on raisins
> *Regulatory:* FDA 21CFR §172.270

Sulfated tallow
> *Uses:* Defoamer for beet sugar and yeast processing
> *Regulatory:* FDA 21CFR §173.340

o-Sulfobenzimide. *See* Saccharin

1-(4-Sulfophenyl)-3-ethylcarboxy-4-(4-sulfonaphthylazo)-5-hydroxypyrazole. *See* Orange B

Sulfur dioxide
> CAS 7446-09-5; EINECS 231-195-2; EEC E220
> *Synonyms:* Sulfurous anhydride; Sulfurous oxide
> *Empirical:* O_2S
> *Formula:* SO_2
> *Properties:* Colorless gas, strong suffocating odor; condenses @ -10 C to colorless liq.; sol. 8.5% in water, 25% in alcohol, 32% in methanol; m.w. 64.06; dens. 1.5 (liq.); m.p. -72 C; b.p. -10 C
> *Precaution:* Nonflamm.; reacts with water or steam to produce toxic and corrosive fumes
> *Toxicology:* TLV:TWA 2 ppm; poison gas; mildly toxic to humans by inh.; corrosive irritant to eyes, respiratory tract; destroys vitamin B_1; experimental tumorigen, teratogen, reproductive effects; human mutagenic data; heated to dec., emits toxic fumes of SO_x
> *Uses:* Bleaching agent, antimicrobial, preservative, dough modifier, vitamin C stabilizer; not for use in meats, sources of vitamin B_1, raw fruits and vegetables, or fresh potatoes
> *Usage level:* up to 0.05%; ADI 0-0.7 mg/kg (EEC)
> *Regulatory:* FDA 21CFR §182.3862, GRAS; BATF 27CFR §240.1051; Japan approved (0.03-5 g/kg); Europe listed; UK approved
> *Manuf./Distrib.:* Pfizer Food Science

Sulfuric acid
> CAS 7664-93-9; EINECS 231-639-5; EEC E513
> *Synonyms:* Hydrogen sulfate; Battery acid; Electrolyte acid; Oil of vitriol
> *Classification:* Inorganic acid
> *Empirical:* H_2O_4S
> *Formula:* H_2SO_4
> *Properties:* Colorless to dark brown dense oily liq.; misc. with water and alcohol; m.w. 98.08; dens. 1.84; m.p. 10.4 C; b.p. 290 C; dec. 340 C
> *Precaution:* Corrosive; powerful acidic oxidizer; ignites or explodes on contact with many materials; reacts with water to produce heat; reactive with oxidizing/reducing materials; heated, emits highly toxic fumes
> *Toxicology:* TLV 1 mg/m³ of air; LD50 (oral, rat) 2.14 g/kg; human poison; mod. toxic by ingestion; strongly corrosive; strong irritant to tissue; can cause severe burns, chronic bronchitis; heated to decomp., emits toxic fumes of SO_x
> *Uses:* Acidulant, pH control agent, processing aid; used in alcoholic beverages, cheese, prod. of caramel and some modified starches
> *Usage level:* Limitation 0.014% (alcoholic beverages), 0.0003% (cheeses)
> *Regulatory:* FDA 21CFR §172.560, 172.892, 173.385, 178.1010, 184.1095, GRAS; BATF 27CFR §240.1051a; Japan restricted; Europe listed; UK approved
> *Manuf./Distrib.:* Rhone-Poulenc

Sulfuric acid, aluminum ammonium salt (2:1:1), dodecahydrate. *See* Ammonium alum
Sulfuric acid, aluminum sodium salt (2:1:1). *See* Sodium alum
Sulfuric acid diammonium salt. *See* Ammonium sulfate
Sulfuric acid, dipotassium salt. *See* Potassium sulfate
Sulfuric acid magnesium salt (1:1). *See* Magnesium sulfate anhyd.
Sulfuric acid monododecyl ester sodium salt. *See* Sodium lauryl sulfate
Sulfuric acid, mono (2-ethylhexyl) ester sodium salt. *See* Sodium octyl sulfate
Sulfuric acid monosodium salt. *See* Sodium bisulfate
Sulfuric acid zinc salt (1:1). *See* Zinc sulfate
Sulfuric acid zinc salt (1:1) heptahydrate. *See* Zinc sulfate heptahydrate
Sulfurous acid, diammonium salt. *See* Ammonium sulfite
Sulfurous acid dipotassium salt. *See* Potassium sulfite
Sulfurous acid disodium salt. *See* Sodium sulfite
Sulfurous acid monosodium salt. *See* Sodium bisulfite

Sulfurous acid potassium salt. *See* Potassium sulfite
Sulfurous acid sodium salt (1:2). *See* Sodium sulfite
Sulfurous anhydride. *See* Sulfur dioxide
Sulfurous oxide. *See* Sulfur dioxide
Sumac wax. *See* Japan wax
Sumatra benzoin. *See* Gum benzoin
Summer savory. *See* Savory
Sunflower oil. *See* Sunflower seed oil
Sunflower seed mono-, di- and tri-glycerides. *See* Sunflower seed oil glycerides

Sunflower seed oil
CAS 8001-21-6; EINECS 232-273-9
Synonyms: Sunflower oil
Definition: Oil expressed from seeds of the sunflower, *Helianthus annuus*
Properties: Amber liq.
Toxicology: Heated to decomp., emits acrid smoke and irritating fumes
Uses: Food ingred., coating agent, emulsifier, formulation aid, texturizer; used in margarine, shortening
Regulatory: FDA 21CFR §175.300, 176.200, GRAS
Manuf./Distrib.: Arista Industries; Penta Mfg.
Trade names: Dewaxed Sunflower Oil 83-070-0; Lipovol SUN

Sunflower seed oil glyceride
Synonyms: Glycerides, sunflower seed mono-
Definition: Monoglyceride derived from sunflower seed oil
Uses: Food emulsifier
Trade names: Monomuls® 90-40; Myverol® 18-92; Myverol® 18-92K; Tegomuls® SB

Sunflower seed oil glycerides
Synonyms: Sunflower seed mono-, di- and tri-glycerides; Glycerides, sunflower seed mono-, di- and tri-
Definition: Mixture of mono, di and triglycerides derived from sunflower seed oil
Uses: Emulsifier, stabilizer, dispersant
Trade names: Monomuls® 60-40

Sunnette. *See* Acesulfame potassium
Sunset Yellow. *See* FD&C Yellow No. 6

Superglycerinated hydrogenated rapeseed oil
Uses: Emulsifier in shortenings or cake mixes
Usage level: Limitation 4% (shortening), 0.5% (cake mix)
Regulatory: FDA 21CFR §184.1555, GRAS

Sweet almond oil
CAS 8007-69-0
Synonyms: Almond oil, sweet
Definition: Fixed oil obtained from the ripe seed kernel of *Prunus amygdalus*, contg. chiefly glyceryl oleate
Properties: Colorless or pale yel. oily liq., almost odorless, bland taste; sl. sol. in alcohol; misc. with benzene,
    chloroform, ether, petrol. ether; insol. in water; dens. 0.910-0.915; iodine no. 93-100; sapon. no. 191-200;
    ref. index 1.4593-1.4646 (40 C)
Precaution: Keep cool, well closed; protect from light
Uses: Food ingred.
Manuf./Distrib.: Arista Industries
Trade names: Lipovol ALM; Sweet Almond Oil BP 73

Sweet basil extract. *See* Basil extract
Sweet basil oil. *See* Basil oil
Sweet bay oil. *See* Laurel leaf oil
Sweet birch oil. *See* Methyl salicylate
Sweet elder. *See* Elder flowers
Sweet fennel extract. *See* Fennel extract
Sweet flag oil. *See* Calamus oil

Sweet marjoram oil
CAS 8015-01-8
Synonyms: Origanum majorana oil
Definition: Volatile oil distilled from leaves of *Origanum majorana*, contg. 40% terpenes, chiefly terpinene; also
    d-terpineol
Properties: Yel. or greenish-yel. liq.; sol. in 2 vols 80% alcohol; sol. in chloroform, ether; insol. in water; dens.

Sweet orange oil

0.888-0.912 (15/15 C)
Uses: Natural flavoring agent
Regulatory: FDA 21CFR §182.20, GRAS

Sweet orange oil. *See* Orange oil
Sweet orange peel extract. *See* Orange peel extract
Sweet oriental gum. *See* Storax
Sweet spirit of nitre. *See* Ethyl nitrite
Sweetwood bark oil. *See* Cascarilla oil

Swertia extract
CAS 97766-44-4
Definition: Extract of flowers, leaves, and stems of various species of *Swertia*
Uses: Natural flavoring
Regulatory: FDA 21CFR §172.510

Synthetic spermaceti wax. *See* Cetyl esters

Synthetic wax
CAS 8002-74-2
Synonyms: Fischer-Tropsch wax; Fischer-Tropsch wax, oxidized
Definition: Hydrocarbon wax derived by Fischer-Tropsch or ethylene polymerization processes
Uses: Masticatory substance in chewing gum base; protective coating on raw fruits and vegetables; defoamer
Regulatory: FDA 21CFR §172.615, 172.888, 173.340, 175.105, 175.250, 176.170, 177.1200, 178.3720

Syringa aldehyde. *See* p-Tolylacetaldehyde
Table salt. *See* Sodium chloride
Table sugar. *See* Sucrose

Tagetes extract
CAS 90131-43-4
Synonyms: Tegetes erecta extract; Marigold extract
Definition: Extract of the flower of *Tagetes erecta*
Uses: Natural flavoring
Regulatory: FDA 21CFR §172.510

Talc
CAS 14807-96-6; EINECS 238-877-9; EEC E553b
Synonyms: Hydrous magnesium silicate; Magnesium hydrogen metasilicate; French chalk; Industrial talc; Cosmetic talc; Platy talc; Talcum; Pigment White 26; CI 77019
Definition: Native, hydrous magnesium silicate sometimes containing small portion of aluminum silicate
Formula: $Mg_3Si_4O_{10}(OH)_2$ or $3MgO \cdot 4SiO_2 \cdot HOH$
Properties: White, apple green, gray powd., pearly or greasy luster, greasy feel; insol. in water, cold acids or in alkalies; dens. 2.7-2.8
Toxicology: TLV:TWA 2 mg/m³, respirable dust; toxic by inhalation; talc with < 1% asbestos is nuisance dust; experimental tumorigen; human skin irritant; prolonged/repeated exposure can produce talc pneumoconiosis
Uses: Anticaking agent; coating agent; texturizing agent, lubricant, release agent, surface-finishing agent; migrating to foods from paper/paperboard, from cotton in dry food pkg.; in chewing gum; filter aid; dusting powd.
Regulatory: FDA 21CFR §73.1550, 175.300, 175.380, 175.390, 176.170, 177.1210, 177.1350, 177.1460, 182.70, 182.90; GRAS; Japan restricted (5000 ppm); Europe listed; UK approved
Manuf./Distrib.: Pfizer

Talcum. *See* Talc
Talin. *See* Thaumatin

Tall oil
CAS 8002-26-4; EINECS 232-304-6
Synonyms: Liquid rosin; Tallol
Definition: Byprod. of wood pulp contg. rosin acids, oleic and linoleic acids, and long chain alcohols
Properties: Dk. brn. liq., acrid odor; dens. 0.95; flash pt. 360 F
Precaution: Combustible when exposed to heat or flame; can react with oxidizing materials
Toxicology: Mild allergen; heated to decomp., emits acrid smoke and irritating fumes
Uses: Drying oil as component of finished resins, migrating from food pkg., in cotton for dry food pkg.
Regulatory: FDA 21CFR §175.105, 175.300, 176.200, 176.210, 177.2600, 177.2800, 181.22, 181.26, 182.70, 186.1557, GRAS as indirect food additive

Tall oil glycerides
   Uses: Used in chewing gum
   Trade names: Zonester® 85

Tallol. *See* Tall oil

Tallow
   CAS 61789-97-7; EINECS 263-099-1
   Synonyms: Beef tallow; Mutton tallow; Bleached deodorized tallow
   Definition: Fat derived from fatty tissue of sheep or cattle; consists primarily of fatty acid glycerides
   Properties: Off-wh. fat
   Toxicology: Heated to decomp., emits acrid smoke and irritating fumes
   Uses: Chewing gum base; coating agent, emulsifier, formulation aid, texturizer; animal feeds
   Regulatory: FDA 21CFR §175.105, 175.210, 176.170, 176.200, 176.210, 177.2800, 182.70, GRAS
   Trade names containing: Top Notch®

Tallow acid
   CAS 61790-37-2; 67701-06-8; EINECS 263-129-3
   Synonyms: Fatty acids, tallow; Acids, tallow
   Definition: Mixture of fatty acids derived from tallow
   Uses: Lubricant, release agent, binder, defoamer in foods; intermediate for food emulsifiers
   Regulatory: FDA 21CFR §175.105, 175.320, 176.200, 176.210, 177.2260, 177.2800, 178.3570, 178.3910
   Trade names: Industrene® 143

Tallow glyceride
   CAS 61789-13-7; EINECS 263-035-2
   Synonyms: Tallow monoglyceride; Glycerides, tallow mono-
   Definition: Monoglyceride derived from tallow
   Uses: Food emulsifier
   Regulatory: FDA 21CFR §175.105, 176.210
   Trade names: Dimodan TH; Monomuls® 90-20; Myverol® 18-30

Tallow glycerides
   CAS 67701-27-3
   Synonyms: Tallow mono, di and tri glycerides; Glycerides, tallow mono-, di- and tri-
   Definition: Mixture of mono, di and triglycerides derived from tallow
   Uses: Emulsifier, stabilizer, dispersant
   Regulatory: FDA 21CFR §175.105, 176.210
   Trade names: Monomuls® 60-20

Tallow, hydrogenated. *See* Hydrogenated tallow
Tallow mono, di and tri glycerides. *See* Tallow glycerides
Tallow monoglyceride. *See* Tallow glyceride

Tamarind
   Definition: Obtained from leguminous tree, *Tamarindus indica*, with pods contg. seeds embedded in brown
      pulp
   Properties: Sweet-sour agreeable taste
   Uses: Natural flavoring agent used in seasonings and curries
   Regulatory: FDA 21CFR §182.20, GRAS; Japan approved, restricted as color
   Manuf./Distrib.: Chart

Tanacetum vulgare extract. *See* Tansy extract
Tanacetum vulgare oil. *See* Tansy oil
Tangantangan oil. *See* Castor oil

Tangerine oil
   FEMA 3041
   Definition: Citrus reticulata
   Uses: Natural flavoring agent for soft drinks, confectionery, other natural fruit flavors
   Regulatory: FDA 21CFR §182.20, GRAS; FEMA GRAS; Europe listed, no restrictions
   Manuf./Distrib.: Florida Treatt

Tannase
   Uses: Natural enzyme
   Regulatory: Japan approved

Tannic acid
   CAS 1401-55-4; EINECS 276-638-0, 215-753-2

Tannin

Synonyms: Glycerite; Tannin; Gallotannic acid; Gallotannin
Classification: Organic acids, mixture
Definition: Occurs in the bark and fruit of many plants, e.g., oak species, sumac
Empirical: $C_{76}H_{52}O_{46}$
Properties: Yellowish-white to lt. brown powd. or flakes, faint char. odor, astringent taste; very sol. in alcohol, acetone; pract. insol. in benzene, chloroform, ether, petrol. ether, carbon disulfide, CCl_4; m.w. 1701.23; dec. 210-215 C
Precaution: Combustible exposed to heat or flame; keep well closed, protect from light; incompat. with salts of heavy metals, alkaloids, gelatin, starch, oxidizers
Toxicology: LD50 (oral, rat) 2260 mg/kg; poison by ingestion, IV, subcutaneous routes; experimental carcinogen, tumorigen, reproductive effects; mutagenic data; may cause liver damage; heated to decomp., emits acrid smoke and irritating fumes
Uses: Food additive, flavoring agent, adjuvant, flavor enhancer, pH control agent, processing aid; clarifying agent for beer and wine; fat rendering aid; boiler water additive for food contact
Usage level: Limitation 0.01% (baked goods/mixes), 0.015% (alcoholic beverages), 0.005% (nonalcoholic beverages), 0.4% (frozen dairy desserts), 0.013% (hard candy), 0.001% (meat prods.), 3 g/L (apple juice, wine), 0.8 g/L (white wine); ADI 0-0.6 mg/kg (FAO/WHO)
Regulatory: FDA 21CFR §173.310, 184.1097, GRAS; 9CFR §318.7; BATF 27CFR §240.1051
Manuf./Distrib.: Aceto; Burlington Bio-Medical; Crompton & Knowles

Tannin. *See* Tannic acid

Tannin extract
Uses: Natural processing aid
Regulatory: Japan approved

Tansy extract
CAS 84961-64-8; EINECS 284-653-9
Synonyms: Tanacetum vulgare extract
Definition: Extract of *Tanacetum vulgare*
Properties: Strong char. aromatic odor, bitter flavor
Uses: Natural flavoring; use has been discontinued
Regulatory: FDA 21CFR §172.510

Tansy oil
Synonyms: Tanacetum vulgare oil
Definition: Volatile oil from leaves and tops of *Tanacetum vulgare*
Properties: Yel. liq.; becomes brn. on exposure to air and light; sol. in alcohol, chloroform, ether; pract. insol. in water; dens. 0.925-0.950
Precaution: Keep cool, well closed; protect from light
Toxicology: Poisonous
Uses: Natural flavoring for alcoholic beverages only
Regulatory: FDA 21CFR §172.510; finished alcoholic beverage must be thujone-free; Japan approved (tansy)
Manuf./Distrib.: Chr. Hansen's

Tapioca. *See* Dextrin

Tapioca starch
Uses: Migrating from cotton in dry food pkg.
Regulatory: FDA 21CFR §182.70

Tara gum
CAS 39300-88-4
Synonyms: Gum-tara
Uses: Natural thickener; stabilizer; film-former
Regulatory: ADI 0-12.5 mg/kg (JECFA)
Regulatory: Japan approved

Taraxacum. *See* Dandelion root
Taraxacum extract. *See* Dandelion extract
Taraxacum officinale extract. *See* Dandelion extract

Tarragon
Synonyms: Estragon
Definition: Dried leaves and flowering tops of the bushy perennial plant, *Artemisia dracunculus*
Uses: Natural anise-like flavoring agent; used to flavor vinegar and pickles
Regulatory: FDA 21CFR §182.10, 182.20, GRAS
Manuf./Distrib.: C.A.L.-Pfizer; Pierre Chauvet (oil, concrete); Florida Treatt

L-Tartaric acid
CAS 87-69-4; EINECS 201-766-0; FEMA 3044; EEC E334
Synonyms: L-(+)-Tartaric acid; L-2,3-Dihydroxybutanedioic acid; d-α,β-Dihydroxysuccinic acid; 2,3-Dihydrosuccinic acid
Classification: Dibasic acid
Empirical: $C_4H_6O_6$
Formula: COOH • CHOH • CHOH • COOH
Properties: Colorless or translucent cryst. or wh. cryst. powd., odorless, acid taste; sol. in water, alcohol, glycerin; m.w. 150.09; dens. 1.0045 (1%, 15/4 C); m.p. 168-170 C
Precaution: Strong organic acid
Toxicology: Mod. toxic by IV route; mildly toxic by ingestion; strong sol'ns. are mildly irritating to humans; heated to decomp., emits acrid smoke and irritating fumes
Uses: Direct food additive, firming agent, flavor enhancer, flavoring agent, humectant, pH control agent, acidulant, antioxidant, sequestrant; mfg. of cream of tartar; effervescent beverages, baking powd., baked goods, gelatin desserts
Usage level: 960 ppm (nonalcoholic beverages), 570 ppm (ice cream), 5400 ppm (candy), 1300 ppm (baked goods), 60 ppm (gelatins, puddings), 3700 ppm (chewing gum), 10,000 ppm (condiments); ADI 0-30 mg/kg (EEC)
Regulatory: FDA 21CFR §184.1099, GRAS; USDA 9CFR §318.7, 381.147; BATF 27CFR §240.1051, 240.364, 240.512; Japan approved; Europe listed; UK approved
Manuf./Distrib.: AB Tech.; Ashland; Browning; Cornelius; CSR Food Ingreds.; Dinoval; Ellis & Everard; Forum; Hays; Int'l. Sourcing; Jungbunzlauer; KWR; Lohmann; North Western; Penta Mfg.; Pfizer; J W Pike; Pointing; Rit-Chem; Siber Hegner; Spice King; Tartaric Chems.

L-(+)-Tartaric acid. *See* L-Tartaric acid
L-Tartaric acid disodium salt. *See* Sodium tartrate

Tartaric acid esters of mono- and diglycerides
EEC E472d
Uses: Emulsifier, stabilizer
Usage level: ADI no limit (EEC)
Regulatory: Europe listed

L-Tartaric acid monopotassium salt. *See* Potassium acid tartrate

Tartrazine. *See also* FD&C Yellow No. 5
CAS 1034 21 0; EINECS 217 600 6; EEC E102
Synonyms: CI 19140; FD&C Yellow No. 5; Acid Yellow 23
Classification: Pyrazole color
Empirical: $C_{16}H9N_4Na_3O_9S_2$
Properties: Bright orange-yel. powd.; sol. in water; m.w. 534.37
Toxicology: Allergen
Uses: Food colorant
Regulatory: Europe listed; UK approved

TBHQ. *See* t-Butyl hydroquinone
TCP. *See* Calcium phosphate tribasic
TDPA. *See* 3,3′-Thiodipropionic acid
TEA. *See* Triethanolamine
Tea extract. *See* Thea sinensis extract
TEC. *See* Triethyl citrate
Tegetes erecta extract. *See* Tagetes extract
Tennecetin. *See* Natamycin

Terpene resin, natural. *See also* Dipentene, Pinene
Toxicology: Heated to decomp., emits acrid smoke and irritating fumes
Uses: Masticatory agent, solv. for elastomers in chewing gum base; color diluent; moisture barrier; component of polypropylene film
Usage level: Limitation 0.07% (of wt. of capsule) 7% (of ascorbic acid and salts)
Regulatory: FDA 21CFR §73.1, 172.280, 172.615, 178.3930
Manuf./Distrib.: Hercules

α-Terpinene
CAS 99-86-5; EINECS 202-795-1
Synonyms: 1-Methyl-4-(1-methylethyl)-1,3-cyclohexadiene; 1-Isopropyl-4-methyl-1,3-cyclohexadiene; p-Mentha-1,3-diene

γ-Terpinene

Empirical: C₁₀H₁₆

Properties: Oil, pleasant odor of lemons; misc. with alcohol, ether; pract. insol. in water; m.w. 136.24; dens. 0.838 (20/4 C); b.p. 173.5-174.8 C; flash pt. 50 C; ref. index 1.4784 (20 C)
Uses: Synthetic flavoring agent
Regulatory: FDA 21CFR §172.515

γ-Terpinene

CAS 99-85-4; EINECS 202-794-6
Synonyms: 1-Isopropyl-4-methyl-1,4-cyclohexadiene; p-Mentha-1,4-diene
Empirical: C₁₀H₁₆
Properties: Oil, herbaceous citrus odor; sol. in alcohol; insol. in water; m.w. 136.24; dens. 0.848 (20/4 C); b.p. 183-186 C; flash pt. 50 C; ref. index 1.474 (20 C)
Precaution: Flamm.
Uses: Synthetic flavoring
Regulatory: FDA 21CFR §172.515
Manuf./Distrib.: Aldrich; Florida Treatt

1-Terpinenol. See p-Menth-3-en-1-ol
4-Terpinenol. See 4-Carvomenthenol
(+)-Terpinen-4-ol. See 4-Carvomenthenol

α-Terpineol

CAS 10482-56-1; FEMA 3045
Synonyms: α,α,4-Trimethyl-3-cyclohexene-1-methanol; 1-Methyl-4-isopropyl-1-cyclohexen-8-ol; p-Menth-1-en-8-ol; p-Menthen-1-en-8-ol
Empirical: C₁₀H₁₈O
Formula: C₁₀H₁₇OH
Properties: Colorless liq.; sol. in propylene glycol; very sl. sol. in water; m.w. 154.24; dens. 0.930; m.p. 40-41 C; f.p. 193 F; b.p. 214-224 C
Uses: Synthetic flavoring agent
Usage level: 5.4 ppm (nonalcoholic beverages), 16 ppm (ice cream), 14 ppm (candy), 19 ppm (baked goods), 12-16 ppm (gelatins, puddings), 40 pmm (chewing gum), 38 ppm (condiments)
Regulatory: FDA 21CFR §172.515; GRAS (FEMA)
Manuf./Distrib.: Aldrich; Hercules; Quest Int'l.

Terpineol anhydrous

CAS 8006-39-1; EINECS 202-680-6; FEMA 3045
Synonyms: p-Menth-1-en-8-ol; Terpineols
Definition: Mixt. of α, β, and γ isomers
Empirical: C₁₀H₁₈O
Properties: Colorless visc. liq., lilac odor; sl. sol. in water, glycerin; m.w. 154.28; dens. 0.930-0.936; b.p. 213-218 C; flash pt. 196 F; ref. index 1.482
Precaution: Combustible liq.
Toxicology: LD50 (oral, rat) 4300 mg/kg; mildly toxic by ingestion; skin irritant; heated to decomp., emits acrid smoke and irritating fumes
Uses: Synthetic flavoring agent
Regulatory: FDA 21CFR §172.515

β-Terpineol

Uses: Synthetic flavoring
Regulatory: FDA 21CFR §172.515

Terpineols. See Terpineol anhydrous

Terpinolene

FEMA 3046
Synonyms: p-Menth-1,4(8)-diene
Uses: Synthetic flavoring
Regulatory: FDA 21CFR §172.515
Manuf./Distrib.: Aldrich

Terpinyl acetate

FEMA 3047
Synonyms: Menthen-1-yl-8 acetate
Empirical: C₁₂H₂₀O₂
Uses: Synthetic flavoring agent
Regulatory: FDA 21CFR §172.515; Japan approved as flavoring

Manuf./Distrib.: Florida Treatt

Terpinyl anthranilate
Uses: Synthetic flavoring
Regulatory: FDA 21CFR §172.515

Terpinyl butyrate
Uses: Synthetic flavoring
Regulatory: FDA 21CFR §172.515

Terpinyl cinnamate
Uses: Synthetic flavoring
Regulatory: FDA 21CFR §172.515

Terpinyl formate
Uses: Synthetic flavoring
Regulatory: FDA 21CFR §172.515

Terpinyl isobutyrate
Uses: Synthetic flavoring
Regulatory: FDA 21CFR §172.515

Terpinyl isovalerate
Uses: Synthetic flavoring
Regulatory: FDA 21CFR §172.515

Terpinyl propionate
Synonyms: Menthen-1-yl-8 propionate
Uses: Synthetic flavoring agent
Regulatory: FDA 21CFR §172.515

Tertiary magnesium phosphate. *See* Magnesium phosphate, tribasic
Tetradecafluorohexane. *See* Perfluorohexane
Tetradecanal. *See* Myristaldehyde
Tetradecanoic acid. *See* Myristic acid
Tetradecanoic acid ethyl ester. *See* Ethyl myristate
Tetradecanoic acid, methyl ester. *See* Methyl myristate
Tetradecanoic acid sodium salt. *See* Sodium myristate
1-Tetradecanol. *See* Myristyl alcohol
Tetradecanol mixed isomers. *See* Myristyl alcohol mixed isomers
n-Tetradecoic acid. *See* Myristic acid
Tetradecyl alcohol mixed isomers. *See* Myristyl alcohol mixed isomers
Tetradecyl aldehyde. *See* Myristaldehyde
Tetradecyl dimethyl benzyl ammonium chloride. *See* Myristalkonium chloride

n-Tetradecyl dimethyl ethylbenzyl ammonium chloride
Uses: Antimicrobial for use in cane-sugar mills
Regulatory: FDA 21CFR §173.320

Tetraglyceryl monooleate. *See* Polyglyceryl-4 oleate
Tetraglyceryl monostearate. *See* Polyglyceryl-4 stearate
Tetraglyceryl pentaoleate. *See* Polyglyceryl-4 pentaoleate
Tetraglyceryl pentastearate. *See* Polyglyceryl-4 pentastearate
Tetraglyceryl tristearate. *See* Polyglyceryl-4 tristearate
Tetrahydro-2-furancarbinol. *See* Tetrahydrofurfuryl alcohol
Tetrahydro-2-furanmethanol. *See* Tetrahydrofurfuryl alcohol

Tetrahydrofurfuryl acetate
CAS 637-64-9; FEMA 3055
Uses: Synthetic flavoring
Regulatory: FDA 21CFR §172.515
Manuf./Distrib.: Aldrich

Tetrahydrofurfuryl alcohol
CAS 97-99-4; EINECS 202-625-6; FEMA 3056
Synonyms: THFA; Tetrahydro-2-furanmethanol; Tetrahydro-2-furancarbinol; Tetrahydro-2-furylmethanol
Empirical: $C_5H_{10}O_2$
Properties: Liq.; hygroscopic; misc. with water, alcohol, ether, acetone, chloroform, benzene; m.w. 102.14; dens. 1.053 (20/4 C); visc. 6.24 cp (20 C); m.p. < -80 C; b.p. 173-177 C; flash pt. 75 C; ref. index 1.453 (20

Tetrahydrofurfuryl butyrate

    C); surf. tens. 37 dyn/cm
    Precaution: Explosive limit 1.5-9.7% by vol.
    Toxicology: Eye irritant; moderately irritating to skin, mucous membranes
    Uses: Synthetic flavoring agent
    Regulatory: FDA 21CFR §172.515
    Manuf./Distrib.: Aldrich

Tetrahydrofurfuryl butyrate
    CAS 92345-48-7; FEMA 3057
    Uses: Synthetic flavoring
    Regulatory: FDA 21CFR §172.515
    Manuf./Distrib.: Aldrich

Tetrahydrofurfuryl propionate
    CAS 637-65-0; FEMA 3056
    Uses: Synthetic flavoring
    Regulatory: FDA 21CFR §172.515
    Manuf./Distrib.: Aldrich

Tetrahydro-2-furylmethanol. *See* Tetrahydrofurfuryl alcohol
Tetrahydrogeraniol. *See* 3,7-Dimethyl-1-octanol

Tetrahydrolinalool
    FEMA 3060
    Synonyms: 3,7-Dimethyloctane-3-ol; 3,7-Dimethyl-3-octanol
    Empirical: $C_{10}H_{22}O_2$
    Properties: Colorless liq., floral odor; sol. in alcohol, fixed oils; insol. in water; m.w. 158.29; dens. 0.923; flash
        pt. 183 F; ref. index 1.431
    Precaution: Combustible liq.
    Toxicology: Heated to decomp., emits acrid smoke and irritating fumes
    Uses: Sweet floral fragrance and flavoring, less herbaceous and woody than linalool
    Regulatory: FDA 21CFR §172.515

Tetrahydro-1,4-oxazine. *See* Morpholine
Tetrahydro-2H-1,4-oxazine. *See* Morpholine

Tetrahydro-pseudo-ionone
    Synonyms: 6,10-Dimethyl-9-undecen-2-one
    Uses: Synthetic flavoring
    Regulatory: FDA 21CFR §172.515

Tetraiodofluorescein sodium salt. *See* FD&C Red No. 3
4-(2,5,6,6-Tetramethyl-2-cyclohexene-1-yl)-3-buten-2-one. *See* α-Irone

Tetramethyl ethylcyclohexenone
    Definition: Mixt. of 5-ethyl-2,3,4,5-tetramethyl-2-cyclohexen-1-one and 5-ethyl-3,4,5,6-tetramethyl-2-
        cyclohexen-1-one
    Uses: Synthetic flavoring
    Regulatory: FDA 21CFR §172.515

3,3,11,15-Tetramethyl-1,6,10,14-hexadecatetraene-3-ol. *See* Geranyl linalool
2-(2,3,5,6-Tetramethylphenoxy)propionic acid. *See* 3,3′-Thiodipropionic acid

2,3,5,6-Tetramethylpyrazine
    CAS 1124-11-4; FEMA 3237
    Empirical: $C_8H_{12}N_2$
    Properties: White crystals or powder
    Uses: Flavoring agent
    Manuf./Distrib.: Aldrich

[2R,4′R,8′R]-2,5,7,8-Tetramethyl-2-(4′,8′,12′-trimethyl-tridecyl)-6-chromanol. *See* d-α-Tocopherol
Tetrapotassium diphosphate. *See* Tetrapotassium pyrophosphate

Tetrapotassium pyrophosphate
    CAS 7320-34-5; EINECS 230-785-7; EEC E450a
    Synonyms: TKPP; Diphosphoric acid tetrapotassium salt; Tetrapotassium diphosphate; Potassium pyrophos-
        phate; Potassium pyrophosphate, normal
    Empirical: $H_4O_7P_2$ • 4K
    Formula: $K_4P_2O_7$ • 3HOH

Properties: Colorless crystals or white powd.; hygroscopic; sol. in water; insol. in alcohol; m.w. 330.4; dens. 2.33; m.p. 1090 C
Toxicology: Nuisance dust
Uses: Sequestrant; emulsifier, suspending agent, texturizer; flume wash water additive for sugar beets; moisture binding in meat treating; alkaline agent, raising agent (Japan)
Usage level: Limitation 5% (pickle for meat prods.), 0.5% (meat prods.), 0.5% (total poultry)
Regulatory: FDA 21CFR §173.315; USDA 9CFR §318.7, 381.147; Japan approved; Europe listed; UK approved
Manuf./Distrib.: Browning; FMC

Tetrapropylene. *See* Dodecene-1
Tetrasodium diphosphate. *See* Tetrasodium pyrophosphate
Tetrasodium edetate. *See* Tetrasodium EDTA

Tetrasodium EDTA
CAS 64-02-8; EINECS 200-573-9
Synonyms: EDTA Na₄; Edetate sodium; Tetrasodium edetate; Ethylene diamine tetraacetic acid, sodium salt
Classification: Substituted amine
Empirical: $C_{10}H_{12}N_2O_8 \cdot 4Na$
Formula: $(NaOOCCH_2)_2NCH_2CH_2N(CH_2COONa)_2$
Properties: White powd.; freely sol. in water; m.w. 380.20; dens. 6.9 lb/gal; m.p. > 300 C
Toxicology: LD50 (IP, mouse) 330 mg/kg; poison by IP route; skin and eye irritant; heated to decomp., emits toxic fumes of NO_x and Na_2O
Uses: Boiler water additive for food contact; flume wash water additive; poultry scald agent
Usage level: Limitation 0.1 ppm (in wash water)
Regulatory: FDA 21CFR §173.310, 173.315, 175.105, 175.125, 175.300, 176.150, 176.170, 176.210, 177.2800, 178.3120, 178.3910; USDA 9CFR §381.147
Manuf./Distrib.: Rhone-Poulenc

Tetrasodium pyrophosphate
CAS 7722-88-5 (anhyd.); EINECS 231-767-1; EEC E450a
Synonyms: TSPP; Tetrasodium diphosphate; Diphosphoric acid tetrasodium salt; Sodium pyrophosphate
Classification: Inorganic salt
Empirical: $Na_4P_2O_7$
Properties: Wh. cryst. powd., gran.; sol. 8 g/100 g water; insol. in alcohol; m.w. 265.91; dens. 2.534; m.p. 988 C
Toxicology: TLV:TWA 5 mg/m³; LD50 (oral, rat) 4000 mg/kg; poison by ingestion, IP, IV, subcutaneous routes; heated to decomp., emits toxic fumes of PO_x and Na_2O; not a cholinesterase inhibitor
Uses: Sequestrant, emulsifier, buffer, dietary supplement, nutrient; in cheeses; dispersant in malted milk; coagulant in instant puddings; crystal formation preventive in tuna fish; boiler water additive; stabilizer migrating from food pkg.
Usage level: up to 3% (cheese); ADI 0-70 mg/kg (EEC, total phosphates)
Regulatory: FDA 21CFR §133.169, 133.173, 133.179, 173.310, 175.210, 175.300, 181.22, 181.29, 182.70, 182.6787, 182.6789, GRAS; Japan approved; Europe listed; UK approved
Manuf./Distrib.: Albright & Wilson Am.; Browning; FMC; Lohmann; Rhone-Poulenc Food Ingreds.
Trade names: Albrite TSPP Food Grade
Trade names containing: Nutrifos® L-50; Wakal® A 601

Tetryl formate. *See* Isobutyl formate
Teucrium scorodonia extract. *See* Germander extract
TGS. *See* Trichlorogalactosucrose

Thaumatin
CAS 53850-34-3
Synonyms: Talin; Katemfe
Classification: Protein
Definition: Purified from African fruit of *Thaumatococcus danielli*
Properties: Sweet taste, licorice aftertaste; strongly cationic; m.w. ≈ 22,000; relatively stable in sol'n. and on heating
Toxicology: No adverse effects in short-term tests; not allergenic, mutagenic, or teratogenic
Uses: Flavor enhancer, nonnutritive sweetener; 750-1600 times sweeter than sucrose on wt. basis; synergistic with saccharin, acesulfame K; for chewing gum, savory flavor, dairy prods., animal feeds, pet foods
Usage level: ADI not specified (WHO)
Regulatory: UK, Japan approved; permitted in U.S. as flavor enhancer in chewing gum

THBP. *See* 2,4,5-Trihydroxybutyrophenone

Thea sinensis extract
CAS 84650-60-2
Synonyms: Camellia sinensis extract; Camelia oleifera extract; Chinese tea extract; Japanese tea extract; Tea extract
Definition: Extract of the leaves of *Thea sinensis* or the seeds of *Carmellia oleifera abal*
Uses: Natural flavoring agent
Regulatory: FDA 21CFR §182.20

Theine. *See* Caffeine
Theobroma cacao. *See* Cacao
Theobroma cacao extract. *See* Cocoa extract
Theobroma oil. *See* Cocoa butter
Thermal black. *See* Carbon black
THFA. *See* Tetrahydrofurfuryl alcohol
Thiaben. *See* Thiabendazole

Thiabendazole
CAS 148-79-8; EEC E233
Synonyms: Thiaben; 2-(Thiazol-4-yl) benzimidazole; 2-(4-Thiazolyl)benzimidazole
Empirical: $C_{10}H_7N_3S$
Properties: Colorless cryst.; sol. in DMF, DMSO; sl. sol. in alcohol, esters, chlorinated hydrocarbons; m.w. 201.25; m.p. 300 C (subl.)
Toxicology: Toxic; LD50 (mice, rats, rabbits) 3.6 g/kg, 3.1 g/kg, > 3.8 g/kg
Uses: Preservative, fungicide
Regulatory: Europe listed; UK approved

Thiamin chloride. *See* Thiamine HCl

Thiamine
EINECS 200-425-3
Synonyms: Vitamin B$_1$
Formula: $C_{12}H_{17}ClN_4OS$
Toxicology: Nontoxic; high oral doses may cause gastric upsets
Uses: Nutrient, enriched flours
Manuf./Distrib.: Hoffmann-La Roche SA; Honeywill & Stein Ltd
Trade names containing: Enrichment R Tablets

Thiamine HCl
CAS 67-03-8; EINECS 200-641-8; FEMA 3322
Synonyms: Thiamine dichloride; Thiamin chloride; Vitamin B$_1$ hydrochloride; Aneurine hydrochloride; Thiamine hydrochloride
Definition: Chloride-hydrochloride salt of thiamine
Empirical: $C_{12}H_{18}N_4OSCl_2$
Formula: $C_{12}H_{17}ClN_4OS \cdot HCl$
Properties: Small white cryst. or cryst. powd., nut-like odor, hygroscopic; sol. in water, glycerol; sl. sol. in alcohol; insol. in ether, benzene; m.w. 337.30; m.p. 248 C (dec.)
Toxicology: LD50 (oral, mouse) 8224 mg/kg; poison by intravenous, intraperitoneal routes; mildly toxic by ingestion; heated to decomp., emits very toxic fumes of HCl, Cl$^-$, SO_x, NO_x
Uses: Direct food additive, flavoring agent, adjuvant, nutrient, dietary supplement; in enriched flours, infant formulas
Usage level: Limitation 0.005 lb/1000 gal (wine)
Regulatory: FDA 21CFR §182.5875, 184.1875, GRAS; BATF 27CFR §240.1051; Japan approved
Manuf./Distrib.: Am. Roland
Trade names: Thiamine Hydrochloride USP, FCC Regular Type No. 601160
Trade names containing: Rich-Pak®; Rich-Pak® Powd. 160

Thiamine dichloride. *See* Thiamine HCl
Thiamine hydrochloride. *See* Thiamine HCl
Thiamine mononitrate. *See* Thiamine nitrate

Thiamine nitrate
CAS 532-43-4; EINECS 208-537-4
Synonyms: 3-[(4-Amino-2-methyl-5-pyrimidinyl)methyl]-4-(2-hydroxyethyl)-4-methylthiazolium nitrate (salt); Thiamine mononitrate; Thiamin nitrate; Vitamin B$_1$ nitrate; Aneurine mononitrate
Classification: Organic compd.
Definition: Mononitrate salt of thiamine

Empirical: $C_{12}H_{17}N_5O_4S$
Formula: $C_{12}H_{17}N_4OS \cdot NO_3$
Properties: Wh. cryst. or cryst. powd., sl. char. odor; pract. nonhygroscopic; sol. 2.7 g/100 ml water; sl. sol. in alcohol, chloroform; m.w. 327.36; m.p. 196-200 C (dec.); pH 6.5-7.1 (2% aq.)
Precaution: Powerful oxidizer
Toxicology: LD50 (IV, rabbit) 113 mg/kg; poison by IV and IP routes; heated to decomp., emits very toxic fumes of NO_x and SO_x
Regulatory: FDA 21CFR §182.5878, 184.1878, GRAS; Japan approved
Manuf./Distrib.: Am. Roland
Trade names: Thiamine Mononitrate USP, FCC Fine Powd. No. 601340
Trade names containing: Rich-Pak® Powd. 160-M; Rocoat® Thiamine Mononitrate 33$^1/_3$% No. 60188

Thiamin nitrate. *See* Thiamine nitrate
2-(Thiazol-4-yl) benzimidazole. *See* Thiabendazole
2-(4-Thiazolyl)benzimidazole. *See* Thiabendazole

2-Thienyl mercaptan
FEMA 3062
Synonyms: 2-Thienylthiol; 2-Mercaptothiophene
Empirical: $C_4H_4S_2$
Properties: Ylsh. or colorless oily liq.; sl. sol. in water; sol. in alcohol; m.w. 116.21; b.p. 166 C
Uses: Synthetic flavoring
Usage level: 0.10 ppm (candy, baked goods)
Regulatory: FDA 21CFR §172.515

2-Thienylthiol. *See* 2-Thienyl mercaptan
Thioallyl ether. *See* Allyl sulfide
Thiobenzyl alcohol. *See* Benzyl mercaptan
1,1′-Thiobisbutane. *See* Butyl sulfide
Thiobis(dodecyl propionate). *See* Dilauryl thiodipropionate
Thiobismethane. *See* Methyl sulfide
3,3′-Thiobispropanoic acid, dioctadecyl ester. *See* Distearyl thiodipropionate
Thiodipropionic acid. *See* 3,3′-Thiodipropionic acid

3,3′-Thiodipropionic acid
CAS 111-17-1; EINECS 203-841-3
Synonyms: TDPA; Bis(2-carboxyethyl) sulfide; Thiodipropionic acid; Diethyl sulfide-2,2′-dicarboxylic acid; 2-(2,3,5,6-Tetramethylphenoxy)propionic acid
Classification: Dicarboxylic acid
Empirical: $C_6H_{10}O_4S$
Formula: HOOCCH$_2$CH$_2$—S—CH$_2$CH$_2$COOH
Properties: Leaflets; very sol. in alcohol, hot water, acetate; sl. sol. in water; m.w. 178.22; m.p. 135 C
Toxicology: LD50 (oral, rat) 3980 mg/kg; irritant; poison by IP, IV routes; mod. toxic by ingestion; skin and eye irritant; heated to decomp., emits toxic fumes of SO_x
Uses: Preservative, antioxidant; food pkg.
Usage level: Use in food restricted to 0.02% of fat and oil content
Regulatory: FDA 21CFR §181.24 (0.005% migrating from food pkg.), 182.3109 (0.02% max. of fat or oil), GRAS
Manuf./Distrib.: Aldrich

Thiodipropionic acid dilauryl ester. *See* Dilauryl thiodipropionate
Thiodipropionic acid, distearyl ester. *See* Distearyl thiodipropionate
Thiophenol. *See* Benzenethiol
Thiosulfuric acid disodium salt. *See* Sodium thiosulfate anhyd.

L-Threonine
CAS 72-19-5; EINECS 200-774-1
Synonyms: α-Amino-β-hydroxybutyric acid; L-2-Amino-3-hydroxybutyric acid
Classification: Essential amino acid
Empirical: $C_4H_9NO_3$
Properties: Colorless cryst. or wh. cryst. powd.; very sol. in hot water; sol. in water; insol. in alcohol, chloroform, ether; m.w. 119.12; m.p. 255-257 C (dec.)
Toxicology: LD50 (IP, rat) 3098 mg/kg; mod. toxic by IP route; heated to decomp., emits toxic fumes of NO_x
Uses: Nutrient, dietary supplement; flavoring
Regulatory: FDA 21CFR §172.320 (5.0% max.); Japan approved
Manuf./Distrib.: Degussa

Thyme extract
CAS 84929-51-1
Synonyms: Thymus vulgaris extract
Definition: Extract of the leaves and flowers of *Thymus vulgaris*
Uses: Natural flavoring agent
Regulatory: FDA 21CFR §172.510, 182.20, GRAS
Manuf./Distrib.: C.A.L.-Pfizer; Chart; Pierre Chauvet

Thyme oil
CAS 8007-46-3; FEMA 3064
Synonyms: Thymus vulgaris oil
Definition: Volatile oil from distillation of flowering plant *Thymus vulgaris*, contg. thymol, carvacrol, cymene, pinene, linalool, bornyl acetate
Properties: Colorless to reddish-brn. liq., pleasant thymol odor, sharp taste; sol. in 2 vols 80% alcohol; very sl. sol. in water; dens. 0.894-0.930; ref. index 1.4830-1.5100 (20 C)
Precaution: Combustible when exposed to heat or flame; keep cool, well closed; protect from light
Toxicology: LD50 (oral, rat) 2840 mg/kg; mod. toxic by ingestion; mutagenic data; allergen; irritant; heated to decomp., emits acrid smoke and irritating fumes
Uses: Natural flavoring agent
Usage level: 0.13-100 ppm
Regulatory: FDA 21CFR §182.10, 182.20, GRAS; 27CFR §21.65, 21.151
Manuf./Distrib.: Chart; Pierre Chauvet; Commodity Services

Thyme oil red
CAS 8007-46-3
Synonyms: Spanish thyme oil
Definition: Oil from *Thymus vulgaris* and *T. zygis*, contg. thymol, carvacrol
Toxicology: LD50 (oral, rat) 4700 mg/kg; mildly toxic by ingestion; severe skin irritant; heated to decomp., emits acrid smoke and irritating fumes
Uses: Natural flavoring agent
Regulatory: FDA 21CFR §182.10, 182.20, GRAS
Manuf./Distrib.: Chr. Hansen's

Thyme, wild
Synonyms: Creeping thyme; Wild thyme
Definition: Thymus serpyllum
Uses: Natural flavoring agent
Regulatory: FDA 21CFR §182.10, 182.20, GRAS

Thymol
CAS 89-83-8; EINECS 201-944-8; FEMA 3066
Synonyms: 5-Methyl-2-(1-methylethyl) phenol; 3-p-Cymenol; 6-Isopropyl-m-cresol; 3-Hydroxy-p-cymene; 2-Isopropyl-5-methylphenol
Classification: Substituted phenol
Empirical: $C_{10}H_{14}O$
Formula: $(CH_3)_2CHC_6H_3(CH_3)OH$
Properties: Colorless translucent cryst., pungent caustic taste; sol. in water, alkali; very sol. in alcohol, ether, chloroform; m.w. 150.24; dens. 0.972; m.p. 51 C; b.p. 233 C; flash pt. 216 F; ref. index 1.523
Precaution: Combustible
Toxicology: Poison by ingestion, intravenous, intraperitoneal, subcutaneous routes; allergen
Uses: Synthetic flavoring agent
Usage level: 2.5-11 ppm (nonalcoholic beverages), 44 ppm (ice cream), 9.4 ppm (candy), 5-6.5 ppm (baked goods), 100 ppm (chewing gum)
Regulatory: FDA 21CFR §172.515, 175.105; 27CFR §21.65, 21.151; GRAS (FEMA)
Manuf./Distrib.: Aldrich

Thymus serpyllum extract. *See* Wild thyme extract
Thymus vulgaris extract. *See* Thyme extract
Thymus vulgaris oil. *See* Thyme oil
Tilia cordata extract. *See* Linden extract
Tin (II) chloride (1:2). *See* Stannous chloride anhyd.
Tin (II) chloride anhydrous. *See* Stannous chloride anhyd.
Tin (II) chloride dihydrate. *See* Stannous chloride dihydrate
Tin crystals. *See* Stannous chloride anhyd.
Tin dichloride. *See* Stannous chloride anhyd.

Tin protochloride. *See* Stannous chloride anhyd.
Tin salt. *See* Stannous chloride anhyd.
Tin stearate. *See* Stannous stearate
Titanic acid anhydride. *See* Titanium dioxide
Titanic anhydride. *See* Titanium dioxide
Titanic earth. *See* Titanium dioxide

Titanium dioxide
CAS 13463-67-7; EINECS 236-675-5; EEC E171
Synonyms: Titanic anhydride; Titanic earth; Titanic acid anhydride; Titanium oxide; Pigment White 6; CI 77891
Classification: Inorganic oxide
Empirical: TiO_2
Properties: White amorphous powd.; insol. in water, HCl, HNO_3, dil. H_2SO_4; sol. in HF, hot conc. H_2SO_4; m.w. 79.90; Anatase: dens. 3.90; Rutile: dens. 4.23
Precaution: Violent or incandescent reaction with metals (e.g., aluminum, calcium, magnesium, potassium, sodium, zinc, lithium)
Toxicology: TLV:TWA 10 mg/m³ of total dust; experimental carcinogen, neoplastigen, tumorigen; human skin irritant; nuisance dust
Uses: White colorant for confectionery panned goods, cheeses, icings; used in paper/paperboard for food pkg.
Usage level: Limitation 1% (in food), 0.5% (canned ham salad spread, creamed canned prods.), 0.5% (poultry salads)
Regulatory: FDA 21CFR §73.575, 73.1575, 73.2575, 175.105, 175.210, 175.300, 175.380, 175.390, 176.170, 177.1200, 177.1210, 177.1350, 177.1400, 177.1460, 177.1650, 177.2260, 177.2600, 177.2800, 181.22, 181.30; USDA 9CFR §318.7, 381.147; Japan restricted as colorant; Europe listed; UK approved; prohibited in Germany
Manuf./Distrib.: Degussa
Trade names: Hilton Davis Titanium Dioxide

Titanium oxide. *See* Titanium dioxide
TKP. *See* Potassium phosphate tribasic
TKPP. *See* Tetrapotassium pyrophosphate
Tocopherol. *See* D-α-Tocopherol, DL-α-Tocopherol
all-rac-α-Tocopherol. *See* DL-α-Tocopherol

D-α-Tocopherol
CAS 59-02-9; EINECS 200-412-2; EEC E306
Synonyms: [2R,4'R,8'R]-2,5,7,8-Tetramethyl-2-(4',8',12'-trimethyl-tridecyl)-6-chromanol; 5,7,8-Trimethyltocol; Vitamin E
Classification: Antioxidant fat-soluble compd.
Empirical: $C_{29}H_{50}O_2$
Properties: Red visc. oil, nearly odorless; freely sol. in oils, fats, acetone, alcohol, chloroform, ether; pract. insol. in water; m.w. 430.72; dens. 0.950 (25/4 C); m.p. 2.5-3.5 C
Toxicology: Experimental reproductive effects; mutagenic data; heated to decomp., emits acrid smoke and irritating fumes
Uses: Nutrient, antioxidant for fats, dietary supplement, preservative; inhibitor of nitrosamine formation in pump-cured bacon; animal feed additive
Usage level: Limitation 0.03% (in rendered animal fat), 30% (in veg. oils for lard or rendered pork fat), 500 ppm (pump cured bacon), 0.03% (poultry based on fat)
Regulatory: FDA 21CFR §182.3890, 182.5890, 182.8890, 184.1890, GRAS; USDA 9CFR §318.7, 381.147; Japan approved; Europe listed; UK approved
Manuf./Distrib.: ADM; Am. Roland

DL-α-Tocopherol
CAS 10191-41-0; EINECS 233-466-0
Synonyms: all-rac-α-Tocopherol; Vitamin E
Classification: Antioxidant fat-soluble compd.
Empirical: $C_{29}H_{50}O_2$
Properties: Pale yel. visc. oil; m.w. 430.72; dens. 0.947-0.958 (25/25 C); ref. index 1.5030-1.5070
Uses: Nutrient, antioxidant for fats, dietary supplement; inhibitor of nitrosamine formation in pump-cured bacon; animal feed additive
Regulatory: FDA 21CFR §182.3890, 182.5890, 182.8890, 184.1890, GRAS; Japan restricted for purpose of antioxidation; UK approved
Manuf./Distrib.: Am. Roland
Trade names: Beta Carotene 1% CWS No. 65659
Trade names containing: 24% Beta Carotene HS-E in Veg. Oil No. 65671; Canthaxanthin 10% Type RVI No. 66523; Canthaxanthin Beadlets 10%; Carotenal Sol'n. #2 No. 66425; Carotenal Sol'n. 4% No. 66424;

Carotenal Sol'n. #73 No. 66428; Dry Beta Carotene Beadlets 10% CWS No. 65633; Dry Beta Carotene Beadlets 10% No. 65661; Dry Canthaxanthin 10% SD No. 66514; Dry Vitamin D_3 Type 100 CWS No. 65242; Palma-Sperse® Type 250-S No. 65322; Palma-Sperse® Type 250A/50 D-S No. 65221; Roxanthin® Red 10WS No. 66515; Vitamin A Palmitate Type 250-CWS No. 65312

Tocopheryl acetate
CAS 1406-70-8; EINECS 231-710-0
Synonyms: D-α Tocopheryl acetate; DL-α Tocopheryl acetate; Vitamin E acetate
Definition: Ester of tocopherol and acetic acid
Empirical: $C_{31}H_{52}O_3$
Uses: Nutrient, dietary supplement
Regulatory: FDA 21CFR §182.5892, 182.5915, 182.8892, GRAS
Trade names: Vitamin E USP, FCC No. 60525; Vitamin E USP, FCC No. 60526; Vitinc® dl-alpha Tocopheryl Acetate USP XXII

d-α-Tocopheryl acetate. *See also* Tocopheryl acetate
CAS 58-95-7
Definition: Obtained from vacuum steam distillation and acetylation of edible vegetable oil prods.
Empirical: $C_{31}H_{52}O_3$
Properties: Cryst., odorless; sol. in alcohol; misc. with acetone, chloroform, ether, vegetable oil; insol. in water; m.w. 472.75; m.p. 25 C
Toxicology: Heated to decomp., emits acrid smoke and irritating fumes
Uses: Dietary supplement, nutrient
Regulatory: FDA 21CFR §182.5892, 182.8892, GRAS

D-α Tocopheryl acetate. *See* Tocopheryl acetate

dl-α-Tocopheryl acetate. *See also* Tocopheryl acetate
CAS 7695-91-2; EINECS 231-710-0
Definition: Obtained from vacuum steam distillation and acetylation of edible vegetable oil prods.
Empirical: $C_{31}H_{52}O_3$
Properties: Colorless to yel. visc. oil, odorless; sol. in alcohol; misc. with acetone, chloroform, ether, vegetable oil; insol. in water; m.w. 472.75; dens. 0.9533 (21.3/4 C); m.p. -27.5 C; b.p. 184 C (0.01 mm); ref. index 1.4950-1.4972
Toxicology: Heated to decomp., emits acrid smoke and irritating fumes
Uses: Dietary supplement, nutrient
Regulatory: FDA 21CFR §182.5892, 182.8892, GRAS
Trade names containing: Dry Vitamin E Acetate 50% SD No. 65356; Dry Vitamin E Acetate 50% Type CWS/F No. 652530001

DL-α Tocopheryl acetate. *See* Tocopheryl acetate
Tocopheryl acid succinate. *See* Tocopheryl succinate

Tocopheryl succinate
CAS 4345-03-3 (d-α); 17407-37-3; EINECS 224-403-8
Synonyms: D-α Tocopheryl succinate; DL-α Tocopheryl succinate; Vitamin E acid succinate; Tocopheryl acid succinate
Definition: Ester of tocopherol and succinic acid
Empirical: $C_{33}H_{54}O_5$
Properties: Colorless to wh. cryst. powd. or needles, odorless, tasteless; very sol. in chloroform; sol. in acetone, alcohol, ether, vegetable oil; pract. insol. in water; m.w. 530.76; m.p. 76-77 C
Toxicology: Heated to decomp., emits acrid smoke and irritating fumes
Uses: Dietary supplement, nutrient
Regulatory: FDA 21CFR §182.5890, 182.8890, GRAS

D-α Tocopheryl succinate. *See* Tocopheryl succinate
DL-α Tocopheryl succinate. *See* Tocopheryl succinate
α-Tolualdehyde. *See* Phenylacetaldehyde
p-Tolualdehyde. *See* p-Tolyl aldehyde

Tolualdehyde glyceryl acetal, mixed o-, m-, p-
Uses: Synthetic flavoring
Regulatory: FDA 21CFR §172.515

Tolualdehydes, mixed o-, m-, p-
CAS 529-20-4 (o-), 620-23-5 (m-), 104-87-0 (p-); FEMA 3068
Synonyms: Methyl benzaldehydes, mixed o-, m-, p-; Tolyl aldehydes, mixed o-, m-, p-
Empirical: C_8H_8O

Properties: Colorless liq., bitter almond odor; sol. in alcohol, ether; sl. sol. in water; m.w. 120.14; b.p. 199-204 C
Uses: Synthetic flavoring agent
Usage level: 11 ppm (nonalcoholic beverages), 16 ppm (ice cream, ices), 25 ppm (candy), 28 ppm (baked goods), 8.3 ppm (gelatins, puddings), 430 ppm (chewing gum), 100 ppm (maraschino cherries)
Regulatory: FDA 21CFR §172.515

α-**Toluenethiol.** *See* Benzyl mercaptan
α-**Toluic acid.** *See* Phenylacetic acid
α-**Toluic aldehyde.** *See* Phenylacetaldehyde
Toluifera balsamam resin. *See* Balsam tolu
Tolu resin. *See* Balsam tolu
p-Toluylaldehyde. *See* p-Tolyl aldehyde

p-Tolylacetaldehyde
FEMA 3071
Synonyms: p-Methyl phenylacetaldehyde; Syringa aldehyde
Empirical: $C_9H_{10}O$
Properties: Colorless oily liq., bitter almond odor, flavor; sol. in most common org. solvs.; insol. in alcohol; m.w. 134.17; dens. 1.010-1.016; b.p. 210 C; flash pt. 70 C; ref. index 1.5300-1.5350
Uses: Synthetic flavoring
Usage level: 2.0 ppm (ice cream, ices, baked goods), 0.03-2.0 ppm (candy)
Regulatory: FDA 21CFR §172.515

Tolyl acetate. *See* Methylbenzyl acetate, mixed o-, m-, p-
o-Tolyl acetate. *See* o-Cresyl acetate
p-Tolyl acetate. *See* p-Cresyl acetate

p-Tolyl aldehyde
CAS 104-87-0, 1334-78-7; EINECS 203-246-9
Synonyms: p-Methyl benzaldehyde; 4-Methylbenzaldehyde; p-Toluylaldehyde; p-Tolualdehyde
Empirical: C_8H_8O
Formula: $CH_3C_6H_4CHO$
Properties: Yel.; sol. in alcohol, ether; sl. sol. in water; m.w. 120.15; dens. 1.016 (20/4 C); b.p. 82-85 C (11 mm); flash pt. 85 C; ref. index 1.515 (20 C)
Precaution: Combustible
Uses: Synthetic flavoring agent

α-**Tolyl aldehyde dimethyl acetal.** *See* Phenylacetaldehyde dimethyl acetal
Tolyl aldehydes, mixed o-, m-, p-. *See* Tolualdehydes, mixed o-, m-, p-

4-(p-Tolyl)-2-butanone
FEMA 3074
Synonyms: p-Methylbenzylacetone
Empirical: $C_{11}H_{14}O$
Properties: Colorless oily liq.; sol. in alcohol; very sl. sol. in water; m.w. 162.23
Uses: Synthetic flavoring
Usage level: 1.0 ppm (nonalcoholic beverages), 1.5 ppm (ice cream, ices), 6.0 ppm (candy, baked goods)
Regulatory: FDA 21CFR §172.515

p-Tolyl dodecanoate. *See* p-Tolyl laurate
p-Tolyl dodecylate. *See* p-Tolyl laurate
α-**Tolylic acid.** *See* Phenylacetic acid

p-Tolyl isobutyrate
CAS 103-93-5; FEMA 3075
Synonyms: p-Cresyl isobutyrate
Definition: Ester of p-cresol and isobutyric acid
Empirical: $C_{11}H_{14}O_2$
Properties: Colorless liq., lily-narcissus odor; sol. in alcohol; insol. in water; m.w. 178.23; dens. 0.993; b.p. 237 C; flash pt. >100 C; ref. index 1.485-1.489
Uses: Synthetic flavoring agent
Usage level: 0.10-4.0 ppm (nonalcoholic beverages), 0.05 ppm (ice cream, ices,etc.), 0.12-7.0 ppm (candy, baked goods)
Regulatory: FDA 21CFR §172.515
Manuf./Distrib.: Aldrich

p-Tolyl laurate
FEMA 3076
Synonyms: p-Cresyl dodecanoate; p-Cresyl laurate; p-Tolyl dodecanoate; p-Tolyl dodecylate
Empirical: $C_{19}H_{30}O_2$
Properties: Colorless oily liq., floral sweet odor, flavor; sol. in alcohol; insol. in water; m.w. 290.45
Uses: Synthetic flavoring
Usage level: 1.0 ppm (nonalcoholic beverages, ice cream, ices), 2.0 ppm (candy, baked goods)
Regulatory: FDA 21CFR §172.515

α-**Tolyl mercaptan.** *See* Benzyl mercaptan
p-Tolyl methyl ether. *See* p-Methylanisole

p-Tolyl phenylacetate
FEMA 3077
Uses: Synthetic flavoring
Regulatory: FDA 21CFR §172.515
Manuf./Distrib.: Aldrich

2-(p-Tolyl) propionaldehyde
Synonyms: p-Methylhydratropic aldehyde
Uses: Synthetic flavoring
Regulatory: FDA 21CFR §172.515

Tonkalide. *See* γ-Hexalactone
Tonquin musk. *See* Musk

Tragacanth gum
CAS 9000-65-1; EINECS 232-552-5; EEC E413
Synonyms: Gum tragacanth; Gum dragon
Definition: Dried gummy exudate from *Astragalus gummifer*
Properties: Wh. powd., wh. to pale yel. translucent, horny pieces, odorless, mucilaginous taste; insol. in alcohol; strongly hydrophilic
Precaution: Combustible when exposed to heat or flame
Toxicology: LD50 (oral, rat) 16,400 mg/kg; mildly toxic by ingestion; mild allergen; may cause intolerance; linked to liver damage in test animals; heated to decomp., emits acrid smoke and irritating fumes
Uses: Stabilizer, thickener, emulsifier, formulation aid, preservative; crystallization inhibitor in confectionery
Usage level: Limitation 0.2% (baked goods), 0.7% (condiments), 1.3% (fats, oils), 0.8% (gravies, sauces), 0.2% (meat prods.), 0.2% (processed fruits, fruit juices), 0.1% (other foods)
Regulatory: FDA 21CFR §133.133, 133.134, 133.162, 133.178, 133.179, 150.141, 150.161, 184.1351, GRAS; Japan approved; Europe listed; UK approved
Manuf./Distrib.: Agrisales; Ashland; Bio-Botanica; Arthur Branwell; Chart; Cornelius; Florexco; Gumix Int'l.; Importers Service; Int'l. Ingreds.; Meer; MLG Enterprises; Penta Mfg.; Quest; Red Carnation; Rhone-Poulenc; Spice King; Thew Arnott; TIC Gums; Welding
Trade names: Nutriloid® Tragacanth; Powdered Gum Tragacanth T-150; Powdered Gum Tragacanth T-200; Powdered Gum Tragacanth T-300; Powdered Gum Tragacanth T-400; Powdered Gum Tragacanth T-500; Powdered Tragacanth Gum Type A/10; Powdered Tragacanth Gum Type E-1; Powdered Tragacanth Gum Type G-3; Powdered Tragacanth Gum Type L; Powdered Tragacanth Gum Type W; TIC Pretested® Tragacanth 440; Tragacanth Flake No. 27; Tragacanth Gum Ribbon No. 1
Trade names containing: Freedom X-PGA; Merecol® FA; Merecol® LK; Ticaloid® S1-102; Ticaloid® S2-102

Triacetin
CAS 102-76-1; EINECS 203-051-9; FEMA 2007
Synonyms: Glyceryl triacetate; Acetin; Enzactin; 1,2,3-Propanetriol triacetate; Triacetyl glycerol; Triacetyl glycerin
Definition: Triester of glycerin and acetic acid
Empirical: $C_9H_{14}O_6$
Formula: $C_3H_5(OCOCH_3)_3$
Properties: Colorless oily liq., sl. fatty odor, bitter taste; sl. sol. in water; sol. in alcohol, ether, other org. solvs.; m.w. 218.20; dens. 1.160 (20 C); m.p. -78 C; b.p. 258-260 C; flash pt. 300 F; ref. index 1.4307 (20 C)
Precaution: Combustible exposed to heat, flame, or powerful oxidizers
Toxicology: LD50 (oral, rat) 3000 mg/kg, (IV, mouse) 1600 ± 81 mg/kg; poison by ingestion; mod. toxic by IP, subcutaneous, IV routes; eye irritant; heated to decomp., emits acrid smoke and irritating fumes
Uses: Direct food additive, nutrient, dietary supplement, synthetic flavoring agent, adjuvant, formulation aid, humectant, solvent, vehicle; plasticizer migrating from food pkg.
Usage level: 190 ppm (nonalcoholic beverages), 60-2000 ppm (ice cream), 560 ppm (candy), 1000 ppm (baked goods), 4100 ppm (chewing gum); ADI not specified (JECFA)

Regulatory: FDA 21CFR §175.300, 175.320, 181.22, 181.27, 184.1901, GRAS
Manuf./Distrib.: Britannia Natural Prods.; Eastman; Unipex

Triacetyl glycerin. *See* Triacetin
Triacetyl glycerol. *See* Triacetin
2,4,6-Triamino-sym-triazine. *See* Melamine
sym-Triaminotriazine. *See* Melamine
2,4,6-Triamino-s-triazine. *See* Melamine
Triatomic oxygen. *See* Ozone
1,3,5-Triazine-2,4,6-triamine. *See* Melamine
1,3,5-Triazine,2,4,6-triamine, polymer with formaldehyde. *See* Melamine/formaldehyde resin
Tribasic calcium phosphate. *See* Calcium phosphate tribasic
Tribasic sodium phosphate dodecahydrate. *See* Sodium phosphate tribasic dodecahydrate
Tributyl acetylcitrate. *See* Acetyl tributyl citrate

Tributyrin
CAS 60-01-5; EINECS 200-451-5; FEMA 2223
Synonyms: Glyceryl tributyrate; Butyrin; Butanoic acid 1,2,3-propanetriyl ester
Definition: Triester of glycerin and butyric acid
Empirical: $C_{15}H_{26}O_6$
Formula: $(C_3H_7COO)_3C_3H_5$
Properties: Colorless oily liq., bitter taste; very sol. in alcohol, ether; insol. in water; m.w. 302.36; dens. 1.032 (20/4 C); m.p. -75 C; b.p. 305-310 C (760 mm), 90C (15 mm); flash pt. 345 F; ref. index 1.4358 (20 C)
Precaution: Combustible liq.
Toxicology: Poison by IV route; mod. toxic by ingestion; experimental tumorigen; heated to decomp., emits acrid smoke and irritating fumes
Uses: Direct food additive, flavoring agent and adjuvant; used in baked goods, alcoholic/nonalcoholic beverages, fats, oils, frozen dairy desserts, gelatins, puddings, fillings, soft candy
Regulatory: FDA 21CFR §172.515, 184.1903, GRAS
Manuf./Distrib.: Aldrich
Trade names containing: Neobee® SL-230

Tricalcium citrate. *See* Calcium citrate
Tricalcium orthophosphate. *See* Calcium phosphate tribasic
Tricalcium phosphate. *See* Calcium phosphate tribasic

Tricalcium silicate
Toxicology: Nuisance dust
Uses: General purpose food additive, anticaking agent in table salt; potatoes
Regulatory: FDA 21CFR §182.2906 (2% max. in table salt), GRAS

Tricaprin
CAS 621-71-6; EINECS 210-702-0
Synonyms: 1,2,3-Propanol tridecanoate; Glyceryl tricaprate; Glycerol tricaprinate; Tridecanoin
Definition: Triester of glycerin and capric acid
Empirical: $C_{33}H_{62}O_6$
Formula: $[CH_3(CH_2)_8COOCH_2]_2CHOCO(CH_2)_8CH_3$
Properties: m.w. 554.86; m.p. 31-32 C
Trade names: Dynasan® 110
Trade names containing: Dynasan® 182

Tricaprylin
CAS 538-23-8; EINECS 208-686-5
Synonyms: Glyceryl tricaprylate; Caprylic acid, 1,2,3-propanetriyl ester; 1,2,3-Propanetriol trioctanoate
Classification: Triester
Definition: Triester of glycerin and caprylic acid
Empirical: $C_{27}H_{50}O_6$
Formula: $[CH_3(CH_2)_6COOCH_2]_2CHOCO(CH_2)_6CH_3$
Properties: m.w. 470.70; dens. 0.954 (20/4 C); m.p. 9-10 C; b.p. 233 C (1 mm); ref. index 1.447
Uses: Used in bakery prods.
Trade names: Captex® 8000; Miglyol® 808

Tricarballylic acid-β-acetoxytributyl ester. *See* Acetyl triethyl citrate
Trichloroethene. *See* Trichloroethylene

Trichloroethylene
CAS 79-01-6; EINECS 201-167-4

Trichlorogalactosucrose

Synonyms: Trichloroethene; 1,1-Dichloro-2-chloroethylene; Ethinyl trichloride; 1-Chloro-2,2-dichloroethylene; Acetylene trichloride
Empirical: C_2HCl_3
Formula: CHCl:CCl$_2$
Properties: Colorless, photoreactive liq., chloroform odor; misc. with common org. solvs.; sl. sol. in water; m.w. 131.40; dens. 1.456-1.462 (25/25 C); m.p. -84.8 C; b.p. 86.7 C (760 mm); ref. index 1.45560 (25 C)
Precaution: High concs. of vapor in high-temp. air can be made to burn mildly under strong flame; heated to decomp., emits toxic fumes of Cl⁻
Toxicology: TLV 50 ppm (air); LD50 (oral, rat) 4.92 ml/kg; mildly toxic to humans by ingestion, inhalation; experimental carcinogen, tumorigen, teratogen, reproductive effects; human systemic effects, mutagenic data; eye, severe skin irritant; carcinogen in mice
Uses: Extraction solvent in mfg. of decaffeinated coffee, spice oleoresins
Usage level: Limitation 25 ppm (decaf coffee), 10 ppm (decaf instant coffee), 30 ppm (spice oleoresins)
Regulatory: FDA 21CFR §173.290
Manuf./Distrib.: Ashland

Trichlorogalactosucrose
Synonyms: TGS

4,1´,6´-Trichlorogalactosucrose. *See* Sucralose

Trichlorotrifluoroethane
CAS 76-13-1; EINECS 200-936-1
Synonyms: CFC 113; 1,1,2-Trichloro-1,2,2-trifluoroethane; Chlorofluorocarbon 113; Fluorocarbon 113
Empirical: $C_2Cl_3F_3$
Formula: ClCF$_2$CCl$_2$F
Properties: Colorless volatile liq., nearly odorless; m.w. 187.4; dens. 1.42 (25 C); b.p. 47.6 C; f.p. -35 C
Toxicology: TLV 1000 ppm in air
Uses: Used with 1% perfluorohexane to quick cool or crust-freeze chickens sealed in intact bags
Regulatory: FDA 21CFR §173.342
Manuf./Distrib.: Aldrich

1,1,2-Trichloro-1,2,2-trifluoroethane. *See* Trichlorotrifluoroethane
l-Tridecanecarboxylic acid. *See* Myristic acid
Tridecanoin. *See* Tricaprin

2-Tridecenal
FEMA 3082
Synonyms: 3-Decylacrolein; Tridecen-2-al-1
Empirical: $C_{13}H_{24}O$
Properties: White or sl. yellowish liq.; oily, citrus odor; sol. in alcohol, most fixed oils; insol. in water; m.w. 196.33; dens. 0.8476; b.p. 232 C; ref. index 1.457-1.460
Uses: Synthetic flavoring agent
Usage level: 0.10-0.30 ppm (nonalcoholic beverages), 1.6-6.0 ppm (ice cream, ices, etc.), 4.0-6.0 ppm (candy, baked goods), 0.10 ppm (chewing gum)
Regulatory: FDA 21CFR §172.515

Tridecen-2-al-1. *See* 2-Tridecenal

Triethanolamine
CAS 102-71-6; EINECS 203-049-8
Synonyms: TEA; 2,2´,2´´-Nitrilotris(ethanol); Trolamine; Trihydroxytriethylamine
Classification: Alkanolamine
Empirical: $C_6H_{15}O_3N$
Formula: N(CH$_2$CH$_2$OH)$_3$
Properties: Colorless visc. liq., sl. ammoniacal odor, very hygroscopic; misc. with water, alcohol; sol. in chloroform; sl. sol. in benzene, ether; m.w. 149.19; dens. 1.126; m.p. 21.2 C; b.p. 335 C; flash pt. (OC) 375 F; ref. index 1.4835
Precaution: Combustible when exposed to heat or flame; can react vigorously with oxidizing materials
Toxicology: LD50 (oral, rat) 8 g/kg; mod. toxic by IP; mildly toxic by ingestion; experimental carcinogen; liver and kidney damage in animals from chronic exposure; human skin irritant; eye irritant; heated to decomp., emits toxic fumes of NO$_x$ and CN⁻
Uses: Flume wash water additive for washing sugar beets prior to slicing
Usage level: Limitation 2 ppm (in wash water)
Regulatory: FDA 21CFR §173.315, 175.105, 175.300, 175.380, 175.390, 176.170, 176.180, 176.200, 176.210, 177.1210, 177.1680, 177.2260, 177.2600, 177.2800, 178.3120, 178.3910

Triethyl acetylcitrate. *See* Acetyl triethyl citrate

Triethyl citrate
    CAS 77-93-0; EINECS 201-070-7; FEMA 3083
    Synonyms: TEC; 2-Hydroxy-1,2,3-propanetricarboxylic acid, triethyl ester; Ethyl citrate
    Definition: Triester of ethyl alchol and citric acid
    Empirical: $C_{12}H_{20}O_7$
    Formula: $C_3H_5O(COOC_2H_5)_3$
    Properties: Colorless mobile oily liq., odorless, bitter taste; sol. 65 g/100 cc water; sol. 0.8g/100 cc oil; m.w.
        276.32; dens. 1.136 (25 C); b.p. 294 C; flash pt. (COC) 303 F; ref. index 1.4420
    Precaution: Combustible liq. when exposed to heat or flame
    Toxicology: LD50 (oral, rat) 5900 mg/kg; mod. toxic by IP route; mildly toxic by ingestion, inh.; heated to
        decomp., emits acrid smoke and irritating fumes
    Uses: Food additive, sequestrant; plasticizer migrating from food pkg.
    Usage level: Limitation 0.25% (in dried egg whites)
    Regulatory: FDA 21CFR §175.300, 175.320, 181.22, 181.27, 182.1911, GRAS; GRAS (FEMA)
    Manuf./Distrib.: H E Daniel; Morflex; Penta Mfg.; Sharon Labs

Triflic acid. *See* Trifluoromethane sulfonic acid

Trifluoromethane sulfonic acid
    CAS 1493-13-6; EINECS 216-087-5
    Synonyms: Triflic acid
    Empirical: CF_3HSO_3
    Formula: CHF_3O_3S
    Properties: Colorless to amber clear liq.; hygroscopic; m.w. 150.08; dens. 1.708 (20/4 C); b.p. 167-180 C; flash
        pt. none; ref. index 1.331 (20 C)
    Precaution: Strong acid; violent reaction with acyl chlorides or aromatic hydrocarbons, evolving toxic hydrogen
        chloride gas
    Toxicology: Corrosive irritant to skin, eyes, mucous membrane; heated to decomp., emits toxic fumes of F^- and
        SO_x
    Uses: Catalyst in prod. of cocoa butter substitute from palm oil
    Usage level: Limitation 0.2% (cocoa butter substitute)
    Regulatory: FDA 21CFR §173.395

Trifolium extract. *See* Clover blossom extract
Trifolium pratense extract. *See* Clover blossom extract
Triglyceryl diisostearate. *See* Polyglyceryl-3 diisostearate
Triglyceryl dioleate. *See* Polyglyceryl-3 dioleate
Triglyceryl distearate. *See* Polyglyceryl-3 distearate
Triglyceryl monolaurate. *See* Polyglyceryl-3 laurate
Triglyceryl oleate. *See* Polyglyceryl-3 oleate
Triglyceryl stearate. *See* Polyglyceryl-3 stearate
3,4,5-Trihydroxybenzoic acid. *See* Gallic acid
3,4,5-Trihydroxybenzoic acid, dodecyl ester. *See* Dodecyl gallate
3,4,5-Trihydroxybenzoic acid, n-propyl ester. *See* Propyl gallate

2,4,5-Trihydroxybutyrophenone
    CAS 1421-63-2; EINECS 215-824-8
    Synonyms: THBP; 2´,4´,5´-Trihydroxybutyrophenone
    Empirical: $C_{10}H_{12}O_4$
    Properties: Yel.-tan cryst.; sol. in alcohol, propylene glycol; very sl. sol. in water; m.w. 196.22; dens. 6 lb/gal
        (20 C); m.p. 149-153 C
    Toxicology: LD50 (IP, mouse) 200 mg/kg; poison by IP route; mutagenic data; heated to decomp., emits acrid
        smoke and irritating fumes
    Uses: Food preservative, antioxidant
    Usage level: Limitation 0.02% (of oil or fat content of food)
    Regulatory: FDA 21CFR §172.190, 181.24 (0.005% migrating from food pkg.)

2´,4´,5´-Trihydroxybutyrophenone. *See* 2,4,5-Trihydroxybutyrophenone
3,7,12-Trihydroxycholanic acid. *See* Cholic acid
2,4a,7-Trihydroxy-1-methyl-8-methylenegibb-3-ene-1,10-carboxylic acid 1-4-lactone. *See* Gibberellic acid
Trihydroxytriethylamine. *See* Triethanolamine

Trilaurin
    CAS 538-24-9; EINECS 208-687-0

Synonyms: Glyceryl trilaurate; Glyceryl tridodecanoate; Lauric acid triglyceride; Dodecanoic acid, 1,2,3-propanetriyl ester; Glycerol trilaurate; 1,2,3-Propanetriol tridodecanoate
Definition: Triester of glycerin and lauric acid
Empirical: $C_{39}H_{74}O_6$
Properties: Wh. solid; m.w. 639.01; m.p. 45-47 C
Uses: Auxiliary in prod. of compressed sweets
Trade names: Dynasan® 112; Massa Estarinum® AM

Trimagnesium phosphate. *See* Magnesium phosphate, tribasic
Trimetaphosphate sodium. *See* Sodium trimetaphosphate

Trimethylamine
CAS 75-50-3; FEMA 3241
Empirical: C_3H_9N
Properties: Fishy oily rancid sweaty odor; m.w. 59.11; dens. 0.932; m.p. -117 C; b.p. 2.9 C; flash pt. 38 F; ref. index 1.3443
Precaution: Flamm. gas
Toxicology: Corrosive
Uses: Synthetic flavoring agent
Manuf./Distrib.: Aldrich

(1S)-1,3,3-Trimethylbicyclo[2.2.1]heptan-2-ol. *See* Fenchyl alcohol
1,7,7-Trimethylbicyclo[2.2.1] heptan-2-one. *See* Camphor
2,6,6-Trimethylbicyclo(3.1.1)-2-hept-2-ene. *See* α-Pinene
α,α,4-Trimethyl-3-cyclohexene-1-methanol. *See* α-Terpineol
4-(2,6,6-Trimethyl cyclohexene-1-yl)-butane-2-one. *See* Dihydro-β-ionone
4-(2,6,6-Trimethyl-1-cyclohexene-1-yl)-3-butene-2-one. *See* β-Ionone
4-(2,6,6-Trimethyl-2-cyclohexene-1-yl)-3-butene-2-one. *See* α-Ionone
1-(2,6,6-Trimethyl-2-cyclohexene-1-yl)-1,6-heptadiene-3-one. *See* Allyl α-ionone
4-(2,6,6-Trimethyl-2-cyclohexen-1-yl)-3-methyl-3-buten-2-one. *See* α-Isomethylionone
5-(2,6,6-Trimethyl-1-cyclohexen-1-yl)-4-penten-3-one. *See* Methyl β-ionone
5-(2,6,6-Trimethyl-2-cyclohexen-1-yl)-4-penten-3-one. *See* Methyl α-ionone
5-(2,6,6-Trimethyl-3-cyclohexen-1-yl)-4-penten-3-one. *See* Methyl δ-ionone

Trimethyldodecatrieneol
CAS 142-50-7 (cis); EINECS 205-540-2 (cis); FEMA 2772
Synonyms: Nerolidol; 3,7,11-Trimethyl-1,6,10-dodecatrien-3-ol; Peruviol
Definition: Found in essential oils from many flowers
Empirical: $C_{15}H_{26}O$
Properties: m.w. 222.36; cis-: Liq.; dens. 0.876 (20/4 C); b.p. 70 C (0.1 mm); flash pt. 96 C; ref. index 1.4775; trans-: Liq.; b.p. 78 C (0.15 mm); ref. index 1.4792
Uses: Synthetic sweetly floral, green, woody, lilly-like fragrance and flavoring
Regulatory: FDA 21CFR §172.515
Manuf./Distrib.: BASF
Trade names: Nerolidol

3,7,11-Trimethyl-1,6,10-dodecatrien-3-ol. *See* Trimethyldodecatrieneol
3,7,11-Trimethyl-2,6,10-dodecatrien-1-ol. *See* Farnesol
4-12,12-Trimethyl-9-methylene-5-oxatricyclo[8.2.2.0.0.4.6]dodecane. *See* β-Caryophyllene oxide
1,3,3-Trimethyl-2-norbornanol. *See* Fenchyl alcohol
d-1,3,3-Trimethyl-2-norbornanone. *See* d-Fenchone
1,3,3-Trimethyl-2-oxabicyclo[2.2.2]octane. *See* Eucalyptol

2,4,5-Trimethyl δ-3-oxazoline
FEMA 3525
Empirical: $C_6H_{11}NO$
Properties: Yellow-orange liq.; nutty odor; sol. in alcohol, water; insol. in most fixed oils; m.w. 113.16; ref. index 1.414-1.435
Uses: Synthetic flavoring agent

2,3,5-Trimethylpyrazine
CAS 14667-55-1; FEMA 3244
Empirical: $C_7H_{10}N_2$
Properties: Colorless to sl. yellow liq.; sweet roasted peanut odor; sol. in water, organic solvents; m.w. 122.17; dens. 0.975; b.p. 171-172 C; flash pt. 130 F; ref. index 1.503-1.507
Uses: Synthetic flavoring agent

Manuf./Distrib.: Aldrich

Trimethyl thiazole
CAS 13623-11-5; FEMA 3325
Classification: Thiazole
Empirical: C_6H_9NS
Properties: m.w. 127.21; dens. 1.013; b.p. 166-167 C (717.5 mm); flash pt. 56 C
Uses: Coffee, nutty, chocolate flavoring
Manuf./Distrib.: Aldrich

5,7,8-Trimethyltocol. *See* d-α-Tocopherol
1,3,7-Trimethylxanthine. *See* Caffeine

Trimyristin
CAS 555-45-3; EINECS 209-099-7
Synonyms: Glyceryl trimyristate; 1,2,3-Propanetriol tritetradecanoate; Myristin
Definition: Triester of glycerin and myristic acid
Empirical: $C_{45}H_{86}O_6$
Properties: White to yellowish-gray solid; insol. in water; sol. in alcohol, benzene, chloroform, ether; m.w.
    723.14; dens. 0.885 (60/4 C); m.p. 56-57 C; ref. index 1.4429 (60 C)
Uses: Binder, lubricant for tablets and compressed confectioneries
Regulatory: FDA 21CFR §177.2800
Trade names: Dynasan® 114

Trioctanoin
CAS 7360-38-5; EINECS 230-896-0
Synonyms: Glyceryl tri(2-ethylhexanoate); Glyceryl trioctanoate; Octanoic acid, 1,2,3-propanetriol ester
Definition: Triester of glycerin and 2-ethylhexanoic acid
Empirical: $C_{27}H_{50}O_6$
Uses: Carrier for flavors; vehicle for vitamins, nutritional prods.

Triolein
CAS 122-32-7; 67701-30-8; EINECS 204-534-7; 266-948-4
Synonyms: Glyceryl trioleate; Olein; 9-Octadecenoic acid, 1,2,3-propanetriyl ester
Definition: Triester of glycerin and oleic acid
Empirical: $C_{57}H_{104}O_6$
Properties: Colorless to yellowish oily liq., tasteless, odorles; pract. insol. in water; sol. in chloroform, ether,
    CCl_4; sl. sol. in alcohol; m.w. 885.40; dens. 0.915 (15/4 C); m.p. -4 to -5 C; b.p. 235-240 C (15 mm); ref.
    index 1.4676 (20 C)
Uses: Solubilizer for flavors, vitamin oils; stabilizer for foods
Regulatory: FDA 21CFR §177.2800
Trade names: Hodag GTO

Tripalmitin
CAS 555-44-2; EINECS 209-098-1
Synonyms: Glyceryl tripalmitate; Palmitin; Hexadecanoic acid, 1,2,3-propanetriyl ester
Definition: Triester of glycerin and palmitic acid
Empirical: $C_{51}H_{98}O_6$
Properties: White crystalline powd. or needles; sol. in ether, chloroform, benzene; pract. insol. in alcohol; insol.
    in water; m.w. 807.29; dens. 0.886 (80/4 C); m.p. 65.5 C; b.p. 310-320 C; ref. index 1.43807 (80 C); sapon.
    no. 208.5
Precaution: Combustible
Uses: Auxiliary in prod. of compressed sweets
Regulatory: FDA 21CFR §177.2800
Trade names: Dynasan® 116

Triphosphoric acid pentapotassium salt. *See* Potassium tripolyphosphate
Triphosphoric acid pentasodium salt. *See* Pentasodium triphosphate
Tripotassium citrate monohydrate. *See* Potassium citrate
Tripotassium orthophosphate. *See* Potassium phosphate tribasic
Tripotassium phosphate. *See* Potassium phosphate tribasic
Tripropionin. *See* Glyceryl tripropanoate
Tris (nicotinato) aluminum. *See* Aluminum nicotinate
Trisodium-3-carboxy-5-hydroxy-1-p-sulfophenyl-4-p-sulfophenylazopyrazole. *See* FD&C Yellow No. 5

Trisodium citrate
CAS 68-04-2 (anhyd.), 6858-44-2 (hydrate), 6132-04-3 (dihydrate); EINECS 200-675-3; EEC E331c

Trisodium dipotassium tripolyphosphate

Synonyms: Citric acid trisodium salt; Sodium citrate tertiary
Empirical: C₆H₅Na₃O₇ (anhyd.), C₆H₅Na₃O₇ • 2H₂O (dihydrate)
Formula: HOC(COONa)(CH₂COONa)₂ (anhyd.), HOC(COONa)(CH₂COONa)₂ • 2H₂O (dihydrate)
Properties: m.w. 258.07 (anhyd.)
Toxicology: Heated to decomp., emits acrid smoke and irritating fumes
Uses: Stabilizer migrating from food pkg.; curing accelerator; buffering salt for pH control in beverages, confectionery; in poultry, meat prods., processed cheese; flavoring, emulsifier (Japan)
Regulatory: FDA 21CFR §131.111, 131.138, 131.146, 131.160, 131.185, 133.112, 133.144, 133.169, 133.173, 133.179, 150.141, 150.161, 175.300, 181.22, 181.29; USDA 9CFR §381.147
Regulatory: Japan approved
Manuf./Distrib.: Jungbunzlauer; Lohmann

Trisodium dipotassium tripolyphosphate
CAS 24315-83-1
Formula: Na₃K₂P₃O₁₀
Properties: m.w. 400.1
Trade names: Nutrifos® SK

Trisodium nitrilotriacetate. *See* Trisodium NTA

Trisodium NTA
CAS 5064-31-3; EINECS 225-768-6
Synonyms: Trisodium nitrilotriacetate; Nitrilotriacetic acid sodium salt; N,N-Bis(carboxymethyl)glycine trisodium salt
Definition: Trisodium salt of nitrilotriacetic acid
Empirical: C₆H₆NO₆ • 3Na
Formula: NaOOCCH₂—N(CH₂COONa)₂
Properties: m.w. 257.10
Toxicology: LD50 (oral, rat) 1100 mg/kg; poison by IP route; mod. toxic by ingestion; experimental neoplastigen, reproductive effects; mutagenic data; heated to decomp., emits toxic fumes of NOₓ and Na₂O
Uses: Boiler water additive for food contact except milk prods.
Regulatory: FDA 21CFR §173.310 (5 ppm max. in boiler feedwater)

Trisodium orthophosphate. *See* Sodium phosphate tribasic
Trisodium phosphate. *See* Sodium phosphate tribasic
Trisodium phosphate dodecahydrate. *See* Sodium phosphate tribasic dodecahydrate

Tristearin
CAS 555-43-1; EINECS 209-097-6
Synonyms: Glyceryl tristearate; Glyceryl monotristearate; Stearin; 1,2,3-Propanetriol trioctadecanoate; Octadecanoic acid, 1,2,3-propanetriyl ester
Definition: Triester of glycerin and stearic acid
Empirical: C₅₇H₁₁₀O₆
Formula: [CH₃(CH₂)₁₆COOCH₂]₂CHOCO(CH₂)₁₆CH₃
Properties: Colorless crystals or powd., odorless, tasteless; insol. in water; sol. in hot alcohol, benzene, chloroform, carbon disulfide; m.w. 891.45; dens. 0.943 (65 C); m.p. 71.6 C; ref. index 1.4385 (80 C)
Precaution: Combustible
Toxicology: Heated to decomp., emits acrid smoke and irritating fumes
Uses: Crystallization accelerator, fermentation aid, formulation aid, lubricant, release agent, emulsifier, surface-finishing agent for foods (imitation chocolate, cocoa, confections, fats, oils)
Usage level: limitation 1% (cocoa), 0.5% (formulation aid, lubricant, release), 3% (confections), 1% (formulation aid in fats, oils)
Regulatory: FDA 21CFR §172.811, 177.2800
Trade names: Dynasan® 118; Neobee® 62; Pationic® 919

Tristearyl citrate
CAS 7775-50-0; EINECS 231-896-3
Synonyms: 2-Hydroxy-1,2,3-propanetricarboxylic acid, trioctadecyl ester
Definition: Triester of stearyl alcohol and citric acid
Empirical: C₆₀H₁₁₆O₇
Toxicology: Heated to decomp., emits acrid smoke and irritating fumes
Uses: Flavor preservative, plasticizer, sequestrant; used in oleomargarine, pkg. materials
Regulatory: FDA 21CFR §175.300, 178.3910, 181.27, 182.6851 (limitation 0.15% as sequestrant)

Triticum aestivum germ oil. *See* Wheat germ oil

Triundecanoin
CAS 13552-80-2; EINECS 236-935-8

Uses: Emollient, solv., fixative, and extender in nutritional applics.; carrier for flavors and fragrances
Trade names: Captex® 8227

Trolamine. *See* Triethanolamine

Trypsin
CAS 9002-07-7; EINECS 232-650-8
Classification: Enzyme
Uses: Enzyme for food use; rennet extender
Regulatory: FDA GRAS; Japan approved
Trade names: Trypsin 1:75; Trypsin 1:80; Trypsin 1:150

DL-α-Tryptophan
CAS 54-12-6; EINECS 200-194-9
Synonyms: (±)-2-Amino-3-(3-indolyl)propionic acid
Classification: Amino acid
Empirical: $C_{11}H_{12}N_2O_2$
Properties: Wh. cryst. or cryst. powd., odorless; sol. in water, dil. acids, alkalies; sl. sol. in alcohol; m.w. 204.23;
     m.p. 295 C (dec.)
Precaution: Photosensitive
Toxicology: Experimental carcinogen; heated to decomp., emits toxic fumes of NO_x
Uses: Nutrient, dietary supplement, cereal enrichment; flavoring
Regulatory: FDA 21CFR §173.320 (1.6% max.); Japan approved
Manuf./Distrib.: Degussa; Penta Mfg.

L-Tryptophan
CAS 73-22-3; EINECS 200-795-6
Synonyms: l-α-Amino-3-indolepropionic acid; Indole-3-alanine
Classification: Amino acid
Empirical: $C_{11}H_{12}N_2O_2$
Properties: Wh. cryst. or cryst. powd., sl. bitter taste; sol. in hot alcohol, alkali hydroxides; sol. 11.4 g/l in water;
     insol. in chloroform; m.w. 204.23; m.p. 289 C (dec.)
Precaution: Photosensitive
Toxicology: Mod. toxic by IP route; experimental tumorigen, teratogen, reproductive effects; human mutagenic
     data; heated to decomp., emits toxic fumes of NO_x
Uses: Nutrient, dietary supplement; flavoring
Regulatory: FDA 21CFR §172.320 (1.6% max.); Japan approved
Manuf./Distrib.: Am. Roland

TSP-12. *See* Sodium phosphate tribasic dodecahydrate
TSP-O. *See* Sodium phosphate tribasic
TSPP. *See* Tetrasodium pyrophosphate

Tuberose
FEMA 3084
Definition: Oil from *Polianthes tuberosa*
Properties: Colorless to very lt. colored oil, intense sweet floral odor; dens. 1.007-1.035 (15 C)
Uses: Natural flavoring agent
Usage level: 0.26 ppm (nonalcoholic beverages), 0.45 ppm (ice cream, ices), 1.5 ppm (candy), 1.7 ppm (baked
     goods)
Regulatory: FDA 21CFR §182.20, GRAS

Tuberyl alcohol. *See* Dihydrocarveol
Tung nut oil. *See* Tung oil

Tung oil
Synonyms: Chinawood oil; Tung nut oil
Definition: Drying oil from seeds of *Aleurites cordata*
Properties: Pale yel. liq., char. disagreeable odor; sol. in chloroform, ether, carbon disulfide, oils; dens. 0.936-
     0.943; iodine no. 163-171; sapon. no. 190-197
Precaution: Combustible when exposed to heat or flame; can react with oxidizing materials
Toxicology: Toxic by ingestion; contact causes dermatitis; ingestion causes nausea, vomiting, cramps,
     diarrhea, dizziness, lethargy, disorientation; large doses can cause fever, tachycardia, respiratory effects
Uses: Drying oil as component of finished resins, migrating from food pkg.
Regulatory: FDA 21CFR §181.26

Turmeric
CAS 458-37-7; EINECS 207-280-5

Turmeric extract

Synonyms: 1,7-Bis(4-hydroxy-3-methoxyphenyl)-1,6-heptadiene-3,5-dione; Natural Yellow 3; CI 75300 (curcumin); Curcumin; Turmeric yellow

Definition: Dried and ground rhizome or bulbous root of *Curcuma longa*, with curcumin as coloring principal

Empirical: $C_{21}H_{20}O_6$ (curcumin)

Properties: Yel. powd., char. odor, sharp taste; Curcumin: Orange-yel. cryst. powd.; sol. in ethanol, glac. acetic acid; insol. in water, ether; m.w. 368.39; m.p. 180-183 C

Toxicology: Human mutagenic data; heated to decomp., emits acrid smoke and irritating fumes

Uses: Natural flavoring, food colorant for use alone or with annatto to shade pickles, mustard, curry powder, spices, margarine, ice cream, cheese, pies, cakes, candies, soups, casings, rendered fats, fish, meat, poultry, rice dishes

Usage level: 0.2-60 ppm (turmeric), 2-640 ppm (oleoresin)

Regulatory: FDA 21CFR §73.600, 182.10, 182.20, GRAS; USDA 9CFR §318.7; Japan approved

Manuf./Distrib.: Chart; MLG Enterprises; Quest Int'l.

Trade names containing: Annatto/Turmeric OSS-1; Annatto/Turmeric WML-1; Annatto/Turmeric WMP-1; Golden Covo® Shortening; Golden Croissant/Danish™ Margarine; Natural Soluble Powder Green Colorant; Shedd's NP Margarine; Turmeric OSO-50; Turmeric WMPE-80; Turmeric WMPE-150; Turmeric WMSE-80

Turmeric extract

CAS 84775-52-0; EINECS 283-882-1; EEC E100

Synonyms: Curcuma domestica extract; Curcuma longa extract

Definition: Extract of the rhizomes of *Curcuma longa*

Uses: Natural flavoring agent

Regulatory: FDA 21CFR §182.20, GRAS

Trade names: Curcumex 1600, 1601; Curcumex Natural Powd. Colorant; Curcumin Extract P8002; Dispersable Curcumex 2600, 2601

Trade names containing: Natural Soluble Powder Q A/L 15 Yellow Colorant; Turmeric DP-300

Turmeric yellow. *See* Curcumin
Turmeric yellow. *See* Turmeric

Turpentine

CAS 8052-14-0

Definition: Mixt. of terpene hydrocarbons obtained from various species of *Pinus*

Uses: Natural flavoring

Regulatory: FDA 21CFR §172.510, 175.105

L-Tyrosine

CAS 60-18-4; EINECS 200-460-4

Synonyms: l-β-(p-Hydroxyphenyl) alanine; 3-(4-Hydroxyphenyl)alanine; p-Tyrosine; (S)-2-Amino-3-(4-hydroxyphenyl)propionic acid

Classification: Nonessential amino acid

Empirical: $C_9H_{11}NO_3$

Properties: Colorless silky need. or wh. cryst. powd.; sol. in water, dil. min. acids, alkaline sol'ns.; sl. sol. in alcohol; m.w. 181.21; dec. 342-344 C

Toxicology: Experimental reproductive effects; heated to decomp., emits acrid smoke and irritating fumes

Uses: Dietary supplement, nutrient; flavoring

Regulatory: FDA 21CFR §172.320 (limitation 4.3%); Japan approved

Manuf./Distrib.: Am. Roland

p-Tyrosine. *See* L-Tyrosine

2,3-Undecadione

FEMA 3090

Synonyms: Acetyl nonyryl; Acetyl nonanoyl; Acetyl pelargonyl

Empirical: $C_{11}H_{20}O_2$

Properties: Yel. oily liq., strong sweet-cream warm odor; sol. in alcohol; very sl. sol. in water; m.w. 184.28; b.p. 109-111 C (10 mm)

Uses: Synthetic flavoring

Usage level: 1.5 ppm (nonalcoholic beverages), 3.0 ppm (ice cream, ices, candy, baked goods)

Regulatory: FDA 21CFR §172.515

γ-Undecalactone

CAS 104-67-6; FEMA 3091

Synonyms: Peach aldehyde; 5-Hydroxyundecanoic acid γ-lactone; Aldehyde C-14 pure

Classification: Heterocyclic compd.

Empirical: $C_{11}H_{20}O_2$
Properties: Colorless to sl. yel. liq., creamy peach-like odor; sol. in alcohol, most fixed oils, propylene glycol; insol. in glycerin, water; m.w. 184.28; dens. 0.943; b.p. 164-166 C (13 mm); flash pt. > 230 F; ref. index 1.430
Precaution: Combustible liq.
Toxicology: Heated to decomp., emits acrid smoke and irritating fumes
Uses: Synthetic flavoring agent
Usage level: 4.4 ppm (nonalcoholic beverages), 3.0 ppm (ice cream, ices, etc.), 11 ppm (candy), 7.1 ppm (baked goods), 90 ppm (gelatins and puddings)
Regulatory: FDA 21CFR §172.515
Manuf./Distrib.: Acme-Hardesty; Aldrich

ς-Undecalactone

CAS 710-04-3; EINECS 203-225-4; FEMA 3294
Synonyms: 5-Heptyldihydro-2(3H)-furanone; Peach aldehyde; Aldehyde C-14 pure
Classification: Heterocyclic compd.
Empirical: $C_{11}H_{20}O_2$
Properties: Colorless to sl. yel. liq., peach odor; sol. in fixed oils, propylene glycol; insol. in glycerin, water; m.w. 184.28; dens. 0.969; b.p. 152-155 C (10.5 mm); flash pt. > 230 F
Precaution: Combustible liq.
Toxicology: Heated to decomp., emits acrid smoke and irritating fumes
Uses: Synthetic flavoring agent for beverages, candy, gelatins, ice cream, puddings
Regulatory: FDA 21CFR §172.515

Undecanal

CAS 112-44-7; FEMA 3092
Synonyms: Aldehyde C-11 undecyclic; n-Undecyl aldehyde; Hendecanal
Empirical: $C_{11}H_{22}O$
Properties: Colorless to sl. yellow liq.; sweet, fatty, floral odor; sol. in most common organic solvents; insol. in water; m.w. 170.30; dens. 0.825; m.p. -4 C; b.p. 118-120 C (20 mm); flash pt. 96 C; ref. index 1.430-1.435
Toxicology: Tends to polymerize unless tightly sealed
Uses: Synthetic flavoring agent
Usage level: 0.95 ppm (nonalcoholic beverages), 3.1 ppm(ice cream, ices, etc.), 2.0 ppm (candy), 2.4 ppm (baked goods), 56 ppm (chewing gum)

1-Undecanol. See Undecyl alcohol

2-Undecanone

CAS 112-12-9; EINECS 203-937-5; FEMA 3093
Synonyms: Methyl nonyl ketone
Empirical: $C_{11}H_{22}O$
Formula: $CH_3(CH_2)_8COCH_3$
Properties: Colorless to sl. yellowish liq.; rue odor; sweet peachy flavor; sol. in most organic solvents; insol. in water; m.w. 170.30; dens. 0.825 (20/4 C); m.p. 11-13 C; b.p. 231-232 C; flash pt. 89 C; ref. index 1.4280-1.4330 (20 C)
Uses: Synthetic flavoring agent
Usage level: 2.8 ppm (nonalcoholic beverages), 0.54 ppm (ice cream, ices, etc.), 2.6 ppm (candy), 3.1 ppm (baked goods), 5.0 ppm (gelatins and puddings)
Regulatory: FDA 21CFR §172.515
Manuf./Distrib.: Aldrich

9-Undecenal

FEMA 3094
Synonyms: Undecenoic aldehyde; Hendecen-9-al; 9-Undecen-1-al
Empirical: $C_{11}H_{20}O$
Properties: Colorless pale yel. oily liq., orange peel-like sweet odor, citrus flavor; sol. in alcohol; insol. in water; m.w. 168.28
Uses: Synthetic flavoring
Usage level: 4.8 ppm (nonalcoholic beverages), 4.2 ppm (ice cream, ices), 4.5 ppm (candy), 4.6 ppm (baked goods)
Regulatory: FDA 21CFR §172.515

9-Undecen-1-al. See 9-Undecenal

10-Undecenal

CAS 112-45-8; FEMA 3095
Synonyms: Aldehyde C-11 undecylenic; Hendecenal; Undecen-10-al; Undecylenaldehyde; Undecylenic

aldehyde
Empirical: $C_{11}H_{20}O$
Properties: Colorless to lt. yel. liq., fatty rose odor on dilution; sol. in fixed oils, propylene glycol; insol. in water, glycerin; m.w. 168.31; dens. 0.840-0.850; b.p. 101-103 C; flash pt. 92 C; ref. index 1.441-1.447
Precaution: Combustible
Toxicology: Skin irritant; heated to decomp., emits acrid smoke and irritating fumes
Uses: Synthetic flavoring agent
Usage level: 0.05-1.0 ppm (nonalcoholic beverages), 0.20 ppm (ice cream, ices, etc.), 0.20 ppm (candy)
Regulatory: FDA 21CFR §172.515
Manuf./Distrib.: Aldrich

Undecen-10-al. *See* 10-Undecenal
Undecenoic acid. *See* Undecylenic acid
10-Undecenoic acid. *See* Undecylenic acid
11-Undecenoic acid. *See* Undecylenic acid
Undecenoic aldehyde. *See* 9-Undecenal

Undecen-1-ol
Synonyms: Undecylenic alcohol
Uses: Synthetic flavoring
Regulatory: FDA 21CFR §172.515

2-Undecenol
Empirical: $C_{11}H_{22}O$
Properties: White to sl. yellow liq.; sweet, floral odor; m.w.170.30; insol. in water; ref. index 1.450-1.452 (22 C)
Uses: Flavoring agent

Undecenyl acetate. *See* 10-Undecen-1-yl acetate

10-Undecen-1-yl acetate
FEMA 3096
Synonyms: Acetate C-11; 10-Hendecenyl acetate; Undecenyl acetate
Empirical: $C_{13}H_{24}O_2$
Properties: Colorless liq., lt. rose-like odor; sol. in most common org. solvs.; m.w. 212.33; dens. 0.8808; b.p. 272 C; ref. index 1.4380-1.4420
Uses: Synthetic flavoring
Usage level: 3.7 ppm (nonalcoholic beverages), 15 ppm (ice cream, ices), 12 ppm (candy, baked goods)
Regulatory: FDA 21CFR §172.515

Undecyl alcohol
CAS 112-42-5; EINECS 203-970-5; FEMA 3097
Synonyms: Hendecanoic alcohol; 1-Hendecanol; Hendecyl alcohol; C-11 primary alcohol; 1-Undecanol; Alcohol C-11
Classification: Aliphatic alcohol
Empirical: $C_{11}H_{24}O$
Formula: $CH_3(CH_2)_9CH_2OH$
Properties: Colorless liq., mild fatty-floral odor; sol. in water, alcohol 60%; m.w. 172.35; dens. 0.822 (35/4 C); m.p. 19 C; b.p. 131 C (15 mm); flash pt. 93.3 C; ref. index 1.4370-1.4430
Precaution: Combustible
Toxicology: LD50 (oral, rat) 3000 mg/kg; mod. toxic by ingestion; low acute inhalation toxicity; severely irritating to eyes; moderately irritating to skin; heated to decomp., emits acrid smoke and irritating fumes
Uses: Synthetic flavoring agent
Usage level: 2.9 ppm (nonalcoholic beverages), 15 ppm (ice cream), 12 ppm (candy), 12 ppm (baked goods)
Regulatory: FDA 21CFR §172.515
Manuf./Distrib.: Aldrich

n-Undecyl aldehyde. *See* Undecanal
Undecylenaldehyde. *See* 10-Undecenal

Undecylenic acid
CAS 112-38-9; EINECS 203-965-8; FEMA 3247
Synonyms: 10-Undecenoic acid; Undecenoic acid; 11-Undecenoic acid
Classification: Aliphatic acid
Empirical: $C_{11}H_{20}O_2$
Formula: $CH_2=CH(CH_2)_8COOH$
Properties: Lt. colored liq., fruity-rosy odor; insol. in water; misc. with alcohol, chloroform, ether, benzene; m.w.

184.28; dens. 0.910-0.913 (25/25 C); m.p. 22 C; b.p. 137 C (2 mm); flash pt. 295 F
Precaution: Combustible
Uses: Synthetic flavoring
Manuf./Distrib.: Aldrich

Undecylenic alcohol. *See* Undecen-1-ol
Undecylenic aldehyde. *See* 10-Undecenal
Unhydrogenated lard. *See* Lard

Urea
CAS 57-13-6; EINECS 200-315-5
Synonyms: Carbamide; Carbonyldiamide; Carbamidic acid; Isourea
Classification: Organic compd.
Empirical: CH_4N_2O
Formula: NH_2CONH_2
Properties: White cryst. or powd., almost odorless; sol. in water, alcohol, benzene; sl. sol. in ether; insol. in chloroform; m.w. 60.06; dens. 1.335; m.p. 132.7 C; b.p. dec.
Precaution: Heated to decomp., emits toxic fumes of NO_x
Toxicology: LD50 (oral, rat) 14,300 mg/kg; mod. toxic by ingestion, IV, subcutaneous routes; experimental carcinogen, neoplastigen, reproductive effects; human reproductive effects by intraplacental route; human mutagenic data; human skin irritant
Uses: Direct food additive, yeast nutrient, formulation aid, fermentation aid used in yeast-raised bakery prods., alcoholic beverages, gelatin prods.
Usage level: Limitation 2 lb/1000 gal (wine)
Regulatory: FDA 21CFR §175.300, 177.1200, 184.1923, GRAS; BATF 27CFR §240.1051
Manuf./Distrib.: Heico
Trade names containing: Carbrea® Tabs; CDC-2001; CDC-2002

Urea amidohydrolase. *See* Urease

Urea-formaldehyde resin
CAS 9011-05-6
Synonyms: Polyoxymethylene urea; Polynoxylin; Urea, polymer with formaldehyde
Classification: Amino resin
Definition: Reaction prod. of urea and formaldehyde
Empirical: $(CH_4N_2O \cdot CH_2O)_x$
Uses: Used in paper/paperboard for food pkg
Regulatory: FDA 21CFR §175.105, 175.300, 177.1200, 177.1650, 177.1900, 181.30
Manuf./Distrib.: Akzo; Cargill; Hercules

Urea, polymer with formaldehyde. *See* Urea-formaldehyde resin

Urease
CAS 9002-13-5; EINECS 232-656-0
Synonyms: Urea amidohydrolase
Definition: Enzyme derived from *Lactobacillus fermentum*; hydrolyzes urea to ammonium carbonate
Properties: Crystals; sol. in water; m.w. ≈ 480,000
Uses: Direct food additive, enzyme to convert urea to ammonia and carbon dioxide, in wine to inhibit formation of ethyl carbamate
Regulatory: FDA 21CFR §184.1924, GRAS; Japan approved

Urotropine. *See* Hexamethylene tetramine

n-Valeraldehyde
CAS 110-62-3; EINECS 203-784-4; FEMA 3098
Synonyms: Aldehyde C-5; n-Pentanal; Amylaldehyde
Empirical: $C_5H_{10}O$
Formula: $CH_3(CH_2)_3CHO$
Properties: Colorless liq.; sol. in alcohol, propylene glycol, oils; m.w. 86.13; dens. 0.81 (20/4 C); m.p -92 C; b.p. 103.4 C; flash pt. 4 C; ref. index 1.3882
Uses: Synthetic pungent fragrance and flavoring
Usage level: 1.3 ppm (nonalcoholic beverages), 5.0 ppm (ice cream, ices, etc.), 4.2 ppm (candy), 5.4 ppm (baked goods)
Regulatory: FDA 21CFR §172.515
Manuf./Distrib.: BASF

Valerian
CAS 8008-88-6

Definition: Dried rhizomes and roots of *Valeriana officinalis*
Uses: Natural flavoring
Regulatory: FDA 21CFR §172.510; Japan approved
Manuf./Distrib.: Chart; Pierre Chauvet (oil)

Valerian extract

CAS 8057-49-6; EINECS 232-501-7
Definition: Extract of rhizomes and roots of *Valeriana officinalis*
Uses: Natural flavoring
Regulatory: FDA 21CFR §172.510
Manuf./Distrib.: Chart

Valerianic acid. *See* n-Valeric acid

n-Valeric acid

CAS 109-52-4; EINECS 203-677-2; FEMA 3101
Synonyms: Carboxylic acid C_5; Valerianic acid; Butanecarboxylic acid; n-Pentanoic acid; Propylacetic acid
Empirical: $C_5H_{10}O_2$
Formula: $CH_3(CH_2)_3COOH$
Properties: Colorless mobile liq., penetrating rancid odor; sol. in water; misc. with alcohol, ether; m.w. 102.14; dens. 0.940 (20/4 C); m.p. -34.5 C; b.p. 186 C; flash pt. 203 F; ref. index 1.405-1.14
Precaution: Combustible liq.; corrosive material
Toxicology: LD50 (oral, mouse) 600 mg/kg; mod. toxic by ingestion, IV, subcutaneous routes; mildly toxic by inh.; corrosive irritant to skin, eyes, mucous membranes; heated to decomp., emits acrid smoke and irritating fumes
Uses: Synthetic flavoring agent
Usage level: 1.2 ppm (nonalcoholic beverages), 1.8 ppm (ice cream), 2.5 ppm (candy), 8 ppm (baked goods)
Regulatory: FDA 21CFR 172.515
Manuf./Distrib.: Aldrich; Hoechst Celanese

4-Valerolactone. *See* γ-Valerolactone

γ-Valerolactone

CAS 108-29-2; EINECS 203-569-5; FEMA 3103
Synonyms: 4-Hydroxypentanoic acid lactone; 4-Valerolactone; 4,5-Dihydro-5-methyl-2(3H)-furanone; 4-Hydroxyvaleric acid lactone; γ-Methyl-γ-butyrolactone; γ-Pentalactone
Empirical: $C_5H_8O_2$
Properties: Colorless mobile liq., sweet herbaceous odor; misc. with alcohol, fixed oils, water; m.w. 100.13; dens. 1.047-1.054; m.p. -31 C; b.p. 205-206.5 C; flash pt. (COC) 205 F; ref. index 1.433
Precaution: Combustible liq. exposed to heat or flame; reactive with oxidizers
Toxicology: LD50 (oral, rat) 8800 mg/kg; mod. toxic by ingestion; skin irritant; heated to decomp., emits acrid smoke and irritating fumes
Uses: Synthetic flavoring agent
Regulatory: FDA 21CFR §172.515
Manuf./Distrib.: Aldrich

Valeryl acetyl. *See* 2,3-Heptanedione

L-Valine

CAS 72-18-4; EINECS 200-773-6
Synonyms: α-Aminoisovaleric acid
Classification: Essential amino acid
Empirical: $C_5H_{11}NO_2$
Properties: Wh. cryst. solid, char. taste; sol. in water; very sl. sol. in alcohol; insol. in ether; m.w. 117.15; dens. 1.230; m.p. 315 C
Toxicology: LD50 (IP, rat) 5390 mg/kg; mutagenic data; heated to decomp., emits toxic fumes of NO_x
Uses: Nutrient, dietary supplement; flavoring
Regulatory: FDA 21CFR §172.320 (limitation 7.4%); Japan approved
Manuf./Distrib.: Degussa

Vanilla

CAS 8024-06-4
Synonyms: Protovanol; Vanilla flavor
Definition: Natural prod. obtained from cured full-grown unripe fruit of *Vanilla planifolia* or *V. tahitensis*
Uses: Natural flavoring
Regulatory: FDA 21CFR §163.111, 163.112, 163.113, 163.114, 163.117, 163.123, 163.130, 163.135, 163.140, 163.145, 163.150, 163.153, 163.155, 182.10, 182.20, GRAS; Japan approved

Manuf./Distrib.: Beck Flavors; Bell Flavours; Berk; Bush Boake Allen; C.A.L.-Pfizer; Consolidated Flavor; Diamalt; Eurovanillin; Frutarom; Givaudan-Roure; Haarmann & Reimer; IFF; V E Kohnstamm; Penta Mfg.; Pointing; Quest; Robertet; Spice King; Synthite; Ungerer
Trade names: Vanilla No. 25450

Vanilla flavor. *See* Vanilla
Vanillal. *See* Ethyl vanillin
Vanillaldehyde. *See* Vanillin
Vanillic aldehyde. *See* Vanillin

Vanillin
CAS 121-33-5; EINECS 204-465-2; FEMA 3107
Synonyms: 4-Hydroxy-m-anisaldehyde; 4-Hydroxy-3-methoxybenzaldehyde; Methylprotocatechualdehyde; Vanillaldehyde; Vanillic aldehyde
Classification: Substituted aromatic aldehyde
Definition: Methyl ether of protocatechuic aldehyde
Empirical: $C_8H_8O_3$
Formula: $(CH_3O)(OH)C_6H_3CHO$
Properties: White cryst. needles, pleasant vanilla odor; sol. in 125 parts water, 20 parts glycerol, 2 parts 95% alcohol, chloroform, ether; m.w. 152.16; dens. 1.056; b.p. 285 C; m.p. 80-81 C
Precaution: Combustible
Toxicology: LD50 (oral, rat) 1580 mg/kg; moderately toxic by ingestion, intraperitoneal, subcutaneous, intravenous routes; experimental reproductive effects; human mutagenic data; heated to decomp., emits acrid smoke and irritating fumes
Uses: Synthetic flavoring agent; migrating to food from paper/paperboard
Usage level: 63 ppm (nonalcoholic beverages), 95 ppm (ice cream), 200 ppm (candy), 220 ppm (baked goods), 120 ppm (gelatins, puddings), 270 ppm (chewing gum), 330-20,000 ppm (syrups), 970 ppm (chocolate), 150 ppm (toppings), 0.2 ppm (margarine)
Regulatory: FDA 21CFR §135.110, 163.111, 163.112, 163.113, 163.114, 163.117, 163.123, 163.130, 163.135, 163.140, 163.145, 163.150, 163.153, 163.155, 182.60, GRAS; Japan approved as flavoring
Manuf./Distrib.: Aldrich; Ashland; Beck Flavors; Chart; Diamalt; Eurovanillin; Forum; Frutarom; Haarmann & Reimer; V E Kohnstamm; Mitsubishi; Penta Mfg.; Pointing; Polarome; Prova; Quest Int'l.; Rhone-Poulenc Food; Rit Chem; Soanchem; Siber Hegner; Spice King
Trade names containing: Vanillin Free Flow

Vanillin acetate
CAS 881-68-5; FEMA 3108
Synonyms: Acetyl vanillin; Vanillyl acetate; 4-Formyl-2-methoxyphenyl acetate
Empirical: $C_{10}H_{12}O_4$
Formula: $CH_3CO_2C_6H_3(CHO)OCH_3$
Properties: Cryst. solid, floral balsamic odor; sol. in alcohol, ether; sl. sol. in water; m.w. 194.19; m.p. 78-79 C
Uses: Synthetic flavoring
Usage level: 11 ppm (nonalcoholic beverages, ice cream, ices), 28 ppm (candy, baked goods)
Regulatory: FDA 21CFR §172.515
Manuf./Distrib.: Aldrich

Vanillin methyl ether. *See* Veratraldehyde
Vanillyl acetate. *See* Vanillin acetate
Vanillylacetone. *See* Zingerone
Vegetable carbon. *See* Carbon black

Vegetable colloids
Uses: Ingredient for pastry coatings
Trade names containing: Danishine™

Vegetable gum. *See* Dextrin
Vegetable lutein. *See* Xanthophyll
Vegetable luteol. *See* Xanthophyll

Vegetable oil
CAS 68956-68-3; 68938-35-2; EINECS 273-313-5
Synonyms: Oils, vegetable
Definition: Expressed oil of vegetable origin consisting primarily of triglycerides of fatty acids
Uses: Shortening, dietary supplements, salad dressings
Manuf./Distrib.: Arista Industries; Karlshamns; A.E. Staley Mfg.
Trade names: Choclin; Coberine; Drewmulse® D-4661

Vegetable oil, hydrogenated

Trade names containing: Amisol™ 785-15K; Annatto OS #2894; Annatto OSL-1; Annatto OSS-1; Annatto/Turmeric OSS-1; Beakin LV3; Beakin LV4; Beakin LV30; Cake Pan Release; Canthaxanthin 10% Type RVI No. 66523; Canthaxanthin Beadlets 10%; Coco-Spred™; Crestawhip H13; Dry Canthaxanthin 10% SD No. 66514; Embanox® 2; Embanox® 5; Kakebake® M/V Cake Icing Shortening; M-C-Thin® ASOL 436; Oxynex® L; Pano-Lube™; Panospray™; Panospray™ LS; Panospray™ SQ; Panospray™ WL; Pristene® 180; Pristene® 181; Pristene® 184; Pristene® 185; Pristene® 186; Pristene® 189; Pristene® RO; Pristene® RW; Pristene® TR; Roxanthin® Red 10WS No. 66515; Sustane® 4A; Sustane® 6; Sustane® 11; Sustane® 15; Sustane® 16; Sustane® 18; Sustane® 20A; Sustane® 20B; Sustane® HW-4; Sustane® T-6; Sustane® W-1; Sustane® W; Tenite®; Tenox® 4A; Tenox® 8; Tenox® 20A; Tenox® 21; Tenox® GT-1; Tenox® GT-2; Turmeric OSO-50; Vykasoid

Vegetable oil, hydrogenated. *See* Hydrogenated vegetable oil
Vegetable pepsin. *See* Papain

Vegetable protein
Uses: Used as a flavoring
Trade names containing: Mayonat V/100

Vegetable protein hydrolysate. *See* Hydrolyzed vegetable protein

Veratraldehyde
CAS 120-14-9; EINECS 204-373-2; FEMA 3109
Synonyms: 3,4-Dimethoxybenzaldehyde; Veratic aldehyde; Vanillin methyl ether; Protocatechualdehyde dimethyl ether; 3,4-Dimethoxybenzenecarbonal
Empirical: $C_9H_{10}O_3$
Properties: Need., vanilla bean odor; freely sol. in alcohol, ether; sl. sol. in hot water; m.w. 166.18; m.p. 41-44 C; b.p. 281 C (760 mm); flash pt. > 230 F
Precaution: Sol'ns. oxidize to veratric acid under influence of light
Toxicology: Skin irritant
Uses: Synthetic flavoring
Usage level: 9.0 ppm (nonalcoholic beverages), 9.2 ppm (ice cream, ices), 32 ppm (candy), 30 ppm (baked goods), 15 ppm (gelatins, puddings)
Regulatory: FDA 21CFR §172.515
Manuf./Distrib.: Aldrich

Veratric aldehyde. *See* Veratraldehyde
Verbascum thapsis extract. *See* Mullein extract

Verbena extract
CAS 84961-67-1
Synonyms: Verbena officinalis extract; Vervain extract
Definition: Extract of *Verbena officinalis*
Uses: Natural flavoring
Regulatory: FDA 21CFR §172.510
Manuf./Distrib.: Chr. Hansen's

Verbena officinalis extract. *See* Verbena extract

Verbenol
Synonyms: 2-Pinen-4-ol
Uses: Synthetic flavoring
Regulatory: FDA 21CFR §172.515

Veronica extract
Synonyms: Veronica officinalis extract; Common speedwell extract; Speedwell extract
Definition: Extract of leaves, flowers, and stems of *Veronica officinalis*
Uses: Natural flavoring
Regulatory: FDA 21CFR §172.510
Manuf./Distrib.: Chr. Hansen's

Veronica officinalis extract. *See* Veronica extract
Vervain extract. *See* Verbena extract

Viburnum
Definition: Natural prod. obtained from *Viburnum* species
Uses: Natural flavoring
Regulatory: FDA 21CFR §172.510

Vinegar acid. *See* Acetic acid

Vinegar naphtha. *See* Ethyl acetate
Vinegar salts. *See* Calcium acetate
Vinyl alcohol polymer. *See* Polyvinyl alcohol
Vinylbenzene. *See* Styrene
Vinyl carbinyl cinnamate. *See* Allyl cinnamate
p-Vinylguaiacol. *See* 2-Methoxy-4-vinylphenol

Vinylidene chloride polymer
  Uses: Used in paper/paperboard for food pkg.
  Regulatory: FDA 21CFR §181.30

cis- and trans-2-Vinyl-2-methyl-5-(1′-hydroxy-1′-methylethyl) tetrahydrofuran. *See* Linalool oxide
3-Vinylpropionic acid. *See* 4-Pentenoic acid
Viola tricolor extract. *See* Pansy extract

Violet
  FEMA 3110
  Definition: Derived from flowers and leaves of *Viola odorata*
  Properties: Pleasant delicate floral odor, sl. bitter taste
  Uses: Natural flavoring agent
  Usage level: 2.3 ppm (nonalcoholic beverages), 8.4 ppm (ice cream, ices), 7.6 ppm (candy), 2.0-2.4 ppm
    (baked goods)
  Regulatory: FDA 21CFR §182.20, GRAS
  Manuf./Distrib.: C.A.L.-Pfizer; Pierre Chauvet (absolute)

Violet leaf alcohol. *See* 2,6-Nonadien-1-ol
Viosterol. *See* Ergocalciferol
Virginia snakeroot. *See* Serpentaria
Vital wheat gluten. *See* Wheat gluten
Vitamin A. *See* Retinol
Vitamin A acetate. *See* Retinyl acetate
Vitamin A alcohol. *See* Retinol
Vitamin A alcohol acetate. *See* Retinyl acetate
Vitamin A palmitate. *See* Retinyl palmitate
Vitamin B₁. *See* Thiamine
Vitamin B₁ hydrochloride. *See* Thiamine HCl
Vitamin B₁ nitrate. *See* Thiamine nitrate
Vitamin B₂. *See* Riboflavin
Vitamin B₂ phosphate sodium. *See* Riboflavin-5′-phosphate sodium
Vitamin B₃. *See* Nicotinic acid
Vitamin B₅, calcium salt. *See* Calcium pantothenate
Vitamin B₆ hydrochloride. *See* Pyridoxine HCl
Vitamin B₇. *See* L-Carnitine
Vitamin B₁₂. *See* Cyanocobalamin
Vitamin Bc. *See* Folicacid
Vitamin C. *See* L-Ascorbic acid
Vitamin C sodium salt. *See* Sodium ascorbate
Vitamin D. *See* Ergocalciferol (Vitamin D₂), Cholecalciferol (Vitamin D₃)
Vitamin D₂. *See* Ergocalciferol
Vitamin D₃. *See* Cholecalciferol
Vitamin E. *See* d-α-Tocopherol
Vitamin E. *See* dl-α-Tocopherol
Vitamin E acetate. *See* Tocopheryl acetate
Vitamin E acid succinate. *See* Tocopheryl succinate
Vitamin G. *See* Riboflavin
Vitamin H. *See* d-Biotin

Vitamin K₁
  Synonyms: Phytonadione; Phylloquinone; Phytodione
  Empirical: $C_{31}H_{46}O_2$
  Properties: m.w. 450.68
  Toxicology: No adverse human effects with prolonged ingestion
  Uses: Infant formula fortifier
  Manuf./Distrib.: Am. Roland
  Trade names: Phytonadione USP No. 61749

Trade names containing: Dry Phytonadione 1% SD No. 61748

Vitamin M. *See* Folicacid
VM&P naphtha. *See* Naphtha

Walnut
Definition: Husks, leaves, and green nuts of *Juglans nigra* or *J. regia*
Uses: Natural flavoring
Regulatory: FDA 21CFR §172.510

Walnut oil
CAS 8024-09-7; 84604-00-2
Definition: Oil derived from the nut meats of walnuts, *Juglans* spp.
Properties: iodine no. 145-155; sapon. no. 190; ref. index 1.4691 (40 C)
Uses: Natural flavoring, food ingred.
Regulatory: FDA 21CFR §172.510, 175.300
Manuf./Distrib.: Arista Industries; Penta Mfg.

Water glass. *See* Sodium silicate
Waxes, microcrystalline. *See* Microcrystalline wax
Waxes, rice bran. *See* Rice bran wax
West Indian lemongrass oil. *See* Lemongrass oil West Indian

Wheat bran
Uses: Dietary fiber supplement
Trade names: Stabilized Red Wheat Bran; Stabilized White Wheat Bran
Trade names containing: Stabilized Cookie Blend

Wheat germ
Uses: Nutritive supplement
Trade names: Stabilized Full-Fat Wheat Germ; Viobin Defatted Wheat Germ #1; Viobin Defatted Wheat Germ #3; Viobin Defatted Wheat Germ #5; Viobin Defatted Wheat Germ #6; Viobin Defatted Wheat Germ #8; Viobin Defatted Wheat Germ #9; Viobin Defatted Wheat Germ #34; Vitinc® Defatted Wheat Germ Nuggets; Vitinc® Defatted Wheat Germ; Vi-Vax
Trade names containing: VCR; Viobin Defatted Wheat Germ #2

Wheat germ oil
CAS 8006-95-9
Synonyms: Triticum aestivum germ oil
Definition: Oil obtained by expression or extraction of wheat germ
Properties: Lt. yel. oil; fat-sol.; sp.gr. 0.9-0.94; iodine no. 120-140; sapon.no. 179-194; ref. index 1.469-1.479
Uses: Food ingred., nutritive additive, dietary supplement
Manuf./Distrib.: Able Prods.; Am. Roland; Arista Industries; S Black; British Arkady; Cornelius; Henry Lamotte; Karlshamns; Penta Mfg.; Quimdis; Spectrum Naturals; Viobin; Westhove
Trade names: Lipovol WGO; Viobin Wheat Germ Oil; Vitinc® Wheat Germ Oil; Wheat Germ Oil
Trade names containing: Rex® Vitamin Fortified Wheat Germ Oil

Wheat gluten
CAS 8002-80-0; EINECS 232-317-7
Synonyms: Gluten; Vital wheat gluten; Devitalized wheat gluten
Definition: Principal protein component of wheat; consists mainly of gliadin and glutenin
Properties: Cream to lt. tan powd.; sol. in alkalies, alcohol
Toxicology: Heated to decomp., emits acrid smoke and irritating fumes
Uses: Direct food additive, dough conditioner/strengthener, formulation aid, nutrient supplement, processing aid, stabilizer, thickener, surface-finishing agent, texturizing agent
Regulatory: FDA 21CFR §184.1322, GRAS
Manuf./Distrib.: Crespel & Deiters GmbH; Goorden
Trade names: Do-Pep Vital Wheat Gluten; Meatbind®-3000; Provim ESP®; SQ®-48; Whetpro®-75; Whetpro®-80
Trade names containing: Lite Pack 350; MG-60; MG-100 K; MG-F; MG-S; Panipower III

Wheat starch
CAS 9005-25-8
Definition: Natural material obtained from wheat, contg. amylose and amylopectin
Uses: Migrating from cotton in dry food pkg.
Regulatory: FDA 21CFR §175.105, 178.3520, 182.70
Trade names: Aytex®-P Food Powd. Wheat Starch; Edigel® 100 Powd. Wheat Starch; Paygel® 290.295 Pregelatinized Food Powd. Wheat Starch; Whetstar® 7 Food Powd. Wheat Starch

Trade names containing: Alcolec® HWS

Whey

Synonyms: Lacto-serum
Definition: Liq. substance obtained by separating coagulum from milk, cream, or skim milk in cheesemaking; as sweet or acid whey, conc. whey, or dried whey
Uses: Food additive used in cheese-making
Regulatory: FDA 21CFR §184.1979, GRAS
Manuf./Distrib.: Blossom Farm Prods.; Borculo
Trade names containing: CFI-10; Medium 7; Medium 10; Medium 55; Medium 700; Medium VS; Precision Base; Qual Guard™ WB; Shedd's Margarine; Shedd's Special 40 Butter Blend Margarine; Shedd's Special 60 Butter Blend Margarine

Whey, dry

Synonyms: Dry whey; Dried whey; Whey powder
Toxicology: Heated to decomp., emits acrid smoke and irritating fumes
Uses: Binder, extender for beef, chili con carne, poultry, sausage, soups, stews
Usage level: Limitation 3.5% (sausage), 8% (chili con carne, pork or beef with BBQ sauce)
Regulatory: FDA 21CFR §184.1979, GRAS; USDA 9CFR §318.7, 381.147
Manuf./Distrib.: Beatrice Cheese; Bio-Isolates; Blossom Farm Prods.; Booths; Byrton Dairy Prods.; DMV Int'l.; EPI Bretagne; G Fiske; Land O'Lakes; Meggle; New Zealand Milk Prods.; J W Pike; J L Priestley; Sorrento Cheese; Unilait France; Van Waters & Rogers

Whey powder. *See* Whey, dry
Whey protein. *See* Whey protein conc.

Whey protein conc.

Synonyms: Whey protein
Definition: Substance obtained by removal of sufficient nonprotein constituents from whey so that finished dry prod. contains ≥ 25% protein; as fluid, conc., or dried
Toxicology: Heated to decomp., emits acrid smoke and irritating fumes
Uses: Food additive, binder; used in beef, chili con carne, sausage, soups, stews
Usage level: Limitation 3.5% (sausages), 8% (chili con carne, pork or beef with BBQ sauce)
Regulatory: FDA 21CFR §184.1979c, GRAS; USDA 9CFR §318.7
Manuf./Distrib.: Adams Food Ingreds.; Am. Casein; Berk; Bio-Isolates; Blossom Farm Prods.; Calaga Food Ingreds.; Cornelius; Dena AG; Domo Food Ingreds.; EPI Bretagne; G Fiske; K&K Greeff; KWR; Meggle; New Zealand Milk Prods.; Pfizer Food; Quest; Unilait France
Trade names: Dairy-Lo™

Whey, reduced lactose

Synonyms: Reduced lactose whey
Definition: Substance obtained by removal of lactose from whey; as fluid, conc., or dried
Uses: Food additive, binder for beef, chili con carne, sausage, soups, stew
Usage level: Limitation 3.5% (sausage), 8% (chili con carne, pork or beef with BBQ sauce)
Regulatory: FDA 21CFR §184.1979a, GRAS; USDA 9CFR §318.7
Manuf./Distrib.: Blossom Farm Prods.

Whey, reduced minerals

Definition: Substance obtained by removal of a portion of minerals from whey; as fluid, conc., or dried
Toxicology: Heated to decomp., emits acrid smoke and irritating fumes
Uses: Food additive, binder for beef, chili con carne, sausage, soups, stews
Usage level: Limitation 3.5% (sausage), 8% (chili con carne, pork or beef with BBQ sauce)
Regulatory: FDA 21CFR §184.1979b, GRAS; USDA 9CFR §318.7

White caustic. *See* Sodium hydroxide
White copperas. *See* Zinc sulfate
White jasmine extract. *See* Jasmine extract
White sandalwood. *See* Sandalwood
White shellac. *See* Shellac
White vitriol. *See* Zinc sulfate
White vitriol. *See* Zinc sulfate heptahydrate
Whole dried egg. *See* Egg powder

Whole fish protein concentrate

Definition: Derived from whole hake and hake-like fish, herring of genera *Clupea*, menhaden, and anchovy (*Engraulis mordax*)
Properties: Faint char. fish odor and taste

Wild bergamot

Uses: Protein supplement in food
Usage level: Limitation 20 g/day (in supplements for children under 8 years)
Regulatory: FDA 21CFR §172.385

Wild bergamot. *See* Horsemint
Wild chamomile extract. *See* Matricaria extract
Wild chamomile oil. *See* Matricaria oil

Wild cherry bark extract
CAS 8000-44-0; 84604-07-9
Synonyms: Prunus serotina extract
Definition: Extract derived from bark of *Prunus serotina*
Uses: Natural flavoring
Regulatory: FDA 21CFR §182.20

Wild marjoram extract
Synonyms: Origanum vulgare extract
Definition: Extract of flowering ends of *Origanum vulgare*
Uses: Natural flavoring
Regulatory: FDA 21CFR §182.20

Wild pennyroyal. *See* Mentha arvensis oil
Wild thyme. *See* Thyme, wild

Wild thyme extract
CAS 84776-98-7
Synonyms: Mother of thyme extract; Thymus serpyllum extract
Definition: Extract of *Thymus serpyllum*
Uses: Natural flavoring
Regulatory: FDA 21CFR §182.20, GRAS

Wilkinite. *See* Bentonite
Wine ether. *See* Ethyl pelargonate
Wine yeast oil. *See* Cognac oil, green or white

Wintergreen
FEMA 3112 (extract), 3113 (oil)
Definition: Gaultheria procumbens
Uses: Natural flavoring agent
Regulatory: Japan approved

Wolf's bane. *See* Arnica
Wood alcohol. *See* Methyl alcohol
Wood creosote. *See* Beechwood creosote
Wood naphtha. *See* Methyl alcohol
Wood pulp, bleached. *See* Cellulose

Woodruff extract
Synonyms: Asperula odorata extract
Definition: Extract of the leaves and flowers of *Asperula odorata*
Uses: Natural flavoring
Regulatory: FDA 21CFR §172.510
Manuf./Distrib.: Chart

Woodruff, sweet
Definition: Derived from *Asperula odorata* L
Properties: Heavy, sweet, tobacco-like flavor, odor
Uses: Natural flavoring agent for alcoholic beverages only
Regulatory: FDA 21CFR §172.510
Manuf./Distrib.: Chart

Wood spirit. *See* Methyl alcohol
Wood sugar. *See* D(+)-Xylose
Wool fat. *See* Lanolin
Wool wax. *See* Lanolin
Wormwood. *See* Artemisia

Wormwood oil
Synonyms: Artemisia absinthium oil

Definition: Volatile oil from leaves and tops of *Artemisia absinthium*, contg. thujyl alcohol and acetate, thujone, phellandrene, cadinene

Properties: Brnsh.-green liq.; sol. in ether, 2 vols 80% aclohol; very sl. sol. in water; dens. 0.925-0.955 (15/15 C); ref. index 1.460-1.4741 (20 C)

Precaution: Keep cool, well closed; protect from light

Uses: Flavoring agent for vermouth; formerly in absinthe

Xanthan. *See* Xanthan gum

Xanthan gum
CAS 11138-66-2; EINECS 234-394-2; EEC E415

Synonyms: Corn sugar gum; Xanthan

Classification: Polysaccharide gum

Definition: High m.w. hetero polysaccharide gum produced by a pure-culture fermentation of a carbohydrate with *Xanthomonas campestris*

Properties: Wh. to cream-colored powd.; very hygroscopic; sol. in water; insol. in oils, most org. solvs.

Toxicology: Heated to decomp., emits acrid smoke and irritating fumes

Uses: Thickener, suspending agent, stabilizer, emulsifier, binder, bodying agent, extender, foam enhancer/ stabilizer; for baked goods, beverages, chili, desserts, gravies, jams, milk prods., poultry, salad dressings, sauces, stews

Usage level: 0.1-0.5%; ADI 0-10 mg/kg (EEC)

Regulatory: FDA 21CFR §133.124, 133.133, 133.134, 133.162, 133.178, 133.179, 172.695, 176.170; USDA 9CFR §318.7 (limitation 8%), 381.147; Japan, JCID, Europe, UK approvals

Manuf./Distrib.: ADM; Am. Roland; D F Anstead; Arthur Branwell; Calaga Food Ingreds.; Cornelius; Courtaulds; Ellis & Everard; G Fiske; Jungbunzlauer; Kelco Int'l.; Meer; Meyhall AG; Penta Mfg.; Rhone-Poulenc; Sanofi; Spice King; TIC Gums; Valmar; Zumbro

Trade names: Alginade XK9; Dariloid® 400; K2B387; K8B249; Keltrol®; Keltrol® 1000; Keltrol® BT; Keltrol® CR; Keltrol® F; Keltrol® GM; Keltrol® RD; Keltrol® SF; Keltrol® T; Keltrol® TF; KOB87; KOB348; KOB349; Merecol® MS; Merezan® 8; Merezan® 20; Nutriloid® Xanthan; Rhodigel®; Rhodigel® 200; Rhodigel® EZ; Rhodigel® Granular; Rhodigel® Supra; Ticaxan® Regular; TIC Pretested® Xanthan 200; TIC Pretested® Xanthan PH

Trade names containing: Avicel® RCN-30; Ches® 500; Dariloid® 100; Dariloid® 300; Dricoid® 200; Dricoid® 280; Emulgel E-21; Emulgel S-32; Freedom X-PGA; GFS®; Kelgum®; Kel-Lite™ CM; Lite Pack 350; Mayonal P3, Merecol® FA, Merecol® GX, Merecol® I, Merecol® RCS, Rhodigum OEH, Rhodigum OEM; Rhodigum WGH; Rhodigum WGL; Rhodigum WGM; Rhodigum WVH; Rhodigum WVM; Thixogum S, Thixogum X, Ticalold® S1-102, Ticalold® S2-102, Vis*Quick™ GX 11, Vis*Quick™ GX 21

Xanthophyll
CAS 127-40-2; EINECS 204-840-0; EEC E161

Synonyms: Lutein; Vegetable lutein; Vegetable luteol

Definition: Carotenoid alcohol found in egg yolk, nettles, algae, and the petals of yellow flowers

Empirical: $C_{40}H_{56}O_2$

Properties: Yel. prisms; insol. in water; sol. in fats and fat solvs.; m.w. 568.85

Uses: Yellow colorant

Regulatory: Europe listed

Manuf./Distrib.: Am. Roland

Xenene. *See* Biphenyl
o-Xenol. *See* o-Phenylphenol

3-(2-Xenolyl)-1,2-epoxypropane
Toxicology: Heated to decomp., emits acrid smoke and irritating fumes

Uses: Plasticizer migrating from food pkg.

Regulatory: FDA 21CFR §181.27

Xylit. *See* Xylitol
Xylite. *See* Xylitol

Xylitol
CAS 87-99-0; EINECS 201-788-0

Synonyms: Xylite; 1,2,3,4,5-Pentahydroxypentane; Xylit

Classification: Pentahydric alcohol

Empirical: $C_5H_{12}O_5$

Properties: Wh. cryst. or cryst. powd., sweet taste with cooling sensation; hygroscopic; sol. in water; sl. sol. in alcohol; m.w. 152.17; m.p. 92-96 C

Toxicology: LD50 (oral, mouse) 22 g/kg; mod. toxic by IV route; mildly toxic by ingestion; heated to decomp.,

D(+)-Xylose

emits acrid smoke and irritating fumes
Uses: Dietary supplement, nutritive sweetener; suitable for diabetic and dietetic foods
Usage level: ADI not specified (WHO)
Regulatory: FDA 21CFR §172.395
Manuf./Distrib.: Am. Roland; Am. Xyrofin; Atomergic Chemetals; F R Benson; Cerestar Int'l.; Food Addit. &
Ingreds.; Forum Chems.; Fruitsource; Melida; Penta Mfg.; Roquette UK; Scanchem; Xyrofin
Trade names: Xylitol C

D(+)-Xylose
CAS 58-86-6; EINECS 200-400-7
Synonyms: Wood sugar
Classification: Complex polysaccharide
Definition: Pentose sugar
Empirical: $C_5H_{10}O_5$
Properties: m.w. 150.13; m.p. 156-158 C
Uses: Sweetener with 40% the sweetness of sugar; diabetic food; savory flavor reducing sugar
Regulatory: Japan approved
Manuf./Distrib.: Am. Xyrofin; F R Benson; Food Additives & Ingreds.; Forum; Molda; Paulaur; Penta Mfg.;
Scanchem; Xyrofin UK

Yarmor. *See* Pine oil

Yarrow
Synonyms: Milfoil; Achillea
Definition: From *Achillea millefolium*
Uses: Natural flavoring agent for beverages only
Usage level: 29 ppm (nonalcoholic beverages), 5-40 ppm (alcoholic beverages)
Regulatory: FDA 21CFR §172.510; finished beverage must be thujone-free

Yarrow extract
CAS 84082-83-7
Synonyms: Achillea millefolium extract; Milfoil extract
Definition: Extract derived from *Achillea millefolium*
Uses: Natural flavoring
Regulatory: FDA 21CFR §172.510
Manuf./Distrib.: Chart

Yarrow oil
CAS 8022-07-9
Synonyms: Achillea millefolium oil
Definition: Oil obtained from the flowering herb *Achillea millefolium*
Uses: Natural flavoring
Regulatory: FDA 21CFR §172.510

Yeast
CAS 68876-77-7
Synonyms: Barm
Definition: Fungus with unicellular growth form; several types are: bakers' yeast, bakers' compressed yeast,
active dry yeast, brewers' yeast
Properties: Ylsh-wh. viscid liq. or soft mass, flakes, or granules, consisting of cells and spores of *Saccharo-
myces cerevisiae*
Uses: Fermentation of sugar, molasses, cereals for alcohol; brewing; baking; dough leavening; food
supplement; source of vitamins, enzymes, nucleic acids
Manuf./Distrib.: Allied Custom Gypsum; Am. Roland; Ashland; Bio Springer; Browning; Croxton & Garry;
Euroma Food Ingreds.; Fermex; Food Additives & Ingreds.; Gist-brocades; Edw. Gittens; Hoechst AG;
Integrated Ingreds.; Novo Nordisk; Provesta; Quest; Red Star
Trade names: Type B Torula Dried Yeast
Trade names containing: BFP White Sour; Hickory Smoked Torula Dried Yeast; Luxor® KB-320; Ry-So

Yeast betaglucan
Definition: Carbohydrate fraction obtained from the hydrolysis of yeast; a component of the yeast cell wall

Yeast extract
CAS 8013-01-2
Synonyms: Extract of yeast
Definition: Preparation of the water-sol. fraction of autolyzed yeast
Properties: Yellowish white liq. or pressed form

Toxicology: Heated to dec., emits acrid smoke and irritating fumes
Uses: Fermentation of sugars, etc., brewing, baking, food supplement; rich source of the B vitamins
Manuf./Distrib.: Ajinomoto; Am. Roland; Berk; Bestoval; Biocon UK; Champlain Industries; Euroma Food Ingreds.; Gist-brocades; Edw. Gittens; W Kündig; Paldena; Provesta; Quest Int'l.; Red Star; Rieber & Son AS; Spicemanns; Universal Flavors Belgium
Trade names: AYE-2000; AYE 2200; AYE 2312; AYE Family; AYS 2311; AYS 2350; Gistex®; Gistex® LS; Gistex® MR; Gistex® X-II; Gistex® Xtra Powd.; Maxarome® Family; Maxarome® MR; Maxarome® Plus; Maxarome® Plus RS; Maxavor™ MYE; Maxavor™ RYE-A, RYE-AS; Maxavor™ RYE-B; Maxavor™ RYE-C; Maxavor™ RYE-CL; Maxavor™ RYE-CR; Maxavor™ RYE-D; Maxavor™ RYE-G; Maxavor™ RYE-PMR; Maxavor™ RYE-T; Maxavor™ RYE Family; Nova-CPLN; Nova-Flav™ 1000; Nova-Flav™ 1001; Nova-Flav™ 1002; Nova-Flav™ 1006; Nova-Flav™ 1020; Nova-Flav™ 1030; Nova-Flav™ 5004; Nova-Flav™ 5006; Nova-Flav™ 5009; Nova-Flav™ 5010; Nova-Flav™ 5030; Nova-Flav™ 5100; Nova-Flav™ 5101; Nova-Flav™ 5102; Nova-Flav™ 5103; Nova-Flav™ 5105; Nova-Flav™ 7000; Nova-Flav™ 7001; Nova-Flav™ 7003; Nova-Flav™ 7004; Nova-Flav™ 7006; Nova-Flav™ 7007; Nova-Flav™ 7009; Nova-Flav™ 7010; Nova-Flav™ 7102; Nova-Flav™ 7105; Nova-Flav™ 7107; Nova-Flav™ 7109; Nova-Flav™ 8002; Nova-Flav™ 8004; Nova-Max 1
Trade names containing: Aromild® S; Medium 7; Medium 10; Medium 55; Medium 700; Medium VS; Natural HVP Replacer

Yellow prussiate of soda. *See* Sodium ferrocyanide
Yellow sandalwood. *See* Sandalwood
Yerba mate extract. *See* Mate extract

Yerba santa extract
CAS 85085-31-0; 68990-14-7; EINECS 285-361-4
Synonyms: Eriodictyon californicum extract; Eriodictyon glutinosum extract
Definition: Extract of leaves of *Eriodictyon californicum*
Uses: Natural flavoring
Regulatory: FDA 21CFR §172.510
Manuf./Distrib.: Chart

Ylang ylang oil
CAS 8006-81-3
Definition: Oil from flowers of *Cananga odorata*
Uses: Natural flavoring
Regulatory: FDA 21CFR §182.20; Japan approved (ylang-ylang)
Manuf./Distrib.: Pierre Chauvet

Yucca
FEMA 3121
Definition: Natural prod. derived from *Yucca* species
Uses: Natural flavoring; foaming agent for soft drinks, cocktail mixes, slush prods.
Regulatory: FDA 21CFR §172.510
Manuf./Distrib.: Chart

Zea mays extract. *See* Corn silk extract
Zea mays oil. *See* Corn oil

Zedoary oil
Definition: Oil obtained from the rhizome of *Curcuma zedoaria*
Uses: Natural flavoring agent
Regulatory: FDA 21CFR §182.10, 182.20, GRAS; Japan approved (zedoary)

Zein
CAS 9010-66-6; EINECS 232-722-9
Definition: Alcohol-sol. protein obtained from corn, *Zea mays*; component of corn gluten
Properties: Powd., odorless; sol. in glycols, glycol ethers; insol. in water, alcohol; dens. 1.226
Precaution: Combustible
Toxicology: Heated to decomp., emits acrid smoke and irritating fumes
Uses: Direct food additive, surface-finishing agent, glaze for confections, grains, nuts, panned goods; processing aid (Japan)
Regulatory: FDA 21CFR §175.105, 184.1984, GRAS; Japan approved
Manuf./Distrib.: Aldrich

Zibet. *See* Civet
Zibeth. *See* Civet
Zibetum. *See* Civet

Zinc chloride
CAS 7646-85-7; EINECS 231-592-0
Empirical: Cl$_2$Zn
Properties: Cubic white deliq. cryst., odorless; sol. in water, alcohol, glycerol, ether; m.w. 136.27; dens. 2.91 (25 C); m.p. 290 C; b.p. 732 C
Toxicology: TLV:TWA 1 mg/m^3 (air); poison by ingestion, intravenous, subcutaneous, intraperitoneal routes; corrosive irritant to skin, eyes, mucous membranes
Uses: Preservative; dietary supplement; nutrient; migrating from cotton in dry food pkg.
Regulatory: FDA 21CFR §182.70, 182.5985, 182.8985, 582.80, GRAS

Zinc dithionite. *See* Zinc hydrosulfite

Zinc gluconate
CAS 4468-02-4 (anhyd.); EINECS 224-736-9
Definition: Zinc salt of gluconic acid
Empirical: C$_{12}$H$_{22}$O$_{14}$Zn
Properties: Wh. gran. or cryst. powd.; sol. in water; very sl. sol. in alcohol; m.w. 455.7
Toxicology: Heated to decomp., emits toxic fumes of ZnO
Uses: Mineral source, nutrient, dietary supplement
Regulatory: FDA 21CFR §182.5988, 182.8988, GRAS; Japan approved with limitations
Manuf./Distrib.: Am. Roland; Lohmann
Trade names: Gluconal® ZN

Zinc hydrosulfite
CAS 7779-86-4
Synonyms: Zinc dithionite
Properties: Wh. amorphous solid; sol. in water
Uses: Migrating to foods from paper/paperboard
Regulatory: FDA 21CFR §182.90, GRAS

Zinc methionine sulfate
CAS 56329-42-1
Toxicology: Heated to decomp., emits very toxic fumes of SO$_x$
Uses: Source of dietary zinc; dietary supplement, nutrient; for vitamin tablets
Regulatory: FDA 21CFR §172.399

Zinc octadecanoate. *See* Zinc stearate

Zinc orthophosphate
CAS 7779-90-0
Synonyms: Zinc phosphate
Formula: Zn$_3$(PO$_4$)$_2$
Properties: Wh. powd.; sol. in acids; insol. in water; dens. 3.998 (15 C); m.p. 900 C
Toxicology: Heated to decomp., emits acrid smoke and irritating fumes
Uses: Stabilizer migrating from food pkg.
Regulatory: FDA 21CFR §181.29 (50 ppm max. in finished food)

Zinc oxide
CAS 1314-13-2; EINECS 215-222-5
Synonyms: Chinese white; Pigment White 4; CI 77947; Zinc white; Flowers of zinc
Classification: Inorganic oxide
Empirical: OZn
Formula: ZnO
Properties: White to gray powd. or crystals, odorless, bitter taste; sol. in dilute acetic or min. acids, alkalies; insol. in water, alcohol; m.w. 81.38; dens. 5.67; m.p. 1975 C; ref. index 2.0041-2.0203; pH 6.95 (Amer. process), 7.37 (French process)
Precaution: Heated to decomp., emits toxic fumes of ZnO
Toxicology: TLV/TWA 5 mg/m^3; LD50 (IP, rat) 240 mg/kg; poison by IP route; fumes may cause metal fume fever with chills, fever, tightness in chest, cough, leukocytes; experimental teratogen; mutagenic data; skin/eye irritant
Uses: Nutrient, dietary supplement
Regulatory: FDA 21CFR §73.1991, 73.2991, 175.300, 177.1460, 182.5991, 182.8991, 582.80, GRAS

Zinc phosphate. *See* Zinc orthophosphate

Zinc resinate
Properties: Powd., clear amber lumps, or ylsh. liq.; sol. in ether, amyl alcohol
Precaution: Combustible

Toxicology: Heated to decomp., emits acrid smoke and irritating fumes
Uses: Stabilizer migrating from food pkg.
Regulatory: FDA 21CFR §181.29 (50 ppm max. in finished food)

Zinc stearate
CAS 557-05-1; EINECS 209-151-9
Synonyms: Zinc octadecanoate; Octadecanoic acid, zinc salt
Definition: Zinc salt of stearic acid
Empirical: $C_{36}H_{70}O_4Zn$
Formula: $Zn(C_{18}H_{35}O_2)_2$
Properties: White powd., faint odor; sol. in acids, common solvs. (hot); insol. in water, alcohol, ether; dec. by dilute acids; m.w. 632.33; dens. 1.095; m.p. 130 C
Precaution: Combustible
Uses: Dietary supplement, nutrient
Regulatory: FDA 21CFR §175.105, 175.300, 176.170, 176.180, 176.200, 176.210, 177.1200, 177.1460, 177.1900, 177.2410, 177.2600, 178.2010, 178.3910, 182.5994, 182.8994, GRAS
Manuf./Distrib.: Aldrich

Zinc sulfate
CAS 7446-02-0 (anhyd.), 7446-19-7 (monohydrate), 7733-02-0; EINECS 231-793-3
Synonyms: Sulfuric acid zinc salt (1:1); White vitriol; White copperas; Zinc vitriol
Classification: Inorganic salt
Empirical: $ZnSO_4$
Properties: Colorless rhombic cryst. or cryst. powd.; sol. in water; almost insol. in alcohol; m.w. 161.43; dens. 3.74 (15 C); m.p. dec. @ 740 C
Toxicology: LD50 (oral, rat) 2949 mg/kg; poison by IP, subcutaneous, IV; mod. toxic by ingestion; experimental tumorigen, teratogen, reproductive effects; human systemic effects; mutagenic data; eye irritant; heated to dec., emits toxic fumes of SO_x and ZnO
Uses: Dietary supplement, nutrient; animal feeds; migrating to food from paper/paperboard
Regulatory: FDA 21CFR §182.90, 182.5997, 182.8997, 582.80, GRAS; Japan approved with limitations

Zinc sulfate heptahydrate
CAS 7446-20-0; EINECS 231-793-3
Synonyms: Sulfuric acid zinc salt (1:1) heptahydrate; White vitriol; Zinc vitriol
Empirical: $O_4SZn \cdot 7H_2O$
Formula: $ZnSO_4 \cdot 7H_2O$
Properties: Colorless cryst. or cryst. powd., odorless, astringent taste; insol. in alcohol, glycerin; m.w. 287.54; dens. 1.97; m.p. 100 C; dec. > 500 C; loses water @ 280 C; pH 4.5
Precaution: Keep well closed
Toxicology: Human poison by unspecified route; poison experimentally by subcutaneous, IV, IP routes; mod. toxic by ingestion; heated to decomp., emits toxic fumes of SO_x and ZnO
Uses: Dietary supplement, nutrient
Regulatory: FDA 21CFR §182.5997, 182.8997, GRAS

Zinc vitriol. *See* Zinc sulfate
Zinc vitriol. *See* Zinc sulfate heptahydrate
Zinc white. *See* Zinc oxide

Zingerone
FEMA 3124
Synonyms: 4-(4-Hydroxy-3-methoxyphenyl)-2-butanone; (4-Hydroxy-3-methoxyphenyl)ethyl methyl ketone; Vanillylacetone; Zingherone; Zingiberone
Empirical: $C_{11}H_{14}O_3$
Properties: Cryst.; sol. in ether, dil. alkalies; sparingly sol. in water, petroleum ether; m.w. 194.22; dens. 1.138-1.139; m.p. 40-41 C; b.p. 187-188 C (14 mm); ref. index 1.5440-1.5450
Uses: Synthetic flavoring
Usage level: 6.9 ppm (nonalcoholic beverages), 7.8 ppm (ice cream, ices), 11 ppm (candy, baked goods), 15 ppm (chewing gum)
Regulatory: FDA 21CFR §172.515

Zingherone. *See* Zingerone
Zingiber officinale oil. *See* Ginger oil
Zingiberone. *See* Zingerone

Part III
Functional Cross-Reference

Functional Cross-Reference

Trade name and generic chemical additives from the first and second parts of this reference are grouped by broad functional areas derived from research and manufacturers' specifications.

Anticaking agents

Trade names: Aerosil® 200; Aldo® MS FG; Aluminum Stearate EA
Cab-O-Sil® L-90
FloAm® 200; FloAm® 200C; FloAm® 210 HC; FloAm® 221; FloAm® 300; FloAm® 400; FloAm® 450; FloAm® System
Justfiber® CL-35-H
Maltrin® M040; Maltrin® QD M440; Miglyol® 812
Qual Flo™; Qual Flo™ C
Radiamuls® MG 2143; Radiamuls® MG 2600; Radiamuls® MG 2602; Radiamuls® MG 2900; Ryoto Sugar Ester B-370
Sipernat® 22; Sipernat® 22S; Sipernat® 50; Sipernat® 50S; Sorbo®; Star-Dri® 1; Star-Dri® 5; Star-Dri® 10; Star-Dri® 15; Star-Dri® 18; Star-Dri® 1005A; Star-Dri® 1015A
Topcithin®

Chemicals: Aluminum calcium silicate; Aluminum caprylate; Aluminum stearate
Calcium gluconate; Calcium oxide; Calcium phosphate tribasic; Calcium silicate; Calcium stearate; Calcium sulfate; Carnauba; Cellulose
Dimethicone
Iron ammonium citrate
Kaolin
Magnesium carbonate; Magnesium carbonate hydroxide; Magnesium oxide; Magnesium silicate; Magnesium silicate hydrate; Magnesium stearate; D-Mannitol; Microcrystalline cellulose
Potassium acid tartrate; Potassium oleate; Potassium stearate; Propylene glycol
Silica; Silica, hydrated; Simethicone; Sodium calcium silicoaluminate, hydrated; Sodium ferrocyanide; Sodium laurate; Sodium methylnaphthalenesulfonate; Sodium myristate; Sodium oleate; Sodium palmitate; Sodium silicoaluminate; Sodium stearate; Sorbitol; Stearic acid
Talc; Tricalcium silicate

Antioxidants

Trade names: Ascorbic Acid USP/FCC, 100 Mesh; Ascorbic Acid USP, FCC Fine Gran. No. 6045655; Ascorbic Acid USP, FCC Fine Powd. No. 6045652; Ascorbic Acid USP, FCC Gran. No. 6045654; Ascorbic Acid USP, FCC Ultra-Fine Powd No. 6045653; Ascorbyl Palmitate NF, FCC No. 60412; Axol® E 66
24% Beta Carotene HS-E in Veg. Oil No. 65671
Calcium Ascorbate FCC No. 60475; CAO®-3; Controx® AT; Controx® KS; Controx® VP; Covi-Ox® T-70; Covipherol T-75; Curafos® 11-2
Eisai Natural Vitamin E Series; Embanox®; Embanox® BHA; Embanox® TBHQ; Eribate®; Evanstab® 18; EverFresh®
Fermcozyme® 1307, BG, BGXX, CBB, CBBXX, M; Flav-R-Keep® FP-51; FloAm® 300; FloAm® 400; FloAm® 450; FloAm® System; Freez-Gard® FP-15
Gamma Oryzanol; Garbefix 31
Lem-O-Fos® 101; Lem-O-Fos® 202; Lowinox® BHA; Lowinox WSP

Antioxidants *(cont'd.)*

Neo-Cebitate®; Nipanox® S-1; Nipanox® Special; Nurupan™; Nutrifos® 088; Nutrifos® B-75; Nutrifos® H-30; Nutrifos® L-50; Nutrifos® Powd; Nutrifos® SK

Oxynex® K; Oxynex® L; Oxynex® LM

Pristene® 180; Pristene® 181; Pristene® 184; Pristene® 185; Pristene® 186; Pristene® 189; Pristene® R20; Pristene® RO; Pristene® RW; Pristene® TR

Sodium Ascorbate USP, FCC Fine Gran. No. 6047709; Sodium Ascorbate USP, FCC Fine Powd. No. 6047708; Sodium Ascorbate USP, FCC Type AG No. 6047710; Sustane® 1-F; Sustane® 3; Sustane® 4A; Sustane® 6; Sustane® 7; Sustane® 7G; Sustane® 8; Sustane® 11; Sustane® 15; Sustane® 16; Sustane® 18; Sustane® 20; Sustane® 20-3; Sustane® 20A; Sustane® 20B; Sustane® 21; Sustane® 27; Sustane® 31; Sustane® BHA; Sustane® BHT; Sustane® CA; Sustane® HW-4; Sustane® P; Sustane® PA; Sustane® PB; Sustane® PG; Sustane® Q; Sustane® S-1; Sustane® T-6; Sustane® TBHQ; Sustane® W; Sustane® W-1

TBHQ®; Tenite®; Tenox® 2; Tenox® 4; Tenox® 4A; Tenox® 4B; Tenox® 5; Tenox® 5B; Tenox® 6; Tenox® 7; Tenox® 8; Tenox® 20; Tenox® 20A; Tenox® 20B; Tenox® 21; Tenox® 22; Tenox® 24; Tenox® 25; Tenox® 26; Tenox® 27; Tenox® A; Tenox® BHA; Tenox® BHT; Tenox® GT-1; Tenox® GT-2; Tenox® PG; Tenox® R; Tenox® S-1; Tenox® TBHQ; Tenox® WD-BHA; TIC Pretested® Arabic FT-1 USP

Vascan; Veltol®; Veltol®-Plus; Vitamin E USP, FCC No. 60525

Wheat Germ Oil

Chemicals: Anoxomer; L-Ascorbic acid; Ascorbyl palmitate; Ascorbyl stearate

BHA; BHT; t-Butyl hydroquinone

Calcium ascorbate; Calcium lactate; Calcium phosphate dibasic; Citric acid; Clove extract; Coffee bean extract

Dilauryl thiodipropionate; Disodium citrate; Disodium EDTA; Distearyl thiodipropionate; Dodecyl gallate

Edetic acid; Erythorbic acid; 6-Ethoxy-1,2-dihydro-2,2,4-trimethylquinoline; Eucalyptus extract

Fumaric acid

Gallic acid; Gentian extract; Guaiac gum

n-Heptyl p-hydroxybenzoate; Heptyl paraben; Hesperetin; 4-Hydroxymethyl-2,6-di-t-butylphenol

Isopropyl citrate

Lecithin

Nordihydroguairetic acid

Octyl gallate; Oryzanol

Phosphatidylcholine; Phosphoric acid; Pimento extract; Potassium citrate; Potassium metabisulfite; Potassium phosphate; Potassium sulfite; Propylene glycol

Rapeseed oil; Rice bran extract

Sage extract; Sodium ascorbate; Sodium carbonate; Sodium citrate; Sodium erythorbate; Sodium hypophosphite; Sodium metabisulfite; Sodium sulfite; Sodium tartrate; Sodium thiosulfate anhyd; Sodium thiosulfate pentahydrate; Stannous chloride anhyd; Stannous chloride dihydrate; Sucrose

L-Tartaric acid; Tocopherol; d-α-Tocopherol; dl-α-Tocopherol; 2,4,5-Trihydroxybutyrophenone

Antispattering agents

Trade names: Asol; Atlas 1400K; Atmul® 84 K; Atmul® 86 K

Bac-N-Fos® Formula 191

Imwitor® 1940

Leciprime™

Myverol® 18-04; Myverol® 18-07

Chemicals: Citric acid esters of mono- and diglycerides of fatty acids

Aromatics

Chemicals: Allyl caproate; Ambergris; p-Anisaldehyde; Anisyl alcohol

Benzyl cinnamate

Aromatics *(cont'd.)*

Citral; Citronellal; β-Citronellol; Cyclohexylethyl acetate
Decanal; Dihydro-β-ionone; Dimethylphenylethylcarbinol
Ethyl butyrate; Ethyl formate; Ethyl isovalerate
Geranyl linalool
Hexanal; 2,3-Hexanedione; 3,4-Hexanedione; Hops oil; Hydrocinnamic alcohol; Hydroxycitronellal; Hydroxycitronellol
α-Ionone; β-Ionone; Isoamyl acetate
Linalool; Linalyl acetate
p-Methylanisole; 3-Methyl-2-buten-1-ol; 3-Methyl-2-buten-1-ol; Methyl cinnamate; Methyl heptenone; Methyl nonyl acetaldehyde
n-Octanal; cis-3-Octen-1-ol
Pentane-2,3-dione; Phenethyl alcohol; Phenylacetaldehyde; 2-Phenylethyl methyl ether; 2-Phenylpropanal; 2-Phenyl propanol-1; 2-Phenylpropionaldehyde dimethylacetal
Tetrahydrolinalool; Trimethyldodecatrieneol
n-Valeraldehyde

Binders

Trade names: Accoline; Alginic Acid FCC; Aluminum Stearate EA; Aqualon® 7HF; Aqualon® 7LF; Aqualon® 9M8F

Benecel® M 043; Benecel® MP 824; Benecel® MP 843; Benecel® MP 874; Binasol™ 15; Binasol™ 81; Britol®; Britol® 6NF; Britol® 7NF; Britol® 9NF; Britol® 20USP; Britol® 35USP; Britol® 50USP

Carbowax® Sentry® PEG 300; Carbowax® Sentry® PEG 400; Carbowax® Sentry® PEG 540 Blend; Carbowax® Sentry® PEG 600; Carbowax® Sentry® PEG 1000; Carbowax® Sentry® PEG 1450; Carbowax® Sentry® PEG 3350; Carbowax® Sentry® PEG 4600; Carbowax® Sentry® PEG 8000; CMC Daicel 1150; CMC Daicel 1160; CMC Daicel 1220; CMC Daicel 1240; CMC Daicel 1260; CMC Daicel 2200; Curavis® 250

Danpro DS; Danpro HV; Danpro S; Danpro S-760; Destab™; Durkote Potassium Chloride/ Hydrog. Veg. Oil; Dynasan® 114

Edicol®; Edifas; Edigel® 100 Powd. Wheat Starch; Egg White Solids Type P-11; Egg White Solids Type P-39; Egg White Solids Type P-110; Egg White Solids Type P-110 High Gel Strength; Egg White Solids Type PF-1; Emulcoll

Fibrim® 1020; Fibrim® 1250; Fibrim® 1255; Fibrim® 1450; Fibrim® 2000; Fri-Bind® 411; Frimulsion 10; Fructodan

Gelatin XF; Genugel® CHP-2; Genugel® CHP-2 Fine Mesh; Genugel® CHP-200; Genugel® MB-51; Genugel® MB-78F; Genuvisco® MP-11; Glass H®; Granular Gum Ghatti #1; Guar Gum

Hexaphos®; Hystrene® 5016 NF FG; Hystrene® 7018 FG; Hystrene® 8718 FG; Hystrene® 9718 NF FG

Iconol NP-7; Industrene® 143; Industrene® 205; Industrene® 225; Industrene® 226 FG; Industrene® 365; Industrene® 4518; Industrene® 5016 NF FG; Industrene® 7018 FG; Industrene® 8718 FG; Industrene® B; Instant Remygel AX-P, AX-2-P; Instant Remyline AX-P; Instant Tender-Jel® 479; IPSO-MR Dispersible

Justfiber® CL-40-H

Kelcogel® IF; Keltose®

Maco-O-Line 091; Maltrin® M150; Maltrin® M180; Maltrin® M200; Maltrin® M250; Maltrin® M510; Maltrin® QD M500; Maltrin® QD M550; Maltrin® QD M600; MB 100; Meatbind®-3000; Methocel® E4M Premium; Methocel® E6 Premium; Methocel® E15 Food Grade; Methocel® E15LV Premium; Methocel® E50LV Premium; Methocel® K4M Premium; Methocel® K100LV Premium; Methocel® K100M Premium; Mira-Cleer® 340

Nutriloid® Arabic

Paygel® 290.295 Pregelatinized Food Powd. Wheat Starch; Pea Protein Conc; Powdered Gum Ghatti #1; Powdered Gum Ghatti #2; Powdered Gum Guar NF Type 80 Mesh B/T; Powdered Gum Guar Type 140 Mesh B/T; Powdered Gum Guar Type ECM; Powdered Gum Guar Type M; Powdered Gum Guar Type MM FCC; Powdered Gum Guar Type MM (HV); Powdered Gum Guar Type MMM $^1/_2$; Powdered Gum Guar Type MMW; Promax® 70; Promax® 70L; Promax® 70LSL; Promax® Plus; Promine® DS; Promine® DS-SL; Pure-Set® B950

Binders *(cont'd.)*

Remygel AX-2; Remygel AX; Remyline AX; Rudol®

Soageena®; Soageena® WX57; Sol-U-Tein EA; Soyamin 70; Soyamin 90; SQ®-48; Stadex® 9; Stadex® 60K; Stadex® 90; Staform P; Sta-Mist® 365; Sta-Mist® 454; Sta-O-Paque®; Star-Dri® 1; Star-Dri® 5; Star-Dri® 10; Star-Dri® 18; Star-Dri® 1005A; Star-Dri® 1015A

Thixogum X; Ticalose® 15; Ticalose® 30; Ticalose® 150 R; Ticalose® 700 R; Ticalose® 2000 R; Ticalose® 5000 R; Ticaxan® Regular; Ticolv

Vascomix 404

Chemicals: Albumin macro aggregates; Aluminum caprylate; Aluminum stearate; Arabinogalactan

Calcium stearate; Caprylic/capric acid; Carboxymethylcellulose sodium; Carboxymethylmethyl-cellulose; Carrageenan; Cellulose

Dextrin

Food starch modified

Gluconolactone

Hydrogenated stearic acid; Hydrogenated vegetable oil

Magnesium stearate; Methoxyethanol; Methylcellulose; Microcrystalline cellulose; Mineral oil

Nonoxynol-7

Oleic acid

Pea protein concentrate; PEG-4; PEG-6; PEG-8; PEG-9; PEG-12; PEG-14; PEG-16; PEG-20; PEG-32; PEG-40; PEG-75; PEG-100; PEG-150; PEG-200; Polyethylene glycol; Potassium oleate; Potassium polymetaphosphate; Potassium stearate; Potassium tripolyphosphate

Rennet

Sodium caseinate; Sodium hexametaphosphate; Sodium laurate; Sodium metaphosphate; Sodium myristate; Sodium oleate; Sodium palmitate; Sodium stearate; Soy acid; Soy protein

Tallow acid; Trimyristin

Whey, dry; Whey protein conc; Whey, reduced lactose; Whey, reduced minerals

Xanthan gum

Bleaching agents • Decolorizing agents

Trade names: ADP; APA; APC

BL™

C™; CAL®; Cane Cal®; Colorsorb; CPG®; CPG® LF

Diaion® HP 10

GW

Marinco® CL

Nutrisoy 7B; Nutrisoy 7B Flour.OL®

PWA™; RC®; SGL®

Chemicals: Acetone peroxides; Ammonium persulfate; Azodicarbonamide

Benzoyl peroxide

Carbon, activated; Catalase; Chlorine dioxide; Chlorine; Chloromethylated aminated styrene-divinylbenzene resin

Dialkyl dimethyl ammonium chloride; Dimethylamine/epichlorohydrin copolymer

Hydrogen peroxide

Lipoxidase

Potassium bromate

Sodium hydrosulfite; Sodium hypochlorite; Sodium metabisulfite; Sodium sulfite; Sulfur dioxide

Bulking agents • Fillers

Trade names: CNP PPRSSRD; CNP PPRSSRDCL; CNP RMDRD18; CNP RSSHM; CNP RSSHMCL; CNP RSSRD; CNP RSSRDCL

Destab™; Diamond D 31

Exafine® 250; Exafine® 500; Exafine® 1000; Exafine® 1000; Exafine® 1000; Exafine® 1000

Fibrex®; Fibrim® 1020; Fibrim® 1250; Fibrim® 1255; Fibrim® 1450; Fibrim® 2000; Fro-Dex® 24 Powd

Jerusalem Arthichoke Flour (JAF); Justfiber® CL-35-H; Justfiber® CL-100-H

Bulking agents • Fillers *(cont'd.)*

Keycel®
Litesse®; Litesse® II; Lo-Dex® 10
Maltrin® M100; Maltrin® M105; Maltrin® M150; Maltrin® M180; Maltrin® M510; Maltrin® M520; Maltrin® M700; Maltrin® QD M500; Maltrin® QD M550; Meatbind®-3000
Phosal® 50 SA; Phosal® 53 MCT; Phosal® 75 SA
Redisol® 248; Rice-Trin® 10; Rice-Trin® 18
Solka-Floc®; Sovital™; SQ®-48; Staley® Maltodextrin 3260; Staley® Tapioca Dextrin 11; Sta-Lite™ 100C; Sta-Lite™ 100CN Neutralized; Sta-Lite™ 100CN Neutralized; Sta-Lite™ 100F; Sta-Lite™ 100CN Neutralized; Star-Dri® 1005A; Star-Dri® 1015A; Sta-Rx®; Sunmalt
Viobin Defatted Wheat Germ #3

Chemicals: Cellulose; Corn starch
Ethylcellulose
Guar gum
Jerusalem artichoke flour
Maltodextrin; D-Mannitol; Methylcellulose; Microcrystalline cellulose
Polydextrose; Polyvinylpolypyrrolidone; Poly(1-vinyl-2-pyrrolidinone) homopolymer
Starch

Catalysts

Trade names: Catalase L
Natural Soluble Powder Green Colorant; Natural Soluble Powder Q A/L 15 Yellow Colorant

Chemicals: Invertase
Lactase
Nickel
Potassium t-butylate; Potassium ethylate; Potassium methylate
Sodium amide; Sodium ethylate; Sodium methylate
Trifluoromethane sulfonic acid

Colorants • Color adjuncts • Color diluents • Color preservatives • Color fixatives • Color improvers • Color retention aids

Trade names: Accoline; Acid Proof Caramel Powd; Albrite STPP-F; Albrite STPP-FC; Allura® Red AC; Annatto ECA-1; Annatto ECH-1; Annatto ECPC; Annatto OS #2894; Annatto OS #2922; Annatto OS #2923; Annatto OSL-1; Annatto OSS-1; Annatto WSL-2; Annatto WSP-75; Annatto WSP-150; Annatto Extract P1003; Annatto Liq. #3968, Acid Proof; Annatto/Turmeric OSS-1; Annatto/Turmeric WML-1; Annatto/Turmeric WMP-1; Anthocyanin Extract P2001; Apo-Carotenal Suspension 20% in Veg. Oil No. 66417
Basic Spice Mix; B&C Caramel Powd; Beetroot Conc. P3003; Beta Carotene 1% CWS No. 65659; 24% Beta Carotene HS-E in Veg. Oil No. 65671; 24% Beta Carotene Semi-Solid Suspension No. 65642; 30% Beta Carotene in Veg. Oil No. 65646; Beta Carotene Emulsion Beverage Type 3.6 No. 65392; 1626 Blue Natural Liq. Colorant
Canthaxanthin 10% Type RVI No. 66523; Canthaxanthin Beadlets 10%; Caramel Color Double Strength; Caramel Color Single Strength; Carbowax® Sentry® PEG 300; Carbowax® Sentry® PEG 400; Carbowax® Sentry® PEG 540 Blend; Carbowax® Sentry® PEG 600; Carbowax® Sentry® PEG 900; Carbowax® Sentry® PEG 1000; Carbowax® Sentry® PEG 1450; Carbowax® Sentry® PEG 3350; Carbowax® Sentry® PEG 4600; Carbowax® Sentry® PEG 8000; Carex for Foods; Carmacid R; Carmacid Y; Carmine 1623; Carmine AS; Carmine Extract P4011; Carmine FG; Carmine Nacarat 40; Carmine PG; Carmine Powd. WS; Carmine XY/UF; Carmisol A; Carmisol NA; Carotenal Sol'n. #2 No. 66425; Carotenal Sol'n. 4% No. 66424; Carotenal Sol'n. #73 No. 66428; Coloreze™; Curcumex 1600, 1601; Curcumex Natural Powd. Colorant; Curcumin Extract P8002
Dascolor Annatoo Enc; Dispersable Curcumex 2600, 2601; Double Strength Acid Proof Caramel Colour; Dry Beta Carotene Beadlets 10% CWS No. 65633; Dry Beta Carotene Beadlets 10%

Colorants *(cont'd.)*

No. 65661; Dry Beta Carotene Beadlets Yellow Type 2.4-S No. 653800100; Dry Canthaxanthin 10% SD No. 66514; Dry β-Carotene Beadlets

Extramalt 10; Extramalt 33; Extramalt 35; Extramalt Dark; Extramalt Light

Goldex II; Grape Skin Extract, 2X #3850; Grape Skin Extract, Double Strength; Grape Skin Extract, Single Strength

Hilton Davis Carmine; Hilton Davis Titanium Dioxide; Hodag PSMO-20

Liponic Sorbitol Powd; Liponic Sorbitol Sol'n. 70% USP; Lucantin® CX; Lucantin® Red; Lucantin® Yellow

Maco-O-Line 091; Maltoferm® A-6001; Maltoferm® MBF CR-40; MicroPlus™

Natural Liquid AP Carmine Colorant; Natural Liquid Carmine Colorant (Type 100, 50, and Simple); Natural Soluble Carmine Powd; Natural Soluble Orange Powd; Natural Yellow Colour Q-500, Q-1000, Q-2000; Neobee® 1053; Neo-Cebitate®; Non-Diastatic Malt Syrup #40600

Oxipur; Oxipur-Mix

P147 Caramel Color; Pie Dough Culture 6450; Powdered Caramel Color, Acid Proof; Powdered Caramel Colour Non-Ammoniated-All Natural T-717; Pure Malt Colorant A6000; Pure Malt Colorant A6001

Red Beet WSL-300; Red Beet WSL-400; Red Beet WSP-300; Red Soluble Powd. Natural Colorant; Rocarna; Roxanthin® Red 10WS No. 66515

Sicovit®; Single Strength Acid Proof Caramel Colour; Spice Mix 614; Spice Mix 808

Turmeric DP-300; Turmeric OSO-50; Turmeric WMPE-80; Turmeric WMPE-150; Turmeric WMSE-80; Tween® 80

Unibix AP (Acid Proof); Unibix CUS; Unibix ENC (Acid Proof); Unibix W; Unisweet Caramel VCR

Wheat-Eze

Yellow Blend No. 3100 Powd; Yellow Blend No. 3300 Powd; Yellow Fat Soluble Liq. G-200

Zea Red

Chemicals: Acetic acid; Acetone; Alganet; Alkanet extract; Aluminum; Amaranth; Ammonium alginate; Ammonium hydroxide; Annatto; Annatto extract; Anthocyanins; L-Ascorbic acid

Barley extract; Beet powder; Bentonite; Benzoin; Brilliant blue FCF; Butyl alcohol

Calcium carbonate; Calcium resinate; Calcium sulfate; Canthaxanthine; Caramel; Carbon black; Carmine; Carmine extract; Carotene; Carotene cochineal; Cetyl alcohol; Chicory extract; Chlorophyll; Citranaxanthine; Citrus Red No. 2; Cochineal; Cottonseed flour, partially defatted, cooked; Curcumin; Cyclohexane

Disodium EDTA; Disodium pyrophosphate

Ethyl acetate; Ethylcellulose

FD&C Blue No. 1; FD&C Blue No. 2; FD&C Green No. 3; FD&C Red No. 3; FD&C Red No. 4; FD&C Red No. 40; FD&C Yellow No. 5; FD&C Yellow No. 6; Ferric oxide; Ferrous gluconate; Flavoxanthin

Gold; Grape color extract; Grape skin extract

Iron sequioxide; Isopropyl alcohol

Magnesium carbonate; Magnesium chloride; Magnesium hydroxide; Malt; Methoxyethanol; Methylene chloride

Naphtha; Nicotinic acid

Orange B

Paprika; 1-(Phenylazo)-2-naphthylamine; Polysorbate 80; Polyvinyl acetate (homopolymer); Polyvinyl alcohol; Poly(1-vinyl-2-pyrrolidinone) homopolymer; Potassium nitrate; Potassium nitrite; PVP

Red beet extract; Riboflavin; Riboflavin tetrabutyrate

Saffron; Shellac; Silica; Sodium ascorbate; Sodium nitrate; Sodium nitrite; Stannous chloride anhyd.

Tartrazine; Terpene resin, natural; Titanium dioxide; Turmeric

Xanthophyll

Cooling agents

Trade names: Frescolat, Type ML
Witarix® 125

Cooling agents *(cont'd.)*

Chemicals: Carbon dioxide
Dichlorodifluoromethane
Menthyl lactate
Perfluorohexane
Trichlorotrifluoroethane

Curing agents • Cure accelerators • Pickling agents

Trade names: Alomine-Mix; Arovas
Citraxine; Culinox® 999 Chemical Grade Salt; Culinox® 999® Food Grade Salt; Curafos® 22-4; Curavis® 150; Curavis® 350
Hamine-ST
Morton® Flour Salt
Neo-Cebitate®
Packers Powd; Purex® All Purpose Salt
Rocarna
Sorbo®
TFC Purex® Salt

Chemicals: Acetic acid
Calcium chloride; Citric acid
Disodium citrate
Gluconolactone
Lactic acid
Nitrites
Potassium nitrate
Sodium ascorbate; Sodium carbonate; Sodium chloride; Sodium citrate; Sodium erythorbate; Sodium nitrate; Sorbitol
Trisodium citrate

Defoamers • Antifoamers • Foam control agents

Trade names: AF 10 FG; AF 30 FG; AF 70; AF 72; AF 75; AF 100 FG; AF 8805 FG; AF 8810 FG; AF 8820 FG; AF 8830 FG; AF 9000; AF 9020; Alcolec® BS; Alcolec® S; Alcolec® SFG; Aldo® MLD FG; Aldo® MOD FG; Aldo® MO FG; Aldo® MS-20 FG; Aldosperse® MO-50 FG; Antispumin ZU; Atlas 3000; Atmos® 300; Atmos® 300 K
Berol 374; Britol® 6NF; Britol® 7NF; Britol® 9NF; Britol® 20USP; Britol® 35USP; Britol® 50USP; Bubble Breaker® 3009-F
Capmul® GMO; Carbowax® Sentry® PEG 300; Carbowax® Sentry® PEG 400; Carbowax® Sentry® PEG 540 Blend; Carbowax® Sentry® PEG 600; Carbowax® Sentry® PEG 900; Carbowax® Sentry® PEG 1000; Carbowax® Sentry® PEG 1450; Carbowax® Sentry® PEG 3350; Carbowax® Sentry® PEG 4600; Carbowax® Sentry® PEG 8000; Cetodan® 90-50; CNC Antifoam 30-FG; Colorin 102, 104, 202; Contraspum 210; Crill 1; Crill 2; Crill 4
Degressal® SD 22; Dehysan Z 4904; Dehysan Z 7225; Dow EP530; Dow Corning® 200 Fluid, Food Grade; Dow Corning® 1920 Powdered Antifoam; Dow Corning® Antifoam 1500; Dow Corning® Antifoam 1510-US; Dow Corning® Antifoam 1520-US; Dow Corning® Antifoam A Compd., Food Grade; Dow Corning® Antifoam AF Emulsion; Dow Corning® Antifoam C Emulsion; Dow Corning® Antifoam FG-10; Drewmulse® D-4661; Drewmulse® POE-SMO; Drewplus® L-523; Drewplus® L-722; Drewplus® L-768; Drewplus® L-790; Drewplus® L-813; Drewplus® L-833; Drewplus® Y-250; Drewpol® 3-1-OK; Drewpol® 8-1-OK; Drewpone® 60K; Drewpone® 80K; Drewsorb® 60K; Dur-Em® 300; Dur-Em® 300K; Durfax® 80
Emargol® KL; Excel 300
Fancol OA-95; Foamaster FGA; Foamaster FLD; Foam Blast 5, 7; Foam Blast 10; Foam Blast 10K; Foam Blast 100; Foam Blast 100 Kosher; Foam Blast 102; Foam Blast 150; Foam Blast 150 Kosher; Foam Blast SPD; Foamkill® 8BA; Foamkill® 8G; Foamkill® 30 Series; Foamkill® 80J Series; Foamkill® 618 Series; Foamkill® 634 Series; Foamkill® 634B-HP; Foamkill®

Defoamers *(cont'd.)*

634C; Foamkill® 634D-HP; Foamkill® 634F-HP; Foamkill® 639J-F; Foamkill® 644 Series; Foamkill® 652H; Foamkill® 652-HF; Foamkill® 663J; Foamkill® 684 Series; Foamkill® 810F; Foamkill® 830F; Foamkill® 836A; Foamkill® 1001 Series; Foamkill® GCP Series; Foamkill® MS Conc; Foamkill® MSC Series; Foamkill® MSF Conc; Foamkill® RP

Glycon® P-45; Glycon® S-65; Glycon® S-70; Glycon® S-90; Glycon® TP; Glycosperse® O-20 FG; Glycosperse® O-20 KFG; GMR-33

Hodag FD Series; Hodag Antifoam CO-350; Hodag Antifoam F-1; Hodag Antifoam F-2; Hodag Antifoam FD-62; Hodag Antifoam FD-82; Hystrene® 5016 NF FG; Hystrene® 7018 FG; Hystrene® 8718 FG; Hystrene® 9718 NF FG

Iconol NP-7; Industrene® 143; Industrene® 205; Industrene® 225; Industrene® 226 FG; Industrene® 365; Industrene® 4518; Industrene® 5016 NF FG; Industrene® 7018 FG; Industrene® 8718 FG; Industrene® B

Kemester® 2050; Kemester® 4516; Kureton 200

Lamesorb® SML; Lamesorb® SML-20; Lamesorb® SMO; Lamesorb® SMO-20; Lamesorb® SMP; Lamesorb® SMP-20; Lamesorb® SMS; Lamesorb® SMS-20; Lamesorb® STS; Lamesorb® STS-20; Lanquell 206, 217; Lipo GMS; Lipo GMS 410; Liposorb O-20; Liposorb S-20; Liposorb TS-20

Masil® SF 5; Masil® SF 10; Masil® SF 20; Masil® SF 50; Masil® SF 100; Masil® SF 200; Masil® SF 350; Masil® SF 350 FG; Masil® SF 500; Masil® SF 1000; Masil® SF 5000; Masil® SF 10,000; Masil® SF 12,500; Masil® SF 30,000; Masil® SF 60,000; Masil® SF 100,000; Masil® SF 300,000; Masil® SF 500,000; Masil® SF 600,000; Masil® SF 1,000,000; Mazol® 300 K; Mazol® GMO; Mazol® GMO K; Mazol® GMS-K; Mazol® PGMSK; Mazu® 10 P Mod 11; Mazu® 22; Mazu® 87-C; Mazu® 150; Mazu® 201; Mazu® 201 A; Mazu® 201 B; Mazu® 201 PM; Mazu® 204; Mazu® 208; Mazu® 280; Mazu® 285; Mazu® 286; Mazu® 287; Mazu® 288; Mazu® 289; Mazu® 307; Mazu® 309; Mazu® 322 A; Mazu® 330 Mod 2; Mazu® 352; Mazu® DF 100S; Mazu® DF 110S; Mazu® DF 130SAV; Mazu® DF 200S; Mazu® DF 200SP; Mazu® DF 210S; Mazu® DF 220S; Mazu® DF 230SP; Monolan® OM Range; Monolan® PK; Monolan® PL; Myrj® 52; Myrj® 52S; Myverol® 18-92; Myverol® 18-99

Nalco® 131; Nalco® 1090

Patco 305 K; Patco 306 K; Patco 307 K; Patco 309 K; Patco 310 K; Patco 315 K; Patco 319 K; Patco 333 K; Patco 337 K; Patco 338 K; Patco 501 K; Patco 502 K; Patco 555 K; Patco 801 K; Patcote® 305; Patcote® 306; Patcote® 307; Patcote® 308; Patcote® 309; Patcote® 310; Patcote® 311; Patcote® 315; Patcote® 319; Patcote® 323; Patcote® 333; Patcote® 337; Patcote® 337K; Patcote® 338K; Patcote® 501K; Patcote® 502K; Patcote® 555K; Patcote® 801K; PD-23; PD-25; Pluracol® W170; Pluracol® W260; Pluracol® W660; Pluracol® W2000; Pluracol® W3520N; Pluracol® W3520N-RL; Pluracol® W5100N; Pluracol® WD1400; Plurafac® LF 1300; Pluronic® PE 6100; Pluronic® PE 6200; Pluronic® PE 6400; Pluronic® PE 8100; Pluronic® RPE 2520; Polyaldo® DGDO KFG; Polycon S60 K; Polycon T60 K; Polycon T80 K; Pronal ST-1

Quso® WR55-FG

Rudol®; Ryoto Sugar Ester S-170; Ryoto Sugar Ester S-270; Ryoto Sugar Ester S-370

Santone® 10-10-O; Serdas GLN; SF18-350; Silicone AF-10 FG; Silicone AF-100 FG; Silicone AF-30 FG; Sipernat® 50S; Sipernat® 50S; S-Maz® 20; S-Maz® 60K; S-Maz® 80K; S-Maz® 85K; S-Maz® 95; Sorgen 30; Sorgen 40; Sorgen 50; Sorgen 90; Sorgen S-30-H; Sorgen S-40-H; Sorgen TW20; Sorgen TW60; Sorgen TW80; Span® 60; Synperonic LF/RA43

Tegomuls® B; Tegomuls® O; Tegomuls® O Spezial; T-Maz® 65K; T-Maz® 80K; Trans-10, -10K; Trans-20, -20K; Trans-25, -25K; Trans-30, -30K; Trans-100; Trans-175, -176; Trans-177S, -177S K; Trans-179; Trans-220; Trans-224; Trans-225S; Trans-1030 K; Trans-134S K; Trans-FG2; Tween® 60; Tween® 65; Tween® 65K; Tween® 80

Wacker Silicone Antifoam Emulsion SE 9; Witafrol® 7420; Witafrol® 7440; Witafrol® 7456; Witafrol® 7480 N; Witafrol® 7490; Witafrol® 7491; Witafrol® 7497 N

Chemicals: Acetylated hydrogenated soybean oil glyceride; Aluminum stearate
BHA
Calcium stearate; Canola oil glyceride; Capric acid; Caprylic acid; Caprylic/capric acid; Caprylic/capric glycerides
Dimethicone
Formaldehyde
Glyceryl mono/dioleate; Glyceryl oleate
Hydrocarbon solvent; Hydrogenated stearic acid; Hydrogenated tallow; Hydrogenated tallow alcohol; Hydroxylated lecithin

Defoamers *(cont'd.)*

Isopropyl alcohol
Lauric acid; Light petroleum hydrocarbons, odorless
Magnesium stearate; Methyl eicosenate; Methyl stearate; Mineral oil; Myristic acid
Oleic acid; Oleyl alcohol; Oxystearin
Palmitic acid; PEG-4; PEG-9; PEG-8 dioleate; PEG-12 dioleate; PEG-6 oleate; PEG-12 ricinoleate; PEG-40 stearate; Petrolatum; Petroleum distillates; Petroleum wax; Poloxamer 105; Poloxamer 108; Poloxamer 122; Poloxamer 123; Poloxamer 124; Poloxamer 181; Poloxamer 182; Poloxamer 183; Poloxamer 184; Poloxamer 185; Poloxamer 188; Poloxamer 212; Poloxamer 215; Poloxamer 217; Poloxamer 231; Poloxamer 234; Poloxamer 235; Poloxamer 237; Poloxamer 238; Poloxamer 282; Poloxamer 284; Poloxamer 288; Poloxamer 331; Poloxamer 333; Poloxamer 334; Poloxamer 335; Poloxamer 338; Poloxamer 401; Poloxamer 402; Poloxamer 403; Poloxamer 407; Polypropylene glycol; Polysorbate 40; Polysorbate 60; Polysorbate 65; Polysorbate 80; Potassium persulfate; Potassium stearate; PPG-5-buteth-7; PPG-12-buteth-16; PPG-20-buteth-30; PPG-28-buteth-35; PPG-33-buteth-45; Propylene glycol alginate; Propylene glycol dicaprylate/dicaprate; Propylene glycol laurate; Propylene glycol oleate; Propylene glycol stearate
Silica; Silica aerogel; Silicone; Silicone emulsions and compounds; Sodium decylbenzene sulfonate; Sorbitan laurate; Sorbitan stearate; Soy acid; Stearic acid; Sulfated tallow; Synthetic wax
Tallow acid

Dough conditioners • Dough strengtheners

Trade names: Adapt; Admul Datem; Admul SSL 2004; Aeromix Baking Powd; Aerophos G; Albrite Monoammonium Phosphate Food Grade; Alcolec® 30-A; Alcolec® HWS; Alcolec® SFG; Aldo® MSD FG; Aldo® MSD KFG; Aldosperse® 40/60; Aldosperse® 40/60 FG; Aldosperse® MS-20 FG; Aldosperse® O-20; Aldosperse® TS-20; Aldosperse® TS-40; Amfal-46; Amidan; Ascorbo-120; Ascorbo-160-2 Powd; Ascorbo-C Tablets; Atlas 90K; Atlas SSL
B-45; B-50; Bagel Base; Basic Natural™; Beta-Tabs; BFP 100; Blendmax 322; Blendmax 322D; Blitz® Danish Conc; Bro 3D 6451; Bro Action 6467; Bromette
Caln's PDC (Pizza Dough Conditioner); Callb; Capmul® EMG; Cap-Shure® AS-125-50; Carbrea® Tabs; CDC-10A; CDC-10A (NB); CDC-50; CDC-79; CDC-79DZ; CDC-2001; CDC-2002; CDC-2500; CDC-3000; C-D Stabilizer; Centrobake® 100L; Centrolex® C; Centrolex® D; Centrolex® P; Centrolex® R; CFI-10
Do Crest Gold; Do Crest Gold Plus; Do Sure; Drewpone® 65K; Drewpone® 80K; Drize-P Powd; Durfax® EOM; Durfax® EOM K
Edicol®; Elasdo Power 70; Emargol® KL; EMG 20; Emulsilac S; Emulsilac S K
FD-6; FD-8
Hodag GMO; Hodag POE (20) GMS
IDX; IDX-20 NB; Imperial FD; Improved Paniplus M
Lactylate Hydrate; Lamegin® DW 8000-Range; Lamegin® DW 8000 HW; Lamegin® DW 8200 VA/HW; Lamegin® DW 9000 HW; Lamemul® K 1000-Range; Lamemul® K 2000-Range; Lamemul® K 3000-Range; Lamesorb® SML; Lamesorb® SML-20; Lamesorb® SMO; Lamesorb® SMO-20; Lamesorb® SMP; Lamesorb® SMP-20; Lamesorb® SMS; Lamesorb® SMS-20; Lamesorb® STS; Lamesorb® STS-20; Lecigran™ 5750; Lecithin L 2000-Range; Lecithin L 3000-Range
Mazol® 80 MGK; Miracle Pie Dough Base; MLO; Morton® Flour Salt; Muffit; Myvatex® Do Control; Myvatex® Do Control K; Myvatex® Mighty Soft®; Myverol® SMG; Myverol® SMG® VK
N'Hance CM; N'Hance (Phase I) Powd; N'Hance (Phase II) Powd; N'Hance (Phase II) Tablets; N'Hance SD; Norfox® Sorbo T-60; Nutrilife®; Nutrisoy 7B
Panaid; Panicrust LC K; Panodan 120 Series; Panodan 140 Series; Panodan 150; Panodan 150 Kosher; Panodan 205; Panodan FDP; Panodan FDP Kosher; Panodan SD; Pationic® 920; Pationic® 925; Pationic® 930; Pationic® 940; Pie Dough-N-Answer; Pita Dough Culture 6448; Pita Soft; Poly E; Prefera® CSL; Prefera® NSL; Proflex; Py-ran®
Reduce®-150; Regular Paniplus; Ryoto Sugar Ester ER-190; Ryoto Sugar Ester ER-290; Ryoto Sugar Ester P-170; Ryoto Sugar Ester P-1570; Ryoto Sugar Ester P-1570S; Ryoto Sugar Ester P-1670; Ryoto Sugar Ester S-1170; Ryoto Sugar Ester S-1170S; Ryoto Sugar Ester S-

Dough conditioners *(cont'd.)*

1570; Ryoto Sugar Ester S-1670
Stadex® 9; Stadex® 60K; Stadex® 90
Tally® 100; Tally® 100 Plus; Tandem 11H; Tandem 930; Terra Alba 114836; Thermolec 68; T-Maz® 60K; Tolerance; Trem-Tabs; Tween® 60K; Tween® 60 VS; Type B Dough Conditioner
Ultra Flex
Vita Choice; Vitalex A; Vitalex P
WaTox-20 P Powd; WaTox-20 Tablets; WaTox 60 Powd; Wheat-Eze
Xpando; Xpando 70
Yelkin SS

Chemicals: Acetone peroxides; Ammonium bicarbonate; Ammonium chloride; Ammonium phosphate; Ammonium sulfate; Azodicarbonamide
Calcium bromate; Calcium carbonate; Calcium iodate; Calcium oxide; Calcium peroxide; Calcium phosphate dibasic; Calcium phosphate monobasic monohydrate; Calcium/sodium stearoyl lactylate; Calcium stearoyl lactylate; L-Cysteine; Cysteine hydrochloride; L-Cystine
Mono- and diglycerides of fatty acids
PEG-20 glyceryl stearate; Poloxamer 105; Poloxamer 108; Poloxamer 123; Poloxamer 124; Poloxamer 181; Poloxamer 182; Poloxamer 183; Poloxamer 184; Poloxamer 185; Poloxamer 188; Poloxamer 212; Poloxamer 215; Poloxamer 217; Poloxamer 231; Poloxamer 234; Poloxamer 235; Poloxamer 237; Poloxamer 238; Poloxamer 282; Poloxamer 284; Poloxamer 288; Poloxamer 331; Poloxamer 333; Poloxamer 334; Poloxamer 335; Poloxamer 338; Poloxamer 401; Poloxamer 402; Poloxamer 403; Poloxamer 407; Polysorbate 40; Polysorbate 60; Potassium bromate; Potassium iodate; Propylene glycol
Sodium chloride; Sodium stearoyl lactylate; Sodium stearyl fumarate; Sorbitan laurate; Succinylated monoglycerides; Succinylated palm oil glyceride; Sucrose erucate; Sucrose stearate; Sulfur dioxide
Wheat gluten

Drying agents

Trade names: Hodag PSMO-20
Ice #12
Myvatex® 8-06; Myvatex® 8-16; Myvatex® MSPS
Polycon T80 K
Ryoto Sugar Ester O-1570; Ryoto Sugar Ester P-1570; Ryoto Sugar Ester P-1670
Sorbo®
Tandem 100 K; Taterfos®; Tween® 65; Tween® 80

Chemicals: Acetylated hydrogenated coconut glycerides; Acetylated hydrogenated soybean oil glyceride; Agar
Calcium sulfate; Castor oil, dehydrated; Cobalt caprylate; Cobalt linoleate; Cobalt naphthenate; Cobalt tallate
Iron caprylate; Iron linoleate; Iron naphthenate; Iron tallate
Linseed oil
Magnesium carbonate; Magnesium hydroxide; Manganese caprylate; Manganese linoleate; Manganese naphthenate; Manganese tallate
Succinic anhydride; Sulfated butyl oleate
Tall oil; Tung oil

Egg replacements

Trade names: Alcolec® 140; Alcolec® EYR SM
Dur-Lec® P
Hentex-10; Hentex-35; Hentex-45; Hentex-81; Hentex-505
Lecigran™ 5750
Nutrisoy 220T
Polypro 5000®
Ultra-Bake® 3100, 3200; Ultra-Bake® 3100K, 3200K

Emulsifiers

Trade names: Ablunol S-20; Ablunol S-40; Ablunol S-60; Ablunol S-80; Ablunol S-85; Ablunol T-20; Ablunol T-40; Ablunol T-60; Ablunol T-80; Acconon 300-MO; Acidan; Actiflo® 68 SB; Actiflo® 68 UB; Actiflo® 70 SB; Actiflo® 70 UB; Admul 1405; Admul Datem; Admul Emulsponge; Admul GLP; Admul GLS; Admul MG 4103; Admul MG 4123; Admul MG 4143; Admul MG 4163; Admul MG 4203; Admul MG 4223; Admul MG 4304; Admul MG 4404; Admul MG 4904; Admul MG 6103; Admul MG 6404; Admul MG 6504; Admul PGE 1405; Admul PGE 1411; Admul PGMP; Admul PGMS; Admul SSL 2003; Admul SSL 2004; Admul WOL 1405; Aerate Cake Emulsifier; Agar Agar NF Flake #1; Albrite MSP Food Grade; Albrite SAPP Food Grade; Alcolec® 30-A; Alcolec® 140; Alcolec® 495; Alcolec® BS; Alcolec® Extra A; Alcolec® F-100; Alcolec® FF-100; Alcolec® Granules; Alcolec® PG; Alcolec® S; Alcolec® XTRA-A; Alcolec® Z-3; Aldo® DC; Aldo® GLP FG; Aldo® HMO FG; Aldo® HMS FG; Aldo® HMS KFG; Aldo® LP FG; Aldo® LP KFG; Aldo® ML; Aldo® MLD FG; Aldo® MOD; Aldo® MOD FG; Aldo® MO FG; Aldo® MS-20 FG; Aldo® MSD FG; Aldo® MS FG; Aldo® MSLG FG; Aldo® MS LG KFG; Aldo® PGHMS KFG; Aldo® PME; Aldo® TC; Aldosperse® 40/60; Aldosperse® 40/60 FG; Aldosperse® 40/60 KFG; Aldosperse® 712 FG; Aldosperse® MO-50 FG; Aldosperse® MS-20 FG; Aldosperse® MS-20 KFG; Aldosperse® O-20; Aldosperse® O-20 FG; Aldosperse® O-20 KFG; Aldosperse® TS-20; Aldosperse® TS-20 FG; Aldosperse® TS-20 KFG; Aldosperse® TS-40; Aldosperse® TS-40 FG; Aldosperse® TS-40 KFG; Alkamuls® PSTO-20; Alkamuls® PSTS-20; Alkamuls® S-6; Alkamuls® SMS; Alkamuls® STS; Alkaquat® DMB-451-50, DMB-451-80; Alpha 500; Alphadim® 90NLK; Alphadim® 90VCK; Aluminum Stearate EA; Amidan; Amidan 500 Series; Amidan 600 Series; Amidan ES; Amidan ES Kosher; Amidan SDM-T; Amidan XTR-A; Amisol™ 329; Amisol™ MS-10; Amisol™ MS-12 BA; Ammonium Alginate Type S; Amoloid HV; Amoloid LV; Ardex D, D Dispersible; Arlacel® 186; Artodan CF 40; Artodan SP 55 Kosher; Asol; Atlas 70K; Atlas 110K; Atlas 800; Atlas 1400K; Atlas 1500; Atlas 2200H; Atlas 3000; Atlas 5000K; Atlas 5520; Atlas A; Atlas B-60; Atlas SSL; Atmos® 150; Atmos® 150 K; Atmos® 300; Atmos® 300 K; Atmos® 378 K; Atmos® 659 K; Atmos® 729; Atmos® 758; Atmos® 758 K; Atmos® 1069; Atmos® 2462; Atmos® 7515 K; Atmul® 27 K; Atmul® 80; Atmul® 82; Atmul® 84; Atmul® 84 K; Atmul® 86 K; Atmul® 122; Atmul® 124; Atmul® 500; Atmul® 601 K; Atmul® 651 K; Atmul® 695; Atmul® 700 H; Atmul® 000; Atmul® 900K, Atmul® 918 K, Atmul® 1003 K, Atmul® 2822 K; Atsurf 594; Axol® C 62; Axol® C 63; Axol® E 41; Axol® E 61; Axol® L 61, L62; Axol® L 626

Dealite™; DFP 30, DFP 04, BFP 64A, BFP 64K; BFP 65; BFP 65 A; BFP 65 C; BFP 65 K; BFP 74; BFP 74A; BFP 74 K; BFP 75; BFP 75A; BFP 75K; BFP 100; BFP L Mono; Bindox-HV-051; Bindox-LV-050; Blendmax 322; Blendmax 322D; Brem

Calsoft L-60; Canamulse 55; Canamulse 70; Canamulse 90K; Canamulse 100; Canamulse 110K; Canamulse 150K; Canamulse 155; Canasperse SBF; Canasperse UBF; Caplube 8350; Capmul® EMG; Capmul® GMO; Capmul® GMS; Capmul® GMVS-K; Capmul® O; Capmul® PGME; Capmul® POE-L; Capmul® POE-O; Capmul® POE-S; Capmul® S; Caprol® 2G4S; Caprol® 3GO; Caprol® 3GS; Caprol® 3GVS; Caprol® 6G2O; Caprol® 6G2S; Caprol® 10G2O; Caprol® 10G4O; Caprol® 10G10O; Caprol® ET; Caprol® JB; Caprol® PGE860; Carsonol® SHS; Cegemett® MZ 490; Cegeprot® Range; Celynol F1; Celynol MSPO-11; Celynol MST-11; Celynol P1M; Celynol TL; Celynol X 8066; Centrocap® 162SS; Centrocap® 162US; Centrocap® 273SS; Centrocap® 273US; Centrol® 2F SB; Centrol® 2F UB; Centrol® 3F SB; Centrol® 3F UB; Centrol® CA; Centrolene® A; Centrolene® S; Centrolex® C; Centrolex® D; Centrolex® F; Centrolex® G; Centrolex® P; Centrolex® R; Centromix® CPS; Centromix® E; Centrophase® C; Centrophase® HR; Centrophase® HR2B; Centrophase® HR2U; Centrophase® HR4B; Centrophase® HR4U; Centrophase® HR6B; Cetodan®; Cetodan® 50-00A; Cetodan® 50-00P Kosher; Cetodan® 70-00A; Cetodan® 70-00P Kosher; Cetodan® 90-40; Cetodan® 90-50; Ches® 500; Chocotop™; Cithrol GDO N/E; Cithrol GDO S/E; Cithrol GDS N/E; Cithrol GDS S/E; Cithrol GML N/E; Cithrol GML S/E; Cithrol GMO N/E; Cithrol GMO S/E; Cithrol GMR N/E; Cithrol GMR S/E; Cithrol GMS Acid Stable; Cithrol GMS N/E; Cithrol GMS S/E; Cithrol PGML N/E; Cithrol PGML S/E; Cithrol PGMO N/E; Cithrol PGMO S/E; Cithrol PGMR N/E; Cithrol PGMR S/E; Cithrol PGMS N/E; Cithrol PGMS S/E; Clearate WDF; Coatingum L; Coconad MT; Coconad PL; Coconad RK; Coconad SK; Colloid 488T; Colloid 602; Colloid 886; Complemix® 50; Complemix® 100; Compritol 888 ATO; Cremodan; Crestawhip; Crestawhip 2; Crestawhip H13; Crestawhip H13A; Crester KZ; Crester L; Crester MG; Crester PR; Crester RA; Crester RB; Crill 1; Crill 2; Crill 3; Crill 35; Crill 41; Crill 43; Crill 45; Crill 50; Crillet 1; Crillet 2; Crillet 3; Crillet 4; Crillet 11; Crillet 31; Crillet 35; Crillet 41; Crillet 45; Crodacreme; Crodascoop; Crodatem L; Crodatem L50; Crodatem T22; Crodatem T25; Crystalsorb B-56

Emulsifiers *(cont'd.)*

Danpro DS; Danpro HV; Danpro S; Danpro S-760; Dariloid®; Dariloid® 100; Dariloid® QH; Datagel; Datamuls® 42; Datamuls® 43; Datamuls® 4720; Datamuls® 4820; Datamuls® 4820 U; Datem esters; Detergent CR; Dimodan BP-T Kosher; Dimodan CP; Dimodan CP Kosher; Dimodan LS Kosher; Dimodan O Kosher; Dimodan P; Dimodan PM; Dimodan PM 300; Dimodan PV; Dimodan PV 300 Kosher; Dimodan PV Kosher; Dimodan PVP Kosher; Dimodan S; Dimodan TH; Dimul DDM K; Dimul S; Dimul S K; DK-Ester; DK-Ester; DK-Ester; DK-Ester; Drewlate 30; Drewlene 10; Drewmulse® 3-1-O; Drewmulse® 3-1-S; Drewmulse® 3-1-SH; Drewmulse® 6-2-S; Drewmulse® 10-4-O; Drewmulse® 10-8-O; Drewmulse® 10-10-S; Drewmulse® 10K; Drewmulse® 70; Drewmulse® 75; Drewmulse® 200; Drewmulse® 200K; Drewmulse® 365; Drewmulse® 700K; Drewmulse® 900K; Drewmulse® 8731-S; Drewmulse® M; Drewmulse® PNO; Drewmulse® POE-SMO; Drewmulse® SMO; Drewmulse® SMS; Drewpol® 3-1-O; Drewpol® 3-1-OK; Drewpol® 3-1-SHK; Drewpol® 3-1-SK; Drewpol® 6-1-O; Drewpol® 6-1-OK; Drewpol® 6-2-O; Drewpol® 6-2-OK; Drewpol® 6-2-SK; Drewpol® 8-1-OK; Drewpol® 10-1-CC; Drewpol® 10-1-CCK; Drewpol® 10-2-OK; Drewpol® 10-4-OK; Drewpol® 10-6-OK; Drewpol® 10-8-OK; Drewpol® 10-10-O; Drewpol® 10-10-OK; Drewpone® 60K; Drewpone® 65K; Drewpone® 80K; Drewsorb® 60K; Drewsorb® 65K; Drewsorb® 80K; Dricoid® 200; Dricoid® 280; Dricoid® KB; Dricoid® KBC; Dur-Em® 114; Dur-Em® 114K; Dur-Em® 117; Dur-Em® 117K; Dur-Em® 204; Dur-Em® 204K; Dur-Em® 207-E; Dur-Em® 207K; Dur-Em® 300; Durfax® 20; Durfax® 60; Durfax® 60K; Durfax® 65; Durfax® 65K; Durfax® 80; Durfax® 80K; Durfax® EOM; Durfax® EOM K; Durlac® 100W; Durlac® 100WK; Dur-Lec® UB; Dur-Lo®; Durpeg® 400MO; Durpro® 107; Durtan® 60; Durtan® 60K; Dynacet® 212

EC-25®; Edifas; Elasdo +; Emalex DISG-2; Emalex DISG-3; Emalex DISG-5; Emalex DSG-2; Emalex DSG-3; Emalex DSG-5; Emalex MSG-2; Emalex MSG-2MA; Emalex MSG-2MB; Emalex MSG-2ME; Emalex MSG-2ML; Emalex TISG-2; Emargol® KL; Emargol® L; EMG 20; Emphos™ D70-31; Emphos™ F27-85; Empilan® GMS LSE32; Empilan® GMS LSE40; Empilan® GMS LSE80; Empilan® GMS MSE40; Empilan® GMS NSE32; Empilan® GMS NSE40; Empilan® GMS NSE90; Empilan® GMS SE40; Empilan® GMS SE70; Emplex; Emrite® 6003; Emrite® 6008; Emrite® 6105; Emrite® 6120; Emrite® 6125; Emulbesto; Emuldan; Emuldan HA 40; Emuldan HA 52; Emuldan HV 40 Kosher; Emuldan HV 52 Kosher; Emuldan HVF 52 K; Emulfluid™; Emulfluid® AS; Emulfluid® E; Emulgator 484; Emulgel E-21; Emulgum BV; Emulpur™ N; Emulpur™ N P-1; Emulsifier D-1; Emulsilac S; Emulsilac S K; Emulthin M-35; Ervol®; Estric™; Excel 122; Excel 124; Excel 150; Excel 200; Excel 84; Excel O-95F; Excel O-95N; Excel O-95R; Excel P-40S; Excel T-95; Excel VS-95; Extramalt 33

Famodan MS Kosher; Freedom X-PGA

G-695; G-991; G-2183; G-2185; Gatodan 415; Glass H®; Glycomul® L; Glycomul® MA; Glycomul® S FG; Glycomul® S KFG; Glycomul® TAO; Glycomul® TS KFG; Glycosperse® HTO-40; Glycosperse® L-20; Glycosperse® O-5; Glycosperse® O-20; Glycosperse® O-20 KFG; Glycosperse® O-20 Veg; Glycosperse® O-20X; Glycosperse® P-20; Glycosperse® S-20; Glycosperse® S-20 FG; Glycosperse® S-20 KFG; Glycosperse® TO-20; Glycosperse® TS-20; Glycosperse® TS-20 FG, TS-20 KFG; Glycosperse® TS-20 KFG; GMR-33; GMS 52; GMS 90; GMS 90 Dbl. Strength; GMS 400; GMS 400V; GMS 402V; Granular Gum Ghatti #1; Guar Gum HV

Hefti GMS-33; Hefti GMS-33-SES; Hefti GMS-99; Hefti GMS-233; Hefti GMS-333; Hefti MO-55-F; Hefti PMS-33; Hercules® AR 160; Hetsorb O-5; Hexaphos®; Hodag GDO-V; Hodag GML; Hodag GMO-D; Hodag GMR; Hodag GMR-D; Hodag GMS; Hodag GMSH; Hodag GTO; Hodag PGO-1010; Hodag PGS; Hodag PGS-1010; Hodag PGSH; Hodag POE (20) GMS; Hodag PSML-20; Hodag PSML-80; Hodag PSMO-5; Hodag PSMO-20; Hodag PSMP-20; Hodag PSMS-20; Hodag PSTO-20; Hodag PSTS-20; Hodag S-35; Hodag SML; Hodag SMO; Hodag SMP; Hodag SMS; Hodag STO; Hodag STS; Hodag SVO-9; Hodag SVO-629; Hodag SVO-1047; Hodag SVS-18; Homodan RD; Hydradan D 42; Hydradan S 21; Hyonic NP-100; Hyonic NP-110; Hyonic NP-120

Ice # 2; Ice #12; Imwitor® 175; Imwitor® 191; Imwitor® 325; Imwitor® 333; Imwitor® 369; Imwitor® 370; Imwitor® 371; Imwitor® 372; Imwitor® 373; Imwitor® 375; Imwitor® 440; Imwitor® 460; Imwitor® 490; Imwitor® 595; Imwitor® 845; Imwitor® 895; Imwitor® 900; Imwitor® 910; Imwitor® 940 K; Imwitor® 945; Imwitor® 960; Imwitor® 1330; Imwitor® 1339; Imwitor® 1940; Imwitor® 2020; Imwitor® 2320; IPSO-FC; IPSO-MR Dispersible

K2B387; K8B243; K8B249; Karaya Gum #1 FCC; Karaya Gum #1 FCC; Karaya Gum #1 FCC; Karaya Gum #1 FCC; Katch® Fish Phosphate; Kelacid®; Kelco® HV; Kelco® LV; Kelcoloid® D; Kelcoloid® DH; Kelcoloid® DO; Kelcoloid® DSF; Kelcoloid® HVF; Kelcoloid® LVF;

Emulsifiers *(cont'd.)*

Kelcoloid® O; Kelcoloid® S; Kelcosol®; Kelgin® F; Kelgin® HV; Kelgin® LV; Kelgin® MV; Kelgin® QL; Kelgin® XL; Kelmar®; Kelmar® Improved; Kelset®; Keltone®; Keltose®; Kelvis®; Kessco® Glycerol Distearate 386F; Kessco® GMC-8; Kirnol® Range; KOB348
Lactodan B 30; Lactodan LW; Lactodan P 22; Lactodan P 22 Kosher; Lamefrost® ES 216 G; Lamefrost® ES 251 G; Lamefrost® ES 315; Lamefrost® ES 375; Lamefrost® ES 379; Lamefrost® ES 424; Lamegin® CSL; Lamegin® DW 8000-Range; Lamegin® DW Range; Lamegin® DWF, DWH, DWP; Lamegin® EE Range; Lamegin® EE 50; Lamegin® EE 70; Lamegin® EE 100; Lamegin® GLP Range; Lamegin® GLP 10, 20; Lamegin® NSL; Lamegin® ZE 30, 60; Lamegin® ZE Range; Lamemul® Range; Lamemul® K 2000-Range; Lamemul® K 3000-Range; Lamephos® Range; Lamesorb® Range; Lamesorb® SML; Lamesorb® SML-20; Lamesorb® SMO; Lamesorb® SMO-20; Lamesorb® SMP; Lamesorb® SMP-20; Lamesorb® SMS; Lamesorb® SMS-20; Lamesorb® STS; Lamesorb® STS-20; Lecidan; Lecigran™ 5750; Lecigran™ 6750; Lecigran™ A; Lecigran™ C; Lecigran™ F; Lecigran™ M; Lecigran™ Super A; Lecigran™ T; Lecimulthin; Leciprime™; Lecithin L 1000-Range; Lecithin L 2000-Range; Lecithin L 3000-Range; Lecithin L 4000-Range; Lecitreme™; Lipal 4-1S; Lipal 70; Lipal 75; Lipal GMS; Lipal M; Lipal PGMS; Lipal PNO; Lipo GMS; Lipo GMS 410; Lipo GMS 450; Lipo GMS 470; Lipodan OM 30; Liposorb L-20; Liposorb S; Liposorb S-20; Liposorb TS; Liposorb TS-20; Lonzest® SMO-20; Lonzest® SMP-20; Lonzest® SMS-20; Lonzest® STO-20; Lonzest® STS-20
Magathin; Maisine; Maltrin® M700; Maltrin® QD M500; Manucol Ester E/PL; Manucol Ester E/RK; Manucol Ester M; Mayodan; Mayonat DF; Mayonat PS; Mayonat V/100; Mazol® 80 MGK; Mazol® 159; Mazol® 300 K; Mazol® GMO K; Mazol® GMS; Mazol® GMS-90; Mazol® GMS-K; Mazol® PGMS; Mazol® PGMSK; Mazol® PGO-31 K; Mazol® PGO-104; M-C-Thin®; M-C-Thin® AF-1 Type DB; M-C-Thin® AF-1 Type SB; M-C-Thin® AF-1 Type UB; M-C-Thin® FWD; M-C-Thin® HL; Merecol® FA; Merecol® FAL; Methocel® A4C Premium; Methocel® A4M Premium; Methocel® A15C Premium; Methocel® A15LV Premium; Methocel® E4M; Methocel® E5 Food Grade; Methocel® E15 Food Grade; Methocel® F4M Premium; Methocel® F50LV Premium; Methocel® K35 Premium; Microlube System 999R; Mira-Cap®; Miranol® J2M Conc; Monomax AH90 B; Monomax VH90 B; Monomuls® Range; Monomuls® 60-10; Monomuls® 60-15; Monomuls® 60-20; Monomuls® 60-25; Monomuls® 60-25/2; Monomuls® 60-25/5; Monomuls® 60-30; Monomuls® 60-35; Monomuls® 60-40; Monomuls® 60-45; Monomuls® 90-10; Monomuls® 90-15; Monomuls® 90-20; Monomuls® 90-25; Monomuls® 90-25/2; Monomuls® 90-25/5; Monomuls® 90-30; Monomuls® 90-35; Monomuls® 90-40; Monomuls® 90-45; Monomuls® 90-O18; Myrj® 52; Myvacet® 5-07; Myvacet® 5-07K; Myvacet® 7-00; Myvacet® 7-07; Myvacet® 9-08; Myvacet® 9-08K; Myvacet® 9-40; Myvacet® 9-45; Myvacet® 9-45K; Myvaplex® 600K; Myvaplex® 600P; Myvatem® 06K; Myvatem® 30; Myvatem® 35K; Myvatem® 92K; Myvatex® 3-50; Myvatex® 3-50K; Myvatex® 7-85; Myvatex® 7-85K; Myvatex® 8-06; Myvatex® 8-06K; Myvatex® 8-16; Myvatex® 8-16K; Myvatex® 8-20; Myvatex® 8-20E; Myvatex® 25-07K; Myvatex® 40-06S; Myvatex® 40-06S K; Myvatex® 90-10K; Myvatex® Do Control; Myvatex® Liquid Lite® (K); Myvatex® Monoset®; Myvatex® Monoset® K; Myvatex® MSPS; Myvatex® Peanut Butter Stabilizer; Myvatex® SSH; Myvatex® Super DO; Myvatex® Texture Lite®; Myvatex® Texture Lite® K; Myverol® 18-00; Myverol® 18-04; Myverol® 18-04K; Myverol® 18-06; Myverol® 18-06K; Myverol® 18-07; Myverol® 18-07K; Myverol® 18-30; Myverol® 18-35; Myverol® 18-35K; Myverol® 18-40; Myverol® 18-50; Myverol® 18-50K; Myverol® 18-85; Myverol® 18-85K; Myverol® 18-92; Myverol® 18-92K; Myverol® 18-98; Myverol® 18-99; Myverol® 18-99K; Myverol® P-06; Myverol® P-06K; Myverol® SMG; Myverol® SMG® VK
New Fresh Conc; Nikkol DGMS; Nikkol MGO; Nikkol MGS-150; Nikkol MGS-A; Nikkol MGS-ASE; Nikkol MGS-B; Nikkol MGS-BSE; Nikkol MGS-DEX; Nikkol MGS-F20; Nikkol MGS-F40; Nikkol MGS-F50; Nikkol MGS-F50SE; Nikkol MGS-F75; Nikkol MGS-TG; Nikkol MGS-TGL; Nikkol SL-10; Nikkol SO-10; Nikkol SO-15; Nikkol SP-10; Nikkol SS-15; Nikkol TO-10; Nikkol TS-10; Nikkol Decaglyn 1-IS; Nikkol Decaglyn 1-L; Nikkol Decaglyn 1-LN; Nikkol Decaglyn 1-M; Nikkol Decaglyn 1-O; Nikkol Decaglyn 1-S; Nikkol Decaglyn 2-O; Nikkol Decaglyn 2-S; Nikkol Tetraglyn 1-O; Nikkol Tetraglyn 1-S; Nikkol Tetraglyn 3-S; Nikkol Tetraglyn 5-O; Nikkol Tetraglyn 5-S; Nissan Nonion CP-08R; Nissan Nonion DN-202; Nissan Nonion DN-203; Nissan Nonion DN-209; Nissan Nonion LP-20R, LP-20RS; Nissan Nonion LT-221; Nissan Nonion MP-30R; Nissan Nonion OP-80R; Nissan Nonion OP-83RAT; Nissan Nonion OP-85R; Nissan Nonion OT-221; Nissan Nonion PP-40R; Nissan Nonion PT-221; Nissan Nonion SP-60R; Nissan Nonion ST-221; Norfox® GMS-FG; Norfox® Sorbo S-60FG; Norfox® Sorbo S-80; Norfox® Sorbo T-60; Norfox® Sorbo T-80; Nurulat; Nurupan™; Nutrifos® 088; Nutrifos® B-75; Nutrifos® H-30; Nutrifos® L-50; Nutrifos® Powd; Nutrifos® SK; Nutriloid®

Emulsifiers *(cont'd.)*

Arabic; Nutrisoft® 55

n-Octenyl Succinic Anhydride; Olicine; Ovothin™; Ovothin™ 120; Ovothin™ 160

Panalite 40; Panalite 40 HVK; Panalite 40 SVK, 50 SA, 50 SVK; Panalite 50; Panalite 50 HVK; Panalite 90D; Panalite 100 K; Panalite EOM-K; Panalite MP Hydrate; Panalite MPH, MPS; Panatex, Panatex HM; Paniplex CK; Paniplex SK; Panodan 120 Series; Panodan 140 Series; Panodan 150; Panodan 205; Panodan 235; Panodan FDP; Panodan FDP Kosher; Panodan SD; Pantex; Pationic® 920; Pationic® 925; PC-35; PGMS 70; Pisane®; Plurol Stearique WL 1009; Polyaldo® 2O10 KFG; Polyaldo® 2P10 KFG; Polyaldo® 10-1-O KFG; Polyaldo® DGDO; Polyaldo® HGDS; Polyaldo® HGDS KFG; Polyaldo® TGMS; Polyaldo® TGMS KFG; Polycon S60 K; Polycon S80 K; Polycon T60 K; Polycon T80 K; Polysorbate 60; Poly-Tergent® P-17A; Poly-Tergent® P-17B; Poly-Tergent® P-17BLF; Poly-Tergent® P-17BX; Poly-Tergent® P-17D; Poly-Tergent® P-32D; Powdered Agar Agar NF M-100 (Gracilaria); Powdered Agar Agar NF MK-80-B; Powdered Agar Agar NF MK-80 (Bacteriological); Powdered Agar Agar NF S-100; Powdered Agar Agar NF S-100-B; Powdered Agar Agar NF S-150; Powdered Agar Agar NF S-150-B; Powdered Gum Ghatti #1; Powdered Gum Ghatti #2; Powdered Gum Karaya Superfine #1 FCC; Powdered Gum Karaya Superfine XXXX FCC; Powdered Tragacanth Gum Type A/10; Powdered Tragacanth Gum Type E-1; Powdered Tragacanth Gum Type G-3; Powdered Tragacanth Gum Type L; Powdered Tragacanth Gum Type W; Prefera® Range; Prefera® SSL 6000 MB; Premium Powdered Gum Ghatti G-1; Promax® 70; Promax® 70L; Promax® 70LSL; Promax® Plus; Promine® DS; Promine® DS-SL; Promine® HV; Promodan SP; Promodan USV; Promodan USV Kosher; Propylene Glycol Alginate HV; Propylene Glycol Alginate LV FCC; Prote-sorb SML; Prote-sorb SMO; Prote-sorb SMP; Prote-sorb SMS; Prote-sorb STO; Prote-sorb STS; Protex™ 20

Qual Flo™; Qual Flo™ C

Radiamuls® Datem 2001; Recodan; Resinogum DD; Rhodacal® LDS-10; Ryoto Sugar Ester B-370; Ryoto Sugar Ester ER-190; Ryoto Sugar Ester ER-290; Ryoto Sugar Ester L-195; Ryoto Sugar Ester L-595; Ryoto Sugar Ester L-1570; Ryoto Sugar Ester L-1695; Ryoto Sugar Ester LN-195; Ryoto Sugar Ester LWA-1570; Ryoto Sugar Ester M-1695; Ryoto Sugar Ester O-170; Ryoto Sugar Ester O-1570; Ryoto Sugar Ester OWA-1570; Ryoto Sugar Ester P-170; Ryoto Sugar Ester P-1570; Ryoto Sugar Ester P-1570S; Ryoto Sugar Ester P-1670; Ryoto Sugar Ester S-070; Ryoto Sugar Ester S-170; Ryoto Sugar Ester S-170 Ac; Ryoto Sugar Ester S-270; Ryoto Sugar Ester S-370; Ryoto Sugar Ester S-370F; Ryoto Sugar Ester S-570; Ryoto Sugar Ester S-770; Ryoto Sugar Ester S-970; Ryoto Sugar Ester S-1170; Ryoto Sugar Ester S-1170S; Ryoto Sugar Ester S-1570; Ryoto Sugar Ester S-1670; Ryoto Sugar Ester S-1670S; Ry-So

Santone® 3-1-S; Santone® 3-1-SH; Santone® 3-1-S XTR; Santone® 8-1-O; Santone® 10-10-O; Servit® Range; Sherbelizer®; Silicone AF-10 FG; Silicone AF-30 FG; S-Maz® 20; S-Maz® 60K; S-Maz® 60KHS; S-Maz® 80K; S-Maz® 95; Sodaphos®; Sodium Alginate HV NF/FCC; Sodium Alginate LV; Sodium Alginate LVC; Sodium Alginate MV NF/FCC; Sodium Citrate USP, FCC Dihydrate Gran. No. 69976; Sodium Citrate USP, FCC Dihydrate Fine Gran. No. 69975; Soft Touch; Sorbax PML-20; Sorbax PMO-5; Sorbax PMO-20; Sorbax PMP-20; Sorbax PMS-20; Sorbax PTO-20; Sorbax PTS-20; Sorbo®; Sorgen 30; Sorgen 40; Sorgen 50; Sorgen 90; Sorgen S-30-H; Sorgen S-40-H; Sorgen TW20; Sorgen TW60; Sorgen TW80; Soyamin 90; Soyapan™; Span® 20; Span® 40; Span® 60; Span® 60K; Span® 65; Span® 65K; Span® 80; Span® 85; Spongolit® Range; Spray Dried Nigerian Gum Arabic; Spraygum C; Spraygum GD; Stabilizer C; Stabolic™ C; Staform P; Star-Dri® 1005A; Starplex® 90; Sucro Ester 7; Sucro Ester 11; Sucro Ester 15; Superfine Ground Mustard; Superior Mustard; Surfactant AR 150; Swedex AR58P-15AC; Swedex SSL-5AC

Tally® 100 Plus; Tandem 5K; Tandem 8; Tandem 9; Tandem 11H K; Tandem 22 H; Tandem 22 H K; Tandem 100 K; Tandem 552; Tandem 552 K; Tegin® C-62 SE; Tegin® C-63 SE; Tegomuls® 19; Tegomuls® 90; Tegomuls® 90 S; Tegomuls® 4070; Tegomuls® 4100; Tegomuls® 4101; Tegomuls® 6070; Tegomuls® 9050; Tegomuls® 9053 S; Tegomuls® 9102; Tegomuls® 9102 S; Tegomuls® B; Tegomuls® M; Tegomuls® P 411; Tegomuls® SB; Tegomuls® SO; Thixogum S; Thixogum X; Ticaxan® Regular; TIC Pretested® Arabic FT Powd; TIC Pretested® Arabic PH-FT; TIC Pretested® Aragum® 9500; TIC Pretested® Saladizer 250; TIC Pretested® Saladizer PH-250; TIC Pretested® Tragacanth 440; TIC Pretested® Xanthan PH; Tinamul®; T-Maz® 20; T-Maz® 28; T-Maz® 60; T-Maz® 60K; T-Maz® 60KHS; T-Maz® 65; T-Maz® 65K; T-Maz® 80; T-Maz® 80K; T-Maz® 80KLM; T-Maz® 81; T-Maz® 95; Toffix® M; Toffix® P; Toffix® S; Topcithin®; Top Mate Kosher; Tragacanth Flake No. 27; Tragacanth Gum Ribbon No. 1; Triodan 20; Triodan 55; Triodan 55 Kosher; Triodan 55/S6; Triodan R 90; Triton® X-45; Triton® X-305-70%; Triton® X-405-70%; Triton®

Emulsifiers *(cont'd.)*

X-705-70%; Trycol® 5888; Trycol® OAL-23; Trycol® SAL-20; Tween® 20; Tween® 20 SD; Tween® 60; Tween® 60K; Tween® 60 VS; Tween® 61; Tween® 65; Tween® 65K; Tween® 65 VS; Tween® 80; Tween® 80K; Tween® 80 VS; Tween® 81; Tween® 85

Ultra Fresh

Vanade; Vanade Kosher; Vanall; Vanall (K); Vanease; Vanease (K); Vanlite; Versa-Whip® 500, 510, 520; Versa-Whip® 520K, 600K, 620K; Vifcoll CCN-40, CCN-40 Powd; Visco Compound™; Vitinc® Cholesterol NF XVII; Vitrafos®; Vrest™; Vrest Plus™; Vykamol 83G; Vykasoid

Wecotop 1013; Witafrol® 7420; Witconol™ 14F; Witconol™ 18F; Witconol™ GMOP

Xpando; Xpando Powd

Chemicals: Acacia; Acetylated hydrogenated coconut glycerides; Acetylated hydrogenated cottonseed glyceride; Acetylated hydrogenated soybean oil glyceride; Acetylated lard glyceride; Acetylated mono- and diglycerides of fatty acids; Acetylated tartaric acid esters of mono- and diglycerides of fatty acids; Acyl lactylates; Agar; Albumen; Algin; Alginic acid; Aluminum caprylate; Aluminum stearate; Ammonium alginate; Ammonium carrageenan; Ammonium furcelleran; Ammonium phosphate, dibasic; Arabinogalactan; Ascorbyl palmitate

Bakers yeast extract; Bentonite

Calcium carrageenan; Calcium citrate; Calcium dihydrogen pyrophosphate; Calcium furcelleran; Calcium lactate; Calcium phosphate monobasic monohydrate; Calcium phosphate tribasic; Calcium/sodium stearoyl lactylate; Calcium stearate; Calcium stearoyl lactylate; Canola oil glyceride; Capric triglyceride; Caprylic/capric triglyceride; Caprylic triglyceride; Carrageenan; Cellulose; Cholesterol; Cholic acid; Citric acid esters of mono- and diglycerides of fatty acids; Coconut oil; Corn glycerides; Corn oil; Cottonseed glyceride; Cottonseed oil

Damar; Diacetyl tartaric acid esters of mono- and diglycerides; Dioctyl sodium sulfosuccinate; Disodium capryloamphodiacetate; Disodium citrate; Disodium phosphate, dihydrate; Disodium pyrophosphate

Ethoxylated mono- and diglycerides

Furcelleran

Glyceryl caprate; Glyceryl caprylate/caprate; Glyceryl citrate/lactate/linoleate/oleate; Glyceryl cocoate; Glyceryl cottonseed oil; Glyceryl dioleate; Glyceryl dioleate SE; Glyceryl distearate; Glyceryl distearate SE; Glyceryl di/tribehenate; Glyceryl lactoesters; Glyceryl lactooleate; Glyceryl lactopalmitate/stearate; Glyceryl laurate; Glyceryl laurate SE; Glyceryl linoleate; Glyceryl mono/dilaurate; Glyceryl mono/dioleate; Glyceryl mono/dioleate; Glyceryl mono/ distearate-palmitate; Glyceryl oleate; Glyceryl oleate SE; Glyceryl palmitate; Glyceryl palmitate lactate; Glyceryl palmitate stearate; Glyceryl ricinoleate; Glyceryl ricinoleate SE; Glyceryl soyate; Glyceryl stearate; Glyceryl stearate citrate; Glyceryl stearate lactate; Glyceryl stearate SE; Guar gum; Gum ghatti

Hydrogenated cottonseed glyceride; Hydrogenated lard glyceride; Hydrogenated lard glycerides; Hydrogenated palm glyceride; Hydrogenated rapeseed oil; Hydrogenated soybean glycerides; Hydrogenated soy glyceride; Hydrogenated tallow glyceride; Hydrogenated tallow glyceride citrate; Hydrogenated tallow glyceride lactate; Hydrogenated tallow glycerides; Hydrogenated vegetable glyceride; Hydrogenated vegetable glycerides; Hydrogenated vegetable oil; Hydroxylated lecithin; Hydroxypropylcellulose; Hydroxypropyl methylcellulose

Karaya gum

Lactic acid esters of mono- and diglycerides of fatty acids; Lactylic esters of fatty acids; Lard; Lard glyceride; Lard glycerides; Lecithin; Locust bean gum

Magnesium stearate; Methylcellulose; Methyl ethyl cellulose; Mono- and diglycerides of fatty acids; Mono- and diglycerides, sodium phosphate derivs

Nonoxynol-10; Nonoxynol-11; Nonoxynol-12

Octenyl succinic anhydride; Octoxynol-5; Octoxynol-30; Octoxynol-40; Octoxynol-70; Oleth-23

Palm glyceride; Palm oil; Palm oil sucroglyceride; Peanut glycerides; Peanut oil; Pea protein concentrate; Pectin; PEG-20 dilaurate; PEG-7 glyceryl cocoate; PEG-20 glyceryl stearate; PEG-6 oleate; PEG-8 oleate; PEG-40 sorbitan hexatallate; PEG-80 sorbitan laurate; PEG-20 sorbitan tritallate; PEG-6 stearate; PEG-8 stearate; PEG-40 stearate; Pentapotassium triphosphate; Phosphatidylcholine; Polyglyceryl-10 decaoleate; Polyglyceryl-10 decastearate; Polyglyceryl-2 diisostearate; Polyglyceryl-3 diisostearate; Polyglyceryl-5 diisostearate; Polyglyceryl-3 dioleate; Polyglyceryl-6 dioleate; Polyglyceryl-10 dioleate; Polyglyceryl-10 dipalmitate; Polyglyceryl-2 distearate; Polyglyceryl-3 distearate; Polyglyceryl-5 distearate; Polyglyceryl-6 distearate; Polyglyceryl-10 distearate; Polyglyceryl-6 hexaoleate; Polyglyceryl-10 hexaoleate; Polyglyceryl-10 isostearate; Polyglyceryl-10 laurate; Polyglyceryl-10 linoleate; Polyglyceryl-10 myristate; Polyglyceryl-2 oleate; Polyglyceryl-3 oleate; Polyglyc-

Emulsifiers *(cont'd.)*

eryl-4 oleate; Polyglyceryl-6 oleate; Polyglyceryl-8 oleate; Polyglyceryl-10 oleate; Polyglyceryl-4 pentaoleate; Polyglyceryl-10 pentaoleate; Polyglyceryl-4 pentastearate; Polyglyceryl polyricinoleate; Polyglyceryl-2 sesquioleate; Polyglyceryl-2 stearate; Polyglyceryl-3 stearate; Polyglyceryl-4 stearate; Polyglyceryl-8 stearate; Polyglyceryl-10 stearate; Polyglyceryl-10 tetraoleate; Polyglyceryl-2 tetrastearate; Polyglyceryl-2 triisostearate; Polyglyceryl-4 tristearate; Polysorbate 20; Polysorbate 21; Polysorbate 40; Polysorbate 60; Polysorbate 61; Polysorbate 65; Polysorbate 80; Polysorbate 81; Polysorbate 85; Potassium alginate; Potassium citrate; Potassium furcelleran; Potassium oleate; Potassium phosphate dibasic; Potassium phosphate tribasic; Potassium polymetaphosphate; Potassium sodium tartrate anhyd; Potassium sodium tartrate tetrahydrate; Potassium tripolyphosphate; Propylene glycol; Propylene glycol alginate; Propylene glycol dicaprylate/dicaprate; Propylene glycol esters of fatty acids; Propylene glycol laurate; Propylene glycol laurate SE; Propylene glycol monodistearate; Propylene glycol oleate; Propylene glycol oleate SE; Propylene glycol palmitate; Propylene glycol ricinoleate; Propylene glycol ricinoleate SE; Propylene glycol stearate; Propylene glycol stearate SE

Rapeseed oil glyceride

Saccharose distearate; Saccharose mono/distearate; Saccharose palmitate; Safflower glyceride; Safflower oil; Sodium acid pyrophosphate; Sodium aluminum phosphate acidic; Sodium aluminum phosphate, basic; Sodium carrageenan; Sodium caseinate; Sodium furcelleran; Sodium hexametaphosphate; Sodium hypophosphite; Sodium laurate; Sodium lauryl sulfate; Sodium metaphosphate; Sodium myristate; Sodium oleate; Sodium palmitate; Sodium phosphate; Sodium phosphate dibasic; Sodium phosphate tribasic; Sodium phosphate tribasic dodecahydrate; Sodium stearate; Sodium stearoyl lactylate; Sodium tartrate; Sorbitan caprylate; Sorbitan laurate; Sorbitan myristate; Sorbitan oleate; Sorbitan palmitate; Sorbitan sesquioleate; Sorbitan sesquistearate; Sorbitan stearate; Sorbitan trioleate; Sorbitan tristearate; Sorbitan tritallate; Soybean oil; Soy protein; Steareth-20; Stearyl-2-lactylic acid; Succinylated monoglycerides; Succistearin; Sucrose dilaurate; Sucrose distearate; Sucrose erucate; Sucrose fatty acid esters; Sucrose laurate; Sucrose myristate; Sucrose oleate; Sucrose palmitate; Sucrose polylaurate; Sucrose polylinoleate; Sucrose polyoleate; Sucrose polystearate; Sucrose stearate; Sucrose tetrastearate triacetate; Sucrose tribehenate; Sucrose tristearate; Sunflower seed oil; Sunflower seed oil glyceride; Sunflower seed oil glycerides; Superglycerinated hydrogenated rapeseed oil

Tallow; Tallow glyceride; Tallow glycerides; Tartaric acid esters of mono- and diglycerides; Tetrapotassium pyrophosphate; Tetrasodium pyrophosphate; Tragacanth gum; Trisodium citrate

Xanthan gum

Encapsulants

Trade names: Alcolec® 140; Aloe Vera Lipo-Quinone Extract™ (AVLQE) Food Grade
Capcithin™; Cap-Shure® AS-165-70; Cap-Shure® BC-140-70; Cap-Shure® BC-900-85; Cap-Shure® C-135-72; Cap-Shure® C-140-72; Cap-Shure® C-140-85; Cap-Shure® C-140E-75; Cap-Shure® C-150-50; Cap-Shure® C-165-63; Cap-Shure® C-165-85; Cap-Shure® C-900-85; Cap-Shure® CLF-165-70; Cap-Shure® F-125-85; Cap-Shure® F-140-63; Cap-Shure® FE-165-50; Cap-Shure® FF-165-60; Cap-Shure® KCL-140-50; Cap-Shure® KCL-165-70; Cap-Shure® LCL-135-50; Capsulec 51-SB; Capsulec 51-UB; Capsulec 56-SB; Capsulec 56-UB; Capsulec 62-SB; Capsulec 62-UB; Cavitron Cyclo-dextrin.™
Descote® Ascorbic Acid 60%; Descote® Sodium Ascorbate 50%
K.L.X; Lo-Dex® 10
Mira-Cap®
Petrolite® C-1035; Polypro 5000®; Polypro 15000® Food Grade; Staley® Maltodextrin 3260

Enzymes

Trade names: AFP 2000; Amflex
Biodiastase; d-Biotin USP, FCC No. 63345; Biozyme L; Biozyme M; Biozyme S; Bromelain 1:10;

Enzymes *(cont'd.)*

Bromelain 150 GDU; Bromelain 600 GDU; Bromelain 1200 GDU; Bromelain 1500 GDU; Bromelain Conc

Catalase L; Celluferm; Cellulase 4000; Cellulase AC; Cellulase AP; Cellulase AP 3; Cellulase L; Cellulase TAP; Cellulase Tr Conc; CG Tase; CK-20L; Clarase® 5,000; Clarase® 40,000; Clarase® Conc; Clarase® L-40,000; Clarex® 5XL; Clarex® L; Clarex® ML; Crackerase

Deamizyme; Dextranase L Amano; Dextranase Novo 25 L; Diazyme® L-200; Dry Phytonadione 1% SD No. 61748

Eversoft Plus Kosher; Extractase L5X, P15X

Fermalpha; Fermcolase®; Fermcozyme® 1307, BG, BGXX, CBB, CBBXX, M; Fermlipase; Fermvertase, 10X, XX; FloAm® 300; Fungal Lactase 100,000; Fungal Protease 31,000; Fungal Protease 60,000; Fungal Protease 500,000; Fungal Protease Conc

Glucanase GV; Glucanex® L-300; Gluczyme; Glutaminase Amano; GNL-3000; Grindamyl

Harina-Ase; Hemi-Cellulase Amano 90; Hemicellulase B 1500; Hemicellulase CE 1500; Hemi-cellulase Conc; Hidelase; HT-Proteolytic® 200; HT-Proteolytic® Conc; HT-Proteolytic® L-175; Hyderase

Lactase AlE; Lactase F Amano; Lipase 8 Powd; Lipase 16 Powd; Lipase 24 Powd; Lipase 30 Powd; Lipase AK; Lipase AP; Lipase AP 6; Lipase AY; Lipase FAP; Lipase G; Lipase GC; Lipase MAP; Lipase N; Lipase PS

MC^2; Microcatalase®; MLO

Naringinase Amano; Neutral-lactase; Newlase; Nutrilife®; Nutrisoy 7B

Optimase® APL-440; Ovazyme, XX

Pancreatic Lipase 250; Pancreatin 3X USP Powd; Pancreatin 4X USP Powd; Pancreatin 5X USP Powd; Pancreatin 6X USP Powd; Pancreatin 8X USP Powd; Pancreatin TA; Pancreatin USP Powd; Pancrelipase USP; Papain 16,000; Papain 30,000; Papain A300; Papain A400; Papain AlE; Papain Conc; Papain M70; Papain P-100; Papain S100; Pearex® L; Pectinase AT; Pectinol® 59L; Pectinol® DL; Pectinol® R10; Pektolase; Pepsin 1:3000 NF XII Powd; Pepsin 1:10,000 Powd. or Gran; Pepsin 1:15,000 Powd; Prevent-O; Proflex; Protoferm; Prozyme 6

Qual Guard™; Qual Guard™ 100; Qual Guard™ WB

Rhozyme® HP-150 Conc; Rhozyme® P11; Rhozyme® P41; Rhozyme® P53, P64

Spark-L® HPG; Spezyme BBA; Spezyme GA; Spezyme IGI; SPL Lipase 30; SPL Lipase CE; SPL Pancreatin 3X USP; SPL Pancreatin 4X USP; SPL Pancreatin 5X USP; SPL Pancreatin 6X USP; SPL Pancreatin 7X USP; SPL Pancreatin 8X USP; SPL Pancreatin USP; SPL Pancrelipase USP; SPL Undiluted Pancreatic Enzyme Conc. (PEC); Super Timeflex

Taka-Sweet®; Taka-Therm® L-340; Tenase® 1200; Tenase® L-340; Tenase® L-1200; Timeflex; Tona Enzyme 201; Tona Papain 14; Tona Papain 90L; Tona Papain 270L; Trem-Tabs; Trypsin 1:75; Trypsin 1:80; Trypsin 1:150

Yeast Lactase L-50,000

Chemicals:
Amylase; α-Amylase; β-Amylase; Amyloglucosidase; Anthocyanase

Bromelain; *Candida guilliermondii; Candida lipolytica*

Carbohydrase; Carbohydrase from *Aspergillus niger*; Carbohydrase from *Rhizopus oryzae*; Carbohydrase-protease; Catalase; Cellulase; Chitinase; Cyclodextrin glucanotransferase

Deaminase; Dextranase; Dismutase superoxide

Esterase; Esterase-lipase

Ficin

α-Galactosidase; Gibberellic acid; Glucanase; Glucoamylase; Glucose isomerase; Glucose oxidase; α-Glucosidase; β-Glucosidase

Hemicellulase

Invertase; Isoamylase

Lactase; Lactase from *Kluyveromyces lactis*; Lipase; Lipoxidase

Malt extract; Milk-clotting enzyme from *Bacillus cereus*; Milk-clotting enzyme from *Endothia parasitica*; Milk-clotting enzyme from *Mucor miehei*; Milk-clotting enzyme from *Mucor pusillus*

Pancreatin; Papain; Pectinase; Pentosanase-hexosanase; Pepsin; Potassium carrageenan; Protease

Rennet

Tannase; Trypsin

Urease

Fat replacers

Trade names: Alcolec® 140; Avicel® CL-611; Avicel® RC-501; Avicel® RC-581; Avicel® RC-591F; Avicel® RCN-10; Avicel® RCN-15; Avicel® RCN-30; Avicel® WC-595
Carcao; CarraFat™; Carralite™; Celish® FD-100F
Dairy-Lo™; Dur-Lo®
Edicol® ULV Series; Enrich®
Fat Replacer 785
Gelcarin® GP-359; Gelcarin® GP-379; Gelcarin® GP-911; Gelcarin® ME-621; Gelcarin® PS-316; Gelcarin® XP-1008; Gelcarin® XP-8004
H.L. PY™
Instant Remygel AX-P, AX-2-P; Instant Remyline AX-P
Kel-Lite™ BK; Kel-Lite™ CM
Lactarin® MV-306; Lactarin® MV-308; Lactarin® MV-406; Lactarin® PS-185X; Lite Pack 350
Maltrin® M040; Maltrin® QD M440; Marine Colloids™ Carrageenan
NatuReal™ Balance; Neobee® C-10; Nutricol® Konjac; Nutrisoy 220T
Optima 7B; Optima 23B; Optima 77IC
Remygel AX-2; Remygel AX; Remyline AX; Rice Complete® 3; Rice Complete® 10; Rice Complete® 18
SeaKem® CM-611; SeaKem® GP-418; SeaKem® IC-611; SeaKem® IC-624; SeaKem® IC-632; SeaKem® IC-912; StarchPlus™ SPR; StarchPlus™ SPW; StarchPlus™ SPW-LP; Sta-Slim™ 142; Sta-Slim™ 143; Sta-Slim™ 150; Sta-Slim™ 171; Stellar®; Summit® 25; Summit® 50
Tem Cote® IE; Tem Cote® II; Tem Crunch™ 93; Tem Crunch™ 110; Tem Freeze™; Tem Plus® 95; Tem Top™; Ticaloid® S1-102; Ticaloid® S2-102; TIC Pretested® Colloid 881 M; TrimChoice™ 5; TrimChoice™ OC
Ultra-Freeze® 400; Ultra-Freeze® 400C
Viobin Wheat Germ Oil; Viscarin® GP-209; Viscarin® ME-389; Viscarin® SD-389; Viscarin® XP-3160; Visco Compound™
Wecobee® FS; Wecobee® FW; Wecobee® HTR; Wecobee® M; Wecobee® R Mono; Wecobee® S; Wecobee® SS; Wecobee® W

Chemicals: Citric acid esters of mono- and diglycerides of fatty acids
Hydrolyzed oat flour
Rice maltodextrin; Rice starch

Fermentation aids • Yeast foods

Trade names: AF 70; AF 72; AF 75; AF HL-36; AFP 2000; Albrite Monoammonium Phosphate Food Grade; Aldosperse® 40/60 FG; Aldosperse® 40/60 KFG; Amflex; Arkady Yeast Food
Best One®; BK-102 Series; Bromelain 1:10
Cap-Shure® SC-165-85FT; Cap-Shure® SC-165X-60FF; Carbrea® Tabs; CDC-2002
Diazyme® L-200; Durkote Glucono-Delta-Lactone/Hydrog. Veg. Oil
EMB
Fermaloid Yeast Food; Fermentase®; Ferment Buffer; Fungal Protease 31,000; Fungal Protease 60,000; Fungal Protease 500,000; Fungal Protease Conc
Hodag Antifoam F-1; Hodag Antifoam F-2; Hodag Antifoam FD-82; HT-Proteolytic® 200; HT-Proteolytic® Conc; HT-Proteolytic® L-175; Hyderase
Lexein® 152D (Dry)
N'Hance (Phase I) Powd; Nissan Anon BL; Nissan Anon LG; Nutrilife®; Nutrimalt®
Papain 16,000; Papain 30,000; Papain Conc.
R.B.-4 Sour
Special C; Super Timeflex
Tenase® 1200; Tenase® L-340; Tenase® L-1200; Terra Alba 114836; Timeflex; 1216 Fast

Chemicals: Ammonium carbonate; Ammonium chloride; Ammonium phosphate; Ammonium sulfate
Bakers yeast extract
Calcium carbonate; Calcium lactate; Calcium oxide; Calcium phosphate monobasic monohydrate; Calcium phosphate tribasic; Choline phosphate; L-Cysteine
Dimethicone
Hydrogenated tallow; Hydrogenated tallow alcohol

Fermentation aids *(cont'd.)*

Magnesium chloride; Magnesium sulfate anhyd; Mineral oil
Phosphoric acid; Potassium bromate; Potassium carbonate; Potassium chloride; Potassium
    gibberellate; Potassium phosphate; Potassium phosphate dibasic
Tristearin
Urea
Yeast; Yeast extract

Film-formers

Trade names: Amisol™ 683 A; Amoloid HV; Amoloid LV; Atlas 1400K; Atlas 1500; Atmos® 150 K; Atmul® 84;
    Atmul® 124
Colloid 488T; Colloid 602; Colloid Stick Tic; Cozeen®
Flavorset® GP-9; Flavorset® GP-10
Kelcogel®; Kelcogel® F; Keltone® LV; Klucel® F Grades
Liquid Fish Gelatin Conc; Lo-Dex® 10
Maltrin® M040; Maltrin® M050; Maltrin® QD M440; Manucol DM; Manucol DMF; Manucol JKT;
    Manucol LB; Manugel C; Manugel DJX; Manugel DMB; Manugel GHB; Manugel JKB;
    Manugel L98; Manugel PTJ; Methocel® E3 Premium; Methocel® E4M Premium; Methocel®
    E6 Premium; Methocel® E15LV Premium; Methocel® E50LV Premium; Methocel® K4M
    Premium; Methocel® K15M Premium; Methocel® K100LV Premium; Mira-Cap®; Myvacet®
    7-00; Myvacet® 7-07; Myvacet® 7-07K
Polycon S60 K; Polycon S80 K; Polypro 5000®; Polypro 15000® Food Grade
Redisol® 78D
Sobalg FD 100 Range; Sol-U-Tein EA; Spray Dried Fish Gelatin; Spray Dried Fish Gelatin/
    Maltodextrin; Spray Dried Hydrolysed Fish Gelatin; Staley® Tapioca Dextrin 12
Ticolv

Chemicals: Acetylated hydrogenated cottonseed glyceride
Calcium lactate, Carboxymethylmethylcellulose
Hydrolyzed gelatin; Hydroxypropylcellulose; Hydroxypropyl methylcellulose
Methylcellulose
Polyacrylamide
Tara gum

Firming agents

Trade names: Py-ran®
Sorbo®
Terra Alba 114836

Chemicals: Acacia; Algin; Aluminum sulfate; Ammonium phosphate, dibasic; Ammonium sulfate
Calcium acetate; Calcium carbonate; Calcium citrate; Calcium gluconate; Calcium hydroxide;
    Calcium lactate; Calcium lactobionate; Calcium phosphate dibasic; Calcium phosphate
    monobasic monohydrate; Calcium sulfate
Guar gum
Magnesium chloride; Magnesium oxide; Magnesium sulfate heptahydrate; D-Mannitol
Potassium alum dodecahydrate
Sodium alum; Sorbitol
L-Tartaric acid

Flavorings • Flavor enhancers • Bittering agents • Sours

Trade names: Accel®; Activera™ 1-1FA (Filtered); Activera™ 1-1 UA (Unfiltered); Adeka Menjust; Almond
    Filling Powd; Alomine 042; Angrex Spices Oleoresins; Aqualon® 7LXF; Aqualon® 9H4XF;
    Aromahop™; Aromild® S; Atmos® 300 K; Autolyzed Type G Torula Dried Yeast; Autolyzed

Flavorings *(cont'd.)*

Type N Torula Dried Yeast; AYE-2000; AYE 2200; AYE 2312; AYE Family; AYS 2311; AYS 2350

Banana Essence 1000 Fold Natural; Basmati Rice Extract; Beef Extract #3041; Beef Spicey No. 25748; Beer Extract; Bittex; BK-102 Series; Blend 424; Bohemian Rye Sour; Brolite 1A; Bro Rye Sour; Bro White Sour; Building Blocks® Beef Extract; Building Blocks® Beef Extract B1; Building Blocks® Dried Flavored Beef Extract B7; Building Blocks® Dried Flavored Beef Extract B8; Building Blocks® Dried Flavored Beef Stock #6; Building Blocks® Dried Flavored Beef Stock B4; Building Blocks® Dried Flavored Meat Stock #6; Building Blocks® Dried Ham Stock H5; Building Blocks® Flavored Beef Extract B7; Building Blocks® Flavored Beef Extract B8; Building Blocks® Flavored Beef Extract B8 LS; Building Blocks® Flavored Beef Stock #6; Building Blocks® Flavored Beef Stock B4; Building Blocks® Flavored Meat Stock #6; Building Blocks® Frozen Roast Beef Juice; Building Blocks® Ham Stock H5; Building Blocks® Pork Stock P1

Cap-Shure® GDL-140-70; Cap-Shure® SC-140X-70; Cap-Shure® SC-165-85FT; Cap-Shure® SC-165X-60FF; Caramel Prep. with Fat No. 75955; Cheddar Harmony #274; Chicken Extract #1083; Citraxine; Citreatt Lemon 3123; Citreatt Lemon 6122; Citreatt Lime 3135; Citreatt Lime 6134; Citreatt Orange 3111; Citreatt Orange 6110; Citric Acid USP FCC Anhyd. Fine Gran. No. 69941; Citrid Acid USP FCC Anhyd. Gran. No. 69942; Citrostabil® NEU; Citrostabil® S; Clam Powd. #60; Clintose® A; Clintose® F; Clintose® L; Clintose® VF; CNP BRSMC35; CNP WRSHG40; CNP WRSHG40CL; CNP WRSHM; CNP WRSHMCL; CNP WRSMC30; CNP WRSMC30CL; CNP WRSRD; CNP WRSRDCL; Cola L; Cola No. 23443; Cola Acid No. 23444; Cola Extra; Concentrated Beef Stock #3021; Concentrated Chicken Broth #1023; Concentrated Chicken Broth #1095; Concentrated Chicken Stock #1021; Concentrated Pork Stock #2021; Concentrated Turkey Stock #4021; Coral®; CornSweet® 95; Crab Extract #41; Culinox® 999 Chemical Grade Salt; Culinox® 999® Food Grade Salt

Dark Rye Sour; DCME; Deli Rye Conc; DMCE; DME Dry Malt Extract; Dried Beef Seasoning #6540; Dried Beef Stock #3024; Dried Chicken Seasoning #1026; Dried Chicken Stock #120; Dried Chicken Stock #1004; Dried Pork Stock #220; Dried Pork Stock #2004; Dried Turkey Stock #410

EMB; Emery® 912; Emery® 916; Emery® 917; Emery® 918; Enzyme Modified Beef #3083; Eromenth; Essiccum®; Exter™ Family; Extract of Whole Grapefruit; Extract of Whole Orange; Extramalt 10; Extramalt 33; Extramalt 35; Extramalt Dark; Extramalt Light

Fancol HON; Fancol Menthol; Fermentase®; Fermented Soy Sauce Powd; Flavor Enhancer Powd; Flav-R-Base® Primary Yeast Autolysate Extracts

Gistex®; Gistex® Family; Gistex® LS; Gistex® MR; Gistex® X-II; Gistex® Xtra Powd; Glutarom® Range; Golden Bake Sour V; Golden Bake Sour Bread Base; Golden Gate Sour; Golden Gate Sour; Golden Gate Sour; Green Pea Powd

Hi Glo™; Hodag PSMO-20; HVP 5-SD; HVP-A; HVP-LS

I+G; Isoamyl Alcohol 95%; Isoamyl Alcohol 99%; Isobutanol HP; Isohop®

Lem-O-Fos® 101; Lem-O-Fos® 202; Lime/Lemon Essence 1574; Luxor® 1517; Luxor® 1576; Luxor® 1626; Luxor® 1639; Luxor® 1658; Luxor® Century™ V; Luxor® CVP 5-SD; Luxor® CVP 1632; Luxor® CVP 1689; Luxor® CVP 1753; Luxor® CVP LS; Luxor® E-40; Luxor® E-50; Luxor® E-610; Luxor® EB-2; Luxor® EB-400; Luxor® FB-10; Luxor® FC-20; Luxor® GR-100; Luxor® GR-150; Luxor® GR-200; Luxor® HVP-A; Luxor® KB-300; Luxor® KB-312; Luxor® KB-320; Luxor® KB-330; Luxor® KB-350; Luxor® KB-400; Luxor® KB-500; Luxor® KB-530; Luxor® KB-600; Luxor® L-625; Luxor® MB-40; Luxor® MB-110; Luxor® MB-120; Luxor® No. 6 Sauce; Luxor® No. 700 Sauce; Luxor® Prozate®; Luxor® R-100; Luxor® Triple-H®

Maltoferm® 10001; Maltoferm® 10001 VDK; Maltoferm® MBF CR-40; Maxarome® Family; Maxarome® MR; Maxarome® Plus; Maxarome® Plus RS; Maxavor™ MYE; Maxavor™ RYE-A, RYE-AS; Maxavor™ RYE-B; Maxavor™ RYE-C; Maxavor™ RYE-CL; Maxavor™ RYE-CR; Maxavor™ RYE-D; Maxavor™ RYE-G; Maxavor™ RYE-PMR; Maxavor™ RYE-T; Maxavor™ RYE Family; MB 100; MB 300; Muffit; Myvatex® Monoset®; Myverol® 18-92; Myverol® 18-99

Natural Flavor Enhancer No. 11.9743; Natural Flavor White Sour; Natural Lipolyzed Butter Oil 500; Natural Pineapple Extract; Natural Starter Distillate Replacer 15X W.S; NatureTones™; ND-201 Syrup; ND-201-C Syrups; ND-305 Syrups; Neral; Nerolidol; Niacet Sodium Acetate 60% FCC; Niacet Sodium Acetate Anhyd. FCC; Niacet Sodium Diacetate FCC; 90/10 Blend; No-GluAce™; Non-Diastatic Malt Syrup #40600; Nova-CPLN; Nova-Chef™ 4005; Nova-Chef™ 4015; Nova-Chef™ 4018; Nova-Chef™ 4021; Nova-Chef™ 4022; Nova-Chef™ 4025; Nova-Chef™ 4026; Nova-Chef™ 4028; Nova-Chef™ 4031; Nova-Chef™ 4032; Nova-

Flavorings *(cont'd.)*

Chef™ 4102; Nova-Chef™ 4108; Nova-Chef™ 4110; Nova-Chef™ 4114; Nova-Chef™ 4120; Nova-Chef™ 4130; Nova-Flav™ 1000; Nova-Flav™ 1001; Nova-Flav™ 1002; Nova-Flav™ 1006; Nova-Flav™ 1020; Nova-Flav™ 1030; Nova-Flav™ 5004; Nova-Flav™ 5006; Nova-Flav™ 5009; Nova-Flav™ 5010; Nova-Flav™ 5030; Nova-Flav™ 5100; Nova-Flav™ 5101; Nova-Flav™ 5102; Nova-Flav™ 5103; Nova-Flav™ 5105; Nova-Flav™ 7000; Nova-Flav™ 7001; Nova-Flav™ 7003; Nova-Flav™ 7004; Nova-Flav™ 7006; Nova-Flav™ 7007; Nova-Flav™ 7009; Nova-Flav™ 7010; Nova-Flav™ 7102; Nova-Flav™ 7105; Nova-Flav™ 7107; Nova-Flav™ 7109; Nova-Flav™ 8002; Nova-Flav™ 8004; Nova-Max 1; Nova-Naturelle™ Anchovy Flavor WONF RC-92; Nova-Naturelle™ Anchovy Flavor WONF RC-93; Nova-Naturelle™ Bacon Type Flavor WONF RC-0027; Nova-Naturelle™ Bacon Type Flavor WONF RC-0038; Nova-Naturelle™ Bacon Type Flavor WONF RC-0056; Nova-Naturelle™ Bacon Type Flavor WONF RC-112; Nova-Naturelle™ Beef Flavor WONF 9101; Nova-Naturelle™ Beef Flavor WONF RC-0034; Nova-Naturelle™ Beef Flavor WONF RC-0035; Nova-Naturelle™ Beef Flavor WONF RC-0044; Nova-Naturelle™ Beef Flavor WONF RC-97; Nova-Naturelle™ Beef Flavor WONF RC-100; Nova-Naturelle™ Beef Flavor WONF RC-200; Nova-Naturelle™ Beef Flavor WONF RC-300; Nova-Naturelle™ Beef Stew Flavor WONF RC-0021; Nova-Naturelle™ Beef Teriyaki Flavor WONF RC-0046; Nova-Naturelle™ Beef Type Flavor WONF RC-0053; Nova-Naturelle™ Beef Type Flavor WONF RC-0058; Nova-Naturelle™ Breakfast Sausage Type Flavor WONF RC-0047; Nova-Naturelle™ Breakfast Sausage Type Flavor WONF RC-0062; Nova-Naturelle™ Chicken Flavor WONF RC-0028; Nova-Naturelle™ Chicken Flavor WONF RC-0031; Nova-Naturelle™ Chicken Flavor WONF RC-0040; Nova-Naturelle™ Chicken Flavor WONF RC-107; Nova-Naturelle™ Chicken Flavor WONF RC-108; Nova-Naturelle™ Chicken Flavor WONF RC-202; Nova-Naturelle™ Chicken Flavor WONF RC-203; Nova-Naturelle™ Chicken Flavor WONF RC-305; Nova-Naturelle™ Chicken Flavor WONF RC-307; Nova-Naturelle™ Clam Flavor WONF RC-95; Nova-Naturelle™ Crab Flavor WONF RC-99; Nova-Naturelle™ Dark Beef Flavor WONF RC-303; Nova-Naturelle™ Fish Flavor WONF RC-89; Nova-Naturelle™ Grilled Beef Flavor WONF RC-0022; Nova-Naturelle™ Grilled Beef Flavor WONF RC-0042; Nova-Naturelle™ Grilled Beef Flavor WONF RC-301; Nova-Naturelle™ Grilled Chicken Flavor WONF RC-0029; Nova-Naturelle™ Grilled Chicken Flavor WONF RC-0049; Nova-Naturelle™ Grilled Chicken Flavor WONF RC-306; Nova-Naturelle™ Grilled Meat Type Flavor WONF RC 0060; Nova-Naturelle™ Grilled Pork Type Flavor RC-0043; Nova-Naturelle™ Grilled Pork Type Flavor WONF RC-0020; Nova-Naturelle™ Hamburger Flavor WONF RC 0023; Nova-Naturelle™ Hamburger Flavor WONF RC-0030; Nova-Naturelle™ Hamburger Type Flavor WONF RC-0052; Nova-Naturelle™ Hamburger Type Flavor WONF RC-0057; Nova-Naturelle™ Hamburger Type Flavor WONF RC-0063; Nova-Naturelle™ Italian Sausage Type Flavor WONF RC-0036; Nova-Naturelle™ Italian Sausage Type Flavor WONF RC-0061; Nova-Naturelle™ Liquid Bacon Flavor O.S. 62093; Nova-Naturelle™ Lobster Flavor WONF RC-90; Nova-Naturelle™ Meat Type Flavor WONF RC-0059; Nova-Naturelle™ Oyster Flavor WONF RC-96; Nova-Naturelle™ Pork Fat Type Flavor WONF RC-0045; Nova-Naturelle™ Pork Flavor WONF RC-304; Nova-Naturelle™ Pork Type Flavor WONF RC-0019; Nova-Naturelle™ Pork Type Flavor WONF RC-0032; Nova-Naturelle™ Pork Type Flavor WONF RC-0033; Nova-Naturelle™ Pork Type Flavor WONF RC-0054; Nova-Naturelle™ Pork Type Flavor WONF RC-106; Nova-Naturelle™ Pork Type Flavor WONF RC-204; Nova-Naturelle™ Roast Beef Flavor WONF 92592; Nova-Naturelle™ Roast Beef Flavor WONF RC-0039; Nova-Naturelle™ Roast Beef Flavor WONF RC-0041; Nova-Naturelle™ Roast Beef Flavor WONF RC-201; Nova-Naturelle™ Roast Beef Flavor WONF RC-302; Nova-Naturelle™ Shrimp Flavor WONF RC-98; Nova-Naturelle™ Turkey Type Flavor WONF RC-0048; Nova-Naturelle™ Turkey Type Flavor WONF RC-110; Nova-Zyme™ MB-0002; Nova-Zyme™ MB-0003; Nova-Zyme™ MB-0004; Nova-Zyme™ MB-5; Nova-Zyme™ MB-6; Nova-Zyme™ MB-7; Nova-Zyme™ MB-8; Nova-Zyme™ MB-9; Nova-Zyme™ MB-10; Nova-Zyme™ MB-11; Nutricol® Konjac; Nutrimalt® Range

Onion Super; 1823 Orange Oil

Pancreatic Lipase 250; Pationic® 1230; Pationic® 1250; Pfico₂Hop®; Phenylacetaldehyde 50; Phenylethyl Alcohol Extra; Pie Dough Culture 6450; Plantex® Family; Polypro 2000® CF 45%; Polypro 2000® UF 45%; Polypro 5000® CF or SF 45%; Powdered Sta-Fudge; Prenol; Pristene® RO; Pristene® RW; Purac®; Purasal® P/USP 60; Purasal® S/SP 60; Purasolv® ELECT; Purasolv® ELS

R.B.-4 Sour; Readi-Glaze Systems; Redihop®; Red Miso Powd; Rhodiarome™; Rye Sour; Rye Sour #4; Ry-Fla-Vor; Ry-So

SeaKem® GP-418; SeasonRite Marinade Systems; Shrimp Extract #51; Shrimp Powd. #50; S-

Flavorings *(cont'd.)*

Maz® 60K; Sorbo®; Sour Base M; Sour Dough Base; Strawberry No. 25820; Strawberry Solid Pack No. 75954; Superfine Ground Mustard; Superior Mustard; Super Vita Rye; Sweet Lime Oil 5 Fold

Tang; Tetrahop®; Tomex®; Toppro; Triple-H; Trophy®

Unisweet EVAN

Vanilla No. 25450; Vanillin Free Flow; Vani-White; VCR; Veg-Juice; Veg-Juice Garlic; Veltol®; Veltol®-Plus; Viobin Defatted Wheat Germ #1; Viobin Defatted Wheat Germ #2; Viobin Defatted Wheat Germ #3; Viobin Defatted Wheat Germ #5; Viobin Defatted Wheat Germ #6; Viobin Defatted Wheat Germ #8; Viobin Defatted Wheat Germ #9; Viobin Defatted Wheat Germ #34; Viobin Pork Liver Fat; V-V Vanilla

White Sour; Whole Egg Solids Type W-7

Yellow Pea Powd

Chemicals: Acacia; Acetal; Acetaldehyde; Acetaldehyde phenethyl propyl acetal; Acetanisole; Acetic acid; Acetisoeugenol; Acetophenone; Acetylacetone; Acetyl butyryl; 3-Acetyl-2,5-dimethylfuran; Acetyl methyl carbinol; 2-Acetylpyrazine; 2-Acetylpyridine; 3-Acetylpyridine; 2-Acetylpyrrole; Aconitic acid; Adipic acid; DL-Alanine; Alfalfa; Alfalfa extract; Algae, brown; Algae, red; Algin; Allspice; Allyl anthranilate; Allyl butyrate; Allyl caproate; Allyl cinnamate; Allyl cyclohexaneacetate; Allyl cyclohexanebutyrate; Allyl cyclohexanehexanoate; Allyl cyclohexanepropionate; Allyl cyclohexanevalerate; Allyl disulfide; Allyl 2-ethylbutyrate; Allyl heptanoate; Allyl α-ionone; Allyl isothiocyanate; Allyl isovalerate; Allyl mercaptan; Allyl nonanoate; Allyl octanoate; Allyl phenoxyacetate; Allyl phenylacetate; Allyl propionate; Allyl sorbate; Allyl sulfide; Allyl tiglate; Allyl 10-undecenoate; Aloe extract; Althea extract; Ambergris; Ambrette seed oil; Ambrettolide; Ammoniated glycyrrhizin; Ammonium chloride; Ammonium glutamate; Ammonium isovalerate; Ammonium sulfide; Amyl acetate; n-Amyl alcohol; Amyl butyrate; α-Amylcinnamaldehyde; α-Amylcinnamaldehyde dimethyl acetal; α-Amylcinnamyl acetate; α-Amylcinnamyl alcohol; α-Amylcinnamyl formate; α-Amylcinnamyl isovalerate; Amyl heptanoate; Amyl hexanoate; Amyl octanoate; Amyl propionate; Amyl salicylate; Amyris oil, West Indian type; Anethole; Angelica; Angelica extract; Angelica root oil; Angelica seed oil; Angola weed; Angostura; p-Anisaldehyde; Anise; Anise extract; Anise oil; Anisole; Anisyl acetate; Anisyl alcohol; Anisyl butyrate; Anisyl formate; Anisyl phenylacetate; Anisyl propionate; Apricot; Apricot kernel oil; L-Arginine; L-Arginine L-glutamate; Arnica; Arnica extract; Artemisia; Artichoke; Artichoke extract; Asafetida; Asafetida extract; L-Ascorbic acid; L-Asparagine; Aspartame; L-Aspartic acid; Autolyzed yeast

Bakers yeast extract; Balm mint; Balm mint extract; Balm mint oil; Balsam Canada; Balsam copaiba; Balsam Peru; Balsam tolu; Barley extract; Basil; Basil, bush; Basil extract; Basil oil; Bay oil; Beechwood creosote; Beeswax; Benzaldehyde; Benzaldehyde dimethyl acetal; Benzaldehyde glyceryl acetal; Benzaldehyde propylene glycol acetal; Benzenethiol; Benzoic acid; Benzoin; Benzoin extract; Benzophenone; Benzyl acetate; Benzyl acetoacetate; Benzyl alcohol; Benzyl benzoate; Benzyl butyl ether; Benzyl butyrate; Benzyl cinnamate; Benzyl 2,3-dimethylcrotonate; Benzyl dipropyl ketone; Benzyl disulfide; Benzyl ether; Benzyl ethyl ether; Benzyl formate; Benzylidene acetone; Benzyl isobutyrate; Benzyl isovalerate; Benzyl mercaptan; Benzyl methoxyethyl acetal; Benzyl phenylacetate; Benzyl propionate; Benzylsalicylate; Bergamot oil; Birch oil; Birch tar oil; Bitter almond extract; Bitter almond oil; Bitter orange extract; Bitter orange oil; Bitter orange peel extract; Blackberry extract; Black caraway; Black currant; Black currant extract; Black pepper oil; Blackthorn berries; Black wattle; Bois de rose; Borneol; Bornyl acetate; Bornyl formate; Bornyl isovalerate; Bornyl valerate; Boronia flowers; β-Bourbonene; Brominated vegetable oil; Buchu leaves; Buckbean extract; Buckbean leaves; 2-Butanol; Butan-3-one-2-yl butyrate; Butter acids; Butter esters; n-Butyl acetate; Butyl acetoacetate; Butyl alcohol; Butylamine; Butyl anthranilate; Butyl butyrate; Butyl butyryllactate; α-Butylcinnamaldehyde; Butyl cinnamate; Butyl 2-decenoate; Butylene glycol; Butyl ethyl malonate; Butyl formate; Butyl heptanoate; Butyl hexanoate; Butyl isobutyrate; n-Butyl isovalerate; Butyl lactate; Butyl laurate; Butyl levulinate; Butylparaben; Butyl phenylacetate; n-Butyl propionate; Butyl stearate; Butyl sulfide; Butyl 10-undecenoate; Butyl valerate; n-Butyraldehyde; n-Butyric acid; Butyrolactone

Cacao; Cade oil; Cadinene; Caffeine; Cajeput oil; Calamus oil; Calcium chloride; Calcium lactate; Calcium stearate; Calendula; Calendula extract; Columba root; Camphene; Camphor; Cananga oil; Canola oil glyceride; Capers; Caproic acid; Caprylic acid; Caprylic alcohol; Caprylic triglyceride; Capsicum; Capsicum extract; Capsicum oleoresin; Caramel; Caraway oil; Cardamom oil; Carob extract; Carrot extract; Carrot juice; Carrot oil; Carrot seed extract; Carvacrol; Carvacryl ethyl ether; Carveol; 4-Carvomenthol; d-Carvone; l-Carvone; cis-Carvone oxide; Carvyl acetate; Carvyl propionate; β-Caryophyllene; Caryophyllene alcohol;

Flavorings *(cont'd.)*

Caryophyllene alcohol acetate; β-Caryophyllene oxide; Cascara extract; Cascara sagraga; Cascarilla oil; Cassie flowers; Castoreum; Castor oil; Catechu, black; Catechu, pale; Cedar leaf oil; Cedar, white; Cedarwood oil alcohols; Cedarwood oil terpenes; Celery seed oil; Centaury; Cetyl alcohol; Chamomile; Chamomile extract; Chamomile oil, Roman; Cherry-laurel oil; Cherry pit extract; Cherry pit oil; Cherry, wild, bark; Chervil; Chestnut leaf extract; Chicory; Chicory extract; Chirata; Chives; Cinchona extract; 1,4-Cineole; Cinnamal; Cinnamaldehyde ethylene glycol acetal; Cinnamic acid; Cinnamon; Cinnamon leaf oil; Cinnamon oil; Cinnamon oil, Ceylon; Cinnamyl acetate; Cinnamyl alcohol; Cinnamyl anthranilate; Cinnamyl benzoate; Cinnamyl butyrate; Cinnamyl cinnamate; Cinnamyl formate; Cinnamyl isobutyrate; Cinnamyl isovalerate; Cinnamyl phenylacetate; Cinnamyl propionate; Citral; Citral diethyl acetate; Citral dimethyl acetal; Citral propylene glycol acetal; Citronella; Citronellal; β-Citronellol; Citronelloxyacetaldehyde; Citronellyl acetate; Citronellyl butyrate; Citronellyl formate; Citronellyl isobutyrate; Citronellyl phenylacetate; Citronellyl propionate; Citronellyl valerate; Citrus extract; Civet; Clary oil; Clove; Clove extract; Clove leaf oil; Clove oil; Clover; Clover blossom extract; Cocoa extract; Coffee bean extract; Cognac oil, green or white; Cola; Cola nut; Coriander oil; Cork, oak; Corn silk; Corn silk extract; Corn syrup solids; Costmary; Costus root oil; p-Cresol; o-Cresyl acetate; p-Cresyl acetate; Cubeb oil; Cuminaldehyde; Cumin extract; Cuminic alcohol; Cumin oil; Curacao orange peel; Cyclamen aldehyde; Cyclohexaneacetic acid; Cyclohexyl acetate; Cyclohexyl anthranilate; Cyclohexyl butyrate; Cyclohexyl cinnamate; Cyclohexylethyl acetate; Cyclohexyl formate; Cyclohexyl isovalerate; Cyclohexyl propionate; p-Cymene; Cysteine hydrochloride; L-Cystine

Damiana; Dandelion; Dandelion extract; Dandelion root; Davana; trans-trans-2,4-Decadienal; Δ-Decalactone; γ-Decalactone; Decanal; Decanal dimethyl acetal; cis-4-Decen-1-al; trans-2-Decenal; Decene-1; 3-Decen-2-one; Decyl acetate; n-Decyl alcohol; Decyl butyrate; Decyl propionate; Denatonium benzoate NF; Diacetyl; Diacetyl tartaric acid esters of mono- and diglycerides; 4,4-Dibutyl-γ-butyrolactone; Dibutyl sebacate; Diethylacetic acid; Diethyl ketone; Diethyl malate; Diethyl malonate; Diethyl sebacate; Diethyl succinate; Diethyl tartrate; 2,5-Diethyltetrahydrofuran; Dihydrocarveol; d-Dihydrocarvone; Dihydrocarvyl acetate; Dihydrocoumarin; Dihydro-β-ionone; Dill; Dill seed oil; Dill seed oil, Indian type; Dillweed oil; m-Dimethoxybenzene; p-Dimethoxybenzene; 2,4-Dimethylacetophenone; Dimethyl anthranilate; Dimethylbenzyl carbinol; Dimethylbenzyl carbinyl acetate; α,α-Dimethylbenzyl isobutyrate; Dimethyl dicarbonate; 2,6-Dimethyl-5-heptenal; 2,6-Dimethyl octanal; 3,7-Dimethyl-1-octanol; α,α-Dimethylphenethyl butyrate; α,α-Dimethylphenethyl formate; Dimethylphenylethylcarbinol; 2,3-Dimethylpyrazine; 2,5-Dimethylpyrazine; 2,6-Dimethylpyrazine; 2,5-Dimethylpyrrole; Dimethyl succinate; 1,3-Diphenyl-2-propanone; Disodium guanylate; Disodium inosinate; Disodium succinate; Dittany; Dittany of Crete; δ-Dodecalactone; γ-Dodecalactone; 2-Dodecenal; Dodecene-1

Elder flowers; Elder tree leaves; Elecampane; Elecampane extract; English oak extract; Enzyme-modified milkfat; Esterase-lipase; Estragole; p-Ethoxybenzaldehyde; Ethyl 2-acetal-3-phenylpropionate; Ethyl acetate; Ethylacetoacetate; Ethyl 2-acetyl-3-phenylpropionate; Ethyl aconitate, mixed esters; Ethyl acrylate; Ethyl-p-anisate; Ethyl anthranilate; Ethyl benzoate; Ethyl benzoylacetate; α-Ethylbenzyl butyrate; 2-Ethylbutyl acetate; 2-Ethylbutyraldehyde; Ethyl butyrate; Ethyl caproate; Ethylcellulose; Ethyl cinnamate; Ethyl crotonate; Ethyl cyclohexanepropionate; Ethyl decanoate; Ethyl-3,5(6)-dimethylpyrazine; Ethylene brassylate; 2-Ethyl fenchol; Ethyl formate; 2-Ethylfuran; Ethyl-2-furanpropionate; 4-Ethylguaiacol; Ethyl heptanoate; 2-Ethyl-2-heptenal; Ethyl isobutyrate; Ethyl isovalerate; Ethyl lactate; Ethyl laurate; Ethyl levulinate; Ethyl maltol; Ethyl 2-methylbutyrate; Ethyl methylphenylglycidate; 2-Ethyl-3-methylpyrazine; Ethyl myristate; Ethyl nitrite; Ethyl 2-nonynoate; Ethyl octanoate; Ethyl oleate; Ethyl oxyhydrate; Ethyl pelargonate; Ethyl phenylacetate; Ethyl-4-phenylbutyrate; Ethyl phenylglycidate; Ethyl-3-phenylpropionate; Ethyl propionate; Ethyl pyruvate; Ethyl salicylate; Ethyl sorbate; Ethyl tiglate; Ethyl undecanoate; Ethyl 10-undecenoate; Ethyl valerate; Ethyl vanillin; Eucalyptol; Eucalyptus extract; Eucalyptus oil; Eugenol; Eugenyl acetate; Eugenyl benzoate; Eugenyl formate; Everlasting extract

Farnesol; Fatty acids; d-Fenchone; Fenchyl alcohol; Fennel; Fennel extract; Fennel oil; Fenugreek; Ferric chloride; Ferric sulfate; Ferrous gluconate; Fir needle oil; Fir needle oil, Siberian type; Formic acid; Fructose; Fumaric acid; Furfural; Furfural mercaptan; Furfuryl alcohol; 2-Furoic acid; (2-Furyl)-2-propanone; Fusel oil refined

Galanga; Garlic extract; Garlic oil; Gentian; Gentian extract; Geraniol; Geranium extract; Geranium oil; Geranyl acetate; Geranyl acetoacetate; Geranyl acetone; Geranyl benzoate; Geranyl butyrate; Geranyl formate; Geranyl hexanoate; Geranyl isobutyrate; Geranyl

Flavorings *(cont'd.)*

isovalerate; Geranyl linalool; Geranyl phenylacetate; Geranyl propionate; Germander extract; Ginger oil; Gluconolactone; D-Glucose anhyd; α-D-Glucose pentaacetate; L-Glutamic acid; L-Glutamine; Glyceryl mono/dioleate; Glyceryl oleate; Glyceryl stearate SE; Glycine; Grapefruit oil; Guaiac extract; Guaiacol; Guaiacyl acetate; Guaiacyl phenylacetate; Guaiene; Guaiol acetate; Guava; Gum benzoin

Heliotropine; γ-Heptalactone; Heptanal; Heptanal dimethyl acetal; Heptanal 1,2-glyceryl acetal; 2,3-Heptanedione; 3-Heptanol; 3-Heptanone; 4-Heptanone; cis-4-Hepten-1-al; Heptyl acetate; Heptyl alcohol; Heptyl butyrate; Heptyl cinnamate; Heptyl formate; Heptyl isobutyrate; Heptyl octanoate; γ-Hexalactone; Hexanal; 2,3-Hexanedione; 3,4-Hexanedione; 2-Hexenal; 2-Hexenol; 3-Hexenol; trans-2-Hexenyl acetate; cis-3-Hexenyl isovalerate; cis-3-Hexenyl 2-methylbutyrate; 3-Hexenyl phenylacetate; Hexyl acetate; 2-Hexyl-4-acetoxytetrahydrofuran; Hexyl alcohol; Hexyl-2-butenoate; Hexyl butyrate; α-Hexylcinnamaldehyde; Hexyl formate; Hexyl hexanoate; 2-Hexylidene cyclopentanone; Hexyl isovalerate; Hexyl 2-methylbutyrate; Hexyl octanoate; Hexyl phenylacetate; Hexyl propionate; Hickory bark; L-Histidine; Histidine hydrochloride monohydrate; Hops extract; Hops oil; Horehound; Horsemint; Horseradish; Hydrocinnamaldehyde; Hydrocinnamic acid; Hydrocinnamic alcohol; Hydrocinnamyl acetate; Hydrolyzed milk protein; Hydrolyzed vegetable protein; Hydroxycitronellal; Hydroxycitronellal diethyl acetal; Hydroxycitronellal dimethyl acetyl; Hydroxycitronellol; 5-Hydroxy-4-octanone; 4-(p-Hydroxyphenyl)-2-butanone; N-Hydroxysuccinic acid; Hyssop

Iceland moss extract; Indole; α-Ionone; β-Ionone; α-Irone; Isoamyl acetate; Isoamyl acetoacetate; Isoamyl alcohol; Isoamyl benzoate; Isoamyl butyrate; Isoamyl cinnamate; Isoamyl formate; Isoamyl 2-furanbutyrate; Isoamyl 2-furanpropionate; Isoamyl hexanoate; Isoamyl isobutyrate; Isoamyl isovalerate; Isoamyl laurate; Isoamyl-2-methylbutyrate; Isoamyl nonanoate; Isoamyl phenylacetate; Isoamyl pyruvate; Isoborneol; Isobornyl acetate; Isobornyl formate; Isobornyl isovalerate; Isobornyl propionate; Isobutyl acetate; Isobutyl acetoacetate; Isobutyl alcohol; Isobutyl angelate; Isobutyl anthranilate; Isobutyl benzoate; Isobutyl butyrate; Isobutyl cinnamate; Isobutyl formate; Isobutyl 2-furanpropionate; Isobutyl heptanoate; Isobutyl hexanoate; Isobutyl isobutyrate; 2-Isobutyl-3-methoxy-pyrazine; α-Isobutylphenethyl alcohol; Isobutyl phenylacetate; Isobutyl propionate; Isobutyl salicylate; 2-Isobutylthiazole; Isobutyl valerate; Isobutyric acid; Isoeugenol; Isoeugenyl benzyl ether; Isoeugenyl ethyl ether; Isoeugenyl formate; Isoeugenyl methyl ether; Isoeugenyl phenylacetate; Isojasmone; L-Isoleucine; α-Isomethylionone; Isopropyl acetate; p-Isopropylacetophenone; Isopropyl alcohol; Isopropyl benzoate; Isopropyl butyrate; Isopropyl cinnamate; Isopropyl formate; Isopropyl hexanoate; Isopropyl isobutyrate; Isopropyl isovalerate; p-Isopropylphenylacetaldehyde; Isopropyl phenylacetate; 3-(p-Isopropylphenyl)-propionaldehyde; Isopropyl propionate; Isopulegol; Isopulegone; Isopulegyl acetate; Isoquinoline; Isovaleraldehyde; Isovaleric acid; Itaconic acid (polymerized)

Jasmine extract; cis-Jasmone; Juniper extract; Juniper oil

Labdanum; Lactic acid; Laurel berries; Laurel leaf oil; Lauric aldehyde; Lauryl acetate; Lauryl alcohol; Lavandin oil; Lavender oil; Lemon extract; Lemongrass extract; Lemongrass oil East Indian; Lemongrass oil West Indian; Lemon juice; Lemon juice extract; Lemon oil; Lemon oil, distilled; Lemon peel extract; Lemon verbena extract; Lepidine; L-Leucine; Levulinic acid; Licorice; Licorice extract; Licorice root extract; Lime oil; d-Limonene; dl-Limonene; l-Limonene; Linaloe wood oil; Linalool; Linalool oxide; Linalyl acetate; Linalyl anthranilate; Linalyl benzoate; Linalyl butyrate; Linalyl cinnamate; Linalyl formate; Linalyl hexanoate; Linalyl isobutyrate; Linalyl isovalerate; Linalyl octanoate; Linalyl propionate; Linden extract; Linden flowers; Linoleic acid; Lovage oil; Lupulin; L-Lysine; L-Lysine L-glutamate; L-Lysine hydrochloride

Mace; Magnesium chloride; Magnesium chloride hexahydrate; Magnesium sulfate anhyd; Magnesium sulfate heptahydrate; Maidenhair fern extract; Malt extract; Maltol; Mandarin orange oil; D-Mannitol; Marjoram oil; Marjoram oil, Spanish; Mate extract; Matricaria extract; Matricaria oil; Mentha arvensis oil; p-Mentha-1,8(10)-dien-9-ol; p-Mentha-1,8(10)-dien-9-yl acetate; Mentha-8-thol-3-one; p-Menth-3-en-1-ol; p-Menth-1-en-9-yl acetate; Menthol; l-Menthone; l-Menthyl acetate; p-Menth-3-yl acetate; Menthyl isovalerate; 2-Mercaptopropionic acid; DL-Methionine; L-Methionine; o-Methoxybenzaldehyde; o-Methoxycinnamaldehyde; 2-Methoxy-4-methylphenol; 4-p-Methoxyphenyl-2-butanone; 1-(4-Methoxyphenyl)-4-methyl-1-penten-3-one; 1-(p-Methoxyphenyl)-1-penten-3-one; 1-(p-Methoxyphenyl)-2-propanone; 2-Methoxy-4-vinylphenol; Methyl acetate; 2-Methylallyl butyrate; Methyl n-amyl ketone; Methyl anisate; o-Methylanisole; p-Methylanisole; Methyl anthranilate; Methyl benzoate; Methylbenzyl acetate, mixed o-, m-, p-; α-Methylbenzyl acetate; α-Methylbenzyl alcohol; α-Methylbenzyl butyrate; α-Methylbenzyl formate; Meth-

Flavorings *(cont'd.)*

ylbenzyl isobutyrate; α-Methylbenzyl propionate; 3-Methyl-2-buten-1-ol; 3-Methyl-2-buten-1-ol; 2-Methyl-3-buten-2-ol; 3-Methyl-2-buten-1-ol; 3-Methyl-2-buten-1-ol; 2-Methylbutyl isovalerate; Methyl p-t-butylphenylacetate; 2-Methylbutyraldehyde; Methyl butyrate; 2-Methylbutyric acid; Methyl caproate; Methyl caprylate; p-Methylcinnamaldehyde; α-Methylcinnamaldehyde; Methyl cinnamate; 2-Methyl-1,3-cyclohexadiene; Methyl cyclopentenolone; Methyl disulfide; Methyl ester of rosin, partially hydrogenated; Methyl ethyl ketone; Methyl eugenol; 6-Methyl-3,5-heptadien-2-one; Methyl heptanoate; 2-Methylheptanoic acid; Methyl-5-hepten-2-ol; Methyl heptenone; Methyl 2-hexanoate; Methyl hexyl ketone; Methyl α-ionone; Methyl β-ionone; Methyl Δ-ionone; Methyl isobutyl ketone; Methyl isobutyrate; Methyl isoeugenol; Methyl isovalerate; Methyl laurate; Methyl mercaptan; Methyl o-methoxybenzoate; Methyl 2-methylbutyrate; Methyl 2-methylthiopropionate; Methyl 3-methylthiopropionate; Methyl 4-methylvalerate; Methyl myristate; Methyl nonanoate; Methyl 2-nonenoate; Methyl nonyl acetaldehyde; Methyl 2-nonynoate; 2-Methyloctanal; Methyl 2-octynoate; Methylparaben; 4-Methyl-2,3-pentanedione; 2-Methylpentanoic acid; Methyl phenylacetate; 3-Methyl-4-phenyl-3-buten-2-one; 2-Methyl-4-phenyl-2-butyl acetate; 2-Methyl-4-phenyl-2-butyl isobutyrate; 3-Methyl-2-phenylbutyraldehyde; Methyl 4-phenylbutyrate; 4-Methyl-1-phenyl-2-pentanone; Methyl 3-phenylpropionate; 2-Methylpropanal; Methyl propionate; 3-Methyl-5-propyl-2-cyclohexen-1-one; Methyl propyl ketone; 2-Methylpyrazine; Methyl salicylate; Methyl sulfide; 5-Methyl-2-thiophene-carboxyaldehyde; 3-Methylthiopropionaldehyde; 2-Methyl-3-tolylpropionaldehyde, mixed o-, m-, p-; Methyl 9-undecenoate; Methyl 2-undecynoate; Methyl valerate; Molasses extract; Monoammonium glycyrrhizinate; Mono- and diglycerides of fatty acids; MSG; Mullein extract; Musk; Mustard, black; Myrcene; Myristaldehyde; Myristyl alcohol; Myrrh extract; Myrrh oil; Myrtle extract

d-Neomenthol; Nerol; Neryl acetate; Neryl butyrate; Neryl formate; Neryl isobutyrate; Neryl isovalerate; Neryl propionate; trans,trans-2,4-Nonadienal; 2,6-Nonadien-1-ol; γ-Nonalactone; Nonanal; 1,3-Nonanediol acetate, mixed esters; Nonanoic acid; 2-Nonanone; 3-Nonanon-1-yl acetate; trans-2-Nonenal; cis-6-Nonen-1-ol; trans-2-Nonen-1-ol; n-Nonyl acetate; Nonyl alcohol; Nonyl isovalerate; Nonyl octanoate; Nootkatone; Nutmeg oil; Nutmeg oil, expressed

Oak bark extract; Oakmoss extract; Ocimene; γ-Octalactone; n-Octanal; Octanal dimethyl acetal; 2-Octanol; 3-Octanol; 3-Octanone; 3-Octanon-1-ol; 1-Octene-3-ol; 1-Octen-3-ol; cis-3-Octen-1-ol; 1-Octen-3-yl acetate; Octyl acetate; 3-Octyl acetate; Octyl butyrate; Octyl formate; Octyl heptanoate; Octyl isobutyrate; Octyl isovalerate; Octyl octanoate; Octyl phenylacetate; Octyl propionate; Olibanum; Olibanum oil; Onion extract; Onion oil; Orange flower extract; Orange flower oil; Orange leaf; Orange oil; Orange oil, distilled; Orange peel extract; Oregano; Origanum oil; Orris root extract

Palmarosa oil; Pansy; Pansy extract; Paprika; Parsley; Parsley extract; Parsley seed oil; Passionflower extract; Patchouli; Peach kernel oil; Peach leaves; PEG-4; PEG-6; PEG-8; PEG-9; PEG-12; PEG-14; PEG-16; PEG-20; PEG-32; PEG-40; PEG-75; PEG-100; PEG-150; PEG-200; Pelargonyl vanillylamide; Pennyroyal extract; Pennyroyal oil; Pennyroyal oil, American; Pentadecalactone; Pentane-2,3-dione; 4-Pentenoic acid; 1-Penten-3-ol; Peppermint extract; Peppermint leaves; Peppermint oil; Perillaldehyde; Perillyl acetate; Perillyl alcohol; Petitgrain oil; α-Phellandrene; Phenethyl alcohol; Phenethyl anthranilate; Phenethyl benzoate; Phenethyl butyrate; Phenethyl cinnamate; Phenethyl formate; Phenethyl isobutyrate; Phenethyl isovalerate; 2-Phenethyl 2-methylbutyrate; Phenethyl phenylacetate; Phenethyl propionate; Phenethyl salicylate; Phenethyl senecioate; Phenethyl tiglate; Phenoxyacetic acid; Phenoxyethyl isobutyrate; Phenylacetaldehyde; Phenylacetaldehyde 2,3-butylene glycol acetal; Phenylacetaldehyde diisobutyl acetal; Phenylacetaldehyde dimethyl acetal; Phenylacetaldehyde glyceryl acetal; Phenylacetic acid; L-Phenylalanine; 4-Phenyl-2-butanol; 4-Phenyl-3-buten-2-ol; 4-Phenyl-3-buten-2-one; 4-Phenyl-2-butyl acetate; 2-Phenylethyl acetate; 2-Phenylethyl methyl ether; 1-Phenyl-3-methyl-3-pentanol; 2-Phenylpropanal; 3-Phenylpropanal; 1-Phenyl-1-propanol; 2-Phenyl propanol-1; 2-Phenylpropionaldehyde dimethylacetal; 2-Phenylpropyl butyrate; 3-Phenylpropyl cinnamate; 3-Phenylpropyl formate; 3-Phenylpropyl hexanoate; 2-Phenylpropyl isobutyrate; 3-Phenylpropyl isobutyrate; 3-Phenylpropyl isovalerate; 3-Phenylpropyl propionate; 2-(3-Phenylpropyl)-tetrahydrofuran; Phosphoric acid; Pimenta leaf oil; Pimenta oil; Pimento extract; Pineapple extract; Pineapple juice; α-Pinene; β-Pinene; Pine needle oil, dwarf; Pine needle oil, Scotch type; Pine oil; Pine tar oil; Pine, white, bark; Pinocarveol; Piperidine; Piperine; Piperitenone; Piperitenone oxide; d-Piperitone; Piperonyl acetate; Piperonyl isobutyrate; Polyethylene glycol; Polylimonene; Polysorbate 20; Polysorbate 60; Polysorbate

Flavorings *(cont'd.)*

80; Poppy seed; Potassium acetate; Potassium carbonate; Potassium caseinate; Potassium chloride; Potassium citrate; Potassium glutamate; Potassium lactate; Potassium sulfate; Propenylguaethol; 2-Propenyl phenylacetate; Propionaldehyde; Propionic acid; Propyl acetate; Propyl alcohol; p-Propyl anisole; Propyl benzoate; Propyl butyrate; Propyl cinnamate; Propyl disulfide; Propylene glycol; Propylene glycol alginate; Propyl formate; Propyl 2-furanacrylate; Propyl gallate; Propyl heptanoate; Propyl hexanoate; 3-Propylidenephthalide; Propyl isobutyrate; Propyl isovalerate; Propyl mercaptan; Propylparaben; α-Propylphenethyl alcohol; Propyl phenylacetate; Propyl propionate; Pulegone; PVP; Pyridine; Pyroligneous acid extract; Pyruvaldehyde; Pyruvic acid

Quassia; Quillaja; Quillaja extract; Quince seed; Quinine; Quinine dihydrochloride; Quinine hydrochloride; Quinine sulfate; Quinoline

Red saunders; Rhodinol; Rhodinyl acetate; Rhodinyl butyrate; Rhodinyl formate; Rhodinyl isobutyrate; Rhodinyl isovalerate; Rhodinyl phenylacetate; Rhodinyl propionate; Rhubarb extract; Rose buds and flowers; Rose geranium oil; Rose hips; Rose leaves; Rosemary oil; Rose oil; Rose water; Rue; Rue oil

Saffron; Safrol; Sage; Sage extract; Sage oil; Sage oil, Dalmatian type; Saint John's bread; Saint John's wort; Salicylaldehyde; Sambucus oil; Sandalwood; Sandalwood extract; Sandalwood oil; Sandarac gum; α and β-Santalol; Santalyl acetate; Santalyl phenylacetate; Sarsaparilla extract; Sassafras extract; Sassafras oil; Savory; Savory extract; Senna extract; L-Serine; Serpentaria; Serpentaria extract; Sesame; Sesame oil; Skatole; Sodium acetate trihydrate; Sodium benzoate; Sodium carbonate; Sodium chloride; Sodium diacetate; Sodium fumarate; Sodium lactate; Sodium metabisulfite; Sodium propionate; Sodium succinate; Sodium sulfoacetate derivs. of mono- and diglycerides; Sorbic acid; Sorbitan stearate; Sorbitol; Soy sauce; Spearmint oil; Spike lavender oil; Spruce; Star anise; Stearic acid; Storax; Styrene; Succinic acid; Sucralose; Sucrose; Sucrose octaacetate; Sweet marjoram oil; Swertia extract

Tagetes extract; Tamarind; Tangerine oil; Tannic acid; Tansy extract; Tansy oil; Tarragon; L-Tartaric acid; α-Terpinene; γ-Terpinene; α-Terpineol; β-Terpineol; Terpineol anhydrous; Terpinolene; Terpinyl acetate; Terpinyl anthranilate; Terpinyl butyrate; Terpinyl cinnamate; Terpinyl formate; Terpinyl isobutyrate; Terpinyl isovalerate; Terpinyl propionate; Tetrahydrofurfuryl acetate; Tetrahydrofurfuryl alcohol; Tetrahydrofurfuryl butyrate; Tetrahydrofurfuryl propionate; Tetrahydrolinalool; Tetrahydro-pseudo-ionone; Tetramethyl ethylcyclohexenone; 2,3,5,6-Tetramethylpyrazine; Thaumatin; Thea sinensis extract; Thiamine HCl; 2-Thienyl mercaptan; L-Threonine; Thyme extract; Thyme oil; Thyme oil red; Thyme, wild; Thymol; Tolualdehyde glyceryl acetal, mixed o-, m-, p-; Tolualdehydes, mixed o-, m-, p-; p-Tolylacetaldehyde; p-Tolyl aldehyde; 4-(p-Tolyl)-2-butanone; p-Tolyl isobutyrate; p-Tolyl laurate; p-Tolyl phenylacetate; 2-(p-Tolyl) propionaldehyde; Triacetin; Tributyl acetylcitrate; Tributyrin; 2-Tridecenal; Trimethylamine; Trimethyldodecatrieneol; 2,4,5-Trimethyl δ-3-oxazoline; 2,3,5-Trimethylpyrazine; Trimethyl thiazole; L-Tryptophan; Tuberose; Turmeric; Turmeric extract; Turpentine; L-Tyrosine

2,3-Undecadione; γ-Undecalactone; ς-Undecalactone; Undecanal; 2-Undecanone; 9-Undecenal; 10-Undecenal; Undecen-1-ol; 2-Undecenol; 10-Undecen-1-yl acetate; Undecyl alcohol; Undecylenic acid

n-Valeraldehyde; Valerian; Valerian extract; n-Valeric acid; γ-Valerolactone; L-Valine; Vanilla; Vanillin; Vanillin acetate; Vegetable protein; Veratraldehyde; Verbena extract; Verbenol; Veronica extract; Viburnum; Violet

Walnut; Walnut oil; Wild cherry bark extract; Wild marjoram extract; Wild thyme extract; Wintergreen; Woodruff extract; Woodruff, sweet; Wormwood oil

Yarrow; Yarrow extract; Yarrow oil; Yerba santa extract; Ylang ylang oil; Yucca

Zedoary oil; Zingerone

Humectants • Moisture barriers

Trade names: Amalty®; Aqualon® 7M8SF; Aqualon® 7MF; Aqualon® 9H4XF; Atlas 3000; Atmos® 300; Avicel® WC-595

Binasol™ 81; Bindox-HV-051; Bindox-LV-050; Britol®; Britol® 6NF; Britol® 7NF; Britol® 9NF; Britol® 20USP; Britol® 35USP; Britol® 50USP

Cake Moist; Cake Sta; CarraFat™; Celish® FD-100F; Cellogen HP-5HS; Cellogen HP-6HS9; Cellogen HP-8A; Cellogen HP-12HS; Cleartic; CNP BRSMC10; CNP BRSMC20; CNP BRSMC35; CNP BRSMC35CL; CNP BRSSHG40; CNP RSSHG40; CNP RSSHG40CL;

Humectants *(cont'd.)*

CNP RSSHM; CNP RSSHMCL; Colloid 451T; Colloid 515MT; Colloid 787; Colloid 1023T; Colloid XC7408; CornSweet® 95; CronSweet® Crystalline Fructose; Crodaglaze; Crodamol GTC/C; Crodyne BY-19; Curafos® 11-2; Curafos® 22-4

Diagum LBG; Durkex 100BHA; Durkex 100BHT; Durkex 100F; Durkex 100TBHQ; Durkex 500; Durkex 500S; Durkote Vitamin B-1/Hydrog. Veg. Oil; Durkote Vitamin C/Hydrog. Veg. Oil

Emery® 912; Emery® 916; Emery® 917; Emery® 918; Emphos™ F27-85; Emulgum™; Extramalt 10; Extramalt 33; Extramalt 35; Extramalt Dark; Extramalt Light

Flav-R-Keep® FP-51; 42/43 Corn Syrup; Freez-Gard® FP-88E

Gelex® Instant Starch; Glycon® G 100; Glycon® G-300

Hodag GMS; Hydex® 100 Coarse Powd; Hydex® 100 Coarse Powd. 35; Hydex® 100 Gran. 206; Hydex® 100 Powd. 60; Hydex® Tablet Grade; Hystar® 3375; Hystar® 4075; Hystar® 5875; Hystar® 6075; Hystar® 7570; Hystar® HM-75

Idealgum 1A; Idealgum 1B; Idealgum 1C; Idealgum 2A; Idealgum 2B; Idealgum 3F; Instant Tender-Jel® 434; Instant Tender-Jel® C

Justfiber® CL-100-H

Karaya Gum #1 FCC; Kel-Lite™ BK; Kelmar® Improved; Kena® FP-28

Liponic Sorbitol Powd; Liponic Sorbitol Sol'n. 70% USP; Litesse®; Locust Bean Gum Speckless Type D-200; Lo-Temp® 452

Maltrin® M040; Maltrin® QD M440; Marine Colloids™ Carrageenan; Merecol® RB; Merecol® RCS; Methocel® A15LV Premium; Methocel® F50LV Premium; Methocel® K100LV Premium; Microlube System 999R; Myvacet® 5-07; Myvacet® 5-07K; Myvacet® 7-00; Myvacet® 7-07

NatuReal™ Essence; 90/10 Blend; Non-Diastatic Malt Syrup #40600; No Stick Emulsifier; Nutrifos® 088; Nutrifos® B-75; Nutrifos® H-30; Nutrifos® L-50; Nutrifos® Powd; Nutrifos® SK; Nutriloid® Xanthan

Oxipur; Oxipur-Mix

Patlac® NAL; Perfect Slice; Pisane®; Polarin® Range; Polypro 5000®; Powdered Gum Karaya Superfine #1 FCC; Powdered Gum Karaya Superfine XXXX FCC; Powdered Locust Bean Gum Type D-200; Powdered Locust Bean Gum Type D-300; Powdered Locust Bean Gum Type P-100; Powdered Locust Bean Gum Type PP-100; Powdered Tragacanth Gum Type A/10; Powdered Tragacanth Gum Type E 1; Powdered Tragacanth Gum Type C 3; Powdered Tragacanth Gum Type L; Powdered Tragacanth Gum Type W; Prevent-O; Procon® 2000; Procon® 20/60; Procon® 20/60 Military; Procon® 20/60 SL; Promax® 70; Promax® 70L; Promax® 70LSL; Promax® Plus; Promine® DS-SL; Promine® HV; Protopet® Alba; Protopet® White 1S; Protopet® White 2L; Protopet® White 3C; Protopet® Yellow 2A; Purasal® P/USP 60; Purasal® S/SP 60

Readi-Glaze Systems; Rice-Trin® 25; Rice-Trin® 35; Rudol®; Ryoto Ester KA

Seal N Glaze Systems for Vegetables and Fruits; 721-A™ Instant Starch; Soalocust®; Sodaphos®; Soft-Set®; Sorbo®; Soyamin 70; Soyamin 90; Staley® Redried Starch A; Staley® Redried Starch B; Staley® Tapioca Dextrin 11; Sta-Lite™ 100C; Sta-Lite™ 100CN Neutralized; Sta-Lite™ 100CN Neutralized; Sta-Lite™ 100F; Sta-Lite™ 100CN Neutralized; Sta-Lite™ 100CN Neutralized; Stamere® CKM FCC; Star; Star-Dri® 55; Super Flex; Superol; Swelite®; Swelite® Micro

Tegomuls® 4100; Tegomuls® M; Tenderbite SF(TB-SF); Ticaloid® 1070; Ticaloid® Lite; Ticaloid® No Fat 102B; Ticalose® 15; Ticalose® 30; Ticalose® 150 R; Ticalose® 700 R; Ticalose® 750; Ticalose® 2000 R; Ticalose® 2500; Ticalose® 4000; Ticalose® 4500; Ticalose® 5000 R; Ticaxan® Regular; TIC Pretested® Arabic FT Powd; TIC Pretested® Arabic PH-FT; TIC Pretested® CMC 2500 S; TIC Pretested® CMC PH-2500; TIC Pretested® Gum Guar 8/22 FCC/NF Powd; TIC Pretested® Locust Bean POR/A; TMP-65; Tragacanth Flake No. 27; Tragacanth Gum Ribbon No. 1; Type B Dough Conditioner

Ultra-Bake® 3100, 3200; Ultra-Bake® 3100K, 3200K; Ultra-Bake® NF

Vascomix 404

Wakal® J; Wakal® JG

Chemicals: Acetylated hydrogenated cottonseed glyceride; Agar; Ammonium alginate
Calcium chloride; Calcium lactate; Carboxymethylmethylcellulose
Glycerin
Hydrolyzed gelatin
Invert sugar; Lactose
Malt extract; D-Mannitol
Pea protein concentrate; Pentapotassium triphosphate; Pentasodium triphosphate; β-Pinene; Polydextrose; Potassium acid tartrate; Potassium lactate; Potassium phosphate; Potassium

Humectants *(cont'd.)*
phosphate dibasic; Potassium polymetaphosphate; Propylene glycol
Sodium lactate; Sodium metaphosphate; Sorbitol
L-Tartaric acid; Terpene resin, natural; Tetrapotassium pyrophosphate; Triacetin

Instantizing agents

Trade names: Alcolec® F-100; Alcolec® FF-100; Alcolec® Granules; Alcolec® Z-3
Beakin LV1; Beakin LV2; Beakin LV3; Beakin LV30
Centrolex® C; Centrolex® D; Centrolex® P; Centrolex® R; Centromix® E; Centrophase® NV
Lecigran™ 5750
M-C-Thin® ASOL; M-C-Thin® ASOL 436; Metarin™; Metarin™ C; Metarin™ CP; Metarin™ F;
Metarin™ P

Leavening agents • Raising agents

Trade names: ABC-Trieb®; actif•8®; actif•8® Hi-Calcium; ADA Tablets; ADM Baking Powd; ADM Cream Acid
Salt; Advitagel; AeroLite LP; Aerophos P; Aerophos X; Ajax®
BL-60®; B.P. Pyro®; B.P. Pyro® Type K
Cap-Shure® GDL-140-70
Donut Pyro®; Donut SAPP; Durkote Citric Acid/Hydrog. Veg. Oil; Levair®
Pationic® 1230; Pationic® 1250; Perfection®; Py-ran®
Regent® 12XX; Royal Baking Powd.
Sapp #4; SAPP 22; SAPP 26; SAPP 28; SAPP 40; SAS Baking Powd; Stabil® 9 High Calcium
Blended Powd.
V-90®; Victor Cream®

Chemicals: Ammonium bicarbonate; Ammonium chloride; Ammonium hydroxide; Ammonium phosphate;
Ammonium phosphate, dibasic; L-Ascorbic acid
Calcium citrate; Calcium dihydrogen pyrophosphate; Calcium lactate; Calcium phosphate
monobasic anhydrous; Calcium phosphate monobasic monohydrate; Calcium phosphate
tribasic; Calcium sulfate; Carbon dioxide
Disodium pyrophosphate
Fumaric acid
Gluconolactone
Lactic acid
Potassium acid tartrate; Potassium bicarbonate; Potassium polymetaphosphate
Sodium acid pyrophosphate; Sodium aluminum phosphate acidic; Sodium bicarbonate; Sodium
fumarate; Sodium metaphosphate; Succinic anhydride
Tetrapotassium pyrophosphate

Lubricants • Release agents • Antisticking agents

Trade names: AA USP; Acconon 300-MO; Admul WOL 1403; Alcolec® 30-A; Alcolec® BS; Alcolec® HWS;
Alcolec® S; Alcolec® SFG; Aldo® MCT KFG; Aldo® MSD FG; Aldo® MSD KFG; Aldo® MS
FG; Amidan; Amisol™ 210-L; Amisol™ 210 LP; Amisol™ 329; Amisol™ 683 A; Amisol™ 785-
15K; Aqualon® 7LF; Asol; Atlas 3000; Atlas 5520; Atmos® 150; Axol® E 61; Axol® E 66
Bake-Well 52; Bake-Well 80/20; Bake-Well Bun Release; Bake-Well Heavy Divider Oil; Bake-
Well High Stability Pan Oil; Bake-Well K Machine Oil; Beakin LV1; Beakin LV2; Beakin LV4;
Beakin LV30; Benol®; Blandol®; Bread Pan Oil; Britol®; Britol® 6NF; Britol® 7NF; Britol®
9NF; Britol® 20USP; Britol® 35USP; Britol® 50USP; Bro-Eze; Bro-Eze III
Cake Pan Release; Cake Sta; Carbowax® Sentry® PEG 300; Carbowax® Sentry® PEG 400;
Carbowax® Sentry® PEG 540 Blend; Carbowax® Sentry® PEG 600; Carbowax® Sentry®
PEG 900; Carbowax® Sentry® PEG 1000; Carbowax® Sentry® PEG 1450; Carbowax®
Sentry® PEG 3350; Carbowax® Sentry® PEG 4600; Carbowax® Sentry® PEG 8000;
Carnation®; Centrolene® A; Centromix® E; Centrophase® 152; Centrophase® C;

Lubricants *(cont'd.)*

Centrophase® HR; Centrophase® HR2B; Centrophase® HR4B; Centrophase® HR4U; Centrophase® HR6B; Centrophase® NV; Centrophil® M; Centrophil® W; Cetodan® 90-50; Cithrol GDO N/E; Coconad RK; Coco-Spred™; Confecto™ Lubes; Confecto No-Stick™; Confecto™ Rubs; Confecto™ Spreds; Crester RT; Crill 35; Crodamol GTC/C

Delios® S; Delios® V; Dimodan PV; Dimodan PV 300 Kosher; Divider Oil 90; Divider Oil 210; Drakeol 5; Drakeol 7; Drakeol 9; Drakeol 13; Drakeol 34; Drewmulse® 10K; Drewmulse® 200; Drewmulse® 365; Drewmulse® 900K; Drewmulse® D-4661; Duratex; Dur-Em® GMO; Durkex 100BHA; Durkex 100BHT; Durkex 100F; Durkex 100L; Durkex 500; Durkex 500S; Dynacet® 285; Dynasan® 114; Dynasan® 119

Eez-Out™; Emphos™ D70-30C; Emphos™ D70-31; Emphos™ F27-85; Emulpur™ N P-1; Emulthin M-35; Ervol®; Estasan GT 8-40 3578; Estasan GT 8-60 3575; Estasan GT 8-60 3580; Estasan GT 8-65 3577; Estasan GT 8-65 3581

Fancol OA-95

Gloria®; Glycon® P-45; Glycon® S-65; Glycon® S-70; Glycon® S-90; Glycon® TP; GMS 400; GMS 400V; GMS 900; Guar Gum

Hodag GMS; Hystrene® 5016 NF FG; Hystrene® 7018 FG; Hystrene® 8718 FG; Hystrene® 9718 NF FG

Iconol NP-7; Industrene® 143; Industrene® 205; Industrene® 225; Industrene® 226 FG; Industrene® 365; Industrene® 4518; Industrene® 5016 NF FG; Industrene® 7018 FG; Industrene® 8718 FG; Industrene® B

Kaydol®; Kemamide® O; Kemamide® S; Kemamide® U; Kemamide® W-39; Kemamide® W-40; Kemamide® W-45

Lamemul® K 1000-Range; Lamemul® K 2000-Range; Lecigran™ 5750; Lecimulthin; Leciprime™; Liposorb L-20; Liposorb S; Lipovol HS; Lipovol PAL; Lipovol SOY

Maltrin® M040; Masil® SF 5; Masil® SF 10; Masil® SF 20; Masil® SF 50; Masil® SF 100; Masil® SF 200; Masil® SF 350; Masil® SF 350 FG; Masil® SF 500; Masil® SF 1000; Masil® SF 5000; Masil® SF 10,000; Masil® SF 12,500; Masil® SF 30,000; Masil® SF 100,000; Masil® SF 300,000; Masil® SF 500,000; Masil® SF 600,000; Masil® SF 1,000,000; Mazol® GMS; Metarin™; Methocel® E3 Premium; Methocel® E4M Premium; Methocel® E6 Premium; Methocel® E15LV Premium; Methocel® E50LV Premium; Methocel® F50LV Premium; Methocel® K4M Premium; Methocel® K15M Premium; Methocel® K100LV Premium; Methocel® K100M Premium; Microlube System 999R; Miglyol® 810; Miglyol® 812; Miglyol® 812 S; Miglyol® 820; Miglyol® 840; Miglyol® 850; Mira-Set® J; Myvacet® 0 08; Myvacet® 0 40; Myvacet® 9-45; Myvacet® 9-45K; Myvaplex® 600; Myvaplex® 600K; Myvaplex® 600P; Myvaplex® 600PK; Myvatex® Mighty Soft®; Myverol® 18-04; Myverol® 18-06; Myverol® 18-07

Neobee® 18; Neobee® M-20; No Stick Emulsifier; Nuodex S-1421 Food Grade; Nuodex S-1520 Food Grade; Nuodex Magnesium Stearate Food Grade

Orzol®

Pano™; Pano-Lube™; Panospray™; Panospray™ LS; Panospray™ SQ; Panospray™ WL; Perlatum® 400; Perlatum® 410; Perlatum® 410 CG; Perlatum® 415; Perlatum® 415 CG; Perlatum® 420; Perlatum® 425; Perlatum® 510; Polyaldo® HGDS KFG; Powdered Gum Guar NF Type 80 Mesh B/T; Powdered Gum Guar Type ECM; Powdered Gum Guar Type M; Powdered Gum Guar Type MM FCC; Powdered Gum Guar Type MM (HV); Powdered Gum Guar Type MMM $^1/_2$; Powdered Gum Guar Type MMW; Protol®; Protopet® Alba; Protopet® White 1S; Protopet® White 2L; Protopet® White 3C; Protopet® Yellow 2A

Quik Release

Rudol®; Ryoto Sugar Ester B-370; Ryoto Sugar Ester L-195; Ryoto Sugar Ester L-595; Ryoto Sugar Ester L-1570; Ryoto Sugar Ester L-1695; Ryoto Sugar Ester LN-195; Ryoto Sugar Ester M-1695; Ryoto Sugar Ester O-170; Ryoto Sugar Ester O-1570; Ryoto Sugar Ester P-1570; Ryoto Sugar Ester P-1570S; Ryoto Sugar Ester P-1670; Ryoto Sugar Ester S-070; Ryoto Sugar Ester S-170; Ryoto Sugar Ester S-170 Ac; Ryoto Sugar Ester S-270; Ryoto Sugar Ester S-370; Ryoto Sugar Ester S-570; Ryoto Sugar Ester S-770; Ryoto Sugar Ester S-970; Ryoto Sugar Ester S-1170; Ryoto Sugar Ester S-1170S; Ryoto Sugar Ester S-1570; Ryoto Sugar Ester S-1670; Ryoto Sugar Ester S-1670S

Santone® 10-10-O; Selin® G deo; Servil® Range; SF18-350; S-Maz® 20; S-Maz® 80K; S-Maz® 85K; S-Maz® 95; Softenol® 3108; Sorbo®; Soy Oil; Special Oil 107; Special Oil 1739; Spezyme BBA; Staley® Dusting Starch; Sterotex® C; Sterotex® HM NF; Sterotex® NF; Super White Fonoline®; Super White Protopet®

Tandem 552; Tegomuls® 90; Tegomuls® 90 K; Tegomuls® 90 S; Tegomuls® 9102; Tegomuls® 9102 S; Thermolec 57; Thermolec 68; Ticaloid® Lite; Ticalose® 15; Ticalose® 30; Trough

Lubricants *(cont'd.)*

Grease
Viscarin® ME-389; Vykacet T/L
Wecobee® FS; Wecobee® FW; Wecobee® HTR; Wecobee® M; Wecobee® R Mono; Wecobee® SS; Witafrol® 7420
Yelkin SS

Chemicals: Acetylated hydrogenated coconut glycerides; Acetylated hydrogenated soybean oil glyceride; Acetylated mono- and diglycerides of fatty acids
Beeswax
C8-10 fatty acid triglyceride; Calcium carbonate; Calcium silicate; Calcium stearate; Candelilla wax; Capric acid; Caprylic acid; Caprylic/capric acid; Carnauba; Castor oil; Coconut oil; Cottonseed oil
Dimethicone
Ethylene distearamide
Glyceryl dioleate; Glyceryl dioleate SE; Glyceryl distearate; Glyceryl distearate SE; Glyceryl laurate SE; Glyceryl oleate SE; Glyceryl ricinoleate SE; Glyceryl stearate; Glyceryl tricaprylate/caprate; Glyceryl trienanthate
Hydrogenated sperm oil; Hydrogenated stearic acid; Hydrogenated vegetable glycerides phosphate; Hydrogenated vegetable oil
Lauric acid; Lecithin; Linoleamide
Magnesium carbonate hydroxide; Magnesium oxide; Magnesium silicate; Magnesium stearate; D-Mannitol; Microcrystalline cellulose; Microcrystalline wax; Mineral oil; Mono- and diglycerides of fatty acids; Mono- and diglycerides, sodium phosphate derivs; Myristic acid
Nonoxynol-7
Oleamide; Oleic acid; Oleyl alcohol; Oxystearin
Palmitamide; Palmitic acid; Palm oil; PEG-4; PEG-6; PEG-8; PEG-9; PEG-12; PEG-14; PEG-16; PEG-20; PEG-32; PEG-40; PEG-75; PEG-100; PEG-150; PEG-200; PEG-6 oleate; PEG-8 oleate; Petrolatum; Polyethylene glycol
Rice bran wax
Sodium glyceryl oleate phosphate; Sorbitol; Soy acid; Soybean oil; Stearamide; Stearic acid; Sucrose dilaurate; Sucrose distearate; Sucrose laurate; Sucrose myristate; Sucrose polylaurate; Sucrose polylinoleate; Sucrose polyoleate; Sucrose stearate; Sucrose tetrastearate triacetate; Sucrose tribehenate
Talc; Tallow acid; Trimyristin; Tristearin

Masticatory aids • Chewing gum ingredients

Trade names: AF 72; AF 75
Be Square® 185; Be Square® 195
Centrolex® F; Centrophil® K; Cetodan®; Cetodan® 90-50
Hercules® Ester Gum 8D; Hercules® Ester Gum 8D-SP; Hercules® Ester Gum 10D
Liponic Sorbitol Powd; Liponic Sorbitol Sol'n. 70% USP
Mekon® White; Monomax AH90 B; Monomax VH90 B; Monomuls® 60-10; Monomuls® 60-15; Monomuls® 60-20; Monomuls® 60-25/2; Monomuls® 60-25/5; Monomuls® 60-30; Monomuls® 60-35; Monomuls® 60-40; Monomuls® 60-45; Monomuls® 90-10; Monomuls® 90-15; Monomuls® 90-20; Monomuls® 90-25; Monomuls® 90-25/2; Monomuls® 90-25/5; Monomuls® 90-30; Monomuls® 90-35; Monomuls® 90-40; Monomuls® 90-45; Multiwax® 180-M; Multiwax® ML-445; Multiwax® W-445; Myvacet® 5-07; Myvacet® 7-07; Myvacet® 9-08; Myverol® 18-06; Myverol® 18-06K; Myverol® 18-07K
Natural Arabic Type Gum Purified, Spray-Dried
Petrolite® C-1035; Piccolyte® C115; Polyaldo® HGDS KFG; Polysar Butyl 101-3; Polywax® 500; Polywax® 655
Ross Candelilla Wax; Ross Carnauba Wax; Ross Rice Bran Wax; Ryoto Sugar Ester ER-190; Ryoto Sugar Ester ER-290; Ryoto Sugar Ester P-170; Ryoto Sugar Ester S-570; Ryoto Sugar Ester S-770; Ryoto Sugar Ester S-970
Spray Dried Gum Arabic NF/FCC CM; Spray Dried Gum Arabic NF/FCC CS (Low Bacteria); Spray Dried Gum Arabic NF/FCC CS-R; Sta-Lite™ 100C; Sta-Lite™ 100CN Neutralized; Sta-Lite™ 100CN Neutralized; Sta-Lite™ 100F; Sta-Lite™ 100CN Neutralized; Sta-Lite™ 100CN Neutralized; Starwax® 100; Staybelite® Ester 5; 07 Stearine; Sustane® 1-F; Sustane® BHA

Masticatory aids *(cont'd.)*

Tenite®; Tenox® BHA; Tenox® BHT; Tenox® GT-1; Tenox® GT-2
Ultraflex®
Victory®
Zea Red; Zonester® 85

Chemicals: Acetylated hydrogenated cottonseed glyceride; Acetylated hydrogenated soybean oil glyceride
Balsam Canada; Balsam copaiba; Beeswax; Butadiene-styrene copolymer; Butadiene-styrene 50/50 rubber; Butadiene-styrene 75/25 Rubber
Calcium phosphate tribasic; Candelilla wax; Carnauba; Chicle; Chilte; Chiquibul; Cocoa butter; Copal resin; Crown gum
Dimethicone
Ester gum
Gelatin; Glycerin; Glyceryl caprylate; Glyceryl ester of tall oil rosin; Glyceryl hydrogenated rosinate; Guaiac gum; Gum benzoin; Gutta percha
Isobutylene/isoprene copolymer
Jojoba wax
Kauri gum; Kelp
Lanolin; Lard
Methyl acetyl ricinoleate
Olibanum; Ozokerite
Paraffin; Pentaerythrityl rosinate; Petroleum wax; Polydipentene; Polyethylene; Polyisobutene; Polyvinyl acetate (homopolymer); Potassium stearate; Potassium tripolyphosphate
Rice bran wax; Rosin
Simethicone; Sodium stearate; Stearic acid; Sucrose distearate; Synthetic wax
Talc; Tall oil glycerides; Tallow; Terpene resin, natural

Nutrients • Dietary supplements • Dietary fiber

Trade names: Aerophos X; Albrite Dicalcium Phosphate Anhyd; Albrite Monoammonium Phosphate Food Grade; Alcolec® Granules; Aldo® MCT KFG; Aloe Con WLG 200; Aloe Vera Gel 1X; Aloe Vera Gel 10X; Aloe Vera Gel 40X; Alomine 042; Anterior Pituitary Substance; Arbran™ Soy Fiber; Arcon F; Arcon G, Fortified Arcon G; Arcon S, Fortified Arcon S; Arcon T, Fortified Arcon T; Arcon VF, Fortified Arcon VF; Ardex D-HD; Ardex DHV Dispersible; Ardex F, F Dispersible; Ardex FR; Ardex R; Ascorbic Acid USP/FCC, 100 Mesh; Ascorbic Acid USP, FCC Fine Gran. No. 6045655; Ascorbic Acid USP, FCC Fine Powd. No. 6045652; Ascorbic Acid USP, FCC Gran. No. 6045654; Ascorbic Acid USP, FCC Type S No. 6045660; Ascorbic Acid USP, FCC Ultra-Fine Powd No. 6045653; Atlas 3000
Bakers Nutrisoy; Beanfeast; Beef Extract Paste #9225; Beef Extract Powd. #9267; Beta Carotene Emulsion Beverage Type 3.6 No. 65392; Bindox-LV-050; Bioblend®; d-Biotin USP, FCC No. 63345; Bitrit-1™ (1% Biotin Trituration No. 65324); Bone Gelatin Type B 200 Bloom; Bone Marrow Powd; Brain Substance
Calcium Ascorbate FCC No. 60475; Calcium Pantothenate USP, FCC Type SD No. 63924; Calfskin Gelatin Type B 200 Bloom; Calfskin Gelatin Type B 225 Bloom; Calfskin Gelatin Type B 250 Bloom; Caliment Dicalcium Phosphate Dihydrate; Cap-Shure® AS-125-50; Cap-Shure® AS-165-70; Cap-Shure® AS-165CR-70; Cap-Shure® FE-165-50; Cap-Shure® FS-165-60; Cap-Shure® KCL-140-50; Cap-Shure® KCL-165-70; Captex® 300; Captex® 350; Captex® 355; Cegeprot® Range; Centex®; CNP RPCCG; CNP RPCXFG; Coated Ascorbic Acid 97.5% No. 60482; Coral Star; Cozeen®; Crodroit CS
Dairy-Lo™; Danpro DS; Danpro HV; Danpro S; Danpro S-760; Desiccated Beef Liver Granular Undefatted; Desiccated Beef Liver Granular Defatted; Desiccated Beef Liver Powd; Desiccated Beef Liver Powd. Defatted; Desiccated Hog Bile; Desiccated Ox Bile; Desiccated Pork Liver Powd; Design C200; Destab™; Dry Vitamin A Palmitate Type 250-SD No. 65378; Dry Vitamin D_3 Beadlets Type 850 No. 652550401, 652550601; Dry Vitamin D_3 Type 100 CWS No. 65242; Dry Vitamin D_3 Type 100-SD No. 65216; Duodenal Substance; Durkote Calcium Carbonate/Starch, Acacia Gum; Durkote Ferrous Fumarate/Hydrog. Veg. Oil; Durkote Ferrous Sulfate/Hydrog. Veg. Oil; Durkote Vitamin B-1/Hydrog. Veg. Oil; Durkote Vitamin C/Hydrog. Veg. Oil; Durkote Vitamin C/Maltodextrin
Edible Beef Gelatin; Edicol® ULV Series; Emulbesto; Enrichment R Tablets; Epamarine®; Epicholin; Epikuron™ 100 P, 100 G; Epikuron™ 100 X; Epikuron™ 130 G; Epikuron™ 130

Nutrients *(cont'd.)*

P; Epikuron™ 130 X; Epikuron™ 135 F; Epikuron™ 145V; Exafine® 250; Exafine® 500; Exafine® 1000; Exafine® 2000; Extract of Hog Bile; Extract of Ox Bile NF XI; Extramalt 10; Extramalt 35; Extramalt Dark; Extramalt Light

Fibrex®; Fibrim® 1020; Fibrim® 1250; Fibrim® 1255; Fibrim® 1450; Fibrim® 2000; Fish Liver Oil; Flavorset® GP-2; Folic Acid 10% Trituration No. 69997; Folic Acid USP, FCC No. 20383; FP 940; Freeze Dried Beef Liver Powd; Freeze Dried Beef Liver Powd. Defatted; Freeze Dried Pork Liver Powd

Gluconal® CA A; Gluconal® CA M; Gluconal® CA M B; Gluconal® CO; Gluconal® CU; Gluconal® FE; Gluconal® K; Gluconal® MG; Gluconal® MN; Gluconal® NA; Gluconal® ZN; Green Pea Fiber; Grits 'N' Bran

Halibut Liver Oil 60,000A/1,000D; Heart Substance; HVP Replacer Powder; Hypothalamus Substance; Hyprol

IPSO-C403; IPSO-FC; IPSO-MR Dispersible; IPSO-NGL; Iron Bile Salts

Jerusalem Arthichoke Flour (JAF)

Kemester® 2050; Kemester® 4516; Kidney Substance

Lecipur™ 95 C; Lecipur™ 95 R; Liver Conc. Paste; Lung Substance; Lutavit® Niacin; Lymphatic Substance

Magnesium Hydroxide USP; Magnesium Hydroxide USP DC; Maltrin® M040; Maltrin® M050; Maltrin® M100; Maltrin® M150; Maltrin® M180; Maltrin® M510; Maltrin® M700; Maltrin® QD M440; Maltrin® QD M500; Maltrin® QD M550; Mammary Substance; Marinco CH; Marinco® CH-Granular; Marinco OH; Merecol® IC; Merecol® SH; Merecol® Y; Micro-White® 10 Codex; Micro-White® 25 Codex; Micro-White® 50 Codex; Micro-White® 100 Codex; Milk Calcium ND (Food Grade); Mix 41; Mix 63; Mix 70; Morton® Flour Salt; Morton® Table Iodized Salt

Natural HVP Replacer; Neobee® 18; Neobee® C-10; Neobee® M-5; Neobee® O; Neobee® SL-110; Neobee® SL-120; Neobee® SL-130; Neobee® SL-140; Neobee® SL-210; Neobee® SL-220; Neobee® SL-230; Neobee® SL-310; Neobee® SL-410; Niacin USP, FCC Fine Granular No. 69901; Niacin USP, FCC No. 69902; Niacinamide Free Flowing No. 69914; Niacinamide USP, FCC No. 69905; Niacinamide USP, FCC Fine Granular No. 69916; Non-Diastatic Malt Syrup #40600; Nuruflakes; Nurugran; Nutriloid® Agar; Nutriloid® Arabic; Nutriloid® Carrageenan; Nutriloid® Guar Special; Nutriloid® Guar Standard; Nutriloid® Locust; Nutriloid® Tragacanth; Nutrisoy 7B Flour; Nutrisoy 220T

Orchic Substance; Ovarian Substance

Pancreas Substance; Panipower III; Parathyroid Substance; Parotid Substance; Pationic® 1230; Pationic® 1250; Pea Protein Conc; Phosal® 50 SA; Phosal® 53 MCT; Phosal® 75 SA; Phospholipon® 90/90 G; Phospholipon® 90/90G; Pineal Substance; Pisane®; Placental Substance; Polypro 5000®; Polypro 5000® CF or SF 45%; Polypro 5000® UF 45%; Polypro 15000® Food Grade; Polypro 15000® Pharmaceutical Grade; Powdered Gum Guar NF Type 80 Mesh B/T; Powdered Gum Guar Type 140 Mesh B/T; Powdered Gum Guar Type ECM; Powdered Gum Guar Type M; Powdered Gum Guar Type MM FCC; Powdered Gum Guar Type MM (HV); Powdered Gum Guar Type MMM $^1/_2$; Powdered Gum Guar Type MMW; Pristene® 180; Procon®; Procon® 2000; Procon® 20/60; Procon® 20/60 Military; Procon® 20/60-SL; Pro Fam 646; Pro Fam 648; Pro Fam 780; Pro Fam 781; Pro Fam 955; Pro Fam 970; Pro Fam 972; Pro Fam 974, 974 Fortified; Promax® 70; Promax® 70L; Promax® 70LSL; Promax® Plus; Promine® DS; Promine® HV; Proplus®; Prostate Substance; Pro-Tein 1550; Pro-Tein SF 1000; Pro-Tein SP 1000; Puracal® PP; Py-ran®; Pyridoxine Hydrochloride USP, FCC Fine Powd. No. 60650

Response®; Rex® Vitamin Fortified Wheat Germ Oil; Riboflavin USP, FCC No. 602940002; Riboflavin-5´-Phosphate Sodium USP, FCC No. 60296; Rice Bran Oil; Rice-Pro® 35W; Rich-Pak®; Rich-Pak® Powd. 160; Rich-Pak® Powd. 160-M; Rocoat® Niacinamide 33$^1/_3$% No. 69907; Rocoat® Niacinamide 33$^1/_3$% Type S No. 69909; Rocoat® Pyridoxine Hydrochloride 33$^1/_3$% No. 60688; Rocoat® Riboflavin 25% No. 60289; Rocoat® Riboflavin 33$^1/_3$ No. 60288; Rocoat® Thiamine Mononitrate 33$^1/_3$% No. 60188

Siverslice; Skipjack Liver Oil 30,000A/20,000D; Skipjack Liver Oil 40,000A/20,000D; Snowite® Oat Fiber; Soageena® MM301; Soalocust®; Sodium Ascorbate USP, FCC Fine Gran. No. 6047709; Sodium Ascorbate USP, FCC Fine Powd. No. 6047708; Sodium Ascorbate USP, FCC Type AG No. 6047710; Soluble Liver Powd; Soluble Trachea CS 16 Substance; Soluble Trachea Substance; Sovital™; Soyamin 50 E; Soyamin 50 T; Soyamin 70; Soyamin 90; Soylec C-6; Soylec C-15; Soylec T-15; Soyoco; Spleen Substance; Spray Dried Hemoglobin; Spray Dried Hydrolysed Fish Gelatin; Stabilized Corn Bran; Stabilized Full-Fat Wheat Germ; Stabilized Micro-Lite Corn Bran; Stabilized Oatex; Stabilized Oat Fiber; Stabilized Red Wheat

Nutrients *(cont'd.)*

Bran; Stabilized White Wheat Bran; Star-Dri® 1; Star-Dri® 5; Star-Dri® 10; Star-Dri® 1005A; Star-Dri® 1015A; Stomach Substance; Superb® 1450 Soy Fiber; Superb® 2000 Soy Fiber; Suprarenal (Adrenal) Substance; Suprarenal Cortex Substance; Supro® 425; Supro® 610; Supro® 661; Supro® 670; Supro® 710; Supro® 760; Supro Plus® 651; Supro Plus® 675; Supro Plus® 2100; Supro Plus® 3000; Swelite®; Swelite® Micro

Tem Crunch™ 93; Tem Crunch™ 110; Tem Freeze™; Tem Plus® 95; Tem Top™; Terra Alba 114836; Terra-Dry™ FD Aloe Vera Powd. Decolorized FDD; Terra-Dry™ FD Aloe Vera Powd. Reg. FDR; Terra-Pulp™ FD Whole Aloe Vera Gel Fillet; Terra-Pure™ FD Non-Preserved Decolorized FDD; Terra-Pure™ FD Non-Preserved Reg. FDR; Terra-Pure™ Non-Preserved Aloe Vera Powd; Terra-Spray™ Spray Dried Aloe Vera Powd. Decolorized SDD; Terra-Spray™ Spray Dried Aloe Vera Powd. Reg. SDR; Textratein; Thiamine Hydrochloride USP, FCC Regular Type No. 601160; Thiamine Mononitrate USP, FCC Fine Powd. No. 601340; Thymus Substance; TIC Pretested® Arabic FT Powd; TIC Pretested® Arabic PH-FT; Tofupro-U; Trachea Substance; TVP, Fortified TVP; Type B Torula Dried Yeast

Uterus Substance

VCR; Viobin Defatted Wheat Germ #2; Viobin Defatted Wheat Germ #3; Viobin Defatted Wheat Germ #5; Viobin Defatted Wheat Germ #6; Viobin Defatted Wheat Germ #8; Viobin Defatted Wheat Germ #9; Viobin Defatted Wheat Germ #34; Viobin Octacosanol; Viobin Wheat Germ Oil; Vitamin A Palmitate Type 250-CWS No. 65312; Vitamin A Palmitate Type PIMO/BH No. 638280100; Vitamin A Palmitate USP, FCC Type P1.7 No. 262090000; Vitamin A Palmitate USP, FCC Type P1.7/BHT No. 63693; Vitamin A Palmitate USP, FCC Type P1.7/E No. 63699; Vitamin B_{12} 0.1% SD No. 65354; Vitamin B_{12} 1% Trituration No. 69992; Vitamin B_{12} 1% Trituration No. 69993; Vitamin B_{12} 1.0% SD No. 65305; Vitamin E USP, FCC No. 60525; Vitamin E USP, FCC No. 60526; Vita-Rite A; Vita-Rite A & D; Vita-Rite A & D-2%; Vita-Rite A & D H-W; Vita-Rite A & D-W; Vita-Rite A2; Vita-Rite A-W; Vita-Rite D3; Vitinc® Defatted Wheat Germ Nuggets; Vitinc® Defatted Wheat Germ; Vitinc® Wheat Germ Oil; Vi-Vax

Wheat Germ Oil; Whitened Pea Fiber; Whole Egg Solids Type W-1; Whole Egg Solids Type W-1-FF; Whole Egg Solids Type W-2; Whole Egg Solids Type W-2-FF; Whole Pituitary Substance

Yelkinol F; Yelkinol G; Yelkinol P; Yellow Pea Fiber

Chemicals. Acacia, N-Acetyl-L-methionine, DL-Alanine, L-Alanine, Aluminum nicotinate, L-Arginine, L-Arginine L-glutamate; L-Arginine monohydrochloride; L-Ascorbic acid; Ascorbyl palmitate; Ascorbyl stearate, L-Asparagine, DL-Aspartic acid, L-Aspartic acid, Attapulgite

Bakers yeast extract; Bakers yeast protein; d-Biotin; Bisbenthiamine

Calcium borogluconate; Calcium carbonate; Calcium caseinate; Calcium chloride; Calcium citrate; Calcium dihydrogen pyrophosphate; Calcium fumarate; Calcium gluconate; Calcium glycerophosphate; Calcium lactate; Calcium oxide; Calcium pantothenate; Calcium pantoth-enate calcium chloride double salt (d or dl); Calcium phosphate dibasic; Calcium phosphate dibasic dihydrate; Calcium phosphate monobasic monohydrate; Calcium phosphate tribasic; Calcium pyrophosphate; Calcium sulfate; L-Carnitine; Carotene; Casein; Cholecalciferol; Choline bitartrate; Choline chloride; Cobalt gluconate; Cod liver oil; Copper; Copper glucon-ate; Copper iodide (ous); Corn bran; Corn gluten; Corn meal; Corn starch; Corn syrup; Cottonseed fiber; Cupric sulfate (anhydrous); Cupric sulfate (pentahydrate); Cyanocobal-amin; L-Cysteine; Cysteine hydrochloride; L-Cystine

Dibenzoyl thiamine; Dibenzoyl thiamine hydrochloride; Disodium phosphate, dihydrate

Egg powder; Ergocalciferol

Ferric chloride; Ferric citrate; Ferric phosphate; Ferric pyrophosphate; Ferric sodium pyrophos-phate; Ferrous ascorbate; Ferrous carbonate; Ferrous citrate; Ferrous fumarate; Ferrous gluconate; Ferrous lactate; Ferrous pyrophosphate; Ferrous sulfate; Ferrous sulfate (dried); Ferrous sulfate heptahydrate; Fish protein isolate; Folicacid; Fumaric acid

Gelatin; D-Gluconic acid; D-Glucose anhyd; L-Glutamic acid; L-Glutamine; Glycine

L-Histidine; Histidine hydrochloride monohydrate; Hydrogenated cottonseed oil; Hydrolyzed gelatin; Hydrolyzed protein

Inositol; Iron; Iron ammonium citrate; L-Isoleucine; L-Isoleucine

Jerusalem artichoke flour

Kelp

Lactose; DL-Leucine; L-Leucine; Linoleic acid; L-Lysine; L-Lysine L-glutamate; L-Lysine hydro-chloride

Magnesium carbonate hydroxide; Magnesium chloride hexahydrate; Magnesium fumarate; Magnesium gluconate; Magnesium hydroxide; Magnesium oxide; Magnesium phosphate dibasic; Magnesium phosphate, tribasic; Magnesium stearate; Magnesium sulfate anhyd;

Nutrients *(cont'd.)*

Magnesium sulfate heptahydrate; Maltose; Manganese chloride tetrahydrate; Manganese citrate; Manganese gluconate; Manganese glycerophosphate; Manganese hypophosphite; Manganese oxide; Manganese sulfate (ous); D-Mannitol; DL-Methionine; L-Methionine; Methyl eicosenate; Methyl stearate; Microcrystalline cellulose; Milk protein; MSG

Niacinamide; Niacinamide ascorbate; Nicotinic acid

Oat fiber; Octacosanol

Pancreatin; D-Panthenol; DL-Panthenol; D-Pantothenamide; Peanut oil; Pea protein concentrate; DL-Phenylalanine; L-Phenylalanine; Potassium bicarbonate; Potassium carbonate; Potassium chloride; Potassium fumarate; Potassium D-gluconate; Potassium glycerophosphate; Potassium iodide; Potassium phosphate dibasic; L-Proline; Pyridoxine HCl; Retinyl acetate; Retinyl palmitate; Riboflavin; Riboflavin tetrabutyrate; Rice bran; Rice protein; Rose hips; DL-Serine; L-Serine; Shark liver oil; Sodium ascorbate; Sodium aspartate; Sodium chloride; Sodium citrate; Sodium ferric pyrophosphate; Sodium fumarate; Sodium gluconate; Sodium hexametaphosphate; Sodium pantothenate; Sodium phosphate; Sodium phosphate dibasic; Sodium phosphate tribasic; Sorbitol; Soybean milk; Soybean oil; Soy fiber; Soy flour; Tetrasodium pyrophosphate; Thiamine; Thiamine HCl; L-Threonine; Tocopherol; d-α-Tocopherol; dl-α-Tocopherol; Tocopheryl acetate; d-α-Tocopheryl acetate; dl-α-Tocopheryl acetate; Tocopheryl succinate; Triacetin; DL-α-Tryptophan; L-Tryptophan; L-Tyrosine; L-Valine; Vitamin B$_6$ hydrochloride; Vitamin K$_1$; Wheat bran; Wheat germ; Wheat germ oil; Wheat gluten; Whole fish protein concentrate

Xylitol

Yeast; Yeast extract

Zinc chloride; Zinc gluconate; Zinc methionine sulfate; Zinc oxide; Zinc stearate; Zinc sulfate; Zinc sulfate heptahydrate

Opacifiers

Trade names:	Avicel® CL-611; Avicel® RC-501; Avicel® RC-581; Avicel® RC-591
	Drewmulse® 200K
	Gelcarin® GP-359; Gelcarin® GP-911; Gelcarin® PS-316; Gelcarin® XP-1008; Glycosperse® S-20 FG; Glycosperse® S-20 KFG
	Hilton Davis Titanium Dioxide; Hodag GML; Hodag GMO; Hodag GMO-D; Hodag GMR; Hodag GMR-D; Hodag GMS; Hodag GTO
	n-Octenyl Succinic Anhydride
	SeaKem® GP-418; S-Maz® 20; S-Maz® 80K; S-Maz® 95; Stabilizer C; Superfine Ground Mustard
	Veri-Lo® 100; Viscarin® SD-389; Visco Compound™
Chemicals:	PEG-6 dilaurate; PEG-20 dilaurate; PEG-32 dilaurate; PEG-75 dilaurate; PEG-150 dilaurate; PEG-6 dioleate; PEG-20 dioleate; PEG-32 dioleate; PEG-75 dioleate; PEG-150 dioleate; PEG-6 distearate; PEG-20 distearate; PEG-32 distearate; PEG-75 distearate; PEG-32 laurate; PEG-32 oleate; PEG-75 oleate; PEG-32 stearate; PEG-75 stearate; Polysorbate 60

Oxidizing agents • Reducing agents

Trade names:	Beta-Tabs; Brew Aid B; Bromette; Bromitabs
	Cap-Shure® AS-125-50; Chocotop™
	Hidelase
	Oxitabs
	Qual Guard™; Qual Guard™ 100; Qual Guard™ WB
	WaTox-20 P Powd; WaTox-20 Tablets
Chemicals:	L-Ascorbic acid
	Benzoyl peroxide
	Calcium peroxide; Catalase; Chlorine; Chlorine dioxide; L-Cysteine
	Hydrogen peroxide
	Potassium hydroxide
	Sodium thiosulfate anhyd.; Sodium thiosulfate pentahydrate; Stannous chloride anhyd.

pH control agents • Acidulants • Alkaline agents • Buffers • Neutralizers

Trade names: Ajax®; Albrite Diammonium Phosphate Food Grade; Albrite Monoammonium Phosphate Food Grade; Albrite MSP Food Grade; Albrite Phosphoric Acid 85% Food Grade; Albrite SAPP Food Grade; Albrite STPP-F; Albrite STPP-FC; Albrite TSPP Food Grade

Brew Aid B

Cap-Shure® C-135-72; Cap-Shure® C-140-72; Cap-Shure® C-140-85; Cap-Shure® C-140E-75; Cap-Shure® C-165-63; Cap-Shure® C-165-85; Cap-Shure® C-900-85; Cap-Shure® CLF-165-70; Cap-Shure® F-125-85; Cap-Shure® F-140-63; Cap-Shure® FF-165-60; Cap-Shure® FS-165-60; Cap-Shure® GDL-140-70; Cap-Shure® LCL-135-50; Citric Acid USP FCC Anhyd. Fine Gran. No. 69941; Citrid Acid USP FCC Anhyd. Gran. No. 69942; Citrocoat® A 1000 HP; Citrocoat® A 2000 HP; Citrocoat® A 4000 TP; Citrocoat® A 4000 TT; Citrostabil® NEU; Citrostabil® S; Cola H; Cola Complex

Destab™; Durkote Citric Acid/Maltodextrin, Acacia Gum; Durkote Fumaric Acid/Hydrog. Veg. Oil; Durkote Malic Acid/Maltodextrin

Essiccum®

Ferment Buffer

Hamine; Hamine-Mix

Levair®; Liquinat®

Magnesium Hydroxide USP; Magnesium Hydroxide USP DC; Marinco CH; Marinco® CH-Granular; Marinco® CL; Marinco H-USP; Marinco OH; Marinco OL

Niacet Calcium Acetate FCC; Niacet Sodium Acetate 60% FCC; Niacet Sodium Acetate Anhyd. FCC; Niacet Sodium Diacetate FCC

Patlac® NAL; Purac®; Purasal® S/SP 60; Py-ran®

Regent® 12XX; Rocarna

Sodium Citrate USP, FCC Dihydrate Gran. No. 69976; Sodium Citrate USP, FCC Dihydrate Fine Gran. No. 69975

Terra Alba 114836

Chemicals: Acetic acid; Adipic acid; Ammonium alum; Ammonium bicarbonate; Ammonium carbonate; Ammonium isovalerate; Ammonium phosphate; Ammonium phosphate, dibasic

Calcium acetate; Calcium chloride; Calcium citrate; Calcium gluconate; Calcium hydroxide; Calcium lactate; Calcium phosphate monobasic monohydrate; Calcium pyrophosphate; Calcium sulfate; Carbon dioxide; Citric acid; Cyclamic acid

Disodium citrate; Disodium phosphate, dihydrate; Disodium succinate

Fumaric acid

D-Gluconic acid; Gluconolactone; Glycine

Hydrochloric acid; N-Hydroxysuccinic acid

α-Ketoglutaric acid

Lactic acid

Magnesium carbonate; Magnesium carbonate hydroxide; Magnesium hydroxide; Magnesium oxide; Magnesium phosphate dibasic; Magnesium phosphate, tribasic

Pentapotassium triphosphate; Phosphoric acid; Potassium acetate; Potassium alum dodecahydrate; Potassium bicarbonate; Potassium carbonate; Potassium chloride; Potassium citrate; Potassium hydroxide; Potassium lactate; Potassium phosphate; Potassium phosphate dibasic; Potassium phosphate tribasic; Potassium polymetaphosphate; Potassium sodium tartrate anhyd; Potassium sodium tartrate tetrahydrate

Sodium acetate trihydrate; Sodium acid pyrophosphate; Sodium alum; Sodium aluminum phosphate acidic; Sodium bicarbonate; Sodium bisulfate; Sodium bisulfate solid; Sodium carbonate; Sodium citrate; Sodium diacetate; Sodium fumarate; Sodium hydroxide; Sodium lactate; Sodium metaphosphate; Sodium phosphate; Sodium phosphate dibasic; Sodium phosphate tribasic; Sodium phosphate tribasic dodecahydrate; Sodium sesquicarbonate; Sodium succinate; Sodium tartrate; Sorbic acid; Succinic acid; Succinic anhydride; Sulfuric acid

Tannic acid; L-Tartaric acid; Tetrapotassium pyrophosphate; Trisodium citrate

Plasticizers

Trade names: Axol® E 61; Axol® E 66

Be Square® 175; Britol® 6NF; Britol® 9NF; Britol® 20USP; Britol® 35USP; Britol® 50USP

Plasticizers *(cont'd.)*

Carbowax® Sentry® PEG 300; Carbowax® Sentry® PEG 400; Carbowax® Sentry® PEG 540 Blend; Carbowax® Sentry® PEG 900; Carbowax® Sentry® PEG 1000; Carbowax® Sentry® PEG 1450; Carbowax® Sentry® PEG 3350; Carbowax® Sentry® PEG 4600; Carbowax® Sentry® PEG 8000; Cetodan®; Cetodan® 90-40
Drakeol 34
Homodan RD
Lamegin® GLP 10, 20; Lecigran™ F
Mazol® GMS; Monomax AH90 B; Monomax VH90 B; Myvacet® 5-07; Myvacet® 7-07; Myvacet® 9-08; Myverol® 18-06
Radiamuls® MG 2602; Radiamuls® MG 2606; Radiamuls® MG 2900; Rudol®
Staybelite® Ester 5
Uniplex 84
Witafrol® 7420

Chemicals: Acetylated hydrogenated cottonseed glyceride; Acetylated hydrogenated soybean oil glyceride; Acetyl tributyl citrate; Acetyl triethyl citrate
p-t-Butylphenyl salicylate; n-Butyl phthalyl-n-butyl glycolate; Butyl stearate
Dibutyl sebacate; Diethyl phthalate; Diisobutyl adipate; Diisooctyl phthalate; Dioctyl phthalate; Diphenyl octyl phosphate; Distearyl citrate
Epoxidized soybean oil; Ethylphthalyl ethyl glycolate
Glycerin; Glyceryl hydrogenated rosinate; Glyceryl oleate
Isopropyl citrate
Lactylic esters of fatty acids
Magnesium glycerophosphate; Magnesium phosphate dibasic
PEG-4; PEG-6; PEG-8; PEG-9; PEG-12; PEG-14; PEG-16; PEG-20; PEG-32; PEG-40; PEG-75; PEG-100; PEG-150; PEG-200; Polyethylene glycol
Rice bran wax
Stearyl citrate
Triacetin; Triethyl citrate; Tristearyl citrate
3-(2-Xenolyl)-1,2-epoxypropane

Preservatives • Antimicrobials • Antistaling agents

Trade names: Adeka Menjust; Admul CSL 2007, CSL 2008; Admul SSL 2003; Admul SSL 2004; Agar Agar NF Flake #1; Alcolec® 30-A; Aldo® MS FG; Alta®; Amfal-46; Arlac P; Arlac S; Ascorbyl Palmitate NF, FCC No. 60412; Atlas 5000K; Atlas 5520; Atmul® 500
Blendmax 322; Blendmax 322D; Britesorb®; Britesorb® A 100
Calcium Ascorbate FCC No. 60475; Cap-Shure® SC-135-63; Cegesterin® Range; CFI-10; Cola H; Cola Complex; Confecto No-Stick™; Cozeen®; CP-2600; CP-3650; Crestawhip H13; Crester KZ; Crodaglaze; Crodatem L; Crodatem L50; Crodatem T22; Crodatem T25; Crolactem; Cyncal®
Delvocid; Drewmulse® 10K; Drewmulse® 200; Drewmulse® 365; Drewmulse® 900K; Durkote Fumaric Acid/Hydrog. Veg. Oil; Durkote Sodium Chloride/Hydrog. Veg. Oil; Durkote Vitamin C/Maltodextrin
Eastman Potassium Sorbate; Eastman Sorbic Acid; Embanox® BHA; Emulgum™; Eribate®; Essiccum®
Flav-R-Keep® FP-51; FloAm® 200; FloAm® 200C; FloAm® 210 HC; FloAm® 221; Fluid Flex; Freez-Gard® FP-88E
Garbefix 31; Gelex® Instant Starch; GMS 305; GMS 402; GMS 402V; GMS 602V; GMS 900; GMS 902; GP-1200; Grindamyl
Hodag GMS; Hodag PGS
Jerusalem Arthichoke Flour (JAF)
Kel-Lite™ BK; Kol Guard® 7413
Lamegin® CSL; Lamemul® Range; Lamemul® K 1000-Range; Lamemul® K 2000-Range; Lamemul® K 3000-Range; Lem-O-Fos® 202; Lexgard® M; Lexgard® P
Merecol® RB; Merecol® RCS; Methocel® K15M Premium; Mira-Cleer® 340; Moldban; Monomuls® 60-10; Monomuls® 60-15; Monomuls® 60-20; Monomuls® 60-25; Monomuls® 60-25/2; Monomuls® 60-25/5; Monomuls® 60-30; Monomuls® 60-35; Monomuls® 60-40; Monomuls® 60-45; Monomuls® 90-10; Monomuls® 90-15; Monomuls® 90-20; Monomuls®

Preservatives *(cont'd.)*

90-25; Monomuls® 90-25/2; Monomuls® 90-25/5; Monomuls® 90-30; Monomuls® 90-35; Monomuls® 90-40; Monomuls® 90-45; Muffin Kote; Myvatex® 7-85; Myvatex® 40-06S; Myvatex® Mighty Soft®; Myvatex® SSH; Myvatex® Texture Lite®; Myverol® 18-50; Myverol® P-06; Myverol® SMG

Natamax™; NatuReal™ Essence; New Fresh Conc; Niacet Calcium Propionate FCC; Niacet Sodium Diacetate FCC; Niacet Sodium Propionate FCC; Nipabenzyl; Nipabutyl; Nipabutyl Potassium; Nipabutyl Sodium; Nipacide® OPP; Nipacide® Potassium; Nipacide® Sodium; Nipacombin PK; Nipagin A; Nipagin A Potassium; Nipagin A Sodium; Nipagin M; Nipagin M Sodium; Nipaheptyl; Nipasept Potassium; Nipasept Sodium; Nipasol M; Nipasol M Potassium; Nipasol M Sodium; Nipastat; Noramium DA.50; Nurupan™; Nutrilife®; Nutrisoft® 55

Ovazyme, XX

Pantex; PAV; PGMS 70; Powdered Agar Agar NF M-100 (Gracilaria); Prefera® Range; Prefera® SSL 6000 MB; Prevent-O; Purac®; Purasal® P/USP 60; Purasal® S/SP 60

Rezista®; Rocarna; Rye Sour; Ryoto Sugar Ester B-370; Ryoto Sugar Ester ER-190; Ryoto Sugar Ester ER-290; Ryoto Sugar Ester L-595; Ryoto Sugar Ester L-1570; Ryoto Sugar Ester L-1695; Ryoto Sugar Ester M-1695; Ryoto Sugar Ester O-1570; Ryoto Sugar Ester P-170; Ryoto Sugar Ester S-1170; Ryoto Sugar Ester S-1170S; Ryoto Sugar Ester S-1570

721-A™ Instant Starch; Sodium Ascorbate USP, FCC Fine Gran. No. 6047709; Sodium Ascorbate USP, FCC Fine Powd. No. 6047708; Sodium Ascorbate USP, FCC Type AG No. 6047710; Sodium Benzoate BP88; Soyamin 50 E; Soyapan™; Special Bro Soft #126 6433; Sporban Regular; Sporban Special; Starplex® 90; Sunmalt; Supercol® GF; Supercol® U; Sustane® 1-F; Sustane® 3; Sustane® 4A; Sustane® 6; Sustane® 8; Sustane® 20; Sustane® 20-3; Sustane® 31; Sustane® BHT; Sustane® HW-4; Sustane® P; Sustane® PA; Sustane® PG; Sustane® Q; Sustane® TBHQ; Sustane® W

Tally® 100; Tally® 100 Plus; Tandem 5K; Tandem 8; Tandem 11H K; Tandem 552; Tandem 552 K; Tegomuls® 9102; Tegomuls® 9102 S; Tenase® 1200; Tenase® L-340; Tenase® L-1200; Tenox® BHA; Ticaloid® Lite; Topcithin®; Tristat; Tristat K; Tristat SDHA

Ultra-Bake® 3100, 3200; Ultra-Bake® 3100K, 3200K; Ultra-Bake® NF; Ultra Fresh

Vascan; Vitrafos® LC; Vykamol 83G

White Sour

Chemicals: Agar; Alcohol; Ammonium persulfate; Anoxomer; L-Ascorbic acid; Ascorbyl palmitate

Benzoic acid; Benzylparaben; BHA; BHT; Biphenyl; Boric acid; Butylparaben

Calcium acetate; Calcium ascorbate; Calcium benzoate; Calcium bromide; Calcium chloride; Calcium citrate; Calcium disodium EDTA; Calcium formate; Calcium propionate; Calcium sorbate; Caprylic acid; Carbon dioxide; Catalase; Cetalkonium chloride; Chlorine; Coriander oil

Dehydroacetic acid; 2,2-Dibromo-3-nitrilopropionamide; Diethyl fumarate; Dilauryl thiodipropionate; Dimethyl dicarbonate; Dimethyl dodecyl betaine; Dimethyl fumarate; Disodium cyanodithioimidocarbonate; Disodium EDTA; Distearyl citrate

Erythorbic acid; 6-Ethoxy-1,2-dihydro-2,2,4-trimethylquinoline; Ethylenediamine; Ethyl fumarate; Ethyl paraben

Formaldehyde; Formic acid

Glucose oxidase; Glutaral; Glyceryl cottonseed oil; Glycine

n-Heptyl p-hydroxybenzoate; Heptyl paraben; Hexamethylene tetramine; Hydrogen peroxide; 4-Hydroxymethyl-2,6-di-t-butylphenol

Imazalil; Isobutyl p-hydroxybenzoate; Isopropyl p-hydroxybenzoate

Lactic acid; Lauralkonium chloride

Methyl fumarate; Myristalkonium chloride

Nabam; Natamycin; Nisin

Ozone

PEG-40 stearate; o-Phenylphenol; Potassium acid tartrate; Potassium benzoate; Potassium bisulfite; Potassium butyl paraben; Potassium ethylparaben; Potassium metabisulfite; Potassium N-methyldithiocarbamate; Potassium nitrate; Potassium nitrite; Potassium propylparaben; Potassium sorbate; Potassium sulfite; Potassium trichlorophenate; Propionic acid; Propylparaben

Quaternium-14

Retinol; Retinyl acetate

Salicylic acid; Sodium acid pyrophosphate; Sodium ascorbate; Sodium benzoate; Sodium bisulfate solid; Sodium bisulfite; Sodium butylparaben; Sodium chloride; Sodium dehydroacetate; Sodium diacetate; Sodium dimethyldithiocarbamate; Sodium formate; Sodium hypophosphite; Sodium metabisulfite; Sodium methylparaben; Sodium nitrate; Sodium

Preservatives (cont'd.)

nitrite; Sodium pentachlorophenate; Sodium o-phenylphenate; Sodium propionate; Sodium propylparaben; Sodium sorbate; Sodium sulfite; Sorbic acid; Stannous chloride anhyd; Stearalkonium chloride; Stearyl citrate; Sucrose; Sucrose erucate; Sulfur dioxide

n-Tetradecyl dimethyl ethylbenzyl ammonium chloride; Thiabendazole; d-α-Tocopherol; Tragacanth gum; 2,4,5-Trihydroxybutyrophenone; Tristearyl citrate

Zinc chloride

Processing aids • Clarifying agents • Clouding agents • Filter aids • Formulation aids • Flocculants

Trade names: Accofloc® A100 PWG; Accofloc® A110 PWG; Accofloc® A120 PWG; Accofloc® A130 PWG; Accofloc® N100 PWG; Accoline; Accoline-Mix; AF 75; AF 100 FG; AF 8805 FG; AF 8810 FG; AF 8820 FG; AF 8830 FG; AFP 2000; Alphadim® 90AB; APA; APC; Aqualon® 7H4F; Atlas SSL

Benecel® MP 843; Bromelain 1:10; Bromelain 150 GDU; Bromelain 600 GDU; Bromelain 1200 GDU; Bromelain 1500 GDU

Calgon® Type RB; Caprol® PGE860; Cleartic; Clouding Agent Powd; Cobee 76; Cobee 92; Cobee 110; Coconut Oil® 76; Coco-Spred™; Colloid 787; Complemix® 50; Complemix® 100

Diahope®-S60; Diaion® WK10; Diazyme® L-200; Divergan® F, R; Dur-Em® 300K; Durkex 100BHA; Durkex 100BHT; Durkex 100F; Durkex 100TBHQ; Durpeg® 400MO

Edicol®

Fermcozyme® 1307, BG, BGXX, CBB, CBBXX, M; Finecol 8000; FishGard® FP-55; Flav-R-Keep®; Freez-Gard® FP-15

Glucanase GV; Glucanex® L-300

Hercules® Ester Gum 8BG; Hodag PGO-103; HT-Proteolytic® 200; HT-Proteolytic® Conc; HT-Proteolytic® L-175

Liponate GC; Liponate PC; Liquid Fish Gelatin Conc; Litesse®

Macol® 5100; Miranol® J2M Conc; Morton® Flour Salt; Myvaplex® 600; Myvaplex® 600K; Myvaplex® 600P; Myvaplex® 600PK

Neobee® 1053; Neobee® M-5; Neobee® O; Norfox® Anionic 27

Optimase® APL-440

Papain 30,000; Papain A300; Papain A400; Papain Conc; Papain M70; Papain P-100; Papain S100; Pastry Wash Powd; Pationic® 920; Pationic® 925; Pearex® L; Pectinase AT; Pektolase; Perfect Slice; Protoferm

Ryoto Sugar Ester LWA-1570; Ryoto Sugar Ester O-1570; Ryoto Sugar Ester OWA-1570

Santone® 8-1-O; Sea-Gard® FP-91; Shur-Fil® 427; Sipernat® 22; Soageena® MM301; Sorbo®; Spa Gelatin; Spark-L® HPG; Spray Dried Fish Gelatin; Spray Dried Fish Gelatin/Maltodextrin; Spray Dried Hydrolysed Fish Gelatin; Sta-Mist® 7415; Sunmalt

Tenase® 1200; Tenase® L-340; Tenase® L-1200; Ticaloid® 1070; TIC Pretested® Aragum® 3000; TIC Pretested® Aragum® 9000 M; TIC Pretested® Xanthan 200; TMP-65; Tona Enzyme 201; Tona Papain 14; Tona Papain 90L; Tona Papain 270L

Veg-X-Ten®; Veri-Lo® 300; Vitiben®

Wacker Silicone Antifoam Emulsion SE 9; Whetstar® 7 Food Powd. Wheat Starch

Chemicals: Acacia; Acetone; Acrylates/acrylamide copolymer; Agar; Albumen; Alcohol; Algin; Alginic acid; Ammonia; Ammonium alum; Ammonium chloride; Ammonium hydroxide; Ammonium phosphate, dibasic

Bentonite; Bromelain

Calcium acetate; Calcium chloride; Calcium gluconate; Calcium hydroxide; Calcium silicate; Calcium/sodium stearoyl lactylate; Calcium sulfate; *Candida guilliermondii*; *Candida lipolytica*; Caramel; Carbohydrase; Carbohydrase-cellilase; Carbohydrase from *Aspergillus niger*; Carbohydrase from *Rhizopus oryzae*; Carbon, activated; Carbon black; Carbon dioxide; Carnauba; Casein; Catalase; Cellulase; Cellulose; Chlorine; Chloromethylated aminated styrene-divinylbenzene resin; Cobalt; Cocamide DEA; Cocoa butter substitute; Coconut oil; Copper; Corn oil; Cottonseed flour, partially defatted, cooked; Cottonseed oil; Cupric sulfate (anhydrous); Cupric sulfate (pentahydrate); Cyclodextrin

Dextrin; Dialkyl dimethyl ammonium chloride; Diatomaceous earth; Diethylaminoethyl cellulose; Dimethylamine/epichlorohydrin copolymer; Dioctyl sodium sulfosuccinate; Divinylbenzene copolymer

Processing aids *(cont'd.)*

Ethanolamine; Ethyl maltol
Ferrous sulfate heptahydrate; Ficin; Fructose
α-Galactosidase; Garlic extract; Gelatin; Glucanase; Glucoamylase; Glyceryl behenate; Glyceryl rosinate; Guar gum
Helium; Hexane; Hydrochloric acid; Hydrogenated coconut oil
Iron
Kaolin; Karaya gum
Lactose; Lard
Magnesium carbonate hydroxide; Magnesium chloride; Magnesium oxide; Magnesium silicate hydrate; Magnesium stearate; Magnesium sulfate anhyd; Magnesium sulfate heptahydrate; D-Mannitol; Mono- and diglycerides of fatty acids
Nickel; Nitrogen
Oregano; Ozone
Palm kernel oil; Palm oil; Papain; Peanut oil; Poloxamer 105; Poloxamer 108; Poloxamer 122; Poloxamer 123; Poloxamer 124; Poloxamer 181; Poloxamer 182; Poloxamer 183; Poloxamer 184; Poloxamer 185; Poloxamer 188; Poloxamer 212; Poloxamer 215; Poloxamer 217; Poloxamer 231; Poloxamer 234; Poloxamer 235; Poloxamer 237; Poloxamer 238; Poloxamer 282; Poloxamer 284; Poloxamer 288; Poloxamer 331; Poloxamer 333; Poloxamer 334; Poloxamer 335; Poloxamer 338; Poloxamer 402; Poloxamer 403; Poloxamer 407; Polyacrylamide; Polyacrylamide resins, modified; Polydextrose; Polyglycerol esters of fatty acids; Polyglyceryl-8 oleate; Polyglyceryl-10 trioleate; Polymaleic acid; Polymaleic acid sodium salt; Polyvinylpolypyrrolidone; Poly(1-vinyl-2-pyrrolidinone) homopolymer; Potassium acid tartrate; Potassium alum dodecahydrate; Potassium bicarbonate; Potassium bromate; Potassium bromide; Potassium carbonate; Potassium hydroxide; Propylene glycol; Propylene glycol alginate; PVP
Rennet
Safflower oil; Silica; Silica aerogel; Sodium carbonate; Sodium caseinate; Sodium dodecylbenzenesulfonate; Sodium ferrocyanide; Sodium hydroxide; Sodium hypochlorite; Sodium lactate; Sodium metasilicate; Sodium methylate; Sodium methylnaphthalenesulfonate; Sodium octyl sulfate; Sodium persulfate; Sodium polyacrylate; Sodium stearoyl lactylate; Sodium sulfate; Sodium thiosulfate pentahydrate; Sorbitol; Soybean oil; Stearic acid; Sulfuric acid; Sunflower seed oil
Talc; Tallow; Tannic acid; Tannin extract; Tragacanth gum; Triacetin; Triethanolamine; Tristearin
Urea
Wheat gluten
Zein

Propellants • Aerating agents • Whipping agents

Trade names: Admul Emulsponge; Alco Whip® 1026; Aldo® HMS FG; Aldo® LP KFG; Aldo® MS FG; Aldo® PGHMS; Aldo® PGHMS KFG; Antelope Aerator; Atlas 800; Atlas 5000K; Atmos® 378 K; Atmos® 659 K; Atmos® 758 K; Atmos® 1069; Atmos® 7515 K; Atmul® 80; Atmul® 500; Atmul® 651 K; Atmul® 918 K
Caprol® 3GS; Cetodan®; Cetodan® 50-00A; Cetodan® 70-00A; Cetodan® 90-40; Crillet 3; Crolactem
Dimodan PV; Dimodan PV 300 Kosher; Dimodan PV Kosher; Dimodan PVP Kosher; Dimul S K; Durlac® 100W
Hodag PGO-1010; Hodag PSMS-20; Hodag SMS; Hyfoama
Lactodan B 30; Lactodan LW; Lactodan P 22; Lactodan P 22 Kosher
Mira-Foam® 100; Mira-Foam® 100K; Mira-Foam® 120V; Mira-Foam® 130H; Myvatex® 3-50; Myvatex® 7-85; Myvatex® 8-16; Myvatex® 40-06S; Myvatex® Texture Lite®; Myverol® 18-06; Myverol® 18-07; Myverol® 18-35; Myverol® 18-50; Myverol® 18-92; Myverol® 18-99; Myverol® P-06K
Pationic® 925; Pationic® 930; Pationic® 940; PGMS 70; Polyaldo® TGMS KFG; Promodan USV
Ryoto Ester SP; Ryoto Sugar Ester ER-190; Ryoto Sugar Ester ER-290; Ryoto Sugar Ester LWA-1570; Ryoto Sugar Ester P-170; Ryoto Sugar Ester S-070
Santone® 3-1-S; Santone® 3-1-SH; Santone® 3-1-S XTR; Soft-Set®; Spongolit® Range
Tandem 5K; Tandem 8; Tandem 11H K; Tandem 22 H; Tandem 100 K; Tegomuls® 4100;

Propellants *(cont'd.)*

Tween® 65

Vanade Kosher; Vanall; Vanease (K); Vegafoom; Versa-Whip® 500, 510, 520; Versa-Whip® 520K, 600K, 620K

Chemicals: Acetylated hydrogenated cottonseed glyceride

Butane

Calcium/sodium stearoyl lactylate; Calcium stearoyl lactylate; Canola oil glyceride; Carbon dioxide; Chloropentafluoroethane

Glyceryl lactopalmitate/stearate

Helium; Hydrogenated vegetable glycerides

Isobutane

Lactylated fatty acid esters of glycerol and propylene glycol

Methyl ethyl cellulose

Nitrogen; Nitrous oxide

Octafluorocyclobutane

Polyglyceryl-3 stearate; Propane

Sodium aluminum phosphate acidic; Sodium bicarbonate; Sodium lauryl sulfate; Sodium stearoyl lactylate; Sucrose erucate; Sucrose oleate; Sucrose palmitate; Sucrose polystearate; Sucrose tristearate

Sequestrants • Chelating agents

Trade names: Albriphos™ Blend 928; Albrite TSPP Food Grade

Citric Acid USP FCC Anhyd. Fine Gran. No. 69941; Citrid Acid USP FCC Anhyd. Gran. No. 69942

Hexaphos®

Katch® Fish Phosphate

Niacet Calcium Acetate FCC; Nutrifos® 088; Nutrifos® B-75; Nutrifos® H-30; Nutrifos® L-50; Nutrifos® Powd; Nutrifos® SK

Py-ran®

Sodaphos®; Sorbo®

Terra Alba 114836

Versene CA; Versene NA; Vitrafos® LC

Chemicals: Ascorbyl palmitate

Calcium acetate; Calcium chloride; Calcium citrate; Calcium diacetate; Calcium disodium EDTA; Calcium gluconate; Calcium hexametaphosphate; Calcium phosphate monobasic monohydrate; Calcium phytate; Citric acid

Disodium citrate; Disodium EDTA; Disodium pyrophosphate; Distearyl citrate

Gluconolactone

Isopropyl citrate

Oxystearin

Pentapotassium triphosphate; Pentasodium triphosphate; Phosphoric acid; Potassium citrate; Potassium D-gluconate; Potassium phosphate; Potassium phosphate tribasic; Potassium sodium tartrate anhyd; Potassium sodium tartrate tetrahydrate

Sodium acid phosphate; Sodium acid pyrophosphate; Sodium carbonate; Sodium citrate; Sodium diacetate; Sodium gluconate; Sodium hexametaphosphate; Sodium metaphosphate; Sodium phosphate; Sodium phosphate dibasic; Sodium phosphate tribasic; Sodium tartrate; Sodium thiosulfate anhyd; Sodium thiosulfate pentahydrate; Sorbitol; Starch; Succinic acid

L-Tartaric acid; Tetrapotassium pyrophosphate; Tetrasodium pyrophosphate; Triethyl citrate; Tristearyl citrate

Solubilizers

Trade names: Alcolec® BS; Alcolec® S; Alcolec® Z-3; Aldo® MLD FG; Aldo® MO FG; Aldosperse® MO-50 FG; Aldosperse® O-20 FG; Arlatone® 650; Arlatone® 827; Axol® E 66

Capmul® POE-L; Capmul® POE-O; Capmul® POE-S; Cavitron Cyclo-dextrin.™; Centrolene® A; Coconut Oil® 76; Coconut Oil® 92; Cremophor® NP 10; Cremophor® NP 14; Crillet 1;

Solubilizers *(cont'd.)*

Crillet 2; Crillet 4; Crillet 11; Crillet 31; Crillet 35; Crillet 41; Crillet 45

DK Ester F-10; DK Ester F-20; DK Ester F-50; DK Ester F 70; DK Ester F 90; DK Ester F-110; DK Ester F-140; DK Ester F-160; Drewmulse® 3-1-O; Drewmulse® 3-1-S; Drewmulse® 6-2-S; Drewmulse® 10-4-O; Drewmulse® 10-8-O; Drewmulse® 10-10-S; Drewmulse® POE-SML; Drewmulse® POE-SMO; Drewmulse® SMO; Drewmulse® SMS; Drewpol® 3-1-OK; Drewpol® 8-1-OK; Drewpol® 10-2-OK; Drewpol® 10-4-O; Drewpol® 10-4-OK; Drewpol® 10-6-OK; Drewpol® 10-8-OK; Drewpol® 10-10-OK; Drewpone® 80K; Dur-Em® 300; Dur-Em® 300K; Dur-Em® GMO; Durfax® 80

Emalex HC-50; Emalex HC-60; Emalex VS-31; Emery® 912; Emery® 916; Emery® 917; Emery® 918; Emulfluid® A; Emulgator 484

Glycosperse® L-20; Glycosperse® O-5; Glycosperse® O-20; Glycosperse® O-20 FG; Glycosperse® O-20 KFG; Glycosperse® O-20 Veg; Glycosperse® O-20X; Glycosperse® P-20; Glycosperse® S-20; Glycosperse® TO-20; Glycosperse® TS-20

Hodag PGO-102; Hodag PGO-1010

Katch® Fish Phosphate; Kessco® GMC-8

Lamesorb® SML; Lamesorb® SML-20; Lamesorb® SMO; Lamesorb® SMO-20; Lamesorb® SMP; Lamesorb® SMP-20; Lamesorb® SMS; Lamesorb® SMS-20; Lamesorb® STS; Lamesorb® STS-20; Lexol® GT-855; Lexol® PG-865; Lonzest® SMO-20; Lonzest® SMP-20; Lonzest® SMS-20; Lonzest® STO-20; Lonzest® STS-20

Maltrin® M700; Maltrin® QD M500; Mazol® PGO-31 K; Mazol® PGO-104; Miglyol® 812

Neobee® 18; Neobee® 20; Neobee® 62; Neobee® 1053; Neobee® 1054; Neobee® M-5; Neobee® M-20; Neobee® O; Nikkol TO-10; Nikkol TS-10; Nikkol Decaglyn 1-IS; Nikkol Decaglyn 1-L; Nikkol Decaglyn 1-LN; Nikkol Decaglyn 1-M; Nikkol Decaglyn 1-O; Nikkol Decaglyn 1-S; Niox KI-29; Norfox® Sorbo T-20; Norfox® Sorbo T-80; Nutrifos® B-75

Pearex® L; Phosal® 50 SA; Phosal® 53 MCT; Phosal® 75 SA; Polyaldo® DGDO KFG

Radiamuls® 157; Ryoto Sugar Ester S-1170; Ryoto Sugar Ester S-1170S; Ryoto Sugar Ester S-1570; Ryoto Sugar Ester S-1670

Santone® 3-1-S; Santone® 10-10-O; S-Maz® 20; S-Maz® 80K; S-Maz® 85K; Sobalg FD 000 Range; Sobalg FD 900 Range; Sorbax PML-20; Sorbax PMO-5; Sorbax PMO-20; Sorbax PMP-20; Sorbax PMS-20; Sorbax PTO-20; Sorbax PTS-20; Span® 60K; Star-Dri® 10; Star-Dri® 15; Star-Dri® 18; Star-Dri® 1005A

Tagat® I; Tagat® I2; Tagat® L; Tagat® L2; Tagat® O; Tagat® O2; Tagat® S; Tagat® S2; Tagat® TO; T-Maz® 20K; T-Maz® 28; T-Maz® 60; T-Maz® 60KHS; T-Maz® 65; T-Maz® 80; T-Maz® 80K; T-Maz® 81; T-Maz® 95; Tween® 20; Tween® 20 SD; Tween® 60; Tween® 61; Tween® 65; Tween® 80; Tween® 80K; Tween® 80 VS; Tween® 81; Tween® 85

Witafrol® 7420

Chemicals: Citric acid esters of mono- and diglycerides of fatty acids

Glyceryl caprylate; Glyceryl caprylate/caprate

Isocetyl alcohol

Monoglyceride citrate

Nonoxynol-10; Nonoxynol-14

PEG-6 dilaurate; PEG-20 dilaurate; PEG-32 dilaurate; PEG-75 dilaurate; PEG-150 dilaurate; PEG-6 dioleate; PEG-20 dioleate; PEG-32 dioleate; PEG-75 dioleate; PEG-150 dioleate; PEG-6 distearate; PEG-20 distearate; PEG-32 distearate; PEG-75 distearate; PEG-20 glyceryl isostearate; PEG-30 glyceryl isostearate; PEG-20 glyceryl laurate; PEG-30 glyceryl laurate; PEG-20 glyceryl oleate; PEG-30 glyceryl oleate; PEG-15 glyceryl ricinoleate; PEG-30 glyceryl stearate; PEG-25 glyceryl trioleate; PEG-50 hydrogenated castor oil; PEG-60 hydrogenated castor oil; PEG-32 laurate; PEG-8 oleate; PEG-32 oleate; PEG-75 oleate; PEG-40 sorbitan diisostearate; PEG-80 sorbitan laurate; PEG-20 sorbitan tritallate; PEG-32 stearate; PEG-75 stearate; Poloxamer 105; Poloxamer 108; Poloxamer 122; Poloxamer 123; Poloxamer 124; Poloxamer 181; Poloxamer 182; Poloxamer 184; Poloxamer 185; Poloxamer 188; Poloxamer 212; Poloxamer 215; Poloxamer 217; Poloxamer 231; Poloxamer 234; Poloxamer 235; Poloxamer 237; Poloxamer 238; Poloxamer 282; Poloxamer 284; Poloxamer 288; Poloxamer 331; Poloxamer 333; Poloxamer 334; Poloxamer 335; Poloxamer 338; Poloxamer 401; Poloxamer 402; Poloxamer 403; Poloxamer 407; Polyglyceryl-10 hexaoleate; Polysorbate 40; Polysorbate 61; Polysorbate 80; Polysorbate 81; Polysorbate 85

Sorbitan laurate; Sucrose distearate; Sucrose stearate

Triolein

Solvents • Vehicles

Trade names: Akofame; Akorex; Akorex B; Aldo® MCT KFG
Captex® 200; Captex® 355; Captex® 800; Captex® 810A; Captex® 810B; Captex® 810C; Captex® 810D; Captex® 910A; Captex® 910B; Captex® 910C; Captex® 910D; Captex® 8000; Captex® 8227
Delios® C; Delios® S; Delios® V
Emery® 917; Estasan GT 8-40 3578; Estasan GT 8-60 3575; Estasan GT 8-60 3580; Estasan GT 8-65 3577; Estasan GT 8-65 3581
Glycerine (Pharmaceutical)
Hodag CC-22; Hodag CC-22-S; Hodag CC-33; Hodag CC-33-F; Hodag CC-33-L; Hodag CC-33-S
Lexol® GT-855; Lexol® PG-800; Lexol® PG-855; Lexol® PG-865; Liponate GC; Liponate PC; Lipovol SES
Miglyol® 810; Miglyol® 812; Miglyol® 840; Myvacet® 9-45K
Neobee® 18; Neobee® 1054; Neobee® 1062; Neobee® M-5; Neobee® M-20; Neobee® O
PD-23; PD-25; Penreco 2251 Oil; Penreco 2257 Oil; Penreco 2263 Oil; Polarin® Range

Chemicals: Acetic acid; Acetone
Butylene glycol
Caprylic alcohol; Caprylic/capric/lauric triglyceride; Caprylic/capric/linoleic triglyceride; Caprylic/capric/oleic triglyceride
Ethyl acetate; Ethylene dichloride
Glycerin; Glyceryl caprylate; Glyceryl oleate; Glyceryl stearate; Glyceryl tricaprylate/caprate
Hexane; Hydrocarbon solvent
Lactic acid
Methyl alcohol; Methylene chloride; Mono- and diglycerides of fatty acids
Naphtha; 2-Nitropropane
PEG-20 dilaurate; 3-Pentanol; Petroleum distillates; Polyglyceryl-10 trioleate; Propylene glycol; Propylene glycol alginate; Propylene glycol dicaprylate/dicaprate; Propylene glycol dioctanoate; Sodium hydroxide
Terpene resin, natural; Triacetin; Trichloroethylene; Trioctanoin; Triundecanoin

Spices

Trade names: Angrex Seasonings Code #02300
Morton® H.G. Blending Prepared Salt; Morton® H.G. Blending Salt; Morton® Lite Salt® Mixt; Morton® Lite Salt® TFC Mixt; Morton® Star Flake® Dendritic ES Salt; Morton® Star Flake® Dendritic Salt; Morton® Table Iodized Salt; Morton® TFC 999® Salt; Morton® TFC H.G. Blending Salt; Morton® Top Flake Coarse Salt; Morton® Top Flake Extra Coarse Salt; Morton® Top Flake Fine Salt; Morton® Top Flake Topping Salt
Seal N' Season Glazes and Seasonings; Super Vita Rye

Chemicals: Garlic juice
Oleoresin angelica seed; Oleoresin anise; Oleoresin basil; Oleoresin black pepper; Oleoresin capsicum; Oleoresin caraway; Oleoresin cardamom; Oleoresin celery; Oleoresin coriander; Oleoresin cumin; Oleoresin dillseed; Oleoresin fennel; Oleoresin ginger; Oleoresin laurel leaf; Oleoresin marjoram; Oleoresin origanum; Oleoresin paprika; Oleoresin parsley leaf; Oleoresin parsley seed; Oleoresin pimenta berries; Oleoresin thyme; Oleoresin turmeric
Rosemary extract

Stabilizers

Trade names: A.B.C. #7; Ablunol S-20; Ablunol S-40; Ablunol S-60; Ablunol S-80; Ablunol S-85; Agar Agar NF Flake #1; Akodel 112; Akodel 118; Albriphos™ Blend 928; Albriphos™ Blend 75-25; Albriphos™ Blend 90-10; Albrite TSPP Food Grade; Alcolec® BS; Alcolec® Granules; Alcolec® S; Aldo® HMS FG; Aldo® HMS KFG; Aldo® MSLG FG; Aldo® MS LG KFG; Alginade DC; Alginade MR; Alginade MRE; Alginade XK9; Alpha W6 HP 0.6; Alpha W6 M1.8;

Stabilizers *(cont'd.)*

Alpha W 6 Pharma Grade; Alphadim® 90AB; Alphadim® 90SBK; Amoloid HV; Amoloid LV; Aqualon® 7H3SF; Aqualon® 7HF; Aqualon® 7MF; Aratex; Arcon S, Fortified Arcon S; Ardex D, D Dispersible; Atlas 1400K; Atlas 1500; Atlas E; Atmos® 150; Atmos® 150 K; Atmul® 84; Atmul® 84 K; Atmul® 124; Atmul® 918 K; Avicel® CL-611; Avicel® RC-501; Avicel® RC-581; Avicel® RC-591F; Avicel® RCN-15; Avicel® WC-595

Bac-N-Fos® Formula 191; Benecel® MP 824; Benecel® MP 843; Benefiber®; Beta W 7; Beta W7 HP 0.9; Beta W7 M1.8; Beta W7 P; BF-46 Icing Powd; BF-50 Icing Powd; Binasol™ 15; Bone Gelatin Type B 200 Bloom; Britesorb®; Britesorb® A 100; Bunge Biscuit Flakes

C-60 Icing Powd; C-249 Icing Powd; C-369; Cake Moist; Calsoft L-60; Cameo Showcase; Cameo Velvet; Cameo Velvet WT; CAO®-3; Caprol® 3GS; Cap-Shure® F-125-85; Cap-Shure® F-140-63; Carralean™ CG-100; Carralean™ CM-70; Carralean™ CM-80; Carralean™ MB-60; Carralean™ MB-93; Carralite™; CarraLizer™ CGB-10; CarraLizer™ CGB-20; CarraLizer™ CGB-40; CarraLizer™ CGB-50; Carudan 000 Range; Carudan 100 Range; Carudan 200 Range; Carudan 300 Range; Carudan 400 Range; Carudan 700 Range; Cavitron Cyclodextrin.™; CC-603; Cegeprot® Range; Celish® FD-100F; Cellogen HP-6HS9; Celynol F1; Celynol MSPO-11; Centrolene® A; Centrolex® C; Centromix® CPS; Cetodan® 50-00A; Cetodan® 70-00A; Cetodan® 90-40; Ches® 500; Cithrol GDO N/E; Cithrol GDO S/E; Cithrol GDS N/E; Cithrol GDS S/E; Cithrol GML N/E; Cithrol GML S/E; Cithrol GMO N/E; Cithrol GMO S/E; Cithrol GMR N/E; Cithrol GMR S/E; Cithrol GMS Acid Stable; Cithrol GMS N/E; Cithrol GMS S/E; Cithrol PGML N/E; Cithrol PGML S/E; Cithrol PGMO N/E; Cithrol PGMO S/E; Cithrol PGMR N/E; Cithrol PGMR S/E; Cithrol PGMS N/E; Cithrol PGMS S/E; CMC Daicel 1150; CMC Daicel 1160; CMC Daicel 1220; CMC Daicel 1240; CMC Daicel 1260; CMC Daicel 2200; Coatingum L; Colloid 886; Concentrated Dariloid®; Concentrated Dariloid® KB; Concentrated Dariloid® XL; Consista®; Controx® AT; Controx® KS; Controx® VP; Cremelite®; Cremodan; Crester KZ; Crester L; Crill 3; Crill 4; Crillet 3; Crodacreme; Crodascoop; Crolactem; Crystallization Inhibitor HL-13788; CS-9 Icing Powd; Curafos® 11-2; Curavis® 250

Dairy-Lo™; Danish Snax® Icing Powd; Dariloid®; Dariloid® 100; Dariloid® 300; Dariloid® 400; Dariloid® QH; Dariloid® XL; Design C200; Design NH; Dimodan PM; Dimodan PV Kosher; Dimodan PVP Kosher; Dimul S K; Divergan® F, R; DK Ester F-10; DK Ester F-20; DK Ester F-50; DK Ester F 70; DK Ester F 90; DK Ester F-110; DK Ester F-140; DK Ester F-160; Dress'n® 300; Drewmulse® 10K; Drewmulse® 200K; Dricoid® 200; Dricoid® 280; Dricoid® KB; Dricoid® KBC; DriRite Glaze Powd; Dur-Em® 207; Durkex 100DS; Durkote Fumaric Acid/Hydrog. Veg. Oil

Edifas; Egg Yolk Solids Type Y-1-FF; Eisai Natural Vitamin E Series; Emulcoll; Emuldan; Emuldan HA 40; Emuldan HA 52; Emuldan HV 40 Kosher; Emuldan HV 52 Kosher; Emuldan HVF 52 K; Emulfluid™; Emulfluid® A; Emulgel E-21; Emulgel S-32; Emulgum BV; Emulsilac S; Emulsilac S K; Enrich®; Essiccum®; Evanstab® 12

Fancol OA-95; Freedom X-PGA; Freezist® M; Frigesa® D 890; Frigesa® F; Frigesa® IC 178; Frigesa® IC 184; Frimulsion 10; Frimulsion 6G; Frimulsion Q8; Frimulsion RA; Frimulsion RF; Frimulsion X5; Frost-N-Wrap™ Icing Powd; Frost-O-Loid Icing Powd; Fruitfil® 1

Gamma W8; Gamma W8 HP0.6; Gamma W8 M1.8; Garbefix 31; Gelatinized Dura-Jel®; Gelcarin® GP-379; Gelcarin® GP-911; Gelcarin® PS-316; Gelcarin® XP-1008; Genu® 18CG-YA; Genu® 104AS-YA; Genu® BB Rapid Set; Genu® JMJ; Genulacta® K-100; Genulacta® KM-5; Genulacta® L-100; Genu® Pectins; GFS®; Glass H®; Glazo™; GMS 400; GMS 400V; GMS 402; GMS 600V; GMS 902; Golden Flake™; Gold Star 6MN Sherbet; Gold Star BMS 3200; Gold Star CDS 3100; Gold Star DD Base; Gold Star HFY; Gold Star LFB; Gold Star MSS 1800; Gold Star MSS 1803; Gold Star NFB; Gold Star SCS 3501; Gold Star SSS 1506; Gold Star WCS 3400; Gold Star Eagle; Gold Star Finch; Gold Star Lark; Gold Star Soft Serve; Gold Star Sparrow; Gold Star Swallow; Gold Star Swan; Granular Gum Arabic NF/FCC C-4010; Granular Gum Ghatti #1; Guar Gum; Guar Gum HV; Gum Arabic NF/FCC Clean Amber Sorts

HBI Icing Powd; Hexaphos®; Hodag GML; Hodag GMO; Hodag GMO-D; Hodag GMR; Hodag GMR-D; Hodag GMS; Hodag GMSH; Hodag GTO; Hodag PGS; Hodag PGSH; Hodag PSMS-20; Homodan RD

Ice # 2; Ice # 81; Idealgum 1A; Idealgum 1B; Idealgum 1C; Idealgum 2A; Idealgum 2B; Idealgum 3F; Imwitor® 370; Imwitor® 371; Imwitor® 372; Imwitor® 373; Imwitor® 375; Instant Remygel AX-P, AX-2-P; Instant Remyline AX-P; Instant Tender-Jel® 480

K2B387; K8B243; K8B249; Karaya Gum #1 FCC; Karaya Gum #1 FCC; Karaya Gum #1 FCC; Karaya Gum #1 FCC; Kelacid®; Kelco® HV; Kelco® LV; Kelcogel®; Kelcogel® F; Kelcoloid® D; Kelcoloid® DH; Kelcoloid® DO; Kelcoloid® DSF; Kelcoloid® HVF; Kelcoloid® LVF;

Stabilizers *(cont'd.)*

Kelcoloid® O; Kelcoloid® S; Kelcosol®; Kelgin® F; Kelgin® HV; Kelgin® LV; Kelgin® MV; Kelgin® QL; Kelgin® XL; Kelmar®; Kelmar® Improved; Kelset®; Keltone®; Keltone® HV; Keltone® LV; Keltose®; Keltrol®; Keltrol® 1000; Keltrol® BT; Keltrol® CR; Keltrol® F; Keltrol® GM; Keltrol® RD; Keltrol® SF; Keltrol® T; Keltrol® TF; Kelvis®; Kena® FP-28; Kimiloid HV; Kimiloid MV; Kimiloid NLS-K; Kimitsu Algin I-1; Kimitsu Algin I-2; Kimitsu Algin I-3; Klucel® F Grades; K.L.X; K.L.X. Flakes; KOB87; KOB348; KOB349; Kol Guard® 7373; Kol Guard® 7413

Lactarin® MV-308; Lactodan P 22; Lamefrost® ES 216 G; Lamefrost® ES 251 G; Lamefrost® ES 315; Lamefrost® ES 375; Lamefrost® ES 379; Lamefrost® ES 424; Lamegin® DW 8000-Range; Lecigran™ 6750; Lecigran™ A; Lecigran™ C; Lecigran™ F; Lecigran™ M; Lecigran™ T; Leciprime™; Lipo GMS 450; Lipo GMS 470; Lipodan CDS Kosher; Lipodan CRE Kosher; Lipodan OM 30; Lipodan SET Kosher; Liponic Sorbitol Powd; Liquid Fish Gelatin Conc; Locust Bean Gum Speckless Type D-200; Lonzest® SMO-20; Lonzest® SMP-20; Lonzest® SMS-20; Lonzest® STO-20; Lonzest® STS-20; L.S.B. Stabl

Maco-O-Line 091; Manucol DH; Manucol DM; Manucol DMF; Manucol JKT; Manucol LB; Manucol SMF; Manucol Ester E/PL; Manucol Ester E/RK; Manucol Ester M; Manugel C; Manugel DJX; Manugel DMB; Manugel GHB; Manugel GMB; Manugel JKB; Manugel L98; Manugel PTB; Manugel PTJ; Manugel PTJA; Marloid® CMS; Maxi-Gel® 7776; Maximaize® DJ; Maximaize® HV; Mayonat DF; Mayonat V/100; M-C-Thin® AF-1 Type DB; M-C-Thin® AF-1 Type SB; M-C-Thin® AF-1 Type UB; Merecol® FA; Merecol® FAL; Merecol® FT; Merecol® G; Merecol® GL; Merecol® GX; Merecol® I; Merecol® IC; Merecol® K; Merecol® LK; Merecol® MS; Merecol® R; Merecol® RB; Merecol® RCS; Merecol® S; Merecol® SH; Merecol® Y; Meringue Stabilizer; Methocel® A4C Premium; Methocel® A4M Premium; Methocel® A15C Premium; Methocel® A15LV Premium; Methocel® E3 Premium; Methocel® E4M; Methocel® E4M Premium; Methocel® E5 Food Grade; Methocel® E6 Premium; Methocel® E15 Food Grade; Methocel® E15LV Premium; Methocel® E50LV Premium; Methocel® F4M Premium; Methocel® F50LV Premium; Methocel® K3 Premium; Methocel® K4M Premium; Methocel® K15M Premium; Methocel® K35 Premium; Methocel® K100LV Premium; Methocel® K100M Premium; Mexpectin HV 400 Range; Mexpectin LA 100 Range; Mexpectin LA 200 Range; Mexpectin LA 400 Range; Mexpectin LC 700 Range; Mexpectin LC 800 Range; Mexpectin LC 900 Range; Mexpectin MRS 300 Range; Mexpectin RS 400 Range; Mexpectin SS 200 Range; Mexpectin XSS 100 Range; Microlube System 999R; Mira-Bake® 505; Mira-Cleer® 187; Mira-Quik® MGL; Mira-Sperse® 623; Mira-Sperse® 626; Mira-Thik® 468; Mira-Thik® 469; Mira-Thik® 603; Mira-Thik® 606; Monomax AH90 B; Monomax VH90 B; Monomuls® 60-10; Monomuls® 60-15; Monomuls® 60-20; Monomuls® 60-25; Monomuls® 60-25/2; Monomuls® 60-25/5; Monomuls® 60-30; Monomuls® 60-35; Monomuls® 60-40; Monomuls® 60-45; Monomuls® 90-10; Monomuls® 90-15; Monomuls® 90-20; Monomuls® 90-25; Monomuls® 90-25/2; Monomuls® 90-25/5; Monomuls® 90-30; Monomuls® 90-35; Monomuls® 90-40; Monomuls® 90-45; Monomuls® 90-O18; Mr. Chips™; Myvacet® 5-07; Myvatex® 90-10K; Myvatex® Monoset®; Myvatex® Monoset® K; Myvatex® Peanut Butter Stabilizer; Myvatex® Texture Lite®; Myverol® 18-04; Myverol® 18-04K; Myverol® 18-07; Myverol® 18-07K; Myverol® 18-50; Myverol® 18-92; Myverol® 18-99; Myverol® P-06; Myverol® P-06K

NatuReal™ Sensoral; NatuReal™ Textural; Neobee® 62; Nikkol MGS-A; Nikkol MGS-ASE; Nikkol MGS-B; Nikkol MGS-BSE; Nikkol MGS-DEX; Nikkol MGS-F20; Nikkol MGS-F40; Nikkol MGS-F50; Nikkol MGS-F75; Nipanox® Special; Norfox® Sorbo T-80; Nu-Col™ 4227; Nustar®; Nutrifos® H-30; Nutriloid® Locust; Nutriloid® Xanthan

O.B. Stabilizer; Oxynex® K

Paniset/Panistay; Panodan 120 Series; Panodan 140 Series; Panodan 150; Panodan 205; Panodan FDP; Panodan SD; Papain 30,000; Papain Conc; Paramount X; Paramount XX; Pationic® 920; Pationic® 925; Pationic® 1230; Pationic® 1250; Patlac® NAL; PB-3 Icing Powd; PB-34 Icing Powd; Pearex® L; Perma-Flo®; Pfico$_2$.Hop®; PGMS 70; PI-64 Icing Powd; Polyaldo® HGDS KFG; Polycon T60 K; Powdered Agar Agar NF M-100 (Gracilaria); Powdered Agar Agar NF MK-80-B; Powdered Agar Agar NF MK-80 (Bacteriological); Powdered Agar Agar NF S-100; Powdered Agar Agar NF S-100-B; Powdered Agar Agar NF S-150; Powdered Agar Agar NF S-150-B; Powdered Gum Arabic NF/FCC G-150; Powdered Gum Arabic NF/FCC Superselect Type NB-4; Powdered Gum Ghatti #1; Powdered Gum Ghatti #2; Powdered Gum Guar NF Type 80 Mesh B/T; Powdered Gum Guar Type 140 Mesh B/T; Powdered Gum Guar Type ECM; Powdered Gum Guar Type M; Powdered Gum Guar Type MM FCC; Powdered Gum Guar Type MM (HV); Powdered Gum Guar Type MMM $^1/_2$; Powdered Gum Guar Type MMW; Powdered Gum Karaya Superfine #1 FCC; Powdered

Stabilizers *(cont'd.)*

Gum Karaya Superfine XXXX FCC; Powdered Locust Bean Gum Type D-200; Powdered Locust Bean Gum Type D-300; Powdered Locust Bean Gum Type P-100; Powdered Locust Bean Gum Type PP-100; Prinza® 452; Prinza® 455; Prinza® Range; Pristene® 180; Pristene® 184; Pristene® 185; Pristene® 186; Pristene® R20; Pristene® RO; Pristene® RW; Pristene® TR; Promax® HV (redesignated Promine® HV); Promine® HV; Propylene Glycol Alginate HV; Propylene Glycol Alginate LV FCC; Pure-Gel® B990; Pure-Gel® B992; Py-ran® Qual Flo™; Qual Flo™ C

Radiamuls® CMG 2930; Readi-Ice Icing Powd; Recodan; Redihop®; Redisol® 78D; Redisol® 88; Redisol® 412; Remygel AX-2; Remygel AX; Remyline AX; Resinogum DD; Rezista®; Rhodigel®; Rhodigel® 200; Rhodigel® EZ; Rhodigel® Supra; Rhodigum OEH; Rhodigum OEM; Rhodigum WGH; Rhodigum WGL; Rhodigum WGM; Rhodigum WVH; Rhodigum WVM; Royal Set

Saladizer® #250PM; Saladizer® #251; Santone® 8-1-O; Santone® 10-10-O; Set-it Powd; Sherbelizer®; Silver Star BMS 3251; Silver Star CDS 3151; Silver Star SCS 3001; Silver Star WCS 3451; Silver Star YS 3302; Sobalg FD 100 Range; Sodaphos®; Sodium Alginate HV NF/FCC; Sodium Alginate LV; Sodium Alginate LVC; Sodium Alginate MV NF/FCC; Sodium Citrate USP, FCC Dihydrate Gran. No. 69976; Sodium Citrate USP, FCC Dihydrate Fine Gran. No. 69975; Soft-Set®; Sorbo®; Span® 20; Span® 40; Span® 60K; Span® 65; Span® 65K; Span® 85; Spray Dried Fish Gelatin; Spray Dried Fish Gelatin/Maltodextrin; Spray Dried Gum Arabic NF/FCC CM; Spray Dried Gum Arabic NF/FCC CS (Low Bacteria); Spray Dried Gum Arabic NF/FCC CS-R; Spray Dried Hydrolysed Fish Gelatin; Spray Dried Nigerian Gum Arabic; Spraygum C; Spraygum GD; Stabilizer C; Stabl Plus; Staform P; Sta-Lite; Sta-Lite; Sta-Lite; Sta-Lite; Stamere® CKM FCC; Stamere® CK-S; Stamere® N-325; Stamere® N-350; Stamere® N-350 S; Stamere® NI; Stamere® NIC FCC; Sta-Nut EE; Sta-Nut P; Sta-Nut R; Starco™ 447; Starco™ 401; Starplex® 90; 27 Stearine; Sunmalt; Supercol® G2S; Super Gel for Glazes and Icings; Super Salox; Super Set; Supersta Icing Powd; Sustane® BHA; #325 Sweet Dough Conc

Tandem 5K; Tandem 8; Tegomuls® 90; Tegomuls® 9102; Tegomuls® 9102 S; Tegomuls® SB; Tegomuls® SO; Tenderbite SF(TB-SF); Tenderfil® 8; Tenderfil® 9; Tenderfil® 428; Tenox® BHA; Tenox® BHT; Tenox® PG; Tenox® TBHQ; Terra Alba 114836; The Edge® C-1600; Thixogum S; Thixogum X; Ticalose® 75; Ticalose® 100; Ticaxan® Regular; TIC Pretested® CMC PH-2500; TIC Pretested® Saladizer 250; TIC Pretested® Saladizer PH-250; TIC Pretested® Xanthan 200; T-Maz® 28; T-Maz® 60; T-Maz® 60KHS; T-Maz® 65; T-Maz® 80; T-Maz® 81; T-Maz® 95; Top Flite™ Icing Powd; Topmulgat; Tween® 60; Tween® 80

Vanease (K); Veltol®; Veltol®-Plus; Vitiben®; Vykamol 83G

Wakal® A 601; Wakal® J; Wakal® JG; Wakal® K-e

Chemicals: Acetylated hydrogenated cottonseed glyceride; Acetylated mono- and diglycerides of fatty acids; Acetylated tartaric acid esters of mono- and diglycerides of fatty acids; Agar; Alginic acid; Aloe extract; Aluminum distearate; Aluminum stearate; Aluminum tristearate; Ammonium alginate; Ammonium carrageenan; Ammonium citrate; Ammonium furcelleran; Ammonium potassium hydrogen phosphate; Arabinogalactan; Ascorbyl palmitate

Bakers yeast extract; Bentonite; Brominated vegetable oil

Calcium acetate; Calcium alginate; Calcium carbonate; Calcium carboxymethyl cellulose; Calcium carrageenan; Calcium chloride; Calcium furcelleran; Calcium glycerophosphate; Calcium lactate; Calcium lignosulfonate; Calcium oleate; Calcium phosphate dibasic; Calcium phosphate dibasic dihydrate; Calcium ricinoleate; Calcium/sodium stearoyl lactylate; Calcium stearate; Calcium sulfate; Canola oil glyceride; Carboxymethylcellulose sodium; Carboxymethylmethylcellulose; Carrageenan; Carrageenan extract; Cassia gum; Chitin; Chitosan; Citric acid esters of mono- and diglycerides of fatty acids

Damar; Dextran; Dextrin; Dipotassium citrate; Disodium citrate; Disodium EDTA

Erythorbic acid; 6-Ethoxy-1,2-dihydro-2,2,4-trimethylquinoline

Ferrous gluconate; Ferrous sulfate heptahydrate; Food starch modified; Furcelleran

Gelatin; Gellan gum; Glucosamine; Glucose oxidase; Glyceryl citrate/lactate/linoleate/oleate; Glyceryl dioleate; Glyceryl dioleate SE; Glyceryl distearate; Glyceryl distearate SE; Glyceryl laurate SE; Glyceryl oleate SE; Glyceryl ricinoleate; Glyceryl ricinoleate SE; Glyceryl stearate SE; Glycine; Guar gum; Gum ghatti

Hydrogenated lard glyceride; Hydrogenated lard glycerides; Hydrogenated palm glyceride; Hydrogenated palm oil; Hydrogenated rapeseed oil; Hydrogenated soybean glycerides; Hydrogenated soy glyceride; Hydrogenated tallow glyceride; Hydrogenated tallow glycerides; Hydrogenated vegetable glycerides; Hydroxypropylcellulose; Hydroxypropyl methylcellulose

Stabilizers *(cont'd.)*

Karaya gum; Kelp extract

Lactic acid esters of mono- and diglycerides of fatty acids; Lactylated fatty acid esters of glycerol and propylene glycol; Lard glycerides; Lecithin; Locust bean gum

Magnesium glycerophosphate; Magnesium phosphate dibasic; Magnesium stearate; Malt extract; D-Mannitol; Methylcellulose; Methyl ethyl cellulose; Microcrystalline cellulose; Mono- and diglycerides of fatty acids

Oleyl alcohol

Palm glyceride; Palm oil sucroglyceride; Pectin; PEG-20 sorbitan tritallate; PEG-8 stearate; Pentapotassium triphosphate; Poloxamer 105; Poloxamer 108; Poloxamer 122; Poloxamer 123; Poloxamer 124; Poloxamer 181; Poloxamer 182; Poloxamer 183; Poloxamer 184; Poloxamer 185; Poloxamer 188; Poloxamer 215; Poloxamer 217; Poloxamer 231; Poloxamer 234; Poloxamer 235; Poloxamer 237; Poloxamer 238; Poloxamer 282; Poloxamer 284; Poloxamer 288; Poloxamer 331; Poloxamer 333; Poloxamer 334; Poloxamer 335; Poloxamer 338; Poloxamer 401; Poloxamer 402; Poloxamer 403; Poloxamer 407; Polyglycerol esters of fatty acids; Polyglyceryl polyricinoleate; Polyglyceryl-3 stearate; Polysorbate 20; Polysorbate 40; Polysorbate 60; Polysorbate 65; Polysorbate 80; Polysorbate 85; Polyvinylpolypyrrolidone; Poly(1-vinyl-2-pyrrolidinone) homopolymer; Potassium acid tartrate; Potassium alginate; Potassium carrageenan; Potassium chloride; Potassium citrate; Potassium furcelleran; Potassium hydroxide; Potassium oleate; Potassium sodium tartrate anhyd; Potassium stearate; Propylene glycol; Propylene glycol alginate; Propylene glycol dicaprylate/dicaprate; Propylene glycol esters of fatty acids; Propylene glycol laurate; Propylene glycol laurate SE; Propylene glycol monodistearate; Propylene glycol oleate; Propylene glycol oleate SE; Propylene glycol ricinoleate; Propylene glycol ricinoleate SE; Propylene glycol stearate; Propylene glycol stearate SE; PVP

Rapeseed oil glyceride; Rennet; Rhamsan gum

Silica; Sodium carrageenan; Sodium caseinate; Sodium citrate; Sodium furcelleran; Sodium hypophosphite; Sodium oleate; Sodium palmitate; Sodium phosphate dibasic; Sodium polyacrylate; Sodium stearate; Sodium stearoyl lactylate; Sodium stearyl fumarate; Sodium tartrate; Sorbitan laurate; Sorbitan oleate; Sorbitan palmitate; Sorbitan tristearate; Sorbitol; Soy glyceride; Soy protein; Stannous stearate; Stearyl glyceridyl citrate; Sucrose fatty acid esters; Sulfur dioxide; Sunflower seed oil glycerides

Tallow glycerides; Tara gum; Tartaric acid esters of mono- and diglycerides; Tetrasodium pyrophosphate; Tragacanth gum; Triolein; Trisodium citrate

Wheat gluten

Xanthan gum

Zinc orthophosphate; Zinc resinate

Starch complexing agents

Trade names: Alphadim® 90AB; Alphadim® 90SBK; Amfal-46; Amidan; Amidan ES Kosher; Artodan SP 55 Kosher

Canamulse 55; CFI-10; Crodatem L; Crodatem L50; Crodatem T22; Crodatem T25

Dimodan PM; Dimodan PV; Dimodan PV 300 Kosher; Dimodan PV Kosher

GMS 305; GMS 402; GMS 402V; GMS 602V; GMS 900; GMS 902

HBF

Monomax AH90 B; Monomax VH90 B; Monomuls® 60-10; Monomuls® 60-15; Monomuls® 60-20; Monomuls® 60-25; Monomuls® 60-25/2; Monomuls® 60-25/5; Monomuls® 60-30; Monomuls® 60-35; Monomuls® 60-40; Monomuls® 60-45; Monomuls® 90-10; Monomuls® 90-15; Monomuls® 90-20; Monomuls® 90-25; Monomuls® 90-25/2; Monomuls® 90-25/5; Monomuls® 90-30; Monomuls® 90-35; Monomuls® 90-40; Monomuls® 90-45; Myvaplex® 600; Myvaplex® 600K; Myvaplex® 600P; Myvaplex® 600PK; Myvatex® Mighty Soft®

Patco® 3

Ryoto Sugar Ester ER-190; Ryoto Sugar Ester ER-290; Ryoto Sugar Ester P-170

Starplex® 90

Tegomuls® 90; Tegomuls® 90 S; Tegomuls® 9053 S

Verv®

Chemicals: Sucrose erucate

Surface-finishing agents • Coatings • Glazes • Waxes

Trade names: AA USP; Akocote 102; Akocote 106; Akocote 109; Akodel 95; Akodel 102; Akodel 108; Akodel 112; Akodel 118; Akomax E; Akomax R; Akopol E-1; Akopol R; Akorine 2A; Akorine 3; Akorine 4; Akorine 9F; Akowesco 1; Akowesco 2; Akowesco 5; Akowesco 7; Akowesco 45 AC; Axol® E 61

Be Square® 195

Carbowax® Sentry® PEG 300; Carbowax® Sentry® PEG 400; Carbowax® Sentry® PEG 540 Blend; Carbowax® Sentry® PEG 900; Carbowax® Sentry® PEG 1000; Carbowax® Sentry® PEG 1450; Carbowax® Sentry® PEG 3350; Carbowax® Sentry® PEG 4600; Carbowax® Sentry® PEG 8000; Castorwax® NF; Cegeskin® Range; Cetodan®; Cetodan® 50-00P Kosher; Cetodan® 70-00P Kosher; Chocolate Flavored Frosteen®; Citrocoat® A 1000 HP; Citrocoat® A 2000 HP; Citrocoat® A 4000 TP; Citrocoat® A 4000 TT; CLSP 399; CLSP 499; CLSP 555; CLSP 874; Coatingum L; Coberine; Cozeen®; Crodaglaze

Danishine™; Delios® C; Delios® S; Delios® V; Descote® Citric Acid 50%; Dimodan PV Kosher; Dimodan PVP Kosher; DK-Ester; DK-Ester; Drakeol 7; Drakeol 9; Drakeol 13; Drewpol® 3-1-SK; Durkex 100BHA; Durkex 100BHT; Durkex 500; Durkex 500S; Durkex Durola; Durkex Gold 77A; Durkex Gold 77F; Durkex Gold 77N; Durlac® 100W; Duromel B108; Durtan® 60; Dynacet® 212; Dynacet® 278; Dynacet® 281; Dynacet® 282; Dynasan® P60

Hi-Glo Wash; Hilton Davis Titanium Dioxide; Hoechst Wax KPS; Hoechst Wax KSL; Hoechst Wax PED 121

Kaokote; Kaola; Kaola D; Kaomax 870; Kaomax-S; Kaomel; Kaoprem-E; Kaydol®; Konut; Koster Keunen Paraffin Wax; Koster Keunen Synthetic Spermaceti

Lactodan P 22 Kosher; Lamegin® EE Range; Lamegin® EE 50; Lamegin® EE 70; Lamegin® EE 100; Luwax OA 5

Maltrin® M200; Maltrin® QD M600; Muffin Kote; Multiwax® X-145A; Myvacet® 5-07; Myvacet® 5-07K; Myvacet® 7-00; Myvacet® 7-07; Myvacet® 7-07K; Myverol® 18-06; Myverol® 18-07; Myverol® 18-35

Nutriloid® Arabic

Orzol®

Paramount X; Paramount XX; Penreco Amber; Penreco Ultima; Petrolite® C-4040; Polyaldo® TGMS KFG; Promise Liq Oil

Ross Rice Bran Wax; Ryoto Sugar Ester ER-190; Ryoto Sugar Ester ER-290; Ryoto Sugar Ester O-1570; Ryoto Sugar Ester P-170; Ryoto Sugar Ester P-1570; Ryoto Sugar Ester P-1670; Ryoto Sugar Ester S-1570

Satina 44; Satina 50; Satina 53NT; Seal N Glaze Systems for Vegetables and Fruits; Seal N' Season Glazes and Seasonings; Sett®; S-Maz® 60K; Snac-Kote; Snac-Kote XTR; Span® 65; Special Fat 168T; Sweet'n'Neat® Oil/Dry Roast

Tem Cote® IE; Tem Cote® II; Ticaloid® 1070; Ticaloid® No Fat 102B; TIC Pretested® Aragum® 1067-T; TIC Pretested® Aragum® 1070; TIC Pretested® Aragum® 2000; T-Maz® 60K; Top Flite™ Icing Powd; Tween® 60

Vykacet T/L; Vykamol 83G

White Frosteen®; Witocan® LM

Chemicals: Acacia; Acetylated hydrogenated tallow glyceride; Acetylated mono- and diglycerides of fatty acids; Agar; Ammonium hydroxide

Beeswax

Calcium chloride; Calcium silicate; Candelilla wax; Carnauba; Castor oil; Cetyl esters; Cocoa butter substitute; Coconut oil; Corn oil; Cottonseed oil; Coumarone-indene resin

Dextrin

Ethylcellulose

Glyceryl stearate; Glyceryl tricaprylate/caprate

Hydrogenated castor oil; Hydrogenated palm oil

Isoparaffinic petroleum hydrocarbons, synthetic

Lanolin; Lard; Light petroleum hydrocarbons, odorless

Magnesium silicate; Methyl caprylate/caprate; Methyl cocoate; Methyl linoleate; Microcrystalline wax; Mineral oil; Mono- and diglycerides of fatty acids

Naphtha

Oleic acid; Oleth-23

1-Palmitoyl-2-oleoyl-3-stearin; Palm kernel oil; Palm oil; Paraffin; Peanut oil; PEG-4; PEG-6; PEG-8; PEG-9; PEG-12; PEG-14; PEG-16; PEG-20; PEG-32; PEG-40; PEG-75; PEG-100; PEG-150; PEG-200; Petrolatum; Petroleum wax; Polyethylene; Polyethylene glycol; Poly-

Surface-finishing agents *(cont'd.)*

ethylene, oxidized; Polyethylene, oxidized; Polyethylene, oxidized; Polyethylene, oxidized; Polysorbate 40; Polysorbate 60; Polyvinyl acetate (homopolymer)

Rice bran wax; Rice maltodextrin; Rosin

Safflower oil; Shellac; Shellac, bleached, wax-free; Sodium succinate; Sorbitan stearate; Sorbitol; Soybean oil; Sucrose fatty acid esters; Sucrose oleate; Sucrose palmitate; Sucrose tristearate; Sunflower seed oil; Synthetic wax

Talc; Tallow; Tristearin

Vegetable colloids

Wheat gluten

Zein

Surfactants • Dispersants • Suspending agents • Foam stabilizing agents

Trade names: Acidan; AF 72; Albrite STPP-F; Albrite STPP-FC; Alcolec® BS; Alcolec® S; Alcolec® Z-3; Aldo® HMS KFG; Aldo® LP KFG; Aldo® MLD FG; Aldo® MO FG; Aldosperse® MO-50 FG; Alkamuls® PSTS-20; Alkaquat® DMB-451-50, DMB-451-80; Alphadim® 90AB; Alphadim® 90SBK; Amisol™ 329; Amisol™ MS-10; Amoloid HV; Amoloid LV; Aqualon® 7H0F; Aqualon® 7HF; Aqualon® 7MF; Aqualon® 9H4XF; Aqualon® 9M31F; Aqualon® 9M31XF; Aqualon® 9M8F; Arlacel® 165; Arlacel® 186; Atlas 2000; Atlas 3000; Atlas G-986K; Atlas MDA; Atlas P-44; Atlas SSL; Atmos® 300 K; Atmul® 400; Atmul® 600H; Atsurf 456K; Atsurf 595; Atsurf 595K; Atsurf 596; Atsurf 596K; Avicel® CL-611; Avicel® RC-501; Avicel® RC-581; Avicel® RC-591F; Axol® C 62

Beakin LV4; Benecel® M 043; Benecel® MP 643; Benecel® MP 824; Benecel® MP 874; Benecel® MP 943; Blendmax 322; Blendmax 322D; Britol® 7NF; Britol® 9NF; Britol® 20USP; Britol® 35USP; Britol® 50USP

Cab-O-Sil® L-90; Calsoft L-60; Canasperse SBF; Canasperse UBF; Capmul® GMO; Capmul® O; Carbowax® Sentry® PEG 300; Carbowax® Sentry® PEG 400; Carbowax® Sentry® PEG 540 Blend; Carbowax® Sentry® PEG 600; Carbowax® Sentry® PEG 900; Carbowax® Sentry® PEG 1000; Carbowax® Sentry® PEG 1450; Carbowax® Sentry® PEG 3350; Carbowax® Sentry® PEG 4600; Carbowax® Sentry® PEG 8000; Carsonol® SHS; Cegepal® TG 126; Cegepal® TG 809; Cegepal® VF 347; Cegepal® VF; Celynol F1; Centrol® 3F SB; Centrolene® A; Centrolene® S; Centrolex® D; Centrolex® F; Centrolex® R; Centromix® E; Centrophase® 152; Centrophase® C; Centrophase® HR; Centrophase® HR2B; Centrophase® HR2U; Centrophase® HR4B; Centrophase® HR4U; Centrophase® HR6B; Centrophase® NV; Centrophil® M; Centrophil® W; Cithrol GDO N/E; Cithrol GDO S/E; Cithrol GDS N/E; Cithrol GDS S/E; Cithrol GML N/E; Cithrol GML S/E; Cithrol GMO N/E; Cithrol GMO S/E; Cithrol GMR N/E; Cithrol GMR S/E; Cithrol GMS Acid Stable; Cithrol GMS N/E; Cithrol GMS S/E; Cithrol PGML N/E; Cithrol PGML S/E; Cithrol PGMO N/E; Cithrol PGMO S/E; Cithrol PGMR N/E; Cithrol PGMR S/E; Cithrol PGMS N/E; Cithrol PGMS S/E; Citric Acid USP FCC Anhyd. Fine Gran. No. 69941; Citrid Acid USP FCC Anhyd. Gran. No. 69942; Clearate Special Extra; CMC Daicel 1150; CMC Daicel 1160; CMC Daicel 1220; CMC Daicel 1240; CMC Daicel 1260; CMC Daicel 2200; Colloid 488T; Colloid 602; Complemix® 50; Complemix® 100; Crill 1; Crill 2; Crill 43; Crill 45; Crill 50; Crillet 1; Crillet 2; Crillet 11; Crillet 31; Crillet 35; Crillet 41; Crillet 45

Dariloid® XL; Detergent CR; Diagum LBG; DK-Ester; DK-Ester; DK Ester F-10; DK Ester F-20; DK Ester F-50; DK Ester F 70; DK Ester F 90; DK Ester F-110; DK Ester F-140; DK Ester F-160; Drewmulse® 3-1-O; Drewmulse® 3-1-S; Drewmulse® 6-2-S; Drewmulse® 10-4-O; Drewmulse® 10-8-O; Drewmulse® 10-10-S; Drewmulse® 10K; Drewmulse® 200; Drewmulse® 365; Drewmulse® 900K; Drewmulse® POE-SMO; Drewmulse® SMO; Drewmulse® SMS; Drewpol® 8-1-OK; Drewpol® 10-2-OK; Drewpol® 10-4-OK; Drewpol® 10-6-OK; Drewpol® 10-8-OK; Drewpol® 10-10-OK; Drewpone® 60K; Drewpone® 65K; Drewpone® 80K; Dur-Em® 300; Dur-Em® 300K; Dur-Em® GMO; Durfax® 65; Durfax® 80; Durtan® 60

Emalex GMS-10SE; Emalex GMS-15SE; Emalex GMS-20SE; Emalex GMS-25SE; Emalex GMS-45RT; Emalex GMS-50; Emalex GMS-55FD; Emalex GMS-195; Emalex GMS-A; Emalex GMS-ASE; Emalex GMS-B; Emalex GMS-P; Emalex GWIS-100; Emalex PGML; Emalex PGMS; Emalex PGO; Emalex PGS; Emphos™ D70-30C; Emphos™ F27-85;

Surfactants *(cont'd.)*

Emulfluid® A; Emulfluid® AS; Emulfluid® E; Emulpur™ N; Emulthin M-35

Fancol OA-95

Genu® 12CG; Genu® DD Extra-Slow Set C, 150 Grade; Genugel® CJ; Genulacta® KM-1; Genulacta® LK-71; Genu® Pectins; Geoldan CL Range; GFS®; Glass H®; Glycosperse® L-20; Glycosperse® O-5; Glycosperse® O-20; Glycosperse® O-20 FG; Glycosperse® O-20 KFG; Glycosperse® O-20 Veg; Glycosperse® O-20X; Glycosperse® P-20; Glycosperse® S-20; Glycosperse® S-20 FG; Glycosperse® S-20 KFG; Glycosperse® TO-20; Glycosperse® TS-20; GMS 300; GMS 402; GMS 402V; GMS 600V; Gum Arabic NF/FCC Clean Amber Sorts; Gum Arabic, Purified, Spray-Dried No. 1834; Gummi Gelatin P-6; Gummi Gelatin P-7; Gummi Gelatin P-8

Hercules® AR 160; Hodag CC-22; Hodag CC-22-S; Hodag CC-33; Hodag CC-33-F; Hodag CC-33-L; Hodag CC-33-S; Hodag GMSH; Hodag PGML; Hodag PGMS; Hodag PGS; Hodag PGS-62; Hodag POE (20) GMS; Hodag PSMO-20; Hodag PSMS-20; Hodag PSTS-20; Hodag SMS; Hyonic NP-60; Hyonic NP-100; Hyonic NP-110; Hyonic NP-120; Hyonic NP-407; Hyonic NP-500

Imwitor® 928; Imwitor® 988

Kelcogel®; Kelcogel® F; Kelcosol®; Kelgin® F; Kelgin® HV; Kelgin® LV; Kelgin® MV; Kelgin® XL; Keltone® HV; Keltone® LV; Keltrol®; Keltrol® CR; Keltrol® F; Keltrol® GM; Keltrol® SF; Keltrol® T; Keltrol® TF; Kelvis®; Klucel® F Grades

Lactodan B 30; Lactodan LW; Lactodan P 22; Lecigran™ 5750; Lecigran™ 6750; Lecigran™ A; Lecigran™ C; Lecigran™ F; Lecigran™ M; Lecigran™ T; Liposorb O-20; Liposorb S; Locust Bean Gum Speckless Type D-200

Manucol DM; Manucol DMF; Manucol JKT; Manucol LB; Manucol Ester B; Manugel C; Manugel DJX; Manugel DMB; Manugel GHB; Manugel JKB; Manugel L98; Manugel PTJ; Mazol® GMO; Mazol® PGMS; Mazol® PGMSK; M-C-Thin®; M-C-Thin® AF-1 Type DB; M-C-Thin® AF-1 Type SB; M-C-Thin® AF-1 Type UB; M-C-Thin® ASOL 436; M-C-Thin® FWD; M-C-Thin® HL; Methocel® E3 Premium; Methocel® E4M Premium; Methocel® E6 Premium; Methocel® E15LV Premium; Methocel® E50LV Premium; Methocel® K3 Premium; Methocel® K4M Premium; Methocel® K15M Premium; Methocel® K100LV Premium; Methocel® K100M Premium; Miglyol® 812; Miglyol® 840; Mira-Bake® 505; Miranol® JB; Mira-Sperse®; Mira-Sperse® 535; Myvatem® 06K; Myvatem® 30; Myvatem® 35K; Myvatem® 92K; Myvatex® 8-06; Myvatex® 8-16; Myverol® 18-35; Myverol® 18-50; Myverol® 18-92

Natural Arabic Type Gum Purified, Spray-Dried; Nikkol TO-10; Nikkol TS-10; Nikkol Decaglyn 1-IS; Nikkol Decaglyn 1-L; Nikkol Decaglyn 1-LN; Nikkol Decaglyn 1-M; Nikkol Decaglyn 1-O; Nikkol Decaglyn 1-S; Norfox® Anionic 27; Nu-Col™ 4227

Panodan 120 Series; Panodan 140 Series; Panodan 150; Panodan 205; Panodan FDP; Panodan SD; Pationic® 920; Pationic® 925; Pegol® 10R8; Pegol® 17R1; Pisane®; Pluracol® E300; Pluracol® E400; Pluracol® E600; Pluracol® E1500; Pluracol® E4000; Polyaldo® 2O10 KFG; Polyaldo® 2P10 KFG; Polyaldo® DGDO KFG; Polyaldo® HGDS KFG; Polycon S60 K; Polycon T60 K; Polycon T80 K; Powdered Gum Arabic NF/FCC G-150; Powdered Gum Arabic NF/FCC Superselect Type NB-4; Powdered Gum Guar NF Type 80 Mesh B/T; Powdered Gum Guar Type ECM; Powdered Gum Guar Type M; Powdered Gum Guar Type MM FCC; Powdered Gum Guar Type MM (HV); Powdered Gum Guar Type MMM ¹/₂; Powdered Gum Guar Type MMW; Powdered Locust Bean Gum Type D-200; Powdered Locust Bean Gum Type D-300; Powdered Locust Bean Gum Type P-100; Powdered Locust Bean Gum Type PP-100; Powdered Tragacanth Gum Type A/10; Powdered Tragacanth Gum Type E-1; Powdered Tragacanth Gum Type G-3; Powdered Tragacanth Gum Type L; Powdered Tragacanth Gum Type W; Promodan USV

Rhodacal® DS-4; Rhodacal® LDS-10; Rhodapon® BOS; Rhodigel®; Rhodigel® 200; Rhodigel® EZ; Rhodigel® Supra; Rudol®; Ryoto Sugar Ester B-370; Ryoto Sugar Ester ER-190; Ryoto Sugar Ester ER-290; Ryoto Sugar Ester L-595; Ryoto Sugar Ester L-1570; Ryoto Sugar Ester L-1695; Ryoto Sugar Ester M-1695; Ryoto Sugar Ester O-1570; Ryoto Sugar Ester P-170; Ryoto Sugar Ester P-1570; Ryoto Sugar Ester P-1570S; Ryoto Sugar Ester P-1670; Ryoto Sugar Ester S-170; Ryoto Sugar Ester S-270; Ryoto Sugar Ester S-370; Ryoto Sugar Ester S-370F; Ryoto Sugar Ester S-570; Ryoto Sugar Ester S-770; Ryoto Sugar Ester S-970; Ryoto Sugar Ester S-1170; Ryoto Sugar Ester S-1170S; Ryoto Sugar Ester S-1570; Ryoto Sugar Ester S-1670; Ryoto Sugar Ester S-1670S

Santone® 3-1-S; Santone® 8-1-O; Shokusen SE; S-Maz® 20; S-Maz® 80K; S-Maz® 85K; Soageena®; Sodaphos®; Sodium Alginate HV NF/FCC; Sodium Alginate LV; Sodium Alginate LVC; Sodium Alginate MV NF/FCC; Sorbanox AL; Span® 60; Spray Dried Gum

Surfactants *(cont'd.)*

Arabic NF/FCC CM; Spray Dried Gum Arabic NF/FCC CS (Low Bacteria); Spray Dried Gum Arabic NF/FCC CS-R; Stabilizer C; Stabolic™ C; Stamere® CK-S; Stamere® N-47; Stamere® N-55; Stamere® N-325; Stamere® N-350; Stamere® N-350 S; Stamere® NI; Star-Dri® 1005A; Sucro Ester 11; Sucro Ester 15; Sul-fon-ate AA-9; Sul-fon-ate AA-10; Sul-fon-ate LA-10; Sulfotex OA; Sunny Safe; Superfine Ground Mustard; Superior Mustard; Surfactant AR 150; Suspengel Elite; Suspengel Micro; Suspengel Ultra

Tagat® I; Tagat® I2; Tagat® L; Tagat® L2; Tagat® O; Tagat® O2; Tagat® S; Tagat® S2; Tagat® TO; Tandem 11H; Tandem 930; Tegomuls® 90 S; Tegomuls® 9053 S; Tegomuls® 9102; Tegomuls® B; Ticalose® 75; Ticalose® 100; Ticalose® 150 R; Ticalose® 700 R; Ticalose® 2000 R; Ticalose® 5000 R; T-Maz® 28; T-Maz® 60; T-Maz® 60KHS; T-Maz® 65; T-Maz® 80; T-Maz® 81; T-Maz® 95; Topcithin®; Tragacanth Flake No. 27; Tragacanth Gum Ribbon No. 1; Triton® X-45; Triton® X-100; Triton® X-102; Triton® X-114; Triton® X-120; Triton® X-305-70%; Triton® X-405-70%; Tween® 20; Tween® 60; Tween® 60K; Tween® 61; Tween® 80; Tween® 81; Tween® 85; Tween®-MOS100K

Visco Compound™

Witafrol® 7420

Yelkinol F; Yelkinol G; Yelkinol P; Yuccafoam™

Chemicals: Algin; Alumina; Ammoniated glycyrrhizin; Ammonium caseinate

Calcium lignosulfonate; Calcium silicate; Calcium/sodium stearoyl lactylate; Carboxymethyl-methylcellulose; Cellulose; Citric acid; Cobalt sulfate (ous)

Dioctyl sodium sulfosuccinate

Ethylene oxide polymer

Glyceryl caprylate; Glyceryl cottonseed oil; Glyceryl dioleate; Glyceryl dioleate SE; Glyceryl distearate; Glyceryl distearate SE; Glyceryl isostearate; Glyceryl laurate; Glyceryl laurate SE; Glyceryl oleate; Glyceryl oleate SE; Glyceryl ricinoleate; Glyceryl ricinoleate SE; Glyceryl stearate SE; Guar gum

Hydrogenated lard glyceride; Hydrogenated lard glycerides; Hydrogenated palm glyceride; Hydrogenated soybean glycerides; Hydrogenated soy glyceride; Hydrogenated tallow glyceride; Hydrogenated tallow glycerides; Hydrogenated vegetable glycerides; Hydroxypropylcellulose; Hydroxypropyl methylcellulose

Lactylated fatty acid esters of glycerol and propylene glycol; Lactylic esters of fatty acids; Lard glycerides; Licorice; Licorice extract; Licorice root extract

Methyl ethyl cellulose; Methyl glucoside-coconut oil ester; Microcrystalline cellulose; Mono- and diglycerides of fatty acids; Mono- and diglycerides, sodium phosphate derivs

Nonoxynol-6; Nonoxynol-10; Nonoxynol-11; Nonoxynol-40; Nonoxynol-50

Octoxynol-5; Octoxynol-8; Octoxynol-9; Octoxynol-13; Octoxynol-30; Octoxynol-40; Oleyl alcohol; Ox bile extract

Palm glyceride; Palm oil sucroglyceride; Pea protein concentrate; PEG-6 dilaurate; PEG-20 dilaurate; PEG-32 dilaurate; PEG-75 dilaurate; PEG-150 dilaurate; PEG-6 dioleate; PEG-20 dioleate; PEG-32 dioleate; PEG-75 dioleate; PEG-150 dioleate; PEG-6 distearate; PEG-20 distearate; PEG-32 distearate; PEG-75 distearate; PEG-32 laurate; PEG-6 oleate; PEG-8 oleate; PEG-32 oleate; PEG-75 oleate; PEG-80 sorbitan laurate; PEG-20 sorbitan tritallate; PEG-32 stearate; PEG-75 stearate; PEG-100 stearate; Phosphatidylcholine; Poloxamer 105; Poloxamer 108; Poloxamer 122; Poloxamer 123; Poloxamer 124; Poloxamer 181; Poloxamer 182; Poloxamer 183; Poloxamer 184; Poloxamer 185; Poloxamer 188; Poloxamer 212; Poloxamer 215; Poloxamer 217; Poloxamer 231; Poloxamer 234; Poloxamer 235; Poloxamer 237; Poloxamer 238; Poloxamer 282; Poloxamer 284; Poloxamer 288; Poloxamer 331; Poloxamer 333; Poloxamer 334; Poloxamer 335; Poloxamer 338; Poloxamer 401; Poloxamer 402; Poloxamer 403; Poloxamer 407; Polyethylene glycol; Polyglyceryl-10 dipalmitate; Polyglyceryl-10 hexaoleate; Polyglyceryl-10 stearate; Polysorbate 20; Polysorbate 40; Polysorbate 60; Polysorbate 80; Polysorbate 81; Polysorbate 85; Poly(1-vinyl-2-pyrrolidinone) homopolymer; Potassium acid tartrate; Potassium persulfate; Potassium tripolyphosphate; Propylene glycol; Propylene glycol alginate; PVP

Quillaja

Simethicone; Sodium acid pyrophosphate; Sodium decylbenzene sulfonate; Sodium glyceryl oleate phosphate; Sodium lauryl sulfate; Sodium stearoyl lactylate; Sorbitan sesquioleate; Sorbitan tritallate; Sucrose dilaurate; Sucrose distearate; Sucrose erucate; Sucrose laurate; Sucrose myristate; Sucrose stearate; Sucrose tribehenate; Sunflower seed oil glycerides

Tallow glycerides; Tetrapotassium pyrophosphate; Tetrasodium pyrophosphate

Xanthan gum

Yucca

Sweeteners

Trade names: Alitame
Barley*Complete® 25; BK-102 Series; BK-305 Series; BK-PR2 Series
Candex®; Clintose® A; Clintose® F; Clintose® L; Clintose® VF; CNP BRSHG40; CNP
  BRSHG40CL; CNP BRSHM; CNP BRSHMCL; CNP BRSMC10; CNP BRSMC20; CNP
  BRSMC35; CNP BRSMC35CL; CNP BRSSHG40; CNP BRSSHG40CL; CNP BRSSHM;
  CNP BRSSHMCL; CNP RSSHG40; CNP RSSHG40CL; CNP RSSHM; CNP RSSHMCL;
  CornSweet® 42; CornSweet® 55; CornSweet® 95; CronSweet® Crystalline Fructose; Corn
  Syrup 36/43; Corn Syrup 42/43; Corn Syrup 42/44; Corn Syrup 52/43; Corn Syrup 62/43; Corn
  Syrup 62/44; Corn Syrup 62/44-1; Corn Syrup 62/44-2; Corn Syrup 97/71
DCME; De-Mol® Molasses; DMCE; DME Dry Malt Extract; Dri-Mol® 604 Molasses; Dri-Mol®
  Molasses
Equal®; Extramalt 10; Extramalt 33; Extramalt 35; Extramalt Dark; Extramalt Light
Finmalt L; 42/43 Corn Syrup; Fructofin® C; FruitSource® Granular; FruitSource® Liquid
  Sweetener; FruitSource® Liquid Sweetener Plus
Glycerine (Pharmaceutical)
Honi-Bake® 705 Honey; Honi-Bake® Honey; HSC Aspartame
Lactitol MC; Liponic Sorbitol Powd; Liponic Sorbitol Sol'n. 70% USP
Maltrin® M205; Maltrin® M250; Maltrin® M255; Maltrin® M365
ND-201 Syrup; ND-201-C Syrups; ND-305 Syrups; Non-Diastatic Malt Syrup #40600;
  NutraSweet®
Rice-Pro® 35W; Rice-Trin® 25; Rice-Trin® 35
62/43 Corn Syrup; Solid Invert Sugar; Sorbo®; Special Liquid Invert Sugar; Sunmalt;
  Sweet'n'Neat® 45; Sweet'n'Neat® 65 Molasses Powd; Sweet'n'Neat® 2000; Sweet'n'Neat®
  3000; Sweet'n'Neat® 4000 Molasses Powd; Sweet'n'Neat® 5000 Molasses Flake;
  Sweet'n'Neat Maple; Sweet'n'Neat® Oil/Dry Roast; Sweet'n'Neat Raisin; Sweet'n'Neat®
  Tack Blend; Syncal® CAS; Syncal® GS; Syncal® GSD; Syncal® S; Syncal® SDI; Syncal®
  SDS; Syncal® US
Unisweet L; Unisweet Lactose; Unisweet MAN; Unisweet SAC; Unisweet SOSAC
Veltol®; Veltol®-Plus
Xylitol C

Chemicals: Acesulfame potassium; N-Acetylglucosamine; Ammonium saccharin; Arabinogalactan; L-Arabi-
  nose; Aspartame
Brown rice syrup
Calcium cyclamate; Calcium saccharin; Corn starch; Corn syrup; Cyclamic acid
Disodium glycyrrhizinate
Fructose
D-Glucose anhyd; D-Glucose monohydrate; Glucose syrup solids; Glycine
Honey; Hydrolyzed barley flour
Invert sugar; Isomalt
Lactitol monohydrate; Lactose
Malt extract; Maltitol; D-Mannitol; Maple; Miraculin; Molasses; Monellin
Rhamnose; Rice syrup
Saccharin; Saccharin sodium; Sodium cyclamate; Sodium saccharin; Sorbitol; Stevioside;
  Sucralose; Sucrose
Thaumatin
Xylitol; D(+)-Xylose

Synergists

Trade names: Benecel® M 042
CornSweet® 95; CronSweet® Crystalline Fructose
Nutrilife®; Nutriloid® Locust
TIC Pretested® CMC 2500 S; TIC Pretested® CMC PH-2500

Chemicals: Ascorbyl palmitate
Calcium chloride; Calcium phosphate dibasic; Calcium sulfate; Citric acid esters of mono- and
  diglycerides of fatty acids; Copper gluconate
Disodium citrate

Synergists *(cont'd.)*

N-Hydroxysuccinic acid
Isopropyl citrate
Magnesium carbonate hydroxide; Monoglyceride citrate
Phosphoric acid; Potassium lactate; Potassium phosphate tribasic; Potassium sodium tartrate
    anhyd; Propyl gallate
Sodium tartrate

Tenderizers • Tissue softening agents

Trade names: Bromelain 1:10; Bromelain 150 GDU; Bromelain 600 GDU; Bromelain 1200 GDU; Bromelain
    1500 GDU; Bromelain Conc; Brosoft 6430
    Fungal Protease 31,000; Fungal Protease 60,000; Fungal Protease 500,000; Fungal Protease
    Conc
    Hodag GMSH
    Kena® FP-28
    Nutricol® Konjac
    Papain 16,000; Papain 30,000; Papain A300; Papain A400; Papain Conc; Papain M70; Papain
    P-100; Papain S100; Protoferm
    Rhozyme® P11
    Tenderbite SF(TB-SF); Ticaloid® 1039; Tona Enzyme 201; Tona Papain 14; Tona Papain 90L;
    Tona Papain 270L
    Viscarin® ME-389; Viscarin® XP-3160; Vitrafos®

Chemicals: Bromelain
    Ficin
    Magnesium chloride
    Papain

Texturizers

Trade names: Albrite STPP-F; Albrite STPP-FC; Albrite TSPP Food Grade; Aldo® HMS FG; Aldo® MS FG;
    Aldo® MS LG KFG; Alomine 223; Aqualon® 7H0F; Aqualon® 7H3SF; Aqualon® 7H3SXF;
    Aqualon® 7H4F; Aqualon® 7H4XF; Aqualon® 7HC4F; Aqualon® 7HCF; Aqualon® 7HF;
    Aqualon® 7HXF; Aqualon® 7LF; Aqualon® 7LXF; Aqualon® 7M8SF; Aqualon® 9M8F; Atlas
    800; Atlas 1400K; Atlas 5000K; Atlas SSL; Atmul® 80; Atmul® 84; Atmul® 84 K; Atmul® 86
    K; Atmul® 500; Atmul® 651 K
    Benecel® MP 643; Benecel® MP 824; Benecel® MP 843; Benecel® MP 872; Bunge Biscuit
    Flakes
    Cap-Shure® SC-140X-70; Cap-Shure® SC-165-85FT; Cap-Shure® SC-165X-60FF;
    CarraLizer™ CGB-10; CarraLizer™ CGB-40; CarraLizer™ CGB-50; Celynol MSPO-11;
    Celynol P1M; CMC Daicel 1220; Colloid 451T; Colloid 862; Consista®; Crester KZ; Curafos®
    11-2; Curavis® 250
    DK-Ester; DK-Ester; Dricoid® 200
    Empruv®; Emulgel S-32
    Famodan MS Kosher; Famodan TS Kosher; Freez-Gard® FP-15; Frimulsion RF; Fructodan
    Golden Covo® Shortening; Guardan 100 Range; Guardan 600 Range; Guardan 700 Range
    Hodag GMSH; Hodag PGS; Hydex® 100 Gran. 206; Hydex® 100 Powd. 60; Hydex® Tablet
    Grade
    Instant Tender-Jel® 419
    Kelcogel®; Kelcogel® F; Kena® FP-28
    Liponic Sorbitol Powd; Liponic Sorbitol Sol'n. 70% USP; Litesse®; Litesse® II; Lo-Temp® 452
    Merecol® GL; Merecol® GX; Merecol® R; Merecol® RB; Merecol® RCS; Methocel® A15LV
    Premium; Methocel® F4M Premium; Methocel® K100LV Premium; Mira-Gel® 463; Mira-
    Thik® 469; Myvatex® 40-06S; Myverol® 18-04; Myverol® 18-07
    Neutral-lactase; Non-Diastatic Malt Syrup #40600; Nutrifos® 088
    Paniplex SK; Pationic® 920; Pationic® 925; Purasal® P/USP 60; Pure-Set™ B965
    Radiamuls® LMG 2950, 2951; Redi-Tex®; Rhodigum OEH; Rhodigum OEM; Rhodigum WGH;

Texturizers *(cont'd.)*

Rhodigum WGL; Rhodigum WGM; Rhodigum WVH; Rhodigum WVM

Saladizer® #228M; Saladizer® #250P; Sorbo®; Staley® Pure Food Powd. (PFP); Sta-Lite™ 100C; Sta-Lite™ 100CN Neutralized; Sta-Lite™ 100F; Star-Dri® 55; Sta-Slim™ 142; Sta-Slim™ 143; Sta-Slim™ 150; Sta-Slim™ 171; Stir & Sperse®; Super White Fonoline®

Tegomuls® 90; Tegomuls® 90 K; Tegomuls® 90 S; Tegomuls® 4100; Tegomuls® 9053 S; Tegomuls® 9102; Tegomuls® 9102 S; Ticaloid® No Fat 102B; Ticalose® 75; Ticalose® 100; TIC Pretested® CMC 2500 S; TIC Pretested® CMC PH-2500; TIC Pretested® Tragacanth 440; Tomex®

Ultra-Freeze® 400; Ultra-Freeze® 400C; Ultra-Freeze® 500

Vascomix 404; Veltol®; Veltol®-Plus; Veri-Lo® 100; Versa-Whip® 500, 510, 520; Versa-Whip® 520K, 600K, 620K; Viscarin® SD-389; Vitinc® Defatted Wheat Germ Nuggets; Vitinc® Defatted Wheat Germ

WaTox-20 P Powd; WaTox-20 Tablets

Chemicals: Acacia; Acetylated mono- and diglycerides of fatty acids; Algin; Ammonium bicarbonate

Bakers yeast extract; Bromelain

Calcium acetate; Calcium chloride; Calcium gluconate; Calcium sulfate; Cellulose; Cocoa butter substitute; Coconut oil; Corn meal; Corn oil; Cottonseed oil

Disodium phosphate, dihydrate

Gelatin; Glyceryl stearate

Hydrolyzed oat flour

Lactose; Lard

Malt extract; Maltodextrin; D-Mannitol; Microparticulated protein prod; Mono- and diglycerides of fatty acids

Palm kernel oil; Palm oil; Palm oil sucroglyceride; Papain; Peanut oil; Pentapotassium triphosphate; Pentasodium triphosphate; Polydextrose; Potassium chloride; Potassium tripolyphosphate; Propylene glycol

Rice maltodextrin

Safflower oil; Sodium metaphosphate; Sodium phosphate dibasic; Sodium stearoyl lactylate; Sorbitol; Soybean oil; Sucrose fatty acid esters; Sucrose oleate; Sucrose palmitate; Sucrose tristearate; Sunflower seed oil

Talo; Tallow; Tetrapotassium pyrophosphate

Wheat gluten

Thickeners • Viscosity modifiers • Bodying agents • Gelling agents • Stiffening agents

Trade names: Ablunol S-20; Ablunol S-40; Ablunol S-60; Ablunol S-80; Ablunol S-85; Actiflo® 68 SB; Actiflo® 68 UB; Actiflo® 70 SB; Actiflo® 70 UB; Admul WOL 1403; Agar Agar NF Flake #1; Akwilox 133; Alcolec® C-150; Alginade DC; Amoloid HV; Amoloid LV; Aqualon® 7H0F; Aqualon® 7H0XF; Aqualon® 7H3SF; Aqualon® 7H3SXF; Aqualon® 7H4F; Aqualon® 7H4XF; Aqualon® 7HF; Aqualon® 7LF; Aqualon® 7MF; Aqualon® 9H4F; Aqualon® 9H4XF; Aqualon® 9M31F; Aqualon® 9M31XF; Aqualon® 9M8XF; Aquasol CSL; Avicel® CL-611; Avicel® RC-501; Avicel® RC-581; Avicel® RC-591F; Avicel® RCN-10; Avicel® RCN-15; Avicel® WC-595

Barley*Complete® 25; Benecel® M 042; Benecel® M 043; Benecel® MP 643; Benecel® MP 824; Benecel® MP 872; Benecel® MP 874; Benecel® MP 943; Binasol™ 81; Binasol™ 90; Bindox-HV-051; Bindox-LV-050; Bone Gelatin Type B 200 Bloom

Carbowax® Sentry® PEG 300; Carbowax® Sentry® PEG 400; Carbowax® Sentry® PEG 540 Blend; Carbowax® Sentry® PEG 600; Carbowax® Sentry® PEG 900; Carbowax® Sentry® PEG 1000; Carbowax® Sentry® PEG 1450; Carbowax® Sentry® PEG 3350; Carbowax® Sentry® PEG 4600; Carbowax® Sentry® PEG 8000; Carralean™ CG-100; Carralean™ CM-70; Carralean™ CM-80; Carralean™ MB-60; Carralean™ MB-93; Carralite™; Carudan 000 Range; Carudan 100 Range; Carudan 200 Range; Carudan 300 Range; Carudan 400 Range; Carudan 700 Range; Celish® FD-100F; Cellogen HP-5HS; Cellogen HP-6HS9; Cellogen HP-12HS; Cellogen HP-SB; Centrocap® 162SS; Centrocap® 162US; Centrocap® 273SS; Centrocap® 273US; Centrol® 2F SB; Centrol® 2F UB; Centrol® 3F SB; Centrol® 3F UB; Centrol® CA; Centrolex® D; Centrolex® F; Centrolex® P; Centrolex® R; Centrophil® K; CMC Daicel 1150; CMC Daicel 1160; CMC Daicel 1240; CMC Daicel 1260; CMC Daicel 2200; CNP

Thickeners *(cont'd.)*

Pregelled Rice Flour; CNP RMDRD05; CNP RMDRD18; Colloid 488T; Colloid 602; Consista®; Crester PR; Crester RT; Crodyne BY-19

Dariloid®; Dariloid® 100; Dariloid® 300; Dariloid® Q; Dariloid® QH; Dariloid® XL; Detergent CR; Diagum LBG; Drewmulse® 200K; Drewmulse® SMO; Drewmulse® SMS; Durlac® 100W; Dynasan® P60

Edicol®; Egg White Solids Type P-39; Egg White Solids Type P-110 High Gel Strength; Egg Yolk Solids Type Y-1-FF; Emery® 912; Emery® 916; Emery® 917; Emery® 918; Emphos™ D70-31; Emulgel S-32

Flavorset® GP-5; Flavorset® GP-6; Flavorset® GP-7; Flavorset® GP-8; Flavorset® GP-9; Flavorset® GP-10; 42/43 Corn Syrup; Freezist® M; Frigesa® D 890; Frigesa® F; Frigesa® IC 178; Frigesa® IC 184; Frimulsion 6G; Frimulsion X5; Fruitfil® 1

Gelatinized Dura-Jel®; Gelcarin® GP-359; Gelodan CC Range; Gelodan CW Range; Gelodan CX Range; Genu® 04CG or 04CB; Genu® 12CG; Genu® 18CG; Genu® 18CG-YA; Genu® 20AS; Genu® 21AS or 21AB; Genu® 22CG; Genu® 102AS; Genu® 104AS; Genu® 104AS-YA; Genu® AA Medium-Rapid Set, 150 Grade; Genu® BA-KING; Genu® BB Rapid Set; Genu® DD Extra-Slow Set; Genu® DD Extra-Slow Set C, 150 Grade; Genu® DD Slow Set; Genu® Type DJ; Genu® VIS; Genugel® CHP-2; Genugel® CHP-2 Fine Mesh; Genugel® CHP-200; Genugel® CJ; Genugel® LC-1; Genugel® LC-4; Genugel® LC-5; Genugel® MB-51; Genugel® MB-78F; Genugel® UE; Genugel® UEU; Genulacta® CP-100; Genulacta® CSM-2; Genulacta® LR-41; Genulacta® LR-60; Genulacta® LRA-50; Genulacta® LRC-21; Genulacta® LRC-30; Genulacta® P-100; Genulacta® PL-93; Genu® Pectins; Genuvisco® CSW-2; Genuvisco® J; Genuvisco® MP-11; Geoldan CL Range; GFS®; Glycon® G 100; Glycon® G-300; GMS 900; Granular Gum Arabic NF/FCC C-4010; Guardan 100 Range; Guardan 600 Range; Guardan 700 Range; Guar Gum; Gum Arabic NF/FCC Clean Amber Sorts; Gum Arabic, Purified, Spray-Dried No. 1834

HG-175; HG-200; Hi-Jel™ S; Hodag PGS; Hodag PGS-62; Hydex® 100 Coarse Powd; Hydex® 100 Coarse Powd. 35; Hystar® 3375; Hystar® 4075; Hystar® 5875; Hystar® 6075; Hystar® 7570; Hystar® HM-75

Ice #12; Idealgum 1A; Idealgum 1B; Idealgum 1C; Idealgum 2A; Idealgum 2B; Idealgum 3F; Instant Remygel AX-P, AX-2-P; Instant Remyline AX-P; Instant Tender-Jel® 419

Jaguar® 6000

Kelacid®; Kelco® HV; Kelco® LV; Kelcogel® BF; Kelcogel® BF10; Kelcogel® CF; Kelcogel® CF10; Kelcogel® F; Kelcogel® IF; Kelcogel® JJ; Kelcogel® PD; Kelcoloid® D; Kelcoloid® DH; Kelcoloid® DO; Kelcoloid® DSF; Kelcoloid® HVF; Kelcoloid® LVF; Kelcoloid® O; Kelcoloid® S; Kelcosol®; Kelgin® F; Kelgin® HV; Kelgin® LV; Kelgin® MV; Kelgin® QL; Kelgin® XL; Kelmar®; Kelmar® Improved; Kelset®; Keltone®; Keltone® HV; Keltone® LV; Keltose®; Keltrol® 1000; Keltrol® BT; Keltrol® CR; Keltrol® GM; Keltrol® RD; Keltrol® SF; Keltrol® T; Keltrol® TF; Kelvis®; Kimiloid HV; Kimiloid MV; Kimiloid NLS-K; Kimitsu Algin I-1; Kimitsu Algin I-2; Kimitsu Algin I-3; Klucel® F Grades; KOB87; Koster Keunen Candelilla

Lactarin® MV-406; Lactarin® PS-185X; Lacticol CFT; Lacticol F336; Lacticol F616; Lamefrost® ES 216 G; Lamefrost® ES 251 G; Lamefrost® ES 315; Lamefrost® ES 375; Lamefrost® ES 379; Lamefrost® ES 424; Lecigran™ F; Leciprime™; Lem-O-Fos® 202; Liposorb TS; Locust Bean Gum Speckless Type D-200

Maco-O-Line 091; Maltrin® M050; Maltrin® M105; Maltrin® M150; Maltrin® M180; Manucol DH; Manucol DM; Manucol DMF; Manucol JKT; Manucol LB; Manucol SMF; Manucol Ester E/PL; Manucol Ester E/RK; Manucol Ester M; Manugel C; Manugel DJX; Manugel DMB; Manugel GHB; Manugel GMB; Manugel JKB; Manugel L98; Manugel PTB; Manugel PTJ; Manugel PTJA; Marine Colloids™ Carrageenan; Marloid® CMS; Maxi-Gel® 542; Maxi-Gel® 7776; Maximaize® DJ; Maximaize® HV; Mazol® GMS; Merecol® FA; Merecol® FAL; Merecol® GL; Merecol® GX; Merecol® I; Merecol® RB; Merecol® RCS; Methocel® A4C Premium; Methocel® A4M Premium; Methocel® A15C Premium; Methocel® A15LV Premium; Methocel® E3 Premium; Methocel® E4M; Methocel® E4M Premium; Methocel® E5 Food Grade; Methocel® E6 Premium; Methocel® E15 Food Grade; Methocel® E15LV Premium; Methocel® E50LV Premium; Methocel® F4M Premium; Methocel® F50LV Premium; Methocel® K4M Premium; Methocel® K15M Premium; Methocel® K35 Premium; Methocel® K100LV Premium; Methocel® K100M Premium; Mexpectin HV 400 Range; Mexpectin LA 100 Range; Mexpectin LA 200 Range; Mexpectin LA 400 Range; Mexpectin LC 700 Range; Mexpectin LC 800 Range; Mexpectin LC 900 Range; Mexpectin MRS 300 Range; Mexpectin RS 400 Range; Mexpectin SS 200 Range; Mexpectin XSS 100 Range; Mira-Bake® 505; Mira-Cleer® 187; Mira-Cleer® 340; Mira-Cleer® 516; Mira-Cleer® 516; Mira-Cleer® 516; Mira-Cleer® 516; Mira-Gel® 463; Mira-Quik® MGL; Mira-Set® B; Mira-Set® J; Mira-Sperse®

Thickeners *(cont'd.)*

623; Mira-Sperse® 626; Mira-Thik® 468; Mira-Thik® 469; Mira-Thik® 603; Mira-Thik® 606; Monomuls® 90-O18; Myvatex® Mighty Soft®; Myvatex® MSPS

Nastar®; Nastar® Instant; Natural Arabic Type Gum Purified, Spray-Dried; NatuReal™ Essence; NatuReal™ Sensoral; NatuReal™ Textural; Niacet Calcium Acetate FCC; Nu-Col™ 4227; Nustar®; Nutrifos® L-50; Nutriloid® Agar; Nutriloid® Carrageenan; Nutriloid® Cellulose Gums; Nutriloid® Guar Special; Nutriloid® Guar Standard; Nutriloid® Locust; Nutriloid® Tragacanth; Nutriloid® Xanthan

n-Octenyl Succinic Anhydride; Optex

Paniset/Panistay; Pationic® 1230; Pationic® 1250; Paygel® 290.295 Pregelatinized Food Powd. Wheat Starch; Perma-Flo®; PF-80; Polyaldo® 2P10 KFG; Powdered Agar Agar NF M-100 (Gracilaria); Powdered Agar Agar NF MK-80-B; Powdered Agar Agar NF MK-80 (Bacteriological); Powdered Agar Agar NF S-100; Powdered Agar Agar NF S-100-B; Powdered Agar Agar NF S-150; Powdered Agar Agar NF S-150-B; Powdered Gum Arabic NF/FCC G-150; Powdered Gum Arabic NF/FCC Superselect Type NB-4; Powdered Gum Guar NF Type 80 Mesh B/T; Powdered Gum Guar Type ECM; Powdered Gum Guar Type M; Powdered Gum Guar Type MM FCC; Powdered Gum Guar Type MM (HV); Powdered Gum Guar Type MMM $^1/_2$; Powdered Gum Guar Type MMW; Powdered Locust Bean Gum Type D-200; Powdered Locust Bean Gum Type P-100; Powdered Locust Bean Gum Type PP-100; Powdered Tragacanth Gum Type A/10; Powdered Tragacanth Gum Type E-1; Powdered Tragacanth Gum Type G-3; Powdered Tragacanth Gum Type L; Powdered Tragacanth Gum Type W; Prime F-25; Prime F-40; Prime F-400; Prime F-600; Prinza® 452; Prinza® 455; Prinza® Range; Pure-Dent® B700; Pure-Gel® B990; Pure-Gel® B992; Pure-Set® B950; Pure-Set™ B965

Redisol® 78D; Redisol® 88; Redisol® 248; Redisol® 412; Remygel AX-2; Remygel AX; Remyline AX; Rezista®; Rhodigel®; Rhodigel® 200; Rhodigel® EZ; Rhodigel® Supra; Rhodigum OEH; Rhodigum OEM; Rhodigum WGH; Rhodigum WGL; Rhodigum WGM; Rhodigum WVH; Rhodigum WVM; Rice Complete® 3; Rice Complete® 10; Rice Complete® 18; Rice Complete® 25; Rice-Trin® 10; Rice-Trin® 18; Rice-Trin® 25; Rice-Trin® 35; Ross Rice Bran Wax

SeaKem® CM-611; Sett®; Sherbelizer®; Soageena®; Soageena® MM501; Soageena® MW321; Soageena® MW351; Soageena® MW371; Soalocust®; Sobalg FD 200 Range; Sobalg FD 300 Range; Sobalg FD 460; Sobalg FD 900 Range; Sodium Alginate LVC; Soft-Set®; Sorbo®; Span® 20; Span® 40; Span® 60, Span® 60K, Span® 65, Span® 65K; Span® 85; Spray Dried Fish Gelatin; Spray Dried Fish Gelatin/Maltodextrin; Spray Dried Gum Arabic NF/FCC CM; Spray Dried Gum Arabic NF/FCC CS (Low Bacteria); Spray Dried Gum Arabic NF/FCC CS-R; Spray Dried Hydrolysed Fish Gelatin; Staform P; Staley® 7025; Staley® 7350 Waxy No. 1 Starch; Staley® Pure Food Powd. (PFP); Staley® Pure Food Powd. Starch Type I; Staley® Pure Food Powd. Starch Type II; Sta-Lite™ 100C; Sta-Lite™ 100CN Neutralized; Sta-Lite™ 100CN Neutralized; Sta-Lite™ 100F; Sta-Lite™ 100CN Neutralized; Sta-Lite™ 100CN Neutralized; Stamere® CK FCC; Stamere® CK-S; Stamere® N-325; Stamere® N-350; Stamere® N-350 E FCC; Stamere® N-350 S; Stamere® NI; Stamere® NK; Starco™ 447; StarchPlus™ SPR; Starco™ 401; Star-Dri® 1; Star-Dri® 5; Star-Dri® 10; Star-Dri® 15; Star-Dri® 18; Star-Dri® 20; Star-Dri® 35F; Star-Dri® 35R; Star-Dri® 42C; Star-Dri® 42F; Star-Dri® 42R; Star-Dri® 1015A; STD-175; Supercol® G2S; Supercol® GF; Supercol® U; Suspengel Elite; Suspengel Micro; Suspengel Ultra; Swelite® Micro

Tenderfil® 8; Tenderfil® 9; Tenderfil® 428; Tenderfil® 473A; Tenderset® M-7; Tenderset® M-8; Tenderset® M-9; Tenderset® MWA-5; Tenderset® MWA-7; Terra Alba 114836; Thin-N-Thik® 99; Ticalose® 15; Ticalose® 30; Ticalose® 75; Ticalose® 100; Ticalose® 150 R; Ticalose® 700 R; Ticalose® 750; Ticalose® 2000 R; Ticalose® 2500; Ticalose® 4000; Ticalose® 4500; Ticalose® 5000 R; Ticaxan® Regular; Ticolv; TIC Pretested® CMC 2500 S; TIC Pretested® CMC PH-2500; TIC Pretested® Colloid 775 Powd; TIC Pretested® Colloid 881 M; TIC Pretested® Gum Agar Agar 100 FCC/NF Powd; TIC Pretested® Gum Guar 8/22 FCC/NF Powd; TIC Pretested® Locust Bean POR/A; TIC Pretested® Pectin HM Slow; TIC Pretested® Saladizer 250; TIC Pretested® Saladizer PH-250; TIC Pretested® Tragacanth 440; TIC Pretested® Xanthan 200; TIC Pretested® Xanthan PH; T-Maz® 28; T-Maz® 60; T-Maz® 60KHS; T-Maz® 65; T-Maz® 80; T-Maz® 81; T-Maz® 95; Tragacanth Flake No. 27; Tragacanth Gum Ribbon No. 1

Uniguar 20; Uniguar 150; Uniguar 200

Viscarin® GP-209; Viscarin® SD-389; Visco Compound™; Vis*Quick™ GX 11

W-300 FG; Wakal® A 601; Wakal® A Range; Wakal® J; Wakal® JG; Wakal® K Range

Thickeners *(cont'd.)*

Chemicals: Agar; Algin; Aloe extract; Aluminum silicate; Ammonium alginate; Ammonium carrageenan; Ammonium furcelleran; Arabinogalactan

Brominated soybean oil

Calcium acetate; Calcium alginate; Calcium carboxymethyl cellulose; Calcium carrageenan; Calcium chloride; Calcium furcelleran; Calcium lactate; Calcium stearate; Calcium sulfate; Carboxymethylcellulose sodium; Carboxymethylmethylcellulose; Carrageenan; Carrageenan extract; Cassia gum; Cellulose; Chitin; Chitosan; Corn starch; Corn syrup

Dextran; Dextrin

Food starch modified; Furcelleran

Gelatin; Gellan gum; Glucosamine; Glycerin; Glyceryl lactopalmitate/stearate; Glyceryl stearate; Glyceryl stearate SE; Guaiac gum; Guar gum; Gum ghatti

Hydrogenated palm oil; Hydrogenated rapeseed oil; Hydrolyzed barley flour; Hydrolyzed oat flour; Hydroxypropylcellulose; Hydroxypropyl methylcellulose

Karaya gum; Kelp extract

Locust bean gum

Malt extract; Maltodextrin; D-Mannitol; Methylcellulose; Methyl ethyl cellulose; Microcrystalline wax; Microparticulated protein prod; Mono- and diglycerides of fatty acids

Octenyl succinic anhydride

Pectin; PEG-4; PEG-6; PEG-8; PEG-9; PEG-12; PEG-14; PEG-16; PEG-20; PEG-32; PEG-40; PEG-75; PEG-100; PEG-150; PEG-200; PEG-6 dilaurate; PEG-20 dilaurate; PEG-32 dilaurate; PEG-75 dilaurate; PEG-150 dilaurate; PEG-6 dioleate; PEG-20 dioleate; PEG-32 dioleate; PEG-75 dioleate; PEG-150 dioleate; PEG-6 distearate; PEG-20 distearate; PEG-32 distearate; PEG-75 distearate; PEG-32 laurate; PEG-8 oleate; PEG-32 oleate; PEG-75 oleate; PEG-20 sorbitan tritallate; PEG-32 stearate; PEG-75 stearate; Polysorbate 81; Poly(1-vinyl-2-pyrrolidinone) homopolymer; Potassium acid tartrate; Potassium alginate; Potassium carrageenan; Potassium chloride; Potassium furcelleran; Potassium hydroxide; Propylene glycol; Propylene glycol alginate; PVP

Rennet; Rhamsan gum; Rice maltodextrin; Rice starch

Silica; Sodium carrageenan; Sodium furcelleran; Sodium polyacrylate; Sorbitol; Starch

Tara gum; Tragacanth gum

Wheat gluten

Xanthan gum

Part IV
Manufacturers Directory

Manufacturers Directory

Aarhus

Aarhus Oliefabrik A/S, Postboks 50, Bruunsgade 27, DK-8100 Aarhus C Denmark (Tel.: 86-12 60 00; Telefax: 86-196252; Telex: 64341)

Aarhus Inc., 131 Marsh St., PO Box 4240, Newark, NJ 07114 USA (Tel.: 201-344-1300)

AB Ingredients Ltd.

Bamway Lodge Farm, Northampton, NN5 7UW UK (Tel.: 44 1604 755522; Telefax: 44 1604 759343; Telex: 311237)

Able Products Ltd.

Tame St., Stalybridge, Cheshire, SK15 1QW UK (Tel.: 44 161 343 1772; Telefax: 44 161 343 1169; Telex: 667014)

Able Sales

1531 Deer Crossing, Diamond Bar, CA 91765 USA (Tel.: 714-861-8671; Telefax: 714-860-5272)

AB Technology Limited

Salthouse Rd, Brackmills, Northampton, NN4 0EX UK (Tel.: 44 1604 768999; Telefax: 44 1604 701503; Telex: 311848)

Aceto

Aceto Chemical Co., Inc., 1 Hollow Lane, Suite 201, Lake Success, NY 11042-1215 USA (Tel.: 516-627-6000; Telefax: 516-627-6093; Telex: FTCC 824609)

Pfaltz & Bauer Inc., Research chemicals subsidiary of Aceto Corp., 172 E. Aurora St., Waterbury, CT 06708 USA (Tel.: 203-574-0075; 800-225-5172; Telefax: 203-574-3181; Telex: 996471)

Acme-Hardesty Co., Div. of Jacob Stern & Sons, Inc.

626 Fox Pavilion, Jenkintown, PA 19046-0831 USA (Tel.: 215-885-3610; 800-223-7054; Telefax: 215-886-2309)

Active Organics, Inc.

11230 Grader St., Dallas, TX 75238 USA (Tel.: 214-348-2015; 800-541-1478; Telefax: 214-348-1557)

Activ International

384 rue de Vaugirard, 75015 Paris France (Tel.: 33 1 4842 2021; Telefax: 33 1 4842 2366; Telex: 201240)

Alfred Adams & Co. Ltd.

PO Box 23, Reliance Works, Church Lane, West Bromwich, W Midlands, B71 1BZ UK (Tel.: 44 121 525 7955; Telex: 339415)

Adams Food Ingredients Ltd.

Prince St., Leek, Staffs, ST13 6DB UK (Tel.: 44 1538 399686; Telex: 36134)

ADM, Div. Archer Daniels Midland Co.

ADM Agri-Industries Ltd., Box 7128, 5550 Maplewood Dr., Windsor, Ontario, N9C 3Z1 Canada (Tel.: 800-265-7315 (C); Telefax: 519-966-7135)

ADM Arkady, 100 Paniplus Rd., Olathe, KS 66061 USA (Tel.: 913-782-8800; 800-633-6919; Telefax: 913-782-1792)

ADM Corn Processing, Box 1470, Decatur, IL 62525 USA (Tel.: 217-424-5200; 800-323-0735; Telefax: 217-424-5978)

ADM Citric Group, Box 1470, Decatur, IL 62525 USA (Tel.: 800-553-8411; Telefax: 217-362-3941)

ADM Ethanol Sales, Box 1470, Decatur, IL 62525 USA (Tel.: 217-424-5200; Telefax: 217-424-5978)

ADM Milling, 4501 College Blvd., Leawood, KS 66211 USA (Tel.: 913-491-9400; Telefax: 913-491-0035)

ADM Milling—Rice Div., Box 7007, Shawnee Mission, KS 66207 USA (Tel.: 913-491-9400; Telefax: 800-879-3291)

ADM Ogilvie, Box 1470, Decatur, IL 62525 USA (Tel.: 217-362-8197; Telefax: 217-362-8194)

ADM Refined Oils, Box 1470, Decatur, IL 62525 USA (Tel.: 217-424-5489; Telefax: 217-424-5467)

ADM Ross & Rowe Lecithin, Box 1470, Decatur, IL 62525 USA (Tel.: 800-637-5843; Telefax: 217-424-5381)

ADM Packaged Oils, Box 1470, Decatur, IL 62525 USA (Tel.: 800-637-1550; Telefax: 217-424-2689)

ADM Protein Specialties, Box 1470, Decatur, IL 62525 USA (Tel.: 217-424-7453; 800-637-5850; Telefax: 217-362-8067)

Fleischmann-Kurth Malting Co., Inc., Box 15067, Minneapolis, MN 55414 USA (Tel.: 612-371-3480; Telefax: 612-371-3417)

Advanced Food Systems, Inc.
69 Veronica Ave., Somerset, NJ 08873 USA (Tel.: 908-828-7878; Telefax: 908-828-5938)

Advanced Sweeteners Ltd.
53 The Green, West Cornforth, Ferry Hill, Co Durham, DL17 9JH UK (Tel.: 44 1740 54955; Telex: 537681)

AECI Aroma & Fine Chemicals (Pty) Ltd., Subsid of AECI Ltd.
Private bag X2, Modderfontein 1620, Rep. of S. Africa (Tel.: 27 11 605 2188; Telefax: 27 11 608 3314)

AFI (Alternative Food Ingredients)
Centre Polytechnique St Louis, 13 Bld de L'Hautil, 95092 Cergy Pontoise Cedex France (Tel.: 33 1 34 24 0411; Telefax: 33 1 30 75 6081)

Agricultural & Chemical Products Ltd.
Unit 2, Zone D, Chelmsford Road Ind. Estate, Geat Dunmow, Essex, CM6 1XG UK (Tel.: 44 1371 875721; Telefax: 44 1371 872014; Telex: 818259)

Agrisales Ltd.
Royal Oak House, 45A Porchester Rd., London, W2 5DP UK (Tel.: 44 171 221 1275; Telefax: 44 171 792 9014; Telex: 266910)

Agrocanet-Socovi
390 Av Cdt Paul-Demarne, 34800 Canet France (Tel.: 33 6796 7009; Telefax: 33 6788 7397; Telex: 485375)

Ajinomoto
Ajinomoto Co., Inc., 15-1, Kyobashi 1-chome, Chuo-ku, Tokyo, 104 Japan (Tel.: (03) 5250-8111; Telex: J22690)

Ajinomoto USA, Inc., Glenpointe Centre West, 500 Frank W. Burr Blvd., Teaneck, NJ 07666-6894 USA (Tel.: 201-488-1212; Telefax: 201-488-6282; Telex: 275425 (AJNJ))

Akzo Chemie
Postfach 100146, Kreuzauer Strasse 46, DW 5160 Duren-Niederau Germany (Tel.: 02421-5951; Telefax: 02421-595380)

Akzo Salt
Akzo Salt Co., Abington Executive Park, Clarks Summit, PA 18411 USA (Tel.: 717-587-5131; Telefax: 717-586-7792; Telex: 756470)

Akzo Salt Europe, PO Box 247, 3800 AE Amersfoort The Netherlands (Tel.: 31 33 676767; Telefax: 31 33 676132; Telex: 79322)

Alba International Inc.
508 Clearwater Dr., N. Aurora, IL 60542 USA (Tel.: 708-897-4200; 800-669-9333; Telefax: 708-377-5330)

Albright & Wilson
Albright & Wilson Ltd., European & Corporate Hdqtrs., PO Box 3, 210-222 Hagley Rd. West, Oldbury, Warley, West Midlands, B68 0NN UK (Tel.: 44 121-429-4942; Telefax: 44 121-420-5151; Telex: 336291)

Albright & Wilson Americas Inc., PO Box 26229, Richmond, VA 23260-6229 USA (Tel.: 804-550-4300; 800-446-3700; Telefax: 804-550-4385)

Albright & Wilson Am. (Canada), 2070 Hadwen Rd., Mississauga, Ontario, L5K 2C9 Canada (Tel.: 905-403-0011; Telefax: 905-403-2231)

Albright & Wilson (Australia) Limited, PO Box 20, Yarraville, Victoria, 3013 Australia (Tel.: 61 3-688-7777; Telefax: 61 3-688-7788)

Albright & Wilson Asia Pacific Pte Ltd., 6 Jalan Besut, Jurong Industrial Estate, Singapore 2261 (Tel.: 011-65-261-2151; Telefax: 011-65-265-1941)

Tenneco, Inc., Div. of Albright & Wilson, PO Box 2511, Houston, TX 77252 USA (Tel.: 713-757-2131; Telefax: 713-757-4257)

Aldrich

Aldrich Flavors & Fragrances, 1101 W. St. Paul Ave., Milwaukee, WI 53233 USA (Tel.: 414-273-3850; 800-227-4563; Telefax: 414-273-5793; Telex: 910-262-3052)

Aldrich Chemical Co. Ltd., The Old Brickyard, New Rd., Gillingham, Dorset, SP8 4JL UK (Tel.: 44 1747 822211, 0800-71 71 81; Telefax: 44 1747 823779; Telex: 417238)

SAF Bulk Chemicals, Kyodo Bldg. Shinkanda, 10-Kanda-Mikuracho, Chiyoda-Ku, Tokyo Japan (Tel.: 81 03 3 258 0155; Telefax: 81 03 3 258 0157)

Alfa Chemicals Ltd.

ARC House, Terrace Rd. South, Binfield, Bracknell, Berks, RG12 5PZ UK (Tel.: 44 1344 861800; Telefax: 44 1344 862010; Telex: 847930)

Algas Marinas SA C/Copenhagen Pectin A/S

DK 4623 Lille-Skensved Denmark (Tel.: 45 5366 9210; Telefax: 45 5366 9446; Telex: 43572)

Alko Ltd.

Koskenkorva Industries, SF-05200 Rajamäki Finland (Tel.: 358 0 13311; Telefax: 358 0 290 1507; Telex: 57 15177 alpro fi)

Alland & Robert

9 Rue de Caintonge, 75003 Paris France (Tel.: 33 1 4272 0055; Telefax: 33 1 4272 5438; Telex: 210063)

Allchem Industries Inc.

4001 Newberry Rd., Suite E-3, Gainesville, FL 32607 USA (Tel.: 904-378-9696; Telefax: 904-338-0400; Telex: 509540 ALLCHEM UD)

Allchem International Ltd.

Broadway House, 21 Broadway, Maidenhead, Berks, SL6 1JE UK (Tel.: 44 1628 776666; Telefax: 44 1628 776591)

Allied Custom Gypsum Inc.

PO Box 69, Lindsay, OK 73052 USA (Tel.: 405-756-9565; Telefax: 405-756-3443)

Allied Mills Ltd.

Kingsgate, 1 King Edward Road, Brentwood, Essex, CM14 4HG UK (Tel.: 44 1277 262525; Telefax: 44 1277 200327; Telex: 996850)

Amano

Amano Int'l. Enzymes, PO Box 1000, Troy, VA 22974 USA (Tel.: 804-589-8278; 800-446-7652; Telex: 822-438)

Amano Enzyme USA Co. Ltd., Rt. 2, Box 1475, Troy, VA 22974 USA (Tel.: 804-589-8278; 800-446-7652; Telefax: 804-589-8270)

Amcan Ingredients International

Amcan Ingredients International, 47 Davies Ave., Toronto, Ontario, M4M 2A9, Canada)

Amcan Ingredients International, 5 Square Channaleilles, 78510 Le Chesnay France (Tel.: 33 3955 5260; Telefax: 33 3954 7268)

American Bio-Synthetics Corp.

710 W. National Ave., Milwaukee, WI 53204 USA (Tel.: 414-384-7017)

American Casein Co.

109 Elbow Lane, Burlington, NJ 08016 USA (Tel.: 609-387-3130; Telefax: 609-387-7204; Telex: 843368)

American Chemical Services
PO Box 190, Griffith, IN 46319 USA (Tel.: 219-924-4370)

American Cyanamid
American Cyanamid/Corporate Headquarters, One Cyanamid Plaza, Wayne, NJ 07470 USA (Tel.: 201-831-3339; 800-922-0187; Telefax: 201-831-2637)

Cyanamid BV, Postbus 1523, NL 3000 BM Rotterdam The Netherlands (Tel.: 010-4116340; Telefax: 010-4136788; Telex: 23554)

American Dairy Products Institute
130 N. Franklin St., Chicago, IL 60606 USA (Tel.: 312-782-4888; Telefax: 312-782-5299)

American Fruit Processors
2121 Lone Star Dr., Dallas, TX 65212 USA (Tel.: 214-634-2121; Telefax: 214-637-8040)

10725 Sutter Ave., PO Box 2596, Pacolma, CA 91331 USA (Tel.: 818-899-9574; Telefax: 818-899-6042)

American Fruits & Flavors
2121 Lone Star Dr., Dallas, TX 75212 USA (Tel.: 214-634-2121)

American Ingredients Co.
3947 Broadway, Kansas City, MO 64111 USA (Tel.: 816-561-9050; 800-821-2250; Telefax: 816-561-0422)

14622 S. Lakeside Ave., Dolton, IL 60419 USA (Tel.: 708-849-8590; 800-821-2250; Telefax: 816-561-0422)

Patco Specialty Chemicals Div., 3947 Broadway, Kansas City, MO 64111 USA (Tel.: 816-561-9050; 800-821-2250; Telefax: 816-561-9909)

American Laboratories Inc.
4410 South 102 St., Omaha, NE 68127 USA (Tel.: 402-339-2494; 800-445-5989; Telefax: 402-339-0801; Telex: 3735593 CACOMA)

American Lecithin. *See* Rhone-Poulenc

American Maize Products Co./Amaizo
1100 Indianapolis Blvd., Hammond, IN 46320-1094 USA (Tel.: 219-659-2000; 800-348-9896; Telefax: 219-473-6601)

American Roland Chemical Corp.
222 Sherwood Ave., Farmingdale, NY 11735-1718 USA (Tel.: 516-694-9090; Telefax: 516-694-9177; Telex: 232771)

American Xyrofin Inc. *See* Xyrofin

Amoco Lubricants Business Unit
MC 1102, 200 East Randolph Dr., Chicago, IL 60601 USA (Tel.: 312-856-4599)

AMPC Inc.
2325 N. Loop Dr., Ames, IA 50010 USA (Tel.: 515-296-7100; Telefax: 515-296-7110)

Amylum NV
Burchstraat 10, 9300 Aalst Belgium (Tel.: 32 53 76333; Telefax: 32 53 771459; Telex: 12363)

Anglia Speciality Oils
King George Dock, Kingston-upon-Hull, North Humberside, HU9 5PX UK (Tel.: 44 1482-701271; Telefax: 44 1482-709447; Telex: 592136)

D.F. Anstead Ltd., Div. of Ellis & Everard
Radford House, Radford Way, Billericay, Essex, CM12 0DE UK (Tel.: 44 1277-63 00 63; Telefax: 44 1277 631356)

Aplin & Barrett Ltd.
Clarks Mill, Stallard St., Trowbridge, Wilts, BA14 8HH UK (Tel.: 44 1225 768406; Telefax: 44 1225 764834; Telex: 44161)

Aprocat SA
Ave. San Julan s/n, 08400 Granoliers Spain (Tel.: 34 849 3555; Telefax: 34 849 5983)

Aqualon
Aqualon Co, A Hercules Inc. Co., 1313 North Market St., Wilmington, DE 19899 USA (Tel.: 302-594-6000; 800-345-8104; Telefax: 302-594-6660; Telex: 4761123)

Aqualon France, 3 Rue Eugene & Armand, Peugeot, 92508 Rueil-Malmaison France (Tel.: 33 1 4751 2919; Telefax: 33 1 4777 0614; Telex: 6314244)

Aquatec Quimica SA
Av. Paulista no. 37-12 andar, 01311-000 Sao Paulo-SP Brazil (Tel.: 011-284-4188; Telefax: 011-288-4431; Telex: 1121312)

Ariake USA Inc.
1977 W. 190th St., Suite 201, Torrance, CA 90504 USA (Tel.: 310-768-8015; Telefax: 310-538-0192)

Arista Industries, Inc.
1082 Post Rd., Darien, CT 06820 USA (Tel.: 203-655-0881; 800-255-6457; Telefax: 203-655-0881; Telex: 996493)

Arizona Chemical Co, Div. of International Paper
1001 E. Business Hwy. 98, Panama City, FL 32401-3633 USA (Tel.: 904-785-6700; 800-526-5294; Telefax: 904-785-2203; Telex: 441695)

Arla Foods AB
PO Box 47, S-59521 Mjolby Sweden (Tel.: 46 142 89000; Telefax: 46 142 10485; Telex: 8155024)

Armor Proteines
35460 St. Brice en Cogles France (Tel.: 33 9998 6381; Telefax: 33 9997 7991)

A. Arnaud SA
68 Ave. du General Michel, Bizot, 75012 Paris France

Arol Chemical Products Co.
649 Ferry St., Newark, NJ 07105 USA (Tel.: 201-344-1510; Telefax: 201-344-7127)

Aromatics Inc.
PO Box 13093, 5240 Gaffin Rd. S.E., Salem, OR 97308 USA (Tel.: 503-363-9494; Telefax: 503-363-3395)

Aromont SA
Route de Reims, PO Box 2, 02340 Montcornet France (Tel.: 33 2321 2121; Telefax: 33 2321 3460)

Arrowhead Mills Inc.
PO Box 3059, Hereford, TX 79045 USA (Tel.: 806-364-0730; Telefax: 806-364-8242)

Asahi Chemical Industry Co., Ltd.
Hibiya Mitsui Bldg., 1-2, Yuraku-cho 1-chome, Chiyoda-ku, Tokyo, 100 Japan (Tel.: (03) 3507-2730; Telefax: (03) 3507-2495; Telex: 222-3518 BEMBRGJ)

Asahi Denka Kogyo K.K.
Furukawa Bldg. 3-14, Nihonbashi Muro-machi 2-chome Chuo-ku, Tokyo, 103 Japan (Tel.: (03) 5255-9017; Telefax: (03) 3270-2463; Telex: 222-2407 TOKADK)

Ashland
Ashland Fine Ingredients Group, PO Box 2219, Columbus, OH 43216 USA (Tel.: 614-889-4530; Telefax: 614-889-3465)

Drew Industrial Div., One Drew Plaza, Boonton, NJ 07005 USA (Tel.: 201-263-7800; 800-526-1015 x7800; Telefax: 201-263-4483; Telex: DREWCHEMS BOON)

Atlantic Gelatin/Kraft General Foods
Hill St., Soburn, MA 01801 USA (Tel.: 617-933-2800; Telefax: 617-935-1566)

Atomergic Chemetals Corp.
222 Sherwood Ave., Farmingdale, NY 11735-1718 USA (Tel.: 516-694-9000; Telefax: 516-694-9177; Telex: 6852289)

Australian Bakels (Pty) Ltd.
PO Box 147, Lidcombe N.S.W. 2141 Australia (Tel.: 648-3833; Telefax: (02) 647-2663; Telex: 23064)

Avebe

Avebe America Inc., 4 Independence Way, Princeton, NJ 08540 USA (Tel.: 609-520-1400; Telefax: 609-520-1473; Telex: 0820713)

Avebe BV, Avebeweg 1, 9607 PT Foxhol The Netherlands (Tel.: 31 5980-42234; Telefax: 5980-97892; Telex: 53018)

Avebe UK Ltd., Thornton Hall, Thornton Curtis, Ulceby, South Humberside, DN39 6XD UK (Tel.: 44 1469-32222; Telefax: 44 1469-31488)

AVOCO

Box 3629, Cobina, CA 91723 USA (Tel.: 818-331-0081; Telefax: 818-915-7153)

Avonmore Foods plc

Ballyragget, Co Kilkenny, Ireland (Tel.: 056 33155; Telefax: 056 33268; Telex: 80663)

Baktech International Ltd.

260 Hillmorton Rd, Rugby, CV22 5BW UK (Tel.: 44 1788 561658; Telefax: 44 1788 565550; Telex: 80663)

Balchem Corp.

PO Box 175, Slate Hill, NY 10973 USA (Tel.: 914-355-2861; Telefax: 914-355-6314)

Barnett & Foster Ltd.

Denington Estate, Wellingborough, Northants, NN8 2QJ UK (Tel.: 44 1933 440022; Telefax: 44 1933 440053; Telex: 311342)

Baromatic Corp.

PO Box 7, Great Neck, NY 11022 USA (Tel.: 516-756-1014; Telefax: 516-756-1015; Telex: 276824)

Bartek Ingredients Inc.

421 Seaman St., Stoney Creek, Ontario, L8E 3J4 Canada (Tel.: 905-662-3292; 800-537-7287 (USA); Telefax: 905-662-8849; Telex: 061-8498)

Barva SA

PO Box 133, Lima, 09, Peru (Tel.: 51 14 673680; Telefax: 51 14 673777)

BASF

BASF AG, ESA/WA-H 201, D-6700 Ludwigshafen Germany (Tel.: 0621-60-99603; Telefax: 0621-60-41787; Telex: 469499-0 BAS D)

BASF Corp./Performance Chemicals, 3000 Continental Dr. North, Mt. Olive, NJ 07828-1234 USA (Tel.: 201-426-2600; 800-669-BASF)

Basildon Chemical Co. Ltd.

Kimber Rd., Abington, Oxon, OX14 1RZ UK (Tel.: 44 1235 526677; Telefax: 44 1235 524334; Telex: 837691)

Beatrice Cheese Inc.

770 N. Springdale Rd., Wakesha, WI 53186 USA (Tel.: 414-782-2750; Telefax: 414-796-0463)

Beck Flavors

411 E. Gano, PO Box 22509, St. Louis, MO 63147 USA (Tel.: 314-436-3133; 800-851-8100; Telefax: 314-436-1049)

Beeta Chemicals

B/19 Nandjyot Industrial Estate, Anher-Kurla Rd., Bombay, 400 072 India

Beldem

Beldem UK, c/o Borm-Reid, Staple Inn Bldgs. South, Staple Inn, London, WC1V 7QE UK (Tel.: 44 171 242 5479; Telex: 21683)

Beldem NV, Industrielaan 10, IIIde Industriezone, B-9320 Erembodegem Belgium (Tel.: 32 53 83 4121; Telefax: 32 53 83 7290; Telex: 11194)

Bell Flavors & Fragrances, Inc.

500 Academy Dr., Northbrook, IL 60062 USA (Tel.: 708-291-8300; 800-323-4387; Telefax: 708-291-1217; Telex: 910-686-0653)

Belovo SC

Zone Industrielle, B6650 Bastogne Belgium (Tel.: 32 61 21 1884; Telefax: 32 61 21 5563; Telex: 41581)

F. R. Benson & Partners Ltd.
Crossroads House, 165 The Parade, High St., Watford, WD1 1NJ UK (Tel.: 44 1923 240560; Telefax: 44 1923 240569)

Berje Chemical Products Inc.
5 Lawrence St., Bloomfield, NJ 07003 USA (Tel.: 201-748-8980; Telefax: 201-680-9618; Telex: 424713)

Berk Natural Products Ltd.
PO Box 56, Priestley Rd., Basingstoke, Hants, RG24 9PU UK (Tel.: 44 1256 29292; Telefax: 44 1256 64711; Telex: 858371)

Berol Nobel AB
Box 11536, S-10061 Stockholm Sweden (Tel.: 8-743-4000; Telefax: 8-644-3955; Telex: 10513 benobls)

David Berryman Ltd.
16 Aldenham Ave., Radlet, Herts, WD7 8HX UK (Tel.: 44 1442 218484; Telefax: 44 1923 857797; Telex: 825771)

Besnier Bridel Alimentaire
Les Placis, 35230 Bourgbarre France (Tel.: 33 9926 6333; Telefax: 33 9926 6684; Telex: 730843)

Bestoval Products Co. Ltd.
49 Ridge Hill, London, NW11 UK (Tel.: 44 181 455 3020)

BFP Goldrei, British Fermenation Products Ltd.
Appold House, 200 North Service Rd., Brentwood, Essex, CM14 4RJ UK (Tel.: 44 1277 233591; Telefax: 44 1277 224902; Telex: 996861)

Biddle Sawyer Foods Ltd.
Unit 76, Lakeside Business Park, Hawley Rd., Dartford, Kent, DA1 1LD UK (Tel.: 44 1322 287797; Telefax: 44 1322 287785; Telex: 893311)

Bio-Botanica, Inc.
75 Commerce Dr., Hauppauge, NY 11788 USA (Tel.: 516-231-5522; 800-645-5720; Telefax: 516-231-7332)

Biocatalysts Ltd.
Main Ave., Treforest Ind. Estate, Pontypridd, Mid Glam, CF37 5UT UK (Tel.: 44 144384 3712; Telefax: 44 144384 1214; Telex: 497126)

Biocon (UK) Ltd.
Dock Rd. South, Bromborough, Wirrl, Merseyside, L62 4SG UK (Tel.: 44 151 645 2060; Telefax: 44 151 643 4400; Telex: 337393)

Bioengineering AG
Sagenrainstrasse 7, 8636 Wald Switzerland (Tel.: 41 55 938 111; Telefax: 41 55 954 964; Telex: 375977)

Bio-Isolates plc
Glendale Business Centre, Deeside Ind. Estate, Welsh Rd., Deeside, Clwyd, CH5 2LR UK (Tel.: 44 1244 830344; Telefax: 44 1244 830054)

Bio Springer/Fould Springer
103 Rue Jean Jaures, 94701 Maisons Alfort France (Tel.: 33 1 4977 1846; Telefax: 33 1 4977 0358; Telex: 264067)

BK-Ladenburg Corp.
2345 Erringer Road 221, Simi Valley, CA 93065 USA (Tel.: 805-581-1979; Telefax: 805-581-2139; Telex: 375977)

S. Black (Import & Export) Ltd.
The Colonnade, High St., Cheshunt, Herts, EN8 0DJ UK (Tel.: 44 1992 30751; Telefax: 44 1992 22838; Telex: 894085)

Blossom Farm Products Co.
12 Rt. 17 N, Paramus, NJ 07652 USA (Tel.: 201-587-1818; 800-729-1818; Telefax: 201-526-0310; Telex: 13-0489)

Boliden Intertrade Inc.
   3400 Peachtree Rd. NE, Suite 401, Atlanta, GA 30326 USA (Tel.: 404-239-6700; 800-241-1912; Telefax: 404-239-6701; Telex: 981036)

Bonlac Foods Ltd.
   566 St. Kilda Rd., Melbourne, Victoria, 3004 Australia (Tel.: 61 3 270 0922; Telefax: 61 3 270 0911; Telex: AA 31082)

Booths Raglan Works
   Methey Rd., Whitwood Mere, Castleford, West Yorks, WF10 1NX UK (Tel.: 44 1977 518515; Telefax: 44 1977 519167)

Boots Chemicals
   D110 Main Office, Beeston, Nottingham, NG2 3AA UK (Tel.: 44 1159 591648; Telefax: 44 1159 593715)

Borculo Whey Products
   Needseweg 23, 7270 AA Borculo The Netherlands (Tel.: 31 5457 56789; Telefax: 31 5457 73275; Telex: 44585)

Bowmans
   Whitley Bridge, Goole, DN14 0LH UK

Bradleys (Hart's Mill Ltd.)
   Mill Lane, Aldington, Ashford, Kent, TN25 7AJ UK (Tel.: 44 1233 720768; Telefax: 44 1233 720007)

Bradshaw-Praeger & Co.
   3248-62 W. 47th Place, Chicago, IL 60632 USA (Tel.: 312-523-2050; Telefax: 312-523-6093)

Arthur Branwell & Co. Ltd.
   Bronte House, 58-62 High St., Epping, Essex, CM16 4AE UK (Tel.: 44 1992 577333; Telefax: 44 1992 575043; Telex: 817158)

Brenntag (UK) Ltd.
   Brenntag House, 45 High St., Hampton Wick, Kingston upon Thames, KT1 4DG UK (Tel.: 44 181 977 3200; Telefax: 44 181 943 4350; Telex: 924629)

Bretagne Chimie Fine
   Boisel, 56140 Pleucadeuc France (Tel.: 33 9726 9121; Telefax: 33 9726 9046; Telex: 951084)

Briess Industries Inc.
   250 W. 5th St., Suite 1020, New York, NY 10107 USA (Tel.: 212-247-0780; Telefax: 212-333-5170)

Britannia Natural Proucts Ltd.
   Unit 5, Woodlands Business Park, Rougham Ind. Estate, Rougham, Suffolk, IP30 9ND UK (Tel.: 44 1359 71461; Telefax: 44 1359 71672)

British Arkady
   British Arkady Co. Ltd., Arkady Soya Mills, Skerton Rd., Old Trafford, Manchester, M16 0NJ UK (Tel.: 44 161 872 7161; Telefax: 44 161 873 8083; Telex: 668488)

   British Arkady Group, 62/70 rue Ivan Tourgueneff, 78380 Bougival France (Tel.: 33 1 3969 7070; Telefax: 33 1 3918 4610; Telex: 689413)

British Bakels Ltd.
   238 Bath Rd., Slough, Berks, SL1 4DU UK (Tel.: 44 1753 526261; Telefax: 44 1753 825455; Telex: 848115)

British Fermentation Products. *See* BFP Goldrei

British Pepper & Spice Co. Ltd.
   Rhosili Rd., Brackmills, Northampton, NN4 0LD UK (Tel.: 44 1604 766461; Telefax: 44 1604 763156; Telex: 312472)

British Sugar plc
   PO Box 26, Oundle Rd., Peterborough, PE2 9QU UK (Tel.: 44 1733 63171; Telefax: 44 1733 63068; Telex: 32273)

Brolite Products Inc.
   2542 N. Elston Ave., Chicago, IL 60647 USA (Tel.: 312-384-0210; Telefax: 312-384-8348)

Bronson & Jacobs Pty. Ltd.

288 Burns Bay Rd., Lane Cove, NSW, 2066 Australia (Tel.: 61 2 427 0066; Telefax: 61 2 428 2845)

Brookside International

2372 Townline Rd., Abbotsford, British Columbia, V2S 1M3 Canada (Tel.: 604-852-5940; Telefax: 604-852-8751)

Browne & Dureau Int'l. Ltd.

Suite 502, Henry Lawson Centre, Birkenhead Point, Drummoyne, NSW, 2047 Australia (Tel.: 61 2 819 7933; Telefax: 61 2 819 6262; Telex: AA 120406)

Browning Chemical Corp./Food Ingredients Div.

707 West Chester Ave., White Plains, NY 10604 USA (Tel.: 914-686-0300; Telefax: 914-686-0310)

Robert Bryce & Co.

1a Queen St., Auburn, NSW 2144 Australia (Tel.: 61 2 646 1777; Telefax: 61 2 646 2904; Telex: 31486)

Chemische Budenheim

Postfach 1147-1149, D-6501 Budenheim Germany (Tel.: 49 6139 89-0; Telefax: 49 6139 89493; Telex: 4187856)

Buffalo Color Corp.

959 Rte. 46 East, Suite 403, Parsippany, NJ 07054 USA (Tel.: 201-316-5600; 800-631-0171; Telefax: 201-316-5828; Telex: 7109885924)

HP Bulmer Pectin Ltd.

Plough Lane, Hereford, HR4 0LE UK (Tel.: 44 1432 352000; Telefax: 44 1432 352081; Telex: 35211)

Bunge Foods

725 N. Kinzie Ave., Bradley, IL 60915 USA (Tel.: 815-937-8129; 800-828-0800; Telefax: 815-939-4289)

Burlington Bio-Medical Corp.

222 Sherwood Ave., Farmingdale, NY 11735-1718 USA (Tel.: 516-694-9000; Telefax: 516-694-9177; Telex: 6852289)

Burton Son & Sanders

Collogo St., Ipowioh, Suffolk, IP4 IDE UK (Tel.: 44 1473 256234; Telefax: 44 1473 211543, Telex. 98224)

Bush Boake Allen

Bush Boake Allen Inc., 7 Mercedes Dr., Montvale, NJ 07645 USA (Tel.: 201-391-9870; Telefax: 201-391-0860; Telex: 98224)

Bush Boake Allen Ltd./GMB Proteins, Blackhorse Lane, Walthamstow, London, E17 5QP UK (Tel.: 44 181-531-4211; Telefax: 44 181-531 6413; Telex: 897808)

Byrnes & Kiefer Co.

131 Kline Ave., Callery, PA 16024 USA (Tel.: 412-538-9200; Telefax: 412-538-9292)

Byrton Dairy Products Inc.

28354 N. Ballard Dr., Lake Forest, IL 60045 USA (Tel.: 708-367-8300; Telefax: 708-367-8332)

Cabot

Cabot Corp./Cab-O-Sil Div., PO Box 188, Tuscola, IL 61953 USA (Tel.: 217-253-3370; 800-222-6745; Telefax: 217-253-4334; Telex: 910-663-2542)

Cabot GmbH/Cab-O-Sil Div., PO Box 1766, D-79607 Rheinfelden Germany (Tel.: 49 7623-9090; Telefax: 49 7623-90932; Telex: 773451)

Cabot Carbon Ltd., Div. of Cabot Corp., Lees Lane, Stanlow, South Wirral, Merseyside L65 4HT UK (Tel.: 44 151-355 36 77; Telefax: 44 151-356-07 12; Telex: 629261 CABLAK G)

Cairn Foods Ltd.

Cairn House, Elgiva Lane, Chesham, Bucks, HP5 2JD UK (Tel.: 44 1494 786066; Telefax: 44 1494 791816; Telex: 837075)

Calaga Food Ingredients

28B Westgate, Grantham, Lincs, NG31 6LX UK (Tel.: 44 1476 590252; Telefax: 44 1476 73436)

Calgene Chemical Inc.
> 7247 North Central Park Ave., Skokie, IL 60076-4093 USA (Tel.: 708-675-3950; 800-432-7187; Telefax: 708-675-3013; Telex: 72-4417)

Calgon
> Calgon Corp., PO Box 717, Pittsburgh, PA 15230-0717 USA (Tel.: 412-787-6700; 800-4-CARBON; Telefax: 4412-787-6713; Telex: 671183CCC)

> Calgon Carbon Corp., PO Box 717, Pittsburgh, PA 15230-0717 USA (Tel.: 412-787-6700; 800-422-7266; Telefax: 412-787-6676; Telex: 6711837 CCC)

California Natural Products
> PO Box 1219, Lathrop, CA 95330 USA (Tel.: 209-858-2525; Telefax: 209-858-4076)

Cal India Foods Int'l.
> 814 N. Del Sol, Diamond Bar, CA 91765 USA (Tel.: 714-860-2188; Telefax: 714-860-4952; Telex: 910 2400197)

J.H. Calo Company Inc.
> 1600 Stewart Ave., PO Box 647, Westbury, NY 11590 USA (Tel.: 516-561-0711; Telefax: 516-561-7758)

C.A.L.-Pfizer. *See* Pfizer

Caminiti Foti & Co. S.r.l.
> Via Calatafimi 2, 98023 Furci Siculo (ME) Italy (Tel.: 39 942 791596; Telefax: 39 942 793832; Telex: 980158)

Canada Packers
> Canada Packers Inc., Edible Oils Div., 30 Weston Rd., Toronto, Ontario, M6N 3P4 Canada (Tel.: 416-761-4172; Telefax: 416-761-4452; Telex: 069-69539)

> Canada Packers Inc., Food Ingredients Group, 30 Weston Rd., Toronto, Ontario, M6N 3P4 Canada (Tel.: 416-761-4231; Telefax: 416-761-4452)

Canadian Harvest
> Canadian Harvest Process, 2 Barrie Blvd., St. Thomas, Ontario, N5P 4B9 Canada (Tel.: 519-633-5030; Telefax: 519-633-3718)

> Canadian Harvest USA, 1001 S. Cleveland St., PO Box 272, Cambridge, MN 55008 USA (Tel.: 612-689-5800; Telefax: 612-689-5949)

Caravan Products Co.
> 100 Adams Dr., Totowa, NJ 07512 USA (Tel.: 201-256-8886; Telefax: 201-256-8395)

Carbery Milk Products Ltd.
> Ballineen, Co Cork, Ireland (Tel.: 353 23 47222; Telefax: 353 23 47541; Telex: 75801)

Carboeshire Ltd. (Wheat Gluten)
> 36 High St., Harrow on the Hill, Middlesex, HA1 3LL UK (Tel.: 44 181 422 2277; Telefax: 44 181 426 8702)

Cargill
> Knowle Hill Park, Fairmile Lane, Cobham, Surrey, KT11 2PD UK (Tel.: 44 1932 861175; Telefax: 44 1932 861286)

Carrageenan Company
> 3830 S. Teakwood St., Santa Ana, CA 92707 USA (Tel.: 714-751-1521; Telefax: 714-850-9865)

CasChem Inc.
> 40 Ave. A, Bayonne, NJ 07002 USA (Tel.: 201-858-7900; 800-CASCHEM; Telefax: 201-437-2728; Telex: 710-729-4466)

CBI
> Beursplein 37, PO Box 30009, 3001 DA Rotterdam The Netherlands (Tel.: 31 1640 82200; Telefax: 31 1640 54489; Telex: 78124)

Ceca SA, Div. of Atochem
> 22, place de l'Iris, La Défense 2, Cedex 54, 92062 Paris-La Défense France (Tel.: 147-96-9090; Telefax: 147-96-9234; Telex: 611444 ckd)

Celite Corp.

PO Box 519, Lompoc, CA 93438-0519 USA (Tel.: 805-735-7791; Telefax: 805-735-5699; Telex: 62776493 ESL UD)

Cellulose Attisholz AG, UK distributors

Wykefold Ltd., 20 Chancellors St., London, W6 9RL UK (Tel.: 44 181 741 8691; Telex: 8813271)

Central Soya

Central Soya Co., Inc./Chemurgy Div., PO Box 2507, Fort Wayne, IN 46801-2507 USA (Tel.: 219-425-5432; 800-348-0960; Telefax: 219-425-5301; Telex: 49609682)

Central Soya Aarhus A/S, Skansevej 2, PO Box 380, DK-8100 Aarhus C Denmark (Tel.: 45 89 31 21 11; Telefax: 45 89 31 21 12; Telex: 64348)

Stern Lecithin & Soja GmbH & Co. KG, Div. of Central Soya, An der Alster 81, D-2000 Hamburg 1 Germany (Tel.: 49 40/2 48 30-01; Telefax: 49 40/280 34 27; Telex: 2 162 052 star d)

Ceratonia SA

PO Box 32, E-43080 Tarragona Spain (Tel.: 34 77 541233; Telefax: 34 77 544606; Telex: 56433)

Cereal & General Products

Broadwall House, 21 Broadwall, London, SE1 9PL UK (Tel.: 44 171 928 8966; Telefax: 44 171 261 9085; Telex: 888220)

Cerestar

Cerestar International Sales, Ave. Louise 149, Bte 13, B-1050 Brussels Belgium (Tel.: 32 2 535 1711; Telefax: 32 2 537 8554; Telex: 22648)

Cerestar UK Ltd., Trafford Park Rd, Trafford Park, Manchester, Lancashire M17 1PA UK (Tel.: 44 161-872 5959; Telefax: 44 161-848-9034; Telex: 667022)

Cesalpinia/Auschem SpA

Via Pinamonte da Brembate N. 3, 24100 Bergamo Italy (Tel.: 39 35 214255; Telefax: 39 35 214022; Telex: 300424)

Cham Foods (Israel) Ltd.

PO Box 299, 38102 Hadera Israel (Tel.: 972 6334755, 972 6336194; Telex: 471794)

Champlain Industries Inc.

PO Box 3055, 25 Styertowne Rd., Clifton, NJ 07012 USA (Tel.: 201-778-4900; 800-222-4904; Telefax: 201-778-0094; Telex: 3725769)

Charabot SA

BP 68, 10 Ave. Y-E Baudoin, 06332 Grasse Cedex France (Tel.: 33 93 09 33 33; Telefax: 33 93 93 33 03; Telex: 470822)

Chart Corporation Inc.

787 E. 27th St., Paterson, NJ 07504 USA (Tel.: 201-345-5554; Telefax: 201-345-2139)

Pierre Chauvet

Pierre Chauvet S.A., 83440 Seillans France (Tel.: 94 76 96 03; Telefax: 94 76 96 16)

Pierre Chauvet, Inc., 3 Reuten Dr., Closter, NJ 07624 USA (Tel.: 201-784-9300; Telefax: 201-784-0604)

Cheam Cheese Ltd.

1-3 Cheam Rd., Ewell, Surrey, KT17 1SP UK (Tel.: 44 181 394 1234; Telefax: 44 181 394 1238)

Cheil Foods

Cheil Foods & Chemicals Inc., 105 Challenger Rd., 9th Floor, Ridgefield Park, NJ 07660-0511 USA (Tel.: 201-229-6037; Telefax: 201-229-6040; Telex: 219176)

Cheil Foods & Chemicals Inc., Samsung House, 3 Riverbank Way, Great West Rd., Brentford, Middx, TW8 9RE UK (Tel.: 44 181 862 9334; Telefax: 44 181 862 0094; Telex: 258237)

Chemax, Inc.

PO Box 6067, Highway 25 South, Greenville, SC 29606 USA (Tel.: 803-277-7000; 800-334-6234; Telefax: 803-277-7807; Telex: 570412 IPM15SC)

Chemcolloids Ltd.

Tunstall Rd., Bosley, Macclesfield, Cheshire, SK11 0PE UK (Tel.: 44 1260 223284; Telefax: 44 1260 223589; Telex: 668002)

Chemische Budenheim. *See under* Budenheim

Cherry Rocher Industries
BP 488, 38312 Bourgoin Jallieu France (Tel.: 33 74 93 3810; Telefax: 33 74 28 4673; Telex: 308208)

Church & Dwight Co. Inc./Specialty Prods. Div.
Box 5297, 469 N. Harrison St., Princeton, NJ 08543-5297 USA (Tel.: 609-683-5900; 800-221-0453; Telefax: 609-497-7176; Telex: 752226)

Ciba-Geigy Corp./Chelate Prods. Group
Box 18300, Greensboro, NC 27419 USA (Tel.: 919-292-7100)

CIG
Unit 37, Slough Business Park, Slough Ave., Silverwater, NSW, 2141 Australia (Tel.: 61 2 741 6600; Telefax: 61 2 741 6666)

Cimbar Performance Minerals
25 Old River Rd. S.E., PO Box 250, Cartersville, GA 30120 USA (Tel.: 404-387-0319; 800-852-6868; Telefax: 404-386-6785)

Classic Flavors & Fragrances
125 E. 23rd St. 400, New York, NY 10010 USA (Tel.: 212-777-0004; Telefax: 212-353-0404)

Claymore Food Ingredients
12 Larchfield, Balerno, Midlothian, EH14 7NN, Scotland (Tel.: 44 131 449 5226; Telefax: 44 131 449 6099)

W.A. Cleary Chemical Corp.
Southview Industrial Park, 178 Route #522 Suite A, Dayton, NJ 08810 USA (Tel.: 908-329-8399; 800-524-1662; Telefax: 908-274-0894)

Clofine Dairy & Food Products Inc.
1407 New Rd., Linwood, NJ 08221 USA (Tel.: 609-653-1000; Telefax: 609-653-0127)

CNC International, Limited Partnership
PO Box 3000, Woonsocket, RI 02895 USA (Tel.: 401-769-6100; Telefax: 401-769-4509)

COFAG Comite des Fabricants d'Acide Glutamique
16 rue Ballu, 75009 Paris France (Tel.: 33 1 4082 3434; Telefax: 33 1 4082 3537; Telex: 285847)

Cofranlait
24 Ave. Hoche, F75008 Paris France (Tel.: 33 1 4413 9720; Telefax: 33 1 4413 9721)

Colloides Naturels International
4 rue Frederic Passy, 92200 Neuilly sur Seine France (Tel.: 33 1 47 47 1850; Telefax: 33 1 47 471891)

Colombus Foods Co.
800 N. Albany Ave., Chicago, IL 60622 USA (Tel.: 312-265-6500; Telefax: 312-265-6985)

Colony Import & Export Corp.
226 Seventh St., Garden City, NY 08807 USA (Tel.: 215-699-7733)

Colorcon
Moyer Blvd., West Point, PA 19486 USA (Tel.: 215-699-7733)

Comet Rice Ingredients/Calbran
10990 Wilshire Blvd., Suite 1800, Los Angeles, CA 90024 USA (Tel.: 310-478-0069; Telefax: 310-312-2077; Telex: 188528)

Commodity Services Int'l. Inc.
114B N. West St., PO Box 1876, Easton, MD 21601 USA (Tel.: 410-820-8880; Telefax: 410-820-8890; Telex: 898099 CSI MD)

Confoco, Inc.
43a Chertsey Rd., Woking, Surrey, GU215AJ UK

Consolidated Flavor Corp.
231 Rock Industrial Dr., Bridgeton, MO 63044 USA (Tel.: 314-291-5444; 800-422-5444; Telefax: 314-291-3289)

Continental Colloids Inc.
245 W. Roosevelt Rd., Bldg. 2, Unit 2, West Chicago, IL 60185 USA (Tel.: 708-231-8650; Telefax: 708-231-8692)

Continental Seasoning Inc.
1700 Palisade Ave., Teaneck, NJ 07666 USA (Tel.: 201-837-6111)

Cooke Aromatics Pty. Ltd.
PO Box 546, Hornsby, NSW, 2077 Australia (Tel.: 61 2 476 1766; Telefax: 61 2 476 1714)

F.D. Copeland & Sons. Ltd.
5 Westfield St., Woolwich, London, SE18 5TL UK (Tel.: 44 181 854 8101; Telefax: 44 181 854 1077; Telex: 896355)

Copenhagen Pectin
DK-4623 Lille Skensved Denmark (Tel.: 45 53 66 9210; Telefax: 45 53 66 9446; Telex: 43572)

Copiaa
87 rue de la Commanderie, 569500 Dovai France (Tel.: 33 27 982828; Telefax: 33 27 918698; Telex: 820837)

Cornelius Chemical Group Ltd.
St. James's House, 27-43 Eastern Road, Romford, Essex, RM1 3NN UK (Tel.: 44 1708 722300; Telefax: 44 1708 768204; Telex: 885589 CORNEL G)

Corn Products/Unit of CPC Int'l.
6500 Archer Rd., Summit-Argo, IL 60501 USA (Tel.: 708-563-2400; Telefax: 708-563-6852; Telex: 708-563-6852)

Cosmark Pty. Ltd.
Suite 204, 30 Risher Rd., Dee Why, NSW, 2099 Australia (Tel.: 61 2 971 0888; Telefax: 61 2 971 0684)

Cosucra
Cosucra BV, Postbus 1288, NL-4700 BG Roosendaal The Netherlands (Tel.: (31) 1650 84333; Telefax: (31) 1650 67796)

Cosucra SA, Haut Vinave 61, B-4350 Momalle Belgium (Tel.: 32 41 50 3108; Telefax: 32 41 50 2365; Telex: 42425)

Courtaulds
Courtaulds plc, Patents Dept, PO Box 111, 72 Lockhurst Lane, Coventry, CV6 5RS UK (Tel.: 44 1203-688771; Telefax: 44 1203-583837)

Courtaulds Fine Chemicals, PO Box 5, Spondon, Derby, DE2 7BP UK (Tel.: 44 1332 661422; Telefax: 44 1332 280610; Telex: 37391)

Courtaulds Water Soluble Polymers, P O Box 5, Spondon, Derby, Derbyshire, DE2 7BP UK (Tel.: 44 1332-661422; Telefax: 44 1332-661078; Telex: 37391 CHMPLS G)

CPC
CPC (UK) Ltd./Food Ingredients Div., Claygate House, Esher, Surrey, KT10 9PN UK (Tel.: 44 1372 462181; Telefax: 44 1372 468780; Telex: 27106)

CPC (UK) Ltd./Bovril Food Ingredients Div., PO Box 18, Wellington Rd., Burton-on-Trent, Staffs, DE14 2AB UK (Tel.: 44 1283 511111; Telefax: 44 1283 510194; Telex: 34322)

CPS Chemical Co Inc.
PO Box 162, Old Bridge, NJ 08857 USA (Tel.: 908-607-2700; Telefax: 908-607-2562; Telex: 844532-CPSOLDB)

Craigmillar Ltd.
Craigmillar House, Stadium Rd., Bromborough, Wirral, Merseyside, L62 3NU UK (Tel.: 44 151 346 1600; Telefax: 44 151 346 1334; Telex: 878805)

Cranfield Brothers Ltd.
PO Box 6, Dock Ruller Mills, Ipswich, Suffolk, IP4 1DB UK (Tel.: 44 1473 252101; Telefax: 44 1473 215355; Telex: 987479)

Crespel & Deiters GmbH
Groner Allee 76, D-4530 Ibbenburen Germany (Tel.: 49 5451 5000-0; Telefax: 49 5451 5000-60; Telex: 94545)

Crest Foods Co. Inc.
PO Box 371, Ashton, IL 61006 USA (Tel.: 815-453-7411)

Croda
Croda Chemicals Ltd., Div. of Croda International plc, Cowick Hall, Snaith Goole, North Humberside, DN14 9AA UK (Tel.: 44 1405-8605551; Telefax: 44 1405-860205; Telex: 57601)

Croda Bakery Services Ltd., Falcon St., Oldham, Lancs, OL8 1JU, UK

Croda Colloids Ltd., Foundry Lane, Ditton Widnes, Cheshire, WA8 8UB UK (Tel.: 44 151 423 3441; Telefax: 44 151 423 3205; Telex: 629586)

Croda Food Products Ltd., Div. of Croda International plc, Cowick Hall, Snaith, Goole, North Humberside, DN14 9AA UK (Tel.: 44 1405 860551; Telefax: 44 1405 860205; Telex: 57601)

Croda International plc, Cowick Hall, Snaith, Goole, North Humberside, DN14 9AA UK (Tel.: 44 1405 860551; Telefax: 44 1405 862253)

Croda Surfactants Ltd., Cowick Hall, Snaith, Goole, North Humberside, DN14 9AA UK (Tel.: 44 1405 860551; Telefax: 44 1405 860205; Telex: 57601)

Croda Universal Ltd., Div. of Croda International plc, Cowick Hall, Snaith, Goole, North Humberside DN14 9AA UK (Tel.: 44 1405 860551; Telefax: 44 1405 860205; Telex: 57601)

Croda Inc., 7 Century Dr., Parsippany, NJ 07054-4698 USA (Tel.: 201-644-4900; Telefax: 201-644-9222)

Croda Singapore PTE Ltd., 20 Chia Ping Rd., S-2261 Singapore (Tel.: 65 261 3008; Telefax: 65 261 2825; Telex: 38943)

Crompton & Knowles Corp./Ingredient Technology Div.
1595 MacArthur Blvd., Mahwah, NJ 07430 USA (Tel.: 201-818-1200; 800-343-4860; Telefax: 201-818-2173)

Crosfield Chemicals, Inc.
101 Ingalls Ave., Joliet, IL 60435 USA (Tel.: 815-727-3651; 800-727-3651; Telefax: 815-727-5312)

Croxton & Garry Ltd.
Curtis Rd. Industrial Estate, Dorking, Surrey, RH4 1XA UK (Tel.: 44 1306-886688; Telefax: 44 1306-887780; Telex: 859567/8 cand g)

CRS Co.
4940 Viking Dr., Minneapolis, MN 55435 USA (Tel.: 612-893-1610)

Crucible Chemical Co. Inc.
PO Box 6786, Donaldson Center, Greenville, SC 29606 USA (Tel.: 803-277-1284; 800-845-8873; Telefax: 803-299-1192)

Crystals International Inc.
1111 W. Haines St., Plant City, FL 33566 USA (Tel.: 813-754-2691; Telefax: 813-757-6448; Telex: ITT 411 8885)

CSR Food Ingredients
Bowman St., Pyrmont, NSW, 2009 Australia (Tel.: 61 2 692 7685; Telefax: 61 2 552 1712)

C&T Refinery Inc.
2000 W. Broad St., Richmond, VA 23220 USA (Tel.: 804-359-5786; Telefax: 804-359-5514)

Curry Co.
551 Wisteria, Bellaire, TX 77401 USA (Tel.: 713-668-3957)

Custom Ingredients Inc.
PO Box 35800, Suite 320, Houston, TX 77235 USA (Tel.: 713-944-7483)

Cyanamid BV. *See* American Cyanamid

Cytec Industries Inc.
Five Garret Mountain Plaza, West Paterson, NJ 07424 USA (Tel.: 201-357-3100; 800-438-5615; Telefax: 201-357-3054; Telex: 130400)

Dafa SA
BP 2, 77313 Marne la Vallé 2 France (Tel.: 33 1 60 06 5525; Telefax: 33 1 64 80 4094; Telex: 691558)

Daicel (USA) Inc., Subsid of Diacel Chem. Industries, Ltd.
One Parker Plaza, 400 Kelby St., Fort Lee, NJ 07024 USA (Tel.: 201-461-4466; Telefax: 201-461-2776)

Dai-ichi Kogyo Seiyaku Co., Ltd.
New Kyoto Center Bldg., 614, Higashishiokoji-cho, Shimokyo-ku, Kyoto, 600 Japan (Tel.: (075) 343-1181; Telefax: (075) 343-1421)

Daiichi Pharmaceutical Co. Ltd.
14-10, Nihonbashi 3-Chome, Chuo-ku, Tokyo, 103 Japan (Tel.: 03-272-0611; Telefax: 03-276-0694; Telex: J22729)

Dairy Crest Ltd. (Dairy Crest Ingredients)
Philpot House, Rayleigh, Essex, SS6 7HH UK (Tel.: 44 1268 775522; Telefax: 44 1268 747666; Telex: 99166)

Dairyfood Milch-U Nahrungsprodukte GmbH
Feldsrabe 5, D2358 Kaltenkirchen Germany (Tel.: 49 4191 88061; Telefax: 49 4191 88051; Telex: 2180159)

Damrow Company Inc./Drying Systems Div.
Fabriksvej 12-14, Kolding 6000 Denmark (Tel.: 45 5 52 67 11; Telefax: 45 5 53 98 70; Telex: 51308)

Danflavour
Ulrikkenborg Pl 10B, Lyngby 2800 Denmark (Tel.: 45 45 930333; Telefax: 45 45 872062)

H.E. Daniel Ltd.
Longfield Rd., North Farm Ind. Estate, Royal Tunbridge Wells, Kent, TN2 3EY UK (Tel.: 44 1892 511444; Telefax: 44 1892 510013; Telex: 957103)

S. Daniels plc
South Quay Plaza, 183 Marsh Wall, London, E14 9SH UK (Tel.: 44 171 538 0188; Telefax: 44 171 410 0067; Telex: 28837)

Danish Dairy Board
Frederiks Alle 22, Aarhus 8000 C Denmark (Tel.: 45 86 13 26 11; Telefax: 45 86 13 26 93; Telex: 64307)

Danmark Protein A/S
Nr Vium, DK 6920 Videbaek Denmark (Tel.: 45 97 17 8244; Telefax: 45 97 17 8206; Telex: 60921)

Daroma STE
ZI de Marcerolles, 26500 Bourg les Valence France (Tel.: 33 7583 4025; Telefax: 33 7583 8547; Telex: 346740)

Dasco Sales Inc.
60 Sunfield Ave., Staten Island, NY 10312 USA (Tel.: 718-984-3600; Telefax: 718-984-5333; Telex: 275757 Dasco UR)

Dawn Foods
Worcester Rd., Evesham, Worcs, WR11 4QU UK (Tel.: 44 1386 41241/3; Telefax: 44 1386 443608; Telex: 338019)

Dawn Foods (UK) Ltd., 27 Perivale Ind. Park, Horsenden Lane South, Greenford, Middx, UB6 7RJ UK (Tel.: 44 181 566 7101; Telefax: 44 181 566 7012; Telex: 338019)

DCL Yeast Limited
Salatin House, 19 Cedar Rd., Sutton, Surrey, SM2 5JG UK (Tel.: 44 181 643 1818; Telefax: 44 181 643 6454; Telex: 8813507)

De Choix Speciality Foods Co., Div. of Amazon Coffee & Tea Co.
58-25 52nd Ave., Woodside, NY 11377 USA (Tel.: 718-507-8080)

Defiance Milling
Cnr Torrence & Wellington, Wooloowin, QLD, 4030 Australia (Tel.: 61 7 864 7133; Telefax: 61 7 857 2115)

Degussa

Degussa AG, Postfach 1345, D-6450 Hanau 1 Germany (Tel.: 6181-59-3983; Telefax: 6181-59-4309; Telex: 415200-25 dwd)

Degussa Ltd., Div. of Degussa AG, Earl Rd, Stanley Green, Handforth, Wilmslow, Cheshire, SK9 3RL UK (Tel.: 44 161-486 6211; Telefax: 44 161-485 6445; Telex: 51665053 DGMCHR G)

Degussa Corp., Wholly owned subsid. of Degussa AG, 65 Challenger Rd., Ridgefield Park, NJ 07660 USA (Tel.: 201-641-6100; Telefax: 201-807-3182; Telex: 221420 degus ur)

Dena AG

Fleher Dieich 3, PO Box 200 260, 4000 Dusseldorf Germany (Tel.: 49 211 15 52 15; Telefax: 49 211 15 60 75; Telex: 8581367)

Descours

St. Barthelemy Le Meil, 07160 Le Cheylard France (Tel.: 33 7529 0144; Telefax: 33 7529 3991; Telex: 345296)

Destilaciones Bordas Chincurreta SA

Apartado 11, 41008 Sevilla Spain (Tel.: 34 5 441 9000; Telefax: 34 5 441 7152; Telex: 72111 72811)

Deutsche Gelatine-Fabriken Stoess AG

Gammelsbacherstr 2, D69402 Eberbach Germany (Tel.: 49 6271 8401; Telefax: 49 6271 842700; Telex: 466240)

Deutsche Nichimen GmbH

Wehrhahn Center, D4000 Dusseldorf Germany (Tel.: 49 211 3551 (278); Telefax: 49 211 362492; Telex: 466240)

De Zaan Inc.

1 Bridge Plaza North, Fort Lee, NJ 07024 USA (Tel.: 201-592-8388; Telefax: 201-592-9247; Telex: 134622)

Daicel (Europa) GmbH, Subsid of Daicel Chem. Industries Ltd.

Ost St. 22, 4000 Düsseldorf 1 Germany (Tel.: (211)369848; Telefax: (211)364429; Telex: (41)8588042 DCELD)

The Dial Corp., A Greyhound Dial Co.

2000 Aucutt Rd., Montgomery, IL 60538 USA (Tel.: 708-892-4381; 800-323-5385)

Diamalt GmbH

Georg-Reismuller-Strasse 32, D80972 500270 Munchen Germany (Tel.: 49 89-81060; Telefax: 49 89-8106513; Telex: 523835)

Dierberger Oleos Essencias S.A.

Sao Paulo, Brazil

Dietary Foods Ltd.

Cumberland House, Brook St., Soham, Ely, Cambs, CB7 5BA UK (Tel.: 44 1353 720791; Telefax: 44 1353 721705; Telex: 817612)

Dietrich's Milk Products Inc./Specialty Prods. Div.

100 McKinley Ave., PO Box 2177, Reading, PA 19605 USA (Tel.: 800-526-8458; Telefax: 215-921-9330)

Dinesen Trading Co.

1809 Shalom Dr., West Bend, WI 53095 USA (Tel.: 414-338-8629; Telefax: 414-375-9120)

Dinoval Chemicals Ltd. *See* Pointing Ltd.

Diversitech Inc.

2411 NW 41st St., Gainesville, FL 32606 USA (Tel.: 904-377-7071; Telefax: 904-377-7073)

DMV

DMV International, NCB-Laan 80, 5462 GE Veghel The Netherlands (Tel.: 31 4130 72222; Telefax: 31 4130 43695; Telex: 74650)

DMV USA, Box 1628, La Crosse, WI 54602 USA (Tel.: 608-781-2345; Telefax: 608-781-3299; Telex: 74650)

Dohler

Dohler GmbH, Reidstrasse 7-9, D-6100 Darmstadt Germany (Tel.: 49 6151 3060; Telefax: 49 6151 306278; Telex: 419545)

Dohler (UK) Ltd., 4 Vincent Ave., Crownhill Business Centre, Milton Keynes, Bucks, MK8 0AB, UK

Dohler America Inc., 561 Jersey Ave., New Brunswick, NJ 08901 USA (Tel.: 908-247-2555; Telefax: 908-246-4210)

Dominion Products Inc.

882 Third Ave., Brooklyn, NY 11232 USA (Tel.: 718-499-3050; Telefax: 718-768-3978; Telex: 425707)

Domo Food Ingredients

Domo Food Ingredients, De Perk 30, 9411 PZ Beilen The Netherlands (Tel.: 31 5930 37171; Telefax: 31 5930 37275; Telex: 53109)

Domo Food Ingredients (UK) Ltd., PO Box 22, South Wirral, L64 6UE UK (Tel.: 44 151 643 9181; Telefax: 44 151 643 1010)

Dow

Dow Chemical U.S.A., 2020 Willard H. Dow Center, Midland, MI 48674 USA (Tel.: 517-636-1000; 800-441-4DOW; Telex: 227455)

Dow Chemical Europe S.A., Bachtobelstrasse 3, CH-8810 Horgen Switzerland (Tel.: 1-728-2111; Telefax: 1-728-2935; Telex: 826940)

Dow Corning Corp.

2200 W. Salzburg Rd., Midland, MI 48686-0994 USA (Tel.: 517-496-6000; 800-248-2481; Telefax: 517-496-5324; Telex: 227450)

PO Box 1593, Midland, MI 48641-1593 USA (Tel.: 800-362-6373; Telefax: 517-46 4821)

Dragoco

Dragoco Gerberding & Co. GmbH, D37601 Holzminden Germany (Tel.: 49 5531 970; Telefax: 49 5531 971391; Telex: 965336)

Dragoco (GB) Ltd., Lady Lane Ind. Estate, Hadleigh, Ipswich, Suffolk, IP7 6AX UK (Tel.: 44 1473 822011; Telefax. 44 1473 024020; Telex: 90420)

Dragoco Inc., 10 Gordon Dr., Totowa, NJ 07512 USA (Tel.: 201-256-3850; Telefax: 201-256-6420; Telex: 130449)

Dragoco Australia Pty. Ltd., 168 South Creek Rd., Dee Why West, NSW, 2099 Australia (Tel.: 61 2 982 7800; Telefax: 61 2 981 2536)

Drew Industrial Div. *See* Ashland

DuCrocq Aromatics International BV

PO Box 6, 1270 AA Huizen (NH) The Netherlands (Tel.: 31 2152 53742; Telefax: 31 2152 53741; Telex: 43369)

J.C. Dudley & Co. Ltd.

Cheyney House, Francis Yard, East St., Chesham, Bucks, HP5 1DG UK (Tel.: 44 1494 792839; Telefax: 44 1494 792875)

DuPont Chemicals

1007 Market St., Wilmington, DE 19898 USA (Tel.: 302-774-2099; 800-441-9442; Telefax: 302-773-4181; Telex: 302-774-7573)

Durkee Industrial Foods Corp.

925 Euclid Ave., Cleveland, OH 44115 USA (Tel.: 216-344-8482; Telefax: 216-344-8485; Telex: 196073)

Dutch Protein & Services BV

Sir Rowland Hillstraat 3, PO Box 6181, 4000 JT Tiel The Netherlands (Tel.: 31 3440 23400; Telefax: 31 3440 24784)

DynaGel Inc.

Wentworth Ave. & Plummer St., Calumet City, IL 60409 USA (Tel.: 708-891-8400; Telefax: 708-891-8432; Telex: 211666)

Eastman

Eastman Chemical Products, Inc., PO Box 431, Kingsport, TN 37662 USA (Tel.: 615-229-2318; 800-EASTMAN; Telefax: 615-229-1196; Telex: 6715569)

Eastman Chemical International AG, Hertizentrum 6, 3263 Zug Switzerland (Tel.: 41 42 23 25 25; Telefax: 41 42 21 12 52; Telex: 868 824)

Eastman Chemical International Ltd., 11 Spring St., Chatswood, NSW, 2067 Australia (Tel.: 61 2 411 3399; Telefax: 61 2 411 6430)

ECC International

5775 Peachtree-Dunwoody Rd. NE, Suite 200G, Atlanta, GA 30342 USA (Tel.: 404-843-1551; 800-843-3222; Telefax: 404-303-4384; Telex: 6827225)

ECC International/Calcium Prods., 5775 Peachtree-Dunwoody Rd, Suite 200G, Atlanta, GA 30342 USA (Tel.: 404-843-1551; 800-251-6327; Telefax: 404-843-8872)

Ecological Chemical Products Co. (Ecochem)

PO Box 299, Adell, WI 53001 USA (Tel.: 414-994-4969; Telefax: 414-994-9103)

The Edlong Company Ltd.

7 Anson Rd., Martlesham Heath, Ipswich, IP5 7RG UK (Tel.: 44 1473 624384/624407; Telefax: 44 1473 626336; Telex: 988864)

Edme Ltd.

Mistley, Manningtree, Essex, CO11 1HG UK (Tel.: 44 1206 393725; Telefax: 44 1206 396699; Telex: 987462)

Egg Processors & Distributors Ltd.

Unit 2, Sovereign Business Park, Bontoft Ave., National Ave., Hull, N. Humberside, HU5 4HF UK (Tel.: 44 1482 449604; Telefax: 44 1482 449612)

Eisai

Eisai Co., Ltd., 6-10, Koishikawa 4-chome, Bunkyo-ku, Tokyo, 112-88 Japan (Tel.: (03) 3817-5015; Telefax: (03) 3811-3305; Telex: J 28859 EISAI TOK J)

Eisai USA, Inc., Glenpointe Centre East, 300 Frank W. Burr Blvd., Teaneck, NJ 07666-6741 USA (Tel.: 201-692-0999; Telefax: 201-692-1972)

Eisai Pharma-Chem Europe Ltd., Commonwealth House, Hammersmith Int'l. Center, 2 Chalkhill Rd., London, W6 8DW UK (Tel.: 44 181 741-1330; Telefax: 44 181 741-5684)

Ellis & Everard Food & Personal Care

Caspian House, East Parade, Bradford, BD1 5EP UK (Tel.: 44 1274 377000; Telefax: 44 1274 725333; Telex: 517464)

Emil Flachsmann AG

Butzenstrasse 60, CH-8038 Zurich Switzerland (Tel.: 41 1 482 1555; Telefax: 41 1 482 34 44; Telex: 815218)

EM Industries, Inc./Fine Chems. Div.

5 Skyline Drive, Hawthorne, NY 10532 USA (Tel.: 914-592-4350; Telefax: 914-592-9469)

Enco Products (London) Ltd.

71-75 Fortess Rd., London, NW5 1AU UK (Tel.: 44 171 485 2217; Telex: 28241)

Enthoven BV

PO Box 4, 2120 AA Bennebroek The Netherlands (Tel.: 31 2502 46741; Telefax: 31 2502 48883; Telex: 41057)

Enzyme Bio-Systems Ltd.

2600 Kennedy Dr., Beloit, WI 53511-3969 USA (Tel.: 608-363-6423; Telefax: 608-365-4526)

Enzyme Development Corp.

2 Penn Plaza, Ste 2439, New York, NY 10121 USA (Tel.: 212-736-1580; Telefax: 212-279-0056; Telex: 427471 BSCE)

EOI Pty. Ltd.

74 Edinburgh Rd., Marrickville, NSW, 2204 Australia (Tel.: 61 2 517 6555; Telefax: 61 2 519 2782)

EPI Bretagne

41 rue des Hauts Chemins, F22360 Langueux France (Tel.: 33 96 627660; Telefax: 33 96 727335)

Erie Foods International Inc.

401 7th Ave., PO Box 648, Erie, IL 61250 USA (Tel.: 309-659-2233; Telefax: 309-659-7270; Telex: 671-1864)

Euroma Food Ingredients

PO Box 4, 8190 AA Wapenveld The Netherlands (Tel.: 31 5206 73550; Telefax: 31 5206 73195)

Eurovanillin

Postfach 330, N-1701 Sarpsborg Norway (Tel.: 47 9 118000; Telefax: 47 9 118640)

Evans Chemetics. *See* Grace

Evernat

BP 4, Z A Les Agreous, 40550 Leon France (Tel.: 33 58 48 7008; Telefax: 33 58 48 7276; Telex: 549276)

Excelpro Inc.

3760 E. 26th St., Los Angeles, CA 90023 USA (Tel.: 213-268-1918; Telefax: 213-268-1993; Telex: 857130)

Exquim SA

Joan Buscala 10, 08190 Saint Cugat de Valles, Barcelona Spain (Tel.: 34 3 589 4622; Telefax: 34 3 589 4502; Telex: 93522)

Extract Oil SA

Finca La Almazara, Santa Ana, 30319 Cartagena Spain (Tel.: 34 68 169254; Telefax: 34 68 169046; Telex: 67222)

Extractos Andinos CA

El Tablon s/n y Av. Maldonado, PO Box 17-19-007-M, Quito, Ecuador (Tel.: 593 2 6131101; Telefax: 593 2 613401; Telex: 22513)

Fairmount Chemical Co., Inc.

117 Blanchard St., Newark, NJ 07105 USA (Tel.: 201-344-5790; 800-872-9999; Telefax: 201-690-5298; Telex: 138905)

Fanning Corp., The

2450 W. Hubbard St., Chicago, IL 60612 USA (Tel.: 312-248-5700; Telefax: 312-248-6810; Telex: 910-221-1335)

Farbest

Farbest Brands, 160 Summit Ave., Montvale, NJ 07645 USA (Tel.: 201-573-4900; Telefax: 201-573-0404)

Farbest-Tallman Foods Corp., 160 Summit Ave., Montvale, NJ 07645 USA (Tel.: 201-573-4900; Telefax: 201-573-0404)

Farmers Rice Cooperative

2525 Natomas Park Dr., Sacramento, CA 95851 USA (Tel.: 916-923-5100; Telefax: 916-920-4295)

Feinkost Ingredient Co., USA

PO Box 1415, Medina, OH 44258 USA (Tel.: 216-948-3006; Telefax: 216-948-3016)

Feinkost Ingredient Co., 103 Billman St., Lodi, OH 44254 USA (Tel.: 216-948-3006; Telefax: 216-948-3016)

Fermex

Fermex International Ltd., E3 Blackpole Trading Estate, (East), Worcester, Worcs, WR3 8SG UK (Tel.: 44 1905 755811; Telefax: 44 1905 754145)

Fermex Australia Pty. Ltd., 1 Ferndell St., Granville, NSW, 2142 Australia (Tel.: 61 2 632 2222; Telefax: 61 2 632 1784)

FFS Flavour & Fragrance Specialties

530 Commerce St., Franklin Lakes, NJ 07147 USA (Tel.: 201-337-8330; Telefax: 201-337-7058)

Fiber Sales & Development Corp.
PO Box 885, Green Brook, NJ 08812-0885 USA (Tel.: 908-968-5024; Telefax: 908-968-5117)

Fibretex Ltd.
16 Dover St., London, W1X 3PB UK (Tel.: 44 171 499 1246; Telefax: 44 171 408 0112; Telex: 261454)

Fibrex
Fibrex AB, S-20504 Malmö Sweden (Tel.: 46 40 53 70 40; Telefax: 46 40 43 21 90)

Norba (UK) Ltd., Distributor for Fibrex AB, Chiltern Lodge, Windsor Lane, Little Kingshill, Great Missenden, Bucks, HP16 0DL UK (Tel.: 010944-494 865004; Telefax: 010944-494 868734)

Fibrisol Service Ltd.
Colville Rd., Acton, London, W3 8TE UK (Tel.: 44 181 993 6291; Telefax: 44 181 993 1033)

Fiddes Paynes Ltd.
The Spice Warehouse, Pepper Alley, Banbury, Oxon, OX16 8JB UK (Tel.: 44 1295 253888; Telefax: 44 1295 269166; Telex: 387333)

Figli Di Guido Lapi SpA
Via Lucchese 164, 500053 Empoli Italy (Tel.: 39 571 581202; Telefax: 39 571 581435; Telex: 575015)

Fina
Fina Chemicals, Div. of Petrofina SA, Nijverheldsstraat, 52 Rue de l'Industrie, B-1040 Brussels Belgium (Tel.: 32 2-2883391; Telefax: 32-2-288-3388; Telex: 21 556 PFINA B)

Fina plc, Fina House, 1 Ashley Ave, Epsom, Surrey, KT18 5AD UK (Tel.: 44 1372-726226; Telefax: 44 1372-744520; Telex: 894317)

Fina Oil and Chemical Co.
8350 North Central Expressway, PO Box 2159, Dallas, TX 75221 USA (Tel.: 214-750-2400; 800-344-FINA)

Finnsugar Bioproducts, Inc.
1400 N. Meacham Rd., Schaumburg, IL 60173-4808 USA (Tel.: 312-843-3200; 800-626-5363; Telefax: 312-843-3368)

Firmenich, Inc.
PO Box 5880, Princeton, NJ 08543-5880 USA (Tel.: 609-452-1000; 800-257-9591; Telefax: 609-921-0719; Telex: 21-99-15)

G. Fiske & Co Ltd.
64 Sheen Rd., Richmond, Surrey, TW9 1UF UK (Tel.: 44 181-948 5811; Telefax: 44 181-948 7059; Telex: 925878)

Flavex Naturextrakte GmbH
PO Box 1140, D66775 Rehlingen Germany (Tel.: 49 6835 4004; Telefax: 49 6835 4005)

Flavo House Inc.
Box 98, San Dimas, CA 91773 USA (Tel.: 909-599-7967; Telefax: 909-599-3517)

Flavor Consortium
2017 Camfield Ave., Commerce, CA 90040 USA (Tel.: 213-724-1010; Telefax: 213-724-3183)

Flavors of North America Inc.
303 Northfield Rd., Northfield, IL 60093 USA (Tel.: 708-441-9740; Telefax: 708-441-9750)

Flavor Specialties Inc.
1521 Commerce St., Corona, CA 91720 USA (Tel.: 909-734-6620; Telefax: 909-734-4214)

Flavours & Essences Private Ltd.
10th KM Hunsur Rd., Belavadi, Mysore, Kamataka, 571 186 India (Tel.: 91 821 42351; Telefax: 91 821 25027; Telex: 846 254)

Fleischmann-Kurth Malting co., Inc. *See* ADM

James Fleming & Co. Ltd.
Newbridge Ind. Estate, Newbridge, Edinburgh, EH28 2PA UK (Tel.: 44 131 333 2323; Telefax: 44 131 333 3991; Telex: 846 254)

Florasynth

Florasynth, Inc., 300 North St., Teterboro, NJ 07608 USA (Tel.: 201-288-3200; Telefax: 201-288-0843, Telex:)

Florasynth Ltd., Wentworth House, 2/4 High St., Chalfont St. Peter, Bucks, SL9 9RA UK (Tel.: 44 1753 890505; Telefax: 44 1753 890370; Telex: 846036)

Florasynth France, 45 Blvd Marcel-Pagnol, F06130 Le Plan de Grasse France (Tel.: 33 93 09 3350; Telefax: 33 93 09 3399; Telex: 970273)

Florexco Inc.

25 Central Park West, Suite 4N, New York, NY 10023 USA (Tel.: 212-586-7588; Telefax: 212-246-7317)

Florida Food Prods., Inc./Aloe Div.

2231 W. Hwy. 44, PO Box 1300, Eustis, FL 32727-1300 USA (Tel.: 904-357-4141; 800-874-2331; Telefax: 904-483-3192)

Florida Treatt. *See* Treatt

Floridin Co.

PO Box 510, 1101 N. Madison St., Quincy, FL 32351-0510 USA (Tel.: 904-627-7688; 800-228-1131; Telefax: 904-875-4408; Telex: 4931835 FLOQYUI)

Florigin Ltd.

Wellington House, 2 Kentwood Hill, Tilehurst, Reading, RG3 6DE UK (Tel.: 44 1734 668850; Telefax: 44 1734 664022; Telex: 848197)

FMC

FMC Corp./Chemical Products Group, 1735 Market St., Philadelphia, PA 19103 USA (Tel.: 215-299-6000; 800-346-5101; Telefax: 215-299-6291; Telex: 685-1326)

FMC Corp./Food Ingredients Div., 1735 Market St., Philadelphia, PA 19103 USA (Tel.: 800-346-5101; Telefax: 215-299-6291; Telex: 6851326 FMC PHA)

FMC Corp./Food Phosphates Marketing, 1735 Market St., Philadelphia, PA 19103 USA (Tel.: 215-299-6884; Telefax: 215-299-6887)

FMC Corp./Marine Colloids Div., 1735 Market St., Philadelphia, PA 19103 USA (Tel.: 215-299-0242; 000-526-3649; Telefax: 215-299-6291; Telex: 6851326)

FMC Corp. N.V., Ave. Louise 480-B9, Brussels 1050 Belgium (Tel.: 322-645 5511; Telefax: 322-640 6350)

FMC International S.A., 4th Floor, Interbank Bldg., 111 Paseo de Roxas, Makati, Metro Manila Phillippines (Tel.: (632) 817 5546; Telefax: (632) 818 1485)

FMC Litex AS, Risingvej 1, D-2665 Vallensbaek Strand Denmark (Tel.: 45 42 73 1122; Telefax: 45 42 73 1225; Telex: 855-33194 DANGEL DK)

Folexco Inc.

25 Davis St., S. Plainfield, NJ 07080 USA (Tel.: 201-769-8400; Telefax: 201-769-8404)

Food Additives & Ingredients Inc.

222 Sherwood Ave., Farmingdale, NY 117535 USA (Tel.: 516-694-9090; Telefax: 516-694-9177; Telex: 6852289)

Food Ingredient Services

17a/21a Woolton St., Liverpool, L25 5NR UK (Tel.: 44 151 428 8732; Telefax: 44 151 421 0029)

Food Ingredient Specialities

PO Box 122, North Hyde Gardens, Hayes, Middlesex, UB3 4SL UK (Tel.: 44 181 569 2069; Telefax: 44 181 569 1633; Telex: 916807)

Food Ingredients Specialities SA

Case Postale 55, CH-1618 Chatel-St-Denis Switzerland (Tel.: 41 21 921 0771; Telefax: 41 21 921 5022)

Food Ingredients Technology Ltd.

Cunningham House, Westfield Lane, Kenton, Harrow, HA3 9ED UK (Tel.: 44 181 907 7278; Telefax: 44 181 909 1053; Telex: 8811603)

Foodmaker Ltd.

16 Princewood Rd., Earlstrees Ind. Estate, Corby, Northants, NN17 2AP UK (Tel.: 44 1536 400560; Telefax: 44 1536 400171)

Foodtech Ltd.
73/75 High St., Hornsey, London, N8 7QW UK (Tel.: 44 181 348 4545; Telefax: 44 181 348 2313; Telex: 24124)

The Foote & Jenkins Corp.
1420 Crestmont Ave., Camden, NJ 08103 USA (Tel.: 609-966-0700; Telefax: 609-966-6132)

Formula One
222 Kensal Rd., London, W10 5BN UK (Tel.: 44 181 969 6807; Telefax: 44 181 969 5337)

Forrester Wood Co. Ltd.
Hawksley Ind. Estate, Heron St., Hollinwood, Oldham, Lancs, OL8 4UJ UK (Tel.: 44 161 620 4124; Telefax: 44 161 627 1050)

Forum Chemicals Ltd.
Forum House, 41-51 Brighton Rd., Redhill, Surrey, RH1 6YS UK (Tel.: 44 1737 773711; Telefax: 44 1737 773116; Telex: 939065)

Framptons Ltd.
Charlton Rd., Shepton Mallet, Somerset, BA4 5PD UK (Tel.: 44 1749 342831; Telefax: 44 1749 344997)

Francexpa
3 rue Moncey, 75009 Paris France (Tel.: 33 1 4463 4826; Telefax: 33 1 4282 1562; Telex: 290448)

Franco-Pacific Food Products Co. Ltd.
44/24 Soi Asoke, Sukhumvit Rd., SO1 21, Bangkok, 10110 Thailand (Tel.: 66 259 4901; Telefax: 66 253 6841; Telex: 82108)

Frec Food
13630 Cimarron Ave., Gardena, CA 90249 USA (Tel.: 310-515-7800; Telefax: 310-515-1570)

Freeman Industries
100 Marbledale Rd., Tuckhoe, NY 10707 USA (Tel.: 914-961-2100; Telefax: 914-961-5793)

Freeman Foods Ltd.
Bourton Mill, Bridge St., Bourton, Gillingham, Dorset, SP8 5BA UK (Tel.: 44 1747 840302; Telefax: 44 1747 840940)

French's Ingredients, Div. of Reckitt & Colman Commercial
1800 Hammons Tower, 901 E. St. Louis, Springfield, MO 65806-2512 USA (Tel.: 417-837-1813; Telefax: 417-837-1801)

Friars Pride Ltd.
Oxney Rd. Ind. Estate, Peterborough, Cambs, PE1 5YW, UK)

Fries & Fries, Div. of Mallinckrodt
1199 Edison Dr., Cincinnati, OH 45216 USA (Tel.: 513-948-8000; 800-543-4643; Telex: 21-4348)

Frigova Produce Ltd.
31 Tooley St., London, SE1 2RY UK (Tel.: 44 171 407 7701; Telefax: 44 171 407 6476; Telex: 888464)

Fruitsource/Treasure Island
1672 Funston, San Francisco, CA 94122 USA (Tel.: 415-665-3553; Telefax: 415-665-2217)

Frutarom
Frutarom USA, 24 Spielman Rd., Fairfield, NJ 07004 USA (Tel.: 201-227-5758; Telefax: 201-808-3948)

Frutarom (UK) Ltd., Northbridge Works, Northbridge Rd., Berkhamsted, Herts, HP4 1EF UK (Tel.: 44 1442 876611; Telefax: 44 1442 876204)

Frutarom Ltd., 25 Hashaish St., 10067, Haifa, 26110 Israel (Tel.: 972 4 725123; Telefax: 972 4 722517; Telex: 45125)

Fuchs Gewürze do Brasil Ltd.
Av das Nacoes, 21 735 Jurubatuba CEP, 07495-100 Sao Paulo-Sp, Brazil)

Fuchs GmbH & Co.
Osterfeldstrabe 2-8, D-4520 Melle 9 Germany (Tel.: 49 5429 2090; Telefax: 49 5429 2290; Telex: 94372)

Fuji Purina Protein, Ltd. *See* Protein Technologies International

Gallard-Schlesinger Industries, Inc.
584 Mineola Ave., Carle Place, NY 11514-1712 USA (Tel.: 516-333-5600; Telefax: 516-333-5628; Telex: 6852390 (WUI))

Galloway-West Co.
325 Tompkins St., Fond Du Lac, WI 54935 USA (Tel.: 414-922-0600; Telefax: 414-922-4702)

Gamay Flavors
2770 S. 171st St., New Berlin, WI 53151 USA (Tel.: 414-789-5104; Telefax: 414-789-5149)

Garuda International Inc.
PO Box 5155, Santa Cruz, CA 95063 USA (Tel.: 408-462-6341; Telefax: 408-462-6355; Telex: 296614 Garuda UR)

Gaspo For Trade & Marketing
50 Youssef Abbas Nasr-City, PO Box 8185, Nasr-City, Cairo Egypt (Tel.: 20 2 2635096; Telefax: 20 2 2635096; Telex: 20682)

Gattefosse
Gattefosse SA, 36 Chemin de Genas, BP 603, F 69804 Saint Priest France (Tel.: 72 22 98 00; Telefax: 78 90 45 67; Telex: 340 240 F)

Gattefosse Corp., 189 Kinderkamack Rd., Westwood, NJ 07675 USA (Tel.: 201-573-1700; Telefax: 201-573-9671)

Gelatine Products Ltd.
Sutton Weaver, Runcorn, Cheshire, WA7 3EH UK (Tel.: 44 1928 716444; Telefax: 44 1928 718325; Telex: 629303)

Gel'Dor SA
Chemin Lebon, Domaine des Terriers, Antibes 06600 France (Tel.: 33 93 33 11 13; Telefax: 33 93 33 27 50; Telex: 470069)

Gemet Industries
44 rue du Louvre, 75001 Paris France (Tel.: 33 1 4508 0717; Telefax: 33 1 4221 0371)

Genarom International Inc.
238 St. Nicholas Ave., So. Plainfield, NJ 07080 USA (Tel.: 201-753-9100; Telefax: 201-753-9635)

Genencor
Sandford Lane, Hurst, Reading, Berks, RG10 0SU UK (Tel.: 44 1734 342328; Telefax: 44 1734 345113)

Genencor International
Genencor Int'l. Inc., 4010 Winnetka Ave., Rolling Meadows, IL 60008 USA (Tel.: 708-870-1030; 800-847-5311; Telefax: 708-870-9977)

Genencor International, Kyllikinportti 2, PO Box 105, SF-002421 Helsinki Finland (Tel.: 358 0 134411; Telefax: 358 0 13441319; Telex: 125737)

General Chemical Corp.
90 East Halsey Rd., PO Box 393, Parsippany, NJ 07054-0393 USA (Tel.: 201-515-0900; 800-631-8050; Telefax: 201-515-4461; Telex: 362262)

General Electric Co./Silicone Products Div.
260 Hudson River Rd., Waterford, NY 12188 USA (Tel.: 518-237-3330; 800-255-8886; Telefax: 518-233-3931)

Geneva Ingredients
2228 Evergreen Rd., Middleton, WI 53562 USA (Tel.: 608-836-7777; Telefax: 608-836-4030)

Ernest George Ltd.
459 Bath Rd., Slough, Berks, SL1 6AA UK (Tel.: 44 16286 67331; Telefax: 44 16286 67052; Telex: 846403)

Germantown Manufacturing Co.
505 Parkway, Box 405, Broomall, PA 19008 USA (Tel.: 215-544-8400; Telefax: 215-544-4490)

Gist-brocades

Gist-brocades, 1 Wateringseweg, PO Box 1, Delft 2600 MA The Netherlands (Tel.: 31 15 793005; Telefax: 31 15 793408; Telex: 38103)

Gist-brocades Food Ingredients, Inc., 2200 Renaissance Blvd., Suite 150, King of Prussia, PA 19406 USA (Tel.: 215-272-4040; 800-662-4478; Telefax: 215-272-5695; Telex: 216902)

Gist-brocades Savoury BV, PO Box 1195, Zaandam 1500 AD The Netherlands (Tel.: 31 75 700041; Telefax: 31 75 161559; Telex: 19115)

Edward Gittens Ltd.

Firwood Works, Firwood Fold, Off Thicketford Rd., Bolton, Lancs, BL2 3AG UK (Tel.: 44 1204 309091; Telefax: 44 1204 309119; Telex: 38103)

Giulini Corp.

105 East Union Ave., Bound Brook, NJ 08805 USA (Tel.: 908-469-6504; Telefax: 908-469-8418; Telex: 700179)

Givaudan-Roure

Givaudan-Roure Corp., 100 Delawanna Ave., Clifton, NJ 07014 USA (Tel.: 201-365-8000; Telefax: 201-777-9304; Telex: 219259 givc ur)

Givaudan-Roure SA, 5 Chemin de la Perfumiere, CH-1214 Vernier-Geneva Switzerland (Tel.: 22-780 91 11; Telefax: 22-780 91 50)

Glidco Organics

PO Box 389, Jacksonville, FL 32201 USA (Tel.: 904-768-5800; Telefax: 904-768-2200; Telex: 441763)

Glorybee Natural Sweeteners Inc.

120 N. Seneca, PO Box 2744, Eugene, OR 97402 USA (Tel.: 503-689-0913)

GMI Products Inc.

2525 Davie Rd., Suite 330, Davie, FL 33317 USA (Tel.: 305-474-9608; Telefax: 305-474-0989)

Goldensun Manufacturing Co.

Ranvir Bldg., 70 Princess St., Bombay, 400 002 India (Tel.: 91 22 205 2745; Telefax: 91 22 201 8054)

Goldschmidt

Goldschmidt AG, Th., Goldschmidtstrasse 100, Postfach 101461, D-4300 Essen 1 Germany (Tel.: 0201-173-01; Telefax: 201-173-2160; Telex: 857170)

Goldschmidt Chemical Corp., 914 E. Randolph Rd., PO Box 1299, Hopewell, VA 23860 USA (Tel.: 804-541-8658; 800-446-1809; Telefax: 804-541-8689; Telex: 710-958-1350)

Good Food Inc.

W. Main St., PO Box 160, Honey Brook, PA 19344 USA (Tel.: 215-273-3776; Telefax: 215-273-2087)

Goorden NV

Melkerijstraat 1, 2900 Schoten Belgium (Tel.: 32 33 243835; Telefax: 32 33 243838)

Gourmet Food Institute A/S

2 Hammervej, DK-2970 Hoersholm Denmark (Tel.: 45 42 866444; Telefax: 45 45 767680; Telex: 37275)

W.R. Grace

W.R. Grace/Davison Chemical Div., PO Box 2117, Baltimore, MD 21203-2117 USA (Tel.: 301-659-9000; Telefax: 410-659-9213; Telex: 192 814 013)

Evans Chemetics Div., 55 Hayden Ave., Lexington, MA 02173 USA (Tel.: 617-861-9700; Telefax: 617-863-8070; Telex: 200076 GRLX UR)

Grain Processing Corp.

1600 Oregon St., Muscatine, IA 52761 USA (Tel.: 319-264-4265; Telefax: 319-264-4289; Telex: 46-8497)

Grayslake Gelatin

PO Box 248, Grayslake, IL 60030 USA (Tel.: 708-223-8141; Telefax: 708-223-8144; Telex: 37275)

Great Lakes Chemical

Great Lakes Chemical Corp., PO Box 2200, W. Lafayette, IN 47906-0200 USA (Tel.: 317-497-6100; 800-428-7947; Telefax: 317-497-6234)

Great Lakes Chemical (Europe) Ltd., P O Box 44, Oil Sites Road, Ellesmere Port, South Wirral, L65 4GD UK (Tel.: 44 151-356 8489; Telefax: 44 151-356 8490)

Great Lakes-QO Chemicals Europe Sarl, 238 route de l'Empereur, F-92500 Rueil Malmaison France (Tel.: 47 32 06 22; Telefax: 47 32 08 88; Telex: 204555)

K & K Greeff Chemicals Ltd., Div. of Beijer Industries AB
Suffolk House, George Streeet, Croydon, Surrey, CR9 3QL UK (Tel.: 44 181-686 0544; Telefax: 44 181 686 4792; Telex: 28386)

Griffith Laboratories
Griffith Laboratories Ltd., Cotes Park, Somercotes, Derbyshire, DE5 4NN UK (Tel.: 44 1773 832171; Telefax: 44 1773 835294; Telex: 37275)

Griffith Laboratories/European Group Office, 58 Rue de Courcelles, 75008 Paris France (Tel.: 33 1 422 57172; Telefax: 33 1 456 38391)

Griffith Laboratories Pty. Ltd., 1 Griffith St., Scoresby, Victoria, 3179 Australia (Tel.: 61 3 763 5300; Telefax: 61 3 763 4887)

Grindsted
Grindsted Products A/S, Edwin Rahrs Vej 38, DK-8220 Brabrand Denmark (Tel.: 45 86-25-3366; Telefax: 45 86-25-1077; Telex: 64177 gvdan dk)

Grindsted Products, Ltd., Northern Way, Bury St. Edmunds, Suffolk, IP32 6NP UK (Tel.: 44 1284 769631; Telefax: 44 1284 760839; Telex: 81203)

Grindsted Products, Inc., 201 Industrial Pkwy., PO Box 26, Industrial Airport, KS 66031 USA (Tel.: 913-764-8100; 800-255-6837; Telefax: 913-764-5407; Telex: 4-37295)

Grindsted Products, Inc., 10 Carlson Court, Suite 580, Rexdale, Ontario, M9W 6L2 Canada (Tel.: 416-674-7340; Telefax: 416-674-7378)

Grindsted de México, S.A. de C.V., Cerrada de las Granjas 623, Col. Jagüey, Delegación Azcapotzalco, 02300 México, D.F. Mexico (Tel.: (5) 352 9102; Telefax: (5) 561 3285)

Nippon Grindsted K.K., Daiichi Nishiwaki Bldg., 1-58-10 Yoyogi, Shibuya-ku, Tokyo, J-151 Japan (Tel.: 3-3375-3481; Telefax: 3-3375-3715)

Grünau GmbH, Chemische Fabrik, A Henkel Group Co.
Postfach 1003, Robert-Hansen-Strasse 1, D-09251 Jllertissen Germany (Tel.: (07303)13-700; Telefax: (07303)13203; Telex: 719114 gruea-d)

Gumix Handels GmbH
Neuer Wall 36, 2000 Hamburg 36 Germany (Tel.: 49 40 362941; Telefax: 49 40 363613; Telex: 215616)

Gumix International Inc.
2160 N. Central Rd., Fort Lee, NJ 07024-7552 USA (Tel.: 201-947-6300; 800-2GU-MIX2; Telefax: 201-947-9265; Telex: 134227)

Gum Technology
PO Box 356, Sta. A, Flushing, NY 11358 USA (Tel.: 718-961-3838; Telefax: 718-961-7297)

F. Gutkind & Co. Ltd.
Chancery House, 53/64 Chancery Lane, London, WC2A 1QX UK (Tel.: 44 171 242 7642; Telefax: 44 171 242 7271; Telex: 24465)

Haarmann & Reimer
Haarmann & Reimer GmbH, Postfach 1253, D-3450 Holzminden Germany (Tel.: 49 5531 7011; Telefax: 49 5531 7016 49; Telex: 965 330)

Haarmann & Reimer Ltd., Fieldhouse Lane, Marlow, Buckinghamshire, SL7 1NA UK (Tel.: 44 1628 472051; Telefax: 44 1628 890795; Telex: 848 859)

Haarmann & Reimer Ltd./Food Ingredients Business Group, Div. of Bayer UK Ltd., Denison Rd, Selby, North Yorkshire, YO8 8EF UK (Tel.: 44 1757 703691; Telefax: 44 1757 701468; Telex: 57852)

Haarmann & Reimer Corp., PO Box 175, 70 Diamond Road, Springfield, NJ 07081 USA (Tel.: 201-467-5600; 800-422-1559; Telefax: 201-912-0499; Telex: 219134 HAR UR)

Haarmann & Reimer Corp./Food Ingreds. Div., A Miles Inc. Company, 1127 Myrtle St., PO Box 932, Elkhart, IN 46515 USA (Tel.: 219-264-8716; 800-348-7414; Telefax: 219-262-6747)

Haarmann & Reimer Australia Pty. Ltd., 9 Garling Rd., Marayong, NSW, 2148 Australia (Tel.: 61 2 671 3444; Telefax: 61 2 621 8086)

Haas Foods Inc.
4550 SW Kruse Way, Suite 305, PO Box 1890, Lake Oswego, OR 97035 USA (FAX 503-636-6434)

Haco Ltd.
3073 Gumligen Switzerland (Tel.: 41 31 950 1111; Telefax: 41 31 951 5050; Telex: 911848)

Hagelin & Co. Inc.
200 Melster Ave., Branchburg, NJ 08876 USA (Tel.: 908-707-4400; Telefax: 908-707-4408)

G.C. Hahn & Co. Ltd.
Broncoed Ind. Park, Wrexham Rd., Mold, Clwyd, CH4 7LR UK (Tel.: 44 1352 758055; Telefax: 44 1352 758086)

Haifa Chemicals Ltd.
PO Box 1809, 31018 Haifa Israel (Tel.: 972 4 469611; Telefax: 972 4 457849; Telex: 45118)

Halcyon Proteins Pty. Ltd.
25 George St., Sandringham, Victoria, 3191 Australia (Tel.: 61 3 598 7900; Telefax: 61 3 597 0842)

Haldin International Inc.
10 Reuten Dr., Closter, NJ 07624 USA (Tel.: 201-784-0044; Telefax: 201-784-2180; Telex: 402512)

Hallmark Foods
Rutland House, Great Chesterford Court, Great Chesterford, Saffron Walden, Essex, CB10 1PF UK (Tel.: 44 1799 31375; Telefax: 44 1799 31356)

M. Hamburger & Sons Ltd.
Tannery House, Tannery Lane, Send, Woking, Surrey, GU23 7HB UK (Tel.: 44 1483 223501; Telefax: 44 1483 224403; Telex: 859296)

Chr. Hansen's Lab
Chr. Hansen's Laboratorium Danmark A/S, 10-12 Boge Allé, DK-2970 Horsholm Denmark (Tel.: 45 45 76 76 76; Telefax: 45 45 76 08 48; Telex: 19184)

Chr Hansen's Lab (Aust), 3/7-9 Newcastle Rd., Bayswater, Victoria, 3153 Australia (Tel.: 61 3 720 8022; Telefax: 61 3 720 7801)

Chr Hansen's Lab Inc., 9015 W. Maple St., Milwaukee, WI 53214 USA (Tel.: 414-476-3630; Telefax: 414-259-9399)

Harcros
Harcros Chemicals UK Ltd./Specialty Chemicals Div., Lankro House, PO Box 1, Eccles, Manchester, M30 0BH UK (Tel.: 44 161-789-7300; Telefax: 44 161-788-7886; Telex: 667725)

Harcros Chemicals Inc./Organics Div., 5200 Speaker Rd., PO Box 2930, Kansas City, KS 66106-1095 USA (Tel.: 913-321-3131; Telefax: 913-621-7718; Telex: 477266)

Harimex-Ligos BV
Kieveen 20, Loenen (Vel.) 7371 GD The Netherlands (Tel.: 31 5765 2351; Telefax: 31 5765 1930; Telex: 36443)

Keith Harris & Co. Ltd.
7 Sefton Rd., Pennant Hills, PO Box 147, Thornleigh, NSW, 2120 Australia (Tel.: 61 2 484 1341; Telefax: 61 2 481 8145; Telex: 26214)

Hartog Trading Corp., Hartog Foods International Inc.
20 E. 46th St., New York, NY 10071 USA (Tel.: 212-687-2000; Telefax: 212-687-2659)

T. Hasegawa Flavors
14017 E. 183rd St., Cerritos, CA 90701 USA (Tel.: 714-522-1900; Telefax: 714-522-6800)

A.C. Hatrick Chemicals Pty Ltd.
49-61 Stephen Road, Botany Bay, Botany, NSW, 2019 Australia (Tel.: 61 2 666-9331; Telefax: 61 2 666 3872)

Hax Ltd.
306 Archway Rd., London, N6 5AU UK (Tel.: 44 181 341 1010; Telefax: 44 181 348 0504; Telex: 27292)

Hayashibara Co., Ltd.
2-3, Shimoishii 1-chome, Okayama, 700 Japan (Tel.: (086)224-4311; Telefax: (086)225-5630; Telex: 5922120 HYSBR J)

Hays Chemicals Distribution Ltd.
215 Tunnel Ave., East Greenwich, London, SE10 0QE UK (Tel.: 44 181 858 8631; Telex: 896173)

Healan Ingredients Ltd.
Limb Rd., Pocklington, York, YO4 2QP UK (Tel.: 44 1759 303303; Telefax: 44 1759 305509; Telex: 57478)

Hefti Ltd. Chemical Products
PO Box 1623, CH-8048 Zurich Switzerland (Tel.: 01-432-1340; Telefax: 01-432-2940; Telex: 822225 hexa ch)

Heico Chemicals, Inc., A Cambrex Co.
Route 611, PO Box 160, Delaware Water Gap, PA 18327-0160 USA (Tel.: 717-420-3900; 800-34-HEICO; Telefax: 717-421-9012)

Henkel
Henkel KGaA/Cospha, Postfach 101100, D-40191, Düsseldorf Germany (Tel.: 0211-797-1; Telefax: 0211-798-7696; Telex: 085817-0)

Henkel Organics, Henkel House, 292-308 Southbury Rd., Enfield, Middlesex, EN1 1TS UK (Tel.: 44 181 804 3343; Telefax: 44 181 443 4392; Telex: 922708)

Henkel Chemicals Ltd., Div. of Henkel KG, Henkel House, 292-308 Southbury Road, Enfield, Middlesex, EN1 1TS UK (Tel.: 44 181-804 3343; Telefax: 44 181 443 2777; Telex: 922708 HENKEL G)

Henkel Belgium NV, Div. of Henkel KG, 66 ave du Port, B-1210 Brussels Belgium (Tel.: 2-423 17 11; Telefax: 2-428 34 67; Telex: 21294 HENKEL B)

Henkel France SA, Div. of Henkel KG, BP 309, 150 rue Gallieni, F-92102 Boulogne Billancourt France (Tel.: 46 84 90 00; Telefax: 46 84 90 90; Telex: 633177 HENKEL F)

Henkel Ireland Ltd., Little Island, Co Cork, Ireland (Tel.: 353 21 354277; Telefax: 353 21 353559; Telex: 75839)

Henkel Corp./Cospha, 300 Brookside Ave., Ambler, PA 19002 USA (Tel.: 215-020-1470, 000-501-0015; Telefax: 215-628-1450; Telex: 125854)

Henkel Corp./Emery Group, 11501 Northlake Dr., Cincinnati, OH 45249 USA (Tel.: 513-530-7300; 800-543-7370; Telefax: 513-530-7581; Telex: 4333016)

Henkel Corp./Fine Chemicals Div./Grünau-Food Tech. Dept., 5325 South 9th Ave., La Grange, IL 60525-3602 USA (Tel.: 708-579-6157; 800-328-6199; Telefax: 312-579-6152)

Henkel Corp./Functional Products, 300 Brookside Ave., Ambler, PA 19002 USA (Tel.: 215-628-1583; 800-654-7588; Telefax: 215-628-1155)

Henkel Corp./Organic Products Div., 300 Brookside Ave., Ambler, PA 19022 USA (Tel.: 215-628-1000; 800-922-0605; Telefax: 215-628-1200; Telex: 6851092)

Henkel Canada Ltd., 2290 Argentia Rd., Mississauga, Ontario, L5N 6H9 Canada (Tel.: 416-542-7554; 800-668-6023; Telefax: 416-542-7588)

Henkel Argentina S.A., Avda. E. Madero Piso 14, 1106 Capital Federal, Argentina)

Henningsen Foods, Inc.
2 Corporate Park Drive, White Plains, NY 10604 USA (Tel.: 914-694-1000; Telefax: 914-694-1221; Telex: 221819 Heno UR)

Henningsen van den Burg
Sluisweg 20, 5145 PE Waalwijk The Netherlands (Tel.: 31 4160 37911; Telefax: 31 4160 35285; Telex: 35114)

Henry Broch & Co.
Three Hawthorn Pkwy., Suite 225, Vernon Hills, IL 60061 USA (Tel.: 708-816-6225; Telefax: 708-816-7415)

Henry Lamotte Import/Export. *See* Lamotte

Herbafood Nahrungsmittel GmbH
Industriestr. 1, D-7556 Otigheim Germany (Tel.: 49 7222 21085; Telefax: 49 722 21088; Telex: 722262)

Herbstreith & Fox KG/Pektin-Fabrik Neuenbürg

PO Box 1261, D-75302 Neuenbürg Germany (Tel.: 49 7082 7913-0; Telefax: 49 7082 20281; Telex: 7245019)

Hercules

Hercules Inc., Hercules Plaza-6205SW, Wilmington, DE 19894 USA (Tel.: 302-594-6500; 800-247-4372; Telefax: 302-594-5400; Telex: 835-479)

Hercules Food Ingredients, Div. of Hercules Inc., 1313 North Market St., PO Box 8740, Wilmington, DE 19899-8740 USA (Tel.: 302-594-5000; 800-654-6529; Telefax: 302-594-6660)

Hercules Inc./PFW Aroma Chemicals, 33 Sprague Ave., Middletown, NY 10940 USA (Tel.: 914-343-1900; Telefax: 914-343-8794; Telex: 283303)

Hercules Ltd., Div. of Hercules Inc., 31 London Road, Reigate, Surrey, RH2 9YA UK (Tel.: 44 1737-242434; Telefax: 44 1737-224288; Telex: 25803)

Hercules BV, 8 Veraartlaan, PO Box 5822, 2280 HV Rijswijk The Netherlands (Tel.: 070-150-000; Telex: 31172)

Hercules BV/Aqualon Div., Postbus 5832, NL-2280 HV Rijswijk The Netherlands (Tel.: 31 70 315 0226; Telefax: 31 70 390 7560)

Hershey Import Co. Inc.

700 E. Lincoln Ave., Rahway, NJ 07065 USA (Tel.: 201-388-9000)

Hesco Inc.

101 W. Kemp Ave., Watertown, SD 57201 USA (Tel.: 605-882-4672; Telefax: 605-882-4985)

Heterene Chemical Co., Inc.

PO Box 247, 795 Vreeland Ave., Paterson, NJ 07543 USA (Tel.: 201-278-2000; Telefax: 201-278-7512; Telex: 883358)

Hilton Davis Chemical Co., A Freedom Chemical Co.

2235 Langdon Farm Rd., Cincinnati, OH 45237 USA (Tel.: 513-841-4000; 800-477-1022; Telefax: 800-477-4565)

Hodgsons Dye Agencies Pty. Ltd.

56 Bay St., Broadway, NSW, 2007 Australia (Tel.: 61 2 211 4366; Telefax: 61 2 212 2151)

Hoechst

Hoechst AG, Entwicklung TH 1, D-6230 Frankfurt am Main 80 Germany (Tel.: 069-305-2298; Telefax: 069-318435; Telex: 6990936)

Hoechst AG, D-65926 Frankfurt/Main Germany (Tel.: 49 69 305-0; Telefax: 49 69 303665/66; Telex: 41234-0)

Hoechst Chemicals (UK) Ltd., Div. of Hoechst AG, Hoechst House, Salisbury Rd., Hounslow, Middlesex, TW4 6JH UK (Tel.: 44 181-570 7712; Telefax: 44 181-577 1854; Telex: 22284)

Hoechst Celanese/Int'l. Headqtrs., 26 Main St., Chatham, NJ 07928 USA (Tel.: 201-635-2600; 800-235-2637; Telefax: 201-635-4330; Telex: 136346)

Hoffmann-La Roche

Hoffmann-La Roche Inc., 340 Kingsland St., Nutley, NJ 07110 USA (Tel.: 201-909-8332; 800-526-0189; Telefax: 201-909-8414)

Hoffmann-La Roche SA, Grenzacherstrasse 124, CH-4002 Basle Switzerland (Tel.: 41 61-688 3451; Telefax: 41 61 688 1680; Telex: 962292 HLR CH)

Holland Company Inc.

Howland Ave., Adams, MA 01220, USA

Holland Sweetener

Holland Sweetener Co. VoF, PO Box 1201, 6201 BE Maastricht The Netherlands (Tel.: 31 43 21 2228; Telefax: 31 43 21 6633; Telex: 56384)

Holland Sweetener North America, Inc., 1100 Circle 75 Parkway, Suite 690, Atlanta, GA 30339-3097 USA (Tel.: 404-956-8443; 800-757-9648; Telefax: 404-956-7102)

Holsum Foods Co.

500 S. Prairie Ave., PO Box 218, Waukesha, WI 53187 USA (Tel.: 414-544-4444)

Honeymead Products Co.
2020 S. River Front Dr., PO Box 3247, Mankato, MN 56002 USA (Tel.: 507-625-7911)

Honeywill & Stein Ltd., Div. of BP Chemicals Ltd.
Times House, Throwley Way, Sutton, Surrey, SM1 4AF UK (Tel.: 44 181-770 7090; Telefax: 44 181-770 7295; Telex: 946560 BPCLGH G)

Honig Merkartikelen BV
Honig Industrie, Lagedijk 3, Koog aan de Zaan 1541 KA The Netherlands (Tel.: 31 75 539222; Telefax: 31 75 701369; Telex: 19102)

Hoogwegt US Inc.
1113 S. Milwaukee Ave., Suite 200, PO Box 459, Llbertyville, IL 60048-0459 USA (Tel.: 708-918-8787; Telefax: 708-918-9189; Telex: 402604)

Hormel & Co., Geo. A.
501 16th Ave. NE, Box 933, Austin, MN 55912 USA (Tel.: 507-437-5676; Telefax: 507-437-5120)

Huijbregts Groep
Vossenbeemd 107, PO Box 165, 5700 AD, 5705 CL Helmond The Netherlands (Tel.: 31 4920 41415; Telefax: 31 4920 50540; Telex: 51518)

Hulman & Co.
PO Box 150, Terre Haute, IN 47808 USA (Tel.: 812-232-9446)

Hüls
Hüls AG, Postfach 1320, D-4370 Marl 1 Germany (Tel.: 02365-49-1; Telefax: 02365-49-2000; Telex: 829211-0)

Hüls AG/Troisdorf Works, PO Box 1347, 5210 Troisdorf Germany (Tel.: 49 2241 85 4321; Telefax: 49 2241 85 4319)

Hüls (UK) Ltd., Featherstone Rd, Wolverton Mill South, Milton Keynes, Buckinghamshire, MK12 5TB UK (Tel.: 44 1908 226 444; Telefax: 44 1908 224 950; Telex: 826500)

Hüls America Inc., PO Box 365, 80 Centennial Ave., Piscataway, NJ 08855-0456 USA (Tel.: 908-980-6800; 800-631-5275; Telefax: 908-980-6970; Telex: 279977 huudex ui)

Humphrey Chemical Co Inc., A Cambrex Co.
Devine St., North Haven, CT 06473-0325 USA (Tel.: 203-230-4945; 800-652-3456; Telefax: 203-287-9197; Telex: 994487)

IAM Sarl
146 rue Paradis, F13294 Marseille Cedex 6 France (Tel.: 33 91 430477; Telefax: 33 91 271650)

N.I. Ibrahim Co.
8 Falaki St., Alexandria Egypt (Tel.: 20 3 4833923; Telex: 54317)

ICI
ICI plc, Imperial Chemical House, 9 Millbank, London, SW11 3JS UK (Tel.: 44 171 834 4444; Telefax: 44 171 834 2040; Telex: 21324)

ICI Surfactants Ltd. (UK), PO Box 90, Wilton, Middlesbrough, Cleveland, TS90 8JE UK (Tel.: 44 1642 454144; Telefax: 44 1642 437374; Telex: 587461)

ICI Surfactants (Belgium), Everslaan 45, B-3078 Everberg Belgium (Tel.: 02-758-9361; Telefax: 02-758-9686; Telex: 26151)

ICI Americas, Inc., Subsidiary of ICI plc, New Murphy Rd. & Concord Pike, Wilmington, DE 19897 USA (Tel.: 302-886-3000; 800-456-3669; Telefax: 302-886-2972; Telex: 4945649)

ICI Atkemix Inc., Div. of ICI, PO Box 1085, 70 Market St., Brantford, Ontario, N3T 5T2 Canada (Tel.: 519-756-6181; Telefax: 519-758-8140)

ICI Specialty Chemicals, Concord Pike & New Murphy Rd., Wilmington, DE 19897 USA (Tel.: 302-886-3000; 800-822-8215; Telefax: 302-886-2972)

ICI Surfactants Americas, Concord Plaza, 3411 Silverside Rd., PO Box 15391, Wilmington, DE 19850 USA (Tel.: 302-886-3000; 800-822-8215; Telefax: 302-887-3525; Telex: 4945649)

Idea Srl
Contrada San Paolo, 96017 Noto (SR) Italy (Tel.: 0931-838730; Telefax: 0931-838030; Telex: 970099 IDEA I)

IDF

2003 E. Sunshine, Suite E, Springfield, MO 65804 USA (Tel.: 417-881-7820; Telefax: 417-881-7274)

IFF International Flavours & Fragrances Ltd.

Duddery Hill, Haverhill, Sulffolk, CB9 8LG UK (Tel.: 44 1440 704488; Telefax: 44 1440 61599; Telex: 818881)

IGI Petroleum Specialities

164 Sheridan St., Perth Amboy, NJ 08861 USA (Tel.: 201-826-0140; Telefax: 201-826-0641)

Igreca

Rue de Bourg de Paille, 49070 Beaucouze France (Tel.: 33 41 482312; Telefax: 33 41 483158; Telex: 720345)

Ikeda Corp.

New Tokyo Bldg., 3-1-Marunouchi, 3-Chome, Chiyoda-ku, Tokyo, 100 Japan (Tel.: 81 3 3212-8791; Telefax: 81 3 3215-5069; Telex: J26370)

The Illes Co.

Box 35412, Dallas, TX 75235 USA (Tel.: 214-631-8350; Telefax: 214-637-3115)

Imarco BV

PO Box 412, 6710 BK Ede The Netherlands (Tel.: 31 8380 78789; Telefax: 31 8380 24025)

Imperial Chemical Industries plc

PO Box 6, Shire Park, Bessemer Road, Welwyn Garden City, Hertfordshire, AL7 1HD UK (Tel.: 44 1707-323400; Telefax: 44 1707-337332; Telex: 94028500 ICIC G)

Importers Service Corp.

233 Suydam Ave., Jersey City, NJ 07304-3399 USA (Tel.: 201-332-6970; Telefax: 201-332-4152)

In-Cide Technologies Inc.

50 N. 41 Ave., Phoenix, AZ 85009 USA (Tel.: 602-233-0756; 800-777-4569; Telefax: 602-272-5864)

Indian Gum Industries Ltd.

51-A, Maker Chamber IV, Nariman Point, Bombay, 400 021 India (Tel.: FAX 91-22-204 0393; Telex: 011-82139 IGUM)

Industrial Proteins Ltd.

405 Lordship Lane, London, SE22 8JN UK (Tel.: 44 181 693 9067; Telefax: 44 181 299 3977; Telex: 898181)

Indústrias Reunidas Jaraguá S.A.

Caixa Postal D-15, Rua Rodolfo Hufenüssler, 755, 89250 Jaraguá do Sul - SC Brazil (Tel.: (0473) 72-2277; Telefax: (0473) 722854; Telex: 474-173 IRJA-BR)

Induvesa

Pricesa 1, Torre de Madrid Pta 14, 28008 Madrid Spain (Tel.: 34 1 541 9603; Telefax: 34 1 547 9167; Telex: 42247)

Induxtra de Suministros Llorella SA

c/. Pere Alsius 2-4, 17820 Banyoles (Girona) Spain (Tel.: 34 972 571058; Telefax: 34 972 575332; Telex: 57071)

Ingredia

51-53 Ave F Lobbedez, BP 946, 62033 Arras Cedex France (Tel.: 33 21 23 8000; Telefax: 33 21 23 8001; Telex: 820530)

Innov'ia

175 Rue de Coureilles, 17000 La Rochelle France (Tel.: 33 46 454511; Telefax: 33 46 448476; Telex: 792167)

Inolex Chemical Co.

Jackson & Swanson Sts., Philadelphia, PA 19148-3497 USA (Tel.: 215-289-9065; 800-521-9891; Telefax: 215-271-2621; Telex: 834617)

Interfood Deutschland GmbH

Louisenstr. 120, 6380 Bad Homburg Germany (Tel.: 49 6172 40100; Telefax: 49 6172 459057)

Integrated Ingredients
1420 Harbor Bay Pkwy., Suite 210, Alameda, CA 94501 USA (Tel.: 510-748-6362; Telefax: 510-748-6375)

International Additives Ltd.
Old Gorsey Lane, Wallasey, Merseyside, LA4 4AH, UK)

International Bioflavors Inc.
404 C. Wilmont Dr., Waukesha, WI 53186 USA (Tel.: 414-521-9200; Telefax: 414-521-9092)

International Bio-Synthetics Inc.
PO Box 241068, Charlotte, NC 28224 USA (Tel.: 704-527-9000; 800-438-1361; Telefax: 704-527-8184)

International Flavors & Fragrances
1515 Hwy 36, Union Beach, NJ 07735 USA (Tel.: 908-264-4500; Telefax: 908-888-2595; Telex: 275284)

International Grain Products Co.
PO Box 677, Wayzata, MN 55391 USA (Tel.: 612-474-6573; Telefax: 612-397-7232)

International Ingredients Co.
City South Business Park, Unit 14, 16-34 Dunning Ave., Rosebery, NSW, 2018 Australia (Tel.: 61 2 663 3999; Telefax: 61 2 633 3933)

International Sourcing Inc.
121 Pleasant Ave., Upper Saddle River, NJ 07458 USA (Tel.: 201-934-8900; 800-772-7672; Telefax: 201-934-8291; Telex: 697-2957 INSOURC)

Irish Dairy Board (Brussels Office)
Radiatorenstraat 1, B-1800 Vilvoorde Belgium (Tel.: 32 2 251 6961; Telefax: 32 2 251 4318; Telex: 23651)

Isnard-Lyraz
11-13 rue de la Loge, BP 100, 94265 Fresnes Cedex France (Tel.: 33 1 40 96 7400; Telefax: 33 1 46 68 1644; Telex: 632700)

Janousek Industriale Srl
Strada per i Laghetti 3, ZI delle Noghere, 340115 Muggia, Trieste Italy (Tel.: 39 40 232691; Telefax: 39 40 232698; Telex: 460437)

Jungbunzlauer
Jungbunzlauer AG, Div. of Montana AG, Postfach 546, Schwarzenbergplatz 16, A-1010 Vienna Austria (Tel.: 222 50200; Telefax: 222 502008; Telex: 133396 JUBA A)

Jungbunzlauer International AG, St. Alban-Vorstadt 90, CH-4002 Basel Switzerland (Tel.: 41 61 295 51 00; Telefax: 41 61 295 51 08; Telex: 964963 jubu ch)

Jungbunzlauer Inc., 75 Wells Ave., Newton Centre, MA 02159-3214 USA (Tel.: 617-969-0900; 800-828-0062; Telefax: 617-964-2921)

Jungbunzlauer Pte Ltd., 62 Cecil St. #06-01, TPI Bldg., Singapore 0104 (Tel.: +65 227-3400; Telefax: +65 227-3600)

Kamena Products Corp.
Kafr Tohormos-Giza, PO Box 2580, Cairo Egypt (Tel.: 20 2 754511; Telefax: 20 2 754391; Telex: 93222)

Kanegrade Ltd.
9 Bowman Estate, Bessemer Dr., Stevenage, Herts, SG1 2DL UK (Tel.: 44 1438 742242; Telefax: 44 1438 742311; Telex: 826862)

Kao Corp. S.A.
Puig dels Tudons, 10, 08210 Barbera Del Valles, Barcelona Spain (Tel.: 3-729-0000; Telefax: 3-718-9829; Telex: 59749)

Kao Corp./Edible Fat & Oil Div.
14-10 Nihonbashi, Kayabacho 1-chome, Chuo-ku, Tokyo, 103 Japan (Tel.: (03) 660-7862; Telefax: (03) 660-7964; Telex: J24816)

Karlshamns
Karlshamns AB, S 37482 Karlshamn Sweden (Tel.: 46 454 82000; Telefax: 46 454 18453; Telex: 4510)

Karlshamns BV, PO Box 17, Koog aan de Zaan 1540 AA The Netherlands (Tel.: 31 75 278400; Telefax: 31 75 285050; Telex: 19268)

Karlshamns Food Ingredients, PO Box 569, Columbus, OH 43216-0569 USA (Tel.: 800-227-2489; Telefax: 614-299-2584)

Karlshamns Lipids for Care, 525 W. First Ave., PO Box 569, Columbus, OH 43216 USA (Tel.: 614-299-3131; 800-848-1340; Telefax: 614-299-2584; Telex: 245494 capctyprdcol)

L. Karp & Sons Inc.
1301 Estes Ave., Elk Grove Village, IL 60007 USA (Tel.: 708-593-5700; Telefax: 708-593-0749)

Kato Worldwide Ltd.
One Bradford Rd., Mt. Vernon, NY 10553 USA (Tel.: 914-664-6200; Telefax: 914-664-0413)

Kelco
Kelco, Div. of Merck & Co Inc., 8355 Aero Drive, PO Box 23576, San Diego, CA 92123-1718 USA (Tel.: 619-292-4900; 800-535-2656; Telefax: 619-467-6520; Telex: 695454)

Kelco International Ltd., Westminster Tower, 3 Albert Embankment, London, SE1 7RZ UK (Tel.: 44 171-735-0333; Telefax: 44 171-735-1363; Telex: 23815 KAILIL G)

Kelco International Ltd., 54-68 Ferndell St., South Granville, Sydney, NSW 2142 Australia (Tel.: 61 2 645-3022; Telefax: 61 2 645-3324)

Kelco International Ltd., c/o MSD Japan Co., Ltd., Seventh Floor, Kowa Bldg. No. 16 Annex, 9-20 Akasaka 1-Chome, Minato-Ku, Tokyo, 107 Japan (Tel.: 81 3 3586 2840; Telefax: 81 3 3586 2889)

Kerry Foods Ltd.
Kerry House, Hillingdon Hill, Uxbridge, Middlesex, UB10 0JH UK (Tel.: 44 181 842 1121; Telefax: 44 181 895 74119; Telex: 9413175)

Kerry Ingredients
Kerry Ingredients/Euro Center, Emanuel-Leutze-Str. 4, W-4000 Dusseldorf 11 Germany (Tel.: 49 211 557210; Telefax: 49 211 5672141; Telex: 8584553)

Kerry Ingredients/Beatreme Div., 352 E. Grand Ave., Beloit, WI 53511 USA (Tel.: 800-328-7517; Telefax: 608-365-6558)

Keylink Ltd.
Blackburn Rd., Rotherham, Lancs, S61 2DR UK (Tel.: 44 1709 550206; Telefax: 44 1709 556867)

Kievit bv Zuivelfabriek de
4A Oliemolenweg, Meppel 7944 HX The Netherlands (Tel.: 31 5220 57233; Telefax: 31 5220 55213; Telex: 42787)

Kikkoman Trading Europe GmbH
Heerdter Lohweg 57-59, D4000 Dusseldorf 11 Germany (Tel.: 49 0211 596087; Telefax: 49 0211 592827; Telex: 8584140)

Kimitsu Chemical Industries Co., Ltd., Distrib. by Unipex
No. 15-4 Uchikanda, 2-Chome, Chiyoda-ku, Tokyo Japan (Tel.: 03 3252 8708; Telefax: 03 3252 8704)

Kimpton Brothers Ltd.
Berkshire House, 168-173 High Holborn, London, WC1V 7AF UK (Tel.: 44 171 379 6422; Telex: 263061)

Kingfood Australia Pty. Ltd.
73 Porters Rd., Kenthurst, NSW, 2156 Australia (Tel.: 61 2 654 2555; Telefax: 61 2 654 2587)

Kiril Mischeff Ltd.
Broadwall House, 21 Broadwall, London, SE1 9PL UK (Tel.: 44 171 928 8966; Telefax: 44 171 261 9085; Telex: 25450)

Kohjin Co., Ltd./Fermentation Div.
Shimbashi 1-1-1, Minato-ku, Tokyo Japan (Tel.: 03-3503-1461; Telefax: 03-3508-7594)

V&E Kohnstamm Inc.
Bldg. #10-Bush Terminal, 882 Third Ave., Brooklyn, NY 11232 USA (Tel.: 718-788-6320; Telefax: 718-768-3978; Telex: 425707)

Konsa Konsantre Sanayi AS

Vatan Cad No. 31, Caglayan, 80340 Istanbul Turkey (Tel.: 90 224 68 40; Telefax: 90 224 68 35; Telex: 39148)

Koster Keunen, Inc.

90 Bourne Blvd., PO Box 447, Sayville, NY 11782 USA (Tel.: 516-589-0456; Telefax: 516-589-0120)

Kowa Europe GmbH

65a Immermannstrasse, D-4000 Dusseldorf 1 Germany (Tel.: 49 211 353444; Telefax: 49 211 161952; Telex: 8587029)

Kraft Food Ingredients

PO Box 398, 6410 Poplar Ave., Memphis, TN 38101 USA (Tel.: 901-766-2100; Telefax: 901-766-2696)

Kulman & Balter Ltd.

Unit 4, Gt. Cambridge Ind. Estate, Lincoln Rd., Enfield, Middlesex, EN1 1SH UK (Tel.: 44 181 804 5565; Telefax: 44 181 804 3755; Telex: 264340)

W. Kündig & Cie AG

Stampfenbachstrasse 38, 8023 Zurich Switzerland (Tel.: 41 1 361 6144; Telefax: 41 1 362 8414; Telex: 815855)

KWR Ltd./Food Ingredients Div.

Priestley Rd., Basingstoke, Hants, RG24 9QH UK (Tel.: 44 1256 810155; Telefax: 44 1256 64711; Telex: 858360)

Kyowa Hakko USA Inc.

599 Lexington Ave., Suite 2780, New York, NY 10022 USA (Tel.: 212-319-5353; Telefax: 212-521-1283)

Lactosan (UK) Ltd.

Lacsan House, 5 Swinbourne Dr., Springwood Ind. Estate, Braintree, Essex, CM7 7UP UK (Tel.: 44 1376 342226; Telefax: 44 1376 342132; Telex: 988709)

Lactovit GmbH

Morsenbroicher Weg 31, 4000 Dusseldorf Germany (Tel.: 49 211 020111, Telefax: 49 211 030723, Telex: 8586899)

Lake States, Div. of Rhinelander Paper Co. Inc.

515 W. Davenport St., Rhinelander, WI 54501 USA (Tel.: 715-369-4356; Telefax: 715-369-4141)

LaMonde Ltd.

500 S. Jefferson St., Placentia, CA 92670 USA (Tel.: 714-993-7700)

Henry Lamotte Import/Export

Auf dem Dreieck 3, 2800 Bremen Germany (Tel.: 49 421 547060; Telefax: 49 421 5470699; Telex: 244144)

Land O'Lakes/Food Ingredients Div.

PO Box 116, Minneapolis, MN 55440 USA (Tel.: 612-481-2064)

La Prosperite Fermiere

BP 946, 51-53 ave. Lobbedez, 62033 Arras Cedex France (Tel.: 33 21 734040; Telefax: 33 21 581633; Telex: 820530)

Latenstein Zetmeel BV

PO Box 392, 6500 AJ Nijmegen The Netherlands (Tel.: 31 80 719811; Telefax: 31 80 786981; Telex: 48725)

Rene Laurent SA

107 ave. Franklin Roosevelt, 06117 Le Cannet Cedex France (Tel.: 33 93 69 2727; Telefax: 33 93 69 1980; Telex: 470709)

Lebermuth Co. Inc.

PO Box 4103, South Bend, IN 46624 USA (Tel.: 219-259-7000; Telefax: 219-258-7450; Telex: 276424)

LeChem Inc.

12537 Scenic Hwy., Baton Rouge, LA 70807 USA (Tel.: 504-775-1801; Telefax: 504-775-5431)

Leeben Color Div./Tricon Colors Inc.
16 Leliarts Lane, Elmwood Park, NJ 07407 USA (Tel.: 201-794-3800)

Leiner USA, Div. of Intergel Corp.
220 White Plains Rd., Tarrytown, NY 10591 USA (Tel.: 800-344-2075; Telefax: 914-631-2505)

Liberty Enterprises
16 S. Oakland Ave., Suite 200, Pasadena, CA 91101 USA (Tel.: 818-792-9706; Telefax: 818-792-9882)

Lifewise Ingredients Inc.
PO Box 7138, Buffalo Grove, IL 60089 USA (Tel.: 708-541-6525; Telefax: 708-541-4428)

Lipo Chemicals, Inc.
207 19th Ave., Paterson, NJ 07504 USA (Tel.: 201-345-8600; Telefax: 201-345-8365; Telex: 130117)

Liquid Sugars Inc. (LSI)
PO Box 96, Oakland, CA 94604 USA (Tel.: 415-420-7100)

Litex Ltd.
New Enterprise House, St. Helen's St., Derby, DE1 3GY UK (Tel.: 44 1332 297959; Telefax: 44 1332 200119; Telex: 377106)

Liveco Inc.
1 Mill St., Burlington, VT 05401 USA (Tel.: 802-864-6818; Telefax: 802-864-6874)

Lochore & Ferguson Ltd.
79 Beardmore Way, Dalmuir, Clydebank, Glasgow, G81 4HA UK (Tel.: 44 141 952 4546; Telefax: 44 141 952 2457)

Lohmann Chemicals, Dr. Paul Lohmann Chemische Frabrik GmbH KG
PO Box 1220, D-3254 Emmerthal 1 Germany (FAX 49 51 55 6 31 18; Telex: 92858)

Lonza Inc.
17-17 Route 208, Fair Lawn, NJ 07410 USA (Tel.: 201-794-2400; 800-777-1875 (tech.); Telefax: 201-703-2028; Telex: 4754539 LONZAF)

Loryma GmbH
Am Falltor 7, Bergstrasse, D6144 Zwingenberg Germany (Tel.: 49 6251 71071; Telex: 468273)

Love Starches
1 Braidwood St., Enfield, NSW, 2136 Australia (Tel.: 61 2 764 8222; Telefax: 61 2 742 5819)

Lowi Chem. Corp.
190 Montrose West Ave. #203, Akron, OH 44321 USA (Tel.: 216-668-6195; Telefax: 216-665-5158)

LSI
LSI/Liquid Sugars, Div. of LSI Inc., 1285 66th St., PO Box 96, Oakland, CA 94604-0096 USA (Tel.: 510-420-7100; Telefax: 510-420-7103)

LSI Specialty Prods., Div. of LSI Inc., Box 96, Oakland, CA 94604 USA (Tel.: 415-420-7100)

The O.C. Lugo Co., Inc.
42 Burd St., Nyack, NY 10960 USA (Tel.: 914-353-7711; Telefax: 914-353-7702)

Luxemburg Cheese Factory Inc.
12495 N. Pleasant Hill Rd., Orangeville, IL 61060 USA (Tel.: 815-789-4227)

MacAndrews & Forbes Co.
3rd St. & Jefferson Ave., Camden, NJ 08104 USA (Tel.: 609-964-8840; Telefax: 609-964-6029; Telex: 845337)

Macphie of Glenbervie Ltd.
Glenbervie, Stonehaven, Kincardineshire, AB3 2YB UK (Tel.: 44 15694 641; Telefax: 44 15694 677; Telex: 73589)

Dr. Madis Laboratories Inc.
375 Huyler St., PO Box 2247, South Hackensack, NJ 07606 USA (Tel.: 201-440-5000; Telefax: 201-342-8000; Telex: 134-200)

Maizecor Foods Ltd.
141 Wincolmlee, Hull, HU2 0HB UK (Tel.: 44 1482 24434; Telefax: 44 1482 212207; Telex: 597256)

Major Products Co.
66 Industrial Ave., Little Ferry, NJ 07643 USA (Tel.: 201-641-5555; Telefax: 201-641-6331)

Mallinckrodt Specialty Chemicals
16305 Swingley Ridge Dr., Chesterfield, MO 63017 USA (Tel.: 314-895-2000; 800-325-7155; Telefax: 314-530-2562)

Malt Products Corp.
PO Box 739, Maywood, NJ 07607 USA (Tel.: 800-526-0180; Telefax: 201-845-0028)

Mane USA
60 Demarest Dr., Wayne, NJ 07470 USA (FAX 201-633-5538)

V. Mane Fils SA
06620 Le Bar Sur Loup France (Tel.: 33 93 09 70 00; Telefax: 33 93 42 54 25; Telex: 470841)

J. Manheimer Inc.
47-22 Pearson Pl., PO Box 4481, Long Island City, NJ 11101 USA (Tel.: 718-392-7800; Telefax: 718-392-7985)

Manna International
128 March St., Sault Ste. Marie, Ontario, P6A 2Z3 Canada (Tel.: 705-946-2662; Telefax: 705-256-6540)

Mantrose-Haeuser Co.
500 Post Road East, Westport, CT 06880 USA (Tel.: 203-454-1800; 800-344-4229; Telefax: 203-227-0558)

Marcel Trading Corp.
926 Araneta Ave., Quezon City, Philippines (Tel.: 201-224-5949 (US); Telefax: 63 2 712 2631)

Dr. Marcus GmbH
Geesthachter Strasse 100-105, 2054 Cocothacht Germany (Tel.: 49 1152 8000 0; Telefax: 49 1152 5479; Telex: 218713)

Mar-Gel Food Products Ltd.
Sudpre House, Worplesdon Rd., Guildford, GU3 3RB UK (Tel.: 44 1483 233001; Telex: 859244)

Marine Colloids, Div. of FMC Corp.
1735 Market St., Philadelphia, PA 19103 USA (Tel.: 215-299-6199)

Marine Magnesium Co.
995 Beaver Grade Rd., Coraopolis, PA 15108 USA (Tel.: 412-264-0200; Telefax: 412-264-9020)

W.H. Marriage & Sons Ltd.
Chelmer Mills, New Street, Chelmsford, Essex, CM1 1PN UK (Tel.: 44 1245 354455; Telefax: 44 1245 261492; Telex: 218713)

Marron Foods Inc.
325 Broadway, Suite 504, New York, NY 10007 USA (Tel.: 212-732-4100; Telefax: 212-962-7300)

Maruzen Fine Chemicals Inc., Div. of Maruzen Kasei Co Ltd.
525 Yale Ave, Pitman, NJ 08071 USA (Tel.: 609-589-4042; Telefax: 609-582-8894; Telex: 333812 MARFINE)

Mastertaste
Draycott Mills, Cam, Dursley, Glos, GL11 5NA UK (Tel.: 44 1453 543272; Telefax: 44 1453 543279; Telex: 449825)

Matthews Foods plc
The Healey Complex, Healey Rd, Ossett, W Yorks, WF5 8NE UK (Tel.: 44 1924 272534; Telefax: 44 1924 277071; Telex: 55483)

Mauri Laboratories
Clarks Mill, Stallard St., Trowbridge, Wilts, BA14 8HH UK (Tel.: 44 1225 768406; Telefax: 44 1225 764834; Telex: 44161)

Gerald McDonald & Co. Ltd.

    1 St. Andrews Hill, London, EC4V 5HA UK (Tel.: 44 171 236 3695; Telefax: 44 171 248 7267; Telex: 8812098)

MD Foods

    Skanderborgvej 277, DK-8260 Viby J Denmark (Tel.: 45 86 281000; Telefax: 45 86 281838; Telex: 68799)

Mears Trading Pty. Ltd.

    41 First Ave., Strathmore, Victoria, 3041 Australia (Tel.: 61 3 379 5597; Telefax: 61 3 379 7901)

Medallion International Inc.

    944 Belmont Ave., PO Box 8208, N. Haledon, NJ 07508 USA (Tel.: 201-427-7781; Telefax: 201-427-0815; Telex: 130159)

Meer Corp.

    PO Box 9006, 9500 Railroad Ave., N. Bergen, NJ 07047-1206 USA (Tel.: 201-861-9500; Telefax: 201-861-9267; Telex: 219130)

Meggle GmbH

    Postfach 40, Megglestrasse 6-12, W-8090 Wasserburg 2 Germany (Tel.: 49 8071-730; Telefax: 49 8071-73444; Telex: 525137)

Melida SpA

    Synthesis, Via Medici del Vascello 40, 20138 Milano Italy (Tel.: 39 2 520 39427; Telefax: 39 2 530 39385; Telex: 3110426)

Mendell Co., Inc., Edward, A Penwest Company

    2981 Rt. 22, Patterson, NY 12563-9970 USA (Tel.: 914-878-3414; 800-431-2457; Telefax: 914-878-3484; Telex: 4971034)

E. Merck

    Postfach 4119, Frankfurter Strasse 250, D-6100 Darmstadt 1 Germany (Tel.: 06151-72-0; Telefax: 06151-72-3684; Telex: 419328-0 em d)

    E. Merck/Fine Chemicals Div., Frankfurter Strasse 250, 6100 Darmstadt Germany (Tel.: 49 6151 72 2108; Telefax: 49 6151 72 7630; Telex: 4193280)

    Merck Pty. Ltd., 207 Colchester Rd., Kilsyth, Victoria, 3137 Australia (Tel.: 61 3 728 5855; Telefax: 61 3 728 1351)

Mero-Roussselot-Satia (Sanofi Bio Industries)

    15 ave. d'Eylau, 75116 Paris France (Tel.: 33 1 4704 6767; Telefax: 33 1 47074 7559; Telex: 614568)

Metarom

    Metarom Corp., 11 School St., Newport, VT 05855 USA (Tel.: 802-334-0117; Telefax: 802-334-7060)

    Metarom Canada Inc., 555 Rutherford, CP 635/PO Box 635, Granby, Quebec, J2G 8W7 Canada (Tel.: 514-378-0841; Telefax: 514-375-7953)

Metayer Aromatiques

    9-11, ave de la Libération, B.P. 31, 94103 St-Maur-des-Fossés Cedex France (Tel.: 48 83 38 75; Telefax: 48 83 10 88)

Metroz Essences SpA

    Via Andrea Doria 40, 20093 Cologno Monzese-Milano Italy (Tel.: 39 2 2549851; Telefax: 39 2 2540105; Telex: 331491)

Metsa-Serla Chemicals BV

    PO Box 31, 6500 AA Nijmegen The Netherlands (Tel.: 31 80 772182; Telefax: 31 788160; Telex: 48071)

Lucas Meyer

    Lucas Meyer GmbH & Co., PO Box 261665, D-2000 Hamburg 26 Germany (Tel.: 40-789-550; Telefax: 40-789-8329; Telex: 2163220 myer d)

    Lucas Meyer GmbH & Co., Ausschlager Elbdeich 62, 2000 Hamburg 26 Germany (Tel.: 49 40 7895 5214; Telefax: 49 40 787173)

    Lucas Meyer (UK) Ltd., Unit 46, Deeside Ind. Park, First Ave., Deeside, Clwyd, CH5 2NU UK (Tel.: 44 1244 281168; Telefax: 44 1244 281169)

Lucas Meyer Inc., 765 E. Pythian Ave., Decatur, IL 62526 USA (Tel.: 217-875-3660; Telefax: 217-877-5046)

Meyhall

Meyhall Chemical AG, Sonnenwiesenstrasse 18, PO Box 862, CH-8280 Kreuzlingen Switzerland (Tel.: 41 72 747576; Telefax: 41 72 752181; Telex: 882222)

Meyhall Chemical UK Ltd., 6E Church Rd., Begington, Wirral, Merseyside, L63 7PG UK (Tel.: 44 151 645 8229; Telefax: 44 151 645 8906; Telex: 627820)

David Michael & Co., Inc.

10801 Decatur Rd., Philadelphia, PA 19154 USA (Tel.: 215-632-3100; Telefax: 215-637-3920)

Mid-America Farms/Food Ingredient Div.

PO Box 1837, Springfield, MO 65801 USA (Tel.: 800-243-2479; Telefax: 417-865-8194)

Mid-America Food Sales Ltd.

3701 Commercial Ave., Suite 11B, Northbrook, IL 60062 USA (Tel.: 708-480-0720; Telefax: 708-480-9392)

Midori Kagaku Co. Ltd.

4F No. 10 Nohagi Bldg., 2-27-8 Minami-Ikebukuro, Toshimaku, Tokyo, 171 Japan (Tel.: 81 3 3980 8808; Telefax: 81 3 3980 8805; Telex: 2324039)

Milei GmbH

Postfach 10 34 28, Sarwey Strasse 5, D70191 Stuttgart Germany (Tel.: 49 711 98178-21; Telefax: 49 711 98178-25; Telex: 723459)

Miles

Miles Inc., One Mellon Center, 500 Grant Street, Pittsburgh, PA 15219-2502 USA (Tel.: 412-394 5500; Telefax: 412-394 5578)

Miles Inc./Polysar Rubber Div., Div. of Bayer AG, 2603 W. Market St., Akron, OH 44313 USA (Tel.: 216-836-0451; Telefax: 216-836-4614)

Miles Ltd./Biotechnology Products Div.

PO Box 37, Stoke Court, Stoke Poges, Slough, SL2 4LY UK (Tel.: 44 12814 5151; Telefax: 44 12814 3893; Telex: 848337)

Milk Products (NZ) Ltd.

Bancroft Place, Reigate, Surrey, RH2 7RP UK (Tel.: 44 1737 221616; Telefax: 44 1737 241288; Telex: 884497)

E.A. Miller Inc.

Hyrum, UT, 84319 USA (Tel.: 801-245-6456)

E.A. Miller/Flavors Div., 410 North 200 West, Hyrum, UT, 84319 USA (Tel.: 800-873-0939; Telefax: 801-245-6634)

Minnesota Dehydrated Veg Inc.

915 Omland Ave. N., Fosston, MN 56542 USA (Tel.: 218-435-1997; Telefax: 218-435-6770)

Mitsubishi International Corp./Fine Chemicals Dept.

520 Madison Ave., New York, NY 10022-4223 USA (Tel.: 212-605-2193; 800-442-6266; Telefax: 212-605-1704)

Mitsubishi Kasei

Mitsubishi Kasei Corp., Mitsubishi Bldg., 5-2, Marunouchi 2-chome, Chiyoda-ku, Tokyo, 100 Japan (Tel.: (03) 3283-6254; Telex: BISICH J 24901)

Mitsubishi-Kasei Foods Corp., Ichikawa Bldg., 13-3, Ginza 5-chome, Chuo-ku, Tokyo, 104 Japan (Tel.: (03) 3542-6525; Telefax: (03) 3545-4860; Telex: BISICHJ 24901 AH.MFC)

Mitsubishi Oil Co., Ltd.

Sanyu Bldg., 2-4, Toranomon 1-chome, Minato-ku, Tokyo, 105 Japan (Tel.: (03) 3595-7663; Telefax: (03) 3508-2521; Telex: 222-4101)

Mitsubishi Petrochemical Co., Ltd.

Mitsubishi Bldg., 5-2, Marunouchi 2-chome, Chiyoda-ku, Tokyo, 100 Japan (Tel.: (03) 3283-5700; Telefax: (03) 3283-5472; Telex: 222-3172)

Mitsubishi Rayon Co., Ltd.

3-19, Kyobashi 2-chome, Chuo-ku, Tokyo, 104 Japan (Tel.: (03) 3272-4321; Telefax: (03) 3245-8781; Telex: MRC TOK J33395)

Mitsubishi Yuka Badische Co., Ltd., Joint venture of Mitsubishi Petrochemical Co., Ltd. and BASF Japan Ltd.

1000, Kawajiri-cho, Yokkaichi, Mie., 510 Japan (Tel.: (0593) 45-7230; Telefax: (0593) 45-7246; Telex: 4948182 DIAPTC)

Mitsui-Cyanamid, Ltd., Joint venture of Mitsui Toatsu Chemicals Inc. and Am. Cyanamid Co., USA

No. 18 Mori Bldg., 2nd Fl., 3-13, Toranomon 2-chome, Minato-ku, Tokyo, 105 Japan (Tel.: (03) 592-1815; Telefax: (03) 581-2424; Telex: 2226952 MTCYAN J)

MLG Enterprises Ltd.

PO Box 52568, Turtle Creek P.O., 1801 Lakeshore Road West, Mississauga, Ontario, L5J 4S6 Canada (Tel.: 905-569-3330; Telefax: 905-569-2133; Telex: 06982341)

Molda UK Ltd.

Crossroads House, 165 The Parade, High St., Watford, Herts, WD1 1NJ UK (Tel.: 44 1923 816188; Telefax: 44 1923 816502)

Molkerei Dahlenburg AG (Molda)

Postfach 1106, Gartenstrasse 13, D-2121 Dahlenburg Germany (Tel.: 49 5851 880; Telefax: 49 5851 7230)

Monsanto Chemical Co.

800 N. Lindbergh Blvd., St. Louis, MO 63167 USA (Tel.: 314-694-1000; 800-325-4330; Telefax: 314-694-7625; Telex: 44-7282)

Moore Fine Foods Ltd.

Hainton House, Hainton Sq., Grimsby, S. Humberside, DN32 9AQ UK (Tel.: 44 1472 240704; Telefax: 44 1472 250075; Telex: 56261)

Morflex, Inc.

2110 High Point Rd., Greensboro, NC 27403 USA (Tel.: 910-292-1781; Telefax: 910-854-4058; Telex: 910240 7846)

Morton International

Morton International Inc., 100 North Riverside Plaza, Randolph Street, Chicago, IL 60606-1598 USA (Tel.: 312-807-2562; Telefax: 312-807-2899; Telex: 25-4433)

Morton Salt, 100 North Riverside Plaza, Chicago, IL 60606-1597 USA (Tel.: 312-807-2562; Telefax: 312-807-2228; Telex: 25-4433)

Moul-Bie (UK) Ltd.

336A Ladbroke Grove, London, W10 5BW UK (Tel.: 44 181 968 1899; Telefax: 44 181 969 5217; Telex: 927713)

Müggenburg Extrakt GmbH

Immenhacken 3, D2359 Henstedt-Ulzburg Germany (Tel.: 49 4193 94033; Telefax: 49 4193 4068; Telex: 2180174)

Multi-Kem Corp.

553 Broad Ave., PO Box 538, Ridgefield, NJ 07657-0538 USA (Tel.: 201-941-4520; 800-462-4425; Telefax: 201-941-5239; Telex: 134 611 multikem)

Munton & Fison plc

Cedars Factory, Stowmarket, Suffolk, IP14 2AG UK (Tel.: 44 1449 612401; Telefax: 44 1449 677800; Telex: 98205)

Nalco Chemical Co.

One Nalco Center, Naperville, IL 60563-1198 USA (Tel.: 708-305-1000; 800-527-7753; Telefax: 708-305-2900)

Bryan W. Nash & Sons Ltd.

15 Rose Hill, Dorking, Surrey, RH4 2ED UK (Tel.: 44 1306 889905; Telefax: 44 1306 740676; Telex: 859912)

Nasoya Foods
   23 Jytek Dr., Leominster, MA 01453 USA (Tel.: 508-537-0713; Telefax: 508-537-9790)

National Cottonseed Products Assn.
   PO Box 172267, Memphis, TN 38187 USA (Tel.: 901-682-0880; Telefax: 901-682-2856)

National Egg Products Corp.
   Box 608, Social Circle, GA 30279 USA (Tel.: 404-464-2652; Telefax: 404-464-2998)

National Stabilizers Inc.
   1846 Business Center Dr., Duarte, CA 91010 USA (Tel.: 818-359-4584)

National Starch & Chemical
   National Starch & Chemical Corp., Box 6500, 10 Finderne Ave., Bridgewater, NJ 08807 USA (Tel.: 908-685-5000; 800-726-0450; Telefax: 908-685-5005)

   National Starch & Chemical, Prestbury Court, Greencourts Business Park, 333 Styal Rd., Manchester, M22 5LW UK (Tel.: 44 161 435 3200; Telefax: 44 161 435 3300)

   National Starch & Chemical (Asia), 107 Neythal Rd., Jurong, S-2262 Singapore (Tel.: 65 2615528; Telefax: 65 264 1870; Telex: 55445)

Natra US Inc.
   1390 S. Dixie Hwy., Suite 2210, Coral Gables, FL 33146 USA (Tel.: 800-262-6216; Telefax: 305-662-1361)

Nattermann Phospholipid GmbH, Div. of Rhone-Poulenc Rorer
   PO Box 350120, Nattermannallee 1, D-5000 Cologne 30 Germany (Tel.: 49 221-509-2267; Telefax: 49 221-509-2816; Telex: 8882663)

Natural Oils International, Inc.
   12350 Montague St., Unit C & D, Pacoima, CA 91331 USA (Tel.: 818-897-0536; Telefax: 818-896-4277; Telex: 371-0352)

Naturex
   BP 152, F84147 Montfavet Cedex France (Tel.: 33 90 239689; Telefax. 33 90 239004)

New England Spice Co. Inc.
   60 Clayton St., Drawer B, Dorchester, MA 01211 USA (Tel.: 617-825-7900)

New Zealand Milk Products
   New Zealand Milk Prods., Inc., 3637 Westwind Blvd., Santa Rosa, CA 95403 USA (Tel.: 707-524-6600; 800-336-1269; Telefax: 707-524-6666)

   New Zealand Milk Products (UK) Ltd., Bancroft Place, 10 Bancroft Rd., Reigate, Surrey, RH2 7RP UK (Tel.: 44 1737 224949; Telefax: 44 1737 241289; Telex: 884497)

   New Zealand Milk Products (Europe) GmbH, Siemensstrasse 6-14, 2084 Rellingen Germany (Tel.: 49 4101 3029 0; Telefax: 49 4101 302920; Telex: 2189074)

   New Zealand Milk Products (Italia) srl, Via Cavalieri Vittorio Veneto 2, 22052 Cernusco Lombardone (CO) Italy (Tel.: 39 39 587351; Telefax: 39 39 587612)

Niacet Corp.
   PO Box 258, 400 47th St., Niagara Falls, NY 14304 USA (Tel.: 716-285-1474; 800-828-1207; Telefax: 716-285-1497; Telex: 6730170)

H.R. Nicholson Co.
   Kenshaw & Oakleaf Ave., Baltimore, MD 21215 USA (Tel.: 800-638-3514; Telefax: 410-764-9125)

Nickstadt-Moeller Inc.
   1169 Edgewater Ave., Ridgefield, NJ 07657 USA (Tel.: 201-943-9300; Telefax: 201-943-9396)

Nidefo
   PO Box 113, 8070 AC Nunspeet The Netherlands (Tel.: 31 3412 57205; Telefax: 31 3412 51274; Telex: 47191)

Thomas Nihal S
   73 Old Kottawa Rd., Mirihana, Nugegoda, Sri Lanka (Tel.: 94 1 552658; Telefax: 94 1 589900; Telex: 23137)

Nihon Emulsion Co., Ltd.
Minami 5-32-7, Koenji, Suginami-ku, Tokyo Japan (Tel.: (03) 314-3211; Telefax: (03) 312-7207; Telex: 2322358 EMALEX J)

Nikken Foods
Nikken Foods Co. Ltd., 37-7-1 Chome, Komaba, Meguro-ku, Tokyo, 153 Japan (Tel.: 81 3 3460 0616; Telefax: 81 3 3460 0683; Telex: 02422597)

Lilar Corp., Distributor for Nikken Foods Co. Ltd., 611 North Tenth St., Suite 375, St. Louis, MO 63101 USA (Tel.: 314-436-5050; Telefax: 314-436-0741)

Nikken Trading Inc.
1374 Clarkson/Clayton Center, Suite 319, St. Louis, MO 63011 USA (Tel.: 314-532-1019; Telefax: 314-527-5057)

Nikko Chemicals Co., Ltd.
4-8, Nihonbashi, Bakuro-cho 1-chome, Chuo-ku, Tokyo, 103 Japan (Tel.: (03) 3661-1677; Telefax: (03) 3664-8620; Telex: 2522744 NIKKOL J)

Nipa Laboratories
Nipa Laboratories, Inc., 3411 Silverside Rd., 104 Hagley Bldg., Wilmington, DE 19810 USA (Tel.: 302-478-1522; Telefax: 302-478-4097; Telex: 905030)

Nipa Laboratories Ltd., Div. of BTP plc, Llanwit Fardre, Pontypridd, Mid Glamorgan, CF38 2SN, Wales UK (Tel.: 44 1443-205311; Telefax: 44 1443-207746; Telex: 497111)

Nippon Oils & Fats Co., Ltd. (NOF Corp.)
Yurakucho Bldg., 10-1, Yarakucho 1-chome Chiyoda-Ku, Tokyo, 100 Japan (Tel.: (03) 3283-7295; Telefax: (03) 3283-7178; Telex: 222-2041 NIPOIL J)

Nippon Rikagakuyakuhin Co., Ltd.
2-12, Nihonbashi-Honcho 4-chome, Chuo-ku, Tokyo, 103 Japan (Tel.: (03) 3241-3557; Telefax: (03) 3242-3345; Telex: 2223651 NITIRJ J)

NIVE BV
Energieweg 9, 8071 DA Nunspeet The Netherlands (Tel.: 31 3412 54774; Telefax: 31 3412 51274; Telex: 47191)

Norba (UK) Ltd. *See* Fibrex

Nordmann Rassmann GmbH & Co.
Kajen 2, D2000 Hamburg 11 Germany (Tel.: 49 40 36 87307; Telefax: 49 40 36 87414; Telex: 212087)

Norman, Fox & Co.
5511 S. Boyle Ave., PO Box 58727, Vernon, CA 90058 USA (Tel.: 213-583-0016; 800-632-1777; Telefax: 213-583-9769)

NorStar Dairy International Ltd.
PO Box 202167, Minneapolis, MN 55420-2167 USA (Tel.: 612-854-1001)

Northville Labs Inc.
1 Vanilla Lane, Northville, MI 48167 USA (Tel.: 313-349-1500)

North Western Bakers Ltd.
74 Roman Way Ind. Estate, Longridge Rd., Preston, Lancs, PR2 5BE UK (Tel.: 44 1772 651616; Telefax: 44 1772 655003)

Nougat Chabert et Guyot
9 rue Charles Chabert, F26200 Montelimar France (Tel.: 33 44 608084; Telefax: 33 44 608076; Telex: 649233)

Novo Nordisk
Novo Nordisk A/S, Bioindustrial Group, Novo Allé, DK-2880 Bagsvaerd Denmark (Tel.: 45 4444-8888; Telefax: 45 4444-5918; Telex: 37173)

Novo Nordisk Bioindustries UK Ltd., 4 St. Georges Yard, Castle St., Farnham, Surrey, GU9 7LW UK (Tel.: 44 1252 711212; Telefax: 44 1252 711187)

Novo Industri A/S, Subsid of Novo Nordisk A/S, Novo Alle, DK-2880 Bagsvaerd Denmark (Tel.: 2-98 23 33; Telefax: 2-98 27 33; Telex: 37173 NOVO DK)

NutraSweet
> NutraSweet AG, Innere Guterstrasse 2-4, 6304 Zug Switzerland (Tel.: 41 42 226622; Telefax: 41 42 214246; Telex: 862339)
>
> The NutraSweet Co., Box 830, 1751 Lake Cook Rd., Deerfield, IL 60015 USA (Tel.: 800-321-7254; Telefax: 708-405-6645)

Nutritonal Research Associates Inc.
> PO Box 354, 407 E. Broad St., S. Whitley, IN 46787 USA (Tel.: 219-723-4931)

Nu-World Amaranth Inc.
> PO Box 2202, Naperville, IL 60567 USA (Tel.: 708-369-6819)

Obipektin AG
> Industriestrasse, CH-9220 Bischofszell Switzerland (Tel.: 41 71 813921; Telefax: 41 71 813625; Telex: 882988)

Ogawa & Co. Ltd., AVRI Companies Inc.
> 1080 Essex Ave., Richmond, CA 94801 USA (Tel.: 510-233-0633; Telefax: 510-233-0636)

O'Laughlin Industries Co. Ltd.
> 20th Floor, Jubilee Centre, 18 Fenwick St., Wanchai Hong Kong (Tel.: 852 527 1031; Telefax: 852 529 0231; Telex: 76724)

Olin Corp.
> 120 Long Ridge Rd., PO Box 1355, Stamford, CT 06904 USA (Tel.: 203-356-3036; 800-243-9171; Telefax: 203-356-3273; Telex: 420202)

OM Ingredients
> PO Box 398, Memphis, TN 38101 USA (Tel.: 800-323-1092; Telefax: 901-766-2696)

Opta Food Ingredients Inc.
> 25 Wiggins Ave., Bedford, MA 01730 USA (Tel.: 617-276-5100; Telefax: 617-276-5101)

Orange-Co Inc.
> PO Box 2158, Bartow, FL 33830 USA (Tel.: 813-533-0551)

Origan
> 26/28 rue Hippolyte Jamot, 95110 Sannois France (Tel.: 33 34 10 6463)

Ottens Flavors
> 1220-34 Hamilton St., Philadelphia, PA 19123 USA (Tel.: 215-627-5030; Telefax: 215-627-0518)

Overseal Foods Ltd.
> Swains Park, Park Rd., Overseal, Swadlincote, Derby, DE12 6JX UK (Tel.: 44 1283 224221; Telefax: 44 1283 222006; Telex: 341345)

Pacific Foods Inc.
> 21612 88th Ave. S., Kent, WA 98031 USA (Tel.: 206-395-9400; Telefax: 206-395-3330)

Paine's plc
> 36 Market Sq., St. Neots, Huntingdon, Cambs, PE19 2AL UK (Tel.: 44 1480 214000; Telex: 32308)

Paldena Ltd.
> 49 Ridge Hill, London, NW11 8PR UK (Tel.: 44 181 455 3020; Telefax: 44 181 458 2064)

Palsgaard Industri A/S
> Palsgaardvej 10, 7130 Juelsminde Denmark (Tel.: 45 75 690122; Telefax: 45 75 690111; Telex: 60656/60799)

Paniplus
> 100 Paniplus Rdwy., Olathe, KS 66061, USA)

Park Tonks Ltd.
> 48 North Rd., Great Abington, Cambridge, CB1 6AS UK (Tel.: 44 1223 891721; Telefax: 44 1223 893571; Telex: 818832)

Parman-Kendall Corp.
> PO Box 157, Goulds, FL 33170 USA (Tel.: 305-258-1628; Telefax: 305-258-2445)

Particle Dynamics Inc.

2503 S. Hanley Rd., St. Louis, MO 63144 USA (Tel.: 314-968-2376; Telefax: 314-968-5208; Telex: 434182)

Par-Way/Tyson Co.

750 W. 17th St., Costa Mesa, CA 92627 USA (Tel.: 714-642-0076; Telefax: 714-642-0611)

Patco Products Div. *See* American Ingredients

Paulaur Corp.

2 Marlen Dr., Trenton, NJ 08691 USA (Tel.: 609-584-0066)

Payan & Bertrand

Ave. Jean XXIII, BP 57, 06332 Grasse France (Tel.: 33 93 40 14 14; Telefax: 33 93 40 10 30; Telex: 470882)

PB Gelatins, Div. of Tessenderlo Chemie

Marius Duchestraat 260, B-1800 Vilvoorde Belgium (Tel.: 32 2 251 3061; Telefax: 32 2 251 6428; Telex: 21131)

Peerless Food Products

Dunnings Bridge Rd., Bootle, Merseyside, L30 6TJ UK (Tel.: 44 151 525 5151; Telefax: 44 151 523 4110; Telex: 629345)

The Peerless Refining Co. (Liverpool) Ltd.

PO Box 15, Dunnings Bridge Rd., Bootle, Merseyside, L30 6TJ UK (Tel.: 44 151 530 1233; Telex: 629345)

Pembroek BV

Industrieweg 3-22, PO Box 34, 1230 AA Loosdrecht The Netherlands (Tel.: 31 215 821355; Telefax: 31 215 826614; Telex: 43215)

Penreco, Div. of Pennzoil Prods. Co.

138 Petrolia St., Box 1, Karns City, PA 16041 USA (Tel.: 412-756-0110; 800-245-3952; Telefax: 412-756-1050; Telex: 1561596)

Penta Manufacturing Co.

PO Box 1448, Fairfield, NJ 07007-1448 USA (Tel.: 201-740-2300; Telefax: 201-740-1839; Telex: 219472 PENT UR)

Pentagon Chemicals Ltd., Div. of Suter plc

Northside, Workington, Cumbria, CA14 1JJ UK (Tel.: 44 1900 604371; Telefax: 44 1900 66943; Telex: 64353 PENTA G)

Penwest Foods Co.

11011 E. Peakview Ave., Englewood, CO 80111-6800 USA (Tel.: 303-649-1900; Telefax: 303-649-1700)

Petrolite Corp./Headquarters & Polymers Div.

6910 E. 14th St., Tulsa, OK 74112 USA (Tel.: 918-836-1601; 800-331-5516; Telefax: 918-834-9718)

Pfaltz & Bauer Inc. *See* Aceto

Pfeifer & Langen

Frankenstrasse 25, PO Box 110320, D4047 Dormagen Germany (Tel.: 49 2106 520; Telefax: 49 2106 52216; Telex: 8517301)

Pfizer

Pfizer Inc./Chem. Div., Minerals, Pigments, Metals, 235 E. 42nd St., New York, NY 10017 USA (Tel.: 212-573-2323; 800-336-9008; Telefax: 212-573-2273; Telex: 420440)

Pfizer/Dairy & Brewery Div., 4215 N. Port Washington Rd., Milwaukee, WI 53212 USA (Tel.: 414-332-3545; 800-231-1590; Telex: ITT 420440)

Pfizer Food Science Group, 235 E. 42nd St., New York, NY 10017 USA (Tel.: 212-573-2323/2548; 800-TECK-SRV; Telefax: 212-573-1166)

Pfizer Canada, PO Box 800, Point Claire/Dorval, Montreal, Quebec, H9R 4V2 Canada (Tel.: 514-695-0500)

Pfizer Europe/Africa/Middle East, 10 Dover Rd., Sandwich, Kent, CT13 0BN UK (Tel.: 44 1304 615518; Telefax: 44 1304 615529; Telex: 966555)

Pfizer Asia/Australia, PO Box 57, West Ryde, NSW, 2114 Australia (Tel.: 61-2-858-9500)

Pfizer K.K., 3-22 Toranomon 2-chome, Minato-ku, Tokyo, 105 Japan (Tel.: 81-3-3503-0441)

C.A.L. Pfizer, 27, Ave. Saint-Lorette, 06130 Grasse France (Tel.: 33 93 36 08 69; Telefax: 33 93 36 81 73)

C.A.L.-Pfizer/Flavor & Fragrance Prod. Group, 230 Brighton Rd., Clifton, NJ 07012 USA (Tel.: 201-470-7892; 800-245-4495; Telefax: 201-470-7895)

Phytone Ltd.

Unit 2B, Boardman Ind. Estate, Hearthcote Rd., Swadlincote, Derbyshire, DE11 9DL UK (Tel.: 44 1283 550338; Telefax: 44 1283 550714)

Tryben Aromatics Inc., U.S.A. agent for Phytone Ltd., 4400 Route 9 South, Freehold, NJ 07728, USA)

J.W. Pike Ltd.

Unit 4 Eley Rd., Eley Estate, Edmonton, London, N18 3BH UK (Tel.: 44 181 807 9924; Telefax: 44 181 803 9972)

Pilot Chemical Co.

11756 Burke St., Santa Fe Springs, CA 90670 USA (Tel.: 213-723-0036; Telefax: 213-945-1877; Telex: 4991200 PILOT)

PLA-Fleischmann

Av. Tlahuac No. 4615, Col El Vergel, 09880 Mexico City Mexico (Tel.: 52 5 656 001 24; Telefax: 52 5 656 02 25)

PMC Specialties Group, Inc., Div. of PMC, Inc.

501 Murray Rd., Cincinnati, OH 45217 USA (Tel.: 513-242-3300; 800-543-2466; Telefax: 513-482-7353; Telex: 5106000948)

PMP Fermentation Products, Inc.

Columbia Center III, 9525 W. Bryn Mawr Ave., Suite 725, Rosemont, IL 60018 USA (Tel.: 708-928-0050; 800-558-1031; Telefax: 708-928-0065)

PMS Foods Inc.

PO Box 1099, Hutchinson, KS 67504 USA (Tel.: 316-663-5711; Telefax: 316-663-7195)

Pointing

Pointing Ltd., Princess Way, Prudhoe, Northumberland, NE42 6NJ UK (Tel.: 44 1661-832621; Telefax: 44 1661-835650; Telex: 537036 POINTX G)

Dinoval Chemicals Ltd., Div. of Pointing Ltd., Stamford House, Stamford New Rd., Altrincham, Cheshire, WA14 1BL UK (Tel.: 44 161-941 4254; Telefax: 44 161-926 8173; Telex: 668839)

Polarome Mfg. Co. Inc.

200 Theodore Conrad Dr., Jersey City, NJ 07305 USA (Tel.: 201-309-4500; Telefax: 201-433-0638; Telex: RCA 233 176)

Pomosin GmbH

Kleine Markstrasse 15, D6050 Offenbach/M, Germany)

PPG

PPG Industries, Inc., One PPG Place, Pittsburgh, PA 15272 USA (Tel.: 412-434-2414; 800-CHEM-PPG; Telex: 86 6570)

PPG Industries, Inc./Specialty Chemicals, 3938 Porett Dr., Gurnee, IL 60031 USA (Tel.: 708-244-3410; 800-323-0856; Telefax: 708-244-9633; Telex: 25-3310)

PPG Canada Inc./Specialty Chem., 2 Robert Speck Pkwy., Suite 900, Mississauga, Ontario, L4Z 1H8 Canada (Tel.: 905-848-2500; Telefax: 905-848-2185; Telex: 38906960351canbizmis)

PPG-Mazer Mexico, S.A. de C.V., Av. Presidente Juarex No. 1978, Tlalnepantla, Edo., C.P. 54090 Mexico (Tel.: 52-5-397-8222; Telefax: 52-5-398-5133)

PPG Industrial do Brazil Ltd.a., Edificio Grande Avenida, Paulista Ave. 1754, Suite 153, 01310 Sao Paulo Brazil (Tel.: 55-011-284-0433; Telefax: 55-011-289-2105; Telex: 011-39104 PPGB BR)

PPG Ouvrie S.A., 64, rue Faldherbe, B.P. 127, 59811 Lesquin Cedex France (Tel.: 33-2087-0510; Telefax: 33-2087-5631; Telex: 131419 F)

PPG Industries Taiwan, Ltd., Suite 601, Worldwide House, No. 131, Ming East Rd., Sec. 3, Taipei, 105, Taiwan, R.O.C. (Tel.: 886-2-514-8052; Telefax: 886-2-514-7957; Telex: 10985 PPGTWN)

PPG Industries-Asia/Pacific Ltd., Takanawa Court, 5th floor, 13-1 Takanawa 3-Chome, Minato-Ku, Tokyo, 108 Japan (Tel.: (03) 3280-2911; Telefax: (03) 3280-2920; Telex: 02-42719 PPGPACJ)

PQ Corp.

PO Box 840, Valley Forge, PA 19482 USA (Tel.: 215-293-7200; 800-944-7411; Telefax: 215-688-3835; Telex: 476 1129 PQCO VAF)

Premier Malt Products, Inc.

PO Box 36359, Grosse Pointe, MI 48236 USA (Tel.: 313-822-2200; 800-521-1057)

J.L. Priestley & Co. Ltd.

Station Rd., Heckington, Sleaford, Lincs, NG34 9NF UK (Tel.: 44 1529 60751; Telefax: 44 1529 60630; Telex: 377189)

Procter & Gamble

Procter & Gamble Co/Chemicals Div., PO Box 599, Cincinnati, OH 45201 USA (Tel.: 513-983-3928; 800-543-1580; Telefax: 513-983-1436; Telex: 21-4185, P&GCIN)

Procter & Gamble Ltd./Europe Div., PO Box 9, 27 Uxbridge Rd., Hayes, Middlesex, UB4 0JD UK (Tel.: 44 181 848 9671; Telex: 936310)

Produits Roche

523 Bd du Parc, F92521 Neuilly/Seine Cedex France (Tel.: 33 46 405000; Telefax: 33 46 405282; Telex: 6121402)

Prolait Euro Proteine

14 Rue d'Inkermann, F79000 Niort France (Tel.: 33 49 240942; Telefax: 33 49 241939; Telex: 790672)

Pronova Biocare AS

Standveien 5, N-1324 Lysaker Norway (Tel.: 47 67 590080; Telefax: 47 67 583450; Telex: 8320351)

Pronova Biopolymer

Pronova Biopolymer A/S, Tomtegt. 36, Postboks 494, N-3002 Drammen Norway (Tel.: 47-32-83 73 00; Telefax: 47-32-83 34 88; Telex: 76594 prota n)

Pronova Biopolymer Ltd., PO Box 8, Alton, Hampshire, GU34 1YL UK (Tel.: 44 1420 82503; Telefax: 44 1420 83360)

Pronova Biopolymer, Inc., 135 Commerce Way, Suite 201, Portsmouth, NH 03801 USA (Tel.: 603-433-1231; 800-223-9030; Telefax: 603-433-1348)

Pronova Biopolymer Asia Ltd., Unit 1401-02 14/F Shun Kwong, Commercial Bldg., 8 Des Voeux Rd. W. Hong Kong (Tel.: 852-517 3028; Telefax: 852 517 3198)

Protan Ltd.

Westbrook House, High St., Alton, Hampshire GU34 1EZ UK (Tel.: 44 1420 82503; Telefax: 44 1420 83360; Telex: 859078)

Protein Foods Scandinavia AS

PO Box 30, 6300 Graasten Denmark (Tel.: 45 74 652065; Telefax: 45 74 652064)

Protein Technologies International

Protein Technologies International, Excelsiorlaan 13, B-1930 Zaventem Belgium (Tel.: 32 2 720 9544; Telefax: 32 2 720 6755; Telex: 64388)

Protein Technologies International, 16 Princewood Rd., Earlstrees Ind. Estate, Corby, Northants, MN17 2AP UK (Tel.: 44 1536 67325; Telefax: 44 1536 61147; Telex: 042546)

Protein Technologies International, One Checkerboard Square, St. Louis, MO 63164 USA (Tel.: 314-982-1983; 800-325-7108; Telefax: 314-982-5057)

Protein Technologies International, 2500 Royal Windsor Dr., Mississauga, Ontario, L5J 1K8 Canada (Tel.: 416-822-1998; Telefax: 416-855-5700)

Fuji Purina Protein, Ltd., SKF Bldg. 6F, 1-9, 1-Chome Shiba Daimon, Minato-ku, Tokyo Japan (Tel.: 81-3-3438-2021; Telefax: 81-3-3438-2654)

Protex

6 Rue Barbes, 92305 Levallois, France

Prova SA

46 rue Colmet Lepinay, 931000 Montreuil France (Tel.: 33 1 42 873676; Telefax: 33 1 42 871013; Telex: 212513)

Provesta Corp.

15 Phillips Bldg., Bartlesville, OK 74004 USA (Tel.: 918-661-5281; Telefax: 918-662-2208)

Puccinelli SpA

Via Manfredini 34, Rovigo 45100 Italy (Tel.: 39 425 35333; Telefax: 39 425 361645; Telex: 480843/434607)

Pulcra SA

Sector E C/42, Barcelona, 08040 Spain (Tel.: 3-323-5914; Telefax: 3-323-6760; Telex: 98301)

Purac

Purac America, Inc., 11 Barclay Blvd., Suite 280, Lincolnshire Corporate Center, Lincolnshire, IL 60069 USA (Tel.: 708-634-6330; Telefax: 708-634-1992; Telex: 280231 PURACINC. ARHT)

Purac Biochem BV, Div. of CSM NV, Postbus 21, Arkelsedijk 46, NL-4200 AA Gorinchem The Netherlands (Tel.: 1830-41799; Telefax: 1830-22741; Telex: 23615 CCA NL)

Purac Biochem (UK), 50-54 St. Paul's Square, Birmingham, B3 1QS UK (Tel.: 44 121 2361828; Telefax: 44 121 2361401)

Purac Far East PTE Ltd., 09-02 Cecil Court, 138 Cecil St., S-0106 Singapore (Tel.: 65 220 6022; Telefax: 65 222 1707; Telex: 23463)

Pure Food Ingredients Inc.

PO Box 265, Verona, WI 53593 USA (Tel.: 608-845-9601)

Pure Malt Products Ltd.

Victoria Bridge, Haddington, East Lothian, EH41 4BD, Scotland (Tel.: 44 162 082 4696; Telefax: 44 162 082 2018; Telex: 728158)

Quadrant Holdings Cambridge Ltd.

Maris Lane, Cambridge, CB2 2SY UK (Tel.: 44 1223 845779; Telefax: 44 1223 842614)

Qualcepts Nutrients Inc.

4940 Viking Dr., Minneapolis, MN 55435 USA (Tel.. 012-093-9970)

Quest International

Quest International Inc., PO Box 630, Woods Corners, Norwich, NY 13815 USA (Tel.: 607-334-9951; Telefax: 607-334-5022; Telex: 646056)

Quest International Flavors & Food Ingreds. Co., 10 Painters Mill Rd., Owings Mills, MD 21117-3686 USA (Tel.: 410-363-2550; Telefax: 410-363-7514; Telex: 87915)

Quest International Fragrances USA, Inc., 400 International Dr., Mt. Olive, NJ 07828 USA (Tel.: 201-691-7100; Telefax: 201-691-7479; Telex: 6714933)

Quest International, Bromborough Port, Wirral, Merseyside, L62 4SU UK (Tel.: 44 151 645 2060; Telefax: 44 151 645 6975; Telex: 627173)

Quest International, Postbus 2, 1400 Bussum The Netherlands (Tel.: 31 2159-99111; Telefax: 31 2159-46067; Telex: 43050 QSTN NL)

Quest International Deutschland GmbH, Postfach 650170, Poppenbütteler Chaussee 36, W-2000 Hamburg 65 Germany (Tel.: 40-607970; Telefax: 40-6079710; Telex: 215196)

Quest International Australasia Pty. Ltd., 12 Britton St., Smithfield, NSW, 2164 Australia (Tel.: 61 2 827 4000; Telefax: 61 2 604 7926)

Quimdis

24 Ave. de Gresillons, 92601 Asnieres France (Tel.: 33 1 47 902580; Telefax: 33 1 47 339096; Telex: 614450)

Quimica Universal SA

Ave. Los Platinos 259, PO Box 2198, 100 Lima, Peru (Tel.: 51 14 706656; Telefax: 51 14 706646; Telex: 21301)

Ragus Sugars

193 Bedford Ave., Slough, Berks, SL1 4RT UK (Tel.: 44 1753 575353; Telefax: 44 1753 691514)

Rank Hovis Ltd.

King Edward House, 27/30 King Edward Court, Windsor, Berks, SL4 1TJ UK (Tel.: 44 1753 840401; Telefax: 44 1753 840401; Telex: 849326)

William Ransom & Son plc

104 Bancroft, Hitchin, Herts, SG5 1LY UK (Tel.: 44 1462 437615; Telefax: 44 1462 420528; Telex: 825631)

Raps & Co.

PO Box 1849, D 8650 Kulmbach Germany (Tel.: 49 9221 807125; Telefax: 49 9221 807100; Telex: 642540)

Raps (UK) Ltd.

6 Oxford Court, St. James Rd., Brackley, Northants, NN13 5XY UK (Tel.: 44 1280 705513; Telefax: 44 1280 705514; Telex: 642540)

Raschig Corp.

5000 Old Osborne Tpke., Box 7656, Richmond, VA 23231 USA (Tel.: 804-222-9516; Telefax: 804-226-1569)

RCR Scientific Inc.

PO Box 340, Goshen, IN 46526 USA (Tel.: 219-533-3351; Telefax: 219-533-3370)

Red Carnation Gums Ltd.

St. John Lyon House, 5 High Timber St., Upper Thames St., London, EC4V 3PA UK (Tel.: 44 171 236 8560; Telefax: 44 171 489 8427; Telex: 8956136)

Redell Industries Inc.

11686 Sheldon St., PO Box 4299, Pacoima, CA 91331 USA (Tel.: 818-767-2038; Telex: 910 337 1285)

Red Star Specialty Products

Red Star Specialty Products, 433 E. Michigan St., Milwaukee, WI 53202 USA (Tel.: 414-347-3936; Telefax: 414-347-3912)

Red Star Specialty Products, Researchpark Hasrode, Interleuvenlaan 37, B-3001 Heverlee Belgium (Tel.: 32 16 40 0465; Telefax: 32 16 40 0565)

Reedy International Corp.

42 First St., Keyport, NJ 07735 USA (Tel.: 908-264-1777; Telefax: 908-264-1189)

Regency Mowbray Co. Ltd.

Regency House, Hixon Ind. Estate, Hixon, Stafford, ST18 0PY UK (Tel.: 44 1889 270554; Telefax: 44 1889 270927)

Reheis Inc.

500 N. Ninth St., PO Box 921, Midlothian, TX 76065 USA (Tel.: 214-775-2307; Telefax: 214-775-3872)

Remy Industries SA, Remylaan 4, B-3018 Wijgmaal-Leuven Belgium (Tel.: 32 16 248511; Telefax: 32 16 440144; Telex: 61392 Rempro B)

A&B Ingredients Inc., U.S. distributor for Remy Industries SA, 24 Spielman Rd., Fairfield, NJ 07004 USA (Tel.: 201-227-1390; Telefax: 201-227-0172)

J.F. Renshaw Ltd.

Crown St., Liverpool, L8 5RF, UK)

J. Rettenmaier & Sohne GmbH & Co.

Faserstoffwerke, D-73494 Rosenberg/Holzmühle Germany (Tel.: 49 7967 1520; Telefax: 49 7967 6111)

RHM Ingredients Ltd.

Owl Lane, Ossett, Wakefield, W Yorkshire, WF5 9AX UK (Tel.: 44 1924 280444; Telefax: 44 1924 281042; Telex: 557491)

Rhone-Poulenc

Rhone-Poulenc Chimie (France)/Secteur Specialites Chimiques, Cedex 29, F-92097 Paris La Défense France (Tel.: 47-68 02 01; Telefax: 47-68 13 31)

Rhone-Poulenc, 25 quai Paul Doumer, 92408 Courbevoie Cedex France (Tel.: 33 1 47 68 12 34; Telex: 610500)

Rhone-Poulenc Chemicals Ltd., Div. of Rhone-Poulenc SA, Oak House, Reeds Crescent, Watford, Herts, WD1 1QH UK (Tel.: 44 181 984 3342; Telefax: 44 181 984 1701; Telex: 28691)

Rhone-Poulenc ABM Brewing & Enzyme Group, Poleacre Lane, Woodley, Stockport, Cheshire, SK6 1PQ UK (Tel.: 44 161 430 4391; Telefax: 44 161 430 8523; Telex: 667835)

Rhone-Poulenc Basic Chemical Co., One Corporate Dr., Box 881, Shelton, CT 06484 USA (Tel.: 203-925-3300; 800-642-4200; Telefax: 203-925-3627)

Rhone-Poulenc Food Ingredients, CN 7500, Prospect Plains Rd., Cranbury, NJ 08512 USA (Tel.: 609-860-4600; 800-253-5052)

Rhone-Poulenc Rorer Pharmaceuticals Inc., 500 Arcola Road, Collegeville, PA 19426-2911 USA (Tel.: 215-454-8000)

Rhone-Poulenc, Inc./Surfactants & Specialties, CN 7500, Prospect Plains Rd., Cranbury, NJ 08512-7500 USA (Tel.: 609-860-8300; 800-922-2189; Telefax: 609-860-7626)

Rhone-Poulenc, Inc./Water Soluble Polymers, CN 7500, Prospect Plains Rd., Cranbury, NJ 08512-7500 USA (Tel.: 609-860-8300; 800-922-2189; Telefax: 609-860-7626)

American Lecithin Co., Div. of Rhone-Poulenc Rorer/Nattermann, 33 Turner Rd., Danbury, CT 06813-1908 USA (Tel.: 203-790-2700; Telefax: 203-790-2705)

Riceland Foods, Inc.
PO Box 8201, Little Rock, AR, 72221 USA (Tel.: 501-225-0936; Telefax: 501-225-9179)

Rieber & Son
Rieber & Son (UK) Ltd., 32 North Hill, Colchester, Essex, CO1 1QR UK (Tel.: 44 1206 762702; Telefax: 44 1206 761931)

Rieber & Son AS, PO Box 987, Nostegaten 58, N-5002 Bergen Norway (Tel.: 47 5 96 7000; Telefax: 47 5 96 7691; Telex: 42052)

Riken Vitamin Oil Co., Ltd.
TDC Bldg., 9-18, Misaki-Cho 2-chome, Chiyoda-Ku, Tokyo, 101 Japan (Tel.: 81 3 5275 5130; Telefax: 81 3 5275 2905; Telex: 2322783)

Ringdex
16 Rue Ballu, F-75009 Paris France (Tel.: 33/1 40 82 35 95; Telefax: 33/1 42 85 17 17; Telex: 650847 F)

R.I.T.A. Corp.
1725 Kilkenny Court, PO Box 585, Woodstock, IL 60098 USA (Tel.: 815-337-2500; 800-426-7759; Telefax: 815-337-2522; Telex: 72-2438)

Rit-Chem Co. Inc.
109 Wheeler Ave., PO Box 435, Pleasantville, NY 10570 USA (Tel.: 914-769-9110; Telefax: 914-769-1408; Telex: 229 639 RTCH)

Riza NV
Metropoolstraat 35, 2120 Schote Antwerp Belgium (Tel.: 32 3 646 7800; Telefax: 32 3 646 4237; Telex: 31486)

Robert Bryce & Co. *See* Bryce

Robertet
Robertet & Cie, 36 ave. Sidi Brahim, BP 100, 06333 Grasse Cedex France (Tel.: 33 93 40 3366; Telefax: 33 93 70 6809; Telex: 470863)

Robertet Flavors Inc., 640 Montrose Ave., PO Box 247, S. Plainfield, NJ 07080 USA (Tel.: 908-561-2181; Telefax: 908-561-7396)

Robin Chemicals Private Ltd.
421 Parekh Market, Opera House, Bombay, 400 004 India (Tel.: 91 91 22 205 2745; Telefax: 91 92 22 2018054)

Roche Vitamins & Fine Chemicals, Div. of Hoffman-La Roche Inc.
340 Kingsland St., Nutley, NJ 07110-1199 USA (Tel.: 201-235-5000; Telefax: 201-535-7606)

Roche Products Ltd./Vitamins & Fine Chemicals Div., Div. of Roche AG
PO Box 8, 40 Broadwater Road, Welwyn Garden City, Herts, AL7 3AY UK (Tel.: 44 1707-328128; Telefax: 44 1707-329587; Telex: 262098 ROCHEW G)

Roehm Ltd./Hexoran Div.
Derwent St., Belper, Derby, DE5 1WQ UK (Tel.: 44 1773 822471; Telefax: 44 1773 820176; Telex: 377716)

Roeper GmbH & Co. CE
Hans-Duncker-Str. 13, D2050 Hamburg 80 Germany (Tel.: 49 40 7341030; Telefax: 49 40 73410335; Telex: 211079)

Rohm GmbH
D64275 Darmstadt Germany (Tel.: 49 6151 1801; Telefax: 49 6151 184193; Telex: 419474)

Roland Industries Inc.
2280 Chaffee Dr., St. Louis, MO 63146 USA (Tel.: 314-567-3800)

Rona, Div. of EM Industries, Inc.
5 Skyline Dr., Hawthorne, NY 10532 USA (Tel.: 914-592-4660; Telefax: 914-592-9469; Telex: 17-8993)

Roquette (UK) Ltd.
Pantiles House, 2 Nevill St., Tunbridge Wells, Kent, TN2 5TT UK (Tel.: 44 1892-540188; Telefax: 44 1892-510872; Telex: 957558 G)

Rose International
PO Box 5020, Santa Rosa, CA 94502 USA (Tel.: 707-576-7050; Telefax: 707-545-7116)

Ross Chemical, Inc.
303 Dale Dr., PO Box 458, Fountain Inn, SC 29644 USA (Tel.: 803-862-4474; 800-521-8246; Telefax: 803-862-2912)

Frank B. Ross Co., Inc.
22 Halladay St., PO Box 4085, Jersey City, NJ 07304-0085 USA (Tel.: 201-433-4512; Telefax: 201-332-3555)

Ross & Rowe Lecithin, Div. Archer-Daniels-Midland. *See under* ADM

Rowse Honey Ltd.
Morton Ave., Wallingford, Oxon, OX10 9DE UK (Tel.: 44 1491 33122; Telefax: 44 1491 25051)

W. Ruitenberg & ZN BV
Postbus 44, 3800 AA Amersfoort The Netherlands (Tel.: 31 33 621364; Telefax: 31 33 633548; Telex: 79153)

SAF Bulk Chemicals. *See* Aldrich

Salga SpA
Via S Cassiano 76, 28069 Trecate Italy (Tel.: 39 321 76001; Telefax: 39 321 71341; Telex: 200003)

Sandoz Nutrition
5320 West 23rd St., Minneapolis, MN 55416 USA (Tel.: 612-925-2100; Telefax: 612-593-2087)

San Giorgio Flavours SpA
Via Fossata 114, 10147 Torino Italy (Tel.: 39 11 216 0991; Telefax: 39 11 221 7273; Telex: 221110)

San-J International Inc.
2880 Sprouse Dr., Richmond, VA 23231 USA (Tel.: 804-226-8333; Telefax: 804-226-8383; Telex: 5101000513)

Sanofi Bio-Industries
Sanofi Bio-Industries, 66 ave. Marceau, 75008 Paris France (Tel.: 33 1 40 73 20 80; Telefax: 33 1 40 73 28 27)

Sanofi UK. *Address unknown*

Sanofi Bio-Industries Inc./N. Amer. Hdq., 8 Neshaminy Interplex, Suite 213, Trevose, PA 19053 USA (Tel.: 215-638-7801; Telefax: 215-638-8168)

Sanofi Bio-Industries Inc./Food Texture Div., 620 Progress Ave., PO Box 1609, Waukesha, WI 53187 USA (Tel.: 414-547-5531)

Sanovo Foods A/S
PO Box 139, DK-5000 Odense C Denmark (Tel.: 45 66 111732; Telefax: 45 66 141132; Telex: 59410)

Sanyo Chemical Industries, Ltd.
11-1, Ikkyo Nomoto-cho, Higashiyama-ku, Kyoto, 605 Japan (Tel.: (075) 541-4311; Telefax: (075) 551-2557; Telex: 05422110)

SAPA
23 rue de la Fraternite, BP 30, 95460 Ezanville France (Tel.: 33 1 39 91 9300; Telefax: 33 1 39 91 1926; Telex: 606040)

Saveur SA
ZI du Pommeret, 35310 Breal sous Monfort France (Tel.: 33 99 60 00760; Telefax: 33 99 60 0842; Telex: 741635)

Scan American Seafood Co. Inc.
1410 80th St. SW, Everett, WA 98203 USA (Tel.: 206-514-0500; Telefax: 206-514-0400)

Scanchem UK Ltd.
16 Jordangate, Macclesfield, Cheshire, SK10 1EW UK (Tel.: 44 1625 511222; Telefax: 44 1625 511391)

Dr. Scholvien GmbH & Co.
Am Schlangengraben 5, D1000 Berlin 20 Germany (Tel.: 49 30 330 8730; Telefax: 49 30 330 8711)

Scientific Protein Laboratories, Div. of Viobin Corp.
PO Box 158, 700 E. Main St., Waunakee, WI 53597 USA (Tel.: 608-849-5944; 800-334-4SPL; Telefax: 608-849-4053; Telex: 26-5479)

SCI Natural Ingredients
4 Kings Rd., Reading, Berks, RG1 3AA UK (Tel.: 44 1734 580247; Telefax: 44 1734 589580; Telex: 847746)

Scottish Pride Quality Dairy Foods
Underwood Rd., Paisley, PA3 1TJ UK (Tel.: 44 141 887 3288; Telefax: 44 141 889 1225; Telex: 779012)

Seafia Inc.
11371 Williamson Rd., Cincinnati, OH 45241 USA (Tel.: 513-489-3331)

Seah International
PO Box 275, 62204 Boulogne sur Mer France (Tel.: 33 21 32 2020; Telofax: 33 21 32 2828; Telex: 110900)

Sefcol (Sales) Ltd.
Whitehouse Ind. Estate, Runcorn, Cheshire, WA7 3BJ UK (Tel.: 44 1928 713121; Telefax: 44 1928 712827)

Selarom Srl
Via Modena 3-5, 20090 Buccinasco (MI) Italy (Tel.: 39 2 4884 3800; Telefax: 39 2 488 3776; Telex: 350864)

Semco Laboratories Inc.
630 E. Keefe Ave., Milwaukee, WI 53212 USA (Tel.: 414-964-3899; Telefax: 414-964-0655)

Seppic, Div. of L'Air Liquide
75 Quai d'Orsay, F-75321 Paris Cedex 07 France (Tel.: 40 62 55 55; Telefax: 40 62 52 53; Telex: 202901 SEPPI F)

Servo Delden B.V.
Postbus 1, NL-7490 AA Delden The Netherlands (Tel.: 5407-63535; Telefax: 5407-64125; Telex: 44347)

Sethness-Greanleaf Inc.
1826 N. Lorel Ave., Chicago, IL 60639 USA (Tel.: 312-889-1400; Telefax: 312-889-0854)

John F. Seyfried & Sons Ltd.
Mickleton, Chipping Campden, Glos, GL55 6SS UK (Tel.: 44 1386 438521; Telefax: 44 1386 438871; Telex: 338137)

S & G Resources Inc.
266 Main St., Medfield, MA 02052 USA (Tel.: 508-359-7771; Telefax: 508-359-7775; Telex: 984586)

Shade Foods Inc.
33063 Western Ave., Union City, CA 94587 USA (Tel.: 415-429-1290; Telefax: 415-429-1629)

Sharon Laboratories Israel
Industrial Zone Ad Halom, Ashdod 77106, Israel

Shemberg USA
PO Box 252, Kidder Rd., Searsport, ME 04974 USA (Tel.: 207-548-2921; Telefax: 207-548-2921)

Shemberg-Hydralco
Bilker Allee 55, 4000 Dusseldorf Germany (Tel.: 49 211 394056; Telefax: 49 211 3982181)

Shimizu Chemical Corp.
2F Foresight Bldg., 3-5-2 Iwamoto-cho Chiyoda-ku, Tokyo, 101 Japan (Tel.: 81 3861 7591; Telefax: 81 3861 7597)

Shin-Etsu Chemical Co., Ltd.
6-1, Otemachi 2-chome, Chiyoda-ku, Tokyo, 100 Japan (Tel.: 81 3 246 5280; Telefax: 81 3 246 5371; Telex: SHINC.HEM J-24790)

Shipton Mill Ltd.
Long Newnton, Tetbury, Glos, GL8 8RP UK (Tel.: 44 1666 505050; Telefax: 44 1666 504666; Telex: 0285 437287)

Showa Ether Co., Inc.
Taiyo Mutual Life Insurance Bldg., 11-2, Nihonbashi 2-chome, Chuo-ku, Tokyo, 103 Japan (Tel.: (03) 3271-2401; Telefax: (03) 3271-2414)

Siber Hegner Ltd.
MacKenzie House, 221-241 Beckenham Rd., Bechenham, Kent, BR3 4UF UK (Tel.: 44 181 659 2345; Telefax: 44 181 659 1292; Telex: 946651)

Silesia (UK) Ltd.
35 The Mall, Ealing, London, W5 3TJ UK (Tel.: 44 181 579 4314; Telefax: 44 181 579 9228; Telex: 935600)

Silesia Gerhard Hanke KG
PF 21 05 54, D41431 Neuss Germany (Tel.: 49 2137/784-0; Telefax: 49 2137/78411; Telex: 8517716)

Silva Ltd.
19 Station Rd., Chinnor, Oxon, OX9 4PU UK (Tel.: 44 1844 354488; Telefax: 44 1844 343499)

Sime Darby Kempas Edible Oil Sdn Bhd
255 Jalan Boon Lay, PO Box 50, Jurong Town Singapore (Tel.: 65 264 3733; Telefax: 65 265 5129; Telex: 24070)

Simplesse Europe
43 rue des Tilleuls, F92100 Boulogne-Billancourt France (Tel.: 33 1 48 25 7878; Telefax: 33 1 48 25 2538)

SKW Chemicals Inc.
4651 Olde Towne Pkwy., Suite 200, Marietta, GA 30068 USA (Tel.: 404-971-1317; Telefax: 404-971-4306)

Soc Ind & Agric de Bretagne
PO Box 17, 35220 Chateaubourg France (Tel.: 33 9962 3823; Telefax: 33 9900 7271)

Soda Aromatic USA Inc.
The Chrysler Bldg., Suite 4920, 405 Lexington Ave., New York, NY 10174 USA (Tel.: 212-557-2071; Telefax: 212-856-9502)

Sofalia
Zone Agro-Industrielle, BP 20, 63720 Ennezat France (Tel.: 33 73 63 4444; Telefax: 33 73 63 4445)

Sogip Protéines Végétales
73/77 rue de Sèvres, BP 72, 92105 Boulogne Billancourt France (Tel.: 33 1 6404 5959; Telefax: 33 1 4825 0754; Telex: 633170)

Solvay Duphar BV
Postbus 900, 1380 DA Weesp The Netherlands (Tel.: 31 2940-77000; Telefax: 31 2940-80253; Telex: 14232)

Solvay Enzymes, Inc.
PO Box 4859, 1230 Randolph St., Elkhart, IN 46514 USA (Tel.: 219-523-3700; 800-487-4704; Telefax: 219-523-3800)

Solvay Biosciences Pty. Ltd.
530 Springvale Rd., Glen Waverley, Victoria, 3150 Australia (Tel.: 61 3 560 8899; Telefax: 61 3 560 8184)

Sorrento Cheese Co.
2375 South Park Ave., Buffalo, NY 14220 USA (Tel.: 716-823-6262; Telefax: 716-823-0448)

Soya Mainz
Dammweg 2, PO Box 3767, D6500 Mainz Germany (Tel.: 49 6131 895-0; Telefax: 49 6131 834104; Telex: 4187761)

Span-Concentres Noel
7 ave. de la République, PO Box 28, F41150 Onzain France (Tel.: 33 5420 8674; Telefax: 33 5420 8563)

SPCI
58 rue du Landy, F93121 Paris France (Tel.: 33 1 4933 3115; Telefax: 33 1 4243 8223)

Specialty Food Ingredients Europa BV
PO Box 5066, 5004 EB Tilburg The Netherlands (Tel.: 31 13 684991; Telefax: 31 13 634109)

Spectrum Chemical Mfg. Corp.
14422 S. San Pedro St., Gardena, CA 90248 USA (Tel.: 213-516-8000; 800-772-8786; Telefax: 213-516-9843; Telex: 182395)

Spectrum Naturals
133 Copeland St., Pentaluma, CA 94952 USA (Tel.: 707-778-8900; Telefax: 707-765-1026)

SPI (Société De Proteines Industrielles)
La Flachec, 56230 Berric France (Tel.: 33 97 670101; Telefax: 33 97 670146; Telex: 950589)

Spice King Corp.
6000 Washington Blvd., Culver City, CA 90232-7488 USA (Tel.: 213-836-7770; Telefax: 213-836-6454; Telex: 664350)

Spicemanns Ltd.
59 Kelvi Ave., Hillingdon Ind. Estate, Glasgow, G52 4LR UK (Tel.: 44 141 883 4707; Telefax: 44 141 810 5242; Telex: 778692)

Spillers Milling Ltd.
180 Aztec West, Almondsbury, Bristol, Avon, BS12 4TH UK (Tel.: 44 1454 201686; Telefax: 44 1454 201686)

New Malden House, 1 Blagden Rd., New Malden, Surrey, KT3 4TB UK (Tel.: 44 181 949 6100; Telefax: 44 181 715 1403; Telex: 896660)

Spillers Premier Products
Station Rd., Cambridge, CB1 2JN UK (Tel.: 44 1223 460666; Telefax: 44 1223 352012)

S.S.T. Corp.
635 Brighton Rd., PO Box 1649, Clifton, NJ 07015-1649 USA (Tel.: 201-473-4300; Telefax: 201-473-4326)

Stabilized Products Inc.
Box 22002, St. Louis, MO 63126 USA (Tel.: 314-677-5764; Telefax: 314-376-5811)

A.E. Staley Manufacturing Co., Subsid of Tate & Lyle PLC
2200 E. Eldorado St., PO Box 151, Decatur, IL 62525 USA (Tel.: 217-423-4411; 800-258-7536; Telefax: 217-421-2881)

Staple Dairy Products Ltd.
Cookhamdene, Manor Park, Chislehurst, Kent, BR7 5QD UK (Tel.: 44 181 467 5505; Telefax: 44 181 467 2350; Telex: 896685)

Star Kay White Inc.
85 Brenner Dr., Congers, NY 10920 USA (Tel.: 914-268-2600; Telefax: 914-268-3572)

Stepan

Stepan Co., 22 West Frontage Rd., Northfield, IL 60093 USA (Tel.: 708-446-7500; 800-745-7837; Telefax: 708-501-2443; Telex: 910-992-1437)

Stepan/Food Ingredients Dept., 100 West Hunter Ave., Maywood, NJ 07607 USA (Tel.: 201-845-3030; Telefax: 201-845-6754; Telex: 710-990-5170)

Stepan Canada, 90 Matheson Blvd. W., Suite 201, Mississauga, Ontario, L5R 3P3 Canada (Tel.: 416-507-1631; Telefax: 416-507-1633)

Stepan Europe, BP127, 38340 Voreppe France (Tel.: 7650-8133; Telefax: 7656-7165; Telex: 320511 F)

Stern-France SARL

40 ave. Gustave Eiffel, F37100 Tours France (Tel.: 33 47 490970; Telefax: 33 47 546633)

Stern Lecithin & Soja GmbH & Co. KG. *See* Central Soya

J. Stewart & Co.

1440 Hicks Rd., Suite C, Rolling Meadows, IL 60008 USA (Tel.: 708-259-9555; Telefax: 708-259-6984)

Stockhausen, Inc.

2408 Doyle St., Greensboro, NC 27406 USA (Tel.: 919-333-3500; Telefax: 919-333-3545; Telex: 574405)

Stork Fibron BV

PO Box 292, 5340 AG Oss, The Netherlands)

Sumitomo Pharmaceuticals Co., Ltd., Subsid of Sumitomo Chemical Co., Ltd.

40, Doshomachi 2-chome, Higashi-ku, Osaka, 541 Japan (Tel.: (06) 229-8925; Telefax: (06) 202-7370; Telex: 522-2115 SUMYAK J)

Sunette Brand Sweetener

25 Worlds Fair Dr., Somerset, NJ 08873 USA (Tel.: 908-271-7220; Telefax: 908-271-7235)

Sutton Laboratories, Inc., Member of the ISP Inc. Group

116 Summit Ave., PO Box 837, Chatham, NJ 07928-0837 USA (Tel.: 201-635-1551; Telefax: 201-635-4964; Telex: 710-999-5607)

Dr. Otto Suwelack Nachf. GmbH & Co.

Josef-Suwelack-Strasse, PO Box 1362, D4425 Billerbeck Germany (Tel.: 49 2543 72-0; Telefax: 49 2543 72-222; Telex: TXX 254340)

SVZ International

Oude Kerkstraat 8, PO Box 27, 4870 AA Etten-Leur The Netherlands (Tel.: 31 1608 27321; Telefax: 31 1608 13321)

Sweeteners Plus Inc.

3239 Rochester Rd., Lakeville, NY 14480 USA (Tel.: 716-346-2318; Telefax: 716-346-2310)

Swiss Valley Farms Co.

Box 4493, Davenport, IA 52808 USA (Tel.: 319-391-3341; Telefax: 319-319-7479)

Synthetic Products Co., Subsid. of Cookson America Inc.

1000 Wayside Rd., Cleveland, OH 44110 USA (Tel.: 216-531-6010; 800-321-4236; Telefax: 216-486-6638)

Synthite Industrial Chemicals Ltd.

Kadayiruppu, 632 311 Kolenchery India (Tel.: 91 484 354616; Telefax: 91 484 370405; Telex: 8856593)

Taiwan Surfactant Corp.

No. 106, 8-1 Floor, Sec. 2, Chung An E. Rd., Taipei, Taiwan, R.O.C. (Tel.: 886-2-507-9155; Telefax: 886-2-507-7011; Telex: 27568 surfact)

Takeda

Takeda Europe GmbH, Domstrasse 17, 20095 Hamburg Germany (Tel.: 49 40 329050; Telefax: 49 40 327506; Telex: 2161408)

Takeda USA, Inc., 8 Corporate Dr., Orangeburg, NY 10962-2614 USA (Tel.: 914-365-2080; 800-825-3328; Telefax: 914-365-2786; Telex: 421149)

Tartaric Chemicals Corp.
515 Madison Ave., New York, NY 10022 USA (Tel.: 212-752-0727; Telefax: 212-207-8037)

Tastemaker
Tastemaker, 1199 Edison Dr., Cincinnati, OH 45216 USA (Tel.: 800-892-1199; Telefax: 513-948-3558)

Tastemaker, Chippenham Dr., Kingston, Milton Keynes, MK10 0AE UK (Tel.: 44 1908 242424; Telefax: 44 1908 282232; Telex: 825989)

Tastemaker BV, PO Box 414, Nijverheidsweg 60, NL-3770 AK Barneveld The Netherlands (Tel.: 31 3420 10411; Telefax: 31 3420 16605; Telex: 70115)

Taxarome Inc.
PO Box 157, Leakey, TX 78873 USA (Tel.: 210-232-6079; Telefax: 210-232-5716)

H.B. Taylor Co.
4830 S. Christiana Ave., Chicago, IL 60632 USA (Tel.: 312-254-4805; Telefax: 312-254-4563)

Richard Taylor Co. Inc.
2485 Huntington Dr. #2, San Marino, CA 91108 USA (Tel.: 818-304-9544; Telefax: 818-304-9542)

Tenneco, Inc. *See* Albright & Wilson

Terry Laboratories, Inc.
390 Wickham Rd. N., Suite F, Melbourne, FL 32935-8647 USA (Tel.: 407-259-1630; 800-367-2563; Telefax: 407-242-0625)

Tesco
Hansaalee 177 c, 4000 Dusseldorf 11 Germany (Tel.: 49 211 596150; Telefax: 49 211 5961530)

Texel
ZA de Buxières, PO Box 10, F86220 Dange St-Romain France (Tel.: 33 49 93 7100; Telefax: 33 49 93 7120; Telex: 791061)

Thew Arnott & Co. Ltd.
Newman Works, 270 London Rd., Wallington, Surrey, SM6 7DJ UK (Tel.: 44 181 669 3131; Telefax: 44 181 669 7747; Telex: 46601)

Thymly Products Inc.
1332 Colora Rd., PO Box 65, Colora, MD 21917 USA (Tel.: 410-658-4820; Telefax: 410-658-4824)

TIC Gums, Inc.
4609 Richlynn Dr., Belcamp, MD 21017-1227 USA (Tel.: 410-273-7300; 800-221-3953; Telefax: 410-273-6469; Telex: 221049)

Tiense Suiderraffinaderij NV
Tervurenlaan 182, 1150 Brussels Belgium (Tel.: 32 2 771 0030; Telefax: 32 2 771-9235; Telex: 24523)

Tipiak
BP 5, 44860 Pont St Martin, Nantes France (Tel.: 33 40 32 1111; Telefax: 33 40 04 0326; Telex: 710005)

Tipiak Inc.
54 Middlebury St., Stamford, CT 06902 USA (Tel.: 203-961-9117; Telefax: 203-975-9081)

Todd's Ltd.
4413 Northeast 14th St., Box 4821, Des Moines, IA 50306 USA (Tel.: 515-266-2276; Telefax: 515-266-1669)

Toho Chemical Industry Co., Ltd.
No. 1-2-5, Ningyo-cho, Nihonbashi, Chuo-ku, Tokyo, 103 Japan (Tel.: (03) 3668-2271; Telefax: (03) 3668-2278; Telex: 252-2332 TOHO K J)

Tokai Bussan Co., Ltd.
7F, T.M.M. Bldg. 1-10-5, Iwamoto-cho, Chiyoda-ku, Tokyo Japan (Tel.: (03) 864-6861)

Toseno GmbH & Co. KG
Am Neuländer Baggerteich 2, D21079 Hamburg Germany (Tel.: 49 40 766 16310; Telefax: 49 40 766 16324; Telex: 2173831)

Tournay Bio-Industries
CD 17 Zone Industrielle, 02990 Fontenoy France (Tel.: 33 23 293222; Telefax: 33 26 500285; Telex: 140839)

Towa Chemical Industry Co., Ltd.
1-2, Ohte-machi 2-chome, Chiyoda-ku, Tokyo, 100 Japan (Tel.: (03) 3243-0045; Telefax: (03) 3242-7407)

Tradimpex JM Thiercelin SA
11-13 Rue Gustave-Eiffel, PO Box 23, 94510 La Queue en Brie France (Tel.: 33 1 4593 0232; Telefax: 33 1 4593 0810; Telex: 262210)

TR-AMC Chemicals
PO Box 296, Hudson Ave., Ridgefield, NJ 07657 USA (Tel.: 201-941-7706; Telefax: 201-941-7702; Telex: 130594)

Trans-Chemco, Inc.
19235 84th St., Bristol, WI 53104 USA (Tel.: 414-857-2363)

Treasure Island Food Co.
1672 Funston, San Francisco, CA 94122 USA (Tel.: 415-665-3553; Telefax: 415-665-3553)

Treatt
R.C. Treatt & Co. Ltd., Northern Way, Bury St. Edmunds, Suffolk, IP32 6NL UK (Tel.: 44 1284 702500; Telefax: 44 1284 752888; Telex: 81583)

Florida Treatt, PO Box 215, Haines City, FL 33845 USA (Tel.: 813-421-4708; 800-866-7704; Telefax: 813-422-5930)

Tricon Colors Inc.
16 Leliarts Ln., Elmwood Park, NJ 07407-3291 USA (Tel.: 201-794-3800; Telefax: 201-797-4660; Telex: 4991537 TRICN)

Tri-K Industries, Inc.
27 Bland St., PO Box 312, Emerson, NJ 07630 USA (Tel.: 201-261-2800; 800-526-0372; Telefax: 201-261-1432; Telex: 215085 TRIK UR)

Tropic Agro Products Ltd.
3 Mountington Park Close off Donnington Rd., Kenton, Harrow, Middlesex, HA3 0NW UK (Tel.: 44 181 907 9428; Telefax: 44 181 909 1661; Telex: 261507)

Trugman Nash Inc.
90 West St., New York, NY 10006 USA (Tel.: 212-964-9350; Telefax: 212-791-1863; Telex: 232098-RCA)

Trustin Foods
Chase Rd., Northern Way, Bury St Edmonds, Suffolk, IP32 6NT UK (Tel.: 44 1284 766265; Telefax: 44 1284 760816; Telex: 81117)

Tsuno Rice Fine Chemicals Co., Ltd.
94 Shinden, Katsuragi-cho, Ito-gun, Wakayama-Ken Japan (Tel.: 81-736-22-0061)

Tunnel Refineries Ltd.
Thames Bank House, Tunnel Ave., Greenwich, London, SE10 0PA UK (Tel.: 44 181 858 3271; Telefax: 44 181 293 4277; Telex: 23455)

Twitchell Flavors Bunge Foods
3582 McCall Pl. NE, Atlanta, GA 30340 USA (Tel.: 800-241-5700; Telefax: 404-986-6282)

Ueno Fine Chemicals Industry, Ltd.
2-8-4 Koraibashi, Chuo-ku, Osaka, 541 Japan (Tel.: (06) 203-0761; Telefax: (06) 222-2413; Telex: J63638 UENOFCI)

UFL Foods Inc.
6320 Northwest Dr., Mississauga, Ontario, L4V 1J7 Canada (Tel.: 416-671-0808; Telefax: 416-671-0809; Telex: 0373902)

Unbar Rothon Ltd.

Radford Way, Billericay, Essex, CM12 0BT UK (Tel.: 44 1277 632211; Telefax: 44 1277 630151; Telex: 995579)

Ungerer & Co.

4 Bridgewater Lane, PO Box U, Lincoln Park, NJ 07035 USA (Tel.: 201-628-0600; Telefax: 201-628-0251; Telex: 4754267)

Unichema North America

4650 S. Racine Ave., Chicago, IL 60609 USA (Tel.: 312-376-9000; 800-833-2864; Telefax: 312-376-0095; Telex: 176068)

Unicorn Food Ingredients

Waveney House, Queen St., Stradbroke, Eye, Suffolk, IP21 5HG UK (Tel.: 44 1379 384650; Telefax: 44 1379 384287)

Unilait France

24 Blvd de l'Hopital, 87005 Paris France (Tel.: 33 1 45354744; Telefax: 33 1 47072591; Telex: 206903)

Union Carbide

Union Carbide Chem. & Plastics Co. Inc./Specialty Chem. Div., 39 Old Ridgebury Rd. H2375, Danbury, CT 06817-0001 USA (Tel.: 203-794-2000; 800-568-4000)

Union Carbide Canada Ltd., 10455 Metropolitan E., Montreal East, Quebec, H1B 1A1 Canada (Tel.: 514-493-2610)

Union Carbide Brazil, Rua Dr. Eduardo De Souza Aranha, 153, Sao Paulo, 04530, Brazil)

Union Carbide Europe S.A., 15 Chemin Louis-Dunant, CH-1211 Geneve 20 Switzerland (Tel.: 22-739-6111; Telefax: 22-739-6545; Telex: 419207)

Union Carbide Japan KK, Toranomon 45 Mori Bldg., 1-5 Toranomon, 5-Chome Minato-Ku, Tokyo, 105 Japan (Tel.: 3431-7281)

Unipektin AG

Bahnhofstrasse 9, 8264 Eschenz Switzerland (Tel.: 41 54 423131; Telefax: 41 54 412763; Telex: 896208)

Unipex

30 Rue du Fort, PO Box 150, 92504 Rueil Malmaison Cedex France (Tel.: 33 1 4732 9293; Telefax: 33 1 4749 0235; Telex: 634022)

Unitex Chemical Corp.

PO Box 16344, 520 Broome Rd., Greensboro, NC 27406 USA (Tel.: 919-378-0965; Telefax: 919-272-4312)

Universal Flavors International Inc., A Universal Foods Co.

5600 West Raymond St., Indianapolis, IN 46241 USA (Tel.: 317-243-3521; Telefax: 317-248-1753)

Universal Flavors UK, Bilton Rd., Bletchley, Milton Keynes, MK1 1HP UK (Tel.: 44 1908 270270; Telefax: 44 1908 270271; Telex: 825533)

Universal Flavors Belgium, Researchpark Haasrode, Interleuvenlaan 37, B3001 Heverlee Belgium (Tel.: 32 16 40 33 20; Telefax: 32 16 40 00 27)

Universal Flavors France, 34/34 Bis Rue de l'Ermitage, 78000 Versailles France (Tel.: 33 13 954 2794; Telefax: 33 13 955 7322; Telex: 696263)

Universe Foods Ltd.

Universe House, 52 Queens Rd., Weybridge, Surrey, KT13 0AN UK (Tel.: 44 1932 840151; Telefax: 44 1932 859912; Telex: 25920)

Universal Preserv-A-Chem Inc./UPI

297 North 7th St., Brooklyn, NY 11211 USA (Tel.: 718-782-7429)

UOP

777 Old Saw Mill River Rd., Tarrytown, NY 10591-6799 USA (Tel.: 914-789-2246; Telefax: 914-789-2279)

UOP Food Antioxidants Dept.

25 East Algonquin Rd., PO Box 5017, Des Plaines, IL 60017-5017 USA (Tel.: 708-391-2425; 800-348-0832; Telefax: 708-391-2097; Telex: 211442)

UPI. *See* Universal Preserv-A-Chem, Inc.

Vaessen-Schoemaker Chemische Industrie B.V.
PO Box 17, 7400 AA Deventer The Netherlands (Tel.: (0)5700-3 25 55; Telefax: 05700-25414; Telex: 49080 vasco nl)

Valmar
ZI de St Mitre, PO Box 539, 13400 Aubagne le Charrel France (Tel.: 33 42 849292; Telefax: 33 42 841079; Telex: 430570)

Vamo-Fuji Specialities
Kuhlmannlaan 36, 9042 Gent Belgium (Tel.: 32 91 430202; Telefax: 32 91 430256; Telex: 11033)

Vamo Mills NV (Group Vandemoortele)
Prins Albertlaan 12, 8870 Izegem Belgium (Tel.: 32 51 332211; Telefax: 32 51 311965; Telex: 81622)

Vandemoortele Professional NV Vamix
Ottergemsesteenweg 806, B-9000 Gent Belgium (Tel.: 32 91 401711; Telefax: 32 91 227264; Telex: 12561)

Van Den Bergh
Van Den Bergh Foods Co., 2200 Cabot Dr., Lisle, IL 60532 USA (Tel.: 708-505-5300; 800-949-7344; Telefax: 708-955-5497)

Van Den Burg Eiprodukten BV, PO Box 220, 5140 AE Waalwijk The Netherlands (Tel.: 31 4160 37911; Telefax: 31 4160 35285)

R.T. Vanderbilt Co Inc.
30 Winfield St, PO Box 5150, Norwalk, CT 06856 USA (Tel.: 203-853-1400; 800-243-6064; Telefax: 203-853-1452; Telex: 6813581 RTVAN)

Jan C. Van de Wetering Import & Export
Schwachhauser Heerstr. 30a, 2800 Bremen 1 Germany (Tel.: 49 421 3499 515; Telefax: 49 421 3499 486; Telex: 246634)

Vanlab Corp.
Box 207, Rochester, NY 14601 USA (Tel.: 716-232-6647; Telefax: 716-232-6168)

Van Waters & Rogers Inc., Subsid of Univar Corp
6100 Carillon Point, Kirkland, WA 98033 USA (Tel.: 206-889-3400; Telefax: 206-889-4133)

Veos NV/Vapran France
Meiboomstraat 1, B-8750 Zweezele Belgium (Tel.: 32 51 623454; Telefax: 32 51 612428; Telex: 81191)

Veriners SA
Route de Quelaines, 53230 Cosse le Vivien France (Tel.: 33 43 98 8127; Telefax: 33 43 98 9365; Telex: 721024)

Vevy Europe SpA
Via P. Semeria 18, PO Box 716, I-16131 Genova Italy (Tel.: 010-5221212; Telefax: 010-5221530; Telex: 281257 Vevy-1)

Vicente Trapani SA
Casilla de Correo 247, 4000 Tucuman, Argentina (Tel.: 54 81 617154; Telefax: 54 81 311381; Telex: 61189)

Viobin Corp., Subsid. of American Home Prods.
226 West Livingston, Monticello, IL 61856 USA (Tel.: 217-762-2561; Telefax: 217-762-2489, Telex:)

PO Box 158, Waunakee, WI 53597 USA (Tel.: 608-849-5944; Telefax: 608-849-4053; Telex: 26-5479)

Virginia Dare
Virginia Dare Extract Co., Inc., 882 Third Ave., Brooklyn, NY 11232 USA (Tel.: 718-788-1776; 800-847-4500 (ex.NY); Telefax: 718-768-3978; Telex: 424708)

Virginia Dare Flavors Inc., 882 Third Ave., Brooklyn, NY 11232 USA (Tel.: 800-847-4500; Telefax: 718-768-3978; Telex: 425707)

Visuvia Chemische-Pharmazeutische Erzeugnisse
Geesthachter Str. 103-105, PO Box 1140, 2054 Geesthacht Germany (Tel.: 49 4152 8000 0; Telefax: 49 4152 5479)

Vitamins, Inc.
200 E. Randolph Dr., Chicago, IL 60601 USA (Tel.: 312-861-0700; Telefax: 312-861-0708; Telex: 25 4717)

Vivolac Cultures Corp.
3862 E. Washington, Indianapolis, IN 46201 USA (Tel.: 317-359-9528; Telefax: 317-356-8450)

Vrymer Commodities
PO Box 545, St. Charles, IL 60174 USA (Tel.: 708-377-2584; Telefax: 708-377-5521)

Vyse Gelatin Co.
5010 N. Rose St., Schiller Park, IL 60176 USA (Tel.: 708-678-4780; Telefax: 708-628-0329)

Wacker
Wacker-Chemie GmbH, Div. S, Hanns-Seidel-Platz 4, D-81737 München Germany (Tel.: (089) 62 79 01; Telefax: (089) 62791771; Telex: 52912156)

Wacker Silicones Corp., Subsid of Wacker-Chemie, 3301 Sutton Rd., Adrian, MI 49221-9397 USA (Tel.: 517-264-8500; 800-248-0063; Telefax: 517-264-8246; Telex: 510-450-2700 sadrnud)

Wander Ltd.
Station Rd., King's Langley, Herts, WD4 8LJ UK (Tel.: 44 1923 266122; Telefax: 44 1923 260038; Telex: 922747)

Warner-Jenkinson
Warner-Jenkinson Co., 2526 Baldwin St., St. Louis, MO 63106 USA (Tel.: 314-658-7469; 800-325-8110; Telefax: 314-658-7431; Telex: 44 7184)

Warner-Jenkinson Europe Ltd., Oldmedow Rd., King's Lynn, Norfolk, PE30 4JJ UK (Tel.: 44 1553 763236; Telefax: 44 1553 770707; Telex: 817144)

Warner-Jenkinson Netherlands, Kleine Koppel 39-40, PO Box 1493, 3800 BL Amersfoort The Netherlands (Tel.: 31 33 673411; Telefax: 31 33 650002; Telex: 70924)

Edgar A. Weber & Co.
PO Box 546, Wheeling, IL 60090 USA (Tel.: 708-215-1980; Telefax: 708-215-2073)

Welding GmbH & Co.
Grosse Theaterstr. 50, 200 Hamburg 36 Germany (Tel.: 49 40 359080; Telefax: 49 40 403870)

Welltep International Inc.
1 Florida Park Dr. S., Suite 324, Palm Coast, FL 32137 USA (Tel.: 904-445-7160; Telefax: 904-445-7169; Telex: 65030 92396)

Wensleydale Foods
Mawson House, The Bridge, Aiskew, Bedale, N Yorks, DL8 1AW UK (Tel.: 44 677 424881; Telefax: 44 677 424588)

Wessanen Meel Fiberland
Noorddijk 70, PO Box 11, 1520 AA Wormerveer The Netherlands (Tel.: 31 75 294294; Telefax: 31 75 294258; Telex: 19071)

Westhove
39 rue Loucheur, BP 73, 62510 Arques France (Tel.: 33 21 38 3316; Telefax: 33 21 98 4437; Telex: 130889)

Westin Inc./Feaster Foods Div.
4727 Center St., Omaha, NE 68106 USA (Tel.: 402-533-3363; Telefax: 402-553-1932)

Peter Whiting (Ingredients)
5 Lord Napier Pl., Upper Mall, London, W6 9UB UK (Tel.: 44 181 741 4025; Telefax: 44 181 741 1737; Telex: 8814670)

The Whole Herb Co.

Box 1203, 19800 8th East St., Sonoma, CA 95476 USA (Tel.: 707-955-1077; Telefax: 727-955-3447; Telex: 275914)

Rudolf Wild of America Inc.

1245 E. Brickyard Rd., Suite 505, Salt Lake City, UT, 84106 USA (Tel.: 801-487-2227; Telefax: 801-487-2229)

D.D. Williamson

D.D. Williamson & Co. Inc., PO Box 6001, Louisville, KY 40206 USA (Tel.: 502-895-2438; Telefax: 502-895-7381)

D.D. Williamson (Ireland), Little Island Ind. Estate, Little Island, Co Cork, Ireland (Tel.: 353 21 353821; Telefax: 353 21 354328; Telex: 75125)

Witco

Witco Corp/Household, Industrial, Personal Care, 520 Madison Ave., New York, NY 10022 USA (Tel.: 212-605-3680; Telefax: 212-486-4198)

Witco Corp/Argus Chem. Div., Bussey Rd., PO Box 1439, Marshall, TX 75671-1439 USA (Tel.: 903-938-5141; 800-431-1413; Telefax: 903-938-2647)

Witco Corp/Petroleum Specialties Group, 520 Madison Ave., New York, NY 10022-4236 USA (Tel.: 212-605-3972; Telefax: 212-754-5676; Telex: 62470)

Wixon/Fontarome

1390 E. Bolivar Ave., St. Francis, WI 53235 USA (Tel.: 414-481-8900; Telefax: 414-481-5570)

Alfred L. Wolff GmbH & Co.

Grosse Baeckerstrasse 13, D20095 Hamburg Germany (Tel.: 49 40 362971; Telefax: 49 40 363912; Telex: 211778)

Woods & Woods

120-124 Carnavon St., Silverwater, NSW, 2141 Australia (Tel.: 61 2 748 2836; Telefax: 61 2 648 3863)

Woodstone Foods Corp.

352 Saulteaux Cr., Winnipeg, Mannitoba, R3J 3T2 Canada (Tel.: 204-831-8702; Telefax: 204-831-8755)

World of Spice Ltd.

Begington Close, Billericay, Essex, CM12 0DT UK (Tel.: 44 1277 633303; Telefax: 44 1277 633036; Telex: 995579)

World Trade Service Singapore

142 Killiney Rd., PT 08-148, Devonshire Court, Singapore 0923)

E.H. Worlee GmbH & Co.

Grusonstrasse 22, 2000 Hamburg 74 Germany (Tel.: 49 40 733 33-0; Telefax: 49 40 733 332 90; Telex: 212384)

Wykefold Ltd.

Britannia House, 9 Glenthorne Rd., Hammersmith, London, W6 0LF UK (Tel.: 44 181 748 9898; Telefax: 44 181 748 8384)

Wynmouth Lehr Ltd.

Kemp House, 158 City Rd., London, EC1V 2PA UK (Tel.: 44 171 253 5871; Telex: 28293)

Xyrofin

Xyrofin (UK) Ltd., A Cultor Company, 41-51 Brighton Rd., Redhill, Surrey, RH1 6YS UK (Tel.: 44 1737 773732; Telefax: 44 1737 773117; Telex: 938830 XYFIN G)

Xyrofin GmbH, Buchenring 53, D-22359 Hamburg Germany (Tel.: 49 40 603 1239; Telefax: 49 40 603 0387)

Xyrofin France SA, 33 Ave. Friedland, 75008 Paris France (Tel.: 33 1 4 053 0909; Telefax: 33 1 4 440 4283)

American Xyrofin Inc., 1101 Perimeter Dr., Suite 475, Schaumburg, IL 60173 USA (Tel.: 708-413-8200; Telefax: 708-413-8282)

American Xyrofin/Cultor Ltd., 1400 N. Meacham Rd., Schaumburg, IL 60173 USA (Tel.: 708-843-3200; Telefax: 708-843-3368)

Xyrofin Far East KK, 4F Towa Kanda-Nishikicho Bldg., 3-4 Kanda-Nishikicho, Chiyoda-ku, Tokyo, 101 Japan (Tel.: 81 3 3295 4011; Telefax: 81 3 3295 5299)

Zeelandia UK

Unit 4, Radford Way, Billericay, Essex, CM12 0DX UK (Tel.: 44 171 248 1212; Telefax: 44 171 489 9033; Telex: 883291)

Zink & Triest Co. Inc.

111 Commerce Dr., Montgomeryville, PA 18936 USA (Tel.: 215-362-1100; Telefax: 215-368-5916; Telex: 846440)

Zschimmer & Schwarz GmbH & Co.

Postfach 2179, D-5420 Lahnstein/Rhein Germany (Tel.: 2621 121; Telefax: 2621-12407; Telex: 869816 ZSO D)

Zumbro Inc.

c/o Garuda Int'l. Inc., PO Box 5155, Santa Cruz, CA 95063 USA (Tel.: 408-462-6341; Telefax: 408-462-6355; Telex: 296614)

Appendices

CAS Number-to-Trade Name Cross-Reference

CAS	Trade name	CAS	Trade name	CAS	Trade name
50-70-4	Hydex® 100 Coarse Powd.		99.5%	72-17-3	Purasal® S/SP 60
50-70-4	Hydex® 100 Coarse Powd. 35	56-81-5	Star	77-62-3	Lowinox® WSP
		56-81-5	Superol	77-90-7	Uniplex 84
50-70-4	Hydex® 100 Gran. 206	57-10-3	Emersol® 6343	77-92-9	Cap-Shure® C-140E-75
50-70-4	Hydex® 100 Powd. 60	57-10-3	Glycon® P-45		
50-70-4	Hydex® Tablet Grade	57-11-4	Emersol® 6320	77-92-9	Citric Acid Anhydrous USP/FCC
50-70-4	Hystar® 7570	57-11-4	Emersol® 6332 NF		
50-70-4	Liponic Sorbitol Powd.	57-11-4	Emersol® 6349	77-92-9	Citric Acid USP FCC Anhyd. Fine Gran. No. 69941
50-70-4	Liponic Sorbitol Sol'n. 70% USP	57-11-4	Emersol® 6351		
		57-11-4	Emersol® 6353		
50-70-4	Sorbo®	57-11-4	Emersol® 7051	77-92-9	Citrid Acid USP FCC Anhyd. Gran. No. 69942
50-70-4	Unisweet 70	57-11-4	Glycon® S-70		
50-81-7	Ascorbic Acid USP/FCC, 100 Mesh	57-11-4	Glycon® S-90		
		57-11-4	Glycon® TP	77-92-9	Citrocoat® A 1000 HP
50-81-7	Ascorbic Acid USP, FCC Fine Gran. No. 6045655	57-11-4	Hystrene® 5016 NF FG	77-92-9	Citrocoat® A 2000 HP
		57-11-4	Hystrene® 7018 FG	77-92-9	Citrocoat® A 4000 TP
		57-11-4	Hystrene® 8718 FG	77-92-9	Citrocoat® A 4000 TT
50-81-7	Ascorbic Acid USP FCC Fine Powd. No. 6045652	57-11-4	Hystrene® 9718 NF FG	77-92-9	Citrostabil® NEU
		57-11-4	Industrene® 4518	77-92-9	Citrostabil® S
		57-11-4	Industrene® 5016 NF FG	77-92-9	Descote® Citric Acid 50%
50-81-7	Ascorbic Acid USP, FCC Gran. No. 6045654	57-11-4	Industrene® 7018 FG	77-92-9	Liquinat®
		57-11-4	Industrene® 8718 FG	78-83-1	Isobutanol HP
50-81-7	Ascorbic Acid USP, FCC Type S No. 6045660	57-55-6	Adeka Propylene Glycol (P)	79-81-2	Vitamin A Palmitate Type PIMO/BH No. 638280100
		57-88-5	Vitinc® Cholesterol NF XVII		
50-81-7	Ascorbic Acid USP, FCC Ultra-Fine Powd No. 6045653	58-56-0	Pyridoxine Hydrochloride USP, FCC Fine Powd. No. 60650	79-81-2	Vitamin A Palmitate USP, FCC Type P1.7 No. 262090000
50-81-7	Ascorbo-120	58-85-5	d-Biotin USP, FCC No. 63345	79-81-2	Vitamin A Palmitate USP, FCC Type P1.7/E No. 63699
50-81-7	Ascorbo-C Tablets				
50-81-7	Cap-Shure® AS-125-50	59-30-3	Folic Acid USP, FCC No. 20383	81-07-2	Syncal® SDI
50-81-7	Cap-Shure® AS-165CR-70	59-67-6	Lutavit® Niacin	81-07-2	Unisweet SAC
		59-67-6	Niacin USP, FCC Fine Granular No. 69901	83-88-5	Riboflavin USP, FCC No. 602940002
50-81-7	Descote® Ascorbic Acid 60%			87-99-0	Xylitol C
		59-67-6	Niacin USP, FCC No. 69902	89-78-1	Fancol Menthol
50-99-7	Candex®			90-43-7	Nipacide® OPP
52-90-4	Do Sure	62-33-9	Versene CA	94-13-3	Lexgard® P
52-90-4	Panicrust LC K	62-54-4	Niacet Calcium Acetate FCC	94-13-3	Nipasol M
56-81-5	Emery® 912			94-18-8	Nipabenzyl
56-81-5	Emery® 916	63-42-3	Unisweet L	94-26-8	Nipabutyl
56-81-5	Emery® 917	63-42-3	Unisweet Lactose	97-64-3	Purasolv® ELECT
56-81-5	Emery® 918	67-03-8	Thiamine Hydrochloride USP, FCC Regular Type No. 601160	97-64-3	Purasolv® ELS
56-81-5	Glycerine (Pharmaceutical)			98-92-0	Niacinamide USP, FCC No. 69905
56-81-5	Glycon® G 100				
56-81-5	Glycon® G-300	68-04-2	Sodium Citrate USP, FCC Dihydrate Fine Gran. No. 69975	98-92-0	Niacinamide USP, FCC Fine Granular No. 69916
56-81-5	Natural Glycerine USP 96%				
56-81-5	Natural Glycerine USP 99%	69-65-8	Unisweet MAN	99-76-3	Lexgard® M
		72-17-3	Arlac S	99-76-3	Nipagin M
56-81-5	Natural Glycerine USP	72-17-3	Patlac® NAL	110-44-1	Tristat

CAS	Trade name	CAS	Trade name	CAS	Trade name
111-03-5	Capmul® GMO	137-66-6	Ascorbyl Palmitate NF,	814-80-2	Pationic® 1240
111-03-5	Cithrol GMO N/E		FCC No. 60412	814-80-2	Pationic® 1250
111-03-5	Mazol® 300 K	139-08-2	Cyncal®	814-80-2	Puracal® PP
111-03-5	Mazol® GMO	139-33-3	Versene NA	996-31-6	Arlac P
111-03-5	Monomuls® 90-O18	141-08-2	Cithrol GMR N/E	996-31-6	Purasal® P/USP 60
111-03-5	Pationic® 1064	141-08-2	Hodag GMR	1066-33-7	ABC-Trieb®
111-03-5	Pationic® 1074	141-08-2	Hodag GMR-D	1107-26-2	Lucantin® Yellow
112-80-1	Emersol® 6313 NF	142-18-7	Cithrol GML N/E	1120-28-1	Kemester® 2050
112-80-1	Emersol® 6321 NF	142-18-7	Grindtek ML 90	1302-78-9	Vitiben®
112-80-1	Emersol® 6333 NF	142-18-7	Hodag GML	1305-79-9	Improved Paniplus M
112-80-1	Emersol® 7021	142-47-2	Asahi Aji®	1305-79-9	Regular Paniplus
112-80-1	Emery® 7021	142-55-2	Cithrol PGML N/E	1309-42-8	Magnesium Hydroxide
112-80-1	Industrene® 205	142-55-2	Emalex PGML		USP
118-71-8	Veltol®	142-55-2	Hodag PGML	1309-42-8	Magnesium Hydroxide
120-47-8	Nipagin A	143-07-7	Emery® 6354		USP DC
121-32-4	Rhodiarome™	143-28-2	Fancol OA-95	1309-42-8	Marinco H-USP
121-32-4	Unisweet EVAN	144-55-8	Sodium Bicarbonate	1309-48-4	Magnesium Oxide USP
121-79-9	Sustane® PG		USP No. 1 Powd.		30 Light
121-79-9	Tenox® PG	144-55-8	Sodium Bicarbonate	1309-48-4	Magnesium Oxide USP
122-32-7	Hodag GTO		USP No. 2 Fine Gran		60 Light
122-78-1	Phenylacetaldehyde 50	144-55-8	Sodium Bicarbonate	1309-48-4	Magnesium Oxide USP
123-28-4	Evanstab® 12		USP No. 5 Coarse		90 Light
123-51-3	Isoamyl Alcohol 95%		Gran	1309-48-4	Magnesium Oxide USP
123-51-3	Isoamyl Alcohol 99%	299-27-4	Gluconal® K		Heavy
123-77-3	ADA Tablets	299-28-5	Gluconal® CA A	1309-48-4	Marinco OH
123-77-3	WaTox-20 Tablets	299-28-5	Gluconal® CA M	1309-48-4	Marinco OL
123-94-4	Drewmulse® 200K	299-29-6	Gluconal® FE	1317-65-3	Atomite®
123-94-4	Drewmulse® 900K	300-92-5	Aluminum Stearate EA	1317-65-3	Carbital® 35
124-07-2	Emersol® 6357	300-92-5	Witco® Aluminum	1317-65-3	Carbital® 50
124-26-5	Kemamide® S		Stearate EA Food	1317-65-3	Carbital® 75
126-13-6	SAIB-SG		Grade	1317-65-3	CC™-101
126-92-1	Rhodapon® BOS	301-02-0	Kemamide® O	1317-65-3	CC™-103
126-96-5	Niacet Sodium	301-02-0	Kemamide® U	1317-65-3	CC™-105
	Diacetate FCC	334-48-5	Emery® 6359	1317-65-3	Micro-White® 10 Codex
127-09-3	Niacet Sodium Acetate	471-34-1	CC™-101	1317-65-3	Micro-White® 25 Codex
	60% FCC	471-34-1	CC™-103	1317-65-3	Micro-White® 50 Codex
127-09-3	Niacet Sodium Acetate	471-34-1	CC™-105	1317-65-3	Micro-White® 100
	Anhyd. FCC	514-78-3	Lucantin® Red		Codex
128-37-0	CAO®-3	527-07-1	Gluconal® NA	1323-39-3	Admul PGMS
128-37-0	Embanox® BHT	527-09-3	Gluconal® CU	1323-39-3	Aldo® PGHMS
128-37-0	Sustane® BHT	532-32-1	ProBenz	1323-39-3	Aldo® PGHMS KFG
128-37-0	Tenox® BHT	532-32-1	Sodium Benzoate BP88	1323-39-3	Canamulse 55
128-44-9	Syncal® GS	532-43-4	Thiamine Mononitrate	1323-39-3	Canamulse 70
128-44-9	Syncal® GSD		USP, FCC Fine Powd.	1323-39-3	Canamulse 90K
128-44-9	Syncal® S		No. 601340	1323-39-3	Cithrol PGMS N/E
128-44-9	Syncal® SDS	538-23-8	Captex® 8000	1323-39-3	Drewlene 10
128-44-9	Syncal® US	538-23-8	Miglyol® 808	1323-39-3	Drewmulse® 10K
128-44-9	Unisweet SOSAC	538-24-9	Dynasan® 112	1323-39-3	Emalex PGMS
130-40-5	Riboflavin-5´-	538-24-9	Massa Estarinum® AM	1323-39-3	Emalex PGS
	Phosphate Sodium	544-63-8	Emery® 6355	1323-39-3	Hefti PMS-33
	USP, FCC No. 60296	546-93-0	Marinco CH	1323-39-3	Hodag PGMS
134-03-2	Descote® Sodium	555-43-1	Dynasan® 118	1323-39-3	Homotex PS-90
	Ascorbate 50%	555-43-1	Neobee® 62	1323-39-3	Lipal PGMS
134-03-2	Sodium Ascorbate	555-43-1	Pationic® 919	1323-39-3	Mazol® PGMS
	USP, FCC Fine Gran.	555-44-2	Dynasan® 116	1323-39-3	Mazol® PGMSK
	No. 6047709	555-45-3	Dynasan® 114	1323-39-3	Myverol® P-06
134-03-2	Sodium Ascorbate	557-04-0	Nuodex Magnesium	1323-39-3	PGMS 70
	USP, FCC Fine Powd.		Stearate Food Grade	1323-39-3	Promodan SP
	No. 6047708	557-04-0	Synpro® Magnesium	1323-83-7	Cithrol GDS N/E
134-03-2	Sodium Ascorbate		Stearate Food Grade	1323-83-7	Kessco® Glycerol
	USP, FCC Type AG	557-61-9	Viobin Octacosanol		Distearate 386F
	No. 6047710	577-11-7	Complemix® 100	1327-36-2	Suspengel Elite
137-08-6	Calcium Pantothenate	585-88-6	Amalty®	1327-36-2	Suspengel Micro
	USP, FCC Type SD No.	585-88-6	Finmalt L	1327-36-2	Suspengel Ultra
	63924	590-00-1	Tristat K	1330-80-9	Cithrol PGMO N/E
137-08-6	Lutavit® Calpan	621-71-6	Dynasan® 110	1330-80-9	Emalex PGO
137-40-6	Niacet Sodium	693-36-7	Evanstab® 18	1330-80-9	G-2185
	Propionate FCC	814-80-2	Pationic® 1230	1335-49-5	Aldo® LP KFG

CAS NUMBER-TO-TRADE NAME CROSS-REFERENCE

CAS	Trade name	CAS	Trade name	CAS	Trade name
1338-39-2	Ablunol S-20	1390-65-4	Natural Soluble Powder	7664-38-2	Albrite Phosphoric Acid
1338-39-2	Crill 1		AP Carmine Colorant		85% Food Grade
1338-39-2	Glycomul® L	1393-63-1	Dascolor Annatoo Enc	7664-38-2	Quaker™ Oatrim 5, 5Q
1338-39-2	Hodag SML	1406-70-8	Vitamin E USP, FCC	7681-93-8	Delvocid
1338-39-2	Lamesorb® SML		No. 60525	7722-76-1	Albrite Monoammonium
1338-39-2	Nissan Nonion LP-20R,	1406-70-8	Vitamin E USP, FCC		Phosphate Food Grade
	LP-20RS		No. 60526	7722-88-5	Albrite TSPP Food
1338-39-2	Prote-sorb SML	1406-70-8	Vitinc® dl-alpha		Grade
1338-39-2	S-Maz® 20		Tocopheryl Acetate	7757-93-9	Albrite Dicalcium
1338-39-2	Span® 20		USP XXII		Phosphate Anhyd
1338-41-6	Ablunol S-60	1592-23-0	Nuodex S-1421 Food	7758-01-2	Bromette
1338-41-6	Alkamuls® SMS		Grade	7758-01-2	Bromitabs
1338-41-6	Atlas 110K	1592-23-0	Nuodex S-1520 Food	7758-16-9	Aerophos P
1338-41-6	Capmul® S		Grade	7758-16-9	Albrite SAPP Food
1338-41-6	Crill 3	1592-23-0	Synpro® Calcium		Grade
1338-41-6	Drewmulse® SMS		Stearate Food Grade	7758-16-9	Antelope Aerator
1338-41-6	Durtan® 60	1948-33-0	Embanox® TBHQ	7758-16-9	B.P. Pyro®
1338-41-6	Durtan® 60K	1948-33-0	Sustane® TBHQ	7758-16-9	B.P. Pyro® Type K
1338-41-6	Famodan MS Kosher	1948-33-0	TBHQ®	7758-16-9	Curacel
1338-41-6	Glycomul® S FG	1948-33-0	Tenox® TBHQ	7758-16-9	Curavis® 150
1338-41-6	Glycomul® S KFG	2277-28-3	Dimodan LS Kosher	7758-16-9	Donut Pyro®
1338-41-6	Hodag SMS	3632-91-5	Gluconal® MG	7758-16-9	Donut SAPP
1338-41-6	Lamesorb® SMS	4075-81-4	Niacet Calcium	7758-16-9	Perfection®
1338-41-6	Liposorb S		Propionate FCC	7758-16-9	Sapp #4
1338-41-6	Nissan Nonion SP-60R	4418-26-2	Tristat SDHA	7758-16-9	SAPP 22
1338-41-6	Norfox® Sorbo S-60FG	4468-02-4	Gluconal® ZN	7758-16-9	SAPP 26
1338-41-6	Polycon S60 K	4940-11-8	Veltol®-Plus	7758-16-9	SAPP 28
1338-41-6	Prote-sorb SMS	25013-16-5	Tenox® BHA	7758-16-9	SAPP 40
1338-41-6	S-Maz® 60K	5026-62-0	Nipagin M Sodium	7758-16-9	Taterfos®
1338-41-6	S-Maz® 60KHS	5743-27-1	Calcium Ascorbate	7758-16-9	Victor Cream®
1338-41-6	Sorgen 50		FCC No. 60475	7758-23-8	Fermaloid Yeast Food
1338-41-6	Span® 60	5743-34-0	Gluconal® CA M B	7758-23-8	Py-ran®
1338-41-6	Span® 60K	5793-94-2	Admul CSL 2007, CSL	7758-23-8	V-90®
1338-41-6	Span® 60 VS		2008	7758-29-4	Albrite STPP-F
1338-43-8	Ablunol S-80	5793-94-2	Artodan OF 40	7758-29-4	Albrite STPP FG
1338-43-8	Capmul® O	5793-94-2	Crolactil CS2L	7758-29-4	Curafos® STP
1338-43-8	Crill 4	5793-94-2	Lamegin® CSL	7758-29-4	Freez-Gard® FP-19
1338-43-8	Crill 50	5793-94-2	Paniplex CK	7758-29-4	Nutrifos® 088
1338-43-8	Drewmulse® SMO	5793-94-2	Pationic® 930	7758-29-4	Nutrifos® Powd.
1338-43-8	Hodag SMO	5793-94-2	Pationic® 940	7758-29-4	Sea-Gard® FP-91
1338-43-8	Lamesorb® SMO	5793-94-2	Prefera® CSL	7758-80-7	Albrite MSP Food
1338-43-8	Nikkol SO-10	5793-94-2	Verv®		Grade
1338-43-8	Nissan Nonion OP-80R	6485-34-3	Syncal® CAS	7783-28-0	Albrite Diammonium
1338-43-8	Norfox® Sorbo S-80	6485-39-8	Gluconal® MN		Phosphate Food Grade
1338-43-8	Prote-sorb SMO	7173-51-5	Querton 210Cl-50	7785-88-8	AeroLite LP
1338-43-8	S-Maz® 80K	7173-51-5	Querton 210Cl-80	7785-88-8	Kasal®
1338-43-8	Sorgen 40	7378-23-6	Neo-Cebitate®	7785-88-8	Levair®
1338-43-8	Sorgen S-40-H	7384-98-7	Captex® 800	7785-88-8	Royal Baking Powd.
1338-43-8	Span® 80	7585-39-9	Alpha W 6 Pharma	7789-77-7	Caliment Dicalcium
1390-65-4	Carmacid Y		Grade		Phosphate Dihydrate
1390-65-4	Carmine 1623	7585-39-9	Beta W 7	7789-80-2	IDX-20 NB
1390-65-4	Carmine AS	7585-39-9	Cavitron Cyclo-	8001-21-6	Dewaxed Sunflower Oil
1390-65-4	Carmine Extract P4011		dextrin.™		83-070-0
1390-65-4	Carmine FG	7585-39-9	Gamma W8	8001-21-6	Lipovol SUN
1390-65-4	Carmine Nacarat 40	7631-86-9	Aerosil® 200	8001-22-7	Archer Soybean Oil 86-
1390-65-4	Carmine PG	7631-86-9	Aerosil® 380		070-0
1390-65-4	Carmine Powd. WS	7647-14-5	Culinox® 999 Chemical	8001-22-7	Capital Soya
1390-65-4	Carmine XY/UF		Grade Salt	8001-22-7	Lipovol SOY
1390-65-4	Carmisol A	7647-14-5	Culinox® 999® Food	8001-22-7	Soybean Oil
1390-65-4	Carmisol NA		Grade Salt	8001-22-7	Soy Oil
1390-65-4	Hilton Davis Carmine	7647-14-5	Morton® H.G. Blending	8001-23-8	Lipovol SAF
1390-65-4	Natural Liquid AP		Salt	8001-23-8	Neobee® 18
	Carmine Colorant	7647-14-5	Morton® TFC 999®	8001-25-0	Pure/Riviera Olive Oil
1390-65-4	Natural Liquid Carmine		Salt		NF
	Colorant (Type 100, 50,	7647-14-5	Purex® All Purpose	8001-29-4	Cottonseed Cooking Oil
	and Simple)		Salt		82-060-0
1390-65-4	Natural Soluble	7647-14-5	Sterling® Purified USP	8001-29-4	Cottonseed Oil
	Carmine Powd.		Salt	8001-29-4	Jewel Oil™

CAS	Trade name	CAS	Trade name	CAS	Trade name
8001-30-7	Corn Oil	8002-43-5	Centrophase® HR2B	8002-43-5	Yelkinol P
8001-30-7	Corn Oil	8002-43-5	Centrophase® HR2U	8002-48-0	DME Dry Malt Extract
8001-30-7	Corn Salad Oil 104-250	8002-43-5	Centrophase® HR4B	8002-48-0	Extramalt Dark
8001-30-7	Liquid Corn Oil 87-	8002-43-5	Centrophase® HR4U	8002-48-0	Maltoferm® 10001
	070-0	8002-43-5	Centrophase® HR6B	8002-48-0	Maltoferm® 10001 VDK
8001-31-8	Cobee 76	8002-43-5	Centrophil® K	8002-48-0	Maltoferm® A-6001
8001-31-8	Coconut Oil® 76	8002-43-5	Chocothin	8002-48-0	ND-201 Syrup
8001-31-8	Coconut Oil® 92	8002-43-5	Chocotop™	8002-48-0	Non-Diastatic Malt
8001-31-8	Konut	8002-43-5	Clearate Special Extra		Syrup #40600
8001-31-8	Pureco® 76	8002-43-5	Clearate WDF	8002-48-0	Nutrimalt® Range
8001-78-3	Castorwax® NF	8002-43-5	Dur-Lec® P	8002-48-0	Pure Malt Colorant
8001-79-4	AA USP	8002-43-5	Dur-Lec® UB		A6000
8002-03-7	Peanut Oil 85-060-0	8002-43-5	Emulbesto	8002-48-0	Pure Malt Colorant
8002-13-9	H.L. 94	8002-43-5	Emulfluid™		A6001
8002-43-5	Actiflo® 68 SB	8002-43-5	Emulfluid® A	8002-50-4	Neobee® SL-310
8002-43-5	Actiflo® 68 UB	8002-43-5	Emulfluid® AS	8002-72-0	Onion Super
8002-43-5	Actiflo® 70 SB	8002-43-5	Emulfluid® E	8002-74-2	Koster Keunen Paraffin
8002-43-5	Actiflo® 70 UB	8002-43-5	Emulgum™		Wax
8002-43-5	Alcolec® 140	8002-43-5	Emulpur™ N	8002-75-3	Lipovol PAL
8002-43-5	Alcolec® 495	8002-43-5	Emulpur™ N P-1	8002-80-0	Do-Pep Vital Wheat
8002-43-5	Alcolec® BS	8002-43-5	Emulthin M-35		Gluten
8002-43-5	Alcolec® Extra A	8002-43-5	Emulthin M-501	8002-80-0	Meatbind®-3000
8002-43-5	Alcolec® F-100	8002-43-5	Epicholin	8002-80-0	Provim ESP®
8002-43-5	Alcolec® FF-100	8002-43-5	Epikuron™ 100 P,	8002-80-0	SQ®-48
8002-43-5	Alcolec® Granules		100 G	8002-80-0	Whetpro®-75
8002-43-5	Alcolec® PG	8002-43-5	Epikuron™ 100 X	8002-80-0	Whetpro®-80
8002-43-5	Alcolec® S	8002-43-5	Epikuron™ 130 G	8006-44-8	Koster Keunen
8002-43-5	Alcolec® SFG	8002-43-5	Epikuron™ 130 P		Candelilla
8002-43-5	Alcolec® XTRA-A	8002-43-5	Epikuron™ 130 X	8006-44-8	Ross Candelilla Wax
8002-43-5	Amisol™ 210-L	8002-43-5	Epikuron™ 135 F	8006-95-9	Lipovol WGO
8002-43-5	Amisol™ 210 LP	8002-43-5	Lecigran™ 5750	8006-95-9	Viobin Wheat Germ Oil
8002-43-5	Amisol™ 329	8002-43-5	Lecigran™ 6750	8006-95-9	Vitinc® Wheat Germ Oil
8002-43-5	Amisol™ 683 A	8002-43-5	Lecigran™ A	8006-95-9	Wheat Germ Oil
8002-43-5	Amisol™ 697	8002-43-5	Lecigran™ F	8007-43-0	Crill 43
8002-43-5	Asol	8002-43-5	Lecigran™ M	8007-43-0	Nikkol SO-15
8002-43-5	Beakin LV1	8002-43-5	Lecimulthin	8007-43-0	Nissan Nonion OP-
8002-43-5	Beakin LV2	8002-43-5	Leciprime™		83RAT
8002-43-5	Blendmax 322	8002-43-5	Lecipur™ 95 C	8007-43-0	Sorgen 30
8002-43-5	Blendmax 322D	8002-43-5	Lecipur™ 95 R	8007-43-0	Sorgen S-30-H
8002-43-5	Canasperse SBF	8002-43-5	Lecithin L 1000-Range	8007-69-0	Lipovol ALM
8002-43-5	Canasperse UBF	8002-43-5	Lecithin L 4000-Range	8007-69-0	Sweet Almond Oil BP
8002-43-5	Canasperse UBF-LV	8002-43-5	Magathin		73
8002-43-5	Canasperse WDF	8002-43-5	M-C-Thin®	8008-26-2	Citreatt Lime 3135
8002-43-5	Capcithin™	8002-43-5	M-C-Thin® AF-1 Type	8008-26-2	Citreatt Lime 6134
8002-43-5	Capsulec 51-SB		DB	8008-26-2	Sweet Lime Oil 5 Fold
8002-43-5	Capsulec 51-UB	8002-43-5	M-C-Thin® AF-1 Type	8008-56-8	Citreatt Lemon 3123
8002-43-5	Capsulec 56-SB		SB	8008-56-8	Citreatt Lemon 6122
8002-43-5	Capsulec 56-UB	8002-43-5	M-C-Thin® AF-1 Type	8008-57-9	Citreatt Orange 3111
8002-43-5	Capsulec 62-SB		UB	8008-57-9	Citreatt Orange 6110
8002-43-5	Capsulec 62-UB	8002-43-5	M-C-Thin® ASOL	8008-74-0	Lipovol SES
8002-43-5	Centrocap® 162SS	8002-43-5	M-C-Thin® FWD	8008-74-0	Vitinc® Sesame Oil NF
8002-43-5	Centrocap® 162US	8002-43-5	Metarin™	8013-01-2	AYE-2000
8002-43-5	Centrocap® 273SS	8002-43-5	Metarin™ C	8013-01-2	AYE 2200
8002-43-5	Centrocap® 273US	8002-43-5	Metarin™ CP	8013-01-2	AYE 2312
8002-43-5	Centrol® 2F SB	8002-43-5	Metarin™ DA 51	8013-01-2	AYE Family
8002-43-5	Centrol® 2F UB	8002-43-5	Metarin™ F	8013-01-2	AYS 2311
8002-43-5	Centrol® 3F SB	8002-43-5	Metarin™ P	8013-01-2	AYS 2350
8002-43-5	Centrol® 3F UB	8002-43-5	Thermolec 57	8013-01-2	Gistex®
8002-43-5	Centrol® CA	8002-43-5	Thermolec 68	8013-01-2	Gistex® LS
8002-43-5	Centrolex® C	8002-43-5	Thermolec 200	8013-01-2	Gistex® MR
8002-43-5	Centrolex® D	8002-43-5	Topcithin®	8013-01-2	Gistex® X-II
8002-43-5	Centrolex® F	8002-43-5	Yelkin DS	8013-01-2	Gistex® Xtra Powd.
8002-43-5	Centrolex® G	8002-43-5	Yelkin Gold	8013-01-2	Maxarome® Family
8002-43-5	Centrolex® P	8002-43-5	Yelkin SS	8013-01-2	Maxarome® MR
8002-43-5	Centrolex® R	8002-43-5	Yelkin T	8013-01-2	Maxarome® Plus
8002-43-5	Centrophase® 152	8002-43-5	Yelkin TS	8013-01-2	Maxarome® Plus RS
8002-43-5	Centrophase® C	8002-43-5	Yelkinol F	8013-01-2	Maxavor™ MYE
8002-43-5	Centrophase® HR	8002-43-5	Yelkinol G	8013-01-2	Maxavor™ RYE-A,

CAS	Trade name	CAS	Trade name	CAS	Trade name
	RYE-AS	8027-32-5	Ointment Base No. 6	8029-44-5	Myverol® 18-85
8013-01-2	Maxavor™ RYE-B	8027-32-5	Penreco Amber	8029-44-5	Myverol® 18-85K
8013-01-2	Maxavor™ RYE-C	8027-32-5	Penreco Blond	8029-76-3	Alcolec® Z-3
8013-01-2	Maxavor™ RYE-CL	8027-32-5	Penreco Cream	8029-76-3	Centrolene® A
8013-01-2	Maxavor™ RYE-CR	8027-32-5	Penreco Lily	8029-76-3	Centrolene® S
8013-01-2	Maxavor™ RYE-D	8027-32-5	Penreco Regent	8029-76-3	M-C-Thin® HL
8013-01-2	Maxavor™ RYE-G	8027-32-5	Penreco Royal	8029-76-3	Thermolec WFC
8013-01-2	Maxavor™ RYE-PMR	8027-32-5	Penreco Snow	8029-76-3	Yelkin 1018
8013-01-2	Maxavor™ RYE-T	8027-32-5	Penreco Super	8029-91-2	Axol® E 61
8013-01-2	Maxavor™ RYE Family	8027-32-5	Penreco Ultima	8029-91-2	Myvacet® 7-00
8013-01-2	Nova-CPLN	8027-32-5	Perlatum® 400	8029-91-2	Tegin® E-61
8013-01-2	Nova-Flav™ 1000	8027-32-5	Perlatum® 410	8029-91-2	Tegin® E-61 NSE
8013-01-2	Nova-Flav™ 1001	8027-32-5	Perlatum® 410 CG	8029-92-3	Cetodan® 90-40
8013-01-2	Nova-Flav™ 1002	8027-32-5	Perlatum® 415	8029-92-3	Myvacet® 9-40
8013-01-2	Nova-Flav™ 1006	8027-32-5	Perlatum® 415 CG	8029-92-3	Tegin® E-66
8013-01-2	Nova-Flav™ 1020	8027-32-5	Perlatum® 420	8029-92-3	Tegin® E-66 NSE
8013-01-2	Nova-Flav™ 1030	8027-32-5	Perlatum® 425	8030-12-4	Special Fat 168T
8013-01-2	Nova-Flav™ 5004	8027-32-5	Perlatum® 510	8040-05-9	Monomuls® 90-15
8013-01-2	Nova-Flav™ 5006	8027-32-5	Protopet® Alba	8042-47-5	Drakeol 5
8013-01-2	Nova-Flav™ 5009	8027-32-5	Protopet® White 1S	8042-47-5	Drakeol 7
8013-01-2	Nova-Flav™ 5010	8027-32-5	Protopet® White 2L	8042-47-5	Drakeol 9
8013-01-2	Nova-Flav™ 5030	8027-32-5	Protopet® White 3C	8042-47-5	Drakeol 10
8013-01-2	Nova-Flav™ 5100	8027-32-5	Protopet® Yellow 2A	8042-47-5	Drakeol 10B
8013-01-2	Nova-Flav™ 5101	8027-32-5	Super White Fonoline®	8042-47-5	Drakeol 13
8013-01-2	Nova-Flav™ 5102	8027-32-5	Super White Protopet®	8042-47-5	Drakeol 15
8013-01-2	Nova-Flav™ 5103	8028-66-8	Fancol HON	8042-47-5	Drakeol 19
8013-01-2	Nova-Flav™ 5105	8028-66-8	Honi-Bake® 705 Honey	8042-47-5	Drakeol 21
8013-01-2	Nova-Flav™ 7000	8028-66-8	Honi-Bake® Honey	8042-47-5	Drakeol 32
8013-01-2	Nova-Flav™ 7001	8028-66-8	Sweet'n'Neat® 45	8042-47-5	Drakeol 34
8013-01-2	Nova-Flav™ 7003	8028-66-8	Sweet'n'Neat® 2000	8042-47-5	Draketex 50
8013-01-2	Nova-Flav™ 7004	8028-66-8	Sweet'n'Neat® 3000	8049-47-6	Pancreatin TA
8013-01-2	Nova-Flav™ 7006	8028-89-5	Acid Proof Caramel	8050-81-5	Dow Corning®
8013-01-2	Nova-Flav™ 7007		Powd.		Antifoam A Compd.,
8013-01-2	Nova-Flav™ 7009	8028-89-5	B&C Caramel Powd.		Food Grade
8013-01-2	Nova-Flav™ 7010	8028-89-5	Caramel Color Double	8050-81-5	Dow Corning®
8013-01-2	Nova-Flav™ 7102		Strength		Antifoam AF Emulsion
8013-01-2	Nova-Flav™ 7105	8028-89-5	Caramel Color Single	8050-81-5	Dow Corning®
8013-01-2	Nova-Flav™ 7107		Strength		Antifoam C Emulsion
8013-01-2	Nova-Flav™ 7109	8028-89-5	Double Strength Acid	8050-81-5	Hodag Antifoam F-1
8013-01-2	Nova-Flav™ 8002		Proof Caramel Colour	8050-81-5	Mazu® DF 200SP
8013-01-2	Nova-Flav™ 8004	8028-89-5	P147 Caramel Color	8050-81-5	Mazu® DF 230SP
8013-01-2	Nova-Max 1	8028-89-5	Powdered Caramel	8050-81-5	Sentry Simethicone NF
8013-17-0	Solid Invert Sugar		Color, Acid Proof	8050-81-5	Wacker Silicone
8013-17-0	Special Liquid Invert	8028-89-5	Powdered Caramel		Antifoam Emulsion
	Sugar		Colour Non-		SE 9
8015-67-6	Annatto Extract P1003		Ammoniated-All Natural	9000-01-5	Coatingum L
8015-67-6	Natural Soluble Orange		T-717	9000-01-5	Emulgum BV
	Powd.	8028-89-5	Single Strength Acid	9000-01-5	Granular Gum Arabic
8015-67-6	Natural Yellow Colour		Proof Caramel Colour		NF/FCC C-4010
	Q-500, Q-1000, Q-2000	8028-89-5	Unisweet Caramel	9000-01-5	Gum Arabic NF/FCC
8015-67-6	Unibix W	8029-43-4	CornSweet® 42		Clean Amber Sorts
8015-86-9	Koster Keunen	8029-43-4	CornSweet® 55	9000-01-5	Gum Arabic, Purified,
	Carnauba	8029-43-4	CornSweet® 95		Spray-Dried No. 1834
8015-86-9	Ross Carnauba Wax	8029-43-4	Corn Syrup 36/43	9000-01-5	Natural Arabic Type
8016-25-9	Pfico₂.Hop®	8029-43-4	Corn Syrup 42/43		Gum Purified, Spray-
8016-60-2	Ross Rice Bran Wax	8029-43-4	Corn Syrup 42/44		Dried
8016-70-4	Akofame	8029-43-4	Corn Syrup 52/43	9000-01-5	Nutriloid® Arabic
8016-70-4	Famous	8029-43-4	Corn Syrup 62/43	9000-01-5	Powdered Gum Arabic
8016-70-4	Lipovol HS	8029-43-4	Corn Syrup 62/44		NF/FCC G-150
8016-70-4	Sterotex® HM NF	8029-43-4	Corn Syrup 62/44-1	9000-01-5	Powdered Gum Arabic
8016-70-4	Witarix® 440	8029-43-4	Corn Syrup 62/44-2		NF/FCC Superselect
8023-79-8	Akowesco 1	8029-43-4	Corn Syrup 97/71		Type NB-4
8024-06-4	Vanilla No. 25450	8029-43-4	42/43 Corn Syrup	9000-01-5	Premium Fine Granular
8024-32-6	Lipovol A	8029-43-4	62/43 Corn Syrup		Gum Arabic
8027-32-5	Fonoline® White	8029-44-5	Dimodan CP	9000-01-5	Premium Granular Gum
8027-32-5	Fonoline® Yellow	8029-44-5	Dimodan CP Kosher		Arabic
8027-32-5	Ointment Base No. 3	8029-44-5	Myvatex® 7-85	9000-01-5	Premium Powdered
8027-32-5	Ointment Base No. 4	8029-44-5	Myvatex® 7-85K		Gum Arabic

CAS	Trade name	CAS	Trade name	CAS	Trade name
9000-01-5	Premium Spray Dried Gum Arabic	9000-07-1	Nutriloid® Carrageenan		Type MM (HV)
9000-01-5	Spray Dried Gum Arabic NF Type CSP	9000-07-1	PF-80	9000-30-0	Powdered Gum Guar Type MMM ¹/₂
9000-01-5	Spray Dried Gum Arabic NF/FCC CM	9000-07-1	Soageena®	9000-30-0	Powdered Gum Guar Type MMW
9000-01-5	Spray Dried Gum Arabic NF/FCC CS (Low Bacteria)	9000-07-1	Soageena® LX22	9000-30-0	Prinza® 452
		9000-07-1	Soageena® ML300		
		9000-07-1	Soageena® MM101	9000-30-0	Prinza® 455
9000-01-5	Spray Dried Gum Arabic NF/FCC CS-R	9000-07-1	Soageena® MM301	9000-30-0	Prinza® Range
		9000-07-1	Soageena® MM330	9000-30-0	Supercol® G2S
9000-01-5	Spray Dried Nigerian Gum Arabic	9000-07-1	Soageena® MM350	9000-30-0	Supercol® GF
		9000-07-1	Soageena® MM501	9000-30-0	Supercol® U
9000-01-5	Spraygum C	9000-07-1	Soageena® MV320	9000-30-0	Ticolv
9000-01-5	Spraygum GD	9000-07-1	Soageena® MW320	9000-30-0	TIC Pretested® Gum Guar 8/22 FCC/NF Powd.
9000-01-5	TIC Pretested® Arabic FT Powd.	9000-07-1	Soageena® MW321		
		9000-07-1	Soageena® MW351	9000-30-0	Uniguar 20
		9000-07-1	Soageena® MW371	9000-30-0	Uniguar 40
9000-01-5	TIC Pretested® Arabic FT-1 USP	9000-07-1	Soageena® WX57	9000-30-0	Uniguar 150
		9000-07-1	Stamere® CK FCC	9000-30-0	Uniguar 200
9000-01-5	TIC Pretested® Arabic PH-FT	9000-07-1	Stamere® CKM FCC	9000-36-6	Karaya Gum #1 FCC
9000-07-1	CarraFat™	9000-07-1	Stamere® CK-S	9000-36-6	Karaya Gum #1 FCC
9000-07-1	Carralean™ CG-100	9000-07-1	Stamere® N-47	9000-36-6	Powdered Gum Karaya Superfine #1 FCC
9000-07-1	Carralean™ CM-70	9000-07-1	Stamere® N-55		
9000-07-1	Carralean™ CM-80	9000-07-1	Stamere® N-325	9000-36-6	Powdered Gum Karaya Superfine XXXX FCC
9000-07-1	Carralean™ MB-60	9000-07-1	Stamere® N-350		
9000-07-1	Carralean™ MB-93	9000-07-1	Stamere® N-350 E FCC	9000-36-6	Premium Powdered Gum Karaya No. 1
9000-07-1	Carralite™				
9000-07-1	CarraLizer™ CGB-10	9000-07-1	Stamere® N-350 S	9000-36-6	Premium Powdered Gum Karaya No. 1 Special
9000-07-1	CarraLizer™ CGB-20	9000-07-1	Stamere® NI		
9000-07-1	CarraLizer™ CGB-40	9000-07-1	Stamere® NIC FCC	9000-36-6	Premium Powdered Gum Karaya No. 2 Special HV
9000-07-1	CarraLizer™ CGB-50	9000-07-1	Stamere® NK		
9000-07-1	Gelodan CC Range	9000-07-1	TIC Pretested® Colloid 775 Powd.	9000-36-6	Premium Powdered Gum Karaya No. 2
9000-07-1	Gelodan CW Range				
9000-07-1	Gelodan CX Range	9000-07-1	TIC Pretested® Colloid 881 M	9000-36-6	Premium Powdered Gum Karaya No. 3
9000-07-1	Genugel® CHP-2	9000-07-1	Wakal® K Range	9000-40-2	Aquasol CSL
9000-07-1	Genugel® CHP-2 Fine Mesh	9000-07-1	Wakal® K-e	9000-40-2	Carudan 000 Range
		9000-30-0	Dycol™ 4000FC	9000-40-2	Carudan 100 Range
9000-07-1	Genugel® CHP-200	9000-30-0	Dycol™ 4500F	9000-40-2	Carudan 200 Range
9000-07-1	Genugel® CJ	9000-30-0	Dycol™ HV400F	9000-40-2	Carudan 300 Range
9000-07-1	Genugel® LC-1	9000-30-0	Edicol®	9000-40-2	Carudan 400 Range
9000-07-1	Genugel® LC-4	9000-30-0	Edicol® ULV Series	9000-40-2	Carudan 700 Range
9000-07-1	Genugel® LC-5	9000-30-0	Emulcoll	9000-40-2	Diagum LBG
9000-07-1	Genugel® MB-51	9000-30-0	Guardan 100 Range	9000-40-2	Hercules® Locust Bean Gum FL 50-40
9000-07-1	Genugel® MB-78F	9000-30-0	Guardan 600 Range		
9000-07-1	Genugel® UE	9000-30-0	Guardan 700 Range	9000-40-2	Hercules® Locust Bean Gum FL 50-50
9000-07-1	Genugel® UEU	9000-30-0	Guar Gum		
9000-07-1	Genulacta® CP-100	9000-30-0	Guar Gum HV	9000-40-2	HG-100
9000-07-1	Genulacta® CSM-2	9000-30-0	Jaguar® 1105	9000-40-2	HG-175
9000-07-1	Genulacta® K-100	9000-30-0	Jaguar® 1110	9000-40-2	HG-200
9000-07-1	Genulacta® KM-1	9000-30-0	Jaguar® 1120	9000-40-2	Idealgum 1A
9000-07-1	Genulacta® KM-5	9000-30-0	Jaguar® 1140	9000-40-2	Idealgum 1B
9000-07-1	Genulacta® L-100	9000-30-0	Jaguar® 2209	9000-40-2	Idealgum 1C
9000-07-1	Genulacta® LK-71	9000-30-0	Jaguar® 2220	9000-40-2	Idealgum 2A
9000-07-1	Genulacta® LR-41	9000-30-0	Jaguar® 2240	9000-40-2	Idealgum 2B
9000-07-1	Genulacta® LR-60	9000-30-0	Jaguar® 6000	9000-40-2	Idealgum 3F
9000-07-1	Genulacta® LRA-50	9000-30-0	Jaguar® Guar Gum	9000-40-2	Locust Bean Gum Speckless Type D-200
9000-07-1	Genulacta® LRC-21	9000-30-0	Nutriloid® Guar Special		
9000-07-1	Genulacta® LRC-30	9000-30-0	Nutriloid® Guar Standard	9000-40-2	Nutriloid® Locust
9000-07-1	Genulacta® P-100			9000-40-2	Powdered Locust Bean Gum Type D-200
9000-07-1	Genulacta® PL-93	9000-30-0	Powdered Gum Guar NF Type 80 Mesh B/T		
9000-07-1	Genuvisco® CSW-2			9000-40-2	Powdered Locust Bean Gum Type D-300
9000-07-1	Genuvisco® J	9000-30-0	Powdered Gum Guar Type 140 Mesh B/T		
9000-07-1	Genuvisco® MP-11	9000-30-0	Powdered Gum Guar Type ECM	9000-40-2	Powdered Locust Bean Gum Type P-100
9000-07-1	Geoldan CL Range				
9000-07-1	Maco-O-Line 091	9000-30-0	Powdered Gum Guar Type M		
9000-07-1	Marine Colloids™ Carrageenan	9000-30-0	Powdered Gum Guar Type MM FCC	9000-40-2	Powdered Locust Bean Gum Type P-100
		9000-30-0	Powdered Gum Guar	9000-40-2	Powdered Locust Bean

CAS NUMBER-TO-TRADE NAME CROSS-REFERENCE

CAS	Trade name	CAS	Trade name	CAS	Trade name
	Gum Type PP-100		Range	9000-92-4	Clarase® 40,000
9000-40-2	Soalocust®	9000-69-5	Mexpectin LC 900	9000-92-4	Clarase® Conc.
9000-40-2	STD-175		Range	9000-92-4	Clarase® L-40,000
9000-40-2	TIC Pretested® Locust	9000-69-5	Mexpectin MRS 300	9000-92-4	Fermalpha
	Bean POR/A		Range	9000-92-4	Spezyme BBA
9000-40-2	Wakal® J	9000-69-5	Mexpectin RS 400	9000-92-4	Taka-Therm® L-340
9000-40-2	Wakal® JG		Range	9000-92-4	Tenase® 1200
9000-65-1	Nutriloid® Tragacanth	9000-69-5	Mexpectin SS 200	9000-92-4	Tenase® L-340
9000-65-1	Powdered Gum		Range	9000-92-4	Tenase® L-1200
	Tragacanth T-150	9000-69-5	Mexpectin XSS 100	9001-05-2	Catalase L
9000-65-1	Powdered Gum		Range	9001-05-2	Fermcolase®
	Tragacanth T-200	9000-69-5	TIC Pretested® Pectin	9001-05-2	Microcatalase®
9000-65-1	Powdered Gum		HM Rapid	9001-37-0	Fermcozyme® 1307,
	Tragacanth T-300	9000-69-5	TIC Pretested® Pectin		BG, BGXX, CBB,
9000-65-1	Powdered Gum		HM Slow		CBBXX, M
	Tragacanth T-400	9000-70-8	Bone Gelatin Type B	9001-37-0	Hidelase
9000-65-1	Powdered Gum		200 Bloom	9001-37-0	Hyderase
	Tragacanth T-500	9000-70-8	Calfskin Gelatin Type B	9001-37-0	Ovazyme, XX
9000-65-1	Powdered Tragacanth		175 Bloom	9001-62-1	Fermlipase
	Gum Type A/10	9000-70-8	Calfskin Gelatin Type B	9001-62-1	Lipase 8 Powd.
9000-65-1	Powdered Tragacanth		200 Bloom	9001-62-1	Lipase 16 Powd.
	Gum Type E-1	9000-70-8	Calfskin Gelatin Type B	9001-62-1	Lipase 24 Powd.
9000-65-1	Powdered Tragacanth		225 Bloom	9001-62-1	Lipase 30 Powd.
	Gum Type G-3	9000-70-8	Calfskin Gelatin Type B	9001-62-1	Lipase AK
9000-65-1	Powdered Tragacanth		250 Bloom	9001-62-1	Lipase AP
	Gum Type L	9000-70-8	Crodyne BY-19	9001-62-1	Lipase AP 6
9000-65-1	Powdered Tragacanth	9000-70-8	Edible Beef Gelatin	9001-62-1	Lipase AY
	Gum Type W	9000-70-8	Edible Gelatins	9001-62-1	Lipase FAP
9000-65-1	TIC Pretested®	9000-70-8	Flavorset® GP-2	9001-62-1	Lipase G
	Tragacanth 440	9000-70-8	Flavorset® GP-3	9001-62-1	Lipase GC
9000-65-1	Tragacanth Flake No.	9000-70-8	Flavorset® GP-4	9001-62-1	Lipase MAP
	27	9000-70-8	Flavorset® GP-5	9001-62-1	Lipase N
9000-65-1	Tragacanth Gum	9000-70-8	Flavorset® GP-6	9001-62-1	Lipase PS
	Ribbon No. 1	9000-70-8	Flavorset® GP-7	9001-62-1	Pancreatic Lipase 250
9000-69-5	Genu® 04CG or 04CB	9000-70-8	Flavorset® GP-8	9001-73-4	Papain A300
9000-69-5	Genu® 12CG	9000-70-8	Flavorset® GP-9	9001-73-4	Papain A400
9000-69-5	Genu® 18CG	9000-70-8	Flavorset® GP-10	9001-73-4	Papain AIE
9000-69-5	Genu® 18CG-YA	9000-70-8	Gelatin XF	9001-73-4	Papain M70
9000-69-5	Genu® 20AS	9000-70-8	Gummi Gelatin P-5	9001-73-4	Papain S100
9000-69-5	Genu® 21AS or 21AB	9000-70-8	Gummi Gelatin P-6	9001-73-4	Tona Enzyme 201
9000-69-5	Genu® 22CG	9000-70-8	Gummi Gelatin P-7	9001-73-4	Tona Papain 14
9000-69-5	Genu® 102AS	9000-70-8	Gummi Gelatin P-8	9001-73-4	Tona Papain 90L
9000-69-5	Genu® 104AS	9000-70-8	Liquid Fish Gelatin	9001-73-4	Tona Papain 270L
9000-69-5	Genu® 104AS-YA		Conc.	9001-75-6	Pepsin 1:3000 NF XII
9000-69-5	Genu® AA Medium-	9000-70-8	Margarine Gelatin P-8		Powd.
	Rapid Set, 150 Grade	9000-70-8	Quickset® D-4	9001-75-6	Pepsin 1:10,000 Powd.
9000-69-5	Genu® BA-KING	9000-70-8	Quickset® D-5		or Gran
9000-69-5	Genu® BB Rapid Set	9000-70-8	Quickset® D-6	9001-75-6	Pepsin 1:15,000 Powd.
9000-69-5	Genu® DD Extra-Slow	9000-70-8	Quickset® D-7	9002-07-7	Trypsin 1:75
	Set	9000-70-8	Quickset® D-8	9002-07-7	Trypsin 1:80
9000-69-5	Genu® DD Extra-Slow	9000-70-8	Quickset® D-9	9002-07-7	Trypsin 1:150
	Set C, 150 Grade	9000-70-8	Quickset® D-10	9002-18-0	Agar Agar NF Flake #1
9000-69-5	Genu® DD Slow Set	9000-70-8	Spa Gelatin	9002-18-0	Cameo Velvet
9000-69-5	Genu® JMJ	9000-70-8	Spray Dried Fish	9002-18-0	Cameo Velvet WT
9000-69-5	Genu® Type DJ		Gelatin	9002-18-0	Nutriloid® Agar
9000-69-5	Genu® VIS	9000-70-8	Spray Dried Hydrolysed	9002-18-0	Powdered Agar Agar
9000-69-5	Genu® Pectins		Fish Gelatin		NF M-100 (Gracilaria)
9000-69-5	Mexpectin HV 400	9000-70-8	Tenderset® M-7	9002-18-0	Powdered Agar Agar
	Range	9000-70-8	Tenderset® M-8		NF MK-60
9000-69-5	Mexpectin LA 100	9000-70-8	Tenderset® M-9	9002-18-0	Powdered Agar Agar
	Range	9000-90-2	Nutrilife®		NF MK-80-B
9000-69-5	Mexpectin LA 200	9000-92-4	Amflex	9002-18-0	Powdered Agar Agar
	Range	9000-92-4	Atlas MDA		NF MK-80 (Bacterio-
9000-69-5	Mexpectin LA 400	9000-92-4	Biodiastase		logical)
	Range	9000-92-4	Biozyme L	9002-18-0	Powdered Agar Agar
9000-69-5	Mexpectin LC 700	9000-92-4	Biozyme M		NF S-100
	Range	9000-92-4	Biozyme S	9002-18-0	Powdered Agar Agar
9000-69-5	Mexpectin LC 800	9000-92-4	Clarase® 5,000		NF S-100-B

CAS	Trade name	CAS	Trade name	CAS	Trade name
9002-18-0	Powdered Agar Agar NF S-150	9004-32-4	Ticalose® 4000	9004-67-5	Benecel® M 043
9002-18-0	Powdered Agar Agar NF S-150-B	9004-32-4	Ticalose® 4500	9004-67-5	Methocel® A4C Premium
		9004-32-4	Ticalose® 5000 R		
9002-18-0	TIC Pretested® Gum Agar Agar 100 FCC/NF Powd.	9004-32-4	TIC Pretested® CMC 2500 S	9004-67-5	Methocel® A4M Premium
		9004-32-4	TIC Pretested® CMC PH-2500	9004-67-5	Methocel® A15C Premium
9002-88-4	Polywax® 500	9004-34-6	Avicel® CL-611	9004-67-5	Methocel® A15LV Premium
9002-88-4	Polywax® 600	9004-34-6	Avicel® RC-501		
9002-88-4	Polywax® 655	9004-34-6	Avicel® RC-581	9004-96-0	Acconon 300-MO
9002-89-5	Elvanol® 71-30	9004-34-6	Avicel® RC-591F	9004-96-0	Durpeg® 400MO
9002-93-1	Triton® X-45	9004-34-6	Avicel® WC-595	9004-98-2	Trycol® OAL-23
9002-93-1	Triton® X-100	9004-34-6	Justfiber® CL-20-H	9004-99-3	Alkamuls® S-6
9002-93-1	Triton® X-102	9004-34-6	Justfiber® CL-35-H	9004-99-3	Myrj® 45
9002-93-1	Triton® X-114	9004-34-6	Justfiber® CL-40-H	9004-99-3	Myrj® 52
9002-93-1	Triton® X-120	9004-34-6	Justfiber® CL-100-H	9004-99-3	Myrj® 52S
9002-93-1	Triton® X-305-70%	9004-34-6	Keycel®	9005-00-9	Trycol® 5888
9002-93-1	Triton® X-405-70%	9004-34-6	Qual Flo™	9005-00-9	Trycol® SAL-20
9002-93-1	Triton® X-705-70%	9004-34-6	Solka-Floc®	9005-25-8	Aytex®-P Food Powd. Wheat Starch
9003-05-8	Accofloc® A100 PWG	9004-53-9	Clinton #600 Dextrin		
9003-05-8	Accofloc® A110 PWG	9004-53-9	Clinton #655 Dextrin	9005-25-8	Edigel® 100 Powd. Wheat Starch
9003-05-8	Accofloc® A120 PWG	9004-53-9	Clinton #656 Dextrin		
9003-05-8	Accofloc® A130 PWG	9004-53-9	Clinton #700 Dextrin	9005-25-8	Mira-Gel® 463
9003-05-8	Accofloc® N100 PWG	9004-53-9	Clinton #721 Dextrin	9005-25-8	Mira-Set® B
9003-11-6	Berol 374	9004-53-9	Stadex® 9	9005-25-8	Paygel® 290.295 Pregelatinized Food Powd. Wheat Starch
9003-73-0	Piccolyte® C115	9004-53-9	Stadex® 60K		
9004-32-4	Aqualon® 7H0F	9004-53-9	Stadex® 126		
9004-32-4	Aqualon® 7H0XF	9004-53-9	Stadex® 128	9005-25-8	Pure-Dent® B700
9004-32-4	Aqualon® 7H3SF	9004-53-9	Staley® Tapioca Dextrin 11	9005-25-8	Pure-Dent® B810
9004-32-4	Aqualon® 7H3SXF			9005-25-8	Pure-Dent® B812
9004-32-4	Aqualon® 7H4F	9004-53-9	Staley® Tapioca Dextrin 12	9005-25-8	Pure-Dent® B815
9004-32-4	Aqualon® 7H4XF			9005-25-8	Pure-Dent® B816
9004-32-4	Aqualon® 7HC4F	9004-64-2	Klucel® F Grades	9005-25-8	Pure-Dent® B880
9004-32-4	Aqualon® 7HCF	9004-65-3	Benecel® MP 643	9005-25-8	Pure Food Powd. Starch 105-A
9004-32-4	Aqualon® 7HF	9004-65-3	Benecel® MP 824		
9004-32-4	Aqualon® 7HXF	9004-65-3	Benecel® MP 843	9005-25-8	Pure Food Powd. Starch 131-C
9004-32-4	Aqualon® 7LF	9004-65-3	Benecel® MP 872		
9004-32-4	Aqualon® 7LXF	9004-65-3	Benecel® MP 874	9005-25-8	Pure Food Starch Bleached 142-A
9004-32-4	Aqualon® 7M8SF	9004-65-3	Benecel® MP 943		
9004-32-4	Aqualon® 7MF	9004-65-3	Methocel® E3 Premium	9005-25-8	Pure-Gel® B990
9004-32-4	Aqualon® 9H4F	9004-65-3	Methocel® E4M	9005-25-8	Staley® 7025
9004-32-4	Aqualon® 9H4XF	9004-65-3	Methocel® E4M Premium	9005-25-8	Staley® 7350 Waxy No. 1 Starch
9004-32-4	Aqualon® 9M31F				
9004-32-4	Aqualon® 9M31XF	9004-65-3	Methocel® E5 Food Grade	9005-25-8	Staley® Moulding Starch
9004-32-4	Aqualon® 9M8F				
9004-32-4	Aqualon® 9M8XF	9004-65-3	Methocel® E6 Premium	9005-25-8	Staley® Pure Food Powd. (PFP)
9004-32-4	Cellogen HP-5HS	9004-65-3	Methocel® E15 Food Grade		
9004-32-4	Cellogen HP-6HS9			9005-25-8	Staley® Pure Food Powd. Starch Type I
9004-32-4	Cellogen HP-8A	9004-65-3	Methocel® E15LV Premium		
9004-32-4	Cellogen HP-12HS			9005-25-8	Staley® Pure Food Powd. Starch Type II
9004-32-4	Cellogen HP-SB	9004-65-3	Methocel® E50LV Premium		
9004-32-4	CMC Daicel 1150			9005-25-8	Staley® Redried Starch A
9004-32-4	CMC Daicel 1160	9004-65-3	Methocel® F4M Premium		
9004-32-4	CMC Daicel 1220			9005-25-8	Staley® Redried Starch B
9004-32-4	CMC Daicel 1240	9004-65-3	Methocel® F50LV Premium		
9004-32-4	CMC Daicel 1260			9005-25-8	Sta-Rx®
9004-32-4	CMC Daicel 2200	9004-65-3	Methocel® K3 Premium	9005-25-8	Stir & Sperse®
9004-32-4	Nutriloid® Cellulose Gums	9004-65-3	Methocel® K4M Premium	9005-25-8	Whetstar® 7 Food Powd. Wheat Starch
		9004-65-3	Methocel® K15M Premium	9005-32-7	Alginic Acid FCC
9004-32-4	Ticalose® 15			9005-32-7	Kelacid®
9004-32-4	Ticalose® 30	9004-65-3	Methocel® K35 Premium	9005-32-7	Kimitsu Acid
9004-32-4	Ticalose® 75			9005-32-7	Sobalg FD 000 Range
9004-32-4	Ticalose® 100	9004-65-3	Methocel® K100LV Premium	9005-34-9	Ammonium Alginate Type S
9004-32-4	Ticalose® 150 R				
9004-32-4	Ticalose® 700 R	9004-65-3	Methocel® K100M Premium	9005-34-9	Amoloid HV
9004-32-4	Ticalose® 750			9005-34-9	Amoloid LV
9004-32-4	Ticalose® 2000 R	9004-67-5	Benecel® M 042	9005-34-9	Sobalg FD 300 Range
9004-32-4	Ticalose® 2500				

CAS	Trade name	CAS	Trade name	CAS	Trade name
9005-35-0	Sobalg FD 460	9005-38-3	Protanal HFC 60	9005-65-6	Hodag PSMO-5
9005-36-1	Kelmar®	9005-38-3	Protanal KC 119	9005-65-6	Hodag PSMO-20
9005-36-1	Kelmar® Improved	9005-38-3	Protanal KP	9005-65-6	Hodag SVO-9
9005-36-1	Sobalg FD 200 Range	9005-38-3	Protanal KPM	9005-65-6	Lamesorb® SMO-20
9005-37-2	Colloid 602	9005-38-3	Protanal LF 5/60	9005-65-6	Liposorb O-20
9005-37-2	Concentrated Dariloid® KB	9005-38-3	Protanal LF 20	9005-65-6	Lonzest® SMO-20
		9005-38-3	Protanal LF 20/40	9005-65-6	Nikkol TO-10
9005-37-2	Dricoid® KB	9005-38-3	Protanal LF 60	9005-65-6	Norfox® Sorbo T-80
9005-37-2	Dricoid® KBC	9005-38-3	Protanal LF 120 M	9005-65-6	Polycon T80 K
9005-37-2	Kelcoloid® D	9005-38-3	Protanal LF 200	9005-65-6	Sorbax PMO-5
9005-37-2	Kelcoloid® DH	9005-38-3	Protanal LFS 40	9005-65-6	Sorbax PMO-20
9005-37-2	Kelcoloid® DO	9005-38-3	Protanal SF 40	9005-65-6	T-Maz® 80
9005-37-2	Kelcoloid® DSF	9005-38-3	Protanal SF 60	9005-65-6	T-Maz® 80K
9005-37-2	Kelcoloid® HVF	9005-38-3	Protanal SF 120	9005-65-6	T-Maz® 80KLM
9005-37-2	Kelcoloid® LVF	9005-38-3	Protanal SF 120 M	9005-65-6	T-Maz® 81
9005-37-2	Kelcoloid® O	9005-38-3	Protanal SP 5 H	9005-65-6	Tween® 80
9005-37-2	Kelcoloid® S	9005-38-3	Protanal VK 687	9005-65-6	Tween® 80K
9005-37-2	Kimiloid HV	9005-38-3	Protanal VK 749	9005-65-6	Tween® 80 VS
9005-37-2	Kimiloid MV	9005-38-3	Protanal VK 805 IMP	9005-65-6	Tween® 81
9005-37-2	Kimiloid NLS-K	9005-38-3	Protanal VK 990	9005-66-7	Crillet 2
9005-37-2	Manucol Ester B	9005-38-3	Protanal VK 998	9005-66-7	Glycosperse® P-20
9005-37-2	Manucol Ester E/PL	9005-38-3	Protanal VPM	9005-66-7	Hodag PSMP-20
9005-37-2	Manucol Ester E/RK	9005-38-3	Protanal VSM	9005-66-7	Lamesorb® SMP-20
9005-37-2	Manucol Ester M	9005-38-3	Sobalg FD 100 Range	9005-66-7	Lonzest® SMP-20
9005-37-2	Propylene Glycol Alginate HV	9005-38-3	Sodium Alginate HV NF/FCC	9005-66-7	Sorbax PMP-20
				9005-67-8	Atlas 70K
9005-37-2	Protanal Ester BI	9005-38-3	Sodium Alginate LV	9005-67-8	Atlas A
9005-37-2	Protanal Ester CF	9005-38-3	Sodium Alginate LVC	9005-67-8	Capmul® POE-S
9005-37-2	Protanal Ester H	9005-38-3	Sodium Alginate MV NF/FCC	9005-67-8	Crillet 3
9005-37-2	Protanal Ester L			9005-67-8	Crillet 31
9005-37-2	Protanal Ester L-25 A/H	9005-38-3	W-300 FG	9005-67-8	Drewpone® 60K
9005-37-2	Protanal Ester PVH-A	9005-64-5	Capmul® POE-L	9005-67-8	Durfax® 60
9005-37-2	Protanal Ester SD-H	9005-64-5	Crillet 1	9005-67-8	Durfax® 60K
9005-38-3	Colloid 488T	9005-64-5	Crillet 11	9005-67-8	Emrite® 6125
0006-38-3	Dariloid® Q	9005-64-5	Drewmulse® POE-SML	9005-67-8	Glycosperse® S-20
9005-38-3	Dariloid® QH	9005-64-5	Durfax® 20	9005-67-8	Glycosperse® S-20 FG
9005-38-3	Kelco® HV	9005-64-5	Glycosperse® L-20	9005-67-8	Glycosperse® S-20 KFG
9005-38-3	Kelco® LV	9005-64-5	Hodag PSML-20		
9005-38-3	Kelcosol®	9005-64-5	Hodag PSML-80	9005-67-8	Hodag PSMS-20
9005-38-3	Kelgin® F	9005-64-5	Lamesorb® SML-20	9005-67-8	Hodag SVS-18
9005-38-3	Kelgin® HV	9005-64-5	Liposorb L-20	9005-67-8	Lamesorb® SMS-20
9005-38-3	Kelgin® LV	9005-64-5	Nissan Nonion LT-221	9005-67-8	Liposorb S-20
9005-38-3	Kelgin® MV	9005-64-5	Norfox® Sorbo T-20	9005-67-8	Lonzest® SMS-20
9005-38-3	Kelgin® QL	9005-64-5	Sorbax PML-20	9005-67-8	Nikkol TS-10
9005-38-3	Kelgin® XL	9005-64-5	Sorgen TW20	9005-67-8	Norfox® Sorbo T-60
9005-38-3	Kelset®	9005-64-5	T-Maz® 20	9005-67-8	Polycon T60 K
9005-38-3	Keltone®	9005-64-5	T-Maz® 20K	9005-67-8	Polysorbate 60
9005-38-3	Keltone® HV	9005-64-5	T-Maz® 28	9005-67-8	Sorbax PMS-20
9005-38-3	Keltone® LV	9005-64-5	Tween® 20	9005-67-8	T-Maz® 60
9005-38-3	Kelvis®	9005-65-6	Atlas E	9005-67-8	T-Maz® 60K
9005-38-3	Kimitsu Algin I-1	9005-65-6	Capmul® POE-O	9005-67-8	T-Maz® 60KHS
9005-38-3	Kimitsu Algin I-2	9005-65-6	Crillet 4	9005-67-8	Tween® 60
9005-38-3	Kimitsu Algin I-3	9005-65-6	Crillet 41	9005-67-8	Tween® 60K
9005-38-3	Manucol DH	9005-65-6	Drewmulse® POE-SMO	9005-67-8	Tween® 60 VS
9005-38-3	Manucol DM			9005-67-8	Tween® 61
9005-38-3	Manucol DMF	9005-65-6	Drewpone® 80K	9005-70-3	Alkamuls® PSTO-20
9005-38-3	Manucol LB	9005-65-6	Durfax® 80	9005-70-3	Crillet 45
9005-38-3	Manugel DJX	9005-65-6	Durfax® 80K	9005-70-3	Glycosperse® TO-20
9005-38-3	Manugel DMB	9005-65-6	Emrite® 6120	9005-70-3	Hodag PSTO-20
9005-38-3	Manugel GHB	9005-65-6	Glycosperse® O-5	9005-70-3	Lonzest® STO-20
9005-38-3	Manugel GMB	9005-65-6	Glycosperse® O-20	9005-70-3	Sorbax PTO-20
9005-38-3	Prime F-25	9005-65-6	Glycosperse® O-20 FG	9005-70-3	Tween® 85
9005-38-3	Prime F-40	9005-65-6	Glycosperse® O-20 KFG	9005-71-4	Alkamuls® PSTS-20
9005-38-3	Prime F-400			9005-71-4	Crillet 35
9005-38-3	Prime F-600	9005-65-6	Glycosperse® O-20 Veg	9005-71-4	Drewpone® 65K
9005-38-3	Proctin BUS			9005-71-4	Durfax® 65
9005-38-3	Protanal 686	9005-65-6	Glycosperse® O-20X	9005-71-4	Durfax® 65K
9005-38-3	Protanal HF 120 M	9005-65-6	Hetsorb O-5	9005-71-4	Glycosperse® TS-20

CAS	Trade name	CAS	Trade name	CAS	Trade name
9005-71-4	Glycosperse® TS-20 FG, TS-20 KFG	9014-01-1	Bromelain 1:10	9050-36-6	Star-Dri® 1
9005-71-4	Glycosperse® TS-20 KFG	9014-01-1	Bromelain Conc.	9050-36-6	Star-Dri® 5
		9014-01-1	Fungal Protease 31,000	9050-36-6	Star-Dri® 10
9005-71-4	Hodag PSTS-20			9050-36-6	Star-Dri® 15
9005-71-4	Lamesorb® STS-20	9014-01-1	Fungal Protease 60,000	9050-36-6	Star-Dri® 18
9005-71-4	Liposorb TS-20			9050-36-6	Star-Dri® 1005A
9005-71-4	Lonzest® STS-20	9014-01-1	Fungal Protease 500,000	9050-36-6	Star-Dri® 1015A
9005-71-4	Sorbax PTS-20			9050-36-6	Wickenol® 550
9005-71-4	T-Maz® 65	9014-01-1	Fungal Protease Conc.	9075-68-7	CK-20L
9005-71-4	T-Maz® 65K	9014-01-1	HT-Proteolytic® 200	9016-45-9	Hyonic NP-120
9005-71-4	Tween® 65	9014-01-1	HT-Proteolytic® Conc.	10031-30-8	Ajax®
9005-71-4	Tween® 65K	9014-01-1	HT-Proteolytic® L-175	10031-30-8	HT® Monocalcium Phosphate, Monohydrate (MCP) Spray Dried Coarse Granular
9005-71-4	Tween® 65 VS	9014-01-1	MC²		
9006-50-2	Egg White Solids Type P-11	9014-01-1	MLO		
		9014-01-1	Newlase		
9006-50-2	Egg White Solids Type P-18G	9014-01-1	Optimase® APL-440	10031-30-8	HT® Monocalcium Phosphate, Monohydrate (MCP) Spray Dried Fines
		9014-01-1	Papain 16,000		
9006-50-2	Egg White Solids Type P-19	9014-01-1	Papain 30,000		
		9014-01-1	Papain Conc.		
9006-50-2	Egg White Solids Type P-20	9014-01-1	Papain P-100	10031-30-8	HT® Monocalcium Phosphate, Monohydrate (MCP) Spray Dried Medium Granular
		9014-01-1	Protoferm		
9006-50-2	Egg White Solids Type P-21	9014-01-1	Prozyme 6		
		9014-01-1	Rhozyme® P11		
9006-50-2	Egg White Solids Type P-25	9014-01-1	Rhozyme® P41	10031-30-8	Regent® 12XX
		9014-01-1	Rhozyme® P53, P64	10124-56-8	Calgon
9006-50-2	Egg White Solids Type P-39	9014-01-1	Wheat-Eze	11042-64-1	Gamma Oryzanol
		9016-45-9	Cremophor® NP 10	11094-60-3	Caprol® 10G10O
9006-50-2	Egg White Solids Type P-110	9016-45-9	Cremophor® NP 14	11094-60-3	Drewpol® 10-10-O
		9016-45-9	Hyonic NP-60	11094-60-3	Drewpol® 10-10-OK
9006-50-2	Egg White Solids Type P-110 High Gel Strength	9016-45-9	Hyonic NP-100	11094-60-3	Hodag PGO-1010
		9016-45-9	Hyonic NP-110	11094-60-3	Polyaldo® DGDO
		9016-45-9	Hyonic NP-407	11094-60-3	Polyaldo® DGDO KFG
9006-50-2	Egg White Solids Type PF-1	9016-45-9	Hyonic NP-500	11094-60-3	Santone® 10-10-O
		9016-45-9	Iconol NP-7	11099-07-3	Mazol® GMS-K
9006-50-2	Hentex Type P-1100	9025-70-1	Dextranase L Amano	11138-66-2	Alginade XK9
9006-50-2	Hentex Type P-1800	9025-70-1	Dextranase Novo 25 L	11138-66-2	Dariloid® 400
9006-50-2	Hentex Type P-2100	9032-08-0	Diazyme® L-200	11138-66-2	K2B387
9006-50-2	Sol-U-Tein EA	9032-08-0	Gluczyme	11138-66-2	K8B249
9007-48-1	Caprol® 3GO	9032-08-0	GNL-3000	11138-66-2	Keltrol®
9007-48-1	Drewmulse® 3-1-O	9032-08-0	Spezyme GA	11138-66-2	Keltrol® 1000
9007-48-1	Drewpol® 3-1-O	9032-75-1	Clarex® 5XL	11138-66-2	Keltrol® BT
9007-48-1	Drewpol® 3-1-OK	9032-75-1	Clarex® L	11138-66-2	Keltrol® CR
9007-48-1	Drewpol® 6-1-O	9032-75-1	Extractase L5X, P15X	11138-66-2	Keltrol® F
9007-48-1	Drewpol® 6-1-OK	9032-75-1	Pearex® L	11138-66-2	Keltrol® GM
9007-48-1	Drewpol® 8-1-OK	9032-75-1	Pectinase AT	11138-66-2	Keltrol® RD
9007-48-1	Hodag PGO	9032-75-1	Pectinol® 59L	11138-66-2	Keltrol® SF
9007-48-1	Hodag PGO-61	9032-75-1	Pectinol® R10	11138-66-2	Keltrol® T
9007-48-1	Hodag PGO-101	9032-75-1	Pektolase	11138-66-2	Keltrol® TF
9007-48-1	Mazol® PGO-31 K	9032-75-1	Spark-L® HPG	11138-66-2	KOB87
9007-48-1	Nikkol Decaglyn 1-O	9038-95-3	Macol® 5100	11138-66-2	KOB348
9007-48-1	Nikkol Tetraglyn 1-O	9038-95-3	Pluracol® W3520N	11138-66-2	KOB349
9007-48-1	Santone® 3-1-SH	9050-36-6	Lo-Dex® 10	11138-66-2	Merecol® I
9007-48-1	Santone® 8-1-O	9050-36-6	Maltrin® M040	11138-66-2	Merecol® MS
9007-48-1	Triodan 20	9050-36-6	Maltrin® M050	11138-66-2	Merezan® 8
9009-48-1	Drewpol® 10-2-OK	9050-36-6	Maltrin® M100	11138-66-2	Merezan® 20
9010-85-9	Polysar Butyl 101-3	9050-36-6	Maltrin® M105	11138-66-2	Nutriloid® Xanthan
9010-85-9	Polysar Butyl 402	9050-36-6	Maltrin® M150	11138-66-2	Rhodigel®
9012-54-8	Celluferm	9050-36-6	Maltrin® M180	11138-66-2	Rhodigel® 200
9012-54-8	Cellulase 4000	9050-36-6	Maltrin® M510	11138-66-2	Rhodigel® EZ
9012-54-8	Cellulase AC	9050-36-6	Maltrin® M520	11138-66-2	Rhodigel® Granular
9102-54-8	Cellulase AP	9050-36-6	Maltrin® M700	11138-66-2	Rhodigel® Supra
9102-54-8	Cellulase AP 3	9050-36-6	Maltrin® QD M440	11138-66-2	Ticaxan® Regular
9102-54-8	Cellulase L	9050-36-6	Maltrin® QD M500	11138-66-2	TIC Pretested® Xanthan 200
9102-54-8	Cellulase TAP	9050-36-6	Maltrin® QD M550		
9012-54-8	Cellulase TRL	9050-36-6	Maltrin® QD M580	11138-66-2	TIC Pretested® Xanthan PH
9012-54-8	Cellulase Tr Conc.	9050-36-6	Microduct®		
9014-01-1	AFP 2000	9050-36-6	Staley® Maltodextrin 3260	12694-22-3	Emalex MSG-2
				12694-22-3	Emalex MSG-2MA

CAS	Trade name	CAS	Trade name	CAS	Trade name
12694-22-3	Emalex MSG-2MB	25383-99-7	Admul SSL 2003		970
12694-22-3	Emalex MSG-2ME	25383-99-7	Admul SSL 2004	27195-16-0	Sucro Ester 7
12694-22-3	Emalex MSG-2ML	25383-99-7	Artodan SP 55 Kosher	31566-31-1	Aldo® HMS FG
12694-22-3	Nikkol DGMS	25383-99-7	Atlas B-60	31566-31-1	Aldo® HMS KFG
12709-64-7	Hodag PGS-103	25383-99-7	Atlas SSL	31566-31-1	Aldo® MS FG
12764-60-2	Hodag PGS-102	25383-99-7	Crolactil SS2L	31566-31-1	Aldo® MSLG FG
12764-60-2	Nikkol Decaglyn 2-S	25383-99-7	Emplex	31566-31-1	Aldo® MS LG KFG
13463-67-7	Hilton Davis Titanium Dioxide	25383-99-7	Emulsilac S	31566-31-1	Cithrol GMS Acid Stable
13552-80-2	Captex® 8227	25383-99-7	Emulsilac S K		
22839-47-0	Equal®	25383-99-7	Lamegin® NSL	31566-31-1	Cithrol GMS N/E
22839-47-0	HSC Aspartame	25383-99-7	Paniplex SK	31566-31-1	Cithrol GMS S/E
22839-47-0	NutraSweet®	25383-99-7	Pationic® 920	31566-31-1	Empilan® GMS NSE40
24315-83-1	Nutrifos® SK	25383-99-7	Prefera® NSL	31566-31-1	Empilan® GMS SE40
25013-16-5	Sustane® 1-F	25383-99-7	Prefera® SSL 6000 MB	31566-31-1	Hodag GMS
25013-16-5	Sustane® BHA	25383-99-7	Swedex SSL-5AC	31566-31-1	Imwitor® 191
25155-30-0	Calsoft L-60	25496-72-4	Aldo® MOD FG	31566-31-1	Lipal GMS
25155-30-0	Rhodacal® DS-4	25496-72-4	Aldo® MO FG	33940-99-7	Caprol® 10G2O
25155-30-0	Sul-fon-ate AA-9	25496-72-4	Caplube 8350	33940-99-7	Hodag PGO-102
25155-30-0	Sul-fon-ate AA-10	25496-72-4	Hodag GMO	33940-99-7	Nikkol Decaglyn 2-O
25155-30-0	Sul-fon-ate LA-10	25496-72-4	Tegomuls® O	34406-66-1	Hodag PGL-101
25168-73-4	Ryoto Sugar Ester S-1170	25496-72-4	Tegomuls® O Spezial	34406-66-1	Nikkol Decaglyn 1-L
		25637-84-7	Cithrol GDO N/E	34424-97-0	Caprol® 6G2S
25168-73-4	Ryoto Sugar Ester S-1170S	25956-17-6	Allura® Red AC	34424-97-0	Drewmulse® 6-2-S
25168-73-4	Ryoto Sugar Ester S-1570	26266-57-9	Ablunol S-40	34424-97-0	Plurol Stearique WL 1009
		26266-57-9	Crill 2		
25168-73-4	Ryoto Sugar Ester S-1670	26266-57-9	Hodag SMP	34424-98-1	Caprol® 10G4O
		26266-57-9	Lamesorb® SMP	34424-98-1	Drewmulse® 10-4-O
25168-73-4	Ryoto Sugar Ester S-1670S	26266-57-9	Nikkol SP-10	34424-98-1	Drewpol® 10-4-O
		26266-57-9	Nissan Nonion PP-40R	34424-98-1	Drewpol® 10-4-OK
25322-68-3	Carbowax® Sentry® PEG 300	26266-57-9	Prote-sorb SMP	34424-98-1	Hodag PGO-104
		26266-57-9	Span® 40	34424-98-1	Mazol® PGO-104
25322-68-3	Carbowax® Sentry® PEG 400	26266-58-0	Ablunol S-85	35285-68-8	Nipagin A Sodium
		26266-58-0	Crill 45	35285-69-9	Nipasol M Sodium
25322-68-3	Carbowax® Sentry® PEG 600	26266-58-0	Hodag STO	36457-19-9	Nipagin A Potassium
		26266-58-0	Nissan Nonion OP-85R	36457-20-2	Nipabutyl Sodium
25322-68-3	Carbowax® Sentry® PEG 900	26266-58-0	Prote-sorb STO	37189-34-7	Bromelain 150 GDU
		26266-58-0	S Mazol® 85K	37189-34-7	Bromelain 600 GDU
		26266-58-0	Span® 85	37189-34-7	Bromelain 1200 GDU
25322-68-3	Carbowax® Sentry® PEG 1000	26402-22-2	Imwitor® 910	37189-34-7	Bromelain 1500 GDU
		26402-26-6	Imwitor® 988	37220-82-9	Mazol® GMO K
25322-68-3	Carbowax® Sentry® PEG 1450	26402-31-3	Cithrol PGMR N/E	37349-34-1	Caprol® 3GS
		26446-38-8	Ryoto Sugar Ester P-170	37349-34-1	Drewmulse® 3-1-S
25322-68-3	Carbowax® Sentry® PEG 3350			37349-34-1	Drewpol® 3-1-SK
		26446-38-8	Ryoto Sugar Ester P-1570	37349-34-1	Hodag PGS
25322-68-3	Carbowax® Sentry® PEG 4600			37349-34-1	Lipal 4-1S
		26446-38-8	Ryoto Sugar Ester P-1570S	37349-34-1	Nikkol Tetraglyn 1-S
25322-68-3	Carbowax® Sentry® PEG 8000			37349-34-1	Polyaldo® TGMS
		26446-38-8	Ryoto Sugar Ester P-1670	37349-34-1	Polyaldo® TGMS KFG
25322-68-3	Dow E300 NF			37349-34-1	Santone® 3-1-S
25322-68-3	Dow E400 NF	26446-38-8	Sucro Ester 15	37349-34-1	Santone® 3-1-S XTR
25322-68-3	Dow E600 NF	26657-96-5	Emalex GMS-P	37349-34-1	Triodan 55
25322-68-3	Dow E900 NF	26658-19-5	Alkamuls® STS	37349-34-1	Witconol™ 18F
25322-68-3	Dow E1000 NF	26658-19-5	Crill 35	38566-94-8	Nipabutyl Potassium
25322-68-3	Dow E1450 NF	26658-19-5	Crill 41	39529-26-5	Caprol® JB
25322-68-3	Dow E3350 NF	26658-19-5	Glycomul® TS KFG	39529-26-5	Drewmulse® 10-10-S
25322-68-3	Dow E4500 NF	26658-19-5	Hodag STS	39529-26-5	Hodag PGS-1010
25322-68-3	Dow E8000 NF	26658-19-5	Lamesorb® STS	51158-08-8	Tagat® S
25322-68-3	Pluracol® E300	26658-19-5	Liposorb TS	51158-08-8	Tagat® S2
25322-68-3	Pluracol® E400	26658-19-5	Prote-sorb STS	51192-09-7	Tagat® O
25322-68-3	Pluracol® E600	26658-19-5	Span® 65	51192-09-7	Tagat® O2
25322-68-3	Pluracol® E4000	26658-19-5	Span® 65K	51248-32-9	Tagat® L
25339-99-5	Ryoto Sugar Ester L-1570	26680-54-6	n-Octenyl Succinic Anhydride	51248-32-9	Tagat® L2
				53124-00-8	Pure-Dent® B890
		27195-16-0	Ryoto Sugar Ester S-570	53124-00-8	Pure-Gel® B992
25339-99-5	Ryoto Sugar Ester L-1695			53637-25-5	Dow EP530
		27195-16-0	Ryoto Sugar Ester S-770	56519-71-2	Lexol® PG-800
				59259-38-0	Frescolat, Type ML
25339-99-5	Ryoto Sugar Ester LWA-1570	27195-16-0	Ryoto Sugar Ester S-	61725-93-7	Drewpol® 6-2-SK

CAS	Trade name	CAS	Trade name	CAS	Trade name
61788-47-4	Emery® 6361	65381-09-1	Liponate GC	68153-28-6	IPSO-FC
61788-85-0	Emalex HC-50	65381-09-1	Miglyol® 810	68153-28-6	IPSO-MR Dispersible
61788-85-0	Emalex HC-60	65381-09-1	Miglyol® 812	68153-28-6	IPSO-NGL
61789-05-7	Drewmulse® 75	65381-09-1	Neobee® 1053	68153-28-6	Mira-Foam® 100
61789-05-7	Imwitor® 928	65381-09-1	Neobee® M-5	68153-28-6	Mira-Foam® 100K
61789-07-9	Myverol® 18-07	65381-09-1	Neobee® O	68153-28-6	Mira-Foam® 120V
61789-07-9	Myverol® 18-07K	65381-09-1	Standamul® 318	68153-28-6	Mira-Foam® 130H
61789-08-0	Myverol® 18-06	65996-62-5	Pure-Bind® B910	68153-28-6	Mix 63
61789-08-0	Myverol® 18-50	65996-62-5	Pure-Bind® B923	68153-28-6	Procon® 2000
61789-08-0	Myverol® 18-50K	65996-63-6	Pure-Set® B950	68153-28-6	Procon® 20/60
61789-09-1	Monomuls® 90-25	65996-63-6	Pure-Set™ B965	68153-28-6	Procon® 20/60 Military
61789-10-4	Dimodan P	66082-42-6	Emalex DISG-3	68153-28-6	Procon® 20/60-SL
61789-10-4	Dimodan S	66988-04-3	Crolactil SISL	68153-28-6	Pro Fam 646
61789-10-4	Monomuls® 90-10	67701-08-0	Industrene® 225	68153-28-6	Pro Fam 648
61789-10-4	Myverol® 18-40	67701-27-3	Monomuls® 60-20	68153-28-6	Pro Fam 780
61789-13-7	Dimodan TH	67701-28-4	Captex® 810A	68153-28-6	Pro Fam 781
61789-13-7	Monomuls® 90-20	67701-28-4	Captex® 810B	68153-28-6	Pro Fam 955
61789-13-7	Myverol® 18-30	67701-28-4	Captex® 810C	68153-28-6	Pro Fam 970
61790-37-2	Industrene® 143	67701-28-4	Captex® 810D	68153-28-6	Pro Fam 972
61790-86-1	Hefti MO-55-F	67701-28-4	Captex® 910A	68153-28-6	Pro Fam 974, 974
61790-95-2	Hefti GMS-333	67701-28-4	Captex® 910B		Fortified
63148-62-9	Masil® SF 350 FG	67701-28-4	Captex® 910C	68153-28-6	Promax® 70
63231-60-7	Be Square® 175	67701-28-4	Captex® 910D	68153-28-6	Promax® 70L
63231-60-7	Be Square® 185	67701-32-0	GMR-33	68153-28-6	Promax® 70LSL
63231-60-7	Be Square® 195	67701-33-1	Alphadim® 90AB	68153-28-6	Promax® Plus
63231-60-7	Mekon® White	67762-36-1	Emery® 6358	68153-28-6	Promine® DS
63231-60-7	Multiwax® 180-M	67762-36-1	Industrene® 365	68153-28-6	Promine® DS-SL
63231-60-7	Multiwax® ML-445	67784-82-1	Polyaldo® 10-1-O KFG	68153-28-6	Promine® HV
63231-60-7	Multiwax® W-445	67784-87-6	Monomuls® 90-35	68153-28-6	Proplus®
63231-60-7	Multiwax® W-835	67784-87-6	Myvatex® 8-16	68153-28-6	Supro® 425
63231-60-7	Multiwax® X-145A	67784-87-6	Myverol® 18-04	68153-28-6	Supro® 610
63231-60-7	Petrolite® C-1035	67938-21-0	Emalex DISG-2	68153-28-6	Supro® 661
63231-60-7	Starwax® 100	68131-37-3	Fro-Dex® 24 Powd.	68153-28-6	Supro® 670
63231-60-7	Ultraflex®	68131-37-3	Maltrin® M200	68153-28-6	Supro® 710
63231-60-7	Victory®	68131-37-3	Maltrin® M205	68153-28-6	Supro® 760
64365-11-3	ADP	68131-37-3	Maltrin® M250	68153-28-6	Textratein
64365-11-3	APA	68131-37-3	Maltrin® M255	68153-28-6	Tofupro-U
64365-11-3	APC	68131-37-3	Maltrin® M365	68153-28-6	Vegafoom
64365-11-3	BL™	68131-37-3	Maltrin® QD M600	68153-28-6	Versa-Whip® 500, 510,
64365-11-3	C™	68131-37-3	Star-Dri® 20		520
64365-11-3	CAL®	68131-37-3	Star-Dri® 35F	68153-28-6	Versa-Whip® 520K,
64365-11-3	Calgon® Type RB	68131-37-3	Star-Dri® 35R		600K, 620K
64365-11-3	Cane Cal®	68131-37-3	Star-Dri® 42C	68153-76-4	Hodag POE (20) GMS
64365-11-3	Colorsorb	68131-37-3	Star-Dri® 42F	68153-76-4	Mazol® 80 MGK
64365-11-3	CPG®	68131-37-3	Star-Dri® 42R	68201-48-9	Monomuls® 60-45
64365-11-3	CPG® LF	68131-37-3	Star-Dri® 42X	68308-54-3	Monomuls® 60-25
64365-11-3	Diahope®-S60	68131-37-3	Star-Dri® 55	68334-00-9	Akoleno D
64365-11-3	GW	68153-28-6	Arcon F	68334-00-9	Akolizer C
64365-11-3	OL®	68153-28-6	Arcon G, Fortified	68334-00-9	Diamond D 75
64365-11-3	PWA™		Arcon G	68334-00-9	Duratex
64365-11-3	RC®	68153-28-6	Arcon S, Fortified	68334-00-9	Duromel
64365-11-3	SGL®		Arcon S	68334-00-9	Duromel B108
64742-14-9	Penreco 2251 Oil	68153-28-6	Arcon T, Fortified	68334-00-9	07 Stearine
64742-46-7	Penreco 2257 Oil		Arcon T	68334-00-9	Sterotex® NF
64742-47-8	Penreco 2263 Oil	68153-28-6	Arcon VF, Fortified	68334-28-1	Lipo SS
65381-09-1	Captex® 300		Arcon VF	68334-28-1	Wecobee® FS
65381-09-1	Captex® 355	68153-28-6	Ardex D, D Dispersible	68334-28-1	Wecobee® FW
65381-09-1	Estasan GT 8-40 3578	68153-28-6	Ardex D-HD	68334-28-1	Wecobee® R Mono
65381-09-1	Estasan GT 8-60 3575	68153-28-6	Ardex DHV Dispersible	68334-28-1	Wecobee® S
65381-09-1	Estasan GT 8-60 3580	68153-28-6	Ardex F, F Dispersible	68334-28-1	Wecobee® SS
65381-09-1	Estasan GT 8-65 3577	68153-28-6	Ardex FR	68334-28-1	Wecobee® W
65381-09-1	Estasan GT 8-65 3581	68153-28-6	Ardex R	68425-17-2	Hystar® 3375
65381-09-1	Hodag CC-33	68153-28-6	Danpro DS	68425-17-2	Hystar® 4075
65381-09-1	Hodag CC-33-F	68153-28-6	Danpro HV	68425-17-2	Hystar® 5875
65381-09-1	Hodag CC-33-L	68153-28-6	Danpro S	68425-17-2	Hystar® 6075
65381-09-1	Hodag CC-33-S	68153-28-6	Danpro S-760	68425-17-2	Hystar® HM-75
65381-09-1	Lexol® GT-855	68153-28-6	FP 940	68425-36-5	Witarix® 450
65381-09-1	Lexol® GT-865	68153-28-6	IPSO-C403	68434-04-4	Litesse®

CAS	Trade name	CAS	Trade name	CAS	Trade name
68434-04-4	Litesse® II	68990-06-7	Lamegin® GLP 10, 20		P8002
68434-04-4	Sta-Lite™ 100C	68990-06-7	Tegin® L 61, L 62	84775-52-0	Dispersable Curcumex
68434-04-4	Sta-Lite™ 100CN	68990-53-4	Canamulse 100		2600, 2601
	Neutralized	68990-53-4	Canamulse 110K	84930-16-5	Nipasol M Potassium
68434-04-4	Sta-Lite™ 100F	68990-53-4	Canamulse 150K	85411-01-4	Emphos™ F27-85
68434-04-4	Sta-Lite™ 100FN	68990-53-4	Canamulse 155	87390-32-7	Nikkol Decaglyn 1-M
	Neutralized	68990-55-6	Axol® E 66	89957-89-1	Beetroot Conc. P3003
68441-17-8	Hoechst Wax PED 121	68990-58-9	Axol® E 41	91052-81-2	Imwitor® 2020
68441-17-8	Luwax OA 5	68990-59-0	Lamegin® ZE 30, 60	91744-38-6	Imwitor® 369
68476-78-8	De-Mol® Molasses	68990-59-0	Tegin® C-62 SE	91744-38-6	Imwitor® 370
68476-78-8	Dri-Mol® 604 Molasses	68990-59-0	Tegin® C-63 SE	94423-19-5	Emalex DSG-3
68476-78-8	Dri-Mol® Molasses	68990-82-9	Akorine 4	97281-47-5	Phospholipon® 90/90 G
68476-78-8	Sweet'n'Neat® 65	68990-82-9	Akowesco 2	97593-29-8	Dimodan PVP Kosher
	Molasses Powd.	68990-82-9	Akowesco 5	97593-31-2	Acidan
68476-78-8	Sweet'n'Neat® 4000	68990-82-9	Akowesco 7	100209-45-8	Exter™ Family
	Molasses Powd.	68990-82-9	Akowesco 45 AC	100209-45-8	HVP-LS
68476-78-8	Sweet'n'Neat® 5000	68990-82-9	Witarix® 212	100209-45-8	Luxor® Century™ V
	Molasses Flake	68991-68-4	Captex® 350	100209-45-8	Luxor® CVP 5-SD
68513-95-1	Bakers Nutrisoy	69005-67-8	Drewsorb® 60K	100209-45-8	Luxor® CVP 1632
68513-95-1	Centex®	69028-36-0	Emuldan HV 40 Kosher	100209-45-8	Luxor® CVP 1689
68513-95-1	Nurupan™	69028-36-0	Emuldan HV 52 Kosher	100209-45-8	Luxor® CVP 1753
68513-95-1	Nutrisoy 220T	69227-21-0	Synperonic LF/RA43	100209-45-8	Luxor® CVP LS
68513-95-1	Soyafluff® 200 W	69468-44-6	Tagat® I	100209-45-8	Luxor® E-610
68513-95-1	Soyamin 50 E	69468-44-6	Tagat® I2	100209-45-8	Luxor® FB-10
68513-95-1	Soyamin 50 T	71010-52-1	Kelcogel®	100209-45-8	Luxor® FC-20
68513-95-1	Soyapan™	71010-52-1	Kelcogel® BF	100209-45-8	Luxor® GR-100
68553-11-7	Aldosperse® MS-20 FG	71010-52-1	Kelcogel® BF10	100209-45-8	Luxor® GR-150
68553-11-7	Aldosperse® MS-20	71010-52-1	Kelcogel® CF	100209-45-8	Luxor® GR-200
	KFG	71010-52-1	Kelcogel® CF10	100209-45-8	Luxor® HVP-A
68553-81-1	Protex™ 20	71010-52-1	Kelcogel® F	100209-45-8	Luxor® KB-312
68583-51-7	Captex® 200	71010-52-1	Kelcogel® IF	100209-45-8	Luxor® L-625
68583-51-7	Neobee® M-20	71010-52-1	Kelcogel® JJ	100209-45-8	Luxor® MB-40
68608-64-0	Miranol® J2M Conc.	71010-52-1	Kelcogel® PD	100209-45-8	Luxor® MB-110
68608-64-0	Miranol® JB	72347-89-8	Caprol® 2G4S	100209-45-8	Luxor® MB-120
00070-77-7	Hickory Smoked Torula	72060-60-3	Lipovol P	100200-45-8	Luxor® No. 6 Sauce
	Dried Yeast	73398-61-5	Aldo® MCT KFG	100209-45-8	Luxor® No. 700 Sauce
68876-77-7	Type B Torula Dried	74623-31-7	Pluracol® W170	100209-45-8	Luxor® Prozate®
	Yeast	74623-31-7	Pluracol® W660	100209-45-8	Luxor® Triple-H®
68915-31-1	Calgon® T Powd. Food	74623-31-7	Pluracol® W2000	100209-45-8	Toppro
	Grade	74623-31-7	Pluracol® W5100N	102051-00-3	Hodag PGO-103
68915-31-1	Vitrafos®	76009-37-5	Caprol® 6G2O	112926-00-8	Sipernat® 22
68915-31-1	Vitrafos® LC	76009-37-5	Drewpol® 6-2-OK	112926-00-8	Sipernat® 22S
68937-16-6	Drewpol® 10-1-CCK	76009-37-5	Hodag PGO-62	112926-00-8	Sipernat® 50
68938-35-2	Drewmulse® D-4661	79777-30-3	Hodag PGS-101	112926-00-8	Sipernat® 50S
68938-37-4	Wecobee® HTR	79777-30-3	Nikkol Decaglyn 1-S	112945-52-5	Cab-O-Sil® HS-5
68938-37-4	Wecobee® M	81025-04-9	Lactitol MC	120486-24-0	Emalex TISG-2
68952-98-7	Akwilox 133	84775-52-0	Curcumex 1600, 1601	133738-23-5	Nikkol Decaglyn 1-IS
68958-64-5	Tagat® TO	84775-52-0	Curcumex Natural	136097-97-7	Koster Keunen
68990-05-6	Axol® C 62		Powd. Colorant		Synthetic Spermaceti
68990-05-6	Axol® C 63	84775-52-0	Curcumin Extract		

CAS Number-to-Chemical
Cross-Reference

CAS	Chemical	CAS	Chemical	CAS	Chemical
50-00-0	Formaldehyde	65-85-0	Benzoic acid	78-84-2	2-Methylpropanal
50-14-6	Ergocalciferol	66-25-1	Hexanal	78-93-3	Methyl ethyl ketone
50-21-5	Lactic acid	67-03-8	Thiamine HCl	78-98-8	Pyruvaldehyde
50-70-4	Sorbitol	67-48-1	Choline chloride	79-01-6	Trichloroethylene
50-81-7	L-Ascorbic acid	67-56-1	Methyl alcohol	79-09-4	Propionic acid
50-99-7	D-Glucose anhyd.	67-63-0	Isopropyl alcohol	79-20-9	Methyl acetate
52-89-1	Cysteine hydrochloride	67-64-1	Acetone	79-31-2	Isobutyric acid
52-90-4	L-Cysteine	67-97-0	Cholecalciferol	79-33-4	L-Lactic acid
54-12-6	DL-α-Tryptophan	68-04-2	Sodium citrate	79-42-5	2-Mercaptopropionic
56-40-6	Glycine	68-04-2	Trisodium citrate anhyd.		acid
56-41-7	L-Alanine	68-19-9	Cyanocobalamin	79-46-9	2-Nitropropane
56-45-1	L-Serine	68-26-8	Retinol	79-69-6	α-Irone
56-81-5	Glycerin	69-65-8	D-Mannitol	79-77-6	β-Ionone
56-84-8	L-Aspartic acid	69-72-7	Salicylic acid	79-78-7	Allyl α-ionone
56-85-9	L-Glutamine	69-79-4	Maltose	79-81-2	Retinyl palmitate
56-86-0	L-Glutamic acid	70-47-3	L-Asparagine anhyd.	79-92-5	Camphene
56-87-1	L-Lysine	71-00-1	L-Histidine	80-56-8	α-Pinene
56-89-3	L-Cystine	71-23-8	Propyl alcohol	81-07-2	Saccharin
57-06-7	Allyl isothiocyanate	71-36-3	Butyl alcohol	81-13-0	D-Panthenol
57-10-3	Palmitic acid	71-41-0	n-Amyl alcohol	81-25-4	Cholic acid
57-11-4	Stearic acid	72-17-3	Sodium lactate	83-34-1	Skatole
57-13-6	Urea	72-18-4	L-Valine	83-88-5	Riboflavin
57-48-7	D-Fructose	72-19-5	L-Threonine	84-66-2	Diethyl phthalate
57-50-1	Sucrose	73-22-3	L-Tryptophan	85-70-1	n-Butyl phthalyl-n-butyl
57-55-6	Propylene glycol	73-32-5	L-Isoleucine		glycolate
57-88-5	Cholesterol	74-79-3	L-Arginine	85-84-7	1-(Phenylazo)-2-
58-08-2	Caffeine	74-93-1	Methyl mercaptan		naphthylamine
58-56-0	Pyridoxine HCl	74-98-6	Propane	85-91-6	Dimethyl anthranilate
58-85-5	d-Biotin	75-07-0	Acetaldehyde	87-18-3	p-t-Butylphenyl
58-86-6	D(+)-Xylose	75-09-2	Methylene chloride		salicylate
58-95-7	d-α-Tocopheryl acetate	75-18-3	Methyl sulfide	87-19-4	Isobutyl salicylate
59-02-9	d-α-Tocopherol	75-28-5	Isobutane	87-20-7	Amyl salicylate
59-30-3	Folicacid	75-50-3	Trimethylamine	87-25-2	Ethyl anthranilate
59-51-8	DL-Methionine	75-71-8	Dichlorodifluoro-	87-29-6	Cinnamyl anthranilate
59-67-6	Nicotinic acid		methane	87-44-5	β-Caryophyllene
60-00-4	Edetic acid	75-85-4	t-Amyl alcohol	87-67-2	Choline bitartrate
60-01-5	Tributyrin	76-13-1	Trichlorotrifluoroethane	87-69-4	L-Tartaric acid
60-12-8	Phenethyl alcohol	76-15-3	Chloropentafluoro-	87-89-8	Inositol
60-18-4	L-Tyrosine		ethane	87-91-2	Diethyl tartrate
60-33-3	Linoleic acid	76-22-2	Camphor	87-99-0	Xylitol
60-93-5	Quinine dihydrochloride	76-49-3	Bornyl acetate	88-09-5	Diethylacetic acid
61-90-5	L-Leucine	77-06-5	Gibberellic acid	88-14-2	2-Furoic acid
62-33-9	Calcium disodium	77-42-9	β-Santalol	89-49-6	Isopulegyl acetate
	EDTA	77-83-8	Ethyl methylphenyl-	89-65-6	Erythorbic acid
62-54-4	Calcium acetate		glycidate	89-74-7	2,4-Dimethylaceto-
63-42-3	Lactose	77-89-4	Acetyl triethyl citrate		phenone
63-68-3	L-Methionine	77-90-7	Acetyl tributyl citrate	89-78-1	Menthol
63-91-2	L-Phenylalanine	77-92-9	Citric acid anhyd.	89-82-7	Pulegone
64-02-8	Tetrasodium EDTA	77-93-0	Triethyl citrate	89-83-8	Thymol
64-17-5	Alcohol	78-35-3	Linalyl isobutyrate	90-02-8	Salicylaldehyde
64-18-6	Formic acid	78-36-4	Linalyl butyrate	90-05-1	Guaiacol
64-19-7	Acetic acid	78-70-6	Linalool	90-43-7	o-Phenylphenol
65-82-7	N-Acetyl-L-methionine	78-83-1	Isobutyl alcohol	90-80-2	Gluconolactone

CAS	Chemical	CAS	Chemical	CAS	Chemical
90-87-9	2-Phenylpropionaldehyde dimethylacetal	100-53-8	Benzyl mercaptan	106-23-0	Citronellal
		100-66-3	Anisole	106-24-1	Geraniol
91-22-5	Quinoline	100-86-7	Dimethylbenzyl carbinol	106-25-2	Nerol
91-53-2	6-Ethoxy-1,2-dihydro-2,2,4-trimethylquinoline	100-88-9	Cyclamic acid	106-27-4	Isoamyl butyrate
		100-97-0	Hexamethylene	106-30-9	Ethyl heptanoate
92-52-4	Biphenyl		tetramine	106-32-1	Ethyl octanoate
93-08-3	2'-Acetonaphthone	101-39-3	α-Methylcinnamaldehyde	106-33-2	Ethyl laurate
93-15-2	Methyl eugenol			106-35-4	3-Heptanone
93-16-3	Methyl isoeugenol	101-41-7	Methyl phenylacetate	106-36-5	Propyl propionate
93-28-7	Eugenyl acetate	101-48-4	Phenylacetaldehyde	106-44-5	p-Cresol
93-29-8	Acetisoeugenol		dimethyl acetal	106-65-0	Dimethyl succinate
93-51-6	2-Methoxy-4-methylphenol	101-86-0	α-Hexylcinnamaldehyde	106-68-3	3-Octanone
				106-70-7	Methyl caproate
93-53-8	2-Phenylpropanal	101-97-3	Ethyl phenylacetate	106-70-7	Methyl 2-hexanoate
93-54-9	1-Phenyl-1-propanol	102-04-5	1,3-Diphenyl-2-propanone	106-72-9	2,6-Dimethyl-5-heptenal
93-58-3	Methyl benzoate			106-73-0	Methyl heptanoate
93-89-0	Ethyl benzoate	102-20-5	Phenethyl phenylacetate	106-97-8	Butane
93-92-5	α-Methylbenzyl acetate			107-02-8	Acrolein
93-92-5	Methylbenzyl acetate, mixed o-, m-, p-	102-71-6	Triethanolamine	107-03-9	Propyl mercaptan
		102-76-1	Triacetin	107-06-2	Ethylene dichloride
94-02-0	Ethyl benzoylacetate	103-25-3	Methyl 3-phenylpropionate	107-15-3	Ethylenediamine
94-13-3	Propylparaben			107-75-5	Hydroxycitronellal
94-18-8	Benzylparaben	103-26-4	Methyl cinnamate	107-87-9	Methyl propyl ketone
94-26-8	Butylparaben	103-28-6	Benzyl isobutyrate	107-88-0	Butylene glycol
94-30-4	Ethyl-p-anisate	103-36-6	Ethyl cinnamate	107-92-6	n-Butyric acid
94-36-0	Benzoyl peroxide	103-37-7	Benzyl butyrate	108-10-1	Methyl isobutyl ketone
94-46-2	Isoamyl benzoate	103-38-8	Benzyl isovalerate	108-21-4	Isopropyl acetate
94-47-3	Phenethyl benzoate	103-41-3	Benzyl cinnamate	108-24-7	Acetic anhydride
94-59-7	Safrol	103-45-7	2-Phenylethyl acetate	108-29-2	γ-Valerolactone
94-62-2	Piperine	103-48-0	Phenethyl isobutyrate	108-30-5	Succinic anhydride
94-86-0	Propenylguaethol	103-50-4	Benzyl ether	108-50-9	2,6-Dimethylpyrazine
96-17-3	2-Methylbutyraldehyde	103-52-6	Phenethyl butyrate	108-64-5	Ethyl isovalerate
96-22-0	Diethyl ketone	103-54-8	Cinnamyl acetate	108-78-1	Melamine
96-33-3	Methyl acrylate polymer	103-58-2	3-Phenylpropyl isobutyrate	108-91-8	Cyclohexylamine
96-48-0	Butyrolactone			108-98-5	Benzenethiol
97-42-7	Carvyl acetate	103-59-3	Cinnamyl isobutyrate	109-08-0	2-Methylpyrazine
97-45-0	Carvyl propionate	103-61-7	Cinnamyl butyrate	109-15-9	Octyl isobutyrate
97-53-0	Eugenol	103-82-2	Phenylacetic acid	109-19-3	n-Butyl isovalerate
97-54-1	Isoeugenol	103-93-5	p-Tolyl isobutyrate	109-21-7	Butyl butyrate
97-61-0	2-Methylpentanoic acid	103-95-7	Cyclamen aldehyde	109-43-3	Dibutyl sebacate
97-62-1	Ethyl isobutyrate	104-20-1	4-p-Methoxyphenyl-2-butanone	109-52-4	n-Valeric acid
97-63-2	Ethyl methacrylate			109-60-4	Propyl acetate
97-64-3	Ethyl lactate	104-46-1	Anethole	109-73-9	Butylamine
97-65-4	Itaconic acid (polymerized)	104-50-7	γ-Octalactone	109-86-4	Methoxyethanol
		104-53-0	Hydrocinnamaldehyde	109-94-4	Ethyl formate
97-67-6	L-Hydroxysuccinic acid	104-54-1	Cinnamyl alcohol	109-95-5	Ethyl nitrite
97-85-8	Isobutyl isobutyrate	104-55-2	Cinnamal	109-97-7	Pyrrole
97-96-1	2-Ethylbutyraldehyde	104-61-0	γ-Nonalactone	110-15-6	Succinic acid
97-99-4	Tetrahydrofurfuryl alcohol	104-65-4	Cinnamyl formate	110-17-8	Fumaric acid
		104-67-6	γ-Undecalactone	110-19-0	Isobutyl acetate
98-00-0	Furfuryl alcohol	104-87-0	p-Tolyl aldehyde	110-30-5	Ethylene distearamide
98-01-1	Furfural	104-93-8	p-Methylanisole	110-38-3	Ethyl decanoate
98-20-2	Furfural mercaptan	105-13-5	Anisyl alcohol	110-39-4	Octyl butyrate
98-85-1	α-Methylbenzyl alcohol	105-21-5	γ-Heptalactone	110-40-7	Diethyl sebacate
98-86-2	Acetophenone	105-37-3	Ethyl propionate	110-41-8	Methyl nonyl acetaldehyde
98-92-0	Niacinamide	105-53-3	Diethyl malonate		
99-48-9	Carveol	105-54-4	Ethyl butyrate	110-43-0	Methyl n-amyl ketone
99-49-0	l-Carvone	105-57-7	Acetal	110-44-1	Sorbic acid
99-76-3	Methylparaben	105-66-8	Propyl butyrate	110-45-2	Isoamyl formate
99-83-2	α-Phellandrene	105-79-3	Isobutyl hexanoate	110-54-3	Hexane
99-85-4	γ-Terpinene	105-85-1	Citronellyl formate	110-62-3	n-Valeraldehyde
99-86-5	α-Terpinene	105-86-2	Geranyl formate	110-74-7	Propyl formate
99-87-6	p-Cymene	105-87-3	Geranyl acetate	110-82-7	Cyclohexane
100-00-0	Acetanisole	105-95-3	Ethylene brassylate	110-86-1	Pyridine
100-37-8	Diethylaminoethanol	106-02-5	Pentadecalactone	110-89-4	Piperidine
100-42-5	Styrene	106-18-3	Butyl laurate	110-91-8	Morpholine
100-51-6	Benzyl alcohol	106-21-8	3,7-Dimethyl-1-octanol	110-93-0	Methyl-5-hepten-2-ol
100-52-7	Benzaldehyde	106-22-9	β-Citronellol	110-93-0	Methyl heptenone

CAS	Chemical	CAS	Chemical	CAS	Chemical
111-03-5	Glyceryl oleate	122-18-9	Cetalkonium chloride		phenate
111-11-5	Methyl caprylate	122-19-0	Stearalkonium chloride	132-27-4	Sodium o-phenyl-
111-12-6	Methyl 2-octynoate	122-32-7	Triolein		phenate
111-13-7	Methyl hexyl ketone	122-40-7	α-Amylcinnamaldehyde	133-18-6	Phenethyl anthranilate
111-17-1	3,3´-Thiodipropionic	122-43-0	Butyl phenylacetate	134-03-2	Sodium ascorbate
	acid	122-57-6	Benzylidene acetone	134-20-3	Methyl anthranilate
111-27-3	Hexyl alcohol	122-59-8	Phenoxyacetic acid	135-02-4	o-Methoxybenzal-
111-30-8	Glutaral	122-67-8	Isobutyl cinnamate		dehyde
111-62-6	Ethyl oleate	122-69-0	Cinnamyl cinnamate	136-77-6	4-Hexylresorcinol
111-70-6	Heptyl alcohol	122-72-5	Hydrocinnamyl acetate	137-08-6	D-Calcium pantothen-
111-71-7	Heptanal	122-78-1	Phenylacetaldehyde		ate
111-76-2	Butoxyethanol	122-91-8	Anisyl formate	137-40-6	Sodium propionate
111-79-5	Methyl 2-nonenoate	122-97-4	Hydrocinnamic alcohol	137-66-6	Ascorbyl palmitate
111-82-0	Methyl laurate	123-11-5	p-Anisaldehyde	138-15-8	L-Glutamic acid
111-87-5	Caprylic alcohol	123-19-3	4-Heptanone		hydrochloride
112-05-0	Nonanoic acid	123-25-1	Diethyl succinate	138-22-7	Butyl lactate
112-06-1	Heptyl acetate	123-28-4	Dilauryl thiodipropio-	138-86-3	dl-Limonene
112-12-9	2-Undecanone		nate	139-06-0	Calcium cyclamate
112-14-1	Octyl acetate	123-29-5	Ethyl pelargonate	139-07-1	Lauralkonium chloride
112-17-4	Decyl acetate	123-32-0	2,5-Dimethylpyrazine	139-08-2	Myristalkonium chloride
112-23-2	Heptyl formate	123-35-3	Myrcene	139-33-3	Disodium EDTA
112-30-1	n-Decyl alcohol	123-38-6	Propionaldehyde	139-43-5	Glyceryl (triacetoxy-
112-31-2	Decanal	123-51-3	Isoamyl alcohol		stearate)
112-32-3	Octyl formate	123-54-6	Acetylacetone	140-03-4	Methyl acetyl ricinoleate
112-38-9	Undecylenic acid	123-66-0	Ethyl caproate	140-10-3	Cinnamic acid
112-41-4	Dodecene-1	123-68-2	Allyl caproate	140-11-4	Benzyl acetate
112-42-5	Undecyl alcohol	123-72-8	n-Butyraldehyde	140-26-1	Phenethyl isovalerate
112-44-7	Undecanal	123-76-2	Levulinic acid	140-39-6	p-Cresyl acetate
112-45-8	10-Undecenal	123-77-3	Azodicarbonamide	140-67-0	Estragole
112-53-8	Lauryl alcohol	123-86-4	n-Butyl acetate	140-88-5	Ethyl acrylate
112-54-9	Lauric aldehyde	123-92-2	Isoamyl acetate	141-01-5	Ferrous fumarate
112-60-7	PEG-4	123-94-4	Glyceryl stearate	141-04-8	Diisobutyl adipate
112-61-8	Methyl stearate	123-95-5	Butyl stearate	141-08-2	Glyceryl ricinoleate
112-63-0	Methyl linoleate	124-04-9	Adipic acid	141-11-7	Rhodinyl acetate
112-66-3	Lauryl acetate	124-06-1	Ethyl myristate	141-12-8	Neryl acetate
112-72-1	Myristyl alcohol	124-07-2	Caprylic acid	141-16-2	Citronellyl butyrate
112-80-1	Oleic acid	124-10-7	Methyl myristate	141-43-5	Ethanolamine
112-92-5	Stearyl alcohol	124-13-0	n-Octanal	141-53-7	Sodium formate
115-18-4	2-Methyl-3-buten-2-ol	124-19-6	Nonanal	141-78-6	Ethyl acetate
115-25-3	Octafluorocyclobutane	124-25-4	Myristaldehyde	141-97-9	Ethylacetoacetate
115-71-9	α-Santalol	124-26-5	Stearamide	142-17-6	Calcium oleate
115-95-7	Linalyl acetate	124-30-1	Stearamine	142-18-7	Glyceryl laurate
115-99-1	Linalyl formate	124-38-9	Carbon dioxide	142-19-8	Allyl heptanoate
117-81-7	Dioctyl phthalate	124-41-4	Sodium methylate	142-30-3	Dimethyl hexynediol
118-58-1	Benzylsalicylate	126-13-6	Sucrose acetate	142-31-4	Sodium octyl sulfate
118-61-6	Ethyl salicylate		isobutyrate	142-47-2	MSG
118-71-8	Maltol	126-14-7	Sucrose octaacetate	142-50-7	cis-Trimethyldodeca-
119-36-8	Methyl salicylate	126-92-1	Sodium octyl sulfate		trieneol
119-53-9	Benzoin	126-95-4	Calcium glycero-	142-55-2	Propylene glycol laurate
119-61-9	Benzophenone		phosphate	142-59-6	Nabam
119-65-3	Isoquinoline	126-96-5	Sodium diacetate	142-62-1	Caproic acid
119-84-6	Dihydrocoumarin		anhyd.	142-92-7	Hexyl acetate
120-14-9	Veratraldehyde	127-08-2	Potassium acetate	143-07-7	Lauric acid
120-24-1	Isoeugenyl phenyl-	127-09-3	Sodium acetate	143-08-8	Nonyl alcohol
	acetate		anhydrous	143-13-5	n-Nonyl acetate
120-45-6	α-Methylbenzyl	127-17-3	Pyruvic acid	143-18-0	Potassium oleate
	propionate	127-40-2	Xanthophyll	143-19-1	Sodium oleate
120-47-8	Ethyl paraben	127-41-3	α-Ionone	143-28-2	Oleyl alcohol
120-51-4	Benzyl benzoate	127-47-9	Retinyl acetate	144-33-2	Disodium citrate
120-57-0	Heliotropine	127-91-3	β-Pinene	144-55-8	Sodium bicarbonate
120-72-9	Indole	128-04-1	Sodium dimethyldithio-	147-81-9	L-Arginine L-glutamate
121-32-4	Ethyl vanillin		carbamate hydrate	147-85-3	L-Proline
121-33-5	Vanillin	128-37-0	BHT	148-18-5	Sodium dimethyldithio-
121-39-1	Ethyl phenylglycidate	128-44-9	Saccharin sodium		carbamate
121-79-9	Propyl gallate	130-40-5	Riboflavin-5´-phosphate	148-79-8	Thiabendazole
122-00-9	4´-Methyl aceto-		sodium	149-91-7	Gallic acid
	phenone	130-95-0	Quinine	150-30-1	DL-Phenylalanine
122-03-2	Cuminaldehyde	131-52-2	Sodium pentachloro-	150-60-7	Benzyl disulfide

CAS	Chemical	CAS	Chemical	CAS	Chemical
150-78-7	p-Dimethoxybenzene	542-55-2	Isobutyl formate		acetate
150-84-5	Citronellyl acetate	544-17-2	Calcium formate	637-65-0	Tetrahydrofurfuryl
151-10-0	m-Dimethoxybenzene	544-40-1	Butyl sulfide		propionate
151-21-3	Sodium lauryl sulfate	544-63-8	Myristic acid	638-11-9	Isopropyl butyrate
298-14-6	Potassium bicarbonate	546-93-0	Magnesium carbonate	646-07-1	4-Methylpentanoic acid
299-27-4	Potassium D-gluconate	547-63-7	Methyl isobutyrate	657-27-2	L-Lysine hydrochloride
299-28-5	Calcium gluconate	554-12-1	Methyl propionate	659-70-1	Isoamyl isovalerate
299-29-6	Ferrous gluconate	555-43-1	Tristearin	692-86-4	Ethyl 10-undecenoate
	anhyd.	555-44-2	Tripalmitin	693-36-7	Distearyl thiodipropio-
300-92-5	Aluminum distearate	555-45-3	Trimyristin		nate
301-02-0	Oleamide	556-24-1	Methyl isovalerate	695-06-7	γ-Hexalactone
302-01-2	Hydrazine	556-82-1	3-Methyl-2-buten-1-ol	705-86-2	Δ-Decalactone
302-72-7	DL-Alanine	557-04-0	Magnesium stearate	706-14-9	γ-Decalactone
302-84-1	DL-Serine	557-05-1	Zinc stearate	710-04-3	ç-Undecalactone
304-59-6	Potassium sodium	557-61-9	Octacosanol	713-95-1	δ-Dodecalactone
	tartrate anhyd.	562-74-3	4-Carvomenthenol	765-70-8	Methyl cyclopenteno-
326-61-4	Piperonyl acetate	563-71-3	Ferrous carbonate		lone
328-50-7	α-Ketoglutaric acid	577-11-7	Dioctyl sodium	771-03-9	Dehydroacetic acid
334-48-5	Capric acid		sulfosuccinate	804-63-7	Quinine sulfate
350-03-8	3-Acetylpyridine	582-25-2	Potassium benzoate	813-94-5	Tricalcium citrate
355-42-0	Perfluorohexane		anhyd.		anhyd.
361-09-1	Ox bile extract	584-02-1	3-Pentanol	814-80-2	Calcium lactate anhyd.
408-35-5	Sodium palmitate	584-08-7	Potassium carbonate	821-55-6	2-Nonanone
409-02-9	Methyl heptenone	585-88-6	Maltitol	822-12-8	Sodium myristate
431-03-8	Diacetyl	589-66-2	Isobutyl-2-butenoate	822-16-2	Sodium stearate
443-79-8	DL-Isoleucine	589-82-2	3-Heptanol	860-22-0	FD&C Blue No. 2
458-37-7	Turmeric	589-98-0	3-Octanol	866-84-2	Potassium citrate
464-49-3	(+)-Camphor	590-00-1	Potassium sorbate		anhyd.
470-67-7	1,4-Cineole	590-01-2	n-Butyl propionate	867-81-2	Sodium pantothenate
470-82-6	Eucalyptol	590-86-3	Isovaleraldehyde	868-14-4	Potassium acid tartrate
471-34-1	Calcium carbonate	591-60-6	Butyl acetoacetate	868-18-8	Sodium tartrate anhyd.
488-10-8	cis-Jasmone	591-80-0	4-Pentenoic acid	868-57-5	Methyl 2-methylbutyrate
491-35-0	Lepidine	592-41-6	1-Hexene	870-23-5	Allyl mercaptan
492-62-6	D-Glucose anhyd.	592-84-7	Butyl formate	872-05-9	Decene-1
497-19-0	Sodium carbonate	592-00-1	Allyl sulfide	881-68-5	Vanillin acetate
	anhyd.	593-29-3	Potassium stearate	915-67-3	Amaranth
499-12-7	Aconitic acid	598-82-3	DL-Lactic acid	927-20-8	Magnesium glycero-
499-75-2	Carvacrol	600-07-7	2-Methylbutyric acid		phosphate anhyd.
500-38-9	Nordihydroguairetic	600-14-6	Pentane-2,3-dione	928-95-0	2-Hexenol
	acid	604-68-2	α-D-Glucose	928-96-1	3-Hexenol
501-52-0	Hydrocinnamic acid		pentaacetate	996-31-6	Potassium lactate
503-74-2	Isovaleric acid	616-25-1	1-Penten-3-ol	1034-01-1	Octyl gallate
506-87-6	Ammonium carbonate	617-35-6	Ethyl pyruvate	1066-33-7	Ammonium bicarbonate
507-70-0	Borneol	617-45-8	DL-Aspartic acid	1072-83-9	2-Acetylpyrrole
512-29-8	Flavoxanthin	617-48-1	DL-Hydroxysuccinic	1107-26-2	β-Apo-8′-carotenal
512-42-5	Sodium methyl sulfate		acid	1119-34-2	L-Arginine mono-
513-86-0	Acetyl methyl carbinol	620-23-5	m-Tolualdehyde		hydrochloride
514-78-3	Canthaxanthine	620-79-1	Ethyl 2-acetal-3-	1120-28-1	Methyl eicosenate
520-36-5	Chamomile		phenylpropionate	1122-62-9	2-Acetylpyridine
520-45-6	Dehydroacetic acid	621-71-6	Tricaprin	1123-85-9	2-Phenyl propanol-1
526-95-4	D-Gluconic acid	621-82-9	Cinnamic acid	1124-11-4	2,3,5,6-Tetramethyl-
527-07-1	Sodium gluconate	622-45-7	Cyclohexyl acetate		pyrazine
527-09-3	Copper gluconate	623-42-7	Methyl butyrate	1125-88-8	Benzaldehyde dimethyl
529-20-4	o-Tolualdehyde	623-70-1	Ethyl crotonate		acetal
532-32-1	Sodium benzoate	623-92-6	Diethyl fumarate	1139-30-6	β-Caryophyllene oxide
532-43-4	Thiamine nitrate	624-24-8	Methyl valerate	1166-52-5	Dodecyl gallate
533-96-0	Sodium sesquicarbon-	624-49-7	Dimethyl fumarate	1185-57-5	Iron ammonium citrate
	ate	624-92-0	Methyl disulfide		green
536-59-4	Perillyl alcohol	625-84-3	2,5-Dimethylpyrrole	1185-57-6	Iron ammonium citrate
536-60-7	Cuminic alcohol	626-11-9	Diethyl malate		brown
538-23-8	Tricaprylin	626-82-4	Butyl hexanoate	1188-02-9	2-Methylheptanoic acid
538-24-9	Trilaurin	627-90-7	Ethyl undecanoate	1241-94-7	Diphenyl octyl
539-82-2	Ethyl valerate	628-63-7	Amyl acetate		phosphate
539-88-8	Ethyl levulinate	629-19-6	Propyl disulfide	1302-42-7	Sodium aluminate
539-90-2	Isobutyl butyrate	629-25-4	Sodium laurate	1302-78-9	Bentonite
540-18-1	Amyl butyrate	629-33-4	Hexyl formate	1303-96-4	Sodium borate
540-42-1	Isobutyl propionate	637-12-7	Aluminum tristearate		decahydrate
541-15-1	L-Carnitine	637-64-9	Tetrahydrofurfuryl	1305-62-0	Calcium hydroxide

CAS	Chemical	CAS	Chemical	CAS	Chemical
1305-78-8	Calcium oxide	1551-44-6	Cyclohexyl butyrate		3(2H)furanone
1305-79-9	Calcium peroxide	1592-23-0	Calcium stearate	3734-33-6	Denatonium benzoate
1309-37-1	Ferric oxide anhyd.	1632-73-1	Fenchyl alcohol		NF
1309-42-8	Magnesium hydroxide	1708-39-0	Benzaldehyde glyceryl	3792-50-5	Sodium aspartate
	anhyd.		acetal	3796-70-1	Geranyl acetone
1309-48-4	Magnesium oxide	1731-84-6	Methyl nonanoate	3844-45-9	Brilliant blue FCF
1310-58-3	Potassium hydroxide	1783-96-6	D-Aspartic acid	3844-45-9	FD&C Blue No. 1
1310-73-2	Sodium hydroxide	1797-74-6	Allyl phenylacetate	3848-24-6	Acetyl butyryl
1314-13-2	Zinc oxide	1866-31-5	Allyl cinnamate	3913-71-1	trans-2-Decenal
1317-65-3	Calcium carbonate	1901-26-4	3-Methyl-4-phenyl-3-	3999-01-7	Linoleamide
1317-65-3	Limestone		buten-2-one	4070-80-8	Sodium stearyl
1319-69-3	Potassium glycero-	1934-21-0	Tartrazine		fumarate
	phosphate anhyd.	1934-21-0	FD&C Yellow No. 5	4075-81-4	Calcium propionate
1320-46-3	Manganese	1948-33-0	t-Butyl hydroquinone		anhyd.
	glycerophosphate	1976-28-9	Aluminum nicotinate	4112-89-4	Guaiacyl phenylacetate
1322-98-1	Sodium decylbenzene	2021-28-5	Ethyl-3-phenyl-	4128-31-8	2-Octanol
	sulfonate		propionate	4151-35-3	Potassium fumarate
1323-38-2	Castor oil	2051-78-7	Allyl butyrate		anhyd.
1323-39-3	Propylene glycol	2052-15-5	Butyl levulinate	4230-97-1	Allyl octanoate
	stearate	2090-05-3	Calcium benzoate	4345-03-3	d-α-Tocopheryl
1323-83-7	Glyceryl distearate	2142-94-1	Neryl formate		succinate
1327-36-2	Aluminum silicate	2179-57-9	Allyl disulfide	4418-26-2	Sodium dehydroacetate
1327-39-5	Aluminum calcium	2198-61-0	Isoamyl hexanoate	4437-51-8	3,4-Hexanedione
	silicate	2216-52-6	d-Neomenthol	4468-02-4	Zinc gluconate anhyd.
1330-43-4	Sodium borate anhyd.	2244-16-8	d-Carvone	4525-33-1	Dimethyl dicarbonate
1330-80-9	Propylene glycol oleate	2277-28-3	Glyceryl linoleate	4548-53-2	FD&C Red No. 4
1332-58-7	Kaolin	2305-05-7	γ-Dodecalactone	4602-84-0	Farnesol
1332-98-5	Iron ammonium citrate	2315-64-2	Octoxynol-5	4606-15-9	Propyl phenylacetate
1333-00-2	Iron ammonium citrate	2315-68-6	Propyl benzoate	4691-65-0	Disodium inosinate
1333-84-2	Alumina, hydrate	2345-24-6	Neryl isobutyrate	4940-11-8	Ethyl maltol
1333-86-4	Carbon black	2353-45-9	FD&C Green No. 3	5001-51-4	Calcium lactobionate
1334-78-7	p-Tolyl aldehyde	2363-89-5	trans-2-Octen-1-al	5026-62-0	Sodium methylparaben
1334-82-3	Amyl furoate	2396-84-1	Ethyl sorbate	5064-31-3	Trisodium NTA
1336-21-6	Ammonium hydroxide	2412-80-8	Methyl 4-methylvalerate	5117-19-1	PEG-8
1336-80-7	Ferric choline citrate	2445-76-3	Hexyl propionate	5292-21-7	Cyclohexaneacetic acid
1337-33-3	Stearyl citrate	2445-77-4	2-Methylbutyl	5392-40-5	Citral
1337-76-4	Attapulgite		isovalerate	5396-89-4	Benzyl acetoacetate
1338-39-2	Sorbitan laurate	2459-05-4	Ethyl fumarate	5408-52-6	L-Lysine L-glutamate
1338-41-6	Sorbitan stearate	2492-26-4	Sodium 2-mercapto-	5454-09-1	Decyl butyrate
1338-43-8	Sorbitan oleate		benzothiazole	5454-19-3	Decyl propionate
1343-88-0	Magnesium silicate	2497-18-9	trans-2-Hexenyl acetate	5461-08-5	Piperonyl isobutyrate
1343-90-4	Magnesium silicate	2611-82-7	Cochineal	5471-51-2	4-(p-Hydroxyphenyl)-2-
	hydrate	2615-15-8	PEG-6		butanone
1343-98-2	Silicic acid	2639-63-6	Hexyl butyrate	5550-12-9	Disodium guanylate
1344-00-9	Sodium silicoaluminate	2705-87-5	Allyl cyclohexane-	5743-27-1	Calcium ascorbate
1344-09-8	Sodium silicate		propionate	5743-34-0	Calcium borogluconate
1344-28-1	Alumina	2783-94-0	FD&C Yellow No. 6	5785-44-4	Tricalcium citrate-4-
1344-43-0	Manganese oxide	2785-89-9	4-Ethylguaiacol		hydrate
1344-95-2	Calcium silicate	2835-39-4	Allyl isovalerate	5794-13-8	L-Asparagine
1369-66-3	Dioctyl sodium	2847-30-5	2-Methoxy-3(5)-		monohydrate
	sulfosuccinate		methylpyrazine	5873-57-4	Sodium fumarate
1390-65-4	Carmine	3012-65-5	Ammonium citrate	5793-94-2	Calcium stearoyl
1390-65-4	Carmine extract	3149-28-8	2-Methoxypyrazine		lactylate
1393-63-1	Annatto	3208-16-0	2-Ethylfuran	5837-78-5	Ethyl tiglate
1398-61-4	Chitin	3268-49-3	3-Methylthiopropion-	5870-93-9	Heptyl butyrate
1401-55-4	Tannic acid		aldehyde	5905-52-2	Ferrous lactate
1406-18-4	Tocopherol	3386-18-3	PEG-9	5910-87-2	trans,trans-2,4-
1406-65-1	Chlorophyll	3391-86-4	1-Octen-3-ol		Nonadienal
1406-70-8	Tocopheryl acetate	3416-24-8	Glucosamine	5910-89-4	2,3-Dimethylpyrazine
1407-03-0	Monoammonium	3460-44-4	α-Methylbenzyl	5934-29-2	L-Histidine hydrochlo-
	glycyrrhizinate		butyrate		ride monohydrate
1414-45-5	Nisin	3522-50-7	Ferric citrate anhyd.	5938-38-5	Sorbitan oleate
1421-63-2	2,4,5-Trihydroxybutyro-	2338-05-8	Ferric citrate	5949-29-1	Citric acid monohydrate
	phenone		monohydrate	5959-89-7	Sorbitan laurate
1493-13-6	Trifluoromethane	3615-41-6	Rhamnose	5968-11-6	Sodium carbonate
	sulfonic acid	3632-91-5	Magnesium gluconate		monohydrate
1504-74-1	o-Methoxycinnamal-		anhyd.	5989-27-5	d-Limonene
	dehyde	3658-77-3	4-Hydroxy-2,5-dimethyl-	5996-10-1	D-Glucose monohy-

CAS	Chemical	CAS	Chemical	CAS	Chemical
	drate	7493-71-2	Allyl tiglate		dibasic
6028-57-5	Aluminum caprylate	7493-72-3	Allyl nonanoate	7757-97-3	D-Pantothenamide
6047-12-7	Ferrous gluconate	7493-74-5	Allyl phenoxyacetate	7758-01-2	Potassium bromate
6100-05-6	Potassium citrate	7512-17-6	N-Acetylglucosamine	7758-02-3	Potassium bromide
	monohydrate	7549-33-9	Anisyl propionate	7758-05-6	Potassium iodate
6119-47-7	Quinine hydrochloride	7549-37-3	Citral dimethyl acetal	7758-09-0	Potassium nitrite
6131-90-4	Sodium acetate	7553-56-2	Iodine	7758-11-4	Potassium phosphate
	trihydrate	7558-16-9	Disodium diphosphate		dibasic
6132-02-1	Sodium carbonate	7558-63-6	Ammonium glutamate	7758-16-9	Sodium acid
	decahydrate	7558-79-4	Sodium phosphate		pyrophosphate
6132-04-3	Trisodium citrate		dibasic anhyd.	7758-23-8	Calcium phosphate
	dihydrate	7558-80-7	Sodium phosphate		monobasic anhydrous
6222-35-1	Cyclohexyl propionate		anhyd.	7758-29-4	Pentasodium
6309-51-9	Isoamyl laurate	7585-39-9	Cyclodextrin		triphosphate
6341-24-8	D-Histidine hydrochlo-	7601-54-9	Sodium phosphate	7758-79-4	Disodium phosphate,
	ride monohydrate		tribasic		dihydrate
6358-53-8	Citrus Red No. 2.	7631-86-9	Diatomaceous earth	7758-87-4	Calcium phosphate
6363-38-8	Calcium pantothenate	7631-86-9	Silica		tribasic
	calcium chloride double	7631-86-9	Silica aerogel	7758-98-7	Cupric sulfate anhyd.
	salt (d or dl).	7631-90-5	Sodium bisulfite	7758-99-8	Cupric sulfate
6363-53-7	Maltose monohydrate	7631-99-4	Sodium nitrate		pentahydrate
6378-65-0	Hexyl hexanoate	7632-00-0	Sodium nitrite	7764-50-3	d-Dihydrocarvone
6381-59-5	Potassium sodium	7646-85-7	Zinc chloride	7772-98-7	Sodium thiosulfate
	tartrate tetrahydrate	7647-01-0	Hydrochloric acid		anhyd.
6381-77-7	Sodium erythorbate	7647-14-5	Sodium chloride	7772-99-8	Stannous chloride
6485-34-3	Calcium saccharin	7660-25-5	Fructose		anhyd.
6485-39-8	Manganese gluconate	7664-38-2	Phosphoric acid	7774-44-9	Cyclohexyl isovalerate
	anhyd.	7664-41-7	Ammonia	7775-14-6	Sodium hydrosulfite
6485-40-1	l-Carvone	7664-93-9	Sulfuric acid	7775-27-1	Sodium persulfate
6728-26-3	2-Hexenal	7681-11-0	Potassium iodide	7775-50-0	Tristearyl citrate
6728-31-0	cis-4-Hepten-1-al	7681-38-1	Sodium bisulfate	7778-18-9	Calcium sulfate anhyd.
6790-09-6	PEG-12	7681-38-1	Sodium bisulfate solid	7778-53-2	Potassium phosphate
6812-78-8	Rhodinol	7681-52-9	Sodium hypochlorite		tribasic
6834-92-0	Sodium metasilicate	7681-53-0	Sodium hypophosphite	7778-77-0	Potassium phosphate
0042-15-5	Dodecene-1	7681-57-4	Sodium metabisulfite	7778-80-5	Potassium sulfate
6858-44-2	Trisodium citrate	7681-65-4	Copper iodide (ous)	7779-70-6	Isoamyl nonanoate
	hydrate	7681-93-8	Natamycin	7779-81-9	Isobutyl angelate
6915-15-7	N-Hydroxysuccinic acid	7695-91-2	dl-α-Tocopheryl	7779-86-4	Zinc hydrosulfite
6916-74-1	Glyceryl behenate		acetate	7779-90-0	Zinc orthophosphate
7047-84-9	Aluminum stearate	7702-01-4	Disodium capryloam-	7782-50-5	Chlorine
7173-51-5	Didecyldimonium		phodiacetate	7782-63-0	Ferrous sulfate
	chloride	7704-71-4	Magnesium fumarate		heptahydrate
7235-40-7	Carotene		anhyd.	7782-75-4	Magnesium phosphate
7320-34-5	Tetrapotassium	7705-08-0	Ferric chloride anhyd.		dibasic trihydrate
	pyrophosphate	7720-78-7	Ferrous sulfate	7782-85-6	Sodium phosphate
7360-38-5	Trioctanoin	7720-78-7	Ferrous sulfate anhyd.		dibasic
7378-23-6	Sodium erythorbate		(dried)	7782-92-5	Sodium amide
7384-98-7	Propylene glycol	7722-76-1	Ammonium phosphate	7783-20-2	Ammonium sulfate
	dioctanoate	7722-84-1	Hydrogen peroxide	7783-28-0	Ammonium phosphate,
7429-90-5	Aluminum	7722-88-5	Tetrasodium		dibasic
7439-89-6	Iron		pyrophosphate anhyd.	7784-24-9	Potassium alum
7440-02-0	Nickel	7727-21-1	Potassium persulfate		dodecahydrate
7440-22-4	Silver	7727-37-9	Nitrogen	7784-25-0	Ammonium alum
7440-48-4	Cobalt	7727-54-0	Ammonium persulfate	7785-26-4	(-)-α-Pinene
7440-50-8	Copper	7727-73-3	Sodium sulfate	7785-70-8	(+)-α-Pinene
7440-57-5	Gold		decahydrate	7785-84-4	Sodium trimetaphos-
7440-59-7	Helium	7733-02-0	Zinc sulfate		phate
7446-02-0	Zinc sulfate anhyd.	7775-39-5	Methylbenzyl	7785-87-7	Manganese sulfate
7446-09-5	Sulfur dioxide		isobutyrate		(ous) anhyd.
7446-19-7	Zinc sulfate monohy-	7757-79-1	Potassium nitrate	7785-88-8	Sodium aluminum
	drate	7757-81-5	Sodium sorbate		phosphate acidic
7446-20-0	Zinc sulfate hepta-	7757-82-6	Sodium sulfate anhyd.	7785-88-8	Sodium aluminum
	hydrate	7757-83-7	Sodium sulfite		phosphate, basic
7447-40-7	Potassium chloride	7757-86-0	Magnesium phosphate	7786-30-3	Magnesium chloride
7452-79-1	Ethyl 2-methylbutyrate		dibasic	7786-58-5	Octyl isovalerate
7487-88-9	Magnesium sulfate	7757-87-1	Magnesium phosphate	7786-67-6	Isopulegol
	anhyd.		tribasic anhyd.	7789-41-5	Calcium bromide
7492-70-8	Butyl butyryllactate	7757-93-9	Calcium phosphate	7789-77-7	Calcium phosphate

CAS	Chemical	CAS	Chemical	CAS	Chemical
	dibasic dihydrate	8007-43-0	Sorbitan sesquioleate	8023-77-6	Capsicum oleoresin
7789-80-2	Calcium iodate	8007-46-3	Thyme oil	8023-79-8	Palm kernel oil
7790-76-3	Calcium pyrophosphate	8007-46-3	Thyme oil red	8023-98-1	Peach kernel oil
7791-18-6	Magnesium chloride	8007-47-4	Balsam Canada	8024-06-4	Vanilla
	hexahydrate	8007-69-0	Sweet almond oil	8024-09-7	Walnut oil
8000-25-7	Rosemary oil	8007-70-3	Anise oil	8024-32-6	Avocado oil
8000-28-0	Lavender oil	8007-75-8	Bergamot oil	8024-37-1	Curcumin
8000-34-8	Clove oil	8007-80-5	Cinnamon oil	8027-32-5	Petrolatum USP
8000-42-8	Caraway oil	8008-26-2	Lime oil	8027-43-8	Birch oil
8000-44-0	Wild cherry bark extract	8008-31-9	Mandarin orange oil	8028-48-6	Bitter orange peel
8000-46-2	Geranium oil	8008-45-5	Nutmeg oil		extract
8000-48-4	Eucalyptus oil	8008-52-4	Coriander oil	8028-66-8	Honey
8000-66-6	Cardamom oil	8008-56-8	Lemon oil	8028-89-5	Caramel
8000-68-8	Parsley seed oil	8008-56-8	Lemon extract	8029-43-4	Corn syrup
8000-73-5	Ammonium carbonate	8008-57-9	Orange oil	8029-44-5	Cottonseed glyceride
8001-21-6	Sunflower seed oil	8008-63-7	Ox bile extract	8029-76-3	Hydroxylated lecithin
8001-22-7	Soybean oil	8008-74-0	Sesame oil	8029-91-2	Acetylated hydroge-
8001-23-8	Safflower oil	8008-79-5	Spearmint oil		nated lard glyceride
8001-25-0	Olive oil	8008-88-6	Valerian	8029-92-3	Acetylated lard
8001-26-1	Linseed oil	8008-94-4	Licorice root extract		glyceride
8001-29-4	Cottonseed oil	8008-99-9	Garlic extract	8030-12-4	Hydrogenated tallow
8001-30-7	Corn oil	8009-03-8	Petrolatum NF	8030-30-6	Naphtha
8001-31-8	Coconut oil	8011-89-0	Balsam tolu	8033-29-2	Hydrogenated palm oil
8001-39-6	Japan wax	8012-89-3	Beeswax, yellow	8040-05-9	Hydrogenated lard
8001-61-4	Balsam copaiba	8012-91-7	Juniper oil		glyceride
8001-67-0	Coffee bean extract	8012-95-1	Mineral oil	8042-47-5	Mineral oil
8001-69-2	Cod liver oil	8013-01-2	Yeast extract	8049-47-6	Pancreatin
8001-78-3	Hydrogenated castor oil	8013-07-8	Epoxidized soybean oil	8050-07-5	Olibanum
8001-79-4	Castor oil	8013-17-0	Invert sugar	8050-09-7	Rosin
8001-88-5	Birch oil	8013-75-0	Fusel oil refined	8050-26-8	Pentaerythrityl rosinate
8002-03-7	Peanut oil	8013-76-1	Bitter almond oil	8050-81-5	Simethicone
8002-05-9	Petroleum distillates	8014-13-9	Cumin oil	8051-30-7	Cocamide DEA
8002-09-3	Pine oil	8014-29-7	Rue oil	8052-10-6	Rosin
8002-13-9	Rapeseed oil	8014-71-9	Balm mint oil	8052-14-0	Turpentine
8002-13-9	Canola oil	8015-01-8	Sweet marjoram oil	8057-49-6	Valerian extract
8002-23-1	Cetyl esters	8015-01-8	Marjoram oil, Spanish	8057-65-6	Arnica extract
8002-26-4	Tall oil	8015-67-6	Annatto extract	8061-51-6	Sodium lignosulfonate
8002-31-1	Cocoa butter	8015-73-4	Basil oil	8061-52-7	Calcium lignosulfonate
8002-43-5	Lecithin	8015-75-6	Bitter almond oil	9000-01-5	Acacia
8002-48-0	Malt extract	8015-80-3	Kukui nut oil	9000-05-9	Gum benzoin
8002-50-4	Menhaden oil	8015-86-9	Carnauba	9000-07-1	Carrageenan
8002-66-2	Matricaria oil	8015-88-1	Carrot oil	9000-16-2	Dammar
8002-74-2	Paraffin	8015-92-7	Chamomile oil, Roman	9000-21-9	Furcelleran
8002-74-2	Petroleum wax	8015-97-2	Clove leaf oil	9000-29-7	Guaiac gum
8002-74-2	Synthetic wax	8016-20-4	Grapefruit oil	9000-30-0	Guar gum
8002-75-3	Palm oil	8016-25-9	Hops extract	9000-36-6	Karaya gum
8002-80-0	Wheat gluten	8016-26-0	Labdanum	9000-40-2	Locust bean gum
8006-39-1	Terpineol anhydrous	8016-31-7	Lovage oil	9000-57-1	Sandarac gum
8006-40-4	Beeswax, white	8016-38-4	Orange flower oil	9000-59-3	Shellac
8006-44-8	Candelilla wax	8016-42-0	Balsam Peru	9000-65-1	Tragacanth gum
8006-54-0	Lanolin anhyd.	8016-55-5	Rhubarb extract	9000-69-5	Pectin
8006-75-5	Dillweed oil	8016-60-2	Rice bran wax	9000-70-8	Gelatin
8006-78-8	Bay oil	8016-63-5	Clary oil	9000-71-9	Casein
8006-80-2	Sassafras oil	8016-70-4	Hydrogenated soybean	9000-71-9	Milk protein
8006-81-3	Ylang ylang oil		oil	9000-90-2	α-Amylase
8006-82-4	Black pepper oil	8020-83-5	Mineral oil	9000-92-4	Amylase
8006-84-6	Fennel oil	8020-84-6	Lanolin hydrous	9001-05-2	Catalase
8006-87-9	Sandalwood oil	8021-29-2	Fir needle oil, Siberian	9001-06-3	Chitinase
8006-90-4	Peppermint oil		type	9001-33-6	Ficin
8006-95-9	Wheat germ oil	8021-55-4	Ozokerite	9001-37-0	Glucose oxidase
8007-00-9	Balsam Peru	8022-07-9	Yarrow oil	9001-62-1	Lipase
8007-01-0	Rose oil	8022-15-9	Lavandin oil	9001-73-4	Papain
8007-02-1	Lemongrass oil West	8022-29-5	Cherry pit oil	9001-75-6	Pepsin
	Indian	8022-37-5	Artemisia	9001-91-3	β-Amylase
8007-04-3	Hops oil	8022-56-8	Sage oil	9001-98-3	Rennet
8007-08-7	Ginger oil	8022-56-8	Sage oil, Dalmatian	9002-07-7	Trypsin
8007-12-3	Nutmeg oil, expressed		type	9002-13-5	Urease
8007-20-3	Cedar leaf oil	8023-62-9	Storax	9002-18-0	Agar

CAS	Chemical	CAS	Chemical	CAS	Chemical
9002-88-4	Polyethylene	9004-65-3	Hydroxypropyl	9006-65-9	Dimethicone
9002-89-5	Polyvinyl alcohol (super		methylcellulose	9007-13-0	Calcium resinate
	and fully hydrolyzed)	9004-67-5	Methylcellulose	9007-48-1	Polyglyceryl-2 oleate
9002-93-1	Octoxynol-5	9004-70-0	Nitrocellulose	9007-48-1	Polyglyceryl-3 oleate
9002-93-1	Octoxynol-8	9004-81-3	PEG-32 laurate	9007-48-1	Polyglyceryl-4 oleate
9002-93-1	Octoxynol-9	9004-87-9	Octoxynol-5	9007-48-1	Polyglyceryl-6 oleate
9002-93-1	Octoxynol-13	9004-87-9	Octoxynol-8	9007-48-1	Polyglyceryl-8 oleate
9002-93-1	Octoxynol-30	9004-87-9	Octoxynol-9	9007-48-1	Polyglyceryl-10 oleate
9002-93-1	Octoxynol-40	9004-87-9	Octoxynol-13	9009-48-1	Polyglyceryl-10 dioleate
9002-93-1	Octoxynol-70	9004-87-9	Octoxynol-30	9010-43-9	Octoxynol-9
9003-04-7	Sodium polyacrylate	9004-87-9	Octoxynol-40	9010-66-6	Zein
9003-05-8	Polyacrylamide	9004-87-9	Octoxynol-70	9010-85-9	Isobutylene/isoprene
9003-08-1	Melamine/formaldehyde	9004-96-0	PEG-6 oleate		copolymer
	resin	9004-96-0	PEG-8 oleate	9011-05-6	Urea-formaldehyde
9003-11-6	Poloxamer 105	9004-96-0	PEG-32 oleate		resin
9003-11-6	Poloxamer 108	9004-96-0	PEG-75 oleate	9012-37-7	Acylase
9003-11-6	Poloxamer 122	9004-97-1	PEG-12 ricinoleate	9012-54-8	Cellulase
9003-11-6	Poloxamer 123	9004-98-2	Oleth-23	9012-54-8	Glucanase from
9003-11-6	Poloxamer 124	9004-99-3	PEG-6 stearate		*Bacillus subtilis*
9003-11-6	Poloxamer 181	9004-99-3	PEG-8 stearate	9012-76-4	Chitosan
9003-11-6	Poloxamer 182	9004-99-3	PEG-32 stearate	9013-34-7	Diethylaminoethyl
9003-11-6	Poloxamer 183	9004-99-3	PEG-40 stearate		cellulose
9003-11-6	Poloxamer 184	9004-99-3	PEG-75 stearate	9013-79-0	Esterase
9003-11-6	Poloxamer 185	9004-99-3	PEG-100 stearate	9014-01-1	Protease
9003-11-6	Poloxamer 188	9005-00-9	Steareth-20	9016-00-6	Dimethicone
9003-11-6	Poloxamer 212	9005-02-1	PEG-6 dilaurate	9016-45-9	Nonoxynol-6
9003-11-6	Poloxamer 215	9005-02-1	PEG-20 dilaurate	9016-45-9	Nonoxynol-7
9003-11-6	Poloxamer 217	9005-02-1	PEG-32 dilaurate	9016-45-9	Nonoxynol-10
9003-11-6	Poloxamer 231	9005-02-1	PEG-75 dilaurate	9016-45-9	Nonoxynol-11
9003-11-6	Poloxamer 234	9005-02-1	PEG-150 dilaurate	9016-45-9	Nonoxynol-12
9003-11-6	Poloxamer 235	9005-07-6	PEG-6 dioleate	9016-45-9	Nonoxynol-14
9003-11-6	Poloxamer 237	9005-07-6	PEG-8 dioleate	9016-45-9	Nonoxynol-40
9003-11-6	Poloxamer 238	9005-07-6	PEG-12 dioleate	9016-45-9	Nonoxynol-50
9003-11-6	Poloxamer 282	9005-07-6	PEG-20 dioleate	9025-70-1	Dextranase
9003-11-6	Poloxamer 284	9005-07-6	PEG-32 dioleate	9029-60-1	Lipoxidase
9003-11-6	Poloxamer 288	9005-07-6	PEG-75 dioleate	9032-08-0	Amyloglucosidase
9003-11-6	Poloxamer 331	9005-07-6	PEG-150 dioleate	9032-08-0	Glucoamylase
9003-11-6	Poloxamer 333	9005-08-7	PEG-6 distearate	9032-75-1	Pectinase
9003-11-6	Poloxamer 334	9005-08-7	PEG-20 distearate	9036-19-5	Octoxynol-5
9003-11-6	Poloxamer 335	9005-08-7	PEG-32 distearate	9036-19-5	Octoxynol-8
9003-11-6	Poloxamer 338	9005-08-7	PEG-75 distearate	9036-19-5	Octoxynol-9
9003-11-6	Poloxamer 401	9005-25-8	Corn starch	9036-19-5	Octoxynol-13
9003-11-6	Poloxamer 402	9005-25-8	Potato starch	9036-19-5	Octoxynol-30
9003-11-6	Poloxamer 403	9005-25-8	Rice starch	9036-19-5	Octoxynol-40
9003-11-6	Poloxamer 407	9005-25-8	Wheat starch	9036-19-5	Octoxynol-70
9003-20-7	Polyvinyl acetate	9005-32-7	Alginic acid	9036-66-2	Arabinogalactan
	(homopolymer)	9005-34-9	Ammonium alginate	9038-95-3	PPG-5-buteth-7
9003-27-4	Polyisobutene	9005-35-0	Calcium alginate	9038-95-3	PPG-12-buteth-16
9003-29-6	Polyisobutene	9005-36-1	Potassium alginate	9038-95-3	PPG-20-buteth-30
9003-39-8	PVP	9005-37-2	Propylene glycol	9038-95-3	PPG-28-buteth-35
9003-54-7	Acrylonitrile-styrene		alginate	9038-95-3	PPG-33-buteth-45
	copolymer	9005-38-3	Algin	9049-05-2	Calcium carrageenan
9003-55-8	Butadiene-styrene	9005-42-9	Ammonium caseinate	9050-36-6	Maltodextrin
	copolymer	9005-46-3	Sodium caseinate	9054-89-1	Dismutase superoxide
9003-56-9	Acrylonitrile-butadiene-	9005-64-5	PEG-80 sorbitan	9061-82-9	Sodium carrageenan
	styrene		laurate	9062-04-8	Propylene glycol
9003-73-0	Polydipentene	9005-64-5	Polysorbate 20		dicaprylate/dicaprate
9004-32-4	Carboxymethylcellulose	9005-64-5	Polysorbate 21	9065-63-8	PPG-5-buteth-7
	sodium	9005-65-5	Polysorbate 81	9065-63-8	PPG-12-buteth-16
9004-34-6	Cellulose	9005-65-6	Polysorbate 80	9065-63-8	PPG-20-buteth-30
9004-34-6	Microcrystalline	9005-66-7	Polysorbate 40	9065-63-8	PPG-28-buteth-35
	cellulose	9005-67-8	Polysorbate 60	9065-63-8	PPG-33-buteth-45
9004-35-7	Cellulose acetate	9005-67-8	Polysorbate 61	9074-99-1	β-Glucanase from
9004-36-3	Sodium caseinate	9005-70-3	Polysorbate 85		*Aspergillus niger*)
9004-53-9	Dextrin	9005-71-4	Polysorbate 65	9075-68-7	Pullulanase
9004-54-0	Dextran	9005-84-9	Starch	9087-61-0	Aluminum starch
9004-57-3	Ethylcellulose	9006-00-2	Pliofilm		octenyl succinate
9004-64-2	Hydroxypropylcellulose	9006-50-2	Albumen	10016-20-3	Cyclodextrin

CAS	Chemical	CAS	Chemical	CAS	Chemical
10024-66-5	Manganese citrate		decaoleate	19473-49-5	Potassium glutamate
10024-97-2	Nitrous oxide	11099-07-3	Glyceryl stearate		monohydrate
10025-69-1	Stannous chloride	11099-07-3	Glyceryl stearate SE	19855-56-2	Calcium fumarate
	dihydrate	11103-57-4	Retinol		anhyd.
10025-77-1	Ferric chloride	11138-49-1	Sodium aluminate	20777-49-5	Dihydrocarvyl acetate
	hexahydrate	11138-66-2	Xanthan gum	21645-51-2	Aluminum hydroxlde
10028-15-6	Ozone	12057-74-8	Magnesium phosphide	21722-83-8	Cyclohexylethyl acetate
10028-22-5	Ferric sulfate anhyd.	12125-02-9	Ammonium chloride	22047-25-2	2-Acetylpyrazine
10028-24-7	Disodium phosphate,	12125-28-9	Magnesium carbonate	22500-92-1	Sorbic acid
	dihydrate		hydroxide	22839-47-0	Aspartame
10031-30-8	Calcium phosphate	12135-76-1	Ammonium sulfide	23383-11-1	Ferrous citrate anhyd.
	monobasic monohy-	12141-46-7	Aluminum silicate	24315-83-1	Trisodium dipotassium
	drate	12167-74-7	Calcium phosphate		tripolyphosphate
10031-82-0	p-Ethoxybenzaldehyde		tribasic	24634-61-5	Potassium sorbate
10031-87-5	2-Ethylbutyl acetate	12694-22-3	Polyglyceryl-2 stearate	24683-00-9	2-Isobutyl-3-methoxy-
10034-96-5	Manganese sulfate	12709-64-7	Polyglyceryl-10		pyrazine
	(ous) monohydrate		tristearate	24817-51-4	2-Phenethyl 2-
10034-99-8	Magnesium sulfate	12764-60-2	Polyglyceryl-10		methylbutyrate
	heptahydrate		distearate	25013-16-5	BHA
10035-04-8	Calcium chloride	13446-34-9	Manganese chloride	25085-02-3	Acrylamide/sodium
	dihydrate		tetrahydrate		acrylate copolymer
10039-32-4	Sodium phosphate	13463-67-7	Titanium dioxide	25086-62-8	Sodium polymethacry-
	dibasic dodecahydrate	13472-35-0	Sodium phosphate		late
10043-52-4	Calcium chloride anhyd.		dihydrate	25152-84-5	trans-trans-2,4-
10043-01-3	Aluminum sulfate	13532-18-8	Methyl 3-methylthio-		Decadienal
10043-35-3	Boric acid		propionate	25155-30-0	Sodium dodecylben-
10043-67-1	Potassium alum	13552-80-2	Triundecanoin		zenesulfonate
	sesquihydrate	13573-18-7	Pentasodium	25168-73-4	Sucrose stearate
10043-84-2	Manganese		triphosphate	25212-19-5	Hydrogenated
	hypophosphite	13601-19-9	Sodium ferrocyanide		vegetable glycerides
10045-86-0	Ferric phosphate	13623-11-5	Trimethyl thiazole		phosphate
	anhyd.	13679-70-4	5-Methyl-2-thiophene-	25322-68-3	PEG-4
10045-87-1	Sodium ferric		carboxyaldehyde	25322-68-3	PEG-6
	pyrophosphate	13845-36-8	Potassium tripolyphos-	25322-68-3	PEG-8
10049-04-4	Chlorine dioxide		phate	25322-68-3	PEG-9
10058-44-3	Ferric pyrophosphate	14073-97-3	l-Menthone	25322-68-3	PEG-12
	anhyd.	14371-10-9	Cinnamal	25322-68-3	PEG-14
10094-36-7	Ethyl cyclohexane-	14536-17-5	Ferrous ascorbate	25322-68-3	PEG-16
	propionate	14667-55-1	2,3,5-Trimethylpyrazine	25322-68-3	PEG-20
10101-39-0	Calcium silicate	14807-96-6	Talc	25322-68-3	PEG-32
10101-41-4	Calcium sulfate	14901-07-6	β-Ionone	25322-68-3	PEG-40
	dihydrate	14940-41-1	Ferric phosphate	25322-68-3	PEG-75
10101-89-0	Sodium phosphate		tetrahydrate	25322-68-3	PEG-100
	tribasic dodecahydrate	15139-76-1	Orange B	25322-68-3	PEG-150
10102-17-7	Sodium thiosulfate	15707-23-0	2-Ethyl-3-methyl-	25322-68-3	PEG-200
	pentahydrate		pyrazine	25322-68-3	PEG-350
10102-71-3	Sodium alum	15892-23-6	2-Butanol	25322-68-3	Polyethylene glycol
10108-28-8	PEG-6 stearate	16409-44-2	Geranyl acetate	25322-69-4	Polypropylene glycol
10117-38-1	Potassium sulfite	16409-45-3	dl-Menthyl acetate	25339-99-5	Sucrose laurate
10124-43-3	Cobalt sulfate (ous)	16409-46-4	Menthyl isovalerate	25383-99-7	Sodium stearoyl
10124-56-8	Sodium hexametaphos-	16423-68-0	Erythrosine		lactylate
	phate	16423-68-0	FD&C Red No. 3	25496-72-4	Glyceryl mono/dioleate
10140-65-5	Sodium phosphate	16485-10-2	DL-Panthenol	25496-72-4	Glyceryl oleate
	dibasic	16721-80-5	Sodium hydrosulfite	25637-84-7	Glyceryl dioleate
10191-41-0	dl-α-Tocopherol	16731-55-8	Potassium metabisulfite	25956-17-6	FD&C Red No. 40
10196-04-0	Ammonium sulfite	16957-70-3	2-Methyl-2-pentenoic	25988-97-0	Dimethylamine/
10222-01-2	2,2-Dibromo-3-		acid		epichlorohydrin
	nitrilopropionamide	17090-93-6	Sodium aspartate		copolymer
10233-87-1	Magnesium phosphate	17407-37-3	Tocopheryl succinate	26027-38-3	Nonoxynol-6
	tribasic pentahydrate	17661-50-6	Cetyl esters	26027-38-3	Nonoxynol-7
10326-41-7	D-Lactic acid	17699-09-1	Dihydrocarveol	26027-38-3	Nonoxynol-10
10361-03-2	Sodium metaphosphate	18016-24-5	Calcium gluconate	26027-38-3	Nonoxynol-11
10361-29-2	Ammonium carbonate		monohydrate	26027-38-3	Nonoxynol-12
10482-56-1	α-Terpineol	18031-40-8	Perillaldehyde	26027-38-3	Nonoxynol-14
10599-70-9	3-Acetyl-2,5-	18368-91-7	2-Ethyl fenchol	26027-38-3	Nonoxynol-40
	dimethylfuran	18640-74-9	2-Isobutylthiazole	26027-38-3	Nonoxynol-50
11042-64-1	Oryzanol	18996-35-5	Sodium citrate	26099-09-2	Polymaleic acid
11094-60-3	Polyglyceryl-10	19089-92-0	Hexyl-2-butenoate	26264-58-4	Sodium methylnaph-

CAS	Chemical	CAS	Chemical	CAS	Chemical
	thalenesulfonate	37205-87-1	Nonoxynol-40	60837-57-2	Anoxomer
26266-57-9	Sorbitan palmitate	37205-87-1	Nonoxynol-50	61332-02-3	Glyceryl isostearate
26266-58-0	Sorbitan trioleate	37220-82-9	Glyceryl oleate	61725-93-7	Polyglyceryl-6
26402-22-2	Glyceryl caprate	37288-57-6	Agarase		distearate
26402-26-6	Caprylic/capric	37349-34-1	Polyglyceryl-3 stearate	61788-47-4	Coconut acid
	glycerides	37349-34-1	Polyglyceryl-4 stearate	61788-59-8	Methyl cocoate
26402-26-6	Glyceryl caprylate	37349-34-1	Polyglyceryl-8 stearate	61788-85-0	PEG-50 hydrogenated
26402-31-3	Propylene glycol	38462-22-5	Mentha-8-thol-3-one		castor oil
	ricinoleate	38566-94-8	Potassium butyl	61788-85-0	PEG-60 hydrogenated
26446-38-8	Sucrose palmitate		paraben		castor oil
26657-96-5	Glyceryl palmitate	39175-72-9	Glyceryl stearate citrate	61789-05-7	Glyceryl cocoate
26658-19-5	Sorbitan tristearate	39300-88-4	Tara gum	61789-07-9	Hydrogenated
26680-54-6	Octenyl succinic	39310-72-0	PEG-15 glyceryl		cottonseed glyceride
	anhydride		ricinoleate	61789-08-0	Glyceryl stearate
26855-43-6	Polyglyceryl-3 stearate	39409-82-0	Magnesium carbonate	61789-08-0	Hydrogenated soy
26855-44-7	Polyglyceryl-4 stearate		hydroxide		glyceride
27176-99-4	Octoxynol-5	39529-26-5	Polyglyceryl-10	61789-08-0	Hydrogenated
27177-01-1	Nonoxynol-6		decastearate		vegetable glyceride
27177-05-5	Nonoxynol-6	42173-90-0	Octoxynol-9	61789-09-1	Hydrogenated tallow
27177-05-5	Nonoxynol-7	49553-76-6	Polyglyceryl-2 oleate		glyceride
27177-08-8	Nonoxynol-10	51142-51-9	PEG-15 glyceryl	61789-10-4	Lard glyceride
27194-74-7	Propylene glycol laurate		ricinoleate	61789-13-7	Tallow glyceride
27195-16-0	Sucrose distearate	51158-08-8	PEG-20 glyceryl	61789-51-3	Cobalt naphthenate
27196-00-5	Myristyl alcohol mixed		stearate	61789-91-1	Jojoba oil
	isomers	51158-08-8	PEG-30 glyceryl	61789-97-7	Tallow
27214-00-2	Calcium glycero-		stearate	61790-37-2	Tallow acid
	phosphate hydrate	51192-09-7	PEG-20 glyceryl oleate	61790-86-1	Polysorbate 80
27215-38-9	Glyceryl laurate SE	51192-09-7	PEG-30 glyceryl oleate	61790-95-2	Polyglyceryl-3 stearate
27321-72-8	Polyglyceryl-3 stearate	51248-32-9	PEG-20 glyceryl laurate	61791-31-9	Cocamide DEA
27479-28-3	Quaternium-14	51248-32-9	PEG-30 glyceryl laurate	63148-62-9	Dimethicone
27554-26-3	Diisooctyl phthalate	52668-97-0	PEG-6 distearate	63231-60-7	Microcrystalline wax
27942-26-3	Nonoxynol-10	52668-97-0	PEG-20 distearate	64365-11-3	Carbon, activated
28069-72-9	2,6-Nonadien-1-ol	52668-97-0	PEG-32 distearate	64742-14-9	Petroleum distillates
29409-82-0	Magnesium carbonate	52668-97-0	PEG-75 distearate	64742-42-3	Microcrystalline wax
30233 64 8	Glyceryl behenate	52688 97 0	PEG 6 dioleate	64742-46-7	Hydrocarbon solvent
30915-61-8	Polymaleic acid sodium	52688-97-0	PEG-8 dioleate	64742-47-8	Petroleum distillates
	salt	52688-97-0	PEG-12 dioleate	64742-49-0	Hexane
31566-31-1	Glyceryl stearate	52688-97-0	PEG-20 dioleate	65072-00-6	Hydrolyzed milk protein
31566-31-1	Glyceryl stearate SE	52688-97-0	PEG-32 dioleate	65381-09-1	Caprylic/capric
31791-00-2	PEG-40 stearate	52688-97-0	PEG-75 dioleate		triglyceride
32057-14-0	Glyceryl isostearate	52688-97-0	PEG-150 dioleate	65996-62-5	Food starch modified
32864-38-3	Butyl ethyl malonate	53022-14-3	Bornyl isovalerate	65996-63-6	Food starch modified
33940-98-6	Polyglyceryl-3 oleate	53124-00-8	Food starch modified	66071-96-3	Corn gluten
33940-99-7	Polyglyceryl-10 dioleate	53637-25-5	Poloxamer 181	66071-96-3	Corn meal
34406-66-1	Polyglyceryl-10 laurate	53850-34-3	Thaumatin	66082-42-6	Polyglyceryl-3
34424-97-0	Polyglyceryl-6	54193-36-1	Sodium polymethacry-		diisostearate
	distearate		late	66085-00-5	Glyceryl isostearate
34424-98-1	Polyglyceryl-10	55031-15-7	2-Ethyl-3,5(6)-	66105-29-1	PEG-7 glyceryl cocoate
	tetraoleate		dimethylpyrazine	66905-25-7	Calcium diacetate
34713-70-7	2-Phenylpropanal	55589-62-3	Acesulfame potassium		anhyd.
35285-68-8	Sodium ethylparaben	55719-85-2	Phenethyl tiglate	66988-04-3	Sodium isostearoyl
35285-69-9	Sodium propylparaben	56038-13-2	Sucralose		lactylate
35554-44-0	Imazalil	56329-42-1	Zinc methionine sulfate	67701-05-7	Coconut acid
35725-46-3	Ferric sodium	56378-72-4	Magnesium carbonate	67701-06-8	Tallow acid
	pyrophosphate anhyd.		hydroxide pentahydrate	67701-08-0	Soy acid
36311-34-9	Isocetyl alcohol	56519-71-2	Propylene glycol	67701-27-3	Hydrogenated tallow
36457-19-9	Potassium ethylpara-		dioctanoate		glycerides
	ben	57576-09-7	Isopulegyl acetate	67701-27-3	Japan wax
36457-20-2	Sodium butylparaben	58748-27-9	Propylene glycol	67701-27-3	Tallow glycerides
36653-82-4	Cetyl alcohol		dicaprylate/dicaprate	67701-28-4	Caprylic/capric/linoleic
37189-34-7	Bromelain	59070-56-3	PEG-20 glyceryl laurate		triglyceride
37200-49-0	Polysorbate 80	59070-56-3	PEG-30 glyceryl laurate	67701-28-4	Caprylic/capric/oleic
37205-87-1	Nonoxynol-6	59259-38-0	Menthyl lactate		triglyceride
37205-87-1	Nonoxynol-7	60177-39-1	Chloromethylated	67701-30-8	Triolein
37205-87-1	Nonoxynol-10		aminated styrene-	67701-32-0	Mono- and diglycerides
37205-87-1	Nonoxynol-11		divinylbenzene resin		of fatty acids
37205-87-1	Nonoxynol-12	60344-26-5	PEG-6 oleate	67701-33-1	Mono- and diglycerides
37205-87-1	Nonoxynol-14	60616-95-7	Sodium carrageenan		of fatty acids

CAS	Chemical	CAS	Chemical	CAS	Chemical
67762-36-1	Caprylic/capric acid		insoluble	75719-57-2	Polyglyceryl-8 stearate
67762-39-4	Methyl caprylate/	68916-04-1	Bitter orange oil	76009-37-5	Polyglyceryl-6 dioleate
	caprate	68916-55-2	Sambucus oil	79665-92-2	Polyglyceryl-6 oleate
67762-40-7	Methyl laurate	68916-88-1	Lemon juice	79665-93-3	Polyglyceryl-10 oleate
67784-82-1	Polyglyceryl-10 oleate	68937-16-6	Polyglyceryl-10	79665-94-4	Polyglyceryl-3 dioleate
67784-87-6	Hydrogenated palm		caprylate	79777-30-3	Polyglyceryl-10 stearate
	glyceride	68937-85-9	Coconut acid	81025-04-9	Lactitol monohydrate
67938-21-0	Polyglyceryl-2	68938-35-2	Vegetable oil	84012-14-6	Artichoke extract
	diisostearate	68938-37-4	Hydrogenated	84012-20-4	Elecampane extract
68002-71-1	Hydrogenated soybean		vegetable oil	84012-25-9	Mullein extract
	oil	68952-98-7	Brominated soybean oil	84012-31-7	Passionflower extract
68002-71-1	Hydrogenated soy	68955-19-1	Sodium lauryl sulfate	84012-33-9	Parsley extract
	glyceride	68955-45-3	Ethylene distearamide	84012-42-0	Pansy extract
68002-72-2	Hydrogenated	68956-68-3	Vegetable oil	84082-36-0	Alfalfa extract
	menhaden oil	68958-64-5	PEG-25 glyceryl	84082-59-7	Mate extract
68004-11-5	Polyglyceryl-4 stearate		trioleate	84082-61-1	Balm mint extract
68037-74-1	Dimethicone (branched)	68988-72-7	Propylene glycol	84082-67-7	Myrtle extract
68081-81-2	Sodium dodecylben-		dicaprylate/dicaprate	84082-70-2	Peppermint extract
	zenesulfonate	68990-05-6	Citric acid esters of	84082-79-1	Sage extract
68130-97-2	Polyethylenimine		mono- and diglycerides	84082-83-7	Yarrow extract
	reaction prod. with 1,2-		of fatty acids	84603-69-0	Juniper extract
	dichloroethane	68990-06-7	Hydrogenated tallow	84603-93-0	Rose oil
68131-37-3	Corn syrup solids		glyceride lactate	84604-00-2	Walnut oil
68153-28-6	Soy protein	68990-14-7	Yerba santa extract	84604-07-9	Wild cherry bark extract
68153-76-4	PEG-20 glyceryl	68990-53-4	Mono- and diglycerides	84604-14-8	Rosemary extract
	stearate		of fatty acids	84625-29-6	Capsicum extract
68201-46-7	PEG-7 glyceryl cocoate	68990-55-6	Acetylated mono- and	84625-32-1	Eucalyptus extract
68201-48-9	Hydrogenated soybean		diglycerides of fatty	84625-39-8	Fennel extract
	glycerides		acids	84649-72-9	Maidenhair fern extract
68308-53-2	Soy acid	68990-58-9	Acetylated hydroge-	84649-86-5	Chamomile extract
68308-54-3	Hydrogenated tallow		nated tallow glyceride	84650-00-0	Coffee bean extract
	glycerides	68990-58-9	Acetylated mono- and	84650-13-5	Guaiac extract
68334-00-9	Hydrogenated		diglycerides of fatty	84650-17-9	Mullein extract
	cottonseed oil		acids	84650-55-5	Cascara extract
68334-00-9	Hydrogenated	68990-59-0	Hydrogenated tallow	84650-59-9	Anise extract
	vegetable oil		glyceride citrate	84650-60-2	Thea sinensis extract
68334-28-1	Hydrogenated	68990-63-6	Shark liver oil	84695-99-8	Chestnut leaf extract
	vegetable oil	68990-67-0	Quillaja	84696-18-4	Benzoin extract
68425-17-2	Hydrogenated starch	68990-82-9	Hydrogenated palm	84696-35-5	Mandarin orange oil
	hydrolysate		kernel oil	84696-37-7	Rice bran oil
68425-36-5	Hydrogenated peanut	68991-68-4	Caprylic/capric/lauric	84775-41-7	Angelica extract
	oil		triglyceride	84775-42-8	Anise extract
68434-04-4	Polydextrose	69005-67-8	Sorbitan stearate	84775-44-0	Serpentaria extract
68441-17-8	Polyethylene, oxidized	69028-36-0	Hydrogenated	84775-51-9	Cumin extract
68476-78-8	Molasses		vegetable glycerides	84775-52-0	Turmeric extract
68513-95-1	Soy flour	69468-44-6	PEG-20 glyceryl	84775-55-3	Dandelion extract
68514-74-9	Hydrogenated palm oil		isostearate	84775-66-6	Licorice extract
68526-79-4	Hexyl alcohol	69468-44-6	PEG-30 glyceryl	84775-71-3	Basil extract
68526-85-2	n-Decyl alcohol		isostearate	84775-98-4	Savory extract
68526-86-3	Lauryl alcohol	70247-90-4	Polymaleic acid sodium	84776-23-8	Calendula extract
68553-11-7	PEG-20 glyceryl		salt	84776-25-0	Iceland moss extract
	stearate	70536-17-3	Albumin macro	84776-28-3	Cinchona extract
68553-81-1	Rice bran		aggregates	84776-64-7	Jasmine extract
68553-81-1	Rice bran oil	70802-40-3	PEG-8 stearate	84776-73-8	Sage oil
68583-51-7	Propylene glycol	71010-52-1	Gellan gum	84776-73-8	Sage oil, Dalmatian
	dicaprylate/dicaprate	71012-10-7	Polyglyceryl-4 oleate		type
68585-47-7	Sodium lauryl sulfate	71789-99-9	Lard	84776-98-7	Wild thyme extract
68603-42-9	Cocamide DEA	72347-89-8	Polyglyceryl-2	84787-69-9	Blackberry extract
68608-64-0	Disodium caproyloam-		tetrastearate	84787-70-2	Sandalwood extract
	phodiacetate	72869-62-6	Sorbitan tristearate	84929-25-9	Cinchona extract
68650-43-1	Chicory extract	72869-69-3	Apricot kernel oil	84929-31-7	Lemon extract
68855-54-9	Diatomaceous earth	73049-62-4	Juniper oil	84929-51-1	Thyme extract
68876-77-7	Yeast	73049-73-7	Hydrolyzed protein	84929-52-2	Linden extract
68889-49-6	PEG-20 glyceryl oleate	73398-61-5	Caprylic/capric	84929-61-3	Carrot extract
68889-49-6	PEG-30 glyceryl oleate		triglyceride	84929-79-3	Benzoin extract
68915-31-1	Sodium hexametaphos-	74623-31-7	PPG-5-buteth-7	84930-16-5	Potassium propylpara-
	phate	74623-31-7	PPG-12-buteth-16		ben
68915-31-1	Sodium metaphosphate	75719-56-1	Polyglyceryl-8 oleate	84961-45-5	Carob extract

CAS	Chemical	CAS	Chemical	CAS	Chemical
84961-64-8	Tansy extract	90028-68-5	Oakmoss extract	97593-29-8	Hydrogenated palm glyceride
84961-67-1	Verbena extract	90028-70-9	Asafetida extract		
85049-36-1	Ethyl oleate	90045-56-0	Everlasting extract	97593-29-8	Lard glyceride
85049-52-1	Bergamot oil	90045-89-9	Orris root extract	97593-31-2	Citric acid esters of mono- and diglycerides of fatty acids
85085-25-2	Clover blossom extract	90045-94-6	Jasmine extract		
85085-31-0	Yerba santa extract	90064-00-9	Pennyroyal extract		
85085-33-2	Fennel extract	90064-13-4	Mullein extract	97676-22-7	Gentian extract
85085-71-8	Senna extract	90106-27-7	Rhubarb extract	97676-23-8	Licorice extract
85117-50-6	Sodium dodecylbenzenesulfonate	90131-43-4	Tagetes extract	97676-24-9	Althea extract
		90131-55-8	Germander extract	97766-44-4	Swertia extract
85186-88-5	Sorbitan trioleate	91744-38-6	Citric acid esters of mono- and diglycerides of fatty acids	99241-24-4	Hydroxypropyl-α-cyclodextrin
85251-77-0	Glyceryl stearate				
85404-84-8	Polyglyceryl-3 diisostearate			99241-25-5	Hydroxypropyl-γ-cyclodextrin
		91744-38-6	Glyceryl stearate citrate		
85409-09-2	Caprylic/capric triglyceride	92345-48-7	Tetrahydrofurfuryl butyrate	100084-96-6	Myrrh extract
				100209-45-8	Hydrolyzed vegetable protein
85411-01-4	Hydrogenated vegetable glycerides phosphate	92666-21-2	Isoeugenyl benzyl ether		
		92797-39-2	Hydrolyzed milk protein	102051-00-3	Polyglyceryl-10 trioleate
		93572-53-3	Hydrogenated menhaden oil	112926-00-8	Silica, hydrated
85507-69-3	Aloe extract			112945-52-5	Silica
85586-21-6	Methyl stearate	94035-02-6	Hydroxypropyl-β-cyclodextrin	120486-24-0	Polyglyceryl-2 triisostearate
85666-92-8	Glyceryl stearate				
85666-92-8	Glyceryl stearate SE	94349-67-4	Barley extract	120962-03-0	Canola oil
85736-49-8	PEG-12 dioleate	94423-19-5	Polyglyceryl-3 distearate	123333-89-1	Sodium methyl sulfate hydrate
86637-84-5	Polyglyceryl-10 pentaoleate				
		95482-05-6	Polyglyceryl-6 hexaoleate	129521-59-1	Palm glycerides
87390-32-7	Polyglyceryl-10 myristate			133738-23-5	Polyglyceryl-10 isostearate
		97281-47-5	Lecithin		
89957-89-1	Beet powder	97281-47-5	Phosphatidylcholine	136097-97-7	Cetyl esters
89998-14-1	Lemongrass extract				

EINECS Number-to-Trade Name Cross-Reference

EINECS	Trade name	EINECS	Trade name	EINECS	Trade name
200-061-5	Hydex® 100 Coarse Powd.		99.5%	200-772-0	Arlac S
200-061-5	Hydex® 100 Coarse Powd. 35	200-289-5	Star	200-772-0	Patlac® NAL
		200-289-5	Superol	200-815-3	Polywax® 500
200-061-5	Hydex® 100 Gran. 206	200-312-9	Emersol® 6343	200-815-3	Polywax® 600
200-061-5	Hydex® 100 Powd. 60	200-312-9	Glycon® P-45	200-815-3	Polywax® 655
200-061-5	Hydex® Tablet Grade	200-313-4	Emersol® 6320	201-067-0	Uniplex 84
200-061-5	Hystar® 7570	200-313-4	Emersol® 6332 NF	201-069-1	Cap-Shure® C-140E-75
200-061-5	Liponic Sorbitol Powd.	200-313-4	Emersol® 6349		
200-061-5	Liponic Sorbitol Sol'n. 70% USP	200-313-4	Emersol® 6351	201-069-1	Citric Acid Anhydrous USP/FCC
200-061-5	Sorbo®	200-313-4	Emersol® 6353	201-069-1	Citric Acid USP FCC Anhyd. Fine Gran. No. 69941
200-061-5	Unisweet 70	200-313-4	Emersol® 7051		
200-066-2	Ascorbic Acid USP/ FCC, 100 Mesh	200-313-4	Glycon® S-70		
200-066-2	Ascorbic Acid USP, FCC Fine Gran. No. 6045655	200-313-4	Glycon® S-90	201-069-1	Citrid Acid USP FCC Anhyd. Gran. No. 69942
		200-313-4	Glycon® TP		
		200-313-4	Hystrene® 5016 NF FG	201-069-1	Citrocoat® A 1000 HP
200-066-2	Ascorbic Acid USP, FCC Fine Powd. No. 6045652	200-313-4	Hystrene® 7018 FG	201-069-1	Citrocoat® A 2000 HP
		200-313-4	Hystrene® 8718 FG	201-069-1	Citrocoat® A 4000 TP
		200-313-4	Hystrene® 9718 NF FG	201-069-1	Citrocoat® A 4000 TT
200-066-2	Ascorbic Acid USP, FCC Gran. No. 6045654	200-313-4	Industrene® 4518	201-069-1	Citrostabil® NEU
		200-313-4	Industrene® 5016 NF FG	201-069-1	Citrostabil® S
200-066-2	Ascorbic Acid USP, FCC Type S No. 6045660	200-313-4	Industrene® 7018 FG	201-069-1	Descote® Citric Acid 50%
		200-313-4	Industrene® 8718 FG		
		200-338-0	Adeka Propylene Glycol (P)	201-069-1	Liquinat®
200-066-2	Ascorbic Acid USP, FCC Ultra-Fine Powd No. 6045653	200-353-2	Vitinc® Cholesterol NF XVII	201-148-0	Isobutanol HP
				201-228-5	Vitamin A Palmitate Type PIMO/BH No. 638280100
200-066-2	Ascorbo-120	200-386-2	Pyridoxine Hydrochloride USP, FCC Powd. No. 60650		
200-066-2	Ascorbo-C Tablets			201-228-5	Vitamin A Palmitate USP, FCC Type P1.7 No. 262090000
200-066-2	Cap-Shure® AS-125-50	200-399-3	d-Biotin USP, FCC No. 63345		
200-066-2	Cap-Shure® AS-165CR-70	200-419-0	Folic Acid USP, FCC No. 20383	201-228-5	Vitamin A Palmitate USP, FCC Type P1.7/E No. 63699
200-066-2	Descote® Ascorbic Acid 60%	200-441-0	Lutavit® Niacin		
		200-441-0	Niacin USP, FCC Fine Granular No. 69901	201-321-0	Syncal® SDI
200-075-1	Candex®			201-321-0	Unisweet SAC
200-158-2	Do Sure	200-441-0	Niacin USP, FCC No. 69902	201-507-1	Riboflavin USP, FCC No. 602940002
200-158-2	Panicrust LC K	200-529-9	Versene CA		
200-289-5	Emery® 912	200-540-9	Niacet Calcium Acetate FCC	201-788-0	Xylitol C
200-289-5	Emery® 916			201-939-0	Fancol Menthol
200-289-5	Emery® 917	200-559-2	Unisweet L	201-993-5	Nipacide® OPP
200-289-5	Emery® 918	200-599-2	Unisweet Lactose	202-307-7	Lexgard® P
200-289-5	Glycerine (Pharmaceutical)	200-641-8	Thiamine Hydrochloride USP, FCC Regular Type No. 601160	202-307-7	Nipasol M
				202-311-9	Nipabenzyl
200-289-5	Glycon® G 100			202-318-7	Nipabutyl
200-289-5	Glycon® G-300	200-675-3	Sodium Citrate USP, FCC Dihydrate Fine Gran. No. 69975	202-713-4	Niacinamide USP, FCC No. 69905
200-289-5	Natural Glycerine USP 96%				
		200-711-8	Unisweet MAN	202-713-4	Niacinamide USP, FCC Fine Granular No. 69916
200-289-5	Natural Glycerine USP 99%	200-716-5	Sunmalt		
200-289-5	Natural Glycerine USP	200-77-20	Purasal® S/SP 60	202-785-7	Lexgard® M
				202-785-7	Nipagin M

EINECS	Trade name	EINECS	Trade name	EINECS	Trade name
203-768-7	Eastman Sorbic Acid	205-455-0	Cithrol GMR N/E	209-406-4	Complemix® 100
203-768-7	Tristat	205-455-0	Hodag GMR	209-567-0	Amalty®
204-007-1	Emersol® 6313 NF	205-455-0	Hodag GMR-D	209-567-0	Finmalt L
204-007-1	Emersol® 6321 NF	205-526-6	Cithrol GML N/E	210-702-0	Dynasan® 110
204-007-1	Emersol® 6333 NF	205-526-6	Grindtek ML 90	211-750-5	Evanstab® 18
204-007-1	Emersol® 7021	205-526-6	Hodag GML	212-406-7	Pationic® 1230
204-007-1	Emery® 7021	205-538-1	Asahi Aji®	212-406-7	Pationic® 1240
204-007-1	Industrene® 205	205-542-3	Cithrol PGML N/E	212-406-7	Pationic® 1250
204-271-8	Veltol®	205-542-3	Emalex PGML	212-406-7	Puracal® PP
204-399-4	Nipagin A	205-542-3	Hodag PGML	213-631-3	Arlac P
204-464-7	Rhodiarome™	205-582-1	Emery® 6354	213-631-3	Purasal® P/USP 60
204-464-7	Unisweet EVAN	205-597-3	Fancol OA-95	213-911-5	ABC-Trieb®
204-498-2	Sustane® PG	205-633-8	Sodium Bicarbonate	214-171-6	Lucantin® Yellow
204-498-2	Tenox® PG		USP No. 1 Powd.	214-304-8	Kemester® 2050
204-534-7	Hodag GTO	205-633-8	Sodium Bicarbonate	215-108-5	Vitiben®
204-574-5	Phenylacetaldehyde 50		USP No. 2 Fine Gran	215-139-4	Improved Paniplus M
204-614-1	Evanstab® 12	205-633-8	Sodium Bicarbonate	215-139-4	Regular Paniplus
204-633-5	Isoamyl Alcohol 95%		USP No. 5 Coarse	215-170-3	Magnesium Hydroxide
204-633-5	Isoamyl Alcohol 99%		Gran		USP
204-650-8	ADA Tablets	206-074-2	Gluconal® K	215-170-3	Magnesium Hydroxide
204-650-8	WaTox-20 Tablets	206-075-8	Gluconal® CA A		USP DC
204-677-5	Emersol® 6357	206-075-8	Gluconal® CA M	215-170-3	Marinco H-USP
204-693-2	Kemamide® S	206-076-3	Gluconal® FE	215-171-9	Magnesium Oxide USP
204-771-6	SAIB-SG	206-101-8	Aluminum Stearate EA		30 Light
204-812-8	Carsonol® SHS	206-101-8	Witco® Aluminum	215-171-9	Magnesium Oxide USP
204-812-8	Norfox® Anionic 27		Stearate EA Food		60 Light
204-812-8	Rhodapon® BOS		Grade	215-171-9	Magnesium Oxide USP
204-812-8	Sulfotex OA	206-103-9	Kemamide® O		90 Light
204-814-9	Niacet Sodium	206-103-9	Kemamide® U	215-171-9	Magnesium Oxide USP
	Diacetate FCC	206-376-4	Emery® 6359		Heavy
204-823-8	Niacet Sodium Acetate	207-439-9	Atomite®	215-171-9	Marinco OH
	60% FCC	207-439-9	Carbital® 35	215-171-9	Marinco OL
204-823-8	Niacet Sodium Acetate	207-439-9	Carbital® 50	215-354-3	Admul PGMS
	Anhyd. FCC	207-439-9	Carbital® 75	215-354-3	Aldo® PGHMS
204-881-4	CAO® 3	207-439-9	CC™-101	215-354-3	Aldo® PGHMS KFG
204-881-4	Embanox® BHT	207-439-9	CC™-103	215-354-3	Canamulse 55
204-881-4	Sustane® BHT	207-439-9	CC™-105	215-354-3	Canamulse 70
204-881-4	Tenox® BHT	207-439-9	Micro-White® 10 Codex	215-354-3	Canamulse 90K
204-886-1	Syncal® GS	207-439-9	Micro-White® 25 Codex	215-354-3	Cithrol PGMS N/E
204-886-1	Syncal® GSD	207-439-9	Micro-White® 50 Codex	215-354-3	Drewlene 10
204-886-1	Syncal® S	207-439-9	Micro-White® 100	215-354-3	Emalex PGMS
204-886-1	Syncal® SDS		Codex	215-354-3	Emalex PGS
204-886-1	Syncal® US	208-187-2	Lucantin® Red	215-354-3	Hefti PMS-33
204-886-1	Unisweet SOSAC	208-407-7	Gluconal® NA	215-354-3	Hodag PGMS
204-988-6	Riboflavin-5´-	208-408-2	Gluconal® CU	215-354-3	Homotex PS-90
	Phosphate Sodium	208-534-8	ProBenz	215-354-3	Lipal PGMS
	USP, FCC No. 60296	208-534-8	Sodium Benzoate BP88	215-354-3	Mazol® PGMS
205-126-1	Descote® Sodium	208-537-4	Thiamine Mononitrate	215-354-3	Mazol® PGMSK
	Ascorbate 50%		USP, FCC Fine Powd.	215-354-3	Myverol® P-06
205-126-1	Sodium Ascorbate		No. 601340	215-354-3	PGMS 70
	USP, FCC Fine Gran.	208-686-5	Captex® 8000	215-354-3	Promodan SP
	No. 6047709	208-686-5	Miglyol® 808	215-359-0	Cithrol GDS N/E
205-126-1	Sodium Ascorbate	208-687-0	Dynasan® 112	215-359-0	Kessco® Glycerol
	USP, FCC Fine Powd.	208-687-0	Massa Estarinum® AM		Distearate 386F
	No. 6047708	208-875-2	Emery® 6355	215-475-1	Suspengel Elite
205-126-1	Sodium Ascorbate	208-915-9	Marinco CH	215-475-1	Suspengel Micro
	USP, FCC Type AG	208-915-9	Marinco® CH-Granular	215-475-1	Suspengel Ultra
	No. 6047710	208-915-9	Marinco® CL	215-549-3	Cithrol PGMO N/E
205-278-9	Calcium Pantothenate	209-097-6	Dynasan® 118	215-549-3	Emalex PGO
	USP, FCC Type SD No.	209-097-6	Neobee® 62	215-549-3	G-2185
	63924	209-097-6	Pationic® 919	215-664-9	Ablunol S-60
205-278-9	Lutavit® Calpan	209-098-1	Dynasan® 116	215-664-9	Alkamuls® SMS
205-290-4	Niacet Sodium	209-099-7	Dynasan® 114	215-664-9	Atlas 110K
	Propionate FCC	209-150-3	Nuodex Magnesium	215-664-9	Capmul® S
205-305-4	Ascorbyl Palmitate NF,		Stearate Food Grade	215-664-9	Crill 3
	FCC No. 60412	209-150-3	Synpro® Magnesium	215-664-9	Drewmulse® SMS
205-352-0	Cyncal®		Stearate Food Grade	215-664-9	Drewsorb® 60K
205-358-3	Versene NA	209-183-3	Elvanol® 71-30	215-664-9	Durtan® 60

EINECS	Trade name	EINECS	Trade name	EINECS	Trade name
215-664-9	Durtan® 60K		PEG 300	231-767-1	Albrite TSPP Food
215-664-9	Famodan MS Kosher	220-045-1	Dow E300 NF		Grade
215-664-9	Glycomul® S FG	220-045-1	Pluracol® E300	231-826-1	Albrite Dicalcium
215-664-9	Glycomul® S KFG	222-848-2	Gluconal® MG		Phosphate Anhyd
215-664-9	Hodag SMS	223-795-8	Niacet Calcium	231-826-1	Caliment Dicalcium
215-664-9	Lamesorb® SMS		Propionate FCC		Phosphate Dihydrate
215-664-9	Liposorb S	224-580-1	Tristat SDHA	231-829-8	Bromette
215-664-9	Nissan Nonion SP-60R	224-736-9	Gluconal® ZN	231-829-8	Bromitabs
215-664-9	Norfox® Sorbo S-60FG	225-714-1	Nipagin M Sodium	231-835-0	Aerophos P
215-664-9	Polycon S60 K	225-856-4	Carbowax® Sentry®	231-835-0	Albrite SAPP Food
215-664-9	Prote-sorb SMS		PEG 400		Grade
215-664-9	S-Maz® 60K	225-856-4	Dow E400 NF	231-835-0	Antelope Aerator
215-664-9	S-Maz® 60KHS	225-856-4	Pluracol® E400	231-835-0	B.P. Pyro®
215-664-9	Sorgen 50	227-335-7	Admul CSL 2007, CSL	231-835-0	B.P. Pyro® Type K
215-664-9	Span® 60		2008	231-835-0	Curacel
215-664-9	Span® 60K	227-335-7	Artodan CF 40	231-835-0	Curavis® 150
215-664-9	Span® 60 VS	227-335-7	Crolactil CS2L	231-835-0	Donut Pyro®
215-665-4	Ablunol S-80	227-335-7	Lamegin® CSL	231-835-0	Donut SAPP
215-665-4	Capmul® O	227-335-7	Paniplex CK	231-835-0	Perfection®
215-665-4	Crill 4	227-335-7	Pationic® 930	231-835-0	Sapp #4
215-665-4	Crill 50	227-335-7	Pationic® 940	231-835-0	SAPP 22
215-665-4	Drewmulse® SMO	227-335-7	Prefera® CSL	231-835-0	SAPP 26
215-665-4	Drewsorb® 80K	227-335-7	Verv®	231-835-0	SAPP 28
215-665-4	Hodag SMO	228-973-9	Eribate®	231-835-0	SAPP 40
215-665-4	Lamesorb® SMO	228-973-9	Neo-Cebitate®	231-835-0	Taterfos®
215-665-4	Nikkol SO-10	229-349-9	Syncal® CAS	231-835-0	Victor Cream®
215-665-4	Nissan Nonion OP-80R	229-350-4	Gluconal® MN	231-837-1	Ajax®
215-665-4	Norfox® Sorbo S-80	229-859-1	Carbowax® Sentry®	231-837-1	HT® Monocalcium
215-665-4	Polycon S80 K		PEG 600		Phosphate, Monohy-
215-665-4	Prote-sorb SMO	229-859-1	Dow E600 NF		drate (MCP) Spray
215-665-4	S-Maz® 80K	229-859-1	Pluracol® E600		Dried Coarse Granular
215-665-4	Sorgen 40	231-449-2	Albrite MSP Food	231-837-1	HT® Monocalcium
215-665-4	Sorgen S-40-H		Grade		Phosphate, Monohy-
215-665-4	Span® 80	231-493-2	Alpha W 6 Pharma		drate (MCP) Spray
215-724-4	Carmacid Y		Grade		Dried Fines
215-724-4	Carmine 1623	231-493-2	Beta W 7	231-837-1	HT® Monocalcium
215-724-4	Carmine AS	231-493-2	Cavitron Cyclo-		Phosphate, Monohy-
215-724-4	Carmine Extract P4011		dextrin.™		drate (MCP) Spray
215-724-4	Carmine FG	231-493-2	Gamma W8		Dried Medium Granular
215-724-4	Carmine Nacarat 40	231-598-3	Culinox® 999 Chemical	231-837-1	Regent® 12XX
215-724-4	Carmine PG		Grade Salt	231-987-8	Albrite Diammonium
215-724-4	Carmine Powd. WS	231-598-3	Culinox® 999® Food		Phosphate Food Grade
215-724-4	Carmine XY/UF		Grade Salt	232-191-3	IDX-20 NB
215-724-4	Carmisol A	231-598-3	Morton® H.G. Blending	232-273-9	Dewaxed Sunflower Oil
215-724-4	Carmisol NA		Salt		83-070-0
215-724-4	Hilton Davis Carmine	231-598-3	Morton® TFC 999®	232-273-9	Lipovol SUN
215-724-4	Natural Liquid AP		Salt	232-274-4	Archer Soybean Oil 86-
	Carmine Colorant	231-598-3	Purex® All Purpose		070-0
215-724-4	Natural Liquid Carmine		Salt	232-274-4	Capital Soya
	Colorant (Type 100, 50,	231-598-3	Sterling® Purified USP	232-274-4	Lipovol SOY
	and Simple)		Salt	232-274-4	Soybean Oil
215-724-4	Natural Soluble	231-633-2	Albrite Phosphoric Acid	232-274-4	Soy Oil
	Carmine Powd.		85% Food Grade	232-276-5	Lipovol SAF
215-724-4	Natural Soluble Powder	231-633-2	Quaker™ Oatrim 5, 5Q	232-276-5	Neobee® 18
	AP Carmine Colorant	231-683-5	Delvocid	232-277-0	Pure/Riviera Olive Oil
215-735-4	Dascolor Annatoo Enc	231-694-5	Albrite STPP-F		NF
216-472-8	Nuodex S-1421 Food	231-694-5	Albrite STPP-FC	232-280-7	Cottonseed Cooking Oil
	Grade	231-694-5	Curafos® STP		82-060-0
216-472-8	Nuodex S-1520 Food	231-694-5	Freez-Gard® FP-19	232-280-7	Cottonseed Oil
	Grade	231-710-0	Vitamin E USP, FCC	232-280-7	Jewel Oil™
216-472-8	Synpro® Calcium		No. 60525	232-281-2	Corn Oil
	Stearate Food Grade	231-710-0	Vitamin E USP, FCC	232-281-2	Corn Oil
217-752-2	Embanox® TBHQ		No. 60526	232-281-2	Corn Salad Oil 104-250
217-752-2	Sustane® TBHQ	231-710-0	Vitinc® dl-alpha	232-281-2	Liquid Corn Oil 87-
217-752-2	TBHQ®		Tocopheryl Acetate		070-0
217-752-2	Tenox® TBHQ		USP XXII	232-282-8	Cobee 76
218-901-4	Dimodan LS Kosher	231-764-5	Albrite Monoammonium	232-282-8	Coconut Oil® 76
220-045-1	Carbowax® Sentry®		Phosphate Food Grade	232-282-8	Coconut Oil® 92

EINECS	Trade name	EINECS	Trade name	EINECS	Trade name
232-282-8	Konut	232-307-2	Clearate WDF	232-310-9	Pure Malt Colorant A6000
232-282-8	Pureco® 76	232-307-2	Dur-Lec® P		
232-292-2	Castorwax® NF	232-307-2	Dur-Lec® UB	232-310-9	Pure Malt Colorant A6001
232-293-8	AA USP	232-307-2	Emulbesto		
232-296-4	Peanut Oil 85-060-0	232-307-2	Emulfluid™	232-311-4	Neobee® SL-310
232-307-2	Actiflo® 68 SB	232-307-2	Emulfluid® A	232-315-6	Koster Keunen Paraffin Wax
232-307-2	Actiflo® 68 UB	232-307-2	Emulfluid® AS		
232-307-2	Actiflo® 70 SB	232-307-2	Emulfluid® E	232-316-1	Lipovol PAL
232-307-2	Actiflo® 70 UB	232-307-2	Emulgum™	232-317-7	Do-Pep Vital Wheat Gluten
232-307-2	Alcolec® 140	232-307-2	Emulpur™ N		
232-307-2	Alcolec® 495	232-307-2	Emulpur™ N P-1	232-317-7	Meatbind®-3000
232-307-2	Alcolec® BS	232-307-2	Emulthin M-35	232-317-7	Provim ESP®
232-307-2	Alcolec® Extra A	232-307-2	Emulthin M-501	232-317-7	SQ®-48
232-307-2	Alcolec® F-100	232-307-2	Epicholin	232-317-7	Whetpro®-75
232-307-2	Alcolec® FF-100	232-307-2	Epikuron™ 100 P, 100 G	232-317-7	Whetpro®-80
232-307-2	Alcolec® Granules			232-347-0	Koster Keunen Candelilla
232-307-2	Alcolec® PG	232-307-2	Epikuron™ 100 X		
232-307-2	Alcolec® S	232-307-2	Epikuron™ 130 G	232-347-0	Ross Candelilla Wax
232-307-2	Alcolec® SFG	232-307-2	Epikuron™ 130 P	232-360-1	Crill 43
232-307-2	Alcolec® XTRA-A	232-307-2	Epikuron™ 130 X	232-360-1	Nikkol SO-15
232-307-2	Amisol™ 210-L	232-307-2	Epikuron™ 135 F	232-360-1	Nissan Nonion OP-83RAT
232-307-2	Amisol™ 210 LP	232-307-2	Lecigran™ 5750		
232-307-2	Amisol™ 329	232-307-2	Lecigran™ 6750	232-360-1	Sorgen 30
232-307-2	Amisol™ 683 A	232-307-2	Lecigran™ A	232-360-1	Sorgen S-30-H
232-307-2	Amisol™ 697	232-307-2	Lecigran™ F	232-370-6	Lipovol SES
232-307-2	Asol	232-307-2	Lecigran™ M	232-370-6	Vitinc® Sesame Oil NF
232-307-2	Beakin LV1	232-307-2	Lecimulthin	232-373-2	Fonoline® White
232-307-2	Beakin LV2	232-307-2	Leciprime™	232-373-2	Fonoline® Yellow
232-307-2	Blendmax 322	232-307-2	Lecipur™ 95 C	232-373-2	Ointment Base No. 3
232-307-2	Blendmax 322D	232-307-2	Lecipur™ 95 R	232-373-2	Ointment Base No. 4
232-307-2	Canasperse SBF	232-307-2	Lecithin L 1000-Range	232-373-2	Ointment Base No. 6
232-307-2	Canasperse UBF	232-307-2	Lecithin L 4000-Range	232-373-2	Penreco Amber
232-307-2	Canasperse UBF-LV	232-307-2	Magathin	232-373-9	Penreco Blond
232-307-2	Canasperse WDF	232-307-2	M-C-Thin®	232-373-2	Penreco Cream
232-307-2	Capcithin™	232-307-2	M-C-Thin® AF-1 Type DB	232-373-2	Penreco Lily
232-307-2	Capsulec 51-SB			232-373-2	Penreco Regent
232-307-2	Capsulec 51 UB	232-307-2	M-C-Thin® AF-1 Type SB	232-373-2	Penreco Royal
232-307-2	Capsulec 56-SB			232-373-2	Penreco Snow
232-307-2	Capsulec 56-UB	232-307-2	M-C-Thin® AF-1 Type UB	232-373-2	Penreco Super
232-307-2	Capsulec 62-SB			232-373-2	Penreco Ultima
232-307-2	Capsulec 62-UB	232-307-2	M-C-Thin® ASOL	232-373-2	Perlatum® 400
232-307-2	Centrocap® 162SS	232-307-2	M-C-Thin® FWD	232-373-2	Perlatum® 410
232-307-2	Centrocap® 162US	232-307-2	Metarin™	232-373-2	Perlatum® 410 CG
232-307-2	Centrocap® 273SS	232-307-2	Metarin™ C	232-373-2	Perlatum® 415
232-307-2	Centrocap® 273US	232-307-2	Metarin™ CP	232-373-2	Perlatum® 415 CG
232-307-2	Centrol® 2F SB	232-307-2	Metarin™ DA 51	232-373-2	Perlatum® 420
232-307-2	Centrol® 2F UB	232-307-2	Metarin™ F	232-373-2	Perlatum® 425
232-307-2	Centrol® 3F SB	232-307-2	Metarin™ P	232-373-2	Perlatum® 510
232-307-2	Centrol® 3F UB	232-307-2	Phospholipon® 90/90 G	232-373-2	Protopet® Alba
232-307-2	Centrol® CA	232-307-2	Thermolec 57	232-373-2	Protopet® White 1S
232-307-2	Centrolex® C	232-307-2	Thermolec 68	232-373-2	Protopet® White 2L
232-307-2	Centrolex® D	232-307-2	Thermolec 200	232-373-2	Protopet® White 3C
232-307-2	Centrolex® F	232-307-2	Topcithin®	232-373-2	Protopet® Yellow 2A
232-307-2	Centrolex® G	232-307-2	Yelkin DS	232-373-2	Super White Fonoline®
232-307-2	Centrolex® P	232-307-2	Yelkin Gold	232-373-2	Super White Protopet®
232-307-2	Centrolex® R	232-307-2	Yelkin SS	232-399-4	Koster Keunen Carnauba
232-307-2	Centrophase® 152	232-307-2	Yelkin T		
232-307-2	Centrophase® C	232-307-2	Yelkin TS	232-399-4	Ross Carnauba Wax
232-307-2	Centrophase® HR	232-307-2	Yelkinol F	232-409-7	Ross Rice Bran Wax
232-307-2	Centrophase® HR2B	232-307-2	Yelkinol G	232-410-2	Akofame
232-307-2	Centrophase® HR2U	232-307-2	Yelkinol P	232-410-2	Akofil A
232-307-2	Centrophase® HR4B	232-310-9	DME Dry Malt Extract	232-410-2	Akofil N
232-307-2	Centrophase® HR4U	232-310-9	Extramalt Dark	232-410-2	Akoleno S
232-307-2	Centrophase® HR6B	232-310-9	Maltoferm® 10001	232-410-2	Akolizer S
232-307-2	Centrophil® K	232-310-9	Maltoferm® 10001 VDK	232-410-2	Akorex
232-307-2	Chocothin	232-310-9	Maltoferm® A-6001	232-410-2	Bunge Biscuit Flakes
232-307-2	Chocotop™	232-310-9	ND-201 Syrup	232-410-2	Bunge Heavy Duty Donut Frying
232-307-2	Clearate Special Extra	232-310-9	Nutrimalt® Range		

EINECS	Trade name	EINECS	Trade name	EINECS	Trade name
	Shortening	232-440-6	Centrolene® A	232-524-2	Genugel® CJ
232-410-2	Clarity	232-440-6	Centrolene® S	232-524-2	Genugel® LC-1
232-410-2	Code 321	232-440-6	M-C-Thin® HL	232-524-2	Genugel® LC-4
232-410-2	Diamond D 31	232-440-6	Thermolec WFC	232-524-2	Genugel® LC-5
232-410-2	Donut Frying	232-440-6	Yelkin 1018	232-524-2	Genugel® MB-51
	Shortening 101-150	232-442-7	Special Fat 168T	232-524-2	Genugel® MB-78F
232-410-2	Durkex 100F	232-519-5	Coatingum L	232-524-2	Genugel® UE
232-410-2	Durkex 500S	232-519-5	Emulgum BV	232-524-2	Genugel® UEU
232-410-2	Durkex Gold 77F	232-519-5	Granular Gum Arabic	232-524-2	Genulacta® CP-100
232-410-2	Famous		NF/FCC C-4010	232-524-2	Genulacta® CSM-2
232-410-2	Kaomax-S	232-519-5	Gum Arabic NF/FCC	232-524-2	Genulacta® K-100
232-410-2	Lipovol HS		Clean Amber Sorts	232-524-2	Genulacta® KM-1
232-410-2	Partially Hydro	232-519-5	Gum Arabic, Purified,	232-524-2	Genulacta® KM-5
	Soybean Oil 86-505-0		Spray-Dried No. 1834	232-524-2	Genulacta® L-100
232-410-2	Shasta®	232-519-5	Natural Arabic Type	232-524-2	Genulacta® LK-71
232-410-2	Soy Flakes		Gum Purified, Spray-	232-524-2	Genulacta® LR-41
232-410-2	17 Stearine		Dried	232-524-2	Genulacta® LR-60
232-410-2	Sterotex® HM NF	232-519-5	Nutriloid® Arabic	232-524-2	Genulacta® LRA-50
232-410-2	Superb® Cookie Bake	232-519-5	Powdered Gum Arabic	232-524-2	Genulacta® LRC-21
	Shortening 101-057		NF/FCC G-150	232-524-2	Genulacta® LRC-30
232-410-2	Superb Oil Hydro	232-519-5	Powdered Gum Arabic	232-524-2	Genulacta® P-100
	Winterized Soybean Oil		NF/FCC Superselect	232-524-2	Genulacta® PL-93
	86-091-0		Type NB-4	232-524-2	Genuvisco® CSW-2
232-410-2	Tem Crunch™ 110	232-519-5	Premium Fine Granular	232-524-2	Genuvisco® J
232-410-2	Tem Plus® 95		Gum Arabic	232-524-2	Genuvisco® MP-11
232-410-2	Witarix® 440	232-519-5	Premium Granular Gum	232-524-2	Geoldan CL Range
232-425-4	Akowesco 1		Arabic	232-524-2	Maco-O-Line 091
232-428-0	Lipovol A	232-519-5	Premium Powdered	232-524-2	Marine Colloids™
232-435-9	Acid Proof Caramel		Gum Arabic		Carrageenan
	Powd.	232-519-5	Premium Spray Dried	232-524-2	Nutriloid® Carrageenan
232-435-9	B&C Caramel Powd.		Gum Arabic	232-524-2	PF-80
232-435-9	Caramel Color Double	232-519-5	Spray Dried Gum	232-524-2	Soageena®
	Strength		Arabic NF Type CSP	232-524-2	Soageena® LX22
232-435-9	Caramel Color Single	232-519-5	Spray Dried Gum	232-524-2	Soageena® ML300
	Strength		Arabic NF/FCC CM	232-524-2	Soageena® MM101
232-435-9	Double Strength Acid	232-519-5	Spray Dried Gum	232-524-2	Soageena® MM301
	Proof Caramel Colour		Arabic NF/FCC CS	232-524-2	Soageena® MM330
232-435-9	P147 Caramel Color		(Low Bacteria)	232-524-2	Soageena® MM350
232-435-9	Powdered Caramel	232-519-5	Spray Dried Gum	232-524-2	Soageena® MM501
	Color, Acid Proof		Arabic NF/FCC CS-R	232-524-2	Soageena® MV320
232-435-9	Powdered Caramel	232-519-5	Spray Dried Nigerian	232-524-2	Soageena® MW321
	Colour Non-		Gum Arabic	232-524-2	Soageena® MW351
	Ammoniated-All Natural	232-519-5	Spraygum C	232-524-2	Soageena® MW371
	T-717	232-519-5	Spraygum GD	232-524-2	Soageena® WX57
232-435-9	Single Strength Acid	232-519-5	TIC Pretested® Arabic	232-524-2	Stamere® CK FCC
	Proof Caramel Colour		FT Powd.	232-524-2	Stamere® CKM FCC
232-435-9	Unisweet Caramel	232-519-5	TIC Pretested® Arabic	232-524-2	Stamere® CK-S
232-436-4	CornSweet® 42		FT-1 USP	232-524-2	Stamere® N-47
232-436-4	CornSweet® 55	232-519-5	TIC Pretested® Arabic	232-524-2	Stamere® N-55
232-436-4	CornSweet® 95		PH-FT	232-524-2	Stamere® N-325
232-436-4	Corn Syrup 36/43	232-524-2	CarraFat™	232-524-2	Stamere® N-350
232-436-4	Corn Syrup 42/43	232-524-2	Carralean™ CG-100	232-524-2	Stamere® N-350 E
232-436-4	Corn Syrup 42/44	232-524-2	Carralean™ CM-70		FCC
232-436-4	Corn Syrup 52/43	232-524-2	Carralean™ CM-80	232-524-2	Stamere® N-350 S
232-436-4	Corn Syrup 62/43	232-524-2	Carralean™ MB-60	232-524-2	Stamere® NI
232-436-4	Corn Syrup 62/44	232-524-2	Carralean™ MB-93	232-524-2	Stamere® NIC FCC
232-436-4	Corn Syrup 62/44-1	232-524-2	Carralite™	232-524-2	Stamere® NK
232-436-4	Corn Syrup 62/44-2	232-524-2	CarraLizer™ CGB-10	232-524-2	TIC Pretested® Colloid
232-436-4	Corn Syrup 97/71	232-524-2	CarraLizer™ CGB-20		775 Powd.
232-436-4	42/43 Corn Syrup	232-524-2	CarraLizer™ CGB-40	232-524-2	TIC Pretested® Colloid
232-436-4	62/43 Corn Syrup	232-524-2	CarraLizer™ CGB-50		881 M
232-438-5	Dimodan CP	232-524-2	Gelodan CC Range	232-524-2	Wakal® K Range
232-438-5	Dimodan CP Kosher	232-524-2	Gelodan CW Range	232-524-2	Wakal® K-e
232-438-5	Myvatex® 7-85	232-524-2	Gelodan CX Range	232-536-8	Dycol™ 4000FC
232-438-5	Myvatex® 7-85K	232-524-2	Genugel® CHP-2	232-536-8	Dycol™ 4500F
232-438-5	Myverol® 18-85	232-524-2	Genugel® CHP-2 Fine	232-536-8	Dycol™ HV400F
232-438-5	Myverol® 18-85K		Mesh	232-536-8	Edicol®
232-440-6	Alcolec® Z-3	232-524-2	Genugel® CHP-200	232-536-8	Edicol® ULV Series

EINECS	Trade name	EINECS	Trade name	EINECS	Trade name
232-536-8	Emulcoll	232-541-5	Idealgum 3F	232-553-0	Mexpectin HV 400
232-536-8	Guardan 100 Range	232-541-5	Locust Bean Gum		Range
232-536-8	Guardan 600 Range		Speckless Type D-200	232-553-0	Mexpectin LA 100
232-536-8	Guardan 700 Range	232-541-5	Nutriloid® Locust		Range
232-536-8	Guar Gum	232-541-5	Powdered Locust Bean	232-553-0	Mexpectin LA 200
232-536-8	Guar Gum HV		Gum Type D-200		Range
232-536-8	Jaguar® 1105	232-541-5	Powdered Locust Bean	232-553-0	Mexpectin LA 400
232-536-8	Jaguar® 1110		Gum Type D-300		Range
232-536-8	Jaguar® 1120	232-541-5	Powdered Locust Bean	232-553-0	Mexpectin LC 700
232-536-8	Jaguar® 1140		Gum Type P-100		Range
232-536-8	Jaguar® 2209	232-541-5	Powdered Locust Bean	232-553-0	Mexpectin LC 800
232-536-8	Jaguar® 2220		Gum Type PP-100		Range
232-536-8	Jaguar® 2240	232-541-5	Soalocust®	232-553-0	Mexpectin LC 900
232-536-8	Jaguar® 6000	232-541-5	STD-175		Range
232-536-8	Jaguar® Guar Gum	232-541-5	TIC Pretested® Locust	232-553-0	Mexpectin MRS 300
232-536-8	Nutriloid® Guar Special		Bean POR/A		Range
232-536-8	Nutriloid® Guar	232-541-5	Wakal® J	232-553-0	Mexpectin RS 400
	Standard	232-541-5	Wakal® JG		Range
232-536-8	Powdered Gum Guar	232-552-5	Nutriloid® Tragacanth	232-553-0	Mexpectin SS 200
	NF Type 80 Mesh B/T	232-552-5	Powdered Gum		Range
232-536-8	Powdered Gum Guar		Tragacanth T-150	232-553-0	Mexpectin XSS 100
	Type 140 Mesh B/T	232-552-5	Powdered Gum		Range
232-536-8	Powdered Gum Guar		Tragacanth T-200	232-553-0	TIC Pretested® Pectin
	Type ECM	232-552-5	Powdered Gum		HM Rapid
232-536-8	Powdered Gum Guar		Tragacanth T-300	232-553-0	TIC Pretested® Pectin
	Type M	232-552-5	Powdered Gum		HM Slow
232-536-8	Powdered Gum Guar		Tragacanth T-400	232-554-6	Bone Gelatin Type B
	Type MM FCC	232-552-5	Powdered Gum		200 Bloom
232-536-8	Powdered Gum Guar		Tragacanth T-500	232-554-6	Calfskin Gelatin Type B
	Type MM (HV)	232-552-5	Powdered Tragacanth		175 Bloom
232-536-8	Powdered Gum Guar		Gum Type A/10	232-554-6	Calfskin Gelatin Type B
	Type MMM ~1/^2	232-552-5	Powdered Tragacanth		200 Bloom
232-536-8	Powdered Gum Guar		Gum Type E-1	232-554-6	Calfskin Gelatin Type B
	Type MMW	232-552-5	Powdered Tragacanth		225 Bloom
232-536-8	Prinza® 452		Gum Type G-3	232-554-6	Calfskin Gelatin Type B
232-536-8	Prinza® 455	232-552-5	Powdered Tragacanth		250 Bloom
232-536-8	Prinza® Range		Gum Type L	232-554-6	Crodyne BY-19
232-536-8	Supercol® G2S	232-552-5	Powdered Tragacanth	232-554-6	Edible Beef Gelatin
232-536-8	Supercol® GF		Gum Type W	232-554-6	Edible Gelatins
232-536-8	Supercol® U	232-552-5	TIC Pretested®	232-554-6	Flavorset® GP-2
232-536-8	Ticolv		Tragacanth 440	232-554-6	Flavorset® GP-3
232-536-8	TIC Pretested® Gum	232-552-5	Tragacanth Flake No.	232-554-6	Flavorset® GP-4
	Guar 8/22 FCC/NF		27	232-554-6	Flavorset® GP-5
	Powd.	232-552-5	Tragacanth Gum	232-554-6	Flavorset® GP-6
232-536-8	Uniguar 20		Ribbon No. 1	232-554-6	Flavorset® GP-7
232-536-8	Uniguar 40	232-553-0	Genu® 04CG or 04CB	232-554-6	Flavorset® GP-8
232-536-8	Uniguar 150	232-553-0	Genu® 12CG	232-554-6	Flavorset® GP-9
232-536-8	Uniguar 200	232-553-0	Genu® 18CG	232-554-6	Flavorset® GP-10
232-541-5	Aquasol CSL	232-553-0	Genu® 18CG-YA	232-554-6	Gelatin XF
232-541-5	Carudan 000 Range	232-553-0	Genu® 20AS	232-554-6	Gummi Gelatin P-5
232-541-5	Carudan 100 Range	232-553-0	Genu® 21AS or 21AB	232-554-6	Gummi Gelatin P-6
232-541-5	Carudan 200 Range	232-553-0	Genu® 22CG	232-554-6	Gummi Gelatin P-7
232-541-5	Carudan 300 Range	232-553-0	Genu® 102AS	232-554-6	Gummi Gelatin P-8
232-541-5	Carudan 400 Range	232-553-0	Genu® 104AS	232-554-6	Liquid Fish Gelatin
232-541-5	Carudan 700 Range	232-553-0	Genu® 104AS-YA		Conc.
232-541-5	Diagum LBG	232-553-0	Genu® AA Medium-	232-554-6	Margarine Gelatin P-8
232-541-5	Hercules® Locust Bean		Rapid Set, 150 Grade	232-554-6	Quickset® D-4
	Gum FL 50-40	232-553-0	Genu® BA-KING	232-554-6	Quickset® D-5
232-541-5	Hercules® Locust Bean	232-553-0	Genu® BB Rapid Set	232-554-6	Quickset® D-6
	Gum FL 50-50	232-553-0	Genu® DD Extra-Slow	232-554-6	Quickset® D-7
232-541-5	HG-100		Set	232-554-6	Quickset® D-8
232-541-5	HG-175	232-553-0	Genu® DD Extra-Slow	232-554-6	Quickset® D-9
232-541-5	HG-200		Set C, 150 Grade	232-554-6	Quickset® D-10
232-541-5	Idealgum 1A	232-553-0	Genu® DD Slow Set	232-554-6	Spa Gelatin
232-541-5	Idealgum 1B	232-553-0	Genu® JMJ	232-554-6	Spray Dried Fish
232-541-5	Idealgum 1C	232-553-0	Genu® Type DJ		Gelatin
232-541-5	Idealgum 2A	232-553-0	Genu® VIS	232-554-6	Spray Dried Hydrolysed
232-541-5	Idealgum 2B	232-553-0	Genu® Pectins		Fish Gelatin

EINECS	Trade name	EINECS	Trade name	EINECS	Trade name
232-554-6	Tenderset® M-7		NF MK-60	232-679-6	Stir & Sperse®
232-554-6	Tenderset® M-8	232-658-1	Powdered Agar Agar	232-680-1	Alginic Acid FCC
232-554-6	Tenderset® M-9		NF MK-80-B	232-680-1	Kelacid®
232-567-7	Amflex	232-658-1	Powdered Agar Agar	232-680-1	Kimitsu Acid
232-567-7	Atlas MDA		NF MK-80 (Bacterio-	232-680-1	Sobalg FD 000 Range
232-567-7	Biodiastase		logical)	232-734-4	Celluferm
232-567-7	Biozyme L	232-658-1	Powdered Agar Agar	232-734-4	Cellulase 4000
232-567-7	Biozyme M		NF S-100	232-734-4	Cellulase AC
232-567-7	Biozyme S	232-658-1	Powdered Agar Agar	232-734-4	Cellulase AP
232-567-7	Clarase® 5,000		NF S-100-B	232-734-4	Cellulase AP 3
232-567-7	Clarase® 40,000	232-658-1	Powdered Agar Agar	232-734-4	Cellulase L
232-567-7	Clarase® Conc.		NF S-150	232-734-4	Cellulase TAP
232-567-7	Clarase® L-40,000	232-658-1	Powdered Agar Agar	232-734-4	Cellulase TRL
232-567-7	Fermalpha		NF S-150-B	232-734-4	Cellulase Tr Conc.
232-567-7	Spezyme BBA	232-658-1	TIC Pretested® Gum	232-752-2	AFP 2000
232-567-7	Taka-Therm® L-340		Agar Agar 100 FCC/NF	232-752-2	Bromelain 1:10
232-567-7	Tenase® 1200		Powd.	232-752-2	Bromelain Conc.
232-567-7	Tenase® L-340	232-674-9	Justfiber® CL-20-H	232-752-2	Fungal Protease
232-567-7	Tenase® L-1200	232-674-9	Justfiber® CL-35-H		31,000
232-577-1	Catalase L	232-674-9	Justfiber® CL-40-H	232-752-2	Fungal Protease
232-577-1	Fermcolase®	232-674-9	Justfiber® CL-100-H		60,000
232-577-1	Microcatalase®	232-674-9	Keycel®	232-752-2	Fungal Protease
232-601-0	Fermcozyme® 1307,	232-674-9	Qual Flo™		500,000
	BG, BGXX, CBB,	232-674-9	Solka-Floc®	232-752-2	Fungal Protease Conc.
	CBBXX, M	232-675-4	Clinton #600 Dextrin	232-752-2	HT-Proteolytic® 200
232-601-0	Hidelase	232-675-4	Clinton #655 Dextrin	232-752-2	HT-Proteolytic® Conc.
232-601-0	Hyderase	232-675-4	Clinton #656 Dextrin	232-752-2	HT-Proteolytic® L-175
232-601-0	Ovazyme, XX	232-675-4	Clinton #700 Dextrin	232-752-2	MC~2
232-619-9	Fermlipase	232-675-4	Clinton #721 Dextrin	232-752-2	MLO
232-619-9	Lipase 8 Powd.	232-675-4	Stadex® 9	232-752-2	Newlase
232-619-9	Lipase 16 Powd.	232-675-4	Stadex® 60K	232-752-2	Optimase® APL-440
232-619-9	Lipase 24 Powd.	232-675-4	Stadex® 126	232-752-2	Papain 16,000
232-619-9	Lipase 30 Powd.	232-675-4	Stadex® 128	232-752-2	Papain 30,000
232-619-9	Lipase AK	232-675-4	Staley® Tapioca	232-752-2	Papain Conc.
232-619-9	Lipase AP		Dextrin 11	232-752-2	Papain P-100
232-619-9	Lipase AP 6	232-675-4	Staley® Tapioca	232-752-2	Protoferm
232-619-9	Lipase AY		Dextrin 12	232-752-2	Prozyme 6
232-619-9	Lipase FAP	232-679-6	Mira-Gel® 463	232-752-2	Rhozyme® P11
232-619-9	Lipase G	232-679-6	Mira-Set® B	232-752-2	Rhozyme® P41
232-619-9	Lipase GC	232-679-6	Pure-Dent® B700	232-752-2	Rhozyme® P53, P64
232-619-9	Lipase MAP	232-679-6	Pure-Dent® B810	232-752-2	Wheat-Eze
232-619-9	Lipase N	232-679-6	Pure-Dent® B812	232-803-9	Dextranase L Amano
232-619-9	Lipase PS	232-679-6	Pure-Dent® B815	232-803-9	Dextranase Novo 25 L
232-619-9	Pancreatic Lipase 250	232-679-6	Pure-Dent® B816	232-877-2	Diazyme® L-200
232-627-2	Papain A300	232-679-6	Pure-Dent® B880	232-877-2	Spezyme GA
232-627-2	Papain A400	232-679-6	Pure Food Powd.	232-885-6	Clarex® 5XL
232-627-2	Papain AIE		Starch 105-A	232-885-6	Clarex® L
232-627-2	Papain M70	232-679-6	Pure Food Powd.	232-885-6	Extractase L5X, P15X
232-627-2	Papain S100		Starch 131-C	232-885-6	Pearex® L
232-627-2	Tona Enzyme 201	232-679-6	Pure Food Starch	232-885-6	Pectinase AT
232-627-2	Tona Papain 14		Bleached 142-A	232-885-6	Pectinol® 59L
232-627-2	Tona Papain 90L	232-679-6	Pure-Gel® B990	232-885-6	Pectinol® R10
232-627-2	Tona Papain 270L	232-679-6	Staley® 7025	232-885-6	Pektolase
232-629-3	Pepsin 1:3000 NF XII	232-679-6	Staley® 7350 Waxy No.	232-885-6	Spark-L® HPG
	Powd.		1 Starch	232-940-4	Lo-Dex® 10
232-629-3	Pepsin 1:10,000 Powd.	232-679-6	Staley® Moulding	232-940-4	Maltrin® M040
	or Gran		Starch	232-940-4	Maltrin® M050
232-629-3	Pepsin 1:15,000 Powd.	232-679-6	Staley® Pure Food	232-940-4	Maltrin® M100
232-650-8	Trypsin 1:75		Powd. (PFP)	232-940-4	Maltrin® M105
232-650-8	Trypsin 1:80	232-679-6	Staley® Pure Food	232-940-4	Maltrin® M150
232-650-8	Trypsin 1:150		Powd. Starch Type I	232-940-4	Maltrin® M180
232-658-1	Agar Agar NF Flake #1	232-679-6	Staley® Pure Food	232-940-4	Maltrin® M510
232-658-1	Cameo Velvet		Powd. Starch Type II	232-940-4	Maltrin® M520
232-658-1	Cameo Velvet WT	232-679-6	Staley® Redried Starch	232-940-4	Maltrin® M700
232-658-1	Nutriloid® Agar		A	232-940-4	Maltrin® QD M440
232-658-1	Powdered Agar Agar	232-679-6	Staley® Redried Starch	232-940-4	Maltrin® QD M500
	NF M-100 (Gracilaria)		B	232-940-4	Maltrin® QD M550
232-658-1	Powdered Agar Agar	232-679-6	Sta-Rx®	232-940-4	Maltrin® QD M580

EINECS NUMBER-TO-TRADE NAME CROSS-REFERENCE

EINECS	Trade name	EINECS	Trade name	EINECS	Trade name
232-940-4	Microduct®	245-261-3	Equal®	247-706-7	Ryoto Sugar Ester P-1570S
232-940-4	Staley® Maltodextrin 3260	245-261-3	HSC Aspartame	247-706-7	Ryoto Sugar Ester P-1670
232-940-4	Star-Dri® 1	245-261-3	NutraSweet®		
232-940-4	Star-Dri® 5	246-376-1	Eastman Potassium Sorbate	247-887-2	Emalex GMS-P
232-940-4	Star-Dri® 10	246-376-1	Tristat K	247-891-4	Alkamuls® STS
232-940-4	Star-Dri® 15	246-563-8	Sustane® 1-F	247-891-4	Crill 35
232-940-4	Star-Dri® 18	246-563-8	Sustane® BHA	247-891-4	Crill 41
232-940-4	Star-Dri® 1005A	246-680-4	Calsoft L-60	247-891-4	Glycomul® TS KFG
232-940-4	Star-Dri® 1015A	246-680-4	Rhodacal® DS-4	247-891-4	Hodag STS
232-940-4	Wickenol® 550	246-680-4	Rhodacal® LDS-10	247-891-4	Lamesorb® STS
232-983-9	CK-20L	246-680-4	Sul-fon-ate AA-9	247-891-4	Liposorb TS
233-343-1	Calgon	246-680-4	Sul-fon-ate AA-10	247-891-4	Prote-sorb STS
233-343-1	Calgon® T Powd. Food Grade	246-680-4	Sul-fon-ate LA-10	247-891-4	Span® 65
233-343-1	Glass H®	246-705-9	Ryoto Sugar Ester S-1170	247-891-4	Span® 65K
233-343-1	Hexaphos®	246-705-9	Ryoto Sugar Ester S-1170S	248-294-1	Cremophor® NP 10
233-343-1	Sodaphos®			248-317-5	Ryoto Sugar Ester S-570
233-343-1	Vitrafos®	246-705-9	Ryoto Sugar Ester S-1570	248-317-5	Ryoto Sugar Ester S-770
233-343-1	Vitrafos® LC	246-705-9	Ryoto Sugar Ester S-1670		
234-316-7	Caprol® 10G10O	246-705-9	Ryoto Sugar Ester S-1670S	248-317-5	Ryoto Sugar Ester S-970
234-316-7	Drewpol® 10-10-O	246-873-3	Ryoto Sugar Ester L-1570	248-403-2	Caprol® 3GS
234-316-7	Drewpol® 10-10-OK			248-403-2	Drewmulse® 3-1-S
234-316-7	Hodag PGO-1010	246-873-3	Ryoto Sugar Ester L-1695	248-403-2	Drewpol® 3-1-SK
234-316-7	Polyaldo® DGDO			248-403-2	Hefti GMS-333
234-316-7	Polyaldo® DGDO KFG	246-873-3	Ryoto Sugar Ester LWA-1570	248-403-2	Hodag PGS
234-316-7	Santone® 10-10-O			248-403-2	Polyaldo® TGMS
234-394-2	Alginade XK9	246-929-7	Admul SSL 2003	248-403-2	Polyaldo® TGMS KFG
234-394-2	Dariloid® 400	246-929-7	Admul SSL 2004	248-403-2	Santone® 3-1-S
234-394-2	K2B387	246-929-7	Artodan SP 55 Kosher	248-403-2	Santone® 3-1-S XTR
234-394-2	K8B249	246-929-7	Atlas B-60	248-403-2	Triodan 55
234-394-2	Keltrol®	246-929-7	Atlas SSL	250-097-0	Compritol 888 ATO
234-394-2	Keltrol® 1000	246-929-7	Atlas SSL	252-011-7	Caprol® 10G4O
234-394-2	Keltrol® BT	246-929-7	Crolactil SS2L	252-011-7	Drewmulse® 10-4-O
234-394-2	Keltrol® CR	246-929-7	Crolactil SS2L	252-011-7	Drewpol® 10-4-O
234-394-2	Keltrol® F	246-929-7	Emplex	252-011-7	Drewpol® 10-4-OK
234-394-2	Keltrol® CM	246-929-7	Emulsilac S	252-011-7	Hodag PGO-104
234-394-2	Keltrol® RD	246-929-7	Emulsilac S K	252-011-7	Mazol® PGO-104
234-394-2	Keltrol® SF	246-929-7	Lamegin® NSL	252-487-6	Nipagin A Sodium
234-394-2	Keltrol® T	246-929-7	Paniplex SK	252-488-1	Nipasol M Sodium
234-394-2	Keltrol® TF	246-929-7	Pationic® 920	253-387-5	Bromelain 150 GDU
234-394-2	KOB87	246-929-7	Prefera® NSL	253-387-5	Bromelain 600 GDU
234-394-2	KOB348	246-929-7	Prefera® SSL 6000 MB	253-387-5	Bromelain 1200 GDU
234-394-2	KOB349	246-929-7	Swedex SSL-5AC	253-387-5	Bromelain 1500 GDU
234-394-2	Merecol® I	247-144-2	Cithrol GDO N/E	254-495-5	Caprol® JB
234-394-2	Merecol® MS	247-568-8	Ablunol S-40	254-495-5	Drewmulse® 10-10-S
234-394-2	Merezan® 8	247-568-8	Crill 2	254-495-5	Hodag PGS-1010
234-394-2	Merezan® 20	247-568-8	Hodag SMP	261-678-3	Frescolat, Type ML
234-394-2	Nutriloid® Xanthan	247-568-8	Lamesorb® SMP	263-027-9	Drewmulse® 75
234-394-2	Rhodigel®	247-568-8	Nikkol SP-10	263-027-9	Imwitor® 928
234-394-2	Rhodigel® 200	247-568-8	Nissan Nonion PP-40R	263-031-0	Monomuls® 90-25
234-394-2	Rhodigel® EZ	247-568-8	Prote-sorb SMP	263-032-6	Dimodan P
234-394-2	Rhodigel® Granular	247-568-8	Span® 40	263-032-6	Dimodan S
234-394-2	Rhodigel® Supra	247-569-3	Ablunol S-85	263-032-6	Monomuls® 90-10
234-394-2	Ticaxan® Regular	247-569-3	Crill 45	263-032-6	Myverol® 18-40
234-394-2	TIC Pretested® Xanthan 200	247-569-3	Hodag STO	263-035-2	Dimodan TH
		247-569-3	Nissan Nonion OP-85R	263-035-2	Monomuls® 90-20
234-394-2	TIC Pretested® Xanthan PH	247-569-3	Prote-sorb STO	263-035-2	Myverol® 18-30
		247-569-3	S-Maz® 85K	263-129-3	Industrene® 143
235-777-7	Emalex MSG-2	247-569-3	Span® 85	264-038-1	Be Square® 175
235-777-7	Emalex MSG-2MA	247-667-6	Imwitor® 910	264-038-1	Be Square® 185
235-777-7	Emalex MSG-2MB	247-668-1	Imwitor® 988	264-038-1	Be Square® 195
235-777-7	Emalex MSG-2ME	247-669-7	Cithrol PGMR N/E	264-038-1	Mekon® White
235-777-7	Emalex MSG-2ML	247-706-7	Ryoto Sugar Ester P-170	264-038-1	Multiwax® 180-M
235-777-7	Nikkol DGMS			264-038-1	Multiwax® ML-445
236-675-5	Hilton Davis Titanium Dioxide	247-706-7	Ryoto Sugar Ester P-1570	264-038-1	Multiwax® W-445
236-935-8	Captex® 8227			264-038-1	Multiwax® W-835

EINECS	Trade name	EINECS	Trade name	EINECS	Trade name
264-038-1	Multiwax® X-145A	269-804-9	Diamond D 75	269-820-6	Sterotex® NF
264-038-1	Petrolite® C-1035	269-804-9	Duratex	269-820-6	Wecobee® FS
264-038-1	Starwax® 100	269-804-9	Duromel	269-820-6	Wecobee® FW
264-038-1	Ultraflex®	269-804-9	Duromel B108	269-820-6	Wecobee® HTR
264-038-1	Victory®	269-804-9	07 Stearine	269-820-6	Wecobee® M
265-724-3	Aldo® MCT KFG	269-820-6	Aratex	269-820-6	Wecobee® R Mono
265-724-3	Delios® S	269-820-6	Cirol	269-820-6	Wecobee® S
265-724-3	Delios® V	269-820-6	CLSP 874	269-820-6	Wecobee® SS
265-724-3	Estasan GT 8-60 3575	269-820-6	Creamtex	269-820-6	Wecobee® W
265-724-3	Estasan GT 8-60 3580	269-820-6	Diamond D 40	270-331-5	Querton 210Cl-50
265-724-3	Estasan GT 8-65 3577	269-820-6	Diamond D 42	270-331-5	Querton 210Cl-80
265-724-3	Estasan GT 8-65 3581	269-820-6	Durkex 500	270-350-9	Witarix® 450
265-724-3	Hodag CC-33	269-820-6	Durlite F	271-397-8	Rice Bran Oil
265-724-3	Hodag CC-33-F	269-820-6	Hydrol 110	271-397-8	Rice Bran Oil SO
265-724-3	Hodag CC-33-L	269-820-6	Kaokote F	273-576-6	Lamegin® GLP 10, 20
265-724-3	Hodag CC-33-S	269-820-6	Kaola	273-576-6	Tegin® L 61, L 62
265-724-3	Lexol® GT-855	269-820-6	Kaola D	273-613-6	Lamegin® ZE 30, 60
265-724-3	Lexol® GT-865	269-820-6	Kaomax 870	273-613-6	Tegin® C-62 SE
265-724-3	Liponate GC	269-820-6	Kaomel	273-613-6	Tegin® C-63 SE
265-724-3	Miglyol® 810	269-820-6	Kaorich	275-117-5	Kelcogel®
265-724-3	Miglyol® 810 N	269-820-6	K.L.X	275-117-5	Kelcogel® BF
265-724-3	Miglyol® 812	269-820-6	K.L.X. Flakes	275-117-5	Kelcogel® BF10
265-724-3	Miglyol® 812 N	269-820-6	Lipo SS	275-117-5	Kelcogel® CF
265-724-3	Miglyol® 812 S	269-820-6	Lipodan SET Kosher	275-117-5	Kelcogel® CF10
265-724-3	Miglyol® 8108	269-820-6	Magna A	275-117-5	Kelcogel® F
265-724-3	Neobee® 1053	269-820-6	Magna B	275-117-5	Kelcogel® IF
265-724-3	Neobee® M-5	269-820-6	Magna C	275-117-5	Kelcogel® JJ
265-724-3	Neobee® O	269-820-6	Optima 7B	275-117-5	Kelcogel® PD
265-724-3	Standamul® 318	269-820-6	Optima 23B	284-283-8	Cobee 92
266-533-8	Crolactil SISL	269-820-6	Optima 77IC	284-283-8	Cobee 110
267-821-6	Emalex DISG-2	269-820-6	Optima 871	284-283-8	Hydrol 92
269-657-0	Industrene® 225	269-820-6	Paramount H	284-283-8	Hydrol 100
269-657-0	Industrene® 226 FG	269-820-6	Paramount X	284-283-8	Pureco® 92
269-658-6	Monomuls® 60-25	269-820-6	Paramount XX	284-283-8	Pureco® 100
269-804-9	Akoleno D	269-820-6	Snac-Kote XTR	284-283-8	Special Fat 42/44
269-804-9	Akolizer C	269-820-6	Sta-Nut R	289-610-8	Beetroot Conc. P3003

EINECS Number-to-Chemical
Cross-Reference

EINECS	Chemical	EINECS	Chemical	EINECS	Chemical
200-001-8	Formaldehyde	200-579-1	Formic acid	201-219-6	α-Irone
200-014-9	Ergocalciferol	200-580-7	Acetic acid	201-224-3	β-Ionone
200-018-0	Lactic acid	200-617-7	N-Acetyl-L-methionine	201-228-5	Retinyl palmitate
200-061-5	Sorbitol	200-618-2	Benzoic acid	201-296-2	L-Lactic acid
200-066-2	L-Ascorbic acid	200-624-5	Hexanal	201-321-0	Saccharin
200-075-1	D-Glucose anhyd.	200-641-8	Thiamine HCl	201-327-3	D-Panthenol
200-075-1	D-Glucose monohy-drate	200-655-4	Choline chloride	201-337-8	Cholic acid
		200-659-6	Methyl alcohol	201-471-7	Skatole
207-757-8	D-Glucose anhyd.	200-661-7	Isopropyl alcohol	201-507-1	Riboflavin
200-157-7	Cysteine hydrochloride	200-662-2	Acetone	201-550-6	Diethyl phthalate
200-158-2	L-Cysteine	200-673-2	Cholecalciferol	201-624-8	n-Butyl phthalyl-n-butyl glycolate
200-194-9	DL-α-Tryptophan	200-675-3	Sodium citrate		
200-272-2	Glycine	200-675-3	Trisodium citrate	201-642-6	Dimethyl anthranilate
200-273-8	L-Alanine	200-680-0	Cyanocobalamin	201-735-1	Ethyl anthranilate
200-274-3	L-Serine	200-683-7	Retinol	201-746-1	β-Caryophyllene
200-289-5	Glycerin	200-711-8	D-Mannitol	201-766-0	L-Tartaric acid
200-291-6	L-Aspartic acid	200-712-3	Salicylic acid	201-781-2	Inositol
200-292-1	L-Glutamine	200-716-5	Maltose	201-783-3	Diethyl tartrate
200-293-7	L-Glutamic acid	200-735-9	L-Asparagine anhyd.	201-788-0	Xylitol
200-294-2	L-Lysine	200-735-9	L-Asparagine monohydrate	201-790-4	Diethylacetic acid
200-296-3	L-Cystine			201-800-4	PVP
200-300-2	Allyl isothiocyanate	200-746-3	L-Histidine	201-803-0	2-Furoic acid
200-312-9	Palmitic acid	200-746-9	Propyl alcohol	201-928-0	Erythorbic acid
200-313-4	Stearic acid	200-751-6	Butyl alcohol	201-935-9	2,4-Dimethylaceto-phenone
200-315-5	Urea	200-752-1	n-Amyl alcohol		
200-333-3	D-Fructose	200-772-0	Sodium lactate	201-939-0	Menthol
200-334-9	Sucrose	200-773-6	L-Valine	201-943-2	Pulegone
200-338-0	Polypropylene glycol	200-774-1	L-Threonine	201-944-8	Thymol
200-338-0	Propylene glycol	200-795-6	L-Tryptophan	201-961-0	Salicylaldehyde
200-353-2	Cholesterol	200-798-2	L-Isoleucine	201-964-7	Guaiacol
200-362-1	Caffeine	200-811-1	L-Arginine	201-993-5	o-Phenylphenol
200-386-2	Pyridoxine HCl	200-815-3	Polyethylene	202-016-5	Gluconolactone
200-399-3	d-Biotin	200-822-1	Methyl mercaptan	202-051-6	Quinoline
200-400-7	D(+)-Xylose	200-827-9	Propane	202-075-7	6-Ethoxy-1,2-dihydro-2,2,4-trimethylquinoline
200-412-2	d-α-Tocopherol	200-836-8	Acetaldehyde		
200-419-0	Folicacid	200-838-9	Methylene chloride	202-163-5	Biphenyl
200-425-3	Thiamine	200-908-9	t-Amyl alcohol	202-216-2	2´-Acetonaphthone
200-432-1	DL-Methionine	200-857-2	Isobutane	202-223-0	Methyl eugenol
200-441-0	Nicotinic acid	201-001-0	Gibberellic acid	202-252-9	2-Methoxy-4-methylphenol
200-449-4	Edetic acid	201-066-5	Acetyl triethyl citrate		
200-451-5	Tributyrin	201-067-0	Acetyl tributyl citrate	202-256-0	1-Phenyl-1-propanol
200-456-2	Phenethyl alcohol	201-069-1	Citric acid	202-259-7	Methyl benzoate
200-460-4	L-Tyrosine	201-070-7	Triethyl citrate	202-284-3	Ethyl benzoate
200-470-9	Linoleic acid	201-134-4	Linalool	202-295-3	Ethyl benzoylacetate
200-522-0	L-Leucine	201-148-0	Isobutyl alcohol	202-307-7	Propylparaben
200-529-9	Calcium disodium EDTA	201-149-6	2-Methylpropanal	202-311-9	Benzylparaben
		201-159-0	Methyl ethyl ketone	202-318-7	Butylparaben
200-540-9	Calcium acetate	201-164-8	Pyruvaldehyde	202-327-6	Benzoyl peroxide
200-559-2	Lactose	201-167-4	Trichloroethylene	202-345-4	Safrol
200-562-9	L-Methionine	201-176-3	Propionic acid	202-348-0	Piperine
200-568-1	L-Phenylalanine	201-185-2	Methyl acetate	202-485-6	2-Methylbutyraldehyde
200-573-9	Tetrasodium EDTA	201-195-7	Isobutyric acid	202-490-3	Diethyl ketone
200-578-6	Alcohol	201-209-1	2-Nitropropane	202-500-6	Methyl acrylate polymer

EINECS	Chemical	EINECS	Chemical	EINECS	Chemical
202-509-5	Butyrolactone	203-378-7	Nerol	203-937-5	2-Undecanone
202-589-1	Eugenol	203-382-9	Ethyl heptanoate	203-942-2	Decyl acetate
202-589-1	Isoeugenol	203-385-5	Ethyl octanoate	203-956-9	n-Decyl alcohol
202-595-4	Ethyl isobutyrate	203-386-0	Ethyl laurate	203-957-4	Decanal
202-597-5	Ethyl methacrylate	203-388-1	3-Heptanone	203-965-8	Undecylenic acid
202-599-6	Itaconic acid	203-389-7	Propyl propionate	203-968-4	Dodecene-1
	(polymerized)	203-398-6	p-Cresol	203-970-5	Undecyl alcohol
202-601-5	N-Hydroxysuccinic acid	203-419-9	Dimethyl succinate	203-982-0	Lauryl alcohol
202-612-5	Isobutyl isobutyrate	203-423-0	3-Octanone	203-983-6	Lauric aldehyde
202-623-5	2-Ethylbutyraldehyde	203-425-1	Methyl caproate	203-989-9	PEG-4
202-625-6	Tetrahydrofurfuryl	203-428-8	Methyl heptanoate	203-990-4	Methyl stearate
	alcohol	203-448-7	Butane	203-993-0	Methyl linoleate
202-626-1	Furfuryl alcohol	203-453-4	Acrolein	204-007-1	Oleic acid
202-627-7	Furfural	203-455-5	Propyl mercaptan	204-017-6	Stearyl alcohol
202-628-2	Furfural mercaptan	203-458-1	Ethylene dichloride	204-068-4	2-Methyl-3-buten-2-ol
202-680-6	Terpineol anhydrous	203-468-6	Ethylenediamine	204-116-4	Linalyl acetate
202-708-7	Acetophenone	203-473-3	Polyethylene glycol	204-211-0	Dioctyl phthalate
202-713-4	Niacinamide	203-518-7	Hydroxycitronellal	204-262-9	Benzylsalicylate
202-785-7	Methylparaben	203-528-1	Methyl propyl ketone	204-265-5	Ethyl salicylate
202-794-6	γ-Terpinene	203-529-7	Butylene glycol	204-271-8	Maltol
202-795-1	α-Terpinene	203-532-3	n-Butyric acid	204-317-7	Methyl salicylate
202-796-7	p-Cymene	203-550-1	Methyl isobutyl ketone	204-331-3	Benzoin
202-845-2	Diethylaminoethanol	203-561-1	Isopropyl acetate	204-337-6	Benzophenone
202-851-5	Styrene	203-564-8	Acetic anhydride	204-341-8	Isoquinoline
202-859-9	Benzyl alcohol	203-569-5	γ-Valerolactone	204-354-9	Dihydrocoumarin
202-860-4	Benzaldehyde	203-570-0	Succinic anhydride	204-373-2	Veratraldehyde
202-862-5	Benzyl mercaptan	203-602-3	Ethyl isovalerate	204-399-4	Ethyl paraben
202-876-1	Anisole	203-615-4	Melamine	204-402-9	Benzyl benzoate
202-905-8	Hexamethylene	203-629-0	Cyclohexylamine	204-409-7	Heliotropine
	tetramine	203-635-3	Benzenethiol	204-420-7	Indole
202-940-9	Methyl phenylacetate	203-656-8	Butyl butyrate	204-442-7	BHA
202-945-6	Phenylacetaldehyde	203-672-5	Dibutyl sebacate	204-464-7	Ethyl vanillin
	dimethyl acetal	203-677-2	n-Valeric acid	204-465-2	Vanillin
202-993-8	Ethyl phenylacetate	203-686-1	Propyl acetate	204-498-2	Propyl gallate
203-000-0	1,3-Diphenyl-2-	203-713-7	Methoxyethanol	204-514-8	4´-Methyl aceto-
	propanone	203-721-0	Ethyl formate		phenone
203-049-8	Triethanolamine	203-724-7	Pyrrole	204-516-9	Cuminaldehyde
203-051-9	Triacetin	203-740-4	Succinic acid	204-526-3	Cetalkonium chloride
203-093-8	Methyl cinnamate	203-743-0	Fumaric acid	204-527-9	Stearalkonium chloride
203-104-6	Ethyl cinnamate	203-745-1	Isobutyl acetate	204-534-7	Triolein
203-109-3	Benzyl cinnamate	203-755-6	Ethylene distearamide	204-555-1	Benzylidene acetone
203-113-5	2-Phenylethyl acetate	203-761-9	Ethyl decanoate	204-556-7	Phenoxyacetic acid
203-118-2	Benzyl ether	203-764-5	Diethyl sebacate	204-574-5	Phenylacetaldehyde
203-121-9	Cinnamyl acetate	203-767-1	Methyl n-amyl ketone	204-587-6	Hydrocinnamic alcohol
203-148-6	Phenylacetic acid	203-768-7	Sorbic acid	204-602-6	p-Anisaldehyde
203-161-7	Cyclamen aldehyde	203-777-6	Hexane	204-608-9	4-Heptanone
203-205-5	Anethole	203-784-4	n-Valeraldehyde	204-612-0	Diethyl succinate
203-211-8	Hydrocinnamaldehyde	203-798-0	Propyl formate	204-614-1	Dilauryl thiodipropio-
203-212-3	Cinnamyl alcohol	203-806-2	Cyclohexane		nate
203-213-9	Cinnamal	203-809-9	Pyridine	204-615-7	Ethyl pelargonate
203-219-1	γ-Nonalactone	203-813-0	Piperidine	204-618-3	2,5-Dimethylpyrazine
203-225-4	ς-Undecalactone	203-815-1	Morpholine	204-623-0	Propionaldehyde
203-246-9	p-Tolyl aldehyde	203-816-7	Methyl heptenone	204-633-5	Isoamyl alcohol
203-253-7	p-Methylanisole	203-827-7	Glyceryl oleate	204-634-0	Acetylacetone
203-273-6	Anisyl alcohol	203-835-0	Methyl caprylate	204-640-3	Ethyl caproate
203-291-4	Ethyl propionate	203-836-6	Methyl 2-octynoate	204-642-4	Allyl caproate
203-305-9	Diethyl malonate	203-837-1	Methyl hexyl ketone	204-646-6	n-Butyraldehyde
203-306-4	Ethyl butyrate	203-841-8	3,3´-Thiodipropionic	204-649-2	Levulinic acid
203-310-6	Acetal		acid	204-650-8	Azodicarbonamide
203-320-0	Propyl butyrate	203-852-3	Hexyl alcohol	204-658-1	n-Butyl acetate
203-341-5	Geranyl acetate	203-856-5	Glutaral	204-662-3	Isoamyl acetate
203-347-8	Ethylene brassylate	203-889-5	Ethyl oleate	204-664-4	Glyceryl stearate
203-354-6	Pentadecalactone	203-897-9	Heptyl alcohol	204-666-5	Butyl stearate
203-370-3	Butyl laurate	203-898-4	Heptanal	204-673-3	Adipic acid
203-374-5	3,7-Dimethyl-1-octanol	203-905-0	Butoxyethanol	204-675-4	Ethyl myristate
203-375-0	β-Citronellol	203-911-3	Methyl laurate	204-677-5	Caprylic acid
203-376-6	Citronellal	203-917-6	Caprylic alcohol	204-680-1	Methyl myristate
203-377-1	Geraniol	203-931-2	Nonanoic acid	204-683-8	n-Octanal

EINECS	Chemical	EINECS	Chemical	EINECS	Chemical
204-688-5	Nonanal	205-749-9	Gallic acid	209-183-3	Polyvinyl alcohol
204-692-7	Myristaldehyde	205-756-7	DL-Phenylalanine	209-235-5	4-Carvomenthenol
204-693-2	Stearamide	205-764-0	Benzyl disulfide	209-406-4	Dioctyl sodium
204-695-3	Stearamine	205-771-9	p-Dimethoxybenzene		sulfosuccinate
204-696-9	Carbon dioxide	205-775-0	Citronellyl acetate	209-481-3	Potassium benzoate
204-771-6	Sucrose acetate	205-783-4	m-Dimethoxybenzene	209-526-7	3-Pentanol
	isobutyrate	205-788-1	Sodium lauryl sulfate	209-529-3	Potassium carbonate
204-772-1	Sucrose octaacetate	206-059-0	Potassium bicarbonate	209-567-0	Maltitol
204-812-8	Sodium octyl sulfate	206-074-2	Potassium D-gluconate	209-661-1	3-Heptanol
204-814-9	Sodium diacetate	206-075-8	Calcium gluconate	209-667-4	3-Octanol
204-822-2	Potassium acetate	206-076-3	Ferrous gluconate	209-691-5	Isovaleraldehyde
204-823-8	Sodium acetate anhyd.	206-101-8	Aluminum distearate	209-732-7	4-Pentenoic acid
204-823-8	Sodium acetate	206-103-9	Oleamide	209-753-1	1-Hexene
	trihydrate	206-126-4	DL-Alanine	209-772-5	Butyl formate
204-824-3	Pyruvic acid	206-130-6	DL-Serine	209-775-1	Allyl sulfide
204-840-0	Xanthophyll	206-156-8	Potassium sodium	209-786-1	Potassium stearate
204-841-6	α-Ionone		tartrate anhyd.	209-954-4	DL-Lactic acid
204-844-2	Retinyl acetate	206-156-8	Potassium sodium	209-982-7	2-Methylbutyric acid
204-876-7	Sodium dimethyldithio-		tartrate tetrahydrate	209-984-8	Pentane-2,3-dione
	carbamate hydrate	206-376-4	Capric acid	210-073-2	α-D-Glucose
204-881-4	BHT	206-496-7	3-Acetylpyridine		pentaacetate
204-886-1	Saccharin sodium	206-585-0	Perfluorohexane	210-472-1	1-Penten-3-ol
204-988-6	Riboflavin-5′-phosphate	206-643-5	Ox bile extract	210-511-2	Ethyl pyruvate
	sodium	206-988-1	Sodium palmitate	210-513-3	DL-Aspartic acid
205-003-2	Quinine	207-069-8	Diacetyl	210-702-0	Tricaprin
205-055-6	Sodium o-phenyl-	207-139-8	DL-Isoleucine	210-736-6	Cyclohexyl acetate
	phenate	207-280-5	Turmeric	210-792-1	Methyl butyrate
205-126-1	Sodium ascorbate	207-355-2	Camphor	210-808-7	Ethyl crotonate
205-132-4	Methyl anthranilate	207-431-5	Eucalyptol	210-819-7	Diethyl fumarate
205-171-7	o-Methoxybenz-	207-439-9	Calcium carbonate	210-838-0	Methyl valerate
	aldehyde	207-734-2	Lepidine	210-871-0	Methyl disulfide
205-278-9	Calcium pantothenate	207-838-8	Sodium carbonate	211-047-3	Amyl acetate
205-290-4	Sodium propionate	207-889-6	Carvacrol	211-079-8	Propyl disulfide
205-305-4	Ascorbyl palmitate	207-903-0	Nordihydroguairetic	211-082-4	Sodium laurate
205-315-9	L-Glutamic acid		acid	211-270-5	Aluminum tristearate
	hydrochloride	207-924-5	Hydrocinnamic acid	211-438-9	L-Histidine hydrochlo-
205-341-0	dl Limonene	207-975-3	Isovaleric acid		ride monohydrate
205-351-5	Lauralkonium chloride	208-174-1	Acetyl methyl carbinol	211-519-9	L-Lysine hydrochloride
205-352-0	Myristalkonium chloride	208-187-2	Canthaxanthine	211-750-5	Distearyl thiodipropio-
205-358-3	Disodium EDTA	208-293-9	Dehydroacetic acid		nate
205-398-1	Cinnamic acid	208-376-3	Calcium formate	211-889-1	D-Decalactone
205-399-7	Benzyl acetate	208-401-4	D-Gluconic acid	212-154-8	Methyl cyclopenteno-
205-438-8	Ethyl acrylate	208-407-7	Sodium gluconate		lone
205-447-7	Ferrous fumarate	208-408-2	Copper gluconate	212-227-4	Dehydroacetic acid
205-450-3	Diisobutyl adipate	208-534-8	Sodium benzoate	212-391-7	Calcium citrate
205-455-0	Glyceryl ricinoleate	208-537-4	Thiamine nitrate	212-406-7	Calcium lactate
205-459-2	Neryl acetate	208-580-9	Sodium sesquicarbon-	212-480-0	2-Nonanone
205-483-3	Ethanolamine		ate	212-487-9	Sodium myristate
205-488-0	Sodium formate	208-639-9	Perillyl alcohol	212-490-5	Sodium stearate
205-500-4	Ethyl acetate	208-686-5	Tricaprylin	212-728-8	FD&C Blue No. 2
205-516-1	Ethylacetoacetate	208-687-0	Trilaurin	212-755-5	Potassium citrate
205-526-6	Glyceryl laurate	208-726-1	Ethyl valerate	212-769-1	Potassium acid tartrate
205-538-1	MSG	208-728-2	Ethyl levulinate	212-773-3	Sodium tartrate
205-540-2	cis-Trimethyldodeca-	208-739-2	Amyl butyrate	212-778-0	Methyl 2-methylbutyrate
	trieneol	208-768-0	L-Carnitine	212-792-7	Allyl mercaptan
205-542-3	Propylene glycol laurate	208-863-7	Calcium formate	212-819-2	Decene-1
205-550-7	Caproic acid	208-870-5	Butyl sulfide	213-149-3	Magnesium glycero-
205-572-7	Hexyl acetate	208-875-2	Myristic acid		phosphate
205-582-1	Lauric acid	208-915-9	Magnesium carbonate	213-191-2	2-Hexenol
205-583-7	Nonyl alcohol	208-929-5	Methyl isobutyrate	213-192-8	3-Hexenol
205-585-8	n-Nonyl acetate	209-060-4	Methyl propionate	213-631-3	Potassium lactate
205-590-5	Potassium oleate	209-097-6	Tristearin	213-853-0	Octyl gallate
205-591-0	Sodium oleate	209-098-1	Tripalmitin	213-911-5	Ammonium bicarbonate
205-597-3	Oleyl alcohol	209-099-7	Trimyristin	214-016-2	2-Acetylpyrrole
205-633-8	Sodium bicarbonate	209-117-3	Methyl isovalerate	214-171-6	β-Apo-8′-carotenal
205-698-2	Potassium sodium	209-141-4	3-Methyl-2-buten-1-ol	214-304-8	Methyl eicosenate
	tartrate tetrahydrate	209-150-3	Magnesium stearate	214-355-6	2-Acetylpyridine
205-702-2	L-Proline	209-151-9	Zinc stearate	214-379-7	2-Phenyl propanol-1

EINECS	Chemical	EINECS	Chemical	EINECS	Chemical
214-413-0	Benzaldehyde dimethyl acetal	218-901-4	Glyceryl linoleate	231-158-0	Cobalt
214-519-7	β-Caryophyllene oxide	219-045-4	Ferric citrate	231-159-6	Copper
214-620-6	Dodecyl gallate	219-091-5	FD&C Green No. 3	231-165-9	Gold
214-686-6	Iron ammonium citrate	219-544-7	Ethyl fumarate	231-168-5	Helium
215-100-1	Sodium aluminate	220-045-1	PEG-6	231-195-2	Sulfur dioxide
215-108-5	Bentonite	220-120-9	Saccharin	231-211-8	Potassium chloride
215-137-3	Calcium hydroxide	220-491-7	FD&C Yellow No. 6	231-225-4	Ethyl 2-methylbutyrate
215-138-9	Calcium oxide	221-146-3	Ammonium citrate	231-298-2	Magnesium sulfate anhyd.
215-139-4	Calcium peroxide	221-882-5	3-Methylthiopropion-aldehyde	231-298-8	Magnesium sulfate heptahydrate
215-168-2	Ferric oxide	222-206-1	PEG-9		
215-170-2	Magnesium hydroxide	222-226-0	1-Octen-3-ol	231-326-3	Butyl butyryllactate
215-171-9	Magnesium oxide	222-848-2	Magnesium gluconate	231-368-2	N-Acetylglucosamine
215-181-3	Potassium hydroxide	223-095-2	Denatonium benzoate NF	231-437-7	Quinine hydrochloride
215-185-5	Sodium hydroxide			231-442-4	Iodine
215-222-5	Zinc oxide	223-269-8	Geranyl acetone	231-448-7	Sodium phosphate dibasic
215-279-6	Limestone	223-339-8	Brilliant blue FCF		
215-291-1	Potassium glycero-phosphate	223-339-8	FD&C Blue No. 1	231-448-7	Disodium phosphate, dihydrate
		223-350-8	Acetyl butyryl		
215-301-4	Manganese glycerophosphate	223-472-1	trans-2-Decenal	231-449-2	Sodium phosphate
		223-644-6	Linoleamide	231-493-2	Cyclodextrin
215-347-5	Sodium decylbenzene sulfonate	223-795-8	Calcium propionate	231-509-8	Sodium phosphate tribasic
		223-938-4	2-Octanol		
215-354-3	Propylene glycol stearate	223-979-8	Potassium fumarate	231-509-8	Sodium phosphate tribasic dodecahydrate
		224-403-8	Tocopheryl succinate		
215-359-0	Glyceryl distearate	224-580-1	Sodium dehydroacetate	231-545-4	Diatomaceous earth
215-475-1	Aluminum silicate	224-651-7	3,4-Hexanedione	231-548-0	Sodium bisulfite
215-540-4	Sodium borate	224-736-9	Zinc gluconate	231-554-3	Sodium nitrate
215-549-3	Propylene glycol oleate	224-859-8	Dimethyl dicarbonate	231-555-9	Sodium nitrite
215-609-9	Carbon black	224-909-9	FD&C Red No. 4	231-592-0	Zinc chloride
215-647-6	Ammonium hydroxide	225-004-1	Farnesol	231-595-7	Hydrochloric acid
215-654-4	Stearyl citrate	225-402-5	2-Dodecenal	231-598-3	Sodium chloride
215-663-3	Sorbitan laurate	225-668-2	Calcium lactobionate	231-633-2	Phosphoric acid
215-664-9	Sorbitan stearate	225-714-1	Sodium methylparaben	231-639-5	Sulfuric acid
215-665-4	Sorbitan oleate	225-768-6	Trisodium NTA	231-659-4	Potassium iodide
215-680-6	Cochineal	225-856-4	PEG-8	231-665-7	Sodium bisulfate
215-681-1	Magnesium silicate	226-132-0	Cyclohexaneacetic acid	231-673-0	Sodium metabisulfite
215-683-2	Silica, hydrated	226-394-6	Citral	231-674-6	Copper iodide (ous)
215-684-8	Sodium silicoaluminate	226-416-4	Benzyl acetoacetate	231-683-5	Natamycin
215-687-4	Sodium silicate	226-516-5	Calcium diacetate	231-694-5	Pentasodium triphosphate
215-691-6	Alumina	227-335-7	Calcium stearoyl lactylate	231-710-0	Tocopheryl acetate
215-710-6	Calcium silicate			231-710-0	dl-α-Tocopheryl acetate
215-724-4	Carmine	227-535-4	Sodium fumarate		
215-724-4	Carmine extract	227-608-0	Ferrous lactate	231-721-0	Disodium capryloam-phodiacetate
215-735-4	Annatto	227-729-9	Sorbitan laurate		
215-744-3	Chitin	227-813-5	d-Limonene	231-724-7	Magnesium fumarate
215-753-2	Tannic acid	227-902-9	Aluminum caprylate	231-729-4	Ferric chloride
215-798-8	Tocopherol	228-733-3	D-Histidine hydrochlo-ride monohydrate	231-753-5	Ferrous sulfate (dried)
215-800-7	Chlorophyll			231-753-5	Ferrous sulfate heptahydrate
215-807-5	Nisin	228-971-8	Ammonium saccharin		
215-824-8	2,4,5-Trihydroxybutyro-phenone	228-973-9	Sodium erythorbate	231-764-5	Ammonium phosphate
		229-349-9	Calcium saccharin	231-765-0	Hydrogen peroxide
216-087-5	Trifluoromethane sulfonic acid	229-350-4	Manganese gluconate	231-767-1	Tetrasodium pyrophosphate
		229-352-5	l-Carvone		
216-472-8	Calcium stearate	229-778-1	2-Hexenal	231-778-1	Bromine
216-639-5	Fenchyl alcohol	229-859-1	PEG-12	231-781-8	Potassium persulfate
217-052-7	Methyl nonanoate	229-912-9	Sodium metasilicate	231-783-9	Nitrogen
217-234-6	D-Aspartic acid	230-325-5	Aluminum stearate	231-786-5	Ammonium persulfate
217-699-5	FD&C Yellow No. 5	230-525-2	Didecyldimonium chloride	231-793-3	Zinc sulfate
217-699-5	Tartrazine			231-793-3	Zinc sulfate hepta-hydrate
217-752-2	t-Butyl hydroquinone	230-636-6	Carotene		
217-966-6	Ethyl-3-phenylpro-pionate	230-785-7	Tetrapotassium pyrophosphate	231-818-8	Potassium nitrate
		230-896-0	Trioctanoin	231-820-9	Sodium sulfate
218-080-2	Amyl salicylate	231-072-3	Aluminum	231-821-4	Sodium sulfite
218-235-4	Calcium benzoate	231-096-4	Iron	231-823-5	Magnesium phosphate dibasic trihydrate
218-548-6	Allyl disulfide	231-111-4	Nickel		
218-691-4	d-Neomenthol	231-131-3	Silver	231-824-0	Magnesium phosphate,
218-827-2	d-Carvone				

EINECS	Chemical	EINECS	Chemical	EINECS	Chemical
	tribasic	232-288-0	Balsam copaiba	232-577-1	Catalase
231-826-1	Calcium phosphate dibasic	232-289-6	Cod liver oil	232-578-7	Chitinase
		232-292-2	Hydrogenated castor oil	232-599-1	Ficin
231-826-1	Calcium phosphate dibasic dihydrate	232-293-8	Castor oil	232-601-0	Glucose oxidase
		232-296-4	Peanut oil	232-619-9	Lipase
231-829-8	Potassium bromate	232-298-5	Petroleum distillates	232-627-2	Papain
231-830-3	Potassium bromide	232-299-0	Rapeseed oil	232-629-3	Pepsin
231-831-9	Potassium iodate	232-304-6	Tall oil	232-650-8	Trypsin
231-832-4	Potassium nitrite	232-307-2	Lecithin	232-656-0	Urease
231-834-5	Potassium phosphate dibasic	232-310-9	Malt extract	232-658-1	Agar
		232-311-4	Menhaden oil	232-674-9	Cellulose
231-835-0	Sodium acid pyrophosphate	232-315-6	Paraffin	232-675-4	Dextrin
		232-316-1	Palm oil	232-677-5	Dextran
231-835-0	Disodium diphosphate	232-317-7	Wheat gluten	232-679-6	Corn starch
231-837-1	Calcium phosphate monobasic monohydrate	232-347-0	Candelilla wax	232-680-1	Alginic acid
		232-348-6	Lanolin	232-686-4	Starch
		232-352-8	Balsam Peru	232-722-9	Zein
231-840-8	Calcium phosphate tribasic	232-360-1	Sorbitan sesquioleate	232-732-3	Acylase
		232-362-2	Balsam Canada	232-734-4	Cellulase
231-847-6	Cupric sulfate anhyd.	232-370-6	Sesame oil	232-734-4	Glucanase from
231-847-6	Cupric sulfate pentahydrate	232-371-1	Garlic oil extract		*Bacillus subtilis*
		232-373-2	Petrolatum	232-752-2	Protease
231-867-5	Sodium thiosulfate anhyd.	232-383-7	Beeswax	232-792-0	α-Galactosidase
		232-384-2	Mineral oil	232-803-9	Dextranase
231-867-5	Sodium thiosulfate pentahydrate	232-391-0	Epoxidized soybean oil	232-853-1	Lipoxidase
		232-395-2	Fusel oil refined	232-877-2	Glucoamylase
231-868-0	Stannous chloride anhyd.	232-399-4	Carnauba	232-885-6	Pectinase
		232-400-8	Cascara sagrada extract	232-940-4	Maltodextrin
231-868-0	Stannous chloride dihydrate			232-943-0	Dismutase superoxide
		232-409-7	Rice bran wax	232-980-2	β-Glucanase from
231-869-6	Manganese chloride tetrahydrate	232-410-2	Hydrogenated soybean oil		*Aspergillus niger*
				232-983-9	Pullulanase
231-890-0	Sodium hydrosulfite	232-425-4	Palm kernel oil	233-007-4	Cyclodextrin
231-892-1	Sodium persulfate	232-428-0	Avocado oil	233-032-0	Nitrous oxide
231-896-3	Tristearyl citrate	232-433-8	Bitter orange peel extract	233-072-0	Ferric sulfate
231-900-3	Calcium sulfate			233-093-3	p-Ethoxybenzaldehyde
231-905-0	Potassium citrate monohydrate	232-435-0	Caramel	233-135-0	Aluminum sulfate
		232-436-4	Corn syrup	233-139-2	Boric acid
231-907-1	Potassium phosphate tribasic	232-438-5	Cottonseed glyceride	233-140-8	Calcium chloride
		232-440-6	Hydroxylated lecithin	233-141-3	Potassium alum dodecahydrate
231-913-4	Potassium phosphate	232-442-7	Hydrogenated tallow		
231-915-5	Potassium sulfate	232-454-2	Ambergris	233-149-7	Ferric phosphate
231-959-5	Chlorine	232-455-8	Mineral oil	233-162-8	Chlorine dioxide
231-971-0	Sodium amide	232-474-1	Olibanum	233-190-0	Ferric pyrophosphate
231-984-1	Ammonium sulfate	232-475-7	Rosin	233-277-3	Sodium alum
231-987-8	Ammonium phosphate, dibasic	232-479-9	Pentaerythrityl rosinate	233-321-1	Potassium sulfite
		232-501-7	Valerian extract	233-343-1	Sodium hexametaphosphate
232-055-3	Ammonium alum	232-519-5	Acacia		
232-077-3	(-)-α-Pinene	232-522-1	Asafetida	233-466-0	dl-α-Tocopherol
232-087-8	(+)-α-Pinene	232-523-7	Gum benzoin	233-484-9	Ammonium sulfite
232-088-3	Sodium trimetaphosphate	232-524-2	Carrageenan	233-713-2	D-Lactic acid
		232-528-4	Dammar	233-786-0	Ammonium carbonate
232-089-9	Manganese sulfate (ous)	232-531-0	Furcelleran	234-059-0	3-Decen-2-one
		232-536-8	Guar gum	234-257-7	Cedarwood oil terpenes
232-094-6	Magnesium chloride	232-539-4	Karaya gum	234-316-7	Polyglyceryl-10 decaoleate
232-094-6	Magnesium chloride hexahydrate	232-541-5	Locust bean gum		
		232-547-8	Sandarac gum	234-325-6	Glyceryl stearate
232-164-6	Calcium bromide	232-549-9	Shellac	234-328-2	Retinol
232-191-3	Calcium iodate	232-550-4	Balsam tolu	234-394-2	Xanthan gum
232-221-5	Calcium pyrophosphate	232-552-5	Tragacanth gum	235-186-4	Ammonium chloride
232-273-9	Sunflower seed oil	232-553-0	Pectin	235-192-7	Magnesium carbonate hydroxide
232-274-4	Soybean oil	232-554-6	Gelatin		
232-276-5	Safflower oil	232-555-1	Milk protein	235-223-4	Ammonium sulfide
232-277-0	Olive oil	232-555-1	Casein	235-777-7	Polyglyceryl-2 stearate
232-278-6	Linseed oil	232-565-6	α-Amylase	236-675-5	Titanium dioxide
232-280-7	Cottonseed oil	232-566-1	β-Amylase	236-883-6	Methyl 3-methylthio-propionate
232-281-2	Corn oil	232-567-7	Amylase		
232-282-8	Coconut oil	232-575-0	Carbohydrase	236-935-8	Triundecanoin

EINECS	Chemical	EINECS	Chemical	EINECS	Chemical
237-081-9	Sodium ferrocyanide	261-678-3	Menthyl lactate	283-461-2	Palmarosa oil extract
237-178-6	5-Methyl-2-thiophene-carboxyaldehyde	262-710-9	Glyceryl isostearate	283-467-5	Chamomile extract
		262-978-7	Coconut acid	283-479-0	Cinnamon leaf oil extract
237-574-9	Potassium tripolyphosphate	262-988-1	Methyl cocoate	283-480-6	Cocoa extract
		263-027-9	Glyceryl cocoate		
238-877-9	Talc	263-031-0	Hydrogenated tallow glyceride	283-481-1	Coffee bean extract
239-018-0	Ferric phosphate tetrahydrate			283-491-6	Geranium extract (G. maculatum extract)
		263-032-6	Lard glyceride		
239-674-8	L-Arginine monohydrochloride	263-035-2	Tallow glyceride	283-492-1	Geranium extract
		263-099-1	Tallow	283-619-0	Chestnut leaf extract
240-029-8	2-Butanol	263-100-5	Lard	283-869-0	Calamus oil extract
240-458-0	Geranyl acetate	263-129-3	Tallow acid	283-871-1	Angelica extract
240-474-8	FD&C Red No. 3	263-163-9	Cocamide DEA	283-872-7	Anise extract
240-795-3	Potassium metabisulfite	264-038-1	Microcrystalline wax	283-873-2	Serpentaria extract
241-155-6	Sodium aspartate	264-705-7	Castor oil, dehydrated	283-880-0	Coriander oil extract
241-640-2	Cetyl esters	265-363-1	Hydrolyzed milk protein	283-881-6	Cumin extract
242-243-7	2-Ethyl fenchol	265-724-3	Caprylic/capric triglyceride	283-882-1	Turmeric extract
242-734-6	Sodium citrate			283-912-3	Dill seed oil extract, Indian type
243-376-3	Calcium fumarate	266-116-0	Corn meal		
244-029-9	Dihydrocarvyl acetate	266-124-4	Glyceryl isostearate	283-949-5	Calendula extract
244-492-7	Aluminum hydroxide	266-533-8	Sodium isostearoyl lactylate	283-952-1	Cinchona extract
245-261-3	Aspartame			283-993-5	Jasmine extract
245-625-1	Ferrous citrate	266-948-4	Triolein	284-283-8	Hydrogenated coconut oil
246-376-1	Potassium sorbate	267-821-6	Polyglyceryl-2 diisostearate		
246-563-8	BHA			284-284-3	Cascarilla oil extract
246-668-9	trans-trans-2,4-Decadienal	269-657-0	Soy acid	284-510-0	Myrrh extract
		269-658-6	Hydrogenated tallow glycerides	284-540-4	Allspice extract
246-680-4	Sodium dodecylbenzenesulfonate			284-545-1	Carrot extract
		269-668-0	Cottonseed flour	284-634-5	Carob extract
246-705-9	Sucrose stearate	269-804-9	Hydrogenated cottonseed oil	284-635-0	Cinnamon oil extract
246-873-3	Sucrose laurate			284-638-7	Clove extract
246-929-7	Sodium stearoyl lactylate	269-820-6	Hydrogenated vegetable oil	284-653-9	Tansy extract
				285-206-0	Ethyl oleate
247-144-2	Glyceryl dioleate	270-350-9	Hydrogenated peanut oil	285-356-7	Clover blossom extract
247-561-6	Sodium methylnaphthalenesulfonate			285-361-4	Yerba santa extract
		271-200-5	Hydrogenated corn oil	286-074-7	Sorbitan trioleate
247-568-8	Sorbitan palmitate	271-397-8	Rice bran oil	286-469-4	Alkanet extract
247-569-3	Sorbitan trioleate	271-792-5	Disodium capryloamphodiacetate	286-476-2	Barley extract
247-667-6	Glyceryl caprate			286-479-9	Costmary extract
247-668-1	Glyceryl caprylate	272-045-6	Chicory extract	286-490-9	Glyceryl stearate
247-669-7	Propylene glycol ricinoleate	272-826-1	Civet	287-824-6	Methyl stearate
		273-277-0	Ethylene distearamide	288-459-5	PEG-12 dioleate
247-706-7	Sucrose palmitate	273-313-5	Vegetable oil	288-920-0	Capsicum extract
247-887-2	Glyceryl palmitate	273-576-6	Hydrogenated tallow glyceride lactate	288-921-6	Caraway oil extract
247-891-4	Sorbitan tristearate			288-922-1	Cardamom oil extract
248-292-0	Nonoxynol-7	273-579-2	Arnica extract	289-561-2	Annatto (tree) extract
248-294-1	Nonoxynol-10	273-612-0	Acetylated hydrogenated tallow glyceride	289-574-3	Serpentaria extract
248-317-5	Sucrose distearate			289-610-8	Beet powder
248-328-5	Calcium glycerophosphate	273-613-6	Hydrogenated tallow glyceride citrate	289-612-9	Bergamot oil extract
				289-620-2	Olibanum oil extract
248-403-2	Polyglyceryl-3 stearate	273-616-2	Shark liver oil	289-668-4	Celery seed oil extract
248-486-5	Quaternium-14	273-620-4	Quillaja	289-695-1	Chives extract
250-097-0	Glyceryl behenate	273-624-6	Dandelion extract	289-707-5	Cinchona extract
250-705-4	Glyceryl stearate	275-117-5	Gellan gum	289-708-0	Cinchona extract
251-269-8	Butyl ethyl malonate	276-638-0	Tannic acid	289-709-6	Cinchona extract
252-011-7	Polyglyceryl-10 tetraoleate	276-951-2	Sorbitan tristearate	289-711-7	Labdanum extract
		277-139-0	Gentian extract	289-720-6	Cola nut extract
252-487-6	Sodium ethylparaben	281-659-3	Artichoke extract	289-741-0	Pumpkin seed oil extract
252-488-1	Sodium propylparaben	281-660-9	Birch oil extract		
252-700-2	Ferric sodium pyrophosphate	281-661-4	Borage seed oil extract	289-752-0	Lemongrass extract
		281-666-1	Elecampane extract	289-752-0	Lemongrass oil West Indian extract
252-964-9	Isocetyl alcohol	281-984-0	Alfalfa extract		
253-149-0	Cetyl alcohol	283-406-2	Eucalyptus extract (E. globulus)	289-766-7	Dittany extract
253-387-5	Bromelain			289-861-3	Oakmoss extract
253-407-2	Glyceryl oleate	283-414-6	Fennel extract	289-870-2	Fir needle oil extract
254-495-5	Polyglyceryl-10 decastearate	283-415-1	Fenugreek extract	289-890-1	Gentian extract
		283-457-0	Maidenhair fern extract (A. capillus-veneris)	289-891-7	Gentian extract
260-364-3	Caryophyllene alcohol			289-892-2	Gentian extract

EINECS NUMBER-TO-CHEMICAL CROSS-REFERENCE

EINECS	Chemical
289-960-1	Jasmine extract
290-249-3	Rhubarb extract
290-367-5	Germander extract
291-060-9	Bitter almond extract
291-068-2	Galanga extract
291-076-6	Amyris oil extract, West Indian type
292-344-5	Chirata extract
294-351-9	Fir needle oil extract, Siberian type
294-354-5	Angostura extract

EINECS	Chemical
294-458-0	Elder flowers (*S. canadensis* extract)
283-259-4	Elder flowers (*S. nigra* extract)
294-926-4	Boronia flowers extract
295-155-6	Davana extract
295-625-0	Glyceryl (triacetoxy-stearate)
296-192-0	Geranium extract
296-432-4	Castoreum (Canada) extract

EINECS	Chemical
296-473-8	Kaolin
297-382-6	Asafetida extract
297-599-6	Capsicum (*C. annuum* extract)
304-454-3	Citrus extract
304-815-5	Cinchona extract
305-181-2	Aloe extract
308-877-4	Black wattle extract
307-497-6	Geranium extract
363-737-7	Potassium glutamate monohydrate

FEMA Number-to-Chemical
Cross-Reference

FEMA	Chemical	FEMA	Chemical	FEMA	Chemical
2002	Acetal	2071	Isoamyl 2-furanpropionate	2153	Bergamot oil
2003	Acetaldehyde	2072	Amyl furoate	2156	Bois de rose
2004	Acetaldehyde phenethyl propyl acetal	2073	Amyl heptanoate	2157	Borneol
		2074	Amyl hexanoate	2158	Isoborneol
2005	Acetanisole	2075	Isoamyl hexanoate	2159	Bornyl acetate
2006	Acetic acid	2077	Isoamyl laurate	2160	Isobornyl acetate
2007	Triacetin	2078	Isoamyl nonanoate	2161	Bornyl formate
2008	Acetyl methyl carbinol	2079	Amyl octanoate	2162	Isobornyl formate
2009	Acetophenone	2081	Isoamyl phenylacetate	2163	Isobornyl propionate
2010	Aconitic acid	2082	Amyl propionate	2165	Bornyl valerate
2011	Adipic acid	2083	Isoamyl pyruvate	2166	Isobornyl isovalerate
2012	Agar	2084	Amyl salicylate	2170	Methyl ethyl ketone
2020	Allyl anthranilate	2085	Isoamyl isovalerate	2171	Butter acids
2021	Allyl butyrate	2086	Anethole	2172	Butter esters
2022	Allyl cinnamate	2096	Star anise oil	2174	n-Butyl acetate
2023	Allyl cyclohexaneacetate	2097	Anisole	2175	Isobutyl acetate
2024	Allyl cyclohexanebutyrate	2098	Anisyl acetate	2176	Butyl acetoacetate
2025	Allyl cyclohexanehexanoate	2099	Anisyl alcohol	2177	Isobutyl acetoacetate
2026	Allyl cyclohexanepropionate	2100	Anisyl butyrate	2178	Butyl alcohol
2027	Allyl cyclohexanevalerate	2101	Anisyl formate	2179	Isobutyl alcohol
2028	Allyl disulfide	2102	Anisyl propionate	2180	Isobutyl angelate
2029	Allyl 2-ethylbutyrate	2106	Asafetida extract	2181	Butyl anthranilate
2031	Allyl heptanoate	2107	Asafetida gum	2182	Isobutyl anthranilate
2032	Allyl caproate	2108	Asafetida oil	2183	BHA
2033	Allyl a-ionone	2109	L-Ascorbic acid	2184	BHT
2034	Allyl isothiocyanate	2126	Beeswax	2185	Isobutyl benzoate
2035	Allyl mercaptan	2127	Bitter almond oil	2186	Butyl butyrate
2036	Allyl nonanoate	2127	Benzaldehyde	2187	Isobutyl butyrate
2037	Allyl octanoate	2128	Benzaldehyde dimethyl acetal	2188	Butyl isobutyrate
2038	Allyl phenoxyacetate			2189	Isobutyl isobutyrate
2039	Allyl phenylacetate	2129	Benzaldehyde glyceryl acetal	2190	Butyl butyryllactate
2040	Allyl propionate			2191	α-Butylcinnamaldehyde
2041	Allyl sorbate	2130	Benzaldehyde propylene glycol acetal	2192	Butyl cinnamate
2042	Allyl sulfide			2193	Isobutyl cinnamate
2043	Allyl tiglate	2131	Benzoic acid	2194	Butyl 2-decenoate
2044	Allyl 10-undecenoate	2132	Benzoin	2195	Butyl ethyl malonate
2045	Allyl isovalerate	2134	Benzophenone	2196	Butyl formate
2048	Althea extract	2135	Benzyl acetate	2197	Isobutyl formate
2053	Ammonium sulfide	2136	Benzyl acetoacetate	2198	Isobutyl 2-furanpropionate
2054	Ammonium isovalerate	2137	Benzyl alcohol	2199	Butyl heptanoate
2055	Isoamyl acetate	2138	Benzyl benzoate	2200	Isobutyl heptanoate
2056	n-Amyl alcohol	2139	Benzyl butyl ether	2201	Butyl hexanoate
2057	Isoamyl alcohol	2140	Benzyl butyrate	2202	Isobutyl hexanoate
2058	Isoamyl benzoate	2141	Benzyl isobutyrate	2203	Butylparaben
2059	Amyl butyrate	2142	Benzyl cinnamate	2205	Butyl lactate
2060	Isoamyl butyrate	2144	Benzyl ethyl ether	2206	Butyl laurate
2061	α-Amylcinnamaldehyde	2145	Benzyl formate	2207	Butyl levulinate
2063	Isoamyl cinnamate	2147	Benzyl mercaptan	2208	α-Isobutylphenethyl alcohol
2065	α-Amylcinnamyl alcohol	2148	Benzyl methoxyethyl acetal	2209	Butyl phenylacetate
2066	α-Amylcinnamyl formate	2149	Benzyl phenylacetate	2210	Isobutyl phenylacetate
2067	α-Amylcinnamyl isovalerate	2150	Benzyl propionate	2211	n-Butyl propionate
2069	Isoamyl formate	2151	Benzylsalicylate	2212	Isobutyl propionate
2070	Isoamyl 2-furanbutyrate	2152	Benzyl isovalerate	2213	Isobutyl salicylate

FEMA	Chemical	FEMA	Chemical	FEMA	Chemical
2214	Butyl stearate	2335	Corn silk	2421	Ethyl anthranilate
2215	Butyl sulfide	2336	Costus root oil	2422	Ethyl benzoate
2216	Butyl 10-undecenoate	2337	p-Cresol	2423	Ethyl benzoylacetate
2217	Butyl valerate	2341	Cuminaldehyde	2424	α-Ethylbenzyl butyrate
2218	n-Butyl isovalerate	2347	Cyclohexaneacetic acid	2425	2-Ethylbutyl acetate
2219	n-Butyraldehyde	2348	Cyclohexylethyl acetate	2426	2-Ethylbutyraldehyde
2220	2-Methylpropanal	2349	Cyclohexyl acetate	2427	Ethyl butyrate
2221	n-Butyric acid	2350	Cyclohexyl anthranilate	2428	Ethyl isobutyrate
2222	Isobutyric acid	2351	Cyclohexyl butyrate	2429	Diethylacetic acid
2223	Tributyrin	2352	Cyclohexyl cinnamate	2430	Ethyl cinnamate
2224	Caffeine	2353	Cyclohexyl formate	2431	Ethyl cyclohexane-
2229	Camphene	2354	Cyclohexyl propionate		propionate
2230	Camphor	2355	Cyclohexyl isovalerate	2432	Ethyl decanoate
2235	Caramel	2356	p-Cymene	2434	Ethyl formate
2245	Carvacrol	2357	Dandelion extract	2435	Ethyl-2-furanpropionate
2246	Carvacryl ethyl ether	2358	Dandelion root	2436	4-Ethylguaiacol
2247	Carveol	2359	Davana	2437	Ethyl heptanoate
2248	4-Carvomenthenol	2360	γ-Decalactone	2438	2-Ethyl-2-heptenal
2249	d-Carvone	2361	Δ-Decalactone	2439	Ethyl caproate
2249	l-Carvone	2362	Decanal	2440	Ethyl lactate
2250	Carvyl acetate	2363	Decanal dimethyl acetal	2441	Ethyl laurate
2251	Carvyl propionate	2365	n-Decyl alcohol	2442	Ethyl levulinate
2252	β-Caryophyllene	2366	trans-2-Decenal	2443	Ethyl 2-methylbutyrate
2253	Cascara sagrada	2367	Decyl acetate	2444	Ethyl methylphenylglycidate
2260	Cassie flowers	2368	Decyl butyrate	2445	Ethyl myristate
2261	Castoreum extract	2369	Decyl propionate	2446	Ethyl nitrite
2262	Castoreum liq.	2370	Diacetyl	2447	Ethyl pelargonate
2263	Castor oil	2371	Benzyl ether	2448	Ethyl 2-nonynoate
2267	Cedar leaf oil	2373	Dibutyl sebacate	2449	Ethyl octanoate
2271	Celery seed oil	2374	Diethyl malate	2450	Ethyl oleate
2276	Cherry, wild, bark	2375	Diethyl malonate	2452	Ethyl phenylacetate
2278	Cherry pit extract	2376	Diethyl sebacate	2454	Ethyl phenylglycidate
2279	Chervil	2377	Diethyl succinate	2455	Ethyl-3-phenylpropionate
2286	Cinnamal	2378	Diethyl tartrate	2456	Ethyl propionate
2287	Cinnamaldehyde ethylene	2379	Dihydrocarveol	2457	Ethyl pyruvate
	glycol acetal	2380	Dihydrocarvyl acetate	2458	Ethyl salicylate
2288	Cinnamic acid	2381	Dihydrocoumarin	2459	Ethyl sorbate
2293	Cinnamyl acetate	2383	Dill seed oil	2460	Ethyl tiglate
2294	Cinnamyl alcohol	2385	m-Dimethoxybenzene	2461	Ethyl 10-undecenoate
2295	Cinnamyl anthranilate	2386	p-Dimethoxybenzene	2462	Ethyl valerate
2296	Cinnamyl butyrate	2387	2,4-Dimethylacetophenone	2463	Ethyl isovalerate
2297	Cinnamyl isobutyrate	2388	α,α-Dimethylbenzyl	2464	Ethyl vanillin
2299	Cinnamyl formate		isobutyrate	2465	Eucalyptol
2300	Cinnamyl phenylacetate	2389	2,6-Dimethyl-5-heptenal	2466	Eucalyptus oil
2301	Cinnamyl propionate	2390	2,6-Dimethyl octanal	2467	Eugenol
2302	Cinnamyl isovalerate	2391	3,7-Dimethyl-1-octanol	2468	Isoeugenol
2303	Citral	2392	Dimethylbenzyl carbinyl	2469	Eugenyl acetate
2304	Citral diethyl acetate		acetate	2470	Acetisoeugenol
2305	Citral dimethyl acetal	2393	Dimethylbenzyl carbinol	2471	Eugenyl benzoate
2306	Citric acid	2394	α,α-Dimethylphenethyl	2472	Isoeugenyl ethyl ether
2307	Citronellal		butyrate	2473	Eugenyl formate
2309	β-Citronellol	2396	Dimethyl succinate	2474	Isoeugenyl formate
2310	Citronelloxyacetaldehyde	2397	1,3-Diphenyl-2-propanone	2475	Methyl eugenol
2311	Citronellyl acetate	2400	γ-Dodecalactone	2476	Methyl isoeugenol
2312	Citronellyl butyrate	2401	δ-Dodecalactone	2476	Isoeugenyl methyl ether
2313	Citronellyl isobutyrate	2402	2-Dodecenal	2477	Isoeugenyl phenylacetate
2314	Citronellyl formate	2411	Estragole	2478	Farnesol
2315	Citronellyl phenylacetate	2413	p-Ethoxybenzaldehyde	2479	d-Fenchone
2316	Citronellyl propionate	2414	Ethyl acetate	2480	Fenchyl alcohol
2317	Citronellyl valerate	2415	Ethylacetoacetate	2481	Fennel extract
2319	Civet	2416	Ethyl 2-acetal-3-	2487	Formic acid
2323	Clove oil (bud)		phenylpropionate	2488	Fumaric acid
2325	Clove oil (leaf)	2416	Ethyl 2-acetyl-3-	2489	Furfural
2326	Clover		phenylpropionate	2491	Furfuryl alcohol
2328	Clove oil (stem)	2417	Ethyl aconitate, mixed	2493	Furfural mercaptan
2331	Cognac oil, green		esters	2497	Fusel oil refined
2332	Cognac oil, white	2418	Ethyl acrylate	2503	Garlic oil
2334	Coriander oil	2420	Ethyl-p-anisate	2507	Geraniol

FEMA	Chemical	FEMA	Chemical	FEMA	Chemical
2508	Geranium oil	2617	Lauryl alcohol		propionate
2509	Geranyl acetate	2618	Lavandin oil	2721	Methyl 4-methylvalerate
2511	Geranyl benzoate	2622	Lavender oil	2722	Methyl myristate
2512	Geranyl butyrate	2624	Lemongrass oil West Indian	2723	2'-Acetonaphthone
2513	Geranyl isobutyrate	2624	Lemongrass oil East Indian	2725	Methyl 2-nonenoate
2515	Geranyl hexanoate	2625	Lemon oil	2726	Methyl 2-nonynoate
2516	Geranyl phenylacetate	2627	Levulinic acid	2728	Methyl caprylate
2517	Geranyl propionate	2631	Lime oil	2729	Methyl 2-octynoate
2518	Geranyl isovalerate	2633	d-Limonene	2730	4-Methyl-2,3-pentanedione
2522	Ginger oil	2634	Linaloe wood oil	2731	Methyl isobutyl ketone
2524	α-D-Glucose pentaacetate	2635	Linalool	2733	Methyl phenylacetate
2530	Grapefruit oil	2636	Linalyl acetate	2734	3-Methyl-4-phenyl-3-buten-2-one
2532	Guaiacol	2637	Linalyl anthranilate		
2533	Guaiac extract	2638	Linalyl benzoate	2740	4-Methyl-1-phenyl-2-pentanone
2535	Guaiacyl phenylacetate	2639	Linalyl butyrate		
2537	Guar gum	2640	Linalyl isobutyrate	2741	Methyl 3-phenylpropionate
2539	γ-Heptalactone	2641	Linalyl cinnamate	2742	Methyl propionate
2540	Heptanal	2642	Linalyl formate	2743	Cyclamen aldehyde
2541	Heptanal dimethyl acetal	2653	Mace oil	2744	Lepidine
2542	Heptanal 1,2-glyceryl acetal	2655	N-Hydroxysuccinic acid	2745	Methyl salicylate
2543	2,3-Heptanedione	2656	Maltol	2746	Methyl sulfide
2544	Methyl n-amyl ketone	2657	Mandarin orange oil	2747	3-Methylthiopropion-aldehyde
2545	3-Heptanone	2664	Perillyl alcohol		
2547	Heptyl acetate	2665	Menthol	2749	Methyl nonyl acetaldehyde
2548	Heptyl alcohol	2666	d-Neomenthol	2752	Methyl valerate
2549	Heptyl butyrate	2667	l-Menthone	2754	2-Methylpentanoic acid
2550	Heptyl isobutyrate	2668	dl-Menthyl acetate	2762	Myrcene
2551	Heptyl cinnamate	2669	Menthyl isovalerate	2763	Myristaldehyde
2552	Heptyl formate	2670	p-Anisaldehyde	2764	Myristic acid
2554	Cetyl alcohol	2671	2-Methoxy-4-methylphenol	2772	Trimethyldodecatrieneol
2556	γ-Hexalactone	2672	4-p-Methoxyphenyl-2-butanone	2773	cis-3,7-Dimethyl-2,6-octadien-1-yl-acetate
2557	Hexanal				
2558	Acetyl butyryl	2674	1-(p-Methoxyphenyl)-2-propanone	2773	Neryl acetate
2558	2,3-Hexanedione			2774	Neryl butyrate
2559	Caproic acid	2676	Methyl acetate	2775	Neryl isobutyrate
2560	2-Hexenal	2677	4'-Methyl acetophenone	2776	Neryl formate
2562	2-Hexenol	2678	2-Methylallyl butyrate	2777	Neryl propionate
2563	3-Hexenol	2681	p-Methylanisole	2778	Neryl isovalerate
2564	trans-2-Hexenyl acetate	2682	Methyl anthranilate	2780	2,6-Nonadien-1-ol
2565	Hexyl acetate	2683	Methyl benzoate	2781	γ-Nonalactone
2566	2-Hexyl-4-acetoxytetra-hydrofuran	2684	α-Methylbenzyl acetate	2782	Nonanal
		2684	Methylbenzyl acetate, mixed o-, m-, p-	2783	1,3-Nonanediol acetate, mixed esters
2567	Hexyl alcohol				
2568	Hexyl butyrate	2685	α-Methylbenzyl alcohol	2784	Nonanoic acid
2569	α-Hexylcinnamaldehyde	2686	α-Methylbenzyl butyrate	2785	2-Nonanone
2570	Hexyl formate	2689	α-Methylbenzyl propionate	2788	n-Nonyl acetate
2572	Hexyl hexanoate	2691	2-Methylbutyraldehyde	2789	Nonyl alcohol
2575	Hexyl octanoate	2692	Isovaleraldehyde	2790	Nonyl octanoate
2576	Hexyl propionate	2693	Methyl butyrate	2793	Nutmeg oil
2577	Hickory bark	2694	Methyl isobutyrate	2796	γ-Octalactone
2581	Horehound	2695	2-Methylbutyric acid	2797	n-Octanal
2582	Horsemint	2697	α-Methylcinnamaldehyde	2799	Caprylic acid
2583	Hydroxycitronellal	2698	Methyl cinnamate	2800	Caprylic alcohol
2584	Hydroxycitronellal diethyl acetal	2699	6-Methylcoumarin	2801	2-Octanol
		2700	Methyl cyclopentenolone	2805	1-Octen-3-ol
2585	Hydroxycitronellal dimethyl acetyl	2705	Methyl heptanoate	2806	Octyl acetate
		2706	2-Methylheptanoic acid	2807	Octyl butyrate
2586	Hydroxycitronellol	2707	Methyl-5-hepten-2-ol	2808	Octyl isobutyrate
2587	5-Hydroxy-4-octanone	2707	Methyl heptenone	2809	Octyl formate
2588	4-(p-Hydroxyphenyl)-2-butanone	2708	Methyl caproate	2813	Octyl propionate
		2708	Methyl 2-hexanoate	2814	Octyl isovalerate
2593	Indole	2710	Methylparaben	2815	Oleic acid
2594	α-Ionone	2714	α-Isomethylionone	2816	Olibanum oil
2595	β-Ionone	2715	Methyl laurate	2821	Orange oil
2597	α-Irone	2716	Methyl mercaptan	2831	Palmarosa oil
2614	Lauric acid	2718	Dimethyl anthranilate	2832	Palmitic acid
2615	Lauric aldehyde	2719	Methyl 2-methylbutyrate	2835	Parsley
2616	Lauryl acetate	2720	Methyl 3-methylthio-	2836	Parsley seed oil

FEMA	Chemical	FEMA	Chemical	FEMA	Chemical
2837	Oleoresin parsley leaf	2939	Isopropyl cinnamate	3071	p-Tolylacetaldehyde
2839	Pennyroyal oil	2943	Propyl formate	3073	p-Cresyl acetate
2840	Pentadecalactone	2944	Isopropyl formate	3074	4-(p-Tolyl)-2-butanone
2841	Pentane-2,3-dione	2945	Propyl 2-furanacrylate	3075	p-Tolyl isobutyrate
2842	Methyl propyl ketone	2948	Propyl heptanoate	3076	p-Tolyl laurate
2843	4-Pentenoic acid	2949	Propyl hexanoate	3077	p-Tolyl phenylacetate
2848	Peppermint oil	2950	Isopropyl hexanoate	3080	Tributyl acetylcitrate
2856	α-Phellandrene	2952	3-Propylidenephthalide	3082	2-Tridecenal
2857	2-Phenylethyl acetate	2953	α-Propylphenethyl alcohol	3083	Triethyl citrate
2858	Phenethyl alcohol	2954	p-Isopropylphenyl-acetaldehyde	3084	Tuberose
2859	Phenethyl anthranilate			3090	2,3-Undecadione
2860	Phenethyl benzoate	2955	Propyl phenylacetate	3091	γ-Undecalactone
2861	Phenethyl butyrate	2956	Isopropyl phenylacetate	3092	Undecanal
2862	Phenethyl isobutyrate	2957	3-(p-Isopropylphenyl)-propionaldehyde	3093	2-Undecanone
2863	Phenethyl cinnamate			3094	9-Undecenal
2864	Phenethyl formate	2958	Propyl propionate	3095	10-Undecenal
2866	Phenethyl phenylacetate	2959	Isopropyl propionate	3096	10-Undecen-1-yl acetate
2868	Phenethyl salicylate	2960	Propyl isovalerate	3097	Undecyl alcohol
2870	Phenethyl tiglate	2961	Isopropyl isovalerate	3098	n-Valeraldehyde
2871	Phenethyl isovalerate	2962	Isopulegol	3101	n-Valeric acid
2872	Phenoxyacetic acid	2963	Pulegone	3102	Isovaleric acid
2873	Phenoxyethyl isobutyrate	2964	Isopulegone	3103	γ-Valerolactone
2874	Phenylacetaldehyde	2965	Isopulegyl acetate	3107	Vanillin
2875	Phenylacetaldehyde 2,3-butylene glycol acetal	2966	Pyridine	3108	Vanillin acetate
		2968	Pyroligneous acid extract	3109	Veratraldehyde
2876	Phenylacetaldehyde dimethyl acetal	2969	Pyruvaldehyde	3110	Violet
		2970	Pyruvic acid	3112	Wintergreen extract
2877	Phenylacetaldehyde glyceryl acetal	2973	Quillaja	3113	Wintergreen oil
		2976	Quinine hydrochloride	3114	Artemisia
2878	Phenylacetic acid	2977	Quinine sulfate	3121	Yucca
2881	Benzylidene acetone	2978	Isoquinoline	3124	Zingerone
2884	1-Phenyl-1-propanol	2980	Rhodinol	3126	2-Acetylpyrazine
2885	Hydrocinnamic alcohol	2981	Rhodinyl acetate	3129	Biphenyl
2886	2-Phenylpropanal	2982	Rhodinyl butyrate	3130	Butylamine
2887	Hydrocinnamaldehyde	2984	Rhodinyl formate	3132	2-Isobutyl-3-methoxy-pyrazine
2888	2-Phenylpropionaldehyde dimethylacetal	2986	Rhodinyl propionate		
		2989	Rose oil	3134	2-Isobutylthiazole
2889	Hydrocinnamic acid	2992	Rosemary oil	3135	trans-trans-2,4-Decadienal
2890	Hydrocinnamyl acetate	2996	Ethyl oxyhydrate	3149	2-Ethyl-3,5(6)-dimethyl-pyrazine
2891	2-Phenylpropyl butyrate	2998	Saffron		
2892	2-Phenylpropyl isobutyrate	3004	Salicylaldehyde	3155	2-Ethyl-3-methylpyrazine
2899	3-Phenylpropyl isovalerate	3005	Sandalwood oil	3168	3,4-Hexanedione
2902	α-Pinene	3006	α and β-Santalol	3174	4-Hydroxy-2,5-dimethyl-3(2H)furanone
2903	β-Pinene	3007	Santalyl acetate		
2904	Pine needle oil, dwarf	3008	Santalyl phenylacetate	3177	Mentha-8-thol-3-one
2905	Fir needle oil, Siberian type	3009	Sarsaparilla extract	3180	2-Mercaptopropionic acid
2906	Pine needle oil, Scotch type	3019	Skatole	3181	o-Methoxycinnamaldehyde
2907	Pine tar oil	3028	Sorbitan stearate	3183	2-Methoxy-3(5)-methylpyrazine
2908	Piperidine	3032	Spearmint oil		
2909	Piperine	3033	Spike lavender oil	3195	2-Methyl-2-pentenoic acid
2910	d-Piperitone	3038	Sucrose octaacetate	3196	cis-Jasmone
2911	Heliotropine	3041	Tangerine oil	3202	2-Acetylpyrrole
2912	Piperonyl acetate	3044	L-Tartaric acid	3209	5-Methyl-2-thiophene-carboxaldehyde
2913	Piperonyl isobutyrate	3045	α-Terpineol		
2922	Propenylguaethol	3045	Terpineol anhydrous	3212	trans,trans-2,4-Nonadienal
2923	Propionaldehyde	3046	Terpinolene	3213	trans-2-Nonenal
2925	Propyl acetate	3047	Terpinyl acetate	3215	trans-2-Octen-1-al
2926	Isopropyl acetate	3055	Tetrahydrofurfuryl acetate	3228	Propyl disulfide
2927	p-Isopropylacetophenone	3056	Tetrahydrofurfuryl alcohol	3237	2,3,5,6-Tetramethylpyrazine
2928	Propyl alcohol	3056	Tetrahydrofurfuryl propionate	3241	Trimethylamine
2929	Isopropyl alcohol			3244	2,3,5-Trimethylpyrazine
2930	p-Propyl anisole	3057	Tetrahydrofurfuryl butyrate	3247	Undecylenic acid
2931	Propyl benzoate	3060	Tetrahydrolinalool	3251	2-Acetylpyridine
2932	Isopropyl benzoate	3062	2-Thienyl mercaptan	3254	Arabinogalactan
2933	Cuminic alcohol	3064	Thyme oil	3263	L-Cysteine
2934	Propyl butyrate	3066	Thymol	3264	cis-4-Decen-1-al
2935	Isopropyl butyrate	3068	Toluadehydes, mixed o-, m-, p-	3271	2,3-Dimethylpyrazine
2937	Isopropyl isobutyrate			3272	2,5-Dimethylpyrazine

FEMA	Chemical	FEMA	Chemical	FEMA	Chemical
3273	2,6-Dimethylpyrazine	3386	Pyrrole		oxazoline
3277	Disodium succinate	3391	3-Acetyl-2,5-dimethylfuran	3536	Methyl disulfide
3277	Sodium succinate	3424	3-Acetylpyridine	3542	Geranyl acetone
3285	L-Glutamic acid	3432	Isobutyl-2-butenoate	3547	3-Heptanol
3286	Glyceryl tripropanoate	3463	4-Methylpentanoic acid	3565	d-Dihydrocarvone
3287	Glycine	3465	cis-6-Nonen-1-ol	3581	3-Octanol
3289	cis-4-Hepten-1-al	3467	cis-3-Octen-1-ol	3584	1-Penten-3-ol
3291	Butyrolactone	3470	Quinoline	3616	Benzenethiol
3294	ç-Undecalactone	3486	Ethyl crotonate	3617	Benzyl disulfide
3302	2-Methoxypyrazine	3487	Ethyl maltol	3632	2-Phenethyl 2-methyl-
3309	2-Methylpyrazine	3491	2-Ethyl fenchol		butyrate
3322	Thiamine HCl	3492	Ethyl undecanoate	3647	3-Methyl-2-buten-1-ol
3325	Trimethyl thiazole	3497	cis-3-Hexenyl 2-	3658	1,4-Cineole
3332	Butan-3-one-2-yl butyrate		methylbutyrate	3673	2-Ethylfuran
3354	Hexyl-2-butenoate	3500	Hexyl isovalerate	3684	L-Glutamine
3379	trans-2-Nonen-1-ol	3506	2-Methylbutyl isovalerate	3698	Isoeugenyl benzyl ether
3384	Phenylacetaldehyde	3521	Propyl mercaptan	7071	2,5-Dimethylpyrrole
	diisobutyl acetal	3525	2,4,5-Trimethyl d-3-		

Tables of Food Regulations

The Tables of Food Regulations are tabular summaries of food additive regulatory information for the United States, Europe, and Japan and include chemical names, functions, usage levels, and limitations. This section is meant to be used as a general guide. Local food legislation should always be consulted regarding all food additive products as they vary from country to country.

U.S. FDA Code of Federal Regulations

21CFR § 73.1
Substance: Diluents in color additive mixt. for food use exempt from certification

21CFR § 137.350
Substance: BHT
Use: Enriched rice

21CFR § 161.175
Use: Frozen raw breaded shrimp

21CFR § 164.110
Use: Mixed nuts

21CFR § 165.175
Use: Soda water (nonalcoholic beverages)

21CFR § 166.110
Use: Margarine

21CFR § 168.110
Substance: Dextrose anhyd.
Use: Sweetener

21CFR § 168.111
Substance: Dextrose monohydrate
Use: Sweetener

21CFR § 168.120
Substance: Glucose syrup
Use: Sweetener

21CFR § 168.121
Substance: Dried glucose syrup (Glucose syrup solids)
Use: Sweetener

21CFR § 168.130
Substance: Cane syrup (Sugar cane syrup)
Use: Sweetener

21CFR § 168.140
Substance: Maple syrup
Use: Sweetener

21CFR § 168.160
Substance: Sorghum syrup (Sorghum)

Use: Sweetener

21CFR § 168.180
Substance: Table syrup
Use: Sweetener

21CFR § 169.175
Substance: Vanilla extract
Use: Flavoring

21CFR § 169.177
Substance: Vanilla flavoring
Use: Flavoring

21CFR § 169.179
Substance: Vanilla powder
Use: Flavoring

21CFR § 169.180
Substance: Vanilla-vanillin extract
Use: Flavoring

21CFR § 169.181
Substance: Vanilla-vanillin flavoring
Use: Flavoring

21CFR § 169.182
Substance: Vanilla-vanillin powder
Use: Flavoring

21CFR § 170
Substance: Food additives

21CFR § 170.45
Substance: Fluorine-contg. compds.
Use: Fluoridation of public or bottled water

21CFR § 170.50
Substance: Glycine (aminoacetic acid)
Limitation: No longer GRAS for use in human food

21CFR § 170.60
Substance: Nitrates and/or nitrites
Use: Curing premixes for meats

21CFR § 171
Food additives petitions

21CFR § 172
Substance: Food additives permitted for direct addition to food for human consumption

21CFR § 172.105
Substance: Anoxomer
Use: Food preservative, antioxidant
Limitation: 5000 ppm on fat and oil content of food

21CFR § 172.110
Substance: BHA
Use: Food preservative
Limitation: 2-1000 ppm

21CFR § 172.115
Substance: BHT
Use: Food preservative
Limitation: 10-200 ppm

21CFR § 172.120
Substance: Calcium disodium EDTA
Use: Food preservative; promotes color, flavor, and texture retention
Limitation: 25-800 ppm

21CFR § 172.130
Substance: Dehydroacetic acid
Use: Food preservative
Limitation: 65 ppm residue on prepared squash

21CFR § 172.133
Substance: Dimethyl dicarbonate
Use: Food preservative
Limitation: 200 ppm in wine

21CFR § 172.135
Substance: Disodium EDTA
Use: Food preservative, sequestrant
Limitation: 36-500 ppm

21CFR § 172.140
Substance: Ethoxyquin (1,2-Dihydro-6-ethoxy-2,2,4-trimethylquinoline)
Use: Food preservative, antioxidant
Limitation: 0.5-5 ppm, zero in milk

21CFR § 172.145
Substance: Heptylparaben (n-Heptyl p-hydroxybenzoate)
Use: Food preservative
Limitation: 12-20 ppm

21CFR § 172.150
Substance: 4-Hydroxy-2,6-di-t-butylphenol
Use: Food preservative, antioxidant
Limitation: 0.02% of oil or fat content

21CFR § 172.155
Substance: Natamycin (Pimaricin)
Use: Food preservative

21CFR § 172.160
Substance: Potassium nitrate
Use: Food preservative, curing agent
Limitation: 200 ppm of finished cod roe

21CFR § 172.165
Substance: Quaternary ammonium chloride combination

Use: Food preservative, antimicrobial

21CFR § 172.170
Substance: Sodium nitrate
Use: Food preservative, color fixative
Limitation: 500 ppm

21CFR § 172.175
Substance: Sodium nitrite
Use: Food preservative, color fixative
Limitation: 10-200 ppm

21CFR § 172.177
Substance: Sodium nitrite used in processing smoked chub
Use: Food preservative
Limitation: 100-200 ppm

21CFR § 172.180
Substance: Stannous chloride
Use: Food preservative, color retention aid
Limitation: 20 ppm (as tin)

21CFR § 172.185
Substance: TBHQ (t-Butyl hydroquinone)
Use: Food preservative, antioxidant
Limitation: 0.02% of oil or fat (total antioxidant content)

21CFR § 172.190
Substance: THBP (2,4,5-Trihydroxybutyrophenone)
Use: Food preservative, antioxidant
Limitation: 0.02% of oil or fat (total antioxidant content)

21CFR § 172.210
Substance: Coatings on fresh citrus fruit

21CFR § 172.215
Substance: Coumarone-indene resin
Use: Coatings
Limitation: 200 ppm residue

21CFR § 172.225
Substance: Methyl and ethyl esters of fatty acids from edible fats and oils
Use: Coatings
Limitation: 200 ppm residue

21CFR § 172.230
Substance: Microcapsules for flavoring substances
Use: Coatings

21CFR § 172.235
Substance: Morpholine
Use: Protective coatings on fruits and vegetables

21CFR § 172.250
Substance: Petroleum naphtha
Use: Solvent in protective coatings on citrus fruits

21CFR § 172.255
Substance: Polyacrylamide
Use: Film-former in imprinting soft-shell gelatin capsules

21CFR § 172.260
Substance: Oxidized polyethylene

Use: Protective coating for fruits and vegetables

21CFR § 172.270
Substance: Sulfated butyl oleate
Use: Dehydrating agent for grapes to produce raisins
Limitation: 100 ppm residue

21CFR § 172.275
Substance: Synthetic paraffin and succinic derivs.
Use: Protective coating for fruits and vegetables

21CFR § 172.280
Substance: Terpene resin (β-Pinene)
Use: Moisture barrier on soft gelatin capsules, powds. of ascorbic acid or its salts
Limitation: 0.07-7%

21CFR § 172.310
Substance: Aluminum nicotinate
Use: Dietary and nutritional additive

21CFR § 172.315
Substance: Nicotinamide-ascorbic acid complex
Use: Dietary and nutritional additive

21CFR § 172.320
Substance: Amino acids
Use: Dietary and nutritional additive

21CFR § 172.325
Substance: Bakers yeast protein
Use: Nutrient supplement

21CFR § 172.330
Substance: Calcium pantothenate, calcium chloride double salt
Use: Dietary and nutritional additive

21CFR § 172.335
Substance: D-Pantothenamide
Use: Dietary and nutritional additive

21CFR § 172.340
Substance: Fish protein isolate
Use: Dietary and nutritional additive

21CFR § 172.345
Substance: Folic acid (Folacin)
Use: Dietary and nutritional additive
Limitation: Daily: 0.4 mg, 0.1 mg (infants), 0.3 mg (children under 4), 0.8 mg (pregnant women)

21CFR § 172.350
Substance: Fumaric acid and its salts
Use: Dietary and nutritional additive

21CFR § 172.365
Substance: Kelp
Use: Dietary and nutritional additive

21CFR § 172.370
Substance: Iron-choline citrate complex
Use: Dietary and nutritional additive

21CFR § 172.372
Substance: N-Acetyl-L-methionine
Use: Dietary and nutritional additive

21CFR § 172.375
Substance: Potassium iodide
Use: Dietary and nutritional additive

21CFR § 172.385
Substance: Whole fish protein conc.
Use: Dietary and nutritional additive

21CFR § 172.395
Substance: Xylitol
Use: Dietary and nutritional additive

21CFR § 172.399
Substance: Zinc methionine sulfate
Use: Dietary and nutritional additive

21CFR § 172.410
Substance: Calcium silicate
Use: Anticaking agent
Limitation: 2% in food, 5% in baking powd.

21CFR § 172.430
Substance: Iron ammonium citrate
Use: Anticaking agent
Limitation: 25 ppm in finished salt

21CFR § 172.480
Substance: Silicon dioxide
Use: Anticaking agent; stabilizer in beer prod.; adsorbent
Limitation: 2%

21CFR § 172.490
Substance: Yellow prussiate of soda (Sodium ferrocyanide decahydrate)
Use: Anticaking agent
Limitation: 13 ppm (as sodium ferrocyanide anhyd.)

21CFR § 172.510
Substance: Natural flavoring substances and natural substances used with flavors
Use: Flavoring agents

21CFR § 172.515
Substance: Synthetic flavoring substances and adjuvants
Use: Flavoring agents

21CFR § 172.520
Substance: Cocoa with dioctyl sodium sulfosuccinate
Use: Flavoring agent for dry beverage mixes
Limitation: 75 ppm of finished beverage (dioctyl sodium sulfosuccinate)

21CFR § 172.530
Substance: Disodium guanylate
Use: Flavor enhancer

21CFR § 172.535
Substance: Disodium inosinate
Use: Flavoring adjuvant

21CFR § 172.540
Substance: DL-Alanine
Use: Flavor enhancer in pickling mixtures
Limitation: 1% of pickling spice added to brine

21CFR § 172.560
Substance: Modified hop extract
Use: Flavoring agent in beer brewing

21CFR § 172.575
Substance: Quinine, hydrochloride salt or sulfate salt
Use: Flavoring agent in carbonated beverages
Limitation: 83 ppm (as quinine)

21CFR § 172.580
Substance: Extract of sassafras, safrole-free
Use: Flavoring agent

21CFR § 172.585
Substance: Sugar beet extract flavor base
Use: Flavoring agent

21CFR § 172.590
Substance: Yeast-malt sprout extract
Use: Flavor enhancer

21CFR § 172.610
Substance: Arabinogalactan
Use: Gum as emulsifier, stabilizer, binder, bodying agent

21CFR § 172.615
Substance: Chewing gum base
Use: Masticatory substances (natural or synthetic) or plasticizers for chewing gum

21CFR § 172.620
Substance: Carrageenan
Use: Emulsifier, stabilizer, thickener

21CFR § 172.623
Substance: Carrageenan with polysorbate 80

21CFR § 172.626
Substance: Carrageenan salts (ammonium, calcium, potassium, or sodium)
Use: Emulsifier, stabilizer, thickener

21CFR § 172.655
Substance: Furcelleran
Use: Emulsifier, stabilizer, thickener

21CFR § 172.660
Substance: Furcelleran salts (ammonium, calcium, potassium, or sodium)
Use: Emulsifier, stabilizer, thickener

21CFR § 172.665
Substance: Gellan gum
Use: Stabilizer, thickener

21CFR § 172.695
Substance: Xanthan gum
Use: Emulsifier, stabilizer, thickener, suspending agent, bodying agent, foam enhancer

21CFR § 172.710
Substance: Adjuvants for pesticide use dilutions

21CFR § 172.715
Substance: Calcium lignosulfonate
Use: Dispersant, stabilizer in pesticides

21CFR § 172.720
Substance: Calcium lactobionate
Use: Firming agent in dry pudding mixes

21CFR § 172.725
Substance: Gibberellic acid and its potassium salt
Use: In malting of barley
Limitation: 2 ppm (as giberellic acid in malt), 0.5 ppm (in finished beverage)

21CFR § 172.730
Substance: Potassium bromate
Use: In malting of barley, prod. of distilled spirits
Limitation: 75 ppm (in malt as Br)

21CFR § 172.735
Substance: Glyceryl ester of wood rosin
Use: Adjusts density of citrus oils in prep. of beverages
Limitation: 100 ppm in finished beverage

21CFR § 172.755
Substance: Stearoyl monoglyceridyl citrate
Use: Emulsion stabilizer in or with shortenings contg. emulsifiers

21CFR § 172.765
Substance: Succistearin (Stearoyl propylene glycol hydrogen succinate)
Use: Emulsion stabilizer in or with shortenings and edible oils

21CFR § 172.770
Substance: Ethylene oxide polymer
Use: Foam stabilizer in fermented malt beverages
Limitation: 300 ppm

21CFR § 172.775
Substance: Methacrylic acid-divinylbenzene copolymer
Use: Carrier of vitamin B_{12} in foods for special dietary use

21CFR § 172.800
Substance: Acesulfame potassium
Use: Sweetening agent

21CFR § 172.802
Substance: Acetone peroxides
Use: Maturing and bleaching agent for flour; dough conditioner

21CFR § 172.804
Substance: Aspartame
Use: Sweetening agent

21CFR § 172.806
Substance: Azodicarbonamide
Use: Aging and bleaching ingred. in cereal flour; dough conditioner
Limitation: 45 ppm

21CFR § 172.808
Substance: Copolymer condensates of ethylene oxide and propylene oxide

Use: Solubilizer, stabilizer in flavor concs.; processing aid, wetting agent; surfactant, defoaming agent in scald baths for poultry defeathering, hog dehairing; dough conditioner

21CFR § 172.810

Substance: Dioctyl sodium sulfosuccinate
Use: Wetting agent, processing aid, solubilizer, emulsifier, dispersant, processing aid
Limitation: 10 ppm (in finished beverage, as dioctyl sodium sulfosuccinate-block copolymer combination)

21CFR § 172.811

Substance: Glyceryl tristearate
Use: Crystallization accelerator, formulation aid, lubricant, release agent, surface-finishing agent, winterization/fractionation aid
Limitation: 0.5-3%

21CFR § 172.812

Substance: Glycine
Use: Masking agent for bitter aftertaste of saccharin in beverages; stabilizer in mono- and diglycerides
Limitation: 0.02% (finished beverage), 0.02% (of mono- and diglycerides)

21CFR § 172.814

Substance: Hydroxylated lecithin
Use: Emulsifier

21CFR § 172.816

Substance: Methyl glucoside-coconut oil ester
Use: Surfactant in molasses
Limitation: 320 ppm

21CFR § 172.818

Substance: Oxystearin
Use: Crystallization inhibitor, release agent in vegetable oils
Limitation: 0.125% (of combined wt. of oil or shortening)

21CFR § 172.820

Substance: Polyethylene glycol (mean m.w. 200-9500)
Use: Coating, binder, plasticizer, lubricant in tablets; flavor adjuvant, bodying agent in nonnutritive sweeteners; dispersant adjuvant for vitamin/mineral preps.; coating on sodium nitrite
Limitation: Zero tolerance in milk for residues

21CFR § 172.822

Substance: Sodium lauryl sulfate
Use: Emulsifier, whipping agent, surfactant, wetting agent
Limitation: 25-1000 ppm, 0.5% (as whipping agent)

21CFR § 172.824

Substance: Sodium mono- and dimethyl naphthalene sulfonates
Use: Crystallization of sodium carbonate; anticaking agent; washing/lye peeling of fruits and vegetables

21CFR § 172.826

Substance: Sodium stearyl fumarate
Use: Dough conditioner, conditioning agent, stabilizer
Limitation: 0.5-1.0% of flour, 0.2% of food

21CFR § 172.828

Substance: Acetylated monoglycerides
Use: Food additive

21CFR § 172.830

Substance: Succinylated monoglycerides
Use: Emulsifier in liq. and plastic shortenings; dough conditioner in bread baking
Limitation: 0.5-3%

21CFR § 172.832

Substance: Monoglyceride citrate
Use: In antioxidants for oils and fats
Limitation: 200 ppm of oil or fat

21CFR § 172.834

Substance: Ethoxylated mono- and diglycerides
Use: Emulsifier, dough conditioner
Limitation: 0.2-0.5%

21CFR § 172.836

Substance: Polysorbate 60
Use: Emulsifier, opacifier, foaming agent, dough conditioner, dispersant, surfactant, wetting agent
Limitation: 0.4-4.5%

21CFR § 172.838

Substance: Polysorbate 65
Use: Emulsifier
Limitation: 0.1-0.66% (combined)

21CFR § 172.840

Substance: Polysorbate 80
Use: Emulsifier, yeast defoamer, solubilizer, dispersant, surfactant, wetting agent
Limitation: Various

21CFR § 172.841

Substance: Polydextrose
Use: Bulking agent, formulation aid, humectant, texturizer

21CFR § 172.842

Substance: Sorbitan monostearate
Use: Emulsifier, rehydration aid
Limitation: Various

21CFR § 172.844

Substance: Calcium stearoyl-2-lactylate
Use: Dough conditioner, whipping agent, conditioning agent
Limitation: Various

21CFR § 172.846

Substance: Sodium stearoyl lactylate
Use: Dough strengthener, emulsifier, processing aid, surface-active agent, stabilizer, formulation aid, texturizer
Limitation: 0.2-0.5%

21CFR § 172.848
Substance: Lactylic esters of fatty acids
Use: Emulsifier, plasticizer, surface-active agent

21CFR § 172.850
Substance: Lactylated fatty acid esters of glycerol
and propylene glycol
Use: Emulsifier, plasticizer, surface-active agent

21CFR § 172.852
Substance: Glyceryl-lacto esters of fatty acids
Use: Emulsifier, plasticizer

21CFR § 172.854
Substance: Polyglycerol esters of fatty acids
Use: Emulsifier; cloud inhibitor in salad oils

21CFR § 172.856
Substance: Propylene glycol mono- and diesters of
fats and fatty acids

21CFR § 172.858
Substance: Propylene glycol alginate
Use: Emulsifier, flavoring adjuvant, formulation aid,
stabilizer, surfactant, thickener
Limitation: 0.3-1.7%

21CFR § 172.859
Substance: Sucrose fatty acid esters
Use: Emulsifier, stabilizer, texturizer, component of
protective coatings on fruits

21CFR § 172.860
Substance: Fatty acids
Use: Lubricant, binder, defoaming agent, compo-
nent in mfg. of food-grade additives

21CFR § 172.861
Substance: Cocoa butter substitute from coconut
and/or palm kernel oils
Use: Coating material for sugar, table salt, vitamins,
citric acid, spices, etc.; in compound coatings,
cocoa creams, chewing sweets, etc.

21CFR § 172.862
Substance: Oleic acid derived from tall oil fatty
acids
Use: Lubricant, binder, defoaming agent, component in mfg. of food-grade additives

21CFR § 172.863
Substance: Salts of fatty acids
Use: Binder, emulsifier, anticaking agent

21CFR § 172.864
Substance: Syn. fatty alcohols
Use: Food additive; as substitute for corresponding
naturally derived fatty alcohol

21CFR § 172.866
Substance: Syn. glycerin produced by hydro-
genolysis of carbohydrates

21CFR § 172.868
Substance: Ethyl cellulose

Use: Food additive, binder, filler; component of
protective coatings for vitamin/mineral tablets;
flavor fixative

21CFR § 172.870
Substance: Hydroxypropyl cellulose
Use: Food additive, emulsifier, film-former, protective colloid, stabilizer, suspending agent, thickener, binder/disintegrant in tablets

21CFR § 172.872
Substance: Methyl ethyl cellulose
Use: Food additive, aerating agent, emulsifier, foam-
ing agent

21CFR § 172.874
Substance: Hydroxypropyl methylcellulose
Use: Emulsifier, film-former, protective colloid, sta-
bilizer, suspending agent, thickener

21CFR § 172.876
Substance: Castor oil
Use: Release agent, antisticking agent, component
of protective coatings for vitamin and mineral
tablets
Limitation: 500 ppm (hard candy)

21CFR § 172.878
Substance: White mineral oil
Use: Release agent, binder, lubricant, defoamer,
float on fermentation fluids to prevent/retard
evaporation and contamination, protective coating, antidusting agent, sealing/polishing agent
Limitation: 0.02-0.6%

21CFR § 172.880
Substance: Petrolatum
Use: Release agent, lubricant, sealing/polishing
agent, protective coating, defoaming agent
Limitation: 0.02-0.2%

21CFR § 172.882
Substance: Syn. isoparaffinic petroleum hydrocarbons
Use: Froth flotation cleaning of vegetables; in insecticide formulations for use on processed
foods; in coatings on fruits, vegetables, egg
shells; float on fermentation fluids

21CFR § 172.884
Substance: Odorless light petroleum hydrocarbons
Use: Coating on egg shells, defoamer, float on
fermentation fluids, in froth flotation cleaning of
vegetables, as component of insecticides

21CFR § 172.886
Substance: Petroleum wax
Use: Masticatory substance in chewing gum base;
protective coating on raw fruits and vegetables;
defoamer; component of microcapsules for spice
flavorings

21CFR § 172.888
Substance: Synthetic petroleum wax

Use: Masticatory substance in chewing gum base; protective coating on raw fruits and vegetables; defoamer

21CFR § 172.890
Substance: Rice bran wax
Use: Coating on candy, fresh fruits and vetables; plasticizing material for chewing gum
Limitation: 50 ppm (as coating), 2.5% (chewing gum)

21CFR § 172.892
Substance: Food starch, modified

21CFR § 172.894
Substance: Modified cottonseed prods. intended for human consumption
Use: Food additive

21CFR § 172.896
Substance: Dried yeasts

21CFR § 172.898
Substance: Bakers yeast glycan
Use: Emulsifier, stabilizer, thickener, texturizer
Limitation: 5% (finished salad dressing), GMP (other foods)

21CFR § 173
Substance: Secondary direct food additives permitted in food for human consumption

21CFR § 173.5
Substance: Acrylate-acrylamide resins
Use: Flocculant in clarification of beet sugar juice, etc.; controls organic and mineral scale in beet or cane sugar juice and liquor
Limitation: 2.5-10 ppm

21CFR § 173.10
Substance: Modified polyacrylamide resin
Use: Flocculant in clarification of beet or cane sugar juice
Limitation: 5 ppm

21CFR § 173.20
Substance: Ion-exchange membranes
Use: In prod. of grapefruit juice to adjust ratio of citric acid to total solids

21CFR § 173.25
Substance: Ion-exchange resins
Use: Purification of foods incl. potable water to remove undesirable ions

21CFR § 173.40
Substance: Molecular sieve resins
Use: As gel filtration media in purification of partially delactosed whey

21CFR § 173.45
Substance: Polymaleic acid and its sodium salt
Use: In processing of beet or cane juice and liquor
Limitation: 0.4 ppm (as the acid)

21CFR § 173.50
Substance: Polyvinylpolypyrrolidone

Use: Clarifying agent in beverages and vinegar

21CFR § 173.55
Substance: Polyvinylpyrrolidone
Use: Clarifying agent for beer, wine, vinegar; tableting adjuvant for vitamin/min., flavor concs. and nonnutritive sweeteners; stabilizer, bodying agent, dispersant
Limitation: 10-60 ppm residual (as clarifying agent); GMP (other)

21CFR § 173.60
Substance: Dimethylamine-epichlorohydrin copolymer
Use: Decolorizing agent, flocculant in clarification of refinery sugar liquors and juices
Limitation: 150 ppm of copolymer by wt. of sugar solids

21CFR § 173.65
Substance: Divinylbenzene copolymer
Use: Removal of organic substances from aq. foods

21CFR § 173.70
Substance: Chloromethylated aminated styrene-divinylbenzene resin
Use: Decolorizing and clarification agent for refinery sugar liquors and juices
Limitation: 500 ppm of additive solids/million parts sugar solids

21CFR § 173.73
Substance: Sodium polyacrylate
Use: Controls mineral scale during evaporation of beet or cane sugar juice
Limitation: 3.6 ppm of raw juice

21CFR § 173.75
Substance: Sorbitan monooleate
Use: Emulsifier in polymer dispersions used to clarify cane or beet sugar juice or liquor
Limitation: 7.5% (in final polymer disp.), 0.7-1.4 ppm (sugar juice/liquor)

21CFR § 173.110
Substance: Amyloglucosidase derived from *Rhizopus niveus*
Use: Enzyme prod. for degrading gelatinized starch into constituent sugars in prod. of distilled spirits and vinegar
Limitation: 0.1% of gelatinized starch

21CFR § 173.120
Substance: Carbohydrate and cellulase derived from *Aspergillus niger*
Use: Enzyme prep. for removal of visceral mass in clams, shells from shrimp

21CFR § 173.130
Substance: Carbohydrase derived from *Rhizopus oryzae*
Use: Enzyme used in prod. of dextrose from starch

21CFR § 173.135
Substance: Catalase derived from *Micrococcus lysodeikticus*

Use: Enzyme for destroying and removing hydrogen peroxide used in cheese mfg.

21CFR § 173.140
Substance: Esterase-lipase derived from *Mucor miehei*
Use: Enzyme as flavor enhancer

21CFR § 173.145
Substance: α-Galactosidase derived from *Mortierella vinaceae*
Use: Enzyme used in prod. of sugar from sugar beets

21CFR § 173.150
Substance: Milk-clotting enzymes, microbial
Use: Enzyme used in prod. of cheese

21CFR § 173.160
Substance: Candida guilliermondii
Use: Enzyme for fermentation prod. of citric acid

21CFR § 173.165
Substance: Candida lipolytica
Use: Enzyme for fermentation prod. of citric acid

21CFR § 173.210
Substance: Acetone
Use: Solvent for extraction of spice oleoresins
Limitation: 30 ppm residual

21CFR § 173.220
Substance: 1,3-Butylene glycol
Use: Solvent for natural and syn. flavoring substances

21CFR § 173.228
Substance: Ethyl acetate
Use: Food additive

21CFR § 173.230
Substance: Ethylene dichloride
Use: Solvent for extraction of spice oleoresins
Limitation: 30 ppm (total residues)

21CFR § 173.240
Substance: Isopropyl alcohol
Use: Solvent for extraction of spice oleoresins, lemon oil, hops
Limitation: 50 ppm (residue in oleoresins), 2% (residue in hops extract)

21CFR § 173.250
Substance: Methyl alcohol residues
Use: Solvent for extraction of spice oleoresins, hops
Limitation: 50 ppm (residue in oleoresins), 2.2% (residue in hops rextract)

21CFR § 173.255
Substance: Methylene chloride
Use: Solvent for extraction of spice oleoresins, hops, caffeine from coffee beans
Limitation: 30 ppm (residue in oleoresins), 2.2% (residue in hops extract), 10 ppm (residue in coffee)

21CFR § 173.270
Substance: Hexane
Use: Solvent for extraction of spice oleoresins, hops
Limitation: 25 ppm (residue in oleoresins), 2.2% (residue in hops extract)

21CFR § 173.275
Substance: Hydrogenated sperm oil
Use: Release agent, lubricant for bakery pans

21CFR § 173.280
Substance: Solvent extraction process for citric acid

21CFR § 173.290
Substance: Trichloroethylene
Use: Solvent for mfg. of spice oleoresins, decaffeinated coffee
Limitation: 10-30 ppm

21CFR § 173.310
Substance: Boiler water additives

21CFR § 173.315
Substance: Chemicals used in washing/lye peeling of fruits and vegetables

21CFR § 173.320
Substance: Chemicals for control of microorganisms in cane and beet sugar mills

21CFR § 173.322
Substance: Chemicals for delinting cottonseed

21CFR § 173.340
Substance: Defoaming agents

21CFR § 173.342
Substance: Chlorofluorocarbon 113 and perfluorohexane
Use: Quick-cool additive

21CFR § 173.345
Substance: Chloropentafluoroethane
Use: Aerating agent for foamed or sprayed food prods.

21CFR § 173.350
Substance: Combustion product gas
Use: For removing/displacing oxygen in processing, storage and pkg. of foods and beverages except fresh meats

21CFR § 173.355
Substance: Dichlorodifluoromethane
Use: Direct-contact freezing agent for foods

21CFR § 173.357
Substance: Materials used as fixing agents in immobilization of enzyme preps.

21CFR § 173.360
Substance: Octafluorocyclobutane
Use: Propellant, aerating agent in foamed or sprayed food prods.

21CFR § 173.385
Substance: Sodium methyl sulfate
Use: May be present in pectin
Limitation: 0.1% of pectin

21CFR § 173.395
Substance: Trifluoromethane sulfonic acid
Use: Catalyst in prod. of cocoa butter substitute from palm oil
Limitation: 0.2 ppm fluoride residual

21CFR § 173.400
Substance: Dimethyldialkylammonium chloride
Use: Food additive, decolorizing agent in clarification of refinery sugar liquors
Limitation: 700 ppm of sugar solids

21CFR § 174
Substance: Indirect food additives: General

21CFR § 175
Substance: Indirect food additives: Adhesives and components of coatings

21CFR § 176
Substance: Indirect food additives: Paper and paperboard components

21CFR § 177
Substance: Indirect food additives: Polymers

21CFR § 178
Substance: Indirect food additives: Adjuvants, production aids, and sanitizers

21CFR§ 170
Substance: Irradiation in the production, processing and handling of food

21CFR § 180
Substance: Food additives permitted in food on an interim basis or in contact with food pending additional study

21CFR § 180.22
Substance: Acrylonitrile copolymers
Use: Food pkg. component

21CFR § 180.25
Substance: Mannitol (1,2,3,4,5,6-Hexanehexol)
Use: Anticaking and free-flow agent
Limitation: 2.5-98%

21CFR § 180.30
Substance: Brominated vegetable oil
Use: Stabilizer for flavoring oils in fruit-flavored beverages
Limitation: 15 ppm (finished beverage)

21CFR § 180.37
Substance: Saccharin, ammonium saccharin, calcium saccharin, sodium saccharin
Use: Sweetening agents; flavor enhancer in chewable vitamin tablets, chewing gum, baked goods

21CFR § 181
Substance: Prior-sanctioned food ingredients

21CFR § 181.22
Substance: Certain substances employed in mfg. of food pkg. materials

21CFR § 181.23
Substance: Antimycotics migrating from food pkg.

21CFR § 181.24
Substance: Antioxidants migrating from food pkg.
Limitation: 0.005%

21CFR § 181.25
Substance: Driers migrating from food pkg.

21CFR § 181.26
Substance: Drying oils as components of finished resins

21CFR § 181.27
Substance: Plasticizers migrating from food pkg.

21CFR § 181.28
Substance: Release agents migrating from food pkg.

21CFR § 181.29
Substance: Stabilizers migrating from food pkg.

21CFR § 181.30
Substance: Substances used in mfg. of paper/paperboard prods. for food pkg.

21CFR § 181.32
Substance: Acrylonitrile copolymers and resins
Use: Films, coatings for food contact use

21CFR § 181.33
Substance: Sodium nitrate and potassium nitrate
Use: USDA prior sanctions for use as source of nitrite in prod. of cured red meat prods. and cured poultry prods.

21CFR § 181.34
Substance: Sodium nitrite and potassium nitrite
Use: USDA prior sanctions for use as color fixatives and preservatives in curing of red meat and poultry prods.

21CFR § 182
Substance: Substances generally recognized as safe (GRAS)

21CFR § 182.1
Substance: Substances that are generally recognized as safe

21CFR § 182.10
Substance: Spices and natural seasonings and flavors

21CFR § 182.20
Substance: Essential oils, oleoresins (solvent-free), and natural extractives (incl. distillates)

21CFR § 182.40
Substance: Natural extractives (solvent-free) used in conjunction with spices, seasonings, and flavorings

21CFR § 182.50
Substance: Certain other spices, seasonings, essential oils, oleoresins, and natural extracts

21CFR § 182.60
Substance: Syn. flavoring substances and adjuvants

21CFR § 182.70
Substance: Substances migrating from cotton/cotton fabrics in dry food pkg.

21CFR § 182.90
Substance: Substances migrating to food from paper/paperboard prods.

21CFR § 182.99
Substance: Adjuvants for pesticide chemicals

21CFR § 182.1033
Substance: Citric acid
Use: GRAS food substance

21CFR § 182.1045
Substance: Glutamic acid
Use: GRAS food substance, salt substitute

21CFR § 182.1047
Substance: Glutamic acid hydrochloride
Use: GRAS food substance, salt substitute

21CFR § 182.1057
Substance: Hydrochloric acid
Use: GRAS food substance, buffer, neutralizing agent

21CFR § 182.1073
Substance: Phosphoric acid
Use: GRAS food substance

21CFR § 182.1087
Substance: Sodium acid pyrophosphate
Use: GRAS food substance

21CFR § 182.1125
Substance: Aluminum sulfate
Use: GRAS food substance

21CFR § 182.1127
Substance: Aluminum ammonium sulfate (Ammonium alum)
Use: GRAS food substance

21CFR § 182.1129
Substance: Aluminum potassium sulfate (Potassium alum)
Use: GRAS food substance

21CFR § 182.1131
Substance: Aluminum sodium sulfate (Sodium alum)
Use: GRAS food substance

21CFR § 182.1180
Substance: Caffeine
Use: GRAS food substance used in cola beverages
Limitation: 0.02%

21CFR § 182.1195
Substance: Calcium citrate
Use: GRAS food substance

21CFR § 182.1217
Substance: Calcium phosphate (mono-, di- and tribasic)
Use: GRAS food substance

21CFR § 182.1235
Substance: Caramel
Use: GRAS food substance

21CFR § 182.1320
Substance: Glycerin
Use: GRAS food substance

21CFR § 182.1480
Substance: Methylcellulose USP
Use: GRAS food substance

21CFR § 182.1500
Substance: Monoammonium glutamate
Use: GRAS food substance

21CFR § 182.1516
Substance: Monopotassium glutamate
Use: GRAS food substance

21CFR § 182.1625
Substance: Potassium citrate
Use: GRAS food substance

21CFR § 182.1711
Substance: Silica aerogel
Use: GRAS food substance as component of anti-foaming agent

21CFR § 182.1745
Substance: Sodium carboxymethylcellulose
Use: GRAS food substance

21CFR § 182.1748
Substance: Sodium caseinate
Use: GRAS food substance

21CFR § 182.1751
Substance: Sodum citrate
Use: GRAS food substance

21CFR § 182.1778
Substance: Sodium phosphate (mono-, di- and tribasic)
Use: GRAS food substance

21CFR § 182.1781
Substance: Sodium aluminum phosphate
Use: GRAS food substance

21CFR § 182.1810
Substance: Sodium tripolyphosphate (Pentasodium triphosphate)
Use: GRAS food substance

21CFR § 182.1866
Substance: High fructose corn syrup
Use: GRAS food substance as nutritive carbohydrate sweetener

21CFR § 182.1911
Substance: Triethyl citrate
Use: GRAS food substance in dried egg whites
Limitation: 0.25%

21CFR § 182.2122
Substance: Aluminum calcium silicate
Use: GRAS food substance as anticaking agent
Limitation: 2%

21CFR § 182.2227
Substance: Calcium silicate
Use: GRAS food substance as anticaking agent
Limitation: 2% (table salt), 5% (baking powd.)

21CFR § 182.2437
Substance: Magnesium silicate
Use: GRAS food substance as anticaking agent
Limitation: 2%

21CFR § 182.2727
Substance: Sodium aluminosilicate
Use: GRAS food substance as anticaking agent
Limitation: 2%

21CFR § 182.2729
Substance: Sodium calcium aluminosilicate, hydrated
Use: GRAS food substance as anticaking agent
Limitation: 2%

21CFR § 182.2906
Substance: Tricalcium silicate
Use: GRAS food substance as anticaking agent
Limitation: 2%

21CFR § 182.3013
Substance: Ascorbic acid
Use: GRAS food substance as preservative

21CFR § 182.3041
Substance: Erythorbic acid
Use: GRAS food substance as preservative

21CFR § 182.3089
Substance: Sorbic acid
Use: GRAS food substance as preservative

21CFR § 182.3109
Substance: Thiodipropionic acid
Use: GRAS food substance as preservative
Limitation: 0.02% of fat or oil (total antioxidants)

21CFR § 182.3149
Substance: Ascorbyl palmitate
Use: GRAS food substance as preservative

21CFR § 182.3169
Substance: Butylated hydroxyanisole (BHA)
Use: GRAS food substance as preservative
Limitation: 0.02% of fat or oil (total antioxidants)

21CFR § 182.3173
Substance: Butylated hydroxytoluene (BHT)
Use: GRAS food substance as preservative
Limitation: 0.02% of fat or oil (total antioxidants)

21CFR § 182.3189
Substance: Calcium ascorbate
Use: GRAS food substance as preservative

21CFR § 182.3225
Substance: Calcium sorbate
Use: GRAS food substance as preservative

21CFR § 182.3280
Substance: Dilauryl thiodipropionate
Use: GRAS food substance as preservative
Limitation: 0.02% of fat or oil (total antioxidants)

21CFR § 182.3616
Substance: Potassium bisulfite
Use: GRAS food substance as preservative
Limitation: Not used in meats, vitamin B_1 sources, raw fruits/vegetables, fresh potatoes

21CFR § 182.3637
Substance: Potassium metabisulfite
Use: GRAS food substance as preservative
Limitation: Not used in meats, vitamin B_1 sources, raw fruits/vegetables, fresh potatoes

21CFR § 182.3640
Substance: Potassium sorbate
Use: GRAS food substance as preservative

21CFR § 182.3731
Substance: Sodium ascorbate
Use: GRAS food substance as preservative

21CFR § 182.3739
Substance: Sodium bisulfite
Use: GRAS food substance as preservative
Limitation: Not used in meats, vitamin B_1 sources, raw fruits/vegetables, fresh potatoes

21CFR § 182.3766
Substance: Sodium metabisulfite
Use: GRAS food substance as preservative
Limitation: Not used in meats, vitamin B_1 sources, raw fruits/vegetables, fresh potatoes

21CFR § 182.3795
Substance: Sodium sorbate
Use: GRAS food substance as preservative

21CFR § 182.3798
Substance: Sodium sulfite
Use: GRAS food substance as preservative
Limitation: Not used in meats, vitamin B_1 sources, raw fruits/vegetables, fresh potatoes

21CFR § 182.3862
Substance: Sulfur dioxide
Use: GRAS food substance as preservative
Limitation: Not used in meats, vitamin B_1 sources, raw fruits/vegetables, fresh potatoes

21CFR § 182.3890
Substance: Tocopherols
Use: GRAS food substance as preservative

21CFR § 182.5013
Substance: Ascorbic acid

Use: GRAS food substance as dietary supplement

21CFR § 182.5065
Substance: Linoleic acid
Use: GRAS food substance as dietary supplement

21CFR § 182.5159
Substance: Biotin
Use: GRAS food substance as dietary supplement

21CFR § 182.5191
Substance: Calcium carbonate
Use: GRAS food substance as dietary supplement

21CFR § 182.5195
Substance: Calcium citrate
Use: GRAS food substance as dietary supplement

21CFR § 182.5201
Substance: Calcium glycerophosphate
Use: GRAS food substance as dietary supplement

21CFR § 182.5210
Substance: Calcium oxide
Use: GRAS food substance as dietary supplement

21CFR § 182.5212
Substance: Calcium pantothenate
Use: GRAS food substance as dietary supplement

21CFR § 182.5217
Substance: Calcium phosphate (mono-, di- and tribasic)
Use: GRAS food substance as dietary supplement

21CFR § 182.5223
Substance: Calcium pyrophosphate
Use: GRAS food substance as dietary supplement

21CFR § 182.5245
Substance: Carotene
Use: GRAS food substance as dietary supplement

21CFR § 182.5250
Substance: Choline bitartrate
Use: GRAS food substance as dietary supplement

21CFR § 182.5252
Substance: Choline chloride
Use: GRAS food substance as dietary supplement

21CFR § 182.5260
Substance: Copper gluconate
Use: GRAS food substance as dietary supplement

21CFR § 182.5301
Substance: Ferric phosphate
Use: GRAS food substance as dietary supplement

21CFR § 182.5304
Substance: Ferric pyrophosphate
Use: GRAS food substance as dietary supplement

21CFR § 182.5306
Substance: Ferric sodium pyrophosphate
Use: GRAS food substance as dietary supplement

21CFR § 182.5308
Substance: Ferrous gluconate

Use: GRAS food substance as dietary supplement

21CFR § 182.5311
Substance: Ferrous lactate
Use: GRAS food substance as dietary supplement

21CFR § 182.5315
Substance: Ferrous sulfate
Use: GRAS food substance as dietary supplement

21CFR § 182.5370
Substance: Inositol
Use: GRAS food substance as dietary supplement

21CFR § 182.5375
Substance: Iron reduced
Use: GRAS food substance as dietary supplement

21CFR § 182.5431
Substance: Magnesium oxide
Use: GRAS food substance as dietary supplement

21CFR § 182.5434
Substance: Magnesium phosphate (di- and tribasic)
Use: GRAS food substance as dietary supplement

21CFR § 182.5443
Substance: Magnesium sulfate
Use: GRAS food substance as dietary supplement

21CFR § 182.5446
Substance: Manganese chloride
Use: GRAS food substance as dietary supplement

21CFR § 182.5449
Substance: Manganese citrate
Use: GRAS food substance as dietary supplement

21CFR § 182.5452
Substance: Manganese gluconate
Use: GRAS food substance as dietary supplement

21CFR § 182.5455
Substance: Manganese glycerophosphate
Use: GRAS food substance as dietary supplement

21CFR § 182.5461
Substance: Manganese sulfate
Use: GRAS food substance as dietary supplement

21CFR § 182.5464
Substance: Manganous oxide
Use: GRAS food substance as dietary supplement

21CFR § 182.5530
Substance: Niacin
Use: GRAS food substance as dietary supplement

21CFR § 182.5535
Substance: Niacinamide
Use: GRAS food substance as dietary supplement

21CFR § 182.5580
Substance: D-Pantothenyl alcohol (Panthenol)
Use: GRAS food substance as dietary supplement

21CFR § 182.5622
Substance: Potassium chloride

Use: GRAS food substance as dietary supplement
Limitation: Preps. contg. > 100 mg potassium/tablet or > 20 mg potassium/ml regulated as drugs

21CFR § 182.5628
Substance: Potassium glycerophosphate
Use: GRAS food substance as dietary supplement

21CFR § 182.5676
Substance: Pyridoxine hydrochloride
Use: GRAS food substance as dietary supplement

21CFR § 182.5695
Substance: Riboflavin
Use: GRAS food substance as dietary supplement

21CFR § 182.5697
Substance: Riboflavin-5-phosphate
Use: GRAS food substance as dietary supplement

21CFR § 182.5772
Substance: Sodium pantothenate
Use: GRAS food substance as dietary supplement

21CFR § 182.5778
Substance: Sodium phosphate (mono-, di- and tribasic)
Use: GRAS food substance as dietary supplement

21CFR § 182.5875
Substance: Thiamine hydrochloride
Use: GRAS food substance as dietary supplement

21CFR § 182.5878
Substance: Thiamine mononitrate
Use: GRAS food substance as dietary supplement

21CFR § 182.5890
Substance: Tocopherols
Use: GRAS food substance as dietary supplement

21CFR § 182.5892
Substance: α-Tocopheryl acetate
Use: GRAS food substance as dietary supplement

21CFR § 182.5930
Substance: Vitamin A (Retinol)
Use: GRAS food substance as dietary supplement

21CFR § 182.5933
Substance: Vitamin A acetate (Retinyl acetate)
Use: GRAS food substance as dietary supplement

21CFR § 182.5936
Substance: Vitamin A palmitate (Retinyl palmitate)
Use: GRAS food substance as dietary supplement

21CFR § 182.5945
Substance: Vitamin B_{12} (Cyanocobalamin)
Use: GRAS food substance as dietary supplement

21CFR § 182.5950
Substance: Vitamin D_2 (Ergocalciferol)
Use: GRAS food substance as dietary supplement

21CFR § 182.5953
Substance: Vitamin D_3 (Cholecalciferol)
Use: GRAS food substance as dietary supplement

21CFR § 182.5985
Substance: Zinc chloride
Use: GRAS food substance as dietary supplement

21CFR § 182.5988
Substance: Zinc gluconate
Use: GRAS food substance as dietary supplement

21CFR § 182.5991
Substance: Zinc oxide
Use: GRAS food substance as dietary supplement

21CFR § 182.5994
Substance: Zinc stearate
Use: GRAS food substance as dietary supplement

21CFR § 182.5997
Substance: Zinc sulfate
Use: GRAS food substance as dietary supplement

21CFR § 182.6033
Substance: Citric acid
Use: GRAS food substance as sequestrant

21CFR § 182.6085
Substance: Sodium acid phosphate
Use: GRAS food substance as sequestrant

21CFR § 182.6195
Substance: Calcium citrate
Use: GRAS food substance as sequestrant

21CFR § 182.6197
Substance: Calcium diacetate
Use: GRAS food substance as sequestrant

21CFR § 182.6203
Substance: Calcium hexametaphosphate
Use: GRAS food substance as sequestrant

21CFR § 182.6215
Substance: Monobasic Calcium phosphate
Use: GRAS food substance as sequestrant

21CFR § 182.6285
Substance: Dipotassium phosphate
Use: GRAS food substance as sequestrant

21CFR § 182.6290
Substance: Disodium phosphate
Use: GRAS food substance as sequestrant

21CFR § 182.6386
Substance: Isopropyl citrate
Use: GRAS food substance as sequestrant
Limitation: 0.02%

21CFR § 182.6511
Substance: Monoisopropyl citrate
Use: GRAS food substance as sequestrant

21CFR § 182.6625
Substance: Potassium citrate
Use: GRAS food substance as sequestrant

21CFR § 182.6751
Substance: Sodium citrate
Use: GRAS food substance as sequestrant

21CFR § 182.6757
Substance: Sodium gluconate
Use: GRAS food substance as sequestrant

21CFR § 182.6760
Substance: Sodium hexametaphosphate
Use: GRAS food substance as sequestrant

21CFR § 182.6769
Substance: Sodium metaphosphate
Use: GRAS food substance as sequestrant

21CFR § 182.6787
Substance: Sodium pyrophosphate
Use: GRAS food substance as sequestrant

21CFR § 182.6789
Substance: Tetrasodium pyrophsophate
Use: GRAS food substance as sequestrant

21CFR § 182.6778
Substance: Sodium phosphate (mono-, di- and tribasic)
Use: GRAS food substance as sequestrant

21CFR § 182.6810
Substance: Sodium tripolyphosphate
Use: GRAS food substance as sequestrant

21CFR § 182.6851
Substance: Stearyl citrate
Use: GRAS food substance as sequestrant
Limitation: 0.15%

21CFR § 182.7255
Substance: Chondrus extract (Carrageenan extract)
Use: GRAS food substance as stabilizer

21CFR § 182.8013
Substance: Ascorbic acid
Use: GRAS food substance as nutrient

21CFR § 182.8159
Substance: Biotin
Use: GRAS food substance as nutrient

21CFR § 182.8195
Substance: Calcium citrate
Use: GRAS food substance as nutrient

21CFR § 182.8217
Substance: Calcium phosphate (mono-, di- and tribasic)
Use: GRAS food substance as nutrient

21CFR § 182.8223
Substance: Calcium pyrophosphate
Use: GRAS food substance as nutrient

21CFR § 182.8250
Substance: Choline bitartrate
Use: GRAS food substance as nutrient

21CFR § 182.8252
Substance: Choline chloride
Use: GRAS food substance as nutrient

21CFR § 182.8458
Substance: Manganese hypophosphite
Use: GRAS food substance as nutrient

21CFR § 182.8778
Substance: Sodium phosphate (mono-, di- and tribasic)
Use: GRAS food substance as nutrient

21CFR § 182.8890
Substance: Tocopherols
Use: GRAS food substance as nutrient

21CFR § 182.8892
Substance: α-Tocopheryl acetate
Use: GRAS food substance as nutrient

21CFR § 182.8985
Substance: Zinc chloride
Use: GRAS food substance as nutrient

21CFR § 182.8988
Substance: Zinc gluconate
Use: GRAS food substance as nutrient

21CFR § 182.8991
Substance: Zinc oxide
Use: GRAS food substance as nutrient

21CFR § 182.8994
Substance: Zinc stearate
Use: GRAS food substance as nutrient

21CFR § 182.8997
Substance: Zinc sulfate
Use: GRAS food substance as nutrient

21CFR § 184
Substance: Direct food substances affirmed as GRAS

21CFR § 184.1005
Substance: Acetic acid
Use: Substance affirmed as GRAS; curing/pickling agent, flavor enhancer, flavoring agent/adjuvant, pH control agent, solvent, vehicle
Limitation: 0.15-9%

21CFR § 184.1007
Substance: Aconitic acid
Use: Substance affirmed as GRAS; flavoring substance and adjuvant
Limitation: 0.0005-0.003%

21CFR § 184.1009
Substance: Adipic acid
Use: Substance affirmed as GRAS; flavoring agent, leavening agent, pH control agent
Limitation: 0.0004-5%

21CFR § 184.1011
Substance: Alginic acid
Use: Substance affirmed as GRAS; emulsifier, formulation aid, stabilizer, thickener

21CFR § 184.1021
Substance: Benzoic acid

Use: Substance affirmed as GRAS; antimicrobial, flavoring agent and adjuvant
Limitation: 0.1%

21CFR § 184.1025
Substance: Caprylic acid
Use: Substance affirmed as GRAS; flavoring agent and adjuvant
Limitation: 0.001-0.04%

21CFR § 184.1027
Substance: Mixed carbohydrase and protease enzyme prod.
Use: Substance affirmed as GRAS; enzyme to hydrolyze proteins or carbohydrates

21CFR § 184.1061
Substance: Lactic acid
Use: Substance affirmed as GRAS; antimicrobial, curing/pickling agent, flavor enhancer, flavoring agent/adjuvant, pH control agent, solvent, vehicle

21CFR § 184.1065
Substance: Linoleic acid
Use: Substance affirmed as GRAS; flavoring agent and adjuvant, nutrient supplement

21CFR § 184.1069
Substance: Malic acid
Use: Substance affirmed as GRAS; flavor enhancer, flavoring agent and adjuvant, pH control agent
Limitation: 0.7-6.9%

21CFR § 184.1077
Substance: Potassium acid tartrate
Use: Substance affirmed as GRAS; anticaking agent, antimicrobial, formulation aid, humectant, leavening agent, pH control agent, processing aid, stabilizer, thickener, surface-active agent

21CFR § 184.1081
Substance: Propionic acid
Use: Substance affirmed as GRAS; antimicrobial, flavoring agent

21CFR § 184.1090
Substance: Stearic acid
Use: Substance affirmed as GRAS; flavoring agent and adjuvant

21CFR § 184.1091
Substance: Succinic acid
Use: Substance affirmed as GRAS; flavor enhancer, pH control agent
Limitation: 0.0061-0.084%

21CFR § 184.1095
Substance: Sulfuric acid
Use: Substance affirmed as GRAS; pH control agent, processing aid
Limitation: 0.0003-0.014%

21CFR § 184.1097
Substance: Tannic acid

Use: Substance affirmed as GRAS; flavoring agent and adjuvant, flavor enhancer, processing aid, pH control agent
Limitation: 0.001-0.04%

21CFR § 184.1099
Substance: Tartaric acid
Use: Substance affirmed as GRAS; firming agent, flavor enhancer, flavoring agent, humectant, pH control agent

21CFR § 184.1101
Substance: Diacetyl tartaric acid esters of mono- and diglycerides (DATEM)
Use: Substance affirmed as GRAS; emulsifier, flavoring agent and adjuvant

21CFR § 184.1115
Substance: Agar-agar
Use: Substance affirmed as GRAS; drying agent, flavoring agent, stabilizer, thickener, surface-finishing agent, formulation aid, humectant
Limitation: 0.8-2%

21CFR § 184.1120
Substance: Brown algae
Use: Substance affirmed as GRAS; flavor enhancer, flavor adjuvant for spices and seasonings

21CFR § 184.1121
Substance: Red algae
Use: Substance affirmed as GRAS; flavor enhancer, flavor adjuvant for spices and seasonings

21CFR § 184.1133
Substance: Ammonium alginate
Use: Substance affirmed as GRAS; stabilizer, thickener, humectant
Limitation: 0.1-0.5%

21CFR § 184.1135
Substance: Ammonium bicarbonate
Use: Substance affirmed as GRAS; dough strengthener, leavening agent, pH control agent, texturizer

21CFR § 184.1137
Substance: Ammonium carbonate
Use: Substance affirmed as GRAS; leavening agent, pH control agent

21CFR § 184.1138
Substance: Ammonium chloride
Use: Substance affirmed as GRAS; dough strengthener, flavor enhancer, leavening agent, processing aid

21CFR § 184.1139
Substance: Ammonium hydroxide
Use: Substance affirmed as GRAS; leavening agent, pH control agent, surface-finishing agent, boiler water additive

21CFR § 184.1141a
Substance: Ammonium phosphate monobasic

Use: Substance affirmed as GRAS; dough strength-ener, pH control agent

21CFR § 184.1141b
Substance: Ammonium phosphate dibasic
Use: Substance affirmed as GRAS; dough strength-ener, firming agent, leavening agent, pH control agent, processing aid

21CFR § 184.1143
Substance: Ammonium sulfate
Use: Substance affirmed as GRAS; dough strength-ener, firming agent, processing aid
Limitation: 0.1-0.15%

21CFR § 184.1155
Substance: Bentonite
Use: Substance affirmed as GRAS; processing aid

21CFR § 184.1157
Substance: Benzoyl peroxide
Use: Substance affirmed as GRAS; bleaching agent

21CFR § 184.1165
Substance: n-Butane and isobutane
Use: Substance affirmed as GRAS; propellant, aerating agent, gas

21CFR § 184.1185
Substance: Calcium acetate
Use: Substance affirmed as GRAS; firming agent, pH control agent, processing aid, sequestrant, stabilizer, thickener, texturizer
Limitation: 0.0001-0.2%

21CFR § 184.1187
Substance: Calcium alginate
Use: Substance affirmed as GRAS; stabilizer, thick-ener
Limitation: 0.002-0.5%

21CFR § 184.1191
Substance: Calcium carbonate
Use: Substance affirmed as GRAS

21CFR § 184.1193
Substance: Calcium chloride
Use: Substance affirmed as GRAS; anticaking agent, antimicrobial, curing/pickling agent, firm-ing agent, flavor enhancer, humectant, nutrient supplement, pH control agent, processing aid, stabilizer, thickener, surface-active agent, syn-ergist, texturizer
Limitation: 0.05-2%

21CFR § 184.1199
Substance: Calcium gluconate
Use: Substance affirmed as GRAS; firming agent, formulation aid, sequestrant, stabilizer, thick-ener, texturizer
Limitation: 0.01-4.5%

21CFR § 184.1201
Substance: Calcium glycerophosphate
Use: Substance affirmed as GRAS; nutrient supple-ment

21CFR § 184.1205
Substance: Calcium hydroxide
Use: Substance affirmed as GRAS

21CFR § 184.1206
Substance: Calcium iodate
Use: Substance affirmed as GRAS; dough strength-ener
Limitation: 0.0075% on flour (bread)

21CFR § 184.1207
Substance: Calcium lactate
Use: Substance affirmed as GRAS; firming agent, flavor enhancer, flavoring agent/adjuvant, leav-ening agent, nutrient supplement, stabilizer, thickener

21CFR § 184.1210
Substance: Calcium oxide
Use: Substance affirmed as GRAS

21CFR § 184.1212
Substance: Calcium pantothenate
Use: Substance affirmed as GRAS; nutrient supple-ment

21CFR § 184.1221
Substance: Calcium propionate
Use: Substance affirmed as GRAS; antimicrobial

21CFR § 184.1229
Substance: Calcium stearate
Use: Substance affirmed as GRAS; flavoring agent and adjuvant, lubricant, release agent, stabi-lizer, thickener

21CFR § 184.1230
Substance: Calcium sulfate
Use: Substance affirmed as GRAS; anticaking agent, color, dough strengthener, drying agent, firming agent, flour treating agent, leavening agent, nutrient supplement, pH control agent, processing aid, stabilizer, thickener, synergist, texturizer

21CFR § 184.1240
Substance: Carbon dioxide
Use: Substance affirmed as GRAS; leavening agent, processing aid, propellant, aerating agent, gas

21CFR § 184.1245
Substance: β-Carotene
Use: Substance affirmed as GRAS; nutrient supple-ment

21CFR § 184.1257
Substance: Clove and its derivs.
Use: Substance affirmed as GRAS; flavoring agent and adjuvant

21CFR § 184.1259
Substance: Cocoa butter substitute primarily from palm oil (1-Palmitoyl-2-oleoyl-3-stearin)
Use: Substance affirmed as GRAS

21CFR § 184.1260
Substance: Copper gluconate

Use: Substance affirmed as GRAS; nutrient supplement, synergist

21CFR § 184.1261
Substance: Copper sulfate (usually pentahydrate form)
Use: Substance affirmed as GRAS; nutrient supplement, processing aid

21CFR § 184.1262
Substance: Corn silk and corn silk extract
Use: Substance affirmed as GRAS; flavoring agent
Limitation: 4-30 ppm

21CFR § 184.1265
Substance: Cuprous iodide
Use: Substance affirmed as GRAS; source of dietary iodine in table salt

21CFR § 184.1271
Substance: L-Cysteine (L-2-Amino-3-mercapto-propanoic acid)
Use: Substance affirmed as GRAS; dough strengthener

21CFR § 184.1272
Substance: L-Cysteine monohydrochloride
Use: Substance affirmed as GRAS; dough strengthener

21CFR § 184.1277
Substance: Dextrin
Use: Substance affirmed as GRAS; formulation aid, processing aid, stabilizer, thickener, surface-finishing agent

21CFR § 184.1278
Substance: Diacetyl
Use: Substance affirmed as GRAS; flavoring agent and adjuvant

21CFR § 184.1282
Substance: Dill and its derivs.
Use: Substance affirmed as GRAS; flavoring agent and adjuvant

21CFR § 184.1287
Substance: Enzyme-modified fats
Use: Substance affirmed as GRAS; flavoring agent and adjuvant

21CFR § 184.1293
Substance: Ethyl alcohol (Ethanol)
Use: Substance affirmed as GRAS; antimicrobial agent on pizza crusts
Limitation: 2% (pizza crusts)

21CFR § 184.1295
Substance: Ethyl formate
Use: Substance affirmed as GRAS; flavoring agent and adjuvant
Limitation: 0.01-0.05%

21CFR § 184.1296
Substance: Ferric ammonium citrate
Use: Substance affirmed as GRAS; nutrient supplement

21CFR § 184.1297
Substance: Ferric chloride
Use: Substance affirmed as GRAS; flavoring agent

21CFR § 184.1298
Substance: Ferric citrate
Use: Substance affirmed as GRAS; nutrient supplement

21CFR § 184.1301
Substance: Ferric phosphate
Use: Substance affirmed as GRAS; nutrient supplement

21CFR § 184.1304
Substance: Ferric pyrophosphate
Use: Substance affirmed as GRAS; nutrient supplement

21CFR § 184.1307
Substance: Ferric sulfate
Use: Substance affirmed as GRAS; flavoring agent

21CFR § 184.1307a
Substance: Ferrous ascorbate
Use: Substance affirmed as GRAS; nutrient supplement

21CFR § 184.1307b
Substance: Ferrous carbonate
Use: Substance affirmed as GRAS; nutrient supplement

21CFR § 184.1307c
Substance: Ferrous citrate
Use: Substance affirmed as GRAS; nutrient supplement

21CFR § 184.1307d
Substance: Ferrous fumarate
Use: Substance affirmed as GRAS; nutrient supplement

21CFR § 184.1308
Substance: Ferrous gluconate
Use: Substance affirmed as GRAS; nutrient supplement

21CFR § 184.1311
Substance: Ferrous lactate
Use: Substance affirmed as GRAS; nutrient supplement

21CFR § 184.1315
Substance: Ferrous sulfate (heptahydrate or dried)
Use: Substance affirmed as GRAS; nutrient supplement, processing aid

21CFR § 184.1317
Substance: Garlic and its derivs.
Use: Substance affirmed as GRAS; flavoring agent and adjuvant

21CFR § 184.1318
Substance: Glucono δ-lactone

Use: Substance affirmed as GRAS; curing/pickling agent, leavening agent, pH control agent, sequestrant

21CFR § 184.1321
Substance: Corn gluten
Use: Substance affirmed as GRAS; nutrient supplement, texturizer

21CFR § 184.1322
Substance: Wheat gluten
Use: Substance affirmed as GRAS; dough strengthener, formulation aid, nutrient supplement, processing aid, stabilizer, thickener, surface-finishing agent, texturizng agent

21CFR § 184.1323
Substance: Glyceryl monooleate
Use: Substance affirmed as GRAS; flavoring agent and adjuvant, solvent, vehicle

21CFR § 184.1324
Substance: Glyceryl monostearate
Use: Substance affirmed as GRAS

21CFR § 184.1328
Substance: Glyceryl behenate
Use: Substance affirmed as GRAS; formulation aid

21CFR § 184.1330
Substance: Acacia (Gum arabic)
Use: Substance affirmed as GRAS; emulsifier, flavoring agent and adjuvant, formulation aid, stabilizer, thickener, humectant, surface-finishing agent, firming agent, texturizer
Limitation: 1-85%

21CFR § 184.1333
Substance: Gum ghatti (Indian gum)
Use: Substance affirmed as GRAS; emulsifier
Limitation: 0.1-0.2%

21CFR § 184.1339
Substance: Guar gum
Use: Substance affirmed as GRAS; emulsifier, formulation aid, stabilizer, thickener, firming agent
Limitation: 0.35-2%

21CFR § 184.1343
Substance: Locust bean gum (Carob bean gum)
Use: Substance affirmed as GRAS; stabilizer, thickener
Limitation: 0.15-0.8%

21CFR § 184.1349
Substance: Karaya gum (Sterculia gum)
Use: Substance affirmed as GRAS; formulation aid, stabilizer, thickener, emulsifier
Limitation: 0.002-0.9%

21CFR § 184.1351
Substance: Gum tragacanth
Use: Substance affirmed as GRAS; emulsifier, formulation aid, stabilizer, thickener
Limitation: 0.1-1.3%

21CFR § 184.1355
Substance: Helium
Use: Substance affirmed as GRAS; processing aid

21CFR § 184.1366
Substance: Hydrogen peroxide
Use: Substance affirmed as GRAS; antimicrobial, oxidizing and reducing agent, bleaching agent; removal/reduction of sulfur dioxide
Limitation: 0.04-1.25%

21CFR § 184.1370
Substance: Inositol
Use: Substance affirmed as GRAS; nutrient supplement, in special dietary foods

21CFR § 184.1372
Substance: Insoluble glucose isomerase enzyme preps.
Use: Substance affirmed as GRAS; enzyme to convert glucose to fructose

21CFR § 184.1375
Substance: Iron, elemental
Use: Substance affirmed as GRAS; nutrient supplement

21CFR § 184.1388
Substance: Lactase enzyme prep. from *Kluyveromyces lactis*
Use: Substance affirmed as GRAS; enzyme to convert lactose to glucose and galactose

21CFR § 184.1400
Substance: Lecithin
Use: Substance affirmed as GRAS

21CFR § 184.1408
Substance: Licorice and licorice derivs.
Use: Substance affirmed as GRAS; flavor enhancer, flavoring agent, surface-active agent
Limitation: 0.05-16%

21CFR § 184.1409
Substance: Ground limestone (essentially calcium carbonate)
Use: Substance affirmed as GRAS

21CFR § 184.1425
Substance: Magnesium carbonate (Magnesium carbonate hydroxide)
Use: Substance affirmed as GRAS; anticaking and free-flow agent, flour treating agent, lubricant, release agent, nutrient supplement, pH control agent, processing aid, synergist

21CFR § 184.1426
Substance: Magnesium chloride (hexahydrate)
Use: Substance affirmed as GRAS; flavoring agent and adjuvant, nutrient supplement

21CFR § 184.1428
Substance: Magnesium hydroxide
Use: Substance affirmed as GRAS; nutrient supplement, pH control agent, processing aid

21CFR § 184.1431
Substance: Magnesium oxide
Use: Substance affirmed as GRAS; anticaking and free-flow agent, firming agent, lubricant, release agent, nutrient supplement, pH control agent

21CFR § 184.1434
Substance: Magnesium phosphate (dibasic and tribasic)
Use: Substance affirmed as GRAS; nutrient supplement, pH control agent

21CFR § 184.1440
Substance: Magnesium stearate
Use: Substance affirmed as GRAS; lubricant, release agent, nutrient supplement, processing aid

21CFR § 184.1443
Substance: Magnesium sulfate (heptahdyrate)
Use: Substance affirmed as GRAS; flavor enhancer, nutrient supplement, processing aid

21CFR § 184.1444
Substance: Maltodextrin
Use: Substance affirmed as GRAS; nonsweet nutrient

21CFR § 184.1445
Substance: Malt syrup (Malt extract)
Use: Substance affirmed as GRAS; flavoring agent and adjuvant

21CFR § 184.1446
Substance: Manganese chloride
Use: Substance affirmed as GRAS; nutrient supplement

21CFR § 184.1449
Substance: Manganese citrate
Use: Substance affirmed as GRAS; nutrient supplement

21CFR § 184.1452
Substance: Manganese gluconate
Use: Substance affirmed as GRAS; nutrient supplement

21CFR § 184.1461
Substance: Manganese sulfate
Use: Substance affirmed as GRAS; nutrient supplement

21CFR § 184.1472
Substance: Hydrogenated and partially hydrogenated menhaden oils
Use: Substance affirmed as GRAS; edible fats and oils

21CFR § 184.1490
Substance: Methylparaben (Methyl p-hydroxybenzoate)
Use: Substance affirmed as GRAS; antimicrobial

21CFR § 184.1498
Substance: Microparticulated protein prod.

Use: Substance affirmed as GRAS; thickener, texturizer

21CFR § 184.1505
Substance: Mono- and diglycerides
Use: Substance affirmed as GRAS; dough strengthener, emulsifier, flavoring agent/adjuvant, formulation aid, lubricant, release agent, solvent, vehicle, stabilizer, thickener, surface-active agent, surface-finishing agent, texturizer

21CFR § 184.1521
Substance: Monosodium phosphate derivs. of mono- and diglycerides
Use: Substance affirmed as GRAS; emulsifier, lubricant, release agent, surface-active agent

21CFR § 184.1530
Substance: Niacin
Use: Substance affirmed as GRAS; nutrient supplement

21CFR § 184.1535
Substance: Niacinamide (nicotinamide)
Use: Substance affirmed as GRAS; nutrient supplement

21CFR § 184.1537
Substance: Nickel
Use: Substance affirmed as GRAS; catalyst; in hydrogenation of fats and oils

21CFR § 184.1538
Substance: Nisin prep.
Use: Substance affirmed as GRAS; antimicrobial agent

21CFR § 184.1540
Substance: Nitrogen
Use: Substance affirmed as GRAS; propellant, aerating agent, gas

21CFR § 184.1545
Substance: Nitrous oxide
Use: Substance affirmed as GRAS; propellant, aerating agent, gas

21CFR § 184.1553
Substance: Peptones
Use: Substance affirmed as GRAS; nutrient supplement, processing aid, surface-active agent

21CFR § 184.1555
Substance: Rapeseed oil (fully hydrog., superglycerinated fully hydrog., or low erucic acid)
Use: Substance affirmed as GRAS; emulsifier
Limitation: Various

21CFR § 184.1560
Substance: Ox bile extract
Use: Substance affirmed as GRAS; surfactant
Limitation: 0.002% (cheese)

21CFR § 184.1563
Substance: Ozone
Use: Substance affirmed as GRAS; antimicrobial
Limitation: 0.4 mg/l (bottled water)

21CFR § 184.1585
Substance: Papain
Use: Substance affirmed as GRAS; enzyme, processing aid, texturizer

21CFR § 184.1588
Substance: Pectins
Use: Substance affirmed as GRAS; emulsifier, stabilizer, thickener

21CFR § 184.1610
Substance: Potassium alginate
Use: Substance affirmed as GRAS; stabilizer, thickener
Limitation: 0.01-0.7%

21CFR § 184.1613
Substance: Potassium bicarbonate
Use: Substance affirmed as GRAS; formulation aid, nutrient supplement, pH control agent, processing aid

21CFR § 184.1619
Substance: Potassium carbonate
Use: Substance affirmed as GRAS; flavoring agent/adjuvant, nutrient supplement, pH control agent, processing aid

21CFR § 184.1622
Substance: Potassium chloride
Use: Substance affirmed as GRAS; flavor enhancer, flavoring agent, nutrient supplement, pH control agent, stabilizer, thickener

21CFR § 184.1631
Substance: Potassium hydroxide
Use: Substance affirmed as GRAS; formulation aid, pH control agent, processing aid, stabilizer, thickener

21CFR § 184.1634
Substance: Potassium iodide
Use: Substance affirmed as GRAS; nutrient supplement, source of dietary iodine in table salt
Limitation: 0.01% (table salt)

21CFR § 184.1635
Substance: Potassium iodate
Use: Substance affirmed as GRAS; dough strengthener
Limitation: 0.0075% on flour (bread)

21CFR § 184.1639
Substance: Potassium lactate
Use: Substance affirmed as GRAS; flavor enhancer, flavroing agent/adjuvant, humectant, pH control agent

21CFR § 184.1643
Substance: Potassium sulfate
Use: Substance affirmed as GRAS; flavoring agent and adjuvant
Limitation: 0.015% (nonalcoholic beverages)

21CFR § 184.1655
Substance: Propane

Use: Substance affirmed as GRAS; propellant, aerating agent, gas

21CFR § 184.1660
Substance: Propyl gallate
Use: Substance affirmed as GRAS; antioxidant
Limitation: 0.02% of fat or oil (total antioxidants)

21CFR § 184.1666
Substance: Propylene glycol
Use: Substance affirmed as GRAS; anticaking agent, antioxidant, dough strengthener, emulsifier, flavor agent, formulation aid, humectant, processing aid, solvent, vehicle, stabilizer, thickener, surface-active agent, texturizer
Limitation: 2.0-97%

21CFR § 184.1670
Substance: Propylparaben (Propyl p-hydroxybenzoate)
Use: Substance affirmed as GRAS; antimicrobial
Limitation: 0.1% in food

21CFR § 184.1676
Substance: Pyridoxine hydrochloride
Use: Substance affirmed as GRAS; nutrient supplement

21CFR § 184.1685
Substance: Rennet (animal-derived) and chymosin (fermentation-derived) prep.
Use: Substance affirmed as GRAS; enzyme, processing aid, stabilizer, thickener

21CFR § 184.1695
Substance: Riboflavin
Use: Substance affirmed as GRAS; nutrient supplement

21CFR § 184.1697
Substance: Riboflavin-5´-phosphate (sodium)
Use: Substance affirmed as GRAS; nutrient supplement

21CFR § 184.1698
Substance: Rue
Use: Substance affirmed as GRAS

21CFR § 184.1699
Substance: Rue oil
Use: Substance affirmed as GRAS; flavoring agent and adjuvant
Limitation: 4-10 ppm

21CFR § 184.1721
Substance: Sodium acetate (anhyd. or trihydrate)
Use: Substance affirmed as GRAS; flavoring agent and adjuvant, pH control agent
Limitation: 0.007-0.6%

21CFR § 184.1724
Substance: Sodium alginate
Use: Substance affirmed as GRAS; texturizer, formulation aid, stabilizer, thickener, firming agent, flavor enhancer, flavor adjuvant, emulsifier, processing aid, surface-active agent

Limitation: 0.3-10%

21CFR § 184.1733
Substance: Sodium benzoate
Use: Substance affirmed as GRAS; antimicrobial agent, flavoring agent and adjuvant
Limitation: 0.1% in food

21CFR § 184.1736
Substance: Sodium bicarbonate
Use: Substance affirmed as GRAS

21CFR § 184.1742
Substance: Sodium carbonate
Use: Substance affirmed as GRAS; antioxidant, curing/pickling agent, flavoring agent and adjuvant, pH control agent, processing aid

21CFR § 184.1754
Substance: Sodium diacetate
Use: Substance affirmed as GRAS; antimicrobial agent, flavoring agent and adjuvant, pH control agent
Limitation: 0.05-0.4%

21CFR § 184.1763
Substance: Sodium hydroxide
Use: Substance affirmed as GRAS; pH control agent, processing aid

21CFR § 184.1764
Substance: Sodium hypophosphite
Use: Substance affirmed as GRAS; emulsifier, stabilizer; in cod liver oil emulsions

21CFR § 184.1768
Substance: Sodium lactate
Use: Substance affirmed as GRAS; emulsifier, flavor enhancer, flavoring agent or adjuvant, humectant, pH control agent

21CFR § 184.1769a
Substance: Sodium metasilicate
Use: Substance affirmed as GRAS; processing aid; in washing/lye peeling of fruits, vegetables, nuts; denuding agent in tripe; hog scald agent; corrosion preventative in bottled water

21CFR § 184.1784
Substance: Sodium propionate
Use: Substance affirmed as GRAS; antimicrobial agent, flavoring agent

21CFR § 184.1792
Substance: Sodium sesquicarbonate
Use: Substance affirmed as GRAS; pH control agent

21CFR § 184.1801
Substance: Sodium tartrate
Use: Substance affirmed as GRAS; emulsifier, pH control agent

21CFR § 184.1804
Substance: Sodium potassium tartrate
Use: Substance affirmed as GRAS; emulsifier, pH control agent

21CFR § 184.1807
Substance: Sodium thiosulfate pentahydrate (Sodium hyposulfite)
Use: Substance affirmed as GRAS; formulation aid, reducing agent
Limitation: 0.00005-0.1%

21CFR § 184.1835
Substance: Sorbitol
Use: Substance affirmed as GRAS; anticaking/free-flow, curing/pickling, drying agent, emulsifier, firming, flavoring, formulation aid, humectant, lubricant, release, nutritive sweetener, sequestrant, stabilizer, thickener, surface-finishing, texturizer
Limitation: 12-99%

21CFR § 184.1845
Substance: Stannous chloride (anhyd. and dihydrate)
Use: Substance affirmed as GRAS; antioxidant
Limitation: 0.0015% in food (as tin)

21CFR § 184.1848
Substance: Starter distillate
Use: Substance affirmed as GRAS; flavoring agent and adjuvant

21CFR § 184.1854
Substance: Sucrose
Use: Substance affirmed as GRAS

21CFR § 184.1857
Substance: Corn sugar
Use: Substance affirmed as GRAS

21CFR § 184.1859
Substance: Invert sugar
Use: Substance affirmed as GRAS

21CFR § 184.1865
Substance: Corn syrup (Glucose syrup)
Use: Substance affirmed as GRAS

21CFR § 184.1875
Substance: Thiamine hydrochloride
Use: Substance affirmed as GRAS; flavoring agent and adjuvant, nutrient supplement

21CFR § 184.1878
Substance: Thiamine mononitrate
Use: Substance affirmed as GRAS; nutrient supplement

21CFR § 184.1890
Substance: α-Tocopherols (d-, dl-)
Use: Substance affirmed as GRAS; inhibitor of nitrosamine formation

21CFR § 184.1901
Substance: Triacetin (Glyceryl triacetate)
Use: Substance affirmed as GRAS; flavoring agent/adjuvant, formulation aid, humectant, solvent, vehicle

21CFR § 184.1903
Substance: Tributyrin (Glyceryl tributyrate)

Use: Substance affirmed as GRAS; flavoring agent and adjuvant

21CFR § 184.1923
Substance: Urea
Use: Substance affirmed as GRAS; formulation aid, fermentation aid

21CFR § 184.1924
Substance: Urease enzyme prep. from *Lactobacillus fermentum*
Use: Substance affirmed as GRAS; enzyme to convert urea to ammonia and carbon dioxide; in wine

21CFR § 184.1930
Substance: Vitamin A (Retinol), Vitamin A acetate (Retinyl acetate), Vitamin A palmitate (Retinyl palmitate)
Use: Substance affirmed as GRAS; nutrient supplement

21CFR § 184.1945
Substance: Vitamin B_{12} (Cyanocobalamin)
Use: Substance affirmed as GRAS; nutrient supplement

21CFR § 184.1950
Substance: Vitamin D_2 (Ergocalciferol), Vitamin D_3 (Cholecalciferol)
Use: Substance affirmed as GRAS; nutrient supplement
Limitation: 42-350 I/100 g

21CFR § 184.1973
Substance: Beeswax (yellow and white)
Use: Substance affirmed as GRAS; flavoring agent and adjuvant, lubricant, surface-finishing agent
Limitation: 0.002-0.1%

21CFR § 184.1976
Substance: Candelilla wax
Use: Substance affirmed as GRAS; lubricant, surface-finishing agent

21CFR § 184.1978
Substance: Carnauba wax
Use: Substance affirmed as GRAS; anticaking agent, formulation aid, lubricant, release agent, surface-finishing agent

21CFR § 184.1979
Substance: Whey (sweet or acid whey, conc. whey, dried whey)
Use: Substance affirmed as GRAS

21CFR § 184.1979a
Substance: Reduced lactose whey
Use: Substance affirmed as GRAS

21CFR § 184.1979b
Substance: Reduced minerals whey
Use: Substance affirmed as GRAS

21CFR § 184.1979c
Substance: Whey protein conc.
Use: Substance affirmed as GRAS

21CFR § 184.1983
Substance: Bakers yeast extract
Use: Substance affirmed as GRAS; flavoring agent and adjuvant

21CFR § 184.1984
Substance: Zein
Use: Substance affirmed as GRAS; surface-finishing agent

21CFR § 186
Substance: Indirect food substances affirmed as GRAS

21CFR § 186.1
Substance: Substances added indirectly to human food affirmed as GRAS

21CFR § 186.1025
Substance: Caprylic acid
Use: Indirect food substance affirmed GRAS; antimicrobial, preservative

21CFR § 186.1093
Substance: Sulfamic acid
Use: Indirect food substance affirmed GRAS

21CFR § 186.1256
Substance: Clay (Kaolin)
Use: Indirect food substance affirmed GRAS; mfg. of food pkg. paper/paperboard

21CFR § 186.1275
Substance: Dextrans
Use: Indirect food substance affirmed GRAS; constituent of food-contact surfaces

21CFR § 186.1316
Substance: Formic acid
Use: Indirect food substance affirmed GRAS; constituent of food pkg. paper/paperboard

21CFR § 186.1330
Substance: Ferric oxide
Use: Indirect food substance affirmed GRAS; constituent of food pkg. paper/paperboard

21CFR § 186.1374
Substance: Iron oxides
Use: Indirect food substance affirmed GRAS; constituent of food pkg. paper/paperboard

21CFR § 186.1551
Substance: Hydrogenated fish oil
Use: Indirect food substance affirmed GRAS; constituent of cotton/cotton fabrics in dry food pkg.

21CFR § 186.1557
Substance: Tall oil
Use: Indirect food substance affirmed GRAS; constituent of cotton/cotton fabrics in dry food pkg.

21CFR § 186.1673
Substance: Pulp
Use: Indirect food substance affirmed GRAS; constituent of food pkg. containers, paper/paperboard

21CFR § 186.1750
Substance: Sodium chlorite
Use: Indirect food substance affirmed GRAS; slimicide in mfg. of paper/paperboard contacting food
Limitation: 125-250 ppm (in paper)

21CFR § 186.1756
Substance: Sodium formate
Use: Indirect food substance affirmed GRAS; constituent of food pkg. paper/paperboard

21CFR § 186.1770
Substance: Sodium oleate
Use: Indirect food substance affirmed GRAS; constituent of food pkg. paper/paperboard; component of lubricants with incidental food contact

21CFR § 186.1771
Substance: Sodium palmitate
Use: Indirect food substance affirmed GRAS; constituent of food pkg. paper/paperboard

21CFR § 186.1797
Substance: Sodium sulfate (Glauber's salt)
Use: Indirect food substance affirmed GRAS; constituent of food pkg. paper/paperboard, cotton/cotton fabric in dry food pkg.

21CFR § 186.1839
Substance: Sorbose
Use: Indirect food substance affirmed GRAS; constituent of cotton, cotton fabrics, paper, paperboard in contact with dry foods

21CFR § 189
Substance: Substances prohibited from use in human food

21CFR § 189.110
Substance: Calamus and derivs.
Use: Substances prohibited from direct addition or use as human food; has been used as a flavoring compound

21CFR § 189.113
Substance: Cinnamyl anthranilate
Use: Substances prohibited from direct addition or use as human food; has been used as flavoring agent in foods

21CFR § 189.120
Substance: Cobaltous salts and derivs.
Use: Substances prohibited from direct addition or use as human food; have been used as foam stabilizer in fermented malt beverages

21CFR § 189.130
Substance: Coumarin
Use: Substances prohibited from direct addition or use as human food; has been used as flavoring compd.

21CFR § 197
Seafood inspection program

EEC E Numbers

E100

Substance: Curcumin
Use: Colorant
Limitation: ADI 0-0.1 mg/kg body wt. (temporary)

E101

Substance: Riboflavin (Vitamin B$_2$; Lactoflavin)
Use: Colorant, vitamin
Limitation: ADI 0-0.5 mg/kg body wt.

E101a

Substance: Riboflavin-5´-phosphate (riboflavin-5´-phosphate sodium)
Use: Colorant, vitamin
Limitation: ADI 0-0.5 mg/kg body wt. (temporary)

E102

Substance: Tartrazine (CI 19140; FD&C Yellow No. 5)
Use: Colorant
Limitation: ADI 0-7.5 mg/kg body wt.; prohibited in Norway, Austria

E104

Substance: Quinoline yellow (CI 47005)
Use: Colorant
Limitation: ADI 0-0.5 mg/kg body wt.; prohibited in Norway, U.S., Australia, Japan

E107

Substance: Yellow 2G (Food Yellow 5)
Use: Colorant
Limitation: no ADI; total ban proposed (EEC); prohibited in EEC (except UK), Japan, U.S.

E110

Substance: Sunset Yellow FCF (CI 15885; FD&C Yellow No. 6)
Use: Colorant
Limitation: ADI 0-2.5 mg/kg body wt.; ADI reduced by 50% by Joint FAO/WHO Expert Comm. on Food Additives

E120

Substance: Cochineal (Carminic acid; CI 75470; Natural Red 4)
Use: Colorant
Limitation: ADI 0-2.5 mg/kg body wt.

E122

Substance: Carmoisine (Azorubine; CI 14720)
Use: Colorant

Limitation: ADI 0-4 mg/kg body wt.; prohibited in U.S., Japan, Norway, Sweden

E123

Substance: Amaranth (CI 16185; FD&C Red No. 2)
Use: Colorant
Limitation: ADI 0-0.75 mg/kg body wt. (temporary); prohibited in U.S., Norway; permitted only in caviar (France)

E124

Substance: Ponceau 4R (CI 16255)
Use: Colorant
Limitation: ADI 0-4 mg/kg body wt.; prohibited in U.S., Norway

E127

Substance: Erythrosine (CI 45430; FD&C Red No. 3)
Use: Colorant
Limitation: ADI 0-1.25 mg/kg body wt.; 0.1 mg/kg body wt. (FAC); prohibited in U.S., Norway

E128

Substance: Red 2G (CI 18050)
Use: Colorant
Limitation: ADI 0-0.1 mg/kg body wt.; FAC recommendations restricted to 20 mg/kg max. in meat prods., veg. protein meat analogs; banned in EEC, U.S., Canada, Japan, Australia

129

Substance: Allura Red (CI 16035; Food Red 17; FD&C Red No. 40)
Use: Colorant
Limitation: ADI 0-0.7 mg/kg body wt.; prohibited in EEC, Japan, Norway, Sweden, Finland, Austria

E131

Substance: Patent Blue V (CI 42051)
Use: Colorant
Limitation: no ADI

E132

Substance: Indigo carmine (Indigotine; CI 73015; FD&C Blue No. 2)
Use: Colorant
Limitation: ADI 0.5 mg/kg body wt.; prohibited in Norway

E133

Substance: Brilliant Blue FCF (CI 42090; FD&C Blue No. 1)
Use: Colorant

Limitation: ADI 0-12.5 mg/kg body wt.; prohibited in several countries

E140
Substance: Chlorophyll (CI 75810)
Use: Colorant
Limitation: ADI not limited

E141
Substance: Copper complexes of chlorophyll/chlorophyllins (CI 75810)
Use: Colorant
Limitation: ADI 0-15 mg/kg body wt. as sum of both complexes

E142
Substance: Green S (Acid Brilliant Green; Food Green S; CI 44090)
Use: Colorant
Limitation: ADI 0-5 mg/kg body wt. (temporary); prohibited in Japan, Canada, U.S., Norway, Sweden, Finland

E150
Substance: Caramel color
Use: Colorant
Limitation: ADI 1000-10,000 mg/kg (FAC)

E150a
Substance: Plain (spirit) caramel
Use: Colorant
Limitation: ADI no limit

E150b
Substance: Caustic sulfite caramel
Use: Colorant

E150c
Substance: Ammonia caramel
Use: Colorant
Limitation: ADI 1-100 mg/kg body wt. (temporary)

E151
Substance: Black PN (Brilliant Black PN; CI 28440)
Use: Colorant
Limitation: ADI 0-1 mg/kg body wt.; prohibited in U.S., Canada, Japan, Norway, Finland

E153
Substance: Carbon black (Vegetable carbon)
Use: Colorant
Limitation: ADI decision postponed; banned by U.S. FDA

E154
Substance: Brown FK (Food brown)
Use: Colorant
Limitation: no ADI; prohibited in EEC (except UK, Irish Rep.), U.S., Canada, Japan, etc.

E155
Substance: Brown HT (CI 20285; Chocolate brown HT)
Use: Colorant
Limitation: ADI 0-0.25 mg/kg body wt. (temporary); prohibited in U.S., France, Norway, Sweden, Germany, etc.

E160a
Substance: α-Carotene, β-Carotene, γ-Carotene
Use: Colorant; becomes vitamin A in the body
Limitation: no ADI

E160b
Substance: Annatto, Bixin, Norbixin (CI 75120)
Use: Colorant
Limitation: ADI 0-0.065 mg/kg body wt. (as bixin)

E160c
Substance: Capsanthin (Capsorubin; Paprika extract; Oleoresin)
Use: Colorant, spice extract

E160d
Substance: Lycopene (CI 75125)
Use: Colorant
Limitation: Proposed for deletion

E160e
Substance: β-Apo-8′-carotenal (C30; β-8′-Apocarotenal)
Use: Colorant
Limitation: ADI 0-5 mg/kg body wt. as sum of the three carotenoids)

E160f
Substance: Ethyl ester of β-apo-8′-carotenoic acid
Use: Colorant

E161
Substance: Xanthophylls
Use: Colorant
Limitation: no ADI

E161a
Substance: Flavoxanthin (CI 75135)
Use: Colorant
Limitation: no ADI

E161b
Substance: Lutein (CI 75135)
Use: Colorant
Limitation: no ADI; proposed for deletion

E161c
Substance: Cryptoxanthin (CI 75135)
Use: Colorant
Limitation: no ADI

E161d
Substance: Rubixanthin (CI 75135)
Use: Colorant
Limitation: no ADI; proposed for deletion

E161e
Substance: Violoxanthin (CI 75135)
Use: Colorant
Limitation: no ADI; proposed for deletion

E161f
Substance: Rhodoxanthin (CI 75135)
Use: Colorant
Limitation: no ADI; proposed for deletion

E161g
Substance: Canthaxanthin (CI 75135)

Use: Colorant
Limitation: ADI 0-25 mg/kg body wt. (1974), expected to be reduced to 0.05 mg/kg

E162

Substance: Beetroot red (Betanin; Betanidin)
Use: Colorant
Limitation: no ADI

E163

Substance: Anthocyanins (Schultz 1394 and 1400)
Use: Colorant

E170

Substance: Calcium carbonate (Chalk; CI 77220)
Use: Colorant, alkali, firming agent, releasing agent, calcium supplement

E171

Substance: Titanium dioxide (CI 77891)
Use: Colorant, opacifier
Limitation: ADI no limit; prohibited in Germany

E172

Substance: Iron oxides, iron hydroxides (CI 77492, red: 77491, brown: 77499)
Use: Colorant
Limitation: ADI 0-0.5 mg/kg body wt.; prohibited in Germany

E173

Substance: Aluminum (CI 77000)
Use: Colorant for surface only
Limitation: no ADI; for external covering only

E174

Substance: Silver (CI 77820)
Use: Surface colorant
Limitation: ADI decision postponed

E175

Substance: Gold (CI 77480)
Use: Surface colorant
Limitation: no ADI

E180

Substance: Pigment rubine (Lithol rubine BK; CI 15850)
Use: Colorant
Limitation: no ADI; solely for coloring rind of hard cheese

E200

Substance: Sorbic acid
Use: Preservative
Limitation: ADI 0-25 mg/kg body wt.

E201

Substance: Sodium sorbate
Use: Preservative

E202

Substance: Potassium sorbate
Use: Antifungal/antibacterial preservative
Limitation: ADI 0-2.5 mg/kg body wt.

E203

Substance: Calcium sorbate
Use: Antifungal/antibacterial preservative
Limitation: ADI 0-2.5 mg/kg body wt.

E210

Substance: Benzoic acid
Use: Antifungal/antibacterial preservative, effective only in acidic media
Limitation: ADI 0-5 mg/kg body wt.

E211

Substance: Sodium benzoate
Use: Antifungal/antibacterial preservative, effective only in sl. acidic media
Limitation: ADI 0-5 mg/kg body wt.

E212

Substance: Potassium benzoate
Use: Antifungal/antibacterial preservative
Limitation: ADI 0-5 mg/kg body wt.

E213

Substance: Calcium benzoate
Use: Antifungal/antibacterial preservative
Limitation: ADI 0-5 mg/kg body wt.

E214

Substance: Ethyl 4-hydroxybenzoate (Ethyl p-hydroxybenzoate; Ethyl paraben)
Use: Antifungal/antibacterial preservative
Limitation: ADI 0-10 mg/kg body wt.

E215

Substance: Ethyl 4-hydroxybenzoate sodium salt
Use: Antifungal/antibacterial preservative

E216

Substance: Propyl 4-hydroxybenzoate (Propyl p-hydroxybenzoate; Propyl paraben)
Use: Antimicrobial preservative
Limitation: ADI 0-10 mg/kg body wt.

E217

Substance: Propyl 4-hydroxybenzoate sodium salt
Use: Antimicrobial preservative

E218

Substance: Methyl 4-hydroxybenzoate (Methyl p-hydroxybenzoate; Methylparaben)
Use: Antimicrobial preservative
Limitation: ADI 0-10 mg/kg body wt.

E219

Substance: Methyl 4-hydroxybenzoate sodium salt
Use: Preservative, effective against fungi and yeasts, less active against bacteria
Limitation: ADI 0-10 mg/kg body wt.

E220

Substance: Sulfur dioxide
Use: Preservative; bleaching agent; improving agent; dough modifier; vitamin C stabilizer
Limitation: ADI 0-0.7 mg/kg body wt.

E221

Substance: Sodium sulfite

Use: Preservative; sterilizer for equip./containers; prevents oxidative discoloration; dough modifier
Limitation: ADI 0-0.7 mg/kg body wt.

E222
Substance: Sodium hydrogen sulfite (Sodium bisulfite; Acid sodium sulfite)
Use: Preservative for alcoholic beverages
Limitation: ADI 0-0.7 mg/kg body wt.

E223
Substance: Sodium metabisulfite (Disodium pyrosulfite)
Use: Antimicrobial preservative; antioxidant; bleaching agent
Limitation: ADI 0-0.7 mg/kg body wt.

E224
Substance: Potassium metabisulfite (Potassium pyrosulfite)
Use: Antimicrobial preservative; antibrowning agent
Limitation: ADI 0-0.7 mg/kg body wt.

E226
Substance: Calcium sulfite
Use: Preservative; firming agent; disinfectant in brewing vats

E227
Substance: Calclum hydrogen sulfite (Calcium bisulfite)
Use: Preservative; firming agent

E230
Substance: Biphenyl (Diphenyl)
Use: Fungistat; preservative
Limitation: ADI 0-0.05 mg/kg body wt.

E231
Substance: 2-Hydroxybiphenyl (o-Phenylphenol)
Use: Antibacterial and antifungal preservative
Limitation: ADI 0-0.02 mg/kg body wt.

E232
Substance: Sodium biphenyl-2-yl oxide (Sodium o-phenylphenol)
Use: Antifungal preservative
Limitation: ADI 0-0.02 mg/kg body wt.

E233
Substance: 2-(Thiazol-4-yl) benzimidazole (Thiabendazole; Thiaben)
Use: Preservative, fungicide

E234
Substance: Nisin
Use: Preservative
Limitation: ADI 33,000 units/kg body wt.

E236
Substance: Formic acid (Methanoic acid)
Use: Preservative, antibacterial; flavor adjunct
Limitation: ADI 0-3 mg/kg body wt.; prohibited in UK

E237
Substance: Sodium formate

Use: Preservative
Limitation: Prohibited in UK

E238
Substance: Calcium formate
Use: Preservative
Limitation: Prohibited in UK

E239
Substance: Hexamine (Hexamethylenetetramine)
Use: Antimicrobial preservative
Limitation: ADI 0-0.15 mg/kg body wt.

E249
Substance: Potassium nitrite
Use: Curing agent, preservative for meat
Limitation: ADI 0-0.2 mg/kg body wt. (temporary)

E250
Substance: Sodium nitrite
Use: Preservative, curing salt, colorant
Limitation: ADI 0-0.2 mg/kg body wt. (temporary)

E251
Substance: Sodium nitrate (Chile saltpetre)
Use: Preservative, curing salt, color fixative
Limitation: ADI 0-5 mg/kg body wt.

E252
Substance: Potassium nitrate (Saltpetre)
Use: Preservative, curing salt, color fixative

E260
Substance: Acetic acid
Use: Antibacterial preservative; acidity stabilizer; diluent for colors; flavoring agent
Limitation: ADI no limit

E261
Substance: Potassium acetate
Use: Preserves natural color of plant/animal tissues; buffer; neutralizer
Limitation: ADI no limit

E262
Substance: Sodium hydrogen diacetate (Sodium diacetate)
Use: Acidity regulator, sequestrant, antimicrobial preservative
Limitation: ADI 0-15 mg/kg body wt.

E262
Substance: Sodium acetate and Sodium acetate anhyd.
Use: Buffer (acid or alkaline stabilizer)
Limitation: ADI no limit

E263
Substance: Calcium acetate
Use: Antimold agent, anti-rope agent; sequestrant; firming agent; stabilizer; buffering agent
Limitation: ADI no limit

E270
Substance: Lactic acid
Use: Preservative; antioxidant synergist; acidulant; flavoring; in malting process

Limitation: ADI no limit

E280
Substance: Propionic acid
Use: Antifungal preservative
Limitation: ADI no limit

E281
Substance: Sodium propionate
Use: Antimicrobial preservative
Limitation: ADI no limit

E282
Substance: Calcium propionate
Use: Antimicrobial preservative
Limitation: ADI no limit

E283
Substance: Potassium propionate
Use: Preservative, mold inhibitor

E290
Substance: Carbon dioxide
Use: Preservative; coolant; freezant (liq. form); packaging gas; aerator

E296
Substance: Malic acid (DL- or L-)
Use: Acid, flavoring

E297
Substance: Fumaric acid
Use: Acidifier; flavoring; raising agent; antioxidant
Limitation: ADI 0-6 mg/kg body wt.

E300
Substance: L-Ascorbic acid (Vitamin C)
Use: Vitamin; browning inhibitor; meat color preservative; antioxidant in brewing

E301
Substance: Sodium L-ascorbate
Use: Vitamin; antioxidant; color preservative

E302
Substance: Calcium L-ascorbate
Use: Vitamin; antioxidant; meat color preservative

E304
Substance: 6-O-Palmitoyl-L-ascorbic acid (Ascorbyl palmitate)
Use: Vitamin; antioxidant; color preservative; antibrowning agent; synergistic with α-tocopherol

E306
Substance: Extracts of natural origin rich in tocopherols (natural Vitamin E)
Use: Vitamin; antioxidant

E307
Substance: Synthetic α-tocopherol (synthetic Vitamin E; DL-α-Tocopherol)
Use: Antioxidant; vitamin

E308
Substance: Synthetic γ-tocopherol (synthetic Vitamin E; DL-γ-Tocopherol)

Use: Antioxidant; vitamin

E309
Substance: Synthetic δ-tocopherol (synthetic Vitamin E; DL-δ-Tocopherol)
Use: Antioxidant; vitamin

E310
Substance: Propyl gallate (Propyl 3,4,5-trihydroxy-benzoate)
Use: Antioxidant in oils and fats; synergistic with BHA and BHT
Limitation: ADI 0-0.5 mg/kg body wt. for total gallates

E311
Substance: Octyl gallate
Use: Antioxidant
Limitation: ADI 0-0.5 mg/kg body wt. for total gallates

E312
Substance: Dodecyl gallate (Dodecyl 3,4,5-trihydroxybenzoate)
Use: Antioxidant
Limitation: ADI 0-0.5 mg/kg body wt. for total gallates

E320
Substance: Butylated hydroxyanisole (BHA)
Use: Antioxidant preventing flavor deterioration, rancidity
Limitation: ADI 0-0.5 mg/kg body wt. (temporary)

E321
Substance: Butylated hydroxytoluene (BHT)
Use: Antioxidant preventing flavor deterioration, rancidity
Limitation: ADI 0-0.5 mg/kg body wt. (temporary)

E322
Substance: Lecithins
Use: Antioxidant; emulsifier; bread improver; protects vitamin A

E325
Substance: Sodium lactate
Use: Humectant; glycerol substitute; synergistic with antioxidants; bodying agent
Limitation: ADI no limit

E326
Substance: Potassium lactate
Use: Synergistic with antioxidants; buffer
Limitation: ADI no limit

E327
Substance: Calcium lactate
Use: Antioxidant; synergistic with other antioxidants; buffer; firming agent; inhibits discoloration; yeast food; dough conditioner
Limitation: ADI no limit

E330
Substance: Citric acid

Use: Synergistic with other antioxidants; inhibits discoloration; stabilizes acidity; sequestrant; flavoring; helps jam to set; in malting, wine prod.
Limitation: ADI no limit

E331
Substance: Sodium citrates
Use: Synergistic with other antioxidants; buffer controlling acidity; emulsifying salt; sequestrant; prevents formation of large milk curds
Limitation: ADI no limit

E331a
Substance: Sodium dihydrogen citrate (Monosodium citrate)
Use: Synergistic with other antioxidants; buffer controlling acidity; emulsifying salt; sequestrant; prevents formation of large milk curds
Limitation: ADI no limit

E331b
Substance: Disodium citrate
Use: Antioxidant; synergistic with other antioxidants; buffer; emulsifying salt

E331c
Substance: Trisodium citrate (Citrosodine)
Use: Antioxidant; buffer; emulsifying salt; sequestrant; stabilizer; used with polyphosphates and flavors to Inject Into chlckens before freezing
Limitation: ADI no limit

E332
Substance: Potassium dihydrogen citrate (Monopotassium citrate)
Use: Buffer; emulsifying salt; yeast food

E332
Substance: Tripotassium citrate (Potassium citrate)
Use: Antioxidant; buffer; emulsifying salt; sequestrant
Limitation: ADI no limit

E333
Substance: Mono-, di-, and tricalcium citrate
Use: Buffer; firming agent; emulsifying salt; sequestrant; improves baking props. of flour
Limitation: ADI no limit

E334
Substance: L-(+)-Tartaric acid
Use: Antioxidant; synergist for other antioxidants; acidity regulator; sequestrant; diluent for food colors; flavor constituent; acid in some baking powds.
Limitation: ADI 0-30 mg/kg body wt.

E335
Substance: Monosodium L-(+)-tartrate and Disodium L-(+)-tartrate
Use: Antioxidant; synergist for other antioxidants; buffer; emulsifying salt; sequestrant
Limitation: ADI 0-30 mg/kg body wt.

E336
Substance: Monopotassium L-(+)-tartrate (Potassium hydrogen tartrate; Cream of tartar)
Use: Acid; buffer; emulsifying salt; raising agent; inverting agent for sugar
Limitation: ADI 0-30 mg/kg body wt. (as L-(+)-tartaric acid)

E336
Substance: Dipotassium L-(+)-tartrate
Use: Antioxidant; synergist for other antioxidants; buffer; emulsifying salt
Limitation: ADI 0-30 mg/kg body wt. (as L-(+)-tartaric acid)

E337
Substance: Potassium sodium L-(+)-tartrate (Sodium potassium tartrate; Rochelle salt)
Use: Buffer; emulsifying salt; stabilizer; synergist for antioxidants
Limitation: ADI 0-30 mg/kg body wt.

E338
Substance: Orthophosphoric acid (Phosphoric acid)
Use: Synergist for antioxidants; acidulant; flavoring agent; in malting process; sequestrant
Limitation: ADI 0-70 mg/kg body wt.

E339a
Substance: Sodium dihydrogen orthophosphate
Use: Texturizer; aids brine penetration; antioxidant synergist; buffer; nutrient; gellant; stabilizer; sugar clarifying agent
Limitation: ADI 0-70 mg/kg body wt.

E339b
Substance: Disodium dihydrogen orthophosphate
Use: Texturizer; aids brine penetration; antioxidant synergist; buffer; nutrient; gellant; stabilizer; sugar clarifying agent
Limitation: ADI 0-70 mg/kg body wt.

E339c
Substance: Trisodium dihydrogen orthophosphate
Use: Texturizer; aids brine penetration; antioxidant synergist; buffer; nutrient; gellant; stabilizer; sugar clarifying agent
Limitation: ADI 0-70 mg/kg body wt.

E340a
Substance: Potassium dihydrogen orthophosphate (Monopotassium phosphate; MKP)
Use: Buffer; sequestrant; emulsifying salt; antioxidant synergist

E340b
Substance: Dipotassium hydrogen orthophosphate (Dipotassium phosphate; DKP)
Use: Buffer; sequestrant; emulsifying salt; antioxidant synergist; yeast food

E340c
Substance: Tripotassium orthophosphate (Potassium phosphate tribasic)
Use: Emulsifying salt; antioxidant synergist; buffer; sequestrant

E341

Substance: Calcium orthophosphates
Use: Dough improver; buffer; firming agent; sequestrant; yeast food; aerator; acidulant; antioxidant synergist; texturizer
Limitation: ADI no limit

E341a

Substance: Calcium tetrahydrogen diorthophosphate (Acid calcium phosphate; MCP)
Use: Dough improver; buffer; firming agent; sequestrant; yeast food; aerator; acidulant; antioxidant synergist; texturizer
Limitation: ADI no limit

E341b

Substance: Calcium hydrogen orthophosphate (Dicalcium phosphate; DCP)
Use: Firming agent; yeast food; nutrient mineral supplement; antioxidant synergist; dough conditioner; abrasive in toothpaste; animal feed supplement

E341c

Substance: Tricalcium diorthophosphate (Tricalcium phosphate)
Use: Anticaking agent; nutrient yeast food; diluent; clarifying agent for sugar syrups

E350

Substance: Sodium malate
Use: Buffer; seasoning agent

E350

Substance: Sodium hydrogen malate
Use: Buffer

E351

Substance: Potassium malate
Use: Buffer

E352

Substance: Calcium malate
Use: Buffer; firming agent; seasoning agent

E352

Substance: Calcium hydrogen malate
Use: Firming agent

E353

Substance: Metatartaric acid
Use: Sequestrant esp. in wine

E355

Substance: Adipic acid (Hexanedioic acid)
Use: Acidulant; buffer; neutralizer; flavoring agent; raising agent in baking powds.
Limitation: ADI 0-5 mg/kg body wt. (free acid basis)

E363

Substance: Succinic acid
Use: Acidulant; buffer; neutralizer

E370

Substance: 1,4-Heptonolactone
Use: Acid; sequestrant

E375

Substance: Nicotinic acid (Niacin; Nicotinamide)
Use: Nutrient; B vitamin; color protector

E380

Substance: Triammonium citrate (Ammonium citrate tribasic)
Use: Buffer; emulsifying salt; softening agent for cheese spreads

E381

Substance: Ammonium ferric citrate (Ferric ammonium citrate)
Use: Dietary iron supplement

E385

Substance: Calcium disodium EDTA
Use: Chelating agent preventing discoloration, rancidity, off-odors; antioxidant
Limitation: ADI 0-2.5 mg/kg body wt.; permitted in UK only in canned fish, shellfish

E400

Substance: Alginic acid
Use: Food additive (rarely used)
Limitation: ADI 0-50 mg/kg body wt.

E401

Substance: Sodium alginate
Use: Stabilizer; suspending agent; thickener; calcium source; gellant; copper fining agent in brewing
Limitation: ADI 0-50 mg/kg body wt.

E402

Substance: Potassium alginate (Potassium polymannuronate)
Use: Emulsifier; stabilizer; boiled water additive; gellant
Limitation: ADI 0-25 mg/kg body wt. (as alginic acid)

E403

Substance: Ammonium alginate
Use: Emulsifier; stabilizer; color diluent; thickener
Limitation: ADI 0-50 mg/kg body wt.

E404

Substance: Calcium alginate
Use: Emulsifier; stabilizer; thickener; gellant
Limitation: ADI 0-25 mg/kg body wt. (as alginic acid)

E405

Substance: Propane-1,2-diol alginate (Propylene glycol alginate)
Use: Emulsifier; stabilizer; thickener; solvent for extracts, flavors, spices; foam stabilizer
Limitation: ADI 0-25 mg/kg body wt.

E406

Substance: Agar (Agar-agar; Japanese isinglass)
Use: Thickening agent, stabilizer, gellant, humectant; copper firming agent in brewing
Limitation: ADI no limit

E407

Substance: Carrageenan (Irish moss)

Use: Stabilizer, thickener, suspending and gelling agent; bodying agent; texturizer

E410
Substance: Locust bean gum (Carob bean gum)
Use: Gellant; stabilizer; emulsifier; thickener; texture modifier; carrageenan modifier

E412
Substance: Guar gum (Guar flour; Jaguar gum)
Use: Thickener; emulsion stabilizer; suspending agent; dietary bulking agent; helps diabetics control blood sugar levels

E413
Substance: Tragacanth (Gum dragon; Gum tragacanth)
Use: Emulsifier; stabilizer; thickener; crystallization inhibitor in sugar confectionery

E414
Substance: Gum arabic (Acacia)
Use: Emulsifier; stabilizer; thickener; crystallization inhibitor in sugar confectionery

E415
Substance: Xanthan gum (Corn sugar gum)
Use: Stabilizer; thickener; emulsifier; gellant
Limitation: ADI 0-10 mg/kg body wt.

E416
Substance: Karaya gum (Sterculia gum; Indian tragacanth)
Use: Stabilizer; thickener; emulsifier; prevents ice crystal formation in ice cream; fat and juice binder in sausages; filler
Limitation: ADI 0-20 mg/kg body wt. (temporary)

E420i
Substance: Sorbitol
Use: Sweetening agent; glycerol substitute; extends shelf life; masks bitter aftertaste of saccharin; humectant; stabilizer; texturizer

E421
Substance: Mannitol (Manna sugar)
Use: Texturizing agent; dietary supplement; humectant; sweetener; anticaking agent; antisticking agent

E422
Substance: Glycerol (Glycerin)
Use: Solvent for oily flavors; humectant; sweetener; bodying agent; plasticizer in edible coatings for meat and cheese

E430
Substance: Polyoxyethylene (8) stearate (PEG-8 stearate)
Use: Emulsifier; stabilizer
Limitation: ADI 0-25 mg/kg body wt.

E431
Substance: Polyoxyethylene (40) stearate (PEG-40 stearate)
Use: Emulsifier; freshness additive in bread

Limitation: ADI 0-25 mg/kg body wt.

E432
Substance: Polyoxyethylene (20) sorbitan laurate (Polysorbate 20)
Use: Emulsifier; stabilizer; dispersant
Limitation: ADI 0-25 mg/kg body wt.

E433
Substance: Polyoxyethylene (20) sorbitan oleate (Polysorbate 80)
Use: Emulsifier, defoamer in sugar beet prod.; stabilizer; aids moistness in rolls, doughnuts; helps sol. of nondairy coffee whiteners
Limitation: ADI 0-25 mg/kg body wt.

E434
Substance: Polyoxyethylene (20) sorbitan palmitate (Polysorbate 40)
Use: Emulsifier; stabilizer; dispersant (esp. for flavors); defoaming agent; wetting agent for powd. processed foods
Limitation: ADI 0-25 mg/kg body wt.

E435
Substance: Polyoxyethylene (20) sorbitan stearate (Polysorbate 60)
Use: Emulsifier; stabilizer; prevents oil and water separation; keeps bread moist; wetting/dispersing agent for powd. processed foods; foaming agent
Limitation: ADI 0-25 mg/kg body wt.

E436
Substance: Polyoxyethylene (20) sorbitan tristearate (Polysorbate 65)
Use: Emulsifier; stabilizer; prevents oil/water separation; keeps bread moist; wetting agent for powd. processed foods; defoamer; flavor dispersant
Limitation: ADI 0-25 mg/kg body wt.

E440a
Substance: Pectin
Use: Emulsifier; gellant; thickener; bodying agent; stabilizer

E440b
Substance: Amidated pectin
Use: Emulsifier; stabilizer; gellant; thickener

E442
Substance: Ammonium phosphatides (Emulsifier YN)
Use: Stabilizer; emulsifier

E450a
Substance: Disodium dihydrogen diphosphate (Disodium pyrophosphate; DSPP)
Use: Buffer; sequestrant; emulsifier; raising agent; color improver; chelating agent
Limitation: ADI 0-70 mg/kg body wt. (total phosphates)

E450a
Substance: Trisodium diphosphate

Use: Buffer; sequestrant; emulsifier; color improver; chelating agent
Limitation: ADI 0-70 mg/kg body wt. (total phosphates)

E450a
Substance: Tetrasodium diphosphate (Tetrasodium pyrophosphate)
Use: Buffer; sequestrant; emulsifier; gellant; stabilizer
Limitation: ADI 0-70 mg/kg body wt. (total phosphates)

E450a
Substance: Tetrapotassium diphosphate
Use: Emulsifying salt; buffer; sequestrant; stabilizer
Limitation: ADI 0-70 mg/kg body wt. (total phosphates)

E450b
Substance: Pentasodium triphosphate (STPP; Sodium tripolyphosphate)
Use: Emulsifying salt; texturizer; buffer; sequestrant; stabilizer
Limitation: ADI 0-70 mg/kg body wt. (total phosphates)

E450b
Substance: Pentapotassium triphosphate (Potassium tripolyphosphate)
Use: Emulsifying salt; texturizer; buffer; sequestrant; stabilizer
Limitation: ADI 0-70 mg/kg body wt. (total phosphates)

E450c
Substance: Sodium polyphosphates
Use: Emulsifying salt; texturizer; sequestrant; stabilizer
Limitation: ADI 0-70 mg/kg body wt. (total phosphates)

E450c
Substance: Potassium polyphosphates
Use: Emulsifying salt; sequestrant; stabilizer
Limitation: ADI 0-70 mg/kg body wt. (total phosphates)

E460
Substance: Microcrystalline cellulose
Use: Nonnutritive bulking agent; binder; anticaking agent; dietary fiber; hydration aid; emulsion/heat stabilizer; binder/disintegrant; carrier; dispersant

E460
Substance: α-Cellulose (Powdered cellulose)
Use: Bulking aid; anticaking agent; binder; dispersant; thickener; filter aid; beer clarifying aid

E461
Substance: Methylcellulose
Use: Emulsifier; stabilizer; thickener; bulking agent; binder; film-former; substitute for water-sol. gums; foods for diabetics; fat barrier

E463
Substance: Hydroxypropylcellulose
Use: Stabilizer; emulsifier; thickener; suspending agent

E464
Substance: Hydroxypropylmethyl cellulose (Hypromellose)
Use: Gellant; suspending agent; emulsifier; stabilizer; thickener; fat barrier

E465
Substance: Ethylmethylcellulose (Methylethylcellulose)
Use: Emulsifier; foam stabilizer; thickener; suspending agent
Limitation: ADI 0-25 mg/kg body wt.

E466
Substance: Carboxymethylcellulose sodium salt (CMC)
Use: Thickener; texture modifier; stabilizer; moisture migration control; gellant; nonnutritive bulking agent; prevents crystal growth, syneresis; foam stabilizer; decreases fat absorption
Limitation: ADI 0-25 mg/kg body wt.

E470
Substance: Sodium, potassium, and calcium salts of fatty acids (soaps)
Use: Emulsifier; stabilizer; anticaking agent

E471
Substance: Mono- and diglycerides of fatty acids (Glyceryl stearate; Glyceryl distearate)
Use: Emulsifier; stabilizer; thickener; in cakes, retains foaming power of egg protein in presence of fat
Limitation: ADI no limit

E472a
Substance: Acetic acid esters of mono- and diglycerides of fatty acids
Use: Emulsifier; stabilizer; coating agent; texture modifying agent; solvent; lubricant
Limitation: ADI no limit

E472b
Substance: Lactic acid esters of mono- and diglycerides of fatty acids
Use: Emulsifier; stabilizer
Limitation: ADI no limit

E472c
Substance: Citric acid esters of mono- and diglycerides of fatty acids
Use: Emulsifier; stabilizer
Limitation: ADI no limit

E472d
Substance: Tartaric acid esters of mono- and diglycerides of fatty acids
Use: Emulsifier; stabilizer
Limitation: ADI no limit

E472e

Substance: Acetylated tartaric acid esters of mono-
and diglycerides of fatty acids
Use: Emulsifier; stabilizer
Limitation: ADI no limit

E473

Substance: Sucrose esters of fatty acids
Use: Emulsifier; stabilizer

E474

Substance: Sucroglycerides
Use: Emulsifier; stabilizer
Limitation: ADI 0-10 mg/kg body wt.

E475

Substance: Polyglycerol esters of fatty acids
Use: Emulsifier; stabilizer
Limitation: ADI 0-25 mg/kg body wt.

E476

Substance: Polyglycerol polyricinoleate
Use: Emulsifier; stabilizer; with lecithin, improves
fluidity of chocolate for coating
Limitation: ADI 0-75 mg/kg body wt.

E477

Substance: Propane-1,2-diol esters of fatty acids
(Propylene glycol esters)
Use: Emulsifier; stabilizer
Limitation: ADI 0-25 mg/kg body wt.

E478

Substance: Lactylated fatty acid esters of glycerol
and propane-1,2-diol
Use: Emulsifier; stabilizer; whipping agent; plasticizers; surface-active agents

E481

Substance: Sodium stearoyl-2-lactylate
Use: Emulsifier; stabilizer
Limitation: ADI 0-20 mg/kg body wt. (temporary)

E482

Substance: Calcium stearoyl-2-lactylate
Use: Emulsifier; stabilizer; whipping aid
Limitation: ADI 0-20 mg/kg body wt.

E483

Substance: Stearyl tartrate
Use: Emulsifier; stabilizer; flour treatment agent
Limitation: ADI 0-500 ppm

E491

Substance: Sorbitan monostearate
Use: Emulsifier; stabilizer; glazing agent
Limitation: ADI 0-25 mg/kg body wt. (total sorbitan
esters)

E492

Substance: Sorbitan tristearate
Use: Emulsifier; stabilizer
Limitation: ADI 0-25 mg/kg body wt. (total sorbitan
esters)

E493

Substance: Sorbitan monolaurate

Use: Emulsifier; stabilizer; antifoam
Limitation: ADI 0-25 mg/kg body wt. (total sorbitan
esters)

E494

Substance: Sorbitan monooleate
Use: Emulsifier; stabilizer
Limitation: ADI 0-25 mg/kg body wt. (total sorbitan
esters)

E495

Substance: Sorbitan monopalmitate
Use: Emulsifier; stabilizer
Limitation: ADI 0-25 mg/kg body wt. (total sorbitan
esters)

E500

Substance: Sodium carbonate
Use: Base; used in beer malting process
Limitation: ADI no limit

E500

Substance: Sodium hydrogen carbonate (Sodium
bicarbonate; Baking soda)
Use: Base; aerating agent; diluent
Limitation: ADI no limit

E500

Substance: Sodium sesquicarbonate
Use: Base

E501

Substance: Potassium carbonate and Potassium
hydrogen carbonate
Use: Base; alkali
Limitation: ADI no limit

E503

Substance: Ammonium carbonate (Hartshorn)
Use: Buffer; neutralizing agent; leavening agent

E503

Substance: Ammonium hydrogen carbonate (Ammonium bicarbonate)
Use: Alkali; buffer; aerating agent; raising agent

E504

Substance: Magnesium carbonate (Magnesite)
Use: Alkali; anticaking agent; acidity regulator;
antibleaching agent
Limitation: ADI no limit

E507

Substance: Hydrochloric acid
Use: Acid; in malting process in beer-making
Limitation: ADI no limit

E508

Substance: Potassium chloride
Use: Gellant; salt substitute; dietary supplement; in
malt process for beer-making

E509

Substance: Calcium chloride
Use: Sequestrant; firming agent; in malt process for
beer-making
Limitation: ADI no limit

E510

Substance: Ammonium chloride
Use: Yeast food, esp. in beer-brewing; flavor
Limitation: ADI no limit

E513

Substance: Sulfuric acid
Use: Acid; in malting process in beer-making; processing aid

E514

Substance: Sodium sulfate (Glauber's salt)
Use: Diluent; in malt process in beer-making

E515

Substance: Potassium sulfate
Use: Salt substitute for dietetic use

E516

Substance: Calcium sulfate (Gypsum; Plaster of Paris)
Use: Firming agent; sequestrant; nutrient; yeast food; inert excipient; in malt process for beer-making
Limitation: ADI no limit

E518

Substance: Magnesium sulfate (Epsom salts)
Use: Dietary supplement; firming agent; in beer-making

E524

Substance: Sodium hydroxide (Caustic soda; Lye)
Use: Base; neutralizer; color solvent; in mfg. of caramel; oxidizer; in malting process in beer-making; processing aid
Limitation: ADI no limit

E525

Substance: Potassium hydroxide
Use: Base; oxidizing agent; processing aid

E526

Substance: Calcium hydroxide
Use: Firming agent; neutralizer; in malting process in beer-making; processing aid
Limitation: ADI no limit

E527

Substance: Ammonium hydroxide
Use: Food coloring diluent and solvent; alkali; processing aid
Limitation: ADI no limit

E528

Substance: Magnesium hydroxide
Use: Alkali; processing aid
Limitation: ADI no limit

E529

Substance: Calcium oxide (Quicklime)
Use: Alkali; nutrient; processing aid
Limitation: ADI no limit

E530

Substance: Magnesium oxide (Periclase)
Use: Anticaking agent; alkali

Limitation: ADI no limit

E535

Substance: Sodium ferrocyanide (Sodium hexacyanoferrate II)
Use: Anticaking agent; crystal modifier
Limitation: ADI 0-0.025 mg/kg body wt.

E536

Substance: Potassium ferrocyanide (Potassium hexacyanoferrate II)
Use: Anticaking agent esp. in table salt; removes excess metals in wine prod.
Limitation: ADI 0-0.25 mg/kg body wt. (as sodium ferrocyanide)

E540

Substance: Dicalcium diphosphate (Dicalcium pyrophosphate)
Use: Neutralizing agent; dietary supplement; buffering agent; yeast food
Limitation: ADI 0-70 mg/kg body wt. (total phosphates)

E541

Substance: Sodium aluminum phosphate acidic (SAP)
Use: Aerator; acidulant; raising agent
Limitation: ADI 0-6 mg/kg body wt. (temporary, as sodium aluminum phosphates)

E541

Substance: Sodium aluminum phosphate basic
Use: Emulsifying salt in processed cheese mfg. in U.S.
Limitation: ADI 0-6 mg/kg body wt. (temporary, as sodium aluminum phosphates)

E542

Substance: Edible bone phosphate
Use: Anticaking agent; mineral supplement; filler in tablet-making

E544

Substance: Calcium polyphosphates
Use: Emulsifying salt; mineral supplement; calcium source; firming agent
Limitation: ADI 0-70 mg/kg body wt. (as phosphorus)

E545

Substance: Ammonium polyphosphates
Use: Emulsifier; emulsifying salt; sequestrant; yeast food; stabilizer
Limitation: ADI 0-70 mg/kg body wt. (as phosphorus)

E551

Substance: Silicon dioxide (Silica)
Use: Suspending agent; anticaking agent; thickener; stabilizer; beer clarifying agent; filtration aid
Limitation: ADI no limit

E552

Substance: Calcium silicate

Use: Anticaking agent; antacid (pharmacology); glazing, polishing, release agent (sweets); dusting agent (chewing gum); coating agent (rice); suspending agent
Limitation: ADI no limit

E553a
Substance: Magnesium silicate synthetic and Magnesium trisilicate
Use: Anticaking agent; tablet excipient; antacid (pharmacology); glazing, polishing, release agent (sweets); dusting agent (chewing gum); coating agent (rice)

E553b
Substance: Talc (French chalk; Magnesium hydrogen metasilicate)
Use: Release agent; anticaking agent; chewing gum component; filtering aid; dusting powd.

E554
Substance: Aluminum sodium silicate (Sodium aluminosilicate)
Use: Anticaking agent
Limitation: ADI no limit

E556
Substance: Aluminum calcium silicate (Calcium aluminum silicate)
Use: Anticaking agent

E558
Substance: Bentonite (Soap clay)
Use: Anticaking agent; clarifying agent (wine); filtration aid; suspending agent; emulsifier

E559
Substance: Kaolin, heavy and light (Aluminum silicate)
Use: Anticaking agent; clarifying agent (wine)
Limitation: ADI no limit

E570
Substance: Stearic acid
Use: Anticaking agent; clarifying agent (wine)

E572
Substance: Magnesium stearate
Use: Anticaking agent; emulsifier; release agent
Limitation: ADI no limit

E575
Substance: D-Glucono-1,5-lactone (Glucono δ-lactone)
Use: Acid; sequestrant; prevents milkstone formation in dairies, beerstone formation in breweries
Limitation: ADI 0-50 mg/kg body wt. (as total gluconic acid)

E576
Substance: Sodium gluconate
Use: Sequestrant; dietary supplement
Limitation: ADI 0-50 mg/kg body wt. (as total gluconic acid)

E577
Substance: Potassium gluconate

Use: Sequestrant
Limitation: ADI 0-50 mg/kg body wt. (as total gluconic acid)

E578
Substance: Calcium gluconate
Use: Buffer; firming agent; sequestrant
Limitation: ADI 0-50 mg/kg body wt. (as total gluconic acid)

E620
Substance: L-Glutamic acid
Use: Dietary supplement; flavor enhancer; salt substitute
Limitation: ADI 0-120 mg/kg body wt.

E621
Substance: Monosodium glutamate (MSG; Sodium hydrogen L-glutamate)
Use: Flavor enhancer; ingred. where reduction of sodium intake desired
Limitation: ADI 0-120 mg/kg body wt.

E622
Substance: Potassium hydrogen L-glutamate (Potassium glutamate)
Use: Flavor enhancer; salt substitute

E623
Substance: Calcium dihydrogen di-L-glutamate
Use: Flavor enhancer; salt substitute
Limitation: ADI 0-120 mg/kg body wt.

E627
Substance: Guanosine 5´-disodium phosphate (Sodium guanylate)
Use: Flavor enhancer

E631
Substance: Inosine 5´-disodium phosphate (Sodium 5´-inosinate)
Use: Flavor enhancer

E635
Substance: Sodium 5´-ribonucleotide
Use: Flavor enhancer

E636
Substance: Maltol
Use: Flavoring agent imparting freshly baked flavor and smell to bread and cakes
Limitation: ADI 0-1 mg/kg body wt.

E637
Substance: Ethyl maltol
Use: Flavoring agent imparting sweet taste; flavor enhancer
Limitation: ADI 0-2 mg/kg body wt.

E900
Substance: Dimethylpolysiloxane (Simethicone; Dimethicone)
Use: Water repellent; antifoaming agent; chewing gum base; anticaking agent; in brewing industry; in fermentation
Limitation: ADI 0-1.5 mg/kg body wt.

E901
Substance: Beeswax, white and yellow
Use: Glazing/polishing agent; release agent

E903
Substance: Carnauba wax
Use: Glazing/polishing agent
Limitation: Permitted only in chocolate prods.

E904
Substance: Shellac
Use: Glazing/polishing agent
Limitation: 0.4% max.

E905
Substance: Mineral hydrocarbons
Use: Polishes; glazing agents; sealing agents; chewing gum component; defoamer in yeast/sugar beet processing; coating; lubricant; binder; release agent
Limitation: 2000 ppm max. (as release agent)

E907
Substance: Refined microcrystalline wax
Use: Chewing gum ingred.; polishing and release agent; stiffening agent; tablet coating
Limitation: May be carcinogenic

E920
Substance: L-Cysteine hydrochloride and L-Cysteine hydrochloride monohydrate
Use: Improving agent for flour; flavor

E924
Substance: Potassium bromate
Use: Flour-maturing or improving agent; in malting process for beer-making
Limitation: ADI 0-75 mg/kg body wt. (temporary)

E925
Substance: Chlorine
Use: Antibacterial and antifungal preservative; bleaching, ageing, and oxidizing agent
Limitation: Powerful irritant

E926
Substance: Chlorine dioxide (Chlorine peroxide)
Use: Bleaching, improving, and oxidizing agent for flour; bleaching agent for oils and fats, beeswax; purification of water; taste and odor control; bactericide; antiseptic
Limitation: Gas is highly irritating; ADI 0-30 mg/kg body wt. (conditional)

E927
Substance: Azodicarbonamide (Azoformamide; Azo bisformamide)

Japanese Food Additive Regulations

Absinthe extract
Use/Usage level: Natural bittering agent

Acetic acid glacial
Use/Usage level: Acidity regulator, food acid

α-Acetolactate decarboxylase
Use/Usage level: Natural enzyme

Acetone
Use/Usage level: Processing aid
Limitation: Restricted to extraction in guarana beverage prep., fractionation of fats/oils; remove prior to prep. of final food

Acetophenone
Use/Usage level: Flavorings
Limitation: Restricted to flavoring use

N-Acetylglucosamine
Use/Usage level: Natural sweetener

Acylase
Use/Usage level: Natural enzyme

Adipic acid
Use/Usage level: Acidity regulator, food acid, raising agent

Agarase
Use/Usage level: Natural enzyme

DL-Alanine
Use/Usage level: Dietary supplement (amino acid); flavorings

L-Alanine
Use/Usage level: Natural dietary supplement; flavorings

Alginic acid
Use/Usage level: Natural thickener/stabilizer

Aliphatic higher alcohols
Use/Usage level: Flavorings
Limitation: Restricted to flavoring use

Aliphatic higher aldehydes
Use/Usage level: Flavorings
Limitation: Except toxic substances; restricted to flavoring use

Aliphatic higher hydrocarbons
Use/Usage level: Flavorings

Limitation: Except toxic substances; restricted to flavoring use

Alkanet
Use/Usage level: Natural flavorings

Alkanet color
Use/Usage level: Color; not permitted in certain foods

Allspice
Use/Usage level: Natural flavorings

Allyl cyclohexylpropionate
Use/Usage level: Flavorings
Limitation: Restricted to flavoring use

Allyl hexanoate
Use/Usage level: Flavorings
Limitation: Restricted to flavoring use

Allyl Isothiocyanate
Use/Usage level: Flavorings
Limitation: Restricted to flavoring use

Almond
Use/Usage level: Natural flavorings

Aloe
Use/Usage level: Natural flavorings

Aloe extract
Use/Usage level: Natural thickener/stabilizer

Aloe vera extract
Use/Usage level: Natural thickener/stabilizer

Aluminum
Use/Usage level: Color; not permitted in certain foods

Aluminum ammonium sulfate
Use/Usage level: Raising agent
Limitation: Not for use in miso

Aluminum potassium sulfate
Use/Usage level: Raising agent
Limitation: Not for use in miso

Ambergris
Use/Usage level: Natural flavorings

Ambrette
Use/Usage level: Natural flavorings

Ammonia
Use/Usage level: Processing aid

Ammonium bicarbonate
Use/Usage level: Raising agent

Ammonium carbonate
Use/Usage level: Raising agent; yeast nutrient

Ammonium chloride
Use/Usage level: Raising agent; yeast nutrient

Ammonium dihydrogen phosphate
Use/Usage level: Emulsifier for cheese food, yeast nutrient, fermentation aid

Ammonium persulfate
Use/Usage level: Flour treatment agent (0.3 g/kg max.)

Ammonium sulfate
Use/Usage level: Yeast nutrient, fermentation aid

α-Amylase
Use/Usage level: Natural enzyme

β-Amylase
Use/Usage level: Natural enzyme

α-Amylcinnamaldehyde
Use/Usage level: Flavorings
Limitation: Restricted to flavoring use

Angelica
Use/Usage level: Natural flavorings

Angostura
Use/Usage level: Natural flavorings

Anisaldehyde
Use/Usage level: Flavorings
Limitation: Restricted to flavoring use

Anise
Use/Usage level: Natural flavorings

Annatto, water-sol.
Use/Usage level: Color

Annatto extract
Use/Usage level: Color; not permitted in certain foods

Anthocyanase
Use/Usage level: Natural enzyme

Apricot
Use/Usage level: Natural flavorings

Arabinogalactan
Use/Usage level: Natural thickener/stabilizer

L-Arabinose
Use/Usage level: Natural sweetener

Areca nut
Use/Usage level: Natural flavorings

L-Arginine
Use/Usage level: Natural dietary supplement; flavorings

L-Arginine L-glutamate
Use/Usage level: Dietary supplement (amino acid); flavorings

Arnica
Use/Usage level: Natural flavorings

Aromatic alcohols
Use/Usage level: Flavorings
Limitation: Restricted to flavoring use

Aromatic aldehydes
Use/Usage level: Flavorings
Limitation: Except toxic substances; restricted to flavoring use

Artemisia
Use/Usage level: Natural flavorings

Artichoke
Use/Usage level: Natural flavorings

Asafetida
Use/Usage level: Natural flavorings

L-Ascorbic acid
Use/Usage level: Antioxidant, dietary supplement, raising agent

L-Ascorbyl palmitate
Use/Usage level: Antioxidant, dietary supplement (vitamin)

L-Ascorbyl stearate
Use/Usage level: Antioxidant, dietary supplement (vitamin)

L-Asparagine
Use/Usage level: Natural dietary supplement; flavorings

Aspartame
Use/Usage level: Sweetener

L-Aspartic acid
Use/Usage level: Natural dietary supplement; flavorings

Basil
Use/Usage level: Natural flavorings

Bay
Use/Usage level: Natural flavorings

Beeswax
Use/Usage level: Natural chewing gum base; glazing agent

Beet red
Use/Usage level: Color; not permitted in certain foods

Bentonite
Use/Usage level: Processing aid (0.5% max. residual)
Limitation: Restricted

Benzaldehyde
Use/Usage level: Flavorings

Limitation: Restricted to flavoring use

Benzoic acid
Use/Usage level: Preservative (0.6-2.5 g/kg as benzoic acid)
Limitation: Limitation with sorbic acid 1 g/kg total

Benzoin
Use/Usage level: Natural flavorings

Benzoin gum
Use/Usage level: Natural chewing gum base

Benzoyl peroxide
Use/Usage level: Flour treatment agent (0.3 g/kg max.)
Limitation: Limited to diluted form

Benzyl acetate
Use/Usage level: Flavorings
Limitation: Restricted to flavoring use

Benzyl alcohol
Use/Usage level: Flavorings
Limitation: Restricted to flavoring use

Benzyl propionate
Use/Usage level: Flavorings
Limitation: Restricted to flavoring use

Bergamot
Use/Usage level: Natural flavorings

BHA
Use/Usage level: Antioxidant (0.2-1.0 g/kg)

BHT
Use/Usage level: Antioxidant (0.2 1.0 g/kg)

Bisbenthiamine
Use/Usage level: Dietary supplement (vitamin)

Blackberry color
Use/Usage level: Color; not permitted in certain foods

Black currant color
Use/Usage level: Color; not permitted in certain foods

Blueberry color
Use/Usage level: Color; not permitted in certain foods

Borage
Use/Usage level: Natural flavorings

d-Borneol
Use/Usage level: Flavorings
Limitation: Restricted to flavoring use

Bromelain
Use/Usage level: Natural enzyme

Butyl acetate
Use/Usage level: Flavorings
Limitation: Restricted to flavoring use

Butyl butyrate
Use/Usage level: Flavorings

Limitation: Restricted to flavoring use

Butyl p-hydroxybenzoate
Use/Usage level: Preservative (0.012-1 g/kg as p-hydroxybenzoic acid)

Butyric acid
Use/Usage level: Flavorings
Limitation: Restricted to flavoring use

Cacao
Use/Usage level: Natural flavorings

Cade
Use/Usage level: Natural flavorings

Caffeine extract
Use/Usage level: Natural bittering agent

Caffeine extract
Use/Usage level: Natural bittering agent

Cajeput
Use/Usage level: Natural flavorings

Calamus
Use/Usage level: Natural flavorings

Calcium carbonate
Use/Usage level: Chewing gum base (2% max. as calcium); dietary supplement (1% max. as calcium); raising agent (1% max. as calcium); yeast nutrient (1% max. as calcium)
Limitation: Restricted

Calcium carboxymethyl cellulose
Use/Usage level: Thickener/stabilizer/gellant (2% max.)
Limitation: Restricted

Calcium chloride
Use/Usage level: Coagulant for tofu (1% max.); dietary supplement (1% max. as calcium)

Calcium citrate
Use/Usage level: Dietary supplement (1% max. as calcium); emulsifier (1% max. as calcium); flavorings for taste-related purposes (1% max. as calcium); raising agent (1% max. as calcium)

Calcium dihydrogen phosphate
Use/Usage level: Dietary supplement (1% max. as calcium); emulsifier (1% max. as calcium); raising agent (1% max. as calcium); yeast nutrient (1% max. as calcium)
Limitation: Restricted

Calcium dihydrogen pyrophosphate
Use/Usage level: Dietary supplement (1% max. as calcium); emulsifier (1% max. as calcium); raising agent (1% max. as calcium)
Limitation: Restricted

Calcium disodium EDTA
Use/Usage level: Antioxidant (0.035 g/kg as calcium disodium EDTA)

Calcium gluconate
Use/Usage level: Dietary supplement (1% max. as calcium)
Limitation: Restricted

Calcium glycerophosphate
Use/Usage level: Dietary supplement (1% max. as calcium)
Limitation: Restricted

Calcium hydroxide
Use/Usage level: Dietary supplement (1% max. as calcium)
Limitation: Restricted

Calcium lactate
Use/Usage level: Dietary supplement (1% max. as calcium); flavorings for taste-related purposes (1% max. as calcium); raising agent (1% max. as calcium)

Calcium monohydrogen phosphate
Use/Usage level: Chewing gum base (1% max. as calcium); dietary supplement (1% max. as calcium); emulsifier (1% max. as calcium); raising agent (1% max. as calcium); yeast nutrient (1% max. as calcium)
Limitation: Restricted

Calcium pantothenate
Use/Usage level: Dietary supplement (1% max. as calcium)

Calcium propionate
Use/Usage level: Preservative (2.5-3 g/kg as propionic acid)
Limitation: Limitation with sorbic acid 3 g/kg total

Calcium 5´-ribonucleotide
Use/Usage level: Flavorings

Calcium stearoyl lactylate
Use/Usage level: Emulsifier for bread (5.5 g/kg max.), cake (8 g/kg max.), confectionery (5 g/kg max.), pasta (4.5 g/kg max.)
Limitation: Restricted

Calcium sulfate
Use/Usage level: Coagulant for tofu (1% max.); dietary supplement (1% max. as calcium); raising agent (1% max. as calcium); yeast nutrient (1% max. as calcium)
Limitation: Restricted

Camomile
Use/Usage level: Natural flavorings

Camphor tree
Use/Usage level: Natural flavorings

Candelilla wax
Use/Usage level: Natural chewing gum base; glazing agent

Caper
Use/Usage level: Natural flavorings

Capsicum
Use/Usage level: Natural flavorings

Caramel color
Use/Usage level: Color; processing aid
Limitation: Not permitted in certain foods

Caraway
Use/Usage level: Natural flavorings

Carbon color
Use/Usage level: Color; not permitted in certain foods

Carbon dioxide
Use/Usage level: Acidity regulator, food acid

Cardamon
Use/Usage level: Natural flavorings

Carnauba wax
Use/Usage level: Natural chewing gum base; glazing agent

Carob
Use/Usage level: Natural flavorings

Carob bean gum
Use/Usage level: Natural thickener/stabilizer

β-Carotene
Use/Usage level: Color; dietary supplement (vitamin)
Limitation: Not permitted in meat, fresh fish/shellfish, etc. (color)

Carrageenan
Use/Usage level: Natural thickener/stabilizer

Cascara
Use/Usage level: Natural flavorings

Cascarilla
Use/Usage level: Natural flavorings

Casein
Use/Usage level: Natural processing aid
Limitation: Not permitted in certain foods

Cassia gum
Use/Usage level: Natural thickener/stabilizer

Cassie
Use/Usage level: Natural flavorings

Castoreum
Use/Usage level: Natural flavorings

Catalase
Use/Usage level: Natural enzyme

Catechu
Use/Usage level: Natural flavorings

Cedar
Use/Usage level: Natural flavorings

Cellulase
Use/Usage level: Natural enzyme

Centaury
Use/Usage level: Natural flavorings

Cherry laurel
Use/Usage level: Natural flavorings

Chervil
Use/Usage level: Natural flavorings

Chicle
Use/Usage level: Natural chewing gum base

Chicory
Use/Usage level: Natural flavorings

Chicory extract
Use/Usage level: Color; not permitted in certain foods

Chilte
Use/Usage level: Natural chewing gum base

Chitin
Use/Usage level: Natural thickener/stabilizer

Chitinase
Use/Usage level: Natural enzyme

Chitosan
Use/Usage level: Natural thickener/stabilizer

Chitosanase
Use/Usage level: Natural enzyme

Chlorine dioxide
Use/Usage level: Flour treatment agent

Chlorophyll
Use/Usage level: Color; not permitted in certain foods

Cholecalciferol
Use/Usage level: Dietary supplement (vitamin)

Cholesterol
Use/Usage level: Natural emulsifier

Cholic acid
Use/Usage level: Natural emulsifier

Choline phosphate
Use/Usage level: Fermentation regulator, taste improver (0.2 g/L in sake compd.)

Chrysanthemum
Use/Usage level: Natural flavorings

1,8-Cineole
Use/Usage level: Flavorings
Limitation: Restricted to flavoring use

Cinnamaldehyde
Use/Usage level: Flavorings
Limitation: Restricted to flavoring use

Cinnamic acid
Use/Usage level: Flavorings
Limitation: Restricted to flavoring use

Cinnamon
Use/Usage level: Natural flavorings

Cinnamon extract
Use/Usage level: Natural processing aid

Limitation: Not permitted in certain foods

Cinnamyl acetate
Use/Usage level: Flavorings
Limitation: Restricted to flavoring use

Cinnamyl alcohol
Use/Usage level: Flavorings
Limitation: Restricted to flavoring use

Citral
Use/Usage level: Flavorings
Limitation: Restricted to flavoring use

Citric acid, anhydrous
Use/Usage level: Acidity regulator, food acid, raising agent

Citric acid, crystal
Use/Usage level: Acidity regulator, food acid, raising agent

Citronella
Use/Usage level: Natural flavorings

Citronellal
Use/Usage level: Flavorings
Limitation: Restricted to flavoring use

Citronellol
Use/Usage level: Flavorings
Limitation: Restricted to flavoring use

Citronellyl acetate
Use/Usage level: Flavorings
Limitation: Restricted to flavoring use

Citronellyl formate
Use/Usage level: Flavorings
Limitation: Restricted to flavoring use

Civet
Use/Usage level: Natural flavorings

Clary sage
Use/Usage level: Natural flavorings

Clove
Use/Usage level: Natural flavorings

Clove extract
Use/Usage level: Natural antioxidant

Clover
Use/Usage level: Natural flavorings

Cobalt
Use/Usage level: Natural processing aid
Limitation: Not permitted in certain foods

Cochineal extract
Use/Usage level: Color; not permitted in certain foods

Coffee bean extract
Use/Usage level: Natural antioxidant

Cola
Use/Usage level: Natural flavorings

Comfrey
Use/Usage level: Natural flavorings

Copaiba balsam
Use/Usage level: Natural chewing gum base

Copal resin
Use/Usage level: Natural chewing gum base

Copper
Use/Usage level: Natural processing aid
Limitation: Not permitted in certain foods

Copper chlorophyll
Use/Usage level: Color (0.0004-0.15 g/kg)

Copper gluconate
Use/Usage level: Dietary supplement (0.6 mg/L as copper in milk)

Coriander
Use/Usage level: Natural flavorings

Crown gum
Use/Usage level: Natural chewing gum base

Cupric sulfate
Use/Usage level: Dietary supplement (0.6 mg/L as copper in milk)

Cyanocobalamin
Use/Usage level: Natural dietary supplement

Cyclodextrin
Use/Usage level: Natural processing aid
Limitation: Not permitted in certain foods

Cyclodextrin glucanotransferase
Use/Usage level: Natural enzyme

Cyclohexyl acetate
Use/Usage level: Flavorings
Limitation: Restricted to flavoring use

Cyclohexyl butyrate
Use/Usage level: Flavorings
Limitation: Restricted to flavoring use

L-Cysteine monohydrochloride
Use/Usage level: Dietary supplement; quality improver

L-Cystine
Use/Usage level: Natural dietary supplement; flavorings

Dammar resin
Use/Usage level: Natural chewing gum base, thickener/stabilizer

Dammar resin
Use/Usage level: Natural thickener/stabilizer

Dandelion
Use/Usage level: Natural flavorings

Dacanal
Use/Usage level: Flavorings
Limitation: Restricted to flavoring use

Decanol
Use/Usage level: Flavorings
Limitation: Restricted to flavoring use

Dextran
Use/Usage level: Natural thickener/stabilizer

Dextranase
Use/Usage level: Natural enzyme

Diammonium hydrogen phosphate
Use/Usage level: Emulsifier for cheese food, yeast nutrient

Diatomaceous earth
Use/Usage level: Processing aid (0.5% max. residual)
Limitation: Restricted

Dibenzoyl thiamine
Use/Usage level: Dietary supplement (vitamin)

Dibenzoyl thiamine hydrochloride
Use/Usage level: Dietary supplement (vitamin)

Dill
Use/Usage level: Natural flavorings

Diphenyl
Use/Usage level: Antimold (preservative) (0.07 g/kg in grapefruit, lemon, oranges)

Dipotassium hydrogen phosphate
Use/Usage level: Acidity regulator, alkaline agent, raising agent, emulsifier for cheese food; flavorings

Disodium 5´-cytidylate
Use/Usage level: Flavorings

Disodium dihydrogen pyrophosphate
Use/Usage level: Acidity regulator, alkaline agent, raising agent, binding agent, emulsifier

Disodium EDTA
Use/Usage level: Antioxidant (0.25 g/kg as calcium disodium EDTA)

Disodium glycyrrhizinate
Use/Usage level: Sweetener

Disodium 5´-guanylate
Use/Usage level: Flavorings

Disodium hydrogen phosphate anhydrous
Use/Usage level: Acidity regulator, alkaline agent, flavoring, raising agent, emulsifier for cheese food; flavorings

Disodium hydrogen phosphate crystal
Use/Usage level: Acidity regulator, alkaline agent, flavoring, raising agent, emulsifier for cheese food; flavorings

Disodium 5´-inosinate
Use/Usage level: Flavorings

Disodium 5´-ribonucleotide
Use/Usage level: Flavorings

Disodium succinate
Use/Usage level: Acidity regulator, food acid, flavoring

Disodium DL-tartrate
Use/Usage level: Acidity regulator, food acid, flavoring

Disodium L-tartrate
Use/Usage level: Acidity regulator, food acid, flavoring

Disodium 5′-uridylate
Use/Usage level: Flavorings

Elder
Use/Usage level: Natural flavorings

Elecampane
Use/Usage level: Natural flavorings

Elemi resin
Use/Usage level: Natural chewing gum base, thickener/stabilizer

Elemi resin
Use/Usage level: Natural thickener/stabilizer

Ergocalciferol
Use/Usage level: Dietary supplement (vitamin)

Erythorbic acid
Use/Usage level: Antioxidant for fish-paste prods., bread
Limitation: Restricted for purpose of antioxidation

Esterase
Use/Usage level: Natural enzyme

Ester gum
Use/Usage level: Chewing gum base

Ethanol
Use/Usage level: Natural processing aid
Limitation: Not permitted in certain foods

Ethyl acetate
Use/Usage level: Flavorings; yeast extract; denaturalization of ethanol; solvent; mfg. of vinegar
Limitation: Restricted; remove prior to prep. of final food

Ethyl acetoacetate
Use/Usage level: Flavorings
Limitation: Restricted to flavoring use

Ethyl butyrate
Use/Usage level: Flavorings
Limitation: Restricted to flavoring use

Ethyl cinnamate
Use/Usage level: Flavorings
Limitation: Restricted to flavoring use

Ethyl decanoate
Use/Usage level: Flavorings
Limitation: Restricted to flavoring use

Ethyl heptanoate
Use/Usage level: Flavorings

Limitation: Restricted to flavoring use

Ethylhexanoate
Use/Usage level: Flavorings
Limitation: Restricted to flavoring use

Ethyl p-hydroxybenzoate
Use/Usage level: Preservative (0.012-1 g/kg as p-hydroxybenzoic acid)

Ethyl isovalerate
Use/Usage level: Flavorings
Limitation: Restricted to flavoring use

Ethyl octanoate
Use/Usage level: Flavorings
Limitation: Restricted to flavoring use

Ethyl phenylacetate
Use/Usage level: Flavorings
Limitation: Restricted to flavoring use

Ethyl propionate
Use/Usage level: Flavorings
Limitation: Restricted to flavoring use

Ethyl vanillin
Use/Usage level: Flavorings
Limitation: Restricted to flavoring use

Eucalyptus
Use/Usage level: Natural flavorings

Eucalyptus leaf extract
Use/Usage level: Natural antioxidant

Eugenol
Use/Usage level: Flavorings
Limitation: Restricted to flavoring use

Fatty acids
Use/Usage level: Flavorings
Limitation: Restricted to flavoring use

Fennel
Use/Usage level: Natural flavorings

Fenugreek
Use/Usage level: Natural flavorings

Ferric ammonium citrate
Use/Usage level: Dietary supplement (mineral)

Ferric chloride
Use/Usage level: Dietary supplement (mineral)

Ferric citrate
Use/Usage level: Dietary supplement (mineral)

Ferric pyrophosphate
Use/Usage level: Dietary supplement (mineral)

Ferrous gluconate
Use/Usage level: Color retention agent (0.15 g/kg max. as iron); dietary supplement

Ferrous gluconate
Use/Usage level: Dietary supplement

Ferrous pyrophosphate
Use/Usage level: Dietary supplement (mineral)

Ferrous sulfate crystal
Use/Usage level: Color retention agent; dietary supplement (mineral); color developer

Ferrous sulfate crystal
Use/Usage level: Dietary supplement (mineral); color developer

Ferrous sulfate exsiccated
Use/Usage level: Color retention agent; dietary supplement (mineral); color developer

Ferrous sulfate exsiccated
Use/Usage level: Dietary supplement (mineral); color developer

Fir balsam
Use/Usage level: Natural chewing gum base

Folic acid
Use/Usage level: Dietary supplement (vitamin)

Food Blue No. 1
Use/Usage level: Color; not permitted in certain foods

Food Blue No. 1 Aluminum Lake.
Use/Usage level: Color; not permitted in certain foods

Food Blue No. 2.
Use/Usage level: Color; not permitted in certain foods

Food Blue No. 2 Aluminum Lake.
Use/Usage level: Color; not permitted in certain foods

Food Green No. 3.
Use/Usage level: Color; not permitted in certain foods

Food Green No. 3 Aluminum Lake.
Use/Usage level: Color; not permitted in certain foods

Food Red No. 2.
Use/Usage level: Color; not permitted in certain foods

Food Red No. 2 Aluminum Lake.
Use/Usage level: Color; not permitted in certain foods

Food Red No. 3.
Use/Usage level: Color; not permitted in certain foods

Food Red No. 3 Aluminum Lake.
Use/Usage level: Color; not permitted in certain foods

Food Red No. 40.
Use/Usage level: Color; not permitted in certain foods

Food Red No. 40 Aluminum Lake.
Use/Usage level: Color; not permitted in certain foods

Food Red No. 102.
Use/Usage level: Color; not permitted in certain foods

Food Red No. 104.
Use/Usage level: Color; not permitted in certain foods

Food Red No. 105.
Use/Usage level: Color; not permitted in certain foods

Food Red No. 106.
Use/Usage level: Color; not permitted in certain foods

Food Yellow No. 4.
Use/Usage level: Color; not permitted in certain foods

Food Yellow No. 4 Aluminum Lake.
Use/Usage level: Color; not permitted in certain foods

Food Yellow No. 5.
Use/Usage level: Color; not permitted in certain foods

Food Yellow No. 5 Aluminum Lake.
Use/Usage level: Color; not permitted in certain foods

Fumaric acid
Use/Usage level: Acidity regulator, food acid, raising agent

Furcellaran
Use/Usage level: Natural thickener/stabilizer

Furfural and derivs.
Use/Usage level: Flavorings
Limitation: Except toxic substances; restricted to flavoring use

α-Galactosidase
Use/Usage level: Natural enzyme

Galanga
Use/Usage level: Natural flavorings

Galbanum
Use/Usage level: Natural flavorings

Gallic acid
Use/Usage level: Natural antioxidant

Gardenia
Use/Usage level: Natural flavorings

Garlic extract
Use/Usage level: Natural processing aid
Limitation: Not permitted in certain foods

Gelatin
Use/Usage level: Natural food additive

Gellan gum
Use/Usage level: Natural thickener/stabilizer

Gentian root extract
Use/Usage level: Natural antioxidant, bittering agent

Geraniol
Use/Usage level: Flavorings
Limitation: Restricted to flavoring use

Geranyl acetate
Use/Usage level: Flavorings
Limitation: Restricted to flavoring use

Geranyl formate
Use/Usage level: Flavorings
Limitation: Restricted to flavoring use

Germander
Use/Usage level: Natural flavorings

Ginger
Use/Usage level: Natural flavorings

Ginger extract
Use/Usage level: Natural processing aid
Limitation: Not permitted in certain foods

Gingko
Use/Usage level: Natural flavorings

Ginseng
Use/Usage level: Natural flavorings

Glucanase
Use/Usage level: Natural enzyme

Glucoamylase
Use/Usage level: Natural enzyme

Gluconic acid
Use/Usage level: Acidity regulator, food acid

Glucono-δ-lactone
Use/Usage level: Acidity regulator, coagulant for tofu, food acid, raising agent

Glucosamine
Use/Usage level: Natural thickener/stabilizer

Glucose isomerase
Use/Usage level: Natural enzyme

Glucose oxidase
Use/Usage level: Natural enzyme

α-Glucosidase
Use/Usage level: Natural enzyme

β-Glucosidase
Use/Usage level: Natural enzyme

L-Glutamic acid
Use/Usage level: Dietary supplement (amino acid); flavorings

L-Glutamine
Use/Usage level: Natural dietary supplement; flavorings

Gluten
Use/Usage level: Natural food additive

Glycerin esters of fatty acids
Use/Usage level: Chewing gum base, emulsifier, plasticizer for chewing gum

Glycerol
Use/Usage level: Plasticizers for chewing gum

Glycine
Use/Usage level: Dietary supplement (amino acid); flavorings

Grape skin color
Use/Usage level: Color; not permitted in certain foods

Grape skin extract
Use/Usage level: Natural processing aid
Limitation: Not permitted in certain foods

Guaiac resin
Use/Usage level: Natural antioxidant (1.0 g/kg max.), chewing gum base

Gualacum
Use/Usage level: Natural flavorings

Guar gum
Use/Usage level: Natural thickener/stabilizer

Gum Arabica
Use/Usage level: Natural thickener/stabilizer

Gum ghatti
Use/Usage level: Natural thickener/stabilizer

Gutta percha
Use/Usage level: Natural chewing gum base

Hemicellulase
Use/Usage level: Natural enzyme

Hesperetin
Use/Usage level: Natural antioxidant

Hesperidine
Use/Usage level: Natural dietary supplement

Hexane
Use/Usage level: Processing aid
Limitation: Restricted for extraction of fats/oils; not permitted in certain foods; remove prior to prep. of final food

Hexane
Use/Usage level: Natural processing aid
Limitation: Not permitted in certain foods

Hexanoic acid
Use/Usage level: Flavorings
Limitation: Restricted to flavoring use

L-Histidine
Use/Usage level: Natural dietary supplement; flavorings

L-Histidine monohydrochloride
Use/Usage level: Dietary supplement (amino acid); flavorings

Honeysuckle
Use/Usage level: Natural flavorings

Hop
Use/Usage level: Natural flavorings

Hop extract
Use/Usage level: Natural food additive

Hyaluronic acid
Use/Usage level: Natural processing aid
Limitation: Not permitted in certain foods

Hydrochloric acid
Use/Usage level: Processing aid
Limitation: Restricted; neutralize or remove prior to prep. of final food

Hydrogen peroxide
Use/Usage level: Sterilizing agent
Limitation: Restricted; decompose or remove prior to prep. of final food

Hydroxycitronellal
Use/Usage level: Flavorings
Limitation: Restricted to flavoring use

Hydroxycitronellal dimethylacetal
Use/Usage level: Flavorings
Limitation: Restricted to flavoring use

L-Hydroxyproline
Use/Usage level: Natural dietary supplement ; flavorings

Hypochlorous acid
Use/Usage level: Sterilizing agent
Limitation: Restricted; not for use in sesame seeds

Hyssop
Use/Usage level: Natural flavorings

Iceland moss
Use/Usage level: Natural flavorings

Imazalil
Use/Usage level: Antimold (preservative) (0.002-0.005 g/kg residual)

Indole and derivs.
Use/Usage level: Flavorings
Limitation: Restricted to flavoring use

Inositol
Use/Usage level: Natural dietary supplement

Invertase
Use/Usage level: Natural enzyme

Ion exchange resin
Use/Usage level: Processing aid
Limitation: Restricted; neutralize or remove prior to prep. of final food

Ionone
Use/Usage level: Flavorings

Limitation: Restricted to flavoring use

Iron
Use/Usage level: Natural processing aid
Limitation: Not permitted in certain foods

Iron lactate
Use/Usage level: Dietary supplement (mineral)

Iron sesquioxide
Use/Usage level: Color for banana; restricted use

Isoamyl acetate
Use/Usage level: Flavorings
Limitation: Restricted to flavoring use

Isoamylase
Use/Usage level: Natural enzyme

Isoamyl butyrate
Use/Usage level: Flavorings
Limitation: Restricted to flavoring use

Isoamyl formate
Use/Usage level: Flavorings
Limitation: Restricted to flavoring use

Isoamyl isovalerate
Use/Usage level: Flavorings
Limitation: Restricted to flavoring use

Isoamyl phenylacetate
Use/Usage level: Flavorings
Limitation: Restricted to flavoring use

Isoamyl propionate
Use/Usage level: Flavorings
Limitation: Restricted to flavoring use

Isobutyl p-hydroxybenzoate
Use/Usage level: Preservative (0.012-1 g/kg as p-hydroxybenzoic acid)

Isobutyl phenylacetate
Use/Usage level: Flavorings
Limitation: Restricted to flavoring use

Isoeugenol
Use/Usage level: Flavorings
Limitation: Restricted to flavoring use

L-Isoleucine
Use/Usage level: Dietary supplement (amino acid); flavorings

Isopropyl citrate
Use/Usage level: Antioxidant (0.1 g/kg max.)

Isopropyl p-hydroxybenzoate
Use/Usage level: Preservative (0.012-1 g/kg as p-hydroxybenzoic acid)

Isothiocyanates
Use/Usage level: Flavorings
Limitation: Except toxic substances; restricted to flavoring use

Itaconic acid
Use/Usage level: Natural food acid

Jasmine
Use/Usage level: Natural flavorings

Jojoba wax
Use/Usage level: Natural chewing gum base

Juniper berry
Use/Usage level: Natural flavorings

Kaolin
Use/Usage level: Processing aid (0.5% max. residual)
Limitation: Restricted

Karaya gum
Use/Usage level: Natural thickener/stabilizer

Kauri gum
Use/Usage level: Natural chewing gum base

Kelp extract
Use/Usage level: Natural thickener/stabilizer

α-Ketoglutaric acid
Use/Usage level: Natural food acid

Ketones
Use/Usage level: Flavorings
Limitation: Restricted to flavoring use

Ketones
Use/Usage level: Flavorings
Limitation: Restricted to flavoring use

Lactase
Use/Usage level: Natural enzyme

Lactic acid
Use/Usage level: Acidity regulator, food acid, raising agent

Lactones
Use/Usage level: Flavorings
Limitation: Except toxic substances; restricted to flavoring use

Lanolin
Use/Usage level: Natural chewing gum base; glazing agent

Laurel
Use/Usage level: Natural flavorings

Lavender
Use/Usage level: Natural flavorings

Lecithin
Use/Usage level: Natural emulsifier

Lemon
Use/Usage level: Natural flavorings

Lemongrass
Use/Usage level: Natural flavorings

L-Leucine
Use/Usage level: Natural dietary supplement; flavorings

Licorice
Use/Usage level: Natural flavorings

Licorice extract
Use/Usage level: Natural sweetener

Lily
Use/Usage level: Natural flavorings

Lime
Use/Usage level: Natural flavorings

Linaloe
Use/Usage level: Natural flavorings

Linalool
Use/Usage level: Flavorings
Limitation: Restricted to flavoring use

Linalyl acetate
Use/Usage level: Flavorings
Limitation: Restricted to flavoring use

Linden
Use/Usage level: Natural flavorings

Linseed gum
Use/Usage level: Natural thickener/stabilizer

Lipase
Use/Usage level: Natural enzyme

Liquid paraffin
Use/Usage level: Mold release/antisticking agent (0.1% max. residual in bread); processing aid

Lotus
Use/Usage level: Natural flavorings

L-Lysine
Use/Usage level: Natural dietary supplement; flavorings

L-Lysine L-aspartate
Use/Usage level: Dietary supplement (amino acid); flavorings

L-Lysine L-glutamate
Use/Usage level: Dietary supplement (amino acid); flavorings

L-Lysine monohydrochloride
Use/Usage level: Dietary supplement (amino acid); flavorings

Magnesium carbonate
Use/Usage level: Anticaking agent (0.5% max.), raising agent (0.5% max. residual)
Limitation: Restricted

Magnesium chloride
Use/Usage level: Coagulant for tofu; processing aid; yeast nutrient

Magnesium oxide
Use/Usage level: Processing aid
Limitation: Restricted

Magnesium sulfate
Use/Usage level: Coagulant for tofu; fermentation aid; yeast nutrient

Maidenhair fern
Use/Usage level: Natural flavorings

Maize
Use/Usage level: Natural flavorings

DL-Malic acid
Use/Usage level: Acidity regulator, food acid, raising agent

Malt
Use/Usage level: Natural flavorings

Maltol
Use/Usage level: Flavorings
Limitation: Restricted to flavoring use

D-Mannitol
Use/Usage level: Antisticking agent (50% max.); flavorings

Marjoram
Use/Usage level: Natural flavorings

Mastic gum
Use/Usage level: Natural chewing gum base

dl-Menthol
Use/Usage level: Flavorings
Limitation: Restricted to flavoring use

l-Menthol
Use/Usage level: Flavorings
Limitation: Restricted to flavoring use

l-Menthyl acetate
Use/Usage level: Flavorings
Limitation: Restricted to flavoring use

DL-Methionine
Use/Usage level: Dietary supplement (amino acid); flavorings

L-Methionine
Use/Usage level: Dietary supplement (amino acid); flavorings

p-Methylacetophenone
Use/Usage level: Flavorings
Limitation: Restricted to flavoring use

Methyl acetyl ricinoleate
Use/Usage level: Chewing gum base

Methyl anthranilate
Use/Usage level: Flavorings
Limitation: Restricted to flavoring use

Methyl cellulose
Use/Usage level: Thickener/stabilizer/gellant (2% max.)
Limitation: Restricted

Methyl cinnamate
Use/Usage level: Flavorings
Limitation: Restricted to flavoring use

Methyl hesperidin
Use/Usage level: Dietary supplement (vitamin)

Methyl N-methylanthranilate
Use/Usage level: Flavorings
Limitation: Restricted to flavoring use

Methyl β-naphthyl ketone
Use/Usage level: Flavorings
Limitation: Restricted to flavoring use

Methyl salicylate
Use/Usage level: Flavorings
Limitation: Restricted to flavoring use

Microcrystalline cellulose
Use/Usage level: Natural processing aid

Microcrystalline wax
Use/Usage level: Natural chewing gum base; glazing agent

Monocalcium di-L-glutamate
Use/Usage level: Flavorings for taste-related purposes (1% max. as calcium)

Monomagnesium di-L-glutamate
Use/Usage level: Flavorings

Monopotassium citrate
Use/Usage level: Flavorings

Monopotassium L-glutamate
Use/Usage level: Flavorings

Monosodium L-aspartate
Use/Usage level: Dietary supplement (amino acid); flavorings

Monosodium fumarate
Use/Usage level: Acidity regulator, food acid, raising agent; flavorings

Monosodium L-glutamate
Use/Usage level: Dietary supplement (amino acid); flavorings

Monosodium succinate
Use/Usage level: Acidity regulator, food acid, flavoring

Montan wax
Use/Usage level: Natural chewing gum base; glazing agent

Morpholine salts of fatty acid
Use/Usage level: Coating material

Mugwort extract
Use/Usage level: Natural bittering agent

Musk
Use/Usage level: Natural flavorings

Mustard
Use/Usage level: Natural flavorings

Myrrh
Use/Usage level: Natural flavorings

Myrrh gum
Use/Usage level: Natural chewing gum base

Myrtle
Use/Usage level: Natural flavorings

Nickel
Use/Usage level: Natural processing aid

Nicotinamide
Use/Usage level: Color retention agent; dietary supplement;
Limitation: Not for use in meat and raw fish/shellfish

Nicotinic acid
Use/Usage level: Color retention agent; diet supplement
Limitation: Not for use in meat and raw fish/shellfish

Nitrogen
Use/Usage level: Natural processing aid

γ-Nonalactone
Use/Usage level: Flavorings
Limitation: Restricted to flavoring use

Nordihydroguaiaretic acid
Use/Usage level: Antioxidant (0.1 g/kg max.)

Nutmeg
Use/Usage level: Natural flavorings

Octanal
Use/Usage level: Flavorings
Limitation: Restricted to flavoring use

Olibanum
Use/Usage level: Natural chewing gum base

Olive
Use/Usage level: Natural flavorings

Opopanax resin
Use/Usage level: Natural chewing gum base

Orange flower
Use/Usage level: Natural flavorings

Oregano extract
Use/Usage level: Natural processing aid

Orris
Use/Usage level: Natural flavorings

γ-Oryzanol
Use/Usage level: Natural antioxidant

Oxalic acid
Use/Usage level: Processing aid
Limitation: Restricted; neutralize or remove prior to prep. of final food

Oxyethylene higher aliphatic alcohols
Use/Usage level: Coating material

Oxygen
Use/Usage level: Natural processing aid

L-Oxyproline
Use/Usage level: Natural dietary supplement; flavorings

Ozokerite
Use/Usage level: Natural chewing gum base

Ozone
Use/Usage level: Natural processing aid

Palladium
Use/Usage level: Natural processing aid

Pancreatin
Use/Usage level: Natural enzyme

Papain
Use/Usage level: Natural enzyme

Paprika
Use/Usage level: Natural food additive

Paprika color
Use/Usage level: Color; not permitted in certain foods

Paraffin wax
Use/Usage level: Natural chewing gum base; glazing agent

Parsley
Use/Usage level: Natural flavorings

Patchouli
Use/Usage level: Natural flavorings

Pectin
Use/Usage level: Natural thickener/stabilizer

Pectinase
Use/Usage level: Natural enzyme

Pepper extract
Use/Usage level: Natural antioxidant

Peppermint
Use/Usage level: Natural flavorings

Pepsin
Use/Usage level: Natural enzyme

Perilla extract
Use/Usage level: Natural processing aid

l-Perillaldehyde
Use/Usage level: Flavorings
Limitation: Restricted to flavoring use

Perlite
Use/Usage level: Processing aid (0.5% max. residual)
Limitation: Restricted

Peroxidase
Use/Usage level: Natural enzyme

Peru balsam
Use/Usage level: Natural flavorings

Petitgrain
Use/Usage level: Natural flavorings

Phenethyl acetate
Use/Usage level: Flavorings
Limitation: Restricted to flavoring use

Phenols
Use/Usage level: Flavorings

Limitation: Except toxic substances; restricted to flavoring use

L-Phenylalanine
Use/Usage level: Dietary supplement (amino acid); flavorings

Phenyl ethers
Use/Usage level: Flavorings
Limitation: Except toxic substances; restricted to flavoring use

o-Phenylphenol
Use/Usage level: Antimold (preservative) (0.01 g/kg residual as o-phenylphenol)

Phospholipase
Use/Usage level: Natural enzyme

Phosphoric acid
Use/Usage level: Acidity regulator, food acid

Phytic acid
Use/Usage level: Natural food acid, processing aid

Pimento extract
Use/Usage level: Natural antioxidant

Piperonal
Use/Usage level: Flavorings
Limitation: Restricted to flavoring use

Piperonyl butoxide
Use/Usage level: Insecticide (0.024 g/kg max. in cereal)

Polybutene
Use/Usage level: Chewing gum base

Polyisobutylene
Use/Usage level: Chewing gum base

Polyvinyl acetate
Use/Usage level: Chewing gum base; coating material

Poppy
Use/Usage level: Natural flavorings

Potassium DL-bitartrate
Use/Usage level: Acidity regulator, flavoring, raising agent

Potassium L-bitartrate
Use/Usage level: Acidity regulator, flavoring, raising agent

Potassium bromate
Use/Usage level: Processing aid (0.03 g/kg max. of flour, as bromic acid)
Limitation: Restricted; remove prior to prep. of final food

Potassium carbonate anhyd.
Use/Usage level: Acidity regulator, yeast nutrient, alkaline agent, raising agent

Potassium chloride
Use/Usage level: Flavorings

Potassium dihydrogen phosphate
Use/Usage level: Acidity regulator, alkaline agent, raising agent, emulsifier for cheese food; flavorings

Potassium metaphosphate
Use/Usage level: Binding agent, emulsifier, alkaline agent, raising agent, emulsifier for cheese food

Potassium nitrate
Use/Usage level: Color fixative (0.007 g/kg max.); fermentation aid (0.2 g/L in cheese, 0.1 g/L in sake)

Potassium norbixin
Use/Usage level: Color

Potassium phosphate
Use/Usage level: Emulsifier for cheese food, alkaline agent

Potassium polyphosphate
Use/Usage level: Binding agent, emulsifier, alkaline agent, raising agent

Potassium polyphosphate
Use/Usage level: Emulsifier for cheese food, alkaline agent, binding agent, raising agent

Potassium pyrophosphate
Use/Usage level: Binding agent, emulsifier, alkaline agent, raising agent

Potassium pyrosulfite
Use/Usage level: Bleaching agent (0.03-5 g/kg residual as sulfur dioxide); emulsifier for cheese food

Potassium sorbate
Use/Usage level: Preservative (0.05-3 g/kg as sorbic acid)
Limitation: Limitation with propionic acid 1-3 g/kg total

L-Proline
Use/Usage level: Natural dietary supplement; flavorings

Propionic acid
Use/Usage level: Flavorings; preservative (2.5-3 g/kg)
Limitation: Restricted to flavoring use; limitation with sorbic acid 3 g/kg total

Propylene glycol
Use/Usage level: Plasticizer for chewing gum (0.6% max.); quality sustainer (0.6-2%)

Propylene glycol alginate
Use/Usage level: Thickener/stabilizer/gellant (1% max.)

Propylene glycol esters of fatty acids
Use/Usage level: Chewing gum base, emulsifier, plasticizer for chewing gum

Propyl gallate
Use/Usage level: Antioxidant (0.1 g/kg max.)

Protease
Use/Usage level: Natural enzyme

Psyllium seed gum
Use/Usage level: Natural thickener/stabilizer

Pullulan
Use/Usage level: Natural thickener/stabilizer

Pullulanase
Use/Usage level: Natural enzyme

Pyridoxine hydrochloride
Use/Usage level: Dietary supplement (vitamin)

Quassia
Use/Usage level: Natural flavorings

Quassia extract
Use/Usage level: Natural bittering agent

Quebracho
Use/Usage level: Natural flavorings

Quercetin
Use/Usage level: Natural antioxidant

Quillaia
Use/Usage level: Natural flavorings

Quillaia extract
Use/Usage level: Natural emulsifier

Quince
Use/Usage level: Natural flavorings

Rape seed oil extract
Use/Usage level: Natural antioxidant

Rennet
Use/Usage level: Natural enzyme

L-Rhamnose
Use/Usage level: Natural sweetener

Rhamsan gum
Use/Usage level: Natural thickener/stabilizer

Rhatany
Use/Usage level: Natural flavorings

Riboflavin
Use/Usage level: Color; dietary supplement (vitamin)

Riboflavin 5´-phosphate sodium
Use/Usage level: Color; dietary supplement (vitamin)

Riboflavin tetrabutyrate
Use/Usage level: Color; dietary supplement (vitamin)

D-Ribose
Use/Usage level: Natural sweetener

Rice bran oil extract
Use/Usage level: Natural antioxidant

Rice bran wax
Use/Usage level: Natural chewing gum base; glazing agent

Rose
Use/Usage level: Natural flavorings

Rosemary
Use/Usage level: Natural flavorings

Rosemary extract
Use/Usage level: Natural antioxidant

Rosin
Use/Usage level: Natural chewing gum base

Rubber
Use/Usage level: Natural chewing gum base

Saccharin
Use/Usage level: Sweetener (0.05 g/kg max. in chewing gum)

Safflower
Use/Usage level: Natural flavorings

Saffron
Use/Usage level: Natural food additive

Saffron color
Use/Usage level: Color; not permitted in certain foods

Sage
Use/Usage level: Natural flavorings

Sage extract
Use/Usage level: Natural antioxidant

Saint John's wort
Use/Usage level: Natural flavorings

Sand
Use/Usage level: Processing aid (0.5% max. residual)
Limitation: Restricted

Sandalwood
Use/Usage level: Natural flavorings

Sandalwood red
Use/Usage level: Color; not permitted in certain foods

Sarsaparilla
Use/Usage level: Natural flavorings

Sassafras
Use/Usage level: Natural flavorings

Seaweed
Use/Usage level: Natural flavorings

L-Serine
Use/Usage level: Natural dietary supplement; flavorings

Sesame
Use/Usage level: Natural flavorings

Sesame seed extract
Use/Usage level: Natural antioxidant

Shellac
Use/Usage level: Natural chewing gum base; glazing agent

Silicon dioxide (fine)
Use/Usage level: Anticaking agent (2% max.)

Silicon dioxide (other than fine)
Use/Usage level: Processing aid
Limitation: Restricted to purpose of filtration aid; remove prior to prep. of final food

Silicone resin
Limitation: Restricted for purpose of antifoaming (0.05 g/kg max.)

Sodium acetate anhyd.
Use/Usage level: Acidity regulator, food acid, flavoring

Sodium acetate crystal
Use/Usage level: Acidity regulator, food acid, flavoring

Sodium alginate
Use/Usage level: Thickener/stabilizer/gelling agent

Sodium L-ascorbate
Use/Usage level: Antioxidant, dietary supplement (vitamin)

Sodium benzoate
Use/Usage level: Preservative (0.6-2.5 g/kg as benzoic acid)
Limitation: Limitation with sorbic acid 1 g/kg total

Sodium bicarbonate
Use/Usage level: Acidity regulator, alkaline agent, raising agent

Sodium carbonate anhyd.
Use/Usage level: Acidity regulator, alkaline agent, raising agent

Sodium carbonate crystal
Use/Usage level: Acidity regulator, alkaline agent, raising agent

Sodium carboxymethyl cellulose
Use/Usage level: Thickener/stabilizer/gellant (2% max.)
Limitation: Restricted

Sodium carboxymethyl starch
Use/Usage level: Thickener/stabilizer/gellant (2% max.)
Limitation: Restricted

Sodium caseinate
Use/Usage level: Processing aid

Sodium chondroitin sulfate
Use/Usage level: Humectant/emulsifier/stabilizer (3-20 g/kg)

Sodium copper chlorophyllin
Use/Usage level: Color (0.0004-0.15 g/kg)

Sodium dehydroacetate
Use/Usage level: Preservative (0.5 g/kg as dehydroacetic acid)

Sodium dihydrogen phosphate anhyd.
Use/Usage level: Acidity regulator, alkaline agent, flavoring, raising agent, emulsifier for cheese food; flavorings

Sodium dihydrogen phosphate crystal
Use/Usage level: Acidity regulator, alkaline agent, flavoring, raising agent, emulsifier for cheese food; flavorings

Sodium erythorbate
Use/Usage level: Antioxidant for fish-paste prods., bread
Limitation: Restricted for purpose of antioxidation

Sodium ferrous citrate
Use/Usage level: Dietary supplement (mineral)

Sodium ferrous citrate
Use/Usage level: Dietary supplement (mineral)

Sodium hydrosulfite
Use/Usage level: Bleaching agent (0.03-5 g/kg residual as sulfur dioxide)
Limitation: Not permitted in certain foods

Sodium hydroxide
Use/Usage level: Processing aid
Limitation: Restricted; neutralize or remove prior to prep. of final food

Sodium hypochlorite
Use/Usage level: Bleaching agent; sterilizing agent (restricted)
Limitation: Restricted; not for use in sesame seeds

Sodium hypochlorite
Use/Usage level: Sterilizing agent
Limitation: Restricted; not for use in sesame seeds

Sodium iron chlorophyllin
Use/Usage level: Color; not permitted in certain foods

Sodium lactate
Use/Usage level: Acidity regulator, food acid, flavoring

Sodium DL-malate
Use/Usage level: Acidity regulator, food acid, flavoring, raising agent

Sodium metaphosphate
Use/Usage level: Binding agent, emulsifier, alkaline agent, raising agent, emulsifier for cheese food

Sodium methoxide
Use/Usage level: Processing aid
Limitation: Restricted; decompose prior to prep. of final food, remove resulting methanol

Sodium nitrate
Use/Usage level: Color fixative (0.007 g/kg max.); fermentation aid (0.2 g/L in cheese, 0.1 g/L in sake)

Sodium nitrite
Use/Usage level: Color fixative (0.005-0.07 g/kg residual as NO_2)

Sodium norbixin
Use/Usage level: Color

Sodium oleate
Use/Usage level: Coating material

Sodium pantothenate
Use/Usage level: Dietary supplement (vitamin)

Sodium o-phenylphenate
Use/Usage level: Antimold (preservative) (0.01 g/kg residual as o-phenylphenol)

Sodium polyacrylate
Use/Usage level: Thickener/stabilizer/gellant (0.2% max.)

Sodium polyphosphate
Use/Usage level: Binding agent, emulsifier, alkaline agent, raising agent, emulsifier for cheese food

Sodium propionate
Use/Usage level: Preservative (2.5-3 g/kg as propionic acid)
Limitation: Limitation with sorbic acid 3 g/kg total

Sodium pyrophosphate anhyd.
Use/Usage level: Binding agent, emulsifier, alkaline agent, raising agent, emulsifier for cheese food

Sodium pyrophosphate crystal
Use/Usage level: Binding agent, emulsifier, alkaline agent, raising agent, emulsifier for cheese food

Sodium pyrosulfite
Use/Usage level: Bleaching agent (0.03-5 g/kg residual as sulfur dioxide)

Sodium saccharin
Use/Usage level: Sweetener (0.1-2 g/kg residual)

Sodium starch phosphate
Use/Usage level: Thickener/stabilizer/gellant (2% max.)
Limitation: Restricted

Sodium sulfate
Use/Usage level: Processing aid

Sodium sulfite (anhydrous)
Use/Usage level: Bleaching agent (0.03-5 g/kg residual as sulfur dioxide)

Sodium sulfite (crystal)
Use/Usage level: Bleaching agent (0.03-5 g/kg residual as sulfur dioxide)

Sorbic acid
Use/Usage level: Preservative (0.05-3 g/kg)
Limitation: Limitation with propionic acid 1-3 g/kg total

Sorbitan esters of fatty acids
Use/Usage level: Chewing gum base, emulsifier, plasticizer for chewing gum

D-Sorbitol
Use/Usage level: Humectant; plasticizers for chewing gum; sweetener

L-Sorbose
Use/Usage level: Natural sweetener

Soybeans
Use/Usage level: Natural flavorings

Spearmint
Use/Usage level: Natural flavorings

Spermaceti wax
Use/Usage level: Natural chewing gum base; glazing agent

Star anise
Use/Usage level: Natural flavorings

Styrax
Use/Usage level: Natural flavorings

Succinic acid
Use/Usage level: Acidity regulator, food acid, flavoring

Sucrose esters of fatty acids
Use/Usage level: Chewing gum base, emulsifier, plasticizer for chewing gum

Sulfur dioxide
Use/Usage level: Bleaching agent (0.03-5 g/kg residual as sulfur dioxide)

Sulfuric acid
Use/Usage level: Processing aid
Limitation: Restricted; neutralize or remove prior to prep. of final food

Superoxide dismutase
Use/Usage level: Natural enzyme

Talc
Use/Usage level: Processing aid (0.5% max. residual, 5% max. in chewing gum); chewing gum base
Limitation: Restricted

Tamarind
Use/Usage level: Natural flavorings

Tamarind color
Use/Usage level: Color; not permitted in certain foods

Tamarind seed gum
Use/Usage level: Natural thickener/stabilizer

Tannase
Use/Usage level: Natural enzyme

Tannin extract
Use/Usage level: Natural processing aid

Tansy
Use/Usage level: Natural flavorings

Tara gum
Use/Usage level: Natural thickener/stabilizer

Tar color preparations
Use/Usage level: Color; not permitted in certain foods

DL-Tartaric acid
Use/Usage level: Acidity regulator; food acid, raising agent

L-Tartaric acid
Use/Usage level: Acidity regulator; food acid; raising agent

Terpenic hydrocarbons
Use/Usage level: Flavorings
Limitation: Restricted to flavoring use

Terpineol
Use/Usage level: Flavorings
Limitation: Restricted to flavoring use

Terpinyl acetate
Use/Usage level: Flavorings
Limitation: Restricted to flavoring use

L-Theanine
Use/Usage level: Dietary supplement (amino acid); flavorings

Theobromine
Use/Usage level: Natural bittering agent

Thiabendazole
Use/Usage level: Antimold (preservative) (0.004-0.01 g/kg)

Thiamine dicetylsulfate
Use/Usage level: Dietary supplement (vitamin)

Thiamine dilaurylsulfate
Use/Usage level: Dietary supplement (vitamin)

Thiamine hydrochloride
Use/Usage level: Dietary supplement (vitamin)

Thiamine mononitrate
Use/Usage level: Dietary supplement (vitamin)

Thiamine naphthalene-1,5-disulfonate
Use/Usage level: Dietary supplement (vitamin)

Thiamine thiocyanate
Use/Usage level: Dietary supplement (vitamin)

Thioethers
Use/Usage level: Flavorings
Limitation: Except toxic substances; restricted to flavoring use

Thiols
Use/Usage level: Flavorings

Limitation: Except toxic substances; restricted to flavoring use

DL-Threonine
Use/Usage level: Dietary supplement (amino acid); flavorings

L-Threonine
Use/Usage level: Dietary supplement (amino acid); flavorings

Thyme
Use/Usage level: Natural flavorings

Titanium dioxide
Use/Usage level: Color; not permitted in certain foods

d-α-Tocopherol
Use/Usage level: Natural antioxidant, dietary supplement

d-δ-Tocopherol
Use/Usage level: Natural antioxidant

d-γ-Tocopherol
Use/Usage level: Natural antioxidant

dl-α-Tocopherol
Limitation: Restricted for purpose of antioxidation

Tolu balsam
Use/Usage level: Natural flavorings

Tonka beans
Use/Usage level: Natural flavorings

Tragacanth gum
Use/Usage level: Natural thickener/stabilizer

Tricalcium phosphate
Use/Usage level: Chewing gum base (1% max. as calcium); dietary supplement (1% max. as calcium); emulsifier (1% max. as calcium); raising agent (1% max. as calcium); yeast nutrient (1% max. as calcium)
Limitation: Restricted

Tripotassium phosphate
Use/Usage level: Flavorings; emulsifier; alkaline agent

Tripotassium citrate
Use/Usage level: Flavorings

Trisodium citrate
Use/Usage level: Acidity regulator, food acid, flavoring, emulsifier for cheese food

Trisodium phosphate anhyd.
Use/Usage level: Emulsifier for cheese food, alkaline agent; flavorings

Trisodium phosphate crystal
Use/Usage level: Emulsifier for cheese food, alkaline agent; flavorings

Trypsin
Use/Usage level: Natural enzyme

DL-Tryptophan
Use/Usage level: Dietary supplement (amino acid); flavorings

L-Tryptophan
Use/Usage level: Dietary supplement (amino acid); flavorings

Turmeric
Use/Usage level: Natural flavorings, food additive

Turmeric
Use/Usage level: Natural food additive

L-Tyrosine
Use/Usage level: Natural dietary supplement; flavorings

δ-Undecalactone
Use/Usage level: Flavorings
Limitation: Restricted to flavoring use

Urease
Use/Usage level: Natural enzyme

Valerian
Use/Usage level: Natural flavorings

L-Valine
Use/Usage level: Dietary supplement (amino acid); flavorings

Vanilla
Use/Usage level: Natural flavorings

Vanillin
Use/Usage level: Flavorings
Limitation: Restricted to flavoring use

Vetiver
Use/Usage level: Natural flavorings

Vitamin A
Use/Usage level: Dietary supplement (vitamin)

Vitamin A esters of fatty acids
Use/Usage level: Dietary supplement (vitamin)

Wasabi
Use/Usage level: Natural flavorings

Wintergreen
Use/Usage level: Natural flavorings

Woodruff
Use/Usage level: Natural flavorings

Wormwood
Use/Usage level: Natural flavorings

Xanthan gum
Use/Usage level: Natural thickener/stabilizer

D-Xylose
Use/Usage level: Sweetener

Ylang-ylang
Use/Usage level: Natural flavorings

Yucca foam extract
Use/Usage level: Natural emulsifier

Zedoary
Use/Usage level: Natural flavorings

Zein
Use/Usage level: Natural processing aid

Zinc gluconate
Use/Usage level: Dietary supplement (6 mg/l as zinc in prepared milk)

Zinc sulfate
Use/Usage level: Dietary supplement (6 mg/l as zinc in prepared milk)

Glossary

absolute (flavoring). Obtained by alcohol extraction of concrete of plant materials; not used extensively in flavors.

Acceptable Daily Intake (ADI). The amount of a food additive that can be taken daily in the diet without risk; expressed as mg/kg body weight. "ADI not specified" or "ADI no limit" means that the total daily intake of the substance due to normal usage in foods does not represent a health hazard.

acid. A compound that may be either organic or inorganic and used in food for taste (to impart a sour or sharp flavor) or to affect the function of other components. A substance that stabilizes the acidity of foods.

acidulant. Any of a number of acids added to food to aid in preservation, to chelate metals, and to modify taste(by offsetting sweetness).

adsorbate. A powdered flavor which is made by coating a liquid flavoring onto the surface of a powdered carrier such as corn starch, salt, or maltodextrin.

alkali. One of a class of chemical compounds which combine with acids to form salts. In water solution, alkalis are bitter; turns litmus blue, and has a pH above 7.0.

allergen. A substance which induces allergy, by acting in the manner of an antigen on coming into contact with body tissues by inhalation, ingestion, or skin adsorption. The allergen causes a specific reagin to be formed in the bloodstream.

allergy. A harmful physiological reaction, caused by an immunologic mechanism.

anticaking agent. A food additive used to prevent or inhibit caking of dry material and thus maintains a free-flowing condition; often used in food products that tend to be hygroscopic, such as cures, salts or seasonings. Examples are: small amounts of anhydrous disodium hydrogen phosphate added to salt or sugar, aluminum calcium silicate or magnesium silicate in table salt, and calcium silicate in baking powder.

antifoaming agent. Substance added to food to prevent excessive frothing on boiling, reduce the formation of scum often caused by the presence of dissolved protein or other stabilizer, or prevent boiling over.

antimicrobial agent. Substance used to preserve foods by destroying or inhibiting the growth of microorganisms and subsequent spoilage .

antioxidant. A substance that retards oxidative rancidity of fat, e.g., propyl gallate octyl gallate, dodecyl gallate, BHA, BHT. Many fats, particularly vegetable oils, contain naturally occurring antioxidants, such as tocopherols, which protect the oils from rancidity for a limited amount of time.

aromatic. A fragrant, usually pleasant, spicy, slightly pungent.

artificial sweetener. Food additive other than sugar that causes a food to have a sweet taste.

antispattering agent. Substance added to fats used in frying, e.g., lecithin, sucrose esters, sodium and sulfoacetate derivatives of mono- and diglycerides. It prevents the coalescence of water droplets.

antistaling agent. Substances that retard the staling of baked products and also soften the crumbs, e.g., sucrose stearate, polyoxyethylene monostearate, glyceryl monostearate and stearoyl tartrate.

azo dyes. Broad series of synthetic dyes having double-bonded nitrogens as the chromophore group. The following are azo dyes: Tartrazine, Yellow 2G, Sunset Yellow FCF, Carmoisine, Amaranth, Ponceau 4R, Red 2G, Brown FK, Chocolate Brown HT black PN, Pigment rubine. A portion of the population is sensitive to azo dyes with reactions that include contractions of the bronchi, watering eyes and nose, and rash.

balsam. The natural exudate from a tree or plant; Canada of balsam from North American balsam fir is a liquid that can be used for flavoring.

bases. Substances added to foods to increase their alkalinity, reduce their acidity, or react with acids to give off carbon dioxide gas for aerating purposes.

binder. A substance that gives a mixture uniform adhesion, solidification, and consistency; absorbs moisture at high temperatures; in the food industry, it is used in reformed meat, fish, and poultry products. They are also used in extruded foods, chewing gums, confections, capsules, and tablets. Examples of binders are starches and salts, dextrins, oils, and gums.

bleaching agent. Substances added to artificially whiten or bleach flour. Examples are nitrogen peroxide and benzoyl peroxide.

buffer. A mixture of compounds that, when added to a solution, protects it from any substantial change in pH. Such mixtures are usually in solution form and contain either a weak acid and its related salt or weak bases. Proteins and amino acids contain both acidic and basic groups and function as buffers.

bulking aids. Nutritive and non-nutritive substances added to food that increase its bulk but not its caloric content. They should not impart any flavor to the product nor should they alter the desired color; also known as fillers.

carcinogen. Cancer-causing agent; any substance that causes the development of cancerous growths in living tissues.

catalyst. Promotes a reaction in which it does not participate; small amounts are needed. Catalysts are used in food processing including hydrogenation of vegetable oils, transesterification of fats (e.g., sodium methoxide), modification of starches, and enzymatic reactions in many food products.

CFR (Code of Federal Regulations). A codification of the general and permanent rules published in the Federal Register by the executive department and agencies of the federal government.

chelating agent. Substances used to remove ions from solutions, e.g., EDTA (ethylenediaminetetraacetic acid).

citrus oil. The essential oil obtained from the peel of a citrus fruit such as lemon, lime, or orange.

clarifying agent. A substance that aids in the removal of small particles that cloud liquids; this can be carried out by filtration, centrifugation, addition of enzymes (proteolytic or pectolytic, or through the addition of flocculating agents; used to prevent chill-hazing in beer, color and flavor changes in fruit juices and wines, and hazes in oils.

clouding agent. A substance that adds a turbid appearance to food products; used in small amounts in soft drinks, citrus-flavored beverages, ice cream, ices, baked goods. Examples are brominated vegetable oils and gums, citrus pulp and peel.

coagulant. A substance capable of removing colloidal material. Coagulants are used in precipitating solids or semisolids from solution, e.g., casein from milk.

coal tar dyes. Dyes which were once made from coal tar but are now made industrially. These include: Quinoline yellow, Erythrosine, Patent blue V, Indigo carmine, Brilliant blue FCF. Azo dyes are also part of this group.

colorant. Any substance that imparts color to another material or mixture; broadly classified colors fall into three groups: natural pigments derived mostly from plant materials , inorganic pigments and lakes (combination of organic coloring materials with metals, and synthetic coal tar dyes. Those colorants permitted as food additives widely vary from country to country.

color adjunct. Substances that enhance the color of a food.

color, certified. Synthetic colorants certified by the FDA for safety and purity for use in foods. May be either a dye (soluble) or a lake (insoluble).

Color Index (CI). Color reference numbers assigned in the Color Index of the Society of Dyers and Colorists.

color, natural. Any of several colors that occur naturally in plant and animal tissues.

concrete. Extract from flowers or plants, obtained by using various types of solvents.

curing agent. Substances imparting a unique flavor or color to a food and usually producing an increase in its shelf-life stability.

decolorizing agent. Any material that removes color by a physical or chemical reaction. Also refers to bleaches involving a chemical reaction for removing color.

defoaming agent. A substance that reduces or inhibits foam formation due to proteins, gases, or nitrogenous materials which may interfere with processing.

degumming agent. Substances used in the refining of fats to remove mucilaginous matter consisting of gum, resin, proteins, and phosphatides, e.g., hydrochloric and phosphoric acids.

dextrose equivalent (DE). A measure of the degree of starch polymer hydrolysis determined by quantitative analysis; defined as reducing sugars calculated as dextrose and expressed as a percentage of the dry substance.

diluents. Substances that dilute or dissolve other additives.

disaccharide. Carbohydrates that are formed when monosaccharide units condense, with the elimination of water.

dispersing agent. A surface-active agent added to a suspending medium to provide uniform and maximum separation of extremely fine, solid particles, often of colloidal size.

distillate. The colorless to pale-colored volatile liquid recovered by condensing the vapors of an extract or press-cake of fruits which is heated to its boiling point in a still.

distilled oil. An essential oil obtained by the distillation of the portion of a botanical material, e.g. peel, leaves, stem, containing the essential oil.

diuretic. A substance that promotes water elimination from the body via kidney function; any substance which increases or stimulates the flow of urine, e.g., beer, coffee, tea.

dough conditioner. A substance that is added to dough to accelerate the aging process or to improve the baking qualities. It produces a stronger more resilient dough. Oxidizing agents such as ammonium persulfate and potassium bromate produce this effect.

drying agent. Substances with the ability to absorb moisture.

emulsifier. A substance that prevents the separation of immiscible substances in an emulsion; helps to distribute evenly one substance in another; used to improve texture, homogeneity, consistency, and stability. Examples are gums, egg yolk, albumin, casein, and lecithin.

emulsifying salts. A mixture, of salts (citrates, phosphates, tartrates) added to cheese when it is melted as part of its processing to prevent the development of stringiness; used in the manufacture of milk powder, evaporated milk, and sterilized cream.

emulsion. A mixture of two immiscible liquids , one being dispersed in the other in the form of fine droplets, e.g., oil in water and water in oil. They are held in suspension by emulsifiers.

enhancer. A substance used to intensify or heighten; a flavor enhancer increases the flavor of a food without contributing any taste of its own.

enzyme. An organic catalyst for metabolic reactions used to improve food processing and the quality of the finished food. Examples are rennet and glucose isomerase.

essence oil. The oil recovered from the distillate that is obtained during the concentration of citrus juices.

essential oil. The active flavoring principles of certain botanicals such as roots, stems, leaves, and buds of spices and herbs, seeds, flowers, citrus fruit skins, and barks of certain trees. The oil is found in small sacs which are distributed throughout the plant and generally constitutes the odoriferous principles of the plant. The oil is inflammable, soluble in alcohol and ether but not in water. Examples are oil of spearmint, oil of bitter almonds, oil of citronella. Essential oils produced only by physical expression are called cold-pressed oils.

E system. Developed by the European Economic Community provides a listing of commonly used additives that are generally recognized as safe within the Common Market and allows food to move from country to country within the Common Market. The list is updated on a regular basis.

European Economic Community (EEC). A federation of European countries organized to promote economic growth and trade; member countries are Belgium, Denmark, France, Germany, Great Britain, Greece, Ireland, Italy, Luxembourg, The Netherlands, Portugal, and Spain.

excipient. An inactive substance used to carry an active substance. Used in the baking industry to denote a carrier substance for additives used in bread.

extract (flavoring). An alcohol or alcohol-water solution containing a flavoring ingredient; less potent than essential oils.

extract-solid. A viscous or semi-solid material obtained by first extracting the botanical material with a water-ethanol solvent and then removing the solvent almost completely. This is a liquid extract which has been concentrated.

fermentation. Any of various aerobic or anaerobic processes used for the manufacture of certain products such as alcohol, acids, vitamins of the B complex, or antibiotics by the action usually of yeast, mold, or bacteria.

fiber. General term given to the indigestible parts of food.

filler. *See* bulking agent.

film former. *See* surface finishing agent.

filter aid. Processing aid used to ease the removal of unwanted substances by filtration, e.g., by breaking down a substance such as cellulose in wines and juices, by physical entrapment and adsorption, or by precipitation (phosphate added to precidpitate proteins in whey).

firming agent. Substance added to precipitate residual pectin, thus strengthening the supporting tissue and preventing its collapse during processing For example, calcium and magnesium salts are used to maintain the natural firmness or crispness of fruits and vegetables and to prevent them from softening during processing.

fixative. A substance used to reduce the overall volatility of flavoring agents.

GLOSSARY

flavor adjunct. Substance added to a flavor but is not an essential part of that flavor; e.g., antioxidants, carriers, emulsifiers, and solvents.

flavor, artificial. A flavor not found in nature; aliphatic, aromatic, and terpene compounds that are made synthetically as opposed to those isolated from natural sources.

flavor enhancer. A substance that will magnify, modify, or supplement the natural or original flavor, taste, or aroma of a food without the substance contributing significantly to that flavor, e.g., monosodium glutamate.

flavor potentiator. A substance that imparts flavor to a food product to a much greater extent than an enhancer. The most important of these are the 5'-nucleotides, which are approved by the FDA. Potentiators do not add any taste of their own, but intensify the taste response to substances already present in the food.

flavoring agent. A substance added to food to give it a specific taste; an extract or essence that imparts its flavor to food.

flavoring, liquid. A flavoring in liquid form which may or may not contain a solvent.

flavoring, powdered. A flavor which is produced either by spray-drying, adsorption, agglomeration, dry-blending, or other such process.

flavor, natural. A flavor derived from a natural animal or plant product and includes spices, herbs, essential oils and their extracts, concentrates and isolates, fruit and fruit juices, animal and vegetable materials and their extracts and aromatic chemicals isolated by physical means from natural products, e.g., citral from oil of lemongrass.

flocculant. A substance that aggregates suspended colloidal particles in such a way that they form small clumps for removal.

foaming agent. A substance that regulates the amount of air in a product.

foam inhibitor. A food additive that prevents the formation of foam in foods during processing.

food. Any substance or mixture that nourishes an organism, builds tissue, and supplies energy.

food additive, direct (intentional). Any substance added purposely to food for technological purposes such as preserving food from bacterial deterioration, protecting it from oxidative changes, and improving its organoleptic characteristics, or texture.

food additive, indirect. Any substance used as a component of articles that contact food.

food additive, unintentional. Chemical substances found in food as a result of environmental or accidental contamination, e.g., insecticide residues, fertilizers.

Food and Agricultural Organization (FAO). Specialized agency of the United Nations concerned with the development of world agriculture, fisheries, and forestry, with the production, processing, preservation, and distribution of food, and with the improvement of nutrition.

Food and Drug Administration (FDA). The U.S. federal agency responsible for enforcement of the Federal Food, Drug, and Cosmetic Act.

food standards. The set of rules defining the criteria that a food must fulfill to be suitable for distribution or sale.

formulation aid. Substances used to promote or produce a desired physical state or texture in food, including carriers, binders, fillers, plasticizers, and film formers.

fumigant. A volatile substance used for controlling insects or pests.

gelling agent. A substance that forms stiff gels when added to water; used in food products for its thickening and water-binding properties.

generally regarded as safe (GRAS). A U.S. FDA term for a group of chemicals that by current knowledge is regarded as safe to use in food.

glazing agent. A substance that imparts a shiny or polished appearance on the surface of the food and may also act as a protective coating; also known as a surface finishing agent.

gum. A sticky substance that can disperse in water to form a viscous mucilaginous mass; this substance is used in food processing to stabilize emulsions, e.g., salad dressings and processed cheese; as a thickener, and in sugar confectionery. Examples of gums extracted from seeds are guar gum, locust, quince, and psyllium; gums from sap or exudate are gum arabic, karaya, tragacanth, ghatti, mesquite, anguo; and gums from seaweed are agar, kelp, alginate, and Irish moss. Gums can be made from starch or cellulose, e.g., dextrins, methyl cellulose, carboxymethylcellulose, or they may be synthetic such as vinyl polymers.

humectant. A liquid that absorbs moisture and is used to maintain the water content of baking products. Examples are glucose syrup, invert sugar, honey, dried whey, glycerol, and sorbitol.

hydrate. A crystalline product made up of salts and closely associated water molecules.

hydrocolloids. Water-soluble gums.

hydrogenation. Any reaction of hydrogen with an organic compound. It may occur as either direct addition of hydrogen to the double bonds of unsaturated molecules, resulting in a saturated product. Cottonseed, maize, and sunflower oils are commonly hydrogenated and used in margarine and cooking fats.

hygroscopic. Descriptive of a liquid or solid material that picks up atmospheric water vapor and thus acts as a drying agent, e.g., calcium chloride and silica gel.

hypersensitivity. When a small amount of a substance produces symptoms that can be objectively verified and repeated.

International Units (IU). Used as a measure of the comparative potency of natural substances, such as vitamins, before they are obtained in sufficiently pure form to measure their weight; it is an arbitrarily defined reproducible standard.

intolerance. When a small amount of a substance produces a response resembling an allergic reaction but immunologic mechanisms are not involved.

intramuscular. Within a muscle.

intraperitoneal. Within the peritoneal cavity.

isolate. An aromatic compound consisting of one ingredient isolated from a natural raw material such as menthol from peppermint oil or citral from lemongrass oil.

JECFA. Joint Expert Committee on Food Additives composed of experts appointed by the Directors General of the World Health Organization and the Food and Agriculture organization. It reviews at regular intervals the safety of additives and establishes values for acceptable daily intakes.

laxative. A substance that accelerates the passage of food through the intestines. If it alters peristaltic activity, it is called a purgative. Cellulose acts as a purgative by retaining water and increasing the volume of the intestinal contents.

leavening agent. A substance that is added to food in order to achieve the following: to produce or stimulate the production of carbon dioxide; to cause fermentation; to make the food porous and light; to cause baked goods to rise, e.g., yeast, yeast foods, and calcium salts.

Lethal Dose Fifty (LD50). A calculated dose of a material which is expected to cause the death of 50% of an entire defined experimental animal population. It is determined from the exposure to the material by any other route than inhalation of a significant number from that population.

liquid freezant. Liquids or liquifiable gases which can freeze food by coming into contact with it and extracting its heat.

lubricants. *See* release agent.

maturing agent. A food additive that accelerates the aging process and improves baking quality. *See* dough conditioner.

mineral hydrocarbons. Substances derived from bitumen (paraffin hydrocarbons). It includes white oil, liquid paraffin, petroleum jelly, microcrystalline wax, and hard paraffin.

monosaccharide. The simplest carbohydrate consisting of one unit of water per carbon atom; monosaccharides have the general formula $(CH_2O)_n$, where n can be three or more; e.g. glucose.

mutagen. A chemical or physical agent that interacts with DNA and causes a mutation.

mutation. A sudden, random, permanent, genetic change; a genetic change within cells that changes its characteristics.

neutralizer. A substance that is used in food to change the acid-alkaline balance.

neutraceutical. Food additives, such as beta carotene, claiming to have disease preventing properties.

nutritional additives. Includes vitamins, minerals, amino acids, and fibers.

oleoresin. The solid, semisolid, or oily, heavy viscous fluid or residue obtained by solvent extraction or percolation of plant matter in preparation of some spices such as pepper, ginger, and capsicum. Oleoresins are stronger in flavor than essential (volatile) oils and are often used to add heat to a product. Also, they have very distinctive odors.

organoleptic. Technical term for taste and smell. Only four tastes can be detected by the tongue (see taste). All other taste perceptions are detected by smell.

oxidizing agent. An agent that causes removal of electrons; an element that gains electrons and is thereby reduced.

polyphosphates. Complex phosphates added to meat products to prevent sausage discoloration, aid in the mixing of fat, speed penetration of brine in curing, cause protein fibers of meat to retain more water and swell, thereby improving texture.

preservative. A substance, either natural or synthetic, that protects food against spoilage, discoloration, or decay; used to retard or prevent microbial or chemical spoilage, e.g., sodium benzoate, sulfur dioxide, sorbic acid, and propionic acid.

processing aid. Substance used as manufacturing aid to enhance the utility or appeal of a food, including clarifying agents, clouding agents, catalysts, flocculants, filter aids, and crystallization inhibitors.

propellant. A liquefied gas used to supply force to expel product.

pungent. Describes an astringent or acrid, sharp odor or flavor.

rancid. Having a rank smell or taste like stale fat, usually from chemical change or decomposition.

reducing agent. A substance that loses electrons and is oxidized and therefore produces a more stable product.

release agent. Substance added to the food product or to the surface of the food product to prevent it from sticking to machinery during food processing; it can also be applied to trays, molds, and tins to make the removal of food packaged in this manner easy, e.g., fatty acid amides, microcrystalline waxes, petrolatums, starch, methylcellulose.

sequestrant. A chemical compound that reacts with trace metal present in the environment to form a complex, thus minimizing the effect of the metal on the food, e.g., citric acid and disodium ethylenediaminetetraacetate.

solubilizer. A substance that will cause dispersion of an insoluble flavorant in water.

solvent. A substance used to dissolve or extract another substance (solute) to form a uniformly dispersed mixture (solution). Examples of permitted solvents in food are: ethyl alcohol, glycerol, propylene glycol.

spice. Any of various aromatic plant materials used to flavor foods or a variety of dried plant products sometimes used in the form of extracts that exhibit pronounced aroma and flavor.

stabilizer. A food additive that thickens, prevents separation, prevents flavor deterioration, retards oxidation by increasing the viscosity, and gives a smoother product; also prevents evaporation and deterioration of volatile flavor.

stimulant. Any agent or drug which temporarily increases action of any organ of the body, e.g., caffeine.

subcutaneous. Beneath the skin; injected beneath the skin.

surface-active agent. Substance used to modify surface properties of liquid food components for a variety of effects other than emulsification; it includes solubilizing agents, dispersants, detergents, wetting agents rehydration enhancers, whipping agents, foaming agents, and defoaming agents.

surface-finishing agent. Substance used to increase palatability, preserve gloss, and inhibit discoloration of foods, including glazes, polishes, waxes, and protective coatings.

suspending agent. Substance that causes particles to mix, but remains undissolved in a liquid or solid.

sweetening agent. A sweet tasting substance used in foods; may be either natural, such as sucrose, fructose, glucose, maltose and lactose; or synthetic and non-nutritive, which usually have a much greater sweetness intensity than sugar , but without the caloric value, e.g., saccharin, aspartame.

synergist. Substance used to act or react with another food ingredient to produce a total effect different or greater than the sum of the effects produced by the individual ingredients. It is used to enhance the functionality of antioxidants, flavors, emulsifiers, and gels. Examples are polyphenolics to enhance antioxidant activity, phosphates which enhance the actions of emulsifiers in dairy products.

taste. Flavor perception of sweet, salty, bitter, or sour.

tenderizer. An enzyme (papain), weak acid (vinegar or lemon juice), salt, or a process used to reduce toughness of meat.

teratogen. Any substance that affects normal development, often causing developmental anomalies; ionizing radiation may have this effect.

texturizing agent. A substance used to impart body, to improve consistency or texture of a food, or to stabilize an emulsion.

thickener. A substance used to impart body, improve the consistency or texture of a food, or to stabilize an emulsion; works by absorbing water.

volatile oil. A substance that will evaporate quickly; responsible for the aroma, odor, and flavor found in the aqueous distillation of organic compounds of flavorings or pure spices.

wetting agent. A surface-active agent which, when added to water causes it to penetrate more easily into, or to spread over the surface of, another material by reducing the surface tension of the water.

World Health Organization (WHO). A branch of the United Nations concerned with international health problems. Its interests are in maintenance of nutrition, wholesomeness of foods, and consumer health.

Bibliography

Aldrich Chemical Company. *Aldrich Catalog Handbook of Fine Chemicals 1994-1995.* 1994

Aldrich Chemical Company. *Aldrich Flavors & Fragrances 1994.* 1993.

Ashurst, P.R. *Food Flavourings.* New York: AVI Publishing Company, Inc., 1991.

Bellanca, Nicolo, and Thomas E. Furia. *Fenaroli's Handbook of Flavor Ingredients.* 2nd Edition, 2 vols. Boca Raton, FL: CRC Press, 1975.

Bender, Arnold E., Ph.D. *Dictionary of Nutrition and Food Technology.* Butterworth & Co. (Publishers) Ltd.. 1975.

Branen, A. Larry, P. Michael Davidson, and Seppo Salminen, Eds. *Food Additives*, New York and Basel: Marcel Dekker, 1990.

Budavari, Susan. *The Merck Index: An Encyclopedia of Chemicals, Drugs, and Biologicals.* 11th Edition. New Jersey: Merck & Co., 1989.

Code of Federal Regulations: Food and Drugs, 21, Parts 170-199. Washington, D.C.: U.S. Government Printing Office, 1993.

Commission of the European Communities. *EINECS, European Inventory of Existing Commercial Chemical Substances, Vol. IV.* Luxenbourg: Office for Official Publications of the European Communities, 1987.

Committee on Codex Specifications. *Food Chemicals Codex.* 3rd Edition. Washington, D.C.: National Academy Press, 1981.

Committee on Food Chemicals Codex. *Food Chemicals Codex, First Supplement to The Third Edition.* Washington, D.C.: National Academy Press, 1983.

Committee on Food Chemicals Codex. *Food Chemicals Codex, Second Supplement to The Third Edition.* Washington, D.C.: National Academy Press, 1986.

Committee on Food Chemicals Codex. *Food Chemicals Codex, Third Supplement to The Third Edition.* Washington, D.C.: National Academy Press, 1992.

Committee on Food Chemicals Codex. *Food Chemicals Codex, Fourth Supplement to The Third Edition.* Washington, D.C.: National Academy Press, 1993.

Conning, D.M., and A.B.G. Lansdown. *Toxic Hazards in Food.* New York : Raven Press, 1983.

Considine, Douglas M. *Foods and Food Production Encyclopedia.* New York: Van Nostrand Reinhold, 1982.

Doull, John. *Cassaret and Doull's Toxicology: The Basic Science of Poisons.* New York: Macmillan, 1980.

Fluka Chemie AG. *Fluka Chemika-Biochemika.* Switzerland, 1993-1994.

Food Ingredients & Processing International. *International Food Ingredients Directory.* Rickmansworth, Hertfordshire, UK: Turret Group plc, 1992.

Freydberg, Nicholas, Ph.D, and Willis A Gorther, Ph.D. *The Food Additives Book.* New York: Bantam Books, 1982

Furia. *CRC Handbook of Food Additives.* 2nd Edition. Vol. 1. 1973.

Grundschober, Friedrich, Dr., and Dr. Jan Stofberg. Consumption Ratio and Food Predominance of Flavoring Materials. *Perfumer & Flavorist.* Vol. 12: August/September 1987: pp. 27-68.

Hanssen, Maurice, and Jill Marsden. *E is for Additives.* London: Thorsons, 1987.

Hathcock, John N. *Nutritional Toxicology.* New York: Academic Press, 1982.

Heath, Henry B., and Gary Reineccius. *Flavor Chemistry and Technology.* Westport, Connecticut: AVI Publishing Company, Inc., 1986.

BIBLIOGRAPHY

Hughes, Christopher C. *The Additives Guide.* New York: John Wiley & Sons, 1987.

Jones, Julie Miller. *Food Safety.* St. Paul, Minnesota: Eagan Press, 1992.

Jukes, DJ. *Food Legislation of the UK: A Concise Guide.* 2nd Edition. Boston: Butterworth & Company, 1987.

Lewis, Richard J., Sr. *Food Additives Handbook.* New York: Van Nostrand Reinhold, 1989.

Lewis, Richard J., Sr. *Hawley's Condensed Chemical Dictionary.* 12th Edition. New York: Van Nostrand Reinhold, 1993.

Morton, I.D. and A.J. MacLeod, eds. *Food Flavours. Part B: The Flavour of Beverages.* New York: Elsevier Science Publishing, 1986.

Ockerman, Herbert. *Food Science Sourcebook.* 2nd Edition. v. 1-2. New York: Van Nostrand Reinhold, 1991.

Secondini, Dr. Olindo. *Handbook of Perfumes and Flavors.* New York: Chemical Publshing Co., 1990

Smith, Jim. *Food Additive User's Handbook.* New York: Van Nostrand Reinhold/Blackie, Inc., 1991.

Smolinske, Susan C. *Handbook of Food, Drug, and Cosmetic Excipients.* Boca Raton, Florida: CRC Press, 1992.

Taylor, Reginald James. *Food Additives.* New York: J. Wiley, 1980.

Vettorazzi, Gaston. *Handbook of International Food Regulatory Toxicology.* New York: SP Medical & Scientific Books, 1980.

Walker, Gibson. *Food Toxicology—Real or Imaginary Problems?* Philadelphia: Taylor & Francis, 1985.

Wenninger, John A. and G.N. McEwen, Jr., Ph.D., J.D., eds. *International Cosmetic Ingredient Dictionary.* 5th Edition, 2 vols. Washington, D.C.: The Cosmetic, Toiletry, and Fragrance Association, 1993.